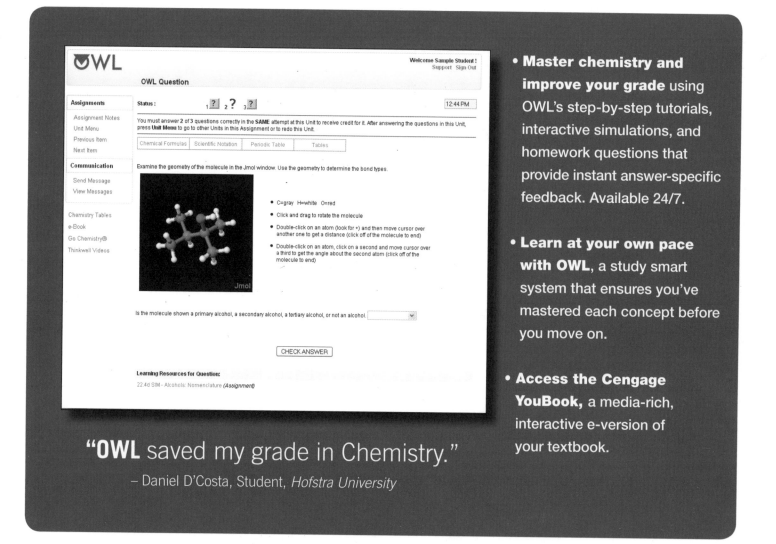

Icons and Colors in Illustrations

The following symbols and colors are used in this text to help in illustrating structures, reactions, and biochemical principles.

Elements:

☐ = Oxygen ☐ = Nitrogen ☐ = Phosphorus ☐ = Sulfur ■ = Carbon ☐ = Chlorine

Inorganic phosphate and pyrophosphate are sometimes symbolized by the following icons:

Ⓟ
Inorganic phosphate (P_i)

ⓅⓅ
Pyrophosphate (PP_i)

Sugars:

Glucose   Galactose   Mannose   Fructose   Ribose

Nucleotides:

☐ = Guanine ☐ = Cytosine ☐ = Adenine ☐ = Thymine ☐ = Uracil

Amino acids:

☐ = Non-polar/hydrophobic ☐ = Polar/uncharged ☐ = Acidic ☐ = Basic

Enzymes:

➕ = Enzyme activation

⊘ = Enzyme inhibition or inactivation

Ⓔ = Enzyme ╱ = Enzyme   Enzyme names are printed in red.

In reactions, blocks of color over parts of molecular structures are used so that discrete parts of the reaction can be easily followed from one intermediate to another, making it easy to see where the reactants originate and how the products are produced.

Some examples:

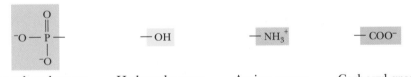

Phosphoryl group   Hydroxyl group   Amino group   Carboxyl group

Red arrows are used to indicate nucleophilic attack. ↶ ↷

These colors are internally consistent within reactions and are generally consistent within the scope of a chapter or treatment of a particular topic.

Biochemistry

FIFTH EDITION

Reginald H. Garrett | Charles M. Grisham

University of Virginia

With molecular graphic images
by Michal Sabat, University of Virginia

BROOKS/COLE
CENGAGE Learning

Australia • Brazil • Japan • Korea • Mexico • Singapore
• Spain • United Kingdom • United States

Biochemistry, **Fifth Edition**
Reginald H. Garrett, Charles M. Grisham

Publisher: Mary Finch

Senior Developmental Editor: Peter McGahey

Assistant Editor: Elizabeth Woods

Senior Media Editor: Lisa Weber

Media Editor: Stephanie Van Camp

Marketing Manager: Nicole Hamm

Marketing Assistant: Julie Stefani

Marketing Communications Manager: Linda Yip

Content Project Manager: Teresa L. Trego

Design Director: Rob Hugel

Art Director: Maria Epes

Manufacturing Planner: Karen Hunt

Rights Acquisitions Specialist: Dean Dauphinais

Production Service: Graphic World Inc.

Text Designer: Lisa Devenish

Photo Researcher: Bill Smith Group

Text Researcher: Ashley Liening

Copy Editor: Graphic World Inc.

OWL Producers: Stephen Battisti, Cindy Stein, David Hart (Center for Educational Software Development, University of Massachusetts, Amherst)

Cover Designer: Beverly Miller

Cover Image: Michal Sabat, University of Virginia

Compositor: Graphic World Inc.

To Georgia, To Rosemary

Reginald H. Garrett was educated in the Baltimore city public schools and at the Johns Hopkins University, where he received his Ph.D. in biology in 1968. Since that time, he has been at the University of Virginia, where he is currently Professor of Biology. He is the author of previous editions of *Biochemistry,* as well as *Principles of Biochemistry* (Cengage, Brooks/Cole), and numerous papers and review articles on the biochemical, genetic, and molecular biological aspects of inorganic nitrogen metabolism. His research interests focused on the pathway of nitrate assimilation in filamentous fungi. His investigations contributed substantially to our understanding of the enzymology, genetics, and regulation of this major pathway of biological nitrogen acquisition. More recently, he has collaborated in systems approaches to the metabolic basis of nutrition-related diseases. His research has been supported by the National Institutes of Health, the National Science Foundation, and private industry. He is a former Fulbright Scholar at the Universität fur Bodenkultur in Vienna, Austria and served as Visiting Scholar at the University of Cambridge on two separate occasions. During the second, he was Thomas Jefferson Visiting Fellow in Downing College. In 2003, he was Professeur Invité at the Université Paul Sabatier/Toulouse III and the Centre National de la Recherche Scientifique, Institute for Pharmacology and Structural Biology in France. He has taught biochemistry at the University of Virginia for more than 40 years. He is a member of the American Society for Biochemistry and Molecular Biology.

Charles M. Grisham was born and raised in Minneapolis, Minnesota, and educated at Benilde High School. He received his B.S. in chemistry from the Illinois Institute of Technology in 1969 and his Ph.D. in chemistry from the University of Minnesota in 1973. Following a postdoctoral appointment at the Institute for Cancer Research in Philadelphia, he joined the faculty of the University of Virginia, where he is Professor of Chemistry. He is the author of previous editions of *Biochemistry* and *Principles of Biochemistry* (Cengage, Brooks/Cole), and numerous papers and review articles on active transport of sodium, potassium, and calcium in mammalian systems, on protein kinase C, and on the applications of NMR and EPR spectroscopy to the study of biological systems. He has also authored *Interactive Biochemistry CD-ROM and Workbook,* a tutorial CD for students. His work has been supported by the National Institutes of Health, the National Science Foundation, the Muscular Dystrophy Association of America, the Research Corporation, the American Heart Association, and the American Chemical Society. He is a Research Career Development Awardee of the National Institutes of Health, and in 1983 and 1984 he was a Visiting Scientist at the Aarhus University Institute of Physiology Denmark. In 1999, he was Knapp Professor of Chemistry at the University of San Diego. He has taught biochemistry, introductory chemistry, and physical chemistry at the University of Virginia for 37 years. He is a member of the American Society for Biochemistry and Molecular Biology.

Charles M. Grisham and Reginald H. Garrett

Georgia Cobb Garrett

Contents in Brief

Detailed Contents

PART III | Metabolism and Its Regulation

ActiveModels Library of Proteins and Nucleic Acids

ActiveModels presents an online library of macromolecular structures with accompanying scripts detailing descriptions of structure-function relationships for the key proteins and nucleic acids mentioned in the book and listed below. The hyperlinked text controls the graphic display and presents a variety of perspectives and features. End-of-chapter problems written specifically to utilize this resource appear in most chapters.

Abl Kinase
Acetylcholinesterase (Recombinant Human)
Acetyl-Coa Carboxylase (Carboxyltransferase Domain)
Acrb Channel
Actin
Acyl-Coa Dehydrogenase
Acyl-Coa Oxidase
Adenylosuccinate Lyase
β-Adrenergic Receptor
Alcohol Dehydrogenase
Aldose Reductase
α-Amylase (salivary)
Amyloid beta-Peptide (A-beta)
Androgen Receptor
Angiotension-1 Converting Enzyme (ACE)
Anthrax Lethal Factor
Apoptosis-Inducing Factor
Aprotinin
Aqp1 Aquaporin (Water Channel)
ASAP3 (ARF GTPase)
ATP/ADP Translocase-1
Atrial Natriuretic Peptide Receptor
Autotransporter (bacterial)
BiP Chaperone
Bone Morphogenetic Protein 2
Bone Morhogentic Protein Domain
BRCA 2
BtuB (Cobalamin transporter)
Calmodulin
Caspase-6
Cat Allergen
Choline Acetyltranferase
Chorismate Mutase
Complex 1 (Electron Transport)
Complex 2 (Electron Transport)
Concanavaline
C-Reactive Protein (human)
Cu,Zn Superoxide Dismutase (human)
Cyclin-Dependent Kinase-2 (human)
Cyclooxygenase
Cystic Fibrosis Transmembrane Conductance Regulator
Cytidine Triphosphate Synthetase
Cytochrome *b* Reductase
Cytochrome *c*
Cytochrome c Oxidase (*Bos tarus*)
Cytochrome c Oxidase (*Rhodobacter sphaeroides*)
D2 Domain of NSf
Dicer I
Di-Heme Cytochrome *c*
3,4-Dihydroxyphenylalanine Decarboxylase
DNA Polymerase n
DrrA Guanine Nucleotide Exchange Factor
Dual Specificity Phosphatase 6
E1 Helicase
E1 Ubiquitin-like Protein Complex
EGFR Kinase Domain
Elastase
Enolase
Enoyl-CoA Hydratase
Estrogen Receptor
Farnesyl Transferase
Fatty Acid Transporter
Ferritin (human) and Bacterioferritin
Ferritin H Chain (human)
Ferrochelatase (human)
Fibrin (human)
FKBP12-Rapamycin Complex
Flavodoxin
Flig (Flagellar Rotor Protein)
Fructose-1-6-Bisphosphatase

GABA Receptor Associated Protein
Galectin 1 (human)
Glucose-6-Phosphate Dehydrogenase
Glutamine Synthetase
Glycogen Phosphorylase
GMP Synthetase
Grb2 Growth Factor Bound Protein 2 Signal Transduction Adaptor
GRD19p
GroEL-GroES Complex
Hemoglobin S
Hepatocyte Nuclear Factor 1b Bound to DNA
Hexokinase (human)
HIV-1 Protease
HIV Reverse Transcriptase with a Rival Purine Inhibitor
HIV Reverse Transcriptase with Inhibitor 7
HLA A2 Class I MHC
HMG-CoA Reductase
Hsp90
Influenza Virus Hemagglutinin
Interleukin 17 Receptor Complex
Interleukin -4 and its Receptor
Ire1 (Transmembrane Serine/Threonine Kinase)
Isocitrate Dehydrogenase
ITK-Sh2 Domain Bound to Phosphopeptide
KIF1A (monomeric kinesin)-Microtubule Complex)
Kinesin
Kinesin (rat)
β-Lactamase
Lactate Dehydrogenase (Malarial)
LDL Receptor
Lipocalin (human)
Luciferase Inhibitor Complex (firefly)
Lysine Gingipain (Kgp) protein
Malonyl-Coa:Acp Transferase
MDM2 (Ubiquitin-Protein Ligase E3)
Metalloprotease
Methemoglobin (horse)
Monoamine Oxidase B (human)
Myoglobin
Myosin 2 - heavy and light chain
Myosin 2 - heavy chain
Myosin 5
Nc6.8 (monoclonal Ab) Fab In Complex With Sweetener Sc45647

Neuropsin (a Serine Protease)
Nicotinic Acetylcholine Receptor
Niemann-Pick C1: Cholesterol
Nitrogenase Reductase
N-Myristoyltransferase With Bound Myristoyl-CoA
NSE/NS4A Protease Apostructure (Hepatitis C Virus)
Nuclear Receptor X Heterodimers
2,3-Oxidosqualene Cylclase with Lanosterol
p53-DNA Complex (p53 DNA-Binding Domain)
Pepsin + DMSO
P-Glycoprotein (MDR)
Phosphofructokinase (Trypsanoma)
Phosphoglucoisomerase (Bacillus)

Laboratory Techniques in Biochemistry

All of our knowledge of biochemistry is the outcome of experiments. For the most part, this text presents biochemical knowledge as established fact, but students should never lose sight of the obligatory connection between scientific knowledge and its validation by observation and analysis. The path of discovery by experimental research is often indirect, tortuous, and confounding before the truth is realized. Laboratory techniques lie at the heart of scientific inquiry, and many techniques of biochemistry are presented within these pages to foster a deeper understanding of the biochemical principles and concepts that they reveal.

Recombinant DNA Techniques

Probing the Function of Biomolecules

Techniques Relevant to Clinical Biochemistry

Isolation/Purification of Macromolecules

Analyzing the Physical and Chemical Properties of Biomolecules

The Fifth Edition

Scientific understanding of the molecular nature of life is growing at an astounding rate. Significantly, society is the prime beneficiary of this increased understanding. Cures for diseases, better public health, remedies for environmental pollution, and the development of cheaper and safer natural products are just a few practical benefits of this knowledge.

In addition, this expansion of information fuels, in the words of Thomas Jefferson, *"the illimitable freedom of the human mind."* Scientists can use the tools of biochemistry and molecular biology to explore all aspects of an organism—from basic questions about its chemical composition, through inquiries into the complexities of its metabolism, its differentiation and development, to analysis of its evolution and even its behavior. *New procedures based on the results of these explorations lie at the heart of the many modern medical miracles.* Biochemistry is a science whose boundaries now encompass all aspects of biology, from molecules to cells, to organisms, to ecology, and *to all aspects of health care.* This fifth edition of *Biochemistry* embodies and reflects the expanse of this knowledge. We hope that this new edition will encourage students to ask questions of their own and to push the boundaries of their curiosity about science.

Making Connections

As the explication of natural phenomena rests more and more on biochemistry, its inclusion in undergraduate and graduate curricula in biology, chemistry, and the health sciences becomes imperative. The challenge to authors and instructors is a formidable one: how to familiarize students with the essential features of modern biochemistry in an introductory course or textbook. Fortunately, the increased scope of knowledge allows scientists to make generalizations connecting the biochemical properties of living systems with the character of their constituent molecules. As a consequence, these generalizations, validated by repetitive examples, emerge in time as principles of biochemistry, principles that are useful in discerning and describing new relationships between diverse biomolecular functions and in predicting the mechanisms underlying newly discovered biomolecular processes. Nevertheless, it is increasingly apparent that students must develop skills in inquiry-based learning, so that, beyond this first encounter with biochemical principles and concepts, students are equipped to explore science on their own. Much of the design of this new edition is meant to foster the development of such skills.

We are both biochemists, but one of us is in a biology department, and the other is in a chemistry department. Undoubtedly, we each view biochemistry through the lens of our respective disciplines. We believe, however, that our collaboration on this textbook represents a melding of our perspectives that will provide new dimensions of appreciation and understanding for all students.

Our Audience

This biochemistry textbook is designed to communicate the fundamental principles governing the structure, function, and interactions of biological molecules to students encountering biochemistry for the first time. We aim to bring an appreciation of biochemistry to a broad audience that includes undergraduates majoring in the life sciences, physical sciences, or premedical programs, as well as medical students and graduate students in the various health sciences for whom biochemistry is an important route to understanding human physiology. To make this subject matter more relevant and interesting to all readers, we emphasize, where appropriate, the biochemistry of humans.

Objectives and Building on Previous Editions

We carry forward the clarity of purpose found in previous editions; namely, to illuminate for students the principles governing the structure, function, and interactions of biological molecules. At the same time, this new edition has been revised to reflect tremendous developments in biochemistry. Significantly, emphasis is placed on the interrelationships of ideas so that students can begin to appreciate the overarching questions of biochemistry.

Features

- **Clarity of Instruction** This edition was re-organized for increased clarity and readability. Many of the lengthier figure legends were shortened and more information was included directly within illustrations. These changes will help the more visual reader.

- **Visual Instruction** The richness of the Protein Data Bank (www.pdb.org) and availability of molecular graphics software has been exploited to enliven this text. Over 340 images of prominent proteins and nucleic acids involved with essential biological functions illustrate and inform the subject matter and were prepared especially for this book.

- **Essential Questions organization** Each chapter in this book is framed around an *Essential Question* that invites students to become actively engaged in their learning and encourages curiosity and imagination about the subject matter. For example, the Essential Question of Chapter 3 asks, "What are the laws and principles of thermodynamics that allow us to describe the flows and interchanges of heat, energy, and matter in biochemical systems?" The section heads then pose *Key Questions* that serve as organizing principles for a lecture such as, "What is the daily human requirement for ATP?" The subheadings are designed to be concept statements that respond to the section headings. The end-of-chapter *Summary* then brings each question together with a synopsis of the answer that summarizes the important concepts and facts to aid students in organizing and understanding the material.

- NEW! **Foundational Biochemistry** At the end of each chapter, a new Foundational Biochemistry feature has been added. These sections provide a comprehensive list of the principal facts and concepts that a student should understand after reading each chapter. Presented as short statements or descriptive phrases, the items of the Foundational Biochemistry list serve as guides to students of the knowledge they have acquired from the chapter and as checklists the students can review in assessing their learning.

- **End-of-Chapter Problems** More than 600 end-of-chapter problems are provided. They serve as meaningful exercises that help students develop problem-solving skills useful in achieving their learning goals. Some problems require students to employ calculations to find mathematical answers to relevant structural or functional questions. Other questions address conceptual problems whose answers require application and integration of ideas and concepts introduced in the chapter. Each set of problems includes MCAT practice questions to aid students in their preparation for standardized examinations such as the MCAT or GRE. Where appropriate a new section presents problems utilizing the ActiveModels database.

- **Further Readings** sections at the end of each chapter make it easy for students to find up-to-date additional information about each topic.

- NEW! **ActiveModels Library of Proteins and Nucleic Acids** An optional online resource for exploration of protein and nucleic acid structures. These molecular documents present macromolecular structures with rich, vivid, state-of-the-art graphics and provide detailed descriptions of structure-function relationships. Each entry was created with ICM Browser Pro software (Molsoft, LLC, La Jolla, CA) by undergraduate biochemistry students at the University of Virginia. Hyperlinks control the graphic display, allowing exploration of structural features in these macromolecules that underlie their biological function. Related questions challenging students to explore macromolecular structure and function are included in the end-of-chapter problems in most of the chapters of the text.

- **Critical Developments in Biochemistry** essays emphasize recent and historical advances in the field.

- **Human Biochemistry** essays emphasize the central role of basic biochemistry in medicine and the health sciences. These essays often present clinically important issues such as diet, diabetes, and cardiovascular health.

- **A Deeper Look** essays expand on the text, highlighting selected topics or experimental observations.

- **Laboratory Techniques** The experimental nature of biochemistry is highlighted, and a list of laboratory techniques found in this book can be seen on page xxx.

New to This Edition

- **ActiveModels Library of Proteins and Nucleic Acids** is an optional online resource for exploration of protein and nucleic acid structures, particularly those described in this text. Each structure is presented as a "Molecular Document" using the ICM Browser Pro modeling program developed by Molsoft, LLC in La Jolla, CA. These molecular documents were created by undergraduate biochemistry students at the University of Virginia and illustrate macromolecular structures with rich, vivid, state-of-the-art graphics accompanied by a script that highlights pertinent structure–function correlations. Hyperlinks in the text window of each entry control the graphic display and present a variety of perspectives and features for each protein or nucleic acid. New end-of-chapter problems challenge students to explore macromolecular structure and function through examination of these molecular documents.

- Presented as short statements or descriptive phrases, the new **Foundational Biochemistry** sections at the end of each chapter provide a short review of the principal facts and concepts that a student should understand after reading the chapter. We anticipate that these lists will serve as guides to students of the knowledge acquired from the chapter and a quick review of the material for assessing learning progress.

- **End-of-chapter problem headings** allow students to place the problem within the context of the subject matter they have learned.

- The *Biochemistry, fifth edition,* **Cengage YouBook** is an interactive and customizable eBook available with OWL. Features of the YouBook include animated figures, the ActiveModels library, and highlighting and note tools for students. See the OWL description below for more details.

Chapter 1 A new Critical Developments in Biochemistry feature on synthetic life: the chemical synthesis of a bacterial genome and its incorporation into host cells to create the first organism with a fully synthetic genome.

Chapter 4 An introduction to the Brainbow technique that enables labeling of many individual neurons simultaneously. This technique will facilitate the analysis of neuronal circuitry on a large scale.

Chapter 6 A new Deeper Look feature on protein sectors—evolutionary units of three-dimensional structure, and a new Deeper Look feature on metamorphic proteins, which exist as an ensemble of structures of similar energies and stabilities.

Chapter 7 A new Deeper Look feature on the carbohydrate chemistry of the *Cucurbitaceae* family, which produces particularly large fruits such as melons, cucumbers, and pumpkins.

Chapter 8 A new Deeper Look feature on glycophospholipids that play a role in formation of plasma membrane signaling microdomains involved in cellular differentiation and maturation. Also a new Human Biochemistry feature on the endocannabinoid signaling system that involves lipid-soluble signals such as anandamide and 2-arachidonoylglycerol.

Chapter 9 A new Human Biochemistry feature on development of inhibitors of N-myristoyltransferase in *T. brucei,* the organism that causes sleeping sickness in Africa. Also revised discussions of the roles of sphingolipid and cholesterol in the formation of membrane rafts and the structures and functions of SNARE proteins and channel proteins.

Chapter 10 An updated introduction to the many roles of small RNAs in the regulation of gene expression: miRNAs and the long, noncoding RNAs (lincRNAs).

Chapter 11 An overview of the emerging technologies to sequence single molecules of DNA. These techniques are at the forefront of "personal genomics": the ability to carry out low-cost sequencing of an individual's genome and the implications of the information obtained on the diagnosis and treatment of disease. Also, new structural models for chromatin at the level of its 30-nm fiber 'secondary structure' give insights into the higher-order structure of chromosomes. This chapter also has a new Human Biochemistry box on DNA methylation, CpG islands, and epigenetics that introduces the concept of heritable changes in the genome without DNA sequence changes.

Chapter 13 A new perspective on the response of enzyme reaction rate to increasing temperature is presented, wherein a temperature-dependent equilibrium between active enzyme and a catalytically inactive but not denatured state of the enzyme affords a deeper understanding of enzyme kinetics.

Chapter 14 A new section on the role of quantum mechanical tunneling in electron and proton transfer reactions of enzymes.

Chapter 15 A new section on the covalent modification of enzymes by reversible acetylation, which has emerged as an important mode of metabolic regulation.

Chapter 16 A revised discussion of P-loops NTPases and their role in molecular motors and a revised section on the contraction cycle of skeletal muscle.

Chapter 18 A new Critical Developments in Biochemistry feature that describes a modern interpretation of the Warburg effect in cancer.

Chapter 19 A new discussion of the structure of pyruvate dehydrogenase complex, a new Deeper Look feature on the role of anaplerosis in insulin secretion, and a new section on the regulation of TCA cycle enzymes by acetylation.

Chapter 20 A new structure (and associated discussion) of Complex I of the electron transport pathway and new information on supercomplexes in electron transport.

Chapters 20 and 21 New information on species variability in the *c*-subunit stoichiometry of F_1F_0- and CF_1CF_0-ATP synthases and the implications of this variability for the energetic cost of ATP formation.

Chapter 22 A new Deeper Look feature on TIGAR, a p53-induced enzyme that mimics fructose-2,6-bisphosphatase and responds to cellular stresses such as oncogenesis and DNA damage events; new information on O-GlcNAc signaling and the hexosamine biosynthetic pathway; and a new Critical Developments in Biochemistry feature describing how consumption of ATP promotes and supports the metabolism of cancer cells.

Chapter 23 A new Deeper Look feature on the biochemistry of obesity describing the role of peroxisome proliferator-activated receptors in regulation of gene expression in fatty acid oxidation and triglyceride metabolism.

Chapter 24 A new Human Biochemistry feature on the role of NPC1 and NPC2 proteins in cholesterol transport in lysosomes and Niemann-Pick type C disease.

Chapter 25 A new discussion of glutamate and its metabolic significance as the most abundant amino acid in human body fluids and tissues. Further, tryptophan catabolism by the kynurenine pathway is presented, because this pathway has been implicated in human neurodegenerative disorders such as Alzheimer's disease.

Chapter 26 Purinosomes as multi-enzyme assemblages of the purine biosynthetic enzymes and the newly solved structure of human ribonucleotide reductase with its revelations regarding regulation by nucleotides are presented.

Chapter 27 AMP-kinase (AMPK) as the cell's energy charge sensor and the newly appreciated protection of AMPK by ADP are discussed. mTORC1 as the integrator of information about nutrient status and as the regulator of cellular synthesis is introduced. The relationships between nutrient intake, AMPK, SIRT1 and protein acetylation and the consequences of these relationships have for caloric intake control and the development of metabolic syndrome. These interactions illuminate the underlying causes of the current obesity epidemic.

Chapter 28 A new section to integrate DNA replication, recombination, and repair as interdependent aspects of DNA metabolism introduces this chapter. Another new feature is an illustration of how homologous recombination helps to prevent cancer.

Chapter 29 An update of eukaryotic translation initiation events in eukaryotes and the emerging science of miRNAs as key regulators of post-transcriptional gene expression are presented, along with new structural and functional information about Mediator and its role as a bridge between enhancers of transcription and RNA polymerase II. The competing concepts of the histone code and histone crosstalk are discussed.

Chapter 30 New features of the G-protein family members, Ef-Tu and EF-G, and their interactions with the ribosome, new structures for the ribosome RF-2 termination complex, and the more richly detailed appreciation of the events in eukaryotic translation initiation highlight this chapter.

Chapter 31 A new Human Biochemistry highlight on autophagy, the process by which cells recycle their materials.

Chapter 32 A new Human Biochemistry feature on neurexins and neuroligins, which function as scaffolding proteins in the formation of synapses and the regulation of synaptic transmission, learning, and memory.

Alternate Versions

Cengage YouBook with OWL

The Cengage YouBook is an interactive and customizable eBook version available with OWL. The instructor text edit feature allows for modification of the narrative by adding notes, re-ordering entire sections and chapters, and hiding content. Additional assets include animated figures, the ActiveModels library, and student highlighting and note tools. See the OWL description below for more details.

eBooks

Cengage Learning textbooks are sold in many eBook formats. Offerings vary over time so be sure to check your favorite eBook retailer if you are interested in a digital version of this text.

Cengage Learning Custom Solutions

Cengage Learning Custom Solutions allows you to develop a personalized version of this textbook to meet your course needs. Match your learning materials to your syllabus and create the perfect learning solution—your customized text will contain the same thought-provoking, scientifically sound content, superior authorship, and stunning art that you've come to expect from Cengage Learning, Brooks/Cole texts, yet in a more flexible format. Visit www.cengage.com/custom to start building your book today.

Supporting Materials

OWL for Biochemistry

OWL with Cengage YouBook 24-Month Instant Access ISBN: 978-1-133-35235-8
OWL with Cengage YouBook 6-Month Instant Access ISBN: 978-1-133-35238-9
By David Gross (University of Massachusetts, Amherst), Susan Young (Hartwick College), and Reginald H. Garrett and Charles M. Grisham (University of Virginia).

OWL Online Web Learning offers more assignable, gradable content and more reliability and flexibility than any other system. OWL's powerful course management tools allow instructors to control due dates, number of attempts, and whether students see answers or receive feedback on how to solve problems. OWL includes the **Cengage** YouBook, an interactive and customizable eBook. Instructors can publish web links, modify the textbook narrative as needed with the text edit tool, quickly re-order entire sections and chapters, and hide any content they don't teach to create an eBook that perfectly matches their syllabus. The Cengage YouBook includes animated figures, video clips, highlighting, notes, the ActiveModels library, and more.

Developed by chemistry instructors for teaching chemistry, OWL is the only system specifically designed to support **mastery learning**, where students work as long as they need to master each chemical concept and skill. OWL has already helped hundreds of thousands of students master chemistry through a wide range of assignment types including tutorials, interactive simulations, and algorithmically generated homework questions that provide instant, answer-specific feedback.

OWL is continually enhanced with online learning tools to address the various learning styles of today's students such as the Jmol molecular visualization program for rotating molecules and measuring bond distances and angles. In addition, when you become an OWL user, you can expect service that goes far beyond the ordinary. To learn more or to see a demo, please contact your Cengage Learning representative or visit us at www.cengage.com/owl.

For the Instructor

PowerLecture Instructor's CD/DVD Package with ExamView®
ISBN: 978-1-133-10850-4

This digital library and presentation tool includes:

- **PowerPoint Lectures**, revised for this text by Reginald H. Garrett and Charles M. Grisham of the University of Virginia. Instructors can import their own lecture slides or other materials to customize the presentation.

- **Image libraries** that contain digital files for figures, photographs, and numbered tables from the text, as well as multimedia animations in a variety of digital formats. Use these files to print transparencies, create your own PowerPoint slides, and supplement your lectures.

- **ExamView Testing Software** enables you to create, deliver, and customize tests using the more than 500 test bank questions written specifically for this text by Scott Lefler of Arizona State University. Digital files of the test banks in both Microsoft Word and PDF format are included for your convenience.

- **Sample chapters of** *Student Solutions Manual, Study Guide and Problems Book*—three chapters are provided. The full book is available for sampling through your Cengage Learning sales representative (see the For the Student section below for a description).

Instructor Companion Site

Supporting materials are available to qualified adopters. Please consult your local Cengage Learning sales representative for details. Go to login.cengage.com, find this textbook, and choose Instructor Companion Site to see samples of these materials, request a desk copy, locate your sales representative, and download the WebCT or Blackboard versions of the Test Bank.

For the Student

Visit CengageBrain.com

To access these and additional course materials or purchase Cengage products, please visit www.cengagebrain.com. At the CengageBrain.com home page, search for this textbook's ISBN (from the back cover of your book). This will take you to the product page where these resources can be found. (Instructors can log in at login.cengage.com.)

Student Solutions Manual, Study Guide and Problems Book by David K. Jemiolo (Vassar College) and Steven M. Theg (University of California, Davis)
ISBN: 978-1-133-10851-1

This comprehensive combination resource contains chapter summaries, important definitions, illustrations of major metabolic pathways, self-tests, detailed solutions to all end-of-chapter problems except those for the ActiveModels library, and additional problems with answers.

Student Companion Website

This site includes a glossary, flashcards, and samples of the *Student Solutions Manual, Study Guide, and Problems Book,* which are all accessible from www.cengagebrain.com.

Acknowledgments

We are indebted to the many experts in biochemistry and molecular biology who carefully reviewed this book at several stages for their outstanding and invaluable advice on how to construct an effective textbook.

Gloria Borgstahl, University of Nebraska–Eppley Institute

Don Bourque, University of Arizona

Richard Calendar, University of California, Berkeley

Robert Dutnall, University of San Diego

Brandt Eichman, Vanderbilt University

Jeff Hansen, Colorado State University

Stuart Haring, North Dakota State University

Christine Martey Ochola, Villanova University

Leisha Mullins, Texas A & M University

Anthony Serianni, University of Notre Dame

Jon Stoltzfus, Michigan State University

We also wish to gratefully acknowledge many other people who assisted and encouraged us in this endeavor. This book remains a legacy of publisher John Vondeling, who originally recruited us to its authorship. We sense his presence still nurturing our book and we are grateful for it. Mary Finch, our new Publisher, has brought enthusiasm and an unwavering emphasis on student learning as the fundamental purpose of our collective enterprise. Sandi Kiselica, Senior Developmental Editor, shepherded the first four editions of this text through conception to production. Any success this fifth edition has will become part of her legacy. Peter McGahey, the Senior Developmental Editor for this edition, has kept us focused on the matters at hand, the urgencies of the schedule, and limits of scale in a textbook's dimensions. He is truly a colleague in these endeavors. We also applaud the unsung but absolutely indispensable contributions by those whose efforts transformed a rough manuscript into this final product: Teresa Trego, Project Manager; Laura Slown, Production Editor at Graphic World Publishing Services; Lisa Weber and Stephanie VanCamp, Media Editors, and Elizabeth Woods, Assistant Editor. If this book has visual appeal and editorial grace, it is due to them. The beautiful illustrations that not only decorate this text but also explain its contents are a testament to the creative work of Eric Hirsch of Graphic World Illustration Studio and to the legacy of John Woolsey and Patrick Lane at J/B Woolsey Associates. Sarah Bonner at the Bill Smith Group assisted in the selection of the wonderful photographs. We are thankful to our many colleagues who provided original art and graphic images for this work, particularly Professor Jane Richardson of Duke University. We are eager to acknowledge the scientific and artistic contributions of Michal Sabat, Senior Scientist in the Department of Chemistry at the University of Virginia. Michal was the creator of most of the PyMOL-based molecular graphics in this book. Much of the visual appeal that you will find in these pages gives testimony to his fine craftsmanship and his unflagging dedication to our purpose. Elizabeth Magnotti, recent B.S. (chemistry) graduate of the University of Virginia, coordinated the student development of the ActiveModels library, a multi-faceted task requiring scientific knowledge and a sense of the important. We owe a special thank you to Rosemary Jurbala Grisham, devoted spouse of Charles and wonderfully tolerant friend of Reg. Also to be acknowledged with love and pride are Georgia Cobb Garrett, spouse of Reg, to whom this book is also dedicated, and our children, Jeffrey, Randal, and Robert Garrett, and David, Emily, and Andrew Grisham. Also to be appreciated is Jasper, the Hungarian Puli whose unseeing eyes view life with an energetic curiosity we all should emulate. Memories of Clancy, a Golden Retriever of epic patience and perspicuity; Jazmine, the alpha dog of the Puli troika; and Jatszi, the recently departed attendant to Charlie's deepest thoughts and daily activities, are a constant presence to our days as authors, our joys in other pursuits, and our reflections on life. We hope this fifth edition of our textbook has captured the growing sense of wonder and imagination that researchers, teachers, and students share as they explore the ever-changing world of biochemistry.

"Imagination is more important than knowledge. For while knowledge defines all we currently know and understand, imagination points to all we might yet discover and create."

—*Albert Einstein*

Reginald H. Garrett   Charles M. Grisham

Charlottesville, VA   *Ivy, VA*

November, 2011

1

The Facts of Life: Chemistry Is the Logic of Biological Phenomena

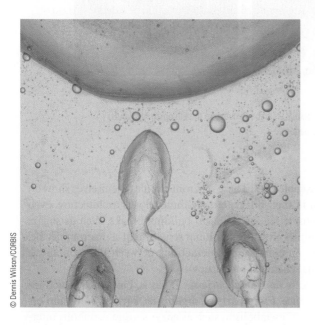

◀ Sperm approaching an egg.

"…everything that living things do can be understood in terms of the jigglings and wigglings of atoms."

Richard P. Feynman
Lectures on Physics, Addison-Wesley, 1963

KEY QUESTIONS

1.1   What Are the Distinctive Properties of Living Systems?

1.2   What Kinds of Molecules Are Biomolecules?

1.3   What Is the Structural Organization of Complex Biomolecules?

1.4   How Do the Properties of Biomolecules Reflect Their Fitness to the Living Condition?

1.5   What Is the Organization and Structure of Cells?

1.6   What Are Viruses?

ESSENTIAL QUESTION

Molecules are lifeless. Yet, the properties of living things derive from the properties of molecules.

Despite the spectacular diversity of life, the elaborate structure of biological molecules, and the complexity of vital mechanisms, are life functions ultimately interpretable in chemical terms?

Molecules are lifeless. Yet, in appropriate complexity and number, molecules compose living things. These living systems are distinct from the inanimate world because they have certain extraordinary properties. They can grow, move, perform the incredible chemistry of metabolism, respond to stimuli from the environment, and most significantly, replicate themselves with exceptional fidelity. The complex structure and behavior of living organisms veil the basic truth that their molecular constitution can be described and understood. The chemistry of the living cell resembles the chemistry of organic reactions. Indeed, cellular constituents or **biomolecules** must conform to the chemical and physical principles that govern all matter. Despite the spectacular diversity of life, the intricacy of biological structures, and the complexity of vital mechanisms, life functions are ultimately interpretable in chemical terms. *Chemistry is the logic of biological phenomena.*

1.1   What Are the Distinctive Properties of Living Systems?

First, the most obvious quality of **living organisms** is that they are *complicated and highly organized* (Figure 1.1). For example, organisms large enough to be seen with the naked eye are composed of many **cells,** typically of many types. In turn, these cells possess subcellular structures, called **organelles,** which are complex assemblies of very large

ʊWL   Online homework and a Student Self Assessment for this chapter may be assigned in OWL

© Dennis Wilson/CORBIS

(a)   (b)

FIGURE 1.1 (a) Gelada *(Theropithecus gelada),* a baboon native to the Ethiopian highlands. (b) Tropical orchid *(Masdevallia norops),* Ecuador.

polymeric molecules, called **macromolecules.** These macromolecules themselves show an exquisite degree of organization in their intricate three-dimensional architecture, even though they are composed of simple sets of chemical building blocks, such as sugars and amino acids. Indeed, the complex three-dimensional structure of a macromolecule, known as its **conformation,** is a consequence of interactions between the monomeric units, according to their individual chemical properties.

Second, *biological structures serve functional purposes.* That is, biological structures play a role in the organism's existence. From parts of organisms, such as limbs and organs, down to the chemical agents of metabolism, such as enzymes and metabolic intermediates, a biological purpose can be given for each component. Indeed, it is this functional characteristic of biological structures that separates the science of biology from studies of the inanimate world such as chemistry, physics, and geology. In biology, it is always meaningful to seek the purpose of observed structures, organizations, or patterns, that is, to ask what functional role they serve within the organism.

Third, *living systems are actively engaged in energy transformations.* Maintenance of the highly organized structure and activity of living systems depends on their ability to extract energy from the environment. The ultimate source of energy is the sun. Solar energy flows from photosynthetic organisms (organisms able to capture light energy by the process of photosynthesis) through food chains to herbivores and ultimately to carnivorous predators at the apex of the food pyramid (Figure 1.2). The biosphere is thus a system through which energy flows. Organisms capture some of this energy, be it from photosynthesis or the metabolism of food, by forming special energized biomolecules, of which **ATP** and **NADPH** are the two most prominent examples (Figure 1.3). (Commonly used abbreviations such as ATP and NADPH are defined on the inside back cover of this book.) ATP and NADPH are energized biomolecules because they represent chemically useful forms of stored energy. We explore the chemical basis of this stored energy in subsequent chapters. For now, suffice it to say that when these molecules react with other molecules in the cell, the energy released can be used to drive energetically unfavorable processes. That is, ATP, NADPH, and related compounds are the power sources that drive the energy-requiring activities of the cell, including biosynthesis, movement, osmotic work against concentration gradients, and in special instances, light emission (bioluminescence). Only upon death does an organism reach equilibrium with its inanimate environment. *The living state is characterized by the flow of energy through the organism.* At the expense of this energy flow, the organism can maintain its intricate order and activity far removed from equilibrium with its surroundings, yet exist in a state of apparent constancy over time. This state of apparent constancy, or so-called **steady state,** is actually a very dynamic condition: Energy and

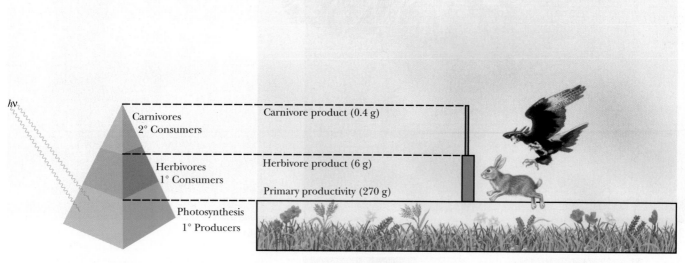

Productivity per square meter of a Tennessee field

FIGURE 1.2 The food pyramid. Photosynthetic organisms at the base capture light energy. Herbivores and carnivores derive their energy ultimately from these primary producers.

material are consumed by the organism and used to maintain its stability and order. In contrast, inanimate matter, as exemplified by the universe in totality, is moving to a condition of increasing disorder or, in thermodynamic terms, maximum entropy.

Fourth, *living systems have a remarkable capacity for self-replication.* Generation after generation, organisms reproduce virtually identical copies of themselves. This self-replication can proceed by a variety of mechanisms, ranging from simple division in bacteria to sexual reproduction in plants and animals; but in every case, it is characterized by an astounding degree of fidelity (Figure 1.4). Indeed, if the accuracy of self-replication were significantly greater, the evolution of organisms would be hampered. This is so because evolution depends upon natural selection operating on individual organisms that vary slightly in their fitness for the environment. The fidelity of self-replication resides ultimately in the chemical nature of the genetic material. This substance consists of polymeric chains of deoxyribonucleic acid, or **DNA,** which are structurally

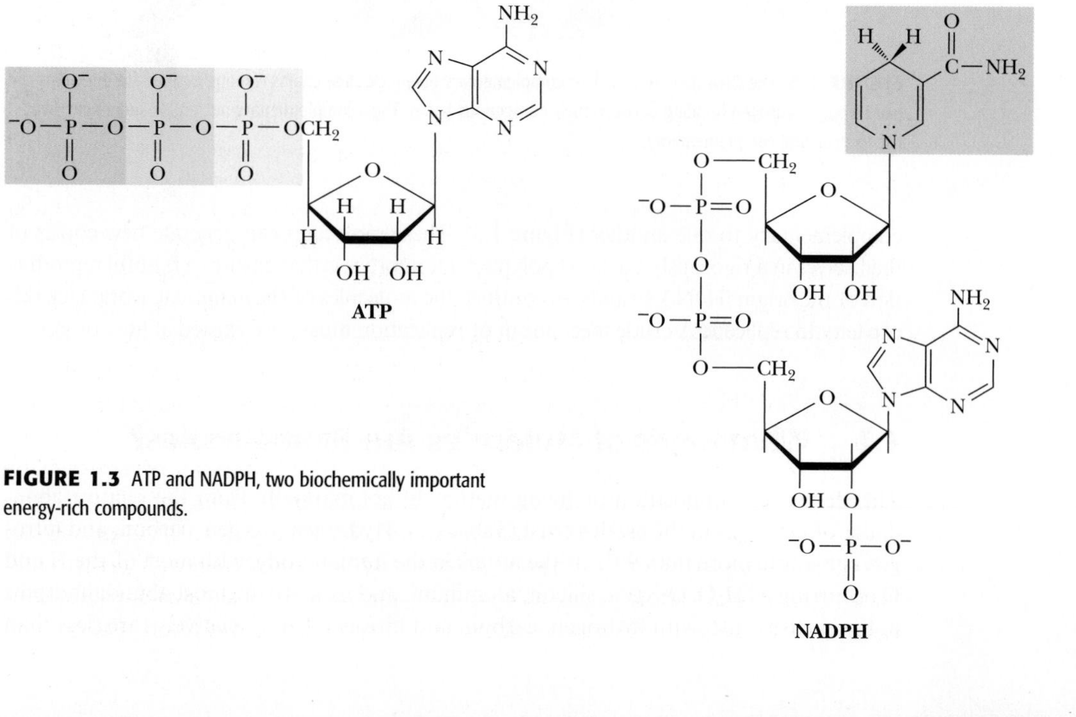

Entropy: is a thermodynamic term used to designate that amount of energy in a system that is unavailable to do work.

FIGURE 1.3 ATP and NADPH, two biochemically important energy-rich compounds.

ATP

NADPH

(a)   (b)

(c)

FIGURE 1.4 Organisms resemble their parents. **(a)** The Garrett guys. Left to right: (seated) Reg Garrett, flanked by grandsons Reggie and Ricky; (standing) grandson Jackson, and sons Jeffrey, Randal, and Robert. **(b)** Orangutan with infant. **(c)** The Grisham family. Left to right: David, Emily, Charles, Rosemary, and Andrew.

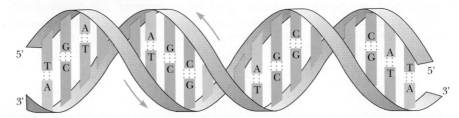

FIGURE 1.5 The DNA double helix. Two complementary polynucleotide chains running in opposite directions can pair through hydrogen bonding between their nitrogenous bases. Their complementary nucleotide sequences give rise to structural complementarity.

complementary to one another (Figure 1.5). These molecules can generate new copies of themselves in a rigorously executed polymerization process that ensures a faithful reproduction of the original DNA strands. In contrast, the molecules of the inanimate world lack this capacity to replicate. A crude mechanism of replication must have existed at life's origin.

1.2 What Kinds of Molecules Are Biomolecules?

The elemental composition of living matter differs markedly from the relative abundance of elements in the earth's crust (Table 1.1). Hydrogen, oxygen, carbon, and nitrogen constitute more than 99% of the atoms in the human body, with most of the H and O occurring as H_2O. Oxygen, silicon, aluminum, and iron are the most abundant atoms in the earth's crust, with hydrogen, carbon, and nitrogen being relatively rare (less than

TABLE 1.1		Composition of the Earth's Crust, Seawater, and the Human Body*			
Earth's Crust		Seawater		Human Body†	
Element	%	Compound	m*M*	Element	%
O	47	Cl⁻	548	H	63
Si	28	Na⁺	470	O	25.5
Al	7.9	Mg²⁺	54	C	9.5
Fe	4.5	SO₄²⁻	28	N	1.4
Ca	3.5	Ca²⁺	10	Ca	0.31
Na	2.5	K⁺	10	P	0.22
K	2.5	HCO₃⁻	2.3	Cl	0.08
Mg	2.2	NO₃⁻	0.01	K	0.06
Ti	0.46	HPO₄²⁻	<0.001	S	0.05
H	0.22			Na	0.03
C	0.19			Mg	0.01

*Figures for the earth's crust and the human body are presented as percentages of the total number of atoms; seawater data are in millimoles per liter. Figures for the earth's crust do not include water, whereas figures for the human body do.
†Trace elements found in the human body serving essential biological functions include Mn, Fe, Co, Cu, Zn, Mo, I, Ni, and Se.

0.2% each). Nitrogen as dinitrogen (N_2) is the predominant gas in the atmosphere, and carbon dioxide (CO_2) is present at a level of 0.04%, a small but critical amount. Oxygen is also abundant in the atmosphere and in the oceans. What property unites H, O, C, and N and renders these atoms so suitable to the chemistry of life? It is their ability to form covalent bonds by electron-pair sharing. Furthermore, H, C, N, and O are among the lightest elements of the periodic table capable of forming such bonds (Figure 1.6). Because the strength of covalent bonds is inversely proportional to the atomic weights of the atoms involved, H, C, N, and O form the strongest covalent bonds. Two other covalent bond-forming elements, phosphorus (as phosphate [$-OPO_3^{2-}$] derivatives) and sulfur, also play important roles in biomolecules.

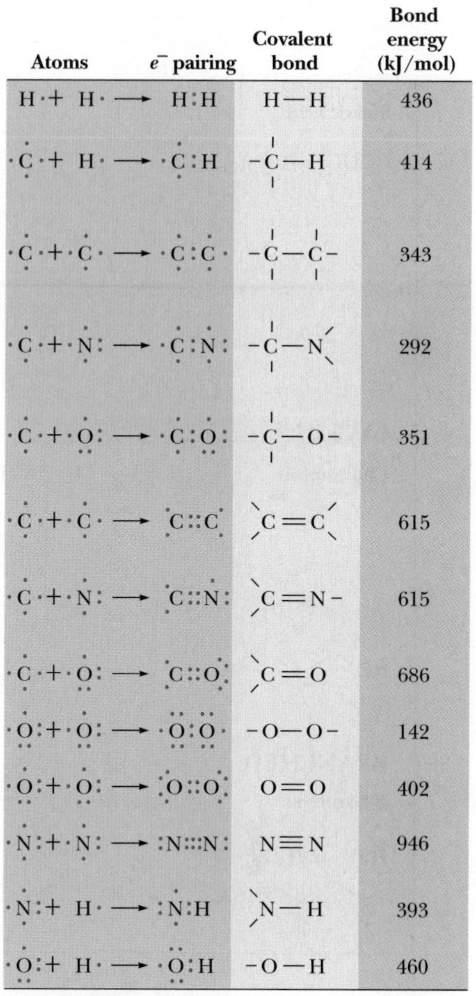

FIGURE 1.6 Covalent bond formation by e^- pair sharing.

Biomolecules Are Carbon Compounds

All biomolecules contain carbon. The prevalence of C is due to its unparalleled versatility in forming stable covalent bonds through electron-pair sharing. Carbon can form as many as four such bonds by sharing each of the four electrons in its outer shell with electrons contributed by other atoms. Atoms commonly found in covalent linkage to C are C itself, H, O, and N. Hydrogen can form one such bond by contributing its single electron to the formation of an electron pair. Oxygen, with two unpaired electrons in its outer shell, can participate in two covalent bonds, and nitrogen, which has three unshared electrons, can form three such covalent bonds. Furthermore, C, N, and O can share two electron pairs to form double bonds with one another within biomolecules, a property that enhances their chemical versatility. Carbon and nitrogen can even share three electron pairs to form triple bonds.

Two properties of carbon covalent bonds merit particular attention. One is the ability of carbon to form covalent bonds with itself. The other is the tetrahedral nature of the four covalent bonds when carbon atoms form only single bonds. Together these properties hold the potential for an incredible variety of linear, branched, and cyclic compounds of C. This diversity is multiplied further by the possibilities for including N, O, and H atoms in these compounds (Figure 1.7). We can therefore envision the ability of C to generate complex structures in three dimensions. These structures, by virtue of appropriately included N, O, and H atoms, can display unique chemistries suitable to the living state. Thus, we may ask, is there any pattern or underlying organization that brings order to this astounding potentiality?

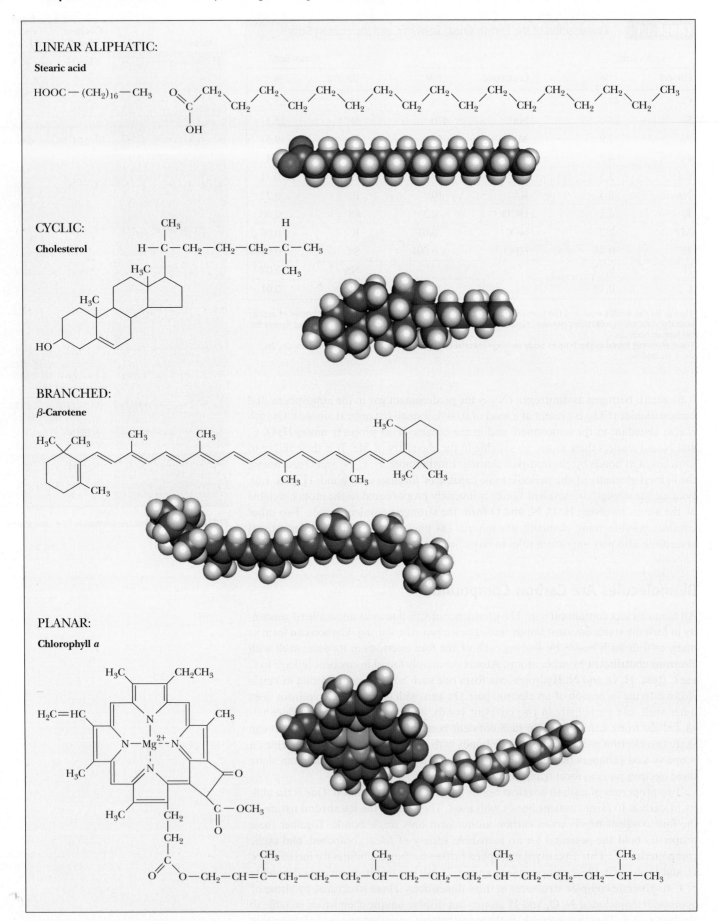

LINEAR ALIPHATIC:

Stearic acid

$HOOC-(CH_2)_{16}-CH_3$

CYCLIC:

Cholesterol

BRANCHED:

β-Carotene

PLANAR:

Chlorophyll *a*

FIGURE 1.7 Examples of the versatility of C—C bonds in building complex structures: linear, cyclic, branched, and planar.

1.3 What Is the Structural Organization of Complex Biomolecules?

Examination of the chemical composition of cells reveals a dazzling variety of organic compounds covering a wide range of molecular dimensions (Table 1.2). As this complexity is sorted out and biomolecules are classified according to the similarities of their sizes and chemical properties, an organizational pattern emerges. The biomolecules are built according to a structural hierarchy: Simple molecules are the units for building complex structures.

The molecular constituents of living matter do not reflect randomly the infinite possibilities for combining C, H, O, and N atoms. Instead, only a limited set of the many possibilities is found, and these collections share certain properties essential to the establishment and maintenance of the living state. The most prominent aspect of biomolecular organization is that macromolecular structures are constructed from simple molecules according to a hierarchy of increasing structural complexity. What properties do these biomolecules possess that make them so appropriate for the condition of life?

Metabolites Are Used to Form the Building Blocks of Macromolecules

The major precursors for the formation of biomolecules are water, carbon dioxide, and three inorganic nitrogen compounds—ammonium (NH_4^+), nitrate (NO_3^-), and dinitrogen (N_2). Metabolic processes assimilate and transform these inorganic precursors through ever more complex levels of biomolecular order (Figure 1.8). In the first step,

TABLE 1.2	Biomolecular Dimensions		

The dimensions of mass* and length for biomolecules are given typically in daltons and nanometers,[†] respectively. One dalton (D) is approximately equal to the mass of one hydrogen atom, 1.66×10^{-24} g. One nanometer (nm) is 10^{-9} m, or 10 Å (angstroms).

Biomolecule	Length (long dimension, nm)	Mass	
		Daltons	Picograms
Water	0.3	18	
Alanine	0.5	89	
Glucose	0.7	180	
Phospholipid	3.5	750	
Ribonuclease (a small protein)	4	12,600	
Immunoglobulin G (IgG)	14	150,000	
Myosin (a large muscle protein)	160	470,000	
Ribosome (bacteria)	18	2,520,000	
Bacteriophage ϕX174 (a very small bacterial virus)	25	4,700,000	
Pyruvate dehydrogenase complex (a multienzyme complex)	60	7,000,000	
Tobacco mosaic virus (a plant virus)	300	40,000,000	6.68×10^{-5}
Mitochondrion (liver)	1,500		1.5
Escherichia coli cell	2,000		2
Chloroplast (spinach leaf)	8,000		60
Liver cell	20,000		8,000

*Molecular mass is expressed in units of daltons (D) or kilodaltons (kD) in this book; alternatively, the dimensionless term *molecular weight*, symbolized by M_r, and defined as the ratio of the mass of a molecule to 1 dalton of mass, is used.
[†]Prefixes used for powers of 10 are

10^6	mega	M	10^{-3}	milli	m
10^3	kilo	k	10^{-6}	micro	μ
10^{-1}	deci	d	10^{-9}	nano	n
10^{-2}	centi	c	10^{-12}	pico	p
			10^{-15}	femto	f

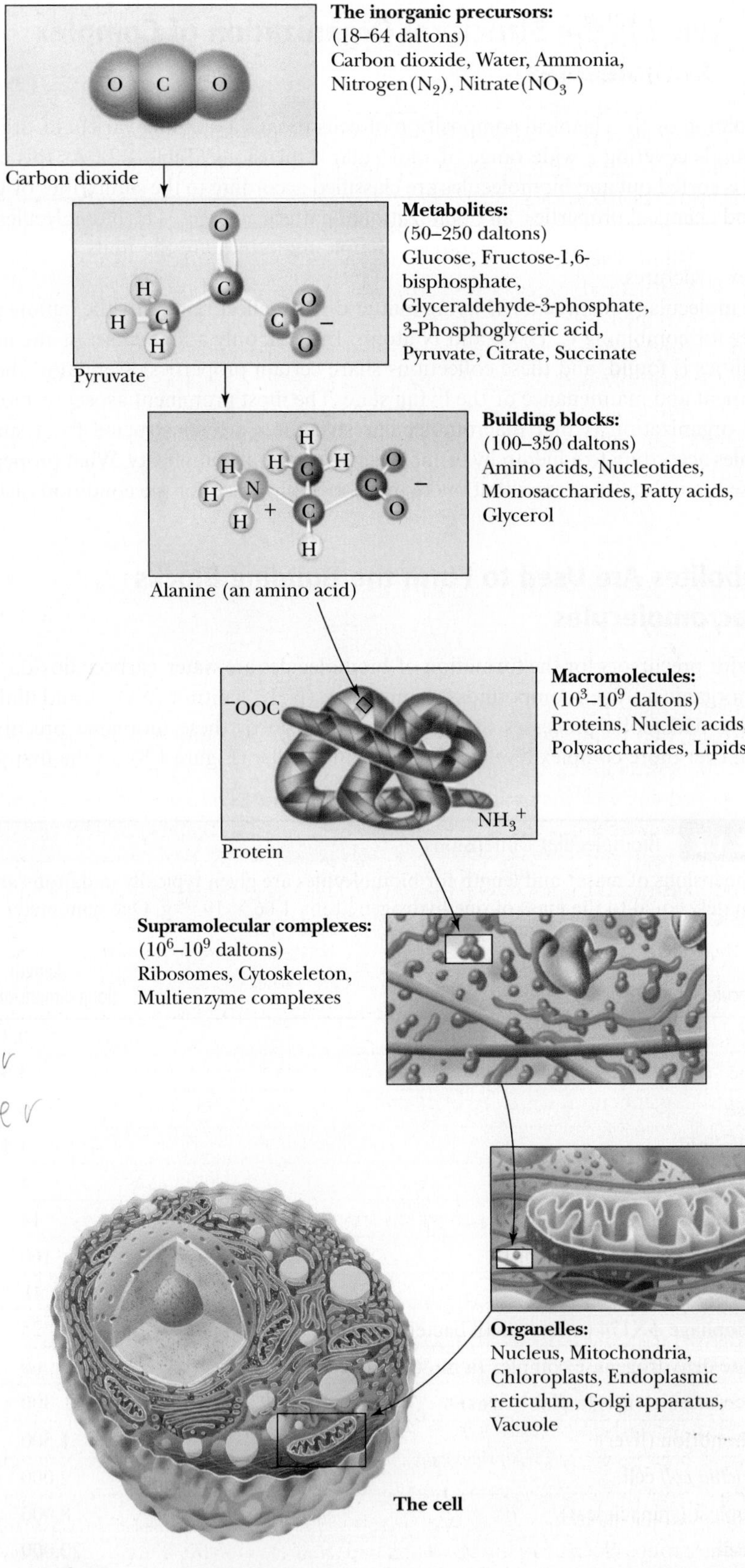

The inorganic precursors:
(18–64 daltons)
Carbon dioxide, Water, Ammonia,
Nitrogen (N_2), Nitrate (NO_3^-)

Carbon dioxide

Metabolites:
(50–250 daltons)
Glucose, Fructose-1,6-
bisphosphate,
Glyceraldehyde-3-phosphate,
3-Phosphoglyceric acid,
Pyruvate, Citrate, Succinate

Pyruvate

Building blocks:
(100–350 daltons)
Amino acids, Nucleotides,
Monosaccharides, Fatty acids,
Glycerol

Alanine (an amino acid)

Macromolecules:
(10^3–10^9 daltons)
Proteins, Nucleic acids,
Polysaccharides, Lipids

^-OOC

NH_3^+

Protein

FIGURE 1.8 Molecular organization in the cell is a hierarchy.

Supramolecular complexes:
(10^6–10^9 daltons)
Ribosomes, Cytoskeleton,
Multienzyme complexes

higher
order

Organelles:
Nucleus, Mitochondria,
Chloroplasts, Endoplasmic
reticulum, Golgi apparatus,
Vacuole

The cell

precursors are converted to **metabolites,** simple organic compounds that are intermediates in cellular energy transformation and in the biosynthesis of various sets of **building blocks:** amino acids, sugars, nucleotides, fatty acids, and glycerol. Through covalent linkage of these building blocks, the **macromolecules** are constructed: proteins, polysaccharides, polynucleotides (DNA and RNA), and lipids. (Strictly speaking, lipids contain relatively few building blocks and are therefore not really polymeric like other macromolecules; however, lipids are important contributors to higher levels of complexity.) Interactions among macromolecules lead to the next level of structural organization, **supramolecular complexes.** Here, various members of one or more of the classes of macromolecules come together to form specific assemblies that serve important subcellular functions. Examples of these supramolecular assemblies are multifunctional enzyme complexes, ribosomes, chromosomes, and cytoskeletal elements. For example, a eukaryotic ribosome contains four different RNA molecules and at least 70 unique proteins. These supramolecular assemblies are an interesting contrast to their components because their structural integrity is maintained by noncovalent forces, not by covalent bonds. These noncovalent forces include hydrogen bonds, ionic attractions, van der Waals forces, and hydrophobic interactions between macromolecules. Such forces maintain these supramolecular assemblies in a highly ordered functional state. Although noncovalent forces are weak (less than 40 kJ/mol), they are numerous in these assemblies and thus can collectively maintain the essential architecture of the supramolecular complex under conditions of temperature, pH, and ionic strength that are consistent with cell life.

Organelles Represent a Higher Order in Biomolecular Organization

The next higher rung in the hierarchical ladder is occupied by the organelles, entities of considerable dimensions compared with the cell itself. Organelles are found only in **eukaryotic cells,** that is, the cells of "higher" organisms (eukaryotic cells are described in Section 1.5). Several kinds, such as mitochondria and chloroplasts, evolved from bacteria that gained entry to the cytoplasm of early eukaryotic cells. Organelles share two attributes: They are cellular inclusions, usually membrane bounded, and they are dedicated to important cellular tasks. Organelles include the nucleus, mitochondria, chloroplasts, endoplasmic reticulum, Golgi apparatus, and vacuoles, as well as other relatively small cellular inclusions, such as peroxisomes, lysosomes, and chromoplasts. The **nucleus** is the repository of genetic information as contained within the linear sequences of nucleotides in the DNA of chromosomes. **Mitochondria** are the "power plants" of cells by virtue of their ability to carry out the energy-releasing aerobic metabolism of carbohydrates and fatty acids, capturing the energy in metabolically useful forms such as ATP. **Chloroplasts** endow cells with the ability to carry out photosynthesis. They are the biological agents for harvesting light energy and transforming it into metabolically useful chemical forms.

Membranes Are Supramolecular Assemblies That Define the Boundaries of Cells

Membranes define the boundaries of cells and organelles. As such, they are not easily classified as supramolecular assemblies or organelles, although they share the properties of both. Membranes resemble supramolecular complexes in their construction because they are complexes of proteins and lipids maintained by noncovalent forces. **Hydrophobic interactions** are particularly important in maintaining membrane structure. Hydrophobic interactions arise because water molecules prefer to interact with each other rather than with nonpolar substances. The presence of nonpolar molecules lessens the range of opportunities for water–water interaction by forcing the water molecules into ordered arrays around the nonpolar groups. Such ordering can be minimized if the individual nonpolar molecules redistribute from a dispersed state in the water into an aggregated organic phase surrounded by water. The spontaneous assembly of membranes in the aqueous environment where life arose and exists is the natural result of the hydrophobic ("water-fearing") character of their lipids and proteins. Hydrophobic interactions are the creative means of membrane formation and the driving force that presumably established the boundary of the first cell. The membranes of organelles, such as nuclei, mitochondria, and chloroplasts, differ from one another, with each having a characteristic protein and lipid composition tailored to the organelle's function. Furthermore, the creation of discrete volumes or **compartments** within cells is not only an inevitable consequence of the presence of membranes but usually an essential condition for proper organellar function.

The Unit of Life Is the Cell

The cell is characterized as the unit of life, the smallest entity capable of displaying the attributes associated uniquely with the living state: growth, metabolism, stimulus response, and replication. In the previous discussions, we explicitly narrowed the infinity of chemical complexity potentially available to organic life and we previewed an organizational arrangement, moving from simple to complex, that provides interesting insights into the functional and structural plan of the cell. Nevertheless, we find no obvious explanation within these features for the living characteristics of cells. Can we find other themes represented within biomolecules that are explicitly chemical yet anticipate or illuminate the living condition?

1.4 : How Do the Properties of Biomolecules Reflect Their Fitness to the Living Condition?

If we consider what attributes of biomolecules render them so fit as components of growing, replicating systems, several biologically relevant themes of structure and organization emerge. Furthermore, as we study biochemistry, we will see that these themes serve as principles of biochemistry. Prominent among them is the *necessity for information and energy in the maintenance of the living state*. Some biomolecules must have the capacity to contain the information, or "recipe," of life. Other biomolecules must have the capacity to translate this information so that the organized structures essential to life are synthesized. Interactions between these structures *are* the processes of life. An orderly mechanism for abstracting energy from the environment must also exist in order to obtain the energy needed to drive these processes. What properties of biomolecules endow them with the potential for such remarkable qualities?

Biological Macromolecules and Their Building Blocks Have a "Sense" or Directionality

The macromolecules of cells are built of units—amino acids in proteins, nucleotides in nucleic acids, and carbohydrates in polysaccharides—that have **structural polarity.** That is, these molecules are not symmetrical, and so they can be thought of as having a "head" and a "tail." Polymerization of these units to form macromolecules occurs by head-to-tail linear connections. Because of this, the polymer also has a head and a tail, and hence, the macromolecule has a "sense" or direction to its structure (Figure 1.9).

Biological Macromolecules Are Informational

Because biological macromolecules have a sense to their structure, the sequential order of their component building blocks, when read along the length of the molecule, has the capacity to specify information in the same manner that the letters of the alphabet can form words when arranged in a linear sequence (Figure 1.10). Not all biological macromolecules are rich in information. Polysaccharides are often composed of the same sugar unit repeated over and over, as in cellulose or starch, which are homopolymers of many glucose units. On the other hand, proteins and polynucleotides are typically composed of building blocks arranged in no obvious repetitive way; that is, their sequences are unique, akin to the letters and punctuation that form this descriptive sentence. In these unique sequences lies meaning. Discerning the meaning, however, requires some mechanism for recognition.

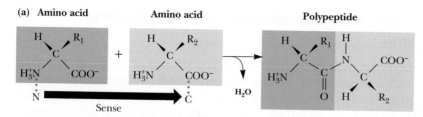

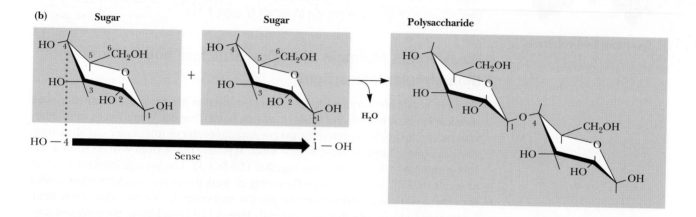

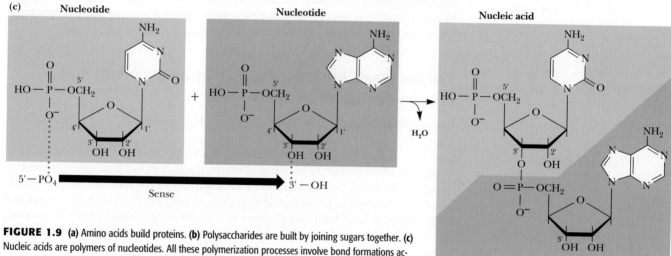

FIGURE 1.9 (a) Amino acids build proteins. **(b)** Polysaccharides are built by joining sugars together. **(c)** Nucleic acids are polymers of nucleotides. All these polymerization processes involve bond formations accompanied by the elimination of water (dehydration synthesis reactions).

A strand of DNA

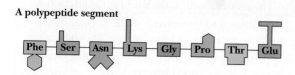

A polypeptide segment

A polysaccharide chain

FIGURE 1.10 The sequence of monomeric units in a biological polymer has the potential to contain information if the diversity and order of the units are not overly simple or repetitive. Nucleic acids and proteins are information-rich molecules; polysaccharides are not.

FIGURE 1.11 Antigen-binding domain of immuno-globulin G (IgG).

Biomolecules Have Characteristic Three-Dimensional Architecture

The structure of any molecule is a unique and specific aspect of its identity. Molecular structure reaches its pinnacle in the intricate complexity of biological macromolecules, particularly the proteins. Although proteins are linear sequences of covalently linked amino acids, the course of the protein chain can turn, fold, and coil in the three dimensions of space to establish a specific, highly ordered architecture that is an identifying characteristic of the given protein molecule (Figure 1.11).

Weak Forces Maintain Biological Structure and Determine Biomolecular Interactions

Covalent bonds hold atoms together so that molecules are formed. In contrast, **weak chemical forces** or **noncovalent bonds** (hydrogen bonds, van der Waals forces, ionic interactions, and hydrophobic interactions) are intramolecular or intermolecular attractions between atoms. None of these forces, which typically range from 4 to 30 kJ/mol, are strong enough to bind free atoms together (Table 1.3). The average kinetic energy of molecules at 25°C is 2.5 kJ/mol, so the energy of weak forces is only several times greater than the dissociating tendency due to thermal motion of molecules. Thus, these weak forces create interactions that are constantly forming and breaking at physiological temperature, unless by cumulative number they impart stability to the structures generated by their collective action. These weak forces merit further discussion because their properties profoundly influence the nature of the biological structures they build.

Van der Waals Attractive Forces Play an Important Role in Biomolecular Interactions

Van der Waals forces are the result of induced electrical interactions between closely approaching atoms or molecules as their negatively charged electron clouds fluctuate instantaneously in time. These fluctuations allow attractions to occur between the positively charged nuclei and the electrons of nearby atoms. Van der Waals attractions operate only over a very limited interatomic distance (0.3 to 0.6 nm) and are an effective bonding interaction at physiological temperatures only when a number of atoms in a molecule can interact with several atoms in a neighboring molecule. For this to occur, the atoms on interacting molecules must pack together neatly. That is, their molecular surfaces must possess a degree of structural complementarity (Figure 1.12).

At best, van der Waals interactions are weak and individually contribute 0.4 to 4.0 kJ/mol of stabilization energy. However, the sum of many such interactions within a macromolecule or between macromolecules can be substantial. Calculations indicate that the attractive van der Waals energy between the enzyme lysozyme and a sugar substrate that it binds is about 60 kJ/mol.

TABLE 1.3	Weak Chemical Forces and Their Relative Strengths and Distances		
Force	Strength (kJ/mol)	Distance (nm)	Description
Van der Waals interactions	0.4–4.0	0.3–0.6	Strength depends on the relative size of the atoms or molecules and the distance between them. The size factor determines the area of contact between two molecules: The greater the area, the stronger the interaction.
Hydrogen bonds	12–30	0.3	Relative strength is proportional to the polarity of the H bond donor and H bond acceptor. More polar atoms form stronger H bonds.
Ionic interactions	20	0.25	Strength also depends on the relative polarity of the interacting charged species. Some ionic interactions are also H bonds: $-NH_3^+ \cdots {}^-OOC-$
Hydrophobic interactions	<40	—	Force is a complex phenomenon determined by the degree to which the structure of water is disordered as discrete hydrophobic molecules or molecular regions coalesce.

(a)

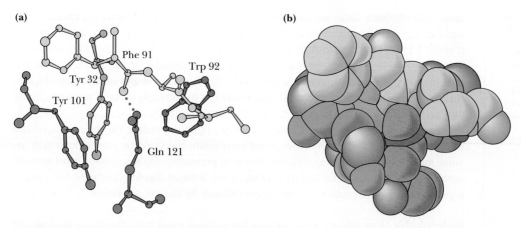

(b)

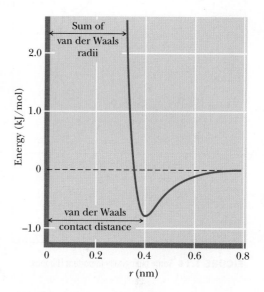

FIGURE 1.12 Van der Waals packing is enhanced in molecules that are structurally complementary. Gln[121], a surface protuberance on lysozyme, is recognized by the antigen-binding site of an antibody against lysozyme. Gln[121] (pink) fits nicely in a pocket formed by Tyr[32] (orange), Phe[91] (light green), Trp[92] (dark green), and Tyr[101] (blue) components of the antibody. (See also Figure 1.16.) **(a)** Ball-and-stick model. **(b)** Space-filling representation. (From Amit, A. G., et al., 1986. Three-dimensional structure of an antigen-antibody complex at 2.8 Å resolution. *Science* 233:747–753, figure 5.)

When two atoms approach each other so closely that their electron clouds interpenetrate, strong *repulsive* van der Waals forces occur, as shown in Figure 1.13. Between the repulsive and attractive domains lies a low point in the potential curve. This low point defines the distance known as the **van der Waals contact distance,** which is the interatomic distance that results if only van der Waals forces hold two atoms together. The limit of approach of two atoms is determined by the sum of their van der Waals radii (Table 1.4).

Hydrogen Bonds Are Important in Biomolecular Interactions

Hydrogen bonds form between a hydrogen atom covalently bonded to an electronegative atom (such as oxygen or nitrogen) and a second electronegative atom that serves as the hydrogen bond acceptor. Several important biological examples are given in Figure 1.14. Hydrogen bonds, at a strength of 12 to 30 kJ/mol, are stronger than van der Waals forces and have an additional property: H bonds are cylindrically symmetrical and tend to be highly directional, forming straight bonds between donor, hydrogen, and acceptor atoms. Hydrogen bonds are also more specific than van der Waals interactions because they require the presence of complementary hydrogen donor and acceptor groups.

FIGURE 1.13 The van der Waals interaction energy profile as a function of the distance, *r*, between the centers of two atoms.

TABLE 1.4	Radii of the Common Atoms of Biomolecules		
Atom	Van der Waals Radius (nm)	Covalent Radius (nm)	Atom Represented to Scale
H	0.1	0.037	
C	0.17	0.077	
N	0.15	0.070	
O	0.14	0.066	
P	0.19	0.096	
S	0.185	0.104	
Half-thickness of an aromatic ring	0.17	—	

H bonds Bonded atoms	Approximate bond length*
O—H---O	0.27 nm
O—H---O⁻	0.26 nm
O—H---N	0.29 nm
N—H---O	0.30 nm
⁺N—H---O	0.29 nm
N—H---N	0.31 nm

*Lengths given are distances from the atom covalently linked to the H to the atom H bonded to the hydrogen:

$$O—H---O$$
$$|\leftarrow 0.27\ nm \rightarrow|$$

Functional groups that are important H-bond donors and acceptors:

Donors   Acceptors

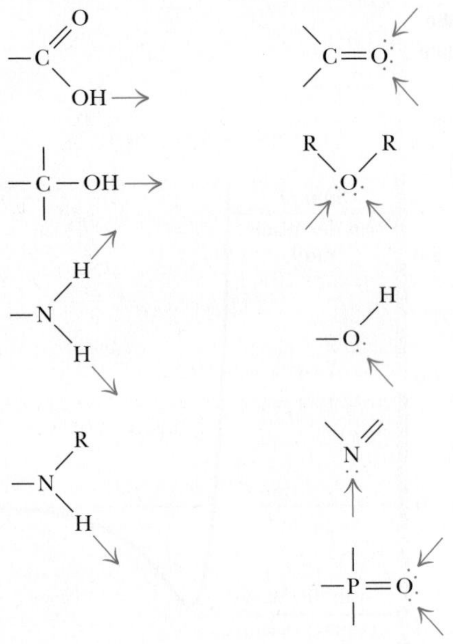

FIGURE 1.14 Some biologically important H bonds.

Ligand: a molecule (or atom) that binds specifically to another molecule (from Latin *ligare, to bind*).

Ionic Interactions **Ionic interactions** are the result of attractive forces between oppositely charged structures, such as negative carboxyl groups and positive amino groups (Figure 1.15). These electrostatic forces average about 20 kJ/mol in aqueous solutions. Typically, the electrical charge is radially distributed, so these interactions may lack the directionality of hydrogen bonds or the precise fit of van der Waals interactions. Nevertheless, because the opposite charges are restricted to sterically defined positions, ionic interactions can impart a high degree of structural specificity.

The strength of electrostatic interactions is highly dependent on the nature of the interacting species and the distance, r, between them. Electrostatic interactions may involve **ions** (species possessing discrete charges), **permanent dipoles** (having a permanent separation of positive and negative charge), or **induced dipoles** (having a temporary separation of positive and negative charge induced by the environment).

Hydrophobic Interactions Hydrophobic interactions result from the strong tendency of water to exclude nonpolar groups or molecules (see Chapter 2). Hydrophobic interactions arise not so much because of any intrinsic affinity of nonpolar substances for one another (although van der Waals forces do promote the weak bonding of nonpolar substances), but because water molecules prefer the stronger interactions that they share with one another, compared to their interaction with nonpolar molecules. Hydrogen-bonding interactions between polar water molecules can be more varied and numerous if nonpolar molecules come together to form a distinct organic phase. This phase separation raises the entropy of water because individual nonpolar molecules are no longer dispersed in the water, and thus, water molecules are no longer arranged in orderly arrays around them. It is these preferential interactions between water molecules that "exclude" hydrophobic substances from aqueous solution and drive the tendency of nonpolar molecules to cluster together. Thus, nonpolar regions of biological macromolecules are often buried in the molecule's interior to exclude them from the aqueous milieu. The formation of oil droplets as hydrophobic nonpolar lipid molecules coalesce in the presence of water is an approximation of this phenomenon. These tendencies have important consequences in the creation and maintenance of the macromolecular structures and supramolecular assemblies of living cells.

The Defining Concept of Biochemistry Is "Molecular Recognition Through Structural Complementarity"

Structural complementarity is the means of recognition in biomolecular interactions. The complicated and highly organized patterns of life depend on the ability of biomolecules to recognize and interact with one another in very specific ways. Such interactions are fundamental to metabolism, growth, replication, and other vital processes. The interaction of one molecule with another, a protein with a metabolite, for example, can be most precise if the structure of one is complementary to the structure of the other, as in two connecting pieces of a puzzle or, in the more popular analogy for macromolecules and their **ligands,** a lock and its key (Figure 1.16). *This principle of structural complementarity is the very essence of biomolecular recognition.* Structural complementarity is the significant clue to understanding the functional properties of biological systems. Biological systems from the macromolecular level to the cellular level operate via specific molecular recognition mechanisms based on structural complementarity: A protein recognizes its specific metabolite, a strand of DNA recognizes its complementary strand, sperm recognize an egg. All these interactions involve structural complementarity between molecules.

Biomolecular Recognition Is Mediated by Weak Chemical Forces

Weak chemical forces underlie the interactions that are the basis of biomolecular recognition. It is important to realize that because these interactions are sufficiently weak, they are readily reversible. Consequently, biomolecular interactions tend to be transient; rigid, static lattices of biomolecules that might paralyze cellular activities are not formed.

Instead, a dynamic interplay occurs between metabolites and macromolecules, hormones and receptors, and all the other participants instrumental to life processes. This interplay is initiated upon specific recognition between complementary molecules and ultimately culminates in unique physiological activities. Biological function is achieved through mechanisms based on structural complementarity and weak chemical interactions.

This principle of structural complementarity extends to higher interactions essential to the establishment of the living condition. For example, the formation of supramolecular complexes occurs because of recognition and interaction between their various macromolecular components, as governed by the weak forces formed between them. If a sufficient number of weak bonds can be formed, as in macromolecules complementary in structure to one another, larger structures assemble spontaneously. The tendency for nonpolar molecules and parts of molecules to come together through hydrophobic interactions also promotes the formation of supramolecular assemblies. Very complex subcellular structures are actually spontaneously formed in an assembly process that is driven by weak forces accumulated through structural complementarity.

Weak Forces Restrict Organisms to a Narrow Range of Environmental Conditions

Because biomolecular interactions are governed by weak forces, living systems are restricted to a narrow range of physical conditions. Biological macromolecules are functionally active only within a narrow range of environmental conditions, such as temperature, ionic strength, and relative acidity. Extremes of these conditions disrupt the weak forces essential to maintaining the intricate structure of macromolecules. The loss of structural order in these complex macromolecules, so-called **denaturation,** is accompanied by loss of function (Figure 1.17). As a consequence, cells cannot tolerate reactions in which large amounts of energy are released, nor can they generate a large energy burst to drive energy-requiring processes. Instead, such transformations take place via sequential series of chemical reactions whose overall effect achieves dramatic energy changes, even though any given reaction in the series proceeds with only modest input or release of energy (Figure 1.18). These sequences of reactions are organized to provide for the release of useful energy to the cell from the breakdown of food or to take such energy and use it to drive the synthesis of biomolecules essential to the living state. Collectively, these reaction sequences constitute cellular **metabolism**—the ordered reaction pathways by which cellular chemistry proceeds and biological energy transformations are accomplished.

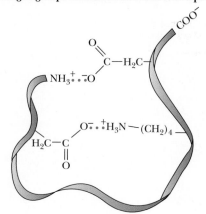

FIGURE 1.15 Ionic bonds in biological molecules.

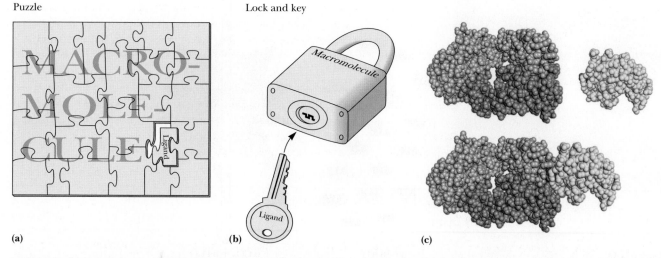

FIGURE 1.16 Structural complementarity: the pieces of a puzzle, the lock and its key, a biological macromolecule and its ligand—an antigen–antibody complex. The antigen on the right (*gold*) is a small protein, lysozyme, from hen egg white. The antibody molecule (IgG) (*left*) has a pocket that is structurally complementary to a surface feature (*red*) on the antigen. (See also Figure 1.12.)

FIGURE 1.17 Denaturation and renaturation of the intricate structure of a protein.

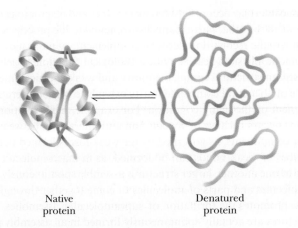

Native
protein

Denatured
protein

Enzymes Catalyze Metabolic Reactions

The sensitivity of cellular constituents to environmental extremes places another constraint on the reactions of metabolism. The rate at which cellular reactions proceed is a very important factor in maintenance of the living state. However, the common ways chemists accelerate reactions are not available to cells; the temperature cannot be raised, acid or base cannot be added, the pressure cannot be elevated, and concentrations cannot be dramatically increased. Instead, biomolecular catalysts mediate cellular reactions. These catalysts, called **enzymes,** accelerate the reaction rates many orders of magnitude and, by selecting the substances undergoing reaction, determine the specific reaction that takes place. Virtually every metabolic reaction is catalyzed by an enzyme (Figure 1.19).

Metabolic Regulation Is Achieved by Controlling the Activity of Enzymes Thousands of reactions mediated by an equal number of enzymes are occurring at any given instant within the cell. Collectively, these reactions constitute cellular metabolism. Metabolism has many branch points, cycles, and interconnections, as subsequent chapters

The combustion of glucose: $C_6H_{12}O_6 + 6\ O_2 \longrightarrow 6\ CO_2 + 6\ H_2O + 2870$ kJ energy

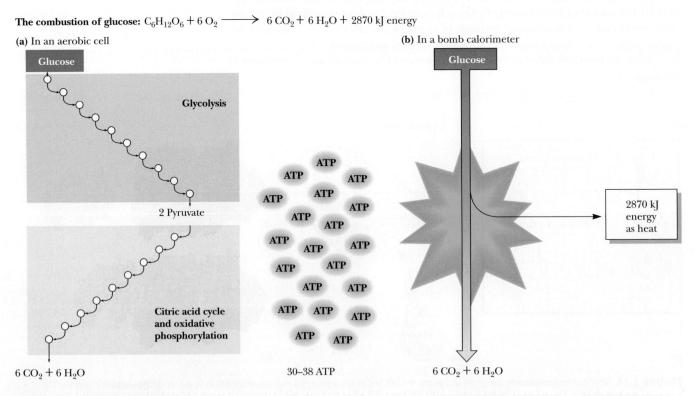

(a) In an aerobic cell

Glucose

Glycolysis

2 Pyruvate

Citric acid cycle
and oxidative
phosphorylation

$6\ CO_2 + 6\ H_2O$

ATP ATP ATP ATP ATP ATP ATP ATP ATP ATP ATP ATP ATP ATP ATP ATP ATP ATP ATP ATP ATP ATP ATP ATP ATP ATP

30–38 ATP

(b) In a bomb calorimeter

Glucose

2870 kJ
energy
as heat

$6\ CO_2 + 6\ H_2O$

FIGURE 1.18 Metabolism is the organized release or capture of small amounts of energy in processes whose overall change in energy is large. **(a)** Cells can release the energy of glucose in a stepwise fashion and the small "packets" of energy appear in ATP. **(b)** Combustion of glucose in a bomb calorimeter results in an uncontrolled, explosive release of energy in its least useful form, heat.

reveal. All these reactions, many of which are at apparent cross-purposes in the cell, must be fine-tuned and integrated so that metabolism and life proceed harmoniously. The need for metabolic regulation is obvious. This metabolic regulation is achieved through controls on enzyme activity so that the rates of cellular reactions are appropriate to cellular requirements.

Despite the organized pattern of metabolism and the thousands of enzymes required, cellular reactions nevertheless conform to the same thermodynamic principles that govern any chemical reaction. Enzymes have no influence over energy changes (the thermodynamic component) in their reactions. Enzymes only influence reaction rates. Thus, cells are systems that take in food, release waste, and carry out complex degradative and biosynthetic reactions essential to their survival while operating under conditions of essentially constant temperature and pressure and maintaining a constant internal environment (**homeostasis**) with no outwardly apparent changes. *Cells are open thermodynamic systems exchanging matter and energy with their environment and functioning as highly regulated isothermal chemical engines.*

The Time Scale of Life

Individual organisms have life spans ranging from a day or less to a century or more, but the phenomena that characterize and define living systems have durations ranging over 33 orders of magnitude, from 10^{-15} sec (electron transfer reactions, photo-excitation in photosynthesis) to 10^{18} sec (the period of evolution, spanning from the first appearance of organisms on the earth more than 3 billion years ago to today) (Table 1.5). Because proteins are the agents of biological function, phenomena involving weak interactions and proteins dominate the shorter times. As time increases, more stable interactions (covalent bonds) and phenomena involving the agents of genetic information (the nucleic acids) come into play.

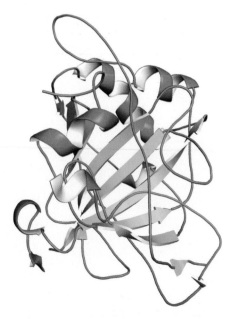

FIGURE 1.19 Carbonic anhydrase, a representative enzyme.

1.5 What Is the Organization and Structure of Cells?

All living cells fall into one of three broad categories—**Archaea, Bacteria** and **Eukarya.** Archaea and Bacteria are referred to collectively as **prokaryotes.** As a group, prokaryotes are single-celled organisms that lack nuclei and other organelles; the

TABLE 1.5	Life Times	
Time (sec)	Process	Example
10^{-15}	Electron transfer	The light reactions in photosynthesis
10^{-13}	Transition states	Transition states in chemical reactions have lifetimes of 10^{-11} to 10^{-15} sec (the reciprocal of the frequency of bond vibrations)
10^{-11}	H-bond lifetimes	H bonds are exchanged between H_2O molecules due to the rotation of the water molecules themselves
10^{-12} to 10^3	Motion in proteins	Fast: tyrosine ring flips, methyl group rotations Slow: bending motions between protein domains
10^{-6} to 10^0	Enzyme catalysis	10^{-6} sec: fast enzyme reactions 10^{-3} sec: typical enzyme reactions 10^0 sec: slow enzyme reactions
10^0	Diffusion in membranes	A typical membrane lipid molecule can diffuse from one end of a bacterial cell to the other in 1 sec; a small protein would go half as far
10^1 to 10^2	Protein synthesis	Some ribosomes synthesize proteins at a rate of 20 amino acids added per second
10^4 to 10^5	Cell division	Prokaryotic cells can divide as rapidly as every hour or so; eukaryotic cell division varies greatly (from hours to years)
10^7 to 10^8	Embryonic development	Human embryonic development takes 9 months (2.4×10^7 sec)
10^5 to 10^9	Life span	Human life expectancy is about 80 years in developed countries (2.5×10^9 sec)
10^{18}	Evolution	The first organisms appeared 3.8×10^9 years ago and evolution has continued since then

word is derived from *pro* meaning "prior to" and *karyot* meaning "nucleus." In traditional biological classification schemes, prokaryotes were grouped together as members of the kingdom Monera. The other four living kingdoms were all Eukarya—the single-celled Protists, such as amoebae, and all multicellular life forms, including the Fungi, Plant, and Animal kingdoms. Eukaryotic cells have true nuclei and other organelles such as mitochondria, with the prefix *eu* meaning "true." Groupings of organisms into kingdoms is useful, but phylogenetic research since 2000 indicates that Eukarya are a more complex domain of life than earlier classification schemes suggest.

The Evolution of Early Cells Gave Rise to Eubacteria, Archaea, and Eukaryotes

For a long time, most biologists believed that eukaryotes evolved from the simpler prokaryotes in some linear progression from simple to complex over the course of geological time. However, contemporary evidence favors the view that present-day organisms are better grouped into the three domains mentioned: eukarya, bacteria, and archaea. All are believed to have evolved from an ancestral communal gene pool shared among primitive cellular entities. Furthermore, contemporary eukaryotic cells are, in reality, composite cells that harbor various bacterial contributions.

Despite great diversity in form and function, cells and organisms share much biochemistry in common. This commonality and diversity has been substantiated by the results of **whole genome sequencing,** the determination of the complete nucleotide sequence within the DNA of an organism. For example, the genome of the metabolically divergent archaea *Methanococcus jannaschii* shows 44% similarity to known genes in eubacteria and eukaryotes, yet 56% of its genes are new to science.

How Many Genes Does a Cell Need?

The genome of the *Mycoplasma genitalium* consists of 523 genes, encoding 484 proteins, in just 580,074 base pairs (Table 1.6). This information sparks an interesting question: How many genes are needed for cellular life? Any **minimum gene set** must encode all the information necessary for cellular metabolism, including the vital functions essential to reproduction. The simplest cell must show at least (1) some degree of metabolism and energy production; (2) genetic replication based on a template molecule that encodes information (DNA or RNA?); and (3) formation and maintenance of a cell boundary (membrane). Top-down studies aim to discover from existing cells what a minimum gene set might be. These studies have focused on

Gene: is a unit of hereditary information, physically defined by a specific sequence of nucleotides in DNA; in molecular terms, a gene is a nucleotide sequence that encodes a protein or RNA product.

> **CRITICAL DEVELOPMENTS IN BIOCHEMISTRY**

Synthetic Life

J. Craig Venter and his colleagues at the J. Craig Venter Institute (JCVI) claim to have created the first synthetic life. They devised a synthetic version of the 1.08×10^6-base pair genome of *Myobacterium mycoides* by designing a thousand segments of DNA, each about 1000 base pairs long, which were then chemically synthesized by the Blue Heron Biotechnology Co., a DNA synthesizing service. The JCVI scientists then assembled these pieces to form a complete synthetic genome that was transferred into the cytoplasm of a related bacterial species, *Mycobacterium capricolum*. The self-replicating cells that grew following this genome transplantation were under the direction of the synthetic genome and produced proteins representative of *M. mycoides*, not *M. capricolum*. Fur-

ther, the only DNA in these cell cultures was the synthetic DNA created by Venter and associates, as evidenced by certain "watermarks" they had designed into their synthetic genome to establish its uniqueness. These cells represent the first artificial living organisms, artificial in the sense that their entire genome was chemically synthesized and not the result of biological evolution. Their significance lies in the possibilities they open for the creation of life forms for specific purposes, such as oil-eating bacteria for environmental remediation or bacteria able to synthesize desired products such as drugs.

Gibson, D. G., 2010. Creation of a bacterial cell controlled by a chemically synthesized genome. *Science* **329**:52–56.

TABLE 1.6	How Many Genes Does It Take To Make An Organism?	
Organism	Number of Cells in Adult*	Number of Genes
Mycobacterium genitalium Pathogenic bacterium	1	523
Methanococcus jannaschii Archaeal methanogen	1	1,800
Escherichia coli K12 Intestinal bacterium	1	4,400
Saccharomyces cereviseae Baker's yeast (eukaryote)	1	6,000
Caenorhabditis elegans Nematode worm	959	19,000
Drosophila melanogaster Fruit fly	10^4	13,500
Arabidopsis thaliana Flowering plant	10^7	27,000
Fugu rubripes Pufferfish	10^{12}	26,700 (est.)
Homo sapiens Human	10^{14}	20,500 (est.)

The first four of the nine organisms in the table are single-celled microbes; the last six are eukaryotes; the last five are multicellular, four of which are animals; the final two are vertebrates. Although pufferfish and humans have roughly the same number of genes, the pufferfish genome, at 0.365 billion nucleotide pairs, is only one-eighth the size of the human genome.
*Numbers for *Arabidopsis thaliana*, the pufferfish, and human are "order-of-magnitude" rough estimates.

simple parasitic bacteria, because parasites often obtain many substances from their hosts and do not have to synthesize them from scratch; thus, they require fewer genes. One study concluded that 206 genes are sufficient to form a minimum gene set. The set included genes for DNA replication and repair, transcription, translation, protein processing, cell division, membrane structure, nutrient transport, metabolic pathways for ATP synthesis, and enzymes to make a small number of metabolites that might not be available, such as pentoses for nucleotides. Yet another study based on computer modeling decided that a minimum gene set might have only 105 protein-coding genes. Bottom-up studies aim to create a minimal cell by reconstruction based on known cellular components. At this time, no such bottom-up creation of an artificial cell has been reported. The simplest functional artificial cell capable of replication would contain an informational macromolecule (presumably a nucleic acid) and enough metabolic apparatus to maintain a basic set of cellular components within a membranelike boundary.

Archaea and Bacteria Have a Relatively Simple Structural Organization

The bacteria form a widely spread group. Certain of them are pathogenic to humans. The archaea, about which we know less, were first discovered growing in unusual environments where other cells cannot survive. Archaea include the **thermoacidophiles** (heat- and acid-loving bacteria) of hot springs, the **halophiles** (salt-loving bacteria) of salt lakes and ponds, and the **methanogens** (bacteria that generate methane from CO_2 and H_2). Archaea are also common in typical microbial habitats, such as soils, seas, and

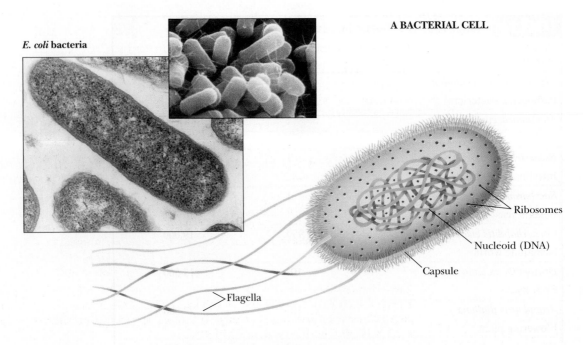

E. coli bacteria

A BACTERIAL CELL

Ribosomes

Nucleoid (DNA)

Capsule

Flagella

FIGURE 1.20 This bacterium is *Escherichia coli,* a member of the coliform group of bacteria that colonize the intestinal tract of humans. (See Table 1.7.) (Photo, Martin Rotker/Phototake, Inc.; inset photo, David M. Phillips/Science Source/Photo Researchers, Inc.)

the guts of animals. Prokaryotes are typically very small, on the order of several microns in length, and are usually surrounded by a rigid **cell wall** that protects the cell and gives it its shape. The characteristic structural organization of one of these cells is depicted in Figure 1.20.

Prokaryotic cells have only a single membrane, the **plasma membrane** or **cell membrane.** Because they have no other membranes, prokaryotic cells contain no nucleus or organelles. Nevertheless, they possess a distinct nuclear area called the **nucleoid** where a single circular chromosome is localized. Some have internal membranous structures derived from and continuous with the cell membrane. Reactions of cellular respiration are localized on these membranes. In **cyanobacteria,** flat, sheetlike membranous structures called **lamellae** are formed from cell membrane infoldings. These lamellae are the sites of photosynthetic activity, but they are not contained within **plastids,** the organelles of photosynthesis found in higher plant cells. Some bacteria have **flagella,** single, long filaments used for motility. Prokaryotes largely reproduce by asexual division, although sexual exchanges can occur. Table 1.7 lists the major features of bacterial cells.

The Structural Organization of Eukaryotic Cells Is More Complex Than That of Prokaryotic Cells

Compared with prokaryotic cells, eukaryotic cells are much greater in size, typically having cell volumes 10^3 to 10^4 times larger. They are also much more complex. These two features require that eukaryotic cells partition their diverse metabolic processes into organized compartments, with each compartment dedicated to a particular function. A system of internal membranes accomplishes this partitioning. A typical animal cell is shown in Figure 1.21 and a typical plant cell in Figure 1.22. Tables 1.8

TABLE 1.7	Major Features of Prokaryotic Cells	
Structure	Molecular Composition	Function
Cell wall	Peptidoglycan: a rigid framework of polysaccharide crosslinked by short peptide chains. Some bacteria possess a lipopolysaccharide- and protein-rich outer membrane.	Mechanical support, shape, and protection against swelling in hypotonic media. The cell wall is a porous nonselective barrier that allows most small molecules to pass.
Cell membrane	The cell membrane is composed of about 45% lipid and 55% protein. The lipids form a bilayer that is a continuous nonpolar hydrophobic phase in which the proteins are embedded.	The cell membrane is a highly selective permeability barrier that controls the entry of most substances into the cell. Important enzymes in the generation of cellular energy are located in the membrane.
Nuclear area or nucleoid	The genetic material is a single, tightly coiled DNA molecule 2 nm in diameter but more than 1 mm in length (molecular mass of *E. coli* DNA is 3×10^9 daltons; 4.64×10^6 nucleotide pairs).	DNA provides the operating instructions for the cell; it is the repository of the cell's genetic information. During cell division, each strand of the double-stranded DNA molecule is replicated to yield two double-helical daughter molecules. Messenger RNA (mRNA) is transcribed from DNA to direct the synthesis of cellular proteins.
Ribosomes	Bacterial cells contain about 15,000 ribosomes. Each is composed of a small (30S) subunit and a large (50S) subunit. The mass of a single ribosome is 2.3×10^6 daltons. It consists of 65% RNA and 35% protein.	Ribosomes are the sites of protein synthesis. The mRNA binds to ribosomes, and the mRNA nucleotide sequence specifies the protein that is synthesized.
Storage granules	Bacteria contain granules that represent storage forms of polymerized metabolites such as sugars or β-hydroxybutyric acid.	When needed as metabolic fuel, the monomeric units of the polymer are liberated and degraded by energy-yielding pathways in the cell.
Cytosol	Despite its amorphous appearance, the cytosol is an organized gelatinous compartment that is 20% protein by weight and rich in the organic molecules that are the intermediates in metabolism.	The cytosol is the site of intermediary metabolism, the interconnecting sets of chemical reactions by which cells generate energy and form the precursors necessary for biosynthesis of macromolecules essential to cell growth and function.

and 1.9 list the major features of a typical animal cell and a higher plant cell, respectively.

Eukaryotic cells possess a discrete, membrane-bounded **nucleus,** the repository of the cell's genetic material, which is distributed among a few or many **chromosomes.** During cell division, equivalent copies of this genetic material must be passed to both daughter cells through duplication and orderly partitioning of the chromosomes by the process known as **mitosis.** Like prokaryotic cells, eukaryotic cells are surrounded by a plasma membrane. Unlike prokaryotic cells, eukaryotic cells are rich in internal membranes that are differentiated into specialized structures such as the **endoplasmic reticulum (ER)** and the **Golgi apparatus.** Membranes also surround certain organelles (**mitochondria** and **chloroplasts,** for example) and various vesicles, including **vacuoles, lysosomes,** and **peroxisomes.** The common purpose of these membranous partitionings is the creation of cellular compartments that have specific, organized metabolic functions, such as the mitochondrion's role as the principal site of cellular energy production. Eukaryotic cells also have a **cytoskeleton** composed of arrays of filaments that give the cell its shape and its capacity to move. Some eukaryotic cells also have long projections on their surface—cilia or flagella—which provide propulsion.

1.6 : What Are Viruses?

Viruses are supramolecular complexes of nucleic acid, either DNA or RNA, encapsulated in a protein coat and, in some instances, surrounded by a membrane envelope (Figure 1.23). Viruses are acellular, but they act as cellular parasites in order to

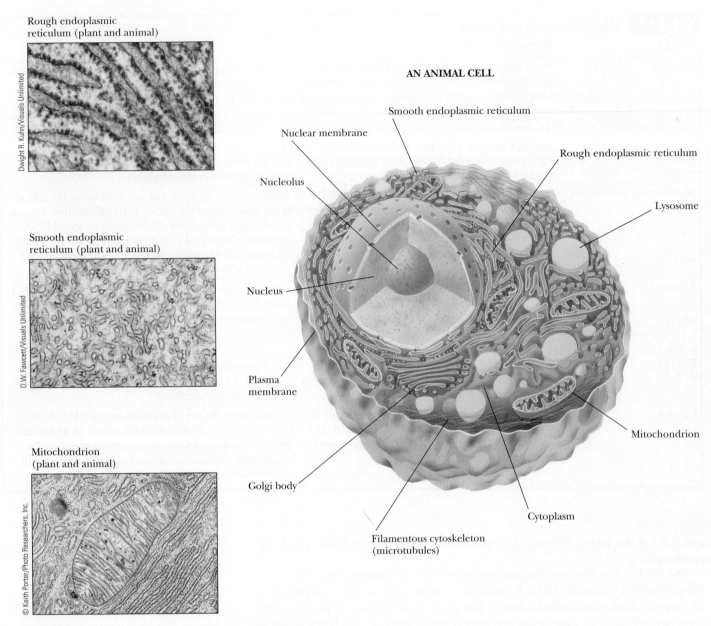

Rough endoplasmic reticulum (plant and animal)

Dwight R. Kuhn/Visuals Unlimited

Smooth endoplasmic reticulum (plant and animal)

D.W. Fawcett/Visuals Unlimited

Mitochondrion (plant and animal)

© Keith Porter/Photo Researchers, Inc.

AN ANIMAL CELL

Smooth endoplasmic reticulum
Nuclear membrane
Rough endoplasmic reticulum
Nucleolus
Lysosome
Nucleus
Plasma membrane
Mitochondrion
Golgi body
Cytoplasm
Filamentous cytoskeleton (microtubules)

FIGURE 1.21 This figure diagrams a rat liver cell, a typical higher animal cell.

reproduce. The bits of nucleic acid in viruses are, in reality, mobile elements of genetic information. The protein coat serves to protect the nucleic acid and allows it to gain entry to the cells that are its specific hosts. Viruses unique for all types of cells are known. Viruses infecting bacteria are called **bacteriophages** ("bacteria eaters"); different viruses infect animal cells and plant cells. Once the nucleic acid of a virus gains access to its specific host, it typically takes over the metabolic machinery of the host cell, diverting it to the production of virus particles. The host metabolic functions are subjugated to the synthesis of viral nucleic acid and proteins. Mature virus particles arise by encapsulating the nucleic acid within a protein coat called the **capsid.** Thus, viruses are supramolecular assemblies that act as parasites of cells (Figure 1.24).

Often, viruses cause disintegration of the cells that they have infected, a process referred to as cell **lysis.** It is their cytolytic properties that are the basis of viral disease.

Chloroplast (plant cell only)

Omikron/Photoresearchers, Inc.

Golgi body (plant and animal)

© Dennis Kunkel Microscopy, Inc./Phototake, NYC

Nucleus (plant and animal)

Biophoto Associates

A PLANT CELL

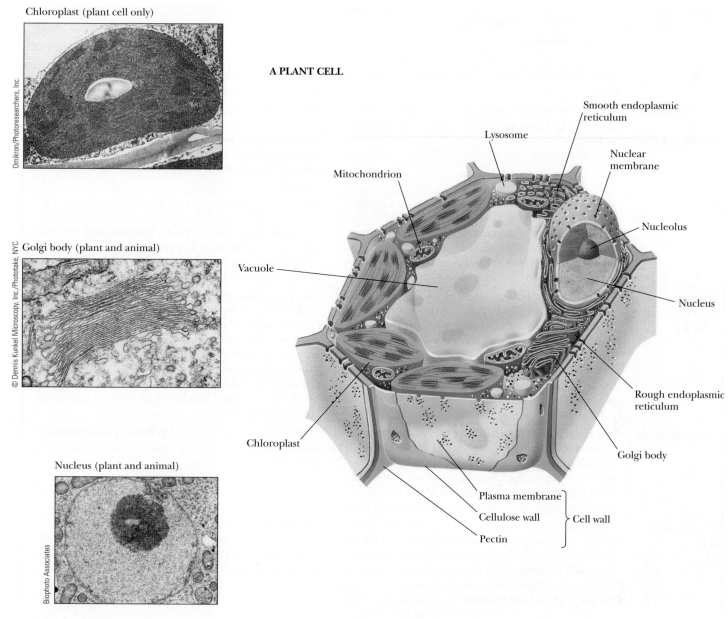

Smooth endoplasmic reticulum

Lysosome

Nuclear membrane

Mitochondrion

Nucleolus

Vacuole

Nucleus

Rough endoplasmic reticulum

Chloroplast

Golgi body

Plasma membrane

Cellulose wall   Cell wall

Pectin

FIGURE 1.22 This figure diagrams a cell in the leaf of a higher plant. The cell wall, membrane, nucleus, chloroplasts, mitochondria, vacuole, endoplasmic reticulum (ER), and other characteristic features are shown.

In certain circumstances, the viral genetic elements may integrate into the host chromosome and become quiescent. Such a state is termed **lysogeny.** Typically, damage to the host cell activates the replicative capacities of the quiescent viral nucleic acid, leading to viral propagation and release. Some viruses are implicated in transforming cells into a cancerous state, that is, in converting their hosts to an unregulated state of cell division and proliferation. Because all viruses are heavily dependent on their host for the production of viral progeny, viruses must have evolved after cells were established. Presumably, the first viruses were fragments of nucleic acid that developed the ability to replicate independently of the chromosome and then acquired the necessary genes enabling protection, autonomy, and transfer between cells. Surprisingly, virus-related DNA makes up almost half of the human genome.

TABLE 1.8	Major Features of a Typical Animal Cell	
Structure	**Molecular Composition**	**Function**
Extracellular matrix	The surfaces of animal cells are covered with a flexible and sticky layer of complex carbohydrates, proteins, and lipids.	This complex coating is cell specific, serves in cell–cell recognition and communication, creates cell adhesion, and provides a protective outer layer.
Cell membrane (plasma membrane)	Roughly 50:50 lipid:protein as a 5-nm-thick continuous sheet of lipid bilayer in which a variety of proteins are embedded.	The plasma membrane is a selectively permeable outer boundary of the cell, containing specific systems—pumps, channels, transporters, receptors—for the exchange of materials with the environment and the reception of extracellular information. Important enzymes are also located here.
Nucleus	The nucleus is separated from the cytosol by a double membrane, the nuclear envelope. The DNA is complexed with basic proteins (histones) to form chromatin fibers, the material from which chromosomes are made. A distinct RNA-rich region, the nucleolus, is the site of ribosome assembly.	The nucleus is the repository of genetic information encoded in DNA and organized into chromosomes. During mitosis, the chromosomes are replicated and transmitted to the daughter cells. The genetic information of DNA is transcribed into RNA in the nucleus and passes into the cytosol, where it is translated into protein by ribosomes.
Endoplasmic reticulum (ER) and ribosomes	Flattened sacs, tubes, and sheets of internal membrane extending throughout the cytoplasm of the cell and enclosing a large interconnecting series of volumes called *cisternae*. The ER membrane is continuous with the outer membrane of the nuclear envelope. Portions of the sheet-like areas of the ER are studded with ribosomes, giving rise to *rough ER*. Eukaryotic ribosomes are larger than prokaryotic ribosomes.	The endoplasmic reticulum is a labyrinthine organelle where both membrane proteins and lipids are synthesized. Proteins made by the ribosomes of the rough ER pass through the ER membrane into the cisternae and can be transported via the Golgi to the periphery of the cell. Other ribosomes unassociated with the ER carry on protein synthesis in the cytosol. The nuclear membrane, ER, Golgi, and additional vesicles are all part of a continuous endomembrane system.
Golgi apparatus	The Golgi is an asymmetrical system of flattened membrane-bounded vesicles often stacked into a complex. The face of the complex nearest the ER is the *cis* face; that most distant from the ER is the *trans* face. Numerous small vesicles found peripheral to the *trans* face of the Golgi contain secretory material packaged by the Golgi.	Involved in the packaging and processing of macromolecules for secretion and for delivery to other cellular compartments.
Mitochondria	Mitochondria are organelles surrounded by two membranes that differ markedly in their protein and lipid composition. The inner membrane and its interior volume—the matrix—contain many important enzymes of energy metabolism. Mitochondria are about the size of bacteria, $\approx 1\ \mu m$. Cells contain hundreds of mitochondria, which collectively occupy about one-fifth of the cell volume.	Mitochondria are the power plants of eukaryotic cells where carbohydrates, fats, and amino acids are oxidized to CO_2 and H_2O. The energy released is trapped as high-energy phosphate bonds in ATP.
Lysosomes	Lysosomes are vesicles 0.2–0.5 μm in diameter, bounded by a single membrane. They contain hydrolytic enzymes such as proteases and nucleases that act to degrade cell constituents targeted for destruction. They are formed as membrane vesicles budding from the Golgi apparatus.	Lysosomes function in intracellular digestion of materials entering the cell via phagocytosis or pinocytosis. They also function in the controlled degradation of cellular components. Their internal pH is about 5, and the hydrolytic enzymes they contain work best at this pH.
Peroxisomes	Like lysosomes, peroxisomes are 0.2–0.5 μm, single-membrane–bounded vesicles. They contain a variety of oxidative enzymes that use molecular oxygen and generate peroxides. They are also formed from membrane vesicles budding from the smooth ER.	Peroxisomes act to oxidize certain nutrients, such as amino acids. In doing so, they form potentially toxic hydrogen peroxide, H_2O_2, and then decompose it to H_2O and O_2 by way of the peroxide-cleaving enzyme catalase.
Cytoskeleton	The cytoskeleton is composed of a network of protein filaments: actin filaments (or microfilaments), 7 nm in diameter; intermediate filaments, 8–10 nm; and microtubules, 25 nm. These filaments interact in establishing the structure and functions of the cytoskeleton. This interacting network of protein filaments gives structure and organization to the cytoplasm.	The cytoskeleton determines the shape of the cell and gives it its ability to move. It also mediates the internal movements that occur in the cytoplasm, such as the migration of organelles and mitotic movements of chromosomes. The propulsion instruments of cells—cilia and flagella—are constructed of microtubules.

TABLE 1.9	Major Features of a Higher Plant Cell: A Photosynthetic Leaf Cell	
Structure	Molecular Composition	Function
Cell wall	Cellulose fibers embedded in a polysaccharide/protein matrix; it is thick (>0.1 μm), rigid, and porous to small molecules.	Protection against osmotic or mechanical rupture. The walls of neighboring cells interact in cementing the cells together to form the plant. Channels for fluid circulation and for cell–cell communication pass through the walls. The structural material confers form and strength on plant tissue.
Cell membrane	Plant cell membranes are similar in overall structure and organization to animal cell membranes but differ in lipid and protein composition.	The plasma membrane of plant cells is selectively permeable, containing transport systems for the uptake of essential nutrients and inorganic ions. A number of important enzymes are localized here.
Nucleus	The nucleus, nucleolus, and nuclear envelope of plant cells are like those of animal cells.	Chromosomal organization, DNA replication, transcription, ribosome synthesis, and mitosis in plant cells are generally similar to the analogous features in animals.
Endoplasmic reticulum, Golgi apparatus, ribosomes, lysosomes, peroxisomes, and cytoskeleton	Plant cells also contain all of these characteristic eukaryotic organelles, essentially in the form described for animal cells.	These organelles serve the same purposes in plant cells that they do in animal cells.
Chloroplasts	Chloroplasts have a double-membrane envelope, an inner volume called the **stroma,** and an internal membrane system rich in thylakoid membranes, which enclose a third compartment, the thylakoid **lumen.** Chloroplasts are significantly larger than mitochondria. Other plastids are found in specialized structures such as fruits, flower petals, and roots and have specialized roles.	Chloroplasts are the site of photosynthesis, the reactions by which light energy is converted to metabolically useful chemical energy in the form of ATP. These reactions occur on the thylakoid membranes. The formation of carbohydrate from CO_2 takes place in the stroma. Oxygen is evolved during photosynthesis. Chloroplasts are the primary source of energy in the light.
Mitochondria	Plant cell mitochondria resemble the mitochondria of other eukaryotes in form and function.	Plant mitochondria are the main source of energy generation in photosynthetic cells in the dark and in non-photosynthetic cells under all conditions.
Vacuole	The vacuole is usually the most obvious compartment in plant cells. It is a very large vesicle enclosed by a single membrane called the **tonoplast.** Vacuoles tend to be smaller in young cells, but in mature cells, they may occupy more than 50% of the cell's volume. Vacuoles occupy the center of the cell, with the cytoplasm being located peripherally around it. They resemble the lysosomes of animal cells.	Vacuoles function in transport and storage of nutrients and cellular waste products. By accumulating water, the vacuole allows the plant cell to grow dramatically in size with no increase in cytoplasmic volume.

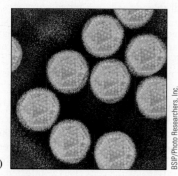
(a)

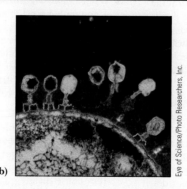
(b)

(c)

FIGURE 1.23 Viruses are genetic elements enclosed in a protein coat. Viruses are not free-living organisms and can reproduce only within cells. Viruses show an almost absolute specificity for their particular host cells, infecting and multiplying only within those cells. Viruses are known for virtually every kind of cell. Shown here are examples of **(a)** an animal virus, adenovirus; **(b)** bacteriophage T₄ on E. coli; and **(c)** a plant virus, tobacco mosaic virus.

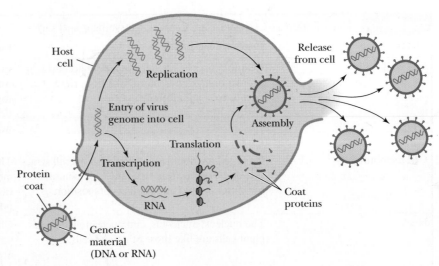

FIGURE 1.24 The virus life cycle. Viruses are mobile bits of genetic information encapsulated in a protein coat. The genetic material can be either DNA or RNA. Once this genetic material gains entry to its host cell, it takes over the host machinery for macromolecular synthesis and subverts it to the synthesis of viral-specific nucleic acids and proteins. These virus components are then assembled into mature virus particles that are released from the cell. Often, this parasitic cycle of virus infection leads to cell death and disease.

SUMMARY

OWL Login to OWL to develop problem-solving skills and complete online homework assigned by your professor.

1.1 What Are the Distinctive Properties of Living Systems? Living systems display an astounding array of activities that collectively constitute growth, metabolism, response to stimuli, and replication. In accord with their functional diversity, living organisms are complicated and highly organized entities composed of many cells. In turn, cells possess subcellular structures known as organelles, which are complex assemblies of very large polymeric molecules, or macromolecules. The monomeric units of macromolecules are common organic molecules (metabolites). Biological structures play a role in the organism's existence. From parts of organisms, such as limbs and organs, down to the chemical agents of metabolism, such as enzymes and metabolic intermediates, a biological purpose can be given for each component. Maintenance of the highly organized structure and activity of living systems requires energy that must be obtained from the environment. Energy is required to create and maintain structures and to carry out cellular functions. In terms of the capacity of organisms to self-replicate, the fidelity of self-replication resides ultimately in the chemical nature of DNA, the genetic material.

1.2 What Kinds of Molecules Are Biomolecules? C, H, N, and O are among the lightest elements capable of forming covalent bonds through electron-pair sharing. Because the strength of covalent bonds is inversely proportional to atomic weight, H, C, N, and O form the strongest covalent bonds. Two properties of carbon covalent bonds merit attention: the ability of carbon to form covalent bonds with itself and the tetrahedral nature of the four covalent bonds when carbon atoms form only single bonds. Together these properties hold the potential for an incredible variety of structural forms, whose diversity is multiplied further by including N, O, and H atoms.

1.3 What Is the Structural Organization of Complex Biomolecules? Biomolecules are built according to a structural hierarchy: Simple molecules are the units for building complex structures. H_2O, CO_2, NH_4^+, NO_3^-, and N_2 are the inorganic precursors for the formation of simple organic compounds from which metabolites are made. These metabolites serve as intermediates in cellular energy transformation and as building blocks (amino acids, sugars, nucleotides, fatty acids, and glycerol) for lipids and for macromolecular synthesis (synthesis of proteins, polysaccharides, DNA, and RNA). The next higher level of structural organization is created when macromolecules come together through noncovalent interactions to form supramolecular complexes, such as multifunctional enzyme complexes, ribosomes, chromosomes, and cytoskeletal elements.

The next higher rung in the hierarchical ladder is occupied by the organelles. Organelles are membrane-bounded cellular inclusions dedicated to important cellular tasks, such as the nucleus, mitochondria, chloroplasts, endoplasmic reticulum, Golgi apparatus, and vacuoles, as well as other relatively small cellular inclusions. At the apex of the biomolecular hierarchy is the cell, the unit of life, the smallest entity displaying those attributes associated uniquely with the living state—growth, metabolism, stimulus response, and replication.

1.4 How Do the Properties of Biomolecules Reflect Their Fitness to the Living Condition? Some biomolecules carry the information of life; others translate this information so that the organized structures essential to life are formed. Interactions between such structures are the processes of life. Properties of biomolecules that endow them with the potential for creating the living state include the following: Biological macromolecules and their building blocks have directionality, and thus biological macromolecules are informational; in addition, biomolecules have characteristic three-dimensional architectures, providing the means for molecular recognition through structural complementarity. Weak forces (H bonds, van der Waals interactions, ionic attractions, and hydrophobic interactions) mediate the interactions between biological molecules and, as a consequence, restrict organisms to the narrow range of environmental conditions where these forces operate.

1.5 What Is the Organization and Structure of Cells? All cells share a common ancestor and fall into one of two broad categories—prokaryotic and eukaryotic—depending on whether the cell has a nucleus. Prokaryotes are typically single-celled organisms and have a rather simple cellular organization. In contrast, eukaryotic cells are structurally more complex, having organelles and various subcellular compartments defined by membranes. Other than the Protists, eukaryotes are multicellular.

1.6 What Are Viruses? Viruses are supramolecular complexes of nucleic acid encapsulated in a protein coat and, in some instances, surrounded by a membrane envelope. Viruses are not alive; they are not even cellular. Instead, they are packaged bits of genetic material that can parasitize cells in order to reproduce. Often, they cause disintegration, or lysis, of the cells they've infected. It is these cytolytic properties that are the basis of viral disease. In certain circumstances, the viral nucleic acid may integrate into the host chromosome and become quiescent, creating a state known as lysogeny. If the host cell is damaged, the replicative capacities of the quiescent viral nucleic acid may be activated, leading to viral propagation and release.

FOUNDATIONAL BIOCHEMISTRY Things You Should Know After Reading Chapter 1.

- Chemistry is the logic of biological phenomena.

- Biological molecules serve functional purposes.

- The living state is characterized by the flow of energy through the organism.

- Biomolecules are compounds of carbon.

- Cellular macromolecules and structures are formed from simple molecules according to a hierarchy of increasing structural complexity.

- Metabolites are used to form the building blocks of macromolecules.

- Membranes are supramolecular assemblies that define the boundaries of cells.

- Biological macromolecules and their building blocks have a "sense," or directionality.

- Biological macromolecules are informational.

- Macromolecules have a defining 3-dimensional architecture.

- Weak forces important in biochemistry include hydrogen bonds, electrostatic (ionic) interactions, van der Waals interactions, and hydrophobic interactions.

- Weak forces maintain biological structure and determine biomolecular interactions.

- The defining concept in biochemistry is "molecular recognition through structural complementarity."

- Biomolecular recognition is mediated by weak chemical forces.

- The importance of weak forces restricts organisms to a narrow range of environmental conditions.

- Enzymes catalyze metabolic reactions.

- Metabolic regulation is achieved by controlling the activity of enzymes.

- The time scale of life ranges from 10^{-15} sec (the speed of electron transfer processes) to 10^{18} sec, the time span that life has been evolving on earth.

- The structural and functional organization of prokaryotic cells.

- The structural and functional organization of animal cells and plant cells, including the functions of the various organelles.

- Viruses are lifeless complexes of nucleic acid and protein that act as cellular parasites in order to reproduce.

PROBLEMS

Answers to all problems are at the end of this book. Detailed solutions are available in the *Student Solutions Manual, Study Guide, and Problems Book.*

1. **The Biosynthetic Capacity of Cells** The nutritional requirements of *Escherichia coli* cells are far simpler than those of humans, yet the macromolecules found in bacteria are about as complex as those of animals. Because bacteria can make all their essential biomolecules while subsisting on a simpler diet, do you think bacteria may have more biosynthetic capacity and hence more metabolic complexity than animals? Organize your thoughts on this question, pro and con, into a rational argument.

2. **Cell Structure** Without consulting the figures in this chapter, sketch the characteristic prokaryotic and eukaryotic cell types and label their pertinent organelle and membrane systems.

3. **The Dimensions of Prokaryotic Cells and Their Constituents** *Escherichia coli* cells are about 2 μm (microns) long and 0.8 μm in diameter.

   a. How many *E. coli* cells laid end to end would fit across the diameter of a pinhead? (Assume a pinhead diameter of 0.5 mm.)

   b. What is the volume of an *E. coli* cell? (Assume it is a cylinder, with the volume of a cylinder given by $V = \pi r^2 h$, where $\pi = 3.14$.)

   c. What is the surface area of an *E. coli* cell? What is the surface-to-volume ratio of an *E. coli* cell?

   d. Glucose, a major energy-yielding nutrient, is present in bacterial cells at a concentration of about 1 mM. What is the concentration of glucose, expressed as mg/mL? How many glucose molecules are contained in a typical *E. coli* cell? (Recall that Avogadro's number = 6.023×10^{23}.)

   e. A number of regulatory proteins are present in *E. coli* at only one or two molecules per cell. If we assume that an *E. coli* cell contains just one molecule of a particular protein, what is the molar concentration of this protein in the cell? If the molecular weight of this protein is 40 kD, what is its concentration, expressed as mg/mL?

   f. An *E. coli* cell contains about 15,000 ribosomes, which carry out protein synthesis. Assuming ribosomes are spherical and have a diameter of 20 nm (nanometers), what fraction of the *E. coli* cell volume is occupied by ribosomes?

   g. The *E. coli* chromosome is a single DNA molecule whose mass is about 3×10^9 daltons. This macromolecule is actually a linear array of nucleotide pairs. The average molecular weight of a nucleotide pair is 660, and each pair imparts 0.34 nm to the length of the DNA molecule. What is the total length of the *E. coli* chromosome? How does this length compare with the overall dimensions of an *E. coli* cell? How many nucleotide pairs does this DNA contain? The average *E. coli* protein is a linear chain of 360 amino acids. If three nucleotide pairs in a gene encode one amino acid in a protein, how many different proteins can the *E. coli* chromosome encode? (The answer to this question is a reasonable approximation of the maximum number of different kinds of proteins that can be expected in bacteria.)

4. **The Dimensions of Mitochondria and Their Constituents** Assume that mitochondria are cylinders 1.5 μm in length and 0.6 μm in diameter.

   a. What is the volume of a single mitochondrion?

   b. Oxaloacetate is an intermediate in the citric acid cycle, an important metabolic pathway localized in the mitochondria of eukaryotic cells. The concentration of oxaloacetate in mitochondria is about 0.03 μM. How many molecules of oxaloacetate are in a single mitochondrion?

5. **The Dimensions of Eukaryotic Cells and Their Constituents** Assume that liver cells are cuboidal in shape, 20 μm on a side.

   a. How many liver cells laid end to end would fit across the diameter of a pinhead? (Assume a pinhead diameter of 0.5 mm.)

   b. What is the volume of a liver cell? (Assume it is a cube.)

   c. What is the surface area of a liver cell? What is the surface-to-volume ratio of a liver cell? How does this compare to the surface-to-volume ratio of an *E. coli* cell (compare this answer with that of problem 3c)? What problems must cells with low surface-to-volume ratios confront that do not occur in cells with high surface-to-volume ratios?

d. A human liver cell contains two sets of 23 chromosomes, each set being roughly equivalent in information content. The total mass of DNA contained in these 46 enormous DNA molecules is 4×10^{12} daltons. Because each nucleotide pair contributes 660 daltons to the mass of DNA and 0.34 nm to the length of DNA, what is the total number of nucleotide pairs and the complete length of the DNA in a liver cell? How does this length compare with the overall dimensions of a liver cell? The maximal information in each set of liver cell chromosomes should be related to the number of nucleotide pairs in the chromosome set's DNA. This number can be obtained by dividing the total number of nucleotide pairs just calculated by 2. What is this value? If this information is expressed in proteins that average 400 amino acids in length and three nucleotide pairs encode one amino acid in a protein, how many different kinds of proteins might a liver cell be able to produce? (In reality, liver cell DNA encodes approximately 20,000 different proteins. Thus, a large discrepancy exists between the theoretical information content of DNA in liver cells and the amount of information actually expressed.)

6. **The Principle of Molecular Recognition Through Structural Complementarity** Biomolecules interact with one another through molecular surfaces that are structurally complementary. How can various proteins interact with molecules as different as simple ions, hydrophobic lipids, polar but uncharged carbohydrates, and even nucleic acids?

7. **The Properties of Informational Macromolecules** What structural features allow biological polymers to be informational macromolecules? Is it possible for polysaccharides to be informational macromolecules?

8. **The Importance of Weak Forces in Biomolecular Recognition** Why is it important that weak forces, not strong forces, mediate biomolecular recognition?

9. **Interatomic Distances in Weak Forces versus Chemical Bonds** What is the distance between the centers of two carbon atoms (their *limit of approach*) that are interacting through van der Waals forces? What is the distance between the centers of two carbon atoms joined in a covalent bond? (See Table 1.4.)

10. **The Strength of Weak Forces Determines the Environmental Sensitivity of Living Cells** Why does the central role of weak forces in biomolecular interactions restrict living systems to a narrow range of environmental conditions?

11. **Cells as Steady-State Systems** Describe what is meant by the phrase "cells are steady-state systems."

12. **A Simple Genome and Its Protein-Encoding Capacity** The genome of the *Mycoplasma genitalium* consists of 523 genes, encoding 484 proteins, in just 580,074 base pairs (Table 1.6). What fraction of the *M. genitalium* genes encode proteins? What do you think the other genes encode? If the fraction of base pairs devoted to protein-coding genes is the same as the fraction of the total genes that they represent, what is the average number of base pairs per protein-coding gene? If it takes 3 base pairs to specify an amino acid in a protein, how many amino acids are found in the average *M. genitalium* protein? If each amino acid contributes on average 120 daltons to the mass of a protein, what is the mass of an average *M. genitalium* protein?

13. **An Estimation of Minimal Genome Size for a Living Cell** Studies of existing cells to determine the minimum number of genes for a living cell have suggested that 206 genes are sufficient. If the ratio of protein-coding genes to non–protein-coding genes is the same in this minimal organism as the genes of *Mycoplasma genitalium,* how many proteins are represented in these 206 genes? How many base pairs would be required to form the genome of this minimal organism if the genes are the same size as *M. genitalium* genes?

14. **An Estimation of the Number of Genes in a Virus** Virus genomes range in size from approximately 3500 nucleotides to approximately 280,000 base pairs. If viral genes are about the same size as *M. genitalium* genes, what is the minimum and maximum number of genes in viruses?

15. **Intracellular Transport of Proteins** The endoplasmic reticulum (ER) is a site of protein synthesis. Proteins made by ribosomes associated with the ER may pass into the ER membrane or enter the lumen of the ER. Devise a pathway by which:

a. a plasma membrane protein may reach the plasma membrane.

b. a secreted protein may be deposited outside the cell.

Preparing for the MCAT® Exam

16. Biological molecules often interact via weak forces (H bonds, van der Waals interactions, etc.). What would be the effect of an increase in kinetic energy on such interactions?

17. Proteins and nucleic acids are informational macromolecules. What are the two minimal criteria for a linear informational polymer?

FURTHER READING

General Biology Textbooks

Campbell, N. A., and Reece, J. B., 2007. *Biology,* 8th ed. San Francisco: Benjamin/Cummings.

Russell, P. J., Hertz P. E., and McMillan, B., 2012. *Biology: The Dynamic Science,* 2nd ed. Pacific Grove, CA: Brooks/Cole.

Cell and Molecular Biology Textbooks

Alberts, B., Johnson, A., Lewis, J., Raff, M., et al., 2007. *Molecular Biology of the Cell,* 5th ed. New York: Garland Press.

Cavicchioli, R., 2007. *Archaea: Cellular and Molecular Biology.* Herndon, VA: ASM Press.

Lewin, B., Cassimeris, L., Plopper, G., and Lingappa, V. R., 2011. *Cells.* Boston, MA: Jones and Bartlett.

Lodish, H., Berk, A., Kaiser, C. A., Kreiger, M., et al., 2008. *Molecular Cell Biology,* 6th ed. New York: W. H. Freeman.

Snyder, L., and Champness, W., 2002. *Molecular Genetics of Bacteria,* 2nd ed. Herndon, VA: ASM Press.

Watson, J. D., Baker, T. A., Bell, S. T., Gann, A., et al., 2007. *Molecular Biology of the Gene,* 6th ed. Menlo Park, CA: Benjamin/Cummings.

Papers on Cell Structure

Gil, R., Silva, F. J., Pereto, J., and Moya, A., 2004. Determination of the core of a minimal bacterial gene set. *Microbiology and Molecular Biology Reviews* **68**:518–537.

Goodsell, D. S., 1991. Inside a living cell. *Trends in Biochemical Sciences* **16**:203–206.

Lewis, P. J., 2004. Bacterial subcellular architecture: Recent advances and future prospects. *Molecular Microbiology* **54**:1135–1150.

Lloyd, C., ed., 1986. Cell organization. *Trends in Biochemical Sciences* **11**:437–485.

Papers on Genomes

Cho, M. K., et al., 1999. Ethical considerations in synthesizing a minimal genome. *Science* **286**:2087–2090.

Gibson, D. G., 2010. Creation of a bacterial cell controlled by a chemically synthesized genome. *Science* **329**:52–56.

Kobayashi, K., Ehrlich, S. D., Albertini, A., Amati, G., et al., 2003. Essential *Bacillus subtilis* genes. *Proceedings of the National Academy of Science, U.S.A.* **100**:4678–4683.

Lartigue, C., Glass, J. I., Alperovich, N., Pieper, R., et al., 2007. Genome transplantation in bacteria: changing one species to another. *Science* **317**:632–638.

Ryan, F., 2010. I, virus: Why you are only half human. *New Scientist* **205**:32–35 (January 27, 2010 issue).

Szathmary, E., 2005. In search of the simplest cell. *Nature* **433**:469–470.

Papers on Early Cell Evolution

Cavalier-Smith, T., 2010. Origin of the cell nucleus, mitosis and sex: roles of intracellular coevolution. *Biology Direct* **5**:7 (78 pages).

Margulis, L., 1996. Archaeal-eubacterial mergers in the origin of Eukarya: Phylogenetic classification of life. *Proceedings of the National Academy of Science, U.S.A.* 93:1071–1076.

Pace, N. R., 2006. Time for a change. *Nature* **441**:289.

Service, R. F., 1997. Microbiologists explore life's rich, hidden kingdoms. *Science* **275**:1740–1742.

Wald, G., 1964. The origins of life. *Proceedings of the National Academy of Science, U.S.A.* **52**:595–611.

Whitfield, J., 2004. Born in a watery commune. *Nature* **427**:674–676.

Woese, C. R., 2002. On the creation of cells. *Proceedings of the National Academy of Science, U.S.A.* **99**:8742–8747.

A Brief History of Life

De Duve, C., 2002. *Life-Evolving: Molecules, Mind, and Meaning.* New York: Oxford University Press.

Morowitz, H., and Smith, E., 2007. Energy flow and the organization of life. *Complexity* **13**:51–59.

Water: The Medium of Life

Where there's water, there's life. ▶

If there is magic on this planet, it is contained in water.

Loren Eisley
(inscribed on the wall of the National Aquarium in Baltimore, Maryland)

KEY QUESTIONS

©Smit/Shutterstock.com

ESSENTIAL QUESTION

Water provided conditions for the origin, evolution, and flourishing of life; water is the medium of life.

What are the properties of water that render it so suited to its role as the medium of life?

⦿WL Online homework and a Student Self Assessment for this chapter may be assigned in OWL

Water is a major chemical component of the earth's surface. It is indispensable to life. Indeed, it is the only liquid that most organisms ever encounter. We are prone to take it for granted because of its ubiquity and bland nature, yet we marvel at its many unusual and fascinating properties. At the center of this fascination is the role of water as the medium of life. Life originated, evolved, and thrives in the seas. Organisms invaded and occupied terrestrial and aerial niches, but none gained true independence from water. Typically, organisms are 70% to 90% water. Indeed, normal metabolic activity can occur only when cells are at least 65% H_2O. This dependency of life on water is not a simple matter, but it can be grasped by considering the unusual chemical and physical properties of H_2O. Subsequent chapters establish that water and its ionization products, hydrogen ions and hydroxide ions, are critical determinants of the structure and function of many biomolecules, including amino acids and proteins, nucleotides and nucleic acids, and even phospholipids and membranes. In yet another essential role, water is an indirect participant—a difference in the concentration of hydrogen ions on opposite sides of a membrane represents an energized condition essential to biological mechanisms of energy transformation. First, let's review the remarkable properties of water.

2.1 | What Are the Properties of Water?

Water Has Unusual Properties

Compared with chemical compounds of similar atomic organization and molecular size, water displays unexpected properties. For example, compare water, the hydride of oxygen, with hydrides of oxygen's nearest neighbors in the periodic table, namely, ammonia (NH_3) and hydrogen fluoride (HF), or with the hydride of its nearest congener, sulfur (H_2S). Water has a substantially higher boiling point, melting point, heat of vaporization, and surface tension. Indeed, all of these physical properties are anomalously high for a substance of this molecular weight that is neither metallic nor ionic. These properties suggest that intermolecular forces of attraction between H_2O molecules are high. Thus, the internal cohesion of this substance is high. Furthermore, water has an unusually high dielectric constant, its maximum density is found in the liquid (not the solid) state, and it has a negative volume of melting (that is, the solid form, ice, occupies more space than does the liquid form, water). It is truly remarkable that so many eccentric properties occur together in this single substance. As chemists, we expect to find an explanation for these apparent eccentricities in the structure of water. The key to its intermolecular attractions must lie in its atomic constitution. Indeed, *the unrivaled ability of water to form hydrogen bonds is the crucial fact to understanding its properties.*

Hydrogen Bonding in Water Is Key to Its Properties

The two hydrogen atoms of water are linked covalently to oxygen, each sharing an electron pair, to give a nonlinear arrangement (Figure 2.1). This "bent" structure of the H_2O molecule has enormous influence on its properties. If H_2O were linear, it would be a nonpolar substance. In the bent configuration, however, the electronegative O atom and the two H atoms form a dipole that renders the molecule distinctly polar. Furthermore, this structure is ideally suited to H-bond formation. Water can serve as both an H donor and an H acceptor in H-bond formation. The potential to form four H bonds per water molecule is the source of the strong intermolecular attractions that endow this substance with its anomalously high boiling point, melting point, heat of vaporization, and surface tension. In ordinary ice, the common crystalline form of water, each H_2O molecule has four nearest neighbors to which it is hydrogen bonded: Each H atom donates an H bond to the O of a neighbor, and the O atom serves as an H-bond acceptor from H atoms bound to two different water molecules (Figure 2.2). A local tetrahedral symmetry results.

Hydrogen bonding in water is cooperative. That is, an H-bonded water molecule serving as an acceptor is a better H-bond donor than an unbonded molecule (and an H_2O molecule serving as an H-bond donor becomes a better H-bond acceptor). Thus, participation in H bonding by H_2O molecules is a phenomenon of mutual reinforcement. The H bonds between neighboring molecules are weak (23 kJ/mol each) relative to the H—O covalent bonds (420 kJ/mol). As a consequence, the hydrogen atoms are situated asymmetrically between the two oxygen atoms along the O-O axis. There is never any ambiguity about which O atom the H atom is chemically bound to, nor to which O it is H bonded.

The Structure of Ice Is Based On H-Bond Formation

In ice, the hydrogen bonds form a space-filling, three-dimensional network. These bonds are directional and straight; that is, the H atom lies on a direct line between the two O atoms. This linearity and directionality mean that the H bonds in ice are strong. In addition, the directional preference of the H bonds leads to an open lattice structure. For example, if the water molecules are approximated as rigid spheres

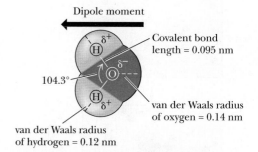

FIGURE 2.1 The structure of water. Two lobes of negative charge formed by the lone-pair electrons of the oxygen atom lie above and below the plane of the diagram. This electron density contributes substantially to the large dipole moment of the water molecule. Note that the H—O—H angle is 104.3°, *not* 109°, the angular value found in molecules with tetrahedral symmetry, such as CH_4. Many of the important properties of water derive from this angular value, such as the decreased density of its crystalline state, ice.

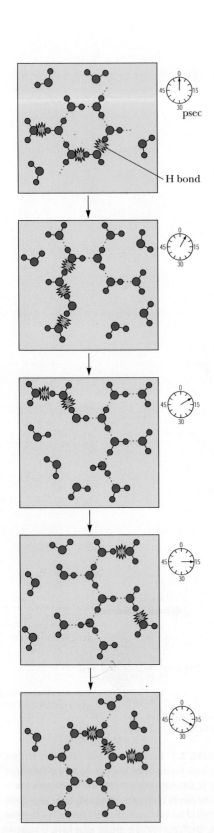

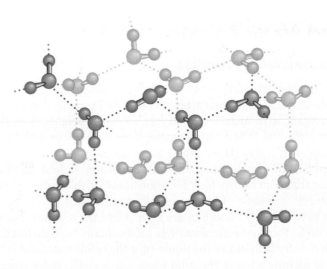

FIGURE 2.2 The structure of normal ice. The smallest number of H_2O molecules in any closed circuit of H-bonded molecules is six, so this structure bears the name hexagonal ice.

centered at the positions of the O atoms in the lattice, then the observed density of ice is actually only 57% of that expected for a tightly packed arrangement of such spheres. The H bonds in ice hold the water molecules apart. Melting involves break-ing some of the H bonds that maintain the crystal structure of ice so that the mole-cules of water (now liquid) can actually pack closer together. Thus, the density of ice is slightly less than that of water. Ice floats, a property of great importance to aquatic organisms in cold climates.

In liquid water, the rigidity of ice is replaced by fluidity and the crystalline period-icity of ice gives way to spatial homogeneity. The H_2O molecules in liquid water form a disordered H-bonded network, with each molecule having an average of 4.4 close neighbors situated within a center-to-center distance of 0.284 nm (2.84 Å). At least half of the hydrogen bonds have nonideal orientations (that is, they are not perfectly straight); consequently, liquid H_2O lacks the regular latticelike structure of ice. The space about an O atom is not defined by the presence of four hydrogens but can be occupied by other water molecules randomly oriented so that the local environment, over time, is essentially uniform. Nevertheless, the heat of melting for ice is but a small fraction (13%) of the heat of sublimation for ice (the energy needed to go from the solid to the vapor state). This fact indicates that the majority of H bonds between H_2O molecules survive the transition from solid to liquid. At 10°C, 2.3 H bonds per H_2O molecule remain and the tetrahedral bond order persists, even though substan-tial disorder is now present.

Molecular Interactions in Liquid Water Are Based on H Bonds

The present interpretation is that water structure at the molecular level is, at any instant, inhomogeneous, consisting of local fluctuations between patches of near-tetrahedral, ordered arrays of water molecules and more asymmetrical ensembles of water mole-cules linked together through distorted H bonds. The participation of each water mol-ecule in an average state of H bonding to its neighbors means that each molecule is connected to every other in a fluid network of H bonds. The average lifetime of an H-bonded connection between two H_2O molecules in water is 9.5 psec (picoseconds, where 1 psec = 10^{-12} sec). Thus, about every 10 psec, the average H_2O molecule moves, reori-ents, and interacts with new neighbors, as illustrated in Figure 2.3.

In summary, pure liquid water consists of H_2O molecules held in a disordered, three-dimensional network that has a local preference for tetrahedral geometry, yet contains a large number of strained or broken hydrogen bonds. The presence of strain creates a ki-netic situation in which H_2O molecules can switch H-bond allegiances; fluidity ensues.

FIGURE 2.3 The fluid network of H bonds linking water molecules in the liquid state. It is revealing to note that, in 10 psec, a photon of light (which travels at 3 × 10^8 m/sec) would move a distance of only 0.003 m.

The Solvent Properties of Water Derive from Its Polar Nature

Because of its highly polar nature, water is an excellent solvent for ionic substances such as salts; nonionic but polar substances such as sugars, simple alcohols, and amines; and carbonyl-containing molecules such as aldehydes and ketones. Although the electrostatic attractions between the positive and negative ions in the crystal lattice of a salt are very strong, water readily dissolves salts. For example, sodium chloride is dissolved because dipolar water molecules participate in strong electrostatic interactions with the Na^+ and Cl^- ions, leading to the formation of **hydration shells** surrounding these ions (Figure 2.4). Although hydration shells are stable structures, they are also dynamic. Each water molecule in the inner hydration shell around a Na^+ ion is replaced on average every 2 to 4 nsec (nanoseconds, where 1 nsec = 10^{-9} sec) by another H_2O. Consequently, a water molecule is trapped only several hundred times longer by the electrostatic force field of an ion than it is by the H-bonded network of water. (Recall that the average lifetime of H bonds between water molecules is about 10 psec.)

Water Has a High Dielectric Constant The attractions between the water molecules interacting with, or **hydrating,** ions are much greater than the tendency of oppositely charged ions to attract one another. Water's ability to surround ions in dipole interactions and diminish their attraction for each other is a measure of its **dielectric constant,** D. Indeed, ionization in solution depends on the dielectric constant of the solvent; otherwise, the strongly attracted positive and negative ions would unite to form neutral molecules. The strength of the dielectric constant is related to the force, F, experienced between two ions of opposite charge separated by a distance, r, as given in the relationship

$$F = e_1e_2/Dr^2$$

where e_1 and e_2 are the charges on the two ions. Table 2.1 lists the dielectric constants of some common liquids. Note that the dielectric constant for water is more than twice that of methanol and more than 40 times that of hexane.

Water Forms H Bonds with Polar Solutes In the case of nonionic but polar compounds such as sugars, the excellent solvent properties of water stem from its ability to readily form hydrogen bonds with the polar functional groups on these compounds, such as hydroxyls, amines, and carbonyls (see Figure 1.14). These polar interactions between solvent and solute are stronger than the intermolecular attractions between solute molecules caused by van der Waals forces and weaker hydrogen bonding. Thus, the solute molecules readily dissolve in water.

Hydrophobic Interactions The behavior of water toward nonpolar solutes is different from the interactions just discussed. Nonpolar solutes (or nonpolar functional groups on biological macromolecules) do not readily H bond to H_2O, and as a result, such compounds tend to be only sparingly soluble in water. The process of dissolving such

TABLE 2.1	Dielectric Constants* of Some Common Solvents at 25°C
Solvent	Dielectric Constant (*D*)
Formamide	109
Water	78.5
Methyl alcohol	32.6
Ethyl alcohol	24.3
Acetone	20.7
Acetic acid	6.2
Chloroform	5.0
Benzene	2.3
Hexane	1.9

*The dielectric constant is also referred to as *relative permitivity* by physical chemists.

FIGURE 2.4 Hydration shells surrounding ions in solution. Water molecules orient so that the electrical charge on the ion is sequestered by the water dipole.

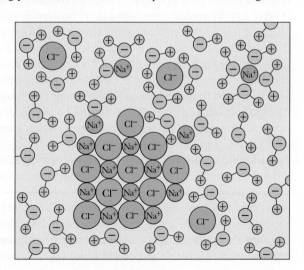

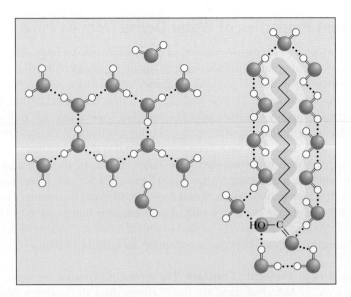

FIGURE 2.5 (*left*) Disordered network of H-bonded water molecules. (*right*) Clathrate cage of ordered, H-bonded water molecules around a nonpolar solute molecule.

substances is accompanied by significant reorganization of the water surrounding the solute so that the response of the solvent water to such solutes can be equated to "structure making." Because nonpolar solutes must occupy space, the random H-bonded network of water must reorganize to accommodate them. At the same time, the water molecules participate in as many H-bonded interactions with one another as the temperature permits. Consequently, the H-bonded water network rearranges toward formation of a local cagelike **(clathrate)** structure surrounding each solute molecule, as shown for a long-chain fatty acid in Figure 2.5. This fixed orientation of water molecules around a hydrophobic "solute" molecule results in a hydration shell. A major consequence of this rearrangement is that the molecules of H_2O participating in the cage layer have markedly reduced options for orientation in three-dimensional space. Water molecules tend to straddle the nonpolar solute such that two or three tetrahedral directions (H-bonding vectors) are tangential to the space occupied by the inert solute. "Straddling" allows the water molecules to retain their H-bonding possibilities because no H-bond donor or acceptor of the H_2O is directed toward the caged solute. The water molecules forming these clathrates are involved in highly ordered structures. That is, clathrate formation is accompanied by significant ordering of structure, or negative entropy.

Multiple nonpolar molecules tend to cluster together, because their joint solvation cage involves less total surface area and thus fewer ordered water molecules than in their separate cages. It is as if the nonpolar molecules had some net attraction for one another. This apparent affinity of nonpolar structures for one another is called **hydrophobic interactions** (Figure 2.6). In actuality, the "attraction" between nonpolar solutes is an entropy-driven process due to a net decrease in order among the H_2O molecules. To be specific, hydrophobic interactions between nonpolar molecules are maintained not so much by direct interactions between the inert solutes themselves as by the increase in entropy when the water cages coalesce and reorganize. Because interactions between nonpolar solute molecules and the water surrounding them are of uncertain stoichiometry and do not share the equality of atom-to-atom participation implicit in chemical bonding, the term *hydrophobic interaction* is more correct than the misleading expression *hydrophobic bond.*

Amphiphilic Molecules Compounds containing both strongly polar and strongly nonpolar groups are called **amphiphilic molecules** (from the Greek *amphi* meaning "both" and *philos* meaning "loving"). Such compounds are also referred to as **amphipathic molecules** (from the Greek *pathos* meaning "passion"). Salts of fatty acids are a typical example that has biological relevance. They have a long nonpolar hydrocarbon tail and a strongly polar carboxyl head group, as in the sodium salt of palmitic acid (Figure 2.7). Their behavior in aqueous solution reflects the combination of the contrasting polar and nonpolar nature of

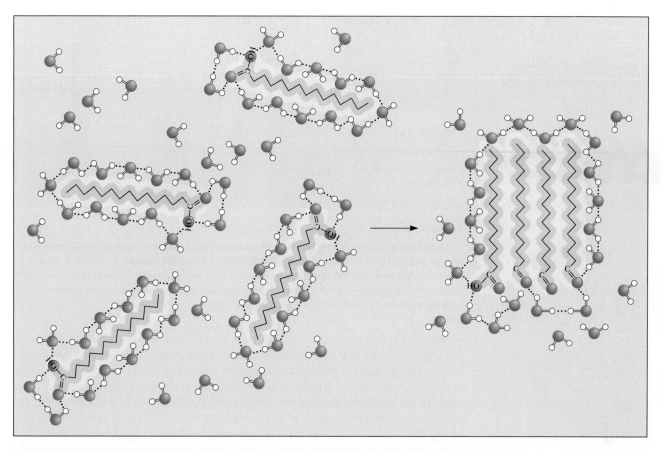

FIGURE 2.6 Hydrophobic interactions between nonpolar molecules (or nonpolar regions of molecules) are due to the increase in entropy of solvent water molecules.

these substances. The ionic carboxylate function hydrates readily, whereas the long hydrophobic tail is intrinsically insoluble. Nevertheless, sodium palmitate and other amphiphilic molecules readily disperse in water because the hydrocarbon tails of these substances are joined together in hydrophobic interactions as their polar carboxylate functions are hydrated in typical hydrophilic fashion. Such clusters of amphipathic molecules are termed **micelles;** Figure 2.7b is a 2-dimensional representation of a spherical micelle.

Influence of Solutes on Water Properties The presence of dissolved substances disturbs the structure of liquid water, thereby changing its properties. The dynamic H-bonding interactions of water must now accommodate the intruding substance. The net effect is that solutes, regardless of whether they are polar or nonpolar, fix nearby water molecules in a more ordered array. Ions, by establishing hydration shells through interactions with the water dipoles, create local order. Hydrophobic substances, for different reasons, make structures within water. To put it another way, by limiting the orientations that neighboring water molecules can assume, solutes give order to the solvent and diminish the dynamic interplay among H_2O molecules that occurs in pure water.

Colligative Properties This influence of the solute on water is reflected in a set of characteristic changes in behavior termed **colligative properties,** or properties related by a common principle. These alterations in solvent properties are related in that they all depend only on the number of solute particles per unit volume of solvent and not on the chemical nature of the solute. These effects include freezing point depression, boiling point elevation, vapor pressure lowering, and osmotic pressure effects. For example, 1 mol of an ideal solute dissolved in 1000 g of water (a 1 *m,* or molal, solution) at 1 atm pressure depresses the freezing point by 1.86°C, raises the boiling point by 0.543°C, lowers the vapor pressure in a temperature-dependent manner, and yields a solution whose osmotic pressure relative to pure water is 22.4 atm (at 25°C). In effect, by imposing local order on the water molecules, solutes make it more difficult for water to assume its crystalline lattice (freeze) or

(a) The sodium salt of palmitic acid: Sodium
palmitate $(Na^+ \ ^-OOC(CH_2)_{14}CH_3)$

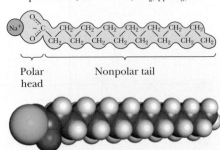

Polar
head

Nonpolar tail

(b)

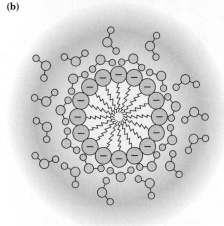

FIGURE 2.7 (a) An amphiphilic molecule: sodium palmitate. **(b)** Micelle formation by amphiphilic molecules in
aqueous solution. Because of their negatively charged surfaces, neighboring micelles repel one another and thereby
maintain a relative stability in solution.

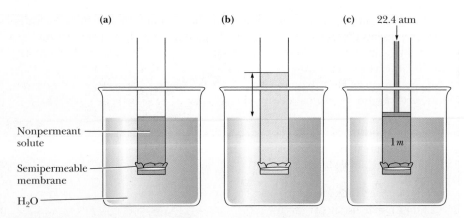

(a)   **(b)**   **(c)**   22.4 atm

Nonpermeant
solute

Semipermeable
membrane

H_2O

1 m

FIGURE 2.8 The osmotic pressure of a 1 molal *(m)* solution is equal to 22.4 atmospheres of pressure. **(a)** If a
nonpermeant solute is separated from pure water by a semipermeable membrane through which H_2O passes freely,
(b) water molecules enter the solution (osmosis) and the height of the solution column in the tube rises. The pressure necessary to push water back through the membrane at a rate exactly equaled by the water influx is the osmotic
pressure of the solution. **(c)** For a 1 *m* solution, this force is equal to 22.4 atm of pressure. Osmotic pressure is directly proportional to the concentration of the nonpermeant solute.

escape into the atmosphere (boil or vaporize). Furthermore, when a solution (such as the
1 *m* solution discussed here) is separated from a volume of pure water by a semipermeable
membrane, the solution draws water molecules across this barrier. The water molecules
are moving from a region of higher effective concentration (pure H_2O) to a region of lower
effective concentration (the solution). This movement of water into the solution dilutes the
effects of the solute that is present. The osmotic force exerted by each mole of solute is so
strong that it requires the imposition of 22.4 atm of pressure to be negated (Figure 2.8).

Osmotic pressure from high concentrations of dissolved solutes is a serious problem for
cells. Bacterial and plant cells have strong, rigid cell walls to contain these pressures. In
contrast, animal cells are bathed in extracellular fluids of comparable osmolarity, so no net
osmotic gradient exists. Also, to minimize the osmotic pressure created by the contents of
their cytosol, cells tend to store substances such as amino acids and sugars in polymeric
form. For example, a molecule of glycogen or starch containing 1000 glucose units exerts
only 1/1000 the osmotic pressure that 1000 free glucose molecules would.

Water Can Ionize to Form H⁺ and OH⁻

Water shows a small but finite tendency to form ions. This tendency is demonstrated by
the electrical conductivity of pure water, a property that clearly establishes the presence
of charged species (ions). Water ionizes because the larger, strongly electronegative oxygen atom strips the electron from one of its hydrogen atoms, leaving the proton to dissociate (Figure 2.9):

$$H\text{—}O\text{—}H \longrightarrow H^+ + OH^-$$

Two ions are thus formed: (1) protons or **hydrogen ions, H^+**, and (2) **hydroxyl ions, OH^-**.
Free protons are immediately hydrated to form **hydronium ions, H_3O^+**:

$$H^+ + H_2O \longrightarrow H_3O^+$$

Indeed, because most hydrogen atoms in liquid water are hydrogen bonded to a neighboring water molecule, this protonic hydration is an instantaneous process and the ion
products of water are H_3O^+ and OH^-:

$$\underset{H}{\overset{H}{\diagdown}}O\cdot\cdot H\text{—}O\underset{H}{\overset{H}{\diagup}} \longrightarrow \underset{H}{\overset{H}{\diagup}}O\text{—}H^+ + OH^-$$

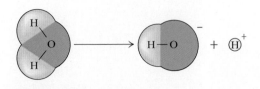

FIGURE 2.9 The ionization of water.

The amount of H_3O^+ or OH^- in 1 L (liter) of pure water at 25°C is 1×10^{-7} mol; the
concentrations are equal because the dissociation is stoichiometric.

Although it is important to keep in mind that the hydronium ion, or hydrated hydrogen ion, represents the true state in solution, the convention is to speak of hydrogen ion concentrations in aqueous solution, even though "naked" protons are virtually nonexistent. Indeed, H_3O^+ itself attracts a hydration shell by H bonding to adjacent water molecules to form an $H_9O_4^+$ species (Figure 2.10) and even more highly hydrated forms. Similarly, the hydroxyl ion, like all other highly charged species, is also hydrated.

K_w, the Ion Product of Water The dissociation of water into hydrogen ions and hydroxyl ions occurs to the extent that 10^{-7} mol of H^+ and 10^{-7} mol of OH^- are present at equilibrium in 1 L of water at 25°C.

$$H_2O \rightleftharpoons H^+ + OH^-$$

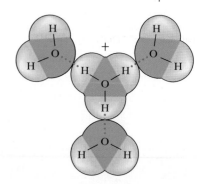

FIGURE 2.10 The hydration of H_3O^+.

The equilibrium constant for this process is

$$K_{eq} = \frac{[H^+][OH^-]}{[H_2O]}$$

where brackets denote concentrations in moles per liter. Because the concentration of H_2O in 1 L of pure water is equal to the number of grams in a liter divided by the gram molecular weight of H_2O, or 1000/18, the molar concentration of H_2O in pure water is 55.5 M (molar). The decrease in H_2O concentration as a result of ion formation ($[H^+]$, $[OH^-] = 10^{-7}M$) is negligible in comparison; thus its influence on the overall concentration of H_2O can be ignored. Thus,

$$K_{eq} = \frac{(10^{-7})(10^{-7})}{55.5} = 1.8 \times 10^{-16} \; M$$

Because the concentration of H_2O in pure water is essentially constant, a new constant, K_w, the **ion product of water,** can be written as

$$K_w = 55.5 \, K_{eq} = 10^{-14} \; M^2 = [H^+][OH^-]$$

This equation has the virtue of revealing the reciprocal relationship between H^+ and OH^- concentrations of aqueous solutions. If a solution is acidic (that is, it has a significant $[H^+]$), then the ion product of water dictates that the OH^- concentration is correspondingly less. For example, if $[H^+]$ is 10^{-2} M, $[OH^-]$ must be 10^{-12} M ($K_w = 10^{-14}$ $M^2 = [10^{-2}][OH^-]$; $[OH^-] = 10^{-12}$ M). Similarly, in an alkaline, or basic, solution in which $[OH^-]$ is great, $[H^+]$ is low.

2.2 What Is pH?

To avoid the cumbersome use of negative exponents to express concentrations that range over 14 orders of magnitude, Søren Sørensen, a Danish biochemist, devised the **pH scale** by defining **pH** as *the negative logarithm of the hydrogen ion concentration*[1]:

$$pH = -\log_{10} [H^+]$$

Table 2.2 gives the pH scale. Note again the reciprocal relationship between $[H^+]$ and $[OH^-]$. Also, because the pH scale is based on negative logarithms, low pH values represent the highest H^+ concentrations (and the lowest OH^- concentrations, as K_w specifies). Note also that

$$pK_w = pH + pOH = 14$$

The pH scale is widely used in biological applications because hydrogen ion concentrations in biological fluids are very low, about 10^{-7} M or 0.0000001 M, a value more easily represented as pH 7. The pH of blood plasma, for example, is 7.4, or 0.00000004 M H^+.

---

[1]To be precise in physical chemical terms, the *activities* of the various components, *not* their molar concentrations, should be used in these equations. The activity *(a)* of a solute component is defined as the product of its molar concentration, *c,* and an *activity coefficient, γ: a = cγ.* Most biochemical work involves dilute solutions, and the use of activities instead of molar concentrations is usually neglected. However, the concentration of certain solutes may be very high in living cells.

TABLE 2.2   pH Scale

The hydrogen ion and hydroxyl ion concentrations are given in moles per liter at 25°C.

pH	[H⁺]		[OH⁻]	
0	(10^0)	1.0	0.00000000000001	(10^{-14})
1	(10^{-1})	0.1	0.0000000000001	(10^{-13})
2	(10^{-2})	0.01	0.000000000001	(10^{-12})
3	(10^{-3})	0.001	0.00000000001	(10^{-11})
4	(10^{-4})	0.0001	0.0000000001	(10^{-10})
5	(10^{-5})	0.00001	0.000000001	(10^{-9})
6	(10^{-6})	0.000001	0.00000001	(10^{-8})
7	$\mathbf{(10^{-7})}$	**0.0000001**	**0.0000001**	$\mathbf{(10^{-7})}$
8	(10^{-8})	0.00000001	0.000001	(10^{-6})
9	(10^{-9})	0.000000001	0.00001	(10^{-5})
10	(10^{-10})	0.0000000001	0.0001	(10^{-4})
11	(10^{-11})	0.00000000001	0.001	(10^{-3})
12	(10^{-12})	0.000000000001	0.01	(10^{-2})
13	(10^{-13})	0.0000000000001	0.1	(10^{-1})
14	(10^{-14})	0.00000000000001	1.0	(10^0)

TABLE 2.3   The pH of Various Common Fluids

Fluid	pH
Household lye	13.6
Bleach	12.6
Household ammonia	11.4
Milk of magnesia	10.3
Baking soda	8.4
Seawater	7.5–8.4
Pancreatic fluid	7.8–8.0
Blood plasma	7.4
Intracellular fluids	
Liver	6.9
Muscle	6.1
Saliva	6.6
Urine	5–8
Boric acid	5.0
Beer	4.5
Orange juice	4.3
Grapefruit juice	3.2
Vinegar	2.9
Soft drinks	2.8
Lemon juice	2.3
Gastric juice	1.2–3.0
Battery acid	0.35

Certain disease conditions may lower the plasma pH level to 6.8 or less, a situation that may result in death. At pH 6.8, the H⁺ concentration is 0.00000016 *M*, four times greater than at pH 7.4.

At pH 7, [H⁺] = [OH⁻]; that is, there is no excess acidity or basicity. The point of **neutrality** is at pH 7, and solutions having a pH of 7 are said to be at **neutral pH.** The pH values of various fluids of biological origin or relevance are given in Table 2.3. Because the pH scale is a logarithmic scale, two solutions whose pH values differ by 1 pH unit have a tenfold difference in [H⁺]. For example, grapefruit juice at pH 3.2 contains more than 12 times as much H⁺ as orange juice at pH 4.3.

Strong Electrolytes Dissociate Completely in Water

Substances that are almost completely dissociated to form ions in solution are called **strong electrolytes.** The term **electrolyte** describes substances capable of generating ions in solution and thereby causing an increase in the electrical conductivity of the solution. Many salts (such as $NaCl$ and K_2SO_4) fit this category, as do strong acids (such as HCl) and strong bases (such as NaOH). Recall from general chemistry that acids are proton donors and bases are proton acceptors. In effect, the dissociation of a strong acid such as HCl in water can be treated as a proton transfer reaction between the acid HCl and the base H_2O to give the **conjugate acid** H_3O^+ and the **conjugate base** Cl^-:

$$HCl + H_2O \longrightarrow H_3O^+ + Cl^-$$

The equilibrium constant for this reaction is

$$K = \frac{[H_3O^+][Cl^-]}{[H_2O][HCl]}$$

Customarily, because the term [H_2O] is essentially constant in dilute aqueous solutions, it is incorporated into the equilibrium constant K to give a new term, K_a, the *acid dissociation constant,* where $K_a = K[H_2O]$. Also, the term [H_3O^+] is often replaced by H⁺, such that

$$K_a = \frac{[H^+][Cl^-]}{[HCl]}$$

For HCl, the value of K_a is exceedingly large because the concentration of HCl in aqueous solution is vanishingly small. Because this is so, the pH of HCl solutions is readily calculated from the amount of HCl used to make the solution:

$$[\text{H}^+] \text{ in solution} = [\text{HCl}] \text{ added to solution}$$

Thus, a 1 M solution of HCl has a pH of 0; a 1 mM HCl solution has a pH of 3. Similarly, a 0.1 M NaOH solution has a pH of 13. (Because $[\text{OH}^-] = 0.1$ M, $[\text{H}^+]$ must be 10^{-13} M.) Viewing the dissociation of strong electrolytes another way, we see that the ions formed show little affinity for each other. For example, in HCl in water, Cl^- has very little affinity for H^+:

$$\text{HCl} \longrightarrow \text{H}^+ + \text{Cl}^-$$

and in NaOH solutions, Na^+ has little affinity for OH^-. The dissociation of these substances in water is effectively complete.

Weak Electrolytes Are Substances That Dissociate Only Slightly in Water

Substances with only a slight tendency to dissociate to form ions in solution are called **weak electrolytes.** Acetic acid, CH_3COOH, is a good example:

$$\text{CH}_3\text{COOH} + \text{H}_2\text{O} \rightleftharpoons \text{CH}_3\text{COO}^- + \text{H}_3\text{O}^+$$

The acid dissociation constant K_a for acetic acid is 1.74×10^{-5} M:

$$K_a = \frac{[\text{H}^+][\text{CH}_3\text{COO}^-]}{[\text{CH}_3\text{COOH}]} = 1.74 \times 10^{-5} \ M$$

K_a is also termed an **ionization constant** because it states the extent to which a substance forms ions in water. The relatively low value of K_a for acetic acid reveals that the un-ionized form, CH_3COOH, predominates over H^+ and CH_3COO^- in aqueous solutions of acetic acid. Viewed another way, CH_3COO^-, the acetate ion, has a high affinity for H^+.

EXAMPLE What is the pH of a 0.1 M solution of acetic acid? In other words, what is the final pH when 0.1 mol of acetic acid (HAc) is added to water and the volume of the solution is adjusted to equal 1 L?

Answer
The dissociation of HAc in water can be written simply as

$$\text{HAc} \rightleftharpoons \text{H}^+ + \text{Ac}^-$$

where Ac^- represents the acetate ion, CH_3COO^-. In solution, some amount x of HAc dissociates, generating x amount of Ac^- and an equal amount x of H^+. Ionic equilibria characteristically are established very rapidly. At equilibrium, the concentration of HAc + Ac^- must equal 0.1 M. So, [HAc] can be represented as $(0.1 - x)$ M, and $[\text{Ac}^-]$ and $[\text{H}^+]$ then both equal x molar. From 1.74×10^{-5} $M =$ $([\text{H}^+][\text{Ac}^-])/[\text{HAc}]$, we get 1.74×10^{-5} $M = x^2/[0.1 - x]$. The solution to quadratic equations of this form ($ax^2 + bx + c = 0$) is $x = (-b \pm \sqrt{b^2 - 4ac})/2a$. For $x^2 + (1.74 \times 10^{-5})x - (1.74 \times 10^{-6}) = 0$, $x = 1.319 \times 10^{-3}$ M, so pH = 2.88. (Note that the calculation of x can be simplified here: Because K_a is quite small, $x << 0.1$ M. Therefore, K_a is essentially equal to $x^2/0.1$. Thus, $x^2 = 1.74 \times 10^{-6}$ M^2, so $x = 1.32 \times 10^{-3}$ M, and pH = 2.88.)

The Henderson–Hasselbalch Equation Describes the Dissociation of a Weak Acid In the Presence of Its Conjugate Base

Consider the ionization of some weak acid, HA, occurring with an acid dissociation constant, K_a. Then,

$$HA \rightleftharpoons H^+ + A^-$$

and

$$K_a = \frac{[H^+][A^-]}{[HA]}$$

Rearranging this expression in terms of the parameter of interest, $[H^+]$, we have

$$[H^+] = \frac{[K_a][HA]}{[A^-]}$$

Taking the logarithm of both sides gives

$$\log [H^+] = \log K_a + \log_{10} \frac{[HA]}{[A^-]}$$

If we change the signs and define $pK_a = -\log K_a$, we have

$$pH = pK_a - \log_{10} \frac{[HA]}{[A^-]}$$

or

$$\mathbf{pH = pK_a + \log_{10} \frac{[A^-]}{[HA]}}$$

This relationship is known as the **Henderson–Hasselbalch equation.** Thus, the pH of a solution can be calculated, provided K_a and the concentrations of the weak acid HA and its conjugate base A^- are known. Note particularly that when $[HA] = [A^-]$, $pH = pK_a$. For example, if equal volumes of 0.1 M HAc and 0.1 M sodium acetate are mixed, then

$$pH = pK_a = 4.76$$

$$pK_a = -\log K_a = -\log_{10}(1.74 \times 10^{-5}) = 4.76$$

(Sodium acetate, the sodium salt of acetic acid, is a strong electrolyte and dissociates completely in water to yield Na^+ and Ac^-.)

The Henderson–Hasselbalch equation provides a general solution to the quantitative treatment of acid–base equilibria in biological systems. Table 2.4 gives the acid dissociation constants and pK_a values for some weak electrolytes of biochemical interest.

EXAMPLE   What is the pH when 100 mL of 0.1 N NaOH is added to 150 mL of 0.2 M HAc if pK_a for acetic acid = 4.76?

Answer
100 mL 0.1 N NaOH = 0.01 mol OH^-, which neutralizes 0.01 mol of HAc, giving an equivalent amount of Ac^-:

$$OH^- + HAc \longrightarrow Ac^- + H_2O$$

0.02 mol of the original 0.03 mol of HAc remains essentially undissociated. The final volume is 250 mL.

$$pH = pK_a + \log_{10} \frac{[Ac^-]}{[HAc]} = 4.76 + \log(0.01\ mol/0.02\ mol)$$

$$pH = 4.76 - \log_{10} 2 = 4.46$$

If 150 mL of 0.2 M HAc had merely been diluted with 100 mL of water, this would leave 250 mL of a 0.12 M HAc solution. The pH would be given by:

$$K_a = \frac{[H^+][Ac^-]}{[HAc]} = \frac{x^2}{0.12\ M} = 1.74 \times 10^{-5}\ M$$

$$x = 1.44 \times 10^{-3} = [H^+]$$

$$pH = 2.84$$

TABLE 2.4	Acid Dissociation Constants and pK_a Values for Some Weak Electrolytes (at 25°C)	
Acid	K_a (M)	pK_a
HCOOH (formic acid)	1.78×10^{-4}	3.75
CH$_3$COOH (acetic acid)	1.74×10^{-5}	4.76
CH$_3$CH$_2$COOH (propionic acid)	1.35×10^{-5}	4.87
CH$_3$CHOHCOOH (lactic acid)	1.38×10^{-4}	3.86
HOOCCH$_2$CH$_2$COOH (succinic acid) pK_1*	6.16×10^{-5}	4.21
HOOCCH$_2$CH$_2$COO$^-$ (succinic acid) pK_2	2.34×10^{-6}	5.63
H$_3$PO$_4$ (phosphoric acid) pK_1	7.08×10^{-3}	2.15
H$_2$PO$_4^-$ (phosphoric acid) pK_2	6.31×10^{-8}	7.20
HPO$_4^{2-}$ (phosphoric acid) pK_3	3.98×10^{-13}	12.40
C$_3$N$_2$H$_5^+$ (imidazole)	1.02×10^{-7}	6.99
C$_6$O$_2$N$_3$H$_{11}^+$ (histidine–imidazole group) pK_R†	9.12×10^{-7}	6.04
H$_2$CO$_3$ (carbonic acid) pK_1	1.70×10^{-4}	3.77
HCO$_3^-$ (bicarbonate) pK_2	5.75×10^{-11}	10.24
(HOCH$_2$)$_3$CNH$_3^+$ (*tris*-hydroxymethyl aminomethane)	8.32×10^{-9}	8.07
NH$_4^+$ (ammonium)	5.62×10^{-10}	9.25
CH$_3$NH$_3^+$ (methylammonium)	2.46×10^{-11}	10.62

*The pK values listed as pK_1, pK_2, or pK_3 are in actuality pK_a values for the respective dissociations. This simplification in notation is used throughout this book.
†pK_R refers to the imidazole ionization of histidine.
Data from *CRC Handbook of Biochemistry*, The Chemical Rubber Co., 1968.

Titration Curves Illustrate the Progressive Dissociation of a Weak Acid

Titration is the analytical method used to determine the amount of acid in a solution. A measured volume of the acid solution is titrated by slowly adding a solution of base, typically NaOH, of known concentration. As incremental amounts of NaOH are added, the pH of the solution is determined and a plot of the pH of the solution versus the amount of OH$^-$ added yields a **titration curve.** The titration curve for acetic acid is

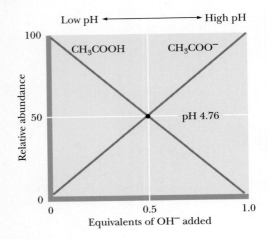

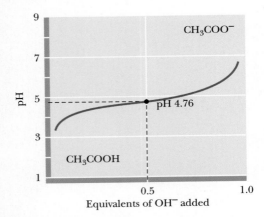

FIGURE 2.11 The titration curve for acetic acid. Note that the titration curve is relatively flat at pH values near the pK_a. In other words, the pH changes relatively little as OH^- is added in this region of the titration curve.

shown in Figure 2.11. In considering the progress of this titration, keep in mind two important equilibria:

$$1. \quad HAc \rightleftharpoons H^+ + Ac^- \qquad K_a = 1.74 \times 10^{-5}$$

$$2. \quad H^+ + OH^- \rightleftharpoons H_2O \qquad K = \frac{[H_2O]}{[K_w]} = 5.55 \times 10^{15}$$

As the titration begins, mostly HAc is present, plus some H^+ and Ac^- in amounts that can be calculated (see the Example on page 39). Addition of a solution of NaOH allows hydroxide ions to neutralize any H^+ present. Note that reaction (2) as written is strongly favored; its apparent equilibrium constant is greater than 10^{15}! As H^+ is neutralized, more HAc dissociates to H^+ and Ac^-. The stoichiometry of the titration is 1:1—for each increment of OH^- added, an equal amount of the weak acid HAc is titrated. As additional NaOH is added, the pH gradually increases as Ac^- accumulates at the expense of diminishing HAc and the neutralization of H^+. At the point where half of the HAc has been neutralized (that is, where 0.5 equivalent of OH^- has been added), the concentrations of HAc and Ac^- are equal and pH = pK_a for HAc. Thus, we have an experimental method for determining the pK_a values of weak electrolytes. These pK_a values lie at the midpoint of their respective titration curves. After all of the acid has been neutralized (that is, when one equivalent of base has been added), the pH rises exponentially.

The shapes of the titration curves of weak electrolytes are identical, as Figure 2.12 reveals. Note, however, that the midpoints of the different curves vary in a way that characterizes the particular electrolytes. The pK_a for acetic acid is 4.76, the pK_a for imidazole is 6.99, and that for ammonium is 9.25. These pK_a values are directly related to the dissociation constants of these substances, or, viewed the other way, to the relative affinities of the conjugate bases for protons. NH_3 has a high affinity for protons compared to Ac^-; NH_4^+ is a poor acid compared to HAc.

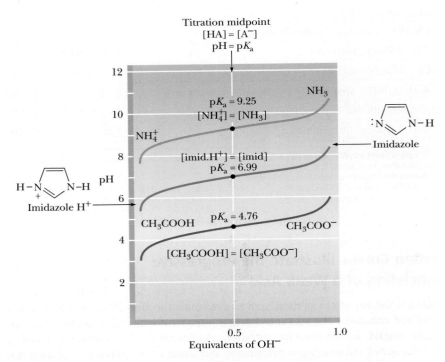

FIGURE 2.12 The titration curves of several weak electrolytes: acetic acid, imidazole, and ammonium.

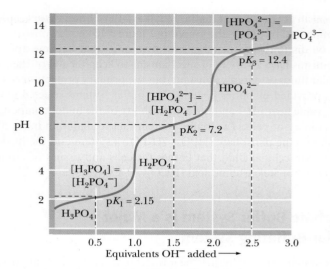

FIGURE 2.13 The titration curve for phosphoric acid.

Phosphoric Acid Has Three Dissociable H$^+$

Figure 2.13 shows the titration curve for phosphoric acid, H_3PO_4. This substance is a polyprotic acid, meaning it has more than one dissociable proton. Indeed, it has three, and thus three equivalents of OH^- are required to neutralize it, as Figure 2.13 shows. Note that the three dissociable H$^+$ are lost in discrete steps, each dissociation showing a characteristic pK_a. Note that pK_1 occurs at pH = 2.15, and the concentrations of the acid H_3PO_4 and the conjugate base $H_2PO_4^-$ are equal. As the next dissociation is approached, $H_2PO_4^-$ is treated as the acid and HPO_4^{2-} is its conjugate base. Their concentrations are equal at pH 7.20, so pK_2 = 7.20. (Note that at this point, 1.5 equivalents of OH^- have been added.) As more OH^- is added, the last dissociable hydrogen is titrated, and pK_3 occurs at pH = 12.4, where $[HPO_4^{2-}] = [PO_4^{3-}]$.

The shape of the titration curves for weak electrolytes has a biologically relevant property: In the region of the pK_a, pH remains relatively unaffected as increments of OH^- (or H$^+$) are added. The weak acid and its conjugate base are acting as a buffer.

2.3 What Are Buffers, and What Do They Do?

Buffers are solutions that tend to resist changes in their pH as acid or base is added. Typically, a buffer system is composed of a weak acid and its conjugate base. A solution of a weak acid that has a pH nearly equal to its pK_a, by definition, contains an amount of the conjugate base nearly equivalent to the weak acid. Note that in this region, the titration curve is relatively flat (Figure 2.14). Addition of H$^+$ then has little effect because it is absorbed by the following reaction:

$$H^+ + A^- \longrightarrow HA$$

Similarly, any increase in $[OH^-]$ is offset by the process

$$OH^- + HA \longrightarrow A^- + H_2O$$

Thus, the pH remains relatively constant. The components of a buffer system are chosen such that the pK_a of the weak acid is close to the pH of interest. It is at the pK_a that the buffer system shows its greatest buffering capacity. At pH values more than 1 pH unit from the pK_a, buffer systems become ineffective because the concentration of one of the components is too low to absorb the influx of H$^+$ or OH^-. The molarity of a buffer is defined as the *sum* of the concentrations of the acid and conjugate base forms.

Maintenance of pH is vital to all cells. The structure and hence the function of proteins, nucleic acids, and many other cellular molecules depends on weak forces such as H bonds and ionic interactions, both of which can be affected by pH. Also, processes

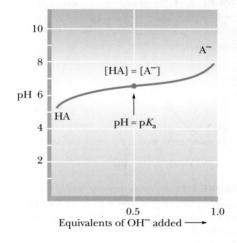

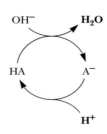

FIGURE 2.14 A buffer system consists of a weak acid, HA, and its conjugate base, A$^-$.

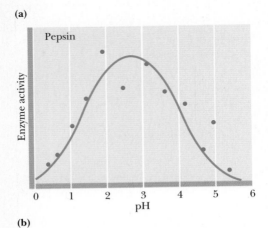

(a)

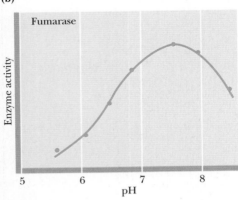

(b)

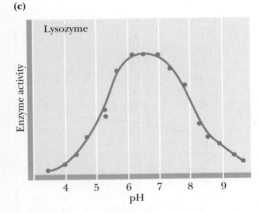

(c)

FIGURE 2.15 pH versus enzymatic activity. Pepsin is a protein-digesting enzyme active in the gastric fluid. Fumarase is a metabolic enzyme found in mitochondria. Lysozyme digests the cell walls of bacteria; it is found in cells and bodily fluids.

such as metabolism are dependent on the activities of enzymes; in turn, enzyme activity is markedly influenced by pH, as the graphs in Figure 2.15 show. Consequently, changes in pH would be disruptive to metabolism for reasons that become apparent in later chapters. Organisms have a variety of mechanisms to keep the pH of their intracellular and extracellular fluids essentially constant, but the primary protection against harmful pH changes is provided by buffer systems. The buffer systems selected reflect both the need for a pK_a value near pH 7 and the compatibility of the buffer components with the metabolic machinery of cells. Two buffer systems act to maintain intracellular pH essentially constant—the phosphate ($HPO_4^{2-}/H_2PO_4^-$) system and the histidine system. The pH of the extracellular fluid that bathes the cells and tissues of animals is maintained by the bicarbonate/carbonic acid (HCO_3^-/H_2CO_3) system.

The Phosphate Buffer System Is a Major Intracellular Buffering System

The **phosphate system** serves to buffer the intracellular fluid of cells at physiological pH because pK_2 lies near this pH value. The intracellular pH of most cells is maintained in the range between 6.9 and 7.4. Phosphate is an abundant anion in cells, both in inorganic form and as an important functional group on organic molecules that serve as metabolites or macromolecular precursors. In both organic and inorganic forms, its characteristic pK_2 means that the ionic species present at physiological pH are sufficient to donate or accept hydrogen ions to buffer any changes in pH, as the titration curve for H_3PO_4 in Figure 2.13 reveals. For example, if the total cellular concentration of phosphate is 20 mM (millimolar) and the pH is 7.4, the distribution of the major phosphate species is given by

$$pH = pK_2 + \log_{10} \frac{\left[HPO_4^{2-}\right]}{\left[H_2PO_4^-\right]}$$

$$7.4 = 7.20 + \log_{10} \frac{\left[HPO_4^{2-}\right]}{\left[H_2PO_4^-\right]}$$

$$\frac{\left[HPO_4^{2-}\right]}{\left[H_2PO_4^-\right]} = 1.58$$

Thus, if $[HPO_4^{2-}] + [H_2PO_4^-] = 20$ mM, then

$$[HPO_4^{2-}] = 12.25 \text{ m}M \quad \text{and} \quad [H_2PO_4^-] = 7.75 \text{ m}M$$

The Imidazole Group of Histidine Also Serves as an Intracellular Buffering System

Histidine is one of the 20 naturally occurring amino acids commonly found in proteins (see Chapter 4). It possesses as part of its structure an imidazole group, a five-membered heterocyclic ring possessing two nitrogen atoms. The pK_a for dissociation of the imidazole hydrogen of histidine is 6.04.

In cells, histidine occurs as the free amino acid, as a constituent of proteins, and as part of dipeptides in combination with other amino acids. Because the concentration of free histidine is low and its imidazole pK_a is more than 1 pH unit removed from prevailing intracellular pH, its role in intracellular buffering is minor. However, protein-bound and

The Bicarbonate Buffer System of Blood Plasma

The important buffer system of blood plasma is the bicarbonate/carbonic acid couple:

$$H_2CO_3 \rightleftharpoons H^+ + HCO_3^-$$

The relevant pK_a, pK_1 for carbonic acid, has a value far removed from the normal pH of blood plasma (pH 7.4). (The pK_1 for H_2CO_3 at 25°C is 3.77 [Table 2.4], but at 37°C, pK_1 is 3.57.) At pH 7.4, the concentration of H_2CO_3 is a minuscule fraction of the HCO_3^- concentration; thus the plasma appears to be poorly protected against an influx of OH^- ions.

$$pH = 7.4 = 3.57 + \log_{10} \frac{[HCO_3^-]}{[H_2CO_3]}$$

$$\frac{[HCO_3^-]}{[H_2CO_3]} = 6761$$

For example, if $[HCO_3^-]$ = 24 mM, then $[H_2CO_3]$ is only 3.55 μM (3.55 × 10^{-6} M), and an equivalent amount of OH^- (its usual concentration in plasma) would swamp the buffer system, causing a dangerous rise in the plasma pH. How, then, can this bicarbonate system function effectively? The bicarbonate buffer system works well because the critical concentration of H_2CO_3 is maintained relatively constant through equilibrium with dissolved CO_2 produced in the tissues and available as a gaseous CO_2 reservoir in the lungs.*

Gaseous CO_2 from the lungs and tissues is dissolved in the blood plasma, symbolized as $CO_2(d)$, and hydrated to form H_2CO_3:

$$CO_2(g) \rightleftharpoons CO_2(d)$$
$$CO_2(d) + H_2O \rightleftharpoons H_2CO_3$$
$$H_2CO_3 \rightleftharpoons H^+ + HCO_3^-$$

Thus, the concentration of H_2CO_3 is itself buffered by the available pools of CO_2. The hydration of CO_2 is actually mediated by an enzyme, *carbonic anhydrase,* which facilitates the equilibrium by rapidly catalyzing the reaction

$$H_2O + CO_2(d) \rightleftharpoons H_2CO_3$$

Under the conditions of temperature and ionic strength prevailing in mammalian body fluids, the equilibrium for this reaction lies far to the left, such that more than 300 CO_2 molecules are present in solution for every molecule of H_2CO_3. Because dissolved CO_2 and H_2CO_3 are in equilibrium, the proper expression for H_2CO_3 availability is $[CO_2(d)] + [H_2CO_3]$, the so-called total carbonic acid pool, consisting primarily of $CO_2(d)$. The overall equilibrium for the bicarbonate buffer system then is

$$CO_2(d) + H_2O \overset{K_h}{\rightleftharpoons} H_2CO_3$$

$$H_2CO_3 \overset{K_1}{\rightleftharpoons} H^+ + HCO_3^-$$

An expression for the ionization of H_2CO_3 under such conditions (that is, in the presence of dissolved CO_2) can be obtained from K_h,

the equilibrium constant for the hydration of CO_2, and from K_1, the first acid dissociation constant for H_2CO_3:

$$K_h = \frac{[H_2CO_3]}{[CO_2(d)]}$$

Thus,

$$[H_2CO_3] = K_h[CO_2(d)]$$

Putting this value for $[H_2CO_3]$ into the expression for the first dissociation of H_2CO_3 gives

$$K_1 = \frac{[H^+][HCO_3^-]}{[H_2CO_3]}$$

$$= \frac{[H^+][HCO_3^-]}{K_h[CO_2(d)]}$$

Therefore, the overall equilibrium constant for the ionization of H_2CO_3 in equilibrium with $CO_2(d)$ is given by

$$K_1 K_h = \frac{[H^+][HCO_3^-]}{[CO_2(d)]}$$

and $K_1 K_h$, the product of two constants, can be defined as a new equilibrium constant, $K_{overall}$. The value of K_h is 0.003 at 37°C and K_1, the ionization constant for H_2CO_3, is $10^{-3.57}$ = 0.000269. Therefore,

$$K_{overall} = (0.000269)(0.003)$$
$$= 8.07 \times 10^{-7}$$
$$pK_{overall} = 6.1$$

which yields the following Henderson–Hasselbalch relationship:

$$pH = pK_{overall} + \log_{10} \frac{[HCO_3^-]}{[CO_2(d)]}$$

Although the prevailing blood pH of 7.4 is more than 1 pH unit away from $pK_{overall}$, the bicarbonate system is still an effective buffer. Note that, at blood pH, the concentration of the acid component of the buffer will be less than 10% of the conjugate base component. One might imagine that this buffer component could be overwhelmed by relatively small amounts of alkali, with consequent disastrous rises in blood pH. However, the acid component is the total carbonic acid pool, that is, $[CO_2(d)] + [H_2CO_3]$, which is stabilized by its equilibrium with $CO_2(g)$. Gaseous CO_2 serves to buffer any losses from the total carbonic acid pool by entering solution as $CO_2(d)$, and blood pH is effectively maintained. Thus, the bicarbonate buffer system is an *open system.* The natural presence of CO_2 gas at a partial pressure of 40 mm Hg in the alveoli of the lungs and the equilibrium

$$CO_2(g) \rightleftharpoons CO_2(d)$$

keep the concentration of $CO_2(d)$ (the principal component of the total carbonic acid pool in blood plasma) in the neighborhood of 1.2 mM. Plasma $[HCO_3^-]$ is about 24 mM under such conditions.

*Well-fed humans exhale about 1 kg of CO_2 daily. Imagine the excretory problem if CO_2 were not a volatile gas.

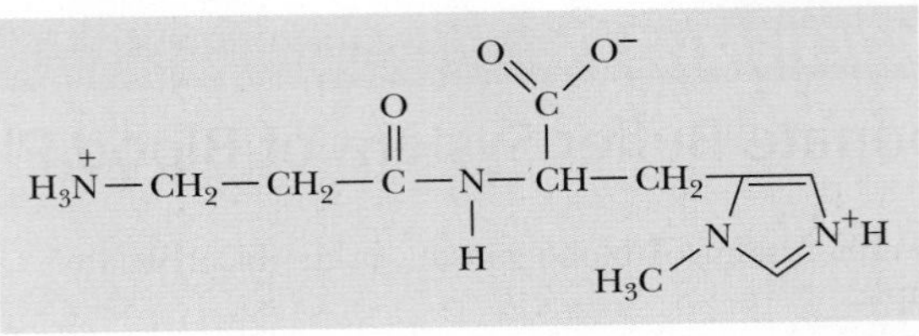

FIGURE 2.16 Anserine (N-β-alanyl-3-methyl-L-histidine) is an important dipeptide buffer in the maintenance of intracellular pH in some tissues. The structure shown is the predominant ionic species at pH 7. pK_1 (COOH) = 2.64; pK_2 (imidazole-N$^+$H) = 7.04; pK_3 (N$^+$H$_3$) = 9.49.

dipeptide histidine may be the dominant buffering system in some cells. In combination with other amino acids, as in proteins or dipeptides, the imidazole pK_a may increase substantially. For example, the imidazole pK_a is 7.04 in **anserine**, a dipeptide containing β-alanine and histidine (Figure 2.16). Thus, this pK_a is near physiological pH, and some histidine peptides are well suited for buffering at physiological pH.

"Good" Buffers Are Buffers Useful Within Physiological pH Ranges

Not many common substances have pK_a values in the range from 6 to 8. Consequently, biochemists conducting in vitro experiments were limited in their choice of buffers effective at or near physiological pH. In 1966, N. E. Good devised a set of synthetic buffers to remedy this problem, and over the years the list has expanded so that a "good" selection is available. Some of these compounds are analogs of *tris*-hydroxymethyl aminomethane (Tris, see end-of-chapter Problem 16), such as triethanolamine (TEA, see end-of-chapter Problem 15) or *N,N-bis* (2-hydroxyethyl) glycine (Bicine, see end-of-chapter Problems 8 and 17). Others are derivatives of *N*-ethane sulfonic acids, such as HEPES (Figure 2.17).

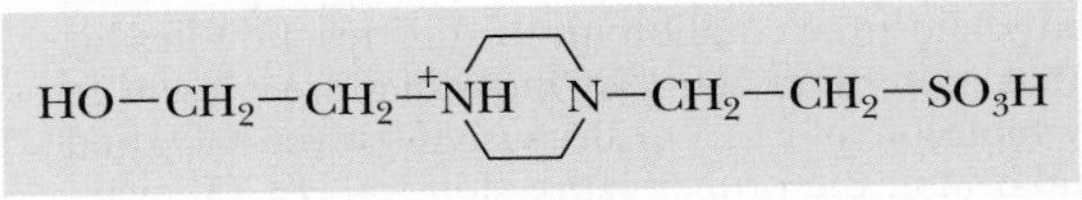

HEPES

FIGURE 2.17 The structure of HEPES, 4-(2-hydroxy)-1-piperazine ethane sulfonic acid, in its fully protonated form. The pK_a of the sulfonic acid group is about 3; the pK_a of the piperazine-N$^+$H is 7.55 at 20°C.

HUMAN BIOCHEMISTRY

Blood pH and Respiration

Hyperventilation, defined as a breathing rate more rapid than necessary for normal CO_2 elimination from the body, can result in an inappropriately low [CO_2(g)] in the blood. Central nervous system disorders such as meningitis, encephalitis, or cerebral hemorrhage, as well as a number of drug- or hormone-induced physiological changes, can lead to hyperventilation. As [CO_2(g)] drops due to excessive exhalation, [H_2CO_3] in the blood plasma falls, followed by a decline in [H^+] and [HCO_3^-] in the blood plasma. Blood pH rises within 20 sec of the onset of hyperventilation, becoming maximal within 15 min. [H^+] can change from its normal value of 40 nM

(pH = 7.4) to 18 nM (pH = 7.74). This rise in plasma pH (increase in alkalinity) is termed **respiratory alkalosis.**

Hypoventilation is the opposite of hyperventilation and is characterized by an inability to excrete CO_2 rapidly enough to meet physiological needs. Hypoventilation can be caused by narcotics, sedatives, anesthetics, and depressant drugs; diseases of the lung also lead to hypoventilation. Hypoventilation results in **respiratory acidosis,** as CO_2(g) accumulates, giving rise to H_2CO_3, which dissociates to form H^+ and HCO_3^-.

2.4 What Properties of Water Give It a Unique Role in the Environment?

The remarkable properties of water render it particularly suitable to its unique role in living processes and the environment, and its presence in abundance favors the existence of life. Let's examine water's physical and chemical properties to see the extent to which they provide conditions that are advantageous to organisms.

As a solvent, water is powerful yet innocuous. No other chemically inert solvent compares with water for the substances it can dissolve. Also, it is very important to life that water is a "poor" solvent for nonpolar substances. Thus, through hydrophobic interactions, lipids coalesce, membranes form, boundaries are created delimiting compartments, and the cellular nature of life is established. Because of its very high dielectric constant, water is a medium for ionization. Ions enrich the living environment in that they enhance the variety of chemical species and introduce an important class of chemical reactions. They provide electrical properties to solutions and therefore to organisms. Aqueous solutions are the prime source of ions.

The thermal properties of water are especially relevant to its environmental fitness. It has great power as a buffer resisting thermal (temperature) change. Its heat capacity, or specific heat (4.1840 J/g°C), is remarkably high; it is ten times greater than iron, five times greater than quartz or salt, and twice as great as hexane. Its heat of fusion is 335 J/g. Thus, at 0°C, it takes a loss of 335 J to change the state of 1 g of H_2O from liquid to solid. Its heat of vaporization (2.24 kJ/g) is exceptionally high. These thermal properties mean that it takes substantial changes in heat content to alter the temperature and especially the state of water. Water's thermal properties allow it to buffer the climate through such processes as condensation, evaporation, melting, and freezing. Furthermore, these properties allow effective temperature regulation in living organisms. For example, heat generated within an organism as a result of metabolism can be efficiently eliminated through evaporation or conduction. The thermal conductivity of water is very high compared with that of other liquids. The anomalous expansion of water as it cools to temperatures near its freezing point is a unique attribute of great significance to its natural fitness. As water cools, H bonding increases because the thermal motions of the molecules are lessened. H bonding tends to separate the water molecules (Figure 2.2), thus decreasing the density of water. These changes in density mean that, at temperatures below 4°C, cool water rises and, most important, ice freezes on the surface of bodies of water, forming an insulating layer protecting the liquid water underneath.

Water has the highest surface tension (75 dyne/cm) of all common liquids (except mercury). Together, surface tension and density determine how high a liquid rises in a capillary system. Capillary movement of water plays a prominent role in the life of plants. Last, consider osmosis as it relates to water and, in particular, the bulk movement of water in the direction from a dilute aqueous solution to a more concentrated one across a semipermeable boundary. Such bulk movements determine the shape and form of living things.

Water is truly a crucial determinant of the fitness of the environment. In a very real sense, organisms are aqueous systems in a watery world.

SUMMARY

OWL Login to OWL to develop problem-solving skills and complete online homework assigned by your professor.

2.1 What Are the Properties of Water? Life depends on the unusual chemical and physical properties of H_2O. Its high boiling point, melting point, heat of vaporization, and surface tension indicate that intermolecular forces of attraction between H_2O molecules are high. Hydrogen bonds between adjacent water molecules are the basis of these forces. Liquid water consists of H_2O molecules held in a random, three-dimensional network that has a local preference for tetrahedral geometry, yet contains a large number of strained or broken hydrogen bonds. The presence of strain creates a kinetic situation in which H_2O molecules can switch H-bond allegiances; fluidity ensues. As kinetic energy decreases (the temperature falls), crystalline water (ice) forms.

The solvent properties of water are attributable to the "bent" structure of the water molecule and polar nature of its O—H bonds. Together these attributes yield a liquid that can form hydration shells around salt ions or dissolve polar solutes through H-bond interactions. Hydrophobic interactions in aqueous environments also arise as a consequence of polar interactions between water molecules. The polarity of the O—H bonds means that water also ionizes to a small but finite extent to release H^+ and OH^- ions. K_w, the ion product of water, reveals that the concentration of $[H^+]$ and $[OH^-]$ at 25°C is $10^{-7}\ M$.

2.2 What Is pH? pH is defined as $-\log_{10}[H^+]$. pH is an important concept in biochemistry because the structure and function of biological molecules depend strongly on functional groups that ionize, or not, depending on small changes in $[H^+]$ concentration. Weak electrolytes are substances that dissociate incompletely in water. The behavior of weak electrolytes determines the concentration of $[H^+]$ and hence, pH. The Henderson–Hasselbalch equation provides a general solution to the quantitative treatment of acid–base equilibria in biological systems.

2.3 What Are Buffers, and What Do They Do? Buffers are solutions composed of a weak acid and its conjugate base. Such solutions can resist changes in pH when acid or base is added to the solution. Maintenance of pH is vital to all cells, and primary protection against harmful pH changes is provided by buffer systems. The buffer systems used by cells reflect a need for a pK_a value near pH 7 and the compatibility of the buffer components with the metabolic apparatus of cells. The phosphate buffer system and the histidine–imidazole system are the two prominent intracellular buffers, whereas the bicarbonate buffer system is the principal extracellular buffering system in animals.

2.4 What Properties of Water Give It a Unique Role in the Environment? Life and water are inextricably related. Water is particularly suited to its unique role in living processes and the environment. As a solvent, water is powerful yet innocuous; no other chemically inert solvent compares with water for the substances it can dissolve. Also, water as a "poor" solvent for nonpolar substances gives rise to hydrophobic interactions, leading lipids to coalesce, membranes to form, and boundaries delimiting compartments to appear.

Water is a medium for ionization. Ions enrich the living environment and introduce an important class of chemical reactions. Ions provide electrical properties to solutions and therefore to organisms.

The thermal properties of water are especially relevant to its environmental fitness. It takes substantial changes in heat content to alter the temperature and especially the state of water. Water's thermal properties allow it to buffer the climate through such processes as condensation, evaporation, melting, and freezing. Furthermore, water's thermal properties allow effective temperature regulation in living organisms.

Osmosis as it relates to water, and in particular, the bulk movement of water in the direction from a dilute aqueous solution to a more concentrated one across semipermeable membranes, determines the shape and form of living things. In large degree, the properties of water define the fitness of the environment. Organisms are aqueous systems in a watery world.

FOUNDATIONAL BIOCHEMISTRY Things You Should Know After Reading Chapter 2.

- Water has unusual properties, and hydrogen bonding between water molecules is the key to understanding these properties.

- Water is an excellent solvent because of its polar nature.

- Hydrophobic interactions between nonpolar molecules in aqueous solutions occur because of an entropy-driven process based on a net decrease in order among water molecules.

- Amphiphilic molecules have both polar and nonpolar chemical groupings.

- Solute molecules affect the properties of the solution through their interactions with water molecules.

- Colligative properties are properties that depend on the number of solute molecules in a solvent, not the particular chemical properties of the solute molecules.

- Water can ionize to a small but finite extent to form H^+ and OH^- ions.

- The ion product of water, K_w, is given by $K_w = [H^+][OH^-]$. $K_w = 10^{-14}\ M^2$ at 25°C.

- pH is defined as the negative logarithm (to the base 10) of the H^+ concentration.

- Strong electrolytes are substances that dissociate completely in water to form ions.

- Strong electrolytes include strong acids, strong bases, and many salts, including the salts of weak acids.

- Weak electrolytes show only a slight tendency to dissociate in water and form ions.

- The ionization constant, K_a, is a measure of the extent to which a substance forms ions in solution.

- The Henderson-Hasselbalch equation relates pH to the pK_a of a weak acid, as determined by the relative concentrations of the weak acid and its conjugate base.

- Titration curves can be used to determine the pK_a values of weak electrolytes.

- Buffers are solutions containing amounts of a weak acid and its conjugate base sufficient to resist large changes in pH when acid or base is added.

- The phosphate buffer system is a major intracellular buffer system.

- The imidazole group of histidine can also serve as an intracellular buffer.

- The bicarbonate/carbonic acid system is a major extracellular buffering system.

- Not many naturally occurring substances have pK_a values in the neutral pH range, so biochemists have devised artificial buffering systems for in vitro work.

- The properties of water give it a significant role in the environment. The fitness of the environment for living systems depends on the properties of water.

PROBLEMS

Answers to all problems are at the end of this book. Detailed solutions are available in the *Student Solutions Manual, Study Guide, and Problems Book.*

1. Calculating pH from [H⁺] Calculate the pH of the following.
   a. $5 \times 10^{-4}\ M$ HCl
   d. $3 \times 10^{-2}\ M$ KOH
   b. $7 \times 10^{-5}\ M$ NaOH
   e. 0.04 mM HCl 0.04×10^{-3}
   c. $2\ \mu M$ HCl 2×10^{-6}
   f. $6 \times 10^{-9} M$ HCl

2. Calculating [H⁺] from pH Calculate the following from the pH values given in Table 2.3.
   a. [H⁺] in vinegar   2.9
   b. [H⁺] in saliva   6.6
   c. [H⁺] in household ammonia   11.4
   d. [OH⁻] in milk of magnesia   10.3
   e. [OH⁻] in beer   4.5
   f. [H⁺] inside a liver cell   6.9

3. Calculating [H⁺] and pK_a from the pH of a Solution of Weak Acid The pH of a 0.02 M solution of an acid was measured at 4.6.
   a. What is the [H⁺] in this solution?
   b. Calculate the acid dissociation constant K_a and pK_a for this acid.

4. Calculating the pH of a Solution of a Weak Acid; Calculating the pH of the Solution after the Addition of Strong Base The K_a for formic acid is $1.78 \times 10^{-4}\ M$.
   a. What is the pH of a 0.1 M solution of formic acid?
   b. 150 mL of 0.1 M NaOH is added to 200 mL of 0.1 M formic acid, and water is added to give a final volume of 1 L. What is the pH of the final solution?

5. Prepare a Buffer by Combining a Solution of Weak Acid with a Solution of the Salt of the Weak Acid Given 0.1 M solutions of acetic acid and sodium acetate, describe the preparation of 1 L of 0.1 M acetate buffer at a pH of 5.4.

6. Calculate the pH in a Muscle Cell from the HPO$_4^{2-}$/H$_2$PO$_4^{-}$ Ratio If the internal pH of a muscle cell is 6.8, what is the [HPO$_4^{2-}$]/[H$_2$PO$_4^{-}$] ratio in this cell?

7. Preparing a Phospate Buffer Solution of pH 7.5 from Solutions of Na$_3$PO$_4$ and H$_3$PO$_4$ Given 0.1 M solutions of Na$_3$PO$_4$ and H$_3$PO$_4$, describe the preparation of 1 L of a phosphate buffer at a pH of 7.5. What are the molar concentrations of the ions in the final buffer solution, including Na$^+$ and H$^+$?

8. Polyprotic Acids: Phosphate Species Abundance at Different pHs What are the approximate fractional concentrations of the following phosphate species at pH values of 0, 2, 4, 6, 8, 10, and 12?
   a. H$_3$PO$_4$
   b. H$_2$PO$_4^{-}$
   c. HPO$_4^{2-}$
   d. PO$_4^{3-}$

9. Polyprotic Acids: Citric Acid Species at Various pHs Citric acid, a tricarboxylic acid important in intermediary metabolism, can be symbolized as H$_3$A. Its dissociation reactions are

$$H_3A \rightleftharpoons H^+ + H_2A^- \qquad pK_1 = 3.13$$
$$H_2A^- \rightleftharpoons H^+ + HA^{2-} \qquad pK_2 = 4.76$$
$$HA^{2-} \rightleftharpoons H^+ + A^{3-} \qquad pK_3 = 6.40$$

If the *total* concentration of the acid *and* its anion forms is 0.02 M, what are the individual concentrations of H$_3$A, H$_2$A$^-$, HA^{2-}, and A^{3-} at pH 5.2?

10. Calculate the pH Change in a Phospate Buffer When Acid or Base is Added
   a. If 50 mL of 0.01 M HCl is added to 100 mL of 0.05 M phosphate buffer at pH 7.2, what is the resultant pH? What are the concentrations of H$_2$PO$_4^-$ and HPO$_4^{2-}$ in the final solution?
   b. If 50 mL of 0.01 M NaOH is added to 100 mL of 0.05 M phosphate buffer at pH 7.2, what is the resultant pH? What are the concentrations of H$_2$PO$_4^-$ and HPO$_4^{2-}$ in this final solution?

11. Explore the Bicarbonate/Carbonic Acid Buffering System of Blood Plasma At 37°C, if the plasma pH is 7.4 and the plasma concentration of HCO$_3^-$ is 15 mM, what is the plasma concentration of H$_2$CO$_3$? What is the plasma concentration of CO$_{2(dissolved)}$? If metabolic activity changes the concentration of CO$_{2(dissolved)}$ to 3 mM and [HCO$_3^-$] remains at 15 mM, what is the pH of the plasma?

12. How to Prepare a Buffer Solution: An Anserine Buffer Draw the titration curve for anserine (Figure 2.16). The isoelectric point of anserine is the pH where the net charge on the molecule is zero; what is the isoelectric point for anserine? Given a 0.1 M solution of anserine at its isoelectric point and ready access to 0.1 M HCl, 0.1 M NaOH and distilled water, describe the preparation of 1 L of 0.04 M anserine buffer solution, pH 7.2.

13. How to Prepare a Buffer Solution: a HEPES Buffer Given a solution of 0.1 M HEPES in its fully protonated form, and ready access to 0.1 M HCl, 0.1 M NaOH and distilled water, describe the preparation of 1 L of 0.025 M HEPES buffer solution, pH 7.8.

14. Determination of the Molecular Weight of a Solute by Freezing Point Depression A 100-g amount of a solute was dissolved in 1000 g of water. The freezing point of this solution was measured accurately and determined to be −1.12°C. What is the molecular weight of the solute?

15. How to Prepare A Triethanolamine Buffer Shown here is the structure of triethanolamine in its fully protonated form:

$$\text{CH}_2\text{CH}_2\text{OH}$$
$$\text{HOCH}_2\text{CH}_2 - \overset{+}{\underset{|}{\text{N}}} - \text{CH}_2\text{CH}_2\text{OH}$$
$$\text{H}$$

Its pK_a is 7.8. You have available at your lab bench 0.1 M solutions of HCl, NaOH, and the uncharged (free base) form of triethanolamine, as well as ample distilled water. Describe the preparation of a 1 L solution of 0.05 M triethanolamine buffer, pH 7.6.

16. How to Prepare a Tris Buffer Solution *Tris*-hydroxymethyl aminomethane (TRIS) is widely used for the preparation of buffers in biochemical research. Shown here is the structure of TRIS in its protonated form:

$$^+\text{NH}_3$$
$$\text{HOCH}_2 - \text{C} - \text{CH}_2\text{OH}$$
$$\text{CH}_2\text{OH}$$

Its acid dissociation constant, K_a, is $8.32 \times 10^{-9}\ M$. You have available at your lab bench a 0.1 M solution of TRIS in its protonated form, 0.1 M solutions of HCl and NaOH, and ample distilled water. Describe the preparation of a 1 L solution of 0.02 M TRIS buffer, pH 7.8.

17. Plot the Titration Curve for Bicine and Calculate How to Prepare a pH 7.5 Bicine Buffer Solution Bicine (*N, N*–bis (2-hydroxyethyl) glycine) is another commonly used buffer in biochemistry labs. The structure of Bicine in its fully protonated form is shown below:

$$\text{CH}_2\text{CH}_2\text{OH}$$
$$\text{H}^+\text{N} - \text{CH}_2\text{COOH}$$
$$\text{CH}_2\text{CH}_2\text{OH}$$

   a. Draw the titration curve for Bicine, assuming the pK_a for its free COOH group is 2.3 and the pK_a for its tertiary amino group is 8.3.

b. Draw the structure of the fully deprotonated form (completely dissociated form) of bicine.

c. You have available a 0.1 M solution of Bicine at its isoelectric point (pH$_I$), 0.1 M solutions of HCl and NaOH, and ample distilled H$_2$O. Describe the preparation of 1 L of 0.04 M Bicine buffer, pH 7.5.

d. What is the concentration of fully protonated form of Bicine in your final buffer solution?

18. Calculate the Concentration of Cl⁻ in Gastric Juice Hydrochloric acid is a significant component of gastric juice. If chloride is the only anion in gastric juice, what is its concentration if pH = 1.2?

19. Calculate the Concentration of Lactate in Blood Plasma at pH 7.4 if [Lactic Acid] = 1.5 μM From the pK_a for lactic acid given in Table 2.4, calculate the concentration of lactate in blood plasma (pH = 7.4) if the concentration of lactic acid is 1.5 μM.

20. Draw the Titration Curve for a Weak Acid and Determine Its pK_a from the Titration Curve When a 0.1 M solution of a weak acid was titrated with base, the following results were obtained:

Equivalents of base added	pH observed
0.05	3.4
0.15	3.9
0.25	4.2
0.40	4.5
0.60	4.9
0.75	5.2
0.85	5.4
0.95	6.0

Plot the results of this titration and determine the pK_a of the weak acid from your graph.

Preparing for the MCAT® Exam

21. The enzyme alcohol dehydrogenase catalyzes the oxidation of ethyl alcohol by NAD⁺ to give acetaldehyde plus NADH and a proton:

$$CH_3CH_2OH + NAD^+ \longrightarrow CH_3CHO + NADH + H^+$$

The rate of this reaction can be measured by following the change in pH. The reaction is run in 1 mL 10 mM TRIS buffer at pH 8.6. If the pH of the reaction solution falls to 8.4 after ten minutes, what is the rate of alcohol oxidation, expressed as nanomoles of ethanol oxidized per mL per sec of reaction mixture?

22. In light of the Human Biochemistry box on page 46, what would be the effect on blood pH if cellular metabolism produced a sudden burst of carbon dioxide?

23. On the basis of Figure 2.12, what will be the pH of the acetate–acetic acid solution when the ratio of [acetate]/[acetic acid] is 10?

a. 3.76

b. 4.76

c. 5.76

d. 11.24

FURTHER READING

Properties of Water

Cooke, R., and Kuntz, I. D., 1974. The properties of water in biological systems. *Annual Review of Biophysics and Bioengineering* **3**:95–126.

Finney, J. L., 2004. Water? What's so special about it? *Philosophical Transactions of the Royal Society, London, Series B* **359**:1145–1165.

Franks, F., ed., 1982. *The Biophysics of Water.* New York: John Wiley & Sons.

Huang, C., Wikfeldt, K. T., Tokushima, T., Nordlund, D., et al., 2009. The inhomogeneous structure of water at ambient conditions. *Proceedings of the National Academy of Sciences USA* **106**:15214–15218.

Stillinger, F. H., 1980. Water revisited. *Science* **209**:451–457.

Tokmakoff, A., 2007. Shining light on the rapidly evolving structure of water. *Science* **317**:54–55.

Properties of Solutions

Cooper, T. G., 1977. *The Tools of Biochemistry,* Chap. 1. New York: John Wiley & Sons.

Segel, I. H., 1976. *Biochemical Calculations,* 2nd ed., Chap. 1. New York: John Wiley & Sons.

Titration Curves

Darvey, I. G., and Ralston, G. B., 1993. Titration curves—misshapen or mislabeled? *Trends in Biochemical Sciences* **18**:69–71.

pH and Buffers

Beynon, R. J., and Easterby, J. S., 1996. *Buffer Solutions: The Basics.* New York: IRL Press: Oxford University Press.

Edsall, J. T., and Wyman, J., 1958. Carbon dioxide and carbonic acid, in *Biophysical Chemistry,* Vol. 1, Chap. 10. New York: Academic Press.

Gillies R. J., and Lynch R. M., 2001. Frontiers in the measurement of cell and tissue pH. *Novartis Foundation Symposium* **240**:7–19.

Kelly, J. A., 2000. Determinants of blood pH in health and disease. *Critical Care* **4**:6–14.

Masoro, E. J., and Siegel, P. D., 1971. *Acid-Base Regulation: Its Physiology and Pathophysiology,* Philadelphia: W.B. Saunders.

Norby, J. G., 2000. The origin and meaning of the little p in pH. *Trends in Biochemical Sciences* **25**:36–37.

Perrin, D. D., 1982. *Ionization Constants of Inorganic Acids and Bases in Aqueous Solution.* New York: Pergamon Press.

Rose, B. D., 1994. *Clinical Physiology of Acid–Base and Electrolyte Disorders,* 4th ed. New York: McGraw-Hill.

Stoll, V. S., and Blanchard, J. S., 2009. Buffers: Principles and practice. *Methods in Enzymology* **463**:43–56.

The Fitness of the Environment

Henderson, L. J., 1913. *The Fitness of the Environment.* New York: Macmillan. (Republished 1970. Gloucester, MA: P. Smith.)

Hille, B., 1992. *Ionic Channels of Excitable Membranes,* 2nd ed., Chap. 10. Sunderland, MA: Sinauer Associates.

Thermodynamics of Biological Systems

© Publiphoto/Photo Researchers, Inc.

◀ The sun is the source of energy for virtually all life. We even harvest its energy in the form of electricity using windmills driven by air heated by the sun.

A theory is the more impressive the greater is the simplicity of its premises, the more different are the kinds of things it relates and the more extended is its range of applicability. Therefore, the deep impression which classical thermodynamics made upon me. It is the only physical theory of universal content which I am convinced, that within the framework of applicability of its basic concepts, will never be overthrown.

Albert Einstein

KEY QUESTIONS

3.1 What Are the Basic Concepts of Thermodynamics?

3.2 What Is the Effect of Concentration on Net Free Energy Changes?

3.3 What Is the Effect of pH on Standard-State Free Energies?

3.4 What Can Thermodynamic Parameters Tell Us About Biochemical Events?

3.5 What Are the Characteristics of High-Energy Biomolecules?

3.6 What Are the Complex Equilibria Involved in ATP Hydrolysis?

3.7 Why Are Coupled Processes Important to Living Things?

3.8 What Is the Daily Human Requirement for ATP?

3.9 What Are Reduction Potentials, and How Are They Used to Account for Free Energy Changes in Redox Reactions?

ESSENTIAL QUESTION

Living things require energy. Movement, growth, synthesis of biomolecules, and the transport of ions and molecules across membranes all demand energy input. All organisms must acquire energy from their surroundings and must utilize that energy efficiently to carry out life processes. To study such bioenergetic phenomena requires familiarity with **thermodynamics.** Thermodynamics also allows us to determine whether chemical processes and reactions occur spontaneously. The student should appreciate the power and practical value of thermodynamic reasoning and realize that this is well worth the effort needed to understand it.

What are the laws and principles of thermodynamics that allow us to describe the flows and interchanges of heat, energy, and matter in biochemical systems?

Even the most complicated aspects of thermodynamics are based ultimately on three rather simple and straightforward laws. These laws and their extensions sometimes run counter to our intuition. However, once truly understood, the basic principles of thermodynamics become powerful devices for sorting out complicated chemical and biochemical problems. Once we reach this milestone in our scientific development, thermodynamic thinking becomes an enjoyable and satisfying activity.

Several basic thermodynamic principles are presented in this chapter, including the analysis of heat flow, entropy production, and free energy functions and the relationship between entropy and information. In addition, some ancillary concepts are considered, including the concept of standard states, the effect of pH on standard-state free energies, the effect of concentration on the net free energy change of a reaction, and the importance of coupled processes in living things. The chapter concludes with a discussion of ATP and other energy-rich compounds.

OWL Online homework and a Student Self Assessment for this chapter may be assigned in OWL

3.1 : What Are the Basic Concepts of Thermodynamics?

In any consideration of thermodynamics, a distinction must be made between the system and the surroundings. The **system** is that portion of the *universe* with which we are concerned. It might be a mixture of chemicals in a test tube, or a single cell, or an entire organism. The **surroundings** include everything else in the universe (Figure 3.1). The nature of the system must also be specified. There are three basic kinds of systems: isolated, closed, and open. An **isolated system** cannot exchange matter or energy with its surroundings. A **closed system** may exchange energy, but not matter, with the surroundings. An **open system** may exchange matter, energy, or both with the surroundings. Living things are typically open systems that exchange matter (nutrients and waste products) and energy (heat from metabolism, for example) with their surroundings.

The First Law: The Total Energy of an Isolated System Is Conserved

It was realized early in the development of thermodynamics that heat could be converted into other forms of energy and moreover that all forms of energy could ultimately be converted to some other form. The **first law of thermodynamics** states that *the total energy of an isolated system is conserved.* Thermodynamicists have formulated a mathematical function for keeping track of heat transfers and work expenditures in thermodynamic systems. This function is called the **internal energy,** commonly designated **E or U,** and it includes all the energies that might be exchanged in physical or chemical processes, including rotational, vibrational, and translational energies of molecules and also the energy stored in covalent and noncovalent bonds. The internal energy depends only on the present state of a system and hence is referred to as a **state function.** The internal energy does not depend on how the system got there and is thus **independent of path.** An extension of this thinking is that we can manipulate the system through any possible pathway of changes, and as long as the system returns to the original state, the internal energy, E, will not have been changed by these manipulations.

The internal energy, E, of any system can change only if energy flows in or out of the system in the form of heat or work. For any process that converts one state (state 1) into another (state 2), the change in internal energy, ΔE, is given as

$$\Delta E = E_2 - E_1 = q + w \qquad (3.1)$$

where the quantity q is the *heat absorbed by the system from the surroundings* and w is *the work done on the system by the surroundings.* **Mechanical work** is defined as *movement through some distance caused by the application of a force.* Both movement and force are required for work to have occurred. Examples of work done in biological systems include the flight of insects and birds, the circulation of blood by a pumping heart, the transmission of an impulse along a nerve, and the lifting of a weight by someone who is exercising. On the other hand, if a person strains to lift a heavy weight but fails to move the weight at all, then, in the thermodynamic sense, no work has been done. (The energy expended in the muscles of the would-be weight lifter is given off in the form of heat.)

Isolated system:
No exchange of matter or energy

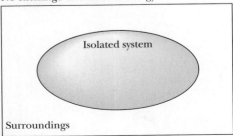

Closed system:
Energy exchange may occur

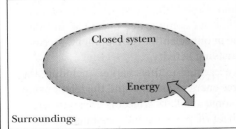

Open system:
Energy exchange and/or matter exchange may occur

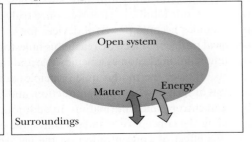

FIGURE 3.1 The characteristics of isolated, closed, and open systems. Isolated systems exchange neither matter nor energy with their surroundings. Closed systems may exchange energy, but not matter, with their surroundings. Open systems may exchange either matter or energy with the surroundings.

In chemical and biochemical systems, work is often concerned with the pressure and volume of the system under study. The mechanical work done on the system is defined as $w = -P\Delta V$, where P is the pressure and ΔV is the volume change and is equal to $V_2 - V_1$. When work is defined in this way, the sign on the right side of Equation 3.1 is positive. (Sometimes w is defined as work done *by* the system; in this case, the equation is $\Delta E = q - w$.) Work may occur in many forms, such as mechanical, electrical, magnetic, and chemical. $\Delta E, q,$ and w must all have the same units. The **calorie,** abbreviated **cal,** and **kilocalorie (kcal)** have been traditional choices of chemists and biochemists, but the SI unit, the **joule,** is now recommended.

Enthalpy Is a More Useful Function for Biological Systems

If the definition of work is limited to mechanical work ($w = -P\Delta V$) and no change in volume occurs, an interesting simplification is possible. In this case, ΔE is merely the *heat exchanged at constant volume.* This is so because if the volume is constant, no mechanical work can be done on or by the system. Then $\Delta E = q$. Thus ΔE is a very useful quantity in constant volume processes. However, chemical and especially biochemical processes and reactions are much more likely to be carried out at constant pressure. In constant pressure processes, ΔE is not necessarily equal to the heat transferred. For this reason, chemists and biochemists have defined a function that is especially suitable for constant pressure processes. It is called the **enthalpy,** $H,$ and it is defined as

$$H = E + PV \tag{3.2}$$

The clever nature of this definition is not immediately apparent. However, if the pressure is constant, then we have

$$\Delta H = \Delta E + P\Delta V = q + w + P\Delta V = q - P\Delta V + P\Delta V = q \tag{3.3}$$

So, ΔE is the heat transferred in a constant volume process, and ΔH is the heat transferred in a constant pressure process.

Often, because biochemical reactions normally occur in liquids or solids rather than in gases, volume changes are typically quite small, and *enthalpy and internal energy are often essentially equal.*

In order to compare the thermodynamic parameters of different reactions, it is convenient to define a *standard state.* **For solutes in a solution, the standard state is normally unit activity (often simplified to 1 M concentration).** Enthalpy, internal energy, and other thermodynamic quantities are often given or determined for standard-state conditions and are then denoted by a superscript degree sign ("°"), as in $\Delta H°, \Delta E°,$ and so on.

Enthalpy changes for biochemical processes can be determined experimentally by measuring the heat absorbed (or given off) by the process in a *calorimeter* (a reaction vessel that can be used to measure the heat evolved by a reaction). Alternatively, for any process A $\rightleftharpoons$ B at equilibrium, the standard-state enthalpy change for the process can be determined from the temperature dependence of the equilibrium constant:

$$\Delta H° = -R\frac{d(\ln K_{eq})}{d(1/T)} \tag{3.4}$$

Here R is the *gas constant,* defined as $R = 8.314$ J/mol $\cdot$ K. A plot of $R(\ln K_{eq})$ versus $1/T$ is called a **van't Hoff plot.** The example below demonstrates how a van't Hoff plot is constructed and how the enthalpy change for a reaction can be determined from the plot itself.

EXAMPLE   In a study[1] of the temperature-induced reversible denaturation of the protein chymotrypsinogen,

$$\text{Native state (N)} \rightleftharpoons \text{denatured state (D)}$$
$$K_{eq} = \text{[D]/[N]}$$

[1]Brandts, J. F., 1964. The thermodynamics of protein denaturation. I. The denaturation of chymotrypsinogen. *Journal of the American Chemical Society* **86:**4291–4301.

(continued)

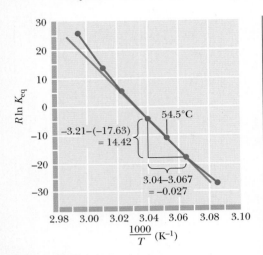

FIGURE 3.2 The enthalpy change, $\Delta H°$, for a reaction can be determined from the slope of a plot of $R \ln K_{eq}$ versus $1/T$. To illustrate the method, the values of the data points on either side of the 327.5 K (54.5°C) data point have been used to calculate $\Delta H°$ at 54.5°C. Regression analysis would normally be preferable. (Adapted from Brandts, J. F., 1964. The thermodynamics of protein denaturation. I. The denaturation of chymotrypsinogen. *Journal of the American Chemical Society* **86**:4291–4301.)

(Example, continued)

John F. Brandts measured the equilibrium constants for the denaturation over a range of pH and temperatures. The data for pH 3:

$T(K)$:	324.4	326.1	327.5	329.0	330.7	332.0	333.8
K_{eq}:	0.041	0.12	0.27	0.68	1.9	5.0	21

A plot of $R(\ln K_{eq})$ versus $1/T$ (a van't Hoff plot) is shown in Figure 3.2. $\Delta H°$ for the denaturation process at any temperature is the negative of the slope of the plot at that temperature. As shown, $\Delta H°$ at 54.5°C (327.5 K) is

$$\Delta H° = -[-3.2 - (-17.6)]/[(3.04 - 3.067) \times 10^{-3}] = +533 \text{ kJ/mol}$$

What does this value of $\Delta H°$ mean for the unfolding of the protein? Positive values of $\Delta H°$ would be expected for the breaking of hydrogen bonds as well as for the exposure of hydrophobic groups from the interior of the native, folded protein during the unfolding process. Such events would raise the energy of the protein solution. The magnitude of this enthalpy change (533 kJ/mol) at 54.5°C is large, compared to similar values of $\Delta H°$ for other proteins and for this same protein at 25°C (Table 3.1). If we consider only this positive enthalpy change for the unfolding process, the native, folded state is strongly favored. As we shall see, however, other parameters must be taken into account.

TABLE 3.1 Thermodynamic Parameters for Protein Denaturation

Protein (and conditions)	$\Delta H°$ kJ/mol	$\Delta S°$ kJ/mol · K	$\Delta G°$ kJ/mol	ΔC_p kJ/mol · K
Chymotrypsinogen (pH 3, 25°C)	164	0.440	31.0	10.9
β-Lactoglobulin (5 M urea, pH 3, 25°C)	−88	−0.300	2.5	9.0
Myoglobin (pH 9, 25°C)	180	0.400	57.0	5.9
Ribonuclease (pH 2.5, 30°C)	240	0.780	3.8	8.4

Adapted from Cantor, C., and Schimmel, P., 1980. *Biophysical Chemistry.* San Francisco: W.H. Freeman; and Tanford, C., 1968. Protein denaturation. *Advances in Protein Chemistry* **23**:121–282.

The Second Law: Systems Tend Toward Disorder and Randomness

The **second law of thermodynamics** has been described and expressed in many different ways, including the following:

1. Systems tend to proceed from *ordered* (*low-entropy* or *low-probability*) states to *disordered* (*high-entropy* or *high-probability*) states.
2. The *entropy* of the system plus surroundings is unchanged by *reversible processes;* the entropy of the system plus surroundings increases for *irreversible processes.*
3. All naturally occurring processes proceed toward **equilibrium,** that is, to a state of minimum potential energy. Energy flows spontaneously so as to become diffused or dispersed or spread out. Energy dispersal results in entropy increase.

Several of these statements of the second law invoke the concept of **entropy,** which is a measure of disorder and randomness in the system (or the surroundings). An organized or ordered state is a low-entropy state, whereas a disordered state is a high-entropy state. All else being equal, reactions involving large, positive entropy changes, ΔS, are more likely to occur than reactions for which ΔS is not large and positive.

$$S = k \ln W \tag{3.5}$$

and

$$\Delta S = k \ln W_{final} - k \ln W_{initial} \tag{3.6}$$

where W_{final} and $W_{initial}$ are the final and initial number of microstates, respectively, and where k is Boltzmann's constant ($k = 1.38 \times 10^{-23}$ J/K).

> **A DEEPER LOOK**

Entropy, Information, and the Importance of "Negentropy"

When a thermodynamic system undergoes an increase in entropy, it becomes more disordered. On the other hand, a decrease in entropy reflects an increase in order. A more ordered system is more highly organized and possesses a greater information content. To appreciate the implications of decreasing the entropy of a system, consider the random collection of letters in the figure. This disorganized array of letters possesses no inherent information content, and nothing can be learned by its perusal. On the other hand, this particular array of letters can be systematically arranged to construct the first sentence of the Einstein quotation that opened this chapter: "A theory is the more impressive the greater is the simplicity of its premises, the more different are the kinds of things it relates and the more extended is its range of applicability."

Arranged in this way, this same collection of 151 letters possesses enormous information content—the profound words of a great scientist. Just as it would have required significant effort to rearrange these 151 letters in this way, so large amounts of energy are required to construct and maintain living organisms. Energy input is required to produce information-rich, organized structures such as proteins and nucleic acids. Information content can be thought of as negative entropy. In 1945 Erwin Schrödinger took time out from his studies of quantum mechanics to publish a delightful book titled *What Is Life?* In it, Schrödinger coined the term *negentropy* to describe the negative entropy changes that confer organization and information content to living organisms. Schrödinger pointed out that organisms must "acquire negentropy" to sustain life.

Seen in this way, *entropy represents* **energy dispersion**—*the dispersion of energy among a large number of molecular motions relatable to quantized states (microstates).* An increase in entropy is just an increase in the number of microstates in any macrostate. On the other hand, if only one microstate corresponds to a given macrostate, then the system has no freedom to choose its microstate—and it has zero entropy. This definition is useful for statistical calculations (in fact, it is a foundation of *statistical thermodynamics*), but a more common form relates entropy to the heat transferred in a process:

$$dS_{\text{reversible}} = \frac{dq}{T} \tag{3.7}$$

where $dS_{\text{reversible}}$ is the entropy change of the system in a reversible[2] process, q is the heat transferred, and T is the temperature at which the heat transfer occurs.

Equation 3.6 says that *entropy change measures the dispersal of energy in a process.* When W_{initial} is less than W_{final}, energy is dispersed from a small number of microstates (W_{initial}) to a larger number of microstates (W_{final}), and ΔS is a positive quantity. That is, dispersal of energy into a larger number of microstates results in an increase in entropy.

The Third Law: Why Is "Absolute Zero" So Important?

The **third law of thermodynamics** states that *the entropy of any crystalline, perfectly ordered substance must approach zero as the temperature approaches 0 K, and, at T = 0 K, entropy is exactly zero.* Based on this, it is possible to establish a quantitative, absolute entropy scale for any substance as

$$S = \int_0^T C_P \, d\ln T \tag{3.8}$$

[2]A reversible process is one that can be reversed by an infinitesimal modification of a variable.

where C_P is the *heat capacity* at constant pressure. The heat capacity of any substance is the amount of heat 1 mole of it can store as the temperature of that substance is raised by 1 degree. For a constant pressure process, this is described mathematically as

$$C_P = \frac{dH}{dT} \tag{3.9}$$

If the heat capacity can be evaluated at all temperatures between 0 K and the temperature of interest, an absolute entropy can be calculated. For biological processes, *entropy changes* are more useful than absolute entropies. The entropy change for a process can be calculated if the enthalpy change and *free energy change* are known.

Free Energy Provides a Simple Criterion for Equilibrium

An important question for chemists, and particularly for biochemists, is, "Will the reaction proceed in the direction written?" J. Willard Gibbs, one of the founders of thermodynamics, realized that the answer to this question lay in a comparison of the enthalpy change and the entropy change for a reaction at a given temperature. The **Gibbs free energy, G,** is defined as

$$G = H - TS \tag{3.10}$$

For any process A $\rightleftharpoons$ B at constant pressure and temperature, the *free energy change* is given by

$$\Delta G = \Delta H - T\Delta S \tag{3.11}$$

If ΔG is equal to 0, the process is at *equilibrium* and there is no net flow either in the forward or reverse direction. When $\Delta G = 0$, $\Delta S = \Delta H/T$ and the enthalpic and entropic changes are exactly balanced. Any process with a nonzero ΔG proceeds spontaneously to a final state of lower free energy. If ΔG is negative, the process proceeds spontaneously in the direction written. If ΔG is positive, the reaction or process proceeds spontaneously in the reverse direction. (The sign and value of ΔG do not allow us to determine *how fast* the process will go.) If the process has a negative ΔG, it is said to be **exergonic,** whereas processes with positive ΔG values are **endergonic.**

The Standard-State Free Energy Change The free energy change, ΔG, for any reaction depends upon the nature of the reactants and products, but it is also affected by the conditions of the reaction, including temperature, pressure, pH, and the concentrations of the reactants and products. As explained earlier, it is useful to define a standard state for such processes. If the free energy change for a reaction is sensitive to solution conditions, what is the particular significance of the standard-state free energy change? To answer this question, consider a reaction between two reactants A and B to produce the products C and D.

$$A + B \rightleftharpoons C + D \tag{3.12}$$

The free energy change for non–standard-state concentrations is given by

$$\Delta G = \Delta G° + RT \ln \frac{[C][D]}{[A][B]} \tag{3.13}$$

At equilibrium, $\Delta G = 0$ and $[C][D]/[A][B] = K_{eq}$. We then have

$$\Delta G° = -RT \ln K_{eq} \tag{3.14}$$

or, in base 10 logarithms,

$$\Delta G° = -2.3RT \log_{10} K_{eq} \tag{3.15}$$

This can be rearranged to

$$K_{eq} = 10^{-\Delta G°/2.3RT} \tag{3.16}$$

In any of these forms, this relationship allows the standard-state free energy change for any process to be determined if the equilibrium constant is known. More importantly, it states that the *point of equilibrium for a reaction in solution is a function of the standard-state free energy change for the process*. That is, $\Delta G°$ is another way of writing an equilibrium constant.

EXAMPLE The equilibrium constants determined by Brandts at several temperatures for the denaturation of chymotrypsinogen (see previous example) can be used to calculate the free energy changes for the denaturation process. For example, the equilibrium constant at 54.5°C is 0.27, so

$$\Delta G° = -(8.314 \text{ J/mol} \cdot \text{K})(327.5 \text{ K}) \ln (0.27)$$
$$\Delta G° = -(2.72 \text{ kJ/mol}) \ln (0.27)$$
$$\Delta G° = 3.56 \text{ kJ/mol}$$

The positive sign of $\Delta G°$ means that the unfolding process is unfavorable; that is, the stable form of the protein at 54.5°C is the folded form. On the other hand, the relatively small magnitude of $\Delta G°$ means that the folded form is only slightly favored. Figure 3.3 shows the dependence of $\Delta G°$ on temperature for the denaturation data at pH 3 (from the data given in the example on page 54).

Having calculated both $\Delta H°$ and $\Delta G°$ for the denaturation of chymotrypsinogen, we can also calculate $\Delta S°$, using Equation 3.11:

$$\Delta S° = -\frac{(\Delta G - \Delta H°)}{T} \qquad (3.17)$$

At 54.5°C (327.5 K),

$$\Delta S° = -(3560 - 533{,}000 \text{ J/mol})/327.5 \text{ K}$$
$$\Delta S° = 1620 \text{ J/mol} \cdot \text{K}$$

Figure 3.4 presents the dependence of $\Delta S°$ on temperature for chymotrypsinogen denaturation at pH 3. A positive $\Delta S°$ indicates that the protein solution has become more disordered as the protein unfolds. Comparison of the value of 1.62 kJ/mol · K with the values of $\Delta S°$ in Table 3.1 shows that the present value (for chymotrypsinogen at 54.5°C) is quite large. The physical significance of the thermodynamic parameters for the unfolding of chymotrypsinogen becomes clear later in this chapter.

3.2 What Is the Effect of Concentration on Net Free Energy Changes?

Equation 3.13 shows that the free energy change for a reaction can be very different from the standard-state value if the concentrations of reactants and products differ significantly from unit activity (1 m (molal)* for solutions). The effects can often be dramatic. Consider the hydrolysis of phosphocreatine:

$$\text{Phosphocreatine} + \text{H}_2\text{O} \longrightarrow \text{creatine} + \text{P}_i \qquad (3.18)$$

This reaction is strongly exergonic, and $\Delta G°$ at 37°C is −42.8 kJ/mol. Physiological concentrations of phosphocreatine, creatine, and inorganic phosphate are normally between 1 and 10 mM. Assuming 1 mM concentrations and using Equation 3.13, the ΔG for the hydrolysis of phosphocreatine is

$$\Delta G = -42.8 \text{ kJ/mol} + (8.314 \text{ J/mol} \cdot \text{K})(310 \text{ K}) \ln \left(\frac{[0.001][0.001]}{[0.001]} \right) \qquad (3.19)$$

$$\Delta G = -60.5 \text{ kJ/mol} \qquad (3.20)$$

At 37°C, the difference between standard-state and 1 mM concentrations for such a reaction is thus approximately −17.7 kJ/mol.

---

*In the dilute solutions typical of biochemical situations, molarity (M) and molality can be used interchangably.

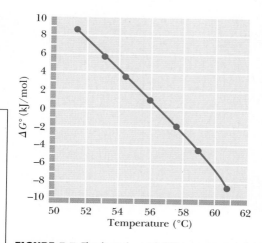

FIGURE 3.3 The dependence of $\Delta G°$ on temperature for the denaturation of chymotrypsinogen. (Adapted from Brandts, J. F., 1964. The thermodynamics of protein denaturation. I. The denaturation of chymotrypsinogen. *Journal of the American Chemical Society* **86:**4291–4301.)

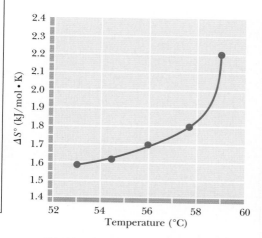

FIGURE 3.4 The dependence of $\Delta S°$ on temperature for the denaturation of chymotrypsinogen. (Adapted from Brandts, J. F., 1964. The thermodynamics of protein denaturation. I. The denaturation of chymotrypsinogen. *Journal of the American Chemical Society* **86:**4291–4301.)

3.3 What Is the Effect of pH on Standard-State Free Energies?

For biochemical reactions in which hydrogen ions (H^+) are consumed or produced, the usual definition of the standard state is awkward. Standard state for the H^+ ion is 1 M, which corresponds to pH 0. At this pH, nearly all enzymes would be denatured and biological reactions could not occur. It makes more sense to use free energies and equilibrium constants determined at pH 7. Biochemists have thus adopted a modified standard state, designated with prime (') symbols, as in $\Delta G^{\circ\prime}$, K_{eq}', $\Delta H^{\circ\prime}$, and so on. For values determined in this way, a standard state of 10^{-7} M H^+ and unit activity (1 M for solutions, 1 atm for gases and pure solids defined as unit activity) for all other components (in the ionic forms that exist at pH 7) is assumed. The two standard states can be related easily. For a reaction in which H^+ is produced,

$$A \longrightarrow B^- + H^+ \tag{3.21}$$

the relation of the equilibrium constants for the two standard states is

$$K_{eq} = \frac{[B^-][H^+]}{[A]} = K_{eq}' \, [H^+] \tag{3.22}$$

and $\Delta G^{\circ\prime}$ is given by

$$\Delta G^{\circ\prime} = \Delta G^\circ + RT \ln [H^+] \tag{3.23}$$

For a reaction in which H^+ is consumed,

$$A^- + H^+ \longrightarrow B \tag{3.24}$$

the equilibrium constants are related by

$$K_{eq} = \frac{K_{eq}'}{[H^+]} \tag{3.25}$$

and $\Delta G^{\circ\prime}$ is given by

$$\Delta G^{\circ\prime} = \Delta G^\circ + RT \ln \left(\frac{1}{[H^+]} \right) = \Delta G^\circ - RT \ln [H^+] \tag{3.26}$$

3.4 What Can Thermodynamic Parameters Tell Us About Biochemical Events?

The best answer to this question is that a single parameter (ΔH or ΔS, for example) is not very meaningful. A positive ΔH° for the unfolding of a protein might reflect either the breaking of hydrogen bonds within the protein or the exposure of hydrophobic groups to water (Figure 3.5). However, *comparison of several thermodynamic parameters can provide meaningful insights about a process.* For example, the transfer of Na^+ and Cl^- ions from the gas phase to aqueous solution involves a very large negative ΔH° (thus a very favorable stabilization of the ions) and a comparatively small ΔS° (Table 3.2). The negative entropy term reflects the ordering of water molecules in the hydration shells of the Na^+ and Cl^- ions. The unfavorable $-T\Delta S$ contribution is more than offset by the large heat of hydration, which makes the hydration of ions a very favorable process overall. The negative entropy change for the dissociation of acetic acid in water also reflects the ordering of water molecules in the ion hydration shells. In this case, however, the enthalpy change is much smaller in magnitude. As a result, ΔG° for dissociation of acetic acid in water is positive, and acetic acid is thus a weak (largely undissociated) acid.

The transfer of a nonpolar hydrocarbon molecule to water is an appropriate model for the exposure of protein hydrophobic groups to solvent when a protein unfolds. The transfer of toluene from liquid toluene to water involves a negative ΔS°, a positive ΔG°, and a ΔH° that is small compared to ΔG° (a pattern similar to that observed for

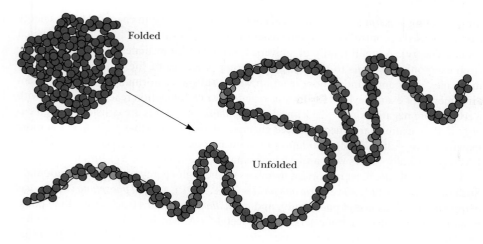

Folded

Unfolded

FIGURE 3.5 Unfolding of a soluble protein exposes significant numbers of nonpolar groups to water, forcing order on the solvent and resulting in a negative $\Delta S°$ for the unfolding process. Orange spheres represent nonpolar groups; blue spheres are polar and/or charged groups.

TABLE 3.2	Thermodynamic Parameters for Several Simple Processes*			
Process	$\Delta H°$ kJ/mol	$\Delta S°$ kJ/mol · K	$\Delta G°$ kJ/mol	ΔC_P kJ/mol · K
Hydration of ions[†] $Na^+(g) + Cl^-(g) \longrightarrow Na^+(aq) + Cl^-(aq)$	−760.0	−0.185	−705.0	
Dissociation of ions in solution[‡] $H_2O + CH_3COOH \longrightarrow H_3O^+ + CH_3COO^-$	−10.3	−0.126	27.26	−0.143
Transfer of hydrocarbon to water[‡] Toluene (in pure toluene) $\longrightarrow$ toluene (aqueous)	1.72	−0.071	22.7	0.265

*All data collected for 25°C.
[†]Berry, R. S., Rice, S. A., and Ross, J., 1980. *Physical Chemistry.* New York: John Wiley.
[‡]Tanford, C., 1980. *The Hydrophobic Effect.* New York: John Wiley.

the dissociation of acetic acid). *What distinguishes these two very different processes is the change in heat capacity* (Table 3.2). These ΔC_P values are dominated by solvent (H$_2$O) effects. A positive heat capacity change for a process indicates that the molecules have acquired new ways to move (and thus to store heat energy). A negative ΔC_P means that the process has resulted in less freedom of motion for the molecules involved. ΔC_P is negative for the dissociation of acetic acid and positive for the transfer of toluene to water. The explanation is that polar and nonpolar molecules *both* induce organization of nearby water molecules, *but in different ways.* The water molecules near a nonpolar solute are *organized but labile.* Hydrogen bonds formed by water molecules near nonpolar solutes rearrange more rapidly than the hydrogen bonds of pure water. On the other hand, the hydrogen bonds formed between water molecules near an ion are less labile (rearrange more slowly) than they would be in pure water. This means that ΔC_P should be negative for the dissociation of ions in solution, as observed for acetic acid (Table 3.2).

3.5 What Are the Characteristics of High-Energy Biomolecules?

Virtually all life on earth depends on energy from the sun. Among life forms, there is a hierarchy of energetics: Certain organisms capture solar energy directly, whereas others derive their energy from this group in subsequent processes. Organisms that absorb light energy directly are called **phototrophic organisms.** These organisms store solar energy in the form of various organic molecules. Organisms that feed on these latter molecules, releasing the stored energy in a series of oxidative reactions, are called **chemotrophic organisms.** Despite these differences, both types of organisms share common mechanisms for generating a useful form of chemical energy. Once captured in chemical form, energy can be released in controlled exergonic reactions to drive a variety of life processes (which require energy). A small family of universal biomolecules mediates the flow of energy from exergonic reactions to the energy-requiring processes of life. These "high-energy molecules" are compounds whose reaction with a substance normally present in the environment (for example, H_2O [hydrolysis reaction] or O_2 [oxidation reaction]) occurs with a $\Delta G°'$ more negative than -25 kJ/mol. Phosphate compounds are considered high energy if they exhibit large negative free energies of hydrolysis (that is, if $\Delta G°'$ is more negative than -25 kJ/mol). Oxidation of reduced coenzymes (NADH, NADPH, and FADH$_2$) typically occurs with a free energy change of -150 to -220 kJ/mol.

Table 3.3 lists the most important members of the high-energy phosphate compounds. Such molecules include *phosphoric anhydrides* (ATP, ADP), an *enol phosphate* (PEP), *acyl phosphates* (such as acetyl phosphate), and *guanidino phosphates* (such as creatine phosphate). Also included are thioesters, such as acetyl-CoA, which do not contain phosphorus, but which have a large negative free energy of hydrolysis. As noted earlier, the exact amount of chemical free energy available from the hydrolysis of such compounds depends on concentration, pH, temperature, and so on, but the $\Delta G°'$ values for hydrolysis of these substances are substantially more negative than those for most other metabolic species. Two important points: First, high-energy phosphate compounds are not long-term energy storage substances. They are transient forms of stored energy, meant to carry energy from point to point, from one enzyme system to another, in the minute-to-minute existence of the cell. (As we shall see in subsequent chapters, other molecules bear the responsibility for long-term storage of energy supplies.) Second, the term *high-energy compound* should not be construed to imply that these molecules are unstable and hydrolyze or decompose unpredictably. ATP, for example, is quite a stable molecule. A substantial *activation energy* must be delivered to ATP to hydrolyze the terminal, or γ, phosphate group. In fact, as shown in Figure 3.6, the activation energy that must be absorbed by the molecule to break the O—P$_γ$ bond is normally 200 to 400 kJ/mol, which is substantially larger than the net

TABLE 3.3	Free Energies of Hydrolysis of Some High-Energy Compounds*		
Compound and Hydrolysis Reaction		$\Delta G°'$ (kJ/mol)	Structure
Phosphoenolpyruvate ⟶ pyruvate + P$_i$		-62.2	Figure 3.12
1,3-Bisphosphoglycerate ⟶ 3-phosphoglycerate + P$_i$		-49.6	Figure 3.10
Creatine phosphate ⟶ creatine + P$_i$		-43.3	Figure 13.21
Acetyl phosphate ⟶ acetate + P$_i$		-43.3	Figure 3.10
Adenosine-5′-triphosphate ⟶ ADP + P$_i$		-35.7[†]	Figure 3.9
Adenosine-5′-triphosphate ⟶ ADP + P$_i$ (with excess Mg^{2+})		-30.5	Figure 3.9
Adenosine-5′-diphosphate ⟶ AMP + P$_i$		-35.7	Figure 3.9
Pyrophosphate ⟶ P$_i$ + P$_i$ (in 5 mM Mg^{2+})		-33.6	Figure 3.8
Adenosine-5′-triphosphate ⟶ AMP + PP$_i$ (excess Mg^{2+})		-32.3	Figure 10.14
Uridine diphosphoglucose ⟶ UDP + glucose		-31.9	Figure 22.14
Acetyl-coenzyme A ⟶ acetate + CoA		-31.5	page 616
S-adenosylmethionine ⟶ methionine + adenosine		-25.6[‡]	Figure 25.28
Glucose-1-phosphate ⟶ glucose + P$_i$		-21.0	Figure 7.13
Sn-Glycerol-3-phosphate ⟶ glycerol + P$_i$		-9.2	Figure 8.5
Adenosine-5′-monophosphate ⟶ adenosine + P$_i$		-9.2	Figure 10.11

*Adapted primarily from *Handbook of Biochemistry and Molecular Biology,* 1976, 3rd ed. In *Physical and Chemical Data,* G. Fasman, ed., Vol. 1, pp. 296–304. Boca Raton, FL: CRC Press.
†From Gwynn, R. W., and Veech, R. L., 1973. The equilibrium constants of the adenosine triphosphate hydrolysis and the adenosine triphosphate-citrate lyase reactions. *Journal of Biological Chemistry* **248**:6966–6972.
‡From Mudd, H., and Mann, J., 1963. Activation of methionine for transmethylation. *Journal of Biological Chemistry* **238**:2164–2170.

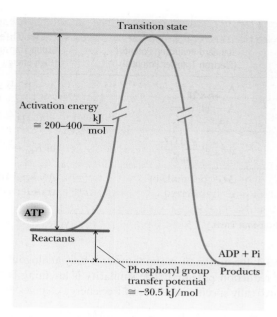

FIGURE 3.6 The activation energies for phosphoryl group transfer reactions (200 to 400 kJ/mol) are substantially larger than the free energy of hydrolysis of ATP (-30.5 kJ/mol).

30.5 kJ/mol released in the hydrolysis reaction. Biochemists are much more concerned with the *net release* of 30.5 kJ/mol than with the activation energy for the reaction (because suitable enzymes cope with the latter). The net release of large quantities of free energy distinguishes the high-energy phosphoric anhydrides from their "low-energy" ester cousins, such as glycerol-3-phosphate (Table 3.3). The next section provides a quantitative framework for understanding these comparisons.

ATP Is an Intermediate Energy-Shuttle Molecule

One last point about Table 3.3 deserves mention. Given the central importance of ATP as a high-energy phosphate in biology, students are sometimes surprised to find that ATP holds an intermediate place in the rank of high-energy phosphates. PEP, 1,3-BPG, creatine phosphate, acetyl phosphate, and pyrophosphate all exhibit higher values of $\Delta G°'$. This is not a biological anomaly. ATP is uniquely situated between the very-high-energy phosphates synthesized in the breakdown of fuel molecules and the numerous lower-energy acceptor molecules that are phosphorylated in the course of further metabolic reactions. ADP can accept both phosphates and energy from the higher-energy phosphates, and the ATP thus formed can donate both phosphates and energy to the lower-energy molecules of metabolism. The ATP/ADP pair is an intermediately placed acceptor/donor system among high-energy phosphates. In this context, ATP functions as a very versatile but intermediate energy-shuttle device that interacts with many different energy-coupling enzymes of metabolism.

Group Transfer Potentials Quantify the Reactivity of Functional Groups

Many reactions in biochemistry involve the transfer of a functional group from a donor molecule to a specific receptor molecule or to water. The concept of **group transfer potential** explains the tendency for such reactions to occur. Biochemists define the group transfer potential as the free energy change that occurs upon hydrolysis, that is, upon transfer of the particular group to water. This concept and its terminology are preferable to the more qualitative notion of *high-energy bonds*.

The concept of group transfer potential is not particularly novel. Other kinds of transfer (of hydrogen ions and electrons, for example) are commonly characterized in terms of appropriate measures of transfer potential (pK_a and reduction potential, $\mathscr{E}_o$, respectively).

TABLE 3.4	Types of Transfer Potential		
	Proton Transfer Potential (Acidity)	Standard Reduction Potential (Electron Transfer Potential)	Group Transfer Potential (High-Energy Bond)
Simple equation	$AH \rightleftharpoons A^- + H^+$	$A \rightleftharpoons A^+ + e^-$	$A \sim P \rightleftharpoons A + P_i$
Equation including acceptor	$AH + H_2O \rightleftharpoons$ $A^- + H_3O^+$	$A + H^+ \rightleftharpoons$ $A^+ + \frac{1}{2}H_2$	$A \sim PO_4^{2-} + H_2O \rightleftharpoons$ $A{-}H + HPO_4^{2-}$
Measure of transfer potential	$pK_a = \dfrac{\Delta G°}{2.303\ RT}$	$\Delta \mathscr{E}_o = \dfrac{-\Delta G°}{n\mathscr{F}}$	$\ln K_{eq} = \dfrac{-\Delta G°}{RT}$
Free energy change of transfer is given by:	$\Delta G°$ per mole of H^+ transferred	$\Delta G°$ per mole of e^- transferred	$\Delta G°$ per mole of phosphate transferred

Adapted from: Klotz, I. M., 1986. *Introduction to Biomolecular Energetics.* New York: Academic Press.

As shown in Table 3.4, the notion of group transfer is fully analogous to those of ionization potential and reduction potential. The similarity is anything but coincidental, because all of these are really specific instances of free energy changes. If we write

$$AH \longrightarrow A^- + H^+ \tag{3.27a}$$

we do not really mean that a proton has literally been removed from the acid AH. In the gas phase at least, this would require the input of approximately 1200 kJ/mol! What we really mean is that the proton has been *transferred* to a suitable acceptor molecule, usually water:

$$AH + H_2O \longrightarrow A^- + H_3O^+ \tag{3.27b}$$

The appropriate free energy relationship is of course

$$pK_a = \frac{\Delta G}{2.303\ RT} \tag{3.28}$$

Similarly, in the case of an oxidation-reduction reaction

$$A \longrightarrow A^+ + e^- \tag{3.29a}$$

we do not really mean that A oxidizes independently. What we really mean (and what is much more likely in biochemical systems) is that the electron is transferred to a suitable acceptor:

$$A + H^+ \longrightarrow A^+ + \tfrac{1}{2}H_2 \tag{3.29b}$$

and the relevant free energy relationship is

$$\Delta \mathscr{E}_o = \frac{-\Delta G°}{n\mathscr{F}} \tag{3.30}$$

where n is the number of equivalents of electrons transferred and $\mathscr{F}$ is **Faraday's constant.**

Similarly, the release of free energy that occurs upon the hydrolysis of ATP and other "high-energy phosphates" can be treated quantitatively in terms of *group transfer*. It is common to write for the hydrolysis of ATP

$$ATP + H_2O \longrightarrow ADP + P_i \tag{3.31}$$

The free energy change, which we henceforth call the *group transfer potential,* is given by

$$\Delta G° = -RT \ln K_{eq} \tag{3.32}$$

where K_{eq} is the equilibrium constant for the group transfer, which is normally written as

$$K_{eq} = \frac{[ADP][P_i]}{[ATP][H_2O]} \tag{3.33}$$

Even this set of equations represents a simplification, because ATP, ADP, and P_i all exist in solutions as a mixture of ionic species. This problem is discussed in a later section. For now, it is enough to note that the free energy changes listed in Table 3.3 are the group transfer potentials observed for transfers to water.

The Hydrolysis of Phosphoric Acid Anhydrides Is Highly Favorable

ATP contains two *pyrophosphoryl* or *phosphoric acid anhydride* linkages, as shown in Figure 3.7. Other common biomolecules possessing phosphoric acid anhydride linkages include ADP, GTP, GDP and the other nucleoside diphosphates and triphosphates, sugar nucleotides such as UDP–glucose, and inorganic pyrophosphate itself. All exhibit large negative free energies of hydrolysis, as shown in Table 3.3. The chemical reasons for the large negative $\Delta G°'$ values for the hydrolysis reactions include destabilization of the reactant due to bond strain caused by electrostatic repulsion, stabilization of the products by ionization and resonance, and entropy factors due to hydrolysis and subsequent ionization.

Destabilization Due to Electrostatic Repulsion Electrostatic repulsion in the reactants is best understood by comparing these phosphoric anhydrides with other reactive anhydrides, such as acetic anhydride. As shown in Figure 3.8a, the electronegative carbonyl oxygen atoms withdraw electrons from the C=O bonds, producing partial negative charges on the oxygens and partial positive charges on the carbonyl carbons. Each of these electrophilic carbonyl carbons is further destabilized by the other acetyl group, which is also electron-withdrawing in nature. As a result, acetic anhydride is unstable with respect to the products of hydrolysis.

The situation with phosphoric anhydrides is similar. The phosphorus atoms of the pyrophosphate anion are electron-withdrawing and destabilize PP_i with respect to its hydrolysis products. Furthermore, the reverse reaction, reformation of the anhydride bond from the two anionic products, requires that the electrostatic repulsion between these anions be overcome (see following).

Stabilization of Hydrolysis Products by Ionization and Resonance The pyrophosphoryl moiety possesses two negative charges at pH values above 7.5 or so (Figure 3.8a). The hydrolysis products, two phosphate esters, each carry about two negative charges at pH values above 7.2. The increased ionization of the hydrolysis products helps stabilize the electrophilic phosphorus nuclei.

Resonance stabilization in the products is best illustrated by the reactant anhydrides (Figure 3.8b). The unpaired electrons of the bridging oxygen atom in acetic anhydride (and phosphoric anhydride) cannot participate in resonance structures with both electrophilic centers at once. This **competing resonance** situation is relieved in the product acetate or phosphate molecules.

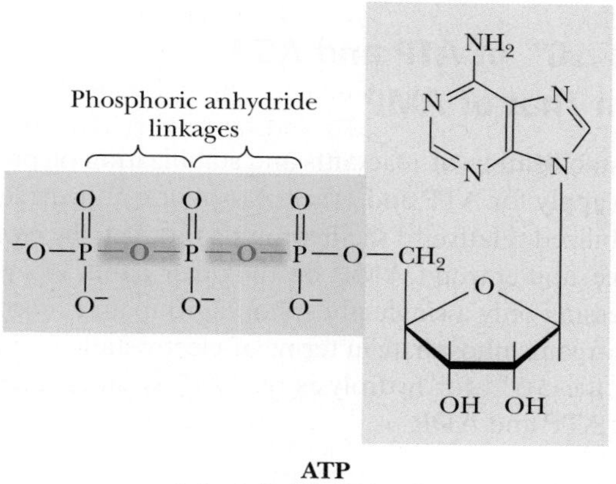

ATP
(adenosine-5'-triphosphate)

FIGURE 3.7 The triphosphate chain of ATP contains two pyrophosphate linkages, both of which release large amounts of energy upon hydrolysis.

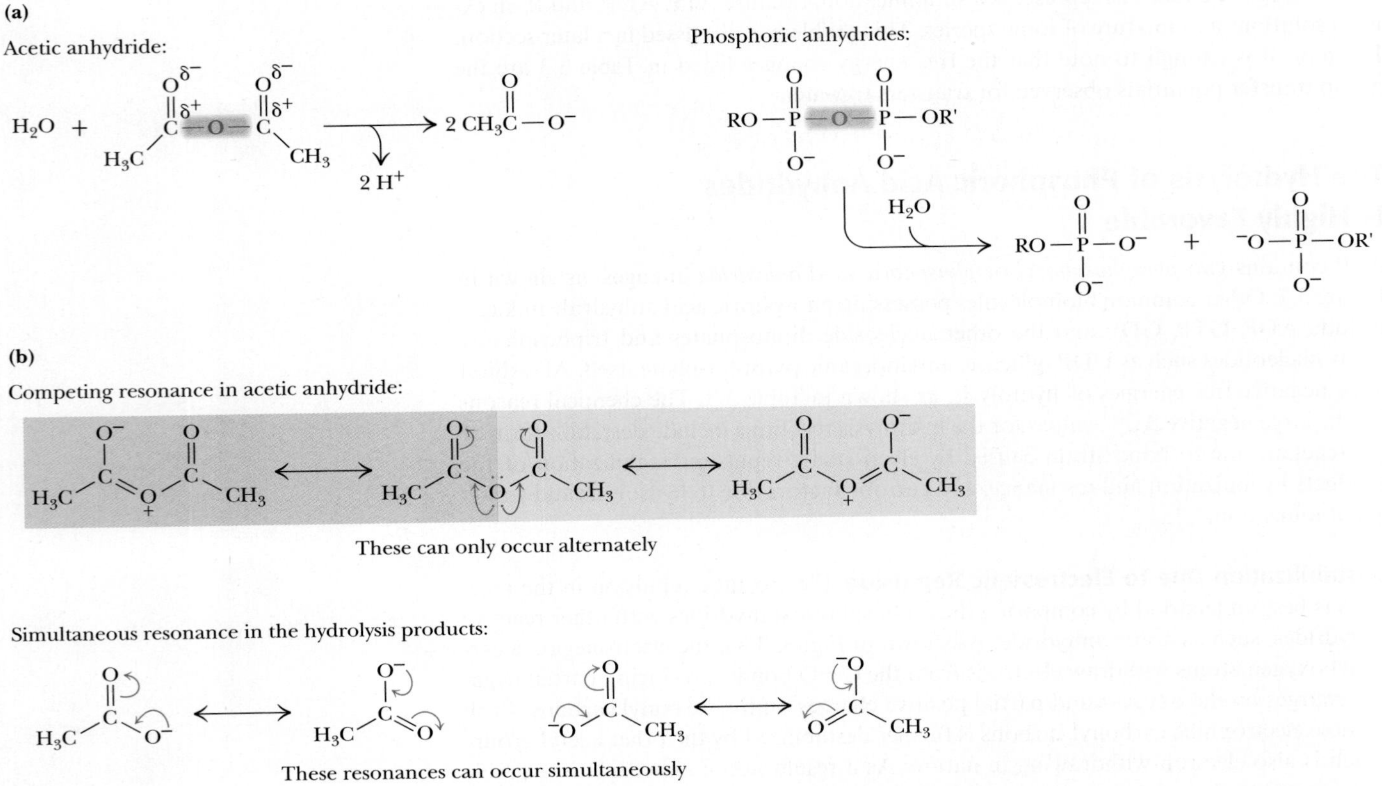

(a)

Acetic anhydride:

Phosphoric anhydrides:

(b)

Competing resonance in acetic anhydride:

These can only occur alternately

Simultaneous resonance in the hydrolysis products:

These resonances can occur simultaneously

FIGURE 3.8 (a) Electrostatic repulsion between adjacent partial positive charges (on carbon and phosphorus, respectively) is relieved upon hydrolysis of the anhydride bonds of acetic anhydride and phosphoric anhydrides. (b) The competing resonances of acetic anhydride and the simultaneous resonance forms of the hydrolysis product, acetate.

Entropy Factors Arising from Hydrolysis and Ionization For the phosphoric anhydrides, and for most of the high-energy compounds discussed here, there is an additional "entropic" contribution to the free energy of hydrolysis. Most of the hydrolysis reactions of Table 3.3 result in an increase in the number of molecules in solution. As shown in Figure 3.9, the hydrolysis of ATP (at pH values above 7) creates three species—ADP, inorganic phosphate (P_i), and a hydrogen ion—from only two reactants (ATP and H_2O). The entropy of the solution increases because the more particles, the more disordered the system.[3] (This effect is ionization-dependent because, at low pH, the hydrogen ion created in many of these reactions simply protonates one of the phosphate oxygens, and one fewer "particle" results from the hydrolysis.)

The Hydrolysis $\Delta G°'$ of ATP and ADP Is Greater Than That of AMP

The concepts of destabilization of reactants and stabilization of products described for pyrophosphate also apply for ATP and other phosphoric anhydrides (Figure 3.9). ATP and ADP are destabilized relative to the hydrolysis products by electrostatic repulsion, competing resonance, and entropy. AMP, on the other hand, is a phosphate ester (not an anhydride) possessing only a single phosphoryl group and is not markedly different from the product inorganic phosphate in terms of electrostatic repulsion and resonance stabilization. Thus, the $\Delta G°'$ for hydrolysis of AMP is much smaller than the corresponding values for ATP and ADP.

[3]Imagine the "disorder" created by hitting a crystal with a hammer and breaking it into many small pieces.

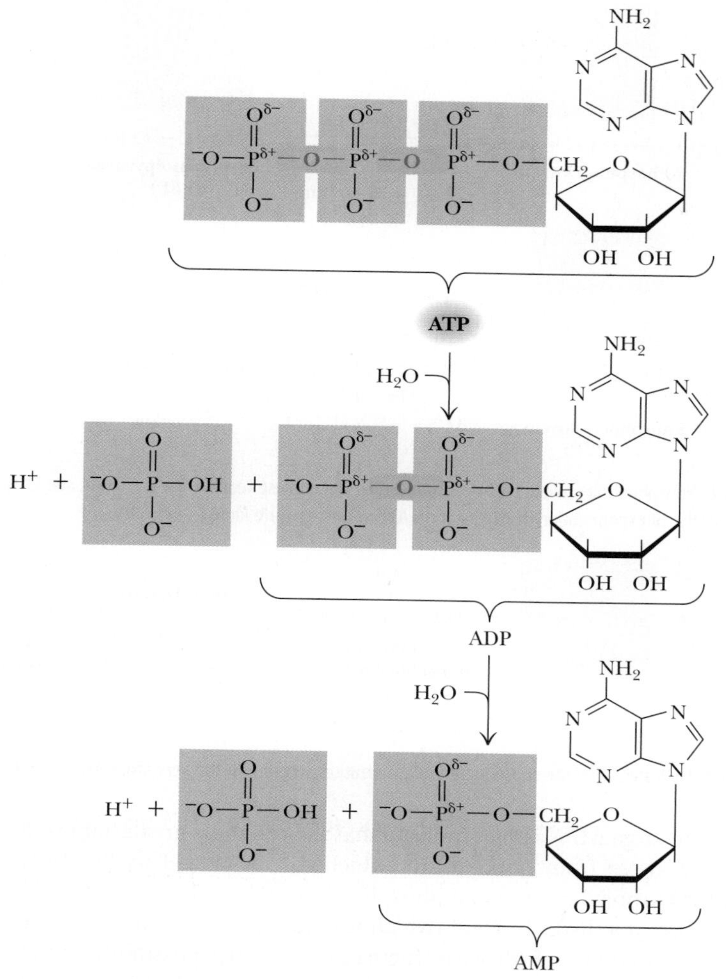

FIGURE 3.9 Hydrolysis of ATP to ADP (and/or of ADP to AMP) leads to relief of electrostatic repulsion.

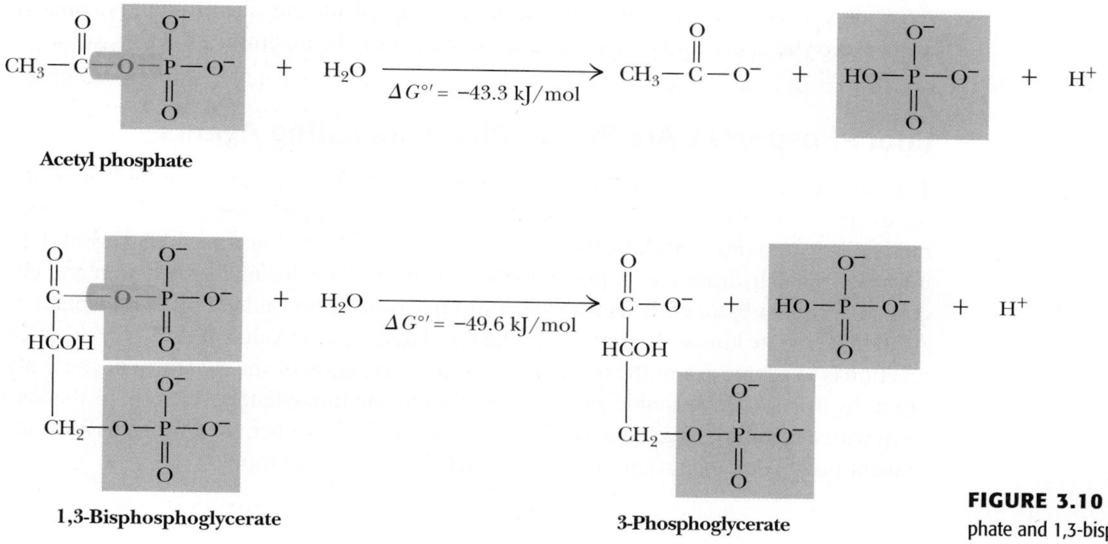

FIGURE 3.10 The hydrolysis reactions of acetyl phosphate and 1,3-bisphosphoglycerate.

Acetyl Phosphate and 1,3-Bisphosphoglycerate Are Phosphoric-Carboxylic Anhydrides

The mixed anhydrides of phosphoric and carboxylic acids, frequently called acyl phosphates, are also energy-rich. Two biologically important acyl phosphates are acetyl phosphate and 1,3-bisphosphoglycerate. Hydrolysis of these species yields acetate and 3-phosphoglycerate, respectively, in addition to inorganic phosphate (Figure 3.10).

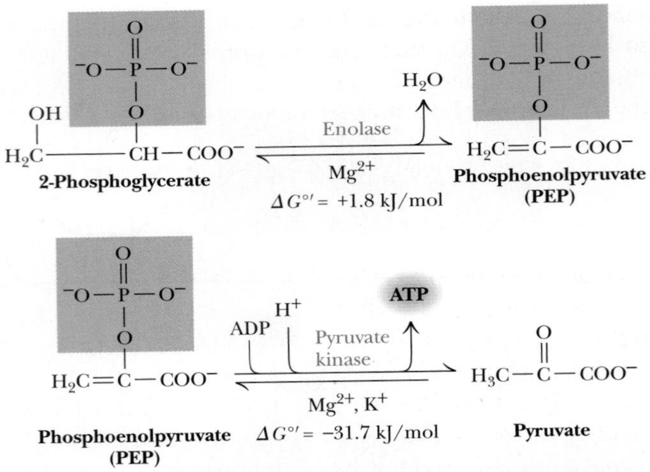

FIGURE 3.11 Phosphoenolpyruvate (PEP) is produced by the enolase reaction (in glycolysis; see Chapter 18) and in turn drives the phosphorylation of ADP to form ATP in the pyruvate kinase reaction.

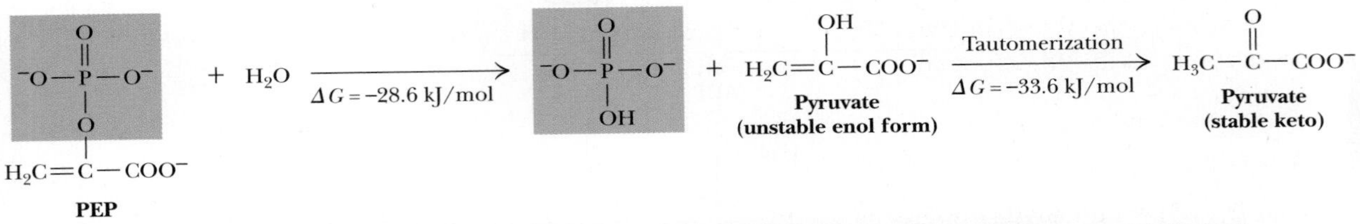

FIGURE 3.12 Hydrolysis and the subsequent tautomerization account for the very large $\Delta G^{\circ\prime}$ of PEP.

Once again, the large $\Delta G^{\circ\prime}$ values indicate that the reactants are destabilized relative to products. This arises from bond strain, which can be traced to the partial positive charges on the carbonyl carbon and phosphorus atoms of these structures. The energy stored in the mixed anhydride bond (which is required to overcome the charge–charge repulsion) is released upon hydrolysis. Increased resonance possibilities in the products relative to the reactants also contribute to the large negative $\Delta G^{\circ\prime}$ values. The value of $\Delta G^{\circ\prime}$ depends on the pK_a values of the starting anhydride and the product phosphoric and carboxylic acids, and of course also on the pH of the medium.

Enol Phosphates Are Potent Phosphorylating Agents

The largest value of $\Delta G^{\circ\prime}$ in Table 3.3 belongs to *phosphoenolpyruvate* or *PEP,* an example of an enolic phosphate. This molecule is an important intermediate in carbohydrate metabolism, and due to its large negative $\Delta G^{\circ\prime}$, it is a potent phosphorylating agent. PEP is formed via dehydration of 2-phosphoglycerate by enolase during fermentation and glycolysis. PEP is subsequently transformed into pyruvate upon transfer of its phosphate to ADP by pyruvate kinase (Figure 3.11). The very large negative value of $\Delta G^{\circ\prime}$ for the latter reaction is to a large extent the result of a secondary reaction of the *enol* form of pyruvate. Upon hydrolysis, the unstable enolic form of pyruvate immediately converts to the keto form with a resulting large negative $\Delta G^{\circ\prime}$ (Figure 3.12). Together, the hydrolysis and subsequent *tautomerization* result in an overall $\Delta G^{\circ\prime}$ of -62.2 kJ/mol.

3.6 What Are the Complex Equilibria Involved in ATP Hydrolysis?

So far, as in Equation 3.34, the hydrolyses of ATP and other high-energy phosphates have been portrayed as simple processes. The situation in a real biological system is far more complex, owing to the operation of several ionic equilibria. First, ATP, ADP, and the other species in Table 3.3 can exist in several different ionization states that must be accounted for in any quantitative analysis. Second, phosphate compounds bind a variety of divalent and monovalent cations with substantial affinity, and the various metal complexes must also be considered in such analyses. Consideration of these special cases makes the quantitative analysis far more realistic. The importance of these multiple equilibria in group transfer reactions is illustrated for the hydrolysis of ATP, but the principles and methods presented are general and can be applied to any similar hydrolysis reaction.

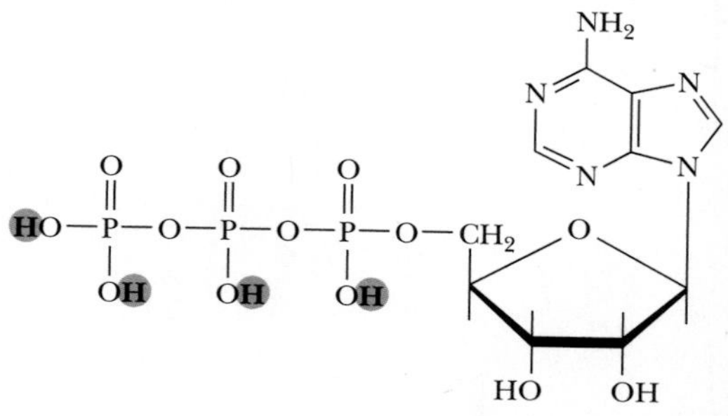

The $\Delta G^{\circ\prime}$ of Hydrolysis for ATP Is pH-Dependent

ATP has four dissociable protons, as indicated in Figure 3.13. Three of the protons on the triphosphate chain dissociate at very low pH. The last proton to dissociate from the triphosphate chain possesses a pK_a of 6.95. At higher pH values, ATP is completely deprotonated. ADP and phosphoric acid also undergo multiple ionizations. These multiple ionizations make the equilibrium constant for ATP hydrolysis more complicated than the simple expression in Equation 3.33. Multiple ionizations must also be taken into account when the pH dependence of ΔG° is considered. The calculations are beyond the scope of this text, but Figure 3.14 shows the variation of ΔG° as a function of pH. The free energy of hydrolysis is nearly constant from pH 4 to pH 6. At higher values of pH, ΔG° varies linearly with pH, becoming more negative by 5.7 kJ/mol for every pH unit of increase at 37°C. Because the pH of most biological tissues and fluids is near neutrality, the effect on ΔG° is relatively small, but it must be taken into account in certain situations.

Metal Ions Affect the Free Energy of Hydrolysis of ATP

Most biological environments contain substantial amounts of divalent and monovalent metal ions, including Mg^{2+}, Ca^{2+}, Na^+, K^+, and so on. What effect do metal ions have on the equilibrium constant for ATP hydrolysis and the associated free energy change? Figure 3.15 shows the change in $\Delta G^{\circ\prime}$ with pMg (that is, $-\log_{10}[Mg^{2+}]$) at pH 7.0 and 38°C. The free energy of hydrolysis of ATP at zero Mg^{2+} is -35.7 kJ/mol, and at 5 mM total Mg^{2+} (the minimum in the plot) the ΔG_{obs}° is approximately -31 kJ/mol. Thus, in most real biological environments (with pH near 7 and Mg^{2+} concentrations of 5 mM or more) the free energy of hydrolysis of ATP is altered more by metal ions than by protons. A widely used "consensus value" for $\Delta G^{\circ\prime}$ of ATP in biological systems is -30.5 **kJ/mol** (Table 3.3). This value, cited in the 1976 *Handbook of Biochemistry and Molecular Biology* (3rd ed., *Physical and Chemical Data*, Vol. 1, pp. 296–304, Boca Raton, FL: CRC Press), was determined at 37°C in the presence of "excess Mg^{2+}." *This is the value we use for metabolic calculations in the balance of this text.*

Why does the $G^{\circ\prime}$ of ATP hydrolysis depend so strongly on Mg^{2+} concentration? The answer lies in the strong binding of Mg^{2+} by the triphosphate oxygens of ATP. As shown in Figure 3.16, the binding of Mg^{2+} to ATP is dependent on Mg^{2+} ion concentration and also on pH. At pH 7 and 1 mM $[Mg^{2+}]$, approximately one Mg^{2+} ion is bound to each ATP. The decrease in binding of Mg^{2+} at low pH is the result of competition by H^+ and Mg^{2+} for the negatively charged oxygen atoms of ATP.

Concentration Affects the Free Energy of Hydrolysis of ATP

Through all these calculations of the effect of pH and metal ions on the ATP hydrolysis equilibrium, we have assumed "standard conditions" with respect to concentrations of all species except for protons. The levels of ATP, ADP, and other high-energy metabolites never even begin to approach the standard state of 1 M. In most cells, the concentrations of these species are more typically 1 to 5 mM or even less. Earlier, we described the effect of concentration on equilibrium constants and free energies in the form of Equation 3.13. For the present case, we can rewrite this as

$$\Delta G = \Delta G^{\circ} + RT \ln \frac{[\Sigma ADP][\Sigma P_i]}{[\Sigma ATP]} \qquad (3.34)$$

where the terms in brackets represent the sum (Σ) of the concentrations of all the ionic forms of ATP, ADP, and P_i.

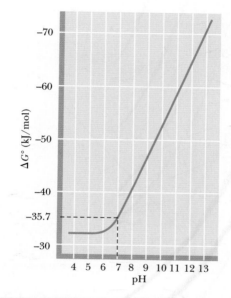

Color indicates the locations of the dissociable protons of ATP

FIGURE 3.13 Adenosine-5′-triphosphate (ATP).

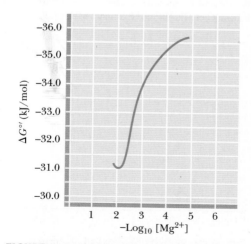

FIGURE 3.14 The pH dependence of the free energy of hydrolysis of ATP. Because pH varies only slightly in biological environments, the effect on ΔG is usually small.

FIGURE 3.15 The free energy of hydrolysis of ATP as a function of total Mg^{2+} ion concentration at 38°C and pH 7.0. (Adapted from Gwynn, R. W., and Veech, R. L., 1973. The equilibrium constants of the adenosine triphosphate hydrolysis and the adenosine triphosphate-citrate lyase reactions. *Journal of Biological Chemistry* **248**:6966–6972.)

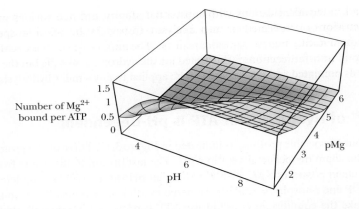

FIGURE 3.16 Number of Mg^{2+} ions bound per ATP as a function of pH and $[Mg^{2+}]$. $pMg = -\log_{10}[Mg^{2+}]$. (*Adapted from: Alberty, R.A., 2003. Thermodynamics of Biochemical Reactions. Hoboken: Wiley-Interscience.*)

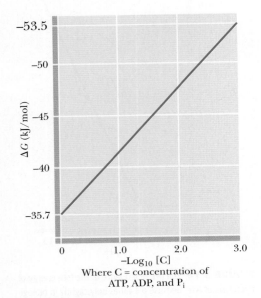

FIGURE 3.17 The free energy of hydrolysis of ATP as a function of concentration at 38°C, pH 7.0. The plot follows the relationship described in Equation 3.34, with the concentrations [C] of ATP, ADP, and P_i assumed to be equal.

It is clear that changes in the concentrations of these species can have large effects on ΔG. The concentrations of ATP, ADP, and P_i may, of course, vary in real biological environments, but if, for the sake of some model calculations, we assume that all three concentrations are equal, then the effect of concentration on ΔG is as shown in Figure 3.17. The free energy of hydrolysis of ATP, which is -35.7 kJ/mol at 1 M, becomes -49.4 kJ/mol at 5 mM (that is, the concentration for which $pC = -2.3$ in Figure 3.17). At 1 mM ATP, ADP, and P_i, the free energy change becomes even more negative at -53.6 kJ/mol. *Clearly, the effects of concentration are much greater than the effects of protons or metal ions under physiological conditions.*

Does the "concentration effect" change ATP's position in the energy hierarchy (in Table 3.3)? Not really. All the other high- and low-energy phosphates experience roughly similar changes in concentration under physiological conditions and thus similar changes in their free energies of hydrolysis. The roles of the very-high-energy phosphates (PEP, 1,3-bisphosphoglycerate, and creatine phosphate) in the synthesis and maintenance of ATP in the cell are considered in our discussions of metabolic pathways. In the meantime, several of the problems at the end of this chapter address some of the more interesting cases.

3.7 Why Are Coupled Processes Important to Living Things?

Many of the reactions necessary to keep cells and organisms alive must run against their **thermodynamic potential,** that is, in the direction of positive ΔG. Among these are the synthesis of adenosine triphosphate (ATP) and other high-energy molecules and the creation of ion gradients in all mammalian cells. These processes are driven in the thermodynamically unfavorable direction via *coupling* with highly favorable processes. Many such *coupled processes* are discussed later in this text. They are crucially important in intermediary metabolism, oxidative phosphorylation, and membrane transport, as we shall see.

We can predict whether pairs of coupled reactions will proceed spontaneously by simply summing the free energy changes for each reaction. For example, consider the reaction from glycolysis (discussed in Chapter 18) involving the conversion of phosphoenolpyruvate (PEP) to pyruvate (Figure 3.18). The hydrolysis of PEP is energetically very favorable, and it is used to drive phosphorylation of adenosine diphosphate (ADP) to form ATP, a process that is energetically unfavorable. Using values of ΔG that would be typical for a human erythrocyte:

$$PEP + H_2O \longrightarrow pyruvate + P_i \qquad \Delta G = -78 \text{ kJ/mol} \qquad (3.35)$$
$$ADP + P_i \longrightarrow ATP + H_2O \qquad \Delta G = +55 \text{ kJ/mol} \qquad (3.36)$$
$$PEP + ADP \longrightarrow pyruvate + ATP \qquad \text{Total } \Delta G = -23 \text{ kJ/mol} \qquad (3.37)$$

The net reaction catalyzed by this enzyme depends upon coupling between the two reactions shown in Equations 3.35 and 3.36 to produce the net reaction shown in Equation

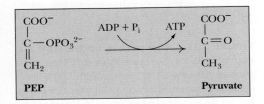

FIGURE 3.18 The pyruvate kinase reaction.

3.37 with a net negative ΔG. Many other examples of coupled reactions are considered in our discussions of intermediary metabolism (see Part 3). In addition, many of the complex biochemical systems discussed in the later chapters of this text involve reactions and processes with positive ΔG values that are driven forward by coupling to reactions with a negative ΔG.

3.8 What Is the Daily Human Requirement for ATP?

We can end this discussion of ATP and the other important high-energy compounds in biology by discussing the daily metabolic consumption of ATP by humans. An approximate calculation gives a somewhat surprising and impressive result. Assume that the average adult human consumes approximately 11,700 kJ (2800 kcal, that is, 2800 Calories) per day. Assume also that the metabolic pathways leading to ATP synthesis operate at a thermodynamic efficiency of approximately 50%. Thus, of the 11,700 kJ a person consumes as food, about 5860 kJ end up in the form of synthesized ATP. As indicated earlier, the hydrolysis of 1 mole of ATP yields approximately 50 kJ of free energy under cellular conditions. This means that the body cycles through 5860/50 = 117 moles of ATP each day. The disodium salt of ATP has a molecular weight of 551 g/mol, so an average person hydrolyzes about

$$(117 \text{ moles}) \frac{551 \text{ g}}{\text{mole}} = 64{,}467 \text{ g of ATP per day}$$

The average adult human, with a typical weight of 70 kg or so, thus consumes approximately 65 kg of ATP per day, an amount nearly equal to his or her own body weight! Fortunately, we have a highly efficient recycling system for ATP/ADP utilization. The energy released from food is stored transiently in the form of ATP. Once ATP energy is used and ADP and phosphate are released, our bodies recycle it to ATP through intermediary metabolism so that it may be reused. The typical 70-kg body contains only about 50 grams of ATP/ADP total. Therefore, each ATP molecule in our bodies must be recycled nearly 1300 times each day! Were it not for this fact, at current commercial prices of about $20 per gram, our ATP "habit" would cost more than $1 million per day! In these terms, the ability of biochemistry to sustain the marvelous activity and vigor of organisms gains our respect and fascination.

> **A DEEPER LOOK**

ATP Changes the K_{eq} by a Factor of 10^8

Consider a process, A $\rightleftharpoons$ B. It could be a biochemical reaction, or the transport of an ion against a concentration gradient, or even a mechanical process (such as muscle contraction). Assume that it is a thermodynamically unfavorable reaction. Let's say, for purposes of illustration, that $\Delta G^{\circ\prime} = +13.8$ kJ/mol. From the equation,

$$\Delta G^{\circ\prime} = -RT \ln K_{eq}$$

we have

$$+13{,}800 = -(8.31 \text{ J/K} \cdot \text{mol})(298 \text{ K}) \ln K_{eq}$$

which yields

$$\ln K_{eq} = -5.57$$

Therefore,

$$K_{eq} = 0.0038 = [B_{eq}]/[A_{eq}]$$

This reaction is clearly unfavorable (as we could have foreseen from its positive $\Delta G^{\circ\prime}$). At equilibrium, there is one molecule of product B for every 263 molecules of reactant A. Not much A was transformed to B.

Now suppose the reaction A $\rightleftharpoons$ B is coupled to ATP hydrolysis, as is often the case in metabolism:

$$A + ATP \rightleftharpoons B + ADP + P_i$$

The thermodynamic properties of this coupled reaction are the same as the sum of the thermodynamic properties of the partial reactions:

A $\rightleftharpoons$ B	$\Delta G^{\circ\prime} = +13.8$ kJ/mol
ATP + H$_2$O $\rightleftharpoons$ ADP + P$_i$	$\Delta G^{\circ\prime} = -30.5$ kJ/mol
A + ATP + H$_2$O $\rightleftharpoons$ B + ADP + P$_i$	$\Delta G^{\circ\prime} = -16.7$ kJ/mol

That is,

$$\Delta G^{\circ\prime}{}_{\text{overall}} = -16.7 \text{ kJ/mol}$$

So

$$-16{,}700 = -RT \ln K_{eq} = -(8.31)(298)\ln K_{eq}$$
$$\ln K_{eq} = -16{,}700/-2476 = 6.75$$
$$K_{eq} = 850$$

(Continued)

Using this equilibrium constant, let's now consider the cellular situation in which the concentrations of A and B are brought to equilibrium in the presence of typical prevailing concentrations of ATP, ADP, and P_i.*

$$K_{eq} = \frac{[B_{eq}][ADP][P_i]}{[A_{eq}][ATP]}$$

$$850 = \frac{[B_{eq}][8 \times 10^{-3}][10^{-3}]}{[A_{eq}][8 \times 10^{-3}]}$$

$$[B_{eq}]/[A_{eq}] = 850{,}000$$

Comparison of the $[B_{eq}]/[A_{eq}]$ ratio for the simple A $\rightleftharpoons$ B reaction with the coupling of this reaction to ATP hydrolysis gives

$$\frac{850{,}000}{0.0038} = 2.2 \times 10^8$$

The equilibrium ratio of B to A is more than 10^8 greater when the reaction is coupled to ATP hydrolysis. A reaction that was clearly unfavorable ($K_{eq} = 0.0038$) has become emphatically spontaneous!

The involvement of ATP has raised the equilibrium ratio of B/A by more than 200 million – fold. It is informative to realize that this

*The concentrations of ATP, ADP, and P_i in a normal, healthy bacterial cell growing at $25\,°C$ are maintained at roughly 8 mM, 8 mM, and 1 mM, respectively. Therefore, the ratio $[ADP][P_i]/[ATP]$ is about 10^{-3}. Under these conditions, ΔG for ATP hydrolysis is approximately -47.6 kJ/mol.

multiplication factor does not depend on the nature of the reaction. Recall that we defined A $\rightleftharpoons$ B in the most general terms. Also, the value of this equilibrium constant ratio, some 2.2×10^8, is not at all dependent on the particular reaction chosen or its standard free energy change, $\Delta G°'$. You can satisfy yourself on this point by choosing some value for $\Delta G°'$ other than $+13.8$ kJ/mol and repeating these calculations (keeping the concentrations of ATP, ADP, and P_i at 8, 8, and 1 mM, as before).

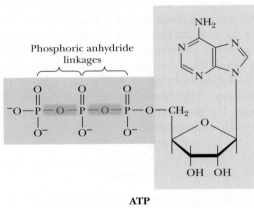

Phosphoric anhydride linkages

ATP
(adenosine-5'-triphosphate)

3.9 What Are Reduction Potentials, and How Are They Used to Account for Free Energy Changes in Redox Reactions?

Earlier, we stressed that NADH and reduced flavoproteins ($[FADH_2]$) are forms of metabolic energy. These reduced coenzymes have a strong tendency to be oxidized—that is, to transfer electrons to other species. Oxidative phosphorylation converts the energy of electron transfer into the energy of phosphoryl transfer stored in the phosphoric anhydride bonds of ATP. Just as the group transfer potential was used in Section 3.5 to quantitate the energy of phosphoryl transfer, the **standard reduction potential,** denoted by $\mathscr{E}_0'$, quantitates the tendency of chemical species to be reduced or oxidized. The standard reduction potential difference describing electron transfer between two species,

$$\text{Reduced donor} \searrow \atop \text{Oxidized donor} \nearrow \quad ne^- \quad \swarrow \text{Oxidized acceptor} \atop \searrow \text{Reduced acceptor}$$

is related to the free energy change for the process by

$$\Delta G°' = -n\mathscr{F}\Delta\mathscr{E}_0' \tag{3.38}$$

where n represents the number of electrons transferred; $\mathscr{F}$ is Faraday's constant, 96,485 J/V · mol; and $\Delta\mathscr{E}_0'$ is the difference in reduction potentials between the donor and acceptor. This relationship is straightforward, but it depends on a standard of reference by which reduction potentials are defined.

Standard Reduction Potentials Are Measured in Reaction Half-Cells

Standard reduction potentials are determined by measuring the voltages generated in **reaction half-cells** (Figure 3.19). A half-cell consists of a solution containing 1 M concentrations of both the oxidized and reduced forms of the substance whose reduction potential is being measured and a simple electrode. (Together, the oxidized and reduced forms of the substance are referred to as a **redox couple.**) Such a **sample half-cell** is connected to a **reference half-cell** and electrode via a conductive bridge (usually a salt-containing agar gel). A sensitive potentiometer (voltmeter) connects the two electrodes so that the electrical potential (voltage) between them can be measured. The reference half-cell normally contains 1 M H^+ in equilibrium with H_2 gas at a pressure of 1 atm. The H^+/H_2 reference half-cell is arbitrarily assigned a standard reduction potential of 0.0 V. The standard reduction potentials of all other redox couples are defined relative to the H^+/H_2 reference half-cell on the basis of the sign and magnitude of the voltage (electromotive force, emf) registered on the potentiometer (Figure 3.19).

If electron flow between the electrodes is toward the sample half-cell, reduction occurs spontaneously in the sample half-cell and the reduction potential is said to be positive. If electron flow between the electrodes is away from the sample half-cell and toward the reference cell, the reduction potential is said to be negative because electron loss (oxidation) is occurring in the sample half-cell. Strictly speaking, the standard reduction potential, $\mathscr{E}_o'$, is the electromotive force generated at 25°C and pH 7.0 by a sample half-cell (containing 1 M concentrations of the oxidized and reduced species) with respect to a reference half-cell. (Note that the reduction potential of the hydrogen half-cell is pH-dependent. The standard reduction potential, 0.0 V, assumes 1 M H^+. The hydrogen half-cell measured at pH 7.0 has an $\mathscr{E}_o'$ of -0.421 V.)

Two Examples Figure 3.19a shows a sample/reference half-cell pair for measurement of the standard reduction potential of the acetaldehyde/ethanol couple. Because electrons flow toward the reference half-cell and away from the sample half-cell, the standard reduction potential is negative, specifically -0.197 V. In contrast, the fumarate/succinate couple (Figure 3.19b) causes electrons to flow from the reference half-cell to the sample half-cell; that is, reduction occurs, and the reduction potential is thus positive. For each half-cell, a **half-cell reaction** describes the reaction taking place. For the fumarate/succinate half-cell coupled to a H^+/H_2 reference half-cell, the reaction occurring is indeed a reduction of fumarate:

$$\text{Fumarate} + 2\,H^+ + 2\,e^- \longrightarrow \text{succinate} \qquad \mathscr{E}_o' = +0.031\ \text{V} \qquad (3.39)$$

However, the reaction occurring in the acetaldehyde/ethanol half-cell is the oxidation of ethanol:

$$\text{Ethanol} \longrightarrow \text{acetaldehyde} + 2\,H^+ + 2\,e^- \qquad \mathscr{E}_o' = -0.197\ \text{V} \qquad (3.40)$$

$\mathscr{E}_o'$ Values Can Be Used to Predict the Direction of Redox Reactions

Some typical half-cell reactions and their respective standard reduction potentials are listed in Table 3.5. Whenever reactions of this type are tabulated, they are uniformly written as reduction reactions, regardless of what occurs in the given half-cell. The sign of the standard reduction potential indicates which reaction really occurs when the given half-cell is combined with the reference hydrogen half-cell. Redox couples that have large positive reduction potentials have a strong tendency to accept electrons, and the oxidized form of such a couple (O_2, for example) is a strong oxidizing agent. Redox couples with large negative reduction potentials have a strong tendency to undergo oxidation (that is, donate electrons), and the reduced form of such a couple (NADPH, for example) is a strong reducing agent.

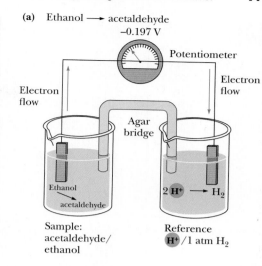

(a) Ethanol $\longrightarrow$ acetaldehyde
-0.197 V

Potentiometer
Electron flow
Electron flow
Agar bridge
Ethanol acetaldehyde
$2\,H^+ \longrightarrow H_2$
Sample: acetaldehyde/ ethanol
Reference H^+/1 atm H_2

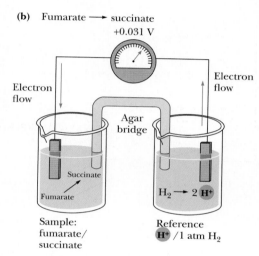

(b) Fumarate $\longrightarrow$ succinate
$+0.031$ V

Electron flow
Electron flow
Agar bridge
Succinate
Fumarate
$H_2 \longrightarrow 2\,H^+$
Sample: fumarate/ succinate
Reference H^+/1 atm H_2

FIGURE 3.19 Experimental apparatus used to measure the standard reduction potential of the indicated redox couples: **(a)** the acetaldehyde/ethanol couple, **(b)** the fumarate/succinate couple.

TABLE 3.5	Standard Reduction Potentials for Several Biological Reduction Half-Reactions	
Reduction Half-Reaction		$\mathscr{E}_0'$ (V)
$\frac{1}{2}O_2 + 2\,H^+ + 2\,e^- \longrightarrow H_2O$		0.816
$Fe^{3+} + e^- \longrightarrow Fe^{2+}$		0.771
Photosystem P700		0.430
$NO_3^- + 2\,H^+ + 2\,e^- \longrightarrow NO_2^- + H_2O$		0.421
Cytochrome f $(Fe^{3+}) + e^- \longrightarrow$ cytochrome f (Fe^{2+})		0.365
Cytochrome $a_3(Fe^{3+}) + e^- \longrightarrow$ cytochrome $a_3(Fe^{2+})$		0.350
Cytochrome $a(Fe^{3+}) + e^- \longrightarrow$ cytochrome $a(Fe^{2+})$		0.290
Rieske Fe-S$(Fe^{3+}) + e^- \longrightarrow$ Rieske Fe-S(Fe^{2+})		0.280
Cytochrome c $(Fe^{3+}) + e^- \longrightarrow$ cytochrome c (Fe^{2+})		0.254
Cytochrome $c_1(Fe^{3+}) + e^- \longrightarrow$ cytochrome $c_1(Fe^{2+})$		0.220
$UQH\cdot + H^+ + e^- \longrightarrow UQH_2$ (UQ = coenzyme Q)		0.190
$UQ + 2\,H^+ + 2\,e^- \longrightarrow UQH_2$		0.060
Cytochrome $b_H(Fe^{3+}) + e^- \longrightarrow$ cytochrome $b_H(Fe^{2+})$		0.050
Fumarate $+ 2\,H^+ + 2\,e^- \longrightarrow$ succinate		0.031
$UQ + H^+ + e^- \longrightarrow UQH\cdot$		0.030
Cytochrome $b_5(Fe^{3+}) + e^- \longrightarrow$ cytochrome b_5 (Fe^{2+})		0.020
[FAD] $+ 2\,H^+ + 2\,e^- \longrightarrow$ [FADH$_2$]		0.003–0.091*
Cytochrome $b_L(Fe^{3+}) + e^- \longrightarrow$ cytochrome $b_L(Fe^{2+})$		−0.100
Oxaloacetate $+ 2\,H^+ + 2\,e^- \longrightarrow$ malate		−0.166
Pyruvate $+ 2\,H^+ + 2\,e^- \longrightarrow$ lactate		−0.185
Acetaldehyde $+ 2\,H^+ + 2\,e^- \longrightarrow$ ethanol		−0.197
FMN $+ 2\,H^+ + 2\,e^- \longrightarrow$ FMNH$_2$		−0.219
FAD $+ 2\,H^+ + 2\,e^- \longrightarrow$ FADH$_2$		−0.219
Glutathione (oxidized) $+ 2\,H^+ + 2\,e^- \longrightarrow$ 2 glutathione (reduced)		−0.230
Lipoic acid $+ 2\,H^+ + 2\,e^- \longrightarrow$ dihydrolipoic acid		−0.290
1,3-Bisphosphoglycerate $+ 2\,H^+ + 2\,e^- \longrightarrow$ glyceraldehyde-3-phosphate $+ P_i$		−0.290
$NAD^+ + 2\,H^+ + 2\,e^- \longrightarrow$ NADH $+ H^+$		−0.320
$NADP^+ + 2\,H^+ + 2\,e^- \longrightarrow$ NADPH $+ H^+$		−0.320
Lipoyl dehydrogenase [FAD] $+ 2\,H^+ + 2\,e^- \longrightarrow$ lipoyl dehydrogenase [FADH$_2$]		−0.340
α-Ketoglutarate $+ CO_2 + 2\,H^+ + 2\,e^- \longrightarrow$ isocitrate		−0.380
$2\,H^+ + 2\,e^- \longrightarrow H_2$		−0.421
Ferredoxin (spinach) $(Fe^{3+}) + e^- \longrightarrow$ ferredoxin (spinach) (Fe^{2+})		−0.430
Succinate $+ CO_2 + 2\,H^+ + 2\,e^- \longrightarrow \alpha$-ketoglutarate $+ H_2O$		−0.670

*Typical values for reduction of bound FAD in flavoproteins such as succinate dehydrogenase (see Bonomi, F., Pagani, S., Cerletti, P., and Giori, C., 1983. Modification of the thermodynamic properties of the electron-transferring groups in mitochondrial succinate dehydrogenase upon binding of succinate. *European Journal of Biochemistry* **134**:439–445).

$\mathscr{E}_0'$ Values Can Be Used to Analyze Energy Changes in Redox Reactions

The half-reactions and reduction potentials in Table 3.5 can be used to analyze energy changes in redox reactions. The oxidation of NADH to NAD^+ can be coupled with the reduction of α-ketoglutarate to isocitrate:

$$NAD^+ + \text{isocitrate} \longrightarrow NADH + H^+ + \alpha\text{-ketoglutarate} + CO_2 \qquad (3.41)$$

This is the isocitrate dehydrogenase reaction of the TCA cycle. Writing the two half-cell reactions, we have

$$NAD^+ + 2\,H^+ + 2\,e^- \longrightarrow NADH + H^+$$
$$\mathscr{E}_o' = -0.32\ V \tag{3.42}$$

$$\alpha\text{-Ketoglutarate} + CO_2 + 2\,H^+ + 2\,e^- \longrightarrow \text{isocitrate}$$
$$\mathscr{E}_o' = -0.38\ V \tag{3.43}$$

In a spontaneous reaction, electrons are donated by (flow away from) the half-reaction with the more negative reduction potential and are accepted by (flow toward) the half-reaction with the more positive reduction potential. Thus, in the present case, isocitrate donates electrons and NAD^+ accepts electrons. The convention defines $\Delta\mathscr{E}_o'$ as

$$\Delta\mathscr{E}_o' = \mathscr{E}_o'\ (\text{acceptor}) - \mathscr{E}_o'\ (\text{donor}) \tag{3.44}$$

In the present case, isocitrate is the donor and NAD^+ the acceptor, so we write

$$\Delta\mathscr{E}_o' = -0.32\ V - (-0.38\ V) = +0.06\ V \tag{3.45}$$

From Equation 20.2, we can now calculate $\Delta G^{\circ\prime}$ as

$$\Delta G^{\circ\prime} = -\ (2)(96.485\ kJ/V \cdot mol)(0.06\ V) \tag{3.46}$$

$$\Delta G^{\circ\prime} = -11.58\ kJ/mol$$

Note that a reaction with a net positive $\Delta\mathscr{E}_o'$ yields a negative $\Delta G^{\circ\prime}$, indicating a spontaneous reaction.

The Reduction Potential Depends on Concentration

We have already noted that the standard free energy change for a reaction, $\Delta G^{\circ\prime}$, does not reflect the actual conditions in a cell, where reactants and products are not at standard-state concentrations ($1\ M$). Equation 3.13 was introduced to permit calculations of actual free energy changes under non-standard-state conditions. Similarly, standard reduction potentials for redox couples must be modified to account for the actual concentrations of the oxidized and reduced species. For any redox couple,

$$ox + ne^- \rightleftharpoons red \tag{3.47}$$

the actual reduction potential is given by

$$\mathscr{E} = \mathscr{E}_o' + (RT/n\mathscr{F}) \ln \frac{[ox]}{[red]} \tag{3.48}$$

Reduction potentials can also be quite sensitive to molecular environment. The influence of environment is especially important for flavins, such as $FAD/FADH_2$ and $FMN/FMNH_2$. These species are normally bound to their respective flavoproteins; the reduction potential of bound FAD, for example, can be very different from the value shown in Table 3.5 for the free $FAD/FADH_2$ couple of $-0.219\ V$.

SUMMARY

The activities of living things require energy. Movement, growth, synthesis of biomolecules, and the transport of ions and molecules across membranes all demand energy input. All organisms must acquire energy from their surroundings and must utilize that energy efficiently to carry out life processes. To study such bioenergetic phenomena requires familiarity with thermodynamics. Thermodynamics also allows us to determine whether chemical processes and reactions occur spontaneously.

3.1 What Are the Basic Concepts of Thermodynamics? The system is that portion of the *universe* with which we are concerned. The surroundings include everything else in the universe. An isolated system cannot exchange matter or energy with its surroundings. A closed system may exchange energy, but not matter, with the surroundings. An open system may exchange matter, energy, or both with the surroundings. Living things are typically open systems. The first law of thermodynamics states that the total energy of an isolated system is conserved. Enthalpy, H, is defined as $H = E + PV$. ΔH is equal to the heat transferred in a constant pressure process. For biochemical reactions in liquids, volume changes are typically quite small, and enthalpy and internal energy are often essentially equal. There are several statements of the second law of thermodynamics, including the following: (1) Systems tend to proceed from ordered (low-entropy or low-probability) states to disordered (high-entropy or high-probability) states. (2) The entropy of the system plus surroundings is unchanged by reversible processes; the entropy of the system plus surroundings increases for irreversible processes. (3) All naturally occurring processes proceed toward equilibrium, that is, to a state of minimum potential energy. The third law of thermodynamics states that the entropy of any crystalline, perfectly ordered substance must approach zero as the temperature approaches 0 K, and, at $T = 0$ K, entropy is exactly zero. The Gibbs free energy, G, defined as $G = H - TS$, provides a simple criterion for equilibrium.

3.2 What Is the Effect of Concentration on Net Free Energy Changes?
The free energy change for a reaction can be very different from the standard-state value if the concentrations of reactants and products differ significantly from unit activity (1 M for solutions). For the reaction A + B $\rightleftharpoons$ C + D, the free energy change for non–standard-state concentrations is given by

$$\Delta G = \Delta G° + RT \ln \frac{[C][D]}{[A][B]}$$

3.3 What Is the Effect of pH on Standard-State Free Energies?
For biochemical reactions in which hydrogen ions (H^+) are consumed or produced, a modified standard state, designated with prime ($'$) symbols, as in $\Delta G°'$, K_{eq}', $\Delta H°'$, may be employed. For a reaction in which H^+ is produced, $\Delta G°'$ is given by

$$\Delta G°' = \Delta G° + RT \ln [H^+]$$

3.4 What Can Thermodynamic Parameters Tell Us About Biochemical Events?
A single parameter (ΔH or ΔS, for example) is not very meaningful, but comparison of several thermodynamic parameters can provide meaningful insights about a process. Thermodynamic parameters can be used to predict whether a given reaction will occur as written and to calculate the relative contributions of molecular phenomena (for example, hydrogen bonding or hydrophobic interactions) to an overall process.

3.5 What Are the Characteristics of High-Energy Biomolecules?
A small family of universal biomolecules mediates the flow of energy from exergonic reactions to the energy-requiring processes of life. These molecules are the reduced coenzymes and the high-energy phosphate compounds. High-energy phosphates are not long-term energy storage substances, but rather transient forms of stored energy.

3.6 What Are the Complex Equilibria Involved in ATP Hydrolysis?
ATP, ADP, and similar species can exist in several different ionization states that must be accounted for in any quantitative analysis. Also, phosphate compounds bind a variety of divalent and monovalent cations with substantial affinity, and the various metal complexes must also be considered in such analyses.

3.7 Why Are Coupled Processes Important to Living Things?
Many of the reactions necessary to keep cells and organisms alive must run against their thermodynamic potential, that is, in the direction of positive ΔG. These processes are driven in the thermodynamically unfavorable direction via coupling with highly favorable processes. Many such coupled processes are crucially important in intermediary metabolism, oxidative phosphorylation, and membrane transport.

3.8 What Is the Daily Human Requirement for ATP?
The average adult human, with a typical weight of 70 kg or so, consumes approximately 2800 calories per day. The energy released from food is stored transiently in the form of ATP. Once ATP energy is used and ADP and phosphate are released, our bodies recycle it to ATP through intermediary metabolism so that it may be reused. The typical 70-kg body contains only about 50 grams of ATP/ADP total. Therefore, each ATP molecule in our bodies must be recycled nearly 1300 times each day.

3.9 What Are Reduction Potentials, and How Are They Used to Account for Free Energy Changes in Redox Reactions?
Just as the group transfer potential is used to quantitate the energy of phosphoryl transfer, the standard reduction potential, denoted by $\mathscr{E}_o'$, quantitates the tendency of chemical species to be reduced or oxidized. Standard reduction potentials are determined by measuring the voltages generated in reaction half-cells. A half-cell consists of a solution containing 1 M concentrations of both the oxidized and reduced forms of the substance whose reduction potential is being measured and a simple electrode.

FOUNDATIONAL BIOCHEMISTRY Things You Should Know After Reading Chapter 3.

- The definitions of isolated, closed, and open systems.

- The first law of thermodynamics—the total energy of an isolated system is conserved.

- The definition of enthalpy (Equations 3.2 and 3.3).

- Enthalpy and internal energy are essentially equal in biochemical reactions.

- The van't Hoff relation (Equation 3.4).

- The second law of thermodynamics—systems tend toward disorder and randomness.

- Entropy represents energy dispersion.

- The third law of thermodynamics—the entropy of a crystalline, perfectly ordered substance is exactly zero at 0 K.

- The definition of free energy (Equations 3.10 and 3.11).

- Free energy depends on reactant and product concentrations (Equation 3.13).

- The equilibrium for a reaction in solution is a function of $\Delta G°'$ (Equation 3.14).

- Free energy is pH-dependent if protons are produced or consumed in a reaction.

- Reduced coenzymes and high-energy phosphates mediate flow of energy from exergonic reactions to endergonic reactions.

- High-energy phosphates include phosphoric anhydrides, enol phosphates, acyl phosphates, and guanidine phosphates.

- High-energy phosphates are not long-term energy storage substances, but rather transient forms of stored energy.

- ATP is an intermediate energy-shuttle molecule.

- The $\Delta G°'$ for ATP hydrolysis used in this text is -30.5 kJ/mol.

- Phospho(enol)pyruvate (PEP) has the largest $\Delta G°'$ of hydrolysis of the high-energy phosphates (-62.2 kJ/mol).

- Group transfer potentials quantify the reactivity of functional groups.

- The favorable $\Delta G°'$ of hydrolysis of phosphoric anhydrides is due to electrostatic, ionization, and resonance effects.

- The $\Delta G°'$ of ATP depends on pH and on the concentration of metal ions.

- The cellular ΔG of ATP hydrolysis depends on the concentrations of ATP, ADP, and P_i.

- Reactions and processes with unfavorable ΔG can be made favorable by coupling with favorable processes, such as the hydrolysis of ATP.

- Humans typically consume their own body weight in ATP each day.

- The typical human body contains about 50 grams of ATP. Recycling of each molecule approximately 1300 times per day satisfies the body's ATP needs.

- $\mathscr{E}_0'$ values can be used to predict the direction of redox reactions.

- $\mathscr{E}_0'$ values can be used to analyze energy changes in redox reactions. The reduction potential depends on concentration.

PROBLEMS

Answers to all problems are at the end of this book. Detailed solutions are available in the Student Solutions Manual, Study Guide, and Problems Book.

1. **Calculating K_{eq} and ΔG from Concentrations** An enzymatic hydrolysis of fructose-1-P,

$$\text{Fructose-1-P} + H_2O \rightleftharpoons \text{fructose} + P_i$$

was allowed to proceed to equilibrium at 25°C. The original concentration of fructose-1-P was 0.2 M, but when the system had reached equilibrium the concentration of fructose-1-P was only 6.52×10^{-5} M. Calculate the equilibrium constant for this reaction and the free energy of hydrolysis of fructose-1-P.

2. **Calculating $\Delta G°$ and $\Delta S°$ from $\Delta H°$** The equilibrium constant for some process A $\rightleftharpoons$ B is 0.5 at 20°C and 10 at 30°C. Assuming that $\Delta H°$ is independent of temperature, calculate $\Delta H°$ for this reaction. Determine $\Delta G°$ and $\Delta S°$ at 20° and at 30°C. Why is it important in this problem to assume that $\Delta H°$ is independent of temperature?

3. **Calculating ΔG from $\Delta G°'$** The standard-state free energy of hydrolysis for acetyl phosphate is $\Delta G° = -42.3$ kJ/mol.

$$\text{Acetyl-P} + H_2O \longrightarrow \text{acetate} + P_i$$

Calculate the free energy change for acetyl phosphate hydrolysis in a solution of 2 mM acetate, 2 mM phosphate, and 3 nM acetyl phosphate.

4. **Understanding State Functions** Define a state function. Name three thermodynamic quantities that are state functions and three that are not.

5. **Calculating the Effect of pH on $\Delta G°$** ATP hydrolysis at pH 7.0 is accompanied by release of a hydrogen ion to the medium

$$ATP^{4-} + H_2O \rightleftharpoons ADP^{3-} + HPO_4^{2-} + H^+$$

If the $\Delta G°'$ for this reaction is -30.5 kJ/mol, what is $\Delta G°$ (that is, the free energy change for the same reaction with all components, including H^+, at a standard state of 1 M)?

6. **Calculating K_{eq} and $\Delta G°$ for Coupled Reactions** For the process A $\rightleftharpoons$ B, K_{eq} (AB) is 0.02 at 37°C. For the process B $\rightleftharpoons$ C, K_{eq} (BC) = 1000 at 37°C.

   a. Determine K_{eq} (AC), the equilibrium constant for the overall process A $\rightleftharpoons$ C, from K_{eq} (AB) and K_{eq} (BC).

   b. Determine standard-state free energy changes for all three processes, and use $\Delta G°$(AC) to determine K_{eq} (AC). Make sure that this value agrees with that determined in part a of this problem.

7. **Understanding Resonance Structures** Draw all possible resonance structures for creatine phosphate and discuss their possible effects on resonance stabilization of the molecule.

8. **Calculating K_{eq} from $\Delta G°'$** Write the equilibrium constant, K_{eq}, for the hydrolysis of creatine phosphate and calculate a value for K_{eq} at 25°C from the value of $\Delta G°'$ in Table 3.3.

9. **Imagining Creatine Phosphate and Glycerol-3-Phosphate as Energy Carriers** Imagine that creatine phosphate, rather than ATP, is the universal energy carrier molecule in the human body. Repeat the calculation presented in Section 3.8, calculating the weight of creatine phosphate that would need to be consumed each day by a typical adult human if creatine phosphate could not be recycled. If recycling of creatine phosphate were possible, and if the typical adult human body contained 20 grams of creatine phosphate, how many times

would each creatine phosphate molecule need to be turned over or recycled each day? Repeat the calculation assuming that glycerol-3-phosphate is the universal energy carrier and that the body contains 20 grams of glycerol-3-phosphate.

10. **Calculating ΔG in a Rat Liver Cell** Calculate the free energy of hydrolysis of ATP in a rat liver cell in which the ATP, ADP, and P_i concentrations are 3.4, 1.3, and 4.8 mM, respectively.

11. **Calculating $\Delta G°'$ and K_{eq} from Cellular Concentrations** Hexokinase catalyzes the phosphorylation of glucose from ATP, yielding glucose-6-P and ADP. Using the values of Table 3.3, calculate the standard-state free energy change and equilibrium constant for the hexokinase reaction.

12. **Evaluating Reactivity of High-Energy Molecules** Would you expect the free energy of hydrolysis of acetoacetyl-coenzyme A (see diagram) to be greater than, equal to, or less than that of acetyl-coenzyme A? Provide a chemical rationale for your answer.

$$\begin{matrix} O & & O \\ \| & & \| \\ CH_3-C-CH_2-C-S-CoA \end{matrix}$$

13. **Assessing the Reactivity of Carbamoyl Phosphate** Consider carbamoyl phosphate, a precursor in the biosynthesis of pyrimidines:

$$\begin{matrix} & O \\ & \| \\ & C \\ H_3N^+ & O-PO_3^{2-} \end{matrix}$$

Based on the discussion of high-energy phosphates in this chapter, would you expect carbamoyl phosphate to possess a high free energy of hydrolysis? Provide a chemical rationale for your answer.

14. **Assessing the Effect of Temperature on Equilibrium** You are studying the various components of the venom of a poisonous lizard. One of the venom components is a protein that appears to be temperature sensitive. When heated, it denatures and is no longer toxic. The process can be described by the following simple equation:

$$\text{T (toxic)} \rightleftharpoons \text{N (nontoxic)}$$

There is only enough protein from this venom to carry out two equilibrium measurements. At 298 K, you find that 98% of the protein is in its toxic form. However, when you raise the temperature to 320 K, you find that only 10% of the protein is in its toxic form.

   a. Calculate the equilibrium constants for the T to N conversion at these two temperatures.

   b. Use the data to determine the $\Delta H°$, $\Delta S°$, and $\Delta G°$ for this process.

15. **Analyzing the Thermodynamics of Protein Denaturation** Consider the data in Figures 3.3 and 3.4. Is the denaturation of chymotrypsinogen spontaneous at 58°C? And what is the temperature at which the native and denatured forms of chymotrypsinogen are in equilibrium?

16. **Assessing the Meaning of Heat Capacity Changes** Consider Tables 3.1 and 3.2, as well as the discussion of Table 3.2 in the text, and discuss the meaning of the positive ΔC_P in Table 3.1.

17. **Assessing the Effect of pH on Metabolic Reactions** The difference between $\Delta G°$ and $\Delta G°'$ was discussed in Section 3.3. Consider the hydrolysis of acetyl phosphate (Figure 3.12) and determine the value of $\Delta G°$ for this reaction at pH 2, 7, and 12. The

value of $\Delta G°'$ for the enolase reaction (Figure 3.13) is 1.8 kJ/mol. What is the value of $\Delta G°$ for enolase at pH 2, 7, and 12? Why is this case different from that of acetyl phosphate?

18. Analyzing the Energetics of Coupled Reactions What is the significance of the magnitude of $\Delta G°'$ for ATP hydrolysis in the calculations in the box on page 69? Repeat these calculations for the case of coupling of a reaction to 1,3-bisphosphoglycerate hydrolysis to see what effect this reaction would have on the equilibrium ratio for components A and B under the conditions stated on this page.

Preparing for the MCAT® Exam

19. The hydrolysis of 1,3-bisphosphoglycerate is favorable, due in part to the increased resonance stabilization of the products of the reaction.

Draw resonance structures for the reactant and the products of this reaction to establish that this statement is true.

20. The acyl-CoA synthetase reaction activates fatty acids for oxidation in cells:

$$R\text{-}COO^- + CoASH + ATP \longrightarrow R\text{-}COSCoA + AMP + pyrophosphate$$

The reaction is driven forward in part by hydrolysis of ATP to AMP and pyrophosphate. However, pyrophosphate undergoes further cleavage to yield two phosphate anions. Discuss the energetics of this reaction both in the presence and absence of pyrophosphate cleavage.

FURTHER READING

General Readings on Thermodynamics

Alberty, R. A., 2003. *Thermodynamics of Biochemical Reactions.* New York: John Wiley.

Alberty, R. A., 2010. Biochemical thermodynamics and rapid-equilibrium kinetics. *Journal of Physical Chemistry B* **114**:17003–17012.

Cantor, C. R., and Schimmel, P. R., 1980. *Biophysical Chemistry.* San Francisco: W. H. Freeman.

Dickerson, R. E., 1969. *Molecular Thermodynamics.* New York: Benjamin Co.

Edsall, J. T., and Gutfreund, H., 1983. *Biothermodynamics: The Study of Biochemical Processes at Equilibrium.* New York: John Wiley.

Edsall, J. T., and Wyman, J., 1958. *Biophysical Chemistry.* New York: Academic Press.

Haynie, D. T., 2008. *Biological Thermodynamics, 2nd ed.* Cambridge: Cambridge University Press.

Klotz, L. M., 1967. *Energy Changes in Biochemical Reactions.* New York: Academic Press.

Lambert, F. L., 2002. Disorder: A cracked crutch for supporting entropy discussions. *Journal of Chemical Education* **79**:187–192.

Lambert, F. L., 2002. Entropy is simple, qualitatively. *Journal of Chemical Education* **79**:1241–1246. (See also http://www.entropysite.com/entropy_is_simple/index.html for a revision of this paper.)

Lehninger, A. L., 1972. *Bioenergetics, 2nd ed.* New York: Benjamin Co.

Morris, J. G., 1968. *A Biologist's Physical Chemistry.* Reading, MA: Addison-Wesley.

Chemistry of Adenosine-5'-Triphosphate

Alberty, R. A., 1968. Effect of pH and metal ion concentration on the equilibrium hydrolysis of adenosine triphosphate to adenosine diphosphate. *Journal of Biological Chemistry* **243**:1337–1343.

Alberty, R. A., 1969. Standard Gibbs free energy, enthalpy, and entropy changes as a function of pH and pMg for reactions involving adenosine phosphates. *Journal of Biological Chemistry* **244**:3290–3302.

Gwynn, R. W., Veech, R. L., 1973. The equilibrium constants of the adenosine triphosphate hydrolysis and the adenosine triphosphate-citrate lyase reactions. *Journal of Biological Chemistry* **248**:6966–6972.

Special Topics

Alberty, R. A., 1994. Legendre transforms in chemical thermodynamics. *Chemical Reviews* **94**:1457-1482.

Brandts, J. F., 1964. The thermodynamics of protein denaturation. I. The denaturation of chymotrypsinogen. *Journal of the American Chemical Society* **86**:4291–4301.

Schneider, E. D., and Sagan, D., 2005. *Into the Cool: Energy Flow, Thermodynamics, and Life.* Chicago: University of Chicago Press.

Schrödinger, E., 1945. *What Is Life?* New York: Macmillan.

Segel, I. H., 1976. *Biochemical Calculations, 2nd ed.* New York: John Wiley.

Tanford, C., 1980. *The Hydrophobic Effect, 2nd ed.* New York: John Wiley.

Amino Acids and the Peptide Bond

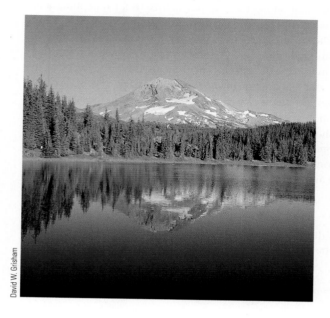

4

David W. Grisham

◀ All objects have mirror images. Like many molecules, amino acids exist in mirror-image forms (stereoisomers) that are not superimposable. Only the L-isomers of amino acids commonly occur in nature. (Three Sisters Wilderness, central Oregon. The Middle Sister, reflected in an alpine lake.)

To hold, as 'twere, the mirror up to nature.

William Shakespeare
Hamlet

ESSENTIAL QUESTION

Proteins are the indispensable agents of biological function, and **amino acids** are the building blocks of proteins. The stunning diversity of the thousands of proteins found in nature arises from the intrinsic properties of only 20 commonly occurring amino acids. These features include (1) the capacity to polymerize, (2) novel acid–base properties, (3) varied structure and chemical functionality in the amino acid side chains, and (4) chirality. This chapter describes each of these properties, laying a foundation for discussions of protein structure (Chapters 5 and 6), enzyme function (Chapters 13–15), and many other subjects in later chapters.

Why are amino acids uniquely suited to their role as the building blocks of proteins?

KEY QUESTIONS

- **4.1** What Are the Structures and Properties of Amino Acids?
- **4.2** What Are the Acid–Base Properties of Amino Acids?
- **4.3** What Reactions Do Amino Acids Undergo?
- **4.4** What Are the Optical and Stereochemical Properties of Amino Acids?
- **4.5** What Are the Spectroscopic Properties of Amino Acids?
- **4.6** How Are Amino Acid Mixtures Separated and Analyzed?
- **4.7** What Is the Fundamental Structural Pattern in Proteins?

4.1 What Are the Structures and Properties of Amino Acids?

Typical Amino Acids Contain a Central Tetrahedral Carbon Atom

The structure of a single typical amino acid is shown in Figure 4.1. Central to this structure is the tetrahedral alpha (α) carbon (C_α), which is covalently linked to both the amino group and the carboxyl group. Also bonded to this α-carbon are a hydrogen and a variable side chain. It is the side chain, the so-called R group, that gives each amino acid its identity. The detailed acid–base properties of amino acids are discussed in the following sections. It is sufficient for now to realize that, in neutral solution (pH 7), the carboxyl group exists as $—COO^-$ and the amino group as $—NH_3^+$. Because the resulting

OWL Online homework and a Student Self Assessment for this chapter may be assigned in OWL.

amino acid contains one positive and one negative charge, it is a neutral molecule called a **zwitterion.** Amino acids are also *chiral* molecules. With four different groups attached to it, the α-carbon is said to be *asymmetric.* The two possible configurations for the α-carbon constitute nonidentical mirror-image isomers or enantiomers. Details of amino acid stereochemistry are discussed in Section 4.4.

Amino Acids Can Join via Peptide Bonds

The crucial feature of amino acids that allows them to polymerize to form peptides and proteins is the existence of their two identifying chemical groups: the amino ($-NH_3^+$) and carboxyl ($-COO^-$) groups, as shown in Figure 4.2. The amino and carboxyl groups of amino acids can react in a head-to-tail fashion, eliminating a water molecule and forming a covalent amide linkage, which, in the case of peptides and proteins, is typically referred to as a **peptide bond.** The equilibrium for this reaction in aqueous solution favors peptide bond hydrolysis. For this reason, biological systems as well as peptide chemists in the laboratory must couple peptide bond formation in an indirect manner or with energy input.

Repetition of the reaction shown in Figure 4.2 produces **polypeptides** and **proteins.** The remarkable properties of proteins, which we shall discover and come to appreciate in later chapters, all depend in one way or another on the unique properties and chemical diversity of the 20 common amino acids found in proteins.

FIGURE 4.1 Anatomy of an amino acid. Except for proline and its derivatives, all of the amino acids commonly found in proteins possess this type of structure.

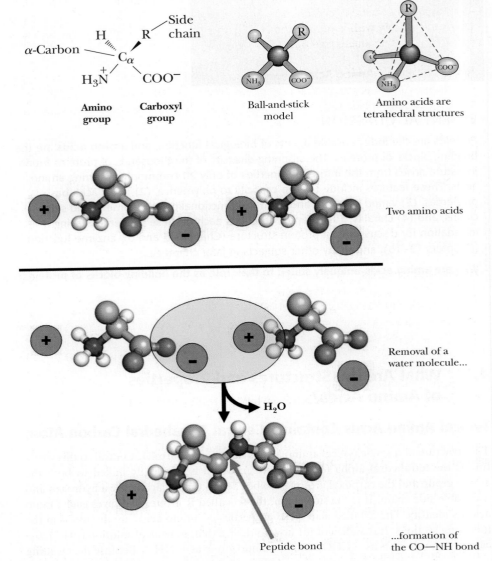

FIGURE 4.2 The α-COOH and α-NH$_3^+$ groups of two amino acids can react with the resulting loss of a water molecule to form a covalent amide bond.

There Are 20 Common Amino Acids

The structures and abbreviations for the 20 amino acids commonly found in proteins are shown in Figure 4.3. All the amino acids except proline have both free α-amino and free α-carboxyl groups (Figure 4.1). There are several ways to classify the common amino acids. The most useful of these classifications is based on the polarity of the side chains. Thus, the structures shown in Figure 4.3 are grouped into the following categories: (1) nonpolar or hydrophobic amino acids, (2) neutral (uncharged) but polar amino acids, (3) acidic amino acids (which have a net negative charge at pH 7.0), and (4) basic amino acids (which have a net positive charge at neutral pH). In later chapters, the importance of this classification system for predicting protein properties becomes clear. Also shown in Figure 4.3 are the three-letter and one-letter codes used to represent the amino acids. These codes are useful when displaying and comparing the sequences of proteins in shorthand form. (Note that several of the one-letter abbreviations are phonetic in origin: arginine = "Rginine" = R, phenylalanine = "Fenylalanine" = F, aspartic acid = "asparDic" = D.)

Nonpolar Amino Acids The nonpolar amino acids (Figure 4.3a) are critically important for the processes that drive protein chains to "fold," that is to form their natural (and functional) structures, as shown in Chapter 6. Amino acids termed *nonpolar* include all those with alkyl chain R groups (alanine, valine, leucine, and isoleucine), as well as proline (with its unusual cyclic structure); methionine (one of the two sulfur-containing amino acids); and two aromatic amino acids, phenylalanine and tryptophan. Tryptophan is sometimes considered a borderline member of this group because it can interact favorably with water via the N—H moiety of the indole ring. Proline, strictly speaking, is not an amino acid but rather an α-imino acid.

Polar, Uncharged Amino Acids The polar, uncharged amino acids (Figure 4.3b), except for glycine, contain R groups that can (1) form hydrogen bonds with water, and (2) play a variety of nucleophilic roles in enzyme reactions. These amino acids are usually more soluble in water than the nonpolar amino acids. The amide groups of asparagine and glutamine; the hydroxyl groups of tyrosine, threonine, and serine; and the sulfhydryl group of cysteine are all good hydrogen bond–forming moieties. Glycine, the simplest amino acid, has only a single hydrogen for an R group, and this hydrogen is not a good hydrogen bond former. Glycine's solubility properties are mainly influenced by its polar amino and carboxyl groups, and thus glycine is best considered a member of the polar, uncharged group. It should be noted that tyrosine has significant nonpolar characteristics due to its aromatic ring and could arguably be placed in the nonpolar group. However, with a pK_a of 10.1, tyrosine's phenolic hydroxyl is a charged, polar entity at high pH.

Acidic Amino Acids There are two acidic amino acids—aspartic acid and glutamic acid—whose R groups contain a carboxyl group (Figure 4.3c). These side-chain carboxyl groups are weaker acids than the α-COOH group but are sufficiently acidic to exist as —COO$^-$ at neutral pH. Aspartic acid and glutamic acid thus have a net negative charge at pH 7. These forms are appropriately referred to as aspartate and glutamate. These negatively charged amino acids play several important roles in proteins. Many proteins that bind metal ions for structural or functional purposes possess metal-binding sites containing one or more aspartate and glutamate side chains. The acid–base chemistry of such groups is considered in detail in Section 4.2.

Basic Amino Acids Three of the common amino acids have side chains with net positive charges at neutral pH: histidine, arginine, and lysine (Figure 4.3d). Histidine contains an imidazole group, arginine contains a guanidino group, and lysine contains a protonated alkyl amino group. The side chains of the latter two amino acids are fully protonated at pH 7, but histidine, with a side-chain pK_a of 6.0, is only 10% protonated at pH 7. With a pK_a near neutrality, histidine side chains play important roles as proton donors and acceptors in many enzyme reactions. Histidine-containing peptides are important biological buffers, as discussed in Chapter 2. Arginine and lysine side chains, which are protonated under physiological conditions, participate in electrostatic interactions in proteins.

(a) Nonpolar (hydrophobic)

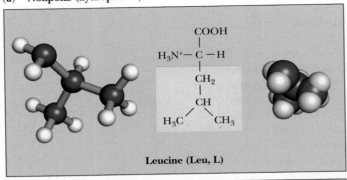

$$
\begin{array}{c}
\text{COOH} \\
| \\
\text{H}_3\text{N}^+ - \text{C} - \text{H} \\
| \\
\text{CH}_2 \\
| \\
\text{CH} \\
\diagup \quad \diagdown \\
\text{H}_3\text{C} \qquad \text{CH}_3
\end{array}
$$

Leucine (Leu, L)

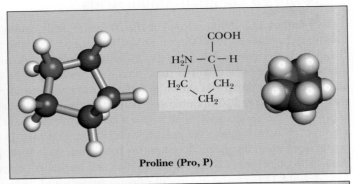

$$
\begin{array}{c}
\text{COOH} \\
| \\
\text{H}_2\text{N}^+ - \text{C} - \text{H} \\
\diagup \qquad \diagdown \\
\text{H}_2\text{C} \qquad \text{CH}_2 \\
\diagdown \quad \diagup \\
\text{CH}_2
\end{array}
$$

Proline (Pro, P)

$$
\begin{array}{c}
\text{COOH} \\
| \\
\text{H}_3\text{N}^+ - \text{C} - \text{H} \\
| \\
\text{CH}_3
\end{array}
$$

Alanine (Ala, A)

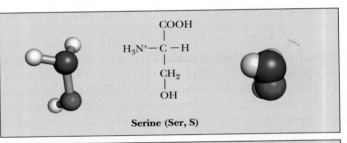

$$
\begin{array}{c}
\text{COOH} \\
| \\
\text{H}_3\text{N}^+ - \text{C} - \text{H} \\
| \\
\text{CH} \\
\diagup \quad \diagdown \\
\text{CH}_3 \qquad \text{CH}_3
\end{array}
$$

Valine (Val, V)

(b) Polar, uncharged

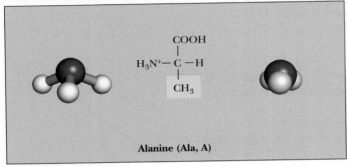

$$
\begin{array}{c}
\text{COOH} \\
| \\
\text{H}_3\text{N}^+ - \text{C} - \text{H} \\
| \\
\text{H}
\end{array}
$$

Glycine (Gly, G)

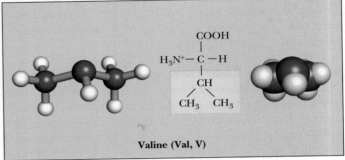

$$
\begin{array}{c}
\text{COOH} \\
| \\
\text{H}_3\text{N}^+ - \text{C} - \text{H} \\
| \\
\text{CH}_2 \\
| \\
\text{OH}
\end{array}
$$

Serine (Ser, S)

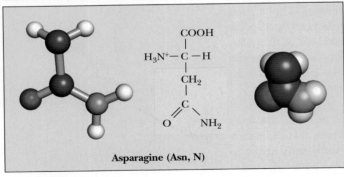

$$
\begin{array}{c}
\text{COOH} \\
| \\
\text{H}_3\text{N}^+ - \text{C} - \text{H} \\
| \\
\text{CH}_2 \\
| \\
\text{C} \\
\diagup \quad \diagdown \\
\text{O} \qquad \text{NH}_2
\end{array}
$$

Asparagine (Asn, N)

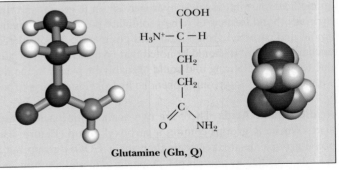

$$
\begin{array}{c}
\text{COOH} \\
| \\
\text{H}_3\text{N}^+ - \text{C} - \text{H} \\
| \\
\text{CH}_2 \\
| \\
\text{CH}_2 \\
| \\
\text{C} \\
\diagup \quad \diagdown \\
\text{O} \qquad \text{NH}_2
\end{array}
$$

Glutamine (Gln, Q)

(c) Acidic

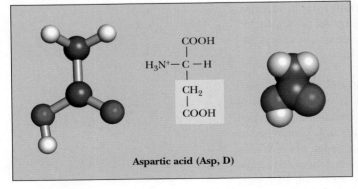

$$
\begin{array}{c}
\text{COOH} \\
| \\
\text{H}_3\text{N}^+ - \text{C} - \text{H} \\
| \\
\text{CH}_2 \\
| \\
\text{COOH}
\end{array}
$$

Aspartic acid (Asp, D)

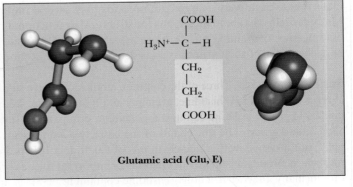

$$
\begin{array}{c}
\text{COOH} \\
| \\
\text{H}_3\text{N}^+ - \text{C} - \text{H} \\
| \\
\text{CH}_2 \\
| \\
\text{CH}_2 \\
| \\
\text{COOH}
\end{array}
$$

Glutamic acid (Glu, E)

FIGURE 4.3 The 20 amino acids that are building blocks of most proteins can be classified as **(a)** nonpolar (hydrophobic); **(b)** polar, neutral; **(c)** acidic; or **(d)** basic.

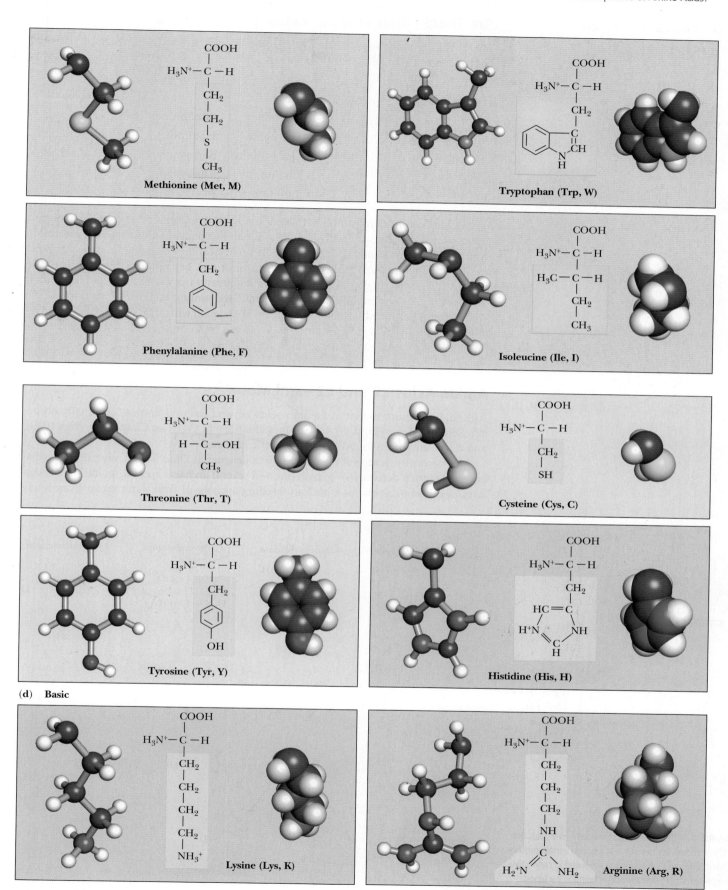

Methionine (Met, M)

Tryptophan (Trp, W)

Phenylalanine (Phe, F)

Isoleucine (Ile, I)

Threonine (Thr, T)

Cysteine (Cys, C)

Tyrosine (Tyr, Y)

Histidine (His, H)

(d) **Basic**

Lysine (Lys, K)

Arginine (Arg, R)

FIGURE 4.3 continued

Are There Other Ways to Classify Amino Acids?

There are alternative ways to classify the 20 common amino acids. For example, it would be reasonable to imagine that the amino acids could be described as hydrophobic, hydrophilic, or amphipathic:

Hydrophobic:		Hydrophilic:		Amphipathic:
Alanine	Proline	Arginine	Glutamine	Lysine
Glycine	Valine	Asparagine	Histidine	Methionine
Isoleucine		Aspartic acid	Serine	Tryptophan
Leucine		Cysteine	Threonine	Tyrosine
Phenylalanine		Glutamic acid		

Lysine can be considered amphipathic because its R group consists of an aliphatic side chain, which can interact with hydrophobic amino acids in proteins, and an amino group, which is normally charged at neutral pH. Methionine is the least polar of the amphipathic amino acids, but its thioether sulfur can be an effective metal ligand in proteins. Cysteine can deprotonate at pH values greater than 7, and the thiolate anion is the most potent nucleophile that can be generated among the 20 common acids. The imidazole ring of histidine has two nitrogen atoms, each with an H. The pK for dissociation of the first of these two H is around 6. However, once one N–H has dissociated, the pK value for the other becomes greater than 10.

Amino Acids 21 and 22—and More?

Although uncommon, natural amino acids beyond the well-known 20 actually do occur. Selenocysteine (Figure 4.4a) was first identified in 1986 (see Chapter 30, page 1049), and it has since been found in a variety of organisms.

More recently, Joseph Krzycki and his colleagues at Ohio State University have discovered a lysine derivative—pyrrolysine—in several archaeal species, including *Methanosarcina barkeri,* found as a bottom-dwelling microbe of freshwater lakes. Pyrrolysine

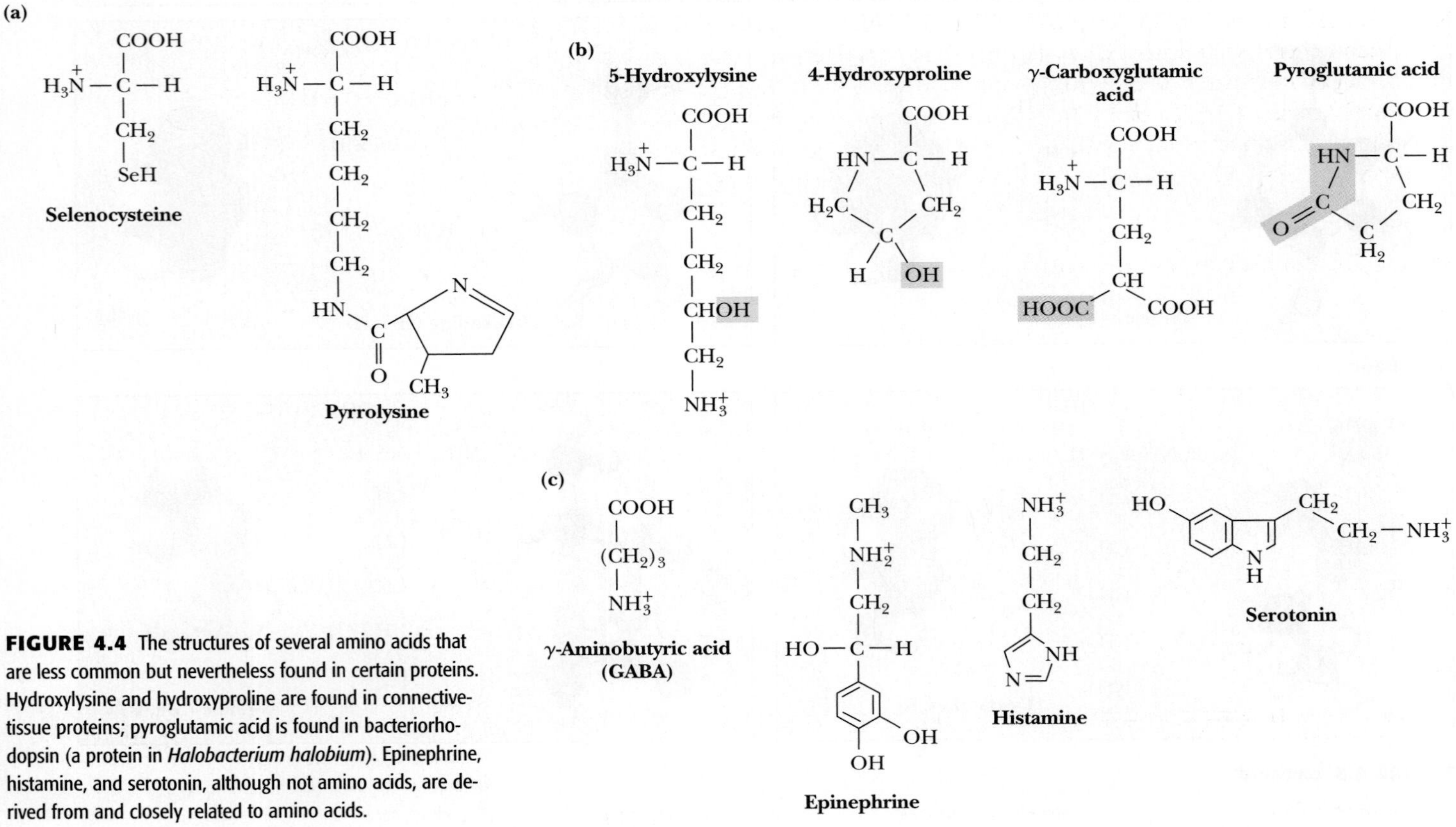

FIGURE 4.4 The structures of several amino acids that are less common but nevertheless found in certain proteins. Hydroxylysine and hydroxyproline are found in connective-tissue proteins; pyroglutamic acid is found in bacteriorhodopsin (a protein in *Halobacterium halobium*). Epinephrine, histamine, and serotonin, although not amino acids, are derived from and closely related to amino acids.

(Figure 4.4a) and selenocysteine both are incorporated naturally into proteins thanks to specific modifications that occur during the reactions of protein synthesis.

Both selenocysteine and pyrrolysine bring novel structural and chemical features to the proteins that contain them. How many more unusual amino acids might be incorporated in proteins in a similar manner?

Several Amino Acids Occur Only Rarely in Proteins

There are several amino acids that occur only rarely in proteins and are produced by modifications of one of the 20 amino acids already incorporated into a protein (Figure 4.4b). These novelties include **hydroxylysine** and **hydroxyproline,** which are found mainly in the collagen and gelatin proteins, **pyroglutamic acid,** which is found in a light-driven proton-pumping protein called bacteriorhodopsin, and **γ-carboxyglutamic acid,** which is found in calcium-binding proteins.

Certain amino acids and their derivatives, although not found in proteins, nonetheless are biochemically important. A few of the more notable examples are shown in Figure 4.4c. **γ-Aminobutyric acid,** or **GABA,** is produced by the decarboxylation of glutamic acid and is a potent neurotransmitter. **Histamine,** which is synthesized by decarboxylation of histidine, and **serotonin,** which is derived from tryptophan, similarly function as neurotransmitters and regulators. **Epinephrine** (also known as **adrenaline**), derived from tyrosine, is an important hormone.

4.2   What Are the Acid–Base Properties of Amino Acids?

Amino Acids Are Weak Polyprotic Acids

From a chemical point of view, the common amino acids are all weak polyprotic acids. The ionizable groups are not strongly dissociating ones, and the degree of dissociation thus depends on the pH of the medium. All the amino acids contain at least two dissociable hydrogens.

Consider the acid–base behavior of glycine, the simplest amino acid. At low pH, both the amino and carboxyl groups are protonated and the molecule has a net positive charge. If the counterion in solution is a chloride ion, this form is referred to as **glycine hydrochloride.** If the pH is increased, the carboxyl group is the first to dissociate, yielding the neutral zwitterionic species Gly^0 (Figure 4.5). A further increase in pH eventually results in dissociation of the amino group to yield the negatively charged **glycinate.** If we denote these three forms as Gly^+, Gly^0, and Gly^-, we can write the first dissociation of Gly^+ as

$$Gly^+ + H_2O \rightleftharpoons Gly^0 + H_3O^+$$

and the dissociation constant K_1 as

$$K_1 = \frac{[Gly^0][H_3O^+]}{[Gly^+]}$$

pH 1   Net charge +1   pH 7   Net charge 0   pH 13   Net charge −1

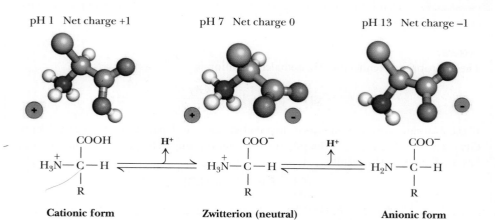

FIGURE 4.5 The ionic forms of the amino acids, shown without consideration of any ionizations on the side chain. The cationic form is the low pH form, and the titration of the cationic species with base yields the zwitterion and finally the anionic form.

TABLE 4.1	pK_a Values of Common Amino Acids		
Amino Acid	α-COOH pK_a	α-NH$_3^+$ pK_a	R group pK_a
Alanine	2.4	9.7	
Arginine	2.2	9.0	12.5
Asparagine	2.0	8.8	
Aspartic acid	2.1	9.8	3.9
Cysteine	1.7	10.8	8.3
Glutamic acid	2.2	9.7	4.3
Glutamine	2.2	9.1	
Glycine	2.3	9.6	
Histidine	1.8	9.2	6.0
Isoleucine	2.4	9.7	
Leucine	2.4	9.6	
Lysine	2.2	9.0	10.5
Methionine	2.3	9.2	
Phenylalanine	1.8	9.1	
Proline	2.1	10.6	
Serine	2.2	9.2	~13
Threonine	2.6	10.4	~13
Tryptophan	2.4	9.4	
Tyrosine	2.2	9.1	10.1
Valine	2.3	9.6	

Values for K_1 for the common amino acids are typically 0.4 to 1.0×10^{-2} M, so that typical values of pK_1 center on values of 2.0 to 2.4 (Table 4.1). In a similar manner, we can write the second dissociation reaction as

$$\text{Gly}^0 + H_2O \rightleftharpoons \text{Gly}^- + H_3O^+$$

and the dissociation constant K_2 as

$$K_2 = \frac{[\text{Gly}^-][H_3O^+]}{[\text{Gly}^0]}$$

Typical values for pK_2 are in the range of 9.0 to 9.8. At physiological pH, the α-carboxyl group of a simple amino acid (with no ionizable side chains) is completely dissociated, whereas the α-amino group has not really begun its dissociation. The titration curve for such an amino acid is shown in Figure 4.6.

EXAMPLE What is the pH of a glycine solution in which the α-NH$_3^+$ group is one-third dissociated?

Answer

The appropriate Henderson–Hasselbalch equation is

$$pH = pK_a + \log_{10} \frac{[\text{Gly}^-]}{[\text{Gly}^0]}$$

If the α-amino group is one-third dissociated, there is 1 part Gly$^-$ for every 2 parts Gly^0. The important pK_a is the pK_a for the amino group. The glycine α-amino group has a pK_a of 9.6. The result is

$$pH = 9.6 + \log_{10} (1/2)$$
$$pH = 9.3$$

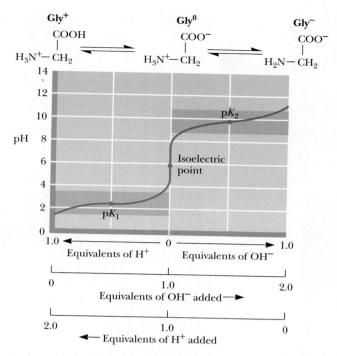

FIGURE 4.6 Titration of glycine, a simple amino acid. The isoelectric point, pI, the pH where glycine has a net charge of 0, can be calculated as $(pK_1 + pK_2)/2$.

Note that the dissociation constants of both the α-carboxyl and α-amino groups are affected by the presence of the other group. The adjacent α-amino group makes the α-COOH group more acidic (that is, it lowers the pK_a), so it gives up a proton more readily than simple alkyl carboxylic acids. Thus, the pK_1 of 2.0 to 2.1 for α-carboxyl groups of amino acids is substantially lower than that of acetic acid ($pK_a = 4.76$), for example. What is the chemical basis for the low pK_a of the α-COOH group of amino acids? The α-NH_3^+ (ammonium) group is strongly electron-withdrawing, and the positive charge of the amino group exerts a strong field effect and stabilizes the carboxylate anion. (The effect of the α-COO^- group on the pK_a of the α-NH_3^+ group is the basis for problem 4 at the end of this chapter.)

Side Chains of Amino Acids Undergo Characteristic Ionizations

As we have seen, the side chains of several of the amino acids also contain dissociable groups. Thus, aspartic and glutamic acids contain an additional carboxyl function, and lysine possesses an aliphatic amino function. Histidine contains an ionizable imidazolium proton, and arginine carries a guanidinium function. Typical pK_a values of these groups are shown in Table 4.1. The β-carboxyl group of aspartic acid and the γ-carboxyl side chain of glutamic acid exhibit pK_a values intermediate to the α-COOH on one hand and typical aliphatic carboxyl groups on the other hand. In a similar fashion, the ϵ-amino group of lysine exhibits a pK_a that is higher than that of the α-amino group but similar to that for a typical aliphatic amino group. These intermediate side-chain pK_a values reflect the slightly diminished effect of the α-carbon dissociable groups that lie several carbons removed from the side-chain functional groups. Figure 4.7 shows typical titration curves for glutamic acid and lysine, along with the ionic species that predominate at various points in the titration. The only other side-chain groups that exhibit any significant degree of dissociation are the *para*-OH group of tyrosine and the —SH group of cysteine. The pK_a of the cysteine sulfhydryl is 8.32, so it is about 5% dissociated at pH 7. The tyrosine *para*-OH group is a very weakly acidic group, with a pK_a of about 10.1. This group is essentially fully protonated and uncharged at pH 7.

It is important to note that side-chain pK_a values for amino acids in proteins can be different from the values shown in Table 4.1. On *average,* values for side chains in proteins

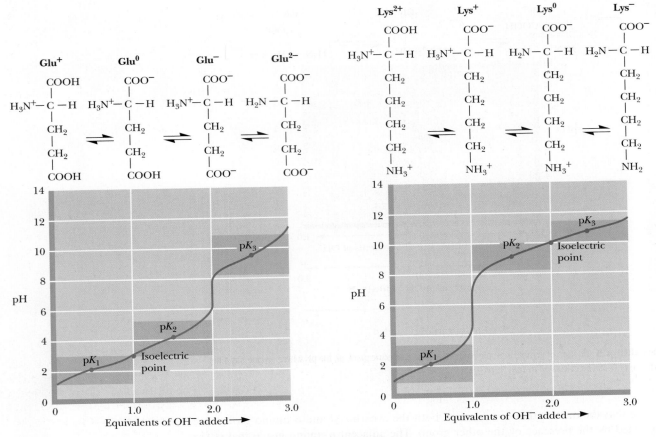

FIGURE 4.7   Titrations of glutamic acid and lysine.

are one pH unit closer to neutrality compared to the free amino acid values. Moreover, environmental effects in the protein can change pK_a values dramatically.

4.3 What Reactions Do Amino Acids Undergo?

A number of reactions of amino acids are noteworthy because they are essential to the degradation, sequencing, and chemical synthesis of peptides and proteins. One of these, the reaction with phenylisothiocyanate, or **Edman reagent,** involves nucleophilic attack by the amino acid α-amino nitrogen, followed by cyclization, to yield a phenylthiohydantoin (PTH) derivative of the amino acid (Figure 4.8a). PTH-amino acids can be easily identified and quantified, as shown in Section 4.6. An important amino acid side-chain reaction is formation of disulfide bonds via reaction between two cysteines. In proteins, cysteine residues form disulfide linkages that stabilize protein structure (Figure 4.8b). Related reactions are discussed in Chapter 5.

4.4 What Are the Optical and Stereochemical Properties of Amino Acids?

Amino Acids Are Chiral Molecules

Except for glycine, all of the amino acids isolated from proteins have four different groups attached to the α-carbon atom. In such a case, the α-carbon is said to be **asymmetric** or **chiral** (from the Greek *cheir*, meaning "hand"), and the two possible configurations for the α-carbon constitute nonsuperimposable mirror-image isomers, or

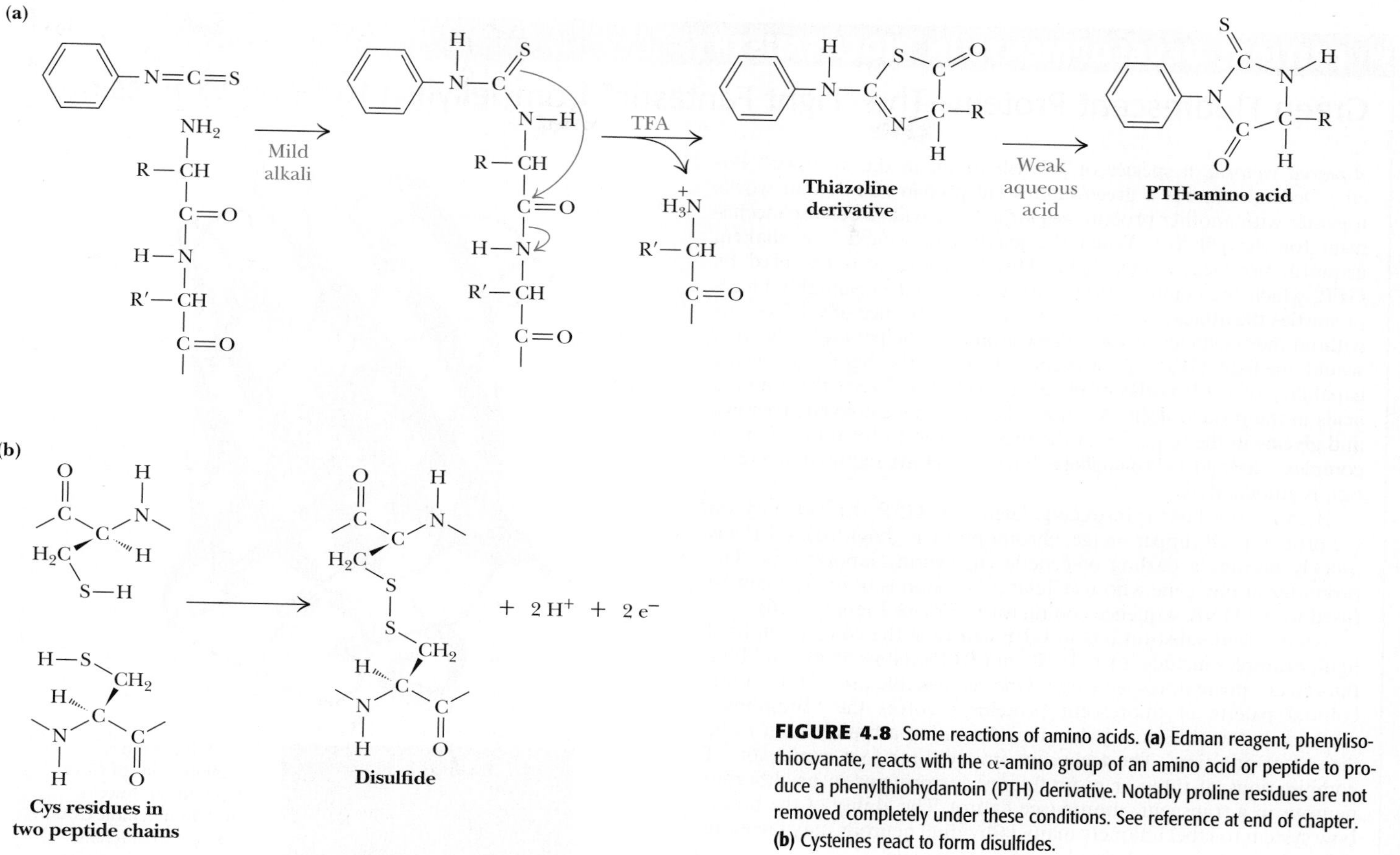

(a)

(b)

$$+ 2 H^+ + 2 e^-$$

Disulfide

**Cys residues in
two peptide chains**

FIGURE 4.8 Some reactions of amino acids. **(a)** Edman reagent, phenyliso-thiocyanate, reacts with the α-amino group of an amino acid or peptide to produce a phenylthiohydantoin (PTH) derivative. Notably proline residues are not removed completely under these conditions. See reference at end of chapter. **(b)** Cysteines react to form disulfides.

enantiomers (Figure 4.9). Enantiomeric molecules display a special property called **optical activity**—the ability to rotate the plane of polarization of plane-polarized light. Clockwise rotation of incident light is referred to as **dextrorotatory** behavior, and counterclockwise rotation is called **levorotatory** behavior. The magnitude and direction of the optical rotation depend on the nature of the amino acid side chain. Some protein-derived amino acids at a given pH are dextrorotatory and others are levorotatory, even though all of them are of the L-configuration. The direction of optical rotation can be specified in the name by using a (+) for dextrorotatory compounds and a (−) for levorotatory compounds, as in L(+)-leucine.

Chiral Molecules Are Described by the D,L and (R,S) Naming Conventions

The discoveries of optical activity and enantiomeric structures (see Critical Developments in Biochemistry, page xx) made it important to develop suitable nomenclature for chiral molecules. Two systems are in common use today: the so-called D,L system and the (R,S) system.

In the **D,L system** of nomenclature, the (+) and (−) isomers of glyceraldehyde are denoted as **D-glyceraldehyde** and **L-glyceraldehyde,** respectively (see Critical Developments in Biochemistry, page xx). Absolute configurations of all other carbon-based molecules are referenced to D- and L-glyceraldehyde. When sufficient care is taken to avoid racemization of the amino acids during hydrolysis of proteins, it is found that all of the amino acids derived from natural proteins are of the L-configuration. Amino acids of the D-configuration are nonetheless found in nature, especially as components of certain peptide antibiotics, such as valinomycin, gramicidin, and actinomycin D, and in the cell walls of certain microorganisms.

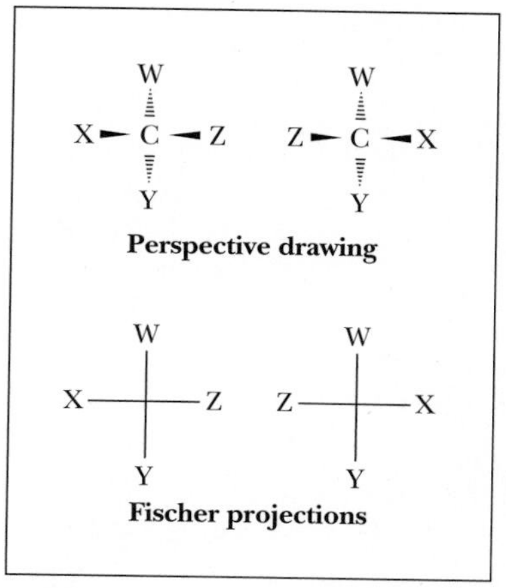

Perspective drawing

Fischer projections

FIGURE 4.9 Enantiomeric molecules based on a chiral carbon atom. Enantiomers are nonsuperimposable mirror images of each other.

CRITICAL DEVELOPMENTS IN BIOCHEMISTRY

Green Fluorescent Protein—The "Light Fantastic" from Jellyfish to Gene Expression

Aquorea victoria, a species of jellyfish found in the northwest Pacific Ocean, contains a **green fluorescent protein (GTP)** that works together with another protein, **aequorin,** to provide a defense mechanism for the jellyfish. When the jellyfish is attacked or shaken, aequorin produces a blue light. This light energy is captured by GFP, which then emits a bright green flash that presumably blinds or startles the attacker. Remarkably, the fluorescence of GFP occurs without the assistance of a **prosthetic group**—a "helper molecule" that would mediate GFP's fluorescence. Instead, the light-transducing capability of GFP is the result of a reaction between three amino acids in the protein itself. As shown below, adjacent **serine, tyrosine,** and **glycine** in the sequence of the protein react to form the pigment complex—termed a **chromophore.** No enzymes are required; the reaction is autocatalytic.

Because the light-transducing talents of GFP depend only on the protein itself (upper image, chromophore highlighted), GFP has quickly become a darling of genetic engineering laboratories. The promoter of any gene whose cellular expression is of interest can be fused to the DNA sequence coding for GFP (see Figure 12.16).

Amino acid substitutions in GFP can tune the color of emitted light; examples include YFP, CFP, and BFP (yellow, cyan, and blue fluorescent protein, respectively). One remarkable use of this multicolored palette of fluorescent proteins involves the "Brainbow" method that enables many distinct cells within a brain circuit to be viewed at one time. The technique involves random expression of combinations of three or more fluorescent proteins in neighboring neurons in a transgenic mouse (see figure). The ability of the brainbow system to label uniquely many individual neurons may facilitate the analysis of neuronal circuitry on a large scale.

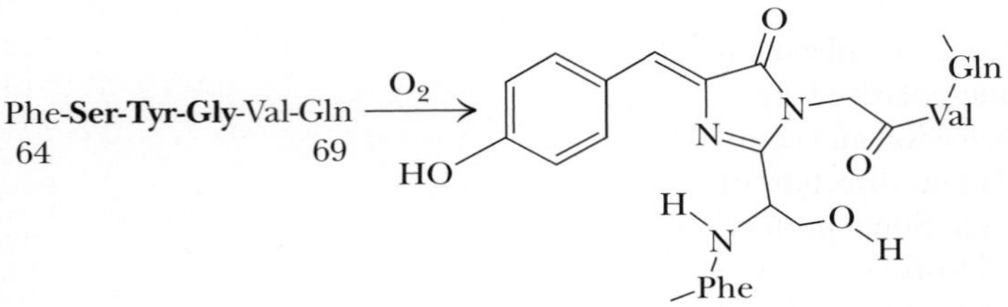

Phe-**Ser-Tyr-Gly**-Val-Gln $\xrightarrow{O_2}$
64   69

◁ Expression of combinations of three different fluorescent proteins in a mouse brain produce 10 different colorations of neurons. Individual neurons in a mouse brain appear in different colors in a fluorescence microscope. Livet, J., Weissman, T. A., Kang, H., Draft, R. W., et al., 2007. Transgenic strategies for combinatorial expression of fluorescent proteins in the nervous system. *Nature* **450:**56–63.

Dentate gyrus

Despite its widespread acceptance, problems exist with the D,L system of nomenclature. For example, this system can be ambiguous for molecules with two or more chiral centers. To address such problems, the **(R,S) system** of nomenclature for chiral molecules was proposed in 1956 by Robert Cahn, Sir Christopher Ingold, and Vladimir Prelog. In this more versatile system, priorities are assigned to each of the groups attached to a chiral center on the basis of atomic number, atoms with higher atomic numbers having higher priorities.

The newer (R,S) system of nomenclature is superior to the older D,L system in one important way: The configuration of molecules with more than one chiral center can be more easily, completely, and unambiguously described with (R,S) notation. Several amino acids, including isoleucine, threonine, hydroxyproline, and hydroxylysine, have two chiral centers. In the (R,S) system, L-threonine is (2S,3R)-threonine.

4.5 What Are the Spectroscopic Properties of Amino Acids?

One of the most important and exciting advances in modern biochemistry has been the application of **spectroscopic methods,** which measure the absorption and emission of energy of different frequencies by molecules and atoms. Spectroscopic studies of

CRITICAL DEVELOPMENTS IN BIOCHEMISTRY

Discovery of Optically Active Molecules and Determination of Absolute Configuration

The optical activity of quartz and certain other materials was first discovered by Jean-Baptiste Biot in 1815 in France, and in 1848 a young chemist in Paris named Louis Pasteur made a related and remarkable discovery. Pasteur noticed that preparations of optically inactive sodium ammonium tartrate contained two visibly different kinds of crystals that were mirror images of each other. Pasteur carefully separated the two types of crystals, dissolved them each in water, and found that each solution was *optically active.* Even more intriguing, the specific rotations of these two solutions were equal in magnitude and of opposite sign. Because these differences in optical rotation were apparent properties of the dissolved molecules, Pasteur eventually proposed that the molecules themselves were mirror images of each other, just like their respective crystals. Based on this and other related evidence, van't Hoff and Le Bel proposed the tetrahedral arrangement of valence bonds to carbon.

In 1888, Emil Fischer decided that it should be possible to determine the *relative* configuration of (+)-glucose, a six-carbon sugar with four asymmetric centers (see figure). Because each of the four C could be either of two configurations, glucose conceivably could exist in any one of 16 possible isomeric structures. It took 3 years to complete the solution of an elaborate chemical and logical puzzle. By 1891, Fischer had reduced his puzzle to a choice between two enantiomeric structures. (Methods for determining *absolute* configuration were not yet available, so Fischer made a simple guess, selecting the structure shown in the figure.) For this remarkable feat, Fischer received the Nobel Prize in Chemistry in 1902. In 1951, J. M. Bijvoet in Utrecht,

the Netherlands, used a new X-ray diffraction technique to show that Emil Fischer's arbitrary guess 60 years earlier had been correct.

It was M. A. Rosanoff, a chemist and instructor at New York University, who first proposed (in 1906) that the isomers of glyceraldehyde be the standards for denoting the stereochemistry of sugars and other molecules. Later, when experiments showed that the configuration of (+)-glyceraldehyde was related to (+)-glucose, (+)-glyceraldehyde was given the designation D. Emil Fischer rejected the **Rosanoff convention,** but it was universally accepted. Ironically, this nomenclature system is often mistakenly referred to as the **Fischer convention.**

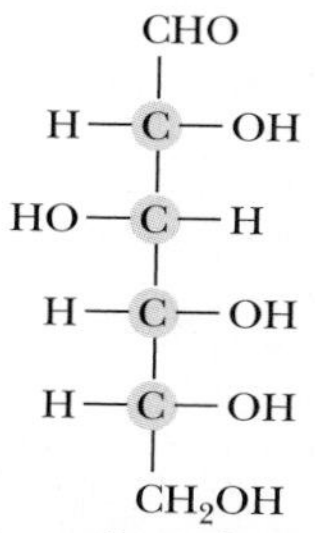

The absolute configuration of (+)-glucose.

*Derewenda, Z. S., 2007. On wine, chirality, and crystallography. *Acta Crystallographica* **A64:**246–258.

proteins, nucleic acids, and other biomolecules are providing many new insights into the structure and dynamic processes in these molecules.

Phenylalanine, Tyrosine, and Tryptophan Absorb Ultraviolet Light

Many details of the structure and chemistry of the amino acids have been elucidated or at least confirmed by spectroscopic measurements. None of the amino acids absorbs light in the visible region of the electromagnetic spectrum. Several of the amino acids, however, do absorb **ultraviolet** radiation, and all absorb in the **infrared** region. The absorption of energy by electrons as they rise to higher-energy states occurs in the ultraviolet/visible region of the energy spectrum. Only the aromatic amino acids phenylalanine, tyrosine, and tryptophan exhibit significant ultraviolet absorption above 250 nm, as shown in Figure 4.10. These strong absorptions can be used for spectroscopic determinations of protein concentration. The aromatic amino acids also exhibit relatively weak fluorescence, and it has recently been shown that tryptophan can exhibit *phosphorescence*—a relatively long-lived emission of light. These fluorescence and phosphorescence properties are especially useful in the study of protein structure and dynamics.

Amino Acids Can Be Characterized by Nuclear Magnetic Resonance

The development in the 1950s of **nuclear magnetic resonance** (NMR), a spectroscopic technique that involves the absorption of radio frequency energy by certain nuclei in the presence of a magnetic field, played an important part in the chemical characterization of amino acids and proteins. Several important principles emerged from these

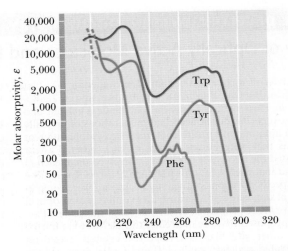

FIGURE 4.10 The ultraviolet absorption spectra of the aromatic amino acids at pH 6. (From Wetlaufer, D. B., 1962. Ultraviolet spectra of proteins and amino acids. *Advances in Protein Chemistry* **17**:303–390.)

studies. First, the **chemical shift**[1] of amino acid protons depends on their particular chemical environment and thus on the state of ionization of the amino acid. Second, the change in electron density during a titration is transmitted throughout the carbon chain in the aliphatic amino acids and the aliphatic portions of aromatic amino acids, as evidenced by changes in the chemical shifts of relevant protons. Finally, the magnitude of the **coupling constants** between protons on adjacent carbons depends in some cases on the ionization state of the amino acid. This apparently reflects differences in

> ## A DEEPER LOOK
>
> ## The Murchison Meteorite—Discovery of Extraterrestrial Handedness
>
> The predominance of L-amino acids in biological systems is one of life's intriguing features. Prebiotic syntheses of amino acids would be expected to produce equal amounts of L- and D-enantiomers. Some kind of enantiomeric selection process must have intervened to select L-amino acids over their D-counterparts as the constituents of proteins. Was it random chance that chose L- over D-isomers?
>
> Analysis of carbon compounds—even amino acids—from extraterrestrial sources might provide deeper insights into this mystery. John Cronin and Sandra Pizzarello have examined the enantiomeric distribution of unusual amino acids obtained from the Murchison meteorite, which struck the earth on September 28, 1969, near
>
> Murchison, Australia. (By selecting unusual amino acids for their studies, Cronin and Pizzarello ensured that they were examining materials that were native to the meteorite and not earth-derived contaminants.) Four α-dialkyl amino acids—α-methylisoleucine, α-methylalloisoleucine, α-methylnorvaline, and isovaline—were found to have an L-enantiomeric excess of 2% to 15%.
>
> This may be the first demonstration that a natural L-enantiomer enrichment occurs in certain cosmological environments. Could these observations be relevant to the emergence of L-enantiomers as the dominant amino acids on the earth? And, if so, could there be life elsewhere in the universe that is based upon the same amino acid handedness?
>
> $$CH_3 - CH_2 - CH - C - COOH$$
> **2-Amino-2,3-dimethylpentanoic acid***
>
> $$CH_3 - CH_2 - C - COOH$$
> **Isovaline**
>
> $$CH_3 - CH_2 - CH_2 - C - COOH$$
> **α-Methylnorvaline**
>
> ◁ Amino acids found in the Murchison meteorite.
>
> *The four stereoisomers of this amino acid include the D- and L-forms of α-methylisoleucine and α-methylalloisoleucine.
> Cronin, J. R., and Pizzarello, S., 1997. Enantiomeric excesses in meteoritic amino acids. *Science* **275**:951–955.
> Pizzarello, S. 2006. The chemistry of life's origin: a carbonaceous meteorite perspective. *Accounts of Chemical Research* **39**:231–237.

[1]The chemical shift for any NMR signal is the difference in resonant frequency between the observed signal and a suitable reference signal. If two nuclei are magnetically coupled, the NMR signals of these nuclei split, and the separation between such split signals, known as the coupling constant, is likewise dependent on the structural relationship between the two nuclei.

CRITICAL DEVELOPMENTS IN BIOCHEMISTRY

Rules for Description of Chiral Centers in the (R,S) System

Naming a chiral center in the (R,S) system is accomplished by viewing the molecule from the chiral center to the atom with the lowest priority. If the other three atoms facing the viewer then decrease in priority in a clockwise direction, the center is said to have the (R) configuration (where R is from the Latin *rectus*, meaning "right"). If the three atoms in question decrease in priority in a counterclockwise fashion, the chiral center is of the (S) configuration (where S is from the Latin *sinistrus*, meaning "left"). If two of the atoms coordinated to a chiral center are identical, the atoms bound to these two are considered for priorities. For such purposes, the

priorities of certain functional groups found in amino acids and related molecules are in the following order:

$$SH > OH > NH_2 > COOH > CHO > CH_2OH > CH_3$$

From this, it is clear that D-glyceraldehyde is (R)-glyceraldehyde and L-alanine is (S)-alanine (see figure). Interestingly, the α-carbon configuration of all the L-amino acids *except for cysteine* is (S). Cysteine, by virtue of its thiol group, is in fact (R)-cysteine.

The assignment of (R) and (S) notation for glyceraldehyde and L-alanine.

the preferred conformations in different ionization states. Proton NMR spectra of two amino acids are shown in Figure 4.11. Because they are highly sensitive to their environment, the chemical shifts of individual NMR signals can detect the pH-dependent ionizations of amino acids. Figure 4.12 shows the ^{13}C chemical shifts occurring in a titration of lysine. Note that the chemical shifts of the carboxyl C, C_α, and C_β carbons of lysine are sensitive to dissociation of the nearby α-COOH and α-NH_3^+ protons

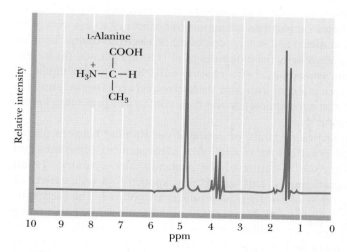

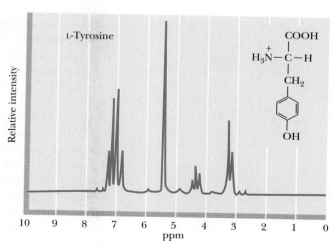

FIGURE 4.11 Proton NMR spectra of several amino acids. Zero on the chemical shift scale is defined by the resonance of tetramethylsilane (TMS). (The large resonance at approximately 5 ppm is due to the normal HDO impurity in the D₂O solvent.) (Adapted from Aldrich Library of NMR Spectra.)

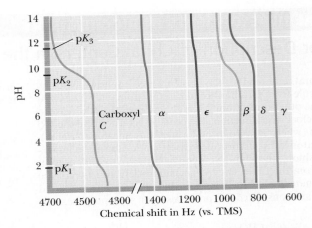

FIGURE 4.12 A plot of chemical shifts versus pH for the carbons of lysine. Changes in chemical shift are most pronounced for atoms near the titrating groups. Note the correspondence between the pK_a values and the particular chemical shift changes. All chemical shifts are defined relative to tetramethylsilane (TMS). (From Suprenant, H., et al., 1980. Carbon-13 NMR studies of amino acids: Chemical shifts, protonation shifts, microscopic protonation behavior. *Journal of Magnetic Resonance* **40**:231–243.)

Kurt Wüthrich received the 2002 Nobel Prize in Chemistry for developing NMR methods for determination of protein structure.

(with pK_a values of about 2 and 9, respectively), whereas the C_δ and C_ϵ carbons are sensitive to dissociation of the ϵ-NH_3^+ group. Such measurements have been very useful for studies of the ionization behavior of amino acid residues in proteins. More sophisticated NMR measurements at very high magnetic fields are also used to determine the three-dimensional structures of peptides and proteins.

4.6 How Are Amino Acid Mixtures Separated and Analyzed?

Amino Acids Can Be Separated by Chromatography

A wide variety of methods is available for the separation and analysis of amino acids (and other biological molecules and macromolecules). All of these methods take advantage of the relative differences in the physical and chemical characteristics of amino acids, particularly ionization behavior and solubility characteristics. Separations of amino acids are usually based on **partition** properties (the tendency to associate with one solvent or phase over another) and separations based on **electrical charge.** In all of the partition methods discussed here, the molecules of interest are allowed (or forced) to flow through a medium consisting of two phases—solid–liquid, liquid–liquid, or gas–liquid. The molecules partition, or distribute themselves, between the two phases in a manner based on their particular properties and their consequent preference for associating with one or the other phase.

In 1903, a separation technique based on repeated partitioning between phases was developed by Mikhail Tswett for the separation of plant pigments (carotenes and chlorophylls). Due to the colorful nature of the pigments thus separated, Tswett called his technique **chromatography.** This term is now applied to a wide variety of separation methods, regardless of whether the products are colored. The success of many chromatography techniques depends on the repeated microscopic partitioning of solutes in a mixture between the available phases. The more frequently this partitioning can be made to occur within a given time span or over a given volume, the more efficient is the resulting separation. Chromatographic methods have advanced rapidly in recent years, due in part to the development of sophisticated new solid-phase materials. Methods important for amino acid separations include ion exchange

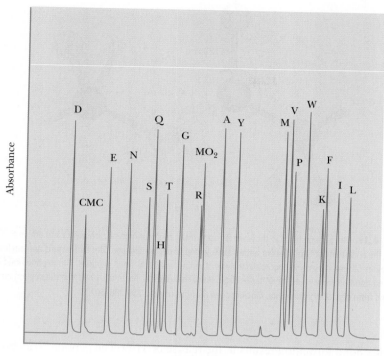

FIGURE 4.13 Gradient separation of common PTH-amino acids, which absorb UV light. Absorbance was monitored at 269 nm. PTH peaks are identified by single-letter notation for amino acid residues and by other abbreviations. D, Asp; CMC, carboxymethyl Cys; E, Glu; N, Asn; S, Ser; Q, Gln; H, His; T, Thr; G, Gly; R, Arg; MO₂, Met sulfoxide; A, Ala; Y, Tyr; M, Met; V, Val; P, Pro; W, Trp; K, Lys; F, Phe; I, Ile; L, Leu. See Figure 4.8a for PTH derivatization. (Adapted from Persson, B., and Eaker, D., 1990. An optimized procedure for the separation of amino acid phenylthiohydantoins by reversed phase HPLC. *Journal of Biochemical and Biophysical Methods* **21**:341–350.)

chromatography, gas chromatography (GC), and high-performance liquid chromatography (HPLC).

A typical HPLC chromatogram using precolumn modification of amino acids to form phenylthiohydantoin (PTH) derivatives is shown in Figure 4.13. HPLC is the chromatographic technique of choice for most modern biochemists. The very high resolution, excellent sensitivity, and high speed of this technique usually outweigh the disadvantage of relatively low capacity.

4.7 What Is the Fundamental Structural Pattern in Proteins?

Chemically, proteins are unbranched polymers of amino acids linked head to tail, from carboxyl group to amino group, through formation of covalent **peptide bonds,** a type of amide linkage (Figure 4.14).

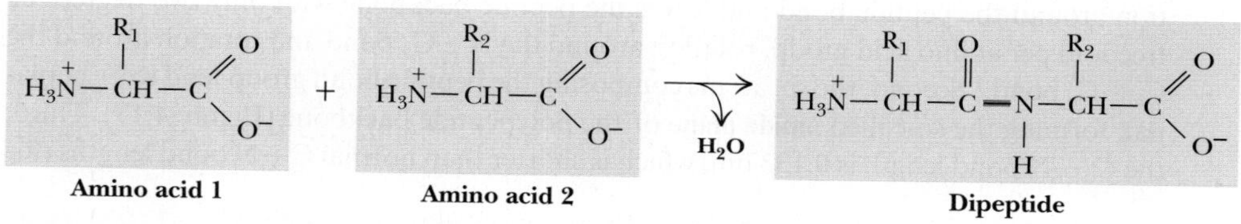

FIGURE 4.14 Peptide formation is the creation of an amide bond between the carboxyl group of one amino acid and the amino group of another amino acid.

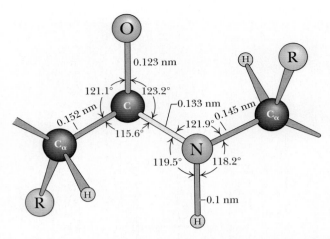

FIGURE 4.15 The peptide bond is shown in its usual *trans* conformation of carbonyl O and amide H. The C_α atoms are the α-carbons of two adjacent amino acids joined in peptide linkage. The dimensions and angles are the average values observed by crystallographic analysis of amino acids and small peptides. The peptide bond is the light-colored bond between C and N. (Adapted from Ramachandran, G. N., et al., 1974. The mean geometry of the peptide unit from crystal structure data. *Biochimica et Biophysica Acta* **359**:298–302.)

Peptide bond formation results in the release of H_2O. The peptide "backbone" of a protein consists of the repeated sequence $—N—C_\alpha—C_o—$, where the N represents the amide nitrogen, the C_α is the α-carbon atom of an amino acid in the polymer chain, and the final C_o is the carbonyl carbon of the amino acid, which in turn is linked to the amide N of the next amino acid down the line. The geometry of the peptide backbone is shown in Figure 4.15. Note that the carbonyl oxygen and the amide hydrogen are *trans* to each other in this figure. This conformation is favored energetically because it results in less steric hindrance between nonbonded atoms in neighboring amino acids. Because the α-carbon atom of the amino acid is a chiral center (in all amino acids except glycine), the polypeptide chain is inherently asymmetric. Only L-amino acids are found in proteins.

The Peptide Bond Has Partial Double-Bond Character

The peptide linkage is usually portrayed by a single bond between the carbonyl carbon and the amide nitrogen (Figure 4.16a). Therefore, in principle, rotation may occur about any covalent bond in the polypeptide backbone because all three kinds of bonds ($N—C_\alpha$, $C_\alpha—C_o$, and the $C_o—N$ peptide bond) are single bonds. In this representation, the C_o and N atoms of the peptide grouping are both in planar sp^2 hybridization and the C_o and O atoms are linked by a π bond, leaving the nitrogen with a lone pair of electrons in a $2p$ orbital. However, another resonance form for the peptide bond is feasible in which the C_o and N atoms participate in a π bond, leaving a lone e^- pair on the oxygen (Figure 4.16b). This structure prevents free rotation about the $C_o—N$ peptide bond because it becomes a double bond. The real nature of the peptide bond lies somewhere between these extremes; that is, it has partial double-bond character, as represented by the intermediate form shown in Figure 4.16c.

Peptide bond resonance has several important consequences. First, it restricts free rotation around the peptide bond and leaves the peptide backbone with only two degrees of freedom per amino acid group: rotation around the $N—C_\alpha$ bond and rotation around the $C_\alpha—C_o$ bond.[1] Second, the six atoms composing the peptide bond group tend to be coplanar, forming the so-called **amide plane** of the polypeptide backbone (Figure 4.17). Third, the $C_o—N$ bond length is 0.133 nm, which is shorter than normal C—N bond lengths (for

[1]The angle of rotation about the $N—C_\alpha$ bond is designated ϕ, phi, whereas the $C_\alpha—C_o$ angle of rotation is designated ψ, psi.

(a)

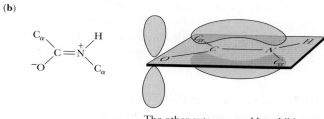

A pure double bond between C and O would permit free rotation around the C—N bond.

(b)

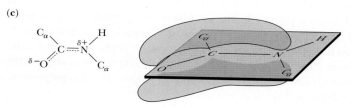

The other extreme would prohibit C—N bond rotation but would place too great a charge on O and N.

(c)

The true electron density is intermediate. The barrier to C—N bond rotation of about 88 kJ/mol is enough to keep the amide group planar.

FIGURE 4.16 The partial double-bond character of the peptide bond. Resonance interactions among the carbon, oxygen, and nitrogen atoms of the peptide group can be represented by two resonance extremes (**a** and **b**). **(a)** The usual way the peptide atoms are drawn. **(b)** In an equally feasible form, the peptide bond is now a double bond; the amide N bears a positive charge and the carbonyl O has a negative charge. **(c)** The actual peptide bond is best described as a resonance hybrid of the forms in **(a)** and **(b)**. Significantly, all of the atoms associated with the peptide group are coplanar, rotation about C_o—N is restricted, and the peptide is distinctly polar. (Illustration: Irving Geis. Rights owned by Howard Hughes Medical Institute. Not to be reproduced without permission.)

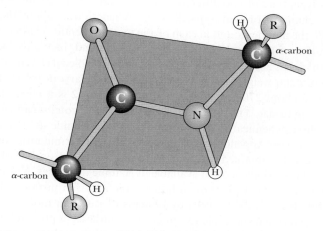

FIGURE 4.17 The coplanar relationship of the atoms in the amide group is highlighted as an imaginary shaded plane lying between two successive α-carbon atoms in the peptide backbone. (Illustration: Irving Geis. Rights owned by Howard Hughes Medical Institute. Not to be reproduced without permission.)

example, the C_α—N bond of 0.145 nm) but longer than typical C=N bonds (0.125 nm). The peptide bond is estimated to have 40% double-bond character.

The Polypeptide Backbone Is Relatively Polar

Peptide bond resonance also causes the peptide backbone to be relatively polar. As shown in Figure 4.16b, the amide nitrogen is in a protonated or positively charged form, and the carbonyl oxygen is a negatively charged atom in this double-bonded resonance state. In actuality, the hybrid state of the partially double-bonded peptide arrangement gives a net positive charge of 0.28 on the amide N and an equivalent net negative charge of 0.28 on the carbonyl O. The presence of these partial charges means that the peptide bond has a permanent dipole. Nevertheless, the peptide backbone is relatively unreactive chemically, and protons are gained or lost by the peptide groups only at extreme pH conditions.

Peptides Can Be Classified According to How Many Amino Acids They Contain

Peptide is the name assigned to short polymers of amino acids. Peptides are classified according to the number of amino acid units in the chain. Each unit is called an **amino acid residue,** the word *residue* denoting what is left after the release of H_2O when an amino acid forms a peptide link upon joining the peptide chain. **Dipeptides** have two amino acid residues, tripeptides have three, tetrapeptides four, and so on. After about 12 residues, this terminology becomes cumbersome, so peptide chains of more than 12 and less than about 20 amino acid residues are usually referred to as **oligopeptides,** and when the chain exceeds several dozen amino acids in length, the term **polypeptide** is used. The distinctions in this terminology are not precise.

Proteins Are Composed of One or More Polypeptide Chains

The terms *polypeptide* and *protein* are used interchangeably in discussing single polypeptide chains. The term **protein** broadly defines molecules composed of one or more polypeptide chains. Proteins with one polypeptide chain are **monomeric proteins.** Proteins composed of more than one polypeptide chain are **multimeric proteins.** Multimeric proteins may contain only one kind of polypeptide, in which case they are **homomultimeric,** or they may be composed of several different kinds of polypeptide chains, in which instance they are **heteromultimeric.** Greek letters and subscripts are used to denote the polypeptide composition of multimeric proteins. Thus, an α_2-type protein is a dimer of identical polypeptide subunits, or a **homodimer.** Hemoglobin (Table 4.2) consists of four polypeptides of two different kinds; it is an $\alpha_2\beta_2$ heteromultimer.

Polypeptide chains of proteins typically range in length from about 100 amino acids to around 2000, the number found in each of the two polypeptide chains of myosin, the contractile protein of muscle. However, exceptions abound, including human cardiac muscle titin, which has 26,926 amino acid residues and a molecular weight of 2,993,497. The average molecular weight of polypeptide chains in eukaryotic cells is about 31,700, corresponding to about 270 amino acid residues. Table 4.2 is a representative list of proteins according to size. The molecular weights (M_r) of proteins can be estimated by a number of physicochemical methods such as polyacrylamide gel electrophoresis or ultracentrifugation (see Appendix to Chapter 5). Precise determinations of protein molecular masses can be obtained by simple calculations based on knowledge of their amino acid sequence, which is often available in genome databases. No simple generalizations correlate the size of proteins with their functions. For instance, the same function may be fulfilled in different cells by proteins of different molecular weight. The *Escherichia coli* enzyme responsible for glutamine synthesis (a protein known as *glutamine synthetase*) has a molecular weight of 600,000, whereas the analogous enzyme in brain tissue has a molecular weight of 380,000.

TABLE 4.2	Size of Protein Molecules*		
Protein	M_r	Number of Residues per Chain	Subunit Organization
Insulin (bovine)	5,733	21 (A) / 30 (B)	$\alpha\beta$
Cytochrome c (equine)	12,500	104	α_1
Ribonuclease A (bovine pancreas)	12,640	124	α_1
Lysozyme (egg white)	13,930	129	α_1
Myoglobin (horse)	16,980	153	α_1
Chymotrypsin (bovine pancreas)	22,600	13 (α) / 132 (β) / 97 (γ)	$\alpha\beta\gamma$
Hemoglobin (human)	64,500	141 (α) / 146 (β)	$\alpha_2\beta_2$
Serum albumin (human)	68,500	550	α_1
Hexokinase (yeast)	96,000	200	α_4
γ-Globulin (horse)	149,900	214 (α) / 446 (β)	$\alpha_2\beta_2$
Glutamate dehydrogenase (liver)	332,694	500	α_6
Myosin (rabbit)	470,000	2,000 (heavy, h) / 190 (α) / 149 (α') / 160 (β)	$h_2\alpha_1\alpha'_2\beta_2$
Ribulose bisphosphate carboxylase (spinach)	560,000	475 (α) / 123 (β)	$\alpha_8\beta_8$
Glutamine synthetase (E. coli)	600,000	468	α_{12}

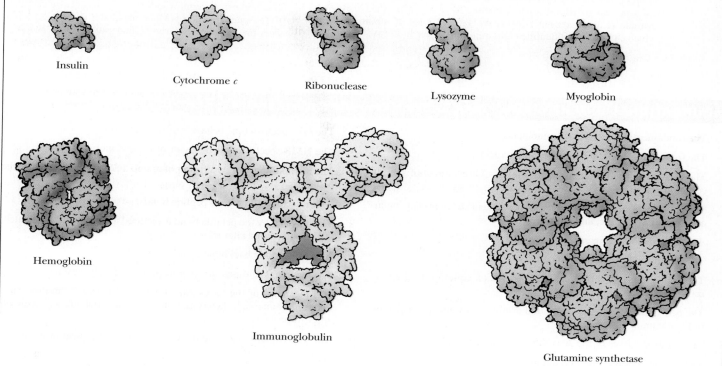

Insulin

Cytochrome c

Ribonuclease

Lysozyme

Myoglobin

Hemoglobin

Immunoglobulin

Glutamine synthetase

*Illustrations of selected proteins listed in Table 4.2 are drawn to constant scale.
Adapted from Goodsell, D. S., and Olson, A. J., 1993. Soluble proteins: Size, shape and function. *Trends in Biochemical Sciences* **18:**65–68.

SUMMARY

OWL Login to OWL to develop problem-solving skills and complete online homework assigned by your professor.

4.1 What Are the Structures and Properties of Amino Acids? The central tetrahedral alpha (α) carbon (C_α) atom of typical amino acids is linked covalently to both the amino group and the carboxyl group. Also bonded to this α-carbon are a hydrogen and a variable side chain. It is the side chain, the so-called R group, that gives each amino acid its identity. In neutral solution (pH 7), the carboxyl group exists as —COO^- and the amino group as —NH_3^+. The amino and carboxyl groups of amino acids can react in a head-to-tail fashion, eliminating a water molecule and forming a covalent amide linkage, which, in the case of peptides and proteins, is typically referred to as a peptide bond. Amino acids are also chiral molecules. With four different groups attached to it, the α-carbon is said to be asymmetric. The two possible configurations for the α-carbon constitute nonidentical mirror-image isomers or enantiomers. The structures of the 20 common amino acids are grouped into the following categories: (1) nonpolar or hydrophobic amino acids, (2) neutral (uncharged) but polar amino acids, (3) acidic amino acids (which have a net negative charge at pH 7.0), and (4) basic amino acids (which have a net positive charge at neutral pH).

4.2 What Are the Acid–Base Properties of Amino Acids? The common amino acids are all weak polyprotic acids. The ionizable groups are not strongly dissociating ones, and the degree of dissociation thus depends on the pH of the medium. All the amino acids contain at least two dissociable hydrogens. The side chains of several of the amino acids also contain dissociable groups. Thus, aspartic and glutamic acids contain an additional carboxyl function, and lysine possesses an aliphatic amino function. Histidine contains an ionizable imidazolium proton, and arginine carries a guanidinium function.

4.3 What Reactions Do Amino Acids Undergo? The reactivities of amino acids are essential to the degradation, sequencing, and chemical synthesis of peptides and proteins. Reaction with phenylisothiocyanate (Edman reagent) forms PTH derivatives of amino acids, which can be easily identified and quantified.

4.4 What Are the Optical and Stereochemical Properties of Amino Acids? Except for glycine, all of the amino acids isolated from proteins are said to be asymmetric or chiral (from the Greek *cheir*, meaning "hand"), and the two possible configurations for the α-carbon constitute nonsuperimposable mirror-image isomers, or enantiomers. Enantiomeric molecules display a special property called optical activity—the ability to rotate the plane of polarization of plane-polarized light. The magnitude and direction of the optical rotation depend on the nature of the amino acid side chain.

4.5 What Are the Spectroscopic Properties of Amino Acids? Many details of the structure and chemistry of the amino acids have been elucidated or at least confirmed by spectroscopic measurements. None of the amino acids absorbs light in the visible region of the electromagnetic spectrum. Several of the amino acids, however, do absorb ultraviolet radiation, and all absorb in the infrared region. Proton NMR spectra of amino acids are highly sensitive to their environment, and the chemical shifts of individual NMR signals can detect the pH-dependent ionizations of amino acids.

4.6 How Are Amino Acid Mixtures Separated and Analyzed? Separation can be achieved on the basis of the relative differences in the physical and chemical characteristics of amino acids, particularly ionization behavior and solubility characteristics. The methods important for amino acids include separations based on partition properties and separations based on electrical charge. HPLC is the chromatographic technique of choice for most modern biochemists. The very high resolution, excellent sensitivity, and high speed of this technique usually outweigh the disadvantage of relatively low capacity.

4.7 What Is the Fundamental Structural Pattern in Proteins? Proteins are linear polymers joined by peptide bonds. The defining characteristic of a protein is the amino acid sequence. The partial double-bonded character of the peptide bond has profound influences on protein conformation. Proteins are also classified according to the length of their polypeptide chains (how many amino acid residues they contain) and the number and kinds of polypeptide chains.

FOUNDATIONAL BIOCHEMISTRY   Things You Should Know After Reading Chapter 4.

- The fundamental structure of amino acids.

- The formation of peptide bonds by amino acids.

- The structures, names, and 1- and 3-letter abbreviations of the 20 common amino acids.

- The classification of the 20 common amino acids as nonpolar, polar and uncharged, acidic, and basic.

- Amino acids that occur rarely in proteins.

- The ionization behavior of amino acids.

- The application of the Henderson-Hasselbalch equation to the ionization behavior of ammo acids.

- The pK_a values of the α-carboxyl and α-amino and side-chain groups of the common amino acids.

- The titration curves of the 20 common amino acids.

- The reactions of the common amino acids.

- How to name amino acids according to the D,L and (R,S) nomenclature schemes.

- The UV spectroscopic behavior of the common amino acids.

- The NMR spectroscopic behavior of the common amino acids.

- The methodology for separation of amino acids on chromatographic columns.

- Amino acids link to form peptide bonds.

- The reason why the peptide bond is pictured as having approximately 40% double bond character.

- The polypeptide backbone is relatively polar.

- The meanings of dipeptide, oligopeptide, and polypeptide.

- The meanings of the terms "monomeric," "dimeric," "multimeric," "homomultimeric," "heteromultimeric," and "polymeric," with respect to proteins.

- The approximate sizes and molecular weights of typical proteins.

PROBLEMS

Answers to all problems are at the end of this book. Detailed solutions are available in the Student Solutions Manual, Study Guide, and Problems Book.

1. **Drawing Fischer Projection Formulas for Amino Acids** Without consulting chapter figures, draw Fischer projection formulas for glycine, aspartate, leucine, isoleucine, methionine, and threonine.

2. **Knowing Abbreviations for Amino Acids** Without reference to the text, give the one-letter and three-letter abbreviations for asparagine, arginine, cysteine, lysine, proline, tyrosine, and tryptophan.

3. **Writing Dissociation Equations for Amino Acids** Write equations for the ionic dissociations of alanine, glutamate, histidine, lysine, and phenylalanine.

4. **Understanding Chemical Effects on Amino Acid pK_a Values** How is the pK_a of the a-NH_3^+ group affected by the presence on an amino acid of the a-COO^-?

5. **Drawing Titration Curves for the Amino Acids** (Integrates with Chapter 2.) Draw an appropriate titration curve for aspartic acid, labeling the axes and indicating the equivalence points and the pK_a values.

6. **Calculating Concentrations of Species in Amino Acid Solutions** (Integrates with Chapter 2.) Calculate the concentrations of all ionic species in a 0.25 M solution of histidine at pH 2, pH 6.4, and pH 9.3.

7. **Calculating pH in Amino Acid Solutions I** (Integrates with Chapter 2.) Calculate the pH at which the g-carboxyl group of glutamic acid is two-thirds dissociated.

8. **Calculating pH in Amino Acid Solutions II** (Integrates with Chapter 2.) Calculate the pH at which the e-amino group of lysine is 20% dissociated.

9. **Calculating pH in Amino Acid Solutions III** (Integrates with Chapter 2.) Calculate the pH of a 0.3 M solution of (a) leucine hydrochloride,
(b) sodium leucinate, and (c) isoelectric leucine.

10. **Understanding Stereochemical Transformations of Amino Acids** Absolute configurations of the amino acids are referenced to D- and L-glyceraldehyde on the basis of chemical transformations that can convert the molecule of interest to either of these reference isomeric structures. In such reactions, the stereochemical consequences for the asymmetric centers must be understood for each reaction step. Propose a sequence of reactions that would demonstrate that L(−)-serine is stereochemically related to L(−)-glyceraldehyde.

11. **Understanding Amino Acid Stereochemistry** Describe the stereochemical aspects of the structure of cystine, the structure that is a disulfide-linked pair of cysteines.

12. **Understanding Amino Acid Reaction Mechanisms** Draw a simple mechanism for the reaction of a cysteine sulfhydryl group with iodoacetamide.

13. **Determining Tyrosine Content of an Unknown Protein** A previously unknown protein has been isolated in your laboratory. Others in your lab have determined that the protein sequence contains 172 amino acids. They have also determined that this protein has no tryptophan and no phenylalanine. You have been asked to determine the possible tyrosine content of this protein. You know from your study of this chapter that there is a relatively easy way to do this. You prepare a pure 50 mM solution of the protein, and you place it in a sample cell with a 1-cm path length, and you measure the absorbance of this sample at 280 nm in a UV-visible spectrophotometer. The absorbance of the solution is 0.372. Are there tyrosines in this protein? How many? (Hint: You will need to use Beer's Law, which is described in any good general chemistry or physical chemistry textbook. You will also find it useful to know that the units of **molar absorptivity** are $M^{-1}cm^{-1}$.)

14. **A Rule of Thumb for Amino Acid Content in Proteins** The simple average molecular weight of the 20 common amino acids is 138, but most biochemists use 110 when estimating the number of amino acids in a protein of known molecular weight. Why do you suppose this is? (Hint: there are two contributing factors to the answer. One of them will be apparent from a brief consideration of the amino acid compositions of common proteins. See, for example, Figure 5.16 of this text.)

15. **Understanding the Chemistry of the Dipeptide Sweetener Aspartame** The artificial sweeteners Equal and Nutrasweet contain aspartame, which has the structure:

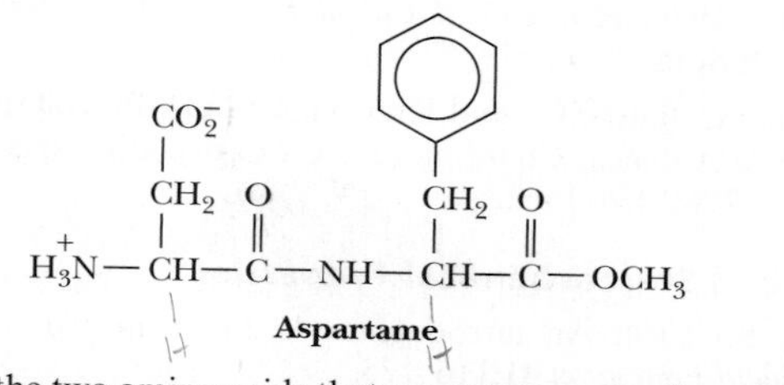

Aspartame

What are the two amino acids that are components of aspartame? What kind of bond links these amino acids? What do you suppose might happen if a solution of aspartame was heated for several hours at a pH near neutrality? Suppose you wanted to make hot chocolate sweetened only with aspartame, and you stored it in a thermos for several hours before drinking it. What might it taste like?

16. **Understanding a Defect of Amino Acid Metabolism** Individuals with phenylketonuria must avoid dietary phenylalanine because they are unable to convert phenylalanine to tyrosine. Look up this condition and find out what happens if phenylalanine accumulates in the body. Would you advise a person with phenylketonuria to consume foods sweetened with aspartame? Why or why not?

17. **Distinguishing Prochiral Isomers** In this chapter, the concept of prochirality was discussed. Citrate (see Figure 19.2) is a prochiral molecule. Describe the process by which you would distinguish between the (R−) and (S−) portions of this molecule and how an enzyme could discriminate between similar but distinct moieties.

18. **Understanding Buffer Capacity of Amino Acids** Amino acids are frequently used as buffers. Describe the pH range of acceptable buffering behavior for the amino acids alanine, histidine, aspartic acid, and lysine.

Preparing for the MCAT® Exam

19. Although the other common amino acids are used as buffers, cysteine is rarely used for this purpose. Why?

20. Draw all the possible isomers of threonine and assign (R,S) nomenclature to each.

FURTHER READING

General Amino Acid Chemistry

Atkins, J. F., and Gesteland, R., 2002. The 22nd amino acid. *Science* **296:**1409–1410.

Barker, R., 1971. *Organic Chemistry of Biological Compounds,* Chap. 4. Englewood Cliffs, NJ: Prentice Hall.

Barrett, G. C., ed., 1985. *Chemistry and Biochemistry of the Amino Acids.* New York: Chapman and Hall.

Greenstein, J. P., and Winitz M., 1961. *Chemistry of the Amino Acids.* New York: John Wiley & Sons.

Herod, D. W., and Menzel, E. R., 1982. Laser detection of latent fingerprints: Ninhydrin. *Journal of Forensic Science* **27:**200–204.

Meister, A., 1965. *Biochemistry of the Amino Acids,* 2nd ed., Vol. 1. New York: Academic Press.

Miyashita, M., Presley, J., et al., 2001. Attomole level protein sequencing by Edman degradation coupled with accelerator mass spectrometry. *Proceedings of the National Academy of Sciences, USA* **98:**4403–4408.

Pizzarello, S., 2006. The chemistry of life's origin: a carbonaceous meteorite perspective. *Accounts of Chemical Research* **39:**231–237.

Segel I. H., 1976. *Biochemical Calculations,* 2nd ed. New York: John Wiley & Sons.

Srinivasan, G., James, C., and Krzycki, J., 2002. Pyrrolysine encoded by UAG in Archaea: Charging of a UAG-decoding specialized tRNA. *Science* **296:**1459–1462.

Optical and Stereochemical Properties

Cahn, R. S., 1964. An introduction to the sequence rule. *Journal of Chemical Education* **41:**116–125.

Iizuke, E., and Yang, J. T., 1964. Optical rotatory dispersion of L-amino acids in acid solution. *Biochemistry* **3:**1519–1524.

Kauffman, G. B., and Priebe, P. M., 1990. The Emil Fischer-William Ramsey friendship. *Journal of Chemical Education* **67:**93–101.

Suprenant, H. L., Sarneski, J. E., Key, R. R., Byrd, J. T., and Reilley, C. N., 1980. Carbon-13 NMR studies of amino acids: Chemical shifts, protonation shifts, microscopic protonation behavior. *Journal of Magnetic Resonance* **40:**231–243.

Separation Methods

Heiser, T., 1990. Amino acid chromatography: The "best" technique for student labs. *Journal of Chemical Education* **67:**964–966.

Mabbott, G., 1990. Qualitative amino acid analysis of small peptides by GC/MS. *Journal of Chemical Education* **67:**441–445.

Moore, S., Spackman, D., and Stein, W. H., 1958. Chromatography of amino acids on sulfonated polystyrene resins. *Analytical Chemistry* **30:**1185–1190.

NMR Spectroscopy

Bovey, F. A., and Tiers, G. V. D., 1959. Proton N.S.R. spectroscopy. V. Studies of amino acids and peptides in trifluoroacetic acid. *Journal of the American Chemical Society* **81:**2870–2878.

de Groot, H. J., 2000. Solid-state NMR spectroscopy applied to membrane proteins. *Current Opinion in Structural Biology* **10:**593–600.

Hinds, M. G., and Norton, R. S., 1997. NMR spectroscopy of peptides and proteins. Practical considerations. *Molecular Biotechnology* **7:**315–331.

James, T. L., Dötsch, V., and Schmitz, U., eds., 2001. *Nuclear Magnetic Resonance of Biological Macromolecules.* San Diego: Academic Press.

Krishna, N. R., and Berliner, L. J., eds., 2003. *Protein NMR for the Millennium.* New York: Kluwer Academic/Plenum.

Opella, S. J., Nevzorov, A., Mesleb, M. F., and Marassi, F. M., 2002. Structure determination of membrane proteins by NMR spectroscopy. *Biochemistry and Cell Biology* **80:**597–604.

Roberts, G. C. K., and Jardetzky, O., 1970. Nuclear magnetic resonance spectroscopy of amino acids, peptides and proteins. *Advances in Protein Chemistry* **24:**447–545.

Amino Acid Analysis

Prata, C., et al., 2001. Recent advances in amino acid analysis by capillary electrophoresis. *Electrophoresis* **22:**4129–4138.

Smith, A. J., 1997. Amino acid analysis. *Methods in Enzymology* **289:**419–426.

Proteins: Their Primary Structure and Biological Functions

Dale Chihuly, Chartreuse Venetian, 1990/Photo by Roger Schreiber

◄ Helices, which sometimes appear as decorative or utilitarian motifs in manmade structures, are a common structural theme in biological macromolecules—proteins, nucleic acids, and even polysaccharides.

…by small and simple things are great things brought to pass.

ALMA 37.6
The Book of Mormon

KEY QUESTIONS

5.1 What Architectural Arrangements Characterize Protein Structure?

5.2 How Are Proteins Isolated and Purified from Cells?

5.3 How Is the Amino Acid Analysis of Proteins Performed?

5.4 How Is the Primary Structure of a Protein Determined?

5.5 What Is the Nature of Amino Acid Sequences?

5.6 Can Polypeptides Be Synthesized in the Laboratory?

5.7 Do Proteins Have Chemical Groups Other Than Amino Acids?

5.8 What Are the Many Biological Functions of Proteins?

ESSENTIAL QUESTIONS

Proteins are polymers composed of hundreds or even thousands of amino acids linked in series by peptide bonds.

What structural forms do these polypeptide chains assume, how can the sequence of amino acids in a protein be determined, and what are the biological roles played by proteins?

Proteins are a diverse and abundant class of biomolecules, constituting more than 50% of the dry weight of cells. Their diversity and abundance reflect the central role of proteins in virtually all aspects of cell structure and function. An extraordinary diversity of cellular activity is possible only because of the versatility inherent in proteins, each of which is specifically tailored to its biological role. The pattern by which each is tailored resides within the genetic information of cells, encoded in a specific sequence of nucleotide bases in DNA. Each such segment of encoded information defines a gene, and expression of the gene leads to synthesis of the specific protein encoded by it, endowing the cell with the functions unique to that particular protein. Proteins are the agents of biological function; they are also the expressions of genetic information.

ᴥOWL Online homework and a Student Self Assessment for this chapter may be assigned in OWL.

5.1 | What Architectural Arrangements Characterize Protein Structure?

Proteins Fall into Three Basic Classes According to Shape and Solubility

As a first approximation, proteins can be assigned to one of three global classes on the basis of shape and solubility: fibrous, globular, or membrane (Figure 5.1). **Fibrous proteins** tend to have relatively simple, regular linear structures. These proteins often serve structural roles in cells. Typically, they are insoluble in water or in dilute salt solutions. In contrast, **globular proteins** are roughly spherical in shape. The polypeptide chain is compactly folded so that hydrophobic amino acid side chains are in the interior of the molecule and the hydrophilic side chains are on the outside exposed to the solvent, water. Consequently, globular proteins are usually very soluble in aqueous solutions. Most soluble proteins of the cell, such as the cytosolic enzymes, are globular in shape. **Membrane proteins** are found in association with the various membrane systems of cells. For interaction with the nonpolar phase within membranes, membrane proteins have hydrophobic amino acid side chains oriented outward. As such, membrane proteins are insoluble in aqueous solutions but can be solubilized in solutions of detergents. Membrane proteins characteristically have fewer hydrophilic amino acids than cytosolic proteins.

Protein Structure Is Described in Terms of Four Levels of Organization

The architecture of protein molecules is quite complex. Nevertheless, this complexity can be resolved by defining various levels of structural organization.

Primary Structure The amino acid sequence is, by definition, the **primary (1°) structure** of a protein, such as that for bovine pancreatic RNase in Figure 5.2, for example.

(a)

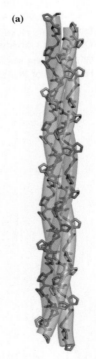

Collagen, a fibrous protein

(b)

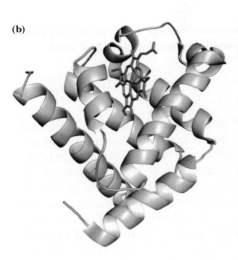

Myoglobin, a globular protein

(c)

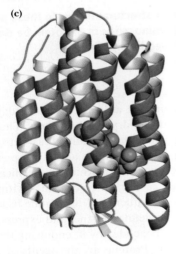

Bacteriorhodopsin, a membrane protein

FIGURE 5.1 **(a)** Collagen, a major component of animal connective tissue, is the most abundant protein in mammals. Proteins having structural roles in cells are typically fibrous and often water insoluble. **(b)** Myoglobin is a globular protein. **(c)** Membrane proteins fold so that hydrophobic amino acid side chains are exposed in their membrane-associated regions. Bacteriorhodopsin binds the light-absorbing pigment, *cis*-retinal, shown here in blue.

Secondary Structure Through hydrogen-bonding interactions between adjacent amino acid residues (discussed in detail in Chapter 6), the polypeptide chain can arrange itself into characteristic helical or pleated segments. These segments constitute structural conformities, so-called **regular structures,** which extend along one dimension, like the coils of a spring. Such architectural features of a protein are designated **secondary (2°) structures** (Figure 5.3). Secondary structures are just one of the higher levels of structure that represent the three-dimensional arrangement of the polypeptide in space.

Tertiary Structure When the polypeptide chains of protein molecules bend and fold in order to assume a more compact three-dimensional shape, the **tertiary (3°) level of structure** is generated (Figure 5.4). It is by virtue of their tertiary structure that proteins adopt a globular shape. A globular conformation gives the lowest surface-to-volume ratio, minimizing interaction of the protein with the surrounding environment.

Quaternary Structure Many proteins consist of two or more interacting polypeptide chains of characteristic tertiary structure, each of which is commonly referred to as a **subunit** of the protein. Subunit organization constitutes another level in the hierarchy of protein structure, defined as the protein's **quaternary (4°) structure** (Figure 5.5). Questions of quaternary structure address the various kinds of subunits within a protein molecule, the number of each, and the ways in which they interact with one another.

Noncovalent Forces Drive Formation of the Higher Orders of Protein Structure

Whereas the primary structure of a protein is determined by the covalently linked amino acid residues in the polypeptide backbone, secondary and higher orders of structure are determined principally by noncovalent forces such as hydrogen bonds and ionic, van der

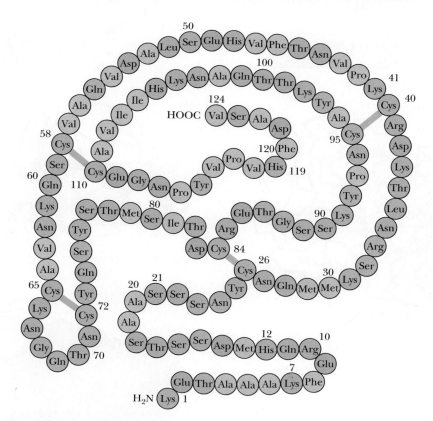

FIGURE 5.2 Bovine pancreatic ribonuclease A contains 124 amino acid residues, none of which are tryptophan. Four intrachain disulfide bridges (S—S) form crosslinks in this polypeptide between Cys[26] and Cys[84], Cys[40] and Cys[95], Cys[58] and Cys[110], and Cys[65] and Cys[72].

α-Helix
Only the N — C_α — C backbone
is represented. The vertical line
is the helix axis.

β-Strand
The N — C_α — C backbone as well
as the C_β of R groups are represented
here. Note that the amide planes
are perpendicular to the page.

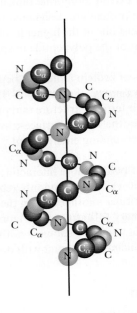

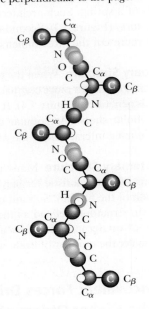

"Shorthand" α-helix

"Shorthand" β-strand

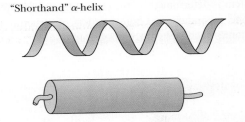

FIGURE 5.3 The α-helix and the β-pleated strand are the two principal secondary structures found in proteins. Simple representations of these structures are a flat, helical ribbon or a cylinder for the α-helix and a flat, wide arrow for β-structures.

(a) Chymotrypsin tertiary structure

(b)

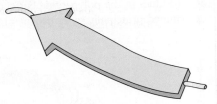

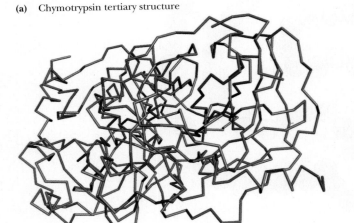

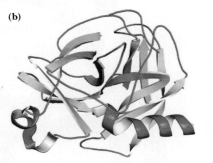

Chymotrypsin ribbon

(c)

Chymotrypsin space-filling model

FIGURE 5.4 Folding of the polypeptide chain into a compact, roughly spherical conformation creates the tertiary level of protein structure. Shown here are three different representations of the tertiary structure of the protein chymotrypsin: **(a)** a tracing showing the position of all of the C_α carbon atoms, **(b)** a ribbon diagram that shows the three-dimensional track of the polypeptide chain, and **(c)** a space-filling representation of the atoms as spheres.

Waals, and hydrophobic interactions. It is important to emphasize that *all the information necessary for a protein molecule to achieve its intricate architecture is contained within its 1° structure,* that is, within the amino acid sequence of its polypeptide chain(s). Chapter 6 presents a detailed discussion of the 2°, 3°, and 4° structure of protein molecules.

A Protein's Conformation Can Be Described as Its Overall Three-Dimensional Structure

The overall three-dimensional architecture of a protein is generally referred to as its **conformation.** This term is not to be confused with **configuration,** which denotes the geometric possibilities for a particular set of atoms (Figure 5.6). In going from one configuration to another, covalent bonds must be broken and rearranged. In contrast, the *conformational possibilities* of a molecule are achieved without breaking any covalent bonds. In proteins, rotations about each of the single bonds along the peptide backbone have the potential to alter the course of the polypeptide chain in three-dimensional space. These rotational possibilities create many possible orientations for the protein chain, referred to as its conformational possibilities. Of the great number of theoretical conformations a given protein might adopt, only a very few are favored energetically under physiological conditions. At this time, the rules that direct the folding of protein chains into energetically favorable conformations are still not entirely clear; accordingly, they are the subject of intensive contemporary research.

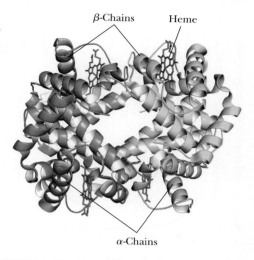

FIGURE 5.5 Hemoglobin is a tetramer consisting of two α and two β polypeptide chains.

5.2 | How Are Proteins Isolated and Purified from Cells?

Cells contain thousands of different proteins. A major problem for protein chemists is to purify a chosen protein so that they can study its specific properties in the absence of other proteins. The starting material for protein purification is usually a cell type or tissue of interest; alternatively, cloned cells engineered to mass produce the protein (see Chapter 12) are used. Proteins can be separated and purified on the basis of their two prominent physical properties: size and electrical charge. A more direct approach is to use **affinity purification** strategies that take advantage of the biological function or specific recognition properties of a protein (see A Deeper Look box: Techniques Used in Protein Purification).

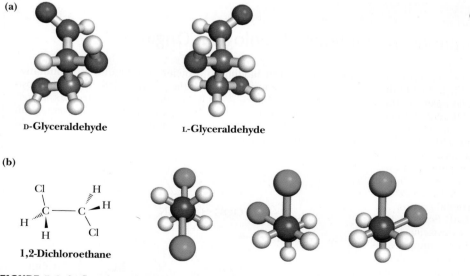

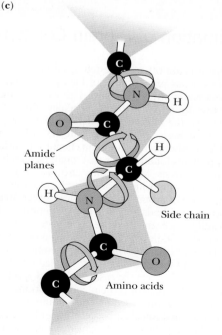

FIGURE 5.6 Configuration and conformation are *not* synonymous. **(a)** Rearrangements between configurational alternatives of a molecule can be achieved only by breaking and remaking bonds, as in the transformation between the D- and L-configurations of glyceraldehyde. **(b)** The intrinsic free rotation around single covalent bonds creates a great variety of three-dimensional conformations, even for relatively simple molecules, such as 1,2-dichloroethane. **(c)** Imagine the conformational possibilities for a protein in which two of every three bonds along its backbone are freely rotating single bonds. (Illustration: Irving Geis. Rights owned by Howard Hughes Medical Institute. Not to be reproduced without permission.)

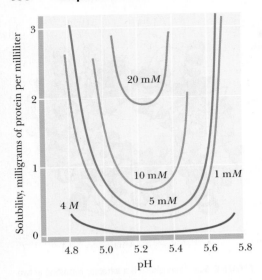

FIGURE 5.7 The solubility of most globular proteins is markedly influenced by pH and ionic strength. This figure shows the solubility of a typical protein as a function of pH and various salt concentrations.

A Number of Protein Separation Methods Exploit Differences in Size and Charge

Separation methods based on size include size exclusion chromatography, ultrafiltration, and ultracentrifugation (see A Deeper Look box: Techniques Used in Protein Purification). The ionic properties of peptides and proteins are determined principally by their complement of amino acid side chains. Furthermore, the ionization of these groups is pH-dependent.

A variety of procedures have been designed to exploit the electrical charges on a protein as a means to separate proteins in a mixture. These procedures include ion exchange chromatography, electrophoresis (see A Deeper Look box: Techniques Used in Protein Purification), and solubility. Proteins tend to be least soluble at their **isoelectric point,** the pH value at which the sum of their positive and negative electrical charges is zero. At this pH, electrostatic repulsion between protein molecules is minimal and they are more likely to coalesce and precipitate out of solution. Ionic strength also profoundly influences protein solubility. Most globular proteins tend to become increasingly soluble as the ionic strength is raised. This phenomenon, the salting-in of proteins, is attributed to the diminishment of electrostatic attractions between protein molecules by the presence of abundant salt ions. Such electrostatic interactions between the protein molecules would otherwise lead to precipitation. However, as the salt concentration reaches high levels (greater than 1 M), the effect may reverse so that the protein is salted out of solution. In such cases, the numerous salt ions begin to compete with the protein for waters of solvation, and as they win out, the protein becomes insoluble. The solubility properties of a typical protein are shown in Figure 5.7.

Although the side chains of nonpolar amino acids in soluble proteins are usually buried in the interior of the protein away from contact with the aqueous solvent, a portion of them may be exposed at the protein's surface, giving it a partially hydrophobic character. Hydrophobic interaction chromatography is a protein purification technique that exploits this hydrophobicity (see A Deeper Look box: Techniques Used in Protein Purification).

> **A DEEPER LOOK**

Estimation of Protein Concentrations in Solutions of Biological Origin

Biochemists are often interested in knowing the protein concentration in various preparations of biological origin. Such quantitative analysis is not straightforward. Cell extracts are complex mixtures that typically contain protein molecules of many different molecular weights, so the results of protein estimations cannot be expressed on a molar basis. Also, aside from the rather unreactive repeating peptide backbone, little common chemical identity is seen among the many proteins found in cells that might be readily exploited for exact chemical analysis. Most of their chemical properties vary with their amino acid composition, for example, nitrogen or sulfur content or the presence of aromatic, hydroxyl, or other functional groups.

Several methods rely on the reduction of Cu^{2+} ions to Cu^+ by readily oxidizable protein components, such as cysteine or the phenols and indoles of tyrosine and tryptophan. For example, *bicinchoninic acid* (BCA) forms a purple complex with Cu^+ in alkaline solution, and the amount of Cu^+ formed upon reaction of a protein sample with Cu^{2+} can be easily measured spectrophotometrically to provide an estimate of protein concentration.

Other assays are based on dye binding by proteins. The Bradford assay is a rapid and reliable technique that uses a dye called *Coomassie Brilliant Blue G-250,* which undergoes a change in its color upon noncovalent binding to proteins. The binding is quantitative and less sensitive to variations in the protein's amino acid composition. The color change is easily measured by a spectrophotometer. A similar, very sensitive method capable of quantifying nanogram amounts of protein is based on the shift in color of colloidal gold upon binding to proteins.

$$Cu^+ + BCA \longrightarrow$$

BCA–Cu$^+$ complex

TABLE 5.1	Example of a Protein Purification Scheme: Purification of an Enzyme from a Cell Extract				
Fraction	Volume (mL)	Total Protein (mg)	Total Activity*	Specific Activity†	Percent Recovery‡
1. Crude extract	3,800	22,800	2,460	0.108	100
2. Salt precipitate	165	2,800	1,190	0.425	48
3. Ion exchange chromatography	65	100	720	7.2	29
4. Molecular sieve chromatography	40	14.5	555	38.3	23
5. Immunoaffinity chromatography§	6	1.8	275	152	11

*The relative enzymatic activity of each fraction is cited as arbitrarily defined units. One unit here is 1 nanomole/mL of substrate converted per minute. The enzyme is xanthine dehydrogenase from the bread mold *Neurospora crassa*.

†The specific activity is the total activity of the fraction divided by the total protein in the fraction. This value gives an indication of the increase in purity attained during the course of the purification as the samples become enriched for the enzyme.

‡The percent recovery of total activity is a measure of the yield of the desired enzyme.

§The last step in the procedure is an affinity method in which antibodies specific for the enzyme are covalently coupled to a chromatography matrix and packed into a glass tube to make a chromatographic column through which fraction 4 is passed. The enzyme is bound by this immunoaffinity matrix while other proteins pass freely out. The enzyme is then recovered by passing through the column a strong salt solution, which dissociates the enzyme-antibody complex.

A Typical Protein Purification Scheme Uses a Series of Separation Methods

Most purification procedures for a particular protein are developed in an empirical manner, the overriding principle being purification of the protein to a homogeneous state with acceptable yield. Table 5.1 presents a summary of a purification scheme for a desired enzyme. Note that the **specific activity** of the enzyme in the immunoaffinity purified fraction (fraction 5) has been increased 152/0.108, or 1407 times the specific activity in the crude extract (fraction 1). Thus, the concentration of this protein has been enriched more than 1400-fold by the purification procedure.

> ## A DEEPER LOOK

Techniques Used in Protein Purification[1]

Dialysis and Ultrafiltration

If a solution of protein is separated from a bathing solution by a semipermeable membrane, small molecules and ions can pass through the semipermeable membrane to equilibrate between the protein solution and the bathing solution, called the *dialysis bath* or *dialysate* (Figure 1). This method is useful for removing small molecules from macromolecular solutions or for altering the composition of the protein-containing solution.

Ultrafiltration is an improvement on the dialysis principle. Filters with pore sizes over the range of biomolecular dimensions are used to filter solutions to select for molecules in a particular size range. Because the pore sizes in these filters are microscopic, high pressures are often required to force the solution through the filter. This technique is useful for concentrating dilute solutions of macromolecules. The concentrated protein can then be diluted into the solution of choice.

Ion Exchange Chromatography Can Be Used to Separate Molecules on the Basis of Charge

Charged molecules can be separated using *ion exchange chromatography,* a process in which the charged molecules of interest *(ions)* are *exchanged* for another ion (usually a salt ion) on a

[1]Although this box is titled *Techniques Used in Protein Purification,* these methods are also applicable to other macromolecules such as nucleic acids.

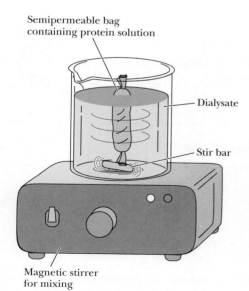

▲ **Figure 1** A dialysis experiment. The solution of macromolecules to be dialyzed is placed in a semipermeable membrane bag, and the bag is immersed in a bathing solution. A magnetic stirrer gently mixes the solution to facilitate equilibrium of diffusible solutes between the dialysate and the solution contained in the bag.

(Continued)

A DEEPER LOOK   Techniques Used in Protein Purification *(Continued)*

charged solid support. In a typical procedure, solutes in a liquid phase, usually water, are passed through a column filled with a porous solid phase composed of synthetic resin particles containing charged groups. Resins containing positively charged groups attract negatively charged solutes and are referred to as *anion exchange resins.* Resins with negatively charged groups are *cation exchangers.* Figure 2 shows several typical anion and cation exchange resins. Weakly acidic or basic groups on ion exchange resins exhibit charges that are dependent on the pH of the bathing solution. Changing the pH will alter the ionic interaction between the resin groups and the bound ions. In all cases, the bare charges on the resin particles must be counterbalanced by oppositely charged ions in solution *(counterions);* salt ions (e.g., Na^+ or Cl^-) usually serve this purpose. The hypothetical separation of a mixture of several proteins on a column of cation exchange resin is illustrated in Figure 3. Increasing the salt concentration in the solution passing through the column leads to competition between the cationic proteins bound to the column and the cations in the salt for binding to the column. Bound cationic proteins that interact weakly with the charged groups on the resin wash out first, and those interacting strongly are washed out only at high salt concentrations.

Size Exclusion Chromatography

Size exclusion chromatography is also known as *gel filtration chromatography* or *molecular sieve chromatography.* In this method, fine, porous beads are packed into a chromatography column. The beads are composed of dextran polymers *(Sephadex),* agarose *(Sepharose),* or polyacrylamide *(Sephacryl* or *BioGel P).* The pore sizes of these beads approximate the dimensions of macromolecules. The total bed volume (Figure 4) of the packed chromatography column, V_t, is equal to the volume outside the porous beads (V_o) plus the volume inside the beads (V_i) plus the volume actually occupied by the bead material (V_g): $V_t = V_o + V_i + V_g$. (V_g is typically less than 1% of V_t and can be conveniently ignored in most applications.)

As a solution of molecules is passed through the column, the molecules passively distribute between V_o and V_i, depending on their ability to enter the pores (that is, their size). If a molecule is too large to enter at all, it is totally excluded from V_i and emerges first from the column at an elution volume, V_e, equal to V_o (Figure 4). If a particular molecule can enter the pores in the gel, its distribution is given by the *distribution coefficient, K_D:*

$$K_D = (V_e - V_o) / V_i$$

where V_e is the molecule's characteristic elution volume (Figure 4). The chromatography run is complete when a volume of solvent equal to V_t has passed through the column.

Electrophoresis

Electrophoretic techniques are based on the movement of ions in an electrical field. An ion of charge q experiences a force F given by $F = Eq/d$, where E is the voltage (or *electrical potential*) and d is the distance between the electrodes. In a vacuum, F would cause the molecule to accelerate. In solution, the molecule experiences *frictional drag, F_f,* due to the solvent:

$$F_f = 6\pi r \eta v$$

where r is the radius of the charged molecule, η is the viscosity of the solution, and v is the velocity at which the charged molecule is moving. So, the velocity of the charged molecule is proportional to its charge q and the voltage E, but inversely proportional to the viscosity of the medium η and d, the distance between the electrodes.

Generally, electrophoresis is carried out *not* in free solution but in a porous support matrix such as polyacrylamide or agarose, which retards the movement of molecules according to their dimensions relative to the size of the pores in the matrix.

SDS-Polyacrylamide Gel Electrophoresis (SDS-PAGE)

SDS is sodium dodecylsulfate (sodium lauryl sulfate) (Figure 5). The hydrophobic tail of dodecylsulfate interacts strongly with polypeptide chains. The number of SDS molecules bound by a polypeptide is proportional to the length (number of amino acid residues) of the polypeptide. Each dodecylsulfate contributes two negative charges. Collectively, these charges overwhelm any intrinsic charge that the protein might have. SDS is also a detergent that

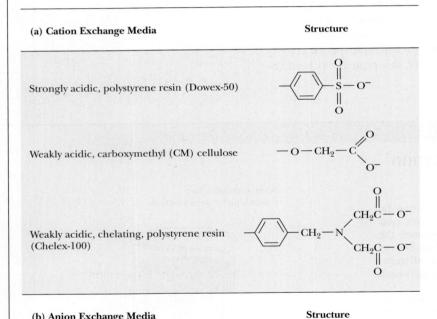

(a) Cation Exchange Media   **Structure**

Strongly acidic, polystyrene resin (Dowex-50)

Weakly acidic, carboxymethyl (CM) cellulose

Weakly acidic, chelating, polystyrene resin (Chelex-100)

(b) Anion Exchange Media   **Structure**

Strongly basic, polystyrene resin (Dowex-1)

Weakly basic, diethylaminoethyl (DEAE) cellulose

▲ **Figure 2** Cation **(a)** and anion **(b)** exchange resins commonly used for biochemical separations.

A DEEPER LOOK Techniques Used in Protein Purification *(Continued)*

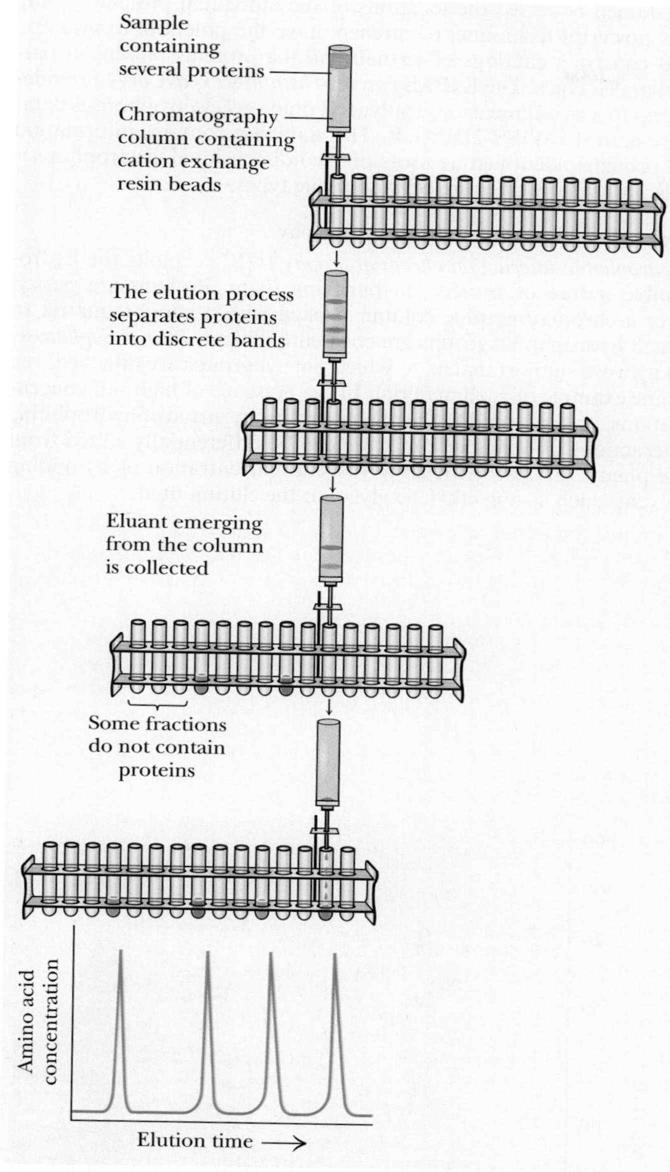

Sample containing several proteins

Chromatography column containing cation exchange resin beads

The elution process separates proteins into discrete bands

Eluant emerging from the column is collected

Some fractions do not contain proteins

Amino acid concentration

Elution time ⟶

▲ **Figure 3** The separation of proteins on a cation exchange column.

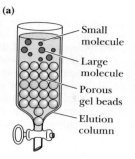

(a)

Small molecule

Large molecule

Porous gel beads

Elution column

(b)

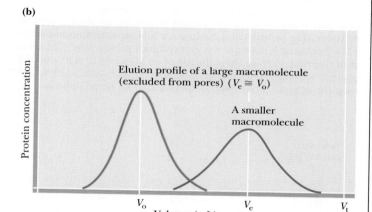

Protein concentration

Elution profile of a large macromolecule (excluded from pores) $(V_e \cong V_o)$

A smaller macromolecule

V_o Volume (mL) ⟶ V_e V_t

▲ **Figure 4** **(a)** A gel filtration chromatography column. Larger molecules are excluded from the gel beads and emerge from the column sooner than smaller molecules, whose migration is retarded because they can enter the beads. **(b)** An elution profile.

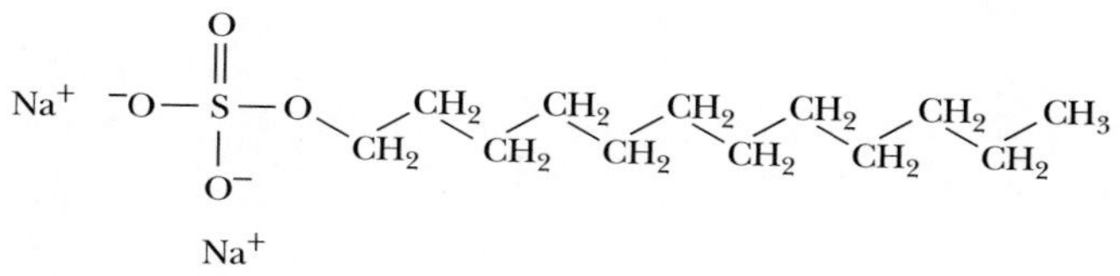

$$Na^+ \ ^-O-\overset{\overset{O}{\|}}{\underset{\underset{O^-}{|}}{S}}-O-\begin{matrix}CH_2 & & CH_2 & & CH_2 & & CH_2 & & CH_2 & & CH_3 \\ & CH_2 & & CH_2 & & CH_2 & & CH_2 & & CH_2 & \end{matrix}$$

$$Na^+$$

▲ **Figure 5** The structure of sodium dodecylsulfate (SDS).

disrupts protein folding (protein 3° structure). SDS-PAGE is usually run in the presence of sulfhydryl-reducing agents such as β-mercaptoethanol so that any disulfide links between polypeptide chains are broken. The electrophoretic mobility of proteins upon SDS-PAGE is inversely proportional to the logarithm of the protein's molecular weight (Figure 6). SDS-PAGE is often used to determine the molecular weight of a protein.

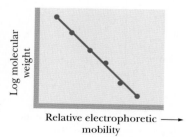

Log molecular weight

Relative electrophoretic mobility ⟶

▲ **Figure 6** A plot of the relative electrophoretic mobility of proteins in SDS-PAGE versus the log of the molecular weights of the individual polypeptides.

(Continued)

Isoelectric Focusing

Isoelectric focusing is an electrophoretic technique for separating proteins according to their *isoelectric points* (pIs). A solution of *ampholytes* (amphoteric electrolytes) is first electrophoresed through a gel, usually contained in a small tube. The migration of these substances in an electric field establishes a pH gradient in the tube. Then a protein mixture is applied to the gel, and electrophoresis is resumed. As the protein molecules move down the gel, they experience the pH gradient and migrate to a position corresponding to their respective pIs. At its pI, a protein has no net charge and thus moves no farther.

Two-Dimensional Gel Electrophoresis

This separation technique uses isoelectric focusing in one dimension and SDS-PAGE in the second dimension to resolve protein mixtures. The proteins in a mixture are first separated according to pI by isoelectric focusing in a polyacrylamide gel in a tube. The gel is then removed and laid along the top of an SDS-PAGE slab, and the proteins are electrophoresed into the SDS polyacrylamide gel, where they are separated according to size (Figure 7). The gel slab can then be stained to reveal the locations of the individual proteins. Using this powerful technique, researchers have the potential to visualize and construct catalogs of virtually *all* the proteins present in particular cell types. The **ExPASy** server *(http://us.expasy.org)* provides access to a two-dimensional polyacrylamide gel electrophoresis database named **SWISS-2DPAGE.** This database contains information on proteins, identified as spots on two-dimensional electrophoresis gels, from many different cell and tissue types.

Hydrophobic Interaction Chromatography

Hydrophobic interaction chromatography (HIC) exploits the hydrophobic nature of proteins in purifying them. Proteins are passed over a chromatographic column packed with a support matrix to which hydrophobic groups are covalently linked. *Phenyl Sepharose,* an agarose support matrix to which phenyl groups are attached, is a prime example of such material. In the presence of high salt concentrations, proteins bind to the phenyl groups by virtue of hydrophobic interactions. Proteins in a mixture can be differentially eluted from the phenyl groups by lowering the salt concentration or by adding solvents such as polyethylene glycol to the elution fluid.

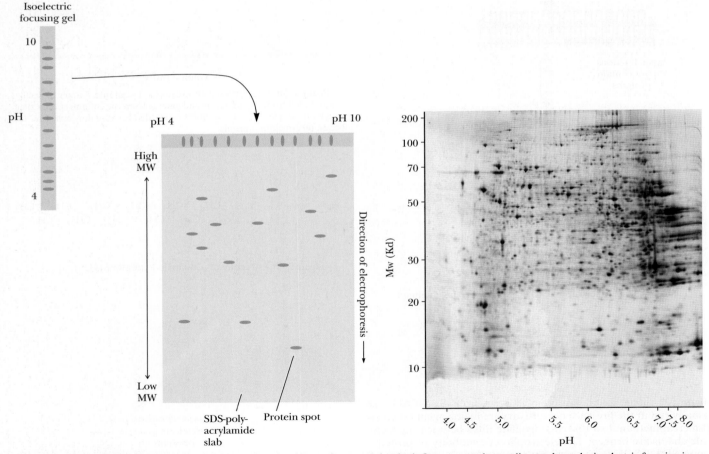

Figure 7 A two-dimensional electrophoresis separation. A mixture of macromolecules is first separated according to charge by isoelectric focusing in a tube gel. The gel containing separated molecules is then placed on top of an SDS-PAGE slab, and the molecules are electrophoresed into the SDS-PAGE gel, where they are separated according to size. Shown on the right is an actual 2D-gel electrophoretogram of mouse liver proteins from the Swiss Institute of Bioinformatics database.

> **A DEEPER LOOK**   Techniques Used in Protein Purification *(Continued)*

High-Performance Liquid Chromatography

The principles exploited in *high-performance* (or high-pressure) *liquid chromatography* (HPLC) are the same as those used in the common chromatographic methods such as ion exchange chromatography or size exclusion chromatography. Very-high-resolution separations can be achieved quickly and with high sensitivity in HPLC using automated instrumentation. *Reverse-phase* HPLC is a widely used chromatographic procedure for the separation of nonpolar solutes. In reverse-phase HPLC, a solution of nonpolar solutes is chromatographed on a column having a nonpolar liquid immobilized on an inert matrix; this nonpolar liquid serves as the *stationary phase*. A more polar liquid that serves as the *mobile phase* is passed over the matrix, and solute molecules are eluted in proportion to their solubility in this more polar liquid.

Affinity Chromatography

Affinity purification strategies for proteins exploit the biological function of the target protein. In most instances, proteins carry out their biological activity through binding or complex formation with specific small biomolecules, or *ligands,* as in the case of an enzyme binding its substrate. If this small molecule can be immobilized through covalent attachment to an insoluble matrix, such as a chromatographic medium like cellulose or polyacrylamide, then the protein of interest, in displaying affinity for its ligand, becomes bound and immobilized itself. It can then be removed from contaminating proteins in the mixture by simple means such as filtration and washing the matrix. Finally, the protein is dissociated or eluted from the matrix by the addition of high concentrations of the free ligand in solution. Figure 8 depicts the protocol for such an *affinity chromatography* scheme. Because this method of purification relies on the biological specificity of the protein of interest, it is a very efficient procedure and proteins can be purified several thousand-fold in a single step.

Ultracentrifugation

Centrifugation methods separate macromolecules on the basis of their characteristic densities. Particles tend to "fall" through a solution if the density of the solution is less than the density of the particle. The velocity of the particle through the medium is proportional to the difference in density between the particle and the solution. The tendency of any particle to move through a solution under centrifugal force is given by the *sedimentation coefficient, S:*

$$S = (\rho_p - \rho_m)V/f$$

where ρ_p is the density of the particle or macromolecule, ρ_m is the density of the medium or solution, V is the volume of the particle, and f is the frictional coefficient, given by

$$f = F_f/v$$

where v is the velocity of the particle and F_f is the frictional drag. Nonspherical molecules have larger frictional coefficients and thus smaller sedimentation coefficients. The smaller the particle and the more its shape deviates from spherical, the more slowly that particle sediments in a centrifuge.

Centrifugation can be used either as a preparative technique for separating and purifying macromolecules and cellular components

A protein interacts with a metabolite. The metabolite is thus a ligand that binds specifically to this protein

Protein   Metabolite

The metabolite can be immobilized by covalently coupling it to an insoluble matrix such as an agarose polymer. Cell extracts containing many individual proteins may be passed through the matrix.

Specific protein binds to ligand. All other unbound material is washed out of the matrix.

Adding an excess of free metabolite that will compete for the bound protein dissociates the protein from the chromatographic matrix. The protein passes out of the column complexed with free metabolite.

Purifications of proteins as much as 1000-fold or more are routinely achieved in a single affinity chromatographic step like this.

Figure 8 Diagram illustrating affinity chromatography.

(see Figure 17.16) or as an analytical technique to characterize the hydrodynamic properties of macromolecules such as proteins and nucleic acids. See also the A Deeper Look box: The Buoyant Density of DNA on page 356.

5.3 How Is the Amino Acid Analysis of Proteins Performed?

Acid Hydrolysis Liberates the Amino Acids of a Protein

Peptide bonds of proteins are hydrolyzed by either strong acid or strong base. Acid hydrolysis is the method of choice for analysis of the amino acid composition of proteins and polypeptides because it proceeds without racemization and with less destruction of certain amino acids (Ser, Thr, Arg, and Cys). Typically, samples of a protein are hydrolyzed with 6 N HCl at 110°C. Tryptophan is destroyed by acid and must be estimated by other means (usually UV spectrophotometry; see Figure 4.10) to determine its contribution to the total amino acid composition. The OH-containing amino acids serine and threonine are slowly destroyed. In contrast, peptide bonds involving hydrophobic residues such as valine and isoleucine are only slowly hydrolyzed in acid. Another complication arises because the β- and γ-amide linkages in asparagine (Asn) and glutamine (Gln) are acid labile. The side-chain amino nitrogen is released as free ammonium, and all of the Asn and Gln residues of the protein are converted to aspartic acid (Asp) and glutamic acid (Glu), respectively. The amount of ammonium released during acid hydrolysis gives an estimate of the total number of Asn and Gln residues in the original protein, but not the amounts of either.

Chromatographic Methods Are Used to Separate the Amino Acids

The complex amino acid mixture in the hydrolysate obtained after digestion of a protein in 6 N HCl can be separated into the component amino acids by using either ion exchange chromatography or reversed-phase high-pressure liquid chromatography (HPLC) (see Figure 4.13 and the A Deeper Look box: Techniques Used in Protein Purification). The amount of each amino acid can then be determined. These methods of separation and analysis are fully automated in instruments called **amino acid analyzers.** Analysis of the amino acid composition of a 30-kD protein by these methods requires less than 1 hour and only 6 μg (0.2 nmol) of the protein.

The Amino Acid Compositions of Different Proteins Are Different

Amino acids almost never occur in equimolar ratios in proteins, indicating that proteins are not composed of repeating arrays of amino acids. There are a few exceptions to this rule. Collagen, for example, contains large proportions of glycine and proline, and much of its structure is composed of (Gly-x-Pro) repeating units, where x is any amino acid. Other proteins show unusual abundances of various amino acids. For example, histones are rich in positively charged amino acids such as arginine and lysine. Histones are a class of proteins found associated with the anionic phosphate groups of eukaryotic DNA.

Amino acid analysis itself does not directly give the number of residues of each amino acid in a polypeptide, but if the molecular weight *and* the exact amount of the protein analyzed are known (or the number of amino acid residues per molecule is known), the molar ratios of amino acids in the protein can be calculated. Amino acid analysis provides no information on the order or sequence of amino acid residues in the polypeptide chain.

5.4 How Is the Primary Structure of a Protein Determined?

The Sequence of Amino Acids in a Protein Is Distinctive

The unique characteristic of each protein is the distinctive sequence of amino acid residues in its polypeptide chain(s). Indeed, it is the **amino acid sequence** of proteins that is encoded by the nucleotide sequence of DNA. This amino acid sequence, then, is a form of genetic information. Because polypeptide chains are unbranched, a polypeptide chain has only two ends, an amino-terminal, or **N-terminal,** end and a carboxy-terminal, or **C-terminal,** end. By convention, the amino acid sequence is read from the N-terminal end of the polypeptide chain through to the C-terminal end. As an example, every molecule of ribonuclease A from bovine pancreas has the same amino acid sequence, beginning with N-terminal lysine at position 1 and ending with C-terminal valine at position 124 (Figure 5.2). Given the possibility of any of the 20 amino acids at each position, the number of unique amino acid sequences is astronomically large. The astounding sequence variation possible within polypeptide chains provides a key insight into the incredible functional diversity of protein molecules in biological systems discussed later in this chapter.

Sanger Was the First to Determine the Sequence of a Protein

In 1953, Frederick Sanger of Cambridge University in England reported the amino acid sequences of the two polypeptide chains composing the protein insulin (Figure 5.8). Not only was this a remarkable achievement in analytical chemistry, but it helped demystify speculation about the chemical nature of proteins. Sanger's results clearly established that all of the molecules of a given protein have a fixed amino acid composition, a defined amino acid sequence, and therefore an invariant molecular weight. In short, proteins are well defined chemically. Today, the amino acid sequences of millions of proteins are known. Although many sequences have been determined from application of the principles first established by Sanger, most are now deduced from knowledge of the nucleotide sequence of the gene that encodes the protein. In addition, in recent years, the application of mass spectrometry to the sequence analysis of proteins has largely superseded the protocols based on chemical and enzymatic degradation of polypeptides that Sanger pioneered. Nevertheless, these chemical and enzymatic protocols not only are useful in confirming gene sequence data, they can also reveal posttranslational modifications of proteins, which often play a critical role in protein function (see Section 5.7). And, instances occur when the amino acid sequence of a protein is desirable information but the gene has not yet been sequenced. Furthur, these protocols inform the student about the nature of proteins, and thus a review of them is worthwhile.

Both Chemical and Enzymatic Methodologies Are Used in Protein Sequencing

The chemical strategy for determining the amino acid sequence of a protein involves six basic steps:

1. If the protein contains more than one polypeptide chain, the chains are separated and purified.
2. Intrachain S—S (disulfide) cross-bridges between cysteine residues in the polypeptide chain are cleaved. (If these disulfides are interchain linkages, then step 2 precedes step 1.)
3. The N-terminal and C-terminal residues are identified.
4. Each polypeptide chain is cleaved into smaller fragments, and the amino acid composition and sequence of each fragment are determined.
5. Step 4 is repeated, using a different cleavage procedure to generate a different and therefore overlapping set of peptide fragments.
6. The overall amino acid sequence of the protein is reconstructed from the sequences in overlapping fragments.

Each of these steps is discussed in greater detail in the following sections.

Step 1. Separation of Polypeptide Chains

If the protein of interest is a **heteromultimer** (composed of more than one type of polypeptide chain), then the protein must be dissociated into its component polypeptide chains, which then must be separated from one another and sequenced individually.

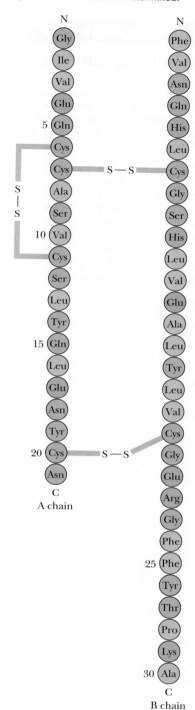

FIGURE 5.8 The hormone insulin consists of two polypeptide chains, A and B, held together by two disulfide cross-bridges (S—S). The A chain has 21 amino acid residues and an intrachain disulfide; the B polypeptide contains 30 amino acids. The sequence shown is for bovine insulin. (Illustration: Irving Geis. Rights owned by Howard Hughes Medical Institute. Not to be reproduced without permission.)

The Virtually Limitless Number of Different Amino Acid Sequences

Given 20 different amino acids, a polypeptide chain of n residues can have any one of 20^n possible sequence arrangements. To portray this, consider the number of tripeptides possible if there were only three different amino acids, A, B, and C (tripeptide $= 3 = n$; $3^n = 3^3 = 27$):

AAA	BBB	CCC
AAB	BBA	CCA
AAC	BBC	CCB
ABA	BAB	CBC
ACA	BCB	CAC
ABC	BAA	CBA
ACB	BCC	CAB
ABB	BAC	CBB
ACC	BCA	CAA

For a polypeptide chain of 100 residues in length, a rather modest size, the number of possible sequences is 20^{100}, or because $20 = 10^{1.3}$, 10^{130} unique possibilities. These numbers are more than astronomical! Because an average protein molecule of 100 residues would have a mass of 12,000 daltons (assuming the average molecular mass of an amino acid residue $= 120$), 10^{130} such molecules would have a mass of 1.2×10^{134} daltons. The mass of the observable universe is estimated to be 10^{80} proton masses (about 10^{80} daltons). Thus, the universe lacks enough material to make just one molecule of each possible polypeptide sequence for a protein only 100 residues in length.

Because subunits in multimeric proteins typically associate through noncovalent interactions, most multimeric proteins can be dissociated by exposure to pH extremes, 8 M urea, 6 M guanidinium hydrochloride, or high salt concentrations. (All of these treatments disrupt polar interactions such as hydrogen bonds both within the protein molecule and between the protein and the aqueous solvent.) Once dissociated, the individual polypeptides can be isolated from one another on the basis of differences in size and/or charge. Occasionally, heteromultimers are linked together by interchain S—S bridges. In such instances, these crosslinks must be cleaved before dissociation and isolation of the individual chains. The methods described under step 2 are applicable for this purpose.

Step 2. Cleavage of Disulfide Bridges

A number of methods exist for cleaving disulfides. An important consideration is to carry out these cleavages so that the original or even new S—S links do not form. Oxidation of a disulfide by performic acid results in the formation of two equivalents of cysteic acid (Figure 5.9a). Because these cysteic acid side chains are ionized SO_3^- groups, electrostatic repulsion (as well as altered chemistry) prevents S—S recombination. Alternatively, sulfhydryl compounds such as 2-mercaptoethanol or dithiothreitol (DTT) readily reduce S—S bridges to regenerate two cysteine—SH side chains, as in a reversal of the reaction shown in Figure 4.8b. However, these SH groups recombine to re-form either the original disulfide link or, if other free Cys—SHs are available, new disulfide links. To prevent this, S—S reduction must be followed by treatment with alkylating agents such as iodoacetate or 3-bromopropylamine, which modify the SH groups and block disulfide bridge formation (Figure 5.9b).

Step 3. N- and C-Terminal Analysis

A. N-Terminal Analysis The amino acid residing at the N-terminal end of a protein can be identified in a number of ways; one method, **Edman degradation,** has become the procedure of choice. This method is preferable because it allows the sequential identification of a series of residues beginning at the N-terminus. In weakly basic solutions, phenylisothiocyanate, or **Edman reagent** (phenyl—N=C=S), combines with the free amino terminus of a protein (see Figure 4.8a), which can be excised from the end of the polypeptide chain and recovered as a PTH derivative. Chromatographic methods can be used to identify this PTH derivative. Importantly, in this

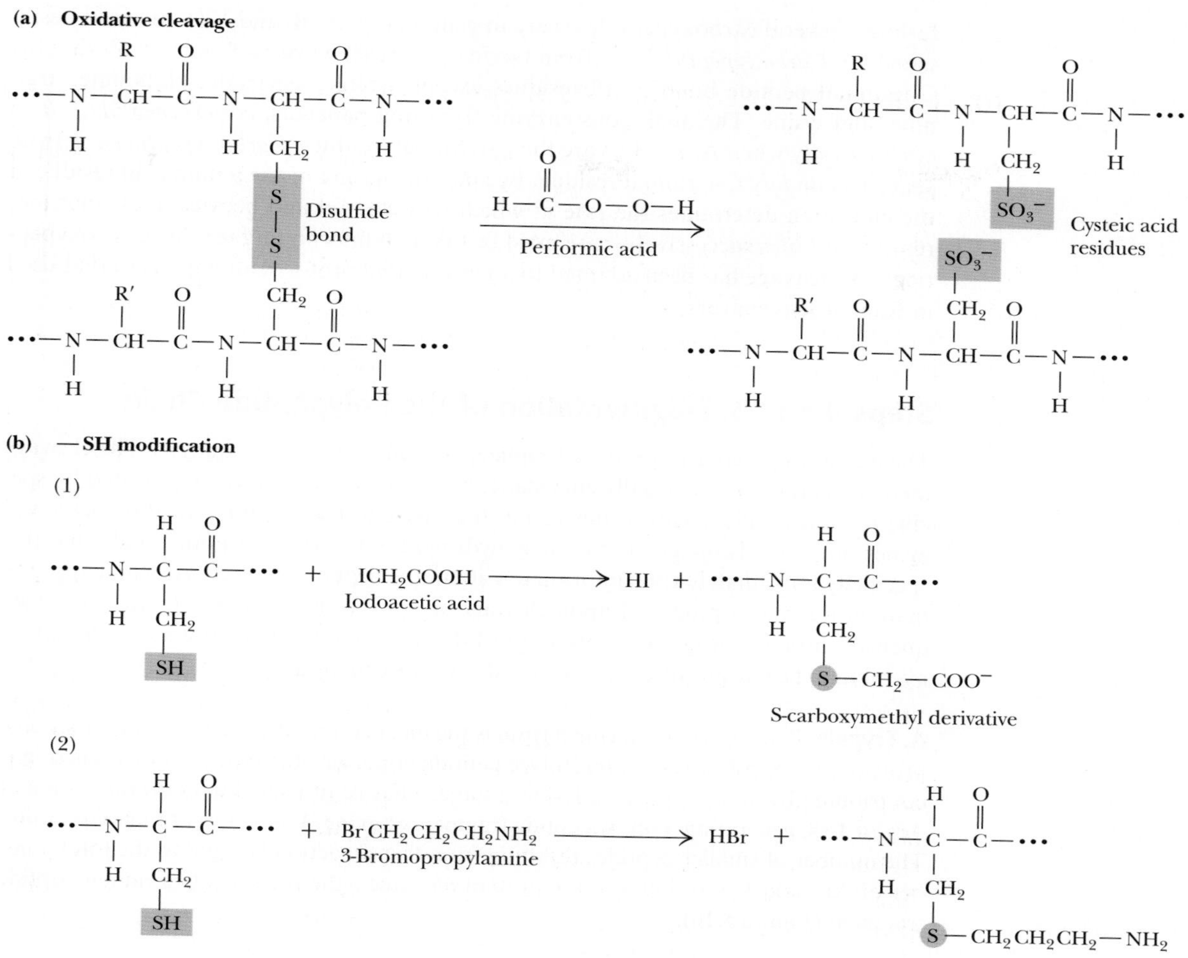

(a) Oxidative cleavage

(b) —SH modification

(1)

S-carboxymethyl derivative

(2)

FIGURE 5.9 Methods for cleavage of disulfide bonds in proteins. **(a)** Oxidative cleavage by reaction with performic acid. **(b)** Disulfide bridges can be broken by reduction with sulfhydryl agents such as β-mercaptoethanol or dithiothreitol (as in the reverse of the reaction shown in Figure 4.8b). Because reaction between the newly reduced —SH groups to reestablish disulfide bonds is a likelihood, S—S reduction must be followed by —SH modification: (1) alkylation with iodoacetate (ICH$_2$COOH) or (2) modification with 3-bromopropylamine (Br—(CH$_2$)$_3$—NH$_2$).

procedure, the rest of the polypeptide chain remains intact and can be subjected to further rounds of Edman degradation to identify successive amino acid residues in the chain. Often, the carboxyl terminus of the polypeptide under analysis is coupled to an insoluble matrix, allowing the polypeptide to be easily recovered by filtration or centrifugation following each round of Edman reaction. Thus, the Edman reaction not only identifies the N-terminal residue of proteins but through successive reaction cycles can reveal further information about sequence. Automated instruments (so-called Edman sequenators) have been designed to carry out repeated rounds of the Edman procedure. In practical terms, as many as 50 cycles of reaction can be accomplished on 50 pmol (about 0.1 µg) of a polypeptide 100 to 200 residues long, revealing the sequential order of the first 50 amino acid residues in the protein. The efficiency with larger proteins is less; a typical 2000–amino acid protein provides only 10 to 20 cycles of reaction.

B. C-Terminal Analysis For the identification of the C-terminal residue of polypeptides, an enzymatic approach is commonly used. Carboxypeptidases are enzymes that cleave amino acid residues from the C-termini of polypeptides in a successive fashion. Several carboxypeptidases are in general use: A, B, and Y (sometimes designated C). *Carboxypeptidase A* (from bovine pancreas) works well in hydrolyzing the C-terminal peptide bond of all residues except proline, aspartate, glutamate, arginine, and lysine. The analogous enzyme from hog pancreas, *carboxypeptidase B,* is effective only when Arg or Lys are the C-terminal residues. *Carboxypeptidase Y* from yeast acts on any C-terminal residue. Because the nature of the amino acid residue at the end often determines the rate at which it is cleaved and because these enzymes remove residues successively, care must be taken in interpreting results. Carboxypeptidase Y cleavage has been adapted to an automated protocol analogous to that used in Edman sequenators.

Steps 4 and 5. Fragmentation of the Polypeptide Chain

The aim at this step is to produce fragments useful for sequence analysis. The cleavage methods employed are usually enzymatic, but proteins can also be fragmented by specific or nonspecific chemical means (such as partial acid hydrolysis). Proteolytic enzymes offer an advantage in that many hydrolyze only specific peptide bonds, and this specificity immediately gives information about the peptide products. As a first approximation, fragments produced upon cleavage should be small enough to yield their sequences through end-group analysis and Edman degradation, yet not so small that an overabundance of products must be resolved before analysis.

A. Trypsin The digestive enzyme *trypsin* is the most commonly used reagent for specific proteolysis. Trypsin will only hydrolyze peptide bonds in which the carbonyl function is contributed by an arginine or a lysine residue. That is, trypsin cleaves on the C-side of Arg or Lys, generating a set of peptide fragments having Arg or Lys at their C-termini. The number of smaller peptides resulting from trypsin action is equal to the total number of Arg and Lys residues in the protein *plus* one—the protein's C-terminal peptide fragment (Figure 5.10).

B. Chymotrypsin *Chymotrypsin* shows a strong preference for hydrolyzing peptide bonds formed by the carboxyl groups of the aromatic amino acids, phenylalanine, tyrosine, and tryptophan. However, over time, chymotrypsin also hydrolyzes amide bonds involving amino acids other than Phe, Tyr, or Trp. For instance, peptide bonds having leucine-donated carboxyls are also susceptible. Thus, the specificity of chymotrypsin is only relative. Because chymotrypsin produces a very different set of products than trypsin, treatment of separate samples of a protein with these two enzymes generates fragments whose sequences overlap. Resolution of the order of amino acid residues in the fragments yields the amino acid sequence in the original protein.

C. Other Endopeptidases A number of other *endopeptidases* (proteases that cleave peptide bonds within the interior of a polypeptide chain) are also used in sequence investigations. These include *clostripain,* which acts only at Arg residues; *endopeptidase Lys-C,* which cleaves only at Lys residues; and *staphylococcal protease,* which acts at the acidic residues, Asp and Glu. Other, relatively nonspecific endopeptidases are handy for digesting large tryptic or chymotryptic fragments. *Pepsin, papain, subtilisin, thermolysin,* and *elastase* are some examples. Papain is the active ingredient in meat tenderizer, soft contact lens cleaner, and some laundry detergents.

D. Cyanogen Bromide Several highly specific chemical methods of proteolysis are available, the most widely used being *cyanogen bromide (CNBr)* cleavage. CNBr acts upon methionine residues (Figure 5.11). The nucleophilic sulfur atom of Met reacts with CNBr, yielding a sulfonium ion that undergoes a rapid intramolecular rearrangement to form a cyclic iminolactone. Water readily hydrolyzes this iminolactone, cleaving the polypeptide and generating peptide fragments having C-terminal homoserine lactone residues at the former Met positions.

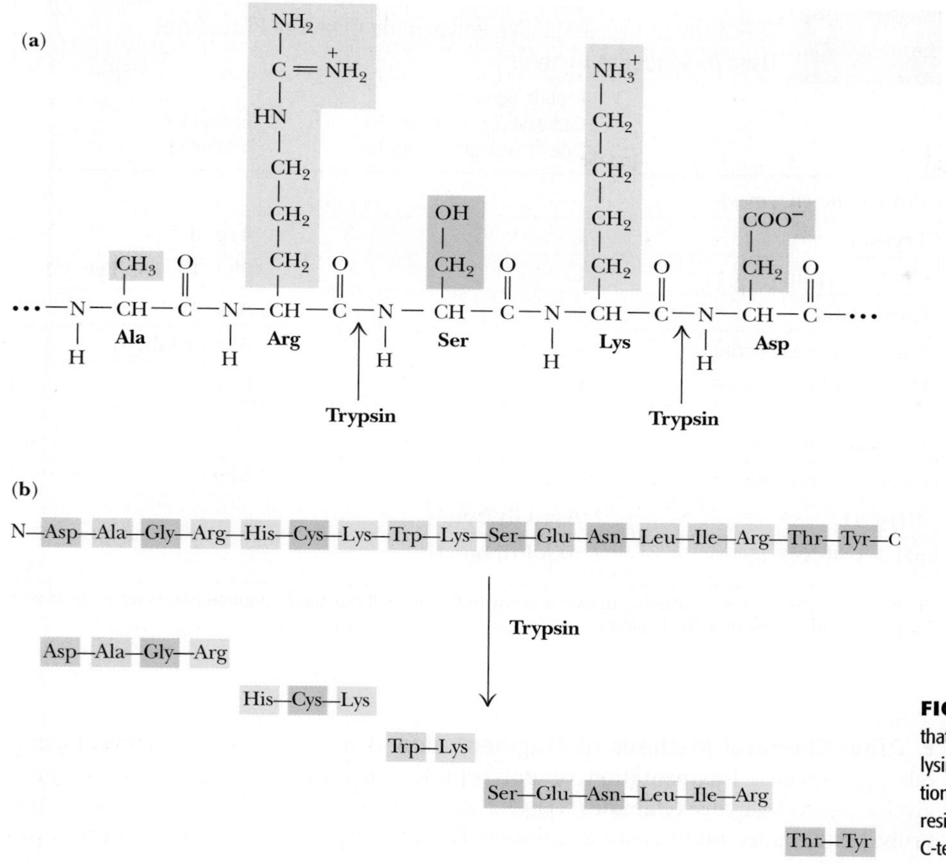

(a)

Trypsin

Trypsin

(b)

N—Asp—Ala—Gly—Arg—His—Cys—Lys—Trp—Lys—Ser—Glu—Asn—Leu—Ile—Arg—Thr—Tyr—C

Trypsin

Asp—Ala—Gly—Arg

His—Cys—Lys

Trp—Lys

Ser—Glu—Asn—Leu—Ile—Arg

Thr—Tyr

FIGURE 5.10 **(a)** Trypsin is a proteolytic enzyme, or *protease,* that specifically cleaves only those peptide bonds in which arginine or lysine contributes the carbonyl function. **(b)** The products of the reaction are a mixture of peptide fragments with C-terminal Arg or Lys residues *and* a single peptide derived from the polypeptide's C-terminal end.

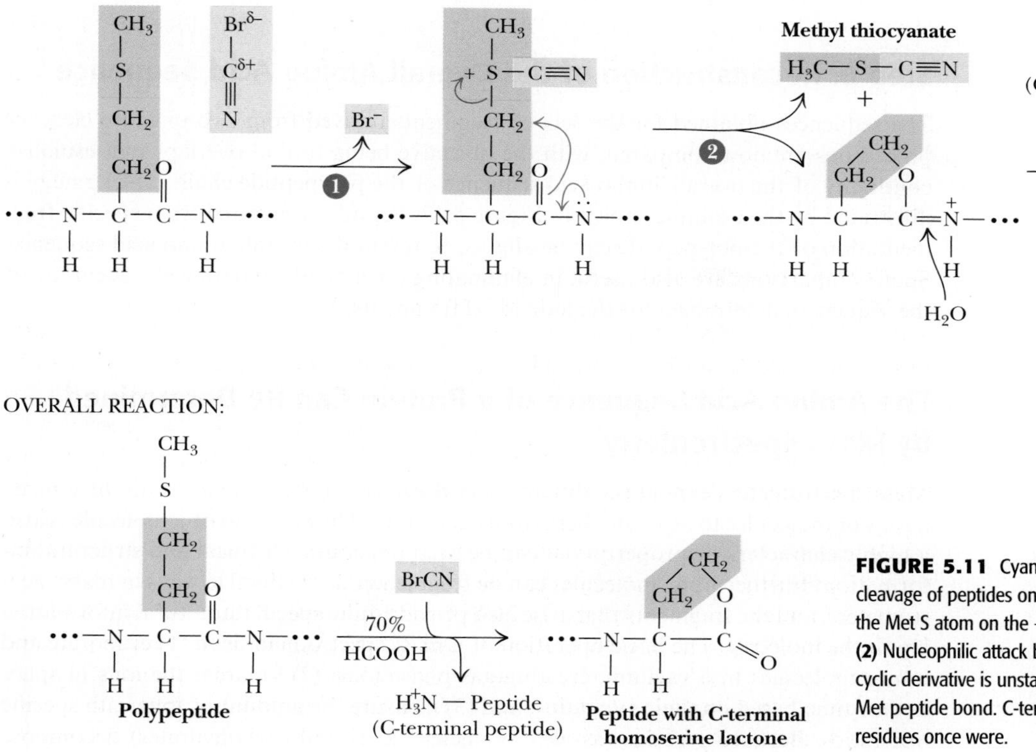

FIGURE 5.11 Cyanogen bromide (CNBr) is a highly selective reagent for cleavage of peptides only at methionine residues. **(1)** Nucleophilic attack of the Met S atom on the —C≡N carbon atom, with displacement of Br. **(2)** Nucleophilic attack by the Met carbonyl oxygen atom on the R group. The cyclic derivative is unstable in aqueous solution. **(3)** Hydrolysis cleaves the Met peptide bond. C-terminal homoserine lactone residues occur where Met residues once were.

OVERALL REACTION:

TABLE 5.2	Specificity of Representative Polypeptide Cleavage Procedures Used in Sequence Analysis	
Method	Peptide Bond on Carboxyl (C) or Amino (N) Side of Susceptible Residue	Susceptible Residue(s)
*Proteolytic enzymes**		
Trypsin	C	Arg or Lys
Chymotrypsin	C	Phe, Trp, or Tyr; Leu
Clostripain	C	Arg
Staphylococcal protease	C	Asp or Glu
Endopeptidase Lys-C	C	Lys
Chemical methods		
Cyanogen bromide	C	Met
NH_2OH	Asn-Gly bonds	
pH 2.5, 40°C	Asp-Pro bonds	

*Some proteolytic enzymes, including trypsin and chymotrypsin, will not cleave peptide bonds where proline is the amino acid contributing the N-atom.

E. Other Chemical Methods of Fragmentation A number of other chemical methods give specific fragmentation of polypeptides, including cleavage at asparagine–glycine bonds by hydroxylamine (NH_2OH) at pH 9 and selective hydrolysis at aspartyl–prolyl bonds under mildly acidic conditions. Table 5.2 summarizes the various procedures described here for polypeptide cleavage. These methods are only a partial list of the arsenal of reactions available to protein chemists. Cleavage products generated by these procedures must be isolated and individually sequenced to accumulate the information necessary to reconstruct the protein's complete amino acid sequence. Peptide sequencing today is most commonly done by Edman degradation of relatively large peptides or by mass spectrometry (see following discussion).

Step 6. Reconstruction of the Overall Amino Acid Sequence

The sequences obtained for the sets of fragments derived from two or more cleavage procedures are now compared, with the objective being to find overlaps that establish continuity of the overall amino acid sequence of the polypeptide chain. The strategy is illustrated by the example shown in Figure 5.12. Peptides generated from specific fragmentation of the polypeptide can be aligned to reveal the overall amino acid sequence. Such comparisons are also useful in eliminating errors and validating the accuracy of the sequences determined for the individual fragments.

The Amino Acid Sequence of a Protein Can Be Determined by Mass Spectrometry

Mass spectrometers exploit the difference in the mass-to-charge *(m/z)* ratio of ionized atoms or molecules to separate them from each other. The *m/z* ratio of a molecule is also a highly characteristic property that can be used to acquire chemical and structural information. Furthermore, molecules can be fragmented in distinctive ways in mass spectrometers, and the fragments that arise also provide quite specific structural information about the molecule. The basic operation of a mass spectrometer is to (1) evaporate and ionize molecules in a vacuum, creating gas-phase ions; (2) separate the ions in space and/or time based on their *m/z* ratios; and (3) measure the amount of ions with specific *m/z* ratios. Because proteins (as well as nucleic acids and carbohydrates) decompose

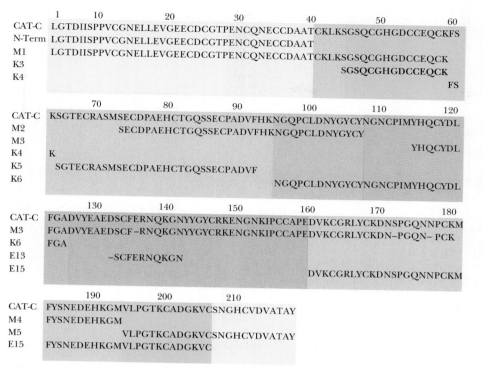

FIGURE 5.12 Summary of the sequence analysis of catrocollastatin-C, a 23.6-kD protein found in the venom of the western diamondback rattlesnake *Crotalus atrox*. Sequences shown are given in the one-letter amino acid code. The overall amino acid sequence (216 amino acid residues long) for catrocollastatin-C as deduced from the overlapping sequences of peptide fragments is shown on the lines headed **CAT-C**. The other lines report the various sequences used to obtain the overlaps. These sequences were obtained from (a) **N-term:** Edman degradation of the intact protein in an automated Edman sequenator; (b) **M:** proteolytic fragments generated by CNBr cleavage, followed by Edman sequencing of the individual fragments (numbers denote fragments M1 through M5); (c) **K:** proteolytic fragments from endopeptidase Lys-C cleavage, followed by Edman sequencing (only fragments K3 through K6 are shown); (d) **E:** proteolytic fragments from *Staphylococcus* protease digestion of catrocollastatin sequenced in the Edman sequenator (only E13 through E15 are shown). (Adapted from Shimokawa, K., et al., 1997. Sequence and biological activity of catrocollastatin-C: A disintegrin-like/cysteine-rich two-domain protein from *Crotalus atrox* venom. *Archives of Biochemistry and Biophysics* **343:**35–43.)

upon heating, rather than evaporating, methods to ionize such molecules for mass spectrometry (MS) analysis require innovative approaches. The two most prominent MS modes for protein analysis are summarized in Table 5.3.

Figure 5.13 illustrates the basic features of electrospray mass spectrometry (ESI MS). In this technique, the high voltage at the electrode causes proteins to pick up protons from the solvent, such that, on average, individual protein molecules acquire about one positive charge (proton) per kilodalton, leading to the spectrum of *m/z* ratios for a

---

TABLE 5.3 The Two Most Common Methods of Mass Spectrometry for Protein Analysis

Electrospray Ionization (ESI-MS)

A solution of macromolecules is sprayed in the form of fine droplets from a glass capillary under the influence of a strong electrical field. The droplets pick up positive charges as they exit the capillary; evaporation of the solvent leaves multiply charged molecules. The typical 20-kD protein molecule will pick up 10 to 30 positive charges. The MS spectrum of this protein reveals all of the differently charged species as a series of sharp peaks whose consecutive *m/z* values differ by the charge and mass of a single proton (see Figure 5.14). Note that decreasing *m/z* values signify increasing number of charges per molecule, *z*. Tandem mass spectrometers downstream from the ESI source (ESI-MS/MS) can analyze complex protein mixtures (such as tryptic digests of proteins or chromatographically separated proteins emerging from a liquid chromatography column), selecting a single *m/z* species for collision-induced dissociation and acquisition of amino acid sequence information.

Matrix-Assisted Laser Desorption Ionization-Time of Flight (MALDI-TOF MS)

The protein sample is mixed with a chemical matrix that includes a light-absorbing substance excitable by a laser. A laser pulse is used to excite the chemical matrix, creating a microplasma that transfers the energy to protein molecules in the sample, ionizing them and ejecting them into the gas phase. Among the products are protein molecules that have picked up a single proton. These positively charged species can be selected by the MS for mass analysis. MALDI-TOF MS is very sensitive and very accurate; as little as attomole (10^{-18} moles) quantities of a particular molecule can be detected at accuracies better than 0.001 atomic mass units (0.001 daltons). MALDI-TOF MS is best suited for very accurate mass measurements.

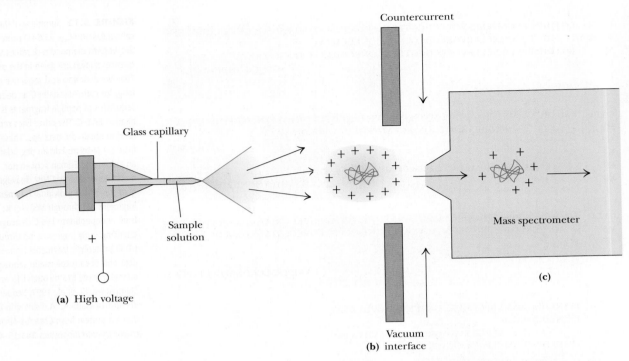

Countercurrent

Glass capillary

Sample
solution

Mass spectrometer

(a) High voltage

Vacuum
(b) interface

(c)

FIGURE 5.13 The three principal steps in electrospray ionization mass spectrometry (ESI-MS). **(a)** Small, highly charged droplets are formed by electrostatic dispersion of a protein solution through a glass capillary subjected to a high electric field; **(b)** protein ions are desorbed from the droplets into the gas phase (assisted by evaporation of the droplets in a stream of hot N_2 gas); and **(c)** the protein ions are separated in a mass spectrometer and identified according to their *m/z* ratios. (Adapted from Figure 1 in Mann, M., and Wilm, M., 1995. Electrospray mass spectrometry for protein characterization. *Trends in Biochemical Sciences* **20:**219–224.)

single protein species (Figure 5.14). Computer analysis can convert these data into a single spectrum that has a peak at the correct protein mass (Figure 5.14, inset).

Sequencing by Tandem Mass Spectrometry *Tandem* MS (or MS/MS) allows sequencing of proteins by hooking two mass spectrometers in tandem. The first mass spectrometer is used as a filter to sort the oligopeptide fragments in a protein digest based on differences in their *m/z* ratios. Each of these oligopeptides can then be selected by the mass spectrometer for further analysis. A selected ionized oligopeptide is directed toward the second mass spectrometer; on the way, this oligopeptide is fragmented by collision with helium or argon gas molecules (a process called *collision-induced dissociation,* or *c.i.d.*), and the fragments are analyzed by the second mass spectrometer (Figure 5.15). Fragmentation occurs primarily at the peptide bonds linking successive amino acids in the oligopeptide. Thus, the products include a series of fragments that represent a nested set of peptides differing in size by one amino acid residue. The various members of this set of fragments differ in mass by 56 atomic mass units [the mass of the peptide backbone atoms (NH—CH—CO)] *plus* the mass of the R group at each position, which ranges from 1 atomic mass unit (Gly) to 130 (Trp). MS sequencing has the advantages of very high sensitivity, fast sample processing, and the ability to work with mixtures of proteins. Subpicomoles (less than 10^{-12} moles) of peptide can be analyzed with these spectrometers. In practice, tandem MS is limited to rather short sequences (no longer than 15 or so amino acid residues). Nevertheless, capillary HPLC-separated peptide mixtures from trypsin digests of proteins can be directly loaded into the tandem MS spectrometer. Furthermore, separation of a complex mixture of proteins from a whole-cell extract by two-dimensional gel electrophoresis (see A Deeper Look: Techniques Used in Protein Purification), followed by trypsin digestion of a specific protein spot on the gel and injection of the digest into the HPLC/tandem MS, gives sequence information that can be used to identify specific proteins. Often, by comparing the mass of tryptic peptides from a protein digest with a

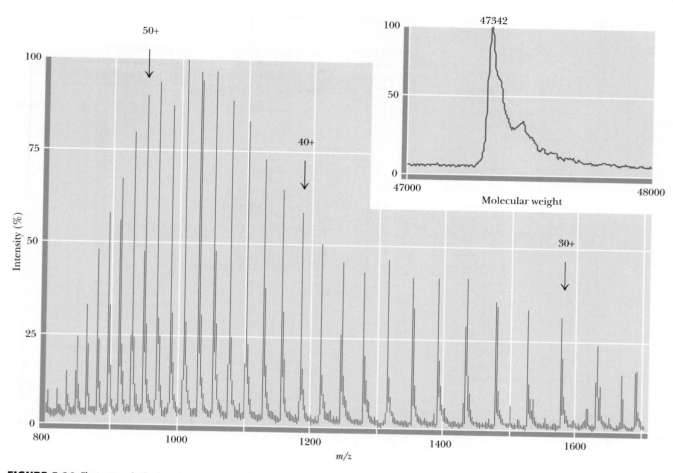

FIGURE 5.14 Electrospray ionization mass spectrum of the protein aerolysin K. The attachment of many protons per protein molecule (from less than 30 to more than 50 here) leads to a series of *m/z* peaks for this single protein. The equation describing each *m/z* peak is: $m/z = [M + n(\text{mass of proton})]/n(\text{charge on proton})$, where M = mass of the protein and n = number of positive charges per protein molecule. Thus, if the number of charges per protein molecule is known and *m/z* is known, M can be calculated. The inset shows a computer analysis of the data from this series of peaks that generates a single peak at the correct molecular mass of the protein. (Adapted from Figure 2 in Mann, M., and Wilm, M., 1995. Electrospray mass spectrometry for protein characterization. *Trends in Biochemical Sciences* **20**:219–224.)

database of all possible masses for tryptic peptides (based on all known protein and DNA sequences), one can identify a protein of interest without actually sequencing it.

Peptide Mass Fingerprinting *Peptide mass fingerprinting* is used to uniquely identify a protein based on the masses of its proteolytic fragments, usually produced by trypsin digestion. MALDI-TOF MS instruments are ideal for this purpose because they yield highly accurate mass data. The measured masses of the proteolytic fragments can be compared to databases (see following discussion) of peptide masses of known sequence. Such information is easily generated from genomic databases: Nucleotide sequence information can be translated into amino acid sequence information, from which very accurate peptide mass compilations are readily calculated. For example, the SWISS-PROT database lists 1197 proteins with a tryptic fragment of *m/z* = 1335.63 (±0.2 D), 16 proteins with tryptic fragments of *m/z* = 1335.63 *and m/z* = 1405.60, but only a single protein (human tissue plasminogen activator [tPA]) with tryptic fragments of *m/z* = 1335.63, *m/z* = 1405.60, *and m/z* = 1272.60.[1] Although the identities of many proteins

[1]The tPA amino acid sequences corresponding to these masses are *m/z* = 1335.63: HEALSPFYSER; *m/z* = 1405.60: ATCYEDQGISYR; and *m/z* = 1272.60: DSKPWCYVFK.

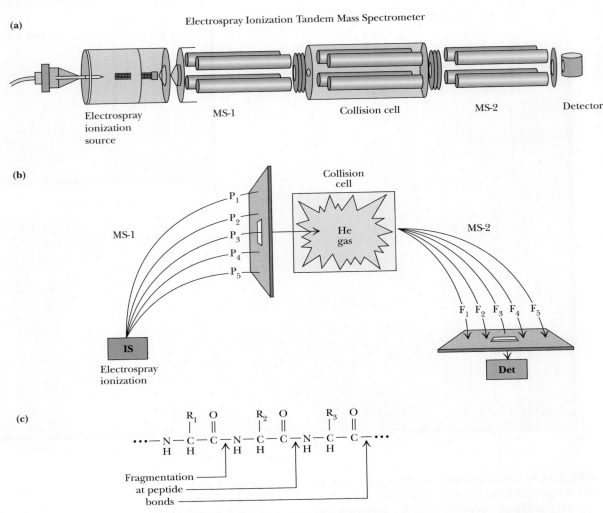

FIGURE 5.15 Tandem mass spectrometry. **(a)** Configuration used in tandem MS. **(b)** Schematic description of tandem MS: Tandem MS involves electrospray ionization of a protein digest (IS in this figure), followed by selection of a single peptide ion mass for collision with inert gas molecules (He) and mass analysis of the fragment ions resulting from the collisions. **(c)** Fragmentation usually occurs at peptide bonds, as indicated. (Adapted from Yates, J. R., 1996. Protein structure analysis by mass spectrometry. *Methods in Enzymology* **271:**351–376; and Gillece-Castro, B. L., and Stults, J. T., 1996. Peptide characterization by mass spectrometry. *Methods in Enzymology* **271:**427–447.)

revealed by genomic analysis remain unknown, peptide mass fingerprinting can assign a particular protein exclusively to a specific gene in a genomic database.

Sequence Databases Contain the Amino Acid Sequences of Millions of Different Proteins

The first protein sequence databases were compiled by protein chemists using chemical sequencing methods. Today, the vast preponderance of protein sequence information has been derived from translating the nucleotide sequences of genes into codons and, thus, amino acid sequences (see Chapter 30). Sequencing the order of nucleotides in cloned genes is a more rapid, efficient, and informative process than determining the amino acid sequences of proteins by chemical methods. Several electronic databases containing continuously updated sequence information are accessible by personal computer. Prominent among these is the SWISS-PROT protein sequence database on the ExPASy (**Ex**pert **P**rotein **A**nalysis **Sy**stem) Molecular Biology server at *http://us.expasy.org* and the PIR (Protein Identification Resource Protein Sequence

Database) at *http://pir.georgetown.edu.* Mass spectrometric proteomics data is freely available from a number of repositories, including PRIDE (PRoteomics IDentification database) at *http://www.ebi.ac.uk/pride/*, PeptideAtlas at *http://peptideatlas.org*, and the Global Proteome Machine at *http://gpmdb.thegpm.org.* Further, protein information from genomic sequences is available in databases such as GenBank, accessible via the National Center for Biotechnology Information (NCBI) Web site located at *http://www.ncbi.nlm.nih.gov.* The protein sequence databases contain millions of entries.

In addition to sequence information, protein structural information can be found in the Worldwide Protein Data Bank (wwPDB), an ensemble of repositories composed of the Research Collaboratory for Structural Bioinformatics Protein Data Bank (RCSB PDB at *http://www.rcsb.org/pdb*), the Protein Data Bank Europe (PDBe), the Protein Data Bank Japan (PDBj), and the BioMagResBank (BMRB), which specifically houses NMR (nuclear magnetic resonance spectroscopy)-derived structural information. The RCSB PDB site contains 3-dimensional structural information on about 70,000 proteins and nucleic acids.

5.5 What Is the Nature of Amino Acid Sequences?

Figure 5.16 illustrates the relative frequencies of the amino acids in proteins. It is very unusual for a globular protein to have an amino acid composition that deviates substantially from these values. Apparently, these abundances reflect a distribution of amino acid polarities that is optimal for protein stability in an aqueous milieu. Membrane proteins tend to have relatively more hydrophobic and fewer ionic amino acids, a condition consistent with their location. Fibrous proteins may show compositions that are atypical with respect to these norms, indicating an underlying relationship between the composition and the structure of these proteins.

Proteins have unique amino acid sequences, and it is this uniqueness of sequence that ultimately gives each protein its own particular personality. Because the number of possible amino acid sequences in a protein is astronomically large, the probability that two proteins will, by chance, have similar amino acid sequences is negligible. Consequently, sequence similarities between proteins imply evolutionary relatedness.

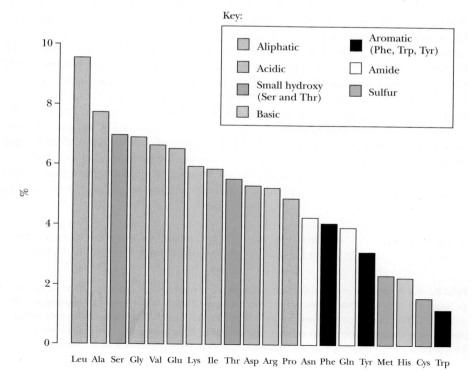

FIGURE 5.16 Amino acid composition: frequencies of the various amino acids in proteins for all the proteins in the SWISS-PROT protein knowledgebase. These data are derived from the amino acid composition of more than 100,000 different proteins (representing more than 40,000,000 amino acid residues). The range is from leucine at 9.55% to tryptophan at 1.18% of all residues.

Homologous Proteins from Different Organisms Have Homologous Amino Acid Sequences

Proteins sharing a significant degree of sequence similarity and structural resemblance are said to be **homologous.** Proteins that perform the same function in different organisms are also referred to as homologous. For example, the oxygen transport protein hemoglobin serves a similar role and has a similar structure in all vertebrates. The study of the amino acid sequences of homologous proteins from different organisms provides very strong evidence for their evolutionary origin within a common ancestor. Homologous proteins characteristically have polypeptide chains that are nearly identical in length, and the degree to which their sequences share identity provides a direct measure of the evolutionary relationship between the species from which they are derived.

Homologous proteins can be further subdivided into **orthologous** and **paralogous** proteins. **Orthologous** proteins are proteins from different species that have homologous amino acid sequences (and often a similar function). Orthologous proteins arose from a common ancestral gene during evolution. **Paralogous** proteins are proteins found within a single species that have homologous amino acid sequences; paralogous proteins arose through gene duplication. For example, the α- and β-globin chains of hemoglobin are paralogs. How is homology revealed?

Computer Programs Can Align Sequences and Discover Homology between Proteins

Protein and nucleic acid sequence databases (see page 122) provide enormous resources for sequence comparisons. If two proteins share homology, it can be revealed through alignment of their sequences using powerful computer programs. In such studies, a given amino acid sequence is used to query the databases for proteins with similar sequences. **BLAST** (**B**asic **L**ocal **A**lignment **S**earch **T**ool) is one commonly used program for rapid searching of sequence databases. The BLAST program detects local as well as global alignments where sequences are in close agreement. Even regions of similarity shared between otherwise unrelated proteins can be detected. Discovery of sequence similarities between proteins can be an important clue to the function of uncharacterized proteins. Similarities are also useful in assigning related proteins to protein families.

The process of sequence alignment is an operation akin to sliding one sequence along another in a search for regions where the two sequences show a good match. Positive scores are assigned everywhere the amino acid in one sequence is similar to or identical with the amino acid in the other; the greater the overall score, the better the match between the two protein sequences. Sometimes two sequences match well at several places along their lengths, but, in one of the proteins, the matching segments are interrupted by a sequence that is dissimilar. When such an interruption is found by the computer program, it inserts a **gap** in the uninterrupted sequence to bring the matching segments of the two sequences into better alignment (Figure 5.17). Because any two sequences would show similarity if a sufficient number of gaps were introduced, a **gap penalty** is imposed for each gap. Gap penalties are negative numbers that lower the overall similarity score. Gaps arise naturally during evolution through insertion and deletion mutations, so-called **indels,** which add or remove residues in the gene and, consequently,

S. acidocaldarius	FPIAKGGTAAIPGPFGSGKTVTLQSLAKWSAAK---VVIYVGCGERGNEMTD
E. coli	CPFAKGGKVGLFGGAGVGKTVNMMELIRNIAIEHSGYSVFAGVGERTREGND

FIGURE 5.17 Alignment of the amino acid sequences of two protein homologs using gaps. Shown are parts of the amino acid sequences of the catalytic subunits from the major ATP-synthesizing enzyme (ATP synthase) in a representative archaea *(Sulfolobus acidocaldarius)* and a bacterium *(Escherichia coli)*. These protein segments encompass the nucleotide-binding site of these enzymes. Identical residues in the two sequences are shown in red. Introduction of a three-residue-long gap in the archaeal sequence optimizes alignment of the two sequences.

the protein. The optimal sequence alignment between two proteins is one that maximizes sequence alignments while minimizing gaps.

Methods for alignment and comparison of protein sequences depend upon some quantitative measure of how similar any two sequences are. One way to measure similarity is to use a matrix that assigns scores for all possible substitutions of one amino acid for another. BLOSUM62 is the substitution matrix most often used with BLAST. This matrix assigns a probability score for each position in an alignment based on the frequency with which that substitution occurs in the consensus sequences of related proteins. **BLOSUM** is an acronym for **Blo**cks **Su**bstitution **M**atrix, a matrix that scores each position on the basis of observed frequencies of different amino acid substitutions within blocks of local alignments in related proteins. In the BLOSUM62 matrix, the most commonly used matrix, the scores are derived using sequences sharing no more than 62% identity (Figure 5.18). BLOSUM substitution scores range from −4 (lowest probability of substitution) to 11 (highest probability of substitution). For example, to look up the value corresponding to the substitution of an asparagine (N) by a tryptophan (W), or vice versa, find the intersection of the "N" column with the "W" row in Figure 5.18. The value −4 means that the substitution of N for W, or vice versa, is not very likely. On the other hand, the substitution of V for I, (BLOSUM score: 3), or vice versa, is very likely. Amino acids whose side chains have unique qualities (such as C, H, P, or W) have high BLOSUM62 scores, because replacing them with any other amino acid may change the protein significantly. Amino acids that are similar (such as R and K, or D and E, or A, V, L, and I) have low scores, since one can replace the other with less likelihood of serious change to the protein structure.

Cytochrome c The electron transport protein **cytochrome c,** found in the mitochondria of all eukaryotic organisms, provides a well-studied example of orthology. Amino acid sequencing of cytochrome c from more than 40 different species has revealed that there are 28 positions in the polypeptide chain where the same amino acid residues are always found (Figure 5.19). These **invariant residues** serve roles crucial to the biological function of this protein, and thus substitutions of other amino acids at these positions

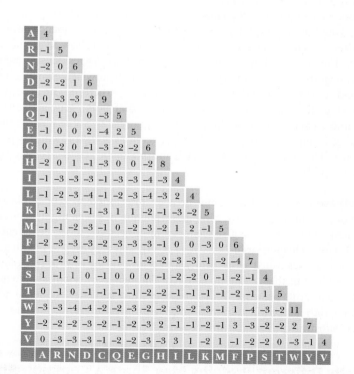

FIGURE 5.18 The BLOSUM62 substitution matrix provides scores for all possible exchanges of one amino acid with another. (From Henikoff, S., and Henikoff, J. G., 1992. Amino acid substitution matrices from protein blocks. *Proceedings of the National Academy of Sciences, USA* **89:**10915–10919.)

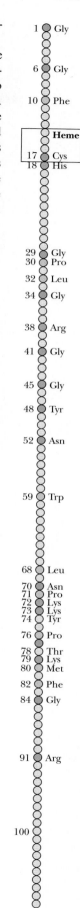

FIGURE 5.19 The sequence of cytochrome c from more than 40 different species reveals that 28 residues are invariant. These invariant residues are scattered irregularly along the polypeptide chain, except for a cluster between residues 70 and 80. All cytochrome c polypeptide chains have a cysteine residue at position 17, and all but one have another Cys at position 14. These Cys residues serve to link the heme prosthetic group of cytochrome c to the protein.

cannot be tolerated. The number of amino acid differences between two cytochrome *c* sequences is proportional to the phylogenetic difference between the species from which they are derived. Cytochrome *c* in humans and in chimpanzees is identical; human and another mammalian (sheep) cytochrome *c* differ at 10 residues. The human cytochrome *c* sequence has 14 variant residues from a reptile sequence (rattlesnake), 18 from a fish (carp), 29 from a mollusc (snail), 31 from an insect (moth), and more than 40 from yeast or higher plants (cauliflower).

The Phylogenetic Tree for Cytochrome *c* Figure 5.20 displays a **phylogenetic tree** (a diagram illustrating the evolutionary relationships among a group of organisms) constructed from the sequences of cytochrome *c*. The tips of the branches are occupied by contemporary species whose sequences have been determined. The tree has been deduced by computer analysis of these sequences to find the minimum number of mutational changes connecting the branches. Other computer methods can be used to infer potential ancestral sequences represented by *nodes,* or branch points, in the tree. Such analysis ultimately suggests a primordial cytochrome *c* sequence lying at the base of the tree. Evolutionary trees constructed in this manner, that is, solely on the basis of amino acid differences occurring in the primary structure of one selected protein, show remarkable agreement with phylogenetic relationships derived from more classic approaches and have given rise to the field of *molecular evolution.*

Related Proteins Share a Common Evolutionary Origin

Amino acid sequence analysis reveals that proteins with related functions often show a high degree of sequence similarity. Such findings suggest a common ancestry for these proteins.

Oxygen-Binding Heme Proteins Myoglobin and the α- and β-globin chains of hemoglobin constitute a set of **paralogous proteins. Myoglobin,** the oxygen-binding heme protein of muscle, consists of a single polypeptide chain of 153 residues. **Hemoglobin,** the oxygen transport protein of erythrocytes, is a tetramer composed of two α-**chains** (141 residues each) and two β-**chains** (146 residues each). These globin paralogs— myoglobin, α-globin, and β-globin—share a strong degree of sequence homology (Figure 5.21). Human myoglobin and the human α-globin chain show 38 amino acid identities, whereas human α-globin and human β-globin have 64 residues in common. The relatedness suggests an evolutionary sequence of events in which chance mutations led to amino acid substitutions and divergence in primary structure. The ancestral myoglobin gene diverged first, after duplication of a primordial globin gene had given rise to its progenitor and an ancestral hemoglobin gene (Figure 5.22). Subsequently, the ancestral hemoglobin gene duplicated to generate the progenitors of the present-day α-globin and β-globin genes. The ability to bind O_2 via a heme prosthetic group is retained by all three of these polypeptides.

Serine Proteases Whereas the globins provide an example of gene duplication giving rise to a set of proteins in which the biological function has been highly conserved, other sets of proteins united by strong sequence homology show more divergent biological functions. **Trypsin, chymotrypsin** (see Chapter 14), and **elastase** are members of a class of proteolytic enzymes called **serine proteases** because of the central role played by specific serine residues in their catalytic activity. **Thrombin,** an essential enzyme in blood clotting, is also a serine protease. These enzymes show sufficient sequence homology to conclude that they arose via duplication of a progenitor serine protease gene, even though their substrate preferences are now quite different.

Apparently Different Proteins May Share a Common Ancestry

A more remarkable example of evolutionary relatedness is inferred from sequence homology between hen egg white **lysozyme** and human milk α-**lactalbumin**, proteins of different biological activity and origin. Lysozyme (129 residues) and α-lactalbumin (123 residues)

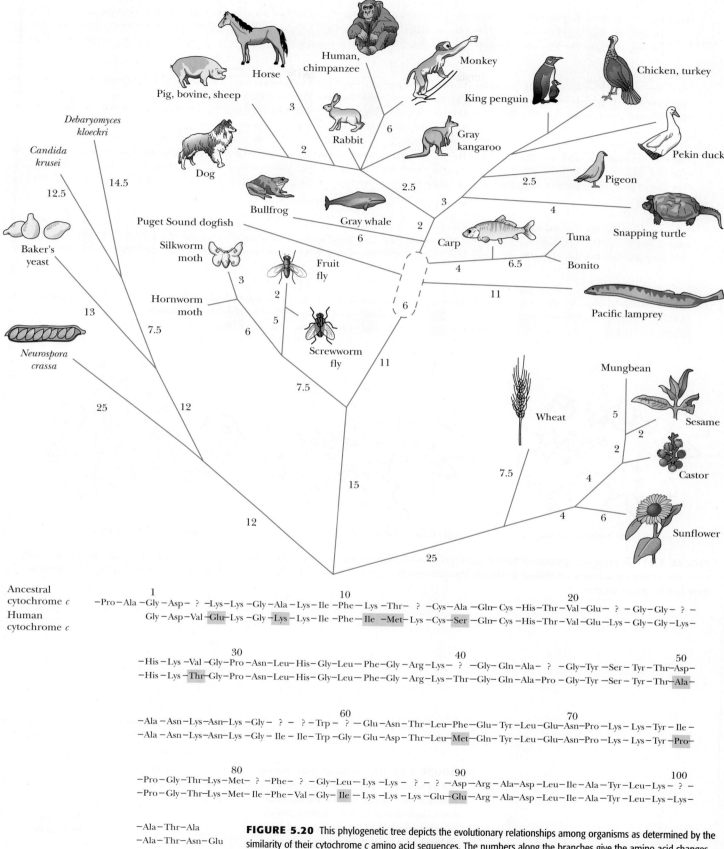

FIGURE 5.20 This phylogenetic tree depicts the evolutionary relationships among organisms as determined by the similarity of their cytochrome *c* amino acid sequences. The numbers along the branches give the amino acid changes between a species and a hypothetical progenitor. Note that extant species are located only at the tips of branches. Below, the sequence of human cytochrome *c* is compared with an inferred ancestral sequence represented by the base of the tree. Uncertainties are denoted by question marks. (Adapted from Creighton, T. E., 1983. *Proteins: Structure and Molecular Properties.* San Francisco: W. H. Freeman.)

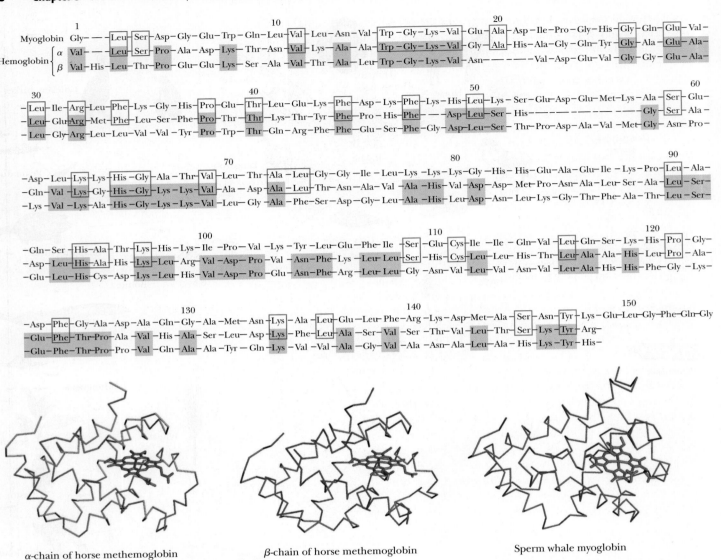

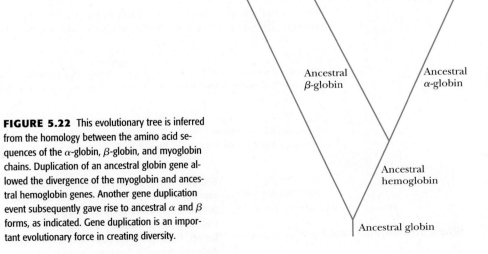

α-chain of horse methemoglobin   β-chain of horse methemoglobin   Sperm whale myoglobin

FIGURE 5.21 The amino acid sequences of the globin chains of human hemoglobin and myoglobin show a strong degree of homology. The α- and β-globin chains share 64 residues of their approximately 140 residues in common. Myoglobin and the α-globin chain have 38 amino acid sequence identities. This homology is further reflected in these proteins' tertiary structure.

FIGURE 5.22 This evolutionary tree is inferred from the homology between the amino acid sequences of the α-globin, β-globin, and myoglobin chains. Duplication of an ancestral globin gene allowed the divergence of the myoglobin and ancestral hemoglobin genes. Another gene duplication event subsequently gave rise to ancestral α and β forms, as indicated. Gene duplication is an important evolutionary force in creating diversity.

are identical at 48 positions. Lysozyme hydrolyzes the polysaccharide wall of bacterial cells, whereas α-lactalbumin regulates milk sugar (lactose) synthesis in the mammary gland. Although both proteins act in reactions involving carbohydrates, their functions show little similarity otherwise. Nevertheless, their tertiary structures are strikingly similar (Figure 5.23). It is conceivable that many proteins are related in this way, but time and the course of evolutionary change erased most evidence of their common ancestry. In contrast to this case, the proteins *G-actin* and *hexokinase* (Figure 5.24) share essentially no sequence homology, yet they have strikingly similar three-dimensional structures, even though their biological roles and physical properties are very different. Actin forms a filamentous polymer that is a principal component of the contractile apparatus in muscle; hexokinase is a cytosolic enzyme that catalyzes the first reaction in glucose catabolism.

A Mutant Protein Is a Protein with a Slightly Different Amino Acid Sequence

Given a large population of individuals, a considerable number of sequence variants can be found for a protein. These variants are a consequence of **mutations** in a gene (base substitutions in DNA) that have arisen naturally within the population. Gene mutations lead to mutant forms of the protein in which the amino acid sequence is altered at one or more positions. Many of these mutant forms are "neutral" in that the functional properties of the protein are unaffected by the amino acid substitution. Others may be nonfunctional (if loss of function is not lethal to the individual), and still others may display a range of aberrations between these two extremes. The severity of the effects on function depends on the nature of the amino acid substitution and its role in the protein. These conclusions are exemplified by the hundreds of human hemoglobin variants that have been discovered to date. Some of these are listed in Table 5.4.

A variety of effects on the hemoglobin molecule are seen in these mutants, including alterations in oxygen affinity, heme affinity, stability, solubility, and subunit interactions between the α-globin and β-globin polypeptide chains. Some variants show no apparent changes, whereas others, such as HbS, sickle-cell hemoglobin (see Chapter 15), result in serious illness. This diversity of response indicates that some amino acid changes are relatively unimportant, whereas others drastically alter one or more functions of a protein.

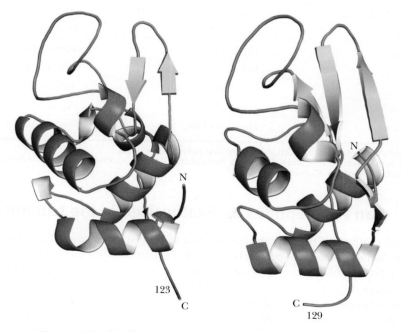

Human milk α-lactalbumin   Hen egg white lysozyme

FIGURE 5.23 The tertiary structures of hen egg white lysozyme and human α-lactalbumin are very similar.

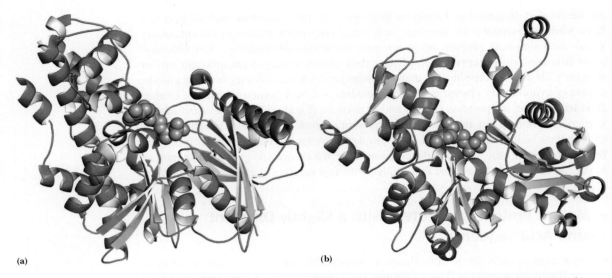

(a)   (b)

FIGURE 5.24 The tertiary structures of **(a)** hexokinase and **(b)** actin; ADP is bound to both proteins (purple).

TABLE 5.4	Some Pathological Sequence Variants of Human Hemoglobin	
Abnormal Hemoglobin*	Normal Residue and Position	Substitution
α-chain		
Torino	Phenylalanine 43	Valine
M$_{Boston}$	Histidine 58	Tyrosine
Chesapeake	Arginine 92	Leucine
G$_{Georgia}$	Proline 95	Leucine
Tarrant	Aspartate 126	Asparagine
Suresnes	Arginine 141	Histidine
β-chain		
S	Glutamate 6	Valine
Riverdale–Bronx	Glycine 24	Arginine
Genova	Leucine 28	Proline
Zurich	Histidine 63	Arginine
M$_{Milwaukee}$	Valine 67	Glutamate
M$_{Hyde Park}$	Histidine 92	Tyrosine
Yoshizuka	Asparagine 108	Aspartate
Hiroshima	Histidine 146	Aspartate

*Hemoglobin variants are often given the geographical name of their origin.
Adapted from Dickerson, R. E., and Geis, I., 1983. *Hemoglobin: Structure, Function, Evolution and Pathology.*
Menlo Park, CA: Benjamin/Cummings.

5.6 Can Polypeptides Be Synthesized in the Laboratory?

Chemical synthesis of peptides and polypeptides of defined sequence can be carried out in the laboratory. Formation of peptide bonds linking amino acids together is not a chemically complex process, but making a specific peptide can be challenging because various functional groups present on side chains of amino acids may also react under the conditions used to form peptide bonds. Furthermore, if correct sequences are to be synthesized, the $α$-COOH group of residue x must be linked to the $α$-NH$_2$ group of neighboring residue y in a way that prevents reaction of the amino group of x with the carboxyl group of y. In essence, any functional groups to be protected from reaction must be blocked while the desired coupling reactions proceed. Also, the blocking groups must be removable later under conditions in which the newly formed peptide bonds are stable. An ingenious synthetic strategy to circumvent these technical problems is *orthogonal synthesis.* An orthogonal system is defined as a set of distinctly different blocking groups—one for side-chain protection, another for α-amino protection, and a third for α-carboxyl protection or anchoring to a solid support (see following discussion). Ideally, any of the three classes of protecting groups can be removed in any order and in the presence of the other two, because the reaction chemistries of the three classes are sufficiently different from one another. In peptide synthesis, all reactions must proceed with high yield if peptide recoveries are to be acceptable. Peptide formation between amino and carboxyl groups is not spontaneous under normal conditions (see Chapter 4), so one or the other of these groups must be activated to facilitate the reaction. Despite these difficulties, biologically active peptides and polypeptides have been recreated by synthetic organic chemistry. Milestones include the pioneering synthesis of the nonapeptide posterior pituitary hormones oxytocin and vasopressin by Vincent du Vigneaud in 1953 and, in later years, larger proteins such as insulin (21 A-chain and 30 B-chain residues), ribonuclease A (124 residues), and HIV protease (99 residues).

Solid-Phase Methods Are Very Useful in Peptide Synthesis

Bruce Merrifield and his collaborators pioneered a clever solution to the problem of recovering intermediate products in the course of a synthesis. The carboxyl-terminal residues of synthesized peptide chains are covalently anchored to an insoluble resin (polystyrene particles) that can be removed from reaction mixtures simply by filtration. After each new residue is added successively at the free amino-terminus, the elongated product is recovered by filtration and readied for the next synthetic step. Because the growing peptide chain is coupled to an insoluble resin bead, the method is called **solid-phase synthesis.** The procedure is detailed in Figure 5.25. This cyclic process is automated and computer controlled so that the reactions take place in a small cup with reagents being pumped in and removed as programmed.

5.7 Do Proteins Have Chemical Groups Other Than Amino Acids?

Many proteins consist of only amino acids and contain no other chemical groups. The enzyme ribonuclease and the contractile protein actin are two such examples. Such proteins are called **simple proteins.** However, many other proteins contain various chemical constituents as an integral part of their structure. Some of these constituents arise through covalent modification of amino acid side chains in proteins after the protein has been synthesized. Such alterations are called **post-translational modifications.** For example, the reaction of two cysteine residues in a protein to form a disulfide linkage (Figure 4.8b) is a post-translational modification. Many of the prominent post-translational modifications, such as those listed in Table 5.5, can act as "on–off

TABLE 5.5	Some Prominent Post-Translational Modifications Found in Proteins		
Name	Nonprotein Part	Amino Acid Side Chain Modified	Examples
Phosphorylation	$-PO_3^{2-}$	S, T, Y	Hormone receptors, regulatory enzymes
Acetylation	$-CH_2COO^-$	K	Histones, metabolic enzymes
Methylation	$-CH_3$	K, R	Histones
Acylation	Palmitic acid	C	G-protein-coupled receptors
Prenylation	Prenyl group	C	Ras p21
ADP-ribosylation	ADP-ribose	H, R	G proteins, eukaryotic elongation factors
Adenylylation	AMP	Y	Glutamine synthetase

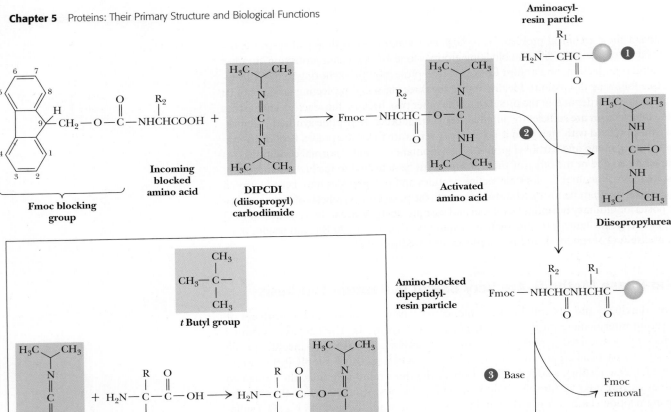

FIGURE 5.25 Solid-phase synthesis of a peptide. The 9-fluorenylmethoxycarbonyl (Fmoc) group is an excellent orthogonal blocking group for the α-amino group of amino acids during organic synthesis because it is readily removed under basic conditions that don't affect the linkage between the insoluble resin and the α-carboxyl group of the growing peptide chain. *(inset)* N,N′-diisopropylcarbodiimide (DIPCDI) is one agent of choice for activating carboxyl groups to condense with amino groups to form peptide bonds. **(1)** The carboxyl group of the first amino acid (the carboxyl-terminal amino acid of the peptide to be synthesized) is chemically attached to an insoluble resin particle (the *aminoacyl-resin particle*). **(2)** The second amino acid, with its amino group blocked by a Fmoc group and its carboxyl group activated with DIPCDI, is reacted with the aminoacyl-resin particle to form a peptide linkage, with elimination of DIPCDI as diisopropylurea. **(3)** Then, basic treatment (with piperidine) removes the N-terminal Fmoc blocking group, exposing the N-terminus of the dipeptide for another cycle of amino acid addition **(4).** Any reactive side chains on amino acids are blocked by addition of acid-labile *tertiary* butyl (*t*Bu) groups as orthogonal protective functions. **(5)** After each step, the peptide product is recovered by collection of the insoluble resin beads by filtration or centrifugation. Following cyclic additions of amino acids, the completed peptide chain is hydrolyzed from linkage to the insoluble resin by treatment with HF; HF also removes any *t*Bu protecting groups from side chains on the peptide.

TABLE 5.6	Some Common Conjugated Proteins		
Name	Nonprotein Part	Association	Examples
Lipoproteins	Lipids	Noncovalent	Blood lipoprotein complexes (HDL, LDL)
Nucleoproteins	RNA, DNA	Noncovalent	Ribosomes, chromosomes
Glycoproteins	Carbohydrate groups	Covalent	Immunoglobulins, LDL receptor
Metalloproteins and metal-activated proteins	Ca^{2+}, K^+, Fe^{2+}, Zn^{2+}, Co^{2+}, others	Covalent to noncovalent	Metabolic enzymes, kinases, phosphatases, among others
Hemoproteins	Heme group	Covalent or noncovalent	Hemoglobin, cytochromes
Flavoproteins	FMN, FAD	Covalent or noncovalent	Electron transfer enzymes

switches" that regulate the function or cellular location of the protein. Approximately 550 post-translational modifications are listed in the RESID database accessible through the National Cancer Institute Frederick Advanced Biomedical Computing Center at *http://www.ncifcrf.gov/RESID.*

A common form of post-translational modification not to be found in such a database or in Table 5.5 is the removal of amino acids from the protein by proteolytic cleavage. Many proteins localized in specific subcellular compartments have N-terminal **signal sequences** that stipulate their proper destination. Such signal sequences typically are clipped off during their journey. Other proteins, such as some hormones or potentially destructive proteases, are synthesized in an inactive form and converted into an active form through proteolytic removal of some of their amino acids.

The general term for proteins containing nonprotein constituents is **conjugated proteins** (Table 5.6). Because association of the protein with the conjugated group does not occur until the protein has been synthesized, these associations are post-translational as well, although such terminology is usually not applied to these proteins (with the possible exception of glycoproteins). As Table 5.6 indicates, conjugated proteins are typically classified according to the chemistry of the nonprotein part. If the nonprotein part participates in the protein's function, it is referred to as a **prosthetic group.** Conjugation of proteins with these different nonprotein constituents dramatically enhances the repertoire of functionalities available to proteins.

5.8 What Are the Many Biological Functions of Proteins?

Proteins are the agents of biological function. Virtually every cellular activity is dependent on one or more particular proteins. Thus, a convenient way to classify the enormous number of proteins is to group them according to the biological roles they serve. Figure 5.26 summarizes the classification of proteins found in the human **proteome** according to their function.

Proteins fill essentially every biological role, with the exception of information storage. The ability to bind other molecules (ligands) is common to many proteins. Binding proteins typically interact noncovalently with their specific ligands. Transport proteins are one class of binding proteins. Transport proteins include membrane proteins that transport substances across membranes, as well as soluble proteins that deliver specific nutrients or waste products throughout the body. Scaffold proteins are a class of binding proteins that uses protein–protein interactions to recruit other proteins into multimeric assemblies whose purpose is to mediate and coordinate the flow of information in cells. Catalytic proteins (enzymes) mediate almost every metabolic reaction. Regulatory proteins that bind to specific nucleotide sequences within DNA control gene expression. Hormones are another kind of regulatory protein in that they convey information about the environment and deliver this information to cells when they bind to specific receptors. Switch proteins such as G-proteins can switch between two conformational states—an "on" state and an "off" state—and act via this conformational switching as regulatory proteins. Structural proteins give form to cells and subcellular structures. The great diversity in function that characterizes biological systems is based on attributes that proteins possess.

Proteome is the complete catalog of proteins encoded by a genome; in cell-specific terms, a proteome is the complete set of proteins found in a particular cell type at a particular time.

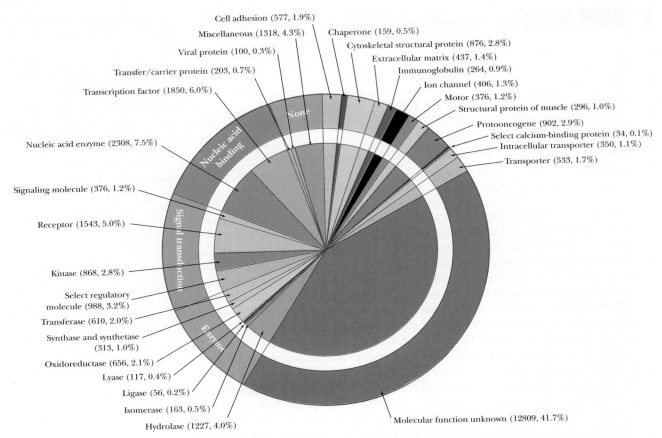

FIGURE 5.26 Proteins of the human genome grouped according to their molecular function, as of 2001. The numbers and percentages within each functional category are enclosed in parentheses. Note that the function of more than 40% of the proteins encoded by the human genome was listed as unknown. Considering those of known function, enzymes (including kinases and nucleic acid enzymes) account for about 20% of the total number of proteins; nucleic acid–binding proteins of various kinds, about 14%, among which almost half are gene-regulatory proteins (transcription factors). Transport proteins collectively constitute about 5% of the total; and structural proteins, another 5%. (Adapted from Figure 15 in Venter, J. C., et al., 2001. The sequence of the human genome. *Science* **291**:1304–1351.)

All Proteins Function through Specific Recognition and Binding of Some Target Molecule Although the classification of proteins according to function has advantages, many proteins are not assigned readily to one of the traditional groupings. Further, classification can be somewhat arbitrary, because many proteins fit more than one category. However, for all categories, the protein always functions through specific recognition and binding of some other molecule, although for structural proteins, it is usually self-recognition and assembly into stable multimeric arrays. Protein behavior provides the cardinal example of *molecular recognition through structural complementarity,* a fundamental principle of biochemistry that was presented in Chapter 1.

Protein Binding The interaction of a protein with its target often can be described in simple quantitative terms. Let's explore the simplest situation in which a protein has a single binding site for the molecule it binds (its **ligand;** Chapter 1). If we treat the interaction between the protein (P) and the ligand (L) as a dissociation reaction:

$$PL \rightleftharpoons P + L.$$

The equilibrium constant for the reaction as written

$$K_{eq} = [P][L]/[PL]$$

is a **dissociation constant,** because it describes the dissociation of the ligand from the protein. Biochemists typically use dissociation constants (K_D) to describe binding

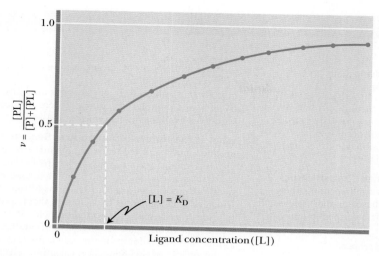

FIGURE 5.27 Saturation curve or binding isotherm.

phenomena. Because brackets ([ ]) denote molar concentrations, dissociation constants have the units of M.

Typically, the ligand concentration is much greater than the protein concentration. Under such conditions, a plot of the moles of ligand bound per mole of protein (defined as $[PL]/([P] + [PL])$) versus $[L]$ yields a hyperbolic curve known as a **saturation curve** or **binding isotherm** (Figure 5.27).

If we define the fractional saturation of P with L, $[PL]/([P] + [PL])$, as ν, a little algebra yields

$$\nu = [L]/(K_D + [L]).$$

Thus, when $\nu = 0.5$,

$$[L] = K_D$$

That is, the concentration of L where half the protein has L bound is equal to the value of K_D. The smaller this number is, the better the ligand binds to the protein; that is, a small K_D means that the protein is half-saturated with L at a low concentration of L. In other words, if K_D is small, the protein binds the ligand avidly. Typical K_D values fall in a range from $10^{-3} M$ to $10^{-12} M$.

The Ligand-Binding Site Ligand binding occurs through noncovalent interactions between the protein and ligand. The lack of covalent interactions means that binding is readily reversible. Proteins display specificity in ligand binding because they possess a specific site, the **binding site,** within their structure that is complementary to the structure of the ligand, its charge distribution, and any H-bond donors or acceptors it might have. Structural complementarity within the binding site is achieved because part of the three-dimensional structure of the protein provides an ensemble of amino acid side chains (and polypeptide backbone atoms) that establish an interactive cavity complementary to the ligand molecule. When a ligand binds to the protein, the protein usually undergoes a conformational change. This new protein conformation provides an even better fit with the ligand than before. Such changes are called **ligand-induced conformational changes,** and the result is an even more stable interaction between the protein and its ligand.

Thus, in a general sense, most proteins are binding proteins because ligand binding is a hallmark of protein function. Catalytic proteins (enzymes) bind substrates; regulatory proteins bind hormones or other proteins or regulatory sequences in genes; structural proteins bind to and interact with each other; and the many types of transport proteins bind ligands, facilitating their movement from one place to another. Many proteins accomplish their function through the binding of other protein molecules, a phenomenon called **protein–protein interaction.** Some proteins engage in protein–protein interactions with proteins that are similar or identical to themselves so that an oligomeric structure is formed, as in hemoglobin. Other proteins engage in protein–protein interactions with proteins that are very different from themselves, as in the anchoring proteins or the scaffolding proteins of signaling pathways.

⚫WL Login to OWL to develop problem-solving skills and complete online homework assigned by your professor.

The primary structure (the amino acid sequence) of a protein is encoded in DNA in the form of a nucleotide sequence. Expression of this genetic information is realized when the polypeptide chain is synthesized and assumes its functional, three-dimensional architecture. Proteins are the agents of biological function.

5.1 What Architectural Arrangements Characterize Protein Structure? Proteins are generally grouped into three fundamental structural classes—soluble, fibrous, and membrane—based on their shape and solubility. In more detail, protein structure is described in terms of a hierarchy of organization:

Primary (1°) structure—the protein's amino acid sequence

Secondary (2°) structure—regular elements of structure (helices, sheets) within the protein created by hydrogen bonds

Tertiary (3°) structure—the folding of the polypeptide chain in three-dimensional space

Quaternary (4°) structure—the subunit organization of multimeric proteins

The three higher levels of protein structure form and are maintained exclusively through noncovalent interactions.

5.2 How Are Proteins Isolated and Purified from Cells? Cells contain thousands of different proteins. A protein of choice can be isolated and purified from such complex mixtures by exploiting two prominent physical properties: size and electrical charge. A more direct approach is to employ affinity purification strategies that take advantage of the biological function or specific recognition properties of a protein. A typical protein purification strategy will use a series of separation methods to obtain a pure preparation of the desired protein.

5.3 How Is the Amino Acid Analysis of Proteins Performed? Acid treatment of a protein hydrolyzes all of the peptide bonds, yielding a mixture of amino acids. Chromatographic analysis of this hydrolysate reveals the amino acid composition of the protein. Proteins vary in their amino acid composition, but most proteins contain at least one of each of the 20 common amino acids. To a very rough approximation, proteins contain about 30% charged amino acids and about 30% hydrophobic amino acids (when aromatic amino acids are included in this number), the remaining being polar, uncharged amino acids.

5.4 How Is the Primary Structure of a Protein Determined? The primary structure (amino acid sequence) of a protein can be determined by a variety of chemical and enzymatic methods. Alternatively, mass spectroscopic methods can also be used. In the chemical and enzymatic protocols, a pure polypeptide chain whose disulfide linkages have been broken is the starting material. Methods that identify the N-terminal and C-terminal residues of the chain are used to determine which amino acids are at the ends, and then the protein is cleaved into defined sets of smaller fragments using enzymes such as trypsin or chymotrypsin or chemical cleavage by agents such as cyanogen bromide. The sequences of these fragments can be obtained by Edman degradation. Edman degradation is a powerful method for stepwise release and sequential identification of amino acids from the N-terminus of the polypeptide. The amino acid sequence of the entire protein can be reconstructed once the sequences of overlapping sets of peptide fragments are known. In mass spectrometry, an ionized protein chain is broken into an array of overlapping fragments. Small differences in the masses of the individual amino acids lead to small differences in the masses of the fragments, and the ability of mass spectrometry to measure mass-to-charge ratios very accurately allows computer deconvolution of the data into an amino acid sequence. The amino acid sequences of more than a million different proteins are known. The vast majority of these amino acid sequences were deduced from nucleotide sequences available in genomic databases.

5.5 What Is the Nature of Amino Acid Sequences? Proteins have unique amino acid sequences, and similarity in sequence between proteins implies evolutionary relatedness. Homologous proteins share sequence similarity and show structural resemblance. These relationships can be used to trace evolutionary histories of proteins and the organisms that contain them, and the study of such relationships has given rise to the field of molecular evolution. Related proteins, such as the oxygen-binding proteins of myoglobin and hemoglobin or the serine proteases, share a common evolutionary origin. Sequence variation within a protein arises from mutations that result in amino acid substitution, and the operation of natural selection on these sequence variants is the basis of evolutionary change. Occasionally, a sequence variant with a novel biological function may appear, upon which selection can operate.

5.6 Can Polypeptides Be Synthesized in the Laboratory? It is possible, although difficult, to synthesize proteins in the laboratory. The major obstacles involve joining desired amino acids to a growing chain using chemical methods that avoid side reactions and the creation of undesired products, such as the modification of side chains or the addition of more than one residue at a time. Solid-state techniques along with orthogonal protection methods circumvent many of these problems, and polypeptide chains having more than 100 amino acid residues have been artificially created.

5.7 Do Proteins Have Chemical Groups Other Than Amino Acids? Although many proteins are composed of just amino acids, other proteins undergo post-translational modifications to certain amino acid side chains. These modifications often regulate the function of the proteins. In addition, many proteins are conjugated with various other chemical components, including carbohydrates, lipids, nucleic acids, metal and other inorganic ions, and a host of novel structures such as heme or flavin. Association with these nonprotein substances dramatically extends the physical and chemical properties that proteins possess, in turn creating a much greater repertoire of functional possibilities.

5.8 What Are the Many Biological Functions of Proteins? Proteins are the agents of biological function. Their ability to bind various ligands is intimately related to their function and thus forms the basis of most classification schemes. Transport proteins bind molecules destined for transport across membranes or around the body. Enzymes bind the reactants unique to the reactions they catalyze. Regulatory proteins are of two general sorts: those that bind small molecules that are physiological or environmental cues, such as hormone receptors, or those that bind to DNA and regulate gene expression, such as transcription activators. These are just a few prominent examples. Indeed, the great diversity in function that characterizes biological systems is based on the attributes that proteins possess. Proteins usually interact noncovalently with their ligands, and often the interaction can be defined in simple quantitative terms by a protein-ligand dissociation constant. Proteins display specificity in ligand binding because the structure of the protein's ligand-binding site is complementary to the structure of the ligand. Some proteins act through binding other proteins. Such protein–protein interactions lie at the heart of many biological functions.

FOUNDATIONAL BIOCHEMISTRY Things You Should Know After Reading Chapter 5.

- Proteins fall into three general classes on the basis of shape and solubility: fibrous, globular, and membrane.

- Protein structure is described in terms of four levels of organization:
  primary (1°)—amino acid sequence
  secondary (2°)—patterns arising by hydrogen bonding
  tertiary (3°)—folding of the polypeptide chain
  quaternary (4°)—subunit organization.

- Noncovalent forces underlie the higher (2°, 3°, 4°) levels of protein structure.

- The overall 3-dimensional structure of a protein is called its conformation.

- Proteins can be purified from cellular sources by exploiting their size, charge, and affinity differences.

- A protein's amino acid composition can be determined following acid hydrolysis and chromatographic analysis of the amino acids released.

- The amino acid compositions of proteins differ, but soluble proteins tend to have about one-third charged amino acid residues, one-third hydrophobic, and one-third polar but uncharged.

- The amino acid sequence of a protein is the defining characteristic of a protein.

- Chemical and enzymatic methodologies can be used to determine the sequence of amino acids along a polypeptide chain.

- The amino acid sequence of a protein can also be determined by mass spectrometric analysis.

- Most of the available amino acid sequence information for proteins was obtained from nucleotide sequence analysis of genes and genomes.

- Homologous proteins share sequence similarity and structural resemblance and often perform similar functions across different organisms.

- Homologous proteins can be subclassified as orthologous or paralogous:
    Orthologous proteins are proteins that perform the same function in different cells.
    Paralogous proteins are proteins found within the same species that have homologous amino acid sequences; paralogous proteins arise through gene duplication.

- Related proteins shared a common evolutionary origin.

- A mutant protein is a protein with a slightly different amino acid sequence.

- Polypeptides of defined sequence can be chemically synthesized; the most effective strategy for chemical synthesis relies on solid-phase methods.

- Post-translational modification of proteins can introduce new properties into proteins.

- Proteins may have other chemical groups attached to them, such as carbohydrates to form glycoproteins.

- Proteins are the agents of biological function.

- All proteins function through recognition and binding of some target molecule.

PROBLEMS

Answers to all problems are at the end of this book. Detailed solutions are available in the *Student Solutions Manual, Study Guide, and Problems Book.*

1. **Calculating the Molecular Weight and Subunit Organization of a Protein From Its Metal Content** The element molybdenum (atomic weight 95.95) constitutes 0.08% of the weight of nitrate reductase. If the molecular weight of nitrate reductase is 240,000, what is its likely quaternary structure?

2. **Solving the Sequence of an Oligopeptide From Sequence Analysis Data** Amino acid analysis of an oligopeptide 7 residues long gave

    Asp   Leu   Lys   Met   Phe   Tyr

   The following facts were observed:
   a. Trypsin treatment had no apparent effect.
   b. The phenylthiohydantoin released by Edman degradation was

   c. Brief chymotrypsin treatment yielded several products, including a dipeptide and a tetrapeptide. The amino acid composition of the tetrapeptide was Leu, Lys, and Met.
   d. Cyanogen bromide treatment yielded a dipeptide, a tetrapeptide, and free Lys.

   What is the amino acid sequence of this heptapeptide?

3. **Solving the Sequence of an Oligopeptide From Sequence Analysis Data** Amino acid analysis of a decapeptide revealed the presence of the following products:

NH_4^+	Asp	Glu	Tyr	Arg
Met	Pro	Lys	Ser	Phe

   The following facts were observed:
   a. Neither carboxypeptidase A or B treatment of the decapeptide had any effect.
   b. Trypsin treatment yielded two tetrapeptides and free Lys.
   c. Clostripain treatment yielded a tetrapeptide and a hexapeptide.
   d. Cyanogen bromide treatment yielded an octapeptide and a dipeptide of sequence NP (using the one-letter codes).
   e. Chymotrypsin treatment yielded two tripeptides and a tetrapeptide. The N-terminal chymotryptic peptide had a net charge of −1 at neutral pH and a net charge of −3 at pH 12.
   f. One cycle of Edman degradation gave the PTH derivative

   What is the amino acid sequence of this decapeptide?

4. **Solving The Sequence of an Oligopeptide from Sequence Analysis Data** Analysis of the blood of a catatonic football fan revealed large concentrations of a psychotoxic octapeptide. Amino acid analysis of this octapeptide gave the following results:
   2 Ala   1 Arg   1 Asp   1 Met   2 Tyr   1 Val   1 NH_4^+

The following facts were observed:

a. Partial acid hydrolysis of the octapeptide yielded a dipeptide of the structure

b. Chymotrypsin treatment of the octapeptide yielded two tetrapeptides, each containing an alanine residue.

c. Trypsin treatment of one of the tetrapeptides yielded two dipeptides.

d. Cyanogen bromide treatment of another sample of the same tetrapeptide yielded a tripeptide and free Tyr.

e. N-terminal analysis of the other tetrapeptide gave Asn.

What is the amino acid sequence of this octapeptide?

5. Solving The Sequence of an Oligopeptide from Sequence Analysis Data Amino acid analysis of an octapeptide revealed the following composition:

2 Arg   1 Gly   1 Met   1 Trp   1 Tyr   1 Phe   1 Lys

The following facts were observed:

a. Edman degradation gave

b. CNBr treatment yielded a pentapeptide and a tripeptide containing phenylalanine.

c. Chymotrypsin treatment yielded a tetrapeptide containing a C-terminal indole amino acid and two dipeptides.

d. Trypsin treatment yielded a tetrapeptide, a dipeptide, and free Lys and Phe.

e. Clostripain yielded a pentapeptide, a dipeptide, and free Phe.

What is the amino acid sequence of this octapeptide?

6. Solving the Sequence of an Oligopeptide from Sequence Analysis Data Amino acid analysis of an octapeptide gave the following results:

1 Ala   1 Arg   1 Asp   1 Gly   3 Ile   1 Val   1 NH_4^+

The following facts were observed:

a. Trypsin treatment yielded a pentapeptide and a tripeptide.

b. Chemical reduction of the free α-COOH and subsequent acid hydrolysis yielded 2-aminopropanol.

c. Partial acid hydrolysis of the tryptic pentapeptide yielded, among other products, two dipeptides, each of which contained C-terminal isoleucine. One of these dipeptides migrated as an anionic species upon electrophoresis at neutral pH.

d. The tryptic tripeptide was degraded in an Edman sequenator, yielding first **A,** then **B:**

What is an amino acid sequence of the octapeptide? Four sequences are possible, but only one suits the authors. Why?

7. Solving the Sequence of an Oligopeptide from Sequence Analysis Data An octapeptide consisting of 2 Gly, 1 Lys, 1 Met, 1 Pro, 1 Arg, 1 Trp, and 1 Tyr was subjected to sequence studies. The following was found:

a. Edman degradation yielded

b. Upon treatment with carboxypeptidases A, B, and Y, only carboxypeptidase Y had any effect.

c. Trypsin treatment gave two tripeptides and a dipeptide.

d. Chymotrypsin treatment gave two tripeptides and a dipeptide. Acid hydrolysis of the dipeptide yielded only Gly.

e. Cyanogen bromide treatment yielded two tetrapeptides.

f. Clostripain treatment gave a pentapeptide and a tripeptide.

What is the amino acid sequence of this octapeptide?

8. Solving the Sequence of an Oligopeptide from Sequence Analysis Data Amino acid analysis of an oligopeptide containing nine residues revealed the presence of the following amino acids:

Arg   Cys   Gly   Leu   Met   Pro   Tyr   Val

The following was found:

a. Carboxypeptidase A treatment yielded no free amino acid.

b. Edman analysis of the intact oligopeptide released

c. Neither trypsin nor chymotrypsin treatment of the nonapeptide released smaller fragments. However, combined trypsin and chymotrypsin treatment liberated free Arg.

d. CNBr treatment of the 8-residue fragment left after combined trypsin and chymotrypsin action yielded a 6-residue fragment containing Cys, Gly, Pro, Tyr, and Val; and a dipeptide.

e. Treatment of the 6-residue fragment with β-mercaptoethanol yielded two tripeptides. Brief Edman analysis of the tripeptide mixture yielded only PTH-Cys. (The sequence of each tripeptide, as read from the N-terminal end, is alphabetical if the one-letter designation for amino acids is used.)

What is the amino acid sequence of this nonapeptide?

9. Describe the Solid-Phase Chemical Synthesis of a Small Peptide Describe the synthesis of the dipeptide Lys-Ala by Merrifield's solid-phase chemical method of peptide synthesis. What pitfalls might be encountered if you attempted to add a leucine residue to Lys-Ala to make a tripeptide?

10. Identify Proteins Using BLAST Searches of Peptide Fragment Sequences Go to the National Center for Biotechnology Information Web site at *http://www.ncbi.nlm.nih.gov/.* From the menu of Popular Resources on the right-hand side, click on "BLAST." Under the Basic BLAST heading on the new page that comes up, click on "protein blast." In the Enter Query Sequence box at the top of the page that comes up, enter the following sequence: NGMMKSSRNLT-KDRCK. Confirm that the database under "Choose Search Set" is set on "nr" (non-redundant protein sequences), then click the BLAST button at the bottom of the page to see the results of your search. Next, enter this sequence from a different protein: SLQTASAPD-VYAIGECA. Identify the protein from which this sequence was derived.

11. Calculate The Mass of a Protein From Mass Spectrometric *M/Z* Values Electrospray ionization mass spectrometry (ESI-MS) of the polypeptide chain of myoglobin yielded a series of *m/z* peaks (similar to those shown in Figure 5.14 for aerolysin K). Two successive peaks had *m/z* values of 1304.7 and 1413.2, respectively. Calculate the mass of the myoglobin polypeptide chain from these data.

12. Phosphorylation of Proteins Introduces New Properties Phosphoproteins are formed when a phosphate group is esterified to an —OH group of a Ser, Thr, or Tyr side chain. At typical cellular pH values, this phosphate group bears two negative charges ($-OPO_3^{2-}$). Compare this side-chain modification to the 20 side chains of the common amino acids found in proteins and comment on the novel properties that it introduces into side-chain possibilities.

13. Using Graphical Analysis to Determine the K_d for the Interaction Between a Protein and its Ligand A quantitative study of the interaction of a protein with its ligand yielded the following results:

Ligand concentration (mM)	1	2	3	4	5	6	9	12
ν (moles of ligand bound per mole of protein)	0.28	0.45	0.56	0.60	0.71	0.75	0.79	0.83

Plot a graph of [L] versus ν. Determine K_D, the dissociation constant for the interaction between the protein and its ligand, from the graph.

Biochemistry on the Web

14. Exploring the ExPASy Proteomics Web Site The human insulin receptor substrate-1 (IRS-1) is designated protein P35568 in the protein knowledge base on the ExPASy Web Site *(http://us.expasy .org/)*. Go to the PeptideMass tool on this Web site and use it to see the results of trypsin digestion of IRS-1. How many amino acids does IRS-1 have? What is the average molecular mass of IRS-1? What is the amino acid sequence of the tryptic peptide of IRS-1 that has a mass of 1741.9629?

Preparing for the MCAT® Exam

15. Proteases such as trypsin and chymotrypsin cleave proteins at different sites, but both use the same reaction mechanism. Based on your knowledge of organic chemistry, suggest a "universal" protease reaction mechanism for hydrolysis of the peptide bond.

16. Table 5.4 presents some of the many known mutations in the genes encoding the α- and β-globin subunits of hemoglobin.

   a. Some of these mutations affect subunit interactions between the subunits. In an examination of the tertiary structure of globin chains, where would you expect to find amino acid changes in mutant globins that affect formation of the hemoglobin $\alpha_2\beta_2$ quaternary structure?

   b. Other mutations, such as the S form of the β-globin chain, increase the tendency of hemoglobin tetramers to polymerize into very large structures. Where might you expect the amino acid substitutions to be in these mutants?

FURTHER READING

General References on Protein Structure and Function

Creighton, T. E., 1992. *Proteins: Structure and Molecular Properties,* 2nd ed. San Francisco: W. H. Freeman and Co.

Creighton, T. E., ed., 1997. *Protein Function—A Practical Approach,* 2nd ed. Oxford: CRI. Press at Oxford University Press.

Fersht, A., 1999. *Structure and Mechanism in Protein Science.* New York: W. H. Freeman and Co.

Lesk, A. M., 2010. *Introduction to Protein Science: Architecture, Function, and Genomics,* 2nd ed. Oxford: Oxford University Press.

Petsko, G. A., and Ringe, D., 2004. *Protein Structure and Function.* Sunderland, MA: Sinauer Associates.

Whitford, D., 2005. *Proteins: Structure and Function.* Hoboken, NJ: John Wiley & Sons Inc.

Protein Purification

Ahmed, H., 2005. *Principles and Reactions of Protein Extraction.* Boca Raton, FL: CRC Press.

Dennison, C., 1999. *A Guide to Protein Isolation.* Norwell, MA: Kluwer Academic Publish.

Linn, S., 2009. Strategies and considerations for protein purifications. *Methods in Enzymology* **463:**9–19. Volume 463 of *Methods in Enzymology* covers many aspects of protein purification. This reference and the next are but two of the relevant chapters.

Noble, J. E., and Bailey, M. J. A., 2009. Quantitation of protein. *Methods in Enzymology* **463:**73–95.

Amino Acid Sequence Analysis

Dayhoff, M. O., 1972–1978. *The Atlas of Protein Sequence and Structure,* Vols. 1–5. Washington, DC: National Medical Research Foundation.

Hsieh, Y. L., et al., 1996. Automated analytical system for the examination of protein primary structure. *Analytical Chemistry* **68:**455–462. An analytical system is described in which a protein is purified by affinity chromatography, digested with trypsin, and its peptides separated by HPLC and analyzed by tandem MS in order to determine its amino acid sequence.

Karger, B. L., and Hancock, W. S., eds. 1996. High resolution separation and analysis of biological macromolecules. Part B: Applications. *Methods in Enzymology* **271.** New York: Academic Press. Sections on liquid chromatography, electrophoresis, capillary electrophoresis, mass spectrometry, and interfaces between chromatographic and electrophoretic separations of proteins followed by mass spectrometry of the separated proteins.

von Heijne, G., 1987. *Sequence Analysis in Molecular Biology: Treasure Trove or Trivial Pursuit?* San Diego: Academic Press.

Mass Spectrometry

Bienvenut, W. V., 2005. Introduction: Proteins analysis using mass spectrometry. In *Acceleration and Improvement of Protein Identification by Mass Spectrometry,* pp. 1–138. Norwell, MA: Springer.

Burlingame, A. L., ed., 2005. Biological mass spectrometry. In *Methods in Enzymology* **405.** New York: Academic Press.

Hamdan, M., and Gighetti, P. G., 2005. *Proteomics Today.* Hoboken, NJ: John Wiley & Sons.

Hernandez, H., and Robinson, C. V., 2001. Dynamic protein complexes: Insights from mass spectrometry. *Journal of Biological Chemistry* **276:**46685–46688. Advances in mass spectrometry open a new view onto the dynamics of protein function, such as protein–protein interactions and the interaction between proteins and their ligands.

Hunt, D. F., et al., 1987. Tandem quadrupole Fourier transform mass spectrometry of oligopeptides and small proteins. *Proceedings of the National Academy of Sciences, U.S.A.* **84:**620–623.

Johnstone, R. A. W., and Rose, M. E., 1996. *Mass Spectrometry for Chemists and Biochemists,* 2nd ed. Cambridge, England: Cambridge University Press.

Kamp, R. M., Cakvete, J. J., and Choli-Papadopoulou, T., eds., 2004. *Methods in Proteome and Protein Analysis.* New York: Springer.

Karger, B. L., and Hancock, W. S., eds. 1996. High resolution separation and analysis of biological macromolecules. Part A: Fundamentals. In *Methods in Enzymology* **270.** New York: Academic Press. Separate sections discussing liquid chromatography, columns and instrumentation, electrophoresis, capillary electrophoresis, and mass spectrometry.

Kinter, M., and Sherman, N. E., 2001. *Protein Sequencing and Identification Using Tandem Mass Spectrometry.* Hoboken, NJ: Wiley-Interscience.

Liebler, D. C., 2002. *Introduction to Proteomics.* Towata, NJ: Humana Press. An excellent primer on proteomics, protein purification methods, sequencing of peptides and proteins by mass spectrometry, and identification of proteins in a complex mixture.

Mann, M., and Wilm, M., 1995. Electrospray mass spectrometry for protein characterization. *Trends in Biochemical Sciences* **20:**219–224. A review of the basic application of mass spectrometric methods to the analysis of protein sequence and structure.

Medzihradszky, K., 2005. Peptide sequence analysis. *Methods in Enzymology* **402:**209–244. Volume 402 of *Methods in Enzymology* has several other chapters on the techniques used in mass spectrometric sequencing of proteins.

Quadroni, M., et al., 1996. Analysis of global responses by protein and peptide fingerprinting of proteins isolated by two-dimensional electrophoresis. Application to sulfate-starvation response of *Escherichia coli. European Journal of Biochemistry* **239:**773–781. This paper describes the use of tandem MS in the analysis of proteins in cell extracts.

Stults, J. T., and Arnott, D., 2005. Proteomics. *Methods in Enzymology* **463:**245–289.

Vestling, M. M., 2003. Using mass spectrometry for proteins. *Journal of Chemical Education* **80:**122–124. A report on the 2002 Nobel Prize in Chemistry honoring the scientists who pioneered the application of mass spectrometry to protein analysis.

Solid-Phase Synthesis of Proteins

Aparicio, F., 2000. Orthogonal protecting groups for *N*-amino and C-terminal carboxyl functions in solid-phase peptide synthesis. *Biopolymers* **55:**123–139.

Fields, G. B., ed., 1997. *Solid-Phase Peptide Synthesis,* Vol. 289, *Methods in Enzymology.* San Diego: Academic Press.

Merrifield, B., 1986. Solid phase synthesis. *Science* **232:**341–347.

Wilken, J., and Kent, S. B. H., 1998. Chemical protein synthesis. *Current Opinion in Biotechnology* **9:**412–426.

6

Proteins: Secondary, Tertiary, and Quaternary Structure

National Archaeological Museum, Athens, Greece/Ancient Art and Architecture Collection Ltd./The Bridgeman Art Library

◀ Like the Greek sea god Proteus, who could assume different forms, proteins act through changes in conformation. Proteins (from the Greek *proteios*, meaning "primary") are the primary agents of biological function. ("*Proteus, Old Man of the Sea, Roman period mosaic, from Thessalonika, 1st century a.d. National Archaeological Museum, Athens/Ancient Art and Architecture Collection Ltd./Bridgeman Art Library, London/New York*)

Growing in size and complexity
Living things, masses of atoms, DNA, protein
Dancing a pattern ever more intricate.
Out of the cradle onto the dry land
Here it is standing
Atoms with consciousness
Matter with curiosity.
Stands at the sea
Wonders at wondering
I
A universe of atoms
An atom in the universe.

Richard P. Feynman (1918–1988)
From "The Value of Science" in Edward Hutchings, Jr., ed. 1958. *Frontiers of Science: A Survey.* New York: Basic Books.

ESSENTIAL QUESTION

Linus Pauling received the Nobel Prize in Chemistry in 1954. The award cited "his research into the nature of the chemical bond and its application to the elucidation of the structure of complex substances." Pauling pioneered the study of secondary structure in proteins.

How do the forces of chemical bonding determine the formation, stability, and myriad functions of proteins?

Nearly all biological processes involve the specialized functions of one or more protein molecules. Proteins function to produce other proteins, control all aspects of cellular metabolism, regulate the movement of various molecular and ionic species across membranes, convert and store cellular energy, and carry out many other activities. Essentially all of the information required to initiate, conduct, and regulate each of these functions must be contained in the structure of the protein itself. The previous chapter described the details of protein primary structure. However, proteins do not normally exist as fully extended polypeptide chains but rather as compact structures that biochemists refer to as "folded." The ability of a particular protein to carry out its function in nature is normally determined by its overall three-dimensional shape, or *conformation*.

This chapter reveals and elaborates upon the exquisite beauty of protein structures. What will become apparent in this discussion is that the three-dimensional structure of proteins and their biological function are linked by several overarching principles:

1. Function depends on structure.
2. Structure depends both on amino acid sequence and on weak, noncovalent forces.
3. The number of protein folding patterns is very large but finite.
4. The structures of globular proteins are marginally stable.
5. Marginal stability facilitates motion.
6. Motion enables function.

KEY QUESTIONS

6.1 What Noncovalent Interactions Stabilize the Higher Levels of Protein Structure?

6.2 What Role Does the Amino Acid Sequence Play in Protein Structure?

6.3 What Are the Elements of Secondary Structure in Proteins, and How Are They Formed?

6.4 How Do Polypeptides Fold into Three-Dimensional Protein Structures?

6.5 How Do Protein Subunits Interact at the Quaternary Level of Protein Structure?

6.1   What Noncovalent Interactions Stabilize the Higher Levels of Protein Structure?

The amino acid sequence (primary structure) of any protein is dictated by covalent bonds, but the higher levels of structure—secondary, tertiary, and quaternary—are formed and stabilized by weak, noncovalent interactions (Figure 6.1). Hydrogen bonds, hydrophobic interactions, electrostatic bonds, and van der Waals forces are all noncovalent in nature, yet they are extremely important influences on protein conformation. The stabilization free energies afforded by each of these interactions may be highly dependent on the local environment within the protein, but certain generalizations can still be made.

Hydrogen Bonds Are Formed Whenever Possible

Hydrogen bonds are generally made wherever possible within a given protein structure. In most protein structures that have been examined to date, component atoms of the peptide backbone tend to form hydrogen bonds with one another. Furthermore, side chains capable of forming H bonds are usually located on the protein surface and form such bonds either with the water solvent or with other surface residues. The strengths of hydrogen bonds depend to some extent on environment. The difference in energy between a side chain hydrogen bonded to water and that same side chain hydrogen bonded to another side chain is usually quite small. On the other hand, a hydrogen bond in the protein interior, away from bulk solvent, can provide substantial stabilization energy to the protein. Although each hydrogen bond may contribute an average of only a few kilojoules per mole in stabilization energy for the protein structure, the number of H bonds formed in the typical protein is very large. For example, in α-helices, the C=O and N—H groups of every interior residue participate in H bonds. The importance of H bonds in protein structure cannot be overstated.

Hydrophobic Interactions Drive Protein Folding

Hydrophobic "bonds," or, more accurately, *interactions,* form because nonpolar side chains of amino acids and other nonpolar solutes prefer to cluster in a nonpolar environment rather than to intercalate in a polar solvent such as water. The forming of hydrophobic "bonds" minimizes the interaction of nonpolar residues with water and is therefore highly favorable. Such clustering is entropically driven, and it is in fact the principal impetus for protein folding. The side chains of the amino acids in the interior or core of the protein structure are almost exclusively hydrophobic. Polar amino acids are much less common in the interior of a protein, but the protein surface may consist of both polar and nonpolar residues.

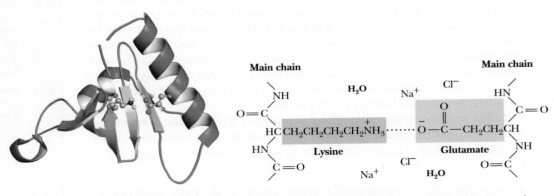

FIGURE 6.1 An electrostatic interaction between the ϵ-amino group of a lysine and the γ-carboxyl group of a glutamate. The protein is IRAK-4 kinase, an enzyme that phosphorylates other proteins (pdb id = 2NRY). The interaction shown is between Lys[213] (left) and Glu[233] (right).

Ionic Interactions Usually Occur on the Protein Surface

Ionic interactions arise either as electrostatic attractions between opposite charges or repulsions between like charges. Chapter 4 discusses the ionization behavior of amino acids. Amino acid side chains can carry positive charges, as in the case of lysine, arginine, and histidine, or negative charges, as in aspartate and glutamate. In addition, the N-terminal and C-terminal residues of a protein or peptide chain usually exist in ionized states and carry positive or negative charges, respectively. All of these may experience ionic interactions in a protein structure. Charged residues are normally located on the protein surface, where they may interact optimally with the water solvent. It is energetically unfavorable for an ionized residue to be located in the hydrophobic core of the protein. Ionic interactions between charged groups on a protein surface are often complicated by the presence of salts in the solution. For example, the ability of a positively charged lysine to attract a nearby negative glutamate may be weakened by dissolved salts such as NaCl (Figure 6.1). The Na^+ and Cl^- ions are highly mobile, compact units of charge, compared to the amino acid side chains, and thus compete effectively for charged sites on the protein. In this manner, ionic interactions among amino acid residues on protein surfaces may be damped out by high concentrations of salts. Nevertheless, these interactions are important for protein stability.

Van der Waals Interactions Are Ubiquitous

Both attractive forces and repulsive forces are included in van der Waals interactions. The attractive forces are due primarily to instantaneous dipole-induced dipole interactions that arise because of fluctuations in the electron charge distributions of adjacent nonbonded atoms. Individual van der Waals interactions are weak ones (with stabilization energies of 0.4 to 4.0 kJ/mol), but many such interactions occur in a typical protein, and by sheer force of numbers, they can represent a significant contribution to the stability of a protein. Peter Privalov and George Makhatadze have shown that, for pancreatic ribonuclease A, hen egg white lysozyme, horse heart cytochrome *c,* and sperm whale myoglobin, van der Waals interactions between tightly packed groups in the interior of the protein are a major contribution to protein stability.

6.2 | What Role Does the Amino Acid Sequence Play in Protein Structure?

It can be inferred from the first section of this chapter that many different forces work together in a delicate balance to determine the overall three-dimensional structure of a protein. These forces operate both within the protein structure itself and between the protein and the water solvent. How, then, does nature dictate the manner of protein folding to generate the three-dimensional structure that optimizes and balances these many forces? *All of the information necessary for folding the peptide chain into its "native" structure is contained in the amino acid sequence of the peptide.*

Just how proteins recognize and interpret the information that is stored in the amino acid sequence is not yet well understood. Certain loci along the peptide chain may act as nucleation points, which initiate folding processes that eventually lead to the correct structures. Regardless of how this process operates, it must take the protein correctly to the final native structure. Along the way, local energy-minimum states different from the native state itself must be avoided. A long-range goal of many researchers in the protein structure field is the prediction of three-dimensional conformation from the amino acid sequence. As the details of secondary and tertiary structure are described in this chapter, the complexity and immensity of such a prediction will be more fully appreciated. This area is one of the greatest intellectual frontiers remaining in molecular biology.

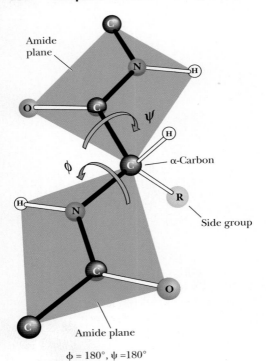

FIGURE 6.2 The amide or peptide bond planes are joined by the tetrahedral bonds of the α-carbon. The rotation parameters are ϕ and ψ. The conformation shown corresponds to $\phi = 180°$ and $\psi = 180°$. Note that positive values of ϕ and ψ correspond to clockwise rotation as viewed from C_α. Starting from 0°, a rotation of 180° in the clockwise direction ($+180°$) is equivalent to a rotation of 180° in the counterclockwise direction ($-180°$). (Illustration: Irving Geis. Rights owned by Howard Hughes Medical Institute. Not to be reproduced without permission.)

6.3 What Are the Elements of Secondary Structure in Proteins, and How Are They Formed?

Any discussion of protein folding and structure must begin with the *peptide bond,* the fundamental structural unit in all proteins. As we saw in Chapter 4, the resonance structures experienced by a peptide bond constrain six atoms—the oxygen, carbon, nitrogen, and hydrogen atoms of the peptide group, as well as the adjacent α-carbons—to lie in a plane. The resonance stabilization energy of this planar structure is approximately 88 kJ/mol, and substantial energy is required to twist the structure about the C—N bond. A twist of θ degrees involves a twist energy of $88 \sin^2\theta$ kJ/mol.

All Protein Structure Is Based on the Amide Plane

The planarity of the peptide bond means that there are only two degrees of freedom per residue for the peptide chain. Rotation is allowed about the bond linking the α-carbon with its carbonyl carbon and also about the bond linking the α-carbon with its amide nitrogen. As shown in Figure 6.2, each α-carbon is the joining point for two planes defined by peptide bonds. The angle about the C_α—N bond is denoted by the Greek letter ϕ (phi), and that about the C_α—C_o is denoted by ψ (psi). For either of these bond angles, a value of 0° corresponds to an orientation with the amide plane bisecting the H—C_α—R (side-chain) angle and a *cis* conformation of the main chain around the rotating bond in question (Figure 6.3).

The entire path of the peptide backbone in a protein is known if the ϕ and ψ rotation angles are all specified. Some values of ϕ and ψ are not allowed due to steric interference between nonbonded atoms. As shown in Figure 6.3, values of $\phi = 180°$ and $\psi = 0°$ are not allowed because of the forbidden overlap of the N—H hydrogens. Similarly, $\phi = 0°$ and $\psi = 180°$ are forbidden because of unfavorable overlap between the carbonyl oxygens.

G. N. Ramachandran and his co-workers in Madras, India, demonstrated that it was convenient to plot ϕ values against ψ values to show the distribution of allowed values

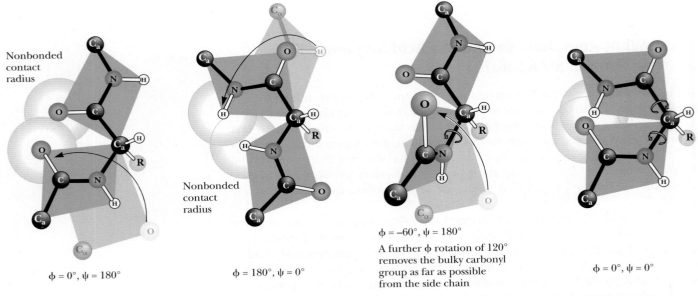

FIGURE 6.3 Many of the possible conformations about an α-carbon between two peptide planes are forbidden because of steric crowding. Several noteworthy examples are shown here.

Note: The formal IUPAC-IUB Commission on Biochemical Nomenclature convention for the definition of the torsion angles ϕ and ψ in a polypeptide chain (*Biochemistry* **9:**3471–3479, 1970) is different from that used here, where the C_α atom serves as the point of reference for both rotations, but the result is the same. (Illustration: Irving Geis. Rights owned by Howard Hughes Medical Institute. Not to be reproduced without permission.)

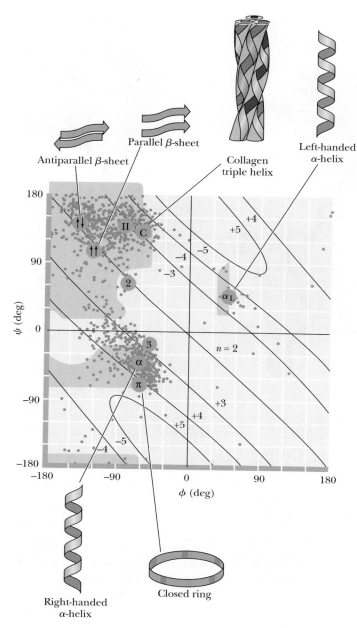

FIGURE 6.4 A Ramachandran diagram showing the sterically reasonable values of the angles ϕ and ψ. The shaded regions indicate particularly favorable values of these angles. Dots in purple indicate actual angles measured for 1000 residues (excluding glycine, for which a wider range of angles is permitted) in eight proteins. The lines running across the diagram (numbered +5 through 2 and −5 through −3) signify the number of amino acid residues per turn of the helix; "+" means right-handed helices; "−" means left-handed helices. (After Richardson, J. S., 1981. The anatomy and taxonomy of protein structure. *Advances in Protein Chemistry* **34**:167–339.)

in a protein or in a family of proteins. A typical **Ramachandran plot** is shown in Figure 6.4. Note the clustering of ϕ and ψ values in a few regions of the plot. Most combinations of ϕ and ψ are sterically forbidden, and the corresponding regions of the Ramachandran plot are sparsely populated. The combinations that are sterically allowed represent the subclasses of structure described in the remainder of this section.

The Alpha-Helix Is a Key Secondary Structure

As noted in Chapter 5, the term *secondary structure* describes local conformations of the polypeptide that are stabilized by hydrogen bonds. In nearly all proteins, the hydrogen bonds that make up secondary structures involve the amide proton of one peptide group and the carbonyl oxygen of another, as shown in Figure 6.5. These structures tend to form in cooperative fashion and involve substantial portions of the peptide chain. When a number of hydrogen bonds form between portions of the peptide chain in this manner, two basic types of structures can result: *α-helices* and *β-pleated sheets*.

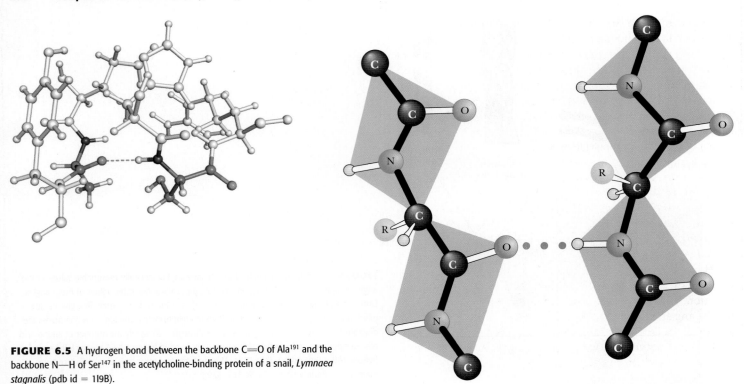

FIGURE 6.5 A hydrogen bond between the backbone C=O of Ala[191] and the backbone N—H of Ser[147] in the acetylcholine-binding protein of a snail, *Lymnaea stagnalis* (pdb id = 1I9B).

> ## A DEEPER LOOK

Knowing What the Right Hand and Left Hand Are Doing

Certain conventions related to peptide bond angles and the "handedness" of biological structures are useful in any discussion of protein structure. To determine the ϕ and ψ angles between peptide planes, viewers should imagine themselves at the C_α carbon looking outward and should imagine starting from the $\phi = 0°$, $\psi = 0°$ conformation. From this perspective, positive values of ϕ correspond to clockwise rotations about the C_α—N bond of the plane that includes the adjacent N—H group. Similarly, positive values of ψ correspond to clockwise rotations about the C_α—C bond of the plane that includes the adjacent C=O group.

Biological structures are often said to exhibit "right-hand" or "left-hand" twists. For all such structures, the sense of the twist can be ascertained by holding the structure in front of you and looking along the polymer backbone. If the twist is clockwise as one proceeds outward and through the structure, it is said to be right-handed. If the twist is counterclockwise, it is said to be left-handed.

The earliest studies of protein secondary structure were those of William Astbury at the University of Leeds. Astbury carried out X-ray diffraction studies on wool and observed differences between unstretched wool fibers and stretched wool fibers. He proposed that the protein structure in unstretched fibers was a helix (which he called the **alpha form**). He also proposed that stretching caused the helical structures to uncoil, yielding an extended structure (which he called the **beta form**). Astbury was the first to propose that hydrogen bonds between peptide groups contributed to stabilizing these structures.

In 1951, Linus Pauling, Robert Corey, and their colleagues at the California Institute of Technology summarized a large volume of crystallographic data in a set of dimensions for polypeptide chains. (A summary of data similar to what they reported is shown in Figure 4.15.) With these data in hand, Pauling, Corey, and Herman Branson proposed a new model for a helical structure in proteins, which they called the α-helix. The report from Caltech was of particular interest to Max Perutz in Cambridge, England, a crystallographer who was also interested in protein structure. By taking into account a critical but previously ignored feature of the X-ray data, Perutz realized that the α-helix existed in keratin, a protein from hair, and also in several other proteins. Since then, the α-helix has proved to be a fundamentally important peptide structure. Several representations of the α-helix are shown in Figure 6.6. One turn of the helix represents 3.6 amino acid residues. (A single turn of the α-helix involves 13 atoms from the O to the H of the H bond. For this reason, the α-helix is sometimes referred to as the 3.6_{13} helix.) This is in fact the feature that most confused crystallographers before the Pauling and Corey α-helix model. Crystallographers were so accustomed to finding twofold, threefold, sixfold, and similar integral axes in simpler molecules that the notion of a nonintegral number of units per turn was never taken seriously before Pauling and Corey's work.

Each amino acid residue extends **1.5 Å (0.15 nm)** along the helix axis. With **3.6 residues per turn,** this amounts to 3.6×1.5 Å or **5.4 Å (0.54 nm)** of travel along the helix axis per turn. This is referred to as the translation distance or the **pitch** of the helix. If one ignores side chains, the helix is about 6 Å in diameter. The side chains, extending outward from the core structure of the helix, are removed from steric interference with the polypeptide backbone. As can be seen in Figure 6.6, *each peptide carbonyl is hydrogen bonded to the peptide N—H group four residues farther up the chain.* Note that all of the H bonds lie parallel to the helix axis and all of the carbonyl groups are pointing in one direction along the helix axis while the N—H groups are pointing in the opposite direction. Recall that the entire path of the peptide backbone can be known if the ϕ and ψ twist angles are specified for each residue. The α-helix is formed if the values of ϕ are approximately $-60°$ and the values of ψ are in the range of -45 to $-50°$. Figure 6.7 shows the structures of two proteins that contain α-helical segments. The number of residues involved in a given α-helix varies from helix to helix and from protein to

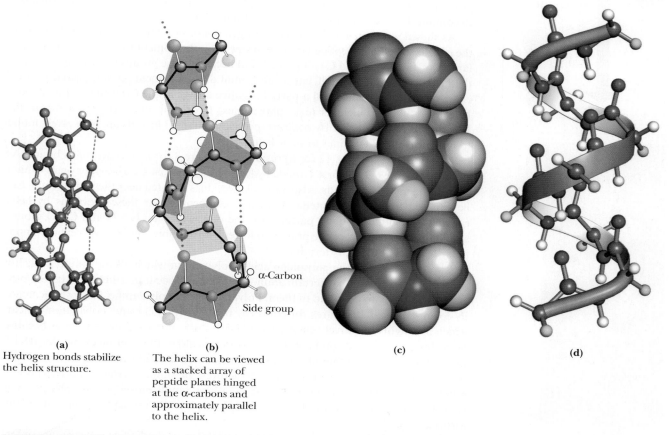

(a)
Hydrogen bonds stabilize the helix structure.

(b)
The helix can be viewed as a stacked array of peptide planes hinged at the α-carbons and approximately parallel to the helix.

α-Carbon

Side group

(c)

(d)

FIGURE 6.6 Four different graphic representations of the α-helix. **(a)** A stick representation with H bonds as dotted lines, as originally conceptualized in Pauling's 1960 *The Nature of the Chemical Bond.* **(b)** Showing the arrangement of peptide planes in the helix. (Illustration: Irving Geis. Rights owned by Howard Hughes Medical Institute. Not to be reproduced without permission.) **(c)** A space-filling computer graphic presentation. **(d)** A "ribbon structure" with an inlaid stick figure, showing how the ribbon indicates the path of the polypeptide backbone.

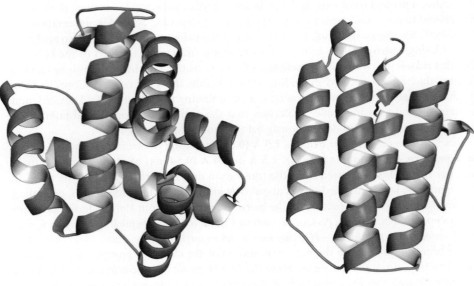

β-Hemoglobin subunit

Myohemerythrin

FIGURE 6.7 The three-dimensional structures of two proteins that contain substantial amounts of α-helix. The helices are represented by the regularly coiled sections of the ribbon drawings. Myohemerythrin is the oxygen-carrying protein in certain invertebrates, including *Sipunculids*, a phylum of marine worm. β-Hemoglobin subunit: pdb id = 1HGA; myohemerythrin pdb id = 1A7D.

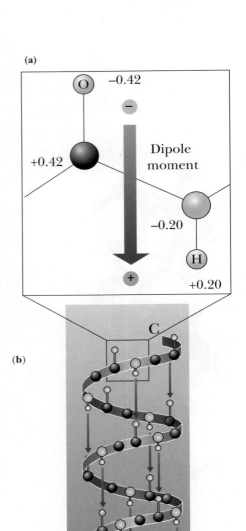

(a)

(b)

FIGURE 6.8 The arrangement of N—H and C=O groups (each with an individual dipole moment) along the helix axis creates a large net dipole for the helix. Numbers indicate fractional charges on respective atoms.

protein. On average, there are about 10 residues per helix. Myoglobin, one of the first proteins in which α-helices were observed, has eight stretches of α-helix that form a box to contain the heme prosthetic group (see Figure 5.1).

As shown in Figure 6.6, all of the hydrogen bonds point in the same direction along the α-helix axis. Each peptide bond possesses a dipole moment that arises from the polarities of the N—H and C=O groups, and because these groups are all aligned along the helix axis, the helix itself has a substantial dipole moment, with a partial positive charge at the N-terminus and a partial negative charge at the C-terminus (Figure 6.8). Negatively charged ligands (e.g., phosphates) frequently bind to proteins near the N-terminus of an α-helix. By contrast, positively charged ligands are only rarely found to bind near the C-terminus of an α-helix.

In a typical α-helix of 12 (or n) residues, there are 8 (or $n - 4$) hydrogen bonds. As shown in Figure 6.9, the first 4 amide hydrogens and the last 4 carbonyl oxygens cannot participate in helix H bonds. Also, nonpolar residues situated near the helix termini can be exposed to solvent. Proteins frequently compensate for these problems by **helix capping**—providing H-bond partners for the otherwise bare N—H and C=O groups and folding other parts of the protein to foster hydrophobic contacts with exposed nonpolar residues at the helix termini.

Careful studies of the **polyamino acids,** polymers in which all the amino acids are identical, have shown that certain amino acids tend to occur in α-helices, whereas others are less likely to be found in them. Polyleucine and polyalanine, for example, readily form α-helical structures. In contrast, polyaspartic acid and polyglutamic acid, which are highly negatively charged at pH 7.0, form only random structures because of strong charge repulsion between the R groups along the peptide chain. At pH 1.5 to 2.5, however, where the side chains are protonated and thus uncharged, these latter species spontaneously form α-helical structures. In similar fashion, polylysine is a random coil at pH values below about 11, where repulsion of positive charges prevents helix formation. At pH 12, where polylysine is a neutral peptide chain, it readily forms an α-helix.

The tendencies of various amino acids to stabilize or destabilize α-helices are different in typical proteins than in polyamino acids. The occurrence of the common amino acids in helices is summarized in Table 6.1. Notably, proline (and hydroxyproline) act as helix breakers due to their unique structure, which limits the value of ϕ (i.e., rotation about the C_α—N bond). Helices can be formed from either D- or L-amino acids, but a

TABLE 6.1	Helix-Forming and Helix-Breaking Behavior of the Amino Acids		
Amino Acid		**Helix Behavior***	
A	Ala	H	(I)
C	Cys	Variable	
D	Asp	Variable	
E	Glu	H	
F	Phe	H	
G	Gly	I	(B)
H	His	H	(I)
I	Ile	H	(C)
K	Lys	Variable	
L	Leu	H	
M	Met	H	
N	Asn	C	(I)
P	Pro	B	
Q	Gln	H	(I)
R	Arg	H	(I)
S	Ser	C	(B)
T	Thr	Variable	
V	Val	Variable	
W	Trp	H	(C)
Y	Tyr	H	(C)

*H = helix former; I = indifferent; B = helix breaker; C = random coil; ( ) = secondary tendency.

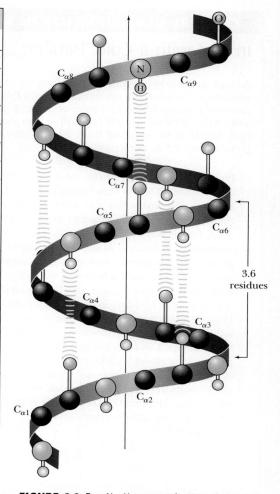

FIGURE 6.9 Four N—H groups at the N-terminal end of an α-helix and four C=O groups at the C-terminal end lack partners for H-bond formation. The formation of H bonds with other nearby donor and acceptor groups is referred to as **helix capping.** Capping may also involve appropriate hydrophobic interactions that accommodate nonpolar side chains at the ends of helical segments.

given helix must be composed entirely of amino acids of one configuration. α-Helices cannot be formed from a mixed copolymer of D- and L-amino acids. An α-helix composed of D-amino acids is left-handed.

The β-Pleated Sheet Is a Core Structure in Proteins

Another type of structure commonly observed in proteins also forms because of local, cooperative formation of hydrogen bonds. That is the pleated sheet, or β-structure, often called the **β-pleated sheet.** This structure was also first postulated by Pauling and Corey in 1951 and has now been observed in many natural proteins. A β-pleated sheet can be visualized by laying thin, pleated strips of paper side by side to make a "pleated sheet" of paper (Figure 6.10). Each strip of paper can then be pictured as a single peptide strand in which the peptide backbone makes a zigzag pattern along the strip, with the α-carbons lying at the folds of the pleats. The pleated sheet can exist in both parallel and antiparallel forms. In the **parallel β-pleated sheet,** adjacent chains run in the same direction. In the **antiparallel β-pleated sheet,** adjacent strands run in opposite directions.

Each single strand of the β-sheet structure can be pictured as a twofold helix, that is, a helix with two residues per turn. The arrangement of successive amide planes has a pleated appearance due to the tetrahedral nature of the C_α atom. It is important to note that the hydrogen bonds in this structure are essentially *inter*strand rather than *intra*strand. Optimum formation of H bonds in the parallel pleated sheet results in a slightly less extended conformation than in the antiparallel sheet. The H bonds thus formed in the parallel β-sheet are bent significantly. The distance between residues is 0.347 nm for the antiparallel pleated sheet, but only 0.325 nm for the parallel pleated sheet. Note that the side chains in the pleated sheet are oriented perpendicular or normal to the plane of the sheet, extending out from the plane on alternating sides.

CRITICAL DEVELOPMENTS IN BIOCHEMISTRY

In Bed with a Cold, Pauling Stumbles onto the α-Helix and a Nobel Prize*

As high technology continues to transform the modern biochemical laboratory, it is interesting to reflect on Linus Pauling's discovery of the α-helix. It involved only a piece of paper, a pencil, scissors, and a sick Linus Pauling, who had tired of reading detective novels. The story is told in the excellent book *The Eighth Day of Creation* by Horace Freeland Judson:

> From the spring of 1948 through the spring of 1951…rivalry sputtered and blazed between Pauling's lab and (Sir Lawrence) Bragg's—over protein. The prize was to propose and verify in nature a general three-dimensional structure for the polypeptide chain. Pauling was working up from the simpler structures of components. In January 1948, he went to Oxford as a visiting professor for two terms, to lecture on the chemical bond and on molecular structure and biological specificity. "In Oxford, it was April, I believe, I caught cold. I went to bed, and read detective stories for a day, and got bored, and thought why don't I have a crack at that problem of alpha keratin." Confined, and still fingering the polypeptide chain in his mind, Pauling called for paper, pencil, and straightedge and attempted to reduce the problem to an almost Euclidean purity. "I took a sheet of paper—I still have this sheet of paper—and drew, rather roughly, the way that I thought a polypeptide chain would look if it were spread out into a plane." The repetitious herringbone of the chain he could stretch across the paper as simply as this—

(a)

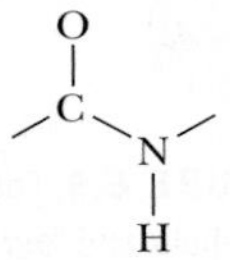

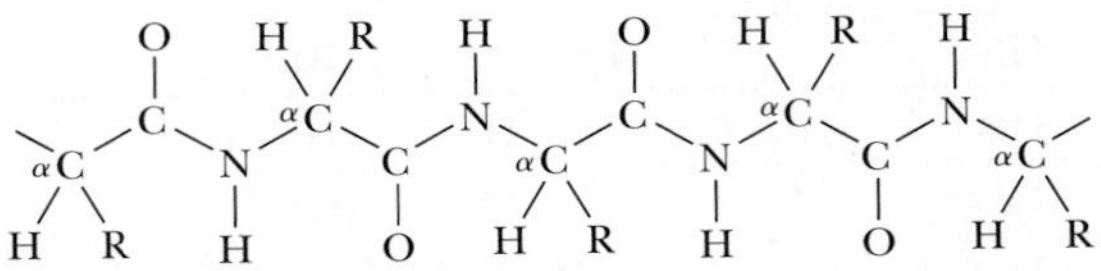

And this meant that the chain could turn corners only at the alpha carbons.…"I creased the paper in parallel creases through the alpha carbon atoms, so that I could bend it and make the bonds to the alpha carbons, along the chain, have tetrahedral value. And then I looked to see if I could form hydrogen bonds from one part of the chain to the next." He saw that if he folded the strip like a chain of paper dolls into a helix, and if he got the pitch of the screw right, hydrogen bonds could be shown to form, N—H⋅⋅O—C, three or four knuckles apart along the backbone, holding the helix in shape. After several tries, changing the angle of the parallel creases in order to adjust the pitch of the helix, he found one where the hydrogen bonds would drop into place, connecting the turns, as straight lines of the right length. He had a model.

—putting in lengths and bond angles from memory.…He knew that the peptide bond, at the carbon-to-nitrogen link, was always rigid:

(b)

*The discovery of the α-helix structure was only one of many achievements that led to Pauling's Nobel Prize in Chemistry in 1954. The official citation for the prize was "for his research into the nature of the chemical bond and its application to the elucidation of the structure of complex substances."

Parallel β-sheets tend to be more regular than antiparallel β-sheets. As can be seen in Figure 6.4, typical ϕ,ψ values for a parallel β-sheet are $\phi = -120°$, $\psi = 105°$, and typical values for an anti-parallel β-sheet are $\phi = -135°$, $\psi = 140°$. However, the range of ϕ and ψ angles for the peptide bonds in parallel sheets is much smaller than that for antiparallel sheets. Parallel sheets are typically large structures; those composed of less than five strands are rare. Antiparallel sheets, however, may consist of as few as two strands. Parallel sheets characteristically distribute hydrophobic side chains on both sides of the sheet, whereas antiparallel sheets are usually arranged with all their hydrophobic residues on one side of the sheet. This requires an alternation of hydrophilic and hydrophobic residues in the primary structure of peptides involved in antiparallel β-sheets because every other side chain projects to the same side of the sheet (Figure 6.10).

Antiparallel pleated sheets are the fundamental structure found in the fabric we know as silk, with the polypeptide chains forming the sheets running parallel to the silk fibers. The silk fibers thus formed have properties consistent with those of the β-sheets that form them. They are quite flexible but cannot be stretched or extended to any appreciable degree.

Helix–Sheet Composites in Spider Silk

Although the intricate designs of spider webs are eye- (and fly-) catching, it might be argued that the composition of web silk itself is even more remarkable. Spider silk (a form of keratin) is synthesized in special glands in the spider's abdomen. The silk strands produced by these glands are both strong and elastic. *Dragline silk* (that from

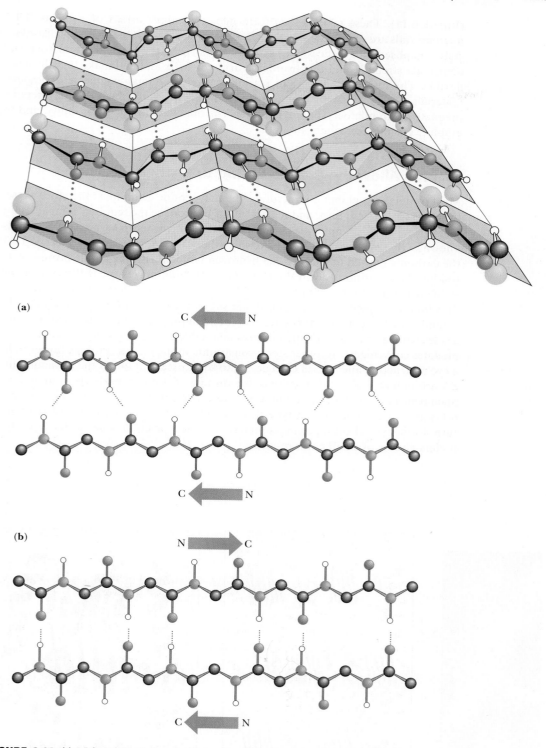

FIGURE 6.10 **(a)** A "pleated sheet" of paper with an antiparallel β-sheet drawn on it. (Illustration: Irving Geis. Rights owned by Howard Hughes Medical Institute. Not to be reproduced without permission.) **(b)** The arrangement of hydrogen bonds in parallel and antiparallel β-pleated sheets.

which the spider hangs) has a tensile strength of 200,000 psi (pounds per square inch)—stronger than steel and similar to Kevlar, the synthetic material used in bulletproof vests! This same silk fiber is also flexible enough to withstand strong winds and other natural stresses. This combination of strength and flexibility derives from the *composite nature* of spider silk. As keratin protein is extruded from the spider's glands, it endures shearing forces that break the H bonds stabilizing keratin α-helices

(Figure 6.11). These regions then form microcrystalline arrays of β-sheets. These microcrystals are surrounded by the keratin strands, which adopt a highly disordered state composed of α-helices and random coil structures. The β-sheet microcrystals contribute strength, and the disordered array of helix and coil make the silk strand flexible. The resulting silk strand resembles modern human-engineered composite materials. Certain tennis racquets, for example, consist of fiberglass polymers impregnated with microcrystalline graphite. The fiberglass provides flexibility, and the graphite crystals contribute strength.

β-Turns Allow the Protein Strand to Change Direction

Most proteins are globular structures. The polypeptide chain must therefore possess the capacity to bend, turn, and reorient itself to produce the required compact, globular structures. A simple structure observed in many proteins is the **β-turn** (also known as the *tight turn* or *β-bend*), in which the peptide chain forms a tight loop with the carbonyl oxygen of one residue hydrogen bonded with the amide proton of the residue three positions down the chain. This H bond makes the β-turn a relatively stable structure. As shown in Figure 6.12, the β-turn allows the protein to reverse the direction of its peptide chain. This figure shows the two major types of β-turns, but a number of less common types are also found in protein structures. Because it lacks a side chain, glycine is sterically the most adaptable of the amino acids, and it accommodates conveniently to other steric constraints in the β-turn. Proline, however, has a cyclic structure and a fixed ϕ angle, so, to some extent, it forces the formation of a β-turn; in many cases this facilitates the turning of a polypeptide chain upon itself. Such bends promote formation of antiparallel β-pleated sheets. Type I turns (Figure 6.12) are more common than type II. Proline fits best in the 3 position of the type I turn. In the type II turn, proline is preferred at the 2 position, whereas the 3 position prefers glycine or small polar residues.

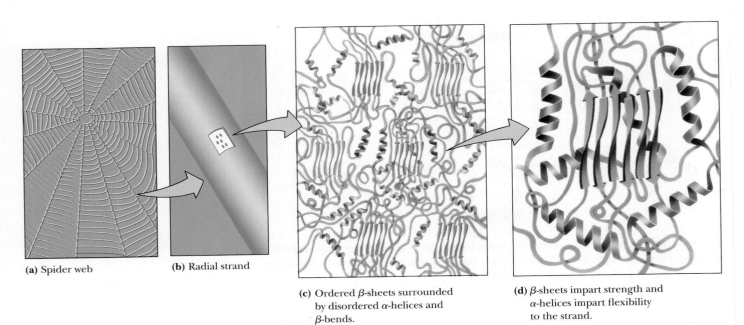

(a) Spider web

(b) Radial strand

(c) Ordered β-sheets surrounded by disordered α-helices and β-bends.

(d) β-sheets impart strength and α-helices impart flexibility to the strand.

FIGURE 6.11 Spider web silks are composites of α-helices and β-sheets. The radial strands of webs must be strong and rigid; silks in radial strands contain a higher percentage of β-sheets. The circumferential strands (termed *capture silk*) must be flexible (to absorb the impact of flying insects); capture silk contains a higher percentage of α-helices. Spiders typically have several different silk gland spinnerets, which secrete different silks of differing composition. Spiders have inhabited the earth for about 470 million years.

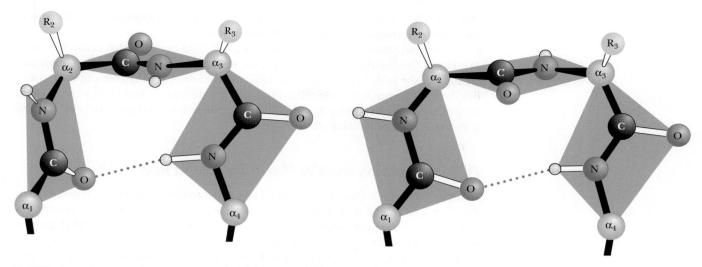

FIGURE 6.12 The structures of two kinds of β-turns (also called tight turns or β-bends). Four residues are required to form a β-turn. Left: Type I; right: Type II. (Illustration: Irving Geis. Rights owned by Howard Hughes Medical Institute. Not to be reproduced without permission.)

6.4 How Do Polypeptides Fold into Three-Dimensional Protein Structures?

The arrangement of all atoms of a single polypeptide chain in three-dimensional space is referred to as its **tertiary structure.** As discussed in Section 6.2 all of the information needed to fold the protein into its native tertiary structure is contained within the primary structure of the peptide chain itself. Sometimes proteins known as **chaperones** assist in the process of protein folding in the cell, but proteins in dilute solution can be unfolded and refolded without the assistance of such chaperones.

The first determinations of the tertiary structure of a protein were by John Kendrew and Max Perutz. Kendrew's structure of myoglobin and Perutz's structure of hemoglobin, reported in the late 1950s, were each the result of more than 20 years of work. Ever since these first protein structures were elucidated, biochemists have sought to understand the principles by which proteins adopt their remarkable structures. Vigorous work in many laboratories has slowly brought important principles to light:

- Secondary structures—helices and sheets—form whenever possible as a consequence of the formation of large numbers of hydrogen bonds.
- α-Helices and β-sheets often associate and pack close together in the protein. No protein is stable as a single-layer structure, for reasons that become apparent later. There are a few common methods for such packing to occur.
- Because the peptide segments between secondary structures in the protein tend to be short and direct, the peptide does not execute complicated twists and knots as it moves from one region of secondary structure to another.
- Proteins generally fold so as to form the most stable structures possible. The stability of most proteins arises from (1) the formation of large numbers of intramolecular hydrogen bonds and (2) the reduction in the surface area accessible to solvent that occurs upon folding.

Two factors lie at the heart of these four "principles." First, proteins are typically a mixture of hydrophilic and hydrophobic amino acids. Why is this important? Imagine a protein that is composed only of polar and charged amino acids. In such a protein, every side chain could hydrogen bond to water. This would leave no reason for the protein to form a compact, folded structure.

Now consider a protein composed of a mixture of hydrophilic and hydrophobic residues. In this case, the hydrophobic side chains cannot form H bonds with water, and their presence will disrupt the hydrogen-bonding structure of water itself. *To minimize* *this, the hydrophobic groups will tend to cluster together. This **hydrophobic effect** induces formation of a compact structure—the folded protein.*

A potential problem with this rather simple folding model is that polar backbone N—H and C=O groups on the hydrophobic residues accompany the hydrophobic side chains into the folded protein interior. This would be energetically costly to the protein, but the actual result is that the polar backbone groups form H bonds with one another, so that the polar backbone N—H and C=O moieties are stabilized in α-helices and β-sheets in the protein interior.

Fibrous Proteins Usually Play a Structural Role

In Chapter 5, we saw that proteins can be grouped into three large classes based on their structure and solubility: *fibrous proteins, globular proteins,* and *membrane proteins.* Fibrous proteins contain polypeptide chains organized approximately parallel along a single axis, producing long fibers or large sheets. Such proteins tend to be mechanically strong and resistant to solubilization in water and dilute salt solutions. Fibrous proteins often play a structural role in nature (see Chapter 5).

α-**Keratin** The α-keratins are the predominant constituents of claws, fingernails, hair, and horns in mammals. As their name suggests, the structure of the α-keratins is dominated by α-helical segments of polypeptide. The amino acid sequence of α-keratin subunits is composed of central α-helix–rich rod domains about 311 to 314 residues in length, flanked by nonhelical N- and C-terminal domains of varying size and composition (Figure 6.13a). The structure of the central rod domain of a typical α-keratin is shown in Figure 6.13b. Pairs of right-handed α-helices wrap around each other to form

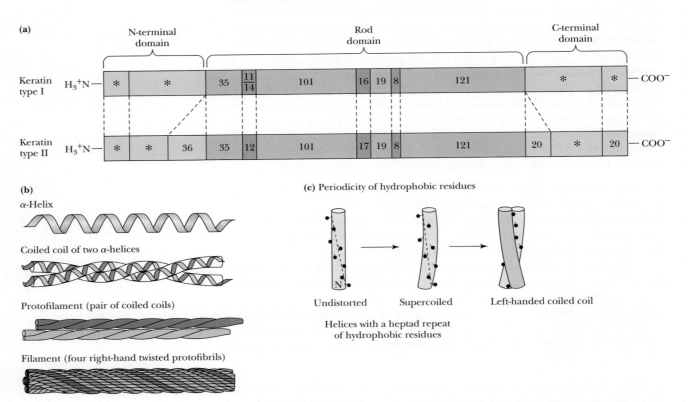

FIGURE 6.13 **(a)** Both type I and type II α-keratin molecules have sequences consisting of long, central rod domains with terminal cap domains. The numbers of amino acid residues in each domain are indicated. Asterisks denote domains of variable length. **(b)** The rod domains form coiled coils consisting of left-twisted right-handed α-helices. These coiled coils form protofilaments that then wind around each other in a right-handed twist. Keratin filaments consist of twisted protofibrils (each a bundle of four coiled coils). **(c)** Periodicity of hydrophobic residues. (Adapted from Steinert, P., and Parry, D., 1985. Intermediate filaments: Conformity and diversity of expression and structure. *Annual Review of Cell Biology* **1**:41–65; and Cohlberg, J., 1993. Textbook error: The structure of alpha-keratin. *Trends in Biochemical Sciences* **18**:360–362.)

▶ **A DEEPER LOOK**

The Coiled-Coil Motif in Proteins

The **coiled-coil** motif was first identified in 1953 by Linus Pauling, Robert Corey, and Francis Crick as the main structural element of fibrous proteins such as keratin and myosin. Since then, many proteins have been found to contain one or more coiled-coil segments or domains. A coiled coil is a bundle of α-helices that are wound into a superhelix. Two, three, or four helical segments may be found in the bundle, and they may be arranged parallel or antiparallel to one another. Coiled coils are characterized by a distinctive and regular packing of side chains in the core of the bundle. This regular meshing of side chains requires that they oc-

cupy equivalent positions turn after turn. This is not possible for undistorted α-helices, which have 3.6 residues per turn. The positions of side chains on their surface shift continuously along the helix surface (see Figure 6.13c). However, giving the right-handed α-helix a left-handed twist reduces the number of residues per turn to 3.5, and because 3.5 × 2 = 7.0, the positions of the side chains repeat after two turns (7 residues). Thus, a **heptad repeat** pattern in the peptide sequence is diagnostic of a coiled-coil structure. The figure shows a sampling of coiled-coil structures (highlighted in color) in various proteins.

(a) Coiled coil

Pitch

(b) Periodicity of hydrophobic residues

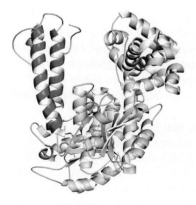

Undistorted   Supercoiled   Left-handed coiled coil

Helices with a heptad repeat
of hydrophobic residues

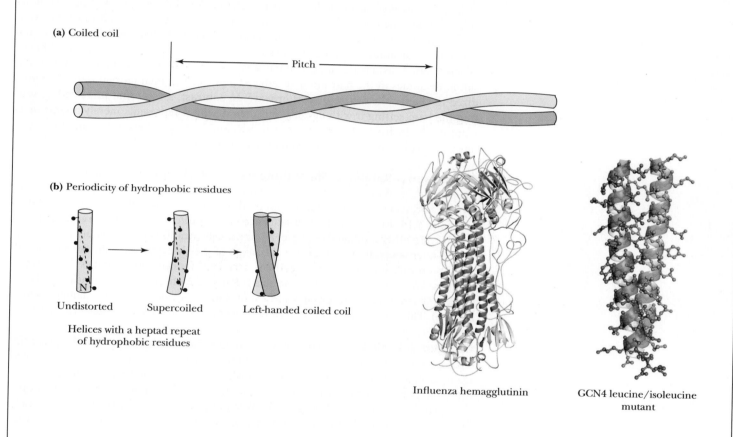

Influenza hemagglutinin   GCN4 leucine/isoleucine
mutant

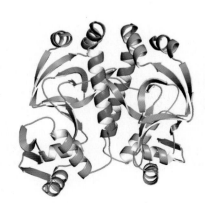

DNA polymerase   Seryl tRNA synthetase   Catabolite activator protein

a left-twisted **coiled coil.** X-ray diffraction data (including the original studies by William Astbury) show that these structures resemble α-helices, but with a pitch of 0.51 nm rather than the expected 0.54 nm. This is consistent with a tilt of the helix relative to the long axis of the coiled coil (and the keratin fiber) in Figure 6.13.

The amino acid sequence of the central rod segments of α-keratin consists of quasi-repeating 7-residue segments of the form $(a\text{-}b\text{-}c\text{-}d\text{-}e\text{-}f\text{-}g)_n$. These units are not true repeats, but residues a and d are usually nonpolar amino acids. In α-helices, with 3.6 residues per turn, these nonpolar residues are arranged in an inclined row or stripe that twists around the helix axis (Figure 6.13c). These nonpolar residues would make the helix highly unstable if they were exposed to solvent, but the association of hydrophobic stripes on two α-helices to form the two-stranded coiled coil effectively buries the hydrophobic residues and forms a highly stable structure (Figure 6.13). The helices clearly sacrifice some stability in assuming this twisted conformation, but they gain stabilization energy from the packing of side chains between the helices. In other forms of keratin, covalent disulfide bonds form between cysteine residues of adjacent molecules, making the overall structure even more rigid, inextensible, and insoluble—important properties for structures such as claws, fingernails, hair, and horns. How and where these disulfides form determines the amount of curling in hair and wool fibers. When a hairstylist creates a permanent wave (simply called a "permanent") in a hair salon, disulfides in the hair are first reduced and cleaved, then reorganized and reoxidized to change the degree of curl or wave. In contrast, a "set" that is created by wetting the hair, setting it with curlers, and then drying it represents merely a rearrangement of the hydrogen bonds between helices and between fibers. (On humid or rainy days, the hydrogen bonds in curled hair may rearrange, and the hair becomes "frizzy.")

Fibroin and β-Keratin: β-Sheet Proteins   The **fibroin** proteins found in silk fibers in the cocoons of the silkworm, *Bombyx mori,* and also in spiderwebs represent another type of fibrous protein. These are composed of stacked antiparallel β-sheets, as shown in Figure 6.14. In the polypeptide sequence of silk proteins, there are large stretches in which every other residue is a glycine. As previously mentioned, the residues of a β-sheet extend alternately above and below the plane of the sheet. As a result, the glycines all end up on one side of the sheet and the other residues (mainly alanines and serines) compose the opposite surface of the sheet. Pairs of β-sheets can then pack snugly together (glycine surface to glycine surface or alanine–serine surface to alanine—serine surface). The β-keratins found in bird feathers are also made up of stacked β-sheets.

Collagen: A Triple Helix   **Collagen** is a rigid, inextensible fibrous protein that is a principal constituent of connective tissue in animals, including tendons, cartilage, bones, teeth, skin, and blood vessels. The high tensile strength of collagen fibers in these structures makes possible the various animal activities such as running and jumping that put severe stresses on joints and skeleton. Broken bones and tendon and cartilage injuries to knees, elbows, and other joints involve tears or hyperextensions of the collagen matrix in these tissues.

The basic structural unit of collagen is **tropocollagen,** which has a molecular weight of 285,000 and consists of three intertwined polypeptide chains, each about 1000 amino acids in length. Tropocollagen molecules are about 300 nm long and only about 1.4 nm in diameter. Several kinds of collagen have been identified. *Type I collagen,* which is the most common, consists of two identical peptide chains designated $\alpha1(I)$ and one different chain designated $\alpha2(I)$. Type I collagen predominates in bones, tendons, and skin. *Type II collagen,* found in cartilage, *and type III collagen,* found in blood vessels, consist of three identical polypeptide chains.

Collagen has an amino acid composition that is unique and is crucial to its three-dimensional structure and its characteristic physical properties. Nearly one residue out of three is a glycine, and the proline content is also unusually high. Three unusual modified amino acids are also found in collagen: 4-hydroxyproline (Hyp), 3-hydroxyproline, and 5-hydroxylysine (Hyl) (Figure 4.4). Proline and Hyp together compose up to 30% of the residues of collagen. Interestingly, these three amino acids are formed from normal proline and lysine *after* the collagen polypeptides are synthesized. The modifications are effected by two enzymes: *prolyl hydroxylase* and *lysyl hydroxylase.* The prolyl hydroxylase

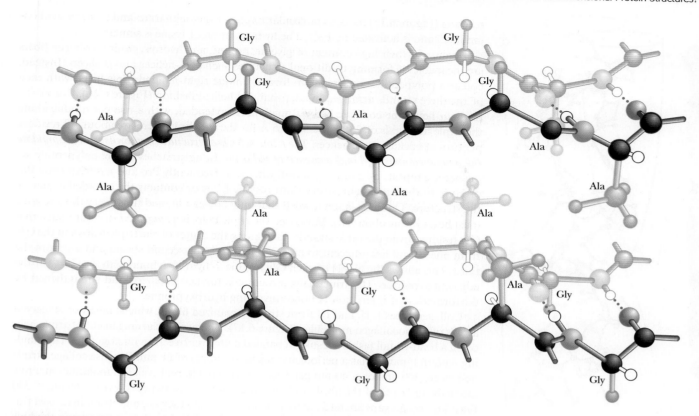

FIGURE 6.14 Silk fibroin consists of a unique stacked array of β-sheets. The primary structure of fibroin molecules consists of long stretches of alternating glycine and alanine or serine residues. When the sheets stack, the more bulky alanine and serine residues on one side of a sheet interdigitate with similar residues on an adjoining sheet. Glycine hydrogens on the alternating faces interdigitate in a similar manner, but with a smaller intersheet spacing.
(Illustration: Irving Geis. Rights owned by Howard Hughes Medical Institute. Not to be reproduced without permission.)

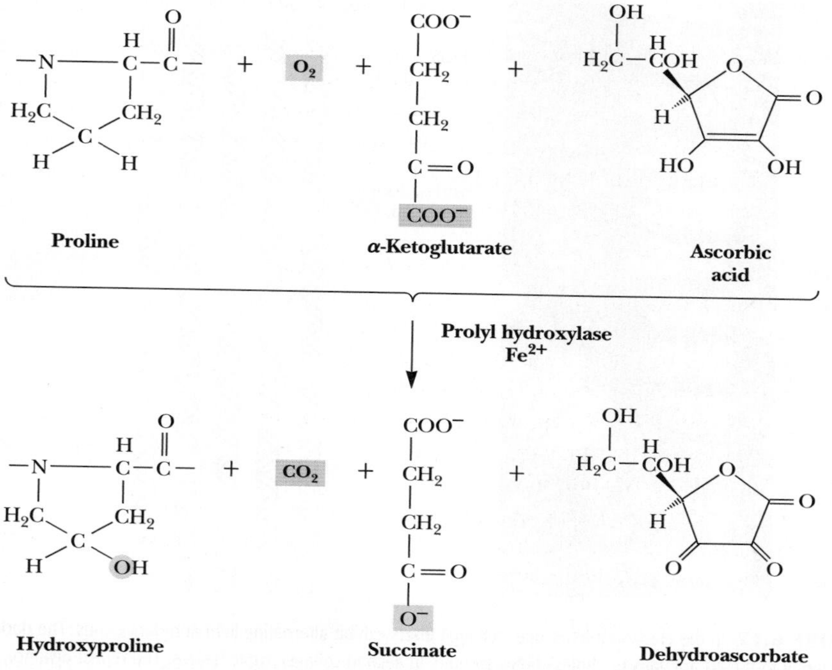

FIGURE 6.15 Hydroxylation of proline residues is catalyzed by prolyl hydroxylase. The reaction requires α-ketoglutarate and ascorbic acid (vitamin C).

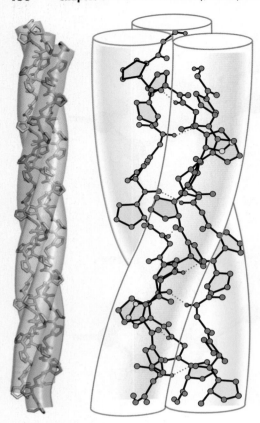

FIGURE 6.16 Poly(Gly-Pro-Pro), a collagenlike right-handed triple helix composed of three left-handed helical chains (pdb id=1K6F). (Adapted from Miller, M. H., and Scheraga, H. A., 1976. Calculation of the structures of collagen models. Role of interchain interactions in determining the triple-helical coiled-coil conformation. I. Poly(glycyl-prolyl-prolyl). *Journal of Polymer Science Symposium* **54:**171–200.)

reaction (Figure 6.15) requires molecular oxygen, α-ketoglutarate, and ascorbic acid (vitamin C) and is activated by Fe^{2+}. The hydroxylation of lysine is similar.

Because of their high content of glycine, proline, and hydroxyproline, collagen fibers are incapable of forming traditional structures such as α-helices and β-sheets. Instead, collagen polypeptides intertwine to form a unique right-handed **triple helix,** with each of the three strands arranged in a left-handed helical fashion (Figure 6.16). Compared to the α-helix, the collagen helix is much more extended, with a rise per residue along the triple helix axis of 2.9 Å (versus 1.5 Å for the α-helix). There are about 3.3 residues per turn of each of these helices. *The triple helix is a structure that forms to accommodate the unique composition and sequence of collagen.* Long stretches of the polypeptide sequence are repeats of a Gly-x-y motif, where x is frequently Pro and y is frequently Pro or Hyp. In the triple helix, every third residue faces or contacts the crowded center of the structure. This area is so crowded that only Gly can fit, and thus every third residue must be a Gly (as observed). Moreover, the triple helix is a *staggered* structure, such that Gly residues from the three strands stack along the center of the triple helix and the Gly from one strand lies adjacent to an x residue from the second strand and a y from the third. This allows the N—H of each Gly residue to hydrogen bond with the C=O of the adjacent x residue. The triple helix structure is further stabilized and strengthened by the formation of interchain H bonds involving hydroxyproline.

Collagen types I, II, and III form strong, organized **fibrils,** which consist of staggered arrays of tropocollagen molecules (Figure 6.17). The periodic arrangement of triple helices in a head-to-tail fashion results in banded patterns in electron micrographs. The banding pattern typically has a periodicity (repeat distance) of 68 nm. Because collagen triple helices are 300 nm long, 40-nm gaps occur between adjacent collagen molecules in a row along the long axis of the fibrils and the pattern repeats every five rows (5×68 nm = 340 nm). The 40-nm gaps are referred to as *hole regions,* and they are important in at least two ways. First, sugars are found covalently attached to 5-hydroxylysine residues in the hole regions of collagen (Figure 6.18). The occurrence of carbohydrate in the hole region has led to the proposal that it plays a role in organizing fibril assembly. Second, the hole regions may play a role in bone formation. Bone consists of microcrystals of **hydroxyapatite,** $Ca_5(PO_4)_3OH$, embedded in a matrix of collagen fibrils. When new bone tissue forms, the formation of new hydroxyapatite crystals occurs at intervals of 68 nm. The hole regions of collagen fibrils may be the sites of nucleation for the mineralization of bone.

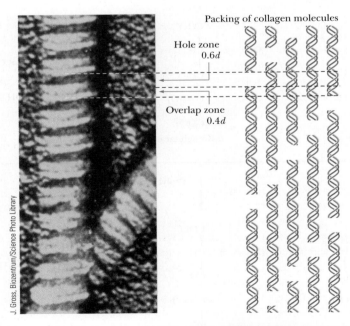

FIGURE 6.17 In the electron microscope, collagen fibers exhibit alternating light and dark bands. The dark bands correspond to the 40-nm gaps or "holes" between pairs of aligned collagen triple helices. The repeat distance, *d*, for the light- and dark-banded pattern is 68 nm. The collagen molecule is 300 nm long, which corresponds to 4.41*d*. The molecular repeat pattern of five staggered collagen molecules corresponds to 5*d*.

Globular Proteins Mediate Cellular Function

Fibrous proteins, although interesting for their structural properties, represent only a small percentage of the proteins found in nature. **Globular proteins,** so named for their approximately spherical shape, are far more numerous. The diversity of protein structures found in nature reflects the remarkable variety of functions performed by proteins—binding, catalysis, regulation, transport, immunity, cellular signaling, and more. The functional diversity and versatility derive in turn from (1) the large number of folded structures that polypeptide chains can adopt and (2) the varied chemistry of the side chains of the 20 common amino acids. Remarkably, this diversity of structure and function derives from a relatively small number of principles and themes of protein folding and design. The balance of Chapter 6 explores and elaborates these principles and themes.

Helices and Sheets Make up the Core of Most Globular Proteins

Globular proteins exist in an enormous variety of three-dimensional structures, but nearly all contain substantial amounts of α-helices and β-sheets folded into a compact structure that is stabilized by both polar and nonpolar groups. A typical example is *bovine ribonuclease A*, a small protein (12.6 kD, 124 residues) that contains a few short α-helices, a broad section of antiparallel β-sheet, a few β-turns, and several peptide segments without defined secondary structure (Figure 6.19). The space between the helices and sheets in the protein interior is filled efficiently and tightly with mostly hydrophobic amino acid side chains. Most polar side chains in ribonuclease face the outside of the protein structure and interact with solvent water. With its hydrophobic core and a hydrophilic surface, ribonuclease illustrates the typical properties of many folded globular proteins.

The helices and sheets that make up the core of most globular proteins probably represent the starting point for protein folding, as shown later in this chapter. Thus, the folding of a globular protein, in its simplest conception, could be viewed reasonably as the condensation of multiple elements of secondary structure. On the other hand, most peptide segments that form helices, sheets, or beta turns in proteins are mostly disordered in small model peptides that contain those amino acid sequences. Thus hydrophobic interactions and other noncovalent interactions with the rest of the protein must stabilize these relatively unstable helices, sheets, and turns in the whole folded protein.

Why should the cores of most globular and membrane proteins consist almost entirely of α-helices and β-sheets? The reason is that the highly polar N—H and C=O moieties of the peptide backbone must be neutralized in the hydrophobic core of the protein. The extensively H-bonded nature of α-helices and β-sheets is ideal for this purpose, and these structures effectively stabilize the polar groups of the peptide backbone in the protein core.

The framework of sheets and helices in the interior of a globular protein is typically constant and conserved in both sequence and structure. The surface of a globular protein is different in several ways. Typically, much of the protein surface is composed of the loops and tight turns that connect the helices and sheets of the protein core, although helices and sheets may also be found on the surface. The result is that the surface of a globular protein is often a complex landscape of different structural elements. These complex surface structures can interact in certain cases with small molecules or even large proteins that have complementary structure or charge (Figure 6.20). These regions of complementary, recognizable structure are formed typically from the peptide segments that connect elements of secondary structure. They are the basis for enzyme–substrate interactions, protein–protein associations in cell signaling pathways, and antigen–antibody interactions, and more.

The segments of the protein that are neither helix, sheet, nor turn have traditionally been referred to as *coil* or *random coil*. Both of these terms are misleading. Most of these "loop" segments are neither coiled nor random, in any sense of the words. These structures are every bit as organized and stable as the defined secondary structures. They just don't conform to any frequently recurring pattern. These so-called coil structures are strongly influenced by side-chain interactions with the rest of the protein.

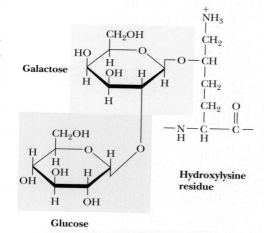

FIGURE 6.18 A disaccharide of galactose and glucose is covalently linked to the 5-hydroxyl group of hydroxylysines in collagen by the combined action of the enzymes galactosyltransferase and glucosyltransferase.

Evan Powers and Jeffery Kelly and co-workers have shown that hydrogen bonds in nonpolar environments (such as the core of a globular protein) can be as much as 5.5 kJ/mole stronger than when they are solvent-exposed.

Gao, J., et al., 2009. Localized thermodynamic coupling between hydrogen bonding and microenvironment polarity substantially stabilizes proteins. *Nature Structural and Molecular Biology* **16:**684–691.

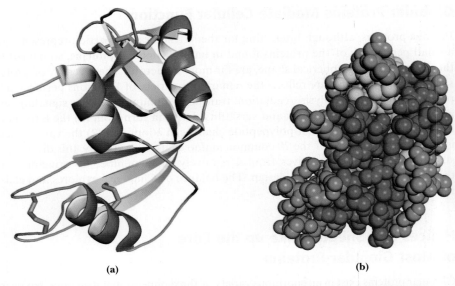

(a)   **(b)**

FIGURE 6.19 The three-dimensional structure of bovine ribonuclease A (pdb id = 1FS3). **(a)** Ribbon diagram; **(b)** space-filling model.

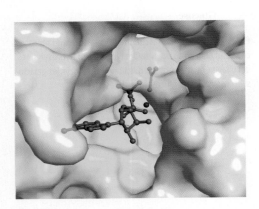

FIGURE 6.20 The surfaces of proteins are complementary to the molecules they bind. PEP carboxykinase (shown here, pdb id = 1K3D) is an enzyme from the metabolic pathway that synthesizes glucose (gluconeogenesis; see Chapter 22). In the so-called active site (yellow) of this enzyme, catalysis depends on complementary binding of substrates. Shown in this image are ADP (brown), a Mg^{2+} ion (blue), and AlF_3^- (a phosphate analog, in green, above the Mg^{2+}).

Waters on the Protein Surface Stabilize the Structure

A globular protein's surface structure also includes water molecules. Many of the polar backbone and side chain groups on the surface of a globular protein make H bonds with solvent water molecules. There are often several such water molecules per amino acid residue, and some are in fixed positions (Figure 6.21). Relatively few water molecules are found inside the protein.

In some globular proteins (Figure 6.22), it is common for one face of an α-helix to be exposed to the water solvent, with the other face toward the hydrophobic interior of the protein. The outward face of such an **amphiphilic helix** consists mainly of polar and charged residues, whereas the inward face contains mostly nonpolar, hydrophobic residues. A good example of such a surface helix is that of residues 153 to 166 of **flavodoxin** from *Anabaena* (Figure 6.22a). Note that the **helical wheel presentation** of this helix readily shows that one face contains four hydrophobic residues and that the other is almost entirely polar and charged.

Less commonly, an α-helix can be completely buried in the protein interior or completely exposed to solvent. **Citrate synthase** is a dimeric protein in which α-helical segments form part of the subunit–subunit interface. As shown in Figure 6.22b, one of these helices (residues 260 to 270) is highly hydrophobic and contains only two polar residues, as would befit a helix in the protein core. On the other hand, Figure 6.22c shows the solvent-exposed helix (residues 74 to 87) of **calmodulin,** which consists of 10 charged residues, 2 polar residues, and only 2 nonpolar residues.

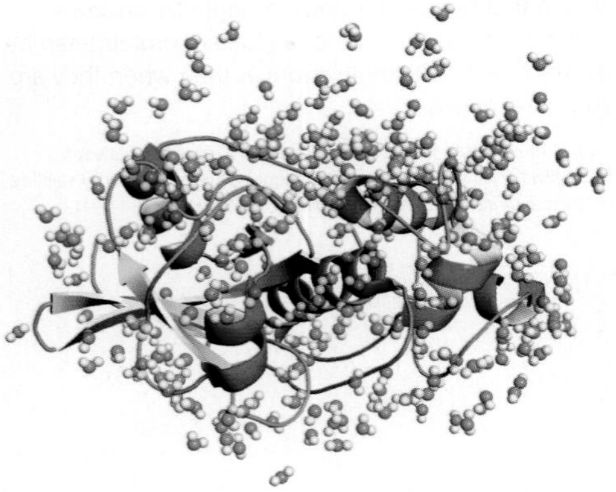

FIGURE 6.21 The surfaces of proteins are ideally suited to form multiple H bonds with water molecules. Shown here are waters (blue and white) associated with actinidin, an enzyme from kiwi fruit that cleaves polypeptide chains at arginine residues (pdb id = 2ACT). The polar backbone atoms and side chain groups on the surface of actinidin are extensively H-bonded with water.

Packing Considerations

The secondary and tertiary structures of ribonuclease A (Figure 6.19) and other globular proteins illustrate the importance of packing in tertiary structures. Secondary structures pack closely to one another and also intercalate with (insert between) extended polypeptide chains. If the sum of the van der Waals volumes of a protein's constituent amino acids is divided by the volume occupied by the protein, packing densities of

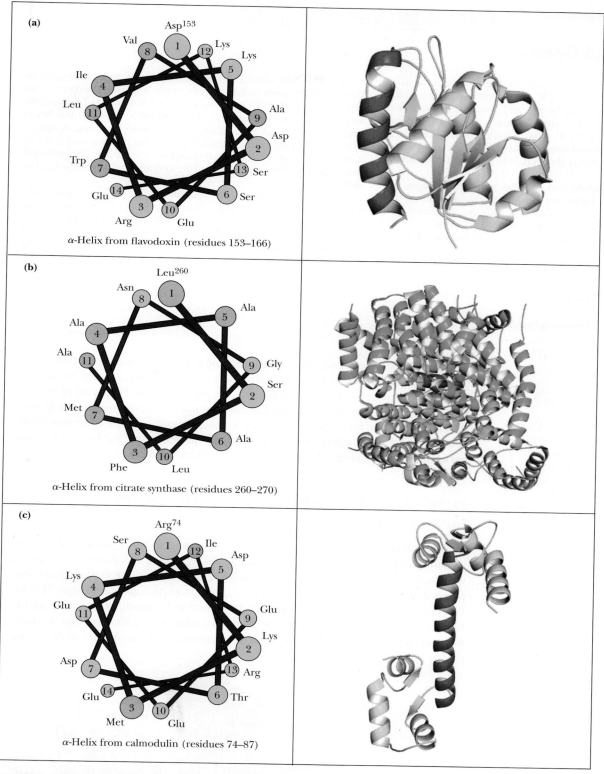

FIGURE 6.22 The so-called helical wheel presentation can reveal the polar or nonpolar character of α-helices. If the helix is viewed end on, and the residues are numbered with residue 1 closest to the viewer, it is easy to see how polar and nonpolar residues are distributed to form a wheel. **(a)** The α-helix consisting of residues 153–166 (red) in flavodoxin from *Anabaena* is a surface helix and is amphipathic (pdb id = 1RCF). **(b)** The two helices (orange and red) in the interior of the citrate synthase dimer (residues 260–270 in each monomer) are mostly hydrophobic (pdb id = 5CSC). **(c)** The exposed helix (residues 74–87—red) of calmodulin is entirely accessible to solvent and consists mainly of polar and charged residues (pdb id = 1CLL).

HUMAN BIOCHEMISTRY

Collagen-Related Diseases

Collagen provides an ideal case study of the molecular basis of physiology and disease. For example, the nature and extent of collagen crosslinking depends on the age and function of the tissue. Collagen from young animals is predominantly un-crosslinked and can be extracted in soluble form, whereas collagen from older animals is highly crosslinked and thus insoluble. The loss of flexibility of joints with aging is probably due in part to increased crosslinking of collagen.

Several serious and debilitating diseases involving collagen abnormalities are known. **Lathyrism** occurs in animals due to the regular consumption of seeds of *Lathyrus odoratus,* the sweet pea, and involves weakening and abnormalities in blood vessels, joints, and bones. These conditions are caused by **β-aminopropionitrile** (see figure), which covalently inactivates lysyl oxidase, preventing intramolecular crosslinking of collagen and causing abnormalities in joints, bones, and blood vessels.

Scurvy results from a dietary vitamin C deficiency and involves the inability to form collagen fibrils properly. This is the result of reduced activity of prolyl hydroxylase, which is vitamin C–dependent, as previously noted. Scurvy leads to lesions in the skin and blood vessels, and in its advanced stages, it can lead to grotesque disfiguration and eventual death. Although rare in the modern world, it was a disease well known to sea-faring explorers in earlier times who did not appreciate the importance of fresh fruits and vegetables in the diet.

A number of rare genetic diseases involve collagen abnormalities, including *Marfan's syndrome* and the *Ehlers–Danlos syndromes,* which result in hyperextensible joints and skin. The formation of *atherosclerotic plaques,* which cause arterial blockages in advanced stages, is due in part to the abnormal formation of collagenous structures in blood vessels.

$$N \equiv C - CH_2 - CH_2 - \overset{+}{N}H_3$$

β-Aminopropionitrile

0.72 to 0.77 are typically obtained. These packing densities are similar to those of a collection of solid spheres. This means that even with close packing, approximately 25% of the total volume of a protein is not occupied by protein atoms. Nearly all of this space is in the form of very small cavities. Cavities the size of water molecules or larger do occasionally occur, but they make up only a small fraction of the total protein volume. It is likely that such cavities provide flexibility for proteins and facilitate conformation changes and a wide range of protein dynamics (discussed later).

Protein Domains Are Nature's Modular Strategy for Protein Design

Proteins range in molecular weight from a thousand to more than a million. It is tempting to think that the size of unique globular, folded structures would increase with molecular weight, but this is not what has been observed. Proteins composed of about 250 amino acids or less often have a simple, compact globular shape. However, larger globular proteins are usually made up of two or more recognizable and distinct structures, termed **domains** or **modules**—compact, folded protein structures that are usually stable by themselves in aqueous solution. Figure 6.23 shows a

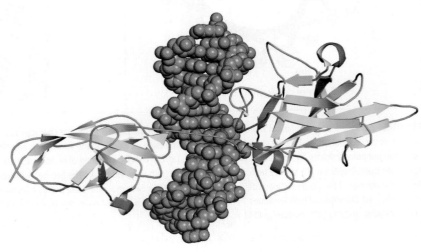

FIGURE 6.23 Ton-EBP is a DNA-binding protein consisting of two distinct domains. The N-terminal domain is shown here on the right, with DNA (orange) in the middle, and the C-terminal domain on the left (pdb id = 1IMH).

two-domain DNA-binding protein, TonEBP, in which the two distinct domains are joined by a short segment of the peptide chain. Most domains consist of a single continuous portion of the protein sequence, but in some proteins the domain sequence is interrupted by a sequence belonging to some other part of the protein that may even form a separate domain (Figure 6.24). In either case, typical domain structures consist of hydrophobic cores with hydrophilic surfaces (as was the case for ribonuclease, Figure 6.19). Importantly, individual domains often possess unique functional behaviors (for example, the ability to bind a particular ligand with high affinity and specificity), and an individual domain from a larger protein often expresses its unique function within the larger protein in which it is found. Multidomain proteins typically possess the sum total of functional properties and behaviors of their constituent domains.

It is likely that proteins consisting of multiple domains (and thus multiple functions) evolved by the fusion of genes that once coded for separate proteins. This would require gene duplication to be common in nature, and analysis of completed genomes has confirmed that approximately 90% of domains in eukaryotes have been duplicated. Thus, the protein domain is a fundamental unit in evolution. Many proteins have been "assembled" by duplicating domains and then combining them in different ways. Many proteins are assemblies constructed from several individual domains, and some proteins contain multiple copies of the same domain. Figure 6.25 shows the tertiary structures of nine domains that are frequently duplicated, and Figure 6.26 presents several proteins that contain multiple copies of one or more of these domains.

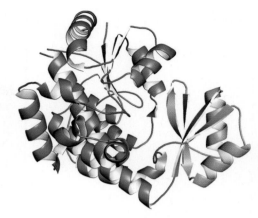

FIGURE 6.24 Malonyl CoA:ACP transacylase (pdb id = 1NM2) is a metabolic enzyme consisting of two subdomains. The large subdomain (blue) includes residues 1–132 and 198–316 and consists of a β-sheet surrounded by 12 α-helices. The small subdomain (gold = residues 133–197) consists of a four-stranded antiparallel β-sheet and two α-helices.

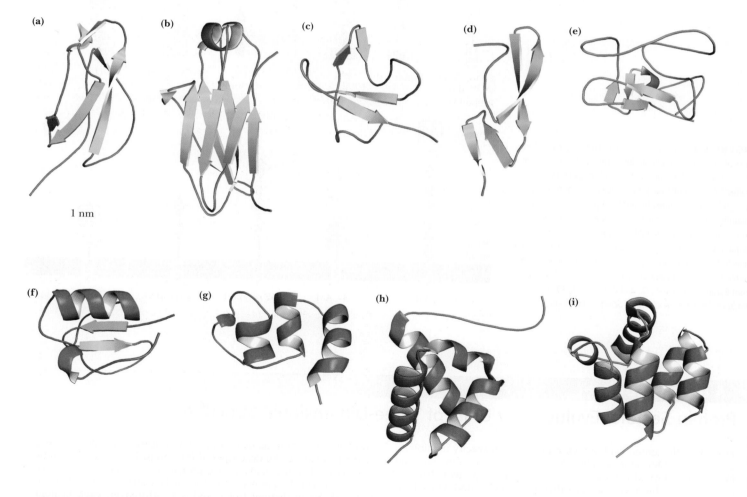

FIGURE 6.25 Ribbon structures of several protein modules used in the construction of complex multimodule proteins. **(a)** The complement control protein module (pdb id = 1HCC). **(b)** The immunoglobulin module (pdb id = 1T89). **(c)** The fibronectin type I module (pdb id = 1Q06). **(d)** The growth factor module (pdb id = 1FSB). **(e)** The kringle module (pdb id = 1HPK). **(f)** The GYF module (pdb id = 1GYF). **(g)** The γ-carboxyglutamate module (pdb id = 1CFI). **(h)** The FF module (pdb id = 1UZC). **(i)** The DED domain (pdb id = 1A1W).

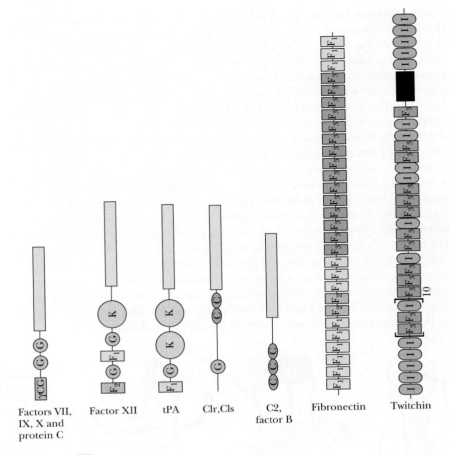

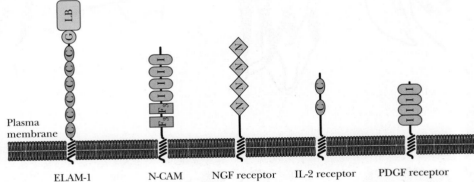

FIGURE 6.26 A sampling of proteins that consist of mosaics of individual protein modules. The modules shown include γCG, a module containing γ-carboxyglutamate residues; G, an epidermal growth factor–like module; K, the "kringle" domain, named for a Danish pastry; C, which is found in complement proteins; F1, F2, and F3, first found in fibronectin; I, the immunoglobulin superfamily domain; N, found in some growth factor receptors; E, a module homologous to the calcium-binding E–F hand domain; and LB, a lectin module found in some cell surface proteins. (Adapted from Baron, M., Norman, D., and Campbell, I., 1991. Protein modules. *Trends in Biochemical Sciences* **16**:13–17.)

> ## ▶ A DEEPER LOOK
>
> # Protein Sectors: Evolutionary Units of Three-Dimensional Structure
>
> How does the amino acid sequence of a protein specify its biological function? Our understanding of protein structure, function, and evolution is normally guided by the hierarchy of protein structure (1°, 2°, 3°, 4°), but there may be other ways to view and understand proteins.
>
> Consider, for example, that the behaviors of amino acids in proteins are **correlated**—that is, an amino acid in a given position in a protein is not independent. Its action and behavior depends on the amino acids nearby. Statistical analysis of correlated evolution between amino acids in a protein indicates that the structure of the protein can be decomposed into quasi-independent groups of correlated amino acids termed **protein sectors.** Such protein sectors are physically connected in the tertiary structure (see figure), and each has a distinct functional role. Moreover, each protein sector constitutes an independent mode of sequence divergence in the protein family. Protein sectors thus appear to represent a structural organization of proteins that reflects their evolutionary histories.

A DEEPER LOOK   Protein Sectors: Evolutionary Units of Three-Dimensional Structure *(Continued)*

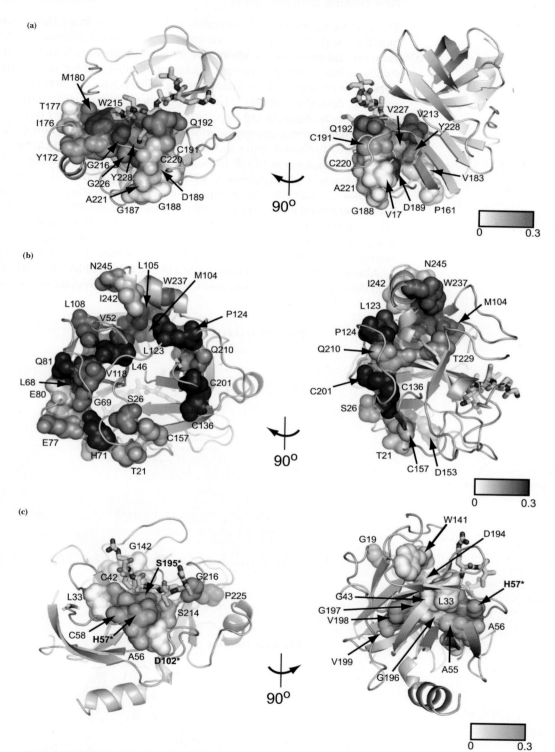

The so-called S1A serine proteases (trypsin, chymotrypsin, elastase, tPA, etc., as described in Chapter 14) are a large family of proteins united by similar structure and function. Several views of rat trypsin show three different sectors in this protein: (red) A contiguous network of amino acids that serves as the primary determinant of substrate specificity; (blue) Another contiguous network of amino acids in the core of the protein; (green) A third contiguous network of amino acids defines the catalytic core of the serine protease family of proteins. (pbd id = 3TGI)

Reference: Halabi, N., Rivoire, O., Leibler, S., and Ranganathan, R., 2009. Protein sectors: evolutionary units of three-dimensional structure. *Cell* **138:**774–786.

Classification Schemes for the Protein Universe Are Based on Domains

The astounding diversity of properties and behaviors in living things can now be explored through the analysis of vast amounts of genomic information. Assessment of sequence and structural data from several million proteins in both protein and genome databases has shown that there is a relatively limited number of structurally distinct domains in proteins. Several comprehensive projects have organized the available information in defined hierarchies or levels of protein structure. The Structural Classification of Proteins database (SCOP, *http://scop.mrc-lmb.cam.ac.uk/scop*) recognizes five overarching classes, which encompass most proteins. SCOP is based on hierarchical levels that embody the evolutionary and structural relationships among known proteins, and protein classification in SCOP is essentially a manual process using visual inspection and comparison of structures. CATH is another hierarchical classification system (*http://www.cathdb.info*) that groups protein domain structures into evolutionary families and structural groupings, depending on sequence and structure similarities. CATH differs from SCOP in that it combines manual analysis with automation based on quantitative algorithms to classify protein structures. Figure 6.27 compares the hierarchical structures of SCOP and CATH and defines the different levels of structure.

Although the hierarchical names in SCOP and CATH differ somewhat, there are common threads shared in these schemes. **Class** is determined from the overall composition of

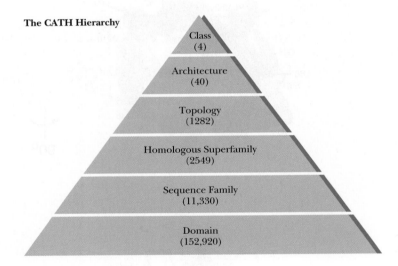

The CATH Hierarchy

Class (4)
Architecture (40)
Topology (1282)
Homologous Superfamily (2549)
Sequence Family (11,330)
Domain (152,920)

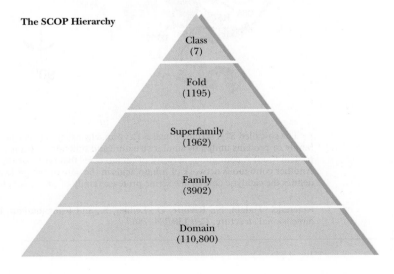

The SCOP Hierarchy

Class (7)
Fold (1195)
Superfamily (1962)
Family (3902)
Domain (110,800)

FIGURE 6.27 SCOP and CATH are hierarchical classification systems for the known proteins. Proteins are classified in SCOP by a manual process, whereas CATH combines manual and automated procedures. Numbers indicate the population of each category.

secondary structure elements in a domain. A **fold** describes the number, arrangement, and connections of these secondary structure elements. A **superfamily** includes domains of similar folds and usually similar functions, thus suggesting a common evolutionary ancestry. A **family** usually includes domains with closely related amino acid sequences (in addition to folding similarities). Although the numbers of unique folds, superfamilies, and families increase as more genomes are known and analyzed, it has become apparent that *the number of protein domains in nature is large but limited*. How many proteins can we expect to identify and understand someday? There are approximately 10^3 to 10^5 genes per organism and approximately 13.6 million species of living organisms on earth (and this latter number is likely an underestimate). Thus, there may be approximately ($10^3 \times 1.36 \times 10^7$) or 10^{10} to 10^{12} different proteins in all organisms on earth. Still, this vast number of proteins may well consist of only about 10^5 sequence domain families (Figure 6.27) and approximately 10^3 protein folds of known structure—a remarkably small number compared to the total number of protein-coding genes (see Table 1.6). It is anticipated that most newly identified proteins will resemble other known proteins and that most structures can be broken into two or more domains, which resemble tertiary structures observed in other proteins.

Because structure depends on sequence, and because function depends on structure, it is tempting to imagine that all proteins of a similar structure should share a common function, but this is not always true. For example, the **TIM barrel** is a common protein fold consisting of eight α-helices and eight β-strands that alternate along the peptide backbone to form a doughnutlike tertiary structure. The TIM barrel is named for triose phosphate isomerase, an enzyme that interconverts ketone and aldehyde substrates in the breakdown of sugars (see Chapter 18). However, other TIM barrel proteins carry out very different functions (Figure 6.28a), including the reduction of aldose sugars and hydrolysis of phosphate esters. Moreover, not all proteins of similar function possess similar domains. Both proteins shown in Figure 6.28b catalyze the same reaction, but they bear little structural similarity to each other.

(a) Same domain type, different functions:

Triose phosphate isomerase

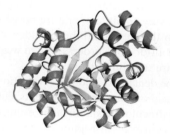
Aldose reductase

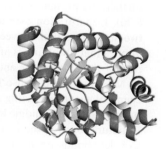
Phosphotriesterase

(b) Same function, different structures:

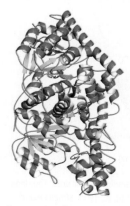
Aspartate aminotransferase

D-amino acid aminotransferase

FIGURE 6.28 **(a)** Some proteins share similar structural features but carry out quite different functions (triose phosphate isomerase, pdb id = 8TIM; aldose reductase, pdb id = 1ADS; phosphotriesterase, pdb id = 1DPM). **(b)** Proteins with quite different structures can carry out similar functions (yeast aspartate aminotransferase, pdb id = 1YAA); D-amino acid aminotransferase, pdb id = 3DAA).

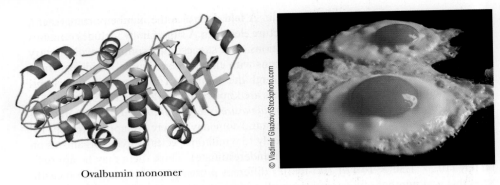

Ovalbumin monomer

FIGURE 6.29 The proteins of egg white are denatured (as evidenced by their precipitation and aggregation) during cooking. More than half of the protein in egg whites is ovalbumin. (Ovalbumin pdb id = 1OVA)

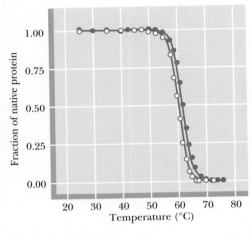

FIGURE 6.30 Proteins can be denatured by heat, with commensurate loss of function. Ribonuclease A (blue) and ribonuclease B (red) lose activity above about 55°C. These two enzymes share identical protein structures, but ribonuclease B possesses a carbohydrate chain attached to Asn[34]. (Adapted from Arnold, U., and Ulbrich-Hofmann, R., 1997. Kinetic and thermodynamic thermal stabilities of ribonuclease A and ribonuclease B. *Biochemistry* **36**:2166-2172.)

Denaturation Leads to Loss of Protein Structure and Function

Whereas the primary structure of proteins arises from covalent bonds, the secondary, tertiary, and quaternary levels of protein structure are maintained by weak, noncovalent forces. The environment of a living cell is exquisitely suited to maintain these weak forces and to preserve the structures of its many proteins. However, a variety of external stresses—for example, heat or chemical treatment—can disrupt these weak forces in a process termed **denaturation**—the loss of protein structure and function.

An everyday example is the denaturation of the protein ovalbumin during the cooking of an egg (Figure 6.29). About 10% of the mass of an egg white is protein, and 54% of that is ovalbumin. When a chicken egg is cracked open, the "egg white" is a nearly transparent, viscous fluid. Cooking turns this fluid to a solid, white mass. The egg white proteins have unfolded and have precipitated out of solution, and the unfolded proteins have aggregated into a solid mass.

As a typical protein solution is heated slowly, the protein remains in its native state until it approaches a characteristic melting temperature, T_m. As the solution is heated further, the protein denatures over a narrow range of temperatures around T_m (Figure 6.30). Denaturation over a very small temperature range such as this is evidence of a **two-state transition** between the native and the unfolded states of the protein, and this implies that unfolding is an all-or-none process: When weak forces are disrupted in one part of the protein, the entire structure breaks down.

Most proteins can also be denatured below the transition temperature by a variety of chemical agents, including acid or base, organic solvents, detergents, and particular denaturing solutes. Guanidine hydrochloride and urea are examples of the latter (Figure 6.31). Denaturation in all these cases involves disruption of the weak forces that stabilize proteins. Covalent bonds are not affected. Acids and bases cause protonation and deprotonation of

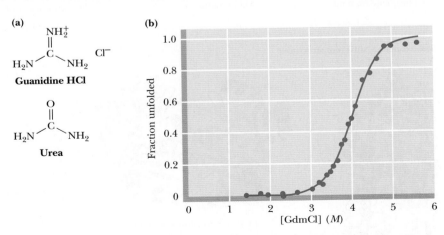

FIGURE 6.31 Proteins can be denatured (unfolded) by high concentrations of guanidine-HCl or urea. The denaturation of chymotrypsin in guanidine-HCl (GdmCl) is plotted here. (Adapted from Fersht, A., 1999. *Structure and Mechanism in Protein Science.* New York, W. H. Freeman.)

dissociable groups on the protein, altering ionic interactions and hydrogen bonds. Organic solvents and detergents disrupt hydrophobic interactions that bury nonpolar groups in the protein interior. The effects of guanidine hydrochloride and urea are more complex. Recent research indicates that these agents denature proteins by both direct effects (binding to hydrophilic groups on the protein) and indirect effects (altering the structure and dynamics of the water solvent). Also, both guanidine hydrochloride and urea are good H-bond donors and acceptors.

Anfinsen's Classic Experiment Proved That Sequence Determines Structure

As noted earlier (Section 6.2), all the information needed to fold a polypeptide into its native structure is contained in the amino acid sequence. This simple but profound truth of protein structure was confirmed in the 1950s by the elegant studies of denaturation and renaturation of proteins by Christian Anfinsen and his co-workers at the National Institutes of Health. For their pivotal studies, they chose the small enzyme ribonuclease A from bovine pancreas, a protein with 124 residues and four disulfide bonds (Figures 6.19 and 6.32). (Ribonuclease cleaves chains of ribonucleic acid. Only ribonuclease in its native

FIGURE 6.32 Ribonuclease can be unfolded by treatment with urea, and β-mercaptoethanol (MCE) cleaves disulfide bonds. If β-mercaptoethanol is then removed (but not urea) under oxidizing conditions, disulfide bonds reform in the still-unfolded protein (one possible hypothetical inactive form is shown). If urea is removed in the presence of a small amount of β-mercaptoethanol with gentle warming, ribonuclease returns to its native structure (with the correct set of disulfide bonds), and full enzymatic activity is restored. This experiment demonstrated that the information required for folding of globular proteins is contained in the primary structure.

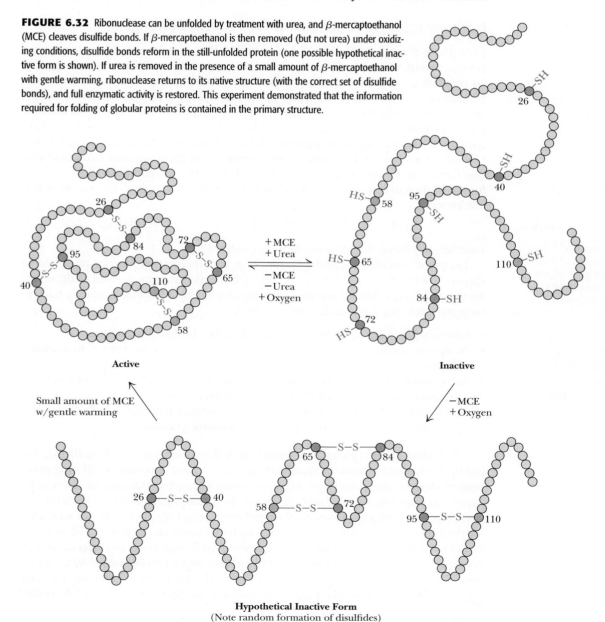

Active

+MCE
+Urea

−MCE
−Urea
+Oxygen

Inactive

Small amount of MCE
w/gentle warming

−MCE
+Oxygen

Hypothetical Inactive Form
(Note random formation of disulfides)

structure posseses enzyme activity, so loss of activity in a denaturation experiment was proof of loss of structure.) They treated solutions of ribonuclease with a combination of urea, which unfolded the protein, and mercaptoethanol, which reduced the disulfide bridges. This treatment destroyed all enzymatic activity of ribonuclease.

Anfinsen discovered that removing the mercaptoethanol but not the urea restored only 1% of the enzyme activity. This was attributed to the formation of random disulfide bridges by the still-denatured protein. With eight Cys residues, there are 105 possible ways to make four disulfide bridges; thus, a residual activity of 1% made sense to Anfinsen. (The first Cys to form a disulfide has seven possible partners, the next Cys has five possible partners, the next has three, and the last Cys has only one choice. $7 \times 5 \times 3 \times 1 = 105$). However, if Anfinsen removed mercaptoethanol *and* urea at the same time, the polypeptide was able to fold into its native structure, the correct set of four disulfides reformed, and full enzyme activity was recovered (Figure 6.32). This experiment demonstrated that the information needed for protein folding resided entirely within the amino acid sequence of the protein itself. Many subsequent experiments with a variety of proteins have confirmed this fundamental postulate. For his studies of the relationship of sequence and structure, Anfinsen shared the 1972 Nobel Prize in Chemistry (with William H. Stein and Stanford Moore).

Is There a Single Mechanism for Protein Folding?

Christian Anfinsen's experiments demonstrated that proteins can fold reversibly. A corollary result of Anfinsen's work is that the native structures of at least some globular proteins are thermodynamically stable states. But the matter of how a given protein achieves such a stable state is a complex one. Cyrus Levinthal pointed out in 1968 that so many conformations are possible for a typical protein that the protein does not have sufficient time to reach its most stable conformational state by sampling all the possible conformations. This argument, termed *Levinthal's paradox,* goes as follows: Consider a protein of 100 amino acids. Assume that there are only two conformational possibilities per amino acid, or $2^{100} = 1.27 \times 10^{30}$ possibilities. Allow 10^{-13} sec for the protein to test each conformational possibility in search of the overall energy minimum:

$$(10^{-13} \text{ sec})(1.27 \times 10^{30}) = 1.27 \times 10^{17} \text{ sec} = 4 \times 10^9 \text{ years}$$

Four billion years is the approximate age of the earth.

Levinthal's paradox led protein chemists to hypothesize that proteins must fold by specific "folding pathways," and many research efforts have been devoted to the search for these pathways. Several consistent themes have emerged from these studies. Each of them may well play a role in the folding process:

- Secondary structures—helices, sheets, and turns—probably form first.
- Nonpolar residues may aggregate or coalesce in a process termed **hydrophobic collapse.**
- Subsequent steps probably involve formation of long-range interactions between secondary structures or other hydrophobic interactions.
- The folding process may involve one or more intermediate states, including transition states and what have become known as **molten globules.**

The folding of most globular proteins may well involve several of these themes. For example, even in the denatured state, many proteins appear to possess small amounts of **residual structure** due to hydrophobic interactions, with strong interresidue contacts between side chains that are relatively distant in the native protein structure. Such interactions, together with small amounts of secondary structure, may act as sites of **nucleation** for the folding process. A bit further in the folding process, the molten globule is postulated to be a flexible but compact form characterized by significant amounts of secondary structure, virtually no precise tertiary structure, and a loosely packed hydrophobic core. Moreover, it is likely that any one of these themes is more important for some proteins than for others. The process of folding is clearly complex, but sophisticated

A collaboration of computer scientists and biochemists led by David Baker at the University of Washington has created *Foldit,* a computer game based on "crowdsourcing" and distributed computing. More than 150,000 *Foldit* players have devoted millions of hours of computer time to solving a variety of protein folding problems. In *Foldit,* players working collaboratively develop a rich assortment of new strategies and algorithms to better understand the folding of globular proteins. *Foldit* can be found at *http://fold.it/portal/*

(Cooper, S., et al., 2009. Predicting protein structures with a multiplayer online game. *Nature* **466:**756–760.)

simulations have already provided reasonable models of folding (and unfolding) pathways for many proteins (Figure 6.33). One school of thought suggests that for any given protein there may be multiple folding pathways. For these cases, Ken Dill has suggested that the folding process can be pictured as a funnel of free energies—an **energy landscape** (Figure 6.34). The rim at the top of the funnel represents the many possible unfolded states for a polypeptide chain, each characterized by high free energy and significant conformational entropy. Polypeptides fall down the wall of the funnel as contacts made between residues establish different folding possibilities. The narrowing of the funnel reflects the smaller number of available states as the protein approaches its final state, and bumps or pockets on the funnel walls represent partially stable intermediates in the folding pathway. The most stable (native) folded state of the protein lies at the bottom of the funnel.

What Is the Thermodynamic Driving Force for Folding of Globular Proteins?

The free energy change for the folding of a globular protein must be negative if the folded state is more stable than the unfolded state. The free energy change for folding depends, in turn, on changes in enthalpy and entropy for the process:

$$\Delta G = \Delta H - T\Delta S$$

When ΔH, $-T\Delta S$, and ΔG are measured separately for the polar side chains and for the nonpolar side chains of the protein, an important insight is apparent. The enthalpy and entropy changes for polar residues largely cancel each other out, and the ΔG of folding for the polar residues is approximately zero.

To understand the behavior of the nonpolar residues, it is helpful to distinguish the ΔH and $-T\Delta S$ contributions for the polypeptide chain and for the water solvent. Both ΔH and $-T\Delta S$ for the nonpolar residues of the peptide chain are positive and thus make unfavorable contributions to the folding free energy. However, large numbers of water molecules restricted and immobilized around nonpolar residues in the unfolded protein are liberated in the folding process. The burying of nonpolar residues in the folded protein's core produces a dramatic entropy increase for these liberated water molecules. This is just enough to make the overall ΔG for folding negative (and thus favorable). The crucial results:

- The largest contribution to the stability of a folded protein is the entropy change for the water molecules associated with the nonpolar residues.
- The overall free energy change for the folding process is not large—typically -20 to -40 kJ/mol.

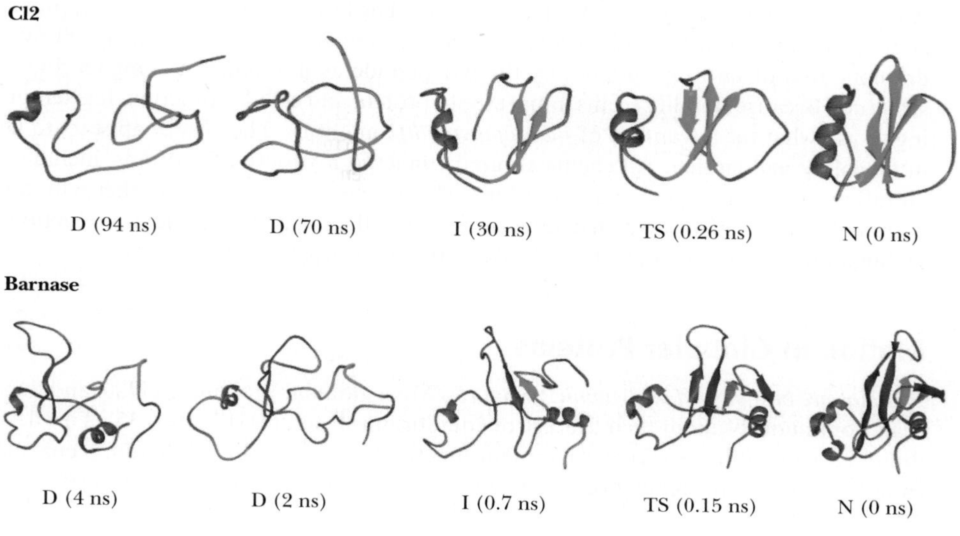

Cl2

D (94 ns)   D (70 ns)   I (30 ns)   TS (0.26 ns)   N (0 ns)

Barnase

D (4 ns)   D (2 ns)   I (0.7 ns)   TS (0.15 ns)   N (0 ns)

FIGURE 6.33 Computer simulations of folding and unfolding of proteins can reveal possible folding pathways. Molecular dynamics simulations of the unfolding of small proteins such as chymotrypsin inhibitor 2 (Cl2) and barnase are presented here on a reversed time scale to show how folding may occur. D = denatured, I = intermediate, TS = transition state, N = native. (Adapted from Daggett, V., and Fersht, A. R., 2003. Is there a unifying mechanism for protein folding? *Trends in Biochemical Sciences* **28**:18–25. Figures provided by Alan Fersht and Valerie Daggett.)

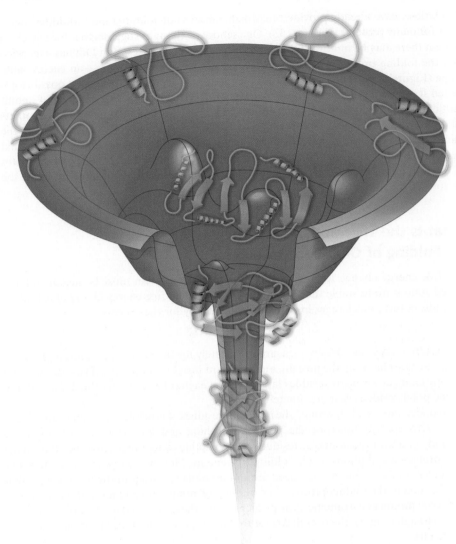

FIGURE 6.34 A model for the steps involved in the folding of globular proteins. The funnel represents a free energy surface or energy landscape for the folding process. The protein folding process is highly cooperative. Rapid and reversible formation of local secondary structures is followed by a slower phase in which establishment of partially folded intermediates leads to the final tertiary structure. Substantial exclusion of water occurs very early in the folding process.

How do proteins shift efficiently and precisely from one conformation to another? Nuclear magnetic resonance measurements by Dorothee Kern and co-workers have shown that transient hydrogen bonds are made in the conversion from one conformation to another in NtrC, a nitrogen regulatory protein.

(Gardino, A., et al., 2010. Transient non-native hydrogen bonds promote activation of a signaling protein. *Cell* **139**:1109–1118.)

Marginal Stability of the Tertiary Structure Makes Proteins Flexible

A typical folded protein is only marginally stable. The hundreds of van der Waals interactions and hydrogen bonds in a folded structure are compensated and balanced by a dramatic loss of entropy suffered by the polypeptide as it assumes a compact folded structure. Because stability seems important to protein and cellular function, it is tempting to ask what the advantage of *marginal stability* might be. The answer appears to lie in **flexibility and motion.** All chemical bonds undergo a variety of motions, including vibrations and (for single bonds) rotations. This propensity to move, together with the marginal stability of protein structures, means that the many noncovalent interactions within a protein can be interrupted, broken, and rearranged rapidly.

Motion in Globular Proteins

Proteins are best viewed as dynamic structures. Most globular proteins oscillate and fluctuate continuously about their average or equilibrium structures (Figure 6.35). This flexibility is essential for a variety of protein functions, including ligand binding, enzyme catalysis, and enzyme regulation, as shown throughout the remainder of this text.

The motions of proteins may be motions of individual atoms, groups of atoms, or even whole sections of the protein. Furthermore, they may arise either from thermal energy or from specific, triggered conformational changes in the protein. **Atomic fluctuations** such as vibrations typically are random, are very fast, and usually occur over small distances, as shown in Table 6.2. These motions arise from the kinetic energy within the protein and are a function of temperature. In the tightly packed interior of the typical protein, atomic movements of an angstrom or less are typical. The closer to the surface of the protein, the more movement can occur, and on the surface atomic movements of several angstroms are possible.

A class of slower motions, which may extend over larger distances, is **collective motions**. These are movements of a group of atoms covalently linked in such a way that the group moves as a unit. Such a group can range from a few atoms to hundreds of atoms. These motions are of two types: (1) those that occur quickly but infrequently, such as tyrosine ring flips, and (2) those that occur slowly, such as the hinge-bending movement between protein domains. For example, the two antigen-binding domains of immunoglobulins move as relatively rigid units to selectively bind separate antigen molecules. These collective motions also arise from thermal energies in the protein and operate on a timescale of 10^{-12} to 10^{-3} sec. It is often important to distinguish the time scale of the motion itself versus the frequency of its occurrence. A tyrosine ring flip takes only a picosecond (1×10^{-12} sec), but such flips occur only about once every millisecond (1×10^{-3} sec).

Conformational changes involve motions of groups of atoms (individual side chains, for example) or even whole sections of proteins. These motions occur on a time scale of 10^{-9} to 10^3 sec, and the distances covered can be as large as 1 nm. These motions may occur in response to specific stimuli or arise from specific interactions within the protein (hydrogen bonding, electrostatic interactions, or ligand binding—see Chapters 14 and 15).

The *cis–trans* **isomerization of proline residues** in proteins (Figure 6.36) occurs over an even longer time scale—typically 10^1 to 10^4 sec. Conversion of even a single proline from its *cis* to its *trans* conformation can alter a protein structure dramatically. Proline *cis–trans* isomerizations sometimes act as switches to activate a protein or open a channel across a membrane (see Chapter 9).

The Folding Tendencies and Patterns of Globular Proteins

Globular proteins adopt the most stable tertiary structure possible. To do this, the peptide chain must both (1) satisfy the constraints inherent in its own structure and (2) fold so as to "bury" the hydrophobic side chains, minimizing their contact with solvent. The polypeptide itself does not usually form simple straight chains. Even in chain segments

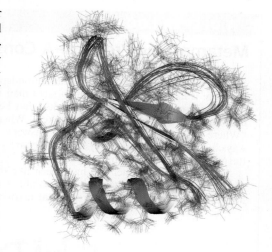

FIGURE 6.35 Proteins are dynamic structures. The marginal stability of a tertiary structure leads to flexibility and motion in the protein. Determination of structures of proteins (such as the SH3 domain of the α-chain of spectrin, shown here) by nuclear magnetic resonance produces a variety of stable tertiary structures that fit the data. Such structural ensembles provide a glimpse into the range of structures that may be accessible to a flexible, dynamic protein (pdb id = 1M8M).

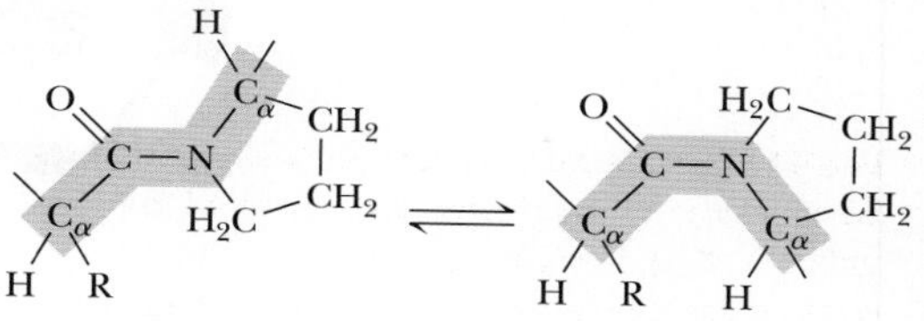

trans   *cis*

FIGURE 6.36 The *cis* and *trans* configurations of proline residues in peptide chains are almost equally stable. Proline *cis-trans* isomerizations, often occurring over relatively long time scales, can alter protein structure significantly.

TABLE 6.2	Motion and Fluctuations in Proteins		
Type of Motion	Spatial Displacement (Å)	Characteristic Time (sec)	Source of Energy
Atomic vibrations	0.01–1	10^{-15}–10^{-11}	Kinetic energy
Collective motions 1. Fast: Tyr ring flips; methyl group rotations 2. Slow: hinge bending between domains	0.01–5 or more	10^{-12}–10^{-3}	Kinetic energy
Triggered conformation changes	0.5–10 or more	10^{-9}–10^3	Interactions with triggering agent
Proline *cis–trans* isomerization	3–10	10^1–10^4	Kinetic energy or enzyme driven

Adapted from Petsko, G. A., and Ringe, D., 1984. Fluctuations in protein structure from X-ray diffraction. *Annual Review of Biophysics and Bioengineering* **13:**331–371.

▶ A DEEPER LOOK

Metamorphic Proteins–A Consequence of Dynamism and Marginal Stability

As our knowledge evolves, we find that our understanding of protein structure is like the story of the inebriated man wandering under a street lamp one night. A passerby asks what he is doing, and the man answers that he is looking for his keys. When the passerby asks "Where did you lose them?", the man answers "Down the street a ways." And when the passerby asks "Why are you looking here?", the man replies, "There's more light here!"

Until recently, biochemists have been like the inebriated man. The proteins that have been most amenable to structural studies have been the more stable proteins that favor a single conforma-

tion. As structural methodologies have evolved, we have come to appreciate that many proteins can exist as an ensemble of structures of more or less similar energies and stabilities. Some of the recent work has been remarkable. For example, it is becoming clear that a few mutations can dramatically change the three-dimensional structure of a protein. Philip Bryan and co-workers have shown that the conversion of *Streptococcus* protein G, a 56-residue protein, from a structure dominated by three α-helices to a structure with four β-strands and an α-helix can occur via a single amino acid substitution—changing residue 45 from Leu to Tyr.

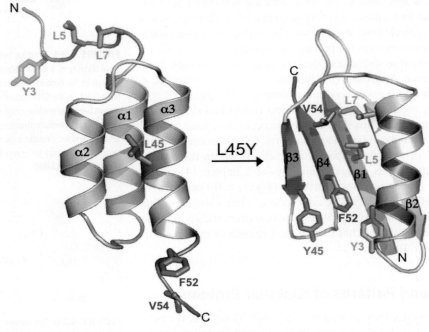

(Left) PDB ID = 2KDL; (Right) PDB ID = 2KDM

Similarly, Sebastian Meier and his co-workers have shown that one mutation (Lys to Pro at residue 21) in a 28-amino acid cysteine-rich domain from nematocyst outer wall antigen (NOWA) results in an

equilibrium between the original folded structure and one that resembles the fold of another domain with which it interacts, Mcol1C. A second mutation (Gly to Val at residue 11) completes this transition.

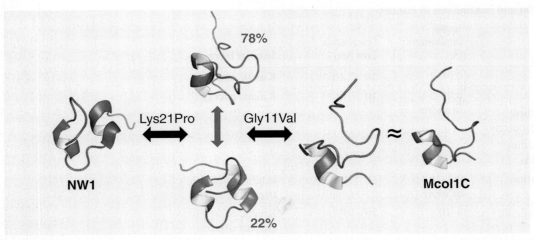

Figure (left: NW1 pdb id = 2HM3; center: NW1 K21P species I pdb id = 2HM4; NW1 K21P species II pdb id = 2HM5; right: NW1 G11V, K21P pdb id = 2HM6)

Metamorphic Proteins—A Consequence of Dynamism and Marginal Stability *(continued)*

Finally, the chemokine* protein lymphotactin has been described as a **metamorphic protein** because it exists naturally in an equilibrium between two diffferent three-dimensional structures under physiological conditions (K_{eq} ~1). This equilibrium is advantageous for lymphotactin, which adopts these different forms in order to interact with multiple proteins in the course of performing its biological function.

*Chemokines are small proteins secreted by cells that induce and regulate cell movement.

References:

Alexander, P., et al., 2009. A minimal sequence code for switching protein structure and function. *Proc. Natl. Acad. Sci.* **106:**21149–21154.
Kuloglu, E., et al., 2001. Monomeric solution structure of the prototypical 'C' chemokine lymphotactin. *Biochemistry* **40:**12486–12496.
Meier, S., et al., 2007. Continuous molecular evolution of protein-domain structures by single amino acid changes. *Current Biology* **17:**173–178.
Murzin, A., 2008. Metamorphic proteins. *Science* **320:**1725–1726.
Tokuriki, N. and Tawfik, D., 2009. Protein dynamism and evolvability. *Science* **324:**203–207.

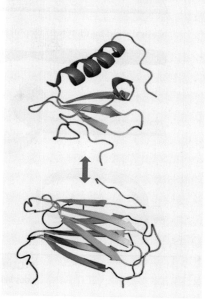

top: pdb id = 1J9O, 1J8I; bottom: pdb id = 2JP1)

where helices and sheets are not formed, an extended peptide chain, being composed of L-amino acids, has a tendency to twist slightly in a right-handed direction. As shown in Figure 6.37, this tendency is apparently the basis for the formation of a variety of tertiary structures having a right-handed sense. Principal among these are the right-handed twists in β-sheets and right-handed cross-overs in parallel β-sheets. Right-handed twisted β-sheets are found at the center of a number of proteins (Figure 6.38) and provide an extended, highly stable structural core.

Connections between β-strands are of two types—hairpins and cross-overs. **Hairpins,** as shown in Figure 6.37, connect adjacent antiparallel β-strands. **Cross-overs** are necessary to connect adjacent (or nearly adjacent) parallel β-strands. Nearly all cross-over structures are right-handed. In many cross-over structures, the cross-over connection itself contains an α-helical segment. This creates a **βαβ-loop.** As shown in Figure 6.37, the strong tendency in nature to form right-handed cross-overs, the wide occurrence of α-helices in the cross-over connection, and the right-handed twists of β-sheets can all be understood as arising from the tendency of an extended polypeptide chain of L-amino acids to adopt a right-handed twist structure. This is a chiral effect. Proteins composed of D-amino acids would tend to adopt left-handed twist structures.

The second driving force that affects the folding of polypeptide chains is the need to bury the hydrophobic residues of the chain, protecting them from solvent water. From a topological viewpoint, then, all globular proteins must have an "inside" where the hydrophobic core can be arranged and an "outside" toward which the hydrophilic groups must be directed. The sequestration of hydrophobic residues away from water is the dominant force in the arrangement of secondary structures and nonrepetitive peptide segments to form a given tertiary structure. Globular proteins can be classified mainly on the basis of the particular kind of core or backbone structure they use to accomplish this goal. The term *hydrophobic core,* as used here, refers to a region in which hydrophobic side chains cluster together, away from the solvent. *Backbone* refers to the polypeptide backbone itself, excluding the particular side chains. Globular proteins can

(a)

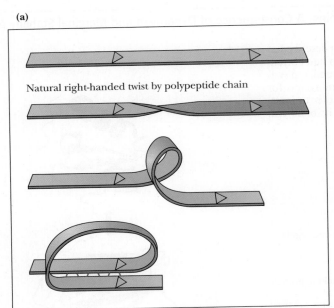

(b)

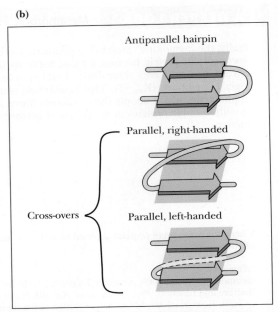

FIGURE 6.37 **(a)** The natural right-handed twist exhibited by polypeptide chains, and **(b)** the types of connections between β-strands.

be pictured as consisting of "layers" of backbone, with hydrophobic core regions between them. More than half the known globular protein structures have two layers of backbone (separated by one hydrophobic core). Roughly one-third of the known structures are composed of three backbone layers and two hydrophobic cores. There are also a few known four-layer structures and at least one five-layer structure. A few structures are not easily classified in this way, but it is remarkable that most proteins fit into one of these classes. Examples of each are presented in Figure 6.38.

Most Globular Proteins Belong to One of Four Structural Classes

In addition to classification based on layer structure, proteins can be grouped according to the type and arrangement of secondary structure (Figure 6.39). There are four such broad groups: **all α proteins** and **all β proteins** (in which the structures are dominated by α-helices and β-sheets, respectively), **α/β proteins** (in which helices and sheets are intermingled), and **α+β proteins** (in which α-helical and β-sheet domains are separated for the most part).

It is important to note that the similarities of tertiary structure within these groups do not necessarily reflect similar or even related functions. Instead, **functional homology** usually depends on structural similarities on a smaller and more intimate scale.

Molecular Chaperones Are Proteins That Help Other Proteins to Fold

To a first approximation, all the information necessary to direct the folding of a polypeptide is contained in its primary structure. On the other hand, the high protein concentration inside cells may adversely affect the folding process because hydrophobic interactions may lead to aggregation of some unfolded or partially folded proteins. Also, it may be necessary to suppress or reverse incorrect or premature folding. A family of proteins, known as **molecular chaperones,** are essential for the correct folding of certain polypeptide chains in vivo; for their assembly into oligomers; and for preventing

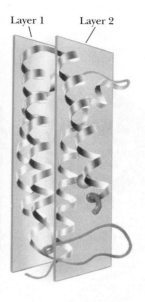

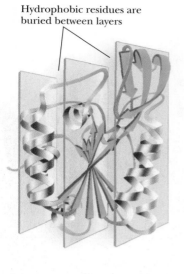

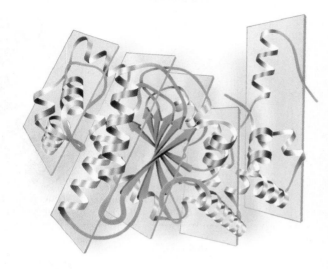

Layer 1   Layer 2

Hydrophobic residues are
buried between layers

(a) Cytochrome c'

(b) Phosphoglycerate kinase
(domain 2)

(c) Phosphorylase
(domain 2)

FIGURE 6.38 Examples of protein domains with different numbers of layers of backbone structure. **(a)** Cytochrome c' with two layers of α-helix. **(b)** Domain 2 of phosphoglycerate kinase, composed of a β-sheet layer between two layers of helix, three layers overall. **(c)** An unusual five-layer structure, domain 2 of glycogen phosphorylase, a β-sheet layer sandwiched between four layers of α-helix. **(d)** The concentric "layers" of β-sheet *(inside)* and α-helix *(outside)* in triose phosphate isomerase. Hydrophobic residues are buried between these concentric layers in the same manner as in the planar layers of the other proteins. The hydrophobic layers are shaded yellow. (Original art courtesy of Jane Richardson.)

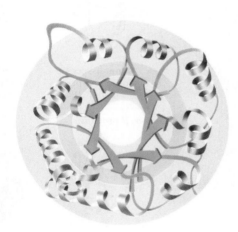

(d) Triosephosphate isomerase

inappropriate liaisons with other proteins during their synthesis, folding, and transport. Many of these proteins were first identified as **heat shock proteins,** which are induced in cells by elevated temperature or other stress. The most thoroughly studied proteins are **Hsp70,** a 70-kD heat shock protein, and the so-called **chaperonins,** also known as **Cpn60s** or **Hsp60s,** a class of 60-kD heat shock proteins. A well-characterized **Hsp60** chaperonin is **GroEL,** an *E. coli* protein that has been shown to affect the folding of several proteins. The mechanism of action of chaperones is discussed in Chapter 31.

Some Proteins Are Intrinsically Unstructured

Remarkably, it is now becoming clear that many proteins exist and function normally in a partially unfolded state. Such proteins, termed **intrinsically unstructured proteins (IUPs)** or **natively unfolded proteins,** do not possess uniform structural properties but are nonetheless essential for basic cellular functions. These proteins are characterized by an almost complete lack of folded structure and an extended conformation with high intramolecular flexibility.

Intrinsically unstructured proteins contact their targets over a large surface area (Figure 6.40). The p27 protein complexed with cyclin-dependent protein kinase 2 (Cdk2) and cyclin A shows that p27 is in contact with its binding partners across its entire length. It binds in a groove consisting of conserved residues on cyclin A. On Cdk2, it binds to the N-terminal domain and also to the catalytic cleft. One of the most appropriate roles for such long-range interactions is assembly of complexes involved in the transcription of DNA into RNA, where large numbers of proteins must be recruited in macromolecular complexes. Thus, the transactivator domain catenin-binding domain (CBD) of TCF3 is bound to several functional domains of β-catenin (Figure 6.40).

All α proteins:

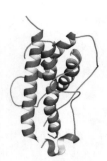

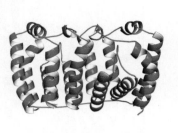

Human growth hormone
(pdb id = 1HGU)

Leucine-rich repeat
variant (pdb id = 1LRV)

Peridinin-chlorophyll protein
(a "solenoid"—pdb id = 1PPR)

Endoglucanase A (an α-helical
barrel—pdb id = 1CEM)

Cat allergen
(pdb id = 1PUO)

All β proteins:

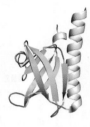

Mannose-specific
aggluttinin (a prism—
(pdb id = 1JPC)

Rieske iron protein
(a 3-layer β-sandwich—
(pdb id = 1RIE)

Hemopexin C-terminal
domain (a 4-bladed
propellor—pdb id = 1HXN)

Lectin from *R. solanacearum*
(a 6-bladed propeller—
pdb id = 1BT9)

Pleckstrin domain of
protein kinase B/AKT
(pdb id = 1UNQ)

α/β proteins:

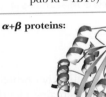

Hevamine (a "TIM barrel"
—pdb id = 2HVM)

Human bactericidal
permeability-increasing
protein (pdb id = 1BP1)

Hepatocyte growth factor
(N-terminal domain
—pdb id = 2HGF)

Prokaryotic ribosomal
protein L9
(pdb id = 1DIV)

α+β proteins:

Equine leucocyte
elastase inhibitor
(pdb id = 1HLE)

RuvA protein
(pdb id = 1CUK)

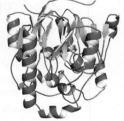

Ribonuclease H
(pdb id = 1RNH)

L-Arginine: glycine
amidinotransferase (a metabolic
enzyme—pdb id = 4JDW)

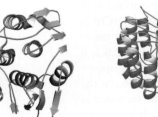

MurA (an α–β prism
—pdb id = 1EYN)

Porcine ribonuclease inhibitor
(a "horseshoe"—pdb id = 2BNH)

Thymidylate synthase
(pdb id = 3TMS)

◀ **FIGURE 6.39** Four major classes of protein structure (as defined in the SCOP database). **(a) All α proteins,** where α-helices dominate the structure; **(b) All β proteins,** in which β-sheets are the primary feature; **(c) α/β proteins,** where α-helices and β-sheets are mixed within a domain; **(d) $\alpha+\beta$ proteins,** in which α-helical and β-sheet domains are separated to at least some extent.

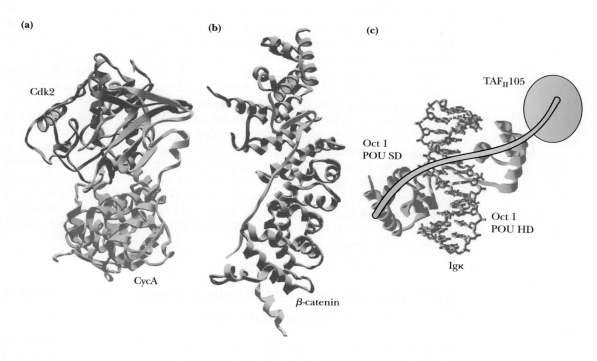

(a) Cdk2 / CycA

(b) β-catenin

(c) TAF$_{II}$105 / Oct 1 POU SD / Oct 1 POU HD / Igκ

FIGURE 6.40 Intrinsically unstructured proteins (IUPs) contact their target proteins over a large surface area. **(a)** p27^{Kip1} (yellow) complexed with cyclin-dependent kinase 2 (Cdk2, blue) and cyclin A (CycA, green). **(b)** The transactivator domain CBD of TCF3 (yellow) bound to β-catenin (blue). **Note:** Part of the β-catenin has been removed for a clear view of the CBD. **(c)** Bob 1 transcriptional coactivator (yellow) in contact with its four partners: TAF$_{II}$105 (green oval), the Oct 1 domains POU SD and POU HD (green), and the Igκ promoter (blue). (From Tompa, P., 2002. Intrinsically unstructured proteins. *Trends in Biochemical Sciences* **27:**527–533.) **(d-g)** Some intrinsically unstructured proteins (in red and yellow) bind to their targets by wrapping around them. Shown here are **(d)** SNAP-25 bound to BoNT/A, **(e)** SARA SBD bound to Smad 2 MH2, **(f)** HIF-1α interaction domain bound to the TAZ1 domain of CBP, and **(g)** HIF-1α interaction domain bound to asparagine hydroxylase FIH. (From *Trends in Biochemical Sciences,* Vol. 27, No. 10, page 530. October 2002.)

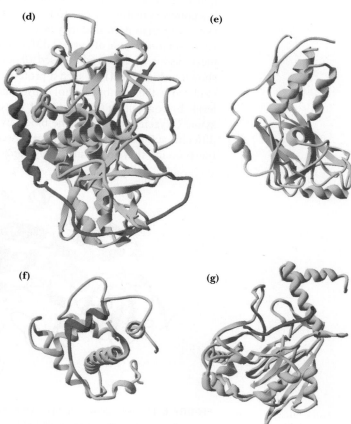

Can amino acid sequence information predict the existence of intrinsically unstructured regions on proteins? Intrinsically unstructured proteins are characterized by a unique combination of high net charge and low overall hydrophobicity. Compared with ordered proteins, IUPs have higher levels of E, K, R, G, Q, S, and P, and low amounts of I, L, V, W, F, Y, C, and N. These features provide a rationale for prediction of regions of disorder from amino acid sequence information, and experimental evidence shows that such predictions are better than 80% accurate. Genomic analysis of disordered proteins indicates that the proportion of the genome encoding IUPs and proteins with substantial regions of disorder tends to increase with the complexity of organisms. Thus, predictive analysis of whole genomes indicates that 2% of archaeal, 4.2% of bacterial, and 33% of eukaryotic proteins probably contain long regions of disorder.

Some proteins are disordered throughout their length, whereas others may contain stretches of 30 to 40 residues or more that are disordered and imbedded in an otherwise folded protein. The prevalence of disordered segments in proteins may reflect two different cellular needs. (1) Disordered proteins are more malleable and thus can adapt their structures to bind to multiple ligands, including other proteins. Each such interaction could provide a different function in the cell. (2) Compared with compact, folded proteins, disordered segments in proteins appear to be able to form larger intermolecular interfaces to which ligands, such as other proteins, could bind (Figure 6.40). Folded proteins might have to be two to three times larger to produce the binding surface possible with a disordered protein. Larger proteins would increase cellular crowding or could increase cell size by 15% to 30%. The flexibility of disordered proteins may thus reduce protein, genome, and cell sizes.

6.5 How Do Protein Subunits Interact at the Quaternary Level of Protein Structure?

Many proteins exist in nature as **oligomers,** complexes composed of (often symmetric) noncovalent assemblies of two or more monomer subunits. In fact, subunit association is a common feature of macromolecular organization in biology. Most intracellular enzymes are oligomeric and may be composed either of a single type of monomer subunit *(homomultimers)* or of several different kinds of subunits *(heteromultimers).* The simplest case is a protein composed of identical subunits. Liver alcohol dehydrogenase, shown in Figure 6.41, is such a protein. Alcohol consumed in a beer or mixed drink is oxidized in the liver by alcohol dehydrogenase. Hormonal signals modulate blood sugar levels by controlling the activity of glycogen phosphorylase, another homodimeric enzyme. Oxygen is carried in the blood by hemoglobin, which contains two each of two different subunits *(heterotetramer).* A counterpoint to these small clusters is made by the proteins that form large *polymeric* aggregates. Proteins are synthesized on large

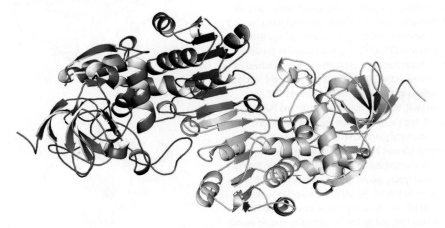

FIGURE 6.41 The quaternary structure of liver alcohol dehydrogenase. Within each subunit is a six-stranded parallel sheet. Between the two subunits is a two-stranded antiparallel sheet (pdb id = 1CDO).

α_1-Antitrypsin—A Tale of Molecular Mousetraps and a Folding Disease

In the human lung, oxygen and CO_2 are exchanged across the walls of **alveoli**—air sacs surrounded by capillaries that connect the pulmonary veins and arteries. The walls of alveoli consist of the elastic protein **elastin.** Inhalation expands the alveoli, and exhalation compresses them. A pair of human lungs contains 300 million alveoli, and the total area of the alveolar walls in contact with capillaries is about 70 m^2—an area about the size of a tennis court! In the lungs, neutrophils (a type of white blood cell) naturally secrete **elastase**, a protein-cleaving enzyme essential to tissue repair. However, elastase also can attack and break down the elastin of the alveolar walls if it spreads from the site of inflammation repair. To prevent this, the liver secretes into the blood α_1-**antitrypsin**—a 52-kD protein belonging to the **serpin** (*ser*ine *p*rotease *in*hibitor) family—which blocks elastase action, preventing alveolar damage.

α_1-Antitrypsin is a molecular mousetrap, with a flexible peptide loop that contains a Met residue as "bait" for the elastase and that can swing like the arm of a mousetrap. When elastase binds to the loop at the Met residue, it cuts the peptide loop. Now free to move, the loop slides into the middle of a large beta sheet (green), at the same time dragging elastase to the opposite side of the α_1-antitrypsin structure. At this new binding site, the elastase structure is distorted, and it cannot complete its reaction and free itself from the α_1-antitrypsin. Cellular scavenger enzymes then attack the elastase–an-

titypsin complex and destroy it. By sacrificing itself in this way, the α_1-antitrypsin has prevented damage to the alveolar elastin.

Defects in α_1-antitrypsin can cause serious lung and liver damage. The gene for α_1-antitrypsin is polymorphic (that is, it occurs as many different sequence variants) and many variants of α_1-antitrypsin are either poorly secreted by the liver or function poorly in the lungs. Even worse, tobacco smoke oxidizes the critical Met residue in the flexible loop of α_1-antitrypsin, and smokers, especially those who carry mutants of this protein, often develop emphysema—the destruction of the elastin connective tissue in the lungs.

The flexible loop of α_1-antitrypsin—its mousetrap spring—is also its Achilles' heel. Mutations in this loop make the protein vulnerable to aberrant conformational changes. The Z-mutation of α_1-antitrypsin is an interesting case, with a Lys in place of Glu at residue 342 (indicated by the arrow in M) at the base of the flexible loop. This causes partial loop insertion in the large β-sheet (M*). This induces the modified β-sheet to accept the flexible loop of another α_1-antitrypsin, forming a dimer. Repetition of these events forms dimers (D), then polymers (P), which are trapped in the liver (often leading to cirrhosis and death). Z variants that manage to make it to the lungs associate so slowly with elastase that they are ineffective in preventing lung damage.

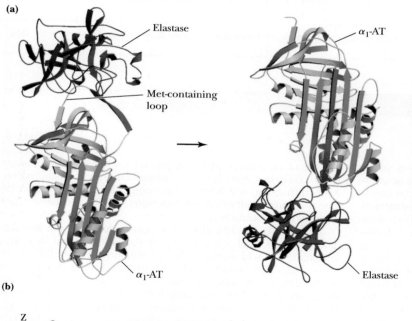

(a)

Elastase

Met-containing loop

α_1-AT

α_1-AT

Elastase

◁ **(a)** Elastase (dark gray) is inactivated by binding to α_1-antitrypsin. When elastase binds, cleaving the flexible loop at a Met residue, the rest of the loop (the red β-strand) rotates more than 180° and inserts into the green β-sheet, swinging the elastase to the other end of the molecule. At this new location, the elastase is distorted and inactivated. **(b)** In the Z-mutant of α_1-antitrypsin, the flexible loop is only partially inserted in the large β-sheet, promoting polymer formation and trapping α_1-antitrypsin at its site of synthesis in the liver. The consequences of this are cirrhosis of the liver, as well as lung damage, since the small amount of α_1-antitrypsin that reaches the lungs is ineffective in preventing lung damage. Individual monomers in the α_1-antitrypsin polymer are colored red, blue, and gold (far right). (*From Lomas, D. A., et al., 2005. Molecular mousetraps and serpinopathies.* Biochem Soc. Transactions *33:321–330. Figure provided by David Lomas.)*

(b)

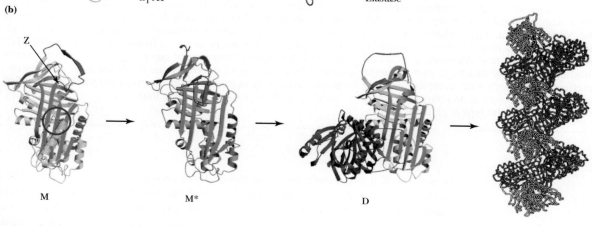

Z

M

M*

D

P

HUMAN BIOCHEMISTRY

Diseases of Protein Folding

A number of human diseases are linked to abnormalities of protein folding. Protein misfolding may cause disease by a variety of mechanisms. For example, misfolding may result in loss of function and the onset of disease. The following table summarizes several other mechanisms and provides an example of each.

Disease	Affected Protein	Mechanism
Alzheimer's disease	β-Amyloid peptide (derived from amyloid precursor protein)	Misfolded β-amyloid peptide accumulates in human neural tissue, forming deposits known as neuritic plaques.
Familial amyloidotic polyneuropathy	Transthyretin	Aggregation of unfolded proteins. Nerves and other organs are damaged by deposits of insoluble protein products.
Cancer	p53	p53 prevents cells with damaged DNA from dividing. One class of p53 mutations leads to misfolding; the misfolded protein is unstable and is destroyed.
Creutzfeldt-Jakob disease (human equivalent of mad cow disease)	Prion	Prion protein with an altered conformation (PrP^{SC}) may seed conformational transitions in normal PrP (PrP^{C}) molecules.
Hereditary emphysema	α_1-Antitrypsin	Mutated forms of this protein fold slowly, allowing its target, elastase, to destroy lung tissue.
Cystic fibrosis	CFTR (cystic fibrosis transmembrane conductance regulator)	Folding intermediates of mutant CFTR forms don't dissociate freely from chaperones, preventing the CFTR from reaching its destination in the membrane.

Aguzzi, A., and O'Connor, T., 2010. Protein aggregation diseases: pathogenicity and therapeutic perspectives. *Nature Reviews Drug Discovery* **9**:237–248.

HUMAN BIOCHEMISTRY

Structural Genomics

The prodigious advances in genome sequencing in recent years, together with advances in techniques for protein structure determination, not only have provided much new information for biochemists but have also spawned a new field of investigation—**structural genomics,** the large-scale analysis of protein structures and functions based on gene sequences. The scale of this new endeavor is daunting: hundreds of thousands of gene sequences are rapidly being determined, and current estimates suggest that there are probably less than 10,000 distinct and stable polypeptide folding patterns in nature. The feasibility of large-scale, high-throughput structure determination programs is being explored in a variety of pilot studies in Europe, Asia, and North America. These efforts seek to add 20,000 or more new protein structures to our collected knowledge in the near future; from this wealth of new information, it should be possible to predict and determine new structures from sequence information alone. This effort will be vastly more complex and more expensive than the Human Genome Project. It presently costs about $100,000 to determine the structure of the typical globular protein,

and one of the goals of structural genomics is to reduce this number to $20,000 or less. Advances in techniques for protein crystallization, X-ray diffraction, and NMR spectroscopy, the three techniques essential to protein structure determination, will be needed to reach this goal in the near future.

The payoffs anticipated from structural genomics are substantial. Access to large amounts of new three-dimensional structural information should accelerate the development of new families of drugs. The ability to scan databases of chemical entities for activities against drug targets will be enhanced if large numbers of new protein structures are available, especially if complexes of drugs and target proteins can be obtained or predicted. The impact of structural genomics will also extend, however, to **functional genomics**—the study of the functional relationships of genomic content—which will enable the comparison of the composite functions of whole genomes, leading eventually to a complete biochemical and mechanistic understanding of all organisms, including humans.

complexes of many protein units and several RNA molecules called *ribosomes.* Muscle contraction depends on large polymer clusters of the protein myosin sliding along filamentous polymers of another protein, actin. The way in which separate folded monomeric protein subunits associate to form the oligomeric protein constitutes the **quaternary structure** of that protein. Table 6.3 lists several proteins and their subunit compositions (see also Table 4.2). Proteins with two to four subunits predominate in nature, but many cases of higher numbers exist.

The subunits of an oligomeric protein typically fold independently and then interact with other subunits. The surfaces at which subunits interact are similar in nature to the interiors of the individual subunits—closely packed with both polar and hydrophobic interactions. Interacting surfaces must therefore possess *complementary* arrangements of polar and hydrophobic groups.

Oligomeric associations of protein subunits can be divided into those between **identical** subunits and those between **nonidentical** subunits. Interactions among identical subunits can be further distinguished as either **isologous** or **heterologous.** In isologous interactions, the interacting surfaces are identical and the resulting structure is necessarily dimeric and closed, with a twofold axis of symmetry (Figure 6.42). If any additional interactions occur to form a trimer or tetramer, these must use different interfaces on the protein's surface. Many proteins, such as transthyretin, form tetramers by means of two sets of isologous interactions (Figure 6.43). Such structures possess three different twofold axes of symmetry. In contrast, heterologous associations among subunits involve nonidentical interfaces. These surfaces must be complementary, but they are generally not symmetric.

There Is Symmetry in Quaternary Structures

Many multimeric proteins are symmetric arrangements of asymmetric objects (the monomer subunits). All of the polypeptide's α-carbons are asymmetric, and the polypeptide nearly always folds to form a low-symmetry structure. (The long helical arrays formed by some synthetic polypeptides are an exception.) Thus, protein subunits do not have mirror reflection planes, points, or axes of inversion. The only symmetry operation possible for protein subunits is a rotation. The most common symmetries observed for multisubunit proteins are cyclic symmetry and dihedral symmetry. In **cyclic symmetry,**

TABLE 6.3	Aggregation Symmetries of Globular Proteins
Protein	Number of Subunits
Alcohol dehydrogenase	2
Malate dehydrogenase	2
Superoxide dismutase	2
Triose phosphate isomerase	2
Glycogen phosphorylase	2
Aldolase	3
Bacteriochlorophyll protein	3
Concanavalin A	4
Glyceraldehyde-3-phosphate dehydrogenase	4
Immunoglobulin	4
Lactate dehydrogenase	4
Prealbumin	4
Pyruvate kinase	4
Phosphoglycerate mutase	4
Hemoglobin	2 + 2
Insulin	6
Aspartate transcarbamoylase	6 + 6
Glutamine synthetase	12
TMV protein disc	17
Apoferritin	24
Coat of tomato bushy stunt virus	180

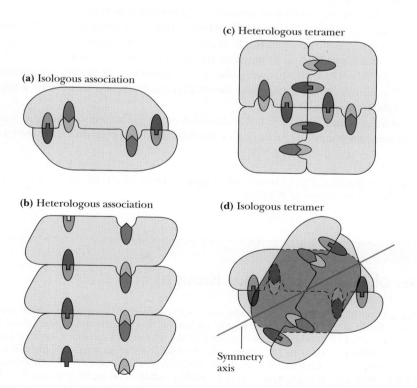

FIGURE 6.42 Isologous and heterologous associations between protein subunits. **(a)** An isologous interaction between two subunits with a twofold axis of symmetry perpendicular to the plane of the page. **(b)** A heterologous interaction that could lead to the formation of a long polymer. **(c)** A heterologous interaction leading to a closed structure—a tetramer. **(d)** A tetramer formed by two sets of isologous interactions.

(a)

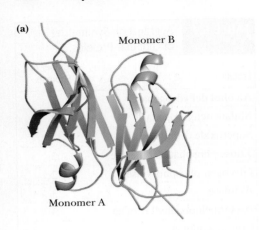

Monomer B

Monomer A

(b)

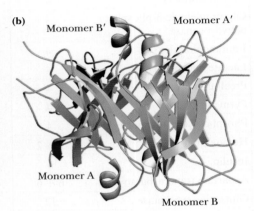

Monomer B′

Monomer A′

Monomer A

Monomer B

FIGURE 6.43 Many proteins form tetramers by means of two sets of isologous interactions. The dimeric **(a)** and tetrameric **(b)** forms of transthyretin (also known as prealbumin) are shown here (pdb id = 1GKE). The monomers of this protein form a dimer in a manner that extends the large monomer β-sheet. The tetramer is formed by isologous interactions between the large β-sheets of two dimers.

the subunits are arranged around a single rotation axis, as shown in Figure 6.44. If there are two subunits, the axis is referred to as a *twofold rotation axis*. Rotating the quaternary structure 180° about this axis gives a structure identical to the original one. With three subunits arranged about a threefold rotation axis, a rotation of 120° about that axis gives an identical structure. **Dihedral symmetry** occurs when a structure possesses at least one twofold rotation axis perpendicular to another *n*-fold rotation axis. This type of subunit arrangement (Figure 6.44) occurs in annexin XII (where $n = 3$).

Quaternary Association Is Driven by Weak Forces

Weak forces stabilize quaternary structures. Typical dissociation constants for simple two-subunit associations range from 10^{-8} to 10^{-16} *M*. These values correspond to free energies of association of about 50 to 100 kJ/mol at 37°C. Dimerization of subunits is accompanied by both favorable and unfavorable energy changes. The favorable interactions include van der Waals interactions, hydrogen bonds, ionic bonds, and hydrophobic interactions. However, considerable entropy loss occurs when subunits interact. When two subunits move as one, three translational degrees of freedom are lost for one subunit because it is constrained to move with the other one. In addition, many residues at the subunit interface, which were previously free to move on the protein surface, now have their movements restricted by the subunit association. This unfavorable energy of association is in the range of 80 to 120 kJ/mol for temperatures of 25° to 37°C. Thus, to achieve stability, the dimerization of two subunits must involve approximately 130 to 220 kJ/mol of favorable interactions.[1] Van der Waals interactions at protein interfaces are numerous, often running to several hundred for a typical monomer–monomer association. This would account for about 150 to 200 kJ/mol of favorable free energy of association. However, when solvent is removed from the protein surface to form the subunit–subunit contacts, nearly as many van der Waals associations are lost as are made. One subunit is simply trading water molecules for residues in the other subunit. As a result, the energy of subunit association due to van der Waals interactions actually contributes little to the stability of the dimer. Hydrophobic interactions at the subunit–subunit interface, however, are generally very favorable. For many proteins, the subunit association process effectively buries as much as 20 nm^2 of surface area previously exposed to solvent, resulting in as much as 100 to 200 kJ/mol of favorable hydrophobic interactions. Together with whatever polar interactions occur at the protein–protein interface, this is sufficient to account for the observed stabilization that occurs when two protein subunits associate.

An additional and important factor contributing to the stability of subunit associations for some proteins is the formation of disulfide bonds between different subunits. All

[1]For example, 130 kJ/mol of favorable interaction minus 80 kJ/mol of unfavorable interaction equals a net free energy of association of 50 kJ/mol.

A DEEPER LOOK

Immunoglobulins—All the Features of Protein Structure Brought Together

The immunoglobulin structure in Figure 6.45 represents the confluence of all the details of protein structure that have been thus far discussed. As for all proteins, the primary structure determines other aspects of structure. There are numerous elements of secondary structure, including β-sheets and tight turns. The tertiary structure consists of 12 distinct domains, and the protein adopts a heterotetrameric quaternary structure. To make matters more interesting, both intrasubunit and intersubunit disulfide linkages act to stabilize the discrete domains and to stabilize the tetramer itself.

One more level of sophistication awaits. As discussed in Chapter 28, the amino acid sequences of both light and heavy immuno-

globulin chains are not constant! Instead, the primary structure of these chains is highly variable in the N-terminal regions (first 108 residues). Heterogeneity of the amino acid sequence leads to variations in the conformation of these variable regions. This variation accounts for antibody diversity and the ability of antibodies to recognize and bind a virtually limitless range of antigens. This full potential of antibody:antigen recognition enables organisms to mount immunological responses to almost any antigen that might challenge the organism. (See also Figure 28.35 and the Special Focus section in Chapter 28.)

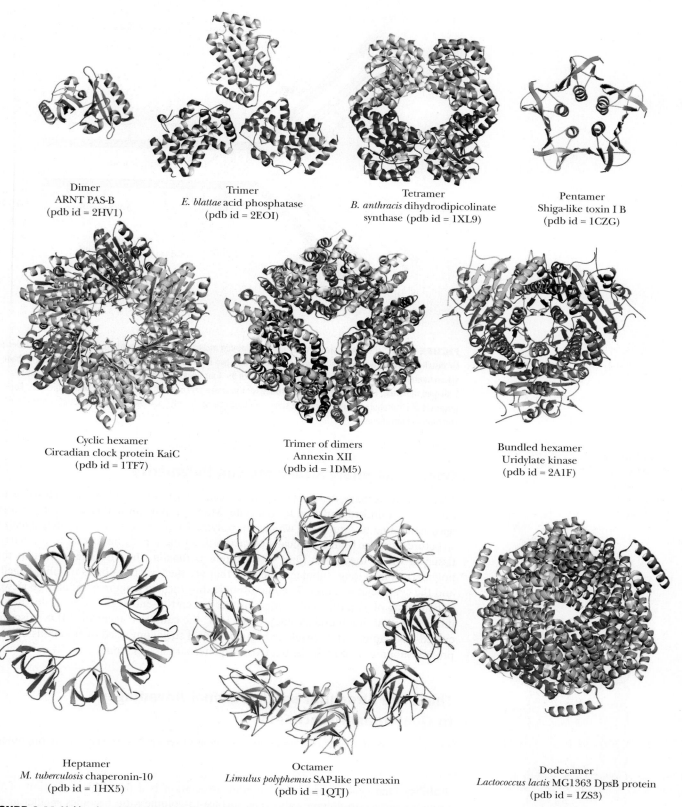

Dimer
ARNT PAS-B
(pdb id = 2HV1)

Trimer
E. blattae acid phosphatase
(pdb id = 2EOI)

Tetramer
B. anthracis dihydrodipicolinate
synthase (pdb id = 1XL9)

Pentamer
Shiga-like toxin I B
(pdb id = 1CZG)

Cyclic hexamer
Circadian clock protein KaiC
(pdb id = 1TF7)

Trimer of dimers
Annexin XII
(pdb id = 1DM5)

Bundled hexamer
Uridylate kinase
(pdb id = 2A1F)

Heptamer
M. tuberculosis chaperonin-10
(pdb id = 1HX5)

Octamer
Limulus polyphemus SAP-like pentraxin
(pdb id = 1QTJ)

Dodecamer
Lactococcus lactis MG1363 DpsB protein
(pdb id = 1ZS3)

FIGURE 6.44 Multimeric proteins are symmetric arrangements of asymmetric objects. A variety of symmetries is displayed in these multimeric structures.

antibodies are $\alpha_2\beta_2$-tetramers composed of two heavy chains (53 to 75 kD) and two light chains (23 kD). In addition to *intrasubunit* disulfide bonds (four per heavy chain, two per light chain), two *intersubunit* disulfide bridges hold the two heavy chains together and a disulfide bridge links each of the two light chains to a heavy chain (Figure 6.45).

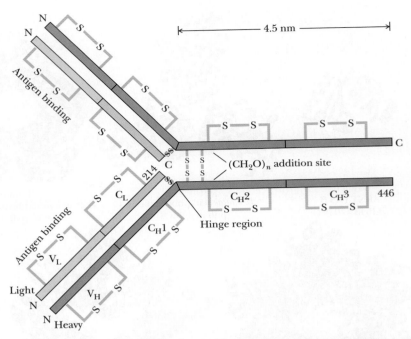

FIGURE 6.45 Schematic drawing of an immunoglobulin molecule, showing the intermolecular and intramolecular disulfide bonds. Two identical L chains are joined with two identical H chains. Each L chain is held to an H chain via an interchain disulfide bond. The variable regions of the four polypeptides lie at the ends of the arms of the Y-shaped molecule. These regions are responsible for the antigen recognition function of antibody molecules. For purposes of illustration, some features are shown on only one or the other L chain or H chain, but all features are common to both chains.

Open Quaternary Structures Can Polymerize

All of the quaternary structures we have considered to this point have been **closed** structures, with a limited capacity to associate. Many proteins in nature associate to form **open** heterologous structures, which can polymerize more or less indefinitely, creating structures that are both esthetically attractive and functionally important to the cells or tissue in which they exist. One such protein is **tubulin,** the $\alpha\beta$-dimeric protein that polymerizes into long, tubular structures that are the structural basis of cilia, flagella, and the cytoskeletal matrix. The microtubule thus formed (Figure 6.46) may be viewed as consisting of 13 parallel filaments arising from end-to-end aggregation of the tubulin dimers. Human immunodeficiency virus, HIV, the causative agent of AIDS (also discussed in Chapter 14), is enveloped by a spherical shell composed of hundreds of coat protein subunits, a large-scale, but closed, quaternary association.

There Are Structural and Functional Advantages to Quaternary Association

There are several important consequences when protein subunits associate in oligomeric structures.

Stability One general benefit of subunit association is a favorable reduction of the protein's surface-to-volume ratio. The surface-to-volume ratio becomes smaller as the radius of any particle or object becomes larger. (This is because surface area is a function of the radius squared and volume is a function of the radius cubed.) Because interactions within the protein usually tend to stabilize the protein and because the interaction of the protein surface with solvent water is often energetically unfavorable, decreased surface-to-volume ratios usually result in more stable proteins. Subunit association may also serve to shield hydrophobic residues from solvent water. Subunits

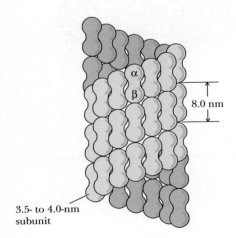

FIGURE 6.46 The structure of a typical microtubule, showing the arrangement of the α- and β-monomers of the tubulin dimer.

> **HUMAN BIOCHEMISTRY**
>
> ## Faster-Acting Insulin: Genetic Engineering Solves a Quaternary Structure Problem
>
> *Insulin* is a peptide hormone secreted by the pancreas that regulates glucose metabolism in the body. Insufficient production of insulin or failure of insulin to stimulate target sites in liver, muscle, and adipose tissue leads to the serious metabolic disorder known as *diabetes mellitus*. Diabetes afflicts millions of people worldwide. Diabetic individuals typically exhibit high levels of glucose in the blood, but insulin injection therapy allows these individuals to maintain normal levels of blood glucose.
>
> Insulin is composed of two peptide chains covalently linked by disulfide bonds (see Figure 5.8). This "monomer" of insulin is the active form that binds to receptors in target cells. However, in solution, insulin spontaneously forms dimers, which themselves aggregate to form hexamers. The surface of the insulin molecule that self-associates to form hexamers is also the surface that binds to insulin receptors in target cells. Thus, hexamers of insulin are inactive.
>
> Insulin released from the pancreas is monomeric and acts rapidly at target tissues. However, when insulin is administered (by injection) to a diabetic patient, the insulin hexamers dissociate slowly and the patient's blood glucose levels typically drop slowly (over several hours).
>
> In 1988, G. Dodson showed that insulin could be genetically engineered to prefer the monomeric (active) state. Dodson and his colleagues used recombinant DNA technology (discussed in Chapter 12) to produce insulin with an aspartate residue replacing a proline at the contact interface between adjacent subunits. The negative charge on the Asp side chain creates electrostatic repulsion between subunits and increases the dissociation constant for the hexamer $\rightleftharpoons$ monomer equilibrium. Injection of this mutant insulin into test animals produced more rapid decreases in blood glucose than did ordinary insulin. This mutant insulin, known as *insulin aspart*, marketed by the Danish pharmaceutical company Novo as NovoLog in the United States and as NovoRapid in Europe, has several advantages over ordinary insulin, in the treatment of diabetes. NovoLog has a faster rate of absorption, a faster onset of action, and a shorter duration of action than regular human insulin. It is particularly suited for mealtime dosing to control postprandial glycemia, the rise in blood sugar following consumption of food. Regular human insulin acts more slowly, so patients must usually administer it 30 minutes before eating.

that recognize either themselves or other subunits avoid any errors arising in genetic translation by binding mutant forms of the subunits less tightly.

Genetic Economy and Efficiency Oligomeric association of protein monomers is genetically economical for an organism. Less DNA is required to code for a monomer that assembles into a homomultimer than for a large polypeptide of the same molecular mass. Another way to look at this is to realize that virtually all of the information that determines oligomer assembly and subunit–subunit interaction is contained in the genetic material needed to code for the monomer. For example, HIV protease, an enzyme that is a dimer of identical subunits, performs a catalytic function similar to homologous cellular enzymes that are single polypeptide chains of twice the molecular mass (see Chapter 14).

Bringing Catalytic Sites Together Many enzymes (see Chapters 13 to 15) derive at least some of their catalytic power from oligomeric associations of monomer subunits. This can happen in several ways. The monomer may not constitute a complete enzyme active site. Formation of the oligomer may bring all the necessary catalytic groups together to form an active enzyme. For example, the active sites of bacterial glutamine synthetase are formed from pairs of adjacent subunits. The dissociated monomers are inactive.

Oligomeric enzymes may also carry out different but related reactions on different subunits. Thus, tryptophan synthase is a tetramer consisting of pairs of different subunits, $\alpha_2\beta_2$. Purified α-subunits catalyze the following reaction:

$$\text{Indoleglycerol phosphate} \rightleftharpoons \text{indole} + \text{glyceraldehyde-3-phosphate}$$

and the β-subunits catalyze this reaction:

$$\text{Indole} + \text{L-serine} \rightleftharpoons \text{L-tryptophan}$$

Indole, the product of the α-reaction and the reactant for the β-reaction, is passed directly from the α-subunit to the β-subunit and cannot be detected as a free intermediate.

Cooperativity There is another, more important consequence when monomer subunits associate into oligomeric complexes. Most oligomeric enzymes regulate catalytic activity by means of subunit interactions, which may give rise to cooperative phenomena.

Multisubunit proteins typically possess multiple binding sites for a given ligand. If the binding of ligand at one site changes the affinity of the protein for ligand at the other binding sites, the binding is said to be **cooperative.** Information transfer in this manner across long distances in proteins is termed **allostery,** literally action "at another site." Increases in affinity at subsequent sites represent positive cooperativity, whereas decreases in affinity correspond to negative cooperativity. The points of contact between protein subunits provide a mechanism for this signal transduction through the protein structure and for communication between the subunits. This in turn provides a way in which the binding of ligand to one subunit can influence the binding behavior at the other subunits. Such cooperative behavior, discussed in greater depth in Chapter 15, is the underlying mechanism for regulation of many biological processes.

SUMMARY

OWL Login to OWL to develop problem-solving skills and complete online homework assigned by your professor.

6.1 What Noncovalent Interactions Stabilize Protein Structure? Several different kinds of noncovalent interactions are of vital importance in protein structure. Hydrogen bonds, hydrophobic interactions, electrostatic bonds, and van der Waals forces are all noncovalent in nature yet are extremely important influences on protein conformations. The stabilization free energies afforded by each of these interactions are highly dependent on the local environment within the protein.

Hydrogen bonds are generally made wherever possible within a given protein structure. Hydrophobic interactions form because nonpolar side chains of amino acids and other nonpolar solutes prefer to cluster in a nonpolar environment rather than to intercalate in a polar solvent such as water. Electrostatic interactions include the attraction between opposite charges and the repulsion of like charges in the protein. Van der Waals interactions involve instantaneous dipoles and induced dipoles that arise because of fluctuations in the electron charge distributions of adjacent nonbonded atoms.

6.2 What Role Does the Amino Acid Sequence Play in Protein Structure? All of the information necessary for folding the peptide chain into its "native" structure is contained in the amino acid sequence of the peptide. Just how proteins recognize and interpret the information that is stored in the polypeptide sequence is not yet well understood. It may be assumed that certain loci along the peptide chain act as nucleation points, which initiate folding processes that eventually lead to the correct structure. Regardless of how this process operates, it must take the protein correctly to the final native structure, without getting trapped in a local energy-minimum state, which, although stable, may be different from the native state itself.

6.3 What Are the Elements of Secondary Structure in Proteins, and How Are They Formed? Secondary structure in proteins forms so as to maximize hydrogen bonding and maintain the planar nature of the peptide bond. Secondary structures include α-helices, β-sheets, and tight turns.

6.4 How Do Polypeptides Fold into Three-Dimensional Protein Structures? First, secondary structures—helices and sheets—form whenever possible as a consequence of the formation of large numbers of hydrogen bonds. Second, α-helices and β-sheets often associate and pack close together in the protein. There are a few common methods for such packing to occur. Third, because the peptide segments between secondary structures in the protein tend to be short and direct, the peptide does not execute complicated twists and knots as it moves from one region of a secondary structure to another. A consequence of these three principles is that protein chains are usually folded so that the secondary structures are arranged in one of a limited number of common patterns. For this reason, there are families of proteins that have similar tertiary structure, with little apparent evolutionary or functional relationship among them. Finally, proteins generally fold so as to form the most stable structures possible. The stability of most proteins arises from (1) the formation of large numbers of intramolecular hydrogen bonds and (2) the reduction in the surface area accessible to solvent that occurs upon folding.

6.5 How Do Protein Subunits Interact at the Quaternary Level of Protein Structure? The subunits of an oligomeric protein typically fold into apparently independent globular conformations and then interact with other subunits. The particular surfaces at which protein subunits interact are similar in nature to the interiors of the individual subunits. These interfaces are closely packed and involve both polar and hydrophobic interactions. Interacting surfaces must therefore possess complementary arrangements of polar and hydrophobic groups.

FOUNDATIONAL BIOCHEMISTRY Things You Should Know After Reading Chapter 6.

- The secondary, tertiary, and quaternary structures of proteins are stabilized by weak forces.

- Protein folding is driven predominantly by hydrophobic interactions.

- Charged residues are found mainly on the surface of proteins; hydrophobic residues are located mainly in the protein interior.

- All the information necessary for folding a polypeptide into its native structure is contained in the amino acid sequence of the peptide.

- The conformation of the peptide backbone is known if the ϕ and ψ angles of each amide plane are known.

- The nature and dimensions of the α–helix and β–sheet and β–turn and how each is stabilized by hydrogen bonds.

- The nature and structure of α– and β–keratin and the collagen helix.

- The reason why helices and sheets comprise the core of most globular proteins.

- The reason why waters on the protein surface stabilize globular structures.

- Packing densities of globular proteins are typically 0.75; most of the space is small cavities that provide flexibility and facilitate protein dynamics.

- Protein domains are nature's modular strategy for protein design.

- Many proteins, especially proteins larger than 250 residues, consist of two or more domains.

- Domains form the basis for the SCOP and CATH classification systems.

- Denaturation leads to loss of structure and function; proteins can be renatured under certain conditions.

- Levinthal's paradox supports the hypothesis that proteins fold by concerted mechanisms.

- The hydrophobic collapse and transient protein states, such as molten globules.

- The largest contribution to the stability of folded proteins is the entropy change for the waters associated with its nonpolar residues.

- Marginal stability contributes to flexibility and motion in proteins.

- Know the types of motion in proteins and the time scale of each.

- Polypeptide chains made from L-amino acids tend to undergo right-handed twists.

- Most globular proteins can be conceptualized as "layer structures."

- The four classes of globular protein structure: all α proteins, all β proteins, α/β proteins, and $\alpha+\beta$ proteins.

- Unstructured proteins can adapt to bind to multiple ligands, thus exerting multiple functions.

- The weak forces that modulate subunit interactions in multimeric proteins are similar to those that stabilize protein monomers.

- Open quaternary proteins structures can polymerize.

- The structural and functional advantages of subunit association include stability, genetic economy and efficiency, juxtaposition of catalytic sites, and cooperativity.

PROBLEMS

Answers to all problems are at the end of this book. Detailed solutions are available in the *Student Solutions Manual, Study Guide, and Problems Book*.

1. **Determining the Length of a Keratin Molecule** The central rod domain of a keratin protein is approximately 312 residues in length. What is the length (in Å) of the keratin rod domain? If this same peptide segment were a true α-helix, how long would it be? If the same segment were a β-sheet, what would its length be?

2. **Calculating the Rate of Collagen Helix Growth** A teenager can grow 4 inches in a year during a "growth spurt." Assuming that the increase in height is due to vertical growth of collagen fibers (in bone), calculate the number of collagen helix turns synthesized per minute.

3. **Assessing the Roles of Weak Forces Between Amino Acids in Proteins** Discuss the potential contributions to hydrophobic and van der Waals interactions and ionic and hydrogen bonds for the side chains of Asp, Leu, Tyr, and His in a protein.

4. **The Role of Proline Residues in β-Turns** Pro is the amino acid least commonly found in α-helices but most commonly found in β-turns. Discuss the reasons for this behavior.

5. **Assessing the Cross-Overs of Flavodoxin** For flavodoxin (pdb id = 5NLL), identify the right-handed cross-overs and the left-handed cross-overs in the parallel β-sheet.

6. **Assessing the Range of ϕ and ψ Angles in Proteins** Choose any three regions in the Ramachandran plot and discuss the likelihood of observing that combination of ϕ and ψ in a peptide or protein. Defend your answer using suitable molecular models of a peptide.

7. **Protein Structure Evaluation Based on Gel Filtration Data** A new protein of unknown structure has been purified. Gel filtration chromatography reveals that the native protein has a molecular weight of 240,000. Chromatography in the presence of 6 M guanidine hydrochloride yields a single peak corresponding to a protein of M_r 60,000. Chromatography in the presence of 6 M guanidine hydrochloride and 10 mM β-mercaptoethanol yields peaks for proteins of M_r 34,000 and 26,000. Explain what can be determined about the structure of this protein from these data.

8. **Understanding the Role of Amino Acids in Oligomerization Behavior** Two polypeptides, A and B, have similar tertiary structures, but A normally exists as a monomer, whereas B exists as a tetramer,

B_4. What differences might be expected in the amino acid composition of A versus B?

9. **Evaluation of α-Helices in Proteins** The hemagglutinin protein in influenza virus contains a remarkably long α-helix, with 53 residues.
   a. How long is this α-helix (in nm)?
   b. How many turns does this helix have?
   c. Each residue in an α-helix is involved in two H bonds. How many H bonds are present in this helix?

10. **Understanding the Role of Gly Residues in Protein Secondary and Tertiary Structure** It is often observed that Gly residues are conserved in proteins to a greater degree than other amino acids. From what you have learned in this chapter, suggest a reason for this observation.

11. **Understanding H-Bond Formation in Proteins** Which amino acids would be capable of forming H bonds with a lysine residue in a protein?

12. **Assessing the pH Dependence of Poly-L-Glutamate Structure** Poly-L-glutamate adopts an α-helical structure at low pH but becomes a random coil above pH 5. Explain this behavior.

13. **Exploring the Dimensions of the α-Helix and Coiled Coils** Imagine that the dimensions of the alpha helix were such that there were exactly 3.5 amino acids per turn, instead of 3.6. What would be the consequences for coiled-coil structures?

14. **(Research Problem) The Nature and Roles of Linear Motifs in Proteins** In addition to domains and modules, there are other significant sequence patterns in proteins—known as *linear motifs*—that are associated with a particular function. Consult the biochemical literature to answer the following questions:
    - What are linear motifs?
    - How are they different from domains?
    - What are their functions?
    - How can they be characterized?

There are several papers that are good starting points for this problem:

Neduva, V., and Russell, R., 2005. Linear motifs: evolutionary interaction switches. *FEBS Letters* **579**:3342–3345.

Gibson, T., 2009. Cell regulation: determined to signal discrete cooperation. *Trends in Biochemical Sciences* **34**:471–482.

Diella, F., Haslam, N., Chica., C. et al., 2009. Understanding eukaryotic linear motifs and their role in cell signaling and regulation. *Frontiers of Bioscience* **13**:6580–6603.

15. (Research Problem) The Nature of Protein-Protein Interactions
How do proteins interact? When one protein binds to another, one or both changes conformation. Two hypotheses have been proposed to describe such binding: In the induced fit model, the interaction between a protein and a ligand induces a conformation change (in the protein or ligand) through a stepwise process. In the conformational selection model, the unliganded protein (in the absence of the ligand) exists as an ensemble of conformations in a dynamic equilibrium. The binding ligand interacts preferentially with one among many of these conformations and shifts the equilibrium in favor of the selected conformation. Three recent papers shed light on this question:

Boehr, D., and Wright, P. E., 2008. How do proteins interact? *Science* **320**:1429–1430.

Gsponer, J., et al., 2008. A coupled equilibrium shift mechanism in calmodulin-mediated signal transduction. *Structure* **16**:736–746.

Lange, O., et al., 2008. Recognition dynamics up to microseconds revealed from an RDC-derived ubiquitin ensemble in solution. *Science* **320**:1471–1475.

Consult these papers and answer the following questions:

- What proteins were studied in these papers?
- What techniques were used, and what time scales of protein motion were studied?
- What were the conclusions of these papers, and how do these results illuminate the choice between induced fit and conformational selection in protein-protein interactions?

16. (Research Problem) Conformational Transitions in Proteins
How do proteins accomplish conformational changes? How is it that proteins convert precisely and efficiently from one conformation to another? Recall from Figure 6.34 that any folding/unfolding transition must involve movement across a free-energy landscape, and try to imagine the nature of a conformational transition. Are bonds formed and broken along the way? What kinds of bonds and interactions might be involved? Suggest how such conformational transitions might occur. One reference that will be useful in this regard is:

Boehr, D., 2009. During transitions proteins make fleeting bonds. *Cell* **139**:1049–1051.

17. (Historical Context) The Third Person of the α-Helix Publication
Who was Herman Branson? What was his role in the elucidation of the structure of the α-helix? Did he receive sufficient credit and

recognition for his contributions? And how did the rest of his career unfold? Do a Google search on Herman Branson to learn about his life, and read the article by David Eisenberg under Further Reading. You may also wish to examine the original paper by Pauling, Corey, and Branson, as well as the following Web site:

http://www.pnas.org/site/misc/classics1.shtml

Pauling, L., Corey, R. B., and Branson, H. R., 1951. The structure of proteins: two hydrogen-bonded helical configurations of the polypeptide chain. *Proceedings of the National Academy of Sciences* **37**:235–240.

Preparing for the MCAT® Exam

18. Consider the following peptide sequences:

EANQIDEMLYNVQCSLTTLEDTVPW
LGVHLDITVPLSWTWTLYVKL
QQNWGGLVVILTLVWFLM
CNMKHGDSQCDERTYP
YTREQSDGHIPKMNCDS
AGPFGPDGPTIGPK

Which of the preceding sequences would be likely to be found in each of the following:
a. A parallel β-sheet
b. An antiparallel β-sheet
c. A tropocollagen molecule
d. The helical portions of a protein found in your hair

19. To fully appreciate the elements of secondary structure in proteins, it is useful to have a practical sense of their structures. On a piece of paper, draw a simple but large zigzag pattern to represent a b-strand. Then fill in the structure, drawing the locations of the atoms of the chain on this zigzag pattern. Then draw a simple, large coil on a piece of paper to represent an α-helix. Then fill in the structure, drawing the backbone atoms in the correction locations along the coil and indicating the locations of the R groups in your drawing.

20. The dissociation constant for a particular protein dimer is 1 micromolar. Calculate the free energy difference for the monomer-to-dimer transition.

ActiveModels Problems

21. Using the ActiveModel for concanavalin A, discuss an example in which a difference in portein primary structure leads to a difference in protein function.

22. Describe the secondary structure of each subdomain of malonyl-CoA: ACP transferase. Explain the difference between parallel and antiparallel beta sheets.

FURTHER READING

General

Branden, C., and Tooze, J., 1991. *Introduction to Protein Structure*. New York: Garland Publishing.

Chothia, C., 1984. Principles that determine the structure of proteins. *Annual Review of Biochemistry* **53**:537–572.

Eisenberg, D., 2003. The discovery of the α-helix and β-sheet, the principal structural features of proteins. *Proceedings of the National Academy of Sciences* **100**:11207–11210.

Fink, A., 2005. Natively unfolded proteins. *Current Opinion in Structural Biology* **15**:35–41.

Greene, L., Lewis, T., Addou, S., et al., 2006. The CATH domain structure database: New protocols and classification levels give a more comprehensive resource for exploring evolution. *Nucleic Acids Research* **35**:D291–D297.

Hardie, D. G., and Coggins, J. R., eds., 1986. *Multidomain Proteins: Structure and Evolution*. New York: Elsevier.

Harper, E., and Rose, G. D., 1993. Helix stop signals in proteins and peptides: The capping box. *Biochemistry* **32**:7605–7609.

Judson, H. F., 1979. *The Eighth Day of Creation*. New York: Simon and Schuster.

Leavitt, M., 2009. Nature of the protein universe. *Proceedings of the National Academy of Sciences* **106**:11079–11084.

Lupas, A., 1996. Coiled coils: New structures and new functions. *Trends in Biochemical Sciences* **21**:375–382.

Perutz, M., 1998. *I Wish I'd Made You Angry Earlier*. Plainview, NY: Coldspring Harbor Laboratory Press.

Petsko, G., and Ringe, D., 2004. *Protein Structure and Function*. London: New Science Press.

Richardson, J. S., 1981. The anatomy and taxonomy of protein structure. *Advances in Protein Chemistry* **34**:167–339.

Schulze, A. J., Huber, R., Bode, W., and Engh, R. A., 1994. Structural aspects of serpin inhibition. *FEBS Letters* **344**:117–124.

Smith, T., 2000. Structural Genomics—special supplement. *Nature Structural Biology* Volume **7**, Issue 11S. This entire supplemental issue is devoted to structural genomics and contains a trove of information about this burgeoning field.

Tompa, P., 2002. Intrinsically unstructured proteins. *Trends in Biochemical Sciences* **27**:527–533.

Tompa, P., Szasz, C., and Buday, L., 2005. Structural disorder throws new light on moonlighting. *Trends in Biochemical Sciences* **30**:484–489.

Uversky, V. N., 2002. Natively unfolded proteins: A point where biology waits for physics. *Protein Science* **11**:739–756.

Webster, D. M., 2000. *Protein Structure Prediction—Methods and Protocols.* New Jersey: Humana Press.

Protein Folding

Aurora, R., Creamer, T., Srinivasan, R., and Rose, G. D., 1997. Local interactions in protein folding: Lessons from the α-helix. *The Journal of Biological Chemistry* **272**:1413–1416.

Baker, D., 2000. A surprising simplicity to protein folding. *Nature* **405**: 39–42.

Creighton, T. E., 1997. How important is the molten globule for correct protein folding? *Trends in Biochemical Sciences* **22**:6–11.

Deber, C. M., and Therien, A. G., 2002. Putting the β-breaks on membrane protein misfolding. *Nature Structural Biology* **9**:318–319.

Dill, K. A., and Chan, H. S., 1997. From Levinthal to pathways to funnels. *Nature Structural Biology* **4**:10–19.

Dinner, A. R., Sali, A., Smith, L. J., Dobson, C. M., and Karplus, M., 2001. Understanding protein folding via free-energy surfaces from theory and experiment. *Trends in Biochemical Sciences* **25**:331–339.

Han, J.-H., Batey, S., Nickson, A., et al., 2007. The folding and evolution of multidomain proteins. *Nature Reviews Molecular Cell Biology* **8**: 319–330.

Kelly, J., 2005. Structural biology: Form and function instructions. *Nature* **437**:486–487.

Mirny, L., and Shakhnovich, E., 2001. Protein folding theory: From lattice to all-atom models. *Annual Review of Biophysics and Biomolecular Structure* **30**:361–396.

Mok, K., Kuhn, L., Goez, M., et al., 2007. A pre-existing hydrophobic collapse in the unfolded state of an ultrafast folding protein. *Nature* **447**:106–109.

Murphy, K. P., 2001. *Protein Structure, Stability, and Folding.* New Jersey: Humana Press.

Myers, J. K., and Oas, T. G., 2002. Mechanisms of fast protein folding. *Annual Review of Biochemistry* **71**:783–815.

Orengo, C., and Thornton, J., 2005. Protein families and their evolution—a structural perspective. *Annual Review of Biochemistry* **74**:867-900.

Radford, S. E., 2000. Protein folding: Progress made and promises ahead. *Trends in Biochemical Sciences* **25**:611-618.

Raschke, T. M., and Marqusee, S., 1997. The kinetic folding intermediate of ribonuclease H resembles the acid molten globule and partially unfolded molecules detected under native conditions. *Nature Structural Biology* **4**:298–304.

Srinivasan, R., and Rose, G. D., 1995. LINUS: A hierarchic procedure to predict the fold of a protein. *Proteins: Structure, Function and Genetics* **22**:81–99.

Worth, C., Gong, S., and Blundell, T., 2009. Structural and functional constraints in the evolution of protein families. *Nature Reviews Molecular and Cell Biology* **10**:709–720.

Secondary Structure

Xiong, H., Buckwalter, B., Shieh, H., and Hecht, M. H., 1995. Periodicity of polar and nonpolar amino acids is the major determinant of secondary structure in self-assembling oligomeric peptides. *Proceedings of the National Academy of Sciences* **92**:6349–6353.

Structural Studies

Boehr, D. D., and Wright, P. E., 2009. How do proteins interact? *Science* **320**:1429–1430.

Bradley, P., Misura, K., and Baker, D., 2005. Toward high-resolution de novo structure prediction for small proteins. *Science* **309**:1868–1871.

Hadley, C., and Jones, D., 1999. A systematic comparison of protein structure classifications: SCOP, CATH, and FSSP. *Structure* **7**:1099–1112.

Lomas, D., Belorgey, D., Mallya, M., et al., 2005. Molecular mousetraps and the serpinopathies. *Biochemical Society Transactions* **33** (part 2): 321–330.

Murzin, A. G., 2008. Metamorphic proteins. *Science* **320**:1725–1726.

Smock, R., and Gierasch, L., 2009. Sending signals dynamically. *Science* **324**:198–203.

Wagner, G., Hyberts, S., and Havel, T., 1992. NMR structure determination in solution: A critique and comparison with X-ray crystallography. *Annual Review of Biophysics and Biomolecular Structure* **21**:167–242.

Wand, A. J., 2001. Dynamic activation of protein function: A view emerging from NMR spectroscopy. *Nature Structural Biology* **8**:926–931.

Diseases of Protein Folding

Bucchiantini, M., et al., 2002. Inherent toxicity of aggregates implies a common mechanism for protein misfolding diseases. *Nature* **416**: 507–511.

Sifers, R. M., 1995. Defective protein folding as a cause of disease. *Nature Structural Biology* **2**:355–367.

Stein, P. E., and Carrell, R. W., 1995. What do dysfunctional serpins tell us about molecular mobility and disease? *Nature Structural Biology* **2**:96–113.

Thomas, P. J., Qu, B-H., and Pedersen, P. L., 1995. Defective protein folding as a basis of human disease. *Trends in Biochemical Sciences* **20**: 456–459.

Carbohydrates and the Glycoconjugates of Cell Surfaces

© Burstein Collection/CORBIS

◀ "The Discovery of Honey"—Piero di Cosimo (1492).

Sugar in the gourd and honey in the horn,
I never was so happy since the hour I was
born.

Turkey in the Straw, *stanza 6*
(classic American folk tune)

KEY QUESTIONS

7.1 How Are Carbohydrates Named?

7.2 What Are the Structure and Chemistry of Monosaccharides?

7.3 What Are the Structure and Chemistry of Oligosaccharides?

7.4 What Are the Structure and Chemistry of Polysaccharides?

7.5 What Are Glycoproteins, and How Do They Function in Cells?

7.6 How Do Proteoglycans Modulate Processes in Cells and Organisms?

7.7 Do Carbohydrates Provide a Structural Code?

ESSENTIAL QUESTION

Carbohydrates are a versatile class of molecules of the formula $(CH_2O)_n$. They are a major form of stored energy in organisms, and they are the metabolic precursors of virtually all other biomolecules. Conjugates of carbohydrates with proteins and lipids perform a variety of functions, including recognition events that are important in cell growth, transformation, and other processes.

What are the structure, chemistry, and biological function of carbohydrates?

Carbohydrates are the single most abundant class of organic molecules found in nature. Energy from the sun captured by green plants, algae, and some bacteria during photosynthesis (see Chapter 21) converts more than 250 billion kilograms of carbon dioxide into carbohydrates every day on earth. In turn, carbohydrates are the metabolic precursors of virtually all other biomolecules. Breakdown of carbohydrates provides the energy that sustains animal life. In addition, carbohydrates are covalently linked with a variety of other molecules. These **glycoconjugates** are important components of cell walls and extracellular structures in plants, animals, and bacteria. In addition to the structural roles such molecules play, they also serve in a variety of processes involving *recognition* between cell types or recognition of cellular structures by other molecules. Recognition events are important in normal cell growth, fertilization, transformation of cells, and other processes.

All of these functions are made possible by the characteristic chemical features of carbohydrates:

- the existence of at least one and often two or more asymmetric centers
- the ability to exist either in linear or ring structures
- the capacity to form polymeric structures via *glycosidic* bonds
- the potential to form multiple hydrogen bonds with water or other molecules in their environment.

ᑌWL Online homework and a Student Self Assessment for this chapter may be assigned in OWL.

7.1 | How Are Carbohydrates Named?

The name *carbohydrate* arises from the basic molecular formula $(CH_2O)_n$, where $n = 3$ or more. $(CH_2O)_n$ can be rewritten $(C \cdot H_2O)_n$ to show that these substances are hydrates of carbon.

Carbohydrates are generally classified into three groups: **monosaccharides** (and their derivatives), **oligosaccharides,** and **polysaccharides.** The monosaccharides are also called **simple sugars** and have the formula $(CH_2O)_n$. Monosaccharides cannot be broken down into smaller sugars under mild conditions. Oligosaccharides derive their name from the Greek word *oligo,* meaning "few," and consist of from two to ten simple sugar residues. Disaccharides are common in nature, and trisaccharides also occur frequently. Four- to six-sugar-unit oligosaccharides are usually bound covalently to other molecules, including glycoproteins. As their name suggests, polysaccharides are polymers of the simple sugars and their derivatives. They may be either linear or branched polymers and may contain hundreds or even thousands of monosaccharide units. Their molecular weights range up to 1 million or more.

Glyco: A generic term relating to sugars.

7.2 | What Are the Structure and Chemistry of Monosaccharides?

Monosaccharides Are Classified as Aldoses and Ketoses

Monosaccharides consist typically of three to seven carbon atoms and are described either as **aldoses** or **ketoses,** depending on whether the molecule contains an aldehyde function or a ketone group. The simplest aldose is glyceraldehyde, and the simplest ketose is dihydroxyacetone (Figure 7.1). These two simple sugars are termed **trioses** because they each contain three carbon atoms. The structures and names of a family of aldoses and ketoses with three, four, five, and six carbons are shown in Figures 7.2 and 7.3. *Hexoses* are the most abundant sugars in nature. Nevertheless, sugars from all these classes are important in metabolism.

Monosaccharides, either aldoses or ketoses, are often given more detailed generic names to describe both the important functional groups and the total number of carbon atoms. Thus, one can refer to *aldotetroses* and *ketotetroses, aldopentoses* and *ketopentoses, aldohexoses* and *ketohexoses,* and so on. Sometimes the ketone-containing monosaccharides are named simply by inserting the letters *-ul-* into the simple generic terms, such as *tetruloses, pentuloses, hexuloses, heptuloses,* and so on. The simplest monosaccharides are water soluble, and most taste sweet.

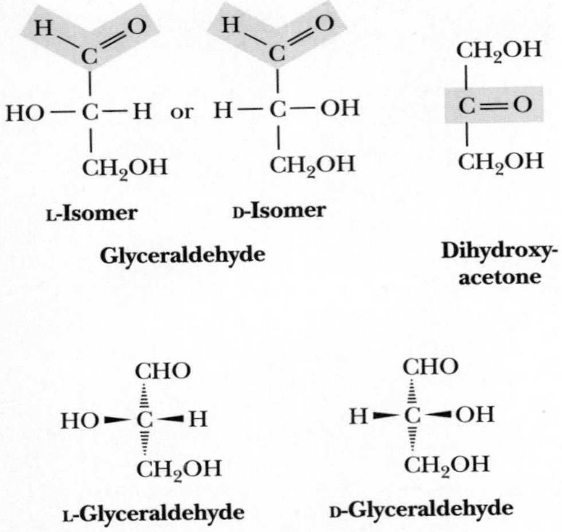

Stereochemistry Is a Prominent Feature of Monosaccharides

Aldoses with at least three carbons and ketoses with at least four carbons contain **chiral centers** (see Chapter 4). The nomenclature for such molecules must specify the **configuration** about each asymmetric center, and drawings of these molecules must be based on a system that clearly specifies these configurations. As noted in Chapter 4, the **Fischer projection** system is used almost universally for this purpose today. The structures shown in Figures 7.2 and 7.3 are Fischer projections. For monosaccharides with two or more asymmetric carbons, the prefix D or L refers to the configuration of the highest numbered asymmetric carbon (the asymmetric carbon farthest from the carbonyl carbon). A monosaccharide is designated D if the hydroxyl group on the highest numbered asymmetric carbon is drawn to the right in a Fischer projection, as in D-glyceraldehyde (Figure 7.1). Note that the designation D or L merely relates the configuration of a given molecule to that of glyceraldehyde and does *not* specify the sign of rotation of plane-polarized light. If the sign of optical rotation is to be specified in the name, the convention of D or L designations may be used along with a + (plus) or − (minus) sign. Thus, D-glucose (Figure 7.2) may also be called D(+)-glucose

FIGURE 7.1 Structure of a simple aldose (glyceraldehyde) and a simple ketose (dihydroxyacetone).

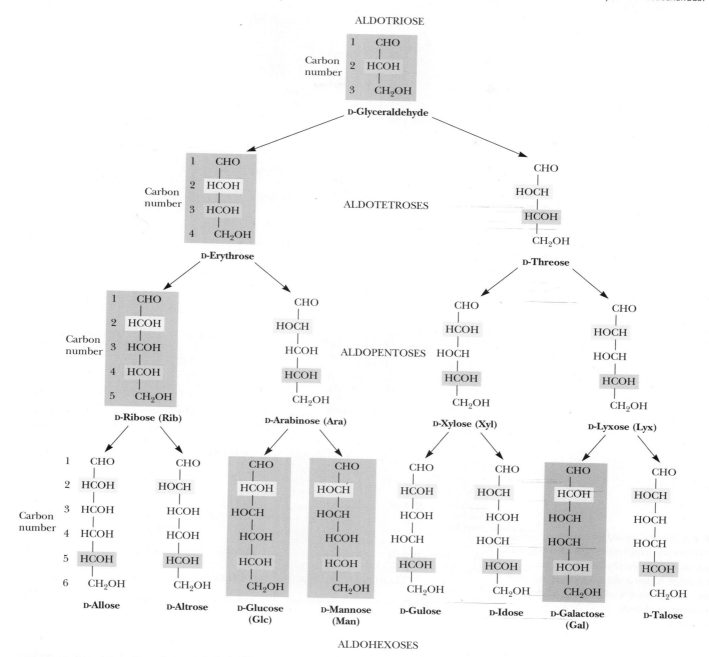

FIGURE 7.2 The structure and stereochemical relationships of D-aldoses with three to six carbons. The configuration in each case is determined by the highest numbered asymmetric carbon (shown in pink). In each row, the "new" asymmetric carbon is shown in yellow. Blue highlights indicate the most common aldoses.

because it is dextrorotatory, whereas D-fructose (Figure 7.3), which is levorotatory, can also be named D(−)-fructose.

All of the structures shown in Figures 7.2 and 7.3 are D-configurations, and the D-forms of monosaccharides predominate in nature, just as L-amino acids do. These preferences, established in apparently random choices early in evolution, persist uniformly in nature because of the stereospecificity of the enzymes that synthesize and metabolize these small molecules. L-Monosaccharides do exist in nature, serving a few relatively specialized roles. L-Galactose is a constituent of certain polysaccharides, and L-arabinose is a constituent of bacterial cell walls.

According to convention, the D- and L-forms of a monosaccharide are *mirror images* of each other, as shown in Figure 7.4 for fructose. Stereoisomers that are mirror images

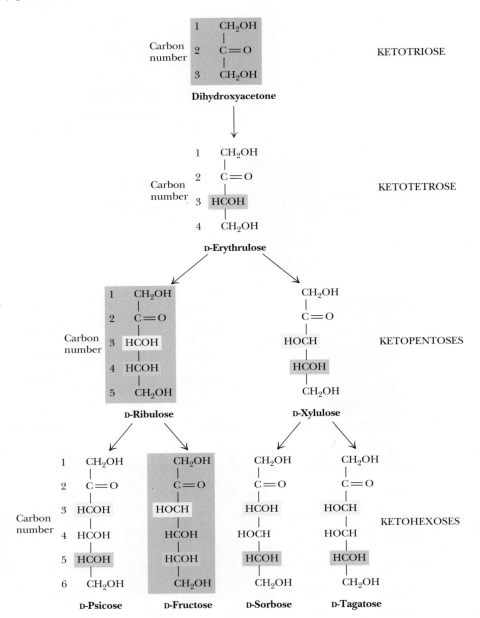

FIGURE 7.3 The structure and stereochemical relationships of D-ketoses with three to six carbons. The configuration in each case is determined by the highest numbered asymmetric carbon (shown in pink). In each row, the "new" asymmetric carbon is shown in yellow. Blue highlights indicate the most common ketoses.

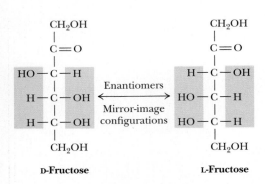

FIGURE 7.4 D-Fructose and L-fructose, an enantiomeric pair. Note that changing the configuration only at C₅ would change D-fructose to L-sorbose.

of each other are called **enantiomers,** or sometimes *enantiomeric pairs.* For molecules that possess two or more chiral centers, more than two stereoisomers can exist. Pairs of isomers that have opposite configurations at one or more of the chiral centers but that are not mirror images of each other are called **diastereomers** or *diastereomeric pairs.* Any two structures in a given row in Figures 7.2 and 7.3 are diastereomeric pairs. Two sugars that differ in configuration at only one chiral center are described as **epimers.** For example, D-mannose and D-talose are epimers and D-glucose and D-mannose are epimers, whereas D-glucose and D-talose are *not* epimers but merely diastereomers.

Monosaccharides Exist in Cyclic and Anomeric Forms

Although Fischer projections are useful for presenting the structures of particular monosaccharides and their stereoisomers, they discount one of the most interesting facets of sugar structure—*the ability to form cyclic structures with formation of an additional asymmetric center.* Alcohols react readily with aldehydes to form **hemiacetals** (Figure 7.5). The British carbohydrate chemist Sir Norman Haworth showed that the

Alcohol   +   **Aldehyde**   ⇌   **Hemiacetal**

D-Glucose

Pyran

Cyclization

α-D-**Glucopyranose**

β-D-**Glucopyranose**

HAWORTH PROJECTION FORMULAS

α-D-**Glucopyranose**

β-D-**Glucopyranose**

FISCHER PROJECTION FORMULAS

FIGURE 7.5 The linear form of D-glucose undergoes an intramolecular reaction to form a cyclic hemiacetal.

β-D-**Glucopyranose**

linear form of glucose (and other aldohexoses) could undergo a similar *intramolecular* reaction to form a *cyclic hemiacetal.* The resulting six-membered, oxygen-containing ring is similar to *pyran* and is designated a **pyranose.** The reaction is catalyzed by acid (H⁺) or base (OH⁻) and is readily reversible.

In a similar manner, ketones can react with alcohols to form **hemiketals.** The analogous intramolecular reaction of a ketose sugar such as fructose yields a *cyclic hemiketal* (Figure 7.6). The five-membered ring thus formed is reminiscent of *furan* and is referred to as a **furanose.** The cyclic pyranose and furanose forms are the preferred structures for monosaccharides in aqueous solution. At equilibrium, the linear aldehyde or ketone structure is only a minor component of the mixture (generally much less than 1%).

When hemiacetals and hemiketals are formed, the carbon atom that carried the carbonyl function becomes an asymmetric carbon atom. Isomers of monosaccharides that differ only in their configuration about that carbon atom are called **anomers,** designated as *α* or *β*, as shown in Figure 7.5, and the carbonyl carbon is thus called the **anomeric carbon.** When the hydroxyl group at the anomeric carbon is on the *same side* of a Fischer projection as the oxygen atom at the highest numbered asymmetric carbon, the configuration at the anomeric carbon is *α*, as in *α*-D-glucose. When the anomeric hydroxyl is on the *opposite side* of the Fischer projection, the configuration is *β*, as in *β*-D-glucopyranose (Figure 7.5).

The addition of this asymmetric center upon hemiacetal and hemiketal formation alters the optical rotation properties of monosaccharides, and the original assignment of the *α* and *β* notations arose from studies of these properties. Early carbohydrate chemists frequently observed that the optical rotation of glucose (and other sugar) solutions could change with time, a process called **mutarotation.** This indicated that a structural change was occurring. It was eventually found that *α*-D-glucose has a specific optical rotation, $[\alpha]_D^{20}$, of 112.2°, and that *β*-D-glucose has a specific optical rotation of 18.7°. Mutarotation involves interconversion of *α*- and *β*-forms of the monosaccharide with intermediate formation of the linear aldehyde or ketone, as shown in Figures 7.5 and 7.6.

Alcohol　　　Ketone　　　　　Hemiketal

α-D-Fructofuranose

D-Fructose

Furan

Cyclization

α-D-Fructofuranose

β-D-Fructofuranose

HAWORTH PROJECTION
FORMULAS

α-D-Fructofuranose

β-D-Fructofuranose

FISCHER PROJECTION
FORMULAS

FIGURE 7.6 The linear form of D-fructose undergoes an intramolecular reaction to form a cyclic hemiketal.

β-D-Fructofuranose

Haworth Projections Are a Convenient Device for Drawing Sugars

Another of Haworth's lasting contributions to the field of carbohydrate chemistry was his proposal to represent pyranose and furanose structures as hexagonal and pentagonal rings lying perpendicular to the plane of the paper, with thickened lines indicating the side of the ring closest to the reader. Such **Haworth projections,** which are now widely used to represent saccharide structures (Figures 7.5 and 7.6), show substituent groups extending either above or below the ring. Substituents drawn to the left in a Fischer projection are drawn above the ring in the corresponding Haworth projection. Substituents drawn to the right in a Fischer projection are below the ring in a Haworth projection. Exceptions to these rules occur in the formation of furanose forms of pentoses and the formation of furanose or pyranose forms of hexoses. In these cases, the structure must be redrawn with a rotation about the carbon whose hydroxyl group is involved in the formation of the cyclic form (Figure 7.7) in order to orient the appropriate hydroxyl group for ring formation. This is merely for illustrative purposes and involves no change in configuration of the saccharide molecule.

The rules previously mentioned for assignment of α- and β-configurations can be readily applied to Haworth projection formulas. For the D-sugars, the anomeric hydroxyl group is below the ring in the α-anomer and above the ring in the β-anomer. For L-sugars, the opposite relationship holds.

As Figure 7.7 implies, in most monosaccharides there are two or more hydroxyl groups that can react with an aldehyde or ketone at the other end of the molecule to form a hemiacetal or hemiketal. Consider the possibilities for glucose, as shown in Figure 7.7. If the C-4 hydroxyl group reacts with the aldehyde of glucose, a five-membered ring is formed, whereas if the C-5 hydroxyl reacts, a six-membered ring is formed. The C-6 hydroxyl does not react effectively because a seven-membered ring is too strained to form a stable hemiacetal. The same is true for the C-2 and C-3 hydroxyls, and thus five- and six-membered rings are by far the most likely to be formed from six-membered monosaccharides. D-Ribose, with five carbons, readily forms either five-membered rings (α- or β-D-ribofuranose) or six-membered rings (α- or β-D-ribopyranose) (Figure 7.7). In general, aldoses and ketoses with five or more carbons can form *either* furanose or pyranose rings, and the more

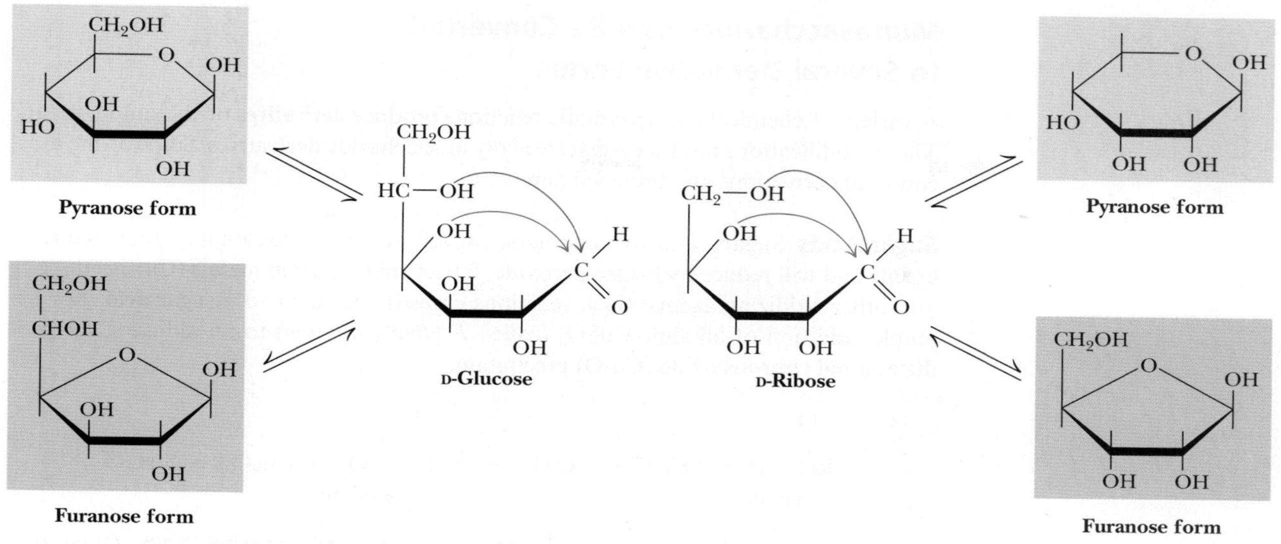

Pyranose form

Furanose form

D-Glucose

D-Ribose

Pyranose form

Furanose form

FIGURE 7.7 D-Glucose, D-ribose, and other simple sugars can cyclize in two ways, forming either furanose or pyranose structures.

stable form depends on structural factors. The nature of the substituent groups on the carbonyl and hydroxyl groups and the configuration about the asymmetric carbon will determine whether a given monosaccharide prefers the pyranose or furanose structure. In general, the pyranose form is favored over the furanose ring for aldohexose sugars, although, as we shall see, furanose structures tend to be more stable for ketohexoses.

Although Haworth projections are convenient for displaying monosaccharide structures, they do not accurately portray the conformations of pyranose and furanose rings. Given C—C—C tetrahedral bond angles of 109° and C—O—C angles of 111°, neither pyranose nor furanose rings can adopt true planar structures. Instead, they take on puckered conformations, and in the case of pyranose rings, the two favored structures are the **chair conformation** and the **boat conformation,** shown in Figure 7.8. Note that the ring substituents in these structures can be **equatorial,** which means approximately coplanar with the ring, or **axial,** that is, parallel to an axis drawn through the ring as shown. Two general rules dictate the conformation to be adopted by a given saccharide unit. First, bulky substituent groups on such rings are more stable when they occupy equatorial positions rather than axial positions, and second, chair conformations are slightly more stable than boat conformations. For a typical pyranose, such as β-D-glucose, there are two possible chair conformations (Figure 7.8). Of all the D-aldohexoses, β-D-glucose is the only one that can adopt a conformation with all its bulky groups in an equatorial position. With this advantage of stability, it may come as no surprise that β-D-glucose is the most widely occurring organic group in nature and the central hexose in carbohydrate metabolism.

(a)

Axis

109°

a = axial bond
e = equatorial bond

Chair

Axis

Boat

(b)

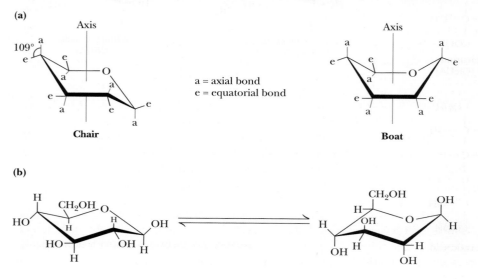

FIGURE 7.8 **(a)** Chair and boat conformations of a pyranose sugar. **(b)** Two possible chair conformations of β-D-glucose.

Monosaccharides Can Be Converted to Several Derivative Forms

A variety of chemical and enzymatic reactions produce **derivatives** of the simple sugars. These modifications produce a diverse array of saccharide derivatives. Some of the most common derivations are discussed here.

Sugar Acids Sugars with free anomeric carbon atoms are reasonably good reducing agents and will reduce hydrogen peroxide, ferricyanide, certain metals (Cu^{2+} and Ag^+), and other oxidizing agents. Such reactions convert the sugar to a **sugar acid.** For example, addition of alkaline $CuSO_4$ (called *Fehling's solution*) to an aldose sugar produces a red cuprous oxide (Cu_2O) precipitate:

$$\underset{\text{Aldehyde}}{RC-H} + 2\ Cu^{2+} + 5\ OH^- \longrightarrow \underset{\text{Carboxylate}}{RC-O^-} + Cu_2O\downarrow + 3\ H_2O$$

and converts the aldose to an **aldonic acid,** such as **gluconic acid** (Figure 7.9). Formation of a precipitate of red Cu_2O constitutes a positive test for an aldehyde. Carbohydrates that can reduce oxidizing agents in this way are referred to as **reducing sugars.** By quantifying the amount of oxidizing agent reduced by a sugar solution, one can accurately

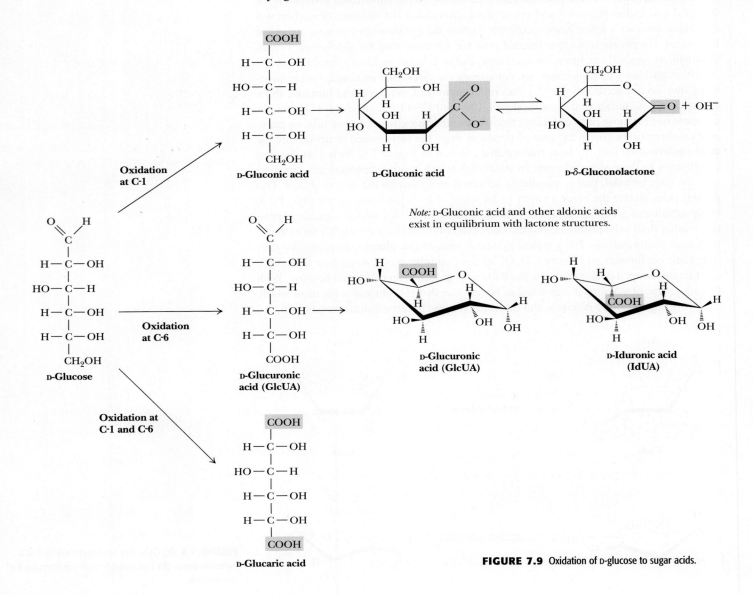

Note: D-Gluconic acid and other aldonic acids exist in equilibrium with lactone structures.

FIGURE 7.9 Oxidation of D-glucose to sugar acids.

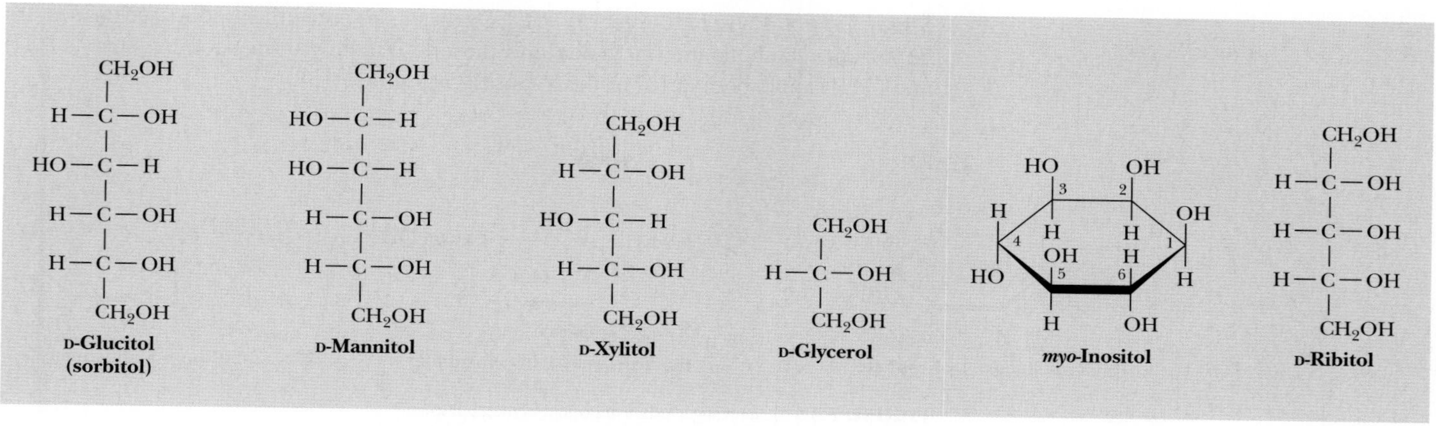

CH$_2$OH
H—C—OH
HO—C—H
H—C—OH
H—C—OH
CH$_2$OH
D-Glucitol
(sorbitol)

CH$_2$OH
HO—C—H
HO—C—H
H—C—OH
H—C—OH
CH$_2$OH
D-Mannitol

CH$_2$OH
H—C—OH
HO—C—H
H—C—OH
CH$_2$OH
D-Xylitol

CH$_2$OH
H—C—OH
CH$_2$OH
D-Glycerol

myo-Inositol

CH$_2$OH
H—C—OH
H—C—OH
H—C—OH
CH$_2$OH
D-Ribitol

FIGURE 7.10 Structures of some sugar alcohols. (Note that *myo*-inositol is a polyhydroxy cyclohexane, not a sugar alcohol.)

determine the concentration of the sugar. *Diabetes mellitus* is characterized by high levels of glucose in urine and blood, and frequent analysis of reducing sugars in diabetic patients is an important part of the diagnosis and treatment of this disease. Over-the-counter kits for the easy and rapid determination of reducing sugars have made this procedure a simple one for diabetic persons.

Monosaccharides can be oxidized enzymatically at C-6, yielding **uronic acids,** such as **D-glucuronic** and **L-iduronic acids** (Figure 7.9). L-Iduronic acid is similar to D-glucuronic acid, except it has an opposite configuration at C-5. Oxidation at both C-1 and C-6 produces **aldaric acids,** such as **D-glucaric acid.**

Sugar Alcohols Sugar alcohols, another class of sugar derivative, can be prepared by the mild reduction (with NaBH$_4$ or similar agents) of the carbonyl groups of aldoses and ketoses. Sugar alcohols, or **alditols,** are designated by the addition of *-itol* to the name of the parent sugar (Figure 7.10). The alditols are linear molecules that cannot cyclize in the manner of aldoses. Nonetheless, alditols are characteristically sweet tasting, and **sorbitol, mannitol,** and **xylitol** are widely used to sweeten sugarless gum and mints (Figure 7.11). Sorbitol buildup in the eyes of diabetic persons is implicated in cataract formation. **Glycerol** and *myo***-inositol,** a cyclic alcohol, are components of lipids (see Chapter 8). There are nine different stereoisomers of inositol; the one shown in Figure 7.10 was first isolated from heart muscle and thus has the prefix *myo*- for muscle. **Ribitol** is a constituent of flavin coenzymes (see Chapter 19).

FIGURE 7.11 Sugar alcohols such as sorbitol, mannitol, and xylitol sweeten many "sugarless" gums and candies.

Deoxy Sugars The **deoxy sugars** are monosaccharides with one or more hydroxyl groups replaced by hydrogens. 2-Deoxy-D-ribose (Figure 7.12), whose systematic name is 2-deoxy-D-erythropentose, is a constituent of DNA in all living things (see Chapter 10). Deoxy sugars also occur frequently in glycoproteins and polysaccharides. L-Fucose and L-rhamnose, both 6-deoxy sugars, are components of some cell walls, and rhamnose is a component of **ouabain,** a highly toxic *cardiac glycoside* found in the bark and root of the ouabaio tree. Ouabain is used by the East African Somalis as an arrow poison. The sugar moiety is not the toxic part of the molecule (see Chapter 9).

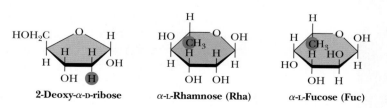

2-Deoxy-α-D-ribose   **α-L-Rhamnose (Rha)**   **α-L-Fucose (Fuc)**

FIGURE 7.12 Several deoxy sugars. Hydrogen and carbon atoms highlighted in red are "deoxy" positions.

Sugar Esters **Phosphate esters** of glucose, fructose, and other monosaccharides are important metabolic intermediates, and the ribose moiety of nucleotides such as ATP and GTP is phosphorylated at the 5′-position (Figure 7.13).

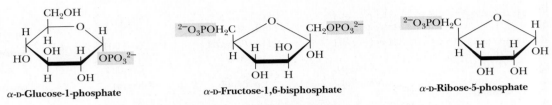

α-ᴅ-**Glucose-1-phosphate**   α-ᴅ-**Fructose-1,6-bisphosphate**   α-ᴅ-**Ribose-5-phosphate**

FIGURE 7.13 Several sugar esters important in metabolism.

A DEEPER LOOK

Honey—An Ancestral Carbohydrate Treat

Honey, the first sweet known to humankind, is the only sweetening agent that can be stored and used exactly as produced in nature. Bees process the nectar of flowers so that their final product is able to survive long-term storage at ambient temperature. Used as a ceremonial material and medicinal agent in earliest times, honey was not regarded as a food until the time of the Greeks and Romans. Only in modern times have cane and beet sugar surpassed honey as the most frequently used sweetener. What is the chemical nature of this magical, viscous substance?

The bees' processing of honey consists of (1) reducing the water content of the nectar (30% to 60%) to the self-preserving range of 15% to 19%, (2) hydrolyzing the significant amount of sucrose in nectar to glucose and fructose by the action of the enzyme invertase, and (3) producing small amounts of gluconic acid from glucose by the action of the enzyme **glucose oxidase.** Most of the sugar in the final product is glucose and fructose, and the final product is supersaturated with respect to these monosaccharides. Honey actually consists of an emulsion of microscopic glucose hydrate and fructose hydrate crystals in a thick syrup. Sucrose accounts for only about 1% of the sugar in the final product, with fructose at about 38% and glucose at 31% by weight.

The accompanying figure shows a ^{13}C nuclear magnetic resonance spectrum of honey from a mixture of wildflowers in southeastern Pennsylvania. Interestingly, five major hexose species contribute to this spectrum. Although most textbooks show fructose exclusively in its furanose form, the predominant form of fructose (67% of total fructose) is β-ᴅ-fructopyranose, with the β- and α-fructofuranose forms

accounting for 27% and 6% of the fructose, respectively. In polysaccharides, fructose invariably prefers the furanose form, but free fructose (and crystalline fructose) is predominantly β-fructopyranose.

Sources: White, J. W., 1978. Honey. *Advances in Food Research* **24**:287–374; and Prince, R. C., Gunson, D. E., Leigh, J. S., and McDonald, G. G., 1982. The predominant form of fructose is a pyranose, not a furanose ring. *Trends in Biochemical Sciences* **7**:239–240.

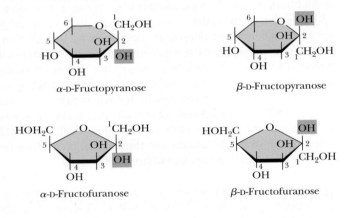

α-ᴅ-Fructopyranose   β-ᴅ-Fructopyranose

α-ᴅ-Fructofuranose   β-ᴅ-Fructofuranose

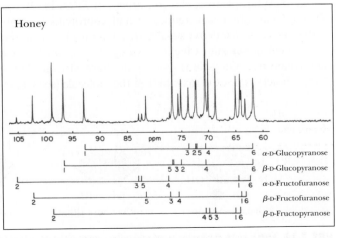

Amino Sugars Amino sugars, including **D-glucosamine** and **D-galactosamine** (Figure 7.14), contain an amino group (instead of a hydroxyl group) at the C-2 position. They are found in many oligosaccharides and polysaccharides, including *chitin,* a polysaccharide in the exoskeletons of crustaceans and insects.

Muramic acid and **neuraminic acid,** which are components of the polysaccharides of cell membranes of higher organisms and also bacterial cell walls, are glycosamines linked to three-carbon acids at the C-1 or C-3 positions. In muramic acid (thus named as an *amine* isolated from bacterial cell wall polysaccharides; *murus* is Latin for "wall"), the hydroxyl group of a lactic acid moiety makes an ether linkage to the C-3 of glucosamine. Neuraminic acid (an *amine* isolated from *neural* tissue) forms a C—C bond between the C-1 of *N*-acetylmannosamine and the C-3 of pyruvic acid (Figure 7.15). The *N*-acetyl and *N*-glycolyl derivatives of neuraminic acid are collectively known as **sialic acids** and are distributed widely in bacteria and animal systems.

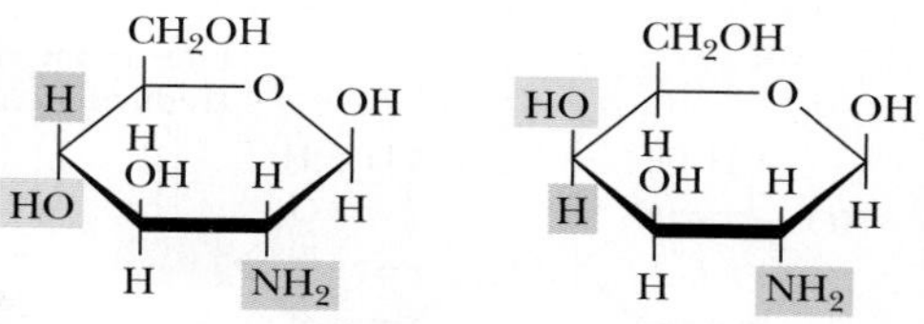

β-D-Glucosamine β-D-Galactosamine

FIGURE 7.14 Structures of D-glucosamine and D-galactosamine.

Acetals, Ketals, and Glycosides Hemiacetals and hemiketals can react with alcohols in the presence of acid to form **acetals** and **ketals,** as shown in Figure 7.16. This reaction is another example of a *dehydration synthesis* and is similar in this respect to the reactions undergone by amino acids to form peptides and nucleotides to form nucleic acids. The pyranose and furanose forms of monosaccharides react with alcohols in this way to form **glycosides** with retention of the α- or β-configuration at the C-1 carbon. The new bond between the anomeric carbon atom and the oxygen atom of the alcohol is called a **glycosidic bond.** Glycosides are named according to the parent monosaccharide. For example, *methyl-β-D-glucoside* (Figure 7.17) can be considered a derivative of β-D-glucose.

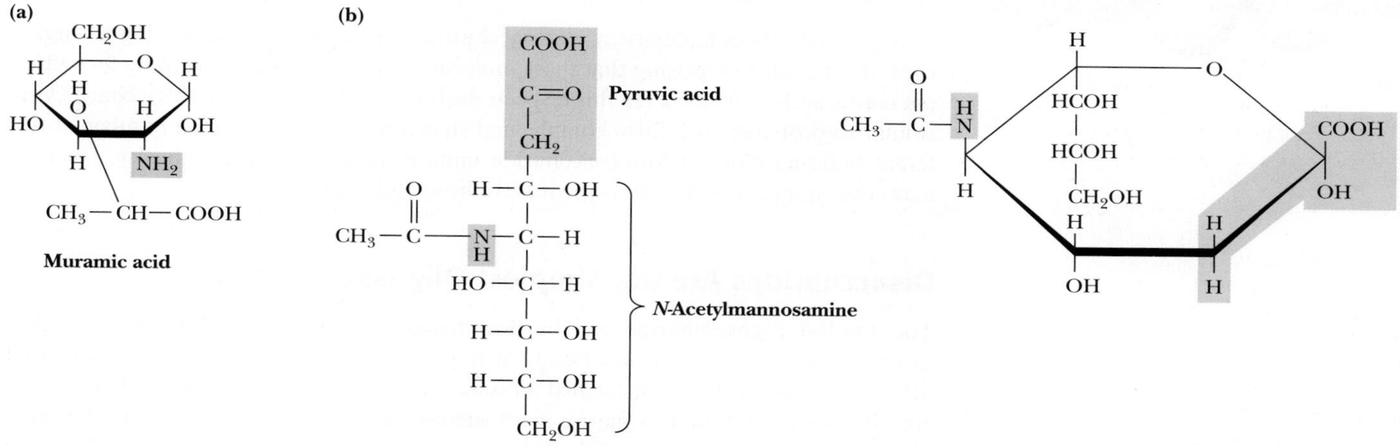

N-Acetyl-D-neuraminic acid (NeuNac, also known as Neu5Ac), a sialic acid

FIGURE 7.15 Structures of **(a)** muramic acid and **(b)** several depictions of a sialic acid.

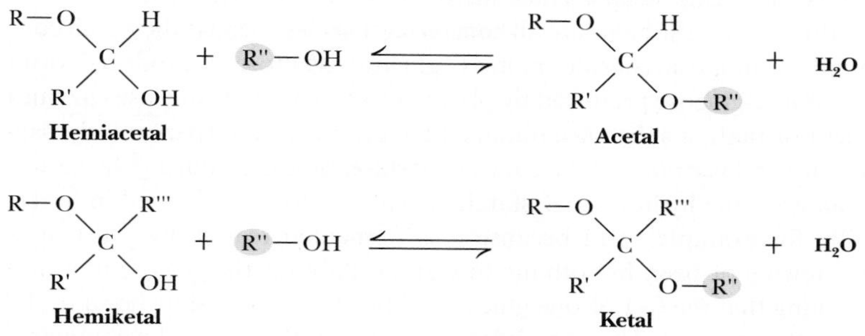

FIGURE 7.16 Acetals and ketals can be formed from hemiacetals and hemiketals, respectively.

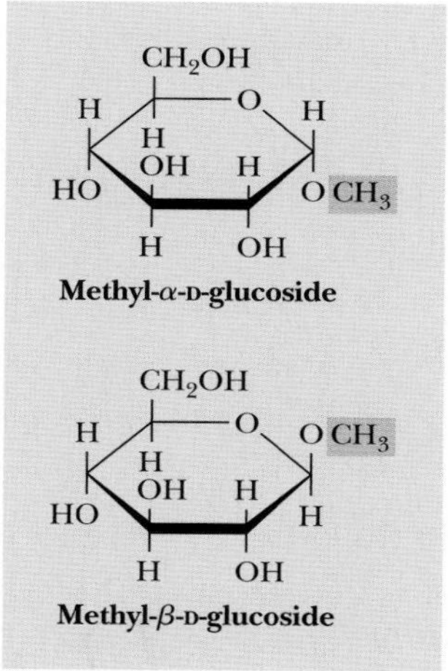

FIGURE 7.17 The anomeric forms of methyl-D-glucoside.

Lactose (galactose-β-1,4-glucose)

Free anomeric carbon
(reducing end)

Sucrose (glucose-1α-β2-fructose)

Maltose (glucose-α-1,4-glucose)

Cellobiose (glucose-β-1,4-glucose)

Isomaltose (glucose-α-1,6-glucose)

Simple sugars	
	Glucose
	Galactose
	Fructose

FIGURE 7.18 The structures of several important disaccharides. Note that the notation "HOH" means that the configuration can be either α or β. If the —OH group is above the ring, the configuration is termed β. The configuration is α if the —OH group is below the ring. Also note that sucrose has no free anomeric carbon atom.

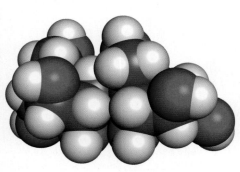

Sucrose

7.3   What Are the Structure and Chemistry of Oligosaccharides?

Given the relative complexity of oligosaccharides and polysaccharides in higher organisms, it is perhaps surprising that these molecules are formed from relatively few different monosaccharide units. (In this respect, the oligosaccharides and polysaccharides are similar to proteins; both form complicated structures based on a small number of different building blocks.) Monosaccharide units include the hexoses glucose, fructose, mannose, and galactose and the pentoses ribose and xylose.

Disaccharides Are the Simplest Oligosaccharides

The simplest oligosaccharides are the **disaccharides,** which consist of two monosaccharide units linked by a glycosidic bond. As in proteins and nucleic acids, each individual unit in an oligosaccharide is termed a *residue.* The disaccharides shown in Figure 7.18 are all commonly found in nature, with **sucrose, maltose,** and **lactose** being the most common. Each is a mixed acetal, with one hydroxyl group provided intramolecularly and one hydroxyl from the other monosaccharide. Except for sucrose, each of these structures possesses one free unsubstituted anomeric carbon atom, and thus each of these disaccharides is a reducing sugar. The end of the molecule containing the free anomeric carbon is called the **reducing end,** and the other end is called the **nonreducing end.** In the case of sucrose, both of the anomeric carbon atoms are substituted, that is, neither has a free —OH group. The substituted anomeric carbons cannot be converted to the aldehyde configuration and thus cannot participate in the oxidation–reduction reactions characteristic of reducing sugars. Thus, sucrose is *not* a reducing sugar.

Maltose, isomaltose, and cellobiose are all **homodisaccharides** because they each contain only one kind of monosaccharide, namely, glucose. Maltose is produced from starch (a polymer of α-D-glucose produced by plants) by the action of amylase enzymes and is a component of malt, a substance obtained by allowing grain (particularly barley) to soften in water and germinate. The enzyme **diastase,** produced during the germination process, catalyzes the hydrolysis of starch to maltose. Maltose is used in beverages (malted milk, for example), and because it is fermented readily by yeast, it is important in the brewing of beer. In both maltose and cellobiose, the glucose units are **1⟶4 linked,** meaning that the C-1 of one glucose is linked by a glycosidic bond to the C-4 oxygen of the other glucose. The only difference between them is in the configuration at the glycosidic bond. Maltose exists in the α-configuration, whereas cellobiose is a β-configuration. **Isomaltose** is obtained in the hydrolysis of some polysaccharides (such as dextran), and **cellobiose** is obtained from the acid hydrolysis of cellulose. Isomaltose also consists of two glucose units in a glycosidic bond, but in this case, C-1 of one glucose is linked to C-6 of the other, and the configuration is α.

The complete structures of these disaccharides can be specified in shorthand notation by using abbreviations for each monosaccharide, α or β, to denote configuration, and appropriate numbers to indicate the nature of the linkage. Thus, cellobiose is Glcβ1–4Glc, whereas isomaltose is Glcα1–6Glc. Often the glycosidic linkage is written with an arrow so that cellobiose and isomaltose would be Glcβ1$\longrightarrow$4Glc and Glcα1$\longrightarrow$6Glc, respectively. Because the linkage carbon on the first sugar is always C-1, a newer trend is to drop the 1– or 1$\longrightarrow$ and describe these simply as Glcβ4Glc and Glcα6Glc, respectively. More complete names can also be used, however; for example, maltose would be O-α-D-glucopyranosyl-(1$\longrightarrow$4)-D-glucopyranose. Cellobiose, because of its β-glycosidic linkage, is formally O-β-D-glucopyranosyl-(1$\longrightarrow$4)-D-glucopyranose.

▶ A DEEPER LOOK

Trehalose—A Natural Protectant for Bugs

Insects use an open circulatory system to circulate **hemolymph** (insect blood). The "blood sugar" is not glucose but rather **trehalose**, an unusual, nonreducing disaccharide (see figure). Trehalose is found typically in organisms that are naturally subject to temperature variations and other environmental stresses—bacterial spores, fungi, yeast, and many insects. (Interestingly, honeybees do not have trehalose in their hemolymph, perhaps because they practice a colonial, rather than solitary, lifestyle. Bee colonies maintain a rather constant temperature of 18°C, protecting the residents from large temperature changes.)

What might explain this correlation between trehalose utilization and environmentally stressful lifestyles? Konrad Bloch* suggests that

trehalose may act as a natural cryoprotectant. Freezing and thawing of biological tissues frequently causes irreversible structural changes, destroying biological activity. High concentrations of polyhydroxy compounds, such as sucrose and glycerol, can protect biological materials from such damage. Trehalose is particularly well suited for this purpose and has been shown to be superior to other polyhydroxy compounds, especially at low concentrations. Support for this novel idea comes from studies by Paul Attfield,[†] which show that trehalose levels in the yeast *Saccharomyces cerevisiae* increase significantly during exposure to high salt and high growth temperatures—the same conditions that elicit the production of heat shock proteins!

*Bloch, K., 1994. *Blondes in Venetian Paintings, the Nine-Banded Armadillo, and Other Essays in Biochemistry.* New Haven: Yale University Press.
[†]Attfield, P. V., 1987. Trehalose accumulates in *Saccharomyces cerevisiae* during exposure to agents that induce heat shock responses. *FEBS Letters* **225**:259.

β-D-Lactose (O-β-D-galactopyranosyl-(1$\longrightarrow$4)-D-glucopyranose) (Figure 7.18) is the principal carbohydrate in milk and is of critical nutritional importance to mammals in the early stages of their lives. It is formed from D-galactose and D-glucose via a β(1$\longrightarrow$4) link, and because it has a free anomeric carbon, it is capable of mutarotation and is a reducing sugar. It is an interesting quirk of nature that lactose cannot be absorbed directly into the bloodstream. It must first be broken down into galactose and glucose by **lactase**, an intestinal enzyme that exists in young, nursing mammals but is not produced in significant quantities in the mature mammal. Most adult humans, with the exception of certain groups in Africa and northern Europe, produce only low levels of lactase. For most individuals, this is not a problem, but some cannot tolerate lactose and experience intestinal pain and diarrhea upon consumption of milk.

Sucrose, in contrast, is a disaccharide of almost universal appeal and tolerance. Produced by many higher plants and commonly known as *table sugar,* it is one of the products of photosynthesis and is composed of fructose and glucose. Sucrose has a specific optical rotation, $[\alpha]_D^{20}$, of $+66.5°$, but an equimolar mixture of its component monosaccharides has a net negative rotation ($[\alpha]_D^{20}$ of glucose is $+52.5°$ and of fructose is $-92°$). Sucrose is hydrolyzed by the enzyme **invertase,** so named for the inversion of optical rotation accompanying this reaction. Sucrose is also easily hydrolyzed by dilute acid. Although sucrose and maltose are important to the human

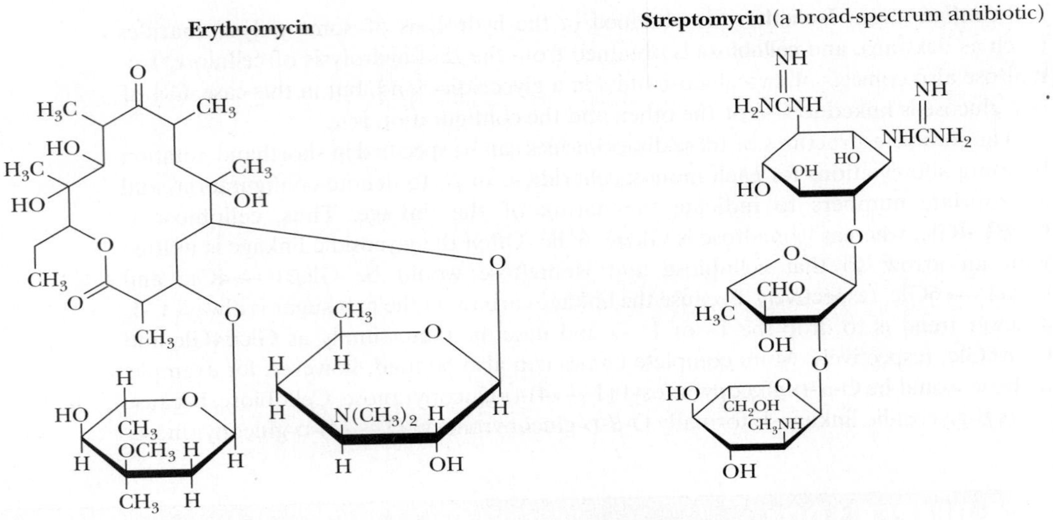

Erythromycin

Streptomycin (a broad-spectrum antibiotic)

FIGURE 7.19 Some antibiotics are oligosaccharides or contain oligosaccharide groups.

diet, they are not taken up directly in the body. In a manner similar to lactose, they are first hydrolyzed by **sucrase** and **maltase,** respectively, in the human intestine.

A Variety of Higher Oligosaccharides Occur in Nature

In addition to the simple disaccharides, many other oligosaccharides are found in both prokaryotic and eukaryotic organisms, either as naturally occurring substances or as hydrolysis products of natural materials.

Oligosaccharides also occur widely as components (via glycosidic bonds) of *antibiotics* derived from various sources. Figure 7.19 shows the structures of two representative carbohydrate-containing antibiotics.

7.4 : What Are the Structure and Chemistry of Polysaccharides?

Nomenclature for Polysaccharides Is Based on Their Composition and Structure

By far the majority of carbohydrate material in nature occurs in the form of polysaccharides. By our definition, polysaccharides include not only those substances composed only of glycosidically linked sugar residues but also molecules that contain polymeric saccharide structures linked via covalent bonds to amino acids, peptides, proteins, lipids, and other structures.

Polysaccharides, also called **glycans,** consist of monosaccharides and their derivatives. If a polysaccharide contains only one kind of monosaccharide molecule, it is a **homopolysaccharide,** or **homoglycan,** whereas those containing more than one kind of monosaccharide are **heteropolysaccharides.** The most common constituent of polysaccharides is D-glucose, but D-fructose, D-galactose, L-galactose, D-mannose, L-arabinose, and D-xylose are also common. Common monosaccharide derivatives in polysaccharides include the amino sugars (D-glucosamine and D-galactosamine), their derivatives (N-acetylneuraminic acid and N-acetylmuramic acid), and simple sugar acids (glucuronic and iduronic acids). Polysaccharides differ not only in the nature of their component monosaccharides but also in the length of their chains and in the amount of chain branching that occurs. Although a given sugar residue has only one anomeric carbon and thus can form only one glycosidic linkage with hydroxyl groups on other

Amylose

Amylopectin

FIGURE 7.20 Amylose and amylopectin are the two forms of starch. Note that the linear linkages are $\alpha(1\longrightarrow4)$ but the branches in amylopectin are $\alpha(1\longrightarrow6)$. Branches in polysaccharides can involve any of the hydroxyl groups on the monosaccharide components. Amylopectin is a highly branched structure, with branches occurring every 12 to 30 residues.

molecules, each sugar residue carries several hydroxyls, one or more of which may be an acceptor of glycosyl substituents (Figure 7.20). This ability to form branched structures distinguishes polysaccharides from proteins and nucleic acids, which occur only as linear polymers.

Polysaccharides Serve Energy Storage, Structure, and Protection Functions

Polysaccharides function as storage materials, structural components, or protective substances. Thus, *starch, glycogen,* and other storage polysaccharides, as readily metabolizable food, provide energy reserves for cells. *Chitin* and *cellulose* provide strong support for the skeletons of arthropods and green plants, respectively. Mucopolysaccharides, such as the *hyaluronic acids,* form protective coats on animal cells. In each of these cases, the relevant polysaccharide is either a homopolymer or a polymer of small repeating units. Recent research indicates, however, that oligosaccharides and polysaccharides with varied structures may also be involved in much more sophisticated tasks in cells, including a variety of cellular recognition and intercellular communication events, as discussed later.

Polysaccharides Provide Stores of Energy

Organisms store carbohydrates in the form of polysaccharides rather than as monosaccharides to lower the osmotic pressure of the sugar reserves. Because osmotic pressures depend only on *numbers of molecules,* the osmotic pressure is greatly reduced by formation of a few polysaccharide molecules out of thousands (or even millions) of monosaccharide units.

Starch By far the most common storage polysaccharide in plants is **starch,** which exists in two forms: **α-amylose** and **amylopectin** (Figure 7.20). Most forms of starch in nature are 10% to 30% α-amylose and 70% to 90% amylopectin. α-Amylose is composed of linear chains of D-glucose in $\alpha(1\longrightarrow4)$ linkages. The chains are of varying length, having molecular weights from several thousand to half a million. As can be seen from the structure in Figure 7.20, the chain has a reducing end and a nonreducing end. Although poorly soluble in water, α-amylose forms micelles in which the polysaccharide chain adopts a helical conformation (Figure 7.21). Iodine reacts with α-amylose to give a

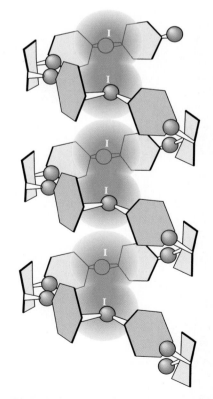

FIGURE 7.21 Suspensions of amylose in water adopt a helical conformation. Iodine (I_2) can insert into the middle of the amylose helix to give a blue color that is characteristic and diagnostic for starch.

characteristic blue color, which arises from the insertion of iodine into the hydrophobic middle of the amylose helix.

Amylopectin is a highly branched chain of glucose units (Figure 7.20). Branches occur in these chains every 12 to 30 residues. The average branch length is between 24 and 30 residues, and molecular weights of amylopectin molecules can range up to 100 million. The linear linkages in amylopectin are $\alpha(1\longrightarrow4)$, whereas the branch linkages are $\alpha(1\longrightarrow6)$. As is the case for α-amylose, amylopectin forms micellar suspensions in water; iodine reacts with such suspensions to produce a red-violet color.

Starch is stored in plant cells in the form of granules in the stroma of plastids (plant cell organelles). When starch is to be mobilized and used by the plant that stored it, it is split into its monosaccharide elements by stepwise phosphorolytic cleavage of glucose units, a reaction catalyzed by **starch phosphorylase** (Figure 7.22). The products are one molecule of glucose-1-phosphate and a starch molecule with one less glucose unit. In α-amylose, this process continues all along the chain until the end is reached.

In animals, digestion and use of plant starches begin in the mouth with **salivary α-amylase** ($\alpha(1\longrightarrow4)$-glucan 4-glucanohydrolase), the major enzyme secreted by the salivary glands. Although the capability of making and secreting salivary α-amylases is widespread in the animal world, some animals (such as cats, dogs, birds, and horses) do not secrete them. Salivary α-amylase is an **endoamylase** that splits $\alpha(1\longrightarrow4)$ glycosidic linkages only within the chain. Raw starch is not very susceptible to salivary endoamylase. However, when suspensions of starch granules are heated, the granules swell, taking up water and causing the polymers to become more accessible to enzymes. Thus, cooked starch is more digestible. Most starch digestion occurs in the small intestine via glycohydrolases.

Glycogen The major form of storage polysaccharide in animals is **glycogen.** Glycogen is found mainly in the liver (where it may amount to as much as 10% of liver mass) and skeletal muscle (where it accounts for 1% to 2% of muscle mass). Liver glycogen consists of granules containing highly branched molecules, with $\alpha(1\longrightarrow6)$ branches occurring every 8 to 12 glucose units. Like amylopectin, glycogen yields a red-violet color with iodine. Glycogen can be hydrolyzed by both α- and β-amylases, yielding glucose and maltose, respectively, as products and can also be hydrolyzed by **glycogen phosphorylase,** an enzyme present in liver and muscle tissue, to release glucose-1-phosphate.

Dextran Another important family of storage polysaccharides is the **dextrans,** which are $\alpha(1\longrightarrow6)$-linked polysaccharides of D-glucose with branched chains found in yeast and bacteria. Because the main polymer chain is $\alpha(1\longrightarrow6)$ linked, the repeating unit is *isomaltose,* Glc$\alpha1\longrightarrow6$Glc. The branch points may be $1\longrightarrow2$, $1\longrightarrow3$, or $1\longrightarrow4$ in various species. The degree of branching and the average chain length between branches depend on the species and strain of the organism. Bacteria growing on the surfaces of teeth produce extracellular accumulations of dextrans, an important component of *dental plaque.*

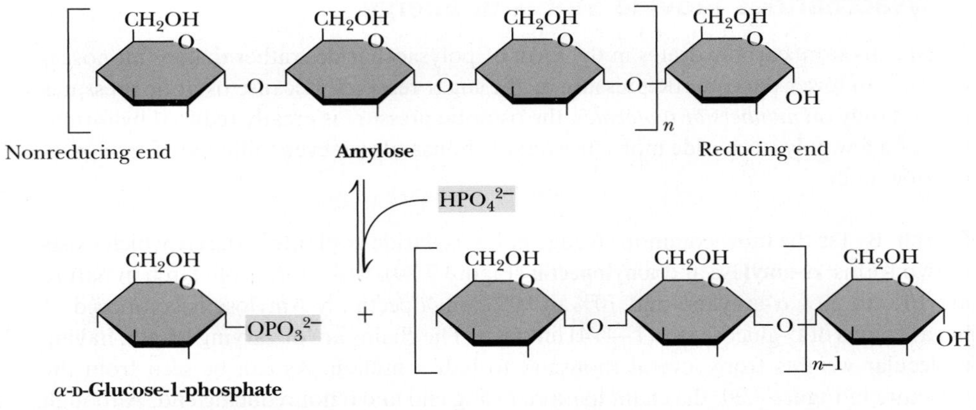

FIGURE 7.22 The starch phosphorylase reaction cleaves glucose residues from amylose, producing α-D-glucose-1-phosphate.

Polysaccharides Provide Physical Structure and Strength to Organisms

Cellulose The **structural polysaccharides** have properties that are dramatically different from those of the storage polysaccharides, even though the compositions of these two classes are similar. The structural polysaccharide **cellulose** is the most abundant natural polymer in the world. Found in the cell walls of nearly all plants, cellulose is one of the principal components providing physical structure and strength. The wood and bark of trees are insoluble, highly organized structures formed from cellulose and also from *lignin* (see Figure 25.35). It is awe-inspiring to look at a large tree and realize the amount of weight supported by polymeric structures derived from sugars and organic alcohols. Cellulose also has its delicate side, however. **Cotton,** whose woven fibers make some of our most comfortable clothing fabrics, is almost pure cellulose. Derivatives of cellulose have found wide use in our society. **Cellulose acetates** are produced by the action of acetic anhydride on cellulose in the presence of sulfuric acid and can be spun into a variety of fabrics with particular properties. Referred to simply as *acetates,* they have a silky appearance, a luxuriously soft feel, and a deep luster and are used in dresses, lingerie, linings, and blouses.

Cellulose is a linear homopolymer of D-glucose units, just as in α-amylose. The structural difference, which completely alters the properties of the polymer, is that in cellulose the glucose units are linked by $\beta(1\longrightarrow4)$-glycosidic bonds, whereas in α-amylose the linkage is $\alpha(1\longrightarrow4)$. The conformational difference between these two structures is shown in Figure 7.23. The $\alpha(1\longrightarrow4)$-linkage sites of amylose are naturally bent, conferring a gradual

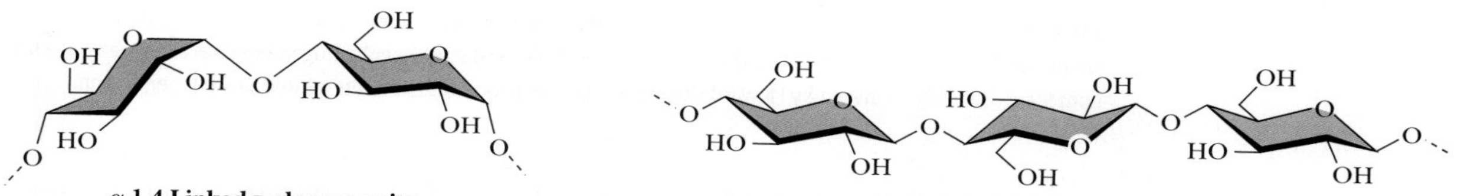

(a) α-1,4-Linked D-glucose units

(b) β-1,4-Linked D-glucose units

FIGURE 7.23 (a) Amylose, composed exclusively of the relatively bent $\alpha(1\longrightarrow4)$ linkages, prefers to adopt a helical conformation, whereas **(b)** cellulose, with $\beta(1\longrightarrow4)$-glycosidic linkages, can adopt a fully extended conformation with alternating 180° flips of the glucose units. The hydrogen bonding inherent in such extended structures is responsible for the great strength of tree trunks and other cellulose-based materials.

> **A DEEPER LOOK**

Billiard Balls, Exploding Teeth, and Dynamite—The Colorful History of Cellulose

Although humans cannot digest it and most people's acquaintance with cellulose is limited to comfortable cotton clothing, cellulose has enjoyed a colorful and varied history of utilization. In 1838, Théophile Pelouze in France found that paper or cotton could be made explosive if dipped in concentrated nitric acid. Christian Schönbein, a professor of chemistry at the University of Basel, prepared "nitrocotton" in 1845 by dipping cotton in a mixture of nitric and sulfuric acids and then washing the material to remove excess acid. In 1860, Major E. Schultze of the Prussian Army used the same material, now called **guncotton,** as a propellant replacement for gunpowder, and its preparation in brass cartridges quickly made it popular for this purpose. The only problem was that it was too explosive and could detonate unpredictably in factories where it was produced. The entire town of Faversham, England, was destroyed in such an accident. In 1868, Alfred Nobel mixed guncotton with ether and alcohol, thus preparing **nitrocellulose,** and in turn mixed this with nitroglycerin and sawdust to produce **dynamite.** Nobel's income from dynamite and also from his profitable development of the Russian oil fields in Baku eventually formed the endowment for the Nobel Prizes.

In 1869, concerned over the precipitous decline (from hunting) of the elephant population in Africa, the billiard ball manufacturers Phelan and Collander offered a prize of $10,000 for production of a substitute for ivory. Brothers Isaiah and John Hyatt in Albany, New York, produced a substitute for ivory by mixing guncotton with camphor, then heating and squeezing it to produce **celluloid.** This product found immediate uses well beyond billiard balls. It was easy to shape, strong, and resilient, and it exhibited a high tensile strength. Celluloid was eventually used to make dolls, combs, musical instruments, fountain pens, piano keys, and a variety of other products. The Hyatt brothers eventually formed the Albany Dental Company to make false teeth from celluloid. Because camphor was used in their production, the company advertised that their teeth smelled "clean," but as reported in the *New York Times* in 1875, the teeth also occasionally exploded!

Portions adapted from Burke, J., 1996. *The Pinball Effect: How Renaissance Water Gardens Made the Carburetor Possible and Other Journeys Through Knowledge.* New York: Little, Brown, & Company.

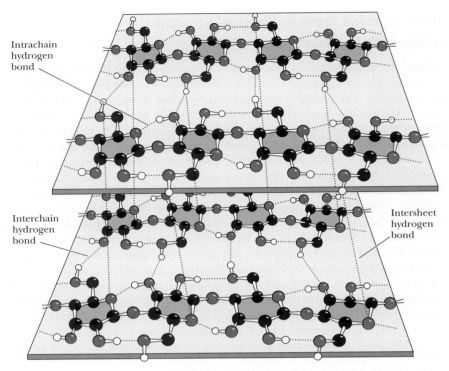

Intrachain hydrogen bond

Interchain hydrogen bond

Intersheet hydrogen bond

FIGURE 7.24 The structure of cellulose, showing the hydrogen bonds (blue) between the sheets, which strengthen the structure. Intrachain hydrogen bonds are in red, and interchain hydrogen bonds are in green. (Illustration: Irving Geis. Rights owned by Howard Hughes Medical Institute. Not to be reproduced without permission.)

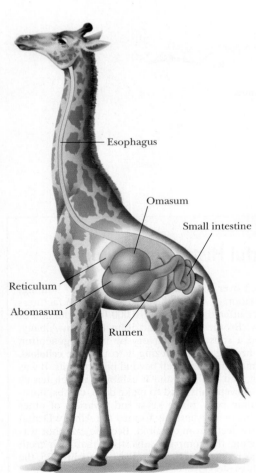

Esophagus

Omasum

Small intestine

Reticulum

Abomasum

Rumen

FIGURE 7.25 Giraffes, cattle, deer, and camels are ruminant animals that are able to metabolize cellulose, thanks to bacterial cellulase in the rumen, a large first compartment in the stomach of a ruminant.

turn to the polymer chain, which results in the helical conformation already described (Figure 7.21). The most stable conformation about the $\beta(1 \longrightarrow 4)$ linkage involves alternating 180° flips of the glucose units along the chain so that the chain adopts a fully extended conformation, referred to as an **extended ribbon.** Juxtaposition of several such chains permits efficient interchain hydrogen bonding, the basis of much of the strength of cellulose.

The structure of one form of cellulose, determined by X-ray and electron diffraction data, is shown in Figure 7.24. The flattened sheets of the chains lie side by side and are joined by hydrogen bonds. These sheets are laid on top of one another in a way that staggers the chains, just as bricks are staggered to give strength and stability to a wall. Cellulose is extremely resistant to hydrolysis, whether by acid or by the digestive tract amylases described earlier. As a result, most animals (including humans) cannot digest cellulose to any significant degree. Ruminant animals, such as cattle, deer, giraffes, and camels, are an exception because bacteria that live in the rumen (Figure 7.25) secrete the enzyme **cellulase,** a β-glucosidase effective in the hydrolysis of cellulose. The resulting glucose is then metabolized in a fermentation process to the benefit of the host animal. Termites and shipworms *(Teredo navalis)* similarly digest cellulose because their digestive tracts also contain bacteria that secrete cellulase.

Chitin A polysaccharide that is similar to cellulose, both in its biological function and its primary, secondary, and tertiary structure, is **chitin.** Chitin is present in the cell walls of fungi and is the fundamental material in the exoskeletons of crustaceans, insects, and spiders. The structure of chitin, an extended ribbon, is identical to that of cellulose, except that the —OH group on each C-2 is replaced by —NHCOCH$_3$, so the repeating units are **N-acetyl-D-glucosamines** in $\beta(1 \longrightarrow 4)$ linkage. Like cellulose (Figure 7.24), the chains of chitin form extended ribbons (Figure 7.26) and pack side by side in a crystalline, strongly hydrogen-bonded form. One significant difference between cellulose and chitin is whether the chains are arranged in **parallel** (all the reducing ends together at one end of a packed bundle and all the nonreducing ends together at the other end) or **antiparallel** (each sheet of chains having the chains arranged oppositely from the sheets above and below). Natural cellulose seems to occur only in parallel arrangements. Chitin, however, can occur in three

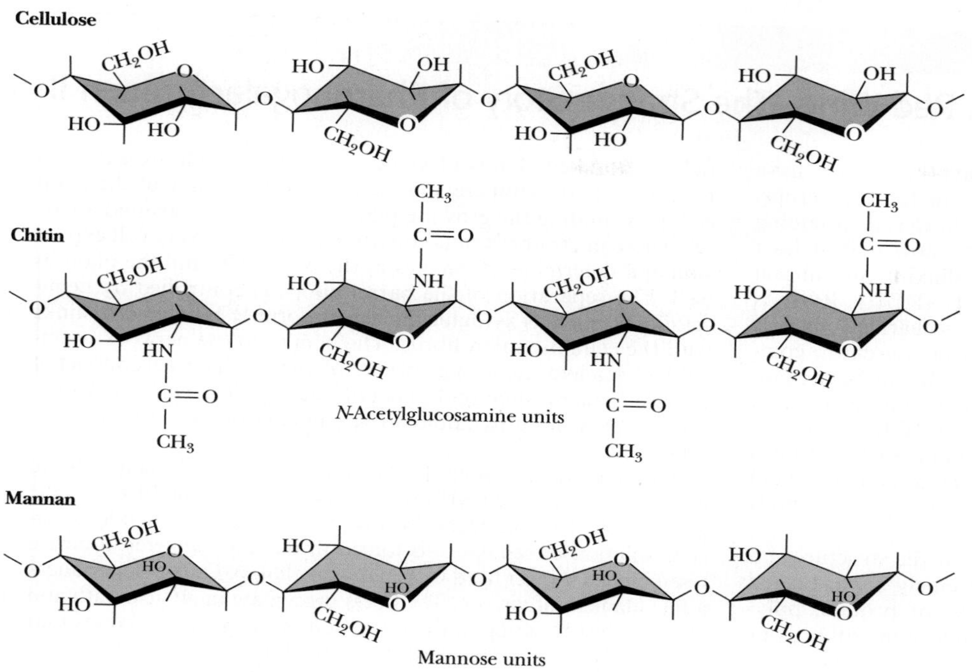

Cellulose

Chitin

N-Acetylglucosamine units

Mannan

Mannose units

FIGURE 7.26 Like cellulose, chitin and mannan form extended ribbons and pack together efficiently, taking advantage of multiple hydrogen bonds.

Agarose

3,6-Anhydro bridge

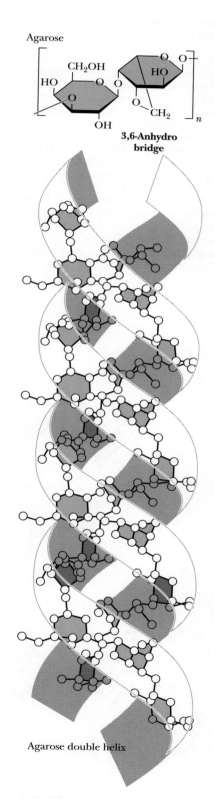

Agarose double helix

FIGURE 7.27 The favored conformation of agarose in water is a double helix with a threefold screw axis.

forms, sometimes all in the same organism. *α-Chitin* is an all-parallel arrangement of the chains, whereas *β-chitin* is an antiparallel arrangement. In *δ-chitin,* the structure is thought to involve pairs of parallel sheets separated by single antiparallel sheets.

Chitin is the earth's second most abundant carbohydrate polymer (after cellulose), and its ready availability and abundance offer opportunities for industrial and commercial applications. Chitin-based coatings can extend the shelf life of fruits, and a chitin derivative that binds to iron atoms in meat has been found to slow the reactions that cause rancidity and flavor loss. Without such a coating, the iron in meats activates oxygen from the air, forming reactive free radicals that attack and oxidize polyunsaturated lipids, causing most of the flavor loss associated with rancidity. Chitin-based coatings coordinate the iron atoms, preventing their interaction with oxygen.

Agarose An important polysaccharide mixture isolated from marine red algae *(Rhodophyceae)* is **agar,** which consists of two components: **agarose** and **agaropectin.** Agarose (Figure 7.27) is a chain of alternating D-galactose and 3,6-anhydro-L-galactose, with side chains of 6-methyl-D-galactose. Agaropectin is similar, but in addition, it contains sulfate ester side chains and D-glucuronic acid. The three-dimensional structure of agarose is a double helix with a threefold screw axis, as shown in Figure 7.27. The central cavity is large enough to accommodate water molecules. Agarose and agaropectin readily form gels containing large amounts (up to 99.5%) of water.

Glycosaminoglycans A class of polysaccharides known as **glycosaminoglycans** is involved in a variety of extracellular (and sometimes intracellular) functions. Glycosaminoglycans consist of linear chains of repeating disaccharides in which one of the monosaccharide units is an amino sugar and one (or both) of the monosaccharide units contains at least one negatively charged sulfate or carboxylate group. The repeating disaccharide structures found commonly in glycosaminoglycans are shown in Figure 7.28. **Heparin,** with the highest net negative charge of the disaccharides shown, is a natural anticoagulant substance. It binds strongly to *antithrombin III* (a protein involved in terminating the clotting process) and inhibits blood clotting. **Hyaluronate** molecules may consist of as many as 25,000 disaccharide units, with molecular weights of up to 10^7. Hyaluronates are important components of the vitreous humor in the eye and of

A DEEPER LOOK

A Complex Polysaccharide in Red Wine—The Strange Story of Rhamnogalacturonan II

For many years, cotton and grape growers and other farmers have known that boron is an essential trace element for their crops. Until recently, however, the role or roles of boron in sustaining plant growth were unknown. Recent reports show that at least one role for boron in plants is that of crosslinking an unusual polysaccharide called rhamnogalacturonan II (RGII). RGII is a low-molecular-weight (5 to 10 kDa) polysaccharide, but it is thought to be the most complex polysaccharide on earth, comprised as it is of 11 different sugar monomers. It can be released from plant cell walls by treatment with a galacturonase, and it is also present in red wine. Part of the structure of RGII is shown in the accompanying figure. The nature of the borate ester crosslinks (also indicated in the figure) was elucidated by Malcolm O'Neill and his colleagues, who used a combination of chemical methods and boron-11 NMR.

Why is rhamnogalacturonan II essential for the structure and growth of plant walls? Plant walls are extremely sophisticated composite materials, composed of networks of protein, polysaccharides, and phenolic compounds. Cellulose microfibrils as strong as steel provide a load-bearing framework for the plant. These microfibrils are tiny wires made of crystalline arrays of β(1⟶4)-linked chains of glucose residues, which are extruded from hexameric spinnerets in the plasma membrane of the plant cell like hoops around a barrel, surrounding the growing plant cell like hoops around a barrel. These microfibrils thus constrain the directions of cell expansion and determine the shapes of the plant cells and the plant as well. The separation of the barrel hoops is controlled by hemicelluloses, such as xyloglucans, which form H-bonded crosslinks with the cellulose microfibrils. The hemicellulose network is embedded in a hydrated gel inside the plant wall. This gel consists of complex galacturonic acid–rich polysaccharides, including RGII. The gel provides a dynamic operating environment for cell wall processes.

It is interesting to note that the tiny spinnerets of plant cells are nature's version of the viscose process, developed in 1910, for the production of rayon fibers. In this process, viscose—literally a *viscous* solution of cellu*lose*—is forced through a spinneret (a device resembling a shower head with many tiny holes). Each hole produces a fine filament of viscose. The fibers precipitate in an acid bath and are stretched resulting in the formation of interchain H bonds that give the filaments the properties essential for use as textile fibers.

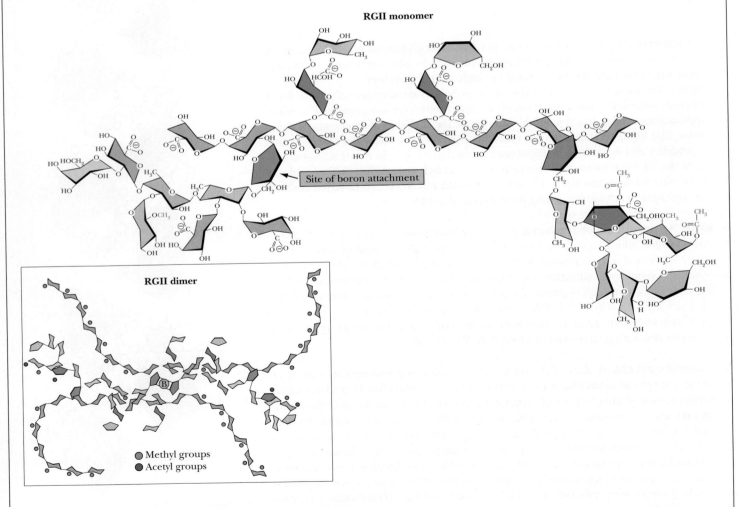

RGII monomer

Site of boron attachment

RGII dimer

● Methyl groups
● Acetyl groups

Source: Höfte, H., 2001. A baroque residue in red wine. *Science* **294**:795–797.

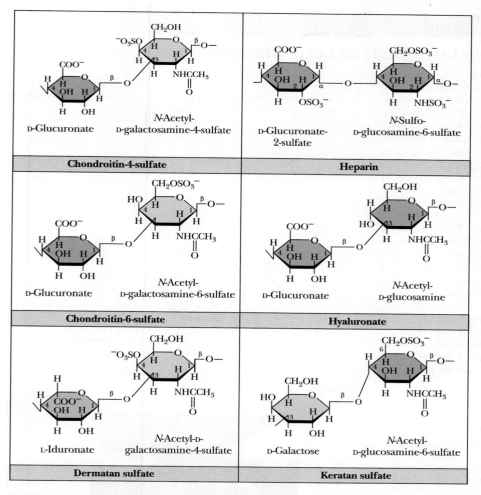

FIGURE 7.28 Glycosaminoglycans are formed from repeating disaccharide arrays. Glycosaminoglycans are components of the proteoglycans.

synovial fluid, the lubricant fluid of joints in the body. The **chondroitins** and **keratan sulfate** are found in tendons, cartilage, and other connective tissue; **dermatan sulfate,** as its name implies, is a component of the extracellular matrix of skin. Glycosaminoglycans are fundamental constituents of *proteoglycans* (discussed later).

Polysaccharides Provide Strength and Rigidity to Bacterial Cell Walls

Some of nature's most interesting polysaccharide structures are found in *bacterial cell walls*. Given the strength and rigidity provided by polysaccharide structures, it is not surprising that bacteria use such structures to provide protection for their cellular contents. Bacteria normally exhibit high internal osmotic pressures and frequently encounter variable, often hypotonic exterior conditions. The rigid cell walls synthesized by bacteria maintain cell shape and size and prevent swelling or shrinkage that would inevitably accompany variations in solution osmotic strength.

Peptidoglycan Is the Polysaccharide of Bacterial Cell Walls

Bacteria are conveniently classified as either **Gram-positive** or **Gram-negative** depending on their response to Gram stain, which identifies multilayered peptidoglycan structures. Despite substantial differences in the various structures surrounding these

▶ A DEEPER LOOK

The Secrets of Phloem and the Large Fruits of *Cucurbitaceae*

The plant family *Cucurbitaceae* includes melons, cucumbers, and pumpkins, all of which produce particularly large fruits. Growth of such large fruits depends on a transport system that can deliver the products of photosynthesis—sugars and other nutrients—from the leaves through the stem and into the "sink tissues" such as the fruits and roots, where they are used in growth and storage. The phloem is the living tissue that is responsible for this transport. However, previously measured values of the sugar content of phloem sap of pumpkins and other cucurbits had been approximately 30 mM—about 30-fold too low to account for photosynthetic output and the requirements of growing large friuts.

New studies by Baichen Zhang and co-workers show that phloem sap used in previous measurements was derived from only one of the two types of phloem—the so-called extrafascicular phloem (the blue strands in the figure). Release of sap from the other type of phloem—the fascicular phloem (pink in the figure)—is prevented by rapid "wound sealing" when the plant tissue is dissected. This wound sealing is achieved by deposition of callose—a special polysaccharide comprised of $\beta(1\longrightarrow 3)$ linked glucose polymers with $\beta(1\longrightarrow 6)$ branches. Using special phloem labeling techniques and video microscopy, Zhang and colleagues have shown that the sugar content of fascicular phloem tissue can reach 1 molar—a concentration sufficient to account for photosynthetic output and the prodigious fruit growth in pumpkins and other cucurbits.

References:
Turgeon, R., and Oparka, K., 2010. The secret phloem of pumpkins. *Proceedings of the National Academy of Sciences* **107**:13201–13202.
Zhang, B., Tolstikov, V., Turnbull, C., Hicks, L., and Fiehn, O., 2010. Divergent metabolome and proteome suggest functional independence of dual phloem transport systems in cucurbits. *Proceedings of the National Academy of Sciences* **107**:13532–13537.

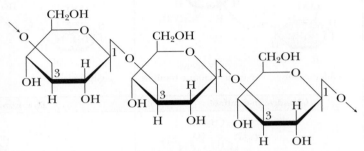

▲ The $\beta(1\longrightarrow 3)$ linked structure of callose. Branches involve $\beta(1\longrightarrow 6)$ linkages.

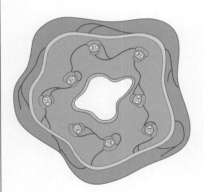

◀ Transverse section of a *C. maxima* stem. The fascicular phloem (pink) occurs on either side of the xylem (gray) in the main vascular bundles of the stem. The extrafascicular phloem (blue) forms a connective network in the pith (pale green) and also in the cortex (dark green). *Adapted from the figure in Turgeon, R., and Oparka, K., 2010. The secret phloem of pumpkins.* Proceedings of the National Academy of Sciences, USA. **107**:13201–13202.

▲ Chris Stevens of New Richmond, Wisconsin, with a record-setting 1,810.5 lb pumpkin at the Stillwater, Minnesota, Harvest Fest.

two types of cells, nearly all bacterial cell walls have a strong, protective peptide–polysaccharide layer called **peptidoglycan.** Gram-positive bacteria have a thick (approximately 25 nm) cell wall consisting of multiple layers of peptidoglycan. This thick cell wall surrounds the bacterial plasma membrane. Gram-negative bacteria, in contrast, have a much thinner (2 to 3 nm) cell wall consisting of a single layer of peptidoglycan sandwiched between the inner and outer lipid bilayer membranes. In either case, peptidoglycan, sometimes called **murein** (from the Latin *murus,* meaning "wall"), is a continuous crosslinked structure—in essence, a single molecule—built around the cell. The structure is shown in Figure 7.29. The backbone is a $\beta(1\longrightarrow 4)$-linked polymer of *N*-acetylglucosamine and *N*-acetylmuramic acid units. This part of the structure is similar to that of chitin, but it is joined to a tetrapeptide, usually L-Ala · D-Glu · L-Lys · D-Ala, in which the L-lysine is linked to the γ-COOH of D-glutamate. The peptide is linked to the *N*-acetylmuramic acid units via its D-lactate moiety. The ε-amino group of lysine in this peptide is linked to the —COOH of

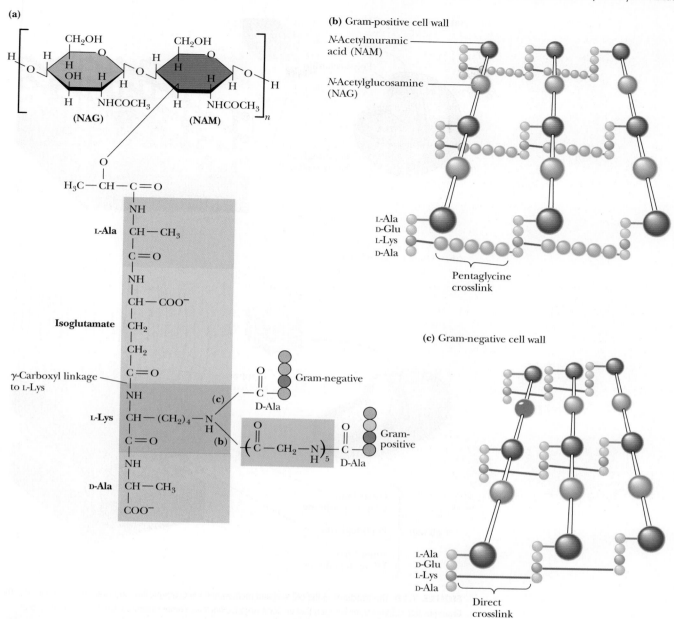

FIGURE 7.29 **(a)** The structure of peptidoglycan. The tetrapeptides linking adjacent backbone chains contain an unusual γ-carboxyl linkage. **(b)** The crosslink in Gram-positive cell walls is a pentaglycine bridge. **(c)** In Gram-negative cell walls, the linkage between the tetrapeptides of adjacent carbohydrate chains in peptidoglycan involves a direct amide bond between the lysine side chain of one tetrapeptide and D-alanine of the other.

D-alanine of an adjacent tetrapeptide. In Gram-negative cell walls, the lysine ε-amino group forms a *direct amide bond* with this D-alanine carboxyl (Figure 7.29). In Gram-positive cell walls, a **pentaglycine chain** bridges the lysine ε-amino group and the D-Ala carboxyl group. Gram-negative cell walls are also covered with highly complex lipopolysaccharides (Figure 7.30).

Cell Walls of Gram-Negative Bacteria

In Gram-negative bacteria, the peptidoglycan wall is the rigid framework around which is built an elaborate membrane structure (Figure 7.30). The peptidoglycan layer encloses the *periplasmic space* and is attached to the outer membrane via a group of **hydrophobic proteins.**

As shown in Figure 7.31, the outer membrane of Gram-negative bacteria is coated with a highly complex **lipopolysaccharide,** which consists of a lipid group (anchored

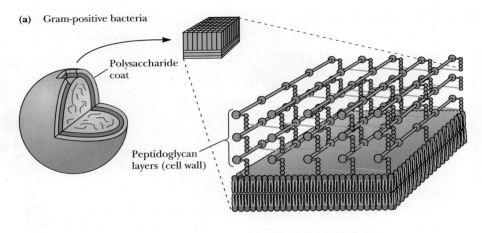

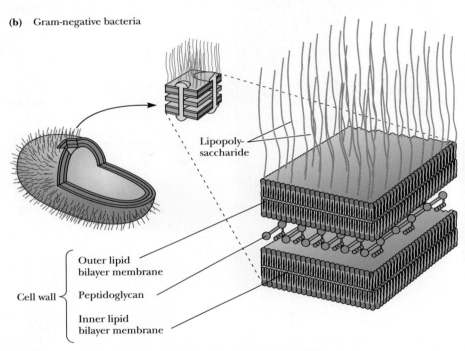

FIGURE 7.30 The structures of the cell wall and membrane(s) in Gram-positive and Gram-negative bacteria. The Gram-positive cell wall is thicker than that in Gram-negative bacteria, compensating for the absence of a second (outer) bilayer membrane.

in the outer membrane) joined to a polysaccharide made up of long chains with characteristic repeating structures (Figure 7.31). Its many different unique units determine the antigenicity of the bacteria; that is, animal immune systems recognize them as foreign substances and raise antibodies against them. As a group, these **antigenic determinants** are called the **O antigens,** and there are thousands of different ones. The *Salmonella* bacteria alone have well over a thousand known O antigens that have been organized into 17 different groups. The great variation in these O antigen structures apparently plays a role in the recognition of one type of cell by another and in evasion of the host immune system.

Cell Walls of Gram-Positive Bacteria In Gram-positive bacteria, the cell exterior is less complex than for Gram-negative cells. Having no outer membrane, Gram-positive cells compensate with a thicker wall. Covalently attached to the peptidoglycan layer are **teichoic**

acids, which often account for 50% of the dry weight of the cell wall. The teichoic acids are polymers of *ribitol phosphate* or *glycerol phosphate* linked by phosphodiester bonds.

Animals Display a Variety of Cell Surface Polysaccharides

Compared to bacterial cells, which are identical within a given cell type (except for O antigen variations), animal cells display a wondrous diversity of structure, constitution, and function. Although each animal cell contains, in its genetic material, the instructions to replicate the entire organism, each differentiated animal cell carefully controls its composition and behavior within the organism. A great part of each cell's uniqueness begins at the cell surface. This surface uniqueness is critical to each animal cell because cells spend their entire life span in intimate contact with other cells and must therefore communicate with one another. That cells are able to pass information among themselves is evidenced by numerous experiments. For example, heart *myocytes,* when grown in culture (in glass dishes), establish *synchrony* when they make contact, so that they "beat" or contract in unison. If they are removed from the culture and separated, they lose their synchronous behavior, but if allowed to reestablish cell-to-cell contact, they spontaneously restore their synchronous contractions.

As these and many other related phenomena show, it is clear that molecular structures on one cell are recognizing and responding to molecules on the adjacent cell or to molecules in the **extracellular matrix,** the complex "soup" of connective proteins and other molecules that exists outside of and among cells. Many of these interactions involve *glycoproteins* on the cell surface and *proteoglycans* in the extracellular matrix. The "information" held in these special carbohydrate-containing molecules is not encoded directly in the genes (as with proteins) but is determined instead by expression of the appropriate enzymes that assemble carbohydrate units in a characteristic way on these molecules. Also, by virtue of the several hydroxyl linkages that can be formed with each carbohydrate monomer, these structures are arguably more information-rich than proteins and nucleic acids, which can form only linear polymers. A few of these glycoproteins and their unique properties are described in the following sections.

7.5 What Are Glycoproteins, and How Do They Function in Cells?

Many proteins found in nature are glycoproteins because they contain covalently linked oligosaccharide and polysaccharide groups. The list of known glycoproteins includes structural proteins, enzymes, membrane receptors, transport proteins, and immunoglobulins, among others. In most cases, the precise function of the bound carbohydrate moiety is not understood.

Carbohydrates on Proteins Can be O-Linked or N-Linked

Carbohydrate groups may be linked to polypeptide chains via the hydroxyl groups of serine, threonine, or hydroxylysine residues (in **O-linked saccharides**) (Figure 7.32a) or via the amide nitrogen of an asparagine residue (in **N-linked saccharides**) (Figure 7.32b). The carbohydrate residue linked to the protein in O-linked saccharides is usually an *N*-acetylgalactosamine, but mannose, galactose, and xylose residues linked to protein hydroxyls are also found (Figure 7.32a). Oligosaccharides O-linked to glycophorin (see Figure 9.10) involve *N*-acetylgalactosamine linkages and are rich in sialic acid residues. N-linked saccharides always have a unique core structure composed of two *N*-acetylglucosamine residues linked to a branched mannose triad (Figure 7.32b, c). Many other sugar units may be linked to each of the mannose residues of this branched core.

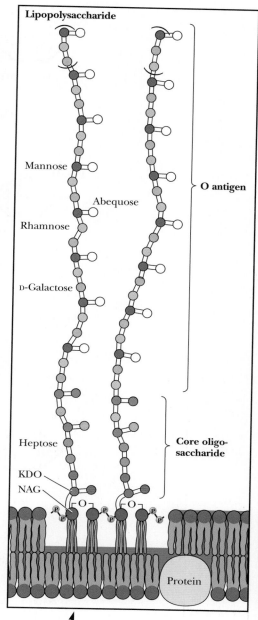

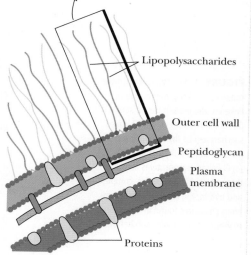

FIGURE 7.31 Lipopolysaccharide (LPS) coats the outer membrane of Gram-negative bacteria. The lipid portion ▶ of the LPS is embedded in the outer membrane and is linked to a complex polysaccharide.

(a) O-linked saccharides

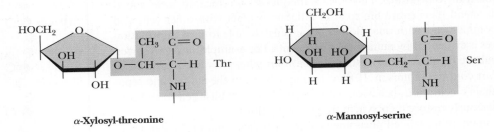

β-Galactosyl-1,3-α-N-acetylgalactosyl-serine

β-N-acetylglucosaminyl-serine
(O-linked GlcNAc)

α-Xylosyl-threonine

α-Mannosyl-serine

(b) Core oligosaccharides in N-linked glycoproteins

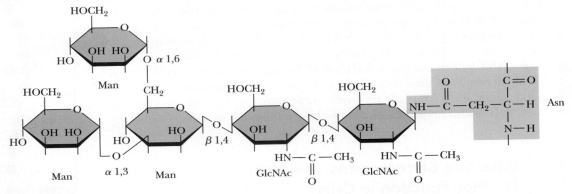

(c) N-linked glycoproteins

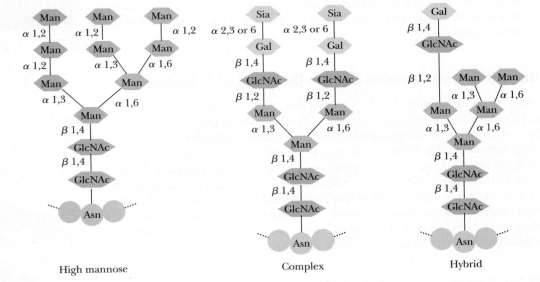

High mannose   Complex   Hybrid

FIGURE 7.32 The carbohydrate moieties of glycoproteins may be linked to the protein via **(a)** serine or threonine residues (in the O-linked saccharides) or **(b)** asparagine residues (in the N-linked saccharides). **(c)** N-linked glycoproteins are of three types: high mannose, complex, and hybrid, the latter of which combines structures found in the high mannose and complex saccharides.

O-GlcNAc Signaling Is Altered in Diabetes and Cancer

O-linked β-N-acetylglucosamine (O-GlcNAc) linked to serine or threonine (Figure 7.32a) on nuclear and cytoplasmic proteins serves as a cellular signaling device, much like phosphorylation at Ser and Thr. O-GlcNAc modifications formed in response to internernal or environmental signals affect transcription, translation, cytoskeletal assembly, signal transduction, and cellular metabolism. Interestingly, O-GlcNAc signaling and its interaction with phosphorylation are both altered in metabolic diseases, including diabetes and cancer.

O-Linked Saccharides Form Rigid Extended Extracellular Structures

O-linked saccharides are often found in cell surface glycoproteins and in **mucins,** the large glycoproteins that coat and protect mucous membranes in the respiratory and gastrointestinal tracts in the body. Certain viral glycoproteins also contain O-linked sugars. O-linked saccharides in glycoproteins are often found clustered in richly glycosylated domains of the polypeptide chain. Physical studies on mucins show that they adopt rigid, extended structures. An individual mucin molecule ($M_r = 10^7$) may extend over a distance of 150 to 200 nm in solution. Inherent steric interactions between the sugar residues and the protein residues in these cluster regions cause the peptide core to adopt an extended and relatively rigid conformation. This interesting effect may be related to the function of O-linked saccharides in glycoproteins. It allows aggregates of mucin molecules to form extensive, intertwined networks, even at low concentrations. These viscous networks protect the mucosal surface of the respiratory and gastrointestinal tracts from harmful environmental agents.

There appear to be two structural motifs for membrane glycoproteins containing O-linked saccharides. Certain glycoproteins, such as **leukosialin,** are O-glycosylated throughout much or most of their extracellular domain (Figure 7.33). Leukosialin, like mucin, adopts a highly extended conformation, allowing it to project great distances above the membrane surface, perhaps protecting the cell from unwanted interactions with

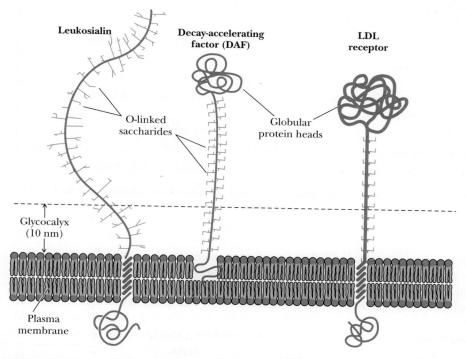

FIGURE 7.33 The O-linked saccharides of glycoproteins appear in many cases to adopt extended conformations that serve to extend the functional domains of these proteins above the membrane surface. (Adapted from Jentoft, N., 1990. Why are proteins O-glycosylated? *Trends in Biochemical Sciences* **15:**291–294.)

A DEEPER LOOK

Drug Research Finds a Sweet Spot

A variety of diseases are being successfully treated with sugar-based therapies. As this table shows, several carbohydrate-based drugs are either on the market or at various stages of clinical trials. Some of these drugs are enzymes, whereas others are glycoconjugates.

Drug	Description	Manufacturer
Tamiflu and Relenza	These "transition state analogs" (see Chapter 14) inhibit neuraminidase, an influenza enzyme that enables new virus particles to bud from a host cell.	GlaxoSmithKline (Relenza) and Genentech (Tamiflu)
Cerezyme (imiglucerase)	This enzyme degrades glycolipids, compensating for an enzyme deficiency that causes Gaucher's disease.	Genzyme Cambridge, MA
Vancocin (vancomycin)	A very potent glycopeptide antibiotic that is typically used against antibiotic-resistant infections. It inhibits synthesis of peptidoglycan in the bacterial cell wall.	Eli Lilly Indianapolis, IN
GMK	A vaccine containing ganglioside GM2; it triggers an immune response against cancer cells carrying GM2.	Progenics Pharmaceuticals Tarrytown, NY
Bimosiamose (TBC1269)	A sugar analog that inhibits selectin-based inflammation in blood vessels.	Revotar PharmAG, Berlin
GCS-100	A sugar that blocks action of galectin-3 (Gal-3), a sugar-binding protein on tumors.	GlycoGenesys Boston
PI-88	A sugar that inhibits growth factor–dependent angiogenesis and enzymes that promote metastasis.	ProgenDarra, Australia
Staphvax	A vaccine that is a protein with a linked bacterial sugar; it is intended to treat *Staphylococcus* infection.	NABI Pharmaceuticals Boca Raton, FL

Adapted from Maeder, T., 2002. Sweet medicines. *Scientific American* **287**:40–47.
Additional References: Alper, J., 2001. Searching for medicine's sweet spot. *Science* **1**:2338–2343. Borman, S., 2007. Sugar medicine. *Chemical & Engineering News* **85**:19–30.

Licensed carbohydrate-based vaccines

Organism Targeted	Vaccine	Manufacturer (Trade Name)
Haemophilus influenzae type b	Glycoconjugate (polysaccharide with tetanus toxin)	Sanofi Pasteur (ActHIB) GlaxoSmithKline (Hiberix)
Nisseria meningitides	Glycoconjugate (meningococcal polysaccharide with DT)	Sanofi Pasteur (Menactra)
Salmonella typhi	Vi capsular polysaccharide	Sanofi Pasteur (TYPHIM Vi)
Streptococcus pneumoniae	Pneumococcal polysaccharide-protein conjugate	Wyeth Pharmaceuticals (Prevnar)
Streptococcus pneumoniae	Pneumococcal polysaccharide	Merck and Co. (Pneumovax 23)

Reference: Astronomo, R. D., and Burton, D. R., 2010. Carbohydrate vaccines: developing sweet solutions to sticky situations? *Nature Reviews Drug Discovery* **9**:308–324.

macromolecules or other cells. The second structural motif is exemplified by the **low-density lipoprotein (LDL) receptor** and by **decay-accelerating factor (DAF).** These proteins contain a highly O-glycosylated stem region that separates the transmembrane domain from the globular, functional extracellular domain. The O-glycosylated stem serves to raise the functional domain of the protein far enough above the membrane surface to make it accessible to the extracellular macromolecules with which it interacts.

Polar Fish Depend on Antifreeze Glycoproteins

A unique family of O-linked glycoproteins permits fish to live in the icy seawater of the Arctic and Antarctic regions, where water temperature may reach as low as −1.9°C.

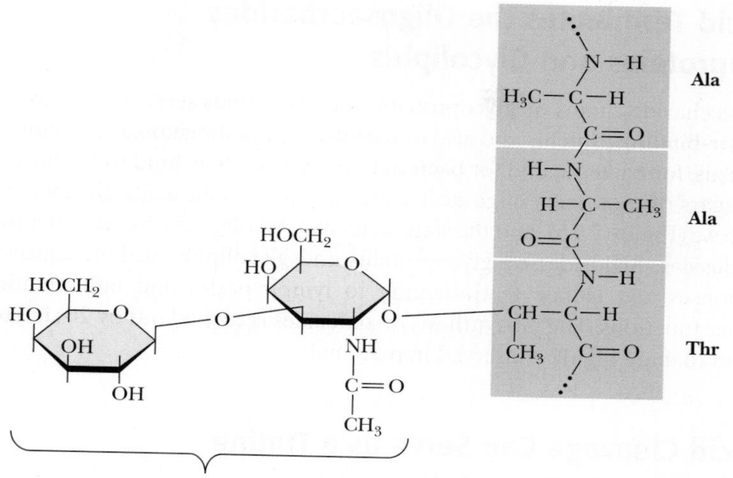

β-Galactosyl-1,3-α-N-acetylgalactosamine

Repeating unit of antifreeze glycoproteins

FIGURE 7.34 The structure of the repeating unit of antifreeze glycoproteins, a disaccharide consisting of β-galactosyl-(1⟶3)-α-N-acetylgalactosamine in glycosidic linkage to a threonine residue.

Antifreeze glycoproteins (AFGPs) are found in the blood of nearly all Antarctic fish and at least five Arctic fish. These glycoproteins have the peptide structure

$$[\text{Ala-Ala-Thr}]_n\text{-Ala-Ala}$$

where n can be 4, 5, 6, 12, 17, 28, 35, 45, or 50. Each of the threonine residues is glycosylated with the disaccharide β-galactosyl-(1⟶3)-α-N-acetylgalactosamine (Figure 7.34). This glycoprotein adopts a **flexible rod** conformation with regions of threefold left-handed helix. The evidence suggests that antifreeze glycoproteins may inhibit the formation of ice in the fish by binding specifically to the growth sites of ice crystals, inhibiting further growth of the crystals.

N-Linked Oligosaccharides Can Affect the Physical Properties and Functions of a Protein

N-linked oligosaccharides are found in many different proteins, including immunoglobulins G and M, ribonuclease B, ovalbumin, and peptide hormones. Many different functions are known or suspected for *N*-glycosylation of proteins. Glycosylation can affect the physical and chemical properties of proteins, altering solubility, mass, and electrical charge. Carbohydrate moieties have been shown to stabilize protein conformations and protect proteins against proteolysis. Eukaryotic organisms use post-translational additions of *N*-linked oligosaccharides to direct selected proteins to various membrane compartments. Recent evidence indicates that N-linked oligosaccharides promote the proper folding of newly synthesized polypeptides in the endoplasmic reticulum (see A Deeper Look below).

> **A DEEPER LOOK**

N-Linked Oligosaccharides Help Proteins Fold

One important effect of N-linked oligosaccharides in eukaryotic organisms may be their contribution to the correct folding of certain globular proteins. This adaptation of saccharide function allows cells to produce and secrete larger and more complex proteins at high levels. Inhibition of glycosylation leads to production of misfolded, aggregated proteins that lack function. Certain proteins are highly dependent on glycosylation, whereas others are much less so, and certain glycosylation sites are more important for protein folding than are others.

Studies with model peptides show that oligosaccharides can alter the conformational preferences near the glycosylation sites. In addition, the presence of polar saccharides may serve to orient that portion of a peptide toward the surface of protein domains. However, it has also been found that saccharides usually are not essential for maintaining the overall folded structure after a glycoprotein has reached its native, folded structure.

Source: Helenius, A., and Aebi, M., 2001. Intracellular functions of N-linked glycans. *Science* **291**:2364–2369.

Sialic Acid Terminates the Oligosaccharides of Glycoproteins and Glycolipids

The oligosaccharide chains of glycoproteins and glycolipids serve as ligands for mammalian sugar-binding proteins and also as receptors for pathogens such as influenza and polyoma virus, for bacteria, and for bacterial toxins when they bind to the host cell. The terminal sugars of most such oligosaccharide chains are sialic acids, the most common being Neu5Ac (Figure 7.15), and the sialic acid–capped oligosaccharides themselves are termed **sialosides.** Sialosides on glycoproteins and glycolipids mediate aspects of immune responses and leukocyte trafficking to lymph nodes and inflammation sites, among other functions. The biosynthesis of sialosides is carried out by 20 **sialyltransferase** enzymes that are highly conserved in mammals.

Sialic Acid Cleavage Can Serve as a Timing Device for Protein Degradation

The slow cleavage of sialic acid residues from N-linked glycoproteins circulating in the blood targets these proteins for degradation by the organism. The liver contains specific receptor proteins that recognize and bind glycoproteins that are ready to be degraded and recycled. Newly synthesized serum glycoproteins contain N-linked **triantennary** (three-chain) oligosaccharides having structures similar to those in Figure 7.35, in which sialic acid residues cap galactose residues. As these glycoproteins circulate, enzymes on the blood vessel walls remove the sialic acid groups, exposing the galactose residues. In the liver, the **asialoglycoprotein receptor** binds the exposed galactose residues of these glycoproteins with very high affinity ($K_D = 10^{-9}$ to 10^{-8} M). The complex of receptor and glycoprotein is then taken into the cell by **endocytosis,** and the glycoprotein is degraded in cellular lysosomes. Highest affinity binding of glycoprotein to the asialoglycoprotein receptor requires three free galactose residues. Oligosaccharides with only one or two exposed galactose residues bind less tightly. This is an elegant way for the body to keep track of how long glycoproteins have been in circulation. Over a period of time—anywhere from a few hours to weeks—the sialic acid groups are cleaved one by

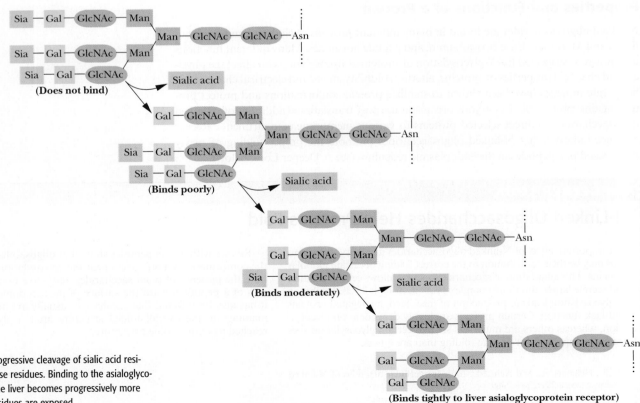

FIGURE 7.35 Progressive cleavage of sialic acid residues exposes galactose residues. Binding to the asialoglycoprotein receptor in the liver becomes progressively more likely as more Gal residues are exposed.

one. The longer the glycoprotein circulates and the more sialic acid residues are removed, the more galactose residues become exposed so that the glycoprotein is eventually bound to the liver receptor.

7.6 How Do Proteoglycans Modulate Processes in Cells and Organisms?

Proteoglycans are a family of glycoproteins whose carbohydrate moieties are predominantly **glycosaminoglycans.** The structures of only a few proteoglycans are known, and even these few display considerable diversity (Figure 7.36). Those known range in size from serglycin, having 104 amino acid residues (10.2 kD), to **versican,** having 2409 residues (265 kD). Each of these proteoglycans contains one or two types of covalently linked glycosaminoglycans. In the known proteoglycans, the glycosaminoglycan units

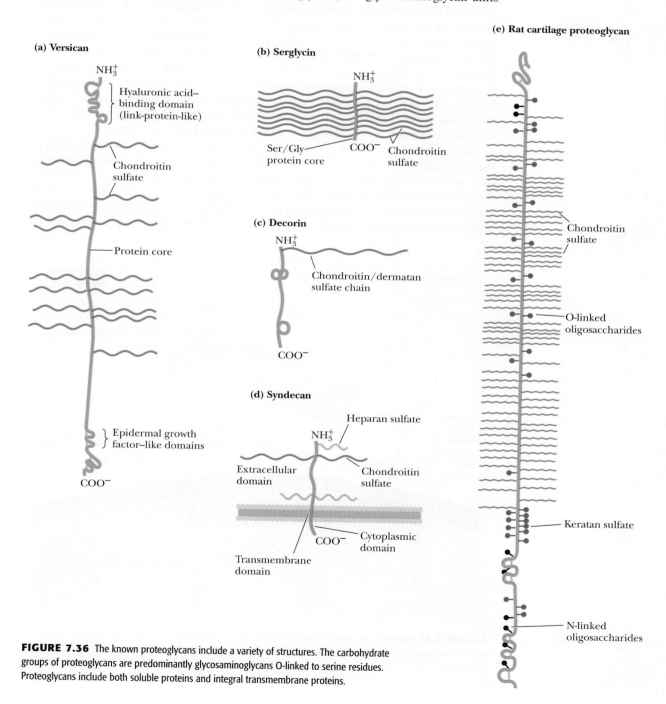

(a) Versican

(b) Serglycin

(c) Decorin

(d) Syndecan

(e) Rat cartilage proteoglycan

FIGURE 7.36 The known proteoglycans include a variety of structures. The carbohydrate groups of proteoglycans are predominantly glycosaminoglycans O-linked to serine residues. Proteoglycans include both soluble proteins and integral transmembrane proteins.

are O-linked to serine residues of Ser-Gly dipeptide sequences. Serglycin is named for a unique central domain of 49 amino acids composed of alternating serine and glycine residues. The **cartilage matrix proteoglycan** contains 117 Ser-Gly pairs to which chondroitin sulfates attach. **Decorin,** a small proteoglycan secreted by fibroblasts and found in the extracellular matrix of connective tissues, contains only three Ser-Gly pairs, only one of which is normally glycosylated. In addition to glycosaminoglycan units, proteoglycans may also contain other N-linked and O-linked oligosaccharide groups.

Functions of Proteoglycans Involve Binding to Other Proteins

Proteoglycans may be *soluble* and located in the extracellular matrix, as is the case for serglycin, versican, and the cartilage matrix proteoglycan, or they may be *integral transmembrane proteins,* such as **syndecan.** Both types of proteoglycan appear to function by interacting with a variety of other molecules through their glycosaminoglycan components and through specific receptor domains in the polypeptide itself. For example, syndecan (from the Greek *syndein,* meaning "to bind together") is a transmembrane proteoglycan that associates intracellularly with the actin cytoskeleton (see Chapter 16). Outside the cell, it interacts with **fibronectin,** an extracellular protein that binds to several cell surface proteins and to components of the extracellular matrix. The ability of syndecan to participate in multiple interactions with these target molecules allows them to act as a sort of "glue" in the extracellular space, linking components of the extracellular matrix, facilitating the binding of cells to the matrix, and mediating the binding of growth factors and other soluble molecules to the matrix and to cell surfaces (Figure 7.37).

Many of the functions of proteoglycans involve the binding of specific proteins to the glycosaminoglycan groups of the proteoglycan. The glycosaminoglycan-binding sites on these specific proteins contain multiple basic amino acid residues. The amino acid sequences BBXB and BBBXXB (where B is a basic amino acid and X is any amino acid) recur repeatedly in these binding domains. Basic amino acids such as lysine and arginine provide charge neutralization for the negative charges of glycosaminoglycan

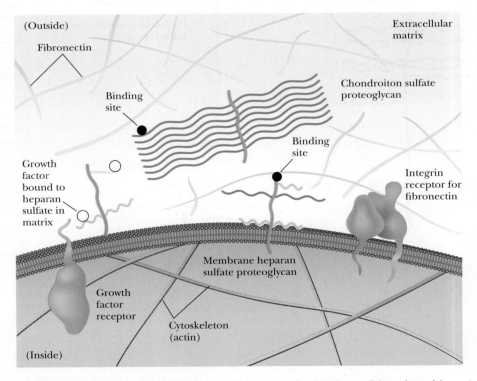

FIGURE 7.37 Proteoglycans serve a variety of functions on the cytoplasmic and extracellular surfaces of the plasma membrane. Many of these functions appear to involve the binding of specific proteins to the glycosaminoglycan groups.

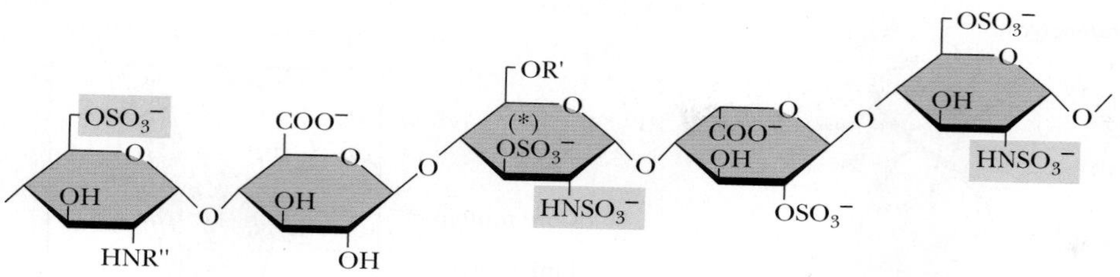

FIGURE 7.38 A portion of the structure of heparin, a carbohydrate having anticoagulant properties. It is used by blood banks to prevent the clotting of blood during donation and storage and also by physicians to prevent the formation of life-threatening blood clots in patients recovering from serious injury or surgery. This sulfated pentasaccharide sequence in heparin binds with high affinity to antithrombin III, accounting for this anticoagulant activity. The 3-O-sulfate marked by an asterisk is essential for high-affinity binding of heparin to antithrombin III.

residues, and in many cases, the binding of extracellular matrix proteins to glycosaminoglycans is primarily charge-dependent. For example, more highly sulfated glycosaminoglycans bind more tightly to fibronectin. However, certain protein–glycosaminoglycan interactions require a specific carbohydrate sequence. A particular pentasaccharide sequence in heparin, for example, binds tightly to antithrombin III (Figure 7.38), accounting for the anticoagulant properties of heparin. Other glycosaminoglycans interact much more weakly.

Proteoglycans May Modulate Cell Growth Processes

Several lines of evidence raise the possibility of modulation or regulation of cell growth processes by proteoglycans. First, heparin and heparan sulfate are known to inhibit cell proliferation in a process involving internalization of the glycosaminoglycan moiety and its migration to the cell nucleus. Second, **fibroblast growth factor** binds tightly to heparin and other glycosaminoglycans, and the heparin–growth factor complex protects the growth factor from degradative enzymes, thus enhancing its activity. There is evidence that binding of fibroblast growth factors by proteoglycans and glycosaminoglycans in the extracellular matrix creates a reservoir of growth factors for cells to use. Third, **transforming growth factor β** has been shown to stimulate the synthesis and secretion of proteoglycans in certain cells. Fourth, several proteoglycan core proteins, including versican and **lymphocyte homing receptor,** have domains similar in sequence to those of **epidermal growth factor** and **complement regulatory factor.** These growth factor domains may interact specifically with growth factor receptors in the cell membrane in processes that are not yet understood.

Proteoglycans Make Cartilage Flexible and Resilient

Cartilage matrix proteoglycan is responsible for the flexibility and resilience of cartilage tissue in the body. In cartilage, long filaments of hyaluronic acid are studded or coated with proteoglycan molecules, as shown in Figure 7.39. The hyaluronate chains can be as long as 4 μm and can coordinate 100 or more proteoglycan units. Cartilage proteoglycan possesses a **hyaluronic acid–binding domain** on the NH_2-terminal portion of the polypeptide, which binds to hyaluronate with the assistance of a **link protein.** The proteoglycan–hyaluronate aggregates can have molecular weights of 2 million or more.

The proteoglycan–hyaluronate aggregates are highly hydrated by virtue of strong interactions between water molecules and the polyanionic complex. When cartilage is compressed (such as when joints absorb the impact of walking or running), water is briefly squeezed out of the cartilage tissue and then reabsorbed when the stress is diminished. This reversible hydration gives cartilage its flexible, shock-absorbing qualities and cushions the joints during physical activities that might otherwise injure the involved tissues.

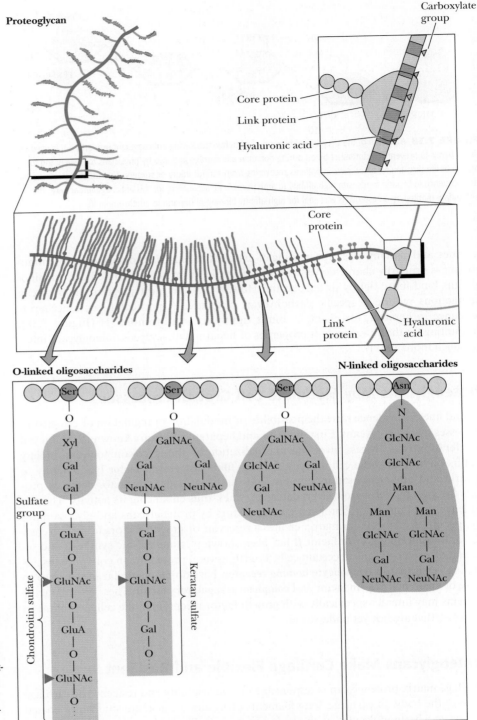

FIGURE 7.39 The human knee joint is a marvel of nature's engineering. Articular cartilage slides with extremely low friction against the cushioning meniscus when the knee bends. Articular cartilage is 68–85% water, adsorbed in a network of collagen fibers (10–20%) and proteoglycans (5–10%). The meniscus is also mostly water (70%). The meniscus's dry weight is 75% collagen and 25% proteoglycan. Hyaluronate (see Figure 7.28) forms the backbone of proteoglycan structures, such as those found in cartilage. The proteoglycan subunits consist of a core protein containing numerous O-linked and N-linked glycosaminoglycans. The glycosaminoglycans in the meniscus proteoglycan include chondroitin sulfate (50–60%), dermatan sulfate (20–30%), and keratan sulfate (15%). (See Figure 7.28 for these structures.) In cartilage, these highly hydrated proteoglycan structures are enmeshed in a network of collagen fibers. Release (and subsequent reabsorption) of water by these structures during compression accounts for the shock-absorbing qualities of cartilaginous tissue.

7.7 | Do Carbohydrates Provide a Structural Code?

The surprisingly low number of genes in the genomes of complex multicellular organisms has led biochemists to consider other explanations for biological complexity and diversity. Oligosaccharides and polysaccharides, endowed with an unsurpassed variability of structures, are information carriers, and **glycoconjugates**—complexes of proteins and lipids with oligosaccharides and polysaccharides—are the mediators of information transfer by these carbohydrate structures. Individual sugar units are the "letters" of **the sugar code**, and the "words" and "sentences" of this code are synthesized by glycosyltransferases, glycosidases, and other enzymes. The total number of permutations for a six-unit polymer formed from an alphabet of 20 hexose monosaccharides is a staggering 1.44×10^{15}, whereas only 6.4×10^7 hexamers can be formed from 20 amino acids and only 4096 hexanucleotides can be formed from the four nucleotides of DNA.

The covalent addition of glycans to proteins and lipids represents not only the most abundant post-translational modification of proteins but also the most structurally diverse. Glycoconjugates respond to and control metabolic states and developmental stages of cells, and they are directly involved in almost every biological process and play a major role in nearly every human disease. The vast array of possible glycan structures adds a **glycomic dimension** to the genomic complexity achieved by protein expression in organisms. The "glycome" is the complete set of glycans and glycoconjugates that are made by a cell or organism under specific conditions, and "glycomics" refers to studies that attempt to define or quantify the glycome of a cell, tissue, or organism.

Studies of glycomics have generated vast amounts of information, which is stored and managed in multiple publicly available carbohydrate databases:
http://www.glycosciences.de/
http://www.genome.jp/kegg/glycan/
http://www.eurocarbdb.org/
http://functionalglycomics.org/

Lectins Translate the Sugar Code

The processes of cell migration, cell–cell interaction, immune response, and blood clotting, along with many other biological processes, depend on information transfer modulated by glycoconjugates. Many of the proteins involved in glycoconjugate formation belong to the **lectins**—a class of proteins that bind carbohydrates with high specificity and affinity. *Lectins are the translators of the sugar code.* Table 7.1 describes a few of the many known lectins, their carbohydrate affinities, and their functions. A few examples of lectin–carbohydrate complexes and their roles in biological information transfer will illustrate the nature of these important and complex interactions.

Selectins, Rolling Leukocytes, and the Inflammatory Response

Human bodies are constantly exposed to a plethora of bacteria, viruses, and other inflammatory substances. To combat these infectious and toxic agents, the body has developed a carefully regulated inflammatory response system. Part of that response is the orderly migration of leukocytes to sites of inflammation. Leukocytes literally roll along the vascular wall and into the tissue site of inflammation. This rolling movement is mediated by reversible adhesive interactions between the leukocytes and the vascular surface.

These interactions involve adhesion proteins called **selectins,** which are found both on the rolling leukocytes and on the endothelial cells of the vascular walls. Selectins have a characteristic domain structure, consisting of an N-terminal extracellular lectin

TABLE 7.1	Specificities and Functions of Some Animal Lectins	
Lectin Family	**Carbohydrate Specificity**	**Function**
Calnexins	Glucose	Ligand-selective molecular chaperones in ER
C-type lectins	Variable	Cell-type specific endocytosis and other functions
ERGIC-53	Mannose	Intracellular routing of glycoproteins and vesicles
Galectins	Galactose/lactose	Cellular growth regulation and cell–matrix interactions
Pentraxins	Variable	Anti-inflammatory action
Selectins	Variable	Cell migration and routing

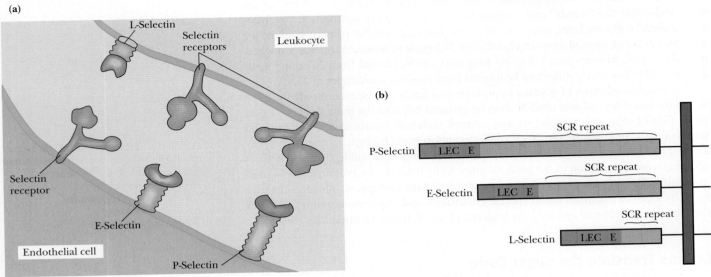

FIGURE 7.40 **(a)** The interactions of selectins with their receptors. **(b)** The selectin family of adhesion proteins.

(LEC) domain, a single epidermal growth factor (E) domain, a series of two to nine short consensus repeat (SCR) domains, a single transmembrane segment, and a short cytoplasmic domain. The lectin domains bind carbohydrates with high affinity and specificity. Selectins of three types are known: E-selectins, L-selectins, and P-selectins. L-Selectin is found on the surfaces of leukocytes, including neutrophils and lymphocytes, and binds to carbohydrate ligands on endothelial cells (Figure 7.40). The presence of L-selectin is a necessary component of leukocyte rolling. P-Selectin and E-selectin are located on the vascular endothelium and bind with carbohydrate ligands on leukocytes. Typical neutrophil cells possess 10,000 to 20,000 P-selectin–binding sites. Selectins are expressed on the surfaces of their respective cells by exposure to inflammatory signal molecules, such as histamine, hydrogen peroxide, and bacterial endotoxins. P-Selectins, for example, are stored in intracellular granules and are transported to the cell membrane within seconds to minutes of exposure to a triggering agent.

Substantial evidence supports the hypothesis that selectin–carbohydrate ligand interactions modulate the rolling of leukocytes along the vascular wall. Studies with L-selectin–deficient and P-selectin–deficient leukocytes show that L-selectins mediate weaker adherence of the leukocyte to the vascular wall and promote faster rolling along the wall. Conversely, P-selectins promote stronger adherence and slower rolling. Thus, leukocyte rolling velocity in the inflammatory response could be modulated by variable exposure of P-selectins and L-selectins at the surfaces of endothelial cells and leukocytes, respectively.

Galectins—Mediators of Inflammation, Immunity, and Cancer

The galectins are a very conserved family of proteins with carbohydrate recognition domains (CRDs) of about 135 amino acids that bind β-galactosides specifically. Galectins occur in both vertebrates and invertebrates, and they participate in processes such as cell adhesion, growth regulation, inflammation, immunity, and cancer metastasis. In humans, one galectin is associated with increased risk of heart attacks and another is implicated in inflammatory bowel disease. Human galectin-1 is a dimer of antiparallel beta-sandwich subunits (Figure 7.41a). Lactose binds at opposite ends of the dimer. Structural studies of the protein in the presence and absence of ligand

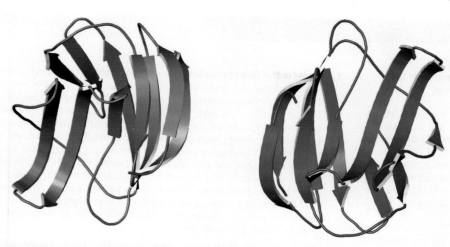

FIGURE 7.41 Structure of the human galectin-1 dimer. The lactose-binding sites are at opposite ends of the dimer.

reveal that the amino acid residues implicated in galactose binding are kept in their proper orientation in the absence of ligand by a hydrogen-bonded network of four water molecules.

C-Reactive Protein—A Lectin That Limits Inflammation Damage

The **pentraxins** are lectins that adopt an unusual quaternary structure in which five identical subunits combine to form a planar ring with a central hole (Figure 7.42a). C-reactive protein is a pentraxin that functions to limit tissue damage, acute inflammation, and autoimmune reactions. C-reactive protein acts by binding to phosphocholine moieties on damaged membranes. Binding of the protein to phosphocholine is apparently mediated through a bound calcium ion and a hydrophobic pocket centered on Phe[66] (Figure 7.42b).

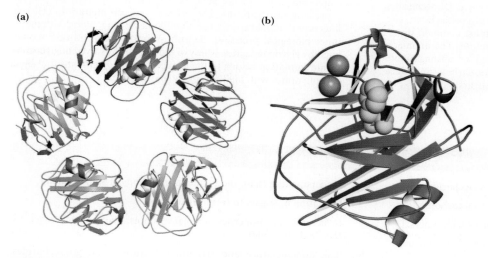

(a)

(b)

FIGURE 7.42 **(a)** The C-reactive protein pentamer. **(b)** The phosphocholine-binding site of C-reactive protein contains two bound Ca^{2+} ions (orange) and a hydrophobic pocket. Phe[66] is shown in yellow.

SUMMARY

⊙**WL** Login to OWL to develop problem-solving skills and complete online homework assigned by your professor.

Carbohydrates are a versatile class of molecules of the formula $(CH_2O)_n$. They are a major form of stored energy in organisms, and they are the metabolic precursors of virtually all other biomolecules. Carbohydrates linked to lipids (glycolipids) are components of biological membranes. Carbohydrates linked to proteins (glycoproteins) are important components of cell membranes and function in recognition between cell types and recognition of cells by other molecules. Recognition events are important in cell growth, differentiation, fertilization, tissue formation, transformation of cells, and other processes.

7.1 How Are Carbohydrates Named? Carbohydrates are classified into three groups: monosaccharides, oligosaccharides, and polysaccharides. Monosaccharides cannot be broken down into smaller sugars under mild conditions. Oligosaccharides consist of from two to ten simple sugar molecules. Polysaccharides are polymers of simple sugars and their derivatives and may be branched or linear. Their molecular weights range up to 1 million or more.

7.2 What Is the Structure and Chemistry of Monosaccharides? Monosaccharides consist typically of three to seven carbon atoms and are described as either aldoses or ketoses. Aldoses with at least three carbons and ketoses with at least four carbons contain chiral centers. The prefixes D- and L- are often used to indicate the configuration of the highest numbered asymmetric carbon. The D- and L-forms of a monosaccharide are mirror images of each other, called enantiomers. Pairs of isomers that have opposite configurations at one or more chiral centers, but are not mirror images of each other, are called diastereomers. Sugars that differ in configuration at only one chiral center are epimers. An interesting feature of carbohydrates is their ability to form cyclic structures with formation of an additional asymmetric center. Aldoses and ketoses with five or more carbons can form either furanose or pyranose rings, and the more stable form depends on structural factors. A variety of chemical and enzymatic reactions produce derivatives of simple sugars, such as sugar acids, sugar alcohols, deoxy sugars, sugar esters, amino sugars, acetals, ketals, and glycosides.

7.3 What Is the Structure and Chemistry of Oligosaccharides? The complex array of oligosaccharides in higher organisms is formed from relatively few different monosaccharide units, particularly glucose, fructose, mannose, galactose, ribose, and xylose. Disaccharides consist of two monosaccharide units linked by a glycosidic bond, and each individual unit is termed a residue. The most common disaccharides in nature are sucrose, maltose, and lactose. The anomeric carbons of oligosaccharides may be substituted or unsubstituted. Disaccharides with a free, unsubstituted anomeric carbon can reduce oxidizing agents and thus are termed reducing sugars.

7.4 What Is the Structure and Chemistry of Polysaccharides? Polysaccharides are formed from monosaccharides and their derivatives. If a polysaccharide consists of only one kind of monosaccharide, it is a homopolysaccharide, whereas those with more than one kind of monosaccharide are heteropolysaccharides. Polysaccharides may function as energy storage materials, structural components of organisms, or protective substances. Starch and glycogen are readily metabolizable and provide energy reserves for cells. Chitin and cellulose provide strong support for the skeletons of arthropods and green plants, respectively. Mucopolysaccharides such as hyaluronic acid form protective coats on animal cells. Peptidoglycan, the strong protective macromolecule of bacterial cell walls, is composed of peptide-linked glycan chains.

7.5 What Are Glycoproteins, and How Do They Function in Cells? Glycoproteins are proteins that contain covalently linked oligosaccharides and polysaccharides. Carbohydrate groups may be linked to proteins via the hydroxyl groups of serine, threonine, or hydroxylysine residues (in O-linked saccharides) or via the amide nitrogen of an asparagine residue (in N-linked saccharides). O-Glycosylated stems of certain proteins raise the functional domain of the protein above the membrane surface and the associated glycocalyx, making these domains accessible to interacting proteins. N-Glycosylation confers a variety of functions to proteins. N-linked oligosaccharides promote the proper folding of newly synthesized polypeptides in the endoplasmic reticulum of eukaryotic cells.

7.6 How Do Proteoglycans Modulate Processes in Cells and Organisms? Proteoglycans are a family of glycoproteins whose carbohydrate moieties are predominantly glycosaminoglycans. Proteoglycans may be soluble and located in the extracellular matrix, as for serglycin, versican, and cartilage matrix proteoglycans, or they may be integral transmembrane proteins, such as syndecan. Both types appear to function by interacting with a variety of other molecules through their glycosaminoglycan components and through specific receptor domains in the polypeptide itself. Proteoglycans modulate cell growth processes and are also responsible for the flexibility and resilience of cartilage tissue in the body.

7.7 Do Carbohydrates Provide a Structural Code? Oligosaccharides and polysaccharides are information carriers, and glycoconjugates are the mediators of information transfer by these carbohydrate structures. The vast array of possible glycan structures adds a glycomic dimension to the genomic complexity achieved by protein expression in organisms. The processes of cell migration, cell–cell interaction, immune response, and blood clotting, along with many other biological processes, depend on information transfer modulated by glycoconjugates. Many of the proteins involved in glycoconjugate formation belong to the lectins—a class of proteins that bind carbohydrates with high specificity and affinity.

FOUNDATIONAL BIOCHEMISTRY   Things You Should Know After Reading Chapter 7.

- Naming conventions for mono-, oligo-, and polysaccharides.
- The structures of simple aldoses and ketoses (3 to 6 carbons).
- The structural features that define and distinguish enantiomers.
- The structural features that define and distinguish diastereomers.
- The structural features that define and distinguish epimers.
- The structural features that define and distinguish anomeric carbons and anomers.
- How to convert linear sugar structures to cyclic Haworth structures.
- Understanding chair and boat conformations of simple sugars.
- Oxidation of sugars to form sugar acids.
- Structures and properties of sugar alcohols, deoxy sugars, sugar esters, and glycosides.
- The nomenclature and structural features of oligosaccharides, including glycosidic bonds and the characteristics of reducing sugars.
- The nomenclature and structural features of storage polysaccharides, including α-amylose, amylopectin, and glycogen.

- The nomenclature and structural features of structural polysaccharides, including cellulose, chitin, agarose, and the glycosaminoglycans.
- The structural features of bacterial peptidoglycans.
- The structural and functional features of O-linked and N-linked oligosaccharides and glycoproteins.
- The structure and function of antifreeze glycoproteins in polar fish.

- The structural and functional features of proteoglycans, especially cartilage matrix proteoglycans.
- The essential features of glycomics, lectins, and the sugar code.
- The role of selectins in the inflammatory response.
- The structural and functional features of galectins, pentraxins, and C-reactive protein.

PROBLEMS

Answers to all problems are at the end of this book. Detailed solutions are available in the *Student Solutions Manual, Study Guide, and Problems Book.*

1. **Drawing Haworth Structures of Sugars** Draw Haworth structures for the two possible isomers of D-altrose (Figure 7.2) and D-psicose (Figure 7.3).

2. **Drawing the Structure of a Glycopeptide** (Integrates with Chapters 4 and 5.) Consider the peptide DGNILSR, where N has a covalently linked galactose and S has a covalently linked glucose. Draw the structure of this glycopeptide, and also draw titration curves for the glycopeptide and for the free peptide that would result from hydrolysis of the two sugar residues.

3. **Separating Glycated Hb From Normal Hb** (Integrates with Chapters 5 and 6.) Human hemoglobin can react with sugars in the blood (usually glucose) to form covalent adducts. The α-amino groups of N-terminal valine in the Hb β-subunits react with the C-1 (aldehyde) carbons of monosaccharides to form aldimine adducts, which rearrange to form very stable ketoamine products. Quantitation of this "glycated hemoglobin" is important clinically, especially for diabetic individuals. Suggest at least three methods by which glycated Hb could be separated from normal Hb and quantitated.

4. **Naming and Characterizing a Disaccharide** Trehalose, a disaccharide produced in fungi, has the following structure:

   a. What is the systematic name for this disaccharide?
   b. Is trehalose a reducing sugar? Explain.

5. **Drawing the Fischer Projection of a Simple Sugar** Draw a Fischer projection structure for L-sorbose (D-sorbose is shown in Figure 7.3).

6. **Calculating the Composition of Anomeric Sugar Mixtures** α-D-Glucose has a specific rotation, $[\alpha]_D^{20}$, of $+112.2°$, whereas β-D-glucose has a specific rotation of $+18.7°$. What is the composition of a mixture of α-D- and β-D-glucose, which has a specific rotation of $83.0°$?

7. **Naming Sugars In the (R,S) System** Use the information in the Critical Developments in Biochemistry box titled "Rules for Description of Chiral Centers in the (R,S) System" (Chapter 4) to name D-galactose using (R,S) nomenclature. Do the same for L-altrose.

8. **Determining the Branch Points and Reducing Ends of Amylopectin** A 0.2-g sample of amylopectin was analyzed to determine the fraction of the total glucose residues that are branch points in the structure. The sample was exhaustively methylated and then digested, yielding 50 μmol of 2,3-dimethylglucose and 0.4 μmol of 1,2,3,6-tetramethylglucose.

   a. What fraction of the total residues are branch points?
   b. How many reducing ends does this sample of amylopectin have?

9. **The Effect of Carbohydrates on Proteolysis of Glycophorin** (Integrates with Chapters 5, 6, and 9.) Consider the sequence of glycophorin (see Figure 9.10), and imagine subjecting glycophorin, and also a sample of glycophorin treated to remove all sugars, to treatment with trypsin and chymotrypsin. Would the presence of sugars in the native glycophorin make any difference to the results?

10. **Assessing the Caloric Content of Protein and Carbohydrate** (Integrates with Chapters 4, 5, and 23.) The caloric content of protein and carbohydrate are quite similar, at approximately 16 to 17 kJ/g, whereas that of fat is much higher, at 38 kJ/g. Discuss the chemical basis for the similarity of the values for carbohydrate and for protein.

11. **Writing a Mechanism for Starch Phosphorylase** Write a reasonable chemical mechanism for the starch phosphorylase reaction (Figure 7.22).

12. **Assessing the Toxicity of Laetrile** Laetrile is a glycoside found in bitter almonds and peach pits. Laetrile treatment is offered in some countries as a cancer therapy. This procedure is dangerous, and there is no valid clinical evidence of its efficacy. Look up the structure of laetrile and suggest at least one reason that laetrile treatment could be dangerous for human patients.

13. **Assessing the Efficacy of Glucosamine and Chondroitin for Arthritis Pain** Treatment with chondroitin and glucosamine is offered as one popular remedy for arthritis pain. Suggest an argument for the efficacy of this treatment, and then comment on its validity, based on what you know of polysaccharide chemistry.

14. **Assessing a Remedy for Flatulence** Certain foods, particularly beans and legumes, contain substances that are indigestible (at least in part) by the human stomach, but which are metabolized readily by intestinal microorganisms, producing flatulence. One of the components of such foods is stachyose.

Stachyose

Beano is a commercial product that can prevent flatulence. Describe the likely breakdown of stachyose in the human stomach and intestines and how Beano could contribute to this process. What would be an appropriate name for the active ingredient in Beano?

15. **Determining the Systematic Name for a Trisaccharide** Give the systematic name for stachyose.

16. Assessing the Formation and Composition of Limit Dextrins Prolonged exposure of amylopectin to starch phosphorylase yields a substance called a limit dextrin. Describe the chemical composition of limit dextrins, and draw a mechanism for the enzyme-catalyzed reaction that can begin the breakdown of a limit dextrin.

17. Assessing the Chemistry and Enzymology of "Light Beer" Biochemist Joseph Owades revolutionized the production of beer in the United States by developing a simple treatment with an enzyme that converted regular beer into "light beer," which was marketed aggressively as a beverage that "tastes great," even though it is "less filling." What was the enzyme-catalyzed reaction that Owades used to modify the fermentation process so cleverly, and how is regular beer different from light beer?

18. Brewing "Light Beer" on a Budget Amateur brewers of beer, who do not have access to the enzyme described in problem 17, have nonetheless managed to brew light beers using a readily available commercial product. What is that product, and how does it work?

19. Assessing the Growth Potential and Enzymology of a Prolific Plant Nuisance Kudzu is a vine that grows prolifically in the southern and southeastern United States. A native of Japan, China, and India, kudzu was brought to the United States in 1876 at the Centennial Exposition in Philadelphia. During the Great Depression of the 1930s, the Soil Conservation Service promoted kudzu for erosion control, and farmers were paid to plant it. Today, however, kudzu is a universal nuisance, spreading rapidly, and covering and destroying trees in large numbers. Already covering 7 to 10 million acres in the U.S., kudzu grows at the rate of a foot per day. Assume that the kudzu vine consists almost entirely of cellulose fibers, and assume that the fibers lie parallel to the vine axis. Calculate the rate of the cellulose synthase reaction that adds glucose units to the growing cellulose molecules. Use the structures in your text to make a reasonable estimate of the unit length of a cellulose molecule (from one glucose monomer to the next).

Preparing for the MCAT® Exam

20. Heparin has a characteristic pattern of hydroxy and anionic functions. Which amino acid side chains on antithrombin III might be the basis for the strong interactions between this protein and the anticoagulant heparin?

21. What properties of hyaluronate, chondroitin sulfate, and keratan sulfate make them ideal components of cartilage?

ActiveModels Problems

22. Using the ActiveModel for Human C-reactive protein (CRP) to characterize the quaternary structure and subunit interactions in this protein.

23. Use examples from the ActiveModel for Human Galectin-1 to describe the hydrophobic effect.

FURTHER READING

Carbohydrate Structure and Chemistry

Breslow, R., and Cheng, Z.-L., 2010. L-Amino acids catalyze the formation of an excess of D-glyceraldehyde, and thus of other D-sugars, under credible prebiotic conditions. *Proceedings of the National Academy of Sciences* **107:**5723–5725.

Collins, P. M., 1987. *Carbohydrates.* London: Chapman and Hall.

Davison, E. A., 1967. *Carbohydrate Chemistry.* New York: Holt, Rinehart and Winston.

Pigman, W., and Horton, D., 1972. *The Carbohydrates.* New York: Academic Press.

Sharon, N., 1980. Carbohydrates. *Scientific American* **243:**90–102.

Polysaccharides

Aspinall, G. O., 1982. *The Polysaccharides,* Vols. 1 and 2. New York: Academic Press.

Höfte, H., 2001. A baroque residue in red wine. *Science* **294:**795–797.

McNeil, M., Darvill, A. G., Fry, S. C., and Albersheim, P., 1984. Structure and function of the primary cell walls of plants. *Annual Review of Biochemistry* **53:**625–664.

O'Neill, M. A., Eberhard, S., Albersheim, P., and Darvill, A. G., 2002. Requirements of borate cross-linking of cell wall rhamnogalacturonan II for *Arabidopsis* growth. *Science* **294:**846–849.

Glycoproteins

Feeney, R. E., Burcham, T. S., and Yeh, Y., 1986. Antifreeze glycoproteins from polar fish blood. *Annual Review of Biophysical Chemistry* **15:**59–78.

Helenius, A., and Aebi, M., 2001. Intracellular functions of N-linked glycans. *Science* **291:**2364–2369.

Jentoft, N., 1990. Why are proteins O-glycosylated? *Trends in Biochemical Sciences* **155:**291–294.

Klein, J., 2010. Repair or replacement: a joint perspective. *Science* **323:**47–48.

Sharon, N., 1984. Glycoproteins. *Trends in Biochemical Sciences* **9:**198–202.

Proteoglycans

Day, A. J., and Prestwich, G. D., 2002. Hyaluronan-binding proteins: Tying up the giant. *Journal of Biological Chemistry* **277:**4585–4588.

Kjellen, L., and Lindahl, U., 1991. Proteoglycans: Structures and interactions. *Annual Review of Biochemistry* **60:**443–475.

Lennarz, W. J., 1980. *The Biochemistry of Glycoproteins and Proteoglycans.* New York: Plenum Press.

Ruoslahti, E., 1989. Proteoglycans in cell regulation. *Journal of Biological Chemistry* **264:**13369–13372.

Glycobiology

Astronomo, R., and Burton, D., 2010. Carbohydrate vaccines: developing sweet solutions to sticky problems. *Nature Review Drug Discovery* **9:**308–324.

Bertozzi, C. R., and Kiessling, L. L., 2001. Chemical glycobiology. *Science* **291:**2357–2363.

Hart, G. W. and Copeland, R. J., 2010. Glycomics hits the big time. *Cell* **143:**672–676.

Liu, F.-T., and Bevins, C. L., 2010. A sweet target for innate immunity. *Nature Medicine* **16:**263–264.

Lodish, H. F., 1991. Recognition of complex oligosaccharides by the multisubunit asialoglycoprotein receptor. *Trends in Biochemical Sciences* **16:**374–377.

Maeder, T., 2002. Sweet medicines. *Scientific American* **287:**40–47.

Miyamoto, S., 2006. Clinical applications of glycomic approaches for the detection of cancer and other diseases. *Current Opinion in Molecular Therapies* **8:**507–513.

Paulsen, J. C., and Rademacher, C., 2010. Glycan terminator. *Nature Structural and Molecular Biology* **16:**1121–1122.

Rademacher, T. W., Parekh, R. B., and Dwek, R. A., 1988. Glycobiology. *Annual Review of Biochemistry* **57:**785–838.

Timmer, M., Stocker, B., and Seeburger, P., 2007. Probing glycomics. *Current Opinion in Chemical Biology* **11:**59–65.

Lipids

Paul Souders/Digital Vision/Jupiter Images

A feast of fat things, a feast of wines on the lees.

Isaiah 25:6

KEY QUESTIONS

8.1 What Are the Structures and Chemistry of Fatty Acids?

8.2 What Are the Structures and Chemistry of Triacylglycerols?

8.3 What Are the Structures and Chemistry of Glycerophospholipids?

8.4 What Are Sphingolipids, and How Are They Important for Higher Animals?

8.5 What Are Waxes, and How Are They Used?

8.6 What Are Terpenes, and What Is Their Relevance to Biological Systems?

8.7 What Are Steroids, and What Are Their Cellular Functions?

8.8 How Do Lipids and Their Metabolites Act as Biological Signals?

8.9 What Can Lipidomics Tell Us about Cell, Tissue, and Organ Physiology?

ESSENTIAL QUESTION

Lipids are a class of biological molecules defined by low solubility in water and high solubility in nonpolar solvents. As molecules that are largely hydrocarbon in nature, lipids represent highly reduced forms of carbon and, upon oxidation in metabolism, yield large amounts of energy. Lipids are thus the molecules of choice for metabolic energy storage.

Lipid molecules are key components of membranes and also serve a myriad of roles as signal molecules in biological systems. Lipids are integrators of cellular function and intercellular communication. Lipid-lipid and lipid-protein interactions regulate cellular physiology.

What are the structure, chemistry, and biological function of lipids?

The lipids found in biological systems are either **hydrophobic** (containing only nonpolar groups) or **amphipathic** (possessing both polar and nonpolar groups). The hydrophobic nature of lipid molecules allows membranes to act as effective barriers to more polar molecules. In this chapter, we discuss the chemical and physical properties of the various classes of lipid molecules. The following chapter considers membranes, whose properties depend intimately on their lipid constituents.

8.1 What Are the Structures and Chemistry of Fatty Acids?

A **fatty acid** is composed of a long hydrocarbon chain ("tail") and a terminal carboxyl group (or "head"). The carboxyl group is normally ionized under physiological conditions. Fatty acids occur in large amounts in biological systems but only rarely in the free,

OWL Online homework and a Student Self Assessment for this chapter may be assigned in OWL.

uncomplexed state. They typically are esterified to glycerol or other backbone structures. Most of the fatty acids found in nature have an even number of carbon atoms (usually 14 to 24). Certain marine organisms, however, contain substantial amounts of fatty acids with odd numbers of carbon atoms. Fatty acids are either **saturated** (all carbon–carbon bonds are single bonds) or **unsaturated** (with one or more double bonds in the hydrocarbon chain). If a fatty acid has a single double bond, it is said to be **monounsaturated,** and if it has more than one, **polyunsaturated.** Fatty acids can be named or described in at least three ways, as shown in Table 8.1. For example, a fatty acid composed of an 18-carbon chain with no double bonds can be called by its systematic name **(octadecanoic acid),** its common name (stearic acid), or its shorthand notation, in which the number of carbons is followed by a colon and the number of double bonds in the molecule (18:0 for stearic acid). The structures of several common fatty acids are given in Figure 8.1. **Stearic acid** (18:0) and **palmitic acid** (16:0) are the most common saturated fatty acids in nature.

Free rotation around each of the carbon–carbon bonds makes saturated fatty acids extremely flexible molecules. Owing to steric constraints, however, the fully extended conformation (Figure 8.1) is the most stable for saturated fatty acids. Nonetheless, the degree of stabilization is slight, and (as will be seen) saturated fatty acid chains adopt a variety of conformations.

Unsaturated fatty acids are slightly more abundant in nature than saturated fatty acids, especially in higher plants. The most common unsaturated fatty acid is **oleic acid,** or $18:1^{\Delta 9}$, with the number following the "superscript Δ" indicating that the double bond is between carbons 9 and 10. The number of double bonds in an unsaturated fatty acid typically varies from one to four, but in the fatty acids found in most bacteria, this number rarely exceeds one.

The double bonds found in fatty acids are nearly always in the *cis* configuration. As shown in Figure 8.1, this causes a bend or "kink" in the fatty acid chain. This bend has very important consequences for the structure of biological membranes. Saturated fatty acid chains can pack closely together to form ordered, rigid arrays under certain conditions, but unsaturated fatty acids prevent such close packing and produce flexible, fluid aggregates.

Some fatty acids are not synthesized by mammals and yet are necessary for normal growth and life. These **essential fatty acids** include **linoleic** and **γ-linolenic acids.** These

TABLE 8.1	Common Biological Fatty Acids			
Number of Carbons	Common Name	Systematic Name	Symbol	Structure
Saturated fatty acids				
12	Lauric acid	Dodecanoic acid	12:0	$CH_3(CH_2)_{10}COOH$
14	Myristic acid	Tetradecanoic acid	14:0	$CH_3(CH_2)_{12}COOH$
16	Palmitic acid	Hexadecanoic acid	16:0	$CH_3(CH_2)_{14}COOH$
18	Stearic acid	Octadecanoic acid	18:0	$CH_3(CH_2)_{16}COOH$
20	Arachidic acid	Eicosanoic acid	20:0	$CH_3(CH_2)_{18}COOH$
22	Behenic acid	Docosanoic acid	22:0	$CH_3(CH_2)_{20}COOH$
24	Lignoceric acid	Tetracosanoic acid	24:0	$CH_3(CH_2)_{22}COOH$
Unsaturated fatty acids (all double bonds are cis*)*				
16	Palmitoleic acid	9-Hexadecenoic acid	16:1*	$CH_3(CH_2)_5CH{=}CH(CH_2)_7COOH$
18	Oleic acid	9-Octadecenoic acid	18:1	$CH_3(CH_2)_7CH{=}CH(CH_2)_7COOH$
18	Linoleic acid	9,12-Octadecadienoic acid	18:2	$CH_3(CH_2)_4(CH{=}CHCH_2)_2(CH_2)_6COOH$
18	α-Linolenic acid	9,12,15-Octadecatrienoic acid	18:3	$CH_3CH_2(CH{=}CHCH_2)_3(CH_2)_6COOH$
18	γ-Linolenic acid	6,9,12-Octadecatrienoic acid	18:3	$CH_3(CH_2)_4(CH{=}CHCH_2)_3(CH_2)_3COOH$
20	Arachidonic acid	5,8,11,14-Eicosatetraenoic acid	20:4	$CH_3(CH_2)_4(CH{=}CHCH_2)_4(CH_2)_2COOH$
24	Nervonic acid	15-Tetracosenoic acid	24:1	$CH_3(CH_2)_7CH{=}CH(CH_2)_{13}COOH$

*Palmitoleic acid can also be described as $16:1^{\Delta 9}$, in a convention used to indicate the position of the double bond.

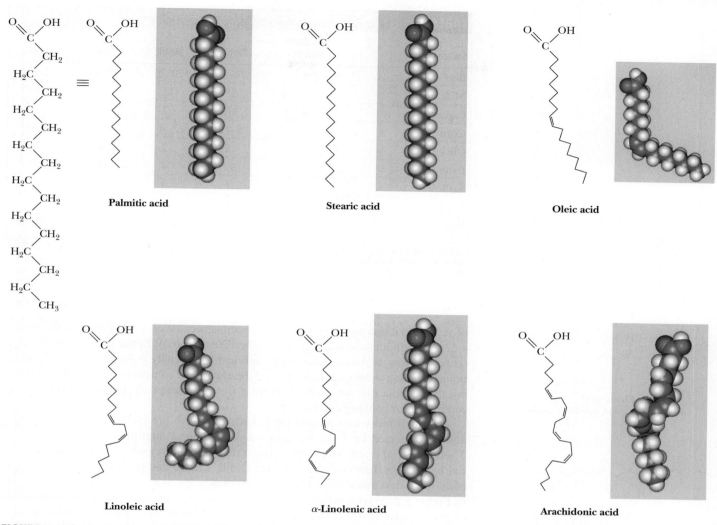

FIGURE 8.1 The structures of some typical fatty acids. Note that most natural fatty acids contain an even number of carbon atoms and that the double bonds are nearly always *cis* and rarely conjugated.

must be obtained by mammals in their diet (specifically from plant sources). **Arachidonic acid,** which is not found in plants, can be synthesized by mammals only from linoleic acid. At least one function of the essential fatty acids is to serve as a precursor for the synthesis of **eicosanoids,** such as prostaglandins, a class of compounds that exert hormonelike effects in many physiological processes (discussed in Chapter 24).

Fats in the modern human diet vary widely in their fatty acid composition (Table 8.2). The incidence of cardiovascular disease is correlated with diets high in saturated fatty acids. By contrast, a diet relatively higher in unsaturated fatty acids (especially polyunsaturated fatty acids) may reduce the risk of heart attacks and strokes. Although vegetable oils usually contain a higher proportion of unsaturated fatty acids than do animal oils and fats, several plant oils are actually high in saturated fats. Palm oil is low in polyunsaturated fatty acids and particularly high in (saturated) **palmitic acid** (hence the name *palmitic*). Coconut oil is particularly high in lauric and myristic acids (both saturated) and contains little unsaturated fatty acid.

Canola oil has been promoted as a healthy dietary oil because it consists primarily of oleic acid (60%), linoleic acid (20%), and α-linolenic acid (9%) with very low saturated fat content (7%). Canola oil is actually rapeseed oil, from the seeds of the rape plant *Brassica rapa* (from the Latin *rapa*, meaning "turnip"), a close relative of mustard, kale, cabbage, and broccoli. Asians and Europeans used rapeseed oil in lamps for hundreds

	TABLE 8.2	Fatty Acid Compositions of Some Dietary Lipids*				
Source	Lauric and Myristic	Palmitic	Stearic	Oleic	Linoleic
Beef	5	24–32	20–25	37–43	2–3
Milk		25	12	33	3
Coconut	74	10	2	7	
Corn		8–12	3–4	19–49	34–62
Olive		9	2	84	4
Palm		39	4	40	8
Safflower		6	3	13	78
Soybean		9	6	20	52
Sunflower		6	1	21	66

Data from *Merck Index,* 10th ed. Rahway, NJ: Merck and Co.; and Wilson, E. D., et al., 1979, *Principles of Nutrition,* 4th ed. New York: Wiley.
*Values are percentages of total fatty acids.

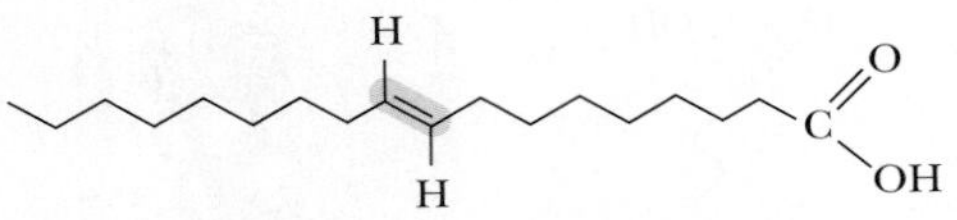

Elaidic acid
trans double bond at 9,10 position

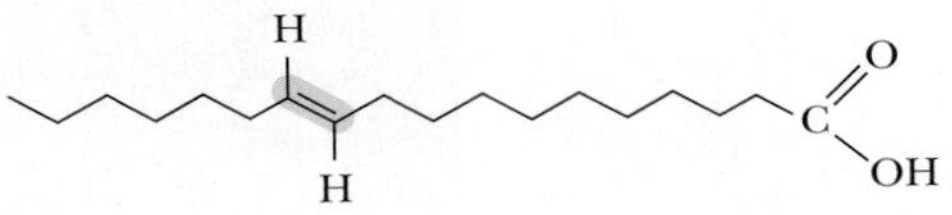

Vaccenic acid
trans double bond at 11,12 position

FIGURE 8.2 Structure of elaidic acid and vacceric acid, two *trans* fatty acids.

of years, but it was not usually considered edible because of its high **erucic acid** content, a $22:1^{\Delta 13}$ monounsaturated fatty acid (often 20% to 60%). In the first half of the 20th century, it was used as a steam engine lubricant (especially in World War II). Conventional breeding techniques have reduced the erucic acid content to less than 1%, producing the "canola oil" (the name is derived from **Can**adian oil, **l**ow **a**cid) now used so commonly for cooking and baking.

Although most unsaturated fatty acids in nature are *cis* fatty acids, *trans* fatty acids are formed by some bacteria via double-bond migration and isomerization. These bacterial reactions produce *trans* fats in ruminant animals (which carry essential bacteria in their rumen), and butter, milk, cheese and the meat of these animals contain modest quantities of *trans* fats (typically 2% to 8% by weight), such as those in Figure 8.2. Margarine and other "processed fats," made by **partial hydrogenation** of polyunsaturated oils (for example, corn, safflower, and sunflower) contain substantial levels of various *trans* fats, and clinical research has shown that chronic consumption of processed foods containing partially hydrogenated vegetable oils can contribute to cardiovascular disease. Diets high in *trans* fatty acids raise plasma low-density lipoprotein (LDL) cholesterol and triglyceride levels while lowering high-density lipoprotein (HDL) cholesterol levels. The effects of *trans* fatty acids on LDL, HDL, and cholesterol levels are similar to those of saturated fatty acids. Diets aimed at reducing the risk of coronary heart disease should be low in both *trans* and saturated fatty acids.

8.2 What Are the Structures and Chemistry of Triacylglycerols?

A significant number of the fatty acids in plants and animals exist in the form of **triacylglycerols** (also called **triglycerides**). Triacylglycerols are a major energy reserve and the principal neutral derivatives of glycerol found in animals. These molecules consist of a glycerol esterified with three fatty acids (Figure 8.3). If all three fatty acid groups are the same, the molecule is called a simple triacylglycerol. Examples include **tristearoylglycerol** (common name tristearin) and **trioleoylglycerol** (triolein). Mixed triacylglycerols contain two or three different fatty acids. Triacylglycerols in animals are found primarily in the adipose tissue (body fat), which serves as a depot or storage site for lipids. Monoacylglycerols and diacylglycerols also exist, but they are far less common than the triacylglycerols. Most natural plant and animal fat is composed of mixtures of simple and mixed triacylglycerols.

FIGURE 8.3 Triacylglycerols are formed from glycerol and fatty acids.

Acylglycerols can be hydrolyzed by heating with acid or base or by treatment with lipases. Hydrolysis with alkali is called **saponification** and yields salts of free fatty acids and glycerol. This is how our ancestors made **soap** (a metal salt of an acid derived from fat). One method used potassium hydroxide (potash) leached from wood ashes to hydrolyze animal fat (mostly triacylglycerols). (The tendency of such soaps to be precipitated by Mg^{2+} and Ca^{2+} ions in hard water makes them less useful than modern detergents.) When the fatty acids esterified at the first and third carbons of glycerol are different, the second carbon is asymmetric. The various acylglycerols are normally soluble in benzene, chloroform, ether, and hot ethanol. Although triacylglycerols are insoluble in water, monoacylglycerols and diacylglycerols readily form organized structures in water (see Chapter 9), owing to the polarity of their free hydroxyl groups.

Triacylglycerols are rich in highly reduced carbons and thus yield large amounts of energy in the oxidative reactions of metabolism. Complete oxidation of 1 g of triacylglycerols yields about 38 kJ of energy, whereas proteins and carbohydrates yield only

A DEEPER LOOK

Polar Bears Prefer Nonpolar Food

The polar bear is magnificently adapted to thrive in its harsh Arctic environment. Research by Malcolm Ramsay (at the University of Saskatchewan in Canada) and others has shown that polar bears eat only during a few weeks out of the year and then fast for periods of 8 months or more, consuming no food or water during that time. Eating mainly in the winter, the adult polar bear feeds almost exclusively on seal blubber (largely composed of triacylglycerols), thus building up its own triacylglycerol reserves. Through the Arctic summer, the polar bear maintains normal physical activity, roaming over long distances, but relies entirely on its body fat for sustenance, burning as much as 1 to 1.5 kg of fat per day. It neither urinates nor defecates for extended periods. All the water needed to sustain life is provided from the metabolism of triacylglycerols (because oxidation of fatty acids yields carbon dioxide and water).

Ironically, the word *Arctic* comes from the ancient Greeks, who understood that the northernmost part of the earth lay under the stars of the constellation Ursa Major, the Great Bear. Although unaware of the polar bear, they called this region *Arktikós,* which means "the country of the great bear."

© Witold Kaszkin/Shutterstock.com

about 17 kJ/g. Also, their hydrophobic nature allows them to aggregate in highly anhydrous forms, whereas polysaccharides and proteins are highly hydrated. For these reasons, triacylglycerols are the molecules of choice for energy storage in animals. Body fat (mainly triacylglycerols) also provides good insulation. Whales and Arctic mammals rely on body fat for both insulation and energy reserves.

8.3 What Are the Structures and Chemistry of Glycerophospholipids?

A 1,2-diacylglycerol that has a phosphate group esterified at carbon atom 3 of the glycerol backbone is a **glycerophospholipid,** also known as a phosphoglyceride or a glycerol phosphatide (Figure 8.4). These lipids form one of the largest and most important classes of natural lipids. They are essential components of cell membranes and are found in small concentrations in other parts of the cell. It should be noted that all glycerophospholipids are members of the broader class of lipids known as **phospholipids.**

The numbering and nomenclature of glycerophospholipids present a dilemma in that the number 2 carbon of the glycerol backbone of a phospholipid is asymmetric. It is possible to name these molecules either as D- or L-isomers. Thus, glycerol phosphate itself can be referred to either as D-glycerol-1-phosphate or as L-glycerol-3-phosphate (Figure 8.5). Instead of naming the glycerol phosphatides in this way, biochemists have adopted the *stereospecific numbering* or *sn-* system. The stereospecific numbering system is based on the concept of prochirality. If a tetrahedral center in a molecule has two identical substituents, it is referred to as **prochiral** because if either of the like substituents is converted to a different group, the tetrahedral center then becomes chiral. Consider glycerol (Figure 8.5): The central carbon of glycerol is prochiral because replacing either of the $-CH_2OH$ groups would make the central carbon chiral. Nomenclature for prochiral centers is based on the (R,S) system (see Chapter 4). To name the otherwise identical substituents of a prochiral center, imagine increasing slightly the priority of one of them (by substituting a deuterium for a hydrogen, for example) as shown in Figure 8.5. The resulting molecule has an (S) configuration about the (now chiral) central carbon atom. The group that contains the deuterium is thus referred to as the *pro-S* group. As a useful exercise, you should confirm that labeling the other CH_2OH group with a deuterium produces the (R) configuration at the central carbon so that this latter CH_2OH group is the *pro-R* substituent.

Now consider the two presentations of glycerol phosphate in Figure 8.5. **In the stereospecific numbering system, the *pro-S* position of a prochiral atom is denoted as the *1-position,* the prochiral atom as the *2-position,* and so on.** When this scheme is used, the prefix *sn-* precedes the molecule name (glycerol phosphate in this case) and distinguishes this nomenclature from other approaches. In this way, the glycerol phosphate in natural phosphoglycerides is named *sn*-glycerol-3-phosphate.

Glycerophospholipids Are the Most Common Phospholipids

Phosphatidic acid, the parent compound for the glycerol-based phospholipids (Figure 8.4), consists of *sn*-glycerol-3-phosphate, with fatty acids esterified at the 1- and 2-positions. Phosphatidic acid is found in small amounts in most natural systems and is an important intermediate in the biosynthesis of the more common glycerophospholipids (Figure 8.6).

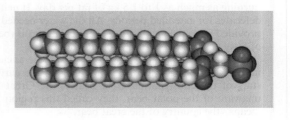

FIGURE 8.4 Phosphatidic acid, the parent compound for glycerophospholipids.

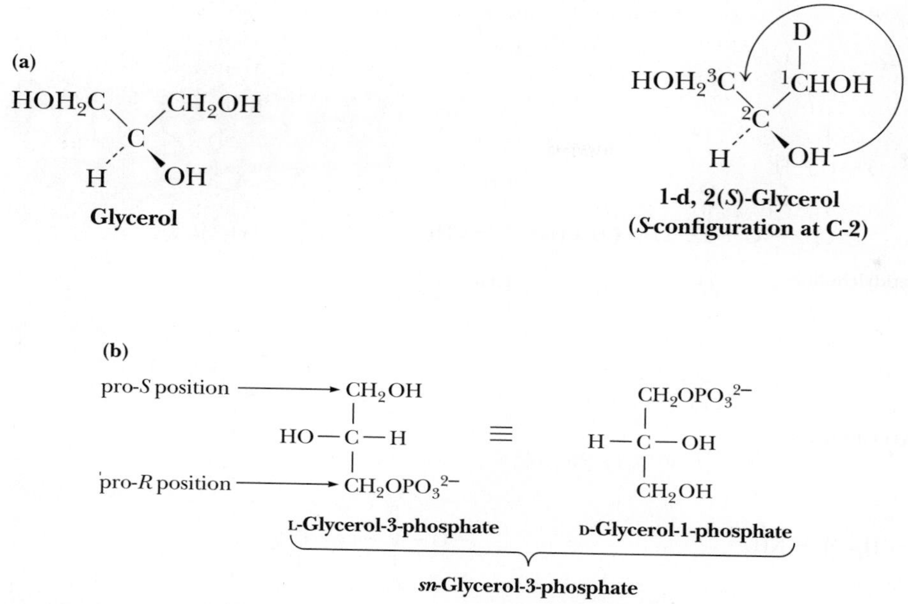

(a)

Glycerol

1-d, 2(S)-Glycerol
(S-configuration at C-2)

(b)

pro-S position $\longrightarrow$ CH$_2$OH

HO — C — H $\equiv$

pro-R position $\longrightarrow$ CH$_2$OPO$_3^{2-}$

L-Glycerol-3-phosphate

CH$_2$OPO$_3^{2-}$

H — C — OH

CH$_2$OH

D-Glycerol-1-phosphate

sn-Glycerol-3-phosphate

FIGURE 8.5 **(a)** The two identical —CH$_2$OH groups on the central carbon of glycerol may be distinguished by imagining a slight increase in priority for one of them (by replacement of an H by a D) as shown. **(b)** The absolute configuration of sn-glycerol-3-phosphate is shown. The pro-R and pro-S positions of the parent glycerol are also indicated.

In these compounds, a variety of polar groups are esterified to the phosphoric acid moiety of the molecule. The phosphate, together with such esterified entities, is referred to as a "head" group. Phosphatides with choline or ethanolamine are referred to as **phosphatidylcholine** (known commonly as **lecithin**) or **phosphatidylethanolamine,** respectively. These phosphatides are two of the most common constituents of biological membranes. Other common head groups found in phosphatides include glycerol, serine, and inositol (Figure 8.6). Another kind of glycerol phosphatide found in many tissues is **diphosphatidylglycerol.** First observed in heart tissue, it is also called **cardiolipin.** In cardiolipin, a phosphatidylglycerol is esterified through the C-1 hydroxyl group of the glycerol moiety of the head group to the phosphoryl group of another phosphatidic acid molecule.

> **A DEEPER LOOK**

Will the Real Glycophospholipid Come Forward?

Phosphatidylinositol is sometimes mistakenly taken to be a glycophospholipid—specifically a phosphatidyl glycoside. However, the head group of phosphatidylinositol is a poly-hydroxy cyclohexane, not a monosaccharide. On the other hand, a true glycophospholipid, namely **phosphatidyl-ß-D-glucopyranoside** (or simply **phosphatidylglucoside**), has been identified in several cell types, including HL-60 cells (a line of human leukemia cells used frequently for studies of blood cell formation). In these cells,

phosphatidylglucoside plays a role in formation of plasma membrane signaling microdomains involved in cellular differentiation and maturation. (See discussion of membrane rafts in Chapter 9.) Interestingly, the phosphatidylglucoside isolated from HL-60 cells is a single molecular species that occurs rarely in mammalian cells. This phosphatidylglucoside is comprised exclusively of stearic acid (18:0) at the sn-C1 position and arachidic acid (20:0) at the sn-C2 position of the glycerol backbone.

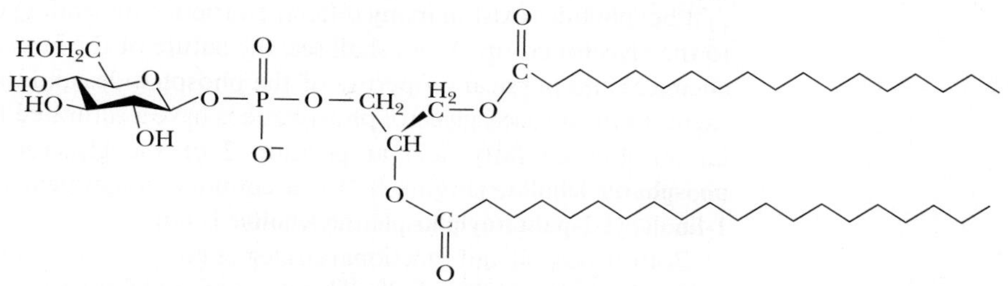

Phosphatidyl-β-D-glucopyranoside

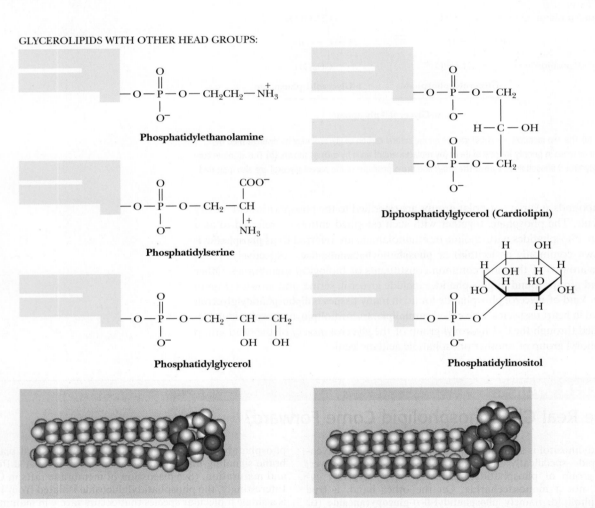

GLYCEROLIPIDS WITH OTHER HEAD GROUPS:

Phosphatidylcholine

Phosphatidylethanolamine

Phosphatidylserine

Phosphatidylglycerol

Diphosphatidylglycerol (Cardiolipin)

Phosphatidylinositol

FIGURE 8.6 Structures of several glycerophospholipids and space-filling models of phosphatidylcholine, phosphatidylglycerol, and phosphatidylinositol.

Phosphatides exist in many different varieties, depending on the fatty acids esterified to the glycerol group. As we shall see, the nature of the fatty acids can greatly affect the chemical and physical properties of the phosphatides and the membranes that contain them. In most cases, glycerol phosphatides have a saturated fatty acid at position 1 and an unsaturated fatty acid at position 2 of the glycerol. Thus, **1-stearoyl-2-oleoyl-phosphatidylcholine** (Figure 8.7) is a common constituent in natural membranes, but **1-linoleoyl-2-palmitoylphosphatidylcholine** is not.

Both structural and functional strategies govern the natural design of the many different kinds of glycerophospholipid head groups and fatty acids. The structural roles of these different glycerophospholipid classes are described in Chapter 9. Certain phospholipids, including phosphatidylinositol and phosphatidylcholine, participate in complex cellular signaling events. These roles are described in Section 8.8 and Chapter 32.

Ether Glycerophospholipids Include PAF and Plasmalogens

Ether glycerophospholipids possess an ether linkage instead of an acyl group at the C-1 position of glycerol (Figure 8.8a). One of the most versatile biochemical signal molecules found in mammals is **platelet-activating factor,** or **PAF,** a unique ether glycerophospholipid (Figure 8.8b). The alkyl group at C-1 of PAF is typically a 16-carbon chain, but the acyl group at C-2 is a 2-carbon acetate unit. By virtue of this acetate group, PAF is much more water soluble than other lipids, allowing PAF to function as a soluble messenger in signal transduction.

Plasmalogens are ether glycerophospholipids in which the alkyl moiety is cis-α,β-unsaturated (Figure 8.9). Common plasmalogen head groups include choline, ethanolamine, and serine. These lipids are referred to as **phosphatidal choline, phosphatidal ethanolamine,** and **phosphatidal serine.**

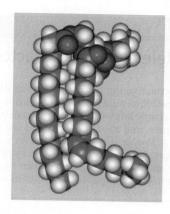

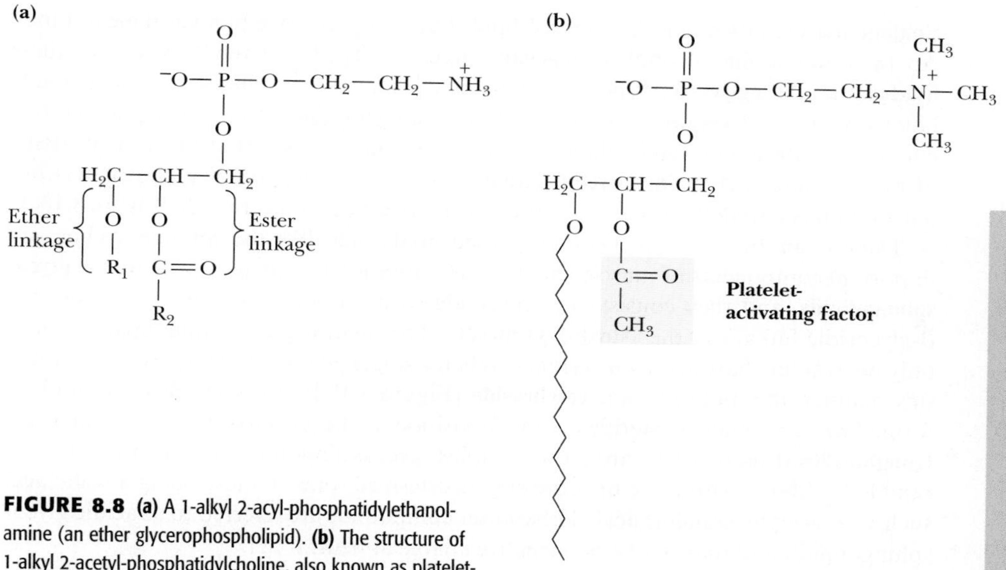

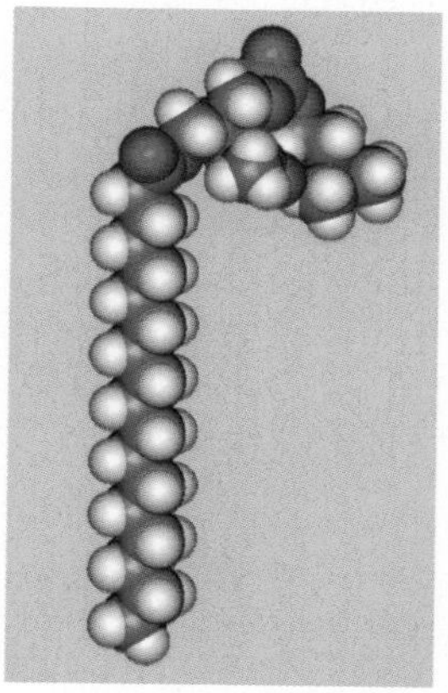

FIGURE 8.7 A space-filling model of 1-stearoyl-2-oleoylphosphatidylcholine.

(a)

(b)

FIGURE 8.8 **(a)** A 1-alkyl 2-acyl-phosphatidylethanolamine (an ether glycerophospholipid). **(b)** The structure of 1-alkyl 2-acetyl-phosphatidylcholine, also known as platelet-activating factor or PAF.

Platelet-activating factor

Choline plasmalogen

The ethanolamine plasmalogens have ethanolamine in place of choline.

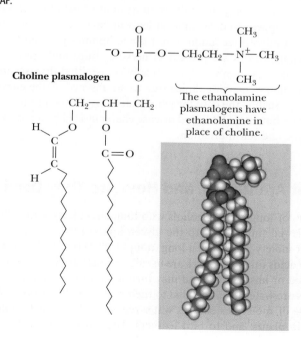

FIGURE 8.9 The structure and a space-filling model of a choline plasmalogen.

HUMAN BIOCHEMISTRY

Platelet-Activating Factor: A Potent Glyceroether Mediator

Platelet-activating factor (PAF) was first identified by its ability (at low levels) to cause platelet aggregation and dilation of blood vessels, but it is now known to be a potent mediator in inflammation, allergic responses, and shock. PAF effects are observed at tissue concentrations as low as $10^{-12}M$. PAF causes a dramatic inflammation of air passages and induces asthmalike symptoms in laboratory animals. **Toxic shock syndrome** occurs when fragments of destroyed bacteria act as toxins and induce the synthesis of PAF. PAF causes a drop in blood pressure and a reduced volume of blood pumped by the heart, which leads to shock and, in severe cases, death.

Beneficial effects have also been attributed to PAF. In reproduction, PAF secreted by the fertilized egg is instrumental in the implantation of the egg in the uterine wall. PAF is produced in significant quantities in the lungs of the fetus late in pregnancy and may stimulate the production of fetal lung surfactant, a protein–lipid complex that prevents collapse of the lungs in a newborn infant.

8.4 What Are Sphingolipids, and How Are They Important for Higher Animals?

Ceramides regulate stress signaling via reorganization of the plasma membrane. The resulting "ceramide-rich platforms" and "rafts," which regulate transmembrane signaling, are discussed in Chapter 9.

Sphingolipids represent another class of lipids frequently found in biological membranes. An 18-carbon amino alcohol, **sphingosine** (Figure 8.10a), forms the backbone of these lipids rather than glycerol. Typically, a fatty acid is joined to a sphingosine via an amide linkage to form a **ceramide** (Figure 8.10b). **Sphingomyelins** represent a phosphorus-containing subclass of sphingolipids and are especially important in the nervous tissue of higher animals. A **sphingomyelin** is formed by the esterification of a phosphorylcholine or a phosphorylethanolamine to the 1-hydroxy group of a ceramide (Figure 8.10c).

There is another class of ceramide-based lipids that, like the sphingomyelins, are important components of muscle and nerve membranes in animals. These are the **glycosphingolipids,** and they consist of a ceramide with one or more sugar residues in a β-glycosidic linkage at the 1-hydroxyl moiety. The neutral glycosphingolipids contain only neutral (uncharged) sugar residues. When a single glucose or galactose is bound in this manner, the molecule is a **cerebroside** (Figure 8.10d). Another class of lipids is formed when a sulfate is esterified at the 3-position of the galactose to make a **sulfatide.** **Gangliosides** (Figure 8.10e) are more complex glycosphingolipids that consist of a ceramide backbone with three or more sugars esterified, one of these being a **sialic acid** such as *N*-**acetylneuraminic acid.** These latter compounds are referred to as acidic glycosphingolipids, and they have a net negative charge at neutral pH.

The glycosphingolipids have a number of important cellular functions, despite the fact that they are present only in small amounts in most membranes. Glycosphingolipids at cell surfaces appear to determine, at least in part, certain elements of tissue and organ specificity. Cell–cell recognition and tissue immunity depend on specific glycosphingolipids. Gangliosides are present in nerve endings and are important in nerve impulse transmission. A number of genetically transmitted diseases involve the accumulation of specific glycosphingolipids due to an absence of the enzymes needed for their degradation. Such is the case for ganglioside G_{M2} in the brains of Tay-Sachs disease victims, a rare but fatal childhood disease characterized by a red spot on the retina, gradual blindness, and self-mutilation.

8.5 What Are Waxes, and How Are They Used?

Waxes are esters of long-chain alcohols with long-chain fatty acids. The resulting molecule can be viewed (in analogy to the glycerolipids) as having a weakly polar head group (the ester moiety itself) and a long, nonpolar tail (the hydrocarbon chains) (Figure 8.11). Fatty acids found in waxes are usually saturated. The alcohols found in waxes may be saturated or unsaturated and may include sterols, such as cholesterol (see later section). Waxes are water insoluble due to their mostly hydrocarbon composition. As a result, this class of molecules confers water-repellant character to animal skin, to the leaves of certain plants, and to bird feathers. The glossy surface of a polished apple

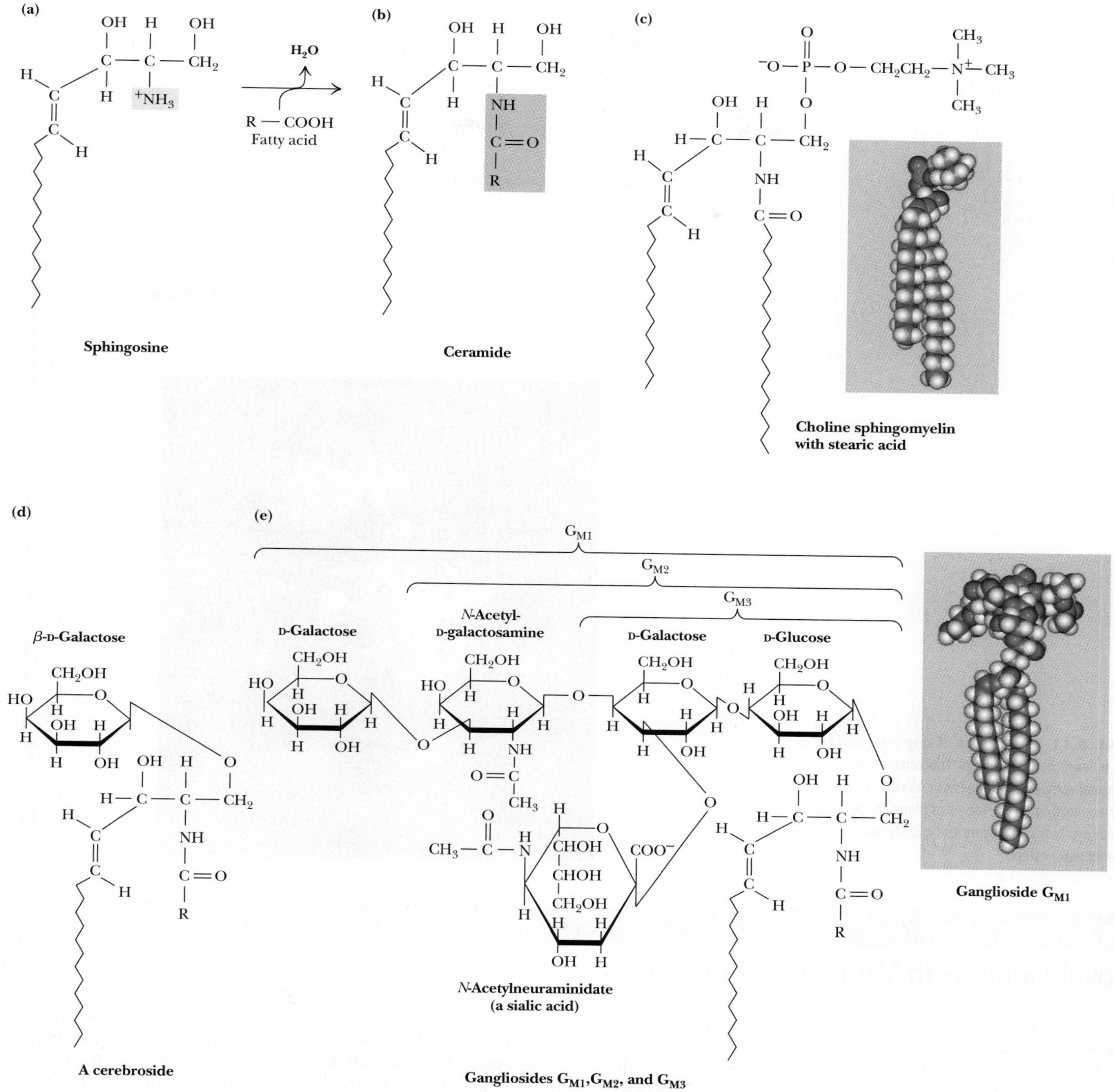

(a)

Sphingosine

(b)

Ceramide

(c)

Choline sphingomyelin with stearic acid

(d)

β-D-Galactose

A cerebroside

(e)

G_{M1}

G_{M2}

G_{M3}

D-Galactose N-Acetyl-D-galactosamine D-Galactose D-Glucose

N-Acetylneuraminidate (a sialic acid)

Gangliosides $G_{M1}, G_{M2},$ and G_{M3}

Ganglioside G_{M1}

FIGURE 8.10 Sphingolipids are based on the structure of sphingosine. A ceramide with a phosphocholine head group is a choline sphingomyelin. A ceramide with a single sugar is a cerebroside. Gangliosides are ceramides with three or more sugars esterified, one of which is a sialic acid.

results from a wax coating. **Carnauba wax,** obtained from the fronds of a species of palm tree in Brazil, is a particularly hard wax used for high-gloss finishes, such as in automobile wax, boat wax, floor wax, and shoe polish. **Lanolin,**[1] a wool wax, is used as a base for pharmaceutical and cosmetic products because it is rapidly assimilated by human skin. The brand name Oil of Olay® was coined by Graham Wulff, a South African chemist who developed it. The name refers to lanolin, a key ingredient.

[1]Lanolin is a complex mixture of waxes with 33 different alcohols esterified to 36 different fatty acids.

Oleoyl alcohol **Stearic acid**

Oleoyl stearate

$$CH_3(CH_2)_{14} - \overset{O}{\overset{\|}{C}} - O - CH_2 - (CH_2)_{16} - CH_3$$

Stearyl palmitate

$$R_1 - \overset{O}{\overset{\|}{C}} - O - R_2$$

General formula of a wax

$$CH_3(CH_2)_{14} - \overset{O}{\overset{\|}{C}} - O - CH_2 - (CH_2)_{28} - CH_3$$

Triacontanol palmitate

FIGURE 8.11 Waxes consist of long-chain alcohols esterified to long-chain fatty acids. Triacontanol palmitate is the principal component of beeswax. Waxes are components of the waxy coating on the leaves of plants, such as jade plants (shown here). Such species typically contain dozens of different waxy esters.

© Yuris/Shutterstock.com

> **A DEEPER LOOK**

Novel Lipids with Valuable Properties

Lipids from many natural sources are useful and highly prized materials for many purposes in our world. When oil from the head of the sperm whale is cooled, **spermaceti,** a translucent wax with a white, pearly luster, crystallizes from the mixture. Spermaceti, which makes up 11% of whale oil, is composed mainly of the wax **cetyl palmitate:**

$$CH_3(CH_2)_{14} - COO - (CH_2)_{15}CH_3$$

as well as smaller amounts of cetyl alcohol:

$$HO - (CH_2)_{15}CH_3$$

In the literary classic *Moby Dick,* Herman Melville describes Ishmael's impressions of spermaceti, when he muses that the waxes "discharged all their opulence, like fully ripe grapes their wine; as I snuffed that uncontaminated aroma—literally and truly, like the smell of spring violets."*

Spermaceti and cetyl palmitate have been widely used in the making of cosmetics, fragrant soaps, and candles. In addition, the low viscosity of sperm whale oil made it an ideal lubricant for a variety of industrial applications until whale populations declined precipi-tously in the 20th century and commercial whaling was outlawed. Since then, other lipids with similar properties have been sought as replacements for whale oil. One example is the lipids found in the shrub *Euonymus alatus* (also known as the "burning bush"), which makes large amounts of 3-acetyl-1,2-diacyl-*sn*-glycerols. These unusual glycerols, with two long acyl chains and one short (2-carbon) chain, possess a much lower viscosity than triglycerides with three long chains. As such, they are appropriate for use in biodiesel fuels.

Durrett, T. P., McClosky, D. D., Tumaney, A. W., Elzinga, D. A., Ohl-rogge, J., and Pollard, M., 2010. A distinct DGAT with *sn*-3 acetyltransfer-ase activity that synthesizes unusual, reduced-viscosity oils in *Euonymus* and transgenic seeds. *Proceedings of the National Academy of Sciences* **107:**9464–9469.
*Melville, H., 1984. *Moby Dick.* London: Octopus Books, p. 205. (Adapted from Waddell, T. G., and Sanderlin, R. R., 1986. Chemistry in *Moby Dick. Journal of Chemical Education* **63:**1019–1020.)

(Continued)

> **A DEEPER LOOK**   Novel Lipids with Valuable Properties *(Continued)*

▲ Sperm whale

▲ Euonymus alatus

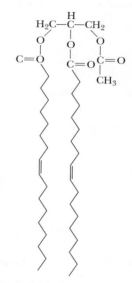

▲ 3-acetyl-1,2-diacyl-*sn*-glycerol

8.6 What Are Terpenes, and What Is Their Relevance to Biological Systems?

The **terpenes** are a class of lipids formed from combinations of two or more molecules of 2-methyl-1,3-butadiene, better known as **isoprene** (a five-carbon unit that is abbreviated C_5). A **monoterpene** (C_{10}) consists of two isoprene units, a **sesquiterpene** (C_{15}) consists of three isoprene units, a **diterpene** (C_{20}) has four isoprene units, and so on. Isoprene units can be linked in terpenes to form straight-chain or cyclic molecules, and the usual method of linking isoprene units is head to tail (Figure 8.12). Monoterpenes occur in all higher plants, whereas sesquiterpenes and diterpenes are less widely known. Several examples of these classes of terpenes are shown in Figure 8.13. The **triterpenes** are C_{30} terpenes and

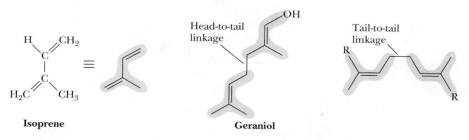

FIGURE 8.12 The structure of isoprene (2-methyl-1,3-butadiene) and the structure of head-to-tail and tail-to-tail linkages. Isoprene itself can be formed by distillation of natural rubber, a linear head-to-tail polymer of isoprene units.

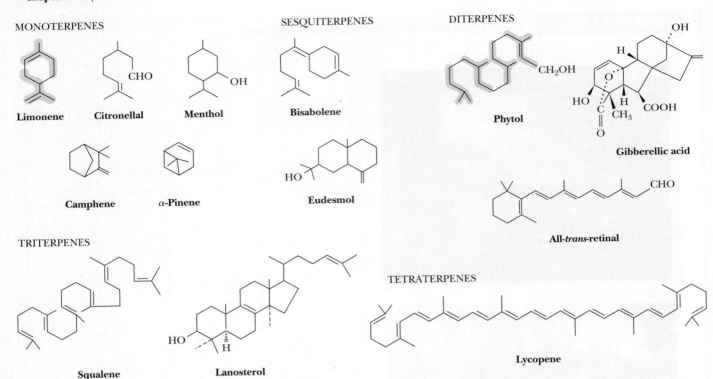

MONOTERPENES

Limonene Citronellal Menthol

Camphene α-Pinene

SESQUITERPENES

Bisabolene

Eudesmol

DITERPENES

Phytol

Gibberellic acid

All-*trans*-retinal

TRITERPENES

Squalene Lanosterol

TETRATERPENES

Lycopene

FIGURE 8.13 Many monoterpenes are readily recognized by their characteristic flavors or odors (limonene in lemons; citronellal in roses, geraniums, and some perfumes; and menthol from peppermint, used in cough drops and nasal inhalers). The diterpenes, which are C_{20} terpenes, include retinal (the essential light-absorbing pigment in rhodopsin, the photoreceptor protein of the eye), and phytol (a constituent of chlorophyll). The triterpene lanosterol is a constituent of wool fat. Lycopene is a carotenoid found in ripe fruit, especially tomatoes.

include **squalene** and **lanosterol,** two of the precursors of cholesterol and other steroids (discussed later). **Tetraterpenes** (C_{40}) are less common but include the carotenoids, a class of colorful photosynthetic pigments. β-Carotene is the precursor of vitamin A, whereas lycopene, similar to β-carotene, is a pigment found in tomatoes.

Long-chain polyisoprenoid molecules with a terminal alcohol moiety are called **polyprenols.** The **dolichols,** one class of polyprenols (Figure 8.14), consist of 16 to 22 isoprene units and, in the form of dolichyl phosphates, function to carry carbohydrate units in the biosynthesis of glycoproteins in animals. Polyprenyl groups serve to anchor certain proteins to biological membranes (discussed in Chapter 9).

The Membranes of Archaea Are Rich in Isoprene-Based Lipids

Archaea are found primarily in harsh environments. Some thrive in the high temperatures of geysers and ocean steam vents, whereas others are found in extremely acidic, cold, or salty

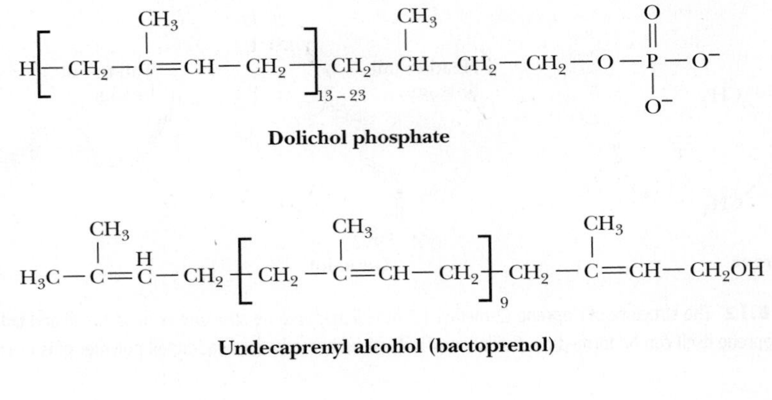

Dolichol phosphate

FIGURE 8.14 Dolichol phosphate is an initiation point for the synthesis of carbohydrate polymers in animals. The analogous alcohol in bacterial systems, *undecaprenol,* also known as *bactoprenol,* consists of 11 isoprene units. Undecaprenyl phosphate delivers sugars from the cytoplasm for the synthesis of cell wall components such as peptidoglycans, lipopolysaccharides, and glycoproteins.

Undecaprenyl alcohol (bactoprenol)

A DEEPER LOOK

Why Do Plants Emit Isoprene?

The Blue Ridge Mountains of Virginia are so named for the misty blue vapor or haze that hangs over them through much of the summer season. This haze is composed in part of isoprene that is produced and emitted by the plants and trees of the mountains. Global emission of isoprene from vegetation is estimated at 3×10^{14} g/yr. Plants frequently emit as much as 15% of the carbon fixed in photosynthesis as isoprene, and Thomas Sharkey, a botanist at the University of Wisconsin, has shown that the kudzu plant can emit as much as 67% of its fixed carbon as isoprene as the result of water stress. Why should plants and trees emit large amounts of isoprene and other hydrocarbons? Sharkey has shown that an isoprene atmosphere or "blanket" can protect leaves from irreversible damage induced by high (summerlike) temperatures. He hypothesizes that isoprene in the air around plants dissolves into leaf-cell membranes, altering the lipid bilayer and/or lipid–protein and protein–protein interactions within the membrane to increase thermal tolerance.

© Jo Crebbin/Shutterstock.com

Blue Ridge Mountains

HUMAN BIOCHEMISTRY

Coumadin or Warfarin—Agent of Life or Death

The isoprene-derived molecule whose structure is shown here is known alternately as **Coumadin** and **warfarin.** By the former name, it is a widely prescribed anticoagulant. By the latter name, it is a component of rodent poisons. How can the same chemical species be used for such disparate purposes? The key to both uses lies in its ability to act as an antagonist of vitamin K in the body.

Vitamin K is necessary for the carboxylation of glutamate residues on certain proteins, including some proteins in the blood-clotting cascade (including **prothrombin, factor VII, factor IX,** and **factor X,** which undergo a Ca^{2+}-dependent conformational change in the course of their biological activity, as well as **protein C** and **protein S,** two regulatory proteins in coagulation). Carboxylation of these coagulation factors is catalyzed by a carboxylase that requires the reduced form of vitamin K (vitamin KH_2), molecular oxygen, and carbon dioxide. KH_2 is oxidized to vitamin K epoxide, which is recycled to KH_2 by the enzymes **vitamin K epoxide reductase** (1) and **vitamin K reductase** (2, 3). Coumadin/warfarin exerts its anticoagulant effect by inhibiting vitamin K epoxide reductase and possibly also vitamin K reductase. This inhibition depletes vitamin KH_2 and reduces the activity of the carboxylase.

Coumadin/warfarin, given at a typical dosage of 4 to 5 mg/day, prevents the deleterious formation in the bloodstream of small blood clots and thus reduces the risk of heart attacks and strokes for individuals whose arteries contain sclerotic plaques. Taken in much larger doses, as for example in rodent poisons, Coumadin/warfarin can cause massive hemorrhages and death.

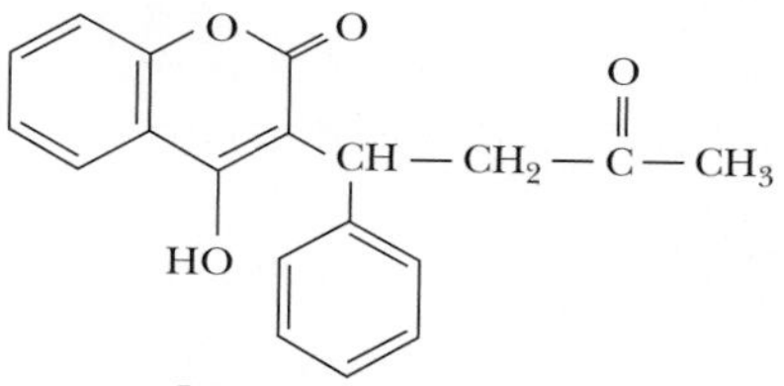

Warfarin (Coumadin)

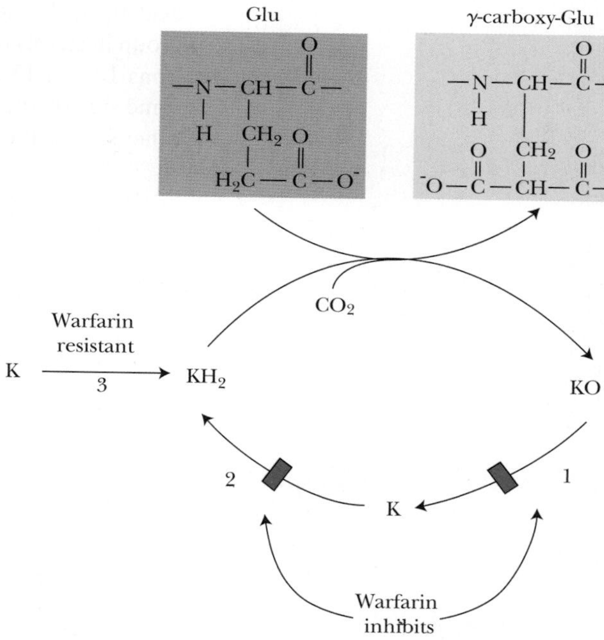

H₂C—O

H—C—O

HOCH₂

Glycerol

CH_2OH

O—C—H

O—CH₂

Glycerol

Isoprene units

Caldarchaeol

FIGURE 8.15 The structure of caldarchaeol, an isoprene-based lipid found in archaea.

environments. Archaea also live in extremes of pH in the digestive tracts of cows, termites, and humans. Archaea are ideally adapted to their harsh environments, and one such adaptation is found in their cell membranes, which contain isoprene-based lipids (Figure 8.15). These isoprene chains are linked at both ends by ether bonds to glycerols. Ether bonds are more stable to hydrolysis than the ester linkages of glycerophospholipids (Figure 8.6). With a length twice that of typical glycerophospholipids, these molecules can completely span a cell membrane, providing additional stability. Interestingly, the glycerols in archaeal lipids are in the (*R*) configuration, whereas glycerolipids of animals, plants, and eubacteria are almost always in the (*S*) configuration.

8.7 What Are Steroids, and What Are Their Cellular Functions?

Cholesterol

A large and important class of terpene-based lipids is the **steroids.** This molecular family, whose members affect an amazing array of cellular functions, is based on a common structural motif of three 6-membered rings and one 5-membered ring all fused together. **Cholesterol** (Figure 8.16) is the most common steroid in animals and the precursor for all other animal steroids. The numbering system for cholesterol applies to all such molecules. Many steroids contain methyl groups at positions 10 and 13 and an 8- to 10-carbon alkyl side chain at position 17. The polyprenyl nature of this compound is particularly evident in the side chain. Many steroids contain an oxygen at C-3, either a hydroxyl group in sterols or a carbonyl group in other steroids. Significantly, the carbons at positions 10 and 13 and the alkyl group at position 17 are nearly always oriented on the same side of the steroid nucleus, the β-orientation. Alkyl groups that extend from the other side of the steroid backbone are in an α-orientation.

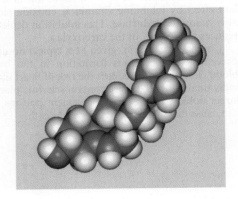

FIGURE 8.16 The structure of cholesterol, shown with steroid ring designations and carbon numbering.

Cholesterol

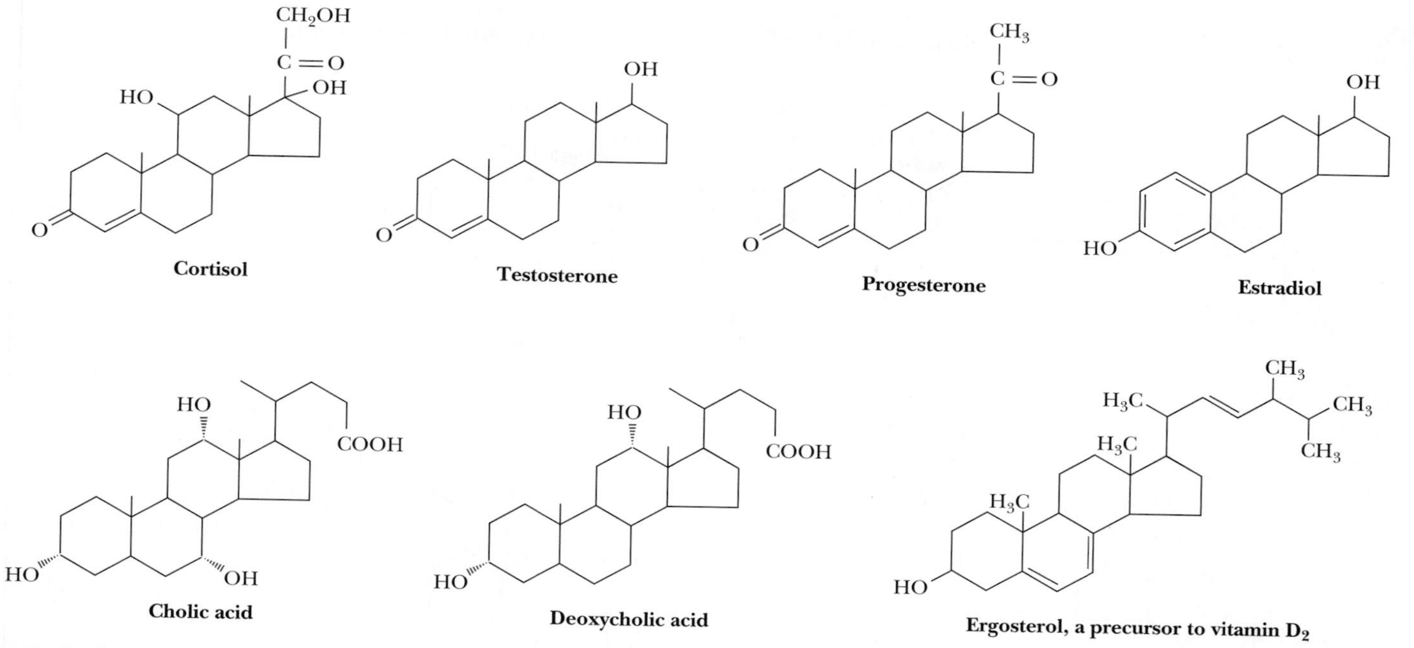

FIGURE 8.17 The structures of several important sterols derived from cholesterol.

Cholesterol is a principal component of animal cell plasma membranes, and smaller amounts of cholesterol are found in the membranes of intracellular organelles. The relatively rigid fused ring system of cholesterol and the weakly polar alcohol group at the C-3 position have important consequences for the properties of plasma membranes. Cholesterol is also a component of lipoprotein complexes in the blood, and it is one of the constituents of plaques that form on arterial walls in atherosclerosis.

Steroid Hormones Are Derived from Cholesterol

Steroids derived from cholesterol in animals include five families of hormones (the androgens, estrogens, progestins, glucocorticoids, and mineralocorticoids) and bile acids (Figure 8.17). **Androgens** such as **testosterone** and **estrogens** such as **estradiol** mediate the development of sexual characteristics and sexual function in animals. The **progestins** such as **progesterone** participate in control of the menstrual cycle and pregnancy. **Glucocorticoids (cortisol,** for example) participate in the control of carbohydrate, protein, and lipid metabolism, whereas the **mineralocorticoids** regulate salt (Na^+, K^+, and Cl^-) balances in tissues. The **bile acids** (including **cholic** and **deoxycholic acid**) are detergent molecules secreted in bile from the gallbladder that assist in the absorption of dietary lipids in the intestine.

HUMAN BIOCHEMISTRY

Plant Sterols and Stanols—Natural Cholesterol Fighters

Dietary guidelines for optimal health call for reducing the cholesterol intake. One strategy involves eating plant sterols and stanols (figure) in place of cholesterol-containing fats such as butter. Despite their structural similarity to cholesterol, minor isomeric differences and the presence of methyl and ethyl groups in the side chains of these substances result in their poor absorption by intestinal mucosal cells. Interestingly, stanols are even less well absorbed than their sterol counterparts. (Stanols are fully reduced sterols.) Both sterols and stanols bind to cholesterol receptors on intestinal cells and block the absorption of cholesterol itself. Stanols esterified with long-chain fatty acids form micelles (see page 263) that are more effectively distributed in the fat phase of the food digest and provide the most effective blockage of cholesterol uptake. Raisio Group, a Finnish company, has developed Benecol, a stanol ester spread that can lower LDL cholesterol by up to 14% if consumed daily (see graph). McNeil Nutritionals has partnered with Raisio Group to market Benecol in the United States.

(Continued)

Plant Sterols and Stanols—Natural Cholesterol Fighters (continued)

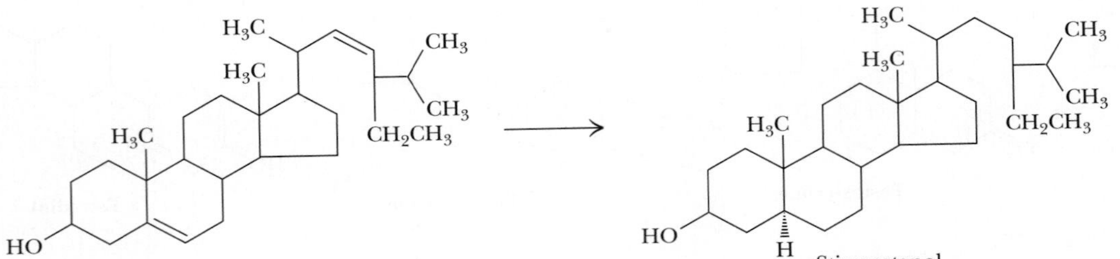

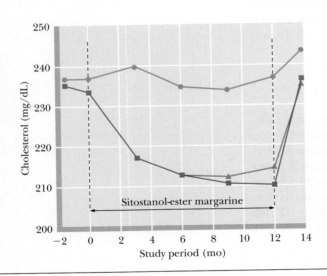

Stigmasterol

Stigmastanol

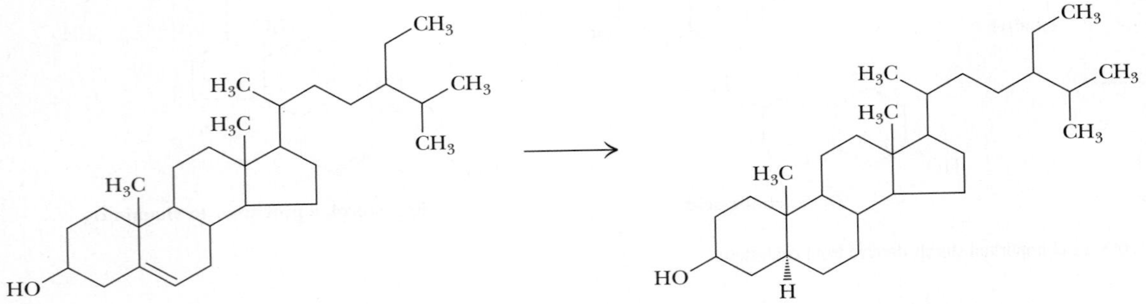

β-Sitosterol

β-Sitostanol

Sitostanol-ester margarine

◀ Serum cholesterol levels before and after the consumption of margarine with and without sitostanol ester for 12 months. Green circles: 0 g/day. Red squares: 2.6 g/day. Blue triangles: 1.8 g/day. Note: The *y*-axis begins at 200 mg cholesterol/dL. (*Adapted from Miettinen, T. A., et al., 1995. Reduction of serum cholesterol with sitostanol-ester margarine in a mildly hypercholesterolemic population.* New England Journal of Medicine **333**:1308–1312.)

17β-Hydroxysteroid Dehydrogenase 3 Deficiency

Testosterone, the principal male sex steroid hormone, is synthesized in five steps from cholesterol, as shown in the following figure. In the last step, five isozymes catalyze the 17β-hydroxysteroid dehydrogenase reaction that interconverts 4-androstenedione and testosterone. Defects in the synthesis or action of testosterone can impair the development of the male phenotype during embryogenesis and cause the disorders of human sexuality termed male pseudohermaphroditism. Specifically, mutations in isozyme 3 of the 17β-hydroxysteroid dehydrogenase in the fetal testes impair the formation of testoster-

one and give rise to genetic males with female external genitalia and blind-ending vaginas. Such individuals are typically raised as females but virilize at puberty, due to an increase in serum testosterone, and develop male hair growth patterns. Fourteen different mutations of 17β-hydroxysteroid dehydrogenase 3 have been identified in 17 affected families in the United States, the Middle East, Brazil, and western Europe. These families account for about 45% of the patients with this disorder reported in scientific literature.

17β-Hydroxysteroid Dehydrogenase 3 Deficiency (continued)

Cholesterol

Desmolase (Mitochondria)

Isocaproic aldehyde

Pregnenolone

(Endoplasmic reticulum)

Progesterone

17α-Hydroxylase

Testosterone

17β-Hydroxysteroid dehydrogenase

4-Androstenedione

17,20-Lyase (Gonads)

17α-Hydroxyprogesterone

8.8 How Do Lipids and Their Metabolites Act as Biological Signals?

Glycerophospholipids and sphingolipids play important structural roles as the principal components of biological membranes (see Chapter 9). However, their modification and breakdown also produce an eclectic assortment of substances that act as powerful **chemical signals** (Figures 8.18 and 8.19). In contrast to the steroid hormones (Figure 8.17), which travel from tissue to tissue in the blood to exert their effects, these lipid metabolites act locally, either within the cell in which they are made or on nearby cells. Signal molecules typically initiate a cascade of reactions with multiple possible effects, and the

Phospholipase D

Phospholipase C

Phospholipase A₂

Phospholipase A₁

(b)

Diamondback rattlesnake

Indian cobra

© Tom Bean/CORBIS

© Joe McDonald/CORBIS

FIGURE 8.18 **(a)** Phospholipases A₁ and A₂ cleave fatty acids from a glycerophospholipid, producing lysophospholipids. Phospholipases C and D hydrolyze on either side of the phosphate in the polar head group. **(b)** Phospholipases are components of the venoms of many poisonous snakes. The pain and physiological consequences of a snake bite partly result from breakdown of cell membranes by phospholipases.

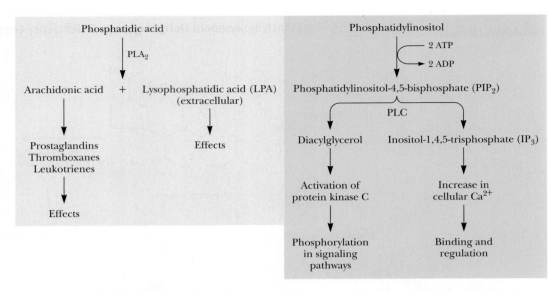

FIGURE 8.19 Modification and breakdown of glycerophospholipids produce a variety of signals and regulatory effects. Phospholipase A$_2$ cleaves a fatty acid from phosphatidic acid to produce lysophosphatidic acid (LPA), which can act as an extracellular signal. If the fatty acid released is arachidonic acid, it can be the substrate for synthesis of prostaglandins, thromboxanes, and leukotrienes. Phospholipase C action on phosphatidylinositol-4,5-bisphosphate produces diacylglycerol and inositol-1,4,5-trisphosphate, two signal molecules that work together to activate protein kinases—enzymes that phosphorylate other proteins in signaling pathways.

> **A DEEPER LOOK**

Glycerophospholipid Degradation: One of the Effects of Snake Venom

The venoms of poisonous snakes contain (among other things) a class of enzymes known as **phospholipases,** enzymes that cause the breakdown of phospholipids. For example, the venoms of the eastern diamondback rattlesnake *(Crotalus adamanteus)* and the Indian cobra *(Naja naja)* both contain phospholipase A$_2$, which catalyzes the hydrolysis of fatty acids at the C-2 position of glycerophospholipids (Figure 8.18). The phospholipid breakdown product of this reaction, *lysolecithin,* acts as a detergent and dissolves the membranes of red blood cells, causing them to rupture. Indian cobras kill several thousand people each year.

> **HUMAN BIOCHEMISTRY**

The Endocannabinoid Signaling System: a Target for Next-Generation Therapeutics

Chemical signals called neurotransmitters are the fundamental mode of intercellular communication in the nervous system. The classic neurotransmitters are small, soluble molecules derived from amino acids (or are amino acids themselves). However, recent research has shown that lipids known as **endocannabinoids** are chemical messengers that operate by a novel retrograde mechanism. This class of chemical signals came to light originally from studies of the physiological effects of **Δ^9-tetrahydrocannabinol** and other metabolites from *Cannabis sativa,* better known as marijuana. Early research showed that the effects of *Cannabis* involved binding of its metabolites to specific receptor proteins in the brain. Further research demonstrated that these receptors are normally activated by endogenous lipid molecules. Two of these are **anandamide** and **2-arachidonoylglycerol (2AG).** The biosyntheses of these signal molecules involves the actions of several different phospholipases and other enzymes, as shown in the figure. Biosynthesis of anandamide from N-arachidonoylphosphatidylethanolamine (NAPE) appears to proceed by at least three different routes in neurons.

Endocannabinoids are termed **retrograde** messengers because they traverse the synaptic cleft in the opposite direction of classical neurotransmitters. Activated endocannabinoid receptors can act to either inhibit or activate neuronal circuits, depending on the type of neurons involved. New research on the physical and functional relationships between endocannabinoid receptors and the proteins with which they interact are providing novel targets for drug discovery.

Δ^9-Tetrahydrocannabinol

Phospholipid

Phospholipase C

Diacylglycerol

Diacylglycerol lipase

2-Arachidonoylglycerol (2AG)

Phosphatidylethanolamine

Phospholipid

Transacylase

N-Arachidonoylphosphatidylethanolamine (NAPE)

Phospholipase C

Phospho-N-arachidonoylethanolamine

Phospholipase D (specific for NAPE)

Hydrolysis of *sn*-1 and *sn*-2 acyl chains, followed by phophodiester cleavage

Phosphatase

N-arachidonoylethanolamine (anandamide)

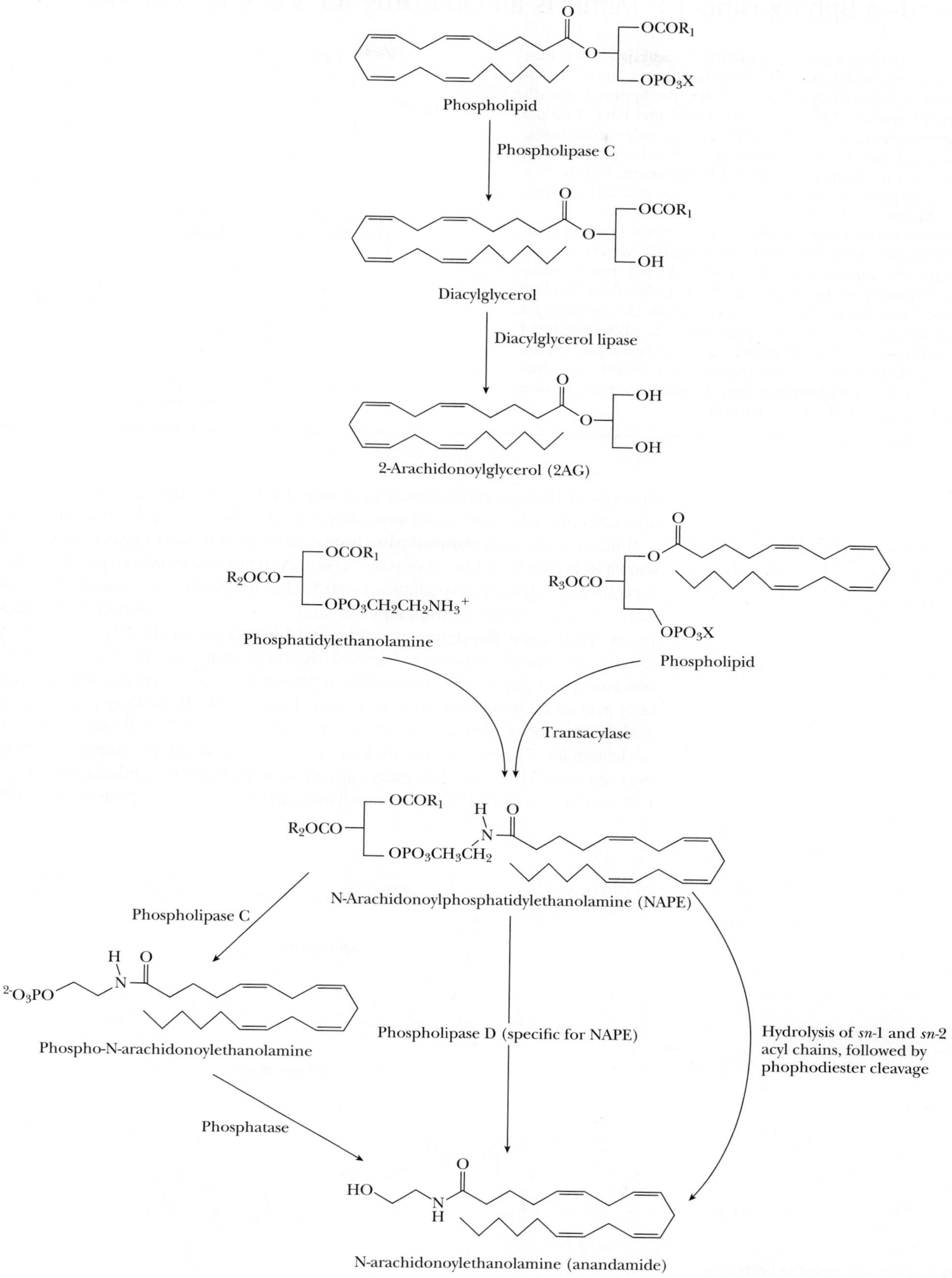

Fingolimod—a Sphingosine-1-P Mimic Is an Oral Drug for Multiple Sclerosis

Multiple sclerosis (MS) is a chronic, potentially disabling illness that is believed to be an autoimmune disorder. In MS, the body's natural defense system mistakenly attacks myelin, the lipid-rich sheath that protects the nerves. About 400,000 Americans have multiple sclerosis. Earlier treatments for MS were all injectable medications, but **fingolimod** has now been approved by the US Food and Drug Administration for relapsing forms of multiple sclerosis. It is the first oral-disease-modifying therapy to be approved for multiple sclerosis in the United States.

In multiple sclerosis, lymph nodes release certain types of lymphocytes (themselves a type of white blood cells), which migrate to and invade the central nervous system, killing neural cells. Lymphocyte release from the lymph nodes is induced by binding of sphingosine-1-phosphate (S1P) to receptors on the lymphocyte surface. Fingolimod, an analog of sphingosine, is phosphorylated by spingosine kinase enzymes, forming *(S)*-fingolimod-phosphate. Binding of fingolimod-1-P to lymphocyte S1P receptors prevents these lymphocytes from migrating to target tissues, averting further lymphocyte-induced neural cell destruction.

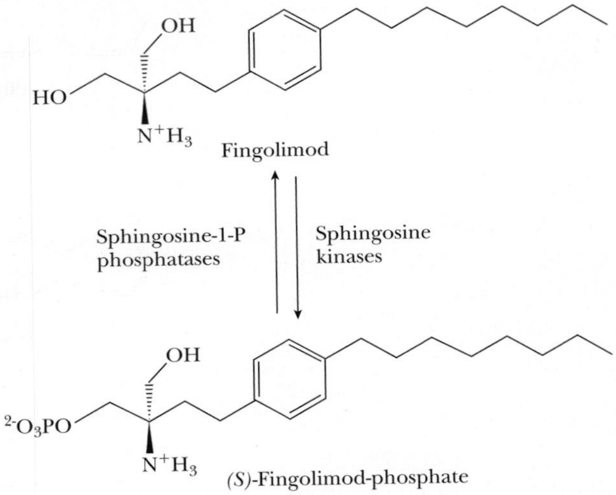

lifetimes of these powerful signals in or near a cell are usually very short. Thus, the creation and breakdown of signal molecules is almost always carefully timed and regulated.

Enzymes known as **phospholipases** hydrolyze the ester bonds of glycerophospholipids as shown in Figure 8.18. Phospholipases A_1 and A_2 remove fatty acid chains from the 1- and 2-positions of glycerophospholipids, respectively. Phospholipases C and D attack the polar head group of a glycerophospholipid. Hydrolysis of inositol phospholipids by phospholipase C produces a **diacylglycerol** and **inositol-1,4,5-trisphosphate (IP_3)** (Figure 8.19), two signal molecules whose combined actions trigger signaling cascades that regulate many cell processes (see Chapter 32). Action of phospholipase A_2 on a phosphatidic acid releases a fatty acid and a **lysophosphatidic acid (LPA,** Figure 8.19). If the fatty acid is arachidonic acid, further chemical modifications can produce a family of 20-carbon compounds—that is, **eicosanoids**. The eicosanoids are local hormones produced as a response to injury and inflammation. They exert their effects on cells near their sites of synthesis (see Chapter 24). LPA produced outside the cell is a signal that can bind to receptor proteins on nearby cells,

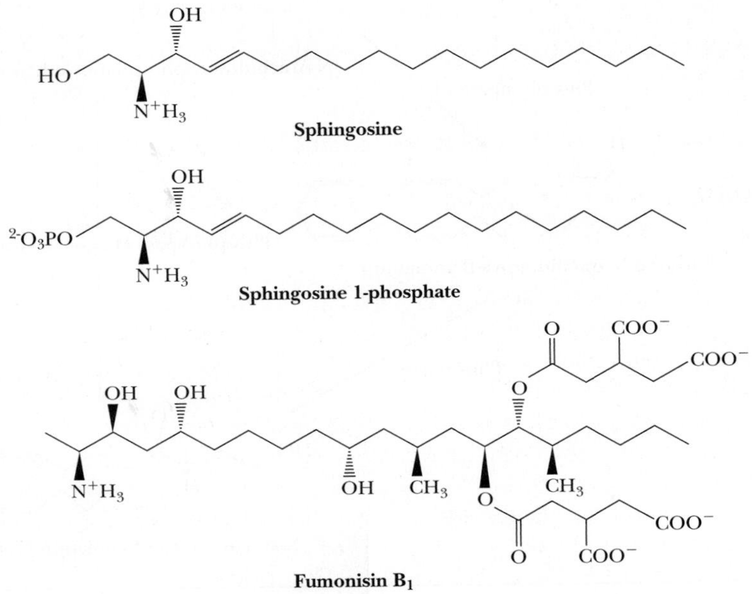

FIGURE 8.20 Structures of sphingosine-1-phosphate (S1P) and fumonisin B_1.

thereby regulating a host of processes, including brain development, cell proliferation and survival, and olfaction (the "sense of smell").

Sphingolipids can also be modified or broken down to produce chemical signals. Sphingosine itself can be phosphorylated to produce **sphingosine-1-phosphate (S1P)** inside cells (Figure 8.20). S1P may either exert a variety of intracellular effects or may be excreted from the cell, where it can bind to membrane receptor proteins, either on adjacent cells or on the cell from which the S1P was released. Excreted S1P binds to many different receptor proteins and provokes many different cell and tissue effects, among them inflammation in allergic reactions, heart rate, and movement and migration of certain cells. Sphingolipid signal molecules are carefully balanced and regulated in organisms, and chemical agents that disturb this balance can be highly toxic. For example, **fumonisin** is a common fungal contaminant of corn and corn-based products that inhibits sphingolipid biosynthesis (Figure 8.20; see also Chapter 24). Fumonisin can trigger esophageal cancer in humans and *leucoencephalomalacia,* a fatal neurological disease in horses.

8.9 What Can Lipidomics Tell Us about Cell, Tissue, and Organ Physiology?

The crucial role of lipids in cells is demonstrated by the large number of human diseases that involve the disruption of lipid metabolic enzymes and pathways. Examples of such diseases include atherosclerosis, diabetes, cancer, infectious diseases, and neurodegenerative diseases. Emerging analytical techniques are making possible the global analysis of lipids and their interacting protein partners in organs, cells, and organelles—an approach termed **lipidomics.** A typical cell may contain more than a thousand different lipids, each with a polar head and a hydrophobic tail or tails. *Various estimates have placed the number of chemical entities within the lipid sphere between 10,000 and 100,000 globally!* It is not clear how and why nature generates this staggering diversity, but there is an increasing awareness of the critical importance of lipids in all aspects of life.

Interestingly, lipid compositions vary in many ways:

- Between species
- Between individuals within a given species
- Within an individual from day to day
- Between tissues within an organism
- Between cell types
- Between cellular compartments
- Between regions of a cellular compartment
- Between parts of a membrane

Complete understanding of lipid function, as well as alteration of such function in disease states, will require the determination of which lipids are present and in what concentrations in every intracellular location. The same knowledge will be needed about each lipid's interaction partners. Mass spectrometric analyses of rat heart muscle reveal that the onset of diabetes results in dramatic changes in triglyceride levels, an increase in phosphatidylinositol levels, and a decrease in phosphatidylethanolamine. On the other hand, mass spectrometric analyses of brain white matter in the very early stages of Alzheimer's disease show a dramatic decrease in one type of plasmalogen and a threefold increase in ceramide levels. Three-dimensional mass spectral imaging can provide 3-D maps of individual lipid species in biological tissues and organs (Figure 8.21).

Cellular lipidomics provides a framework for understanding the myriad roles of lipids, which include (but are not limited to) membrane transport (see Chapter 9), metabolic regulation (see Chapters 18–27), and cell signaling (see Chapter 32). For example, six different classes of lipids have been shown to modulate systems important in the regulation of pain responses. Each of these classes of lipids exerts its action by interacting with one or more receptor proteins. True understanding of the molecular basis for diseases and physiologic conditions may require comprehensive and simultaneous analyses of many lipid species and their respective receptors.

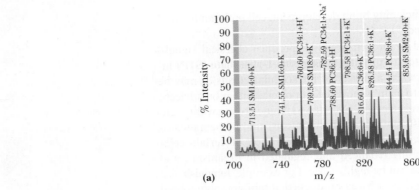

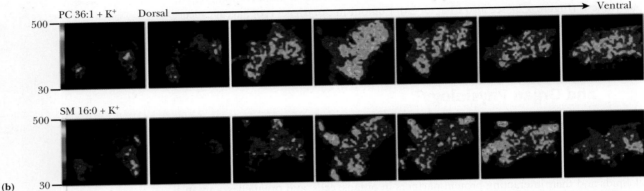

FIGURE 8.21 Identification and 3D mapping of lipids in the crab (*Cancer borealis*) brain. (a) Representative mass spectrometry (MS) spectrum of lipids from crab brain slice with the masses and identities of each peak labeled; **(b)** Comparison of 3D distribution of two different species of lipids PC 36:1 (*upper*) and SM 16:0 (*lower*). (Adapted from Chen, R., Hui, L., Sturm, R. M., and Li, L., 2009. Three dimensional mapping of neuropeptides and lipids in crustacean brain by mass spectral imaging. *J. Amer. Soc. Mass Spectrom.* **20**:1068–1077.)

SUMMARY

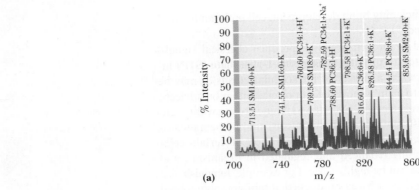 Login to OWL to develop problem-solving skills and complete online homework assigned by your professor.

Lipids are a class of biological molecules defined by low solubility in water and high solubility in nonpolar solvents. As molecules that are largely hydrocarbon in nature, lipids represent highly reduced forms of carbon and, upon oxidation in metabolism, yield large amounts of energy. Lipids are thus the molecules of choice for metabolic energy storage. The lipids found in biological systems are either hydrophobic (containing only nonpolar groups) or amphipathic (containing both polar and nonpolar groups). The hydrophobic nature of lipid molecules allows membranes to act as effective barriers to more polar molecules.

8.1 What Are the Structures and Chemistry of Fatty Acids? A fatty acid is composed of a long hydrocarbon chain ("tail") and a terminal carboxyl group ("head"). The carboxyl group is normally ionized under physiological conditions. Fatty acids occur in large amounts in biological systems but only rarely in the free, uncomplexed state. They typically are esterified to glycerol or other backbone structures.

8.2 What Are the Structures and Chemistry of Triacylglycerols? A significant number of the fatty acids in plants and animals exist in the form of triacylglycerols (also called triglycerides). Triacylglycerols are a major energy reserve and the principal neutral derivatives of glycerol found in animals. These molecules consist of a glycerol esterified with three fatty acids. Triacylglycerols in animals are found primarily in the adipose tissue (body fat), which serves as a depot or storage site for lipids. Monoacylglycerols and diacylglycerols also exist, but they are far less common than the triacylglycerols.

8.3 What Are the Structures and Chemistry of Glycerophospholipids? A 1,2-diacylglycerol that has a phosphate group esterified at carbon atom 3 of the glycerol backbone is a glycerophospholipid, also known as a phosphoglyceride or a glycerol phosphatide. This substance is the parent compound for one of the largest and most important classes of natural lipids. They are essential components of cell membranes and are found in small concentrations in other parts of the cell. All glycerophospholipids are members of the broader class of lipids known as phospholipids.

8.4 What Are Sphingolipids, and How Are They Important for Higher Animals? Sphingolipids represent another class of lipids in biological membranes. An 18-carbon amino alcohol, sphingosine, forms the backbone of these lipids rather than glycerol. Typically, a fatty acid is joined to a sphingosine via an amide linkage to form a ceramide. Sphingomyelins are a phosphorus-containing subclass of sphingolipids especially important in the nervous tissue of higher animals. A sphingomyelin is formed by the esterification of a phosphorylcholine or a phosphorylethanolamine to the 1-hydroxy group of a ceramide. Glycosphingolipids are another class of ceramide-based lipids that, like the sphingomyelins, are important components of muscle and nerve membranes in animals. Glycosphingolipids consist of a ceramide with one or more sugar residues in a β-glycosidic linkage at the 1-hydroxyl moiety.

8.5 What Are Waxes, and How Are They Used? Waxes are esters of long-chain alcohols with long-chain fatty acids. The resulting molecule can be viewed (in analogy to the glycerolipids) as having a weakly polar head group (the ester moiety itself) and a long, non-polar tail (the hydrocarbon chains). Fatty acids found in waxes are usually saturated. The alcohols found in waxes may be saturated or unsaturated and may include sterols, such as cholesterol. Waxes are water insoluble due to their predominantly hydrocarbon nature.

8.6 What Are Terpenes, and What Is Their Relevance to Biological Systems?
The terpenes are a class of lipids formed from combinations of two or more molecules of 2-methyl-1,3-butadiene, better known as isoprene (a five-carbon unit abbreviated C_5). A monoterpene (C_{10}) consists of two isoprene units, a sesquiterpene (C_{15}) consists of three isoprene units, a diterpene (C_{20}) has four isoprene units, and so on. Isoprene units can be linked in terpenes to form straight-chain or cyclic molecules, and the usual method of linking isoprene units is head to tail. Monoterpenes occur in all higher plants, whereas sesquiterpenes and diterpenes are less widely known.

8.7 What Are Steroids, and What Are Their Cellular Functions?
A large and important class of terpene-based lipids is the steroids. This molecular family, whose members affect an amazing array of cellular functions, is based on a common structural motif of three 6-membered rings and one 5-membered ring all fused together. Cholesterol is the most common steroid in animals and the precursor for all other animal steroids. The numbering system for cholesterol applies to all such molecules. The polyprenyl nature of this compound is particularly evident in the side chain. Many steroids contain an oxygen at C-3, either a hydroxyl group in sterols or a carbonyl group in other steroids. The methyl groups at positions 10 and 13 and the alkyl group at position 17 are usually oriented on the same side of the steroid nucleus, the β-orientation. Alkyl groups that extend from the other side of the steroid backbone are in an α-orientation. Cholesterol is a principal component of animal cell plasma membranes. Steroids derived from cholesterol in animals include five families of hormones (the androgens, estrogens, progestins, glucocorticoids, and mineralocorticoids) and bile acids.

8.8 How Do Lipids and Their Metabolites Act as Biological Signals?
Modification and breakdown of cellular lipids produce an eclectic assortment of substances that act as powerful chemical signals. Signal molecules typically initiate a cascade of reactions with multiple possible effects. The creation and breakdown of signal molecules is almost always carefully timed and regulated. Phospholipases initiate the production of a variety of lipid signals, including arachidonic acid (the precursor to eicosanoids), lysophosphatidic acid, inositol-1,4,5-trisphosphate, and diacylglycerol.

8.9 What Can Lipidomics Tell Us about Cell, Tissue, and Organ Physiology?
The comprehensive analysis of lipids and their interacting protein partners in organs, cells, and organelles is termed lipidomics. A typical cell may contain more than a thousand different lipids. Complete understanding of lipid function, as well as alteration of such function in disease states, will require the determination of which lipids are present and in what concentrations in every intracellular location. The same knowledge will be needed about each lipid's interaction partners.

FOUNDATIONAL BIOCHEMISTRY Things You Should Know After Reading Chapter 8.

- The structures and chemical features of the saturated and unsaturated fatty acids

- The identities and structures of the essential fatty acids

- The fat compositions of common dietary oils

- The structures and chemical features of representative *trans*-fatty acids

- The structures and chemical features of triacylglycerols

- The structures and chemical features of glycerophospholipids

- The prochirality of glycerol and the *sn*-numbering system for glycerophospholipids

- The structures of ether glycerophospholipids, especially plasmalogens

- The structures and chemical features of sphingolipids, including ceramides, sphingomyelins, cerebrosides, and gangliosides

- The structures and properties of waxes

- The structure of isoprene and the structural and chemical features of terpenes

- The chemical features and cellular properties of steroids, including cholesterol, the androgens, the progestins, the corticoids, and the bile acids

- The roles of lipids and their metabolites (including sphingosine-1-phosphate, IP_3, and diacylglycerol) as biological signals

- The identities and chemical and biological actions of phospholipases A_1, A_2, C, and D

- The role of lipidomics and lipidomic analyses in elucidation of the functions of lipids in cells, tissues, and organs in complex organisms

---

PROBLEMS

Answers to all problems are at the end of this book. Detailed solutions are available in the *Student Solutions Manual, Study Guide, and Problems Book*.

1. **Drawing Structures of Triacylglycerols** Draw the structures of (a) all the possible triacylglycerols that can be formed from glycerol with stearic and arachidonic acid and (b) all the phosphatidylserine isomers that can be formed from palmitic and linolenic acids.

2. **Distinguishing structures of Glycerolipids and Sphingolipids** Describe in your own words the structural features of
   a. a ceramide and how it differs from a cerebroside.
   b. a phosphatidylethanolamine and how it differs from a phosphatidylcholine.
   c. an ether glycerophospholipid and how it differs from a plasmalogen.
   d. a ganglioside and how it differs from a cerebroside.
   e. testosterone and how it differs from estradiol.

3. **Identifying Glycerophospholipids** From your memory of the structures, name the glycerophospholipids that
   a. carry a net positive charge at pH 7.
   b. carry a net negative charge at pH 7.
   c. have zero net charge at pH 7.

4. **Comparing Health Effects of Cholesterol and Plant Sterols** Compare and contrast two individuals, one whose diet consists largely

of meats containing high levels of cholesterol and the other whose diet is rich in plant sterols. Are their risks of cardiovascular disease likely to be similar or different? Explain your reasoning.

5. **Evaluating Organic Secretions of Trees** James G. Watt, Secretary of the Interior (1981–1983) in Ronald Reagan's first term, provoked substantial controversy by stating publicly that trees cause significant amounts of air pollution. Based on your reading of this chapter, evaluate Watt's remarks.

6. **Evaluating Nutritional Benefits of Foods** In a departure from his usual and highly popular westerns, author Louis L'Amour wrote a novel in 1987, *Last of the Breed* (Bantam Press), in which a military pilot of Native American ancestry is shot down over the former Soviet Union and is forced to use the survival skills of his ancestral culture to escape his enemies. On the rare occasions when he is able to trap and kill an animal for food, he selectively eats the fat, not the meat. Based on your reading of this chapter, what is his reasoning for doing so?

7. **Exploring the Action of Phospholipase A₂** As you read Section 8.7, you might have noticed that phospholipase A$_2$, the enzyme found in rattlesnake venom, is also the enzyme that produces essential and beneficial lipid signals in most organisms. Explain the differing actions of phospholipase A$_2$ in these processes.

8. **Evaluating the Action of Warfarin** Visit a grocery store near you, stop by the rodent poison section, and examine a container of warfarin or a related product. From what you can glean from the packaging, how much warfarin would a typical dog (40 lbs) have to consume to risk hemorrhages and/or death?

9. **Understanding Isoprene Structures** Refer to Figure 8.13 and draw each of the structures shown and try to identify the isoprene units in each of the molecules. (Note that there may be more than one correct answer for some of these molecules, unless you have the time and facilities to carry out ^{14}C labeling studies with suitable organisms.)

10. **Evaluating the Caloric Content of Stored Fat** (Integrates with Chapter 3.) As noted in the Deeper Look box on polar bears, a polar bear may burn as much as 1.5 kg of fat resources per day. What weight of seal blubber would you have to ingest if you were to obtain all your calories from this energy source?

11. **Assessing the Fat Composition of Desserts** If you are still at the grocery store working on problem 8, stop by the cookie shelves and choose your three favorite cookies from the shelves. Estimate how many calories of fat, and how many other calories from other sources, are contained in 100 g of each of these cookies. Survey the ingredients listed on each package, and describe the contents of the package in terms of (a) saturated fat, (b) cholesterol, and (c) *trans* fatty acids. (Note that food makers are required to list ingredients in order of decreasing amounts in each package.)

12. **Evaluating the Structures of Sterols** Describe all of the structural differences between cholesterol and stigmasterol.

13. **Understanding the Functions of Steroid Hormones** Describe in your own words the functions of androgens, glucocorticoids, and mineralocorticoids.

14. **Understanding Structures and Properties of Terpenes** Look through your refrigerator, your medicine cabinet, and your cleaning solutions shelf or cabinet, and find at least three commercial products that contain fragrant monoterpenes. Identify each one by its scent and then draw its structure.

15. **Understanding the Chemistry of Soaps** Our ancestors kept clean with homemade soap (page 237), often called "lye soap." Go to *http://www.wikihow.com/Make-Your-Own-Soap* and read the procedure for making lye soap from vegetable oils and lye (sodium hydroxide). What chemical process occurs in the making of lye soap? Draw reactions to explain. How does this soap work as a cleaner?

16. **Understanding the Chemistry of Common Foods** Mayonnaise is mostly vegetable oil and vinegar. So what's the essential difference between oil and vinegar salad dressing and mayonnaise? Learn for yourself: Combine a half cup of pure vegetable oil (olive oil will work) with two tablespoons of vinegar in a bottle, cap it securely, and shake the mixture vigorously. What do you see? Now let the mixture sit undisturbed for an hour. What do you see now? Add one egg yolk to the mixture, and shake vigorously again. Let the mixture stand as before. What do you see after an hour? After two hours? Egg yolk is rich in phosphatidylcholine. Explain why the egg yolk caused the effect you observed.

17. **Evaluating the Physiological Benefits of Stanol Esters** The cholesterol-lowering benefit of stanol-ester margarine is only achieved after months of consumption of stanol esters (see graph, page 253). Suggest why this might be so. Suppose dietary sources represent approximately 25% of total serum cholesterol. Based on the data in the graph, how effective are stanol esters at preventing uptake of dietary cholesterol?

18. **Comparing Therapies for Cholesterol Lowering** Statins are cholesterol-lowering drugs that block cholesterol synthesis in the human liver (see Chapter 24). Would you expect the beneficial effects of stanol esters and statins to be duplicative or additive? Explain.

19. **Assessing the Sterol and Stanol Content of Food Products** If most plant-derived food products contain plant sterols and stanols, would it be as effective (for cholesterol-lowering purposes) to simply incorporate plant fats in one's diet as to use a sterol- or stanol-fortified spread like Benecol? Consult a suitable reference (for example, *http://lpi.oregonstate.edu/infocenter/phytochemicals/sterols/#sources* at the Linus Pauling Institute) to compose your answer.

20. **Evaluating the History and Culture of Anabolic Steroids** Tetrahydrogestrinone is an anabolic steroid. It was banned by the U.S. Food and Drug Administration in 2003, but it has been used illegally since then by athletes to increase muscle mass and strength. Nicknamed "The Clear," it has received considerable attention in high-profile steroid-abuse cases among athletes such as baseball player Barry Bonds and track star Marion Jones. Use your favorite Web search engine to learn more about this illicit drug. How is it synthesized? Who is "the father of prohormones" who first synthesized it? Why did so many prominent athletes use The Clear (and its relative, "The Cream") when less expensive and more commonly available anabolic steroids are in common use? (Hi nt: There are at least two answers to this last question.)

Tetrahydrogestrinone

Preparing for the MCAT® Exam

21. Make a list of the advantages polar bears enjoy from their nonpolar diet. Why wouldn't juvenile polar bears thrive on an exclusively nonpolar diet?

22. Snake venom phospholipase A$_2$ causes death by generating membrane-soluble anionic fragments from glycerophospholipids. Predict the fatal effects of such molecules on membrane proteins and lipids.

ActiveModels Problem

23. Examine the ActiveModel for N-myristoyltransferase and explain the mechanism of N-myristoylation.

FURTHER READING

General

Robertson, R. N., 1983. *The Lively Membranes.* Cambridge: Cambridge University Press.

Seachrist, L., 1996. A fragrance for cancer treatment and prevention. *The Journal of NIH Research* **8**:43.

Vance, D. E., and Vance, J. E. (eds.), 1985. *Biochemistry of Lipids and Membranes.* Menlo Park, CA: Benjamin/Cummings.

Sterols

Anderson, S., Russell, D. W., and Wilson, J. D., 1996. 17β-Hydroxysteroid dehydrogenase 3 deficiency. *Trends in Endocrinology and Metabolism* **7**:121–126.

DeLuca, H. F., and Schneos, H. K., 1983. Vitamin D: Recent advances. *Annual Review of Biochemistry* **52**:411–439.

Denke, M. A., 1995. Lack of efficacy of low-dose sitostanol therapy as an adjunct to a cholesterol-lowering diet in men with moderate hypercholesterolemia. *American Journal of Clinical Nutrition* **61**:392–396.

Thompson, G., and Grundy, S., 2005. History and development of plant sterol and stanol esters for cholesterol-lowering purposes. *American Journal of Cardiology* **96**:3D–9D.

Vanhanen, H. T., Blomqvist, S., Ehnholm, C., et al., 1993. Serum cholesterol, cholesterol precursors, and plant sterols in hypercholesterolemic subjects with different apoE phenotypes during dietary sitostanol ester treatment. *Journal of Lipid Research* **34**:1535–1544.

Isoprenes and Prenyl Derivatives

Dowd, P., Ham, S.-W., Naganathan, S., and Hershline, R., 1995. The mechanism of action of vitamin K. *Annual Review of Nutrition* **15**:419–440.

Hirsh, J., Dalen, J. E., Deykin, D., Poller, L., and Bussey, H., 1995. Oral anticoagulants: Mechanism of action, clinical effectiveness, and optimal therapeutic range. *Chest* **108**:231S–246S.

Sharkey, T. D., 1995. Why plants emit isoprene. *Nature* **374**:769.

Sharkey, T. D., 1996. Emission of low molecular-mass hydrocarbons from plants. *Trends in Plant Science* **1**:78–82.

Eicosanoids

Chakrin, L. W., and Bailey, D. M., 1984. *The Leukotrienes—Chemistry and Biology.* Orlando: Academic Press.

Keuhl, F. A., and Egan, R. W., 1980. Prostaglandins, arachidonic acid and inflammation. *Science* **210**:978–984.

Trans Fatty Acids

Katan, M. B., Zock, P. L., and Mensink, R. P., 1995. *Trans* fatty acids and their effects on lipoproteins in humans. *Annual Review of Biochemistry* **15**:473–493.

Lipids of Archaea

Hanford, M., and Peebles, T., 2002. Archaeal tetraetherlipids: Unique structures and applications. *Applied Biochemistry and Biotechnology* **97**:45–62.

Lipid Alterations in Disease States

Malan, T. P., and Porreca, F., 2005. Lipid mediators regulating pain sensitivity. *Prostaglandins and Other Lipid Mediators* **77**:123–130.

Smith, L. E. H., and Connor, K. M., 2005. A radically twisted lipid regulates vascular death. *Nature Medicine* **11**:1275–1276.

Lipidomics

German, J., Gillies, L., Smilowitz, J., Zivkovic, A., and Watkins, S., 2007. Lipidomics and lipid profiling in metabolomics. *Current Opinion in Lipidology* **18**:66–71.

Muralikrishna, R., Hatcher, J., and Dempsey, R., 2006. Lipids and lipidomics in brain injury and diseases. *AAPS Journal* **8**:E314–E321.

Shevchenko, A., and Simons, K., 2010. Lipidomics: coming to grips with lipid diversity. *Nature Reviews Molecular Cell Biology* **11**:593–598.

Van Meer, G., 2005. Cellular lipidomics. *EMBO Journal* **24**:3159–3165.

Weak, M. R., 2005. The emerging field of lipidomics. *Nature Reviews Drug Discovery* **4**:594–610.

Wenk, M. R., 2010. Lipidomics: New tools and applications. *Cell* **143**:888–895.

Lipids as Signaling Molecules

Brinkman, V., Billich, A., Baumruker, T., Heining, P., Schmouder, R., Francis, G., Aradhye, S., and Burtin, P., 2010. Fingolimod (FTY720): discovery and development of an oral drug to treat multiple sclerosis. *Nature Reviews Drug Discovery* **9**:883–897.

Connor, M., Vaughn, C. W., and Vandenberg, R. J., 2010. N-Acyl amino acids and N-acyl neurotransmitter conjugates: neuromodulators and probes for new drug targets. *British Journal of Pharmacology* **160**:1857–1871.

Eyster, K., 2007. The membrane and lipids as integral participants in signal transduction: Lipid signal transduction for the non-lipid biochemist. *Advances in Physiology Education* **31**:5–16.

Fernandis, A., and Wenk, M., 2007. Membrane lipids as signaling molecules. *Current Opinion in Lipidology* **18**:121–128.

Maccarone, M., Gasperi, V., Catani, M., Diep, T., Dainese, E., and Hansen, H. S., 2010. The endocannabinoid system and its relevance for nutrition. *Annual Review of Nutrition* **30**:423–440.

Sphingolipids

Breslow, D. S., and Weissman, J. S., 2010. Membranes in balance: Mechanisms of sphingolipid homeostasis. *Molecular Cell* **40**:267–279.

Fyrst, H., and Saba, J. D., 2010. An update on sphingosine-1-phosphate and other sphingolipid mediators. *Nature Chemical Biology* **6**:489–498.

Pyne, N. J., and Pyne, S., 2010. Sphingosine-1-phosphate and cancer. *Nature Reviews Cancer* **10**:489–503.

Stancevic, B., and Kolesnick, R., 2010. Ceramide-rich platforms in transmembrane signaling. *FEBS Letters* **584**:1728–1740.

Walther, T. C., 2010. Keeping sphingolipid levels nORMal. *Proceedings of the National Academy of Sciences, USA* **107**:5701–5702.

9

Membranes and Membrane Transport

Frog eggs are macroscopic facsimiles of microscopic cells. All cells are surrounded by a thin, ephemeral yet stable membrane. ▶

It takes a membrane to make sense out of disorder in biology.

Lewis Thomas
The World's Biggest Membrane,
The Lives of a Cell (1974)

Dale Chihuly, Chartreuse Venetian, 1990/Photo by Roger Schreiber

OWL Online homework and a Student Self Assessment for this chapter may be assigned in OWL.

ESSENTIAL QUESTION

Membranes serve a number of essential cellular functions. They constitute the boundaries of cells and intracellular organelles, and they provide a surface where many important biological reactions and processes occur. Membranes have proteins that mediate and regulate the transport of metabolites, macromolecules, and ions. Hormones and many other biological signal molecules and regulatory agents exert their effects via interactions with membranes. Photosynthesis, electron transport, oxidative phosphorylation, muscle contraction, and electrical activity all depend on membranes and membrane proteins. As an index of the importance of membranes, note that 30 percent of the genes of *Mycoplasma genitalium* are thought to encode membrane proteins.

What are the properties and characteristics of biological membranes that account for their broad influence on cellular processes and transport?

Membranes are key structural and functional elements of cells. All cells have a cytoplasmic membrane, or *plasma membrane,* that functions (in part) to separate the cytoplasm from the surroundings. The plasma membrane is also responsible for (1) the exclusion of certain toxic ions and molecules from the cell, (2) the accumulation of cell nutrients, and (3) energy transduction. It functions in (4) cell locomotion, (5) reproduction, (6) signal transduction processes, and (7) interactions with molecules or other cells in the vicinity.

Even the plasma membranes of prokaryotic cells are complex (Figure 9.1). With no intracellular organelles to divide and organize the work, bacteria carry out processes either at the plasma membrane or in the cytoplasm itself. Eukaryotic cells, however, contain numerous intracellular organelles that perform specialized tasks. Nucleic acid

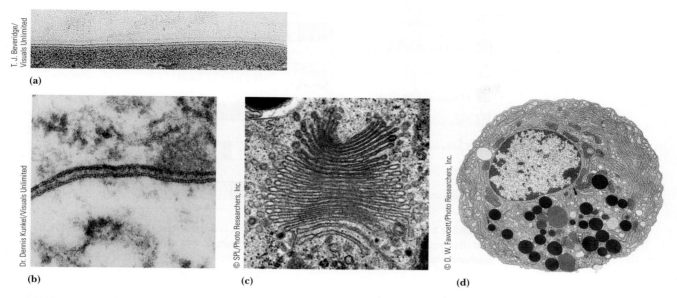

FIGURE 9.1 Electron micrographs of several different membrane structures: **(a)** Plasma membrane of *Menoidium,* a protozoan. **(b)** Two plasma membranes from adjacent neurons in the central nervous system. **(c)** Golgi apparatus. **(d)** Many membrane structures are evident in pancreatic acinar cells.

biosynthesis is handled in the nucleus; mitochondria are the site of electron transport, oxidative phosphorylation, fatty acid oxidation, and the tricarboxylic acid cycle; and secretion of proteins and other substances is handled by the endoplasmic reticulum (ER) and the Golgi apparatus. This partitioning of labor is not the only contribution of the membranes in these cells. Many of the processes occurring in these organelles (or in the prokaryotic cell) actively involve membranes. Thus, some of the enzymes involved in nucleic acid metabolism are membrane associated. The electron transfer chain and its associated system for ATP synthesis are embedded in the mitochondrial membrane. Many enzymes responsible for aspects of lipid biosynthesis are located in the ER membrane.

This chapter discusses the composition, structure, and dynamic processes of biological membranes.

9.1 What Are the Chemical and Physical Properties of Membranes?

Water's tendency to form hydrogen bonds and share in polar interactions, and the hydrophobic effect, which promotes self-association of lipids in water to maximize entropy, are the basis for the interactions of lipids and proteins to form membranes. These forces drive amphiphilic glycerolipids, sphingolipids, and sterols to form membrane structures in water, and these forces facilitate the association of proteins (and thus myriad biological functions) with membranes. A symphony of molecular events over a range of times from picoseconds to many seconds results in the movement of lipids and proteins across and between membranes; coordinates reactions at or in the membrane and the transport of ions, sugars, and amino acids across membranes; and organizes and directs hundreds of cell signaling events.

The Composition of Membranes Suits Their Functions

Biological membranes may contain as much as 75% to 80% protein (and only 20% to 25% lipid) or as little as 15% to 20% protein. Membranes that carry out many enzyme-catalyzed reactions and transport activities (the inner mitochondrial membrane,

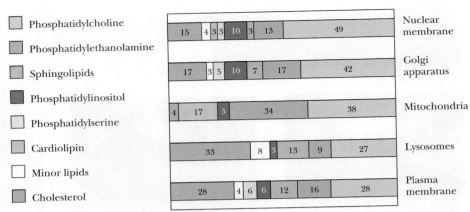

FIGURE 9.2 The lipid composition of rat liver cell membranes, in weight percent. (Adapted from Andreoli, T. E., 1987. *Membrane Physiology, 2nd ed.* Chapter 27, Table II, and Daum, G., 1985. Lipids of mitochondria. *Biochimica et Biophysica Acta* **822**:1–42.)

chloroplast membranes, and the plasma membrane of *Escherichia coli*, for example) are typically richer in protein, whereas membranes that carry out fewer protein-related functions (myelin sheaths, the protective coating around neurons, for example) are richer in lipid.

Cellular mechanisms adjust lipid composition to functional needs. Thus, for example, the lipid makeup of red blood cell membranes is consistent across species, whereas the lipid complement of different (specialized) membranes within a particular cell type (rat liver, Figure 9.2) reflects differences of function. Plasma membranes are enriched in cholesterol but do not contain diphosphatidylglycerol (cardiolipin), whereas mitochondria contain considerable amounts of cardiolipin (essential for some mitochondrial proteins) and no cholesterol. The protein components of membranes vary even more greatly than their lipid compositions.

Lipids Form Ordered Structures Spontaneously in Water

Monolayers and Micelles Amphipathic lipids spontaneously form a variety of structures when added to aqueous solution. All these structures form in ways that minimize contact between the hydrophobic lipid chains and the aqueous milieu. For example, when small amounts of a fatty acid are added to an aqueous solution, a monolayer is formed at the air–water interface, with the polar head groups in contact with the water surface and the hydrophobic tails in contact with the air (Figure 9.3). Few lipid molecules are found as monomers in solution.

Further addition of fatty acid eventually results in the formation of micelles. **Micelles** formed from an amphipathic lipid in water position the hydrophobic tails in the center of the lipid aggregation with the polar head groups facing outward. Amphipathic molecules that form micelles are characterized by a unique **critical micelle concentration, or CMC.** Below the CMC, individual lipid molecules predominate. Nearly all the lipid added above the CMC, however, spontaneously forms micelles. Micelles are the preferred form of aggregation in water for detergents and soaps. Some typical CMC values are listed in Figure 9.4.

Lipid Bilayers **Lipid bilayers** consist of back-to-back arrangements of monolayers (Figure 9.3). The nonpolar portions of the lipids face the middle of the bilayer, with the polar head groups arrayed on the bilayer surface. Phospholipid bilayers form rapidly and spontaneously when phospholipids are added to water, and they are stable structures in aqueous solution. As opposed to micelles, which are small, self-limiting structures of a few hundred molecules, bilayers may form spontaneously over large areas (10^8 nm^2 or more). Because exposure of the edges of the bilayer to solvent is highly

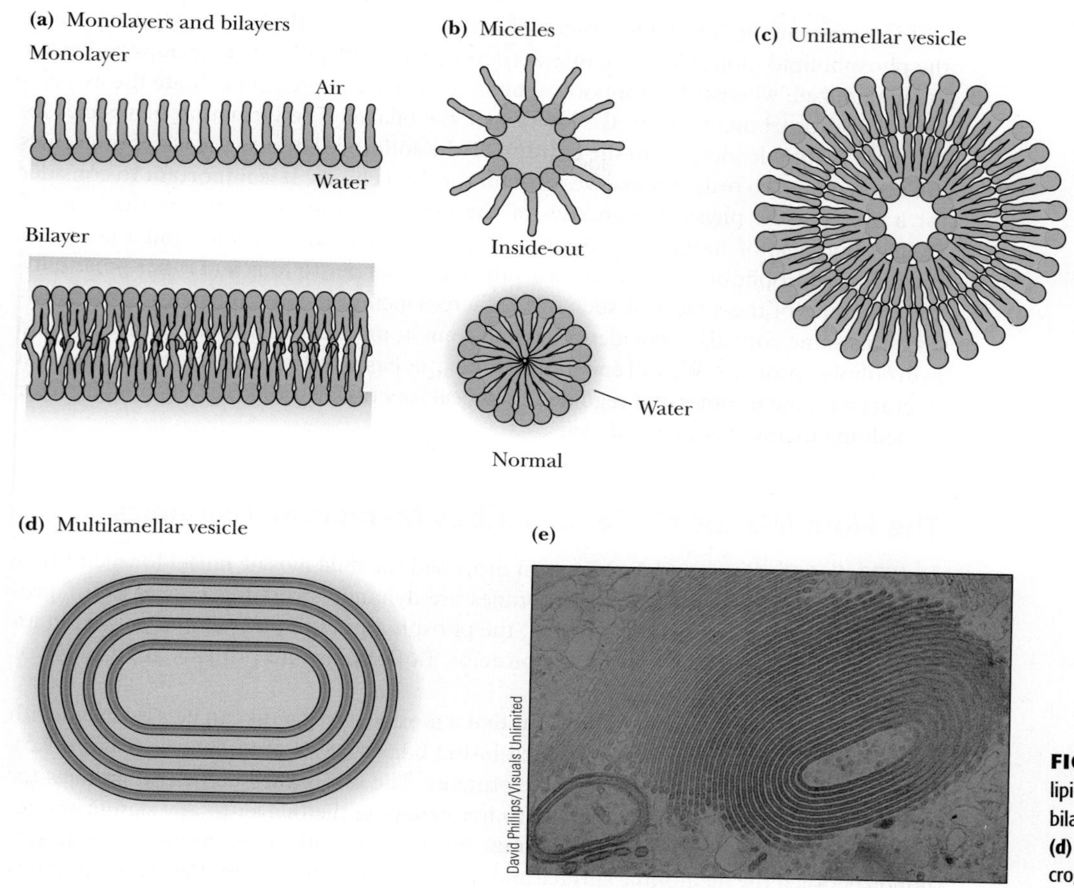

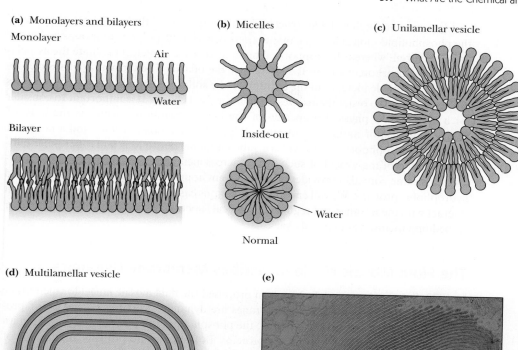

(a) Monolayers and bilayers

Monolayer

Air

Water

Bilayer

(b) Micelles

Inside-out

Water

Normal

(c) Unilamellar vesicle

(d) Multilamellar vesicle

(e)

David Phillips/Visuals Unlimited

FIGURE 9.3 Several spontaneously formed lipid structures. Drawings of **(a)** monolayers and bilayers, **(b)** micelles, **(c)** a unilamellar vesicle, **(d)** a multilamellar vesicle, and **(e)** an electron micrograph of a multilamellar Golgi structure.

Structure	M_r	CMC	Micelle M_r
Triton X-100			
$CH_3-C(CH_3)_2-CH_2-C(CH_3)_2-\phi-(OCH_2CH_2)_{10}-OH$	625	0.24 mM	90–95,000
Octyl glucoside			
glucose—$O-(CH_2)_7-CH_3$	292	25 mM	
$C_{12}E_8$ (Dodecyl octaoxyethylene ether)			
$C_{12}H_{25}-(OCH_2CH_2)_8-OH$	538	0.071 mM	

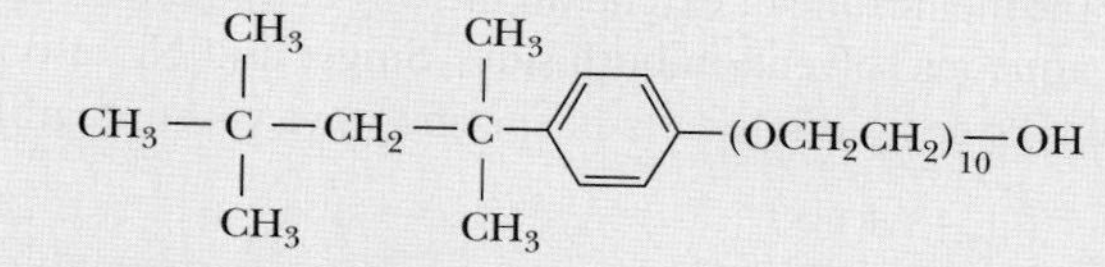

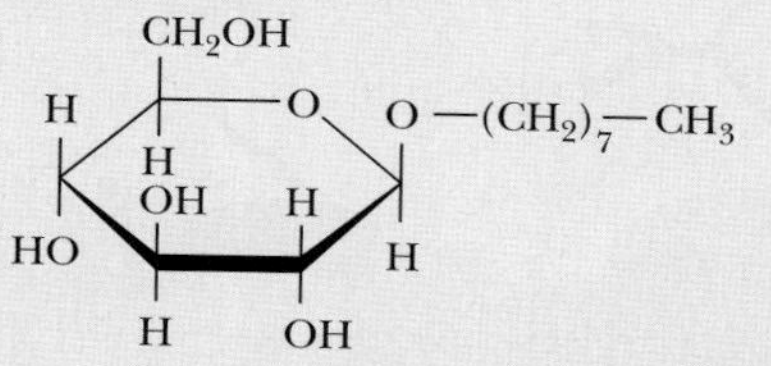

FIGURE 9.4 The structures of some common detergents and their physical properties. Micelles formed by detergents can be quite large. Triton X-100, for example, typically forms micelles with a total molecular mass of 90 to 95 kD. This corresponds to approximately 150 molecules of Triton X-100 per micelle.

unfavorable, extensive bilayers normally wrap around themselves and form closed vesicles (Figure 9.3). The nature and integrity of these vesicle structures are very much dependent on the lipid composition. Physico-chemical studies in the laboratory with these substances reveal that phospholipids can form either *unilamellar vesicles* (with a single lipid bilayer), known as *liposomes,* or *multilamellar vesicles.* These latter structures are reminiscent of the layered structure of onions.

Liposomes are highly stable structures, a consequence of the amphipathic nature of the phospholipid molecule. Ionic interactions between the polar head groups and water are maximized, whereas hydrophobic interactions (see Chapter 2) facilitate the association of hydrocarbon chains in the interior of the bilayer. The formation of vesicles results in a favorable increase in the entropy of the solution, because the water molecules are not required to order themselves around the lipid chains. It is important to consider for a moment the physical properties of the bilayer membrane, which is the basis of vesicles and also of natural membranes. Bilayers have a polar surface and a nonpolar core. This hydrophobic core provides a substantial barrier to ions and other polar entities. The rates of movement of such species across membranes are thus quite slow. However, this same core also provides a favorable environment for nonpolar molecules and hydrophobic proteins. We will encounter numerous cases of hydrophobic molecules that interact with membranes and regulate biological functions in some way by binding to or embedding themselves in membranes.

The Fluid Mosaic Model Describes Membrane Dynamics

In 1972, S. J. Singer and G. L. Nicolson proposed the **fluid mosaic model** for membrane structure, which suggested that membranes are dynamic structures composed of proteins and phospholipids. In this model, the phospholipid bilayer is a *fluid* matrix, in essence, a two-dimensional solvent for proteins. Both lipids and proteins are capable of rotational and lateral movement.

Singer and Nicolson also pointed out that a mosaic of proteins can be associated with the surface of this bilayer or embedded in the bilayer to varying degrees (Figure 9.5). They defined two classes of membrane proteins. The first, called **peripheral proteins** (or **extrinsic proteins**), includes those that do not penetrate the bilayer to any significant degree and are associated with the membrane by virtue of ionic interactions and hydrogen bonds between the membrane surface and the surface of the protein. Peripheral proteins can be dissociated from the membrane by treatment with salt solutions or by changes in pH (treatments that disrupt hydrogen bonds and ionic interactions). **Integral proteins** (or **intrinsic proteins**), in contrast, possess hydrophobic surfaces that can readily penetrate the lipid bilayer itself, as well as surfaces that prefer contact with the aqueous medium. These proteins can either insert into the membrane or extend all the way across the membrane and expose themselves to the aqueous solvent on both sides. Singer and Nicolson also suggested that a portion of the bilayer lipid interacts in specific ways with integral

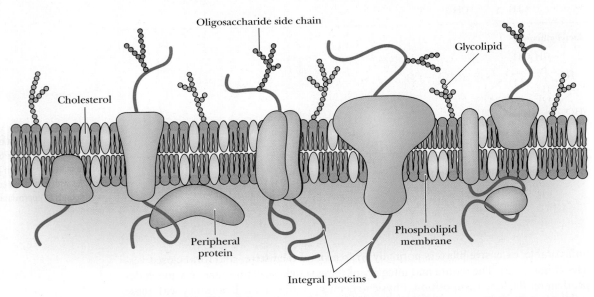

FIGURE 9.5 The fluid mosaic model of membrane structure proposed by S. J. Singer and G. L. Nicolson. In this model, the lipids and proteins are assumed to be mobile; they can diffuse laterally in the plane of the membrane. Transverse motion may also occur, but it is much slower.

membrane proteins and that these interactions might be important for the function of certain membrane proteins. Because of these intimate associations with membrane lipid, integral proteins can be removed from the membrane only by agents such as detergents or organic solvents that are capable of breaking up the hydrophobic interactions within the lipid bilayer itself. The fluid mosaic model became the paradigm for modern studies that have advanced our understanding of membrane structure and function.

The Thickness of a Membrane Depends on Its Components Electron micrographs of typical cellular membranes show the thickness of the entire membrane—including lipid bilayer and embedded protein—to be 50 Å or more. Electron microscopy, NMR, and X-ray and neutron diffraction measurements have shown that membrane thickness is influenced by the particular lipids and proteins in the membrane. The thickness of a phospholipid bilayer made from dipalmitoyl phosphatidylcholine, measured as the phosphorus-to-phosphorus spacing, is about 37 Å, and the hydrophobic phase of such membranes is approximately 26 Å thick. Natural membranes are thicker overall than simple lipid bilayers because many membrane proteins extend out of the bilayer significantly.

Among the known membrane protein structures, there is considerable variation in the hydrophobic surface perpendicular to the membrane plane. If the hydrophobic surface of the protein is larger or smaller than the lipid bilayer, the thickness of the lipid bilayer must be increased or decreased. The change in bilayer thickness due to membrane proteins can be as much as 5 Å.

Lipid Chains May Bend and Tilt in the Membrane The long hydrocarbon chains of lipids are typically portrayed as more or less perpendicular to the membrane plane (Figure 9.3). In fact, the hydrocarbon tails of phospholipids may tilt and bend and adopt a variety of orientations. Typically, the portions of a lipid chain near the membrane surface lie most nearly perpendicular to the membrane plane, and lipid chain ordering decreases toward the end of the chain (toward the middle of the bilayer).

Membranes Are Crowded with Many Different Proteins Membranes are crowded places, with a large number of proteins either embedded or associated in some way. The *E. coli* genome codes for more than a thousand membrane proteins. Moreover, as more membrane protein structures are determined (Figure 9.6), it has become apparent that many membrane proteins have large structures extending outside the lipid bilayer that share steric contacts and other interactions. Donald Engelman has suggested that most membranes are more crowded than first portrayed in Singer and Nicolson's model (Figure 9.7).

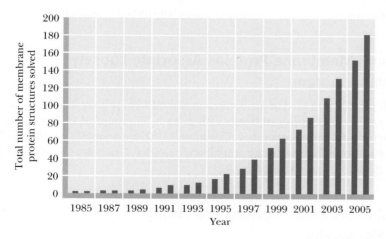

FIGURE 9.6 Membrane protein structures, by year published. By August, 2011, the number had reached 302. (Data from the Web site Membrane Proteins of Known 3D Structure at the laboratory of Stephen White, http://blanco.biomol.uci.edu/mpstruc/listAll/list, and from the Web site of Hartmut Michel, http://www.mpibp-frankfurt.mpg.de/michel/public/memprotstruct.html.)

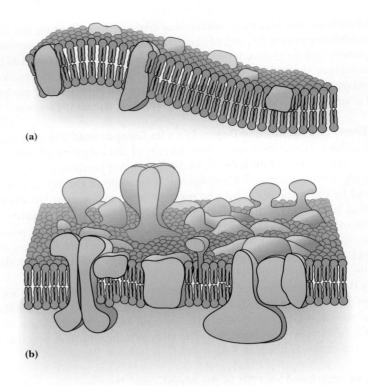

FIGURE 9.7 Comparison of (a) the Singer-Nicolson model with (b) an updated model for membrane structure, as proposed by Donald Engelman. (Adapted from Engelman, D., 2005. Membranes are more mosaic than fluid. *Nature* **438:**578–580.)

9.2 What Are the Structure and Chemistry of Membrane Proteins?

Although the lipid bilayer constitutes the fundamental structural unit of all biological membranes, proteins carry out essentially all of the active functions of membranes. Singer and Nicolson defined peripheral proteins as globular proteins that interact with the membrane mainly through electrostatic and hydrogen-bonding interactions, and integral proteins as those that are strongly associated with the lipid bilayer. Another class of proteins not anticipated by Singer and Nicolson, the **lipid-anchored proteins,** is important in a variety of functions in different cells and tissues. These proteins associate with membranes by means of a variety of covalently linked lipid anchors.

Peripheral Membrane Proteins Associate Loosely with the Membrane

Peripheral proteins can bind to membranes in several ways (Figure 9.8). They may form ionic interactions and hydrogen bonds with polar head groups of membrane lipids or with other (integral) proteins, or they may interact with the nonpolar membrane core by inserting a hydrophobic loop or an amphipathic α-helix. Examples of each of these interaction types are shown in Figure 9.9.

Integral Membrane Proteins Are Firmly Anchored in the Membrane

Hundreds of structures of integral membrane proteins are now available in the Protein Data Bank, and the number of membrane protein structures is doubling about every 3 years. The known structures show a surprising diversity, but in all cases the portions

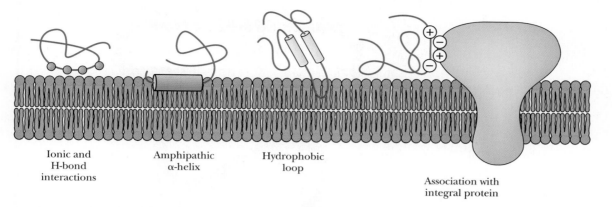

FIGURE 9.8 Four possible modes for the binding of peripheral membrane proteins.

Ionic and H-bond interactions

Amphipathic α-helix

Hydrophobic loop

Association with integral protein

of the protein in contact with the nonpolar core of the lipid bilayer are dominated by α-helices or β-sheets, because these secondary structures neutralize the highly polar N—H and C=O functions of the peptide backbone through H-bond formation.

Proteins with a Single Transmembrane Segment In proteins that are anchored by a single hydrophobic segment, that segment typically takes the form of an α-helix. One of the best examples is **glycophorin.** Most of glycophorin's mass is oriented on the outside surface of the red blood cell, exposed to the aqueous milieu (Figure 9.10). Hydrophilic oligosaccharide units are attached to this extracellular domain. These oligosaccharide groups constitute the ABO and MN blood group antigenic specificities of the red cell. Glycophorin has a total molecular weight of about 31,000 and is approximately 40% protein and 60% carbohydrate. The glycophorin primary structure consists of a segment of 19 hydrophobic amino acid residues with a short hydrophilic sequence on one end and a longer hydrophilic sequence on the other end. The 19-residue sequence is just the right length to span the cell membrane if it is coiled in the shape of an α-helix.

Monoamine oxidase from the mitochondrial outer membrane is another typical single transmembrane–segment protein (Figure 9.11); this enzyme is the target for many antidepressant and neuroprotective drugs. Each monomer of the dimeric protein binds to the membrane through a C-terminal transmembrane α-helix. Residues in two loops (Pro-109 and Ile-110 in the 99–112 loop and Phe-481, Leu-482, Leu-486, and Pro-487 in the 481–488 loop) also provide nonpolar residues that participate in membrane binding.

Approximately 10% to 30% of transmembrane proteins have a single helical transmembrane segment. In animals, many of these function as cell surface receptors for extracellular signaling molecules or as recognition sites that allow the immune system to recognize and distinguish cells of the host organism from invading foreign cells or viruses. The proteins that represent the *major transplantation antigens H2* in mice (Figure 9.11) and *human leukocyte associated (HLA) proteins* in humans are members of this class. Other such proteins include the *surface immunoglobulin receptors* on B lymphocytes and the *spike proteins* of many membrane viruses. The function of many of these proteins depends primarily on their extracellular domain; thus, the segment facing the intracellular surface is often a shorter one.

Proteins with Multiple Transmembrane Segments Most integral transmembrane proteins cross the lipid bilayer more than once. These **multi-spanning** membrane proteins typically have 2 to 12 transmembrane segments, and they carry out a variety of cellular functions (Figure 9.12). A well-characterized example of such a protein is **bacteriorhodopsin** (Figure 9.13), which clusters in purple patches in the membrane of the archaeon *Halobacterium halobium.* The name *Halobacterium* refers to the fact that this prokaryote thrives in solutions having high concentrations of sodium chloride, such as the salt ponds of San Francisco Bay. *Halobacterium* carries out a light-driven proton transport by means of bacteriorhodopsin, named in reference to its spectral similarities to rhodopsin in the rod outer segments of the mammalian retina. The amino acid

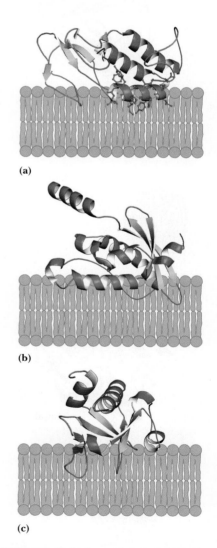

(a)

(b)

(c)

FIGURE 9.9 Models for membrane association of peripheral proteins. **(a)** Bee venom phospholipase A₂ (pdb id = 1POC), **(b)** p40 phox PX domain of NADH oxidase (pdb id = 1H6H), and **(c)** PH domain of phospholipase Cδ (pdb id = 1MAI).

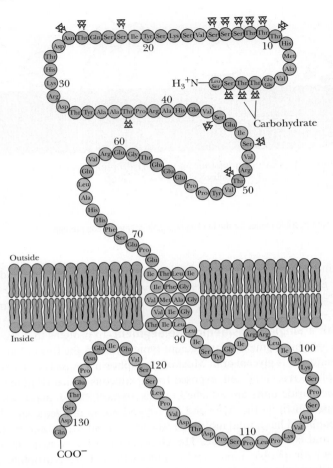

FIGURE 9.10 Glycophorin A spans the membrane of the human erythrocyte via a single α-helical transmembrane segment. The C-terminus of the peptide faces the cytosol of the erythrocyte; the N-terminal domain is extracellular. Points of attachment of carbohydrate groups are indicated by triangles.

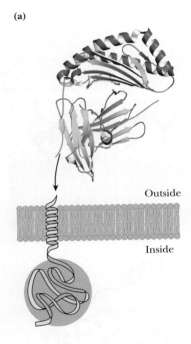

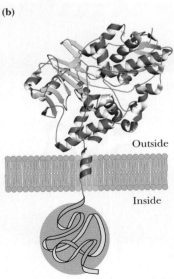

FIGURE 9.11 (a) Major histocompatibility antigen HLA-A2 (pdb id = 1JF1) and (b) monoamine oxidase (pdb id = 1GOS) are membrane-associated proteins with a single transmembrane helical segment.

sequence of bacteriorhodopsin contains seven different segments, each about 20 nonpolar residues in length—just the right size for an α-helix that could span a bilayer membrane. (Twenty residues times 1.5 Å per residue equals 30 Å.)

Bacteriorhodopsin clusters in symmetric, repeating arrays in the purple membrane patches of *Halobacterium,* and it was this orderly, repeating arrangement of proteins in the membrane that enabled Nigel Unwin and Richard Henderson in 1975 to determine the bacteriorhodopsin structure. The polypeptide chain crosses the membrane seven times, in seven α-helical segments, with very little of the protein exposed to the aqueous milieu. The bacteriorhodopsin structure became a model of globular membrane protein structure. Many other integral membrane proteins contain numerous hydrophobic sequences that, like those of bacteriorhodopsin, form α-helical transmembrane segments.

Membrane Protein Topology Can Be Revealed by Hydropathy Plots The **topology** of a membrane protein is a specification of the number of transmembrane segments and their orientation across the membrane. The topology of a transmembrane helical protein can be revealed by a **hydropathy plot** based on its amino acid sequence. If a measure of hydrophobicity is assigned to each amino acid (Table 9.1), then the overall hydrophobicity of a segment of a polypeptide chain can be estimated. The **hydropathy index** for any segment is an average of the hydrophobicity values for its residues.

The hydropathy index can be calculated at any residue in a sequence by averaging the hydrophobicity values for a segment surrounding that residue. Typically, segment sizes for such calculations can be 7 to 21 residues. With a 7-residue segment size, the calculation of hydropathy index at residue 10 would average the values for residues 7 through

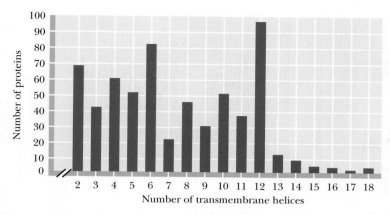

FIGURE 9.12 Most membrane proteins possess 2 to 12 transmembrane segments. Those involved in transport functions have between 6 and 12 transmembrane segments. (Adapted from von Heijne, G., 2006. Membrane-protein topology. *Nature Reviews Molecular Cell Biology* **7**:909–918.)

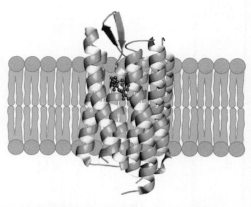

FIGURE 9.13 Bacteriorhodopsin is composed of seven transmembrane α-helical segments connected by short loops (pdb id = 1M0M). Nearly all of this protein is embedded in the membrane. Only the short loops connecting helices are exposed to solvent. A retinal chromophore (a light-absorbing molecule, shown in blue) lies approximately parallel to the membrane and between the helical segments. A proline residue (red) induces a kink in one of the helical segments (green).

13. The calculation for a 21-residue segment around residue 100 would include residues 90 to 110. A polypeptide segment approximately 20 residues long with a high hydropathy index is likely to be an α-helical transmembrane segment. A hydropathy plot for glycophorin (Figure 9.14a) reveals a single region of high hydropathy index between residues 73 and 93, the location of the α-helical segment in this transmembrane protein (Figure 9.10). A hydropathy plot for rhodopsin (Figure 9.14b) reveals the locations of its seven α-helical transmembrane segments. Rhodopsin, the light-absorbing pigment protein of the eye, is a member of the **G-protein–coupled receptor (GPCR)** family of membrane proteins (see Chapter 32).

Proline Residues Can Bend a Transmembrane α-Helix Transmembrane α-helices often contain distortions and "kinks"—more so than for water-soluble proteins. As more integral membrane protein structures have been determined, it has become clear that most transmembrane α-helices contain significant distortions from ideal helix geometry. Helix distortions may have evolved in membrane proteins because (1) helices, even distorted ones, are highly stable in the membrane environment, and (2) helix distortions may be one way to create structural diversity.

About 60% of known membrane helix distortions are kinks at proline residues (Figure 9.13). Proline distorts the ideal α-helical geometry because of its fixed ϕ value, steric conflict with the preceding residue, and because of the loss of a backbone H bond. Proline-induced kinks create weak points in the helix, which may facilitate movements required for transmembrane transport channels.

Amino Acids Have Preferred Locations in Transmembrane Helices Transmembrane protein sequences and structures are adapted to the transition from water on one side of the membrane, to the hydrocarbon core of the membrane, and then to water on the other side of the membrane. The amino acids that make up transmembrane segments reflect these transitions. Hydrophobic amino acids (Ala, Val, Leu, Ile, and Phe) are found most often in the hydrocarbon interior, where charged and polar amino acids almost never reside (Figure 9.15b). Charged residues (Figure 9.15a) occur commonly at the lipid-water interface, but positively charged residues are found more often on the cytoplasmic face of transmembrane proteins. Gunnar von Heijne has termed this the **positive inside rule.** Tryptophan, histidine, and tyrosine are special cases (Figure 9.15c). These residues have a mixed character, with nonpolar aromatic rings that also contain polar parts (the ring N—H of Trp and the substituent —OH of Tyr). As such, Trp and Tyr are found commonly at the lipid–water interface of transmembrane proteins.

The amino acids Lys and Arg frequently behave in novel ways at the lipid–water interface. Both of these residues possess long aliphatic side chains with positively charged

TABLE 9.1	Hydropathy Scale for Amino Acid Side Chains in Proteins*
Side Chain	Hydropathy Index
Isoleucine	4.5
Valine	4.2
Leucine	3.8
Phenylalanine	2.8
Cysteine	2.5
Methionine	1.9
Alanine	1.8
Glycine	−0.4
Threonine	−0.7
Serine	−0.8
Tryptophan	−0.9
Tyrosine	−1.3
Proline	−1.6
Histidine	−3.2
Glutamic acid	−3.5
Glutamine	−3.5
Aspartic acid	−3.5
Asparagine	−3.5
Lysine	−3.9
Arginine	−4.5

*From Kyte, J., and Doolittle, R., 1982. A simple method for displaying the hydropathic character of a protein. *Journal of Molecular Biology* **157**:105–132.

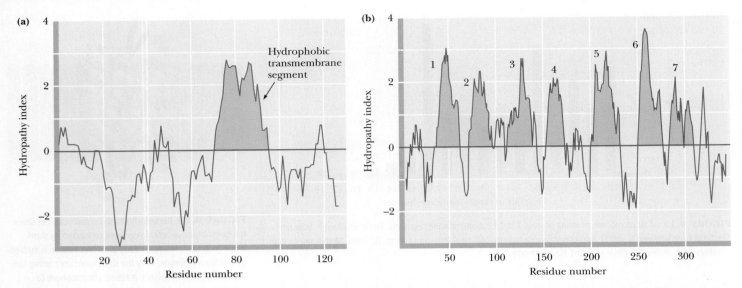

FIGURE 9.14 Hydropathy plots for **(a)** glycophorin and **(b)** rhodopsin. Hydropathy index is plotted versus residue number. At each position in the polypeptide chain, the average of hydropathy indices for a certain number of adjacent residues (eight, in this case) is calculated and plotted on the y-index, and the number of the residue in the middle of this "window" is shown on the x-axis.

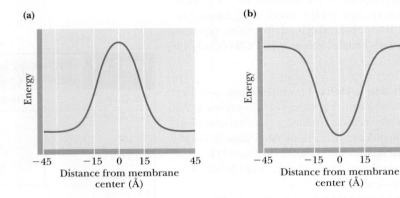

FIGURE 9.15 Amino acids have distinct preferences for different parts of the membrane. The graphs show relative stabilization energies as a function of location in the membrane for **(a)** Arg, Asp, Glu, Lys, Asn, Gln, and Pro; **(b)** Ala, Gly, Ile, Leu, Met, Phe, and Val; and **(c)** His, Tyr, and Trp. Polar and charged residues are less stable in the membrane interior, whereas nonpolar residues tend to be more stable in the membrane interior. The stability profiles for His, Tyr, and Trp are more complex. (Adapted from von Heijne, G., 2006. Membrane-protein topology. *Nature Reviews Molecular and Cell Biology* **7:**909–918.)

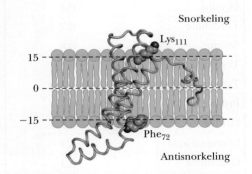

FIGURE 9.16 Snorkeling and antisnorkeling behavior in membrane proteins. The SdhC subunit of succinate dehydrogenase (pdb id = 1NEK). Lys[111] snorkels away from the membrane core and Phe[72] antisnorkels toward the membrane core. (Adapted from Liang, J., Adamian, L., and Jackups, R., Jr., 2005. The membrane-water interface region of membrane proteins: Structural bias and the anti-snorkeling effect. *Trends in Biochemical Sciences* **30:**355–357.)

groups at the end. In many membrane proteins, the aliphatic chain of Lys or Arg is associated with the hydrophobic portion of the bilayer, with the positively charged groups (amino or guanidinium) extending beyond to associate with negatively charged phosphate groups. This behavior, with the side chain pointing up out of the membrane core, has been termed *snorkeling* (Figure 9.16). If a Phe residue occurs near the lipid–water interface, it is typically arranged with the aromatic ring oriented toward the membrane core. This is termed *antisnorkeling*.

Membrane Protein Structures Show Many Variations on the Classical Themes

Although it revealed many insights of membrane protein structure, bacteriorhodopsin gave a relatively limited view of the structural landscape. Many membrane protein structures obtained since bacteriorhodopsin (and a few others) have provided a vastly more complex picture to biochemists. For example, the structures of a homodimeric chloride ion transport protein and a glutamate transport protein show several novel structural features (Figure 9.17). In addition to several transmembrane helices that lie perpendicular to the membrane plane (like those of bacteriorhodopsin), these structures each contain several *long, severely tilted helices* that span the membrane. Both these proteins also contain several **reentrant loops,** consisting of a pair of short α-helices and a connecting loop that together penetrate part way into the membrane core. There are also regions of **nonhelical polypeptide** deep in the membrane core of these proteins, with helical segments on either side that extend to the membrane surface (Figure 9.17).

Finally, most membrane protein structures are relatively stable; that is, transmembrane helices do not flip in and out of the membrane, and they do not flip across the lipid bilayer, inverting their orientation. However, a few membrane proteins can in fact change their membrane orientation. Aquaporin-1 is a protein that functions normally with six transmembrane α-helices. When this protein is first inserted into its membrane, it has only four transmembrane α-helices (Figure 9.18a). One of these, the third transmembrane helix (TM3), reorients across the membrane, pulling helices 2 and 4 into the membrane. Similarly, a glycoprotein of the hepatitis B virus is initially inserted into the viral membrane with its N-terminal domain lying outside. During the viral maturation process, about half of these glycoproteins rearrange (Figure 9.18b), with the N-terminal segment moving across the membrane as TM4 creates a new transmembrane segment.

Some Proteins Use β-Strands and β-Barrels To Span the Membrane The α-helix is not the only structural motif by which a protein can cross a membrane. Some integral transmembrane proteins use structures built from β-strands and β-sheets to diminish the polar character of the peptide backbone as it crosses the nonpolar membrane core. These **β-barrel** structures (Figure 9.19) maximize hydrogen bonding and are highly stable. The barrel interior is large enough to accommodate water molecules and often structures as large as peptide chains, and most barrels are literally water filled.

How does the β-barrel structure tolerate water on one surface (the inside) and the nonpolar membrane core on the other? In all transmembrane β-barrels, polar and nonpolar residues alternate along the β-strands, with polar residues facing the center of the barrel and nonpolar residues facing outward, where they can interact with the hydrophobic lipid milieu of the membrane.

Porin proteins found in the outer membranes (OMs) of Gram-negative bacteria such as *Escherichia coli,* and also in the outer mitochondrial membranes of eukaryotic cells, span their respective membranes with large β-barrels. A good example is **maltoporin,** also known as **LamB protein** or **lambda receptor,** which participates in the entry of maltose

(a)   (b)

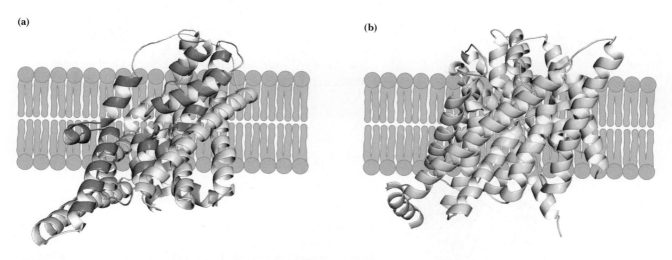

FIGURE 9.17 Not all the embedded segments of membrane proteins are transmembrane and oriented perpendicular to the membrane plane. **(a)** The glutamate transporter homolog (pdb id = 1XFH). "Reentrant" helices (orange) and interrupted helices (red) are shown. Several of the transmembrane helices deviate significantly from the perpendicular. **(b)** The *E. coli* ClC chloride transporter (pdb id = 1KPK). Few of the transmembrane helices are perpendicular to the membrane plane.

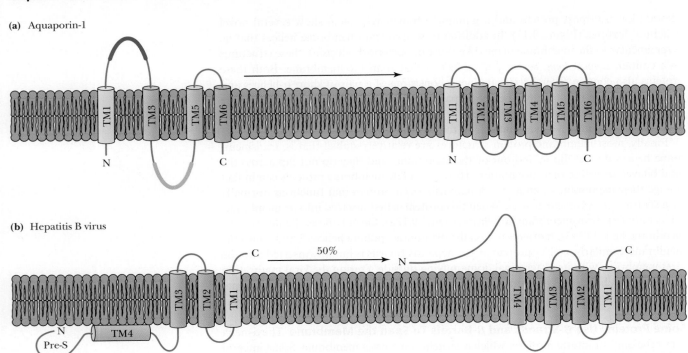

(a) Aquaporin-1

(b) Hepatitis B virus

FIGURE 9.18 Dynamic insertion of helical segments of membrane proteins. **(a)** Aquaporin-1. The second and fourth transmembrane helices insert properly across the membrane only after reorientation of the third transmembrane helix. **(b)** The large envelope glycoprotein of the hepatitis B virus. The N-terminal "pre-S" domain translocates across the endoplasmic reticulum membrane in a slow process in 50% of the molecules. (Adapted from von Heijne, G., 2006. *Nature Reviews Molecular and Cell Biology* **7**:909–918.)

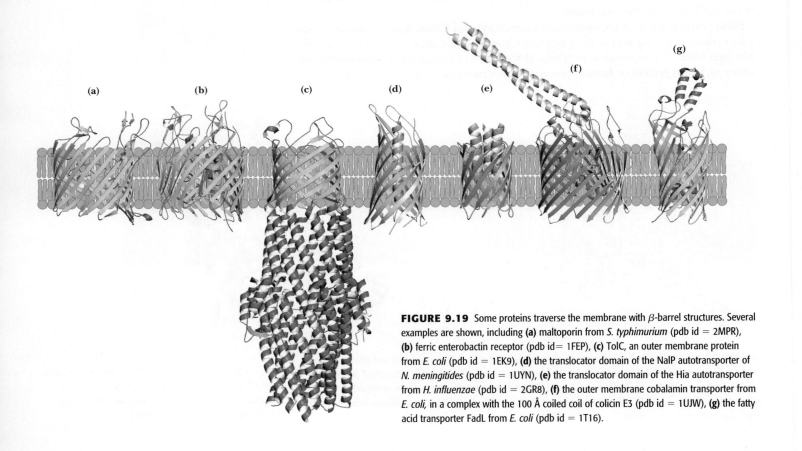

FIGURE 9.19 Some proteins traverse the membrane with β-barrel structures. Several examples are shown, including **(a)** maltoporin from *S. typhimurium* (pdb id = 2MPR), **(b)** ferric enterobactin receptor (pdb id= 1FEP), **(c)** TolC, an outer membrane protein from *E. coli* (pdb id = 1EK9), **(d)** the translocator domain of the NalP autotransporter of *N. meningitides* (pdb id = 1UYN), **(e)** the translocator domain of the Hia autotransporter from *H. influenzae* (pdb id = 2GR8), **(f)** the outer membrane cobalamin transporter from *E. coli,* in a complex with the 100 Å coiled coil of colicin E3 (pdb id = 1UJW), **(g)** the fatty acid transporter FadL from *E. coli* (pdb id = 1T16).

and maltodextrins into *E. coli*. Maltoporin is active as a trimer. The 421-residue monomer forms an 18-strand β-barrel with antiparallel β-strands connected to their nearest neighbors either by long loops or by β-turns (Figure 9.20; see also Figure 9.19a). The long loops are found at the end of the barrel that is exposed to the cell exterior, whereas the turns are located on the intracellular face of the barrel. Three of the loops fold into the center of the barrel.

β-barrels can also be constructed from multiple subunits. The α-hemolysin toxin (Figure 9.21) forms a 14-stranded β-barrel with seven identical subunits that each contribute two antiparallel β-strands connected by a short loop. *Staphylococcus aureus* secretes monomers of this toxin, which bind to the plasma membranes of host blood cells. Upon binding, the monomers oligomerize to form the 7-subunit structure. The channel thus formed facilitates uncontrolled permeation of water, ions, and small molecules, destroying the host cell.

Why have certain proteins evolved to use β-strands instead of α-helices as membrane-crossing devices? Among other reasons, there is an advantage of genetic economy in the use of β-strands to traverse the membrane instead of α-helices. An α-helix requires 21 to 25 amino acid residues to span a typical biological membrane; a β-strand can cross the same membrane with 9 to 11 residues. Therefore, a given amount of genetic information could encode a larger number of membrane-spanning segments using a β-strand motif instead of α-helical arrays.

Transmembrane Barrels Can also Be Formed with α-Helices Many bacteria, including *E. coli*, produce extracellular polysaccharides, some of which form a discrete structural layer—the **capsule,** which shields the cell, allowing it to evade or counteract host immune systems. In *E. coli*, the components of this polysaccharide capsule are synthesized inside the cell and then transported outward through an octameric outer membrane protein called **Wza.** To cross the outer membrane, Wza uses a novel **α-helical barrel** (Figure 9.22). Wza is composed of three novel domains that, with the α-helical barrel, form a large central cavity that accommodates the transported polysaccharides.

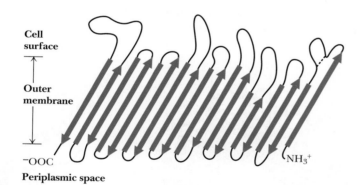

FIGURE 9.20 The arrangement of the peptide chain in maltoporin from *E. coli*.

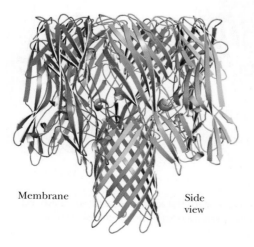

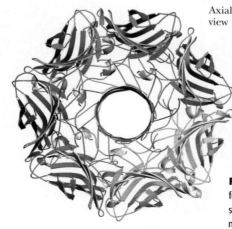

FIGURE 9.21 The structure of the heptameric channel formed by *Staphylococcus aureus* α-hemolysin. Each of the seven subunits contributes a β-sheet hairpin to the transmembrane channel (pdb id = 7AHL).

The transmembrane α-helices of Wza are amphiphilic, with hydrophobic outer surfaces that face the lipid bilayer and hydrophilic inner surfaces that face the water-filled pore.

Lipid-Anchored Membrane Proteins Are Switching Devices

Certain proteins are found to be covalently linked to lipid molecules. For many of these proteins, covalent attachment of lipid is required for association with a membrane. The lipid moieties can insert into the membrane bilayer, effectively **anchoring** their linked proteins to the membrane. Some proteins with covalently linked lipid normally behave as soluble proteins; others are integral membrane proteins and remain membrane associated even when the lipid is removed. Covalently bound lipid in these latter proteins can play a role distinct from membrane anchoring. In many cases, attachment to the membrane via the lipid anchor serves to modulate the activity of the protein.

Another interesting facet of lipid anchors is that they are transient. Lipid anchors can be reversibly attached to and detached from proteins. This provides a "switching device" for altering the affinity of a protein for the membrane. Reversible lipid anchoring is one factor in the control of **signal transduction pathways** in eukaryotic cells (see Chapter 32).

Four different types of lipid-anchoring motifs have been found to date. These are **amide-linked myristoyl** anchors, **thioester-linked fatty acyl** anchors, **thioether-linked prenyl** anchors, and **amide-linked glycosyl phosphatidylinositol** anchors. Each of these anchoring motifs is used by a variety of membrane proteins, but each nonetheless exhibits a characteristic pattern of structural requirements.

Amide-Linked Myristoyl Anchors Myristic acid may be linked via an amide bond to the α-amino group of the N-terminal glycine residue of selected proteins (Figure 9.23a). The reaction is referred to as **N-myristoylation** and is catalyzed by *myristoyl–CoA:protein N-myristoyltransferase*, known simply as **NMT**. N-Myristoyl–anchored proteins include the catalytic subunit of *cAMP-dependent protein kinase*, the *pp60^{src} tyrosine kinase*, the phosphatase known as *calcineurin B*, the α-subunit of *G proteins* (involved in GTP-dependent transmembrane signaling events), and the *gag proteins* of certain retroviruses (including the HIV-1 virus that causes AIDS).

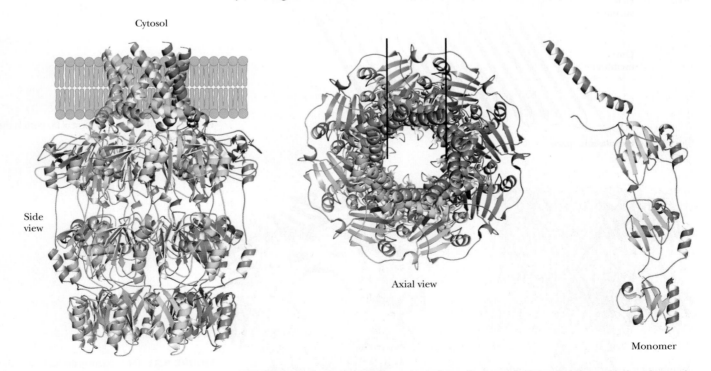

Cytosol

Side view

Axial view

Monomer

FIGURE 9.22 The structure of Wza, an octameric membrane protein that anchors the peptidoglycan layer and the outer membrane of Gram-negative bacteria. The structure contains a central barrel constructed from α-helical segments (pdb id = 2J58).

"Fat-Free Proteins" May Point the Way to Drugs for Sleeping Sickness

Human African trypanosomiasis (HAT, also known as sleeping sickness) is an invariably fatal disease caused by the protozoan **Trypanosoma brucei** and related organisms. No safe and effective drugs for treatment of sleeping sickness exist, but the research of Paul G. Wyatt and colleagues has taken a first step on a path that may yield useful medications.

    T. brucei is a unicellular parasite that shares many biological similarities with its human hosts, but also displays differences that could provide a basis for therapeutic strategies. Wyatt and colleagues focused on the *N*-myristoyltransferase (NMT) that attaches myristic acid anchors to a small number of essential cellular proteins. In a screening of 62,000 chemical structures, they found several moderately active NMT inhibitors, then synthesized more than 200 additional compounds based on the structures of the most promising. One of these, DDD85646, inhibits trypanosome NMT at concentrations 200-fold lower than those that inhibit human NMT. DDD85646, administered orally, effectively cures mice of stage 1 trypanosomiasis infections. Although this drug is not effective in stage 2, when the parasite has invaded the central nervous system, Wyatt and colleagues may have opened a door to development of a drug that could prevent up to 30,000 deaths annually from sleeping sickness in Africa.

DDD85646, an NMT inhibitor

Reference: Cross, G. A. M., 2010. Fat-free proteins kill parasites. *Nature* **464**:689–690.

Thioester-Linked Fatty Acyl Anchors A variety of cellular and viral proteins contain fatty acids covalently bound via ester linkages to the side chains of cysteine and sometimes to serine or threonine residues within a polypeptide chain (Figure 9.23b). This type of fatty acyl chain linkage has a broader fatty acid specificity than *N*-myristoylation. Myristate, palmitate, stearate, and oleate can all be esterified in this way, with the C_{16} and C_{18} chain lengths being most commonly found. Proteins anchored to membranes via fatty acyl thioesters include *G-protein–coupled receptors,* the *surface glycoproteins* of several viruses, the *reggie* proteins of nerve axons, and the *transferrin receptor* protein.

Thioether-Linked Prenyl Anchors As noted in Chapter 8, polyprenyl (or simply prenyl) groups are long-chain polyisoprenoid groups derived from isoprene units. Prenylation of proteins destined for membrane anchoring can involve either **farnesyl** or **geranylgeranyl** groups (Figure 9.23c and d). The addition of a prenyl group typically occurs at the cysteine residue of a carboxy-terminal CAAX sequence of the target protein, where C is cysteine, A is any aliphatic residue, and X can be any amino acid. As shown in Figure 9.23c and d, the result is a thioether-linked farnesyl or geranylgeranyl group. Once the prenylation reaction has occurred, a specific protease cleaves the three carboxy-terminal residues, and the carboxyl group of the now terminal Cys is methylated to produce an ester. All of these modifications appear to be important for subsequent activity of the prenyl-anchored protein. Proteins anchored to membranes via prenyl groups include *yeast mating factors,* the *p21^ras protein* (the protein product of the *ras* oncogene; see Chapter 32), and the *nuclear lamins,* structural components of the lamina of the inner nuclear membrane.

Glycosyl Phosphatidylinositol Anchors Glycosyl phosphatidylinositol, or **GPI,** groups are structurally more elaborate membrane anchors than fatty acyl or prenyl groups. GPI groups modify the carboxy-terminal amino acid of a target protein via an ethanolamine residue linked to an oligosaccharide, which is linked in turn to the inositol moiety of a phosphatidylinositol (Figure 9.23e). The oligosaccharide typically consists of a conserved tetrasaccharide core of three mannose residues and a glucosamine, which can be altered by addition of galactosyl side chains of various sizes and extra phosphoethanolamines, *N*-acetylgalactose, or mannosyl residues (Figure 9.23e). The inositol moiety can also be modified by an additional fatty acid, and a variety of fatty acyl groups are found linked to the glycerol group. GPI groups anchor a wide variety of *surface antigens, adhesion molecules,* and *cell surface hydrolases* to plasma membranes in various eukaryotic organisms. GPI anchors have not yet been observed in prokaryotic organisms or plants.

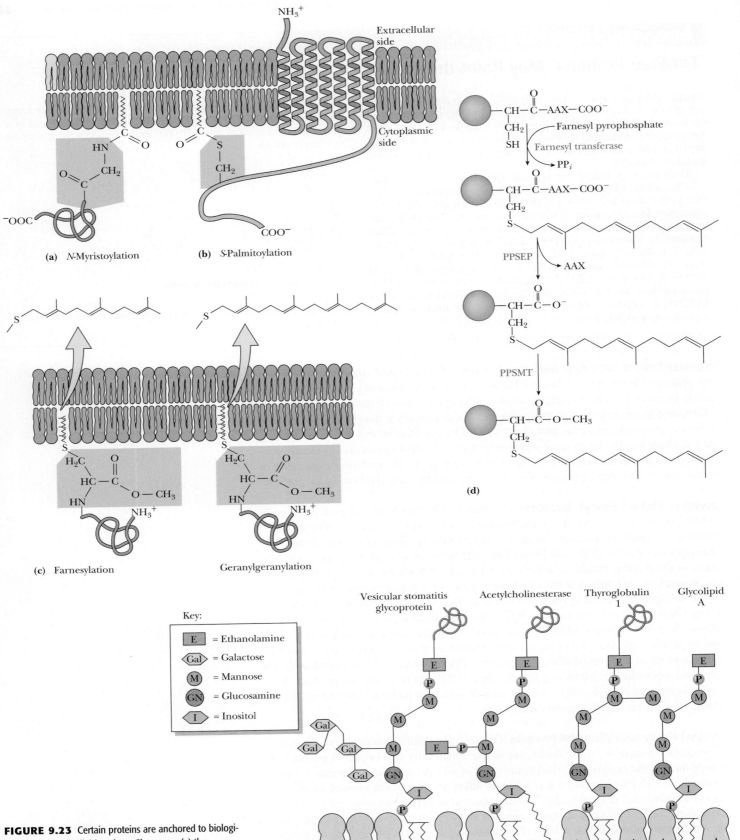

FIGURE 9.23 Certain proteins are anchored to biological membranes by lipid anchors. Shown are **(a)** the N-myristoyl motif, **(b)** the S-palmitoyl motif, **(c)** the farnesyl and geranylgeranyl motifs, **(d)** the prenylation and subsequent processing of prenyl-anchored proteins (PPSEP = prenyl protein-specific endoprotease, and PPSMT = prenyl protein-specific methyltransferase) and **(e)** several cases of the glycosyl phosphatidylinositol (GPI) motif.

9.3 How Are Biological Membranes Organized?

Membranes Are Asymmetric and Heterogeneous Structures

Biological membranes are **asymmetric** and heterogeneous structures. The two monolayers of the lipid bilayer have different lipid compositions and different complements of proteins. The membrane composition is also different from place to place across the plane of the membrane. There are clusters of particular kinds of lipids, particular kinds of proteins, and a variety of specific lipid-protein associations and aggregates, all of which serve the functional needs of the cell. We say that both the lipids and the proteins of membranes exhibit lateral heterogeneity and transverse asymmetry. **Lateral heterogeneity** arises when lipids or proteins of particular types cluster in the plane of the membrane. **Transverse asymmetry** refers to different lipid or protein compositions in the two leaflets or monolayers of a bilayer membrane.

Many properties of a membrane depend on its two-sided nature. Properties that are a consequence of membrane "sidedness" include membrane transport that is driven in one direction only; the effects of hormones at the outsides of cells; and the recognition reactions that occur between cells (necessarily involving only the outside surfaces of the cells). The proteins involved in these and other interactions must be arranged asymmetrically in the membrane.

Lipid transverse asymmetry can be seen in the typical animal cell, where the amine-containing phospholipids are enriched in the cytoplasmic leaflet of the plasma membrane, and the choline-containing phospholipids and sphingolipids are enriched in the outer leaflet (Figure 9.24). In the erythrocyte, for example, phosphatidylcholine (PC) comprises about 29% of the total phospholipid in the membrane. Of this amount, 76% is found in the outer monolayer and 24% is found in the inner monolayer.

Asymmetric lipid distributions are important to cells in several ways. The carbohydrate groups of glycolipids (and of glycoproteins) always face the outside of plasma membranes, where they participate in cell recognition phenomena. Asymmetric lipid distributions may also be important to various integral membrane proteins, which may prefer particular lipid classes in the inner and outer monolayers.

Loss of transverse lipid asymmetry has dramatic (and often severe) consequences for cells and organisms. For example, appearance of PS in the outer leaflet of the plasma membrane triggers apoptosis, the programmed death of the cell. Similarly, aging erythrocytes and platelets slowly externalize PS, culminating in engulfment by macrophages. Many disease states, including diabetes and malaria, involve microvascular occlusions that may result in part from alterations of transverse lipid asymmetry.

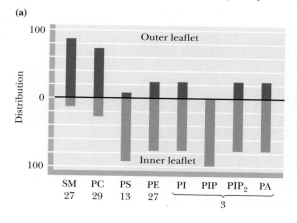

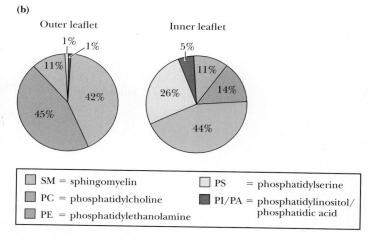

FIGURE 9.24 Phospholipids are distributed asymmetrically in most membranes, including the human erythrocyte membrane, as shown here. **(a)** The distribution of phospholipids across the inner and outer leaflets of human erythrocytes. The *x*-axis values show, for each lipid type, its percentage of the total phospholipid in the membrane. **(b)** The phospholipid compositions of the inner and outer leaflets. Note that the bar chart and pie charts provide complementary information. All percentages in **(a)** and **(b)** are weight percentages. (Adapted from Zachowski, A., 1993. Phospholipids in animal eukaryotic membranes: Transverse asymmetry and movement. *Biochemical Journal* **294**:1–14; and from Andreoli, T. E., 1987. *Membrane Physiology,* 2nd ed. Chapter 27, Table I. New York: Springer.)

9.4 What Are the Dynamic Processes That Modulate Membrane Function?

Lipids and Proteins Undergo a Variety of Movements in Membranes

Motions of lipids and proteins in membranes underlie many cell functions. Lipid movements (Figure 9.25) range from bond vibrations (at 10^{12} per sec), to bilayer undulations (1 to 10^6 per sec), to transverse motion—called "flip-flop" (roughly one per day to one per sec). Lateral movement of lipids (in the plane of the membrane) is rapid. Adjacent lipids can change places with each other on the order of 10^7/sec. Thus, a typical phospholipid can diffuse laterally in a membrane at a linear rate of several microns per second. At that rate, a phospholipid molecule travels from one end of a bacterial cell to the other in less than a second or traverses a typical animal cell in a few minutes.

Many membrane proteins move laterally (through the plane of the membrane) at a rate of a few microns per minute. On the other hand, some integral membrane proteins are more restricted in their lateral movement, with diffusion rates of about 10 nm per sec or even slower. Slower protein motion is likely for proteins that associate and bind with each other and for proteins that are anchored to the cytoskeleton, a complex latticelike structure that maintains the cell's shape and assists in the controlled movement of various substances through the cell.

Flippases, Floppases, and Scramblases: Proteins That Redistribute Lipids Across the Membrane Proteins that can "flip" and "flop" phospholipids from one side of a bilayer to the other have also been identified in several tissues (Figure 9.26). Three classes of such proteins are known:

1. ATP-dependent **flippases** that transport PS, and to a lesser extent PE, from the outer leaflet to the inner leaflet of the plasma membrane
2. ATP-dependent **floppases** that transport a variety of amphiphilic lipids, especially cholesterol, PC, and sphigomyelin from the inner leaflet to the outer leaflet
3. bidirectional, Ca^{2+}-activated (but ATP-independent) **scramblases** that function to randomize lipids and thus degrade transverse asymmetry

One class of ATP-driven flippases, the subfamily IV P-type ATPases (also known as P4-ATPases), share structural similarities with other P-type ATPases, including the plasma membrane Na,K-ATPase and the sarcoplasmic reticulum Ca-ATPase. Both these latter transport enzymes are discussed in Section 9.8.

(Puts, C. F., and Holthuis, J. C. M., 2009. Mechanism and significance of P4 ATPase-catalyzed lipid transport: Lessons from a Na+/K+-ATPase. Biochimica et Biophysica Acta **1791**:603–611.)

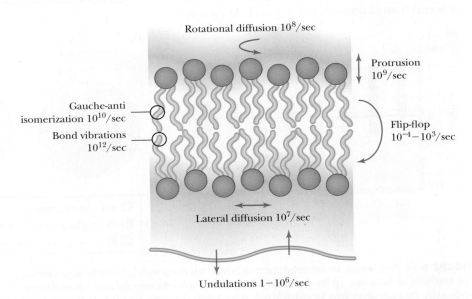

FIGURE 9.25 Lipid motions in the membrane and their characteristic frequencies. (Adapted from Gawrisch, K., 2005. The dynamics of membrane lipids. In *The Structure of Biological Membranes*, Chapter 4, Figure 4.1, Yeagle, P. L., ed., 2005. Boca Raton: CRC Press.)

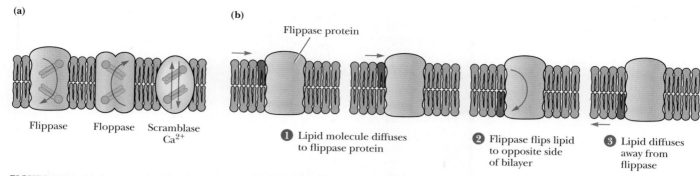

(a)

Flippase Floppase Scramblase Ca^{2+}

(b)

Flippase protein

① Lipid molecule diffuses to flippase protein

② Flippase flips lipid to opposite side of bilayer

③ Lipid diffuses away from flippase

FIGURE 9.26 **(a)** Phospholipids can be flipped, flopped, or scrambled across a bilayer membrane by the action of flippase, floppase, and scramblase proteins. **(b)** When, by normal diffusion through the bilayer, the lipid encounters one of these proteins, it can be moved quickly to the other face of the bilayer.

These proteins reduce the half-time for phospholipid movement across a membrane from days to a few minutes or less. Approximately one ATP is consumed per lipid transported by flippases and floppases. Energy-dependent lipid flippase activity is essential for the creation and maintenance of transverse lipid asymmetries. A number of diseases have been linked to defects in flippases and floppases. Tangier disease causes accumulation of high concentrations of cholesterol in various tissues and leads to cardiovascular problems. Infants with respiratory distress syndrome produce low amounts of lung surfactant (a mix of lipids) and typically die a few days after birth. Both of these diseases involve flippase or floppase defects.

Membrane Lipids Can Be Ordered to Different Extents

The phospholipids and sterols of membranes can adopt different structures depending on the exact lipid and protein composition of the membrane and on the temperature. At low temperatures, bilayer lipids are highly ordered, forming a **gel phase** with the acyl chains nearly perpendicular to the plane of the membrane plane (Figure 9.27). In this state—called the **solid-ordered state** (or **S_o state**)—the lipid chains are tightly packed and undergo relatively little motion. The lipid chains are in their fully extended conformation, the surface area per lipid is minimal, and the bilayer thickness is maximal. At higher temperatures, the acyl chains undergo much more motion, with rotations around the acyl chain C–C bonds and significant large-scale disordering of the acyl chains. The membrane is then said to be in a **liquid crystalline** phase or **liquid-disordered state** (**L_d state**) (Figure 9.27). In this less ordered state, the surface area per lipid increases and the bilayer thickness decreases by 10% to 15%. Under most conditions, the transition from the gel phase to the liquid crystalline phase is a true **phase transition**, and the temperature at which this change occurs is referred to as a **transition temperature** or **melting temperature (T_m)**.

The sharpness of the transition in pure lipid preparations shows that the phase change is a cooperative behavior. This is to say that the behavior of one or a few molecules affects the behavior of many other molecules in the vicinity. The sharpness of the transition then reflects the number of molecules that are acting in concert. Sharp transitions involve large numbers of molecules all "melting" together.

Phase transitions have been characterized in a number of different pure and mixed lipid systems. Table 9.2 shows a comparison of the transition temperatures observed for several different phosphatidylcholines with different fatty acyl chain compositions. General characteristics of bilayer phase transitions include the following:

1. The transitions are always endothermic; heat is absorbed as the temperature increases through the transition (Figure 9.27).
2. Particular phospholipids display characteristic transition temperatures (T_m). As shown in Table 9.2, T_m increases with chain length, decreases with unsaturation, and depends on the nature of the polar head group.

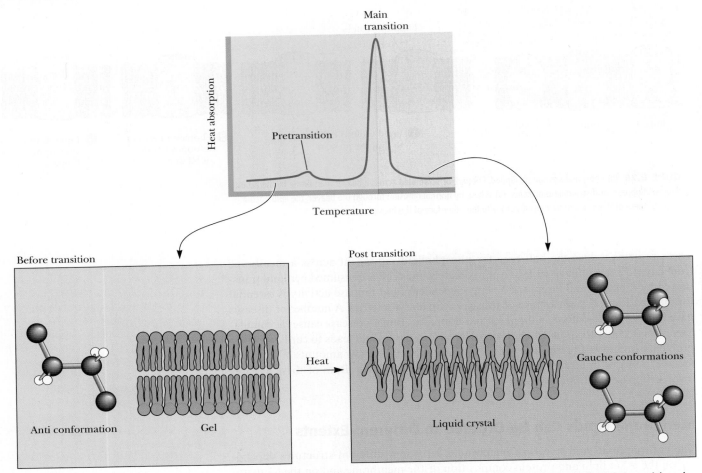

FIGURE 9.27 The gel-to-liquid crystalline phase transition, which occurs when a membrane is warmed through the transition temperature, T_m. In the transition, the surface area increases, the membrane thickness decreases, and the mobility of the lipid chains increases dramatically. Membrane phase transitions can be characterized by measuring the rate of heat absorption by a membrane sample in a calorimeter. Pure, homogeneous bilayers (containing only a single lipid component) give sharp calorimetric peaks. As membrane heterogeneity increases, the calorimetric peaks broaden. Below phase transitions, lipid chains primarily adopt the anti conformation. Above the phase transition, lipid chains have absorbed a substantial amount of heat. This is reflected in the adoption of higher-energy conformations, including the gauche conformations shown.

TABLE 9.2	Phase Transition Temperatures for Phospholipids in Water
Phospholipid	**Transition Temperature (T_m), °C**
Dimyristol phosphatidylcholine (Di 14:0)	23.6
Dipalmitoyl phosphatidylcholine (Di 16:0)	41.4
Distearoyl phosphatidylcholine (Di 18:0)	58
1-Stearoyl-2-oleoyl-phosphatidylcholine (1-18:0, 2-18:1 PC)	3
Dioleoyl phosphatidylcholine (Di 18:1 PC)	−22
Egg phosphatidylcholine (Egg PC)	−15
Dipalmitoyl phosphatidic acid (Di 16:0 PA)	67
Dipalmitoyl phosphatidylethanolamine (Di 16:0 PE)	63.8
Dipalmitoyl phosphatidylglycerol (Di 16:0 PG)	41.0

Adapted from Jain, M., and Wagner, R. C., 1980. *Introduction to Biological Membranes.* New York: John Wiley and Sons; and Martonosi, A., ed., 1982. *Membranes and Transport,* Vol. 1. New York: Plenum Press.

3. For pure phospholipid bilayers, the transition occurs over a narrow temperature range. The phase transition for dimyristoyl lecithin has a peak width of about 0.2°C.
4. Native biological membranes also display characteristic phase transitions, but these are broad and strongly dependent on the lipid and protein composition of the membrane.
5. With certain lipid bilayers, a change of physical state referred to as a *pretransition* occurs 5° to 15°C below the phase transition itself. These pretransitions involve a tilting of the hydrocarbon chains.
6. A volume change is usually associated with phase transitions in lipid bilayers.
7. Bilayer phase transitions are sensitive to the presence of solutes that interact with lipids, including multivalent cations, lipid-soluble agents, peptides, and proteins.

Cells adjust the lipid composition of their membranes to maintain proper fluidity as environmental conditions change.

The Evidence for Liquid Ordered Domains and Membrane Rafts In addition to the solid ordered (S_o) and liquid disordered (L_d) states, model lipid bilayers can exhibit a third structural phase if the membrane contains sufficient sphingolipid and cholesterol. The **liquid-ordered (L_o) state** is characterized by a high degree of acyl chain ordering (like the S_o state) but has the translational disorder characteristic of the L_d state. Lipid diffusion in the L_o phase is about twofold to threefold slower than in the L_d phase.

Biological membranes are hypothesized to contain regions equivalent to the L_o phase of model membranes. These **microdomains** are postulated to be aggregates of specific proteins, cholesterol, and glycosphingolipids with long, saturated fatty acyl chains (particularly ceramides and gangliosides), and they are termed **membrane rafts,** an apt description because these protein-rich lipid aggregates move together laterally in the "sea" of phospholipid. (Sphingolipids are found only in the extracellular leaflet of the plasma membrane, whereas cholesterol is believed to occur in roughly equal proportions in both leaflets.) Up to 50% of the plasma membrane may consist of rafts. Given that sphingolipids typically contain long, largely saturated acyl chains, they pack more tightly together, thus giving the lipids of rafts a much higher melting temperature than for other membrane domains. The physical evidence for membrane rafts is indirect; thus, their existence is a matter of debate among membrane biochemists. Direct measurements of rafts are difficult, because they are small (with postulated diameters of 10 to 50 nm) and because they are apparently transient, with lifetimes from a tenth of a millisecond or less to a few seconds or more. The most likely scenario, based on existing data, is shown in Figure 9.28.

Many of the proteins that appear to associate with and stabilize rafts are cell surface receptor proteins and other proteins involved in cell signaling processes (see Chapter 32). Association in rafts may be advantageous for the functioning of these proteins. The spontaneous association of sphingolipid with cholesterol together with the apparent specificity of protein association with rafts, may lead to selective modulation of membrane bioactivity.

Lateral Membrane Diffusion Is Restricted by Barriers and Fences A variety of studies of lateral diffusion rates in membranes have shown that membrane proteins and lipids in plasma membranes diffuse laterally at a rate 5 to 50 times slower than those of artificial lipid membranes. Why should this be? Part of the answer has come from **single particle tracking** experiments (Figure 9.29), which reveal that lipids and at least some membrane proteins tend to undergo **hop diffusion,** such that they can diffuse freely within a membrane "compartment" for a time and then hop to an adjacent compartment, where the process repeats. Akihiro Kusumi and colleagues have proposed the **membrane–skeleton fence** model to explain this behavior, suggesting that certain proteins that comprise the **cytoskeleton**—a network of proteins on the cytoplasmic face of the plasma membrane—restrict the lateral diffusion of other membrane proteins (Figure 9.30). The "fence" proteins may include *spectrin,* a filamentous cytoskeletal protein in red blood cells, and *actin,* a cytoskeletal protein that is ubiquitous in eukaryotic cells. The single particle tracking experiments show that lipid molecules are typically confined within fenced compartments for approximately 13 to 15 msec, whereas transmembrane proteins are typically confined for 45 to 65 msec. Even lipids in the outer leaflet of the plasma membrane undergo hop diffusion, leading Kusumi and colleagues to postulate that transmembrane proteins act as

(a)

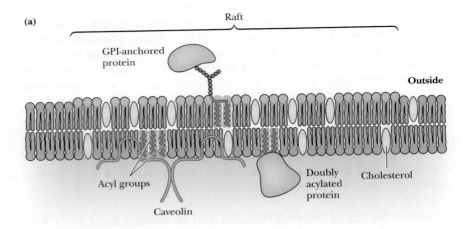

(b)

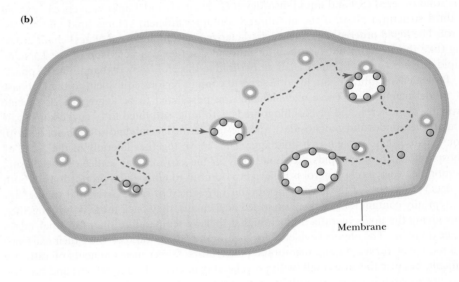

FIGURE 9.28 **(a)** A model for a membrane raft. Relative to other parts of the membrane, rafts are presumed to be enriched in cholesterol, fatty acyl-anchored proteins, and GPI-anchored proteins. Sphingolipids are found predominantly in the outer leaflet of the raft bilayer. **(b)** Rafts are postulated to "grow" by accumulation of these components as they diffuse through the plane of the membrane. They become increasingly stable as they grow. Green circles represent GPI-anchored proteins, which accumulate in lipid rafts as they grow in size. (Adapted from Hancock, J. F., 2006. Lipid rafts: Contentious only from simplistic standpoints. *Nature Reviews Molecular Cell Biology* **7:**456–462, and Parton, R. G., and K. Simons, 2007. The multiple faces of caveolae. *Nature Reviews Molecular Cell Biology* **8:**185–194.)

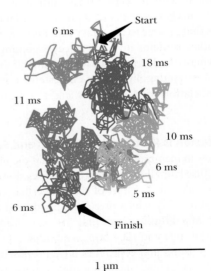

FIGURE 9.29 Motions of a single (fluorescently labeled) lipid molecule on the surface of a cell can be measured by video fluorescence microscopy. Video recording at 40,000 frames per second yields a time resolution of 25 microseconds. Data collected over 62 milliseconds (a total of 2500 frames) show that a lipid diffuses rapidly within small domains (defined by colors) and occasionally jumps or hops to an adjacent region (shown as a color change). (Adapted from Kusumi, A., et al., 2005. Paradigm shift of the plasma membrane concept from the two-dimensional continuum fluid to the partitioned fluid: High-speed single-molecule tracking of membrane molecules. *Annual Review of Biophysics and Biomolecular Structure* **34:**351–378.)

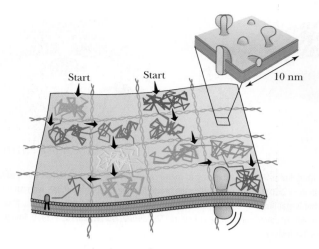

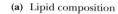

FIGURE 9.30 Studies of single lipid molecule movement in membranes (Figure 9.29) are consistent with a "compartmentalized" model for the membrane, in which lipids and proteins undergo short-term diffusion within "fenced" compartments, with occasional hops to adjacent compartments. Akihiro Kusumi has suggested that elements of the cytoskeleton may define the fence boundaries at the membrane.

rows of "pickets" extending across both monolayers in these membrane fences. Fences thus define regions of relatively unrestricted lipid diffusion.

Lipids and Proteins Direct Dynamic Membrane Remodeling and Curvature

The complex shapes of cells and organelles are the result of forces that operate on their membranes, and these forces in turn are orchestrated by lipids and proteins. Membranes change their shapes in special ways during movement, cell division, and other cellular events. This dynamic membrane remodeling is also accomplished by the interplay of lipids and proteins. The various membrane subdomains with particular curvatures have precise and specialized biological properties and functions.

There are several ways to induce curvature in a membrane (Figure 9.31). Lipids can influence or accommodate membrane curvature, either because of lipid molecule geometry or because of an imbalance in the number of lipids in the inner and outer leaflets of the bilayer. (In a liposome of 50-nm diameter, there is 56% more lipid in the outer leaflet than in the inner leaflet.) Integral membrane proteins with conical shapes can promote membrane curvature. The structure of a voltage-gated K^+ channel is an example of a shape conducive to membrane curvature (see Figure 9.41). Proteins of the cytoskeleton, such as actin, typically contact the plasma membrane and can generate curvature by rearrangements of their own structure. Moreover, motor proteins (see Chapter 16) moving along filaments of the cytoskeleton can generate curvature in the membrane. **Scaffolding proteins,** which can bind on either side of the plasma membrane, can influence membrane curvature in many ways. For example, BAR domains are dimeric, banana-shaped structures (Figure 9.32) that bind preferentially to and stabilize curved regions of the plasma membrane. Finally, amphipathic α-helices can insert into bilayers, parallel to the membrane surface, thus forcing curvature on the membrane. N-BAR domains are BAR domains that have an N-terminal α-helix preceding the BAR domain. The helix typically inserts to induce curvature, and the BAR domain binds to stabilize the curved structure. Harvey McMahon and his colleagues have proposed a structure (Figure 9.32b) for N-BAR domain-mediated membrane curvature by *endophilin-A1*, a protein found at synapses and implicated in the formation of synaptic vesicles. Membrane curvature is essential to a variety of cellular functions, including cell division, viral budding, and the processes of endocytosis and exocytosis, described in the next section.

Caveolins and Caveolae Respond to Plasma Membrane Changes

Caveolae are flask-shaped indentations in plasma membranes. Caveolae (Figure 9.33) are rich in cholesterol, sphingolipids, and **caveolin,** an integral membrane protein of 22,000 MW. There are three members of the caveolin family. CAV1 and CAV2 are found in endothelial,

(a) Lipid composition

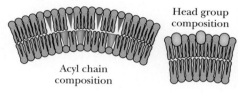

Head group composition

Acyl chain composition

(b) Membrane proteins

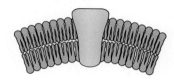

(c) Amphipathic helix insertion

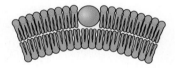

(d) Scaffolding

(e) Cytoskeleton

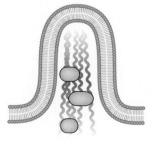

FIGURE 9.31 Membrane curvature can occur by several different mechanisms, including **(a)** changes in lipid composition, **(b)** insertion of membrane proteins that have intrinsic curvature or that oligomerize, **(c)** insertion of amphipathic helices into one leaflet of the bilayer, **(d)** interaction of the bilayer with molecular scaffolding proteins, or **(e)** changes induced by the cytoskeletal filaments inside the cell. (Adapted from McMahon, H. T., and Gallop, J. L., 2005. Membrane curvature and mechanisms of dynamic cell membrane remodeling. *Nature* **438:**590–596.)

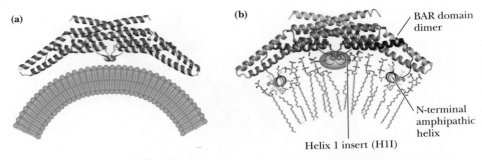

FIGURE 9.32 Model of BAR domain binding to membranes. **(a)** The classical model for binding of BAR domains to membranes (pdb id = 2C08). **(b)** A model for membrane binding of endophilin-A1, with amphiphilic helices inducing curvature that is stabilized by BAR domain binding. (Adapted from Gallop, J. L., et al., 2006. Mechanism of endophilin N-BAR domain-mediated membrane curvature. *The EMBO Journal* **25**:2898–2910. Image in (b) kindly provided by Harvey T. McMahon.)

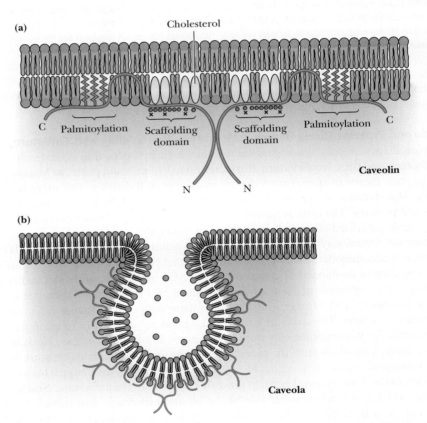

FIGURE 9.33 **(a)** Caveolin (pink) possesses a central hydrophobic segment flanked by three covalently bound fatty acyl anchors on the C-terminal side and a scaffolding domain on the N-terminal side. **(b)** Approximately 144 molecules of caveolin combine to force curvature in the lipid bilayer and form a caveolar structure. A caveola may contain as many as 20,000 cholesterol molecules. (Adapted from Parton, R. G., and K. Simons, 2007. The multiple faces of caveolae. *Nature Reviews Molecular Cell Biology* **8**:185–194.)

fibrous, and adipose (fat) tissue, whereas CAV3 is unique to skeletal muscle. Caveolins form homodimers in the plasma membrane, with both N- and C-termini oriented toward the cytosolic face of the membrane. The C-terminal domain has several palmitoyl lipid anchors and is separated from the N-terminal oligomerization domain by a 33-residue intramembrane hairpin domain. A typical caveolar structure consists of about 144 caveolin molecules, with up to 20,000 cholesterol molecules. Caveolae participate in **mechanosensation,** the detection and sensing of mechanical forces at the membrane, and **mechanotransduction,** the conversion of mechanical forces into biochemical signals that result in cell responses that regulate cell growth, differentiation, cell shape, and cell death.

Vesicle Formation and Fusion Are Essential Membrane Processes The membranes of cells are not static. Normal cell function requires that the various membrane-enclosed compartments of the cell constantly reorganize and exchange proteins, lipids, and other materials. These processes, and others such as cell division, exocytosis, endocytosis, and viral infection, all involve either fusion of one membrane with another or budding and separation of a vesicle from a membrane. Eukaryotic organelles communicate with one another by the exchange of "trafficking vesicles." Vesicles are generated at a precursor membrane, transported to their destination, and then fused with the target compartment (Figure 9.34). Although the organelles involved in these processes are diverse—endoplasmic reticulum, Golgi, endosomes, and others—the basic reactions of budding and fusion are accomplished by protein families and multiprotein complexes that have been conserved throughout eukaryotic evolution. The molecular events of exocytotic release of neurotransmitters into the synapses of nerve cells are good examples of such processes.

Neurons communicate with one another by converting electrical signals into chemical signals and back again. When electrical signals arrive at the synapse, vesicles containing neurotransmitters (such as acetylcholine—see Chapter 32) fuse with the plasma membrane, releasing the neurotransmitters into the synaptic cleft. Binding of neurotransmitters to receptors on an adjacent neuron generates an electrical signal that is passed along. The fusion of vesicles with the plasma membrane is directed by **SNAREs**—a family of proteins that "snare" vesicles to initiate the fusion process. (The acronym, a somewhat strained effort to describe their function cleverly, stands for *s*oluble *N*-ethylmaleimide–sensitive factor *a*ttachment protein *re*ceptor.) SNAREs are small proteins with a simple domain structure (Figure 9.35) that includes a **SNARE motif,** consisting of 60 to 70 residues of classical 7-residue repeats (see Chapter 6). These SNARE domains form a four-helix bundle that bridges two membranes, drawing them together in the first stage of membrane fusion. The N-terminal domains are variable across the SNARE family, but

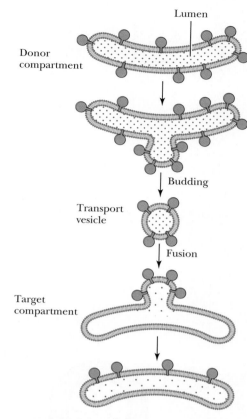

FIGURE 9.34 Vesicle-mediated transport in cells involves budding of vesicles from a donor membrane, followed by fusion of the vesicle membrane with the membrane of a target compartment, a process that transfers the contents of the donor compartment, as well as selected membrane proteins. (Adapted from Alberts, B., 2007. *Molecular Biology of the Cell,* 5th ed., New York: Garland Science.)

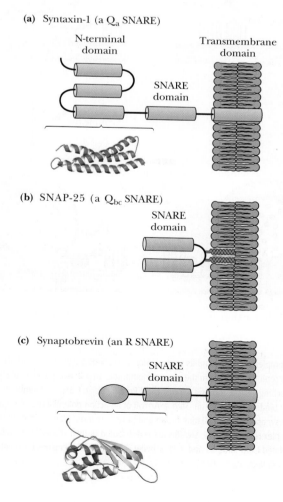

(a) Syntaxin-1 (a Q_a SNARE)

(b) SNAP-25 (a Q_{bc} SNARE)

(c) Synaptobrevin (an R SNARE)

FIGURE 9.35 (a) The domain structure of the SNARE protein families. A variety of N-terminal domains are found in Q_a SNARE proteins, including the three-helix bundle of syntaxin-1 (pdb id = 1BR0); **(b)** Q_{bc} SNAREs (such as SNAP-25) are anchored in the membrane by palmitic acid lipid anchors; **(c)** Many R SNAREs contain small globular N-terminal domains such as Vam7, a PX-homology domain (pdb id = 1OCS). (Adapted from Jahn, R., and R. H. Scheller, 2006. SNAREs—engines for membrane fusion. *Nature Reviews Molecular Cell Biology* **7:**631–643.)

at their C-termini most SNAREs have a single transmembrane domain joined to the SNARE motif by a short linker. **Syntaxin-1** (known as a **Qa SNARE,** for a conserved glutamine) has an N-terminal three-helix bundle, whereas **SNAP-25** (known as a **Q$_{bc}$ SNARE)** consists of two SNARE domains joined by a linker with two palmitoyl lipid anchors (Figure 9.35). **Synaptobrevin** (known as an **R SNARE,** for a conserved arginine (R)) consists of a small N-terminal domain, in addition to its SNARE and transmembrane domains.

The fusion of a vesicle with the plasma membrane is a multistep process (Figure 9.36). When a neurotransmitter-laden vesicle approaches the plasma membrane, syntaxin-1 and SNAP-25 on the plasma membrane join with synaptobrevin on the vesicle to form a four-helix bundle that bridges the two membranes, drawing them together in the first stage of membrane fusion, termed **Priming Stage I.** The binding of **complexin,** a small helical protein, "clamps" the complex so that it is poised for membrane fusion but is unable to complete the process. Arrival of an action potential (electrical signal—see Chapter 32) triggers flow of Ca^{2+} ions into the cell through channel proteins, and binding of Ca^{2+} ion to **synaptotagmin** displaces complexin and promotes joining of the membranes (to form **Priming Stage II**) and the formation of a fusion pore. The complexin clamp (Figure 9.36) ensures that neurotransmitter release can occur in an instant following Ca^{2+} influx, because the slow steps of SNARE assembly have already been completed.

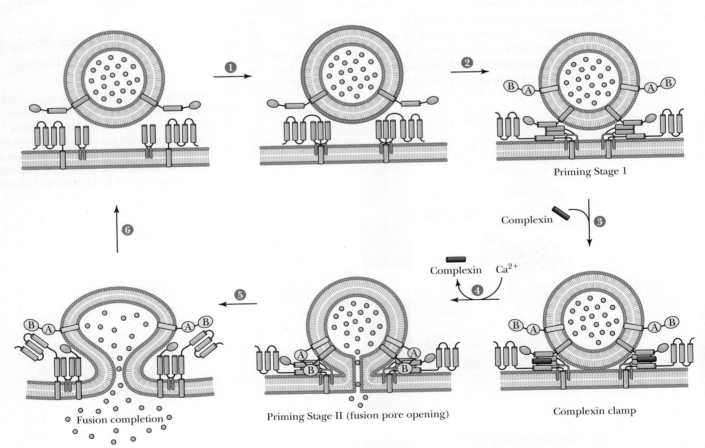

FIGURE 9.36 SNARE complex assembly and its control. *Step 1:* Q SNAREs (syntaxin-1 and SNAP-25), organized in clusters, assemble into acceptor complexes in the plasma membrane. *Step 2:* Acceptor complexes interact with synaptobrevin in an approaching vesicle, forming a four-helical Priming Stage I. *Step 3:* Membrane fusion and pore formation are prevented by binding of complexin. *Step 4:* Arrival of an action potential (nerve impulse) triggers displacement of complexin by synaptotagmin, initiating fusion and pore formation. *Step 5:* Upon completion of the fusion process, the *trans* complex relaxes. *Step 6:* SNAREs are redistributed to their respective membrane domains and vesicles are reformed. (Adapted from Jahn, R., and R. H. Scheller, 2006. SNAREs—engines for membrane fusion. *Nature Reviews Molecular Cell Biology* **7:**631–643.)

9.5 | How Does Transport Occur Across Biological Membranes?

Transport processes are vitally important to all life forms, because all cells must exchange materials with their environment. Cells obviously must have ways to bring nutrient molecules into the cell and ways to send waste products and toxic substances out. Also, inorganic electrolytes must be able to pass in and out of cells and across organelle membranes. All cells maintain **concentration gradients** of various metabolites across their plasma membranes and also across the membranes of intracellular organelles. By their very nature, cells maintain a very large amount of potential energy in the form of such concentration gradients. Sodium and potassium ion gradients across the plasma membrane mediate the transmission of nerve impulses and the normal functions of the brain, heart, kidneys, and liver, among other organs. Storage and release of calcium from cellular compartments controls muscle contraction, as well as the response of many cells to hormonal signals. High acid concentrations in the stomach are required for the digestion of food. Extremely high hydrogen ion gradients are maintained across the plasma membranes of the mucosal cells lining the stomach in order to maintain high acid levels in the stomach.

We shall consider the molecules and mechanisms that mediate these transport activities. In nearly every case, the molecule or ion transported is water soluble, yet moves across the hydrophobic, impermeable lipid membrane at a rate high enough to serve the metabolic and physiological needs of the cell. This perplexing problem is solved in each case by a specific transport protein. The transported species either diffuses through a channel-forming protein or is carried by a carrier protein. Transport proteins are all classed as **integral membrane proteins.**

From a thermodynamic and kinetic perspective, there are only three types of membrane transport processes: *passive diffusion, facilitated diffusion,* and *active transport.* To be thoroughly appreciated, membrane transport phenomena must be considered in terms of thermodynamics. Some of the important kinetic considerations also will be discussed.

9.6 | What Is Passive Diffusion?

Passive diffusion is the simplest transport process. In passive diffusion, the transported species moves across the membrane in the thermodynamically favored direction without the help of any specific transport system/molecule. For an uncharged molecule, passive diffusion is an entropic process, in which movement of molecules across the membrane proceeds until the concentration of the substance on both sides of the membrane is the same. For an uncharged molecule, the free energy difference between side 1 and side 2 of a membrane (Figure 9.37) is given by

$$\Delta G = G_2 - G_1 = RT \ln \frac{[C_2]}{[C_1]} \tag{9.1}$$

The difference in concentrations, $[C_2] - [C_1]$, is termed the *concentration gradient,* and ΔG here is the *chemical potential difference.*

Charged Species May Cross Membranes by Passive Diffusion

For a charged species, the situation is slightly more complicated. In this case, the movement of a molecule across a membrane depends on its **electrochemical potential.** This is given by

$$\Delta G = G_2 - G_1 = RT \ln \frac{[C_2]}{[C_1]} + Z\mathscr{F}\Delta\psi \tag{9.2}$$

where Z is the **charge** on the transported species, $\mathscr{F}$ is **Faraday's constant** (the charge on 1 mole of electrons = 96,485 coulombs/mol = 96,485 joules/volt·mol, since

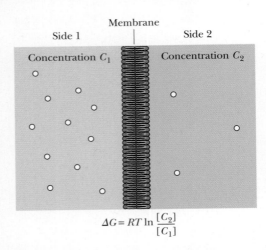

Membrane
Side 1 | Side 2

Concentration C_1 Concentration C_2

$$\Delta G = RT \ln \frac{[C_2]}{[C_1]}$$

FIGURE 9.37 Passive diffusion of an uncharged species across a membrane depends only on the concentrations (C_1 and C_2) on the two sides of the membrane.

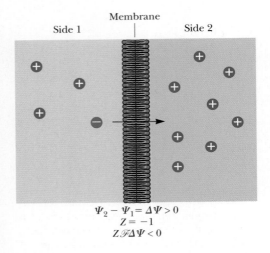

Membrane
Side 1 | Side 2

$$\Psi_2 - \Psi_1 = \Delta\Psi > 0$$
$$Z = -1$$
$$Z\mathcal{F}\Delta\Psi < 0$$

FIGURE 9.38 The passive diffusion of a charged species across a membrane depends on the concentration and also on the charge of the particle, Z, and the electrical potential difference across the membrane, $\Delta\psi$.

1 volt = 1 joule/coulomb), and $\Delta\psi$ is the electric potential difference (that is, voltage difference) across the membrane. The second term in the expression thus accounts for the movement of a charge across a potential difference. Note that the effect of this second term on ΔG depends on the magnitude and the sign of both Z and $\Delta\psi$.

For example, as shown in Figure 9.38, if side 2 has a higher potential than side 1 (so that $\Delta\psi$ is positive), for a negatively charged ion the term $Z\mathcal{F}\Delta\psi$ makes a negative contribution to ΔG. In other words, the negative charge is spontaneously attracted to the more positive potential—and ΔG is negative. In any case, if the sum of the two terms on the right side of Equation 9.2 is a negative number, transport of the ion in question from side 1 to side 2 would occur spontaneously. The driving force for passive transport is the ΔG term for the transported species itself.

9.7 : How Does Facilitated Diffusion Occur?

The transport of many substances across simple lipid bilayer membranes via passive diffusion is far too slow to sustain life processes. On the other hand, the transport rates for many ions and small molecules across actual biological membranes are much higher than anticipated from passive diffusion alone. This difference is due to specific proteins in the cell membranes that **facilitate** transport of these species across the membrane. Proteins capable of effecting **facilitated diffusion** of a variety of solutes are present in essentially all natural membranes. Such proteins have two features in common: (1) They facilitate net movement of solutes only in the thermodynamically favored direction (that is, $\Delta G < 0$), and (2) they display a measurable affinity and specificity for the transported solute. Consequently, facilitated diffusion rates display **saturation behavior** similar to that observed with substrate binding by enzymes (see Chapter 13). Such behavior provides a simple means for distinguishing between passive diffusion and facilitated diffusion experimentally. The dependence of transport rate on solute concentration takes the form of a rectangular hyperbola (Figure 9.39), so the transport rate approaches a limiting value, V_{max}, at very high solute concentration. Figure 9.39 also shows the graphical behavior exhibited by simple passive diffusion. Because passive diffusion does not involve formation of a specific solute:protein complex, the plot of rate versus concentration is linear, not hyperbolic.

Membrane Channel Proteins Facilitate Diffusion

The structures of hundreds of membrane proteins have been determined by X-ray diffraction and NMR spectroscopy. Many of these proteins are membrane transport channels that carry out facilitated diffusion. In contrast to active transport systems (or "pumps") like Na$^+$, K$^+$-ATPase and Ca^{2+}-ATPase, channels simply enable the (energetically passive) downhill movement of ions and other molecules. However, active pumps and most channels share one fundamental property: an ability to transport species in a selective manner. Molecular selectivity is crucial to the operation of both pumps and channels.

The membrane channel structures determined to date have revealed some of nature's strategies for moving ions and molecules across biological membranes. Channel composition can take several forms. A single channel pore can be formed from dimers, trimers, tetramers, or pentamers of protein subunits (for example, channels for Na$^+$, K$^+$, Mg^{2+}, and glutamate; see Table 9.3). On the other hand, multimeric assemblies in which each subunit has its own pore are known (in channels for Cl$^-$, NH$_3$, water, and glycerol). Several recurring themes are apparent from the channel structures studied to date:

- Some channels are **selective** for a particular ion or molecule. Others select either for cations or for anions, and some are nonselective.
- Channels can be "gated"—that is, they can open or close in response to a triggering signal. Some channels are **voltage-gated** and open and close in response to a change

in membrane electrical potential (i.e., voltage). Other channels are **ligand-gated,** opening or closing in response to binding of an ion, a small organic molecule, or even a protein.

- Channels that are ion-selective usually have pores lined with (oppositely) charged amino acids. Cation channels are lined with negatively charged residues; anion channels are lined with positively charged residues.
- Ions may flow through channels either in a hydrated or unhydrated state, depending on the width of the channel. Wider pores are often unselective and accept ions in their hydrated states. Narrower pores lined with charged amino acids are more likely to be highly selective.

The key questions for any channel protein are:

- How does the channel open and close?
- How does gating occur—that is, how is a voltage change sensed or how does a triggering ligand bind?
- How is voltage sensing or ligand binding **coupled** to channel opening and closing?

These recurring themes and key questions are illustrated particularly well by the K$^+$ channels characterized largely by Roderick MacKinnon and his colleagues.

Potassium Channels Combine High Selectivity with High Conduction Rates

Potassium transport (that is, conduction) is essential for many cell processes, including regulation of cell volume, electrical impulse formation (in electrically excitable cells, such as neurons), and secretion of hormones; all cells thus conduct K$^+$ ions across the cell membrane. Potassium channels are facilitated diffusion devices, conducting K$^+$ down the electrochemical gradient for K$^+$. Whether found in bacteria, archaea, plants, or animals, all known potassium channels are members of a single protein family. Potassium channels have two important characteristics: They are highly selective for K$^+$ ions over Na$^+$ ions, and they conduct K$^+$ ions at very high rates (almost as fast as any entity can diffuse in water—the so-called diffusion limit).

The structure of **KcsA,** from *Streptococcus lividans,* is typical of the simpler prokaryotic K$^+$ channels (Figure 9.40). The structure consists of four identical subunits, and, facing the cytosol, it has a water-filled pore that traverses more than half of the membrane bilayer, ending at the selectivity filter. A hydrated K$^+$ ion is suspended in the center of the pore. Each subunit consists of two transmembrane helices (S1 and S2, the outer and inner helices, respectively), with an intervening reentrant loop consisting of a ten-residue pore helix and a 5-residue β-strand that is one quarter of the selectivity filter. The inner (S2) helices form an inverted "teepee" with the shorter pore helices arranged within.

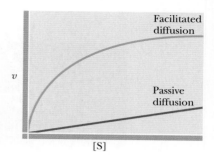

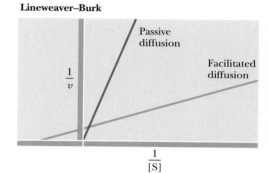

Lineweaver–Burk

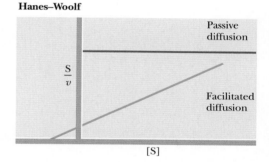

Hanes–Woolf

FIGURE 9.39 Passive diffusion and facilitated diffusion may be distinguished graphically. The plots for facilitated diffusion are similar to plots of enzyme-catalyzed processes (see Chapter 13), and they display saturation behavior.

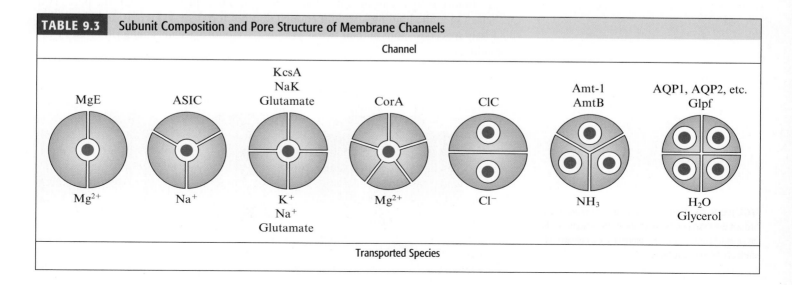

TABLE 9.3	Subunit Composition and Pore Structure of Membrane Channels

Channel

MgE	ASIC	KcsA NaK Glutamate	CorA	ClC	Amt-1 AmtB	AQP1, AQP2, etc. Glpf
Mg^{2+}	Na$^+$	K$^+$ Na$^+$ Glutamate	Mg^{2+}	Cl$^-$	NH$_3$	H$_2$O Glycerol

Transported Species

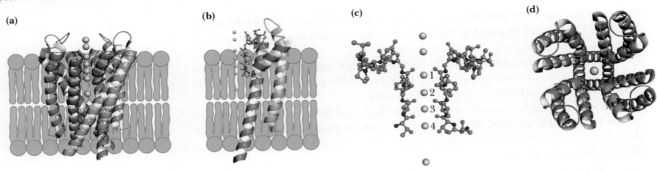

(a) **(b)** **(c)** **(d)**

FIGURE 9.40 Structure of the KcsA potassium channel from *Streptococcus lividans* (pdb id = 1K4C). **(a)** The four identical subunits of the channel, which surround a central pore, are shown in different colors. **(b)** Each subunit contributes three α-helices (blue, green, red) to the tetramer structure. **(c)** The selectivity filter is made from loops from each of the subunits, two of which are shown here. **(d)** The tetrameric channel, as viewed through the pore.

In contrast to the simple prokaryotic K⁺ channels, the mammalian voltage-gated potassium channels involved in nerve and muscle function are more complex, with each tetramer subunit consisting of six transmembrane helical segments (S1-S6). Helices S1-S4 form the voltage sensor domain, whereas the S5-S6 domain is analogous to the transmembrane helices of KcsA and forms the transport channel.

The selectivity filter in KcsA (Figure 9.41) consists of four pentapeptides, one from each subunit, with the sequence TVGYG. The backbone carbonyls of the first four residues and the threonine side-chain oxygen—evenly spaced—face the center of the pore. These oxygens create four possible K⁺-binding sites. In each site, a bound, dehydrated K⁺ is surrounded by eight oxygens from the protein: four above and four below. The arrangement of protein oxygens at each site is very similar to the arrangement of water molecules around a hydrated K⁺. This simple structure is strikingly selective for K⁺. The physical basis for selection between K⁺ and Na⁺ is the atomic radius—1.33 Å for K⁺ and 0.95 Å for Na⁺. Still, K⁺ channels select for K⁺ over Na⁺ by a factor of more than a thousand!

As K⁺ moves through the KcsA channel, there are, on average, two K⁺ ions bound in the selectivity filter at any given time, either in positions 1 and 3 or positions 2 and 4 (with water molecules occupying the other positions). Ions can move in either direction across the channel, depending on the existing electrochemical gradient. One K⁺ enters the channel from one side as a different ion exits on the other side. The cycle of steps for inward or outward movement is shown in Figure 9.41.

High selectivity, along with high conduction rates, seems at first paradoxical. If K⁺ ions bind too tightly in the filter, they could not move quickly through the pore. Two factors keep the binding just tight enough, but not too tight: (1) repulsion between the closely spaced K⁺ ions at their two sites and (2) a conformational change induced by K⁺ binding.

FIGURE 9.41 Model for outward and inward transport through the KcsA potassium channel. The selectivity filter in the channel contains four K⁺-binding sites, only two of which are filled at any time.

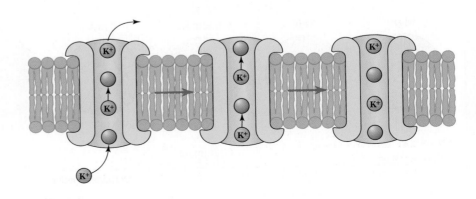

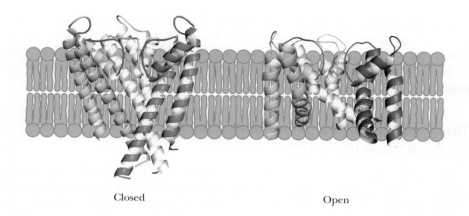

Closed   Open

FIGURE 9.42 Comparison of the closed (pdb id = 1K4C) and open (pdb id = 1LNQ) states of the potassium channel. The inner helices obstruct the central pore in the closed conformation. However, bending at a glycine residue near the center of the membrane splays the inner helices outward from the channel center, allowing free access for ions between the cytosol and the selectivity filter. This critical Gly residue is conserved in most K⁺ channel sequences, making this a likely gating mechanism for most K⁺ channels.

The KcsA channel is gated by intracellular pH. It is closed at neutral pH and above, and it opens at acidic pH. What is the conformational change that opens this and other K⁺ channels? After comparing the closed pore conformation of KcsA with the open pore conformation of the related MthK channel (Figure 9.42), MacKinnon has proposed that helix bending and rearrangement deep in the membrane opens K⁺ channels.

The mammalian K⁺ channel is voltage-gated. (See Chapter 32 for details.) In this more complex structure (Figure 9.43), helix S4 contains a series of five positively

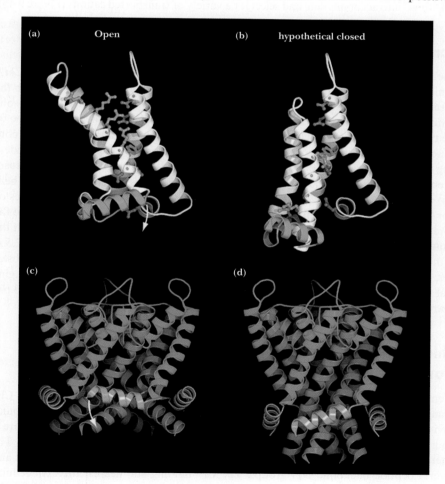

FIGURE 9.43 Model for voltage-dependent gating in mammalian K⁺ channels. See pdb id = 1J4C. Parts (a) and (b) show a single subunit of the channel; (c) and (d) show the native tetrameric channel. Parts (a) and (c) represent the open channel conformation; (b) and (d) show the closed state. Photo courtesy of Roderick MacKinnon.

Sequence of selectivity filter	Selective for
T V G Y G D L Y P	K^+
T V G D G N F S P	Na^+, K^+
L T G E D W N S V	Ca^{2+}

FIGURE 9.44 Ion selectivity in cation channels is a function of peptide sequence of the selectivity filter. Conserved glycines in the TVGYG motif of the KcsA potassium channel are shown in red. Amino acids that are chemically similar are yellow. From top to bottom, the selectivity for K^+ over Na^+ decreases and the selectivity for Ca^{2+} increases. (Adapted from Zagotta, W. N., 2006. Permutations of permeability. *Nature* **440:**427–428.)

(a)

Extracellular

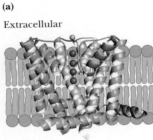

Intracellular

(b)

(c)

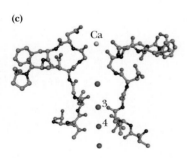

FIGURE 9.45 Structure of the channel from *Bacillus cereus* (pdb id = 2AHY), which has equal preference for Na^+ and K^+. **(a)** One subunit of the tetramer has been removed to reveal the five ion binding sites in the center of the channel. A Ca^{2+} ion is bound at the extracellular entrance to the channel (aqua, top), and K^+ is bound to the other four sites. **(b)** The tetrameric channel, as viewed through the pore. **(c)** Substitution of D for Y in the selectivity filter eliminates binding sites 1 and 2, leaving a pore vestibule that binds a K^+ with low affinity. Sites 3 and 4 are preserved, but bind Na^+ and K^+ equally well. The bottom site contains a fully hydrated K^+, in a manner similar to the KcsA channel (Figure 9.40).

charged Arg residues, spaced every three residues. These five residues all face the outside surface of the membrane in the open state of the channel. Roderick MacKinnon and co-workers have postulated that channel closing involves movement of the S4 helix inward, which pushes down on the short S4-S5 linker helix, which induces a conformation change in S5 and S6, thus closing the channel.

The *B. cereus* NaK Channel Uses a Variation on the K^+ Selectivity Filter

Could the K^+ channel selectivity filter be modified to accommodate other ions, for instance Na^+? Comparison of amino acid sequences from a variety of ion channels (Figure 9.44) shows that this is indeed the case. Variations on the TVGYG filter sequence are found in ion channels with a range of selectivities for K^+, Na^+, and even Ca^{2+}. *Bacillus cereus* contains an ion channel with equal preference for Na^+ and K^+ that is similar to the **transient receptor potential (TRP)** channels found widely in eukaryotes. The structure of this channel (Figure 9.45) is similar in many ways to the K^+ channels, but the selectivity filter sequence of this **NaK channel** is TVGDG. The substitution of D for Y changes the filter in several ways. Binding sites 1 and 2, the sites most selective for K^+, are eliminated, leaving a "pore vestibule" that can accommodate an ion but not bind it tightly. The remaining sites, binding sites 3 and 4, bind Na^+ and K^+ equally well. In addition, the D for Y substitution creates a Ca^{2+}-binding site at the extracellular entrance to the selectivity filter (see Figure 9.45c). It appears likely that variations of the selectivity filter sequence can "tune" it to accommodate and select for a variety of transported cations (Figure 9.45).

CorA Is a Pentameric Mg^{2+} Channel

The transport of Mg^{2+} in bacteria and archaea is accomplished primarily by the **CorA** family of membrane channels. Its pentameric structure (Figure 9.46) contrasts with the tetrameric K^+ and NaK channels in several ways. With a large N-terminal cytosolic domain and C-terminal transmembrane domain, it resembles a funnel or cone. One of the two transmembrane α-helices extends 100 Å into the cytosol and is the longest continuous α-helix in any known protein. Remarkable features of the structure include a ring of 20 lysine residues around the outside of the structure near the membrane–cytosol interface and a cluster of 50 aspartate and glutamate residues (on so-called willow helices) adjacent to the lysines and extending into the cytosol. The available structures of CorA are closed-pore structures. A ring of five Asn[314] residues blocks the opening to the pore on the periplasmic face, a ring of Met[291] residues narrows the pore to 3.3 Å at the center of the membrane, and a ring of Leu[294] residues reduce the pore diameter to 2.5 Å at the cytosolic face of the membrane. Gating of the pore must overcome these obstacles, as well as the repulsive ring of positive Lys side chains. It is tempting to imagine that the long α-helix and the negatively charged willow helices could act as a lever to pry apart the repulsive ring of lysines and open the Mg^{2+} pore.

Chloride, Water, Glycerol, and Ammonia Flow Through Single-Subunit Pores

Membrane channels can also be formed within a single subunit of a protein. The **ClC channels** (ubiquitous in cells from bacteria to animals) are homodimeric, each subunit having two similar halves, with opposite orientation in the membrane. The ClC pore is hourglass-shaped, with a 15-Å-long selectivity filter in the middle (Figure 9.47a). The filter contains 3 Cl^--binding sites, with coordination from Tyr and Ser hydroxyls and several peptide backbone NH groups. The Cl^- binding site nearest the extracellular solution can be occupied either by a Cl^- ion or by a glutamate carboxyl group. With the glutamate carboxyl in place, the pore is closed, but an increase in Cl^- concentration can displace the Glu side chain, with Cl^- binding to this position and opening the pore. Thus, this Cl^- channel is chloride-gated.

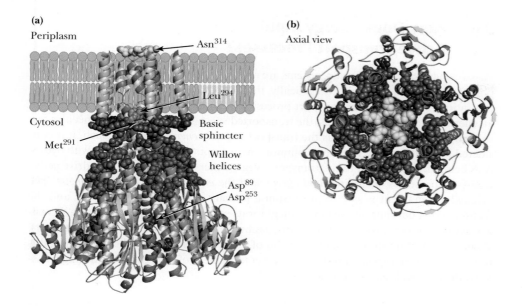

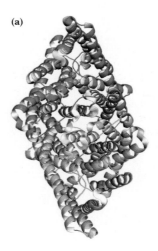

FIGURE 9.46 The structure of the pentameric CorA Mg^{2+} channel from *Thermus maritima* (pdb id = 2HN2). The transmembrane pore is formed from five short α-helices (red) and stabilized by four longer α-helices (aqua). The pore entrance from the periplasm is gated closed by a ring of 5 Asn residues (gold), and a ring of 5 Leu (orange) and 5 Met (beige) residues narrows the pore to 2.5 Å. The large cytosolic domain includes a basic sphincter of 20 Lys residues (blue), and a ring of 50 Asp and Glu residues (green) at the tips of the willow helices. Asp^{89} and Asp^{253} (purple) participate in Mg^{2+} binding.

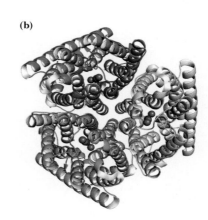

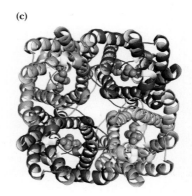

Channel proteins often solve chemical and thermodynamic problems in innovative ways. Ion selectivity, for example, requires that ions be dehydrated in the channel, and dehydration is energetically expensive. Binding sites have to compensate for the energetic cost of dehydration by providing favorable compensatory interactions between the ion and the binding amino acid residues. The ammonia transport channel solves a different problem. Ammonia is a gas, but the protonated ammonium ion, NH_4^+, is the species that diffuses to the channel opening. The transport channel in this case is a hydrophobic pore 20 Å in length (Figure 9.47b). The hydrophobic character of the channel lowers the pK_a of ammonium from its normal 9.25 to less than 6, facilitating the transport of NH_3 but not the monovalent cation, NH_4^+. In the narrow hydrophobic channel, His^{168} and His^{318} line the pore and stabilize three NH_3 molecules through hydrogen bonding. On either side of the narrow pore, broad vestibules contain NH_3 in equilibrium with NH_4^+.

Another example of adaptation to the transported species is the tetrameric glycerol channel GlpF from *E. coli,* with a transport pore in each monomer. Six transmembrane helices and two half-membrane-spanning helices form a right-handed helical bundle around each channel (Figure 9.47c). Glycerol molecules taken up by an *E. coli* cell first enter a 15-Å-wide vestibule in the transport protein, becoming progressively dehydrated before entry into a 28-Å amphipathic channel and selectivity filter. The channel accommodates three glycerol molecules, lined up in a single file, with their alkyl backbones wedged against a hydrophobic corner and their hydroxyl groups forming hydrogen bonds with the side chains of channel residues.

The aquaporin water channels are closely related to the GlpF glycerol channel, with tetrameric structures and similar right-handed helical bundles forming transport channels. Selection for water or glycerol in these proteins is based on subtle differences in the selectivity filters within the transport channels. For example, the Phe and Trp residues that comprise the hydrophobic corner surrounding the alkyl moiety of the middle glycerol site in GlpF are replaced by a His residue in the corresponding water-binding site in aquaporin Apq1.

FIGURE 9.47 Structures of channels for **(a)** chloride, **(b)** ammonia, and **(c)** glycerol. All structures are axial views. **(a)** The ClC chloride channel from *E. coli* (pdb id = 1OTS). **(b)** The AmtB ammonia channel from *E. coli* with four bound NH_4^+ (pdb id = 1U7G). **(c)** The GlpF glycerol channel from *E. coli,* with bound glycerol (pdb id = 1FX8).

9.8 How Does Energy Input Drive Active Transport Processes?

Passive and facilitated diffusion systems are relatively simple, in the sense that the transported species flow downhill energetically, that is, from high concentration to low concentration. However, other transport processes in biological systems must be *driven* in an energetic sense. In these cases, the transported species move from low concentration to high concentration, and thus the transport requires *energy input*. As such, it is considered **active transport.** The most common energy input is **ATP hydrolysis,** with hydrolysis being *tightly coupled* to the transport event. Other energy sources also drive active transport processes, including *light energy* and the *energy stored in ion gradients*. The original ion gradient is said to arise from a **primary active transport** process, and the transport that depends on the ion gradient for its energy input is referred to as a **secondary active transport** process (see later discussion of the *E. coli* proton–drug exchanger). When transport results in a net movement of electric charge across the membrane, it is referred to as **electrogenic transport.** If no net movement of charge occurs during transport, the process is electrically neutral.

All Active Transport Systems Are Energy-Coupling Devices

Hydrolysis of ATP is essentially a chemical process, whereas movement of species across a membrane is a mechanical process (that is, movement). An active transport process that depends on ATP hydrolysis thus couples chemical free energy to mechanical (translational) free energy. The bacteriorhodopsin protein in *Halobacterium halobium* couples light energy and mechanical energy. Oxidative phosphorylation (see Chapter 20) involves coupling between electron transport, proton translocation, and the capture of chemical energy in the form of ATP synthesis. Similarly, the overall process of photosynthesis (see Chapter 21) amounts to a coupling between captured light energy, proton translocation, and chemical energy stored in ATP.

Many Active Transport Processes are Driven by ATP

Monovalent Cation Transport: Na+,K+-ATPase All animal cells actively extrude Na^+ ions and accumulate K^+ ions. These two transport processes are driven by **Na+,K+-ATPase,** also known as the **sodium pump,** an integral protein of the plasma membrane. Most animal cells maintain cytosolic concentrations of Na^+ and K^+ of 10 mM and 100 mM, respectively. The extracellular milieu typically contains about 100 to 140 mM Na^+ and 5 to 10 mM K^+. Potassium is required within the cell to activate a variety of processes, whereas high intracellular sodium concentrations are inhibitory. The transmembrane gradients of Na^+ and K^+ and the attendant gradients of Cl^- and other ions provide the means by which neurons communicate. They also serve to regulate cellular volume and shape. Animal cells also depend upon these Na^+ and K^+ gradients to drive transport processes involving amino acids, sugars, nucleotides, and other substances. In fact, maintenance of these Na^+ and K^+ gradients consumes large amounts of energy in animal cells—20% to 40% of total metabolic energy in many cases and up to 70% in neural tissue.

The Na^+- and K^+-dependent ATPase comprises three subunits: an α-subunit of 1016 residues (120 kD), a 35-kD β-subunit, and a 6.5-kD γ-subunit. The sodium pump actively pumps three Na^+ ions out of the cell and two K^+ ions into the cell per ATP hydrolyzed:

$$ATP^{4-} + H_2O + 3\ Na^+(inside) + 2\ K^+(outside) \longrightarrow ADP^{3-} + H_2PO_4^- +$$
$$3\ Na^+(outside) + 2\ K^+(inside) \quad (9.3)$$

ATP hydrolysis occurs on the cytoplasmic side of the membrane (Figure 9.48), and the net movement of one positive charge outward per cycle makes the sodium pump electrogenic in nature.

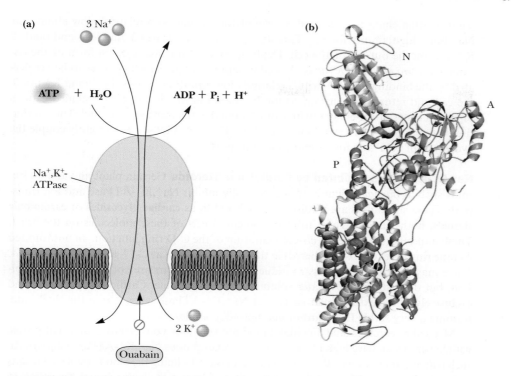

FIGURE 9.48 **(a)** A schematic diagram of the Na^+,K^+-ATPase of the mammalian plasma membrane. ATP hydrolysis occurs on the cytoplasmic side of the membrane, Na^+ ions are transported out of the cell, and K^+ ions are transported in. The transport stoichiometry is 3 Na^+ out and 2 K^+ in per ATP hydrolyzed. Ouabain and other cardiac glycosides inhibit Na^+,K^+-ATPase by binding on the extracellular surface of the pump protein. **(b)** Structure of the Na^+,K^+-ATPase, showing the α-subunit, residues 28–73 of the β-subunit (gray) and the transmembrane helix (residues 23–51, yellow) of the γ-subunit (pdb id = 3B8E).

The α-subunit of Na^+,K^+-ATPase consists of ten transmembrane α-helices, with three cytoplasmic domains, denoted A, P, and N. A large cytoplasmic loop between transmembrane helices 4 and 5 forms the P (phosphorylation) and N (nucleotide-binding) domains. The enzyme is covalently phosphorylated at Asp^{369} during ATP hydrolysis. The crystal structure of the enzyme reveals two rubidium ions bound to putative K^+-binding sites in the center of the protein (Figure 9.48).

A minimal mechanism for Na^+,K^+-ATPase postulates that the enzyme cycles between two principal conformations, denoted E_1 and E_2 (Figure 9.49). E_1 has a high affinity for Na^+ and ATP and is rapidly phosphorylated in the presence of Mg^{2+} to form E_1-P, a state that *contains three occluded Na^+ ions* (occluded in the sense that they are tightly bound and not easily dissociated from the enzyme in this conformation). A

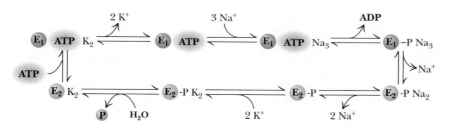

FIGURE 9.49 A mechanism for Na^+,K^+-ATPase. The model assumes two principal conformations, E_1 and E_2. Binding of Na^+ ions to E_1 is followed by phosphorylation and release of ADP. Na^+ ions are transported and released, and K^+ ions are bound before dephosphorylation of the enzyme. Transport and release of K^+ ions complete the cycle.

conformation change yields E_2-P, a form of the enzyme with relatively low affinity for Na^+ but a high affinity for K^+. This state presumably releases 3 Na^+ ions and binds 2 K^+ ions on the outside of the cell. Dephosphorylation leaves E_2K_2, a form of the enzyme with *two occluded K^+ ions.* A conformation change, which appears to be accelerated by the binding of ATP (with a relatively low affinity), releases the bound K^+ inside the cell and returns the enzyme to the E_1 state. Enzyme forms with occluded cations represent states of the enzyme with cations bound in the transport channel. The alternation between high and low affinities for Na^+, K^+, and ATP serves to tightly couple the hydrolysis of ATP and ion binding and transport.

Na^+,K^+-ATPase Is Inhibited by Cardiotonic Steroids Certain plant and animal steroids such as *ouabain* (Figure 9.50) specifically inhibit Na^+,K^+-ATPase and ion transport. These substances are traditionally referred to as **cardiac glycosides** or **cardiotonic steroids,** both names derived from the potent effects of these molecules on the heart. These molecules all possess a *cis*-configuration of the C-D ring junction, an unsaturated lactone ring (five- or six-membered) in the β-configuration at C-17, and a β-OH at C-14. There may be one or more sugar residues at C-3. The sugars are not required for inhibition, but do contribute to water solubility of the molecule. Cardiotonic steroids bind exclusively to the extracellular surface of Na^+,K^+-ATPase when it is in the E_2-P state, forming a very stable E_2-P(cardiotonic steroid) complex.

Medical researchers studying high blood pressure have consistently found that people with hypertension have high blood levels of an endogenous Na^+,K^+-ATPase inhibitor. In such patients, inhibition of the sodium pump in the cells lining the blood vessel wall results in accumulation of sodium and calcium in these cells and the narrowing of the vessels to create hypertension. An 8-year study aimed at the isolation and identification of the agent responsible for these effects by researchers at the University of Maryland Medical School and the Upjohn Laboratories in Michigan yielded a surprising result. Mass spectrometric analysis of compounds isolated from many hundreds of gallons of blood plasma has revealed that the hypertensive agent is ouabain itself or a closely related molecule!

Calcium Transport: Ca^{2+}-ATPase Calcium, an ion acting as a cellular signal in virtually all cells (see Chapter 32), plays a special role in muscles. It is the signal that stimulates muscles to contract (see Chapter 16). In the resting state, the levels of Ca^{2+} near the muscle fibers are very low (approximately 0.1 μM), and nearly all of the calcium ion in muscles is sequestered inside a complex network of vesicles called the **sarcoplasmic reticulum,** or **SR** (see Figure 16.1). Nerve impulses induce the SR membrane to quickly release large amounts of Ca^{2+}, with cytosolic levels rising to approximately 10 μM. At these levels, Ca^{2+} stimulates contraction. Relaxation of the muscle requires that cytosolic Ca^{2+} levels be reduced to their resting values. This is accomplished by an ATP-driven Ca^{2+} transport protein known as the **Ca^{2+}-ATPase,** which bears many similarities to the Na^+,K^+-ATPase. It has an α-subunit of the same approximate size, it forms

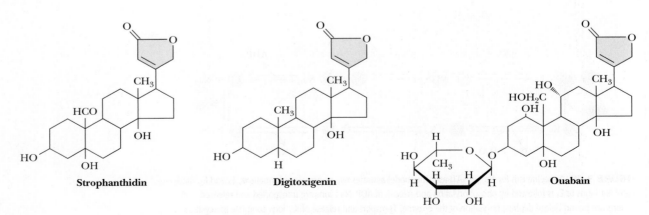

FIGURE 9.50 The structures of several cardiotonic steroids. The lactone rings are yellow.

a covalent E-P intermediate during ATP hydrolysis, and its mechanism of ATP hydrolysis and ion transport is similar in many ways to that of the sodium pump.

The structure of the Ca^{2+}-ATPase includes a transmembrane (M) domain consisting of ten α-helical segments and a large cytoplasmic domain that itself consists of a nucleotide-binding (N) domain, a phosphorylation (P) domain, and an actuator (A) domain (Figure 9.51).

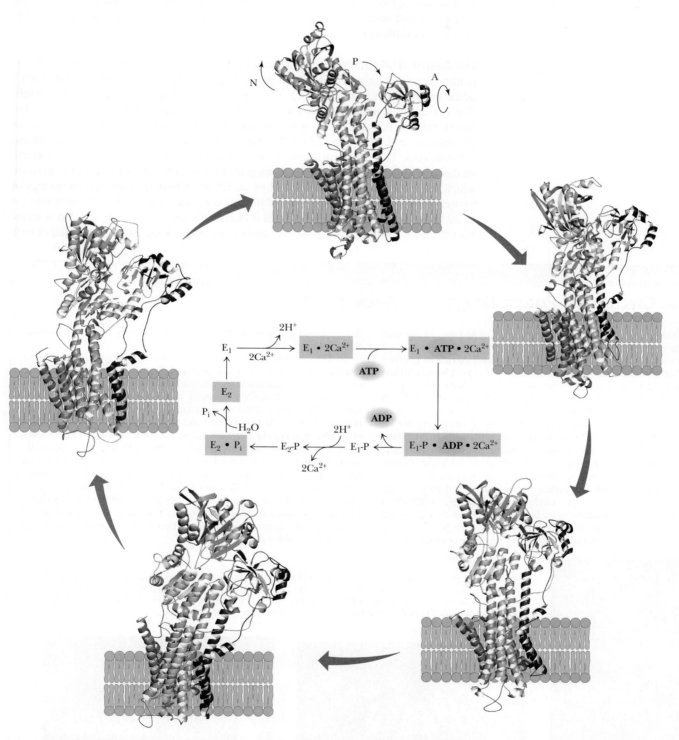

FIGURE 9.51 The transport cycle of the sarcoplasmic reticulum Ca^{2+}-ATPase involves at least five different conformations of the protein. The states shown here are $E_1 \cdot 2Ca^{2+}$ (pdb id = 3B9B); $E_1 \cdot ATP \cdot 2Ca^{2+}$ (pdb id = 1SU4); E_1-P·ADP·2Ca^{2+} (pdb id = 1T5C); $E_2 \cdot P_i$ (pdb id = 2ZBD); and E_2 (pdb id = 2EAR). Blue-shaded states in the reaction sequence correspond to adjacent structures.

The calcium transport cycle begins with binding of two Ca^{2+} ions. Subsequent ATP binding causes a 90° rotation of N and a 30° rotation of A, thus joining all three cytoplasmic domains (N, A, and P), and pulling a transmembrane helix partly out of the membrane. Phosphorylation of Asp^{351}, dissociation of ADP, and conversion of the E_1-P state to E_2-P induce a 110° rotation of A and a rearrangement of the transmembrane domain, which acts like a piston to release Ca^{2+} inside the SR. A TGES sequence in A (residues 181 to 184) then guides nucleophilic attack of water on E_2-P, releasing phosphate and restoring the original structures of both the transmembrane and the cytoplasmic domains of the enzyme.

The Gastric H^+,K^+-ATPase Production of protons is a fundamental activity of cellular metabolism, and proton production plays a special role in the stomach. The highly acidic environment of the stomach is essential for the digestion of food in all animals. The pH of the stomach fluid is normally 0.8 to 1. The pH of the parietal cells of the gastric mucosa in mammals is approximately 7.4. This represents a **pH gradient** across the mucosal cell membrane of 6.6, the largest known transmembrane gradient in eukaryotic cells. This enormous gradient must be maintained constantly so that food can be digested in the stomach without damage to the cells and organs adjacent to the stomach. The gradient of H^+ is maintained by an **H^+,K^+-ATPase,** which uses the energy of hydrolysis of ATP to pump H^+ out of the mucosal cells and into the stomach interior in exchange for K^+ ions. This transport is electrically neutral, and the K^+ that is transported into the mucosal cell is subsequently pumped back out of the cell together with

---

> **A DEEPER LOOK**

Cardiac Glycosides: Potent Drugs from Ancient Times

The cardiac glycosides have a long and colorful history. Many species of plants producing these agents grow in tropical regions and have been used by natives in South America and Africa to prepare poisoned arrows used in fighting and hunting. Zulus in South Africa, for example, have used spears tipped with cardiac glycoside poisons. The sea onion, found commonly in southern Europe and northern Africa, was used by the Romans and the Egyptians as a cardiac stimulant, diuretic, and expectorant. The Chinese have long used a medicine made from the skins of certain toads for similar purposes. Cardiac glycosides are also found in several species of domestic plants, including the foxglove, lily of the valley, oleander (figure part a), and milkweed plants. Monarch butterflies (figure part b) acquire these compounds by feeding on milkweed and then storing the cardiac glycosides in their exoskeletons. Cardiac glycosides deter predation of monarch butterflies by birds, which learn by experience not to feed on monarchs. Viceroy butterflies (figure part c) mimic monarchs in overall appearance. Although viceroys contain no car-

diac glycosides and are edible, they are avoided by birds that mistake them for monarchs.

In 1785, the physician and botanist William Withering described the medicinal uses for agents derived from the foxglove plant. In modern times, **digitalis** (a preparation of dried leaves prepared from the foxglove, *Digitalis purpurea*) and other purified cardiotonic steroids have been used to increase the contractile force of heart muscle, to slow the rate of beating, and to restore normal function in hearts undergoing fibrillation (a condition in which heart valves do not open and close rhythmically but rather remain partially open, fluttering in an irregular and ineffective way). Inhibition of the cardiac sodium pump increases the intracellular Na^+ concentration, leading to stimulation of the Na^+-Ca^{2+} exchanger, which extrudes sodium in exchange for inward movement of calcium. Increased intracellular Ca^{2+} stimulates muscle contraction. Careful use of digitalis drugs has substantial therapeutic benefit for patients with heart problems.

◁ **(a)** Cardiac glycoside inhibitors of Na^+,K^+-ATPase are produced by many plants, including foxglove, lily of the valley, milkweed, and oleander (shown here). **(b)** The monarch butterfly, which concentrates cardiac glycosides in its exoskeleton, is shunned by predatory birds. **(c)** Predators also avoid the viceroy, even though it contains no cardiac glycosides, because it is similar in appearance to the monarch.

(a) Oleander   **(b)** Monarch butterfly   **(c)** Viceroy butterfly

Cl^- in a second electroneutral process (Figure 9.52). Thus, the net transport effected by these two systems is the movement of HCl into the interior of the stomach. (Only a small amount of K^+ is needed, because it is recycled.) The H^+,K^+-ATPase bears many similarities to the plasma membrane Na^+,K^+-ATPase and the SR Ca^{2+}-ATPase described earlier. It has a similar molecular weight, it forms an E-P intermediate, and many parts of its peptide sequence are homologous with the Na^+,K^+-ATPase and Ca^{2+}-ATPase.

Bone Remodeling by Osteoclast Proton Pumps Other proton-translocating ATPases exist in eukaryotic and prokaryotic systems. **Vacuolar ATPases** (V-type ATPases) are found in vacuoles, lysosomes, endosomes, Golgi, chromaffin granules, and coated vesicles. Various H^+-transporting ATPases occur in yeast and bacteria as well. H^+-transporting ATPases found in **osteoclasts** (multinucleate cells that break down bone during normal bone remodeling) provide a source of circulating calcium for soft tissues such as nerves and muscles. About 5% of bone mass in the human body undergoes remodeling at any given time. Once growth is complete, the body balances formation of new bone tissue by cells called **osteoblasts** with resorption of existing bone matrix by osteoclasts. Osteoclasts possess proton pumps—which are in fact V-type ATPases—on the portion of the plasma membrane that attaches to the bone. This region of the osteoclast membrane is called the ruffled border. The osteoclast attaches to the bone in the manner of a cup turned upside down on a saucer (Figure 9.53), leaving an extracellular space between the bone surface and the cell. The H^+-ATPases in the ruffled border pump protons into this space, creating an acidic solution that dissolves the bone mineral matrix. Bone mineral consists mainly of poorly crystalline hydroxyapatite $[Ca_{10}(PO_4)_6(OH)_2]$ with some carbonate (HCO_3^-) replacing OH^- or PO_4^{3-} in the crystal lattice. Transport of protons out of the osteoclasts lowers the pH of the extracellular space near the bone to about 4, dissolving the hydroxyapatite.

ABC Transporters Use ATP to Drive Import and Export Functions and Provide Multidrug Resistance

The word *cell* is from the Latin *cella,* meaning a "small room." Cells, just like humans, must keep their rooms neat and tidy, and they do this with special membrane transporters known as **multidrug resistance (MDR)** efflux pumps, often referred to as "molecular vacuum cleaners." MDR pumps export cellular waste molecules and toxins, as well as drugs that find their way into cells in various ways. Bacteria also have influx pumps, which bring essential nutrients (for example vitamin B_{12}) into the cell (Figure 9.54). At least five families of influx and efflux pumps are known, among them the **ABC transporters**. In eukaryotes, ABC transporters are problematic because they export potentially

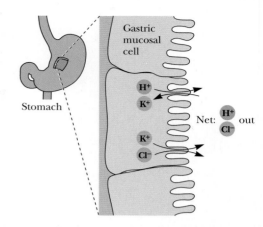

FIGURE 9.52 The H^+,K^+-ATPase of gastric mucosal cells mediates proton transport into the stomach. Potassium ions are recycled by means of an associated K^+/Cl^- cotransport system. The action of these two pumps results in net transport of H^+ and Cl^- into the stomach.

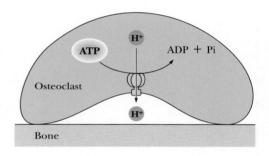

FIGURE 9.53 Proton pumps cluster on the ruffled border of osteoclast cells and function to pump protons into the space between the cell membrane and the bone surface. High proton concentration in this space dissolves the mineral matrix of the bone.

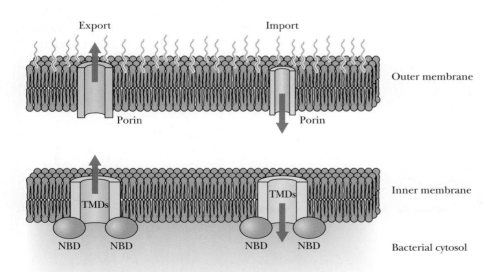

FIGURE 9.54 Influx pumps in the inner membrane of Gram-negative bacteria bring nutrients into the cell, whereas efflux pumps export cellular waste products and toxins. (Adapted from Garmory, H. S., and Titball, R. W. 2004. ATP-binding cassette transporters are targets for the development of antibacterial vaccines and therapies. *Infection and Immunity* **72**:6757–6763.)

therapeutic drugs (Figure 9.55) from cancer cells, so chemotherapy regimens must be changed often to avoid rejection of the beneficial drugs.

All ABC transporters consist of two transmembrane domains (TMDs), which form the transport pore, and two cytosolic nucleotide-binding domains (NBDs) that bind and hydrolyze ATP. The TMDs and NBDs are separate subunits (thus composing a tetramer) in bacterial ABC **importers** (Figure 9.56). Bacterial **exporters,** on the other hand, are homodimers, with each monomer made up of an N-terminal TMD and a C-terminal NBD. Eukaryotic ABC exporters are monomeric, with all four necessary domains in a single polypeptide chain.

The NBDs of ABC transporters from nearly all sources are similar in size, sequence, and structure. The TMDs, on the other hand, vary considerably in sequence, architecture, and number of transmembrane helices. ABC exporters contain a conserved core of 12 transmembrane helices, whereas ABC importers can have between 10 and 20 transmembrane helices. A variety of studies show that human MDR ATPases are similar to Sav1866, an exporting ABC transporter from *S. aureus*, and Sav1866 is considered to be a good model for the architecture of all ABC exporters.

The structures of several ABC transporters, in different stages of the transport cycle, provide a picture of how ATP binding and hydrolysis by the NBDs might be coupled to import and export of molecules (Figure 9.56). The TMDs can cycle from inward-facing to outward-facing conformations and back again, whereas the NBDs alternate between open and closed states. In all ABC transporters, a short "coupling helix" lies at the interface between each NBD and its corresponding TMD. Binding of ATP induces "closing," or joining of the NBD domains, bringing the coupling helices 10 to 15 Å closer to each other than in the ATP-free state. The merger of the coupling helices in turn triggers a flip-flop of the TMDs from the inward-facing to the outward-facing conformation. In this state, ABC exporters release bound drugs to the extracellular environment, whereas ABC importers accept substrate molecules from their associated substrate-binding proteins. Following ATP hydrolysis, release of ADP and inorganic phosphate allows the TMD to revert to its inward-facing conformation, where importers can release their substrates into the cytosol and exporters can bind new substrates to be exported.

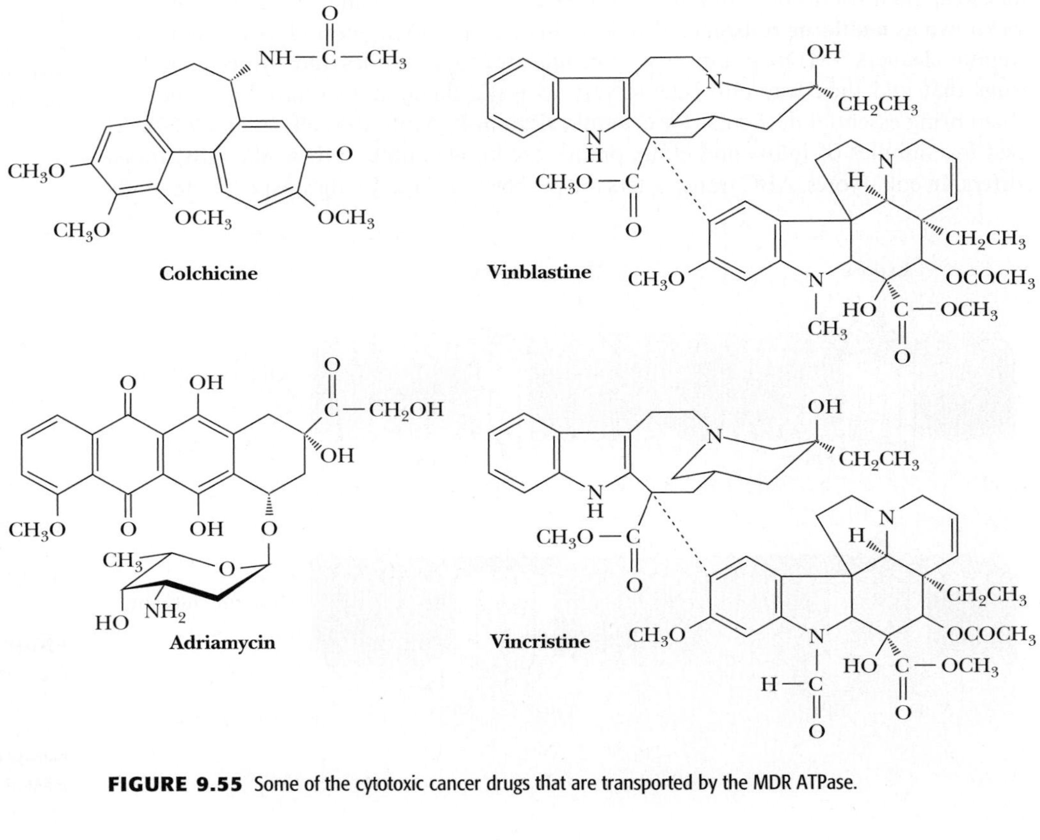

FIGURE 9.55 Some of the cytotoxic cancer drugs that are transported by the MDR ATPase.

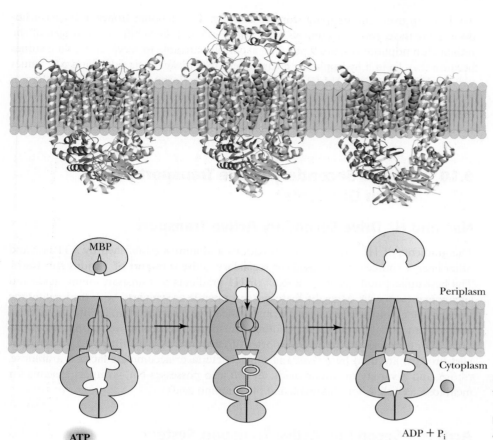

FIGURE 9.56 Several ABC transporters are shown in different stages of their transport cycles. Left to right: pdb id = 1L7V, pdb id = 2QI9, pdb id = 2NQ2. MBP is a multidrug binding protein, which binds molecules to be transported and delivers them to the transport channel. It is shown bound to the transport channel in the middle structures.

9.9 How Are Certain Transport Processes Driven by Light Energy?

As noted previously, certain biological transport processes are driven by light energy rather than by ATP. Two well-characterized systems are **bacteriorhodopsin,** the light-driven H^+-pump, and **halorhodopsin,** the light-driven Cl^- pump, of *Halobacterium halobium,* an archaeon that thrives in high-salt media. *H. halobium* grows optimally at an NaCl concentration of 4.3 *M*. It was extensively characterized by Walther Stoeckenius, who found it growing prolifically in the salt pools near San Francisco Bay, where salt is commercially extracted from seawater. *H. halobium* carries out normal respiration if oxygen and metabolic energy sources are plentiful. However, when these substrates are lacking, *H. halobium* survives by using bacteriorhodopsin to capture light energy. In oxygen- and nutrient-deficient conditions, **purple patches** appear on the surface of *H. halobium.* These purple patches of membrane are 75% protein, the only protein being **bacteriorhodopsin (bR).** The purple color arises from a retinal molecule that is covalently bound in a Schiff base linkage with an ϵ-NH_2 group of Lys[216] on each bacteriorhodopsin protein (Figure 9.57). Bacteriorhodopsin is a 26-kD transmembrane protein that packs so densely in the membrane that it naturally forms a two-dimensional crystal in the plane of the membrane. The retinal moiety lies parallel to the membrane plane, about 1 nm below the membrane's outer surface (Figure 9.13).

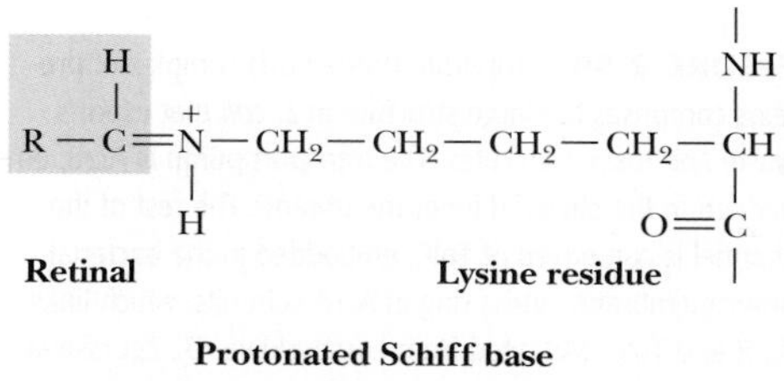

Bacteriorhodopsin Uses Light Energy to Drive Proton Transport

Light energy drives transport of protons (H^+) through bacteriorhodopsin, providing energy for the bacterium in the form of a transmembrane proton gradient. Protons hop from site to site across bacteriorhodopsin, just as a person crossing a creek

FIGURE 9.57 The Schiff base linkage between the retinal chromophore and Lys[216].

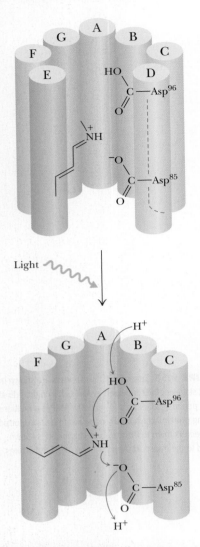

FIGURE 9.58 The mechanism of proton transport by bacteriorhodopsin. Asp[85] and Asp[96] on the third transmembrane segment (C) and the Schiff base of bound retinal serve as stepping stones for protons driven across the membrane by light-induced conformation changes. The hydrophobic environments of Asp[85] and Asp[96] raise the pK$_a$ values of their side-chain carboxyl groups, making it possible for these carboxyls to accept protons as they are transported across the membrane.

would jump from one stepping stone to another. The stepping stones in bacteriorhodopsin are the carboxyl groups of Asp[85] and Asp[96] and the Schiff base nitrogen of the retinal chromophore (Figure 9.58). The aspartates are able to serve as stepping stones because they lie in a hydrophobic environment that makes their side-chain pK$_a$ values very high (more than 11). Light absorption in bacteriorhodopsin is different than in rhodopsin in that it converts retinal from all-*trans* to the 13-*cis* configuration, triggering conformation changes that induce pK$_a$ changes and thus facilitate H$^+$ transfers (between Asp[96], the Schiff base, and Asp[85]) and net H$^+$ transport across the membrane.

9.10 How Is Secondary Active Transport Driven by Ion Gradients?

Na$^+$ and H$^+$ Drive Secondary Active Transport

The gradients of H$^+$, Na$^+$, and other cations and anions established by ATPases and other energy sources can be used for **secondary active transport** of various substrates. The best-understood systems use Na$^+$ or H$^+$ gradients to transport amino acids and sugars in certain cells. Many of these systems operate as **symports**, with the ion and the transported amino acid or sugar moving in the same direction (that is, into the cell). In **antiport** processes, the ion and the other transported species move in opposite directions. (For example, the anion transporter of erythrocytes is an antiport.) **Proton symport** proteins are used by *E. coli* and other bacteria to accumulate lactose, arabinose, ribose, and a variety of amino acids. *E. coli* also possesses Na$^+$-symport systems for melibiose, as well as for glutamate and other amino acids.

AcrB Is a Secondary Active Transport System

The ABC transporters described in Section 9.8 are just one of five different families of multidrug resistance transporters. **AcrB,** the major MDR transporter in *E. coli,* is responsible for pumping a variety of molecules including drugs such as erythromycin, tetracycline, and the β-lactams (for example, penicillin). AcrB is part of a large **tripartite complex** that bridges the *E. coli* inner and outer membranes and spans the entire periplasmic space (Figure 9.59). AcrB works with its partners, **AcrA** and **TolC,** to transport drugs and other toxins from the cytoplasm across the entire cell envelope and into the extracellular medium.

AcrB is a secondary active transport system and an **H$^+$-drug antiporter.** As protons flow spontaneously inward through AcrB in the *E. coli* inner membrane, drug

FIGURE 9.59 A tripartite (three-part) complex of proteins comprises the large structure in *E. coli* that exports waste and toxin molecules. The transport pump is AcrB, embedded in the bacterial inner membrane. The rest of the channel is composed of TolC, embedded in the bacterial outer membrane, and a ring of AcrA subunits, which links AcrB and TolC. (Adapted from Lomovskaya, O., Zgurskaya, H. I., et al., 2007. Waltzing transporters and 'the dance macabre' between humans and bacteria. *Nature Reviews Drug Discovery* **6**:56–65.)

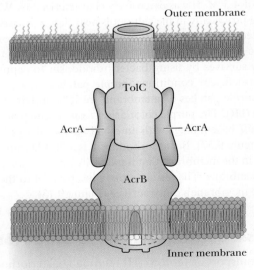

molecules are driven outward. AcrB is a homotrimer of large, 1100-residue subunits. Remarkably, the three identical subunits adopt slightly different conformations, denoted loose (L), tight (T), and open (O). Transported drug molecules enter AcrB through a tunnel that starts in the periplasmic space, about 15 Å above the inner membrane, and ends at the trimer center (Figure 9.60). The three conformations of the AcrB monomers are three consecutive states of a transport cycle. As each monomer cycles through the L, T, and O states, drug molecules enter the tunnel, are bound, and then are exported (Figure 9.61). Poetically, this three-step rotation has been likened to a Viennese waltz, and AcrB has been dubbed a "waltzing pump" by Olga Lomovskaya and her co-workers.

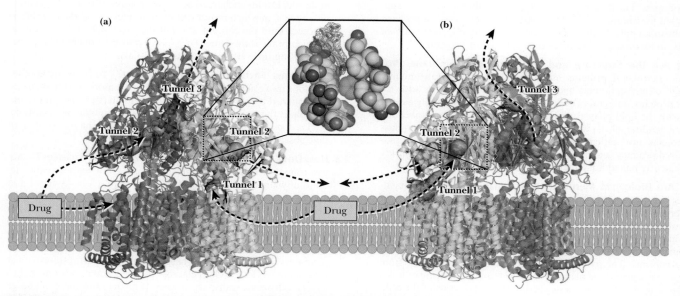

FIGURE 9.60 In the AcrB trimer, the three identical subunits adopt three different conformations. The "loose" L state (blue), the "tight" T state (yellow), and the "open" O (orange) state are indicated. Possible transport paths of drugs through the tunnels are shown in green. Tunnel 1 is lined with hydrophobic residues and is the likely point of entrance for drugs in the membrane bilayer. Tunnel 2 may serve either as an entrance port for water-soluble drugs or as an exit channel for nonsubstrates. Tunnel 3 is the exit pathway. Tunnels 1 and 2 converge at the hydrophobic substrate binding pocket, where minocyclin (an antibiotic similar to tetracycline) is bound in a hydrophobic pocket defined by phenylalanines 136, 178, 610, 615, 617, and 628; valines 139 and 612; isoleucines 277 and 626; and tyrosine 327. (Inset—all shown in spacefill. Minocyclin is shown in stick and wireframe.) Panels A and B represent one step in a L-T-O (or T-L-O) transport cycle. (Image kindly provided by Klass Martinus Pos.)

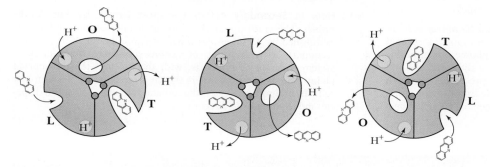

FIGURE 9.61 A model for drug transport by AcrB involves three possible conformations—loose (L, blue), tight (T, green), and open (O, pink)—for each of the three identical monomer subunits of the complex. The lateral grooves in L and T indicate low affinity and high affinity binding of drugs, respectively. The circle in the O state indicates that there is no drug binding in this state. Drugs to be transported (such as acridine, shown here) bind first to the L state. A conformational change to the T state moves the drug deeper into the tunnel, and a second conformation change opens the tunnel to the opposite side of the membrane, followed by release of the drug molecule. Binding, transport, and release of H^+ drives the drug transport cycle. (Adapted from Seeger, M., Schiefner, A., et al., 2006. Structural asymmetry of AcrB trimer suggests a peristaltic pump mechanism. *Science* **313**:1295–1298.)

SUMMARY

OWL Login to OWL to develop problem-solving skills and complete online homework assigned by your professor.

Membranes constitute the boundaries of cells and intracellular organelles, and they provide an environment where many important biological reactions and processes occur. Membranes have proteins that mediate and regulate the transport of metabolites, macromolecules, and ions.

9.1 What Are the Chemical and Physical Properties of Membranes? Amphipathic lipids spontaneously form a variety of structures when added to aqueous solution, including micelles and lipid bilayers. The fluid mosaic model for membrane structure suggests that membranes are dynamic structures composed of proteins and phospholipids. In this model, the phospholipid bilayer is a *fluid* matrix, in essence, a two-dimensional solvent for proteins.

9.2 What Are the Structure and Chemistry of Membrane Proteins? Peripheral proteins interact with the membrane mainly through electrostatic and hydrogen-bonding interactions with integral proteins. Integral proteins are those that are strongly associated with the lipid bilayer, with a portion of the protein embedded in, or extending all the way across, the lipid bilayer. Another class of proteins not anticipated by Singer and Nicolson, the lipid-anchored proteins, associate with membranes by means of a variety of covalently linked lipid anchors.

9.3 How Are Biological Membranes Organized? Biological membranes are asymmetric structures, and the lipids and proteins of membranes exhibit both lateral and transverse asymmetries. The two monolayers of the lipid bilayer have different lipid compositions and different complements of proteins. Loss of transverse lipid asymmetry has dramatic (and often severe) consequences for cells and organisms. The membrane composition is also different from place to place across the plane of the membrane. Clustering of lipids and proteins in specific ways serves the functional needs of the cell.

9.4 What Are the Dynamic Processes That Modulate Membrane Function? Motions of lipids and proteins in membranes underlie many cell functions. Lipid bilayers typically undergo gel-to-liquid crystalline phase transitions, with the transition temperature being dependent upon bilayer composition. Lipids and proteins undergo a variety of movements in membranes, including bond vibrations, rotations, and lateral and transverse motion, with a range of characteristic times. These motions modulate a variety of membrane processes, including lipid phase transitions, raft formation, membrane curvature, membrane remodeling, caveolae formation, and membrane fusion events that regulate vesicle trafficking.

9.5 How Does Transport Occur Across Biological Membranes? In most biological transport processes, the molecule or ion transported is water soluble, yet moves across the hydrophobic, impermeable lipid membrane at a rate high enough to serve the metabolic and physiological needs of the cell. Most of these processes occur with the assistance of specific transport protein. The transported species either diffuses through a channel-forming protein or is carried by a carrier protein. Transport proteins are all classed as integral membrane proteins. From a thermodynamic and kinetic perspective, there are only three types of membrane transport processes: *passive diffusion, facilitated diffusion, and active transport.*

9.6 What Is Passive Diffusion? In passive diffusion, the transported species moves across the membrane in the thermodynamically favored direction without the help of any specific transport system or molecule. For an uncharged molecule, passive diffusion is an entropic process, in which movement of molecules across the membrane proceeds until the concentration of the substance on both sides of the membrane is the same. The passive transport of charged species depends on their electrochemical potentials.

9.7 How Does Facilitated Diffusion Occur? Certain metabolites and ions move across biological membranes more readily than can be explained by passive diffusion alone. In all such cases, a protein that binds the transported species is said to facilitate its transport. Facilitated diffusion rates display saturation behavior similar to that observed with substrate binding by enzymes.

9.8 How Does Energy Input Drive Active Transport Processes? Active transport involves the movement of a given species against its thermodynamic potential. Such systems require energy input and are referred to as active transport systems. Active transport may be driven by the energy of ATP hydrolysis, by light energy, or by the potential stored in ion gradients. The original ion gradient arises from a primary active transport process, and the transport that depends on the ion gradient for its energy input is referred to as a secondary active transport process. When transport results in a net movement of electric charge across the membrane, it is referred to as an electrogenic transport process. If no net movement of charge occurs during transport, the process is electrically neutral. The Na^+,K^+-ATPase of animal plasma membranes, the Ca^{2+}-ATPase of muscle sarcoplasmic reticulum, the gastric ATPase, the osteoclast proton pump, and the multidrug transporter all use the free energy of hydrolysis of ATP to drive transport processes.

9.9 How Are Certain Transport Processes Driven by Light Energy? Light energy drives a series of conformation changes in the transmembrane protein bacteriorhodopsin that drive proton transport. The transport involves the *trans-13–cis* isomerization of retinal in Schiff base linkage to the protein via a lysine residue.

9.10 How Is Secondary Active Transport Driven by Ion Gradients? The gradients of H^+, Na^+, and other cations and anions established by ATPases and other energy sources can be used for secondary active transport of various substrates. Many of these systems operate as symports, with the ion and the transported amino acid or sugar moving in the same direction (that is, into the cell). In antiport processes, the ion and the other transported species move in opposite directions.

FOUNDATIONAL BIOCHEMISTRY Things You Should Know After Reading Chapter 9.

- The variation of membrane composition in tissues and subcellular organelles.

- The spontaneous formation of stable structures by lipids in water.

- The meaning of critical micelle concentrations for amphiphilic molecules.

- The essential features of the Singer-Nicolson fluid mosaic model.

- The relationship between membrane composition and membrane thickness.

- The characteristics of peripheral and integral membrane proteins.

- The structures and essential features of membrane proteins with single transmembrane segments and multiple transmembrane segments.

- The role of hydropathy plots in analysis of membrane proteins.

- The role prolines play in transmembrane helical segments.
- The location preferences of amino acids in transmembrane helices.
- The roles of reentrant loops and nonhelical segments in membrane proteins.
- The characteristics and functions of β-barrel proteins in membranes.
- The characteristics and functions of lipid-anchored membrane proteins.
- The asymmetric and heterogeneous nature of biological membranes.
- The types and time scales of motions in biological membranes.
- The characteristics and functions of flippases, floppases, and scramblases.
- The structural and functional consequences of lipid ordering in membranes.
- The structural and function features of lipid rafts.
- The essential features of the hop-diffusion model for lipids in membranes.
- The ways in which lipids and proteins accomplish curvature of membranes.
- The essential features of caveolins and caveolae.
- The essential features of SNARE proteins.

- The current model for SNARE complex assembly and synaptic vesicle fusion.
- The definitions of passive and facilitated diffusion and active transport.
- The calculation of the chemical potential for a membrane gradient involving an uncharged molecule.
- The calculation of the electrochemical potential for a membrane gradient involving an ionic or charged or molecular species.
- The graphical methods for distinguishing passive and facilitated diffusion.
- The structural and functional characteristics of membrane channels.
- The energy sources that are responsible for biological active transport.
- The structural and functional features of Na,K-ATPase and SR Ca-ATPase.
- The structural and functional features of the gastric and osteoclast H^+-ATPases.
- The structural and functional features of MDR efflux pumps.
- The essential features of light-driven proton transport by bacteriorhodopsin.
- The essential features of secondary active transport driven by ion gradients.

PROBLEMS

Answers to all problems are at the end of this book. Detailed solutions are available in the *Student Solutions Manual, Study Guide,* and *Problems Book.*

1. **Understanding the Occurrence of Natural Phospholipids** In problem 1 (b) in Chapter 8 (page 257), you were asked to draw all the possible phosphatidylserine isomers that can be formed from palmitic and linolenic acids. Which of the PS isomers are not likely to be found in biological membranes?

2. **Calculation of Phospholipid-to-Protein Ratios** The purple patches of the *Halobacterium halobium* membrane, which contain the protein bacteriorhodopsin, are approximately 75% protein and 25% lipid. If the protein molecular weight is 26,000 and an average phospholipid has a molecular weight of 800, calculate the phospholipid-to-protein mole ratio.

3. **Understanding the Densities of Membrane Components** Sucrose gradients for separation of membrane proteins must be able to separate proteins and protein–lipid complexes having a wide range of densities, typically 1.00 to 1.35 g/mL.

   a. Consult reference books (such as the *CRC Handbook of Biochemistry*) and plot the density of sucrose solutions versus percent sucrose by weight (g sucrose per 100 g solution), and versus percent by volume (g sucrose per 100 mL solution). Why is one plot linear and the other plot curved?

   b. What would be a suitable range of sucrose concentrations for separation of three membrane-derived protein–lipid complexes with densities of 1.03, 1.07, and 1.08 g/mL?

4. **Understanding Diffusion of Phospholipids in Membranes** Phospholipid lateral motion in membranes is characterized by a diffusion coefficient of about 1×10^{-8} cm²/sec. The distance traveled in two dimensions (across the membrane) in a given time is $r = (4Dt)^{1/2}$, where r is the distance traveled in centimeters, D is the diffusion coefficient, and t is the time during which diffusion occurs. Calculate the distance traveled by a phospholipid across a bilayer in 10 msec (milliseconds).

5. **Understanding Diffusion of Proteins in Membranes** Protein lateral motion is much slower than that of lipids because proteins

are larger than lipids. Also, some membrane proteins can diffuse freely through the membrane, whereas others are bound or anchored to other protein structures in the membrane. The diffusion constant for the membrane protein fibronectin is approximately 0.7×10^{-12} cm²/sec, whereas that for rhodopsin is about 3×10^{-9} cm²/sec.

   a. Calculate the distance traversed by each of these proteins in 10 msec.

   b. What could you surmise about the interactions of these proteins with other membrane components?

6. **Understanding the Phase Transitions of Membrane Phospholipids** Discuss the effects on the lipid phase transition of pure dimyristoyl phosphatidylcholine vesicles of added (a) divalent cations, (b) cholesterol, (c) distearoyl phosphatidylserine, (d) dioleoyl phosphatidylcholine, and (e) integral membrane proteins.

7. **Determining the Free Energy of a Galactose Gradient** Calculate the free energy difference at 25°C due to a galactose gradient across a membrane, if the concentration on side 1 is 2 mM and the concentration on side 2 is 10 mM.

8. **Determining the Electrochemical Potential of a Sodium Ion Gradient** Consider a phospholipid vesicle containing 10 mM Na$^+$ ions. The vesicle is bathed in a solution that contains 52 mM Na$^+$ ions, and the electrical potential difference across the vesicle membrane $\Delta\psi = \psi_{outside} - \psi_{inside} = -30$ mV. What is the electrochemical potential at 25°C for Na$^+$ ions?

9. **Assessing the Nature of Transmembrane Histidine Transport** Transport of histidine across a cell membrane was measured at several histidine concentrations:

[Histidine], μM	Transport, μmol/min
2.5	42.5
7	119
16	272
31	527
72	1220

   Does this transport operate by passive diffusion or by facilitated diffusion?

10. **Determining the Concentration Limits of an Active Transport System** (Integrates with Chapter 3.) Fructose is present outside a cell at 1 μM concentration. An active transport system in the plasma membrane transports fructose into this cell, using the free energy of ATP hydrolysis to drive fructose uptake. What is the highest intracellular concentration of fructose that this transport system can generate? Assume that one fructose is transported per ATP hydrolyzed; that ATP is hydrolyzed on the intracellular surface of the membrane; and that the concentrations of ATP, ADP, and P_i are 3 mM, 1 mM, and 0.5 mM, respectively. T = 298 K. (*Hint:* Refer to Chapter 3 to recall the effects of concentration on free energy of ATP hydrolysis.)

11. **Assessing the Energy Coupling of a Transport Process** In this chapter, we have examined coupled transport systems that rely on ATP hydrolysis or on primary gradients of Na^+ or H^+. Suppose you have just discovered an unusual strain of bacteria that transports rhamnose across its plasma membrane. Suggest experiments that would test whether it was linked to any of these other transport systems.

12. **Characterization of a Myristoyl Lipid Anchor** Which of the following peptides would be the most likely to acquire an N-terminal myristoyl lipid anchor?
    a. VLIHGLEQN
    b. THISISIT
    c. RIGHTHERE
    d. MEMEME
    e. GETREAL

13. **Characterization of a Prenyl Lipid Anchor** Which of the following peptides would be the most likely to acquire a prenyl anchor?
    a. RIGHTCALL
    b. PICKME
    c. ICANTICANT
    d. AINTMEPICKA
    e. None of the above

14. **Creating and Analyzing a Hydropathy Plot Online** What would the hydropathy plot of a soluble protein look like, compared to those in Figure 9.14? Find out by creating a hydropathy plot at *www.expasy .ch*. In the search box at the top of the page, type in "bovine pancreatic ribonuclease" and click "Go." The search engine should yield UniProtKB/Swiss-Prot entry P61823. Scroll to the bottom of the page and click "ProtScale" under Sequence Analysis Tools. On the next page, select the radio button for "Hphob. / Kyte and Doolittle," then scroll to the bottom of the page, and click "Submit." On the next page, scroll to the bottom of the page and click "Submit" again. At the bottom of the next page, after a few seconds, you should see a hydropathy plot. How does the plot for ribonuclease compare to those in Figure 9.14? You should see a large positive peak at the left side of the plot. This is the signal sequence portion of the polypeptide. You can read about signal sequences on page 1095.

15. **Assessing the Nature of Proline in Transmembrane α-Helices** Proline residues are almost never found in short α-helices; nearly all transmembrane α-helices that contain proline are long ones (about 20 residues). Suggest a reason for this observation.

16. **Analyzing the Structure of Proline-Containing α-Helices** As described in this chapter, proline introduces kinks in transmembrane α-helices. What are the molecular details of the kink, and why does it form? A good reference for this question is von Heijne, G., 1991. Proline kinks in transmembrane α-helices. *Journal of Molecular Biology* **218:**499–503. Another is Barlow, D. J., and Thornton, J. M., 1988. Helix geometry in proteins. *Journal of Molecular Biology* **201:**601–619.

17. **Comparing Membrane Barrel Structures** Compare the porin proteins, which have transmembrane pores constructed from β-barrels, with the Wza protein, which has a transmembrane pore constructed from a ring of α-helices. How many amino acids are required to form the β-barrel of a porin? How many would be required to form the same-sized pore from α-helices?

18. **Assessing the Structural Consequences of the Hop-Diffusion Model** The hop-diffusion model of Akihiro Kusumi suggests that lipid molecules in natural membranes diffuse within "fenced" areas before hopping the molecular fence to an adjacent area. Study Figure 9.29 and estimate the number of phospholipid molecules that would be found in a typical fenced area of local diffusion. For the purpose of calculations, you can assume that the surface area of a typical phospholipid is about 60 $Å^2$.

19. **Assessing the Energetic Consequences of Snorkeling in Membrane Proteins** What are the energetic consequences of snorkeling for a charged amino acid? Consider the lysine residue shown in Figure 9.16. If the lysine side chain was reoriented to extend into the center of the membrane, how far from the center would the positive charge of the lysine be? The total height of the peak for the lysine plot in Figure 9.15 is about $4kT$, where k is Boltzmann's constant. If the lysine side chain in Figure 9.16 was reoriented to face the membrane center, how much would its energy increase? How does this value compare with the classical value for the average translational kinetic energy of a molecule in an ideal gas ($3/2kT$)?

20. **Assessing the Dissociation Behavior of Aspartic Acid Residues in a Membrane** As described in the text, the pK_a values of Asp[85] and Asp[96] of bacteriorhodopsin are shifted to high values (more than 11) because of the hydrophobic environment surrounding these residues. Why is this so? What would you expect the dissociation behavior of aspartate carboxyl groups to be in a hydrophobic environment?

21. **Assessing the Dissociation Behavior of Lysine and Arginine Residues in a Membrane** Extending the discussion from problem 20, how would a hydrophobic environment affect the dissociation behavior of the side chains of lysine and arginine residues in a protein? Why?

22. **Analyzing the Nature of Light-Driven Proton Transport** In the description of the mechanism of proton transport by bacteriorhodopsin, we find that light-driven conformation changes promote transmembrane proton transport. Suggest at least one reason for this behavior. In molecular terms, how could a conformation change facilitate proton transport?

Preparing for the MCAT® Exam

23. Singer and Nicolson's fluid mosaic model of membrane structure presumed all of the following statements to be true EXCEPT:
    a. The phospholipid bilayer is a fluid matrix.
    b. Proteins can be anchored to the membrane by covalently linked lipid chains.
    c. Proteins can move laterally across a membrane.
    d. Membranes should be about 5 nm thick.
    e. Transverse motion of lipid molecules can occur occasionally.

ActiveModels Problems

24. Examine the ActiveModel for the KcsA potassium channel and explain the concept of a selectivity filter. Explain how the structural features of the KcsA selectivity filter enable its function.

25. Study the ActiveModel for aquaporin 1 and propose a reason for the different water transport rates for AQP0 and AQP1.

26. Consider the ActiveModel for the P-glycoprotein and describe how the primary structure of the Walker A region of this protein facilitates ATP binding.

FURTHER READING

Membrane Composition and Structure

Engelman, D. M., 2005. Membranes are more mosaic than fluid. *Nature* **438**:578–580.

Gallop, J., Jao, C., et al., 2006. Mechanism of endophilin N-BAR domain–mediated membrane curvature. *EMBO Journal* **25**: 2898–2910.

Granseth, E., Von Heijne, G., et al., 2004. A study of the membrane–water interface region of membrane proteins. *Journal of Molecular Biology* **346**:377–385.

Joh, N. H., Oberai, A., Yang, D., Whitelegge, J. P., Bowie, J. U., 2010. Similar energetic contributions of packing in the core of membrane and water-soluble proteins. *Journal of the American Chemical Society* **131**:10846–10847.

Kucerka, N., Marquardt, D., Harroun, T. A., Nieh, M.-P., Wassall, S. R., and Katsaras, J., 2009. Functional significance of lipid diversity: Orientation of cholesterol in bilayers is determined by lipid species. *Journal of the American Chemical Society* **131**:16368–16359.

Kusumi, A., Nadaka, C., et al., 2005. Paradigm shift of the plasma membrane concept from the two-dimensional continuum fluid to the partitioned fluid. *Annual Review of Biophysics and Biomolecular Structure* **34**:351–378.

MacKinnon, R., and von Heijne, G., 2006. Membranes. *Current Opinion in Structural Biology* **16**:431.

Pan, Y., and Konermann, L., 2010. Membrane protein structural insights from chemical labeling and mass spectrometry. *Analyst* **135**:1191–1200.

Phillips, R., Ursell, T., Wiggins, P., and Sens, P., 2010. Emerging roles for lipids in shaping membrane-protein function. *Nature* **459**:379–385.

Singer, S. J., and Nicolson, G. L., 1972. The fluid mosaic model of the structure of cell membranes. *Science* **175**:720–731.

Suzuki, K., Ritchie, K., et al., 2005. Rapid hop diffusion of a G-protein–coupled receptor in the plasma membrane as revealed by single-molecule techniques. *Biophysical Journal* **88**:3659–3680.

van Meer, G., and Vaz, W., 2005. Membrane curvature sorts lipids. *EMBO Reports* **6**(5):418–419.

Membrane Rafts

Chidlow, J. H., and Sessa, W. C., 2010. Caveolae, caveolins, and cavins: Complex control of cellular signaling and inflammation. *Cardiovascular Research* **86**:219–225.

Leventa, I., Grzybek, M., and Simons, K., 2010. Greasing their way: Lipid modifications determine protein association with membrane rafts. *Biochemistry* **49**:6305–6316.

Lingwood, D., and Simons, K., 2009. Lipid rafts as a membrane-organizing principle. *Science* **327**:46–50.

Simons, K., and Gerl, M. J., 2010. Revitalizing membrane rafts: new tools and insights. *Nature Reviews Molecular Cell Biology* **11**:688–699.

Membrane Proteins

Baker, M., 2010. Making membrane proteins for structures: A trillion tiny tweaks. *Nature Methods* **7**:429–434.

Bowie, J. U., 2006. Flip-flopping membrane proteins. *Nature Structural and Molecular Biology* **13**(2):94–96.

Cross, G. A. M., 2010. Fat-free proteins kill parasites. *Nature* **464**:689–690.

Daley, D. O., 2008. The assembly of membrane proteins into complexes. *Current Opinion in Structural Biology* **18**:420–424.

Dong, C., Beis, K., et al., 2006. Wza the translocon for *E. coli* capsular polysaccharides defines a new class of membrane protein. *Nature* **444**:226–229.

Elofsson, A., and von Heijne, G., 2007. Membrane protein structure: Prediction versus reality. *Annual Review of Biochemistry* **76**:125–140.

Fischer, F., Wolters, D., et al., 2006. Toward the complete membrane proteome. *Molecular and Cellular Proteomics* **5**(3):444–453.

Frearson, J. A., Brand, S., McElroy, S. P., Cleghorn, L. A., et al., 2010. N-Myristoyltransferase inhibitors as new leads to treat sleeping sickness. *Nature* **464**:728–732.

Hunte, C., and Richers, S., 2008. Lipids and membrane protein structures. *Current Opinion in Structural Biology* **18**:406–411.

Lee, J. K., and Stroud, R. M., 2010. Unlocking the eukaryotic membrane protein structural proteome. *Current Opinion in Structural Biology* **20**:464–470.

Liang, J., Adamian, L., et al., 2006. The membrane–water interface region of membrane proteins: Structural bias and the anti-snorkeling effect. *Trends in Biochemical Sciences* **30**:355–357.

Lindahl, E., and Sansom, M. S. P., 2008. Membrane proteins: molecular dynamics simulations. *Current Opinion in Structural Biology* **18**:425–431.

Rosenbaum, D. M., Rasmussen, S. G. F., and Kobilka, B. K., 2009. The structure and function of G-protein-coupled receptors. *Nature* **469**:356–363.

Skach, W. R., 2009. Cellular mechanisms of membrane protein folding. *Nature Structural and Molecular Biology* **16**:606–612.

Vinothkumar, K. R., and Henderson, R., 2010. Structures of membrane proteins. *Quarterly Review of Biophysics* **43**:65–158.

von Heijne, G., 2006. Membrane–protein topology. *Nature Reviews Molecular Cell Biology* **7**:909–918.

von Heijne, G., 2007. The membrane protein universe: What's out there and why bother? *Journal of Internal Medicine* **261**:543–557.

White, S. H., 2007. Membrane protein insertion: The biology–physics nexus. *Journal of General Physiology* **129**(5):363–369.

White, S. H., 2010. Biophysical dissection of membrane proteins. *Nature* **459**:344–346.

Zimmerberg, J., and Kozlov, M. M., 2006. How proteins produce cellular membrane curvature. *Nature Reviews Molecular Cell Biology* **7**:9–19.

Flippases

Baby-Periyanayaki, M., Natarajan, P., Zhou, N., and Graham, T. R., 2009. Linking phospholipid flippases to vesicle-mediated protein transport. *Biochimica et Biophysica Acta* **1791**:612–619.

Daleke, D. L., 2007. Phospholipid flippases. *Journal of Biological Chemistry* **282**:821–825.

Devaux, P. F., Herrmann, A., Ohlwein, N., and Kozlov, M. M., 2008. How lipid flippases can modulate membrane structure. *Biochimica et Biophysica Acta* **1778**:1592–1600.

Pomorski, T., and Menon, A. K., 2006. Lipid flippases and their biological functions. *Cellular and Molecular Life Sciences* **63**:2908–2921.

Sanyal, S., and Menon, A. K., 2010. Flipping Lipids: Why an' what's the reason for? *ACS Chemical Biology* **4**:895–909.

Van der Velden, L. M., Van de Graaf, S. F. J., and Klomp, L. W.O., 2010. Biochemical and cellular functions of P4 ATPases. *Biochemical Journal* **431**:1–11.

Active Transport Systems

Bublitz, M., Poulson, H., Morth, J., Nissen, P., 2010. In and out of the cation pumps: P-type ATPase structure revisited. *Current Opinion in Structural Biology* **20**:431–439.

Fang, Y., Jayaram, H., Shane, T., Kolmakova-Partensky, L., et al., 2009. Structure of a prokaryotic virtual proton pump at 3.2 Å resolution. *Nature* **460**:1040–1043.

Gadsby, D. C., 2009. Ion channels versus ion pumps: The principal difference, in principle. *Nature Reviews Molecular and Cell Biology* **10**:344–352.

Giacomini, K. M., Huang, S. M., Tweedie, D. J., Benet, L. Z., et al., 2010. Membrane transporters in drug development. *Nature Reviews Drug Discovery* **9**:215–236.

Hollenstein, K., Dawson, R. J., et al., 2007. Structure and mechanism of ABC transporter proteins. *Current Opinion in Structural Biology* **17**: 412–418.

Hvorup, R. N., Goetz, B., et al., 2007. Asymmetry in the structure of the ABC transporter–binding protein complex BtuCD-BtuF. *Science* **317**: 1387–1390.

Lingrel, J. B., 2010. The physiological significance of the cardiotonic steroid/ouabain-binding site of Na,K-ATPase. *Annual Review of Physiology* **72**:395–412.

Lomovskaya, O., Zgurskaya, H. I., Totrov, M., and Watkins, W. J., 2007. Waltzing transporters and "the dance macabre" between humans and bacteria. *Nature Reviews Drug Discovery* **6**:56–65.

Moller, J., Nissen, P., et al., 2005. Transport mechanism of the sarcoplasmic reticulum Ca^{2+}-ATPase pump. *Current Opinion in Structural Biology* **15**:387–393.

Morth, J., Pedersen, B., et al., 2007. Crystal structure of the sodium–potassium pump. *Nature* **450**:1043–1050.

Parcej, D., and Tampe, R., 2007. Caught in the act: An ABC transporter on the move. *Structure* **15**:1028–1030.

Parcej, D., and Tampé, R., 2010. ABC proteins in antigen translocation and inhibition. *Nature Chemical Biology* **6**:572–581.

Seeger, M., Schiefner, A., et al., 2006. Structural asymmetry of AcrB trimer suggests a peristaltic pump mechanism. *Science* **313**:1295–1298.

Shinoda, T., Ogawa, H., Cornelius, F., and Toyoshima, C., 2009. Crystal structure of the sodium-potassium pump at 2.4 Å resolution. *Nature* **459**:446–451.

Toyoshima, C., and Mizutani, T., 2004. Crystal structure of the calcium pump with a bound ATP analogue. *Nature* **430**:529–535.

Tsuda, T., and Toyoshima, C., 2009. Nucleotide recognition by CopA, a Cu^{2+}-transporting P-type ATPase. *The EMBO Journal* **28**:1782–1791.

Facilitated Diffusion and Membrane Channels

Bocquet, N., Nury, H., Baaden, M., Le Poupon, C., et al., 2009. X-ray structure of a pentameric ligand-gated ion channel in an apparently open conformation. *Nature* **457**:111–114.

Dutzler, R., 2006. The ClC family of chloride channels and transporters. *Current Opinion in Structural Biology* **16**:439–446.

Elvington, S. M., Liu, C. W., and Maduke, M. C., 2009. Substrate-driven conformational changees in ClC-ec1 observed by fluorine NMR. *The EMBO Journal* **28**:3090–3102.

Gouaux, E., and MacKinnon, R., 2005. Principles of selective ion transport in channels and pumps. *Science* **310**:1461–1465.

Hartzell, H. C., 2008. CaCl-ing channels get the last laugh. *Science* **322**:532–535.

Liu, Z., Gandhi, C. S., and Rees, D., 2009. Structure of a tetrameric MscL in an expanded intermediate state. *Nature* **461**:120–126.

Lunin, V. V., Dobrovetsky, E., et al., 2006. Crystal structure of the CorA Mg^{2+} transporter. *Nature* **440**:833–837.

Maguire, M., 2006. The structure of CorA: A Mg^{2+}-selective channel. *Current Opinion in Structural Biology* **16**:432–438.

Shi, N., Ye, S., et al., 2006. Atomic structure of a Na^+- and K^+-conducting channel. *Nature* **440**:570–574.

Tao, X., Avalos, J. L., Chen, J., and MacKinnon, R., 2009. Crystal structure of the eukaryotic strong inward-rectifier K^+ Channel Kir2.2 at 3.1 Å resolution. *Science* **326**:1668–1674.

Vásquez, V., and Perozo, E., 2009. A channel with a twist. *Nature* **461**:47–49.

Waight, A. B., Love, J., and Wang, D.-N., 2010. Structure and mechanism of a pentameric formate channel. *Nature Structural and Molecular Biology* **17**:31–38.

Wu, B., Steinbronn, C., Alsterfjord, M., Zeuthen, T., and Beitz, E., 2009. Concerted action of two cation filters in the aquaporin water channel. *The EMBO Journal* **28**:2188–2194.

Zagotta, W. N., 2006. Permutations of permeability. *Nature* **440**:42–429.

Zeth, K., and Thein, M., 2010. Porins in prokaryotes and eukaryotes: Common themes and variations. *Biochemical Journal* **431**:13–22.

Vesicles, Caveolae, and Membrane Fusion

Jahn, R., and Scheller, R. H., 2006. SNAREs: Engines for membrane fusion. *Nature Reviews Molecular Cell Biology* **7**:631–643.

Langer, J. D., Stoops, E. H., et al., 2007. Conformational changes of coat proteins during vesicle formation. *FEBS Letters* **581**:2083–2088.

Melia, T. J., 2007. Putting the clamps on membrane fusion: How complexin sets the stage for calcium-mediated exocytosis. *FEBS Letters* **581**:2131–2139.

Pang, Z. P., and Südhof, T. C., 2010. Cell biology of Ca^{2+}-triggered exocytosis. *Current Opinion in Cell Biology* **22**:496–505.

Parton, R. G., and Simons, K., 2007. The multiple faces of caveolae. *Nature Reviews Molecular Cell Biology* **8**:185–194.

Rizo, J., 2010. Synaptotagmin-SNARE coupling enlightened. *Nature Structural and Molecular Biology* **17**:260–262.

Rizo, J., and Rosenmund, C., 2008. Synaptic vesicle fusion. *Nature Structural & Molecular Biology* **15**:665–674.

Nucleotides and Nucleic Acids

© Barrington Brown/Photo Researchers, Inc.

◄ Francis Crick *(right)* and James Watson *(left)* point out features of their model for the structure of DNA.

We have discovered the secret of life!

Proclamation by Francis H. C. Crick to patrons of the Eagle, a pub in Cambridge, England (1953)

KEY QUESTIONS

10.1 What Are the Structure and Chemistry of Nitrogenous Bases?

10.2 What Are Nucleosides?

10.3 What Are the Structure and Chemistry of Nucleotides?

10.4 What Are Nucleic Acids?

10.5 What Are the Different Classes of Nucleic Acids?

10.6 Are Nucleic Acids Susceptible to Hydrolysis?

ESSENTIAL QUESTION

Nucleotides and nucleic acids are compounds containing nitrogen bases (aromatic cyclic structures possessing nitrogen atoms) as part of their structure. Nucleotides are essential to cellular metabolism, and nucleic acids are the molecules of genetic information storage and expression.

What are the structures of the nucleotides? How are nucleotides joined together to form nucleic acids? How is information stored in nucleic acids? What are the biological functions of nucleotides and nucleic acids?

Nucleotides are biological molecules that possess a heterocyclic nitrogenous base, a five-carbon sugar (pentose), and phosphate as principal components of their structure. The biochemical roles of nucleotides are numerous; they participate as essential intermediates in virtually all aspects of cellular metabolism. Serving an even more central biological purpose are the **nucleic acids,** the elements of heredity and the agents of genetic information transfer. Just as proteins are linear polymers of amino acids, nucleic acids are linear polymers of nucleotides. Like the letters in this sentence, the orderly sequence of nucleotide residues in a nucleic acid can encode information. The two basic kinds of nucleic acids are **deoxyribonucleic acid (DNA)** and **ribonucleic acid (RNA).** The five-carbon sugar in DNA is 2-deoxyribose; in RNA, it is ribose. (See Chapter 7 for a detailed discussion of sugars and other carbohydrates.) DNA is the repository of genetic information in cells, whereas RNA serves in the expression of this information through the processes of **transcription** and **translation** (Figure 10.1). An interesting exception to this rule is that some viruses have their genetic information stored as RNA.

This chapter describes the chemistry of nucleotides and the major classes of nucleic acids. Chapter 11 presents methods for determination of nucleic acid primary structure (nucleic acid sequencing) and describes the higher orders of nucleic acid structure. Chapter 12

ⵁWL Online homework and a Student Self Assessment for this chapter may be assigned in OWL.

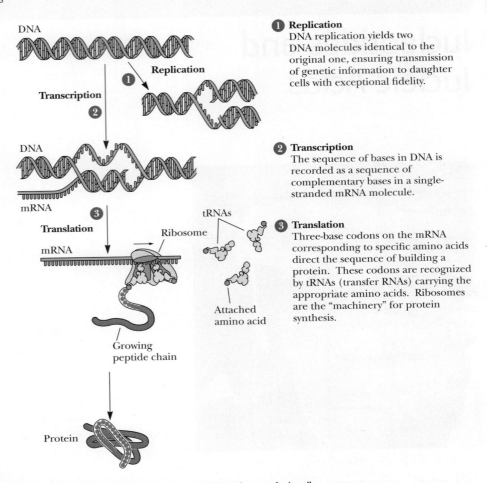

1 Replication
DNA replication yields two DNA molecules identical to the original one, ensuring transmission of genetic information to daughter cells with exceptional fidelity.

2 Transcription
The sequence of bases in DNA is recorded as a sequence of complementary bases in a single-stranded mRNA molecule.

3 Translation
Three-base codons on the mRNA corresponding to specific amino acids direct the sequence of building a protein. These codons are recognized by tRNAs (transfer RNAs) carrying the appropriate amino acids. Ribosomes are the "machinery" for protein synthesis.

FIGURE 10.1 The fundamental process of information transfer in cells.

introduces the *molecular biology of recombinant DNA:* the construction and uses of novel DNA molecules assembled by combining segments from different DNA molecules.

10.1 | What Are the Structure and Chemistry of Nitrogenous Bases?

The bases of nucleotides and nucleic acids are derivatives of either **pyrimidine** or **purine.** Pyrimidines are six-membered heterocyclic aromatic rings containing two nitrogen atoms (Figure 10.2a). The atoms are numbered in a clockwise fashion, as shown in Figure 10.2. The purine ring system consists of two rings of atoms: one resembling the pyrimidine ring and another resembling the imidazole ring (Figure 10.2b). The nine atoms in this fused ring system are numbered according to the convention shown.

The pyrimidine ring system is planar, whereas the purine system deviates somewhat from planarity in having a slight pucker between its imidazole and pyrimidine portions. Both are relatively insoluble in water, as might be expected from their pronounced aromatic character.

Three Pyrimidines and Two Purines Are Commonly Found in Cells

The common naturally occurring pyrimidines are **cytosine, uracil,** and **thymine** (5-methyluracil) (Figure 10.3). Cytosine and thymine are the pyrimidines typically found in DNA, whereas cytosine and uracil are common in RNA. Note that the

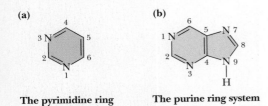

(a)

The pyrimidine ring

(b)

The purine ring system

FIGURE 10.2 **(a)** The pyrimidine ring system; by convention, atoms are numbered as indicated. **(b)** The purine ring system, atoms numbered as shown.

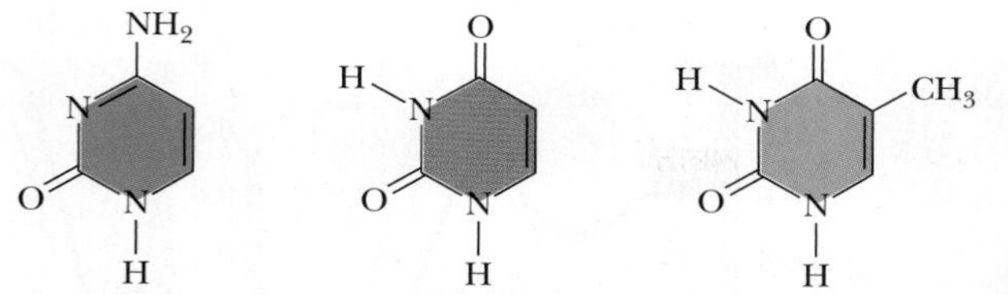

Cytosine (2-oxy-4-amino pyrimidine) Uracil (2-oxy-4-oxy pyrimidine) Thymine (2-oxy-4-oxy 5-methyl pyrimidine)

FIGURE 10.3 The common pyrimidine bases—cytosine, uracil, and thymine—in the tautomeric forms predominant at pH 7.

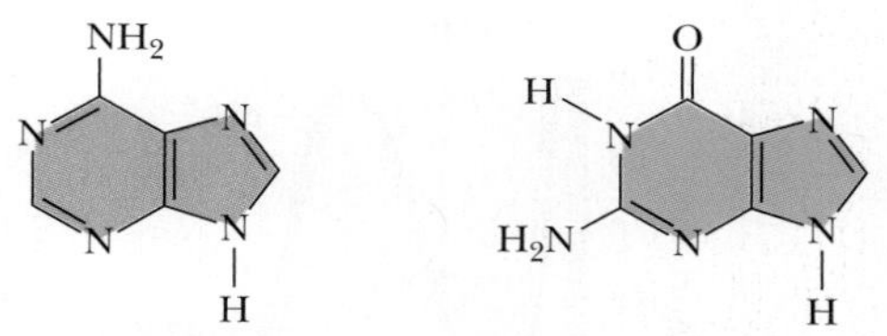

Adenine (6-amino purine) Guanine (2-amino-6-oxy purine)

FIGURE 10.4 The common purine bases—adenine and guanine—in the tautomeric forms predominant at pH 7.

5-methyl group of thymine is the only thing that distinguishes it from uracil. Various pyrimidine derivatives, such as dihydrouracil, are present as minor constituents in certain RNA molecules.

Adenine (6-amino purine) and **guanine** (2-amino-6-oxy purine), the two common purines, are found in both DNA and RNA (Figure 10.4). Other naturally occurring purine derivatives include **hypoxanthine, xanthine,** and **uric acid** (Figure 10.5). Hypoxanthine and xanthine are found only rarely as constituents of nucleic acids. Uric acid, the most oxidized state for a purine derivative, is never found in nucleic acids.

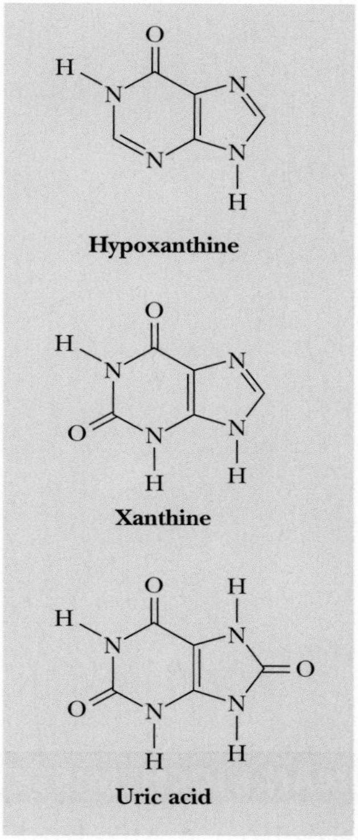

FIGURE 10.5 Other naturally occurring purine derivatives—hypoxanthine, xanthine, and uric acid.

The Properties of Pyrimidines and Purines Can Be Traced to Their Electron-Rich Nature

The aromaticity of the pyrimidine and purine ring systems and the electron-rich nature of their carbonyl and ring nitrogen substituents endow them with the capacity to undergo **keto–enol tautomeric shifts.** That is, pyrimidines and purines exist as tautomeric pairs, as shown in Figure 10.6 for uracil and Figure 10.7 for guanine. The keto tautomers of uracil, thymine, and guanine vastly predominate at neutral pH. In other words, pK_a values for ring nitrogen atoms 1 and 3 in uracil (Figure 10.6) are greater than 8 (the pK_a value for N-3 is 9.5). In contrast, the enamine form of cytosine predominates at pH 7 and the pK_a value for N-3 in this pyrimidine is 4.5. Similarly, for guanine (Figure 10.7), the pK_a value is 9.4 for N-1 and less than 5 for N-3. These pK_a values specify whether protons are associated with the various ring nitrogens at neutral pH. As such, they are important in determining

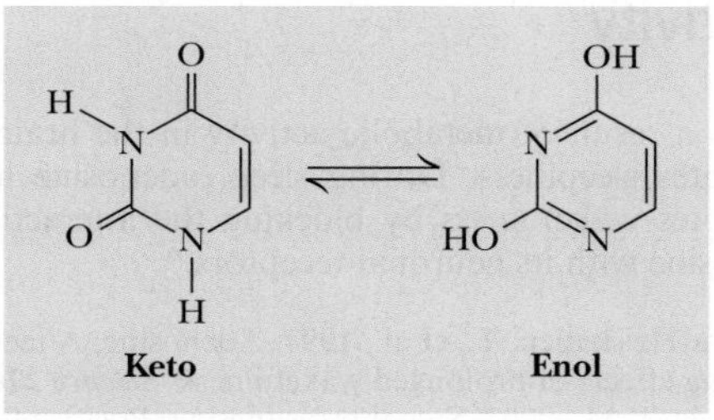

FIGURE 10.6 The keto–enol tautomerization of uracil.

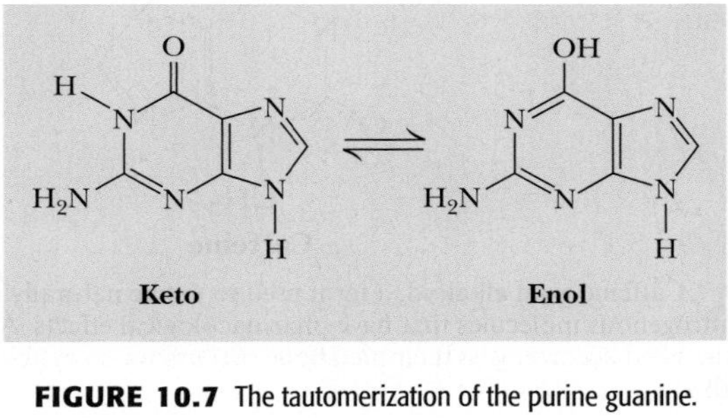

FIGURE 10.7 The tautomerization of the purine guanine.

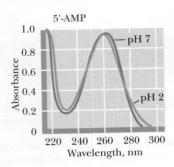
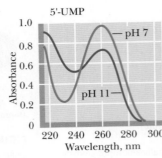
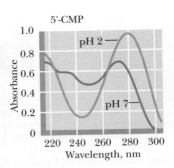
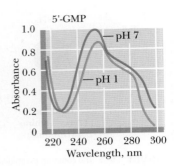

FIGURE 10.8 The UV absorptioan spectra of the common ribonucleotides.

whether these nitrogens serve as H-bond donors or acceptors. Hydrogen bonding between purine and pyrimidine bases is fundamental to the biological functions of nucleic acids, as in the formation of the double-helix structure of DNA (see Section 10.5). The important functional groups participating in H-bond formation are the amino groups of cytosine, adenine, and guanine; the ring nitrogens at position 3 of pyrimidines and position 1 of purines; and the strongly electronegative oxygen atoms attached at position 4 of uracil and thymine, position 2 of cytosine, and position 6 of guanine (see Figure 10.17).

Another property of pyrimidines and purines is their strong absorbance of ultraviolet (UV) light, which is also a consequence of the aromaticity of their heterocyclic ring structures. Figure 10.8 shows characteristic absorption spectra of several of the common bases of nucleic acids—adenine, uracil, cytosine, and guanine—in their nucleotide forms: AMP, UMP, CMP, and GMP (see Section 10.3). This property is particularly useful in quantitative and qualitative analysis of nucleotides and nucleic acids.

10.2 What Are Nucleosides?

Nucleosides are compounds formed when a base is linked to a sugar. The sugars of nucleosides are **pentoses** (five-carbon sugars, see Chapter 7). **Ribonucleosides** contain the pentose D-ribose, whereas 2-deoxy-D-ribose is found in **deoxyribonucleosides.** In both instances, the

---

HUMAN BIOCHEMISTRY

Adenosine: A Nucleoside with Physiological Activity

For the most part, nucleosides have no biological role other than to serve as component parts of nucleotides. Adenosine is a rare exception. In mammals, adenosine functions as an **autacoid,** or "local hormone," and as a neuromodulator. This nucleoside circulates in the bloodstream, acting locally on specific cells to influence such diverse physiological phenomena as blood vessel dilation, smooth muscle contraction, neuronal discharge, neurotransmitter release, and metabolism of fat. For example, when muscles work hard, they release adenosine, causing the surrounding blood vessels to dilate, which in turn increases the flow of blood and its delivery of O_2 and nutrients to the muscles. In a different autacoid role, adenosine acts in regulating heartbeat. The natural rhythm of the heart is controlled by a pacemaker, the sinoatrial node, which cyclically sends a wave of electrical excitation to the heart muscles. By blocking the flow of electrical current, adenosine slows the heart rate. *Supraventricular tachycardia* is a heart condition characterized by a rapid heartbeat. Intravenous injection of adenosine causes a momentary interruption of the rapid cycle of contraction and restores a normal heart rate. Adenosine is licensed and marketed as *Adenocard* to treat supraventricular tachycardia.

In addition, adenosine is implicated in sleep regulation. During periods of extended wakefulness, extracellular adenosine levels rise as a result of metabolic activity in the brain, and this increase promotes sleepiness. During sleep, adenosine levels fall. Caffeine promotes wakefulness by blocking the interaction of extracellular adenosine with its neuronal receptors.*

*Porrka-Heiskanen, T., et al., 1997. Adenosine: A mediator of the sleep-inducing effects of prolonged wakefulness. *Science* **276:**1265–1268; and Vaugeois, J.-M., 2002. Signal transduction: Positive feedback from coffee. *Nature* **418:**734–736.

Caffeine

Caffeine is an alkaloid, a term used to define naturally occurring nitrogenous molecules that have pharmacological effects. Alkaloids are classified according to their metabolic precursors, so caffeine is a purine alkaloid.

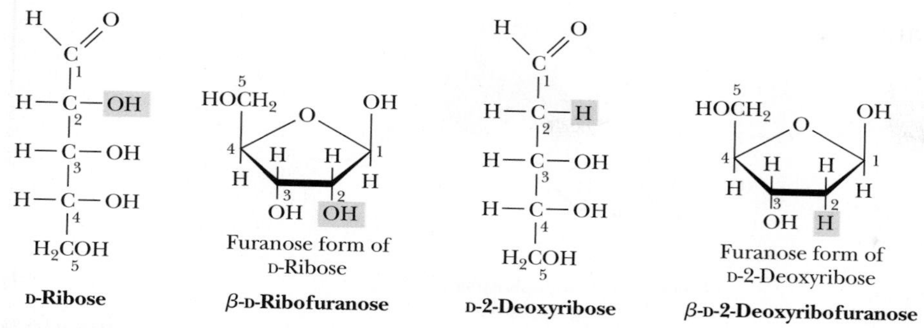

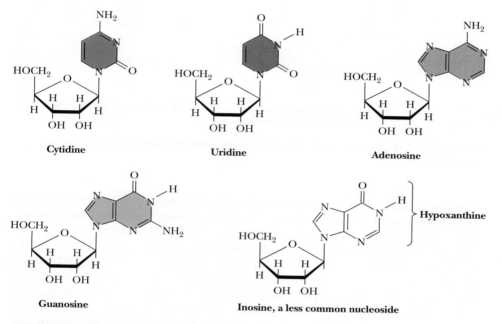

FIGURE 10.9 Furanose structures–ribose and deoxyribose.

FIGURE 10.10 The common ribonucleosides–cytidine, uridine, adenosine, and guanosine–and the less-common inosine. (Purine nucleosides and nucleotides usually adopt the anti conformation, where the purine ring is not above the ribose, as it would be in the syn conformation. Pyrimidines are always anti, never syn, because the 2-O atom of pyrimidines sterically hinders the ring from a position above the ribose.)

pentose is in the five-membered ring form furanose: D-ribofuranose for ribonucleosides and 2-deoxy-D-ribofuranose for deoxyribonucleosides (Figure 10.9). In nucleosides, these ribofuranose atoms are numbered as 1′, 2′, 3′, and so on to distinguish them from the ring atoms of the nitrogenous bases. The base is linked to the sugar via a **glycosidic bond.** Glycosidic bonds in nucleosides (and nucleotides, see following discussion) are always of the β-configuration. Nucleosides are named by adding the ending *-idine* to the root name of a pyrimidine or *-osine* to the root name of a purine. The common nucleosides are thus **cytidine, uridine, thymidine, adenosine,** and **guanosine** (Figure 10.10). Nucleosides are more water soluble than the free bases, because of the hydrophilicity of the pentose.

10.3 What Are the Structure and Chemistry of Nucleotides?

A **nucleotide** results when phosphoric acid is esterified to a sugar —OH group of a nucleoside. The nucleoside ribose ring has three —OH groups available for esterification, at C-2′, C-3′, and C-5′ (although 2′-deoxyribose has only two). The vast majority of monomeric nucleotides in the cell are **ribonucleotides** having 5′-phosphate groups. Figure 10.11

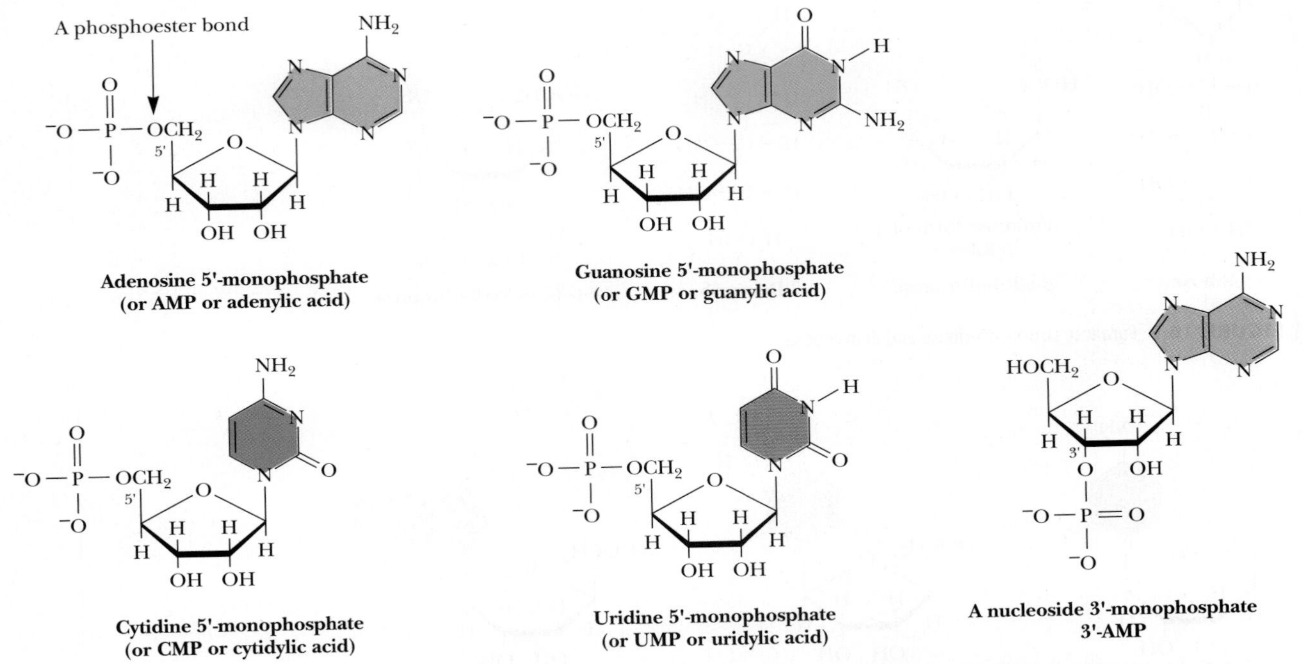

Adenosine 5'-monophosphate
(or AMP or adenylic acid)

Guanosine 5'-monophosphate
(or GMP or guanylic acid)

Cytidine 5'-monophosphate
(or CMP or cytidylic acid)

Uridine 5'-monophosphate
(or UMP or uridylic acid)

A nucleoside 3'-monophosphate
3'-AMP

FIGURE 10.11 Structures of the four common ribonucleotides—AMP, GMP, CMP, and UMP. Also shown is the nucleotide 3'-AMP.

shows the structures of the common four *ribonucleotides,* whose formal names are **adenosine 5'-monophosphate, guanosine 5'-monophosphate, cytidine 5'-monophosphate,** and **uridine 5'-monophosphate.** These compounds are more often referred to by their abbreviations: **5'-AMP, 5'-GMP, 5'-CMP,** and **5'-UMP,** or even more simply as **AMP, GMP, CMP,** and **UMP.** Because the pK_a value for the first dissociation of a proton from the phosphoric acid moiety is 1.0 or less, the nucleotides have acidic properties. This acidity is implicit in the other names by which these substances are known—**adenylic acid, guanylic acid, cytidylic acid,** and **uridylic acid.** The pK_a value for the second dissociation, pK_2, is about 6.0, so at neutral pH or above, the net charge on a nucleoside monophosphate is -2. Nucleic acids, which are polymers of nucleoside monophosphates, derive their name from the acidity of these phosphate groups.

Cyclic Nucleotides Are Cyclic Phosphodiesters

Nucleoside monophosphates in which the phosphoric acid is esterified to *two* of the available ribose hydroxyl groups (Figure 10.12) are found in all cells. Forming two such ester linkages with one phosphate results in a cyclic phosphodiester structure. **3',5'-cyclic AMP,** often abbreviated **cAMP,** and its guanine analog **3',5'-cyclic GMP,** or **cGMP,** are important regulators of cellular metabolism (see Parts 3 and 4).

Nucleoside Diphosphates and Triphosphates Are Nucleotides with Two or Three Phosphate Groups

Additional phosphate groups can be linked to the phosphoryl group of a nucleotide through the formation of phosphoric anhydride linkages, as shown in Figure 10.13. Addition of a second phosphate to AMP creates **adenosine 5'-diphosphate,** or **ADP,** and adding a third yields **adenosine 5'-triphosphate,** or **ATP.** The respective phosphate groups are designated by the Greek letters α, β, and γ, starting with the α-phosphate as the one linked directly to the pentose. The abbreviations **GTP, CTP,** and **UTP** represent the other corresponding nucleoside 5'-triphosphates. Like the nucleoside 5'-monophosphates, the nucleoside 5'-diphosphates and 5'-triphosphates all occur in the free state in the cell,

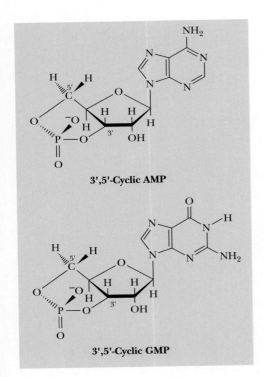

3',5'-Cyclic AMP

3',5'-Cyclic GMP

FIGURE 10.12 The cyclic nucleotides cAMP and cGMP.

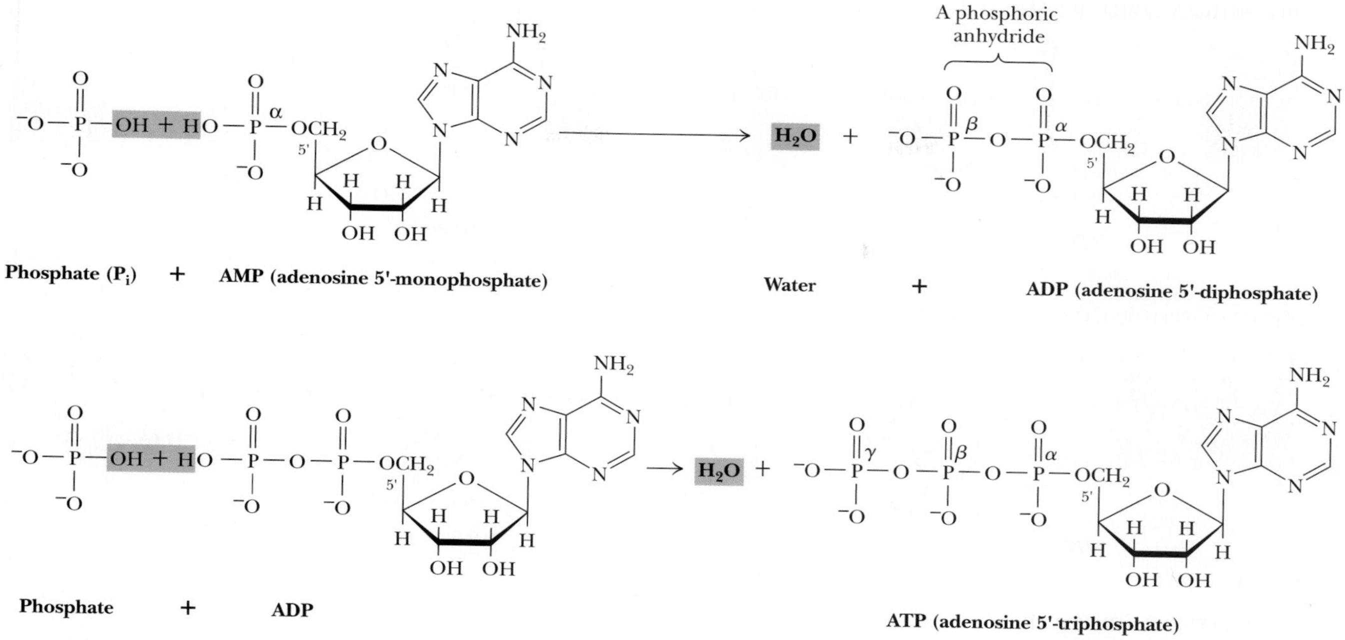

FIGURE 10.13 Formation of ADP and ATP by the successive addition of phosphate groups via phosphoric anhydride linkages. Note that the reaction is a dehydration synthesis reaction.

as do their deoxyribonucleoside phosphate counterparts, represented as dAMP, dADP, and dATP; dGMP, dGDP, and dGTP; dCMP, dCDP, and dCTP; dUMP, dUDP, and dUTP; and dTMP, dTDP, and dTTP.

NDPs and NTPs Are Polyprotic Acids

Nucleoside 5′-diphosphates (NDPs) and **nucleoside 5′-triphosphates (NTPs)** are relatively strong *polyprotic acids* in that they dissociate three and four protons, respectively, from their phosphoric acid groups. The resulting phosphate anions on NDPs and NTPs form stable complexes with divalent cations such as Mg^{2+} and Ca^{2+}. Because Mg^{2+} is present at high concentrations (as much as 40 mM) intracellularly, NDPs and NTPs occur primarily as Mg^{2+} complexes in the cell.

Nucleoside 5′-Triphosphates Are Carriers of Chemical Energy

Nucleoside 5′-triphosphates are indispensable agents in metabolism because the phosphoric anhydride bonds they possess are a prime source of chemical energy to do biological work.

Virtually all of the biochemical reactions of nucleotides involve either *phosphoryl* or *pyrophosphoryl group transfer:* the release of a phosphoryl group from an NTP to give an NDP, the release of a pyrophosphoryl group to give an NMP unit, or the acceptance of a phosphoryl group by an NMP or an NDP to give an NDP or an NTP (Figure 10.14). The pentose and the base are *not* directly involved in this chemistry. A "division of labor" directs ATP to serve as the primary nucleotide in central pathways of energy metabolism, whereas GTP is used to drive protein synthesis. CTP is essential for phospholipid synthesis (see Chapter 24), and UTP forms activated intermediates with sugars that go on to serve as substrates in the biosynthesis of complex carbohydrates and polysaccharides (see Chapter 22). Thus, the various nucleotides are channeled in appropriate metabolic directions through specific recognition of the base of the nucleotide. The bases of nucleotides never participate directly in the covalent bond chemistry that goes on. To complete the picture, the four NTPs and their dNTP counterparts are the substrates for synthesis of the remaining great class of biomolecules—the nucleic acids.

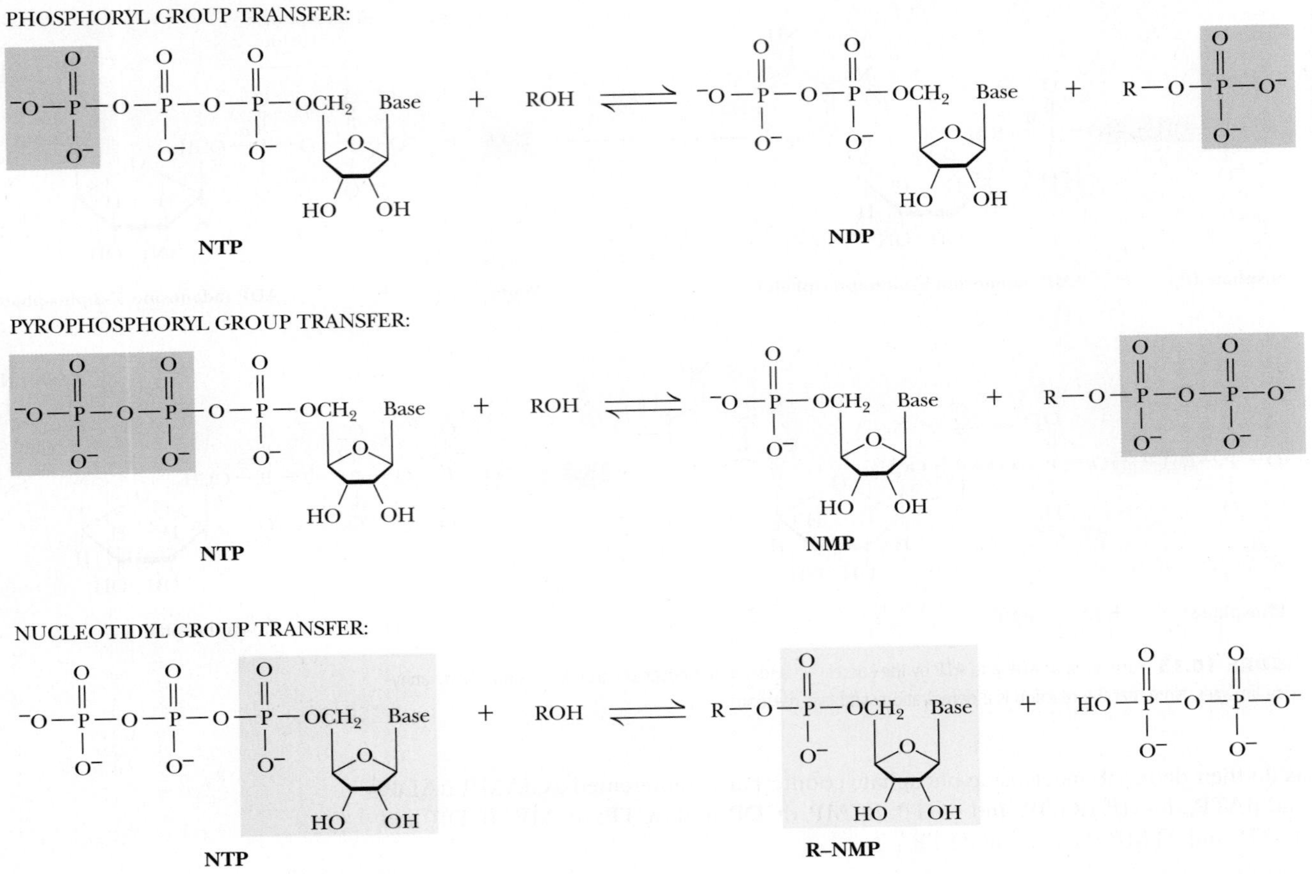

PHOSPHORYL GROUP TRANSFER:

PYROPHOSPHORYL GROUP TRANSFER:

NUCLEOTIDYL GROUP TRANSFER:

FIGURE 10.14 Phosphoryl, pyrophosphoryl, and nucleotidyl group transfer, the major biochemical reactions of nucleotides.

10.4 : What Are Nucleic Acids?

Nucleic acids are **polynucleotides:** linear polymers of nucleotides linked 3′ to 5′ by **phosphodiester bridges** (Figure 10.15). They are formed as 5′-nucleoside monophosphates are successively added to the 3′-OH group of the preceding nucleotide, a process that gives the polymer a directional sense. Polymers of ribonucleotides are named **ribonucleic acid,** or **RNA.** Deoxyribonucleotide polymers are called **deoxyribonucleic acid,** or **DNA.** Because C-1′ and C-4′ in deoxyribonucleotides are involved in furanose ring formation and because there is no 2′-OH, only the 3′- and 5′-hydroxyl groups are available for internucleotide phosphodiester bonds. In the case of DNA, a polynucleotide chain may contain hundreds of millions of nucleotide units. *The convention in all notations of nucleic acid structure is to read the polynucleotide chain from the 5′-end of the polymer to the 3′-end.* Note that this reading direction actually passes through each phosphodiester from 3′ to 5′ (Figure 10.15). A repetitious uniformity exists in the covalent backbone of polynucleotides.

The Base Sequence of a Nucleic Acid Is Its Defining Characteristic

The only significant variation that commonly occurs in the chemical structure of nucleic acids is the nature of the base at each nucleotide position. These bases are not part of the sugar–phosphate backbone but instead serve as distinctive side chains, much like the R groups of amino acids along a polypeptide backbone. They give the polymer its unique identity. A simple notation for nucleic acid structures is merely to list the order of bases in the polynucleotide using single capital letters—A, G, C, and U (or T).

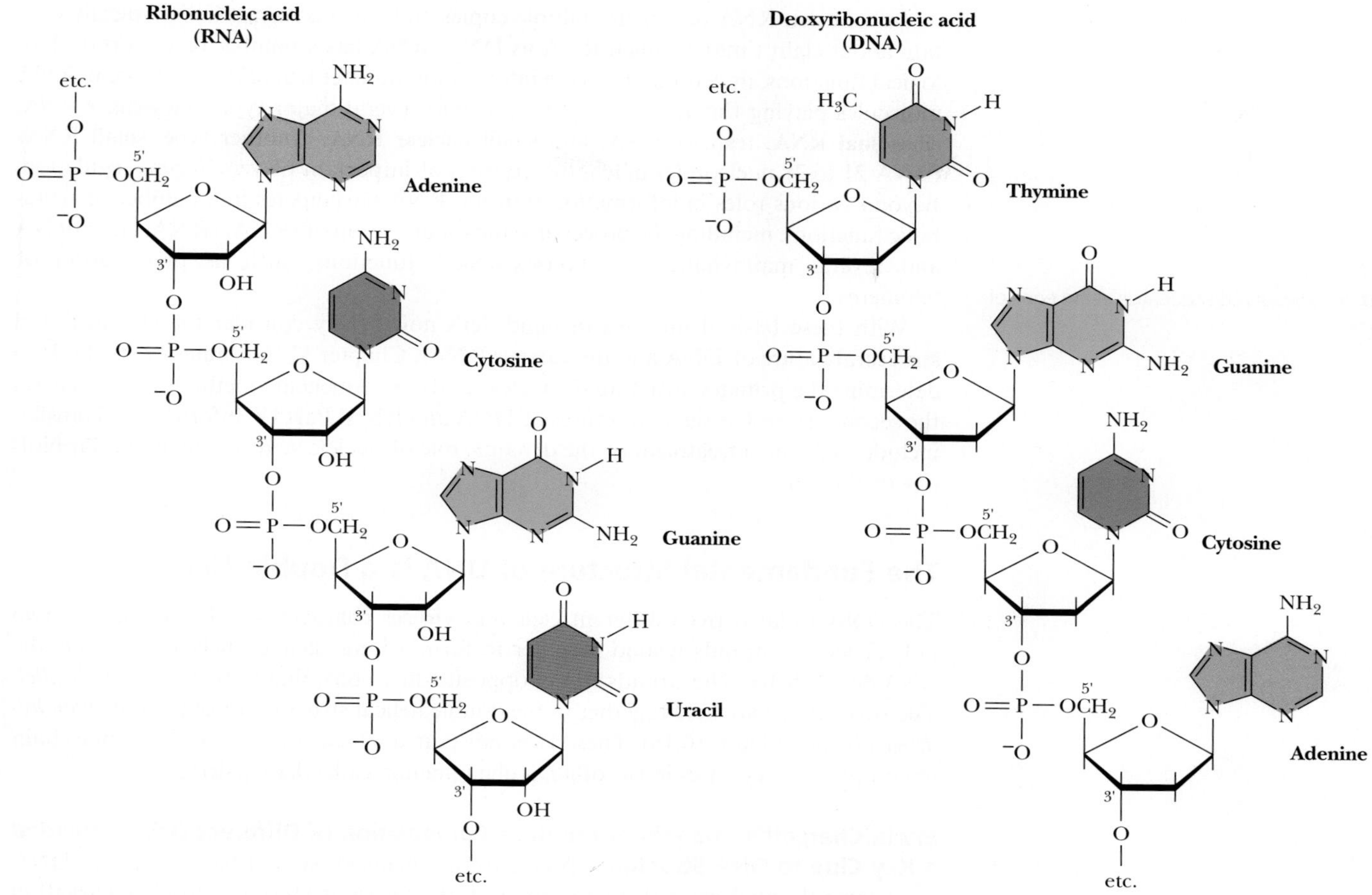

FIGURE 10.15 3′, 5′-phosphodiester bridges link nucleotides together to form polynucleotide chains. The 5′-ends of the chains are at the top; the 3′-ends are at the bottom.

Occasionally, a lowercase "p" is written between each successive base to indicate the phosphodiester bridge, as in GpApCpGpUpA.

To distinguish between RNA and DNA sequences, DNA sequences may be preceded by a lowercase "d" to denote deoxy, as in d-GACGTA. From a simple string of letters such as this, any biochemistry student should be able to draw the unique chemical structure of, for example, a pentanucleotide, even though it may contain more than 200 atoms.

10.5 What Are the Different Classes of Nucleic Acids?

The two major classes of nucleic acids are DNA and RNA. DNA has only one biological role, but it is the more central one. The information to make all the functional macromolecules of the cell (even DNA itself) is preserved in DNA and accessed through transcription of the information into RNA copies. Coincident with DNA's singular purpose, simple life forms such as viruses or bacteria usually contain only a single DNA molecule (or "chromosome"). Such DNA molecules must be quite large in order to embrace enough information for making the macromolecules necessary to maintain a living cell. The *Escherichia coli* chromosome has a molecular mass of 2.9×10^9 D and contains more than 9 million nucleotides. Eukaryotic cells have many chromosomes, and DNA is found principally in two copies in the diploid chromosomes of the nucleus. DNA is also found in mitochondria and in chloroplasts, where it encodes some of the proteins and RNAs unique to these organelles.

In contrast, RNA occurs in multiple copies and various forms. Cells typically contain about eight times as much RNA as DNA. RNA has a number of important biological functions, its central one being information transfer from DNA to protein. RNA molecules playing this role are categorized into several major types: **messenger RNA, ribosomal RNA, transfer RNA,** and **small nuclear RNA.** Another type, **small RNAs** (RNA 21 to 28 nucleotides in length), consists of important players in gene regulation. Beyond various roles in information transfer, RNA participates in a number of metabolic functions, including the processing and modification of tRNA, rRNA, and mRNA and several maintenance or "housekeeping" functions, such as preservation of telomeres.

Telomeres are specialized sequences at the ends of chromosomes.

With these basic definitions in mind, let's now briefly consider the chemical and structural nature of DNA and the various RNAs. Chapter 11 elaborates on methods to determine the primary structure of nucleic acids by sequencing methods and discusses the secondary and tertiary structures of DNA and RNA. Part IV, Information Transfer, includes a detailed treatment of the dynamic role of nucleic acids in the molecular biology of the cell.

The Fundamental Structure of DNA Is a Double Helix

The DNA isolated from different cells and viruses characteristically consists of two polynucleotide strands wound together to form a long, slender, helical molecule, the **DNA double helix.** The strands run in opposite directions; that is, they are *antiparallel*. The two strands are held together in the double helical structure through *interchain hydrogen bonds* (Figure 10.16). These H bonds pair the bases of nucleotides in one chain to complementary bases in the other, a phenomenon called **base pairing.**

Erwin Chargaff's Analysis of the Base Composition of Different DNAs Provided a Key Clue to DNA Structure   A clue to the chemical basis of base pairing in DNA came from the analysis of the base composition of various DNAs by Erwin Chargaff in the late 1940s. His data showed that the four bases commonly found in DNA (A, C, G, and T) do not occur in equimolar amounts and that the relative amounts of each vary from species to species (Table 10.1). Nevertheless, Chargaff noted that certain pairs of bases, namely, adenine and thymine, and guanine and cytosine, are always found in a 1:1 ratio and that the number of pyrimidine residues always equals the number of purine residues. These findings are known as *Chargaff's rules:* [A] = [T]; [C] = [G]; [pyrimidines] = [purines].

Table 10.1	**Molar Ratios Leading to the Formulation of Chargaff's Rules**				
Source	Adenine to Guanine	Thymine to Cytosine	Adenine to Thymine	Guanine to Cytosine	Purines to Pyrimidines
Ox	1.29	1.43	1.04	1.00	1.1
Human	1.56	1.75	1.00	1.00	1.0
Hen	1.45	1.29	1.06	0.91	0.99
Salmon	1.43	1.43	1.02	1.02	1.02
Wheat	1.22	1.18	1.00	0.97	0.99
Yeast	1.67	1.92	1.03	1.20	1.0
Haemophilus influenzae	1.74	1.54	1.07	0.91	1.0
E. coli K-12	1.05	0.95	1.09	0.99	1.0
Avian tubercle bacillus	0.4	0.4	1.09	1.08	1.1
Serratia marcescens	0.7	0.7	0.95	0.86	0.9
Bacillus schatz	0.7	0.6	1.12	0.89	1.0

Source: After Chargaff, E., 1951. Structure and function of nucleic acids as cell constituents. *Federation Proceedings* **10:**654–659.

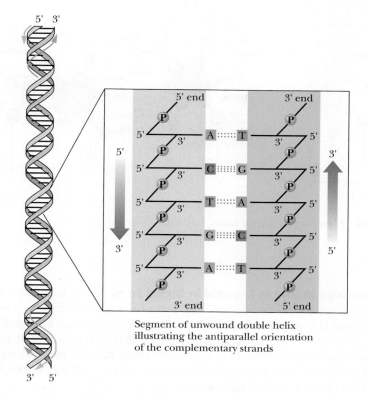

Segment of unwound double helix illustrating the antiparallel orientation of the complementary strands

FIGURE 10.16 The antiparallel nature of the DNA double helix.

Watson and Crick's Postulate of the DNA Double Helix Became the Icon of DNA Structure James Watson and Francis Crick, working in the Cavendish Laboratory at Cambridge University in 1953, took advantage of Chargaff's results and the data obtained by Rosalind Franklin and Maurice Wilkins in X-ray diffraction studies on the structure of DNA to conclude that DNA was a *complementary double helix*. Two strands of deoxyribonucleic acid (sometimes referred to as the *Watson strand* and the *Crick strand*) are held together by the bonding interactions between unique base pairs, always consisting of a purine in one strand and a pyrimidine in the other. Base pairing is very specific: If the purine is adenine, the pyrimidine must be thymine. Similarly, guanine pairs only with cytosine (Figure 10.17). Thus, if an A occurs in one strand of the helix, T must occupy the complementary position in the opposing strand. Likewise, a G in one dictates a C in the other. Because of this exclusive pairing of A only with T and G only with C, these pairs are taken as the standard or accepted law, and the A:T and G:C base pairs are often referred to as canonical. As Watson recognized from testing various combinations of bases using structurally accurate models, the A:T pair and the G:C pair form spatially equivalent units (Figure 10.17). The backbone-to-backbone distance of an A:T pair is 1.11 nm, virtually identical to the 1.08 nm chain separation in G:C base pairs.

Base pairing in the DNA molecule not only conforms to Chargaff results and Watson and Crick's rules but also has a profound property relating to heredity: *The sequence of bases in one strand has a complementary relationship to the sequence of bases in the other strand.* That is, the information contained in the sequence of one strand is conserved in the sequence of the other. Therefore, separation of the two strands and faithful replication of each, through a process in which base pairing specifies the nucleotide sequence in the newly synthesized strand, leads to two progeny molecules identical in every respect to the parental double helix (Figure 10.18). Elucidation of the double helical structure of DNA represented one of the most significant events in the history of science. This discovery more than any other marked the beginning of molecular biology. Indeed, upon solving the structure of DNA, Crick proclaimed in The Eagle, a pub just across from the Cavendish lab, "We have discovered the secret of life!"

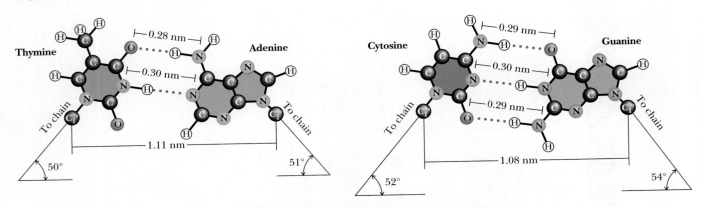

FIGURE 10.17 The Watson–Crick base pairs A : T and G : C.

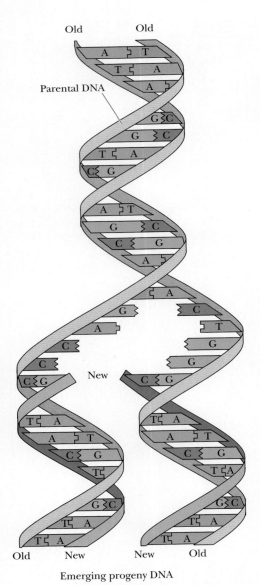

FIGURE 10.18 Replication of DNA gives identical progeny molecules because base pairing is the mechanism that determines the nucleotide sequence of each newly synthesized strand.

The Information in DNA Is Encoded in Digital Form In this digital age, we are accustomed to electronic information encoded in the form of extremely long arrays of just two digits: ones (1s) and zeros (0s). DNA uses four digits to encode biological information: A, C, G, and T. A significant feature of the DNA double helix is that virtually any base sequence (encoded information) is possible: Other than the base-pairing rules, no structural constraints operate to limit the potential sequence of bases in DNA.

DNA contains two kinds of information:

1. The base sequences of genes that encode the amino acid sequences of proteins and the nucleotide sequences of functional RNA molecules such as rRNA and tRNA (see following discussion)
2. The gene regulatory networks that control the expression of protein-encoding (and functional RNA-encoding) genes (see Chapter 29)

DNA Is in the Form of Enormously Long, Threadlike Molecules Because of the double helical nature of DNA molecules, their size can be represented in terms of the numbers of paired nucleotides (or **base pairs**) they contain. For example, the *E. coli* chromosome consists of 4.64×10^6 base pairs (abbreviated bp) or 4.64×10^3 kilobase pairs (kbp). DNA is a threadlike molecule. The diameter of the DNA double helix is only 2 nm, but the length of the DNA molecule forming the *E. coli* chromosome is over 1.6×10^6 nm (1.6 mm). The long dimension of an *E. coli* cell is only 2000 nm (0.002 mm), so its chromosome must be highly folded. Because of their long, threadlike nature, DNA molecules are easily sheared into shorter fragments during isolation procedures, and it is difficult to obtain intact chromosomes even from the simple cells of prokaryotes.

DNA in Cells Occurs in the Form of Chromosomes DNA occurs in various forms in different cells. The single chromosome of prokaryotic cells (Figure 10.19) is typically a circular DNA molecule. Proteins are associated with prokaryotic DNA, but unlike eukaryotic chromosomes, prokaryotic chromosomes are not uniformly organized into ordered nucleoprotein arrays. The DNA molecules of eukaryotic cells, each of which defines a chromosome, are linear and richly adorned with proteins. A class of arginine- and lysine-rich basic proteins called **histones** interact ionically with the anionic phosphate groups in the DNA backbone to form **nucleosomes,** structures in which the DNA double helix is wound around a protein "core" composed of pairs of four different histone polypeptides (see Section 11.5 in Chapter 11). Chromosomes also contain a varying mixture of other proteins, so-called **nonhistone chromosomal proteins,** many of which are involved in regulating which genes in DNA are transcribed at any given moment. The amount of DNA in a diploid mammalian cell is typically more than 1000 times that found in an *E. coli* cell. Some higher plant cells contain more than 50,000 times as much.

Various Forms of RNA Serve Different Roles in Cells

Unlike DNA, cellular RNA molecules are almost always single-stranded. However, all of them typically contain double-stranded regions formed when stretches of nucleotides with complementary base sequences align in an antiparallel fashion and form canonical A:U and G:C base pairs. (Compare Figures 10.3 and 10.17 to convince yourself that U would pair with A in the same manner T does.) Such base pairing creates secondary structure.

Messenger RNA Carries the Sequence Information for Synthesis of a Protein

Messenger RNA (**mRNA**) serves to carry the information or "message" that is encoded in genes to the sites of protein synthesis in the cell, where this information is translated into a polypeptide sequence. That is, mRNA molecules are transcribed copies of the protein-coding genetic units of DNA. Prokaryotic mRNAs have from 75 to 3,000 nucleotides; mRNA constitutes about 2% of total prokaryotic RNA.

Messenger RNA is synthesized during transcription, an enzymatic process in which an RNA copy is made of the sequence of bases along one strand of DNA. This mRNA then directs the synthesis of a polypeptide chain as the information that is contained within its nucleotide sequence is translated into an amino acid sequence by the protein-synthesizing machinery of the ribosomes. Ribosomal RNA and tRNA molecules are also synthesized by transcription of DNA sequences, but unlike mRNA molecules, these RNAs are not subsequently translated to form proteins. In prokaryotes, a single mRNA may contain the information for the synthesis of several polypeptide chains

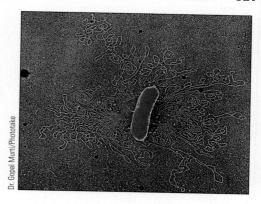

Dr. Gopal Murti/Phototake

FIGURE 10.19 If the cell walls of bacteria such as *Escherichia coli* are partially digested and the cells are then osmotically shocked by dilution with water, the contents of the cells are extruded to the exterior. In electron micrographs, the most obvious extruded component is the bacterial chromosome, shown here surrounding the cell.

> **A DEEPER LOOK**

Do the Properties of DNA Invite Practical Applications?

The molecular recognition between one DNA strand and its complementary partner not only leads to formation of a double-stranded DNA, but it also creates a molecule with mechanical properties distinctly different from single-stranded DNA. DNA double helices are relatively rigid rods. Single-stranded DNA molecules are flexible strands. These features are of interest to the practitioners of **nanotechnology,** the new branch of applied science that aims to manipulate matter at the molecular, or nanometer, level for practical purposes. Nanotechnology seeks to create **nanodevices,** nanoscale devices with simple machinelike qualities.

DNA chains have been used to construct nanomachines capable of simple movements such as rotation or pincerlike motions, and more elaborate DNA-based devices can even act as motors that walk along DNA tracks. To illustrate the principles, consider the DNA "tweezers" composed of three DNA strands (Q [red strand], S1, and S2) with regions of partial sequence complementarity. The 40-nucleotide Q strand is hybridized with two different 42-nucleotide-long DNA strands, S1 (blue/purple) and S2 (green/purple). Terminal segments of S1 and S2 are designed to be complementary to 18-nucleotide stretches at the opposite ends of Q. Base pairing between Q, S1, and S2 forms a V-shaped supramolecular structure, the tweezers, in an open conformation (1). Both S1 and S2 have 24-nucleotide-long ends that remain unpaired. The DNA tweezers can be driven into a closed conformation by the "fuel," a 56-nucleotide DNA strand (F) that has 24 bases complementary to the unpaired region of S1 and 24 bases complementary to the unpaired region of S2. Hybridization of the "fuel" to the unpaired segments of S1 and S2 (blue and green, respectively) closes the "tweezers" (2). The F strand has eight unpaired nucleotides remaining at its end; this region is called the "toehold." The "toehold" serves as the hybridization site for a fifth DNA strand, the R or "removal" strand. R is complementary to F along its length. Hybridization of the end of R to the complementary eight-base "toehold" region of F results in the unzipping of F from S2 and then S1 as it zips up with R. Removal of

F returns the DNA tweezers to the open conformation (1). The F:R duplex is the "waste" generated by the operation of the DNA nanomachine. Thus, this DNA tweezers nanomachine consumes "fuel" and generates "waste," as many common machines do.

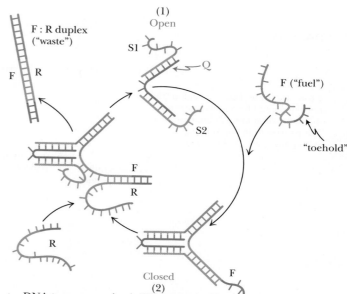

DNA tweezers—a simple DNA nanomachine. (*Adapted from Yurke, B., Turberfield, A. J., Mills, A. P., Jr., Simmel, F. C., and Neumann, J. L. 2000. A DNA-fuelled molecular machine made of DNA. Nature* **406:**605–608, *as discussed in an article in the whimsically named nanoscience journal* Small *by Simmel, F. C., and Dittmer W. U., 2006. DNA nanodevices.* Small *1:284–299*).

within its nucleotide sequence (Figure 10.20). In contrast, eukaryotic mRNAs encode only one polypeptide but are more complex in that they are synthesized in the nucleus in the form of much larger precursor molecules called **heterogeneous nuclear RNA,** or **hnRNA.** hnRNA molecules contain stretches of nucleotide sequence that have no protein-coding capacity. These noncoding regions are called **intervening sequences** or **introns** because they intervene between coding regions, which are called **exons.** Introns interrupt the continuity of the information specifying the amino acid sequence of a protein and must be spliced out before the message can be translated. In addition, eukaryotic hnRNA and mRNA molecules have a run of 100 to 200 adenylic acid residues attached at their 3′-ends, so-called **poly(A) tails.** This polyadenylation occurs after transcription has been completed and is essential for efficient translation and stability of the mRNA. The properties of mRNA molecules as they move through transcription and translation in prokaryotic versus eukaryotic cells are summarized in Figure 10.20.

Ribosomal RNA Provides the Structural and Functional Foundation for Ribosomes Ribosomes, the supramolecular assemblies where protein synthesis occurs, are about 65% RNA of the ribosomal RNA type. Ribosomal RNA **(rRNA)** molecules fold into characteristic secondary structures as a consequence of intramolecular base-pairing interactions (Figure 10.21). The different species of rRNA are generally referred to

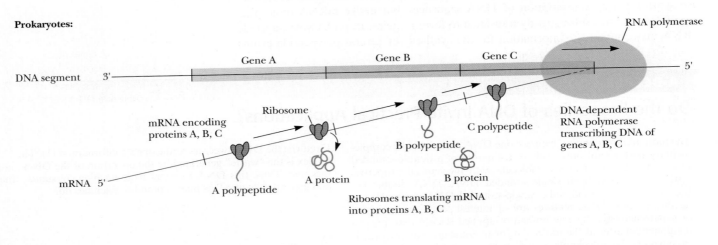

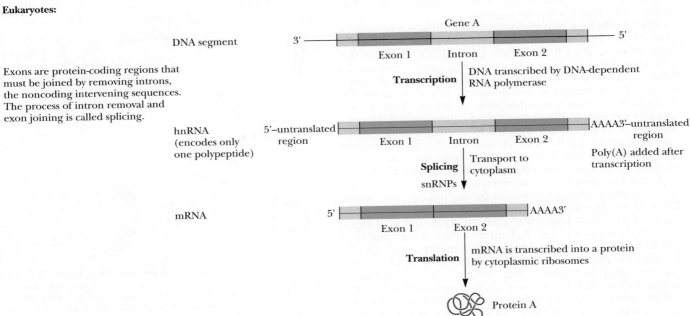

FIGURE 10.20 Transcription and translation of mRNA molecules in prokaryotic versus eukaryotic cells.

according to their **sedimentation coefficients**[1] (see the Chapter 5 A Deeper Look box Techniques Used in Protein Purification.), which are a rough measure of their relative size (Figure 10.22).

Ribosomes are composed of two subunits of different sizes that dissociate from each other if the Mg^{2+} concentration is below 10^{-3} M. Each subunit is a supramolecular assembly of proteins and RNA and has a total mass of 10^6 D or more. *E. coli* ribosomal subunits have sedimentation coefficients of 30S (the small subunit) and 50S (the large subunit). Eukaryotic ribosomes are somewhat larger than prokaryotic ribosomes, consisting of 40S and 60S subunits. More than 80% of total cellular RNA is represented by the various forms of rRNA.

Ribosomal RNAs characteristically contain a number of specially modified nucleotides, including **pseudouridine** residues, **ribothymidylic acid,** and methylated bases (Figure 10.23). The central role of ribosomes in the biosynthesis of proteins is treated in detail in Chapter 30. Here we briefly note the significant point that genetic information in the nucleotide sequence of an mRNA is translated into the amino acid sequence of a polypeptide chain by ribosomes.

Transfer RNAs Carry Amino Acids to Ribosomes for Use in Protein Synthesis

Transfer RNAs **(tRNAs)** serve as the carrier of amino acids for protein synthesis (see Chapter 30). tRNA molecules also fold into a characteristic secondary structure (Figure 10.24). tRNAs are small RNA molecules, containing 73 to 94 residues, a substantial number of which are methylated or otherwise unusually modified. Each of the 20 amino acids in proteins has at least one unique tRNA species dedicated to chauffeuring its delivery to ribosomes for insertion into growing polypeptide chains, and some amino acids are served by several tRNAs. In eukaryotes, there are even discrete sets of tRNA

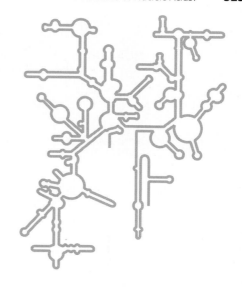

FIGURE 10.21 Ribosomal RNA has a complex secondary structure due to many intrastrand hydrogen bonds. The gray line in this figure traces a polynucleotide chain consisting of more than 1000 nucleotides. Aligned regions represent H-bonded complementary base sequences.

[1]Sedimentation coefficients are a measure of the velocity with which a particle sediments in a centrifugal force field. Sedimentation coefficients are expressed in *Svedbergs* (symbolized S), named to honor The Svedberg, developer of the ultracentrifuge. One S$=10^{-13}$ sec.

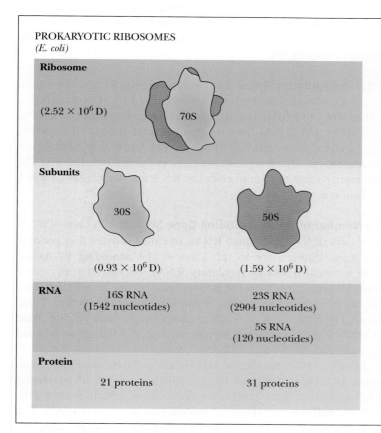

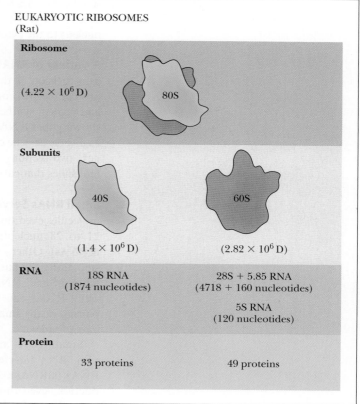

PROKARYOTIC RIBOSOMES
(*E. coli*)

Ribosome		
$(2.52 \times 10^6$ D)	70S	
Subunits	30S	50S
	$(0.93 \times 10^6$ D)	$(1.59 \times 10^6$ D)
RNA	16S RNA (1542 nucleotides)	23S RNA (2904 nucleotides)
		5S RNA (120 nucleotides)
Protein	21 proteins	31 proteins

EUKARYOTIC RIBOSOMES
(Rat)

Ribosome		
$(4.22 \times 10^6$ D)	80S	
Subunits	40S	60S
	$(1.4 \times 10^6$ D)	$(2.82 \times 10^6$ D)
RNA	18S RNA (1874 nucleotides)	28S + 5.85 RNA (4718 + 160 nucleotides)
		5S RNA (120 nucleotides)
Protein	33 proteins	49 proteins

FIGURE 10.22 The organization and composition of prokaryotic and eukaryotic ribosomes.

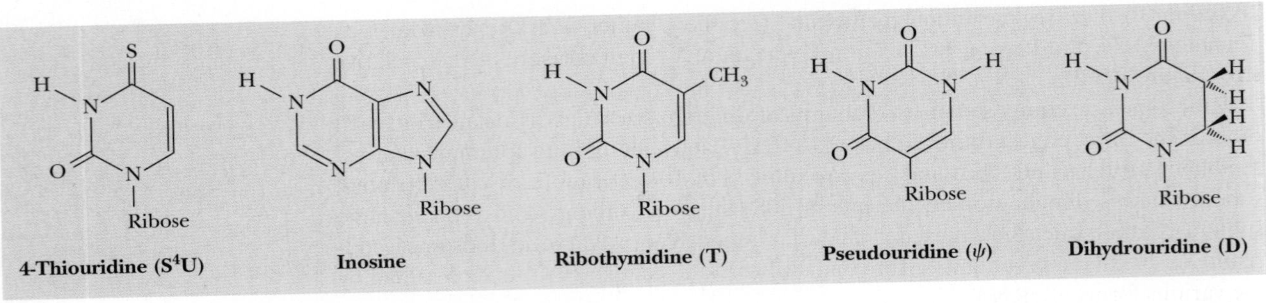

4-Thiouridine (S⁴U)   **Inosine**   **Ribothymidine (T)**   **Pseudouridine (ψ)**   **Dihydrouridine (D)**

FIGURE 10.23 Unusual bases in RNA.

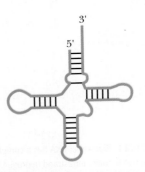

FIGURE 10.24 Transfer RNA also has a complex secondary structure due to many intrastrand hydrogen bonds. The black lines represent base-paired nucleotides in the sequence.

molecules for each site of protein synthesis—the cytoplasm, the mitochondrion, and in plant cells, the chloroplast. All tRNA molecules possess a 3′-terminal nucleotide sequence that reads **-CCA,** and the amino acid is carried to the ribosome attached as an acyl ester to the free 3′-OH of the terminal A residue. These **aminoacyl-tRNAs** are the substrates of protein synthesis, the amino acid being transferred to the carboxyl end of a growing polypeptide. The peptide bond–forming reaction is a catalytic process intrinsic to ribosomes.

Small Nuclear RNAs Mediate the Splicing of Eukaryotic Gene Transcripts (hnRNA) into mRNA Small nuclear RNAs, or snRNAs, are a class of RNA molecules found in eukaryotic cells, principally in the nucleus. They are neither tRNA nor small rRNA molecules, although they are similar in size to these species. They contain from 100 to about 200 nucleotides, some of which, like tRNA and rRNA, are methylated or otherwise modified. No snRNA exists as naked RNA. Instead, snRNA is found in stable complexes with specific proteins forming **small nuclear ribonucleoprotein particles,** or **snRNPs,** which are about 10S in size. Their occurrence in eukaryotes, their location in the nucleus, and their relative abundance (1% to 10% of the number of ribosomes) are significant clues to their biological purpose: snRNPs are important in the processing of eukaryotic gene transcripts (hnRNA) into mature messenger RNA for export from the nucleus to the cytoplasm (Figure 10.20).

A Variety of RNAs Serve Regulatory Roles The 21st century has witnessed a significant revision of the dogma of molecular biology as summarized by "DNA→RNA→ protein," which confined RNA to information transfer (mRNA), processing (snRNPs), and protein synthesis (rRNA and tRNA). This revision has been driven by discoveries showing that RNA molecules act in a multitude of ways to regulate biological processes as diverse as gene expression, chromatin organization, and programmed cell death. The next two sections summarize these newfound roles for RNA and the classes of RNA molecules that carry them out.

Small RNAs Serve a Number of Roles, Including Gene Silencing A class of RNA molecules even smaller than tRNAs is the **small RNAs,** so-called because they are only 21 to 28 nucleotides long. (Some refer to this class as the **noncoding RNAs** [or **ncRNAs**]. Others refer to small RNAs as **regulatory RNAs,** because virtually every step along the pathway of gene expression can be regulated by one or another small RNA.) Small RNAs are involved in a number of novel biological functions. These small RNAs can target DNA or RNA through complementary base pairing. Base pairing of the small RNA with particular nucleotide sequences in the target is called *direct readout.*

Small RNAs are classified into a number of subclasses on the basis of their function. **RNA interference (RNAi)** is mediated by one subclass, the **small interfering RNAs (siRNAs).** siRNAs disrupt gene expression by blocking specific protein production, even though the mRNA encoding the protein has been synthesized. The 21- to 23-nucleotide-long siRNAs act by base pairing with complementary sequences within a particular mRNA to form regions of double-stranded RNA (dsRNA).

The RNA World and Early Evolution

Proteins are encoded by nucleotide sequences in DNA. DNA replication depends on the activity of protein enzymes. These two statements form a "chicken and egg" paradox: Which came first in evolution—DNA or protein? Neither, it seems. The 1989 Nobel Prize in Chemistry was awarded to Thomas Cech and Sidney Altman for their discovery that RNA molecules are not only informational but also may be catalytic. This discovery gave evidence to earlier speculation by Carl Woese, Francis Crick, and Leslie Orgel that prebiotic evolution (that is, early evolution before cells arose) depended on self-replicating and catalytic RNAs, with proteins and DNA appearing later. Three basic assumptions about the prebiotic RNA world are (1) RNA replication maintained information-carrying RNAs, (2) Watson–Crick base pairing was essential to RNA replication, and (3) genetically encoded proteins were unavailable as catalysts. The challenge shifts to explaining the origin of nucleotides and their polymerization to form RNA.

Adenine exists in outer space and is found in comets and meteorites. A likely source is conversion of aminoimidazolecarbonitrile to adenine. (Aminoimidazolecarbonitrile is a tetramer of HCN; adenine is a pentamer.)

Aminoimidazolecarbonitrile   **Adenine**

▲ *Adapted from Glaser, R., et al., 2007. Adenine synthesis in interstellar space: Mechanisms of prebiotic pyrimidine-ring formation of monocyclic HCN-pentamers.* Astrobiology 7:455–470.

Glycolaldehyde can combine with other simple compounds to form ribose (and glucose). Glycolaldehyde has been detected in a gas cloud at the center of the Milky Way, our galaxy.

D-Ribose
$C_5H_{10}O_5$

Glycolaldehyde
$C_2H_4O_2$

α-D-Glucopyranose
$C_6H_{12}O_6$

(Acetic acid and methyl formate have the same eight atoms as glycolaldehyde; these two useful precursor molecules have also been detected in intergalactic clouds.) Inorganic phosphate, the remaining ingredient in nucleotides, is a common component in naturally occurring aqueous solutions. Its negative charge allows it to interact readily with positively charged mineral surfaces, upon which the first nucleotides may have spontaneously assembled. Cyclic nucleotides such as cAMP (3', 5'-cyclic AMP) and cGMP (3', 5'-cyclic GMP) are formed spontaneously and accumulate under mild conditions in the presence of formamide. Recent studies show that the chemical formation of short RNA polynucleotides from such cyclic nucleotides is thermodynamically favorable in the absence of enzymes. These tantalizing facts are bright spots along the dim thread that connects us to our distant past. The RNA world is an attractive hypothesis.

References: Gesteland, R. F., Cech, T. R., Atkins, J. F., eds., 2006. *The RNA World: The Nature of Modern RNA Suggests a Prebiotic RNA World,* 3rd ed. Cold Spring Harbor, NY: Cold Spring Harbor Laboratory Press.

Costanzo, G., Pino, S., Ciciriello, F., and Di Mauro, E., 2009. 3', 5'-cyclic AMP or 3', 5'-cyclic GMP generation of long RNA chains in water. *The Journal of Biological Chemistry* **284**:33206–33216.

These dsRNA regions are then specifically degraded, eliminating the mRNA from the cell (see Chapter 12). Thus, RNAi is a mechanism to silence the expression of specific genes, even after they have been transcribed, a phenomenon referred to as **gene silencing.** RNAi is also implicated in modifying the structure of chromatin and causing large-scale influences in gene expression. Another subclass, the **micro RNAs (miRNAs)** control developmental timing by base pairing with and preventing the translation of certain mRNAs, thus blocking synthesis of specific proteins. Thus, miRNAs also act in gene silencing. However, unlike siRNAs, miRNAs (22 nucleotides long) do not cause mRNA degradation. A third subclass is the **small nucleolar RNAs (snoRNAs).** snoRNAs (60 to 300 nucleotides long) are catalysts that accomplish some of the chemical modifications found in tRNA, rRNA, and even DNA (see Figure 10.23, for example). *Small RNAs* in bacteria (known by the acronym **sRNAs**) play an important role altering gene expression in response to stressful environmental situations.

Long Noncoding RNAs Are Another Class of Regulatory RNAs Long noncoding RNAs, or **lincRNAs,** are RNA transcripts greater than 5 kb in size that do not code for proteins. They are transcribed from DNA regions lying between genes and thus are sometimes referred to as "long intergenic noncoding RNAs" or "long intervening noncoding RNAs" (lincRNAs). RNAs of this type play roles in the regulation of chromatin structure and gene repression. Interestingly, some encode short peptides (11-32 amino acids long) that serve as the agents by which the lincRNA regulates gene expression. (Such small peptides are below the threshold for detection of protein-coding genes by conventional computer algorithms for protein prediction.)

The Chemical Differences Between DNA and RNA Have Biological Significance

Two fundamental chemical differences distinguish DNA from RNA:

1. DNA contains 2-deoxyribose instead of ribose.
2. DNA contains thymine instead of uracil.

What are the consequences of these differences, and do they hold any significance in common? An argument can be made that, because of these differences, DNA is chemically more stable than RNA. The greater stability of DNA over RNA is consistent with the respective roles these macromolecules have assumed in heredity and information transfer.

Consider first why DNA contains thymine instead of uracil. The key observation is that *cytosine deaminates to form uracil* at a finite rate in vivo (Figure 10.25). Because C in one DNA strand pairs with G in the other strand, whereas U would pair with A, conversion of a C to a U could potentially result in a heritable change of a C:G pair to a U:A pair. Such a change in nucleotide sequence would constitute a *mutation* in the DNA. To prevent this C deamination from leading to permanent changes in nucleotide sequence, a cellular repair mechanism "proofreads" DNA, and when a U arising from C deamination is encountered, it is treated as inappropriate and is replaced by a C. If DNA normally contained U rather than T, this repair system could not readily distinguish U formed by C deamination from U correctly paired with A. However, the U in DNA is "5-methyl-U" or, as it is conventionally known, thymine (Figure 10.26). That is, the 5-methyl group on T labels it as if to say "this U belongs; do not replace it."

The other chemical difference between RNA and DNA is that the ribose 2'-OH group on each nucleotide in RNA is absent in DNA. Consequently, the ubiquitous 3'-O of polynucleotide backbones lacks a vicinal hydroxyl neighbor in DNA. This difference leads to a greater resistance of DNA to alkaline hydrolysis, examined in detail in the following section. To view it another way, RNA is less stable than DNA because its vicinal 2'-OH group makes the 3'-phosphodiester bond susceptible to nucleophilic cleavage (Figure 10.27). For just this reason, it is selectively advantageous for the heritable form of genetic information to be DNA rather than RNA.

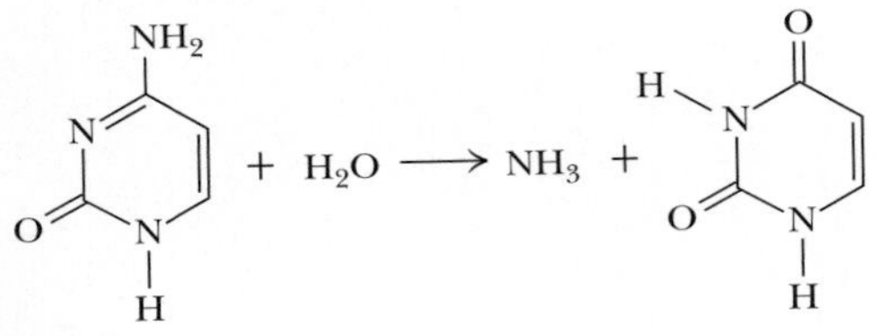

FIGURE 10.25 Deamination of cytosine forms uracil.

Cytosine + H_2O ⟶ NH_3 + **Uracil**

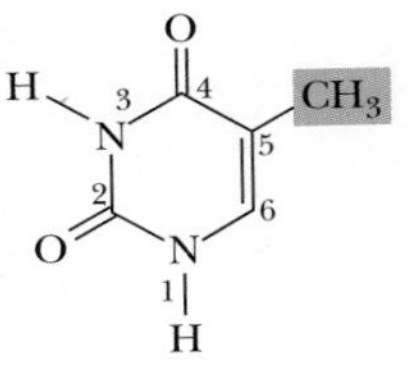

FIGURE 10.26 The 5-methyl group on thymine labels it as a special kind of uracil.

10.6 : Are Nucleic Acids Susceptible to Hydrolysis?

Most reactions of nucleic acid hydrolysis break phosphodiester bonds in the polynucleotide backbone even though such bonds are among the most stable chemical bonds found in biological molecules. In the laboratory, hydrolysis of polynucleotides will generate smaller fragments that are easier to manipulate and study.

RNA Is Susceptible to Hydrolysis by Base, but DNA Is Not

RNA is relatively resistant to the effects of dilute acid, but gentle treatment of DNA with 1 mM HCl leads to hydrolysis of purine glycosidic bonds and the loss of purine bases from the DNA. The glycosidic bonds between pyrimidine bases and 2'-deoxyribose are not affected, and in this case, the polynucleotide's sugar–phosphate backbone remains intact. The purine-free polynucleotide product is called **apurinic acid.**

DNA is not susceptible to alkaline hydrolysis. On the other hand, RNA is alkali labile and is readily hydrolyzed by hydroxide ions (Figure 10.27). DNA has no 2'-OH; therefore, DNA is alkali stable.

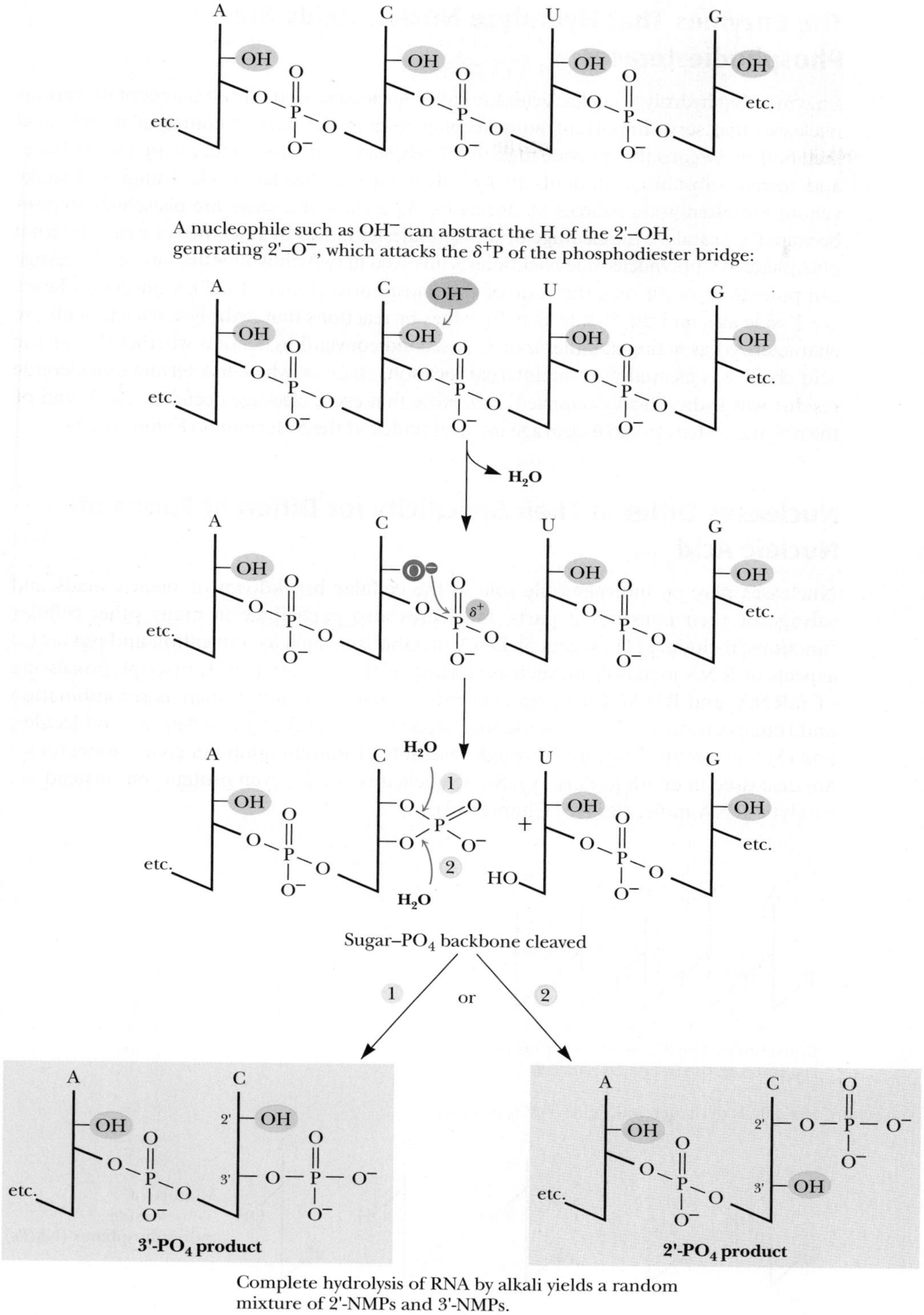
A nucleophile such as OH^- can abstract the H of the 2'–OH, generating 2'–O^-, which attacks the δ^+P of the phosphodiester bridge:
H_2O
Sugar–PO_4 backbone cleaved
or
3'-PO_4 product
2'-PO_4 product
Complete hydrolysis of RNA by alkali yields a random mixture of 2'-NMPs and 3'-NMPs.

FIGURE 10.27 Alkaline hydrolysis of RNA. The vertical lines represent ribose; the diagonals the phosphodiester linkages joining successive nucleotides. Nucleophilic attack by OH^- on the P atom leads to 5′-phosphoester cleavage and random hydrolysis of the cyclic 2′,3′-phosphodiester intermediate to give a mixture of 2′- and 3′-nucleoside monophosphate products.

The Enzymes That Hydrolyze Nucleic Acids Are Phosphodiesterases

Enzymes that hydrolyze nucleic acids are called **nucleases.** Virtually all cells contain various nucleases that serve important housekeeping roles in the normal course of nucleic acid metabolism. Organs that provide digestive fluids, such as the pancreas, are rich in nucleases and secrete substantial amounts to hydrolyze ingested nucleic acids. Fungi and snake venom are often good sources of nucleases. As a class, nucleases are **phosphodiesterases** because they catalyze the cleavage of phosphodiester bonds by H_2O. Because each internal phosphate in a polynucleotide backbone is involved in two phosphoester linkages, cleavage can potentially occur on either side of the phosphorus (Figure 10.28). Convention labels the 3'-side as a and the 5'-side as b. Enzymes or reactions that hydrolyze nucleic acids are characterized as acting at either a or b. A second convention denotes whether the nucleic acid chain was cleaved at some internal location, *endo,* or whether a terminal nucleotide residue was hydrolytically removed, *exo.* Note that exo a cleavage occurs at the 3'-end of the polymer, whereas exo b cleavage involves attack at the 5'-terminus (Figure 10.28).

Nucleases Differ in Their Specificity for Different Forms of Nucleic Acid

Nucleases play an indispensable role in the cellular breakdown of nucleic acids and salvage of their constituent parts. Nucleases also participate in many other cellular functions, including (1) aspects of DNA metabolism, such as replication and repair; (2) aspects of RNA metabolism, such as splicing of the primary gene transcript, processing of mRNA, and RNAi; (3) rearrangements of genetic material, such as recombination and transposition; (4) host defense mechanisms against foreign nucleic acid molecules; and (5) the immune response, through assembly of immunoglobulin genes (these topics are discussed in depth in Part IV). Some nucleases are not even proteins but instead are catalytic RNA molecules (see Chapter 13).

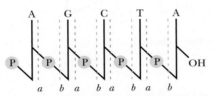

Convention: The 3'-side of each phosphodiester is termed a; the 5'-side is termed b.

(a)   Hydrolysis of the a bond yields 5'-PO_4 products:

Mixture of 5'-nucleoside monophosphates (NMPs)

(b)   Hydrolysis of the b bond yields 3'-PO_4 products:

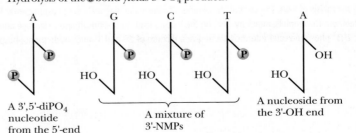

A 3',5'-diPO_4 nucleotide from the 5'-end

A mixture of 3'-NMPs

A nucleoside from the 3'-OH end

FIGURE 10.28 Cleavage in polynucleotide chains. **(a)** Cleavage on the a side leaves the phosphate attached to the 5'-position of the adjacent nucleotide, while **(b)** b-side hydrolysis yields 3'-phosphate products.

Like most enzymes (see Chapter 13), nucleases exhibit selectivity or *specificity* for the nature of the substance on which they act. That is, some nucleases act only on DNA **(DNases),** whereas others are specific for RNA (the **RNases**). Still others are nonspecific and are referred to simply as **nucleases**. Nucleases may also show specificity for only single-stranded nucleic acids or may act only on double helices. Some display a decided preference for acting only at certain bases in a polynucleotide, or as we shall see for *restriction endonucleases*, act only at a particular nucleotide sequence four to eight nucleotides (or more) in length. To the molecular biologist, nucleases are the surgical tools for the dissection and manipulation of nucleic acids in the laboratory.

Restriction Enzymes Are Nucleases That Cleave Double-Stranded DNA Molecules

Restriction endonucleases are enzymes, isolated chiefly from bacteria, that have the ability to cleave double-stranded DNA. The term *restriction* comes from the capacity of prokaryotes to defend against or "restrict" the possibility of takeover by foreign DNA that might gain entry into their cells. Prokaryotes degrade foreign DNA by using their unique restriction enzymes to chop it into relatively large but noninfective fragments. Restriction enzymes are classified into three types: I, II, or III. Types I and III require ATP to hydrolyze DNA and can also catalyze chemical modification of DNA through addition of methyl groups to specific bases. Type I restriction endonucleases cleave DNA randomly, whereas type III recognize specific nucleotide sequences within ds-DNA and cut the DNA at or near these sites.

Type II Restriction Endonucleases Are Useful for Manipulating DNA in the Lab

Type II restriction enzymes have received widespread application in the cloning and sequencing of DNA molecules. Their hydrolytic activity is not ATP-dependent, and they do not modify DNA by methylation or other means. Most important, they cut DNA within or near particular nucleotide sequences that they specifically recognize. These recognition sequences are typically four or six nucleotides in length and have a twofold axis of symmetry. For example, *E. coli* has a restriction enzyme, *Eco*RI, that recognizes the hexanucleotide sequence GAATTC:

$$5'\text{---}N\text{---}N\text{---}N\text{---}N\text{---}G\text{---}A\text{---}A\text{---}T\text{---}T\text{---}C\text{---}N\text{---}N\text{---}N\text{---}N\text{---}3'$$
$$3'\text{---}N\text{---}N\text{---}N\text{---}N\text{---}C\text{---}T\text{---}T\text{---}A\text{---}A\text{---}G\text{---}N\text{---}N\text{---}N\text{---}N\text{---}5'$$

Note the twofold symmetry: the sequence read $5' \rightarrow 3'$ is the same in both strands.

When *Eco*RI encounters this sequence in dsDNA, it causes a staggered, double-stranded break by hydrolyzing each chain between the G and A residues:

$$5'\text{---}N\text{---}N\text{---}N\text{---}N\text{---}G \quad A\text{---}A\text{---}T\text{---}T\text{---}C\text{---}N\text{---}N\text{---}N\text{---}N\text{---}3'$$
$$3'\text{---}N\text{---}N\text{---}N\text{---}N\text{---}C\text{---}T\text{---}T\text{---}A\text{---}A \quad G\text{---}N\text{---}N\text{---}N\text{---}N\text{---}5'$$

Staggered cleavage results in fragments with protruding single-stranded 5'-ends:

$$5'\text{---}N\text{---}N\text{---}N\text{---}N\text{---}G$$
$$3'\text{---}N\text{---}N\text{---}N\text{---}N\text{---}C\text{---}T\text{---}T\text{---}A\text{---}A\ 5'$$

$$5'\ A\text{---}A\text{---}T\text{---}T\text{---}C\text{---}N\text{---}N\text{---}N\text{---}N\text{---}3'$$
$$G\text{---}N\text{---}N\text{---}N\text{---}N\text{---}5'$$

Because the protruding termini of *Eco*RI fragments have complementary base sequences, they can form base pairs with one another.

Table 10.2 Restriction Endonucleases

About 1000 restriction enzymes have been characterized. They are named by italicized three-letter codes; the first is a capital letter denoting the genus of the organism of origin, and the next two letters are an abbreviation of the particular species. Because prokaryotes often contain more than one restriction enzyme, the various representatives are assigned letter and number codes as they are identified. Thus, *Eco*RI is the initial restriction endonuclease isolated from *Escherichia coli*, strain R. With one exception (*Nci*I), all known type II restriction endonucleases generate fragments with 5′-PO$_4$ and 3′-OH ends.

Enzyme	Common Isoschizomers	Recognition Sequence	Compatible Cohesive Ends
*Alu*I		AG↓CT	Blunt
*Apy*I	*Atu*I, *Eco*RII	CC↓G(^A_T)GG	
*Asu*II		TT↓CGAA	*Cla*I, *Hpa*II, *Taq*I
*Ava*I		G↓PyCGPuG	*Sal*I, *Xho*I, *Xma*I
*Avr*II		C↓CTAGG	
*Bal*I		TGG↓CCA	Blunt
*Bam*HI		G↓GATCC	*Bcl*I, *Bgl*II, *Mbo*I, Sau3A, *Xho*II
*Bcl*I		T↓GATCA	*Bam*HI, *Bgl*II, *Mbo*I, Sau3A, *Xho*II
*Bgl*II		A↓GATCT	*Bam*HI, *Bcl*I, *Mbo*I, Sau3A, *Xho*II
*Bst*EII		G↓GTNACC	
*Bst*XI		CCANNNNN↓NTGG	
*Cla*I		AT↓CGAT	*Acc*I, *Acy*I, *Asu*II, *Hpa*II, *Taq*I
*Dde*I		C↓TNAG	
*Eco*RI		G↓AATTC	
*Eco*RII	*Atu*I, *Apy*I	↓CC(^A_T)GG	
*Fnu*DII	*Tha*I	CG↓CG	Blunt
*Hae*I		(^A_T)GG↓CC(^A_T)	Blunt
*Hae*II		PuGCGC↓Py	
*Hae*III		GG↓CC	Blunt
*Hinc*II		GTPy↓PuAC	Blunt
*Hind*III		A↓AGCTT	
*Hpa*I		GTT↓AAC	Blunt
*Hpa*II		C↓CGG	*Acc*I, *Acy*I, *Asu*II, *Cla*I, *Taq*I
*Kpn*I		GGTAC↓C	
*Mbo*I	Sau3A	↓GATC	*Bam*HI, *Bcl*I, *Bgl*II, *Xho*II
*Msp*I		C↓CGG	
*Mst*I		TGC↓GCA	Blunt
*Not*I		GC↓GGCCGC	
*Pst*I		CTGCA↓G	
*Sac*I	*Sst*I	GAGCT↓C	
*Sal*I		G↓TCGAC	*Ava*I, *Xho*I
*Sau*3A		↓GATC	*Bam*HI, *Bcl*I, *Bgl*II, *Mbo*I, *Xho*II
*Sfi*I		GGCCNNNN↓NGGCC	
*Sma*I	*Xma*I	CCC↓GGG	Blunt
*Sph*I		GCATG↓C	
*Sst*I	*Sac*I	GAGCT↓C	
*Taq*I		T↓CGA	*Acc*I, *Acy*I, *Asu*II, *Cla*I, *Hpa*II
*Xba*I		T↓CTAGA	
*Xho*I		C↓TCGAG	*Ava*I, *Sal*I
*Xho*II		(^A_G)↓GATC(^C_T)	*Bam*HI, *Bcl*I, *Bgl*II, *Mbo*I, Sau3A
*Xma*I	*Sma*I	C↓CCGGG	*Ava*I

$$
\begin{array}{c}
\text{——N—N—N—N—G—A—A—T—T—C—N—N—N—N——} \\
\text{——N—N—N—N—C—T—T—A—A—G—N—N—N—N——}
\end{array}
$$

Therefore, DNA restriction fragments having such "sticky" ends can be joined together to create new combinations of DNA sequence. If fragments derived from DNA molecules of different origin are combined, novel recombinant forms of DNA are created.

*Eco*RI leaves staggered 5′-termini. Other restriction enzymes, such as *Pst*I, which recognizes the sequence 5′-CTGCAG-3′ and cleaves between A and G, produce cohesive staggered 3′-ends. Still others, such as *Bal*I, act at the center of the twofold symmetry axis of their recognition site and generate blunt ends that are noncohesive. *Bal*I recognizes 5′-TGGCCA-3′ and cuts between G and C.

Table 10.2 lists many of the commonly used restriction endonucleases and their recognition sites. Different restriction enzymes sometimes recognize and cleave within identical target sequences. Such enzymes are called **isoschizomers,** meaning that they cut at the same site; for example, *Mbo*I and *Sau*3A are isoschizomers.

Restriction Fragment Size Assuming random distribution and equimolar proportions for the four nucleotides in DNA, a particular tetranucleotide sequence should occur every 4^4 nucleotides, or every 256 bases. Therefore, the fragments generated by a restriction enzyme that acts at a four-nucleotide sequence should average about 250 bp in length. "Six-cutters," enzymes such as *Eco*RI or *Bam*HI, will find their unique hexanucleotide sequences on the average once in every 4096 (4^6) bp of length. Because the genetic code is a triplet code with three successive bases in a DNA strand specifying one amino acid in a polypeptide sequence, and because polypeptides typically contain at most 1000 amino acid residues, the fragments generated by six-cutters are approximately the size of prokaryotic genes. This property makes these enzymes useful in the construction and cloning of genetically useful recombinant DNA molecules. For the isolation of even larger nucleotide sequences, such as those of genes encoding large polypeptides (or those of eukaryotic genes that are disrupted by large introns), partial or limited digestion of DNA by restriction enzymes can be employed. However, restriction endonucleases that cut only at specific nucleotide sequences 8 or even 13 nucleotides in length are also available, such as *Not*I and *Sfi*I.

Restriction Endonucleases Can Be Used to Map the Structure of a DNA Fragment

The application of these sequence-specific nucleases to problems in molecular biology is considered in detail in Chapter 12, but one prominent application is described here. Because restriction endonucleases cut dsDNA at unique sites to generate large fragments, they provide a means for mapping DNA molecules that are many kilobase pairs in length. Restriction digestion of a DNA molecule is in many ways analogous to proteolytic digestion of a protein by an enzyme such as trypsin (see Chapter 5): The restriction endonuclease acts only at its specific sites so that a discrete set of nucleic acid fragments is generated. This action is analogous to trypsin cleavage only at Arg and Lys residues to yield a particular set of tryptic peptides from a given protein. The restriction fragments represent a unique collection of different-sized DNA pieces. Fortunately, this complex mixture can be resolved by *electrophoresis* (see the Chapter 5 A Deeper Look box Techniques Used in Protein Purification). Electrophoresis of DNA molecules on gels of restricted pore size (as formed in agarose or polyacrylamide media) separates them according to size, the largest being retarded in their migration through the gel pores while the smallest move relatively unhindered. Figure 10.29 shows a hypothetical electrophoretogram obtained for a DNA molecule treated with two different restriction nucleases, alone and in combination. Just as cleavage of a protein with different proteases to generate overlapping fragments allows an ordering of the peptides, restriction fragments can be ordered or "mapped" according to their sizes, as deduced from the patterns depicted in Figure 10.29.

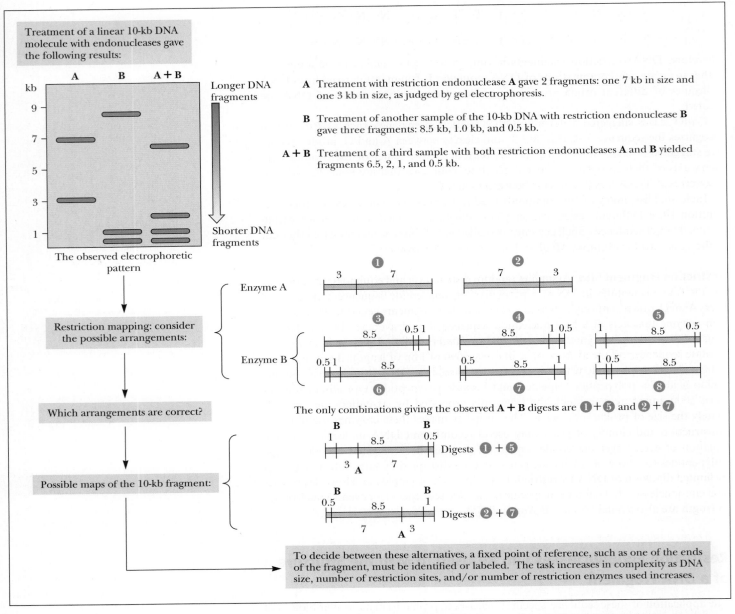

FIGURE 10.29 Restriction mapping of a DNA molecule as determined by an analysis of the electrophoretic pattern obtained for different restriction endonuclease digests. (Keep in mind that a dsDNA molecule has a unique nucleotide sequence and therefore a definite polarity; thus, fragments from one end are distinctly different from fragments derived from the other end.)

SUMMARY

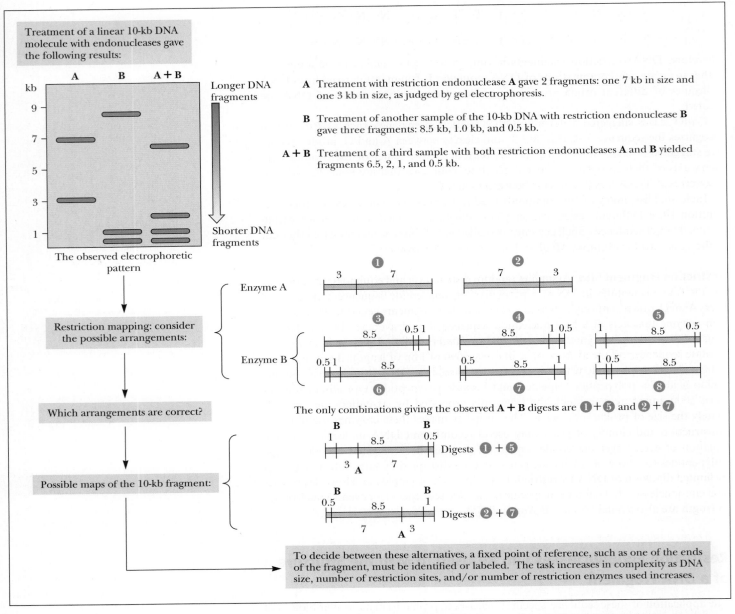 Login to OWL to develop problem-solving skills and complete online homework assigned by your professor.

Nucleotides and nucleic acids possess heterocyclic nitrogenous bases as principal components of their structure. Nucleotides participate as essential intermediates in virtually all aspects of cellular metabolism. Nucleic acids are the substances of heredity (DNA) and the agents of genetic information transfer (RNA).

10.1 What Are the Structure and Chemistry of Nitrogenous Bases? The bases of nucleotides and nucleic acids are derivatives of either pyrimidine (cytosine, uracil, and thymine) or purine (adenine and guanine). The aromaticity of the pyrimidine and purine

ring systems and the electron-rich nature of their —OH and ring nitrogen substituents allow them to undergo keto–enol tautomeric shifts and endow them with the capacity to absorb UV light.

10.2 What Are Nucleosides? Nucleosides are formed when a base is linked to a sugar. The usual sugars of nucleosides are pentoses; ribonucleosides contain the pentose D-ribose, whereas 2-deoxy-D-ribose is found in deoxyribonucleosides. Nucleosides are more water soluble than free bases.

10.3 What Are the Structure and Chemistry of Nucleotides? A nucleotide results when phosphoric acid is esterified to a sugar —OH group of a nucleoside. Successive phosphate groups can be linked to the phosphoryl group of a nucleotide through phosphoric anhydride linkages. Nucleoside 5′-triphosphates, as carriers of chemical energy, are indispensable agents in metabolism because phosphoric anhydride bonds are a prime source of chemical energy to do biological work. Virtually all of the biochemical reactions of nucleotides involve either *phosphoryl* or *pyrophosphoryl group transfer*. The bases of nucleotides serve as "information symbols."

10.4 What Are Nucleic Acids? Nucleic acids are polynucleotides: linear polymers of nucleotides linked 3′ to 5′ by phosphodiester bridges. The only significant variation in the chemical structure of nucleic acids is the particular base at each nucleotide position. These bases are not part of the sugar–phosphate backbone but instead serve as distinctive side chains.

10.5 What Are the Different Classes of Nucleic Acids? The two major classes of nucleic acids are DNA and RNA. Two fundamental chemical differences distinguish DNA from RNA: The nucleotides in DNA contain 2-deoxyribose instead of ribose as their sugar component, and DNA contains the base thymine instead of uracil. These differences confer important biological properties on DNA.

DNA consists of two antiparallel polynucleotide strands wound together to form a long, slender, double helix. The strands are held together through specific base pairing of A with T and C with G. The information in DNA is encoded in digital form in terms of the sequence of bases along each strand. Because base pairing is specific, the information in the two strands is complementary. DNA molecules may contain tens or even hundreds of millions of base pairs. In eukaryotic cells, DNA is complexed with histone proteins to form a nucleoprotein complex known as chromatin.

RNA occurs in multiple forms in cells, almost all of which are single-stranded. Nevertheless, the presence of complementary nucleotide sequences within the strand gives rise to multiple double-stranded regions in RNA molecules. Messenger RNA (mRNA) molecules are direct copies of the base sequences of protein-coding genes. Ribosomal RNA (rRNA) molecules provide the structural and functional foundations for ribosomes, the agents for translating mRNAs into proteins. In protein synthesis, the amino acids are delivered to the ribosomes in the form of aminoacyl-tRNA (transfer RNA) derivatives. Small nuclear RNAs (snRNAs) are characteristic of eukaryotic cells and are necessary for processing the RNA transcripts of protein-coding genes into mature mRNA molecules. Small RNAs are a recently discovered class of regulatory RNA molecules. A prominent role of small RNAs is gene silencing, particularly in the phenomenon of RNA interference (RNAi).

10.6 Are Nucleic Acids Susceptible to Hydrolysis? Like all biological polymers, nucleic acids are susceptible to hydrolysis, particularly hydrolysis of the phosphoester bonds in the polynucleotide backbone. RNA is susceptible to hydrolysis by base: DNA is not. Nucleases are hydrolytic enzymes that cleave the phosphoester linkages in the sugar–phosphate backbone of nucleic acids. Nucleases abound in nature, with varying specificity for RNA or DNA, single- or double-stranded nucleic acids, endo versus exo action, and 3′- versus 5′-cleavage of phosphodiesters. Restriction endonucleases of the type II class are sequence-specific endonucleases useful in mapping the structure of DNA molecules.

FOUNDATIONAL BIOCHEMISTRY Things You Should Know After Reading Chapter 10.

- Nucleotides are composed of either a pyrimidine or a purine base, a ribose, and phosphate.
- Nucleotides participate in virtually all aspects of metabolism.
- Nucleic acids are polynucleotides.
- The two kinds of nucleic acid are DNA and RNA.
- DNA is the repository of genetic information in cells.
- RNA is involved in the expression of genetic information through the processes of transcription and translation.
- Pyrimidines and purines are heterocyclic aromatic substances containing two (pyrimidines) or four (purines) nitrogen atoms in their ring structures.
- The three principal pyrimidines are cytosine, uracil, and thymine.
- The two principal purines are adenine and guanine.
- Purines and pyrimidines undergo keto-enol tautomeric shifts.
- The particular keto or enol form of a nitrogenous base affects its H-bonding possibilities.
- The aromaticity of pyrimidines and purines endows them with UV absorbance properties.
- Nucleosides are formed when ribose (or deoxyribose) is attached to a nitrogenous base via a β–N-glycosidic bond.
- Nucleotides are formed when phosphoric acid is esterified to an —OH group of the ribose of a nucleoside.
- Cyclic nucleotides result when the phosphate group forms ester linkages with two ribose —OH groups.

- Nucleoside diphosphates (NDPs) and nucleoside triphosphates (NTPs) are nucleotides with two or three phosphoryl groups.
- NDPs and NTPs are polyprotic acids.
- NTPs are the carriers of chemical energy.
- Nucleic acids are polymers of nucleotides linked 3′ to 5′ by phosphodiester bridges to form a sugar-phosphate backbone from which the bases project.
- The sequence of bases along the sugar-phosphate backbone is the defining characteristic of a nucleic acid.
- DNA is found in chromosomes as the genetic material.
- RNA has as its central role the transfer of information from DNA to protein.
- The fundamental structure of DNA is a double helix that is maintained by base pairing between a purine base in one strand and a pyrimidine base in the other.
- The canonical base pairs are A:T and G:C.
- The information of DNA is encoded in digital form.
- Messenger RNA (mRNA) carries the sequence information for synthesis of a protein.
- Ribosomal RNA (rRNA) provides both the structural framework and the protein-synthesizing function of ribosomes.
- Transfer RNAs (tRNAs) deliver amino acids in the form of aminoacyl-tRNAs for use in protein synthesis.
- Small, nuclear ribonucleoproteins (snRNPs) mediate the splicing of enkaryotic gene transcripts to form mature mRNAs.

- Small RNAs and long noncoding RNAs serve a variety of regulatory roles in cells.

- The chemical differences between DNA and RNA – thymine vs. uracil, presence or absence of 2′-OH group – have important biological consequences.

- DNA is acid-labile; RNA is not.

- RNA is susceptible to hydrolysis by base because of its 2′-OH group; DNA lacks this —OH and is not susceptible to alkaline hydrolysis.

- Enzymes that hydrolyze the phosphoester bonds of nucleic acids are called nucleases.

- Nucleases differ in their specificity for the different forms of nucleic acid, such as DNA vs. RNA or single-stranded vs. double-stranded.

- Type II restriction endonucleases are nucleases that cleave double-stranded DNA at specific base sequences. These enzymes are useful for in vitro manipulation of DNA molecules.

- Restriction endonucleases can be used to map the structure of DNA molecules.

PROBLEMS

Answers to all problems are at the end of this book. Detailed solutions are available in the *Student Solutions Manual, Study Guide, and Problems Book.*

1. **The Structure and Ionization Properties of Nucleotides** Draw the principal ionic species of 5′-GMP occurring at pH 2.

2. **Oligonucleotide Structure** Draw the chemical structure of pACG.

3. **Chargaff's Rules for the Base Composition of DNA** Chargaff's results (Table 10.1) yielded a molar ratio of 1.29 for A to G in ox DNA, 1.43 for T to C, 1.04 for A to T, and 1.00 for G to C. Given these values, what are the approximate mole fractions of A, C, G, and T in ox DNA?

4. **Abundance of the Different Bases in the Human Genome** Results on the human genome published in *Science* (*Science* **291**: 1304–1350 [2001]) indicate that the haploid human genome consists of 2.91 gigabase pairs (2.91×10^9 base pairs) and that 27% of the bases in human DNA are A. Calculate the number of A, T, G, and C residues in a typical human cell.

5. **The Base Sequence in the Two Polynucleotide Chains of a DNA Double Helix Is Complementary** Adhering to the convention of writing nucleotide sequences in the 5′→3′ direction, what is the nucleotide sequence of the DNA strand that is complementary to d-ATCGCAACTGTCACTA?

6. **The Relationship Between the Nucleotide Sequence of an mRNA and the DNA Strand from Which It Is Transcribed** Messenger RNAs are synthesized by RNA polymerases that read along a DNA template strand in the 3′→5′ direction, polymerizing ribonucleotides in the 5′→3′ direction (see Figure 10.20). Give the nucleotide sequence (5′→3′) of the DNA template strand from which the following mRNA segment was transcribed: 5′-UAGUGA-CAGUUGCGAU-3′.

7. **The Sequence Relationship Between an Antisense RNA Strand and Its Template DNA Strand** The DNA strand that is complementary to the template strand copied by RNA polymerase during transcription has a nucleotide sequence identical to that of the RNA being synthesized (except T residues are found in the DNA strand at sites where U residues occur in the RNA). An RNA transcribed from this nontemplate DNA strand would be complementary to the mRNA synthesized by RNA polymerase. Such an RNA is called antisense RNA because its base sequence is complementary to the "sense" mRNA. One strategy to thwart the deleterious effects of genes activated in disease states (such as cancer) is to generate antisense RNAs in affected cells. These antisense RNAs would form double-stranded hybrids with mRNAs transcribed from the activated genes and prevent their translation into protein. Suppose transcription of a cancer-activated gene yielded an mRNA whose sequence included the segment 5′-UACGGUCUAAGCUGA. What is the corresponding nucleotide sequence (5′→3′) of the template strand in a DNA duplex that might be introduced into these cells so that an antisense RNA could be transcribed from it?

8. **Restriction Endonuclease Mapping of a DNA Fragment** A 10-kb DNA fragment digested with restriction endonuclease *Eco*RI yielded fragments 4 kb and 6 kb in size. When digested with *Bam*HI, fragments 1, 3.5, and 5.5 kb were generated. Concomitant digestion with both *Eco*RI and *Bam*HI yielded fragments 0.5, 1, 3, and 5.5 kb in size. Give a possible restriction map for the original fragment.

9. **Design of DNA Sequences with Overlapping Restriction Endonuclease Sites** Based on the information in Table 10.2, describe two different 20-base nucleotide sequences that have restriction sites for *Bam*H1, *Pst*I, *Sal*I, and *Sma*I. Give the sequences of the *Sma*I cleavage products of each.

10. **Calculate the Free Energy Change for Synthesis of a Polynucleotide** (Integrates with Chapter 3.) The synthesis of RNA can be summarized by the reaction:

$$n\,\text{NTP} \longrightarrow (\text{NMP})_n + n\,\text{PP}_i$$

What is the $\Delta G^{\circ\prime}_{\text{overall}}$ for synthesis of an RNA molecule 100 nucleotides in length, assuming that the $\Delta G^{\circ\prime}$ for transfer of an NMP from an NTP to the 3′-O of polynucleotide chain is the same as the $\Delta G^{\circ\prime}$ for transfer of an NMP from an NTP to H_2O? (Use data given in Table 3.3.)

11. **Protein-DNA Interactions** Gene expression is controlled through the interaction of proteins with specific nucleotide sequences in double-stranded DNA.

    a. List the kinds of noncovalent interactions that might take place between a protein and DNA.

    b. How do you suppose a particular protein might specifically interact with a particular nucleotide sequence in DNA? That is, how might proteins recognize specific base sequences within the double helix?

12. **The Properties Restriction Endonucleases Must Have** Restriction endonucleases also recognize specific base sequences and then act to cleave the double-stranded DNA at a defined site. Speculate on the mechanisms by which this sequence recognition and cleavage reaction might occur by listing a set of requirements for the process to take place.

13. **The Properties That Carbohydrates Contribute to Nucleosides, Nucleotides, and Nucleic Acids** A carbohydrate group is an integral part of a nucleoside.

    a. What advantage does the carbohydrate provide?
    Polynucleotides are formed through formation of a sugar–phosphate backbone.

    b. Why might ribose be preferable for this backbone instead of glucose?

    c. Why might 2-deoxyribose be preferable to ribose in some situations?

14. **How Does the Presence of Phosphate Affect the Properties of Nucleotides?** Phosphate groups are also integral parts of nucleotides, with the second and third phosphates of a nucleotide linked through phosphoric anhydride bonds, an important distinction in terms of the metabolic role of nucleotides.

a. What property does a phosphate group have that a nucleoside lacks?

b. How are phosphoric anhydride bonds useful in metabolism?

c. How are phosphate anhydride bonds an advantage to the energetics of polynucleotide synthesis?

15. Calculate the Frequency of Occurrence of an RNAi Target Sequence The RNAs acting in RNAi are about 21 nucleotides long. To judge whether it is possible to uniquely target a particular gene with a RNA of this size, consider the following calculation: What is the expected frequency of occurrence of a specific 21-nucleotide sequence?

16. Calculate the Length of a Nucleotide Sequence Whose Expected Frequency Is Once per Haploid Human Genome The haploid human genome consists of 3×10^9 base pairs. Using the logic in problem 15, one can calculate the minimum length of a unique DNA sequence expected to occur by chance just once in the human genome. That is, what is the length of an oligonucleotide whose expected frequency of occurrence is once every 3×10^9 bp?

17. Design a Sequencing Strategy for Nucleic Acids Snake venom phosphodiesterase is an *a*-specific exonuclease (Figure 10.28) that acts equally well on single-stranded RNA or DNA. Design a protocol based on snake venom phosphodiesterase that would allow you to determine the base sequence of an oligonucleotide. Hint: Adapt the strategy for protein sequencing by Edman degradation, as described on pages 10–11 and 114–115.

18. Calculate the Mass of DNA in a Human Cell From the answer to problem 4 and the molecular weights of dAMP (331 D), dCMP (307 D), dGMP (347 D), and dTMP (322 D), calculate the mass (in daltons) of the DNA in a typical human cell.

Preparing for the MCAT® Exam

19. The bases of nucleotides and polynucleotides are "information symbols." Their central role in providing information content to DNA and RNA is clear. What advantages might bases as "information symbols" bring to the roles of nucleotides in metabolism?

20. Structural complementarity is the key to molecular recognition, a lesson learned in Chapter 1. The principle of structural complementarity is relevant to answering problems 5, 6, 7, 11, 12, and 19. The quintessential example of structural complementarity in all of biology is the DNA double helix. What features of the DNA double helix exemplify structural complementarity?

FURTHER READING

Nucleic Acid Biochemistry and Molecular Biology

Adams, R. L. P., Knowler, J. T., and Leader, D. P., 1992. *The Biochemistry of the Nucleic Acids,* 11th ed. New York: Chapman and Hall (Methuen and Co., distrib.).

Watson, J. D., Baker, T. A., Bell, S. T., Gann, A., et al., 2007. *The Molecular Biology of the Gene,* 6th ed. Menlo Park, CA: Benjamin/Cummings.

The History of Discovery of the DNA Double Helix

Judson, H. F., 1979. *The Eighth Day of Creation.* New York: Simon and Schuster.

DNA as Information

Hood, L., and Galas, D., 2003. The digital code of DNA. *Nature* **421:** 444–448.

DNA Nanomachines

Liu, H., and Liu, D., 2009. DNA nanomachines and their functional evolution. *Chemical Communications* **19:**2625–2635.

Seeman, N. D., 2010. Nanomaterials based on DNA. *Annual review of Biochemistry* **79:**65–87.

The Catalytic Properties of RNA and Its Role in Early Evolution

Caprara, M. G., and Nilsen, T. W., 2000. RNA: Versatility in form and function. *Nature Structural Biology* **7:**831–833.

Gray, M. W., and Cedergren, R., eds., 1993. The new age of RNA. *The FASEB Journal* **7:**4–239. A collection of articles emphasizing the new appreciation for RNA in protein synthesis, in evolution, and as a catalyst.

Small RNAs and Long Noncoding RNAs and Their Novel Biological Roles

Barsotti, A. M., and Prives, C., 2010. Noncoding RNAs: The missing "linc" in p53-mediated repression. *Cell* **142:**358–360.

Cartthrew, R. W., 2006. Gene regulation by microRNAs. *Current Opinion in Genetics & Development* **18:**203–208.

Hannon, G. J., 2002. RNA interference. *Nature* **418:**244–251. A review of RNAi, a widely conserved biological response to the intracellular presence of double-stranded RNA. RNAi provides an experimental method for manipulating gene expression as well as a mechanism to investigate specific gene function at the whole genome level.

Kondo, T., Plaza, S., Zanet, J., et al., 2010. Small peptides switch the transcriptional activity of Shavenbaby during *Drosophila* embryogenesis. *Science* **329:**336–339.

Liu, Q., and Paroo, Z., 2010. Biochemical principles of small RNA pathways. *Annual Review of Biochemistry* **79:**295–319.

Pillai, R. S., et al., 2007. Repression of protein synthesis by miRNAs: How many mechanisms? *Trends in Cell Biology* **17:**118–126.

Storz, G., Altuvia, A., and Wassarman, K. M., 2005. An abundance of RNA regulators. *Annual Review of Biochemistry* **74:**199–217.

Tsai, M.-C., Manor, O., Wan, Y., et al., 2010. Long noncoding RNA as modular scaffold of histone modification complexes. *Science* **329:**689–693.

Tuschi, T., 2003. RNA sets the standard. *Nature* **421:**220–221. Overview of the use of RNA interference to inactivate all the genes in a model organism *(Caenorhabditis elegans)* as a means of identifying gene function.

Zmora, P. D., and Haley, B., 2005. Ribo-gnome: The big world of small RNAs. *Science* **309:**1519–1524. This review in the September 2, 2005, issue of *Science* is accompanied by a series of articles on the various noncoding RNA types.

Nucleases and DNA Manipulation

Linn, S. M., Lloyd, R. S., and Roberts, R. J., 1993. *Nucleases,* 2nd ed. Long Island, NY: Cold Spring Harbor Laboratory Press.

Mishra, N. C., 2002. *Nucleases: Molecular Biology and Applications.* Hoboken, NJ: Wiley-Interscience.

Sambrook, J., and Russell, D., 2000. *Molecular Cloning: A Laboratory Manual,* 3rd ed. Cold Spring Harbor, NY: Cold Spring Harbor Laboratory.

Structure of Nucleic Acids

Renaissance masons created this double helix adorning the cathedral in Orvieto, Italy, some 500 years ago. What do you suppose they might have thought had they known that a double helix lies at the heart of heredity?

The Structure of DNA: "A melody for the eye of the intellect, with not a note wasted."

Horace Freeland Judson
The Eighth Day of Creation

KEY QUESTIONS

Reginald H. Garrett

ESSENTIAL QUESTION

The nucleotide sequence—the primary structure—of DNA not only determines its higher-order structure but it is also the physical representation of genetic information in organisms. RNA sequences, as copies of specific DNA segments, direct both the higher-order structure and the function of RNA molecules in information transfer processes.

What are the higher-order structures of DNA and RNA, and what methodologies have allowed scientists to probe these structures and the functions that derive from them?

Chapter 10 presented the structure and chemistry of nucleotides and how these units are joined via phosphodiester bonds to form nucleic acids, the biological polymers for information storage and transmission. In this chapter, we investigate biochemical methods that reveal this information by determining the sequential order of nucleotides in a polynucleotide, the so-called primary structure of nucleic acids. Then, we consider the higher orders of structure in the nucleic acids: the secondary and tertiary levels. Although the focus here is primarily on the structural and chemical properties of these macromolecules, it is fruitful to keep in mind the biological roles of these remarkable substances. The sequence of nucleotides in nucleic acids is the embodiment of genetic information (see Part IV). We can anticipate that the cellular mechanisms for accessing this information, as well as reproducing it with high fidelity, will be illuminated by knowledge of the chemical and structural qualities of these polymers.

OWL Online homework and a Student Self Assessment for this chapter may be assigned in OWL.

11.1 | How Do Scientists Determine the Primary Structure of Nucleic Acids?

Determining the primary structure of nucleic acids (the nucleotide sequence) would seem to be a more formidable problem than amino acid sequencing of proteins, simply because nucleic acids contain only 4 unique monomeric units (A, C, G, and T) whereas proteins have 20. With only four, there are *apparently* fewer specific sites for selective cleavage, distinctive sequences are more difficult to recognize, and the likelihood of ambiguity is greater. The much greater number of monomeric units in most polynucleotides as compared to polypeptides is a further difficulty. However, two simple tools make nucleic acid sequencing substantially easier than polypeptide sequencing. One of these tools is the set of type II *restriction endonucleases* that cleave DNA at specific oligonucleotide sites, generating unique fragments of manageable size (see Chapter 10). The second is *gel electrophoresis*, a method capable of separating nucleic acid fragments that differ from one another in length by just a single nucleotide.

The Nucleotide Sequence of DNA Can Be Determined from the Electrophoretic Migration of a Defined Set of Polynucleotide Fragments

The traditional protocol for nucleic acid sequencing has been the **chain termination** or **dideoxy method** of Frederick Sanger, which relies on enzymatic replication of the DNA to be sequenced. Very sensitive analytical techniques that can detect the newly synthesized DNA chains following electrophoretic separation are available, so Sanger sequencing could be carried out on as little as 1 attomole (amol, 10^{-18} mol) of DNA contained in less than 0.1 μL volume. (10^{-18} moles of DNA are roughly equivalent to 10^{-12} grams (pg) of 1-kb sized DNA molecules.) These analytical techniques typically rely on fluorescent detection of the DNA products.

Sanger's Chain Termination or Dideoxy Method Uses DNA Replication to Generate a Defined Set of Polynucleotide Fragments

To appreciate the rationale of the chain termination or dideoxy method, we first must briefly examine the biochemistry of DNA replication. DNA is a double helical molecule. In the course of its replication, the sequence of nucleotides in one strand is copied in a complementary fashion to form a new second strand by the enzyme **DNA polymerase.** Each original strand of the double helix serves as a **template** for the biosynthesis that yields two daughter DNA duplexes from the parental double helix (Figure 11.1). DNA polymerase will carry out this reaction *in vitro* in the presence of the four deoxynucleotide monomers and copy single-stranded DNA, provided a double-stranded region of DNA is artificially generated by adding a **primer.** This primer is merely an oligonucleotide capable of forming a short stretch of dsDNA by base pairing with the ssDNA (Figure 11.2). The primer must have a free 3′-OH end from which the new polynucleotide

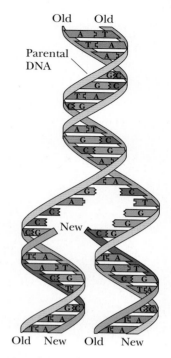

FIGURE 11.1 DNA replication yields two daughter DNA duplexes identical to the parental DNA molecule.

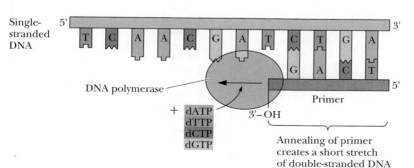

$$(dNMP)_n\text{-}3'OH + dNTP \rightarrow (dNMP)_{n+1}\text{-}3'OH + PP_i + H^+$$

The DNA polymerase reaction: dNMP addition to the 3′-OH end of a DNA strand that is base-paired with a template strand

FIGURE 11.2 Primed synthesis of a DNA template by DNA polymerase, using the four deoxynucleoside triphosphates as the substrates.

chain can grow as the first residue is added in the initial step of the polymerization process. DNA polymerases synthesize new strands by adding successive nucleotides in the 5'→3' direction.

The Chain Termination Protocol In the chain termination method of DNA sequencing, a DNA fragment of unknown sequence serves as a template in a polymerization reaction using some type of DNA polymerase, usually a genetically engineered version that lacks all traces of exonuclease activity that might otherwise degrade the DNA. (DNA polymerases usually have an intrinsic exonuclease activity that allows proofreading and correction of the DNA strand being synthesized; see Chapter 28.) The primer requirement is met by an appropriate oligonucleotide (this method is also known as the **primed synthesis method** for this reason). The reaction is run in the presence of all four deoxynucleoside triphosphates dATP, dGTP, dCTP, and dTTP, which are the substrates for DNA polymerase (Figure 11.3). In addition, the reaction mixture contains the four corresponding 2′,3′-**di**deoxynucleotides (ddATP, ddGTP, ddCTP, and ddTTP); it is these dideoxynucleotides that give the method its name.

Because dideoxynucleotides lack 3'-OH groups, they cannot serve as acceptors for 5'-nucleotide addition in the polymerization reaction; thus, the chain is terminated where they become incorporated. The concentrations of the deoxynucleotides in each reaction mixture are significantly greater than the concentrations of the dideoxynucleotides, so incorporation of a dideoxynucleotide is infrequent. Therefore, base-specific premature chain termination is only a random, occasional event, and a population of new strands of varying length is synthesized. Nevertheless, termination, although random, occurs everywhere

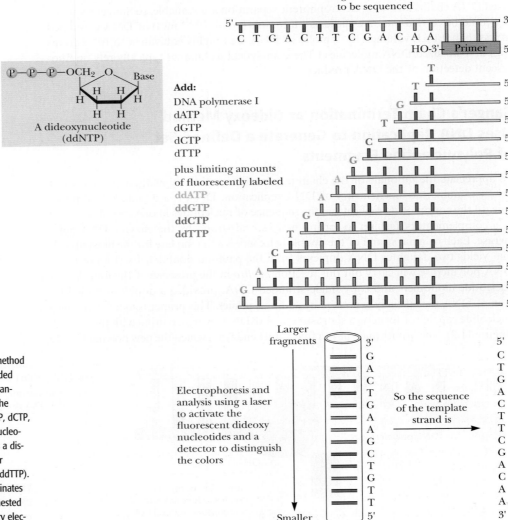

FIGURE 11.3 The chain termination or dideoxy method of DNA sequencing. A template DNA (the single-stranded DNA to be sequenced) with a complementary primer annealed at its 3'-end is copied by DNA polymerase in the presence of the four deoxynucleotide substrates (dATP, dCTP, dGTP, dTTP) and small amounts of the four dideoxynucleotide analogs of these substrates, each of which carries a distinctive fluorescence tag (illustrated here as orange for ddATP, blue for ddCTP, green for ddGTP, and red for ddTTP). Occasional incorporation of a dideoxynucleotide terminates further synthesis of that complementary strand. The nested set of terminated strands can be separated by capillary electrophoresis and identified by laser fluorescence spectroscopy.

in the sequence. Thus, the population of newly synthesized DNAs forms a nested set of molecules that differ in length by just one nucleotide. Each newly synthesized strand has a dideoxynucleotide at its 3′-end, and each of the four dideoxynucleotides used in Sanger sequencing is distinctive because each bears a fluorescent tag of a different color. (These fluorescent tags are attached to the 5-position of pyrimidine dideoxynucleotides or the 7-position of purine dideoxynucleotides, where these tags do not impair the ability of DNA polymerase to add them to a growing polynucleotide chain.) The color of a particular fluorescence (as in orange for ddA, blue for ddC, green for ddG, and red for ddT) reveals which base was specified by the template and incorporated by DNA polymerase at that spot.

Reading Dideoxy Sequencing Gels The sequencing products are visualized by fluorescence spectroscopy following their separation according to size by capillary electrophoresis (Figure 11.3). Because the smallest fragments migrate fastest upon electrophoresis and because fragments differing by only a single nucleotide in length are readily resolved, the sequence of nucleotides in the set of newly synthesized DNA fragments is given by the order of the fluorescent colors emerging from the capillary. Thus, the gel in Figure 11.3 is read TTGTCGAAGTCAG (5′→3′). Because of the way DNA polymerase acts, this observed sequence is complementary to the corresponding unknown template sequence. Knowing this, the template sequence now can be written CTGACTTCGACAA (5′→3′).

Sanger sequencing has been fully automated. Automation is achieved through the use of robotics for preparing the samples, running the DNA sequencing reactions, loading the chain-terminated DNA fragments onto capillary electrophoresis tubes, performing the electrophoresis, and imaging the results for computer analysis. These advances made it feasible to sequence the entire genomes of organisms (see Chapter 12). Celera Genomics, the private enterprise that reported a sequence for the 2.91 billion–bp human genome in 2001, used 300 automated DNA sequencers/analyzers to sequence more than 1 billion bases every month. Today, the more tedious aspect of DNA sequencing is the isolation and preparation of DNA fragments of interest, such as cloned genes; automated sequencing makes the rest routine.

High-Throughput DNA Sequencing by the Light of Fireflies

The enormous significance of DNA sequence information to fundamental questions in biology, medicine, and personal health is a compelling force for the development of more rapid and efficient DNA sequencing technologies, so-called ultra-high-throughput sequencing, or UHT, methods. One important UHT advance is **454 Technology**, a methodology developed by 454 Life Sciences, a division of Roche Company. Like Sanger sequencing, 454 Technology relies on DNA polymerase-catalyzed copying of a primed single-stranded DNA. (However, because 454 Technology does not rely on chain termination or creation of a nested set of DNA fragments, dideoxynucleotide terminators are not needed.) Multiple copies of unique single-stranded template DNA molecules paired with primer strands are immobilized on microscopic beads that can be loaded into micro-microtiter wells at a scale of 1.6 million different wells on a 6 cm × 6 cm platform (see Figure 11.4). Each well receives a unique DNA template. The reagents for primed synthesis are passed over the platform in sequential order: First, a reaction mixture with DNA polymerase plus dTTP (but no other dNTPs), a wash, then a reaction mixture with enzymes but only dATP, a wash, then the dGTP-specific mixture, a wash, and finally the dCTP mixture and a wash. Such cycles are repeated up to 100 times over an 8-hour period. Up to 500 cycles are possible in one run. A fiber-optic array to monitor light emission from each well is aligned with the platform.

The methodology is based on detection of DNA polymerase action through light emission. To do this, the technology exploits an overlooked product of the polymerase reaction, namely, the pyrophosphate released each time a dNTP contributes the correct complementary dNMP in the polymerase reaction (see Figure 11.2). Pyrophosphate release is coupled to light emission through two reactions (Figure 11.4). The first is catalyzed by **ATP sulfurylase,** which uses PP_i plus **adenosine-5′-phosphosulfate (APS)** to form ATP. The second reaction, catalyzed by the ATP-dependent firefly enzyme **luciferase,** oxidizes luciferin to form oxyluciferin with the emission of light.

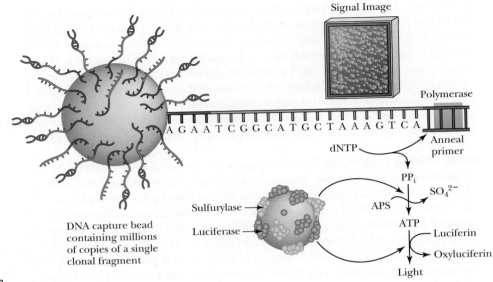

FIGURE 11.4 Ultra-high-throughput DNA sequencing by the 454 method. Light emission signals nucleotide incorporation in a DNA polymerase-catalyzed primed synthesis reaction.

Reaction 1:

$$PP_i^{2-} + APS \rightarrow ATP + SO_4^{2-}$$

(This is the reverse of the ATP sulfurylase reaction shown as reaction 1 in Figure 25.34.)

Reaction 2:

$$ATP + luciferin + O_2 \rightarrow AMP + PP_i + CO_2 + oxyluciferin + light$$

Luciferin   **Oxyluciferin**

Light detection confirms that addition of a dNMP by primed synthesis has occurred. Using computer recording of light emission to keep track of when in each cycle each well emitted a pulse of light allows reconstruction of sequence information for each of 1.6 million templates. Using this methodology, the 580,069-nucleotide sequence of *Mycoplasma genitalium* was confirmed in one run on the 454 Genome Sequencer. An order-of-magnitude cheaper approach developed by a company called Ion Torrents and named *PostLight* measures the pH change due to proton release when a deoxynucleotide is added in the DNA polymerase primed synthesis reaction (see Figure 11.2). This strategy obviates the need for the expensive ATP sulfurylase/luciferase reagents. The pH change measurement relies on a semiconductor-based ion sensor to observe H^+ release as each deoxynucleotide is added.

Emerging Technologies to Sequence DNA Are Based on Single-Molecule Sequencing Strategies

The rich information available from genome sequencing has immediate significance to medicine, cell biology, and evolution, to name only several disciplines greatly impacted by this technology. Growing demand for sequence information is driving the development of faster and cheaper methods. The most promising emerging technologies are based upon single-molecule DNA sequencing strategies that do not rely on Sanger-based primed synthesis of strands complementary to prepared DNA samples. Instead, several of these new technologies seek to determine directly the sequence of single molecules of single-stranded DNA. One new approach is to measure the change in the electrical properties of nanopores as single strands of DNA are pulled through them, one base at a time. As each base passes through such a nanopore, it alters the electrical conductance of the pore in subtly different ways, depending on whether it is an A, C, G, or T. Read-out of these electrical changes

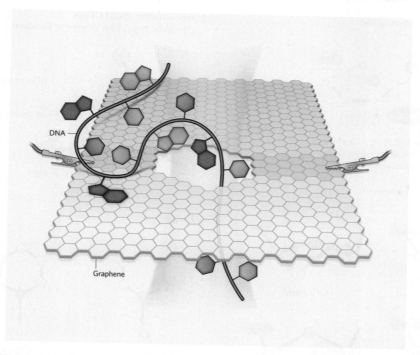

FIGURE 11.5 A graphene monolayer is one possible platform for nanopore-based DNA sequencing. An electrical potential is placed across a graphene monolayer punctured with nanopores. As single-stranded DNA is pulled through a nanopore, the fluctuations in ion flow (shown as yellow shading) reveal which base (A, C, G, or T) is momentarily occupying the pore (adapted from Figure 1 in Bayley, H., 2010. Nanotechnology: Holes with an edge. *Nature* **467:**164–165).

reveals the nucleotide sequence of the DNA directly, without the need for expensive reagents or chemical or enzymatic reactions (Figure 11.5). The challenge is to create a platform that can read single-stranded DNA sequences rapidly, accurately, and reliably.

11.2 What Sorts of Secondary Structures Can Double-Stranded DNA Molecules Adopt?

Conformational Variation in Polynucleotide Strands

Polynucleotide strands are inherently flexible. Each deoxyribose–phosphate segment of the backbone has six degrees of freedom (Figure 11.6a) as a consequence of the six successive single bonds per segment along the chain. Furanose rings of pentoses are not planar but instead adopt puckered conformations, four of which are shown in Figure 11.6b. A seventh degree of freedom per nucleotide unit arises because of free rotation about the C1'-N glycosidic bond. This freedom allows the plane of the base to rotate relative to the path of the polynucleotide strand (Figure 11.6c).

DNA Usually Occurs in the Form of Double-Stranded Molecules

Double-stranded DNA molecules adopt one of three secondary structures, termed A, B, and Z. In a moment, we will address the "ABZs of DNA secondary structure"; first we must consider some general features of DNA double helices. Fundamentally, double-

(a) The six degrees of freedom in the
sugar–PO₄ backbone:

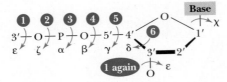

Rotation about bonds 1, 2, 3, 4, 5, and 6
correspond to 6 degrees of freedom designated
α, β, γ, δ, ε, and ζ as indicated.

(b) Four puckered conformations of
furanose rings:

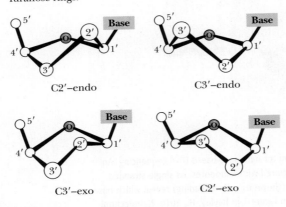

(c) Free rotation about C1′–N
glycosidic bond (7th degree of freedom):

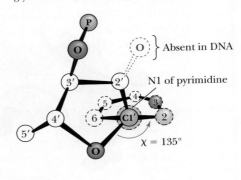

Pyrimidine:

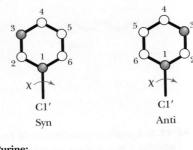

Purine:

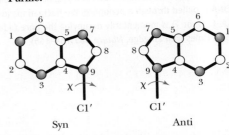

FIGURE 11.6 **(a)** The six degrees of freedom in the deoxyribose–PO₄ units of the polynucleotide chain. **(b)** Four puckered conformations of the furanose rings. **(c)** Free rotation about the C1′–N glycosidic bond.

stranded DNA is a regular two-chain structure with hydrogen bonds formed between opposing bases on the two chains (see Chapter 10). Such H bonding is possible only when the two chains are antiparallel. The polar sugar–phosphate backbones of the two chains are on the outside. The bases are stacked on the inside of the structure; these heterocyclic bases, as a consequence of their π-electron clouds, are hydrophobic on their flat sides. One purely hypothetical conformational possibility for a two-stranded arrangement would be a ladderlike structure (Figure 11.7) in which the base pairs are fixed at 0.6 nm apart because this is the distance between adjacent sugars along a polynucleotide strand. Because H₂O molecules could fit into the spaces between the hydrophobic surfaces of the bases, this conformation is energetically unfavorable. This ladderlike structure converts to a double helix when given a simple right-handed twist. Helical twisting brings the base-pair rungs of the ladder closer together, stacking them 0.34 nm apart, without affecting the sugar–sugar distance of 0.6 nm. Because this helix repeats itself approximately every 10 bp, its **pitch** is 3.4 nm. This is the major conformation of DNA in solution, and it is called **B-DNA.**

Watson–Crick Base Pairs Have Virtually Identical Dimensions

As indicated in Chapter 10, the base pairing in DNA is size complementary: Large bases (purines) pair with small bases (pyrimidines). Hydrogen bond formation between purines and pyrimidines dictates that the purine adenine pairs with the pyrimidine thymine; the purine guanine pairs with the pyrimidine cytosine. Size complementarity means that the A:T pair and G:C pair have virtually identical dimensions (Figure 11.8). Watson and Crick realized that units of such structural equivalence could serve as spatially invariant substructures to build a polymer whose exterior dimensions would be uniform along its length, regardless of the sequence of bases. That is, the pairing of smaller pyrimidines with larger purines everywhere across the double-stranded molecule allows the two polynucleotide strands to assume essentially identical helical conformations.

The DNA Double Helix Is a Stable Structure

Several factors account for the stability of the double helical structure of DNA.

H Bonds Although it has long been emphasized that the two strands of DNA are held together by H bonds formed between the complementary purines and pyrimidines, two in an A:T pair and three in a G:C pair (Figure 11.8), the H bonds between base pairs impart little net stability to the double-stranded structure compared to the separated strands in solution. When the two strands of the double helix are separated, the H bonds between base pairs are replaced by H bonds between individual bases and surrounding water molecules. Polar atoms in the sugar–phosphate backbone do form external H bonds with surrounding water molecules, but these form with separated strands as well.

Electrostatic Interactions A prominent feature of the backbone of a DNA strand is the repeating array of negatively charged phosphate groups. These arrays of negative charge along the strands repel each other so that their sugar–phosphate backbones are kept apart and the two strands come together through Watson–Crick base pairing. As a consequence, the negative charges are situated on the exterior surface of the double helix, such that repulsive effects are minimized. Further these charges become electrostatically shielded from one another because divalent cations, particularly Mg^{2+}, bind strongly to the anionic phosphates.

Van der Waals and Hydrophobic Interactions The core of the helix consists of the base pairs, and these base pairs stack together through π, π-electronic interactions (a form of van der Waals interaction), and hydrophobic forces. These base-pair stacking interactions range from -16 to -51 kJ/mol (expressed as the energy of interaction between adjacent base pairs), contributing significantly to the overall stabilizing energy.

A stereochemical consequence of the way A:T and G:C base pairs form is that the sugars of the respective nucleotides have opposite orientations. This is why the sugar–phosphate backbones of the two chains run in opposite or "antiparallel" directions. Furthermore, the two glycosidic bonds holding the bases in each base pair are not directly across the helix from each other, defining a common diameter (Figure 11.9). Consequently, the sugar–phosphate backbones of the helix are not equally spaced along the helix axis and the grooves between them are not the same size. Instead, the intertwined chains create a **major groove** and a **minor groove** (Figure 11.9). The edges of the base pairs have a specific relationship to these grooves. The "top" edges of the base pairs ("top" as defined by placing the glycosidic bond at the bottom, as in Figure 11.9) are exposed along the interior surface or "floor" of the major groove; the base-pair edges nearest to the glycosidic bond form the interior surface of the minor groove. Some proteins that bind to DNA can actually recognize specific nucleotide sequences by "reading" the pattern of H-bonding possibilities presented by the edges of the bases in these grooves. Such DNA–protein interactions provide one step toward understanding how cells regulate the expression of genetic information encoded in DNA (see Chapter 29).

Double Helical Structures Can Adopt a Number of Stable Conformations

In solution, DNA ordinarily assumes the familar structure we have been discussing: B-DNA. However, nucleic acids also occur naturally in other double helical forms. The base-pairing arrangement remains the same, but the inherently flexible sugar–phosphate backbone can adopt different conformations. Base-pair rotations are another kind of conformational variation. **Helical twist** is the rotation (around the axis

(a) Ladder

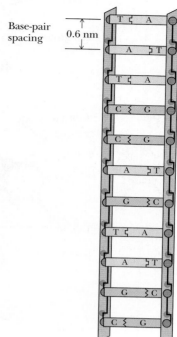

Base-pair spacing   0.6 nm

(b) Helix

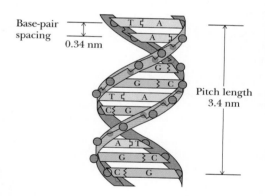

Base-pair spacing   0.34 nm

Pitch length 3.4 nm

FIGURE 11.7 (a) Double-stranded DNA as an imaginary ladderlike structure. **(b)** A simple right-handed twist converts the ladder to a helix.

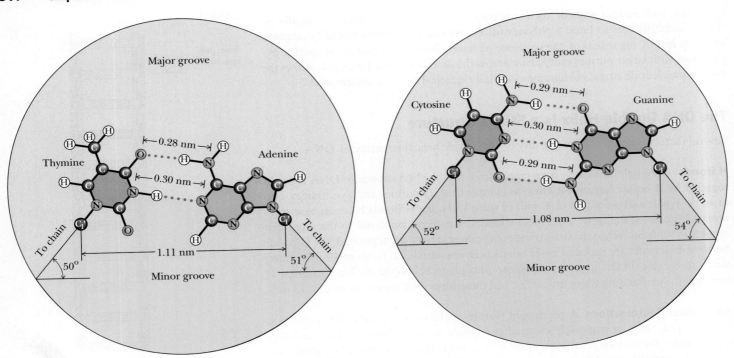

FIGURE 11.8 Watson–Crick A:T and G:C base pairs. All H bonds in both base pairs are straight.

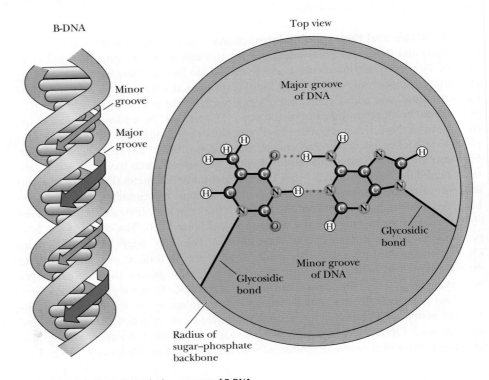

FIGURE 11.9 The major and minor grooves of B-DNA.

(a)

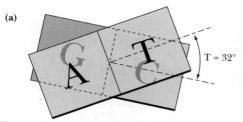

Two base pairs with 32° of right-handed helical twist: the *minor-groove edges are drawn with heavy shading.*

(b)

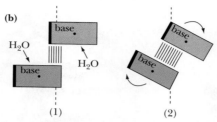

(1)   (2)

Propellor twist, as in (2), allows greater overlap of successive bases along the same strand and reduces the area of contact between the bases and water.

(c)

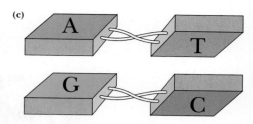

Propellor-twisted base pairs. Note how the hydrogen bonds between bases are distorted by this motion, yet remain intact. The minor-groove edges of the bases are shaded gray.

FIGURE 11.10 Helical twist and propellor twist in DNA. **(a)** Successive base pairs in B-DNA show a rotation with respect to each other. **(b)** Rotation in a different dimension—**propellor twist**—allows the hydrophobic surfaces of bases to overlap better. Dots represent axes perpendicular to the helix axis. The view is from the sugar–P backbone. **(c)** Each of the bases in a base pair shows positive propellor twist (a clockwise rotation from the horizontal, as viewed along the *N*-glycosidic bond, from the pentose C1′ to the base). (Adapted from Figure 3.4 in Callandine, C. R., and Drew, H. R., 1992. *Understanding DNA: The Molecule and How It Works.* London: Academic Press.)

of the double helix) of one base pair relative to the next (Figure 11.10). Successive base pairs in B-DNA show a mean rotation of 36° with respect to each other. **Propellor twist** involves rotation around a different axis, namely, an axis perpendicular to the helix axis (Figure 11.10b). Propellor twist allows greater overlap between successive bases along a strand of DNA and diminishes the area of contact between bases and solvent water.

A-Form DNA Is an Alternative Form of Right-Handed DNA

An alternative form of the right-handed double helix is **A-DNA.** A-DNA molecules differ from B-DNA molecules in a number of ways. The pitch, or distance required to complete one helical turn, is different. In B-DNA, it is 3.4 nm, whereas in A-DNA it is 2.46 nm. One turn in A-DNA requires 11 bp to complete. Depending on local sequence, 10 to 10.6 bp define one helical turn in B-form DNA. In A-DNA, the base pairs are no longer nearly perpendicular to the helix axis but instead are tilted 19° with respect to this axis. Successive base pairs occur every 0.23 nm along the axis, as opposed to 0.332 nm in B-DNA. The B-form of DNA is thus longer and thinner than the short, squat A-form, which has its base pairs displaced around, rather than centered on, the helix axis. Figure 11.11 and Table 11.1 show the relevant structural characteristics of the A- and B-forms of DNA. (Z-DNA, another form of DNA to be discussed shortly, is also depicted in Figure 11.11 and Table 11.1.) A comparison of the structural properties of A-, B-, and Z-DNA is summarized in Table 11.1.

Relatively dehydrated DNA fibers can adopt the A-conformation, and DNA may be in the A-form in dehydrated structures, such as bacterial and fungal spores. The pentose conformation in A-DNA is 3′-endo, as opposed to 2′-endo in B-DNA. Double helical DNA:RNA hybrids have an A-like conformation. The 2′-OH in RNA sterically prevents double helical regions of RNA chains from adopting the B-form helical arrangement. Importantly, double-stranded regions in RNA chains often assume an A-like conformation, with their bases strongly tilted with respect to the helix axis.

Z-DNA Is a Conformational Variation in the Form of a Left-Handed Double Helix

Z-DNA was first discovered when X-ray analysis of crystals of the synthetic deoxynucleotide dCpGpCpGpCpG revealed an antiparallel double helix of unexpected conformation. The alternating pyrimidine–purine (Py–Pu) sequence of this oligonucleotide is

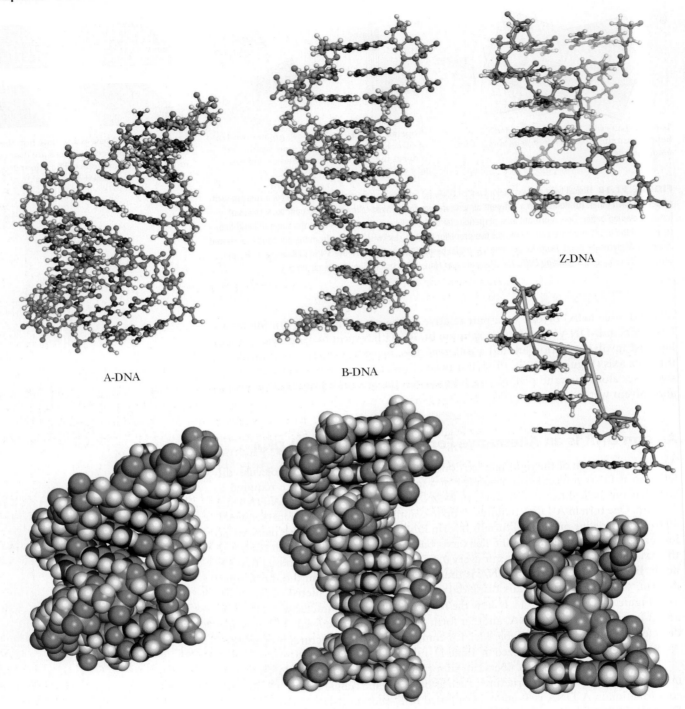

A-DNA   B-DNA

Z-DNA

FIGURE 11.11 Comparison of the A-, B-, and Z-forms of the DNA double helix. The A- and B-structures show 12 bp of DNA; the Z-structures, 6 bp. The middle Z-structure shows just one strand of a Z-DNA double helix to illustrate better the left-handed zigzag path of the polynucleotide backbones in Z-DNA. (The light blue line was added to show the imaginary zigzag path.) A-DNA: pdb id = 2D47, B-DNA: pdb id = 355D, Z-DNA: pdb id = 1DCG.

the key to its unusual properties. The *N*-glycosyl bonds of G residues in this alternating copolymer are rotated 180° with respect to their conformation in B-DNA, so now the purine ring is in the syn rather than the anti conformation (Figure 11.12). The C residues remain in the anti form. Because the G ring is "flipped," the C ring must also flip to maintain normal Watson–Crick base pairing. However, pyrimidine nucleosides do

TABLE 11.1 Comparison of the Structural Properties of A-, B-, and Z-DNA			
	Double Helix Type		
	A	**B**	**Z**
Overall proportions	Short and broad	Longer and thinner	Elongated and slim
Rise per base pair	2.3 Å	3.32 Å ± 0.19 Å	3.8 Å
Helix packing diameter	25.5 Å	23.7 Å	18.4 Å
Helix rotation sense	Right-handed	Right-handed	Left-handed
Base pairs per helix repeat	1	1	2
Base pairs per turn of helix	~11	~10	12
Mean rotation per base pair	33.6°	35.9° ± 4.2°	$-60°/2$
Pitch per turn of helix	24.6 Å	33.2 Å	45.6 Å
Base-pair tilt from the perpendicular	+19°	$-1.2° ± 4.1°$	$-9°$
Base-pair mean propeller twist	+18°	+16° ± 7°	~0°
Helix axis location	Major groove	Through base pairs	Minor groove
Major groove proportions	Extremely narrow but very deep	Wide and with intermediate depth	Flattened out on helix surface
Minor groove proportions	Very broad but shallow	Narrow and with intermediate depth	Extremely narrow but very deep
Glycosyl bond conformation	anti	anti	anti at C, syn at G

Adapted from Dickerson, R. L., et al., 1983. Helix geometry and hydration in A-DNA, B-DNA, and Z-DNA. *Cold Spring Harbor Symposium on Quantitative Biology* **47**:13–24.

not readily adopt the syn conformation because it creates steric interference between the pyrimidine C-2 oxy substituent and atoms of the pentose. Because the cytosine ring does not rotate relative to the pentose, the whole C nucleoside (base and sugar) must flip 180° (Figure 11.13). It is topologically possible for the G to go syn and the C nucleoside to undergo rotation by 180° without breaking and re-forming the G:C hydrogen bonds. In other words, the B-to-Z structural transition can take place without disrupting the bonding relationships among the atoms involved.

Because alternate nucleotides assume different conformations, the repeating unit on a given strand in the Z-helix is the dinucleotide. That is, for any number of bases, *n*, along one strand, *n* − 1 dinucleotides must be considered. For example, a GpCpGpC subset of sequence along one strand is composed of *three* successive dinucleotide units: GpC, CpG, and GpC. (In A- and B-DNA, the nucleotide conformations are essentially uniform and the repeating unit is the mononucleotide.) It follows that the CpG sequence is distinct conformationally from the GpC sequence along the alternating copolymer chains in the Z-double helix. The conformational alterations going from B to Z realign

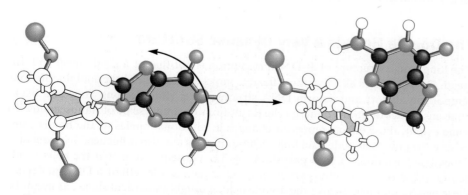

Deoxyguanosine in B-DNA (anti position)

Deoxyguanosine in Z-DNA (syn position)

FIGURE 11.12 Comparison of the deoxyguanosine conformation in B- and Z-DNA.

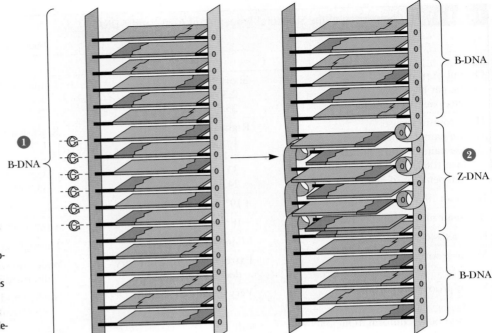

FIGURE 11.13 The change in topological relationships of base pairs from B- to Z-DNA. A six-base-pair GCGCGC segment of B-DNA (**1**) is converted to Z-DNA (**2**) through rotation of the base pairs, as indicated by the curved arrows. The purine rings (green) of the deoxyguanosine nucleosides rotate via an anti to syn change in the conformation of the guanine–deoxyribose glycosidic bond; the pyrimidine rings (blue) are rotated by flipping the entire deoxycytosine nucleoside (base *and* deoxyribose).

the sugar–phosphate backbone along a zigzag course that has a left-handed orientation (Figure 11.11), thus the designation Z-DNA. Note that in any GpCpGp subset, the sugar–phosphates of GpC form the horizontal "zig" while the CpG backbone segment forms the vertical "zag." The mean rotation angle circumscribed around the helix axis is $-15°$ for a CpG step and $-45°$ for a GpC step (giving $-60°$ for the dinucleotide repeat). The minus sign denotes a left-handed or counterclockwise rotation about the helix axis. Z-DNA is more elongated and slimmer than B-DNA.

Cytosine Methylation and Z-DNA The Z-form can arise in sequences that are not strictly alternating Py–Pu. For example, the hexanucleotide ^{m5}CGAT^{m5}CG, a Py-Pu-Pu-Py-Py-Pu sequence containing two 5-methylcytosines (^{m5}C), crystallizes as Z-DNA. Indeed, the in vivo methylation of C at the 5-position is believed to favor a B-to-Z switch because, in B-DNA, these hydrophobic methyl groups would protrude into the aqueous environment of the major groove, a destabilizing influence. In Z-DNA, the same methyl groups can form a stabilizing hydrophobic patch. It is likely that the Z-conformation naturally occurs in specific regions of cellular DNA, which otherwise is predominantly in the B-form. Furthermore, because methylation is implicated in gene regulation, the occurrence of Z-DNA may affect the expression of genetic information (see Part 4).

The Double Helix Is a Very Dynamic Structure

The long-range structure of B-DNA in solution is not that of a rigid, linear rod. Instead, DNA behaves as a dynamic, flexible molecule. Localized thermal fluctuations temporarily distort and deform DNA structure over short regions. Base and backbone ensembles of atoms undergo elastic motions on a time scale of nanoseconds. To some extent, these effects represent changes in rotational angles of the bonds comprising the polynucleotide backbone. These changes are also influenced by sequence-dependent variations in base-pair stacking. The consequence is that the helix bends gently. When these variations are summed over the great length of a DNA molecule, these bending influences give the double helix a roughly spherical shape, as might be expected for a long, semirigid rod undergoing apparently random coiling. It is also

DNA Methylation, CpG Islands, and Epigenetics

In Chapter 10, we saw that rRNA and tRNA often contain chemically modified bases (Figure 10.23); a common form of modification is methylation. Methylation of eukaryotic DNA occurs only at cytosine residues, forming 5-methylcytosine.

5-Methylcytosine

In mammals, cytosine methylation is essential for normal embryonic development. A general rule of thumb is that **cytosine methylation** switches genes to an "off" state. Then, the information they encode is not expressed. In somatic cells (cells other than gametes), most cytosine methylation occurs at cytosine residues next to Gs, that is, within CpG dinucleotides. (Note that the complementary strand would have a 5'-CpG-3' as well.) CpG dinucleotides are rare in mammalian genomes, appearing at just 20% of their statistically expected numbers.

However, some regions of the genome contain CpG dinucleotides at expected frequencies. Small regions of DNA (less than 500 bp or so) that contain expected amounts of CpG dinucleotides are called **CpG islands** (CGIs). CGIs are G-C rich (about 65% G-C content,

compared to the average 40% G-C content of mammalian genomes). The Cs in CGIs are usually unmethylated. Those that are methylated tend to lie in gene regulatory regions and shut down gene expression.

Epigenetics is the study of heritable changes in the genome that occur without a change in nucleotide sequence. Epigenetic changes are reversible and include modifications that occur in DNA (such as methylation) and in histones, the principal proteins found in chromatin, the DNA-protein complex in the eukaryotic nucleus. Epigenetic changes can influence expression of the information encoded by the genome.

Hypermethylation of CGIs in tumor suppressor genes can shut down the expression of these genes, leading to cancer. Further, deamination of 5-methylcytosine yields thymine, and thus mutation of a C:G base pair in DNA to a T:A base pair. Accumulation of somatic mutations can cause cancer.

Branciamore, S., Chen, Z.-X., Riggs, A. D., and Rodin, S. N., 2010. CpG island clusters and pro-epigenetic selection for CpGs in protein-coding exons of HOX and other transcription factors. Proceedings of the National Academy of Sciences, USA *107*:15485–15490.

Goll, M. G., and Bestor, T. H., 2005. Eukaryotic cytosine methyltransferases. Annual Review of Biochemistry *74*:481–514.

Riddihough, G., and Zahn, L. M., 2010. What is epigenetics? Science *330*:611. This October 29, 2010, issue of Science features a number of articles on epigenetics.

worth noting that, on close scrutiny, the surface of the double helix is *not* that of a totally featureless, smooth, regular "barber pole" structure. Different base sequences impart their own special signatures to the molecule by subtle influences on such factors as the groove width, the angle between the helix axis and base planes, and the mechanical rigidity. Certain regulatory proteins bind to specific DNA sequences and participate in activating or suppressing expression of the information encoded therein. These proteins bind at unique sites by virtue of their ability to recognize novel structural characteristics imposed on the DNA by the local nucleotide sequence.

Intercalating Agents Distort the Double Helix Aromatic macrocycles, flat hydrophobic molecules composed of fused, heterocyclic rings, such as **ethidium bromide, acridine orange,** and **actinomycin D** (Figure 11.14), can slip between the stacked base pairs of DNA. The base pairs move apart to accommodate these so-called **intercalating agents,** causing an unwinding of the helix to a more ladderlike structure. The deoxyribose–phosphate backbone is almost fully extended as successive base pairs are displaced 0.7 nm from one another, and the rotational angle about the helix axis between adjacent base pairs is reduced from 36° to 10°.

Dynamic Nature of the DNA Double Helix in Solution Intercalating substances insert with ease into the double helix, indicating that the van der Waals stacking interactions that they share with the bases sandwiching them are more favorable than similar interactions between the bases themselves. Furthermore, the fact that these agents slip in suggests that the double helix must momentarily unwind and present gaps for these agents to occupy. That is, the DNA double helix in solution must be represented by a set of metastable alternatives to the standard B-conformation. These alternatives constitute a flickering repertoire of dynamic structures.

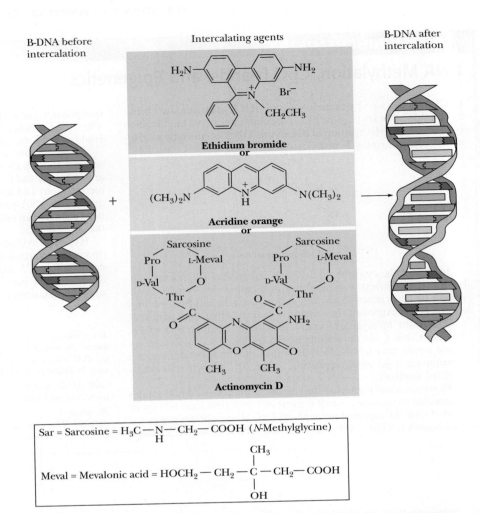

FIGURE 11.14 The structures of ethidium bromide, acridine orange, and actinomycin D, three intercalating agents, and their effects on DNA structure.

Alternative Hydrogen-Bonding Interactions Give Rise to Novel DNA Structures: Cruciforms, Triplexes, and Quadruplexes

Cruciform Structures Arise from Inverted Repeats **Inverted repeats** (Figure 11.15) are duplex DNA sequences showing twofold symmetry (the 5′→3′ sequence is identical in both strands). **Palindromes** are words, phrases, or sentences that read the same backward or forward, such as "radar," "sex at noon taxes," "Madam, I'm Adam," and "a man, a plan, a canal, Panama." Inverted repeats are sometimes referred to as palindromes (despite the inaccuracy of this description). Inverted repeats have the potential to adopt **cruciform** (meaning "cross-shaped") structures if the normal interstrand base pairing is replaced by intrastrand pairing. In effect, each strand forms a hairpin structure through alignment and pairing of the self-complementary sequences along the strand. Cruciforms are never as stable as normal DNA duplexes because an unpaired segment must exist in the loop region. Cruciforms potentially create novel structures that can serve as distinctive recognition sites for specific DNA-binding proteins.

Hoogsteen Base Pairs and DNA Multiplexes The A:T and G:C base pairs first seen by Watson (Figure 11.8) are the canonical building blocks for DNA structures. However, Karst Hoogsteen found that adenine and thymine do not pair in this way when crystallized from aqueous solution. Instead, they form two H bonds in a different arrangement (Figure 11.16). Further, Hoogsteen observed that, in mildly acidic solutions, guanine and cytosine form base pairs different from Watson–Crick G:C base pairs.

These Hoogsteen base pairs depend upon protonation of cytosine N-3 (Figure 11.16) and have only two H bonds, not three. In both A:T and G:C Hoogsteen base pairs, the purine N-7 atom is an H-bond acceptor. The functional groups of adenine and guanine that participate in Watson–Crick H bonds remain accessible in Hoogsteen base pairs. Thus, base triplets can form, as shown in Figure 11.17, giving rise to TAT and a C⁺GC triplets, where each purine interacts with one of its pyrimidine partners through Hoogsteen base pairing and the other through Watson–Crick base pairing.

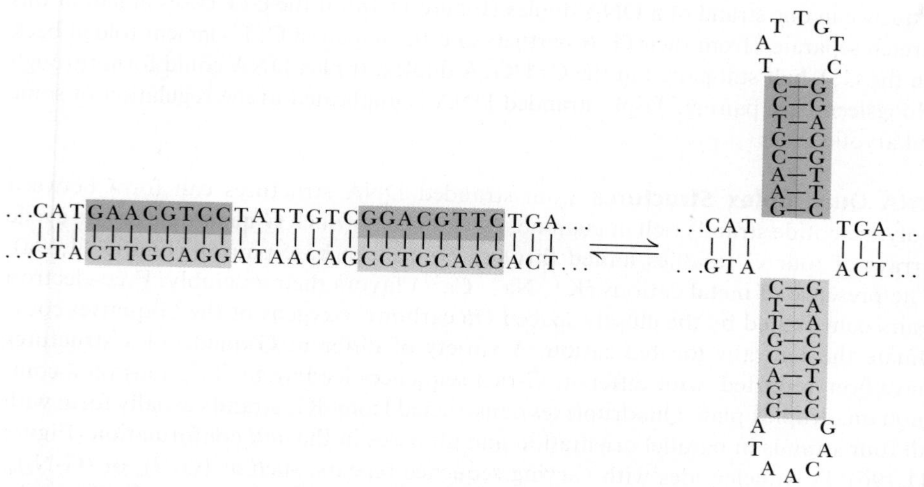

FIGURE 11.15 Self-complementary inverted repeats can rearrange to form hydrogen-bonded cruciform stem-loop structures.

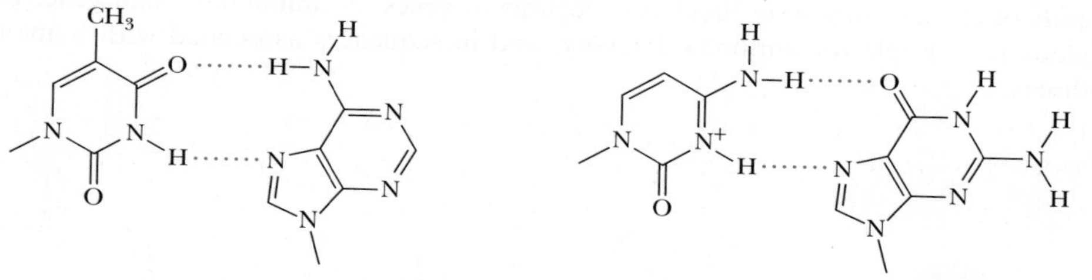

FIGURE 11.16 Hoogsteen base pairs: A:T *(left)* and C⁺:G *(right)*.

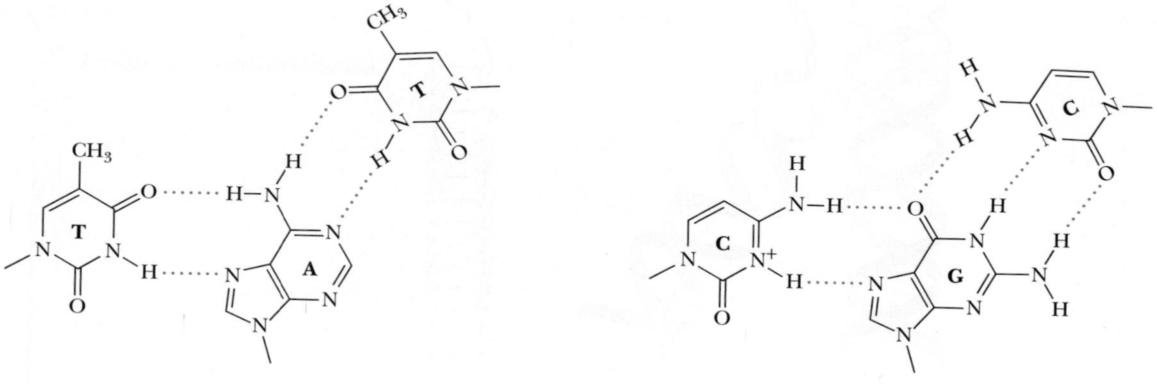

FIGURE 11.17 Base triplets formed when a purine interacts with one pyrimidine by Hoogsteen base pairing and another by Watson–Crick base pairing.

H-DNA Is Triplex DNA Under certain conditions, triple-stranded DNA structures can form. In H-DNA, two of the strands are pyrimidine-rich and the third is purine-rich. One pyrimidine-rich strand is hydrogen bonded to the purine-rich strand via Watson–Crick base pairing, and the other pyrimidine-rich strand is hydrogen bonded to the purine-rich strand by Hoogsteen base pairing. Such structures were originally referred to as H-DNA, because protonation of the cytosine N-3 atom was necessary, but the name also fits because a hinge is present between double- and triple-stranded DNA regions when H-DNA forms. Consider, for example, a long stretch of alternating C:T sequence in one strand of a DNA duplex (Figure 11.18). If the C:T bases in half of this stretch separated from their G:A partners and the unpaired C:T segment folded back on the C:T half still paired in the C:T/G:A duplex, triplex DNA could form through Hoogsteen base pairing. Triple-stranded DNA is implicated in the regulation of some eukaryotic genes.

DNA Quadruplex Structures Four-stranded DNA structures can form between polynucleotide strands rich in guanine. At the heart of such **G-quadruplexes** are cyclic arrays of four G residues united through Hoogsteen base pairing (Figure 11.19a). The presence of metal cations (K^+, Na^+, Ca^{2+}) favors their assembly. Free-electron pairs contributed by the closely spaced O6 carbonyl oxygens of the G-quartet coordinate the centrally located cation. A variety of different G-quadruplex structures have been reported, with different G-rich sequences leading to variations on a common quadruplex plan. Quadruplexes constructed from dG_n strands usually form with all four strands in parallel orientation and all bases in the *anti* conformation (Figure 11.19b). Polynucleotides with varying sequence repeats, such as $(G_3N)_n$ or $(G_2N_2)_n$, form G-quadruplexes with variations on the dG_n structural theme, such as the $(dG_4T_4G_4)_2$ structure in which two such strands pair in antiparallel fashion to form the G-quadruplex (Figure 11.19c and 11.19d). G-quadruplex structures have biological significance because they have been found in telomeres (structures that define the ends of chromosomes), in regulatory regions of genes, in immunoglobulin gene regions responsible for antibody diversity, and in sequences associated with human diseases.

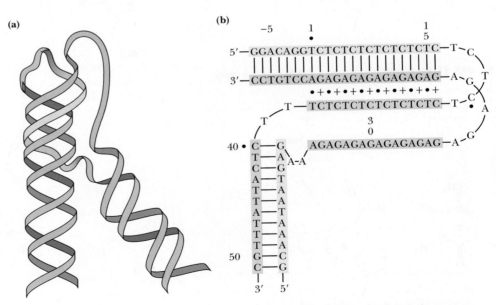

FIGURE 11.18 H-DNA. **(a)** The pyrimidine-rich strands of the duplex regions are blue, and the purine-rich strands are green. The Hoogsteen base-paired pyrimidine-rich strand in the triplex (H-DNA) structure is yellow. **(b)** Nucleotide sequence representation of H-DNA formation. T:A Hoogsteen base pairing leading to triplex formation is shown by dots; C⁺-G Hoogsteen base pairing leading to triplex formation is shown by + signs. (Adapted from Htun, H., and Dahlberg, J. E., 1989. Topology and formation of triple-stranded H-DNA. *Science* **243**:1571–1576.)

(a)

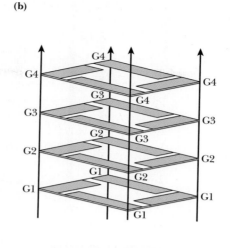

(b)

(c)

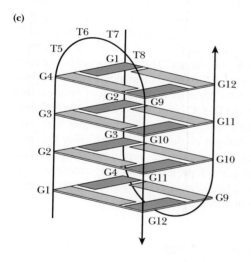

(d)

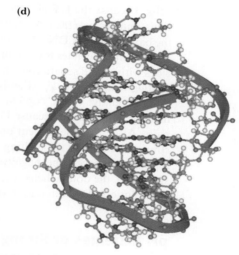

FIGURE 11.19 (a) G-quadruplex showing the cyclic array of guanines linked by Hoogsteen hydrogen bonding. **(b)** Four G-rich polynucleotide strands in parallel alignment with all bases in anti conformation. **(c)** Antiparallel dimeric hairpin quadruplex formed from d(G$_4$T$_4$G$_4$)$_2$. **(d)** Structure of d(G$_4$T$_4$G$_4$)$_2$K$^+$ solved by X-ray crystallography. Two d(G$_4$T$_4$G$_4$) strands come together as hairpins to form a G-quadruplex . The backbones of the two strands are traced in violet. (Adapted from Keniry, M. A., 2001. Quadruplex structures in nucleic acids. *Biopolymers* **56:**123–146.)

11.3 Can the Secondary Structure of DNA Be Denatured and Renatured?

Thermal Denaturation of DNA Can Be Observed by Changes in UV Absorbance

When duplex DNA molecules are subjected to conditions of pH, temperature, or ionic strength that disrupt base-pairing interactions, the strands are no longer held together. That is, the double helix is **denatured,** and the strands separate as individual random coils. If temperature is the denaturing agent, the double helix is said to *melt.* The course of this dissociation can be followed spectrophotometrically because the relative absorbance of the DNA solution at 260 nm increases as much as 40% as the bases unstack. This absorbance increase, or **hyperchromic shift,** is due to the fact that the aromatic

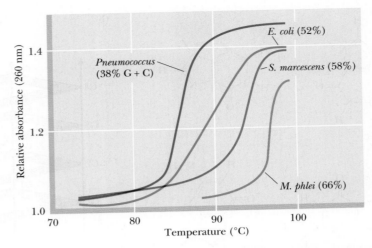

FIGURE 11.20 Heat denaturation of DNA from various sources, so-called melting curves. (From Marmur, J., 1959. Heterogenity in deoxyribonucleic acids. *Nature* **183:**1427–1429.)

bases in DNA interact via their π-electron clouds when stacked together in the double helix. Because the UV absorbance of the bases is a consequence of π-electron transitions, and because the potential for these transitions is diminished when the bases stack, the bases in duplex DNA absorb less 260-nm radiation than expected for their numbers. Unstacking alleviates this suppression of UV absorbance. The rise in absorbance coincides with strand separation, and the midpoint of the absorbance increase is termed the **melting temperature,** T_m (Figure 11.20). DNAs differ in their T_m values because they differ in relative G + C content. The higher the G + C content of a DNA, the higher its melting temperature because G:C pairs have higher base stacking energies than A:T pairs. Also, T_m is dependent on the ionic strength of the solution; the lower the ionic strength, the lower the melting temperature. Because cations suppress the electrostatic repulsion between the negatively charged phosphate groups in the complementary strands of the double helix, the double-stranded form of DNA is more stable in dilute salt solutions. DNA in pure water melts even at room temperature.

pH Extremes or Strong H-Bonding Solutes also Denature DNA Duplexes

At pH values greater than 10, the bases of DNA become deprotonated, which destroys their base-pairing potential, thus denaturing the DNA duplex. Extensive protonation of the bases below pH 2.3 also disrupts base pairing. Alkali is the preferred denaturant because, unlike acid, it does not hydrolyze the glycosidic bonds linking purine bases to the sugar–phosphate backbone. Small solutes that readily form H bonds can also denature duplex DNA at temperatures below T_m. If present in sufficiently high concentrations, such small solutes will form H bonds with the bases, thereby disrupting H-bonding interactions between the base pairs. Examples include formamide and urea.

Single-Stranded DNA Can Renature to Form DNA Duplexes

Denatured DNA will **renature** to re-form the duplex structure if the denaturing conditions are removed (that is, if the solution is cooled, the pH is returned to neutrality, or the denaturants are diluted out). Renaturation requires reassociation of the DNA strands into a double helix, a process termed **reannealing.** For this to occur, the strands must realign themselves so that their complementary bases are once again in register and the helix can be zipped up (Figure 11.21). Renaturation is dependent on both DNA concentration and time. Many of the realignments are imperfect, and thus the strands must dissociate again to allow for proper pairings to be formed. The process

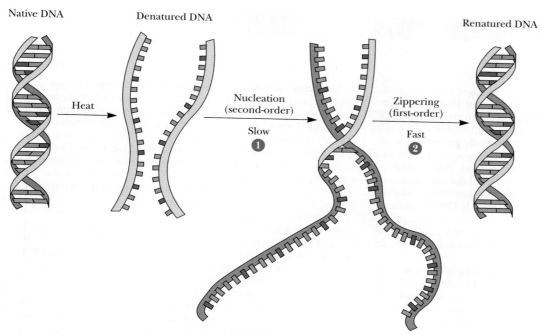

FIGURE 11.21 Steps in the thermal denaturation and renaturation of DNA. The nucleation phase of the reaction is a second-order process depending on sequence alignment of the two strands (**1**). This process takes place slowly because it takes time for complementary sequences to encounter one another in solution and then align themselves in register. Once the sequences are aligned, the strands zipper up quickly (**2**).

occurs more quickly if the temperature is warm enough to promote diffusion of the large DNA molecules but not so warm as to cause melting.

The Rate of DNA Renaturation Is an Index of DNA Sequence Complexity

The renaturation rate of DNA is an excellent indicator of the sequence complexity of DNA. For example, the DNA of bacteriophage T4 contains 2×10^5 base pairs; an *Escherichia coli* chromosome contains more than ten times as much (4.64×10^6 base pairs). *E. coli* DNA is considerably more complex in that it encodes more information. Expressed in another way, for any fixed amount of single-stranded DNA (in grams), the nucleotide sequences represented in an *E. coli* sample will show greater sequence variation than those in an equal weight of phage T4 DNA. Thus, it will take longer for the *E. coli* DNA strands to find their complementary partners and reanneal. Because the rate of DNA duplex formation depends on complementary DNA sequences encountering one another and beginning the process of sequence alignment and reannealing, the time necessary for reconstituting double-stranded DNA molecules is an excellent index of the degree of sequence complementarity in a DNA sample.

Nucleic Acid Hybridization: Different DNA Strands of Similar Sequence Can Form Hybrid Duplexes

If DNA from two different species are mixed, denatured, and allowed to cool slowly so that reannealing can occur, **hybrid duplexes** may form, provided the DNA from one species is similar in nucleotide sequence to the DNA of the other. The degree of hybridization is a measure of the sequence similarity or *relatedness* between the two species. Depending on the conditions of the experiment, about 25% of the DNA from a human forms hybrids with mouse DNA, implying that some of the nucleotide sequences (genes) in humans are very similar to those in mice (Figure 11.22). Mixed RNA:DNA hybrids

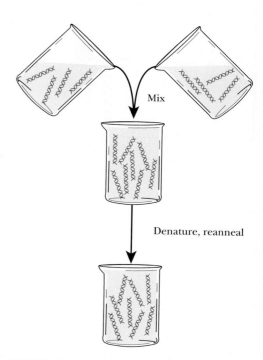

FIGURE 11.22 Solutions of human DNA (red) and mouse DNA (blue) are mixed and denatured, and the single strands are allowed to reanneal. About 25% of the human DNA strands form hybrid duplexes (one red and one blue strand) with mouse DNA.

The Buoyant Density of DNA

Density gradient ultracentrifugation is a variant of the basic technique of ultracentrifugation (discussed in Chapter 5). The densities of DNAs are about the same as those of concentrated solutions of cesium chloride, CsCl (1.6 to 1.8 g/mL). Centrifugation of CsCl solutions at very high rotational speeds, where the centrifugal force becomes 10^5 times stronger than the force of gravity, causes the formation of a density gradient within the solution. This gradient is the result of a balance that is established between the sedimentation of the salt ions toward the bottom of the tube and their diffusion upward toward regions of lower concentration. If DNA is present in the centrifuged CsCl solution, it moves to a position of equilibrium in the gradient equivalent to its buoyant density (as shown in the figure). For this reason, this technique is also called **isopycnic centrifugation.**

Cesium chloride centrifugation is an excellent means of removing RNA and proteins in the purification of DNA. The density of DNA is typically slightly greater than 1.7 g/cm³, whereas the density of RNA is more than 1.8 g/cm³. Proteins have densities less than 1.3 g/cm³. In CsCl solutions of appropriate density, the DNA bands near the center of the tube, RNA pellets to the bottom, and the proteins float near the top. Single-stranded DNA is denser than double helical DNA. The irregular structure of randomly coiled ssDNA allows the atoms to pack together through van der Waals interactions. These interactions compact the molecule into a smaller volume than that occupied by a hydrogen-bonded double helix.

The net movement of solute particles in an ultracentrifuge is the result of two processes: diffusion (from regions of higher concentration to regions of lower concentration) and sedimentation due to centrifugal force (in the direction away from the axis of rotation). In general, diffusion rates for molecules are inversely proportional to their molecular weight—larger molecules diffuse more slowly than smaller ones. On the other hand, sedimentation rates increase with increasing molecular weight. A macromolecular species that has reached its position of equilibrium in isopycnic centrifugation has formed a concentrated band of material.

Essentially three effects are influencing the movement of the molecules in creating this concentration zone: (1) diffusion away to regions of lower concentration, (2) sedimentation of molecules situated at positions of slightly lower solution density in the density gradient, and (3) flotation (buoyancy or "reverse sedimentation") of molecules that have reached positions of slightly greater solution density in the gradient. The consequence of the physics of these effects is that, at equilibrium, *the width of the concentration band established by the macromolecular species is inversely proportional to the square root of its molecular weight.* That is, a population of large molecules will form a concentration band that is narrower than the band formed by a population of small molecules. For example, the bandwidth formed by dsDNA will be less than the bandwidth formed by the same DNA when dissociated into ssDNA.

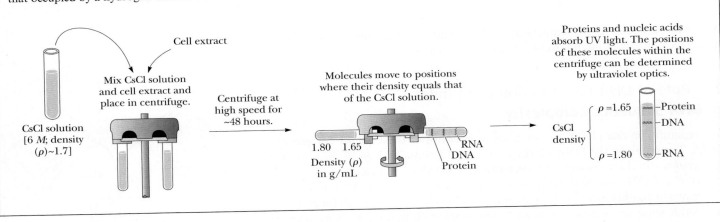

can be created in vitro if single-stranded DNA is allowed to anneal with RNA copies of itself, such as those formed when genes are transcribed into mRNA molecules.

Nucleic acid hybridization is a commonly employed procedure in molecular biology. First, it can reveal evolutionary relationships. Second, it gives researchers the power to identify specific genes selectively against a vast background of irrelevant genetic material: An appropriately labeled oligonucleotide or polynucleotide, referred to as a **probe,** is constructed so that its sequence is complementary to a target gene. The probe specifically base pairs with the target gene, allowing identification and subsequent isolation of the gene. Also, the quantitative expression of genes (in terms of the amount of mRNA synthesized) can be assayed by hybridization experiments.

11.4 Can DNA Adopt Structures of Higher Complexity?

DNA can adopt regular structures of higher complexity in several ways. For example, many DNA molecules are circular. Most, but not all, bacterial chromosomes are covalently closed, circular DNA duplexes, as are most plasmid DNAs. **Plasmids** are naturally occurring, self-replicating, extrachromosomal DNA molecules found in bacteria; plasmids carry genes specifying novel metabolic capacities advantageous to the host bacterium. Various animal virus DNAs are circular as well.

Supercoils Are One Kind of Structural Complexity in DNA

In duplex DNA, the two strands are wound about each other once every 10 bp or so, that is, once every turn of the helix. Double-stranded circular DNA (or linear DNA duplexes whose ends are not free to rotate) form **supercoils** if the strands are underwound *(negatively supercoiled)* or overwound *(positively supercoiled)* (Figure 11.23). Underwound duplex DNA has fewer than the normal number of turns, whereas overwound DNA has more. DNA supercoiling is analogous to twisting or untwisting a two-stranded rope so that it is torsionally stressed. Negative supercoiling introduces a torsional stress that favors unwinding of the right-handed B-DNA double helix, whereas positive supercoiling overwinds such a helix. Both forms of supercoiling compact the DNA so that it sediments faster upon ultracentrifugation or migrates more rapidly in an electrophoretic gel in comparison to **relaxed DNA** (DNA that is not supercoiled). Cellular DNA is almost always negatively supercoiled (underwound).

Linking Number   The basic parameter characterizing supercoiled DNA is the **linking number** (L). This is the number of times the two strands are intertwined, and provided both strands remain covalently intact, L cannot change. In a relaxed circular DNA duplex of 400 bp, L is 40 (assuming 10 bp per turn in B-DNA). The linking number for relaxed DNA is usually taken as the reference parameter and is written as L_0. L can be equated to the **twist** (T) and **writhe** (W) of the duplex, where twist is the number of helical turns and writhe is the number of supercoils:

$$L = T + W$$

Figure 11.24 shows the values of T and W for a simple striped circular tube in various supercoiled forms. In any closed, circular DNA duplex that is relaxed, $W = 0$. A relaxed circular DNA of 400 bp has 40 helical turns, $T = L = 40$. This linking number can be changed only by breaking one or both strands of the DNA, winding them tighter or looser, and rejoining the ends. Enzymes capable of carrying out such reactions are called **topoisomerases** because they change the topological state of DNA. Topoisomerases fall into two basic classes: I and II. Topoisomerases of the I type cut one strand of a DNA double helix, pass the other strand through, and then rejoin the cut ends. Topoisomerase II enzymes cut both strands of a dsDNA, pass a region of the DNA duplex between

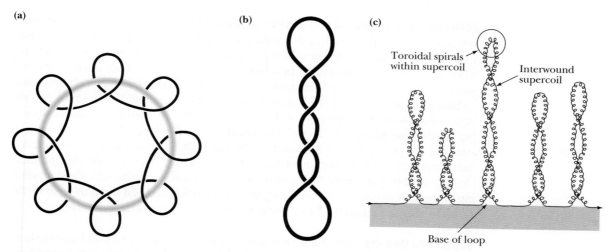

(a)   (b)   (c)

Toroidal spirals within supercoil

Interwound supercoil

Base of loop

FIGURE 11.23   Toroidal and interwound varieties of supercoiling. **(a)** The DNA is coiled in a spiral fashion about an imaginary toroid (yellow circle). **(b)** The DNA interwinds and wraps about itself. **(c)** Supercoils in long, linear DNA arranged into loops whose ends are restrained—a model for chromosomal DNA. (Adapted from Figures 6.1 and 6.2 in Calladine, C. R., and Drew, H. R., 1992. *Understanding DNA: The Molecule and How It Works.* London: Academic Press.)

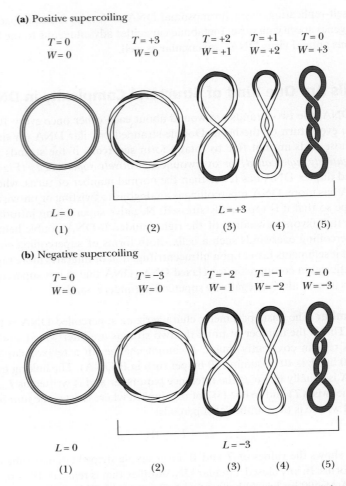

FIGURE 11.24 Supercoil topology for a simple circular tube with a single stripe along it. (Adapted from Figures 6.5 and 6.6 in Callandine, C. R., and Drew, H. R., 1992. *Understanding DNA: The Molecule and How It Works.* London: Academic Press.)

the cut ends, and then rejoin the ends (Figure 11.25). Topoisomerases are important players in DNA replication (see Chapter 28).

DNA Gyrase The bacterial enzyme **DNA gyrase** is a topoisomerase that introduces negative supercoils into DNA in the manner shown in Figure 11.25. Suppose DNA gyrase puts four negative supercoils into the 400-bp circular duplex, then $W = -4$, T remains the same, and $L = 36$ (Figure 11.26). In actuality, the negative supercoils cause a torsional stress on the molecule, so T tends to decrease; that is, the helix becomes a bit unwound, so base pairs are separated. The extreme would be that T would decrease by 4 and the supercoiling would be removed ($T = 36$, $L = 36$, and $W = 0$). That is, negative supercoiling has the potential to cause localized unwinding of the DNA double helix so that single-stranded regions (or bubbles) are created (Figure 11.26). Usually the real situation is a compromise in which the negative value of W is reduced, T decreases slightly, and these changes are distributed over the length of the circular duplex so that no localized unwinding of the helix ensues. Nevertheless, negative supercoiling makes it easier to separate the DNA strands and access the information encoded by the nucleotide sequence.

Superhelix Density The difference between the linking number of a DNA and the linking number of its relaxed form is $\Delta L = (L - L_0)$. In our example with four negative supercoils, $\Delta L = -4$. The **superhelix density** or **specific linking difference** is defined as $\Delta L / L_0$ and is sometimes termed *sigma*, σ. For our example, $\sigma = -4/40$, or -0.1. As a ratio, σ is a measure of supercoiling that is independent of length. Its sign reflects whether the

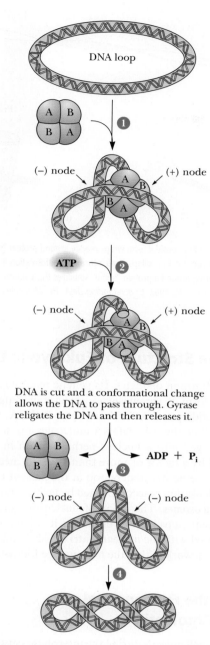

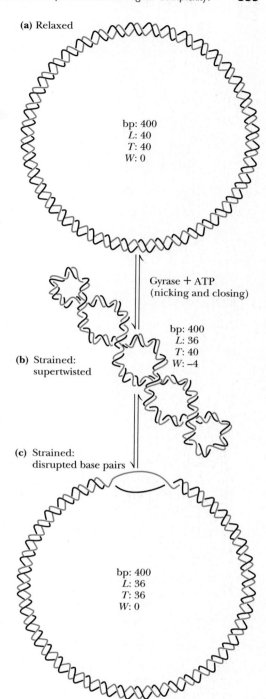

FIGURE 11.25 A model for the action of bacterial DNA gyrase (topoisomerase II). The A-subunits cut the DNA duplex (**1**) and then hold onto the cut ends (**2**). Conformational changes in the enzyme allow an intact region of the DNA duplex to pass between the cut ends. The cut ends are religated (**3**), and the covalently complete DNA duplex is released from the enzyme. The circular DNA now contains two negative supercoils (**4**).

FIGURE 11.26 A 400-bp circular DNA molecule in different topological states: (**a**) relaxed, (**b**) negative supercoils distributed over the entire length, and (**c**) negative supercoils creating a localized single-stranded region.

supercoiling tends to unwind *(negative σ)* or overwind *(positive σ)* the helix. In other words, the superhelix density states the number of supercoils per 10 bp, which also is the same as the number of supercoils per B-DNA repeat. Circular DNA isolated from natural sources is always found in the underwound, negatively supercoiled state.

Toroidal Supercoiled DNA Negatively supercoiled DNA can arrange into a toroidal state (Figure 11.27). The toroidal state of negatively supercoiled DNA is stabilized by wrapping around proteins that serve as spools for the DNA "ribbon." This toroidal conformation of DNA is found in protein–DNA interactions that are the basis of phenomena as diverse as chromosome structure (see Figure 11.30) and gene expression.

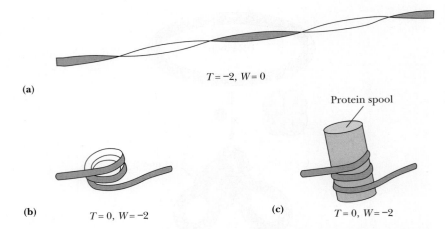

FIGURE 11.27 Supercoiled DNA in a toroidal form wraps readily around protein "spools." A twisted segment of linear DNA with two negative supercoils **(a)** can collapse into a toroidal conformation if its ends are brought closer together **(b)**. Wrapping the DNA toroid around a protein "spool" stabilizes this conformation **(c)**. (Adapted from Figure 6.6 in Callandine, C. R., and Drew, H. R., 1992. *Understanding DNA: The Molecule and How It Works.* London: Academic Press.)

11.5 What Is the Structure of Eukaryotic Chromosomes?

A typical human cell is 20 μm in diameter. Its genetic material consists of 23 pairs of dsDNA molecules in the form of **chromosomes,** the average length of which is 3×10^9 bp/23 or 1.3×10^8 nucleotide pairs. At 0.34 nm/bp in B-DNA, this represents a DNA molecule 5 cm long. Together, these 46 dsDNA molecules amount to more than 2 m of DNA that must be packaged into a nucleus perhaps 5 μm in diameter! Clearly, the DNA must be condensed by a factor of more than 10^5. The mechanisms by which this condensation is achieved are poorly understood at the present time, but it is clear that the first stage of this condensation is accomplished by neatly wrapping the DNA around protein spools called **nucleosomes.** The string of nucleosomes is then coiled to form a helical filament. Subsequent steps are less clear, but it is believed that this filament is arranged in loops associated with the **nuclear matrix,** a skeleton or scaffold of proteins providing a structural framework within the nucleus (see following discussion).

Nucleosomes Are the Fundamental Structural Unit in Chromatin

The DNA in a eukaryotic cell nucleus during the interphase between cell divisions exists as a nucleoprotein complex called **chromatin.** The proteins of chromatin fall into two classes: **histones** and **nonhistone chromosomal proteins.** Histones are abundant and play an important role in chromatin structure. In contrast, the nonhistone class is defined by a great variety of different proteins, all of which are involved in genetic regulation; typically, there are only a few molecules of each per cell. Five distinct histones are known: **H1, H2A, H2B, H3,** and **H4** (Table 11.2). All five are relatively small, positively charged, arginine- or lysine-rich proteins that interact via ionic bonds with the negatively charged phosphate groups on the polynucleotide backbone. Pairs of histones H2A, H2B, H3, and H4 aggregate to form octameric core structures, and the DNA helix is wound about these core octamers, creating **nucleosomes.**

If chromatin is swelled suddenly in water and prepared for viewing in the electron microscope, the nucleosomes are evident as "beads on a string," dsDNA being the string. The structure of the histone octamer core wrapped with DNA has been solved by T. J. Richmond and collaborators (Figure 11.28). The core octamer has surface landmarks that guide the course of the DNA; 147 bp of B-DNA in a flat, left-handed superhelical conformation make 1.6 turns around the histone core (Figure 11.28), which itself is a protein superhelix consisting of a spiral array of the four histone dimers. Histone H1, a three-domain protein, organizes an additional 29–43 bp of DNA and links consecutive nucleosomes.

TABLE 11.2	Properties of Histones		
Histone	Ratio of Lysine to Arginine	Size (kD)	Copies per Nucleosome
H1	59/3	21.2	1 (not in core)
H2A	13/13	14.1	2
H2B	20/8	13.1	2
H3	13/17	15.1	2
H4	11/14	11.4	2

(a)   **(b)**

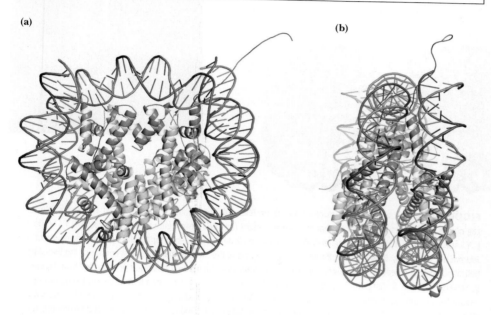

FIGURE 11.28 The nucleosome core particle wrapped with 1.65 turns of DNA (147 bp). The DNA is shown as a blue and orange double helix. The four types of core histones are shown as different colors. *(left)* View down the axis of the nucleosome; *(right)* view perpendicular to the axis (pdb id = 1AOI). (Adapted from Luger, K., et al., 1997. Crystal structure of the nucleosome core particle at 2.8 Å resolution. *Nature* **389**:251–260. Photos courtesy of T. J. Richmond, ETH-Hönggerberg, Zurich, Switzerland.)

Each complete nucleosome unit contains 176–190 bp of DNA. The N-terminal tails of histones H3 and H4 are accessible on the surface of the nucleosome. Lysine and serine residues in these tails can be covalently modified in myriad ways (lysines may be acetylated, methylated, or ubiquitinated; serines may be phosphorylated). These modifications play an important role in chromatin dynamics and gene expression (see Chapter 29).

Higher-Order Structural Organization of Chromatin Gives Rise to Chromosomes

Details about the structural organization of chromatin beyond the level of the nucleosome have remained elusive despite intensive study. If we think of nucleosomes in their characteristic beads-on-a-string motif as the "primary" structure of chromatin, the "secondary" level of chromatin structure is the so-called **30-nm fiber.** Recent models suggest that the 30-nm fiber is formed when the array of nucleosomes adopts a zig-zag pattern and wraps as a two-start helix through interactions between alternating nucleosomes (Figure 11.29). (A two-start helix can be formed from a single-stranded structure whose alternating units are wrapped around an axis to form a "double" helix.) Higher levels of chromatin structural organization are achieved when the 30-nm fiber forms long loops of variable length, each containing on average between 60,000 and 150,000 bp. Electron microscopic analysis of human chromosome 4 suggests that 18 such loops are then arranged radially about the circumference of a single turn to form a **miniband unit** of the chromosome. According to this model, approximately 10^6 of these minibands are arranged along a central axis in each of the chromatids of human chromosome 4 that form at mitosis (Figure 11.30). An essential feature of chromosomes is their structural plasticity, as for example chromosomes condense prior to mitosis and disperse afterward, or as genes are transcribed or not. Such plasticity has a tendency to make structural conclusions less precise.

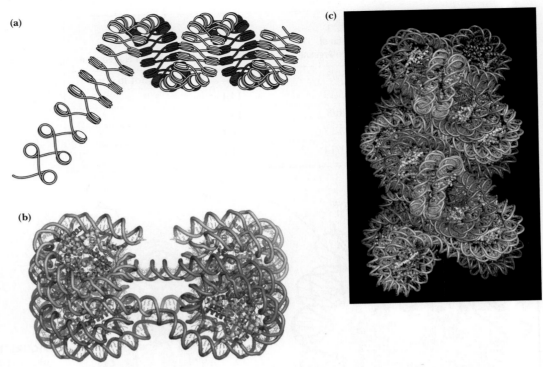

FIGURE 11.29 A suggested structure for the 30-nm fiber. **(a)** The "beads on a string" motif arranges into a zig-zag ribbon that adopts a two-start left-handed helical pattern (adapted from Figure 8 in Woodcock, C. L. F., Frado, L.-L. Y., and Rattner, J. B., 1984. The higher-order structure of chromatin: Evidence for a helical ribbon arrangement. *Journal of Cell Biology* **99**:42–52. **(b)** Packing of nucleosomes in the 30-nm fiber as a consequence of the two-start helical pattern, as revealed by the X-ray crystallographic structure of a tetranucleosome unit (adapted from Figure 1 in Schlach, T., Duda, S., Sargent, D. F., and Richmond, T. J., 2005. X-ray structure of a tetranucleosome and its implications for the chromatin fibre. *Nature* **436**:138–141. **(c)** An idealized model for the 30-nm fiber based in the X-ray structure of the tetranucleosome (adapted from Figure 3 in Schlach, T., Duda, S., Sargent, D. F., and Richmond, T. J., 2005. X-ray structure of a tetranucleosome and its implications for the chromatin fibre. *Nature* **436**:138–141. Images (b) and (c) courtesy of T. J. Richmond and with permission of *Nature*.)

SMC Proteins Establish Chromosome Organization and Mediate Chromosome Dynamics

Although the details remain a mystery, we know that the process of chromatin organization into chromosomes involves **SMC proteins**. SMC stands for structural maintenance of chromosomes. SMC proteins are members of the nonhistone chromosomal protein class. SMC proteins form a large superfamily of ATPases involved in higher-order chromosome organization and dynamics. SMC protein representatives are found in all forms of life—archaea, bacteria, and eukaryotes. Chromosomal dynamics includes chromosome condensation, sister chromatid cohesion, genetic recombination, and DNA repair, as well as other phenomena. SMC proteins have a characteristic five-domain organization, consisting of an N-terminal globular ATP-binding domain, a rodlike dimerization domain involved in coiled coil formation, a flexible hinge region, another rodlike and coiled coil–forming region, and finally a C-terminal globular domain termed *DA* for its DNA-binding and ATPase abilities (Figure 11.31). Five subgroups of SMC proteins are found in eukaryotes, and functional SMC proteins are heterodimers. SMC2/SMC4 heterodimers are essential for chromatin condensation as part of **condensin** complexes; SMC1/SMC3 heterodimers act in sister chromatid cohesion as part of **cohesin** complexes. Current models of SMC protein function suggest that V-shaped heterodimers bind to DNA through their DA domains and mediate chromosomal dynamics in an ATP-dependent manner. The flexible hinge region of each SMC subunit is located at the point of the V, and hinge-bending motions allow the DNA-binding parts of the two globular heads to move closer together, compacting the DNA into a higher-order structure (Figure 11.31).

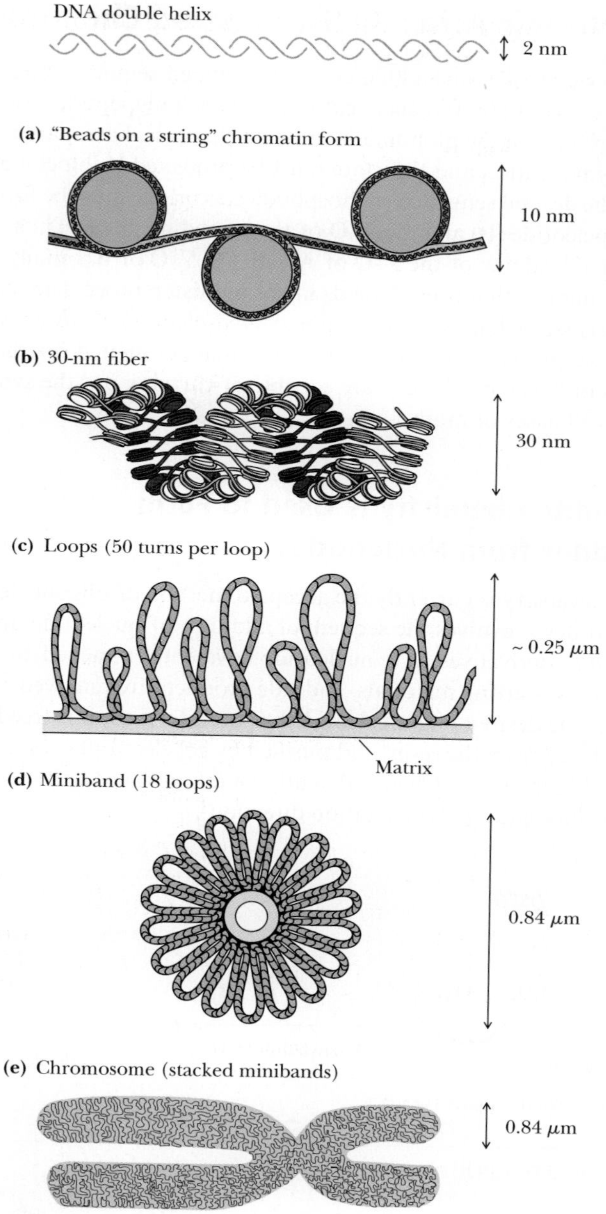

DNA double helix

‡ 2 nm

(a) "Beads on a string" chromatin form

10 nm

(b) 30-nm fiber

30 nm

(c) Loops (50 turns per loop)

~ 0.25 μm

Matrix

(d) Miniband (18 loops)

0.84 μm

(e) Chromosome (stacked minibands)

0.84 μm

FIGURE 11.30 A model for chromosome structure, human chromosome 4.

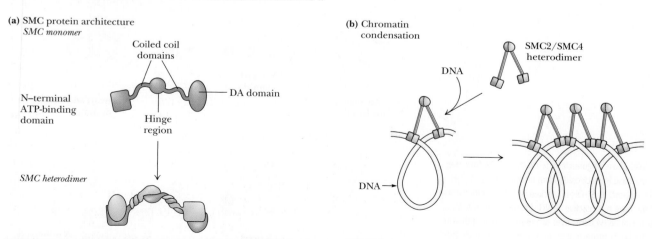

(a) SMC protein architecture

SMC monomer

Coiled coil domains

N–terminal ATP-binding domain

DA domain

Hinge region

SMC heterodimer

(b) Chromatin condensation

SMC2/SMC4 heterodimer

DNA

DNA

FIGURE 11.31 SMC protein architecture and function. **(a)** SMC protein architecture. SMC proteins range from 115 to 165 kD in size. **(b)** SMC protein function. SMC proteins are reminiscent of motor proteins. Illustrated in **(b)** is a condensation of DNA into a coiled arrangement through SMC2/SMC4-mediated interactions.

11.6 : Can Nucleic Acids Be Synthesized Chemically?

Laboratory synthesis of oligonucleotide chains of defined sequence presents some of the same problems encountered in chemical synthesis of polypeptides (see Chapter 5). First, functional groups on the monomeric units (in this case, bases) are reactive under conditions of polymerization and therefore must be protected by blocking agents. Second, to generate the desired sequence, a phosphodiester bridge must be formed between the 3′-O of one nucleotide (B) and the 5′-O of the preceding one (A) in a way that precludes the unwanted bridging of the 3′-O of A with the 5′-O of B. Finally, recoveries at each step must be high so that overall yields in the multistep process are acceptable. As in peptide synthesis (see Chapter 5), orthogonal *solid-phase methods* are used to overcome some of these problems. Commercially available automated instruments, called **DNA synthesizers** or "gene machines," are capable of carrying out the synthesis of oligonucleotides of 150 bases or more.

Phosphoramidite Chemistry Is Used to Form Oligonucleotides from Nucleotides

Phosphoramidite chemistry is currently the accepted method of oligonucleotide synthesis. The general strategy involves the sequential addition of nucleotide units as *nucleoside phosphoramidite* derivatives to a nucleoside covalently attached to the insoluble resin. Excess reagents, starting materials, and side products are removed after each step by filtration. After the desired oligonucleotide has been formed, it is freed of all blocking groups, hydrolyzed from the resin, and purified by gel electrophoresis. The four-step cycle is shown in Figure 11.32. Chemical synthesis takes place in the 3′→5′ direction (the reverse of the biological polymerization direction).

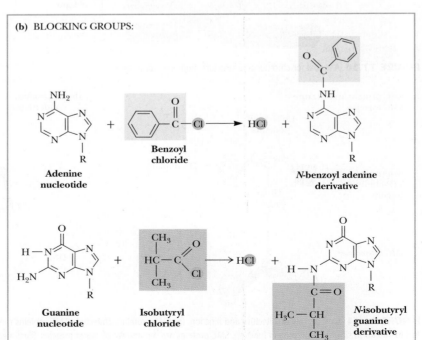

FIGURE 11.32 Solid-phase oligonucleotide synthesis. The four-step cycle starts with the first base in nucleoside form attached by its 3′-OH group to an insoluble support. Its 5′-OH is blocked with a dimethoxytrityl (DMTr) group **(a)**. If the base has reactive —NH₂ functions, as in A, G, or C, then N-benzoyl or N-isobutyryl derivatives are used to prevent their reaction **(b)**. In step 1, the DMTr protecting group is removed by trichloroacetic acid treatment. Step 2 is the coupling step: The second base is added in the form of a nucleoside phosphoramidite derivative whose 5′-OH bears a DMTr blocking group so it cannot polymerize with itself **(c)**.

(c)

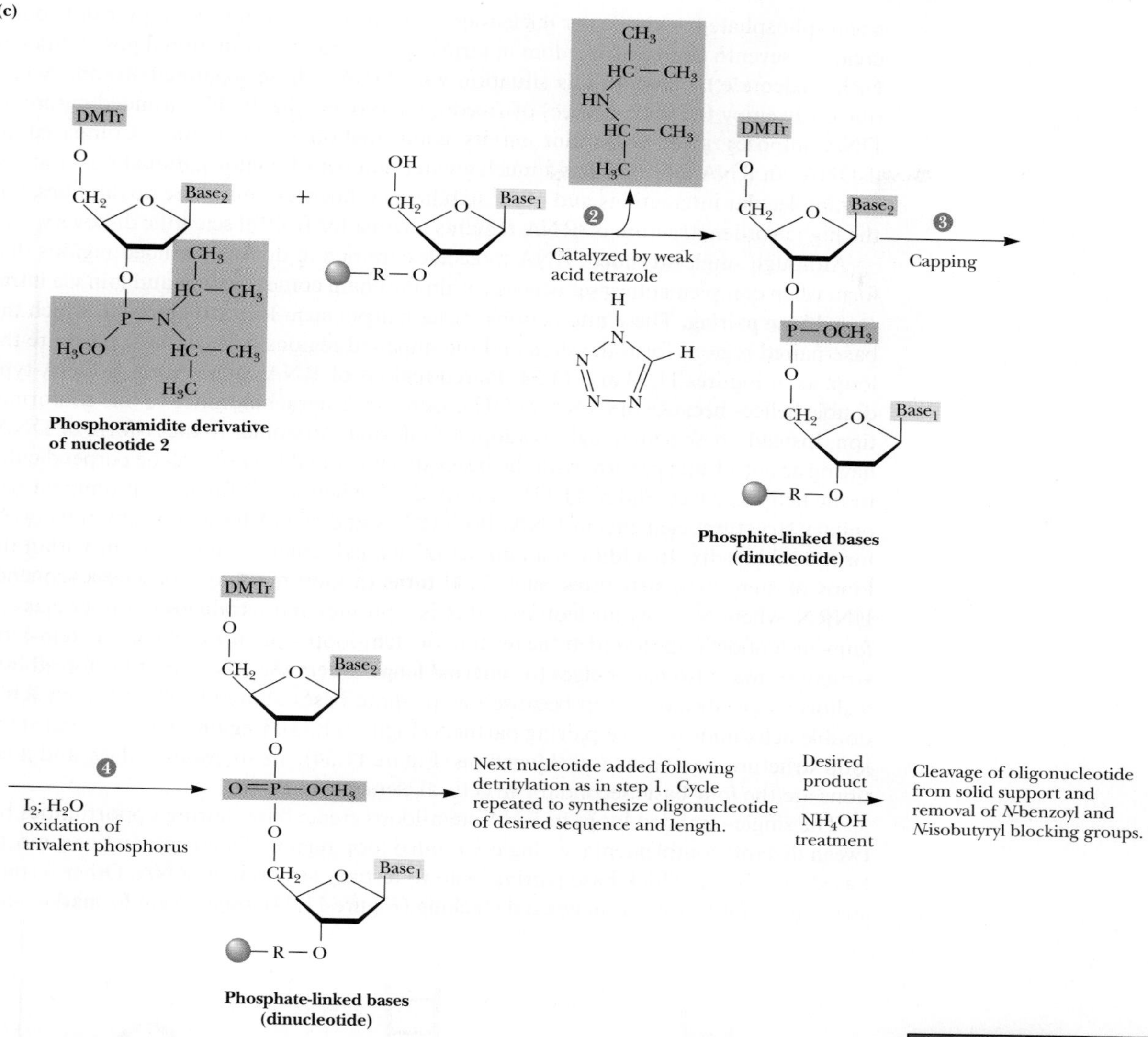

FIGURE 11.32 continued In step 2, the presence of a weak acid, such as tetrazole, activates the phosphoramidite, and it rapidly reacts with the free 5′-OH of N-1, forming a dinucleotide linked by a phosphite group. Unreacted free 5′-OHs of N-1 are blocked from further participation in the polymerization process by acetylation with acetic anhydride in step 3, referred to as *capping*. In step 4, the phosphite linkage between N-1 and N-2 is oxidized by aqueous iodine (I_2) to form the desired more stable phosphate group. Subsequent cycles add successive residues to the resin-immobilized chain. When the chain is complete, it is cleaved from the support with NH_4OH, which also removes the *N*-benzoyl– and *N*-isobutyryl–protecting groups from the amino functions on the A, G, and C residues.

Genes Can Be Synthesized Chemically

It is possible to synthesize genes using phosphoramidite chemistry (Table 11.3). Because protein-coding genes are characteristically much larger than the 150-bp practical limit on oligonucleotide synthesis, gene synthesis involves joining a series of oligonucleotides to assemble the overall sequence.

11.7 What Are the Secondary and Tertiary Structures of RNA?

RNA molecules (see Chapter 10) are typically single-stranded. The course of a single-stranded RNA in three-dimensional space conceivably would have six degrees of freedom per nucleotide, represented by rotation about each of the six single bonds along the

TABLE 11.3	Some Chemically Synthesized Genes
Gene	Size (bp)
tRNA	126
α-Interferon	542
Secretin	81
γ-Interferon	453
Rhodopsin	1057
Proenkephalin	77
Connective tissue activating peptide III	280
Lysozyme	385
Tissue plasminogen activator	1610
c-Ha-ras	576
RNase T1	324
Cytochrome b_5	330
Bovine intestinal Ca-binding protein	298
Hirudin	226
RNase A	375

sugar–phosphate backbone per nucleotide unit. (Rotation about the β-glycosidic bond creates a seventh degree of freedom in terms of the total conformational possibilities at each nucleotide.) Compare this situation with DNA, whose separated strands would obviously enjoy the same degrees of freedom. However, the double-stranded nature of DNA imposes great constraint on its conformational possibilities. Compared to dsDNA, an RNA molecule has a much greater number of conformational possibilities. Intramolecular interactions and other stabilizing influences limit these possibilities, but the higher-order structure of RNA remains an area for fruitful scientific discovery.

Although single-stranded, RNA molecules are rich in double-stranded regions that form when complementary sequences within the chain come together and join via **intrastrand base pairing.** These interactions create hairpin **stem-loop structures,** in which the base-paired regions form the stem and the unpaired regions between base pairs are the loop, as in Figures 11.33 and 11.34. Paired regions of RNA cannot form B-DNA-type double helices because the RNA 2′-OH groups are a steric hindrance to this conformation. Instead, these paired regions adopt a conformation similar to the A-form of DNA, having about 11 bp per turn, with the bases strongly tilted from the plane perpendicular to the helix axis (see Figure 11.11). A-form double helices are the most prominent secondary structural elements in RNA. Both tRNA and rRNA have large amounts of A-form double helix. In addition, a number of defined structural motifs recur within the loops of stem-loop structures, such as **U-turns** (a loop motif of consensus sequence UNRN, where N is any nucleotide and R is a purine) and **tetraloops** (another class of four-nucleotide loops found at the termini of stem-loop structures). Stems of stem-loop structures may also have **bulges** (or **internal loops**) where the RNA strand is forced into a short single-stranded loop because one or more bases along one strand in an RNA double helix finds no base-pairing partners (Figure 11.33). Regions where several stem-loop structures meet are termed **junctions** (Figure 11.34). Stems, loops, bulges, and junctions are the four basic secondary structural elements in RNA.

The single-stranded loops in RNA stem-loops create base-pairing opportunities between distant, complementary, single-stranded loop regions. These interactions, mostly based on Watson–Crick base pairing, lead to tertiary structure in RNA. Other tertiary structural motifs arise from **coaxial stacking** (Figure 11.34), **pseudoknot formation,** and

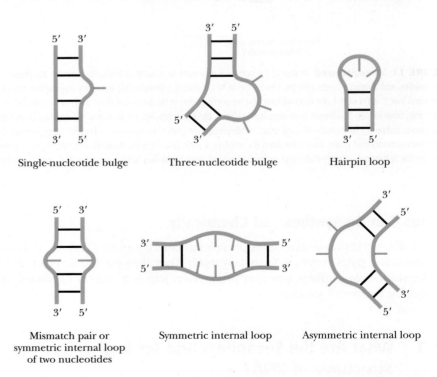

Single-nucleotide bulge   Three-nucleotide bulge   Hairpin loop

Mismatch pair or
symmetric internal loop
of two nucleotides   Symmetric internal loop   Asymmetric internal loop

FIGURE 11.33 Bulges and loops formed in RNA when aligned sequences are not fully complementary. (Adapted from Appendix Figure 1 in Gesteland, R. F., Cech, T. R., and Atkins, J. F., eds. *The RNA World,* 2nd ed. New York: Cold Spring Harbor Press.)

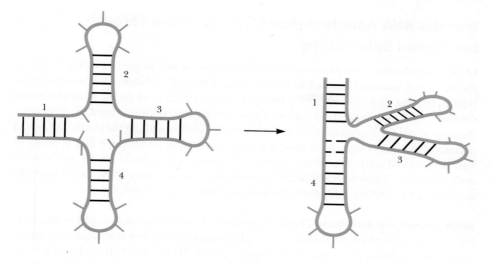

FIGURE 11.34 Junctions and coaxial stacking in RNA. Stem junctions (or multibranched loops) are another type of RNA secondary structure. Coaxial stacking of stems or stem-loops (as in stacking of stem 1 on stem-loop 4) is a tertiary structural feature found in many RNAs. (Adapted from Figure 1 in Tyagi, R., and Matthews, D. H., 2007. Predicting co-axial stacking in multibranch loops. *RNA* **13**:1–13.)

ribose zippers. In coaxial stacking, the blunt, nonloop ends of stem-loops situated next to one another in the RNA sequence stack upon each other to create an uninterrupted stack of base pairs. A good example of coaxial stacking is found in the tertiary structure of tRNAs, where the acceptor end of the L-shaped tRNA is formed by coaxial stacking of the acceptor stem on the TΨC stem-loop and the anticodon end is formed by coaxial stacking of the dihydrouracil stem-loop on the anticodon stem-loop (Figures 11.36 and 11.38). Pseudoknots occur when bases in the loops of stem-loop structures form a short double helix by base pairing with nearby single-stranded regions in the RNA (Figure 11.35). Ribose zippers are found when two antiparallel, single-stranded regions of RNA align as an H-bonded network forms between the 2′-OH groups of the respective strands, the O at the 2′-OH position of one strand serving as the H-bond acceptor while the H on the 2′-OH of the other strand is the H-bond donor. Ribose zippers and the other RNA structures mentioned here are well represented by many examples in the SCOR (Structural Classification of RNA) database at http://scor.berkeley.edu and NDB (Nucleic Acid Database) at http://ndbserver.rutgers.edu/.

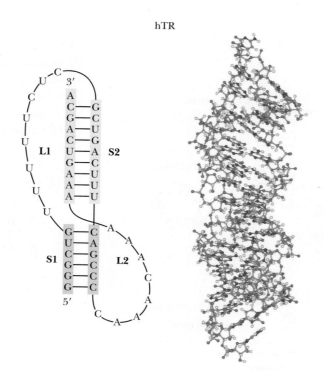

FIGURE 11.35 RNA pseudoknots are formed when a single-stranded region of RNA folds to base-pair with a hairpin loop. Loops L1 and L2, as shown on the sequence representation of human telomerase RNA (hTR) on the left, form a pseudoknot. The three-dimensional structure of an hTR pseudoknot is shown on the right (pdb id = 1YMO). (Adapted from Figure 2 in Staple, D. W., and Butcher, S. E., 2005. Pseudoknots: RNA structures with diverse functions. *PLoS Biology* **3**:e213.)

Transfer RNA Adopts Higher-Order Structure Through Intrastrand Base Pairing

In tRNA molecules, which contain 73 to 94 nucleotides in a single chain, a majority of the bases are hydrogen bonded to one another. Figure 11.36 shows the structure that typifies tRNAs. *Hairpin turns* bring complementary stretches of bases in the chain into contact so that double helical regions form, creating stem-loop secondary structures. Because of the arrangement of the complementary stretches along the chain, the overall pattern of base pairing can be represented as a *cloverleaf*. Each cloverleaf consists of four base-paired segments—three loops and the stem where the 3'- and 5'-ends of the molecule meet. These four segments are designated the ***acceptor stem,*** the ***D loop***, the ***anticodon loop,*** and the ***TψC loop*** (the latter two are U-turn motifs).

tRNA Secondary Structure The *acceptor stem* is where the amino acid is linked to form the aminoacyl-tRNA derivative, which serves as the amino acid–donating species in protein synthesis; this is the physiological role of tRNA. The carboxyl group of an amino acid is linked to the 3'-OH of the 3'-terminal A nucleotide, thus forming an aminoacyl ester (Figure 11.36). The 3'-end of tRNA is invariantly CCA-3'-OH. The *D loop* is so named because this tRNA loop often contains dihydrouridine, or D, residues. In addition to dihydrouridine, tRNAs characteristically contain a number of unusual bases, including inosine, thiouridine, pseudouridine, and hypermethylated purines (see Figure 10.23). The *anticodon stem-loop* consists of a double helical segment and seven unpaired bases, three of which are the **anticodon**—a three-nucleotide unit that recognizes and base pairs with a particular mRNA **codon**, a complementary three-base unit in mRNA providing the genetic information that specifies an amino acid. In the 5'→3' direction beyond the anticodon stem-loop lies a loop that varies from tRNA to tRNA in the number of residues that it has, the so-called **extra** or **variable loop.** The last loop in the tRNA, reading 5'→3', is within the **TψC stem-loop.** It contains seven unpaired bases, including the sequence TψC, where ψ is the symbol for **pseudouridine.** Most of the invariant residues common to tRNAs lie within the non–hydrogen-bonded regions of the cloverleaf structure.

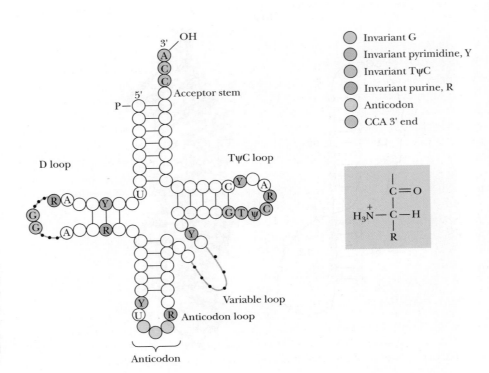

FIGURE 11.36 A general diagram for the structure of tRNA. The positions of invariant bases as well as bases that seldom vary are shown in color. R = purine; Y = pyrimidine. Dotted lines denote sites in the D loop and variable loop regions where varying numbers of nucleotides are found in different tRNAs. Inset: An aminoacyl group can add to the 3'-OH to create an aminoacyl-tRNA.

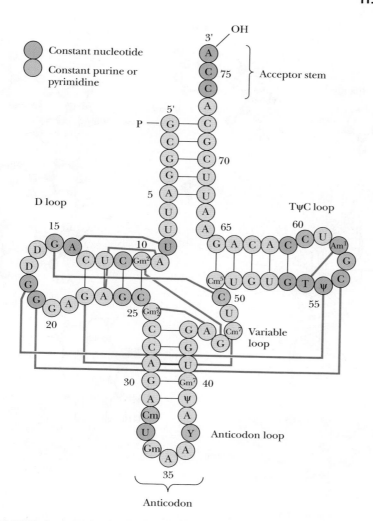

FIGURE 11.37 Tertiary interactions in yeast phenylalanine tRNA. The molecule is presented in the conventional cloverleaf secondary structure generated by intrastrand hydrogen bonding. Solid lines connect bases that are hydrogen bonded when this cloverleaf pattern is folded into the characteristic tRNA tertiary structure (see also Figure 11.38).

tRNA Tertiary Structure Tertiary structure in tRNA arises from base-pairing interactions between bases in the D loop with bases in the variable and TψC loops, as shown for yeast phenylalanine tRNA in Figure 11.37. Note that these base-pairing interactions involve the invariant nucleotides of tRNAs. These interactions fold the D and TψC arms together and bend the cloverleaf into the stable L-shaped tertiary form (Figure 11.38). Many of these base-pairing interactions involve base pairs that are not canonical A:T or G:C pairings, as illustrated around the central ribbon diagram of the tRNA in Figure 11.38. Note that three of the interactions involve three bases. The amino acid acceptor stem (highlighted in green) is at one end of the inverted, backward L shape, separated by 7 nm or so from the anticodon at the opposite end of the L. The D and TψC loops form the corner of the L. Hydrophobic stacking interactions between the flat faces of the bases contributes significantly to L-form stabilization.

Ribosomal RNA Also Adopts Higher-Order Structure Through Intrastrand Base Pairing

rRNA Secondary Structure A large degree of *intrastrand sequence complementarity* is found in all ribosomal RNA strands, and all assume a highly folded pattern that allows base pairing between these complementary segments, giving rise to multiple stem-loop

(a)

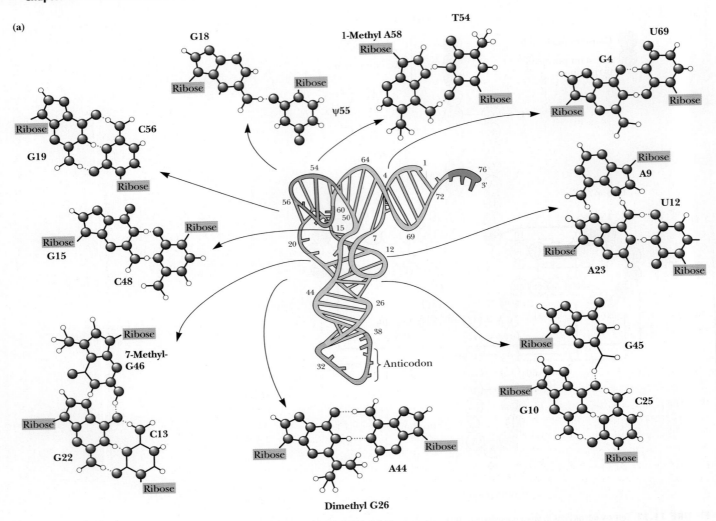

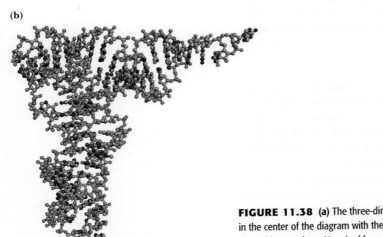

FIGURE 11.38 **(a)** The three-dimensional structure of yeast phenylalanine tRNA. The tertiary folding is illustrated in the center of the diagram with the ribose–phosphate backbone presented as a continuous ribbon; H bonds are indicated by crossbars. Unpaired bases are shown as short, unconnected rods. The anticodon loop is at the bottom and the -CCA 3′-OH acceptor end is at the top right. **(b)** A space-filling model of the molecule (pdb id = 6TNA).

structures. Furthermore, the loop regions of stem-loops contain the characteristic structural motifs, such as U-turns, tetraloops, and bulges. Figure 11.39 shows the secondary structure of several 16S rRNAs, based on computer alignment of each nucleotide sequence into optimal H-bonding segments. The reliability of these alignments is then tested through a comparative analysis of whether very similar secondary structures are

(a) *E. coli* (a eubacterium)   **(b)** *H. volcanii* (an archaebacterium)   **(c)** *S. cerevisiae* (yeast, a lower eukaryote)

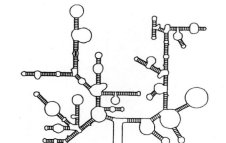
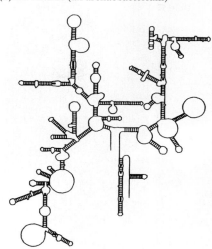
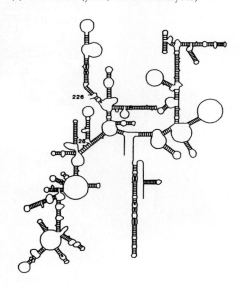

FIGURE 11.39 Comparison of secondary structures of 16S-like rRNAs from **(a)** a bacterium *(E. coli)*, **(b)** an archaeon *(H. volcanii)*, and **(c)** a eukaryote (*S. cerevisiae*, a yeast).

observed. If so, then such structures are apparently conserved. The approach is based on the thesis that because ribosomal RNA species (regardless of source) serve common roles in protein synthesis, it may be anticipated that they share structural features. These secondary structures resemble one another, even though the nucleotide sequences of these 16S rRNAs exhibit little sequence similarity. Apparently, evolution is acting at the level of rRNA secondary structure, not rRNA nucleotide sequence. Similar conserved folding patterns are seen for the 5S-like rRNAs and 23S-like rRNAs that reside in the large ribosomal subunits of various species. An insightful conclusion may be drawn regarding the persistence of such strong secondary structure conservation despite the millennia that have passed since these organisms diverged: *All ribosomes are similar, and all function in a similar manner.* As usual with RNAs, the single-stranded regions of rRNA create the possibility of base-pairing opportunities with distant, complementary, single-stranded regions. Such interactions are the driving force for tertiary structure formation in RNAs.

rRNA Tertiary Structure Recently, the detailed structure of ribosomes has been revealed through X-ray crystallography and cryoelectron microscopy of ribosomes (see Chapter 30). These detailed images not only disclose the tertiary structure of the rRNAs but also the quaternary interactions that must occur when ribosomal proteins combine with rRNAs and when the ensuing ribonucleoprotein complexes, the small and large subunits, come together to form the complete ribosome. Only the rRNAs of the 50S ribosomal subunit are shown in Figure 11.40; *no* ribosomal proteins are shown. Note that the overall anatomy of the 50S ribosomal subunit (shown diagrammatically in Figure 10.22) is essentially the same as that of the rRNA molecules within this subunit, even though these rRNAs account for only 65% of the mass of this particle. An assortment of tertiary structural features are found in the rRNAs, including coaxial stacks, pseudoknots, and ribose zippers. We will consider the role of rRNA in ribosome structure and function in Chapter 30.

Aptamers Are Oligonucleotides Specifically Selected for Their Ligand-Binding Ability

Aptamers are synthetic oligonucleotides, usually RNA, which fold into very specific three-dimensional structures that selectively bind ligands with high affinity. Ligand binding by aptamers is based on the fundamental principle of structural complementarity.

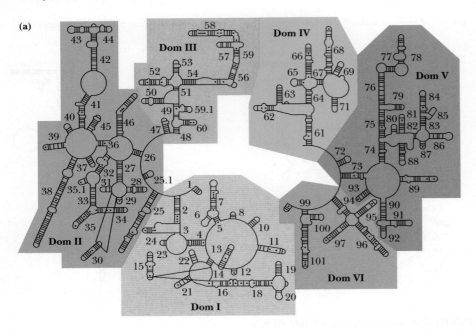

(a)

23S rRNA 5′ end   **3′ end**

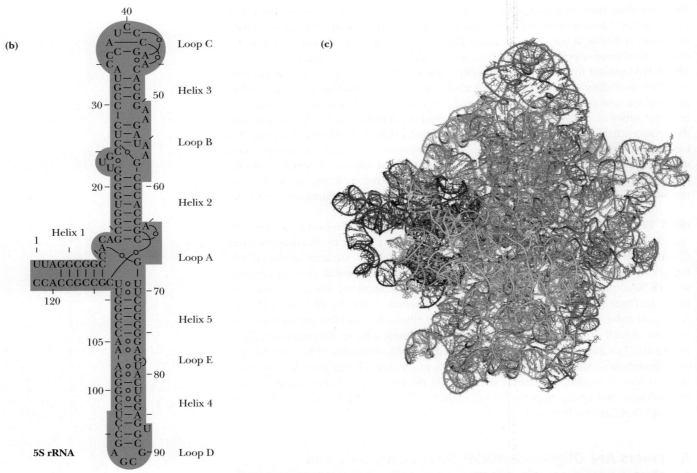

(b)

5S rRNA

(c)

FIGURE 11.40 The secondary and tertiary structures of rRNAs in the 50S ribosomal subunit from the archaeon *Haloarcula marismortui* (pdb id = 1FFk). **(a)** Secondary structure of the 23S rRNA, with various domains color-coded. **(b)** Secondary structure of 5S rRNA. **(c)** Tertiary structure of the 5S and 23S rRNAs within the 50S ribosomal subunit. The 5S rRNA (red) lies atop the 23S rRNA. (Adapted from Figure 4 in Ban, N., et al., 2000. The complete atomic structure of the large ribosomal subunit at 2.4 Å resolution. *Science* **289**:905–920.)

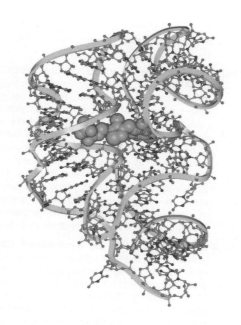

FIGURE 11.41 Structure of the thiamine pyrophosphate (TPP) riboswitch, a conserved region within the mRNA that encodes enzymes for synthesis of this coenzyme (pdb id = 2CKY). TTP, a pyrimidine-containing compound, is shown in orange. (From Figure 1b in Thore, S., Leibundgdut, M., and Ban, N., 2006. Structure of the eukaryotic thiamine pyrophosphate riboswitch with its regulatory ligand. *Science* **312**:1208–1211.)

The rich array of interactive possibilities presented by the four bases and the sugar–phosphate backbone, coupled with the inherent flexibility of polynucleotide chains, make nucleic acids very good ligand-binding candidates. The bases project polar amino and carbonyl functionalities, and their π-electron density gives them nonpolar properties. The sugar–phosphate backbone presents polar —OH groups and regularly spaced, negatively charged phosphate groups. These phosphate groups can coordinate cations and thus provide foci of positive charge. Synthetic aptamers designed to target a selected protein can be potent inhibitors of protein function; they are of interest in drug development.

Riboswitches, a naturally occurring class of aptamers, are conserved regions of mRNAs that reversibly bind specific metabolites and coenzymes and usually act as regulators of gene expression. Riboswitches are usually buried within the 5′- or 3′-untranslated regions of the mRNAs whose expression they regulate. Binding of the metabolite to the riboswitch typically blocks expression of the mRNA. Figure 11.41 shows the structure of the thiamine pyrophosphate riboswitch.

SUMMARY

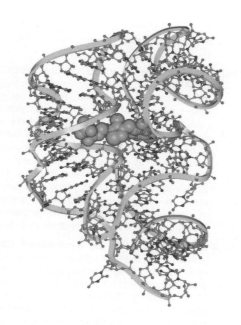 **OWL** Login to OWL to develop problem-solving skills and complete online homework assigned by your professor.

11.1 How Do Scientists Determine the Primary Structure of Nucleic Acids? The traditional protocol for nucleic acid sequencing is Sanger's chain termination (also called the dideoxy or the primed synthesis) method. A DNA fragment of unknown sequence serves as template in a polymerization reaction using DNA polymerase. Polymerization depends on an oligonucleotide primer base-paired to the template. All four DNA polymerase deoxynucleotide substrates—dATP, dGTP, dCTP, and dTTP—are present. In addition, the reaction mixture contains the four corresponding 2′,3′-**di**deoxynucleotides (ddATP, ddGTP, ddCTP, and ddTTP). As synthesis proceeds, a deoxynucleotide is usually added to the 3′-OH end of the growing chain as the newly formed strand is extended in the 5′→3′ direction. Occasionally, however, a dideoxynucleotide is added and, because it lacks a 3′-OH group, it cannot serve as a deoxynucleotide acceptor in chain extension. Then synthesis is terminated. This base-specific premature chain termination is only a random, occasional event, and a population of new strands of varying length is synthesized. The population of

newly synthesized DNAs forms a nested set of molecules differing in length by just one nucleotide. Each has a dideoxynucleotide at its 3′-end. Because each of the four dideoxynucleotides bears a different fluorescent tag, the particular fluorescence (as in orange for ddA, blue for ddC, green for ddG, and red for ddT) indicates which base was specified by the template and incorporated by DNA polymerase at that spot. The sequencing products are visualized by fluorescence spectroscopy following capillary electrophoresis, revealing the sequence of the newly synthesized strands. This observed sequence is complementary to the corresponding unknown template sequence. Sanger sequencing has been fully automated.

Contemporary technology such as the 454 ultra-high-throughput DNA sequencing system also exploits primer-dependent DNA polymerase synthesis of a DNA strand of unknown sequence, but uses an enzyme-coupled reaction to emit light when pyrophosphate is released as each deoxynucleotide is added to the growing chain. The related Ion Torrents technology simply records the pH change as a proton is released upon deoxynucleotide addition. Such systems can simultaneously sequence over a million different strands and report over 500 successive nucleotides for each strand in one run. Even more rapid and inexpensive sequencing methodologies are under development, aimed at lowering the cost of sequencing a person's genome for costs as low as $1000.

11.2 What Sorts of Secondary Structures Can Double-Stranded DNA Molecules Adopt?

DNA typically occurs as a double helical molecule, with the two DNA strands running antiparallel to one another, bases inside, sugar–phosphate backbone outside. The double helical arrangement dramatically curtails the conformational possibilities otherwise available to single-stranded DNA. DNA double helices can be in a number of stable conformations, with the three predominant forms termed A-, B-, and Z-DNA. B-DNA, has about 10.5 base pairs per turn, each contributing about 0.332 nm to the length of the double helix. The base pairs in B-DNA are nearly perpendicular to the helix axis. In A-DNA, the pitch is 2.46 nm, with 11 bp per turn. A-DNA has its base pairs displaced around, rather than centered on, the helix axis. Z-DNA has four distinctions: It is left-handed, it is G:C-rich, the repeating unit on a given strand is the dinucleotide, and the sugar–phosphate backbone follows a zigzag course. Alternative hydrogen-bonding interactions between A:T and G:C gives rise to Hoogsteen base pairs. Interstrand Hoogsteen base pairing creates novel multiplex structures composed of three or four DNA strands. These multiplex structures occur naturally and have biological implications.

11.3 Can the Secondary Structure of DNA Be Denatured and Renatured?

When duplex DNA is subjected to conditions that disrupt base-pairing interactions, the double helix is denatured and the two DNA strands separate as individual random coils. Denatured DNA will renature to re-form a duplex structure if the denaturing conditions are removed. The rate of DNA renaturation is an index of DNA sequence complexity.

If DNA from two different species are mixed, denatured, and allowed to anneal, artificial hybrid duplexes may form, provided the DNA from one species is similar in nucleotide sequence to the DNA of the other. Nucleic acid hybridization can reveal evolutionary relationships, and it can be exploited to identify specific DNA sequences.

11.4 Can DNA Adopt Structures of Higher Complexity?

Supercoils are one kind of DNA tertiary structure. In relaxed, B-form DNA, the two strands wind about each other once every 10 bp or so (once every turn of the helix). DNA duplexes form supercoils if the strands are underwound (*negatively supercoiled*) or overwound (*positively supercoiled*). The basic parameter characterizing supercoiled DNA is the linking number, L. L can be equated to the twist (T) and writhe (W), where twist is the number of helical turns and writhe is the number of supercoils: $L = T + W$. L can be changed only if one or both strands of the DNA are broken, the strands are wound tighter or looser, and their ends are rejoined. DNA gyrase is a topoisomerase that introduces negative supercoils into bacterial DNA.

11.5 What Is the Structure of Eukaryotic Chromosomes?

The DNA in a eukaryotic cell exists as chromatin, a nucleoprotein complex mostly composed of DNA wrapped around a protein core consisting of eight histone polypeptide chains—two copies each of histones H2A, H2B, H3, and H4. This DNA:histone core structure is termed a nucleosome, the fundamental structural unit of chromosomes. A higher order of chromatin structure is created when the array of nucleosomes arranges into a two-start helical structure known as the 30-nm filament. This 30-nm filament then is formed into long DNA loops, and loops are arranged radially about the circumference of a single turn to form a miniband unit of a chromosome. SMC proteins mediate chomosomal dynamics, including chromatin condensation and chromosome formation.

11.6 Can Nucleic Acids Be Synthesized Chemically?

Laboratory synthesis of oligonucleotide chains of defined sequence is accomplished through orthogonal solid-phase methods based on phosphoramidite chemistry. Chemical synthesis takes place in the $3'{\rightarrow}5'$ direction (the reverse of the biological polymerization direction). Commercially available automated instruments called DNA synthesizers can synthesize oligonucleotide chains with 150 bases or more.

11.7 What Are the Secondary and Tertiary Structures of RNA?

Compared to double-stranded DNA, single-stranded RNA has many more conformational possibilities, but intramolecular interactions and other stabilizing influences limit these possibilities. RNA molecules have many double-stranded regions formed via intrastrand hydrogen bonding. Such double-stranded regions give rise to hairpin stem-loop structures. A number of defined structural motifs recur within the loops of stem-loop structures, such as U-turns and tetraloops.

Single-stranded loops in RNA stem-loops create base-pairing opportunities between distant, complementary, single-stranded loop regions. Other tertiary structural motifs arise from coaxial stacking, pseudoknot formation, and ribose zippers.

In tRNAs, the formation of stem-loops leads to a cloverleaf pattern of secondary structure formed from four base-paired segments: the acceptor stem, the D loop, the anticodon loop, and the TψC loop. Base-pairing interactions between bases in the D and TψC loops give rise to tertiary structure by bending the cloverleaf into the stable L-shaped form.

Substantial intrastrand sequence complementarity also is found in ribosomal RNA molecules, leading to a highly folded pattern based on base pairing between complementary segments. The complete three-dimensional structure of rRNAs has revealed an assortment of the tertiary structural features common to RNAs, including coaxial stacks, pseudoknots, and ribose zippers.

FOUNDATIONAL BIOCHEMISTRY **Things You Should Know After Reading Chapter 11.**

- DNA sequencing: Sanger's chain termination protocol based on dideoxynucleotides uses DNA replication to generate a defined set of polynucleotide fragments.

- The nucleotide sequence of DNA can be determined from the electrophoretic migration of a defined set of polynucleotide fragments.

- Current ultra-high-throughput methodologies for DNA sequencing also rely on DNA polymerase-driven primer extension to copy unknown DNA sequences.

- Ultra-high-throughput sequencing technologies use light emission or pH change to follow DNA polymerase-catalyzed nucleotide addition and note which base was added. These reactions can be carried out on millions of different DNA strands at the same time, all of which are localized on a 6-cm^2 platform.

- DNA structure: The sugar-phosphate backbone of nucleic acids has six degrees of freedom. Rotation about the C_1'-N glycosidic bond creates a seventh degree of freedom.

- DNA usually occurs in the form of double-stranded molecules.

- The Watson–Crick base pairs (A-T; G-C) have virtually identical dimensions.

- The DNA double helix is stabilized by H bonds, electrostatic interactions, and the van der Waals and hydrophobic interactions arising from base-pair stacking.

- The DNA double helix has a major groove and a minor groove.

- DNA double helices can occur as A-, B-, or Z-DNA conformations; the B-DNA conformation is the canonical form for double-helical DNA.

- Z-DNA is a left-handed double helix.

- RNA double helices and hybrid DNA:RNA helices adopt the A-conformation due to steric hindrance caused by the RNA 3′-OH groups.

- Alternative H-bonding interactions give rise to novel DNA structures, such as cruciforms, triplexes, and quadruplexes.

- Hoogsteen base pairs, an alternative mode of base-pair formation, arise in triplex DNA structures known as H-DNA.

- Thermal denaturation of DNA can be followed by observing increases in UV absorbance.

- Single-stranded DNA can renature (re-anneal) to form DNA duplexes.

- The rate of DNA renaturation is a measure of DNA sequence complexity.

- Supercoils are one kind of structural complexity in DNA. Supercoils alter the topological state of DNA.

- Supercoils can be positive (overwound DNA) or negative (underwound DNA), relative to B-DNA.

- The linking number, L, of double-helical DNA is given by $L = T + W$, where T = twist (the number of helical turns in a DNA molecule) and W = writhe (the number of supercoils).

- Topoisomerases are enzymes that change the topological state of DNA.

- Eukaryotic DNA: Nucleosomes are the fundamental structural unit of chromatin.

- Higher-order structural organization of chromatin gives rise to chromosomes.

- SMC proteins establish chromosome organization and mediate chromosome dynamics.

- Nucleic acids can be created synthetically, using an orthogonal procedure and phosphoramidite chemistry.

- RNA molecules are rich in secondary structures formed through intrastrand H-bonding.

- RNAs contain a variety of structural motifs, such as stem-loops, U-turns, tetraloops, bulges, junctions, and coaxial stacks.

- Transfer RNA adopts higher-order structure through base-pairing interactions between the single-stranded regions of stem-loop structures.

- Ribosomal RNA also adopts higher-order structure through base-pairing interactions between the single-stranded regions of stem-loop structures.

- Ribosomal RNAs adopt an evolutionarily conserved secondary structure.

- Aptamers are oligonucleotides with highly selective ligand-binding ability.

- Riboswitches are naturally occurring aptamers involved in regulating the expression of genetic information.

PROBLEMS

Answers to all problems are at the end of this book. Detailed solutions are available in the *Student Solutions Manual, Study Guide, and Problems Book.*

1. **Predicting a Sanger Sequencing Pattern** The oligonucleotide d-AGATGCCTGACT was subjected to sequencing by Sanger's dideoxy method, using fluorescent-tagged dideoxynucleotides and capillary electrophoresis, essentially as shown in Figure 11.3. Draw a diagram of the gel-banding pattern within the capillary.

2. **Deducing DNA Sequence from Sanger Sequencing Results** The output of an automated DNA sequence determination by the Sanger dideoxy chain termination method, performed as illustrated in Figure 11.3, is displayed at right. What is the sequence of the original oligonucleotide?

3. **Helical Parameters for a Double-Stranded DNA** X-ray diffraction studies indicate the existence of a double-stranded DNA helical conformation in which ΔZ (the rise per base pair) = 0.32 nm and P (the pitch) = 3.36 nm. What are the other parameters of this novel helix: (a) the number of base pairs per turn, (b) $\Delta\phi$ (the mean rotation per base pair), and (c) c (the true repeat)?

4. **B- and Z-DNA Helical Parameters I** A 41.5-nm-long duplex DNA molecule in the B-conformation adopts the A-conformation upon dehydration. How long is it now? What is its approximate number of base pairs?

5. **B- and Z-DNA Helical Parameters II** If 80% of the base pairs in a duplex DNA molecule (12.5 kbp) are in the B-conformation and 20% are in the Z-conformation, what is the length of the molecule?

6. **DNA Supercoiling Parameters** A "relaxed," circular, double-stranded DNA molecule (1600 bp) is in a solution where conditions favor 10 bp per turn. What is the value of L_0 for this DNA molecule? Suppose DNA gyrase introduces 12 negative supercoils into this molecule. What are the values of L, W, and T now? What is the superhelical density, σ?

7. **B- and Z-DNA Supercoiling Parameters** Suppose one double helical turn of a superhelical DNA molecule changes conformation from B- to Z-form. What are the changes in L, W, and T? Why do you suppose the transition of DNA from B- to Z-form is favored by negative supercoiling?

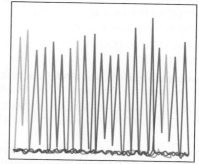

8. **Calculate the Number of Nucleosomes in a Human Diploid Cell** Assume that there is one nucleosome for every 200 bp of eukaryotic DNA. How many nucleosomes are there in a diploid human cell? Nucleosomes can be approximated as disks 11 nm in diameter and 6 nm long. If all the DNA molecules in a diploid human cell are in the B-conformation, what is the sum of their lengths? If this DNA is now arrayed on nucleosomes in the beads-on-a-string motif, what would be the approximate total height of the nucleosome column if these disks were stacked atop one another?

9. **The Linear Arrangement of Complementary Sequences along the tRNA Strand** The characteristic secondary structures of tRNA and rRNA molecules are achieved through intrastrand hydrogen bonding. Even for the small tRNAs, remote regions of the nucleotide sequence interact via H bonding when the molecule adopts the cloverleaf pattern. Using Figure 11.36 as a guide, draw the primary structure of a tRNA and label the positions of its various self-complementary regions.

10. **Using Chargaff's Results to Order DNAs According to Their T_ms** Using the data in Table 10.1, arrange the DNAs from the following sources in order of increasing T_m: human, salmon, wheat, yeast, E. coli.

11. **Calculating T_ms and Separating DNA Molecules That Differ in G:C Content** At 0.2 M Na$^+$, the melting temperature of double-stranded DNA is given by the formula, $T_m = 69.3 + 0.41$ (% G + C). The DNAs from mice and rats have (G + C) contents of 44% and 40%, respectively. Calculate the T_ms for these DNAs in 0.2 M NaCl. If samples of these DNAs were inadvertently mixed, how might they be separated from one another?

12. **Calculating the Density of DNAs That Differ in G:C Content** The buoyant density of DNA is proportional to its (G + C) content. (G:C base pairs have more atoms per volume than A:T base pairs.) Calculate the density (ρ) of avian tubercle bacillus DNA from the data presented in Table 10.1 and the equation $\rho = 1.660 + 0.098$(GC), where (GC) is the mole fraction of (G + C) in DNA.

13. **Draw a ψ:G Base Pair** (Integrates with Chapter 10.) Pseudouridine (ψ) is an invariant base in the TψC loop of tRNA; ψ is also found in strategic places in rRNA. (Figure 10.23 shows the structure of pseudouridine.) Draw the structure of the base pair that ψ might form with G.

14. **The Dimensions of Closed Circular DNA Molecules** The plasmid pBR322 is a closed circular dsDNA containing 4363 base pairs. What is the length in nm of this DNA (that is, what is its circumference if it were laid out as a perfect circle)? The E. coli K12 chromosome is a closed circular dsDNA of about 4,639,000 base pairs. What would be the circumference of a perfect circle formed from this chromosome? What is the diameter of a dsDNA molecule? Calculate the ratio of the length of the circular plasmid pBR322 to the diameter of the DNA of which it's made. Do the same for the E. coli chromosome.

15. **Identifying DNA Structural and Functional Elements from Nucleotide Sequence Information** Listed below are four DNA sequences. Which one contains a type-II restriction endonuclease ("six-cutter") hexanucleotide site? Which one is likely to form a cruciform structure? Which one is likely to be found in Z-DNA? Which one represents the 5′-end of a tRNA gene? Which one is most likely to be found in a triplex DNA structure?
   a. CGCGCGCCGCGCACGCGCTCGCGCGCCGC
   b. GAACGTCGTATTCCCGTACGACGTTC
   c. CAGGTCTCTCTCTCTCTCTCTC
   d. TGGTGCGAATTCTGTGGAT
   e. ATCGGAATTCATCG

16. **Draw the Cloverleaf Structure of a tRNA from RNA T_ms Sequence Information** The nucleotide sequence of E. coli tRNA^{Gln} is as follows:

    UGGGGUAUCG$_{10}$CCAAGC−GGU$_{20}$AAGGCACCGG$_{30}$AU-UCUGA$\Psi\Psi$C$_{40}$CGGCAUUCCG$_{50}$AGGTΨCGAAU$_{60}$CCUCGUA-CCC$_{70}$CAGCCA$_{76}$

    From this primary structure information, draw the secondary structure (cloverleaf) of this RNA and identify its anticodon.

17. **Use the Tools of the Protein Data Bank to Explore Ribosome Structure** The Protein Data Bank (PDB) is also a repository for nucleic acid structures. Go to the PDB at www.rcsb.org and enter pdb id = 1YI2. 1YI2 is the PDB ID for the structure of the H. marismortui 50S ribosomal subunit with erythromycin bound. Erythromycin is an antibiotic that acts by inhibiting bacterial protein synthesis. In the list of the display options under the image of the 50S subunit, click on one of the viewing options to view the structure. Using the tools of the viewer, zoom in and locate erythromycin within this structure. If the 50S ribosomal subunit can be compared to a mitten, where in the mitten is erythromycin?

18. **Use the Human Genome Database to Explore the Relationship Between Genes, Proteins, and Diseases** Online resources provide ready access to detailed information about the human genome. Go the National Center for Biotechnology Information (NCBI) genome database at http://www.ncbi.nlm.nih.gov/Genomes/index.html and click on Homo sapiens in the Map Viewer genome annotation updates list to access the chromosome map and organization of the human genome. Next, go to http://www.ncbi.nlm.nih.gov/genome/. In the "Search For" box, type in the following diseases to discover the chromosomal location of the affected gene and, by exploring links highlighted by the search results, discover the name of the protein affected by the disease:
   a. Sickle cell anemia
   b. Tay Sachs disease
   c. Leprechaunism
   d. Hartnup disorder

Preparing for the MCAT® Exam

19. (Integrates with Chapter 10.) Erwin Chargaff did not have any DNA samples from thermoacidophilic bacteria such as those that thrive in the geothermal springs of Yellowstone National Park. (Such bacteria had not been isolated by 1951 when Chargaff reported his results.) If he had obtained such a sample, what do you think its relative G:C content might have been? Why?

20. Think about the structure of DNA in its most common B-form double helical conformation and then list its most important structural features (deciding what is "important" from the biological role of DNA as the material of heredity). Arrange your answer with the most significant features first.

ActiveModels Problems

21. Examine the ActiveModel for the thiamine-PP riboswitch (pbd id = 2CKY, shown in Figure 11.38). The RNA recognizes its TPP ligand via conserved residues located within two "sensor" helices. Describe the interactions between TPP and the RNA that account for its binding and recognition.

22. Examine the ActiveModel for yeast phenylalanine-tRNA (pbd id = 6TNA), and describe the interactions that convert its cloverleaf secondary structure into its folded, tertiary conformation.

23. Examine the ActiveModel for an H-DNA (pbd id = 1B4Y), and describe the interactions that lead to triplex formation.

FURTHER READING

General References

Adams, R. L. P., Knowler, J. T., and Leader, D. P., 1992. *The Biochemistry of the Nucleic Acids,* 11th ed. London: Chapman and Hall.

Gesteland, R. F., et al., eds. 2006. *The RNA World,* 3rd ed. Cold Spring Harbor, NY: Cold Spring Harbor Laboratory Press.

Kornberg, A., and Baker, T. A., 1991. *DNA Replication,* 2nd ed. New York: W. H. Freeman.

Sinden, R. R., 1994. *DNA Structure and Function.* St. Louis: Elsevier/Academic Press.

Watson, J. D., et al., 2007. *The Molecular Biology of the Gene,* 6th ed. Menlo Park, CA: Pearson/Benjamin Cummings.

DNA Sequencing

Fox, S., Filichkin, S., and Mocker, T. C., 2009. Applications of ultra-high-throughput sequencing. *Plant Systems Biology: Methods in Molecular Biology* **553:**79–108.

Meldrum, D., 2000. Automation for genomics, Part One: Preparation for sequencing. *Genome Research* **10:**1081–1092.

Meldrum, D., 2000. Automation for genomics, Part Two: Sequencers, microarrays, and future trends. *Genome Research* **10:**1288–1303.

Nunnally, B. K., 2005. *Analytical Techniques in DNA Sequencing.* Boca Raton, FL: CRC Group, Taylor and Francis.

Xu, M., Fujita, D., and Hanagata, N, 2009. Perspectives and challenges of emerging single-molecule DNA sequencing technologies. *Small* **5:**2638–2649.

Ziebolz, B., and Droege, M. 2007. Toward a new era in sequencing. *Biotechnology Annual Review* **13:**1–26.

Higher-Order DNA Structure

Bates, A. D., and Maxwell, A., 1993. *DNA Topology.* New York: IRL Press at Oxford University Press.

Benner, S. A., 2004. Redesigning genetics. *Science* **306:**625–626.

Callandine, C. R., et al., 2004. *Understanding DNA: The Molecule and How It Works,* 3rd ed. London: Academic Press.

Frank-Kamenetskii, M. D., and Mirkin, S. A. M., 1995. Triplex DNA structures. *Annual Review of Biochemistry* **64:**65–95.

Fry, M., 2007. Tetraplex DNA and its interacting proteins. *Frontiers in Biosciences* **12:**4336–4351.

Htun, H., and Dahlberg, J. E., 1989. Topology and formation of triple-stranded H-DNA. *Science* **243:**1571–1576.

Keniry, M. A., 2001. Quadruplex structures in nucleic acids. *Biopolymers* **56:**123–146.

Rich, A., 2003. The double helix: A tale of two puckers. *Nature Structural Biology* **10:**247–249.

Rich, A., Nordheim, A., and Wang, A. H.-J., 1984. The chemistry and biology of left-handed Z-DNA. *Annual Review of Biochemistry* **53:**791–846.

Watson, J. D., ed., 1983. *Structures of DNA. Cold Spring Harbor Symposia on Quantitative Biology,* Volume XLVII. New York: Cold Spring Harbor Laboratory.

Wells, R. D., 1988. Unusual DNA structures. *Journal of Biological Chemistry* **263:**1095–1098.

Zain, R., and Sun, J.-S., 2003. So natural triple-helical structures occur and function in vivo? *Cellular and Molecular Life Sciences* **60:**862–870.

Nucleosomes

Cobbe, N., and Heck, M. M. S., 2000. Review: SMCs in the world of chromosome biology—from prokaryotes to higher eukaryotes. *Journal of Structural Biology* **129:**123–143.

Hirano, T., 2005. SMC proteins and chromosome mechanics: From bacteria to humans. *Philosophical Transactions of the Royal Society London, Series B* **360:**507–514.

Luger, C., et al., 1997. Crystal structure of the nucleosome core particle at 2.8 Å resolution. *Nature* **389:**251–260.

Rhodes, D., 1997. The nucleosome core all wrapped up. *Nature* **389:** 231–233.

Chromosome Structure

Pienta, K. J., and Coffey, D. S., 1984. A structural analysis of the role of the nuclear matrix and DNA loops in the organization of the nucleus and chromosomes. In Cook, P. R., and Laskey, R. A., eds., Higher order structure in the nucleus. *Journal of Cell Science Supplement* **1:**123–135.

Schlach, T., Duda, S., Sargent, D. F., and Richmond, T. J., 2005. X-ray structure of a tetranucleosome and its implications for the chromatin fibre. *Nature* **436:**138–141.

Sumner, A. T., 2003. *Chromosomes: Organization and Function.* Malden, MA: Blackwell Science.

Tremethick, D. J., 2007. Higher-order structures of chromatin: The elusive 30 nm fiber. *Cell* **128:**651-654.

Woodcock, C. L. F., Frado, L.-L. Y., and Rattner, J. B., 1984. The higher-order structure of chromatin: Evidence for a helical ribbon arrangement. *Journal of Cell Biology* **99:**42–52.

Woodcock, C. L., and Ghosh, R. P., 2010. Chromatin higher-order structure and dynamics. *Cold Spring Harbor Laboratory Perspectives in Biology* **2:**a000596.

Telomeres

Axelrod, N., 1996. Of telomeres and tumors. *Nature Medicine* **2:**158–159.

Feng, J., Funk, W. D., Wang, S.-S., Weinrich, S. L., et al., 1995. The RNA component of human telomerase. *Science* **269:**1236–1241.

Chemical Synthesis of Genes

Ferretti, L., Karnik, S. S., Khorana, H. G., Nassal, M., and Oprian, D. D., 1986. Total synthesis of a gene for bovine rhodopsin. *Proceedings of the National Academy of Sciences U.S.A.* **83:**599–603.

Higher-Order RNA Structure

Ban, N., et al., 2000. The complete atomic structure of the large ribosomal subunit at 2.4 Å resolution. *Science* **289:**905–920.

Gray, M. W., and Cedergren, R., eds., 1993. The new age of RNA. *The FASEB Journal* **7:**4–239. A collection of articles emphasizing the new appreciation for RNA in protein synthesis, in evolution, and as a catalyst.

Holbrook, S. R., 2005. RNA structure: The long and the short of it. *Current Opinion in Structural Biology* **15:**302–308.

Klosterman, P. S., et al., 2005. Three-dimensional motifs from the SCOR, structural classification of RNA database: Extruded strands, base triples, tetraloops, and U-turns. *Nucleic Acids Research* **32:**2342–2352.

Nilsen, T. W., 2007. RNA 1997–2007: A remarkable decade of discovery. *Molecular Cell* **28:**715–720.

Recombinant DNA: Cloning and Creation of Chimeric Genes

The Chimera of Arezzo, of Etruscan origin and probably from the fifth century B.C., was found near Arezzo, Italy, in 1553. Chimeric animals existed only in the imagination of the ancients. But the ability to create chimeric DNA molecules is a very real technology that has opened up a whole new field of scientific investigation.

…how many vain chimeras have you created?…
Go and take your place with the seekers after
gold.

Leonardo da Vinci
The Notebooks (1508–1518), Volume II, Chapter 25

KEY QUESTIONS

12.1 What Does It Mean "To Clone"?

12.2 What Is a DNA Library?

12.3 Can the Cloned Genes in Libraries Be Expressed?

12.4 What Is the Polymerase Chain Reaction (PCR)?

12.5 How Is RNA Interference Used to Reveal the Function of Genes?

12.6 Is It Possible to Make Directed Changes in the Heredity of an Organism?

Scala/Art Resource, NY

ESSENTIAL QUESTIONS

Using techniques for the manipulation of nucleic acids in the laboratory, scientists can join together different DNA segments from different sources. Such manmade products are called recombinant DNA molecules, and the use of such molecules to alter the genetics of organisms is termed genetic engineering.

What are the methods that scientists use to create recombinant DNA molecules; can scientists create genes from recombinant DNA molecules; and can scientists modify the heredity of an organism using recombinant DNA?

In the early 1970s, technologies for the laboratory manipulation of nucleic acids emerged. In turn, these technologies led to the construction of DNA molecules composed of nucleotide sequences taken from different sources. The products of these innovations, **recombinant DNA molecules**,[1] opened exciting new avenues of investigation in molecular biology and genetics, and a new field was born—**recombinant DNA technology. Genetic engineering** is the application of this technology to the manipulation of genes. These advances were made possible by methods for **amplification** of any particular DNA segment, regardless of source, within bacterial host cells. Or, in the language of recombinant DNA technology, the **cloning** of virtually any DNA sequence became feasible.

[1]The advent of molecular biology, like that of most scientific disciplines, generated a jargon all its own. Learning new fields often requires gaining familiarity with a new vocabulary. We will soon see that many words—*vector, amplification,* and *insert* are but a few examples—have been bent into new meanings to describe the marvels of molecular biology.

12.1 | What Does It Mean "To Clone"?

In classical biology, a *clone* is a population of identical organisms derived from a single parental organism. For example, the members of a colony of bacterial cells that arise from a single cell on a petri plate are a clone. Molecular biology has borrowed the term to mean a collection of molecules or cells all identical to an original molecule or cell. So, if a single bacterial cell harboring a recombinant DNA molecule in the form of a plasmid grows and multiplies on a petri plate to form a colony, the plasmids within the millions of cells in the bacterial colony represent a clone of the original DNA molecule, and these molecules can be isolated and studied. Furthermore, if the cloned DNA molecule is a gene (or part of a gene)—that is, it encodes a functional product—a new avenue to isolating and studying this product has opened. Recombinant DNA methodology offers exciting new vistas in biochemistry.

Plasmids Are Very Useful in Cloning Genes

Plasmids are naturally occurring, circular, extrachromosomal DNA molecules (see Chapter 11). Natural strains of the common colon bacterium *Escherichia coli* isolated from various sources contain diverse plasmids. Often these plasmids carry genes specifying novel metabolic activities that are advantageous to the host bacterium. These activities range from catabolism of unusual organic substances to metabolic functions that endow the host cells with resistance to antibiotics, heavy metals, or bacteriophages. Plasmids that are able to perpetuate themselves in *E. coli,* the bacterium favored by bacterial geneticists and molecular biologists, are the workhorses of recombinant DNA technology. Because restriction endonuclease digestion of plasmids can generate fragments with overlapping or "sticky" ends, artificial plasmids can be constructed by ligating different fragments together. Such artificial plasmids were among the earliest recombinant DNA molecules. These recombinant molecules can be autonomously replicated, and hence propagated, in suitable bacterial host cells, provided they still possess a site signaling where DNA replication can begin (a so-called **origin of replication** or *ori* sequence).

Plasmids as Cloning Vectors The idea arose that "foreign" DNA sequences could be inserted into artificial plasmids and that these foreign sequences would be carried into *E. coli* and propagated as part of the plasmid. That is, these plasmids could serve as **cloning vectors** to carry genes. (The word *vector* is used here in the sense of "a vehicle or carrier.") Plasmids useful as cloning vectors possess three common features: **a replicator, a selectable marker,** and **a cloning site** (Figure 12.1). A *replicator* is an origin of replication, or *ori.* The *selectable marker* is typically a gene conferring resistance to an antibiotic. Only cells containing the cloning vector will grow in the presence of the antibiotic. Therefore, growth on antibiotic-containing media "selects for" plasmid-containing cells. Typically, the *cloning site* is a sequence of nucleotides representing one or more restriction endonuclease cleavage sites. Cloning sites are located where the insertion of foreign DNA neither disrupts the plasmid's ability to replicate nor inactivates essential markers.

Virtually Any DNA Sequence Can Be Cloned Nuclease cleavage at a restriction site opens, or *linearizes,* the circular plasmid so that a foreign DNA fragment can be inserted. The ends of this linearized plasmid are joined to the ends of the fragment so that the circle is closed again, creating a recombinant plasmid (Figure 12.2). **Recombinant plasmids** are hybrid DNA molecules consisting of plasmid DNA sequences plus inserted DNA elements (called *inserts*). Such hybrid molecules are also called **chimeric constructs** or **chimeric plasmids.** (The term *chimera* is borrowed from mythology and refers to a beast composed of the body and head of a lion, the heads of a goat and a snake, and the wings of a bat.) The presence of foreign DNA sequences does not adversely affect replication of the plasmid, so chimeric plasmids can be propagated in bacteria just like the original plasmid. Bacteria often harbor several hundred copies of common cloning vectors per cell. Hence, large amounts of a cloned DNA sequence can be recovered from bacterial cultures. The enormous power of recombinant DNA technology stems in part

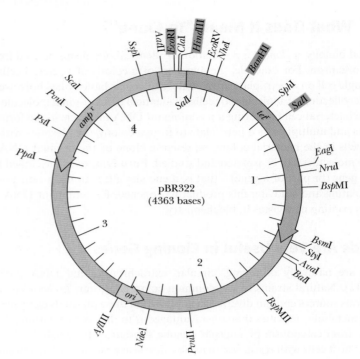

FIGURE 12.1 One of the first widely used cloning vectors, the plasmid pBR322. This 4363-bp plasmid contains an *ori* and genes for resistance to the drugs ampicillin (*amp*ʳ) and tetracycline (*tet*ʳ). The locations of restriction endonuclease cleavage sites are indicated.

from the fact that *virtually any DNA sequence can be selectively cloned and amplified in this manner.* DNA sequences that are difficult to clone include inverted repeats, origins of replication, centromeres, and telomeres. The only practical limitation is the size of the foreign DNA segment: Most plasmids with inserts larger than about 10 kbp are not replicated efficiently. However, bacteriophages such as bacteriophage λ can be manipulated so that DNA sequences as large as 40 kbp can be inserted into the bacteriophage genome. Such recombinant phage DNA molecules lack essential λ genes and replicate in *E. coli* as plasmids.

Construction of Chimeric Plasmids Creation of chimeric plasmids requires joining the ends of the foreign DNA insert to the ends of a linearized plasmid. This ligation is facilitated if the ends of the plasmid and the insert have complementary, single-stranded overhangs. Then these ends can base-pair with one another, annealing the two molecules together. One way to generate such ends is to cleave the DNA with restriction enzymes that make staggered cuts; many such restriction endonucleases are available (see Table 10.2). For example, if the sequence to be inserted is an *Eco*RI fragment and the plasmid is cut with *Eco*RI, the single-stranded sticky ends of the two DNAs can anneal (Figure 12.2). The interruptions in the sugar–phosphate backbone of DNA can then be sealed with DNA ligase to yield a covalently closed, circular chimeric plasmid. DNA ligase is an enzyme that covalently links adjacent 3′-OH and 5′-PO₄ groups. An inconvenience of this strategy is that *any* pair of *Eco*RI sticky ends can anneal with each other. So, plasmid molecules can reanneal with themselves, as can the foreign DNA restriction fragments. These DNAs can be eliminated by selection schemes designed to identify only those bacteria containing chimeric plasmids.

Blunt-end ligation is an alternative method for joining different DNAs. The most widely used DNA ligase, **bacteriophage T4 DNA ligase,** is an ATP-dependent enzyme that can even ligate two DNA fragments whose ends lack overhangs (blunt-ended DNAs). Many restriction endonucleases cut double-stranded DNA so that blunt ends are formed.

A great number of variations on these basic themes have emerged. For example, short synthetic DNA duplexes whose nucleotide sequence consists of little more than a restriction site can be blunt-end ligated onto any DNA. These short DNAs are known as **linkers.** Cleavage of the ligated DNA with the restriction enzyme then leaves tailor-made sticky ends useful in cloning reactions (Figure 12.3). Similarly, many vectors contain a **polylinker** cloning site, a short region of DNA sequence bearing numerous restriction sites.

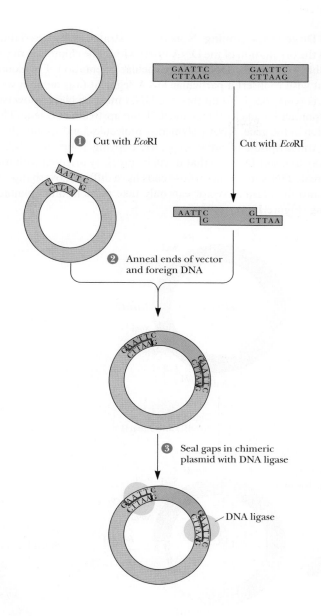

FIGURE 12.2 An *Eco*RI restriction fragment of foreign DNA can be inserted into a plasmid having an *Eco*RI cloning site by **(1)** cutting the plasmid at this site with *Eco*RI, **(2)** annealing the linearized plasmid with the *Eco*RI foreign DNA fragment, and **(3)** sealing the nicks with DNA ligase.

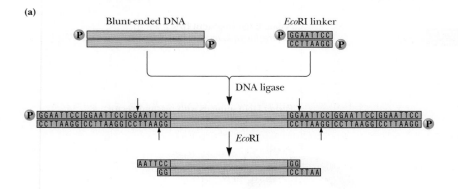

(b) A vector cloning site containing multiple restriction sites, a so-called polylinker.

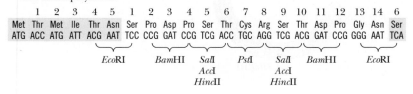

FIGURE 12.3 **(a)** The use of linkers to create tailor-made ends on cloning fragments. Note that the ligation reaction can add multiple linkers on each end of the blunt-ended DNA. *Eco*RI digestion removes all but the terminal one, leaving the desired 5′-overhangs. **(b)** Cloning vectors often have polylinkers consisting of a multiple array of restriction sites at their cloning sites, so restriction fragments generated by a variety of endonucleases can be incorporated into the vector. Note that the polylinker is engineered not only to have multiple restriction sites but also to have an uninterrupted sequence of codons, so this region of the vector has the potential for translation into protein (see Figure 12.15). (Adapted from Figure 1.14.2 in Greenwich, D., and Brent, R., 2003. *UNIT 1.14 Introduction to Vectors Derived from Filamentous Phages,* in *Current Protocols in Molecular Biology,* Ausubel, F. M., Brent, R., Kingston, R. E., Moore, D. D., Seidman, J. G., Smith, J. A., and Struhl, K., eds. New York: John Wiley and Sons.)

Promoters and Directional Cloning Note that the strategies discussed thus far create hybrids in which the orientation of the DNA insert within the chimera is random. Sometimes it is desirable to insert the DNA in a particular orientation. For example, an experimenter might wish to insert a particular DNA segment (a gene) in a vector so that its gene product is synthesized. To do this, the DNA must be placed downstream from a **promoter.** A promoter is a nucleotide sequence lying upstream of a gene. The promoter controls expression of the gene. RNA polymerase molecules bind specifically at promoters and initiate transcription of adjacent genes, copying template DNA into RNA products. One way to insert DNA so that it will be properly oriented with respect to the promoter is to create DNA molecules whose ends have different overhangs. Ligation of such molecules into the plasmid vector can only take place in one orientation to give **directional cloning** (Figure 12.4).

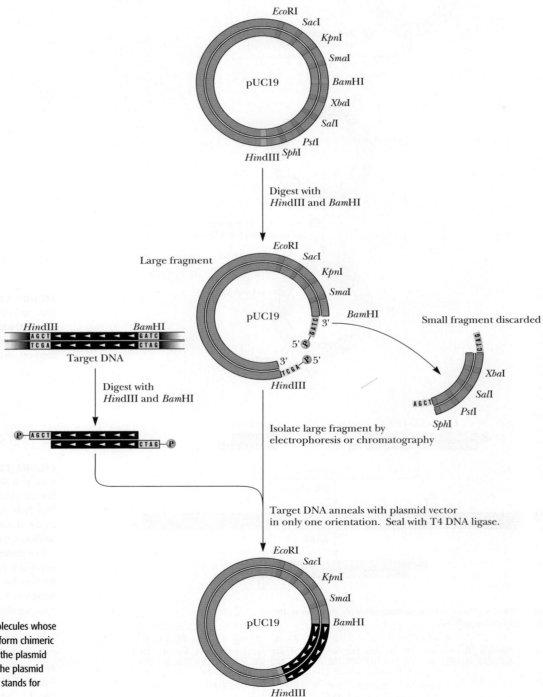

FIGURE 12.4 Directional cloning. DNA molecules whose ends have different overhangs can be used to form chimeric constructs in which the foreign DNA can enter the plasmid in only one orientation. The foreign DNA and the plasmid are digested with the same two enzymes. pUC stands for universal cloning plasmid.

Biologically Functional Chimeric Plasmids The first biologically functional chimeric DNA molecules constructed in vitro were assembled from parts of different plasmids in 1973 by Stanley Cohen, Annie Chang, Herbert Boyer, and Robert Helling. These plasmids were used to **transform** recipient *E. coli* cells (*transformation* means the uptake and replication of exogenous DNA by a recipient cell). To facilitate transformation, the bacterial cells were rendered somewhat permeable to DNA by Ca^{2+} treatment and a brief 42°C heat shock. Although less than 0.1% of the Ca^{2+}-treated bacteria became competent for transformation, transformed bacteria could be selected by their resistance to certain antibiotics (Figure 12.5). Consequently, the chimeric plasmids must have been

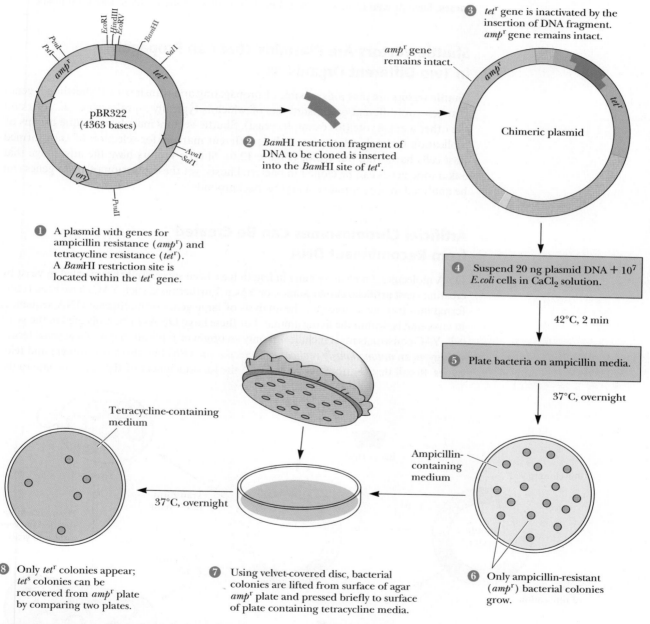

① A plasmid with genes for ampicillin resistance (*amp*^r) and tetracycline resistance (*tet*^r). A *Bam*HI restriction site is located within the *tet*^r gene.

② *Bam*HI restriction fragment of DNA to be cloned is inserted into the *Bam*HI site of *tet*^r.

③ *tet*^r gene is inactivated by the insertion of DNA fragment. *amp*^r gene remains intact.

④ Suspend 20 ng plasmid DNA + 10⁷ *E.coli* cells in $CaCl_2$ solution.

42°C, 2 min

⑤ Plate bacteria on ampicillin media.

37°C, overnight

⑥ Only ampicillin-resistant (*amp*^r) bacterial colonies grow.

⑦ Using velvet-covered disc, bacterial colonies are lifted from surface of agar *amp*^r plate and pressed briefly to surface of plate containing tetracycline media.

⑧ Only *tet*^r colonies appear; *tet*^s colonies can be recovered from *amp*^r plate by comparing two plates.

FIGURE 12.5 A typical bacterial transformation experiment. Here the plasmid pBR322 is the cloning vector. (1) Cleavage of pBR322 with *Bam*HI, followed by (2) annealing and ligation of inserts generated by *Bam*HI cleavage of some foreign DNA, (3) creates a chimeric plasmid. (4) The chimeric plasmid is then used to transform Ca^{2+}-treated heat-shocked *E. coli* cells, and the bacterial sample is plated on a petri plate. (5) Following incubation of the petri plate overnight at 37°C, (6) colonies of *amp*^r bacteria are evident. (7) Replica plating of these bacteria on plates of tetracycline-containing media (8) reveals which colonies are *tet*^r and which are tetracycline sensitive (*tet*^s). Only the *tet*^s colonies possess plasmids with foreign DNA inserts.

biologically functional in at least two aspects: They replicated stably within their hosts, and they expressed the drug resistance markers they carried.

In general, plasmids used as cloning vectors are engineered to be small (2.5 kbp to about 10 kbp in size) so that the size of the insert DNA can be maximized. These plasmids have only a single origin of replication, so the time necessary for complete replication depends on the size of the plasmid. Under selective pressure in a growing culture of bacteria, overly large plasmids are prone to delete any nonessential "genes," such as any foreign inserts. Such deletion would thwart the purpose of most cloning experiments. The useful upper limit on cloned inserts in typical plasmids is about 10 kbp. Many eukaryotic genes exceed this size. F′, an atypical plasmid, can carry foreign DNA inserts of 350 kb or more. Recombinant F′ constructs, called **BACs,** for **bacterial artificial chromosomes,** have proven to be very useful in cloning eukaryotic DNA in bacterial hosts.

Shuttle Vectors Are Plasmids That Can Propagate in Two Different Organisms

Shuttle vectors are plasmids capable of propagating and transferring ("shuttling") genes between two different organisms, one of which is typically a prokaryote *(E. coli)* and the other a eukaryote (for example, yeast). Shuttle vectors must have unique origins of replication for each cell type as well as different markers for selection of transformed host cells harboring the vector (Figure 12.6). Shuttle vectors have the advantage that eukaryotic genes can be cloned in bacterial hosts, yet the expression of these genes can be analyzed in appropriate eukaryotic backgrounds.

Artificial Chromosomes Can Be Created from Recombinant DNA

DNA molecules 2 megabase pairs in length have been successfully propagated in yeast by creating **yeast artificial chromosomes,** or **YACs.** Furthermore, such YACs have been transferred into transgenic mice for the analysis of large genes or multigenic DNA sequences in vivo, that is, within the living animal. For these large DNAs to be replicated in the yeast cell, YAC constructs must include not only an origin of replication (known in yeast terminology as an *autonomously replicating sequence* or *ARS*) but also a centromere and telomeres. Recall that centromeres provide the site for attachment of the chromosome to the

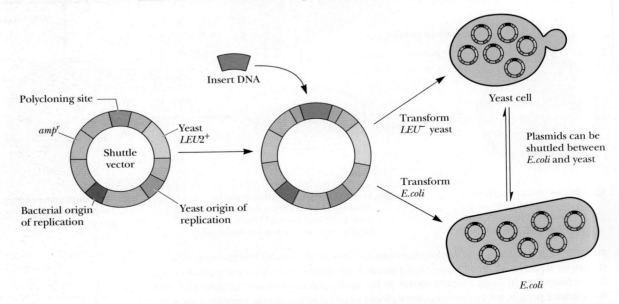

FIGURE 12.6 A typical shuttle vector. *LEU2*⁺ is a gene in the yeast pathway for leucine biosynthesis. The recipient yeast cells are *LEU2*⁻ (defective in this gene) and thus require leucine for growth. *LEU2*⁻ yeast cells transformed with this shuttle vector can be selected on medium lacking any leucine supplement.

spindle during mitosis and meiosis, and telomeres are nucleotide sequences defining the ends of chromosomes. Telomeres are essential for proper replication of the chromosome.

12.2 | What Is a DNA Library?

A DNA library is a set of cloned fragments that collectively represent the genes of a specific organism. Particular genes can be isolated from DNA libraries, much as books can be obtained from conventional libraries. The secret is knowing where and how to look.

Genomic Libraries Are Prepared from the Total DNA in an Organism

Any particular gene constitutes only a small part of an organism's genome. For example, if the organism is a mammal whose entire genome exceeds 10^6 kbp and the gene is 10 kbp, then the gene represents less than 0.001% of the total nuclear DNA. It is impractical to attempt to recover such rare sequences directly from isolated nuclear DNA because of the overwhelming amount of extraneous DNA sequences. Instead, a **genomic library** is prepared by isolating total DNA from the organism, digesting it into fragments of suitable size, and cloning the fragments into an appropriate vector. This approach is called *shotgun cloning* because the strategy has no way of targeting a particular gene but instead seeks to clone all the genes of the organism at one time. The intent is that at least one recombinant clone will contain at least part of the gene of interest. Usually, the isolated DNA is only partially digested by the chosen restriction endonuclease so that not every

---

> ### CRITICAL DEVELOPMENTS IN BIOCHEMISTRY

Combinatorial Libraries

Specific recognition and binding of other molecules is a defining characteristic of any protein or nucleic acid. Often, target ligands of a particular protein are unknown, or in other instances, a unique ligand for a known protein may be sought in the hope of blocking the activity of the protein or otherwise perturbing its function. Or, the hybridization of nucleic acids with each other according to base-pairing rules, as an act of specific recognition, can be exploited to isolate or identify pairing partners. **Combinatorial libraries** are the products of strategies to facilitate the identification and characterization of macromolecules (proteins, DNA, RNA) that interact with small-molecule ligands or with other macromolecules. Unlike genomic libraries, combinatorial libraries consist of synthetic oligomers. Arrays of synthetic oligonucleotides printed as tiny dots on miniature solid supports are known as **DNA chips.** (See the section titled "DNA Microarrays *(Gene Chips)* Are Arrays of Different Oligonucleotides Immobilized on a Chip.")

Specifically, combinatorial libraries contain very large numbers of chemically synthesized molecules (such as peptides or oligonucleotides) with randomized sequences or structures. Such libraries are designed and constructed with the hope that one molecule among a vast number will be recognized as a ligand by the protein (or nucleic acid) of interest. If so, perhaps that molecule will be useful in a pharmaceutical application. For instance, the synthetic oligomer may serve as a drug to treat a disease involving the protein to which it binds.

An example of this strategy is the preparation of a **synthetic combinatorial library** of hexapeptides. The maximum number of sequence combinations for hexapeptides is 20^6, or 64,000,000. One approach to simplify preparation and screening possibilities for such a library is to specify the first two amino acids in the hexapeptide while the next four are randomly chosen. In this approach, 400 libraries (20^2) are synthesized, each of which is unique in terms of the amino acids at positions 1 and 2 but random at the other four positions (as in AAXXXX, ACXXXX, ADXXXX, etc.), so each of the 400 libraries contains 20^4, or 160,000, different sequence combinations. Screening these libraries with the protein of interest reveals which of the 400 libraries contains a ligand with high affinity. Then, this library is expanded systematically by specifying the first three amino acids (knowing from the chosen 1-of-400 libraries which amino acids are best as the first two); only 20 synthetic libraries (each containing 20^3, or 8000, hexapeptides) are made here (one for each third-position possibility, the remaining three positions being randomized). Selection for ligand binding, again with the protein of interest, reveals the best of these 20, and this particular library is then varied systematically at the fourth position, creating 20 more libraries (each containing 20^2, or 400, hexapeptides). This cycle of synthesis, screening, and selection is repeated until all six positions in the hexapeptide are optimized to create the best ligand for the protein. A variation on this basic strategy using synthetic oligonucleotides rather than peptides identified a unique 15-mer (sequence GGTTGGTGTGGTTGG) with high affinity ($K_D = 2.7$ nM) toward thrombin, a serine protease in the blood coagulation pathway. Thrombin is a major target for the pharmacological prevention of clot formation in coronary thrombosis.

From Cortese, R., 1996. *Combinatorial Libraries: Synthesis, Screening and Application Potential.* Berlin: Walter de Gruyter.

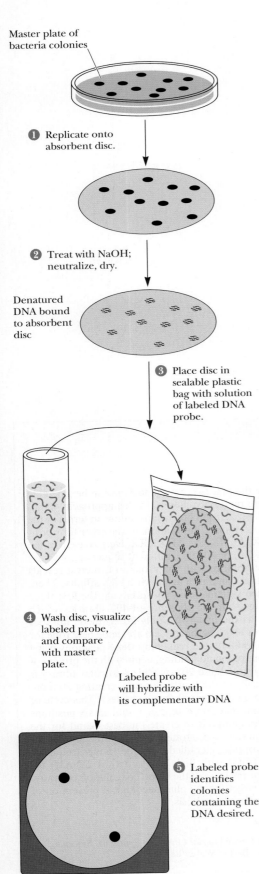

Master plate of
bacteria colonies

1 Replicate onto
absorbent disc.

2 Treat with NaOH;
neutralize, dry.

Denatured
DNA bound
to absorbent
disc

3 Place disc in
sealable plastic
bag with solution
of labeled DNA
probe.

4 Wash disc, visualize
labeled probe,
and compare
with master
plate.

Labeled probe
will hybridize with
its complementary DNA

5 Labeled probe
identifies
colonies
containing the
DNA desired.

Labeling reveals position(s) of
colonies bearing the desired clone

restriction site is cleaved in every DNA molecule. Then, even if the gene of interest contains a susceptible restriction site, some intact genes might still be found in the digest. Genomic libraries have been prepared from thousands of different species.

Many clones must be created to be confident that the genomic library contains the gene of interest. The probability, P, that some number of clones, N, contains a particular fragment representing a fraction, f, of the genome is

$$P = 1 - (1 - f)^N$$

Thus,

$$N = \frac{\ln(1 - P)}{\ln(1 - f)}$$

For example, if the library consists of 10-kbp fragments of the *E. coli* genome (4640 kbp total), more than 2000 individual clones must be screened to have a 99% probability ($P = 0.99$) of finding a particular fragment. Since $f = 10/4640 = 0.0022$ and $P = 0.99$, $N = 2093$. For a 99% probability of finding a particular sequence within the 3×10^6 kbp human genome, N would equal almost 1.4 million if the cloned fragments averaged 10 kbp in size. The need for cloning vectors capable of carrying very large DNA inserts becomes obvious from these numbers.

Libraries Can Be Screened for the Presence of Specific Genes

A common method of screening genomic libraries is to carry out a **colony hybridization experiment.** In a typical experiment, host bacteria containing a plasmid-based library are plated out on a petri dish and allowed to grow overnight to form colonies (Figure 12.7). A replica of the bacterial colonies is then obtained by overlaying the plate with a flexible, absorbent disc. The disc is removed, treated with alkali to dissociate bound DNA duplexes into single-stranded DNA, dried, and placed in a sealed bag with labeled probe (see the Critical Developments in Biochemistry box on page 388). If the probe DNA is duplex DNA, it must be denatured by heating at 70°C. The probe and target DNA complementary sequences must be in a single-stranded form if they are to hybridize with one another. Any DNA sequences complementary to probe DNA will be revealed by the location of hybridized labeled probe on the absorbent disc. Bacterial colonies containing clones bearing target DNA can be recovered from the master plate.

Probes for Southern Hybridization Can Be Prepared in a Variety of Ways

Clearly, specific probes are essential reagents if the goal is to identify a particular gene against a background of innumerable DNA sequences. Usually, the probes that are used to screen libraries are nucleotide sequences that are complementary to some part of the target gene. Making useful probes requires some information about the gene's nucleotide sequence. Sometimes such information is available. Alternatively, if the amino acid sequence of the protein encoded by the gene is known, it is possible to work backward

FIGURE 12.7 Screening a genomic library by colony hybridization. Host bacteria transformed with a plasmid-based genomic library are plated on a petri plate and incubated overnight to allow bacterial colonies to form. A replica of the colonies is obtained by overlaying the plate with a flexible disc composed of absorbent material (such as nitrocellulose or nylon) **(1)**. Nitrocellulose strongly binds nucleic acids; single-stranded nucleic acids are bound more tightly. Once the disc has taken up an impression of the bacterial colonies, it is removed and the petri plate is set aside and saved. The disc is treated with 2 *M* NaOH, neutralized, and dried **(2)**. NaOH both lyses any bacteria (or phage particles) and dissociates the DNA strands. When the disc is dried, the DNA strands become immobilized on the filter. The dried disc is placed in a sealable plastic bag, and a solution containing heat-denatured (single-stranded), labeled probe is added. (Probes can be labeled in a variety of ways: fluorescent or colored dyes, radioactive isotopes, and more.) **(3)**. The bag is incubated to allow annealing of the probe DNA to any target DNA sequences that might be present on the disc. The filter is then washed, dried, and prepared for visualization of the presence of labeled probe **(4)**. The position of any spots reveals where the labeled probe has hybridized with target DNA **(5)**. The location of these spots can be used to recover the genomic clone from the bacteria on the original petri plate.

through the genetic code to the DNA sequence (Figure 12.8). Because the genetic code is *degenerate* (that is, several codons may specify the same amino acid; see Chapter 30), probes designed by this approach are usually **degenerate oligonucleotides** about 17 to 50 residues long (such oligonucleotides are so-called 17- to 50-mers). The oligonucleotides are synthesized so that different bases are incorporated at sites where degeneracies occur in the codons. The final preparation thus consists of a mixture of equal-length oligonucleotides whose sequences vary to accommodate the degeneracies. Presumably, one oligonucleotide sequence in the mixture will hybridize with the target gene. These oligonucleotide probes are at least 17-mers because shorter degenerate oligonucleotides might hybridize with sequences unrelated to the target sequence.

A piece of DNA from the corresponding gene in a related organism can also be used as a probe in screening a library for a particular gene. Such probes are termed **heterologous probes** because they are not derived from the homologous (same) organism.

Problems arise if a complete eukaryotic gene is the cloning target; eukaryotic genes can be tens or even hundreds of kilobase pairs in size. Genes this size are fragmented in most cloning procedures. Thus, the DNA identified by the probe may represent a clone that carries only part of the desired gene. However, most cloning strategies are based on a partial digestion of the genomic DNA, a technique that generates an overlapping set of genomic fragments. This being so, DNA segments from the ends of the identified clone can now be used to probe the library for clones carrying DNA sequences that flanked the original isolate in the genome. Repeating this process ultimately yields the complete gene among a subset of overlapping clones.

cDNA Libraries Are DNA Libraries Prepared from mRNA

cDNAs are DNA molecules copied from mRNA templates. cDNA libraries are constructed by synthesizing cDNA from purified cellular mRNA. These libraries present an alternative strategy for gene isolation, especially eukaryotic genes. Because most eukaryotic mRNAs carry 3′-poly(A) tails, mRNA can be selectively isolated from preparations of total cellular RNA by oligo(dT)-cellulose chromatography (Figure 12.9).

Known amino acid sequence:

Phe	Met	Glu	Trp	His	Lys	Asn

Possible mRNA sequence:

UUU	AUG	GAA	UGG	CAU	AGG	AAU
UUC		GAG		CAC	AAA	AAC

① Absorbent disc replica of bacterial colonies carrying different DNA fragments

② Synthesize 32 possible DNA oligonucleotides and attach suitable label

③ Incubate disc with probe solution in plastic bag

④ Hybridization of the correct oligonucleotide to the DNA

⑤ Visualize location of labeled probe on disc

FIGURE 12.8 Cloning genes using oligonucleotide probes designed from a known amino acid sequence. A labeled set of DNA (degenerate) oligonucleotides representing all possible mRNA coding sequences is synthesized and is used to probe the genomic library by colony hybridization (see Figure 12.7). (Adapted from Fig. 19-18, in Watson, J. D., et al., 1987. *Molecular Biology of the Gene* 4/e. Menlo Park, Calif.: Benjamin/Cummings. Reprinted with permission.)

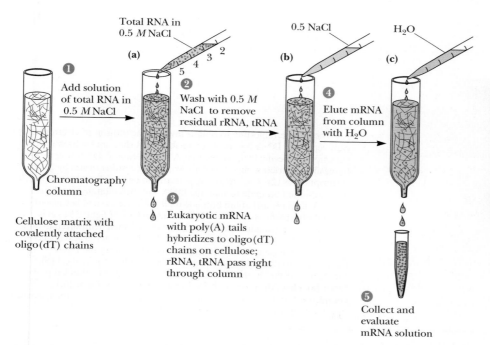

① Add solution of total RNA in 0.5 *M* NaCl

Chromatography column

Cellulose matrix with covalently attached oligo(dT) chains

Total RNA in 0.5 *M* NaCl

(a)

② Wash with 0.5 *M* NaCl to remove residual rRNA, tRNA

③ Eukaryotic mRNA with poly(A) tails hybridizes to oligo(dT) chains on cellulose; rRNA, tRNA pass right through column

0.5 NaCl

(b)

④ Elute mRNA from column with H₂O

H₂O

(c)

⑤ Collect and evaluate mRNA solution

FIGURE 12.9 Isolation of eukaryotic mRNA via oligo(dT)-cellulose chromatography. **(a)** In the presence of 0.5 *M* NaCl, the poly(A) tails of eukaryotic mRNA anneal with short oligo(dT) chains covalently attached to an insoluble chromatographic matrix such as cellulose. Other RNAs, such as rRNA (green), pass right through the chromatography column. **(b)** The column is washed with more 0.5 *M* NaCl to remove residual contaminants. **(c)** Then the poly(A) mRNA (red) is recovered by washing the column with water because the base pairs formed between the poly(A) tails of the mRNA and the oligo(dT) chains are unstable in solutions of low ionic strength.

Identifying Specific DNA Sequences by Southern Blotting (Southern Hybridization)

Any given DNA fragment is unique solely by virtue of its specific nucleotide sequence. The only practical way to find one particular DNA segment among a vast population of different DNA fragments (such as you might find in genomic DNA preparations) is to exploit its sequence specificity to identify it. In 1975, E. M. Southern invented a technique capable of doing just that.

Electrophoresis

Southern first fractionated a population of DNA fragments according to size by gel electrophoresis (see step 2 in figure). The electrophoretic mobility of a nucleic acid is inversely proportional to its molecular mass. Polyacrylamide gels are suitable for separation of nucleic acids of 25 to 2000 bp. Agarose gels are better if the DNA fragments range up to 10 times this size. Most preparations of genomic DNA show a broad spectrum of sizes, from less than 1 kbp to more than 20 kbp. Typically, no discrete-size fragments are evident following electrophoresis, just a "smear" of DNA throughout the gel.

Blotting

Once the fragments have been separated by electrophoresis (step 3), the gel is soaked in a solution of NaOH. Alkali denatures duplex DNA, converting it to single-stranded DNA. After the pH of the gel is adjusted to neutrality with buffer, a sheet of absorbent material soaked in a concentrated salt solution is then placed over the gel, and salt solution is drawn through the gel in a direction perpendicular to the direction of electrophoresis (step 4). The salt solution is pulled through the gel in one of three ways: capillary action *(blotting)*, suction *(vacuum blotting)*, or electrophoresis *(electroblotting)*. The movement of salt solution through the gel carries the DNA to the absorbent sheet, which binds the single-stranded DNA molecules very tightly, effectively immobilizing them in place on the sheet. Note

that the distribution pattern of the electrophoretically separated DNA is maintained when the single-stranded DNA molecules bind to the absorbent sheet (step 5 in figure). The sheet is then dried. Next, in the *prehybridization step,* the sheet is incubated with a solution containing protein (serum albumin, for example) and/or a detergent such as sodium dodecylsulfate. The protein and detergent molecules saturate any remaining binding sites for DNA on the absorbent sheet, so no more DNA can bind nonspecifically.

Hybridization

To detect a particular DNA within the electrophoretic smear of countless DNA fragments, the prehybridized sheet is incubated in a sealed plastic bag with a solution of specific probe molecules (step 6 in figure). A **probe** is usually a single-stranded DNA of defined sequence that is distinctively labeled, either with a radioactive isotope (such as ^{32}P) or some other easily detectable tag. The nucleotide sequence of the probe is designed to be complementary to the sought-for or *target* DNA fragment. The single-stranded probe DNA **anneals** with the single-stranded target DNA bound to the sheet through specific base pairing to form a DNA duplex. This annealing, or **hybridization** as it is usually called, labels the target DNA, revealing its position on the sheet (step 7 in figure). For example, if the probe is ^{32}P-labeled, its location can be detected by autoradiographic exposure of a piece of X-ray film laid over the sheet.

Southern's procedure has been extended to the identification of specific RNA and protein molecules. In a play on Southern's name, the identification of particular RNAs following separation by gel electrophoresis, blotting, and probe hybridization is called **Northern blotting.** The analogous technique for identifying protein molecules is termed **Western blotting.** In Western blotting, the probe of choice is usually an antibody specific for the target protein.

▶ The Southern blotting technique involves the transfer of electrophoretically separated DNA fragments to an absorbent sheet and subsequent detection of specific DNA sequences. A preparation of DNA fragments [typically a restriction digest, **(1)**] is separated according to size by gel electrophoresis **(2)**. The separation pattern can be visualized by soaking the gel in ethidium bromide to stain the DNA and then illuminating the gel with UV light **(3)**. Ethidium bromide molecules intercalated between the hydrophobic bases of DNA are fluorescent under UV light. The gel is soaked in strong alkali to denature the DNA and then neutralized in buffer. Next, the gel is placed on a sheet of DNA-binding material and concentrated salt solution is passed through the gel **(4)** to carry the DNA fragments out of the gel where they are bound tightly to the sheet **(5)**. Incubation of the sheet with a solution of labeled, single-stranded probe DNA **(6)** allows the probe to hybridize with target DNA sequences complementary to it. The location of these target sequences is then revealed by an appropriate means of visualization **(7)**.

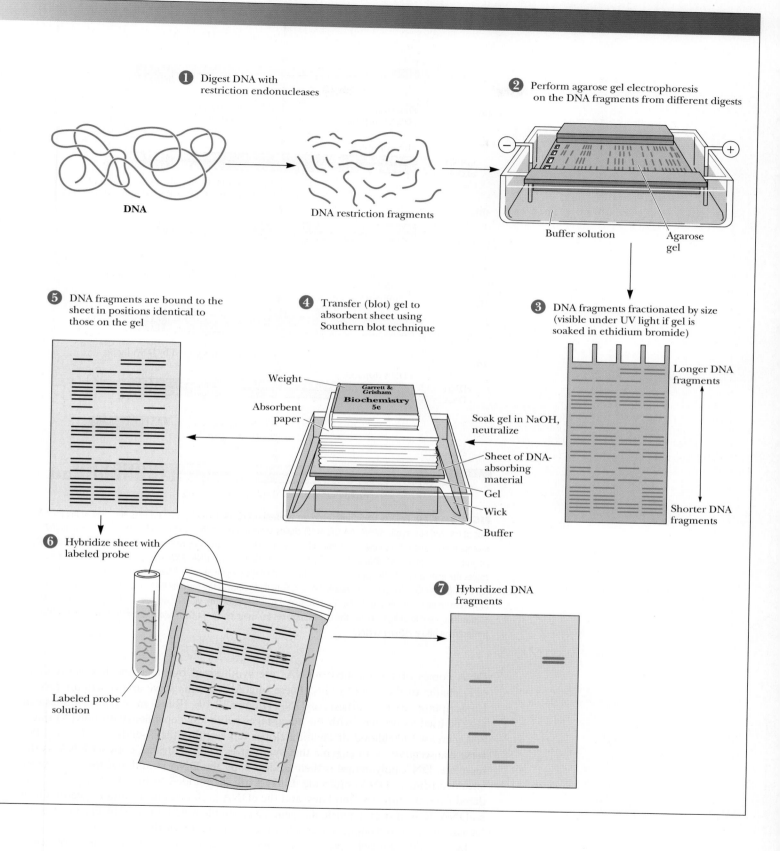

① Digest DNA with restriction endonucleases

DNA

DNA restriction fragments

② Perform agarose gel electrophoresis on the DNA fragments from different digests

Buffer solution

Agarose gel

③ DNA fragments fractionated by size (visible under UV light if gel is soaked in ethidium bromide)

Longer DNA fragments

Shorter DNA fragments

Soak gel in NaOH, neutralize

④ Transfer (blot) gel to absorbent sheet using Southern blot technique

Weight

Absorbent paper

Garrett & Grisham
Biochemistry
5e

Sheet of DNA-absorbing material

Gel

Wick

Buffer

⑤ DNA fragments are bound to the sheet in positions identical to those on the gel

⑥ Hybridize sheet with labeled probe

Labeled probe solution

⑦ Hybridized DNA fragments

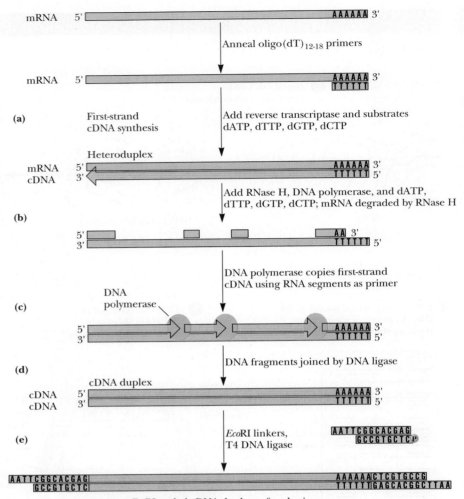

FIGURE 12.10 Reverse transcriptase–driven synthesis of cDNA from oligo(dT) primers annealed to the poly(A) tails of purified eukaryotic mRNA. **(a)** Oligo(dT) chains serve as primers for synthesis of a DNA copy of the mRNA by reverse transcriptase. Following completion of first-strand cDNA synthesis by reverse transcriptase, RNase H and DNA polymerase are added **(b)**. RNase H specifically digests RNA strands in DNA : RNA hybrid duplexes. DNA polymerase copies the first-strand cDNA, using as primers the residual RNA segments after RNase H has created nicks and gaps **(c)**. DNA polymerase has a $5' \rightarrow 3'$ exonuclease activity that removes the residual RNA as it fills in with DNA. The nicks remaining in the second-strand DNA are sealed by DNA ligase **(d)**, yielding duplex cDNA. EcoRI adapters with 5'-overhangs are then ligated onto the cDNA duplexes **(e)** using phage T4 DNA ligase to create EcoRI-ended cDNA for insertion into a cloning vector.

DNA copies of the purified mRNAs are synthesized by first annealing short oligo (dT) chains to the poly(A) tails. These oligo(dT) chains serve as primers for reverse transcriptase–driven synthesis of DNA (Figure 12.10). [Random oligonucleotides can also be used as primers, with the advantages being less dependency on poly(A) tracts and increased likelihood of creating clones representing the 5′-ends of mRNAs.] **Reverse transcriptase** is an enzyme that synthesizes a DNA strand, copying RNA as the template. DNA polymerase is then used to copy the DNA strand and form a double-stranded (duplex DNA) molecule. Linkers are then added to the DNA duplexes rendered from the mRNA templates, and the cDNA is cloned into a suitable vector. Once a cDNA derived from a particular gene has been identified, the cDNA becomes an effective probe for screening genomic libraries for isolation of the gene itself.

Because different cell types in eukaryotic organisms express selected subsets of genes, RNA preparations from cells or tissues in which genes of interest are selectively transcribed are enriched for the desired mRNAs. cDNA libraries prepared from such mRNA are representative of the pattern and extent of gene expression that uniquely define particular

HUMAN BIOCHEMISTRY

The Human Genome Project

Completed in 2003, the Human Genome Project was a 13-year collaborative international, government- and private-sponsored effort to map and sequence the entire human genome, some 3 billion base pairs distributed among the two sex chromosomes (**X** and **Y**) and 22 **autosomes** (chromosomes that are not sex chromosomes). A primary goal was to identify and map at least 3000 genetic **markers** (genes or other recognizable loci on the DNA), which were evenly distributed throughout the chromosomes at roughly 100-kb intervals. At the same time, determination of the entire nucleotide sequence of the human genome was undertaken. J. Craig Venter and colleagues working at Celera, a private corporation, took an alternative approach based on computer alignment of sequenced human DNA fragments. A working draft of the human genome was completed in June 2000 and published in February 2001. An ancillary part of the project focused on sequencing the genomes of other species (such as yeast, *Drosophila melanogaster* [the fruit fly], mice, and *Arabidopsis thaliana* [a plant]) to reveal comparative aspects of genetic and sequence organization (Table 12.1). Information about whole genome sequences of organisms has created a new branch of science called **bioinformatics:** the study of the nature and organization of biological information. Bioinformatics includes such approaches as **functional genomics** and **proteomics.** *Functional genomics* addresses global issues of gene expression, such as looking at *all* the genes that are activated during major metabolic shifts (as from growth under aerobic to growth under anaerobic conditions) or during embryogenesis and development of organisms. **Transcriptome** is the word used in functional genomics to define the entire set of genes expressed (as mRNAs transcribed from DNA) in a particular cell or tissue under defined conditions. Functional genomics also provides new insights into evolutionary relationships between organisms. *Proteomics* is the study of all the proteins expressed by a certain cell or tissue under specified conditions. Typically, this set of proteins, the **proteome,** is revealed by running two-dimensional polyacrylamide gel electrophoresis on a cellular extract or by coupling protein separation techniques to mass spectrometric analysis.

The Human Genome Project has proven to be beneficial to medicine. Many human diseases have been traced to genetic defects whose position within the human genome has been identified. As of 2010, the Human Gene Mutation Database (HGMD) listed more than 102,000 mutations in more than 3800 nuclear genes associated with human disease. Among these are

cystic fibrosis gene

the *breast cancer* genes, BRCA1 and BRCA2

Duchenne muscular dystrophy gene* (at 2.4 megabases, one of the largest known genes in any organism)

Huntington's disease gene

neurofibromatosis gene

neuroblastoma gene (a form of brain cancer)

amyotrophic lateral sclerosis gene (Lou Gehrig's disease)

melanocortin-4 receptor gene (obesity and binge eating)

fragile X-linked mental retardation gene*

as well as genes associated with the development of diabetes, a variety of other cancers, and affective disorders such as *schizophrenia* and *bipolar affective disorder* (manic depression).

TABLE 12.1	Completed Genome Nucleotide Sequences[1]		
Genome		Genome Size[2]	Year Completed
Bacteriophage φX174		0.0054	1977
Bacteriophage λ		0.048	1982
Marchantia[3] chloroplast genome		0.187	1986
Vaccinia virus		0.192	1990
Marchantia[3] mitochondrial genome		0.187	1992
Variola (smallpox) virus		0.186	1993
Haemophilus influenzae[4] (Gram-negative bacterium)		1.830	1995
Mycobacterium genitalium (mycobacterium)		0.58	1995
Escherichia coli (Gram-negative bacterium)		4.64	1996
Saccharomyces cerevisiae (yeast)		12.1	1996
Methanococcus jannaschii (archaeon)		1.66	1998
Arabidopsis thaliana (green plant)		115	2000
Caenorhabditis elegans (simple animal: nematode worm)		88	1998
Drosophila melanogaster (fruit fly)		117	2000
Homo sapiens (human)		3038	2001
Pan troglodytes (chimpanzee)		3109	2005
Homo sapiens neanderthalensis (Neanderthal)		3200	2010

[1]Data available from the National Center for Biotechnology Information at the National Library of Medicine. Website: *http://www.ncbi.nlm.nih.gov/*
[2]Genome size is given as millions of base pairs (mb).
[3]*Marchantia* is a bryophyte (a nonvascular green plant).
[4]The first complete sequence for the genome of a free-living organism.

*X-chromosome–linked gene. As of 2010, more than 784 disease-related genes have been mapped to the X chromosome (source: the *GeneCards* website at the Weizmann Institute of Science, Israel.)

kinds of differentiated cells. cDNA libraries of many normal and diseased human cell types are commercially available, including cDNA libraries of many tumor cells. Comparison of normal and abnormal cDNA libraries, in conjunction with two-dimensional gel electrophoretic analysis (see A Deeper Look: Techniques Used in Protein Purification in Chapter 5) of the proteins produced in normal and abnormal cells, is a useful strategy in clinical medicine to understand disease mechanisms.

Expressed Sequence Tags When a cDNA library is prepared from the mRNAs synthesized in a particular cell type under certain conditions, these cDNAs represent the nucleotide sequences (genes) that have been expressed in this cell type under these conditions. **Expressed sequence tags (ESTs)** are relatively short (~200 nucleotides or so) sequences obtained by determining a portion of the nucleotide sequence for each insert in randomly selected cDNAs. An EST represents part of a gene that is being expressed. Probes derived from ESTs can be labeled, radioactively or otherwise, and used in hybridization experiments to identify which genes in a genomic library are being expressed in the cell. For example, labeled ESTs can be hybridized to a *gene chip* (see following discussion).

DNA Microarrays *(Gene Chips)* Are Arrays of Different Oligonucleotides Immobilized on a Chip

Robotic methods can be used to synthesize combinatorial libraries of DNA oligonucleotides directly on a solid support, such that the completed library is a two-dimensional array of different oligonucleotides (see the Critical Developments in Biochemistry box on combinatorial libraries, page 385). Synthesis is performed by phosphoramidite chemistry (Figure 11.32) adapted into a photochemical process that can be controlled by light. Computer-controlled masking of the light allows chemistry to take place at some spots in the two-dimensional array of growing oligonucleotides and not at others, so each spot on the array is a population of identical oligonucleotides of unique sequence. The final products of such procedures are referred to as "gene chips" because the oligonucleotide sequences synthesized upon the chip represent the sequences of chosen genes. Typically, the oligonucleotides are up to 25 nucleotides long (there are more than 10^{15} possible sequence arrangements for 25-mers made from four bases), and as many as 100,000 different oligonucleotides can be arrayed on a chip 1 cm square. The oligonucleotides on such gene chips are used as the probes in a hybridization experiment to reveal gene expression patterns. Figure 12.11 shows one design for gene chip analysis of gene expression.

12.3 | Can the Cloned Genes in Libraries Be Expressed?

Expression Vectors Are Engineered So That the RNA or Protein Products of Cloned Genes Can Be Expressed

Expression vectors are engineered so that any cloned insert can be transcribed into RNA, and, in many instances, even translated into protein. cDNA expression libraries can be constructed in specially designed vectors. Proteins encoded by the various cDNA clones within such expression libraries can be synthesized in the host cells, and if suitable assays are available to identify a particular protein, its corresponding cDNA clone can be identified and isolated. Expression vectors designed for RNA expression or protein expression, or both, are available.

RNA Expression A vector for in vitro expression of DNA inserts as RNA transcripts can be constructed by putting a highly efficient promoter adjacent to a versatile cloning site. Figure 12.12 depicts such an expression vector. Linearized recombinant vector DNA is transcribed in vitro using SP6 RNA polymerase. Large amounts of RNA product can be obtained in this manner; if radioactive or fluorescent-labeled ribonucleotides are used as substrates, labeled RNA molecules useful as probes are made.

Protein Expression Because cDNAs are DNA copies of mRNAs, cDNAs are uninterrupted copies of the exons of expressed genes. Because cDNAs lack introns, it is feasible to express these cDNA versions of eukaryotic genes in prokaryotic hosts that cannot process the complex primary transcripts of eukaryotic genes. To express a eukaryotic

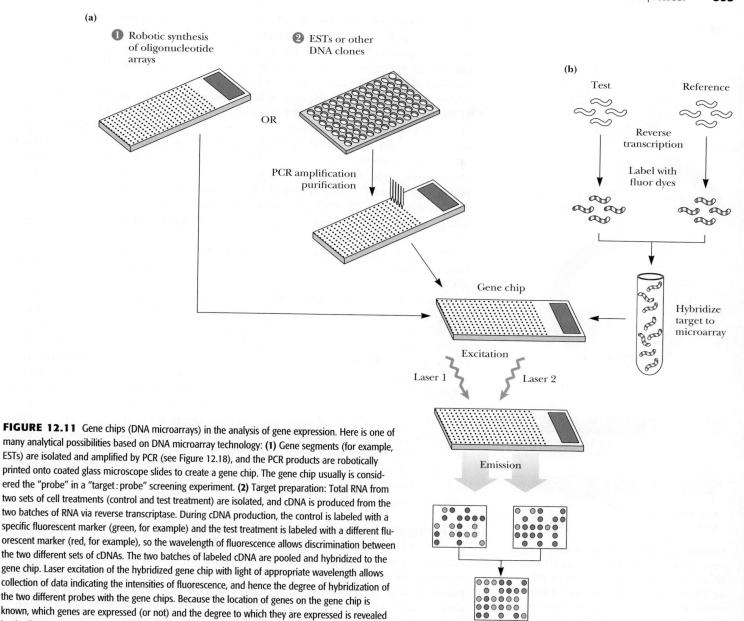

(a)

① Robotic synthesis of oligonucleotide arrays

② ESTs or other DNA clones

OR

PCR amplification purification

Gene chip

Excitation

Laser 1   Laser 2

Emission

Computer analysis

(b)

Test   Reference

Reverse transcription

Label with fluor dyes

Hybridize target to microarray

FIGURE 12.11 Gene chips (DNA microarrays) in the analysis of gene expression. Here is one of many analytical possibilities based on DNA microarray technology: **(1)** Gene segments (for example, ESTs) are isolated and amplified by PCR (see Figure 12.18), and the PCR products are robotically printed onto coated glass microscope slides to create a gene chip. The gene chip usually is considered the "probe" in a "target:probe" screening experiment. **(2)** Target preparation: Total RNA from two sets of cell treatments (control and test treatment) are isolated, and cDNA is produced from the two batches of RNA via reverse transcriptase. During cDNA production, the control is labeled with a specific fluorescent marker (green, for example) and the test treatment is labeled with a different fluorescent marker (red, for example), so the wavelength of fluorescence allows discrimination between the two different sets of cDNAs. The two batches of labeled cDNA are pooled and hybridized to the gene chip. Laser excitation of the hybridized gene chip with light of appropriate wavelength allows collection of data indicating the intensities of fluorescence, and hence the degree of hybridization of the two different probes with the gene chips. Because the location of genes on the gene chip is known, which genes are expressed (or not) and the degree to which they are expressed is revealed by the fluorescent patterns. (Adapted from Figure 1 in Duggan, D. J., et al., 1999. Expression profiling using cDNA microarrays. *Nature Genetics* **21** supplement:10–14.)

protein in *E. coli,* the eukaryotic cDNA must be cloned in an *expression vector* that contains regulatory signals for both transcription and translation. Accordingly, a *promoter* where RNA polymerase initiates transcription as well as a *ribosome-binding site* to facilitate translation are engineered into the vector just upstream from the restriction site for inserting foreign DNA. The AUG initiation codon that specifies the first amino acid in the protein (the *translation start site*) is contributed by the insert (Figure 12.13).

Strong promoters have been constructed that drive the synthesis of foreign proteins to levels equal to 30% or more of total *E. coli* cellular protein. An example is the hybrid promoter, p_{tac}, which was created by fusing part of the promoter for the *E. coli* genes encoding the enzymes of lactose metabolism (the *lac* promoter) with part of the promoter for the genes encoding the enzymes of tryptophan biosynthesis (the *trp* promoter) (Figure 12.14). In cells carrying p_{tac} expression vectors, the p_{tac}

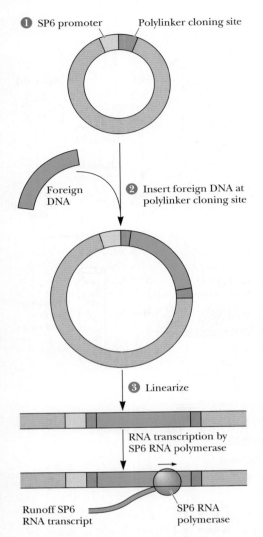

❶ SP6 promoter   Polylinker cloning site

Foreign DNA

❷ Insert foreign DNA at polylinker cloning site

❸ Linearize

RNA transcription by SP6 RNA polymerase

Runoff SP6 RNA transcript

SP6 RNA polymerase

FIGURE 12.12 Expression vectors carrying the promoter recognized by the RNA polymerase of bacteriophage SP6 are useful for the production of multiple RNA copies of any DNA inserted at the polylinker. Before transcription is initiated, the circular expression vector is linearized by a single cleavage at or near the end of the insert so that transcription terminates at a fixed point.

promoter is not induced to drive transcription of the foreign insert until the cells are exposed to *inducers* that lead to its activation. Analogs of lactose (a β-galactoside) such as *isopropyl-β-thiogalactoside,* or *IPTG,* are excellent inducers of p_{tac}. Thus, expression of the foreign protein is easily controlled. (See Chapter 29 for detailed discussions of inducible gene expression.)

A widely used protein expression system is based on the pET plasmid. Transcription of the cloned gene insert is under the control of the bacteriophage T7 RNA polymerase promoter in pET. This promoter is not recognized by the *E. coli* RNA polymerase, so transcription can only occur if the T7 RNA polymerase is present in host cells. Host *E. coli* cells are engineered so that the T7 RNA polymerase gene is inserted in the host chromosome under the control of the *lac* promoter. IPTG induction triggers T7 RNA polymerase production and subsequent transcription and translation of the pET insert. The bacteriophage T7 RNA polymerase is so active that most of the host cell's resources are directed into protein expression and levels of expressed protein approach 50% of total cellular protein.

The bacterial production of valuable eukaryotic proteins represents one of the most important uses of recombinant DNA technology. For example, human insulin for the clinical treatment of diabetes is now produced in bacteria.

Analogous systems for expression of foreign genes in eukaryotic cells include vectors carrying promoter elements derived from mammalian viruses, such as *simian virus 40 (SV40),* the *Epstein–Barr virus,* and the human *cytomegalovirus (CMV).* A system for high-level expression of foreign genes uses insect cells infected with the *baculovirus* expression vector. **Baculoviruses** infect *lepidopteran* insects (butterflies and moths). In engineered baculovirus vectors, the foreign gene is cloned downstream of the promoter for **polyhedrin,** a major viral-encoded structural protein, and the recombinant vector is incorporated into insect cells grown in culture. Expression from the polyhedrin promoter can lead to accumulation of the foreign gene product to levels as high as 500 mg/L.

Technologies for the expression of recombinant proteins in mammalian cell cultures are commercially available. These technologies have the advantage that the unique post-translational modifications of proteins (such as glycosylation; see Chapter 31) seen in mammalian cells take place in vivo so that the expressed protein is produced in its naturally occurring form.

Screening cDNA Expression Libraries with Antibodies Antibodies that specifically cross-react with a particular protein of interest are often available. If so, these antibodies can be used to screen a cDNA expression library to identify and isolate cDNA clones encoding the protein. The cDNA library is introduced into host bacteria, which are plated out and grown overnight, as in the colony hybridization scheme previously described. DNA-binding nylon membranes are placed on the plates to obtain a replica of the bacterial colonies. The nylon membrane is then incubated under conditions that

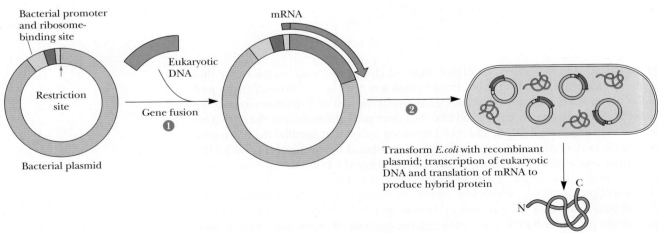

Bacterial promoter and ribosome-binding site

mRNA

Eukaryotic DNA

Restriction site

Bacterial plasmid

Gene fusion ❶

❷

Transform *E.coli* with recombinant plasmid; transcription of eukaryotic DNA and translation of mRNA to produce hybrid protein

N   C

Recovery of eukaryotic protein product

FIGURE 12.13 A typical expression-cloning vector. (Adapted from Fig. 19-19, in Watson, J. D., et al., 1987. *Molecular Biology of the Gene* 4/e. Menlo Park, Calif.: Benjamin/Cummings. Reprinted with permission.)

induce protein synthesis from the cloned cDNA inserts, and the cells are treated to release the synthesized protein. The synthesized protein binds tightly to the nylon membrane, which can then be incubated with the specific antibody. Binding of the antibody to its target protein product reveals the position of any cDNA clones expressing the protein, and these clones can be recovered from the original plate. Like other libraries, expression libraries can be screened with oligonucleotide probes, too.

Fusion Protein Expression Some expression vectors carry cDNA inserts cloned directly into the coding sequence of a vector-borne protein-coding gene (Figure 12.15). Translation of the recombinant sequence leads to synthesis of a *hybrid protein* or *fusion protein.* The N-terminal region of the fused protein represents amino acid sequences encoded in the vector, whereas the remainder of the protein is encoded by the foreign insert. Keep in mind that the triplet codon sequence within the cloned insert must be in phase with codons contributed by the vector sequences to make the right protein. The N-terminal protein sequence contributed by the vector can be chosen to suit purposes. Furthermore, adding an N-terminal signal sequence that targets the hybrid protein for secretion from the cell simplifies recovery of the fusion protein. A variety of gene fusion systems have been developed to facilitate isolation of a specific protein encoded by a cloned insert. The isolation procedures are based on affinity chromatography purification of the fusion protein through exploitation of the unique ligand-binding properties of the vector-encoded protein (Table 12.2).

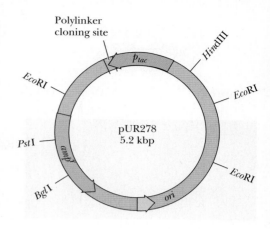

FIGURE 12.14 A p_{tac} protein expression vector contains the hybrid promoter p_{tac} derived from fusion of the *lac* and *trp* promoters. Isopropyl-β-D-thiogalactoside, or IPTG, induces expression from p_{tac}.

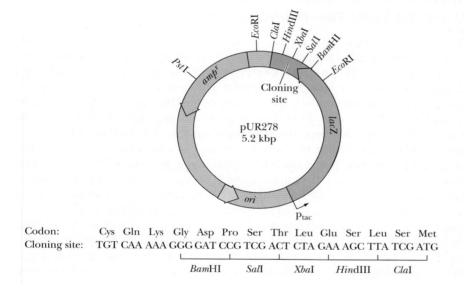

Codon:	Cys	Gln	Lys	Gly	Asp	Pro	Ser	Thr	Leu	Glu	Ser	Leu	Ser	Met
Cloning site:	TGT	CAA	AAA	GGG	GAT	CCG	TCG	ACT	CTA	GAA	AGC	TTA	TCG	ATG

| | *Bam*HI | *Sal*I | *Xba*I | *Hind*III | *Cla*I |

FIGURE 12.15 A typical expression vector for the synthesis of a hybrid protein. The cloning site is located at the end of the coding region for the protein β-galactosidase. Insertion of foreign DNAs at this site fuses the foreign sequence to the β-galactosidase coding region (the *lacZ* gene). IPTG induces the transcription of the *lacZ* gene from its promoter p_{lac}, causing expression of the fusion protein. (Adapted from Figure 2, Rüther, U., and Müller-Hill, B., 1983. *EMBO Journal* **2**:1791–1794.)

TABLE 12.2	Gene Fusion Systems for Isolation of Cloned Fusion Proteins	
Fusion Protein	**Secreted?***	**Affinity Ligand**
β-Galactosidase	No	*p*-Aminophenyl-β-D-thiogalactoside (APTG)
Protein A	Yes	Immunoglobulin G (IgG)
Chloramphenicol acetyltransferase (CAT)	Yes	Chloramphenicol
Streptavidin	Yes	Biotin
Glutathione-S-transferase (GST)	No	Glutathione
Maltose-binding protein (MBP)	Yes	Starch
Hexahistidine tag	No	Nickel or cobalt
Hemagglutinin (HA) peptide	No	HA-peptide antibody

*This indicates whether combined secretion–fusion gene systems have led to secretion of the protein product from the cells, which simplifies its isolation and purification.

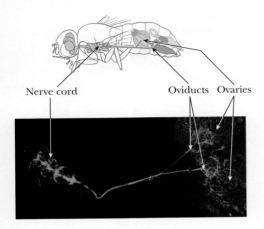

Nerve cord   Oviducts Ovaries

FIGURE 12.16 Green fluorescent protein (GFP) as a reporter gene. In the experiment here, GFP expression depends on the promoter for the *Drosophilia melanogaster Tdc2* gene. *Tdc2* encodes a neuronal tyrosine decarboxylase (TDC) whose expression is necessary for egg laying in fruit flies. *(Bottom)* Green fluorescence highlights neuronal projections expressing the *Tdc2* gene. *(Top)* Diagram of a fly, its nervous system, and ovaries. Note that *Tdc2* neurons innervate the ovaries and oviducts of flies. (See Cole, S. H., et al., 2005. Two functional but noncomplementing *Drosophila* tyrosine decarboxylase genes. *Journal of Biological Chemistry* **280:**14948–14955. GFP image courtesy of Shannon H. Cole and Jay Hirsh, the University of Virginia. Fly image derived from the *Atlas of* Drosophila *Development* by Volker Hartenstein, http://flybase.bio.indiana.edu/allied-data/lk/interactive-fly/atlas/00contents.htm.)

Reporter Gene Constructs Are Chimeric DNA Molecules Composed of Gene Regulatory Sequences Positioned Next to an Easily Expressible Gene Product

Potential regulatory regions of genes (such as promoters) can be investigated by placing these regulatory sequences into plasmids upstream of a gene, called a **reporter gene,** whose expression is easy to measure. Such chimeric plasmids are then introduced into cells of choice (including eukaryotic cells) to assess the potential function of the nucleotide sequence in regulation because expression of the reporter gene serves as a report on the effectiveness of the regulatory element. A number of different genes have been used as reporter genes. A reporter gene with many inherent advantages is that encoding the **green fluorescent protein** (or **GFP**), described in Chapter 4. Unlike the protein expressed by other reporter gene systems, GFP does not require any substrate to measure its activity, nor is it dependent on any cofactor or prosthetic group. Detection of GFP requires only irradiation with near-UV or blue light (400-nm light is optimal), and the green fluorescence (light of 500 nm) that results is easily observed with the naked eye, although it can also be measured precisely with a fluorometer. Figure 12.16 demonstrates the use of GFP as a reporter gene. EGFP is an engineered version of GFP that shows enhanced fluorescent properties.

Specific Protein–Protein Interactions Can Be Identified Using the Yeast Two-Hybrid System

Specific interactions between proteins (so-called protein–protein interactions) lie at the heart of many essential biological processes. One method to identify specific protein–protein interactions in vivo is through expression of a reporter gene whose transcription is dependent on a functional transcriptional activator, the *GAL4* protein. The *GAL4* protein consists of two domains: a DNA-binding (or **DB**) domain and a transcriptional activation (or **TA**) domain. Even if expressed as separate proteins, these two domains will still work, provided they can be brought together. The method depends on two separate plasmids encoding two hybrid proteins, one consisting of the *GAL4* DB domain fused to protein X and the other consisting of the *GAL4* TA domain fused to protein Y (Figure 12.17a). If proteins X and Y interact in a specific protein–protein interaction, the *GAL4* DB and TA domains are brought together so that transcription of a reporter gene driven by the *GAL4* promoter can take place (Figure 12.17b). Protein X, fused to the *GAL4*-DNA–binding domain (DB), serves as the "bait" to fish for the protein Y "target" and its fused *GAL4* TA domain. This method can be used to screen cells for protein "targets" that interact specifically with a particular "bait" protein. To do so, cDNAs encoding proteins from the cells of interest are inserted into the TA-containing plasmid to create fusions of the cDNA coding sequences with the *GAL4* TA domain coding sequences, so a fusion protein library is expressed. Identification of a target of the "bait" protein by this method also yields directly a cDNA version of the gene encoding the "target" protein.

Identifying Protein–Protein Interactions Through Immunoprecipitation If antibodies against one protein of a multiprotein complex are available, the entire complex can be immunoprecipitated and its composition analyzed. Attachment of such antibodies to glass or agarose beads, which easily sediment in a centrifuge, makes recovery of the complex very simple. Because antibodies against it are commercially available, the hemagglutinin (HA) peptide, sequence YPYDVPDYA, is a useful protein fusion tag, not only for fusion protein purification (Table 12.2) but also for analysis of protein–protein interactions. Expressing an HA-tagged protein in vivo, followed by immunoprecipitation, allows the isolation of protein complexes of which the HA-tagged protein is a member. The other members of the complex can then be identified to establish the various interacting partners within the multiprotein complex.

(a)

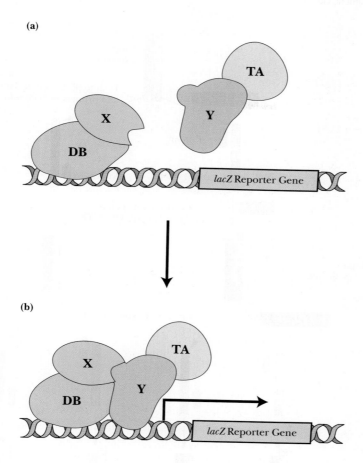

(b)

FIGURE 12.17 The yeast two-hybrid system for identifying protein–protein interactions. If proteins X and Y interact, the *lacZ* reporter gene is expressed. Cells expressing *lacZ* exhibit β-galactosidase activity.

12.4 What Is the Polymerase Chain Reaction (PCR)?

Polymerase chain reaction, or **PCR,** is a technique for dramatically amplifying the amount of a specific DNA segment. A preparation of denatured DNA containing the segment of interest serves as template for DNA polymerase, and two specific oligonucleotides serve as primers for DNA synthesis (as in Figure 12.18). These primers, designed to be complementary to the two 3′-ends of the specific DNA segment to be amplified, are added in excess amounts of 1000 times or greater (Figure 12.18). They prime the DNA polymerase–catalyzed synthesis of the two complementary strands of the desired segment, effectively doubling its concentration in the solution. Then the DNA is heated to dissociate the DNA duplexes and then cooled so that primers bind to both the newly formed and the old strands. Another cycle of DNA synthesis follows. The protocol has been automated through the invention of **thermal cyclers** that alternately heat the reaction mixture to 95°C to dissociate the DNA, followed by cooling, annealing of primers, and another round of DNA synthesis. The isolation of heat-stable DNA polymerases from thermophilic bacteria (such as the *Taq* DNA polymerase from *Thermus aquaticus*) has made it unnecessary to add fresh enzyme for each round of synthesis. Because the amount of target DNA theoretically doubles each round, 25 rounds would increase its concentration about 33 million times. In practice, the increase is actually more like a million times, which is more than ample for gene isolation. Thus, starting with a tiny amount of total genomic DNA, a particular sequence can be produced in quantity in a few hours.

PCR amplification is an effective cloning strategy if sequence information for the design of appropriate primers is available. Because DNA from a single cell can be used as a template, the technique has enormous potential for the clinical diagnosis of infectious diseases and genetic abnormalities. With PCR techniques, DNA from a single hair or sperm can be analyzed to identify particular individuals in criminal cases without

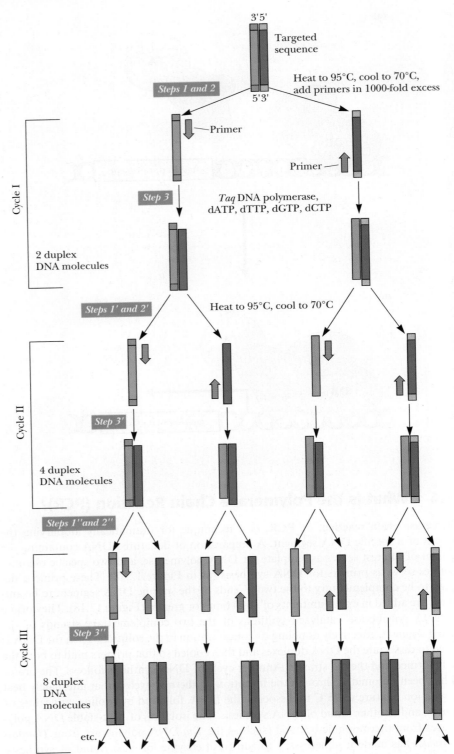

FIGURE 12.18 Polymerase chain reaction (PCR).

ambiguity. **RT-PCR,** a variation on the basic PCR method, is useful when the nucleic acid to be amplified is an RNA (such as mRNA). Reverse transcriptase (RT) is used to synthesize a cDNA strand complementary to the RNA, and this cDNA serves as the template for further cycles of PCR. (RT-PCR is also used to refer to yet another variation on PCR whose full name is *real-time PCR*. Real-time PCR uses PCR amplification to measure the relative amounts of mRNAs expressed in vivo.)

In Vitro Mutagenesis

The advent of recombinant DNA technology has made it possible to clone genes, manipulate them in vitro, and express them in a variety of cell types under various conditions. The function of any protein is ultimately dependent on its amino acid sequence, which in turn can be traced to the nucleotide sequence of its gene. The introduction of purposeful changes in the nucleotide sequence of a cloned gene represents an ideal way to make specific structural changes in a protein. The effects of these changes on the protein's function can then be studied. Such changes constitute *mutations* introduced in vitro into the gene. In vitro **mutagenesis** makes it possible to alter the nucleotide sequence of a cloned gene systematically, as opposed to the chance occurrence of mutations in natural genes.

One efficient technique for in vitro mutagenesis is **PCR-based mutagenesis.** Mutant primers are added to a PCR reaction in which the gene (or segment of a gene) is undergoing amplification. The *mutant primers* are primers whose sequence has been specifically altered to introduce a directed change at a particular place in the nucleotide sequence of the gene being amplified (Figure 12.19). Mutant versions of the gene can then be cloned and expressed to determine any effects of the mutation on the function of the gene product.

12.5 : How Is RNA Interference Used to Reveal the Function of Genes?

RNA interference (**RNAi**) has emerged as a method of choice in eukaryotic gene inactivation. RNAi leads to targeted destruction of a selected gene's transcript. The consequences following loss of gene function reveal the role of the gene product in cell metabolism. Inactivation of gene expression by RNAi is sometimes referred to as **gene knockdown.** (Gene knockdown is a term that contrasts the method with **gene knockout,** a procedure that inactivates a gene by disrupting its nucleotide sequence; see Chapter 28.)

Procedures for silencing gene expression via RNAi depend on the introduction of double-stranded RNA (dsRNA) molecules into target cells by transfection, viral infection, or artificial expression. One strand of the dsRNA is designed to be an antisense RNA, in that its nucleotide sequence is complementary to the RNA transcript of the gene selected for silencing. An ATP-dependent endogenous cellular protein system known as **Dicer** processes the dsRNA. Dicer is an RNase III family member that catalyzes endonucleolytic cleavage of both strands of dsRNA molecules to produce a double-stranded **small interfering RNA (siRNA)** 21 to 23 nucleotides long and having 2-nucleotide-long 3′-overhangs on each strand (Figure 12.20). The siRNA is then passed to another protein complex known as **RNA-induced silencing complex (RISC).** In an ATP-dependent process, RISC unwinds the double-stranded siRNA and selects the antisense strand, which is referred to as the **guide strand.** The other strand, referred to as the **passenger strand,** is discarded. RISC pairs the single-stranded guide strand with a complementary region on the targeted gene transcript. RISC then carries out its "slicer function" by cleaving the RNA transcript between nucleotides 10 and 11 of the mRNA region that is base-paired with the guide strand. Such cleavage prevents expression of the product encoded by the mRNA. The guide strand remains associated with RISC, and RISC can use it for multiple cycles of mRNA cleavage and post-transcriptional gene silencing.

12.6 : Is It Possible to Make Directed Changes in the Heredity of an Organism?

Recombinant DNA technology is a powerful tool for the genetic modification of organisms. The strategies and methodologies described in this chapter are but an overview of the repertoire of experimental approaches that have been devised by molecular biologists

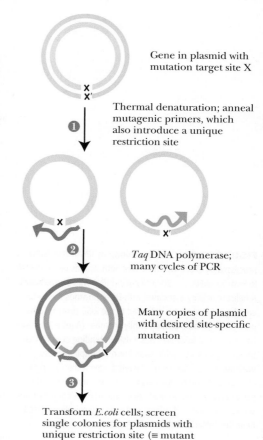

Gene in plasmid with mutation target site X

Thermal denaturation; anneal mutagenic primers, which also introduce a unique restriction site

Taq DNA polymerase; many cycles of PCR

Many copies of plasmid with desired site-specific mutation

Transform *E. coli* cells; screen single colonies for plasmids with unique restriction site (≡ mutant gene)

FIGURE 12.19 One method of PCR-based site-directed mutagenesis. **(1)** Template DNA strands are separated and amplified by PCR using mutagenic primers (represented as bent arrows) whose sequences introduce a single base substitution at site X (and its complementary base X′; thus, the desired amino acid change in the protein encoded by the gene). Ideally, the mutagenic primers also introduce a unique restriction site into the plasmid that was not present before. **(2)** Following many cycles of PCR, the DNA product can be used to transform *E. coli* cells. **(3)** The plasmid DNA can be isolated and screened for the presence of the mutation by screening for the presence of the unique restriction site by restriction endonuclease cleavage. For example, the nucleotide sequence GGATCT within a gene codes for amino acid residues Gly-Ser. Using mutagenic primers of nucleotide sequence AGATCT (and its complement AGATCT) changes the amino acid sequence from Gly-Ser to Arg-Ser and creates a *Bgl*II restriction site (see Table 10.2). Gene expression of the isolated mutant plasmid in *E. coli* allows recovery and analysis of the mutant protein.

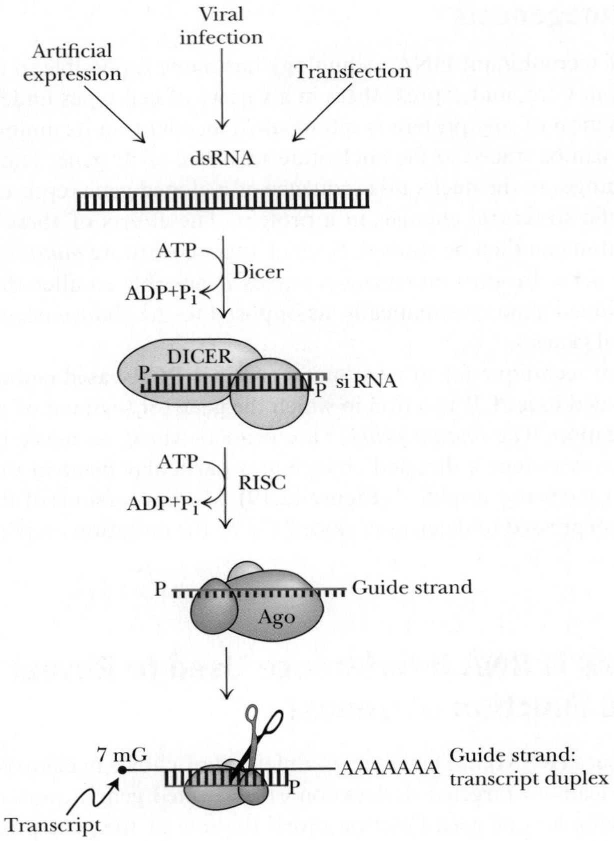

FIGURE 12.20 Gene knockdown by RNAi. The dsRNA is processed by Dicer, which cleaves both strands of the dsRNA to form an siRNA, a ~20-nucleotide dsRNA with 3′-overhangs. A helicase activity associated with Dicer unwinds the siRNA, and the guide strand is delivered to the RISC protein complex. An Argonaute protein family member (Ago) is the catalytic subunit of RISC. Ago has a dsRNA-binding domain that brings together the guide strand and a complementary nucleotide sequence on the targeted gene transcript. Ago also has a RNase H-type catalytic domain that cleaves the gene transcript, rendering it incapable of translation by ribosomes. This RNase H activity of Ago is whimsically referred to as the "slicer" function in RNAi.

in order to manipulate DNA and the information inherent in it. The enormous success of recombinant DNA technology means that the molecular biologist's task in searching genomes for genes is now akin to that of a lexicographer compiling a dictionary, a dictionary in which the "letters" (the nucleotide sequences), spell out not words but rather genes and what they mean. Molecular biologists have no index or alphabetic arrangement to serve as a guide through the vast volume of information in a genome; nevertheless, this information and its organization is rapidly being disclosed by the imaginative efforts and diligence of these scientists and their growing arsenal of analytical schemes.

Recombinant DNA technology now verges on the ability to engineer at will the heredity (or genetic makeup) of organisms for desired ends. The commercial production of therapeutic biomolecules in microbial cultures is already established (for example, the production of human insulin in quantity in *E. coli* cells). Agricultural crops with desired attributes, such as enhanced resistance to herbicides or elevated vitamin levels, are in cultivation. Transgenic mice are widely used as experimental animals to investigate models of human disease and physiology (see Chapter 28). Already, transgenic versions of domestic animals such as pigs, sheep, and even fish have been developed for human benefit. Perhaps most important, in a number of instances, clinical trials have been approved for **gene replacement therapy** (or, more simply, *gene therapy*) to correct particular human genetic disorders.

Human Gene Therapy Can Repair Genetic Deficiencies

Human gene therapy seeks to repair the damage caused by a genetic deficiency through introduction of a functional version of the defective gene. To achieve this end, a cloned variant of the gene must be incorporated into the organism in such a manner that it is expressed only at the proper time *and* only in appropriate cell types. At this time, these conditions impose serious technical and clinical difficulties. Many gene therapies have received approval from the National Institutes of Health for trials in human patients, including the introduction of gene constructs into patients. Among these are constructs designed to cure ADA⁻ SCID (**s**evere **c**ombined **i**mmuno**d**eficiency due to adenosine deaminase [ADA] deficiency), neuroblastoma, or cystic fibrosis or to treat cancer through expression of the *E1A* and *p53* tumor suppressor genes.

A basic strategy in human gene therapy involves incorporation of a functional gene into target cells. The gene is typically in the form of an **expression cassette** consisting of a cDNA version of the gene downstream from a promoter that will drive expression of the gene in one of two ways. One way, the ex-vivo route, is to introduce a vector carrying the expression cassette into cells isolated from a patient and cultured in the laboratory. The modified cells are then reintroduced into the patient. The other way involves direct incorporation of the gene by treating the patient with a viral vector carrying the expression cassette.

Retroviruses are RNA viruses that replicate their RNA genome by first making a DNA intermediate. Because retroviruses can transfer their genetic information directly into the genome of host cells, retroviruses provide a route for permanent modification of host cells ex vivo. A replication-deficient mutant of *Maloney murine leukemia virus* (MMLV) can be generated by deleting the *gag, pol,* and *env* genes. This mutant retrovirus can introduce expression cassettes up to 9 kb (Figure 12.21). In the cytosol of the patient's cells, a DNA copy of the viral RNA is synthesized by viral reverse transcriptase. This DNA is then randomly integrated into the host cell genome, where its expression leads to synthesis of the expression cassette gene product (Figure 12.21).

In 2000, scientists at the Pasteur Institute in Paris used such an ex vivo approach to successfully treat infants with X-linked SCID. The gene encoding the γc cytokine receptor

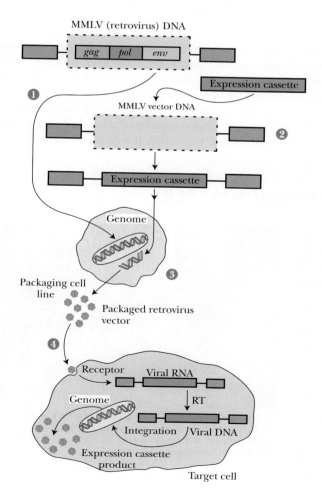

FIGURE 12.21 Retrovirus-mediated gene delivery ex vivo using MMLV. Deletion of the essential genes *gag, pol,* and *env* from MMLV **(1)** creates a space for insertion of an expression cassette **(2)**. The modified MMLV acts as a vector for the expression cassette. A second virus (the packaging cell line) that carries intact *gag, pol,* and *env* genes allows the modified MMLV to reproduce **(3)**, and the packaged recombinant viruses can be collected and used to infect a patient **(4)**. (Adapted from Figure 1 in Crystal, R. G., 1995. Transfer of genes to humans: Early lessons and obstacles to success. *Science* **270:**404.)

FIGURE 12.22 Adenovirus-mediated gene delivery in vivo. Adenoviruses are DNA viruses. The adenovirus genome (**1**). Adenovirus vectors are generated by deleting gene E1 (and sometimes E3 if more space for an expression cassette is needed) (**2**). Insertion of an expression cassette (**3**). Adenovirus progeny from the complementing cell line can be isolated and used to infect a patient (**4**). The recombinant viral DNA gains access to the cell nucleus (**5**), where the gene carried by the cassette is expressed (**6**). (Adapted from Figure 2 in Crystal, R. G., 1995. Transfer of genes to humans: Early lessons and obstacles to success. *Science* **270**:404.)

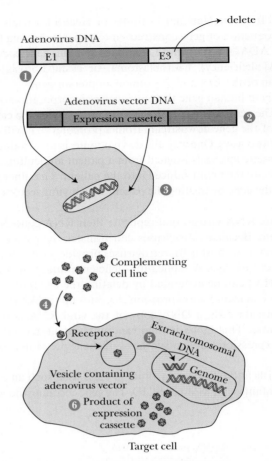

subunit gene was defective in these infants, and gene therapy was used to deliver a functional *γc* cytokine receptor subunit gene to stem cells harvested from the infants. Transformed stem cells were reintroduced into the patients, who were then able to produce functional lymphocytes and lead normal lives. This achievement represents the first successful outcome in human gene therapy. Unfortunately, this therapeutic approach had to be abandoned after several patients developed leukemia (attributed to the chance insertion of the gene into the genome at places that altered expression of nearby oncogenes).

HUMAN BIOCHEMISTRY

The Biochemical Defects in Cystic Fibrosis and ADA⁻ SCID

The gene defective in cystic fibrosis codes for CFTR (cystic fibrosis transmembrane conductance regulator), a membrane protein that pumps Cl^- out of cells. If this Cl^- pump is defective, Cl^- ions remain in cells, which then take up water from the surrounding mucus by osmosis. The mucus thickens and accumulates in various organs, including the lungs, where its presence favors infections such as pneumonia. Left untreated, children with cystic fibrosis seldom survive past the age of 5 years.

ADA⁻ SCID (adenosine deaminase–defective severe combined immunodeficiency) is a fatal genetic disorder caused by defects in the gene that encodes ADA. The consequence of ADA deficiency is accumulation of adenosine and 2′-deoxyadenosine, substances toxic to lymphocytes, important cells in the immune response. 2′-Deoxyadenosine is particularly toxic because its presence leads to accumulation of its nucleotide form, dATP, an essential substrate in DNA synthesis. Elevated levels of dATP actually block DNA replication and cell division by inhibiting synthesis of the other deoxynucleoside 5′-triphosphates (see Chapter 26). Accumulation of dATP also leads to selective depletion of cellular ATP, robbing cells of energy. Children with ADA⁻ SCID fail to develop normal immune responses and are susceptible to fatal infections, unless kept in protective isolation.

© Bettmann / Corbis

▲ David, the Boy in the Bubble. David was born with SCID and lived all 12 years of his life inside a sterile plastic "bubble" to protect him from germs common in the environment. He died in 1984 following an unsuccessful bone marrow transplant.

More recently, successful human gene therapy has been achieved using the ex vivo approach with recombinant lentiviral vectors (lentiviruses are also retroviruses and include HIV, the virus that causes AIDS). Successes include: (1) treatment of X-linked adrenoleukodystrophy (ALD, a fatal central nervous system disease) by infection of hematopoietic stem cells (HSCs) with an HIV-1-derived lentiviral vector carrying a functional ALD gene (2009); and (2) treatment of β–thalassemia patients via ex vivo transduction of their HSCs with a recombinant lentiviral vector carrying the gene for β–globin (2010). Sufferers of β–thalassemia cannot produce functional β–globin chains, lack an effective population of red blood cells, and suffer life-threatening anemia. To date, neither of these successful lentiviral-based human gene therapies has resulted in cancer.

Transduction: The introduction of foreign DNA into another cell via a viral vector.

Adenovirus vectors, which can carry expression cassettes up to 7.5 kb, are a possible in vivo approach to human gene therapy (Figure 12.22). Adenoviruses are DNA viruses. The 36-kb adenovirus genome is divided into *early genes* (E1 to E4) and *late genes* (L1 to L5). Deletion of E1 renders the adenovirus incapable of replication unless introduced into a complementing cell line carrying the E1 gene. The complementing cell line produces adenovirus particles that can be used to infect patients. The recombinant adenoviruses enter the patient's cells via specific receptors on the target cell surface; the transferred genetic information is expressed directly from the adenovirus recombinant DNA and is never incorporated into the host cell genome. No adenovirus-based gene therapy has been successful to date. Although many problems remain to be solved, human gene therapy has become a feasible clinical strategy.

SUMMARY

⊙WL Login to OWL to develop problem-solving skills and complete online homework assigned by your professor.

12.1 What Does It Mean "To Clone"? A clone is a collection of molecules or cells all identical to an original molecule or cell. Plasmids (naturally occurring, circular, extrachromosomal DNA molecules) are very useful in cloning genes. Artificial plasmids can be created by ligating different DNA fragments together. In this manner, "foreign" DNA sequences can be inserted into artificial plasmids, carried into *E. coli,* and propagated as part of the plasmid. Recombinant plasmids are hybrid DNA molecules consisting of plasmid DNA sequences plus inserted DNA elements. A great number of cloning strategies have emerged to make recombinant plasmids and other vectors for use in prokaryotic and eukaryotic cells, including those from humans.

12.2 What Is a DNA Library? A DNA library is a set of cloned fragments representing all the genes of an organism. Particular genes can be isolated from DNA libraries, even though a particular gene constitutes only a small part of an organism's genome. Genomic libraries have been prepared from thousands of different species. Libraries can be screened for the presence of specific genes. A common method of screening plasmid-based genomic libraries is colony hybridization. Making useful probes requires some information about the gene's nucleotide sequence (or the amino acid sequence of a protein whose gene is sought). DNA from the corresponding gene in a related organism can also be used as a probe in screening a library for a particular gene.

cDNA libraries are DNA libraries prepared from mRNA. Because different cell types in eukaryotic organisms express selected subsets of genes, cDNA libraries prepared from such mRNA are representative of the pattern and extent of gene expression that uniquely define particular kinds of differentiated cells.

Expressed sequence tags (ESTs) are relatively short (~200 nucleotides or so) sequences derived from determining a portion of the nucleotide sequence for each insert in randomly selected cDNAs. ESTs can be used to identify which genes in a genomic library are being expressed in the cell. For example, labeled ESTs can be hybridized to DNA microarrays *(gene chips)*. DNA microarrays are arrays of different oligonucleotides immobilized on a solid support, or *chip*. The oligonucleotides on the chip represent a two-dimensional array of different oligonucleotides. Such gene chips are used to reveal gene expression patterns.

12.3 Can the Cloned Genes in Libraries Be Expressed? Expression vectors are engineered so that any cloned insert can be transcribed into RNA and, in many instances, translated into protein. Strong promoters have been constructed that drive the synthesis of foreign proteins to levels equal to 30% or more of total *E. coli* cellular protein. Similarly efficient systems have been developed for protein expression in eukaryotic cells. cDNA expression libraries can also be screened with antibodies to identify and isolate cDNA clones encoding a particular protein.

Reporter gene constructs are chimeric DNA molecules composed of gene regulatory sequences positioned next to an easily expressible gene product, such as green fluorescent protein. Reporter gene constructs introduced into cells of choice (including eukaryotic cells) can reveal the function of nucleotide sequences involved in regulation.

12.4 What Is the Polymerase Chain Reaction (PCR)? PCR is a technique for dramatically amplifying the amount of a specific DNA segment. Denatured DNA containing the segment of interest serves as template for DNA polymerase, and two specific oligonucleotides serve as primers for DNA synthesis. The protocol has been automated through the invention of thermal cyclers that alternately heat the reaction mixture to 95°C to dissociate the DNA, followed by cooling, annealing of primers, and another round of DNA synthesis. Because DNA from a single cell can be used as a template, the technique has enormous potential for the clinical diagnosis of infectious diseases and genetic abnormalities.

Recombinant DNA technology makes it possible to clone genes, manipulate them in vitro, and express them in a variety of cell types under various conditions. The introduction of changes in the nucleotide sequence of a cloned gene represents an ideal way to make specific structural changes in a protein; such changes constitute mutations introduced in vitro into the gene. One efficient technique for in vitro mutagenesis is PCR-based mutagenesis.

12.5 How Is RNA Interference Used to Reveal the Function of Genes? RNAi can be used to selectively inactivate the expression of a target gene in a host cell (gene knockdown). Such inactivation reveals the function of the gene. RNAi relies on processing an introduced double-stranded RNA molecule (dsRNA), one of whose strands (the guide strand) is complementary to a region of the RNA transcript made from the gene destined for knockdown. The dsRNA is processed by the host cell Dicer protein complex to yield a ~20-nucleotide-long siRNA, followed by delivery of the siRNA guide strand sequence to the RISC protein complex. RISC then aligns the guide strand with its complemen-

tary RNA transcript and cleaves the RNA transcript between nucleotides 10 and 11 of the region that is base-paired with the guide strand. Transcript cleavage causes post-transcriptional gene silencing because the cleaved transcript cannot be translated into protein.

12.6 Is It Possible to Make Directed Changes in the Heredity of an Organism? Recombinant DNA technology now verges on the ability to engineer at will the heredity (or genetic makeup) of organisms for desired ends. Human gene therapy seeks to repair the damage caused by a genetic deficiency through the introduction of a functional version of the defective gene. In a number of instances, human gene therapy has been successful in correcting particular inherited human diseases.

FOUNDATIONAL BIOCHEMISTRY   Things You Should Know After Reading Chapter 12.

- The laboratory manipulation of nucleic acid molecules has led to the fields of recombinant DNA technology and genetic engineering.

- A "clone" in molecular biology refers to a population of molecules or cells identical to an original molecule or cell.

- Plasmids are very useful as vectors for cloning genes.

- Virtually any DNA sequence can be cloned.

- Chimeric plasmids can be constructed by inserting foreign DNA into a plasmid that has been cleaved by a restriction endonuclease.

- Directional cloning allows the insertion of foreign DNA into a plasmid or other vector in the desired orientation with respect to adjacent DNA elements such as promoters.

- Biologically functional chimeric plasmids have been used to transform cells, and the gene(s) they carry have been expressed in the recipient cells.

- Shuttle vectors are plasmids that can be propagated in two different organisms.

- Artificial chromosomes can be created from recombinant DNA.

- Genomic libraries are prepared from the total DNA in an organism.

- Genomic libraries can be screened for the presence of specific genes.

- Southern hybridization is a method for identifying specific DNA sequences in a large population of different DNA fragments.

- Probes for Southern hybridization can be prepared in a variety of ways.

- Northern hybridization is a technique for finding specific RNA sequences among a large population of RNA molecules.

- cDNA libraries are libraries prepared from mRNA using reverse transcriptase.

- DNA microarrays or "gene chips" are arrays of different oligonucleotides immobilized on a chip.

- Expression vectors are designed so that the RNA transcripts or protein products encoded by specific genes can be expressed in host cells.

- cDNA libraries can be screened for the presence of specific protein-encoding cDNAs by using antibodies to the protein.

- Expression of proteins from cDNA libraries where the coding sequences of the cDNAs have been fused to easily isolated proteins or peptides allows rapid purification of the protein encoded by the cDNA.

- Reporter gene constructs are chimeric DNA molecules composed of gene regulatory sequences positioned next to an easily expressible gene product, such as GFP.

- Specific protein–protein interactions can be identified by the yeast two-hybrid system.

- Protein–protein interactions can also be revealed by immunoprecipitation using antibodies against one of the proteins.

- PCR (polymerase chain reaction) is a technique for dramatically amplifying a particular DNA sequence.

- In vitro mutagenesis makes it possible to alter the nucleotide sequence of a cloned gene in a systematic way.

- RNAi (RNA interference) can be used to reveal the function of genes.

- It is possible to make directed changes in the heredity of organisms through gene therapy.

- Successful human gene therapy has been achieved using the ex vivo approach with recombinant lentiviral vectors.

- Inherited genetic diseases successfully treated by human gene therapy thus far include X-linked severe combined immunodeficiency (X-linked SCID), X-linked adrenoleukodystrophy (ALD), and β–thalassemia.

PROBLEMS

Answers to all problems are at the end of this book. Detailed solutions are available in the *Student Solutions Manual, Study Guide, and Problems Book.*

1. **Construction of a Recombinant Plasmid** A DNA fragment isolated from an *Eco*RI digest of genomic DNA was combined with a plasmid vector linearized by *Eco*RI digestion so that the sticky ends could anneal. Phage T4 DNA ligase was then added to the mixture. List all possible products of the ligation reaction.

2. **Overlapping Restriction Sites in a Plasmid Vector Polylinker** The nucleotide sequence of a polylinker in a particular plasmid vector is

   -GAATTCCCGGGGGATCCTCTAGAGTCGACCTGCAGGCATGC-

This polylinker contains restriction sites for *Bam*HI, *Eco*RI, *Pst*I, *Sal*I, *Sma*I, *Sph*I, and *Xba*I. Indicate the location of each restriction site in this sequence. (See Table 10.2 for the cleavage sites of restriction enzymes.)

3. **Design a Vector Polylinker for Directional Cloning** A vector has a polylinker containing restriction sites in the following order: *Hind*III, *Sac*I, *Xho*I, *Bgl*II, *Xba*I, and *Cla*I.

   a. Give a possible nucleotide sequence for the polylinker.

   b. The vector is digested with *Hind*III and *Cla*I. A DNA segment contains a *Hind*III restriction site fragment 650 bases upstream from a *Cla*I site. This DNA fragment is digested with *Hind*III and *Cla*I, and the resulting *Hind*III–*Cla*I fragment is directionally cloned into the *Hind*III–*Cla*I-digested vector. Give the nucleotide sequence at each end of the vector and the insert and show that the insert can be cloned into the vector in only one orientation.

4. The Number of Clones Needed to Screen a Yeast Genomic Library at 99% Confidence Yeast *(Saccharomyces cerevisiae)* has a genome size of 1.21×10^7 bp. If a genomic library of yeast DNA was constructed in a vector capable of carrying 16-kbp inserts, how many individual clones would have to be screened to have a 99% probability of finding a particular fragment?

5. The Number of Clones Needed to Screen a Very Large Genomic Library at 99% Confidence The South American lungfish has a genome size of 1.02×10^{11} bp. If a genomic library of lungfish DNA was constructed in a vector capable of carrying inserts averaging 45 kbp in size, how many individual clones would have to be screened to have a 99% probability of finding a particular DNA fragment?

6. Designing Primers for PCR Amplification of a DNA Sequence Given the following short DNA duplex of sequence $(5'\rightarrow 3')$

ATGCCGTAGTCGATCATTACGATAGCATAGCACAGGGATCCA-CATGCACACACATGACATAGGACAGATAGCAT

what oligonucleotide primers (17-mers) would be required for PCR amplification of this duplex?

7. A Polylinker for Expression of a β–Galactosidase Fusion Protein Figure 12.3 shows a polylinker that falls within the β-galactosidase coding region of the *lac*Z gene. This polylinker serves as a cloning site in a fusion protein expression vector where the closed insert is expressed as a β-galactosidase fusion protein. Assume the vector polylinker was cleaved with *Bam*HI and then ligated with an insert whose sequence reads

GATCCATTTATCCACCGGAGAGCTGGTATCCCCAAAAGACG-GCC . . .

What is the amino acid sequence of the fusion protein? Where is the junction between β-galactosidase and the sequence encoded by the insert? (Consult the genetic code table on the inside front cover to decipher the amino acid sequence.)

8. Using PCR for the In Vitro Mutagenesis of a Protein Coding Sequence The amino acid sequence across a region of interest in a protein is

Asn-Ser-Gly-Met-His-Pro-Gly-Lys-Leu-Ala-Ser-Trp-Phe-Val-Gly-Asn-Ser

The nucleotide sequence encoding this region begins and ends with an *Eco*RI site, making it easy to clone out the sequence and amplify it by the polymerase chain reaction (PCR). Give the nucleotide sequence of this region. Suppose you wished to change the middle Ser residue to a Cys to study the effects of this change on the protein's activity. What would be the sequence of the mutant oligonucleotide you would use for PCR amplification?

9. Combinatorial Libraries: Calculating the Number of Sequence Possibilities for Oligonucleotides and Peptides Combinatorial chemistry can be used to synthesize polymers such as oligopeptides or oligonucleotides. The number of sequence possibilities for a polymer is given by x^y, where x is the number of different monomer types (for example, 20 different amino acids in a protein or 4 different nucleotides in a nucleic acid) and y is the number of monomers in the oligomers.

  a. Calculate the number of sequence possibilities for RNA oligomers 15 nucleotides long.

  b. Calculate the number of amino acid sequence possibilities for pentapeptides.

10. Using the Yeast Two-Hybrid System to Discover Protein–Protein Intereactions Imagine that you are interested in a protein that interacts with proteins of the cytoskeleton in human epithelial cells.

Describe an experimental protocol based on the yeast two-hybrid system that would allow you to identify proteins that might interact with your protein of interest.

11. Preparing cDNA Libraries from Different Cells Describe an experimental protocol for the preparation of two cDNA libraries, one from anaerobically grown yeast cells and the second from aerobically grown yeast cells.

12. Using Gene Chips to Study Differential Gene Expression Describe an experimental protocol based on DNA microarrays (gene chips) that would allow you to compare gene expression in anaerobically grown yeast versus aerobically grown yeast.

13. Using Antibodies to Screen cDNA Libraries You have an antibody against yeast hexokinase A (hexokinase is the first enzyme in the glycolytic pathway). Describe an experimental protocol using the cDNA libraries prepared in problem 11 that would allow you to identify and isolate the cDNA for hexokinase. Consulting Chapter 5 for protein analysis protocols, describe an experimental protocol to verify that the protein you have identified is hexokinase A.

14. Design of a Reporter Gene Construct to Study Promoter Function In your experiment in problem 12, you discover a gene that is strongly expressed in anaerobically grown yeast but turned off in aerobically grown yeast. You name this gene *nox* (for "no oxygen"). You have the "bright idea" that you can engineer a yeast strain that senses O_2 levels if you can isolate the *nox* promoter. Describe how you might make a reporter gene construct using the *nox* promoter and how the yeast strain bearing this reporter gene construct might be an effective oxygen sensor.

Biochemistry on the Web

15. Search the National Center for Biotechnology Information (NCBI) website at *http://www.ncbi.nlm.nih.gov/sites/entrez?db=Genome* to discover the number of organisms whose genome sequences have been completed. Explore the rich depository of sequence information available here by selecting one organism from the list and browsing through the contents available.

Preparing for the MCAT® Exam

16. Figure 12.1 shows restriction endonuclease sites for the plasmid pBR322. You want to clone a DNA fragment and select for it in transformed bacteria by using resistance to tetracycline and sensitivity to ampicillin as a way of identifying the recombinant plasmid. What restriction endonucleases might be useful for this purpose?

17. Suppose in the amino acid sequence in Figure 12.8, tryptophan was replaced by cysteine. How would that affect the possible mRNA sequence? (Consult the inside front cover of this textbook for amino acid codons.) How many nucleotide changes are necessary in replacing Trp with Cys in this coding sequence? What is the total number of possible oligonucleotide sequences for the mRNA if Cys replaces Trp?

ActiveModels Problems

18. Using the ActiveModel for Dicer, explain the conformational change that occurs in the PAZ domains after the binding of double-stranded RNA.

19. Examine the ActiveModel for green fluorescent protein (GFP), identify the residues that form the chromophore, and draw the structure of the chromophore itself.

FURTHER READING

Cloning Manuals and Procedures

Ausubel, F. M., Brent, R., Kingston, R. E., Moore, D. D., Seidman, J. G., Smith, J. A., and Struhl, K., eds., 2003. *Current Protocols in Molecular Biology,* New York: John Wiley and Sons. Constantly updated online at http://onlinelibrary.wiley.com/10.1002/0471142727.

Brown, T. A., 2006. *Gene Cloning and DNA Analysis,* 5th ed. Malden, MA: Blackwell Publishing.

Cohen, S. N., Chang, A. C. Y., Boyer, H. W., and Helling, R. B., 1973. Construction of biologically functional bacterial plasmids in vitro. *Proceedings of the National Academy of Sciences U.S.A.* **70:**3240–3244. The classic paper on the construction of chimeric plasmids.

Peterson, K. R., et al., 1997. Production of transgenic mice with yeast artificial chromosomes. *Trends in Genetics* **13:**61–66.

Sambrook, J., and Russell, D., 2006. The Condensed Protocols from *Molecular Cloning, A Laboratory Manual,* Long Island, NY: Cold Spring Harbor Laboratory Press.

Expression and Screening of DNA Libraries

Glorioso, J. C., and Schmidt, M. C., eds., 1999. Expression of recombinant genes in eukaryotic cells. *Methods in Enzymology* **306:**1–403.

Hillier, L., et al., 1996. Generation and analysis of 280,000 human expressed sequence tags. *Genome Research* **6:**807–828.

Southern, E. M., 1975. Detection of specific sequences among DNA fragments separated by gel electrophoresis. *Journal of Molecular Biology* **98:**503–517. The classic paper on the identification of specific DNA sequences through hybridization with unique probes.

Thorner, J., and Emr, S., eds., 2000. Applications of chimeric genes and hybrid proteins. *Methods in Enzymology* **328:**1–690.

Weissman, S., ed., 1999. cDNA preparation and display. *Methods in Enzymology* **303:**1–575.

Young, R. A., and Davis, R. W., 1983. Efficient isolation of genes using antibody probes. *Proceedings of the National Academy of Sciences U.S.A.* **80:**1194–1198. Using antibodies to protein expression libraries to isolate the structural gene for a specific protein.

Combinatorial Libraries and Microarrays

Bowtell, D., MacCallum, P., and Sambrook, J., 2003. *DNA Microarrays: A Molecular Cloning Manual,* 2nd ed. Long Island, NY: Cold Spring Harbor Laboratory Press. Techniques used in preparing microarrays and using them in genomic analysis and bioinformatics.

Duggan, D. J., et al., 1999. Expression profiling using cDNA microarrays. *Nature Genetics* **21:**10–14. This is one of a number of articles published in a special supplement of *Nature Genetics* **21** devoted to the use of DNA microarrays to study global gene expression.

Geysen, H. M., et al., 2003. Combinatorial compound libraries for drug discovery: An ongoing challenge. *Nature Reviews Drug Discovery* **2:**222–230.

MacBeath, G., and Schreiber, S. L., 2000. Printing proteins as microarrays for high-throughput function determination. *Science* **289:**1760–1763. This paper describes robotic construction of protein arrays (functionally active proteins immobilized on a solid support) to study protein function.

Southern, E. M., 1996. DNA chips: Analysing sequence by hybridization to oligonucleotides on a large scale. *Trends in Genetics* **12:**110–115.

Stoughton, R. B., 2005. Applications of DNA microarrays in biology. *Annual Review of Biochemistry* **74:**53–83.

Genomes

Collins, F., and the International Human Genome Consortium, 2001. Initial sequencing and analysis of the human genome. *Nature* **409:**860–921.

Ewing, B., and Green, P., 2002. Analysis of expressed sequence tags indicates 35,000 human genes. *Nature Genetics* **25:**232–234.

Lander, E., Page, D., and Lifton, R., eds. 2000–present. *Annual Review of Genomics and Human Genetics,* Vols. 1–3. Palo Alto, CA: Annual Reviews, Inc. A review series on genomics and human diseases.

Venter, J. C., et al., 2001. The sequence of the human genome. *Science* **291:**1304–1351.

The Two-Hybrid System

Chien, C.-T., et al., 1991. The two-hybrid system: A method to identify and clone genes for proteins that interact with a protein of interest. *Proceedings of the National Academy of Sciences U.S.A.* **88:**9578–9582.

Golemis, E., and Adams, P., eds., 2005. *Protein–Protein Interactions: A Molecular Cloning Manual,* 2nd ed. Long Island, NY: Cold Spring Harbor Laboratory Press.

Uetz, P., et al., 2000. A comprehensive analysis of protein–protein interactions in *Saccharomyces cerevisiae. Nature* **403:**623–627.

Reporter Gene Constructs

Chalfie, M., et al., 1994. Green fluorescent protein as a marker for gene expression. *Science* **263:**802–805.

Polymerase Chain Reaction (PCR)

Saiki, R. K., Gelfand, D. H., Stoeffel, B., et al., 1988. Primer-directed amplification of DNA with a thermostable DNA polymerase. *Science* **239:**487–491. Discussion of the polymerase chain reaction procedure.

Timmer, W. C., and Villalobos, J. M., 1993. The polymerase chain reaction. *Journal of Chemical Education* **70:**273–280.

RNAi

Filipowicz, W., 2005. RNAi: The nuts and bolts of the RISC machine. *Cell* **122:**17–20.

Filipowicz, W., et al., 2005. Post-transcriptional gene silencing by siRNAs and miRNAs. *Current Opinion in Structural Biology* **15:**331–341.

Mohr, S., Bakal, C., and Perrimon, N., 2010. Genomic screening with RNAi: results and challenges. *Annual Review of Biochemistry* **79:**37–64.

Gene Therapy

Cartier, N., Hacein-Bey-Abina, S., Bartholomae, C. C., et al., 2009. Hemapoietic stem cell gene therapy with a lentiviral vector in X-linked adrenoleukodystrophy. *Science* **236:**818–323.

Cavazzana-Calvo, M., et al., 2000. Gene therapy of human severe combined immunodeficiency (SCID)-X1 disease. *Science* **288:**669–672.

Cavazzana-Calvo, M., Payen, E., Negre, O., et al., 2010. Transfusion independence and *HmGA2* activation after gene therapy of human β–thalassaemia. *Nature* **467:**318–322.

Crystal, R. G., 1995. Transfer of genes to humans: Early lessons and obstacles to success. *Science* **270:**404–410.

Gaspar, H. B., Cooray, S., Gilmour, K. C., Parsley, K. L., et al., 2011. Long-term persistence of a polyclonal T cell repertoire after gene therapy for X-linked severe combined immunodeficiency. *Science Translational Medicine* **3:**97ra79.

Gaspar, H. B., Cooray, S., Gilmour, K. C., Parsley, K. L., et al., 2011. Hematopoietic stem cell gene therapy for adenosine deaminase-deficient severe combined immunodeficiency leads to a long-term immunological recovery and metabolic correction. *Science Translational Medicine* **3:**97ra80.

Lyon, J., and Gorner, P., 1995. *Altered Fates. Gene Therapy and the Retooling of Human Life.* New York: Norton.

Verma, I. M., and Weitzman, M. D., 2005. Gene therapy: Twenty-first century medicine. *Annual Review of Biochemistry* **74:**711–738.

Enzymes—Kinetics and Specificity

13

© Mark M. Lawrence/CORBIS

◄ The space shuttle must accelerate from zero velocity to a velocity of more than 18,000 miles per hour in order to enter earth orbit. Its three main engines are powered by energy released in the reaction: $2H_2 + O_2 \rightarrow 2 H_2O$. The rate of O_2 consumption per engine is 13,000 moles per second.

There is more to life than increasing its speed.

Mahatma Gandhi (1869–1948)

KEY QUESTIONS

13.1 What Characteristic Features Define Enzymes?

13.2 Can the Rate of an Enzyme-Catalyzed Reaction Be Defined in a Mathematical Way?

13.3 What Equations Define the Kinetics of Enzyme-Catalyzed Reactions?

13.4 What Can Be Learned from the Inhibition of Enzyme Activity?

13.5 What Is the Kinetic Behavior of Enzymes Catalyzing Bimolecular Reactions?

13.6 How Can Enzymes Be So Specific?

13.7 Are All Enzymes Proteins?

13.8 Is It Possible to Design an Enzyme to Catalyze Any Desired Reaction?

ESSENTIAL QUESTIONS

At any moment, thousands of chemical reactions are taking place in any living cell. Enzymes are essential for these reactions to proceed at rates fast enough to sustain life.

What are enzymes, and what do they do?

Living organisms seethe with metabolic activity. Thousands of chemical reactions are proceeding very rapidly at any given instant within all living cells. Virtually all of these transformations are mediated by **enzymes**—proteins (and occasionally RNA) specialized to catalyze metabolic reactions. The substances transformed in these reactions are often organic compounds that show little tendency for reaction outside the cell. An excellent example is glucose, a sugar that can be stored indefinitely on the shelf with no deterioration. Most cells quickly oxidize glucose, producing carbon dioxide and water and releasing lots of energy:

$$C_6H_{12}O_6 + 6\,O_2 \longrightarrow 6\,CO_2 + 6\,H_2O + 2870 \text{ kJ of energy}$$

(-2870 kJ/mol is the standard-state free energy change $[\Delta G°']$ for the oxidation of glucose.) In chemical terms, 2870 kJ is a large amount of energy, and glucose can be viewed as an energy-rich compound even though at ambient temperature it is not readily reactive with oxygen outside of cells. Stated another way, glucose represents **thermodynamic potentiality:** Its reaction with oxygen is strongly exergonic, but it doesn't occur under normal conditions. On the other hand, enzymes can catalyze such thermodynamically favorable reactions, causing them to proceed at extraordinarily rapid rates (Figure 13.1). In glucose oxidation and countless other instances, enzymes provide cells with the ability to exert *kinetic control over thermodynamic potentiality.* That is, living systems use enzymes to accelerate and control the rates of vitally important biochemical reactions.

OWL Online homework and a Student Self Assessment for this chapter may be assigned in OWL.

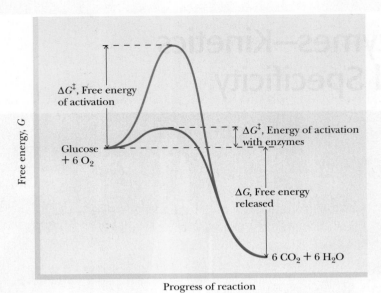

FIGURE 13.1 Reaction profile showing the large $\Delta G^{\ddagger}$ for glucose oxidation. Enzymes lower $\Delta G^{\ddagger}$, thereby accelerating rate.

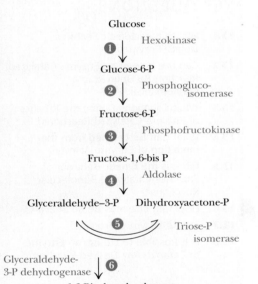

FIGURE 13.2 The breakdown of glucose by *glycolysis* provides a prime example of a metabolic pathway.

Enzymes Are the Agents of Metabolic Function

Acting in sequence, enzymes form metabolic pathways by which nutrient molecules are degraded, energy is released and converted into metabolically useful forms, and precursors are generated and transformed to create the literally thousands of distinctive biomolecules found in any living cell (Figure 13.2). Situated at key junctions of metabolic pathways are specialized **regulatory enzymes** capable of sensing the momentary metabolic needs of the cell and adjusting their catalytic rates accordingly. The responses of these enzymes ensure the harmonious integration of the diverse and often divergent metabolic activities of cells so that the living state is promoted and preserved.

13.1 : What Characteristic Features Define Enzymes?

Enzymes are remarkably versatile biochemical catalysts that have in common three distinctive features: **catalytic power, specificity,** and **regulation.**

Catalytic Power Is Defined as the Ratio of the Enzyme-Catalyzed Rate of a Reaction to the Uncatalyzed Rate

Enzymes display enormous catalytic power, accelerating reaction rates as much as 10^{21} over uncatalyzed levels, which is far greater than any synthetic catalysts can achieve, and enzymes accomplish these astounding feats in dilute aqueous solutions under mild conditions of temperature and pH. For example, the enzyme jack bean *urease* catalyzes the hydrolysis of urea:

$$\overset{\overset{\text{O}}{\|}}{H_2N-C-NH_2} + 2\,H_2O + H^+ \longrightarrow 2\,NH_4^+ + HCO_3^-$$

At 20°C, the rate constant for the enzyme-catalyzed reaction is 3×10^4/sec; the rate constant for the uncatalyzed hydrolysis of urea is 3×10^{-10}/sec. Thus, 10^{14} is the ratio of the catalyzed rate to the uncatalyzed rate of reaction. Such a ratio is defined as the relative **catalytic power** of an enzyme, so the catalytic power of urease is 10^{14}.

Specificity Is the Term Used to Define the Selectivity of Enzymes for Their Substrates

A given enzyme is very selective, both in the substances with which it interacts and in the reaction that it catalyzes. The substances upon which an enzyme acts are traditionally called **substrates.** In an enzyme-catalyzed reaction, none of the substrate is diverted into nonproductive side reactions, so no wasteful by-products are produced. It follows then that the products formed by a given enzyme are also very specific. This situation can be contrasted with your own experiences in the organic chemistry laboratory, where yields of 50% or even 30% are viewed as substantial accomplishments (Figure 13.3). The selective qualities of an enzyme are collectively recognized as its **specificity.** Intimate interaction between an enzyme and its substrates occurs through molecular recognition based on structural complementarity; such mutual recognition is the basis of specificity. The specific site on the enzyme where substrate binds and catalysis occurs is called the **active site.**

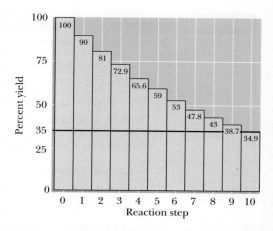

FIGURE 13.3 A 90% yield over 10 steps, for example, in a metabolic pathway, gives an overall yield of 35%. Therefore, yields in biological reactions *must be substantially greater;* otherwise, unwanted by-products would accumulate to unacceptable levels.

Regulation of Enzyme Activity Ensures That the Rate of Metabolic Reactions Is Appropriate to Cellular Requirements

Regulation of enzyme activity is essential to the integration and regulation of metabolism. Enzyme regulation is achieved in a variety of ways, ranging from controls over the amount of enzyme protein produced by the cell to more rapid, reversible interactions of the enzyme with metabolic inhibitors and activators. Chapter 15 is devoted to discussions of this topic. Because most enzymes are proteins, we can anticipate that the functional attributes of enzymes are due to the remarkable versatility found in protein structures.

Enzyme Nomenclature Provides a Systematic Way of Naming Metabolic Reactions

Traditionally, enzymes were named by adding the suffix *-ase* to the name of the substrate upon which they acted, as in *urease* for the urea-hydrolyzing enzyme or *phosphatase* for enzymes hydrolyzing phosphoryl groups from organic phosphate compounds. Other enzymes acquired names bearing little resemblance to their activity, such as the peroxide-decomposing enzyme *catalase* or the proteolytic enzymes *(proteases)* of the digestive tract, *trypsin* and *pepsin.* Because of the confusion that arose from these trivial designations, an International Commission on Enzymes was established to create a systematic basis for enzyme nomenclature. Although common names for many enzymes remain in use, all enzymes now are classified and formally named according to the reaction they catalyze. Six classes of reactions are recognized (Table 13.1). Within each class are subclasses, and under each subclass are sub-subclasses within which individual enzymes are listed. Classes, subclasses, sub-subclasses, and individual entries are each numbered so that a series of four numbers serves to specify a particular enzyme. A systematic name, descriptive of the reaction, is also assigned to each entry. To illustrate, consider the enzyme that catalyzes this reaction:

$$ATP + \text{D-glucose} \longrightarrow ADP + \text{D-glucose-6-phosphate}$$

A phosphate group is transferred from ATP to the C-6-OH group of glucose, so the enzyme is a *transferase* (class 2, Table 13.1). Subclass 7 of transferases is *enzymes transferring phosphorus-containing groups,* and sub-subclass 1 covers those *phosphotransferases with an alcohol group as an acceptor.* Entry 2 in this sub-subclass is **ATP:D-glucose-6-phosphotransferase,** and its classification number is **2.7.1.2.** In use, this number is written preceded by the letters **E.C.,** denoting the Enzyme Commission. For example, entry 1 in the same sub-subclass is E.C.2.7.1.1, ATP:D-hexose-6-phosphotransferase, an ATP-dependent enzyme that transfers a phosphate to the 6-OH of hexoses (that is, it is nonspecific regarding its hexose acceptor). These designations can be cumbersome, so in

TABLE 13.1	Systematic Classification of Enzymes According to the Enzyme Commission			
E.C. Number	**Systematic Name and Subclasses**		**E.C. Number**	**Systematic Name and Subclasses**
1	*Oxidoreductases* (oxidation–reduction reactions)		4	*Lyases* (bond cleavage by means other than hydrolysis or oxidation)
1.1	Acting on CH—OH group of donors		4.1	C—C lyases
1.1.1	With NAD or NADP as acceptor		4.1.1	Carboxy lyases
1.1.3	With O_2 as acceptor		4.1.2	Aldehyde lyases
1.2	Acting on the $>$C$=$O group of donors		4.2	C—O lyases
			4.2.1	Hydrolases
1.2.3	With O_2 as acceptor		4.3	C—N lyases
1.3	Acting on the CH—CH group of donors		4.3.1	Ammonia lyases
1.3.1	With NAD or NADP as acceptor		5	*Isomerases* (isomerization reactions)
2	*Transferases* (transfer of functional groups)		5.1	Racemases and epimerases
2.1	Transferring C-1 groups		5.1.3	Acting on carbohydrates
2.1.1	Methyltransferases		5.2	*Cis-trans* isomerases
2.1.2	Hydroxymethyltransferases and formyltransferases		6	*Ligases* (formation of bonds with ATP cleavage)
2.1.3	Carboxyltransferases and carbamoyltransferases		6.1	Forming C—O bonds
			6.1.1	Amino acid–RNA ligases
2.2	Transferring aldehydic or ketonic residues		6.2	Forming C—S bonds
2.3	Acyltransferases		6.3	Forming C—N bonds
2.4	Glycosyltransferases		6.4	Forming C—C bonds
2.6	Transferring N-containing groups		6.4.1	Carboxylases
2.6.1	Aminotransferases			
2.7	Transferring P-containing groups			
2.7.1	With an alcohol group as acceptor			
3	*Hydrolases* (hydrolysis reactions)			
3.1	Cleaving ester linkage			
3.1.1	Carboxylic ester hydrolases			
3.1.3	Phosphoric monoester hydrolases			
3.1.4	Phosphoric diester hydrolases			

everyday usage, trivial names are commonly used. The glucose-specific enzyme E.C.2.7.1.2 is called *glucokinase,* and the nonspecific E.C.2.7.1.1 is known as *hexokinase. Kinase* is a trivial term for enzymes that are ATP-dependent phosphotransferases.

Coenzymes and Cofactors Are Nonprotein Components Essential to Enzyme Activity

Many enzymes carry out their catalytic function relying solely on their protein structure. Many others require nonprotein components, called **cofactors** (Table 13.2). Cofactors may be metal ions or organic molecules referred to as **coenzymes.** Coenzymes and cofactors provide proteins with chemically versatile functions not found in amino acid side chains. Many coenzymes are vitamins or contain vitamins as part of their structure. Usually coenzymes are actively involved in the catalytic reaction of the enzyme, often serving as intermediate carriers of functional groups in the conversion of substrates to products. In most cases, a coenzyme is firmly associated with its enzyme, perhaps even by covalent bonds, and it is difficult to separate the two. Such tightly bound coenzymes are referred to as **prosthetic groups** of the enzyme. The catalytically active complex of protein and prosthetic group is called the **holoenzyme.** The protein without the prosthetic group is called the **apoenzyme;** it is catalytically inactive.

TABLE 13.2	Enzyme Cofactors: Some Metal Ions and Coenzymes and the Enzymes with Which They Are Associated			
Metal Ions and Some Enzymes That Require Them		Coenzymes Serving as Transient Carriers of Specific Atoms or Functional Groups		
Metal Ion	Enzyme	Coenzyme	Entity Transferred	Representative Enzymes Using Coenzymes
Fe^{2+} or Fe^{3+}	Cytochrome oxidase	Thiamine pyrophosphate (TPP)	Aldehydes	Pyruvate dehydrogenase
	Catalase	Flavin adenine dinucleotide (FAD)	Hydrogen atoms	Succinate dehydrogenase
	Peroxidase	Nicotinamide adenine dinucleotide (NAD)	Hydride ion ($:H^-$)	Alcohol dehydrogenase
Cu^{2+}	Cytochrome oxidase			
Zn^{2+}	DNA polymerase	Coenzyme A (CoA)	Acyl groups	Acetyl-CoA carboxylase
	Carbonic anhydrase	Pyridoxal phosphate (PLP)	Amino groups	Aspartate aminotransferase
	Alcohol dehydrogenase			
Mg^{2+}	Hexokinase	5′-Deoxyadenosylcobalamin (vitamin B_{12})	H atoms and alkyl groups	Methylmalonyl-CoA mutase
	Glucose-6-phosphatase			
Mn^{2+}	Arginase	Biotin (biocytin)	CO_2	Propionyl-CoA carboxylase
K^+	Pyruvate kinase (also requires Mg^{2+})	Tetrahydrofolate (THF)	Other one-carbon groups, such as formyl and methyl groups	Thymidylate synthase
Ni^{2+}	Urease			
Mo	Nitrate reductase			
Se	Glutathione peroxidase			

13.2 Can the Rate of an Enzyme-Catalyzed Reaction Be Defined in a Mathematical Way?

Kinetics is the branch of science concerned with the rates of reactions. The study of **enzyme kinetics** addresses the biological roles of enzymatic catalysts and how they accomplish their remarkable feats. In enzyme kinetics, we seek to determine the maximum reaction velocity that the enzyme can attain and its binding affinities for substrates and inhibitors. Coupled with studies on the structure and chemistry of the enzyme, analysis of the enzymatic rate under different reaction conditions yields insights regarding the enzyme's mechanism of catalytic action. Such information is essential to an overall understanding of metabolism.

Significantly, this information can be exploited to control and manipulate the course of metabolic events. The science of pharmacology relies on such a strategy. **Pharmaceuticals,** or **drugs,** are often special inhibitors specifically targeted at a particular enzyme in order to overcome infection or to alleviate illness. A detailed knowledge of the enzyme's kinetics is indispensable to rational drug design and successful pharmacological intervention.

Chemical Kinetics Provides a Foundation for Exploring Enzyme Kinetics

Before beginning a quantitative treatment of enzyme kinetics, it will be fruitful to review briefly some basic principles of chemical kinetics. **Chemical kinetics** is the study of the rates of chemical reactions. Consider a reaction of overall stoichiometry:

$$A \longrightarrow P$$

Although we treat this reaction as a simple, one-step conversion of A to P, it more likely occurs through a sequence of elementary reactions, each of which is a simple molecular process, as in

$$A \longrightarrow I \longrightarrow J \longrightarrow P$$

where I and J represent intermediates in the reaction. Precise description of all of the elementary reactions in a process is necessary to define the overall reaction mechanism for A⟶P.

Let us assume that A⟶P *is* an elementary reaction and that it is spontaneous and essentially irreversible. Irreversibility is easily assumed if the rate of P conversion to A is very slow *or* the concentration of P (expressed as [P]) is negligible under the conditions chosen. The **velocity,** *v*, or **rate,** of the reaction A⟶P is the amount of P formed or the amount of A consumed per unit time, *t*. That is,

$$v = \frac{d[P]}{dt} \quad \text{or} \quad v = \frac{-d[A]}{dt} \tag{13.1}$$

The mathematical relationship between reaction rate and concentration of reactant(s) is the **rate law.** For this simple case, the rate law is

$$v = \frac{-d[A]}{dt} = k[A] \tag{13.2}$$

From this expression, it is obvious that the rate is proportional to the concentration of A, and *k* is the proportionality constant, or **rate constant.** *k* has the units of $(\text{time})^{-1}$, usually $\sec^{-1}$. *v* is a function of [A] to the first power, or in the terminology of kinetics, *v* is first-order with respect to A. For an elementary reaction, the **order** for any reactant is given by its exponent in the rate equation. The number of molecules that must simultaneously interact is defined as the **molecularity** of the reaction. Thus, the simple elementary reaction of A⟶P is a **first-order reaction.** Figure 13.4 portrays the course of a first-order reaction as a function of time. The rate of decay of a radioactive isotope, like ^{14}C or ^{32}P, is a first-order reaction, as is an intramolecular rearrangement, such as A⟶P. Both are **unimolecular reactions** (the molecularity equals 1).

Bimolecular Reactions Are Reactions Involving Two Reactant Molecules

Consider the more complex reaction, where two molecules must react to yield products:

$$A + B \longrightarrow P + Q$$

Assuming this reaction is an elementary reaction, its molecularity is 2; that is, it is a **bimolecular reaction.** The velocity of this reaction can be determined from the rate of disappearance of either A or B, or the rate of appearance of P or Q:

$$v = \frac{-d[A]}{dt} = \frac{-d[B]}{dt} = \frac{d[P]}{dt} = \frac{d[Q]}{dt} \tag{13.3}$$

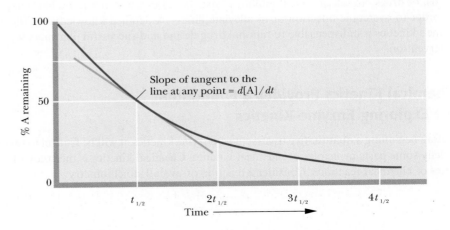

FIGURE 13.4 Plot of the course of a first-order reaction. The half-time, $t_{1/2}$, is the time for one-half of the starting amount of A to disappear.

The rate law is

$$v = k[A][B] \tag{13.4}$$

Since A and B must collide in order to react, the rate of their reaction will be proportional to the concentrations of both A and B. Because it is proportional to the product of two concentration terms, the reaction is **second-order** overall, first-order with respect to A and first-order with respect to B. (Were the elementary reaction $2A \longrightarrow P + Q$, the rate law would be $v = k[A]^2$, second-order overall and second-order with respect to A.) Second-order rate constants have the units of $(concentration)^{-1}(time)^{-1}$, as in $M^{-1} sec^{-1}$.

Molecularities greater than 2 are rarely found (and greater than 3, never). (The likelihood of simultaneous collision of three molecules is very, very small.) When the overall stoichiometry of a reaction is greater than two (for example, as in $A + B + C \longrightarrow$ or $2A + B \longrightarrow$), the reaction almost always proceeds via unimolecular or bimolecular elementary steps, and the overall rate obeys a simple first- or second-order rate law.

At this point, it may be useful to remind ourselves of an important caveat that is the first principle of kinetics: *Kinetics cannot prove a hypothetical mechanism.* Kinetic experiments can only rule out various alternative hypotheses because they do not fit the data. However, through thoughtful kinetic studies, a process of elimination of alternative hypotheses leads ever closer to the reality.

Catalysts Lower the Free Energy of Activation for a Reaction

In a first-order chemical reaction, the conversion of A to P occurs because, at any given instant, a fraction of the A molecules has the energy necessary to achieve a reactive condition known as the **transition state.** In this state, the probability is very high that the particular rearrangement accompanying the $A \longrightarrow P$ transition will occur. This transition state sits at the apex of the energy profile in the energy diagram describing the energetic relationship between A and P (Figure 13.5). The average free energy of A molecules defines the initial state, and the average free energy of P molecules is the final state along the reaction coordinate. The rate of any chemical reaction is proportional to the concentration of reactant molecules (A in this case) having this transition-state energy. Obviously, the higher this energy is above the average energy, the smaller the fraction of molecules that will have this energy and the slower the reaction will proceed. The height of this energy barrier is called the **free energy of activation, $\Delta G^{\ddagger}$.** Specifically, $\Delta G^{\ddagger}$ is the energy required to raise the average energy of 1 mol of reactant (at a given temperature)

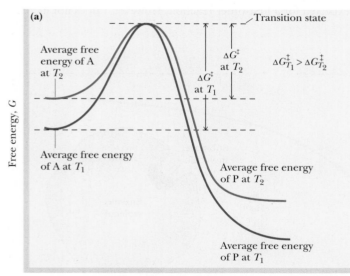
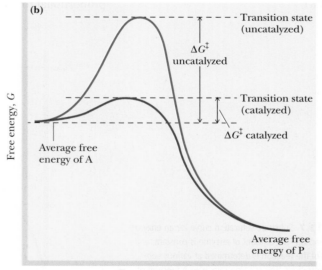

FIGURE 13.5 Energy diagram for a chemical reaction ($A \longrightarrow P$) and the effects of **(a)** raising the temperature from T_1 to T_2 or **(b)** adding a catalyst.

to the transition-state energy. The relationship between activation energy and the rate constant of the reaction, k, is given by the **Arrhenius equation:**

$$k = Ae^{-\Delta G^{\ddagger}/RT} \tag{13.5}$$

where A is a constant for a particular reaction (not to be confused with the reactant species, A, that we're discussing). Another way of writing this is $1/k = (1/A)e^{\Delta G^{\ddagger}/RT}$. That is, k is inversely proportional to $e^{\Delta G^{\ddagger}/RT}$. Therefore, if the energy of activation decreases, the reaction rate increases.

Decreasing $\Delta G^{\ddagger}$ Increases Reaction Rate

We are familiar with two general ways that rates of chemical reactions may be accelerated. First, the temperature can be raised. This will increase the kinetic energy of reactant molecules, and more reactant molecules will possess the energy to reach the transition state (Figure 13.5a). In effect, increasing the average energy of reactant molecules makes the energy difference between the average energy and the transition-state energy smaller. (Also note that the equation $k = Ae^{-\Delta G^{\ddagger}/RT}$ demonstrates that k increases as T increases.) The rates of many chemical reactions are doubled by a 10°C rise in temperature. Second, the rates of chemical reactions can also be accelerated by catalysts. Catalysts work by lowering the energy of activation rather than by raising the average energy of the reactants (Figure 13.5b). Catalysts accomplish this remarkable feat by combining transiently with the reactants in a way that promotes their entry into the reactive, transition-state condition. Two aspects of catalysts are worth noting: (1) They are regenerated after each reaction cycle (A⟶P), and therefore can be used over and over again; and (2) catalysts have *no* effect on the overall free energy change in the reaction, the free energy difference between A and P (Figure 13.5b).

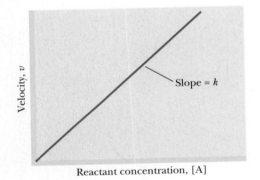

FIGURE 13.6 A plot of *v* versus [A] for the unimolecular chemical reaction, A⟶P, yields a straight line having a slope equal to *k*.

13.3 | What Equations Define the Kinetics of Enzyme-Catalyzed Reactions?

Examination of the change in reaction velocity as the reactant concentration is varied is one of the primary measurements in kinetic analysis. Returning to A⟶P, a plot of the reaction rate as a function of the concentration of A yields a straight line whose slope is k (Figure 13.6). The more A that is available, the greater the rate of the reaction, v. Similar analyses of enzyme-catalyzed reactions involving only a single substrate yield remarkably different results (Figure 13.7). At low concentrations of the substrate S, v is proportional to [S], as expected for a first-order reaction. However, v does not increase

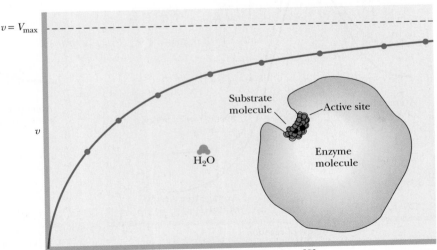

FIGURE 13.7 Substrate saturation curve for an enzyme-catalyzed reaction. The amount of enzyme is constant, and the velocity of the reaction is determined at various substrate concentrations. The reaction rate, *v*, as a function of [S] is described mathematically by a rectangular hyperbola. The H_2O molecule provides a rough guide to scale.

proportionally as [S] increases, but instead begins to level off. At high [S], v becomes virtually independent of [S] and approaches a maximal limit. The value of v at this limit is written V_{max}. Because rate is no longer dependent on [S] at these high concentrations, the enzyme-catalyzed reaction is now obeying **zero-order kinetics;** that is, the rate is independent of the reactant (substrate) concentration. This behavior is a **saturation effect:** When v shows no increase even though [S] is increased, the system is saturated with substrate. Such plots are called **substrate saturation curves.** The physical interpretation is that every enzyme molecule in the reaction mixture has its substrate-binding site occupied by S. Indeed, such curves were the initial clue that an enzyme interacts directly with its substrate by binding it.

The Substrate Binds at the Active Site of an Enzyme

An enzyme molecule is often (but not always) orders of magnitude larger than its substrate. In any case, its **active site,** that place on the enzyme where S binds, comprises only a portion of the overall enzyme structure. The conformation of the active site is structured to form a special pocket or cleft whose three-dimensional architecture is complementary to the structure of the substrate. The enzyme and the substrate molecules "recognize" each other through this structural complementarity. The substrate binds to the enzyme through relatively weak forces—H bonds, ionic bonds (salt bridges), and van der Waals interactions between sterically complementary clusters of atoms.

The Michaelis–Menten Equation Is the Fundamental Equation of Enzyme Kinetics

Leonor Michaelis and Maud L. Menten proposed a general theory of enzyme action in 1913 consistent with observed enzyme kinetics. Their theory was based on the assumption that the enzyme, E, and its substrate, S, associate reversibly to form an enzyme–substrate complex, ES:

$$E + S \underset{k_{-1}}{\overset{k_1}{\rightleftharpoons}} ES \tag{13.6}$$

This association/dissociation is assumed to be a rapid equilibrium, and K_S is the *enzyme : substrate dissociation constant.* At equilibrium,

$$k_{-1}[ES] = k_1[E][S] \tag{13.7}$$

and

$$K_s = \frac{[E][S]}{[ES]} = \frac{k_{-1}}{k_1} \tag{13.8}$$

Product, P, is formed in a second step when ES breaks down to yield E + P.

$$E + S \underset{k_{-1}}{\overset{k_1}{\rightleftharpoons}} ES \xrightarrow{k_2} E + P \tag{13.9}$$

E is then free to interact with another molecule of S.

Assume That [ES] Remains Constant During an Enzymatic Reaction

The interpretations of Michaelis and Menten were refined and extended in 1925 by Briggs and Haldane, who assumed the concentration of the enzyme–substrate complex ES quickly reaches a constant value in such a dynamic system. That is, ES is formed as rapidly from E + S as it disappears by its two possible fates: dissociation to regenerate

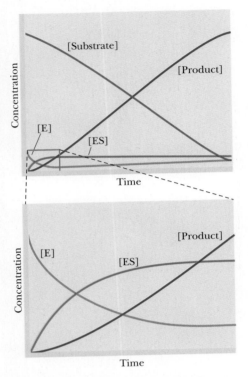

FIGURE 13.8 Time course for a typical enzyme-catalyzed reaction obeying the Michaelis–Menten, Briggs–Haldane models for enzyme kinetics. The early stage of the time course is shown in greater magnification in the bottom graph.

E + S and reaction to form E + P. This assumption is termed the **steady-state assumption** and is expressed as

$$\frac{d[ES]}{dt} = 0 \qquad (13.10)$$

That is, the change in concentration of ES with time, t, is 0. Figure 13.8 illustrates the time course for formation of the ES complex and establishment of the steady-state condition.

Assume That Velocity Measurements Are Made Immediately After Adding S

One other simplification will be advantageous. Because enzymes accelerate the rate of the reverse reaction as well as the forward reaction, it would be helpful to ignore any back reaction by which E + P might form ES. The velocity of this back reaction would be given by $v = k_{-2}[E][P]$. However, if we observe only the *initial velocity* for the reaction immediately after E and S are mixed in the absence of P, the rate of any back reaction is negligible because its rate will be proportional to [P], and [P] is essentially 0. Given such simplification, we now analyze the system described by Equation 13.9 in order to describe the initial velocity v as a function of [S] and amount of enzyme.

The total amount of enzyme is fixed and is given by the formula

$$\text{Total enzyme, } [E_T] = [E] + [ES] \qquad (13.11)$$

where [E] is free enzyme and [ES] is the amount of enzyme in the enzyme–substrate complex. From Equation 13.9, the rate of [ES] formation is

$$v_f = k_1([E_T] - [ES])[S]$$

where

$$[E_T] - [ES] = [E] \qquad (13.12)$$

From Equation 13.9, the rate of [ES] disappearance is

$$v_d = k_{-1}[ES] + k_2[ES] = (k_{-1} + k_2)[ES] \qquad (13.13)$$

At steady state, $d[ES]/dt = 0$, and therefore, $v_f = v_d$. So,

$$k_1([E_T] - [ES])[S] = (k_{-1} + k_2)[ES] \qquad (13.14)$$

Rearranging gives

$$\frac{([E_T] - [ES])[S]}{[ES]} = \frac{(k_{-1} + k_2)}{k_1} \qquad (13.15)$$

The Michaelis Constant, K_m, Is Defined as $(k_{-1} + k_2)/k_1$

The ratio of constants $(k_{-1} + k_2)/k_1$ is itself a constant and is defined as the **Michaelis constant**, K_m

$$K_m = \frac{(k_{-1} + k_2)}{k_1} \qquad (13.16)$$

Note from Equation 13.15 that K_m is given by the ratio of two concentrations $(([E_T] - [ES])$ and [S]) to one ([ES]), so K_m has the units of *molarity*. (Also, because the units of k_{-1} and k_2 are in time^{-1} and the units of k_1 are M^{-1}time^{-1}, it becomes obvious that the units of K_m are M.) From Equation 13.15, we can write

$$\frac{([E_T] - [ES])[S]}{[ES]} = K_m \qquad (13.17)$$

which rearranges to

$$[ES] = \frac{[E_T][S]}{K_m + [S]} \qquad (13.18)$$

Now, the most important parameter in the kinetics of any reaction is the **rate of product formation.** This rate is given by

$$v = \frac{d[\text{P}]}{dt} \tag{13.19}$$

and for this reaction

$$v = k_2[\text{ES}] \tag{13.20}$$

Substituting the expression for [ES] from Equation 13.18 into Equation 13.20 gives

$$v = \frac{k_2[\text{E}_T][\text{S}]}{K_m + [\text{S}]} \tag{13.21}$$

The product $k_2[\text{E}_T]$ has special meaning. When [S] is high enough to saturate all of the enzyme, the velocity of the reaction, v, is maximal. At saturation, the amount of [ES] complex is equal to the total enzyme concentration, E_T, its maximum possible value. From Equation 13.20, the initial velocity v then equals $k_2[\text{E}_T] = V_{\text{max}}$. Written symbolically, when $[\text{S}] >> [\text{E}_T]$ (and K_m), $[\text{E}_T] = [\text{ES}]$ and $v = V_{\text{max}}$. Therefore,

$$V_{\text{max}} = k_2[\text{E}_T] \tag{13.22}$$

Substituting this relationship into the expression for v gives the **Michaelis–Menten equation:**

$$v = \frac{V_{\text{max}}[\text{S}]}{K_m + [\text{S}]} \tag{13.23}$$

This equation says that the initial rate of an enzyme-catalyzed reaction, v, is determined by two constants, K_m and V_{max}, and the initial concentration of substrate.

When [S] = K_m, v = V_{max}/2

We can provide an operational definition for the constant K_m by rearranging Equation 13.23 to give

$$K_m = [\text{S}] \left(\frac{V_{\text{max}}}{v} - 1 \right) \tag{13.24}$$

Then, at $v = V_{\text{max}}/2$, $K_m = [\text{S}]$. That is, K_m is defined by the substrate concentration that gives a velocity equal to one-half the maximal velocity. Table 13.3 gives the K_m values of some enzymes for their substrates.

Plots of v Versus [S] Illustrate the Relationships Between V_{max}, K_m, and Reaction Order

The Michaelis–Menten equation (Equation 13.23) describes a curve known from analytical geometry as a *rectangular hyperbola.* In such curves, as [S] is increased, v approaches the limiting value, V_{max}, in an asymptotic fashion. V_{max} can be approximated experimentally from a substrate saturation curve (Figure 13.7), and K_m can be derived from $V_{\text{max}}/2$, so the two constants of the Michaelis–Menten equation can be obtained from plots of v versus [S]. Note, however, that actual estimation of V_{max}, and consequently K_m, is only approximate from such graphs. That is, according to Equation 13.23, to get $v = 0.99\ V_{\text{max}}$, [S] must equal $99\ K_m$, a concentration that may be difficult to achieve in practice.

From Equation 13.23, when $[\text{S}] >> K_m$, then $v = V_{\text{max}}$. That is, v is no longer dependent on [S], so the reaction is obeying zero-order kinetics. Also, when $[\text{S}] < K_m$, then $v \approx (V_{\text{max}}/K_m)[\text{S}]$. That is, the rate, v, approximately follows a first-order rate equation, $v = k'[\text{A}]$, where $k' = V_{\text{max}}/K_m$.

TABLE 13.3	K_m Values for Some Enzymes	
Enzyme	Substrate	K_m (mM)
Carbonic anhydrase	CO_2	12
Chymotrypsin	N-Benzoyltyrosinamide	2.5
	Acetyl-L-tryptophanamide	5
	N-Formyltyrosinamide	12
	N-Acetyltyrosinamide	32
	Glycyltyrosinamide	122
Hexokinase	Glucose	0.15
	Fructose	1.5
β-Galactosidase	Lactose	4
Glutamate dehydrogenase	NH_4^+	57
	Glutamate	0.12
	α-Ketoglutarate	2
	NAD^+	0.025
	NADH	0.018
Aspartate aminotransferase	Aspartate	0.9
	α-Ketoglutarate	0.1
	Oxaloacetate	0.04
	Glutamate	4
Threonine deaminase	Threonine	5
Arginyl-tRNA synthetase	Arginine	0.003
	tRNA^{Arg}	0.0004
	ATP	0.3
Pyruvate carboxylase	HCO_3^-	1.0
	Pyruvate	0.4
	ATP	0.06
Penicillinase	Benzylpenicillin	0.05
Lysozyme	Hexa-N-acetylglucosamine	0.006

K_m and V_{max}, once known explicitly, define the rate of the enzyme-catalyzed reaction, *provided:*

1. The reaction involves only one substrate, *or* if the reaction is multisubstrate, the concentration of only one substrate is varied while the concentrations of all other substrates are held constant.
2. The reaction ES$\longrightarrow$E + P is irreversible, *or* the experiment is limited to observing only initial velocities where [P] = 0.
3. [S]$_0$ > [E$_T$] and [E$_T$] is held constant.
4. All other variables that might influence the rate of the reaction (temperature, pH, ionic strength, and so on) are held constant.

Turnover Number Defines the Activity of One Enzyme Molecule

The **turnover number** of an enzyme, k_{cat}, is a measure of its maximal catalytic activity. k_{cat} is defined as the number of substrate molecules converted into product per enzyme molecule per unit time when the enzyme is saturated with substrate. The turnover number is also referred to as the **molecular activity** of the enzyme. For the simple Michaelis–Menten reaction (Equation 13.9) under conditions of initial velocity measurements,

$k_2 = k_{cat}$. Provided the concentration of enzyme, $[E_T]$, in the reaction mixture is known, k_{cat} can be determined from V_{max}. At saturating $[S]$, $v = V_{max} = k_2[E_T]$. Thus,

$$k_2 = \frac{V_{max}}{[E_T]} = k_{cat} \qquad (13.25)$$

The term k_{cat} represents the kinetic efficiency of the enzyme. Table 13.4 lists turnover numbers for some representative enzymes. Catalase has the highest turnover number known; each molecule of this enzyme can degrade 40 million molecules of H_2O_2 in 1 second! At the other end of the scale, lysozyme requires 2 seconds to cleave a glycosidic bond in its glycan substrate.

In many situations, the actual molar amount of the enzyme is not known. However, its amount can be expressed in terms of the activity observed. The International Commission on Enzymes defines **one international unit** as the amount that catalyzes the formation of 1 micromole of product in 1 minute. (Because enzymes are very sensitive to factors such a pH, temperature, and ionic strength, the conditions of assay must be specified.) In the process of purifying enzymes from cellular sources, many extraneous proteins may be present. Then, the units of enzyme activity are expressed as enzyme units per mg protein, a term known as **specific activity** (see Table 5.1).

TABLE 13.4	Values of k_{cat} (Turnover Number) for Some Enzymes
Enzyme	k_{cat} (sec^{-1})
Catalase	40,000,000
Carbonic anhydrase	1,000,000
Acetylcholinesterase	14,000
Penicillinase	2,000
Lactate dehydrogenase	1,000
Chymotrypsin	100
DNA polymerase I	15
Lysozyme	0.5

The Ratio, k_{cat}/K_m, Defines the Catalytic Efficiency of an Enzyme

Under physiological conditions, $[S]$ is seldom saturating and k_{cat} itself is not particularly informative. That is, the in vivo ratio of $[S]/K_m$ usually falls in the range of 0.01 to 1.0, so active sites often are not filled with substrate. Nevertheless, we can derive a meaningful index of the efficiency of Michaelis–Menten–type enzymes under these conditions by using the following equations. As presented in Equation 13.23, if

$$v = \frac{V_{max}[S]}{K_m + [S]}$$

and $V_{max} = k_{cat}[E_T]$, then

$$v = \frac{k_{cat}[E_T][S]}{K_m + [S]} \qquad (13.26)$$

When $[S] \ll K_m$, the concentration of free enzyme, $[E]$, is approximately equal to $[E_T]$, so

$$v = \left(\frac{k_{cat}}{K_m}\right)[E][S] \qquad (13.27)$$

That is, k_{cat}/K_m is an apparent second-order rate constant for the reaction of E and S to form product. Because K_m is inversely proportional to the affinity of the enzyme for its substrate and k_{cat} is directly proportional to the kinetic efficiency of the enzyme, k_{cat}/K_m provides an index of the catalytic efficiency of an enzyme operating at substrate concentrations substantially below saturation amounts.

An interesting point emerges if we restrict ourselves to the simple case where $k_{cat} = k_2$. Then

$$\frac{k_{cat}}{K_m} = \frac{k_1 k_2}{k_{-1} + k_2} \qquad (13.28)$$

But k_1 must always be greater than or equal to $k_1 k_2/(k_{-1} + k_2)$. That is, the reaction can go no faster than the rate at which E and S come together. Thus, k_1 sets the upper limit for k_{cat}/K_m. In other words, *the catalytic efficiency of an enzyme cannot exceed the diffusion-controlled rate of combination of E and S to form ES.* In H_2O, the rate constant for such diffusion is approximately $10^9/M \cdot sec$ for small substrates (for example, glyceraldehyde 3-P) and an order of magnitude smaller ($\approx 10^8/M \cdot sec$) for substrates the size of nucleotides. Those enzymes that are most efficient in their catalysis have k_{cat}/K_m ratios

TABLE 13.5	Enzymes Whose k_{cat}/K_m Approaches the Diffusion-Controlled Rate of Association with Substrate			
Enzyme	Substrate	k_{cat} (sec^{-1})	K_m (M)	k_{cat}/K_m (M^{-1} sec^{-1})
Acetylcholinesterase	Acetylcholine	1.4×10^4	9×10^{-5}	1.6×10^8
Carbonic anhydrase	CO_2	1×10^6	0.012	8.3×10^7
	HCO_3^-	4×10^5	0.026	1.5×10^7
Catalase	H_2O_2	4×10^7	1.1	4×10^7
Crotonase	Crotonyl-CoA	5.7×10^3	2×10^{-5}	2.8×10^8
Fumarase	Fumarate	800	5×10^{-6}	1.6×10^8
	Malate	900	2.5×10^{-5}	3.6×10^7
Triosephosphate isomerase	Glyceraldehyde-3-phosphate*	4.3×10^3	1.8×10^{-5}	2.4×10^8
β-Lactamase	Benzylpenicillin	2×10^3	2×10^{-5}	1×10^8

*K_m for glyceraldehyde-3-phosphate is calculated on the basis that only 3.8% of the substrate in solution is unhydrated and therefore reactive with the enzyme.
Adapted from Fersht, A., 1985. *Enzyme Structure and Mechanism,* 2nd ed. New York: W. H. Freeman.

approaching this value. Their catalytic velocity is limited only by the rate at which they encounter S; enzymes this efficient have achieved so-called catalytic perfection. All E and S encounters lead to reaction because such "catalytically perfect" enzymes can channel S to the active site, regardless of where S hits E. Table 13.5 lists the kinetic parameters of several enzymes in this category. Note that k_{cat} and K_m both show a substantial range of variation in this table, even though their ratio falls around $10^8/M \cdot$ sec.

Linear Plots Can Be Derived from the Michaelis–Menten Equation

Because of the hyperbolic shape of v versus [S] plots, V_{max} can be determined only from an extrapolation of the asymptotic approach of v to some limiting value as [S] increases indefinitely (Figure 13.7); and K_m is derived from that value of [S] giving $v = V_{max}/2$. However, several rearrangements of the Michaelis–Menten equation transform it into a straight-line equation. The best known of these is the **Lineweaver–Burk double-reciprocal plot:**

Taking the reciprocal of both sides of the Michaelis–Menten equation, Equation 13.23, yields the equality

$$\frac{1}{v} = \left(\frac{K_m}{V_{max}}\right)\left(\frac{1}{[S]}\right) + \frac{1}{V_{max}} \tag{13.29}$$

This conforms to $y = mx + b$ (the equation for a straight line), where $y = 1/v$; m, the slope, is K_m/V_{max}; $x = 1/[S]$; and $b = 1/V_{max}$. Plotting $1/v$ versus $1/[S]$ gives a straight line whose x-intercept is $-1/K_m$, whose y-intercept is $1/V_{max}$, and whose slope is K_m/V_{max} (Figure 13.9).

The **Hanes–Woolf plot** is another rearrangement of the Michaelis–Menten equation that yields a straight line:

Multiplying both sides of Equation 13.29 by [S] gives

$$\frac{[S]}{v} = [S]\left(\frac{K_m}{V_{max}}\right)\left(\frac{1}{[S]}\right) + \frac{[S]}{V_{max}} = \frac{K_m}{V_{max}} + \frac{[S]}{V_{max}} \tag{13.30}$$

and

$$\frac{[S]}{v} = \left(\frac{1}{V_{max}}\right)[S] + \frac{K_m}{V_{max}} \tag{13.31}$$

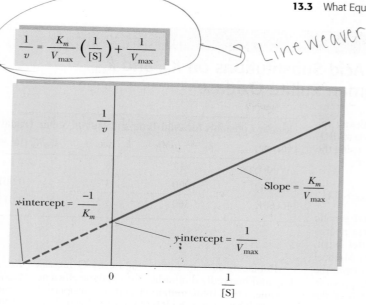

$$\frac{1}{v} = \frac{K_m}{V_{max}} \left(\frac{1}{[S]}\right) + \frac{1}{V_{max}}$$

→ Lineweaver

FIGURE 13.9 The Lineweaver–Burk double-reciprocal plot.

Graphing [S]/v versus [S] yields a straight line where the slope is $1/V_{max}$, the y-intercept is K_m/V_{max}, and the x-intercept is $-K_m$, as shown in Figure 13.10. The Hanes–Woolf plot has the advantage of not overemphasizing the data obtained at low [S], a fault inherent in the Lineweaver–Burk plot. The common advantage of these plots is that they allow both K_m and V_{max} to be accurately estimated by extrapolation of straight lines rather than asymptotes. Computer fitting of v versus [S] data to the Michaelis–Menten equation is more commonly done than graphical plotting.

Nonlinear Lineweaver–Burk or Hanes–Woolf Plots Are a Property of Regulatory Enzymes

If the kinetics of the reaction disobey the Michaelis–Menten equation, the violation is revealed by a departure from linearity in these straight-line graphs. We shall see in the next chapter that such deviations from linearity are characteristic of the kinetics of regulatory enzymes known as **allosteric enzymes.** Such regulatory enzymes are very important in the overall control of metabolic pathways.

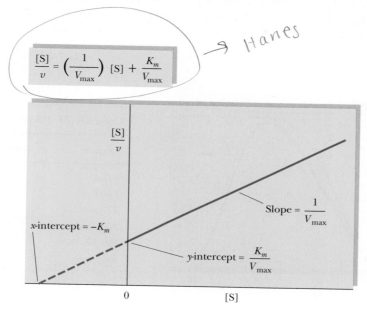

$$\frac{[S]}{v} = \left(\frac{1}{V_{max}}\right)[S] + \frac{K_m}{V_{max}}$$

→ Hanes

FIGURE 13.10 A Hanes–Woolf plot of [S]/v versus [S].

A DEEPER LOOK

An Example of the Effect of Amino Acid Substitutions on K_m and k_{cat}: Wild-Type and Mutant Forms of Human Sulfite Oxidase

Mammalian sulfite oxidase is the last enzyme in the pathway for degradation of sulfur-containing amino acids. Sulfite oxidase (SO) catalyzes the oxidation of sulfite (SO_3^{2-}) to sulfate (SO_4^{2-}), using the heme-containing protein, cytochrome c, as electron acceptor:

$$SO_3^{2-} + 2 \text{ cytochrome } c_{oxidized} + H_2O \rightleftharpoons$$
$$SO_4^{2-} + 2 \text{ cytochrome } c_{reduced} + 2 H^+$$

Isolated sulfite oxidase deficiency is a rare and often fatal genetic disorder in humans. The disease is characterized by severe neurological abnormalities, revealed as convulsions shortly after birth. R. M. Garrett and K. V. Rajagopalan at Duke University Medical Center have isolated the human cDNA for sulfite oxidase from the cells of normal *(wild-type)* and SO-deficient individuals. Expression of these SO cDNAs in transformed *Escherichia coli* cells allowed the isolation and kinetic analysis of wild-type and mutant forms of SO, including one (designated R160Q) in which the Arg at position 160 in the polypeptide chain is replaced by Gln. A genetically engineered version of SO (designated R160K) in which Lys replaces Arg[160] was also studied.

Kinetic Constants for Wild-Type and Mutant Sulfite Oxidase

Enzyme	K_m^{sulfite} (μM)	k_{cat} (sec^{-1})	k_{cat}/K_m (10^6 M^{-1} sec^{-1})
Wild-type	17	18	1.1
R160Q	1900	3	0.0016
R160K	360	5.5	0.015

Replacing R[160] in sulfite oxidase by Q increases K_m, decreases k_{cat}, and markedly diminishes the catalytic efficiency (k_{cat}/K_m) of the enzyme. The R160K mutant enzyme has properties intermediate between wild-type and the R160Q mutant form. The substrate, SO_3^{2-}, is strongly anionic, and R[160] is one of several Arg residues situated within the SO substrate-binding site. Positively charged side chains in the substrate-binding site facilitate SO_3^{2-} binding and catalysis, with Arg being optimal in this role.

Enzymatic Activity Is Strongly Influenced by pH

Enzyme–substrate recognition and the catalytic events that ensue are greatly dependent on pH. An enzyme possesses an array of ionizable side chains and prosthetic groups that not only determine its secondary and tertiary structure but may also be intimately involved in its active site. Furthermore, the substrate itself often has ionizing groups, and one or another of the ionic forms may preferentially interact with the enzyme. Enzymes in general are active only over a limited pH range, and most have a particular pH at which their catalytic activity is optimal. These effects of pH may be due to effects on K_m or V_{max} or both. Figure 13.11 illustrates the relative activity of four enzymes as a function of pH. Trypsin, an intestinal protease, has a slightly alkaline pH optimum, whereas pepsin, a gastric protease, acts in the acidic confines of the stomach and has a pH optimum near 2. Papain, a protease found in papaya, is relatively insensitive to pHs between 4 and 8. Cholinesterase activity is pH-sensitive below pH 7 but not between pH 7 and 10. The cholinesterase activity-pH profile suggests that an ionizable group with a pK_a near 6 is essential to its activity. Might this group be a histidine side chain within its

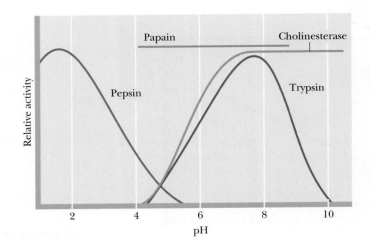

FIGURE 13.11 The pH activity profiles of four different enzymes.

Optimum pH of Some Enzymes

Enzyme	Optimum pH
Pepsin	1.5
Catalase	7.6
Trypsin	7.7
Fumarase	7.8
Ribonuclease	7.8
Arginase	9.7

active site? Although the pH optimum of an enzyme often reflects the pH of its normal environment, the optimum may not be precisely the same. This difference suggests that the pH-activity response of an enzyme may be a factor in the intracellular regulation of its activity.

The Response of Enzymatic Activity to Temperature Is Complex

Like most chemical reactions, the rates of enzyme-catalyzed reactions generally increase with increasing temperature. However, at temperatures above 50° to 60°C, enzymes typically show a decline in activity (Figure 13.12). Several effects are operating here: (1) the characteristic increase in reaction rate with temperature; (2) a temperature-dependent equilibrium between active enzyme (E_{active}) and the catalytically inactive but not denatured state of the enzyme (E_{inact}); and (3) thermal denaturation of protein structure at higher temperatures. An equilibrium model of these effects proposed by Roy M. Daniel and Michael J. Danson fits well with experimentally observed responses of enzyme activity versus temperature,

$$E_{active} \underset{}{\overset{K_{eq}}{\rightleftharpoons}} E_{inact} \xrightarrow{k_{inact}} X$$

Where X designates the irreversibly inactivated form of the enzyme and k_{inact} is the first-order rate constant for thermal denaturation. Most enzymatic reactions double in rate for every 10°C rise in temperature (that is, $Q_{10} = 2$, where Q_{10} is defined as *the ratio of activities at two temperatures 10° apart*) as long as the enzyme is stable and fully active. Some enzymes, those catalyzing reactions having very high activation energies, show proportionally greater Q_{10} values. The increasing rate with increasing temperature is ultimately offset by the temperature-dependent equilibrium between active and inactive states of the native enzyme, and, ultimately, the denaturation of protein structure at elevated temperatures. The fit of this model with the experimental data supports the existence of E_{inact}. Indeed, E_{inact} reflects the intrinsic flexibility, and, hence, greater thermal sensitivity of the active site, compared to the rest of the enzyme's structure. Not all enzymes are quite so thermally labile. For example, the enzymes of thermophilic prokaryotes (*thermophilic* = "heat-loving") found in geothermal springs retain full activity at temperatures in excess of 85°C.

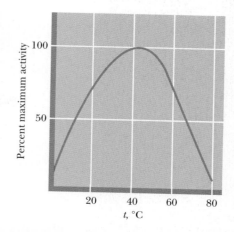

FIGURE 13.12 The effect of temperature on enzyme activity.

13.4 What Can Be Learned from the Inhibition of Enzyme Activity?

If the velocity of an enzymatic reaction is decreased or **inhibited** by some agent, the kinetics of the reaction obviously have been perturbed. Systematic perturbations are a basic tool of experimental scientists; much can be learned about the normal workings of any system by inducing changes in it and then observing the effects of the change. The study of enzyme inhibition has contributed significantly to our understanding of enzymes.

Enzymes May Be Inhibited Reversibly or Irreversibly

Enzyme inhibitors are classified in several ways. The inhibitor may interact either reversibly or irreversibly with the enzyme. **Reversible inhibitors** interact with the enzyme through noncovalent association/dissociation reactions. In contrast, **irreversible inhibitors** usually form covalent bonds with side chains or prosthetic groups in the enzyme. That is, the consequence of irreversible inhibition is a decrease in the concentration of active enzyme. The kinetics observed are consistent with this interpretation, as we shall see later.

Reversible Inhibitors May Bind at the Active Site or at Some Other Site

Reversible inhibitors fall into three major categories: competitive, noncompetitive, and uncompetitive. **Competitive inhibitors** are characterized by the fact that the substrate and inhibitor compete for the same binding site on the enzyme, the so-called **active site** or **substrate-binding site.** Thus, increasing the concentration of S favors the likelihood of S binding to the enzyme instead of the inhibitor, I. That is, high [S] can overcome the effects of I. The effects of the other major types, noncompetitive and uncompetitive inhibition, cannot be overcome by increasing [S]. The three types can be distinguished by the particular patterns obtained when the kinetic data are analyzed in linear plots, such as double-reciprocal (Lineweaver–Burk) plots. A general formulation for common inhibitor interactions in our simple enzyme kinetic model would include

$$E + I \rightleftharpoons EI \quad \text{and/or} \quad I + ES \rightleftharpoons IES \tag{13.32}$$

Competitive Inhibition Consider the following system:

$$E + S \underset{k_{-1}}{\overset{k_1}{\rightleftharpoons}} ES \overset{k_2}{\longrightarrow} E + P \qquad E + I \underset{k_{-3}}{\overset{k_3}{\rightleftharpoons}} EI \tag{13.33}$$

where an inhibitor, I, binds *reversibly* to the enzyme at the same site as S. S-binding and I-binding are mutually exclusive, *competitive* processes. Formation of the ternary complex, IES, where both S and I are bound, is physically impossible. This condition leads us to anticipate that S and I must share a high degree of structural similarity because they bind at the same site on the enzyme. Also notice that, in our model, EI does not react to give rise to E + P. That is, I is not changed by interaction with E. The rate of the product-forming reaction is $v = k_2[ES]$.

It is revealing to compare the equation for the uninhibited case, Equation 13.23 (the Michaelis–Menten equation) with Equation 13.43 for the rate of the enzymatic reaction in the presence of a fixed concentration of the competitive inhibitor, [I]

$$v = \frac{V_{max}[S]}{K_m + [S]}$$

$$v = \frac{V_{max}[S]}{[S] + K_m\left(1 + \dfrac{[I]}{K_I}\right)}$$

(see also Table 13.6). The K_m term in the denominator in the inhibited case is increased by the factor $(1 + [I]/K_I)$; thus, v is less in the presence of the inhibitor, as expected. Clearly, in the absence of I, the two equations are identical. Figure 13.13 shows a Lineweaver–Burk plot of competitive inhibition. Several features of competitive inhibition are evident. First, at a given [I], v decreases ($1/v$ increases). When [S] becomes infinite, $v = V_{max}$ and is unaffected by I because all of the enzyme is in the ES form. Note that the value of the $-x$-intercept decreases as [I] increases. This $-x$-intercept is often termed the *apparent* K_m (or K_{mapp}) because it is the K_m apparent under these conditions.

TABLE 13.6	The Effect of Various Types of Inhibitors on the Michaelis–Menten Rate Equation and on Apparent K_m and Apparent V_{max}		
Inhibition Type	Rate Equation	Apparent K_m	Apparent V_{max}
None	$v = V_{max}[S]/(K_m+[S])$	K_m	V_{max}
Competitive	$v = V_{max}[S]/([S]+K_m(1+[I]/K_I))$	$K_m(1+[I]/K_I)$	V_{max}
Noncompetitive	$v = (V_{max}[S]/(1+[I]/K_I))/(K_m+[S])$	K_m	$V_{max}/(1+[I]/K_I)$
Mixed	$v = V_{max}[S]/((1+[I]/K_I)K_m+(1+[I]/K_I'[S]))$	$K_m(1+[I]/K_I)/(1+[I]/K_I')$	$V_{max}/(1+[I]/K_I')$
Uncompetitive	$v = V_{max}[S]/(K_m+[S](1+[I]/K_I'))$	$K_m/(1+[I]/K_I')$	$V_{max}/(1+[I]/K_I')$

K_I is defined as the enzyme:inhibitor dissociation constant $K_I=[E][I]/[EI]$; K_I' is defined as the enzyme–substrate complex:inhibitor dissociation constant $K_I'=[ES][I]/[IES]$.

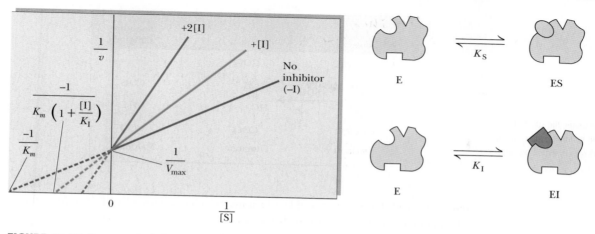

FIGURE 13.13 Lineweaver–Burk plot of competitive inhibition, showing lines for no I, [I], and 2[I]. Note that when [S] is infinitely large (1/[S] ≈ 0), V_{max} is the same, whether I is present or not.

The diagnostic criterion for competitive inhibition is that V_{max} is unaffected by I; that is, all lines share a common y-intercept. This criterion is also the best experimental indication of binding at the same site by two substances. Competitive inhibitors resemble S structurally.

Succinate Dehydrogenase—A Classic Example of Competitive Inhibition The enzyme *succinate dehydrogenase (SDH)* is competitively inhibited by malonate. Figure 13.14 shows the structures of succinate and malonate. The structural similarity between them is obvious and is the basis of malonate's ability to mimic succinate and bind at the

---

> **A DEEPER LOOK**

The Equations of Competitive Inhibition

Given the relationships between E, S, and I described previously and recalling the steady-state assumption that $d[ES]/dt = 0$, from Equations (13.14) and (13.16) we can write

$$ES = \frac{k_1[E][S]}{(k_2 + k_{-1})} = \frac{[E][S]}{K_m} \tag{13.34}$$

Assuming that $E + I \rightleftharpoons EI$ reaches rapid equilibrium, the rate of EI formation, $v_f' = k_3[E][I]$, and the rate of disappearance of EI, $v_d' = k_{-3}[EI]$, are equal. So,

$$k_3[E][I] = k_{-3}[EI] \tag{13.35}$$

Therefore,

$$[EI] = \frac{k_3}{k_{-3}}[E][I] \tag{13.36}$$

If we define K_I as k_{-3}/k_3, an enzyme-inhibitor dissociation constant, then

$$[EI] = \frac{[E][I]}{K_I} \tag{13.37}$$

knowing $[E_T] = [E] + [ES] + [EI]$. Then

$$[E_T] = [E] + \frac{[E][S]}{K_m} + \frac{[E][I]}{K_I} \tag{13.38}$$

Solving for [E] gives

$$[E] = \frac{K_I K_m [E_T]}{(K_I K_m + K_I[S] + K_m[I])} \tag{13.39}$$

Because the rate of product formation is given by $v = k_2[ES]$, from Equation 13.34 we have

$$v = \frac{k_2[E][S]}{K_m} \tag{13.40}$$

So,

$$v = \frac{(k_2 K_I[E_T][S])}{(K_I K_m + K_I[S] + K_m[I])} \tag{13.41}$$

Because $V_{max} = k_2[E_T]$,

$$v = \frac{V_{max}[S]}{K_m + [S] + \dfrac{K_m[I]}{K_I}} \tag{13.42}$$

or

$$v = \frac{V_{max}[S]}{[S] + K_m\left(1 + \dfrac{[I]}{K_I}\right)} \tag{13.43}$$

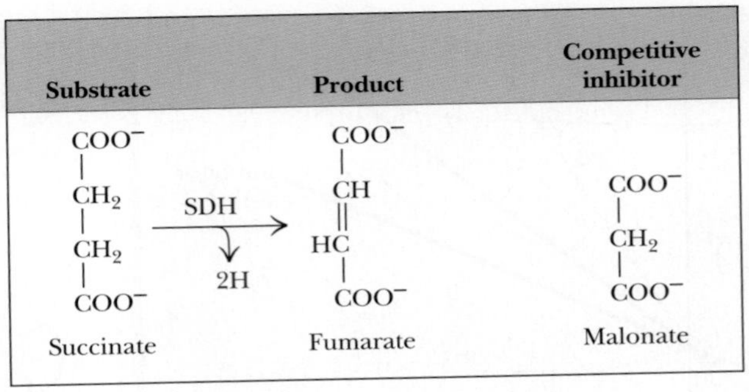

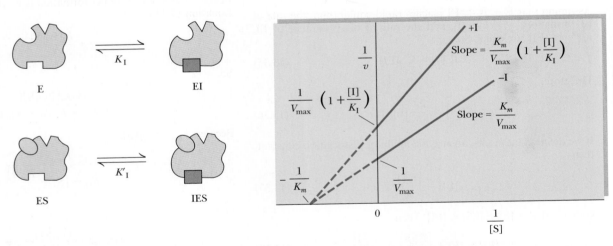

FIGURE 13.14 Structures of succinate, the substrate of succinate dehydrogenase (SDH), and malonate, the competitive inhibitor. Fumarate (the product of SDH action on succinate) is also shown.

active site of SDH. However, unlike succinate, which is oxidized by SDH to form fumarate, malonate cannot lose two hydrogens; consequently, it is unreactive.

Noncompetitive Inhibition Noncompetitive inhibitors interact with both E and ES (or with S and ES, but this is a rare and specialized case). Obviously, then, the inhibitor is not binding to the same site as S, and the inhibition cannot be overcome by raising [S]. There are two types of noncompetitive inhibition: pure and mixed.

Pure Noncompetitive Inhibition In this situation, the binding of I by E has no effect on the binding of S by E. That is, S and I bind at different sites on E, and binding of I does not affect binding of S. Consider the system

$$E + I \overset{K_I}{\rightleftharpoons} EI \qquad ES + I \overset{K_I'}{\rightleftharpoons} IES \qquad (13.44)$$

Pure noncompetitive inhibition occurs if $K_I = K_I'$. This situation is relatively uncommon; the Lineweaver–Burk plot for such an instance is given in Figure 13.15. Note that K_m is unchanged by I (the x-intercept remains the same, with or without I). Note also that the apparent V_{max} decreases. A similar pattern is seen if the amount of enzyme in the experiment is decreased. Thus, it is as if I lowered [E].

Mixed Noncompetitive Inhibition In this situation, the binding of I by E influences the binding of S by E. Either the binding sites for I and S are near one another or conformational changes in E caused by I affect S binding. In this case, K_I and K_I', as defined previously, are not equal. Both the apparent K_m and the apparent V_{max} are altered by the presence of I, and K_m/V_{max} is not constant (Figure 13.16). This inhibitory pattern is commonly encountered. A reasonable explanation is that the inhibitor is binding at a site distinct from the active site yet is influencing the binding of S at the active site.

FIGURE 13.15 Lineweaver–Burk plot of pure noncompetitive inhibition. Note that I does not alter K_m but that it decreases V_{max}.

(a) $K_I < K_I'$

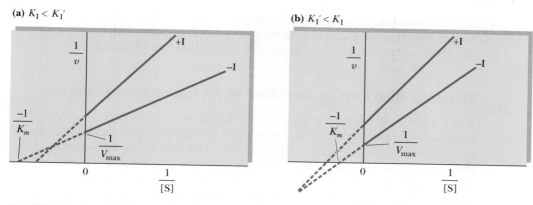

(b) $K_I' < K_I$

FIGURE 13.16 Lineweaver–Burk plot of mixed noncompetitive inhibition. Note that both intercepts and the slope change in the presence of I. **(a)** When K_I is less than K_I'; **(b)** when K_I is greater than K_I'.

Presumably, these effects are transmitted via alterations in the protein's conformation. Table 13.6 includes the rate equations and apparent K_m and V_{max} values for both types of noncompetitive inhibition.

Uncompetitive Inhibition Completing the set of inhibitory possibilities is uncompetitive inhibition. Unlike competitive inhibition (where I combines only with E) or noncompetitive inhibition (where I combines with E and ES), in uncompetitive inhibition, I combines only with ES.

$$ES + I \xrightleftharpoons{K_I'} IES \tag{13.45}$$

Because IES does not lead to product formation, the observed rate constant for product formation, k_2, is uniquely affected. In simple Michaelis–Menten kinetics, k_2 is the only rate constant that is part of both V_{max} and K_m. The pattern obtained in Lineweaver–Burk plots is a set of parallel lines (Figure 13.17). A clinically important example is the action of lithium in alleviating manic depression; Li^+ ions are uncompetitive inhibitors

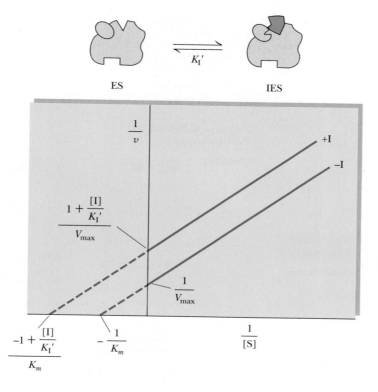

FIGURE 13.17 Lineweaver–Burk plot of uncompetitive inhibition. Note that both intercepts change but the slope (K_m/V_{max}) remains constant in the presence of I.

of *myo*-inositol monophosphatase. Some pesticides are also uncompetitive inhibitors, such as Roundup, an uncompetitive inhibitor of 3-enolpyruvylshikimate-5-P synthase, an enzyme essential to aromatic amino acid biosynthesis (see Chapter 25).

Enzymes Also Can Be Inhibited in an Irreversible Manner

If the inhibitor combines irreversibly with the enzyme—for example, by covalent attachment—the kinetic pattern seen is like that of noncompetitive inhibition, because the net effect is a loss of active enzyme. Usually, this type of inhibition can be distinguished from the noncompetitive, reversible inhibition case because the reaction of I with E (and/or ES) is not instantaneous. Instead, there is a *time-dependent decrease in enzymatic activity* as E + I⟶EI proceeds, and the rate of this inactivation can be followed. Also, unlike reversible inhibitions, dilution or dialysis of the enzyme:inhibitor solution does not dissociate the EI complex and restore enzyme activity.

Suicide Substrates—Mechanism-Based Enzyme Inactivators **Suicide substrates** are inhibitory substrate analogs designed so that, via normal catalytic action of the enzyme, a very reactive group is generated. This reactive group then forms a covalent bond with a nearby functional group within the active site of the enzyme, thereby causing irreversible inhibition. Suicide substrates, also called *Trojan horse substrates,* are a type of **affinity label.** As substrate analogs, they bind with specificity and high affinity to the enzyme active site; in their reactive form, they become covalently bound to the enzyme. This covalent link effectively labels a particular functional group within the active site, identifying the group as a key player in the enzyme's catalytic cycle.

Penicillin—A Suicide Substrate Several drugs in current medical use are mechanism-based enzyme inactivators. For example, the antibiotic **penicillin** exerts its effects by covalently reacting with an essential serine residue in the active site of *glycopeptide transpeptidase,* an enzyme that acts to crosslink the peptidoglycan chains during synthesis of bacterial cell walls (Figure 13.18). Penicillin consists of a thiazolidine ring fused to

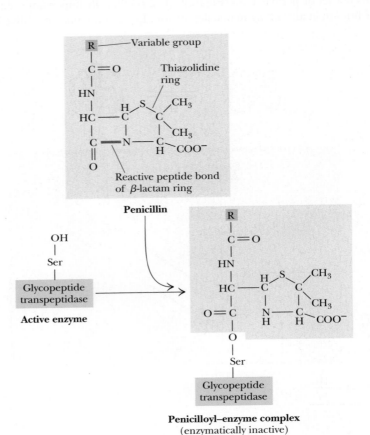

FIGURE 13.18 Penicillin is an irreversible inhibitor of the enzyme *glycopeptide transpeptidase*, also known as *glycoprotein peptidase*, which catalyzes an essential step in bacterial cell wall synthesis.

a β-lactam ring to which a variable R group is attached. A reactive peptide bond in the β-lactam ring covalently attaches to a serine residue in the active site of the glycopeptide transpeptidase. (The conformation of penicillin around its reactive peptide bond resembles the transition state of the normal glycopeptide transpeptidase substrate.) The penicillinoyl–enzyme complex is catalytically inactive. Once cell wall synthesis is blocked, the bacterial cells are very susceptible to rupture by osmotic lysis and bacterial growth is halted.

13.5 What Is the Kinetic Behavior of Enzymes Catalyzing Bimolecular Reactions?

Thus far, we have considered only the simple case of enzymes that act upon a single substrate, S. This situation is not common. Usually, enzymes catalyze reactions in which two (or even more) substrates take part.

Consider the case of an enzyme catalyzing a reaction involving two substrates, A and B, and yielding the products P and Q:

$$A + B \underset{\text{enzyme}}{\rightleftharpoons} P + Q \tag{13.46}$$

Enzymatic reactions involving two substrates are called **bisubstrate reactions.** In general, bisubstrate reactions proceed by one of two possible routes:

1. Both A and B are bound to the enzyme and then reaction occurs to give P + Q:

$$E + A + B \longrightarrow AEB \longrightarrow PEQ \longrightarrow E + P + Q \tag{13.47}$$

HUMAN BIOCHEMISTRY

Viagra—An Unexpected Outcome in a Program of Drug Design

Prior to the accumulation of detailed biochemical information on metabolism, enzymes, and receptors, drugs were fortuitous discoveries made by observant scientists; the discovery of penicillin as a bacteria-killing substance by Fleming is an example. Today, **drug design** is the rational application of scientific knowledge and principles to the development of pharmacologically active agents. A particular target for therapeutic intervention is identified (such as an enzyme or receptor involved in illness), and chemical analogs of its substrate or ligand are synthesized in hopes of finding an inhibitor (or activator) that will serve as a drug to treat the illness. Sometimes the outcome is unanticipated, as the story of **Viagra** (sildenafil citrate) reveals.

When the smooth muscle cells of blood vessels relax, blood flow increases and blood pressure drops. Such relaxation is the result of decreases in intracellular $[Ca^{2+}]$ triggered by increases in intracellular [cGMP] (which in turn is triggered by nitric oxide, NO; see Chapter 32). Cyclic GMP (cGMP) is hydrolyzed by *phosphodiesterases* to form 5′-GMP, and the muscles contract again. Scientists at Pfizer reasoned that, if phosphodiesterase inhibitors could be found, they

might be useful drugs to treat *angina* (chest pain due to inadequate blood flow to heart muscle) or *hypertension* (high blood pressure). The phosphodiesterase (PDE) prevalent in vascular muscle is PDE 5, one of at least nine different subtypes of PDE in human cells. The search was on for substances that inhibit PDE 5, but not the other prominent PDE types, and Viagra was found. Disappointingly, Viagra showed no significant benefits for angina or hypertension, but some men in clinical trials reported penile erection. Apparently, Viagra led to an increase in [cGMP] in penile vascular tissue, allowing vascular muscle relaxation, improved blood flow, and erection. A drug was born.

In a more focused way, detailed structural data on enzymes, receptors, and the ligands that bind to them has led to **rational drug design,** in which *computer modeling of enzyme-ligand interactions* replaces much of the initial chemical synthesis and clinical prescreening of potential therapeutic agents, saving much time and effort in drug development.

cGMP

Viagra

Note the structural similarity between cGMP *(left)* and Viagra *(right)*.

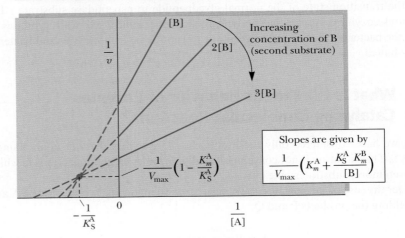

Double-reciprocal form of the rate equation:

$$\frac{1}{v} = \frac{1}{V_{max}}\left(K_m^A + \frac{K_S^A \; K_m^B}{[B]}\right)\left(\frac{1}{[A]} + \frac{1}{V_{max}}\left(1 + \frac{K_m^B}{[B]}\right)\right)$$

FIGURE 13.19 Single-displacement bisubstrate mechanism.

Reactions of this type are defined as **sequential** or **single-displacement reactions.** They can be either of two distinct classes:

a. **random,** where either A or B may bind to the enzyme first, followed by the other substrate, or

b. **ordered,** where A, designated the *leading substrate,* must bind to E first before B can be bound.

Both classes of single-displacement reactions are characterized by lines that intersect to the left of the $1/v$ axis in Lineweaver–Burk plots where the rates observed with different fixed concentrations of one substrate (B) are graphed versus a series of concentrations of A (Figure 13.19).

2. The other general possibility is that one substrate, A, binds to the enzyme and reacts with it to yield a chemically modified form of the enzyme (E′) plus the product, P. The second substrate, B, then reacts with E′, regenerating E and forming the other product, Q.

$$E + A \longrightarrow EA \longrightarrow E'P \searrow E' \nearrow E'B \longrightarrow EQ \longrightarrow E + Q$$

$$P \qquad B \qquad (13.48)$$

Reactions that fit this model are called **ping-pong** or **double-displacement reactions.** Two distinctive features of this mechanism are the obligatory formation of a modified enzyme intermediate, E′, and the pattern of parallel lines obtained in double-reciprocal plots of the rates observed with different fixed concentrations of one substrate (B) versus a series of concentrations of A (see Figure 13.22).

The Conversion of AEB to PEQ Is the Rate-Limiting Step in Random, Single-Displacement Reactions

In this type of sequential reaction, all possible binary enzyme–substrate complexes (AE, EB, PE, EQ) are formed rapidly and reversibly when the enzyme is added to a reaction mixture containing A, B, P, and Q:

$$A + E \rightleftharpoons AE \searrow \qquad \nearrow EP \rightleftharpoons P + E$$
$$AEB \rightleftharpoons PEQ$$
$$E + B \rightleftharpoons EB \nearrow \qquad \searrow QE \rightleftharpoons E + Q \qquad (13.49)$$

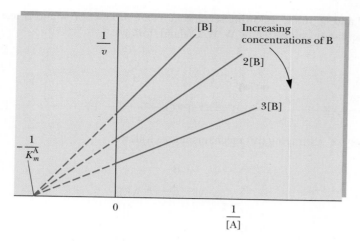

FIGURE 13.20 Random, single-displacement bisubstrate mechanism where A does not affect B binding, and vice versa.

The rate-limiting step is the reaction AEB⟶PEQ. It does not matter whether A or B binds first to E, or whether Q or P is released first from QEP. Sometimes, reactions that follow this random order of addition of substrates to E can be distinguished from reactions obeying an ordered, single-displacement mechanism. If A has *no* influence on the binding constant for B (and vice versa) and the mechanism is purely random, the lines in a Lineweaver–Burk plot intersect at the 1/[A] axis (Figure 13.20).

Creatine Kinase Acts by a Random, Single-Displacement Mechanism An example of a random, single-displacement mechanism is seen in the enzyme creatine kinase, a phosphoryl transfer enzyme that uses ATP as a phosphoryl donor to form creatine phosphate (CrP) from creatine (Cr). Creatine-P is an important reservoir of phosphate-bond energy in muscle cells (Figure 13.21).

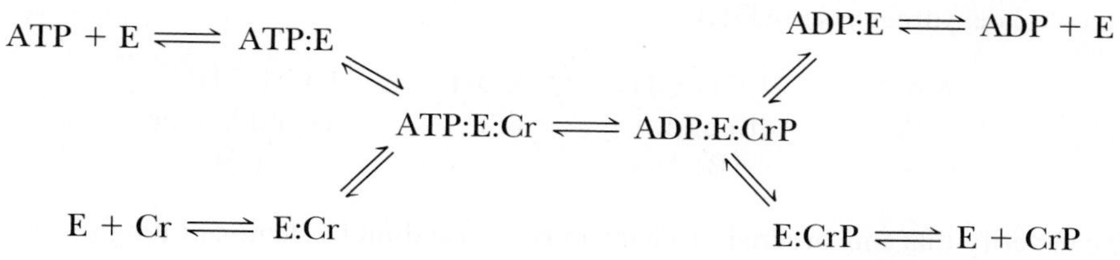

$$ATP + E \rightleftharpoons ATP{:}E$$
$$ADP{:}E \rightleftharpoons ADP + E$$
$$ATP{:}E{:}Cr \rightleftharpoons ADP{:}E{:}CrP$$
$$E + Cr \rightleftharpoons E{:}Cr$$
$$E{:}CrP \rightleftharpoons E + CrP$$

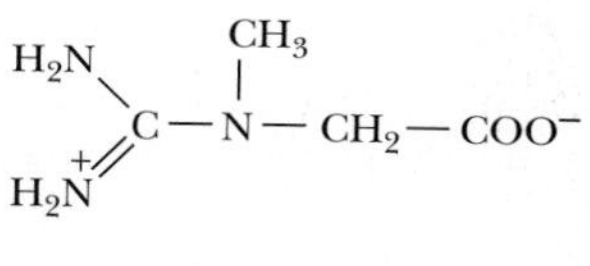

The overall direction of the reaction will be determined by the relative concentrations of ATP, ADP, Cr, and CrP and the equilibrium constant for the reaction. The enzyme can be considered to have two sites for substrate (or product) binding: an adenine nucleotide site, where ATP or ADP binds, and a creatine site, where Cr or CrP is bound. In such a mechanism, ATP and ADP compete for binding at their unique site while Cr and CrP compete at the specific Cr/CrP-binding site. Note that no modified enzyme form (E′), such as an E-P intermediate, appears here. The reaction is characterized by rapid and reversible binary ES complex formation, followed by addition of the remaining substrate, and the rate-determining reaction taking place within the ternary complex.

Creatine

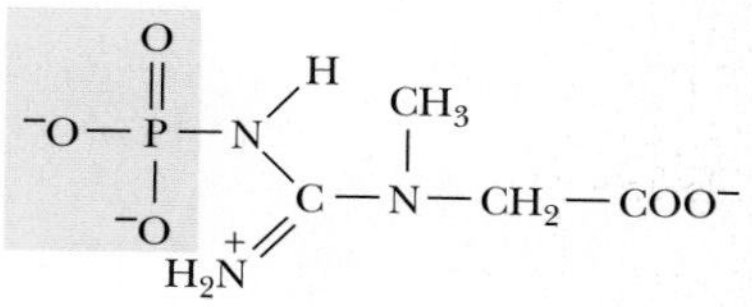

Creatine-P

FIGURE 13.21 The structures of creatine and creatine phosphate, guanidinium compounds that are important in muscle energy metabolism.

In an Ordered, Single-Displacement Reaction, the Leading Substrate Must Bind First

In an ordered reaction, the **leading substrate,** A (also called the **obligatory** or **compulsory substrate**), must bind first. Then the second substrate, B, binds. Strictly speaking, B cannot bind to free enzyme in the absence of A. Reaction between A and B occurs in the ternary complex and is usually followed by an ordered release of the products of the reaction, P and Q. In the following schemes, P is the product of A and is released last. One representation, suggested by W. W. Cleland, follows:

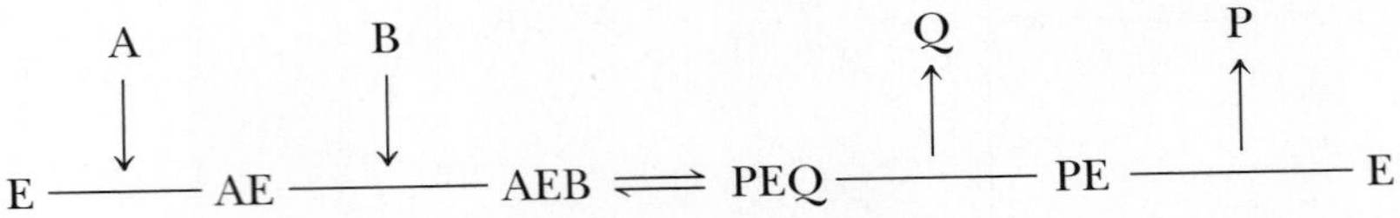

$$E \underset{}{\overset{A}{\downarrow}} AE \underset{}{\overset{B}{\downarrow}} AEB \rightleftharpoons PEQ \overset{Q}{\uparrow} PE \overset{P}{\uparrow} E \quad (13.50)$$

Another way of portraying this mechanism is as follows:

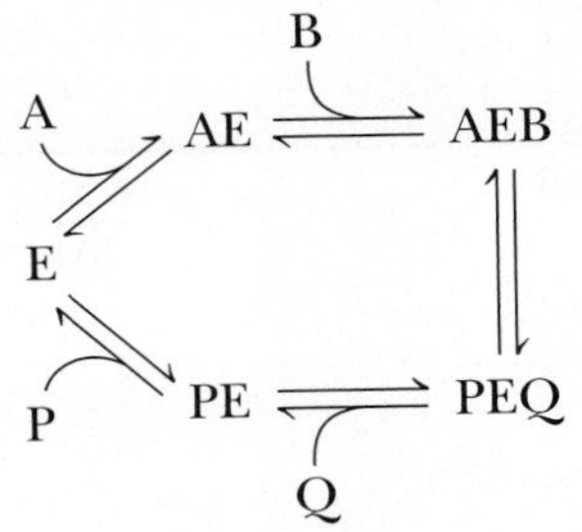

Note that A and P are competitive for binding to the free enzyme, E, but not A and B (or P and B).

NAD$^+$-Dependent Dehydrogenases Show Ordered Single-Displacement Mechanisms *Nicotinamide adenine dinucleotide (NAD$^+$)-dependent dehydrogenases* are enzymes that typically behave according to the kinetic pattern just described. A general reaction of these dehydrogenases is

$$NAD^+ + BH_2 \rightleftharpoons NADH + H^+ + B$$

The leading substrate (A) is nicotinamide adenine dinucleotide (NAD$^+$), and NAD$^+$ and NADH (product P) compete for a common site on E. A specific example is offered by alcohol dehydrogenase (ADH):

$$\underset{\text{(A)}}{NAD^+} + \underset{\substack{\text{ethanol} \\ \text{(B)}}}{CH_3CH_2OH} \rightleftharpoons \underset{\text{(P)}}{NADH} + H^+ + \underset{\substack{\text{acetaldehyde} \\ \text{(Q)}}}{CH_3CHO}$$

We can verify that this ordered mechanism is not random by demonstrating that no B (ethanol) is bound to E in the absence of A (NAD$^+$).

Double-Displacement (Ping-Pong) Reactions Proceed Via Formation of a Covalently Modified Enzyme Intermediate

Double-displacement reactions are characterized by a pattern of parallel lines when $1/v$ is plotted as a function of $1/[A]$ at different concentrations of B, the second substrate (Figure 13.22). Reactions conforming to this kinetic pattern are characterized by the fact that the product of the enzyme's reaction with A (called P in the following schemes) is released *prior* to reaction of the enzyme with the second substrate, B. As a result of this process, the enzyme, E, is converted to a modified form, E′, which then reacts with B to give the second product, Q, and regenerate the unmodified enzyme form, E:

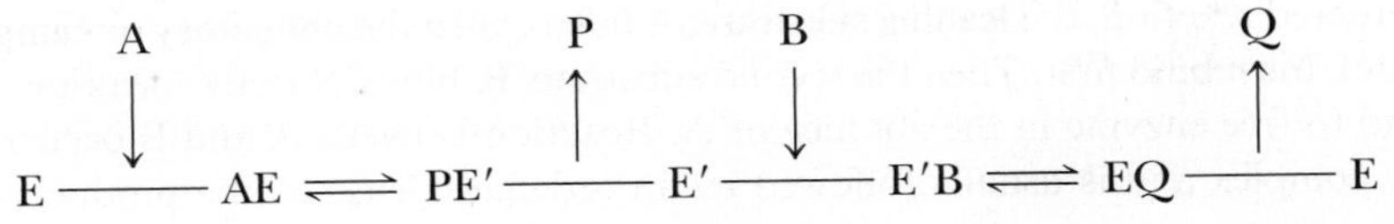

$$E \underset{}{\overset{A}{\downarrow}} AE \rightleftharpoons PE' \overset{P}{\uparrow} E' \underset{}{\overset{B}{\downarrow}} E'B \rightleftharpoons EQ \overset{Q}{\uparrow} E$$

Double-reciprocal form of the rate equation:

$$\frac{1}{v} = \frac{K_m^A}{V_{max}}\left(\frac{1}{[A]}\right) + \left(1 + \frac{K_m^B}{[B]}\right)\left(\frac{1}{V_{max}}\right)$$

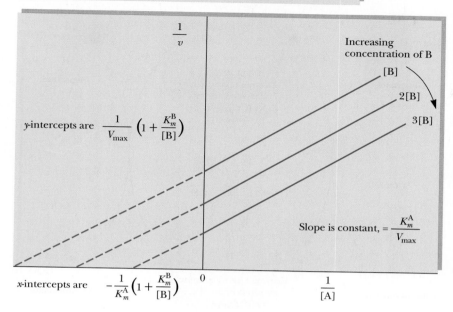

$\frac{1}{v}$

Increasing concentration of B

[B]

2[B]

3[B]

y-intercepts are $\dfrac{1}{V_{max}}\left(1 + \dfrac{K_m^B}{[B]}\right)$

Slope is constant, $= \dfrac{K_m^A}{V_{max}}$

x-intercepts are $-\dfrac{1}{K_m^A}\left(1 + \dfrac{K_m^B}{[B]}\right)$

$\frac{1}{[A]}$

FIGURE 13.22 Double-displacement (ping-pong) bisubstrate mechanisms are characterized by parallel lines.

or

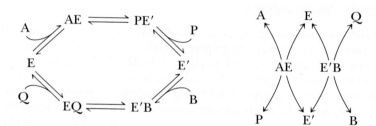

Note that these schemes predict that A and Q compete for the free enzyme form, E, while B and P compete for the modified enzyme form, E′. A and Q do not bind to E′, nor do B and P combine with E.

Aminotransferases Show Double-Displacement Catalytic Mechanisms

One class of enzymes that follow a ping-pong–type mechanism are *aminotransferases* (previously known as transaminases). These enzymes catalyze the transfer of an amino group from an amino acid to an α-keto acid. The products are a new amino acid and the keto acid corresponding to the carbon skeleton of the amino donor:

$$\text{amino acid}_1 + \text{keto acid}_2 \longrightarrow \text{keto acid}_1 + \text{amino acid}_2$$

A specific example would be *glutamate : aspartate aminotransferase.* Figure 13.23 depicts the scheme for this mechanism. Note that glutamate and aspartate are competitive for E and that oxaloacetate and α-ketoglutarate compete for E′. In glutamate : aspartate aminotransferase, an enzyme-bound coenzyme, *pyridoxal phosphate* (a vitamin B_6 derivative), serves as the amino group acceptor/donor in the enzymatic reaction. The unmodified enzyme, E, has the coenzyme in the aldehydic pyridoxal form, whereas in the modified enzyme, E′, the coenzyme is actually pyridoxamine phosphate (Figure 13.23). Not all enzymes displaying ping-pong–type mechanisms require coenzymes as carriers for the chemical substituent transferred in the reaction.

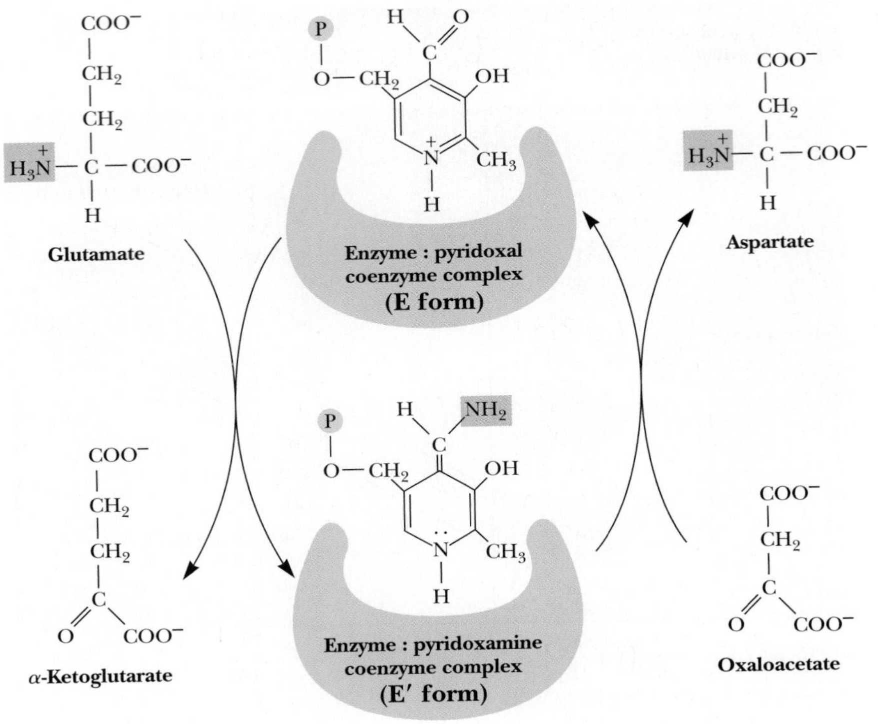

FIGURE 13.23 *Glutamate : aspartate aminotransferase,* an enzyme conforming to a double-displacement bisubstrate mechanism.

Exchange Reactions Are One Way to Diagnose Bisubstrate Mechanisms

Kineticists rely on a number of diagnostic tests for the assignment of a reaction mechanism to a specific enzyme. One is the graphic analysis of the kinetic patterns observed. It is usually easy to distinguish between single- and double-displacement reactions in this manner, and examining competitive effects between substrates aids in assigning reactions to random versus ordered patterns of S binding. A second diagnostic test is to determine whether the enzyme catalyzes an **exchange reaction.** Consider as an example the two enzymes *sucrose phosphorylase* and *maltose phosphorylase.* Both catalyze the phosphorolysis of a disaccharide and both yield glucose-1-phosphate and a free hexose:

$$\text{Sucrose} + P_i \rightleftharpoons \text{glucose-1-phosphate} + \text{fructose}$$

$$\text{Maltose} + P_i \rightleftharpoons \text{glucose-1-phosphate} + \text{glucose}$$

Interestingly, in the absence of sucrose and fructose, sucrose phosphorylase will catalyze the exchange of inorganic phosphate, P_i, into glucose-1-phosphate. This reaction can be followed by using $^{32}P_i$ as a radioactive tracer and observing the incorporation of ^{32}P into glucose-1-phosphate:

$$^{32}P_i + \text{G-1-P} \rightleftharpoons P_i + \text{G-1-}^{32}P$$

Maltose phosphorylase cannot carry out a similar reaction. The ^{32}P exchange reaction of sucrose phosphorylase is accounted for by a double-displacement mechanism where E′ is E-glucose:

$$\text{Sucrose} + E \rightleftharpoons \text{E-glucose} + \text{fructose}$$

$$\text{E-glucose} + P_i \rightleftharpoons E + \text{glucose-1-phosphate}$$

Thus, in the presence of just $^{32}P_i$ and glucose-1-phosphate, sucrose phosphorylase still catalyzes the second reaction and radioactive P_i is incorporated into glucose-1-phosphate over time.

Maltose phosphorylase proceeds via a single-displacement reaction that necessarily requires the formation of a ternary maltose : E : P_i (or glucose : E : glucose-1-phosphate) complex for any reaction to occur. Exchange reactions are a characteristic of enzymes that obey double-displacement mechanisms at some point in their catalysis.

Multisubstrate Reactions Can Also Occur in Cells

Thus far, we have considered enzyme-catalyzed reactions involving one or two substrates. How are the kinetics described in those cases in which more than two substrates participate in the reaction? An example might be the glycolytic enzyme *glyceraldehyde-3-phosphate dehydrogenase* (see Chapter 18):

$$NAD^+ + glyceraldehyde\text{-}3\text{-}P + P_i \rightleftharpoons NADH + H^+ + 1,3\text{-}bisphosphoglycerate$$

Many other multisubstrate examples abound in metabolism. In effect, these situations are managed by realizing that the interaction of the enzyme with its many substrates can be treated as a series of unisubstrate or bisubstrate steps in a multistep reaction pathway. Thus, the complex mechanism of a multisubstrate reaction is resolved into a sequence of steps, each of which obeys the single- and double-displacement patterns just discussed.

13.6 How Can Enzymes Be So Specific?

The extraordinary ability of an enzyme to catalyze only one particular reaction is a quality known as **specificity.** Specificity means an enzyme acts only on a specific substance, its substrate, invariably transforming it into a specific product. That is, an enzyme binds only certain compounds, and then, only a specific reaction ensues. Some enzymes show absolute specificity, catalyzing the transformation of only one specific substrate to yield a unique product. Other enzymes carry out a particular reaction but act on a class of compounds. For example, *hexokinase* (ATP:hexose-6-phosphotransferase) will carry out the ATP-dependent phosphorylation of a number of hexoses at the 6-position, including glucose. Specificity studies on enzymes entail an examination of the rates of the enzymatic reaction obtained with various **structural analogs** of the substrate. By determining which functional and structural groups within the substrate affect binding or catalysis, enzymologists can map the properties of the active site, analyzing questions such as: Can the active site accommodate sterically bulky groups? Are ionic interactions between E and S important? Are H bonds formed?

The "Lock and Key" Hypothesis Was the First Explanation for Specificity

Pioneering enzyme specificity studies at the turn of the 20th century by the great organic chemist Emil Fischer led to the notion of an enzyme resembling a **"lock"** and its particular substrate the **"key."** This analogy captures the essence of the specificity that exists between an enzyme and its substrate, but enzymes are not rigid templates like locks.

The "Induced Fit" Hypothesis Provides a More Accurate Description of Specificity

Enzymes are highly flexible, conformationally dynamic molecules, and many of their remarkable properties, including substrate binding and catalysis, are due to their structural pliancy. Realization of the conformational flexibility of proteins led Daniel Koshland to hypothesize that the binding of a substrate by an enzyme is an interactive process. That is, the shape of the enzyme's active site is actually modified upon binding S, in a process of dynamic recognition between enzyme and substrate aptly called **induced fit.** In essence, substrate binding alters the conformation of the protein, so that the protein and the substrate "fit" each other more precisely. The process is truly interactive in

New ideas do not always gain immediate acceptance: "Although we did many experiments that in my opinion could only be explained by the induced-fit theory, gaining acceptance for the theory was still an uphill fight. One (journal) referee wrote, 'The Fischer Key-Lock theory has lasted 100 years and will not be overturned by speculation from an embryonic scientist.'" *Daniel Koshland, 1996. How to get paid for having fun.* Annual Review of Biochemistry *65:1–13.*

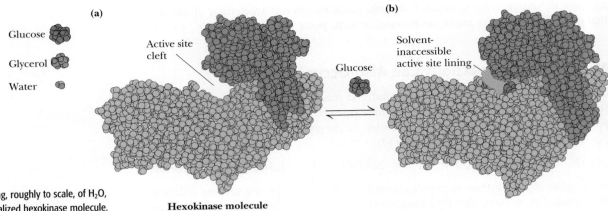

FIGURE 13.24 A drawing, roughly to scale, of H_2O, glycerol, glucose, and an idealized hexokinase molecule.

that the conformation of the substrate also changes as it adapts to the conformation of the enzyme.

This idea also helps explain some of the mystery surrounding the enormous catalytic power of enzymes: In enzyme catalysis, precise orientation of catalytic residues comprising the active site is necessary for the reaction to occur; substrate binding induces this precise orientation by the changes it causes in the protein's conformation.

"Induced Fit" Favors Formation of the Transition State

The catalytically active enzyme substrate complex is an interactive structure in which the enzyme causes the substrate to adopt a form that mimics the transition state of the reaction. Thus, a poor substrate would be one that was less effective in directing the formation of an optimally active enzyme:transition state conformation. This active conformation of the enzyme molecule is thought to be relatively unstable in the absence of substrate, and free enzyme thus reverts to a different conformational state.

Specificity and Reactivity

Consider, for example, why hexokinase catalyzes the ATP-dependent phosphorylation of hexoses but not smaller phosphoryl-group acceptors such as glycerol, ethanol, or even water. Surely these smaller compounds are not sterically forbidden from approaching the active site of hexokinase (Figure 13.24). Indeed, water should penetrate the active site easily and serve as a highly effective phosphoryl-group acceptor. Accordingly, hexokinase should display high ATPase activity. It does not. Only the binding of hexoses induces hexokinase to assume its fully active conformation. The hexose-binding site of hexokinase is located between two protein domains. Binding of glucose in the active site induces a conformational change in hexokinase that causes the two domains to close upon one another, creating the catalytic site.

In Chapter 14, we explore in greater detail the factors that contribute to the remarkable catalytic power of enzymes and examine specific examples of enzyme reaction mechanisms.

13.7 | Are All Enzymes Proteins?

RNA Molecules That Are Catalytic Have Been Termed "Ribozymes"

It was long assumed that all enzymes are proteins. However, several decades ago, instances of biological catalysis by RNA molecules were discovered. Catalytic RNAs, or **ribozymes,** satisfy several enzymatic criteria: They are substrate specific, they enhance the reaction rate, and they emerge from the reaction unchanged. Most ribozymes act in RNA processing, cutting the phosphodiester backbone at specific sites and religating needed segments to form functional RNA strands while discarding extraneous pieces. For example, bacterial RNase P is a ribozyme involved in the formation of mature tRNA molecules from longer RNA transcripts. RNase P requires an RNA component as well as a protein subunit for its activity in the cell. In vitro, the protein alone is incapable of catalyzing the maturation reaction, but the RNA component by itself can carry out the reaction under appropriate conditions. As another example, the introns within some rRNAs and mRNAs are ribozymes that can catalyze their own excision from large RNA transcripts by a process known as **self-splicing.** For instance, in the ciliated protozoan *Tetrahymena,* formation of mature ribosomal RNA from a pre-rRNA precursor involves the removal of an internal RNA segment and the joining of the two ends. The excision of this intron and ligation of the exons is catalyzed by the intron itself, in the presence of Mg^{2+} and a free molecule of guanosine nucleoside or nucleotide (Figure 13.25). In vivo, the intervening sequence RNA probably acts only in splicing itself out; in vitro, however, it can act many times, turning over like a true enzyme.

The Ribosome Is a Ribozyme A particularly significant case of catalysis by RNA occurs in protein synthesis. The **peptidyl transferase reaction,** which is the reaction of peptide bond formation during protein synthesis, is catalyzed by the 23S rRNA of the 50S subunit of ribosomes (see Chapters 10 and 30). The substrates for the peptidyl transferase reaction are two tRNA molecules, one bearing the growing peptide chain (the

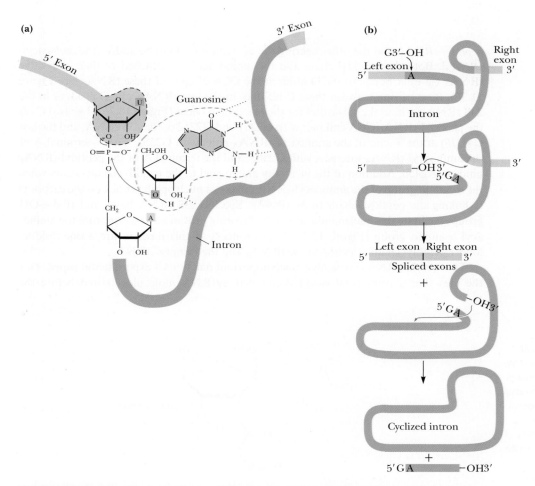

FIGURE 13.25 RNA splicing in *Tetrahymena* rRNA maturation: **(a)** the guanosine-mediated reaction involved in the autocatalytic excision of the *Tetrahymena* rRNA intron and **(b)** the overall splicing process. The cyclized intron is formed via nucleophilic attack of the freed 3'-OH end of the intron on the phosphodiester bond that is 15 nucleotides from the 5'-GA end of the spliced-out intron. Cyclization frees a linear 15-mer with a 5'-GA end.

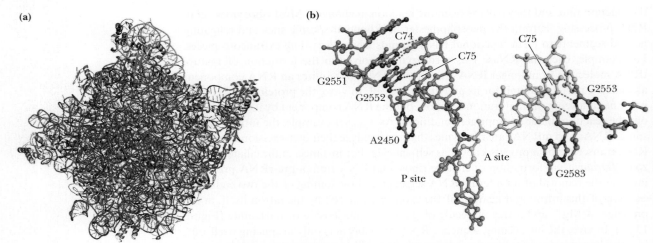

(a)

(b)

FIGURE 13.26 **(a)** The 50S subunit from *H. marismortui* (pdb id = 1FFK). Ribosomal proteins are shown in blue, the 23S rRNA backbone in brown, the 5S rRNA backbone in olive, and a tRNA substrate analog in red. The tRNA analog identifies the peptidyl transferase catalytic center of the 50S subunit. **(b)** The aminoacyl-tRNA$_A$ (yellow) and the peptidyl-tRNA$_P$ (orange) in the peptidyl transferase active site. Bases of the 23S rRNA shown in green and labeled according to their position in the 23S rRNA sequence. Interactions between the tRNAs and the 23S rRNA are indicated by dotted lines. The α-amino group of the aminoacyl-tRNA$_A$ (blue) is positioned for the attack on the carbonyl-C (green) of the peptidyl-tRNA$_P$. (Adapted from Figure 2 in Beringer, M., and Rodnina, M. V., 2007. The ribosomal peptidyl transferase. *Molecular Cell* **26:**311–321.)

peptidyl-tRNA$_P$) and the other bearing the next amino acid to be added (the *aminoacyl-tRNA$_A$*). Both the peptidyl chain and the amino acid are attached to their respective tRNAs via ester bonds to the O atom at the CCA-3′ ends of these tRNAs (see Figure 11.36). Base-pairing between these C residues in the two tRNAs and G residues in the 23S rRNA position the substrates for the reaction to occur (Figure 13.26). The two Cs at the peptidyl-tRNA$_P$ CCA end pair with G2251 and G2252 of the 23S rRNA, and the last C (C75) at the 3′-end of the aminoacyl-tRNA$_A$ pairs with G2553. The 3′-terminal A of the aminoacyl-tRNA$_A$ interacts with G2583, and the terminal A of the peptidyl-tRNA$_P$ binds to A2450. Addition of the incoming amino acid to the peptidyl chain occurs when the α-amino group of the aminoacyl-tRNA$_A$ makes a nucleophilic attack on the carbonyl C linking the peptidyl chain to its tRNA$_P$. Specific 23S rRNA bases and ribose-OH groups facilitate this nucleophilic attack by favoring proton abstraction from the aminoacyl α-amino group (Figure 13.27). The products of this reaction are a one-residue-longer peptidyl chain attached to the tRNA$_A$ and the "empty" tRNA$_P$.

The fact that RNA can catalyze such important reactions is experimental support for the idea that a primordial world dominated by RNA molecules existed before the

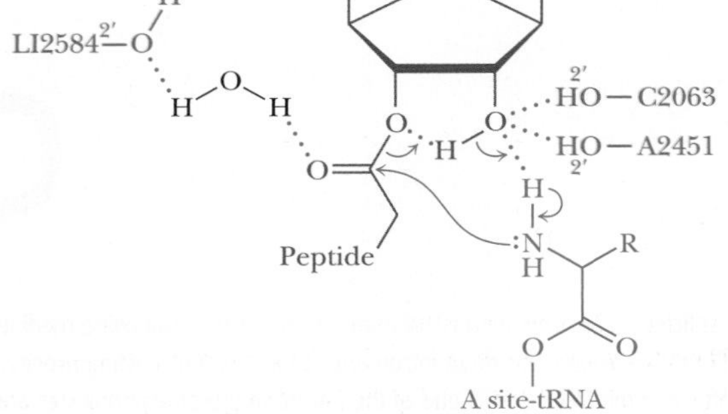

FIGURE 13.27 The peptidyl transferase reaction. Abstraction of an amide proton from the α-amino group of the aminoacyl-tRNA$_A$ (shown in red) by the 2′-O of the terminal A of the peptidyl-tRNA$_P$ (blue) is aided by hydrogen-bonding interactions with neighboring 23S rRNA nucleotides (green). These interactions facilitate nucleophilic attack by the α-amino group of the aminoacyl-tRNA$_A$ on the carbonyl C of the peptidyl-tRNA$_P$ and peptide bond formation between the incoming amino acid and the growing peptide chain to give a one-residue-longer peptide chain attached to the tRNA$_A$. (Adapted from Figure 3 in Beringer, M., and Rodnina, M. V., 2007. The ribosomal peptidyl transferase. *Molecular Cell* **26:**311–321.)

evolution of DNA and proteins. Sidney Altman and Thomas R. Cech shared the 1989 Nobel Prize in Chemistry for their discovery of the catalytic properties of RNA.

Antibody Molecules Can Have Catalytic Activity

Antibodies are *immunoglobulins,* which, of course, are proteins. Catalytic antibodies are antibodies with catalytic activity (catalytic antibodies are also called **abzymes,** a word created by combining "Ab," the abbreviation for antibody, with "enzyme.") Like other antibodies, catalytic antibodies are elicited in an organism in response to immunological challenge by a foreign molecule called an **antigen** (see Chapter 28 for discussions on the molecular basis of immunology). In this case, however, the antigen is purposefully engineered to be *an analog of the transition state in a reaction.* The strategy is based on the idea that a protein specific for binding the transition state of a reaction will promote entry of the normal reactant into the reactive, transition-state conformation. Thus, a catalytic antibody facilitates, or catalyzes, a reaction by forcing the conformation of its substrate in the direction of its transition state. (A prominent explanation for the remarkable catalytic power of conventional enzymes is their great affinity for the transition state in the reactions they catalyze; see Chapter 14.)

One proof of this principle has been to prepare ester analogs by substituting a phosphorus atom for the carbon in the ester group (Figure 13.28). The phosphonate compound mimics the natural transition state of ester hydrolysis, and antibodies elicited against these analogs act like enzymes in accelerating the rate of ester hydrolysis as much as 1000-fold. Abzymes have been developed for a number of other classes of reactions, including C—C bond formation via aldol condensation (the reverse of the aldolase reaction [see Figure 13.2, reaction 4, and Chapter 18]) and the pyridoxal 5′-P– dependent aminotransferase reaction shown in Figure 13.23. This biotechnology offers the real possibility of creating specially tailored enzymes designed to carry out specific catalytic processes.

Catalytic antibodies apparently occur naturally. Autoimmune diseases are diseases that arise because an individual begins to produce antibodies against one of their own cellular constituents. Multiple sclerosis (MS), one such autoimmune disease, is characterized by gradual destruction of the myelin sheath surrounding neurons throughout the brain and spinal cord. Blood serum obtained from some MS patients contains antibodies capable of carrying out the proteolytic destruction of myelin basic protein (MBP). That is, these antibodies were MBP-destructive proteases. Similarly, hemophilia A is a blood-clotting disorder due to lack of the factor VIII, an essential protein for formation of a

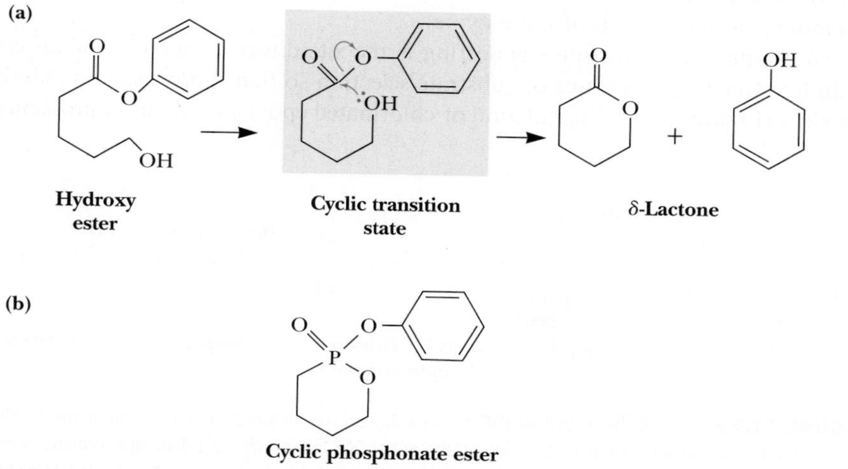

FIGURE 13.28 (a) The intramolecular hydrolysis of a hydroxy ester to yield as products a δ-lactone and the alcohol phenol. Note the cyclic transition state. (b) The cyclic phosphonate ester analog of the cyclic transition state.

blood clot. Serum from some sufferers of hemophilia A contained antibodies with proteolytic activity against factor VIII. Thus, some antibodies may be proteases.

13.8 Is It Possible to Design an Enzyme to Catalyze Any Desired Reaction?

Enzymes have evolved to catalyze metabolic reactions with high selectivity, specificity, and rate enhancements. Given these remarkable attributes, it would be very desirable to have the ability to create **designer enzymes** individually tailored to catalyze any imaginable reaction, particularly those that might have practical uses in industrial chemistry, the pharmaceutical industry, or environmental remediation. To this end, several approaches have been taken to create a desired enzyme de novo (*de novo:* literally "anew"; colloquially "from scratch." In biochemistry, the synthesis of some end product from simpler precursors.) Most approaches begin with a known enzyme and then engineer it by using in vitro mutagenesis (see Chapter 12) to replace active-site residues with a new set that might catalyze the desired reaction. This strategy has the advantage that the known protein structure provides a stable scaffold into which a new catalytic site can be introduced. As pointed out in Chapter 6, despite the extremely large number of possible amino acid sequences for a polypeptide chain, the number of stable tertiary structures that proteins can adopt is rather limited. Yet proteins have an extraordinary range of functional possibilities. So, this approach is rational. A second, more difficult, approach attempts the completely new design of a protein with the desired structure and activity. Often, this approach relies on in silico methods, where the folded protein structure and the spatial and reactive properties of its putative active site are modeled, refined, and optimized via computer. Although this approach has fewer limitations in terms of size and shape of substrates, it brings other complications, such as protein folding and stability, to the problem, to say nothing of the difficulties of going from the computer model (in silico) to a real enzyme in a cellular environment (in vivo).

Enzymes have shown adaptability over the course of evolution. New enzyme functions have appeared time and time again, as mutation and selection according to Darwinian principles operate on existing enzymes. Some enzyme designers have coupled natural evolutionary processes with rational design using in vitro mutagenesis. Expression of mutated versions of the gene encoding the enzyme in bacteria, followed by rounds of selection for bacteria producing an enzyme with even better catalytic properties, takes advantage of naturally occurring processes to drive further mutation and selection for an optimal enzyme. This dual approach is whimsically referred to as *semirational design* because it relies on the rational substitution of certain amino acids with new ones in the active site, followed by directed evolution (selection for bacteria expressing more efficient versions of the enzyme).

An example of active-site engineering is the site-directed mutation of an epoxide hydrolase to change its range of substrate selection so that it now acts on chlorinated epoxides (Figure 13.29). Degradation of chlorinated epoxides is a major problem in the

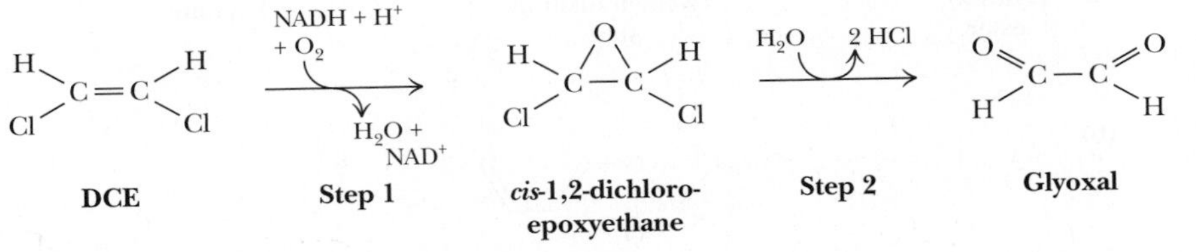

FIGURE 13.29 *cis*-1,2-Dichloroethylene (DCE) is an industrial solvent that poses hazards to human health; DCE occurs as a pollutant in water sources. Bacterial metabolism of DCE to form *cis*-1,2-dichloroepoxyethane (step 1) occurs readily, but enzymatic degradation of the epoxide to glyoxal and chloride ions (step 2) is limited. Microbial detoxification of DCE in ground water requires enzymes for both steps 1 and 2. Genetic engineering of an epoxide hydrolase to create an enzyme capable of using *cis*-1,2-dichloroepoxyethane as a substrate and catalyzing step 2 is a practical example of de novo enzyme design.

removal of toxic pollutants from water resources. Mutation of a bacterial epoxide hydrolase at three active-site residues (F[108], I[219], and C[248]) and selection in bacteria for enhanced chlorinated epoxide hydrolase activity yielded an F108L, I219L, C248I mutant enzyme that catalyzed the conversion of *cis*-dichloroepoxyethane to Cl$^-$ ions and glyoxal with a dramatically increased V_{max}/K_m ratio.

SUMMARY

OWL Login to OWL to develop problem-solving skills and complete online homework assigned by your professor.

Living systems use enzymes to accelerate and control the rates of vitally important biochemical reactions. Enzymes provide kinetic control over thermodynamic potentiality: Reactions occur in a timeframe suitable to the metabolic requirements of cells. Enzymes are the agents of metabolic function.

13.1 What Characteristic Features Define Enzymes? Enzymes can be characterized in terms of three prominent features: catalytic power, specificity, and regulation. The site on the enzyme where substrate binds and catalysis occurs is called the active site. Regulation of enzyme activity is essential to the integration and regulation of metabolism.

13.2 Can the Rate of an Enzyme-Catalyzed Reaction Be Defined in a Mathematical Way? Enzyme kinetics can determine the maximum reaction velocity that the enzyme can attain, its binding affinities for substrates and inhibitors, and the mechanism by which it accomplishes its catalysis. The kinetics of simple chemical reactions provides a foundation for exploring enzyme kinetics. Enzymes, like other catalysts, act by lowering the free energy of activation for a reaction.

13.3 What Equations Define the Kinetics of Enzyme-Catalyzed Reactions? A plot of the velocity of an enzyme-catalyzed reaction v versus the concentration of the substrate S is called a substrate saturation curve. The Michaelis–Menten equation is derived by assuming that E combines with S to form ES and then ES reacts to give E + P. Rapid, reversible combination of E and S and ES breakdown to yield P reach a steady-state condition where [ES] is essentially constant. The Michaelis–Menten equation says that the initial rate of an enzyme reaction, v, is determined by two constants, K_m and V_{max}, and the initial concentration of substrate. The turnover number of an enzyme, k_{cat}, is a measure of its maximal catalytic activity (the number of substrate molecules converted into product per enzyme molecule per unit time when the enzyme is saturated with substrate). The ratio k_{cat}/K_m defines the catalytic efficiency of an enzyme. This ratio, k_{cat}/K_m, cannot exceed the diffusion-controlled rate of combination of E and S to form ES.

Several rearrangements of the Michaelis–Menten equation transform it into a straight-line equation, a better form for experimental determination of the constants K_m and V_{max} and for detection of regulatory properties of enzymes.

13.4 What Can Be Learned from the Inhibition of Enzyme Activity? Inhibition studies on enzymes have contributed significantly to our understanding of enzymes. Inhibitors may interact either reversibly or irreversibly with an enzyme. Reversible inhibitors bind to the enzyme through noncovalent association/dissociation reactions. Irreversible inhibitors typically form stable, covalent bonds with the enzyme. Reversible inhibitors may bind at the active site of the enzyme (competitive inhibition) or at some other site on the enzyme (noncompetitive inhibition). Uncompetitive inhibitors bind only to the ES complex.

13.5 What Is the Kinetic Behavior of Enzymes Catalyzing Bimolecular Reactions? Usually, enzymes catalyze reactions in which two (or even more) substrates take part, so the reaction is bimolecular. Several possibilities arise. In single-displacement reactions, both substrates, A and B, are bound before reaction occurs. In double-displacement (or ping-pong) reactions, one substrate (A) is bound and reaction occurs to yield product P and a modified enzyme form, E′. The second substrate (B) then binds to E′ and reaction occurs to yield product Q and E, the unmodified form of enzyme. Graphical methods can be used to distinguish these possibilities. Exchange reactions are another way to diagnose bisubstrate mechanisms.

13.6 How Can Enzymes Be So Specific? Early enzyme specificity studies by Emil Fischer led to the hypothesis that an enzyme resembles a "lock" and its particular substrate the "key." However, enzymes are not rigid templates like locks. Koshland noted that the conformation of an enzyme is dynamic and hypothesized that the interaction of E with S is also dynamic. The enzyme's active site is actually modified upon binding S, in a process of dynamic recognition between enzyme and substrate called induced fit. Hexokinase provides a good illustration of the relationship between substrate binding, induced fit, and catalysis.

13.7 Are All Enzymes Proteins? Not all enzymes are proteins. Catalytic RNA molecules ("ribozymes") play important cellular roles in RNA processing and protein synthesis, among other things. Catalytic RNAs give support to the idea that a primordial world dominated by RNA molecules existed before the evolution of DNA and proteins.

Antibodies that have catalytic activity ("abzymes") can be elicited in an organism in response to immunological challenge with an analog of the transition state for a reaction. Such antibodies are catalytic because they bind the transition state of a reaction and promote entry of the normal substrate into the reactive, transition-state conformation.

13.8 Is It Possible to Design an Enzyme to Catalyze Any Desired Reaction? Several approaches have been taken to create **designer enzymes** individually tailored to catalyze any imaginable reaction. One rational approach is to begin with a known enzyme and then engineer it using in vitro mutagenesis to replace active-site residues with a new set that might catalyze the desired reaction. A second, more difficult approach uses computer modeling to design a protein with the desired structure and activity. A third approach is to couple natural evolutionary processes with rational design using in vitro mutagenesis. Expression of mutated versions of the gene encoding the enzyme in bacteria is followed by selection for bacteria producing an enzyme with even better catalytic properties. This dual approach is sometimes called semirational design, because it relies on the rational substitution of certain amino acids with new ones in the active site, followed by directed evolution. Active-site engineering and site-directed mutation have been used to modify an epoxide hydrolase so that it now acts on chlorinated epoxides, substances that are serious pollutants in water resources.

FOUNDATIONAL BIOCHEMISTRY Things You Should Know After Reading Chapter 13.

- Enzymes are the agents of metabolic function.

- Enzymes exert kinetic control over thermodynamic potentiality.

- Catalytic power is defined as the ratio of the enzyme-catalyzed rate of a reaction to the uncatalyzed rate.

- The substance upon which an enzyme acts is termed the substrate of the enzyme.

- Specificity is the term used to define the selectivity of enzymes for their substrates.

- The specific site on the enzyme to which the substrate binds and where catalysis occurs is called the active site.

- Regulation of enzyme activity ensures that the rate of a metabolic reaction is appropriate to cellular requirements.

- Enzyme nomenclature provides a systematic way of naming metabolic reactions.

- Coenzymes and cofactors are nonprotein components of enzymes that are essential for enzyme activity.

- Tightly bound coenzymes are termed prosthetic groups; the protein without the prosthetic group is called an apoenzyme.

- Enzyme kinetics is the branch of science that studies the properties of enzyme-catalyzed reactions.

- Enzymes, like other catalysts, lower the free energy of activation for a reaction.

- Decreasing the free energy of activation, $\Delta G^{\ddagger}$, increases reaction rate.

- The rate of a simple enzyme-catalyzed reaction can be described by the Michaelis-Menten equation.

- The Briggs-Haldane steady-state assumption ($d[ES]/dt = 0$) provides an improved interpretation of the Michaelis-Menten equation.

- K_m, the Michaelis constant, is defined as $(k_{-1} + k_2)/k_1$.

- The velocity of the reaction, v, is given by $v = d[P]/dt = k_2[ES]$.

- The maximal velocity of the reaction, V_{max}, is achieved when all of the enzyme, E, is in the ES form. $V_{max} = k_2[E_T]$.

- When $[S] = K_m$, $v = V_{max}/2$.

- The turnover number, or molecular activity, of an enzyme is defined as k_{cat}. $k_{cat} = V_{max}[E_T]$.

- The catalytic efficiency of an enzyme is defined as k_{cat}/K_m.

- The catalytic efficiency of an enzyme cannot exceed the diffusion-controlled rate of encounters between E and S. Thus, the upper limit on k_{cat}/K_m is about 10^9 $M^{-1}sec^{-1}$.

- Lineweaver-Burk double-reciprocal ($1/v$ vs. $1/[S]$) plots and Hanes-Woolf ($[S]/v$ vs. $[S]$) plots convert the hyperbolic Michaelis-Menten v vs. $[S]$ response into a straight line.

- Enzyme activity is affected by pH, temperature, and ionic strength, all of which can affect protein conformation.

- Enzyme inhibitors are useful probes of the properties of enzymatic reactions.

- Many drugs and other pharmacological agents are enzyme inhibitors.

- Enzyme inhibitors are classified as competitive (compete with S for binding to the active site, noncompetitive (bind elsewhere on the enzyme) or uncompetitive (bind only to the ES complex).

- Irreversible inhibitors bind irreversibly to or form covalent bonds with the enzyme.

- Bisubstrate reaction mechanisms may be single-displacement (both substrates bind and then the reaction occurs) or double-displacement (the first substrate binds, reaction occurs and product is released and then the second substrate binds, further reaction occurs and the second product is released).

- Single-displacment reactions may be random (it does not matter which substrate binds first) or ordered (one substrate, called the leading substrate, binds first; only when it is bound can the second substrate bind).

- Fischer's "lock and key hypothesis" was the original explanation for enzyme specificity.

- Koshland's "induced fit hypothesis" is a better explanation of enzyme specificity because it considers the conformational change induced in the enzyme (and the substrate) when S is bound by E.

- Not only protein molecules but RNA molecules may be catalytic; catalytic RNA molecules are termed ribozymes.

- The ribosome is a ribozyme.

- Antibodies that have catalytic activity are called abzymes.

- Designer enzymes are enzymes designed to catalyze a reaction for which no naturally occurring enzyme exists.

- Several approaches have been taken to create designer enzymes: rational enzyme design (altering the active site of a known enzyme so that it now catalyzes a new reaction) or in silico design (using computers to model an active site that will catalyze a desired reaction and then attempting to create a protein that will have such an active site).

- Designer enzymes have potential applications in organic synthesis (such as the synthesis of new drugs) and environmental remediation (to remove pollutants).

PROBLEMS

Answers to all problems are at the end of this book. Detailed solutions are available in the *Student Solutions Manual, Study Guide, and Problems Book.*

1. **Exploring the Michaelis–Menten Equation - I** According to the Michaelis–Menten equation, what is the v/V_{max} ratio when $[S] = 4\ K_m$?

2. **Exploring the Michaelis–Menten Equation - II** If $V_{max} = 100\ \mu mol/mL \cdot sec$ and $K_m = 2\ mM$, what is the velocity of the reaction when $[S] = 20\ mM$?

3. **Exploring the Michaelis–Menten Equation - III** For a Michaelis–Menten reaction, $k_1 = 7 \times 10^7/M \cdot sec$, $k_{-1} = 1 \times 10^3/sec$, and $k_2 = 2 \times 10^4/sec$. What are the values of K_S and K_m? Does substrate binding approach equilibrium, or does it behave more like a steady-state system?

4. **Graphing the Results from Kinetics Experiments with Enzyme Inhibitors** The following kinetic data were obtained for an enzyme in the absence of any inhibitor (1), and in the presence of two different

inhibitors (2) and (3) at 5 mM concentration. Assume [E_T] is the same in each experiment.

[S] (mM)	(1) v (μmol/ mL · sec)	(2) v (μmol/ mL · sec)	(3) v (μmol/ mL · sec)
1	12	4.3	5.5
2	20	8	9
4	29	14	13
8	35	21	16
12	40	26	18

Graph these data as Lineweaver–Burk plots and use your graph to find answers to a. and b.

a. Determine V_{max} and K_m for the enzyme.

b. Determine the type of inhibition and the K_I for each inhibitor.

5. **How Varying the Amount of Enzyme or the Addition of Inhibitors Affects v Versus [S] Plots** Using Figure 13.7 as a model, draw curves that would be obtained in v versus [S] plots when

a. twice as much enzyme is used.

b. half as much enzyme is used.

c. a competitive inhibitor is added.

d. a pure noncompetitive inhibitor is added.

e. an uncompetitive inhibitor is added.

For each example, indicate how V_{max} and K_m change.

6. **Using Graphical Methods to Derive the Kinetic Constants for an Ordered, Single-Displacement Reaction** The general rate equation for an ordered, single-displacement reaction where A is the leading substrate is

$$v = \frac{V_{max}[A][B]}{(K_S^A K_m^B + K_m^A[B] + K_m^B[A] + [A][B])}$$

Write the Lineweaver–Burk (double-reciprocal) equivalent of this equation and from it calculate algebraic expressions for the following:

a. The slope

b. The y-intercepts

c. The horizontal and vertical coordinates of the point of intersection when 1/v is plotted versus 1/[B] at various *fixed* concentrations of A

7. **Interpreting Kinetics Experiments from Graphical Patterns** The following graphical patterns obtained from kinetic experiments have several possible interpretations depending on the nature of the experiment and the variables being plotted. Give at least two possibilities for each.

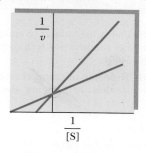

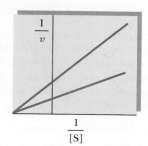

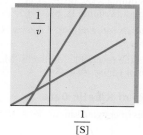

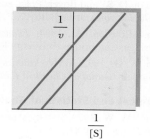

8. **Using the Equations of Enzyme Kinetics to Treat Methanol Intoxication** Liver alcohol dehydrogenase (ADH) is relatively nonspecific and will oxidize ethanol or other alcohols, including methanol. Methanol oxidation yields formaldehyde, which is quite toxic, causing, among other things, blindness. Mistaking it for the cheap wine he usually prefers, my dog Clancy ingested about 50 mL of windshield washer fluid (a solution 50% in methanol). Knowing that methanol would be excreted eventually by Clancy's kidneys if its oxidation could be blocked, and realizing that, in terms of methanol oxidation by ADH, ethanol would act as a competitive inhibitor, I decided to offer Clancy some wine. How much of Clancy's favorite vintage (12% ethanol) must he consume in order to lower the activity of his ADH on methanol to 5% of its normal value if the K_m values of canine ADH for ethanol and methanol are 1 millimolar and 10 millimolar, respectively? (The K_I for ethanol in its role as competitive inhibitor of methanol oxidation by ADH is the same as its K_m.) Both the methanol and ethanol will quickly distribute throughout Clancy's body fluids, which amount to about 15 L. Assume the densities of 50% methanol and the wine are both 0.9 g/mL.

9. **Quantitative Relationships Between Rate Constants to Calculate K_m, Kinetic Efficiency (k_{cat}/K_m) and V_{max} - I** Measurement of the rate constants for a simple enzymatic reaction obeying Michaelis–Menten kinetics gave the following results:
$k_1 = 2 \times 10^8 \, M^{-1} \, sec^{-1}$
$k_{-1} = 1 \times 10^3 \, sec^{-1}$
$k_2 = 5 \times 10^3 \, sec^{-1}$

a. What is K_S, the dissociation constant for the enzyme–substrate complex?

b. What is K_m, the Michaelis constant for this enzyme?

c. What is k_{cat} (the turnover number) for this enzyme?

d. What is the catalytic efficiency (k_{cat}/K_m) for this enzyme?

e. Does this enzyme approach "kinetic perfection"? (That is, does k_{cat}/K_m approach the diffusion-controlled rate of enzyme association with substrate?)

f. If a kinetic measurement was made using 2 nanomoles of enzyme per mL and saturating amounts of substrate, what would V_{max} equal?

g. Again, using 2 nanomoles of enzyme per mL of reaction mixture, what concentration of substrate would give $v = 0.75 \, V_{max}$?

h. If a kinetic measurement was made using 4 nanomoles of enzyme per mL and saturating amounts of substrate, what would V_{max} equal? What would K_m equal under these conditions?

10. **Quantitative Relationships Between Rate Constants to Calculate K_m, Kinetic Efficiency (k_{cat}/K_m) and V_{max} - II** Triose phosphate isomerase catalyzes the conversion of glyceraldehyde-3-phosphate to dihydroxyacetone phosphate.

$$\text{Glyceraldehyde-3-P} \rightleftharpoons \text{dihydroxyacetone-P}$$

The K_m of this enzyme for its substrate glyceraldehyde-3-phosphate is $1.8 \times 10^{-5} \, M$. When [glyceraldehydes-3-phosphate] = 30 μM, the rate of the reaction, v, was 82.5 μmol mL^{-1} sec^{-1}.

a. What is V_{max} for this enzyme?

b. Assuming 3 nanomoles per mL of enzyme was used in this experiment ([E_{total}] = 3 nanomol/mL), what is k_{cat} for this enzyme?

c. What is the catalytic efficiency (k_{cat}/K_m) for triose phosphate isomerase?

d. Does the value of k_{cat}/K_m reveal whether triose phosphate isomerase approaches "catalytic perfection"?

e. What determines the ultimate speed limit of an enzyme-catalyzed reaction? That is, what is it that imposes the physical limit on kinetic perfection?

11. **Quantitative Relationships Between Rate Constants to Calculate K_m, Kinetic Efficiency (k_{cat}/K_m) and V_{max} - III** The citric acid cycle enzyme *fumarase* catalyzes the conversion of fumarate to form malate.

Fumarate + $H_2O \rightleftharpoons$ malate

The turnover number, k_{cat}, for fumarase is 800/sec. The K_m of fumarase for its substrate fumarate is 5 μM.

a. In an experiment using 2 nanomole/mL of fumarase, what is V_{max}?
b. The cellular concentration of fumarate is 47.5 μM. What is v when [fumarate] = 47.5 μM?
c. What is the catalytic efficiency of fumarase?
d. Does fumarase approach "catalytic perfection"?

12. **Quantitative Relationships Between Rate Constants to Calculate K_m, Kinetic Efficiency (k_{cat}/K_m) and V_{max} - IV** Carbonic anhydrase catalyzes the hydration of CO_2:

$$CO_2 + H_2O \rightleftharpoons H_2CO_3$$

The K_m of carbonic anhydrase for CO_2 is 12 mM. Carbonic anhydrase gave an initial velocity v_o = 4.5 μmol H_2CO_3 formed/mL · sec when [CO_2] = 36 mM.

a. What is V_{max} for this enzyme?
b. Assuming 5 pmol/mL (5×10^{-12} moles/mL) of enzyme were used in this experiment, what is k_{cat} for this enzyme?
c. What is the catalytic efficiency of this enzyme?
d. Does carbonic anhydrase approach "catalytic perfection"?

13. **Quantitative Relationships Between Rate Constants to Calculate K_m, Kinetic Efficiency (k_{cat}/K_m) and V_{max} - V** Acetylcholinesterase catalyzes the hydrolysis of the neurotransmitter acetylcholine:

Acetylcholine + $H_2O \longrightarrow$ acetate + choline

The K_m of acetylcholinesterase for its substrate acetylcholine is 9×10^{-5} M. In a reaction mixture containing 5 nanomoles/mL of acetylcholinesterase and 150 μM acetylcholine, a velocity v_o = 40 μmol/mL · sec was observed for the acetylcholinesterase reaction.

a. Calculate V_{max} for this amount of enzyme.
b. Calculate k_{cat} for acetylcholinesterase.
c. Calculate the catalytic efficiency (k_{cat}/K_m) for acetylcholinesterase.
d. Does acetylcholinesterase approach "catalytic perfection"?

14. **Quantitative Relationships Between Rate Constants to Calculate K_m, Kinetic Efficiency (k_{cat}/K_m) and V_{max} - VI** The enzyme catalase catalyzes the decomposition of hydrogen peroxide:

$$2 H_2O_2 \rightleftharpoons 2 H_2O + O_2$$

The turnover number (k_{cat}) for catalase is 40,000,000 sec^{-1}. The K_m of catalase for its substrate H_2O_2 is 0.11 M.

a. In an experiment using 3 nanomole/L of catalase, what is V_{max}?
b. What is v when [H_2O_2] = 0.75 M?
c. What is the catalytic efficiency of fumarase?
d. Does catalase approach "catalytic perfection"?

15. **The Effect of the Accumulation of P, the Reaction Product, on the Michaelis–Menten Equation** Equation 13.9 presents the simple Michaelis–Menten situation where the reaction is considered to be irreversible ([P] is negligible). Many enzymatic reactions are reversible, and P does accumulate.

a. Derive an equation for v, the rate of the enzyme-catalyzed reaction S$\longrightarrow$P in terms of a modified Michaelis–Menten model that incorporates the reverse reaction that will occur in the presence of product, P.
b. Solve this modified Michaelis–Menten equation for the special situation when v = 0 (that is, S$\rightleftharpoons$P is at equilibrium, or in other words, K_{eq} = [P]/[S]).

(J. B. S. Haldane first described this reversible Michaelis–Menten modification, and his expression for K_{eq} in terms of the modified M–M equation is known as the Haldane relationship.)

Preparing for the MCAT® Exam

16. **Enzyme A follows simple Michaelis–Menten kinetics.**

a. The K_m of enzyme A for its substrate S is K_m^S=1 mM. Enzyme A also acts on substrate T and its K_m^T=10 mM. Is S or T the preferred substrate for enzyme A?
b. The rate constant k_2 with substrate S is 2×10^4 sec^{-1}; with substrate T, $k_2 = 4 \times 10^5$ sec^{-1}. Does enzyme A use substrate S or substrate T with greater catalytic efficiency?

17. **Use Figure 13.12 to answer the following questions.**

a. Is the enzyme whose temperature versus activity profile is shown in Figure 13.12 likely to be from an animal or a plant? Why?
b. What do you think the temperature versus activity profile for an enzyme from a thermophilic prokaryote growing in an 80°C pool of water would resemble?

ActiveModels Problems

18. Examine the ActiveModel for glutamate-oxaloacetate aminotransferase (pdb id = GOT1) and identify the amino acid side chains involved in binding pyridoxal-5′-phosphate

19. Using the ActiveModel for hexokinase, explain the conformational change that occurs upon substrate binding.

FURTHER READING

Enzymes in General

Bennett, B. D., Kimball, E. H., Osterhout, R., Van Dien, S. J., and Rabinowitz, J. D., 2009. Absolute metabolite concentrations and implied enzyme active site occupancy in *Escherichia coli*. *Nature Chemical Biology* **5**:593–599.

Creighton, T. E., 1997. *Protein Structure: A Practical Approach* and *Protein Function: A Practical Approach.* Oxford: Oxford University Press.

Fersht, A., 1999. *Structure and Mechanism in Protein Science.* New York: Freeman & Co. A guide to protein structure, chemical catalysis, enzyme kinetics, enzyme regulation, protein engineering, and protein folding.

Petsko, G. A., and Ringe, D., 2004. *Protein Structure and Function.* Sunderland, MA: New Science Press, LTD; Sinauer Associates, Inc.

Catalytic Power

Miller, B. G., and Wolfenden, R., 2002. Catalytic proficiency: The unusual case of OMP decarboxylase. *Annual Review of Biochemistry* **71**:847–885.

Wolfenden, R., 2011. Benchmark reaction rates, the stability of biological molecules in water, and the evolution of catalytic power in enzymes. *Annual Review of Biochemistry* **80**:645–667.

Wolfenden, R., and Snider, M. J., 2001. The depth of chemical time and the power of enzymes as catalysts. *Accounts of Chemical Research* **34**:938–945.

General Reviews of Enzyme Kinetics

Cleland, W. W., 1990. Steady-state kinetics. In *The Enzymes,* 3rd ed. Sigman, D. S., and Boyer, P. D., eds. Volume XIX, pp. 99–158. See also, *The Enzymes,* 3rd ed. Boyer, P. D., ed., Volume II, pp. 1–65, 1970.

Cornish-Bowden, A., 1994. *Fundamentals of Enzyme Kinetics.* Cambridge: Cambridge University Press.

Smith, W. G., 1992. In vivo kinetics and the reversible Michaelis–Menten model. *Journal of Chemical Education* **12**:981–984.

Graphical and Statistical Analysis of Kinetic Data

Cleland, W. W., 1979. Statistical analysis of enzyme kinetic data. *Methods in Enzymology* **82**:103–138.

Naqui, A., 1986. Where are the asymptotes of Michaelis–Menten? *Trends in Biochemical Sciences* **11**:64–65. A proof that the Michaelis–Menten equation describes a rectangular hyperbola.

Rudolph, F. B., and Fromm, H. J., 1979. Plotting methods for analyzing enzyme rate data. *Methods in Enzymology* **63**:138–159. A review of the various rearrangements of the Michaelis–Menten equation that yield straight-line plots.

Segel, I. H., 1976. *Biochemical Calculations,* 2nd ed. New York: John Wiley & Sons. An excellent guide to solving problems in enzyme kinetics.

Effect of Active Site Amino Acid Substitutions on k_{cat}/K_m

Garrett, R. M., et al., 1998. Human sulfite oxidase R160Q: Identification of the mutation in a sulfite oxidase-deficient patient and expression and characterization of the mutant enzyme. *Proceedings of the National Academy of Sciences U.S.A.* **95**:6394–6398.

Garrett, R. M., and Rajagopalan, K. V., 1996. Site-directed mutagenesis of recombinant sulfite oxidase. *Journal of Biological Chemistry* **271**: 7387–7391.

Enzymes and Rational Drug Design

Cornish-Bowden, A., and Eisenthal, R., 1998. Prospects for antiparasitic drugs: The case of *Trypanosoma brucei,* the causative agent of African sleeping sickness. *Journal of Biological Chemistry* **273**:5500–5505. An analysis of why drug design strategies have had only limited success.

Kling, J., 1998. From hypertension to angina to Viagra. *Modern Drug Discovery* **1**:31–38. The story of the serendipitous discovery of Viagra in a search for agents to treat angina and high blood pressure.

Enzyme Inhibition

Cleland, W. W., 1979. Substrate inhibition. *Methods in Enzymology* **63**: 500–513.

Pollack, S. J., et al., 1994. Mechanism of inositol monophosphatase, the putative target of lithium therapy. *Proceedings of the National Academy of Sciences U.S.A.* **91**:5766–5770.

Silverman, R. B., 1988. *Mechanism-Based Enzyme Inactivation: Chemistry and Enzymology,* Vols. I and II. Boca Raton, FL: CRC Press.

Catalytic RNA

Altman, S., 2000. The road to RNase P. *Nature Structural Biology* **7**: 827–828.

Cech, T. R., and Bass, B. L., 1986. Biological catalysis by RNA. *Annual Review of Biochemistry* **55**:599–629. A review of the early evidence that RNA can act like an enzyme.

Doherty, E. A., and Doudna, J. A., 2000. Ribozyme structures and mechanisms. *Annual Review of Biochemistry* **69**:597–615.

Frank, D. N., and Pace, N. R., 1998. Ribonuclease P: Unity and diversity in a tRNA processing ribozyme. *Annual Review of Biochemistry* **67**: 153–180.

Narlikar, G. J., and Herschlag, D., 1997. Mechanistic aspects of enzymatic catalysis: Comparison of RNA and protein enzymes. *Annual Review of Biochemistry* **66**:19–59. A comparison of RNA and protein enzymes that addresses fundamental principles in catalysis and macromolecular structure.

Nissen, P., et al., 2000. The structural basis of ribosome activity in peptide bond synthesis. *Science* **289**:920–930. Peptide bond formation by the ribosome: the ribosome is a ribozyme.

Schimmel, P., and Kelley, S. O., 2000. Exiting an RNA world. *Nature Structural Biology* **7**:5–7. Review of the in vitro creation of an RNA capable of catalyzing the formation of an aminoacyl-tRNA. Such a ribozyme would be necessary to bridge the evolutionary gap between a primordial RNA world and the contemporary world of proteins.

Watson, J. D., ed., 1987. Evolution of catalytic function. *Cold Spring Harbor Symposium on Quantitative Biology* **52**:1–955. Publications from a symposium on the nature and evolution of catalytic biomolecules (proteins and RNA) prompted by the discovery that RNA could act catalytically.

Wilson, D. S., and Szostak, J. W., 1999. In vitro selection of functional nucleic acids. *Annual Review of Biochemistry* **68**:611–647. Screening libraries of random nucleotide sequences for catalytic RNAs.

Catalytic Antibodies

Hilvert, D., 2000. Critical analysis of antibody catalysis. *Annual Review of Biochemistry* **69**:751–793. A review of catalytic antibodies that were elicited with rationally designed transition-state analogs.

Janda, K. D., 1997. Chemical selection for catalysis in combinatorial antibody libraries. *Science* **275**:945.

Lacroix-Desmazes, S., et al., 2002. The prevalence of proteolytic antibodies against factor VIII in Hemophilia A. *New England Journal of Medicine* **346**:662–667.

Ponomarenko, N. A., 2006. Autoantibodies to myelin basic protein catalyze site-specific degradation of their antigen. *Proceedings of the National Academy of Sciences U S A* **103**:281–286.

Wagner, J., Lerner, R. A., and Barbas, C. F., III, 1995. Efficient adolase catalytic antibodies that use the enamine mechanism of natural enzymes. *Science* **270**:1797–1800.

Designer Enzymes

Chica, R. A., Doucet, N., and Pelletier, J. N., 2005. Semi-rational approaches to engineering enzyme activity: Combining the benefits of directed evolution and rational design. *Current Opinion in Biotechnology* **16**:378–384.

Jiang, L., Althoff, E. A., Clemente, F. R., et al., 2008. De novo computational design of retro-aldol enzymes. *Science* **319**:1387–1391.

Kaplan, J., and DeGrado, W. F., 2004. *De novo* design of catalytic proteins. *Proceedings of the National Academy of Sciences U S A* **101**:11566–11570.

Lippow, S. M., and Tidor, B., 2007. Progress in computational protein design. *Current Opinion in Biotechnology* **18**:305–311.

Rui, L., Cao L., Chen W., et al., 2004. Active site engineering of the epoxide hydrolase from *Agrobacterium radiobacter* AD1 to enhance aerobic mineralization of *cis*-1,2,-dichloroethylene in cells expressing an evolved toluene *ortho*-monooxygenase. *The Journal of Biological Chemistry* **279**:46810–46817.

Siegel, J. B., Zanghellini, A., Lovick, H. M., et al., 2010. Computational design of an enzyme catalyst for a stereoselective bimolecular Diels-Alder reaction. *Science* **329**:309–313. See also the overview by Lutz, S., 2010. Reengineering enzymes. *Science* **309**:285–287.

Walter, K. U., Vamvaca, K., and Hilvert, D., 2005. An active enzyme constructed from a 9-amino acid alphabet. *The Journal of Biological Chemistry* **280**:37742–37746.

Woycechowsky, K. L., et al., 2007. Novel enzymes through design and evolution. *Advances in Enzymology and Related Areas of Molecular Biology* **75**:241–294.

Specificity

Jencks, W. P., 1975. Binding energy, specificity, and enzymic catalysis: The Circe effect. *Advances in Enzymology* **43**:219–410. Enzyme specificity stems from the favorable binding energy between the active site and the substrate and unfavorable binding or exclusion of nonsubstrate molecules.

Johnson, K. A., 2008. Role of induced fit in enzyme specificity: A molecular forward/reverse switch. *The Journal of Biological Chemistry* **283**: 26297–26301.

The Effect of Temperature on Enzyme Activity

Daniel, R. M., and Danson, M. J., 2010. A new understanding of how temperature affects the catalytic activity of enzymes. *Trends in Biochemical Sciences* **35**:584–591.

Mechanisms of Enzyme Action

David W. Grisham

◀ Like the workings of machines, the details of enzyme mechanisms are at once complex and simple.

No single thing abides but all things flow.
Fragment to fragment clings and thus they grow
Until we know them by name.
Then by degrees they change and are no more the things we know.

Lucretius (ca. 94 B.C.–50 B.C.)

KEY QUESTIONS

14.1 What Are the Magnitudes of Enzyme-Induced Rate Accelerations?

14.2 What Role Does Transition-State Stabilization Play in Enzyme Catalysis?

14.3 How Does Destabilization of ES Affect Enzyme Catalysis?

14.4 How Tightly Do Transition-State Analogs Bind to the Active Site?

14.5 What Are the Mechanisms of Catalysis?

14.6 What Can Be Learned from Typical Enzyme Mechanisms?

ESSENTIAL QUESTION

Although the catalytic properties of enzymes may seem almost magical, it is simply chemistry—the breaking and making of bonds—that gives enzymes their prowess. This chapter will explore the unique features of this chemistry. The mechanisms of thousands of enzymes have been studied in at least some detail. In this chapter, it will be possible to examine only a few of these.

What are the universal chemical principles that influence the mechanisms of enzymes and allow us to understand their enormous catalytic power?

14.1 What Are the Magnitudes of Enzyme-Induced Rate Accelerations?

Enzymes are powerful catalysts. Enzyme-catalyzed reactions are typically 10^7 to 10^{15} times faster than their uncatalyzed counterparts (Table 14.1). The most impressive reaction acceleration known is that of fructose-1,6-bisphosphatase, an enzyme found in liver and kidneys that is involved in the synthesis of glucose (see Chapter 22).

These large rate accelerations correspond to substantial decreases in the free energy of activation for the reaction in question. The urease reaction, for example,

$$H_2N-\overset{\overset{\textstyle O}{\|}}{C}-NH_2 + 2\ H_2O + H^+ \longrightarrow 2\ NH_4^+ + HCO_3^-$$

shows an energy of activation 84 kJ/mol smaller than that for the corresponding uncatalyzed reaction. To fully understand any enzyme reaction, it is important to account for the rate acceleration in terms of the structure of the enzyme and its mechanism of action.

In all chemical reactions, the reacting atoms or molecules pass through a state that is intermediate in structure between the reactant(s) and the product(s). Consider the transfer of a proton from a water molecule to a chloride anion:

$$\underset{\textbf{Reactants}}{H-O-H + Cl^-} \ \rightleftharpoons \ \underset{\textbf{Transition state}}{H-O^{\delta-}\cdots H\cdots Cl^{\delta-}} \ \rightleftharpoons \ \underset{\textbf{Products}}{HO^- + H-Cl}$$

OWL Online homework and a Student Self Assessment for this chapter may be assigned in OWL.

TABLE 14.1	A Comparison of Enzyme-Catalyzed Reactions and Their Uncatalyzed Counterparts			
Reaction	Enzyme	Uncatalyzed Rate, v_u (sec^{-1})	Catalyzed Rate, v_e (sec^{-1})	v_e/v_u
Fructose-1,6-bisP $\longrightarrow$ fructose-6-P + P$_i$	Fructose-1,6-bisphosphatase	2×10^{-20}	21	1.05×10^{21}
(Glucose)$_n$ + H$_2$O $\longrightarrow$ (glucose)$_{n-2}$ + maltose	β-amylase	1.9×10^{-15}	1.4×10^3	7.2×10^{17}
DNA, RNA cleavage	Staphylococcal nuclease	7×10^{-16}	95	1.4×10^{17}
CH$_3$—O—PO$_3^{2-}$ + H$_2$O $\longrightarrow$ CH$_3$OH + HPO$_4^{2-}$	Alkaline phosphatase	1×10^{-15}	14	1.4×10^{16}
$\overset{\text{O}}{\overset{\|}{\text{H}_2\text{N—C—NH}_2}}$ + 2 H$_2$O + H$^+$ $\longrightarrow$ 2 NH$_4^+$ + HCO$_3^-$	Urease	3×10^{-10}	3×10^4	1×10^{14}
$\overset{\text{O}}{\overset{\|}{\text{R—C—O—CH}_2\text{CH}_3}}$ + H$_2$O $\longrightarrow$ RCOOH + HOCH$_2$CH$_3$	Chymotrypsin	1×10^{-10}	1×10^2	1×10^{12}
Glucose + ATP $\longrightarrow$ Glucose-6-P + ADP	Hexokinase	$<1 \times 10^{-13}$	1.3×10^{-3}	$>1.3 \times 10^{10}$
CH$_3$CH$_2$OH + NAD$^+$ $\longrightarrow$ $\overset{\text{O}}{\overset{\|}{\text{CH}_3\text{CH}}}$ + NADH + H$^+$	Alcohol dehydrogenase	$<6 \times 10^{-12}$	2.7×10^{-5}	$>4.5 \times 10^6$
CO$_2$ + H$_2$O $\longrightarrow$ HCO$_3^-$ + H$^+$	Carbonic anhydrase	10^{-2}	10^5	1×10^7
Creatine + ATP $\longrightarrow$ Cr-P + ADP	Creatine kinase	$<3 \times 10^{-9}$	4×10^{-5}	$>1.33 \times 10^4$

Adapted from Koshland, D., 1956. Molecular geometry in enzyme action. *Journal of Cellular Comparative Physiology*, Supp. 1, **47**:217; and Wolfenden, R., 2006. Degrees of difficulty of water-consuming reactions in the absence of enzymes. *Chemical Reviews* **106**:3379–3396.

In the middle structure, the proton undergoing transfer is shared equally by the hydroxyl and chloride anions. This structure represents, as nearly as possible, the transition between the reactants and products, and it is known as the **transition state.**[1] All transition states contain at least one partially formed bond.

Linus Pauling was the first to suggest (in 1946) that the active sites of enzymes bind the transition state more readily than the substrate and that, by doing so, they stabilize the transition state and lower the activation energy of the reaction. Many subsequent studies have shown that this idea is essentially correct, but it is just the beginning in understanding what enzymes do. Reaction rates can also be accelerated by destabilizing (raising the energy of) the enzyme–substrate complex. Chemical groups arrayed across the active site actually guide the entering substrate toward the formation of the transition state. Thus, the enzyme active site is said to be "preorganized." Enzymes are dynamic, and fluctuations in the active-site structure are presumed to organize the initial enzyme–substrate complex into a reactive conformation and on to the transition state. Along the way, electrostatic and hydrophobic interactions between the enzyme and the substrate mediate and direct these changes that make the reaction possible. Often, catalytic groups provided by the enzyme participate directly in proton transfers and other bond-making and bond-breaking events.

This chapter describes and elaborates on each of these contributions to the catalytic prowess of enzymes and then illustrates the lessons learned by looking closely at the mechanisms of three well-understood enzymes.

14.2 | What Role Does Transition-State Stabilization Play in Enzyme Catalysis?

Chemical reactions in which a substrate (S) is converted to a product (P) can be pictured as involving a transition state (which we henceforth denote as X‡), a species intermediate in structure between S and P (Figure 14.1). As seen in Chapter 13, the catalytic role

[1]It is important to distinguish **transition states** from **intermediates.** A transition state is envisioned as an extreme distortion of a bond, and thus the lifetime of a typical transition state is viewed as being on the order of the lifetime of a bond vibration, typically 10^{-13} sec. Intermediates, on the other hand, are longer lived, with lifetimes in the range of 10^{-13} to 10^{-3} sec.

(a)

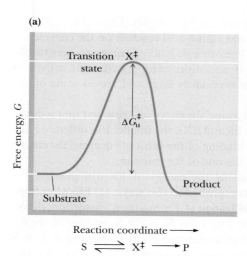

(b)

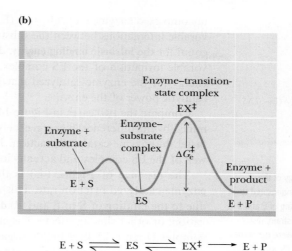

FIGURE 14.1 Enzymes catalyze reactions by lowering the activation energy. Here the free energy of activation for **(a)** the uncatalyzed reaction, $\Delta G_u^{\ddagger}$, is larger than that for **(b)** the enzyme-catalyzed reaction, $\Delta G_e^{\ddagger}$.

of an enzyme is to reduce the energy barrier between substrate and transition state. This is accomplished through the formation of an **enzyme–substrate complex** (ES). This complex is converted to product by passing through a transition state, $EX^{\ddagger}$ (Figure 14.1). As shown, the energy of $EX^{\ddagger}$ is clearly lower than that of $X^{\ddagger}$. One might be tempted to conclude that this decrease in energy explains the rate enhancement achieved by the enzyme, but there is more to the story.

The energy barrier for the uncatalyzed reaction (Figure 14.1) is of course the difference in energies of the S and $X^{\ddagger}$ states. Similarly, the energy barrier to be surmounted in the enzyme-catalyzed reaction, assuming that E is saturated with S, is the energy difference between ES and $EX^{\ddagger}$. *Reaction rate acceleration by an enzyme means simply that the energy barrier between ES and $EX^{\ddagger}$ is less than the energy barrier between S and $X^{\ddagger}$.* In terms of the free energies of activation, $\Delta G_e^{\ddagger} < \Delta G_u^{\ddagger}$.

There are important consequences for this statement. The enzyme must stabilize the transition-state complex, $EX^{\ddagger}$, more than it stabilizes the substrate complex, ES. Put another way, enzymes bind the transition-state structure more tightly than the substrate (or the product). The dissociation constant for the enzyme–substrate complex is

$$K_S = \frac{[E][S]}{[ES]} \tag{14.1}$$

and the corresponding dissociation constant for the transition-state complex is

$$K_T = \frac{[E][X^{\ddagger}]}{[EX^{\ddagger}]} \tag{14.2}$$

Enzyme catalysis requires that $K_T < K_S$. According to **transition-state theory** (see references at the end of this chapter), the rate constants for the enzyme-catalyzed (k_e) and uncatalyzed (k_u) reactions can be related to K_S and K_T by

$$k_e/k_u \approx K_S/K_T \tag{14.3}$$

Thus, the enzymatic rate enhancement is approximately equal to the ratio of the dissociation constants of the enzyme–substrate and enzyme–transition-state complexes, at least when E is saturated with S.

14.3   How Does Destabilization of ES Affect Enzyme Catalysis?

How is it that $X^{\ddagger}$ is stabilized more than S at the enzyme active site? To understand this, we must dissect and analyze the formation of the enzyme–substrate complex, ES. There are a number of important contributions to the free energy difference between the

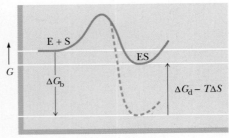

FIGURE 14.2 The intrinsic binding energy of the enzyme–substrate (ES) complex (ΔG_b) is compensated to some extent by entropy loss due to the binding of E and S ($T\Delta S$) and by destabilization of ES (ΔG_d) by strain, distortion, desolvation, and similar effects. If ΔG_b were not compensated by $T\Delta S$ and ΔG_d, the formation of ES would follow the dashed line.

uncomplexed enzyme and substrate (E + S) and the ES complex (Figure 14.2). The favorable interactions between the substrate and amino acid residues on the enzyme account for the **intrinsic binding energy, ΔG_b.** The intrinsic binding energy ensures the favorable formation of the ES complex, but if uncompensated, it makes the activation energy for the enzyme-catalyzed reaction unnecessarily large and wastes some of the catalytic power of the enzyme.

Compare the two cases in Figure 14.3. Because the enzymatic reaction rate is determined by the difference in energies between ES and EX‡, the smaller this difference, the faster the enzyme-catalyzed reaction. Tight binding of the substrate deepens the energy well of the ES complex and actually lowers the rate of the reaction.

The message of Figure 14.3 is that raising the energy of ES will increase the enzyme-catalyzed reaction rate. This is accomplished in two ways: (1) **loss of entropy** due to the binding of S to E and (2) **destabilization of ES** by strain, distortion, desolvation, or other similar effects. The entropy loss arises from the formation of the ES complex (Figure 14.4), a highly organized (low-entropy) entity compared to E + S in solution (a disordered, high-entropy situation). Because ΔS is negative for this process, the term $-T\Delta S$ is a positive quantity, and *the intrinsic binding energy of ES is compensated to some extent by the entropy loss that attends the formation of the complex.*

Destabilization of the ES complex can involve **structural strain, desolvation,** or **electrostatic effects.** Destabilization by strain or distortion is usually just a consequence of the fact (noted previously) that *the enzyme is designed to bind the transition state more strongly than the substrate.* When the substrate binds, the imperfect nature of the "fit" results in distortion or strain in the substrate, the enzyme, or both.

Solvation of charged groups on a substrate in solution releases energy, making the charged substrate more stable. When a substrate with charged groups moves from water into an enzyme active site (Figure 14.4), the charged groups are often desolvated to some extent, becoming less stable and therefore more reactive.

Similarly, when a substrate enters the active site, charged groups may be forced to interact (unfavorably) with charges of like sign, resulting in **electrostatic destabilization** (Figure 14.4). The reaction pathway acts in part to remove this stress. If the charge on the substrate is diminished or lost in the course of reaction, electrostatic destabilization can result in rate acceleration.

Whether by strain, desolvation, or electrostatic effects, destabilization raises the energy of the ES complex, and this increase is summed in the term ΔG_d, the free energy of destabilization (Figures 14.2 and 14.3).

(a)

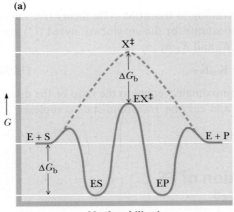

No destabilization,
thus no catalysis

(b)

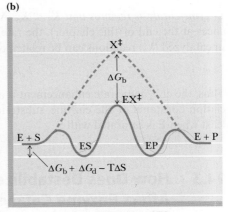

Destabilization of ES
facilitates catalysis

FIGURE 14.3 (a) Catalysis does not occur if the ES complex and the transition state for the reaction are stabilized to equal extents. (b) Catalysis *will* occur if the transition state is stabilized to a greater extent than the ES complex *(right).* Entropy loss and destabilization of the ES complex ΔG_d ensure that this will be the case.

(a)

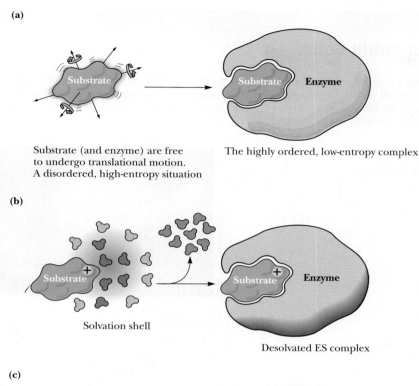

Substrate (and enzyme) are free to undergo translational motion. A disordered, high-entropy situation

The highly ordered, low-entropy complex

(b)

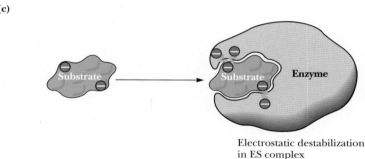

Solvation shell

Desolvated ES complex

(c)

Electrostatic destabilization in ES complex

FIGURE 14.4 (a) Formation of the ES complex results in entropy loss. Before binding, E and S are free to undergo translational and rotational motion. The ES complex is a more highly ordered, low-entropy complex. **(b)** Substrates typically lose waters of hydration in the formation of the ES complex. Desolvation raises the energy of the ES complex, making it more reactive. **(c)** Electrostatic destabilization of a substrate may arise from juxtaposition of like charges in the active site. If such charge repulsion is relieved in the course of the reaction, electrostatic destabilization can result in a rate increase.

14.4 | How Tightly Do Transition-State Analogs Bind to the Active Site?

Although not apparent at first, there are other important implications of Equation 14.3. It is important to consider the magnitudes of K_S and K_T. The ratio k_e/k_u may even exceed 10^{16}, as noted previously. Given a ratio of 10^{16} and a typical K_S of 10^{-4} M, the value of K_T should be 10^{-20} M. The value of K_T for fructose-1,6-bisphosphatase (see Table 14.1) is an astounding 7×10^{-26} M! This is the dissociation constant for the transition state from the enzyme, and this very low value corresponds to very tight binding of the transition state by the enzyme.

It is unlikely that such tight binding in an enzyme transition state will ever be determined in a direct equilibrium measurement, however, because the transition state itself is a "moving target." It exists only for about 10^{-14} to 10^{-13} sec, less than the time required for a bond vibration. On the other hand, the nature of the elusive transition state can be explored using **transition-state analogs**, stable molecules that are chemically and structurally similar to the transition state. Such molecules should bind more strongly than a substrate and more strongly than competitive inhibitors that bear no significant similarity to the transition state. Hundreds of examples of such behavior have been reported. For example, Robert Abeles studied a series of inhibitors of

▶ **A DEEPER LOOK**

Transition-State Analogs Make Our World Better

Enzymes (human, plant, and bacterial) are often targets for drugs and other beneficial agents. Transition-state analogs (TSAs), with very high affinities for their enzyme-binding sites, often make ideal enzyme inhibitors, and TSAs have become ubiquitous therapeutic agents that improve the lives of millions and millions of people. A few applications of transition-state analogs for human health and for agriculture are shown here.

Enalapril and Aliskiren Lower Blood Pressure

High blood pressure is a significant risk factor for cardiovascular disease, and drugs that lower blood pressure reduce the risk of heart attacks, heart failure, strokes, and kidney disease. Blood pressure is partly regulated by aldosterone, a steroid synthesized in the adrenal cortex and released in response to angiotensin II, a peptide produced from angiotensinogen in two proteolytic steps by renin (an aspartic protease) and angiotensin-converting enzyme (ACE). Vasotec (enalapril) manufactured by Merck and Biovail is an ACE inhibitor. Novartis and Speedel have developed Tekturna (aliskiren) as a renin inhibitor. Both are TSAs.

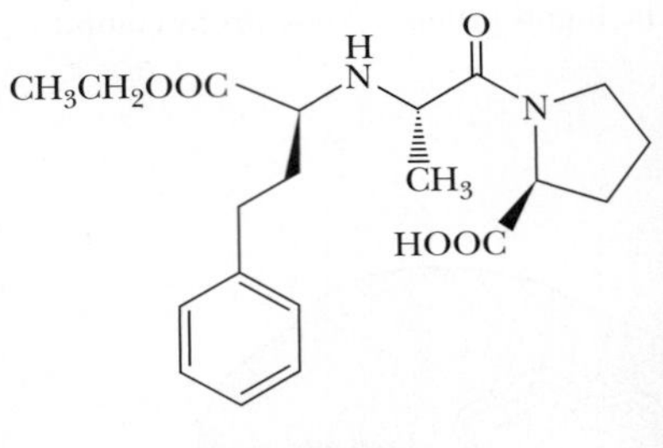

Enalapril

Courtesy of James Gathany/CDC

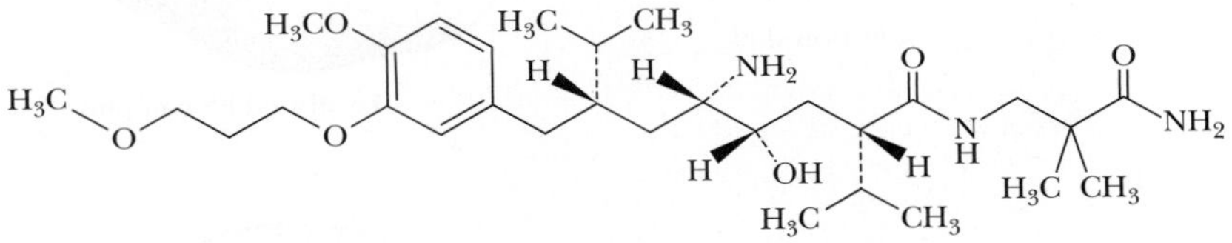

Aliskiren

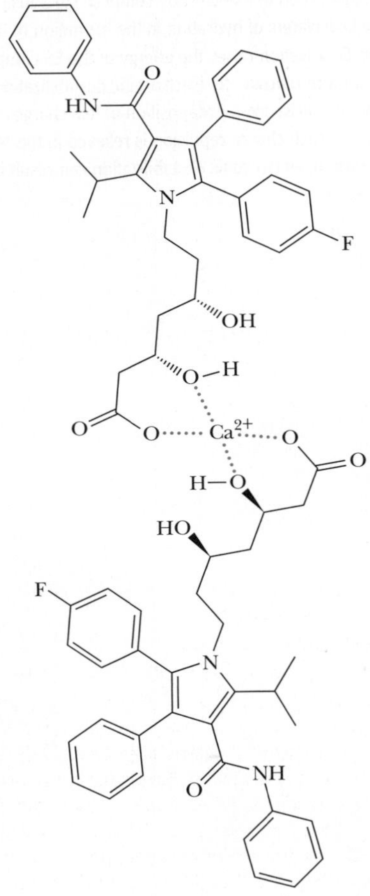

Statins Lower Serum Cholesterol

Statins such as Lipitor are powerful cholesterol-lowering drugs, because they are transition-state analog inhibitors of HMG-CoA reductase, a key enzyme in the biosynthetic pathway for cholesterol (discussed in Chapter 24).

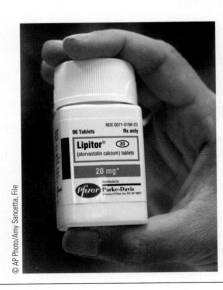

© AP Photo/Amy Sancetta, File

Atorvastatin (Lipitor)

Protease Inhibitors Are AIDS Drugs

Crixivan (indinavir) by Merck, Invirase (saquinavir) by Roche, and similar "protease inhibitor" drugs are transition-state analogs for the HIV-1 protease, discussed on pages 470–471.

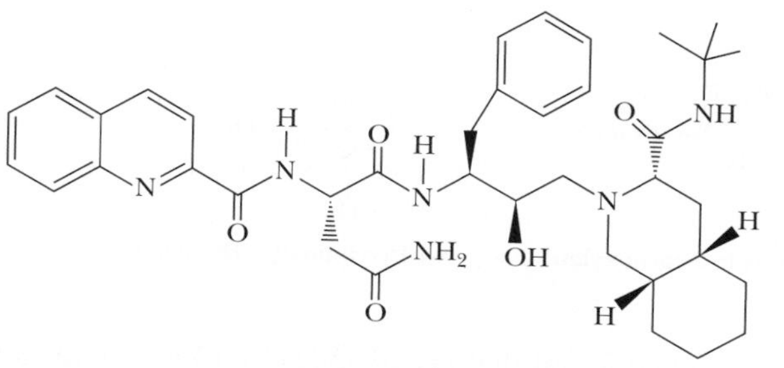

Saquinavir

Tamiflu Is a Viral Neuraminidase Inhibitor

Influenza is a serious respiratory illness that affects 5% to 15% of the earth's population each year and results in 250,000 to 500,000 deaths annually, mostly among children and the elderly. Protection from influenza by vaccines is limited by the antigenic variation of the influenza virus. Neuraminidase is a major glycoprotein on the influenza virus membrane envelope that is essential for viral replication and infectivity. Tamiflu is a neuraminidase inhibitor and antiviral agent based on the transition state of the neuraminidase reaction.

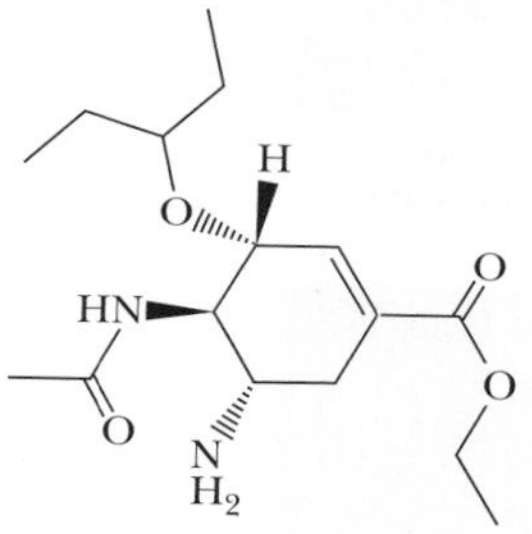

Tamiflu

Courtesy of the Otis Historical Archives/National Museum of Health and Medicine

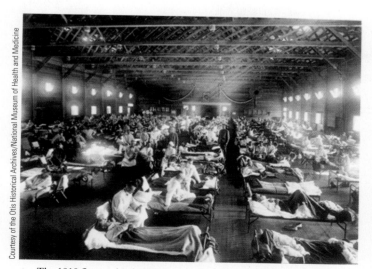

The 1918 flu pandemic killed more than 20 million people worldwide.

Juvenile Hormone Esterase Is a Pesticide Target

Insects have significant effects on human health, being the transmitting agents (vectors) for diseases such as malaria, West Nile virus, and viral encephalitis, all carried by mosquitoes, and Lyme disease and Rocky Mountain spotted fever, carried by ticks. One strategy for controlling insect populations is to alter the actions of **juvenile hormone**, a terpene-based substance that regulates insect life cycle processes. Levels of juvenile hormone are controlled by **juvenile hormone esterase (JHE),** and inhibition of JHE is toxic to insects. OTFP (figure) is a potent transition state analog inhibitor of JHE.

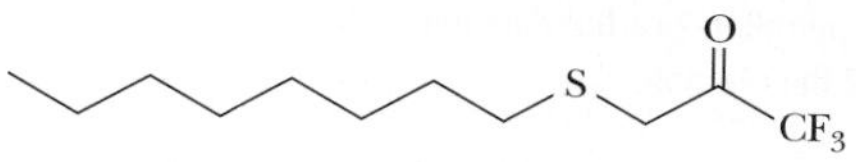

3-Octylthio-1,1,1-trifluoropropan-2-one (OTFP)

© Darwin Dale/Photo Researchers, Inc.

Courtesy of James Gathany/CDC

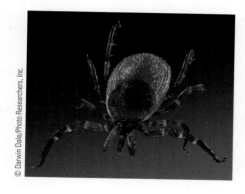

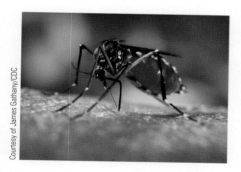

How Many Other Drug Targets Might There Be?

If the human genome contains approximately 20,000 genes, how many of these might be targets for drug therapy? Andrew Hopkins has proposed the term "druggable genome" to conceptualize the subset of human genes that might express proteins able to bind drug-like molecules. The DrugBank database (http://drugbank.ca) contains more than 1400 FDA-approved small molecule drugs. Nearly half of these are directed specifically to enzymes. More than 6800 experimental drugs are presently under study and testing. It is easy to imagine that thousands more drugs will eventually be developed, with many of these designed as transition-state analogs for enzyme reactions.

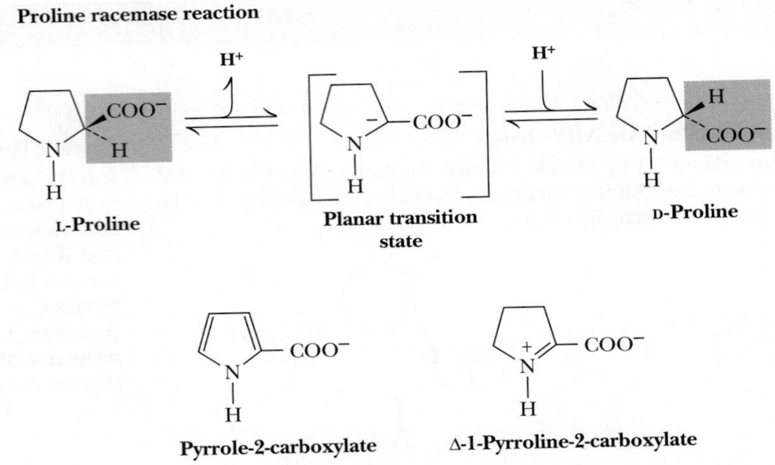

Proline racemase reaction

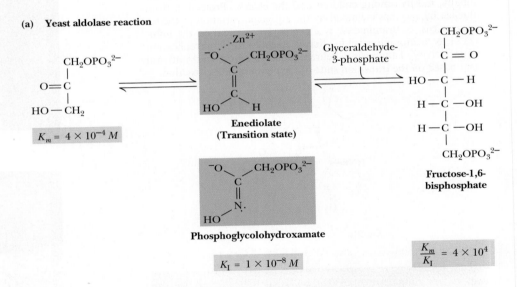

FIGURE 14.5 The proline racemase reaction. Pyrrole-2-carboxylate and Δ-1-pyrroline-2-carboxylate mimic the planar transition state of the reaction.

proline racemase (Figure 14.5) and found that *pyrrole-2-carboxylate* bound to the enzyme 160 times more tightly than L-proline, the normal substrate. This analog binds so tightly because it is planar and is similar in structure to the planar transition state for the racemization of proline. Two other examples of transition-state analogs are shown in Figure 14.6. *Phosphoglycolohydroxamate* binds 40,000 times more tightly to yeast

FIGURE 14.6 **(a)** Phosphoglycolohydroxamate is an analog of the enediolate transition state of the yeast aldolase reaction. **(b)** Purine riboside, a potent inhibitor of the calf intestinal adenosine deaminase reaction, binds to adenosine deaminase as the 1,6-hydrate. The hydrated form of purine riboside is an analog of the proposed transition state for the reaction.

aldolase than the substrate dihydroxyacetone phosphate. Even more remarkable, the 1,6-hydrate of *purine ribonucleoside* has been estimated to bind to adenosine deaminase with a K_I of 3×10^{-13} M!

It should be noted that transition-state analogs are only approximations of the transition state itself and will never bind as tightly as would be expected for the true transition state. These analogs are, after all, stable molecules and cannot be expected to resemble a true transition state too closely.

14.5 What Are the Mechanisms of Catalysis?

Enzymes Facilitate Formation of Near-Attack Conformations

Exquisite and beautiful details of enzyme active-site structure and dynamics have emerged from X-ray crystal structures of enzymes and computer simulations of molecular conformation and motion at the active site. Importantly, these studies have shown that the reacting atoms and catalytic groups are precisely positioned for their roles. This "preorganization" of the active site allows it to select and stabilize conformations of the substrate(s) in which *the reacting atoms are in van der Waals contact and at an angle resembling the bond to be formed in the transition state.* Thomas Bruice has termed such arrangements **near-attack conformations (NACs),** and he has proposed that NACs are the precursors to transition states of reactions (Figure 14.7). In the absence of an enzyme, potential reactant molecules adopt a NAC only about 0.0001% of the time. On the other hand, NACs have been shown to form in enzyme active sites from 1% to 70% of the time.

The **alcohol dehydrogenase (ADH)** reaction provides a good example of a NAC on the pathway to the reaction transition state (Figure 14.8). The ADH reaction converts a primary alcohol to an aldehyde (through an ordered, single-displacement mechanism, see page 432). The reaction proceeds by a proton transfer to water followed by a hydride transfer to NAD^+. In the enzyme active site, Ser^{48} accepts the proton from the alcohol substrate, the resulting negative charge is stabilized by a zinc ion, and the substrate *pro-R* hydrogen is poised above the NAD^+ ring prior to hydride transfer (Figure 14.8). Computer simulations of the enzyme–substrate complex involving the deprotonated alcohol show that this intermediate exists as a NAC 60% of the time. The kinetic advantage of such an enzymatic reaction, compared to its nonenzymatic counterpart, is the *ease of formation of the NAC and the favorable free energy difference between the NAC and the transition state* (Figure 14.7).

Protein Motions Are Essential to Enzyme Catalysis

Proteins are constantly moving. As noted in Chapter 6 (Table 6.2), bonds vibrate, side chains bend and rotate, backbone loops wiggle and sway, and whole domains move with respect to each other. Enzymes depend on such motions to provoke and direct catalytic

(a)

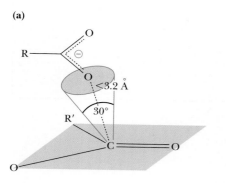

(b)

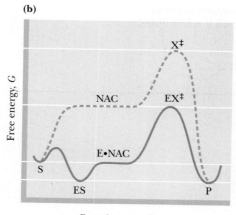

Reaction coordinate

FIGURE 14.7 **(a)** For reactions involving bonding between O, N, C, and S atoms, NACs are characterized as having reacting atoms within 3.2 Å and an approach angle of $\pm15°$ of the bonding angle in the transition state. **(b)** In an enzyme active site, the enzyme–substrate complex and the NAC are separated by a small energy barrier, and NACs form readily. In the absence of the enzyme, the energy gap between the substrate and the NAC is much greater and NACs are rarely formed. The energy separation between the NAC and the transition state is approximately the same in the presence and absence of the enzyme. (Adapted from Bruice, T., 2002. A view at the millennium: The efficiency of enzymatic catalysis. *Accounts of Chemical Research* 35:139–148.)

> **A DEEPER LOOK**

How to Read and Write Mechanisms

The custom among chemists and biochemists of writing chemical reaction mechanisms with electron dots and curved arrows began with two of the greatest chemists of the 20th century. Gilbert Newton Lewis was the first to suggest that a covalent bond consists of a shared pair of electrons, and Sir Robert Robinson was the first to use curved arrows to illustrate a mechanism in a paper in the *Journal of the Chemical Society* in 1922.

Learning to read and write reaction mechanisms should begin with a review of Lewis dot structures in any good introductory chemistry text. It is also important to understand valence electrons and "formal charge." The formal charge of an atom is calculated as the number of valence electrons minus the "electrons owned" by an atom. More properly

Formal charge = group number −
$$\text{nonbonding electrons} - (1/2 \text{ shared electrons})$$

Students of mechanisms should also appreciate electronegativity—the tendency of an atom to attract electrons. Electronegativities of the atoms important in biochemistry go in the order

$$F > O > N > C > H$$

Thus, in a C–N bond, the N should be viewed as more electron-rich and C as electron-deficient. An electron-rich atom is termed *nucleophilic* and will have a tendency to react with an electron-deficient (electrophilic) atom.

In written mechanisms, a curved arrow shows the movement of an electron pair from its original position to a new one. The tail of the arrow shows where the electron pair comes from, and the head of the arrow shows where the electron pair is going. Thus, the arrow represents the actual movement of a pair of electrons from a filled orbital into an empty one. By convention, an arrow with a full arrowhead ⌒ represents movement of an electron pair, whereas a half arrowhead ⌒ represents a single electron (for example, in a free radical reaction). For a bond-breaking event, the arrow begins in the middle of the bond, and the arrowhead points at the atom that will accept the electrons:

$$A \overset{\frown}{-} B \longrightarrow A^+ + B^-$$

For a bond-making event, the arrow begins at the source of the electrons (for example, a nonbonded pair), and the arrowhead points to the atom where the new bond will be formed:

$$A^+ \overset{\frown}{} :B^- \longrightarrow A-B$$

It has been estimated that 75% of the steps in enzyme reaction mechanisms are proton (H⁺) transfers. If the proton is donated or accepted by a group on the enzyme, it is often convenient (and traditional) to represent that group as B, for "base," even if B is protonated and behaving as an acid:

$$B \overset{\frown}{-} H \overset{\frown}{} :N{<} \longrightarrow B^- + H \overset{+}{-}N{<}$$

It is important to appreciate that a proton transfer can change a nucleophile into an electrophile, and vice versa. Thus, it is necessary to consider (1) the protonation states of substrate and active-site residues and (2) how pK_a values can change in the environment of the active site. For example, an active-site histidine, which might normally be protonated, can be deprotonated by another group and then act as a base, accepting a proton from the substrate:

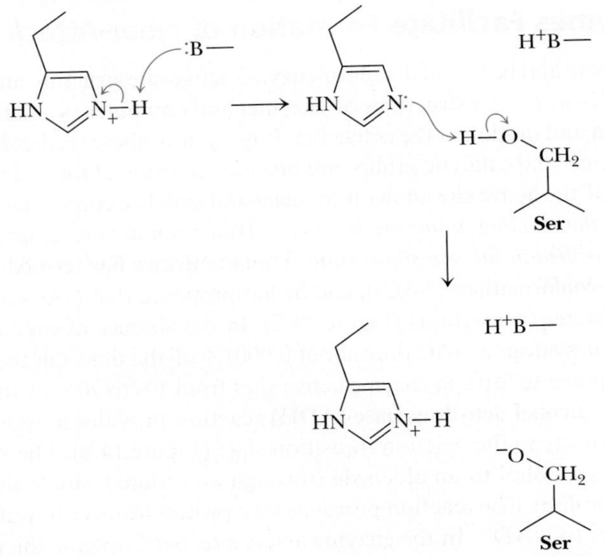

Water can often act as an acid or base at the active site through proton transfer with an assisting active-site residue:

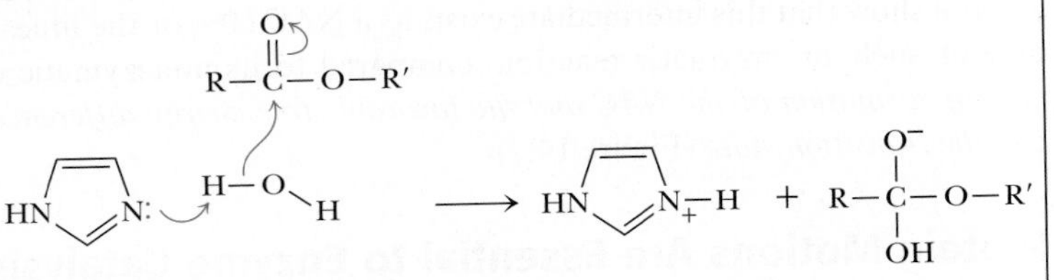

These concepts provide a sense of what is reasonable and what makes good chemical logic in a reaction. Practice and experience are essential to building skills for reading and writing enzyme mechanisms. Excellent Web sites are available where such skills can be built (http://www.abdn.ac.uk/curly-arrows).

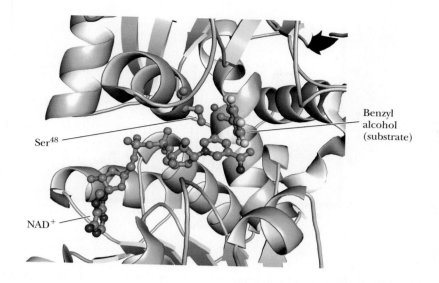

Ser⁴⁸

Benzyl alcohol (substrate)

NAD⁺

FIGURE 14.8 The complex of horse liver ADH with benzyl alcohol illustrates the approach to a near-attack conformation. Computer simulations by Bruice and co-workers show that the side-chain oxygen of Ser⁴⁸ approaches within 1.8 Å of the hydroxyl hydrogen of the substrate, benzyl alcohol, and that the *pro-R* hydrogen of benzyl alcohol lies 2.75 Å from the C-4 carbon of the nicotinamide ring. The reaction mechanism involves hydroxyl proton abstraction by Ser⁴⁸ and hydride transfer from the substrate to C-4 of the NAD⁺ nicotinamide ring (pdb id = 1HLD).

events. Protein motions may support catalysis in several ways: Active site conformation changes can

- assist substrate binding
- bring catalytic groups into position around a substrate
- induce formation of a NAC
- assist in bond making and bond breaking
- facilitate conversion of substrate to product

A good example of protein motions facilitating catalysis is human **cyclophilin A,** which catalyzes the interconversion between *cis* and *trans* conformations of proline in peptides (Figure 14.9). NMR studies of cyclophilin A have provided direct

(a)

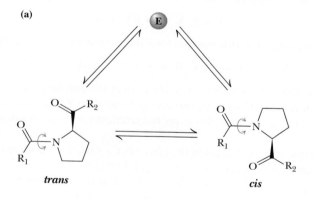

trans

cis

(b)

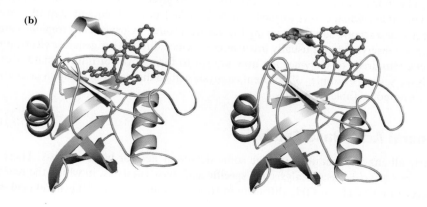

FIGURE 14.9 **(a)** Human cyclophilin A is a **prolyl isomerase,** which catalyzes the interconversion between *trans* and *cis* conformations of proline in peptides. **(b)** The active site of cyclophilin with a bound peptide containing proline in *cis* and *trans* conformations (pdb id = 1RMH).

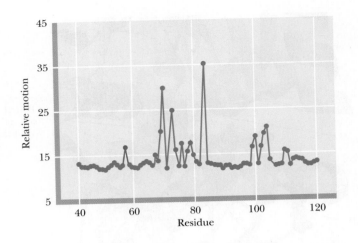

FIGURE 14.10 Catalysis in enzyme active sites depends on motion of active-site residues. NMR studies by Dorothee Kern and her co-workers show that several cyclophilin active-site residues, including Arg[55] (red dot) and Lys[82], Leu[98], Ser[99], Ala[101], Gln[102], Ala[103], and Gly[109] (green dots), undergo greater motion during catalysis than residues elsewhere in the protein. (Adapted from Eisenmesser, E., et al., 2002. Enzyme dynamics during catalysis. *Science* **295:**1520–1523.)

measurements of the active-site motions occurring in this enzyme. Certain active-site residues (Lys[82], Leu[98], Ser[99], Ala[101], Gln[102], Ala[103], and Gly[109]) of the enzyme undergo conformation changes during substrate binding, whereas Arg[55] is involved directly in the *cis*-to-*trans* interconversion itself (Figure 14.10).

The protein motions that assist catalysis may be quite complex. Stephen Benkovic and Sharon Hammes-Schiffer have characterized an extensive network of coupled protein motions in dihydrofolate reductase. This network extends from the active site to the surface of the protein, and the motions in this network span time scales of femtoseconds (10^{-15} sec) to milliseconds. Such extensive networks of motion make it likely that the entire folded structure of the protein may be involved in catalysis at the active site.

Covalent Catalysis

Some enzyme reactions derive much of their rate acceleration from the formation of **covalent bonds** between enzyme and substrate. Consider the reaction:

$$BX + Y \longrightarrow BY + X$$

and an enzymatic version of this reaction involving formation of a **covalent intermediate:**

$$BX + Enz \longrightarrow E:B + X + Y \longrightarrow Enz + BY$$

If the enzyme-catalyzed reaction is to be faster than the uncatalyzed case, the acceptor group on the enzyme must be a better attacking group than Y and a better leaving group than X. Note that most enzymes that carry out covalent catalysis have ping-pong kinetic mechanisms.

The side chains of amino acids in proteins offer a variety of **nucleophilic** centers for catalysis, including amines, carboxylates, aryl and alkyl hydroxyls, imidazoles, and thiol groups. These groups readily attack electrophilic centers of substrates, forming covalently bonded enzyme–substrate intermediates. Typical electrophilic centers in substrates include phosphoryl groups, acyl groups, and glycosyl groups (Figure 14.11). The covalent intermediates thus formed can be attacked in a subsequent step by a water molecule or a second substrate, giving the desired product. **Covalent electrophilic catalysis** is also observed, but it usually involves coenzyme adducts that generate electrophilic centers. Hundreds of enzymes are now known to form covalent intermediates during catalysis. Several examples of covalent catalysis will be discussed in detail in later chapters, as noted in Table 14.2.

General Acid–Base Catalysis

Nearly all enzyme reactions involve some degree of acid or base catalysis. There are two types of acid–base catalysis: (1) **specific acid–base catalysis,** in which the reaction is accelerated by H^+ or OH^- diffusing in from the solution, and (2) **general acid–base**

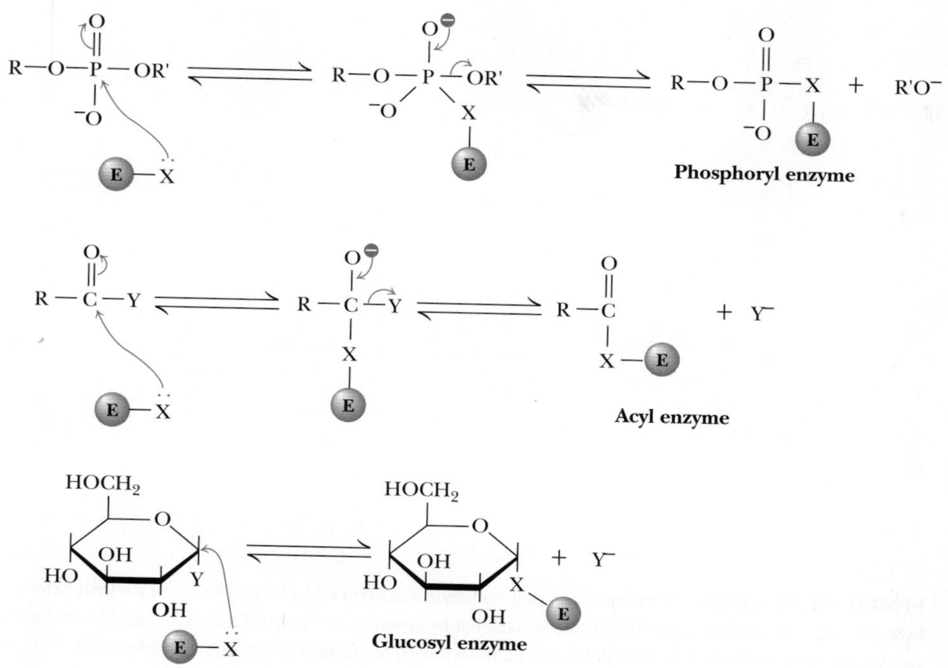

Phosphoryl enzyme

Acyl enzyme

Glucosyl enzyme

FIGURE 14.11 Examples of covalent bond formation between enzyme and substrate. In each case, a nucleophilic center (X:) on an enzyme attacks an electrophilic center on a substrate.

TABLE 14.2	Enzymes That Form Covalent Intermediates	
Enzyme	Reacting Group	Covalent Intermediate
Trypsin Chymotrypsin (pages 465–466)	Serine	Acyl-Ser
Glyceraldehyde-3-P dehydrogenase (page 589)	Cysteine	Acyl-Cys
Phosphoglucomutase (page 493)	Serine	Phospho-Ser
Phosphoglycerate mutase (page 591) Succinyl-CoA synthetase (page 623)	Histidine	Phospho-His
Aldolase (page 587) Pyridoxal phosphate enzymes (pages 434, 856, and 881)	Lysine and other amino groups	Schiff base

catalysis, in which H^+ or OH^- is created *in the transition state* by another molecule or group, which is termed the general acid or general base, respectively. *By definition, general acid–base catalysis is catalysis in which a proton is transferred in the transition state.* Consider the hydrolysis of *p*-nitrophenylacetate by specific base catalysis or with imidazole acting as a general base (Figure 14.12). In the specific base mechanism, hydroxide diffuses into the reaction from solution. In the general base mechanism, the hydroxide that catalyzes the reaction is generated from water in the transition state. The water has been made more nucleophilic without generation of a high concentration of OH^- or without the formation of unstable, high-energy species. General acid or general base catalysis may increase reaction rates 10- to 100-fold. In an enzyme, ionizable groups on the protein provide the H^+ transferred in the transition state. Clearly, an ionizable group will be most effective as an H^+ transferring agent at or near its pK_a. Because the pK_a of the histidine side chain is near 7, histidine is often the most effective general acid or base. Descriptions of several cases of general acid–base catalysis in typical enzymes follow.

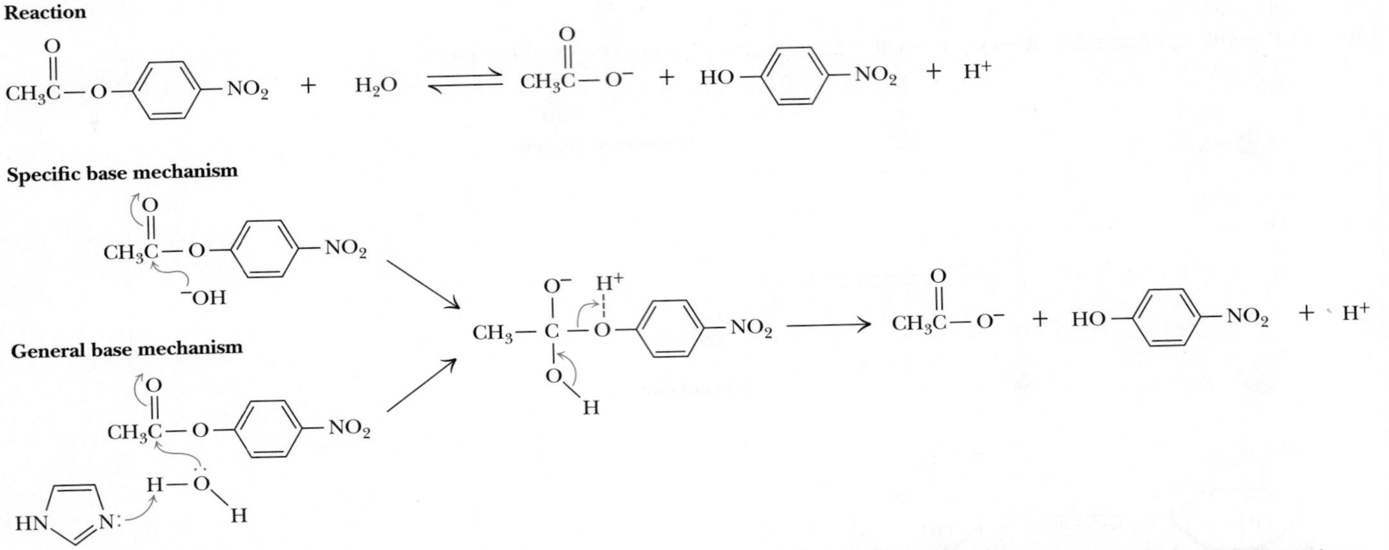

Reaction

$$CH_3\overset{\displaystyle O}{\overset{\displaystyle \|}{C}}-O-\underset{}{\bigcirc}-NO_2 \;+\; H_2O \;\rightleftharpoons\; CH_3\overset{\displaystyle O}{\overset{\displaystyle \|}{C}}-O^- \;+\; HO-\underset{}{\bigcirc}-NO_2 \;+\; H^+$$

Specific base mechanism

General base mechanism

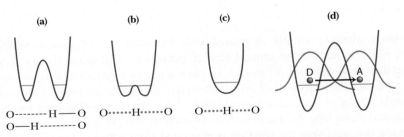

FIGURE 14.12 Catalysis of *p*-nitrophenylacetate hydrolysis can occur either by specific base hydrolysis (where hydroxide from the solution is the attacking nucleophile) or by general base catalysis (in which a base like imidazole can promote hydroxide attack on the substrate carbonyl carbon by removing a proton from a nearby water molecule).

Low-Barrier Hydrogen Bonds

As previously noted, the typical strength of a hydrogen bond is 10 to 30 kJ/mol. For an O—H—O hydrogen bond, the O—O separation is typically 0.28 nm and the H bond is a relatively weak electrostatic interaction. The hydrogen is firmly linked to one of the oxygens at a distance of approximately 0.1 nm, and the distance to the other oxygen is thus about 0.18 nm, which corresponds to a bond order of about 0.07. Not all hydrogen bonds are weak, however. As the distance between heteroatoms becomes smaller, the overall bond becomes stronger, the hydrogen becomes centered, and the bond order approaches 0.5 for both O—H interactions (Figure 14.13). These interactions are more nearly covalent in nature, and the stabilization energy is much higher. Notably, the barrier that the hydrogen atom must surmount to exchange oxygens becomes lower as the O—O separation decreases (Figure 14.13). When the barrier to hydrogen exchange has dropped to the point that it is at or below the zero-point energy level of hydrogen, the interaction is referred to as a **low-barrier hydrogen bond (LBHB).** The hydrogen is now

Bond order refers to the number of electron pairs in a bond. (For a single bond, the bond order is 1.)

(a)   (b)   (c)   (d)

O--------H—O   O••••••H••••••O   O••••H••••O
O—H--------O

FIGURE 14.13 Comparison of conventional (weak) hydrogen bonds **(a)** and low-barrier hydrogen bonds **(b** and **c)**. The horizontal line in each case is the zero-point energy of hydrogen. **(a)** shows an O—H—O hydrogen bond of length 0.28 nm, with the hydrogen attached to one or the other of the oxygens. The bond order for the stronger O—H interaction is approximately 1.0, and the weaker O—H interaction is 0.07. As the O-O distance decreases, the hydrogen bond becomes stronger, and the bond order of the weakest interaction increases. In **(b)**, the O-O distance is 0.25 nm, and the barrier is equal to the zero-point energy. In **(c)**, the O-O distance is 0.23 to 0.24 nm, and the bond order of each O—H interaction is 0.5. **(d)** If the distance for particle (proton or electron) transfer is sufficiently small, overlap of probability functions (red curves) permit efficient quantum mechanical tunneling between donor (D) and acceptor (A) groups.

free to move anywhere between the two oxygens (or, more generally, two heteroatoms). The stabilization energy of LBHBs may approach 100 kJ/mol in the gas phase and 60 kJ/mol or more in solution. LBHBs require matched pK_as for the two electronegative atoms that share the hydrogen. As the two pK_a values diverge, the stabilization energy of the LBHB is decreased. Widely divergent pK_a values thus correspond to ordinary, weak hydrogen bonds.

How may low-barrier hydrogen bonds affect enzyme catalysis? A weak hydrogen bond in an enzyme ground state may become an LBHB in a transient intermediate, or even in the transition state for the reaction. In such a case, the energy released in forming the LBHB is used to help the reaction that forms it, lowering the activation barrier for the reaction. Alternatively, the purpose of the LBHB may be to redistribute electron density in the reactive intermediate, achieving rate acceleration by facilitation of "hydrogen tunneling." Enzyme mechanisms that will be examined later in this chapter (the serine proteases and aspartic proteases) appear to depend upon one or the other of these effects.

Quantum Mechanical Tunneling in Electron and Proton Transfers

The fundamental premise of transition-state theory, as described in this chapter, is that reactants cannot become products until they acquire enough energy to distort their structures to that of the transition state. However, **quantum mechanical tunneling** offers a path around this energy barrier and many enzymes exploit it. According to quantum theory, if an atom or electron is transferred in a chemical reaction from one site to another across an activation barrier, there is a finite probability that the particle will appear (as part of the product) on the other side of the energy barrier, even though it cannot achieve sufficient energy to reach the transition state (Figure 14.1). The likelihood of tunneling depends on the distance over which the particle must move and the shape (width and height) of the potential energy barrier that separates reactant from product. Wave/particle duality in quantum theory implies that a wavelength λ can be calculated for a particle of mass m with a given kinetic energy E, according to the de Broglie equation:

$$\lambda = \frac{h}{\sqrt{2mE}} \tag{14.4}$$

where h is Planck's constant. Tunneling can only play a significant role in a reaction when the wavelength of the transferring particle is similar to the distance over which it is transferred. In such cases, the overlap of probability functions for the particle in the reactant and product states is sufficient to permit effective tunneling (Figure 14.13).

In enzyme reactions, only electrons and hydrogen (H^+, $H\cdot$, or $H:$) satisfy this condition. Consider the low likelihood of tunneling for atoms larger than hydrogen. For example, the de Broglie wavelength of a carbon atom, with a mass of 12 daltons, is approximately 0.25 Å, a distance too short for tunneling in any reaction involving carbon transfer. Note also that tunneling permits transfer of electrons over substantial molecular distances. Assuming a kinetic energy of 10 kJ/mole, Equation 14.4 yields de Broglie wavelengths for the electron and hydrogen of 38 Å and 0.9 Å, respectively. The latter wavelength is on the order of distances involved in proton, hydrogen, and hydride transfers in enzyme reactions, and it its likely that tunneling is a contributing factor in most, if not all, hydrogen transfer reactions.

Metal Ion Catalysis

Many enzymes require metal ions for maximal activity. If the enzyme binds the metal very tightly or requires the metal ion to maintain its stable, native state, it is referred to as a **metalloenzyme.** Enzymes that bind metal ions more weakly, perhaps only during the catalytic cycle, are referred to as **metal activated.** One role for metals in metal-activated enzymes and metalloenzymes is to act as electrophilic catalysts, stabilizing the increased

How Do Active-Site Residues Interact to Support Catalysis?

Only about half of the common amino acid residues (that is, His, Cys, Asp, Glu, Arg, Lys, Tyr, Ser, Thr, Asn, and Gln) engage directly in catalytic effects in enzyme active sites. These polar and charged residues provide a relatively limited range of catalytic capabilities. They can act as nucleophiles, facilitate substrate binding, and stabilize transition states. It has been estimated that up to 75% of the steps in enzyme mechanisms involve a simple proton transfer. Is this enough to explain the dramatic catalytic power of enzymes? Or might there be other phenomena at work?

Janet Thornton and Alex Gutteridge have analyzed residue interactions at the active sites of 191 different enzymes. In this group of enzymes, each polar catalytic residue interacts with (on average) 2.3 other polar residues in the active site, whereas noncatalytic, buried polar residues have, on average, interactions with only 1.2 other polar residues. This suggests that some of the interactions between catalytic and noncatalytic residues are functional in some way. At the same time, in only 88 of the enzymes does the key catalytic residue have a direct interaction with a second catalytic residue, indicating that most catalytic residues do not require direct interactions with other catalytic residues to be active.

The catalytic capacities of polar and charged residues can be influenced by other polar and charged residues at the active site and even by hydrophobic residues. The so-called secondary, or noncatalytic, residues at the active site play interesting roles:

- *Raising or lowering catalytic residue pK_a values* through electrostatic or hydrophobic interactions. In aldoketoreductase, an Asp–Lys pair facilitates general acid–base catalysis, with Lys[84] lowering the pK_a of Tyr[58] so that it can donate a proton to the substrate. On the other hand, nearby hydrophobic residues can provide a nonpolar environment that tends to raise the pK_a values of acidic residues (such as Asp or Glu) and to lower the pK_a values of basic residues (such as lysine and arginine). Hydrophobic environments can change pK_a values by as much as 5 or 6 pH units.
- *Orientation* of catalytic residues, as will be seen in the serine proteases, where Asp[102] orients His[57] (see Figure 14.21).

- *Charge stabilization*, as will be seen in chorismate mutase, where active-site arginines stabilize negatively charged carboxyl groups on the substrate (see Figures 14.31 and 14.33).
- *Proton transfers via hydrogen tunneling.* In such quantum mechanical tunneling, the proton transfer is accomplished by molecular motions that lead to degeneracy of a pair of localized proton vibrational states (Figure 14.13). Proton tunneling can be facilitated by nearby molecular motions of secondary residues coupled to the motion and vibration of the bonds in question. David Leys has shown that aromatic amine dehydrogenase probably accomplishes catalysis by coupling local motions (of two secondary residues, C[171] and T[172]) to the vibrational states involved in a proton transfer reaction with D[128], as shown here.

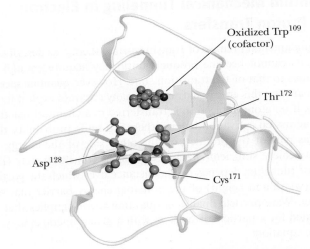

Closeup of the crystal structure of aromatic amine dehydrogenase, showing the relationship of Asp[128], Thr[172], and Cys[171]. N atoms are blue; O atoms are red; C atoms are green; S atom is gold (pdb id = 2AH1).

electron density or negative charge that can develop during reactions. Among the enzymes that function in this manner (Figure 14.14) is thermolysin. Another potential function of metal ions is to provide a powerful nucleophile at neutral pH. Coordination to a metal ion can increase the acidity of a nucleophile with an ionizable proton:

$$M^{2+} + NucH \rightleftharpoons M^{2+} (NucH) \rightleftharpoons M^{2+} (Nuc^-) + H^+$$

The reactivity of the coordinated, deprotonated nucleophile is typically intermediate between that of the un-ionized and ionized forms of free nucleophile. Carboxypeptidase (see Chapter 5) contains an active site Zn^{2+}, which facilitates deprotonation of a water molecule in this manner.

FIGURE 14.14 Thermolysin is an endoprotease (that is, it cleaves polypeptides in the middle of the chain) with a catalytic Zn^{2+} ion in the active site. The Zn^{2+} ion stabilizes the buildup of negative charge on the peptide carbonyl oxygen, as a glutamate residue deprotonates water, promoting hydroxide attack on the carbonyl carbon. Thermolysin is found in certain laundry detergents, where it is used to remove protein stains from fabrics.

Initial step of thermolysin reaction

14.6 What Can Be Learned from Typical Enzyme Mechanisms?

The balance of this chapter will be devoted to several classic and representative enzyme mechanisms, including the serine proteases, the aspartic proteases, and chorismate mutase. Both the serine proteases and the aspartic proteases use general acid–base catalysis chemistry; the serine proteases also employ a covalent catalysis strategy. Chorismate mutase, on the other hand, uses neither of these and depends instead on the formation of a NAC to carry out its reaction. These particular cases are well understood, because the three-dimensional structures of the enzymes and the bound substrates are known at atomic resolution and because great efforts have been devoted to kinetic and mechanistic studies. They are important because they represent reaction types that appear again and again in living systems and because they demonstrate many of the catalytic principles cited previously. Enzymes are the catalytic machines that sustain life, and what follows is an intimate look at the inner workings of the machinery.

Serine Proteases

Serine proteases are a class of proteolytic enzymes whose catalytic mechanism is based on an active-site serine residue. Serine proteases are one of the best-characterized families of enzymes. This family includes *trypsin, chymotrypsin, elastase, thrombin, subtilisin, plasmin, tissue plasminogen activator,* and other related enzymes. The first three of these are digestive enzymes and are synthesized in the pancreas and secreted into the digestive tract as inactive **proenzymes,** or **zymogens.** Within the digestive tract, the zymogen is converted into the active enzyme form by cleaving off a portion of the peptide chain. Thrombin is a crucial enzyme in the blood-clotting cascade, subtilisin is a bacterial protease, and plasmin breaks down the fibrin polymers of blood clots. Tissue plasminogen activator (TPA) specifically cleaves the proenzyme *plasminogen,* yielding plasmin. Owing to its ability to stimulate breakdown of blood clots, TPA can minimize the harmful consequences of a heart attack, if administered to a patient within 30 minutes of onset. Finally, although not itself a protease, *acetylcholinesterase is a serine esterase* and is related mechanistically to the serine proteases. It degrades the neurotransmitter acetylcholine in the synaptic cleft between neurons.

The Digestive Serine Proteases

Trypsin, chymotrypsin, and elastase all carry out the same reaction—the cleavage of a peptide chain—and although their structures and mechanisms are quite similar, they display very different specificities. Trypsin cleaves peptides on the carbonyl side of the basic amino acids, arginine or lysine (see Table 5.2). Chymotrypsin prefers to cleave on the carbonyl side of aromatic residues, such as phenylalanine and tyrosine. Elastase is not as specific as the other two; it mainly cleaves peptides on the carbonyl side of small, neutral residues. These three enzymes all possess molecular weights in the range of 25,000, and all have similar sequences (Figure 14.15) and three-dimensional structures. The structure of chymotrypsin is typical (Figure 14.16). The molecule is ellipsoidal in shape and contains an α-helix at the C-terminal end (residues 230 to 245) and several β-sheet domains. Most of the aromatic and hydrophobic residues are buried in the interior of the protein, and most of the charged or hydrophilic residues are on the surface. Three polar residues—His[57], Asp[102], and Ser[195]—form what is known as a **catalytic triad** at the active site (Figure 14.17). These three residues are conserved in trypsin and elastase as well. The active site is actually a depression on the surface of the enzyme, with a pocket that the enzyme uses to identify the residue for which it is specific (Figure 14.18). Chymotrypsin, for example, has a pocket surrounded by hydrophobic residues and large enough to accommodate an aromatic side chain. The pocket in trypsin has a negative charge (Asp[189]) at its bottom, facilitating the binding of positively charged arginine and lysine residues. Elastase, on the other hand, has a shallow pocket with bulky threonine

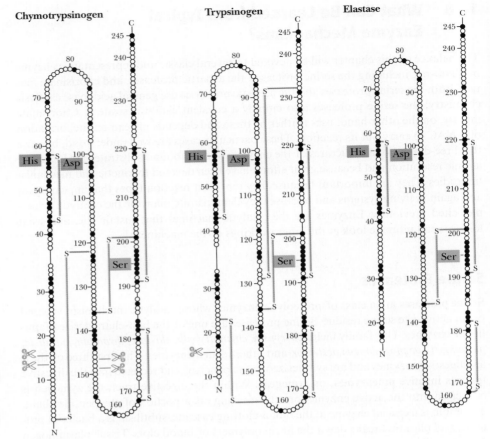

Chymotrypsinogen **Trypsinogen** **Elastase**

FIGURE 14.15 Comparison of the amino acid sequences of chymotrypsinogen, trypsinogen, and elastase. Each circle represents one amino acid. Numbering is based on the sequence of chymotrypsinogen. Filled circles indicate residues that are identical in all three proteins. Disulfide bonds are indicated in orange. The positions of the three catalytically important active-site residues (His^{57}, Asp^{102}, and Ser^{195}) are indicated.

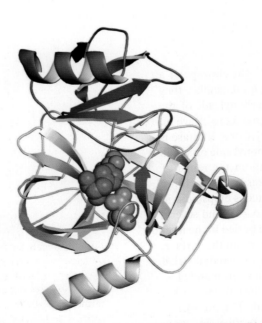

FIGURE 14.16 Structure of chymotrypsin (white) in a complex with eglin C (blue ribbon structure), a target protein. The residues of the catalytic triad (His^{57}, Asp^{102}, and Ser^{195}) are highlighted. His^{57} (red) is flanked by Asp^{102} (gold) and by Ser^{195} (green). The catalytic site is filled by a peptide segment of eglin. Note how close Ser^{195} is to the peptide that would be cleaved in the chymotrypsin reaction (pdb id = 1ACB).

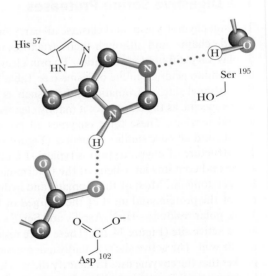

FIGURE 14.17 The catalytic triad of chymotrypsin.

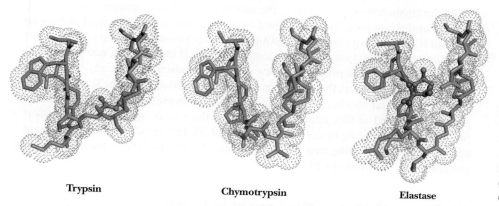

Trypsin   **Chymotrypsin**   **Elastase**

FIGURE 14.18 The substrate-binding pockets of trypsin (pdb id = 2CMY), chymotrypsin (pdb id = 1ACB), and elastase (pdb id = 3EST). Asp[189] (aqua) coordinates Arg and Lys residues of peptide substrates in the trypsin binding pocket. Val[216] (purple) and Thr[226] (green) make the elastase binding pocket shallow and able to accommodate only small, nonbulky residues.

and valine residues at the opening. Only small, nonbulky residues can be accommodated in its pocket. The backbone of the peptide substrate is hydrogen bonded in anti-parallel fashion to residues 215 to 219 and bent so that the peptide bond to be cleaved is bound close to His[57] and Ser[195].

The Chymotrypsin Mechanism in Detail: Kinetics

Much of what is known about the chymotrypsin mechanism is based on studies of the hydrolysis of artificial substrates—simple organic esters, such as p-nitrophenylacetate (Figure 14.19). p-Nitrophenylacetate is an especially useful model substrate, because the nitrophenolate product is easily observed, owing to its strong absorbance at 400 nm. When large amounts of chymotrypsin are used in kinetic studies with this substrate, a **rapid initial burst** of p-nitrophenolate is observed (in an amount approximately equal to the enzyme concentration), followed by a much slower, linear rate of nitrophenolate release (Figure 14.20). Observation of a burst, followed by slower, steady-state product release, is strong evidence for a multistep mechanism, with a fast first step and a slower second step.

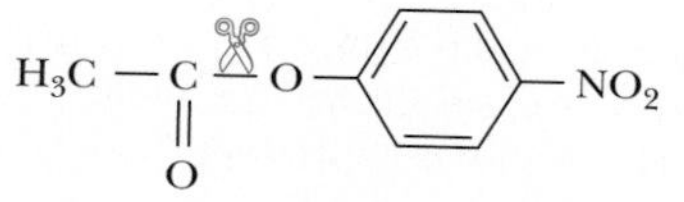

$$H_3C - C \overset{\curvearrowleft}{\underset{\parallel}{\overset{\,}{O}}} O - \bigcirc - NO_2$$

***p*-Nitrophenylacetate**

FIGURE 14.19 Chymotrypsin cleaves simple esters, in addition to peptide bonds. p-Nitrophenylacetate has been used in studies of the chymotrypsin mechanism.

(a)

Acetate or pNO$_2^-$ phenolate release (vertical axis)

Steady-state release

p-Nitrophenolate

Acetate

Burst

Lag

Time

(b)

E—Ser[195]—OH + [p-nitrophenylacetate ring with NO$_2$ and O—C—CH$_3$, O] →(Fast step) NO$_2$ ring with O$^-$ + H$^+$ → E—Ser[195]—O—C—CH$_3$, O →(Slow step, H$_2$O, H$^+$) $^-$O—C—CH$_3$, O

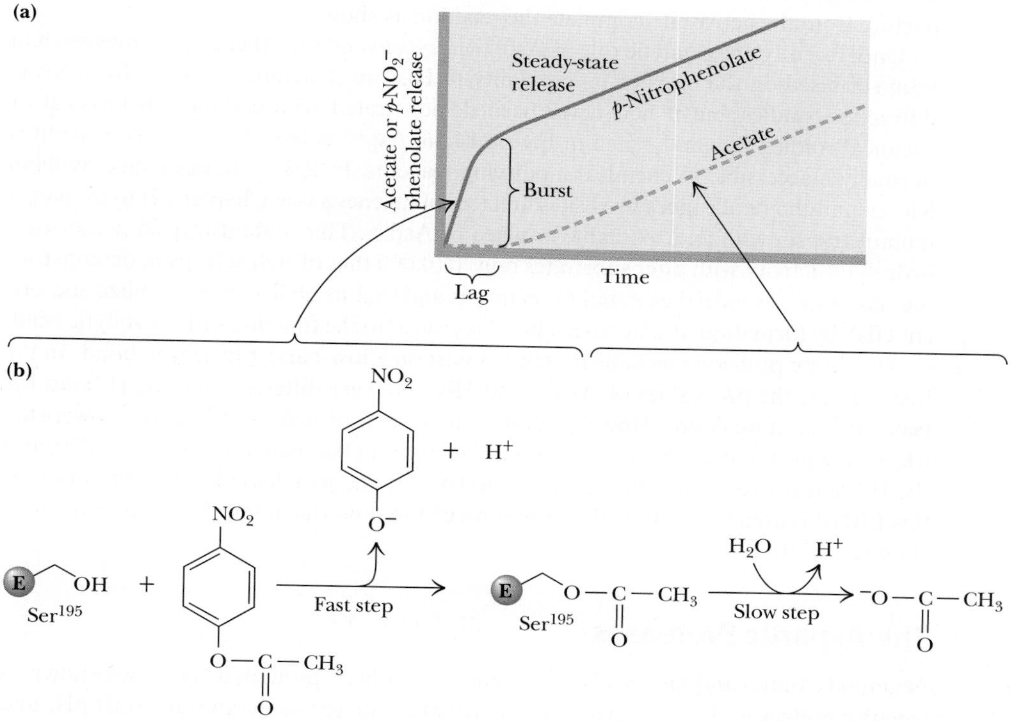

FIGURE 14.20 Burst kinetics observed in the chymotrypsin reaction **(a)**. A burst of nitrophenolate (**b,** first step) is followed by a slower, steady-state release. After an initial lag period, acetate release (**b,** second step) is observed. This kinetic pattern is consistent with rapid formation of an acyl-enzyme intermediate (and the burst of nitrophenolate). The slower, steady-state release of products corresponds to rate-limiting breakdown of the acyl-enzyme intermediate.

In the chymotrypsin mechanism, the nitrophenylacetate combines with the enzyme to form an ES complex. This is followed by a rapid step in which an **acyl-enzyme intermediate** is formed, with the acetyl group covalently bound to the very reactive Ser[195]. The nitrophenyl moiety is released as nitrophenolate (Figure 14.20), accounting for the burst of nitrophenolate product. Attack of a water molecule on the acyl-enzyme intermediate yields acetate as the second product in a subsequent, slower step. The enzyme is now free to bind another molecule of p-nitrophenylacetate, and the p-nitrophenolate product produced at this point corresponds to the slower, steady-state formation of product in the upper right portion of Figure 14.20. In this mechanism, the release of acetate is the **rate-limiting step** and accounts for the observation of **burst kinetics**—the pattern shown in Figure 14.20.

The Serine Protease Mechanism in Detail: Events at the Active Site

A likely mechanism for peptide hydrolysis is shown in Figure 14.21. As the backbone of the substrate peptide binds adjacent to the catalytic triad, the specific side chain fits into its pocket. Asp[102] of the catalytic triad positions His[57] and immobilizes it through a hydrogen bond as shown. In the first step of the reaction, His[57] acts as a general base to withdraw a proton from Ser[195], facilitating nucleophilic attack by Ser[195] on the carbonyl carbon of the peptide bond to be cleaved. This is probably a *concerted step,* because proton transfer prior to Ser[195] attack on the acyl carbon would leave a relatively unstable negative charge on the serine oxygen. In the next step, donation of a proton from His[57] to the peptide's amide nitrogen creates a protonated amine on the covalent, tetrahedral intermediate, facilitating the subsequent bond breaking and dissociation of the amine product. The negative charge on the peptide oxygen is unstable; the tetrahedral intermediate is short lived and rapidly breaks down to expel the amine product. The acyl-enzyme intermediate that results is reasonably stable; it can even be isolated using substrate analogs for which further reaction cannot occur. With normal peptide substrates, however, subsequent nucleophilic attack at the carbonyl carbon by water generates another transient tetrahedral intermediate (Figure 14.21g). His[57] acts as a general base in this step, accepting a proton from the attacking water molecule. The subsequent collapse of the tetrahedral intermediate is assisted by proton donation from His[57] to the serine oxygen in a concerted manner. Deprotonation of the carboxyl group and its departure from the active site complete the reaction as shown.

Until recently, the catalytic role of Asp[102] in trypsin and the other serine proteases had been surmised on the basis of its proximity to His[57] in structures obtained from X-ray diffraction studies, but it had never been demonstrated with certainty in physical or chemical studies. As can be seen in Figure 14.16, Asp[102] is buried at the active site; it is normally inaccessible to chemical modifying reagents. In 1987, Charles Craik, William Rutter, and their colleagues used site-directed mutagenesis (see Chapter 12) to prepare a mutant trypsin with an asparagine in place of Asp[102]. This mutant trypsin possessed a hydrolytic activity with ester substrates only 1/10,000 that of native trypsin, demonstrating that Asp[102] is indeed essential for catalysis and that its ability to immobilize and orient His[57] by formation of a hydrogen bond is crucial to the function of the catalytic triad.

The serine protease mechanism relies in part on a low-barrier hydrogen bond. In the free enzyme, the pK_a values of Asp[102] and His[57] are very different, and the H bond between them is a weak one. However, donation of the proton of Ser[195] to His[57] lowers the pK_a of the protonated imidazole ring so it becomes a close match to that of Asp[102], and the H bond between them becomes an LBHB. The energy released in the formation of this LBHB is used to facilitate the formation of the subsequent tetrahedral intermediate (Figure 14.21c, g).

The Aspartic Proteases

Mammals, fungi, and higher plants produce a family of proteolytic enzymes known as **aspartic proteases.** These enzymes are active at acidic (or sometimes neutral) pH, and each possesses two aspartic acid residues at the active site. Aspartic proteases carry out

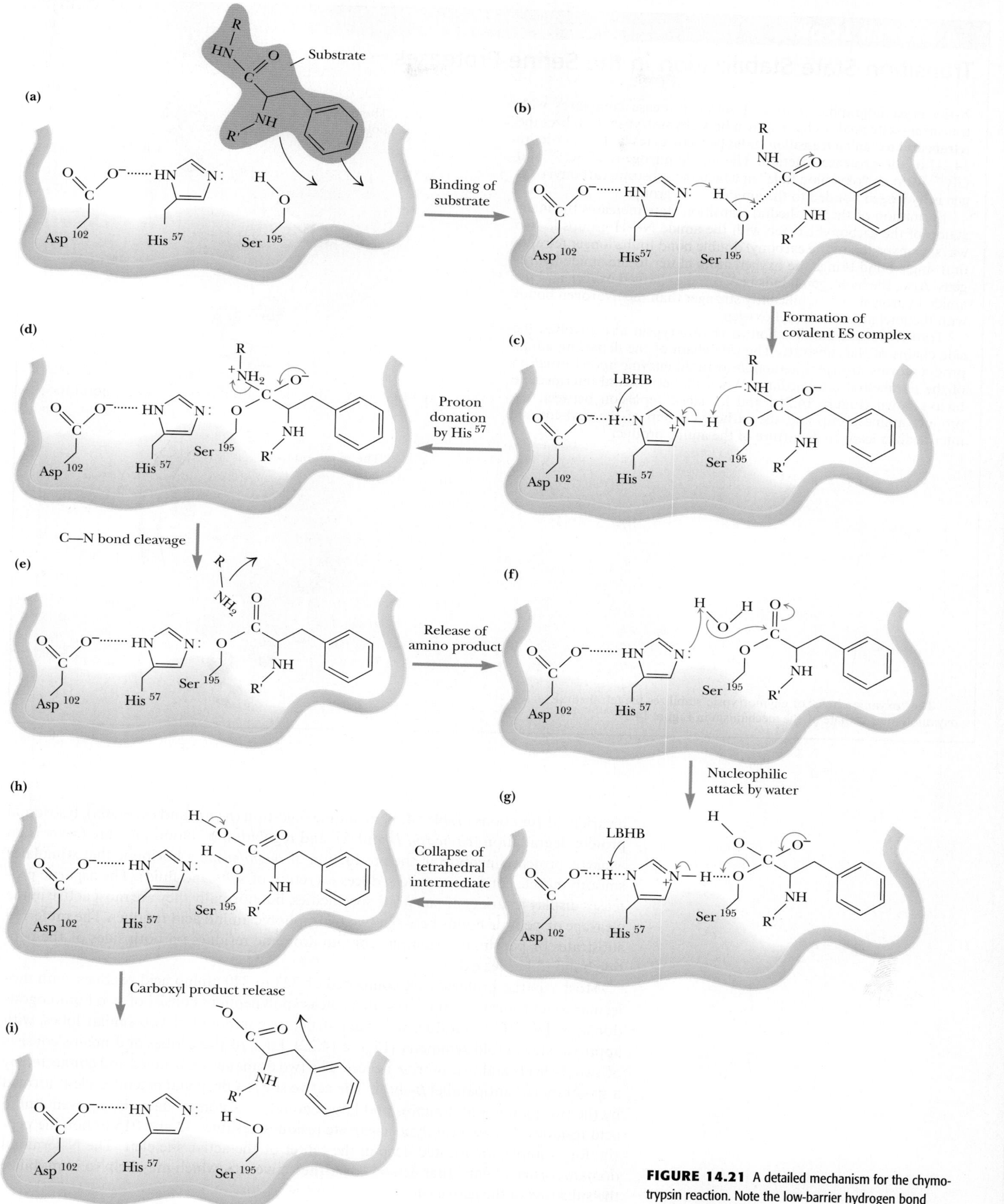

FIGURE 14.21 A detailed mechanism for the chymotrypsin reaction. Note the low-barrier hydrogen bond (LBHB) in **(c)** and **(g)**.

> **A DEEPER LOOK**

Transition-State Stabilization in the Serine Proteases

X-ray crystallographic studies of serine protease complexes with transition-state analogs have shown how chymotrypsin stabilizes the **tetrahedral oxyanion transition states** [structures (c) and (g) in Figure 14.21] of the protease reaction. The amide nitrogens of Ser[195] and Gly[193] form an "oxyanion hole" in which the substrate carbonyl oxygen is hydrogen bonded to the amide N—H groups.

Formation of the tetrahedral transition state increases the interaction of the carbonyl oxygen with the amide N—H groups in two ways. Conversion of the carbonyl double bond to the longer tetrahedral single bond brings the oxygen atom closer to the amide hydrogens. Also, the hydrogen bonds between the charged oxygen and the amide hydrogens are significantly stronger than the hydrogen bonds with the uncharged carbonyl oxygen.

Transition-state stabilization in chymotrypsin also involves the side chains of the substrate. The side chain of the departing amine product forms stronger interactions with the enzyme upon formation of the tetrahedral intermediate. When the tetrahedral intermediate breaks down (Figure 14.21d and h), steric repulsion between the product amine group and the carbonyl group of the acyl-enzyme intermediate leads to departure of the amine product.

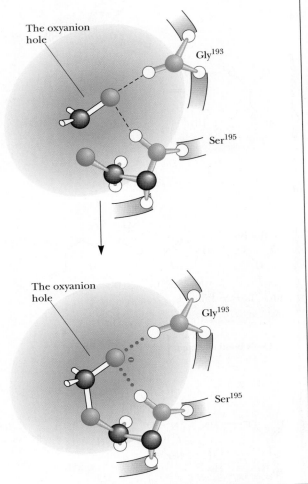

▶ The "oxyanion hole" of chymotrypsin stabilizes the tetrahedral oxyanion intermediate of the mechanism in Figure 14.21.

a variety of functions (Table 14.3), including digestion (*pepsin* and *chymosin*), lysosomal protein degradation (*cathepsin D* and *E*), and regulation of blood pressure (*renin* is an aspartic protease involved in the production of *angiotensin,* a hormone that stimulates smooth muscle contraction and reduces excretion of salts and fluid). The aspartic proteases display a variety of substrate specificities, but normally they are most active in the cleavage of peptide bonds between two hydrophobic amino acid residues. The preferred substrates of pepsin, for example, contain aromatic residues on both sides of the peptide bond to be cleaved.

Most aspartic proteases are composed of 323 to 340 amino acid residues, with molecular weights near 35,000. Aspartic protease polypeptides consist of two homologous domains that fold to produce a tertiary structure composed of two similar lobes, with approximate twofold symmetry (Figure 14.22). Each of these lobes or domains consists of two β-sheets and two short α-helices. The two domains are bridged and connected by a six-stranded, antiparallel β-sheet. The active site is a deep and extended cleft, formed by the two juxtaposed domains and large enough to accommodate about seven amino acid residues. The two catalytic aspartate residues, residues 32 and 215 in porcine pepsin, for example, are located deep in the center of the active site cleft. The N-terminal domain forms a "flap" that extends over the active site, which may help to immobilize the substrate in the active site.

TABLE 14.3	Some Representative Aspartic Proteases	
Name	Source	Function
Pepsin*	Stomach	Digestion of dietary protein
Chymosin†	Stomach	Digestion of dietary protein
Cathepsin D	Spleen, liver, and many other animal tissues	Lysosomal digestion of proteins
Renin‡	Kidney	Conversion of angiotensinogen to angiotensin I; regulation of blood pressure
HIV-protease§	AIDS virus	Processing of AIDS virus proteins

*The second enzyme to be crystallized (by John Northrop in 1930). Even more than urease before it, pepsin study by Northrop established that enzyme activity comes from proteins.
†Also known as rennin, it is the major pepsinlike enzyme in gastric juice of fetal and newborn animals. It has been used for thousands of years, in a gastric extract called rennet, in the making of cheese.
‡A drop in blood pressure causes release of renin from the kidneys, which converts more angiotensinogen to angiotensin.
§A dimer of identical monomers, homologous to pepsin

(a)

(b)

FIGURE 14.22 Structures of **(a)** HIV-1 protease, a dimer (pdb id = 7HVP), and **(b)** pepsin, a monomer. Pepsin's N-terminal half is shown in red; C-terminal half is shown in blue (pdb id = 5PEP).

On the basis, in part, of comparisons with chymotrypsin, trypsin, and the other serine proteases, it was at first hypothesized that aspartic proteases might function by formation of covalent enzyme–substrate intermediates involving the active-site aspartate residues. However, all attempts to trap or isolate a covalent intermediate failed, and a mechanism (see following section) favoring noncovalent enzyme–substrate intermediates and general acid–general base catalysis is now favored for aspartic proteases.

The Mechanism of Action of Aspartic Proteases

A crucial datum supporting the general acid–general base model is the pH dependence of protease activity (Figure 14.23). For many years, enzymologists hypothesized that the aspartate carboxyl groups functioned alternately as general acid and general base. This model requires that one of the aspartate carboxyls be protonated and one be deprotonated when substrate binds. (This made sense, because X-ray diffraction data on aspartic proteases had shown that the active-site structure in the vicinity of the two aspartates is highly symmetric.) However, Stefano Piana and Paolo Carloni reported in 2000 that molecular dynamics simulations of aspartic proteases were consistent with a

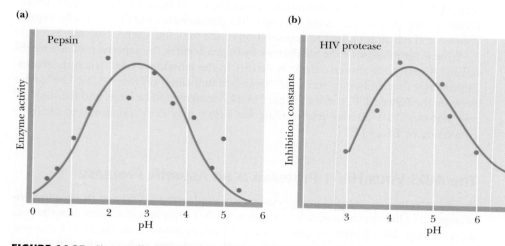

FIGURE 14.23 pH-rate profiles for **(a)** pepsin and **(b)** HIV protease. (Adapted from Denburg, J., et al., 1968. The effect of pH on the rates of hydrolysis of three acylated dipeptides by pepsin. *Journal of the American Chemical Society* **90**:479–486; and Hyland, J., et al., 1991. Human immunodeficiency virus-1 protease. 2. Use of pH rate studies and solvent kinetic isotope effects to elucidate details of chemical mechanism. *Biochemistry* **30**:8454–8463.)

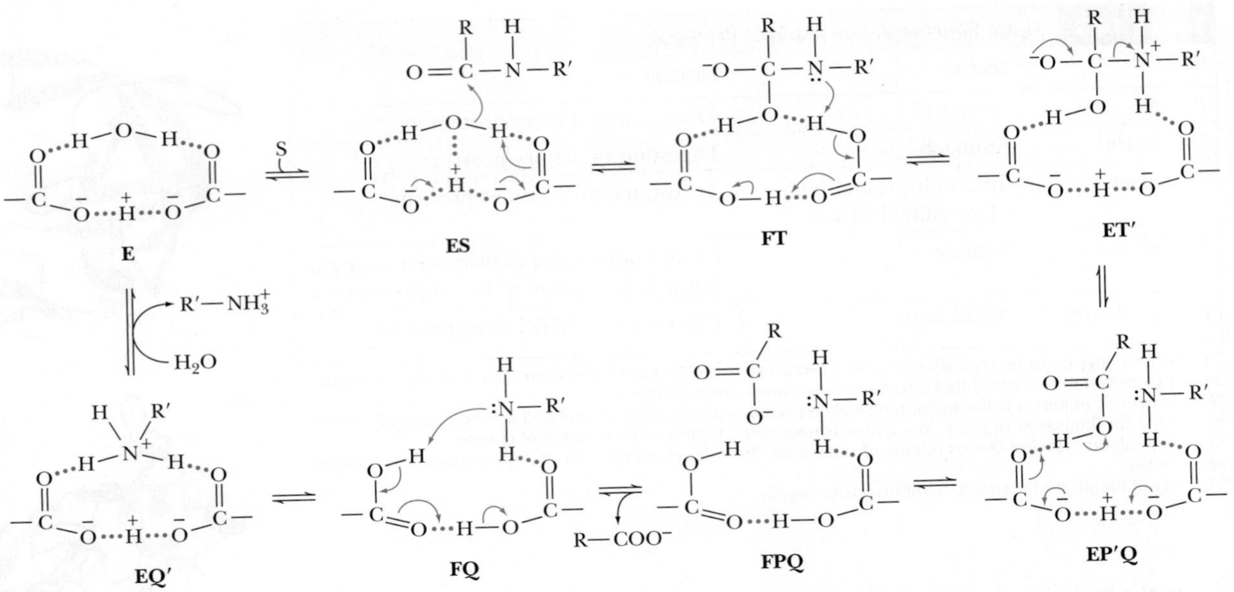

FIGURE 14.24 Mechanism for the aspartic proteases. The letter titles describe the states as follows: E represents the enzyme form with a low-barrier hydrogen bond between the catalytic aspartates, F represents the enzyme form with both aspartates protonated and joined by a conventional hydrogen bond, S represents bound substrate, T represents a tetrahedral amide hydrate intermediate, P represents bound carboxyl product, and Q represents bound amine product. This mechanism is based in part on a mechanism proposed by Dexter Northrop, a distant relative of John Northrop, who had first crystallized pepsin in 1930. (Northrop, D. B., 2001. Follow the protons: A low-barrier hydrogen bond unifies the mechanisms of the aspartic proteases. *Accounts of Chemical Research* **34**:790–797.) The mechanism is also based on data of Thomas Meek. (Meek, T. D., Catalytic mechanisms of the aspartic proteinases. In Sinnott, M., ed, *Comprehensive Biological Catalysis: A Mechanistic Reference,* San Diego: Academic Press, 1998.)

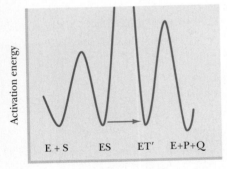

FIGURE 14.25 Energy level diagram for the aspartic protease reaction, showing ground-state hydrogen tunneling (arrow), with consequent rate acceleration.

low-barrier hydrogen bond involving the two active-site aspartates. This led to a new mechanism for the aspartic proteases (Figure 14.24) that begins with Piana and Carloni's model of the LBHB structure of the free enzyme (state E). In this model, the LBHB holds the twin aspartate carboxyls in a coplanar conformation, with the catalytic water molecule on the opposite side of a ten-atom cyclic structure.

Following substrate binding, a counterclockwise flow of electrons moves two protons clockwise and creates a tetrahedral intermediate bound to a diprotonated enzyme form (FT). Then a clockwise movement of electrons moves two protons counterclockwise and generates the zwitterion intermediate bound to a monoprotonated enzyme form (ET'). Collapse of the zwitterion cleaves the C—N bond of the substrate. Dissociation of one product leaves the enzyme in the diprotonated FQ form. Finally, deprotonation and rehydration lead to regeneration of the ten-atom cyclic structure, E.

What is the purpose of the low-barrier hydrogen bond in the aspartic protease mechanism? It may be to disperse electron density in the ten-atom cyclic structure, accomplishing rate acceleration by means of "hydrogen tunneling" (Figure 14.25). The barrier between the ES and ET' states of Figure 14.24 is imagined to be large, and the state FT may not exist as a discrete intermediate but rather may exist transiently to facilitate conversion of ES and ET'.

The AIDS Virus HIV-1 Protease Is an Aspartic Protease

Recent research on acquired immunodeficiency syndrome (AIDS) and its causative viral agent, the human immunodeficiency virus (HIV-1), has brought a new aspartic protease to light. **HIV-1 protease** cleaves the polyprotein products of the HIV-1 genome, producing several proteins necessary for viral growth and cellular infection (Figure 14.26). HIV-1 protease cleaves several different peptide linkages. For example, the protease cleaves between the Tyr and Pro residues of the sequence Ser-Gln-Asn-Tyr-Pro-Ile-Val, which joins the p17 and p24 HIV-1 proteins.

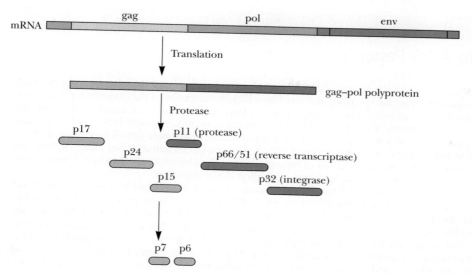

FIGURE 14.26 HIV mRNA provides the genetic information for synthesis of a polyprotein. Proteolytic cleavage of this polyprotein by HIV protease produces the individual proteins required for viral growth and cellular infection.

The HIV-1 protease is a remarkable viral imitation of mammalian aspartic proteases: It is a **dimer of identical subunits** that mimics the two-lobed monomeric structure of pepsin and other aspartic proteases. The HIV-1 protease subunits are 99-residue polypeptides that are homologous with the individual domains of the monomeric proteases. Structures determined by X-ray diffraction studies reveal that the active site of HIV-1 protease is formed at the interface of the homodimer and consists of two aspartate residues, designated Asp^{25} and $Asp^{25'}$, one contributed by each subunit (Figure 14.27). In the homodimer, the active site is covered by two identical "flaps," one from each subunit, in contrast to the monomeric aspartic proteases, which possess only a single active-site flap. Enzyme kinetic measurements by Thomas Meek and his collaborators at SmithKline Beecham Pharmaceuticals have shown that the mechanism of HIV-1 protease is very similar to those of other aspartic proteases.

Chorismate Mutase: A Model for Understanding Catalytic Power and Efficiency

Direct comparison of an enzyme reaction with the analogous uncatalyzed reaction is usually difficult, if not impossible. There are several problems: First, many enzyme-catalyzed reactions do not proceed at measurable rates in the absence of the enzyme. Second, many enzyme-catalyzed reactions involve formation of a covalent intermediate between the enzyme and the substrate. Third, a reaction occurring in an enzyme active site might proceed through a different transition state than the corresponding solution reaction. **Chorismate mutase** is a happy exception to all these potential problems. First, although the rate of this reaction is more than a million times faster on the enzyme, the

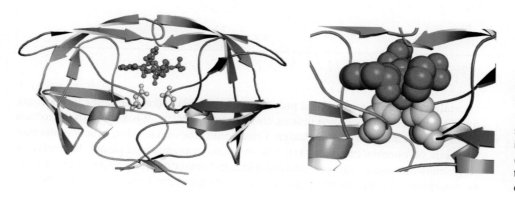

FIGURE 14.27 *(left)* HIV-1 protease complexed with the inhibitor Crixivan (red) made by Merck. The flaps (residues 46–55 from each subunit) covering the active site are shown in green, and the active-site aspartate residues involved in catalysis are shown in light purple. *(right)* The close-up of the active site shows the interaction of Crixivan with the carboxyl groups (yellow) of the essential aspartate residues (pdb id = 1HSG).

HUMAN BIOCHEMISTRY

Protease Inhibitors Give Life to AIDS Patients

Infection with HIV was once considered a death sentence, but the emergence of a new family of drugs called protease inhibitors has made it possible for some AIDS patients to improve their overall health and extend their lives. These drugs are all specific inhibitors of the HIV protease. By inhibiting the protease, they prevent the development of new virus particles in the cells of infected patients. A combination of drugs—including a protease inhibitor together with a reverse transcriptase inhibitor like AZT—can reduce the human immunodeficiency virus (HIV) to undetectable levels in about 40% to 50% of infected individuals. Patients who respond successfully to this combination therapy have experienced dramatic improvement in their overall health and a substantially lengthened life span.

Four of the protease inhibitors approved for use in humans by the U.S. Food and Drug Administration are shown below: Crixivan by Merck, Invirase by Hoffman-LaRoche, Norvir by Abbott, and Viracept by Agouron. These drugs were all developed from a "structure-based" design strategy; that is, the drug molecules were designed to bind tightly to the active site of the HIV-1 protease. The backbone OH-group in all these substances inserts between the two active-site carboxyl groups of the protease.

In the development of an effective drug, it is not sufficient merely to show that a candidate compound can cause the desired biochemi-

cal effect. It must also be demonstrated that the drug can be effectively delivered in sufficient quantities to the desired site(s) of action in the organism and that the drug does not cause undesirable side effects. The HIV-1 protease inhibitors shown here fulfill all of these criteria. Other drug candidates have been found that are even better inhibitors of HIV-1 protease in cell cultures, but many of these fail the test of bioavailability—the ability of a drug to be delivered to the desired site(s) of action in the organism.

Candidate protease inhibitor drugs must be relatively specific for the HIV-1 protease. Many other aspartic proteases exist in the human body and are essential to a variety of body functions, including digestion of food and processing of hormones. An ideal drug thus must strongly inhibit the HIV-1 protease, must be delivered effectively to the lymphocytes where the protease must be blocked, and should not adversely affect the activities of the essential human aspartic proteases.

A final but important consideration is viral mutation. Certain mutant HIV strains are resistant to one or more of the protease inhibitors, and even for patients who respond initially to protease inhibitors it is possible that mutant viral forms may eventually arise and thrive in infected individuals. The search for new and more effective protease inhibitors is ongoing.

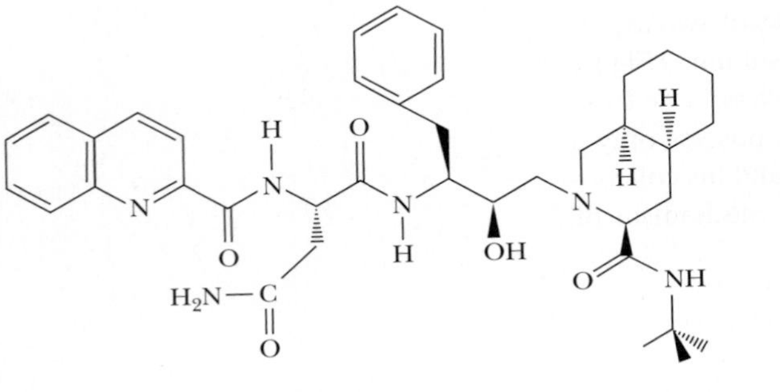

Invirase (saquinavir)

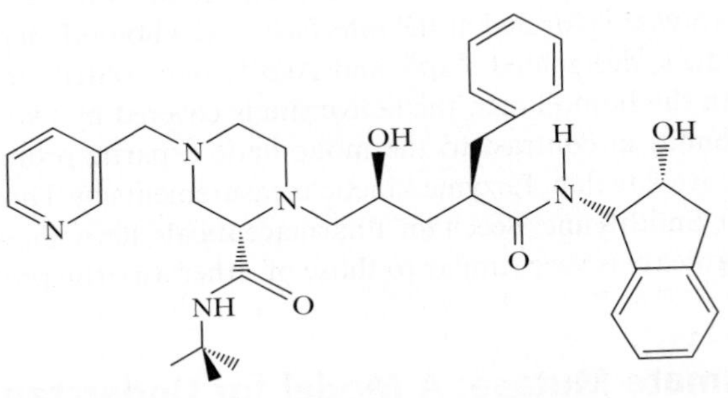

Crixivan (indinavir)

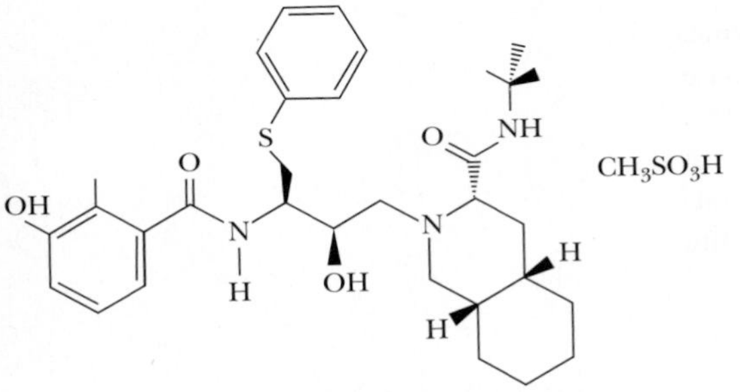

Viracept (nelfinavir mesylate)

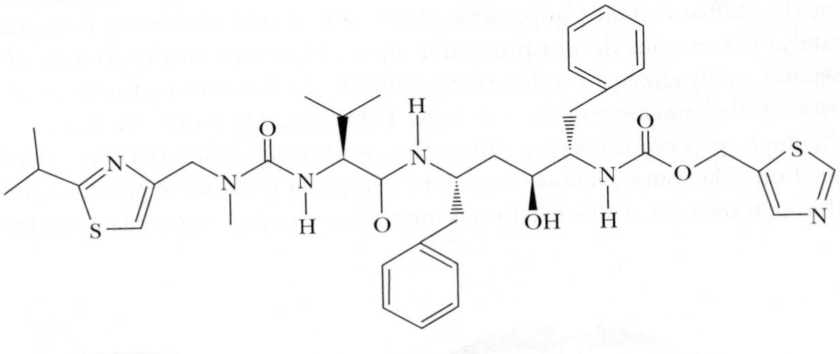

Norvir (ritonavir)

uncatalyzed solution reaction still proceeds at reasonable and measurable rates. Second, the enzyme reaction does not employ a covalent intermediate. What about the transition states for the catalyzed and uncatalyzed reactions? Chorismate mutase acts in the biosynthesis of phenylalanine and tyrosine in microorganisms and plants. It involves a single substrate and catalyzes a concerted intramolecular rearrangement of chorismate to prephenate. In this simple reaction, one carbon-oxygen bond is broken, and one

(a)   Chorismate mutase reaction

(b)   Classic Claisen rearrangement

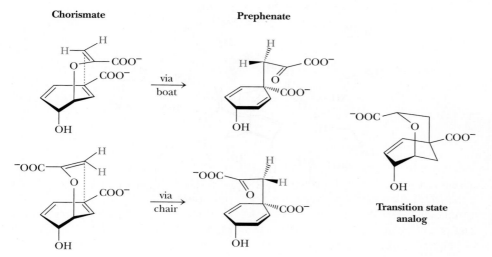

Allyl phenyl ether   Cyclohexadienone   2-Allyl phenol
                          intermediate

FIGURE 14.28 (a) The chorismate mutase reaction converts chorismate to prephenate. **(b)** A classic Claisen rearrangement. Conversion of allyl phenyl ether to 2-allyl alcohol proceeds through a cyclohexadienone intermediate, which then undergoes a keto-enol tautomerization.

carbon-carbon bond is formed. It is an example of a **Claisen rearrangement,** familiar to any student of organic chemistry (Figure 14.28). There are two possible transition states, one involving a chair conformation and the other a boat (Figure 14.29). Jeremy Knowles and his co-workers have shown that both the enzymatic and the solution reactions take place via a chair transition state, and a transition-state analog of this state has been characterized (Figure 14.29).

The Chorismate Mutase Active Site Lies at the Interface of Two Subunits The chorismate mutase from *E. coli* is the N-terminal portion (109 residues) of a bifunctional enzyme, termed the **P protein,** which also has a C-terminal prephenate dehydrogenase activity. The N-terminal portion of the P protein has been prepared as a separate protein by recombinant DNA techniques, and this engineered protein is a fully functional chorismate mutase. The structure shown in Figure 14.30 is a homodimer, each monomer consisting of three α-helices (denoted H1, H2, and H3) connected by short loops. The two monomers are dovetailed in the dimer structure, with the H1 helices paired and the H3 helices overlapping significantly. The long, ten-turn H1 helices form an antiparallel coiled coil, with leucines at positions 10, 17, 24, and 31 in a classic 7-residue repeat pattern (see Chapter 6).

Chorismate   **Prephenate**

via boat

via chair

Transition state analog

FIGURE 14.29 The conversion of chorismate to prephenate could occur (in principle) through a boat transition state or a chair transition state. The difference can be understood by imagining two different isotopes of hydrogen (blue and green) at carbon 9 of chorismate and the products that would result in each case. Knowles and co-workers have shown that both the uncatalyzed reaction and the reaction on chorismate mutase occur through a chair transition state. The molecule shown at right is a transition state analog for the chorismate mutase reaction.

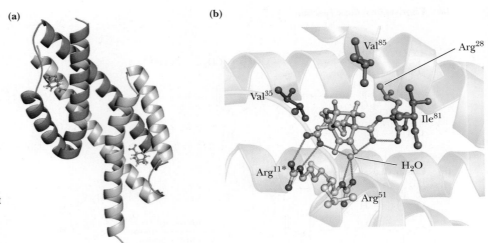

(a)   **(b)**

FIGURE 14.30 Chorismate mutase is a symmetric homodimer, each monomer consisting of three α-helices connected by short loops. **(a)** The dimer contains two equivalent active sites, each formed from portions of both monomers (pdb id = 4CSM). **(b)** A close-up of the active site, showing the bound transition-state analog (pink, see Figure 14.29).

The chorismate mutase dimer contains two equivalent active sites, each formed from portions of both monomers. The structure shown in Figure 14.20 contains a bound transition-state analog (Figure 14.29) stabilized by 12 electrostatic and hydrogen-bonding interactions (Figure 14.31). Arg[28] from one subunit and Arg[11*] from the other coordinate the carboxyl groups of the analog, and a third arginine (Arg[51]) coordinates a water molecule, which in turn coordinates both carboxyls of the analog. Each oxygen of the analog is coordinated by two groups from the active site. In addition, there are hydrophobic residues surrounding the analog, especially Val[35] on one side and Ile[81] and Val[85] on the other.

The Chorismate Mutase Active Site Favors a Near-Attack Conformation The chorismate mutase reaction mechanism requires that the carboxyvinyl group fold over the chorismate ring to facilitate the Claisen rearrangement (Figure 14.32). This implies the formation of a NAC on the way to the transition state. Bruice and his co-workers have carried out extensive molecular dynamics simulations of the chorismate mutase reaction. Their calculations show that, in the nonenzymatic reaction, only 0.0001% of chorismate in solution exists in the NAC required for reaction. Similar calculations show that, in the enzyme active site, chorismate adopts a NAC 30% of the time. The computer-simulated NAC in the chorismate mutase active site (Figure 14.33) is similar in many ways to the chorismate mutase-TSA complex, with Arg[28] and Arg[11*] coordinating the two carboxylate groups of chorismate so as to position the carboxyvinyl group

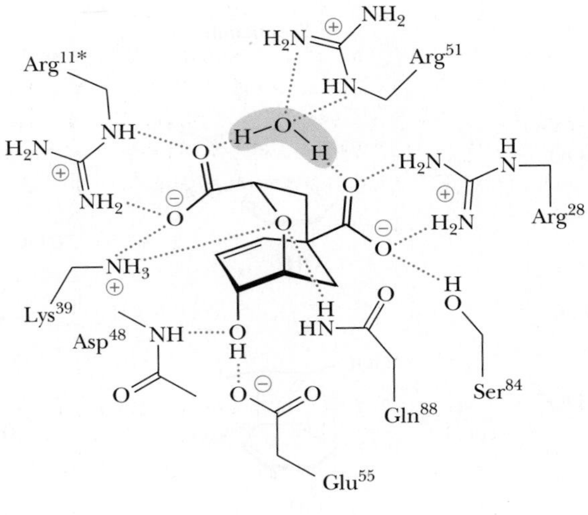

FIGURE 14.31 In the chorismate mutase active site, the transition-state analog is stabilized by 12 electrostatic and hydrogen-bonding interactions. (Adapted from Lee, A., et al., 1995. Atomic structure of the buried catalytic pocket of *Escherichia coli* chorismate mutase. *Journal of the American Chemical Society* **117**:3627–3628.)

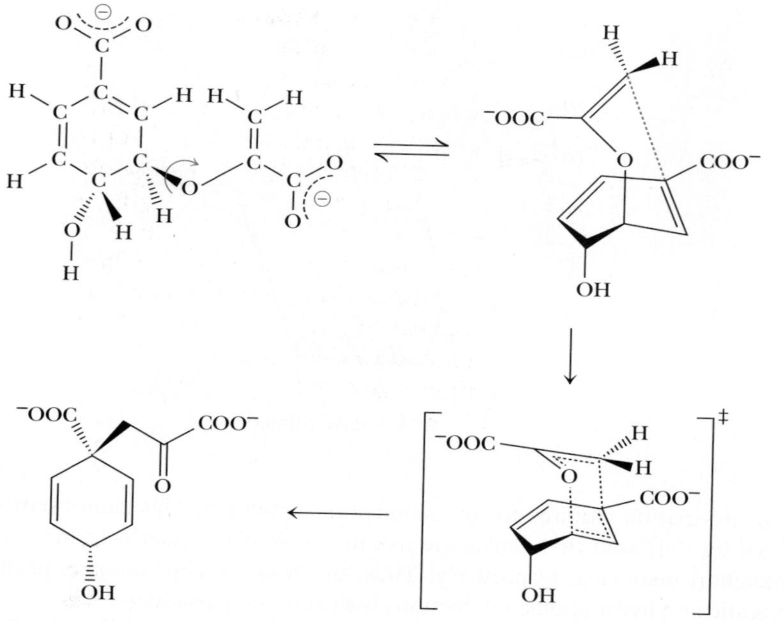

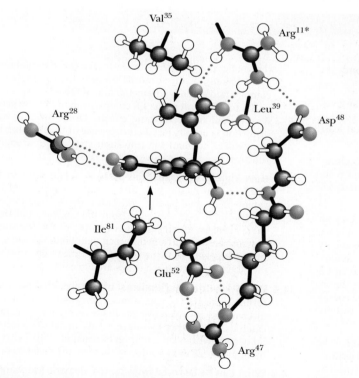

FIGURE 14.32 The mechanism of the chorismate mutase reaction. The carboxyvinyl group folds up and over the chorismate ring, and the reaction proceeds via an internal rearrangement.

FIGURE 14.33 Chorismate bound to the active site of chorismate mutase in a structure that resembles a NAC. Arrows indicate hydrophobic interactions, and red dotted lines indicate electrostatic interactions. (Adapted from Hur, S., and Bruice, T., 2003. The near attack conformation approach to the study of the chorismate to prephenate reaction. *Proceedings of the National Academy of Sciences USA* **100**:12015–12020.)

FIGURE 14.34 The energetic profile of the chorismate mutase reaction. Computer simulations by Bruice and his co-workers show that the NAC and the E-S complex are separated by only 0.42 kJ/mol, meaning that the NAC forms much more readily in the enzyme active site than it does in the absence of enzyme. The NAC and the reaction transition state are separated by similar energy barriers in either the presence or the absence of the enzyme. Thus, the catalytic prowess of the enzyme lies in its ability to form the NAC at a very small energetic cost. (Adapted from Bruice, T., 2002. A view at the millennium: The efficiency of enzymatic catalysis. *Accounts of Chemical Reactions* **35**:139–148.)

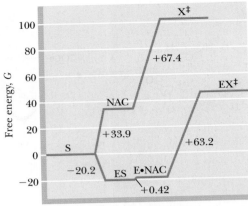

in the conformation required for transition-state formation. This conformation is also stabilized by Val[35] and Ile[85], which are in van der Waals contact with the vinyl group and the chorismate ring, respectively. Thus, the NAC of chorismate is promoted by electrostatic and hydrophobic interactions with active-site residues.

The energetics of the chorismate mutase reaction are revealing (Figure 14.34). Computer simulations by Bruice and his co-workers show that formation of a NAC in the absence of the enzyme is energetically costly, whereas formation of the NAC in the enzyme active site is facile, with only a modest energy cost. On the other hand, the energy required to move from the NAC to the transition state is about the same for the solution and the enzyme reactions. Clearly, the catalytic advantage of chorismate mutase is the ease of formation of a NAC in the active site.

SUMMARY

OWL Login to OWL to develop problem-solving skills and complete online homework assigned by your professor.

It is simply chemistry—the breaking and making of bonds—that gives enzymes their prowess. This chapter explores the unique features of this chemistry. The mechanisms of thousands have been studied in at least some detail.

14.1 What Are the Magnitudes of Enzyme-Induced Rate Accelerations? Enzymes are powerful catalysts. Enzyme-catalyzed reactions are typically 10^7 to 10^{14} times faster than their uncatalyzed counterparts and may exceed 10^{16}.

14.2 What Role Does Transition-State Stabilization Play in Enzyme Catalysis? The energy barrier for the uncatalyzed reaction is the difference in energies of the S and $X^‡$ states. Similarly, the energy barrier to be surmounted in the enzyme-catalyzed reaction, assuming that E is saturated with S, is the energy difference between ES and $EX^‡$. Reaction rate acceleration by an enzyme means simply that the energy barrier between ES and $EX^‡$ is less than the energy barrier between S and $X^‡$. In terms of the free energies of activation, $\Delta G_e^‡ < \Delta G_u^‡$.

14.3 How Does Destabilization of ES Affect Enzyme Catalysis? The favorable interactions between the substrate and amino acid residues on the enzyme account for the intrinsic binding energy, ΔG_b. The intrinsic binding energy ensures the favorable formation of the ES complex, but if uncompensated, it makes the activation energy for the enzyme-catalyzed reaction unnecessarily large and wastes some of the catalytic power of the enzyme. Because the enzymatic reaction rate is determined by the difference in energies between ES and $EX^‡$,

the smaller this difference, the faster the enzyme-catalyzed reaction. Tight binding of the substrate deepens the energy well of the ES complex and actually lowers the rate of the reaction.

Destabilization of the ES complex can involve structural strain, desolvation, or electrostatic effects. Destabilization by strain or distortion is usually just a consequence of the fact that the enzyme has evolved to bind the transition state more strongly than the substrate.

14.4 How Tightly Do Transition-State Analogs Bind to the Active Site? Given a ratio k_e/k_u of 10^{12} and a typical K_S of 10^{-3} M, the value of K_T would be 10^{-15} M. This is the dissociation constant for the transition-state complex from the enzyme, and this very low value corresponds to very tight binding of the transition state by the enzyme. It is unlikely that such tight binding in an enzyme transition state will ever be measured experimentally, however, because the lifetimes of transition states are typically 10^{-14} to 10^{-13} sec.

14.5 What Are the Mechanisms of Catalysis? Enzymes facilitate formation of NACs (near-attack conformations). Enzyme reaction mechanisms involve covalent bond formation, general acid–base catalysis, low-barrier hydrogen bonds, metal ion effects, and proximity and favorable orientation of reactants. Most enzymes display involvement of two or more of these in any given reaction.

14.6 What Can Be Learned from Typical Enzyme Mechanisms? The enzymes examined in this chapter—serine proteases, aspartic proteases, and chorismate mutase—provide representative examples of catalytic mechanisms; all embody two or more of the rate enhancement contributions.

FOUNDATIONAL BIOCHEMISTRY Things You Should Know After Reading Chapter 14.

- The approximate magnitudes of catalyzed and uncatalyzed reactions described in Section 14.1.
- The role that transition-state stabilization plays in enzyme catalysis.
- The manner in which destabilization of ES affects enzyme catalysis.
- The nature, lifetime, and binding characteristics of the transition state of an enzyme reaction.
- The utility of transition state analogs as therapeutic agents.
- How to read and write chemical reaction mechanisms.
- The nature of near-attack complexes in enzyme reactions.
- The role of protein motions in enzyme catalysis.
- The role of covalent intermediates in enzyme catalysis.
- The role of general acid–base catalysis in enzyme reactions.
- The characteristic features of low-barrier hydrogen bonds and their role in enzyme catalysis.
- The role of metal ion catalysis in enzyme reactions.

- The essential features of the serine proteases.
- The mechanistic basis of burst kinetics in an enzyme reaction.
- The mechanism of serine proteases.
- The essential features of the aspartic proteases.
- The mechanism of aspartic proteases.
- The roles of low-barrier hydrogen bonds in the serine and aspartic proteases.
- The structural contrasts between mammalian aspartic proteases and the HIV-1 protease.
- The significance of protease inhibitors as AIDS drugs.
- The essential features of chorismate mutase.
- The mechanism of action of chorismate mutase.
- The role of a near-attack complex in the chorismate mutase reaction.

PROBLEMS

Answers to all problems are at the end of this book. Detailed solutions are available in the *Student Solutions Manual, Study Guide, and Problems Book.*

1. **Characterizing a Covalent Enzyme Inhibitor** Tosyl-L-phenylalanine chloromethyl ketone (TPCK) specifically inhibits chymotrypsin by covalently labeling His[57].

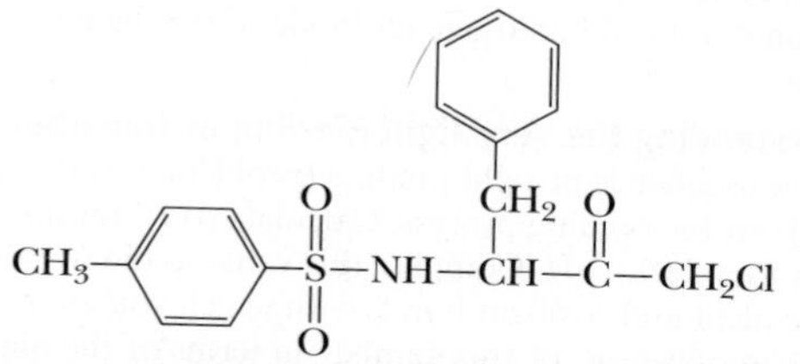

Tosyl-L-phenylalanine chloromethyl ketone (TPCK)

   a. Propose a mechanism for the inactivation reaction, indicating the structure of the product(s).
   b. State why this inhibitor is specific for chymotrypsin.
   c. Propose a reagent based on the structure of TPCK that might be an effective inhibitor of trypsin.

2. **Using Site-Directed Mutants to Understand an Enzyme Mechanism** In this chapter, the experiment in which Craik and Rutter replaced Asp[102] with Asn in trypsin (reducing activity 10,000-fold) was discussed.

   a. On the basis of your knowledge of the catalytic triad structure in trypsin, suggest a structure for the "uncatalytic triad" of Asn-His-Ser in this mutant enzyme.
   b. Explain why the structure you have proposed explains the reduced activity of the mutant trypsin.
   c. See the original journal articles (Sprang, et al., 1987. *Science* **237:** 905–909; and Craik, et al., 1987. *Science* **237:**909–913) to see Craik and Rutter's answer to this question.

3. **Assessing the Action of an Aspartic Protease Inhibitor** Pepstatin (see below) is an extremely potent inhibitor of the monomeric aspartic proteases, with K_I values of less than 1 nM.

   a. On the basis of the structure of pepstatin, suggest an explanation for the strongly inhibitory properties of this peptide.
   b. Would pepstatin be expected to also inhibit the HIV-1 protease? Explain your answer.

4. **Deriving an Expression for Catalytic Power** Based on the following reaction scheme, derive an expression for k_e/k_u, the ratio of the rate constants for the catalyzed and uncatalyzed reactions,

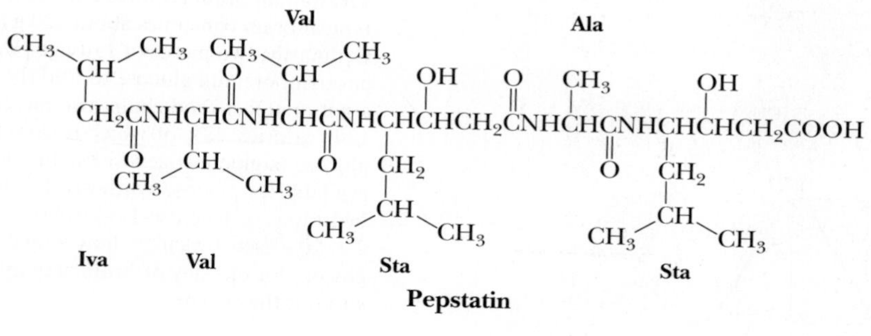

Pepstatin

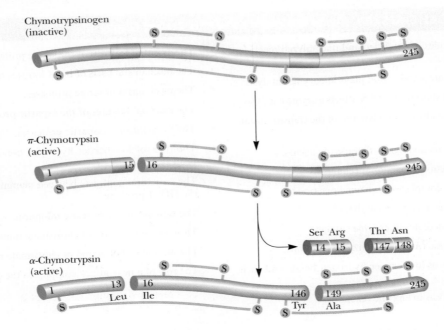

Chymotrypsinogen
(inactive)

π-Chymotrypsin
(active)

Ser Arg   Thr Asn
14 15   147 148

α-Chymotrypsin
(active)

13 16
Leu Ile

146 149   245
Tyr Ala

respectively, in terms of the free energies of activation for the catalyzed ($\Delta G_e^{\ddagger}$) and the uncatalyzed ($\Delta G_u^{\ddagger}$) reactions.

$$S \underset{}{\overset{K_u}{\rightleftharpoons}} X^{\ddagger} \overset{k_u'}{\longrightarrow} P$$

$$E \downarrow\uparrow K_S \qquad\qquad \downarrow\uparrow E$$

$$ES \underset{}{\overset{K_e}{\rightleftharpoons}} EX^{\ddagger} \overset{k_e'}{\longrightarrow} EP$$

5. Comparison of Enzymatic and Nonenzymatic Rate Constants
The k_{cat} for alkaline phosphatase–catalyzed hydrolysis of methylphosphate is approximately 14/sec at pH 8 and 25°C. The rate constant for the uncatalyzed hydrolysis of methylphosphate under the same conditions is approximately 10^{-15}/sec. What is the difference in the free energies of activation of these two reactions?

6. Understanding the Actions of Proteolytic Enzymes Active α-chymotrypsin is produced from chymotrypsinogen, an inactive precursor, as shown in the color figure at the top of this page. The first intermediate—π-chymotrypsin—displays chymotrypsin activity. Suggest proteolytic enzymes that might carry out these cleavage reactions effectively.

7. Understanding the Mechanism of Carboxypeptidase A Consult the classic paper by William Lipscomb (1982. *Accounts of Chemical Research* **15**:232–238), and on the basis of this article write a simple mechanism for the enzyme carboxypeptidase A.

8. Calculation of Rate Enhancement from Energies of Activation
The relationships between the free energy terms defined in the solution to Problem 4 above are shown in the following figure:

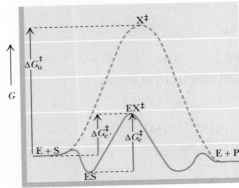

Reaction coordinate

If the energy of the ES complex is 10 kJ/mol lower than the energy of E + S, the value of $\Delta G_e^{\ddagger}$ is 20 kJ/mol, and the value of $\Delta G_u^{\ddagger}$ is 90 kJ/mol, what is the rate enhancement achieved by an enzyme in this case?

9. Understanding the Implications of Transition State Stabilization
As noted on page 451, a true transition state can bind to an enzyme active site with a K_T as low as 7×10^{-26} M. This is a remarkable number, with interesting consequences. Consider a hypothetical solution of an enzyme in equilibrium with a ligand that binds with a K_D of 10^{-27} M. If the concentration of free enzyme, [E], is equal to the concentration of the enzyme–ligand complex, [EL], what would [L], the concentration of free ligand, be? Calculate the volume of solution that would hold one molecule of free ligand at this concentration.

10. Understanding the Very Tight Binding of Transition States Another consequence of tight binding (problem 9) is the free energy change for the binding process. Calculate $\Delta G°'$ for an equilibrium with a K_D of 10^{-27} M. Compare this value to the free energies of the noncovalent and covalent bonds with which you are familiar. What are the implications of this number, in terms of the binding of a transition state to an enzyme active site?

11. Assessing the Metabolic Consequences of Life Without Enzymes The incredible catalytic power of enzymes can perhaps best be appreciated by imagining how challenging life would be without just one of the thousands of enzymes in the human body. For example, consider life without fructose-1,6-bisphosphatase, an enzyme in the gluconeogenesis pathway in liver and kidneys (see Chapter 22), which helps produce new glucose from the food we eat:

Fructose-1,6-bisphosphate + H_2O → Fructose-6-P + P_i

The human brain requires glucose as its only energy source, and the typical brain consumes about 120 g (or 480 calories) of glucose daily. Ordinarily, two pieces of sausage pizza could provide more than enough potential glucose to feed the brain for a day. According to a national fast-food chain, two pieces of sausage pizza provide 1340 calories, 48% of which is from fat. Fats cannot be converted to glucose in gluconeogenesis, so that leaves 697 calories potentially available for glucose synthesis. The first-order rate constant for the hydrolysis of fructose-1,6-bisphosphate in the absence of enzyme is 2×10^{-20}/sec. Calculate how long it would take to provide enough glucose for one day of brain activity from two pieces of sausage pizza without the enzyme.

Preparing for the MCAT® Exam

The following graphs show the temperature and pH dependencies of four enzymes, A, B, X, and Y. Problems 12 through 18 refer to these graphs.

(a)

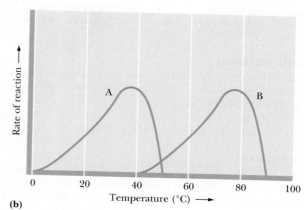

(b)

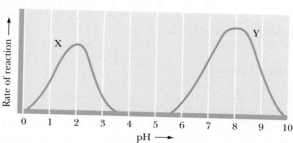

12. Enzymes X and Y in the figure are both protein-digesting enzymes found in humans. Where would they most likely be at work?

a. X is found in the mouth, Y in the small intestine.

b. X in the small intestine, Y in the mouth.

c. X in the stomach, Y in the small intestine.

d. X in the small intestine, Y in the stomach.

13. Which statement is true concerning enzymes X and Y?

a. They could not possibly be at work in the same part of the body at the same time.

b. They have different temperature ranges at which they work best.

c. At a pH of 4.5, enzyme X works slower than enzyme Y.

d. At their appropriate pH ranges, both enzymes work equally fast.

14. What conclusion may be drawn concerning enzymes A and B?

a. Neither enzyme is likely to be a human enzyme.

b. Enzyme A is more likely to be a human enzyme.

c. Enzyme B is more likely to be a human enzyme.

d. Both enzymes are likely to be human enzymes.

15. At which temperatures might enzymes A and B both work?

a. Above 40°C

b. Below 50°C

c. Above 50°C and below 40°C

d. Between 40° and 50°C

16. An enzyme–substrate complex can form when the substrate(s) bind(s) to the active site of the enzyme. Which environmental condition might alter the conformation of an enzyme to the extent that its substrate is unable to bind?

a. Enzyme A at 40°C

b. Enzyme B at pH 2

c. Enzyme X at pH 4

d. Enzyme Y at 37°C

17. At 35°C, the rate of the reaction catalyzed by enzyme A begins to level off. Which hypothesis best explains this observation?

a. The temperature is too far below optimum.

b. The enzyme has become saturated with substrate.

c. Both A and B.

d. Neither A nor B.

18. In which of the following environmental conditions would digestive enzyme Y be unable to bring its substrate(s) to the transition state?

a. At any temperature below optimum

b. At any pH where the rate of reaction is not maximum

c. At any pH lower than 5.5

d. At any temperature higher than 37°C

19. Review the mechanisms of the serine and aspartic proteases, and compare these two mechanisms carefully. Are there steps in the mechanisms that are similar? How are they similar? How are they different? Suggest experiments that could support or refute your hypotheses.

ActiveModels Problems

20. Examine the ActiveModel for porcine pepsin, then identify and explain the purpose of the functionally important residues in this enzyme's active site.

21. Using specific examples from the ActiveModel for angiotensin converting enzyme (ACE), explain how N-glycosylation stabilizes ACE.

22. As noted in its ActiveModel, the flap domain of HIV-1 protease is responsible for ligand-binding interactions. Why do the flap domains have glycine-rich ends?

23. According to its ActiveModel, where is the HIV-1 protease cleavage site?

FURTHER READING

General

Alegre-Cebollada, J., Perez-Jimenez, R., Kosuri, P., and Fernandez, J. M., 2010. Single-molecule force spectroscopy approach to enzyme catalysis. *Journal of Biological Chemistry* **285:**18961–18966.

Allen, K. N., and Dunaway-Mariano, D., 2010. Markers of fitness in a successful enzyme superfamily. *Current Opinion in Structural Biology* **19:**658–665.

Cleland, W. W., 2005. The use of isotope effects to determine enzyme mechanisms. *Archives of Biochemistry and Biophysics* **433:**2–12.

Eigen, M., 1964. Proton transfer, acid–base catalysis, and enzymatic hydrolysis. *Angewandte Chemie International Edition* **3:**1–72.

Hammes-Schiffer, S., and Stuchebrukhov, A. A., 2010. Coupled electron and proton transfer reactions. *Chemical Reviews* **110:**6939–6960.

Kamerlin, S. C. L., and Warshel, A., 2010. At the dawn of the 21st century: is dynamics the missing link for understanding enzyme catalysis. *Proteins* **78:**1339–1375.

Ringe, D., and Petsko, G. A., 2008. How enzymes work. *Science* **320:**1428–1429.

Schramm, V., 2011. Chemical mechanisms in biochemical reactions. *Journal of the American Chemical Society* **133:**13207–13212.

Stockbridge, R. B., Lewis, Jr., C. A., Yuan, Y., and Wolfenden, R., 2010. Impact of structure on the time required for the establishment of primordial biochemistry, and for the evolution of enzymes. *Proceedings of the National Academy of Sciences* **107:**22102–22105.

Tang, Q., and Leyh, T. S., 2010. Precise, facile, initial rate measurements. *Journal of Physical Chemistry B* **114:**16131–16136.

Warshel, A., Sharma, P. K., Kato, M., Xiang, Y., Liu, H., and Olsson, M., 2006. Electrostatic basis for enzyme catalysis. *Chemical Reviews* **106**:3210–3235.

Wolfenden, R., 2006. Degree of difficulty of water-consuming reactions in the absence of enzymes. *Chemical Reviews* **106**:3379–3397.

Wolfenden, R., and Snider, M. T., 2001. The depth of chemical time and the power of enzymes as catalysts. *Accounts of Chemical Research* **34**:938–945.

Zhang, X., and Houk, K. N., 2005. Why enzymes are proficient catalysts: Beyond the Pauling paradigm. *Accounts of Chemical Research* **38**:379–385.

Transition-State Stabilization and Transition-State Analogs

Hopkins, A. L., and Groom, C. R., 2002. The druggable genome. *Nature Reviews Drug Discovery* **1**:727–730.

Overington, J. P., Al-Lazikani, B., and Hopkins, A. L., 2006. How many drug targets are there? *Nature Reviews Drug Discovery* **5**:993–996.

Schramm, V. L., 2007. Enzymatic transition state theory and transition state analogue design. *Journal of Biological Chemistry* **282**:28297–28300.

Schwartz, S. D., and Schramm, V. L., 2010. Enzymatic transition states and dynamic motion in barrier crossing. *Nature Chemical Biology* **5**:551–558.

Wogulis, M., Wheelock, C. E., et al., 2006. Structural studies of a potent insect maturation inhibitor bound to the juvenile hormone esterase of *Manduca sexta*. *Biochemistry* **45**:4045–4057.

Near-Attack Conformations

Bruice, T. C., 2002. A view at the millennium: The efficiency of enzymatic catalysis. *Accounts of Chemical Research* **35**:139–148.

Bruice, T. C., 2006. Computational approaches: reaction trajectories, structures, and atomic motions. Enzyme reactions and proficiency. *Chemical Reviews* **106**:3119–3139.

Hur, S., and Bruice, T., 2003. The near attack conformation approach to the study of the chorismate to prephenate reaction. *Proceedings of the National Academy of Sciences USA* **100**:12015–12020.

Luo, J., and Bruice, T. C., 2007. Low frequency normal modes in horse liver alcohol dehydrogenase and motions of residues involved in the enzymatic reaction. *Biophysical Chemistry* **126**:80–85.

Schowen, R. L., 2003. How an enzyme surmounts the activation energy barrier. *Proceedings of the National Academy of Sciences USA* **100**:11931–11932.

Motion in Enzymes

Baldwin, A. J., and Kay, L. E., 2009. NMR spectroscopy brings invisible protein states into focus. *Nature Chemical Biology* **5**:808–814.

Benkovic, S. J., and Hammes-Schiffer, S., 2006. Enzyme motions inside and out. *Science* **312**:208–209.

Boehr, D. D., Dyson, H. J., and Wright, P. E., 2006. An NMR perspective on enzyme dynamics. *Chemical Reviews* **106**:3055–3079.

Hammes-Schiffer, S., and Benkovic, S. J., 2006. Relating protein motion to catalysis. *Annual Review of Biochemistry* **75**:519–541.

Hirschi, J. S., Arora, K., Brooks III, C. L., and Schramm, V. L., 2010. Conformational dynamics in human purine nucleoside phosporylase with reactants and transition-state analogues. *Journal of Physical Chemistry B* **114**:16263–16272.

Nashine, V. C., Hammes-Schiffer, S., and Benkovic, S. J., 2010. Coupled motions in enzyme catalysis. *Current Opinion in Chemical Biology* **14**:644–651.

Low-Barrier Hydrogen Bonds

Cleland, W. W., 2000. Low barrier hydrogen bonds and enzymatic catalysis. *Archives of Biochemistry and Biophysics* **382**:1–5.

Yamaguchia, S., Kamikuboa, H., Kuriharab, K., Kurokib, R., et al., 2009. Low barrier hydrogen bond in photoactive yellow protein. *Proceedings of the National Academy of Sciences* **106**:440–444.

Serine Proteases

Bruice, T. C., and Bruice, P. Y., 2005. Covalent intermediates and enzyme proficiency. *Journal of the American Chemical Society* **127**:12478–12479.

Craik, C. S., Roczniak, S., et al. 1987. The catalytic role of the active site aspartic acid in serine proteases. *Science* **237**:909–913.

Sprang, S., and Standing, T., 1987. The three dimensional structure of Asn[102] mutant of trypsin: Role of Asp[102] in serine protease catalysis. *Science* **237**:905–909.

Aspartic Proteases

Northrop, D. B., 2001. Follow the protons: A low-barrier hydrogen bond unifies the mechanisms of the aspartic proteases. *Accounts of Chemical Research* **34**:790–797.

Wolfe, M. S., 2010. Structure, mechanism and inhibition of γ-secretase and presenilin-like proteases. *Biology and Chemistry* **391**:839–847.

HIV-1 Protease

Hyland, L., et al., 1991. Human immunodeficiency virus-1 protease 1: Initial velocity studies and kinetic characterization of reaction intermediates by [18]O isotope exchange. *Biochemistry* **30**:8441–8453.

Hyland, L., Tomaszek, T., and Meek, T., 1991. Human immunodeficiency virus-1 protease 2: Use of pH rate studies and solvent isotope effects to elucidate details of chemical mechanism. *Biochemistry* **30**:8454–8463.

Chorismate Mutase

Bartlett, P. A., and Johnson, C. R., 1985. An inhibitor of chorismate mutase resembling the transition state conformation. *Journal of the American Chemical Society* **107**:7792–7793.

Copley, S. D., and Knowles, J. R., 1985. The uncatalyzed Claisen rearrangement of chorismate to prephenate prefers a transition state of chairlike geometry. *Journal of the American Chemical Society* **107**:5306–5308.

Hur, S., and Bruice, T. C., 2003. The near attack conformation approach to the study of the chorismate to prephenate reaction. *Proceedings of the National Academy of Sciences USA* **100**:12015–12020.

Lee, A., Karplus, A., et al., 1995. Atomic structure of the buried catalytic pocket of *Escherichia coli* chorismate mutase. *Journal of the American Chemical Society* **117**:3627–3628.

Sasso, S., Okvist, M., Rodere, K., et al., 2009. Structure and function of a complex between chorismate mutase and DAHP synthase: efficiency boost for the junior partner. *The EMBO Journal* **28**:2128–2142.

Sogo, S. G., Widlanski, T. S., et al., 1984. Stereochemistry of the rearrangement of chorismate to prephenate: Chorismate mutase involves a chair transition state. *Journal of the American Chemical Society* **106**:2701–2703.

Zhang, X., Zhang, X., et al., 2005. A definitive mechanism for chorismate mutase. *Biochemistry* **44**:10443–10448.

Tunneling in Enzyme Mechanisms

Hammes-Schiffer, 2010. Introduction: Proton-Coupled Electron Transfer. *Chemical Reviews* **110**:6937–6938.

Machleder, S. Q., Pineda, E. T., and Schwartz, S. D., 2010. On the orgin of the chemical barrier and tunneling in enzymes. *Journal of Physical Organic Chemistry* **23**:690–695.

Nagel, Z. D., and Klinman, J. P., 2009. A 21st century revisionist's view at a turning point in enzymology. *Nature Chemical Biology* **5**:543–550.

Nagel, Z. D., and Klinman, J. P., 2010. Tunneling and dynamics in enzymatic hydride transfer. *Chemical Reviews* **110**:PR41–PR67.

Pudney, C. R., Johannissen, L. O., Sutcliffe, M. J., Hay, S., and Scrutton, N. S., 2010. Direct analysis of donor-acceptor distance and relationship to isotope effects and the force constant for barrier compression in enzymatic H-tunneling reactions. *Journal of the American Chemical Society* **132**:11329–11335.

Enzyme Regulation

© Christie's Images/CORBIS © 2011 Calder Foundation, New York/Artists Rights Society (ARS), New York.

◀ Metabolic regulation is achieved through an exquisitely balanced interplay among enzymes and small molecules.

Allostery is a key chemical process that makes possible intracellular and intercellular regulation: ". . . the molecular interactions which ensure the transmission and interpretation of (regulatory) signals rest upon (allosteric) proteins endowed with discriminatory stereospecific recognition properties."

Jacques Monod
Chance and Necessity

KEY QUESTIONS

15.1 What Factors Influence Enzymatic Activity?

15.2 What Are the General Features of Allosteric Regulation?

15.3 Can Allosteric Regulation Be Explained by Conformational Changes in Proteins?

15.4 What Kinds of Covalent Modification Regulate the Activity of Enzymes?

15.5 Is the Activity of Some Enzymes Controlled by Both Allosteric Regulation and Covalent Modification?

Special Focus: Is There an Example in Nature That Exemplifies the Relationship Between Quaternary Structure and the Emergence of Allosteric Properties? Hemoglobin and Myoglobin—Paradigms of Protein Structure and Function

ESSENTIAL QUESTION

Enzymes catalyze essentially all of the thousands of metabolic reactions taking place in cells. Many of these reactions are at cross-purposes: Some enzymes catalyze the breakdown of substances, whereas others catalyze synthesis of the same substances; many metabolic intermediates have more than one fate; and energy is released in some reactions and consumed in others. At key positions within the metabolic pathways, regulatory enzymes sense the momentary needs of the cell and adjust their catalytic activity accordingly. Regulation of these enzymes ensures the harmonious integration of the diverse and often divergent reactions of metabolism.

What are the properties of regulatory enzymes? How do regulatory enzymes sense the momentary needs of cells? What molecular mechanisms are used to regulate enzyme activity?

15.1 | What Factors Influence Enzymatic Activity?

The activity displayed by enzymes is affected by a variety of factors, some of which are essential to the harmony of metabolism. Two of the more obvious ways to regulate the amount of activity at a given time are (1) to increase or decrease the number of enzyme molecules and (2) to increase or decrease the activity of each enzyme molecule. Although these ways are obvious, the cellular mechanisms that underlie them are complex and varied, as we shall see. A general overview of factors influencing enzyme activity includes the following considerations.

OWL Online homework and a Student Self Assessment for this chapter may be assigned in OWL.

The Availability of Substrates and Cofactors Usually Determines How Fast the Reaction Goes

The availability of substrates and cofactors typically determines the enzymatic reaction rate. In general, enzymes have evolved such that their K_m values approximate the prevailing in vivo concentration of their substrates. (It is also true that the concentration of some enzymes in cells is within an order of magnitude or so of the concentrations of their substrates.)

As Product Accumulates, the Apparent Rate of the Enzymatic Reaction Will Decrease

The enzymatic rate, $v = d[\text{P}]/dt$, "slows down" as product accumulates and equilibrium is approached. The apparent decrease in rate is due to the conversion of P to S by the reverse reaction as [P] rises. Once $[\text{P}]/[\text{S}] = K_{eq}$, no further reaction is apparent. K_{eq} defines thermodynamic equilibrium. Enzymes have no influence on the thermodynamics of a reaction. Also, product inhibition can be a kinetically valid phenomenon: Some enzymes are actually inhibited by the products of their action.

Genetic Regulation of Enzyme Synthesis and Decay Determines the Amount of Enzyme Present at Any Moment

The amounts of enzyme synthesized by a cell are determined by transcription regulation (see Chapter 29). If the gene encoding a particular enzyme protein is turned on or off, changes in the amount of enzyme activity soon follow. **Induction,** which is the activation of enzyme synthesis, and **repression,** which is the shutdown of enzyme synthesis, are important mechanisms for the regulation of metabolism. By controlling the amount of an enzyme that is present at any moment, cells can either activate or terminate various metabolic routes. Genetic controls over enzyme levels have a response time ranging from minutes in rapidly dividing bacteria to hours (or longer) in higher eukaryotes. Once synthesized, the enzyme may also be degraded, either through normal turnover of the protein or through specific decay mechanisms that target the enzyme for destruction. These latter mechanisms are discussed in detail in Chapter 31.

Enzyme Activity Can Be Regulated Allosterically

Enzymatic activity can also be activated or inhibited through noncovalent interaction of the enzyme with small molecules (metabolites) other than the substrate. This form of control is termed **allosteric regulation,** because the activator or inhibitor binds to the enzyme at a site *other* than (*allo* means "other") the active site. Furthermore, such allosteric regulators, or **effector molecules,** are often quite different sterically from the substrate. Because this form of regulation results simply from reversible binding of regulatory ligands to the enzyme, the cellular response time can be virtually instantaneous.

Enzyme Activity Can Be Regulated Through Covalent Modification

Enzymes can be regulated by **covalent modification,** the reversible covalent attachment of a chemical group. Enzymes susceptible to such regulation are called **interconvertible enzymes,** because they can be reversibly converted between two forms. Thus, a fully active enzyme can be converted into an inactive form simply by the covalent attachment of a functional group. For example, **protein kinases** are enzymes that act in covalent modification by attaching a phosphoryl moiety to target proteins (Figure 15.1). Protein kinases catalyze the ATP-dependent phosphorylation of —OH groups on Ser, Thr, or

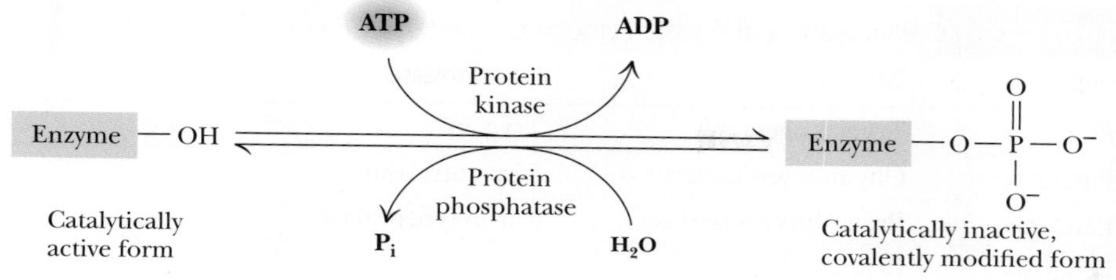

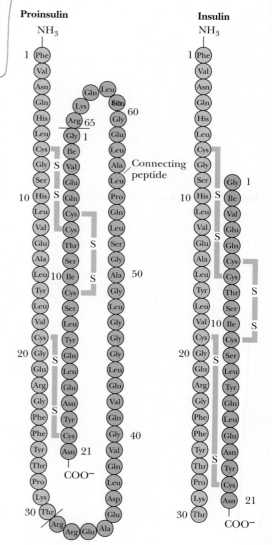

FIGURE 15.1 Enzyme regulation by reversible covalent modification.

Tyr side chains (or in some cases, the imidazole-N atoms in histidine residues). Removal of the phosphate group by a **phosphoprotein phosphatase** returns the enzyme to its original state. In contrast to the example in the figure, some enzymes exist in an inactive state unless specifically converted into the active form through covalent addition of a functional group. Covalent modification reactions are catalyzed by special **converter enzymes,** which are themselves subject to metabolic regulation. (Protein kinases are one class of converter enzymes.) Although covalent modification represents a stable alteration of the enzyme, a different converter enzyme operates to remove the modification, so when the conditions that favored modification of the enzyme are no longer present, the process can be reversed, restoring the enzyme to its unmodified state. Because covalent modification events are catalyzed by enzymes, they occur very quickly, with response times of seconds or even less for significant changes in metabolic activity.

Regulation of Enzyme Activity Also Can Be Accomplished in Other Ways

Enzyme regulation is an important matter to cells, and evolution has provided a variety of additional options, including zymogens, isozymes, and modulator proteins. We will discuss these options first and then return to the major topics of this chapter—enzyme regulation through allosteric mechanisms and covalent modification.

Zymogens Are Inactive Precursors of Enzymes

Most proteins become fully active as their synthesis is completed and they spontaneously fold into their native, three-dimensional conformations. Some proteins, however, are synthesized as inactive precursors, called zymogens or **proenzymes,** that acquire full activity only upon specific proteolytic cleavage of one or several of their peptide bonds. Unlike allosteric regulation or covalent modification, zymogen activation by specific proteolysis is an irreversible process. Activation of enzymes and other physiologically important proteins by specific proteolysis is a strategy frequently exploited by biological systems to switch on processes at the appropriate time and place, as the following examples illustrate.

Insulin Some protein hormones are synthesized in the form of inactive precursor molecules, from which the active hormone is derived by proteolysis. For instance, **insulin,** an important metabolic regulator, is generated by proteolytic excision of a specific peptide from **proinsulin** (Figure 15.2).

Proteolytic Enzymes of the Digestive Tract Enzymes of the digestive tract that serve to hydrolyze dietary proteins are synthesized in the stomach and pancreas as zymogens (Table 15.1). Only upon proteolytic activation are these enzymes able to form a catalytically active substrate-binding site. The activation of chymotrypsinogen is an interesting example (Figure 15.3). **Chymotrypsinogen** is a 245-residue polypeptide chain crosslinked by five disulfide bonds. Chymotrypsinogen is converted to an enzymatically active form called π-chymotrypsin when trypsin cleaves the peptide bond joining Arg[15] and Ile[16]. The enzymatically active π-chymotrypsin acts upon other π-chymotrypsin molecules, excising two dipeptides: Ser[14]–Arg[15] and Thr[147]–Asn[148]. The end product of

FIGURE 15.2 Proinsulin is an 86-residue precursor to insulin (the sequence shown here is human proinsulin). Proteolytic removal of residues 31 to 65 yields insulin. Residues 1 through 30 (the B chain) remain linked to residues 66 through 87 (the A chain) by a pair of interchain disulfide bridges.

TABLE 15.1	Pancreatic and Gastric Zymogens	
Origin	Zymogen	Active Protease
Pancreas	Trypsinogen	Trypsin
Pancreas	Chymotrypsinogen	Chymotrypsin
Pancreas	Procarboxypeptidase	Carboxypeptidase
Pancreas	Proelastase	Elastase
Stomach	Pepsinogen	Pepsin

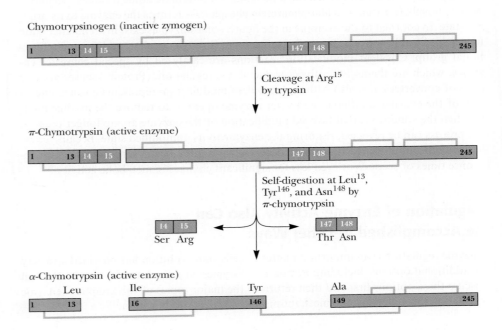

FIGURE 15.3 The proteolytic activation of chymotrypsinogen.

this processing pathway is the mature protease **α-chymotrypsin,** in which the three peptide chains, A (residues 1 through 13), B (residues 16 through 146), and C (residues 149 through 245), remain together because they are linked by two disulfide bonds, one from A to B and one from B to C.

Blood Clotting The formation of blood clots is the result of a series of zymogen activations (Figure 15.4). The amplification achieved by this cascade of enzymatic activations allows blood clotting to occur rapidly in response to injury. Seven of the clotting factors in their active form are serine proteases: **kallikrein, XII$_a$, XI$_a$, IX$_a$, VII$_a$, X$_a$,** and **thrombin.** Two routes to blood clot formation exist. The **intrinsic pathway** is instigated when the blood comes into physical contact with abnormal surfaces caused by injury; the **extrinsic pathway** is initiated by factors released from injured tissues. The pathways merge at factor X and culminate in clot formation. Thrombin excises peptides rich in negative charge from **fibrinogen,** converting it to **fibrin,** a molecule with a different surface charge distribution. Fibrin readily aggregates into ordered fibrous arrays that are subsequently stabilized by covalent crosslinks. Thrombin specifically cleaves Arg–Gly peptide bonds and is homologous to trypsin, which is also a serine protease (recall that trypsin acts only at Arg and Lys residues).

Isozymes Are Enzymes with Slightly Different Subunits

A number of enzymes exist in more than one quaternary form, differing in their relative proportions of structurally equivalent but catalytically distinct polypeptide subunits. A classic example is mammalian **lactate dehydrogenase (LDH),** which exists in two

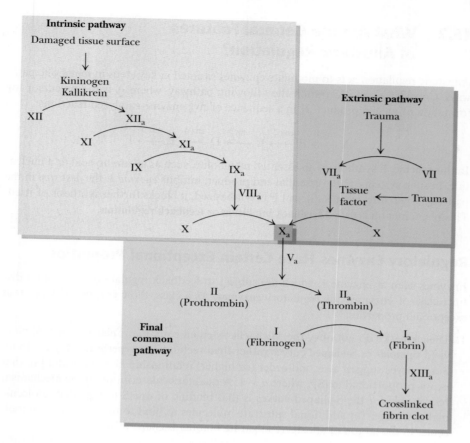

FIGURE 15.4 The cascade of activation steps leading to blood clotting. The intrinsic and extrinsic pathways converge at factor X, and the final common pathway involves the activation of thrombin and its conversion of fibrinogen into fibrin, which aggregates into ordered filamentous arrays that become crosslinked to form the clot.

prominent oligomeric forms, M and H. LDH M is an A_4 homotetramer (composed of four subunits encoded by the LDH A gene). LDH H is also a homotetramer, but its subunits are products of the LDH B gene, so its subunit composition is B_4 (Figure 15.5). Heterotetrameric isozymes (A_3B, A_2B_2, AB_3) can also form, depending on the relative expression of the LDH A and LDH B genes in different tissues. The kinetic properties of the various LDH isozymes differ. Different tissues express different isozyme forms, as appropriate to their particular metabolic needs. By regulating the relative amounts of A and B subunits they synthesize, the cells of various tissues control which isozymic forms are likely to assemble and thus which kinetic parameters prevail.

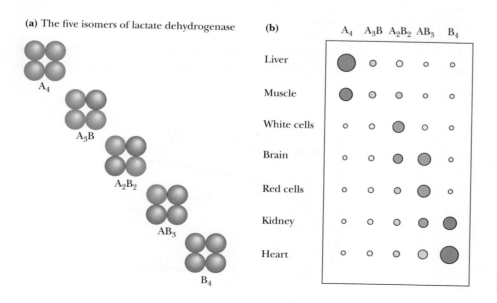

(a) The five isomers of lactate dehydrogenase

FIGURE 15.5 The isozymes of lactate dehydrogenase (LDH). Active skeletal muscle tissue becomes anaerobic and produces pyruvate from glucose via glycolysis (see Chapter 18). It needs LDH to regenerate NAD^+ from NADH so that glycolysis can continue. The lactate produced is released into the blood. LDH M works best in the NAD^+-regenerating direction. Heart tissue is aerobic and uses lactate as a fuel, converting it to pyruvate via LDH and using the pyruvate to fuel the citric acid cycle to obtain energy. LDH H works best in the lactate-to-pyruvate direction.

15.2 What Are the General Features of Allosteric Regulation?

Allosteric regulation acts to modulate enzymes situated at key steps in metabolic pathways. Consider as an illustration the following pathway, where A is the precursor for formation of an end product, F, in a sequence of five enzyme-catalyzed reactions:

$$A \xrightarrow{\text{enz 1}} B \xrightarrow{\text{enz 2}} C \xrightarrow{\text{enz 3}} D \xrightarrow{\text{enz 4}} E \xrightarrow{\text{enz 5}} F$$

In this scheme, F symbolizes an essential metabolite, such as an amino acid or a nucleotide. In such systems, F, the essential end product, inhibits *enzyme 1*, the *first step* in the pathway. Therefore, when sufficient F is synthesized, it blocks further synthesis of itself. This phenomenon is called **feedback inhibition** or **feedback regulation.**

Regulatory Enzymes Have Certain Exceptional Properties

Enzymes such as enzyme 1, which are subject to feedback regulation, represent a distinct class of enzymes, the **regulatory enzymes.** As a class, these enzymes have certain exceptional properties:

1. Their kinetics do not obey the Michaelis–Menten equation. Their v versus [S] plots yield **sigmoid-** or **S-shaped** curves rather than rectangular hyperbolas (Figure 15.6). Such curves suggest a second-order (or higher) relationship between v and [S]; that is, v is proportional to $[S]^n$, where $n > 1$. A qualitative description of the mechanism responsible for the S-shaped curves is that binding of one S to a protein molecule makes it easier for additional substrate molecules to bind to the same protein molecule. In the jargon of allostery, substrate binding is **cooperative.**

2. Inhibition of a regulatory enzyme by a feedback inhibitor does not conform to any normal inhibition pattern, and the feedback inhibitor F bears little structural similarity to A, the substrate for the regulatory enzyme. F apparently acts at a binding site distinct from the substrate-binding site. The term *allosteric* is apt, because F is sterically dissimilar and, moreover, acts at a site other than the site for S. Its effect is called **allosteric inhibition.**

3. Regulatory or allosteric enzymes like enzyme 1 are, in some instances, regulated by activation. That is, whereas some effector molecules such as F exert negative effects on enzyme activity, other effectors show stimulatory, or positive, influences on activity.

4. Allosteric enzymes typically have an oligomeric organization. They are composed of more than one polypeptide chain (subunit), and each subunit has a binding site for substrate, as well as a distinct binding site for allosteric effectors. Thus, allosteric enzymes typically have more than one S-binding site and more than one effector-binding site per enzyme molecule.

5. The working hypothesis is that, by some means, interaction of an allosteric enzyme with effectors alters the distribution of conformational possibilities or subunit

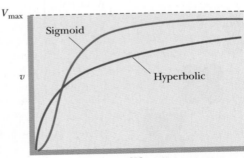

FIGURE 15.6 Sigmoid v versus [S] plot. The blue line represents the hyperbolic plot characteristic of normal Michaelis–Menten-type enzyme kinetics.

interactions available to the enzyme. That is, the regulatory effects exerted on the enzyme's activity are achieved by conformational changes occurring in the protein when effector metabolites bind.

In addition to enzymes, noncatalytic proteins may exhibit many of these properties; hemoglobin is the classic example. The allosteric properties of hemoglobin are the subject of a Special Focus at the end of this chapter.

15.3 Can Allosteric Regulation Be Explained by Conformational Changes in Proteins?

The Symmetry Model for Allosteric Regulation Is Based on Two Conformational States for a Protein

Various models have been proposed to account for the behavior of allosteric proteins. All of them note that proteins can exist in different conformational states. Models usually propose a small number of conformations (two or, at most, three) for a given protein. For example, the model for allosteric behavior of Jacques Monod, Jeffries Wyman, and Jean-Pierre Changeux **(the MWC model)** proposes two conformational states for an allosteric protein: the R (relaxed) state and the T (taut) state. The MWC model is sometimes referred to as the **symmetry model** because all subunits in an oligomer are assumed to have the same conformation, whether it is R or T. R-state and T-state protein molecules are in equilibrium, with the T conformation greatly favored over the R ($[T] \gg [R]$), under conditions in which no ligands are present. This model further suggests that substrate and allosteric activators (positive effectors) bind only to the R state and allosteric inhibitors (negative effectors) bind only to the T state. Figure 15.7 illustrates such a model for a dimeric protein, each monomer of which has a substrate-binding site and an effector-binding site. Because substrate (S) binds only to the R state, S binding perturbs the R $\rightleftharpoons$ T equilibrium in favor of more R-state conformers and thus more S binding. That is, S binding is cooperative. The concentration of ligand giving half-maximal response is defined as $K_{0.5}$. (Like K_m, the units of $K_{0.5}$ are molarity; K_m cannot be used to describe these constants, because the protein does not conform to the Michaelis–Menten model for enzyme kinetics.) The MWC model accounts for the action of allosteric effectors. Positive effectors bind only to the R state and thus cause a shift of the R $\rightleftharpoons$ T equilibrium in favor of more R and thus easier S binding. Negative effectors do the opposite; they perturb the R $\rightleftharpoons$ T equilibrium in favor of T, the conformation that cannot bind S. Note that positive effectors (allosteric activators) cause a decline in the $K_{0.5}$ for S (signifying easier binding of S) and negative effectors raise $K_{0.5}$ for S (Figure 15.7). Note that the MWC model assumes an equilibrium between conformational states, but ligand binding does not alter the conformation of the protein.

The Sequential Model for Allosteric Regulation Is Based on Ligand-Induced Conformational Changes

An alternative model proposed by Daniel Koshland, George Nemethy, and David Filmer **(the KNF model)** relies on the well-accepted idea that ligand binding triggers a change in the conformation of a protein. And, if the protein is oligomeric, ligand-induced conformational changes in one subunit may lead to changes in the conformation of its neighbors. Such ligand-induced conformational change could cause the subunits of an oligomeric protein to shift from a low-affinity state to a high-affinity state. For example, S binding to one monomer may cause the other monomers to adopt conformations with higher affinity for S (Figure 15.8). Interestingly, the KNF model also explains how ligand-induced conformational changes could cause subunits of a protein to adopt conformations with little or no affinity for the ligand, a phenomenon referred to as **negative cooperativity**. The KNF model is termed the **sequential model** because

The notable difference between the MWC and KNF models:

In the MWC model, the different conformations have different affinities for the various ligands, and the concept of ligand-induced conformational changes is ignored. In contrast, the KNF model is based on ligand-induced conformational changes.

A dimeric protein that can exist in either of two states: R_0 or T_0.
This protein can bind three ligands:
1) Substrate (S): Binds only to R at site S
2) Activator (A): A positive effector that binds only to R at site F
3) Inhibitor (I): A negative effector that binds only to T at site F
Substrate
R_0
T_0
Activator
Inhibitor
R
$R_{1(S)}$
Activator
R
$R_{1(A)}$
T
$T_{1(I)}$
Substrate
R
$R_{1(A,S)}$
1.0
+A
No A or I
+I
Y_S 0.5
$K_{0.5}$
0
0
1.0
2.0
[S]
Effects of A:
$A + R_0 \rightarrow R_{1(A)}$
Increase in number of R-conformers shifts $R_0 \rightleftharpoons T_0$ so that $T_0 \rightarrow R_0$
(1) More binding sites for S made available.
(2) Decrease in cooperativity of substrate saturation curve.
Effects of I:
$I + T_0 \rightarrow T_{1(I)}$
Increase in number of T-conformers (decrease in R_0 as $R_0 \rightarrow T_0$ to restore equilibrium)
Thus, I inhibits association of S and A with R by lowering R_0 level. I increases cooperativity of substrate saturation curve.

FIGURE 15.7 Allosteric effects: A and I binding to R and T, respectively. The linked equilibria lead to changes in the relative amounts of R and T and, therefore, shifts in the substrate saturation curve. The parameters of such a system are that (1) S and A (or I) have different affinities for R and T and (2) A (or I) modifies the apparent $K_{0.5}$ for S by shifting the relative R versus T population.

(a) Binding of S induces a conformational change.
S
Symmetric protein dimer
Asymmetric protein dimer
(b)
Transmitted conformational change
If the relative affinities of the various conformations for S are
positive cooperativity ensues.
If the relative affinities of the various conformations for S are
negative cooperativity ensues.
(c)
1.0
0.8
0.6
0.4
0.2
v/V_{max}
0
3
6
9
$[S]/K_{0.5}$
Positive cooperativity
No cooperativity
Negative cooperativity

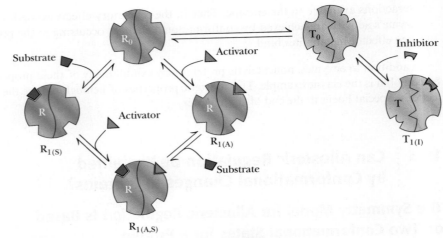

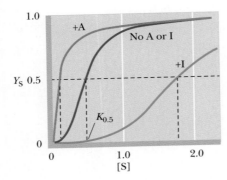

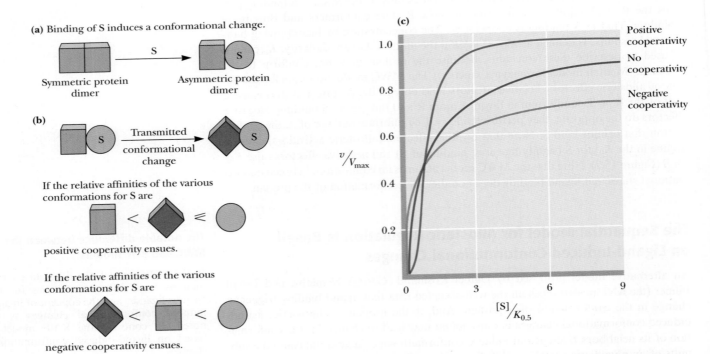

FIGURE 15.8 The Koshland–Nemethy–Filmer sequential model for allosteric behavior. **(a)** S binding can, by induced fit, cause a conformational change in the subunit to which it binds. **(b)** If subunit interactions are tightly coupled, binding of S to one subunit may cause the other subunit to assume a conformation having a greater or lesser affinity for S. That is, the ligand-induced conformational change in one subunit can affect the adjoining subunit. Such effects could be transmitted between neighboring peptide domains by changing alignments of nonbonded amino acid residues. **(c)** Theoretical curves for the binding of a ligand to a protein having four identical subunits, each with one binding site for the ligand. The fraction of maximal velocity (v/V_{max}) is plotted as a function of $[S]/K_{0.5}$.

subunits undergo sequential changes in conformation due to ligand binding. A comparison of the response of velocity to substrate concentration for positive versus negative cooperativity is shown in Figure 15.8c.

Changes in the Oligomeric State of a Protein Can Also Give Allosteric Behavior

Although the MWC and KNF models are the best-known paradigms for allosteric protein behavior, other models have been put forward. For example, instead of R and T, consider a monomer–oligomer equilibrium for an allosteric protein, where only the oligomer binds S and [monomer] $\gg$ [oligomer]. This model strongly resembles the MWC model. In yet another model, we might have a monomeric protein with distinct binding sites for several different ligands. In this case, binding of ligand A to its site might cause a conformational change such that the protein shows much greater affinity for S than it would in the absence of A. Or, binding of ligand I might result in a conformational change in the protein such that its affinity for S is abolished. Although the binding of other ligands may affect the affinity of the monomer for S, S binding cannot show cooperativity in monomeric proteins, because, unlike oligomers, the monomer has only one binding site for S.

It is important to realize that all of these various models are attempts to use simple concepts to explain the complex behavior of a protein. Although these models provide reasonable approximations and useful insights, the molecular mechanisms underlying allostery cannot be expected to conform rigidly to any one of these models. Shortly, we explore the regulated behavior of a real protein (glycogen phosphorylase) with these models in mind.

15.4 : What Kinds of Covalent Modification Regulate the Activity of Enzymes?

Covalent Modification Through Reversible Phosphorylation

As we saw in Figure 15.1, enzyme activity can be regulated through reversible phosphorylation; indeed it is the most prominent form of covalent modification in cellular regulation. Phosphorylation is accomplished by protein kinases that target specific enzymes for modification. Phosphoprotein phosphatases operate in the reverse direction to remove the phosphate group through hydrolysis of the side-chain phosphoester bond. Because protein kinases and phosphoprotein phosphatases work in opposing directions, regulation must be imposed on these converter enzymes so that their interconvertible enzyme targets are locked in the desired state (active versus inactive) and a wasteful cycle of ATP hydrolysis is avoided. Thus, converter enzymes are themselves the targets of allosteric regulation or covalent modification.

Protein Kinases: Target Recognition and Intrasteric Control

Protein kinases are converter enzymes that catalyze the ATP-dependent phosphorylation of serine, threonine, or tyrosine hydroxyl groups in target proteins (Table 15.2). Phosphorylation introduces a bulky group bearing two negative charges, causing conformational changes that alter the target protein's function. (Unlike a phosphoryl group, no amino acid side chain can provide *two* negative charges.) Protein kinases represent a protein superfamily whose members are widely diverse in terms of size, subunit structure, and subcellular localization. Nevertheless, all share a common catalytic mechanism based on a conserved catalytic core/kinase domain of approximately 260 amino acid residues (Figure 15.9). Protein kinases are classified as Ser/Thr and/or Tyr specific. They also differ in terms of the target proteins that they recognize and phosphorylate; target selection depends on the presence of an amino acid sequence within the target

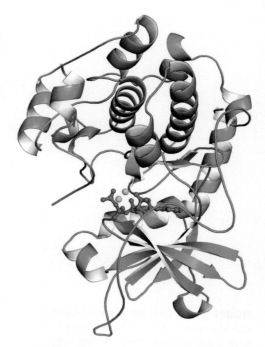

FIGURE 15.9 Protein kinase A is shown complexed with a pseudosubstrate peptide (orange). This complex also includes ATP (red) and two Mn²⁺ ions (yellow) bound at the active site (pdb id = 1ATP).

TABLE 15.2	Classification of Protein Kinases		
Protein Kinase Class		Target Sequence*	Activators
I. Ser/Thr protein kinases			
A. Cyclic nucleotide–dependent			
cAMP-dependent (PKA)		—R(R/K)X(S*/T*)—	cAMP
cGMP-dependent		—(R/K)KKX(S*/T*)—	cGMP
B. Ca²⁺-calmodulin (CaM)–dependent			
Phosphorylase kinase (PhK)		—KRKQIS*VRGL—	Phosphorylation by PKA
Myosin light-chain kinase (MLCK)		—KKRPQRATS*NV—	Ca²⁺-CaM
C. Protein kinase C (PKC)			Ca²⁺, diacylglycerol
D. Mitogen-activated protein kinases (MAP kinases)		—PXX(S*/T*)P—	Phosphorylation by MAPK kinase
E. G-protein–coupled receptors			
β-Adrenergic receptor kinase (BARK)			
Rhodopsin kinase			
II. Ser/Thr/Tyr protein kinases			
MAP kinase kinase (MAPK kinase)		—TEY—	Phosphorylation by *Raf* (a protein kinase)
III. Tyr protein kinases			
A. Cytosolic tyrosine kinases (*src, fgr, abl*, etc.)			
B. Receptor tyrosine kinases (RTKs)			
Plasma membrane receptors for hormones such as *epidermal growth factor* (EGF) or *platelet-derived growth factor* (PDGF)			

*X denotes any amino acid.

protein that is recognized by the kinase. For example, cAMP-dependent protein kinase **(PKA)** phosphorylates proteins having Ser or Thr residues within an R(R/K)X(S*/T*) target consensus sequence (* denotes the residue that becomes phosphorylated). That is, PKA phosphorylates Ser or Thr residues that occur in an Arg-(Arg or Lys)-(any amino acid)-(Ser or Thr) sequence segment (Table 15.2).

Targeting of protein kinases to particular consensus sequence elements within proteins creates a means to regulate these kinases by **intrasteric control.** Intrasteric control occurs when a regulatory subunit (or protein domain) has a **pseudosubstrate sequence** that mimics the target sequence but lacks an OH-bearing side chain at the right place. For example, the cAMP-binding regulatory subunits of PKA (R subunits in Figure 15.10) possess the pseudosubstrate sequence RRGA*I, and this sequence binds to the active site of PKA catalytic subunits, blocking their activity. This pseudosubstrate sequence has an alanine residue (A*) where serine occurs in the PKA target sequence; Ala is sterically similar to serine but lacks a phosphorylatable OH group. When these PKA regulatory subunits bind cAMP, they undergo a conformational change and dissociate from the catalytic (C) subunits, and the active site of PKA is free to bind and phosphorylate its targets. In other protein kinases, the pseudosubstrate sequence involved in intrasteric control and the kinase domain are part of the same polypeptide chain. In these cases, binding of an allosteric effector (like cAMP) induces a conformational change in the protein that releases the pseudosubstrate sequence from the active site of the kinase domain.

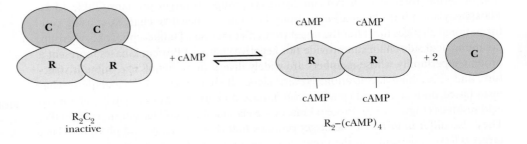

FIGURE 15.10 Cyclic AMP–dependent protein kinase (also known as *PKA*) is a 150- to 170-kD R_2C_2 tetramer in mammalian cells. The two R (regulatory) subunits bind cAMP ($K_D = 3 \times 10^{-8}$ *M*); cAMP binding releases the R subunits from the C (catalytic) subunits. C subunits are enzymatically active as monomers.

The abundance of many protein kinases in cells is an indication of the great importance of protein phosphorylation in cellular regulation. Exactly 113 protein kinase genes have been recognized in yeast, and 868 putative protein kinase genes have been identified in the human genome. **Tyrosine kinases** (protein kinases that phosphorylate Tyr residues) occur only in multicellular organisms (yeast has no tyrosine kinases). Tyrosine kinases are components of signaling pathways involved in cell–cell communication (see Chapter 32).

Phosphorylation Is Not the Only Form of Covalent Modification That Regulates Protein Function

Several hundred different chemical modifications of proteins have been discovered thus far, ranging from carboxylation (addition of a carboxyl group), acetylation (addition of an acetyl group, see Figure 29.30), prenylation (see Figure 9.23), and glycosylation (see Figures 7.32–7.37) to covalent attachment of a polypeptide to the protein (addition of ubiquitin to free amino groups on proteins; see Figure 31.8), to name just a few. A compilation of known protein modifications can be found in RESID, the European Bioinformatics Institute online database *(http://www.ebi.ac.uk/RESID/)*. Only a small number of these covalent modifications are used to achieve metabolic regulation through *reversible conversion of an enzyme between active and inactive forms.* Table 15.3 presents a few examples.

Note that three of these types of covalent modification require nucleoside triphosphates (ATP, UTP) that are related to cellular energy status; another relies on reducing potential within the cell, which also reflects cellular energy status.

Acetylation Is a Prominent Modification for the Regulation of Metabolic Enzymes

Reversible acetylation/deacetylation of histone lysine residues is a well-recognized way to regulate chromatin organization in eukaryotes (Figure 29.30). Acetylation of an ϵ–NH_3^+ group on a K residue changes it from a positively charged amino group to a neutral amide, with all the implications that has for protein structure and thus function. The enzymes responsible are **acetyl-CoA-dependent lysine acetyltransferases** or **KATs.** Over 30 different KATs are known in mammals. Proteomics studies show that acetylation of metabolic enzymes is an important mechanism for regulating the flow of metabolic substrates (carbohydrates, fats) through the central metabolic pathways of

TABLE 15.3	Additional Examples of Regulation by Covalent Modification	
Reaction	Amino Acid Side Chain	Reaction (see figure indicated)
Adenylylation	Tyrosine	Transfer of AMP from ATP to Tyr-OH (Figure 25.16)
Uridylylation	Tyrosine	Transfer of UMP from UTP to Tyr-OH (Figure 25.17)
ADP-ribosylation	Arginine	Transfer of ADP-ribose from NAD^+ to Arg (Figure 25.8)
Methylation	Glutamate	Transfer of methyl group from S-adenosylmethionine to Glu γ-carboxyl group
Oxidation-reduction	Cysteine (disulfide)	Reduction of Cys-S—S-Cys to Cys-SH HS-Cys (Figure 21.27)
Acetylation	Lysine	Transfer of acetyl group from acetyl-CoA to Lys ϵ-amino group (Figure 29.30)

FIGURE 15.11 Activation of the tricarboxylic acid cycle enzyme, malate dehydrogenase (MDH) by acetylation. **(a)** Relative MDH activity from human liver cells in the absence (−) and presence (+) of inhibitors of lysine deacetylating enzymes. **(b)** Activation of MDH in human liver cells treated with glucose. The relative MDH activity in human liver cells exposed to a range (0, 6.25, 12.5, and 25 mM) glucose concentrations is shown. Glucose enhances MDH acetylation. Adapted from Figure 2 in Zhao, S., Xu, W., Jiang, W., et al. 2010. Regulation of cellular metabolism by protein lysine acetylation. *Science* **327**:1000–1004.

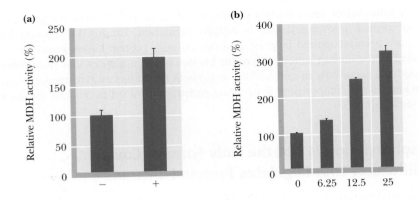

carbohydrate and fat metabolism. Acetylation activates some enzymes and inhibits others. Cellular levels of major metabolic fuels such as glucose, fatty acids, and amino acids influence the degree of acetylation (Figure 15.11). The in vivo stability of some enzymes may also be regulated by acetylation. Deacetylation by **KDACs (lysine deacetylases)** reverses the effects of acetylation. KDACs include **sirtuins,** a class of NAD⁺-dependent protein deacetylating enzymes. Sirtuins are implicated in the regulation of energy metabolism and longevity (Figure 27.13).

15.5 : Is the Activity of Some Enzymes Controlled by Both Allosteric Regulation and Covalent Modification?

Glycogen phosphorylase, the enzyme that catalyzes the release of glucose units from glycogen, serves as an excellent example of the many enzymes regulated both by allosteric controls and by covalent modification.

The Glycogen Phosphorylase Reaction Converts Glycogen into Readily Usable Fuel in the Form of Glucose-1-Phosphate

The cleavage of glucose units from the nonreducing ends of glycogen molecules is catalyzed by **glycogen phosphorylase,** an allosteric enzyme. The enzymatic reaction involves phosphorolysis of the bond between C-1 of the departing glucose unit and the glycosidic oxygen, to yield *glucose-1-phosphate* and a glycogen molecule that is shortened by one residue (Figure 15.12). (Because the reaction involves attack by phosphate instead of H₂O, it is referred to as a **phosphorolysis** rather than a hydrolysis.) Phosphorolysis

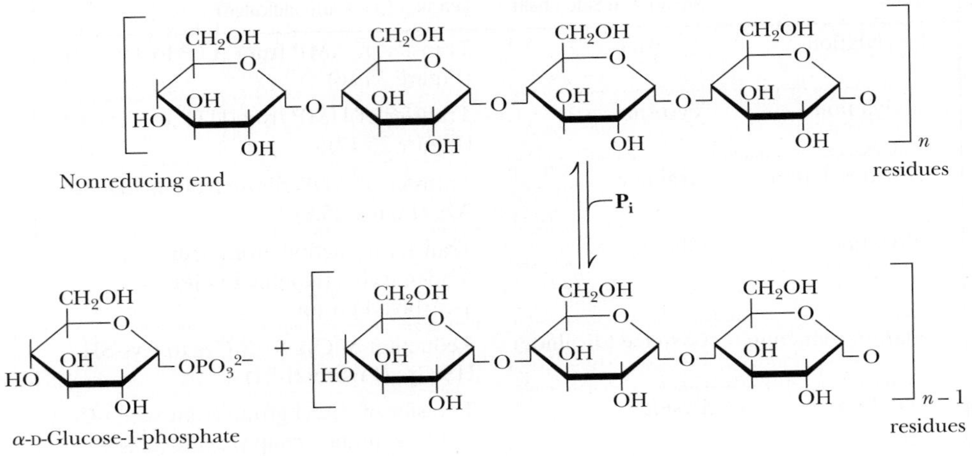

FIGURE 15.12 The glycogen phosphorylase reaction.

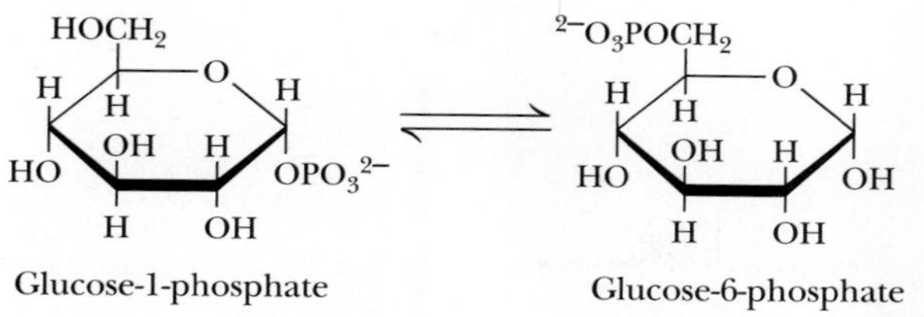

Glucose-1-phosphate ⇌ Glucose-6-phosphate

FIGURE 15.13 The phosphoglucomutase reaction.

produces a phosphorylated sugar product, glucose-1-P, which is converted to the glycolytic substrate, glucose-6-P, by *phosphoglucomutase* (Figure 15.13). In muscle, glucose-6-P proceeds into glycolysis, providing needed energy for muscle contraction. In the liver, hydrolysis of glucose-6-P yields glucose, which is exported to other tissues via the circulatory system.

Glycogen Phosphorylase Is a Homodimer

Muscle glycogen phosphorylase is a dimer of two identical subunits (842 residues, 97.44 kD). Each subunit contains an active site (at the center of the subunit) and an allosteric effector site near the subunit interface (Figure 15.14). In addition, a regulatory phosphorylation site is located at Ser^{14} on each subunit. A glycogen-binding site on each subunit facilitates prior association of glycogen phosphorylase with its substrate and also exerts regulatory control on the enzymatic reaction.

Each subunit contributes a tower helix (residues 262 to 278) to the subunit–subunit contact interface in glycogen phosphorylase. In the phosphorylase dimer, the tower helices extend from their respective subunits and pack against each other in an antiparallel manner.

Glycogen Phosphorylase Activity Is Regulated Allosterically

Muscle Glycogen Phosphorylase Shows Cooperativity in Substrate Binding The binding of the substrate *inorganic phosphate* (P_i) to muscle glycogen phosphorylase is highly cooperative (Figure 15.15a), which allows the enzyme activity to increase markedly over a rather narrow range of substrate concentration.

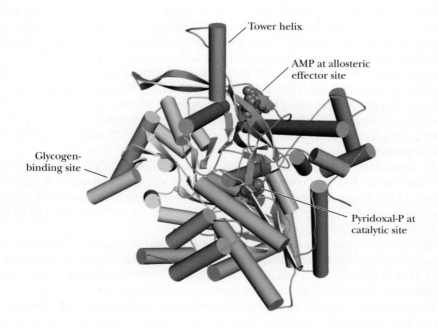

Tower helix

AMP at allosteric effector site

Glycogen-binding site

Pyridoxal-P at catalytic site

FIGURE 15.14 Structure of the glycogen phosphorylase monomer (pdb id = 8GPB).

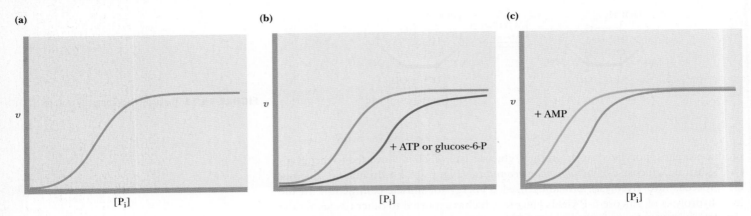

FIGURE 15.15 *v* versus S curves for glycogen phosphorylase. **(a)** The response to the concentration of the substrate phosphate (P_i). **(b)** ATP and glucose-6-P are feedback inhibitors. **(c)** AMP is a positive effector. It binds at the same site as ATP.

ATP and Glucose-6-P Are Allosteric Inhibitors of Glycogen Phosphorylase ATP can be viewed as the "end product" of glycogen phosphorylase action, in that the glucose-1-P liberated by glycogen phosphorylase is degraded in muscle via metabolic pathways whose purpose is energy (ATP) production. Glucose-1-P is readily converted into glucose-6-P to feed such pathways. (In the liver, glucose-1-P from glycogen is converted to glucose and released into the bloodstream to raise blood glucose levels.) Thus, feedback inhibition of glycogen phosphorylase by ATP and glucose-6-P provides a very effective way to regulate glycogen breakdown. Both ATP and glucose-6-P act by decreasing the affinity of glycogen phosphorylase for its substrate P_i (Figure 15.15b). Because the binding of ATP or glucose-6-P has a negative effect on substrate binding, these substances act as *negative effectors*. Note in Figure 15.15b that the substrate saturation curve is displaced to the right in the presence of ATP or glucose-6-P, and a higher substrate concentration is needed to achieve half-maximal velocity ($V_{max}/2$). When concentrations of ATP or glucose-6-P accumulate to high levels, glycogen phosphorylase is inhibited; when [ATP] and [glucose-6-P] are low, the activity of glycogen phosphorylase is regulated by availability of its substrate, P_i.

AMP Is an Allosteric Activator of Glycogen Phosphorylase AMP also provides a regulatory signal to glycogen phosphorylase. It binds to the same site as ATP, but it stimulates glycogen phosphorylase rather than inhibiting it (Figure 15.15c). AMP acts as a *positive effector*, meaning that it enhances the binding of substrate to glycogen phosphorylase. Significant levels of AMP indicate that the energy status of the cell is low and that more energy (ATP) should be produced. Reciprocal changes in the cellular concentrations of ATP and AMP and their competition for binding to the same site (the *allosteric site*) on glycogen phosphorylase, with opposite effects, allow these two nucleotides to exert *rapid and reversible control* over glycogen phosphorylase activity. Such reciprocal regulation ensures that the production of energy (ATP) is commensurate with cellular needs.

To summarize, muscle glycogen phosphorylase is allosterically activated by AMP and inhibited by ATP and glucose-6-P; caffeine can also act as an allosteric inhibitor (Figure 15.16). When ATP and glucose-6-P are abundant, glycogen breakdown is inhibited. When cellular energy reserves are low (i.e., high [AMP] and low [ATP] and [G-6-P]), glycogen catabolism is stimulated.

Glycogen phosphorylase conforms to the MWC model of allosteric transitions, with the active form of the enzyme designated the **R state** and the inactive form denoted as the **T state** (Figure 15.16). Thus, AMP promotes the conversion to the active R state, whereas ATP, glucose-6-P, and caffeine favor conversion to the inactive T state.

X-ray diffraction studies of glycogen phosphorylase in the presence of allosteric effectors have revealed the molecular basis for the T $\rightleftharpoons$ R conversion. Although the

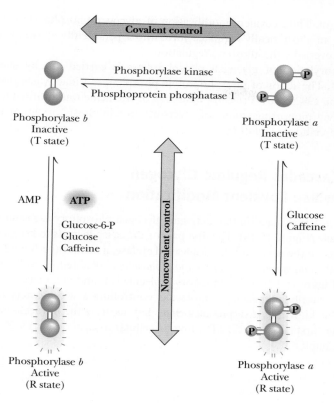

FIGURE 15.16 The mechanism of covalent modification and allosteric regulation of glycogen phosphorylase.

structure of the central core of the phosphorylase subunits is identical in the T and R states, a significant change occurs at the subunit interface between the T and R states. This conformation change at the subunit interface is linked to a structural change at the active site that is important for catalysis. In the T state, the negatively charged carboxyl group of Asp^{283} faces the active site, so binding of the anionic substrate phosphate is unfavorable. In the conversion to the R state, Asp^{283} is displaced from the active site and replaced by Arg^{569}. The exchange of negatively charged aspartate for positively charged arginine at the active site provides a favorable binding site for phosphate anion. These allosteric controls serve as a mechanism for adjusting the activity of glycogen phosphorylase to meet normal metabolic demands. However, in crisis situations in which abundant energy (ATP) is needed immediately, these controls can be overridden by covalent modification of glycogen phosphorylase. Covalent modification through phosphorylation of Ser^{14} in glycogen phosphorylase converts the enzyme from a less active, allosterically regulated form (the *b* form) to a more active, allosterically unresponsive form (the *a* form).

Covalent Modification of Glycogen Phosphorylase Trumps Allosteric Regulation

As early as 1938, it was known that glycogen phosphorylase existed in two forms: the less active **phosphorylase *b*** and the more active **phosphorylase *a*.** In 1956, Edwin Krebs and Edmond Fischer reported that a "converting enzyme" could convert phosphorylase *b* to phosphorylase *a*. Three years later, Krebs and Fischer demonstrated that the conversion of phosphorylase *b* to phosphorylase *a* involved covalent phosphorylation, as shown in Figure 15.16.

Phosphorylation of Ser^{14} causes a dramatic conformation change in phosphorylase. Upon phosphorylation, the amino-terminal end of the protein (including residues 10 through 22) swings through an arc of 120°, moving into the subunit interface (Figure 15.17). This conformation change moves Ser^{14} by more than 3.6 nm. The phosphorylated or *a* form of glycogen phosphorylase is much less sensitive to allosteric regulation

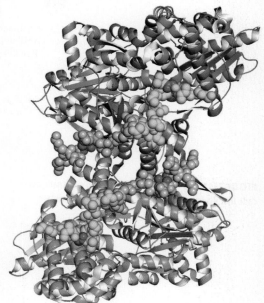

FIGURE 15.17 The major conformational change that occurs in the N-terminal residues upon phosphorylation of Ser^{14}. Ser^{14} is shown in red. N-terminal conformation of unphosphorylated enzyme (phosphorylase *b*): yellow; N-terminal conformation of phosphorylated enzyme (phosphorylase *a*): cyan. (Molecular graphic created from pdb id = 8GPB and pdb id = 1GPA.)

than the *b* form. Thus, covalent modification of glycogen phosphorylase converts this enzyme from an allosterically regulated form into a persistently active form. Covalent modification overrides the allosteric regulation.

Dephosphorylation of glycogen phosphorylase is carried out by **phosphoprotein phosphatase 1.** The action of phosphoprotein phosphatase 1 inactivates glycogen phosphorylase. The 1992 Nobel Prize in Physiology or Medicine was awarded to Krebs and Fischer for their pioneering studies of reversible protein phosphorylation as an important means of cellular regulation.

Enzyme Cascades Regulate Glycogen Phosphorylase Covalent Modification

The phosphorylation reaction that activates glycogen phosphorylase is mediated by an **enzyme cascade** (Figure 15.18). The first part of the cascade leads to hormonal stimulation (described in the next section) of **adenylyl cyclase,** a membrane-bound enzyme that converts ATP to *adenosine-3',5'-cyclic monophosphate,* denoted as *cyclic AMP* or simply *cAMP* (Figure 15.19). This regulatory molecule is found in all eukaryotic cells and acts as an intracellular messenger molecule, controlling a wide variety of processes. Cyclic AMP is known as a **second messenger** because it is the intracellular agent of a hormone (the "first messenger"). (The myriad cellular roles of cyclic AMP are described in detail in Chapter 32.)

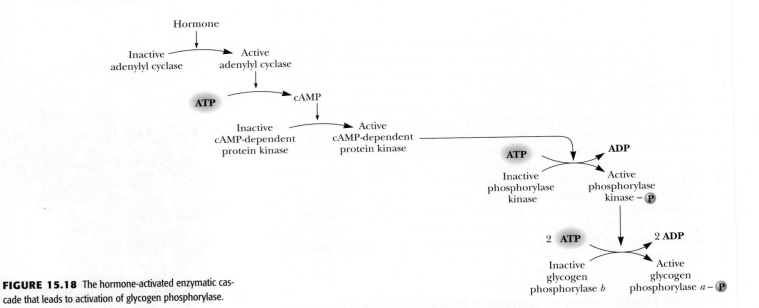

FIGURE 15.18 The hormone-activated enzymatic cascade that leads to activation of glycogen phosphorylase.

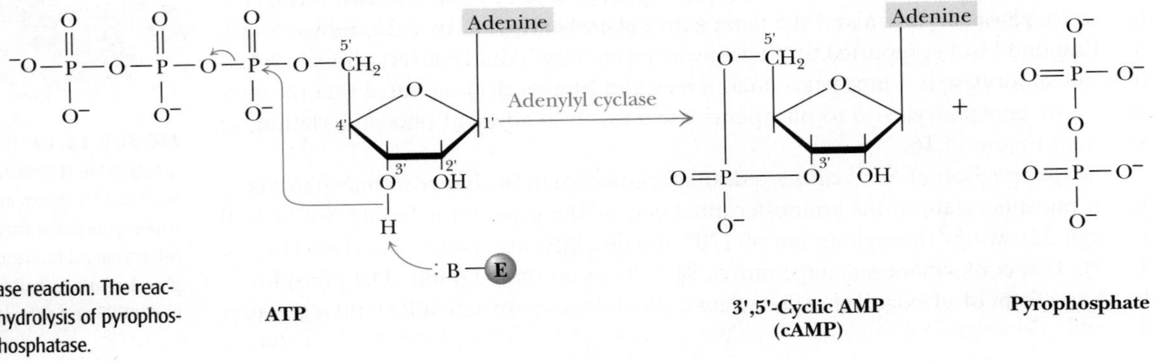

FIGURE 15.19 The adenylyl cyclase reaction. The reaction is driven forward by subsequent hydrolysis of pyrophosphate by the enzyme inorganic pyrophosphatase.

The hormonal stimulation of adenylyl cyclase is effected by a transmembrane signaling pathway consisting of three components, all membrane associated. Binding of hormone to the external surface of a hormone receptor causes a conformational change in this transmembrane protein, which in turn stimulates a **GTP-binding protein** (abbreviated **G protein**). G proteins are heterotrimeric proteins consisting of α- (45–47 kD), β- (35 kD), and γ- (7–9 kD) subunits. The α-subunit binds GDP or GTP and has an intrinsic, slow GTPase activity. In the inactive state, the $G_{\alpha\beta\gamma}$ complex has GDP at the nucleotide site. When a G protein is stimulated by a hormone–receptor complex, GDP dissociates and GTP binds to G_{α}, causing it to dissociate from $G_{\beta\gamma}$ and to associate with adenylyl cyclase (Figure 15.20). *Binding of G_{α} (GTP) activates adenylyl cyclase to form cAMP from ATP.* However, the intrinsic GTPase activity of G_{α} eventually hydrolyzes GTP to GDP, leading to dissociation of G_{α} (GDP) from adenylyl cyclase and reassociation with $G_{\beta\gamma}$ to form the inactive $G_{\alpha\beta\gamma}$ complex. This cascade amplifies the hormonal signal because a single hormone–receptor complex can activate many G proteins before the hormone dissociates from the receptor, and because the G_{α}-activated adenylyl cyclase can synthesize many cAMP molecules before bound GTP is hydrolyzed by G_{α}. More than 100 different G-protein–coupled receptors and at least 21 distinct G_{α} proteins are known (see Chapter 32).

Cyclic AMP is an essential activator of *cAMP-dependent protein kinase (PKA)*. Binding of cyclic AMP to the regulatory subunits induces a conformation change that causes the dissociation of the C monomers from the R dimer (Figure 15.10). The free C subunits are active and can phosphorylate other proteins. One of the many proteins phosphorylated by PKA is *phosphorylase kinase* (Figure 15.18). Phosphorylase kinase is inactive in the unphosphorylated state and active in the phosphorylated form. As its name implies, phosphorylase kinase functions to phosphorylate (and activate) glycogen phosphorylase. Thus, hormonal activation of adenylyl cyclase leads to activation of glycogen breakdown.

Is There an Example in Nature That Exemplifies the Relationship Between Quaternary Structure and the Emergence of Allosteric Properties? Hemoglobin and Myoglobin—Paradigms of Protein Structure and Function

SPECIAL FOCUS

Ancient life forms evolved in the absence of oxygen and were capable only of anaerobic metabolism. As the earth's atmosphere changed over time, so too did living things. Indeed, the production of O_2 by photosynthesis was a major factor in altering the atmosphere. Evolution to an oxygen-based metabolism was highly beneficial. Aerobic metabolism of sugars, for example, yields far more energy than corresponding anaerobic processes. Two important oxygen-binding proteins appeared in the course of evolution so that aerobic metabolic processes were no longer limited by the solubility of O_2 in water. These proteins are represented in animals as **hemoglobin (Hb)** in blood and **myoglobin (Mb)** in muscle. Because hemoglobin and myoglobin are two of the most-studied proteins in nature, they have become paradigms of protein structure and function. Moreover, hemoglobin is a model for protein quaternary structure and allosteric function. The binding of O_2 by hemoglobin, and its modulation by effectors such as protons, CO_2, and 2,3-bisphosphoglycerate, depend on interactions between subunits in the Hb tetramer. Subunit–subunit interactions in Hb reveal much about the functional significance of quaternary associations and allosteric regulation.

The Comparative Biochemistry of Myoglobin and Hemoglobin Reveals Insights into Allostery

A comparison of the properties of hemoglobin and myoglobin offers insights into allosteric phenomena, even though these proteins are *not* enzymes. Hemoglobin displays sigmoid-shaped O_2-binding curves (Figure 15.21). The unusual shape of these curves

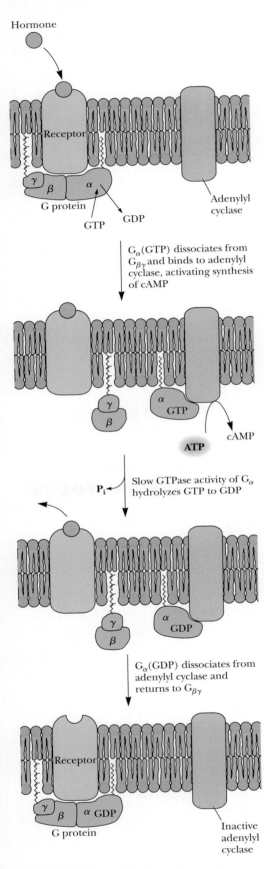

Hormone

Receptor

γ β α

G protein

GTP GDP

Adenylyl cyclase

G_α(GTP) dissociates from $G_{\beta\gamma}$ and binds to adenylyl cyclase, activating synthesis of cAMP

α GTP

γ β

ATP cAMP

Slow GTPase activity of G_α hydrolyzes GTP to GDP

P_i

α GDP

γ β

G_α(GDP) dissociates from adenylyl cyclase and returns to $G_{\beta\gamma}$

Receptor

γ β α GDP

G protein

Inactive adenylyl cyclase

FIGURE 15.20 Hormone binding to its receptor leads via G-protein activation to cAMP synthesis. Adenylyl cyclase and the hormone receptor are integral plasma membrane proteins; G_α and $G_{\beta\gamma}$ are membrane-anchored proteins.

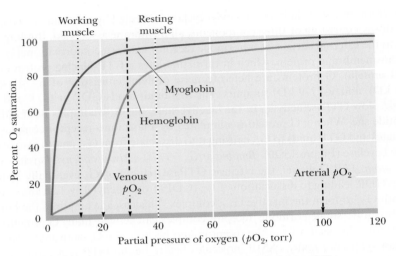

FIGURE 15.21 O_2-binding curves for hemoglobin and myoglobin.

was once a great enigma in biochemistry. Such curves closely resemble allosteric enzyme:substrate saturation graphs (see Figure 15.6). In contrast, myoglobin's interaction with oxygen obeys classical Michaelis–Menten-type substrate saturation behavior.

Before examining myoglobin and hemoglobin in detail, let us first encapsulate the lesson: Myoglobin is a compact globular protein composed of a single polypeptide chain 153 amino acids in length; its molecular mass is 17.2 kD (Figure 15.22). It contains **heme,** a porphyrin ring system complexing an iron ion, as its prosthetic group (Figure 15.23). Oxygen binds to Mb via its heme. Hemoglobin (Hb) is also a compact globular protein, but Hb is a tetramer. It consists of four polypeptide chains, each of which is very similar structurally to the myoglobin polypeptide chain, and each bears a heme group. Thus, a hemoglobin molecule can bind four O_2 molecules. In adult human Hb, there are two identical chains of 141 amino acids, the α-chains, and two identical β-chains, each of 146 residues. The human Hb molecule is an $\alpha_2\beta_2$-type tetramer of molecular mass 64.45 kD.

The myoglobin polypeptide chain and the α- and β-chains of hemoglobin are composed of 8 α-helical segments denoted by the letters A through H. The short, unordered regions that connect the helices are named for the segments they connect, as in the AB region or the EF region. In an amino acid numbering system unique to globin chains, successive residues in the helices are numbered, such as the histidine at position 8 in the F helix, known as His F8.

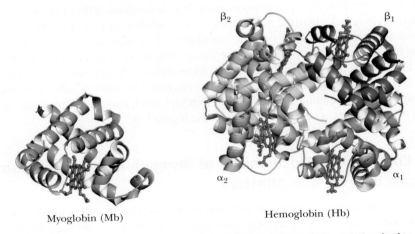

Myoglobin (Mb)   Hemoglobin (Hb)

FIGURE 15.22 The myoglobin (pdb id = 2MM1) and hemoglobin (pdb id = 2HHB) molecules.

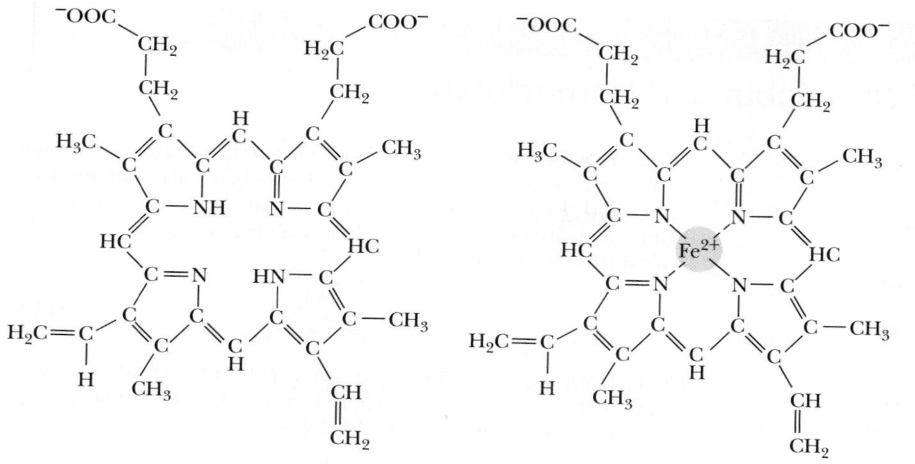

Protoporphyrin IX

Heme
(Fe-protoporphyrin IX)

FIGURE 15.23 Heme is formed when protoporphyrin IX binds Fe^{2+}.

The tetrameric nature of Hb is crucial to its biological function: *When a molecule of O_2 binds to a heme in Hb, the heme Fe ion is drawn into the plane of the porphyrin ring. This slight movement sets off a chain of conformational events that are transmitted to adjacent subunits, dramatically enhancing the affinity of their heme groups for O_2.* That is, the binding of O_2 to one heme of Hb makes it easier for the Hb molecule to bind additional equivalents of O_2. Hemoglobin is a marvelously constructed molecular machine. Let us dissect its mechanism, beginning with its monomeric counterpart, the myoglobin molecule.

Myoglobin Is an Oxygen-Storage Protein

Myoglobin is the oxygen-storage protein of muscle. The muscles of diving mammals such as seals and whales are especially rich in this protein, which serves as a store for O_2 during the animal's prolonged periods underwater. Myoglobin is abundant in skeletal and cardiac muscle of nondiving animals as well. Myoglobin is the cause of the characteristic red color of muscle.

O_2 Binds to the Mb Heme Group

Iron prefers to interact with six ligands, four of which share a common plane. The fifth and sixth ligands lie above and below this plane (see Figure 15.24). In heme, four of the ligands are provided by the nitrogen atoms of the four pyrroles. A fifth ligand is donated by the imidazole side chain of amino acid residue His F8. When myoglobin binds O_2 to become **oxymyoglobin,** the O_2 molecule adds to the heme iron ion as the sixth ligand (Figure 15.24). O_2 adds end on to the heme iron, but it is not oriented perpendicular to the plane of the heme. Rather, it is tilted about 60° with respect to the perpendicular.

O_2 Binding Alters Mb Conformation

What happens when the heme group of myoglobin binds oxygen? X-ray crystallography has revealed that a crucial change occurs in the position of the iron atom relative to the plane of the heme. In deoxymyoglobin, the ferrous ion actually lies 0.055 nm above the plane of the heme, in the direction of His F8. The iron–porphyrin complex is therefore dome-shaped. When O_2 binds, the iron atom is pulled back toward the porphyrin plane and is now displaced from it by only 0.026 nm. The consequences of this small motion are trivial as far as the biological role of myoglobin is concerned. However, as we shall soon see, this slight movement profoundly affects the properties of hemoglobin.

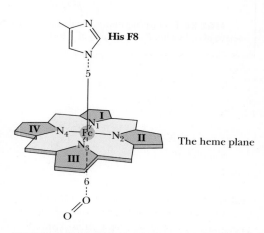

His F8

The heme plane

FIGURE 15.24 The six liganding positions of an iron ion.

> **A DEEPER LOOK**

The Oxygen-Binding Curves of Myoglobin and Hemoglobin

Myoglobin

The reversible binding of oxygen to myoglobin,

$$MbO_2 \rightleftharpoons Mb + O_2$$

can be characterized by the equilibrium dissociation constant, K.

$$K = \frac{[Mb][O_2]}{[MbO_2]} \tag{15.1}$$

If Y is defined as the **fractional saturation** of myoglobin with O_2, that is, the fraction of myoglobin molecules having an oxygen molecule bound, then

$$Y = \frac{[MbO_2]}{[MbO_2] + [Mb]} \tag{15.2}$$

The value of Y ranges from 0 (no myoglobin molecules carry an O_2) to 1.0 (all myoglobin molecules have an O_2 molecule bound). Substituting from Equation 15.1, $([Mb][O_2])/K$ for $[MbO_2]$ gives

$$Y = \frac{\left(\dfrac{[Mb][O_2]}{K}\right)}{\left(\dfrac{[Mb][O_2]}{K} + [Mb]\right)} = \frac{\left(\dfrac{[O_2]}{K}\right)}{\left(\dfrac{[O_2]}{K} + 1\right)} = \frac{[O_2]}{[O_2] + K} \tag{15.3}$$

and, if the concentration of O_2 is expressed in terms of the partial pressure (in torr) of oxygen gas in equilibrium with the solution of interest, then

$$Y = \frac{pO_2}{pO_2 + K} \tag{15.4}$$

(In this form, K has the units of torr.) The relationship defined by Equation 15.4 plots as a hyperbola. That is, the MbO_2 saturation

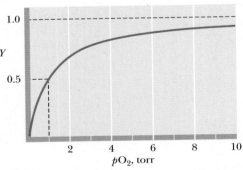

FIGURE 1 Oxygen saturation curve for myoglobin in the form of Y versus pO_2 showing P_{50} is at a pO_2 of 1 torr.

curve resembles an enzyme:substrate saturation curve. For myoglobin, a partial pressure of 1 torr for pO_2 is sufficient for half-saturation (Figure 1). We can define P_{50} as the partial pressure of O_2 at which 50% of the myoglobin molecules have a molecule of O_2 bound (that is, $Y = 0.5$), then

$$0.5 = \frac{pO_2}{pO_2 + P_{50}} \tag{15.5}$$

(Note from Equation 15.1 that when $[MbO_2] = [Mb]$, $K = [O_2]$, which is the same as saying when $Y = 0.5$, $K = P_{50}$.) The general equation for O_2 binding to Mb becomes

$$Y = \frac{pO_2}{pO_2 + P_{50}} \tag{15.6}$$

The ratio of the fractional saturation of myoglobin, Y, to free myoglobin, $1 - Y$, depends on pO_2 and K according to the equation

$$\frac{Y}{1 - Y} = \frac{pO_2}{K} \tag{15.7}$$

Hemoglobin

New properties emerge when four heme-containing polypeptides come together to form a tetramer. The O_2-binding curve of hemoglobin is sigmoid rather than hyperbolic (see Figure 15.21), and Equation 15.4 does not describe such curves. Of course, each hemoglobin molecule has four hemes and can bind up to four oxygen molecules. Suppose for the moment that O_2 binding to hemoglobin is an "all-or-none" phenomenon, where Hb exists either free of O_2 or with four O_2 molecules bound. This supposition represents the extreme case for cooperative binding of a ligand by a protein with multiple binding sites. In effect, it says that if one ligand binds to the protein molecule, then all other sites are immediately occupied by ligand. Or, to say it another way for the case in hand, suppose that four O_2 molecules bind to Hb simultaneously:

$$Hb + 4\,O_2 \rightleftharpoons Hb(O_2)_4$$

Then the dissociation constant, K, would be

$$K = \frac{[Hb][O_2]^4}{[Hb(O_2)_4]} \tag{15.8}$$

By analogy with Equation 15.4, the equation for fractional saturation of Hb is given by

$$Y = \frac{[pO_2]^4}{[pO_2]^4 + K} \tag{15.9}$$

Cooperative Binding of Oxygen by Hemoglobin Has Important Physiological Significance

The relative oxygen affinities of hemoglobin and myoglobin reflect their respective physiological roles (see Figure 15.21). Myoglobin, as an oxygen storage protein, has a greater affinity for O_2 than hemoglobin at all oxygen pressures. Hemoglobin, as the oxygen carrier, becomes saturated with O_2 in the lungs, where the partial pressure of O_2 (pO_2) is about 100 torr.[1] In the capillaries of tissues, pO_2 is typically 40 torr, and oxygen is

[1] The **torr** is a unit of pressure named for Evangelista Torricelli, inventor of the barometer. One torr corresponds to 1 mm Hg (1/760th of an atmosphere).

A plot of Y versus pO_2 according to Equation 15.9 is presented in Figure 2. This curve has the characteristic sigmoid shape seen for O_2 binding by Hb. Half-saturation is set to be a pO_2 of 26 torr. Note that when pO_2 is low, the fractional saturation, Y, changes very little as pO_2 increases. The interpretation is that Hb has little affinity for O_2 at these low partial pressures of O_2. However, as pO_2 reaches some threshold value and the first O_2 is bound, Y, the fractional saturation, increases rapidly. Note that the slope of the curve is steepest in the region where $Y = 0.5$. The sigmoid character of this curve is diagnostic of the fact that the binding of O_2 to one site on Hb strongly enhances binding of additional O_2 molecules to the remaining vacant sites on the same Hb molecule, a phenomenon aptly termed **cooperativity**. (If each O_2 bound independently, exerting no influence on the affinity of Hb for more O_2 binding, this plot would be hyperbolic.)

The experimentally observed oxygen-binding curve for Hb does not fit the graph given in Figure 2 exactly. If we generalize Equation 15.9 by replacing the exponent 4 with n, we can write the equation as

$$Y = \frac{[pO_2]^n}{[pO_2]^n + K} \qquad (15.10)$$

Rearranging yields

$$\frac{Y}{1 - Y} = \frac{[pO_2]^n}{K} \qquad (15.11)$$

This equation states that the ratio of oxygenated heme groups (Y) to O_2-free heme ($1 - Y$) is equal to the nth power of the pO_2 divided by the apparent dissociation constant, K.

Archibald Hill demonstrated in 1913, well before any knowledge about the molecular organization of Hb existed, that the O_2-binding behavior of Hb could be described by Equation 15.11. If a value of 2.8 is taken for n, Equation 15.11 fits the experimentally observed O_2-binding curve for Hb very well (Figure 3). If the binding of O_2 to Hb were an all-or-none phenomenon, n would equal 4, as discussed previously. If the O_2-binding sites on Hb were completely noninteracting, that is, if the binding of one O_2 to Hb had no influence on the binding of additional O_2 molecules to the same Hb, n would equal 1. Figure 3 compares these extremes. Obviously, the real situation falls between the extremes of $n = 1$ or 4. The qualitative answer is that O_2 binding by Hb is highly cooperative, and the binding of the first O_2 markedly enhances the binding of subsequent O_2 molecules. However, this binding is not quite an all-or-none phenomenon.

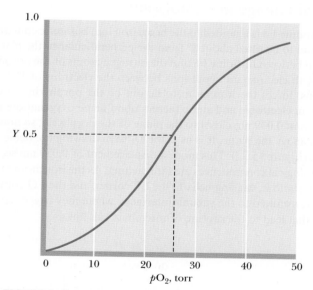

FIGURE 2 Oxygen saturation curve for Hb in the form of Y versus pO_2, assuming $n = 4$ and $P_{50} = 26$ torr. The graph has the characteristic experimentally observed sigmoid shape.

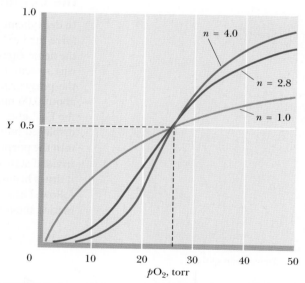

FIGURE 3 A comparison of the experimentally observed O_2 curve for Hb yielding a value for n of 2.8 (blue), the hypothetical curve if $n = 4$ (red), and the curve if $n = 1$ (noninteracting O_2-binding sites, purple).

released from Hb. In muscle, some of it can be bound by myoglobin, to be stored for use in times of severe oxygen deprivation, such as during strenuous exercise.

Hemoglobin Has an $\alpha_2\beta_2$ Tetrameric Structure

As noted, hemoglobin is an $\alpha_2\beta_2$ tetramer. Each of the four subunits has a conformation virtually identical to that of myoglobin (Figure 15.22). The subunits pack in a tetrahedral array, creating a roughly spherical molecule $6.4 \times 5.5 \times 5.0$ nm. The four heme groups, nestled within the easily recognizable cleft formed between the E and F helices of each polypeptide, are exposed at the surface of the molecule. The heme

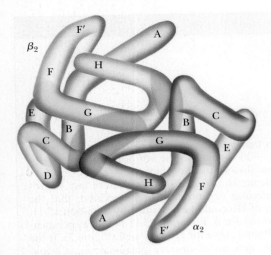

FIGURE 15.25 Side view of one of the two $\alpha\beta$-dimers in Hb, with α-β packing contacts indicated in blue. The sliding contacts made with the other dimer are shown in yellow. The changes in these sliding contacts are shown in Figure 15.26. (Illustration: Irving Geis. Rights owned by Howard Hughes Medical Institute. Not to be reproduced without permission.)

groups are quite far apart; 2.5 nm separates the closest iron ions, those of hemes α_1 and β_2, and those of hemes α_2 and β_1. The subunit interactions are mostly between dissimilar chains: Each of the α-chains is in contact with both β-chains, but there are few α–α or β–β interactions.

Oxygenation Markedly Alters the Quaternary Structure of Hb

Crystals of deoxyhemoglobin shatter when exposed to O_2. Furthermore, X-ray crystallographic analysis reveals that oxyhemoglobin and deoxyhemoglobin differ markedly in quaternary structure. In particular, specific $\alpha\beta$-subunit interactions change. The $\alpha\beta$ contacts are of two kinds. The $\alpha_1\beta_1$ and $\alpha_2\beta_2$ contacts involve helices B, G, and H and the GH corner. These contacts are extensive and important to subunit packing; they remain unchanged when hemoglobin goes from its deoxy to its oxy form. The $\alpha_1\beta_2$ and $\alpha_2\beta_1$ contacts are called **sliding contacts.** They principally involve helices C and G and the FG corner (Figure 15.25). When hemoglobin undergoes a conformational change as a result of ligand binding to the heme, these contacts are altered (Figure 15.26). Hemoglobin, as a conformationally dynamic molecule, consists of two dimeric halves, an $\alpha_1\beta_1$-subunit pair and an $\alpha_2\beta_2$-subunit pair. Each $\alpha\beta$-dimer moves as a rigid body, and the two halves of the molecule slide past each other upon oxygenation of the heme. The two halves rotate some 15° about an imaginary pivot passing through the $\alpha\beta$-subunits; some atoms at the interface between $\alpha\beta$-dimers are relocated by as much as 0.6 nm.

Movement of the Heme Iron by Less Than 0.04 nm Induces the Conformational Change in Hemoglobin

In deoxyhemoglobin, histidine F8 is liganded to the heme iron ion, but steric constraints force the Fe^{2+} : His-N bond to be tilted about 8° from the perpendicular to the plane of the heme. Steric repulsion between histidine F8 and the nitrogen atoms of the porphyrin ring system, combined with electrostatic repulsions between the electrons of Fe^{2+} and the porphyrin π-electrons, forces the iron atom to lie out of the porphyrin plane by about 0.06 nm. Changes in electronic and steric factors upon heme oxygenation allow the Fe^{2+} atom to move about 0.039 nm closer to the plane of the porphyrin, so now it is displaced only 0.021 nm above the plane. It is as if the O_2 were drawing the heme Fe^{2+} into the porphyrin plane (Figure 15.27). This modest displacement of 0.039 nm seems a trivial distance, but its biological consequences are far reaching. As the iron atom moves, it drags histidine F8 along with it, causing helix F, the EF corner, and the FG corner to follow. These shifts are transmitted to the subunit interfaces, where they trigger conformational readjustments that lead to the rupture of interchain salt links.

(a) Deoxyhemoglobin

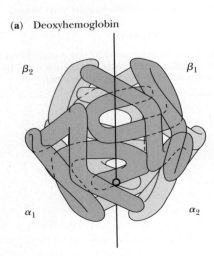

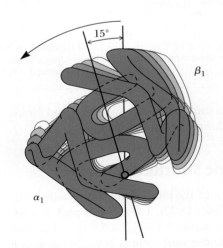

(b) Oxyhemoglobin

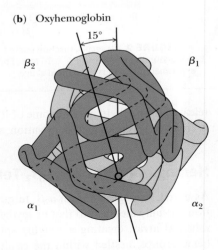

FIGURE 15.26 Subunit motion in hemoglobin when the molecule goes from the **(a)** deoxy to the **(b)** oxy form. (Illustration: Irving Geis. Rights owned by Howard Hughes Medical Institute. Not to be reproduced without permission.)

> ▶ **A DEEPER LOOK**
>
> # The Physiological Significance of the Hb:O_2 Interaction
>
> We can determine quantitatively the physiological significance of the sigmoid nature of the hemoglobin oxygen-binding curve, or, in other words, the biological importance of cooperativity. The equation
>
> $$\frac{Y}{1 - Y} = \frac{[pO_2]^n}{P_{50}}$$
>
> describes the relationship between pO_2, the affinity of hemoglobin for O_2 (defined as P_{50}, the partial pressure of O_2 giving half-maximal saturation of Hb with O_2), and the fraction of hemoglobin with O_2 bound, Y, versus the fraction of Hb with no O_2 bound, $(1 - Y)$ (see A Deeper Look: The Oxygen-Binding Curves of Myoglobin and Hemoglobin on pages 500–501). The coefficient n is the Hill coefficient, an index of the cooperativity (sigmoidicity) of the hemoglobin
>
> oxygen-binding curve. Taking pO_2 in the lungs as 100 torr, P_{50} as 26 torr, and n as 2.8, the fractional saturation of the hemoglobin heme groups with O_2, is 0.98. If pO_2 were to fall to 10 torr within the capillaries of an exercising muscle, Y would drop to 0.06. The oxygen delivered under these conditions would be proportional to the difference, $Y_{lungs} - Y_{muscle}$, which is 0.92. That is, virtually all the oxygen carried by Hb would be released. Suppose instead that hemoglobin binding of O_2 were not cooperative; in that case, the hemoglobin oxygen-binding curve would be hyperbolic, and $n = 1.0$. Then Y in the lungs would be 0.79 and Y in the capillaries, 0.28; the difference in Y values would be 0.51. Thus, under these conditions, the cooperativity of oxygen binding by Hb means that 0.92/0.51 or 1.8 times as much O_2 can be delivered.

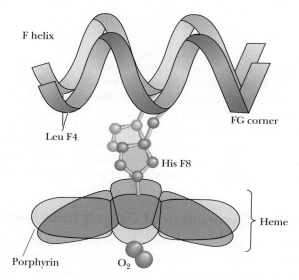

FIGURE 15.27 Changes in the position of the heme iron atom upon oxygenation lead to conformational changes in the hemoglobin molecule.

The Oxy and Deoxy Forms of Hemoglobin Represent Two Different Conformational States

Hemoglobin resists oxygenation (see Figure 15.21) because the deoxy form is stabilized by specific hydrogen bonds and salt bridges (ion-pair bonds) (Figure 15.28). All of these interactions are broken in oxyhemoglobin, as the molecule stabilizes into a new conformation. The shift in helix F upon oxygenation leads to rupture of the Tyr β145:Val β98 hydrogen bond. In deoxyhemoglobin, with these interactions intact, the C-termini of the four subunits are restrained, and this conformational state is termed **T**, the **tense** or **taut form.** In oxyhemoglobin, these C-termini have almost complete freedom of rotation, and the molecule is now in its **R, or relaxed, form.**

The Allosteric Behavior of Hemoglobin Has Both Symmetry (MWC) Model and Sequential (KNF) Model Components

Oxygen is accessible only to the heme groups of the α-chains when hemoglobin is in the T conformational state. Max Perutz has pointed out that the heme environment of β-chains in the T state is virtually inaccessible because of steric hindrance by amino acid

(a)

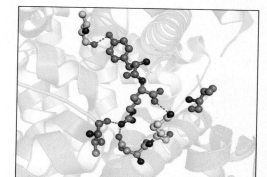

(b)

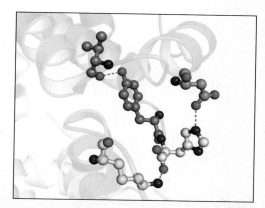

FIGURE 15.28 Salt bridges between different subunits in human deoxyhemoglobin. These noncovalent, electrostatic interactions are disrupted upon oxygenation. **(a)** A focus on those salt bridges and hydrogen bonds involving interactions between N-terminal and C-terminal residues in the α-chains. Residues in the lower center are Arg α_1 141 (green) with Val α_1 1 (purple), Asp α_2 126 (orange), Lys α_2 127 (yellow), and Val β_2 34 (olive); residues at top are Val α_1 93 (yellow) with Tyr α_1 140 (purple). **(b)** A focus on those salt bridges and hydrogen bonds involving C-terminal residues of β-chains: Val β_2 78 (olive) with Tyr β_2 145 (purple); His β_2 146 (light blue) with Asp β_2 94 (orange) and Lys α_1 40 (yellow) (pdb id = 2HHB).

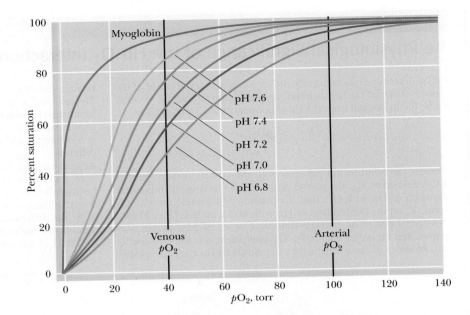

FIGURE 15.29 The oxygen saturation curves for myoglobin and for hemoglobin at five different pH values: 7.6, 7.4, 7.2, 7.0, and 6.8.

residues in the E helix. This hindrance disappears when the hemoglobin molecule undergoes transition to the R conformational state. Binding of O_2 to the β-chains is thus dependent on a T-to-R conformational shift, and this shift is triggered by the subtle changes that occur when O_2 binds to the α-chain heme groups. Together these observations lead to a model that is partially MWC and partially KNF: O_2 binding to one α-subunit and then the other leads to sequential changes in conformation, followed by a switch in quaternary structure at the $Hb : 2O_2$ state from T to R. Thus, the real behavior of this protein is an amalgam of the two prominent theoretical models for allosteric behavior.

H⁺ Promotes the Dissociation of Oxygen from Hemoglobin

Protons, carbon dioxide, and chloride ions, as well as the metabolite 2,3-bisphosphoglycerate (or BPG), all affect the binding of O_2 by hemoglobin. Their effects have interesting ramifications, which we shall see as we discuss them in turn. Deoxyhemoglobin has a higher affinity for protons than oxyhemoglobin. Thus, as the pH decreases, dissociation of O_2 from hemoglobin is enhanced. In simple symbolism, ignoring the stoichiometry of O_2 or H^+ involved:

$$HbO_2 + H^+ \rightleftharpoons HbH^+ + O_2$$

Expressed another way, H^+ is an antagonist of oxygen binding by Hb, and the saturation curve of Hb for O_2 is displaced to the right as acidity increases (Figure 15.29). This phenomenon is called the **Bohr effect,** after its discoverer, the Danish physiologist

> **A DEEPER LOOK**

Changes in the Heme Iron upon O₂ Binding

In deoxyhemoglobin, the six d electrons of the heme Fe^{2+} exist as four unpaired electrons and one electron pair, and five ligands can be accommodated: the four N-atoms of the porphyrin ring system and histidine F8. In this electronic configuration, the iron atom is paramagnetic and in the **high-spin state.** When the heme binds O_2 as a sixth ligand, these electrons are rearranged into three e^- pairs and the iron changes to the **low-spin state** and is diamagnetic. This change in spin state allows the bond between the Fe^{2+} ion and histidine F8 to become perpendicular to the heme plane and to shorten. In addition, interactions between the porphyrin N atoms and the iron strengthen. Also, high-spin Fe^{2+} has a greater atomic volume than low-spin Fe^{2+} because its four unpaired e^- occupy four orbitals rather than two when the electrons are paired in low-spin Fe^{2+}. So, low-spin iron is less sterically hindered and able to move nearer to the porphyrin plane.

Christian Bohr (the father of Niels Bohr, the atomic physicist). The effect has important physiological significance because actively metabolizing tissues produce acid, promoting O_2 release where it is most needed. About two protons are taken up by deoxyhemoglobin. The N-termini of the two α-chains and the His $\beta146$ residues have been implicated as the major players in the Bohr effect. (The pK_a of a free amino terminus in a protein is about 8.0, but the pK_a of a protein histidine imidazole is around 6.5.) Neighboring carboxylate groups of Asp $\beta94$ residues help stabilize the protonated state of the His $\beta146$ imidazoles that occur in deoxyhemoglobin. However, when Hb binds O_2, changes in the conformation of β-chains upon Hb oxygenation move the negative Asp function away, and dissociation of the imidazole protons is favored.

CO$_2$ Also Promotes the Dissociation of O$_2$ from Hemoglobin

Carbon dioxide has an effect on O_2 binding by Hb that is similar to that of H^+, partly because it produces H^+ when it dissolves in the blood:

$$CO_2 + H_2O \xrightleftharpoons{\text{carbonic anhydrase}} \underset{\text{carbonic acid}}{H_2CO_3} \rightleftharpoons H^+ + \underset{\text{bicarbonate}}{HCO_3^-}$$

The enzyme *carbonic anhydrase* promotes the hydration of CO_2. Many of the protons formed upon ionization of carbonic acid are picked up by Hb as O_2 dissociates. The bicarbonate ions are transported with the blood back to the lungs. When Hb becomes oxygenated again in the lungs, H^+ is released and reacts with HCO_3^- to re-form H_2CO_3, from which CO_2 is liberated. The CO_2 is then exhaled as a gas.

In addition, some CO_2 is directly transported by hemoglobin in the form of *carbamate* ($—NHCOO^-$). Free α-amino groups of Hb react with CO_2 reversibly:

$$R—NH_2 + CO_2 \rightleftharpoons R—NH—COO^- + H^+$$

This reaction is driven to the right in tissues by the high CO_2 concentration; the equilibrium shifts the other way in the lungs where $[CO_2]$ is low. Thus, carbamylation of the N-termini converts them to anionic functions, which then form salt links with the cationic side chains of Arg $\alpha141$ that stabilize the deoxy or T state of hemoglobin.

In addition to CO_2, Cl^- and BPG also bind better to deoxyhemoglobin than to oxyhemoglobin, causing a shift in equilibrium in favor of O_2 release. These various effects are demonstrated by the shift in the oxygen saturation curves for Hb in the presence of one or more of these substances (Figure 15.30). Note that the O_2-binding curve for Hb + BPG + CO_2 fits that of whole blood very well.

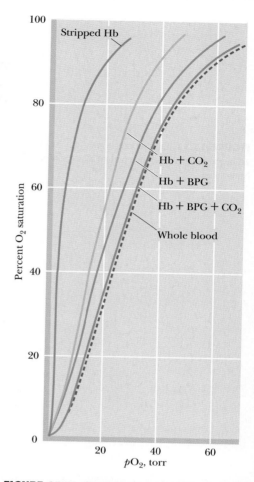

FIGURE 15.30 Oxygen-binding curves of blood and of hemoglobin in the absence and presence of CO_2 and BPG. From left to right: stripped Hb, Hb + CO_2, Hb + BPG, Hb + BPG + CO_2, and whole blood.

2,3-Bisphosphoglycerate Is an Important Allosteric Effector for Hemoglobin

The binding of 2,3-bisphosphoglycerate (BPG) to Hb promotes the release of O_2 (Figure 15.30). Erythrocytes (red blood cells) normally contain about 4.5 mM BPG, a concentration equivalent to that of tetrameric hemoglobin molecules. Interestingly, this equivalence is maintained in the Hb:BPG binding stoichiometry because the tetrameric Hb molecule has but one binding site for BPG. This site is situated within the central cavity formed by the association of the four subunits. The strongly negative BPG molecule (Figure 15.31) is electrostatically bound via interactions with the positively charged functional groups of each Lys $\beta82$, His $\beta2$, His $\beta143$, and the NH_3^+-terminal group of each β-chain. These positively charged residues are arranged to form an electrostatic pocket complementary to the conformation and charge distribution of BPG (Figure 15.32). In effect, BPG crosslinks the two β-subunits. The ionic bonds between BPG and the two β-chains aid in stabilizing the conformation of Hb in its deoxy form, thereby favoring the dissociation of oxygen. In oxyhemoglobin, this central cavity is too small for BPG to fit. Or, to put it another way, the conformational changes in the Hb molecule that accompany O_2 binding perturb the BPG-binding site so that BPG can no

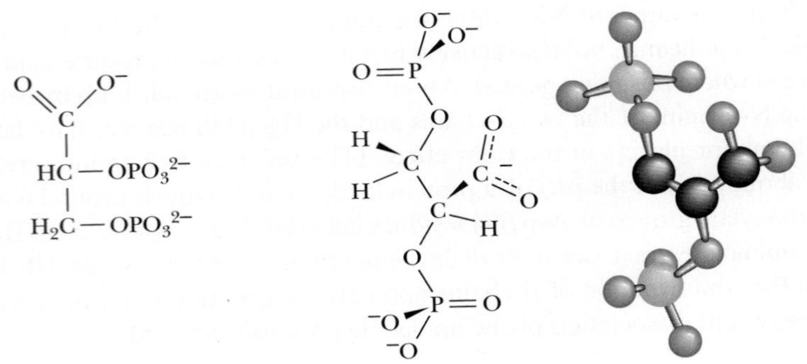

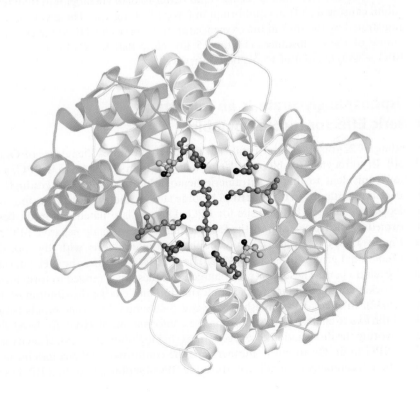

FIGURE 15.31 The structure, in ionic form, of BPG or 2,3-bisphosphoglycerate, an important allosteric effector for hemoglobin.

longer be accommodated. Thus, BPG and O_2 are mutually exclusive allosteric effectors for Hb, even though their binding sites are physically distinct.

BPG Binding to Hb Has Important Physiological Significance

The importance of the BPG effect is evident in Figure 15.30. Hemoglobin stripped of BPG is virtually saturated with O_2 at a pO_2 of only 20 torr, and it cannot release its oxygen within tissues, where the pO_2 is typically 40 torr. BPG shifts the oxygen saturation curve of Hb to the right, making Hb an O_2 delivery system eminently suited to the needs of the organism. BPG serves this vital function in humans, most primates, and a number of other mammals. However, the hemoglobins of cattle, sheep, goats, deer, and other animals have an intrinsically lower affinity for O_2, and these Hbs are relatively unaffected by BPG.

Fetal Hemoglobin Has a Higher Affinity for O_2 Because It Has a Lower Affinity for BPG

The fetus depends on its mother for an adequate supply of oxygen, but its circulatory system is entirely independent. Gas exchange takes place across the placenta. Ideally then, fetal Hb should be able to absorb O_2 better than maternal Hb so that an effective

FIGURE 15.32 The ionic binding of BPG to the two β-subunits of Hb. BPG lies at center of the cavity between the two β-subunits. The highlighted residues are N-terminal Val β_1 and Val β_2 (yellow), His β_1 2, His β_2 2, His β_1 143, and His β_2 143 (purple), Lys β_1 82 and Lys β_2 82 (green) (pdb id = 1B86).

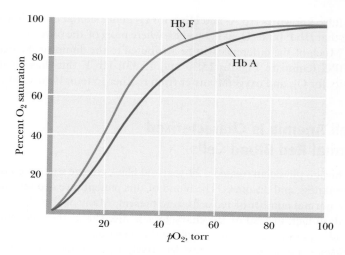

FIGURE 15.33 Comparison of the oxygen saturation curves of Hb A and Hb F under similar conditions of pH and [BPG].

transfer of oxygen can occur. Fetal Hb differs from adult Hb in that the β-chains are replaced by very similar, but not identical, 146-residue subunits called γ-chains (gamma chains). Fetal Hb is thus $\alpha_2\gamma_2$. Recall that BPG functions through its interaction with the β-chains. BPG binds less effectively with the γ-chains of fetal Hb (also called Hb F). (Fetal γ-chains have Ser instead of His at position 143 and thus lack two of the positive charges in the central BPG-binding cavity.) Figure 15.33 compares the relative affinities

---

HUMAN BIOCHEMISTRY

Hemoglobin and Nitric Oxide

Nitric oxide (NO $\cdot$) is a simple gaseous molecule whose many remarkable physiological functions are still being discovered. For example, NO $\cdot$ is known to act as a neurotransmitter and as a second messenger in signal transduction (see Chapter 32). Furthermore, **endothelial relaxing factor** (ERF, also known as endothelium-derived relaxing factor, or EDRF), a once-elusive hormonelike agent that acts to relax the musculature of the walls (**endothelium**) of blood vessels and lower blood pressure, has been identified as NO $\cdot$. It has long been known that NO $\cdot$ is a high-affinity ligand for Hb, binding to its heme-Fe^{2+} atom 10,000 times better than O_2. An enigma thus arises: Why isn't NO $\cdot$ instantaneously bound by Hb within human erythrocytes and prevented from exerting its vasodilation properties?

The reason that Hb doesn't block the action of NO $\cdot$ is due to a unique interaction between Cys 93β of Hb and NO $\cdot$ discovered by Li Jia, Celia and Joseph Bonaventura, and Johnathan Stamler at Duke University. Nitric oxide reacts with the sulfhydryl group of Cys 93β, forming an S-nitroso derivative:

$$-CH_2-S-N=O$$

This S-nitroso group is in equilibrium with other S-nitroso compounds formed by reaction of NO $\cdot$ with small-molecule thiols such as free cysteine or glutathione (an isoglutamylcysteinylglycine tripeptide):

$$H_3{}^+N-\underset{\underset{COO^-}{|}}{\overset{\overset{H}{|}}{C}}-CH_2-CH_2-\overset{\overset{O}{\|}}{C}-\underset{\underset{H}{|}}{N}-\underset{\underset{CH_2-S-N=O}{|}}{\overset{\overset{H}{|}}{C}}-\overset{\overset{O}{\|}}{C}-\underset{\underset{H}{|}}{N}-CH_2-COO^-$$

S-nitrosoglutathione

These small-molecule thiols serve to transfer NO $\cdot$ from erythrocytes to endothelial receptors, where it acts to relax vascular tension. NO $\cdot$ itself is a reactive free-radical compound whose biological half-life is very short (1–5 sec). S-nitrosoglutathione has a half-life of several hours.

The reactions between Hb and NO $\cdot$ are complex. NO $\cdot$ forms a ligand with the heme-Fe^{2+} that is quite stable in the absence of O_2. However, in the presence of O_2, NO $\cdot$ is oxidized to $NO_3{}^-$ and the heme-Fe^{2+} of Hb is oxidized to Fe^{3+}, forming methemoglobin. Fortunately, the interaction of Hb with NO $\cdot$ is controlled by the allosteric transition between R-state Hb (oxyHb) and T-state Hb (deoxyHb). Cys 93β is more exposed and reactive in R-state Hb than in T-state Hb, and binding of NO $\cdot$ to Cys 93β precludes reaction of NO $\cdot$ with heme iron. Upon release of O_2 from Hb in tissues, Hb shifts conformation from R state to T state, and binding of NO $\cdot$ at Cys 93β is no longer favored. Consequently, NO $\cdot$ is released from Cys 93β and transferred to small-molecule thiols for delivery to endothelial receptors, causing capillary vasodilation. This mechanism also explains the puzzling observation that free Hb produced by recombinant DNA methodology for use as a whole-blood substitute causes a transient rise of 10 to 12 mm Hg in diastolic blood pressure in experimental clinical trials. (Conventional whole-blood transfusion has no such effect.) It is now apparent that the "synthetic" Hb, which has no bound NO $\cdot$, is binding NO $\cdot$ in the blood and preventing its vasoregulatory function.

In the course of hemoglobin evolution, the only invariant amino acid residues in globin chains are His F8 (the obligatory heme ligand) and a Phe residue acting to wedge the heme into its pocket. However, in mammals and birds, Cys 93β is also invariant, no doubt due to its vital role in NO $\cdot$ delivery.

Adapted from Jia, L., et al., 1996. S-Nitrosohaemoglobin: A dynamic activity of blood involved in vascular control. *Nature* **380**:221–226.

of adult Hb (also known as Hb A) and Hb F for O_2 under similar conditions of pH and [BPG]. Note that Hb F binds O_2 at pO_2 values where most of the oxygen has dissociated from Hb A. Much of the difference can be attributed to the diminished capacity of Hb F to bind BPG (compare Figures 15.30 and 15.33); Hb F thus has an intrinsically greater affinity for O_2, and oxygen transfer from mother to fetus is ensured.

Sickle-Cell Anemia Is Characterized by Abnormal Red Blood Cells

In 1904, a Chicago physician treated a 20-year-old black college student complaining of headache, weakness, and dizziness. The blood of this patient revealed serious anemia—only half the normal number of red cells were present. Many of these cells were abnormally shaped; in fact, instead of the characteristic disc shape, these erythrocytes were elongated and crescentlike in form, a feature that eventually gave name to the disease **sickle-cell anemia.** These sickle cells pass less freely through the capillaries, impairing circulation and causing tissue damage. Furthermore, these cells are more fragile and rupture more easily than normal red cells, leading to anemia.

Sickle-Cell Anemia Is a Molecular Disease

A single amino acid substitution in the β-chains of Hb causes sickle-cell anemia. Replacement of the glutamate residue at position 6 in the β-chain by a valine residue marks the only chemical difference between Hb A and sickle-cell hemoglobin, Hb S.

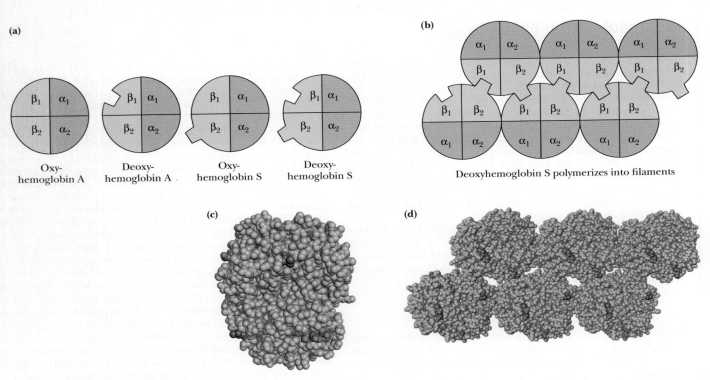

FIGURE 15.34 The polymerization of Hb S via the interactions between the hydrophobic Val side chains at position $\beta6$ and the hydrophobic pockets in the EF corners of β-chains in neighboring Hb molecules. **(a)** The protruding "block" on Oxy S represents the Val hydrophobic protrusion. The complementary hydrophobic pocket in the EF corner of deoxy β-chains is represented by a square-shaped indentation. (This indentation is probably present in Hb A also.) Only the β_2 Val protrusions and the β_1 EF pockets are shown. (The β_1 Val protrusions and the β_2 EF pockets are not involved, although they are present.) **(b)** The polymerization of Hb S via β_2 Val6 insertions into neighboring β_1 pockets. **(c)** Molecular graphic of an Hb S dimer of tetramers. β_2 Val residues are highlighted in blue; heme is shown in red (pdb id = 2HBS). **(d)** Molecular graphic of the Hb S filament (pdb id = 2HBS).

The amino acid residues at position $\beta 6$ lie at the surface of the hemoglobin molecule. In Hb A, the ionic R groups of the Glu residues are suited to this environment. In contrast, the aliphatic side chains of the Val residues in Hb S create hydrophobic protrusions where none existed before. To the detriment of individuals who carry this trait, a hydrophobic pocket forms in the EF corner of each β-chain of Hb when it is in the deoxy state, and this pocket nicely accommodates the Val side chain of a neighboring Hb S molecule (Figure 15.34). This interaction leads to the aggregation of Hb S molecules into long, chainlike polymeric structures. The obvious consequence is that deoxyHb S is less soluble than deoxyHb A. The concentration of hemoglobin in red blood cells is high (about 150 mg/mL), so even in normal circumstances it is on the verge of crystallization. The formation of insoluble deoxyHb S fibers distorts the red cell into the elongated sickle shape characteristic of the disease.[2]

[2]In certain regions of Africa, the sickle-cell trait is found in 20% of the people. Why does such a deleterious heritable condition persist in the population? Individuals with this trait are less susceptible to the most virulent form of malaria. The geographic distribution of malaria and the sickle-cell trait are positively correlated.

SUMMARY

OWL Login to OWL to develop problem-solving skills and complete online homework assigned by your professor.

15.1 What Factors Influence Enzymatic Activity? The two prominent ways to regulate enzyme activity are (1) to increase or decrease the number of enzyme molecules or (2) to increase or decrease the intrinsic activity of each enzyme molecule. Changes in enzyme amounts are typically regulated via gene expression and protein degradation. Changes in the intrinsic activity of enzyme molecules are achieved principally by allosteric regulation or covalent modification.

15.2 What Are the General Features of Allosteric Regulation? Allosteric enzymes show a sigmoid response of velocity, v, to increasing [S], indicating that binding of S to the enzyme is cooperative. Allosteric enzymes often are susceptible to feedback inhibition. Allosteric enzymes may also respond to allosteric activation. Allosteric activators signal a need for the end product of the pathway in which the allosteric enzyme functions. As a general rule, allosteric enzymes are oligomeric, with each monomer possessing a substrate-binding site and an allosteric site where effectors bind. Interaction of one subunit of an allosteric enzyme with its substrate (or its effectors) is communicated to the other subunits of the enzyme through intersubunit interactions. These interactions can lead to conformational transitions that make it easier (or harder) for additional equivalents of ligand (S, A, or I) to bind to the enzyme.

15.3 Can Allosteric Regulation Be Explained by Conformational Changes in Proteins? Monod, Wyman, and Changeux postulated that the subunits of allosteric enzymes can exist in two conformational states (R and T), that all subunits in any enzyme molecule are in the same conformational state *(symmetry)*, that equilibrium strongly favors the T conformational state, and that S binds preferentially ("only") to the R state. Sigmoid binding curves result, provided that $[T_0] \gg [R_0]$ in the absence of S and that S binds "only" to R. Positive or negative effectors influence the relative T/R equilibrium by binding preferentially to T (negative effectors) or R (positive effectors), and the substrate saturation curve is shifted to the right (negative effectors) or left (positive effectors).

In an alternative allosteric model suggested by Koshland, Nemethy, and Filmer (the KNF model), S binding leads to conformational changes in the enzyme. The altered conformation of the enzyme may display higher affinity for the substrate (positive cooperativity) or lower affinity for the substrate or other ligand (negative cooperativity). Negative cooperativity is not possible within the MWC model. Reversible changes in the oligomeric state of a protein can also yield allosteric behavior. For example, a monomer–oligomer equilibrium for an allosteric protein, where only the oligomer binds S and [monomer] $\gg$ [oligomer], would show cooperative substrate binding.

15.4 What Kinds of Covalent Modification Regulate the Activity of Enzymes? Reversible phosphorylation is the most prominent form of covalent modification in cellular regulation. Phosphorylation is accomplished by protein kinases; phosphoprotein phosphatases act in the reverse direction to remove the phosphate group. Regulation must be imposed on these converter enzymes so that their enzyme targets adopt the metabolically appropriate state (active versus inactive). Thus, these converter enzymes are themselves the targets of allosteric regulation or covalent modification. Although several hundred chemical modifications of proteins have been described, only a few are used for reversible conversion of enzymes between active and inactive forms. Besides phosphorylation, these regulatory types include acetylation, adenylylation, uridylylation, ADP-ribosylation, methylation, and oxidation-reduction of protein disulfide bonds.

15.5 Is the Activity of Some Enzymes Controlled by Both Allosteric Regulation and Covalent Modification? Some enzymes are subject to both allosteric regulation and regulation by covalent modification. A prime example is glycogen phosphorylase. Glycogen phosphorylase exists in two forms, *a* and *b*, which differ only in whether or not Ser[14]-OH is phosphorylated *(a)* or not *(b)*. Glycogen phosphorylase *b* shows positive cooperativity in binding its substrate, phosphate. In addition, glycogen phosphorylase *b* is allosterically activated by the positive effector AMP. In contrast, ATP and glucose-6-P are negative effectors for glycogen phosphorylase *b*. Covalent modification of glycogen phosphorylase *b* by phosphorylase kinase converts it from a less active, allosterically regulated form to the more active *a* form that is less responsive to allosteric regulation. Glycogen phosphorylase is both activated and freed from allosteric control by covalent modification.

Special Focus: Is There an Example in Nature That Exemplifies the Relationship Between Quaternary Structure and the Emergence of Allosteric Properties? Hemoglobin and Myoglobin—Paradigms of Protein Structure and Function Myoglobin and hemoglobin have illuminated our understanding of protein structure and function. Myoglobin is monomeric, whereas hemoglobin has a quaternary structure. Myoglobin functions as an oxygen-storage protein in muscle; Hb is an

O_2-transport protein. When Mb binds O_2, its heme iron atom is drawn within the plane of the heme, slightly shifting the position of the F helix of the protein. Hemoglobin shows cooperative binding of O_2 and allosteric regulation by H^+, CO_2, and 2,3-bisphosphoglycerate. The allosteric properties of Hb can be traced to the movement of the F helix upon O_2 binding to Hb heme groups and the effects of F-helix movement on interactions between the protein's subunits that alter the intrinsic affinity of the other subunits for O_2. The allosteric transitions in Hb partially conform to the MWC model in that a concerted conformational change from a T-state,

low-affinity conformation to an R-state, high-affinity form takes place after 2 O_2 are bound (by the 2 Hb α-subunits). However, Hb also behaves somewhat according to the KNF model of allostery in that oxygen binding leads to sequential changes in the conformation and O_2 affinity of hemoglobin subunits. Sickle-cell anemia is a molecular disease traceable to a tendency for Hb S to polymerize as a consequence of having a βE6V amino acid substitution that creates a "sticky" hydrophobic patch on the Hb surface.

FOUNDATIONAL BIOCHEMISTRY Things You Should Know After Reading Chapter 15.

- The availability of substrates and cofactors usually determines the rate of an enzymatic reaction.

- As the product of the reaction accumulates, the apparent rate of product formation will slow down due to the increasing rate of the reverse reaction, which is directly dependent on [P].

- Genetic regulation of enzyme synthesis and decay determines the amount of enzyme present at any moment.

- Enzymes can be regulated allosterically through reversible binding of metabolic effectors at sites other than the active site.

- Enzyme activity can also be regulated through covalent modification, the reversible attachment of chemical groups to amino acid side chains in the enzyme.

- Enzymes susceptible to reversible covalent modification are called interconvertible enzymes.

- The enzymes that catalyze reversible covalent modification are called converter enzymes, and such converter enzymes are subject to metabolic regulation.

- Zymogens are inactive precursors from which active enzymes can be generated by proteolytic cleavage.

- Isozymes are enzymes with slightly different subunit composition. The different isozyme subunits have different kinetic properties.

- Allosteric enzymes catalyze committed steps in metabolic pathways.

- Many allosteric enzymes are susceptible to feedback inhibition by the end product of the metabolic pathway.

- The kinetic behavior of allosteric enzymes does not conform to the Michaelis–Menten model for enzyme kinetics.

- Substrate saturation curves for allosteric enzymes are sigmoid (S-shaped).

- Allosteric effectors interact with allosteric enzymes at binding sites distinct from the substrate-binding (active) site.

- The activity of an allosteric enzyme can be activated or inhibited, depending on the nature of the allosteric effector it binds.

- Allosteric enzymes are oligomers; each subunit in an allosteric enzyme has a substrate-binding site and an effector-binding site.

- The interaction of allosteric enzymes with their substrates or effectors alters the conformation of the subunits.

- Conformational changes in allosteric enzymes are the basis of their changing affinity for the various ligands.

- The Monod–Wyman–Changeux (MWC) model of allosteric behavior postulates that
  1. Allosteric enzymes are oligomeric.
  2. The allosteric enzyme can exist in (at least) two conformational states, called R and T.

3. All subunits in any molecule of enzyme are in the same conformation (either all R or all T).
4. The different conformations of the allosteric enzyme have different affinities for the various ligands.
5. L is the equilibrium constant for the R $\rightleftharpoons$ T equilibrium defined as $L = T_0/R_0$. If L is large, and S binds "only" to R, substrate binding is cooperative.
6. The concentration of ligand giving half-maximal saturation is defined as $K_{0.5}$.

- The Monod–Wyman–Changeux (MWC) model of allosteric behavior is based on linked equilibria between conformational states and ligand-binding properties of allosteric proteins.

- The Koshland–Nemethy–Filmer (KNF) model for allosteric regulation is based on ligand-induced conformational changes in allosteric proteins.

- Reversible changes in the oligomeric state of an allosteric protein can also give rise to allosteric behavior.

- Covalent modification through reversible phosphorylation is a prominent means of metabolic regulation.

- Protein kinases catalyze the ATP-dependent phosphorylation of target proteins.

- Protein kinases phosphorylate certain Ser, Thr, or Tyr-OH side chains.

- Target recognition: The specificity of protein kinases is determined by the sequence context in which the Ser, Thr, or Tyr residue is found.

- The activity of protein kinases is regulated by intrasteric control: A pseudosubstrate sequence occupies and blocks the protein kinase active site.

- Binding of metabolic regulators (or phosphorylation) leads to the dissociation of the pseudosubstrate sequence from the protein kinase active site and activation of its protein kinase function.

- Phosphorylation is not the only form of covalent modification that regulates protein function.

- Acetylation, adenylylation, uridylylation, ADP-ribosylation, methylation, and disulfide oxidation-reduction are other forms of covalent modification acting in metabolic regulation.

- Glycogen phosphorylase is a paradigm of allosteric regulation and covalent modification through reversible phosphorylation.

- AMP allosterically activates glycogen phosphorylase; ATP and glucose allosterically inhibit glycogen phosphorylase.

- Glycogen phosphorylase is also regulated through reversible phosphorylation of residue Ser^{14}.

- Glycogen phosphorylase b, the unphosphorylated form of the enzyme, is intrinsically less active and more susceptible to allosteric regulation than glycogen phosphorylase *a*, the phosphorylated form.

- Glycogen phosphorylase behavior reveals an important theme in metabolic regulation: covalent modification overrides allosteric regulation.

- Whether glycogen phosphorylase is covalently modified is determined by an enzyme cascade initiated by signal transduction in response to hormones.

- The signal transduction/enzyme cascade culminating in glycogen phosphorylase phosphorylation and activation involves a hormone receptor, G protein, adenylyl cyclase transmembrane signaling pathway and activation of a series of protein kinases.

- The comparative biochemistry of the oxygen-binding proteins myoglobin and hemoglobin reveals insights into allostery.

- Myoglobin (Mb) is an oxygen-storage protein; hemoglobin (Hb) is an oxygen-transport protein.

- Cooperative binding of O_2 by hemoglobin has important physiological significance in oxygen delivery to tissues.

- Oxygenation markedly alters the conformation of hemoglobin's quaternary structure.

- Deoxy- and oxy-Hb represent two conformational states of the protein.

- The allosteric behavior of Hb has attributes of both the MWC and KNF models.

- H^+ promotes the dissociation of O_2 from Hb, as does CO_2 and 2,3-bisphosphoglycerate (BPG).

- Fetal hemoglobin has a higher affinity for O_2 than adult Hb because it has a lower affinity for BPG.

- Sickle-cell anemia is a molecular disease due to a mutation in the β–globin gene that replaces the glutamate residue at position 6 with a valine residue.

- The valine residue at position 6 in the β–globin chains of HbS (sickle cell Hb) leads to the polymerization of the deoxy form of HbS into filaments that distort the shape of red blood cells.

PROBLEMS

Answers to all problems are at the end of this book. Detailed solutions are available in the *Student Solutions Manual, Study Guide, and Problems Book*.

1. **General Controls Over Enzyme Activity** List six general ways in which enzyme activity is controlled.

2. **Why Zymogens Are Advantageous** Why do you suppose proteolytic enzymes are often synthesized as inactive zymogens?

3. **Graphical Analysis of MWC Allosteric Enzyme Kinetics** (Integrates with Chapter 13.) Draw both Lineweaver–Burk plots and Hanes–Woolf plots for an MWC allosteric enzyme system, showing separate curves for the kinetic response in (a) the absence of any effectors, (b) the presence of allosteric activator A, and (c) the presence of allosteric inhibitor I.

4. **Graphical Analysis of Negative Cooperativity in KNF Allosteric Enzyme Kinetics** The KNF model for allosteric transitions includes the possibility of negative cooperativity. Draw Lineweaver–Burk and Hanes–Woolf plots for the case of negative cooperativity in substrate binding. (As a point of reference, include a line showing the classic Michaelis–Menten response of *v* to [S].)

5. **The Quantitative Advantage of Allosteric Behavior for O_2-Transporting Heme Proteins**

   The equation $\dfrac{Y}{(1 - Y)} = \left(\dfrac{pO_2}{P_{50}}\right)^n$

   allows the calculation of *Y* (the fractional saturation of hemoglobin with O_2), given P_{50} and *n* (see box on pages 500–501). Let $P_{50} = 26$ torr and $n = 2.8$. Calculate *Y* in the lungs, where $pO_2 = 100$ torr, and *Y* in the capillaries, where $pO_2 = 40$ torr. What is the efficiency of O_2 delivery under these conditions (expressed as $Y_{lungs} - Y_{capillaries}$)? Repeat the calculations, but for $n = 1$. Compare the values for $Y_{lungs} - Y_{capillaries}$ for $n = 2.8$ versus $Y_{lungs} - Y_{capillaries}$ for $n = 1$ to determine the effect of cooperative O_2 binding on oxygen delivery by hemoglobin.

6. **Pedict the Effect of Caffeine Consumption on Glycogen Phosphorylase Activity** The cAMP formed by adenylyl cyclase (Figure 15.19) does not persist because 5′-phosphodiesterase activity prevalent in cells hydrolyzes cAMP to give 5′-AMP. Caffeine inhibits 5′-phosphodiesterase activity. Describe the effects on glycogen phosphorylase activity that arise as a consequence of drinking lots of caffeinated coffee.

7. **Predict the O_2-Binding Properties of Stored Blood Whose Hemoglobin Is BPG-Depleted** If no precautions are taken, blood that has been stored for some time becomes depleted in 2,3-BPG. What happens if such blood is used in a transfusion?

8. **Estimate the Quantitative Effects of Allosteric Regulators on Glycogen Phosporylase Activity** Enzymes have evolved such that their K_m values (or $K_{0.5}$ values) for substrate(s) are roughly equal to the in vivo concentration(s) of the substrate(s). Assume that glycogen phosphorylase is assayed at $[P_i] \approx K_{0.5}$ in the absence and presence of AMP or ATP. Estimate from Figure 15.15 the relative glycogen phosphorylase activity when (a) neither AMP or ATP is present, (b) AMP is present, and (c) ATP is present. (*Hint:* Use a ruler to get relative values for the velocity *v* at the appropriate midpoints of the saturation curves.)

9. **Describe the Effects on cAMP and Glycogen Levels in Cells Exposed to Cholera Toxin** Cholera toxin is an enzyme that covalently modifies the G_α-subunit of G proteins. (Cholera toxin catalyzes the transfer of ADP-ribose from NAD^+ to an arginine residue in G_α, an ADP-ribosylation reaction.) Covalent modification of G_α inactivates its GTPase activity. Predict the consequences of cholera toxin on cellular cAMP and glycogen levels.

10. **One Way Negative Cooperativity Might Make Metabolic Sense** Allosteric enzymes that sit at branch points leading to several essential products sometimes display negative cooperativity for feedback inhibition (allosteric inhibition) by one of the products. What might be the advantage of negative cooperativity instead of positive cooperativity in feedback inhibitor binding by such enzymes?

11. **Protein Kinase Specificity Via Consensus Target Sequences and Intrasteric Control** Consult Table 15.2 and

    a. Suggest a consensus amino acid sequence within phosphorylase kinase that makes it a target of protein kinase A (the cAMP-dependent protein kinase).

    b. Suggest an effective amino acid sequence for a regulatory domain pseudosubstrate sequence that would exert intrasteric control on phosphorylase kinase by blocking its active site.

12. **Allosteric Regulation Versus Covalent Modification** What are the relative advantages (and disadvantages) of allosteric regulation versus covalent modification?

13. **Potential Treatment of Sickle-Cell Anemia by Drugs Targeted to HbS** You land a post as scientific investigator with a pharmaceutical company that would like to develop drugs to treat people with

sickle-cell anemia. They want ideas from you! What molecular properties of Hb S might you suggest as potential targets of drug therapy?

14. Nitric Oxide Is a Metabolic Regulator, But What Kind? Under appropriate conditions, nitric oxide (NO ·) combines with Cys 93β in hemoglobin and influences its interaction with O_2. Is this interaction an example of allosteric regulation or covalent modification?

15. Is Lactate an Allosteric Effector for Mb? Lactate, a metabolite produced under anaerobic conditions in muscle, lowers the affinity of myoglobin for O_2. This effect is beneficial, because O_2 dissociation from Mb under anaerobic conditions will provide the muscle with oxygen. Lactate binds to Mb at a site distinct from the O_2-binding site at the heme. In light of this observation, discuss whether myoglobin should be considered an allosteric protein.

16. Morpheeins—a Model for Allosteric Regulation Based on the Reversible Association-Dissociation of a Monomeric-Oligomeric Protein An allosteric model based on multiple oligomeric states of a protein has been proposed by E. K. Jaffe (2005. Morpheeins: A new structural paradigm for allosteric regulation. *Trends in Biochemical Sciences* 30:490–497). This model coins the term **morpheeins** to describe the different forms of a protein that can assume more than one conformation, where each distinct conformation assembles into an oligomeric structure with a fixed number of subunits. For example, conformation A of the protein monomer forms trimers, whereas conformation B of the monomer forms tetramers. If trimers and tetramers have different kinetic properties (K_m and k_{cat} values), as in low-activity trimers and high-activity tetramers, then the morpheein ensemble behaves like an allosterically regulated enzyme. Drawing on the traditional MWC model as an analogy, diagram a simple morpheein model in which wedge-shaped protein monomers assemble into trimers but the alternative conformation for the monomer (a square shape) forms tetramers. Further, the substrate, S, or allosteric regulator, A, binds "only" to the square conformation, and its binding prevents the square from adopting the wedge conformation. Describe how your diagram yields allosteric behavior.

17. Draw Substrate Saturation Curves for the Substrates of CTP Synthetase That Show Both Positive and Negative Cooperativity CTP synthetase catalyzes the synthesis of CTP from UTP:

$$UTP + ATP + glutamine \rightleftharpoons CTP + glutamate + ADP + P_i$$

The substrates UTP and ATP show positive cooperativity in their binding to the enzyme, which is an α_4-type homotetramer. However, the other substrate, glutamine, shows negative cooperativity. Draw substrate saturation curves of the form v versus $[S]/K_{0.5}$ for each of these three substrates that illustrate these effects.

18. Draw a Model for the Conformational States of Glyceraldehyde-3-P Dehydrogenase, a Tetrameric Enzyme Displaying Negative Cooperativity Glyceraldehyde-3-phosphate dehydrogenase catalyzes the synthesis of 1,3-bisphosphoglycerate:

$$Glyceraldehyde\text{-}3\text{-}P + P_i + NAD^+ \rightleftharpoons 1,3\text{-}BPG + NADH + H^+$$

The enzyme is a tetramer. NAD^+ binding shows negative cooperativity. Draw a diagram of possible conformational states for this tetrameric enzyme and its response to NAD^+ binding that illustrates negative cooperativity.

19. Proteomics as a Tool to Study Metabolic Regulation Proteomics studies have revealed that protein acetylation is an important mechanism for regulation of metabolic enzymes. Describe how proteomics might have led to this discovery.

Preparing for the MCAT® Exam

20. On the basis of the graphs shown in Figures 15.29 and 15.30 and the relationship between blood pH and respiration (Chapter 2), predict the effect of hyperventilation and hypoventilation on Hb:O_2 affinity.

21. Figure 15.18 traces the activation of glycogen phosphorylase from hormone to phosphorylation of the *b* form of glycogen phosphorylase to the *a* form. These effects are reversible when hormone disappears. Suggest reactions by which such reversibility is achieved.

ActiveModels Problems

22. Examine the ActiveModel for ferrochelatase, then describe the function of the ligands in the active site. How is enzyme activity in ferrochelatase regulated?

23. Using the ActiveModel for thrombin, give an example of regulation through covalent modification that occurs at exosite II.

24. Use the ActiveModel for glycogen phosphorylase to describe how this enzyme is regulated by allosteric and covalent controls.

FURTHER READING

General References

Fersht, A., 1999. *Structure and Mechanism in Protein Science: A Guide to Enzyme Catalysis and Protein Folding.* New York: W. H. Freeman.

Protein Kinases

Johnson, L., 2007. Protein kinases and their therapeutic exploitation. *Transactions* 35:7–11.

Kee, J.-M., Villany, B., Carpenter, L. R., and Muir, T. W., 2010. Development of stable phosphohistidine analogues. *Journal of the American Chemical Society* 132:14327–14329.

Manning, G., et al., 2002. The protein kinase complement of the human genome. *Science* 298:1912–1934. A catalog of the protein kinase genes identified within the human genome. About 2% of all eukaryotic genes encode protein kinases.

Taylor, S. S., and Kornev, A. P., 2011. Protein kinases: Evolution of dynamic regulatory proteins. *Trends in Biochemical Sciences* 36:65–77.

Regulation of Metabolic Enzymes by Acetylation

Wang, Q., Zhang, Y., Xu, W., Yang, C., et al., 2010. Acetylation of metabolic enzymes coordinates carbon source utilization and metabolic flux. *Science* 327:1004–1007.

Zhao, S., Xu, W., Jiang, W., et al., 2010. Regulation of cellular metabolism by protein lysine acetylation. *Science* 327:1000–1004.

Allosteric Regulation

Changeux, J.-P., and Edelstein, S. J., 2005. Allosteric mechanisms of signal transduction. *Science* 308:1424–1428.

Helmstaedt, K., Krappman, S., and Braus, G. H., 2001. Allosteric regulation of catalytic activity. *Escherichia coli* aspartate transcarbamoylase versus yeast chorismate mutase. *Microbiology and Molecular Biology Reviews* 65:404–421. The authors present evidence to show that the MWC two-state model is oversimplified, as Monod, Wyman, and Changeux themselves originally stipulated.

Koshland, D. E., Jr., and Hamadani, K., 2002. Proteomics and models for enzyme cooperativity. *Journal of Biological Chemistry* 277:46841–46844. An overview of both the MWC and the KNF models for allostery and a discussion of the relative merits of these models. The fact that the number of allosteric enzymes showing negative cooperativity is about the same as the number showing positive cooperativity is an important focus of this review.

Koshland, D. E., Jr., Nemethy, G., and Filmer, D., 1966. Comparison of experimental binding data and theoretical models in proteins containing subunits. *Biochemistry* 5:365–385. The KNF model.

Kuriyan, J., and Eisenberg, D., 2007. The origin of protein interactions and allostery in colocalization. *Nature* **450:**983–990.

Monod, J., Wyman, J., and Changeux, J.-P., 1965. On the nature of allosteric transitions: A plausible model. *Journal of Molecular Biology* **12:**88–118. The classic paper that provided the first theoretical analysis of allosteric regulation.

Schachman, H. K., 1990. Can a simple model account for the allosteric transition of aspartate transcarbamoylase? *Journal of Biological Chemistry* **263:**18583–18586. Tests of the postulates of the allosteric models through experiments on aspartate transcarbamoylase.

Swain, J. F., and Gierasch, L. M., 2006. The changing landscape of protein allostery. *Current Opinion in Structural Biology* **16:**102–108.

Glycogen Phosphorylase

Johnson, L. N., 2009. The regulation of protein phosphorylation. *Biochemical Society Transactions* **37:**627–641.

Johnson, L. N., and Barford, D., 1993. The effects of phosphorylation on the structure and function of proteins. *Annual Review of Biophysics and Biomolecular Structure* **22:**199–232.

Johnson, L. N., and Barford, D., 1994. Electrostatic effects in the control of glycogen phosphorylase by phosphorylation. *Protein Science* **3:**1726–1730.

Lin, K., et al., 1996. Comparison of the activation triggers in yeast and muscle glycogen phosphorylase. *Science* **273:**1539–1541.

Lin, K., et al., 1997. Distinct phosphorylation signals converge at the catalytic center in glycogen phosphorylases. *Structure* **5:**1511–1523.

Rath, V. L., et al., 1996. The evolution of an allosteric site in phosphorylase. *Structure* **4:**463–473.

Hemoglobin

Ackers, G. K., 1998. Deciphering the molecular code of hemoglobin allostery. *Advances in Protein Chemistry* **51:**185–253.

Dickerson, R. E., and Geis, I., 1983. *Hemoglobin: Structure, Function, Evolution and Pathology.* Menlo Park, CA: Benjamin/Cummings.

Henry, E. R., et al., 2002. A tertiary two-state allosteric model for hemoglobin. *Biophysical Chemistry* **98:**149–164.

Weiss, J. N., 1997. The Hill equation revisited: Uses and abuses. *The FASEB Journal* **11:**835–841.

Molecular Motors

© Bettmann/CORBIS

◀ Michelangelo's *David* epitomizes the musculature of the human form.

Buying bread from a man in Brussels
He was six foot four and full of muscles
I said "Do you speak-a my language?"
He just smiled and gave me a Vegemite
sandwich.

Colin Hay and Ron Strykert,
lyrics from *Down Under*

KEY QUESTIONS

16.1 What Is a Molecular Motor?

16.2 What Is the Molecular Mechanism of Muscle Contraction?

16.3 What Are the Molecular Motors That Orchestrate the Mechanochemistry of Microtubules?

16.4 How Do Molecular Motors Unwind DNA?

16.5 How Do Bacterial Flagella Use a Proton Gradient to Drive Rotation?

ESSENTIAL QUESTION

Movement is an intrinsic property associated with all living things. Within cells, molecules undergo coordinated and organized movements, and cells themselves may move across a surface. At the tissue level, **muscle contraction** allows higher organisms to carry out and control crucial internal functions, such as peristalsis in the gut and the beating of the heart. Muscle contraction also enables the organism to perform organized and sophisticated movements, such as walking, running, flying, and swimming.

How can biological macromolecules, carrying out conformational changes on the molecular level, achieve these feats of movement that span the microscopic and macroscopic worlds?

16.1 What Is a Molecular Motor?

Motor proteins, also known as **molecular motors,** use chemical energy (ATP) to orchestrate movements, transforming ATP energy into the mechanical energy of motion. In all cases, ATP hydrolysis is presumed to drive and control protein conformational changes that result in sliding or walking movements of one molecule relative to another. To carry out directed movements, molecular motors must be able to associate and dissociate reversibly with a polymeric protein array, a surface or substructure in the cell. ATP hydrolysis drives the process by which the motor protein ratchets along the protein array or surface. As fundamental and straightforward as all this sounds, elucidation of these basically simple processes has been extremely challenging for biochemists, involving the application of many sophisticated chemical and physical methods in many different laboratories. This chapter describes the structures and chemical functions of molecular motor proteins and some of the experiments by which we have come to understand them.

OWL Online homework and a Student Self Assessment for this chapter may be assigned in OWL.

Molecular motors may be **linear** or **rotating.** Linear motors crawl or creep along a polymer lattice, whereas rotating motors consist of a rotating element (the "rotor") and a stationary element (the "stator"), in a fashion much like a simple electrical motor. The linear motors we will discuss include **kinesins** and **dyneins** (which crawl along microtubules), **myosin** (which slides along actin filaments in muscle), and **DNA helicases** (which move along a DNA lattice, unwinding duplex DNA to form single-stranded DNA). Rotating motors include the flagellar motor complex, described in this chapter, and the ATP synthase, which will be described in Chapter 20.

16.2 What Is the Molecular Mechanism of Muscle Contraction?

Muscle Contraction Is Triggered by Ca²⁺ Release from Intracellular Stores

Muscle contraction is the result of interactions between myosin and **actin,** the two predominant muscle proteins. Thick filaments of myosin slide along thin filaments of actin to cause contraction.

The cells of skeletal muscle are long and multinucleate and are referred to as **muscle fibers.** Skeletal muscles in higher animals consist of 100-μm-diameter fibers, some as long as the muscle itself. Each of these muscle fibers contains hundreds of **myofibrils** (Figure 16.1), each of which spans the length of the fiber and is about 1 to 2 μm in diameter. Myofibrils are linear arrays of cylindrical **sarcomeres,** the basic structural units of muscle contraction. Each myofibril is surrounded by a specialized endoplasmic reticulum called the **sarcoplasmic reticulum (SR).** The SR contains high concentrations of Ca^{2+}, and the release of Ca^{2+} from the SR and its interactions within the sarcomeres trigger muscle contraction. The muscle fiber is surrounded by the **sarcolemma,** a specialized plasma membrane. Extensions of the sarcolemma, called **transverse tubules,** or

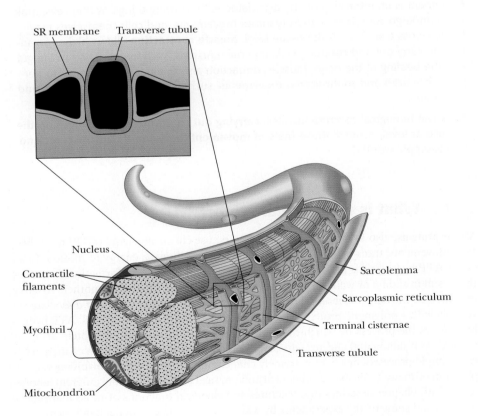

FIGURE 16.1 The structure of a skeletal muscle cell, showing the manner in which transverse tubules enable the sarcolemmal membrane to extend into the interior of the fiber. T-tubules and sarcoplasmic reticulum (SR) membranes are juxtaposed at structures termed triad junctions (inset).

Smooth Muscle Effectors Are Useful Drugs

Not all vertebrate muscle is skeletal muscle. Vertebrate organisms employ **smooth muscle** for long, slow, and involuntary contractions in various organs, including large blood vessels, intestinal walls, the gums of the mouth, and in the female, the uterus. Smooth muscle contraction is triggered by Ca^{2+}-activated phosphorylation of myosin by myosin light-chain kinase (MLCK). The action of epinephrine and related agents forms the basis of therapeutic control of smooth muscle contraction. Breathing disorders, including asthma and various allergies, can result from excessive contraction of bronchial smooth muscle tissue. Treatment with epinephrine, whether by tablets or aerosol inhalation, inhibits MLCK and relaxes bronchial muscle tissue. More specific **bronchodilators,** such as **albuterol**

(see accompanying figure), act more selectively on the lungs and avoid the undesirable side effects of epinephrine on the heart. Albuterol is also used to prevent premature labor in pregnant women because of its relaxing effect on uterine smooth muscle. Conversely, **oxytocin,** known also as **Pitocin,** stimulates contraction of uterine smooth muscle. This natural secretion of the pituitary gland is often administered to induce labor.

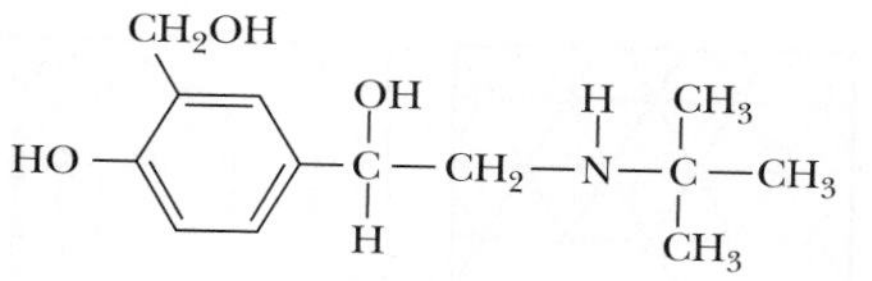

Albuterol

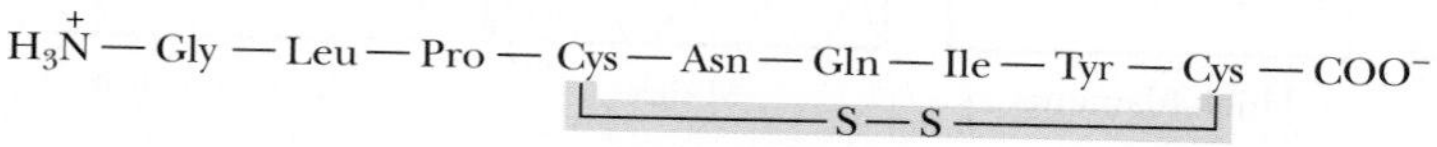

$H_3\overset{+}{N}$ — Gly — Leu — Pro — Cys — Asn — Gln — Ile — Tyr — Cys — COO^-

|_____ S — S _____|

Oxytocin (Pitocin)

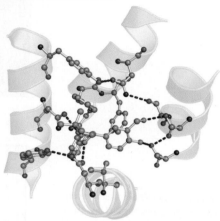

A model of albuterol bound to the β_2-adrenergic receptor. PDB file from Katritch, V., et al., 2009. Analysis of full and partial agonist binding to β-2-adrenergic receptor suggests a role of transmembrane helix V in agonist-specific conformational changes. *Journal of Molecular Recognition* **22:**307–318.

t-tubules, reach deep into the muscle fiber, enabling the sarcolemmal membrane to be in contact with each myofibril.

Skeletal muscle contractions are initiated by nerve stimuli that act directly on the muscle. Nerve impulses produce an electrochemical signal (see Chapter 32) called an action potential that spreads over the sarcolemmal membrane and into the fiber along the t-tubule network. This signal induces the release of Ca^{2+} ions from the SR. These Ca^{2+} ions bind to proteins within the muscle fibers and induce contraction.

The Molecular Structure of Skeletal Muscle Is Based on Actin and Myosin

Examination of myofibrils in the electron microscope reveals a banded or striated structure. The bands are traditionally identified by letters (Figure 16.2). Regions of high electron density, denoted **A bands,** alternate with regions of low electron density, the **I bands.** Small, dark **Z lines** lie in the middle of the I bands, marking the ends of the sarcomere. Each A band has a central region of slightly lower electron density called the **H zone,** which contains a central **M disc** (also called an **M line**). Electron micrographs of cross sections of each of these regions reveal molecular details. The H zone shows a regular, hexagonally arranged array of thick filaments of myosin (15 nm diameter), whereas the I band shows a regular, hexagonal array of thin filaments of actin, together with proteins known as **troponin** and **tropomyosin** (7 nm diameter). In the dark regions at the ends of each A band, the thin and thick filaments interdigitate, as shown in Figure 16.2. The thin and thick filaments are joined by **cross-bridges.** These cross-bridges are actually extensions of the myosin molecules, and muscle contraction is accomplished by the sliding of the cross-bridges along the thin filaments, a mechanical movement driven by the free energy of ATP hydrolysis.

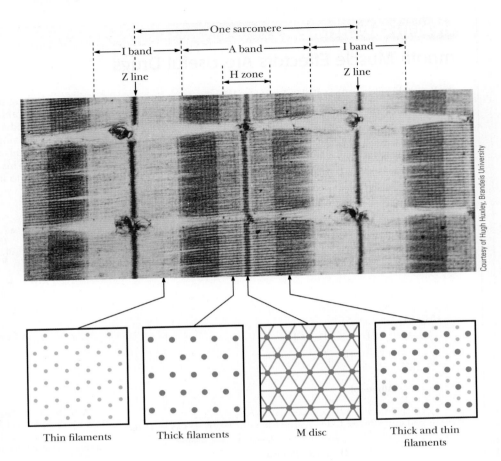

FIGURE 16.2 Electron micrograph of a skeletal muscle myofibril (in longitudinal section). The length of one sarcomere is indicated, as are the A and I bands, the H zone, the M disc, and the Z lines. Cross sections from the H zone show a hexagonal array of thick filaments, whereas the I band cross section shows a hexagonal array of thin filaments.

Thin filaments   Thick filaments   M disc   Thick and thin filaments

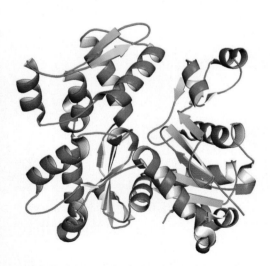

FIGURE 16.3 The three-dimensional structure of an actin monomer from skeletal muscle. This view shows the two domains (left and right) of actin (pdb id = 1J6Z).

The Composition and Structure of Thin Filaments Actin, the principal component of thin filaments, is found in substantial amounts in most eukaryotic cells. At low ionic strength, actin exists as a 42-kD globular protein, denoted **G-actin** (Figure 16.3). Under physiological conditions (higher ionic strength), G-actin polymerizes to form a fibrous form of actin, called **F-actin.** As shown in Figure 16.4, F-actin is a right-handed helical structure, with a helix pitch of about 72 nm per turn. The F-actin helix is the core of the thin filament, to which **tropomyosin** and the **troponin complex** also add. Tropomyosin winds around actin filaments and prevents myosin binding in resting muscle. When a nerve impulse arrives at the sarcolemmal membrane, Ca^{2+} ions released from the sarcoplasmic reticulum bind to the troponin complex, inducing a conformation change that allows myosin to bind to actin, initiating contraction. In nonmuscle cells, actin filaments are the highways across which a variety of cellular cargo is transported.

The Composition and Structure of Thick Filaments Myosin, the principal component of muscle thick filaments, is a large protein consisting of six polypeptides, with an aggregate molecular weight of approximately 540 kD. As shown in Figure 16.5, the six peptides include two 230-kD **heavy chains,** as well as two pairs of different 20-kD **light chains,** denoted **LC1** and **LC2.** The heavy chains consist of globular amino-terminal **myosin heads,** joined to long α-helical carboxy-terminal segments, the **tails.** These tails are intertwined to form a left-handed coiled coil approximately 2 nm in diameter and 130 to 150 nm long. Each of the heads in this dimeric structure is associated with an LC1 and an LC2. The myosin heads exhibit **ATPase activity,** and hydrolysis of ATP by the myosin heads drives muscle contraction. LC1 is also known as the **essential light chain,** and LC2 is designated the **regulatory light chain.** Both light chains are homologous to

(a)

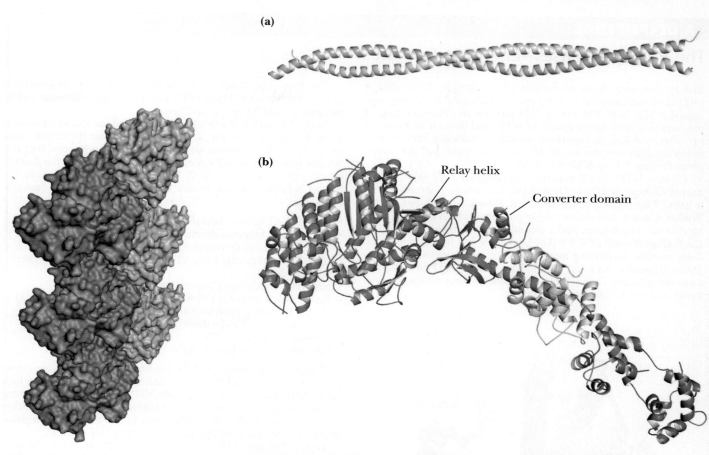

(b)

Relay helix

Converter domain

FIGURE 16.4 A molecular model of an actin polymer, based on the actin monomer structure shown in Figure 16.3 (pdb id = 1A5X). Actin consists of two strands of actin polymer wrapped around each other.

FIGURE 16.5 (a) A schematic drawing of a myosin hexamer, showing the two heavy chains and four light chains. The tail is a coiled coil of intertwined α-helices extending from the two globular heads. One of each of the myosin light-chain proteins, LC1 and LC2, is bound to each of the globular heads. (b) A ribbon diagram shows the structure of the myosin head. The head and neck domains of the heavy chain are red; the essential light chain is yellow and the regulatory light chain is blue (pdb id = 1B7T).

calmodulin and troponin C. Dissociation of LC1 from the myosin heads in vitro results in loss of the myosin ATPase activity.

The myosin head consists of a globular domain, where ATP is bound and hydrolyzed, and a long α-helical neck, to which the light chains are bound. The most prominent feature of the globular head is the actin-binding cleft between the so-called upper and lower domains. The N-terminal domain and the upper domain together form a seven-stranded β-sheet. The ATP-binding site is partially defined by three loops: switch 1, switch 2, and the P-loop. Conformation changes driven by ATP hydrolysis cause a rotation of the converter domain (Figure 16.5) and the long-neck helix—the fundamental event in contraction.

Repeating Structural Elements Are the Secret of Myosin's Coiled Coils A number of key features of the myosin sequence are responsible for the α-helical coiled coils formed by myosin tails. Several orders of repeating structure are found in all myosin tails, including 7-residue, 28-residue, and 196-residue repeating units. Large stretches of the tail domain are composed of 7-residue repeating segments. The first and fourth residues of these 7-residue units are generally small, hydrophobic amino acids, whereas the second, third, and sixth are likely to be charged residues. The consequence of this arrangement is shown in Figure 16.6. Seven residues form two turns of an α-helix, and in the coiled coil structure of the myosin tails, the first and fourth residues face the

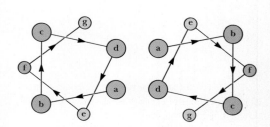

FIGURE 16.6 An axial view of the two-stranded, α-helical coiled coil of a myosin tail. Hydrophobic residues a and d of the 7-residue repeat sequence align to form a hydrophobic core. Residues b, c, and f face the outer surface of the coiled coil and are typically ionic.

The P-Loop NTPases—Energy to Run the Motors

Energy obtained from hydrolysis of nucleoside triphosphates (NTPs) is fundamental to a myriad of biological processes, and indeed to life itself. The **P-loop NTPases** are an abundant class of enzymes that harness the energy of NTPs and direct it to a vast array of critical cellular functions, including molecular motors. Genomic analysis reveals that 10–18% of prokaryotic and eukaryotic genes contain P-loop NTP domains!

P-Loop NTPases all contain a polypeptide loop with the sequence GxxxxGK(S/T) or GxxGK(S/T) between a β-strand and an α-helix. This sequence, the so-called **P-loop,** also known as the **Walker A motif,** coordinates the triphosphate chain of the NTP to be cleaved. The P-loop, and a second, more variable region called the **Walker B motif** ($\Psi\Psi\Psi\Psi$[D/E], where Ψ is a hydrophobic residue), provide the binding interactions with the nucleoside triphosphate (typically ATP or GTP). The P-loop NTPases share a common $\alpha\beta\alpha$ core, consisting of a parallel β-sheet sandwiched between two sets of α-helices (see figure).

The P-loop NTPases can be divided into two major structural groups: the **kinase-GTPase (KG)** group and the **additional strand catalytic E (ASCE)** group (see figure).

In the KG group, the Walker B strand and the strand connected to the P-loop are adjacent, so that the β-sheet has the strand order 54132. In the ASCE group, an additional strand is inserted between Walker B and the strand connected to the P-loop, and the β-sheet strand order changes to 51432. Within these two large groups, the P-loop NTPases can be divided into numerous distinct lineages, including:

- the ABC ATPases (Chapter 9)
- the myosins, kinesins, dyneins, helicases, and AAA+ ATPases, described later in this chapter
- the ATP-synthesizing F_1F_0-ATP synthase (a rotary motor; see Chapter 20)
- adenylate kinase (see Chapter 27)
- RecA, which mediates DNA strand exchange (see Chapter 28)
- the GTP-binding proteins known as G-proteins (see Figure 15.27, as well as EF-Tu discussed in Chapter 30 and Ras discussed in Chapter 32)
- the proteasome (Chapter 31)

The essential features of NTP binding and hydrolysis are represented in the F_1F_0-ATP synthase described in Chapter 20, which binds ADP and P_i and synthesizes ATP at the interface between two RecA-type domains. The ATP binding site consists of Walker A (P-loop) and Walker B motifs from one subunit and an **arginine finger** that extends from the adjacent subunit (see figure).

Here ADP and AlF_4^- bound in the NTP site mimic the pentavalent transition state for ATP hydrolysis, with Al in place of the γ-phosphorus of ATP and the F atoms in place of oxygens.

The binding and hydrolysis of NTP between adjacent RecA or AAA+ domains is a common behavior of many P-loop NTPases, as we will see with the helicases described later in this chapter.

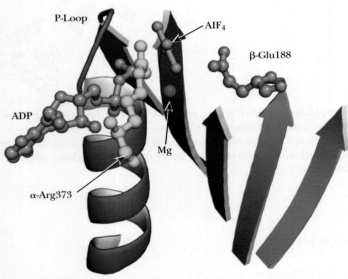

▲ P-loop NTPase (pdb id = 1H8E). The Walker A motif (P-loop) is shown in red, the Walker B motif in aqua-blue (with E^{189} in orange), ADP in salmon, and the arginine finger in yellow.

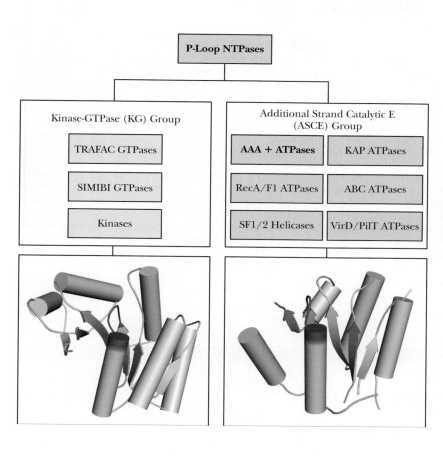

The major divisions and representative structures of the KG and ASCE structural groups of P-loop NTPases. The KG group structures are built on a 54132 β-sheet core. The ASCE group structures are built on a 51432 β-sheet core. The KG structure shown is the GTPase domain of the signal recognition particle from *Thermus aquaticus* (pdb id = 1FFH). The ASCE structure shown is the nucleotide-binding domain of the DNA clamp loader complex of *Saccharomyces cerevisiae* (pdb id = 1SXJ).

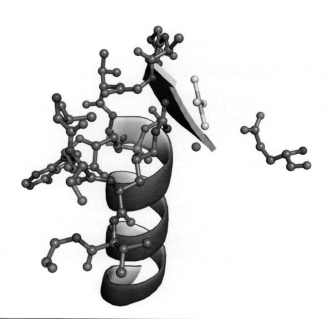

The P loop (ball and stick) coordinates the triphosphate of NTPs (pdb id = 1H8E). The view here is the same as on the previous page.

The Molecular Defect in Duchenne Muscular Dystrophy Involves an Actin-Anchoring Protein

Duchenne muscular dystrophy is a degenerative and fatal disorder of muscle affecting approximately 1 in 3500 boys. Victims of Duchenne dystrophy show early abnormalities in walking and running. By the age of 5, the victim cannot run and has difficulty standing, and by early adolescence, walking is difficult or impossible. The loss of muscle function progresses upward in the body, affecting next the arms and the diaphragm. Respiratory problems or infections usually result in death by the age of 30. Louis Kunkel and his coworkers identified the Duchenne muscular dystrophy gene in 1986. This gene produces a protein called **dystrophin,** which is highly homologous to α-actinin and spectrin. A defect in dystrophin is responsible for the muscle degeneration of Duchenne dystrophy.

Dystrophin is located on the cytoplasmic face of the muscle plasma membrane, linked to the plasma membrane via an integral membrane glycoprotein. Dystrophin has a high molecular mass (427 kD) but constitutes less than 0.01% of the total muscle pro-

tein. It folds into four principal domains (see accompanying figure, part a), including an N-terminal domain similar to the actin-binding domains of actinin (in muscle) and spectrin (in red blood cells), a long repeat domain, a β-dystroglycan-binding domain, and a C-terminal domain that is unique to dystrophin. The repeat domain consists of 24 triple-helical repeat units of approximately 109 residues each. "Spacer sequences" high in proline content, which do not align with the repeat consensus sequence, occur at the beginning and end of the repeat domain. Spacer segments are found between repeat elements 3 and 4 and 19 and 20. The high proline content of the spacers suggests that they may represent hinge domains. The spacer/hinge segments are sensitive to proteolytic enzymes, indicating that they may represent more exposed regions of the polypeptide. The N-terminal actin-binding domain appears capable of binding to 24 actin monomers in a polymerized actin filament.

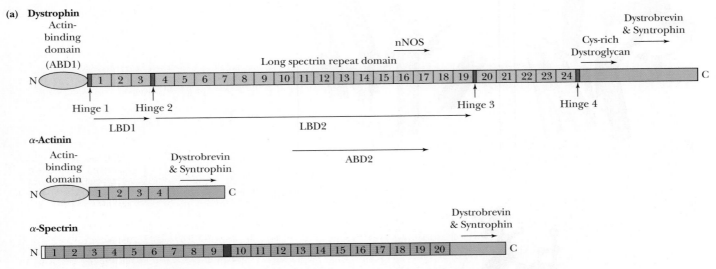

A comparison of the amino acid sequence of dystrophin, α-actinin, and spectrin. The potential hinge segments in the dystrophin structure are indicated. Dystrophin molecule and its identified partners. H1-4 denotes hinges 1–4; ABD1 and ABD2, actin-binding domains 1 and 2; LBD1 and LBD2, lipid-binding domains 1 and 2.

interior contact region of the coiled coil. Residues b, c, and f (2, 3, and 6) of the 7-residue repeat face the periphery, where charged residues can interact with the water solvent. At the 28 (4 × 7) residue and 196 (28 × 7) residue levels, specialized amino acid sequence patterns promote packing of large numbers of myosin tails in offset or staggered arrays (Figure 16.7).

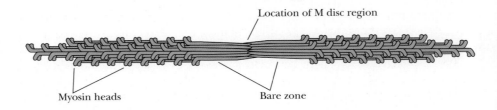

FIGURE 16.7 The packing of myosin molecules in a thick filament. Adjoining molecules are offset by approximately 14 nm, a distance corresponding to 98 residues of the coiled coil.

Dystrophin itself appears to be part of an elaborate protein–glycoprotein complex that bridges the inner cytoskeleton (actin filaments) and the extracellular matrix (via a matrix protein called laminin) (see figure). It is now clear that defects in one or more of the proteins in this complex are responsible for many of the other forms of muscular dystrophy. The glycoprotein complex is composed of two subcomplexes, the dystroglycan complex and the sarcoglycan complex. The dystroglycan complex consists of α-dystroglycan, an extracellular protein that binds to merosin, a laminin subunit and component of the extracellular matrix, and β-dystroglycan, a transmembrane protein that binds the C-terminal domain of dystrophin inside the cell (see figure). The sarcoglycan complex is composed of α-, β-, and γ-sarcoglycans, all of which are transmembrane glycoproteins. Alterations of the sarcoglycan proteins are linked to limb-girdle muscular dystrophy and autosomal recessive muscular dystrophy. Mutations in the gene for merosin, which binds to α-dystroglycan, are linked to severe congenital muscular dystrophy, yet another form of the disease.

(b)

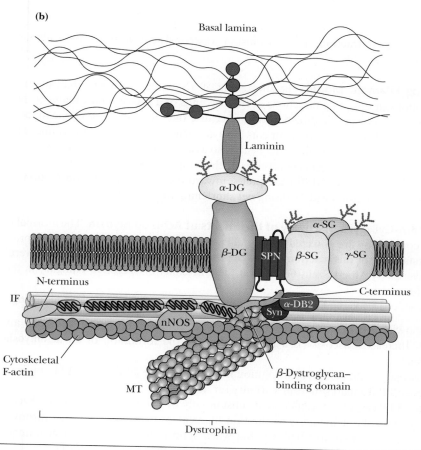

◁ A model for the actin–dystrophin–glycoprotein complex in skeletal muscle. Lipid-binding domains closely attach dystrophin to the sarcolemmal membrane by interactions with the membrane phospholipids. Actin-binding domains interact with the cytoskeleton via γ-actin (green), and the Cys-rich domain and C-terminal end interact with β-dystroglycan (βDG), α-dystrobrevin-2 (αDB2), and syntrophins (Syn). Abbreviations: αDG, α-dystroglycan; SGC, sarcoglycan; SPN, sarcospan complex; MT, microtubules; IF, intermediate filaments; nNOS, neuronal nitric oxide synthase. This dystrophin-anchored complex may function to stabilize the sarcolemmal membrane during contraction–relaxation cycles, link the contractile force generated in the cell (fiber) with the extracellular environment, or maintain local organization of key proteins in the membrane. The dystrophin-associated membrane proteins (dystroglycans, DGs, and sarcoglycans, SGs) range from 25 to 154 kD.

The Mechanism of Muscle Contraction Is Based on Sliding Filaments

When muscle fibers contract, the thick myosin filaments slide or walk along the thin actin filaments. The basic elements of the **sliding filament model** were first described in 1954 by two different research groups: Hugh Huxley and his colleague Jean Hanson, and Andrew Huxley and his colleague Ralph Niedergerke. Several key discoveries paved the way for this model. Electron microscopic studies of muscle revealed that sarcomeres decreased in length during contraction and that this decrease was due to decreases in the width of both the I band and the H zone (Figure 16.8). At the same time, the width of the A band (which is the length of the thick filaments) and the distance from the Z discs to the nearby H zone (that is, the length of the thin filaments) did not change. These observations made it clear that the lengths of both the thin and thick filaments were

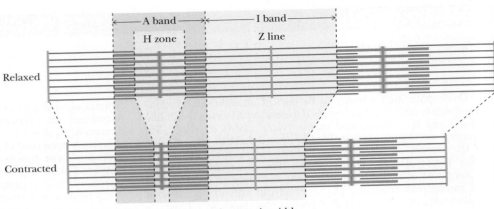

FIGURE 16.8 The sliding filament model of skeletal muscle contraction. The decrease in sarcomere length is due to decreases in the width of the I band and H zone, with no change in the width of the A band. These observations mean that the lengths of both the thick and thin filaments do not change during contraction. Rather, the thick and thin filaments slide along one another.

constant during contraction. This conclusion was consistent with a sliding filament model.

The Sliding Filament Model The shortening of a sarcomere (Figure 16.8) involves sliding motions in opposing directions at the two ends of a myosin thick filament. Net sliding motions in a specific direction occur because the thin and thick filaments both have **directional character.** Actin filaments always extend outward from the Z lines in a uniform manner. The myosin thick filaments also assemble in a directional manner. The polarity of myosin thick filaments reverses at the M disc, which means that actin filaments on either side of the M disc are pulled toward the M disc during contraction by the sliding of the myosin heads, causing net shortening of the sarcomere.

Albert Szent-Györgyi's Discovery of the Effects of Actin on Myosin The molecular events of contraction are powered by the ATPase activity of myosin. Much of our present understanding of this reaction and its dependence on actin can be traced to several key discoveries by Albert Szent-Györgyi at the University of Szeged in Hungary in the early 1940s.

In a series of elegant and insightful experiments, Szent-Györgyi showed the following:

- Solution viscosity is increased dramatically when solutions of myosin and actin are mixed. Increased viscosity is a manifestation of the formation of an **actomyosin complex.**
- The viscosity of an actomyosin solution is lowered by the addition of ATP, indicating that ATP decreases myosin's affinity for actin.
- Myosin ATPase activity is increased substantially by actin. (For this reason, Szent-Györgyi gave the name *actin* to the thin filament protein.) The ATPase turnover number of pure myosin is 0.05/sec, but when actin is added, the turnover number increases to about 10/sec, a number more like that of intact muscle fibers.

Szent-Györgyi's experiments demonstrated that ATP hydrolysis and the association and dissociation of actin and myosin are coupled. It is this coupling that enables ATP hydrolysis to power muscle contraction.

The Coupling Mechanism: ATP Hydrolysis Drives Conformation Changes in the Myosin Heads The only remaining piece of the puzzle is this: How does the close coupling of actin-myosin binding and ATP hydrolysis result in the shortening of myofibrils? Put another way, how are ATP hydrolysis and the sliding filament model related? The answer to this puzzle is shown in Figure 16.9. The free energy of ATP hydrolysis is translated into conformation changes in the myosin head, so dissociation of myosin and actin, hydrolysis of ATP, and rebinding of myosin and actin occur with stepwise movement of the myosin head along the actin filament. The conformation changes in the myosin head are driven by the binding and hydrolysis of ATP.

> **CRITICAL DEVELOPMENTS IN BIOCHEMISTRY**

Molecular "Tweezers" of Light Take the Measure of a Muscle Fiber's Force

The optical trapping experiment involves the attachment of myosin molecules to silica beads that are immobilized on a microscope coverslip (see accompanying figure). Actin filaments are then prepared such that a polystyrene bead is attached to each end of the filament. These beads can be "caught" and held in place in solution by a pair of "optical traps"—two high-intensity infrared laser beams, one focused on the polystyrene bead at one end of the actin filament and the other focused on the bead at the other end of the actin filament. The force acting on each bead in such a trap is proportional to the position of the bead in the "trap," so displacement and forces acting on the bead (and thus on the actin filament) can both be measured. When the "trapped" actin filament is brought close to the myosin-coated silica bead, one or a few myosin molecules may interact with sites on the actin and ATP-induced interactions of individual myosin molecules with the trapped actin filament can be measured and quantitated. Such optical trapping experiments have shown that *a single cycle or turnover of a single myosin molecule along an actin filament involves an average movement of 4 to 11 nm (40–110 Å) and generates an average force of 1.7 to 4 × 10⁻¹² newton (1.7–4 piconewtons [pN]).*

The magnitudes of the movements observed in the optical trapping experiments are consistent with the movements predicted by cryoelectron microscopy imaging data. Can the movements and forces detected in a single contraction cycle by optical trapping also be related to the energy available from hydrolysis of a single ATP molecule? The energy required for a contraction cycle is defined by the "work" accomplished by contraction, and work *(w)* is defined as force *(F)* times distance *(d)*:

$$w = F \cdot d$$

For a movement of 4 nm against a force of 1.7 pN, we have

$$w = (1.7 \text{ pN}) \cdot (4 \text{ nm}) = 0.68 \times 10^{-20} \text{ J}$$

For a movement of 11 nm against a force of 4 pN, the energy requirement is larger:

$$w = (4 \text{ pN}) \cdot (11 \text{ nm}) = 4.4 \times 10^{-20} \text{ J}$$

If the cellular free energy of hydrolysis of ATP is taken as −50 kJ/mol, the free energy available from the hydrolysis of a single ATP molecule is

$$\Delta G = (-50 \text{ kJ/mol})/(6.02 \times 10^{23} \text{ molecules/mol}) = 8.3 \times 10^{-20} \text{ J}$$

Thus, the free energy of hydrolysis of a single ATP molecule is sufficient to drive the observed movements against the forces that have been measured.

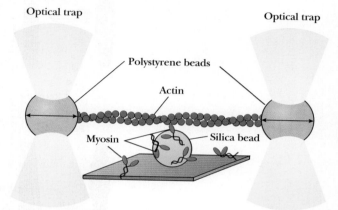

▲ Movements of single myosin molecules along an actin filament can be measured by means of an optical trap consisting of laser beams focused on polystyrene beads attached to the ends of actin molecules. *(Adapted from Finer, J. T., et al., 1994. Single myosin molecule mechanics: Piconewton forces and nanometre steps. Nature **368**:113–119. See also Block, S. M., 1995. Macromolecular physiology. Nature **378**:132–133.)*

As shown in the cycle in Figure 16.9, the myosin heads—with the hydrolysis products ADP and P_i bound—are mainly dissociated from the actin filaments in resting muscle. When the signal to contract is presented (see following discussion), the myosin heads move out from the thick filaments to bind to actin on the thin filaments (step 1). Actin binding closes the cleft in the myosin head, which causes a twist in the large β-sheet. The twist causes switch 1 and the P-loop to "open," both of them moving away from the bound ADP and P_i (Figure 16.10). The β-sheet twist also straightens the kink in the relay helix. These conformation changes result in the **top-of-power stroke** state shown in Figure 16.9, but this state is transient. The power stroke occurs almost immediately, accompanied by dissociation of P_i and then ADP. The power stroke consists of a 60° rotation of the converter domain and the long α-helical neck into the down position relative to the myosin head (Figure 16.9)—a movement that results in a 100-Å movement of the end of the neck in the direction of contraction.

The end of the power stroke is termed the **rigor-like** state because without access to additional ATP, the actin–myosin pair would be locked together, unable to dissociate. However, binding of another ATP causes dissociation of myosin from actin, as first noticed by Szent-Györgyi. This dissociation occurs with "opening" of switch 2, in which the switch 2 segment (the lower part of strand 5 and a short following loop) moves out of the plane of the seven-stranded β-sheet (Figure 16.10).

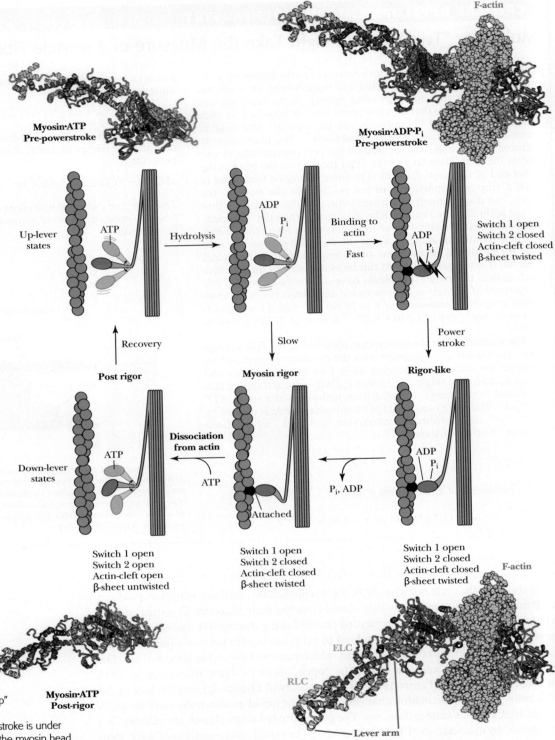

Myosin·ATP Pre-powerstroke

Myosin·ADP·P$_i$ Pre-powerstroke

F-actin

Up-lever states

ATP

Hydrolysis

ADP

P$_i$

Binding to actin

Fast

ADP

P$_i$

Switch 1 open
Switch 2 closed
Actin-cleft closed
β-sheet twisted

Recovery

Slow

Power stroke

Post rigor

Myosin rigor

Rigor-like

Down-lever states

ATP

Dissociation from actin

ATP

Attached

P$_i$, ADP

ADP

P$_i$

Switch 1 open
Switch 2 open
Actin-cleft open
β-sheet untwisted

Switch 1 open
Switch 2 closed
Actin-cleft closed
β-sheet twisted

Switch 1 open
Switch 2 closed
Actin-cleft closed
β-sheet twisted

F-actin

ELC

RLC

Lever arm

Myosin·ATP Post-rigor

The conversion from the detached "up" state of the myosin head to the actin-bound state that precedes the powerstroke is under kinetic control! That is, the binding of the myosin head to actin while in the "up" position is not the thermodynamically favored process, but it occurs faster than the conversion of the myosin head to the "down state." Otherwise, binding to actin would form the rigor state (the center path of Figure 16.9), which would occur without a powerstroke and thus waste a cycle of ATP hydrolysis. (Malnasi-Csizmadia, A., and Kovacs, M., 2010. Emerging complex pathways of the actomyosin powerstroke. *Trends in Biochemical Sciences* **35:**684–690; and Sweeney, H. L., and Houdusse, A., 2010. Structural and functional insights into the myosin motor mechanism. *Annual Review of Biophysics* **39:**539–557.)

FIGURE 16.9 The mechanism of skeletal muscle contraction. ATP binding to the myosin head dissociates the strongly bound actomyosin "rigor" complex. During the recovery step, which occurs with ATP bound, the myosin head can move freely up and down. ATP hydrolysis occurs only in the "up" state. At this point, there are two possibilities. If the myosin head returns to a "down" state, the ATP hydrolysis cycle is completed without any contraction work being done. However, if the myosin head binds to the actin filament while still in an "up" position, the free energy of ATP hydrolysis drives a conformational change in the myosin head, and an effective powerstroke can occur, resulting in net movement of the myosin head along the actin filament. Experiments indicate that this latter possibility is favored kinetically but not thermodynamically. (Adapted from Malnasi-Csizmadia, A., and Kovacs, M., 2010. Emerging complex pathways of the actomyosin powerstroke. *Trends in Biochemical Sciences* **35:**684–690; and Sweeney, H. L., and Houdusse, A., 2010. Structural and functional insights into the myosin motor mechanism. *Annual Review of Biophysics* **39:**539–557.)

▶ **FIGURE 16.10** Details of the switch domains, the relay helix and the converter domain are shown for **(a)** the post-rigor state and **(b)** the pre-power stroke state of skeletal muscle myosin. ATP hydrolysis drives these conformation changes. Actin binding induces a twist in the large β-sheet of the myosin head, causing the switch 1 and P-loop segments to "open." (Adapted from Geeves, M., and Holmes, K., 2005. The molecular mechanism of muscle contraction. *Advances in Protein Chemistry* **71**:161–193. Figure provided by Kenneth Holmes, Max Plank Institute for Medical Research, Heidelberg.)

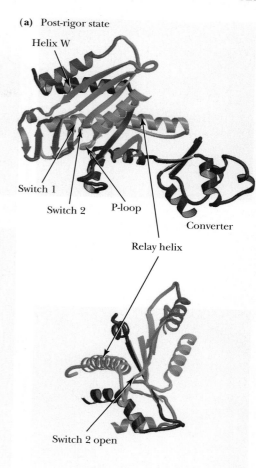

(a) Post-rigor state

Helix W

Switch 1

Switch 2   P-loop

Converter

Relay helix

Switch 2 open

ATP forms a strong interaction with switch 1 in the myosin active site, inducing switch 1 to close (moving toward the β- and γ-phosphates of bound ATP), presumably also causing switch 2 to close (with the lower part of strand 5 moving back into the plane of the seven-stranded β-sheet). Switch 2 closing induces formation of the kink in the relay helix, causing a 60° rotation of the converter domain and neck helix into the up position relative to the myosin head. This movement completes the cycle of contraction as it "primes" the motor, preparing it for the next power stroke. This cycle is repeated at rates up to 5/sec in a typical skeletal muscle contraction. The conformational changes occurring in this cycle are the secret of the energy coupling that allows ATP binding and hydrolysis to drive muscle contraction.

16.3 | What Are the Molecular Motors That Orchestrate the Mechanochemistry of Microtubules?

Filaments of the Cytoskeleton Are Highways That Move Cellular Cargo

Most eukaryotic cells contain elaborate networks of protein fibers collectively termed the **cytoskeleton.** The cytoskeleton is a dynamic, three-dimensional structure (Figure 16.11) that fills the cytoplasm and functions to:

- Establish cell shape
- Provide mechanical strength
- Facilitate cell movement
- Support intracellular transport of organelles and other cellular cargo
- Guide chromosome separation during mitosis and meiosis

Three types of fibers comprise the cytoskeleton: microfilaments of actin (with a diameter of 3 to 6 nm), microtubules made from tubulin (20 to 25 nm diameter), and intermediate filaments formed from a variety of proteins (about 10 nm diameter). All of these have dynamic properties that facilitate the movement of organelles and other molecular cargo through the cell.

Intermediate filaments provide a supporting network that allows cells to resist mechanical stress and deformation. Intermediate filaments are dynamic, and short filament segments (termed "squiggles" by Robert Goldman and his co-workers) can be transported across cells by motor proteins riding on microfilaments and microtubules. Polymeric actin microfilaments serve at least two functions in cells: They form networks just beneath the plasma membrane that link transmembrane proteins to cytoplasmic proteins, and they provide transcellular tracks on which organelles can be transported by myosin-like proteins. Microtubules are the best understood components of the cytoskeleton, and they are the focus of this section.

Microtubules are hollow, cylindrical structures, approximately 30 nm in diameter, formed from **tubulin,** a dimeric protein composed of two similar 55-kD subunits known as *α-tubulin* and *β-tubulin.* Eva Nogales, Sharon Wolf, and Kenneth Downing have determined the structure of the bovine tubulin αβ-dimer to 3.7 Å resolution (Figure 16.12a). Tubulin dimers polymerize as shown in Figure 16.12b to form microtubules, which are essentially helical structures, with 13 tubulin monomer "residues" per turn. Microtubules grown in vitro are dynamic structures that are constantly being assembled

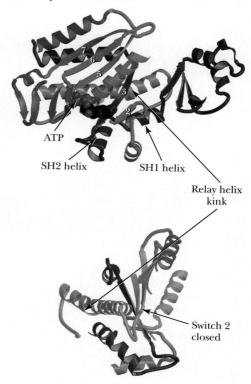

(b) Pre-power stroke state

ATP

SH2 helix   SH1 helix

Relay helix kink

Switch 2 closed

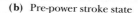

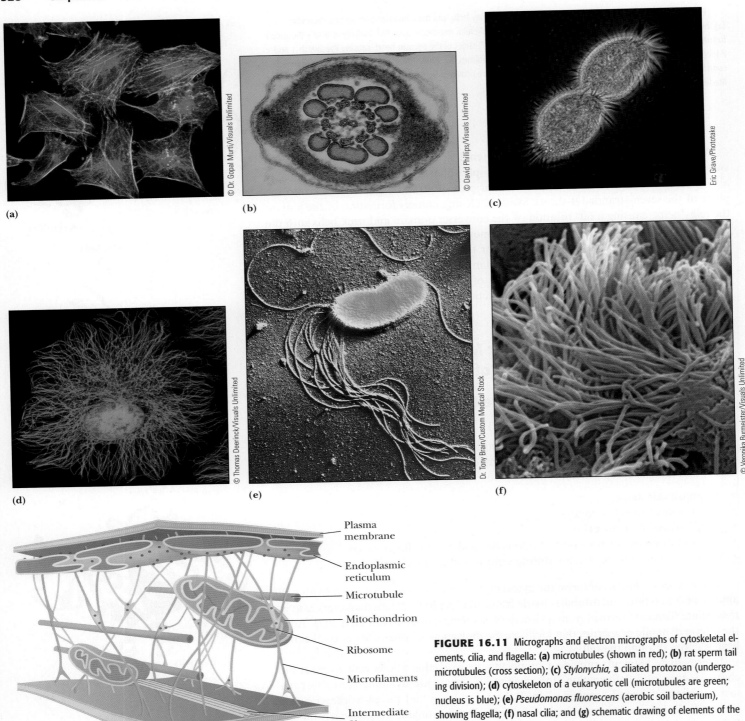

FIGURE 16.11 Micrographs and electron micrographs of cytoskeletal elements, cilia, and flagella: **(a)** microtubules (shown in red); **(b)** rat sperm tail microtubules (cross section); **(c)** *Stylonychia,* a ciliated protozoan (undergoing division); **(d)** cytoskeleton of a eukaryotic cell (microtubules are green; nucleus is blue); **(e)** *Pseudomonas fluorescens* (aerobic soil bacterium), showing flagella; **(f)** nasal cilia; and **(g)** schematic drawing of elements of the cytoskeleton.

and disassembled. Because all tubulin dimers in a microtubule are oriented similarly, microtubules are polar structures. The end of the microtubule at which growth occurs is the **plus end,** and the other is the **minus end.** Microtubules in vitro carry out a GTP-dependent process called **treadmilling,** in which tubulin dimers are added to the plus end at about the same rate at which dimers are removed from the minus end (Figure 16.13).

Although composed only of 55-kD tubulin subunits, microtubules can grow sufficiently large to span a eukaryotic cell or to form large structures such as cilia and flagella. Inside cells, networks of microtubules play many functions, including formation

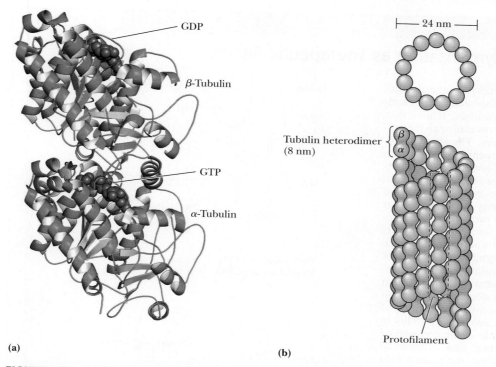

(a)

(b)

FIGURE 16.12 **(a)** The structure of the tubulin $\alpha\beta$-heterodimer (pdb id = 1JFF). **(b)** Microtubules may be viewed as consisting of 13 parallel, staggered protofilaments of alternating α-tubulin and β-tubulin subunits. The sequences of the α- and β-subunits of tubulin are homologous, and the $\alpha\beta$-tubulin dimers are quite stable if Ca^{2+} is present. The dimer is dissociated only by strong denaturing agents.

of the mitotic spindle that segregates chromosomes during cell division, the movement of organelles and various vesicular structures through the cell, and the variation and maintenance of cell shape. In most cells, microtubules are oriented with their minus ends toward the centrosome and their plus ends toward the cell periphery. This consistent orientation is important for mechanisms of intracellular transport.

Three Classes of Motor Proteins Move Intracellular Cargo

Three principal classes of motor proteins move organelles and other cellular cargo on cytoskeletal filament highways in both eukaryotic and prokaryotic cells. In addition to the myosins, most cells contain kinesins and dyneins. Humans possess 40 genes for myosins, 45 for kinesins, and at least 14 for dyneins (Table 16.1). The large number of genes

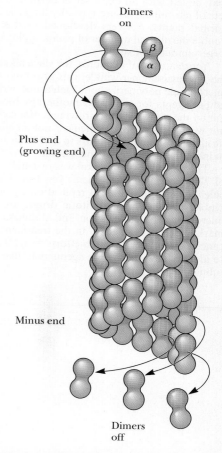

FIGURE 16.13 A model of the GTP-dependent treadmilling process. Both α- and β-tubulin possess two different binding sites for GTP. The polymerization of tubulin to form microtubules is driven by GTP hydrolysis in a process that is only beginning to be understood in detail.

TABLE 16.1	Genes for Molecular Motors		
	Number of Genes		
Genome	**Kinesins**	**Dyneins**	**Myosins**
Giardia lamblia (protozoan parasite)	25	10	0
Saccharomyces cerevisiae (yeast)	25	1	5
Drosophila melanogaster (fruit fly)	25	13	13
Caenorhabditis elegans (roundworm)	20	2	17
Arabidopsis thaliana (flowering plant)	61	0	17
Homo sapiens (human)	45	14–16	40

Effectors of Microtubule Polymerization as Therapeutic Agents

Microtubules in eukaryotic cells are important for the maintenance and modulation of cell shape and the disposition of intracellular elements during the growth cycle and mitosis. It may thus come as no surprise that the inhibition of microtubule polymerization can block many normal cellular processes. The alkaloid **colchicine** (see accompanying figure), a constituent of the swollen, underground stems of the autumn crocus *(Colchicum autumnale)* and meadow saffron, inhibits the polymerization of tubulin into microtubules. This effect blocks the mitotic cycle of plants and animals. Colchicine also inhibits cell motility and intracellular transport of vesicles and organelles (which in turn blocks secretory processes of cells). Colchicine has been used for hundreds of years to alleviate some of the acute pain of gout and rheumatism. In gout, white cell lysosomes surround and engulf small crystals of uric acid. The subsequent rupture of the lysosomes and the attendant lysis of the white cells initiate an inflammatory response that causes intense pain. The mechanism of pain alleviation by colchicine is not known for certain, but appears to involve inhibition of white cell movement in tissues. Interestingly, colchicine's ability to inhibit mitosis has given it an important role in the commercial development of new varieties of agricultural and ornamental plants. When mitosis is blocked by colchicine, the treated cells may be left with an extra set of chromosomes. Plants with extra sets of chromosomes are typically larger and more vigorous than normal plants. Flowers developed in this way may grow with double the normal number of petals, and fruits may produce much larger amounts of sugar.

Another class of alkaloids, the **vinca alkaloids** from *Vinca rosea,* the Madagascar periwinkle, can also bind to tubulin and inhibit microtubule polymerization. **Vinblastine** and **vincristine** are used as potent agents for cancer chemotherapy because of their ability to inhibit the growth of fast-growing tumor cells. For reasons that are not well understood, colchicine is not an effective chemotherapeutic agent, although it appears to act similarly to the vinca alkaloids in inhibiting tubulin polymerization.

The antitumor drug **taxol** was originally isolated from the bark of *Taxus brevifolia,* the Pacific yew tree. Like vinblastine and colchicine, taxol inhibits cell replication by acting on microtubules. Unlike these other antimitotic drugs, however, taxol stimulates microtubule polymerization and stabilizes microtubules. The remarkable success of taxol in the treatment of breast and ovarian cancers stimulated research efforts to synthesize taxol directly and to identify new antimitotic agents that, like taxol, stimulate microtubule polymerization.

Vinblastine: R = CH$_3$
Vincristine: R = CHO

Colchicine

Taxol

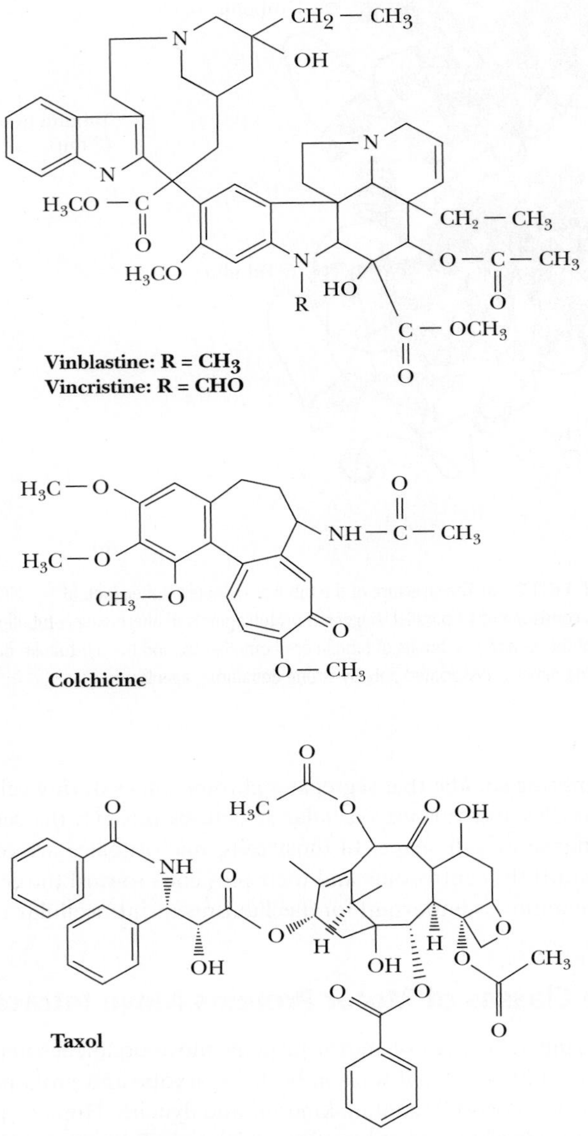

The structures of vinblastine, vincristine, colchicine, and taxol.

in each class reflects specialized structures required for a variety of functions. This diversity notwithstanding, these three classes of motor proteins share remarkable similarities of structure and function, as we shall see.

Kinesin 1, also called conventional kinesin, is a tetramer consisting of a dimer of heavy chains (110 kD) associated with two light chains (65 kD). The heavy chains contain an N-terminal motor domain, a long coiled-coil stalk with a central hinge, and a globular C-terminal tail domain where the light chains bind (Figure 16.14). The motor domain binds to tubulin in microtubules, and the globular tail domain associates with the intended cellular cargo, for example, an organelle, an mRNA molecule, or an intermediate filament. In different kinesin families, the motor domain is located in different places in the sequence, depending on the function of the specific family.

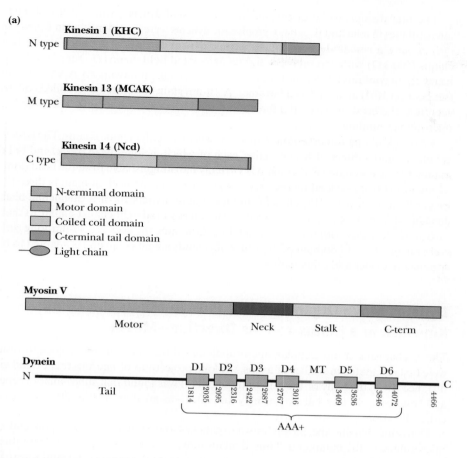

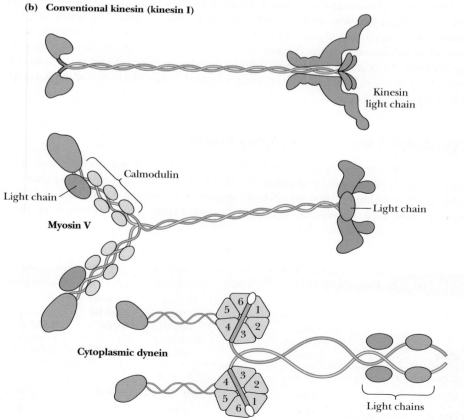

FIGURE 16.14 (a) Domain structure of kinesins, myosin V, and cytoplasmic dynein. **(b)** Molecular models of kinesin 1, myosin V, and cytoplasmic dynein. (Adapted from Vale, R., 2003. The molecular motor toolbox for intracellular transport. *Cell* **112:**467–480.)

The first dyneins to be discovered were **axonemal dyneins,** which cause sliding of microtubules in cilia and flagella. **Cytoplasmic dyneins** were first identified in *Caenorhabditis elegans*, a nematode worm. Cytoplasmic dynein consists of a dimer of two heavy chains (500 kD) with several other tightly associated light chains (Figure 16.14). Each heavy chain contains a large motor domain (380 kD) encompassing six AAA+ domains (see Section 16.4) arranged as a hexamer. A 10-nm stalk composed of a coiled coil projects from the head, between the fourth and the fifth AAA+ domains. The stalk is the microtubule-binding domain.

Myosin V is a multimeric protein that consists of 16 polypeptide chains. The structure is built around a dimer of heavy chains, each of which includes head, neck, and tail domains. The heavy chain head domain is virtually indistinguishable from the head domain of myosin II from skeletal muscle (see Figure 16.5), but the neck domain is three times longer than the myosin II neck helix and it contains six repeats of a **calmodulin-binding domain.** Myosin V is normally associated with an essential light chain (similar to that of myosin II), together with several calmodulins. Adjoining the neck is a 30-nm-long coiled-coil domain. The tail domain of myosin V also binds a light chain and is adapted to bind specific organelles and other cargoes.

Dyneins Move Organelles in a Plus-to-Minus Direction; Kinesins, in a Minus-to-Plus Direction—Mostly

The mechanisms of intracellular, microtubule-based transport of organelles and vesicles were first elucidated in studies of **axons,** the long projections of neurons that extend great distances from the body of the cell. In these cells, it was found that subcellular organelles and vesicles could travel at surprisingly fast rates—as great as 1000 to 2000 nm/sec—in either direction.

Cytosolic dyneins specifically move organelles and vesicles from the plus end of a microtubule to the minus end. Thus, dyneins move vesicles and organelles from the cell periphery toward the centrosome (or, in an axon, from the synaptic termini toward the cell body). Most **kinesins,** on the other hand, assist the movement of organelles and vesicles from the minus end to the plus end of microtubules, resulting in outward movement of organelles and vesicles (Figure 16.15). Certain unconventional kinesins move in the opposite direction, transporting cargo in the plus-to-minus direction on microtubules. These kinesins have their motor domain located at the C-terminus of the polypeptide (see Figure 16.14).

Cytoskeletal Motors Are Highly Processive

The motors that move organelles and other cellular cargo on microtubules and actin filaments must be processive, meaning that they must make many steps along their cellular journey without letting go of their filamentous highway. Dyneins, nonskeletal myosins, and most kinesins are processive motors (Table 16.2). Motors in all these

TABLE 16.2	**Processivity of Motor Proteins**				
Motor	Rate of Movement (nm/sec)	Step Size (nm)	Distance Traveled Before Dissociating (nm)	Processivity (average number of steps before dissociating)	% Chance of Dissociating in One Step
Kinesin 1	600	8	800–1200	100–120	~1
Myosin V	1000	36	700–2100	20–60	~2–5
Dynein (cytoplasmic)	600	24–32	1000	30–40	~2–3

(a)

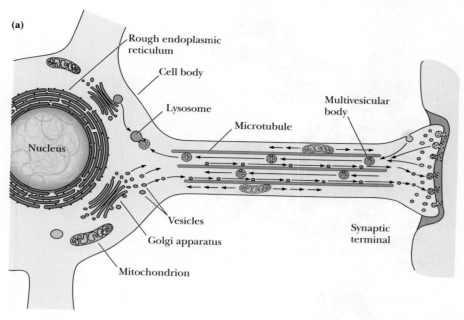

(b)

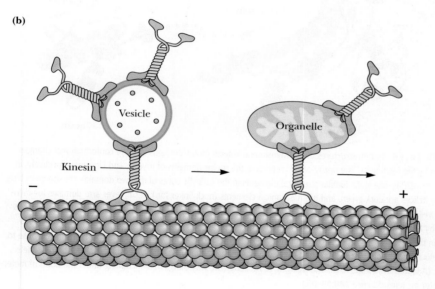

FIGURE 16.15 **(a)** Rapid axonal transport along microtubules permits the exchange of material between the synaptic terminal and the body of the nerve cell. **(b)** Vesicles, multivesicular bodies, and mitochondria are carried through the axon by this mechanism. (Adapted from a drawing by Ronald Vale.)

classes can carry cargo over roughly similar distances (700 to 2100 nm) before dissociating.[1] The step size of kinesin 1 is smaller than those of myosin V and cytoplasmic dynein; thus, its overall processivity (the average number of steps made before dissociating) is necessarily higher. Moving at rates of 600 to 1000 nm/sec, these motors can carry their cargoes for a second or more before dissociating from their filaments.

ATP Binding and Hydrolysis Drive Hand-over-Hand Movement of Kinesin

Kinesin movement along microtubules is driven by the cycle of ATP binding and hydrolysis. The molecular details are similar in some ways to those of the skeletal muscle myosin–actin motor but are quite different in other ways. Kinesins, like skeletal muscle myosin, contain switch 1 and switch 2 domains that open and close in response to ATP

[1]Compare these distances with the dimensions of typical cells in Table 1.2.

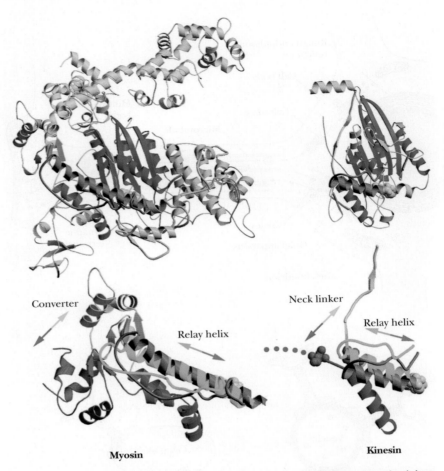

FIGURE 16.16 Ribbon structures of the myosin and kinesin motor domains and the conformational changes triggered by the relay helix. The upper panels represent the motor domains of myosin and kinesin, respectively, in the ATP- or ADP-P_i–like state. Similar structural elements in the catalytic cores of the two domains are shown in blue, the relay helices are dark green, and the mechanical elements (neck linker for kinesin, lever arm domains for myosin) are yellow. The nucleotide is shown as a white space-filling model. The similarity of the conformation changes caused by the relay helix in going from the ATP/ADP-P_i–bound state to the ADP-bound or nucleotide-free state is shown in the lower panels. In both cases, the mechanical elements of the protein shift their positions in response to relay helix motion. Note that the direction of mechanical element motion is nearly perpendicular to the relay helix motion. (Adapted from Vale, R. D., and Milligan, R. A., 2000. The way things move: Looking under the hood of molecular motor proteins. *Science* **288**:88–95.)

binding and hydrolysis. Together these switches act as a "γ-phosphate sensor," which can detect the presence or absence of the γ-phosphate of an adenine nucleotide in the active site. The switch movements between the ATP-bound and the ADP-bound states thus induce conformation changes that are propagated through a relay helix to a neck linker that rotates, in ways similar to skeletal muscle myosin (Figure 16.16). *Thus, just as in skeletal myosin, small movements of the γ-phosphate sensor at the ATP site are translated into piston-like movement of a relay helix and then into rotations of the neck linker that result in motor movement.*

Here the kinesin and myosin models diverge, however, because the dimer of kinesin heavy chains translates these ATP-induced conformation changes into a hand-over-hand movement of its motor domain heads along the microtubule filament. Ronald Vale and Ronald Milligan have likened this movement of kinesin heads along a microtubule to a judo expert throwing an opponent with a forward swing of the arm.

A model of kinesin motor movement is shown in Figure 16.17. Kinesin heads in solution (that is, not attached to a microtubule) contain tightly bound ADP. Binding of

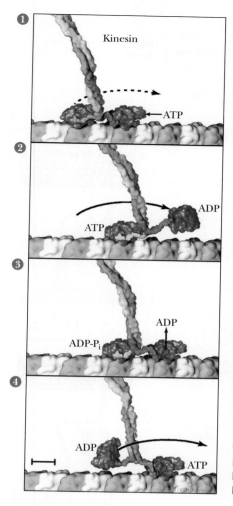

FIGURE 16.17 A model for the motility cycle of kinesin. The two heads of the kinesin dimer work together to move processively along a microtubule. **Frame 1:** Each kinesin head is bound to the tubulin surface. The heads are connected to the coiled coil by "neck linker" segments (one shown in orange/yellow, and the other in red). **Frame 2:** Conformation changes in the neck linkers flip the trailing head by 160°, over and beyond the leading head and toward the next tubulin binding site. **Frame 3:** The new leading head binds to a new site on the tubulin surface (with ADP dissociation), completing an 80 Å movement of the coiled coil and the kinesin's cargo. During this time, the trailing head hydrolyzes ATP to ADP and P_i. **Frame 4:** ATP binds to the leading head, and P_i dissociates from the trailing head, completing the cycle. (Adapted from Vale, R., and Milligan, R., 2000. The way things move: Looking under the hood of molecular motor proteins. *Science* **288**:88–95.)

one head of a kinesin multimer to a microtubule causes dissociation of ADP from that head. ATP binds rapidly, triggering the neck linker to rotate or ratchet forward, throwing the second head forward as well and bringing it near the next binding site on the microtubule, 8 nm farther along the filament. The trailing head then hydrolyzes ATP and releases inorganic phosphate (but retains ADP), inducing its neck linker to return to its original orientation relative to the head. Exchange of ADP for ATP on the forward head begins the cycle again. The structure of the kinesin–microtubule complex (Figure 16.18) shows the switch 2 helix of kinesin in intimate contact with the microtubule at the junction of the α- and β-subunits of a tubulin dimer.

The Conformation Change That Leads to Movement Is Different in Myosins and Dyneins

The movement of myosin motors on cytoskeletal actin filaments is presumed to be similar to the myosin–actin interaction in skeletal muscle. Clearly, however, the different structure of the dynein hexameric motor domain and its associated coiled-coil stalk (see Figure 16.14) must represent a different motor mechanism. ATP-dependent conformation changes in the ring of AAA+ modules must be translated into movements of the tip of the coiled-coil stalk along a microtubule. A proposed mechanism for dynein movement (Figure 16.19) suggests that the events of ATP binding and hydrolysis and ADP and P_i release at an AAA+ module swing a linker that joins the AAA+ domain and the dynein tail.

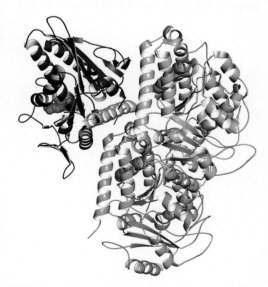

FIGURE 16.18 In the kinesin–microtubule complex, the switch 2 helix (yellow) of kinesin (*left*) lies in contact with the microtubule (*right*) at the subunit interface of a tubulin dimer (pdb id = 2HXH).

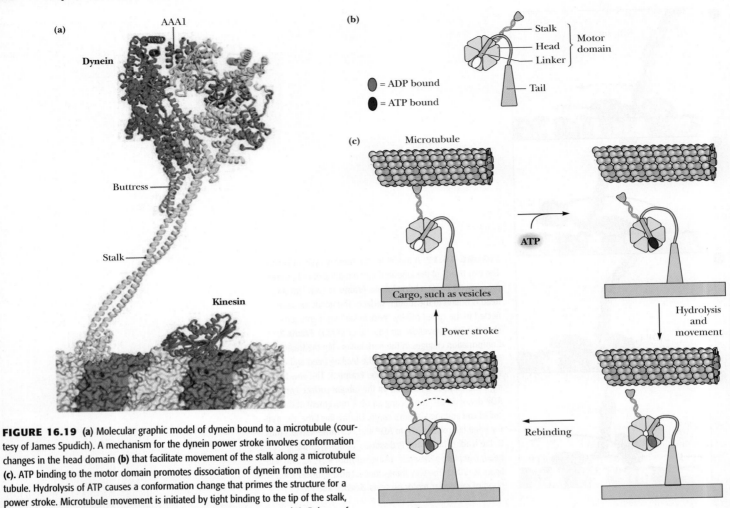

FIGURE 16.19 **(a)** Molecular graphic model of dynein bound to a microtubule (courtesy of James Spudich). A mechanism for the dynein power stroke involves conformation changes in the head domain **(b)** that facilitate movement of the stalk along a microtubule **(c)**. ATP binding to the motor domain promotes dissociation of dynein from the microtubule. Hydrolysis of ATP causes a conformation change that primes the structure for a power stroke. Microtubule movement is initiated by tight binding to the tip of the stalk, which promotes a conformation change in the head ring (the power stroke). Release of ADP and P$_i$ from the catalytic site causes tilting of the stalk at the end of the cycle. (Adapted from Oiwa, K., and Sakakibara, H., 2005. Recent progress in dynein structure and mechanism. *Current Opinion in Cell Biology* **17**:98–103.)

16.4 | How Do Molecular Motors Unwind DNA?

The ability of proteins to move in controlled ways along nucleic acid chains is important to many biological processes. For example, when DNA is to be replicated, the strands of the double helix must be unwound and separated to expose single-stranded DNA templates. Similarly, histone octamers (Figure 11.28) slide along DNA strands in chromatin remodeling, Holliday junctions (see Figure 28.18) move, and nucleic acids move in and out of viral capsids. The motor proteins that move directionally along nucleic acid strands and accomplish these many functions are called **translocases.** The translocases that unwind DNA or RNA duplex substrates are termed **helicases.** Thus, all helicases are translocases, but not vice versa.

All translocases and helicases are members of six protein "superfamilies" (Table 16.3 and Figure 16.20), all of them related evolutionarily to **RecA,** a DNA-binding protein (See A Deeper Look box on pages 520–521). **Superfamily 1 (SF1) and superfamily 2 (SF2)** helicases are monomeric and consist of two RecA domains in a tandem repeat. SF1 and SF2 helicases participate in several diverse DNA and RNA manipulations. The other helicases (SF3–SF6), which often participate at the DNA replication fork (Chapter 29), form hexameric rings, each monomer built from a single RecA domain. Each superfamily possesses characteristic conserved residues and sequence elements (Table 16.3), most of which are shared between several superfamilies. All members of a given superfamily move in the same direction on a DNA or RNA

TABLE 16.3	Helicase Superfamilies				
Superfamily	Subunit Structure		Quaternary Structure	Direction of Movement	Representative Motor
SF1A		Core domains	Monomeric	A*	Rep
SF1B		Accessory domains		B	RecD
	DNA binding DNA binding Unknown				
SF2			Monomeric	A or B	NS3
	Protease Protein:protein?				
SF3			Hexameric	A	BPV E1
	Origin binding				
SF4			Hexameric	B	T7 gp4
	Primase				
SF5			Hexameric	B	Rho
	Origin-binding domain				
SF6			Hexameric	A or B	FtsK
	Zn binding				MCM

*A: helicase moves 3′→5′ on nucleic acid. B: helicase moves 5′→3′.
Adapted from table in Singleton, M. R., Dillingham, M. S., and Wigley, D. B., 2007. Structure and mechanism of helicases and nucleic acid translocases. *Annual Review of Biochemistry* **76**:23–50.

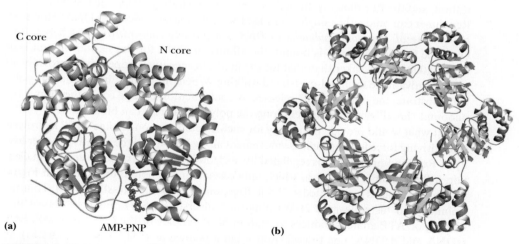

(a) C core   N core   AMP-PNP

(b)

FIGURE 16.20 (a) Translocase and helicase motors of SF1 and SF2 are monomers that consist of two RecA domains in a tandem repeat (pdb id = 1QHG). (b) Motor peptides of SF3 through SF6 associate to form hexamers (as shown) or dodecamers (pdb id = 1CR0).

template (either 5′ to 3′ or 3′ to 5′). As noted in the A Deeper Look box on pages 520–521, all of these helicases bind and hydrolyze NTP at the interface between two recA-like domains. The SF1 and SF2 helicases contain two recA-like domains coupled by a short linker, and the ATP binding and hydrolysis site is located at the interface of these two domains. In the hexameric helicases, the ATP site consists of elements derived from adjacent monomers in the complex. The hexameric motor proteins of the SF3 and

SF6 superfamilies are members of the ancient **AAA+ ATPase family.** (AAA stands for "**A**TPases **a**ssociated with various cellular **a**ctivities," and the "+" sign refers to an expanded definition of the family characteristics.)

Translocases and helicases, like other molecular motors, require energy for their function. The energy for movement along a nucleic acid strand, as well as for separation of the strands of a duplex (DNA or RNA), is provided by hydrolysis of ATP. Translocases and helicases move on nucleic acid strands at rates of a few base pairs to several thousand base pairs per second. These movements are carefully regulated by accessory proteins in nearly all cases. Translocases and helicases typically move along the DNA or RNA lattice for long distances without dissociating. This is termed **processive movement,** and helicases are said to have a high processivity. For example, the *E. coli* BCD helicase, which is involved in recombination processes, can unwind 33,000 base pairs before it dissociates from the DNA lattice. Processive movement is essential for helicases involved in DNA replication, where millions of base pairs must be replicated rapidly.

Helicases have evolved at least two structural and functional strategies for achieving high processivity. The hexameric helicases (of the SF3 through SF6 superfamilies) form ringlike structures that completely encircle at least one of the strands of a DNA duplex. The SF1 and SF2 helicases, notably **Rep helicase** from *E. coli,* are monomeric or homodimeric and move processively along the DNA helix by means of a "hand-over-hand" movement that is remarkably similar to that of kinesin's movement along microtubules. A key feature of hand-over-hand movement of a dimeric motor protein along a polymer is that at least one of the motor subunits must be bound to the polymer at any moment.

Negative Cooperativity Facilitates Hand-over-Hand Movement

How does hand-over-hand movement of a motor protein along a polymer occur? Clues have come from the structures of Rep helicase and its complexes with DNA. The Rep helicase from *E. coli* is a 76-kD protein that is monomeric in the absence of DNA. Binding of Rep helicase to either single-stranded or double-stranded DNA induces dimerization, and the Rep dimer is the active species in unwinding DNA. Each subunit of the Rep dimer can bind either single-stranded (ss) or double-stranded (ds) DNA. However, *the binding of Rep dimer subunits to DNA is negatively cooperative* (see Chapter 15). Once the first Rep subunit is bound, the affinity of DNA for the second subunit is at least 10,000 times weaker than that for the first! This negative cooperativity provides an obvious advantage for hand-over-hand walking. When one "hand" has bound the polymer substrate, the other "hand" releases. A conformation change could then move the unbound "hand" one step farther along the polymer where it can bind again.

But what would provide the energy for such a conformation change? ATP hydrolysis is the driving force for Rep helicase movement along DNA, and the negative cooperativity of Rep binding to DNA is regulated by nucleotide binding. In the absence of nucleotide, a Rep dimer is favored, in which only one subunit is bound to ssDNA. In Figure 16.21a, this state is represented as P_2S [a Rep dimer (P_2) bound to ssDNA (S)]. Timothy Lohman and his colleagues at Washington University in St. Louis have shown that binding of ATP analogs induces formation of a complex of the Rep dimer with both ssDNA and dsDNA, one to each Rep subunit (shown as P_2SD in Figure 16.21a). In their model, unwinding of the dsDNA and ATP hydrolysis occur at this point, leaving a P_2S_2 state in which both Rep subunits are bound to ssDNA. Dissociation of ADP and P_i leave the P_2S state again (Figure 16.21a).

Work by Lohman and his colleagues has shown that coupling of ATP hydrolysis and hand-over-hand movement of Rep over the DNA involves the existence of the Rep dimer in an asymmetric state. A crystal structure of the Rep dimer in complex with ssDNA and ADP shows that the two Rep monomers are in different conformations (Figure 16.21b). The two conformations differ by a 130° rotation about a hinge region between two subdomains within the monomer subunit. The hand-over-hand walking of the Rep dimer along the DNA surface may involve alternation of each subunit between these two conformations, with coordination of the movements by nucleotide binding and hydrolysis.

(a)

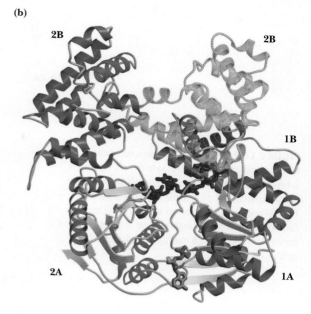

FIGURE 16.21 **(a)** A hand-over-hand model for movement along (and unwinding of) DNA by *E. coli* Rep helicase. The P_2S state consists of a Rep dimer bound to ssDNA. The P_2SD state involves one Rep monomer bound to ssDNA and the other bound to dsDNA. The P_2S_2 state has ssDNA bound to each Rep monomer. ATP binding and hydrolysis control the interconversion of these states and walking along the DNA substrate. **(b)** Crystal structure of the *E. coli* Rep helicase monomer with bound ssDNA (dark blue, ball and stick) and ADP (red). The monomer consists of four domains designated 1A (residues 1–84 and 196–276), 1B (residues 85–195), 2A (residues 277–373 and 543–670), and 2B (residues 374–542). The open (purple) and closed (green) conformations of the 2B domain are superimposed in this figure (pdb id = 1UAA). (From Korolev, S., Hsieh, J., Gauss, G., Lohman, T. L., and Waksman, G., 1997. Major domain swiveling revealed by the crystal structures of complexes of *E. coli* Rep helicase bound to single-stranded DNA and ADP. *Cell* **90**:635–647. Reprinted by permission of Cell Press.)

Papillomavirus E1 Helicase Moves along DNA on a Spiral Staircase

Papillomaviruses are tumor viruses that cause both cancerous and benign lesions in a host. Replication of papillomaviral DNA within a host cell requires the multifunctional 605-residue viral **E1 protein.** Monomers of E1 assemble at a replication origin on DNA and form a pair of hexamers that wrap around a single strand of DNA. These assemblies are helicases that operate bidirectionally in the replication of viral DNA. The N-terminal half of the E1 protein includes a regulatory domain and a sequence-specific DNA-binding domain, whereas the helicase activity is located in the C-terminal half of the protein. The C-terminal helicase domain (Figure 16.22) includes a segment involved in oligomer formation (residues 300 to 378) and an AAA+ domain (residues 379 to 605).

AAA+ domains are found in proteins of many functions, including motor activity by dyneins (see Section 16.3) and helicases, protein degradation by proteasomes (see Chapter 31), and disassembly of SNARE complexes (see Chapter 9). This ubiquitous module consists of two subdomains: an N-terminal segment known as an α/β **Rossman** fold, and a C-terminal α-helical domain (Figure 16.23). The Rossman fold is wedge-shaped and has a β-sheet of parallel strands in a $\beta5$-$\beta1$-$\beta4$-$\beta3$-$\beta2$ pattern. Key features of this fold include a Lys residue in the Walker A motif, an Asp–Asp or Asp–Glu pair in the Walker B motif, and a crucial Arg residue in a structure called an **arginine finger.** These three motifs are essential for ATP binding and hydrolysis. In an AAA+ hexamer, the ATP-binding sites lie at the interface between any two subunits, involving the arginine finger of any given subunit and the Walker A and Walker B motifs of the adjacent subunit.

The structure of a large fragment (residues 306 to 577) of the papillomavirus E1 protein bound to a segment of ssDNA (Figure 16.24) reveals the remarkable mechanism by which this helicase traverses a DNA chain. The oligomerization domains form a symmetric hexamer, but the six AAA+ domains each display a unique conformation. The DNA strand is bound in the center pore of the AAA+ hexamer, with six nucleotides of the DNA chain each bound to residues from each of the protein subunits. The crucial nucleotide-binding residues include Lys[506] and His[507] on a hairpin loop and Phe[464], all of which face the center pore of the protein (Figure 16.25). Lys[506] interacts with one DNA phosphate oxygen, and His[507] forms a hydrogen bond with the phosphate of an adjacent nucleotide in the DNA chain. The aliphatic portion of Lys[506] and the aromatic groups of Phe[464] and His[507] share van der Waals interactions with the

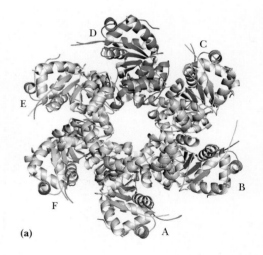

(a)

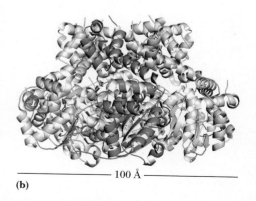

— 100 Å —

(b)

FIGURE 16.22 **(a)** The papillomavirus E1 protein is a 605-residue monomer that forms hexameric assemblies at specific sites on single-stranded DNA (pdb id = 2V9P). **(b)** The C-terminal helicase domain shown here includes an oligomerization domain (magenta) and the AAA+ domain (blue).

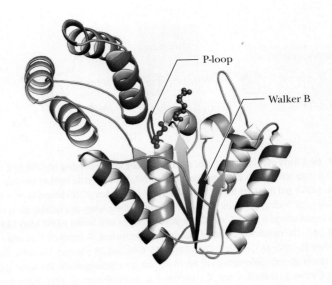

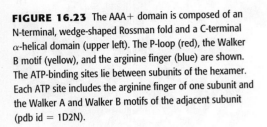

FIGURE 16.23 The AAA+ domain is composed of an N-terminal, wedge-shaped Rossman fold and a C-terminal α-helical domain (upper left). The P-loop (red), the Walker B motif (yellow), and the arginine finger (blue) are shown. The ATP-binding sites lie between subunits of the hexamer. Each ATP site includes the arginine finger of one subunit and the Walker A and Walker B motifs of the adjacent subunit (pdb id = 1D2N).

DNA sugar moiety linking these two phosphates. *The hairpin loops of the six protein subunits form a spiral staircase, following the ssDNA as it threads through the central pore of the hexamer* (Figure 16.26).

Eric Enemark and Leemor Joshua-Tor have suggested that a central hairpin loop from one of the AAA+ subunits coordinates each DNA nucleotide as it enters the helicase pore. Then, as each AAA+ domain proceeds through the intermediate states of ATP binding and hydrolysis, its hairpin loop steps down through the six conformations of the staircase, maintaining continuous contact with its nucleotide, as it escorts it through the pore, finally releasing the nucleotide as it exits the pore. Following release, the hairpin moves back to the top of the staircase, picks up the next available nucleotide, and begins another journey down the staircase. For one full cycle of the hexamer, each subunit hydrolyzes one ATP, releases one ADP, and translocates one nucleotide through the central pore. A full cycle thus translocates six nucleotides with associated hydrolysis of six ATPs and release of six ADPs.

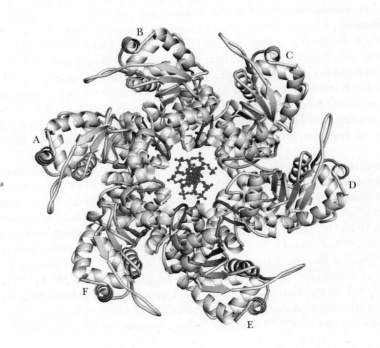

FIGURE 16.24 A view of the E1 helicase along the ssDNA axis, showing the DNA-binding hairpin loops from each monomer (colored) interacting with the phosphates of DNA. DNA is shown as a ball-and-stick model in the center of the structure (pdb id = 2GXA).

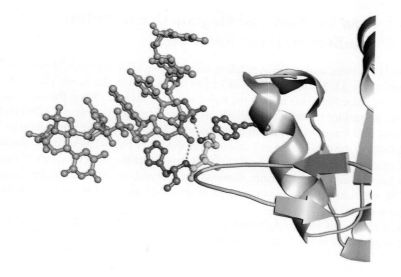

(a)

FIGURE 16.25 The hairpin loops of each subunit in the E1 helicase interact with two adjacent nucleotides in the DNA chain. Interactions include an ionic bond between Lys[506] (yellow) and a DNA phosphate oxygen, a hydrogen bond between His[507] and the phosphate of an adjacent nucleotide, and van der Waals interactions between the aromatic rings of Phe[464] (purple) and His[507] (olive) and the aliphatic chain of Lys[506], with the sugar linking the two phosphates (pdb id = 2GXA).

(b)

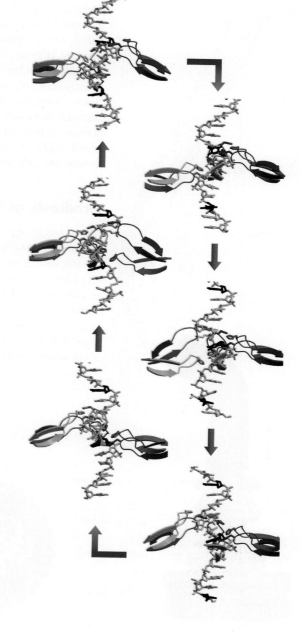

FIGURE 16.26 The hairpin loops of the E1 helicase hexamer are arranged in a spiral staircase that winds around the DNA strand (a) axial view (pdb id = 2GXA) and (b) the coordinated escort mechanism for DNA translocation by sequential ATP hydrolysis. As the helicase moves along the strand, the hairpin loop of one protein monomer binds each nucleotide as it enters the central cavity of the helicase. The loop adopts six conformations as the helicase moves along the DNA, preserving the loop–nucleotide interaction until the nucleotide exits the cavity. The released protein loop then returns to the other end of the cavity to bind a new, incoming nucleotide. Every sixth nucleotide of DNA is colored black and is escorted by the purple hairpin. (b) Courtesy of Leemore Joshua-Tor.

16.5 How Do Bacterial Flagella Use a Proton Gradient to Drive Rotation?

Bacterial cells swim and move by rotating their flagella. The flagella of *E. coli* are helical filaments up to 15,000 nm (15 μm) in length and 15 nm in diameter. The direction of rotation of these filaments affects the movements of the cell. When the half-dozen flagella on the surface of the bacterial cell rotate in a counterclockwise (CCW) direction, they twist and bundle together in a left-handed helical structure and rotate in a concerted fashion, propelling the cell through the medium. Every few seconds, the flagellar motor reverses, the helical bundle of filaments (now turning clockwise, or CW) unwinds into a jumble, and the bacterium somersaults or tumbles. Alternating between CCW and CW rotations, the bacterium can move toward food sources, such as amino acids and sugars.

The rotations of bacterial flagellar filaments are the result of the rotation of motor protein complexes in the bacterial plasma membrane.

The Flagellar Rotor Is a Complex Structure

The flagellum is built from at least 25 proteins and comprises three parts: a rotary motor anchored in the bacterial inner membrane, a long filament that serves as a helical propellor, and a "hook" that functions as a universal joint that connects the motor with the filament (Figure 16.27). The rotary motor includes several rings of protein subunits, including the C ring, the MS ring, the P ring, and the L ring. The MS ring is built from 26 copies of the protein FliF. The C ring is attached to the MS ring and includes three "rotor" proteins—FliG, FliM, and FliN—involved in rotation of the motor. The C ring includes 26 copies of FliG, 34 copies of FliM, and 34 × 4 = 136 copies of FliN. The stationary portion of the motor—the "stator"—is formed from the proteins motA and motB. Eight motA$_4$–motB$_2$ complexes are embedded in the bacterial inner membrane around the MS ring.

Gradients of H⁺ and Na⁺ Drive Flagellar Rotors

What energy source drives the flagellar motor? Gradients of protons and Na⁺ ions exist across bacterial inner membranes, typically with more H⁺ and Na⁺ outside the cell. In *E. coli*, spontaneous inward flow of protons through the motA–motB complexes drives the rotation of the motor (Figure 16.28). In *Vibrio cholerae*, inward Na⁺ ion flow powers the motor. Flagellar motors are thus energy conversion devices. In *E. coli*, each motA–motB complex passes 70 H⁺ per revolution of the motor. With a full complement of eight motA–motB complexes, a motor conducts about 560 protons per revolution. The H⁺-driven flagellar rotors reach top rotational speeds of about 360 Hz (corresponding to 21,600 rpm). Thus, the overall rate of proton flow for the motor is

Filament (Flagellin)

Cap FliD

Hook FlgE

LP rings
FlgI, FlgH

MS rings
FliF, FliG

C ring
FliM, FliN
and FliG

L

P

S

M

Outer membrane

Peptidogylcan layer (cell wall)

Cell membrane

Cytoplasm

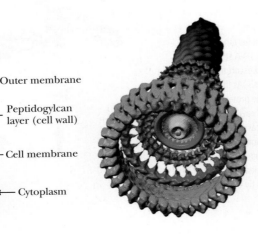

FIGURE 16.27 A model of the *E. coli* flagellar motor. The motor is anchored by interactions of stationary motA and motB proteins in the M and S rings with the inner membrane. Spontaneous flow of protons through the motA–motB complexes and into the cell drives the rotation of the motor. Flow rates of 200,000 protons per second drive the motor at speeds approaching 22,000 rpm. (Adapted from Thomas, D. R., Morgan, D. G., Francis, N. R., and DeRosier, D. J., 2007. Bit by bit The structure of the complete flagellar hook/basal body complex. *Microscopy and Microanalysis* **13**:34–35. Image provided by David J. DeRosier, Brandeis University.)

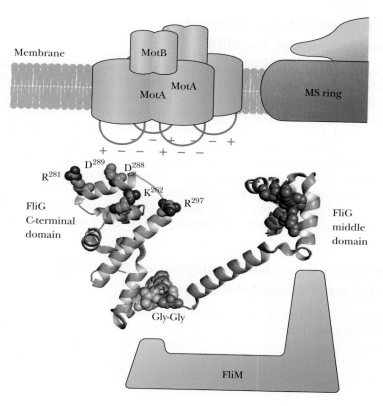

FIGURE 16.28 Interactions between the stationary motA–motB complexes and the rotating FliG ring drive the flagellar motor. Proton flow through the motA–motB complexes is presumably coupled to conformation changes that alter ionic interactions between charged residues at the motA–motB and FliG interface, driving rotation of the FliG ring. Other conserved features include a hydrophobic patch (light green), a Gly-Gly motif (purple), and a EHPQR motif (blue, in the middle domain (pdb id = 1LKV).

approximately 200,000 H$^+$/sec! Flagellar motors driven by Na$^+$ ions are even faster, with rotational rates of 1700 Hz (100,000 rpm) observed in *Vibrio*.

The motA–motB complexes work with FliG in the C ring to transfer protons across the membrane. FliG contains 335 residues, and most of the FliG protein structure (residues 104 to 335) consists of two compact domains joined by an α-helix (Figure 16.28). A ridge on the C-terminal domain contains five charged residues that interact with motA and are important for motor rotation. Asp^{32} of motB is essential for rotation of the motor and is probably involved in proton transfer. David Blair has proposed a model for creation of two membrane channels from the transmembrane segments of the motA$_4$–motB$_2$ complex. Blair has suggested that each encounter of a motA–motB complex with a FliG subunit as the motor turns results in movement of one proton through each of these channels. The passage of about 70 H$^+$ through each motA–motB complex in one revolution of the motor (which would involve encounters with 34 FliG subunits) is consistent with this suggestion (70/34 = ~2).

The Flagellar Rotor Self-Assembles in a Spontaneous Process

Flagellar rotors are true masterpieces of biological self-assembly. The ring of FliF subunits, within the MS ring, is the first to assemble in the plasma membrane. Other proteins then attach to this ring one after another, from the base to the tip, to construct the motor structure. Once the motor has formed, the flexible hook and the flagellar filament are assembled. Precise recognition of the existing template structure allows this highly ordered self-assembly process to proceed without error. The flagellar filament is made from 20,000 to 30,000 copies of **flagellin** polymerized into a hollow helical tube structure. Each turn of the helical filament contains about 5000 flagellin subunits and is about 2300 nm long. A complete flagellum can have up to six full helical turns. Flagellin molecules are transported through the long, narrow, central channel of the motor and flagellum from the cell interior to the far end of the flagellum, where they self-assemble with the help of a pentameric complex of **FliD**, a capping protein (see Figure 16.27). The FliD complex has a plate and five leg domains. It rotates in a stepping fashion at the end of the filament, exposing one binding site at a time and guiding the binding of newly arriving flagellin molecules in a helical pattern.

Flagellar Filaments Are Composed of Protofilaments of Flagellin

Each cylindrical flagellar filament is composed of 11 fibrils or **protofilaments** that form the cylinder, with each fibril lying at a slight tilt to the cylinder axis (Figure 16.29a). An end-on view of the filament shows 11 subunits, each representing the end of a protofilament (Figure 16.29b). The flagellin protein of *Salmonella typhimurium* contains 494 residues and consists of four domains, denoted D0 through D3 (Figure 16.29c). D0 and D1 are composed of α-helices, whereas D2 and D3 consist primarily of β-strands. The N-terminus of the peptide chain lies at the base of D0. The peptide runs from D0 to D3 and then reverses and returns to D0, where the N- and C-termini are juxtaposed. The structure resembles a Greek capital gamma (Γ), with a height of 140 Å and a width of 110 Å. Each flagellin protein is arranged with D0 inside the filament and D3 facing the outside. The central pore, 20 Å in diameter, is lined by the α-helices of D0.

The conformational change from the L to the R form involves a change of orientation in the D1 domain relative to the D0 domain and a conformational switch in the D1 domain itself. (Maki-Yonekura, S., Yonekura, K., and Namba, K., 2010. Conformational change of flagellin for polymorphic supercoiling of the flagellar filament. *Nature Structural and Molecular Biology* **17:**417–423.)

Motor Reversal Involves Conformation Switching of Motor and Filament Proteins

The flagellar motor reverses direction every few seconds so that the bacterium can change its swimming direction to seek better environments. Motor reversal involves conformation changes both in motor proteins and also in the filament itself. In the

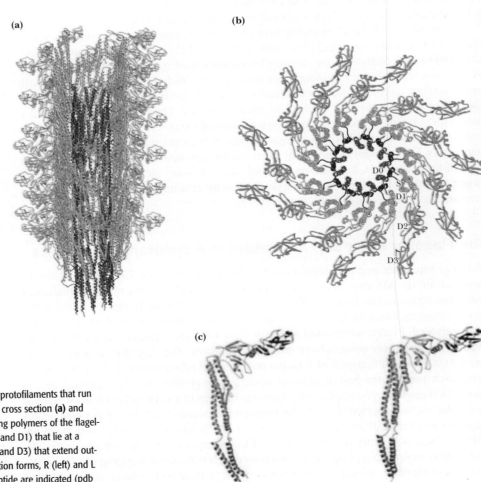

(a)

(b)

(c)

FIGURE 16.29 The *E. coli* flagellum is composed of 11 protofilaments that run the length of the flagellar filament. The filament is shown in cross section **(a)** and perpendicular to the filament **(b).** The protofilaments are long polymers of the flagellin protein **(c),** which consists of two α-helical domains (D0 and D1) that lie at a slight tilt to the filament axis and two β-sheet domains (D2 and D3) that extend outward from the filament. Flagellin is shown in two conformation forms, R (left) and L (right; pdb id = 3A5X). The N- and C-termini of the polypeptide are indicated (pdb id = 1UCU). (Parts (a) and (b) courtesy of Keiichi Namba, Osaka University, Japan.)

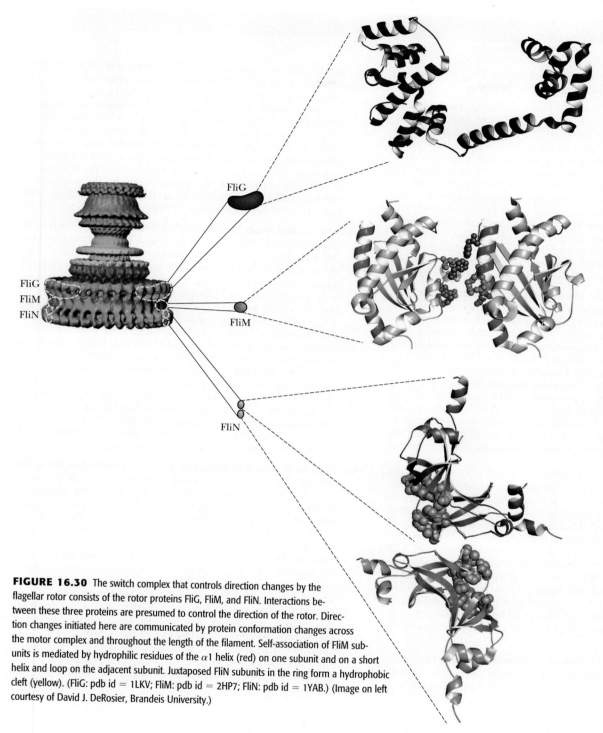

FIGURE 16.30 The switch complex that controls direction changes by the flagellar rotor consists of the rotor proteins FliG, FliM, and FliN. Interactions between these three proteins are presumed to control the direction of the rotor. Direction changes initiated here are communicated by protein conformation changes across the motor complex and throughout the length of the filament. Self-association of FliM subunits is mediated by hydrophilic residues of the α1 helix (red) on one subunit and on a short helix and loop on the adjacent subunit. Juxtaposed FliN subunits in the ring form a hydrophobic cleft (yellow). (FliG: pdb id = 1LKV; FliM: pdb id = 2HP7; FliN: pdb id = 1YAB.) (Image on left courtesy of David J. DeRosier, Brandeis University.)

motor structure, the rotor proteins FliG, FliM, and FliN work together to control direction changes of the motor, and they are known collectively as the **switch complex.** FliN appears to lie at the base of the C ring, FliG lies at the top of the C ring, and FliM resides in the middle, contacting both FliN and FliG (Figure 16.30).

Reversal of the flagellar motor causes the long filament to switch from a left-handed (L) helical structure to a right-handed (R) helical form. This makes the bundle of flagella fall apart, causing the bacterium to tumble. This left-to-right switch in the filament is caused by a conformational change that occurs in the flagellin subunits in some protofilaments. Interestingly, the driving force for these conformation changes is probably the torque applied to D0 and D1 of flagellin subunits along the filament when the motor itself reverses.

⊘WL Login to OWL to develop problem-solving skills and complete online homework assigned by your professor.

16.1 What Is a Molecular Motor? Motor proteins, also known as molecular motors, use chemical energy (ATP) to orchestrate different movements, transforming ATP energy into the mechanical energy of motion. In all cases, ATP hydrolysis is presumed to drive and control protein conformational changes that result in sliding or walking movements of one molecule relative to another. To carry out directed movements, molecular motors must be able to associate and dissociate reversibly with a polymeric protein array, a surface, or substructure in the cell. ATP hydrolysis drives the process by which the motor protein ratchets along the protein array or surface. Molecular motors may be linear or rotating. Linear motors crawl or creep along a polymer lattice, whereas rotating motors consist of a rotating element (the "rotor") and a stationary element (the "stator"), in a fashion much like a simple electrical motor.

16.2 What Is the Molecular Mechanism of Muscle Contraction? Examination of myofibrils in the electron microscope reveals a banded or striated structure. The so-called H zone shows a regular, hexagonally arranged array of thick filaments, whereas the I band shows a regular, hexagonal array of thin filaments. In the dark regions at the ends of each A band, the thin and thick filaments interdigitate. The thin filaments are composed primarily of three proteins called actin, troponin, and tropomyosin. The thick filaments consist mainly of a protein called myosin. The thin and thick filaments are joined by cross-bridges. These cross-bridges are actually extensions of the myosin molecules, and muscle contraction is accomplished by the sliding of the cross-bridges along the thin filaments, a mechanical movement driven by the free energy of ATP hydrolysis.

Myosin, the principal component of muscle thick filaments, is a large protein consisting of six polypeptides, including light chains and heavy chains. The heavy chains consist of globular amino-terminal myosin heads, joined to long α-helical carboxy-terminal segments, the tails. These tails are intertwined to form a left-handed coiled coil approximately 2 nm in diameter and 130 to 150 nm long. The myosin heads exhibit ATPase activity, and hydrolysis of ATP by the myosin heads drives muscle contraction.

The free energy of ATP hydrolysis is translated into a conformation change in the myosin head, so dissociation of myosin and actin, hydrolysis of ATP, and rebinding of myosin and actin occur with stepwise movement of the myosin S1 head along the actin filament. The conformation change in the myosin head is driven by the hydrolysis of ATP.

16.3 What Are the Molecular Motors That Orchestrate the Mechanochemistry of Microtubules? Microtubules are hollow, cylindrical structures, approximately 30 nm in diameter, formed from tubulin, a dimeric protein composed of two similar 55-kD subunits known as α-tubulin and β-tubulin. Tubulin dimers polymerize to form microtubules, which are essentially helical structures, with 13 tubulin monomer "residues" per turn. Microtubules are, in fact, a significant part of the cytoskeleton, a sort of intracellular scaffold formed of microtubules, intermediate filaments, and microfilaments. In most cells, microtubules are oriented with their minus ends toward the centrosome and their plus ends toward the cell periphery. This consistent orientation is important for mechanisms of intracellular transport. Microtubules are also the fundamental building blocks of eukaryotic cilia and flagella. Microtubules also mediate the intracellular motion of organelles and vesicles.

16.4 How Do Molecular Motors Unwind DNA? When DNA is replicated or repaired, the strands of the double helix must be unwound and separated to form single-stranded DNA intermediates. This separation is carried out by molecular motors known as DNA helicases that move along the length of the DNA lattice, sequentially destabilizing the hydrogen bonds between complementary base pairs. The movement along the lattice and the separation of the DNA strands are coupled to the hydrolysis of nucleoside 5'-triphosphates. The *E. coli* BCD helicase, which is involved in recombination processes, can unwind 33,000 base pairs before it dissociates from the DNA lattice. Processive movement is essential for helicases involved in DNA replication, where millions of base pairs must be replicated rapidly. Certain hexameric helicases form ring-like structures that completely encircle at least one of the strands of a DNA duplex. Other helicases, notably Rep helicase from *E. coli*, are homodimeric and move processively along the DNA helix by means of a "hand-over-hand" movement that is remarkably similar to that of kinesin's movement along microtubules.

16.5 How Do Bacterial Flagella Use a Proton Gradient to Drive Rotation? Bacterial cells swim and move by rotating their flagella. The direction of rotation of these flagella affects the movements of the cell. When the half-dozen flagella on the surface of the bacterial cell rotate in a counterclockwise direction, they twist and bundle together and rotate in a concerted fashion, propelling the cell through the medium. The rotations of bacterial flagellar filaments are the result of the rotation of motor protein complexes in the bacterial plasma membrane. The flagellar motor consists of multiple rings (including the MS ring and the C ring). The rings are surrounded by a circular array of membrane proteins. In all, at least 40 genes appear to code for proteins involved in this magnificent assembly. One of these proteins, motB, lies on the edge of the M ring, where it interacts with the motA protein, located in the membrane protein array and facing the M ring. In contrast to the many other motor proteins described in this chapter, a proton gradient, not ATP hydrolysis, drives the flagellar motor.

FOUNDATIONAL BIOCHEMISTRY Things You Should Know After Reading Chapter 16.

- The essential features of molecular motors.

- The differences between linear and rotating motors.

- The morphology of skeletal muscle (Figure 16.1).

- The actions of bronchodilators on smooth muscle.

- The triggering action of Ca^{2+} ions on skeletal muscle.

- The composition and structure of skeletal muscle thick filaments.

- The composition and structure of skeletal muscle thin filaments.

- The repeating characteristics of myosin coiled coils.

- The characteristics of the P-loop in motor ATPases.

- The essential features of dystrophin and its role in muscular dystrophies.

- The essential features of the sliding filament model of muscle contraction.

- Szent-Györgyi's experiments that revealed the coupling of ATP hydrolysis and actin-myosin association-dissociation.

- The mechanism of skeletal muscle contraction, including the role of ATP hydrolysis in myosin head conformational changes.

- The optical trapping experiments that measure molecular forces in muscle contraction.

- The functions of the cytoskeleton in eukaryotic cells.
- The types of fibers that comprise the cytoskeleton.
- The structural features of microtubules and tubulin.
- The essential structural and functional features of kinesins and dyneins.
- The effects of colchicine, vinblastine, vincristine, and taxol on microtubules.
- The essential features of organelle and vesicle transport by dyneins and kinesins.
- The definition of processivity and the processivities of kinesins and dyneins.
- The hand-over-hand mechanism of kinesin movement and how it is driven by ATP.
- The mechanism of the dynein power stroke and how it is driven by ATP.

- The essential features of the helicase superfamilies.
- The essential features of helicase processive movement.
- The hand-over-hand movement mechanism of the Rep helicase.
- The essential structural and functional features of papillomavirus E1 helicase.
- The essential structural features of the *E. coli* flagellar rotor.
- The mechanism by which proton transport drives the *E. coli* flagellar rotor.
- The mechanism of self-assembly of the *E. coli* flagellar rotor.
- The essential structural features of flagellin protein.
- The mechanism of flagellar motor reversal.

PROBLEMS

Answers to all problems are at the end of this book. Detailed solutions are available in the *Student Solutions Manual, Study Guide, and Problems Book.*

1. **Assessing the Performance Consequences of Skeletal Muscle Anatomy** The cheetah is generally regarded as nature's fastest mammal, but another amazing athlete in the animal kingdom (and almost as fast as the cheetah) is the pronghorn antelope, which roams the plains of Wyoming. Whereas the cheetah can maintain its top speed of 70 mph for only a few seconds, the pronghorn antelope can run at 60 mph for about an hour! (It is thought to have evolved to do so in order to elude now-extinct ancestral cheetahs that lived in North America.) What differences would you expect in the muscle structure and anatomy of pronghorn antelopes that could account for their remarkable speed and endurance?

2. **Analyzing the Action of a Muscle Contraction Inhibitor** An ATP analog, β,γ-methylene-ATP, in which a —CH_2— group replaces the oxygen atom between the β- and γ-phosphorus atoms, is a potent inhibitor of muscle contraction. At which step in the contraction cycle would you expect β,γ-methylene-ATP to block contraction?

3. **Understanding the Role of Phosphocreatine** ATP stores in muscle are augmented or supplemented by stores of phosphocreatine. During periods of contraction, phosphocreatine is hydrolyzed to drive the synthesis of needed ATP in the creatine kinase reaction:

$$\text{Phosphocreatine} + \text{ADP} \longrightarrow \text{creatine} + \text{ATP}$$

Muscle cells contain two different isozymes of creatine kinase, one in the mitochondria and one in the sarcoplasm. Explain.

4. **Understanding the Biochemistry of Rigor Mortis** *Rigor* is a muscle condition in which muscle fibers, depleted of ATP and phosphocreatine, develop a state of extreme rigidity and cannot be easily extended. (In death, this state is called *rigor mortis,* the rigor of death.) From what you have learned about muscle contraction, explain the state of rigor in molecular terms.

5. **Assessing the Contraction Potential of Human Muscle** Skeletal muscle can generate approximately 3 to 4 kg of tension or force per square centimeter of cross-sectional area. This number is roughly the same for all mammals. Because many human muscles have large cross-sectional areas, the force that these muscles can (and must) generate is prodigious. The gluteus maximus (on which you are probably sitting as you read this) can generate a tension of 1200 kg! Estimate the cross-sectional area of all of the muscles in your body and the total force that your skeletal muscles could generate if they all contracted at once.

6. **Understanding the Structure of Tubulin** Calculate a diameter for a tubulin monomer, assuming that the monomer MW is 55,000, that

the monomer is spherical, and that the density of the protein monomer is 1.3 g/mL. How does the number that you calculate compare to the dimension portrayed in Figure 16.12?

7. **Assessing the Tubulin Content of a Liver Cell** Use the number you obtained in problem 6 to calculate how many tubulin monomers would be found in a microtubule that stretched across the length of a liver cell. (See Table 1.2 for the diameter of a liver cell.)

8. **Understanding Axonal Vesicle Transport** The giant axon of the squid may be up to 4 inches in length. Use the value cited in this chapter for the rate of movement of vesicles and organelles across axons to determine the time required for a vesicle to traverse the length of this axon.

9. **Understanding the Molecular Structure of Myosin** As noted in this chapter, the myosin molecules in thick filaments of muscle are offset by approximately 14 nm. How many residues of a coiled-coil structure are needed to span 14nm?

10. **Assessing the Energetics of Sarcoplasmic Reticulum Calcium Transport** Use the equations of Chapter 9 to determine the free energy difference represented by a Ca^{2+} gradient across the sarcoplasmic reticulum membrane if the luminal (inside) concentration of Ca^{2+} is 1 mM and the concentration of Ca^{2+} in the solution bathing the muscle fibers is 1 μM.

11. **Understanding the Energetic Driving Force of the Ca-ATPase** Use the equations of Chapter 3 to determine the free energy of hydrolysis of ATP by the sarcoplasmic reticulum Ca-ATPase if the concentration of ATP is 3 mM, the concentration of ADP is 1 mM, and the concentration of P_i is 2 mM.

12. **Understanding the Stoichiometry of Sarcoplasmic Reticulum Calcium Transport** Under the conditions described in problems 10 and 11, what is the maximum number of Ca^{2+} ions that could be transported per ATP hydrolyzed by the Ca-ATPase?

13. **Understanding the Energetics of Motor Proteins** For each of the motor proteins in Table 16.2, calculate the force exerted over the step size given, assuming that the free energy of hydrolysis of ATP under cellular conditions is −50 kJ/mol.

14. **Relating Myosin Function and Stoichiometry to Weight Lifting** When you go to the gym to work out, you not only exercise many muscles but also involve many myosins (and actins) in any given exercise activity. Suppose you lift a 10-kg weight a total distance of 0.4 m. Using the data in Table 16.2 for myosin, calculate the minimum number of myosin heads required to lift this weight and the number of sliding steps these myosins must make along their associated actin filaments.

15. **Understanding Human Smooth Muscle** In which of the following tissues would you expect to find smooth muscle?

    a. Arteries

    b. Stomach

    c. Urinary bladder

    d. Diaphragm

    e. Uterus

    f. The gums in your mouth

16. **Analyzing the Mechanism of Muscle Contraction** When an action potential (nerve impulse) arrives at a muscle membrane (sarcolemma), in what order do the following events occur?

    a. Release of Ca^{2+} ions from the sarcoplasmic reticulum

    b. Hydrolysis of ATP, with release of energy

    c. Detachment of myosin from actin

    d. Sliding of myosin along actin filament

    e. Opening of switch 1 and switch 2 on myosin head

17. **Essay Question on the Functional Relevance of Molecular Motors** (Essay question.) You are invited by the National Science Foundation to attend a scientific meeting to set the agenda for funding of basic research related to molecular motors for the next 10 years. Only basic research will be funded, ruling out studies on human subjects. You are asked to suggest the research area most worthy of scientific research. Your presentation must include (1) a brief background on what we currently know about the subject; (2) identification of a key research topic about which more needs to be known; and (3) a justification of why additional knowledge in this area is critical for advancing the field (that is, why investigations in this area are especially important). You are not being asked to provide the methods or experiments that might be used to address the problem—only the concept. Base your presentation on what you have learned in this chapter (you may consult and include references from the Further Reading section), and limit your presentation to 300 words.

Preparing for the MCAT® Exam

18. Consult Figure 16.17 and use the data in problem 8 to determine how many steps a kinesin motor must take to traverse the length of the squid giant axon.

19. When athletes overexert themselves on hot days, they often suffer immobility from painful muscle cramps. Which of the following is a reasonable hypothesis to explain such cramps?

    a. Muscle cells do not have enough ATP for normal muscle relaxation.

    b. Excessive sweating has affected the salt balance within the muscles.

    c. Prolonged contractions have temporarily interrupted blood flow to parts of the muscle.

    d. All of the above.

20. Duchenne muscular dystrophy is a sex-linked recessive disorder associated with severe deterioration of muscle tissue. The gene for the disease:

    a. is inherited by males from their mothers.

    b. should be more common in females than in males.

    c. both a and b.

    d. neither a nor b.

ActiveModels Problems

21. Examine the ActiveModel for myosin II, and explain the function of each subunit.

22. Using the ActiveModel for the kinesin protein K1F1A, identify the function of switch 1, and explain why K1F1A cannot move cargo when switch 1 is bound to AMPPNP.

23. Using the ActiveModel for E1 helicase, describe the features of AAA+ domains.

FURTHER READING

P-Loop NTPases

Snider, J., and Houry, W. A., 2008. AAA+ proteins: diversity in function, similarity in structure. *Biochemical Society Transactions* **36**:72–77.

Thomsen, N. D., and Berger, J. M., 2008. Structural frameworks for considering microbial protein- and nucleic acid-dependent motor ATPases. *Molecular Microbiology* **69**:1071–1090.

Ye, J., Osborne, A. R., Groll, M., and Rapoport, T. A., 2004. RecA-like motor ATPases–lessons from structures. *Biochimica et Biophysica Acta* **1659**:1–18.

Muscle Contraction

Bagshaw, C. R., 2007. Myosin mechanochemistry. *Structure* **15**:511–512.

Carlsson, A. E., 2010. Actin dynamics: from nanoscale to microscale. *Annual Review of Biophysics* **39**:91–110.

Geeves, M. A., and Holmes, K. C., 2005. The molecular mechanism of muscle contraction. *Advances in Protein Chemistry* **71**:161–193.

Malnasi-Csizmadia, A., and Kovacs, M., 2010. Emerging complex pathways of the actomyosin powerstroke. *Trends in Biochemical Sciences* **36**:684–690.

Sun, Y., Sato, O., Ruhnow, F., Arsenault, M. E., Ikebe, M., and Goldman, Y. E., 2010. Single-molecule stepping and structural dynamics of myosin X. *Nature Structural and Molecular Biology* **17**:485–492.

Sweeney, H. L., and Houdusse, A., 2010. Myosin VI rewrites the rules for myosin motors. *Cell* **141**:573–582.

Sweeney, H. L., and Houdusse, A., 2010. Structural and functional insights into the myosin motor mechanism. *Annual Review of Biophysics* **39**:539–557.

Thomas, D. D., Kast, D., and Korman, V. L., 2009. Site-directed spectroscopic probes of actomyosin structural dynamics. *Annual Review of Biophysics* **38**:347–369.

Uribe, R., and Jay, D., 2009. A review of actin binding proteins: new perspectives. *Molecular Biology Reports* **36**:121–125.

Dystrophin and Muscular Dystrophy

Davies, K. E., and Nowak, K. J., 2006. Molecular mechanisms of muscular dystrophies: Old and new players. *Nature Reviews Molecular Cell Biology* **7**:762–773.

Le Rumeur, E., Winder, S. J., and Jubert, J.-F., 2010. Dystrophin: more than just the sum of its parts. *Biochimica et Biophysica Acta* **1804**:1713–1722.

Kinesins

Carter, N. J., and Cross, R. A., 2005. Mechanics of the kinesin step. *Nature* **435**:308–312.

Carter, N. J., and Cross, R. A., 2006. Kinesin's moonwalk. *Current Opinion in Structural Biology* **18**:61–67.

Guydosh, N. R., and Block, S. M., 2009. Direct observation of the binding state of the kinesin head to the microtubule. *Nature* **461**:125–129.

Hirokawa, N., Noda, Y., Tanaka, Y., and Niwa, S., 2009. Kinesin superfamily motor proteins and intracellular transport. *Nature Reviews Molecular Cell Biology* **10**:682–696.

Khalil, A. S., Appleyard, D. C., Labno, A. K., Georges, A., et al., 2009. Kinesin's cover-neck bundle folds forward to generate force. *Proceedings of the National Academy of Sciences* **105**:19247–19252.

Kikkawa, M., 2008. The role of microtubules in kinesin movement. *Trends in Cell Biology* **18**:128–135.

Marx, A., Hoenger, A., and Mandelkow, E., 2009. Structures of kinesin motor proteins. *Cell Motility and the Cytoskeleton* **66**:958–966.

Marx, A., Muller, J., et al., 2005. The structure of microtubule motor proteins. *Advances in Protein Chemistry* **71**:299–344.

Skowronek, K. J., Kocik, E., et al., 2007. Subunits interactions in kinesin motors. *European Journal of Cell Biology* **86**:559–568.

Verhey, K. J., and Hammond, J. W., 2009. Traffic control: regulation of kinesin motors. *Nature Reviews Molecular Cell Biology* **10**:765–777.

Yildiz, A., and Selvin, P. R., 2005. Kinesin: Walking, crawling or sliding along? *Trends in Cell Biology* **15**:112–120.

Dyneins

Carter, A. P., Garbarino, J. E., Wilson-Kubalek. E. M., Shipley, W. E., et al., 2008. Structure and functional role of dynein's microtubule-binding domain. *Science* **322**:1691–1695.

Carter, A. P., and Vale, R. D., 2010. Communication between the AAA+ ring and microtubule-binding domain of dynein. *Biochemistry and Cell Biology* **88**:15–21.

Gennerich, A., and Vale, R. D., 2009. Walking the walk: how kinesin and dynein coordinate their steps. *Current Opinion in Cell Biology* **21**: 59–67.

Gross, S. P., Vershinin, M., et al., 2007. Cargo transport: Two motors are sometimes better than one. *Current Biology* **17**:R478–R486.

Kardon. J. R., and Vale, R. D., 2009. Regulators of the cytoplasmic dynein motor. *Nature Reviews Molecular Cell Biology* **10**:854–865.

Kon, T., Imamula, K., Roberts, A. J., Ohkura, R., et al., 2009. Helix sliding in the stalk coiled coil of dynein couples ATPase and microtubule binding. *Nature Structural and Molecular Biology* **16**:325–333.

Serohijos, A. W. R., Chen, Y., et al., 2006. A structural model reveals energy transduction in dynein. *Proceedings of the National Academy of Sciences U.S.A.* **103**:18540–18545.

Intermediate Filaments

Caviston, J. P., and Holzbaur, E. L., 2006. Microtubule motors at the intersection of trafficking and transport. *Trends in Cell Biology* **16**:530–537.

Chou, Y.-H., Flitney, F. W., et al., 2007. The motility and dynamic properties of intermediate filaments and their constituent proteins. *Experimental Cell Research* **313**:2236–2243.

Goldman, R. D., Grin, B., Mendez, M. G., and Kuczmarski, E. R., 2008. Intermediate filaments: versatile building blocks of cell structure. *Current Opinion in Cell Biology* **20**:28–34.

Hirokawa, N., 2006. mRNA transport in dendrites: RNA granules, motors, and tracks. *Journal of Neuroscience* **26**:7139–7142.

Hirokawa, N., and Takemura, R., 2005. Molecular motors and mechanisms of directional transport in neurons. *Nature Reviews Neuroscience* **6**:201–214.

Michie, K., and Lowe, J., 2006. Dynamic filaments of the bacterial cytoskeleton. *Annual Review of Biochemistry* **75**:467–492.

Styers, M., Kawalczyk, A. P., et al., 2007. Intermediate filaments and vesicular membrane traffic: The odd couple's first dance? *Traffic* **6**:359–365.

Toivola, M., Strnad, P., Habtezion, A., and Omary, M. B., 2010. Intermediate filaments take the heat as stress proteins. *Trends in Cell Biology* **20**:79–91.

Vale, R. D., 2003. The molecular motor toolbox for intracellular transport. *Cell* **112**:467–480.

Vale, R. D., and Milligan, R. A., 2000. The way things move: Looking under the hood of molecular motor proteins. *Science* **288**:88–95.

Verhey, K. J., and Gaertig, J., 2007. The tubulin code. *Cell Cycle* **6**:2152–2160.

Helicases

Blainey, P. C., Luo, G., Kou, S. C., Mangel, W. F., et al., 2009. Nonspecifically bound proteins spin while diffusing along DNA. *Nature Structural and Molecular Biology* **16**:1224–1230.

Castella, S., Bingham, G., et al., 2006. Common determinants in DNA melting and helicase-catalyzed DNA unwinding by papillomavirus replication protein E1. *Nucleic Acids Research* **34**:3008–3019.

Enemark, E. J., and Joshua-Tor, L., 2006. Mechanism of DNA translocation in a replicative hexameric helicase. *Nature* **442**:270–275.

Enemark, E. J., and Joshua-Tor, L., 2008. On helicases and other motor proteins. *Current Opinion in Structural Biology* **18**:243–257.

Georgescu, R. E., and O'Donnell, M., 2007. Getting DNA to unwind. *Science* **317**:1181–1182.

Ha, T., 2007. Need for speed: Mechanical regulation of a replicative helicase. *Cell* **129**:1249–1250.

Johnson, D. S., Bai, L., et al., 2007. Single-molecule studies reveal dynamics of DNA unwinding by the ring-shaped T7 helicase. *Cell* **129**:1299–1309.

Okorokov, A. L., Waugh, A., et al., 2007. Hexameric ring structure of human MCM10 DNA replication factor. *EMBO Reports* **8**:925–930.

Raney, K. D., 2006. A helicase staircase. *Nature Structural and Molecular Biology* **13**:671–672.

Singleton, M. R., Dillingham, M. S., et al., 2007. Structure and mechanism of helicases and nucleic acid translocases. *Annual Review of Biochemistry* **76**:23–50.

Yoshimoto, K., Arora, K., and Brooks III, C. L., 2010. Hexameric helicase deconstructed: Interplay of conformational changes and substrate coupling. *Biophysical Journal* **98**:1449–1457.

Flagellar Rotor

Brown, P. N., Terrazas, M., et al., 2007. Mutational analysis of the flagellar protein FliG: Sites of interaction with FliM and implications for organization of the switch complex. *Journal of Bacteriology* **189**:305–312.

Calladine, C. R., 2010. New twists for bacterial flagella. *Nature Structural and Molecular Biology* **17**:395–396.

Dyer, C., and Dahlquist, F. W., 2006. Switched or not? The structure of unphosphorylated CheY bound to the N-terminus of FliM. *Journal of Bacteriology* **188**:7354–7363.

Fujii, T., Kato, T., and Namba, K., 2010. Specific arrangement of α-helical coiled coils in the core domain of the bacterial flagellar hook for the universal joint function. *Structure* **17**:1485–1493.

Hosking, E. R., Vogt, C., et al., 2006. The *Escherichia coli* motAB proton channel unplugged. *Journal of Molecular Biology* **364**:921–937.

Maki-Yonekura, S., Yonekura, K., and Namba, K., 2010. Conformational change of flagellin for polymorphic supercoiling of the flagellar filament. *Nature Structural and Molecular Biology* **17**:417–423.

Nakamura, S., Kami-Ike, N., Yokata, J. P., Minamino, T., and Namba, K., 2010. Evidence for symmetry in the elementary process of bidirectional torque generation by the bacterial flagellar motor. *Proceedings of the National Academy of Sciences* **107**:17616–17620.

Van den Heuvel, M. G. L., and Dekker, C., 2007. Motor proteins at work for nanotechnology. *Science* **317**:333–336.

Waters, R. C., O'Toole, P. W., et al., 2007. The FliK protein and flagellar hook-length control. *Protein Science* **16**:769–780.

Yakushi, T., Yang, J., et al., 2006. Roles of charged residues of rotor and stator in flagellar rotation: Comparative study using H^+-driven and Na^+-driven motors in *Escherichia coli*. *Journal of Bacteriology* **188**:1466–1472.

Metabolism: An Overview

17

© Gray Hardel/CORBIS

◀ Anise swallowtail butterfly *(Papilio zelicans)* with its pupal case. Metamorphosis of butterflies is a dramatic example of metabolic change.

All is flux, nothing stays still.
Nothing endures but change.

Heraclitus (c. 540–c. 480 B.C.)

KEY QUESTIONS

17.1 Is Metabolism Similar in Different Organisms?

17.2 What Can Be Learned from Metabolic Maps?

17.3 How Do Anabolic and Catabolic Processes Form the Core of Metabolic Pathways?

17.4 What Experiments Can Be Used to Elucidate Metabolic Pathways?

17.5 What Can the Metabolome Tell Us about a Biological System?

17.6 What Food Substances Form the Basis of Human Nutrition?

ESSENTIAL QUESTION

The word *metabolism* derives from the Greek word for "change." **Metabolism** represents the sum of the chemical changes that convert **nutrients,** the "raw materials" necessary to nourish living organisms, into energy and the chemically complex finished products of cells. Metabolism consists of literally hundreds of enzymatic reactions organized into discrete pathways. These pathways proceed in a stepwise fashion, transforming substrates into end products through many specific chemical **intermediates.** Metabolism is sometimes referred to as intermediary metabolism to reflect this aspect of the process.

What are the anabolic and catabolic processes that satisfy the metabolic needs of the cell?

17.1   Is Metabolism Similar in Different Organisms?

One of the great unifying principles of modern biology is that organisms show marked similarity in their major pathways of metabolism. Given the almost unlimited possibilities within organic chemistry, this generality would appear most unlikely. Yet it's true, and it provides strong evidence that all life has descended from a common ancestral form. All forms of nutrition and almost all metabolic pathways evolved in early prokaryotes prior to the appearance of eukaryotes 1 billion years ago. For example, **glycolysis,** the metabolic pathway by which energy is released from glucose and captured in the form of ATP under anaerobic conditions, is common to almost every cell. It is believed to be the most ancient of metabolic pathways, having arisen prior to the appearance of oxygen in abundance in the atmosphere. All organisms, even those that can synthesize their own glucose, are capable of glucose degradation and ATP synthesis via glycolysis. Other prominent pathways are also virtually ubiquitous among organisms.

OWL Online homework and a Student Self Assessment for this chapter may be assigned in OWL.

TABLE 17.1	Metabolic Classification of Organisms According to Their Carbon and Energy Requirements				
Classification	Carbon Source	Energy Source	Electron Donors		Examples
Photoautotrophs	CO_2	Light	H_2O, H_2S, S, other inorganic compounds		Green plants, algae, cyanobacteria, photosynthetic bacteria
Photoheterotrophs	Organic compounds	Light	Organic compounds		Nonsulfur purple bacteria
Chemoautotrophs	CO_2	Oxidation–reduction reactions	Inorganic compounds: H_2, H_2S, NH_4^+, NO_2^-, Fe^{2+}, Mn^{2+}		Nitrifying bacteria; hydrogen, sulfur, and iron bacteria
Chemoheterotrophs	Organic compounds	Oxidation–reduction reactions	Organic compounds (e.g., glucose)		All animals, most microorganisms, nonphotosynthetic plant tissue such as roots, photosynthetic cells in the dark

Living Things Exhibit Metabolic Diversity

Although most cells have the same basic set of central metabolic pathways, different cells (and, by extension, different organisms) are characterized by the alternative pathways they might express. These pathways offer a wide diversity of metabolic possibilities. For instance, organisms are often classified according to the major metabolic pathways they exploit to obtain carbon or energy. Classification based on carbon requirements defines two major groups: autotrophs and heterotrophs. **Autotrophs** are organisms that can use carbon dioxide as their only source of carbon. **Heterotrophs** require an organic form of carbon, such as glucose, in order to synthesize other essential carbon compounds.

Classification based on energy sources also gives two groups: phototrophs and chemotrophs. **Phototrophs** are *photosynthetic organisms,* which use light as a source of energy. **Chemotrophs** use organic compounds such as glucose or, in some instances, oxidizable inorganic substances such as Fe^{2+}, NO_2^-, NH_4^+, or elemental sulfur as sole sources of energy. Typically, the energy is extracted through oxidation–reduction reactions. Based on these characteristics, every organism falls into one of four categories (Table 17.1).

Metabolic Diversity Among the Five Kingdoms Prokaryotes (the kingdom Monera—archaea and bacteria) show a greater metabolic diversity than all the four eukaryotic kingdoms (Protista, Fungi, Plants, and Animals) put together. Prokaryotes are variously chemoheterotrophic, photoautotrophic, photoheterotrophic, or chemoautotrophic. No protists are chemoautotrophs; fungi and animals are exclusively chemoheterotrophs; plants are characteristically photoautotrophs, although some are heterotrophic in their mode of carbon acquisition.

Oxygen Is Essential to Life for Aerobes

A further metabolic distinction among organisms is whether or not they can use oxygen as an electron acceptor in energy-producing pathways. Those that can are called **aerobes** or *aerobic organisms;* others, termed **anaerobes,** can subsist without O_2. Organisms for which O_2 is obligatory for life are called **obligate aerobes;** humans are an example. Some species, the so-called **facultative anaerobes,** can adapt to anaerobic conditions by substituting other electron acceptors for O_2 in their energy-producing pathways; *Escherichia coli* is an example. Yet others cannot use oxygen at all and are even poisoned by it; these are the **obligate anaerobes.** *Clostridium botulinum,* the bacterium that produces botulin toxin, is representative.

The Flow of Energy in the Biosphere and the Carbon and Oxygen Cycles Are Intimately Related

The primary source of energy for life is the sun. Photoautotrophs utilize light energy to drive the synthesis of organic molecules, such as carbohydrates, from atmospheric CO_2 and water (Figure 17.1). Molecular oxygen (O_2) is a by-product of this process.

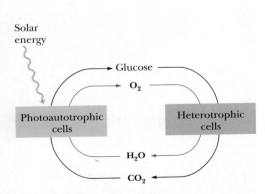

FIGURE 17.1 The flow of energy in the biosphere is coupled primarily to the carbon and oxygen cycles.

▶ **A DEEPER LOOK**

Calcium Carbonate—A Biological Sink for CO_2

A major biological sink for CO_2 that is often overlooked is the calcium carbonate shells of corals, molluscs, and crustacea. These invertebrate animals deposit $CaCO_3$ in the form of protective exoskeletons. In some invertebrates, such as the *scleractinians* (hard corals) of tropical seas, photosynthetic dinoflagellates (marine eukaryotic algae) live within the animal cells as **endosymbionts** called *zooxanthellae*. These phototrophic cells use light to drive the resynthesis of organic molecules from CO_2 released (as bicarbonate ion) by the animal's metabolic activity. In the presence of Ca^{2+}, the photosynthetic CO_2 fixation "pulls" the deposition of $CaCO_3$, as summarized in the following coupled reactions:

$$Ca^{2+} + 2\ HCO_3^- \rightleftharpoons CaCO_{3(s)}\downarrow + H_2CO_3$$
$$H_2CO_3 \rightleftharpoons H_2O + CO_2$$
$$H_2O + CO_2 \longrightarrow carbohydrate + O_2$$

Heterotrophic cells then use these organic products of photosynthetic cells both as fuels and as building blocks, or precursors, for the biosynthesis of their own unique complement of biomolecules. O_2 is reduced to water as a consequence of energy-releasing electron-transfer reactions that underlie the use of these organic fuels. Ultimately, CO_2 is the end product of heterotrophic carbon metabolism, and CO_2 is returned to the atmosphere for reuse by the photoautotrophs. In effect, solar energy is converted to the chemical energy of organic molecules by photoautotrophs, and heterotrophs recover this energy by oxidizing the organic substances. The flow of energy in the biosphere is thus conveyed within the carbon and oxygen cycles, and the impetus driving these cycles is light energy.

17.2 | What Can Be Learned from Metabolic Maps?

Metabolic maps (Figure 17.2) portray the principal reactions of the intermediary metabolism of carbohydrates, lipids, amino acids, nucleotides, and their derivatives. These maps are very complex at first glance and would seem virtually impossible to learn easily. Despite their appearance, these maps become easy to follow once the major metabolic routes are known and their functions are understood. The underlying order of metabolism and the important interrelationships between the various pathways then appear as simple patterns against the seemingly complicated background.

The Metabolic Map Can Be Viewed as a Set of Dots and Lines

One interesting transformation of the intermediary metabolism map is to represent each intermediate as a black dot and each enzyme as a line (Figure 17.3). Then, the more than 1000 different enzymes and substrates are represented by just two symbols. This chart has about 520 dots (intermediates). Table 17.2 lists the numbers of dots that have one or two or more lines (enzymes) associated with them. Thus, this table classifies intermediates by the number of enzymes that act upon them. A dot connected to just a single line must be either a nutrient, a storage form, an end product, or an excretory product of metabolism. Also, because many pathways tend to proceed in only one direction (that is, they are essentially irreversible under physiological conditions), a dot connected to just two lines is probably an intermediate in only one pathway and has only one fate in metabolism. If three lines are connected to a dot, that intermediate has at least two possible metabolic fates; four lines, three fates; and so on. Note that about 80% of the intermediates connect to only one or two lines and thus have only a single role in the cell. However, intermediates at branch points are subject to a variety of fates. In such instances, the pathway followed is an important regulatory choice. Indeed, whether any substrate is routed down a particular metabolic pathway is the consequence of a regulatory decision made in response to the cell's (or organism's) momentary requirements for energy or nutrition. The regulation of metabolism is an interesting and important subject to which we will return often.

TABLE 17.2	Number of Dots (Intermediates) in the Metabolic Map of Figure 17.2, and the Number of Lines Associated with Them
Lines	Dots
1 or 2	410
3	71
4	20
5	11
6 or more	8

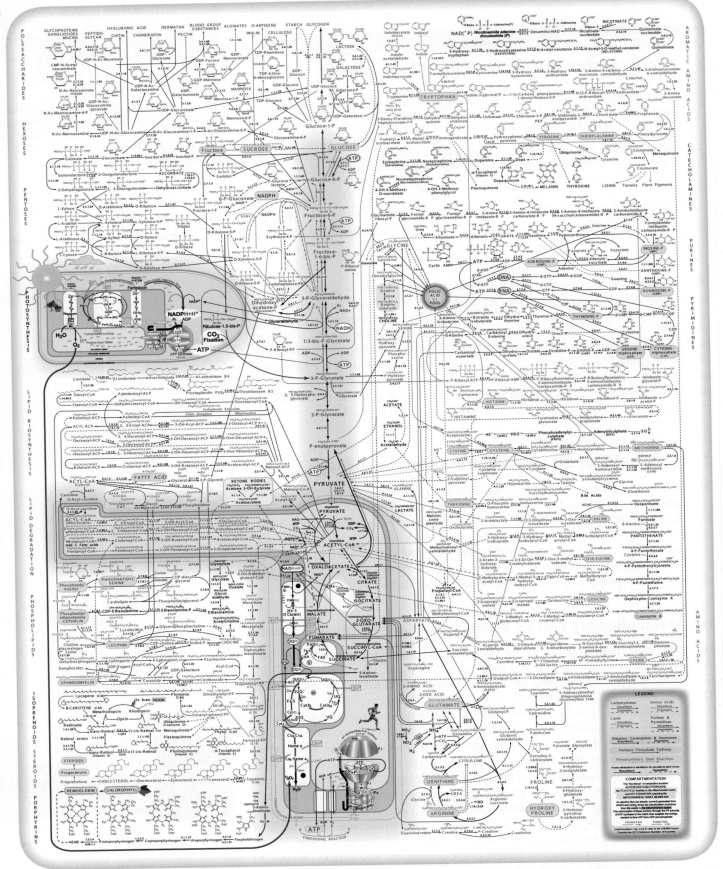

FIGURE 17.2 A metabolic map, indicating the reactions of intermediary metabolism and the enzymes that catalyze them. More than 500 different chemical intermediates, or metabolites, and a greater number of enzymes are represented here. (Source: From Donald Nicholson, Map #22, © International Union of Biochemistry and Molecular Biology.)

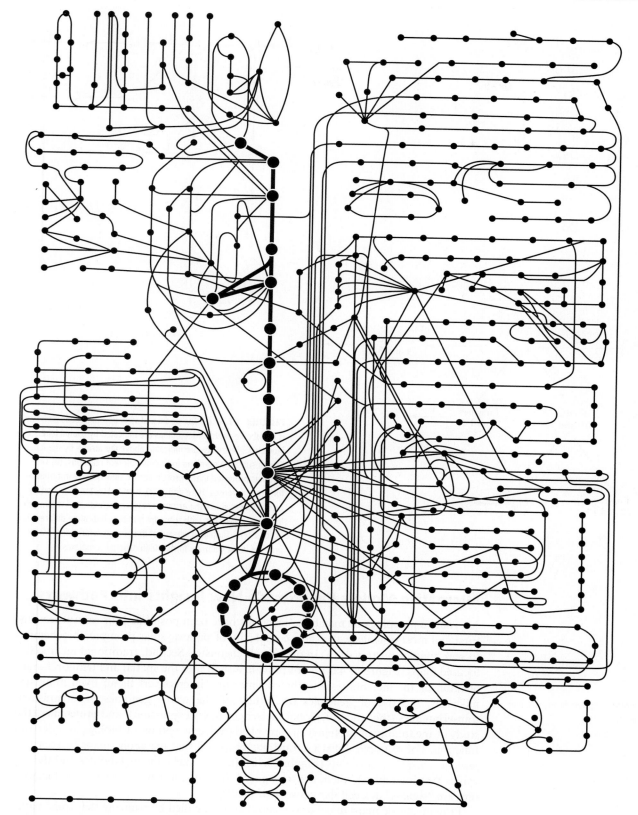

FIGURE 17.3 The metabolic map as a set of dots and lines. The heavy dots and lines trace the central energy-releasing pathways known as glycolysis and the citric acid cycle. (Adapted from Alberts, B., et al., 1989. *Molecular Biology of the Cell,* 2nd ed. New York: Garland Publishing Co.)

(a)

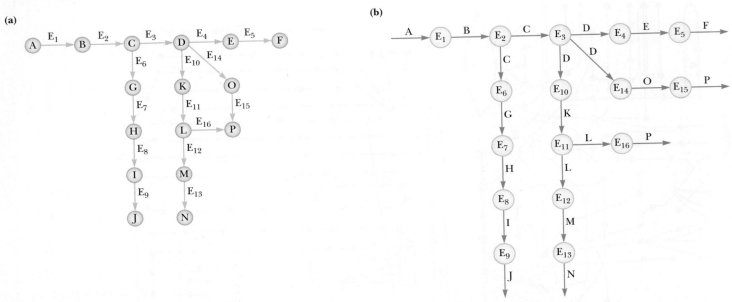

(c)

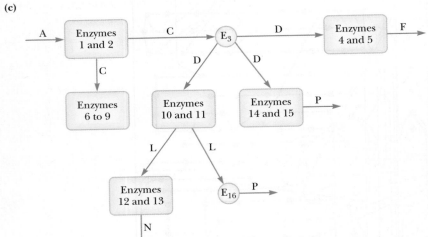

Metabolite: Any small organic molecule involved in or produced by metabolism

FIGURE 17.4 **(a)** The traditional view of a metabolic pathway is metabolite-centric. Metabolites are indicated by capital letters in blue circles; enzymes are indicated above the arrows by the letter E, with subscript numbers indicating different reactions. **(b)** Gerrard has proposed that a protein-centric view is more informative for some purposes. Here and in **(c)**, metabolites are indicated by capital letters above the arrows; enzymes are indicated within yellow circles or boxes, with subscript numbers indicating different reactions. **(c)** A simplified version of the protein-centric view where proteins in the pathway form multifunctional complexes.

Alternative Models Can Provide New Insights into Pathways

Alternative mappings of metabolic reactions have been postulated for several reasons. First and most obviously, the sheer complexity of pathways has prompted biochemists to seek simpler portrayals of an organism's chemistry. Second, traditional metabolite-focused maps (Figure 17.4a) do not provide insight into the spatial and temporal organization of the metabolites and the enzymes that interconvert them. Even more significantly, the rise of **genomics** (the study of the whole genomes of organisms), **transcriptomics** (the study of global messenger RNA expression), and **proteomics** (the study of the totality of proteins) has provoked fresh conceptions of biological order and function. For example, Juliet Gerrard has proposed that metabolic maps be recast in *protein-centric* presentations (Figure 17.4b). In such maps, the metabolites and the enzymes that interconvert them are transposed, revealing a new emphasis—the metabolites are "signals" in a cellular network of proteins.

Protein-centric maps may be condensed and simplified by realizing that some pathway enzymes are clustered in *multienzyme complexes* and that metabolites are literally passed from enzyme to enzyme within such clusters (Figure 17.4c). The result is a simplified representation of metabolic networks, containing only the essential signaling information. Metabolic maps are representations of large amounts of information. Conceptualizing them in different formats enables biochemists to analyze vast amounts of information in new and insightful ways.

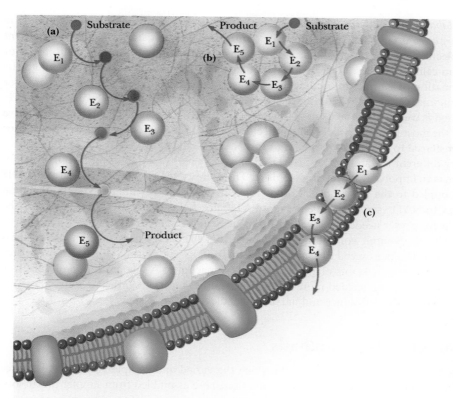

FIGURE 17.5 Schematic representation of types of multienzyme systems carrying out a metabolic pathway: **(a)** Physically separate, soluble enzymes with diffusing intermediates. **(b)** A multienzyme complex. Substrate enters the complex and becomes bound and then sequentially modified by enzymes E_1 to E_5 before product is released. No intermediates are free to diffuse away. **(c)** A membrane-bound multienzyme system.

Multienzyme Systems May Take Different Forms

The individual metabolic pathways of anabolism and catabolism consist of sequential enzymatic steps (Figure 17.5). Several types of organization are possible. The enzymes of some multienzyme systems may exist as physically separate, soluble entities, with diffusing intermediates (Figure 17.5a). In other instances, the enzymes of a pathway are collected to form a discrete multienzyme complex, and the substrate is sequentially modified as it is passed along from enzyme to enzyme (Figure 17.5b). This type of organization has the advantage that intermediates are not lost or diluted by diffusion. In a third pattern of organization, the enzymes common to a pathway reside together as a *membrane-bound system* (Figure 17.5c). In this case, the enzyme participants (and perhaps the substrates as well) must diffuse in just the two dimensions of the membrane to interact with their neighbors.

As research reveals the ultrastructural organization of the cell in ever greater detail, more and more of the so-called soluble enzyme systems are found to be physically united into functional complexes. Thus, in many (perhaps all) metabolic pathways, the consecutively acting enzymes interact with one another to form stable aggregations that are sometimes referred to as **metabolons,** a word meaning "units of metabolism." Further, in more highly evolved organisms such as birds and mammals, evolutionary processes have led to the fusion of genes encoding enzymes for several sequential steps in a metabolic pathway. Expression of the fused gene produces a single **multifunctional polypeptide** capable of catalyzing a series of steps in a pathway.

17.3 How Do Anabolic and Catabolic Processes Form the Core of Metabolic Pathways?

Metabolism serves two fundamentally different purposes: the generation of energy to drive vital functions and the synthesis of biological molecules. To achieve these ends, metabolism consists largely of two contrasting processes: catabolism and anabolism.

Catabolic pathways are characteristically energy yielding, whereas anabolic pathways are energy requiring. **Catabolism** involves the oxidative degradation of complex nutrient molecules (carbohydrates, lipids, and proteins) obtained either from the environment or from cellular reserves. The breakdown of these molecules by catabolism leads to the formation of simpler molecules such as lactic acid, ethanol, carbon dioxide, urea, or ammonia. Catabolic reactions are usually exergonic, and often the chemical energy released is captured in the form of ATP (see Chapter 3). Because catabolism is oxidative for the most part, part of the chemical energy may be conserved as energy-rich electrons transferred to the coenzymes NAD^+ and $NADP^+$. These two reduced coenzymes have very different metabolic roles: NAD^+ reduction is part of catabolism; NADPH oxidation is an important aspect of anabolism. The energy released upon oxidation of NADH is coupled to the phosphorylation of ADP in aerobic cells, and so NADH oxidation back to NAD^+ serves to generate more ATP; in contrast, NADPH is the source of the reducing power needed to drive reductive biosynthetic reactions.

Thermodynamic considerations demand that the energy necessary for biosynthesis of any substance exceed the energy available from its catabolism. Otherwise, organisms could achieve the status of perpetual motion machines: A few molecules of substrate whose catabolism yielded more ATP than required for its resynthesis would allow the cell to cycle this substance and harvest an endless supply of energy.

Anabolism Is Biosynthesis

Anabolism is a synthetic process in which the varied and complex biomolecules (proteins, nucleic acids, polysaccharides, and lipids) are assembled from simpler precursors. Such biosynthesis involves the formation of new covalent bonds, and an input of chemical energy is necessary to drive such endergonic processes. The ATP generated by catabolism provides this energy. Furthermore, NADPH is an excellent donor of high-energy electrons for the reductive reactions of anabolism. Despite their divergent roles, anabolism and catabolism are interrelated in that the products of one provide the substrates of the other (Figure 17.6). Many metabolic intermediates are shared between the two processes, and the precursors needed by anabolic pathways are found among the products of catabolism.

Anabolism and Catabolism Are Not Mutually Exclusive

Interestingly, anabolism and catabolism occur simultaneously in the cell. The conflicting demands of concomitant catabolism and anabolism are managed by cells in two ways. First, the cell maintains tight and separate regulation of both catabolism and

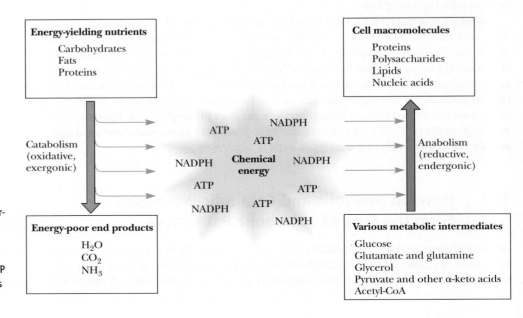

FIGURE 17.6 Energy relationships between the pathways of catabolism and anabolism. Oxidative, exergonic pathways of catabolism release free energy and reducing power that are captured in the form of ATP and NADPH, respectively. Anabolic processes are endergonic, consuming chemical energy in the form of ATP and using NADPH as a source of high-energy electrons for reductive purposes.

anabolism, so metabolic needs are served in an immediate and orderly fashion. Second, competing metabolic pathways are often localized within different cellular compartments. Isolating opposing activities within distinct compartments, such as separate organelles, avoids interference between them. For example, the enzymes responsible for catabolism of fatty acids, the *fatty acid oxidation pathway*, are localized within mitochondria. In contrast, *fatty acid biosynthesis* takes place in the cytosol. In subsequent chapters, we shall see that the particular molecular interactions responsible for the regulation of metabolism become important for an understanding and appreciation of metabolic biochemistry.

The Pathways of Catabolism Converge to a Few End Products

If we survey the catabolism of the principal energy-yielding nutrients (carbohydrates, lipids, and proteins) in a typical heterotrophic cell, we see that the degradation of these substances involves a succession of enzymatic reactions. In the presence of oxygen *(aerobic catabolism)*, these molecules are degraded ultimately to carbon dioxide, water, and ammonia. Aerobic catabolism consists of three distinct stages. In **stage 1,** the nutrient macromolecules are broken down into their respective building blocks. Despite the great diversity of macromolecules, these building blocks represent a rather limited number of products. Proteins yield up their 20 component amino acids, polysaccharides give rise to carbohydrate units that are convertible to glucose, and lipids are broken down into glycerol and fatty acids (Figure 17.7).

In **stage 2,** the collection of product building blocks generated in stage 1 is further degraded to yield an even more limited set of simpler metabolic intermediates. The deamination of amino acids leaves α-keto acid carbon skeletons. Several of these α-keto acids are citric acid cycle intermediates and are fed directly into stage 3 catabolism via this cycle. Others are converted either to the three-carbon α-keto acid *pyruvate* or to the acetyl groups of *acetyl-coenzyme* A (acetyl-CoA). Glucose and the glycerol from lipids also generate pyruvate, whereas the fatty acids are broken into two-carbon units that appear as *acetyl-CoA*. Because pyruvate also gives rise to acetyl-CoA, we see that the degradation of macromolecular nutrients converges to a common end product, acetyl-CoA (Figure 17.7).

The combustion of the acetyl groups of acetyl-CoA by the citric acid cycle and *oxidative phosphorylation* to produce CO_2 and H_2O represents **stage 3** of catabolism. The end products of the citric acid cycle, CO_2 and H_2O, are the ultimate waste products of aerobic catabolism. As we shall see in Chapter 19, the oxidation of acetyl-CoA during stage 3 metabolism generates most of the energy produced by the cell.

Anabolic Pathways Diverge, Synthesizing an Astounding Variety of Biomolecules from a Limited Set of Building Blocks

A rather limited collection of simple precursor molecules is sufficient to provide for the biosynthesis of virtually any cellular constituent, be it protein, nucleic acid, lipid, or polysaccharide. All of these substances are constructed from appropriate building blocks via the pathways of anabolism. In turn, the building blocks (amino acids, nucleotides, sugars, and fatty acids) can be generated from metabolites in the cell. For example, amino acids can be formed by amination of the corresponding α-keto acid carbon skeletons, and pyruvate can be converted to hexoses for polysaccharide biosynthesis.

Amphibolic Intermediates Play Dual Roles

Certain of the central pathways of intermediary metabolism, such as the citric acid cycle, and many metabolites of other pathways have dual purposes—they serve in both catabolism and anabolism. This dual nature is reflected in the designation of such pathways as **amphibolic** rather than solely catabolic or anabolic. In any event, in contrast to catabolism—which converges to the common intermediate, acetyl-CoA—the pathways

Amphi is from the Greek for "on both sides."

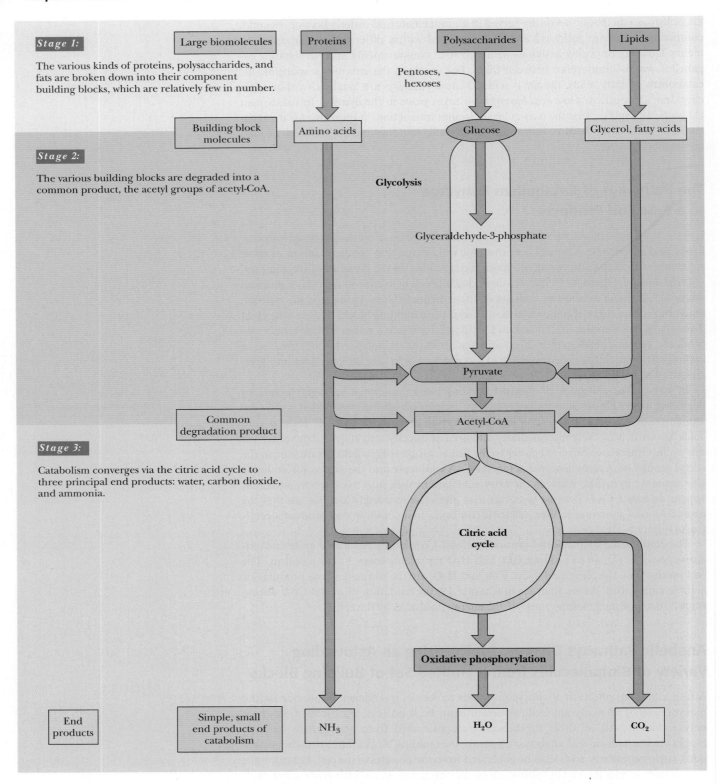

Stage 1:

The various kinds of proteins, polysaccharides, and fats are broken down into their component building blocks, which are relatively few in number.

Stage 2:

The various building blocks are degraded into a common product, the acetyl groups of acetyl-CoA.

Stage 3:

Catabolism converges via the citric acid cycle to three principal end products: water, carbon dioxide, and ammonia.

Large biomolecules

Building block molecules

Common degradation product

End products

Simple, small end products of catabolism

Proteins

Amino acids

Polysaccharides

Pentoses, hexoses

Glucose

Glycolysis

Glyceraldehyde-3-phosphate

Pyruvate

Acetyl-CoA

Lipids

Glycerol, fatty acids

Citric acid cycle

Oxidative phosphorylation

NH_3

H_2O

CO_2

FIGURE 17.7 The three stages of catabolism. **Stage 1:** Proteins, polysaccharides, and lipids are broken down into their component building blocks, which are relatively few in number. **Stage 2:** The various building blocks are degraded into the common product, the acetyl groups of acetyl-CoA. **Stage 3:** Catabolism converges to three principal end products: water, carbon dioxide, and ammonia.

of anabolism diverge from a small group of simple metabolic intermediates to yield a spectacular variety of cellular constituents.

Corresponding Pathways of Catabolism and Anabolism Differ in Important Ways

The anabolic pathway for synthesis of a given end product usually does not match precisely the pathway used for catabolism of the same substance. Some of the intermediates may be common to steps in both pathways, but different enzymatic reactions and unique metabolites characterize other steps. A good example of these differences is found in a comparison of the catabolism of glucose to pyruvic acid by the pathway of glycolysis and the biosynthesis of glucose from pyruvate by the pathway called *gluconeogenesis*. The glycolytic pathway from glucose to pyruvate consists of 10 enzymes. Although it may seem efficient for glucose synthesis from pyruvate to proceed by a reversal of all 10 steps, gluconeogenesis uses only seven of the glycolytic enzymes in reverse, replacing those remaining with four enzymes specific to glucose biosynthesis. In similar fashion, the pathway responsible for degrading proteins to amino acids differs from the protein synthesis system, and the oxidative degradation of fatty acids to two-carbon acetyl-CoA groups does not follow the same reaction path as the biosynthesis of fatty acids from acetyl-CoA.

Metabolic Regulation Requires Different Pathways for Oppositely Directed Metabolic Sequences A second reason for different pathways serving in opposite metabolic directions is that such pathways must be independently regulated. If catabolism and anabolism passed along the same set of metabolic tracks, equilibrium considerations would dictate that slowing the traffic in one direction by inhibiting a particular enzymatic reaction would necessarily slow traffic in the opposite direction. Independent regulation of anabolism and catabolism can be accomplished only if these two contrasting processes move along different routes or, in the case of shared pathways, the rate-limiting steps serving as the points of regulation are catalyzed by enzymes that are unique to each opposing sequence (Figure 17.8). Note that opposing pathways are reciprocally regulated: Activation of one is accompanied by inhibition of the other. Reciprocal regulation means that the opposing metabolic sequences function in only one

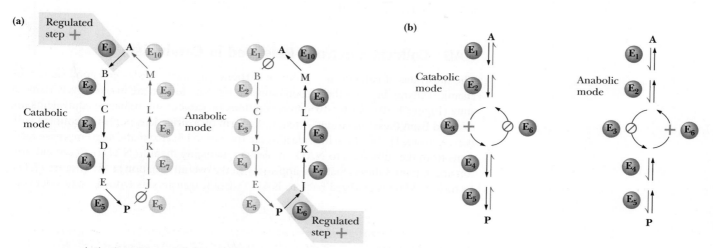

FIGURE 17.8 Parallel pathways of catabolism and anabolism must differ in at least one metabolic step in order that they can be regulated independently. Shown here are two possible arrangements of opposing catabolic and anabolic sequences between A and P. **(a)** The parallel sequences proceed via independent routes. **(b)** Only one reaction has two different enzymes, a catabolic one (E_5) and its anabolic counterpart (E_6). These provide sites for regulation.

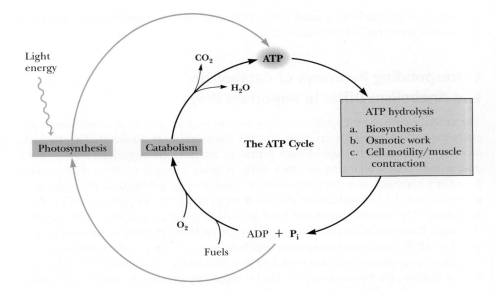

direction at any moment in time. The direction in which they operate is determined by allosteric effectors acting on key steps in the opposing pathways.

ATP Serves in a Cellular Energy Cycle

We saw in Chapter 3 that ATP is the energy currency of cells. In phototrophs, ATP is one of the two energy-rich primary products resulting from the transformation of light energy into chemical energy. (The other is NADPH; see the following discussion.) In heterotrophs, the pathways of catabolism have as their major purpose the release of free energy that can be captured in the form of energy-rich phosphoric anhydride bonds in ATP. In turn, ATP provides the energy that drives the manifold activities of all living cells—the synthesis of complex biomolecules, the osmotic work involved in transport-ing substances into cells, the work of cell motility, and the work of muscle contraction. These diverse activities are all powered by energy released in the hydrolysis of ATP to ADP and P_i. Thus, there is an energy cycle in cells where ATP serves as the vessel carry-ing energy from photosynthesis or catabolism to the energy-requiring processes unique to living cells (Figure 17.9).

NAD$^+$ Collects Electrons Released in Catabolism

The substrates of catabolism—proteins, carbohydrates, and lipids—are good sources of chemical energy because the carbon atoms in these molecules are in a relatively reduced state (Figure 17.10). In the oxidative reactions of catabolism, reducing equivalents are released from these substrates, often in the form of **hydride ions** (a proton coupled with two electrons, $H:^-$). These hydride ions are transferred in enzymatic **dehydrogenase** reac-tions from the substrates to NAD$^+$ molecules, reducing them to NADH. A second pro-ton accompanies these reactions, appearing in the overall equation as H^+ (Figure 17.11). In turn, NADH is oxidized back to NAD$^+$ when it transfers its reducing equivalents to

FIGURE 17.10 Comparison of the state of reduction of carbon atoms in biomolecules: $—CH_2—$ (fats) $>$ $—CHOH—$ (carbohydrates) $>C=O$ (carbonyls) $>$ $—COOH$ (carboxyls) $> CO_2$ (carbon dioxide, the final product of catabolism).

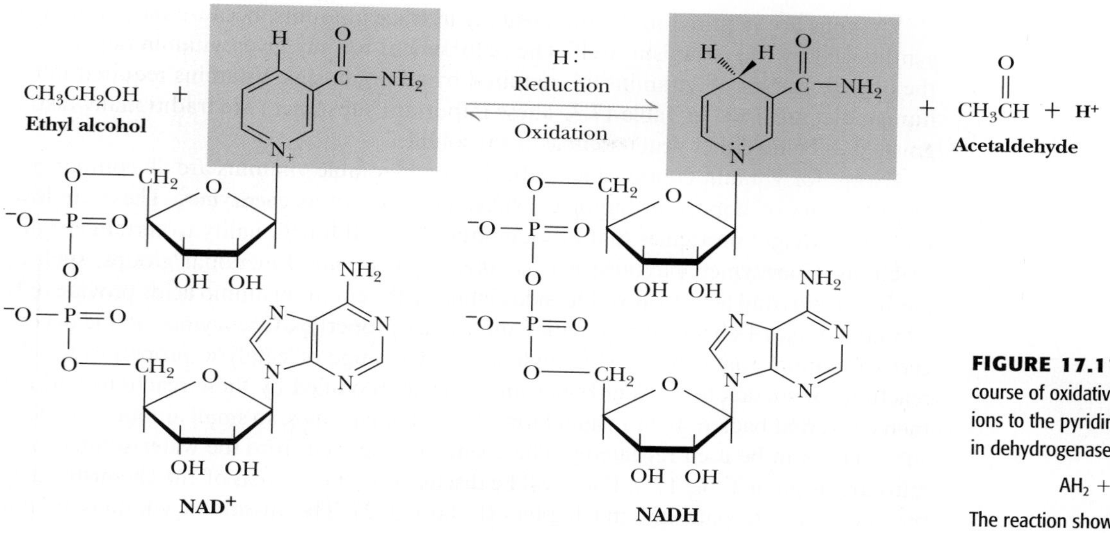

CH_3CH_2OH +
Ethyl alcohol

NAD$^+$

$\xrightarrow[\text{Oxidation}]{\begin{array}{c}H:^-\\ \text{Reduction}\end{array}}$

NADH

+ CH_3CH + H^+
Acetaldehyde

FIGURE 17.11 Hydrogen and electrons released in the course of oxidative catabolism are transferred as hydride ions to the pyridine nucleotide, NAD$^+$, to form NADH + H$^+$ in dehydrogenase reactions of the type

$$AH_2 + NAD^+ \longrightarrow A + NADH + H^+$$

The reaction shown is catalyzed by alcohol dehydrogenase.

electron acceptor systems that are part of the metabolic apparatus of the mitochondria. The ultimate oxidizing agent (e^- acceptor) is O_2, becoming reduced to H_2O.

Oxidation reactions are exergonic, and the energy released is coupled with the formation of ATP in a process called **oxidative phosphorylation.** The NAD$^+$–NADH system can be viewed as a *shuttle* that carries the electrons released from catabolic substrates to the mitochondria, where they are transferred to O_2, the ultimate electron acceptor in catabolism. In the process, the free energy released is trapped in ATP. The NADH cycle is an important player in the transformation of the chemical energy of carbon compounds into the chemical energy of phosphoric anhydride bonds. Such transformations of energy from one form to another are referred to as **energy transduction.** Oxidative phosphorylation is one cellular mechanism for energy transduction. Chapter 20 is devoted to electron transport reactions and oxidative phosphorylation.

NADPH Provides the Reducing Power for Anabolic Processes

Whereas catabolism is fundamentally an oxidative process, anabolism is, by its contrasting nature, reductive. The biosynthesis of the complex constituents of the cell begins at the level of intermediates derived from the degradative pathways of catabolism; or, less commonly, biosynthesis begins with oxidized substances available in the inanimate environment, such as carbon dioxide. When the hydrocarbon chains of fatty acids are assembled from acetyl-CoA units, activated hydrogens are needed to reduce the carbonyl (C=O) carbon of acetyl-CoA into a —CH$_2$— at every other position along the chain. When glucose is synthesized from CO_2 during photosynthesis in plants, reducing power is required. These reducing equivalents are provided by NADPH, the usual source of high-energy hydrogens for reductive biosynthesis. NADPH is generated when NADP$^+$ is reduced with electrons in the form of hydride ions. In heterotrophic organisms, these electrons are removed from fuel molecules by NADP$^+$-specific dehydrogenases. In these organisms, NADPH can be viewed as the carrier of electrons from catabolic reactions to anabolic reactions (Figure 17.12). In photosynthetic organisms, the energy of light is used to pull electrons from water and transfer them to NADP$^+$; O_2 is a by-product of this process.

Coenzymes and Vitamins Provide Unique Chemistry and Essential Nutrients to Pathways

In addition to NAD$^+$ and NADPH, a variety of other small molecules are essential to metabolism. Some of these are essential nutrients called **vitamins.** (The name was coined by Kazimierz Funk, who discovered thiamine as a cure for beriberi in 1912 and termed it a "vital amine." He later proposed that other diseases might be cured by "vitamins.")

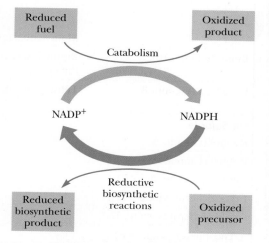

FIGURE 17.12 Transfer of reducing equivalents from catabolism to anabolism via the NADPH cycle.

Vitamins are required in the diet, usually in trace amounts, because they cannot be synthesized by the organism itself. The requirement for any given vitamin depends on the organism. Not all vitamins are required by all organisms. Vitamins required in the human diet are listed in Table 17.3. These important substances are traditionally distinguished as being either water soluble or fat soluble.

Except for vitamin C (ascorbic acid), the water-soluble vitamins are all components or precursors of important biological substances known as **coenzymes.** These are low-molecular-weight molecules that bring unique chemical functionality to certain enzyme reactions. Coenzymes may also act as *carriers* of specific functional groups, such as methyl groups and acyl groups. The side chains of the common amino acids provide only a limited range of chemical reactivities and carrier properties. Coenzymes, acting in concert with appropriate enzymes, provide a broader range of catalytic properties for the reactions of metabolism. Coenzymes are typically modified by these reactions and are then converted back to their original forms by other enzymes, so small amounts of these substances can be used repeatedly. The coenzymes derived from the water-soluble vitamins are listed in Table 17.3. Each will be discussed in the context of the chemistry they provide to specific pathways in Chapters 18 through 27. The fat-soluble vitamins are not

TABLE 17.3	Vitamins and Coenzymes		
Vitamin	Coenzyme Form	Function	Discussed in Chapter
Water-Soluble			
Thiamine (vitamin B_1)	Thiamine pyrophosphate	Decarboxylation of α-keto acids and formation and cleavage of α-hydroxyketones	19, 22
Niacin (vitamin B_3, nicotinic acid)	Nicotinamide adenine dinucleotide (NAD^+)	Hydride transfer	18–27
	Nicotinamide adenine dinucleotide phosphate ($NADP^+$)	Hydride transfer	21, 22, 24–26
Riboflavin (vitamin B_2)	Flavin adenine dinucleotide (FAD)	One- and two-electron transfer	19, 20, 23, 26
	Flavin mononucleotide (FMN)	One- and two-electron transfer	20
Pantothenic acid (vitamin B_5)	Coenzyme A	Activation of acyl groups for transfer by nucleophilic attack, and activation of the α-hydrogen of the acyl group for abstraction as a proton	19, 23, 24, 27
Pyridoxal, pyridoxine, pyridoxamine (vitamin B_6)	Pyridoxal phosphate	Formation of stable Schiff base (aldimine) adducts with α-amino groups of amino acids; serving as an electron sink to stabilize reaction intermediates	25
Cobalamin (vitamin B_{12})	5'-Deoxyadenosylcobalamin	Intramolecular rearrangement, reduction of ribonucleotides to deoxyribonucleotides, and methyl group transfer	23
	Methylcobalamin		
Biotin (vitamin B_7)	Biotin–lysine complexes (biocytin)	Carrier of carboxyl groups in carboxylation reactions	22, 24
Folic acid (vitamin B_9)	Tetrahydrofolate	Acceptor and donor of 1-C units for all oxidation levels of carbon except that of CO_2	25, 26
Fat-Soluble			
Retinol (vitamin A)			
Retinal (vitamin A)			
Retinoic acid (vitamin A)			
Ergocalciferol (vitamin D_2)			
Cholecalciferol (vitamin D_3)			
α-Tocopherol (vitamin E)			
Menaquinone (vitamin K)			

directly related to coenzymes, but they play essential roles in a variety of critical biological processes, including vision, maintenance of bone structure, and blood coagulation. The mechanisms of action of fat-soluble vitamins are not as well understood as their water-soluble counterparts, but modern research efforts are gradually closing this gap.

17.4 What Experiments Can Be Used to Elucidate Metabolic Pathways?

Armed with the knowledge that metabolism is organized into pathways of successive reactions, we can appreciate by hindsight the techniques employed by early biochemists to reveal their sequence. A major intellectual advance took place at the end of the 19th century when Eduard Buchner showed that the fermentation of glucose to yield ethanol and carbon dioxide can occur in extracts of broken yeast cells. Until this discovery, many thought that metabolism was a vital property, unique to intact cells; even the eminent microbiologist Louis Pasteur, who contributed so much to our understanding of fermentation, was a *vitalist,* one of those who believed that the processes of living substance transcend the laws of chemistry and physics. After Buchner's revelation, biochemists searched for intermediates in the transformation of glucose and soon learned that inorganic phosphate was essential to glucose breakdown. This observation gradually led to the discovery of a variety of phosphorylated organic compounds that serve as intermediates along the fermentative pathway.

An important tool for elucidating the steps in the pathway was the use of *metabolic inhibitors.* Adding an enzyme inhibitor to a cell-free extract caused an accumulation of intermediates in the pathway prior to the point of inhibition (Figure 17.13). Each inhibitor was specific for a particular site in the sequence of metabolic events. As the arsenal of inhibitors was expanded, the individual steps in metabolism were revealed.

Mutations Create Specific Metabolic Blocks

Genetics provides an approach to the identification of intermediate steps in metabolism that is somewhat analogous to inhibition. Mutation in a gene encoding an enzyme often results in an inability to synthesize the enzyme in an active form. Such a defect leads to a block in the metabolic pathway at the point where the enzyme acts, and the enzyme's substrate accumulates. Such genetic disorders are lethal if the end product of the pathway is essential or if the accumulated intermediates have toxic effects. In microorganisms, however, it is often possible to manipulate the growth medium so that essential end

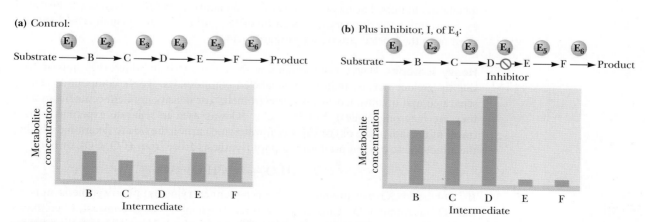

FIGURE 17.13 The use of inhibitors to reveal the sequence of reactions in a metabolic pathway. **(a) Control:** Under normal conditions, the steady-state concentrations of a series of intermediates will be determined by the relative activities of the enzymes in the pathway. **(b) Plus inhibitor:** In the presence of an inhibitor (in this case, an inhibitor of *enzyme 4*), intermediates upstream of the metabolic block (B, C, and D) accumulate, revealing themselves as intermediates in the pathway. The concentration of intermediates lying downstream (E and F) will fall.

TABLE 17.4	Properties of Radioactive and Stable "Heavy" Isotopes Used as Tracers in Metabolic Studies			
Isotope	Type	Radiation Type	Half-Life	Relative Abundance*
^{2}H	Stable			0.0154%
^{3}H	Radioactive	β^-	12.1 years	
^{13}C	Stable			1.1%
^{14}C	Radioactive	β^-	5700 years	
^{15}N	Stable			0.365%
^{18}O	Stable			0.204%
^{24}Na	Radioactive	β^-, γ	15 hours	
^{32}P	Radioactive	β^-	14.3 days	
^{35}S	Radioactive	β^-	87.1 days	
^{36}Cl	Radioactive	β^-	310,000 years	
^{42}K	Radioactive	β^-	12.5 hours	
^{45}Ca	Radioactive	β^-	152 days	
^{59}Fe	Radioactive	β^-, γ	45 days	
^{131}I	Radioactive	β^-, γ	8 days	

*The relative natural abundance of a stable isotope is important because, in tracer studies, the amount of stable isotope is typically expressed in terms of atoms percent excess over the natural abundance of the isotope.

products are provided. Then the biochemical consequences of the mutation can be investigated. Studies on mutations in genes of the filamentous fungus *Neurospora crassa* led G. W. Beadle and E. L. Tatum to hypothesize in 1941 that genes are units of heredity that encode enzymes (a principle referred to as the "one gene–one enzyme" hypothesis).

Isotopic Tracers Can Be Used as Metabolic Probes

Another widely used approach to the elucidation of metabolic sequences is to "feed" cells a substrate or metabolic intermediate labeled with a particular isotopic form of an element that can be traced. Two sorts of isotopes are useful in this regard: radioactive isotopes, such as ^{14}C, and stable "heavy" isotopes, such as ^{18}O or ^{15}N (Table 17.4). Because the chemical behavior of isotopically labeled compounds is rarely distinguishable from that of their unlabeled counterparts, isotopes provide reliable "tags" for observing metabolic changes. The metabolic fate of a radioactively labeled substance can be traced by determining the presence and position of the radioactive atoms in intermediates derived from the labeled compound (Figure 17.14).

Heavy Isotopes Heavy isotopes endow the compounds in which they appear with slightly greater masses than their unlabeled counterparts. These compounds can be separated and quantitated by mass spectrometry (or density gradient centrifugation, if they are macromolecules). For example, ^{18}O was used in separate experiments as a tracer of the fate of the oxygen atoms in water and carbon dioxide to determine whether the atmospheric oxygen produced in photosynthesis arose from H_2O, CO_2, or both:

$$CO_2 + H_2O \longrightarrow (CH_2O) + O_2$$

If ^{18}O-labeled CO_2 was presented to a green plant carrying out photosynthesis, none of the ^{18}O was found in O_2. Curiously, it was recovered as $H_2^{18}O$. In contrast, when plants fixing CO_2 were equilibrated with $H_2^{18}O$, $^{18}O_2$ was evolved. These latter labeling experiments established that photosynthesis is best described by the equation

$$C^{16}O_2 + 2\ H_2^{18}O \longrightarrow (CH_2^{16}O) + {}^{18}O_2 + H_2^{16}O$$

That is, in the process of photosynthesis, the two oxygen atoms in O_2 come from two H_2O molecules. One O is lost from CO_2 and appears in H_2O, and the other O of CO_2 is

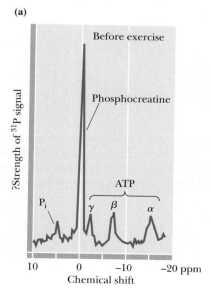

FIGURE 17.14 One of the earliest experiments using a radioactive isotope as a metabolic tracer. Cells of *Chlorella* (a green alga) synthesizing carbohydrate from carbon dioxide were exposed briefly (5 sec) to ^{14}C-labeled CO_2. The products of CO_2 incorporation were then quickly isolated from the cells, separated by two-dimensional paper chromatography, and observed via autoradiographic exposure of the chromatogram. Such experiments identified radioactive 3-phosphoglycerate (PGA) as the primary product of CO_2 fixation. The 3-phosphoglycerate was labeled in the 1-position (in its carboxyl group). Radioactive compounds arising from the conversion of 3-phosphoglycerate to other metabolic intermediates included phosphoenolpyruvate (PEP), malic acid, triose phosphate, alanine, and sugar phosphates and diphosphates.

retained in the carbohydrate product. Two of the four H atoms are accounted for in (CH_2O), and two reduce the O lost from CO_2 to H_2O.

NMR Spectroscopy Is a Noninvasive Metabolic Probe

A technology analogous to isotopic tracers is provided by **nuclear magnetic resonance (NMR) spectroscopy.** The atomic nuclei of certain isotopes, such as ^1H and ^{31}P, have *magnetic moments.* The resonance frequency of a magnetic moment depends on the local chemical environment. That is, the NMR signal of the nucleus is influenced in an identifiable way by the chemical nature of its neighboring atoms in the compound. Such variation in an atom's nuclear magnetic resonance frequency as a function of the electronic environment in its neighborhood is called **chemical shift.** Chemical shifts reveal a great deal of structural information about the environment around the atom and thus the nature of the compound containing the atom. Transformations of substrates and metabolic intermediates labeled with magnetic nuclei can be traced by following changes in NMR spectra. Furthermore, NMR spectroscopy is a noninvasive procedure. Whole-body NMR spectrometers are being used today in hospitals to directly observe the metabolism (and clinical condition) of living subjects (Figure 17.15). NMR is a useful tool

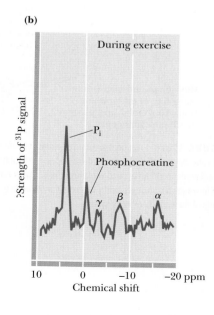

FIGURE 17.15 With NMR spectroscopy, one can observe the metabolism of a living subject in real time. These NMR spectra show the changes in ATP, creatine-P (phosphocreatine), and P_i levels in the forearm muscle of a human subjected to 19 minutes of exercise. Note that the three P atoms of ATP (α, β, and γ) have different chemical shifts, reflecting their different chemical environments.

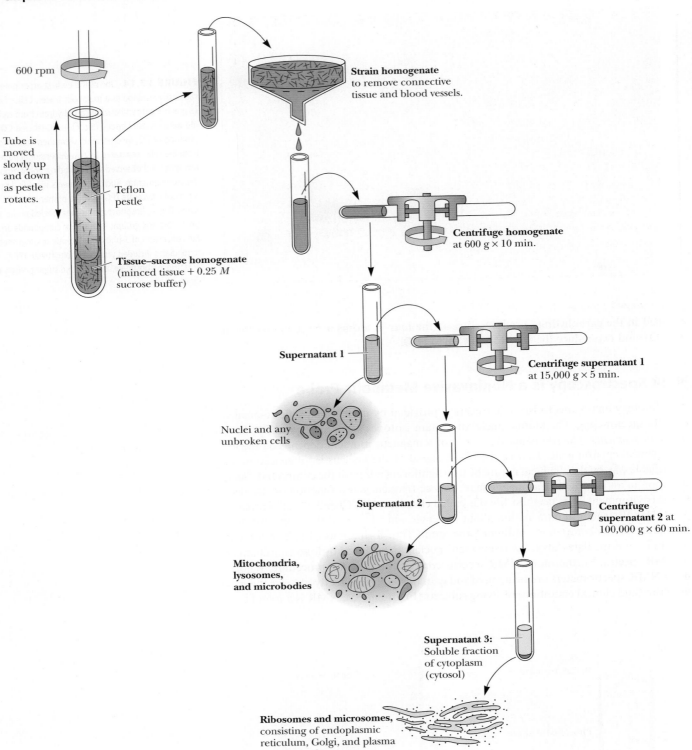

600 rpm

Tube is moved slowly up and down as pestle rotates.

Teflon pestle

Tissue–sucrose homogenate (minced tissue + 0.25 *M* sucrose buffer)

Strain homogenate to remove connective tissue and blood vessels.

Centrifuge homogenate at 600 g × 10 min.

Supernatant 1

Centrifuge supernatant 1 at 15,000 g × 5 min.

Nuclei and any unbroken cells

Supernatant 2

Centrifuge supernatant 2 at 100,000 g × 60 min.

Mitochondria, lysosomes, and microbodies

Supernatant 3: Soluble fraction of cytoplasm (cytosol)

Ribosomes and microsomes, consisting of endoplasmic reticulum, Golgi, and plasma membrane fragments

FIGURE 17.16 Fractionation of a cell extract by differential centrifugation. It is possible to separate organelles and subcellular particles in a centrifuge because their inherent size and density differences give them different rates of sedimentation in an applied centrifugal field. Nuclei are pelleted in relatively weak centrifugal fields and mitochondria in somewhat stronger fields, whereas very strong centrifugal fields are necessary to pellet ribosomes and fragments of the endomembrane system.

for clinical diagnosis and for the investigation of metabolism in situ (literally "in site," meaning, in this case, "where and as it happens").

Metabolic Pathways Are Compartmentalized Within Cells

Although the interior of a prokaryotic cell is not subdivided into compartments by internal membranes, the cell still shows some segregation of metabolism. For example, certain metabolic pathways, such as phospholipid synthesis and oxidative phosphorylation, are localized in the plasma membrane. Protein biosynthesis is carried out on ribosomes.

In contrast, eukaryotic cells are extensively compartmentalized by an endomembrane system. Each of these cells has a true nucleus bounded by a double membrane called the *nuclear envelope*. The nuclear envelope is continuous with the endomembrane system, which is composed of differentiated regions: the endoplasmic reticulum; the Golgi complex; various membrane-bounded vesicles such as lysosomes, vacuoles, and microbodies; and, ultimately, the plasma membrane itself. Eukaryotic cells also possess mitochondria and, if they are photosynthetic, chloroplasts. Disruption of the cell membrane and fractionation of the cell contents into the component organelles have allowed an analysis of their respective functions (Figure 17.16). Each compartment is dedicated to specialized metabolic functions, and the enzymes appropriate to these specialized functions are confined together within the organelle. In many instances, the enzymes of a metabolic sequence occur together within the organellar membrane. Thus, *the flow of metabolic intermediates in the cell is spatially as well as chemically segregated.* For example, the 10 enzymes of glycolysis are found in the cytosol, but pyruvate, the product of glycolysis, is fed into the mitochondria. These organelles contain the citric acid cycle enzymes, which oxidize pyruvate to CO_2. The great amount of energy released in the process is captured by the oxidative phosphorylation system of mitochondrial membranes and used to drive the formation of ATP (Figure 17.17).

Temporal Compartmentalization of Metabolic Pathways The spatial compartmentalization of metabolic pathways within cells provides important advantages, one of which is isolating competing pathways from one another. Cells and organisms also exhibit temporal compartmentalization of their metabolic pathways. That is, metabolic pathways may be turned on and off in a time-dependent and/or cyclic fashion. For

temporal: Relating to time as distinguished from space; chronological

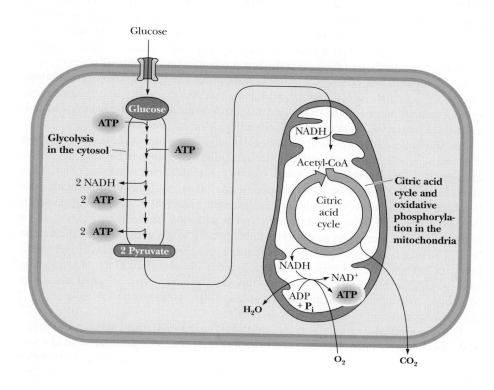

FIGURE 17.17 Compartmentalization of glycolysis, the citric acid cycle, and oxidative phosphorylation.

circadian: About a day; from the Latin *circa* = approximately, and *diem* = day

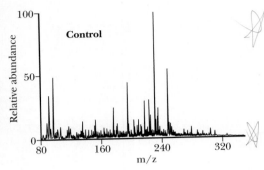

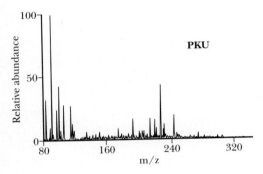

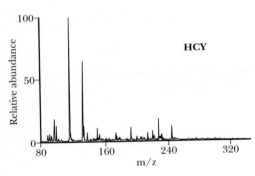

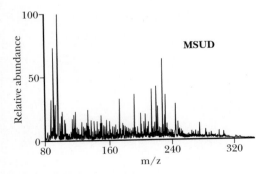

FIGURE 17.18 Mass spectrometry offers remarkable sensitivity for metabolomic analyses. Shown here are desorption electrospray ionization mass spectra for urine samples from individuals with inborn errors of metabolism. PKU = phenylketonuria; HCY = homocystinuria; MSUD = maple syrup urine disease. (Adapted from Pan, Z., Gu, H., et al., 2007. Principal component analysis of urine metabolites by NMR and DESI-MS in patients with inborn errors of metabolism. *Analytical and Bioanalytical Chemistry* **387**:539–549.)

example, the metabolism of many organisms—microbes, animals, and plants—is regulated in synchrony with the 24-hour cycle of day and night, a pattern called **circadian rhythmicity** and often referred to as the **biological clock.** Why? Because light and/or varying nutrient availability represent key signals regarding the transitory nature of the environment, and organisms have evolved and adapted to exploit the information in such signals. Photosynthesis and carbon dioxide fixation in plants typically is active during daylight hours and down-regulated during the dark. Significantly, recent studies have demonstrated that mammalian genes involved in lipid metabolism, glycolysis, gluconeogenesis, electron transport, mitochondrial ATP synthesis, and other metabolic pathways show a circadian pattern of regulated expression. In particular, the genes encoding the rate-limiting enzymes in many of these pathways are under circadian regulation. Feeding time (nutrient availability) is an important external cue that sets the phase of these metabolic gene expression cycles. In the liver, gluconeogenesis and glycogen breakdown are active during the sleep/fasting phase of the circadian rhythm, apparently to maintain blood glucose levels. On the other hand, glycolysis and cholesterol synthesis are active during the wake/feeding phase when substrates become available. Notably, disruptions of the circadian clock result in elevated fat accumulation and other metabolic abnormalities. Yeast cells *(Saccharomyces cerevisiae)* maintained under continuous nutrient-limited conditions adopt a 4- to 5-hour cycle of respiratory activity, as demonstrated by oscillations in O_2 consumption. Over half of the genes in the yeast genome are expressed rhythmically during this yeast metabolic cycle, with genes encoding proteins of similar metabolic function showing synchronous oscillations in expression. Those genes involved with energy metabolism show particularly robust metabolic cycles.

17.5 What Can the Metabolome Tell Us about a Biological System?

Metabolomics is a newly emerged field in biochemistry and an obvious next step in the –omics series of genomics, transcriptomics, and proteomics. The **metabolome** is the complete set of low-molecular-weight substances involved in metabolism or produced by a cell or organism under specified physiological conditions. Metabolomics is the systematic identification and quantitation of all of these substances. Moving from the genome, which specifies all the genetic information available to an organism as encoded in its DNA, to the metabolome takes one from the **genotype** of the organism (its complete genetic potential) to its **phenotype** (the observed traits of an organism achieved through expression of its genes and the interaction of those gene products with the environment). As the ultimate expression of an organism's genetic possibilities and its responses to its environment, the metabolome is the best predictor of an organism's phenotype. As such, metabolomics is occupying a prominent position in the rise of **systems biology,** where comprehensive information about the genome, the transcriptome, the proteome, and the metabolome combine to provide insightful descriptions of the behavior of cells and organisms, as well as detailed understanding of many diseases.

Metabolomic analyses present daunting challenges, even for simple organisms. More than 500 metabolites are represented in the metabolic map shown in Figure 17.2, but far more exist in a typical cell. For example, the 40 or so fatty acids occurring in a cell can alone account for more than a thousand phosphatidic acids. (Phosphatidic acids, with two fatty acids esterified to *sn*-glycerol-3-phosphate create $40 \times 40 = 1600$ species by themselves!) The Human Metabolomics Database (www.hmdb.ca) provides data on more than 7900 metabolites in cells and fluids (blood, urine, and so on) of the human body. The metabolomes of plants are even more complex, with estimates suggesting as many as 200,000 different metabolites across the plant kingdom. Metabolomic assays must be able to resolve and discriminate this extraordinary array of small molecules. Moreover, concentrations of metabolites in a cell vary widely, from less than $10^{-12}\ M$ (for many hormones) to 0.1 M (for Na^+ ions).

The principal tools of metabolomics are NMR spectroscopy and mass spectrometry. Mass spectrometry offers unmatched sensitivity for detection of metabolites at low concentrations (Figure 17.18), and NMR spectroscopy can provide exceptional resolution and discrimination of metabolites in complex mixtures (Figure 17.19). Coupling these

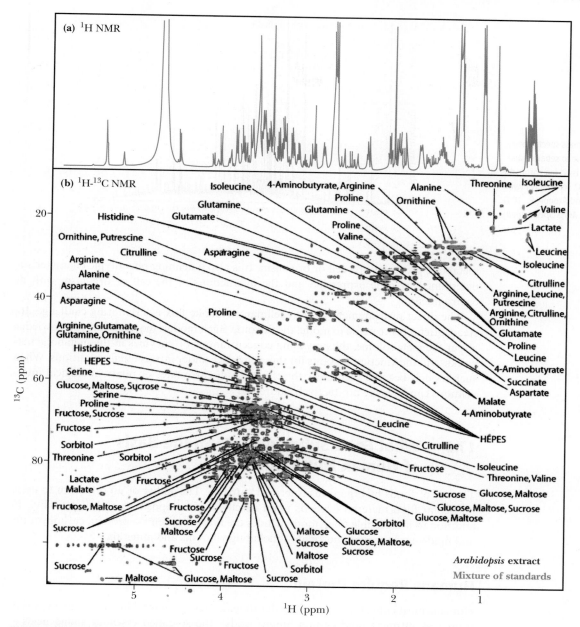

FIGURE 17.19 **(a)** One-dimensional ^1H NMR spectrum of an equimolar mixture of 26 small-molecule standards. **(b)** Two-dimensional NMR spectrum of the same mixture (red) overlaid onto a spectrum of aqueous whole-plant extract from *Arabidopsis thaliana,* a model organism for the study of plant molecular biology and biochemistry. (Adapted from Lewis, I., Schommer, S., et al., 2007. Method for determining molar concentrations of metabolites in complex solutions from two-dimensional ^1H-^{13}C NMR spectra. *Analytical Chemistry* **79**:9385–9390.)

techniques with a variety of chromatographic separation protocols (as in the example shown in Figure 17.20) makes it possible to rapidly analyze thousands of metabolites in biological samples at low cost.

Fluxomics The true phenotype of a cell depends not only on knowledge of all the metabolites in a cell at any given moment, but also on information about the flow of metabolites through its metabolic networks. The quantitative study of metabolite flow, or **flux,** is termed **fluxomics.** The fluxome is defined by all of the metabolic fluxes in a metabolic network, and thus represents systems-level information on those cellular processes defined by the metabolic network. As such, the fluxome depends on the metabolome and the proteome, with the proteome representing those enzymes that mediate the reactions comprising the metabolic network. Thus, the fluxome is a dynamic view of the

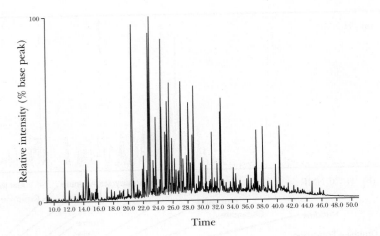

FIGURE 17.20 The combination of mass spectrometry and gas chromatography makes it possible to separate and identify hundreds of metabolites. Shown is an ion chromatogram of a human urine sample, with 1582 separately identified peaks. (Adapted from Dettmer, K., and Aronov, P., 2007. Mass spectrometry-based metabolomics. *Mass Spectrometry Reviews* **26**:51–79.)

metabolic processes in a cell or organism. It is also a defining phenotypic characteristic determined by gene expression and the interactions of the cell or organism with its environment.

The direct measurement of intracellular metabolic flux is a daunting challenge. Instead, attempts are made to model the fluxome. Such models require detailed knowledge about the metabolome, the prevailing concentrations of enzymes in the metabolic networks, and the kinetic constants for these enzymes under intracellular conditions. When such models become predictive, the effects of disease states or pharmacological interventions can be more productively addressed.

17.6 What Food Substances Form the Basis of Human Nutrition?

The use of foods by organisms is termed **nutrition.** The ability of an organism to use a particular food material depends upon its chemical composition and digestibility. In addition to indigestible fiber, food includes the macronutrients—protein, carbohydrate, and lipid—and the micronutrients—including vitamins and minerals.

Humans Require Protein

Humans must consume protein in order to make new proteins. Dietary protein is a rich source of nitrogen, and certain amino acids—the so-called **essential amino acids**—cannot be synthesized by humans (and various animals) and can be obtained only in the diet. The average adult in the United States consumes far more protein than required for synthesis of essential proteins. Excess dietary protein is then merely a source of metabolic energy. Some of the amino acids (termed **glucogenic**) can be converted into glucose, whereas others, the **ketogenic** amino acids, can be converted to fatty acids and/or keto acids. If fat and carbohydrate are already adequate for the energy needs of the individual, then both kinds of amino acids will be converted to triacylglycerol and stored in adipose tissue.

An individual's protein undergoes a constant process of degradation and resynthesis, a cycle known as **protein turnover.** Together with dietary protein, this recycled protein material participates in a nitrogen equilibrium. For most adults, the amount of protein synthesis is equaled by the amount of protein degradation, and the dietary amount of protein is equalled by the amount of nitrogen excreted in the urine (as urea), in the feces, and in sweat. Persons whose nitrogen intake is balanced by their nitrogen excretion are in **zero nitrogen balance.** A positive nitrogen balance occurs whenever there is a net increase in the organism's protein content, such as during periods of growth. Nitrogen balance is positive in healthy individuals from birth through adolescence, in pregnant women, and in people recovering from illness or protein insufficiency. A negative nitrogen balance exists when dietary intake of nitrogen is insufficient to meet the demands for new protein synthesis, as in people who are starving or who need to replace tissue due to illness or injury.

Carbohydrates Provide Metabolic Energy

The principal purpose of carbohydrate in the diet is production of metabolic energy. Simple sugars are metabolized in the glycolytic pathway (see Chapter 18). Complex carbohydrates are degraded into simple sugars, which then can enter the glycolytic pathway. Carbohydrates are also essential components of nucleotides, nucleic acids, glycoproteins, and glycolipids. Human metabolism can adapt to a wide range of dietary carbohydrate levels, but the brain requires glucose for fuel. When dietary carbohydrate consumption exceeds the energy needs of the individual, excess carbohydrate is converted to triacylglycerols and glycogen for long-term energy storage. On the other hand, when dietary carbohydrate intake is low, **ketone bodies** are formed from acetate units to provide metabolic fuel for the brain and other organs.

Lipids Are Essential, But in Moderation

Fatty acids and triacylglycerols can be used as fuel by many tissues in the human body, and phospholipids are essential components of all biological membranes. Even though the human body can tolerate a wide range of fat intake levels, there are disadvantages in either extreme. Excess dietary fat is stored as triacylglycerols in adipose tissue, but high levels of dietary fat can also increase the risk of atherosclerosis and heart disease. Moreover, high dietary fat levels are also correlated with increased risk for colon, breast, and prostate cancers. When dietary fat consumption is low, there is a risk of **essential fatty acid** deficiencies. As will be seen in Chapter 24, the human body cannot synthesize linoleic and linolenic acids, so these must be acquired in the diet. In addition, arachidonic acid can by synthesized in humans only from linoleic acid, so it too is classified as essential. The essential fatty acids are key components of biological membranes, and arachidonic acid is the precursor to prostaglandins, which mediate a variety of processes in the body.

SUMMARY

OWL Login to OWL to develop problem-solving skills and complete online homework assigned by your professor.

17.1 Is Metabolism Similar in Different Organisms? One of the great unifying principles of modern biology is that organisms show marked similarity in their major pathways of metabolism. Given the almost unlimited possibilities within organic chemistry, this generality would appear most unlikely. Yet it's true, and it provides strong evidence that all life has descended from a common ancestral form. All forms of nutrition and almost all metabolic pathways evolved in early prokaryotes prior to the appearance of eukaryotes 1 billion years ago. All organisms, even those that can synthesize their own glucose, are capable of glucose degradation and ATP synthesis via glycolysis. Other prominent pathways are also virtually ubiquitous among organisms.

17.2 What Can Be Learned from Metabolic Maps? Metabolism represents the sum of the chemical changes that convert nutrients, the "raw materials" necessary to sustain living organisms, into energy and the chemically complex finished products of cells. Metabolism consists of literally hundreds of enzymatic reactions organized into discrete pathways. Metabolic maps portray the principal reactions of the intermediary metabolism of carbohydrates, lipids, amino acids, and their derivatives. In such maps, arrows connect metabolites and represent the enzyme reactions that interconvert the metabolites. Alternative mappings of biochemical pathways have been proposed in a response to the emergence of genomic, transcriptomic, and proteomic perspectives on the complexity of biological systems.

17.3 How Do Anabolic and Catabolic Processes Form the Core of Metabolic Pathways? *Catabolism* involves the oxidative degradation of complex nutrient molecules (carbohydrates, lipids, and proteins) obtained either from the environment or from cellular reserves. The breakdown of these molecules by catabolism leads to the formation of simpler molecules such as lactic acid, ethanol, carbon dioxide, urea, or ammonia. Catabolic reactions are usually exergonic, and often the chemical energy released is captured in the form of ATP. *Anabolism* is a synthetic process in which the varied and complex biomolecules (proteins, nucleic acids, polysaccharides, and lipids) are assembled from simpler precursors. Such biosynthesis involves the formation of new covalent bonds, and an input of chemical energy is necessary to drive such endergonic processes. The ATP generated by catabolism provides this energy. Furthermore, NADPH is an excellent donor of high-energy electrons for the reductive reactions of anabolism.

17.4 What Experiments Can Be Used to Elucidate Metabolic Pathways?

An important tool for elucidating the steps in the pathway is the use of *metabolic inhibitors*. Adding an enzyme inhibitor to a cell-free extract causes an accumulation of intermediates in the pathway prior to the point of inhibition. Each inhibitor is specific for a particular site in the sequence of metabolic events. Genetics provides an approach to the identification of intermediate steps in metabolism that is somewhat analogous to inhibition. Mutation in a gene encoding an enzyme often results in an inability to synthesize the enzyme in an active form. Such a defect leads to a block in the metabolic pathway at the point where the enzyme acts, and the enzyme's substrate accumulates. Such genetic disorders are lethal if the end product of the pathway is essential or if the accumulated intermediates have toxic effects. In microorganisms, however, it is often possible to manipulate the growth medium so that essential end products are provided. Then the biochemical consequences of the mutation can be investigated. Isotopic tracers can also be used to follow metabolic changes. NMR spectroscopy is a noninvasive metabolic probe that allows real-time observation of metabolism. Results of many studies established that metabolic pathways are compartmentalized within cells.

17.5 What Can the Metabolome Tell Us about a Biological System?

Rapid advances in chemical analysis have made it possible to carry out comprehensive studies of the many metabolites in a living organism. The metabolome is the complete set of low-molecular-weight molecules involved in metabolism or produced by a cell or organism under a given set of physiological conditions. Metabolomics is the systematic identification and quantitation of all these metabolites in a given organism or sample. Mass spectrometry offers unmatched sensitivity for detection of metabolites at very low concentrations, whereas NMR spectroscopy can provide remarkable resolution and discrimination of metabolites in complex mixtures. Whereas the metabolome provides information about all the metabolites in a cell at any given moment, the real phenotype of a cell or organism is represented by the *fluxome*, defined as all of the metabolic fluxes in a metabolic network. The fluxome represents important systems biology information, as the final step in the genome, transcriptome, metabolome, fluxome progression.

17.6 What Food Substances Form the Basis of Human Nutrition?

In addition to indigestible fiber, the food that human beings require includes the macronutrients—protein, carbohydrate, and lipid—and the micronutrients—including vitamins and minerals.

FOUNDATIONAL BIOCHEMISTRY Things You Should Know After Reading Chapter 17.

- Organisms show marked similarity in their metabolic pathways.

- Living things exhibit metabolic diversity.

- Autotrophs are organisms that can subsist on carbon dioxide as the only source of carbon; heterotrophs require an organic form of carbon.

- Phototrophs are photosynthetic organisms; chemotrophs use organic or inorganic compounds as a source of energy.

- Prokaryotes show greater metabolic diversity than found in all the eukaryotic organisms.

- Oxygen is essential for aerobic organisms.

- The flow of energy in the biosphere is intimately related to the carbon and oxygen cycles.

- Metabolic maps portray the principal reactions comprising the intermediary metabolism of carbohydrates, lipids, amino acids, nucleotides, and their derivatives.

- Traditional metabolic maps emphasize metabolites, but alternative models focused on the enzymes or enzyme complexes can provide new insights.

- Multienzyme systems may take several different forms.

- Multifunctional polypeptides are proteins that can catalyze several of the reactions in a metabolic pathway.

- Catabolic pathways are characteristically energy yielding, whereas anabolic pathways are energy requiring.

- Catabolism involves the oxidative degradation of complex nutrient molecules, such as carbohydrates, lipids, and proteins.

- The energy released in catabolic reactions is often captured in ATP.

- During catabolism, energy-rich electrons are transferred to the coenzymes NAD^+ and $NADP^+$.

- Energy released upon oxidation of NADH is coupled to ATP synthesis.

- NADPH is the source of reducing potential to drive reductive biosynthetic reactions.

- Anabolism is biosynthesis; as such, it is endergonic.

- Anabolism and catabolism are not mutually exclusive.

- The pathways of catabolism converge to a few end products, such as lactate, ethanol, CO_2, H_2O, and NH_4^+.

- Anabolic pathways diverge, synthesizing an astounding variety of biomolecules from a limited set of metabolic intermediates that serve as building blocks.

- Amphibolic intermediates play dual roles, serving roles in both catabolic and anabolic pathways.

- Corresponding pathways of catabolism and anabolism, such as glycolysis and gluconeogenesis or fatty acid oxidation and fatty acid biosynthesis, differ in important ways.

- Metabolic regulation requires different pathways for oppositely directed metabolic sequences.

- Opposing metabolic pathways are reciprocally regulated.

- ATP serves in a cellular energy cycle, being produced in catabolism (or photosynthesis) and consumed in biosynthesis, or osmotic work, or motility.

- NAD^+ collects electrons released in catabolism.

- The energy released in NADH oxidation is coupled with the formation of ATP; the coupled process is called oxidative phosphorylation.

- Transformation of the energy released in an oxidation-reduction reaction into the chemical energy of a phosphoric anhydride bond is a form of energy transduction.

- Coenzymes provide unique chemistry to metabolic pathways.

- Water-soluble vitamins are precursors for coenzyme synthesis.

- Metabolic pathways have been elucidated by a variety of experimental approaches.

- Metabolic inhibitors block pathways at their site of inhibition.

- Mutations create specific metabolic blocks.

- Either radioactive isotopes or stable "heavy" isotopes can be used to trace the course of metabolic reactions.

- NMR spectroscopy is a noninvasive metabolic probe.

- Metabolic pathways are compartmentalized with cells.

- Cells and organisms also exhibit temporal compartmentalization of their metabolic pathways.

- The metabolism of many organisms shows circadian rhythmicity.

- The metabolome is the complete set of metabolites involved in metabolism or produced by a cell or organism under specified physiological conditions.

- The metabolome is an excellent predictor of a cell or organism's phenotype.

- The metabolome is an important aspect of systems biology.

- NMR spectroscopy and mass spectrometry are the principal tools of metabolomic analyses.

- Fluxomics is the quantitative study of metabolic flux.

- Human nutrition depends on carbohydrates, lipids, and proteins for energy and replenishment of metabolic building blocks.

- Vitamins and minerals are also essential to human nutrition.

- Proteins undergo a continuous cycle of synthesis and degradation termed protein turnover.

- Most adults excrete as much nitrogen as they consume and thus are in zero nitrogen balance.

- Carbohydrates and triglycerides are important sources of nutritional energy.

- Essential fatty acids cannot be synthesized by humans and must be obtained in the diet.

PROBLEMS

Answers to all problems are at the end of this book. Detailed solutions are available in the *Student Solutions Manual, Study Guide, and Problems Book*.

1. **Global Carbon Dioxide Cycling Expressed as Human Equivalents** If 3×10^{14} kg of CO_2 are cycled through the biosphere annually, how many human equivalents (70-kg people composed of 18% carbon by weight) could be produced each year from this amount of CO_2?

2. **The Different Modes of Carbon and Energy Metabolism** Define the differences in carbon and energy metabolism between *photoautotrophs* and *photoheterotrophs* and between *chemoautotrophs* and *chemoheterotrophs*.

3. **Where Do the O Atoms in Organisms Come From?** Name three principal inorganic sources of oxygen atoms that are commonly available in the inanimate environment and readily accessible to the biosphere.

4. **How Do Catabolism and Anabolism Differ?** What are the features that generally distinguish pathways of catabolism from pathways of anabolism?

5. **How Are the Enzymes of Metabolic Pathways Organized?** Name the four principal modes of enzyme organization in metabolic pathways.

6. **Why Do Anabolic and Catabolic Pathways Differ?** Why is the pathway for the biosynthesis of a biomolecule at least partially different from the pathway for its catabolism? Why is the pathway for the biosynthesis of a biomolecule inherently more complex than the pathway for its degradation?

7. **What Are Primary Metabolic Roles of the Three Principal Nucleotides?** (Integrates with Chapters 1 and 3.) What are the metabolic roles of ATP, NAD^+, and NADPH?

8. **How Is Metabolism Regulated?** (Integrates with Chapter 15.) Metabolic regulation is achieved via regulating enzyme activity in three prominent ways: allosteric regulation, covalent modification, and enzyme synthesis and degradation. Which of these three modes of regulation is likely to be the quickest; which the slowest? For each of these general enzyme regulatory mechanisms, cite conditions in which cells might employ that mode in preference to either of the other two.

9. **What Are the Advantages of Metabolic Compartmentalization?** What are the advantages of compartmentalizing particular metabolic pathways within specific organelles?

10. **Why Are Metabolomic Analyses Challenging?** Name and discuss four challenges associated with metabolomic measurements in biological systems.

11. **Which Is "Better": NMR or MS?** Compare and contrast mass spectrometry and NMR in terms of their potential advantages and disadvantages for metabolomic analysis.

12. **How Do Vitamin-Derived Coenzymes Aid Metabolism?** What chemical functionality is provided to enzyme reactions by pyridoxal phosphate (see Chapter 13)? By coenzyme A (see Chapter 19)? By vitamin B_{12} (see Chapter 23)? By thiamine pyrophosphate (see Chapter 19)?

13. **What Are the Features of the Series of -omes?** Define the following terms:
    a. Genome
    b. Transcriptome
    c. Proteome
    d. Metabolome
    e. Fluxome

14. **What Is the Alcohol Dehydrogenase Enzyme Mechanism?** The alcohol dehydrogenase reaction, described in Figure 17.11, interconverts ethyl alcohol and acetaldehyde and involves hydride transfer to and from NAD^+ and NADH, respectively. Write a reasonable mechanism for the conversion of ethanol to acetaldehyde by alcohol dehydrogenase.

15. **Where Are Various Metabolic Pathways Localized and What Are Their Inputs?** For each of the following metabolic pathways, describe where in the cell it occurs and identify the starting material and end product(s):
    a. Citric acid cycle
    b. Glycolysis
    c. Oxidative phosphorylation
    d. Fatty acid synthesis

16. **A Possible Solution to Global Warming** Many solutions to the problem of global warming have been proposed. One of these involves strategies for carbon sequestration—the removal of CO_2 from the earth's atmosphere by various means. From your reading of this chapter, suggest and evaluate a strategy for carbon sequestration in the ocean.

17. **Radioactive Isotopes as Metabolic Tracers** Consult Table 17.4, and consider the information presented for ^{32}P and ^{35}S. Write reactions for the decay events for these two isotopes, indicating clearly the products of the decays, and calculate what percentage of each would remain from a sample that contained both and decayed for 100 days.

Preparing for the MCAT® Exam

18. Which statement is most likely to be true concerning obligate anaerobes?

a. These organisms can use oxygen if it is present in their environment.

b. These organisms cannot use oxygen as their final electron acceptor.

c. These organisms carry out fermentation for at least 50% of their ATP production.

d. Most of these organisms are vegetative fungi.

19. Which of the following experimental approaches to metabolic analysis cannot be used directly with living cells?

a. NMR spectroscopy

b. Metabolic inhibitors

c. Mass spectroscopy

d. Isotopic tracers

ActiveModels Problem

20. Examine the ActiveModel for alcohol dehydrogenase and identify the amino acid side chains involved in binding the NAD coenzyme.

FURTHER READING

Metabolism

Atkinson, D. E., 1977. *Cellular Energy Metabolism and Its Regulation.* New York, Academic Press.

Chatterjee, R., and Yuan, L., 2006. Directed evolution of metabolic pathways. *Trends in Biotechnology* **24**(1):28–38.

Frayn, K., 2003. *Metabolic Regulation.* New York: Wiley.

Gerrard, J. A., Sparrow, A. D., et al., 2001. Metabolic databases: What next? *Trends in Biochemical Sciences* **26**:137–140.

Gropper, S., and Smith, J. L., 2008. *Advanced Nutrition and Human Metabolism.* Belmont, CA: Wadsworth Publishing.

Hosler, J. P., Ferguson-Miller, S., et al., 2006. Energy transduction: Proton transfer through the respiratory complexes. *Annual Review of Biochemistry* **75**:165–187.

Metzger, R. P., 2006. Thoughts on the teaching of metabolism. *Biochemistry and Molecular Biology Education* **34**:78–87.

Nicholls, D. G., and Ferguson, S. J., 2007. *Bioenergetics 3.* New York: Academic Press.

Nicholson, D. E., 2003. *Metabolic Pathways,* 22nd ed. St. Louis: Sigma-Aldrich.

Nicholson, D. E., 2005. From metabolic pathways charts to animaps in 50 years. *Biochemistry and Molecular Biology Education* **33**:156–158.

Smith, E. and Morowitz, H. J., 2004. Universality in intermediary metabolism. *Proceedings of the National Academy of Sciences U.S.A.* **101**:13168–13173.

Teichmann, S. A., Rison, S. C. G., et al., 2001. Small-molecule metabolism: An enzyme mosaic. *Trends in Biotechnology* **19**:482–486.

Tu, B. P., and McKnight, S. L., 2006. Metabolic cycles as an underlying basis of biological oscillations. *Nature Reviews Molecular Cell Biology* **7**:696–701.

Metabolic Cycles

Bass, J., and Takahashi, J. S., 2010. Circadian integration of metabolism and energetics. *Science* **330**:1349–1354.

Green, C. B., Takahashi, J. S., and Bass, J., 2008. The meter of metabolism. *Cell* **134**:728–742.

Tu, B. P., Kudlicki, A., Rowicka, M., and McKnight, S. L., 2005. Logic of the yeast metabolic cycle. *Science* **310**:1152–1158.

Metabolomics and Metabonomics

Alexander, P. B., Wang, J., and McKnight, S. L., 2011. Targeted killing of a mammalian cell based upon its specialized metabolic state. *Proceedings of the National Academy of Sciences, U. S. A.* **108**:15828–15833.

Almaas, E., Kovacs, B., Visek, T., Oltval, Z. N., and Barabasi, A.-L., 2004. Global organization of metabolic fluxes in the bacterium *Escherichia coli. Nature* **427**:839–843.

Beckonert, O., Keun, H. C., et al., 2007. Metabolic profiling, metabolomic and metabonomic procedures for NMR spectroscopy of urine, plasma, serum, and tissue extracts. *Nature Protocols* **2**: 2692–2703.

Breitling, R., Pitt, A. R., et al., 2006. Precision mapping of the metabolome. *Trends in Biotechnology* **24**:543–548.

Cascante, M., and Marin, S., 2008. Metabolomics and fluxomics approaches. *Essays in Biochemistry* **45**:67–81.

Dettmer, K., Aronov, P. A., and Hammock, B. D., 2007. Mass spectrometry-based metabolomics. *Mass Spectrometry Reviews* **26**:51–78.

Fiehn, O., 2002. Metabolomics: The link between genotypes and phenotypes. *Plant Molecular Biology* **48**:155–171.

Griffen, J. L., 2006. The Cinderella story of metabolic profiling: Does metabolomics get to go to the functional genomics ball? *Philosophical Transactions of the Royal Society B* **361**:147–161.

Idle, J. R., and Gonzalez, F. J., 2007. Metabolomics. *Cell Metabolism* **6**:348–351.

Kell, D. B., 2004. Metabolomics and systems biology: Making sense of the soup. *Current Opinion in Microbiology* **7**:296–307.

Lane, A. N., Fan, T. W.-M., et al., 2008. Isotopomer-based metabolomic analysis by NMR and mass spectrometry. *Methods in Cell Biology* **84:** 541–588.

Lewis, I. A., Schommer, S. C., et al., 2007. Method for determining molar concentrations of metabolites in complex solutions from two-dimensional ^{1}H-^{13}C NMR spectra. *Analytical Chemistry* **79**:9385–9390.

Li, X., Gianoulis, T. A., Yip, K. Y., Gerstein, M., and Snyder, M., 2010. Extensive in vivo metabolite-protein interactions revealed by large-scale systematic analyses. *Cell* **143**:1–12.

Nicholson, J. K., and Lindon, J. C., 2008. Metabolomics. *Nature* **455**:1054–1056.

Pan, Z., and Raftery, D., 2007. Comparing and combining NMR spectroscopy and mass spectrometry in metabolomics. *Analytical Biochemistry and Chemistry* **387**:525–527.

Pearson, H., 2007. Meet the human metabolome. *Nature* **446**:8.

Wu, H., Southam, A. D., et al., 2007. High-throughput tissue extraction protocol for NMR- and MS-based metabolomics. *Analytical Biochemistry* **372**:204–212.

Systems Biology

Doolittle, R. F., 2005. Evolutionary aspects of whole-genome biology. *Current Opinion in Structural Biology* **15**:248–253.

Kell, D. B., Brown, M., et al., 2005. Metabolic footprinting and systems biology: The medium is the message. *Nature Reviews Microbiology* **3**:557–565.

Vitamins

Abelson, J. N., Simon, M. I., et al., 1997. *Vitamins and Coenzymes,* Part I. New York: Academic Press.

Dennis, E. A., Simon, M. I., et al., 1997. *Vitamins and Coenzymes,* Part L. New York: Academic Press.

Glycolysis

© the food passionates/Corbis

◄ Louis Pasteur's scientific investigations into fermentation of grape sugar were pioneering studies of glycolysis.

Living organisms, like machines, conform to the law of conservation of energy, and must pay for all their activities in the currency of catabolism.

Ernest Baldwin
Dynamic Aspects of Biochemistry (1952)

KEY QUESTIONS

18.1 What Are the Essential Features of Glycolysis?

18.2 Why Are Coupled Reactions Important in Glycolysis?

18.3 What Are the Chemical Principles and Features of the First Phase of Glycolysis?

18.4 What Are the Chemical Principles and Features of the Second Phase of Glycolysis?

18.5 What Are the Metabolic Fates of NADH and Pyruvate Produced in Glycolysis?

18.6 How Do Cells Regulate Glycolysis?

18.7 Are Substrates Other Than Glucose Used in Glycolysis?

18.8 How Do Cells Respond to Hypoxic Stress?

ESSENTIAL QUESTION

Nearly every living cell carries out a catabolic process known as **glycolysis**—the stepwise degradation of glucose (and other simple sugars). Glycolysis is a paradigm of metabolic pathways. Carried out in the cytosol of cells, it is basically an anaerobic process; its principal steps occur with no requirement for oxygen. Living things first appeared in an environment lacking O₂, and glycolysis was an early and important pathway for extracting energy from nutrient molecules. It played a central role in anaerobic metabolic processes during the first 2 billion years of biological evolution on earth. Contemporary organisms still employ glycolysis to provide precursor molecules for aerobic catabolic pathways (such as the tricarboxylic acid cycle) and as a short-term energy source when oxygen is limiting. A complex array of molecular signals fine-tune and adapt glycolysis to a variety of metabolic needs.

What are the chemical basis and logic for this central pathway of metabolism; that is, how does glycolysis work?

18.1 What Are the Essential Features of Glycolysis?

In the glycolysis pathway (Figure 18.1), a molecule of glucose is converted in 10 enzyme-catalyzed steps to two molecules of 3-carbon pyruvate. Most of the details of this pathway (the first metabolic pathway to be elucidated) were worked out in the first half of the 20th century by the German biochemists Otto Warburg, Gustav Embden, and Otto Fritz Meyerhof. In fact, the sequence of reactions in Figure 18.1 is often referred to as the **Embden–Meyerhof pathway.**

Why is glycolysis so important to organisms? There are several reasons. For some tissues (such as brain, kidney medulla, and rapidly contracting skeletal muscles) and for some cells (such as erythrocytes and sperm cells), glucose is the only source of metabolic

ᗜWL Online homework and a Student Self Assessment for this chapter may be assigned in OWL.

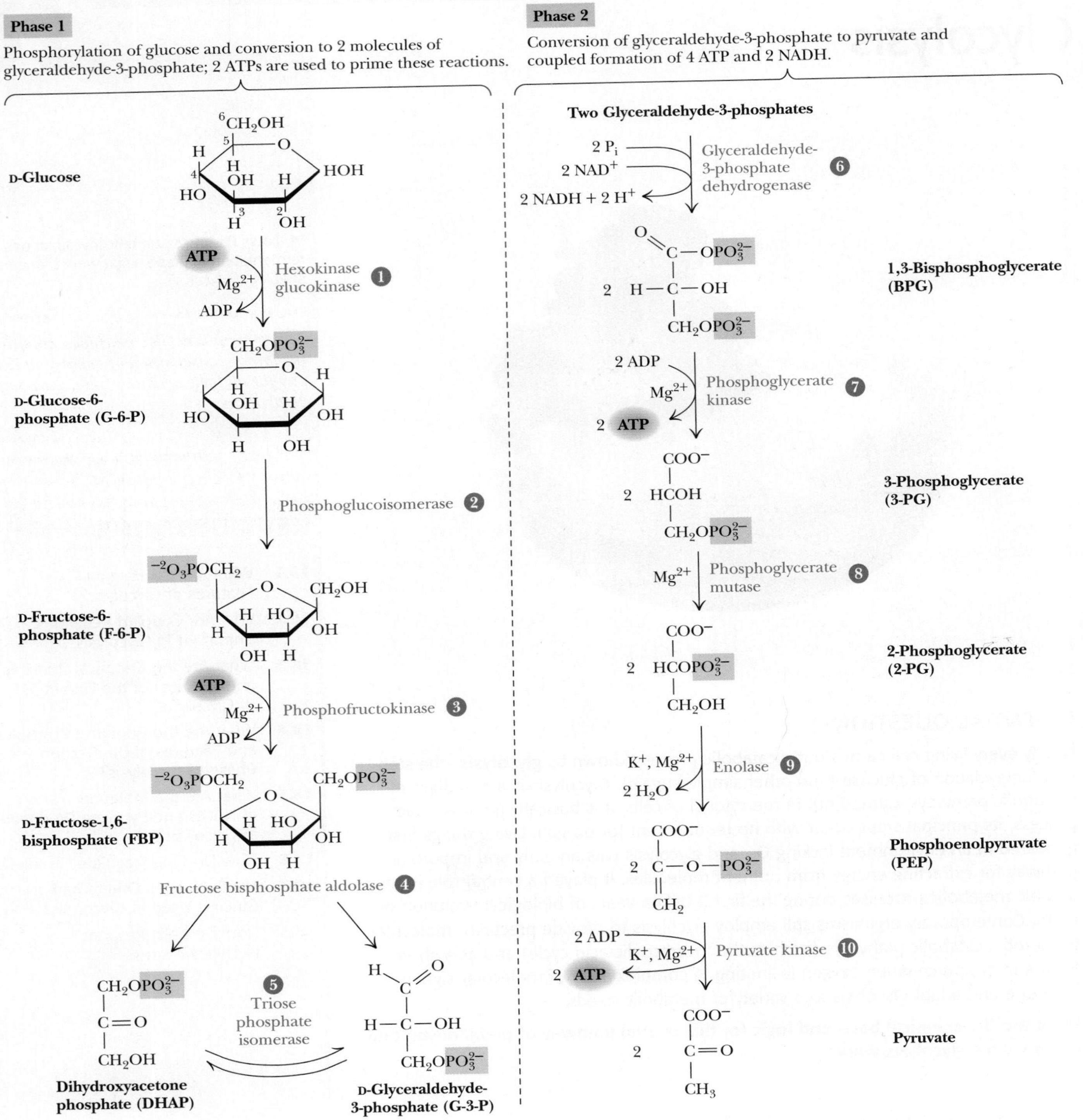

Phase 1

Phosphorylation of glucose and conversion to 2 molecules of glyceraldehyde-3-phosphate; 2 ATPs are used to prime these reactions.

Phase 2

Conversion of glyceraldehyde-3-phosphate to pyruvate and coupled formation of 4 ATP and 2 NADH.

Two Glyceraldehyde-3-phosphates

D-Glucose

ATP

Hexokinase glucokinase ❶

Mg^{2+}

ADP

D-Glucose-6-phosphate (G-6-P)

$CH_2OPO_3^{2-}$

Phosphoglucoisomerase ❷

D-Fructose-6-phosphate (F-6-P)

$^{-2}O_3POCH_2$

ATP

Mg^{2+} Phosphofructokinase ❸

ADP

D-Fructose-1,6-bisphosphate (FBP)

Fructose bisphosphate aldolase ❹

❺ Triose phosphate isomerase

Dihydroxyacetone phosphate (DHAP)

D-Glyceraldehyde-3-phosphate (G-3-P)

$2 P_i$
$2 NAD^+$
$2 NADH + 2 H^+$

Glyceraldehyde-3-phosphate dehydrogenase ❻

1,3-Bisphosphoglycerate (BPG)

$2 ADP$
Mg^{2+} Phosphoglycerate kinase ❼
2 ATP

3-Phosphoglycerate (3-PG)

Mg^{2+} Phosphoglycerate mutase ❽

2-Phosphoglycerate (2-PG)

K^+, Mg^{2+} Enolase ❾
$2 H_2O$

Phosphoenolpyruvate (PEP)

$2 ADP$
K^+, Mg^{2+} Pyruvate kinase ❿
2 ATP

Pyruvate

FIGURE 18.1 The glycolytic pathway.

energy. In addition, the product of glycolysis—pyruvate—is a versatile metabolite that can be used in several ways. In most tissues, when oxygen is plentiful (aerobic conditions), pyruvate is oxidized (with loss of the carboxyl group as CO_2), and the remaining two-carbon unit becomes the acetyl group of acetyl-coenzyme A (acetyl-CoA) (Figure 18.2). This acetyl group is metabolized in the tricarboxylic acid (TCA) cycle (and fully oxidized) to yield CO_2 (see Chapter 19).

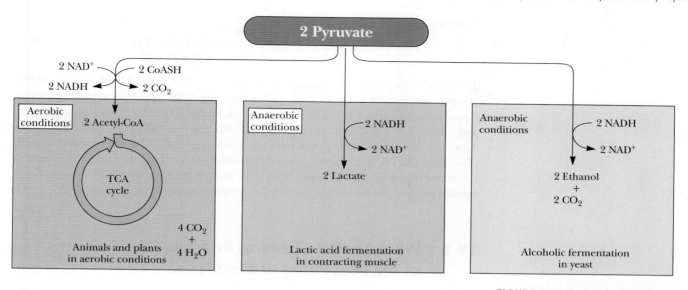

Alternatively, in the absence of oxygen (anaerobic conditions), pyruvate can be reduced to lactate through oxidation of NAD$^+$ to NAD$^+$—a process termed **lactic acid fermentation.** In microorganisms such as brewer's yeast, and in certain plant tissues, pyruvate can be reduced to ethanol, again with oxidation of NADH to NAD$^+$. Most students will recognize this process as **alcoholic fermentation.**

Glycolysis consists of two phases. In the first phase, a series of five reactions, glucose is broken down to two molecules of glyceraldehyde-3-phosphate. In the second phase, five subsequent reactions convert these two molecules of glyceraldehyde-3-phosphate into two molecules of pyruvate. Phase 1 consumes two molecules of ATP (Figure 18.1). The later stages of glycolysis result in the production of four molecules of ATP. The net is $4 - 2 = 2$ molecules of ATP produced per molecule of glucose. Microorganisms, plants, and animals (including humans) carry out the 10 reactions of glycolysis in more or less similar fashion, although the rates of the individual reactions and the means by which they are regulated differ from species to species.

18.2   Why Are Coupled Reactions Important in Glycolysis?

The process of glycolysis converts some, but not all, of the metabolic energy of the glucose molecule into ATP. The free energy change for the conversion of glucose to two molecules of lactate (the anaerobic route in contracting muscle) is -183.6 kJ/mol:

$$C_6H_{12}O_6 \rightarrow 2\ H_3C-CHOH-COO^- + 2\ H^+ \qquad (18.1)$$
$$\Delta G^{\circ\prime} = -183.6 \text{ kJ/mol}$$

This process occurs with no net oxidation or reduction. Although several individual steps in the pathway involve oxidation or reduction, these steps compensate each other exactly. Thus, the conversion of a molecule of glucose to two molecules of lactate involves simply a rearrangement of bonds, with no net loss or gain of electrons. The energy made available through this rearrangement is a relatively small part of the total energy obtainable from glucose (see Figure 1.18).

The production of two molecules of ATP in glycolysis is an energy-requiring process:

$$2\ ADP + 2\ P_i \longrightarrow 2\ ATP + 2\ H_2O \qquad (18.2)$$
$$\Delta G^{\circ\prime} = 2 \times 30.5 \text{ kJ/mol} = 61.0 \text{ kJ/mol}$$

Glycolysis couples these two reactions:

$$\text{Glucose} + 2\,\text{ADP} + 2\,\text{P}_i \rightarrow 2\,\text{lactate} + 2\,\text{ATP} + 2\,\text{H}^+ + 2\,\text{H}_2\text{O} \qquad (18.3)$$
$$\Delta G^{\circ\prime} = -183.6 + 61 = -122.6 \text{ kJ/mol}$$

Thus, under standard-state conditions, $(61/183.6) \times 100\%$, or 33%, of the free energy released is preserved in the form of ATP in these reactions. However, as we discussed in Chapter 3, the various solution conditions, such as pH, concentration, ionic strength, and presence of metal ions, can substantially alter the free energy change for such reactions. Under actual cellular conditions, the free energy change for the synthesis of ATP (Equation 18.2) is much larger, and approximately 50% of the available free energy is converted into ATP. Clearly, then, more than enough free energy is available in the conversion of glucose into lactate to drive the synthesis of two molecules of ATP.

18.3 What Are the Chemical Principles and Features of the First Phase of Glycolysis?

In the first phase of glycolysis, glucose will be phosphorylated at C-1 and C-6, and the six-carbon skeleton of glucose will be cleaved to yield two three-carbon molecules of glyceraldehyde-3-phosphate. Phosphorylation and cleavage reorganize the glucose molecule so that molecules of ATP can be produced in the second phase of glycolysis.

Reaction 1: Glucose Is Phosphorylated by Hexokinase or Glucokinase—The First Priming Reaction

The initial reaction of the glycolysis pathway involves phosphorylation of glucose at carbon atom 6 by either hexokinase or glucokinase. (Recall that "kinases" are enzymes that transfer the γ-phosphate of ATP to nucleophilic acceptors.) Phosphorylation activates glucose for the subsequent reactions in the pathway. However, the formation of such a phosphoester is thermodynamically unfavorable and requires energy input to operate in the forward direction (see Chapter 3). The energy comes from ATP, a requirement that at first seems counterproductive. Glycolysis is designed to *make* ATP, not consume it. However, the hexokinase or glucokinase reaction (Figure 18.1) is one of two **priming reactions** in the pathway. Just as old-fashioned, hand-operated water pumps (Figure 18.3) have to be primed with a small amount of water to deliver more water to the thirsty pumper, the glycolysis pathway requires two priming ATP molecules to start the sequence of reactions and delivers four molecules of ATP in the end.

The complete reaction for the first step in glycolysis is

$$\alpha\text{-D-Glucose} + \text{ATP}^{4-} \longrightarrow \alpha\text{-D-glucose-6-phosphate}^{2-} + \text{ADP}^{3-} + \text{H}^+ \qquad (18.4)$$
$$\Delta G^{\circ\prime} = -16.7 \text{ kJ/mol}$$

The hydrolysis of ATP makes 30.5 kJ/mol available in this reaction, and the phosphorylation of glucose "costs" 13.8 kJ/mol (Table 8.1). Thus, the reaction liberates 16.7 kJ/mol under standard-state conditions (1 M concentrations), and the equilibrium of the reaction lies far to the right ($K_{eq} = 850$ at 25°C; see Table 18.1). Under cellular conditions (Table 18.2), this first reaction of glycolysis is even more favorable than at standard state, with a ΔG of -33.9 kJ/mol (see Table 18.1).

The Cellular Advantages of Phosphorylating Glucose The incorporation of a phosphate into glucose in this energetically favorable reaction is important for several reasons. First, phosphorylation keeps the substrate in the cell. Glucose is a neutral molecule and could diffuse across the cell membrane, but phosphorylation confers a negative charge on glucose and the plasma membrane is essentially impermeable to glucose-6-phosphate (Figure 18.4). Moreover, rapid conversion of glucose to glucose-6-phosphate keeps the *intracellular* concentration of glucose low, favoring facilitated diffusion of

FIGURE 18.3 Just as a water pump must be "primed" with water to get more water out, the glycolytic pathway is primed with ATP in steps 1 and 3 in order to achieve net production of ATP in the second phase of the pathway.

TABLE 18.1	Reactions and Thermodynamics of Glycolysis				
Reaction	Enzyme	$\Delta G°'$ (kJ/mol)	K_{eq} at 25°C	ΔG (kJ/mol)
α-D-Glucose + ATP^{4-} $\rightleftharpoons$ glucose-6-phosphate^{2-} + ADP^{3-} + H$^+$	Hexokinase Glucokinase	−16.7	850	−33.9*
Glucose-6-phosphate^{2-} $\rightleftharpoons$ fructose-6-phosphate^{2-}	Phosphoglucoisomerase	+1.67	0.51	−2.92
Fructose-6-phosphate^{2-} + ATP^{4-} $\rightleftharpoons$ fructose-1,6-bisphosphate^{4-} + ADP^{3-} + H$^+$	Phosphofructokinase	−14.2	310	−18.8
Fructose-1,6-bisphosphate^{4-} $\rightleftharpoons$ dihydroxyacetone-P^{2-} + glyceraldehyde-3-P^{2-}	Fructose bisphosphate aldolase	+23.9	6.43×10^{-5}	−0.23
Dihydroxyacetone-P^{2-} $\rightleftharpoons$ glyceraldehyde-3-P^{2-}	Triose phosphate isomerase	+7.56	0.0472	+2.41
Glyceraldehyde-3-P^{2-} + P$_i^{2-}$ + NAD$^+$ $\rightleftharpoons$ 1,3-bisphosphoglycerate^{4-} + NADH + H$^+$	Glyceraldehyde-3-P dehydrogenase	+6.30	0.0786	−1.29
1,3-Bisphosphoglycerate^{4-} + ADP^{3-} $\rightleftharpoons$ 3-P-glycerate^{3-} + ATP^{4-}	Phosphoglycerate kinase	−18.9	2060	+0.1
3-Phosphoglycerate^{3-} $\rightleftharpoons$ 2-phosphoglycerate^{3-}	Phosphoglycerate mutase	+4.4	0.169	+0.83
2-Phosphoglycerate^{3-} $\rightleftharpoons$ phosphoenolpyruvate^{3-} + H$_2$O	Enolase	+1.8	0.483	+1.1
Phosphoenolpyruvate^{3-} + ADP^{3-} + H$^+$ $\rightleftharpoons$ pyruvate$^-$ + ATP^{4-}	Pyruvate kinase	−31.7	3.63×10^5	−23.0
Pyruvate$^-$ + NADH + H$^+$ $\rightleftharpoons$ lactate$^-$ + NAD$^+$	Lactate dehydrogenase	−25.2	2.63×10^4	−14.8

*ΔG values calculated for 310K (37°C) using the data in Table 18.2 for metabolite concentrations in erythrocytes. $\Delta G°'$ values are assumed to be the same at 25° and 37°C.

glucose *into* the cell. In addition, because regulatory control can be imposed only on reactions not at equilibrium, the favorable thermodynamics of this first reaction makes it an important site for regulation.

The Isozymes of Hexokinase In most animal, plant, and microbial cells, the enzyme that phosphorylates glucose is **hexokinase**. Magnesium ion (Mg^{2+}) is required for this reaction, as for the other kinase enzymes in the glycolytic pathway. The true substrate for the hexokinase reaction is MgATP^{2-}. There are four isozymes of hexokinase in most animal tissues. Hexokinase I is the principal form in the brain. Hexokinase in skeletal muscle is a mixture of types I (70% to 75%) and II (25% to 30%). The K_m for glucose is 0.03 mM for type I and 0.3 mM for type II; thus, hexokinase operates efficiently at normal blood glucose levels of 4 mM or so. The animal isozymes are allosterically inhibited by the product, glucose-6-phosphate. High levels of glucose-6-phosphate inhibit

TABLE 18.2	Steady-State Concentrations of Glycolytic Metabolites in Erythrocytes
Metabolite	mM
Glucose	5.0
Glucose-6-phosphate	0.083
Fructose-6-phosphate	0.014
Fructose-1,6-bisphosphate	0.031
Dihydroxyacetone phosphate	0.14
Glyceraldehyde-3-phosphate	0.019
1,3-Bisphosphoglycerate	0.001
2,3-Bisphosphoglycerate	4.0
3-Phosphoglycerate	0.12
2-Phosphoglycerate	0.030
Phosphoenolpyruvate	0.023
Pyruvate	0.051
Lactate	2.9
ATP	1.85
ADP	0.14
P$_i$	1.0

Adapted from Minakami, S., and Yoshikawa, H., 1965. Thermodynamic considerations on erythrocyte glycolysis. *Biochemical and Biophysical Research Communications* **18:**345.

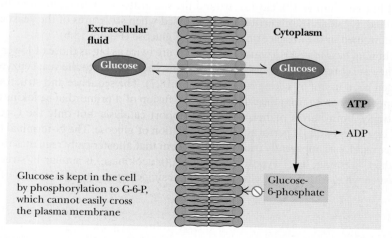

FIGURE 18.4 Phosphorylation of glucose to glucose-6-phosphate by ATP creates a charged molecule that cannot easily cross the plasma membrane.

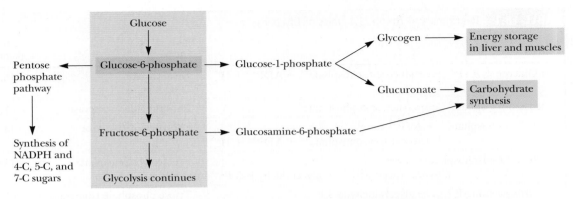

FIGURE 18.5 Glucose-6-phosphate is the branch point for several metabolic pathways.

(a)

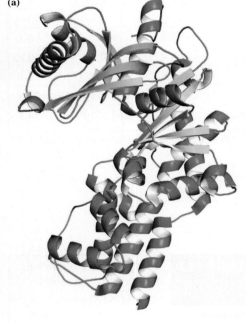

(b)

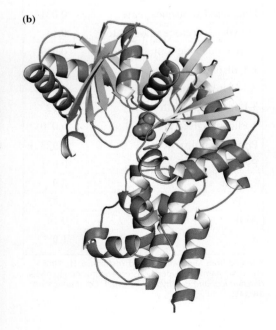

hexokinase activity until consumption by glycolysis lowers its concentration. The hexokinase reaction is one of three points in the glycolysis pathway that are *regulated*. As the generic name implies, hexokinase can phosphorylate a variety of hexose sugars, including glucose, mannose, and fructose.

The type IV isozyme of hexokinase, called **glucokinase,** is found predominantly in the liver and pancreas. Type IV is highly specific for D-glucose, has a much higher K_m for glucose (approximately 10 mM), and is not product inhibited. With such a high K_m for glucose, glucokinase becomes important metabolically only when liver glucose levels are high (for example, when the individual has consumed large amounts of carbohydrates). When glucose levels are low, hexokinase is primarily responsible for phosphorylating glucose. However, when glucose levels are high, glucose is converted by glucokinase to glucose-6-phosphate and is eventually stored in the liver as glycogen. Glucokinase is an *inducible* enzyme—the amount present in the liver is controlled by *insulin* (secreted by the pancreas). (Patients with **diabetes mellitus** produce insufficient insulin. They have low levels of glucokinase, cannot tolerate high levels of blood glucose, and produce little liver glycogen.) Because glucose-6-phosphate is common to several metabolic pathways (Figure 18.5), it occupies a branch point in glucose metabolism.

Hexokinase Binds Glucose and ATP with an Induced Fit In most organisms, hexokinase occurs in a single form: a two-lobed 50-kD monomer that resembles a clamp, with a large groove in one side (Figure 18.6; see also Figure 13.24). Daniel Koshland predicted, years before structures were available, that hexokinase would undergo an induced fit (see Chapter 13), closing around the substrates ATP and glucose when they were bound. Koshland's prediction was confirmed when structures of the yeast enzyme were determined in the absence and presence of glucose (Figure 18.6).

The human hexokinase isozymes I, II, and III are twice as big as those of lower organisms. They are composed of two separate domains, each similar to the yeast enzyme, and connected head to tail by a long α-helix (Figure 18.7). The sequence and structure similarity apparently arose from the duplication and fusion of a primordial hexokinase gene. Interestingly, both halves of hexokinase II support catalysis, but only the C-terminal half of isozymes I and III performs phosphorylation of glucose. The N-terminal half, on the other hand, has apparently evolved into a form that allosterically regulates the activity of the C-terminal half! Type IV hexokinase (glucokinase) is similar in structure to the yeast enzyme, with a single clamp domain, a single active site, and a mass of 50 kD (Figure 18.7).

Glucose

FIGURE 18.6 The **(a)** open and **(b)** closed states of yeast hexokinase. Binding of glucose (green) induces a conformation change that closes the active site, as predicted by Koshland (a: pdb id = 1IG8; b: pdb id = 1BDG).

(a)

(b)

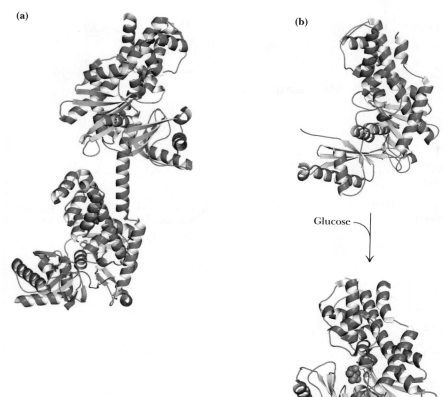

Glucose

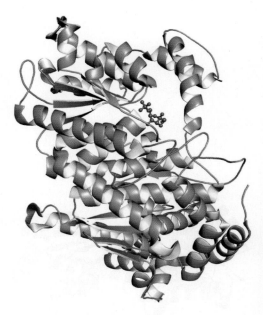

FIGURE 18.7 (a) Mammalian hexokinase I (pdb id = 1CZA) contains an N-terminal domain (top) and a C-terminal domain (bottom) joined by a long α-helix. Each of these domains is similar in sequence and structure to yeast hexokinase. **(b)** Human glucokinase undergoes an induced fit upon binding glucose (green). (Top: pdb id = 1V4T; bottom: pdb id = 1V4S).

Reaction 2: Phosphoglucoisomerase Catalyzes the Isomerization of Glucose-6-Phosphate

The second step in glycolysis is a common type of metabolic reaction: the isomerization of a sugar. In this particular case, the carbonyl oxygen of glucose-6-phosphate is shifted from C-1 to C-2. This amounts to isomerization of an aldose (glucose-6-phosphate) to a ketose—fructose-6-phosphate (Figure 18.8). The reaction is necessary for two reasons. First, the next step in glycolysis is phosphorylation at C-1, and the hemiacetal —OH of glucose would be more difficult to phosphorylate than a simple primary hydroxyl. Second, the isomerization to fructose (with a carbonyl group at position 2 in the linear form) activates C-3, facilitating C-C bond cleavage in the fourth step of glycolysis. The enzyme responsible for this isomerization is **phosphoglucoisomerase,** also known as **phosphoglucose isomerase** and **glucose phosphate isomerase.** In humans, the enzyme requires Mg^{2+} for activity and is highly specific for glucose-6-phosphate. The $\Delta G^{\circ\prime}$ is 1.67 kJ/mol, and the value of ΔG under cellular conditions (Table 18.1) is −2.92 kJ/mol. This small value means that the reaction operates near equilibrium in the cell and is readily reversible. Phosphoglucoisomerase proceeds through an *enediol* intermediate, as shown in Figure 18.8. Although the predominant forms of glucose-6-phosphate and fructose-6-phosphate in solution are the ring forms, the isomerase interconverts the open-chain form of G-6-P with the open-chain form of F-6-P.

Phosphoglucoisomerase, with fructose-6-P (blue) bound (pdb id = 1HOX).

FIGURE 18.8 The phosphoglucoisomerase mechanism involves opening of the pyranose ring (step 1), proton abstraction leading to enediol formation (step 2), and proton addition to the double bond (step 3), followed by ring closure.

$\Delta G^\sigma = -14.2$ kJ/mol
$\Delta G_{\text{erythrocyte}} = -18.8$ kJ/mol

Fructose-6-phosphate

Fructose-1,6-bisphosphate

Reaction 3: ATP Drives a Second Phosphorylation by Phosphofructokinase—The Second Priming Reaction

The action of phosphoglucoisomerase, "moving" the carbonyl group from C-1 to C-2, creates a new primary alcohol function at C-1 (see Figure 18.8). The next step in the glycolytic pathway is the phosphorylation of this group by **phosphofructokinase.** Once again, the substrate that provides the phosphoryl group is ATP. Like the hexokinase/glucokinase reaction, the phosphorylation of fructose-6-phosphate is a priming reaction and is endergonic:

$$\text{Fructose-6-P} + \text{P}_i \longrightarrow \text{fructose-1,6-bisphosphate} \qquad (18.5)$$
$$\Delta G^{\circ\prime} = 16.3 \text{ kJ/mol}$$

When coupled (by phosphofructokinase) with phosphoryl transfer from ATP, the overall reaction becomes exergonic:

$$\text{Fructose-6-P} + \text{ATP} \longrightarrow \text{fructose-1,6-bisphosphate} + \text{ADP} \qquad (18.6)$$
$$\Delta G^{\circ\prime} = -14.2 \text{ kJ/mol}$$
$$\Delta G \text{ (in erythrocytes)} = -18.8 \text{ kJ/mol}$$

At pH 7 and 37°C, the phosphofructokinase reaction equilibrium lies far to the right. Just as the hexokinase reaction commits the cell to taking up glucose, *the phosphofructokinase reaction commits the cell to metabolizing glucose* rather than converting it to

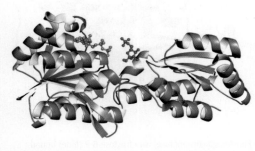

Phosphofructokinase with ADP (in orange) and fructose-6-phosphate (in red) (pdb id = 4PFK).

another sugar or storing it. Similarly, just as the large free energy change of the hexokinase reaction makes it a likely candidate for regulation, so the phosphofructokinase reaction is an important site of regulation—indeed, the most important site in the glycolytic pathway.

Regulation of Phosphofructokinase Phosphofructokinase is the "valve" controlling the rate of glycolysis. In addition to its role as a substrate, ATP is also an allosteric inhibitor of this enzyme. Thus, phosphofructokinase has two distinct binding sites for ATP; a high-affinity substrate site and a low-affinity regulatory site. In the presence of high ATP concentrations, phosphofructokinase behaves cooperatively, plots of enzyme activity versus fructose-6-phosphate are sigmoid, and the K_m for fructose-6-phosphate is increased (Figure 18.9). Thus, when ATP levels are sufficiently high in the cytosol, glycolysis "turns off." Under most cellular conditions, however, the ATP concentration does not vary over a large range. The ATP concentration in muscle during vigorous exercise, for example, is only about 10% lower than that during the resting state. The rate of glycolysis, however, varies much more. A large range of glycolytic rates cannot be directly accounted for by only a 10% change in ATP levels.

AMP reverses the inhibition due to ATP, and AMP levels in cells *can* rise dramatically when ATP levels decrease, due to the action of the enzyme *adenylate kinase,* which catalyzes the reaction

$$ADP + ADP \rightleftharpoons ATP + AMP$$

with the equilibrium constant:

$$K_{eq} = \frac{[ATP][AMP]}{[ADP]^2} = 0.44 \qquad (18.7)$$

Adenylate kinase rapidly interconverts ADP, ATP, and AMP to maintain this equilibrium. ADP levels in cells are typically 10% of ATP levels, and AMP levels are often less than 1% of the ATP concentration. Due to the nature of the adenylate kinase equilibrium, a small decrease in ATP concentration due to ATP consumption results in a much larger relative increase in the AMP levels because of adenylate kinase activity.

Clearly, the activity of phosphofructokinase depends on both ATP and AMP levels and is a function of the cellular energy status. Phosphofructokinase activity is increased when the energy status falls and is decreased when the energy status is high. The rate of glycolytic activity thus decreases when ATP is plentiful and increases when more ATP is needed.

Glycolysis and the citric acid cycle (to be discussed in Chapter 19) are coupled via phosphofructokinase, because *citrate,* an intermediate in the citric acid cycle, is an allosteric inhibitor of phosphofructokinase. When the citric acid cycle reaches saturation,

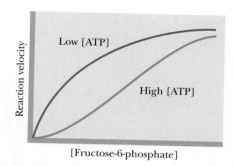

FIGURE 18.9 At high [ATP], phosphofructokinase (PFK) behaves cooperatively and the plot of enzyme activity versus [fructose-6-phosphate] is sigmoid. High [ATP] thus inhibits PFK, decreasing the enzyme's affinity for fructose-6-phosphate.

▶ A DEEPER LOOK

Phosphoglucoisomerase—A Moonlighting Protein

When someone has a day job but also works at night (that is, under the moon) at a second job, they are said to be "moonlighting." Similarly, a number of proteins have been found to have two or more different functions, and Constance Jeffery at Brandeis University has dubbed these "moonlighting proteins." Phosphoglucoisomerase catalyzes the second step of glycolysis but also moonlights as a nerve growth factor outside animal cells. In fact, outside the cell, this protein is known as neuroleukin (NL), autocrine motility factor (AMF), and differentiation and maturation mediator (DMM). Neuroleukin is secreted by (immune system) T cells and promotes the survival of certain spinal neurons and sensory nerves. AMF is secreted by tumor cells and stimulates cancer cell migration. DMM causes certain leukemia cells to differentiate.

How phosphoglucoisomerase is secreted by the cell for its moonlighting functions is unknown, but there is evidence that the organism itself may be harmed by this secretion. Diane Mathis and Christophe Benoist at the University of Strasbourg have shown that, in mice with disorders similar to rheumatoid arthritis, the immune system recognizes extracellular phosphoglucoisomerase as an antigen—that is, a protein that is "nonself." That a protein can be vital to metabolism inside the cell and also function as a growth factor and occasionally act as an antigen outside the cell is indeed remarkable.

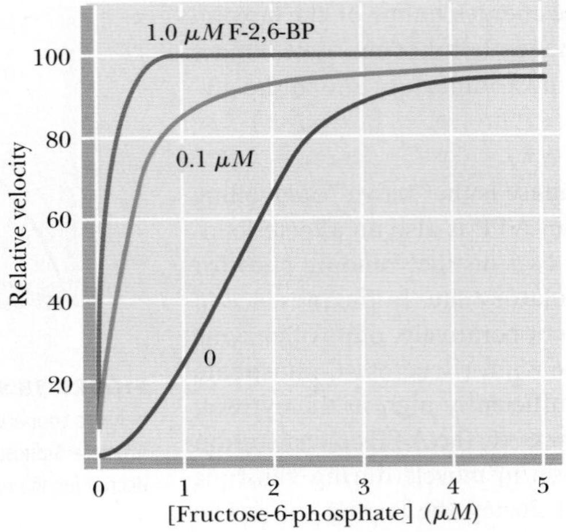

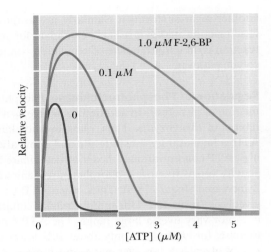

FIGURE 18.10 Fructose-2,6-bisphosphate activates phosphofructokinase, increasing the affinity of the enzyme for fructose-6-phosphate and restoring the hyperbolic dependence of enzyme activity on substrate concentration.

FIGURE 18.11 Fructose-2,6-bisphosphate decreases the inhibition of phosphofructokinase due to ATP.

glycolysis (which "feeds" the citric acid cycle under aerobic conditions) slows down. The citric acid cycle directs electrons into the electron-transport chain (for the purpose of ATP synthesis in oxidative phosphorylation) and also provides precursor molecules for biosynthetic pathways. Inhibition of glycolysis by citrate ensures that glucose will not be committed to these activities if the citric acid cycle is already saturated.

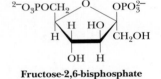

Fructose-2,6-bisphosphate

Phosphofructokinase is also regulated by β-D-fructose-2,6-bisphosphate, a potent allosteric activator that increases the affinity of phosphofructokinase for the substrate fructose-6-phosphate (Figure 18.10). Stimulation of phosphofructokinase is also achieved by decreasing the inhibitory effects of ATP (Figure 18.11). Fructose-2,6-bisphosphate increases the net flow of glucose through glycolysis by stimulating phosphofructokinase and, as we shall see in Chapter 22, by inhibiting fructose-1,6-bisphosphatase, the enzyme that catalyzes this reaction in the opposite direction.

Reaction 4: Cleavage by Fructose Bisphosphate Aldolase Creates Two 3-Carbon Intermediates

Fructose bisphosphate aldolase cleaves fructose-1,6-bisphosphate between the C-3 and C-4 carbons to yield two triose phosphates. The products are dihydroxyacetone phosphate (DHAP) and glyceraldehyde-3-phosphate. The reaction has an equilibrium constant of approximately 10^{-4} *M*, and a corresponding $\Delta G°'$ of +23.9 kJ/mol. These values might imply that the reaction does not proceed effectively from left to right as written. However, the reaction makes two molecules (glyceraldehyde-3-P and dihydroxyacetone-P) from one molecule (fructose-1,6-bisphosphate), and the

D-Fructose-1,6-bisphosphate (FBP) | **Dihydroxyacetone phosphate (DHAP)** | **D-Glyceraldehyde 3-phosphate (G-3-P)**

$\Delta G°' = 23.9$ kJ/mol

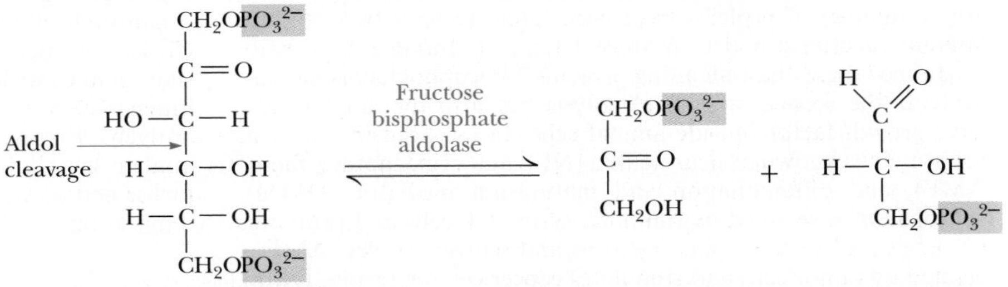

equilibrium is thus greatly influenced by concentration. The value of ΔG in erythrocytes is actually -0.23 kJ/mol (see Table 18.1). At physiological concentrations, the reaction is essentially at equilibrium.

Two classes of aldolase enzymes are found in nature. Animal tissues produce a Class I aldolase, characterized by the formation of a covalent Schiff base intermediate between an active-site lysine and the carbonyl group of the substrate. Class I aldolases do not require a divalent metal ion. Class II aldolases are produced mainly in bacteria and fungi and do not form a covalent E-S intermediate, but they contain an active-site metal (normally zinc, Zn^{2+}). Cyanobacteria and some other simple organisms possess both classes of aldolase.

The aldolase reaction is merely the reverse of the **aldol condensation** well known to organic chemists. The latter reaction involves an attack by a nucleophilic enolate anion of an aldehyde or ketone on the carbonyl carbon of an aldehyde. The opposite reaction, aldol cleavage, begins with removal of a proton from the β-hydroxyl group, which is followed by the elimination of the enolate anion. A mechanism for the aldol cleavage reaction of fructose-1,6-bisphosphate in the Class I–type aldolases is shown in Figure 18.12a. In Class II aldolases, an active-site metal such as Zn^{2+} behaves as an electrophile, polarizing the carbonyl group of the substrate and stabilizing the enolate intermediate (Figure 18.12b).

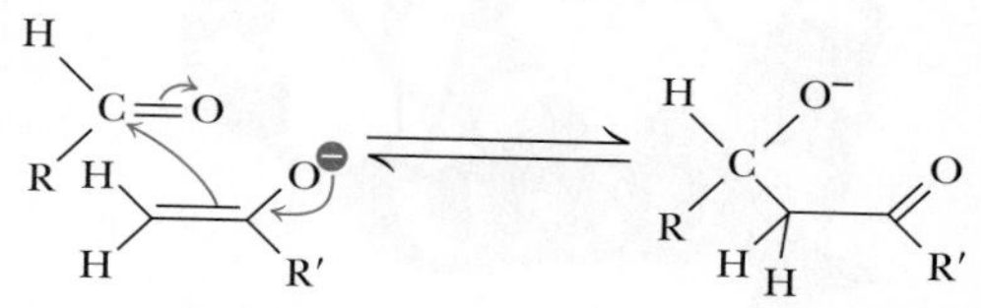

R′ = H (aldehyde)
R′ = alkyl, etc. (ketone)

Aldol condensation

Reaction 5: Triose Phosphate Isomerase Completes the First Phase of Glycolysis

Of the two products of the aldolase reaction, only glyceraldehyde-3-phosphate goes directly into the second phase of glycolysis. The other triose phosphate, dihydroxyacetone phosphate, must be converted to glyceraldehyde-3-phosphate by the enzyme

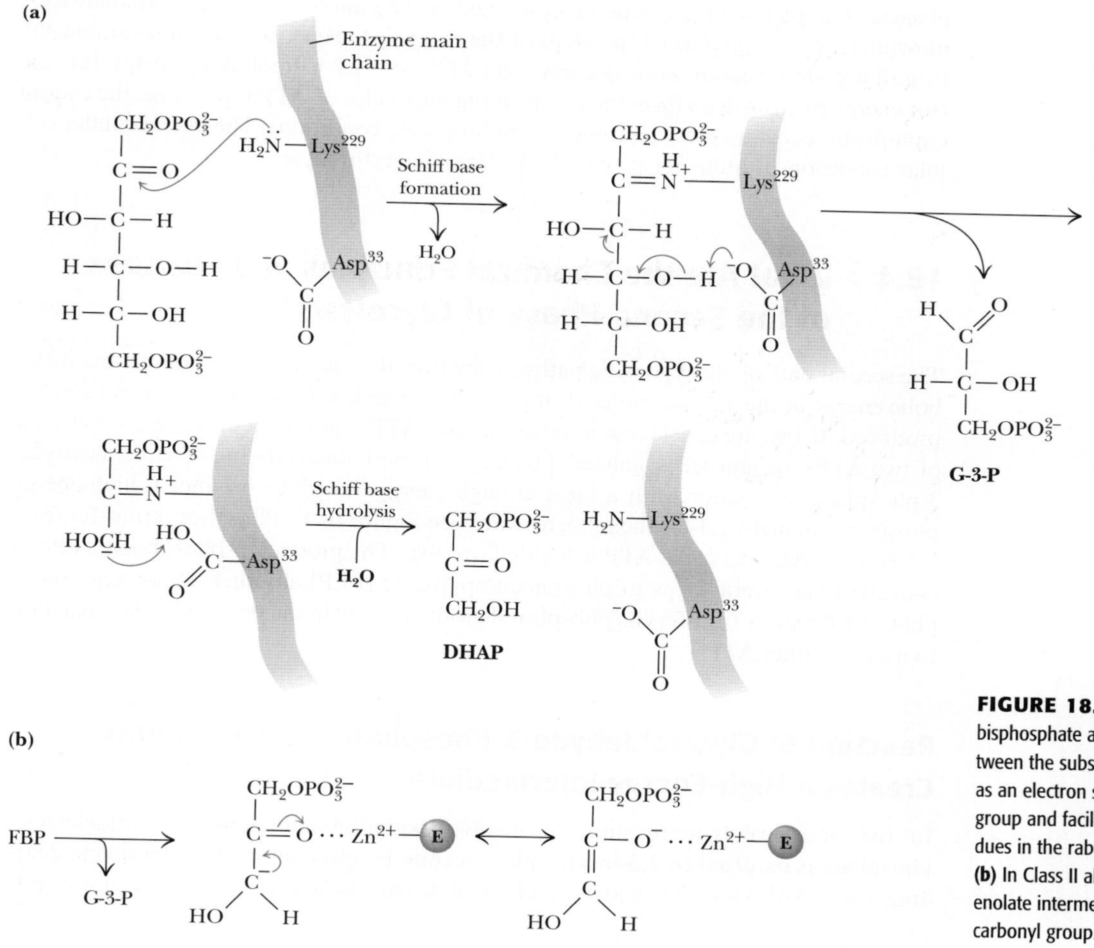

(a)

(b)

FIGURE 18.12 (a) A mechanism for the fructose-1,6-bisphosphate aldolase reaction. The Schiff base formed between the substrate carbonyl and an active-site lysine acts as an electron sink, increasing the acidity of the β-hydroxyl group and facilitating cleavage as shown. The catalytic residues in the rabbit muscle enzyme are Lys[229] and Asp[33]. **(b)** In Class II aldolases, an active-site Zn^{2+} stabilizes the enolate intermediate, leading to polarization of the substrate carbonyl group.

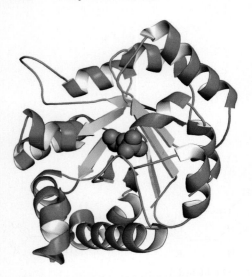

Triose phosphate isomerase with substrate analog 2-phosphoglycerate shown in cyan (pdb id = 1YPI).

$$CH_2OH \atop |$$
$$C=O \atop |$$
$$CH_2OPO_3{}^{2-}$$

DHAP

Triose phosphate isomerase →

$$H \quad O \atop \diagdown C \diagup$$
$$| \atop HCOH$$
$$| \atop CH_2OPO_3{}^{2-}$$

G-3-P

$\Delta G°' = +7.56$ kJ/mol

FIGURE 18.13 A reaction mechanism for triose phosphate isomerase. In the enzyme from yeast, the catalytic residue is Glu^{165}.

Enediol intermediate

Glyceraldehyde-3-P

triose phosphate isomerase. This reaction thus permits both products of the aldolase reaction to continue in the glycolytic pathway and in essence makes the C-1, C-2, and C-3 carbons of the starting glucose molecule equivalent to the C-6, C-5, and C-4 carbons, respectively. The reaction mechanism involves an enediol intermediate that can donate either of its hydroxyl protons to a basic residue on the enzyme and thereby become either dihydroxyacetone phosphate or glyceraldehyde-3-phosphate (Figure 18.13). Triose phosphate isomerase is one of the enzymes that have evolved to a state of "catalytic perfection," with a turnover number near the diffusion limit (see Table 13.5).

The triose phosphate isomerase reaction completes the first phase of glycolysis, each glucose that passes through being converted to two molecules of glyceraldehyde-3-phosphate. Although the last two steps of the pathway are energetically unfavorable, the overall five-step reaction sequence has a net $\Delta G°'$ of +2.2 kJ/mol ($K_{eq} \approx 0.43$). It is the free energy of hydrolysis from the two priming molecules of ATP that brings the overall equilibrium constant close to 1 under standard-state conditions. The net ΔG under cellular conditions is quite negative (−53.4 kJ/mol in erythrocytes).

18.4 What Are the Chemical Principles and Features of the Second Phase of Glycolysis?

The second half of the glycolytic pathway involves the reactions that convert the metabolic energy in the glucose molecule into ATP. Altogether, four new ATP molecules are produced. If two are considered to offset the two ATPs consumed in phase 1, a net yield of two ATPs per glucose is realized. Phase 2 starts with the oxidation of glyceraldehyde-3-phosphate, a reaction with a large enough energy "kick" to produce a high-energy phosphate, namely, 1,3-bisphosphoglycerate (see Figure 18.1). Phosphoryl transfer from 1,3-BPG to ADP to make ATP is highly favorable. The product, 3-phosphoglycerate, is converted via several steps to phosphoenolpyruvate (PEP), another high-energy phosphate. PEP readily transfers its phosphoryl group to ADP in the pyruvate kinase reaction to make another ATP.

Reaction 6: Glyceraldehyde-3-Phosphate Dehydrogenase Creates a High-Energy Intermediate

In the first glycolytic reaction to involve oxidation–reduction, glyceraldehyde-3-phosphate is oxidized to 1,3-bisphosphoglycerate by **glyceraldehyde-3-phosphate dehydrogenase**. Although the oxidation of an aldehyde to a carboxylic acid is a highly

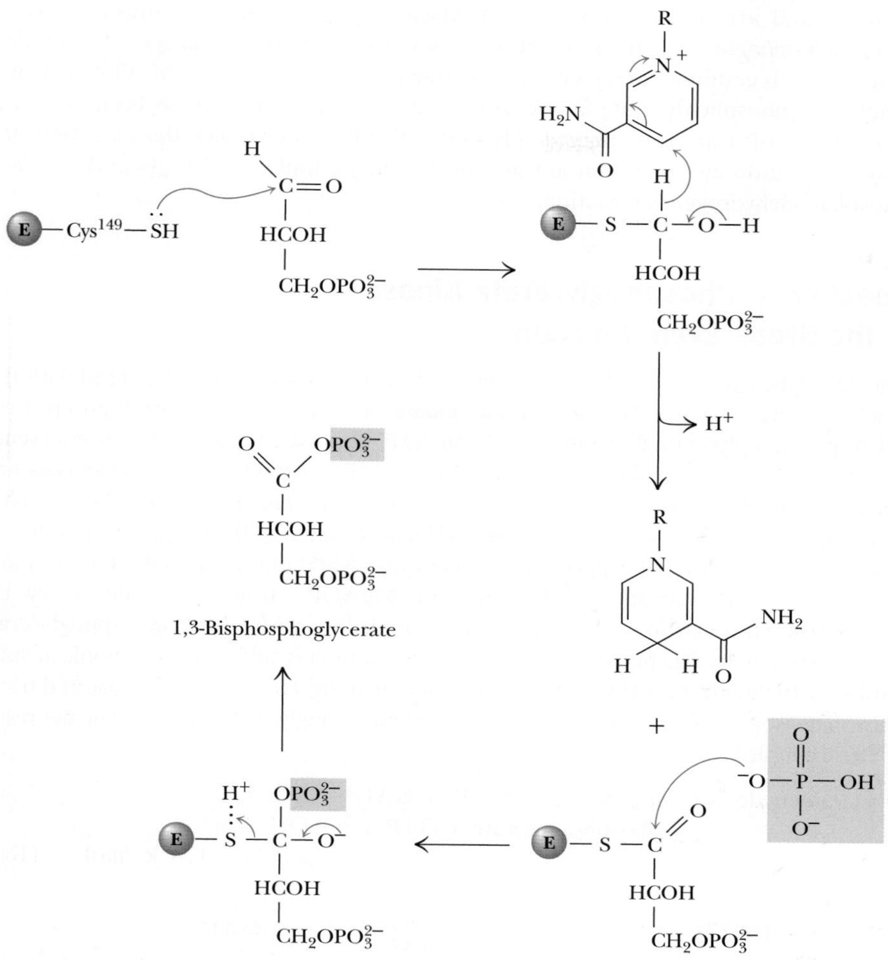

FIGURE 18.14 A mechanism for the glyceraldehyde-3-phosphate dehydrogenase reaction. Reaction of an enzyme sulfhydryl with the carbonyl carbon of glyceraldehyde-3-P forms a thiohemiacetal, which loses a hydride to NAD$^+$ to become a thioester. Phosphorolysis of this thioester releases 1,3-bisphosphoglycerate. In the enzyme from rabbit muscle, the catalytic residue is Cys^{149}.

exergonic reaction, the overall reaction involves both formation of a carboxylic–phosphoric anhydride and the reduction of NAD$^+$ to NADH and is therefore slightly endergonic at standard state, with a $\Delta G°'$ of $+6.30$ kJ/mol. The free energy that might otherwise be released as heat in this reaction is directed into the formation of a high-energy phosphate compound, 1,3-bisphosphoglycerate, and the reduction of NAD$^+$. The reaction mechanism involves nucleophilic attack by a cysteine —SH group on the carbonyl carbon of glyceraldehyde-3-phosphate to form a hemithioacetal (Figure 18.14). The hemithioacetal intermediate decomposes by hydride (H:$^-$) transfer to NAD$^+$ to form a high-energy thioester. Nucleophilic attack by phosphate displaces the product, 1,3-bisphosphoglycerate, from the enzyme. The enzyme can be inactivated by reaction with iodoacetate, which reacts with and blocks the essential cysteine sulfhydryl.

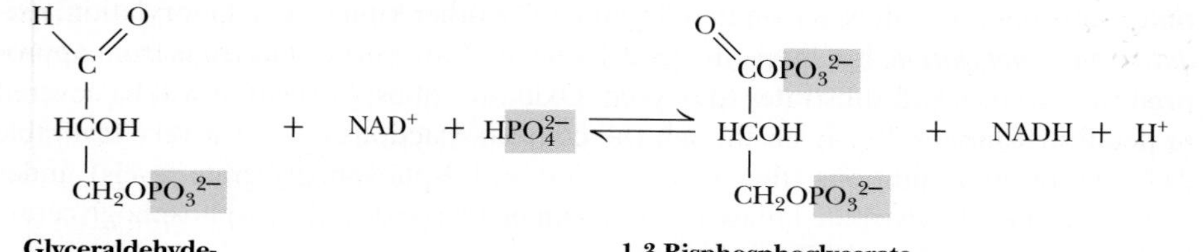

Glyceraldehyde-3-phosphate (G-3-P)

$+$ NAD$^+$ $+$ HPO$_4^{2-}$ $\rightleftharpoons$

1,3-Bisphosphoglycerate (1,3-BPG)

$+$ NADH $+$ H$^+$

$\Delta G°' = +6.3$ kJ/mol

The glyceraldehyde-3-phosphate dehydrogenase reaction is the site of action of *ar-senate* (AsO$_4^{3-}$), an anion analogous to phosphate. Arsenate is an effective substrate in this reaction, forming *1-arseno-3-phosphoglycerate*, but acyl arsenates are quite

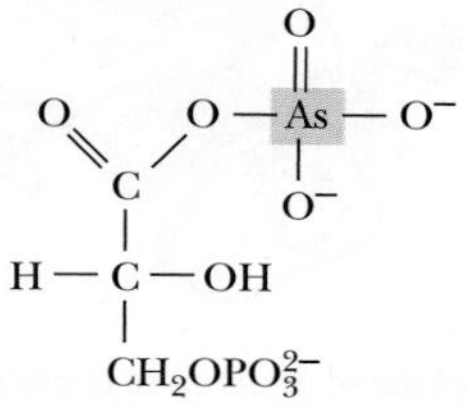

1-Arseno-3-phosphoglycerate

unstable and are rapidly hydrolyzed. 1-Arseno-3-phosphoglycerate breaks down to yield *3-phosphoglycerate,* the product of the seventh reaction of glycolysis. The result is that glycolysis continues in the presence of arsenate, but the molecule of ATP formed in reaction 7 (phosphoglycerate kinase) is not made because this step has been bypassed. The lability of 1-arseno-3-phosphoglycerate effectively uncouples the oxidation and phosphorylation events, which are normally tightly coupled in the glyceraldehyde-3-phosphate dehydrogenase reaction.

Reaction 7: Phosphoglycerate Kinase Is the Break-Even Reaction

The glycolytic pathway breaks even in terms of ATPs consumed and produced with this reaction. The enzyme **phosphoglycerate kinase** transfers a phosphoryl group from 1,3-bisphosphoglycerate to ADP to form an ATP. Because each glucose molecule sends two molecules of glyceraldehyde-3-phosphate into the second phase of glycolysis and because two ATPs were consumed per glucose in the first-phase reactions, the phosphoglycerate kinase reaction "pays off" the ATP debt created by the priming reactions. As might be expected for a phosphoryl transfer enzyme, Mg^{2+} ion is required for activity and the true nucleotide substrate for the reaction is $MgADP^{-}$. It is appropriate to view the sixth and seventh reactions of glycolysis as a coupled pair, with 1,3-bisphosphoglycerate as an intermediate. The phosphoglycerate kinase reaction is sufficiently exergonic at standard state to pull the G-3-P dehydrogenase reaction along. (In fact, the aldolase and triose phosphate isomerase are also pulled forward by phosphoglycerate kinase.) The net result of these coupled reactions is

$$\text{Glyceraldehyde-3-phosphate} + ADP + P_i + NAD^+ \longrightarrow$$
$$\text{3-phosphoglycerate} + ATP + NADH + H^+$$
$$\Delta G^{\circ\prime} = -12.6 \text{ kJ/mol} \qquad (18.8)$$

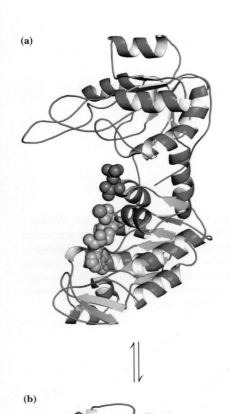

$\Delta G^{\circ\prime} = -18.9$ kJ/mol

Another reflection of the coupling between these reactions lies in their values of ΔG under cellular conditions (Table 18.1). Despite its strongly negative $\Delta G^{\circ\prime}$, the phosphoglycerate kinase reaction operates at equilibrium in the erythrocyte ($\Delta G = 0.1$ kJ/mol). In essence, the free energy available in the phosphoglycerate kinase reaction is used to bring the three previous reactions closer to equilibrium. Viewed in this context, it is clear that ADP has been phosphorylated to form ATP at the expense of a substrate, namely, glyceraldehyde-3-phosphate. This is an example of **substrate-level phosphorylation,** a concept that will be encountered again. (The other kind of phosphorylation, *oxidative phosphorylation,* is driven energetically by the transport of electrons from appropriate coenzymes and substrates to oxygen. Oxidative phosphorylation will be covered in detail in Chapter 20.) Even though the coupled reactions exhibit a very favorable $\Delta G^{\circ\prime}$, there are conditions (that is, high ATP and 3-phosphoglycerate levels) under which the phosphoglycerate kinase reaction can be reversed so that 3-phosphoglycerate is phosphorylated from ATP.

An important regulatory molecule, 2,3-bisphosphoglycerate, is synthesized and metabolized by a pair of reactions that make a detour around the phosphoglycerate kinase reaction. 2,3-BPG, which stabilizes the deoxy form of hemoglobin and is primarily responsible for the cooperative nature of oxygen binding by hemoglobin (see Chapter 15), is formed from 1,3-bisphosphoglycerate by **bisphosphoglycerate mutase** (Figure 18.15). Interestingly, 3-phosphoglycerate is required for this reaction, which

The open **(a)** and closed **(b)** forms of phosphoglycerate kinase. ATP (cyan), 3-phosphoglycerate (purple), and Mg^{2+} (gold) (a: pdb id = 3PGK; b: pdb id = 1VPE).

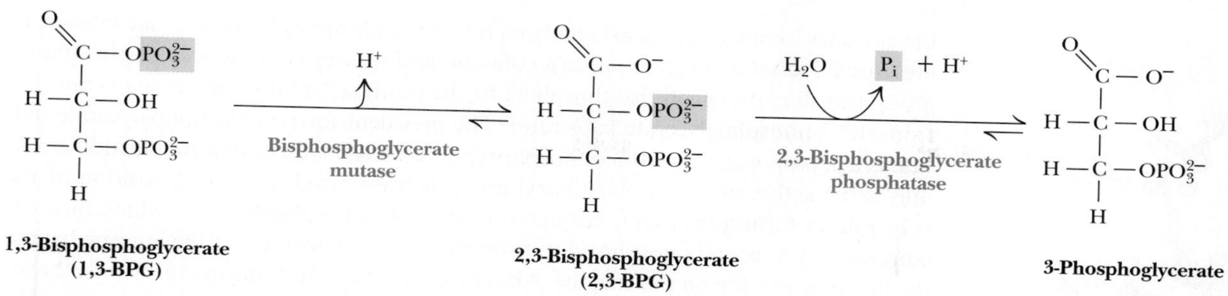

FIGURE 18.15 Formation and decomposition of 2,3-bisphosphoglycerate.

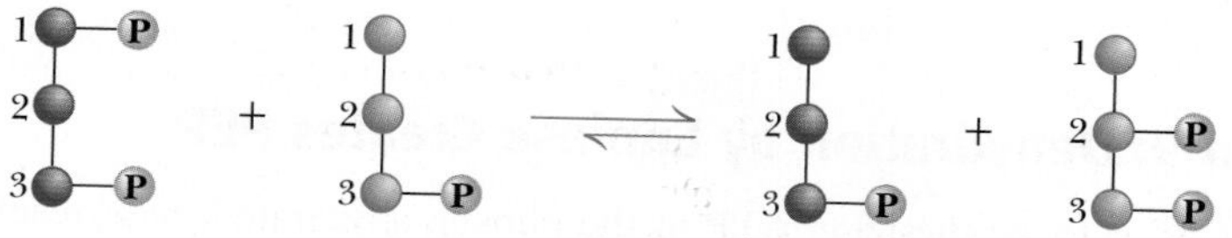

FIGURE 18.16 The mutase that forms 2,3-BPG from 1,3-BPG requires 3-phosphoglycerate. The reaction is actually an intermolecular phosphoryl transfer from C-1 of 1,3-BPG to C-2 of 3-PG.

involves phosphoryl transfer from the C-1 position of 1,3-bisphosphoglycerate to the C-2 position of 3-phosphoglycerate (Figure 18.16). Hydrolysis of 2,3-BPG is carried out by *2,3-bisphosphoglycerate phosphatase*. Although other cells contain only a trace of 2,3-BPG, erythrocytes typically contain 4 to 5 mM 2,3-BPG.

Reaction 8: Phosphoglycerate Mutase Catalyzes a Phosphoryl Transfer

The remaining steps in the glycolytic pathway prepare for synthesis of the second ATP equivalent. This begins with the **phosphoglycerate mutase** reaction, in which the phosphoryl group of 3-phosphoglycerate is moved from C-3 to C-2. (The term *mutase* is applied to enzymes that catalyze migration of a functional group within a substrate molecule.) The free energy change for this reaction is very small under cellular conditions ($\Delta G = 0.83$ kJ/mol in erythrocytes). Phosphoglycerate mutase enzymes isolated from different sources exhibit different reaction mechanisms. As shown in Figure 18.17,

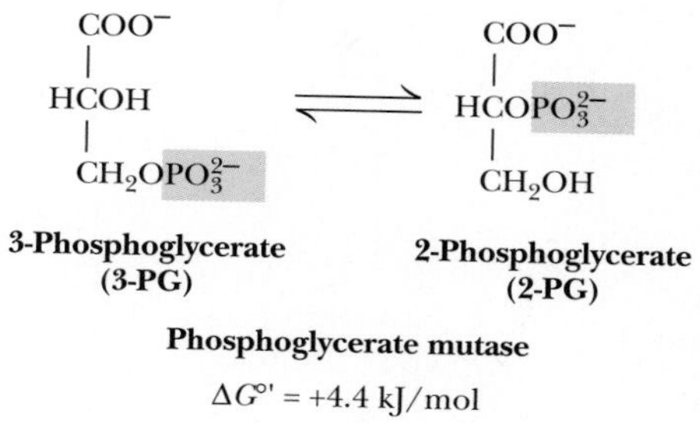

3-Phosphoglycerate (3-PG) 2-Phosphoglycerate (2-PG)

Phosphoglycerate mutase

$\Delta G°' = +4.4$ kJ/mol

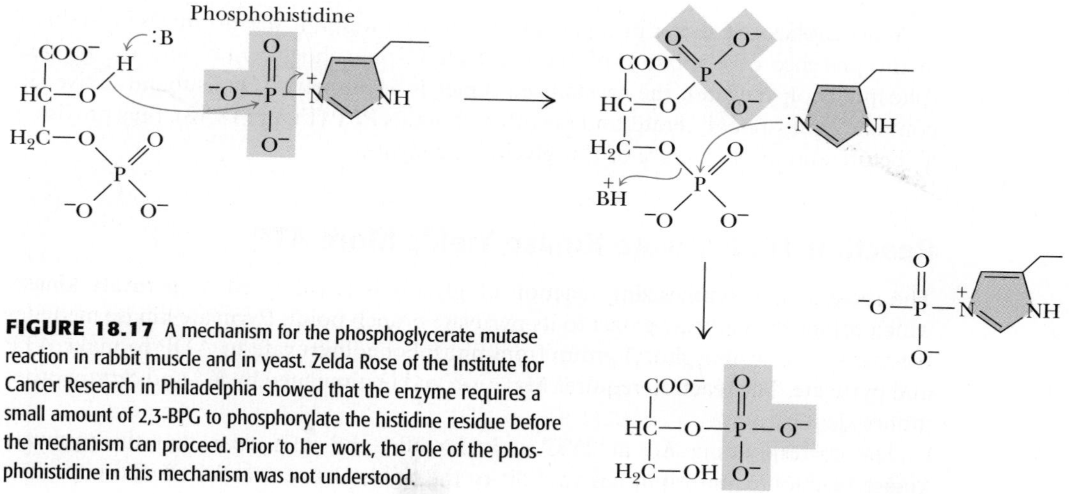

FIGURE 18.17 A mechanism for the phosphoglycerate mutase reaction in rabbit muscle and in yeast. Zelda Rose of the Institute for Cancer Research in Philadelphia showed that the enzyme requires a small amount of 2,3-BPG to phosphorylate the histidine residue before the mechanism can proceed. Prior to her work, the role of the phosphohistidine in this mechanism was not understood.

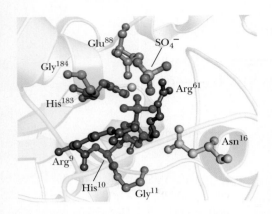

The catalytic histidine (His[183]) at the active site of *Escherichia coli* phosphoglycerate mutase (pdb id = 1E58). Note that His[10] is phosphorylated.

the enzymes isolated from yeast and from rabbit muscle form *phosphoenzyme* intermediates, use *2,3-bisphosphoglycerate* as a cofactor, and undergo *inter*molecular phosphoryl group transfers (in which the phosphate of the product 2-phosphoglycerate is not that from the 3-phosphoglycerate substrate). The prevalent form of phosphoglycerate mutase is a *phosphoenzyme,* with a phosphoryl group covalently bound to a histidine residue at the active site. This phosphoryl group is transferred to the C-2 position of the substrate to form a transient, enzyme-bound 2,3-bisphosphoglycerate, which then decomposes by a second phosphoryl transfer from the C-3 position of the intermediate to the histidine residue on the enzyme. About once in every 100 enzyme turnovers, the intermediate, 2,3-bisphosphoglycerate, dissociates from the active site, leaving an inactive, unphosphorylated enzyme. The unphosphorylated enzyme can be reactivated by binding 2,3-BPG. For this reason, maximal activity of phosphoglycerate mutase requires the presence of small amounts of 2,3-BPG.

Reaction 9: Dehydration by Enolase Creates PEP

Recall that prior to synthesizing ATP in the phosphoglycerate kinase reaction, it was necessary to first make a substrate having a high-energy phosphate. Reaction 9 of glycolysis similarly makes a high-energy phosphate in preparation for ATP synthesis. **Enolase** catalyzes the formation of *phosphoenolpyruvate* from 2-phosphoglycerate. The reaction involves the removal of a water molecule to form the enol structure of PEP. The $\Delta G^{\circ\prime}$ for this reaction is relatively small at 1.8 kJ/mol ($K_{eq} = 0.5$); and, under cellular conditions, ΔG is very close to zero. In light of this condition, it may be difficult at first to understand how the enolase reaction transforms a substrate with a relatively low free energy of hydrolysis into a product (PEP) with a very high free energy of hydrolysis. This puzzle is clarified by realizing that 2-phosphoglycerate and PEP contain about the same amount of *potential* metabolic energy, with respect to decomposition to P_i, CO_2, and H_2O. What the enolase reaction does is rearrange the substrate into a form from which more of this potential energy can be released upon hydrolysis. The enzyme is strongly inhibited by fluoride ion in the presence of phosphate. Thomas Nowak has shown that fluoride, phosphate, and a divalent cation form a transition-state–like complex in the enzyme active site, with fluoride apparently mimicking the hydroxide ion nucleophile in the enolase reaction.

$$\underset{\substack{\textbf{2-Phosphoglycerate}\\\textbf{(2-PG)}}}{\overset{\displaystyle COO^-}{\underset{\displaystyle CH_2OH}{\overset{|}{\underset{|}{HC-O-PO_3{}^{2-}}}}}} \quad \xrightleftharpoons{Mg^{2+}} \quad \underset{\substack{\textbf{Phosphoenolpyruvate}\\\textbf{(PEP)}}}{\overset{\displaystyle COO^-}{\underset{\displaystyle CH_2}{\overset{|}{\underset{\|}{C-O-PO_3{}^{2-}}}}}} \quad + \quad H_2O$$

$$\Delta G^{\circ\prime} = +1.8 \text{ kJ/mol}$$

Yeast enolase is a dimer of identical subunits. However, if the enzyme is crystallized in the presence of a mixture of the substrate (2-phosphoglycerate) and the product (phosphoenolpyruvate), the crystallized dimer is asymmetric! One subunit active site contains 2-phosphoglycerate, and the other contains PEP (Figure 18.18), thus providing a "before-and-after" picture of this glycolytic enzyme.

Reaction 10: Pyruvate Kinase Yields More ATP

The second ATP-synthesizing reaction of glycolysis is catalyzed by **pyruvate kinase,** which brings the pathway at last to its pyruvate branch point. Pyruvate kinase mediates the transfer of a phosphoryl group from phosphoenolpyruvate to ADP to make ATP and pyruvate. The reaction requires Mg^{2+} ion and is stimulated by K^+ and certain other monovalent cations.

The corresponding K_{eq} at 25°C is 3.63×10^5, and it is clear that the pyruvate kinase reaction equilibrium lies very far to the right. Concentration effects reduce the

(a)   **(b)**

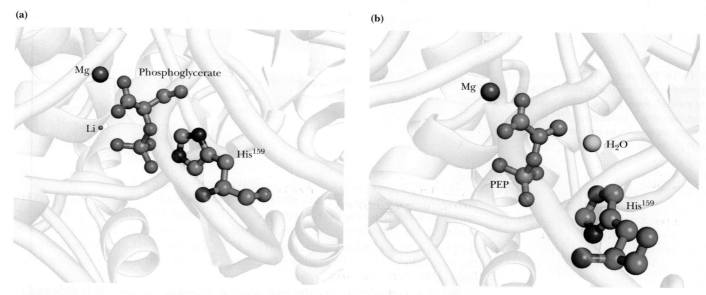

FIGURE 18.18 The yeast enolase dimer is asymmetric. The active site of one subunit **(a)** contains 2-phosphoglycerate, the enolase substrate. Also shown are a Mg^{2+} ion (blue), a Li^+ ion (purple), and His^{159}, which participates in catalysis. The other subunit **(b)** binds phosphoenolpyruvate, the product of the enolase reaction. An active-site water molecule (yellow), Mg^{2+} (blue), and His^{159} are also shown (pdb id = 2ONE).

$$\underset{\textbf{PEP}}{\overset{\displaystyle COO^-}{\underset{\displaystyle CH_2}{\overset{\displaystyle |}{\underset{\displaystyle |}{\overset{\displaystyle C}{||}}-O-PO_3{}^{2-}}}}} + H^+ + ADP^{3-} \underset{K^+}{\overset{Mg^{2+}}{\rightleftharpoons}} \underset{\textbf{Pyruvate}}{\overset{\displaystyle COO^-}{\underset{\displaystyle CH_3}{\overset{\displaystyle |}{\underset{\displaystyle |}{\overset{\displaystyle C}{=}O}}}}} + \textbf{ATP}^{4-}$$

$$\Delta G^{\circ\prime} = -31.7 \ kJ/mol$$

magnitude of the free energy change somewhat in the cellular environment, but the ΔG in erythrocytes is still quite favorable at -23.0 kJ/mol. The high free energy change for the conversion of **PEP** to pyruvate is due largely to the highly favorable and spontaneous conversion of the enol tautomer of pyruvate to the more stable keto form (Figure 18.19) following the phosphoryl group transfer step.

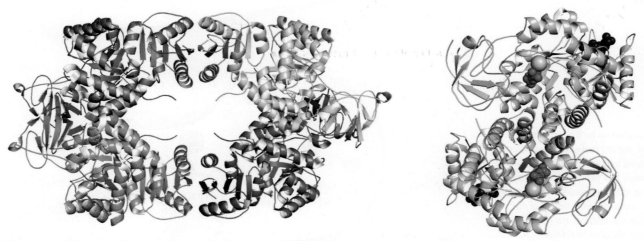

The structure of the pyruvate kinase tetramer is sensitive to bound ligands. The inactive *E. coli* enzyme in the absence of ligands (left, pdb id = 1E0U). The active form of the yeast dimer, with fructose-1,6-bisphosphate (an allosteric regulator, *blue*), substrate analog (*red*), and K^+ (*gold*) (pdb id = 1A3W).

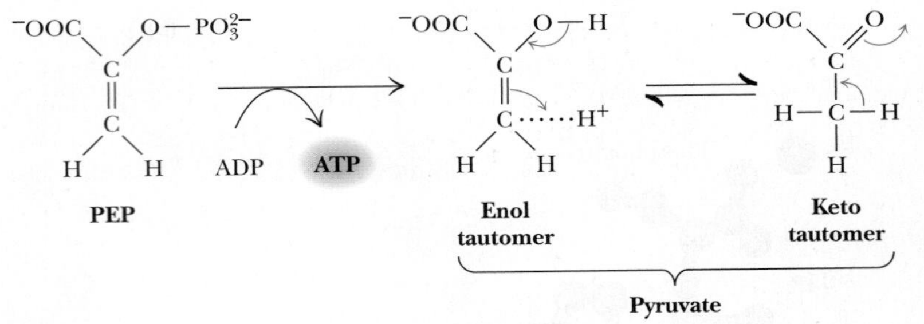

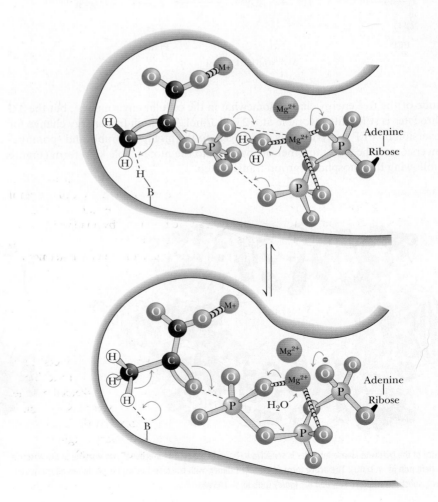

FIGURE 18.19 The conversion of phosphoenolpyruvate (PEP) to pyruvate may be viewed as involving two steps: phosphoryl transfer followed by an enol–keto tautomerization. The tautomerization is spontaneous ($\Delta G^{\circ\prime} \approx$ −35–40 kJ/mol) and accounts for much of the free energy change for PEP hydrolysis.

The large negative ΔG of this reaction makes pyruvate kinase a suitable target site for regulation of glycolysis. For each glucose molecule in the glycolysis pathway, two ATPs are made at the pyruvate kinase stage (because two triose molecules were produced per glucose in the aldolase reaction). Because the pathway broke even in terms of ATP at the phosphoglycerate kinase reaction (two ATPs consumed and two ATPs produced), the two ATPs produced by pyruvate kinase represent the "payoff" of glycolysis—a net yield of two ATP molecules.

Pyruvate kinase possesses allosteric sites for numerous effectors. It is activated by AMP and fructose-1,6-bisphosphate and inhibited by ATP, acetyl-CoA, and alanine. (Note that alanine is the α-amino acid counterpart of the α-keto acid, pyruvate.) Furthermore, liver pyruvate kinase is regulated by covalent modification. Hormones such as *glucagon* activate cAMP-dependent protein kinase, which transfers a phosphoryl group from ATP to the enzyme. The phosphorylated form of pyruvate kinase is more strongly inhibited by ATP and alanine and has a higher K_m for PEP, so in the presence of physiological levels of PEP, the enzyme is inactive. Then PEP is used as a substrate

FIGURE 18.20 A mechanism for the pyruvate kinase reaction, based on NMR and EPR studies by Albert Mildvan and colleagues. Phosphoryl transfer from phosphoenolpyruvate (PEP) to ADP occurs in four steps: (1) A water on the Mg^{2+} ion coordinated to ADP is replaced by the phosphoryl group of PEP, (2) Mg^{2+} dissociates from the α-P of ADP, (3) the phosphoryl group is transferred, and (4) the enolate of pyruvate is protonated. (Adapted from Mildvan, A., 1979. The role of metals in enzyme-catalyzed substitutions at each of the phosphorus atoms of ATP. *Advances in Enzymology* **49**:103–126.)

for glucose synthesis in the *gluconeogenesis* pathway (to be described in Chapter 22), instead of going on through glycolysis and the citric acid cycle (or fermentation routes). A suggested active-site geometry for pyruvate kinase, based on NMR and EPR studies by Albert Mildvan and colleagues, is presented in Figure 18.20. The carbonyl oxygen of pyruvate and the γ-phosphorus of ATP lie within 0.3 nm of each other at the active site, consistent with direct transfer of the phosphoryl group without formation of a phosphoenzyme intermediate.

18.5 What Are the Metabolic Fates of NADH and Pyruvate Produced in Glycolysis?

In addition to ATP, the products of glycolysis are NADH and pyruvate. Their processing depends upon other cellular pathways. NADH must be recycled to NAD^+, lest NAD^+ become limiting in glycolysis. NADH can be recycled by both aerobic and anaerobic paths, either of which results in further metabolism of pyruvate. What a given cell does with the pyruvate produced in glycolysis depends in part on the availability of oxygen. Under aerobic conditions, pyruvate can be sent into the citric acid cycle (also known as the TCA cycle; see Chapter 19), where it is oxidized to CO_2 with the production of additional NADH (and $FADH_2$). Under aerobic conditions, the NADH produced in glycolysis and the citric acid cycle is reoxidized to NAD^+ in the mitochondrial electron-transport chain (see Chapter 20).

Anaerobic Metabolism of Pyruvate Leads to Lactate or Ethanol

Under anaerobic conditions, the pyruvate produced in glycolysis is processed differently. In yeast, it is reduced to ethanol; in other microorganisms and in animals, it is reduced to lactate. These processes are examples of **fermentation**—the production of ATP energy by reaction pathways in which organic molecules function as donors and acceptors of electrons. In either case, reduction of pyruvate provides a means of reoxidizing the NADH produced in the glyceraldehyde-3-phosphate dehydrogenase reaction of glycolysis (Figure 18.21). In yeast, alcoholic fermentation is a two-step process. Pyruvate is decarboxylated to acetaldehyde by **pyruvate decarboxylase** in an essentially irreversible reaction. Thiamine pyrophosphate (see page 614) is a required cofactor for this enzyme. The second step, the reduction of acetaldehyde to ethanol by NADH, is catalyzed by **alcohol dehydrogenase** (Figure 18.21). At pH 7, the reaction equilibrium strongly favors ethanol. The end products of alcoholic fermentation are thus ethanol and carbon dioxide. Alcoholic fermentations are the basis for the brewing of beers and the fermentation of grape sugar in wine making. Lactate produced by anaerobic microorganisms during lactic acid fermentation is responsible for the taste of sour milk and for the characteristic taste and fragrance of sauerkraut, which in reality is fermented cabbage.

Lactate Accumulates Under Anaerobic Conditions in Animal Tissues

In animal tissues experiencing anaerobic conditions, pyruvate is reduced to lactate. Pyruvate reduction occurs in tissues that normally experience minimal access to blood flow (for example, the cornea of the eye) and also in rapidly contracting skeletal muscle. When skeletal muscles are exercised strenuously, the available tissue oxygen is consumed and the pyruvate generated by glycolysis can no longer be oxidized in the TCA cycle. Instead, excess pyruvate is reduced to lactate by **lactate dehydrogenase** (Figure 18.21). The rate of anaerobic glycolysis in skeletal muscle can increase up to 2000-fold almost instantaneously, for example, to support the intense demands of a sprinting animal. Large amounts of ATP are generated rapidly, at the expense of lactate accumulation. In

(a) Alcoholic fermentation

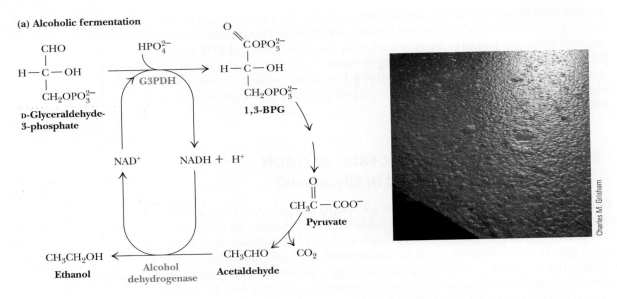

D-Glyceraldehyde-3-phosphate

1,3-BPG

G3PDH

NAD^+ → $NADH + H^+$

Pyruvate

CH_3CH_2OH ← Alcohol dehydrogenase ← CH_3CHO → CO_2

Ethanol

Acetaldehyde

(b) Lactic acid fermentation

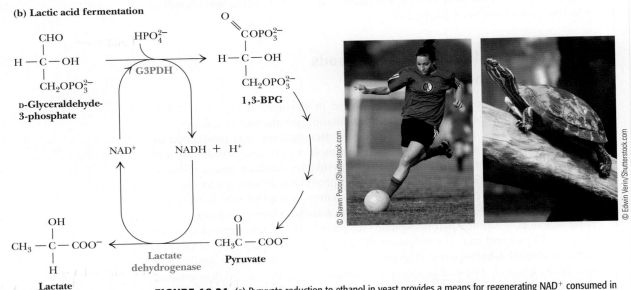

D-Glyceraldehyde-3-phosphate

G3PDH

1,3-BPG

NAD^+ → $NADH + H^+$

Pyruvate

Lactate

Lactate dehydrogenase

FIGURE 18.21 **(a)** Pyruvate reduction to ethanol in yeast provides a means for regenerating NAD^+ consumed in the glyceraldehyde-3-P dehydrogenase reaction. Photo: Fermentation at a bourbon distillery. A "mash" of corn and other grains is fermented by yeast, producing ethanol and carbon dioxide, which can be seen bubbling to the surface. Woodford Reserve Distillery, Versailles, KY. **(b)** In oxygen-depleted muscle, for example, during strenuous exercise (photo), NAD^+ is regenerated in the lactate dehydrogenase reaction. Hibernating fresh-water turtles, trapped beneath ice and lying in mud, become "anoxic," and convert glucose reserves mainly to lactate. Their shells release minerals to buffer the lactate throughout the period of hibernation (photo).

anaerobic muscle tissue, lactate represents the end of glycolysis. Anyone who exercises to the point of depleting available muscle oxygen stores knows the cramps and muscle fatigue associated with the buildup of lactic acid in the muscle. Most of this lactate must be carried out of the muscle by the blood and transported to the liver, where it can be resynthesized into glucose in gluconeogenesis (see Chapter 22). Moreover, because glycolysis generates only a fraction of the total energy available from the breakdown of glucose (the rest is generated by the TCA cycle and oxidative phosphorylation), the onset of anaerobic conditions in skeletal muscle also means a reduction in the energy available from the breakdown of glucose.

CRITICAL DEVELOPMENTS IN BIOCHEMISTRY

The Warburg Effect and Cancer

Otto Warburg observed in 1924 that rapidly proliferating cancer cells metabolize glucose in a manner distinct from that of quiescent (non-dividing) tissues. Warburg found that cancer cells tend to convert glucose into lactate, even in the presence of oxygen sufficient to support mitochondrial oxidative phosphorylation. (Because this occurs even in an abundance of oxygen, it is referred to as "aerobic glycolysis.") This is at first surprising, because conversion of glucose to lactate produces only 2 molecules of ATP per molecule of glucose whereas complete metabolism of glucose in oxidative phosphorylation produces more than 30 molecules of ATP. Why do proliferating cells switch to an apparently less efficient metabolism?

Lewis Cantley and co-workers have suggested that proliferating cells need more than just ATP—they must synthesize large amounts of nucleotides, amino acids, and lipids. For these purposes, cells require carbon for "biomass" and NADPH ("reducing equivalents" for anabolic metabolism). These needs cannot be met merely by converting glucose to CO_2 and ATP via oxidative phosphorylation. Some glucose must be diverted to macromolecular precursors such as glycolytic intermediates for nonessential amino acids, ribose for nucleotides, and acetate units for fatty acids. C-13 NMR measurements have shown in fact that glioblastoma (a type of cancer) cells convert as much as 90% of acquired glucose and 60% of acquired glutamine into lactate or alanine, producing large amounts of NADPH. The lactate and alanine produced in this way can be recycled by nearby specialized nonproliferating cells, and also by the liver.

Numerous signaling proteins and pathways regulate the glycolytic pathway. Phosphoinositide 3-kinase (PI3K) activates glucose transporter expression, enchancing glucose capture by hexokinase. PI3K also stimulates PFK. In contrast to quiescent cells, proliferating cells selectively express the M2 (embryonic) isoform of pyruvate kinase (PK-M2), which is regulated by tyrosine-phosphorylated proteins.

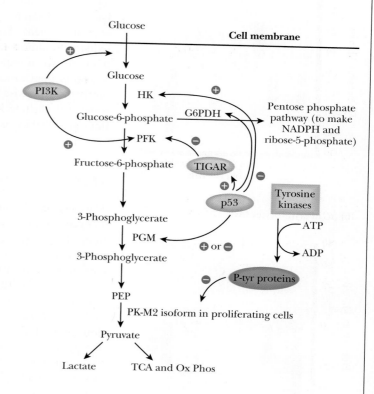

△ Otto Warburg received the Nobel Prize in Physiology or Medicine in 1931 with the citation *"for his discovery of the nature and mode of action of the respiratory enzyme."* Warburg's "respiratory enzyme" is cytochrome oxidase, the last enzyme in the electron transport pathway (see Chapter 20). Warburg, who studied under the great Emil Fischer, mentored three future recipients of the Nobel Prize, including Hans Krebs, discoverer of the tricarboxylic acid cycle (see Chapter 19). A skilled equestrian, Warburg served with a cavalry unit at the front during World War I, and he received the Iron Cross for his outstanding service.

The activity of PK-M2 is reduced by binding phosphotyrosine-containing proteins, allowing PK-M2 to act as a gatekeeper to control the flow of carbon into biosynthetic pathways versus catabolism for ATP production. p53, the protein encoded by a prominent tumor suppressor gene, is implicated in the regulation of cellular responses to stress. The p53 gene is also mutated in most cancer cells. TIGAR, a p53-activated protein, is a fructose bisphosphatase. TIGAR inhibits PFK by lowering fructose-2,6,bisphosphate levels, thereby redirecting glucose to the pentose phosphate pathway (Chapter 22). The pentose phosphate pathway produces NADPH and a variety of sugars essential to anabolic pathways. p53 has additional effects on glycolysis and related pathways. For example, p53 inhibits the expression of glucose transporter proteins (which bring glucose into the cell) and also regulates expression of hexokinase and phosphoglycerate mutase. These effects of p53 occur in the nucleus, but cytosolic p53 directly inhibits glucose-6-P dehydrogenase (which inhibits the shunting of glucose-6-P into the pentose phosphate pathway, forestalling NADPH production for fatty acid synthesis and ribose-5-P formation for nucleotide and nucleic acid biosynthesis: Chapters 22 and 26 and references below).

The emerging evidence highlights the role of glycolysis as a dynamically regulated system that is programmed to meet the specific needs of quiescent differentiated tissues or fit the requirements of rapidly proliferating cells.

References

Gottlieb, E., 2011. p53 guards the metabolic pathway less travelled. *Nature Cell Biology* **13**:195–197.

Gottlieb, E., and Vousden, K., 2009. p53 regulation of metabolism. *Cold Spring Harbor Perspectives in Biology* **2**:1–11.

Vander Heiden, M. G., Cantley, L. C., and Thompson, C. B., 2009. Understanding the Warburg effect: The metabolic requirements of cell proliferation. *Science* **324**:1029–1033.

(a) ΔG at standard state $(\Delta G°)$

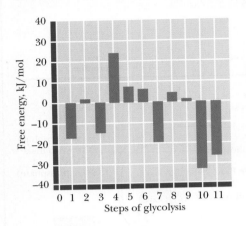

(b) ΔG in erythrocytes (ΔG)

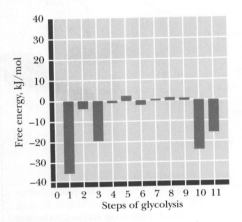

FIGURE 18.22 A comparison of free energy changes for the reactions of glycolysis (step 1 = hexokinase) under **(a)** standard-state conditions and **(b)** actual intracellular conditions in erythrocytes. The values of $\Delta G°'$ provide little insight into the actual free energy changes that occur in glycolysis. On the other hand, under intracellular conditions, seven of the glycolytic reactions operate near equilibrium (with ΔG near zero). The driving force for glycolysis lies in the hexokinase (1), phosphofructokinase (3), and pyruvate kinase (10) reactions. The lactate dehydrogenase (step 11) reaction also exhibits a large negative ΔG under cellular conditions.

The Old Shell Game—How Turtles Survive the Winter

Freshwater turtles (Figure 18.21b) use a remarkable adaptation of their form to survive the harsh conditions of winter, under frozen lakes for months, without breathing. Donald Jackson has shown that turtles in such conditions become "anoxic" (subsisting without oxygen). Their heart beats only once every ten minutes, and brain activity is reduced to almost nothing. They metabolize only a limited store of glycogen, rather than fat (which requires oxygen to be metabolized). The glycogen is converted to glucose, and then to lactate via glycolysis. The turtle's shell releases mineral reserves to buffer the otherwise fatal buildup of lactate in the turtle's blood, and the shell also traps some of the lactate, keeping it out of the bloodstream. Without the biochemical adaptations provided by its shell, the turtle could not survive the winter.

18.6 How Do Cells Regulate Glycolysis?

The elegance of nature's design for the glycolytic pathway may be appreciated through an examination of Figure 18.22. The standard-state free energy changes for the 10 reactions of glycolysis and the lactate dehydrogenase reaction (Figure 18.22a) are variously positive and negative and, taken together, offer little insight into the coupling that occurs in the cellular milieu. On the other hand, the values of ΔG under cellular conditions (Figure 18.22b) fall into two distinct classes. For reactions 2 and 4 through 9, ΔG is very close to zero, meaning these reactions operate essentially at equilibrium. Small changes in the concentrations of reactants and products could "push" any of these reactions either forward or backward. By contrast, the hexokinase, phosphofructokinase, and pyruvate kinase reactions all exhibit large negative ΔG values under cellular conditions. These reactions are thus the sites of glycolytic regulation. When these three enzymes are active, glycolysis proceeds and glucose is readily metabolized to pyruvate or lactate. Inhibition of the three key enzymes by allosteric effectors brings glycolysis to a halt. When we consider **gluconeogenesis**—the biosynthesis of glucose—in Chapter 22, we will see that different enzymes are used to carry out reactions 1, 3, and 10 in reverse, effecting the net synthesis of glucose. The maintenance of reactions 2 and 4 through 9 at or near equilibrium permits these reactions (and their respective enzymes!) to operate effectively in *either* the forward or reverse direction.

18.7 Are Substrates Other Than Glucose Used in Glycolysis?

The glycolytic pathway described in this chapter begins with the breakdown of glucose, but other sugars, both simple and complex, can enter the cycle if they can be converted by appropriate enzymes to one of the intermediates of glycolysis. Figure 18.23 shows the routes by which several simple metabolites can enter the glycolytic pathway. **Fructose,** for example, which is produced by breakdown of sucrose, may participate in glycolysis by at least two different routes. In the liver, fructose is phosphorylated at C-1 by the enzyme **fructokinase:**

$$\text{D-Fructose} + \text{ATP}^{4-} \longrightarrow \text{D-fructose-1-phosphate}^{2-} + \text{ADP}^{3-} + \text{H}^+ \qquad (18.9)$$

Subsequent action by **fructose-1-phosphate aldolase** cleaves fructose-1-P in a manner like the fructose bisphosphate aldolase reaction to produce dihydroxyacetone phosphate and D-glyceraldehyde:

$$\text{D-Fructose-1-P}^{2-} \longrightarrow \text{D-glyceraldehyde} + \text{dihydroxyacetone phosphate}^{2-} \qquad (18.10)$$

Dihydroxyacetone phosphate is of course an intermediate in glycolysis. D-Glyceraldehyde can be phosphorylated by **triose kinase** in the presence of ATP to form D-glyceraldehyde-3-phosphate, another glycolytic intermediate.

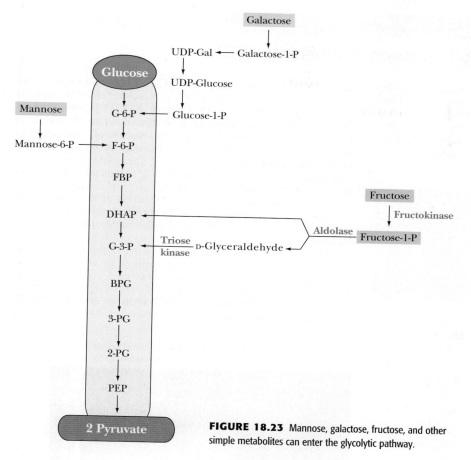

FIGURE 18.23 Mannose, galactose, fructose, and other simple metabolites can enter the glycolytic pathway.

In the kidney and in muscle tissues, fructose is readily phosphorylated by hexokinase, which, as pointed out previously, can utilize several different hexose substrates. The free energy of hydrolysis of ATP drives the reaction forward:

$$\text{D-Fructose} + \text{ATP}^{4-} \longrightarrow \text{D-fructose-6-phosphate}^{2-} + \text{ADP}^{3-} + \text{H}^+ \qquad (18.11)$$

Fructose-6-phosphate generated in this way enters the glycolytic pathway directly in step 3, the second priming reaction. This is the principal means for channeling fructose into glycolysis in adipose tissue, which contains high levels of fructose.

Mannose Enters Glycolysis in Two Steps

Another simple sugar that enters glycolysis at the same point as fructose is **mannose,** which occurs in many glycoproteins, glycolipids, and polysaccharides (see Chapter 7). Mannose is also phosphorylated from ATP by hexokinase, and the mannose-6-phosphate thus produced is converted to fructose-6-phosphate by **phosphomannoisomerase.**

$$\text{D-Mannose} + \text{ATP}^{4-} \longrightarrow \text{D-mannose-6-phosphate}^{2-} + \text{ADP}^{3-} + \text{H}^+ \qquad (18.12)$$

$$\text{D-Mannose-6-phosphate}^{2-} \longrightarrow \text{D-fructose-6-phosphate}^{2-} \qquad (18.13)$$

Galactose Enters Glycolysis via the Leloir Pathway

A somewhat more complicated route into glycolysis is followed by **galactose,** another simple hexose sugar. The process, called the **Leloir pathway** after Luis Leloir, its discoverer, begins with phosphorylation from ATP at the C-1 position by **galactokinase:**

$$\text{D-Galactose} + \text{ATP}^{4-} \longrightarrow \text{D-galactose-1-phosphate}^{2-} + \text{ADP}^{3-} + \text{H}^+ \qquad (18.14)$$

HUMAN BIOCHEMISTRY

Tumor Diagnosis Using Positron Emission Tomography (PET)

More than 70 years ago, Otto Warburg at the Kaiser Wilhelm Institute of Biology in Germany demonstrated that most animal and human tumors displayed a very high rate of glycolysis compared to that of normal tissue (see also the Critical Developments in Biochemistry box on page 597). This observation from long ago is the basis of a modern diagnostic method for tumor detection called positron emission tomography, or **PET**. PET uses molecular probes that contain a neutron-deficient, radioactive element such as carbon-11 or fluorine-18. An example is 2-[^{18}F]fluoro-2-deoxy-glucose (FDG), a molecular mimic of glucose. The ^{18}F nucleus is unstable and spontaneously decays by emission of a positron (an antimatter* particle) from a proton, thus converting a proton to a neutron and transforming the ^{18}F to ^{18}O. The emitted positron typically travels a short distance (less than a millimeter) and collides with an electron, annihilating both particles and creating a pair of high-energy photons—gamma rays. Detection of the gamma rays with special cameras can be used to construct three-dimensional models of the location of the radiolabeled molecular probe in the tissue of interest.

FDG is taken up by human cells and converted by hexokinase to 2-[^{18}F]fluoro-2-deoxy-glucose-6-phosphate in the first step of glycolysis. Cells of a human brain, for example, accumulate FDG in direct proportion to the amount of glycolysis occuring in those cells. Tumors can be identified in PET scans as sites of unusually high FDG accumulation.

(a)

CH$_2$OH

O

OH   HOH

HO

^{18}F

2-[^{18}F]Fluoro-2-deoxyglucose

(b)

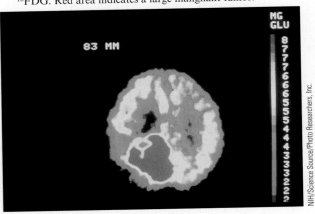

(c) PET image of human brain following administration of ^{18}FDG. Red area indicates a large malignant tumor.

*The existence of antimatter in the form of positrons was first postulated by Robert Oppenheimer, the father of the atomic bomb.

Galactose-1-phosphate is then converted into *UDP-galactose* (a sugar nucleotide) by **galactose-1-phosphate uridylyltransferase** (Figure 18.24), with concurrent production of glucose-1-phosphate and consumption of a molecule of UDP-glucose. The uridylyltransferase reaction (Figure 18.25) proceeds via a "ping-pong" mechanism (see Chapter 13, page 430) with a covalent enzyme-UMP intermediate. The glucose-1-phosphate produced by the transferase reaction is a substrate for the **phosphoglucomutase** reaction (Figure 18.24), which produces glucose-6-phosphate, a glycolytic substrate. The other transferase product, UDP-galactose, is converted to UDP-glucose by **UDP-glucose-4-epimerase.** The combined action of the uridylyltransferase and epimerase thus produces glucose-1-P from galactose-1-P, with regeneration of UDP-glucose.

A rare hereditary condition known as **galactosemia** involves defects in galactose-1-P uridylyltransferase that render the enzyme inactive. Toxic levels of galactose accumulate in afflicted individuals, causing cataracts and permanent neurological disorders. These problems can be prevented by removing galactose and lactose from the diet. In adults, the toxicity of galactose appears to be less severe, due in part to the metabolism of galactose-1-P by **UDP-glucose pyrophosphorylase,** which apparently can accept galactose-1-P in place of glucose-1-P (Figure 18.26). The levels of this enzyme may increase in galactosemic individuals in order to accommodate the metabolism of galactose.

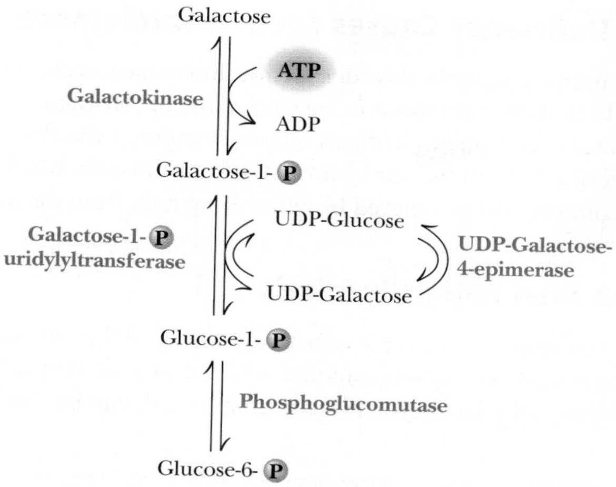

FIGURE 18.24 Galactose metabolism via the Leloir pathway.

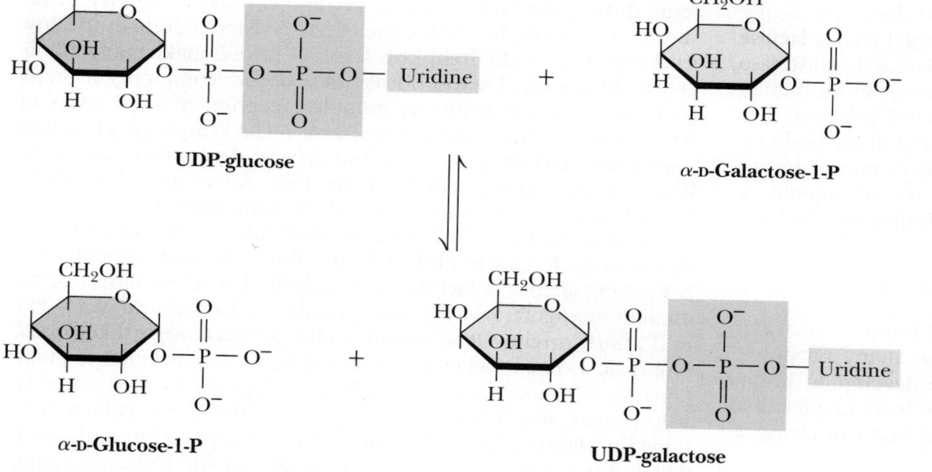

FIGURE 18.25 The galactose-1-phosphate uridylyltransferase reaction involves a "ping-pong" kinetic mechanism.

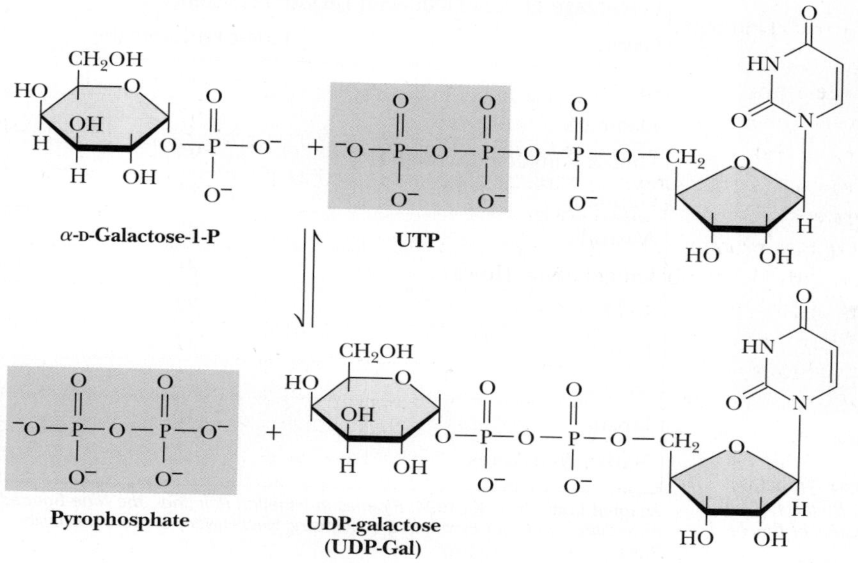

FIGURE 18.26 The UDP-glucose pyrophosphorylase reaction also works with galactose-1-P.

An Enzyme Deficiency Causes Lactose Intolerance

A much more common metabolic disorder, **lactose intolerance,** occurs commonly in most parts of the world (notable exceptions being some parts of Africa and northern Europe). Lactose intolerance is an inability to digest lactose because of the absence of the enzyme **lactase** in the intestines of adults. The symptoms of this disorder, which include diarrhea and general discomfort, can be relieved by eliminating milk from the diet.

Glycerol Can Also Enter Glycolysis

Glycerol is the last important simple substance whose ability to enter the glycolytic pathway must be considered. This metabolite, which is produced in substantial amounts by the decomposition of triacylglycerols (see Chapter 23), can be converted to glycerol-

---

HUMAN BIOCHEMISTRY

Lactose—From Mother's Milk to Yogurt—and Lactose Intolerance

Lactose is an interesting sugar in many ways. In placental mammals, it is synthesized only in the mammary gland, and then only during late pregnancy and lactation. The synthesis is carried out by **lactose synthase,** a dimeric complex of two proteins: galactosyl transferase and α-lactalbumin. Galactosyl transferase is present in all human cells, and it is normally involved in incorporation of galactose into glycoproteins. In late pregnancy, the pituitary gland in the brain releases a protein hormone, prolactin, which triggers production of α-lactalbumin by certain cells in the breast. α-Lactalbumin, a 123-residue protein, associates with galactosyl transferase to form lactose synthase, which catalyzes the reaction:

$$\text{UDP-galactose} + \text{glucose} \longrightarrow \text{lactose} + \text{UDP}$$

Lactose breakdown by **lactase** in the small intestine provides newborn mammals with essential galactose for many purposes, including the synthesis of gangliosides in the developing brain. Lactase is a **β-galactosidase** that cleaves lactose to yield galactose and glucose—in fact, the only human enzyme that can cleave a β-glycosidic linkage:

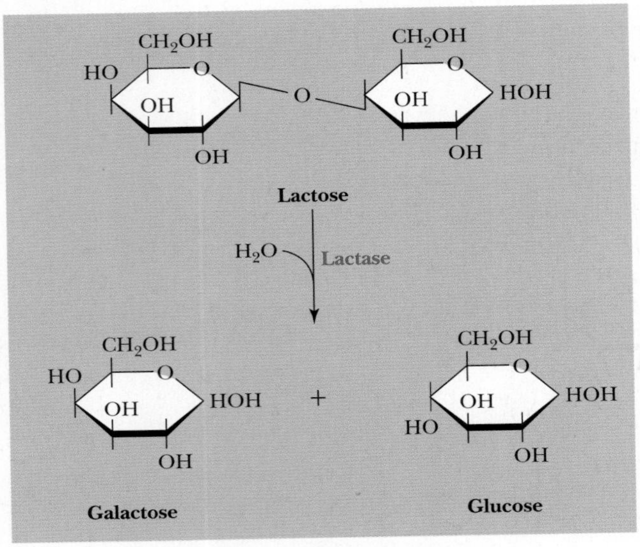

Lactase is an inducible enzyme in mammals, and it appears in the fetus during the late stages of gestation. Lactase activity peaks shortly after birth, but by the age of 3 to 5 years, it declines to a low level in nearly all children. Low levels of lactase make many adults **lactose intolerant.** Lactose intolerance occurs commonly in most parts of the world (with the notable exception of some parts of Africa and northern Europe; see table). The symptoms of lactose intolerance, including diarrhea and general discomfort, can be relieved by eliminating milk from the diet. Alternatively, products containing β-galactosidase are available commercially.

Certain bacteria, including several species of *Lactobacillus,* thrive on the lactose in milk and carry out lactic acid fermentation, converting lactose to lactate via glycolysis. This is the basis of production of yogurt, which is now popular in the Western world but of Turkish origin. Other cultures also produce yogurtlike foods. Nomadic Tatars in Siberia and Mongolia used camel milk to make *koumiss,* which was used for medicinal purposes. In the Caucasus, *kefir* is made much like yogurt, except that the starter culture contains (in addition to *Lactobacillus*) *Streptococcus lactis* and yeast, which convert some of the glucose to ethanol and CO_2, producing an effervescent and slightly intoxicating brew.

Breakdown of lactose to galactose and glucose by lactase.

Percentage of Population with Lactase Persistence	
Country	Lactase Persistence (%)
Sweden	99
Denmark	97
United Kingdom (Scotland)	95
Germany	88
Australia	82
United States (Iowa)	81
Spain	72
Jordan	21
Greece	12
Japan	10
Native Americans	5

Portions adapted from Hill, R., and Brew, K., 1975. Lactose synthetase. *Advances in Enzymology* **43:**411–485; and Bloch, K., 1994. *Blondes in Venetian Paintings, the Nine-Banded Armadillo, and Other Essays in Biochemistry.* New Haven, CT: Yale University Press.

Adapted from Bloch, K., 1994. *Blondes in Venetian Paintings, the Nine-Banded Armadillo, and Other Essays in Biochemistry.* New Haven, CT: Yale University Press.

Glycerol kinase reaction

$$\underset{\text{Glycerol}}{\text{HOCH}\begin{array}{c}\text{CH}_2\text{OH}\\|\\\text{CH}_2\text{OH}\end{array}} + \text{ATP} \xrightarrow{\text{Mg}^{2+}} \underset{sn\text{-Glycerol-3-phosphate}}{\text{HOCH}\begin{array}{c}\text{CH}_2\text{OH}\\|\\\text{CH}_2\text{OPO}_3^{2-}\end{array}} + \text{ADP}$$

$$\underset{sn\text{-Glycerol-3-phosphate}}{\text{H\,OCH}\begin{array}{c}\text{CH}_2\text{OH}\\|\\\text{CH}_2\text{OPO}_3^{2-}\end{array}} + \text{NAD}^+ \longrightarrow \underset{\substack{\text{Dihydroxyacetone}\\\text{phosphate}}}{\text{C}=\text{O}\begin{array}{c}\text{CH}_2\text{OH}\\|\\\text{CH}_2\text{OPO}_3^{2-}\end{array}} + \text{NADH} + \text{H}^+$$

Glycerol-P dehydrogenase reaction

3-phosphate by the action of **glycerol kinase** and then oxidized to dihydroxyacetone phosphate by the action of **glycerol phosphate dehydrogenase,** with NAD^+ as the required coenzyme. The dihydroxyacetone phosphate thereby produced enters the glycolytic pathway as a substrate for triose phosphate isomerase.

18.8 How Do Cells Respond to Hypoxic Stress?

Glycolysis is an **anaerobic** pathway—it does not require oxygen. But as noted in Figure 18.1, operation of the TCA cycle (the subject of Chapter 19) depends on oxygen, so it is **aerobic.** When oxygen is abundant, cells prefer aerobic metabolism, which yields more energy per glucose consumed. However, as Louis Pasteur first showed, when oxygen is limited, cells adapt to make the most of glycolysis, the less energetic, anaerobic alternative. In mammalian tissues, **hypoxia** (oxygen limitation) can cause changes in gene expression that result in increased angiogenesis (the growth of new blood vessels), increased synthesis of red blood cells, and increased levels of some glycolytic enzymes (and thus a higher rate of glycolysis).

What is the molecular basis for the increased expression of glycolytic enzymes? One of the triggers for this expression is a DNA-binding protein called **hypoxia inducible factor (HIF).** HIF is a heterodimer of a constitutive, stable nuclear β subunit (HIF-1β) and an inducible, unstable α-subunit. Both subunits are basic helix-loop-helix transcription factors that bind to hypoxia-inducible genes, and both subunits exist as a series of isoforms (for example, HIF-1α, HIF-2α, and HIF-3α). HIF-α subunit regulation is a multistep process that includes gene splicing, phosphorylation, acetylation, and hydroxylation. HIF-1α is the best-studied HIF-α isoform. Under normal oxygen levels, the α-subunits are continually synthesized but quickly degraded. When oxygen is plentiful, HIF-1α is hydroxylated by oxygen-dependent **prolyl hydroxylases (PHDs)** at Pro[402] and Pro[564]. These hydroxylations ensure its binding to ubiquitin E3 ligase, which leads to rapid proteolysis by the 26S proteasome (see Chapter 31). HIF-1α binding to the ligase is also promoted by acetylation of Lys[532] by the ARD1 acetyltransferase. In addition, the presence of oxygen induces the hydroxylation of HIF-1α Asn[803] by the **hydroxylase factor–inhibiting HIF (FIH-1).** Figure 18.27 shows the structure of FIH bound to a fragment of HIF-1α.

Because PHDs and FIH-1 both are oxygen-dependent, lowering oxygen concentration means that HIF-1α avoids degradation and is available to promote gene transcription (Figure 18.28). Phosphorylation of HIF-1α by a protein kinase promotes binding of HIF-1α to HIF-1β, which enhances transcription. HIF-1α–HIF-1β dimers bind to **hypoxia responsive elements (HREs),** activating transcription of HRE-regulated genes, including hexokinase and lactate dehydrogenase, and the genes for glucose and lactate transporters (Figure 18.28). Pasteur observed more than 100 years ago that fermentation amounted to "life without air." The "Pasteur effect" depends on HIF-mediated

HIF is induced also by inflammatory and oxidative stresses, and also in tumors.

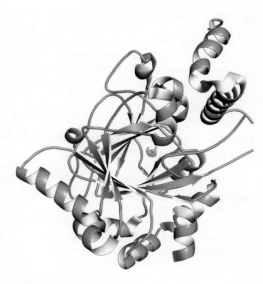

FIGURE 18.27 FIH (green) bound to HIF.

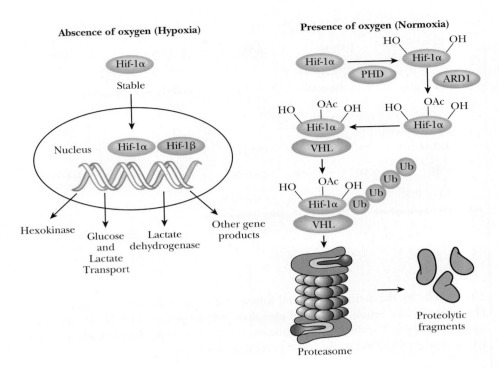

FIGURE 18.28 The HIF transcription factor is composed of two subunits: a ubiquitous HIF-1β subunit and a hypoxia-responsive HIF-1α subunit. In response to hypoxia, inactivation of the PHDs allows HIF-1α stabilization, dimerization with HIF-1β, binding of the dimer to the hypoxia responsive element (HRE) of HIF target genes, and activation of the transcription of these genes. VHL is the von Hippel Lindau subunit of the ubiquitin E3 ligase that targets proteins for proteasome degradation.

activation of the genes encoding glycolytic enzymes in the absence of oxygen. The linking of glycolytic activity to oxygen level is the result of an exquisite dance of oxygen-sensitive enzymes with proteins, which undergo covalent modifications that control protein–protein and protein–DNA interactions, a dance that Pasteur could hardly have anticipated.

SUMMARY

OWL Login to OWL to develop problem-solving skills and complete online homework assigned by your professor.

Nearly every living cell carries out a catabolic process known as glycolysis—the stepwise degradation of glucose (and other simple sugars). Glycolysis is a paradigm of metabolic pathways. Localized in the cytosol of cells, it is basically an anaerobic process; its principal steps occur with no requirement for oxygen.

18.1 What Are the Essential Features of Glycolysis? Glycolysis consists of two phases. In the first phase, a series of five reactions, glucose is broken down to two molecules of glyceraldehyde-3-phosphate. In the second phase, five subsequent reactions convert these two molecules of glyceraldehyde-3-phosphate into two molecules of pyruvate. Phase 1 consumes two molecules of ATP. The later stages of glycolysis result in the production of four molecules of ATP. The net is $4 - 2 = 2$ molecules of ATP produced per molecule of glucose.

18.2 Why Are Coupled Reactions Important in Glycolysis? Coupled reactions permit the energy of glycolysis to be used for generation of ATP. Conversion of glucose to pyruvate in glycolysis drives the production of two molecules of ATP.

18.3 What Are the Chemical Principles and Features of the First Phase of Glycolysis? In the first phase of glycolysis, glucose is converted into two molecules of glyceraldehyde-3-phosphate. First, glucose is phosphorylated to glucose-6-P, which is isomerized to fructose-6-P. Another phosphorylation and then cleavage yields two 3-carbon intermediates. One of these is glyceraldehyde-3-P, and the other, dihydroxyacetone-P, is converted to glyceraldehyde-3-P.

Energy released from this high-energy molecule in the second phase of glycolysis is then used to synthesize ATP.

18.4 What Are the Chemical Principles and Features of the Second Phase of Glycolysis? The second half of the glycolytic pathway involves the reactions that convert the metabolic energy in the glucose molecule into ATP. Phase 2 starts with the oxidation of glyceraldehyde-3-phosphate, a reaction with a large enough energy "kick" to produce a high-energy phosphate, namely, 1,3-bisphosphoglycerate. Phosphoryl transfer from 1,3-BPG to ADP to make ATP is highly favorable. The product, 3-phosphoglycerate, is converted via several steps to phosphoenolpyruvate (PEP), another high-energy phosphate. PEP readily transfers its phosphoryl group to ADP in the pyruvate kinase reaction to make another ATP.

18.5 What Are the Metabolic Fates of NADH and Pyruvate Produced in Glycolysis? In addition to ATP, the products of glycolysis are NADH and pyruvate. Their processing depends upon other cellular pathways. NADH must be recycled to NAD^+, lest NAD^+ become limiting in glycolysis. NADH can be recycled by both aerobic and anaerobic paths, either of which results in further metabolism of pyruvate. What a given cell does with the pyruvate produced in glycolysis depends in part on the availability of oxygen. Under aerobic conditions, pyruvate can be sent into the citric acid cycle, where it is oxidized to CO_2 with the production of additional NADH (and $FADH_2$). Under aerobic conditions, the NADH produced in glycolysis and the citric acid cycle is reoxidized to NAD^+ in the mitochondrial electron-transport chain.

Under anaerobic conditions, the pyruvate produced in glycolysis is not sent to the citric acid cycle. Instead, it is reduced to ethanol in yeast; in other microorganisms and in animals, it is reduced to lactate. These processes are examples of fermentation—the production of ATP energy by reaction pathways in which organic molecules function as donors and acceptors of electrons. In either case, reduction of pyruvate provides a means of reoxidizing the NADH produced in the glyceraldehyde-3-phosphate dehydrogenase reaction of glycolysis.

18.6 How Do Cells Regulate Glycolysis? The standard-state free energy changes for the 10 reactions of glycolysis are variously positive and negative and, taken together, offer little insight into the coupling that occurs in the cellular milieu. On the other hand, the values of ΔG under cellular conditions fall into two distinct classes. For reactions 2 and 4 through 9, ΔG is very close to zero, meaning these reactions operate essentially at equilibrium. Small changes in the concentrations of reactants and products could "push" any of these reactions either forward or backward. By contrast, the hexokinase, phosphofructokinase, and pyruvate kinase reactions all exhibit large negative ΔG values under cellular conditions. These reactions are thus the sites of glycolytic regulation.

18.7 Are Substrates Other Than Glucose Used in Glycolysis? Fructose enters glycolysis by either of two routes. Mannose, galactose, and glycerol enter via reactions that are linked to the glycolytic pathway.

18.8 How Do Cells Respond to Hypoxic Stress? Glycolysis is an anaerobic pathway, but it normally feeds pyruvate into aerobic metabolic pathways. However, when oxygen is limited, cells adapt to make the most of glycolysis. In mammalian tissues, oxygen limitation (hypoxia) can cause changes in gene expression that result in increased angiogenesis, red blood cell synthesis, and elevated levels of some glycolytic enzymes. One of the triggers for this expression is a DNA-binding protein, HIF, which binds to hypoxia-inducible genes. HIF-α regulation is a multistep process that includes gene splicing, phosphorylation, acetylation, and hydroxylation. The Pasteur effect depends on HIF-mediated activation of the genes encoding glycolytic enzymes in the absence of oxygen.

FOUNDATIONAL BIOCHEMISTRY Things You Should Know After Reading Chapter 18.

- The essential features of glycolysis.
- The definitions of lactic acid fermentation and alcoholic fermentation.
- The ten steps of the glycolysis pathway.
- Why coupled reactions are important in glycolysis.
- The chemical principles and features of the first phase of glycolysis.
- What is meant by a priming reaction.
- The cellular advantages of phosphorylating glucose.
- The role of induced fit in the binding of ATP and glucose to hexokinase.
- The chemical logic of the phosphoglucoisomerase reaction in glycolysis.
- What it means to say that the phosphofructokinase reaction commits the cell to metabolizing glucose.
- The nature of regulation of phosphofructokinase.
- The mechanisms of class I and class II aldolase reactions.
- How the aldolase reaction is similar to an aldol condensation.
- The role and mechanism of the triose phosphate isomerase reaction.
- The chemical principles and features of the second phase of glycolysis.
- Why the glyceraldehyde-3-phosphate dehydrogenase reaction is considered an oxidation/reduction reaction.
- The role of NAD+ in the glyceraldehyde-3-phosphate dehydrogenase reaction.

- The mechanism of the glyceraldehyde-3-phosphate dehydrogenase reaction.
- Why the phosphoglycerate kinase reaction is considered the break-even point of glycolysis.
- Why the phosphoglycerate kinase reaction is termed a substrate-level phosphorylation.
- Why the phosphoglycerate mutase reaction is important in glycolysis.
- Why the phosphoglycerate mutase reaction requires a small amount of 2,3-bisphosphoglycerate.
- The mechanism of the phosphoglycerate mutase reaction.
- How the enolase reaction makes a "high-energy" product from a "low-energy" reactant.
- The two energetic driving forces that make the pyruvate kinase reaction possible (see Figure 3.12 for a hint).
- The possible metabolic fates of pyruvate following glycolysis.
- Which steps of glycolysis are regulatory and why.
- The means by which fructose, mannose, galactose, and glycerol enter glycolysis.
- The meaning and origins of lactose intolerance.
- The mechanism by which cells respond to hypoxic stress.
- The nature of the Warburg effect in proliferating (esp. cancerous) cells.

PROBLEMS

Answers to all problems are at the end of this book. Detailed solutions are available in the *Student Solutions Manual, Study Guide, and Problems Book.*

1. **Characterizing Glycolysis** List the reactions of glycolysis that
    a. are energy consuming (under standard-state conditions).
    b. are energy yielding (under standard-state conditions).
    c. consume ATP.
    d. yield ATP.
    e. are strongly influenced by changes in concentration of substrate and product because of their molecularity.
    f. are at or near equilibrium in the erythrocyte (see Table 18.2).

2. **Radiotracer Labeling of Pyruvate from Glucose** Determine the anticipated location in pyruvate of labeled carbons if glucose molecules labeled (in separate experiments) with ^{14}C at each position of the carbon skeleton proceed through the glycolytic pathway.

3. **Effects of Changing Metabolite Concentrations on Glycolysis** In an erythrocyte undergoing glycolysis, what would be the effect of a sudden increase in the concentration of

 a. ATP?   b. AMP?

 c. fructose-1,6-bisphosphate? d. fructose-2,6-bisphosphate?

 e. citrate?   f. glucose-6-phosphate?

4. **Understanding NADH Cycling in Glycolysis** Discuss the cycling of NADH and NAD$^+$ in glycolysis and the related fermentation reactions.

5. **Understanding the Reaction Mechanisms of Glycolysis** For each of the following reactions, name the enzyme that carries out this reaction in glycolysis, and write a suitable mechanism for the reaction.

6. **The Reactions and Mechanisms of the Leloir Pathway** Write the reactions that permit galactose to be utilized in glycolysis. Write a suitable mechanism for one of these reactions.

7. **The Effect of Iodoacetic Acid on the Glyceraldehyde-3-P Dehydrogenase Reaction** (Integrates with Chapters 4 and 14.) How might iodoacetic acid affect the glyceraldehyde-3-phosphate dehydrogenase reaction in glycolysis? Justify your answer.

8. **Radiotracer Studies with ^{32}P in Glycolysis** If ^{32}P-labeled inorganic phosphate were introduced to erythrocytes undergoing glycolysis, would you expect to detect ^{32}P in glycolytic intermediates? If so, describe the relevant reactions and the ^{32}P incorporation you would observe.

9. **Comparing Glycolysis Entry Points for Sucrose** Sucrose can enter glycolysis by either of two routes:

Sucrose phosphorylase:
Sucrose + P$_i$ $\rightleftharpoons$ fructose + glucose-1-phosphate

Invertase:
Sucrose + H$_2$O $\rightleftharpoons$ fructose + glucose

Would either of these reactions offer an advantage over the other in the preparation of hexoses for entry into glycolysis?

10. **Assessing the Role of Mg^{2+} in Glycolysis** What would be the consequences of a Mg^{2+} ion deficiency for the reactions of glycolysis?

11. **Energetics of the Triose Phosphate Isomerase Reaction** (Integrates with Chapter 3.) Triose phosphate isomerase catalyzes the conversion of dihydroxyacetone-P to glyceraldehyde-3-P. The standard free energy change, $\Delta G^{\circ\prime}$, for this reaction is +7.6 kJ/mol. However, the observed free energy change (ΔG) for this reaction in erythrocytes is +2.4 kJ/mol.

 a. Calculate the ratio of [dihydroxyacetone-P]/[glyceraldehyde-3-P] in erythrocytes from ΔG.

 b. If [dihydroxyacetone-P] = 0.2 mM, what is [glyceraldehyde-3-P]?

12. **Energetics of the Enolase Reaction** (Integrates with Chapter 3.) Enolase catalyzes the conversion of 2-phosphoglycerate to phospho-

enolpyruvate + H$_2$O. The standard free energy change, $\Delta G^{\circ\prime}$, for this reaction is +1.8 kJ/mol. If the concentration of 2-phosphoglycerate is 0.045 mM and the concentration of phosphoenolpyruvate is 0.034 mM, what is ΔG, the free energy change for the enolase reaction, under these conditions?

13. **Energetics of PEP Hydrolysis** (Integrates with Chapter 3.) The standard free energy change ($\Delta G^{\circ\prime}$) for hydrolysis of phosphoenol-pyruvate (PEP) is −62.2 kJ/mol. The standard free energy change ($\Delta G^{\circ\prime}$) for ATP hydrolysis is −30.5 kJ/mol.

 a. What is the standard free energy change for the pyruvate kinase reaction:

$$\text{ADP} + \text{phosphoenolpyruvate} \longrightarrow \text{ATP} + \text{pyruvate}$$

 b. What is the equilibrium constant for this reaction?

 c. Assuming the intracellular concentrations of [ATP] and [ADP] remain fixed at 8 mM and 1 mM, respectively, what will be the ratio of [pyruvate]/[phosphoenolpyruvate] when the pyruvate kinase reaction reaches equilibrium?

14. **Energetics of Fructose-1,6-bisP Hydrolysis** (Integrates with Chapter 3.) The standard free energy change ($\Delta G^{\circ\prime}$) for hydrolysis of fructose-1,6-bisphosphate (FBP) to fructose-6-phosphate (F-6-P) and P$_i$ is −16.7 kJ/mol:

$$\text{FBP} + \text{H}_2\text{O} \longrightarrow \text{fructose-6-P} + \text{P}_i$$

The standard free energy change ($\Delta G^{\circ\prime}$) for ATP hydrolysis is −30.5 kJ/mol:

$$\text{ATP} + \text{H}_2\text{O} \longrightarrow \text{ADP} + \text{P}_i$$

 a. What is the standard free energy change for the phosphofructokinase reaction:

$$\text{ATP} + \text{fructose-6-P} \longrightarrow \text{ADP} + \text{FBP}$$

 b. What is the equilibrium constant for this reaction?

 c. Assuming the intracellular concentrations of [ATP] and [ADP] are maintained constant at 4 mM and 1.6 mM, respectively, in a rat liver cell, what will be the ratio of [FBP]/[fructose-6-P] when the phosphofructokinase reaction reaches equilibrium?

15. **Energetics of 1,3-Bisphosphoglycerate Hydrolysis** (Integrates with Chapter 3.) The standard free energy change ($\Delta G^{\circ\prime}$) for hydrolysis of 1,3-bisphosphoglycerate (1,3-BPG) to 3-phosphoglycerate (3-PG) and P$_i$ is −49.6 kJ/mol:

$$\text{1,3-BPG} + \text{H}_2\text{O} \longrightarrow \text{3-PG} + \text{P}_i$$

The standard free energy change ($\Delta G^{\circ\prime}$) for ATP hydrolysis is −30.5 kJ/mol:

$$\text{ATP} + \text{H}_2\text{O} \longrightarrow \text{ADP} + \text{P}_i$$

 a. What is the standard free energy change for the phosphoglycerate kinase reaction:

$$\text{ADP} + \text{1,3-BPG} \longrightarrow \text{ATP} + \text{3-PG}$$

 b. What is the equilibrium constant for this reaction?

 c. If the steady-state concentrations of [1,3-BPG] and [3-PG] in an erythrocyte are 1 μM and 120 μM, respectively, what will be the ratio of [ATP]/[ADP], assuming the phosphoglycerate kinase reaction is at equilibrium?

16. **Energetics of the Hexokinase Reaction** The standard-state free energy change, $\Delta G^{\circ\prime}$, for the hexokinase reaction is −16.7 kJ/mol. Use the values in Table 18.2 to calculate the value of ΔG for this reaction in the erythrocyte at 37°C.

17. **Analyzing the Concentration Dependence of the Adenylate Kinase Reaction** Taking into consideration the equilibrium constant for the adenylate kinase reaction (page 585), calculate the change in concentration in AMP that would occur if 8% of the ATP in an erythrocyte (red blood cell) were suddenly hydrolyzed to ADP. In addition to the concentration values in Table 18.2, it may be useful to assume that the initial concentration of AMP in erythrocytes is 5 μM.

18. **Distinguishing the Mechanisms of Class I and Class I Aldolases** Fructose bisphosphate aldolase in animal muscle is a class I aldolase, which forms a Schiff base intermediate between substrate (for example, fructose-1,6-bisphosphate or dihydroxyacetone phosphate) and a lysine at the active site (see Figure 18.12). The chemical evidence for this intermediate comes from studies with aldolase and the reducing agent sodium borohydride, $NaBH_4$. Incubation of the enzyme with dihydroxyacetone phosphate and $NaBH_4$ inactivates the enzyme. Interestingly, no inactivation is observed if $NaBH_4$ is added to the enzyme in the absence of substrate. Write a mechanism that explains these observations and provides evidence for the formation of a Schiff base intermediate in the aldolase reaction.

19. **Understanding the Ping-Pong Mechanism of Gal-1-P Uridylyl-transferase** As noted on page 600, the galactose-1-phosphate uridylyltransferase reaction proceeds via a ping-pong mechanism. Consult Chapter 13, page 430, to refresh your knowledge of ping-pong mechanisms, and draw a diagram to show how a ping-pong mechanism would proceed for the uridylyltransferase.

20. **Understanding the Mechanism of Hemolytic Anemia** Genetic defects in glycolytic enzymes can have serious consequences for humans. For example, defects in the gene for pyruvate kinase can result in a condition known as hemolytic anemia. Consult a reference to learn about hemolytic anemia, and discuss why such genetic defects lead to this condition.

Preparing for the MCAT® Exam

21. Regarding phosphofructokinase, which of the following statements is true:

   a. Low [ATP] stimulates the enzyme, but fructose-2,6-bisphosphate inhibits.

   b. High [ATP] stimulates the enzyme, but fructose-2,6-bisphosphate inhibits.

   c. High [ATP] stimulates the enzyme, and fructose-2,6-bisphosphate activates.

   d. The enzyme is more active at low [ATP] than at high, and fructose-2,6-bisphosphate activates the enzyme.

   e. [ATP] and fructose-2,6-bisphosphate both inhibit the enzyme.

22. Based on your reading of this chapter, what would you expect to be the most immediate effect on glycolysis if the steady-state concentration of glucose-6-P were 8.3 mM instead of 0.083 mM?

ActiveModels Problems

23. Examine the ActiveModel for alcohol dehydrogenase and describe the structure and function of the catalytic zinc center.

24. Based on your knowledge of the structure of NAD^+ and an assumption that coenzyme dissociation is the rate limiting step of the alcohol dehydrogenase mechanism, hypothesize why a N249W mutation at the coenzyme binding site would increase the rate of catalysis.

25. Using the ActiveModel for phosphofructokinase *(Trypanosoma)*, describe the difference between the APO1, APO2, and holoenzyme conformations.

FURTHER READING

General

Fothergill-Gilmore, L., 1986. The evolution of the glycolytic pathway. *Trends in Biochemical Sciences* **11**:47–51.

Kim, J-W., and Dang, C. V., 2006. Cancer's molecular sweet tooth and the Warburg effect. *Cancer Research* **66**:8927–8930.

Sparks, S., 1997. The purpose of glycolysis. *Science* **277**:459–460.

Waddell, T. G., 1997. Optimization of glycolysis: A new look at the efficiency of energy coupling. *Biochemical Education* **25**:204–205.

Welberg, L., 2010. Spotlight on aerobic glycolysis. *Nature Reviews Neuroscience* **11**:729.

Enzymes of Glycolysis

Aleshin, A. E., Kirby, C., et al., 2000. Crystal structures of mutant monomeric hexokinase I reveal multiple ADP-binding sites and conformational changes relevant to allosteric regulation. *Journal of Molecular Biology* **296**:1001–1015.

Choi, K. H., Shi, J., et al., 2001. Snapshots of catalysis: The structure of fructose-1,6-(bis)phosphate aldolase covalently bound to the substrate dihydroxyacetone phosphate. *Biochemistry* **40**:13868–13875.

Didierjean, C., Corbier, C., et al., 2003. Crystal structure of two ternary complexes of phosphorylating glyceraldehyde-3-phosphate dehydrogenase from *Bacillus sterothermophilus* with NAD and D-glyceraldehyde 3-phosphate. *Journal of Biological Chemistry* **278**:12968–12976.

Jeffery, C. J., 1999. Moonlighting proteins. *Trends in Biochemical Sciences* **24**:8–11.

Jeffery, C. J., 2004. Molecular mechanisms for multitasking: Recent crystal structures of moonlighting proteins. *Current Opinion in Structural Biology* **14**:663–668.

Kim, J-W., and Dang, C. V, 2005. Multifaceted roles of glycolytic enzymes. *Trends in Biochemical Sciences* **30**:142–150.

Lee, J. H., Chang, K. Z., et al., 2001. Crystal structure of rabbit phosphoglucose isomerase complexed with its substrate D-fructose 6-phosphate. *Biochemistry* **40**:7799–7805.

Lolis, E., and Petsko, G., 1990. Crystallographic analysis of the complex between triosephosphate isomerase and 2-phosphoglycolate at 2.5 Å resolution: Implications for catalysis. *Biochemistry* **29**:6619–6625.

Schirmer, T., and Evans, P. R., 1999. Structural basis of the allosteric behaviour of phosphofructokinase. *Nature* **343**:140–145.

Valentini, G., Chiarelli, L., et al., 2000. The allosteric regulation of pyruvate kinase. *Journal of Biological Chemistry* **275**:18145–18152.

Wilson, J. E., 2003. Isozymes of mammalian hexokinase: Structure, subcellular localization and metabolic function. *Journal of Experimental Biology* **206**:2049–2057.

Zhang, E., Brewer, J. M., et al., 1997. Mechanism of enolase: The crystal structure of asymmetric dimer enolase-2-phospho-D-glycerate/enolase-phosphoenopyruvate at 2.0 Å resolution. *Biochemistry* **36**:12526–12534.

Muscle Biochemistry

Green, H. J., 1997. Mechanisms of muscle fatigue in intense exercise. *Journal of Sports Sciences* **15**:247–256.

HIF and Glycolysis

Cramer, T., Yamanishi, Y., et al., 2003. HIF-1α is essential for myeloid cell-mediated inflammation. *Cell* **112**:645–657.

Dang, C. V., Kim, J.-W., Gao, P., and Yustein, J., 2008. The interplay between MYC and HIF in cancer. *Nature Reviews Cancer* **8**:51–56.

Kroemer, G., and Pouyssegur, J., 2008. Tumor cell metabolism: Cancer's Achilles' heel. *Cancer Cell* **13**:472–482.

Melillo, G., 2006. Inhibiting hypoxia-inducible factor 1 for cancer therapy. *Molecular Cancer Research* **4**:601–605.

Melillo, G., 2007. Targeting hypoxia cell signaling for cancer therapy. *Cancer and Metastasis Reviews* **26**:341–352.

North, S., Moenner, M., et al., 2005. Recent developments in the regulation of the angiogenic switch by cellular stress factors in tumors. *Cancer Letters* **218**:1–14.

Glycolysis and Cancer

Cairns, R. A., Harris, I. S., and Mak, T. W., 2011. Regulation of cancer cell metabolism. *Nature Reviews Cancer* **11**:85–95.

Ferreira, L. M. R., 2010. Cancer metabolism: The Warburg effect today. *Experimental and Molecular Pathology* **89**:372–380.

Fritz, V., and Fajas, L., 2010. Metabolism and proliferation share common regulatory pathways in cancer cells. *Oncogene* **29**:4369–4377.

Furata, E., Okuda, H., Kobayashi, A., and Watabe, K., 2010. Metabolic genes in cancer: Their roles in tumor progression and clinical implications. *Biochimica et Biophysica Acta* **1805**:141–152.

Levine, A. J., and Puzio-Kuter, A. M., 2010. The control of the metabolic switch in cancers by oncogenes and tumor suppressor genes. *Science* **330**:1340–1344.

McKnight, S., 2010. On geting there from here. *Science* **330**:1338–1339.

Pelicano, H., Martin, D. S., Xu, R.-H., and Huang, P., 2006. Glycolysis inhibition for anticancer treatment. *Oncogene* **25**:4633–4646.

Tennant, D. A., Duran, Raul V., and Gottlieb, E., 2011. Targeting metabolic transformation for cancer therapy. *Nature Reviews Cancer* **10**:267–277.

Metabolic Adaptations of Turtles

Jackson, D. C., and Ultsch, G. R., 2010. Physiology of hibernation under the ice by turtles and frogs. *Journal of Experimental Zoology* **313A**:311–327.

Warren, D. E., and Jackson, D.C., 2008. Lactate metabolism in anoxic turtles: An integrative review. *Journal of Comparative Physiology B* **178**:133–148.

Signaling in Glycolysis

Buchakjian, M. R., and Kornbluth, S., 2010. The engine driving the ship: Metabolic steering of cell proliferation and death. *Nature Reviews Molecular Cell Biology* **11**:715–727.

Christofk, H. R., Vander Heiden, M. G., Wu, N., Asara, J. M., Cantley, L. C., 2008. Pyruvate kinase M2 is a phosphotyrosine-binding protein. *Nature* **452**:181–186.

Corcoran, C. A., Huang, Y., and Sheikh, M. S., 2006. The regulation of energy generating metabolic pathways by p53. *Cancer Biology and Therapy* **5**:1610–1613.

Feng, Z., and Levine, A. J., 2010. The regulation of energy metabolism and the IGF-1/mTOR pathways by the p53 protein. *Trends in Biochemical Sciences* **20**:427–434.

Gottlieb, E., 2011. p53 guards the metabolic pathway less travelled. *Nature Cell Biology* **13**:195–197.

Gottlieb, E., and Vousden, K.H., 2009. p53 regulation of metabolic pathways. *Cold Spring Harbor Perspectives in Biology* **2**:1–11.

Li, H., and Jogl, G., 2009. Structural and biochemical studies of TIGAR (TP53-induced glycolysis and apoptosis regulator). *Journal of Biological Chemistry* **284**:1748–1754.

Olovnikov, I. A., Kravchenko, J. E., and Chumakov, P.M., 2009. Homeostatic functions of the p53 tumor suppressor: Regulation of energy metabolism and antioxidant defense. *Seminars in Cancer Biology* **19**:32–41.

Salminen, A., and Kaarniranta, K., 2010. Glycolysis links p53 function with NF-κB signaling: Impact on cancer and aging process. *Journal of Cellular Physiology* **224**:1–6.

Vander Heiden, M. G., Cantley, L. C. and Thompson, C. B., 2009. Understanding the Warburg effect: The metabolic requirements of cell proliferation. *Science* **324**:1029–1033.

Vousden, K. H., and Ryan, K. M., 2010. p53 and metabolism. *Nature Reviews Cancer* **9**:691–700.

The Tricarboxylic Acid Cycle

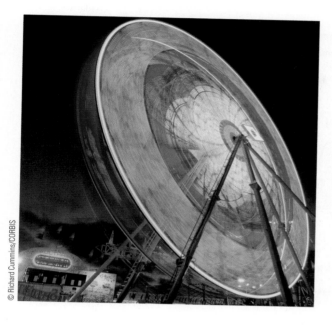

© Richard Cummins/CORBIS

◀ A time-lapse photograph of a ferris wheel at night. Aerobic cells use a metabolic wheel—the tricarboxylic acid cycle—to generate energy by acetyl-CoA oxidation.

Thus times do shift, each thing his turn does hold;
New things succeed, as former things grow old.

Robert Herrick
Hesperides (1648), "Ceremonies for Christmas Eve"

KEY QUESTIONS

19.1 What Is the Chemical Logic of the TCA Cycle?

19.2 How Is Pyruvate Oxidatively Decarboxylated to Acetyl-CoA?

19.3 How Are Two CO_2 Molecules Produced from Acetyl-CoA?

19.4 How Is Oxaloacetate Regenerated to Complete the TCA Cycle?

19.5 What Are the Energetic Consequences of the TCA Cycle?

19.6 Can the TCA Cycle Provide Intermediates for Biosynthesis?

19.7 What Are the Anaplerotic, or "Filling Up," Reactions?

19.8 How Is the TCA Cycle Regulated?

19.9 Can Any Organisms Use Acetate as Their Sole Carbon Source?

ESSENTIAL QUESTION

The glycolytic pathway converts glucose to pyruvate and produces two molecules of ATP per glucose—only a small fraction of the potential energy available from glucose. Under anaerobic conditions, pyruvate is reduced to lactate in animals and to ethanol in yeast, and much of the potential energy of the glucose molecule remains untapped. In the presence of oxygen, however, a much more interesting and thermodynamically complete story unfolds.

How is pyruvate oxidized to CO_2 under aerobic conditions, and what is the chemical logic that dictates how this process occurs?

Under aerobic conditions, pyruvate from glycolysis is converted to acetyl-coenzyme A (acetyl-CoA) and oxidized to CO_2 in the **tricarboxylic acid (TCA) cycle** (also called the **citric acid cycle**). The electrons liberated by this oxidative process are passed via NADH and $FADH_2$ through an elaborate, membrane-associated **electron-transport pathway** to O_2, the final electron acceptor. Electron transfer is coupled to creation of a proton gradient across the membrane. Such a gradient represents an energized state, and the energy stored in this gradient is used to drive the synthesis of many equivalents of ATP.

ATP synthesis as a consequence of electron transport is termed **oxidative phosphorylation;** the complete process is diagrammed in Figure 19.1. Aerobic pathways permit the production of 30 to 38 molecules of ATP per glucose oxidized. Although two molecules of ATP come from glycolysis and two more directly out of the TCA cycle, most of the ATP arises from oxidative phosphorylation. Specifically, reducing equivalents released in the oxidative reactions of glycolysis, pyruvate decarboxylation, and the TCA cycle are captured in the form of NADH and enzyme-bound $FADH_2$, and these reduced coenzymes fuel the electron-transport pathway and oxidative phosphorylation.

Complete oxidation of glucose to CO_2 involves the removal of 24 electrons—that is, it is a 24-electron oxidation. In glycolysis, 4 electrons are removed as NADH, and

OWL Online homework and a Student Self Assessment for this chapter may be assigned in OWL.

(a)

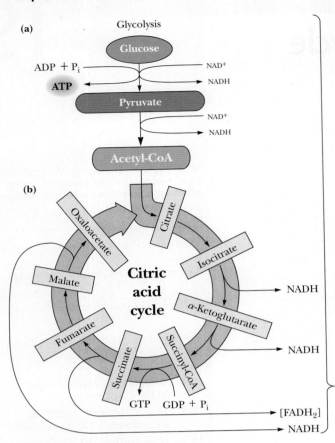

(b)

FIGURE 19.1 (a) Pyruvate produced in glycolysis is oxidized in (b) the tricarboxylic acid (TCA) cycle. (c) Electrons liberated in this oxidation flow through the electron-transport chain and drive the synthesis of ATP in oxidative phosphorylation. In eukaryotic cells, this overall process occurs in mitochondria.

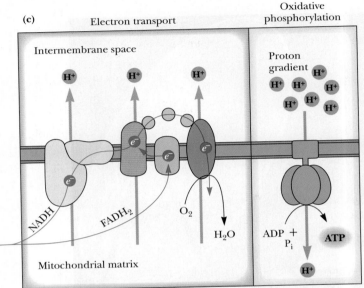

4 more exit as two more NADH in the decarboxylation of two molecules of pyruvate to two acetyl-CoA (Figure 19.1). For each acetyl-CoA oxidized in the TCA cycle, 8 more electrons are removed (as three NADH and one $FADH_2$):

$$H_3CCOO^- + 2H_2O + H^+ \longrightarrow 2CO_2 + 8H$$

In the electron-transport pathway, these 8 electrons combine with oxygen to form water:

$$8H + 2O_2 \longrightarrow 4H_2O$$

So, the net reaction for the TCA cycle and electron-transport pathway is

$$H_3CCOO^- + 2O_2 + H^+ \longrightarrow 2CO_2 + 2H_2O$$

As German biochemist Hans Krebs showed in the 1930s, the eight-electron oxidation of acetate by the TCA cycle is accomplished with the help of oxaloacetate. (In his honor, the TCA cycle is often referred to as the **Krebs cycle**.) Beginning with acetate, a series of five reactions produces two molecules of CO_2, with four electrons extracted in the form of NADH and four electrons passed to oxaloacetate to produce a molecule of succinate. The pathway becomes a cycle by three additional reactions that accomplish a four-electron oxidation of succinate back to oxaloacetate. *This special trio of reactions is used repeatedly in metabolism: first, oxidation of a single bond to a double bond, then addition of the elements of water across the double bond, and finally oxidation of the resulting alcohol to a carbonyl.* We will see it again in fatty acid oxidation (see Chapter 23), in reverse in fatty acid synthesis (see Chapter 24), and in amino acid synthesis and breakdown (see Chapter 25).

19.1 ⋮ What Is the Chemical Logic of the TCA Cycle?

The entry of new carbon units into the cycle is through acetyl-CoA. This entry metabolite can be formed either from pyruvate (from glycolysis) or from oxidation of fatty acids (discussed in Chapter 23). Transfer of the two-carbon acetyl group from acetyl-CoA to the four-carbon oxaloacetate to yield six-carbon citrate is catalyzed by citrate synthase. A dehydration–rehydration rearrangement of citrate yields isocitrate. Two successive decarboxylations produce α-ketoglutarate and then succinyl-CoA, a CoA conjugate of a four-carbon unit. Several steps later, oxaloacetate is regenerated and can combine with another two-carbon unit of acetyl-CoA. Thus, carbon enters the cycle as acetyl-CoA and exits as CO_2. In the process, metabolic energy is captured in the form of ATP, NADH, and enzyme-bound $FADH_2$ (symbolized as $[FADH_2]$).

The TCA Cycle Provides a Chemically Feasible Way of Cleaving a Two-Carbon Compound

The cycle shown in Figure 19.2 at first appears to be a complicated way to oxidize acetate units to CO_2, but there is a chemical basis for the apparent complexity. Oxidation of an acetyl group to a pair of CO_2 molecules requires C—C cleavage:

$$CH_3COO^- \longrightarrow CO_2 + CO_2$$

In many instances, C—C cleavage reactions in biological systems occur between carbon atoms α and β to a carbonyl group:

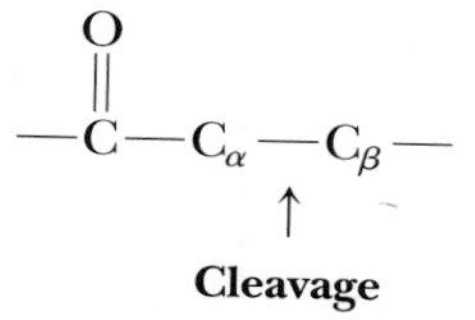

$$-\overset{\overset{\displaystyle O}{\|}}{C}-C_\alpha-C_\beta-$$
$$\uparrow$$
Cleavage

A good example of such a cleavage is the fructose bisphosphate aldolase reaction (see Figure 18.12).

Another common type of C—C cleavage is α-cleavage of an α-hydroxyketone:

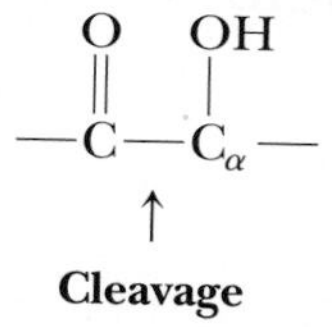

$$-\overset{\overset{\displaystyle O}{\|}}{C}-\overset{\overset{\displaystyle OH}{|}}{C_\alpha}-$$
$$\uparrow$$
Cleavage

(We will see this type of cleavage in the transketolase reaction described in Chapter 22.) Neither of these cleavage strategies is suitable for acetate. It has no β-carbon, and the second method would require hydroxylation—not a favorable reaction for acetate. Instead, living things have evolved the clever chemistry of condensing acetate with oxaloacetate and then carrying out a β-cleavage. The TCA cycle combines this β-cleavage reaction with oxidation to form CO_2, regenerate oxaloacetate, and capture the liberated metabolic energy in NADH and ATP.

19.2 ⋮ How Is Pyruvate Oxidatively Decarboxylated to Acetyl-CoA?

Pyruvate produced by glycolysis is a significant source of acetyl-CoA for the TCA cycle. Because, in eukaryotic cells, glycolysis occurs in the cytoplasm, whereas the TCA cycle reactions and all subsequent steps of aerobic metabolism take place in the mitochondria, pyruvate must first enter the mitochondria to enter the TCA cycle. The oxidative decarboxylation of pyruvate to acetyl-CoA

$$Pyruvate + CoA + NAD^+ \longrightarrow acetyl\text{-}CoA + CO_2 + NADH$$

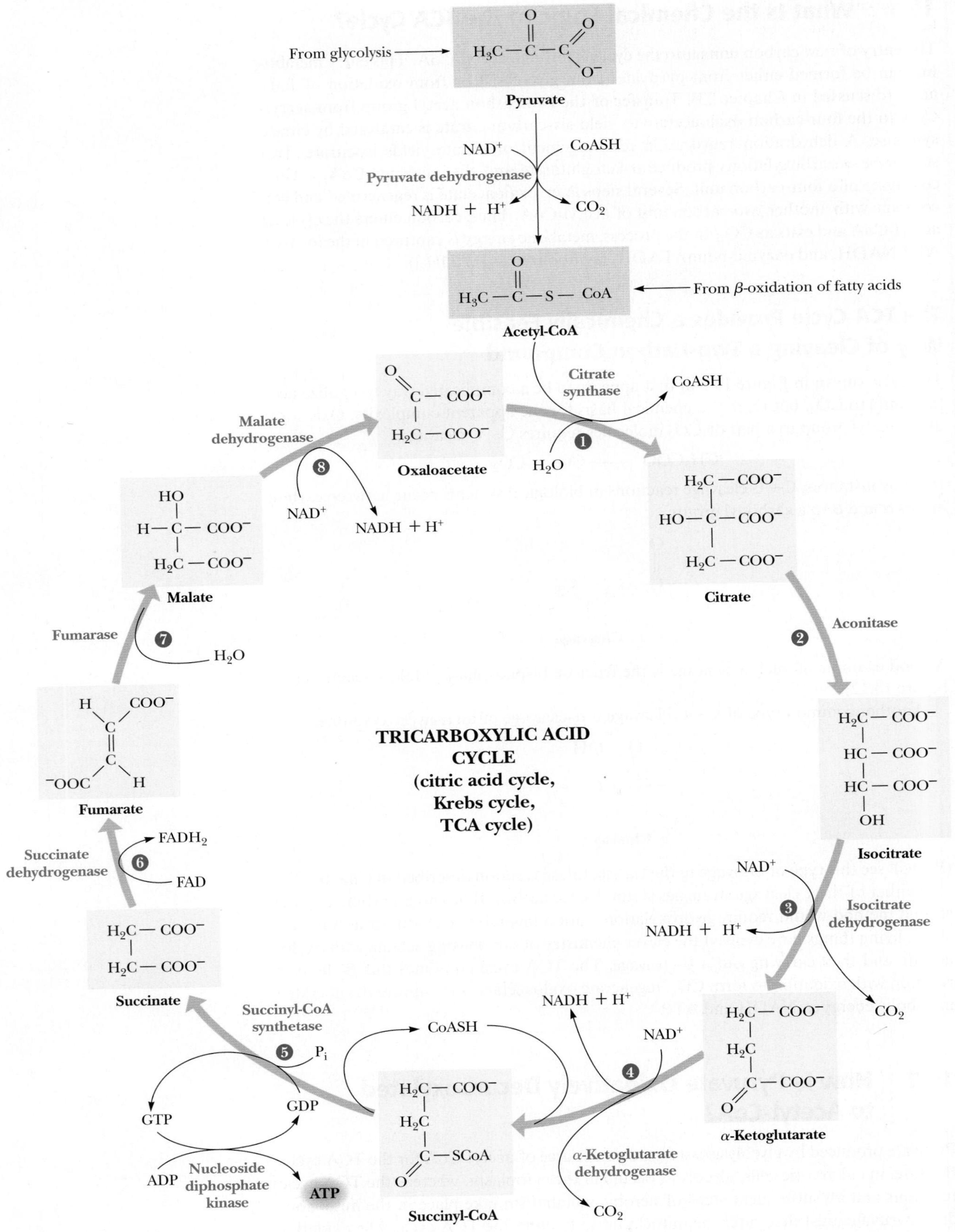

FIGURE 19.2 The TCA cycle.

is the connecting link between glycolysis and the TCA cycle. The reaction is catalyzed by pyruvate dehydrogenase, a multienzyme complex.

The **pyruvate dehydrogenase complex (PDC)** is formed from multiple copies of three enzymes: *pyruvate dehydrogenase* (E1), *dihydrolipoamide acetyltransferase* (E2), and *dihydrolipoamide dehydrogenase* (E3). All are involved in the conversion of pyruvate to acetyl-CoA. In addition, eukaryotic PDC contains a unique subunit, termed E3BP, that has no known enzymatic function. The active sites of all three enzymes are not far removed from one another, and the product of the first enzyme is passed directly to the second enzyme, and so on, without diffusion of substrates and products through the solution. Eukaryotic PDC, one of the largest-known multienzyme complexes (with a diameter of approximately 500 Å) is a 9.5-megadalton assembly.

The central core of PDC consists of E2 subunits (and E3BP in eukaryotes), forming either a 24-meric octahedron (in Gram-negative bacteria) or a 60-meric pentagonal dodecahedron (in the mitochondria of eukaryotes and Gram-positive bacteria). In mitochondria, six E3BP dimers and 48 E2 subunits comprise in the core structure, and binding of E3 is mediated by these substituted E3BPs. Thirty E1 heterotetramers and 12 E3 homodimers are bound to the E2/E3BP core. Eukaryotic E2 subunits consist of an inner catalytic (IC) domain and an outer globular domain, itself comprised of two lipoic-acid binding domains (L1 and L2) and a small E1-binding domain, connected by Ala- and Pro-rich hinge regions 20-30 residues in length (Figure 19.3a). Human E3BP

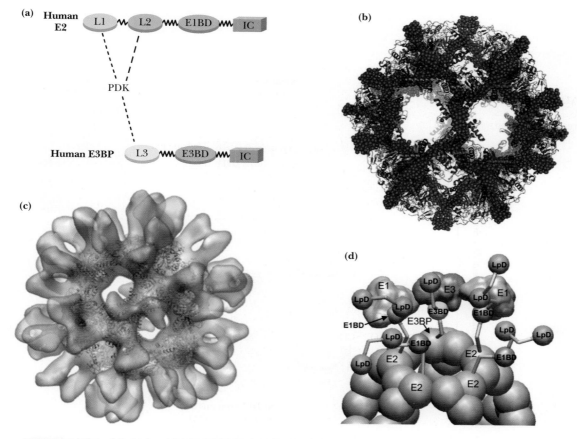

FIGURE 19.3 Models of human pyruvate dehydrogenase. **(a)** Domain structures of human E2 and E3BP. **(b)** Model of a truncated version of human E2. **(c)** Model of the human E2/E3BP:E3 core complex. **(d)** Model of human pyruvate dehydrogenase complex. Six E3BP dimers (brown) substitute for 12 E2 subunits (purple) to form the E2/E2BP core. Around each pentagonal opening, one E3 and two E1 subunits interact to form a shell connected to the E2/E3BP core through linker segments. (Adapted from Vijayakrishnan, S., Kelly, S. M., Gilber, R. J. C., Callow, P., Bhella, D., Forsyth, T., Lindsay, J. G., and Byron, O., 2010. Solution structure and characterization of the human pyruvate dehydrogenase complex core assembly. *Journal of Molecular Biology* **399**:71–93; and Yu, X., Hiromasa, Y., Tsen, H., Stoops, J. K., Roche, T. E., and Zhou, Z. H., 2007. Structures of the human pyruvate dehydrogenase complex cores: A highly conserved catalytic center with flexible N-terminal domains. *Structure* **16**:104–114.)

A DEEPER LOOK

The Coenzymes of the Pyruvate Dehydrogenase Complex

Coenzymes are small molecules that bring unique chemistry to enzyme reactions. Five coenzymes are used in the pyruvate dehydrogenase reaction.

Thiamine Pyrophosphate

TPP assists in the decarboxylation of α-keto acids (here) and in the formation and cleavage of α-hydroxy ketones (as in the transketolase reaction, see Chapter 22).

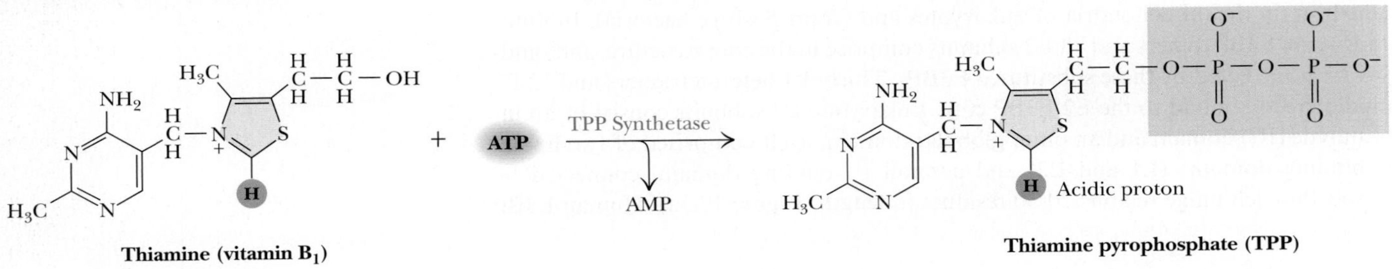
ATP
TPP Synthetase
AMP
Acidic proton
Thiamine (vitamin B1)
Thiamine pyrophosphate (TPP)

The Nicotinamide Coenzymes

NAD⁺/NADH and NADP⁺/NADPH carry out hydride (H:⁻) transfer reactions. All reactions involving these coenzymes are two-electron transfers.

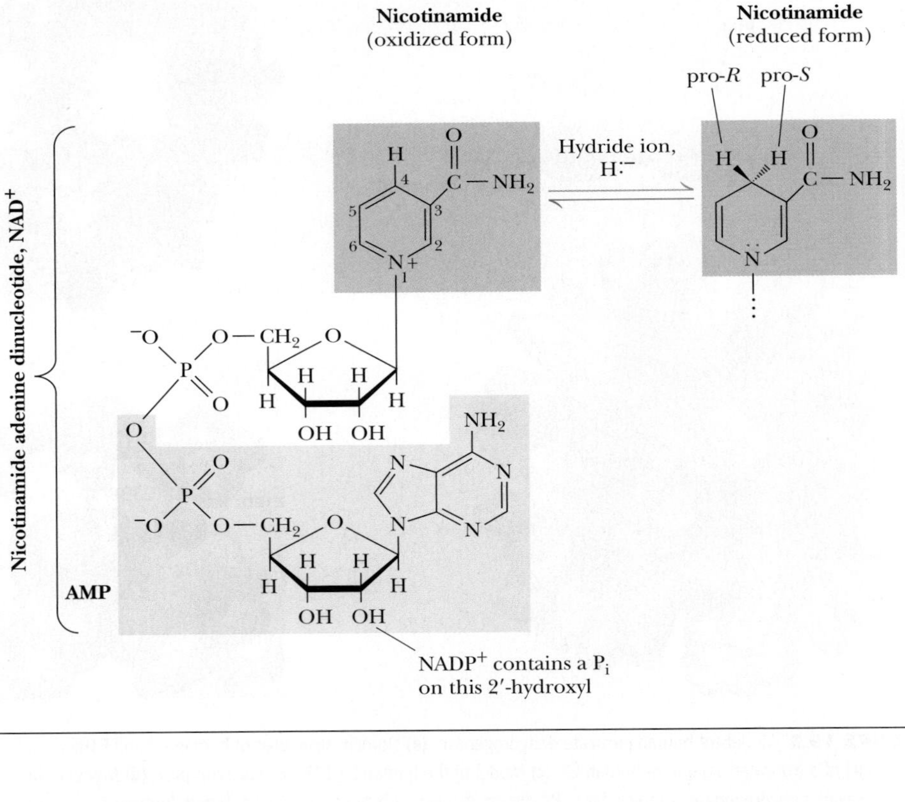
Nicotinamide
(oxidized form)
Nicotinamide
(reduced form)
pro-R pro-S
Hydride ion, H:⁻
Nicotinamide adenine dinucleotide, NAD⁺
AMP
NADP⁺ contains a Pi on this 2′-hydroxyl

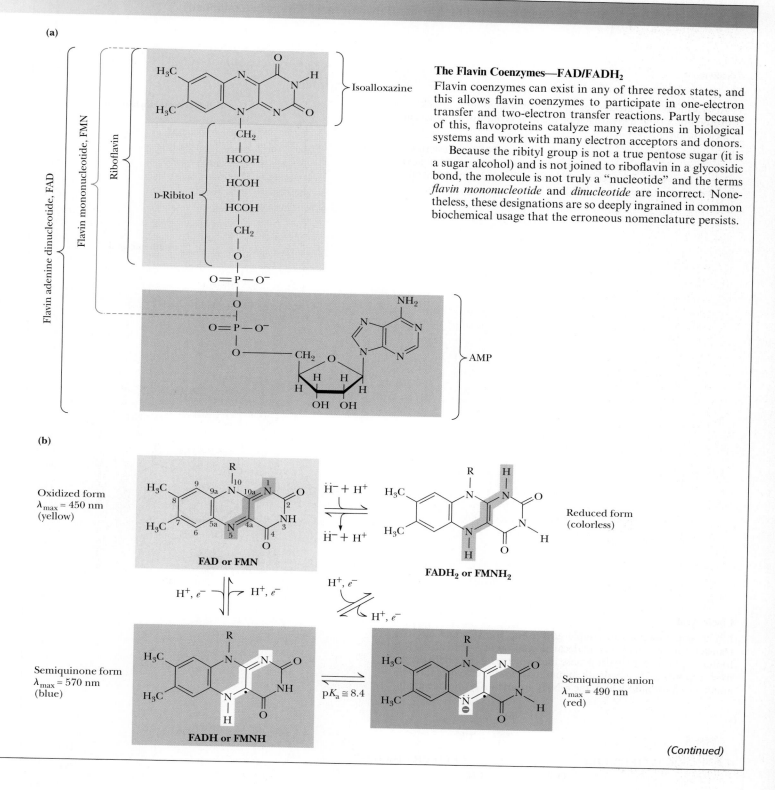

(a)

The Flavin Coenzymes—FAD/FADH$_2$

Flavin coenzymes can exist in any of three redox states, and this allows flavin coenzymes to participate in one-electron transfer and two-electron transfer reactions. Partly because of this, flavoproteins catalyze many reactions in biological systems and work with many electron acceptors and donors.

Because the ribityl group is not a true pentose sugar (it is a sugar alcohol) and is not joined to riboflavin in a glycosidic bond, the molecule is not truly a "nucleotide" and the terms *flavin mononucleotide* and *dinucleotide* are incorrect. Nonetheless, these designations are so deeply ingrained in common biochemical usage that the erroneous nomenclature persists.

(b)

Oxidized form
$\lambda_{max} = 450$ nm
(yellow)

Reduced form
(colorless)

Semiquinone form
$\lambda_{max} = 570$ nm
(blue)

Semiquinone anion
$\lambda_{max} = 490$ nm
(red)

(Continued)

> **A DEEPER LOOK** The Coenzymes of the Pyruvate Dehydrogenase Complex (continued)

Coenzyme A

The two main functions of CoA are:

1. Activation of acyl groups for transfer by nucleophilic attack
2. Activation of the α-hydrogen of the acyl group for abstraction as a proton

The reactive sulfhydryl group on CoA mediates both of these functions. The sulfhydryl group forms **thioester** linkages with acyl groups. The two main functions of CoA are illustrated in the citrate synthase reaction (see Figure 19.6).

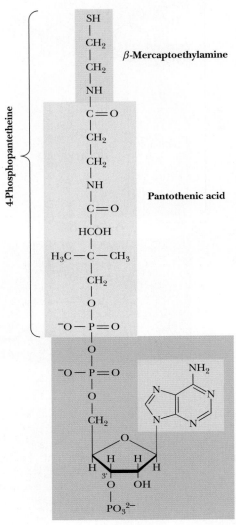

3′,5′–ADP

Lipoic Acid

Lipoic acid functions to couple acyl-group transfer and electron transfer during oxidation and decarboxylation of α-keto acids. It is found in pyruvate dehydrogenase and α-ketoglutarate dehydrogenase. Lipoic acid is covalently bound to relevant enzymes through amide bond formation with the ϵ-NH$_2$ group of a lysine side chain.

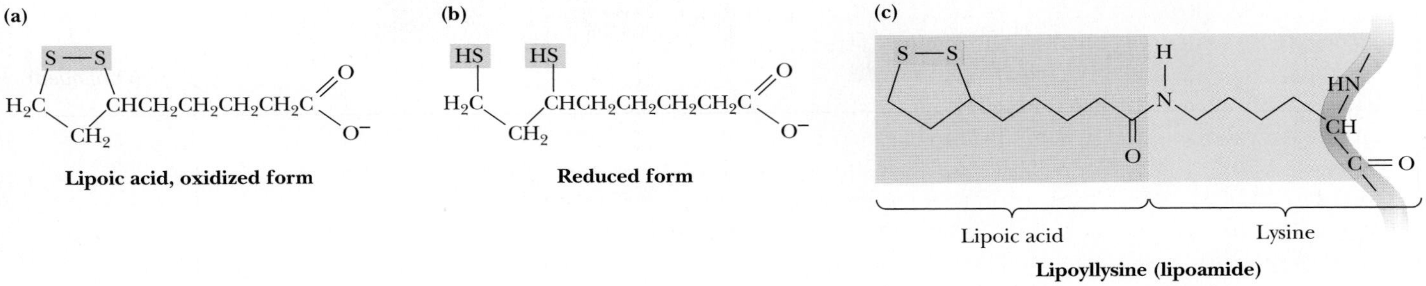

(a)

Lipoic acid, oxidized form

(b)

Reduced form

(c)

Lipoic acid   Lysine

Lipoyllysine (lipoamide)

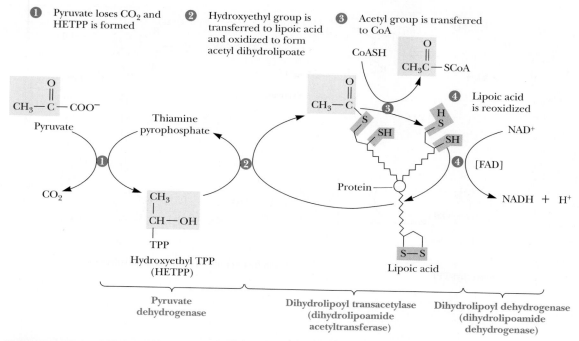

FIGURE 19.4 The reaction mechanism of the pyruvate dehydrogenase complex. Decarboxylation of pyruvate occurs with formation of hydroxyethyl-TPP (step 1). Transfer of the two-carbon unit to lipoic acid in step 2 is followed by formation of acetyl-CoA in step 3. Lipoic acid is reoxidized in step 4 of the reaction.

is composed of three linker-connected domains in a configuration similar to that of human E2, and its IC domain has 50% sequence identity with E2. The flexible linker segments in E2 and E3BP impart the flexibility that allows the lipoic acid groups to visit all three active sites during catalysis.

The pyruvate dehydrogenase reaction (Figure 19.4) is a tour de force of mechanistic chemistry, involving as it does a total of three enzymes and five different coenzymes. The first step of this reaction, decarboxylation of pyruvate and transfer of the acetyl group to lipoic acid, depends on accumulation of negative charge on the transferred two-carbon fragment, as facilitated by the quaternary nitrogen on the thiazolium group of thiamine pyrophosphate (TPP). As shown in Figure 19.5, this cationic imine nitrogen plays two distinct roles in TPP-catalyzed reactions:

1. It provides electrostatic stabilization of the thiazole carbanion formed upon removal of the C-2 proton. (The sp^2 hybridization and the availability of vacant d orbitals on the adjacent sulfur probably also facilitate proton removal at C-2.)
2. TPP attack on pyruvate leads to decarboxylation. The TPP cationic imine nitrogen can act as an effective electron sink to stabilize the negative charge that must develop on the carbon that has been attacked. This stabilization takes place by resonance interaction through the double bond to the nitrogen atom.

This resonance-stabilized intermediate can be protonated to give **hydroxyethyl-TPP**. The reaction of hydroxyethyl-TPP with the oxidized form of lipoic acid yields the energy-rich acetyl-thiol ester of reduced lipoic acid through oxidation of the hydroxyl-carbon of the two-carbon substrate unit. Nucleophilic attack by CoA on the carbonyl carbon (a characteristic feature of CoA chemistry) results in transfer of the acetyl group from lipoic acid to CoA. The subsequent oxidation of lipoic acid is catalyzed by the FAD-dependent dihydrolipoyl dehydrogenase, and NAD$^+$ is reduced.

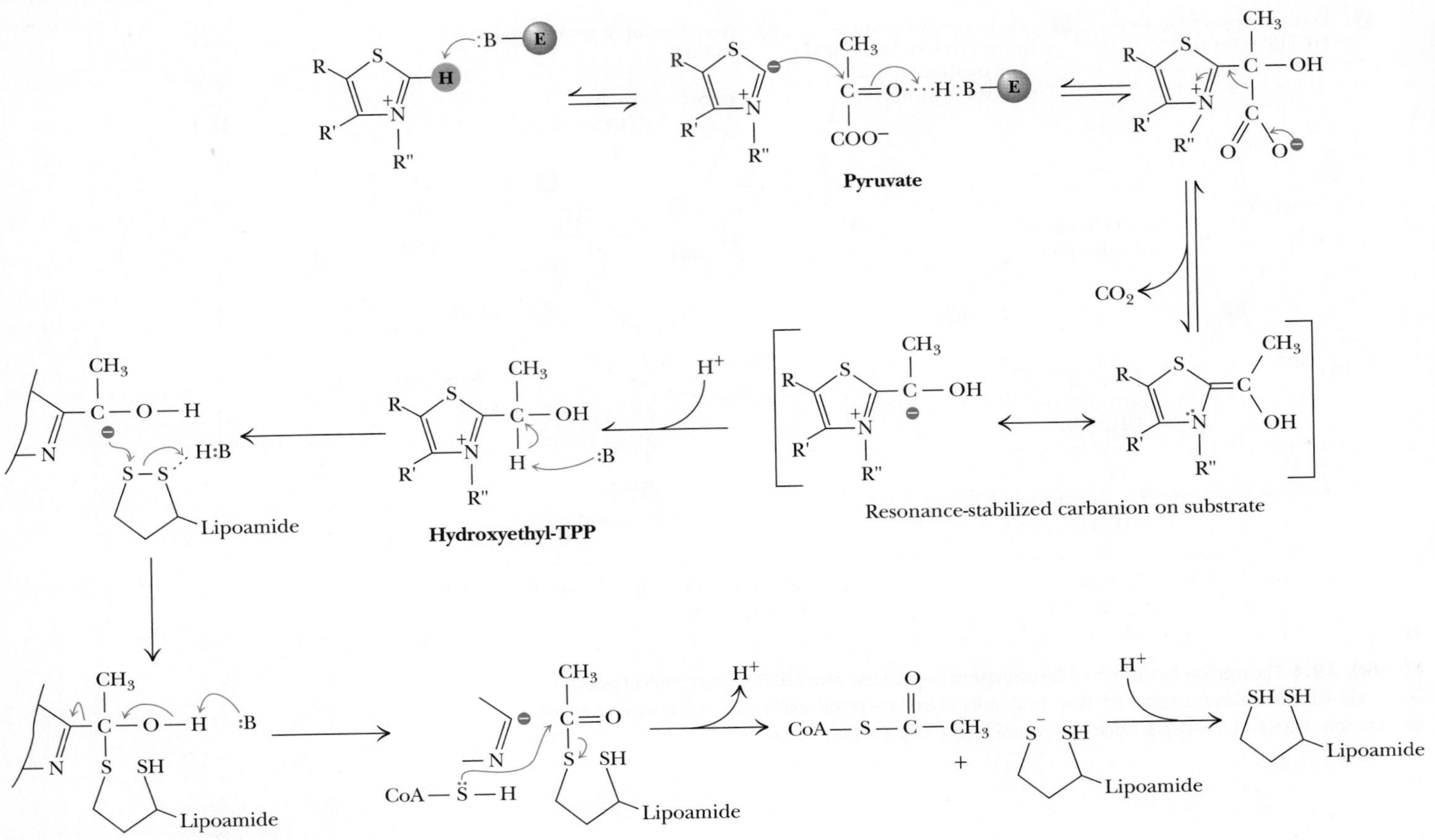

FIGURE 19.5 The mechanistic details of the first three steps of the pyruvate dehydrogenase complex reaction.

19.3 How Are Two CO_2 Molecules Produced from Acetyl-CoA?

The Citrate Synthase Reaction Initiates the TCA Cycle

The first reaction within the TCA cycle, the one by which carbon atoms are introduced, is the **citrate synthase reaction** (Figure 19.6). Here acetyl-CoA reacts with oxaloacetate in a **Perkin condensation** (a carbon–carbon condensation between a ketone or aldehyde and an ester). The acyl group is activated in two ways in an acyl-CoA molecule: The carbonyl carbon is activated for attack by nucleophiles, and the C_α carbon is more acidic and can be deprotonated to form a carbanion. The citrate synthase reaction depends upon the latter mode of activation. As shown in Figure 19.6, a general base on the

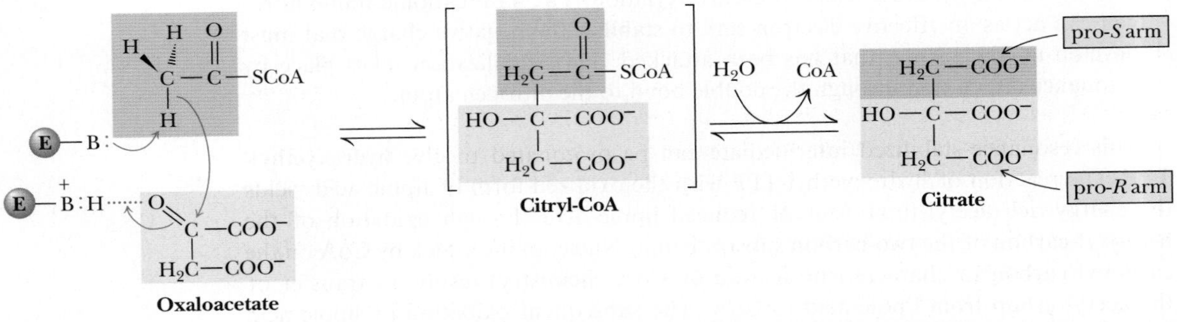

FIGURE 19.6 Citrate is formed in the citrate synthase reaction from oxaloacetate and acetyl-CoA. The mechanism involves nucleophilic attack by the carbanion of acetyl-CoA on the carbonyl carbon of oxaloacetate, followed by thioester hydrolysis.

TABLE 19.1	The Enzymes and Reactions of the TCA Cycle			
Reaction		**Enzyme**	$\Delta G^{\circ\prime}$ (kJ/mol)	ΔG (kJ/mol)
1. Acetyl-CoA + oxaloacetate + H_2O $\rightleftharpoons$ CoASH + citrate		Citrate synthase	−31.4	−53.9
2. Citrate $\rightleftharpoons$ isocitrate		Aconitase	+6.7	+0.8
3. Isocitrate + NAD^+ $\rightleftharpoons$ α-ketoglutarate + NADH + CO_2		Isocitrate dehydrogenase	−8.4	−17.5
4. α-Ketoglutarate + CoASH + NAD^+ $\rightleftharpoons$ succinyl-CoA + NADH + CO_2		α-Ketoglutarate dehydrogenase complex	−30	−43.9
5. Succinyl-CoA + GDP + P_i $\rightleftharpoons$ succinate + GTP + CoASH		Succinyl-CoA synthetase	−3.3	≈0
6. Succinate + [FAD] $\rightleftharpoons$ fumarate + [FADH$_2$]		Succinate dehydrogenase	+0.4	≠0
7. Fumarate + H_2O $\rightleftharpoons$ L-malate		Fumarase	−3.8	≈0
8. L-Malate + NAD^+ $\rightleftharpoons$ oxaloacetate + NADH + H^+		Malate dehydrogenase	+29.7	≈0

ΔG values from Newsholme, E. A., and Leech, A. R., 1983. *Biochemistry for the Medical Sciences.* New York: Wiley.

enzyme accepts a proton from the methyl group of acetyl-CoA, producing a stabilized α-carbanion of acetyl-CoA. This strong nucleophile attacks the α-carbonyl of oxaloacetate, yielding citryl-CoA. This part of the reaction has an equilibrium constant near 1, but the overall reaction is driven to completion by the subsequent hydrolysis of the high-energy thioester to citrate and free CoA. The overall $\Delta G^{\circ\prime}$ is −31.4 kJ/mol, and under standard conditions the reaction is essentially irreversible. Although the mitochondrial concentration of oxaloacetate is very low (much less than 1 μM—see example in Section 19.4), the strong, negative $\Delta G^{\circ\prime}$ drives the reaction forward.

Citrate Synthase Is a Dimer Citrate synthase in mammals is a dimer of 49-kD subunits (Table 19.1). On each subunit, oxaloacetate and acetyl-CoA bind to the active site, which lies in a cleft between two domains and is surrounded mainly by α-helical segments (Figure 19.7). Binding of oxaloacetate induces a conformational change that facilitates the binding of acetyl-CoA and closes the active site so that the reactive carbanion of acetyl-CoA is protected from protonation by water.

NADH Is an Allosteric Inhibitor of Citrate Synthase Citrate synthase is the first step in this metabolic pathway, and as stated the reaction has a large negative $\Delta G^{\circ\prime}$. As might be expected, it is a highly regulated enzyme. NADH, a product of the TCA cycle, is an allosteric inhibitor of citrate synthase, as is succinyl-CoA, the product of the fifth step in the cycle (and an acetyl-CoA analog).

Citrate Is Isomerized by Aconitase to Form Isocitrate

Citrate itself poses a problem: It is a poor candidate for further oxidation because it contains a tertiary alcohol, which could be oxidized only by breaking a carbon–carbon bond. An obvious solution to this problem is to isomerize the tertiary alcohol to a secondary alcohol, which the cycle proceeds to do in the next step.

Citrate is isomerized to isocitrate by **aconitase** in a two-step process involving aconitate as an intermediate (Figure 19.8). In this reaction, the elements of water are first abstracted from citrate to yield aconitate, which is then rehydrated with H— and HO— adding back in opposite positions to produce isocitrate. The net effect is the conversion of a tertiary alcohol (citrate) to a secondary alcohol (isocitrate). Oxidation of the secondary alcohol of isocitrate involves breakage of a C—H bond, a simpler matter than the C—C cleavage required for the direct oxidation of citrate.

Inspection of the citrate structure shows a total of four chemically equivalent hydrogens, but only one of these—the pro-*R* H atom of the pro-*R* arm of citrate—is abstracted by aconitase, which is quite stereospecific. Formation of the double bond of aconitate following proton abstraction requires departure of hydroxide ion from the

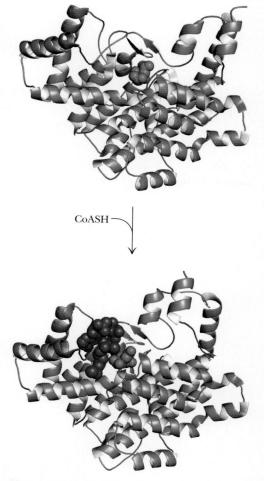

CoASH

FIGURE 19.7 Citrate synthase. In the monomer shown here, citrate is shown in blue, and CoA is red. (Top: pdb id = 1CTS; bottom: pdb id = 2CTS.)

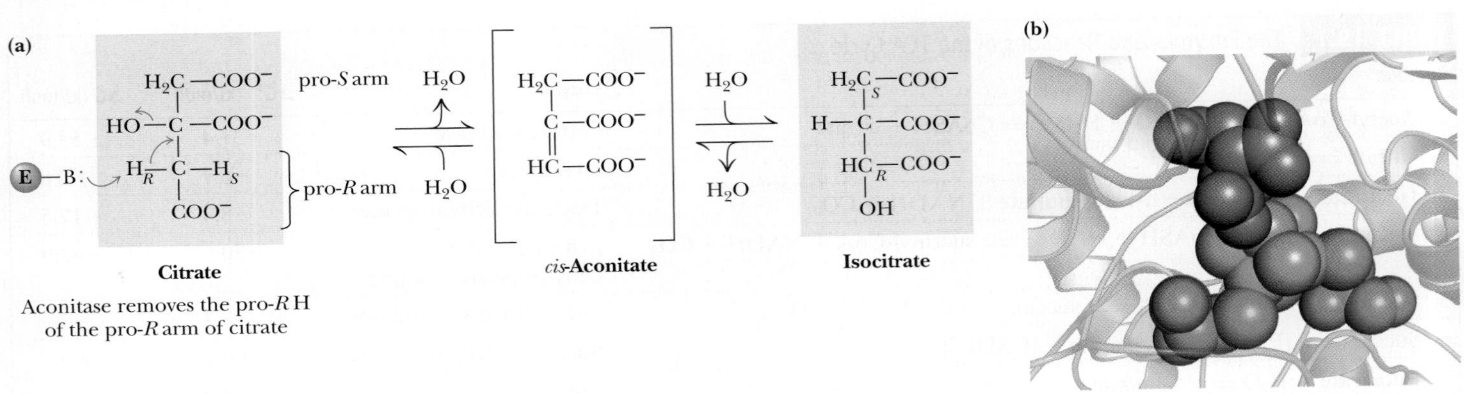

(a)

Citrate

Aconitase removes the pro-*R* H
of the pro-*R* arm of citrate

cis-Aconitate

Isocitrate

(b)

FIGURE 19.8 **(a)** The aconitase reaction converts citrate to *cis*-aconitate and then to isocitrate. Aconitase is stereospecific and removes the pro-*R* hydrogen from the pro-*R* arm of citrate. **(b)** The active site of aconitase. The iron–sulfur cluster (pink) is coordinated by cysteines (orange) and isocitrate (purple) (pdb id = 1B0J).

C-3 position. Hydroxide is a relatively poor leaving group, and its departure is facilitated in the aconitase reaction by coordination with an iron atom in an iron–sulfur cluster.

Aconitase Utilizes an Iron–Sulfur Cluster Aconitase contains an **iron–sulfur cluster** consisting of three iron atoms and four sulfur atoms in a near-cubic arrangement (Figure 19.9). Cysteine residues from the enzyme coordinate the three iron atoms. In the inactive state of the enzyme, one corner of the cube is vacant. Binding of iron (as Fe^{2+}) to this position activates aconitase. The iron atom in this position can coordinate the C-3 carboxyl and hydroxyl groups of citrate. This iron atom thus acts as a Lewis acid, accepting an unshared pair of electrons from the hydroxyl, making it a better leaving group. The equilibrium for the aconitase reaction favors citrate, and an equilibrium mixture typically contains about 90% citrate, 4% *cis*-aconitate, and 6% isocitrate. The $\Delta G^{\circ\prime}$ is +6.7 kJ/mol.

Fluoroacetate Blocks the TCA Cycle Fluoroacetate is an extremely poisonous agent that blocks the TCA cycle in vivo, although it has no apparent effect on any of the isolated enzymes. Its LD_{50}, the lethal dose for 50% of animals consuming it, is 0.2 mg per kilogram of body weight; it has been used as a rodent poison. The action of fluoroacetate has been traced to aconitase, which is inhibited in vivo by fluorocitrate, which is formed from fluoroacetate in two steps.

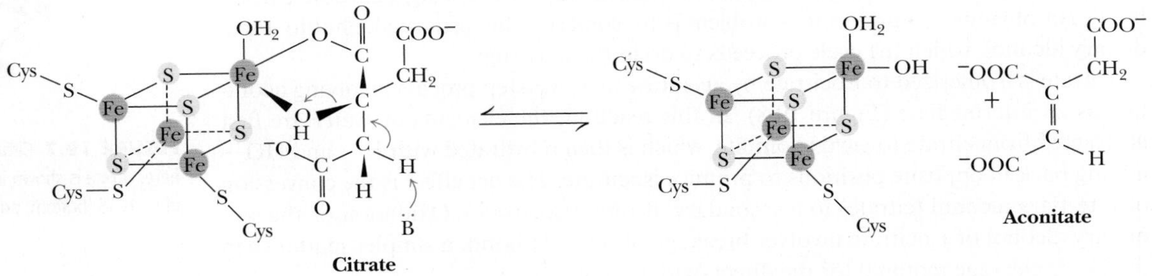

Citrate

Aconitate

FIGURE 19.9 The iron–sulfur cluster of aconitase. Binding of Fe^{2+} to the vacant position of the cluster activates aconitase. The added iron atom coordinates the C-3 carboxyl and hydroxyl groups of citrate and acts as a Lewis acid, accepting an electron pair from the hydroxyl group and making it a better leaving group.

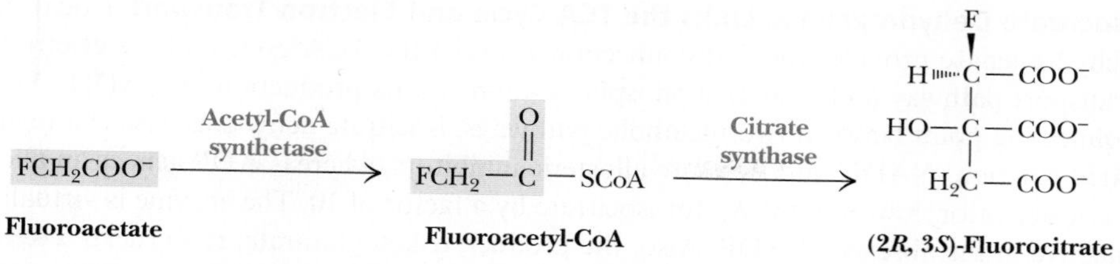

Fluoroacetate readily crosses both the cellular and the mitochondrial membranes, and in mitochondria it is converted to fluoroacetyl-CoA by acetyl-CoA synthetase. Fluoroacetyl-CoA is a substrate for citrate synthase, which condenses it with oxaloacetate to form fluorocitrate. Fluoroacetate may thus be viewed as a **trojan horse inhibitor.** Analogous to the giant Trojan horse of legend—which the soldiers of Troy took into their city, not knowing that Greek soldiers were hidden inside it and waiting to attack—fluoroacetate enters the TCA cycle innocently enough, in the citrate synthase reaction. Citrate synthase converts fluoroacetate to inhibitory fluorocitrate for its TCA cycle partner, aconitase, blocking the cycle.

Isocitrate Dehydrogenase Catalyzes the First Oxidative Decarboxylation in the Cycle

In the next step of the TCA cycle, isocitrate is oxidatively decarboxylated to yield α-ketoglutarate, with concomitant reduction of NAD^+ to NADH in the isocitrate dehydrogenase reaction (Figure 19.10). The reaction has a net $\Delta G^{\circ\prime}$ of -8.4 kJ/mol, and it is sufficiently exergonic to pull the aconitase reaction forward. This two-step reaction involves (1) oxidation of the C-2 alcohol of isocitrate to form oxalosuccinate, followed by (2) a β-decarboxylation reaction that expels the central carboxyl group as CO_2, leaving the product α-ketoglutarate. Oxalosuccinate, the β-keto acid produced by the initial dehydrogenation reaction, is unstable and thus is readily decarboxylated.

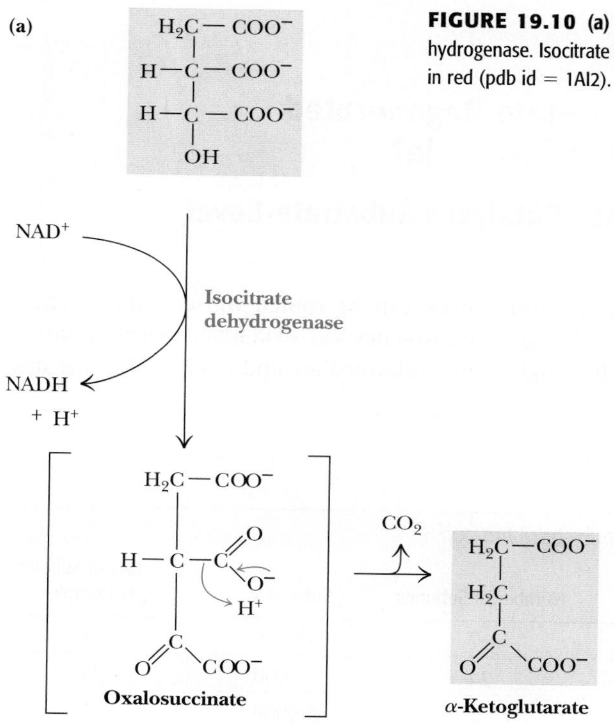

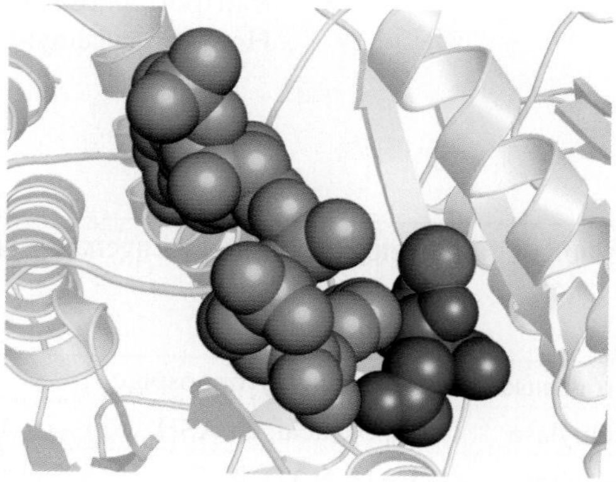

FIGURE 19.10 **(a)** The isocitrate dehydrogenase reaction. **(b)** The active site of isocitrate dehydrogenase. Isocitrate is shown in blue, $NADP^+$ (as an NAD^+ analog) is shown in gold, with Ca^{2+} in red (pdb id = 1AI2).

Isocitrate Dehydrogenase Links the TCA Cycle and Electron Transport Isocitrate dehydrogenase provides the first connection between the TCA cycle and the electron-transport pathway and oxidative phosphorylation, via its production of NADH. As a connecting point between two metabolic pathways, isocitrate dehydrogenase is a regulated reaction. NADH and ATP are allosteric inhibitors, whereas ADP acts as an allosteric activator, lowering the K_m for isocitrate by a factor of 10. The enzyme is virtually inactive in the absence of ADP. Also, the product, α-ketoglutarate, is a crucial α-keto acid for aminotransferase reactions (see Chapters 13 and 25), connecting the TCA cycle (that is, carbon metabolism) with nitrogen metabolism.

α-Ketoglutarate Dehydrogenase Catalyzes the Second Oxidative Decarboxylation of the TCA Cycle

A second oxidative decarboxylation occurs in the α-ketoglutarate dehydrogenase reaction.

α-Ketoglutarate   Succinyl-CoA

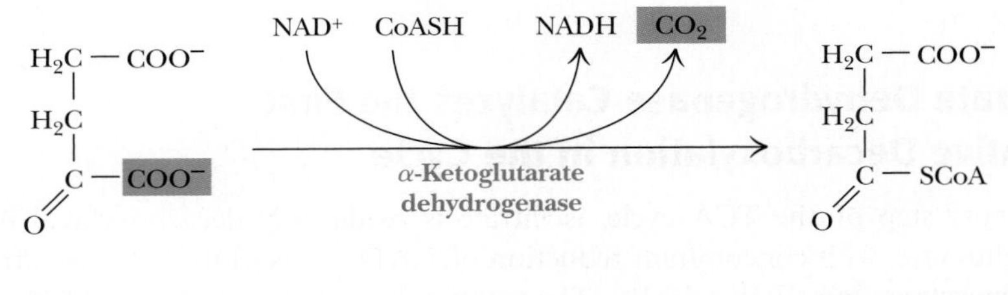

Like the pyruvate dehydrogenase complex, α-ketoglutarate dehydrogenase is a multienzyme complex—consisting of *α-ketoglutarate dehydrogenase, dihydrolipoyl transsuccinylase,* and *dihydrolipoyl dehydrogenase*—that employs five different coenzymes (Table 19.2). The dihydrolipoyl dehydrogenase in this reaction is identical to that in the pyruvate dehydrogenase reaction. The mechanism is analogous to that of pyruvate dehydrogenase. As with the pyruvate dehydrogenase reaction, this reaction produces NADH and a thioester product—in this case, succinyl-CoA. Succinyl-CoA and NADH products are energy-rich species that are important sources of metabolic energy in subsequent cellular processes.

19.4 How Is Oxaloacetate Regenerated to Complete the TCA Cycle?

Succinyl-CoA Synthetase Catalyzes Substrate-Level Phosphorylation

The NADH produced in the foregoing steps can be routed through the electron-transport pathway to make high-energy phosphates via oxidative phosphorylation. However, succinyl-CoA is itself a high-energy intermediate and is utilized in the next

TABLE 19.2	Composition of the α-Ketoglutarate Dehydrogenase Complex from *Escherichia coli*				
Enzyme	Coenzyme	Enzyme M_r	Number of Subunits	Subunit M_r	Number of Subunits per Complex
α-Ketoglutarate dehydrogenase	Thiamine pyrophosphate	192,000	2	96,000	24
Dihydrolipoyl transsuccinylase	Lipoic acid, CoASH	1,700,000	24	70,000	24
Dihydrolipoyl dehydrogenase	FAD, NAD$^+$	112,000	2	56,000	12

step of the TCA cycle to drive the phosphorylation of GDP to GTP (in mammals) or ADP to ATP (in plants and bacteria).

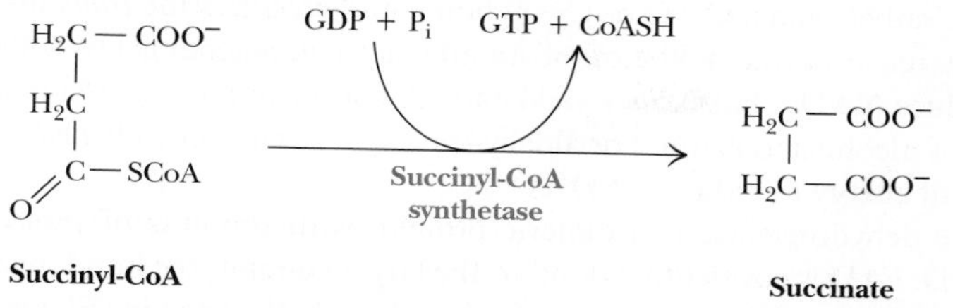

Succinyl-CoA   **Succinate**

The reaction is catalyzed by **succinyl-CoA synthetase,** sometimes called **succinate thiokinase.** The free energies of hydrolysis of succinyl-CoA and GTP or ATP are similar, and the net reaction has a $\Delta G°'$ of -3.3 kJ/mol. Succinyl-CoA synthetase provides another example of a **substrate-level phosphorylation** (see Chapter 18), in which a substrate, rather than an electron-transport chain or proton gradient, provides the energy for phosphorylation. It is the only such reaction in the TCA cycle. The GTP produced by mammals in this reaction can exchange its terminal phosphoryl group with ADP via the **nucleoside diphosphate kinase reaction:**

$$GTP + ADP \xrightleftharpoons[\text{kinase}]{\text{Nucleoside diphosphate}} ATP + GDP$$

The Mechanism of Succinyl-CoA Synthetase Involves a Phosphohistidine The mechanism of succinyl-CoA synthetase is postulated to involve displacement of CoA by phosphate, forming succinyl phosphate at the active site, followed by transfer of the phosphoryl group to an active-site histidine (making a phosphohistidine intermediate) and release of succinate. The phosphoryl moiety is then transferred to GDP to form GTP (Figure 19.11). This sequence of steps "preserves" the energy of the thioester bond of succinyl-CoA in a series of high-energy intermediates that lead to a molecule of ATP:

$$\text{Thioester} \longrightarrow [\text{succinyl-P}] \longrightarrow [\text{phosphohistidine}] \longrightarrow GTP \longrightarrow ATP$$

The First Five Steps of the TCA Cycle Produce NADH, CO$_2$, GTP (ATP), and Succinate This is a good point to pause in our trip through the TCA cycle and see what has happened. A two-carbon acetyl group has been introduced as acetyl-CoA and linked to oxaloacetate, and two CO$_2$ molecules have been liberated. The cycle has produced two molecules of NADH and one of GTP or ATP and has left a molecule of succinate.

The TCA cycle can now be completed by converting succinate to oxaloacetate. This latter process represents a net oxidation. The TCA cycle breaks it down into (consecutively) an oxidation step, a hydration reaction, and a second oxidation step. The oxidation steps are accompanied by the reduction of an [FAD] and an NAD$^+$. The reduced coenzymes, [FADH$_2$] and NADH, subsequently provide reducing power in the electron-transport chain. (It will be seen in Chapter 23 that virtually the same chemical strategy is used in β-oxidation of fatty acids.)

Succinate Dehydrogenase Is FAD-Dependent

The oxidation of succinate to fumarate is carried out by **succinate dehydrogenase,** a membrane-bound enzyme that is actually part of the electron-transport chain.

Succinate   **Fumarate**

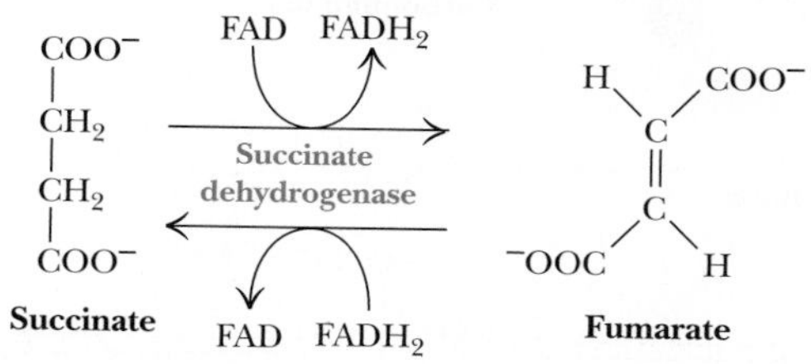

As will be seen in Chapter 20, succinate dehydrogenase is identical with the succinate–coenzyme Q reductase of the electron-transport chain. In contrast with all of the other enzymes of the TCA cycle, which are soluble proteins found in the mitochondrial matrix,

Condensation: A reaction between two or more molecules that results in formation of a larger molecule, with elimination of some simpler molecule, such as water (as in dehydration synthesis).

Synthase: A condensation reaction that does not require a nucleoside triphosphate as an energy source.

Synthetase: A condensation reaction that requires a nucleoside triphosphate (often ATP) as an energy source.

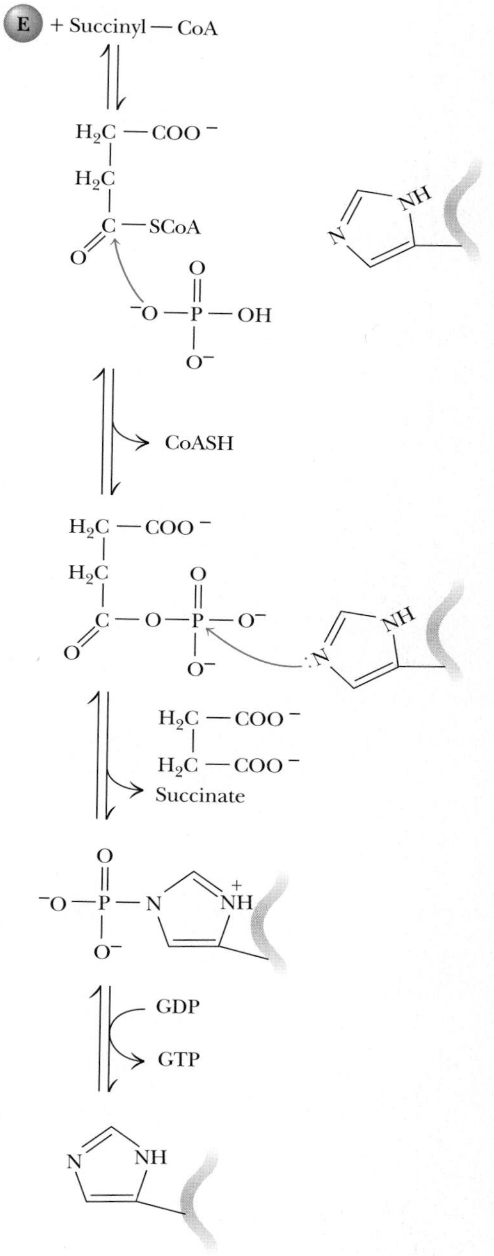

FIGURE 19.11 The mechanism of the succinyl-CoA synthetase reaction.

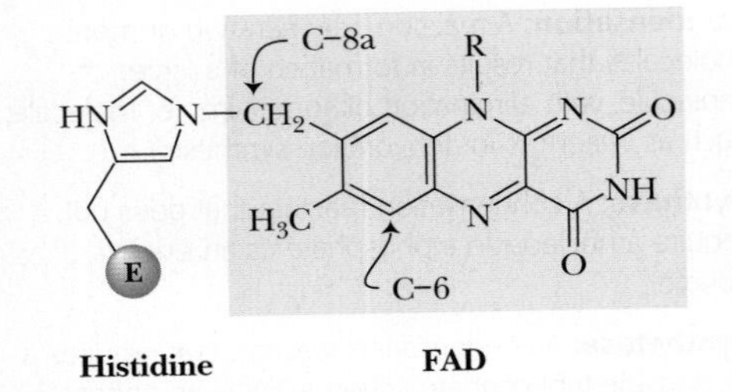

FIGURE 19.12 The covalent bond between FAD and succinate dehydrogenase involves the C-8a methylene group of FAD and the N-3 of a histidine residue on the enzyme.

succinate dehydrogenase is an integral membrane protein tightly associated with the inner mitochondrial membrane. Succinate oxidation involves removal of H atoms across a C—C bond, rather than a C—O or C—N bond, and produces the *trans*-unsaturated fumarate. This reaction (the oxidation of an alkane to an alkene) is not sufficiently exergonic to reduce NAD$^+$, but it does yield enough energy to reduce [FAD]. (By contrast, oxidations of alcohols to ketones or aldehydes are more energetically favorable and provide sufficient energy to reduce NAD$^+$.)

Succinate dehydrogenase is a dimeric protein, with subunits of molecular masses 70 and 27 kD. FAD is covalently bound to the larger subunit; the bond involves a methylene group of C-8a of FAD and N-3 of a histidine on the protein (Figure 19.12). Succinate dehydrogenase also contains three different iron–sulfur clusters: a 4Fe-4S cluster, a 3Fe-4S cluster, and a 2Fe-2S cluster, shown below.

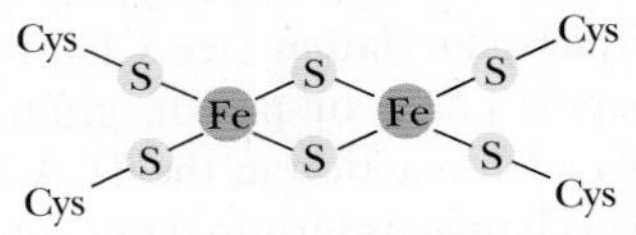

Viewed from either end of the succinate molecule, the reaction involves dehydrogenation α,β to a carbonyl (actually, a carboxyl) group. The dehydrogenation is stereospecific, with the pro-S hydrogen removed from one carbon atom and the pro-R hydrogen removed from the other. The electrons captured by [FAD] in this reaction are passed directly into the iron–sulfur clusters of the enzyme and on to coenzyme Q (UQ). The covalently bound FAD is first reduced to [FADH$_2$] and then reoxidized to form [FAD] and the reduced form of coenzyme Q, UQH$_2$. Electrons captured by UQH$_2$ then flow through the rest of the electron-transport chain in a series of events that will be discussed in detail in Chapter 20.

Note that flavin coenzymes can carry out either one-electron or two-electron transfers. The succinate dehydrogenase reaction represents a net two-electron reduction of FAD.

Fumarase Catalyzes the *Trans*-Hydration of Fumarate to Form L-Malate

Fumarate is hydrated in a stereospecific reaction by fumarase to give L-malate.

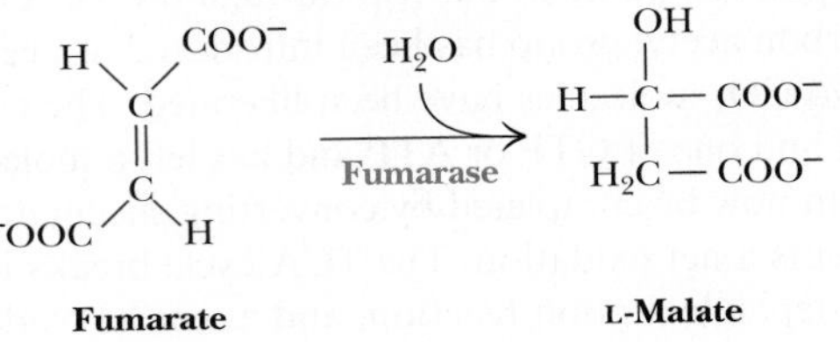

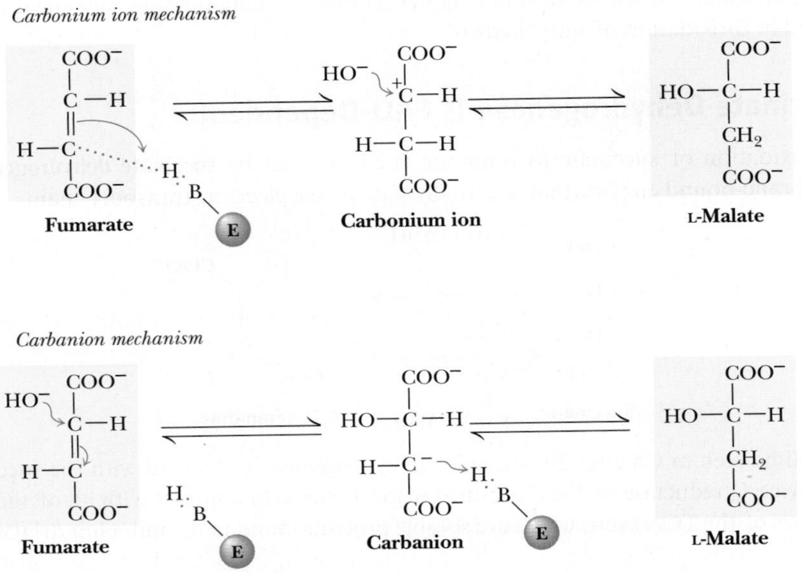

FIGURE 19.13 Two possible mechanisms for the fumarase reaction.

The reaction involves *trans*-addition of the elements of water across the double bond. Recall that aconitase carries out a similar reaction and that *trans*-addition of —H and —OH occurs across the double bond of *cis*-aconitate. Although the exact mechanism is uncertain, it may involve protonation of the double bond to form an intermediate carbonium ion (Figure 19.13) or possibly attack by water or OH⁻ anion to produce a carbanion, followed by protonation.

Malate Dehydrogenase Completes the Cycle by Oxidizing Malate to Oxaloacetate

In the last step of the TCA cycle, L-malate is oxidized to oxaloacetate by malate dehydrogenase. This reaction is very endergonic, with a $\Delta G^{\circ\prime}$ of $+30$ kJ/mol.

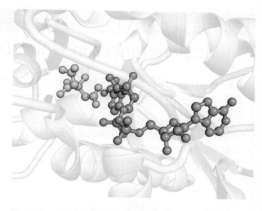

FIGURE 19.14 The active site of malate dehydrogenase. Malate is shown in yellow; NAD⁺ is pink (pdb id = 1EMD).

Consequently, the concentration of oxaloacetate in the mitochondrial matrix is usually quite low. The reaction, however, is pulled forward by the favorable citrate synthase reaction. Oxidation of malate is coupled to reduction of yet another molecule of NAD⁺, the third one of the cycle. Counting the [FAD] reduced by succinate dehydrogenase, this makes the fourth coenzyme reduced through oxidation of a single acetate unit.

Malate dehydrogenase is structurally and functionally similar to other dehydrogenases, notably lactate dehydrogenase (Figure 19.14). Both consist of alternating β-sheet and α-helical segments. Binding of NAD⁺ causes a conformational change in the 20-residue segment that connects the D and E strands of the β-sheet. The change is triggered by an interaction between the adenosine phosphate moiety of NAD⁺ and an arginine residue in this loop region. Such a conformational change is consistent with an ordered single-displacement mechanism for NAD⁺-dependent dehydrogenases (see Chapter 13).

19.5 What Are the Energetic Consequences of the TCA Cycle?

The net reaction accomplished by the TCA cycle, as follows, shows two molecules of CO_2, one ATP, and four reduced coenzymes produced per acetate group oxidized. The cycle is exergonic, with a net $\Delta G^{\circ\prime}$ for one pass around the cycle of approximately -40 kJ/mol. Table 19.1 compares the $\Delta G^{\circ\prime}$ values for the individual reactions with the overall $\Delta G^{\circ\prime}$ for the net reaction.

Acetyl-CoA + 3 NAD⁺ + [FAD] + ADP + P$_i$ + 2 H$_2$O $\longrightarrow$
$$2\ CO_2 + 3\ NADH + 3\ H^+ + [FADH_2] + ATP + CoASH$$
$$\Delta G^{\circ\prime} = -40\ \text{kJ/mol}$$

Glucose metabolized via glycolysis produces two molecules of pyruvate and thus two molecules of acetyl-CoA, which can enter the TCA cycle. Combining glycolysis and the TCA cycle gives the net reaction shown:

Glucose + 2 H$_2$O + 10 NAD⁺ + 2 [FAD] + 4 ADP + 4 P$_i$ $\longrightarrow$
$$6\ CO_2 + 10\ NADH + 10\ H^+ + 2\ [FADH_2] + 4\ ATP$$

All six carbons of glucose are liberated as CO_2, and a total of four molecules of ATP are formed thus far in substrate-level phosphorylations. The 12 reduced coenzymes produced up to this point can eventually produce as many as 34 molecules of ATP in the

▶ **A DEEPER LOOK**

Steric Preferences in NAD⁺-Dependent Dehydrogenases

The enzymes that require nicotinamide coenzymes are stereospecific and transfer hydride to either the pro-*R* or the pro-*S* positions selectively.

What accounts for this stereospecificity? It arises from the fact that the enzymes (and especially the active sites of enzymes) are inherently asymmetric structures. The nicotinamide coenzyme (and the substrate) fit the active site in only one way. Malate dehydroge-

nase, the citric acid cycle enzyme, transfers hydride to the H_R position of NADH, but glyceraldehyde-3-P dehydrogenase in the glycolytic pathway transfers hydride to the H_S position, as shown in the accompanying figure. Dehydrogenases are stereospecific with respect to the substrates as well. Note that alcohol dehydrogenase removes hydrogen from the pro-*R* position of ethanol and transfers it to the pro-*R* position of NADH.

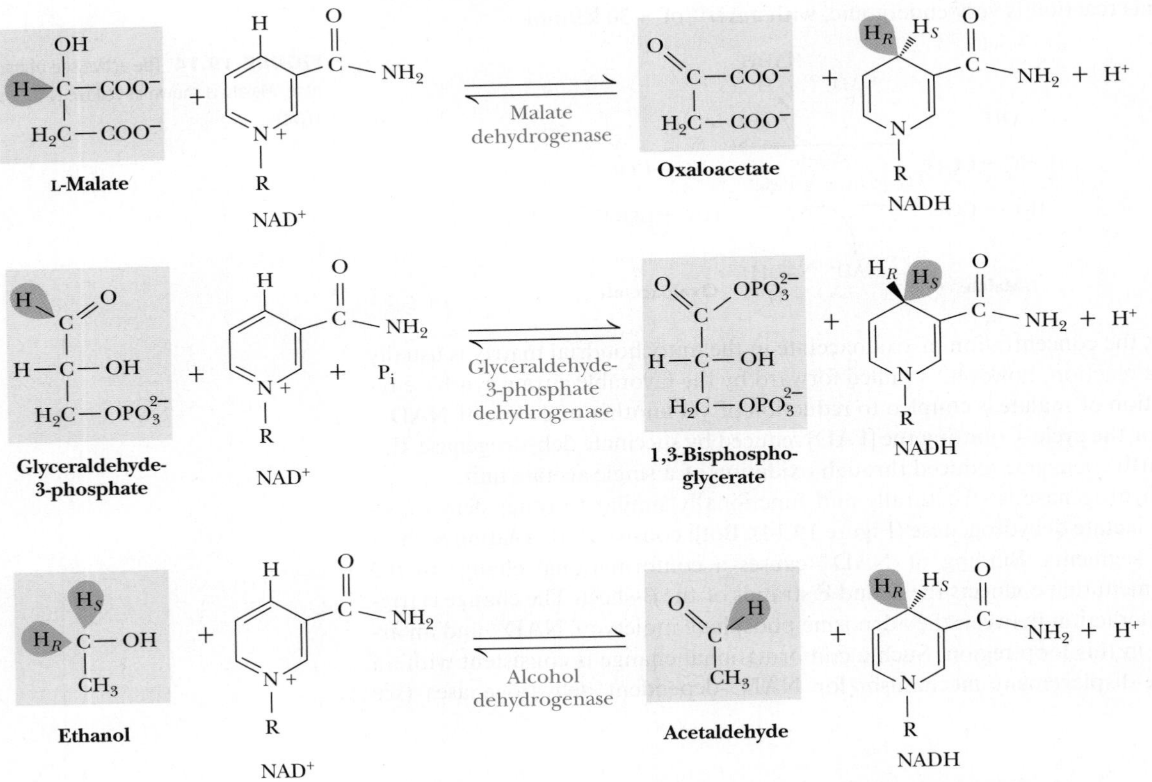

electron-transport and oxidative phosphorylation pathways. A stoichiometric relationship for these subsequent processes would be

$$NADH + H^+ + \tfrac{1}{2}O_2 + 3\ ADP + 3\ P_i \longrightarrow NAD^+ + 3\ ATP + 4\ H_2O$$
$$[FADH_2] + \tfrac{1}{2}O_2 + 2\ ADP + 2\ P_i \longrightarrow [FAD] + 2\ ATP + 3\ H_2O$$

Thus, a total of 3 ATP per NADH and 2 ATP per FADH₂ may be produced through the processes of electron transport and oxidative phosphorylation, but see Chapter 20 for a detailed discussion of this stoichiometry.

The Carbon Atoms of Acetyl-CoA Have Different Fates in the TCA Cycle

It is instructive to consider how the carbon atoms of a given acetate group are routed through several turns of the TCA cycle. As shown in Figure 19.15, neither of the carbon atoms of a labeled acetate unit is lost as CO_2 in the first turn of the cycle. The CO_2 evolved in any turn of the cycle derives from the carboxyl groups of the oxaloacetate acceptor (produced in the previous turn), not from incoming acetyl-CoA. On the other hand, succinate labeled on one end from the original labeled acetate forms two different

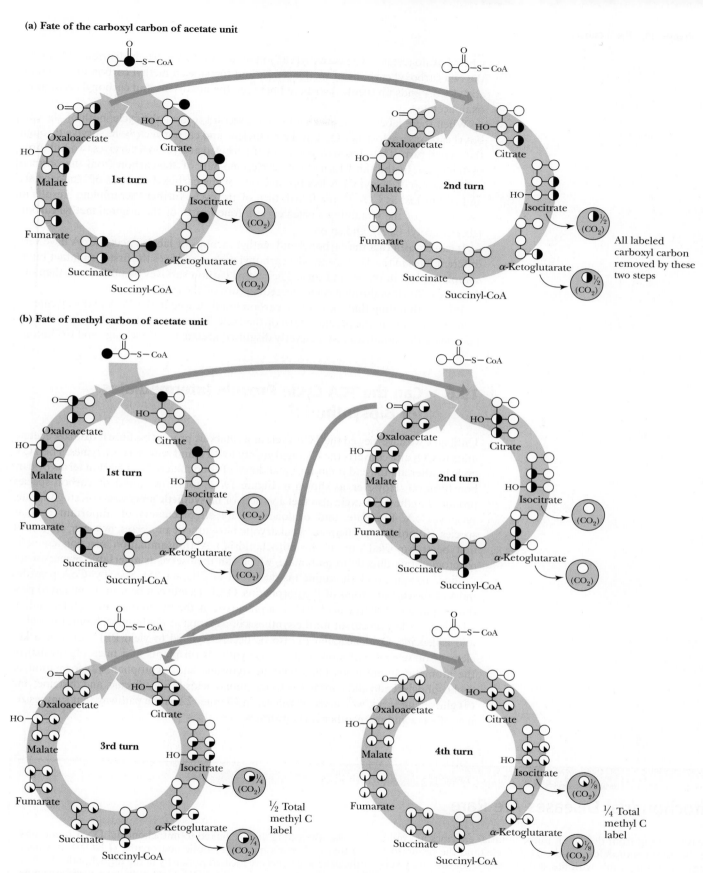

FIGURE 19.15 The fate of the carbon atoms of acetate in successive TCA cycles. Assume at the start, labeled acetate is added to cells containing unlabeled metabolites. **(a)** The carbonyl carbon of acetyl-CoA is fully retained through one turn of the cycle but is lost completely in a second turn of the cycle. **(b)** The methyl carbon of a labeled acetyl-CoA survives two full turns of the cycle but becomes equally distributed among the four carbons of oxaloacetate by the end of the second turn. In each subsequent turn of the cycle, one-half of this carbon (the original labeled methyl group) is lost.

labeled oxaloacetates. The carbonyl carbon of acetyl-CoA is evenly distributed between the two carboxyl carbons of oxaloacetate, and the labeled methyl carbon of incoming acetyl-CoA ends up evenly distributed between the methylene and carbonyl carbons of oxaloacetate.

When these labeled oxaloacetates enter a second turn of the cycle, both of the carboxyl carbons are lost as CO_2, but the methylene and carbonyl carbons survive through the second turn. Thus, the methyl carbon of a labeled acetyl-CoA survives two full turns of the cycle. In the third turn of the cycle, one-half of the carbon from the original methyl group of acetyl-CoA has become one of the carboxyl carbons of oxaloacetate and is thus lost as CO_2. In the fourth turn of the cycle, further "scrambling" results in loss of half of the remaining labeled carbon (one-fourth of the original methyl carbon label of acetyl-CoA), and so on.

It can be seen that the carbonyl and methyl carbons of labeled acetyl-CoA have very different fates in the TCA cycle. The carbonyl carbon survives the first turn intact but is completely lost in the second turn. The methyl carbon survives two full turns, then undergoes a 50% loss through each succeeding turn of the cycle.

It is worth noting that the carbon–carbon bond cleaved in the TCA pathway entered as an acetate unit in the previous turn of the cycle. Thus, the oxidative decarboxylations that cleave this bond are just a cleverly disguised acetate C—C cleavage and oxidation.

19.6 Can the TCA Cycle Provide Intermediates for Biosynthesis?

Until now we have viewed the TCA cycle as a catabolic process because it oxidizes acetate units to CO_2 and converts the liberated energy to ATP and reduced coenzymes. The TCA cycle is, after all, the end point for breakdown of food materials, at least in terms of carbon turnover. However, as shown in Figure 19.16, four-, five-, and six-carbon species produced in the TCA cycle also fuel a variety of **biosynthetic processes.** α-Ketoglutarate, succinyl-CoA, fumarate, and oxaloacetate are all precursors of important cellular species. (In order to participate in eukaryotic biosynthetic processes, however, they must first be transported out of the mitochondria.) A transamination reaction converts α-ketoglutarate directly to glutamate, which can then serve as a versatile precursor for proline, arginine, and glutamine (as described in Chapter 25). Succinyl-CoA provides most of the carbon atoms of the porphyrins. Oxaloacetate can be transaminated to produce aspartate. Aspartic acid itself is a precursor of the pyrimidine nucleotides and, in addition, is a key precursor for the synthesis of asparagine, methionine, lysine, threonine, and isoleucine. Oxaloacetate can also be decarboxylated to yield PEP, which is a key element of several pathways, namely (1) synthesis (in plants and microorganisms) of the aromatic amino acids phenylalanine, tyrosine, and tryptophan; (2) formation of 3-phosphoglycerate and conversion to the amino acids serine, glycine, and cysteine; and (3) gluconeogenesis, which, as we will see in Chapter 22, is the pathway that synthesizes new glucose and many other carbohydrates.

HUMAN BIOCHEMISTRY

Mitochondrial Diseases Are Rare

Diseases arising from defects in mitochondrial enzymes are quite rare, because major defects in the TCA cycle (and the respiratory chain) are incompatible with life and affected embryos rarely survive to birth. Even so, about 150 different hereditary mitochondrial diseases have been reported. Even though mitochondria carry their own DNA, many of the reported diseases map to the nuclear genome, because most of the mitochondrial proteins are imported from the cytosol.

An interesting disease linked to mitochondrial DNA mutations is that of Leber's hereditary optic neuropathy (LHON), in which the genetic defects are located primarily in the mitochondrial DNA coding for the subunits of NADH–CoQ reductase, also known as Complex I of the electron-transport chain (see Chapter 20). Leber's disease is the most common form of blindness in otherwise healthy young men and occurs less often in women.

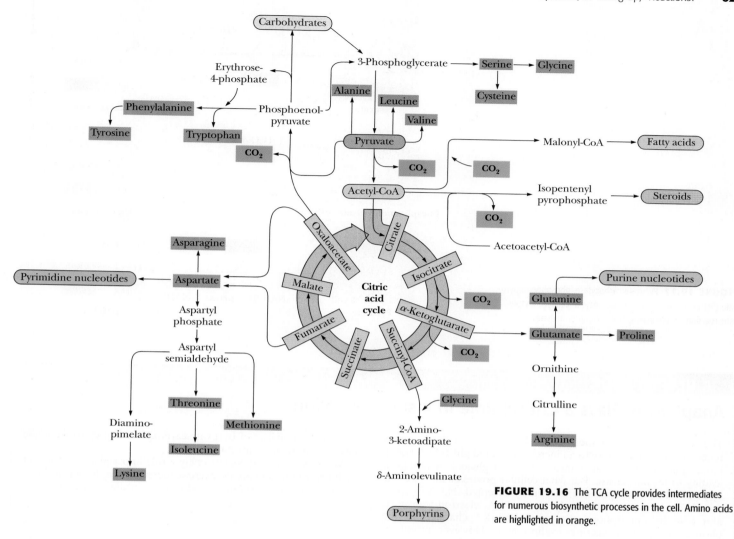

FIGURE 19.16 The TCA cycle provides intermediates for numerous biosynthetic processes in the cell. Amino acids are highlighted in orange.

Finally, citrate can be exported from the mitochondria and then broken down by **ATP–citrate lyase** to yield oxaloacetate and acetyl-CoA, a precursor of fatty acids. Oxaloacetate produced in this reaction is rapidly reduced to malate, which can then be processed in either of two ways: It may be transported into mitochondria, where it is reoxidized to oxaloacetate, or it may be oxidatively decarboxylated to pyruvate by **malic enzyme,** with subsequent mitochondrial uptake of pyruvate. This cycle permits citrate to provide acetyl-CoA for biosynthetic processes, with return of the malate and pyruvate byproducts to the mitochondria.

19.7 What Are the Anaplerotic, or "Filling Up," Reactions?

In a sort of reciprocal arrangement, the cell also feeds many intermediates back into the TCA cycle from other reactions. Because such reactions replenish the TCA cycle intermediates, Hans Kornberg proposed that they be called **anaplerotic reactions** (literally, the "filling up" reactions). Thus, **PEP carboxylase** and **pyruvate carboxylase** synthesize oxaloacetate from pyruvate (Figure 19.17).

Pyruvate carboxylase is the most important of the anaplerotic reactions. It exists in the mitochondria of animal cells but not in plants, and it provides a direct link between glycolysis and the TCA cycle. The enzyme is tetrameric and contains covalently bound

FIGURE 19.17 Pyruvate carboxylase, phosphoenolpyruvate (PEP) carboxylase, and malic enzyme catalyze anaplerotic reactions, replenishing TCA cycle intermediates.

> **A DEEPER LOOK**

Anaplerosis Plays a Critical Role in Insulin Secretion

The so-called β-cells of the islets of Langerhans in the pancreas release insulin in response to an increase of blood glucose. A rise in blood glucose induces an increase in β-cell glucose metabolism, leading to insulin release. But what cellular processes mediate this response? For many years, it was widely accepted that ATP (produced from the processes of glycolysis, TCA, electron transport, and oxidative phosphorylation) activated K^+ channels in the plasma membrane, triggering insulin release. However, recent research has shown that anaplerotic enzymes feed alternative pathways that produce cytosolic signal molecules that also support insulin secretion.

Pancreatic β-cells maintain high levels of the anaplerotic enzyme pyruvate carboxylase. In these cells, only about half of the pyruvate from glycolysis enters the TCA cycle via pyruvate dehydrogenase, leading to oxidative processes that generate ATP. The rest of the pyruvate is converted by pyruvate carboxylase to oxaloacetate, much of which is converted to malate and exported to the cytosol, as part of a **pyruvate/malate cycle** (see figure). NADPH and other metabolites generated from this cycle act as intracellular messengers that appear to be as significant as ATP in provoking insulin secretion.

In support of this new model for insulin secretion, Jamie Joseph and his colleagues have shown that β-cells that lack hypoxia-inducible factor-1 beta (HIF-1β) display reduced anaplerotic activity, reduced activities of pyruvate carboxylase and cytosolic malic enzyme, and reduced insulin secretion.

Anaplerosis also appears to play a role in peripheral tissues. In insulin-resistant individuals, exercise increases anaplerotic activity, increases fatty acid oxidation, and restores insulin sensitivity.

References

Jensen, M., Joseph, J. W., Ronnebaum, S. M., Burgess, S. C., Sherry, A. D., and Newgard, C. B., 2008. Metabolic cycling in control of glucose-stimulated insulin secretion. *American Journal of Physiology and Endocrine Metabolism* **295**:E1287–E1297.

Pillai, R., Huypens, P., Huang, M., Schaefer, S., Sheinin, T., Wettig, S. D., and Joseph, J. W., 2011. Aryl hydrocarbon receptor nuclear translocator/hypoxia-inducible factor-1β plays a critical role in maintaining glucose-stimulated anaplerosis and insulin release from pancreatic β-cells. *Journal of Biological Chemistry* **286**:1014–1024.

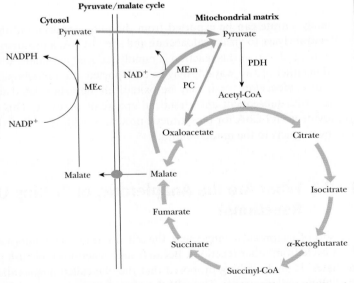

The pyruvate/malate cycle in pancreatic β-cells. Oxaloacetate produced by high levels of pyruvate carboxylase is converted to malate, then exported to the cytosol. Cytosolic malic enzyme (MEc) converts malate to pyruvate, which is returned to the mitochondrial matrix. NADPH and other cytosolic signals produced by this and related cycles provoke insulin secretion.

> ► **A DEEPER LOOK**

Fool's Gold and the Reductive Citric Acid Cycle—The First Metabolic Pathway?

How did life arise on the planet Earth? It was once supposed that a reducing atmosphere, together with random synthesis of organic compounds, gave rise to a prebiotic "soup" in which the first living things appeared. However, certain key compounds, such as arginine, lysine, and histidine; the straight-chain fatty acids; porphyrins; and essential coenzymes, have not been convincingly synthesized under simulated prebiotic conditions. This and other problems have led researchers to consider other models for the evolution of life.

One of these alternative models, postulated by Günter Wächtershäuser, involves an archaic version of the TCA cycle running in the reverse (reductive) direction. Reversal of the TCA cycle results in assimilation of CO_2 and fixation of carbon as shown. For each turn of the reversed cycle, two carbons are fixed in the formation of isocitrate and two more are fixed in the reductive transformation of acetyl-CoA to oxaloacetate. Thus, for every succinate that enters the reversed cycle, two succinates are returned, making the cycle highly autocatalytic. Because TCA cycle intermediates are involved in many biosynthetic pathways (see Section 19.6), a reversed TCA cycle would be a bountiful and broad source of metabolic substrates.

A reversed, reductive TCA cycle would require energy input to drive it. What might have been the thermodynamic driving force for such a cycle? Wächtershäuser hypothesizes that the anaerobic reaction of FeS and H_2S to form insoluble FeS_2 (pyrite, also known as fool's gold) in the prebiotic milieu could have been the driving reaction:

$$FeS + H_2S \longrightarrow FeS_2 \text{ (pyrite)} \downarrow + H_2$$

This reaction is highly exergonic, with a standard-state free energy change ($\Delta G°'$) of -38 kJ/mol. Under the conditions that might have existed in a prebiotic world, this reaction would have been sufficiently exergonic to drive the reductive steps of a reversed TCA cycle.

Wächtershäuser has also suggested that early metabolic processes first occurred on the surface of pyrite and other related mineral materials. The iron–sulfur chemistry that prevailed on these mineral surfaces may have influenced the evolution of the iron–sulfur proteins that control and catalyze many reactions in modern pathways (including the succinate dehydrogenase and aconitase reactions of the TCA cycle). This reductive citric acid cycle has been shown to occur in certain extant archaea and bacteria, where it serves all their carbon needs.

A reductive, reversed TCA cycle.

biotin and an Mg^{2+} site on each subunit. (It is examined in greater detail in our discussion of gluconeogenesis in Chapter 22.) Pyruvate carboxylase has an absolute allosteric requirement for acetyl-CoA. Thus, when acetyl-CoA levels exceed the oxaloacetate supply, allosteric activation of pyruvate carboxylase by acetyl-CoA raises oxaloacetate levels, so the excess acetyl-CoA can enter the TCA cycle.

PEP carboxylase occurs in yeast, bacteria, and higher plants, but not in animals. The enzyme is specifically inhibited by aspartate, which is produced by transamination of oxaloacetate. Thus, organisms utilizing this enzyme control aspartate production by regulation of PEP carboxylase. Malic enzyme is found in the cytosol or mitochondria of many animal and plant cells and is an NADPH-dependent enzyme.

It is worth noting that the reaction catalyzed by **PEP carboxykinase** could also function as an anaplerotic reaction, were it not for the particular properties of the enzyme.

CO_2 binds weakly to PEP carboxykinase, whereas oxaloacetate binds very tightly ($K_D = 2 \times 10^{-6}$ M), and, as a result, the enzyme favors formation of PEP from oxaloacetate.

The catabolism of amino acids provides pyruvate, acetyl-CoA, oxaloacetate, fumarate, α-ketoglutarate, and succinate, all of which may be oxidized by the TCA cycle. In this way, proteins may serve as excellent sources of nutrient energy, as seen in Chapter 25.

19.8 | How Is the TCA Cycle Regulated?

Situated as it is between glycolysis and the electron-transport chain, the TCA cycle must be carefully controlled. If the cycle were permitted to run unchecked, large amounts of metabolic energy could be wasted in overproduction of reduced coenzymes and ATP; conversely, if it ran too slowly, ATP would not be produced rapidly enough to satisfy the needs of the cell. Also, as just seen, the TCA cycle is an important source of precursors for biosynthetic processes and must be able to provide them as needed.

What are the sites of regulation in the TCA cycle? Based on our experience with glycolysis (see Figure 18.22), we might anticipate that some of the reactions of the TCA cycle would operate near equilibrium under cellular conditions (with $\Delta G < 0$), whereas others—the sites of regulation—would be characterized by large negative ΔG values. Estimates for the values of ΔG in mitochondria, based on mitochondrial concentrations of metabolites, are summarized in Table 19.1. Three reactions of the cycle—citrate synthase, isocitrate dehydrogenase, and α-ketoglutarate dehydrogenase—operate with large negative ΔG values under mitochondrial conditions and are thus the primary sites of regulation in the cycle.

The regulatory actions that control the TCA cycle are shown in Figure 19.18. As one might expect, the principal regulatory "signals" are the concentrations of acetyl-CoA, ATP, NAD$^+$, and NADH, with additional effects provided by several other metabolites. The main sites of regulation are pyruvate dehydrogenase, citrate synthase, isocitrate

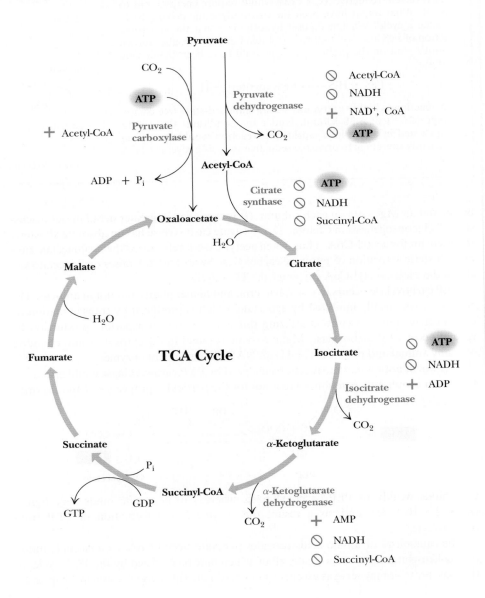

FIGURE 19.18 Regulation of the TCA cycle.

dehydrogenase, and α-ketoglutarate dehydrogenase. All of these enzymes are inhibited by NADH, so when the cell has produced all the NADH that can conveniently be turned into ATP, the cycle shuts down. For similar reasons, ATP is an inhibitor of pyruvate dehydrogenase and isocitrate dehydrogenase. The TCA cycle is turned on, however, when either the ADP/ATP or NAD$^+$/NADH ratio is high, an indication that the cell has run low on ATP or NADH. Regulation of the TCA cycle by NADH, NAD$^+$, ATP, and ADP thus reflects the energy status of the cell. On the other hand, succinyl-CoA is an intracycle regulator, inhibiting citrate synthase and α-ketoglutarate dehydrogenase. Acetyl-CoA acts as a signal to the TCA cycle that glycolysis or fatty acid breakdown is producing two-carbon units. Acetyl-CoA activates pyruvate carboxylase, the anaplerotic reaction that provides oxaloacetate, the acceptor for increased flux of acetyl-CoA into the TCA cycle.

Pyruvate Dehydrogenase Is Regulated by Phosphorylation/Dephosphorylation

As we shall see in Chapter 22, most organisms can synthesize sugars such as glucose from pyruvate. However, animals cannot synthesize glucose from acetyl-CoA. For this reason, the pyruvate dehydrogenase complex, which converts pyruvate to acetyl-CoA, plays a pivotal role in metabolism. Conversion to acetyl-CoA commits nutrient carbon atoms either to oxidation in the TCA cycle or to fatty acid synthesis (see Chapter 24). Because this choice is so crucial to the organism, pyruvate dehydrogenase is a carefully regulated enzyme. It is subject to product inhibition and is further regulated by nucleotides. Finally, activity of pyruvate dehydrogenase is regulated by phosphorylation and dephosphorylation of the enzyme complex itself.

High levels of either product, acetyl-CoA or NADH, allosterically inhibit the pyruvate dehydrogenase complex. Acetyl-CoA specifically blocks dihydrolipoyl transacetylase, and NADH acts on dihydrolipoyl dehydrogenase. The mammalian pyruvate dehydrogenase is also regulated by covalent modifications. As shown in Figure 19.19, a Mg^{2+}-dependent **pyruvate dehydrogenase kinase** is associated with the enzyme in mammals. This kinase is allosterically activated by NADH and acetyl-CoA, and when levels of these metabolites rise in the mitochondrion, they stimulate phosphorylation of a serine residue on the pyruvate dehydrogenase α-subunit, blocking the first step of the pyruvate dehydrogenase reaction, the decarboxylation of pyruvate. Inhibition of the dehydrogenase in this manner eventually lowers the levels of NADH and acetyl-CoA in the

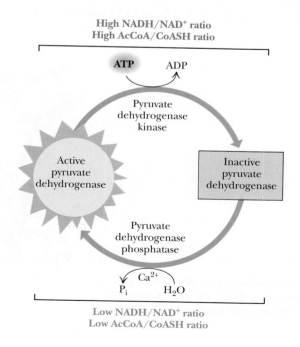

FIGURE 19.19 Regulation of the pyruvate dehydrogenase reaction.

matrix of the mitochondrion. Reactivation of the enzyme is carried out by **pyruvate dehydrogenase phosphatase,** a Ca^{2+}-activated enzyme that binds to the dehydrogenase complex and hydrolyzes the phosphoserine moiety on the dehydrogenase subunit. At low ratios of NADH to NAD^+ and low acetyl-CoA levels, the phosphatase maintains the dehydrogenase in an activated state, but a high level of acetyl-CoA or NADH once again activates the kinase and leads to the inhibition of the dehydrogenase. Insulin and Ca^{2+} ions activate dephosphorylation, and pyruvate inhibits the phosphorylation reaction.

Pyruvate dehydrogenase is also sensitive to the energy status of the cell. AMP activates pyruvate dehydrogenase, whereas GTP inhibits it. High levels of AMP are a sign that the cell may be energy-poor. Activation of pyruvate dehydrogenase under such conditions commits pyruvate to energy production.

Isocitrate Dehydrogenase Is Strongly Regulated

The mechanism of regulation of isocitrate dehydrogenase is in some respects the reverse of pyruvate dehydrogenase. The mammalian isocitrate dehydrogenase is subject only to allosteric activation by ADP and NAD^+ and to inhibition by ATP and NADH. Thus, high NAD^+/NADH and ADP/ATP ratios stimulate isocitrate dehydrogenase and TCA cycle activity.

It may seem surprising that isocitrate dehydrogenase is strongly regulated, because it is not an apparent branch point within the TCA cycle. However, the citrate/isocitrate ratio controls the rate of production of cytosolic acetyl-CoA, because acetyl-CoA in the cytosol is derived from citrate exported from the mitochondrion. (Breakdown of cytosolic citrate produces oxaloacetate and acetyl-CoA, which can be used in a variety of biosynthetic processes.) Thus, isocitrate dehydrogenase activity in the mitochondrion favors catabolic TCA cycle activity over anabolic utilization of acetyl-CoA in the cytosol.

Regulation of TCA Cycle Enzymes by Acetylation

Lysine acetylation is a dynamic, reversible, and regulatory post-translational modification that enables cells to sense energy status and respond to environmental changes and stresses (Chapter 15). All of the TCA cycle enzymes are acetylated in organisms from bacteria to plants to mammals, and acetylation inhibits TCA enzyme activity in the cases that have been studied, particularly isocitrate dehydrogenase, succinate dehydrogenase, and malate dehydrogenase. SIRT3 is a mitochondrial NAD^+-dependent deacetylase of the sirtuin family (Chapter 15) that can deacetylate and activate all these TCA enzymes (Figure 19.20 and Figure 27.13). SIRT3 expression is a function of exercise

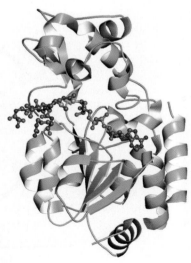

FIGURE 19.20 Structure of human SIRT3 with a bound peptide substrate analog (shown in ball-and-stick). The small domain of SIRT3 is at the top; the large domain is at the bottom (pdb id = 3GLT).

status in humans. Adults who exercise vigorously for an hour six days a week express SIRT3 at levels nearly three times those of sedentary adults. Elevated levels of SIRT3 presumably activate TCA cycle enzymes, and studies have shown that increased levels of SIRT3 are correlated with extended lifespans in humans.

Two Covalent Modifications Regulate *E. coli* Isocitrate Dehydrogenase

Interestingly, the *Escherichia coli* isocitrate dehydrogenase is regulated by two different covalent modifications. Serine residues on each subunit of the dimeric enzyme are phosphorylated by the protein kinase activity of **AceK,** a bifunctional kinase/phosphatase, causing inhibition of the isocitrate dehydrogenase activity. Activity is restored by the action of the phosphatase activity of AceK (Figure 19.20). When TCA cycle and glycolytic intermediates—such as isocitrate, 3-phosphoglycerate, pyruvate, PEP, and oxaloacetate—are high, the kinase is inhibited, the phosphatase is activated, and the TCA cycle operates normally. When levels of these intermediates fall, the kinase is activated, isocitrate dehydrogenase is inhibited, and isocitrate is diverted to the glyoxylate pathway, as explained in the next section. Interestingly, the kinase and phosphatase activities of AceK depend on acetylation of AceK by lysine acetyltransferases and removal of these acetyl groups by deacetylases. Acetylated AceK displays a reduced ability to inactivate isocitrate dehydrogenase (by phosphorylation) but at the same time an increased ability to activate isocitrate dehydrogenase (via dephosphoryation by the phosphatase).

19.9 Can Any Organisms Use Acetate as Their Sole Carbon Source?

Plants (particularly seedlings, which cannot yet accomplish efficient photosynthesis), as well as some bacteria and algae, can use acetate as the only source of carbon for all the carbon compounds they produce. Although we saw that the TCA cycle can supply intermediates for some biosynthetic processes, the cycle gives off 2 CO_2 for every two-carbon acetate group that enters and cannot effect the net synthesis of TCA cycle intermediates. Thus, it would not be possible for the cycle to produce the massive amounts of biosynthetic intermediates needed for acetate-based growth unless alternative reactions were available. In essence, the TCA cycle is geared primarily to energy production, and it "wastes" carbon units by giving off CO_2. Modification of the cycle to support acetate-based growth would require eliminating the CO_2-producing reactions and enhancing the net production of four-carbon units (that is, oxaloacetate). Plants and bacteria employ a modification of the TCA cycle called the **glyoxylate cycle** to produce four-carbon dicarboxylic acids (and eventually even sugars) from two-carbon acetate units. The glyoxylate cycle bypasses the two oxidative decarboxylations of the TCA cycle and instead routes isocitrate through the **isocitrate lyase** and **malate synthase** reactions (Figure 19.21). Glyoxylate produced by isocitrate lyase reacts with a second molecule of acetyl-CoA to form L-malate. The net effect is to conserve carbon units, using two acetyl-CoA molecules per cycle to generate oxaloacetate. Some of this is converted to PEP and then to glucose by pathways discussed in Chapter 22.

The Glyoxylate Cycle Operates in Specialized Organelles

The enzymes of the glyoxylate cycle in plants are contained in **glyoxysomes,** organelles devoted to this cycle. Yeast and algae carry out the glyoxylate cycle in the cytoplasm. The enzymes common to both the TCA and glyoxylate pathways exist as isozymes, with spatially and functionally distinct enzymes operating independently in the two cycles.

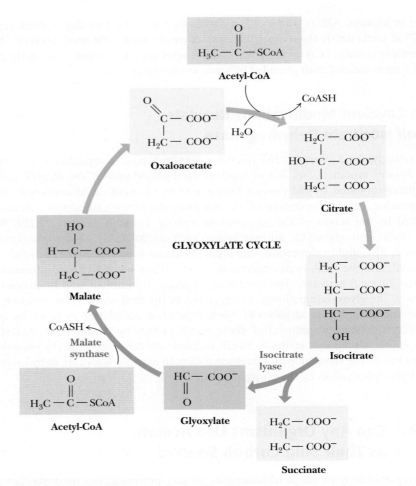

FIGURE 19.21 The glyoxylate cycle. The first two steps are identical to TCA cycle reactions. The third step bypasses the CO_2-evolving steps of the TCA cycle to produce succinate and glyoxylate. The malate synthase reaction forms malate from glyoxylate and another acetyl-CoA. The result is that one turn of the cycle consumes one oxaloacetate and two acetyl-CoA molecules but produces two molecules of oxaloacetate. (Succinate produced in the isocitrate lyase reaction is converted to oxaloacetate by TCA cycle reactions.) The net for this cycle is one oxaloacetate from two acetyl-CoA molecules.

Isocitrate Lyase Short-Circuits the TCA Cycle by Producing Glyoxylate and Succinate

The **isocitrate lyase** reaction (Figure 19.22) produces succinate, a four-carbon product of the cycle, as well as glyoxylate, which can then combine with a second molecule of acetyl-CoA. Isocitrate lyase catalyzes an aldol cleavage and is similar to the reaction mediated by aldolase in glycolysis. The **malate synthase** reaction (see Figure 19.21), a Claisen condensation of acetyl-CoA with the aldehyde of glyoxylate to yield malate, is quite similar to the citrate synthase reaction. Compared with the TCA cycle, the

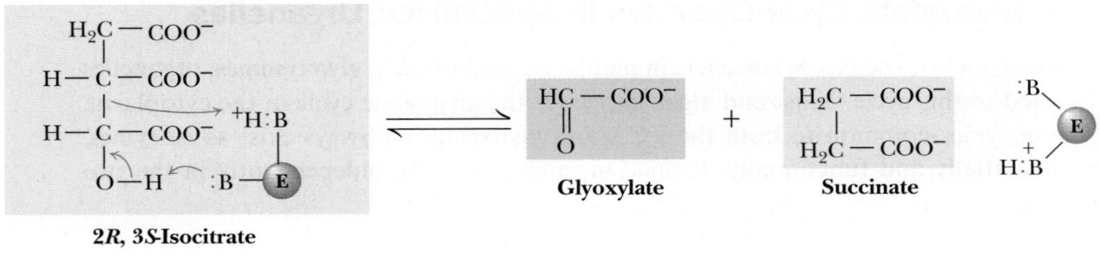

FIGURE 19.22 The isocitrate lyase reaction.

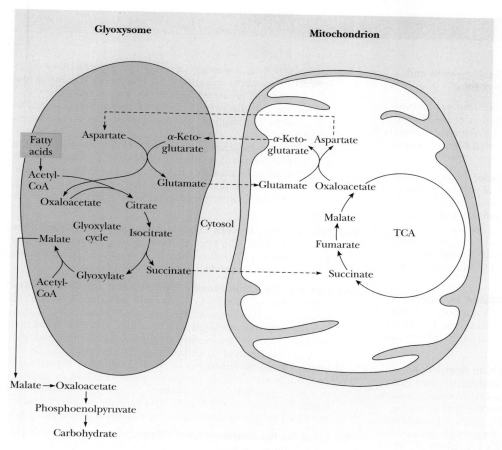

FIGURE 19.23 Glyoxysomes lack three of the enzymes needed to run the glyoxylate cycle. Succinate dehydrogenase, fumarase, and malate dehydrogenase are all "borrowed" from the mitochondria in a shuttle in which succinate and glutamate are passed to the mitochondria and α-ketoglutarate and aspartate are passed to the glyoxysome.

glyoxylate cycle (1) contains only five steps (as opposed to eight), (2) lacks the CO_2-liberating reactions, (3) consumes two molecules of acetyl-CoA per cycle, and (4) produces four-carbon units (oxaloacetate) as opposed to one-carbon units.

The Glyoxylate Cycle Helps Plants Grow in the Dark

The existence of the glyoxylate cycle explains how certain seeds grow underground (or in the dark), where photosynthesis is impossible. Many seeds (peanuts, soybeans, and castor beans, for example) are rich in lipids, and as we will see in Chapter 23, most organisms degrade the fatty acids of lipids to acetyl-CoA. Glyoxysomes form in seeds as germination begins, and the glyoxylate cycle uses the acetyl-CoA produced in fatty acid oxidation to provide large amounts of oxaloacetate and other intermediates for carbohydrate synthesis. Once the growing plant begins photosynthesis and can fix CO_2 to produce carbohydrates (see Chapter 21), the glyoxysomes disappear.

Glyoxysomes Must Borrow Three Reactions from Mitochondria

Glyoxysomes do not contain all the enzymes needed to run the glyoxylate cycle: Succinate dehydrogenase, fumarase, and malate dehydrogenase are absent. Consequently, glyoxysomes must cooperate with mitochondria to run their cycle (Figure 19.23). Succinate travels from the glyoxysomes to the mitochondria, where it is converted to oxaloacetate. Transamination to aspartate follows because oxaloacetate cannot be transported out of the mitochondria. Aspartate formed in this way then moves from the mitochondria back to the glyoxysomes, where a reverse transamination with α-ketoglutarate forms oxaloacetate, completing the shuttle. Finally, to balance the transaminations, glutamate shuttles from glyoxysomes to mitochondria.

OWL Login to OWL to develop problem-solving skills and complete online homework assigned by your professor.

The glycolytic pathway converts glucose to pyruvate and produces two molecules of ATP per glucose—only a small fraction of the potential energy available from glucose. In the presence of oxygen, pyruvate is oxidized to CO_2, releasing the rest of the energy available from glucose via the TCA cycle and oxidative phosphorylation.

19.1 What Is the Chemical Logic of the TCA Cycle? The new carbon units enter the cycle as acetyl-CoA. Transfer of the two-carbon acetyl group from acetyl-CoA to the four-carbon oxaloacetate to yield six-carbon citrate is catalyzed by citrate synthase. A dehydration–rehydration rearrangement of citrate yields isocitrate. Two successive decarboxylations produce α-ketoglutarate and then succinyl-CoA, a CoA conjugate of a four-carbon unit. Several steps later, oxaloacetate is regenerated and can combine with another two-carbon unit of acetyl-CoA.

19.2 How Is Pyruvate Oxidatively Decarboxylated to Acetyl-CoA? The pyruvate dehydrogenase complex (PDC) is a noncovalent assembly of three different enzymes and five coenzymes operating in concert to catalyze successive steps in the conversion of pyruvate to acetyl-CoA.

19.3 How Are Two CO_2 Molecules Produced from Acetyl-CoA? Citrate synthase combines acetyl-CoA with oxaloacetate in a Perkin condensation (a carbon–carbon condensation between a ketone or aldehyde and an ester). A general base on the enzyme accepts a proton from the methyl group of acetyl-CoA, producing a stabilized α-carbanion of acetyl-CoA. This strong nucleophile attacks the α-carbonyl of oxaloacetate, yielding citryl-CoA, which is hydrolyzed to citrate and CoASH.

Citrate is isomerized to isocitrate by aconitase in a two-step process involving aconitate as an intermediate. The elements of water are first abstracted from citrate to yield aconitate, which is then rehydrated with H— and HO— adding back in opposite positions to produce isocitrate. The net effect is the conversion of a tertiary alcohol (citrate) to a secondary alcohol (isocitrate).

The two-step isocitrate dehydrogenase reaction involves (1) oxidation of the C-2 alcohol of isocitrate to form oxalosuccinate, followed by (2) a β-decarboxylation reaction that expels the central carboxyl group as CO_2, leaving the product α-ketoglutarate. Oxalosuccinate, the β-keto acid produced by the initial dehydrogenation reaction, is unstable and thus is readily decarboxylated.

α-Ketoglutarate dehydrogenase is a multienzyme complex—consisting of α-ketoglutarate dehydrogenase, dihydrolipoyl transsuccinylase, and dihydrolipoyl dehydrogenase—that employs five different coenzymes. The dihydrolipoyl dehydrogenase in this reaction is identical to that in the pyruvate dehydrogenase reaction. The mechanism is an oxidative decarboxylation analogous to that of pyruvate dehydrogenase. Succinyl-CoA is the product.

19.4 How Is Oxaloacetate Regenerated to Complete the Cycle? Succinyl-CoA synthetase catalyzes a substrate-level phosphorylation: Succinyl-CoA is a high-energy intermediate and is used to drive the phosphorylation of GDP to GTP (in mammals) or ADP to ATP (in plants and bacteria).

Succinate dehydrogenase (succinate–coenzyme Q reductase of the electron-transport chain) catalyzes removal of H atoms across a C—C bond and produces the *trans*-unsaturated fumarate.

Fumarate is hydrated in a stereospecific reaction by fumarase to give L-malate. The reaction involves *trans*-addition of the elements of water across the double bond.

Malate dehydrogenase completes the TCA cycle. This reaction is very endergonic, with a $\Delta G°'$ of $+30$ kJ/mol. Consequently, the concentration of oxaloacetate in the mitochondrial matrix is usually quite low. Oxidation of malate to oxaloacetate is coupled to reduction of yet another molecule of NAD^+, the third one of the cycle.

19.5 What Are the Energetic Consequences of the TCA Cycle? The cycle is exergonic, with a net $\Delta G°'$ for one pass around the cycle of approximately -40 kJ/mol. Three NADH, one $[FADH_2]$, and one ATP equivalent are produced in each turn of the cycle.

19.6 Can the TCA Cycle Provide Intermediates for Biosynthesis? α-Ketoglutarate, succinyl-CoA, fumarate, and oxaloacetate are all precursors of important cellular species. A transamination reaction converts α-ketoglutarate directly to glutamate, which can then serve as a precursor for proline, arginine, and glutamine. Succinyl-CoA provides most of the carbon atoms of the porphyrins. Oxaloacetate can be transaminated to produce aspartate. Aspartic acid itself is a precursor of the pyrimidine nucleotides and, in addition, is a key precursor for the synthesis of asparagine, methionine, lysine, threonine, and isoleucine. Oxaloacetate can also be decarboxylated to yield PEP, which is a key element of several pathways.

19.7 What Are the Anaplerotic, or "Filling Up," Reactions? Anaplerotic reactions replenish the TCA cycle intermediates. Examples include PEP carboxylase and pyruvate carboxylase, both of which synthesize oxaloacetate from pyruvate.

19.8 How Is the TCA Cycle Regulated? The main sites of regulation are pyruvate dehydrogenase, citrate synthase, isocitrate dehydrogenase, and α-ketoglutarate dehydrogenase. All of these enzymes are inhibited by NADH. ATP is an inhibitor of pyruvate dehydrogenase and isocitrate dehydrogenase. The TCA cycle is turned on, however, when either the ADP/ATP or NAD^+/NADH ratio is high. Regulation of the TCA cycle by NADH, NAD^+, ATP, and ADP thus reflects the energy status of the cell. Succinyl-CoA is an intracycle regulator, inhibiting citrate synthase and α-ketoglutarate dehydrogenase. Acetyl-CoA activates pyruvate carboxylase, the anaplerotic reaction that provides oxaloacetate, the acceptor for acetyl-CoA entry into the TCA cycle.

19.9 Can Any Organisms Use Acetate as Their Sole Carbon Source? Plants and bacteria employ a modification of the TCA cycle called the glyoxylate cycle to produce four-carbon dicarboxylic acids (and eventually even sugars) from two-carbon acetate units. The glyoxylate cycle bypasses the two oxidative decarboxylations of the TCA cycle and instead routes isocitrate through the isocitrate lyase and malate synthase reactions. Glyoxylate produced by isocitrate lyase reacts with a second molecule of acetyl-CoA to form L-malate. The net effect is to conserve carbon units, using two acetyl-CoA molecules per cycle to generate oxaloacetate.

FOUNDATIONAL BIOCHEMISTRY Things You Should Know After Reading Chapter 19.

- The essential features and reactions of the tricarboxylic acid cycle.
- How the TCA cycle is linked to electron transport and oxidative phosphorylation.
- The chemical logic of the TCA cycle as a way to cleave and oxidize a two-carbon unit.
- The chemical steps and reactions of the tricarboxylic acid cycle.
- The essential features, structure, and mechanism of the pyruvate dehydrogenase reaction.
- The identities and functional characteristics of the coenzymes of pyruvate dehydrogenase.
- The energetics of the reactions of the tricarboxylic acid cycle.
- The reaction, structure, mechanism, and regulation of citrate synthase.
- The involvement of iron–sulfur clusters in aconitase.
- The reaction and mechanism of aconitase.
- The means by which fluoracetate blocks the TCA cycle.
- The reaction, mechanism, and regulation of isocitrate dehydrogenase.
- The reaction, mechanism, and regulation of α–ketoglutarate dehydrogenase.

- The reaction and mechanism of the succinyl-CoA synthetase reaction.
- The role of substrate-level phosphorylation in succinyl-CoA synthetase.
- The difference between a synthase and a synthetase.
- The reaction and mechanism of fumarase.
- The reaction and mechanism of malate dehydrogenase.
- The energetic consequences of the TCA cycle.
- The fates of the carbon atoms of acetyl-CoA in the TCA cycle.
- The role of TCA in providing intermediates for biosynthesis.
- The identity and nature of the anaplerotic reactions.
- The regulation of the pyruvate dehydrogenase reaction by phosphorylation.
- The regulation of isocitrate dehydrogenase by phosphorylation and acetylation.
- The importance of the glyoxylate cycle in plants and bacteria.
- The reactions of the glyoxylate cycle and their mechanisms.
- The compartmentation of the reactions of the glyoxylate cycle.

PROBLEMS

Answers to all problems are at the end of this book. Detailed solutions are available in the *Student Solutions Manual, Study Guide, and Problems Book.*

1. **Radiolabeling with ¹⁴C-Glutamate** Describe the labeling pattern that would result from the introduction into the TCA cycle of glutamate labeled at C_γ with ^{14}C.

2. **Assessing the Effects of Substrate Concentrations on the TCA Cycle** Describe the effect on the TCA cycle of (a) increasing the concentration of NAD^+, (b) reducing the concentration of ATP, and (c) increasing the concentration of isocitrate.

3. **Assessing the Effect of Active-Site Phosphorylation on Enzyme Activity** (Integrates with Chapter 15.) The serine residue of isocitrate dehydrogenase that is phosphorylated by protein kinase lies within the active site of the enzyme. This situation contrasts with most other examples of covalent modification by protein phosphorylation, where the phosphorylation occurs at a site remote from the active site. What direct effect do you think such active-site phosphorylation might have on the catalytic activity of isocitrate dehydrogenase? (See Barford, D., 1991. Molecular mechanisms for the control of enzymic activity by protein phosphorylation. *Biochimica et Biophysica Acta* **1133**:55–62.)

4. **Understanding the Mechanism of the α-Ketoglutarate Dehydrogenase Reaction** The first step of the α-ketoglutarate dehydrogenase reaction involves decarboxylation of the substrate and leaves a covalent TPP intermediate. Write a reasonable mechanism for this reaction.

5. **Understanding the Action of Fluoroacetate on the TCA Cycle** In a tissue where the TCA cycle has been inhibited by fluoroacetate, what difference in the concentration of each TCA cycle metabolite would you expect, compared with a normal, uninhibited tissue?

6. **Designing an Assay for Succinate Dehydrogenase** On the basis of the descriptions of the physical properties of FAD and $FADH_2$, suggest a method for the measurement of the enzyme activity of succinate dehydrogenase.

7. **Understanding the Oxidative Processes in the TCA Cycle** Starting with citrate, isocitrate, α-ketoglutarate, and succinate, state which of the individual carbons of the molecule undergo oxidation in the next step of the TCA cycle. Which molecules undergo a net oxidation?

8. **Designing Substrate Analog Inhibitors for the TCA Cycle** In addition to fluoroacetate, consider whether other analogs of TCA cycle metabolites or intermediates might be introduced to inhibit other, specific reactions of the cycle. Explain your reasoning.

9. **Understanding the Mechanism of the Pyruvate Dehydrogenase Reaction** Based on the action of thiamine pyrophosphate in catalysis of the pyruvate dehydrogenase reaction, suggest a suitable chemical mechanism for the pyruvate decarboxylase reaction in yeast:

$$\text{Pyruvate} \longrightarrow \text{acetaldehyde} + CO_2$$

10. **Understanding the Energetics of the Aconitase Reaction** (Integrates with Chapter 3.) Aconitase catalyzes the citric acid cycle reaction:

$$\text{Citrate} \rightleftharpoons \text{isocitrate}$$

The standard free energy change, $\Delta G^{\circ\prime}$, for this reaction is +6.7 kJ/mol. However, the observed free energy change (ΔG) for this reaction in pig heart mitochondria is +0.8 kJ/mol. What is the ratio of [isocitrate]/[citrate] in these mitochondria? If [isocitrate] = 0.03 mM, what is [citrate]?

11. **Radiolabeling with ¹⁴CO₂ in the Pyruvate Carboxylase** Describe the labeling pattern that would result if $^{14}CO_2$ were incorporated into the TCA cycle via the pyruvate carboxylase reaction.

12. **Consequences of Radiolabeling With ¹⁴CO₂ in the Reductive TCA Cycle** Describe the labeling pattern that would result if the reductive, reversed TCA cycle (see A Deeper Look on page 631) operated with $^{14}CO_2$.

13. **Consequences of Radiolabeling with ¹⁴CH₃-Acetyl-CoA in the Glyoxylate Cycle** Describe the labeling pattern that would result in the glyoxylate cycle if a plant were fed acetyl-CoA labeled at the —CH_3 carbon.

14. Understanding the Mechanism of the Malate Synthase Reaction The malate synthase reaction, which produces malate from acetyl-CoA and glyoxylate in the glyoxylate pathway, involves chemistry similar to the citrate synthase reaction. Write a mechanism for the malate synthase reaction, and explain the role of CoA in this reaction.

15. Understanding How Cells Provide Acetate Units for Fatty Acid Synthesis in the Cytosol In most cells, fatty acids are synthesized from acetate units in the cytosol. However, the primary source of acetate units is the TCA cycle in mitochondria, and acetate cannot be transported directly from the mitochondria to the cytosol. Cells solve this problem by exporting citrate from the mitochondria and then converting citrate to acetate and oxaloacetate. Then, because cells cannot transport oxaloacetate into mitochondria directly, they must convert it to malate or pyruvate, both of which can be taken up by mitochondria. Draw a complete pathway for citrate export, conversion of citrate to malate and pyruvate, and import of malate and pyruvate by mitochondria.

   a. Which of the reactions in this cycle might require energy input?

   b. What would be the most likely source of this energy?

   c. Do you recognize any of the enzyme reactions in this cycle?

   d. What coenzymes might be required to run this cycle?

16. Assessing the Equilibrium Concentrations in the Malate Dehydrogenase Reaction A typical intramitochondrial concentration of malate is 0.22 mM. If the ratio of NAD^+ to $NADH$ in mitochondria is 20, and if the malate dehydrogenase reaction is at equilibrium, calculate the concentration of oxaloacetate in the mitochondrion at 20°C. A typical mitochondrion can be thought of as a cylinder 1 μm in diameter and 2 μm in length. Calculate the number of molecules of oxaloacetate in a mitochondrion. In analogy with pH (the negative logarithm of $[H^+]$), what is pOAA?

17. Understanding the Oxidation of Glucose and Its Products in the TCA Cycle Glycolysis, the pyruvate dehydrogenase reaction, and the TCA cycle result in complete oxidation of a molecule of glucose to CO_2. Review the calculation of oxidation numbers for individual atoms in any molecule, and then calculate the oxidation numbers of the carbons of glucose, pyruvate, the acetyl carbons of acetyl-CoA, and the metabolites of the TCA cycle to convince yourself that complete oxidation of glucose involves removal of 24 electrons and that each acetyl-CoA through the TCA cycle gives up 8 electrons.

18. A Simple Way to Understand and Remember the Reactions of the TCA Cycle Recalling all reactions of the TCA cycle can be a challenging proposition. One way to remember these is to begin with the simplest molecule—succinate, which is a symmetric four-carbon molecule. Begin with succinate, and draw the eight reactions of the TCA cycle. Remember that succinate $\longrightarrow$ oxaloacetate is accomplished by a special trio of reactions: oxidation of a single bond to a double bond, hydration across the double bond, and oxidation of an alcohol to a ketone. From there, a molecule of acetyl-CoA is added. If you remember the special function of acetyl-CoA (see A Deeper Look, page 616), this is an easy reaction to draw. From there, you need only isomerize, carry out the two oxidative decarboxylations, and remove the CoA molecule to return to succinate.

19. Understanding the Mechanism-Based Inactivation of Aconitase by Fluoroacetate Aconitase is rapidly inactivated by 2R, 3R-fluorocitrate, which is produced from fluoroacetate in the citrate synthase reaction. Interestingly, inactivation by fluorocitrate is accompanied by stoichiometric release of fluoride ion (i.e., one F^- ion is lost per aconitase active site). This observation is consistent with "mechanism-based inactivation" of aconitase by fluorocitrate. Suggest a mechanism for this inactivation, based on formation of 4-hydroxy-$trans$-aconitate, which remains tightly bound at the active site. To assess your answer, consult this reference: Lauble, H., Kennedy, M., et al., 1996. The reaction of fluorocitrate with aconitase and the crystal structure of the enzyme-inhibitor complex. *Proceedings of the National Academy of Sciences* **93**:13699–13703.

Preparing for the MCAT® Exam

20. Complete oxidation of a 16-carbon fatty acid can yield 129 molecules of ATP. Study Figure 19.2 and determine how many ATP molecules would be generated if a 16-carbon fatty acid were metabolized solely by the TCA cycle, in the form of 8 acetyl-CoA molecules.

21. Study Figure 19.18 and decide which of the following statements is false.

   a. Pyruvate dehydrogenase is inhibited by NADH.

   b. Pyruvate dehydrogenase is inhibited by ATP.

   c. Citrate synthase is inhibited by NADH.

   d. Succinyl-CoA activates citrate synthase.

   e. Acetyl-CoA activates pyruvate carboxylase.

ActiveModels Problem

22. Examine the ActiveModel for isocitrate dehydrogenase, and identify the α-helices, 3_{10} helices, and β-sheets in this structure. Locate the enzyme residues involved in catalysis.

FURTHER READING

General

Bodner, G. M., 1986. The tricarboxylic acid (TCA), citric acid or Krebs cycle. *Journal of Chemical Education* **63**:673–677.

Dalsgaard, M. K., 2006. Fuelling cerebral activity in exercising man. *Journal of Cerebral Blood Flow and Metabolism* **26**:731–750.

Fisher, C. R., and Girguis, P., 2007. A proteomic snapshot of life at a vent. *Science* **315**:198–199.

Gibala, M. J., Young, M. E., et al., 2000. Anaplerosis of the citric acid cycle: Role in energy metabolism of heart and skeletal muscle. *Acta Physiologica Scandinavica* **168**:657–665.

Holmes, F. L., 1993. *Hans Krebs: Architect of Intermediary Metabolism, 1933-1937*, Vol. 2. Oxford: Oxford University Press.

Hu, Y., and Holden, J. F., 2006. Citric acid cycle in the hyperthermophilic archaeon *Pyrobaculum islandicum* grown autotrophically, heterotrophically, and mixotrophically with acetate. *Journal of Bacteriology* **188**:4350–4355.

Krebs, H. A., 1981. *Reminiscences and Reflections*. Oxford: Oxford University Press.

Newsholme, E. A., and Leech, A. R., 1983. *Biochemistry for the Medical Sciences*. New York: John Wiley and Sons.

Schurr, A., 2006. Lactate: The ultimate cerebral oxidative energy substrate? *Journal of Cerebral Blood Flow and Metabolism* **26**:142–152.

Smith, E., and Morowitz, H. J., 2004. Universality in intermediary metabolism. *Proceedings of the National Academy of Sciences U.S.A.* **101**:13168–13173.

Enzymes of the TCA Cycle

Fraser, M. E., James, M. N. G., et al., 1999. A detailed structural description of *Escherichia coli* succinyl-CoA synthetase. *Journal of Molecular Biology* **285**:1633–1653.

Perham, R. N., 2000. Swinging arms and swinging domains in multifunctional enzymes: Catalytic machines for multistep reactions. *Annual Review of Biochemistry* **69**:961–1004.

St. Maurice, M., Reinhardt, L., et al., 2007. Domain architecture of pyruvate carboxylase, a biotin-dependent multifunctional enzyme. *Science* **317**:1076–1079.

Diseases of the TCA Cycle

Briere, J.-J., Favier, J., et al., 2006. Tricarboxylic acid cycle dysfunction as a cause of human diseases and tumor formation. *American Journal of Physiology and Cellular Physiology* **291**:C1114–C1120.

Pithukpakorn, M., 2005. Disorders of pyruvate metabolism and the tricarboxylic acid cycle. *Molecular Genetics and Metabolism* **85**:243–246.

Regulation of the TCA Cycle

Atkinson, D. E., 1977. *Cellular Energy Metabolism and Its Regulation.* New York: Academic Press.

Bott, M., 2007. Offering surprises: TCA cycle regulation in *Corynebacterium glutamicum. Trends in Microbiology* **15**:417–425.

Chen, J.-Q., Brown, T. R., and Russo, J., 2009. Regulation of energy metabolism pathways by estrogens and estrogenic chemicals and potential implications in obesity associated with increased exposure to endocrine disrupters. *Biochimica et Biophysica Acta* **1793**:1128–1143.

Gibson, D., and Harris, R., 2001. *Metabolic Regulation in Mammals.* New York: Taylor and Francis.

McKnight, S. L., 2010. On getting there from here. *Science* **330**:1338–1339.

Pyruvate Dehydrogenase

Brautigam, C. A., Wynn, R. M., et al., 2006. Structural insight into interactions between dihydrolipoamide dehydrogenase (E3) and E3-binding protein of human pyruvate dehydrogenase complex. *Structure* **14**:611–621.

Harris, R. A., Bowker-Kinley, M. M., et al., 2002. Regulation of the activity of the pyruvate dehydrogenase complex. *Advances in Enzyme Regulation* **42**:249–259.

Jordan, F., Arjunan, P., Kale, S., Nemeria, N. S., and Furey, W., 2009. Multiple roles of active center loops in the E1 component of the *Escherichia coli* pyruvate dehydrogenase complex—Linkage of protein dynamics to catalysis. *Journal of Molecular Catalysis B Enzymology* **61**:14–22.

Milne, J. L. S., Shi, D., et al., 2002. Molecular architecture and mechanism of an icosahedral pyruvate dehydrogenase complex: A multifunctional catalytic machine. *EMBO Journal* **21**:5587–5598.

Song, J., Park, Y.-H., Nemeria, N. S., Kale, S., Kakalis, L., and Jordan, F., 2010. Nuclear magnetic resonance evidence for the role of the flexible regions of the E1 component of the pyruvate dehydrogenase complex from Gram-negative bacteria. *Journal of Biological Chemistry* **285**:4680–4694.

Sugden, M. C., and Holness, M. J., 2006. Mechanisms underlying regulation of the expression and activities of the mammalian pyruvate dehydrogenase kinases. *Archives of Physiology and Biochemistry* **112**:139–149.

Vijayakrishnan, S., Kelly, S. M., Gilber, R. J. C., Callow, P., Bhella, D., Forsyth, T., Lindsay, J. G., and Byron, O., 2010. Solution structure and characterization of the human pyruvate dehydrogenase complex core assembly. *Journal of Molecular Biology* **399**:71–93.

Yu, X., Hiromasa, Y., Tsen, H., Stoops, J. K., Roche, T. E., and Zhou, Z. H., 2007. Structures of the human pyruvate dehydrogenase complex cores: A highly conserved catalytic center with flexible N-terminal domains. *Structure* **16**:104–114.

Zhou, Z. H., McCarthy, D. B., et al., 2001. The remarkable structural and functional organization of the eukaryotic pyruvate dehydrogenase complexes. *Proceedings of the National Academy of Sciences U.S.A.* **98**:14802–14807.

Glyoxylate Cycle

Eastmond, P. J., and Graham, I. A., 2001. Re-examining the role of the glyoxylate cycle in oilseeds. *Trends in Plant Science* **6**:72–77.

TCA and Cancer

Dang, C. V., Hamaker, M., Sun, P., and Le, A., 2011. Therapeutic targeting of cancer cell metabolism. *Journal of Molecular Medicine* **89**:205–212.

Dang, C. V., White, D. W., Gross, S., et al., 2009. Cancer-associated IDH1 mutations produce 2-hydroxyglutarate. *Nature* **462**:739–746.

Diaz-Ruiz, R., Uribe-Carvajal, S., Devin, A., and Rigoulet, M., 2009. Tumor cell energy metabolism and its common features with yeast metabolism. *Biochimica et Biophysica Acta* **1796**:252–265.

Frezza, C., Pollard, P. J., and Gottlieb, E., 2011. Inborn and acquired metabolic defects in cancer. *Journal of Molecular Medicine* **89**:213–220.

Linehan, W. M., Srinivasan, R., and Schmidt, L. S., 2010. The genetic basis of kidney cancer: A metabolic disease. *Nature Reviews Urology* **7**:277–285.

Vander Heiden, M. G., Cantley, L. C., and Thompson, C. B., 2009. Understanding the Warburg effect: The metabolic requirements of cell proliferation. *Science* **324**:1029–1033.

Weinberg, F., and Chandel, N. S., 2009. Mitochondrial metabolism and cancer. *New York Academy of Sciences* **1177**:66–73.

Lysine Acetylation and SIRT3 in TCA

Arif, M., Senapati, P., Shandilya, J., and Kundu, T. K., 2010. Protein lysine acetylation in cellular function and its role in cancer manifestation. *Biochimica et Biophysica Acta* **1799**:702–716.

Schlicker, C., Gertz, M., Paptheodorouo, P., Kachholz, B., Becker, C. F. W., and Steegborn, C., 2008. *Journal of Molecular Biology* **382**:790–801.

Smith, B. C., Settles, B., Hallows, W. C., Craven, M. W., and Denu, J. M., 2011. SIRT3 substrate specificity determined by peptide arrays and machine learning. *ACS Chemical Biology* **6**:146–157.

Verdin, E., Hirschey, M. D., Finley, L. W. S., and Haigis, M. C., 2010. Sirtuin regulation of mitochondria: Energy production, apoptosis, and signaling. *Trends in Biochemical Sciences* **35**:669–674.

Wang, Q., Zhang, Y., Yang, C., Xiong, H., et al., 2010. Acetylation of metabolic enzymes coordinates carbon source utilization and metabolic flux. *Science* **327**:1004–1007.

Wu, X., Oh, M.-H., Schwarz, E. M., Larue, C. T., et al., 2011. Lysine acetylation is a widespread protein modification for diverse proteins in *Arabidopsis. Plant Physiology* **155**:1769–1778.

Zhang, D., Liu, Y., and Chen, D., 2011. SIRT–ain relief from age-inducing stress. *Aging* **3**:158–161.

Zhao, S., Xu, W., Jiang, W., et al., 2010. Regulation of cellular metabolism by protein lysine acetylation. *Science* **327**:1000–1004.

Zheng, J. and Jia, Z., 2010. Structure of the bifunctional isocitrate dehydrogenase kinase/phosphatase. *Nature* **465**:961–966.

The TCA Metabolome

Meyer, F. M., Gerwig, J., Hammer, E., Herzberg, C., et al., 2011. Physical interactions between tricarboxylic acid cycle enzymes in *Bacillus subtilis:* Evidence for a metabolon. *Metabolic Engineering* **13**:18–27.

Sadykov, M. R., Zhang, B., Halouska, S., Nelson, J. L., et al., 2010. Using NMR metabolomics to investigate tricarboxylic acid cycle-dependent signal transduction in *Staphylococcus epidermidis. Journal of Biological Chemistry* **285**:36616–36624.

Anaplerosis and Diabetes

Jensen, M., Joseph, J. W., Ronnebaum, S. M., Burgess, S. C., et al., 2008. Metabolic cycling in control of glucose-stimulated insulin secretion. *American Journal of Physiology and Endocrine Metabolism* **295**:E1287–E1297.

Pillai, R., Huypens, P., Huang, M., Schaefer, S., et al., 2011. Aryl hydrocarbon receptor nuclear translocator/hypoxia-inducible factor-1β plays a critical role in maintaining glucose–stimulated anaplerosis and insulin release from pancreatic β-cells. *Journal of Biological Chemistry* **286**:1014–1024.

Electron Transport and Oxidative Phosphorylation

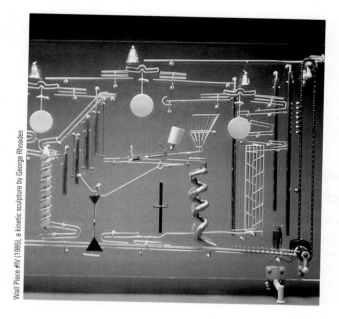

Wall Piece #IV (1985), a kinetic sculpture by George Rhoades

◀ Wall Piece #IV (1985), a kinetic sculpture by George Rhoads. This complex mechanical art form can be viewed as a metaphor for the molecular apparatus underlying electron transport and ATP synthesis by oxidative phosphorylation.

In all things of nature there is something of the marvelous.

Aristotle (384–322 B.C.)

KEY QUESTIONS

20.1 Where in the Cell Do Electron Transport and Oxidative Phosphorylation Occur?

20.2 How Is the Electron-Transport Chain Organized?

20.3 What Are the Thermodynamic Implications of Chemiosmotic Coupling?

20.4 How Does a Proton Gradient Drive the Synthesis of ATP?

20.5 What Is the P/O Ratio for Mitochondrial Oxidative Phosphorylation?

20.6 How Are the Electrons of Cytosolic NADH Fed into Electron Transport?

20.7 How Do Mitochondria Mediate Apoptosis?

ESSENTIAL QUESTION

Living cells save up metabolic energy predominantly in the form of fats and carbohydrates, and they "spend" this energy for biosynthesis, membrane transport, and movement. In both directions, energy is exchanged and transferred in the form of ATP. In Chapters 18 and 19 we saw that glycolysis and the TCA cycle convert some of the energy available from stored and dietary sugars directly to ATP. However, most of the metabolic energy that is obtainable from substrates entering glycolysis and the TCA cycle is funneled via oxidation–reduction reactions into NADH and reduced flavoproteins, the latter symbolized by [FADH₂].

How do cells oxidize NADH and [FADH₂] and convert their reducing potential into the chemical energy of ATP?

Whereas ATP made in glycolysis and the TCA cycle is the result of substrate-level phosphorylation, NADH-dependent ATP synthesis is the result of **oxidative phosphorylation.** Electrons stored in the form of the reduced coenzymes, NADH or [FADH₂], are passed through an elaborate and highly organized chain of proteins and coenzymes, the so-called **electron-transport chain,** finally reaching O_2 (molecular oxygen), the terminal electron acceptor. Each component of the chain can exist in (at least) two oxidation states, and each component is successively reduced and reoxidized as electrons move through the chain from NADH (or [FADH₂]) to O_2. In the course of electron transport, a proton gradient is established across the inner mitochondrial membrane. It is the energy of this proton gradient that drives ATP synthesis.

OWL Online homework and a Student Self Assessment for this chapter may be assigned in OWL.

20.1 | Where in the Cell Do Electron Transport and Oxidative Phosphorylation Occur?

The processes of electron transport and oxidative phosphorylation are **membrane associated.** Prokaryotes are the simplest life form, and prokaryotic cells typically consist of a single cellular compartment surrounded by a plasma membrane and a more rigid cell wall. In such a system, the conversion of energy from NADH and [FADH$_2$] to the energy of ATP via electron transport and oxidative phosphorylation is carried out at (and across) the plasma membrane.

In eukaryotic cells, electron transport and oxidative phosphorylation are localized in mitochondria, which are also the sites of TCA cycle activity and (as we shall see in Chapter 23) fatty acid oxidation. Mammalian cells contain 800 to 2500 mitochondria; other types of cells may have as few as one or two or as many as half a million mitochondria. Human erythrocytes, whose purpose is simply to transport oxygen to tissues, contain no mitochondria at all. The typical mitochondrion is about 0.5 ± 0.3 micron in diameter and from 0.5 micron to several microns long; its overall shape is sensitive to metabolic conditions in the cell.

Mitochondrial Functions Are Localized in Specific Compartments

Mitochondria are surrounded by a simple **outer membrane** and a more complex **inner membrane** (Figure 20.1). The space between the inner and outer membranes is referred to as the **intermembrane space.** Several enzymes that utilize ATP (such as creatine kinase and adenylate kinase) are found in the intermembrane space. The smooth outer membrane is about 30% to 40% lipid and 60% to 70% protein and has a relatively high concentration of phosphatidylinositol. The outer membrane contains significant amounts of **porin**—a transmembrane protein, rich in β-sheets, that forms large channels across the membrane, permitting free diffusion of molecules with molecular weights of about 10,000 or less. The outer membrane plays a prominent role in maintaining the shape of the mitochondrion. The inner membrane is richly packed with proteins, which account for nearly 80% of its weight; thus, its density is higher than that of the outer membrane. The fatty acids of inner membrane lipids are highly unsaturated. Cardiolipin and diphosphatidylglycerol (see

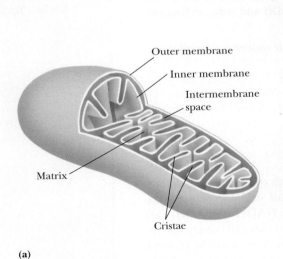

Outer membrane

Inner membrane

Intermembrane space

Matrix

Cristae

(a)

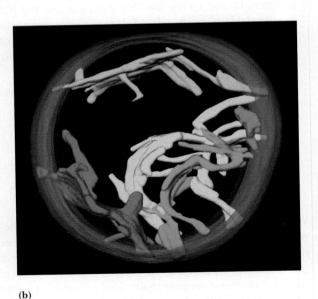

(b)

FIGURE 20.1 **(a)** A drawing of a mitochondrion with components labeled. **(b)** Tomography of a rat liver mitochondrion. The tubular structures in red, yellow, green, purple, and aqua represent individual cristae formed from the inner mitochondrial membrane. (b, Frey, T. G., and Mannella, C. A., 2000. The internal structure of mitochondria. *Trends in Biochemical Sciences* **25**:319–324.)

Chapter 8) are abundant. The inner membrane lacks cholesterol and is quite impermeable to molecules and ions. Species that must cross the mitochondrial inner membrane—ions, substrates, fatty acids for oxidation, and so on—are carried by specific transport proteins in the membrane. Notably, the inner membrane is extensively folded (Figure 20.1). The folds, known as **cristae,** provide the inner membrane with a large surface area in a small volume. During periods of active respiration, the inner membrane appears to shrink significantly, leaving a comparatively large intermembrane space.

The Mitochondrial Matrix Contains the Enzymes of the TCA Cycle

The space inside the inner mitochondrial membrane is called the **matrix,** and it contains most of the enzymes of the TCA cycle and fatty acid oxidation. (An important exception, succinate dehydrogenase of the TCA cycle, is located in the inner membrane itself.) In addition, mitochondria contain circular DNA molecules, along with ribosomes and the enzymes required to synthesize proteins coded within the mitochondrial genome. Although some of the mitochondrial proteins are made this way, most are encoded by nuclear DNA and synthesized by cytosolic ribosomes.

20.2 How Is the Electron-Transport Chain Organized?

As we have seen, the metabolic energy from oxidation of food materials—sugars, fats, and amino acids—is funneled into formation of reduced coenzymes (NADH) and reduced flavoproteins ([FADH$_2$]). The electron-transport chain reoxidizes the coenzymes and channels the free energy obtained from these reactions into the creation of a proton gradient. This reoxidation process involves the removal of both protons and electrons from the coenzymes. Electrons move from NADH and [FADH$_2$] to molecular oxygen, O_2, which is the terminal acceptor of electrons in the chain. The reoxidation of NADH,

$$\text{NADH (reductant)} + H^+ + O_2 \text{ (oxidant)} \longrightarrow NAD^+ + H_2O \qquad (20.1)$$

involves the following half-reactions:

$$NAD^+ + 2\,H^+ + 2\,e^- \longrightarrow NADH + H^+ \qquad \mathscr{E}_0' = -0.32 \text{ V} \qquad (20.2)$$

$$\tfrac{1}{2}\,O_2 + 2\,H^+ + 2\,e^- \longrightarrow H_2O \qquad \mathscr{E}_0' = +0.816 \text{ V} \qquad (20.3)$$

Here, half-reaction 20.3 is the electron acceptor and half-reaction 20.2 is the electron donor. Then

$$\Delta\mathscr{E}_0' = 0.816 - (-0.32) = 1.136 \text{ V} \qquad (20.4)$$

and, according to Equation 3.38, the standard-state free energy change, $\Delta G^{\circ\prime}$, is -219 kJ/mol. Molecules along the electron-transport chain have reduction potentials between the values for the NAD^+/NADH couple and the oxygen/H_2O couple, so electrons move down the energy scale toward progressively more positive reduction potentials (Figure 20.2).

Although electrons move from more negative to more positive reduction potentials in the electron-transport chain, it should be emphasized that the electron carriers do not operate in a simple linear sequence. This will become evident when the individual components of the electron-transport chain are discussed in the following paragraphs.

The Electron-Transport Chain Can Be Isolated in Four Complexes

The electron-transport chain involves several different molecular species, including:

1. **Flavoproteins,** which contain tightly bound FMN or FAD as prosthetic groups and which may participate in one- or two-electron transfer events.

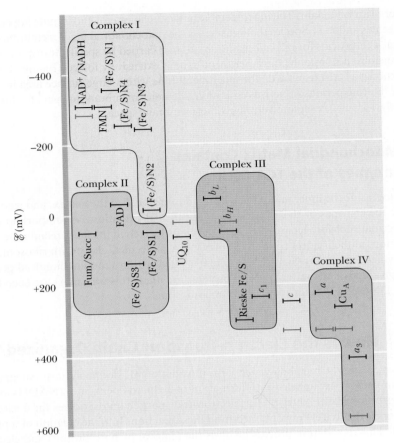

FIGURE 20.2 $\mathscr{E}_0'$ and $\mathscr{E}$ values for the components of the mitochondrial electron-transport chain. Values indicated are consensus values for animal mitochondria. Black bars represent $\mathscr{E}_0'$; red bars, $\mathscr{E}$.

2. **Coenzyme Q,** also called **ubiquinone** (and abbreviated **CoQ** or **UQ**) (see Figure 20.4), which can function in either one- or two-electron transfer reactions.
3. Several **cytochromes** (proteins containing heme prosthetic groups [see Chapter 5], which function by carrying or transferring electrons), including cytochromes b, c, c_1, a, and a_3. Cytochromes are one-electron transfer agents in which the heme iron is converted from Fe^{2+} to Fe^{3+} and back.
4. A number of **iron–sulfur proteins,** which participate in one-electron transfers involving the Fe^{2+} and Fe^{3+} states.
5. Protein-bound **copper,** a one-electron transfer site that converts between Cu^+ and Cu^{2+}.

All these intermediates except for cytochrome c are membrane associated (either in the mitochondrial inner membrane of eukaryotes or in the plasma membrane of prokaryotes). Three types of proteins involved in this chain—flavoproteins, cytochromes, and iron–sulfur proteins—possess electron-transferring **prosthetic groups.**

The components of the electron-transport chain can be purified from the mitochondrial inner membrane. Solubilization of the membranes containing the electron-transport chain results in the isolation of four distinct protein complexes, and the complete chain can thus be considered to be composed of four parts: (I) **NADH–coenzyme Q reductase,** (II) **succinate–coenzyme Q reductase,** (III) **coenzyme Q–cytochrome c reductase,** and (IV) **cytochrome c oxidase** (Figure 20.3). Complex I accepts electrons from NADH, serving as a link between glycolysis, the TCA cycle, fatty acid oxidation, and the electron-transport chain. Complex II includes succinate dehydrogenase and thus forms a direct link between the TCA cycle and electron transport. Complexes I and II produce a common product, reduced coenzyme Q (UQH_2), which is the substrate for coenzyme Q–cytochrome c reductase (Complex III). As shown in Figure 20.3, there are two other ways to feed electrons to UQ: the **electron-transferring flavoprotein,** which transfers electrons from the flavoprotein-linked step of fatty acyl-CoA dehydrogenase, and **sn-glycerophosphate dehydrogenase.** Complex III oxidizes UQH_2 while reducing

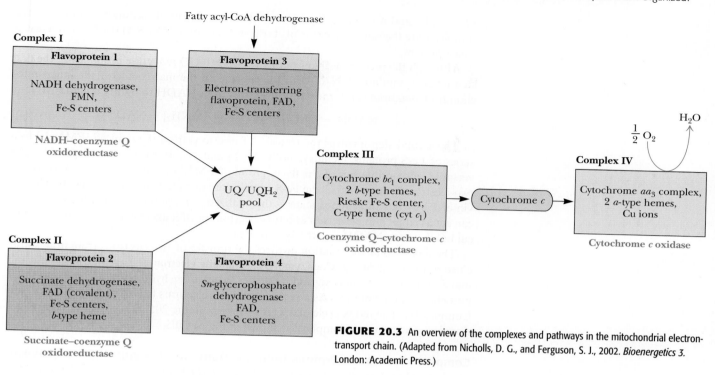

FIGURE 20.3 An overview of the complexes and pathways in the mitochondrial electron-transport chain. (Adapted from Nicholls, D. G., and Ferguson, S. J., 2002. *Bioenergetics 3*. London: Academic Press.)

cytochrome *c,* which in turn is the substrate for Complex IV, cytochrome *c* oxidase. Complex IV is responsible for reducing molecular oxygen. Each of the complexes shown in Figure 20.3 is a large multisubunit complex embedded within the inner mitochondrial membrane.

Complex I Oxidizes NADH and Reduces Coenzyme Q

As its name implies, this complex transfers a pair of electrons from NADH to coenzyme Q, a small, hydrophobic, yellow compound. Another common name for this enzyme complex is *NADH dehydrogenase.* The complex from bovine heart mitochondria (with an estimated mass of 980 kD) involves at least 45 polypeptide chains, one molecule of flavin mononucleotide (FMN), and eight or nine Fe-S clusters, together containing a total of 20 to 26 iron atoms (Table 20.1). Bacterial complex I is much simpler, with just

TABLE 20.1	Protein Complexes of the Mitochondrial Electron-Transport Chain			
Complex	Mass (kD)	Subunits	Prosthetic Group	Binding Site for:
NADH–UQ reductase	980	$\geq$45	FMN Fe-S	NADH (matrix side) UQ (lipid core)
Succinate–UQ reductase	140	4	FAD Fe-S	Succinate (matrix side) UQ (lipid core)
UQ–Cyt *c* reductase	250	9–10	Heme b_L Heme b_H Heme c_1 Fe-S	Cyt *c* (intermembrane space side)
Cytochrome *c*	13	1	Heme *c*	Cyt c_1 Cyt *a*
Cytochrome *c* oxidase	162	13	Heme *a* Heme a_3 Cu_A Cu_B	Cyt *c* (intermembrane space side)

14 subunits and a mass of 500 kD, but the cofactors in mitochondrial and bacterial complex I are the same. By virtue of its dependence on FMN, NADH–UQ reductase is a flavoprotein.

Although the precise mechanism of the NADH–UQ reductase is unknown, the first step involves binding of NADH to the enzyme on the matrix side of the inner mitochondrial membrane and transfer of electrons from NADH to tightly bound FMN:

$$NADH + [FMN] + H^+ \longrightarrow [FMNH_2] + NAD^+ \qquad (20.5)$$

The second step involves the transfer of electrons from the reduced $[FMNH_2]$ to a series of Fe-S proteins, including both 2Fe-2S and 4Fe-4S clusters (see page 651). The versatile redox properties of the flavin group of FMN are important here. NADH is a two-electron donor, whereas the Fe-S proteins are one-electron transfer agents. The flavin of FMN has three redox states—the oxidized, semiquinone, and reduced states. It can act as either a one-electron or a two-electron transfer agent and may serve as a critical link between NADH and the Fe-S proteins.

The final step of the reaction involves the transfer of two electrons from iron–sulfur clusters to coenzyme Q. Coenzyme Q is a **mobile electron carrier.** Its isoprenoid tail makes it highly hydrophobic, and it diffuses freely in the hydrophobic core of the inner mitochondrial membrane. As a result, it shuttles electrons from Complexes I and II to Complex III. The redox cycle of UQ is shown in Figure 20.4. The structural and functional organization of Complex I is shown in Figure 20.5.

Complex I Transports Protons from the Matrix to the Cytosol The oxidation of one NADH and the reduction of one UQ by NADH–UQ reductase results in the net transport of protons from the matrix side to the cytosolic side of the inner membrane. The cytosolic side, where H^+ accumulates, is referred to as the **P** (for positive) face; similarly, the matrix side is the **N** (for negative) face. Some of the energy liberated by the flow of electrons through this complex is used in a coupled process to drive the transport of protons across the membrane. (This is an example of active transport, a phenomenon examined in detail in Chapter 9.) Available experimental evidence suggests a stoichiometry of four H^+ transported per two electrons passed from NADH to UQ.

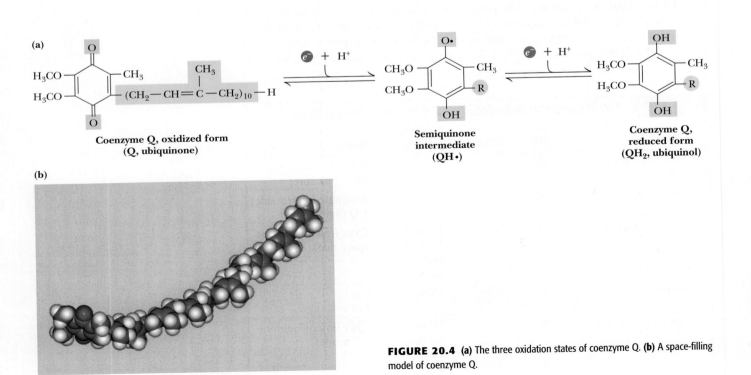

FIGURE 20.4 (a) The three oxidation states of coenzyme Q. (b) A space-filling model of coenzyme Q.

HUMAN BIOCHEMISTRY

Solving a Medical Mystery Revolutionized Our Treatment of Parkinson's Disease

A tragedy among illegal drug users was the impetus for a revolutionary treatment of Parkinson's disease. In 1982, several mysterious cases of paralysis came to light in southern California. The victims, some of them teenagers, were frozen like living statues, unable to talk or move. The case was baffling at first, but it was soon traced to a batch of synthetic heroin that contained MPTP (1-methyl-4-phenyl-1,2,3,6-tetrahydropyridine) as a contaminant. MPTP is rapidly converted in the brain to MPP$^+$ (1-methyl-4-phenylpyridine) by the enzyme monoamine oxidase B. MPP$^+$ is a potent inhibitor of mitochondrial Complex I (NADH–UQ reductase), and it acts preferentially in the *substantia nigra,* an area of the brain that is essential to movement and also the region of the brain that deteriorates slowly in Parkinson's disease.

Parkinson's disease results from the inability of the brain to produce sufficient quantities of dopamine, a neurotransmitter. Neurologist J. William Langston, asked to consult on the treatment of some of these patients, recognized that the symptoms of this drug-induced disorder were in fact similar to those of parkinsonism. He began treatment of the patients with L-dopa, which is decarboxylated in the brain to produce dopamine. The treated patients immediately regained movement. Langston then took a bold step. He implanted fetal brain tissue into the brains of several of the affected patients, prompting substantial recovery from the Parkinson-like symptoms. Langston's innovation sparked a revolution in the use of tissue implantation for the treatment of neurodegenerative diseases.

Other toxins may cause similar effects in neural tissue. Timothy Greenmyre at Emory University has shown that rats exposed to the pesticide rotenone (see Figure 20.26) over a period of weeks experience a gradual loss of function in dopaminergic neurons and then develop symptoms of parkinsonism, including limb tremors and rigidity. This finding supports earlier research that links long-term pesticide exposure to Parkinson's disease.

A Helical Piston Drives the Proton Pump of Complex I The structure of Complex I is elegantly suited to its biological function. This multi-subunit complex consists of a large hydrophilic domain extending into the mitochondrial matrix and a large hydrophobic domain in the inner mitochondrial membrane. The hydrophilic domain is essentially a superhighway for electrons, from the FMN cofactor site through a chain of seven Fe-S clusters that extends approximately 95 Å through the protein to a UQ site deep in the membrane (Figure 20.5). The hydrophobic domain is composed of multiple subunits in an extended array with a total of 55 (in *E. coli*) or 63 (in *Thermus thermophilus*) transmembrane α-helical segments, with an unusual 110 Å helix lying parallel to the membrane. In both these organisms, the three largest transmembrane subunits are very similar in structure to antiporter proteins that transport Na$^+$ and H$^+$ across membranes. Each of these subunits contains a **discontinuous helix** consisting of two helical segments interrupted by a short extended peptide (Figure 20.5) that is essential for ion transport in these transporters and channel proteins. Leonid Sazanov and his colleagues have suggested that the 110 Å helix lying across these subunits acts as a connecting rod or piston, sliding back and forth to open or close H$^+$ channels in complex I in response to conformational changes induced by electron flow through the hydrophilic domain of the complex.

Complex II Oxidizes Succinate and Reduces Coenzyme Q

Complex II is perhaps better known by its other name—succinate dehydrogenase, the only TCA cycle enzyme that is an integral membrane protein in the inner mitochondrial membrane. This complex (Figure 20.6) has a mass of 124 kD and is composed of two hydrophilic subunits, a flavoprotein (Fp, 68 kD) and an iron–sulfur protein (Ip, 29 kD), and two hydrophobic, membrane-anchored subunits (15 kD and 11 kD), which contain one heme *b* and provide the binding site for UQ. Fp contains an FAD covalently bound to a His residue (see Figure 19.12), and Ip contains three Fe-S centers: a 4Fe-4S cluster, a 3Fe-4S cluster, and a 2Fe-2S cluster. When succinate is converted to fumarate in the

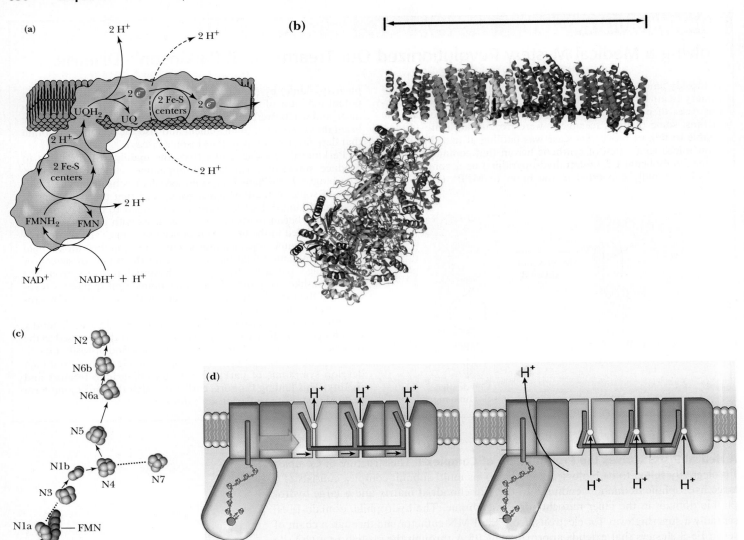

FIGURE 20.5 **(a)** Structural organization of mammalian Complex I, based on electron microscopy, showing functional relationships within the L-shaped complex. Electron flow from NADH to UQH$_2$ in the membrane pool is indicated. **(b)** Structure of the hydrophilic domain of Complex I from *Thermus thermophilus* is shown on a model of the membrane-associated complex (pdb id = 2FUG). The locations of individual subunits are indicated. **(c)** Arrangement of the redox centers in Complex I. The various iron–sulfur centers of Complex I are designated by capital N. **(d)** Blue arrows indicate the pathway of electron transfer from NADH to coenzyme Q. The energy of electron transfer allows a proton to cross Complex I near the interface between its hydrophilic and membrane domains. Conformational changes accompanying these events push the long helical rod (magenta), reorienting residues in the NuoL, NuoM, and NuoN subunits, so that protons bound there can cross into the intermembrane space. (Part c adapted from Figure 1 of Sazanov, L., and Hinchliffe, P., 2006. Structure of the hydrophilic domain of respiratory Complex I from *Thermus thermophilus*. *Science* **311**:1430–1436; part d adapted from Figure 1 of Ohnishi, T., 2010. Piston drives a proton pump. *Nature* **465**:428–429.)

TCA cycle, concomitant reduction of bound [FAD] to [FADH$_2$] occurs in succinate dehydrogenase. This [FADH$_2$] transfers its electrons immediately to Fe-S centers, which pass them on to UQ. Electron flow from succinate to UQ,

$$\text{Succinate} \longrightarrow \text{fumarate} + 2\,\text{H}^+ + 2\,e^- \tag{20.6}$$

$$\text{UQ} + 2\,\text{H}^+ + 2\,e^- \longrightarrow \text{UQH}_2 \tag{20.7}$$

$$\text{Net rxn: Succinate} + \text{UQ} \longrightarrow \text{fumarate} + \text{UQH}_2 \qquad \Delta\mathscr{E}_o' = 0.029\ \text{V} \tag{20.8}$$

(a)

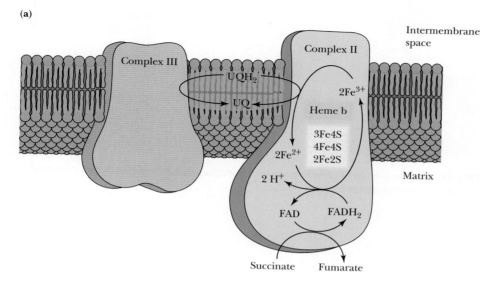

(b)

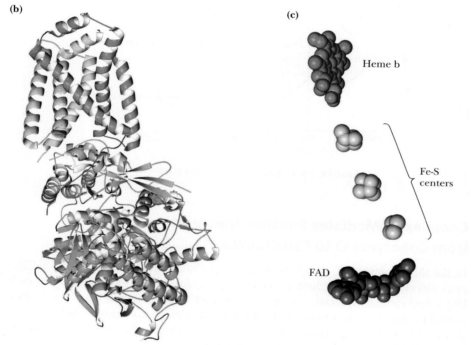

(c)

FIGURE 20.6 **(a)** A scheme for electron flow in Complex II. Oxidation of succinate occurs with reduction of [FAD]. Electrons are then passed to Fe-S centers and then to co-enzyme Q (UQ). Proton transport does not occur in this complex. **(b)** The structure of Complex II from pig heart (pdb id = 1ZOY). **(c)** The arrangement of redox centers. Electron flow is from bottom to top.

yields a net reduction potential of 0.029 V. (Note that the first half-reaction is written in the direction of the e^- flow. As always, $\Delta\mathscr{E}_o'$ is calculated according to Equation 3.44.) The small free energy change of this reaction does not contribute to the transport of protons across the inner mitochondrial membrane.

This is a crucial point because (as we will see) proton transport is coupled with ATP synthesis. Oxidation of one [FADH$_2$] in the electron-transport chain results in synthesis of approximately two molecules of ATP, compared with the approximately three ATPs produced by the oxidation of one NADH. Other flavoproteins can also supply electrons to UQ, including mitochondrial sn-glycerophosphate dehydrogenase, an inner membrane-bound shuttle enzyme, and the fatty acyl-CoA dehydrogenases, three soluble matrix enzymes involved in fatty acid oxidation (Figure 20.7; also see Chapter 23). The path of electrons from succinate to UQ is shown in Figure 20.6.

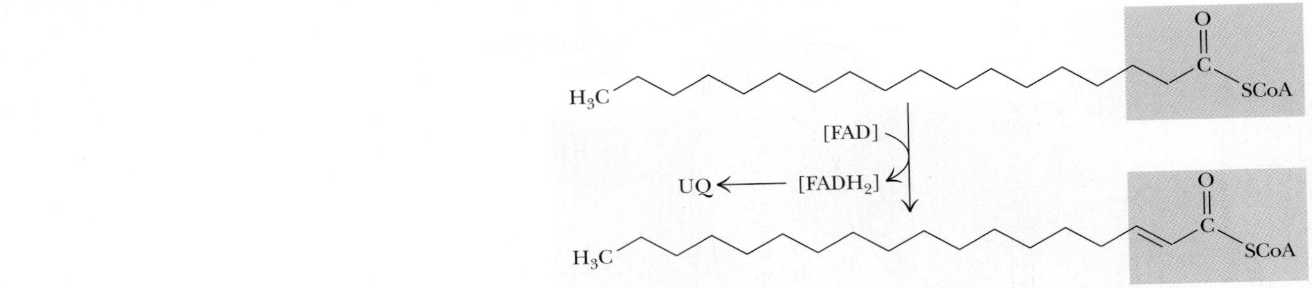

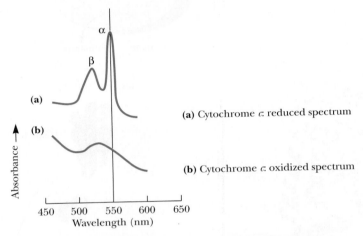

FIGURE 20.7 The fatty acyl-CoA dehydrogenase reaction, emphasizing that the reaction involves reduction of enzyme-bound FAD (indicated by brackets).

(a) Cytochrome c: reduced spectrum

(b) Cytochrome c: oxidized spectrum

FIGURE 20.8 Visible absorption spectra of cytochrome c.

Complex III Mediates Electron Transport from Coenzyme Q to Cytochrome c

In the third complex of the electron-transport chain, reduced coenzyme Q (UQH_2) passes its electrons to cytochrome c via a unique redox pathway known as the **Q cycle.** UQ–cytochrome c reductase (UQ–cyt c reductase), as this complex is known, involves three different cytochromes and an Fe-S protein. In the cytochromes of these and similar complexes, the iron atom at the center of the porphyrin ring cycles between the reduced Fe^{2+} (ferrous) and oxidized Fe^{3+} (ferric) states.

Cytochromes were first named and classified on the basis of their absorption spectra (Figure 20.8), which depend upon the structure and environment of their heme groups. The **b cytochromes** contain *iron protoporphyrin IX* (Figure 20.9), the same heme found in hemoglobin and myoglobin. The **c cytochromes** contain *heme c,* derived from iron protoporphyrin IX by the covalent attachment to cysteine residues from the associated protein. (One other heme variation, heme a, contains a 15-carbon isoprenoid chain on a modified vinyl group and a formyl group in place of one of the methyls [see Figure 20.9]. **Cytochrome a** is found in two forms in Complex IV of the electron-transport chain, as we shall see.) UQ–cyt c reductase (Figure 20.10) contains a b-type cytochrome, of 30 to 40 kD, with two different heme sites and one c-type cytochrome. The two hemes on the b cytochrome polypeptide in UQ–cyt c reductase are distinguished by their reduction potentials and the wavelength (λ_{max}) of the so-called α-band. One of these hemes, known as b_L or b_{566}, has a standard reduction potential, $\mathscr{E}_o'$, of -0.100 V and a wavelength of maximal absorbance (λ_{max}) of 566 nm. The other, known as b_H or b_{562}, has a standard reduction potential of $+0.050$ V and a λ_{max} of 562 nm. (H and L here refer to high and low reduction potential.)

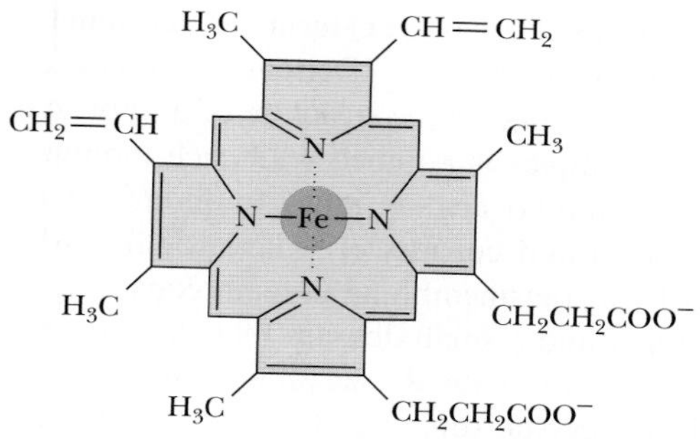

H_3C $CH = CH_2$

$CH_2 = CH$ CH_3

N
$N - Fe - N$
N

H_3C $CH_2CH_2COO^-$

H_3C $CH_2CH_2COO^-$

Iron protoporphyrin IX
(found in cytochrome *b*,
myoglobin, and hemoglobin)

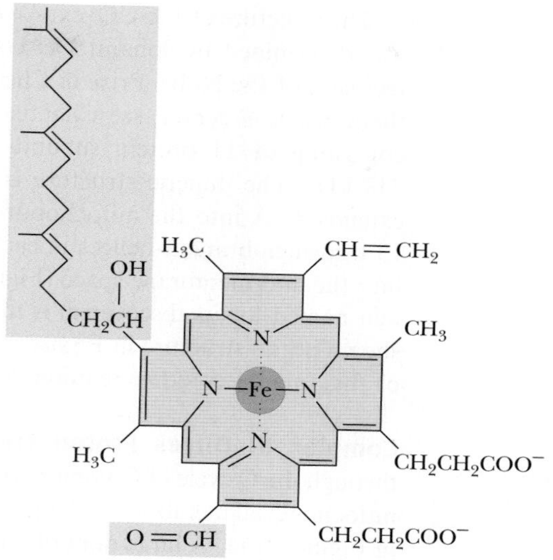

S
$CHCH_3$

H_3C

S
CH_3CH

N CH_3
$N - Fe - N$
N

H_3C

H_3C $CH_2CH_2COO^-$

H_3C $CH_2CH_2COO^-$

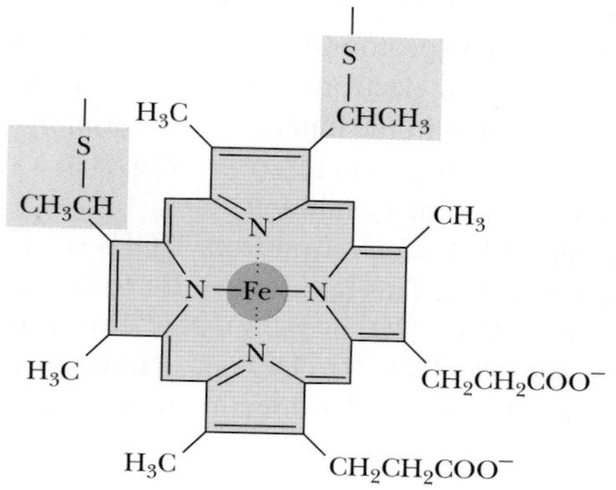

Heme *c*
(found in cytochrome *c*)

OH
CH_2CH

H_3C $CH = CH_2$

N CH_3
$N - Fe - N$
N

H_3C $CH_2CH_2COO^-$

$O = CH$ $CH_2CH_2COO^-$

Heme *a*
(found in cytochrome *a*)

FIGURE 20.9 The structures of iron protoporphyrin IX,
heme *c*, and heme *a*.

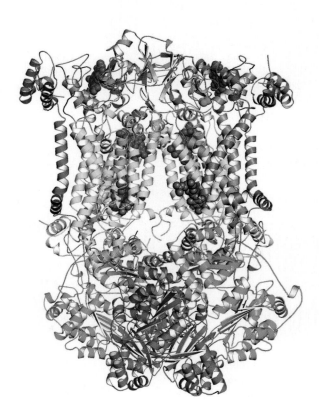

FIGURE 20.10 The structure of UQ–cyt *c* reductase,
also known as the cytochrome *bc₁* complex. The α-helical
bundle near the top of the structure defines the transmembrane domain of the protein (pdb id = 1BCC).

The structure of the UQ–cyt c reductase, also known as the **cytochrome bc_1 complex,** was determined by Johann Deisenhofer and his colleagues. (Deisenhofer was a co-recipient of the Nobel Prize in Chemistry for his work on the structure of a photosynthetic reaction center; see Chapter 21). The complex is a dimer, with each monomer consisting of 11 protein subunits and 2165 amino acid residues (monomer mass, 248 kD). The dimeric structure is pear-shaped and consists of a large domain that extends 75 Å into the mitochondrial matrix, a transmembrane domain consisting of 13 transmembrane α-helices in each monomer and a small domain that extends 38 Å into the intermembrane space (Figure 20.10). Most of the **Rieske protein** (an Fe-S protein named for its discoverer) is mobile in the crystal (only 62 of its 196 residues are shown in the structure in Figure 20.10), and Deisenhofer has postulated that mobility of this subunit could be required for electron transfer in the function of this complex.

Complex III Drives Proton Transport As with Complex I, passage of electrons through the Q cycle of Complex III is accompanied by proton transport across the inner mitochondrial membrane. The postulated pathway for electrons in this system is shown in Figure 20.11. A large pool of UQ and UQH_2 exists in the inner mitochondrial membrane. The Q cycle is initiated when a molecule of UQH_2 from this pool diffuses to a site (called Q_p) on Complex III near the cytosolic face of the membrane.

Oxidation of this UQH_2 occurs in two steps. First, an electron from UQH_2 is transferred to the Rieske protein and then to cytochrome c_1. This releases two H^+ to the cytosol and leaves $UQ \cdot^-$, a semiquinone anion form of UQ, at the Q_p site. The second electron is then transferred to the b_L heme, converting $UQ \cdot^-$ to UQ. The Rieske protein and cytochrome c_1 are similar in structure; each has a globular domain and is anchored

(a) First half of Q cycle

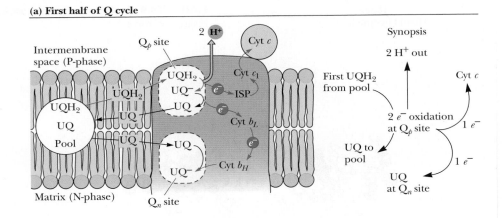

(b) Second half of Q cycle

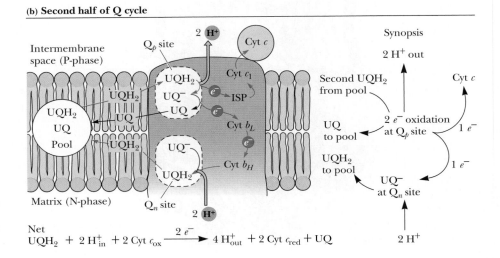

FIGURE 20.11 The Q cycle in mitochondria. **(a)** The electron-transfer pathway following oxidation of the first UQH_2 at the Q_p site near the cytosolic face of the membrane. **(b)** The pathway following oxidation of a second UQH_2.

Net
$$UQH_2 + 2\,H_{in}^+ + 2\,\text{Cyt}\,c_{ox} \xrightarrow{2\,e^-} 4\,H_{out}^+ + 2\,\text{Cyt}\,c_{red} + UQ$$

to the inner mitochondrial membrane by a hydrophobic segment. However, the hydrophobic segment is N-terminal in the Rieske protein and C-terminal in cytochrome c_1.

The electron on the b_L heme facing the cytosolic side of the membrane is now passed to the b_H heme on the matrix side of the membrane. This electron transfer occurs against a membrane potential of 0.15 V and is driven by the loss of redox potential as the electron moves from b_L ($\mathscr{E}_o' = -0.100$ V) to b_H ($\mathscr{E}_o' = +0.050$ V). The electron is then passed from b_H to a molecule of UQ at a second quinone-binding site, $\mathbf{Q}_n$, converting this UQ to $UQ \cdot^-$. The resulting $UQ \cdot^-$ remains firmly bound to the Q_n site. This completes the first half of the Q cycle (Figure 20.11a).

The second half of the cycle (Figure 20.11b) is similar to the first half, with a second molecule of UQH_2 oxidized at the Q_p site, one electron being passed to cytochrome c_1 and the other transferred to heme b_L and then to heme b_H. In this latter half of the Q cycle, however, the b_H electron is transferred to the semiquinone anion, $UQ \cdot^-$, at the Q_n site. With the addition of two H^+ from the mitochondrial matrix, this produces a molecule of UQH_2, which is released from the Q_n site and returns to the coenzyme Q pool, completing the Q cycle.

The Q Cycle Is an Unbalanced Proton Pump Why has nature chosen this rather convoluted path for electrons in Complex III? First of all, Complex III takes up two protons on the matrix side of the inner membrane and releases four protons on the cytoplasmic side for each pair of electrons that passes through the Q cycle. The other significant feature of this mechanism is that it offers a convenient way for a two-electron carrier, UQH_2, to interact with the b_L and b_H hemes, the Rieske protein Fe-S cluster, and cytochrome c_1, all of which are one-electron carriers.

Cytochrome c Is a Mobile Electron Carrier Electrons traversing Complex III are passed through cytochrome c_1 to **cytochrome c.** Cytochrome c is the only one of the mitochondrial cytochromes that is water soluble. Its structure (Figure 20.12) is globular; the planar heme group lies near the center of the protein, surrounded predominantly by hydrophobic amino acid residues. The iron in the porphyrin ring is coordinated both to a histidine nitrogen and to the sulfur atom of a methionine residue. Coordination with ligands in this manner on both sides of the porphyrin plane precludes the binding of oxygen and other ligands, a feature that distinguishes cytochrome c from hemoglobin (see Chapter 15).

Cytochrome c, like UQ, is a mobile electron carrier. It associates loosely with the inner mitochondrial membrane (in the intermembrane space on the cytosolic side of the inner membrane) to acquire electrons from the Fe-S–cyt c_1 aggregate of Complex III, and then it migrates along the membrane surface in the reduced state, carrying electrons to cytochrome c oxidase, the fourth complex of the electron-transport chain.

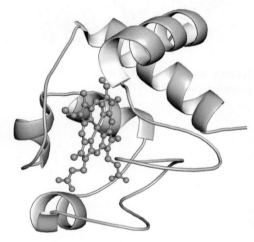

FIGURE 20.12 The structure of mitochondrial cytochrome c. The heme is shown at the center of the structure. It is covalently linked to the protein via two sulfur atoms. A third sulfur from a methionine residue coordinates the iron (pdb id = 2B4Z).

Complex IV Transfers Electrons from Cytochrome c to Reduce Oxygen on the Matrix Side

Complex IV is called cytochrome c oxidase because it accepts electrons from cytochrome c and directs them to the four-electron reduction of O_2 to form H_2O:

$$4 \text{ cyt } c \text{ (Fe}^{2+}) + 4 \text{ H}^+ + O_2 \longrightarrow 4 \text{ cyt } c \text{ (Fe}^{3+}) + 2 \text{ H}_2O \tag{20.9}$$

Thus, cytochrome c oxidase and O_2 are the final destination for the electrons derived from the oxidation of food materials. In concert with this process, cytochrome c oxidase also drives transport of protons across the inner mitochondrial membrane. The combined processes of oxygen reduction and proton transport involve a total of $8H^+$ in each catalytic cycle—four H^+ for O_2 reduction and four H^+ transported from the matrix to the intermembrane space.

The total number of subunits in cytochrome c oxidase varies from 2–4 (in bacteria) to 13 (in mammals). Three subunits (I, II, and III) are common to most organisms (Figure 20.13). This minimal complex, which contains two hemes (termed a and a_3) and

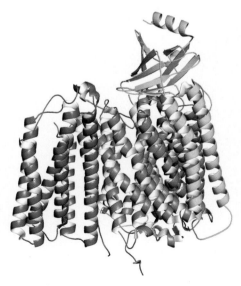

FIGURE 20.13 Bovine cytochrome c oxidase consists of 13 subunits. The 3 largest subunits—I (purple), II (yellow), and III (blue)—contain the proton channels and the redox centers (pdb id = 2EIJ).

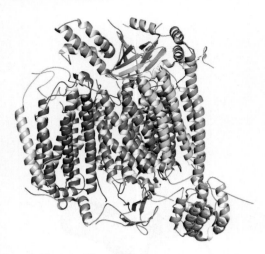

FIGURE 20.14 The complete structure of bovine cytochrome *c* oxidase (pdb id = 2EIJ).

three copper ions (two in the Cu_A center and one in the Cu_B site), is sufficient to carry out both oxygen reduction and proton transport.

The total mass of the protein in mammalian Complex IV (Figure 20.14) is 204 kD. In mammals, subunits I through III, the largest ones, are encoded by mitochondrial DNA, synthesized in the mitochondrion, and inserted into the inner membrane from the matrix side. The 10 smaller subunits are coded by nuclear DNA, are synthesized in the cytosol, and are presumed to play regulatory roles in the complex.

In the bovine structure, subunit I is cylindrical in shape and consists of 12 transmembrane helices, without any significant extramembrane parts. Hemes *a* and a_3, which lie perpendicular to the membrane plane, and Cu_B are cradled by the helices of subunit I (Figure 20.15). Subunits II and III lie on opposite sides of subunit I and do not contact each other (see Figure 20.13). Subunit II has an extramembrane domain on the outer face of the inner mitochondrial membrane. This domain consists of a 10-strand β-barrel that holds the two copper ions of the Cu_A site 7 Å from the nearest surface atom of the subunit. Subunit III consists of seven transmembrane helices with no significant extramembrane domains.

Electron Transfer in Complex IV Involves Two Hemes and Two Copper Sites

Electron transfer through Complex IV begins with binding of cytochrome *c* to the β-barrel of subunit II. Four electrons are transferred sequentially (one each from four molecules of cytochrome *c*) first to the Cu_A center, next to heme *a*, and finally to the Cu_B/heme a_3 active site, where O_2 is reduced to H_2O (Figure 20.15):

$$\text{Cyt } c \longrightarrow Cu_A \longrightarrow \text{heme } a \longrightarrow Cu_B/\text{heme } a_3 \longrightarrow O_2 \qquad (20.10)$$

A tryptophan residue, which lies 5Å above the Cu_A site (Figure 20.16a), is the entry point for electrons from cytochrome *c*. It lies in a hydrophobic patch on subunit II, surrounded by a ring of negatively charged Asp and Glu residues. Electrons flow rapidly from Cu_A to

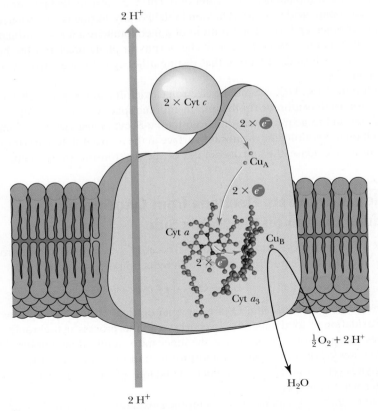

FIGURE 20.15 The electron-transfer pathway for cytochrome oxidase. Cytochrome *c* binds on the cytosolic side, transferring electrons through the copper and heme centers to reduce O_2 on the matrix side of the membrane (pdb id = 2EIJ).

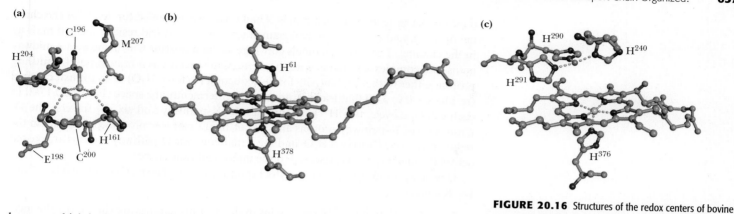

FIGURE 20.16 Structures of the redox centers of bovine cytochrome c oxidase. **(a)** The Cu_A site, **(b)** the heme a site, and **(c)** the binuclear Cu_B/heme a_3 site (pdb id = 2EIJ).

heme a, which is coordinated by a pair of His residues (Figure 20.16b), and then to the Cu_B/heme a_3 complex. The Fe atom in heme a_3 is five coordinate (Figure 20.16c), with four ligands from the heme plane and one from His^{376}. This leaves a sixth position free, and this is the catalytic site where O_2 binds and is reduced. Cu_B is about 5Å from the Fe atom of heme a_3 and is coordinated by three histidine ligands, including His^{240}, His^{290}, and His^{291} (Figure 20.16c). An unusual crosslink between His^{240} and Tyr^{244} lowers the pK_a of the Tyr hydroxyl so that it can participate in proton transport across the membrane.

Proton Transport Across Cytochrome c Oxidase Is Coupled to Oxygen Reduction

Proton transport in *R. sphaeroides* cytochrome c oxidase takes place via two channels denoted the D- and K-pathways (Figure 20.17a). Both these channels contain water molecules, and they are lined with polar residues that can either protonate and

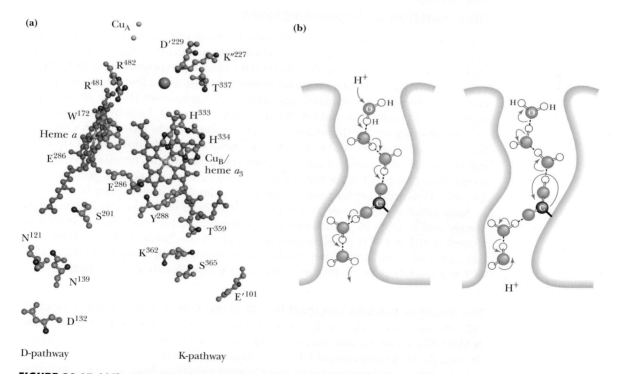

FIGURE 20.17 **(a)** The proton channels of cytochrome c oxidase from *R. sphaeroides*. Functional residues in the D- and K-pathways are indicated. The D- and K-pathways converge at the Cu_B/heme a_3 center. The proton exit channel is lined by residues 320 to 340 of subunit I (pdb id = 1M56). **(b)** Protons are presumed to "hop" along arrays of water molecules in the proton transport channels of cytochrome c oxidase. Such a chain of protonation and deprotonation events means that the proton eventually released from the exit channel is far removed from the proton that entered the D-pathway and initiated the cascade.

deprotonate or form hydrogen bonds. The D-pathway is named for Asp[132] at the channel opening, and the K-pathway is named for Lys[362], a critical residue located midway in the channel. These two channels converge at the binuclear Cu_B/heme a_3 site midway across the complex and the membrane. Here, Glu[286] serves as a branch point, shuttling protons either to the catalytic site for O_2 reduction (to form H_2O) or to the exit channel (residues 320 to 340) that leads protons to the intermembrane space (Figure 20.17a). In each catalytic cycle, two H^+ pass through the K-pathway and six H^+ traverse the D-pathway. The K-pathway protons and two of the D-pathway protons participate in the reduction of one O_2 to two H_2O, and the remaining four D-pathway protons are passed across the membrane and released to the intermembrane space.

How are protons driven across cytochrome c oxidase? The mechanism involves three key features:

- The pK_a values of protein side chains in the proton channels are shifted (by the local environment) to make them effective proton donors or acceptors during transport. For example, the pK_a of Glu[286] is unusually high at 9.4. (This is similar to the behavior of Asp[85] and Asp[96] in bacteriorhodopsin; see page 302, Chapter 9.)
- Electron-transfer events induce conformation changes that control proton transport. For example, redox events at the Cu_B/heme a_3 site are sensed by Glu[286] and an adjacent proton-gating loop (residues 169 to 175), controlling H^+ binding and release by Glu[286] and proton movement through the exit channel.
- Protons are "transported" via chains of hydrogen-bonded water molecules in the proton channels (Figure 20.17b). Sequential hopping of protons along these "proton wires" essentially transfers a "positive charge" between distant residues in the channel. (Note that the H^+ that arrives at an accepting residue is not the same proton that left the donating residue.)

The Complexes of Electron Transport May Function as Supercomplexes

For many years, the complexes of the electron-transport chain were thought to exist and function independently in the mitochondrial inner membrane, because each complex could be isolated in a pure yet functional state. Growing experimental evidence, however, supports the existence of multimeric **supercomplexes** of the four electron-transport complexes. These supercomplexes, also known as **respirasomes** (Figure 20.18a), may represent functional states, and association of two or more of the individual complexes may be advantageous for the organism. Supercomplexes can be identified by single particle electron microscopy, and kinetic measurements support the operation of the respiratory chain as one functional unit. Mutation studies show that Complex III is required to maintain Complex I in the mitochondria of human and mouse cells in culture, and Complex IV is necessary for proper assembly of Complex I in mouse fibroblasts.

Substantial evidence points to a requirement for cardiolipin (see page 240) in the assembly and stability of mitochondrial supercomplexes. For example, association of Complexes III and IV requires the presence of cardiolipin. Studies by Carola Hunte and colleagues show that cardiolipin stabilizes supercomplexes by neutralizing the charges of lysine residues at the binding interface between Complex III and Complex IV.

The Model of Electron Transport Is a Dynamic One The model that emerges for electron transport is shown in Figure 20.18b. Coenzyme Q collects electrons from NADH–UQ reductase and succinate–UQ reductase and delivers them (by diffusion through the membrane core) to UQ–cyt c reductase. Cytochrome c is water soluble and moves freely in the intermembrane space, carrying electrons from UQ–cyt c reductase to cytochrome c oxidase. In the process of these electron transfers, protons are driven across the inner membrane (from the matrix side to the intermembrane space). The proton gradient generated by electron transport represents an enormous source of potential energy. As seen in the next section, this potential energy is used to synthesize ATP as protons flow back into the matrix.

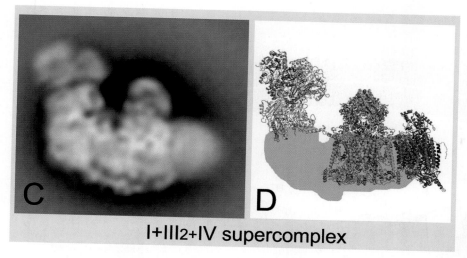

I+III₂+IV supercomplex

Reprinted from *Biochimica et Biophysica Acta* (BBA)—Bioenergetics, Vol. 1797, Natalya V. Dudkina, Roman Kouril, Katrin Peters, Hans-Peter Braun, Egbert J. Boekema, "Structure and function of mitochondrial supercomplexes," pp. 664–670, © 2010, with permission from Elsevier.

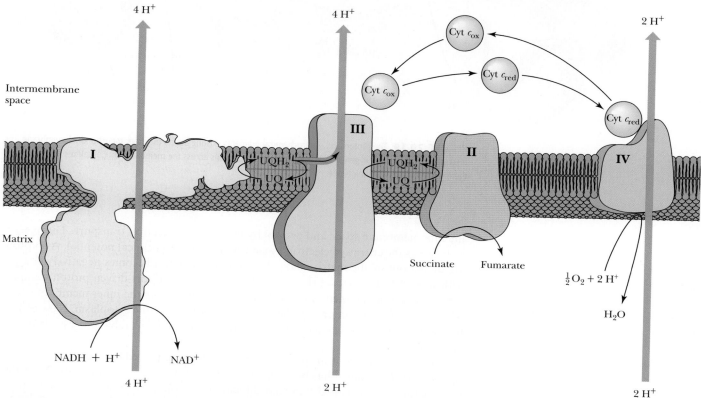

FIGURE 20.18 **(a)** An electron microscopy image of a supercomplex formed from Complex I, Complex III, and Complex IV, and a model of this complex. The hydrophilic portion of Complex I is yellow, the hydrophobic portion is solid beige, the Complex III dimer is green, and Complex IV is purple. From Dudkina, N., et al., 2010. Structure and function of mitochondrial supercomplexes. *Biochimica et Biophysica Acta* **1797:**664–670. **(b)** A model for the electron-transport pathway in the mitochondrial inner membrane. UQ/UQH₂ and cytochrome *c* are mobile electron carriers and function by transferring electrons between the complexes. The proton transport driven by Complexes I, III, and IV is indicated.

Electron Transfer Energy Stored in a Proton Gradient: The Mitchell Hypothesis

In 1961, Peter Mitchell, a British biochemist, proposed that the energy stored in a proton gradient across the inner mitochondrial membrane by electron transport drives the synthesis of ATP in cells. The proposal became known as **Mitchell's chemiosmotic hypothesis.** In this hypothesis, protons are driven across the membrane from the matrix to

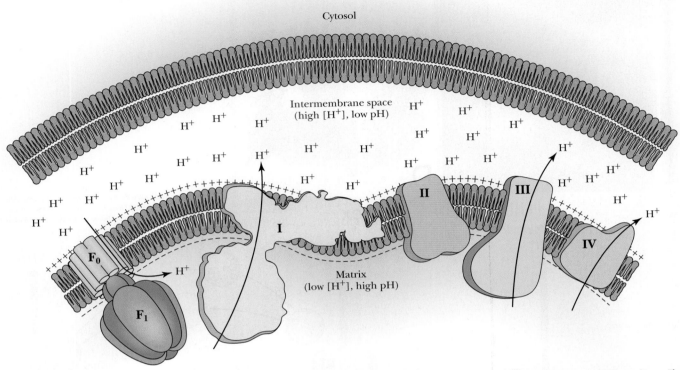

FIGURE 20.19 The proton and electrochemical gradients existing across the inner mitochondrial membrane. The electrochemical gradient is generated by the transport of protons across the membrane by Complexes I, III, and IV in the inner mitochondrial membrane.

the intermembrane space and cytosol by the events of electron transport. This mechanism stores the energy of electron transport in an electrochemical potential. As protons are driven out of the matrix, the pH rises and the matrix becomes negatively charged with respect to the cytosol (Figure 20.19). Electron transport-driven proton pumping thus creates a pH gradient and an electrical gradient across the inner membrane, both of which tend to attract protons back into the matrix from the cytoplasm. Flow of protons down this electrochemical gradient, an energetically favorable process, drives the synthesis of ATP.

The ratio of protons transported per pair of electrons passed through the chain—the so-called $H^+/2e^-$ ratio—has been an object of great interest for many years. Nevertheless, the ratio has remained extremely difficult to determine. The consensus estimate for the electron-transport pathway from succinate to O_2 is $6H^+/2e^-$. The ratio for Complex I by itself remains uncertain, but recent best estimates place it as high as $4H^+/2e^-$. On the basis of this value, the stoichiometry of transport for the pathway from NADH to O_2 is $10H^+/2e^-$. Although this is the value assumed in Figure 20.18, it is important to realize that this represents a consensus drawn from many experiments.

20.3 What Are the Thermodynamic Implications of Chemiosmotic Coupling?

Mitchell's chemiosmotic hypothesis revolutionized our thinking about the energy coupling that drives ATP synthesis by means of an electrochemical gradient. How much energy is stored in this electrochemical gradient? For the transmembrane flow of protons across the inner membrane (from inside [matrix] to outside), we could write

$$H^+_{in} \longrightarrow H^+_{out}$$

(20.11)

The free energy difference for protons across the inner mitochondrial membrane includes a term for the concentration difference and a term for the electrical potential. This is expressed as

$$\Delta G = RT \ln \frac{[c_2]}{[c_1]} + Z \mathcal{F} \Delta \psi \qquad (20.12)$$

where c_1 and c_2 are the proton concentrations on the two sides of the membrane, Z is the charge on a proton, $\mathcal{F}$ is Faraday's constant, and $\Delta \psi$ is the potential difference across the membrane. For the case at hand, this equation becomes

$$\Delta G = RT \ln \frac{[H_{out}^+]}{[H_{in}^+]} + Z \mathcal{F} \Delta \psi \qquad (20.13)$$

In terms of the matrix and cytoplasm pH values, the free energy difference is

$$\Delta G = -2.303 \, RT(pH_{out} - pH_{in}) + \mathcal{F} \Delta \psi \qquad (20.14)$$

Reported values for $\Delta \psi$ and ΔpH vary, but the membrane potential is always found to be positive outside and negative inside, and the pH is always more acidic outside and more basic inside. Taking typical values of $\Delta \psi = 0.18$ V and $\Delta pH = 1$ unit, the free energy change associated with the movement of one mole of protons from inside to outside is

$$\Delta G = 2.3 \, RT + \mathcal{F}(0.18 \text{ V}) \qquad (20.15)$$

With $\mathcal{F} = 96.485$ kJ/V · mol, the value of ΔG at 37°C is

$$\Delta G = 5.9 \text{ kJ} + 17.4 \text{ kJ} = 23.3 \text{ kJ} \qquad (20.16)$$

which is the free energy change for movement of a mole of protons across an inner membrane. Note that the free energy terms for both the pH difference and the potential difference are unfavorable for the outward transport of protons, with the latter term making the greater contribution. On the other hand, the ΔG for inward flow of protons is -23.3 kJ/mol. It is this energy that drives the synthesis of ATP, in accord with Mitchell's model. Peter Mitchell was awarded the Nobel Prize in Chemistry in 1978.

20.4 : How Does a Proton Gradient Drive the Synthesis of ATP?

The great French chemist Antoine Lavoisier showed in 1777 that foods undergo combustion in the body. Since then, chemists and biochemists have wondered how energy from food oxidation is captured by living things. Mitchell paved the way by suggesting that a proton gradient across the inner mitochondrial membrane could drive the synthesis of ATP. But how could the proton gradient be coupled to ATP production? The answer lies in a mitochondrial complex called ATP synthase, or sometimes F_1F_0- ATPase (for the reverse reaction it catalyzes). The F_1 portion of the ATP synthase was first identified in early electron micrographs of mitochondrial preparations as spherical, 8.5-nm projections or particles on the inner membrane. The purified particles catalyze ATP hydrolysis, the reverse reaction of the ATP synthase. Stripped of these particles, the membranes can still carry out electron transfer but cannot synthesize ATP. In one of the first reconstitution experiments with membrane proteins, Efraim Racker showed that adding the particles back to stripped membranes restored electron transfer-dependent ATP synthesis.

ATP Synthase Is Composed of F_1 and F_0

ATP synthase is a remarkable molecular machine. It is an enzyme, a proton pump, and a rotating molecular motor. Nearly all the ATP that fuels our cellular processes is made by this multifaceted molecular superstar. The spheres observed in electron micrographs make up the **F_1 unit,** which catalyzes ATP synthesis (Figure 20.20). These F_1 spheres are

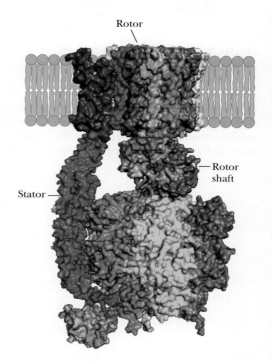

FIGURE 20.20 The ATP synthase, a rotating molecular motor. The c-, γ-, and ϵ-subunits constitute the rotating portion (the rotor) of the motor. Flow of protons from the a-subunit through the c-subunits turns the rotor and drives the cycle of conformational changes in α and β that synthesize ATP (pdb id = 1C17; 1E79; 2A7U; 2CLY; and 2BO5).

(a)

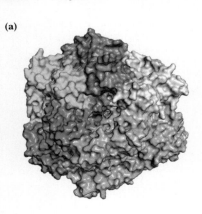

(b)

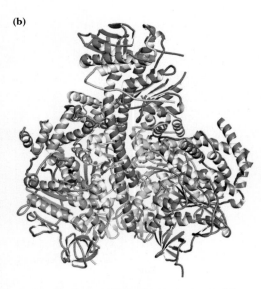

FIGURE 20.21 (a) An axial view of the F$_1$ unit of the F$_1$F$_0$-ATP synthase, showing alternating α- and β-subunits in a hexameric array, with the γ-subunit (purple) visible in the center of the structure. (b) A side view of the F1 unit, with one α-subunit and one β-subunit removed to show how the γ-subunit (red) extends through the center of the $\alpha_3\beta_3$ hexamer. Also shown are the δ-subunit (aqua) and the ϵ-subunit (pink), which link the γ-subunit to the F$_0$ unit (pdb id = 1E79).

TABLE 20.2	Yeast F$_1$F$_0$–ATP Synthase Subunit Organization		
Complex	Protein Subunit Function	Mass (kD)	Stoichiometry
F$_1$	α	55.4	3 Stator
	β	51.3	3 Stator
	γ	30.6	1 Rotor
	δ	14.6	1 Rotor†
	ϵ	6.6	1 Rotor
F$_0$	a	27.9	1 Stator
	b	23.3	1 Stator
	c	7.8	8–15* Rotor
	d	19.7	1 Stator
	h	10.4	1 Stator
	OSCP	20.9	1 Stator

*The number of c-subunits varies among organisms: bovine mitochondria, 8; yeast mitochondria and *E coli*, 10; *Ilyobacter tartaricus*, 11; spinach chloroplasts, 14; *Spirulina platensis*, 15.
†The subunit nomenclature can be confusing. *E. coli* ATP synthase lacks a δ-subunit in its rotor; its δ-subunit is analogous structrually and functionally to the mitochondrial OSCP. Similarly, subunit F6 in the bovine ATP synthase is weakly homologous to subunit h in yeast.

attached to an integral membrane protein aggregate called the **F$_0$ unit.** F$_1$ consists of five polypeptide chains named α, β, γ, δ, and ϵ, with a subunit stoichiometry $\alpha_3\beta_3\gamma\delta\epsilon$ (Table 20.2 and Figure 20.21). F$_0$ includes three hydrophobic subunits denoted by a, b, and c, with an apparent stoichiometry of $a_1b_2c_{8-15}$. F$_0$ forms the transmembrane pore or channel through which protons move to drive ATP synthesis. The F$_0$ structures in prokaryotic organisms and plants have 10 to 15 c-subunits, whereas c-rings of only 8 subunits are found in the F$_0$ structures of vertebrates and all or most invertebrates (Table 20.2).

The a- and b-subunits of F$_0$ form part of the **stator**—a stationary component anchored in the membrane—and a ring of 8 to 15 c-subunits (see Table 20.3) constitutes a major component of the **rotor** of the motor. Protons flowing through the a–c complex cause the c-ring to rotate in the membrane. Each c-subunit is a folded pair of α-helices joined by a short loop, whereas the a-subunit is presumed to be a cluster of α-helices. The b-subunit, together with the d- and h-subunits and the oligomycin sensitivity-conferring protein (OSCP), form a long, slender **stalk** that connects F$_0$ in the membrane with F$_1$, which extends out into the matrix. The b-, d-, and h-subunits form long α-helical segments that comprise the stalk, and OSCP adds a helical bundle cap that sits at the bottom of an α-subunit of F$_1$ (Figure 20.20). The stalk is a stable link between F$_0$ and F$_1$, essentially joining the two, both structurally and functionally.

Are there structural accommodations in the mitochondrial membrane that suit different numbers of subunits in the circular ring of c-subunits? Nobelist John Walker and colleagues have shown that Lys-43 of the c-subunits in bovine mitochondria are completely trimethylated, forming a bulky quaternary amino group on each subunit, adjacent to the phospholipid headgroups of the inner membrane. This bulky group would clash with phospholipid headgroups, but Walker has postulated that cardiolipin, which is essential to ATP synthase activity and which contains two phosphatidylglycerol groups but no headgroup, can accommodate the trimethylamino groups and link adjacent c-subunits in the relatively small 8-subunit c-ring. No organism with c-rings of more than 8 subunits are trimethylated at Lys^{43}.

The Catalytic Sites of ATP Synthase Adopt Three Different Conformations

The F$_1$ structure appears at first to be a symmetric hexamer of α- and β-subunits. However, it is asymmetric in several ways. The α- and β-subunits, arranged in an alternating pattern in the hexamer, are similar but not identical. The hexamer contains six ATP-

In the absence of a proton gradient:

$$ADP + P_i \rightleftharpoons [ATP] \rightleftharpoons ADP + \;^{18}O-P-^{18}OH$$

Enzyme bound

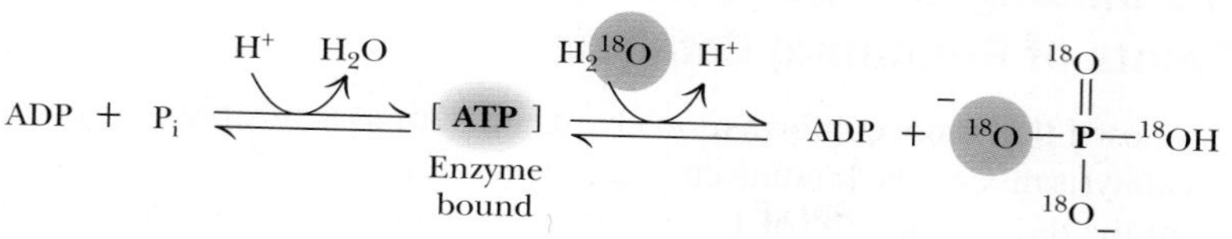

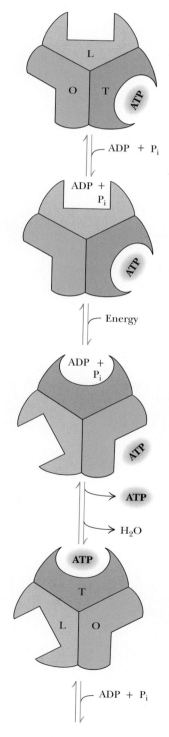

FIGURE 20.22 ATP–ADP exchange in the absence of a proton gradient. Exchange leads to incorporation of ^{18}O in phosphate as shown. Boyer's experiments showed that ^{18}O could be incorporated into all four positions of phosphate, demonstrating that the free energy change for ATP formation from enzyme-bound ADP + P_i is close to zero. (From Parsons, D. F., 1963. *Science* **140**:985.)

binding sites, each of them arranged at the interface of adjacent subunits. Three of these, each located mostly on a β-subunit but with some residues contributed by an α-subunit, are catalytic sites for ATP synthesis. The other three, each located mostly on an α-subunit but with residues contributed by a β-subunit, are noncatalytic and inactive. The noncatalytic α-sites have similar structures, but *the three catalytic β-sites have three quite different conformations.* In the crystal structure first characterized by John Walker, one of the β-subunit ATP sites contains AMP-PNP (a nonhydrolyzable analog of ATP), another contains ADP, and the third site is empty.

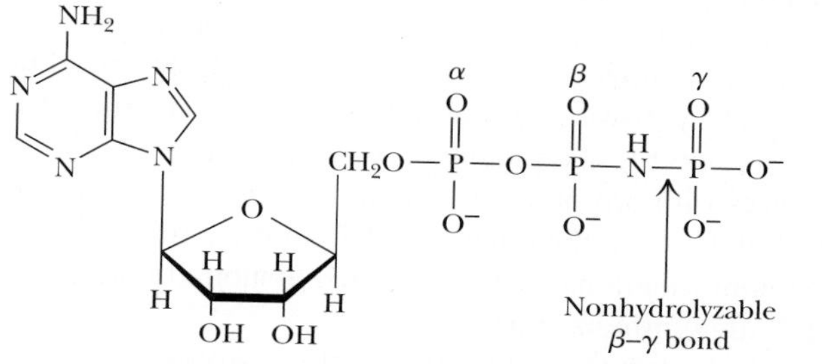

(20.17)

Nonhydrolyzable β–γ bond

β, γ-Imidoadenosine 5'-triphosphate (AMP-PNP)

Walker's work provided structural verification for a novel hypothesis first advanced by Paul Boyer, the **binding change mechanism** for ATP synthesis. Walker and Boyer, whose efforts provided complementary insights into the workings of this molecular motor, shared in the Nobel Prize for Chemistry in 1997.

Boyer's ^{18}O Exchange Experiment Identified the Energy-Requiring Step

The elegant studies by Boyer of ^{18}O exchange in ATP synthase provided important insights into the mechanism of the enzyme. Boyer and his colleagues studied the ability of the synthase to incorporate labeled oxygen from $H_2^{18}O$ into P_i. This reaction (Figure 20.22) occurs via synthesis of ATP from ADP and P_i, followed by hydrolysis of ATP with incorporation of oxygen atoms from the solvent. Although net production of ATP requires coupling with a proton gradient, Boyer observed that this exchange reaction occurs readily, even in the absence of a proton gradient. The exchange reaction was so facile that, eventually, all four oxygens of phosphate were labeled with ^{18}O. This important observation indicated that the formation of enzyme-bound ATP does not require energy. The experiments that followed, by Boyer, Harvey Penefsky, and others, showed clearly that the energy-requiring step in the ATP synthase was actually the release of newly synthesized ATP from the enzyme (Figure 20.23). Eventually, it would be shown that flow of protons through F_0 drives the enzyme conformational changes that result in the binding of substrates on ATP synthase, ATP synthesis, and the release of products.

FIGURE 20.23 The binding change mechanism for ATP synthesis by ATP synthase. This model assumes that F_1 has three interacting and conformationally distinct active sites: an open (O) conformation with almost no affinity for ligands, a loose (L) conformation with low affinity for ligands, and a tight (T) conformation with high affinity for ligands. Synthesis of ATP is initiated (step 1) by binding of ADP and P_i to an L site. In the second step, an energy-driven conformational change converts the L site to a T conformation and converts T to O and O to L. In the third step, ATP is synthesized at the T site and released from the O site. Two additional passes through this cycle produce two more ATPs and return the enzyme to its original state.

Cycle repeats

Boyer's Binding Change Mechanism Describes the Events of Rotational Catalysis

Boyer proposed that these conformation changes occurred in a rotating fashion. His rotational catalysis model, the binding change mechanism (Figure 20.23), suggested that at any instant the three β subunits of F_1 existed in three different conformations, that these different states represented the three steps of ATP synthesis, and that each site stepped through the three states to make ATP. A site beginning with ADP and phosphate bound (the first state) would synthesize ATP (producing the second state) and then release ATP, leaving an empty site (the third state). In the binding change mechanism, the three catalytic sites thus cycle concertedly through the three intermediate states of ATP synthesis.

Proton Flow Through F_0 Drives Rotation of the Motor and Synthesis of ATP

How might the cycling proposed by Boyer's binding change mechanism occur? Important clues have emerged from several experiments that show that the γ-subunit rotates with respect to the $\alpha\beta$ complex. How such rotation might be linked to transmembrane proton flow and ATP synthesis is shown in Figure 20.24. The ring of c-subunits is a rotor that turns with respect to the a-subunit, a stator component consisting of five transmembrane α-helices with proton access channels on either side of the membrane. The γ-subunit is the link between the functions of F_1 and F_0. In one complete rotation, the γ-subunit drives conformational changes in each β-subunit that lead to ATP synthesis. Thus, three ATPs are synthesized per turn.

But how does the F_0 complex couple the events of proton transport and ATP synthesis? The a-subunit contains two half-channels, a proton inlet channel that opens to the intermembrane space and a proton outlet channel that opens to the matrix. The c-subunits are proton carriers that transfer protons from the inlet channel to the outlet channel only by rotation of the c-ring. Each c-subunit contains a protonatable residue, Asp^{61}. Protons flowing from the intermembrane space through the inlet half-channel protonate the Asp^{61} of a passing c-subunit and ride the rotor around the ring until they reach the outlet channel and flow out into the matrix.

The molecular details of this process are shown in Figure 20.24. Each c-subunit in the c-ring has an inner helix and an outer helix. Asp^{61} is located midway along the outer α-helix. When protonated, the Asp carboxyl faces into the adjacent subunit. Rotation of the entire outer α-helix exposes Asp^{61} to the outside when it is deprotonated. Arg^{210}, located midway on a transmembrane helix of the a-subunit, forms hydrogen bonds with Asp^{61} residues on two adjacent c-subunits. The half-channels of the a-subunit extend up and down from Arg^{210}. The inlet channel terminates in Asn^{214}, whereas the outlet channel terminates at Ser^{206}.

The structure of the c-subunit complex is exquisitely suited for proton transport. When a proton enters the a-subunit inlet channel and is transferred from a-subunit Asn^{214} to c-subunit Asp^{61}, the α-helix of that c-subunit rotates clockwise to bury the Asp carboxyl group (Figure 20.24c). Each Asp^{61} remains protonated once it leaves the a-subunit interface, because the hydrophobic environment of the membrane interior makes deprotonation (and charge formation) highly unfavorable. However, when a protonated Asp residue approaches the a-subunit outlet channel, the proton is transferred to Ser^{206} and exits through the outlet channel. The a-subunit Arg^{210} side chain orients adjacent Asp^{61} groups and promotes transfers of entering protons from a-subunit Asn^{214} to Asp^{61} and transfers of exiting protons from Asp^{61} to a-subunit Ser^{206}. Arg^{210}, because it is protonated, also prevents direct proton transfer from Asn^{214} to Ser^{206}, which would circumvent ring rotation and motor function.

ATP synthesis occurs in concert with rotation of the c-ring, because the γ-subunit is anchored to the rotating c-ring and rotates with it. Rotation causes the γ-subunit to turn relative to the three β-subunit nucleotide sites of F_1, changing the conformation of each in sequence, so ADP is first bound, then phosphorylated, then released, according to Boyer's binding change mechanism.

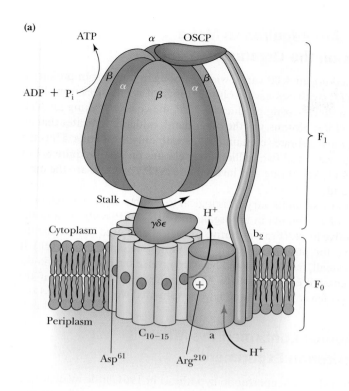

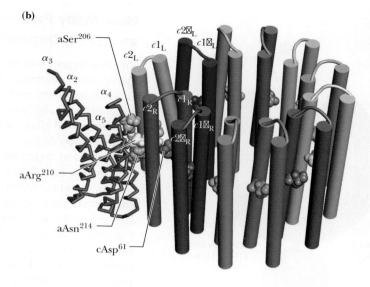

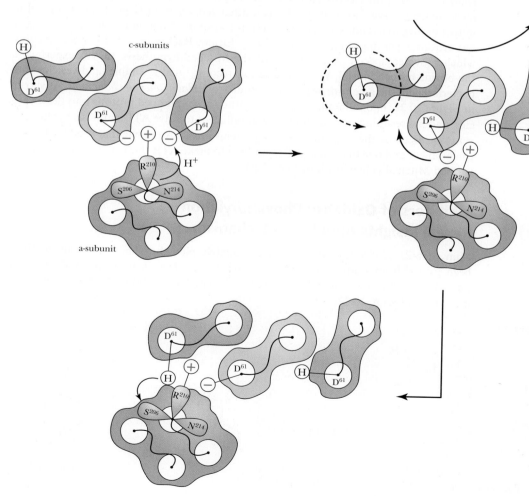

FIGURE 20.24 (a) Protons entering the inlet half-channel in the *a*-subunit are transferred to binding sites on *c*-subunits. Rotation of the *c*-ring delivers protons to the outlet half-channel in the *a*-subunit. Flow of protons through the structure turns the rotor and drives the cycle of conformational changes in β that synthesize ATP. (b) Arg[210] on the *a*-subunit lies between the end of the inlet half-channel (Asn[214]) and the end of the outlet half-channel (Ser[206]) (pdb id = 1C17). (c) A view looking down into the plane of the membrane. Transported protons flow from the inlet half-channel to Asp[61] residues on the *c*-ring, around the ring, and then into the outlet half-channel. When Asp[61] is protonated, the outer helix of the *c*-subunit rotates clockwise to bury the protonated carboxyl group for its trip around the *c*-ring. Counterclockwise ring rotation then brings another protonated Asp[61] to the *a*-subunit, where an exiting proton is transferred to the outlet half-channel.

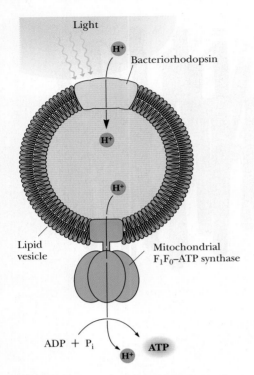

FIGURE 20.25 The reconstituted vesicles containing ATP synthase and bacteriorhodopsin used by Stoeckenius and Racker to confirm the Mitchell chemiosmotic hypothesis. (Note that both proteins are in reverse inside-out orientation relative to their normal orientation in biological membranes.)

How Many Protons Are Required to Make an ATP? It Depends on the Organism

The bioenergetic cost of making an ATP varies widely among organisms. In prokarya, chloroplasts, and fungi, ATP synthases have evolved a range of c-ring sizes from 10 to 15 subunits. Since one turn of the c-ring produces 3 ATP, the cost of making an ATP varies from 3.3 (10/3) to 5 (15/3) protons. On the other hand, evidence indicates that all vertebrates and all or most invertebrates contain c-rings with only 8 subunits. Thus the approximately 50,000 vertebrates and two million invertebrates on earth require 8/3 or 2.7 protons to make one ATP. Vertebrate and invertebrate ATP synthases are the most efficient that have been found.

Why might the number of c subunits vary? It is logical to assume an inverse relationship between the number of c subunits in an organism's F_1F_0-ATP synthase and the strength of the proton-motive force. (Proton-motive force is a term used to describe the free energy difference, ΔG, for protons across a membrane [see equation 20.12].) If the proton-motive force is small, more H^+ would have to traverse the membrane via the c-subunit ring to deliver enough energy to drive ATP synthesis, and of course, if the proton-motive force is large, fewer protons (and fewer c subunits) would be needed.

Racker and Stoeckenius Confirmed the Mitchell Model in a Reconstitution Experiment

When Mitchell first described his chemiosmotic hypothesis in 1961, little evidence existed to support it and it was met with considerable skepticism by the scientific community. Eventually, however, considerable evidence accumulated to support this model. It is now clear that the electron-transport chain generates a proton gradient, and careful measurements have shown that ATP is synthesized when a pH gradient is applied to mitochondria that cannot carry out electron transport. Even more relevant is a simple but crucial experiment reported in 1974 by Efraim Racker and Walther Stoeckenius, which provided specific confirmation of the Mitchell hypothesis. In this experiment, the bovine mitochondrial ATP synthase was reconstituted in simple lipid vesicles with **bacteriorhodopsin,** a light-driven proton pump from *Halobacterium halobium.* As shown in Figure 20.25, upon illumination, bacteriorhodopsin pumped protons into these vesicles, and the resulting proton gradient was sufficient to drive ATP synthesis by the ATP synthase. Because the only two kinds of proteins present were one that produced a proton gradient and one that used such a gradient to make ATP, this experiment essentially verified Mitchell's chemiosmotic hypothesis.

Inhibitors of Oxidative Phosphorylation Reveal Insights About the Mechanism

Many details of electron transport and oxidative phosphorylation mechanisms have been gained from studying the effects of particular electron transport and oxidative phosphorylation inhibitors (Figure 20.26). The sites of inhibition by these agents are indicated in Figure 20.27.

FIGURE 20.26 The structures of several inhibitors of electron transport and oxidative phosphorylation.

Rotenone

Amytal (amobarbital)

Demerol (meperidine)

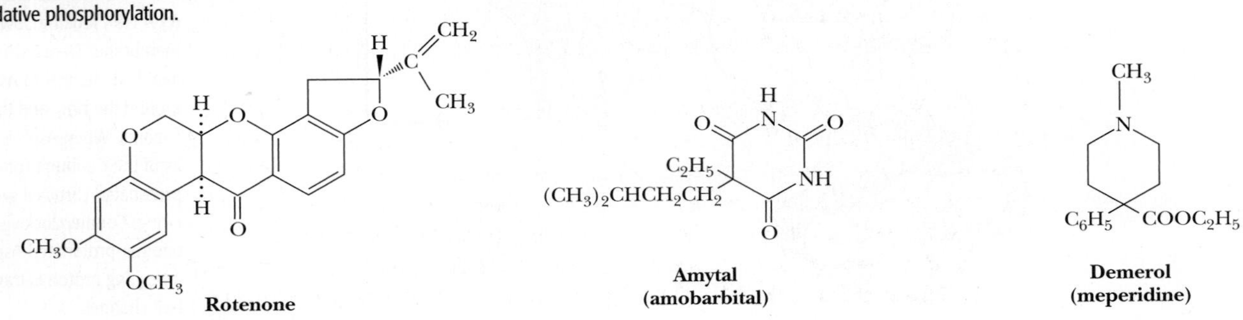

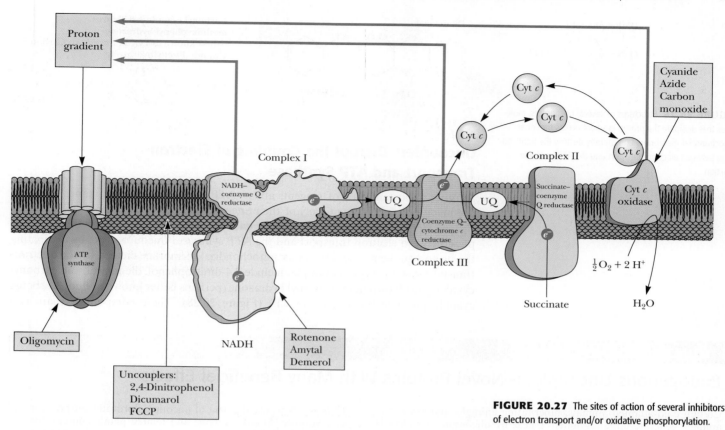

FIGURE 20.27 The sites of action of several inhibitors of electron transport and/or oxidative phosphorylation.

Inhibitors of Complexes I, II, and III Block Electron Transport Rotenone is a common insecticide that strongly inhibits the NADH–UQ reductase. Rotenone is obtained from the roots of several species of plants. Natives in certain parts of the world have made a practice of beating the roots of trees along riverbanks to release rotenone into the water, where it paralyzes fish and makes them easy prey. Amytal and other barbiturates and the widely prescribed painkiller Demerol also inhibit Complex I. All these substances appear to inhibit reduction of coenzyme Q and the oxidation of the Fe-S clusters of NADH–UQ reductase.

Cyanide, Azide, and Carbon Monoxide Inhibit Complex IV Complex IV, the cytochrome c oxidase, is specifically inhibited by cyanide, azide, and carbon monoxide (Figure 20.27). Cyanide and azide bind tightly to the ferric form of cytochrome a_3, whereas carbon monoxide binds only to the ferrous form. The inhibitory actions of cyanide and azide at this site are very potent, whereas the principal toxicity of carbon monoxide arises from its affinity for the iron of hemoglobin. Herein lies an important distinction between the poisonous effects of cyanide and carbon monoxide. Because animals (including humans) carry many, many hemoglobin molecules, they must inhale a large quantity of carbon monoxide to die from it. These same organisms, however, possess comparatively few molecules of cytochrome a_3. Consequently, a limited exposure to cyanide can be lethal. The sudden action of cyanide attests to the organism's constant and immediate need for the energy supplied by electron transport.

Oligomycin Is an ATP Synthase Inhibitor Inhibitors of ATP synthase include oligomycin. Oligomycin is a polyketide antibiotic that acts directly on ATP synthase by binding to the OSCP subunit of F_0. Oligomycin also blocks the movement of protons through F_0.

Dinitrophenol

Dicumarol

Carbonyl cyanide-*p*-trifluoro-methoxyphenyl hydrazone
—best known as **FCCP**; for **F**luoro **C**arbonyl **C**yanide **P**henylhydrazone

$F_3C - O$

$C \equiv N$

$C \equiv N$

FIGURE 20.28 Structures of several uncouplers, molecules that dissipate the proton gradient across the inner mitochondrial membrane and thereby destroy the tight coupling between electron transport and the ATP synthase reaction.

Uncouplers Disrupt the Coupling of Electron Transport and ATP Synthase

Another important class of reagents affects ATP synthesis, but in a manner that does not involve direct binding to any of the proteins of the electron-transport chain or the F_1F_0–ATPase. These agents are known as **uncouplers** because they disrupt the tight coupling between electron transport and the ATP synthase. Uncouplers act by dissipating the proton gradient across the inner mitochondrial membrane created by the electron-transport system. Typical examples include 2,4-dinitrophenol, dicumarol, and carbonyl cyanide-*p*-trifluoro-methoxyphenyl hydrazone (perhaps better known as fluorocarbonyl cyanide phenylhydrazone, or FCCP) (Figure 20.28). These compounds share two

---

HUMAN BIOCHEMISTRY

Endogenous Uncouplers—Novel Proteins With Many Beneficial Effects

Certain cold-adapted animals, hibernating animals, and newborn animals generate large amounts of heat by uncoupling oxidative phosphorylation. These organisms have a type of fat known as brown adipose tissue, so called for the color imparted by the many mitochondria this adipose tissue contains. The inner membrane of brown adipose tissue mitochondria contains large amounts of an endogenous protein called **thermogenin** (literally, "heat maker") or **uncoupling protein 1 (UCP1)**. UCP1 creates a passive proton channel through which protons flow from the cytosol to the matrix and UCP1 is a universal temperature regulator for animals. Mice that lack UCP1 cannot maintain their body temperature in cold conditions, whereas normal animals produce larger amounts of UCP1 when they are cold-adapted.

Two other uncoupling proteins, UCP2 and UCP3, show 60% sequence identity with UCP1 and, like UCP1, they can also dissipate the mitochondrial proton gradient, but they do so in the service of very different physiological functions. UCP2 is found mainly in kidney, pancreas, spleen, immune cells and the central nervous system, and it is involved in a variety of processes, including regulation of food intake, insulin secretion, and immunity. In particular, it plays a role in regulation of the generation of **reactive oxygen species (ROS)**, which include hydrogen peroxide, hypochlorite ion, hydroxyl radical, and superoxide anion (see figure).

$HO - OH$
Hydrogen peroxide

$:\ddot{O}:H$
Hydroxyl radical

$:\ddot{O}:\ddot{C}l:{}^{\ominus}$
Hypochlorite ion

$\cdot\ddot{O}:\ddot{O}{}^{\ominus}$
Superoxide anion

Mitochondrial ROS production correlates exponentially with mitochondrial proton-motive force (Equation 20.12 and page 661). (The larger the proton-motive force, the more likely that the respiratory electron carriers will donate electrons to oxygen to form ROS.) Dissipation of the proton-motive force by UCP2 thus attenuates ROS formation and reduces oxidative stress, which has been linked to diabetes and atherosclerosis. On the other hand, UCP3 may play a similar role in attenuation of ROS during fatty acid oxidation. It also appears to be essential for adaptation to fasting.

Certain plants use the heat of uncoupled proton transport for a special purpose. Skunk cabbage and related plants contain floral spikes that are maintained as much as 20° above ambient temperature in this way. The warmth of the spikes serves to vaporize odiferous molecules, which attract insects that fertilize the flowers. Red tomatoes have very small mitochondrial membrane proton gradients compared with green tomatoes—evidence that uncouplers are more active in red tomatoes.

Source: Azzu, V., and Brand, M., 2010. The on-off switches of the mitochondrial uncoupling proteins. *Trends in Biochemical Sciences* **35:**298–307.

common features: hydrophobic character and a dissociable proton. As uncouplers, they function by carrying protons across the inner membrane. Their tendency is to acquire protons on the cytosolic surface of the membrane (where the proton concentration is high) and carry them to the matrix side, thereby destroying the proton gradient that couples electron transport and the ATP synthase. In mitochondria treated with uncouplers, electron transport continues and protons are driven out through the inner membrane. However, they leak back in so rapidly via the uncouplers that ATP synthesis does not occur. Instead, the energy released in electron transport is dissipated as heat.

ATP–ADP Translocase Mediates the Movement of ATP and ADP Across the Mitochondrial Membrane

ATP, the cellular energy currency, must exit the mitochondria to carry energy throughout the cell, and ADP must be brought into the mitochondria for reprocessing. Neither of these processes occurs spontaneously because the highly charged ATP and ADP molecules do not readily cross biological membranes. Instead, these processes are mediated by a single transport system, the **ATP–ADP translocase.** This protein tightly couples the exit of ATP with the entry of ADP so that the mitochondrial nucleotide levels remain approximately constant. For each ATP transported out, one ADP is transported into the matrix. The translocase is the most abundant protein in the inner mitochondrial membrane and accounts for approximately 14% of the total mitochondrial membrane protein. It is a homodimer of 30-kD subunits. The structure of the bovine translocase consists of six transmembrane α-helices. The helices are all tilted with respect to the membrane, and the first, third, and fifth helices are bent or kinked at proline residues in the middle of the membrane (Figure 20.29).

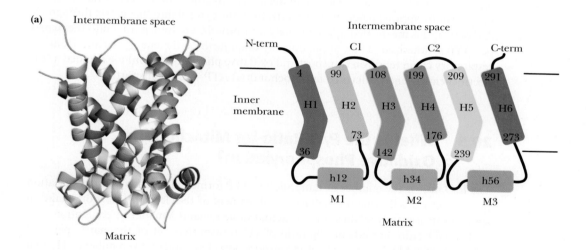

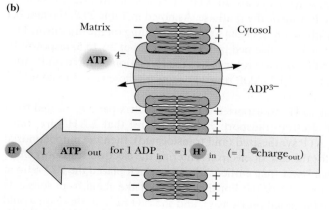

FIGURE 20.29 **(a)** The bovine ATP–ADP translocase (pdb id = 2C3E). **(b)** Outward transport of ATP (via the ATP–ADP translocase) is favored by the membrane electrochemical potential.

Transport occurs via a single nucleotide-binding site, which alternately faces the matrix and the intermembrane space. It binds ATP on the matrix side, reorients to face the outside, and exchanges ATP for ADP, with subsequent rearrangement to face the matrix side of the inner membrane. Transport of ADP and ADP is regulated by phosphorylation of Tyr[194] on transmembrane helix number 5 (Figure 20.29). The relatively large size and charge of the ATP and ADP substrates, compared to the relatively small size of the translocase itself, may require large conformation changes during transport. Export of ATP by the translocase is coupled to the ATP/ADP ratio and the proton-motive force across the inner mitochondrial membrane, ensuring that the energy needs of the cell will be met.

Outward Movement of ATP Is Favored over Outward ADP Movement The charge on ATP at pH 7.2 or so is about -4, and the charge on ADP at the same pH is about -3. Thus, net exchange of an ATP (out) for an ADP (in) results in the net movement of one negative charge from the matrix to the cytosol. (This process is equivalent to the movement of a proton from the cytosol to the matrix.) Recall that the inner membrane is positive outside (see Figure 20.19), and it becomes clear that outward movement of ATP is favored over outward ADP transport, ensuring that ATP will be transported out (Figure 20.29). Inward movement of ADP is favored over inward movement of ATP for the same reason. Thus, the membrane electrochemical potential itself controls the specificity of the ATP–ADP translocase. However, the electrochemical potential is diminished by the ATP–ADP translocase cycle and therefore operates with an energy cost to the cell. The cell must compensate by passing yet more electrons down the electron-transport chain.

What is the cost of ATP–ADP exchange relative to the energy cost of ATP synthesis itself? We already noted that moving one ATP out and one ADP in is the equivalent of one proton moving from the cytosol to the matrix. Synthesis of an ATP results from the movement of approximately three protons from the cytosol into the matrix through F_0. Altogether, this means that approximately four protons are transported into the matrix per ATP synthesized. Thus, approximately one-fourth of the energy derived from the respiratory chain (electron transport and oxidative phosphorylation) is expended as the electrochemical energy devoted to mitochondrial ATP–ADP transport.

20.5 What Is the P/O Ratio for Mitochondrial Oxidative Phosphorylation?

The **P/O ratio** is the number of molecules of ATP formed in oxidative phosphorylation per two electrons flowing through a defined segment of the electron-transport chain. In spite of intense study of this ratio, its actual value remains a matter of contention.

The P/O ratio depends on the ratio of H^+ transported out of the matrix per $2\,e^-$ passed from NADH to O_2 in the electron-transport chain and on the number of H^+ that pass through the ATP synthase to synthesize an ATP. The latter number depends on the number of c-subunits in the F_0 ring of the synthase. As noted in Table 20.2, the number of c-subunits in the ATP synthase ranges from 8 to 15, depending on the organism. This would correspond to ratios of H^+ consumed per ATP from about 2.7 to 5, respectively, since each rotation of the ATP synthase rotor drives the formation of three ATP. Adding one H^+ for the action of the ATP–ADP translocase raises these values to about 3.7 and 6, respectively.

If we accept the value of 10 H^+ transported out of the matrix per $2\,e^-$ passed from NADH to O_2 through the electron-transport chain, and agree that 3.7 H^+ are transported into the matrix per ATP synthesized (and translocated), then the mitochondrial P/O ratio is 10/3.7, or 2.7, for the case of electrons entering the electron-transport chain as NADH. This is somewhat lower than earlier estimates, which placed the P/O ratio at 3 for mitochondrial oxidation of NADH. For the portion of the chain from succinate to O_2, the $H^+/2e^-$ ratio is 6 (as noted previously), and the P/O ratio in this case would be 6/3.7, or 1.6; earlier estimates placed this number at 2. The consensus of more recent

$$\left(\frac{1\,\text{ATP}}{3.7\,H^+}\right)\left(\frac{10\,H^+}{2\,e^-\,[\text{NADH}\rightarrow\frac{1}{2}O_2]}\right) = \frac{10}{3.7} = \frac{P}{O}$$

experimental measurements of P/O ratios for these two cases has been closer to the values of 2.7 and 1.6. Many chemists and biochemists, accustomed to the integral stoichiometries of chemical and metabolic reactions, were once reluctant to accept the notion of nonintegral P/O ratios. At some point, as we learn more about these complex coupled processes, it may be necessary to reassess the numbers.

20.6 How Are the Electrons of Cytosolic NADH Fed into Electron Transport?

Most of the NADH used in electron transport is produced in the mitochondrial matrix, an appropriate site because NADH is oxidized by Complex I on the matrix side of the inner membrane. Furthermore, the inner mitochondrial membrane is impermeable to NADH. Recall, however, that NADH is produced in glycolysis by glyceraldehyde-3-P dehydrogenase in the cytosol. If this NADH were not oxidized to regenerate NAD^+, the glycolytic pathway would cease to function due to NAD^+ limitation. Eukaryotic cells have a number of **shuttle systems** that harvest the electrons of cytosolic NADH for delivery to mitochondria without actually transporting NADH across the inner membrane (Figures 20.30 and 20.31).

The Glycerophosphate Shuttle Ensures Efficient Use of Cytosolic NADH

In the **glycerophosphate shuttle,** two different **glycerophosphate dehydrogenases,** one in the cytosol and one on the outer face of the mitochondrial inner membrane, work together to carry electrons into the mitochondrial matrix (see Figure 20.30). NADH produced in the cytosol transfers its electrons to dihydroxyacetone phosphate, thus reducing it to glycerol-3-phosphate. This metabolite is reoxidized by the FAD^+-dependent mitochondrial membrane enzyme to reform dihydroxyacetone phosphate and enzyme-bound $FADH_2$. The two electrons of $[FADH_2]$ are passed directly to UQ, forming UQH_2. Thus, via this shuttle, cytosolic NADH can be used to produce mitochondrial $[FADH_2]$ and, subsequently, UQH_2. As a result, cytosolic NADH oxidized via this shuttle route yields only 1.6 molecules of ATP. The cell "pays" with a potential ATP molecule for the convenience of getting cytosolic NADH into the mitochondria. Although this may seem wasteful, there is an important payoff. The glycerophosphate shuttle is essentially irreversible, and even when NADH levels are very low relative to NAD^+, the cycle operates effectively.

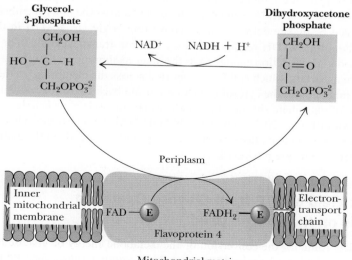

FIGURE 20.30 The glycerophosphate shuttle (also known as the glycerol phosphate shuttle) couples the cytosolic oxidation of NADH with mitochondrial reduction of [FAD].

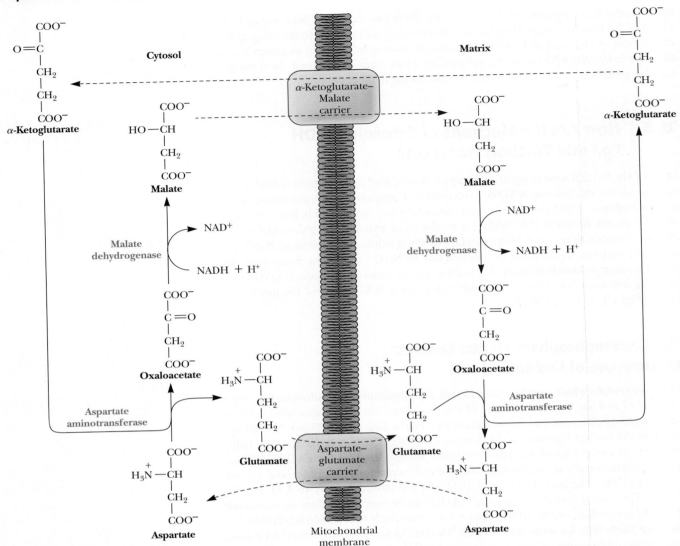

FIGURE 20.31 The malate (oxaloacetate)–aspartate shuttle, which operates across the inner mitochondrial membrane.

The Malate–Aspartate Shuttle Is Reversible

The second electron shuttle system, called the **malate–aspartate shuttle,** is shown in Figure 20.31. Oxaloacetate is reduced in the cytosol, acquiring the electrons of NADH (which is oxidized to NAD^+). Malate is transported across the inner membrane, where it is reoxidized by malate dehydrogenase, converting NAD^+ to NADH in the matrix. This mitochondrial NADH readily enters the electron-transport chain. The oxaloacetate produced in this reaction cannot cross the inner membrane and must be transaminated to form aspartate, which can be transported across the membrane to the cytosolic side. Transamination in the cytosol recycles aspartate back to oxaloacetate. In contrast to the glycerol phosphate shuttle, the malate–aspartate cycle is reversible, and it operates as shown in Figure 20.31 only if the NADH/NAD^+ ratio in the cytosol is higher than the ratio in the matrix. Because this shuttle produces NADH in the matrix, the full 2.7 ATPs per NADH are recovered.

The Net Yield of ATP from Glucose Oxidation Depends on the Shuttle Used

The complete route for the conversion of the metabolic energy of glucose to ATP has now been described in Chapters 18 through 20. Assuming appropriate P/O ratios, the number of ATP molecules produced by the complete oxidation of a molecule of glucose

TABLE 20.3	Yield of ATP from Glucose Oxidation		
		ATP Yield per Glucose	
Pathway		Glycerol–Phosphate Shuttle	Malate–Aspartate Shuttle
Glycolysis: glucose to pyruvate (cytosol)			
Phosphorylation of glucose		-1	-1
Phosphorylation of fructose-6-phosphate		-1	-1
Dephosphorylation of 2 molecules of 1,3-BPG		$+2$	$+2$
Dephosphorylation of 2 molecules of PEP		$+2$	$+2$
Oxidation of 2 molecules of glyceraldehyde-3-phosphate yields 2 NADH			
Pyruvate conversion to acetyl-CoA (mitochondria)			
2 NADH			
Citric acid cycle (mitochondria)			
2 molecules of GTP from 2 molecules of succinyl-CoA		$+2$	$+2$
Oxidation of 2 molecules each of isocitrate, α-ketoglutarate, and malate yields 6 NADH			
Oxidation of 2 molecules of succinate yields 2 [FADH$_2$]			
Oxidative phosphorylation (mitochondria)			
2 NADH from glycolysis yield 1.6 ATPs each if NADH is oxidized by glycerol–phosphate shuttle; 2.7 ATP by malate–aspartate shuttle		$+3.2$	$+5.4$
Oxidative decarboxylation of 2 pyruvate to 2 acetyl-CoA:			
2 NADH produce 2.7 ATPs each		$+5.4$	$+5.4$
2 [FADH$_2$] from each citric acid cycle produce 1.6 ATPs each		$+3.2$	$+3.2$
6 NADH from citric acid cycle produce 2.7 ATPs each		$+16.2$	$+16.2$
Net Yield		32.0	34.2

Note: These P/O ratios of 2.7 and 1.6 for mitochondrial oxidation of NADH and [FADH$_2$] are "consensus values." Because they may not reflect actual values and because these ratios may change depending on metabolic conditions, these estimates of ATP yield from glucose oxidation are approximate.

can be estimated. Keeping in mind that P/O ratios must be viewed as approximate, for all the reasons previously cited, we will assume the values of 2.7 and 1.6 for the mitochondrial oxidation of NADH and succinate, respectively. In eukaryotic cells, the combined pathways of glycolysis, the TCA cycle, electron transport, and oxidative phosphorylation then yield a net of approximately 32 to 34 molecules of ATP per molecule of glucose oxidized, depending on the shuttle route used (Table 20.3).

The net stoichiometric equation for the oxidation of glucose, using the glycerol phosphate shuttle, is

$$\text{Glucose} + 6\,O_2 + {\sim}32\,\text{ADP} + {\sim}32\,P_i \longrightarrow$$
$$6\,CO_2 + {\sim}32\,\text{ATP} + {\sim}38\,H_2O \qquad (20.18)$$

Because the 2 NADH formed in glycolysis are "transported" by the glycerol phosphate shuttle in this case, they each yield only 1.6 ATP, as already described. On the other hand, if these 2 NADH take part in the malate–aspartate shuttle, each yields 2.7 ATP, giving a total (in this case) of 34.2 ATP formed per glucose oxidized. Most of the ATP—28 out of 32 or 30 out of 34—is produced by oxidative phosphorylation; only 4 ATP molecules result from direct synthesis during glycolysis and the TCA cycle.

The situation in bacteria is somewhat different. Prokaryotic cells need not carry out ATP–ADP exchange. Thus, bacteria have the potential to produce as many as 38 ATP per glucose.

3.5 Billion Years of Evolution Have Resulted in a Very Efficient System

Hypothetically speaking, how much energy does a eukaryotic cell extract from the glucose molecule? Taking a value of 50 kJ/mol for the hydrolysis of ATP under cellular conditions (see Chapter 3), the production of 34 ATPs per glucose oxidized yields 1700 kJ/mol of glucose. The cellular oxidation (combustion) of glucose yields $\Delta G = -2937$ kJ/mol. We can calculate an efficiency for the pathways of glycolysis, the TCA cycle, electron transport, and oxidative phosphorylation of $1700/2937 \times 100\% = 58\%$.

20.7 How Do Mitochondria Mediate Apoptosis?

ROS: Reactive oxygen species, such as oxygen ions, free radicals, and peroxides.

Mitochondria not only are the home of the TCA cycle and oxidative phosphorylation but also are a crossroads for several cell signaling pathways. Mitochondria take up Ca^{2+} ions released from the endoplasmic reticulum, thus helping control intracellular Ca^{2+} signals. They produce reactive oxygen species (ROS) that play signaling roles in cells, although ROS can also cause cellular damage. Mitochondria also participate in the programmed death of cells, a process known as **apoptosis** (the second "p" is silent in this word).

Apoptosis is a mechanism through which certain cells are eliminated from higher organisms. It is central to the development and homeostasis of multicellular organisms, and it is the route by which unwanted or harmful cells are eliminated. Under normal circumstances, apoptosis is suppressed through compartmentation of the involved activators and enzymes. Mitochondria play a major role in this subcellular partitioning of the apoptotic activator molecules. One such activator is cytochrome c, which normally resides in the intermembrane space, bound tightly to a lipid chain of cardiolipin in the membrane (Figure 20.32). A variety of triggering agents, including Ca^{2+}, ROS, certain lipid molecules, a family of **caspases** (proteases with a catalytic Cys residue that cleave next to Asp residues in targets), and certain protein kinases, can induce the opening of pores in the mitochondrial membrane. For example, using hydrogen peroxide as a substrate, cytochrome c can oxidize its bound cardiolipin chain, releasing itself from the membrane. When the outer membrane is made permeable by other apoptotic signals, cytochrome c can enter the cytosol. This is termed **mitochondrial outer membrane permeabilization (MOMP)**.

Permeabilization events, which occur at points where outer and inner mitochondrial membranes are in contact, involve association of the ATP–ADP translocase in the inner membrane and the **voltage-dependent anion channel (VDAC)** in the outer membrane. This interaction leads to the opening of protein-permeable pores. Cytochrome c, as well as several other proteins, can pass through these pores. Pore formation is carefully regulated by the BCL-2 family of proteins, which includes both proapoptotic members (proteins known as Bax, Bid, and Bad) and antiapoptotic members (BCL-2 itself, as well as BCL-X_L and BCL-W). The BCL-2 proteins also activate a series of events that trigger **autophagy**, a catabolic process involving the degradation of the cell's own components through the lysosomal machinery (Chapter 31). The cell's choice of apoptosis or autophagy is regulated in part by a membrane protein called **RAGE**, the **receptor for advanced glycation endproducts** (see Chapter 22 for more about advanced glycation endproducts).

Cytochrome c Triggers Apoptosome Assembly

But how is the release of cytochrome c translated into the activation of the death cascade, a point of no return for the cell? The answer lies in the assembly, in the cytosol, of a signaling platform called the **apoptosome** (Figure 20.33). The function of the

(a)

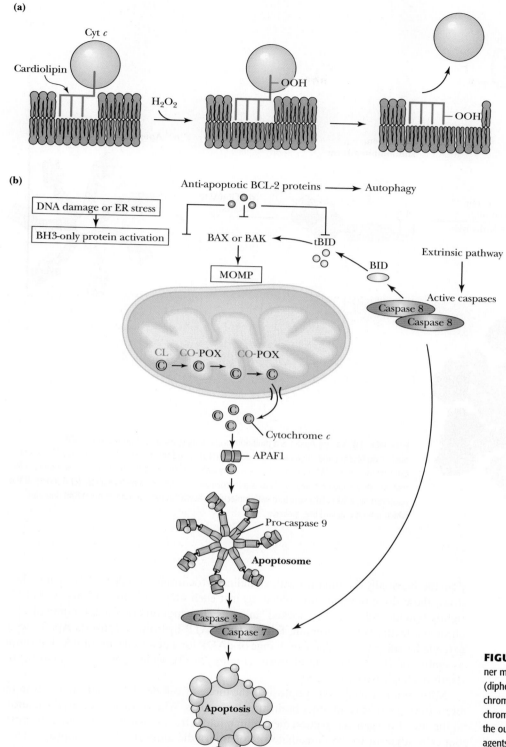

(b)

FIGURE 20.32 **(a)** Cytochrome *c* is anchored at the inner mitochondrial membrane by association with cardiolipin (diphosphatidylglycerol). The peroxidase activity of cytochrome *c* oxidizes a cardiolipin lipid chain, releasing cytochrome *c* from the membrane. **(b)** The opening of pores in the outer membrane, induced by a variety of triggering agents, releases cytochrome *c* to the cytosol, where it initiates the events of apoptosis.

apoptosome is to activate a cascade of proteases called **caspases.** (Here, "c" is for cysteine and "asp" is for aspartic acid. Caspases have Cys at the active site and cleave their peptide substrates after Asp residues.) The apoptosome is a wheel-shaped, heptameric platform that looks like (and in some ways behaves like) an earth-orbiting space station. It is assembled from seven subunits of the **apoptotic protease-activating factor 1 (Apaf-1),** a multidomain protein. Apaf-1 contains an ATPase domain (which prefers dATP over ATP in some organisms), a **caspase-recruitment domain (CARD),** and a **WD40 repeat**

(a) Apaf-1

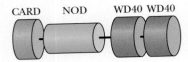

(b) Locked form

FIGURE 20.33 **(a)** Apaf-1 is a multidomain protein, consisting of an N-terminal CARD, a nucleotide-binding and oligomerization domain (NOD), and several WD40 domains. **(b)** Binding of cytochrome *c* to the WD40 domains and ATP hydrolysis unlocks Apaf-1 to form the semiopen conformation. Nucleotide exchange leads to oligomerization and apoptosome formation. **(c)** A model of the apoptosome, a wheellike structure with molecules of cytochrome *c* bound to the WD40 domains, which extend outward like spokes.

domain. Normally (before the death-signaling cytochrome *c* is released from mitochondria), these three domains are folded against each other (Figure 20.33b), with dATP tightly bound, and Apaf-1 is "locked" in an inactive monomeric state. Binding of cytochrome *c* to the WD40 domain, followed by dATP hydrolysis, converts Apaf-1 to an extended conformation. Then, exchange of dADP for a new molecule of dATP prompts assembly of the heptameric platform (Figure 20.33), which goes on to activate the death-dealing caspase cascade.

Mitochondria-mediated apoptosis is the mode of cell death for many neurons in the brain during strokes and other brain-trauma injuries. When a stroke occurs, the neurons at the site of oxygen deprivation die within minutes by a nonspecific process of necrosis, but cells adjacent to the immediate site of injury die more slowly by apoptosis. These latter cells have been saved by a variety of therapeutic interventions that suppress apoptosis in laboratory studies, raising the hope that strokes and other neurodegenerative conditions may someday be treated clinically in similar ways.

SUMMARY

OWL Login to OWL to develop problem-solving skills and complete online homework assigned by your professor.

20.1 Where in the Cell Do Electron Transport and Oxidative Phosphorylation Occur?
The processes of electron transport and oxidative phosphorylation are membrane associated. In prokaryotes, the conversion of energy from NADH and [FADH$_2$] to the energy of ATP via electron transport and oxidative phosphorylation is carried out at (and across) the plasma membrane. In eukaryotic cells, electron transport and oxidative phosphorylation are localized in mitochondria. Mitochondria are surrounded by a simple outer membrane and a more complex inner membrane. The space between the inner and outer membranes is referred to as the intermembrane space.

20.2 How Is the Electron-Transport Chain Organized?
The components of the electron-transport chain can be purified from the mitochondrial inner membrane as four distinct protein complexes: (I) NADH–coenzyme Q reductase, (II) succinate–coenzyme Q reductase, (III) coenzyme Q–cytochrome c reductase, and (IV) cytochrome c oxidase. In complexes I, III, and IV, electron transfer drives the movement of protons from the mitochondrial matrix to the intermembrane space.

Complex I (NADH dehydrogenase) involves more than 45 polypeptide chains, 1 molecule of flavin mononucleotide (FMN), and as many as nine Fe-S clusters, together containing a total of 20 to 26 iron atoms. The complex transfers electrons from NADH to FMN, then to a series of FeS proteins, and finally to coenzyme Q.

Complex II (succinate dehydrogenase) oxidizes succinate to fumarate, with concomitant reduction of bound FAD to FADH$_2$. This FADH$_2$ transfers its electrons immediately to Fe-S centers, which pass them on to UQ. Electrons flow from succinate to UQ.

Complex III drives electron transport from coenzyme Q to cytochrome c via a unique redox pathway known as the Q cycle. UQ–cytochrome c reductase (UQ–cyt c reductase), as this complex is known, involves three different cytochromes and an Fe-S protein. In the cytochromes of these and similar complexes, the iron atom at the center of the porphyrin ring cycles between the reduced Fe^{2+} (ferrous) and oxidized Fe^{3+} (ferric) states.

Complex IV transfers electrons from cytochrome c to reduce oxygen on the matrix side. Complex IV (cytochrome c oxidase) accepts electrons from cytochrome c and directs them to the four-electron reduction of O$_2$ to form 2H$_2$O via Cu$_A$ sites, the heme iron of cytochrome a, Cu$_B$, and the heme iron of a_3.

20.3 What Are the Thermodynamic Implications of Chemiosmotic Coupling?
Peter Mitchell was the first to propose that electron transport leads to formation of a proton gradient that drives ATP synthesis. The free energy difference for protons across the inner mitochondrial membrane includes a term for the concentration difference and a term for the electrical potential. It is this energy that drives the synthesis of ATP, in accord with Mitchell's model.

20.4 How Does a Proton Gradient Drive the Synthesis of ATP?
The mitochondrial complex that carries out ATP synthesis is ATP synthase (F$_1$F$_0$–ATPase). ATP synthase consists of two principal complexes, designated F$_1$ and F$_0$. Protons taken up from the cytosol by one of the proton access channels in the a-subunit of F$_0$ ride the rotor of c-subunits until they reach the other proton access channel on a, from which they are released into the matrix. Such rotation causes the γ-subunit of F$_1$ to turn relative to the three β-subunit nucleotide sites of F$_1$, changing the conformation of each in sequence, so ADP is first bound, then phosphorylated, then released, according to Boyer's binding change mechanism.

The inhibitors of oxidative phosphorylation include rotenone, a common insecticide that strongly inhibits the NADH–UQ reductase. Complex IV is specifically inhibited by cyanide (CN$^-$), azide (N$_3^-$), and carbon monoxide (CO). Cyanide and azide bind tightly to the ferric form of cytochrome a_3, whereas carbon monoxide binds only to the ferrous form.

Uncouplers disrupt the coupling of electron transport and ATP synthase. Uncouplers share two common features: hydrophobic character and a dissociable proton. They function by carrying protons across the inner membrane, acquiring protons on the outer surface of the membrane (where the proton concentration is high) and carrying them to the matrix side. Uncouplers destroy the proton gradient that couples electron transport and the ATP synthase.

ATP–ADP translocase mediates the movement of ATP and ADP across the mitochondrial membrane. The ATP–ADP translocase is an inner membrane protein that tightly couples the exit of ATP with the entry of ADP so that the mitochondrial nucleotide levels remain approximately constant. For each ATP transported out, one ADP is transported into the matrix. ATP–ADP translocase binds ATP on the matrix side, reorients to face the intermembrane space, and exchanges ATP for ADP, with subsequent reorientation back to the matrix face of the inner membrane.

20.5 What Is the P/O Ratio for Mitochondrial Oxidative Phosphorylation?
The P/O ratio is the number of molecules of ATP formed in oxidative phosphorylation per two electrons flowing through a defined segment of the electron-transport chain. The consensus value for the mitochondrial P/O ratio is 10/3.7, or 2.7, for electrons entering the electron-transport chain as NADH. For succinate to O$_2$, the P/O ratio in this case would be 6/3.7, or 1.6.

20.6 How Are the Electrons of Cytosolic NADH Fed into Electron Transport?
Eukaryotic cells have a number of shuttle systems that collect the electrons of cytosolic NADH for delivery to mitochondria without actually transporting NADH across the inner membrane. In the glycerophosphate shuttle, two different glycerophosphate dehydrogenases, one in the cytosol and one on the outer face of the mitochondrial inner membrane, work together to carry electrons into the mitochondrial matrix. In the malate–aspartate shuttle, oxaloacetate is reduced in the cytosol, acquiring the electrons of NADH (which is oxidized to NAD$^+$). Malate is transported across the inner membrane, where it is reoxidized by malate dehydrogenase, converting NAD$^+$ to NADH in the matrix.

20.7 How Do Mitochondria Mediate Apoptosis?
Mitochondria are a crossroads for several cell signaling pathways. Mitochondria take up Ca^{2+} ions released from the endoplasmic reticulum, helping control intracellular Ca^{2+} signals. They produce ROS that play signaling roles in cells. They also participate in apoptosis, the programmed death of cells. Triggering agents, including Ca^{2+}, ROS, and certain lipid molecules and protein kinases, can induce the opening of pores in the mitochondrial membrane, releasing cytochrome c, which then binds to the WD40 domain of Apaf-1, activating formation of the heptameric apoptosome platform. Mitochondria-mediated apoptosis is the mode of cell death of many neurons in the brain during strokes and other brain-trauma injuries, and interventions that suppress apoptosis may eventually be useful in clinical settings.

- The essential features of mitochondrial structure.

- The energetics and organization of the four complexes of the electron-transport chain.

- The structure and function of Complex I, the NADH dehydrogenase.

- The structure and function of Complex II, the succinate dehydrogenase.

- The structure and function of Complex III, the UQ-cytochrome c reductase.

- The mechanism of electron and proton transport in the Q cycle in Complex III.

- The structure and function of Complex IV, the cytochrome c oxidase.

- The essential features of supramolecular complexes in the mitochondrial inner membrane.

- The details of the Mitchell chemiosmotic hypothesis.

- The thermodynamic consequences of chemiosmotic coupling.

- The mechanism by which a proton gradient drives synthesis of ATP in mitochondria.

- The structure and function of the mitochondrial F_1F_0-ATP synthase.

- The essential features of the Boyer binding change mechanism for the mitochondrial F_1F_0-ATP synthase.

- The details of the Racker-Stoeckenius experiment that affirmed the Mitchell chemiosmotic hypothesis.

- The inhibitors of electron transport and oxidative phosphorylation and the proteins to which they bind.

- The mechanism of action of uncouplers of electron transport and oxidative phosphorylation.

- The essential features of the mitochondrial ATP-ADP translocase.

- The meaning of the P/O ratio for mitochondrial oxidative phosphorylation.

- The essential features of the shuttle systems that feed electrons from cytosolic NADH into electron transport.

- The details of the calculation of ATP yield from glucose oxidation.

- The processes and molecules in mitochondria that mediate apoptosis.

PROBLEMS

Answers to all problems are at the end of this book. Detailed solutions are available in the *Student Solutions Manual, Study Guide, and Problems Book.*

1. **Understanding Redox Couples** For the following reaction,

$$[FAD] + 2\ cyt\ c\ (Fe^{2+}) + 2\ H^+ \longrightarrow [FADH_2] + 2\ cyt\ c\ (Fe^{3+})$$

determine which of the redox couples is the electron acceptor and which is the electron donor under standard-state conditions, calculate the value of $\Delta\mathscr{E}_o'$, and determine the free energy change for the reaction.

2. **Determining the Redox Potential of the Glyceraldehyde-3-Phosphate Dehydrogenase Reaction** Calculate the value of $\Delta\mathscr{E}_o'$ for the glyceraldehyde-3-phosphate dehydrogenase reaction, and calculate the free energy change for the reaction under standard-state conditions.

3. **Understanding the pH Dependence of NAD⁺ Reduction** For the following redox reaction,

$$NAD^+ + 2\ H^+ + 2\ e^- \longrightarrow NADH + H^+$$

suggest an equation (analogous to Equation 3.48) that predicts the pH dependence of this reaction, and calculate the reduction potential for this reaction at pH 8.

4. **Understanding an Antidote for Cyanide Poisoning** Sodium nitrite $(NaNO_2)$ is used by emergency medical personnel as an antidote for cyanide poisoning (for this purpose, it must be administered immediately). Based on the discussion of cyanide poisoning in Section 20.4, suggest a mechanism for the lifesaving effect of sodium nitrite.

5. **Assessing the Mechanism of Uncoupler Action** A wealthy investor has come to you for advice. She has been approached by a biochemist who seeks financial backing for a company that would market dinitrophenol and dicumarol as weight-loss medications. The biochemist has explained to her that these agents are uncouplers and that they would dissipate metabolic energy as heat. The investor wants to know if you think she should invest in the biochemist's company. How do you respond?

6. **Assessing the [ATP]/[ADP][Pᵢ] Ratio and ATP Synthesis** Assuming that 3 H⁺ are transported per ATP synthesized in the mitochondrial matrix, the membrane potential difference is 0.18 V (negative inside), and the pH difference is 1 unit (acid outside, basic inside),

calculate the largest ratio of [ATP]/[ADP][Pᵢ] under which synthesis of ATP can occur.

7. **Analyzing the Succinate Dehydrogenase Reaction** Of the dehydrogenase reactions in glycolysis and the TCA cycle, all but one use NAD⁺ as the electron acceptor. The lone exception is the succinate dehydrogenase reaction, which uses covalently bound FAD of a flavoprotein as the electron acceptor. The standard reduction potential for this bound FAD is in the range of 0.003 to 0.091 V (see Table 3.5). Compared with the other dehydrogenase reactions of glycolysis and the TCA cycle, what is unique about succinate dehydrogenase? Why is bound FAD a more suitable electron acceptor in this case?

8. **Analysis of the NADH-CoQ Reductase Reaction**
   a. What is the standard free energy change ($\Delta G^{o\prime}$) for the reduction of coenzyme Q by NADH as carried out by Complex I (NADH–coenzyme Q reductase) of the electron-transport pathway if $\mathscr{E}_o'$ (NAD⁺/NADH) = −0.320 V and $\mathscr{E}_o'$ (CoQ/CoQH₂) = +0.060 V.
   b. What is the equilibrium constant (K_{eq}) for this reaction?
   c. Assume that (1) the actual free energy release accompanying the NADH–coenzyme Q reductase reaction is equal to the amount released under standard conditions (as calculated in part a), (2) this energy can be converted into the synthesis of ATP with an efficiency = 0.75 (that is, 75% of the energy released upon NADH oxidation is captured in ATP synthesis), and (3) the oxidation of 1 equivalent of NADH by coenzyme Q leads to the phosphorylation of 1 equivalent of ATP.

   Under these conditions, what is the maximum ratio of [ATP]/[ADP] attainable by oxidative phosphorylation when [Pᵢ] = 1 mM? (Assume $\Delta G^{o\prime}$ for ATP synthesis = +30.5 kJ/mol.)

9. **Determining the Impact of the Succinate Dehydrogenase Reaction on Oxidative Phosphorylation** Consider the oxidation of succinate by molecular oxygen as carried out via the electron-transport pathway

$$Succinate + \tfrac{1}{2}O_2 \longrightarrow fumarate + H_2O$$

   a. What is the standard free energy change ($\Delta G^{o\prime}$) for this reaction if $\mathscr{E}_o'$ (Fum/Succ) = +0.031 V and $\mathscr{E}_o'$ ($\tfrac{1}{2}O_2/H_2O$) = +0.816 V?

b. What is the equilibrium constant (K_{eq}) for this reaction?

c. Assume that (1) the actual free energy release accompanying succinate oxidation by the electron-transport pathway is equal to the amount released under standard conditions (as calculated in part a), (2) this energy can be converted into the synthesis of ATP with an efficiency = 0.7 (that is, 70% of the energy released upon succinate oxidation is captured in ATP synthesis), and (3) the oxidation of 1 succinate leads to the phosphorylation of 2 equivalents of ATP.

Under these conditions, what is the maximum ratio of [ATP]/[ADP] attainable by oxidative phosphorylation when [P_i] = 1 mM? (Assume $\Delta G^{\circ\prime}$ for ATP synthesis = +30.5 kJ/mol.)

10. Determining the Effects of NADH Oxidation on Oxidative Phosphorylation Consider the oxidation of NADH by molecular oxygen as carried out via the electron-transport pathway

$$NADH + H^+ + \tfrac{1}{2}O_2 \longrightarrow NAD^+ + H_2O$$

a. What is the standard free energy change ($\Delta G^{\circ\prime}$) for this reaction if $\mathscr{E}_0{}'$ (NAD$^+$/NADH) = -0.320 V and $\mathscr{E}_0{}'$ (O$_2$/H$_2$O) = $+0.816$ V.

b. What is the equilibrium constant (K_{eq}) for this reaction?

c. Assume that (1) the actual free energy release accompanying NADH oxidation by the electron-transport pathway is equal to the amount released under standard conditions (as calculated in part a), (2) this energy can be converted into the synthesis of ATP with an efficiency = 0.75 (that is, 75% of the energy released upon NADH oxidation is captured in ATP synthesis), and (3) the oxidation of 1 NADH leads to the phosphorylation of 3 equivalents of ATP.

Under these conditions, what is the maximum ratio of [ATP]/[ADP] attainable by oxidative phosphorylation when [P_i] = 2 mM? (Assume $\Delta G^{\circ\prime}$ for ATP synthesis = +30.5 kJ/mol.)

11. Determining the Effect of the Cytochrome Oxidase Reaction on Oxidative Phosphorylation Write a balanced equation for the reduction of molecular oxygen by reduced cytochrome c as carried out by Complex IV (cytochrome oxidase) of the electron-transport pathway.

a. What is the standard free energy change ($\Delta G^{\circ\prime}$) for this reaction if

$$\Delta \mathscr{E}_0{}' \text{ cyt } c(Fe^{3+})/cyt \ c(Fe^{2+}) = +0.254 \text{ volts and}$$
$$\mathscr{E}_0{}' \ (\tfrac{1}{2}O_2/H_2O) = 0.816 \text{ volts}$$

b. What is the equilibrium constant (K_{eq}) for this reaction?

c. Assume that (1) the actual free energy release accompanying cytochrome c oxidation by the electron-transport pathway is equal to the amount released under standard conditions (as calculated in part a), (2) this energy can be converted into the synthesis of ATP with an efficiency = 0.6 (that is, 60% of the energy released upon cytochrome c oxidation is captured in ATP synthesis), and (3) the reduction of 1 molecule of O_2 by reduced cytochrome c leads to the phosphorylation of 2 equivalents of ATP.

Under these conditions, what is the maximum ratio of [ATP]/[ADP] attainable by oxidative phosphorylation when [P_i] = 3 mM? (Assume $\Delta G^{\circ\prime}$ for ATP synthesis = +30.5 kJ/mol and T = 298 K.)

12. Understanding the Energetics of the Lactate Dehydrogenase Reaction The standard reduction potential for (NAD$^+$/NADH) is -0.320 V, and the standard reduction potential for (pyruvate/lactate) is -0.185 V.

a. What is the standard free energy change ($\Delta G^{\circ\prime}$) for the lactate dehydrogenase reaction:

$$NADH + H^+ + pyruvate \longrightarrow lactate + NAD^+$$

b. What is the equilibrium constant (K_{eq}) for this reaction?

c. If [pyruvate] = 0.05 mM and [lactate] = 2.9 mM and ΔG for the lactate dehydrogenase reaction = -15 kJ/mol in erythrocytes, what is the [NAD$^+$]/[NADH] ratio under these conditions?

13. Assessing Proton Transport and Free Energy Change in Bacterial ATP Synthesis Assume that the free energy change (ΔG) associated with the movement of 1 mole of protons from the outside to the

inside of a bacterial cell is -23 kJ/mol and 3 H$^+$ must cross the bacterial plasma membrane per ATP formed by the bacterial F_1F_0–ATP synthase. ATP synthesis thus takes place by the coupled process:

$$3 \ H^+{}_{out} + ADP + P_i \Longleftrightarrow 3 \ H^+{}_{in} + ATP + H_2O$$

a. If the overall free energy change ($\Delta G_{overall}$) associated with ATP synthesis in these cells by the coupled process is -21 kJ/mol, what is the equilibrium constant (K_{eq}) for the process?

b. What is $\Delta G_{synthesis}$, the free energy change for ATP synthesis, in these bacteria under these conditions?

c. The standard free energy change for ATP hydrolysis ($\Delta G^{\circ\prime}{}_{hydrolysis}$) is -30.5 kJ/mol. If [P_i] = 2 mM in these bacterial cells, what is the [ATP]/[ADP] ratio in these cells?

14. Describing the Path of Electrons Through the Q Cycle of Complex III Describe in your own words the path of electrons through the Q cycle of Complex III.

15. Describing the Path of Electrons Through Complex IV Describe in your own words the path of electrons through the copper and iron centers of Complex IV.

16. Understanding Oxidation of Unsaturated Lipids In the course of events triggering apoptosis, a fatty acid chain of cardiolipin undergoes peroxidation to release the associated cytochrome c. Lipid peroxidation occurs at a double bond. Suggest a mechanism for the reaction of hydrogen peroxide with an unsaturation in a lipid chain, and identify a likely product of the reaction.

17. Determining the Number of Protons in a Typical Mitochondrion In problem 18 at the end of Chapter 19, you might have calculated the number of molecules of oxaloacetate in a typical mitochondrion. What about protons? A typical mitochondrion can be thought of as a cylinder 1 μm in diameter and 2 μm in length. If the pH in the matrix is 7.8, how many protons are contained in the mitochondrial matrix?

18. Understanding the Unique Requirement for FAD in the Succinate Dehydrogenase Reaction Considering that all other dehydrogenases of glycolysis and the TCA cycle use NADH as the electron donor, why does succinate dehydrogenase, a component of the TCA cycle and the electron-transfer chain, use FAD as the electron acceptor from succinate, rather than NAD$^+$? Note that there are two justifications for the choice of FAD here—one based on energetics and one based on the mechanism of electron transfer for FAD versus NAD$^+$.

Preparing for the MCAT® Exam

19. Based on your reading on the F_1F_0–ATPase, what would you conclude about the mechanism of ATP synthesis:

a. The reaction proceeds by nucleophilic substitution via the S_N2 mechanism.

b. The reaction proceeds by nucleophilic substitution via the S_N1 mechanism.

c. The reaction proceeds by electrophilic substitution via the E1 mechanism.

d. The reaction proceeds by electrophilic substitution via the E2 mechanism.

20. Imagine that you are working with isolated mitochondria and you manage to double the ratio of protons outside to protons inside. In order to maintain the overall ΔG at its original value (whatever it is), how would you have to change the mitochondria membrane potential?

ActiveModels Problems

21. Using the ActiveModel for Complex I of the electron transport chain: Describe the interactions of different arms of the L-shape. Describe Nqo1's Rossman fold. Determine the specific interactions of each residue of the catalytic binding site with NADH.

22. Using the ActiveModel for ADP-ATP translocase, explain the formation of the kinked α–helices.

FURTHER READING

Bioenergetics

Belogrudov, G. I., 2009. Recent advances in structure–functional studies of mitochondrial factor B. *Journal of Bioenergetics and Biomembranes* **41:**137–143.

Fulda, S., Galluzzi, L., and Kroemer, G., 2010. Targeting mitochondria for cancer therapy. *Nature Reviews Drug Discovery* **9:**447–464.

Hinkle, P. C., 2005. P/O ratios of mitochondrial oxidative phosphorylation. *Biochimica et Biophysica Acta* **1706:**1–11.

Merz, S., Hammermeister, M., et al., 2007. Molecular machinery of mitochondrial dynamics in yeast. *Biological Chemistry* **388:**917–926.

Nicholls, D. G., 2008. Forty years of Mitchell's proton circuit: From little grey books to little grey cells. *Biochimica et Biophysica Acta* **1777:**550–556.

Ubarretxena-Belandia, I., and Stokes, D. L., 2010. Present and future of membrane protein structure determination by electron crystallography. *Advances in Protein Chemistry and Structural Biology* **81:**33–60.

Electron Transfer

Belevich, I., and Verkhovsky, M. I., 2008. Molecular mechanism of proton translocation by cytochrome *c* oxidase. *Antioxidants and Redox Signaling* **10:**1–29.

Boekema, E. J., and Braun, H-P., 2007. Supramolecular structure of the mitochondrial oxidative phosphorylation system. *Journal of Biological Chemistry* **282:**1–4.

Busenlehner, L. S., Branden, G., et al., 2008. Structural elements involved in proton translocation by cytochrome *c* oxidase as revealed by backbone amide hydrogen–deuterium exchange of the E286H mutant. *Biochemistry* **47:**73–83.

Cramer, W. A., Hasan, S. S., and Yamashita, E., 2011. The Q cycle of cytochrome *bc* complexes: A structure perspective. *Biochimica et Biophysica Acta* **1807:**788–802.

Crofts, A. R., Holland, J. T., Victoria, D., Kolling, D., Dikanov, S. A., Gilbreth, R., Lhee, S., Kuras, R., and Kuras, M. G., 2008. The Q–cycle reviewed: How well does a monomeric mechanism of the *bc*₁ complex accounts for the function of a dimeric complex? *Biochimica et Biophysica Acta* **1777:**1001–1019.

Eframov, R. G., Baradaran, R., and Sazanov, L. A., 2010. The architecture of respiratory complex I. *Nature* **465:**441–447.

Maklashina, E., and Cecchini, G., 2010. The quinone–binding and catalytic site of complex II. *Biochimica et Biophysica Acta* **1797:**1877–1882.

Seibold, S. A., Mills, D. A., et al., 2005. Water chain formation and possible proton pumping routes in *Rhodobacter sphaeroides* cytochrome *c* oxidase: A molecular dynamics comparison of the wild type and R481K mutant. *Biochemistry* **44:**10475–10485.

Yoshikawa, S., Muramoto, K., et al., 2006. Reaction mechanism of bovine heart cytochrome *c* oxidase. *Biochimica et Biophysica Acta* **1757:**395–400.

F₁F₀–ATP Synthase

Bason, J. V., Runswick, M. J., Fearnley, I. M., and Walker, J. E., 2011. Binding of the inhibitor protein IF₁ to bovine F₁-ATPase. *Journal of Molecular Biology* **406:**443–453.

Dickson, V. K., Silvester, J. A., et al., 2006. On the structure of the stator of the mitochondrial ATP synthase. *EMBO Journal* **25:**2911–2918.

Düser, M. G., Zarribi, N., Cipriano, D. J., Ernst, S., Glick, G. D., Dunn, S. D., and Börsch, M., 36° step size of proton-driven *c*-ring rotation in ATP synthase. *EMBO Journal* **28:**2689–2696.

Von Ballmoos, C., Cook, G. M., and Dimroth, P., 2008. Unique Rotary ATP synthase and its biological diversity. *Annual Review of Biophysics* **37:**43–64.

Von Ballmoos, C., Wiedenmann, A., and Dimroth, P., 2009. Essentials for ATP synthesis by F₁F₀ ATP synthases. *Annual Review of Biochemistry* **78:**649–672.

Wächter, A., Bi, Y., Dunn, S. D., Cain, B. D., Sielaff, H., Wintermann, F., Engelbrecht, S., and Junge, W., 2011. Two rotary motors in F-ATP synthase are elastically coupled by a flexible rotor and a stiff stator stalk. *Proceedings of the National Academy of Sciences* **108:**3924–3929.

Watt, I. N., Montgomery, M. G., Runswick, M. J., Leslie, A. G., and Walker, J. E., 2010. Bioenergetic cost of making an adenosine triphosphate molecule in animal mitochondria. *Proceedings of the National Academy of Sciences* **107:**16823–16827.

Weber, J., 2007. ATP synthase: The structure of the stator stalk. *Trends in Biochemical Sciences* **32:**53–55.

Supramolecular Complexes

Bulltema, J. B., Braun, H.-P., Boekema, E. J., and Kouril, R., 2008. Organization of the oxidative phosphorylation system by structural analysis of respiratory supercomplexes from potato. *Biochimica et Biophysica Acta* **1787:**60–67.

Dudkina, N. V., Kouril, R., Peters, K., Braun, H.-P., and Boekema, E. J., 2010. Structure and function of mitochondrial supercomplexes. *Biochimica et Biophysica Acta* **1797:**664–670.

Lenaz, G., and Genova, M. L., 2010. Structure and organization of mitochondrial respiratory complexes: New understanding of an old subject. *Antioxidants and Redox Signaling* **12:**961–1008.

Mitochondrial Carriers

Azzu, V., and Brand, M. D., 2010. The on-off switches of the mitochondrial uncoupling proteins. *Trends in Biochemical Sciences* **35:**298–307.

Claypool, S. M., 2009. Cardiolipin, a critical determinant of mitochondrial carrier protein assembly. *Biochimica et Biophysica Acta* **1788:**2059–2068.

Feng, J., Lucchinetti, E., Enkavi, G., Wang, Y., et al., 2010. Tyrosine phosphorylation by Src within the cavity of the adenine nucleotide translocase 1 regulates ADP/ATP exchange in mitochondria. *American Journal of Physiology and Cell Physiology* **298:**C740–C748.

Klingenberg, M., 2008. The ADP and ATP transport in mitochondria and its carrier. *Biochimica et Biophysica Acta* **1778:**1978–2021.

Klingenberg, M., 2009. Cardiolipin and mitochondrial carriers. *Biochimica et Biophysica Acta* **1788:**2048–2058.

Kunji, E. R. S., and Crichton, P. G., 2010. Mitochondrial carriers function as monomers. *Biochimica et Biophysica Acta* **1797:**817–831.

Mookerjee, S. A., Divakaruni, A. S., Jastroch, M., and Brand, M. D., 2010. Mitochondrial uncoupling and lifespan. *Mechanisms of Ageing and Development* **131:**463–472.

Nath, S., 2010. Beyond the chemiosmotic theory: Analysis of key fundamental aspects of energy coupling in oxidative phosphorylation in the light of a torsional mechanism for energy transduction and ATP Synthesis. *Journal of Bioenergetics and Biomembranes* **42:**293–300.

Nury, H., Blesneac, I., Ravaud, S., and Pebay-Peyroula, E., 2010. Structural approaches of the mitochondrial carrier family. *Methods in Molecular Biology* **654:**105–117.

Nury, H., Dahout-Gonzalez, C., et al., 2006. Relations between structure and function of the mitochondrial ADP/ATP carrier. *Annual Review of Biochemistry* **75:**713–741.

Rey, M., Man, P., Clémençon, B., Trezeguet, V., Brandolin, G., and Forest, E., 2010. Conformational dynamics of the bovine mitochondrial ADP/ATP carrier isoform 1 revealed by hydrogen/deuterium exchange coupled to mass spectrometry. *Journal of Biological Chemistry* **285:**34981–34990.

Apoptosis

Cerveny, K. L., Tamura, Y., et al., 2007. Regulation of mitochondrial fusion and division. *Trends in Cell Biology* **17:**563–569.

Kang, R., Tang, D., Lotze, M. T., and Zeh, H. J., 2011. Apoptosis to autophagy switch triggered by the MHC class III–encoded receptor for advanced glycation endproducts (RAGE). *Autophagy* **7:**92–93.

Ledgerwood, E. C., and Morison, I. M., 2009. Targeting the apoptosome for cancer therapy. *Clinical Cancer Research* **15:**420–424.

Mace, P. D., and Riedl, S. J., 2010. Molecular cell death platforms and assemblies. *Current Opinion in Cell Biology* **22:**828–836.

Orrenius, S., 2007. Reactive oxygen species in mitochondria-mediated cell death. *Drug Metabolism Reviews* **39:**443–455.

Riedl, S. J., and Salvesen, G. S., 2007. The apoptosome: Signalling platform of cell death. *Nature Reviews Molecular Cell Biology* **8:**405–413.

Strappazzon, F., Vietri-Rudan, M., Campello, S., et al., 2011. Mitochondrial BCL-2 inhibitis AMBRA1-induced autophagy. *The EMBO Journal* **30:**1195–1206.

Tait, S. W. G., and Green, D. R., 2010. Mitochondria and cell death: Outer membrane permeabilization and beyond. *Nature Reviews Molecular and Cell Biology* **11:**621–632.

Yuan, S., Yu, X., Topf, M., Ludtke, S. J., Wang, X., and Akey, C. W., 2010. Structure of an apoptosome-procaspase-9 CARD complex. *Structure* **18:**571–583.

Zhou, F., Yang, Y., and Xing, D., 2010. Bcl-2 and Bcl-xL play important roles in the crosstalk between autophagy and apoptosis. *FEBS Journal* **278:**403–413.

Photosynthesis

© Richard Hamilton Smith/CORBIS

In a sun-flecked lane,
Beside a path where cattle trod,
Blown by wind and rain,
Drawing substance from air and sod;

In ruggedness, it stands aloof,
The ragged grass and puerile leaves,
Lending a hand to fill the woof
In the pattern that beauty weaves.

What mystery this, hath been wrought;
Beauty from sunshine, air, and sod!
Could we thus gain the ends we sought-
Tell us thy secret, Goldenrod.

Rosa Staubus
Oklahoma pioneer (1886–1966)

KEY QUESTIONS

21.1 What Are the General Properties of Photosynthesis?

21.2 How Is Solar Energy Captured by Chlorophyll?

21.3 What Kinds of Photosystems Are Used to Capture Light Energy?

21.4 What Is the Molecular Architecture of Photosynthetic Reaction Centers?

21.5 What Is the Quantum Yield of Photosynthesis?

21.6 How Does Light Drive the Synthesis of ATP?

21.7 How Is Carbon Dioxide Used to Make Organic Molecules?

21.8 How Does Photorespiration Limit CO_2 Fixation?

ESSENTIAL QUESTIONS

Photosynthesis is the primary source of energy for all life forms (except chemolithotrophic prokaryotes). Much of the energy of photosynthesis is used to drive the synthesis of organic molecules from atmospheric CO_2.

How is solar energy captured and transformed into metabolically useful chemical energy? How is the chemical energy produced by photosynthesis used to create organic molecules from carbon dioxide?

The vast majority of energy consumed by living organisms stems from solar energy captured by the process of photosynthesis. Only chemolithotrophic prokaryotes are independent of this energy source. Of the 1.5×10^{22} kJ of energy reaching the earth each day from the sun, 1% is absorbed by photosynthetic organisms and transduced into chemical energy.[1] This energy, in the form of biomolecules, becomes available to other members of the biosphere through food chains. The transduction of solar, or light, energy into chemical energy is often expressed in terms of **carbon dioxide fixation,** in which hexose is formed from carbon dioxide and oxygen is evolved:

$$6\, CO_2 + 6\, H_2O \xrightarrow{\text{Light}} C_6H_{12}O_6 + 6\, O_2 \qquad (21.1)$$

Estimates indicate that 10^{11} tons of carbon dioxide are fixed globally per year, of which one-third is fixed in the oceans, primarily by photosynthetic marine microorganisms.

Although photosynthesis is traditionally equated with CO_2 fixation, light energy (or rather the chemical energy derived from it) drives all endergonic processes in phototrophic cells. The assimilation of inorganic forms of nitrogen and sulfur into organic molecules

[1]Of the remaining 99%, two-thirds is absorbed by the earth and oceans, thereby heating the planet; the remaining one-third is lost as light reflected back into space.

OWL Online homework and a Student Self Assessment for this chapter may be assigned in OWL.

(see Chapter 25) represents two other metabolic conversions closely coupled to light energy in green plants. Our previous considerations of aerobic metabolism (Chapters 18 through 20) treated cellular respiration (precisely the reverse of Equation 21.1) as the central energy-releasing process in life. It necessarily follows that the formation of hexose from carbon dioxide and water, the products of cellular respiration, must be endergonic. The necessary energy comes from light. Note that in the carbon dioxide fixation reaction described, light is used to drive a chemical reaction against its thermodynamic potential.

21.1 What Are the General Properties of Photosynthesis?

Photosynthesis Occurs in Membranes

Organisms capable of photosynthesis are very diverse, ranging from simple prokaryotic forms to the largest organisms of all, *Sequoia gigantea,* the giant redwood trees of California. Despite this diversity, we find certain generalities regarding photosynthesis. An important one is that *photosynthesis occurs in membranes.* In photosynthetic prokaryotes, the photosynthetic membranes fill up the cell interior; in photosynthetic eukaryotes, the photosynthetic membranes are localized in large organelles known as **chloroplasts** (Figures 21.1 and 21.2). Chloroplasts are one member in a family of related plant-specific organelles known as **plastids.** Chloroplasts themselves show a range of diversity, from the single, spiral chloroplast that gives *Spirogyra* its name to the multitude of ellipsoidal plastids typical of higher plant cells (Figure 21.3).

Characteristic of all chloroplasts, however, is the organization of the inner membrane system, the so-called **thylakoid membrane.** The thylakoid membrane is organized into paired folds that extend throughout the organelle, as in Figure 21.2. These paired folds, or **lamellae,** give rise to flattened sacs or discs, **thylakoid vesicles** (from the Greek *thylakos,* meaning "sack"), which occur in stacks called **grana.** A single stack, or **granum,** may contain dozens of thylakoid vesicles, and different grana are joined by lamellae that run through the soluble portion, or **stroma,** of the organelle. Chloroplasts thus possess three membrane-bound aqueous compartments: the intermembrane space, the stroma, and the interior of the thylakoid vesicles, the so-called **thylakoid space** (also known as the **thylakoid lumen**). As we shall see, this third compartment serves an important function in the transduction of light energy into ATP formation. The thylakoid membrane has a highly characteristic lipid composition and, like the inner membrane of the mitochondrion, is impermeable to most ions and molecules. Chloroplasts, like their mitochondrial counterparts, possess DNA, RNA, and ribosomes and consequently display a considerable amount of autonomy. However, many critical chloroplast components are encoded by nuclear genes, so autonomy is far from absolute.

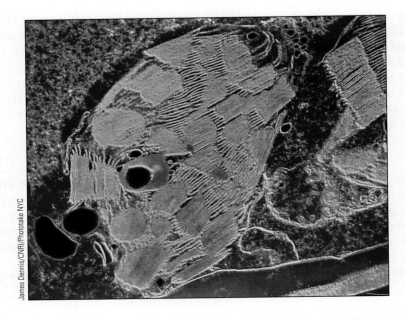

FIGURE 21.1 Electron micrograph of a representative chloroplast.

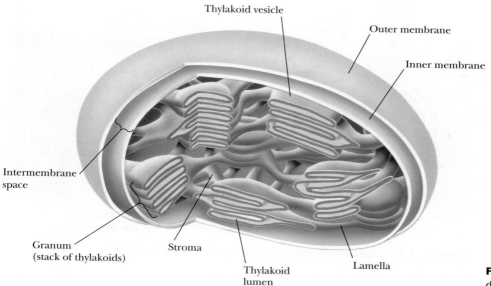

FIGURE 21.2 Schematic diagram of an idealized chloroplast.

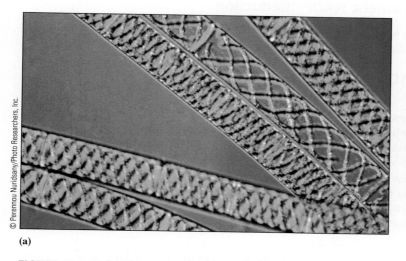

(a)

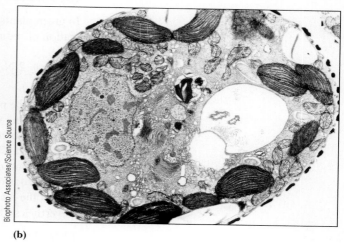

(b)

FIGURE 21.3 **(a)** *Spirogyra*—a freshwater green alga. **(b)** A higher plant cell.

Photosynthesis Consists of Both Light Reactions and Dark Reactions

If a chloroplast suspension is illuminated in the absence of carbon dioxide, oxygen is evolved. Furthermore, if the illuminated chloroplasts are now placed in the dark and supplied with CO_2, net hexose synthesis can be observed (Figure 21.4). Thus, the evolution of oxygen can be temporally separated from CO_2 fixation and also has a light dependency that CO_2 fixation lacks. The **light reactions** of photosynthesis, of which O_2 evolution is only one part, are associated with the thylakoid membranes. In contrast, the light-independent reactions, or so-called **dark reactions,** notably CO_2 fixation, are located in the stroma. A concise summary of the photosynthetic process is that radiant electromagnetic energy (light) is transformed by a specific photochemical system located in the thylakoids to yield chemical energy in the form of reducing potential (NADPH) and high-energy phosphate (ATP). NADPH and ATP can then be used to drive the endergonic process of hexose formation from CO_2 by a series of enzymatic reactions found in the stroma (see Equation 21.3, which follows).

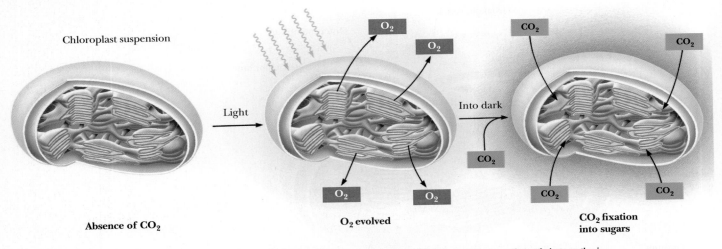

Chloroplast suspension

Light

Into dark

Absence of CO$_2$

O$_2$ evolved

CO$_2$ fixation into sugars

FIGURE 21.4 The light-dependent and light-independent reactions of photosynthesis.

Water Is the Ultimate e^- Donor for Photosynthetic NADP$^+$ Reduction

In green plants, water serves as the ultimate electron donor for the photosynthetic generation of reducing equivalents. The reaction sequence

$$2\ H_2O + 2\ NADP^+ + x\ ADP + x\ P_i \xrightarrow{nh\upsilon}$$
$$O_2 + 2\ NADPH + 2\ H^+ + x\ ATP + x\ H_2O \qquad (21.2)$$

describes the process, where $nh\upsilon$ symbolizes light energy (n is some number of photons of energy $h\upsilon$, where h is Planck's constant and υ is the frequency of the light). Light energy is necessary to make the unfavorable reduction of NADP$^+$ by H$_2$O ($\Delta \mathscr{E}_0' = -1.136$ V; $\Delta G^{\circ\prime} = +219$ kJ/mol NADP$^+$) thermodynamically favorable. Thus, the light energy input, $nh\upsilon$, must exceed 219 kJ/mol NADP$^+$. The stoichiometry of ATP formation depends on the pattern of photophosphorylation operating in the cell at the time and on the ATP yield in terms of the chemiosmotic ratio, ATP/H$^+$, as we will see later. Nevertheless, the stoichiometry of the metabolic pathway of CO$_2$ fixation is certain:

$$12\ NADPH + 12\ H^+ + 18\ ATP + 6\ CO_2 + 12\ H_2O \longrightarrow$$
$$C_6H_{12}O_6 + 12\ NADP^+ + 18\ ADP + 18\ P_i \qquad (21.3)$$

A More Generalized Equation for Photosynthesis In 1931, comparative study of photosynthesis in bacteria led van Niel to a more general formulation of the overall reaction:

$$\underset{\substack{\textbf{Hydrogen} \\ \textbf{acceptor}}}{CO_2} + \underset{\substack{\textbf{Hydrogen} \\ \textbf{donor}}}{2\ H_2A} \xrightarrow{\text{Light}} \underset{\substack{\textbf{Reduced} \\ \textbf{acceptor}}}{(CH_2O)} + \underset{\substack{\textbf{Oxidized} \\ \textbf{donor}}}{2A} + H_2O \qquad (21.4)$$

In photosynthetic bacteria, H$_2$A is variously H$_2$S (photosynthetic green and purple sulfur bacteria), isopropanol, or some similar oxidizable substrate. [(CH_2O) symbolizes a carbohydrate unit.]

$$CO_2 + 2\ H_2S \longrightarrow (CH_2O) + H_2O + 2\ S$$

$$CO_2 + 2\ CH_3\!-\!CHOH\!-\!CH_3 \longrightarrow (CH_2O) + H_2O + 2\ CH_3\!-\!\overset{\overset{\displaystyle O}{\displaystyle \|}}{C}\!-\!CH_3$$

In cyanobacteria and the eukaryotic photosynthetic cells of algae and higher plants, H$_2$A is H$_2$O, as implied earlier, and 2 A is O$_2$. The accumulation of O$_2$ to constitute 21% of the earth's atmosphere is the direct result of eons of global oxygenic photosynthesis.

21.2 How Is Solar Energy Captured by Chlorophyll?

Photosynthesis depends on the photoreactivity of **chlorophyll.** Chlorophylls are magnesium-containing substituted tetrapyrroles whose basic structure is reminiscent of heme, the iron-containing porphyrin (see Chapters 5 and 20). Chlorophylls differ from heme in a number of properties: Magnesium instead of iron is coordinated in the center of the planar conjugated ring structure; a long-chain alcohol, **phytol,** is esterified to a pyrrole ring substituent; and the methine bridge linking pyrroles III and IV is substituted and crosslinked to ring III, leading to the formation of a fifth five-membered ring. The structures of chlorophyll *a* and *b* are shown in Figure 21.5a.

Chlorophylls are excellent light absorbers because of their aromaticity. That is, they possess delocalized π electrons above and below the planar ring structure. The energy differences between electronic states in these π orbitals correspond to the energies of visible light photons. When light energy is absorbed, an electron is promoted to a higher orbital, enhancing the potential for transfer of this electron to a suitable acceptor. Loss of such a photoexcited electron to an acceptor is an oxidation–reduction reaction. The net result is the transduction of light energy into the chemical energy of a redox reaction.

Chlorophylls and Accessory Light-Harvesting Pigments Absorb Light of Different Wavelengths

The absorption spectra of chlorophylls *a* and *b* (Figure 21.5b) differ somewhat. Plants that possess both chlorophylls can harvest a wider spectrum of incident energy. Other pigments in photosynthetic organisms, so-called **accessory light-harvesting pigments** (Figure 21.6), increase the possibility for absorption of incident light of wavelengths not absorbed by the chlorophylls. Carotenoids and phycocyanobilins, like chlorophyll, possess many conjugated double bonds and thus absorb visible light. Carotenoids have two

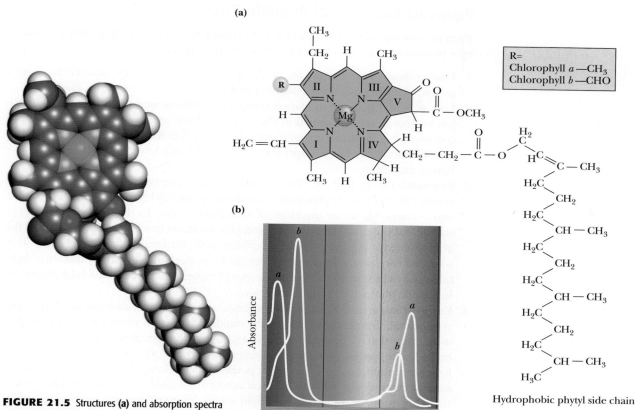

FIGURE 21.5 Structures **(a)** and absorption spectra **(b)** of chlorophyll *a* and *b*. The phytyl side chain of ring IV provides a hydrophobic tail to anchor the chlorophyll in membrane protein complexes.

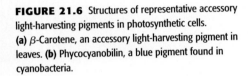

(a)

β-Carotene

(b)

Phycocyanobilin

FIGURE 21.6 Structures of representative accessory light-harvesting pigments in photosynthetic cells. **(a)** β-Carotene, an accessory light-harvesting pigment in leaves. **(b)** Phycocyanobilin, a blue pigment found in cyanobacteria.

primary roles in photosynthesis—light harvesting and photoprotection through destruction of reactive oxygen species that arise as byproducts of photoexcitation.

The Light Energy Absorbed by Photosynthetic Pigments Has Several Possible Fates

Each photon represents a quantum of light energy. A quantum of light energy absorbed by a photosynthetic pigment has four possible fates (Figure 21.7):

1. **Loss as heat.** The energy can be dissipated as heat through redistribution into atomic vibrations within the pigment molecule.
2. **Loss as light.** Energy of excitation reappears as **fluorescence** (light emission); a photon of fluorescence is emitted as the e^- drops down to a lower orbital. This fate is common only in saturating light intensities. For thermodynamic reasons, the photon of fluorescence has a longer wavelength and hence lower energy than the quantum of excitation.
3. **Resonance energy transfer.** The excitation energy can be transferred by resonance energy transfer to a neighboring molecule if the energy level difference between the two corresponds to the quantum of excitation energy. In this process, the energy transferred raises an electron in the receptor molecule to a higher energy state as the photoexcited e^- in the original absorbing molecule returns to ground state. This so-called *Förster resonance energy transfer* is the mechanism whereby quanta of light falling anywhere within an array of pigment molecules can be transferred ultimately to specific photochemically reactive sites.
4. **Energy transduction.** The energy of excitation, in raising an electron to a higher energy orbital, dramatically changes the standard reduction potential, $\mathscr{E}_o'$, of the pigment such that it becomes a much more effective electron donor. That is, the excited-state species, by virtue of having an electron at a higher energy level through light absorption, has become a more potent electron donor. Reaction of this excited-state electron donor with an electron acceptor situated in its vicinity leads to the transformation, or **transduction,** of light energy (photons) to chemical energy (reducing power, the potential for electron-transfer reactions). *Transduction of light energy into chemical energy, the photochemical event, is the essence of photosynthesis.*

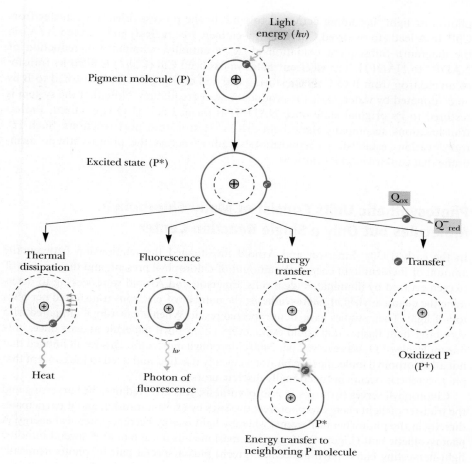

FIGURE 21.7 Possible fates of the quantum of light energy absorbed by photosynthetic pigments.

The Transduction of Light Energy into Chemical Energy Involves Oxidation–Reduction

The diagram presented in Figure 21.8 illustrates the fundamental transduction of light energy into chemical energy (an oxidation–reduction reaction) that is the basis of photosynthesis. Chlorophyll (Chl) resides in a membrane in close association with molecules competent in e^- transfer, symbolized here as A and B. Chl absorbs a

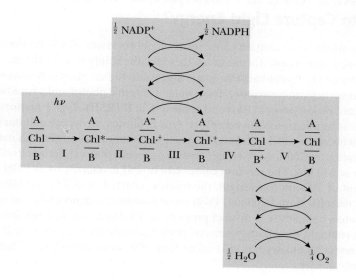

FIGURE 21.8 Model for light absorption by chlorophyll and transduction of light energy into an oxidation–reduction reaction. **I:** Photoexcitation of Chl creates Chl*. **II:** Electron transfer from Chl* to A yields oxidized Chl (Chl$^+$) and reduced A (A$^-$) **III:** An electron-transfer pathway from A$^-$ to NADP$^+$ leads to NADPH formation and restoration of oxidized A (A). **IV:** Chl$^+$ accepts an electron from B, restoring Chl and generating oxidized B (B$^+$). **V:** B$^+$ is reduced back to B by an electron originating in H$_2$O. Water oxidation is the source of O$_2$ formation.

photon of light, becoming activated to Chl* in the process. Electron transfer from Chl* to A leads to oxidized Chl (Chl·$^+$, a **cationic free radical**) and reduced A (A$^-$ in the diagram). Subsequent oxidation of A$^-$ eventually culminates in reduction of NADP$^+$ to NADPH. The **electron "hole"** in oxidized Chl (Chl·$^+$) is filled by transfer of an electron from B to Chl·$^+$, restoring Chl and creating B$^+$. B$^+$ is restored to B by an e^- donated by water. O$_2$ is the product of water oxidation. Note that the system is restored to its original state once NADPH is formed and H$_2$O is oxidized. Proton translocations accompany these light-driven electron-transport reactions. Such H$^+$ translocations establish a chemiosmotic gradient across the photosynthetic membrane that can drive ATP synthesis.

Photosynthetic Units Consist of Many Chlorophyll Molecules but Only a Single Reaction Center

In the early 1930s, Emerson and Arnold investigated the relationship between the amount of incident light energy, the amount of chlorophyll present, and the amount of oxygen evolved by illuminated algal cells. Emerson and Arnold were seeking to determine the **quantum yield of photosynthesis:** the number of electrons transferred per photon of light. Their studies gave an unexpected result: When algae were illuminated with very brief light flashes that could excite every chlorophyll molecule at least once, only one molecule of O$_2$ was evolved per 2400 chlorophyll molecules. This result implied that not all chlorophyll molecules are photochemically reactive, and it led to the concept that photosynthesis occurs in functionally discrete units.

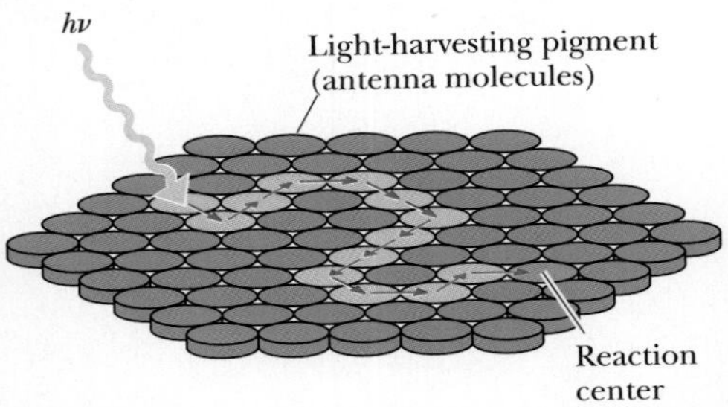

Chlorophyll serves two roles in photosynthesis. It is involved in light harvesting and the transfer of light energy to photoreactive sites by exciton transfer, and it participates directly in the photochemical events whereby light energy becomes chemical energy. A **photosynthetic unit** (Figure 21.9) can be envisioned as an antenna of several hundred light-harvesting chlorophyll molecules (green) plus a special pair of photochemically reactive chlorophyll *a* molecules called the **reaction center** (orange). The purpose of the vast majority of chlorophyll in a photosynthetic unit is to harvest light incident within the unit and funnel it, via resonance energy transfer, to the reaction center chlorophyll dimers that are photochemically active. Most chlorophyll thus acts as a large light-collecting antenna, and it is at the reaction centers that the photochemical event occurs. Oxidation of chlorophyll leaves a **cationic free radical,** Chl·$^+$, whose properties as an electron acceptor have important consequences for photosynthesis. Note that the Mg^{2+} ion does not change in valence during these redox reactions.

hv

Light-harvesting pigment
(antenna molecules)

Reaction
center

FIGURE 21.9 Schematic diagram of a photosynthetic unit.

21.3 What Kinds of Photosystems Are Used to Capture Light Energy?

All photosynthetic cells contain some form of photosystem. Photosynthetic bacteria have only one photosystem; furthermore, they lack the ability to use light energy to split H$_2$O and release O$_2$. Cyanobacteria, green algae, and higher plants are *oxygenic phototrophs* because they can generate O$_2$ from water. Oxygenic phototrophs have two distinct photosystems: **photosystem I (PSI)** and **photosystem II (PSII).** Type I photosystems use ferredoxins as terminal electron acceptors; type II photosystems use quinones as terminal electron acceptors. PSI is defined by reaction center chlorophylls with maximal red light absorption at 700 nm; PSII uses reaction centers that exhibit maximal red light absorption at 680 nm. The reaction center Chl of PSI is referred to as **P700** because it absorbs light of 700-nm wavelength; the reaction center Chl of PSII is called **P680** for analogous reasons. Both P700 and P680 are chlorophyll *a* dimers situated within specialized protein complexes. A distinct property of PSII is its role in light-driven O$_2$ evolution. Interestingly, the photosystems of photosynthetic bacteria are type II photosystems that resemble eukaryotic PSII more than PSI, even though these bacteria lack O$_2$-evolving capacity.

Chlorophyll Exists in Plant Membranes in Association with Proteins

Detergent treatment of a suspension of thylakoids dissolves the membranes, releasing complexes containing both chlorophyll and protein. These chlorophyll–protein complexes represent integral components of the thylakoid membrane, and their organization reflects their roles as either **light-harvesting complexes (LHC), PSI complexes,** or **PSII complexes.** All chlorophyll is apparently localized within these three macromolecular assemblies.

PSI and PSII Participate in the Overall Process of Photosynthesis

What are the roles of the two photosystems, and what is their relationship to each other? PSI provides reducing power in the form of NADPH. PSII splits water, producing O_2, and feeds the electrons released into an electron-transport chain that couples PSII to PSI. Electron transfer between PSII and PSI pumps protons for chemiosmotic ATP synthesis. As summarized by Equation 21.2, photosynthesis involves the reduction of $NADP^+$, using electrons derived from water and activated by light, $h\nu$. ATP is generated in the process. The standard reduction potential for the $NADP^+/NADPH$ couple is -0.32 V. Thus, a strong reductant with an $\mathscr{E}_o'$ more negative than -0.32 V is required to reduce $NADP^+$ under standard conditions. By similar reasoning, a very strong oxidant will be required to oxidize water to oxygen because $\mathscr{E}_o'(\frac{1}{2}O_2/H_2O)$ is $+0.82$ V. Separation of the oxidizing and reducing aspects of Equation 21.2 is accomplished in nature by devoting PSI to $NADP^+$ reduction and PSII to water oxidation. PSI and PSII are linked via an electron-transport chain so that the weak reductant generated by PSII can provide an electron to reduce the weak oxidant side of P700 (Figure 21.10). Thus, *electrons flow from H_2O to $NADP^+$,* driven by light energy absorbed at the reaction centers. Oxygen is a by-product of the **photolysis,** literally "light-splitting," of water. Accompanying electron flow is production of a proton gradient and ATP synthesis (see Section 21.6). This light-driven phosphorylation is termed *photophosphorylation.*

The Pathway of Photosynthetic Electron Transfer Is Called the Z Scheme

Photosystems I and II contain unique complements of electron carriers, and these carriers mediate the stepwise transfer of electrons from water to $NADP^+$. When the individual redox components of PSI and PSII are arranged as an e^- transport chain according to their standard reduction potentials, the zigzag result resembles the letter Z laid sideways (Figure 21.11). The various electron carriers are indicated as follows: "Mn complex" symbolizes the manganese-containing oxygen-evolving complex; D is its e^- acceptor and the immediate e^- donor to $P680^+$; Q_A and Q_B represent special plastoquinone molecules (see Figure 21.13), and PQ, the plastoquinone pool; Fe-S stands for the Rieske iron–sulfur center, and cyt f, cytochrome f. PC is the abbreviation for plastocyanin,

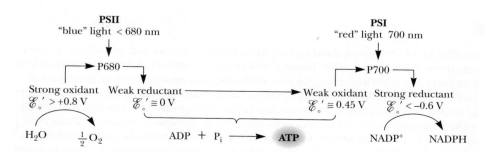

FIGURE 21.10 Roles of the two photosystems, PSI and PSII.

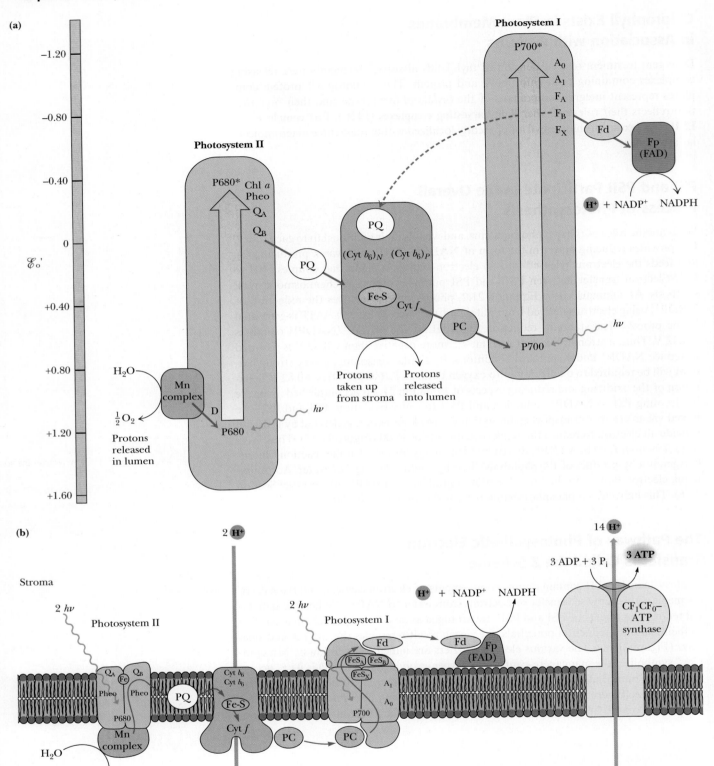

(a)

(b)

FIGURE 21.11 The Z scheme of photosynthesis. **(a)** The Z scheme is a diagrammatic representation of photosynthetic electron flow from H_2O to $NADP^+$. The energy relationships can be derived from the $\mathscr{E}_0'$ scale beside the Z diagram. Energy input as light is indicated by two broad arrows, one photon appearing in P680 and the other in P700. P680* and P700* represent photoexcited states. The three supramolecular complexes (PSI, PSII, and the cytochrome b_6f complex) are in shaded boxes. Proton translocations that establish the proton-motive force driving ATP synthesis are illustrated as well. **(b)** The functional relationships among PSII, the cytochrome b_6f complex, PSI, and the photosynthetic CF_1CF_0–ATP synthase within the thylakoid membrane. The functional relationships for photosynthetic electron transport through PSII, cytochrome b_6f, and PSI are based on 4 photons (2 per photosystem) driving the passage of an electron pair from water to $NADP^+$. ATP synthesis is based on an H^+/ATP ratio of 14/3.

the immediate e^- donor to P700$^+$; and F_A, F_B, and F_X represent the membrane-associated ferredoxins downstream from A_0 (a specialized Chl a) and A_1 (a specialized PSI quinone). Fd is the soluble ferredoxin pool that serves as the e^- donor to the flavoprotein (Fp), called **ferredoxin–NADP$^+$ reductase**, which catalyzes reduction of NADP$^+$ to NADPH. Cyt(b_6)$_N$,(b_6)$_P$ symbolizes the cytochrome b_6 moieties of the cytochrome b_6f complex. PQ and the cytochrome b_6f complex also serve to transfer e^- from F_A/F_B back to P700$^+$ during cyclic photophosphorylation (the pathway symbolized by the dashed arrow).

Overall photosynthetic electron transfer is accomplished by three **membrane-spanning supramolecular complexes** composed of intrinsic and extrinsic polypeptides (shown as shaded boxes bounded by solid black lines in Figure 22.11). These complexes are the PSII complex, the cytochrome b_6f complex, and the PSI complex. The PSII complex is aptly described as a light-driven **water : plastoquinone oxidoreductase;** it is the enzyme system responsible for photolysis of water, and as such, it is also referred to as the **oxygen-evolving complex,** or **OEC.** PSII possesses a metal cluster containing 4 Mn^{2+} atoms that coordinate two water molecules. As P680 undergoes four cycles of light-induced oxidation, four protons and four electrons are removed from the two water molecules and their O atoms are joined to form O_2. A tyrosyl side chain of the PSII complex (see following discussion) mediates electron transfer between the Mn^{2+} cluster and P680. The O_2-evolving reaction requires Ca^{2+} and Cl$^-$ ions in addition to the (Mn^{2+})$_4$ cluster.

Oxygen Evolution Requires the Accumulation of Four Oxidizing Equivalents in PSII

When isolated chloroplasts that have been held in the dark are illuminated with very brief flashes of light, O_2 evolution reaches a peak on the third flash and every fourth flash thereafter (Figure 21.12a). The oscillation in O_2 evolution dampens over repeated flashes and converges to an average value. These data are interpreted to mean that the P680 reaction center complex cycles through five different oxidation states, numbered S_0 to S_4. One electron and one proton are removed photochemically in each step. When S_4 is attained, an O_2 molecule is released (Figure 21.12b) as PSII returns to oxidation state S_0 and two new water molecules bind. (The reason the first pulse of O_2 release occurred on the third flash [Figure 21.12a] is that the PSII reaction centers in the isolated chloroplasts were already poised at S_1 reduction level.)

Electrons Are Taken from H$_2$O to Replace Electrons Lost from P680

The events intervening between H_2O and P680 involve D, the name assigned to a specific protein tyrosine residue that mediates e^- transfer from H_2O via the Mn complex to P680$^+$ (see Figure 21.11). The oxidized form of D is a tyrosyl free radical species, $D\cdot^+$. To begin the cycle, an exciton of energy excites P680 to P680*, whereupon P680* transfers an electron to a nearby Chl a molecule, which is the direct electron acceptor from P680*. This Chl a then reduces a molecule of **pheophytin,** symbolized by "Pheo" in Figure 21.11. Pheophytin is like chlorophyll a, except 2 H$^+$ replace the centrally coordinated Mg^{2+} ion. Loss of an electron from P680* creates P680$^+$, the electron acceptor for D. Electrons flow from Pheo via specialized molecules of **plastoquinone,** represented by "Q" in Figure 21.11, to a pool of plastoquinone (PQ) within the membrane. Because of its lipid nature, plastoquinone is mobile within the membrane and hence serves to shuttle electrons from the PSII supramolecular complex to the cytochrome b_6f complex. Alternate oxidation–reduction of plastoquinone to its hydroquinone form involves the uptake of protons (Figure 21.13). The asymmetry of the thylakoid membrane is designed to exploit this proton uptake and release so that protons (H$^+$) accumulate within the lumen of thylakoid vesicles, establishing an electrochemical gradient. Note that plastoquinone is an analog of coenzyme Q, the mitochondrial electron carrier (see Chapter 20).

Ferredoxin (Fd): A generic term for small proteins possessing iron-sulfur clusters that participate in various electron-transfer reactions.

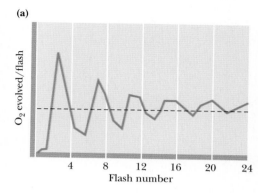

(a)

(b)

FIGURE 21.12 Oxygen evolution requires the accumulation of four oxidizing equivalents in PSII. **(a)** O_2 evolution after brief light flashes. **(b)** The cycling of the PSII reaction center through five different oxidation states, S_0 to S_4. One e^- is removed photochemically at each light flash, moving the reaction center successively through S_1, S_2, S_3, and S_4. S_4 decays spontaneously to S_0 by oxidizing 2 H_2O to O_2.

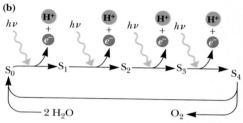

Plastoquinone A

Plastohydroquinone A

FIGURE 21.13 The structures of plastoquinone A and its reduced form, plastohydroquinone (or plastoquinol). Plastoquinone A has nine isoprene units and is the most abundant plastoquinone in plants and algae.

Electrons from PSII Are Transferred to PSI via the Cytochrome b_6f Complex

The cytochrome b_6f or **plastoquinol : plastocyanin oxidoreductase** is a large (210 kD) multimeric protein possessing 26 transmembrane α-helices. This membrane protein complex is structurally and functionally homologous to the cytochrome bc_1 complex (Complex III) of mitochondria (see Chapter 20). It includes the two heme-containing electron transfer proteins for which it is named, as well as *iron–sulfur clusters*, which also participate in electron transport. The purpose of this complex is to mediate the transfer of electrons from PSII to PSI and to pump protons across the thylakoid membrane via a plastoquinone-mediated Q cycle, analogous to that found in mitochondrial e^- transport (see Chapter 20). Cytochrome f (f from the Latin *folium,* meaning "foliage") is a c-type cytochrome, with a reduction potential of $+0.365$ V. Cytochrome b_6 in two forms (low- and high-potential) participates in the oxidation of plastoquinol and the Q cycle of the b_6f complex. The cytochrome b_6f complex can also serve in an alternative **cyclic electron-transfer pathway.** Under certain conditions, electrons derived from P700* are not passed on to $NADP^+$ but instead cycle down an alternative path, whereby reduced ferredoxin contributes its electron to PQ. This electron is then passed to the cytochrome b_6f complex, and then back to $P700^+$. This cyclic flow yields no O_2 evolution or $NADP^+$ reduction but can lead to ATP synthesis via so-called cyclic photophosphorylation, discussed later.

Plastocyanin Transfers Electrons from the Cytochrome b_6f Complex to PSI

Plastocyanin (**PC** in Figure 21.11) is an electron carrier capable of diffusion along the inside of the thylakoid and migration in and out of the membrane, aptly suited to its role in shuttling electrons between the cytochrome b_6f complex and PSI. Plastocyanin is a low-molecular-weight (10.4 kD) protein containing a single copper atom. PC functions as a single-electron carrier ($\mathscr{E}_o' = +0.32$ V) as its copper atom undergoes alternate oxidation–reduction between the cuprous (Cu^+) and cupric (Cu^{2+}) states. PSI is a light-driven **plastocyanin : ferredoxin oxidoreductase.** When P700, the specialized chlorophyll a dimer of PSI, is excited by light and oxidized by transferring its e^- to an adjacent chlorophyll a molecule that serves as its immediate e^- acceptor, $P700^+$ is formed. (The standard reduction potential for the $P700^+/P700$ couple is about $+0.45$ V.) $P700^+$ readily gains an electron from plastocyanin.

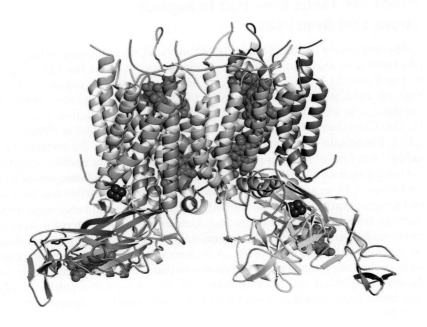

Structure of the cyanobacterial cytochrome b_6f complex. The heme groups of cytochromes b_6N, b_6P, and f are shown in red; the iron-sulfur clusters are blue (pdb id = 1BF5). The upper bundle of $\propto$-helices defines the transmembrane domain.

The immediate electron acceptor for P700* is a special molecule of chlorophyll. This unique Chl a (A_0) rapidly passes the electron to a specialized quinone (A_1), which in turn passes the e^- to the first in a series of *membrane-bound ferredoxins*. This Fd series ends with a soluble form of ferredoxin, Fd_s, which serves as the immediate electron donor to the flavoprotein (Fp) that catalyzes $NADP^+$ reduction, namely, **ferredoxin : $NADP^+$ reductase.**

21.4 What Is the Molecular Architecture of Photosynthetic Reaction Centers?

What molecular architecture couples the absorption of light energy to rapid electron-transfer events, in turn coupling these e^- transfers to proton translocations so that ATP synthesis is possible? Part of the answer to this question lies in the membrane-associated nature of the photosystems. A major breakthrough occurred in 1984, when Johann Deisenhofer, Hartmut Michel, and Robert Huber reported the first X-ray crystallographic analysis of a membrane protein. To the great benefit of photosynthesis research, this protein was the reaction center from the photosynthetic purple bacterium *Rhodopseudomonas viridis*. This research earned these three scientists the 1984 Nobel Prize in Chemistry.

The *R. viridis* Photosynthetic Reaction Center Is an Integral Membrane Protein

R. viridis is a photosynthetic prokaryote with a single photosystem that resembles PSII (even though it lacks an OEC and the capacity to oxidize water). The reaction center (145 kD) of the *R. viridis* photosystem is localized in the plasma membrane of these photosynthetic bacteria and is composed of four different polypeptides, designated *L* (273 amino acid residues), *M* (323 residues), *H* (258 residues), and *cytochrome* (333 amino acid residues) (Figure 21.14a). *L* and *M* each consist of five membrane-spanning α-helical segments; *H* has one such helix, the majority of the protein forming a globular domain in the cytoplasm (Figure 21.14b). The cytochrome subunit contains four heme groups; the N-terminal amino acid of this protein is cysteine. This cytochrome is anchored to the periplasmic face of the membrane via the hydrophobic chains of two fatty acid groups that are esterified to a glyceryl moiety joined via a thioether bond to the Cys (Figure 21.14a). *L* and *M* each bear two *bacteriochlorophyll* molecules (the bacterial version of Chl) and one *bacteriopheophytin*. *L* also has a bound quinone molecule, Q_A. Together, *L* and *M* coordinate an Fe atom. The photochemically active species of the *R. viridis* reaction center, **P870,** is composed of two bacteriochlorophylls, one contributed by *L* and the other by *M*.

Photosynthetic Electron Transfer by the *R. viridis* Reaction Center Leads to ATP Synthesis

The prosthetic groups of the *R. viridis* reaction center (P870, BChl, BPheo, and the bound quinones) are fixed in a spatial relationship to one another that favors photosynthetic e^- transfer (Figure 21.14a,c). Photoexcitation of P870 (creation of P870*) leads to e^- loss (P870$^+$) via electron transfer to the nearby bacteriochlorophyll (BChl). The e^- is then transferred via the *L* bacteriopheophytin (BPheo) to Q_A, which is also an *L* prosthetic group. The corresponding site on *M* is occupied by a loosely bound quinone, Q_B, and electron transfer from Q_A to Q_B takes place. An interesting aspect of the system is that *no* electron transfer occurs through *M*, even though it has components apparently symmetric to and identical to the *L* e^- transfer pathway.

The reduced quinone formed at the Q_B site is free to diffuse to a neighboring *cytochrome bc_1* membrane complex, where its oxidation is coupled to H^+ translocation and,

(a)

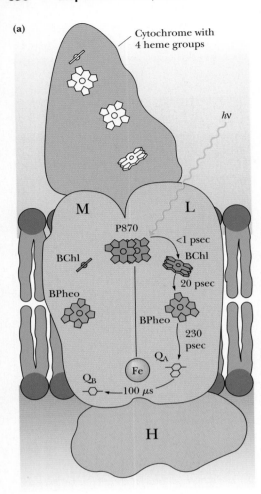

Note: The cytochrome subunit is membrane associated via a diacylglycerol moiety on its N-terminal Cys residue:

(b)   **(c)**

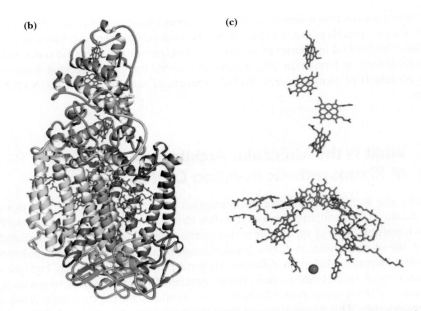

FIGURE 21.14 The *R. viridis* reaction center (RC). **(a)** Diagram of the RC showing light activation and path of e^- transfer. **(b)** Molecular graphic of the *R. viridis* RC. M and L are yellow and blue; H is orange; the cytochrome is green. **(c)** Deletion of the *R. viridis* RC protein chains reveals the spatial relationship between its heme, chlorophyll, and quinone prosthetic groups. The iron atom is represented by a sphere (pdb id = 1PRC).

hence, ultimately to ATP synthesis. The use of light energy to drive ATP synthesis by the concerted action of these membrane proteins is called **photophosphorylation** (Figure 21.15).

Cytochrome c_2, a periplasmic protein, serves to cycle electrons back to P870$^+$ via the four hemes of the reaction center cytochrome subunit. A specific tyrosine residue of *L* (Tyr^{162}) is situated between P870 and the closest cytochrome heme. This Tyr is the immediate e^- donor to P870$^+$ and completes the light-driven electron transfer cycle.

The Molecular Architecture of PSII Resembles the *R. viridis* Reaction Center Architecture

Some have argued that PSII is the most important enzyme system in the history of life. Its appearance in cyanobacteria about 2.5 billion years ago dramatically altered the trajectory of evolution. PSII accomplishes the light-driven formation of O_2 from 2 H_2O, and it is directly responsible for all of the oxygen in the atmosphere. The availability of molecular oxygen for highly exergonic oxidative reactions allowed the evolution of energy-intensive organisms such as animals. So, we are here because of PSII.

Type II photosystems of higher plants, green algae, and cyanobacteria contain more than 20 subunits and are considerably more complex than the *R. viridis* reaction center. The structure of PSII from the thermophilic cyanobacterium *Thermosynechoccus elongatus* has been revealed by X-ray crystallography, providing insight into PSII structures in general. Interestingly, both type II and type I photosystems show significant similarity to the *R. viridis* reaction center, thus establishing a strong evolutionary connection between reaction centers.

T. elongatus PSII is a homodimeric structure. Each "monomer" has a mass of almost 350 kD and 23 different protein subunits, the 4 largest being the reaction center pair of subunits (**D1** and **D2**) and two chlorophyll-containing inner antenna subunits (**CP43** and **CP47**) that bracket D1 and D2 (Figure 21.16). Together, CP43 and CP47 have a total of 26 Chl *a* molecules, and exciton energy is collected and transferred from them

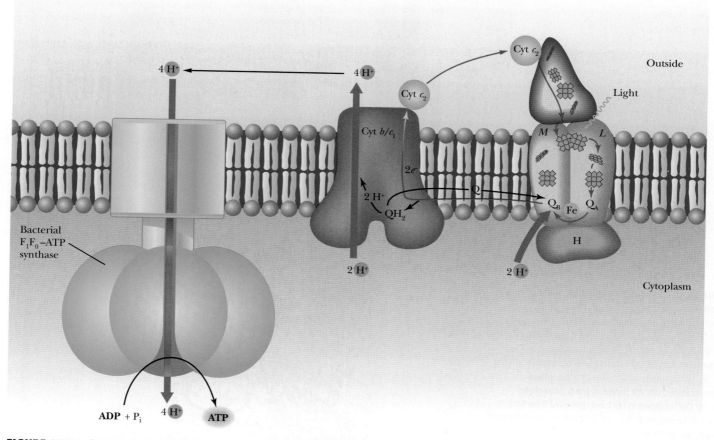

FIGURE 21.15 Photophosphorylation. Photoexcitation of the *R. viridis* RC leads to reduction of a quinone, Q, to form QH₂. Oxidation of QH₂ by the cytochrome *bc₁* complex leads to H⁺ translocation for ATP synthesis by the *R. viridis* F₁F₀–ATP synthase.

to P680. Collectively, the protein subunits in a PSII "monomer" have at least 34 transmembrane α-helical segments, 22 of which are found in the D1-D2-CP43-CP47 "core" structure. D1 and D2 each have five membrane-spanning α-helices. Structurally and functionally, these two subunits are a direct counterpart of the *L* and *M* subunits of the *R. viridis* reaction center. P680 consists of a pair of Chl *a* molecules, with D1 and D2 each contributing one. D1 and D2 each have two other Chl *a* molecules, one near each P680 (**Chl$_{D1}$** and **Chl$_{D2}$**, respectively) and another that interacts with CP43/CP47 (Chl$_{Z\text{-}D1}$ and Chl$_{Z\text{-}D2}$, respectively) (Figure 21.16). Two equivalents of pheophytin (Pheo) are located on D1 and D2. The tyrosine species D is Tyr[161] in the D1 amino acid sequence. Complexed to D2 is a tightly bound plastoquinone molecule, Q$_A$. Electrons flow from P680* to Chl$_{D1}$ and on to Pheo$_{D1}$. Pheo$_{D1}$ then transfers the electron to Q$_A$ on D2, where it then moves to a second plastoquinone situated in the Q$_B$ site on D1 (Figure 21.16). Electron transfer from Q$_A$ and Q$_B$ is assisted by the iron atom located between them. Each plastoquinone that enters the Q$_B$ site accepts two electrons derived from water and two H⁺ from the stroma before it is released into the membrane as the hydroquinone PQH₂. Thus, two photons are required to reduce each PQ that enters the Q$_B$ site. The stoichiometry of the overall reaction catalyzed by PSII is 2 H₂O + 2 PQ + 4 *hυ* $\longrightarrow$ O₂ + 2 PQH₂.

Electrons withdrawn from water are the source of the electrons needed to restore oxidized P680 (P680⁺) back to the P680 state for another cycle of light-driven electron transport. The oxidation of two H₂O molecules results in O₂ formation and is carried out by the **oxygen evolving complex (OEC)** located on the lumenal side of the PSII D1 subunit. These electrons pass from the OEC to P680⁺ via Tyr[161] (also known as Y$_Z$) of

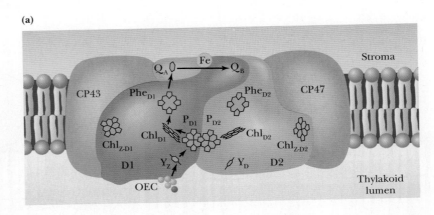

(a)

(b)

FIGURE 21.16 Molecular architecture of the *Thermosynechoccus elongatus* PSII dimer. **(a)** The arrow shows the path of electron transfer from P680* to Chl$_{D1}$ to Pheo$_{D1}$ to Q$_A$ on D2 and then, via the Fe atom, to Q$_B$ on D1. The Tyr^{161} residue of D1, symbolized by Y$_z$, is situated between P680 and the (Mn)$_4$ cluster. **(b)** Structure of *T. elongatus* PSII (pdb id = 3BZ1). Chlorophylls of the reaction center and electron-transfer path are shown in green; pheophytins, in blue. The OEC is shown in brick red. (Adapted from Barber, J., 2003. Photosystem II: The engine of life. *Quarterly Review of Biophysics* **36**:71–89.)

the D1 subunit. The water-binding site of the OEC is a cluster of 4 manganese and 1 calcium atom (Mn$_4$Ca); an associated Cl$^-$ ion is essential to water oxidation. Protons liberated upon oxidation of these H$_2$O molecules are passed directly into the thylakoid lumen.

How Does PSII Generate O$_2$ from H$_2$O?

PSII catalyzes what is probably the most thermodynamically difficult reaction in nature, the light-driven oxidation of water to molecular oxygen. The protons and electrons released in this reaction are used to reduce NADP$^+$ to NADPH and to establish a proton gradient across the photosynthetic membrane that can be tapped to drive chemiosmotic ATP synthesis (see Figure 21.20). Accumulation of molecular oxygen in the atmosphere as a byproduct of the photo-oxidation of water has transformed the planet since the evolutionary appearance of this reaction some 2 billion years ago in cyanobacteria. How does PSII evolve oxygen?

The Structure of the Oxygen-Evolving Complex The architecture of the *T. elongatus* photosynthetic OEC reveals a large globular protein domain juxtaposed on the lumenal side of the D1 subunit of PSII (Figure 21.16). The active site of the OEC is identifiable by the elongated Mn$_4$Ca cluster, whose Mn atoms are bridged by 5 O atoms, as shown in Figure 21.17. This metal cluster is held by Glu^{189}, Asp^{342}, His^{332}, and His^{337} of the PSII D1 subunit and Glu^{354} of CP43. The chloride ion (Cl$^-$) required for O$_2$ evolution is believed to be a Ca^{2+} ligand. Note also that Tyr^{161} (Y$_Z$) is situated near the metal cluster, ideally poised to serve in electron transfer between H$_2$O and P680$^+$. When four e^- have been removed from the cluster (one from each Mn atom) through e^- transfer to PSII via Tyr^{161}, two H$_2$O molecules provide the four e^- needed to re-reduce the Mn atoms, and O$_2$ is evolved. The four H$^+$ released contribute to the proton gradient.

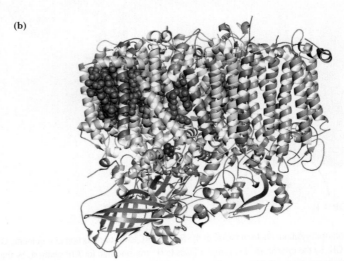

FIGURE 21.17 Structure of the PSII oxygen-evolving complex (OEC). Four Mn atoms (red, lettered A–D) and a Ca atom (green) form the water-splitting metal cluster of the OEC. Bridging O atoms are purple. (Adapted from Figure 4 in Yano, J., et al., 2006. Where water is oxidized to dioxygen: Structure of the photosynthetic Mn$_4$Ca cluster. *Science* **314**:821–825.)

The Molecular Architecture of PSI Resembles the *R. viridis* Reaction Center and PSII Architecture

The structure of PSI from the thermophilic cyanobacterium *T. elongatus* also has been solved by X-ray crystallography, completing our view of reaction center structure and confirming the fundamental similarities in organization that exist in these energy-transducing integral membrane proteins. Because of direct correlations with information about eukaryotic PSI, this cyanobacterial PSI provides a general model for all P700-dependent photosystems.

T. elongatus PSI exists as a cloverleaf-shaped trimeric structure. Each "monomer" (356 kD) consists of 12 different protein subunits and 127 cofactors: 96 chlorophyll *a* molecules, 2 phylloquinones, 3 Fe_4S_4 clusters, 22 carotenoids, and 4 lipids that are an intrinsic part of the protein complex (Figure 21.18). All of the electron-transferring prosthetic groups essential to PSI function are localized to just three polypeptides: **PsaA, PsaB,** and **PsaC.** PsaA and PsaB (83 kD each) compose the reaction center heterodimer, a structural pattern now seen as universal in photosynthetic reaction centers. PsaA and PsaB each have 11 transmembrane α-helices, with the 5 most C-terminal α-helices of each serving as the scaffold for the reaction center photosynthetic electron-transfer apparatus. PsaC interacts with the stromal face of the PsaA–PsaB heterodimer. PsaC carries two of the Fe_4S_4 clusters, F_A and F_B, and interacts with **PsaD.** Together they provide a docking site for ferredoxin. The electron-transfer system of PSI consists of three pairs of chlorophyll molecules: **P700** (a heterodimer of Chl *a* and an epimeric form, Chl *a'*) and two additional Chl *a* pairs (symbolized by A_0) that mediate e^- transfer to the quinone acceptor. The *T. elongatus* quinone acceptor (A_1) is **phylloquinone** (also known as vitamin K_1). The **Fx** Fe_4S_4 cluster bridges PsaA and PsaB; two of its four cysteine ligands come from PsaA, the other two from PsaB. Photochemistry begins with exciton absorption at P700, creating P700*, which has the most negative reduction potential known in nature. Almost instantaneous electron transfer and charge separation (to form $P700^+{:}A_0^-$) is followed by transfer of the electron from A_0 to A_1, then to F_A, to F_B, and then on to a ferredoxin molecule located at the "stromal" side of the membrane.

(a)

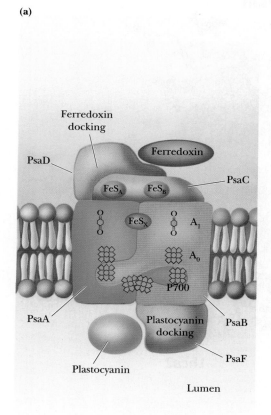

(b)

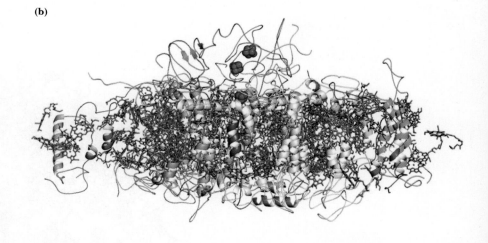

FIGURE 21.18 The molecular architecture of PSI. **(a)** Subunit organization. (Adapted from Golbeck, J. H., 1992. Structure and function of photosystem I. *Annual Review of Plant Physiology and Plant Molecular Biology* **43**:293–324; and Fromme, P., Jordan, P., and Krausse, N., 2001. Structure of photosystem I. *Biochimica et Biophysica Acta* **1507**:5–31.) **(b)** Molecular graphic PSI. PsaA is salmon; PsaB is salmon; chlorophylls are green. The iron-sulfur clusters are red (pdb id = 2WSC).

The positive charge at P700$^+$ and the e^- at F_A/F_B represent a charge separation across the membrane, an energized condition created by light. Plastocyanin (or in cyanobacteria, a lumenal cytochrome designated *cytochrome c$_6$*) delivers an electron to fill the electron hole in P700$^+$.

How Do Green Plants Carry Out Photosynthesis?

Do the higher plant photosystems follow the structural and functional pattern first revealed in the bacterial RC and recapitulated in the cyanobacterial PSI and PSII? With new structural information for the higher plant PSI (from *Pisum sativum,* garden pea) and PSII (from spinach), the fundamental organization pattern for photosystems seen earlier is confirmed for higher plants. Further, the structure of a plant membrane protein supercomplex consisting of the PSI reaction center and its light-harvesting antenna **LHC1** (light-harvesting complex 1) has been described. This supercomplex exists as a "monomer" composed of 17 distinct protein subunits and 193 prosthetic groups, including 173 chlorophylls, 2 phylloquinones, and 3 Fe$_4$S$_4$ clusters (Figure 21.19). The four LHC1 subunits form an arc around one side of the PSI RC. A second light-harvesting complex (LHC2) binds to another side. This plant PSI system, like all photosystems, is remarkably efficient, showing a quantum efficiency of nearly 1 (one electron transferred per photon falling anywhere within the supercomplex). The many Chl and other light-harvesting molecules of the supercomplex form an integrated network for highly efficient transfer of light energy into P700.

21.5 | What Is the Quantum Yield of Photosynthesis?

The quantum yield of photosynthesis can be defined as the amount of product formed per equivalent of light input. In terms of exciton delivery to reaction center Chl dimers and subsequent e^- transfer, the quantum yield of photosynthesis typically approaches

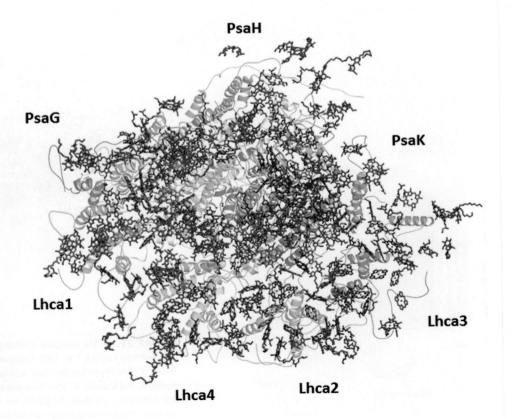

FIGURE 21.19 View of the plant PSI-LHC1 supercomplex, from the stromal side of the thylakoid membrane. Chl molecules are shown in green and carotenoids and lipids in red. The 16 protein subunits are shown as ribbon diagrams in the background, with the positions of PsaG, PsaH, PsaK, and LHC1-4 subunits indicated. (pdb id = 3LW5).

the theoretical limit of 1. The quantum yield of photosynthesis can also be expressed as the ratio of CO_2 fixed or O_2 evolved per photon absorbed. Interestingly, Figure 21.11b indicates an overall stoichiometry of 7 H^+ contributed to the proton gradient across the thylakoid membrane for each pair of electrons moving from H_2O to $NADP^+$: 2 into the lumen upon oxidation of a water molecule, 4 into the lumen from oxidation of PQH_2 by the cytochrome b_6f complex, and 1 H^+ disappearing from the stroma as $NADP^+$ is reduced to NADPH. More appropriately, 4 hv per photosystem (8 quanta) total would drive the evolution of 1 O_2, the reduction of 2 $NADP^+$, and a contribution of 14 H^+ to the proton gradient. The H^+/ATP ratio depends on the number of c-subunits in the F_1F_0-ATP synthase possessed by a cell or organism (chloroplast F_1F_0-ATP synthase is termed CF_1CF_0-ATP synthase, C signifying "chloroplast," see below). As we have seen, the number of c-subunits in ATP synthase can vary from organism to organism. Chloroplast CF_1CF_0-ATP synthase has 14 c-subunits; thus, 3 ATP are formed per 14 H^+, so 3 ATP would be synthesized in chloroplasts from an input of 8 quanta.

The energy of a photon depends on its wavelength, according to the equation $E = hv = hc/\lambda$, where E is energy, c is the speed of light, and λ is its wavelength. Expressed in molar terms, an einstein is the amount of energy in Avogadro's number of photons: $E = Nhc/\lambda$. Light of 700-nm wavelength is the longest-wavelength and the lowest-energy light acting in the eukaryotic photosystems. An einstein of 700-nm light is equivalent in energy to approximately 170 kJ. Eight einsteins of this light, 1360 kJ, theoretically generate 2 moles of NADPH, 3 moles of ATP, and 1 mole of O_2.

Calculation of the Photosynthetic Energy Requirements for Hexose Synthesis Depends on H^+/hv and ATP/H^+ Ratios

The fixation of carbon dioxide to form hexose, the dark reactions of photosynthesis, requires considerable energy. The overall stoichiometry of this process (see Equation 21.3) involves 12 NADPH and 18 ATPs. Thus, the ATP/NADPH ratio for CO_2 fixation is 1.5. To generate 12 equivalents of NADPH and 18 equivalents of ATP necessitates the input of 48 einsteins of light, minimally 170 kJ each, or 8160 kJ, from which one mole of hexose would be synthesized. The standard free energy change, $\Delta G^{\circ\prime}$, for hexose formation from carbon dioxide and water (the exact reverse of cellular respiration) is +2870 kJ/mol. Note that many assumptions underlie these calculations, including assumptions about the ATP/H^+ ratio, the H^+/e^- ratio, and ultimately, the relationship between quantum input and overall yields of NADPH and ATP.

21.6 How Does Light Drive the Synthesis of ATP?

Light-driven ATP synthesis, or photophosphorylation, is a fundamental part of the photosynthetic process. The conversion of light energy to chemical energy results in electron-transfer reactions, which lead to the generation of reducing power (reduced quinones or NADPH). Coupled with these electron transfers, protons are driven across the thylakoid membranes from the stromal side to the lumenal side. These proton translocations occur in a manner analogous to the proton translocations accompanying mitochondrial electron transport that provide the driving force for oxidative phosphorylation (see Chapter 20). Figure 21.11 indicates that proton translocations can occur at a number of sites. For example, protons are produced in the thylakoid lumen upon photolysis of water by PSII. The oxidation–reduction events as electrons pass through the plastoquinone pool and the Q cycle are another source of proton translocations. The proton transfer accompanying $NADP^+$ reduction also can be envisioned as protons being taken from the stromal side of the thylakoid vesicle. The current view is that three protons are translocated for each electron that flows from H_2O to $NADP^+$. Because this electron transfer requires two photons, one falling at PSII and one at PSI, the overall yield is 1.5 protons per quantum of light.

The Mechanism of Photophosphorylation Is Chemiosmotic

The thylakoid membrane is asymmetrically organized, or "sided," like the mitochondrial membrane. It also shares the property of being a barrier to the passive diffusion of H^+ ions. Photosynthetic electron transport thus establishes an electrochemical gradient, or proton-motive force, across the thylakoid membrane with the interior, or lumen, side accumulating H^+ ions relative to the stroma of the chloroplast. Like oxidative phosphorylation, the mechanism of photophosphorylation is chemiosmotic.

A proton-motive force of approximately -250 mV is needed to achieve ATP synthesis. This proton-motive force, Δp, is composed of a membrane potential, $\Delta \psi$, and a pH gradient, ΔpH (see Chapter 20). The proton-motive force is defined as the free energy difference, ΔG, divided by $\mathscr{F}$, Faraday's constant:

$$\Delta p = \Delta G/\mathscr{F} = \Delta \psi - (2.3 \; RT/\mathscr{F})\Delta \text{pH} \tag{21.5}$$

In chloroplasts, the value of $\Delta \psi$ is typically -50 to -100 mV and the pH gradient is equivalent to about 3 pH units, so $-(2.3 \; RT/\mathscr{F})\Delta$pH $= -170$ mV. This situation contrasts with the mitochondrial proton-motive force, where the membrane potential contributes relatively more to Δp than does the pH gradient.

CF$_1$CF$_0$–ATP Synthase Is the Chloroplast Equivalent of the Mitochondrial F$_1$F$_0$–ATP Synthase

The transduction of the electrochemical gradient into the chemical energy represented by ATP is carried out by the chloroplast ATP synthase, which is highly analogous to the mitochondrial F_1F_0–ATP synthase. The chloroplast enzyme complex is called **CF$_1$CF$_0$–ATP synthase**, "C" symbolizing chloroplast. Like the mitochondrial complex, CF$_1$CF$_0$–ATP synthase is a heteromultimer of α-, β-, γ-, δ-, ϵ-, a-, b-, and c-subunits (see Chapter 20), consisting of a knoblike structure some 9 nm in diameter (CF$_1$) attached to a stalked base (CF$_0$) embedded in the thylakoid membrane. The mechanism of action of CF$_1$CF$_0$–ATP synthase in coupling ATP synthesis to the collapse of the pH gradient is similar to that of the mitochondrial ATP synthase described in Chapter 20. However, spinach chloroplast CF$_1$CF$_0$ ATP synthase has 14 c-subunits in its F_0 rotor, whereas the ATP synthase from the cyanobacterium *Spirulina platensis* has 15 c-subunits. Thus, one turn of the spinach F_0 c-subunit ring would require 14 H^+, and one turn of the *S. platensis* rotor, 15. All proton-translocating ATP synthases have 3 ATP-synthesizing sites, so the H^+/ATP ratio for spinach chloroplasts is 14/3 or 4.67; the H^+/ATP ratio for the *S. platensis* enzyme is 5. The mechanism of photophosphorylation is summarized schematically in Figure 21.20. The structure of the spinach choroplast CF$_1$CF$_0$–ATP synthase c-subunit assembly is shown in Figure 21.21.

Photophosphorylation Can Occur in Either a Noncyclic or a Cyclic Mode

Photosynthetic electron transport, which pumps H^+ into the thylakoid lumen, can occur in two modes, both of which lead to the establishment of a transmembrane proton-motive force. Thus, both modes are coupled to ATP synthesis and are considered alternative mechanisms of photophosphorylation, even though they are distinguished by differences in their electron-transfer pathways. The two modes are cyclic and noncyclic photophosphorylation. **Noncyclic photophosphorylation** has been the focus of our discussion and is represented by the scheme in Figure 21.20, where electrons activated by quanta at PSII and PSI flow from H_2O to $NADP^+$, with concomitant establishment of the proton-motive force driving ATP synthesis. Note that in noncyclic photophosphorylation, O_2 is evolved and $NADP^+$ is reduced.

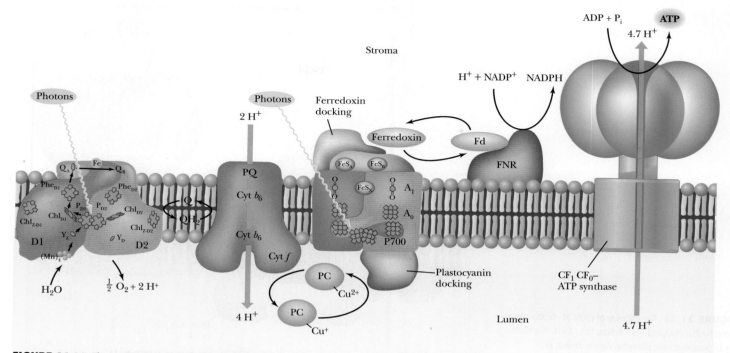

FIGURE 21.20 The mechanism of photophosphorylation. Photosynthetic electron transport establishes a proton gradient that is tapped by the CF_1CF_0–ATP synthase to drive ATP synthesis. The value, 4.7 H^+/ATP, is the estimated ratio for the spinach chloroplast CF_1CF_0–ATP synthase.

Cyclic Photophosphorylation Generates ATP but Not NADPH or O_2

In cyclic photophosphorylation, the "electron hole" in $P700^+$ created by electron loss from P700 is filled *not* by an electron derived from H_2O via PSII but by a cyclic pathway in which the photoexcited electron returns ultimately to $P700^+$. This pathway is

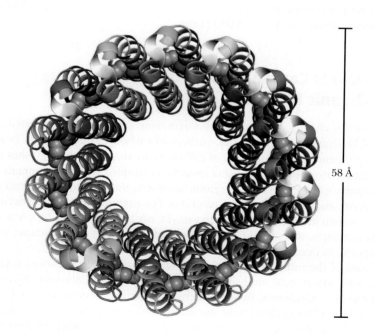

FIGURE 21.21 The *c*-subunit assembly of the spinach chloroplast CF_1CF_0–ATP synthase consists of a cylindrical array of 14 identical subunits. Each subunit reversibly binds an H^+ through protonation/deprotonation of its Glu^{61} side chain (pdb id = 2W5J).

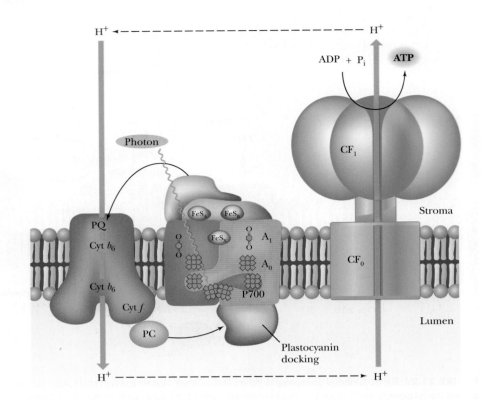

FIGURE 21.22 The pathway of cyclic photophosphory-lation by PSI. (Adapted from Arnon, D. I., 1984. The discovery of photosynthetic phosphorylation. *Trends in Biochemical Sciences* **9:**258–262.)

schematically represented in Figure 21.11 by the dashed line connecting ferredoxin (Fd) and plastoquinone (PQ) within the membrane. This pathway diverts the activated e^- lost from PSI back through the PQ pool, the cytochrome b_6f complex, and plastocyanin to re-reduce $P700^+$ (Figure 21.22).

Proton translocations accompany these cyclic electron transfer events so that ATP synthesis can be achieved. In cyclic photophosphorylation, ATP is the sole product of energy conversion. No NADPH is generated, and because PSII is not involved, no oxygen is evolved. Cyclic photophosphorylation theoretically yields 2 H^+ per e^- (2 $H^+/h\nu$) from the operation of the cytochrome b_6f complex. Thus, cyclic photophosphorylation provides a mechanism for supplementing ATP production. Estimates indicate that cyclic photophosphorylation may contribute about 10% of total chloroplast ATP synthesis and thereby elevate the ATP/NADPH ratio.

21.7 How Is Carbon Dioxide Used to Make Organic Molecules?

As we began this chapter, we saw that photosynthesis traditionally is equated with the process of CO_2 fixation, that is, the net synthesis of carbohydrate from CO_2. Indeed, the capacity to perform net accumulation of carbohydrate from CO_2 distinguishes the phototrophic (and autotrophic) organisms from heterotrophs. Although animals possess enzymes capable of linking CO_2 to organic acceptors, they cannot achieve a net accumulation of organic material by these reactions. For example, fatty acid biosynthesis is primed by covalent attachment of CO_2 to acetyl-CoA to form malonyl-CoA (see Chapter 24). Nevertheless, this "fixed CO_2" is liberated in the very next reaction, so no net CO_2 incorporation occurs.

Elucidation of the pathway of CO_2 fixation represents one of the earliest applications of radioisotope tracers to the study of biology. In 1945, Melvin Calvin and his colleagues at the University of California, Berkeley, were investigating photosynthetic CO_2 fixation in *Chlorella*. Using $^{14}CO_2$, they traced the incorporation of radioactive ^{14}C into organic products and found that the earliest labeled product was **3-phosphoglycerate** (see Figure

17.14). Although this result suggested that the CO_2 acceptor was a two-carbon compound, further investigation revealed that, in reality, 2 equivalents of 3-phosphoglycerate were formed following addition of CO_2 to a five-carbon (pentose) sugar:

$$CO_2 + \text{5-carbon acceptor} \longrightarrow \text{[6-carbon intermediate]} \longrightarrow$$
$$\text{two 3-phosphoglycerates}$$

Ribulose-1,5-Bisphosphate Is the CO_2 Acceptor in CO_2 Fixation

The five-carbon CO_2 acceptor was identified as **ribulose-1,5-bisphosphate (RuBP),** and the enzyme catalyzing this key reaction of CO_2 fixation is **ribulose bisphosphate carboxylase/ oxygenase,** or, in the jargon used by workers in this field, **rubisco.** The name *ribulose bisphosphate carboxylase/oxygenase* reflects the fact that rubisco catalyzes the reaction of either CO_2 or, alternatively, O_2 with RuBP. Rubisco is found in the chloroplast stroma. It is a very abundant enzyme, constituting more than 15% of the total chloroplast protein. Given the preponderance of plant material in the biosphere, rubisco is probably the world's most abundant protein. Rubisco is large: In higher plants, rubisco is a 550-kD heteromultimeric $(\alpha_8\beta_8)$ complex consisting of eight identical large subunits (55 kD) and eight small subunits (15 kD) (Figure 21.23). The large subunit is the catalytic unit of the enzyme. It binds both substrates (CO_2 and RuBP) and Mg^{2+} (a divalent cation essential for enzymatic activity). The small subunit modulates the catalytic efficiency of the enzyme, increasing k_{cat} more than 100-fold. The rubisco large subunit is encoded by a gene within the chloroplast DNA, whereas the small subunit is encoded by a multigene family in the nuclear DNA. Assembly of active rubisco heteromultimers occurs within chloroplasts following transit of the small subunit polypeptide across the chloroplast membrane.

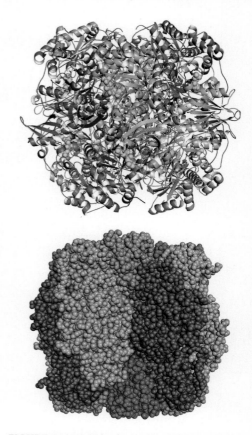

FIGURE 21.23 Molecular graphic of ribulose bisphosphate carboxylase. The enzyme consists of 8 equivalents each of two subunits. Clusters of four small subunits (orange and red) are located at each end of the symmetric octamer formed by the *L* subunits (light and dark green). The active sites are revealed in the ribbon diagram by bound ribulose-1,5-bisphosphate (yellow) (pdb id = 1RXO).

2-Carboxy-3-Keto-Arabinitol Is an Intermediate in the Ribulose-1,5-Bisphosphate Carboxylase Reaction

The addition of CO_2 to ribulose-1,5-bisphosphate results in the formation of an enzyme-bound intermediate, **2-carboxy-3-keto-arabinitol** (Figure 21.24, II). This intermediate arises when CO_2 adds to the enediol intermediate generated from ribulose-1,5-bisphosphate. Hydrolysis of the C_2—C_3 bond of the intermediate generates two molecules of 3-phosphoglycerate. The CO_2 ends up as the carboxyl group of one of the two molecules.

Ribulose-1,5-Bisphosphate Carboxylase Exists in Inactive and Active Forms

Rubisco exists in three forms: an inactive form, designated *E;* a carbamylated, but inactive, form, designated *EC;* and an active form, *ECM,* which is carbamylated and has Mg^{2+} at its active sites as well. Carbamylation of rubisco takes place by addition of CO_2

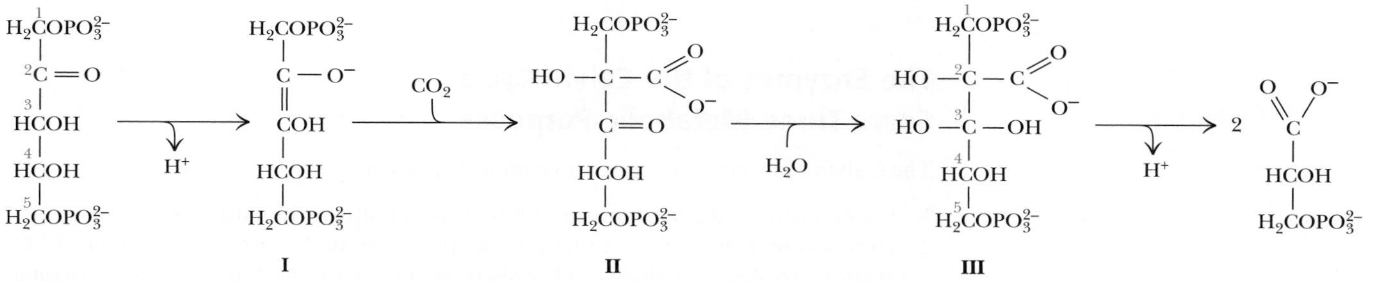

FIGURE 21.24 The ribulose bisphosphate carboxylase reaction. Mg^{2+} at the active site aids in stabilizing the 2,3-enediol transition state **(I)** for CO_2 addition and in facilitating the carbon–carbon bond cleavage that leads to product formation.

to its Lys^{201} ϵ-NH$_2$ groups (to give ϵ—NH—COO$^-$ derivatives). The CO$_2$ molecules used to carbamylate Lys residues do not become substrates. The carbamylation reaction occurs spontaneously at slightly alkaline pH (pH 8). Carbamylation of rubisco completes the formation of a binding site for the Mg^{2+} that participates in the catalytic reaction. Once Mg^{2+} binds to *EC,* rubisco achieves its active *ECM* form. Activated rubisco displays a K_m for CO$_2$ of 10 to 20 μM. The relative abundance of CO$_2$ in the atmosphere is low, about 0.03%. The concentration of CO$_2$ dissolved in aqueous solutions equilibrated with air is about 10 μM.

Substrate RuBP binds much more tightly to the inactive *E* form of rubisco (K_D = 20 nM) than to the active *ECM* form (K_m for RuBP = 20 μM). Thus, RuBP is also a potent inhibitor of rubisco activity. Release of RuBP from the active site of rubisco is mediated by **rubisco activase**, a member of the AAA+ ATPase family of proteins involved in conformational remodeling of other proteins (see Chapter 31). Rubisco activase is a *regulatory protein;* it binds to *E*-form rubisco near its catalytic site and, in an ATP-dependent reaction, promotes the release of RuBP. Rubisco then becomes activated by carbamylation and Mg^{2+} binding. Rubisco activase itself is activated in an indirect manner by light. Thus, light is the ultimate activator of rubisco.

CO$_2$ Fixation into Carbohydrate Proceeds Via the Calvin–Benson Cycle

The immediate product of CO$_2$ fixation, 3-phosphoglycerate, must undergo a series of transformations before the net synthesis of carbohydrate is realized. Among carbohydrates, hexoses (particularly glucose) occupy center stage. Glucose is the building block for both cellulose and starch synthesis. These plant polymers constitute the most abundant organic material in the living world, and thus, the central focus on glucose as the ultimate end product of CO$_2$ fixation is amply justified. Also, sucrose (α-D-glucopyranosyl-(1→2)-β-D-fructofuranoside) is the major carbon form translocated out of leaves to other plant tissues. In nonphotosynthetic tissues, sucrose is metabolized via glycolysis and the TCA cycle to produce ATP.

The set of reactions that transforms 3-phosphoglycerate into hexose is named the **Calvin–Benson cycle** (often referred to simply as the Calvin cycle) for its discoverers. The reaction series is indeed cyclic because not only must carbohydrate appear as an end product, but the five-carbon acceptor, RuBP, must be regenerated to provide for continual CO$_2$ fixation. Balanced equations that schematically represent this situation are

$$6(1) + 6(5) \longrightarrow 12(3)$$
$$12(3) \longrightarrow 1(6) + 6(5)$$

$$\textit{Net: } 6(1) \longrightarrow 1(6)$$

Each number in parentheses represents the number of carbon atoms in a compound, and the number preceding the parentheses indicates the stoichiometry of the reaction. Thus, 6(1), or 6 CO$_2$, condense with 6(5) or 6 RuBP to give 12 3-phosphoglycerates. These 12(3)s are then rearranged in the Calvin cycle to form one hexose, 1(6), and regenerate the six 5-carbon (RuBP) acceptors.

The Enzymes of the Calvin Cycle Serve Three Metabolic Purposes

The Calvin cycle enzymes serve three important ends:

1. They constitute the predominant CO$_2$ fixation pathway in nature.
2. They accomplish the reduction of 3-phosphoglycerate, the primary product of CO$_2$ fixation, to glyceraldehyde-3-phosphate so that carbohydrate synthesis becomes feasible.
3. They catalyze reactions that transform three-carbon compounds into four-, five-, six-, and seven-carbon compounds.

Most of the enzymes mediating the reactions of the Calvin cycle also participate in either glycolysis (see Chapter 18) or the pentose phosphate pathway (see Chapter 22). The aim of the Calvin scheme is to account for hexose formation from 3-phosphoglycerate. In the course of this metabolic sequence, the NADPH and ATP produced in the light reactions are consumed, as indicated earlier in Equation 21.3.

The Calvin cycle of reactions starts with *ribulose bisphosphate carboxylase* catalyzing formation of 3-phosphoglycerate from CO_2 and RuBP and concludes with **ribulose-5-phosphate kinase** (also called *phosphoribulose kinase*), which forms RuBP (Figure 21.25 and Table 21.1). The carbon balance is given at the right side of Table 21.1. Several features of the reactions in this table merit discussion. Note that a total of 18 equivalents of ATP consumed in hexose formation are expended (reactions 2 and 15): 12 to form 12 equivalents of 1,3-bisphosphoglycerate from 3-phosphoglycerate by a reversal of the normal glycolytic reaction catalyzed by **3-phosphoglycerate kinase** and six to phosphorylate Ru-5-P to regenerate 6 RuBP. All 12 NADPH equivalents are used in reaction 3. Plants possess an **NADPH-specific glyceraldehyde-3-phosphate dehydrogenase,** which contrasts with its glycolytic counterpart in its specificity for NADP over NAD and in the direction in which the reaction normally proceeds.

The Calvin Cycle Reactions Can Account for Net Hexose Synthesis

When carbon rearrangements are balanced to account for net hexose synthesis, five of the glyceraldehyde-3-phosphate molecules are converted to dihydroxyacetone phosphate (DHAP). Three of these DHAPs then condense with three glyceraldehyde-3-P via the aldolase reaction to yield three hexoses in the form of fructose-1,6-bisphosphate (Figure 21.25). (Recall that the $\Delta G°'$ for the aldolase reaction in the glycolytic direction

---

TABLE 21.1	The Calvin Cycle Series of Reactions

Reactions 1 through 15 constitute the cycle that leads to the formation of one equivalent of glucose. The enzyme catalyzing each step, a concise reaction, and the overall carbon balance are given. Numbers in parentheses show the numbers of carbon atoms in the substrate and product molecules. Prefix numbers indicate in a stoichiometric fashion how many times each step is carried out in order to provide a balanced net reaction.

Reaction	Carbon balance
1. Ribulose bisphosphate carboxylase: $6\ CO_2 + 6\ H_2O + 6\ RuBP \longrightarrow 12\ 3\text{-PG}$	$6(1) + 6(5) \longrightarrow 12(3)$
2. 3-Phosphoglycerate kinase: $12\ 3\text{-PG} + 12\ ATP \longrightarrow 12\ 1,3\text{-BPG} + 12\ ADP$	$12(3) \longrightarrow 12(3)$
3. NADP$^+$-glyceraldehyde-3-P dehydrogenase: $12\ 1,3\text{-BPG} + 12\ NADPH \longrightarrow 12\ NADP^+ + 12\ G\text{-3-P} + 12\ P_i$	$12(3) \longrightarrow 12(3)$
4. Triose-P isomerase: $5\ G\text{-3-P} \longrightarrow 5\ DHAP$	$5(3) \longrightarrow 5(3)$
5. Aldolase: $3\ G\text{-3-P} + 3\ DHAP \longrightarrow 3\ FBP$	$3(3) + 3(3) \longrightarrow 3(6)$
6. Fructose bisphosphatase: $3\ FBP + 3\ H_2O \longrightarrow 3\ F\text{-6-P} + 3\ P_i$	$3(6) \longrightarrow 3(6)$
7. Phosphoglucoisomerase: $1\ F\text{-6-P} \longrightarrow 1\ G\text{-6-P}$	$1(6) \longrightarrow 1(6)$
8. Glucose phosphatase: $1\ G\text{-6-P} + 1\ H_2O \longrightarrow 1\ GLUCOSE + 1\ P_i$ The remainder of the pathway involves regenerating six RuBP acceptors ($= 30$ C) from the leftover two F-6-P (12 C), four G-3-P (12 C), and two DHAP (6 C).	$1(6) \longrightarrow 1(6)$
9. Transketolase: $2\ F\text{-6-P} + 2\ G\text{-3-P} \longrightarrow 2\ Xu\text{-5-P} + 2\ E\text{-4-P}$	$2(6) + 2(3) \longrightarrow 2(5) + 2(4)$
10. Aldolase: $2\ E\text{-4-P} + 2\ DHAP \longrightarrow 2$ sedoheptulose-1,7-bisphosphate (SBP)	$2(4) + 2(3) \longrightarrow 2(7)$
11. Sedoheptulose bisphosphatase: $2\ SBP + 2\ H_2O \longrightarrow 2\ S\text{-7-P} + 2\ P_i$	$2(7) \longrightarrow 2(7)$
12. Transketolase: $2\ S\text{-7-P} + 2\ G\text{-3-P} \longrightarrow 2\ Xu\text{-5-P} + 2\ R\text{-5-P}$	$2(7) + 2(3) \longrightarrow 4(5)$
13. Phosphopentose epimerase: $4\ Xu\text{-5-P} \longrightarrow 4\ Ru\text{-5-P}$	$4(5) \longrightarrow 4(5)$
14. Phosphopentose isomerase: $2\ R\text{-5-P} \longrightarrow 2\ Ru\text{-5-P}$	$2(5) \longrightarrow 2(5)$
15. Phosphoribulose kinase: $6\ Ru\text{-5-P} + 6\ ATP \longrightarrow 6\ RuBP + 6\ ADP$	$6(5) \longrightarrow 6(5)$
Net: $6\ CO_2 + 18\ ATP + 12\ NADPH + 12\ H^+ + 12\ H_2O \longrightarrow$ glucose $+ 18\ ADP + 18\ P_i + 12\ NADP^+$	$6(1) \longrightarrow 1(6)$

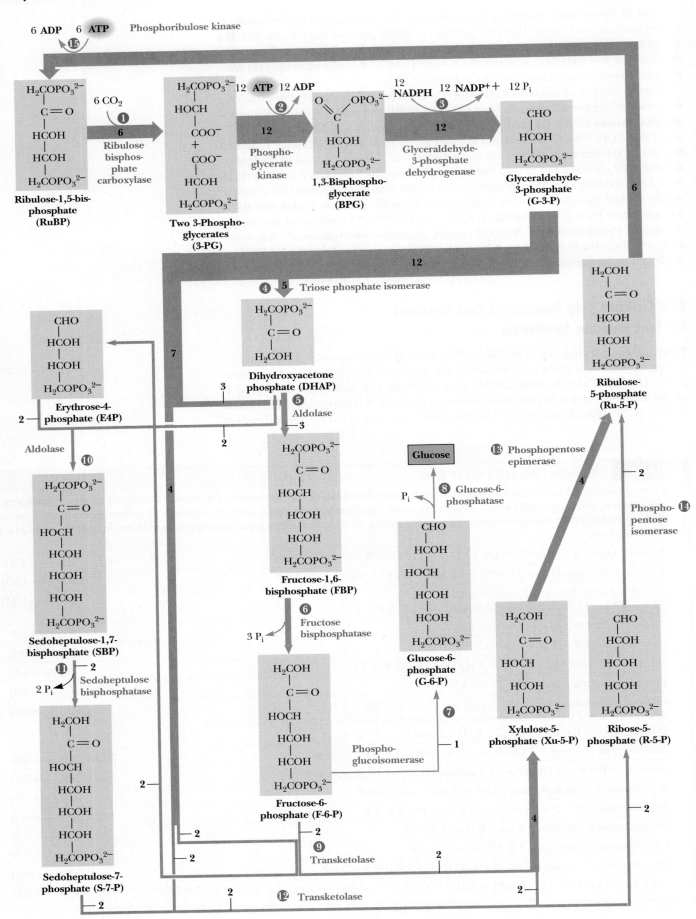

FIGURE 21.25 The Calvin–Benson cycle of reactions. The number associated with the arrow at each step indicates the number of molecules reacting in a turn of the cycle that produces one molecule of glucose. Reactions are numbered as in Table 21.1.

is +23.9 kJ/mol. Thus, the aldolase reaction running "in reverse" in the Calvin cycle would be thermodynamically favored under standard-state conditions.) Taking one FBP to glucose, the desired product of this scheme, leaves 30 carbons, distributed as 2 fructose-6-phosphates, 4 glyceraldehyde-3-phosphates, and 2 DHAP. These 30 Cs are reorganized into 6 RuBP by reactions 9 through 15. Step 9 and steps 12 through 14 involve carbohydrate rearrangements like those in the pentose phosphate pathway (see Chapter 22). Reaction 11 is mediated by **sedoheptulose-1,7-bisphosphatase.** This phosphatase is unique to plants; it generates sedoheptulose-7-P, the seven-carbon sugar serving as the transketolase substrate. Likewise, **phosphoribulose kinase** carries out the unique plant function of providing RuBP from Ru-5-P (reaction 15). The net conversion accounts for the fixation of 6 equivalents of carbon dioxide into 1 hexose at the expense of 18 ATP and 12 NADPH.

The Carbon Dioxide Fixation Pathway Is Indirectly Activated by Light

Plant cells contain mitochondria and can carry out cellular respiration (glycolysis, the citric acid cycle, and oxidative phosphorylation) to provide energy in the dark. Futile cycling of carbohydrate to CO_2 by glycolysis and the citric acid cycle in one direction and CO_2 to carbohydrate by the CO_2 fixation pathway in the opposite direction is thwarted through regulation of the Calvin cycle (Figure 21.26). In this regulation the activities of key Calvin cycle enzymes are coordinated with the output of photosynthesis. In effect, these enzymes respond indirectly to **light activation.** Thus, when light energy is available to generate ATP and NADPH for CO_2 fixation, the Calvin cycle proceeds. In the dark, when ATP and NADPH cannot be produced by photosynthesis, fixation of CO_2 ceases. The light-induced changes in the chloroplast, which regulate key Calvin cycle enzymes, include (1) *changes in stromal pH,* (2) *generation of reducing power,* and (3) *Mg^{2+} efflux from the thylakoid lumen.*

Light Induces pH Changes in Chloroplast Compartments As discussed in Section 21.6, illumination of chloroplasts leads to light-driven pumping of protons into the thylakoid lumen, which causes pH changes in both the stroma and the thylakoid lumen (Figure 21.27). The stromal pH rises, typically to pH 8. Because rubisco and rubisco activase are more active at pH 8, CO_2 fixation is activated as stromal pH rises. *Fructose-1,6-bisphosphatase, ribulose-5-phosphate kinase,* and *glyceraldehyde-3-phosphate dehydrogenase* all have alkaline pH optima. Thus, their activities increase as a result of the light-induced pH increase in the stroma.

Light Energy Generates Reducing Power Illumination of chloroplasts initiates photosynthetic electron transport, which generates reducing power in the form of reduced ferredoxin. In turn, Fd_{red} leads to reduced thioredoxin via *ferredoxin–thioredoxin reductase* (FTR) (Figure 21.28). Thioredoxin is a small (12 kD) protein possessing in its

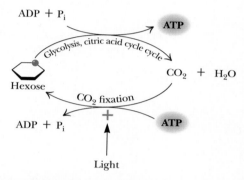

FIGURE 21.26 Light regulation of CO_2 fixation prevents a substrate cycle between cellular respiration and hexose synthesis by CO_2 fixation.

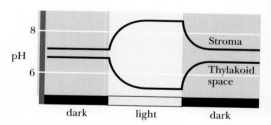

FIGURE 21.27 Light-induced pH changes in chloroplast compartments. These pH changes modulate the activity of key Calvin cycle enzymes.

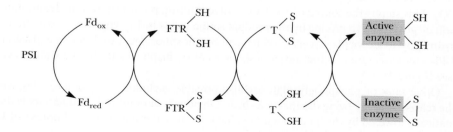

FIGURE 21.28 The pathway for light-regulated reduction of Calvin cycle enzymes. Light-generated reducing power (Fd_{red} = reduced ferredoxin) provides e^- for reduction of thioredoxin (T) by FTR.

reduced state a pair of sulfhydryls (—SH HS—), which upon oxidation form a disulfide bridge (—S—S—). Several Calvin cycle enzymes have pairs of cysteine residues that are involved in a disulfide–sulfhydryl transition between an inactive (—S—S—) and an active (—SH HS—) form. These enzymes include *fructose-1,6-bisphosphatase* (residues Cys^{174} and Cys^{179}), *NADP$^+$-malate dehydrogenase* (residues Cys^{10} and Cys^{15}), and *ribulose-5-phosphate kinase* (residues Cys^{16} and Cys^{55}). Thus, light activates these key enzymes through this ferredoxin-dependent, thioredoxin-mediated pathway (Figure 21.28).

Light Induces Movement of Mg^{2+} Ions from the Thylakoid Vesicles into the Stroma When light-driven proton pumping across the thylakoid membrane occurs, a concomitant efflux of Mg^{2+} ions from vesicles into the stroma is observed. This efflux of Mg^{2+} somewhat counteracts the charge accumulation due to H$^+$ influx and is one reason why the membrane potential change in response to proton pumping is less in chloroplasts than in mitochondria (see Equation 21.5). Both *ribulose bisphosphate carboxylase* and *fructose-1,6-bisphosphatase* are Mg^{2+}-activated enzymes, and Mg^{2+} flux into the stroma as a result of light-driven proton pumping stimulates the CO$_2$ fixation pathway at these key steps. Activity measurements have indicated that fructose bisphosphatase may be the rate-limiting step in the Calvin cycle. The recurring theme of fructose bisphosphatase as the target of the light-induced changes in the chloroplasts implicates this enzyme as a key point of control in the Calvin cycle.

Protein–Protein Interactions Mediated by an Intrinsically Unstructured Protein Also Regulate Calvin–Benson Cycle Activity

The 8.5-kD chloroplast protein CP12 is an instrinsically unstructured protein (see Chapter 6) that interacts with ribulose-5-P kinase (R5PK) and glyceraldehyde-3-P dehydrogenase (GAPDH) to form a complex. When complexed with CP12, R5PK and GAPDH are relatively inactive. Note that the first of these enzymes is ATP-dependent and the second is NADPH-dependent. Thus, formation of this complex constrains the use of ATP and NADPH, the principal products of the light reactions of photosynthesis. Reduced thioredoxin, through interactions with CP12, leads to the dissociation of the GAPDH–CP12–R5PK complex. Uncomplexed R5PK and GAPDH are inherently more active, and operation of the Calvin–Benson cycle is enhanced.

21.8 How Does Photorespiration Limit CO$_2$ Fixation?

As indicated, ribulose bisphosphate carboxylase/oxygenase catalyzes an alternative reaction in which O$_2$ replaces CO$_2$ as the substrate added to RuBP (Figure 21.29a). The *ribulose-1,5-bisphosphate oxygenase* reaction diminishes plant productivity because it leads to loss of RuBP, the essential CO$_2$ acceptor. The K_m for O$_2$ in this oxygenase reaction is about 200 μM. Given the relative abundance of CO$_2$ and O$_2$ in the atmosphere and their relative K_m values in these rubisco-mediated reactions, the ratio of carboxylase to oxygenase activity in vivo is about 3 or 4 to 1.

The products of ribulose bisphosphate oxygenase activity are *3-phosphoglycerate* and *phosphoglycolate*. Dephosphorylation and oxidation convert phosphoglycolate to **glyoxylate,** the α-keto acid of glycine (Figure 21.29b). Transamination yields glycine. In mitochondria, two glycines from photorespiration are converted into one serine and CO$_2$. This step is the source of the CO$_2$ evolved during photorespiration. Transamination of glyoxylate to glycine by the product serine yields hydroxypyruvate; reduction of hydroxypyruvate yields glycerate, which can be phosphorylated to 3-phosphoglycerate. 3-Phosphoglycerate can fuel resynthesis of ribulose bisphosphate by the Calvin cycle (see Figure 21.25).

Other fates of phosphoglycolate are also possible, including oxidation to CO$_2$, with the released energy being dissipated as heat. Obviously, agricultural productivity is dramatically lowered by this phenomenon, which, because it is a light-related uptake of O$_2$

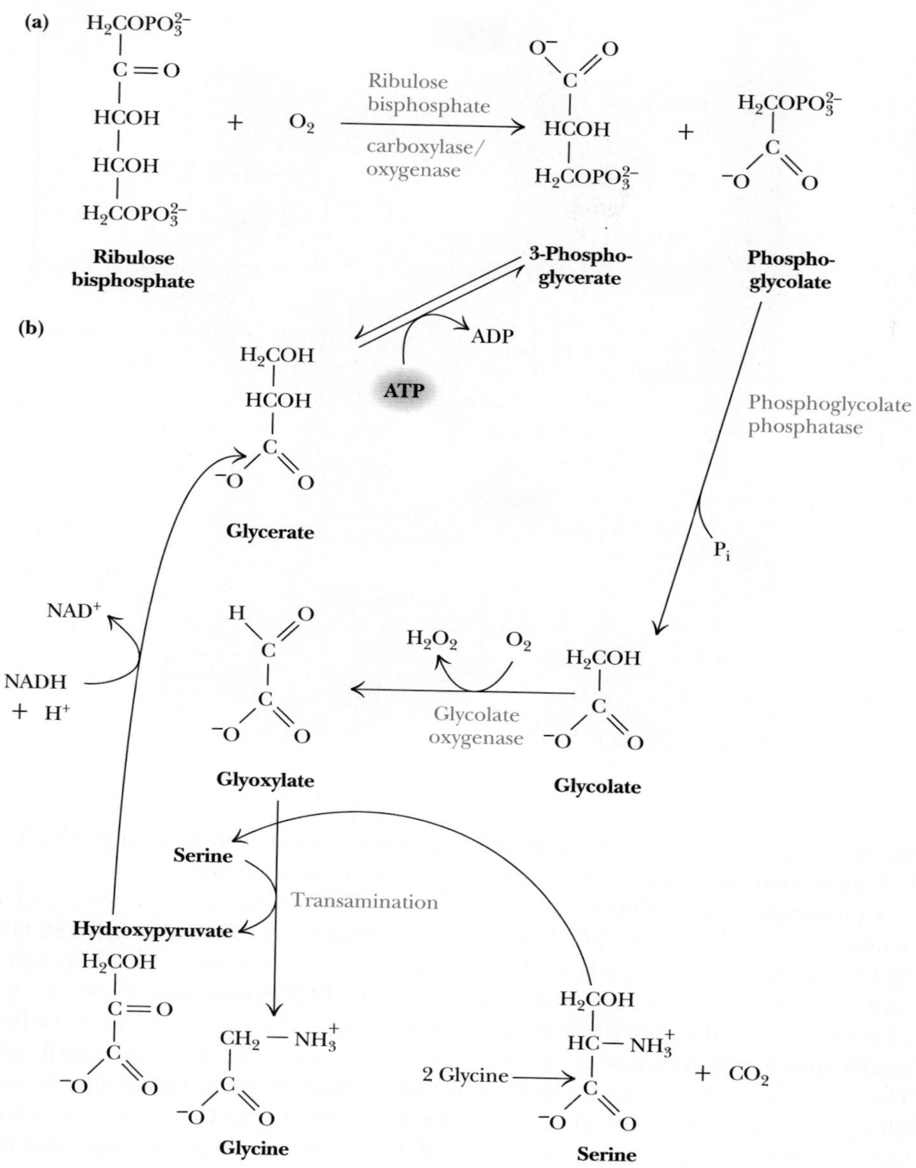

(a)

(b)

FIGURE 21.29 The oxygenase reaction of rubisco. **(a)** The reaction of ribulose bisphosphate carboxylase with O$_2$ and ribulose bisphosphate yields 3-phosphoglycerate and phosphoglycolate. **(b)** Conversion of two phosphoglycolates to serine + CO$_2$.

and release of CO$_2$, is termed **photorespiration.** As we shall see, certain plants, particularly tropical grasses, have evolved means to circumvent photorespiration. These plants are more efficient users of light for carbohydrate synthesis.

Tropical Grasses Use the Hatch–Slack Pathway to Capture Carbon Dioxide for CO$_2$ Fixation

Tropical grasses are less susceptible to the effects of photorespiration, as noted earlier. Studies using ^{14}CO$_2$ as a tracer indicated that the first organic intermediate labeled in these plants was not a three-carbon compound but a four-carbon compound. M. D. Hatch and C. R. Slack, two Australian biochemists, first discovered this C-4 product of CO$_2$ fixation, and the C-4 pathway of CO$_2$ incorporation is named the **Hatch–Slack pathway** after them. The C-4 pathway is not an alternative to the Calvin cycle series of reactions or even a net CO$_2$ fixation scheme. Instead, it functions as a CO$_2$ delivery system, carrying CO$_2$ from the relatively oxygen-rich surface of the leaf to interior cells, where oxygen is lower in concentration and, hence, less effective in competing with CO$_2$ in the rubisco reaction. Thus, the C-4 pathway is a means of avoiding photorespiration

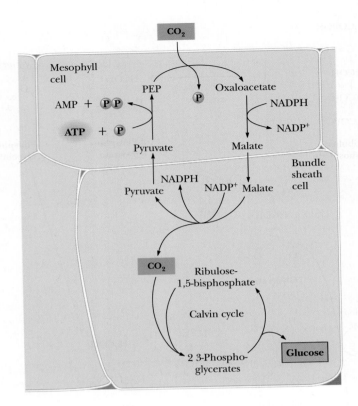

FIGURE 21.30 Essential features of the compartmentation and biochemistry of the Hatch–Slack pathway of carbon dioxide uptake in C4 plants.

by sheltering the rubisco reaction in a cellular compartment away from high $[O_2]$. The C-4 compounds serving as CO_2 transporters are malate or *aspartate.*

Compartmentation of these reactions to prevent photorespiration involves the interaction of two cell types: *mesophyll cells* and *bundle sheath cells.* The mesophyll cells take up CO_2 at the leaf surface, where O_2 is abundant, and use it to carboxylate phosphoenolpyruvate to yield OAA in a reaction catalyzed by **PEP carboxylase** (Figure 21.30). This four-carbon dicarboxylic acid is then either reduced to malate by an **NADPH-specific malate dehydrogenase** or transaminated to give *aspartate* in the mesophyll cells. The 4-C CO_2 carrier (malate or aspartate) is then transported to the bundle sheath cells, where it is decarboxylated to yield CO_2 and a 3-C product. The CO_2 is then fixed into organic carbon by the Calvin cycle localized within the bundle sheath cells, and the 3-C product is returned to the mesophyll cells, where it is reconverted to PEP in preparation to accept another CO_2 (Figure 21.30). Plants that use the C-4 pathway are termed **C4 plants,** in contrast to those plants with the conventional pathway of CO_2 uptake **(C3 plants).**

Intercellular Transport of Each CO_2 Via a C-4 Intermediate Costs 2 ATPs The transport of each CO_2 requires the expenditure of two high-energy phosphate bonds. The energy of these bonds is expended in the phosphorylation of pyruvate to PEP (phosphoenolpyruvate) by the plant enzyme **pyruvate-P_i dikinase;** the products are PEP, AMP, and pyrophosphate (PP_i). This represents a unique phosphotransferase reaction in that both the β- and γ-phosphates of a single ATP are used to phosphorylate the two substrates, pyruvate and P_i. The reaction mechanism involves an enzyme phosphohistidine intermediate. The γ-phosphate of ATP is transferred to P_i, whereas formation of E-His-P occurs by addition of the β-phosphate from ATP:

$$E\text{—His} + AMP_\alpha\text{—}P_\beta\text{—}P_\gamma + P_i \longrightarrow E\text{—His—}P_\beta + AMP_\alpha + P_\gamma P_i$$
$$E\text{—His—}P_\beta + \text{pyruvate} \longrightarrow PEP + E\text{—His}$$

Net: $ATP + \text{pyruvate} + P_i \longrightarrow AMP + PEP + PP_i$

Pyruvate-P_i dikinase is regulated by reversible phosphorylation of a threonine residue, the nonphosphorylated form being active. Interestingly, ADP is the phosphate donor in this interconvertible regulation. Despite the added metabolic expense of two phosphodiester bonds for each equivalent of carbon dioxide taken up, CO_2 fixation is more efficient in C4 plants, provided that light intensities and temperatures are both high. (As temperature rises, photorespiration in C3 plants rises and efficiency of CO_2 fixation falls.) Tropical grasses that are C4 plants include sugarcane, maize, and crabgrass. In terms of photosynthetic efficiency, cultivated fields of sugarcane represent the pinnacle of light-harvesting efficiency. Approximately 8% of the incident light energy on a sugarcane field appears as chemical energy in the form of CO_2 fixed into carbohydrate. This efficiency compares dramatically with the estimated photosynthetic efficiency of 0.2% for uncultivated plant areas. Research on photorespiration is actively pursued in hopes of enhancing the efficiency of agriculture by controlling this wasteful process. Only 1% of the 300,000 or so different plant species known are C4 plants; most are in hot climates.

Cacti and Other Desert Plants Capture CO_2 at Night

In contrast to C4 plants, which have separated CO_2 uptake and fixation into distinct cells in order to minimize photorespiration, succulent plants native to semiarid and tropical environments separate CO_2 uptake and fixation in time. Carbon dioxide (as well as O_2) enters the leaf through microscopic pores known as **stomata,** and water vapor escapes from plants via these same openings. In nonsucculent plants, the stomata are open during the day, when light can drive photosynthetic CO_2 fixation, and closed at night. Succulent plants, such as the *Cactaceae* (cacti) and *Crassulaceae,* cannot open their stomata during the heat of day because any loss of precious H_2O in their arid habitats would doom them. Instead, these plants open their stomata to take up CO_2 only at night, when temperatures are lower and water loss is less likely. This carbon dioxide is immediately incorporated into PEP to form OAA by PEP carboxylase; OAA is then reduced to malate by malate dehydrogenase and stored within vacuoles until morning. During the day, the malate is released from the vacuoles and decarboxylated to yield CO_2 and a 3-C product. The CO_2 is then fixed into organic carbon by rubisco and the reactions of the Calvin cycle. Because this process involves the accumulation of organic acids (OAA, malate) and is common to succulents of the *Crassulaceae* family, it is referred to as *crassulacean acid metabolism,* and plants capable of it are called *CAM plants.*

SUMMARY

⊌WL Login to OWL to develop problem-solving skills and complete online homework assigned by your professor.

21.1 What Are the General Properties of Photosynthesis? Photosynthesis takes place in membranes. In photosynthetic eukaryotes, the photosynthetic membranes form an inner membrane system within chloroplasts that is called the thylakoid membrane system. Photosynthesis is traditionally broken down into two sets of reactions: the light reactions, whereby light energy is used to generate NADPH and ATP concomitant with O_2 evolution, and the dark reactions in which NADPH and ATP provide the chemical energy for fixation of CO_2 into glucose. Water is the ultimate e^- donor for $NADP^+$ reduction.

21.2 How Is Solar Energy Captured by Chlorophyll? Chlorophyll and various accessory light-harvesting pigments absorb light throughout the visible spectrum and use the light energy to initiate electron-transfer reactions. The absorption of a photon of light by a pigment molecule promotes an electron of the pigment molecule to a higher orbital (and higher energy level). As a result, the pigment molecule is a much better electron donor. Photosynthetic units consist of arrays of hundreds of chlorophyll molecules and accessory light-harvesting pigments, but only a single reaction center. The reaction center is formed from a pair of Chl molecules.

21.3 What Kinds of Photosystems Are Used to Capture Light Energy? Photosynthetic bacteria have a single photosystem, but eukaryotic phototrophs have two distinct photosystems. Type I photosystems use proteins with Fe_4S_4 clusters as terminal e^- acceptors; type II photosystems reduce quinones, such as plastoquinone or phylloquinone. In oxygenic phototrophs (cyanobacteria, green algae, and higher plants), photosystem II (PSII) generates a strong oxidant that functions in O_2 evolution through the photolysis of water and a weak reductant that reduces plastoquinone to plastohydroquinone (PQH_2). Photosystem I (PSI) generates a weak oxidant that accepts electrons from plastohydroquinone via the cytochrome b_6f complex and a strong reductant capable of reducing $NADP^+$ to NADPH. Overall photosynthetic electron transfer is accomplished by three supramolecular membrane-spanning complexes: PSII, the cytochrome b_6f complex, and PSI. Oxygen evolution requires the accumulation of four oxidizing equivalents in PSII. The electrons withdrawn from water are used to re-reduce $P680^+$ back to P680, restoring its ability to absorb another photon, become P680*, and transfer an e^- once again. Electrons from P680* traverse PSII and reduce plastoquinone. Plastohydroquinone is oxidized via the cytochrome b_6f complex, with plastocyanin serving as e^- acceptor. The cytochrome b_6f complex catalyzes a Q cycle: It translocates 4 H^+ from the stroma to the thylakoid lumen for each molecule of

PQH_2 that it oxidizes. PSI is a light-driven plastocyanin:ferredoxin oxidoreductase having P700 as its reaction center Chl dimer. Electrons from P700* are transferred to the Fe_4S_4 cluster of ferredoxin. Reduced ferredoxin reduces $NADP^+$ via the ferredoxin:$NADP^+$ reductase flavoprotein. The electron "hole" in $P700^+$ is filled by reduced plastocyanin.

21.4 What Is the Molecular Architecture of Photosynthetic Reaction Centers?
All known photosynthetic reaction centers have a universal molecular architecture. The "core" structure is a pair of protein subunits having (at least) five transmembrane α-helical segments that provide a scaffold upon which the reaction center Chl pair and its associated chain of electron transfer cofactors are arrayed in a characteristic spatial pattern that facilitates rapid removal of an electron from the photoactivated RC and efficient transfer of the e^- across the membrane to a terminal acceptor (such as a quinone or a ferredoxin molecule). Photosynthetic electron transport is always coupled to H^+ translocation across the membrane, creating the potential for ATP synthesis by F_1F_0-type ATP synthases.

21.5 What Is the Quantum Yield of Photosynthesis?
The absorption of light energy by the photosynthetic apparatus is very efficient. The quantum yield of chemical energy, either in the form of ATP and NADPH, or in the form of glucose, depends on a number of factors that are still subject to investigation, including the H^+/e^- ratio and the ATP/H^+ ratio.

21.6 How Does Light Drive the Synthesis of ATP?
Photosynthetic electron transport leads to proton translocation across the photosynthetic membrane and creation of an H^+ gradient that can be used by an F_1F_0-type ATP synthase to drive ATP formation from ADP and P_i. Photophosphorylation occurs by either of two modes: noncyclic and cyclic. Noncyclic photophosphorylation depends on both PSI and PSII and leads to O_2 evolution, $NADP^+$ reduction, and ATP synthesis. In cyclic photophosphorylation, only PSI is used, no $NADP^+$ is reduced, and no O_2 is evolved. However, the electron-transfer events of cyclic photophosphorylation lead to H^+ translocation and ATP synthesis.

21.7 How Is Carbon Dioxide Used to Make Organic Molecules?
Ribulose-1,5-bisphosphate is the CO_2 acceptor in the key reaction for conversion of carbon dioxide into organic compounds. The reaction is catalyzed by rubisco (ribulose bisphosphate carboxylase/oxygenase); the products of CO_2 fixation by the rubisco reaction are 2 equivalents of 3-phosphoglycerate.

The Calvin–Benson cycle is a series of reactions that converts the 3-phosphoglycerates formed by rubisco into carbohydrates such as glyceraldehyde-3-P, dihydroxyacetone-P, and glucose. CO_2 fixation is activated by light through a variety of mechanisms, including changes in stromal pH, generation of reducing power in the form of ferredoxin by photosynthetic electron transport, and increased Mg^{2+} efflux from the thylakoid lumen to the stroma.

21.8 How Does Photorespiration Limit CO_2 Fixation?
When O_2 replaces CO_2 in the rubisco, or ribulose bisphosphate carboxylase/oxygenase, reaction, ribulose-1,5-bisP is destroyed through conversion into 3-phosphoglyerate and phosphoglycolate. Phosphoglycolate is oxidized to form CO_2, with loss of organic substance from the cell. Because O_2 is taken up and CO_2 is released in these reactions, the process is called *photorespiration*.

Tropical grasses carry out the Calvin–Benson cycle of reactions in cells shielded from high O_2 levels. CO_2 is first incorporated into PEP by PEP carboxylase to form oxaloacetate (OAA) in the mesophyll cells on the leaf surface. OAA is then reduced to malate and transported to bundle sheath cells. There, CO_2 is released and taken up by the rubisco reaction to form 3-phosphoglyerate, initiating the Calvin–Benson cycle. Plants capable of doing this are called C4 plants.

Succulent plants of semiarid and tropical regions such as *Cactaceae* and *Crassulaceae* exchange gases through their stomata only at night in order to avoid precious water loss. CO_2 taken up at night is added to PEP by PEP carboxylase to form OAA, which is then reduced to malate in the dark. During the day, malate is decarboxylated to yield CO_2, which then enters the Calvin–Benson cycle. This metabolic variation is referred to as *crassulacean acid metabolism*.

FOUNDATIONAL BIOCHEMISTRY Things You Should Know After Reading Chapter 21.

- The vast majority of energy used by living organisms can be traced to solar energy captured by photosynthesis.

- Photosynthesis is often expressed in terms of CO_2 fixation into carbohydrates.

- The capture of light energy in photosynthesis occurs in membranes.

- Chloroplasts are plant organelles devoted principally to photosynthesis.

- The capture of light energy during photosynthesis is localized to the inner membrane of chloroplasts, the thylakoid membrane.

- Chloroplasts possess an inner compartment not found in mitochondria, the thylakoid lumen.

- Photosynthesis consists of light-dependent ("light") reactions and light-independent ("dark") reactions.

- The "light" reactions are responsible for O_2 evolution, NADPH reduction, and ATP synthesis and are associated with the chloroplast thylakoid membranes.

- The "dark" reactions carry out the ATP- and NADPH-dependent process of CO_2 fixation and occur in the stroma of the chloroplast.

- Water is the ultimate electron donor for photosynthetic $NADP^+$ reduction.

- All of the O_2 in the Earth's atmosphere was produced by photosynthesis.

- Photosynthesis depends on the photoreactivity of chlorophyll (Chl).

- When light energy is absorbed by Chl, an electron in this pigment is excited to a higher orbital, making the loss of this electron by an oxidation-reduction reaction more likely.

- Chlorophylls and accessory light-harvesting pigments absorb light of differing wavelengths, allowing photosynthetic organisms to capture energy across the visible light spectrum.

- Light energy absorbed by photosynthetic pigments has several possible fates—loss as heat, loss through fluorescence, resonance energy transfer to a neighboring pigment molecule, or transduction into the chemical energy of an oxidation-reduction reaction.

- Photosynthetic units consist of many Chl molecules but only a single reaction-center Chl dimer.

- Oxygenic phototrophs have two Chl-based photosystems, PSI and PSII.

- P700 is the PSI reaction-center Chl dimer.

- P60 is the PSII reaction-center Chl dimer.

- Chl exists in photosynthetic membranes in association with proteins.

- The Chl-containing protein complexes are PSI, PSII, and light-harvesting complexes (LHC).

- PSI and PSII participate in photosynthesis, with PSII catalyzing the light-driven oxidation of water to O_2 and the generation of a weak reductant, and PSI catalyzing the light-driven reduction of $NADP^+$.

- Together, PSI and PSII carry out the light-driven transfer of electrons from H_2O to $NADP^+$.

- Transfer of reducing equivalents from PSII to PSI through the cytochrome $b_6 f$ complex leads to proton translocations that contribute to ATP synthesis.

- Synthesis of ATP by means of a proton gradient established by light-driven electron transport is called photophosphorylation.

- The pathway of photosynthetic electron transfer is called the Z scheme.

- Three membrane-spanning supramolecular complexes achieve the Z scheme of electron transfer: PSII, cytochrome $b_6 f$ and PSI.

- Plastoquinone and plastocyanin mediate the transfer of electrons from PSII to cytochrome $b_6 f$ and from cytochrome $b_6 f$ to PSI, respectively.

- O_2 evolution requires the accumulation of 4 oxidizing equivalents in PSII.

- Electrons taken from H_2O replace electrons lost upon photo-oxidation of P680.

- X-ray crystallographic solution of the *Rhodopseudomonas viridis* reaction center (RC) established the molecular architecture of reaction centers.

- The *R. viridis* RC is an integral membrane protein containing a reaction-center bacteriochlorophyll dimer (P870), bacteriochlorophyll, bacteriopheophytin, quinone, cytochromes, and a Fe atom.

- The molecular architectures of PSII and PSI resemble the molecular architecture of the *R. viridis* RC.

- The oxygen-evolving complex (OEC) of PSII possesses 4 Mn^{2+} and 1 Ca^{2+}.

- The plant PSI complex has a quantum efficiency that is near-perfect: Essentially every photon of light absorbed leads to the transfer of an electron.

- The quantum yield of photosynthesis is expressed in terms of O_2 evolution, H^+ translocation, $NADP^+$ reduction and ATP synthesis.

- The photosynthetic energy requirements for hexose synthesis depend on the $H^+/h\nu$ and the ATP/H^+ ratios.

- The mechanism of photophosphorylation, like oxidative phosphorylation, is chemiosmotic.

- $CF_1 CF_0$-ATP synthase is the chloroplast equivalent of mitochondrial $F_1 F_0$-ATP synthase.

- The chloroplast ATP synthase c-subunit rotor has 14 c-subunits, so the H^+/ATP ratio for ATP synthesis in chloroplasts = 14/3 = 4.67.

- Photophosphorylation can occur in either a noncyclic or a cyclic mode.

- Noncyclic photophosphorylation yields ATP, O_2, and NADPH.

- Cyclic photophosphorylation yields only ATP.

- Ribulose-1,5-bisphosphate is the CO_2 acceptor in the Calvin–Benson carbon dioxide fixation pathway.

- Ribulose bisphosphate carboxylase/oxygenase (rubisco) catalyzes the formation of two 3-phosphoglycerate molecules from ribulose-1,5-bisphosphate + CO_2.

- 2-Carboxy-3-keto-arabinitol is the key intermediate in the rubisco reaction.

- Rubisco exists in an inactive state that is activated through carbamoylation of Lys^{201}, coordination of Mg^{2+}, and removal of tightly bound substrate by rubisco activase.

- The Calvin–Benson cycle accounts for the vast preponderance of CO_2 fixation in the biosphere, carbohydrate synthesis, and the production of 4-, 5-, 6-, and 7- carbon sugars.

- The conversion of 6 CO_2 to one hexose requires 12 NADPH and 18 ATP.

- The Calvin–Benson cycle is indirectly activated by light through light-driven pH changes in the chloroplast compartments, reduction of disulfides in key enzymes to pairs of —SH groups, and Mg^{2+} movement from the thylakoid lumen into the stroma.

- Protein-protein interactions mediated by an IUP also regulate the Calvin–Benson cycle activity.

- Photorespiration limits CO_2 fixation through the wasteful consumption of RuBP by the oxygenase activity of rubisco.

- Tropical grasses use the Hatch–Slack pathway to capture carbon dioxide for CO_2 fixation.

- In tropical grasses, carbon dioxide capture occurs in mesophyll cells and fixation occurs in bundle sheath cells.

- Pyruvate-P_i kinase is a key enzyme in PEP synthesis for CO_2 capture.

- Cacti and other succulent plants of arid climates capture CO_2 at night.

PROBLEMS

Answers to all problems are at the end of this book. Detailed solutions are available in the *Student Solutions Manual, Study Guide, and Problems Book*.

1. **P700* Has the Most Negative Standard Reduction Potential Found in Nature** In photosystem I, P700 in its ground state has an $\mathscr{E}_o' = +0.4$ V. Excitation of P700 by a photon of 700-nm light alters the $\mathscr{E}_o'$ of P700* to -0.6 V. What is the efficiency of energy capture in this light reaction of P700?

2. **PSII Has the Most Positive Standard Reduction Potential Found in Nature** What is the $\mathscr{E}_o'$ for the light-generated primary oxidant of photosystem II if the light-induced oxidation of water (which leads to O_2 evolution) proceeds with a $\Delta G^{o'}$ of -25 kJ/mol?

3. **Equating the Energy of an Oxidation-Reduction Potential Difference and the Energy for ATP Synthesis** (Integrates with Chapters 3 and 20.) Assuming that the concentrations of ATP, ADP,

and P_i in chloroplasts are 3, 0.1, and 10 mM, respectively, what is the ΔG for ATP synthesis under these conditions? Photosynthetic electron transport establishes the proton-motive force driving photophosphorylation. What redox potential difference is necessary to achieve ATP synthesis under the foregoing conditions, assuming 1.3 ATP equivalents are synthesized for each electron pair transferred?

4. **The Q Cycle in Photosynthesis** (Integrates with Chapter 20.) Write a balanced equation for the Q cycle as catalyzed by the cytochrome $b_6 f$ complex of chloroplasts.

5. **The Relative Efficiency of ATP Synthesis in Noncyclic versus Cyclic Photophosphorylation** If noncyclic photosynthetic electron transport leads to the translocation of 7 $H^+/2e^-$ and cyclic photosynthetic electron transport leads to the translocation of 2 H^+/e^-, what is the relative photosynthetic efficiency of ATP synthesis (expressed as the number of photons absorbed per ATP synthesized) for noncyclic

versus cyclic photophosphorylation? (Assume that the CF_1CF_0–ATP synthase yields 3 ATP/14 H^+.)

6. ΔpH and $\Delta\psi$ in the Chloroplast Proton-Motive Force (Integrates with Chapter 20.) In mitochondria, the membrane potential ($\Delta\psi$) contributes relatively more to Δp (proton-motive force) than does the pH gradient (ΔpH). The reverse is true in chloroplasts. Why do you suppose that the proton-motive force in chloroplasts can depend more on ΔpH than mitochondria can? Why is ($\Delta\psi$) less in chloroplasts than in mitochondria?

7. The Role of Tyr[161] in PSII Predict the consequences of a Y161F mutation in the amino acid sequence of the D1 subunit of PSII.

8. H^+/ATP Ratio in Photosynthetic Bacteria (Integrates with Chapter 20.) Calculate (in Einsteins and in kJ/mol) how many photons would be required by the *Rhodopseudomonas viridis* photophosphorylation system to synthesize 3 ATPs. (Assume that the *R. viridis* F_1F_0–ATP synthase *c*-subunit rotor contains 12 *c*-subunits and that the *R. viridis* cytochrome bc_1 complex translocates 2 H^+/e^-.)

9. The Standard-State Free Energy Change for NADP-Glyceraldehyde-3-P Dehydrogenase (Integrates with Chapters 18 and 20.) Calculate $\Delta G°'$ for the $NADP^+$-specific glyceraldehyde-3-P dehydrogenase reaction of the Calvin–Benson cycle.

10. Photosynthesis of Glucose Write a balanced equation for the synthesis of a glucose molecule from ribulose-1,5-bisphosphate and CO_2 that involves the first three reactions of the Calvin cycle and subsequent conversion of the two glyceraldehyde-3-P molecules into glucose.

11. Tracing the Fate of CO_2 During Photosynthesis ^{14}C-labeled carbon dioxide is administered to a green plant, and shortly thereafter the following compounds are isolated from the plant: 3-phosphoglycerate, glucose, erythrose-4-phosphate, sedoheptulose-1,7-bisphosphate, and ribose-5-phosphate. In which carbon atoms will radioactivity be found?

12. Regulation of CO_2 Fixation During Photosynthesis The photosynthetic CO_2 fixation pathway is regulated in response to specific effects induced in chloroplasts by light. What is the nature of these effects, and how do they regulate this metabolic pathway?

13. The Formation of Phosphoglycerate During Photorespiration Write a balanced equation for the conversion of phosphoglycolate to glycerate-3-P by the reactions of photorespiration. Does this balanced equation demonstrate that photorespiration is a wasteful process?

14. The Source of the Oxygen Atoms in Photosynthetic O_2 Evolution The overall equation for photosynthetic CO_2 fixation is

$$6\ CO_2 + 6\ H_2O \longrightarrow C_6H_{12}O_6 + 6\ O_2$$

All the O atoms evolved as O_2 come from water; *none* comes from carbon dioxide. But 12 O atoms are evolved as 6 O_2, and only 6 O atoms appear as 6 H_2O in the equation. Also, 6 CO_2 have 12 O atoms, yet there are only 6 O atoms in $C_6H_{12}O_6$. How can you account for

these discrepancies? (*Hint:* Consider the partial reactions of photosynthesis: ATP synthesis, $NADP^+$ reduction, photolysis of water, and the overall reaction for hexose synthesis in the Calvin–Benson cycle.)

15. The Influence of *c*-Subunit Stoichiometry on the Efficiency of ATP Synthesis The number of *c*-subunits in F_1F_0-type ATP synthases shows some variation from organism to organism. For example, the yeast ATP synthase contains 10 *c*-subunits, the spinach CF_1CF_0–ATP synthase has 14, and the cyanobacterium *Spirulina platensis* enzyme apparently has 15.

    a. What is the consequence of *c*-subunit stoichiometry for the H^+/ATP ratio?

    b. What is the relationship between *c*-subunit stoichiometry and the magnitude of Δp (the proton-motive force)?

16. The Energetics of Hydroquinone Reduction by Photosynthetic Reaction Centers The reduction of membrane-associated quinones, such as coenzyme Q and plastoquinones, is a common feature of photosystems (see Figures 21.15, 21.20, and 21.22). Assume $\mathscr{E}_0'$ for PQ/PQH$_2$ = 0.07 V and the potential of the ground-state chlorophyll molecule = 0.5 V, calculate ΔG for the reduction of plastoquinone by

    a. 870-nm light.

    b. 700-nm light.

    c. 680-nm light.

17. The Energetics of Proton Translocation by the Cytochrome b_6f Complex Plastoquinone oxidation by cytochrome bc_1 and cytochrome b_6f complexes apparently leads to the translocation of $4^+/2e^-$. If $\mathscr{E}_0'$ for cytochrome f = 0.365 V (Table 20.1) and E_0' for PQ/PQH$_2$ = 0.07 V, calculate ΔG for the coupled reaction:

$$2\ h\nu + 4\ H^+_{in} \longrightarrow 4\ H^+_{out}$$

(Assume a value of 23 kJ/mol for the free energy change (ΔG) associated with moving protons from inside to outside.)

18. The Overall Free Energy Change for Photosynthetic $NADP^+$ Reduction What is the overall free energy change (ΔG) for noncyclic photosynthetic electron transport?

$$4\ (700\text{-nm photons}) + 4\ (680\text{-nm photons}) + 2\ H_2O + 2\ NADP^+ \longrightarrow$$
$$O_2 + 2\ NADPH + 2\ H^+$$

Preparing for the MCAT® Exam

19. From Figure 21.5, predict the spectral properties of accessory light-harvesting pigments found in plants.

20. Draw a figure analogous to Figure 21.27, plotting [Mg^{2+}] in the stroma and thylakoid lumen on the *y*-axis and *dark-light-dark* on the *x*-axis.

ActiveModels Problems

21. Using the ActiveModel for Photosystem II, explain why dimerization occurs, and identify the residues that drive dimerization.

22. Look at the Ramachandran plot in the ActiveModel for Photosystem II, and determine the ϕ and ψ angles of the α-helices in Photosystem II. What are typical ϕ and ψ angles for α-helices?

FURTHER READING

General References

Blankenship, R. E., 2002. *Molecular Mechanisms of Photosynthesis*. Malden, MA: Blackwell Science.

Björn, L. O., Papageorgiou, G. C., Blankenship, R. E., and Govindjee, 2009. A viewpoint: Why chlorophyll a? *Photosynthesis Research* **99:**85–98.

Buchanan, B. B., Gruissem, W., and Jones, R. I., 2000. *Biochemistry and Molecular Biology of Plants*. Rockville, MD: American Society of Plant Physiologists.

Cramer, W. A., and Knaff, D. B., 1990. *Energy Transduction in Biological Membranes—A Textbook of Bioenergetics*. New York: Springer-Verlag.

Harold, F. M., 1987. *The Vital Force: A Study of Bioenergetics*. Chapter 8: Harvesting the Light. San Francisco: Freeman & Company.

Heathcote, P., Fyfe, P. K., and Jones, M. R., 2002. Reaction centers: The structure and evolution of biological solar power. *Trends in Biochemical Sciences* **27:**79–87.

Photosynthetic Pigments

Glazer, A. N., 1983. Comparative biochemistry of photosynthetic light-harvesting pigments. *Annual Review of Biochemistry* **52**:125–157.

Green, B. R., and Durnford, D. G., 1996. The chlorophyll-carotenoid proteins of oxygenic photosynthesis. *Annual Review of Plant Physiology and Plant Molecular Biology* **47**:685–714.

Hoffman, E., et al., 1996. Structural basis of light harvesting by carotenoids: Peridinin-chlorophyll protein from *Amphidinium carterae*. *Science* **272**:1788–1791.

Properties of the Thylakoid Membranes

Anderson, J. M., 1986. Photoregulation of the composition, function and structure of the thylakoid membrane. *Annual Review of Plant Physiology* **37**:93–136.

Anderson, J. M., and Anderson, B., 1988. The dynamic photosynthetic membrane and regulation of solar energy conversion. *Trends in Biochemical Sciences* **13**:351–355.

Photosynthetic Reaction Centers of Photosynthetic Bacteria

Deisenhofer, J., and Michel, H., 1989. The photosynthetic reaction center from the purple bacterium *Rhodopseudomonas viridis*. *Science* **245**: 1463–1473.

Deisenhofer, J., Michel, H., and Huber, R., 1985. The structural basis of light reactions in bacteria. *Trends in Biochemical Sciences* **10**:243–248.

Deisenhofer, J., et al., 1985. Structure of the protein subunits in the photosynthetic reaction center of *Rhodopseudomonas viridis* at 3 Å resolution. *Nature* **318**:618–624; also *Journal of Molecular Biology* (1984) **180**:385–398.

Structure and Function of Photosystems I and II and the Cytochrome b_6f Complex

Amunts, A., and Nelson, N., 2008. Functional organization of plant photosystem I: Evolution of a highly efficient photochemical machine. *Plant Physiology and Biochemistry* **46**:228–237.

Amunts, A., Toporik, H., Borovikova, A., and Nelson, N., 2010. Structure determination and improved model of plant photosystem I. *Journal of Biological Chemistry* **285**:3478–3486.

Armstrong, F. A., 2008. Why did Nature choose manganese to make oxygen? *Philosophical Transactions of the Royal Society B* **363**:1263–1270.

Barber, J., 2008. Photosynthetic generation of oxygen. *Philosophical Transactions of the Royal Society B* **363**:2665–2675.

Cramer, W. A., et al., 2006. Transmembrane traffic in the cytochrome b_6f complex. *Annual Review of Biochemistry* **75**:769–790.

Guskov, A., Kern, J., Gabdulkhakov, A., Broser, M., et al., 2009. Cyanobacterial photosystem II at 2.9-Å resolution and the role of quinones, lipids, channels, and chloride. *Nature Structural and Molecular Biology* **16**:334–342.

Jensen, P. E., Bassi, R., Boekema,, E. J., Dekker, J. P., et al., 2007. Structure, function, and regulation of plant photosystem I. *Biochimica et Biophysica Acta* **1767**:335–352.

Jordan, P., Fromme, P., Witt, H. T., Klukas, O. , et al., 2001. Three dimensional structure of cyanobacterial photosystem I at 2.5 Å resolution. *Nature* **411**:909–17.

Merchant, S., and Sawaya, M. R., 2005. The light reactions: A guide to recent acquisitions for the picture gallery. *Plant Cell* **17**:648–663.

Nelson, N., and Yocum, C. F., 2006. Structure and function of photosystems I and II. *Annual Review of Plant Biology* **57**:521–566.

Yano, J., et al., 2006. Where water is oxidized to dioxygen: Structure of the photosynthetic Mn_4Ca cluster. *Science* **314**:821–825.

Zouni, A., et al., 2001. Crystal structure of photosystem II from *Synechococcus elongatus* at 3.8 Å resolution. *Nature* **409**:739–743.

Photophosphorylation

Allen, J. F., 2002. Photosynthesis of ATP—electrons, proton pumps, rotors, and poise. *Cell* **110**:273–276.

Arnon, D. I., 1984. The discovery of photosynthetic phosphorylation. *Trends in Biochemical Sciences* **9**:258–262.

Avenson, T. J., et al., 2005. Regulating the proton budget of higher plant photosynthesis. *Proceedings of the National Academy of Sciences U.S.A.* **102**:9709–9713.

Jagendorf, A. T., and Uribe, E., 1966. ATP formation caused by acid-base transition of spinach chloroplasts. *Proceedings of the National Academy of Sciences U.S.A.* **55**:170–177.

Pogoryelov, D., Yidiz, O., Faraldo-Gómez, J. D., and Meier, T., 2009. High-resolution structure of the rotor ring of a proton-dependent ATP synthase. *Nature Structural & Molecular Biology* **16**:1068–1073.

Remy, A., and Gerwert, K., 2003. Coupling of light-induced electron transfer to proton uptake in photosynthesis. *Nature Structural Biology* **10**:637–644.

Seelert, H., Dencher, N., and Müller, D. J., 2003. Fourteen protomers compose the oligomer III of the proton-rotor in spinach chloroplast ATP synthase. *Journal of Molecular Biology* **333**:337–344.

Shikanai, T., 2007. Cyclic electron transport around photosystem I: Genetic approaches. *Annual Review of Plant Biology* **58**:199–217.

Vollmar, M., Schlieper, D., Winn, M., Büchner, C., and Groth, G., 2009. Structure of the c_{14} rotor ring of the proton translocating chloroplast ATP synthase. *Journal of Biological Chemistry* **284**:18228–18235.

Carbon Dioxide Fixation

Andersson, I., and Backlund, A., 2008. Structure and function of Rubisco. *Plant Physiology and Biochemistry* **46**:275–291.

Burnell, J. N., and Hatch, M. D., 1985. Light–dark modulation of leaf pyruvate, P_1 dikinase. *Trends in Biochemical Sciences* **10**:288–291.

Cushman, J. C., and Bohnert, H. J., 1999. Crassulacean acid metabolism: Molecular genetics. *Annual Review of Plant Physiology and Plant Molecular Biology* **50**:305–332.

Graciet, E., Lebreton, S., and Gontero, B, 2004. Emergence of new regulatory mechanisms in the Benson–Calvin pathway via protein–protein interactions: A glyceraldehyde-3-phosphate dehydrogenase/CP12/phosphoribulokinase complex. *Journal of Experimental Botany* **55**:1245–1254.

Hatch, M. D., 1987. C_4 photosynthesis: A unique blend of modified biochemistry, anatomy, and ultrastructure. *Biochimica et Biophysica Acta* **895**:81–106.

Kaplan, A., and Reinhold, L., 1999. CO_2-concentrating mechanisms in photosynthetic organisms. *Annual Review of Plant Physiology and Plant Molecular Biology* **50**:539–570.

Knaff, D. B., 1989. The regulatory role of thioredoxin in chloroplasts. *Trends in Biochemical Sciences* **14**:433–434.

Portis, A. R., Jr., 1992. Regulation of ribulose-1,5-bisphosphate carboxylase/oxygenase activity. *Annual Review of Plant Physiology and Plant Molecular Biology* **43**:415–437.

Spreitzer, R. J., and Salvucci, M. E., 2002. Rubisco: Structure, regulatory interactions, and possibilities for a better enzyme. *Annual Review of Plant Biology* **53**:449–475.

Wingler, A., et al., 2000. Photorespiration: Metabolic pathways and their role in stress protection. *Philosophical Transactions of the Royal Society of London B* **355**:1517–1529.

Gluconeogenesis, Glycogen Metabolism, and the Pentose Phosphate Pathway

© Jason Alan/iStockphoto.com

Con pan y vino se anda el camino.
(With bread and wine you can walk your road.)

Spanish proverb

KEY QUESTIONS

22.1 What Is Gluconeogenesis, and How Does It Operate?

22.2 How Is Gluconeogenesis Regulated?

22.3 How Are Glycogen and Starch Catabolized in Animals?

22.4 How Is Glycogen Synthesized?

22.5 How Is Glycogen Metabolism Controlled?

22.6 Can Glucose Provide Electrons for Biosynthesis?

ESSENTIAL QUESTIONS

As shown in Chapters 18 and 19, the metabolism of sugars is an important source of energy for cells. Animals, including humans, typically obtain significant amounts of glucose and other sugars from the breakdown of starch in their diets. Glucose can also be supplied via breakdown of cellular reserves of glycogen (in animals) or starch (in plants).

What is the nature of *gluconeogenesis,* the pathway that synthesizes glucose from noncarbohydrate precursors; how is glycogen synthesized from glucose; and how are electrons from glucose used in biosynthesis?

22.1 : What Is Gluconeogenesis, and How Does It Operate?

The ability to synthesize glucose from common metabolites is very important to most organisms. Human metabolism, for example, consumes about 160 ± 20 grams of glucose per day, about 75% of this in the brain. Body fluids carry only about 20 grams of free glucose, and glycogen stores normally can provide only about 180 to 200 grams of free glucose. Thus, the body carries only a little more than a 1-day supply of glucose. If glucose is not obtained in the diet, the body must produce new glucose from noncarbohydrate precursors. The term for this activity is **gluconeogenesis,** which means the generation *(genesis)* of new *(neo)* glucose.

Furthermore, muscles consume large amounts of glucose via glycolysis, producing large amounts of pyruvate. In vigorous exercise, muscle cells become anaerobic and pyruvate is converted to lactate. Gluconeogenesis salvages this pyruvate and lactate and reconverts it to glucose.

Another pathway of glucose catabolism, the *pentose phosphate pathway,* is the primary source of NADPH, the reduced coenzyme essential to most reductive biosynthetic processes. For example, NADPH is crucial to the biosynthesis of fatty acids (see Chapter 24) and amino acids (see Chapter 25). The pentose phosphate pathway also results in the production of ribose-5-phosphate, an essential component of ATP, NAD^+, FAD, coenzyme A, and particularly DNA and RNA. This important pathway will also be considered in this chapter.

The Substrates for Gluconeogenesis Include Pyruvate, Lactate, and Amino Acids

In addition to pyruvate and lactate, other noncarbohydrate precursors can be used as substrates for gluconeogenesis in animals. These include most of the amino acids, as well as glycerol and all the TCA cycle intermediates. On the other hand, fatty acids are not substrates for gluconeogenesis in animals, because most fatty acids yield only acetyl-CoA upon degradation, and animals cannot carry out net synthesis of sugars from acetyl-CoA. Lysine and leucine are the only amino acids that are not substrates for gluconeogenesis. These amino acids produce only acetyl-CoA upon degradation. Note also that acetyl-CoA can be a substrate for gluconeogenesis in plants when the glyoxylate cycle is operating (see Chapter 19).

Nearly All Gluconeogenesis Occurs in the Liver and Kidneys in Animals

Interestingly, the mammalian organs that consume the most glucose, namely, brain and muscle, carry out very little glucose synthesis. The major sites of gluconeogenesis are the liver and kidneys, which account for about 90% and 10% of the body's gluconeogenic activity, respectively. Glucose produced by gluconeogenesis in the liver and kidneys is released into the blood and is subsequently absorbed by brain, heart, muscle, and red

---

HUMAN BIOCHEMISTRY

The Chemistry of Glucose Monitoring Devices

Individuals with diabetes must measure their serum glucose concentration frequently, often several times a day. The advent of computerized, automated devices for glucose monitoring has made this necessary chore easier, far more accurate, and more convenient than it once was. These devices all use a simple chemical scheme for glucose measurement that involves oxidation of glucose to gluconic acid by glucose oxidase. This reaction produces two molecules of hydrogen peroxide per molecule of glucose oxidized. The H_2O_2 is then used to oxidize a dye, such as o-dianisidine, to a colored product that can be measured:

Glucose + 2 H_2O + O_2 $\longrightarrow$ gluconic acid + 2 H_2O_2
o-dianisidine (colorless) + H_2O_2 $\longrightarrow$ oxidized o-anisidine (colored) + H_2O

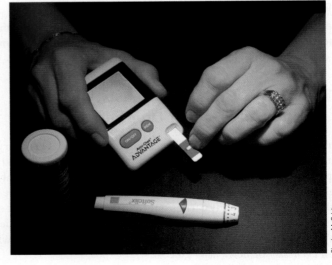

Charles M. Grisham

The amount of colored dye produced is directly proportional to the amount of glucose in the sample.

The patient typically applies a drop of blood (from a fingerprick*) to a plastic test strip that is then inserted into the glucose monitor. Within half a minute, a digital readout indicates the blood glucose value. Modern glucose monitors store several days of glucose measurements, and the data can be easily transferred to a computer for analysis and graphing.

*How does the monitor deal with getting just the right amount of blood? The blood flows up an absorbent "wick" by capillary action. It is impossible to overfill this device, but the monitor will give an error signal if not enough blood flows up the strip.

blood cells to meet their metabolic needs. In turn, pyruvate and lactate produced in these tissues are returned to the liver and kidneys to be used as gluconeogenic substrates.

Gluconeogenesis Is Not Merely the Reverse of Glycolysis

In some ways, gluconeogenesis is the reverse, or antithesis, of glycolysis. Glucose is synthesized, not catabolized; ATP is consumed, not produced; and NADH is oxidized to NAD^+, rather than the other way around. However, gluconeogenesis cannot be merely the reversal of glycolysis, for two reasons. First, glycolysis is exergonic, with a $\Delta G^{\circ\prime}$ of approximately -74 kJ/mol. If gluconeogenesis were merely the reverse, it would be a strongly endergonic process and could not occur spontaneously. Somehow the energetics of the process must be augmented so that gluconeogenesis can proceed spontaneously. Second, the processes of glycolysis and gluconeogenesis must be regulated in a reciprocal fashion so that when glycolysis is active, gluconeogenesis is inhibited, and when gluconeogenesis is proceeding, glycolysis is turned off. Both of these limitations are overcome by having unique reactions within the routes of glycolysis and gluconeogenesis, rather than a completely shared pathway.

Gluconeogenesis—Something Borrowed, Something New

The complete route of gluconeogenesis is shown in Figure 22.1, side by side with the glycolytic pathway. Gluconeogenesis employs four different reactions, catalyzed by four different enzymes, for the three steps of glycolysis that are highly exergonic (and highly regulated). In essence, seven of the ten steps of glycolysis are merely reversed in gluconeogenesis. The six reactions between fructose-1,6-bisphosphate and PEP are shared by the two pathways, as is the isomerization of glucose-6-P to fructose-6-P. The three exergonic, regulated reactions—the hexokinase (glucokinase), phosphofructokinase, and pyruvate kinase reactions—are replaced by alternative reactions in the gluconeogenic pathway.

The conversion of pyruvate to PEP that initiates gluconeogenesis is accomplished by two unique reactions. **Pyruvate carboxylase** catalyzes the first, converting pyruvate to oxaloacetate. Then, **PEP carboxykinase** catalyzes the conversion of oxaloacetate to PEP. Conversion of fructose-1,6-bisphosphate to fructose-6-phosphate is catalyzed by a specific phosphatase, **fructose-1,6-bisphosphatase.** The final step to produce glucose, hydrolysis of glucose-6-phosphate, is mediated by **glucose-6-phosphatase.** Each of these steps is considered in detail in the following paragraphs. The overall conversion of pyruvate to PEP by pyruvate carboxylase and PEP carboxykinase has a $\Delta G^{\circ\prime}$ close to zero but is pulled along by subsequent reactions. The conversion of fructose-1,6-bisphosphate to glucose in the last three steps of gluconeogenesis is strongly exergonic, with a $\Delta G^{\circ\prime}$ of about -30.5 kJ/mol. This sequence of two phosphatase reactions separated by an isomerization accounts for most of the free energy release that makes the gluconeogenesis pathway spontaneous.

Four Reactions Are Unique to Gluconeogenesis

1. Pyruvate Carboxylase—A Biotin-Dependent Enzyme
Initiation of gluconeogenesis occurs in the **pyruvate carboxylase reaction**—the conversion of pyruvate to oxaloacetate.

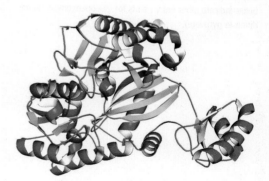

$$\underset{\text{Pyruvate}}{CH_3\overset{\overset{\displaystyle O}{\|}}{C}-COO^-} + \underset{\text{Bicarbonate}}{HCO_3^-} + ATP \longrightarrow \underset{\text{Oxaloacetate}}{{}^-OOC-CH_2\overset{\overset{\displaystyle O}{\|}}{C}-COO^-} + ADP + P_i$$

The reaction takes place in two discrete steps, involves ATP and bicarbonate as substrates, and utilizes biotin as a coenzyme and acetyl-CoA as an allosteric activator. Pyruvate carboxylase is a tetrameric enzyme (with a molecular mass of about 500 kD). Each monomer possesses a biotin covalently linked to the ϵ-amino group of a lysine residue at the active site (Figure 22.2). The first step of the reaction involves nucleophilic

In most organisms, pyruvate carboxylase is a homotetramer of 130-kD subunits, with each subunit composed of three functional domains named biotin carboxylase, carboxyl transferase, and biotin carboxyl carrier protein. Shown here is the biotin carboxylase domain of pyruvate carboxylase from *Bacillus thermodenitrificans*. (pdb id = 2DZD).

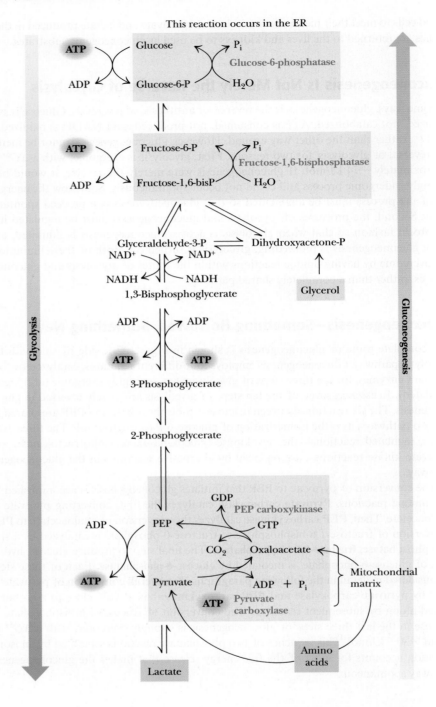

FIGURE 22.1 The pathways of gluconeogenesis and glycolysis. Species in blue, green, and peach-colored shaded boxes indicate other entry points for gluconeogenesis (in addition to pyruvate).

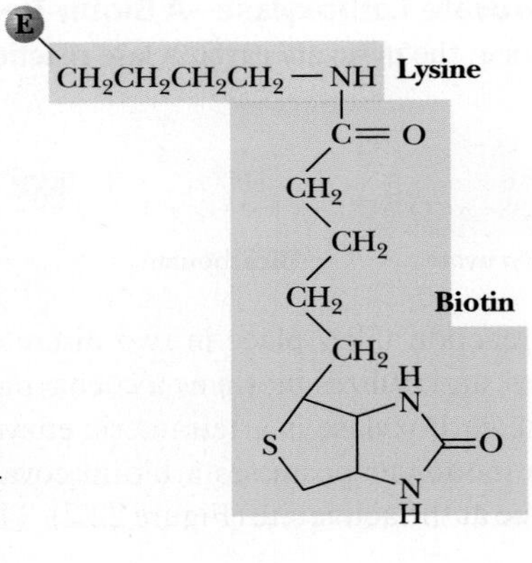

FIGURE 22.2 Covalent linkage of biotin to an active-site lysine in pyruvate carboxylase.

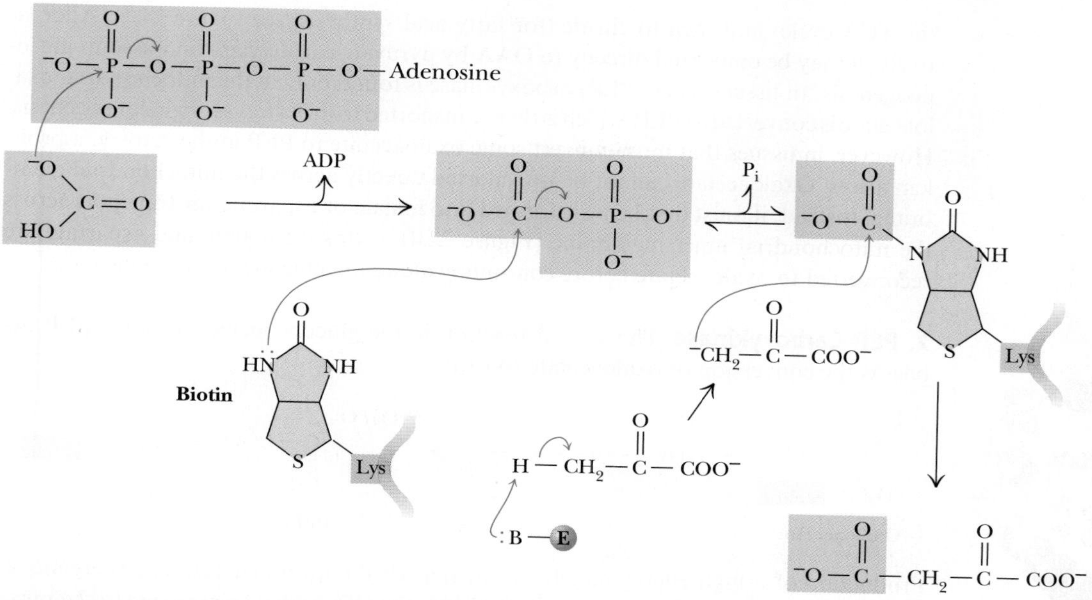

FIGURE 22.3 A mechanism for the pyruvate carboxylase reaction. Bicarbonate must be activated for attack by the pyruvate carbanion. This activation is driven by ATP and involves formation of a carbonylphosphate intermediate—a mixed anhydride of carbonic and phosphoric acids. (*Carbonylphosphate* and *carboxyphosphate* are synonyms.)

attack of a bicarbonate oxygen at the γ-P of ATP to form **carbonylphosphate,** an activated form of CO_2, and ADP (Figure 22.3). Reaction of carbonylphosphate with biotin occurs rapidly to form N-carboxybiotin, liberating inorganic phosphate. The third step involves abstraction of a proton from the C-3 of pyruvate, forming a carbanion that can attack the carbon of N-carboxybiotin to form oxaloacetate.

Pyruvate Carboxylase Is Allosterically Activated by Acetyl-Coenzyme A Two particularly interesting aspects of the pyruvate carboxylase reaction are (1) allosteric activation of the enzyme by acyl-CoA derivatives and (2) compartmentation of the reaction in the mitochondrial matrix. The carboxylation of biotin requires the presence (at an allosteric site) of acetyl-CoA or other acylated CoA derivatives. The second half of the carboxylase reaction—the attack by pyruvate to form oxaloacetate—is not affected by CoA derivatives.

Activation of pyruvate carboxylase by acetyl-CoA provides an important physiological regulation. Acetyl-CoA is the primary substrate for the TCA cycle, and oxaloacetate (formed by pyruvate carboxylase) is an important intermediate in both the TCA cycle and the gluconeogenesis pathway. If levels of ATP and/or acetyl-CoA (or other acyl-CoAs) are low, pyruvate is directed primarily into the TCA cycle, which eventually promotes the synthesis of ATP. If ATP and acetyl-CoA levels are high, pyruvate is converted to oxaloacetate and consumed in gluconeogenesis. Clearly, high levels of ATP and CoA derivatives are signs that energy is abundant and that metabolites will be converted to glucose (and perhaps even glycogen). If the energy status of the cell is low (in terms of ATP and CoA derivatives), pyruvate is consumed in the TCA cycle. Also, as noted in Chapter 19, pyruvate carboxylase is an important anaplerotic enzyme. Its activation by acetyl-CoA leads to oxaloacetate formation, replenishing the level of TCA cycle intermediates.

Compartmentalized Pyruvate Carboxylase Depends on Metabolite Conversion and Transport The second interesting feature of pyruvate carboxylase is that it is found only in the matrix of the mitochondria. By contrast, the next enzyme in the gluconeogenic pathway, PEP carboxykinase, may be localized in the cytosol, in the mitochondria, or both. For example, rabbit liver PEP carboxykinase is predominantly mitochondrial, whereas the rat liver enzyme is strictly cytosolic. In human liver, PEP carboxykinase is found both in the cytosol and in the mitochondria. Pyruvate is transported into the mitochondrial matrix (Figure 22.4), where it can be converted to acetyl-CoA (for use in

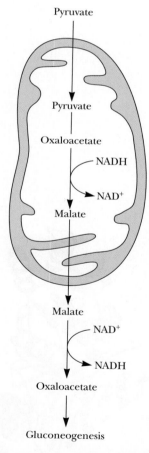

FIGURE 22.4 Pyruvate carboxylase is a compartmentalized reaction. Pyruvate is converted to oxaloacetate in the mitochondria. Because oxaloacetate cannot be transported across the mitochondrial membrane, it must be reduced to malate, which is then transported to the cytosol and oxidized back to oxaloacetate before gluconeogenesis can continue.

the TCA cycle) and then to citrate (for fatty acid synthesis; see Figure 24.1). Alternatively, it may be converted directly to OAA by pyruvate carboxylase and used in gluconeogenesis. In tissues where PEP carboxykinase is found only in the mitochondria, oxaloacetate is converted to PEP, which is then transported to the cytosol for gluconeogenesis. However, in tissues that must convert some oxaloacetate to PEP in the cytosol, a problem arises. Oxaloacetate cannot be transported directly across the mitochondrial membrane. Instead, it must first be transformed into malate or aspartate for transport across the mitochondrial inner membrane (Figure 22.4). Cytosolic malate and aspartate are reconverted to oxaloacetate before continuing along the gluconeogenic route.

2. PEP Carboxykinase The second reaction in the gluconeogenic pyruvate–PEP bypass is the conversion of oxaloacetate to PEP.

$$\underset{\textbf{Oxaloacetate}}{\overset{\displaystyle O}{\underset{\displaystyle H_2C-\boxed{COO^-}}{\overset{\displaystyle \|}{C}-COO^-}}} + GTP \xrightarrow{\text{PEP carboxykinase}} \underset{\textbf{PEP}}{\overset{\displaystyle ^{2-}O_3PO}{\underset{\displaystyle CH_2}{C-COO^-}}} + GDP + \boxed{CO_2}$$

Production of a high-energy metabolite such as PEP requires energy. The energetic requirements are handled in two ways here. First, the CO_2 added to pyruvate in the pyruvate carboxylase step is removed in the PEP carboxykinase reaction. Decarboxylation is a favorable process and helps drive the formation of the very high-energy enol phosphate in PEP. This decarboxylation drives a reaction that would otherwise be highly endergonic. Note the inherent metabolic logic in this pair of reactions: Pyruvate carboxylase consumed an ATP to drive a carboxylation so that the PEP carboxykinase could use the decarboxylation to facilitate formation of PEP. Second, another high-energy phosphate is consumed by the carboxykinase. Mammals and several other species use GTP in this reaction, rather than ATP. The use of GTP here is equivalent to the consumption of an ATP, due to the activity of the **nucleoside diphosphate kinase** (see Figure 19.2). The substantial free energy of hydrolysis of GTP is crucial to the synthesis of PEP in this step. The overall ΔG for the pyruvate carboxylase and PEP carboxykinase reactions under physiological conditions in the liver is −22.6 kJ/mol. Once PEP is formed in this way, the phosphoglycerate mutase, phosphoglycerate kinase, glyceraldehyde-3-P dehydrogenase, aldolase, and triose phosphate isomerase reactions act to eventually form fructose-1,6-bisphosphate, as shown in Figure 22.1.

3. Fructose-1,6-Bisphosphatase The hydrolysis of fructose-1,6-bisphosphate to fructose-6-phosphate,

$$\underset{\textbf{Fructose-1,6-bisphosphate}}{\text{F-1,6-BP}} + H_2O \xrightarrow[\text{bisphosphatase}]{\text{Fructose-1,6-}} \underset{\textbf{Fructose-6-phosphate}}{\text{F-6-P}} + \text{P}$$

$$\Delta G^{\circ\prime} = -16.7 \text{ kJ/mol}$$

like all phosphate ester hydrolyses, is a thermodynamically favorable (exergonic) reaction under standard-state conditions ($\Delta G^{\circ\prime} = -16.7$ kJ/mol). Under physiological conditions in the liver, the reaction is also exergonic ($\Delta G = -8.6$ kJ/mol). **Fructose-1,6-bisphosphatase** is an allosterically regulated enzyme. Citrate stimulates bisphosphatase activity, but *fructose-2,6-bisphosphate* is a potent allosteric inhibitor. AMP also inhibits the bisphosphatase; the inhibition by AMP is enhanced by fructose-2,6-bisphosphate.

4. Glucose-6-Phosphatase The final step in the gluconeogenesis pathway is the conversion of glucose-6-phosphate to glucose by the action of **glucose-6-phosphatase** (Figure 22.5). This enzyme is present in the membranes of the endoplasmic reticulum of liver and kidney cells but is absent in muscle and brain. For this reason, gluconeogenesis is not carried out in muscle and brain. The glucose-6-phosphatase system includes the phosphatase itself and three transport proteins, T1, T2, and T3. The glucose-6-phosphate

PEP carboxykinase from *Escherichia coli* with ADP (blue), pyruvate (purple), and Mg^{2+} (pdb id = 1OS1).

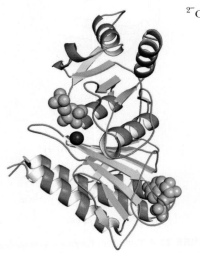

Fructose-1,6-bisphosphatase from pig with fructose-6-phosphate (orange), AMP (blue), and Mg^{2+} (dark blue) (pdb id = 1FBP).

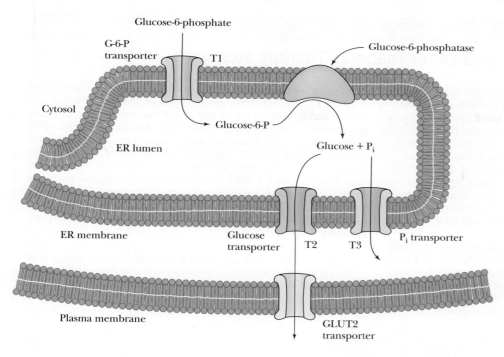

FIGURE 22.5 Glucose-6-phosphatase is localized in the endoplasmic reticulum. Conversion of glucose-6-phosphate to glucose occurs following transport into the endoplasmic reticulum. Glucose-6-phosphatase and the three transporters, T1, T2, and T3, are known collectively as the glucose-6-phosphatase system.

transporter (T1) takes glucose-6-phosphate into the endoplasmic reticulum, where it is hydrolyzed to glucose and P_i. The T2 and T3 transporters export glucose and P_i, respectively, to the cytosol, and glucose is then exported (to the circulation) by the GLUT2 transporter. The glucose-6-phosphatase reaction involves a phosphorylated enzyme intermediate, phosphohistidine (Figure 22.6). The ΔG for the glucose-6-phosphatase reaction in liver is -5.1 kJ/mol.

Coupling with Hydrolysis of ATP and GTP Drives Gluconeogenesis The net reaction for the conversion of pyruvate to glucose in gluconeogenesis is

$$2\ \text{Pyruvate} + 4\ \text{ATP} + 2\ \text{GTP} + 2\ \text{NADH} + 2\ \text{H}^+ + 6\ \text{H}_2\text{O} \longrightarrow$$
$$\text{glucose} + 4\ \text{ADP} + 2\ \text{GDP} + 6\ \text{P}_i + 2\ \text{NAD}^+$$

The net free energy change, $\Delta G^{\circ\prime}$, for this conversion is -37.7 kJ/mol. The consumption of a total of six nucleoside triphosphates drives this process forward. If glycolysis were merely reversed to achieve the net synthesis of glucose from pyruvate, the net reaction would be

$$2\ \text{Pyruvate} + 2\ \text{ATP} + 2\ \text{NADH} + 2\ \text{H}^+ + 2\ \text{H}_2\text{O} \longrightarrow$$
$$\text{glucose} + 2\ \text{ADP} + 2\ \text{P}_i + 2\ \text{NAD}^+$$

and the overall $\Delta G^{\circ\prime}$ would be about $+74$ kJ/mol. Such a process would be highly endergonic and therefore thermodynamically unfeasible. Hydrolysis of four additional high-energy phosphate bonds makes gluconeogenesis thermodynamically favorable.

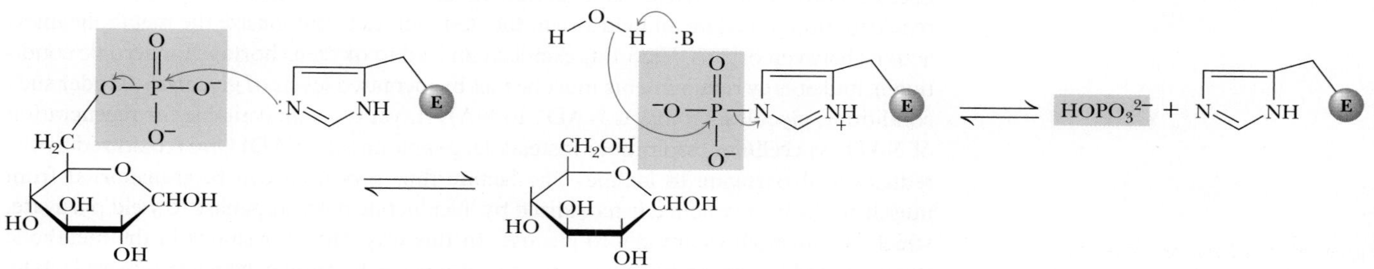

FIGURE 22.6 The glucose-6-phosphatase reaction involves formation of a phosphohistidine intermediate.

HUMAN BIOCHEMISTRY

Gluconeogenesis Inhibitors and Other Diabetes Therapy Strategies

Diabetes, the inability to assimilate and metabolize blood glucose, afflicts millions of people. People with type 1 diabetes are unable to synthesize and secrete insulin. On the other hand, people with type 2 diabetes make sufficient insulin, but the molecular pathways that respond to insulin are defective. Many type 2 diabetic people exhibit a condition termed **insulin resistance** even before the onset of diabetes. **Metformin** (see accompanying figure) is a drug that improves sensitivity to insulin, primarily by stimulating glucose uptake by glucose transporters in peripheral tissues. It also increases binding of insulin to insulin receptors, stimulates tyrosine kinase activity (see Chapter 32) of the insulin receptor, and inhibits gluconeogenesis in the liver.

Metformin may have several specific actions in the body. Many lines of evidence point to its activation of AMP-dependent kinase (AMPK), which in turn modulates a host of effects in cells (Chapter 27). Raymond Ochs and colleagues have shown that the primary effect of metformin is probably inhibition of AMP deaminase (Chapter 26), which raises cellular AMP levels, activating AMPK.

Gluconeogenesis inhibitors may be the next wave in diabetes therapy. Drugs that block gluconeogenesis without affecting glycolysis would need to target one of the enzymes unique to gluconeogenesis. 3-Mercaptopicolinate and hydrazine inhibit PEP carboxykinase, and **chlorogenic acid,** a natural product found in the skin of peaches, inhibits the transport activity of the glucose-6-phosphatase system (but not the glucose-6-phosphatase enzyme activity). The drug S-3483, a derivative of chlorogenic acid, also inhibits the glucose-6-phosphatase transport activity and binds a thousand times more tightly to the transporter than chlorogenic acid. Drugs of this type may be useful in the treatment of type 2 diabetes.

Ouyang, J., Parakhia, R. A., and Ochs, R., 2011. Metformin activates AMP kinase through inhibition of AMP deaminase. *The Journal of Biological Chemistry* **286:**1–11.

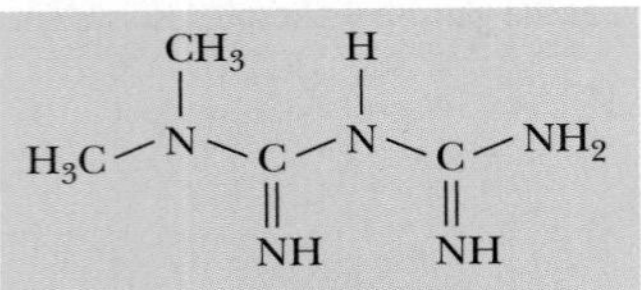

Metformin

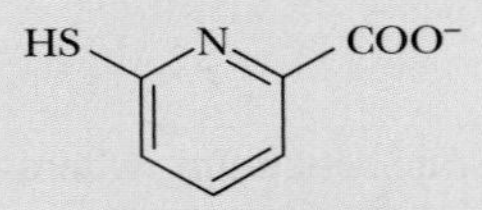

3-Mercaptopicolinate

Hydrazine

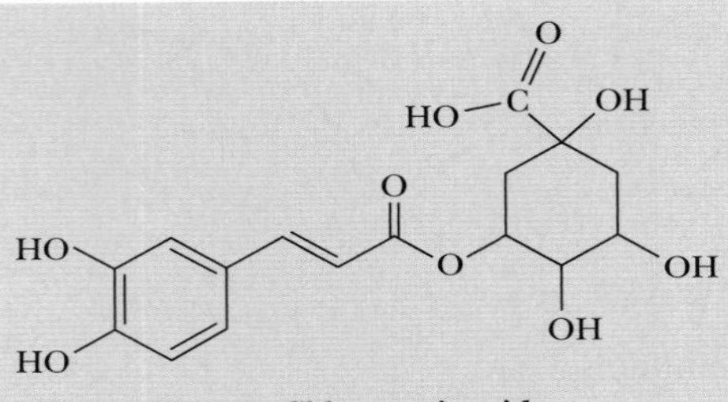

Chlorogenic acid

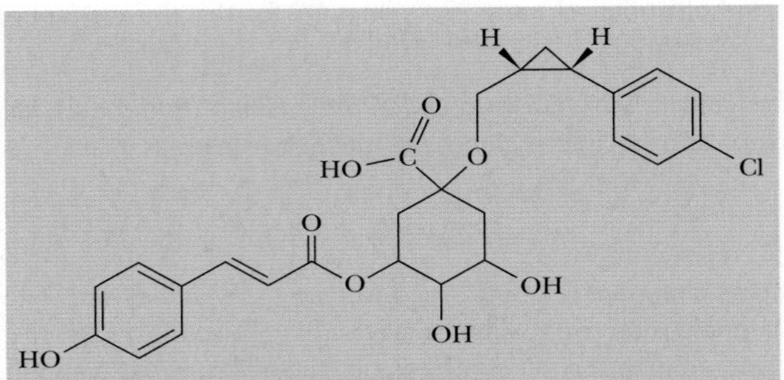

S-3483

Under physiological conditions, however, gluconeogenesis is somewhat less favorable than at standard state, with an overall ΔG of -15.6 kJ/mol for the conversion of pyruvate to glucose.

Lactate Formed in Muscles Is Recycled to Glucose in the Liver A final point on the redistribution of lactate and glucose in the body serves to emphasize the metabolic interactions between organs. Vigorous exercise can lead to oxygen shortage (anaerobic conditions), and energy requirements must be met by increased levels of glycolysis. Under such conditions, glycolysis converts NAD^+ to NADH, yet O_2 is unavailable for regeneration of NAD^+ via cellular respiration. Instead, large amounts of NADH are reoxidized by the reduction of pyruvate to lactate. The lactate thus produced can be transported from muscle to the liver, where it is reoxidized by liver lactate dehydrogenase to yield pyruvate, which is converted eventually to glucose. In this way, the liver shares in the metabolic stress created by vigorous exercise. It exports glucose to muscle, which produces lactate, and lactate from muscle can be processed by the liver into new glucose. This is referred

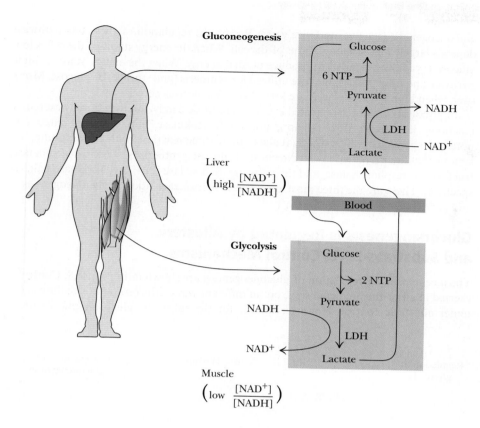

FIGURE 22.7 The Cori cycle.

to as the Cori cycle (Figure 22.7). Liver, with a typically high $NAD^+/NADH$ ratio (about 700), readily produces more glucose than it can use. Muscle that is vigorously exercising will enter anaerobiosis and show a decreasing $NAD^+/NADH$ ratio, which favors reduction of pyruvate to lactate.

22.2 How Is Gluconeogenesis Regulated?

Nearly all of the reactions of glycolysis and gluconeogenesis take place in the cytosol. If metabolic control were not exerted over these reactions, glycolytic degradation of glucose and gluconeogenic synthesis of glucose could operate simultaneously, with no net benefit to the cell and with considerable consumption of ATP. This is prevented by a sophisticated system of **reciprocal control,** which inhibits glycolysis when

The Cori cycle is named for Carl and Gerty Cori, who received the Nobel Prize in Physiology or Medicine in 1947 for their studies of glycogen metabolism and blood glucose regulation. Carl Ferdinand Cori and Gerty Theresa Radnitz were both born in Prague (then in Austria). They earned medical degrees from the German University of Prague in 1920 and were married later that year. They joined the faculty of the Washington University School of Medicine in St. Louis in 1931. Their remarkable collaboration resulted in many fundamental advances in carbohydrate and glycogen metabolism. They were credited with the discovery of glucose-1-phosphate, also known at the time as the "Cori ester." They also showed that glucose-6-phosphate was produced from glucose-1-P by the action of phosphoglucomutase. They isolated and crystallized glycogen phosphorylase and elucidated the pathway of glycogen breakdown. In 1952, they showed that absence of glucose-6-phosphatase in the liver was the enzymatic defect in von Gierke's disease, an inherited glycogen-storage disease. Six eventual Nobel laureates received training in their laboratory. Gerty Cori was the first American woman to receive a Nobel Prize. Carl Cori said of their remarkable collaboration: "Our efforts have been largely complementary and one without the other would not have gone so far…"

gluconeogenesis is active, and vice versa. Reciprocal regulation of these two pathways depends largely on the energy status of the cell. When the energy status of the cell is low, glucose is rapidly degraded to produce needed energy. When the energy status is high, pyruvate and other metabolites are utilized for synthesis (and storage) of glucose. Moreover, when blood glucose levels are low, gluconeogenesis is active.

In glycolysis, the three regulated enzymes are those catalyzing the strongly exergonic reactions: hexokinase (glucokinase), phosphofructokinase, and pyruvate kinase. As noted, the gluconeogenic pathway replaces these three reactions with corresponding reactions that are exergonic in the direction of glucose synthesis: glucose-6-phosphatase, fructose-1,6-bisphosphatase, and the pyruvate carboxylase–PEP carboxykinase pair, respectively. These are the three most appropriate sites of regulation in gluconeogenesis.

Gluconeogenesis Is Regulated by Allosteric and Substrate-Level Control Mechanisms

The mechanisms of regulation of gluconeogenesis are shown in Figure 22.8. Control is exerted at all of the predicted sites, but in different ways. Glucose-6-phosphatase is not under allosteric control. However, the K_m for the substrate, glucose-6-phosphate, is

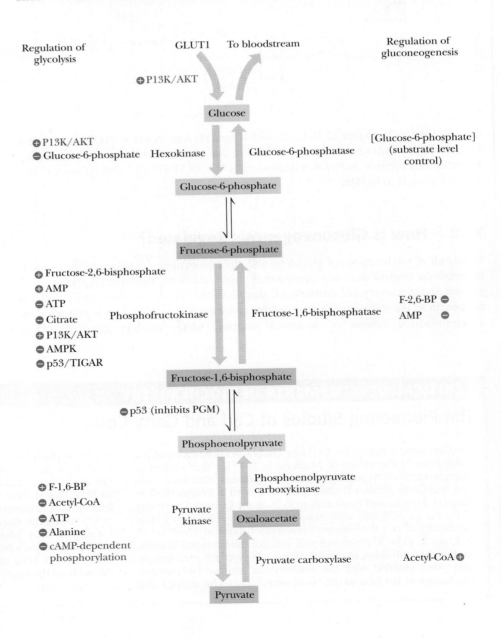

FIGURE 22.8 The principal regulatory mechanisms in glycolysis and gluconeogenesis. Activators are indicated by plus signs and inhibitors by minus signs. Carbohydrate metabolism is also modulated by a variety of cellular signals, including the protein kinases PI3K, AKT, and AMPK, the tumor suppressor p53, and TIGAR (see A Deeper Look, page 730).

considerably higher than the normal range of substrate concentrations. As a result, glucose-6-phosphatase displays a near-linear dependence of activity on substrate concentrations and is thus said to be under **substrate-level control** by glucose-6-phosphate.

Acetyl-CoA is a potent allosteric effector of glycolysis and gluconeogenesis. It allosterically inhibits pyruvate kinase (as noted in Chapter 18) and activates pyruvate carboxylase. Because it also allosterically inhibits pyruvate dehydrogenase (the enzymatic link between glycolysis and the TCA cycle), the cellular fate of pyruvate is strongly dependent on acetyl-CoA levels. A rise in [acetyl-CoA] indicates that cellular energy levels are high and that carbon metabolites can be directed to glucose synthesis and storage. When acetyl-CoA levels drop, the activities of pyruvate kinase and pyruvate dehydrogenase increase and flux through the TCA cycle increases, providing needed energy for the cell.

Fructose-1,6-bisphosphatase is another important site of gluconeogenic regulation. This enzyme is inhibited by AMP and activated by citrate. These effects by AMP and citrate are the opposites of those exerted on phosphofructokinase in glycolysis, providing another example of reciprocal regulatory effects. When AMP levels increase, gluconeogenic activity is diminished and glycolysis is stimulated. An increase in citrate concentration signals that TCA cycle activity can be curtailed and that pyruvate should be directed to sugar synthesis instead.

Fructose-2,6-Bisphosphate—Allosteric Regulator of Gluconeogenesis

Emile Van Schaftingen and Henri-Géry Hers demonstrated in 1980 that fructose-2,6-bisphosphate is a potent stimulator of phosphofructokinase (see Chapter 18). Cognizant of the reciprocal nature of regulation in glycolysis and gluconeogenesis, Van Schaftingen and Hers also considered the possibility of an opposite effect—inhibition—for fructose-1,6-bisphosphatase. In 1981, they reported that fructose-2,6-bisphosphate was indeed a powerful inhibitor of fructose-1,6-bisphosphatase (Figure 22.9). Inhibition occurs in either the presence or absence of AMP, and the effects of AMP and fructose-2,6-bisphosphate are synergistic.

Cellular levels of fructose-2,6-bisphosphate are controlled by **phosphofructokinase-2 (PFK-2)**, an enzyme distinct from the phosphofructokinase of the glycolytic pathway, and by **fructose-2,6-bisphosphatase (F-2,6-BPase)**. Remarkably, these two enzymatic activities are both found in the same protein molecule, which is an example of a **bifunctional**, or **tandem, enzyme** (Figure 22.10). The opposing activities of this bifunctional enzyme are themselves regulated in two ways. First, fructose-6-phosphate, the substrate of phosphofructokinase and the product of fructose-1,6-bisphosphatase, allosterically

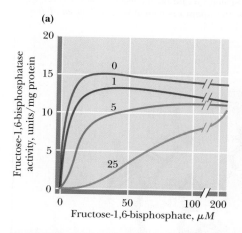

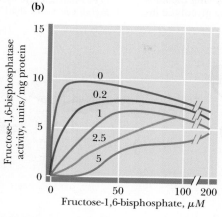

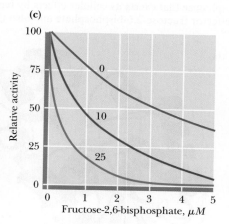

Fructose-2,6-bisphosphate

FIGURE 22.9 Inhibition of fructose-1,6-bisphosphatase by fructose-2,6-bisphosphate in the **(a)** absence and **(b)** presence of 25 m*M* AMP. In (a) and (b), enzyme activity is plotted against substrate (fructose-1,6-bisphosphate) concentration. Concentrations of fructose-2,6-bisphosphate (in m*M*) are indicated above each curve. **(c)** The effect of AMP (0, 10, and 25 m*M*) on the inhibition of fructose-1,6-bisphosphatase by fructose-2,6-bisphosphate. Activity was measured in the presence of 10 m*M* fructose-1,6-bisphosphate. (Adapted from Van Schaftingen, E., and Hers, H-G., 1981. Inhibition of fructose-1,6-bisphosphatase by fructose-2,6-bisphosphate. *Proceedings of the National Academy of Science, U.S.A.* **78**:2861–2863.)

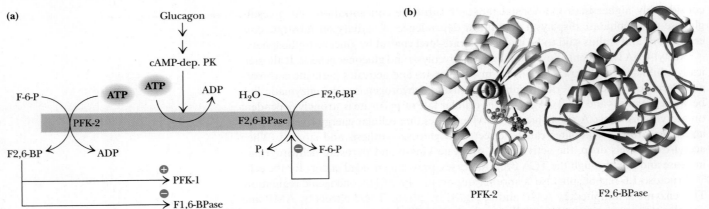

(a)

(b)

PFK-2

F2,6-BPase

FIGURE 22.10 **(a)** Synthesis and degradation of fructose-2,6-bisphosphate are catalyzed by the same bifunctional enzyme. **(b)** The structure of PFK-2/F-2,6-BPase from rat liver. PFK-2 activity resides in the N-terminal portion of the protein (*left*), and the C-terminal domain (*right*) contains F-2,6-BPase activity. The PFK-2 domain has a bound ATP analog; the F-2,6-BPase has two phosphates bound (pdb id = 1BIF).

activates PFK-2 and inhibits F-2,6-BPase. Second, the phosphorylation by **cAMP-dependent protein kinase** of a single Ser (Ser^{32}) residue on the liver enzyme exerts reciprocal control of the PFK-2 and F-2,6-BPase activities. Phosphorylation inhibits PFK-2 activity (by increasing the K_m for fructose-6-phosphate) and stimulates F-2,6-BPase activity.

> **A DEEPER LOOK**

TIGAR—a p53-Induced Enzyme That Mimics Fructose-2,6-Bisphosphatase

Cellular stresses, such as oncogenesis and DNA damage, provoke the activation of target genes by the tumor suppressor p53. A principal target of p53 is **TIGAR (TP53-induced glycolysis and apoptosis regulator).** TIGAR inhibits glycolysis, reduces cellular levels of reactive oxygen species (ROS) and protects cells from apoptosis. Hui Li and Gerwald Jogl have shown that TIGAR's structure is nearly identical to that of the fructose-2,6-bisphosphatase domain of the bifunctional PFK-2/F-2,6-BPase (see figure). TIGAR is a fructose bisphosphatase that exerts its cellular effects by removing the allosteric effector fructose-2,6-bisphosphate and also the glycolytic in-

termediate fructose-1,6-bisphosphate. **(a)** TIGAR (pbd id = 3E9D) is orange; F-2, 6-BPase (pbd id = 2BIF) is blue. **(b)** Active-site comparison, with phosphates in yellow/red.

References

Green, D.R., and Chipuk, J. E., 2006. p53 and metabolism: Inside the TIGAR. *Cell* **126**:30–32

Li, H., and Jogl, G., 2011. Structural and biochemical studies of TIGAR (TP53-induced glycolysis and apoptosis regulator. *The Journal of Biological Chemistry* **284**:1748–1754.

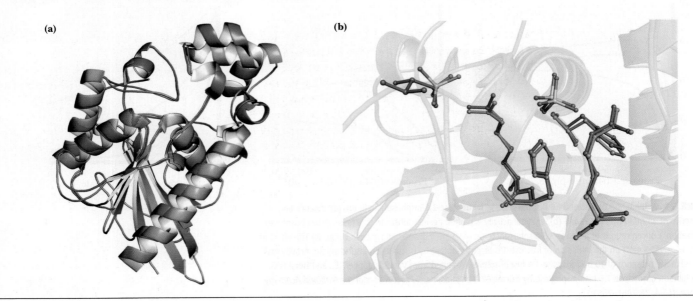

(a)

(b)

Substrate Cycles Provide Metabolic Control Mechanisms

If fructose-1,6-bisphosphatase and phosphofructokinase acted simultaneously, they would constitute a **substrate cycle** in which fructose-1,6-bisphosphate and fructose-6-phosphate became interconverted with net consumption of ATP:

$$\text{Fructose-1,6-BP} + H_2O \longrightarrow \text{fructose-6-P} + P_i$$
$$\text{Fructose-6-P} + \text{ATP} \longrightarrow \text{fructose-1,6-BP} + \text{ADP}$$

Net: $\qquad \text{ATP} + H_2O \longrightarrow \text{ADP} + P_i$

Because substrate cycles such as this appear to operate with no net benefit to the cell, they were once regarded as metabolic quirks and were referred to as futile cycles. More recently, substrate cycles have been recognized as important devices for controlling metabolite concentrations.

The three steps in glycolysis and gluconeogenesis that differ constitute three such substrate cycles, each with its own particular metabolic raison d'être. Consider, for example, the regulation of the fructose-1,6-BP–fructose-6-P cycle by fructose-2,6-bisphosphate. As already noted, fructose-1,6-bisphosphatase is subject to allosteric inhibition by fructose-2,6-bisphosphate, whereas phosphofructokinase is allosterically activated by fructose-2,6-bisphosphate. The combination of these effects should permit either phosphofructokinase or fructose-1,6-bisphosphatase (but not both) to operate at any one time and should thus prevent futile cycling. For instance, in the **fasting state,** when food (that is, glucose) intake is zero, phosphofructokinase (and therefore glycolysis) is inactive due to the low concentration of fructose-2,6-bisphosphate. In the liver, gluconeogenesis operates to provide glucose for the brain. However, in the fed state, up to 30% of fructose-1,6-bisphosphate formed from phosphofructokinase is recycled back to fructose-6-P (and then to glucose). Because the dependence of fructose-1,6-bisphosphatase activity on fructose-1,6-bisphosphate is sigmoidal in the presence of fructose-2,6-bisphosphate (see Figure 22.9), substrate cycling occurs only at relatively high levels of fructose-1,6-bisphosphate. Substrate cycling in this case prevents the accumulation of excessively high levels of fructose-1,6-bisphosphate.

22.3 How Are Glycogen and Starch Catabolized in Animals?

Dietary Starch Breakdown Provides Metabolic Energy

As noted earlier, well-fed adult human beings normally metabolize about 160 grams of carbohydrates each day. A balanced diet easily provides this amount, mostly in the form of starch. If too little carbohydrate is supplied by the diet, glycogen reserves in liver and muscle tissue can also be mobilized. The reactions by which ingested starch and glycogen are digested are shown in Figure 22.11. The enzyme known as **α-amylase** is an important component of saliva and pancreatic juice. (**β-Amylase** is found in plants. The α- and β-designations for these enzymes serve only to distinguish the two and do not refer to glycosidic linkage nomenclature.) α-Amylase is an endoglycosidase that hydrolyzes α(1⟶4) linkages of amylopectin and glycogen at random positions, eventually producing a mixture of maltose, maltotriose [with three α(1⟶4)-linked glucose residues], and other small oligosaccharides. α-Amylase can cleave on either side of an amylopectin branch point, but activity is reduced in highly branched regions of the polysaccharide and stops four residues from any branch point.

The highly branched polysaccharides that are left after extensive exposure to α-amylase are called **limit dextrins.** These structures can be further degraded by the action of a **debranching enzyme,** which carries out two distinct reactions. The first of these, known as **oligo(α1,4⟶α1,4) glucanotransferase** activity, removes a trisaccharide unit and transfers this group to the end of another, nearby branch (Figure 22.12). This leaves a single glucose residue in α(1⟶6) linkage to the main chain. The **α(1⟶6) glucosidase** activity of the debranching enzyme then cleaves this residue from the chain, leaving a

(a)

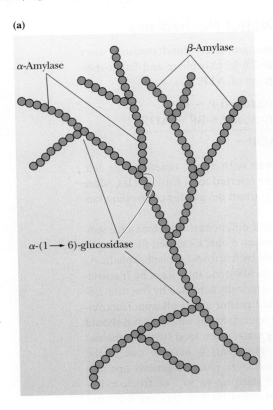

(b)

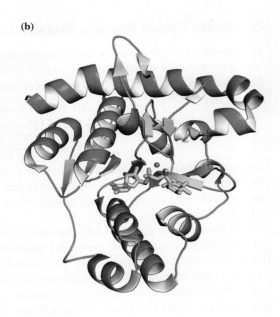

FIGURE 22.11 **(a)** The sites of hydrolysis of starch by α- and β-amylase are indicated. **(b)** Glycogenin is a glycosyltransferase that initiates eukaryotic glycogen synthesis from glucose. Bound UDP-glucose (blue) and Mn^{2+} ion (purple) are shown (pdb id = 1LL2).

polysaccharide chain with one branch fewer. Repetition of this sequence of events leads to complete degradation of the polysaccharide.

β-Amylase is an exoglycosidase that cleaves maltose units from the free, nonreducing ends of amylopectin branches, as in Figure 22.11. Like α-amylase, however, β-amylase does not cleave either the $\alpha(1\longrightarrow6)$ bonds at the branch points or the $\alpha(1\longrightarrow4)$ linkages near the branch points.

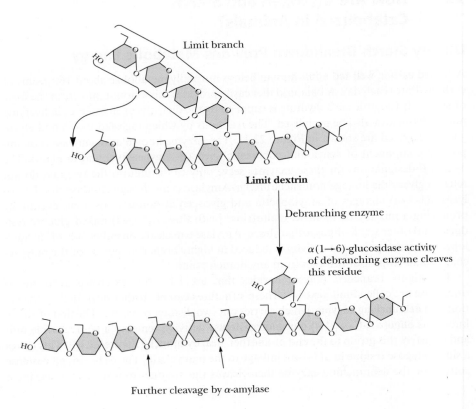

FIGURE 22.12 The reactions of debranching enzyme. Transfer of a group of three $\alpha(1\longrightarrow4)$-linked glucose residues from a limit branch to another branch is followed by cleavage of the $\alpha(1\longrightarrow6)$ bond of the residue that remains at the branch point.

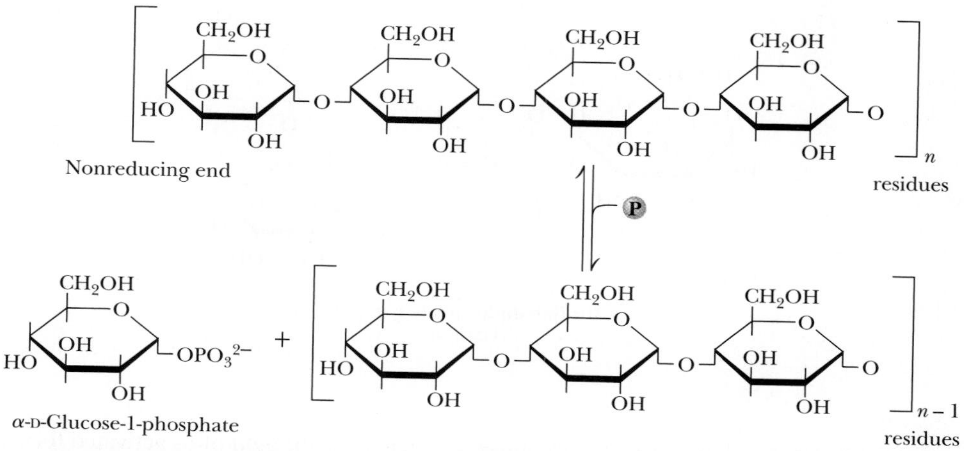

FIGURE 22.13 The glycogen phosphorylase reaction.

Metabolism of Tissue Glycogen Is Regulated

Digestion itself is a highly efficient process in which almost 100% of ingested food is absorbed and metabolized. Digestive breakdown of starch is an unregulated process. On the other hand, tissue glycogen represents an important reservoir of potential energy, and it should be no surprise that the reactions involved in its degradation and synthesis are carefully controlled and regulated. Glycogen reserves in liver and muscle tissue are stored in the cytosol as granules exhibiting a molecular weight range from 6×10^6 to 1600×10^6. These granular aggregates contain the enzymes required to synthesize and catabolize the glycogen, as well as all the enzymes of glycolysis.

The principal enzyme of glycogen catabolism is **glycogen phosphorylase,** a highly regulated enzyme that was discussed extensively in Chapter 15. The glycogen phosphorylase reaction (Figure 22.13) involves phosphorolysis at a nonreducing end of a glycogen polymer. The standard-state free energy change for this reaction is +3.1 kJ/mol, but the intracellular ratio of [P_i] to [glucose-1-P] approaches 100, and thus the actual ΔG in vivo is approximately −6 kJ/mol. There is an energetic advantage to the cell in this phosphorolysis reaction. If glycogen breakdown were hydrolytic and yielded glucose as a product, it would be necessary to phosphorylate the product glucose (with the expenditure of a molecule of ATP) to initiate its glycolytic degradation.

The glycogen phosphorylase reaction degrades glycogen to produce limit dextrins, which are further degraded by debranching enzyme, as already described.

22.4 | How Is Glycogen Synthesized?

Animals synthesize and store glycogen when glucose levels are high, but the synthetic pathway is not merely a reversal of the glycogen phosphorylase reaction. High levels of phosphate in the cell favor glycogen breakdown and prevent the phosphorylase reaction from synthesizing glycogen in vivo, despite the fact that $\Delta G^{\circ\prime}$ for the phosphorylase reaction actually favors glycogen synthesis. Hence, another reaction pathway must be employed in the cell for the net synthesis of glycogen. In essence, this pathway must activate glucose units for transfer to glycogen chains.

Glucose Units Are Activated for Transfer by Formation of Sugar Nucleotides

We are familiar with several examples of chemical activation as a strategy for group transfer reactions. Acetyl-CoA is an activated form of acetate; biotin and tetrahydrofolate activate one-carbon groups for transfer; and ATP is an activated form of phosphate. Luis Leloir, a biochemist in Argentina, showed in the 1950s that glycogen

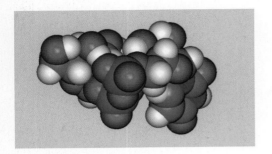

Uridine diphosphate glucose (UDP–glucose).

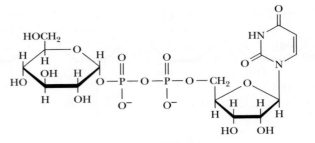

**Uridine diphosphate glucose
(UDPG)**

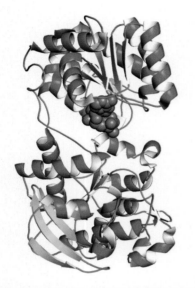

Glycogen synthase from *Agrobacterium tumefaciens* consists of two Rossman folds (see Chapter 16) separated by a deep cleft that includes the active site (shown here with bound ADP, purple) (pdb id = 1RZU).

synthesis depended upon **sugar nucleotides,** which may be thought of as activated forms of sugar units. For example, formation of an ester linkage between the C-1 hydroxyl group and the β-phosphate of UDP activates the glucose moiety of **UDP–glucose.**

UDP–Glucose Synthesis Is Driven by Pyrophosphate Hydrolysis

Sugar nucleotides are formed from sugar-1-phosphates and nucleoside triphosphates by specific **pyrophosphorylase** enzymes (Figure 22.14). For example, **UDP–glucose pyrophosphorylase** catalyzes the formation of UDP–glucose from glucose-1-phosphate and uridine 5′-triphosphate:

$$\text{Glucose-1-P} + \text{UTP} \longrightarrow \text{UDP–glucose} + \text{pyrophosphate}$$

FIGURE 22.14 The UDP–glucose pyrophosphorylase reaction is a phosphoanhydride exchange, with a phosphoryl oxygen of glucose-1-P attacking the α-phosphorus of UTP to form UDP–glucose and pyrophosphate.

The reaction proceeds via attack by a phosphate oxygen of glucose-1-phosphate on the α-phosphorus of UTP, with departure of the pyrophosphate anion. The reaction is a reversible one, but—as is the case for many biosynthetic reactions—it is driven forward by subsequent hydrolysis of pyrophosphate:

$$\text{Pyrophosphate} + \text{H}_2\text{O} \longrightarrow 2\ \text{P}_i$$

The net reaction for sugar nucleotide formation (combining the preceding two equations) is thus

$$\text{Glucose-1-P} + \text{UTP} + \text{H}_2\text{O} \longrightarrow \text{UDP–glucose} + 2\ \text{P}_i$$

Sugar nucleotides of this type act as donors of sugar units in the biosynthesis of oligosaccharides and polysaccharides. In animals, UDP–glucose is the donor of glucose units for glycogen synthesis, but ADP–glucose is the glucose source for starch synthesis in plants.

Glycogen Synthase Catalyzes Formation of $\alpha(1\longrightarrow4)$ Glycosidic Bonds in Glycogen

The very large glycogen polymer is built around a tiny protein core. The first glucose residue is covalently joined to the protein **glycogenin** (see Figure 22.11b) via an acetal linkage to a tyrosine–OH group on the protein. Sugar units can then be added by the action of **glycogen synthase.** The reaction involves transfer of a glucosyl unit from UDP–glucose to the C-4 hydroxyl group at a nonreducing end of a glycogen strand. The mechanism proceeds by cleavage of the C—O bond between the glucose moiety and the β-phosphate of UDP–glucose, leaving an oxonium ion intermediate, which is rapidly attacked by the C-4 hydroxyl oxygen of a terminal glucose unit on glycogen (Figure 22.15). The reaction is exergonic and has a $\Delta G^{\circ\prime}$ of -13.3 kJ/mol.

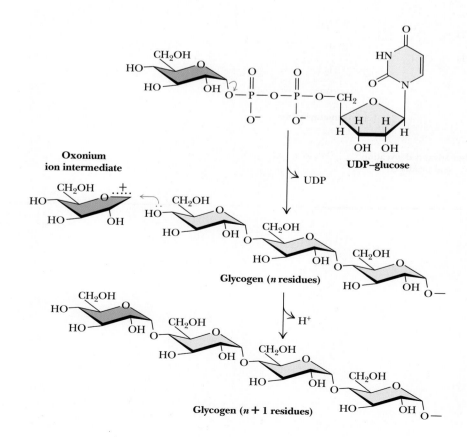

FIGURE 22.15 The glycogen synthase reaction. Cleavage of the C—O bond of UDP–glucose yields an oxonium intermediate. Attack by the hydroxyl oxygen of the terminal residue of a glycogen molecule completes the reaction.

Advanced Glycation End Products—A Serious Complication of Diabetes

Covalent linkage of sugars to proteins to form glycoproteins normally occurs through the action of enzymes that use sugar nucleotides as substrates. However, sugars may also react nonenzymatically with proteins. The C-1 carbonyl group of glucose forms Schiff base linkages with lysine side chains of proteins. These Schiff base adducts undergo Amadori rearrangements and subsequent oxidations to form irreversible "glycation" products, including carboxymethyllysine and pentosidine derivatives (see accompanying figure). These **advanced glycation end products (AGEs)** can alter the function of the protein. Such AGE-dependent changes are thought to contribute to circulation, joint, and vision problems in people with diabetes.

Nonenzymatic glycation of hemoglobin is a better diagnostic yardstick for type 2 diabetes than serum glucose levels. Red blood cells have an average life expectancy of about 4 months. By measuring the concentration of "glycated hemoglobin" in a patient, it is possible to determine the average glucose concentration in the blood over the past several months.

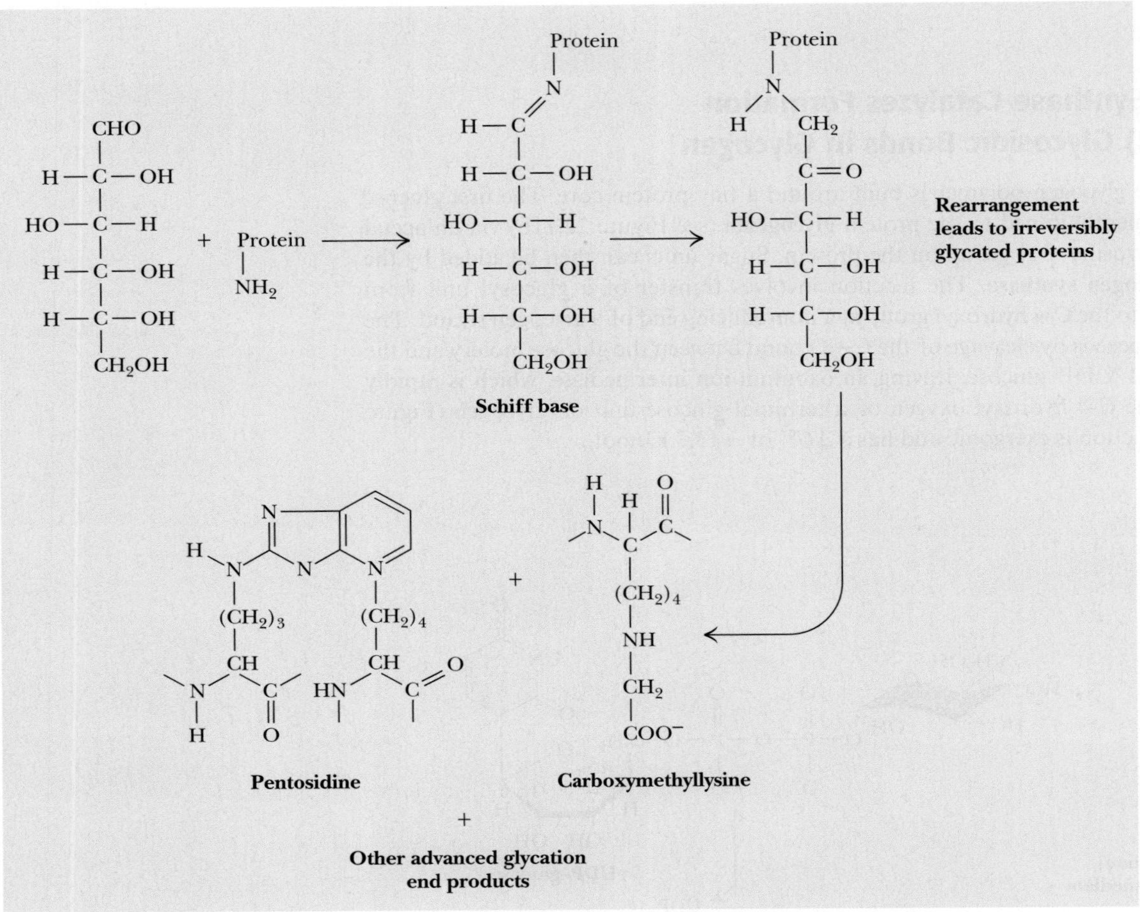

The cell surface receptor that binds **a**dvanced **g**lycation **e**nd products **(RAGE)** is a multifunctional protein of the innate immune system that plays pivotal roles in diabetes, chronic inflammation, neurodegenerative diseases, T-lymphocyte proliferation, and cancer. RAGE is a single-transmembrane segment receptor expressed mainly in immune cells, neurons, endothelial and vascular smooth muscle cells, bone-forming cells, and a variety of cancer cells. RAGE consists of three extracellular immunoglobulin domains, a transmembrane segment and a small cytoplasmic domain (figure, part a). The first two extracellular domains of RAGE contain a basic patch (figure, part b) and a hydrophobic patch. Binding of its ligands (AGE-modified proteins, amyloid-β, and S100 proteins)

depends on interactions with these surfaces, particularly the large basic patch comprised of Arg and Lys side chains. Binding of ligands induces dimerization of RAGE and triggers a variety of intracellular responses related to inflammation and tumorigenesis. Understanding these cellular responses will be essential in light of the developing worldwide epidemic of type 2 diabetes.

References

Park, H., and Boyington, J. C., 2010. The 1.5Å crystal structure of human receptor for advanced glycation end products (RAGE) ectodomains reveals unique features determining ligand binding. *The Journal of Biological Chemistry* **285**:40762–40770.

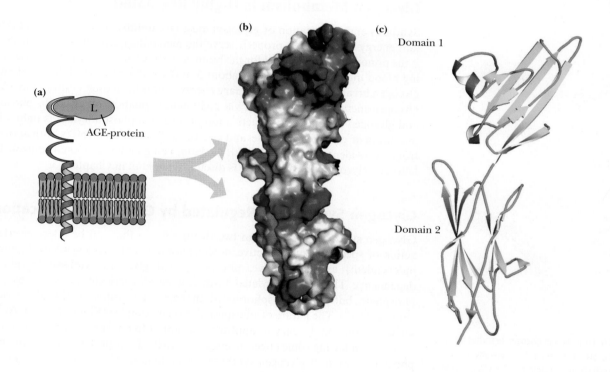

(a)
L
AGE-protein

(b)

(c)
Domain 1

Domain 2

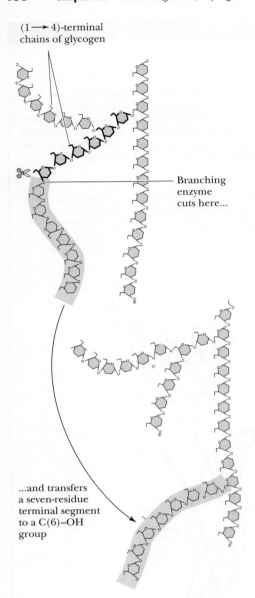

(1⟶4)-terminal
chains of glycogen

Branching
enzyme
cuts here...

...and transfers
a seven-residue
terminal segment
to a C(6)–OH
group

FIGURE 22.16 Formation of glycogen branches by the branching enzyme. Six- or seven-residue segments of a growing glycogen chain are transferred to the C-6 hydroxyl group of a glucose residue on the same or a nearby chain.

Glycogen Branching Occurs by Transfer of Terminal Chain Segments

Glycogen is a branched polymer of glucose units. The branches arise from $\alpha(1\longrightarrow6)$ linkages, which occur every 8 to 12 residues. As noted in Chapter 7, the branches provide multiple sites for rapid degradation or elongation of the polymer and also increase its solubility. Glycogen branches are formed by **amylo-(1,4⟶1,6)-transglycosylase,** also known as *branching enzyme*. The reaction involves the transfer of a 6- or 7-residue segment from the nonreducing end of a linear chain at least 11 residues in length to the C-6 hydroxyl of a glucose residue of the same chain or another chain (Figure 22.16). For each branching reaction, the resulting polymer has gained a new terminus at which growth can occur.

22.5 How Is Glycogen Metabolism Controlled?

Glycogen Metabolism Is Highly Regulated

Synthesis and degradation of glycogen must be carefully controlled so that this important energy reservoir can properly serve the metabolic needs of the organism. Glucose is the principal metabolic fuel for the brain, and the concentration of glucose in circulating blood must be maintained at about 5 mM for this purpose. Glucose derived from glycogen breakdown is also a primary energy source for muscle contraction. Control of glycogen metabolism is effected via reciprocal regulation of glycogen phosphorylase and glycogen synthase. Thus, activation of glycogen phosphorylase is tightly linked to inhibition of glycogen synthase, and vice versa. Regulation involves both allosteric control and covalent modification, with the latter being under hormonal control. The regulation of glycogen phosphorylase is discussed in detail in Chapter 15.

Glycogen Synthase Is Regulated by Covalent Modification

Glycogen synthase also exists in two distinct forms that can be interconverted by the action of specific enzymes: active, dephosphorylated **glycogen synthase I** (glucose-6-P–independent) and less active, phosphorylated **glycogen synthase D** (glucose-6-P–dependent). The phosphorylated form can be allosterically activated by glucose-6-phosphate, but the unphosphorylated enzyme is insensitive to this allosteric effector (Figure 22.17). The nature of phosphorylation is complex (Figure 22.17a). At least nine serine residues on the enzyme appear to be subject to phosphorylation, each site's phosphorylation having some effect on enzyme activity. Four protein kinases are involved in phosphorylation of glycogen synthase: casein kinase, AMP-dependent protein kinase, protein kinase A, and glycogen synthase kinase 3 (GSK3).

Dephosphorylation of both glycogen phosphorylase and glycogen synthase is carried out by **phosphoprotein phosphatase-1 (PP1).** The action of PP1 inactivates glycogen phosphorylase and activates glycogen synthase.

Hormones Regulate Glycogen Synthesis and Degradation

Storage and utilization of tissue glycogen, maintenance of blood glucose concentration, and other aspects of carbohydrate metabolism are meticulously regulated by hormones, including *insulin, glucagon, epinephrine,* and the *glucocorticoids.*

Insulin Is a Response to Increased Blood Glucose The primary hormone responsible for conversion of glucose to glycogen is **insulin** (see Figure 5.8). Insulin is secreted by the β-cells in the pancreas within the **islets of Langerhans.** *Secretion of insulin is a response to increased glucose in the blood.* When blood glucose levels rise (after a meal, for example), insulin is secreted from the pancreas into the *pancreatic vein,* which

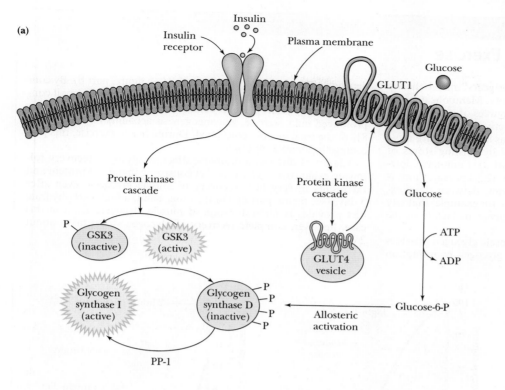

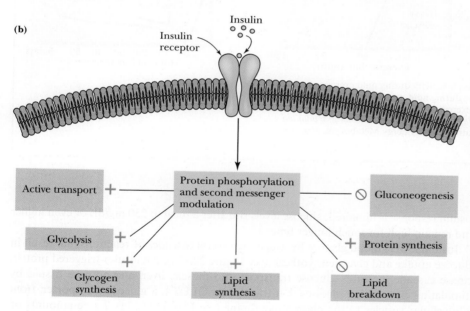

FIGURE 22.17 (a) Binding of insulin to plasma membrane receptors in the liver and muscles triggers protein kinase cascades that stimulate glycogen synthesis. Insulin's effects include inactivation of GSK3 and stimulation of PP1, both actions activating glycogen synthase, as well as recruitment of GLUT4 to the plasma membrane. Glucose uptake provides substrate for glycogen synthesis and glucose-6-phosphate, which allosterically activates the otherwise inactive form of glycogen synthase. (b) The metabolic effects of insulin are mediated through protein phosphorylation and second messenger modulation.

empties into the **portal vein system** (Figure 22.18), so insulin traverses the liver before it enters the systemic blood supply. Insulin acts to rapidly lower blood glucose concentration in several ways. Insulin stimulates glycogen synthesis and inhibits glycogen breakdown in liver and muscle.

Insulin Triggers Glycogen Synthesis When Blood Glucose Rises The action of insulin when blood glucose rises is immediate and powerful. During periods between meals, typical human blood glucose levels are 70 to 90 mg/dL. Glucose levels normally rise to about 150 mg/dL within the first hour following a carbohydrate-rich meal (Figure 22.19) and then return to normal within 2 to 3 hours. (For diabetic subjects, whose

▶ **A DEEPER LOOK**

Carbohydrate Utilization in Exercise

Animals have a remarkable ability to "shift gears" metabolically during periods of strenuous exercise or activity. Metabolic adaptations allow the body to draw on different sources of energy (all of which produce ATP) for different types of activity. During periods of short-term, high-intensity exercise (such as a 100-m dash), most of the required energy is supplied directly by existing stores of ATP and creatine phosphate (see figure, part a). Long-term, low-intensity exercise (such as a 10-km run or a 42.2-km marathon) is fueled almost entirely by aerobic metabolism. Between these extremes is a variety of activities (an 800-m run, for example) that rely on anaerobic glycolysis—conversion of glucose to lactate in the muscles and utilization of the Cori cycle.

For all these activities, breakdown of muscle glycogen provides much of the needed glucose. The rate of glycogen consumption depends on the intensity of the exercise (see figure, part b). By contrast, glucose derived from gluconeogenesis makes only small contributions to total glucose consumed during exercise. During prolonged mild exercise, gluconeogenesis accounts for only about 8% of the total glucose consumed. During heavy exercise, this percentage becomes even lower.

Choice of diet has a dramatic effect on glycogen recovery following exhaustive exercise. A diet consisting mainly of protein and fat results in very little recovery of muscle glycogen, even after 5 days (see figure, part c). On the other hand, a high-carbohydrate diet provides faster restoration of muscle glycogen. Even in this case, however, complete recovery of glycogen stores takes about 2 days.

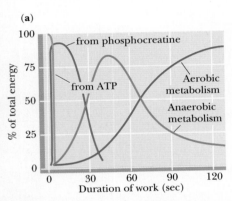

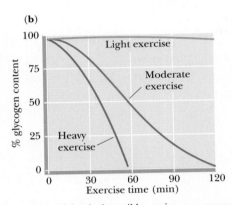

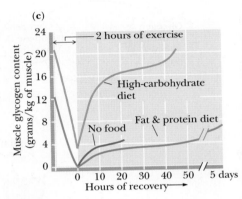

▲ (a) Contributions of the various energy sources to muscle activity during mild exercise.
(b) Consumption of glycogen stores in fast-twitch muscles during light, moderate, and heavy exercise.
(c) Rate of glycogen replenishment following exhaustive exercise. *(a and c adapted from Rhodes, R., and Pflanzer, R. G., 1992.* Human Physiology. *Philadelphia: Saunders College Publishing; b adapted from Horton, E. S., and Terjung, R. L., 1988.* Exercise, Nutrition and Energy Metabolism. *New York: Macmillan.)*

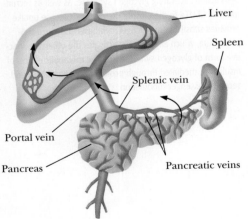

FIGURE 22.18 The portal vein system carries pancreatic secretions such as insulin and glucagon to the liver and then into the rest of the circulatory system.

insulin response is impaired, glucose levels rise after a meal to 250 mg/dL or even higher and remain high for much longer times.)

Insulin lowers blood glucose by triggering several cascades of reactions that result in glucose uptake and glycogen synthesis (see Figure 22.17a). An insulin-triggered protein kinase cascade increases glucose transport into muscle, liver, and adipose tissues by stimulating exocytotic processes that translocate GLUT4, a glucose transporter, from intracellular vesicles to the plasma membrane (see Figure 22.17a). Large amounts of glucose thus transported into the cell are converted to glucose-6-P, which can be directed to glycogen synthesis (by conversion to glucose-1-P). Also, glucose-6-P is the allosteric effector that activates the otherwise inactive, phosphorylated form of glycogen synthase.

Binding of insulin to the plasma membrane, in either liver or muscle cells, triggers another protein kinase cascade (see Figure 15.17 and Chapter 32) that results in phosphorylation and inactivation of **glycogen synthase kinase 3 (GSK3).** This kinase normally phosphorylates and inactivates glycogen synthase. Inhibition of GSK3 means that more of the cell's glycogen synthase will remain in the unphosphorylated, active state (see Figure 22.17a). Insulin also stimulates PP1, which dephosphorylates (and activates) glycogen synthase.

Several other physiological effects of insulin also serve to lower blood and tissue glucose levels (see Figure 22.17b). Insulin increases cellular utilization of glucose by

HUMAN BIOCHEMISTRY

von Gierke Disease—A Glycogen-Storage Disease

In 1929, the physician Edgar von Gierke treated a patient with a very enlarged abdomen. The patient's liver and kidneys were severely enlarged due to massive accumulations of glycogen, and von Gierke appropriately called the condition "hepato-nephromegalia glycogenica." Now termed **von Gierke's disease,** or Type Ia glycogen storage disease, this condition results from the absence of glucose-6-phosphatase activity in the affected organs. This simple genetic defect causes a host of difficult complications, including a striking elevation of serum triglycerides, excess adipose tissue in the cheeks, thin extremities, short stature, excessive curvature of the lumbar spine, and delay of puberty.

The absence of glucose-6-phosphatase activity in the liver blocks the last steps of glycogenolysis and gluconeogenesis, interrupting the recycling of glucose and causing affected individuals to be hypoglycemic. The accumulation of glucose-6-phosphate in the liver leads to greatly increased glycolytic activity, with consequent elevation of lactic acid, a condition known more commonly as **lactic acidosis.** Large amounts of uric acid and lipids are produced, and the high rates of glycolysis produce excess NADH.

The treatment of von Gierke's disease consists of trying to maintain normal levels of glucose in the patient's serum. This often requires oral administration of large amounts of glucose, in its various forms, including, for example, uncooked cornstarch, which acts as a slow-release form of glucose.

---

inducing the synthesis of several important glycolytic enzymes, namely, glucokinase, phosphofructokinase, and pyruvate kinase. In addition, insulin acts to inhibit several enzymes of gluconeogenesis. These various actions enable the organism to respond quickly to increases in blood glucose levels.

Glucagon and Epinephrine Stimulate Glycogen Breakdown Catabolism of tissue glycogen is triggered by the actions of the hormones epinephrine and glucagon (Figure 22.20). *In response to decreased blood glucose,* glucagon is released from the α-cells in pancreatic islets of Langerhans. This peptide hormone travels through the blood to specific receptors on liver cell membranes. (Glucagon acts on liver and adipose tissue but not other tissues.) Similarly, signals from the central nervous system cause release of *epinephrine*—also known as adrenaline—from the adrenal glands into the bloodstream. Epinephrine acts on liver and muscles. When either hormone binds to its receptor on the outside surface of the cell membrane, a cascade is initiated that activates glycogen phosphorylase and inhibits glycogen synthase (Figure 22.20). The result of these actions is *tightly coordinated stimulation of glycogen breakdown and inhibition of glycogen synthesis.*

The Phosphorylase Cascade Amplifies the Hormonal Signal Stimulation of glycogen breakdown involves consumption of molecules of ATP at three different steps in the hormone-sensitive adenylyl cyclase cascade (see Figure 15.18). Note that the cascade mechanism is a means of chemical amplification, because the binding of just a few molecules of epinephrine or glucagon results in the synthesis of many molecules of cyclic AMP, which, through the action of cAMP-dependent protein kinase, can activate many more molecules of phosphorylase kinase and even more molecules of phosphorylase. For example, an extracellular level of 10^{-10} to 10^{-8} M epinephrine prompts the formation of 10^{-6} M cyclic AMP, and for each protein kinase activated by cyclic AMP, approximately 30 phosphorylase kinase molecules are activated; these in turn activate some 800 molecules of phosphorylase. Each of these catalyzes the formation of many molecules of glucose-1-P.

The Difference Between Epinephrine and Glucagon Although both epinephrine and glucagon exert glycogenolytic effects, they do so for quite different reasons. Epinephrine is secreted as a response to anger or fear and may be viewed as an alarm or danger signal for the organism. Called the "fight or flight" hormone, it prepares the organism for mobilization of large amounts of energy. Among the many physiological changes elicited by epinephrine, one is the initiation of the enzyme cascade, as in Figure

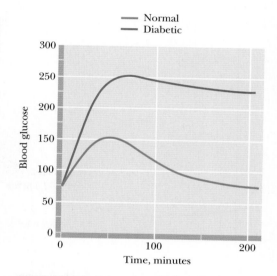

FIGURE 22.19 A glucose tolerance test involves ingestion of a glucose solution followed by measurements of blood glucose for about 3 hours. Normal subjects exhibit a rise in blood glucose to about 150 mg/dL, followed by a decline to normal values over a 3-hour period. In diabetic subjects, blood glucose rises to higher values and remains high for longer periods.

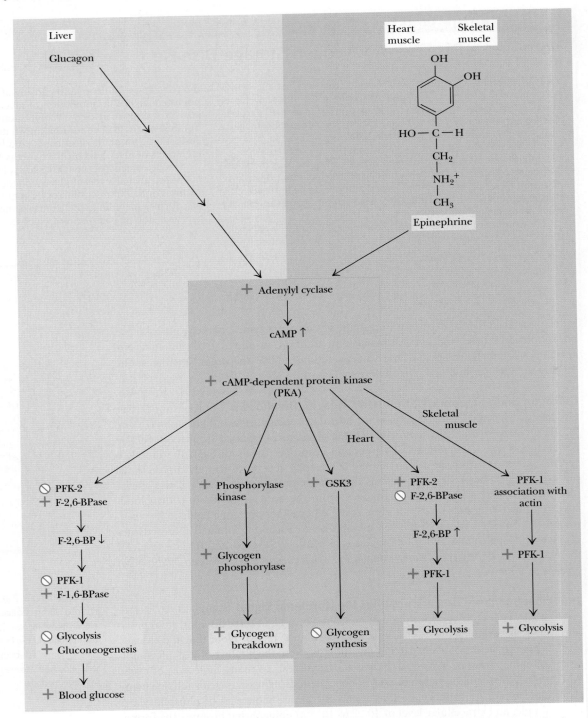

FIGURE 22.20 Glucagon and epinephrine each activate a cascade of reactions that stimulate glycogen breakdown and inhibit glycogen synthesis in liver and muscles, respectively. The effects of these hormones on other metabolic pathways depend on the tissue. In liver, glucagon inhibits glycolysis and stimulates gluconeogenesis, facilitating export of glucose into the bloodstream. In muscles, epinephrine stimulates glycolysis to provide energy for contraction. These effects all depend on protein phosphorylations by cAMP-dependent protein kinase. Note that the liver and heart isoforms of PFK-2/F-2,6-BPase respond oppositely to phosphorylation by PKA. Glucagon is a 29-residue peptide with the sequence H_3^+N-HSEGTFTSDYSKYLDSRRAQDFVQWLMNT-COO$^-$.

15.18, which leads to rapid breakdown of glycogen, inhibition of glycogen synthesis, stimulation of glycolysis, and production of energy. The burst of energy produced is the result of a 2000-fold amplification of the rate of glycolysis. Because a fear or anger response must include generation of energy (in the form of glucose)—both immediately in localized sites (the muscles) and eventually throughout the organism (as supplied by the liver)—epinephrine must be able to activate glycogenolysis in both liver and muscles.

Glucagon is involved in the long-term maintenance of steady-state levels of glucose in the blood. It performs this function by stimulating the liver to release glucose from glycogen stores into the bloodstream. To further elevate glucose levels, glucagon also

---

▶ CRITICAL DEVELOPMENTS IN BIOCHEMISTRY

O-GlcNAc Signaling and the Hexosamine Biosynthetic Pathway

O-linked β-N-acetylglucosamine (O-GlcNAc) is a post-translational modification consisting of a single N-acetylglucosamine moiety attached in O-β-glycosidic linkage to Ser and Thr residues. Modification with a single O-GlcNAc (as opposed to a complex oligosaccharide) is a signaling device similar to phosphorylation. Emerging research reveals that O-GlcNAc signaling and its crosstalk with phosphorylation modulate a wide array of cellular events. Such signaling and crosstalk are altered in diseases such as diabetes and cancer.

Uridine diphosphate N-acetylglucosamine (UDP-GlcNAc), the high-energy donor substrate for O-GlcNAcylation, is synthesized in the **hexosamine biosynthetic pathway (HBP**—see figure), a nexus of glucose, nitrogen, fatty acid, and nucleic acid metabolic pathways which consumes approximately 2–3% of total cellular glucose. The amide N of glutamine is transferred to fructose-6-P to form glucosamine-6-P by **glutamine:fructose-6-P aminotransferase (GFAT)**, the rate-limiting step in the HBP. O-GlcNac-dependent glycosylation (O-GlcNacylation) of proteins is catalyzed by **O-GlcNAc transferase (OGT)** and O-GlcNAc groups are removed from protein targets by **O-GlcNAcase (OGA).**

O-GlcNAcylation is intertwined with cellular metabolism; sugars can be attached or removed dynamically in reponse to changes in the cellular environment triggered by stress, hormones, or nutri-ents. In some cases, the same Ser or Thr residues may be either phosphorylated or modified by O-GlcNAc, provoking different cellular responses. Moreover, O-GlcNAcylated and phosphorylated residues can be in proximity to each and either can sterically impair the attachment of the other.

O-GlcNAcylation plays a direct role in insulin signaling and cancer. Cellular stimulation by insulin provokes OGT association with the insulin receptor, followed by OGT phosphorylation, and an increase in OGT activity, resulting in O-GlcNAcylation of several components of the insulin signaling pathway. In cancer cells, expression of the transcription factor c-Myc is typically enhanced. c-Myc is O-GlcNAcylated at a Thr residue that is also a phosphorylation site, and these modifications have opposing effects on the activity of c-Myc. Similar competition between O-GlcNAcylation and phosphorylation, with consequent effects on activity, are observed for several other transcription factors, oncogenes, and tumor suppressors that play essential roles in tumorigenesis and cell proliferation.

Reference

Slawson, C., Copeland, R. J., and Hart, G. W., 2010. O-GlcNAc signaling: A metabolic link between diabetes and cancer? *Trends in Biochemical Sciences* **35**:547–555.

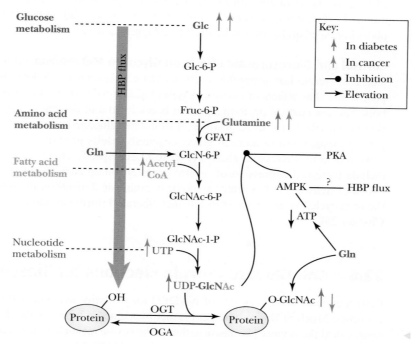

◀ Adapted from Slawson et al.

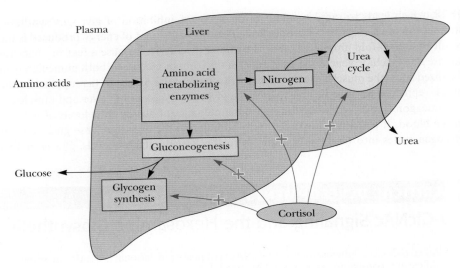

FIGURE 22.21 The effects of cortisol on carbohydrate and protein metabolism in the liver.

stimulates liver gluconeogenesis by activating F-2,6-BPase activity (see Figure 22.10). It is important to note, however, that stabilization of blood glucose levels is managed almost entirely by the liver. Glucagon does not activate the phosphorylase cascade in muscle (muscle membranes do not contain glucagon receptors). Muscle glycogen breakdown occurs only in response to epinephrine release, and muscle tissue does not participate in maintenance of steady-state glucose levels in the blood.

Glucagon and epinephrine both trigger glycogen breakdown and inhibit glycogen synthesis (in liver and muscles, respectively), but their other effects on metabolic pathways are adapted exquisitely to the needs of the tissues involved. The liver must export glucose to the bloodstream to support other tissues. Thus, in the liver, PFK-2 is phosphorylated and inhibited by protein kinase A (PKA), lowering [fructose-2,6-bisphosphate], inhibiting glycolysis, and activating gluconeogenesis (see Figure 22.20). In muscles, the glucose provided by glycogen breakdown is used immediately and locally to provide ATP energy for contraction. Therefore, glycolysis should be activated in concert with glycogen breakdown in muscles. Activation of glycolysis is accomplished in different ways, depending on the muscle. Heart muscle PFK-2 is activated upon phosphorylation at Ser^{466} and Ser^{483} by PKA in response to epinephrine, thus activating glycolysis (see Figure 22.20). Skeletal muscle PFK-2, however, is not a substrate for PKA. Instead, skeletal muscle PFK-1 is phosphorylated and activated by PKA (see Figure 22.20).

Cortisol and Glucocorticoid Effects on Glycogen Metabolism Glucocorticoids are a class of steroid hormones that exert distinct effects on liver, skeletal muscle, and adipose tissue. The effects of cortisol, a typical glucocorticoid, are best described as catabolic because cortisol promotes protein breakdown and decreases protein synthesis in skeletal muscle. In the liver, however, it stimulates gluconeogenesis and increases glycogen synthesis. Cortisol-induced gluconeogenesis results primarily from increased conversion of amino acids into glucose (Figure 22.21). Specific effects of cortisol in the liver include increased expression of several genes encoding enzymes of the gluconeogenic pathway, activation of enzymes involved in amino acid metabolism, and stimulation of the urea cycle, which disposes of nitrogen liberated during amino acid catabolism (see Chapter 25).

22.6 Can Glucose Provide Electrons for Biosynthesis?

Cells require a constant supply of NADPH for reductive reactions vital to biosynthetic purposes. Much of this requirement is met by a glucose-based metabolic sequence variously called the **pentose phosphate pathway,** the **hexose monophosphate shunt,** or the **phosphogluconate pathway.** In addition to providing NADPH for biosynthetic processes, this pathway produces *ribose-5-phosphate,* which is essential for nucleic acid synthesis. Several metabolites of the pentose phosphate pathway can also be shuttled into glycolysis.

The Pentose Phosphate Pathway Operates Mainly in Liver and Adipose Cells

The pentose phosphate pathway begins with glucose-6-phosphate, a six-carbon sugar, and produces three-, four-, five-, six-, and seven-carbon sugars (Figure 22.22). As we will see, two successive oxidations lead to the reduction of $NADP^+$ to NADPH and the release of CO_2. Five subsequent nonoxidative steps produce a variety of carbohydrates, some of which may enter the glycolytic pathway. The enzymes of the pentose phosphate pathway are particularly abundant in the cytoplasm of liver and adipose cells. These enzymes are largely absent in muscle, where glucose-6-phosphate is utilized primarily for energy production via glycolysis and the TCA cycle. These pentose phosphate pathway enzymes are located in the cytosol, which is the site of fatty acid synthesis, a pathway heavily dependent on NADPH for reductive reactions.

The Pentose Phosphate Pathway Begins with Two Oxidative Steps

1. Glucose-6-Phosphate Dehydrogenase The pentose phosphate pathway begins with the oxidation of glucose-6-phosphate. The products of the reaction are a cyclic ester (the lactone of phosphogluconic acid) and NADPH (Figure 22.23). **Glucose-6-phosphate dehydrogenase (G6PDH),** which catalyzes this reaction, is highly specific for $NADP^+$. As the first step of a major pathway, the reaction is irreversible and highly regulated. Glucose-6-phosphate dehydrogenase is strongly inhibited by the product coenzyme, NADPH, and also by fatty acid esters of coenzyme A (which are intermediates of fatty acid biosynthesis). Inhibition due to NADPH depends upon the cytosolic $NADP^+$/NADPH ratio, which in the liver is about 0.015 (compared to about 725 for the NAD^+/NADH ratio in the cytosol).

2. Gluconolactonase The gluconolactone produced in step 1 is hydrolytically unstable and readily undergoes a spontaneous ring-opening hydrolysis, although an enzyme, gluconolactonase, accelerates this reaction (Figure 22.24). The linear product, the sugar acid 6-phospho-D-gluconate, is further oxidized in step 3.

3. 6-Phosphogluconate Dehydrogenase The oxidative decarboxylation of 6-phosphogluconate by **6-phosphogluconate dehydrogenase** yields D-ribulose-5-phosphate and another equivalent of NADPH. There are two distinct steps in this reaction (Figure 22.25): The initial $NADP^+$-dependent dehydrogenation yields a β-keto acid, 3-keto-6-phosphogluconate, which is very susceptible to decarboxylation (the second step). The resulting product, D-ribulose-5-P, is the substrate for the nonoxidative reactions composing the rest of this pathway.

There Are Four Nonoxidative Reactions in the Pentose Phosphate Pathway

This portion of the pathway begins with an isomerization and an epimerization, and it leads to the formation of either D-ribose-5-phosphate or D-xylulose-5-phosphate. These intermediates can then be converted into glycolytic intermediates or directed to biosynthetic processes.

4. Phosphopentose Isomerase This enzyme interconverts ribulose-5-P and ribose-5-P via an enediol intermediate (Figure 22.26). The reaction (and mechanism) is quite similar to the phosphoglucoisomerase reaction of glycolysis, which interconverts

G6PDH is the rate-limiting step of the pentose phosphate pathway. Mian Wu and Xiaolu Wang have shown that G6PDH is directly inhibited by the tumor suppressor p53.

Gottlieb, E., 2010. p53 guards the metabolic pathway less travelled. *Nature Cell Biology* **13:**195–197.

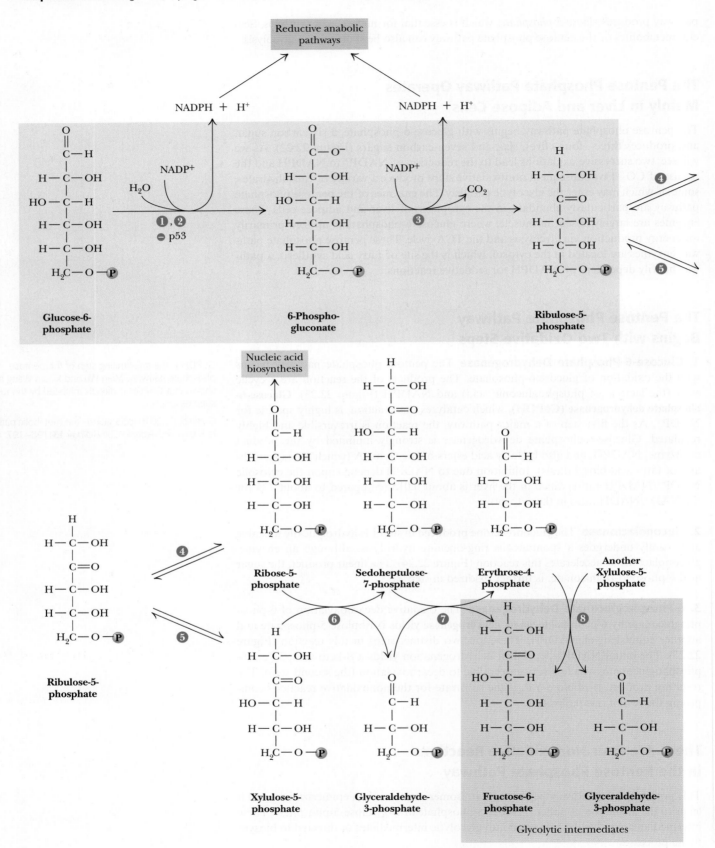

FIGURE 22.22 The pentose phosphate pathway. The numerals in the blue circles indicate the steps discussed in the text. The glucose-6-phosphate dehydrogenase reaction is inhibited by the tumor suppressor p53.

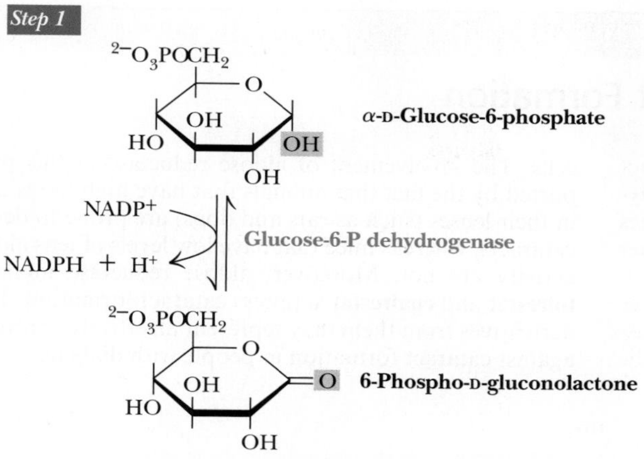

Step 1

α-D-**Glucose-6-phosphate**

NADP$^+$

NADPH + H$^+$

Glucose-6-P dehydrogenase

6-Phospho-D-gluconolactone

FIGURE 22.23 The glucose-6-phosphate dehydrogenase reaction is the committed step in the pentose phosphate pathway.

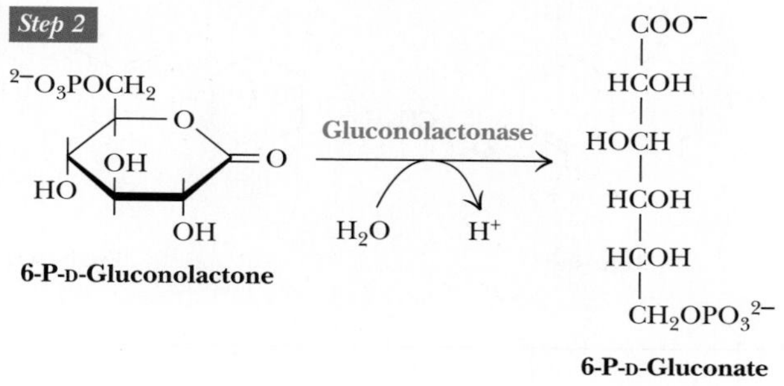

Step 2

6-P-D-Gluconolactone

Gluconolactonase

H$_2$O   H$^+$

COO$^-$
|
HCOH
|
HOCH
|
HCOH
|
HCOH
|
CH$_2$OPO$_3{}^{2-}$

6-P-D-Gluconate

FIGURE 22.24 The gluconolactonase reaction.

Step 3

COO$^-$
|
HCOH
|
HOCH
|
HCOH
|
HCOH
|
CH$_2$OPO$_3{}^{2-}$

6-P-D-Gluconate

NADP$^+$ NADPH

H$^+$
+

6-Phosphogluconate
dehydrogenase

$\Bigg[$
COO$^-$
|
HCOH
|
C=O
|
HCOH
|
HCOH
|
CH$_2$OPO$_3{}^{2-}$
$\Bigg]$

3-Keto-6-P-D-Gluconate

H$^+$ CO$_2$

CH$_2$OH
|
C=O
|
HCOH
|
HCOH
|
CH$_2$OPO$_3{}^{2-}$

D-Ribulose-5-P

FIGURE 22.25 The 6-phosphogluconate dehydrogenase reaction.

Step 4

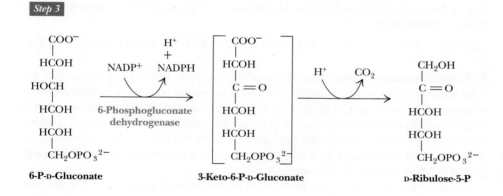

CH$_2$OH
|
C=O
|
HCOH
|
HCOH
|
CH$_2$OPO$_3{}^{2-}$

D-Ribulose-5-P (ketose)

$\Bigg[$
HC—OH
‖
C—OH
|
HCOH
|
HCOH
|
CH$_2$OPO$_3{}^{2-}$
$\Bigg]$

Enediol

HC=O
|
HCOH
|
HCOH
|
HCOH
|
CH$_2$OPO$_3{}^{2-}$

Ribose-5-P (aldose)

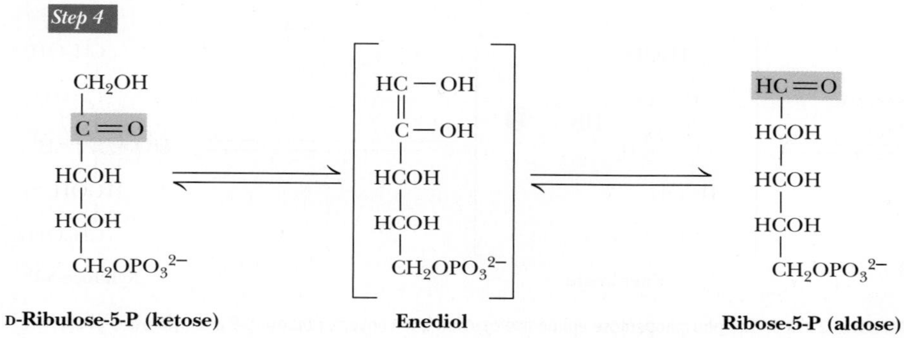

FIGURE 22.26 The phosphopentose isomerase reaction involves an enediol intermediate.

HUMAN BIOCHEMISTRY

Aldose Reductase and Diabetic Cataract Formation

The complications of diabetes include a high propensity for cataract formation in later life, both in type 1 and type 2 diabetics. Hyperglycemia is the suspected cause, but by what mechanism? Several lines of evidence point to the **polyol pathway,** in which glucose and other simple sugars are reduced in NADPH-dependent reactions. Glucose, for example, is reduced by *aldose reductase* to sorbitol (see accompanying figure), which accumulates in lens fiber cells, increasing the intracellular osmotic pressure and eventually rupturing the

cells. The involvement of aldose reductase in this process is supported by the fact that animals that have high levels of this enzyme in their lenses (such as rats and dogs) are prone to develop diabetic cataracts, whereas mice that have low levels of lens aldose reductase activity are not. Moreover, aldose reductase inhibitors such as **tolrestat** and **epalrestat** suppress cataract formation. These drugs or derivatives from them may represent an effective preventive therapy against cataract formation in people with diabetes.

(a)

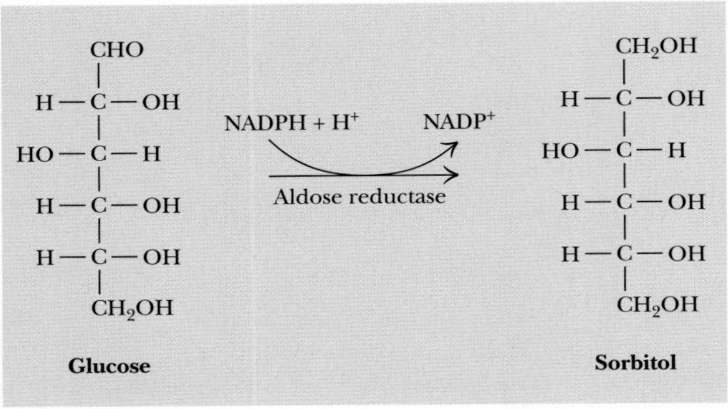

(b)

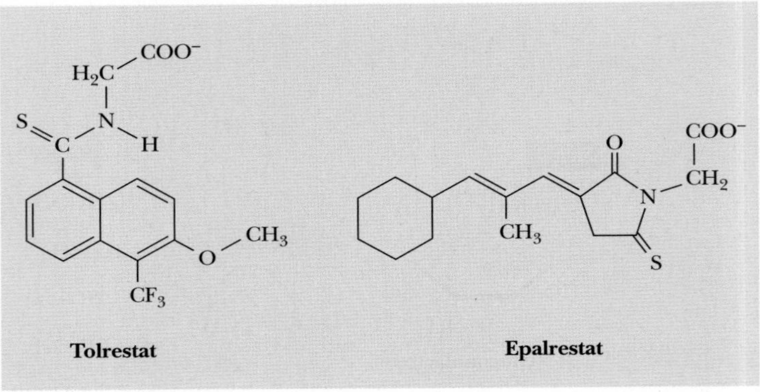

glucose-6-P and fructose-6-P. The ribose-5-P produced in this reaction is utilized in the biosynthesis of coenzymes (including NADH, NADPH, FAD, and B_{12}), nucleotides, and nucleic acids (DNA and RNA). The net reaction for the first four steps of the pentose phosphate pathway is

$$\text{Glucose-6-P} + 2 \text{ NADP}^+ + H_2O \longrightarrow \text{ribose-5-P} + 2 \text{ NADPH} + 2 \text{ H}^+ + CO_2$$

5. Phosphopentose Epimerase This reaction converts ribulose-5-P to another ketose, namely, xylulose-5-P. This reaction also proceeds by an enediol intermediate but involves an inversion at C-3 (Figure 22.27). In the reaction, an acidic proton located α- to a carbonyl carbon is removed to generate the enediolate, but the proton is added back to the same carbon from the opposite side. Note the distinction in nomenclature here. Interchange of groups on a single carbon is an epimerization, and interchange of groups between carbons is an isomerization.

To this point, the pathway has generated a pool of pentose phosphates. The $\Delta G^{\circ\prime}$ for each of the last two reactions is small, and the three pentose-5-phosphates coexist at

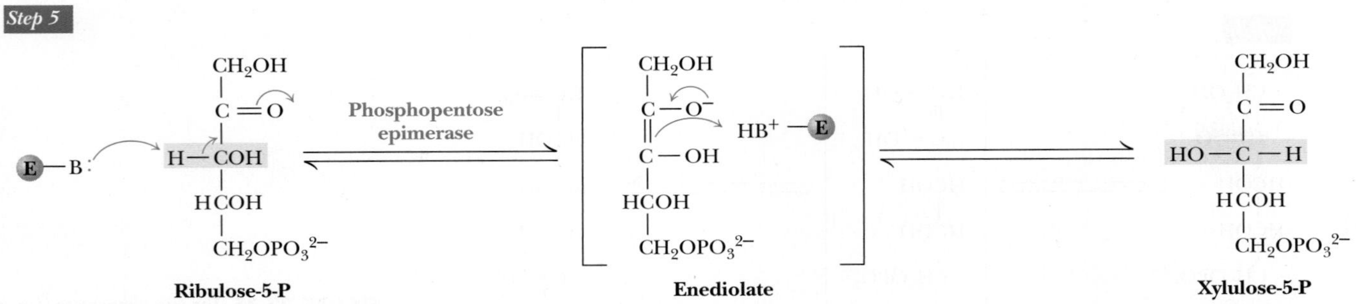

FIGURE 22.27 The phosphopentose epimerase reaction interconverts ribulose-5-P and xylulose-5-phosphate. The mechanism involves an enediol intermediate and occurs with inversion at C-3.

equilibrium. The pathway has also produced two molecules of NADPH for each glucose-6-P converted to pentose-5-phosphate. The next three steps rearrange the five-carbon skeletons of the pentoses to produce three-, four-, six-, and seven-carbon units, which can be used for various metabolic purposes. Why should the cell do this? Very often, the cellular need for NADPH is considerably greater than the need for ribose-5-phosphate. The next three steps thus return some of the five-carbon units to glyceraldehyde-3-phosphate and fructose-6-phosphate, which can enter the glycolytic pathway. The advantage of this is that the cell has met its needs for NADPH and ribose-5-phosphate in a single pathway, yet at the same time it can return the excess carbon metabolites to glycolysis.

6 and 8. Transketolase The transketolase enzyme acts at both steps 6 and 8 of the pentose phosphate pathway. In both cases, the enzyme catalyzes the transfer of two-carbon units. In these reactions (and also in step 7, the transaldolase reaction, which transfers three-carbon units), the donor molecule is a ketose and the recipient is an aldose. In step 6, xylulose-5-phosphate transfers a two-carbon unit to ribose-5-phosphate to form glyceraldehyde-3-phosphate and sedoheptulose-7-phosphate (Figure 22.28). Step 8 involves a two-carbon transfer from xylulose-5-phosphate to erythrose-4-phosphate to produce another glyceraldehyde-3-phosphate and a fructose-6-phosphate (Figure 22.29). Three of these products enter directly into the glycolytic pathway. (The sedoheptulose-7-phosphate is taken care of in step 7, as we shall see.) Transketolase is a thiamine pyrophosphate–dependent enzyme, and the mechanism (Figure 22.30) involves abstraction of the acidic thiazole proton of TPP, attack by the resulting carbanion at the carbonyl carbon of the ketose phosphate substrate, expulsion of the glyceraldehyde-3-phosphate product, and transfer of the two-carbon unit. Transketolase can process a variety of 2-keto sugar phosphates in a similar manner. It is specific for ketose substrates with the configuration shown but can accept a variety of aldose phosphate substrates.

7. Transaldolase The transaldolase functions primarily to make a useful glycolytic substrate from the sedoheptulose-7-phosphate produced by the first transketolase reaction. This reaction (Figure 22.31) is quite similar to the aldolase reaction of glycolysis, involving

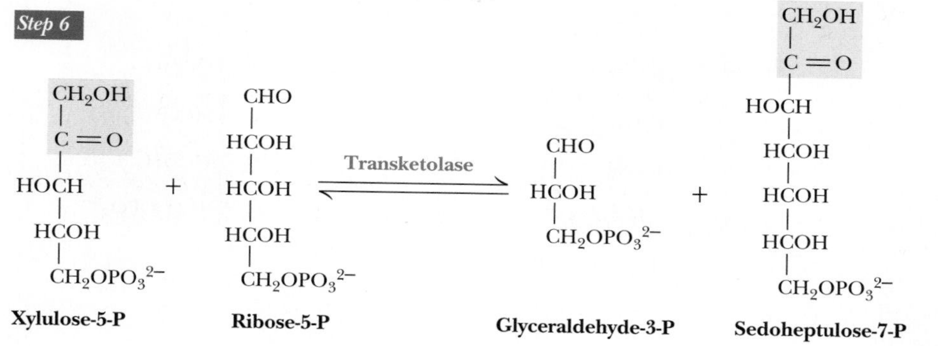

FIGURE 22.28 The transketolase reaction of step 6 in the pentose phosphate pathway.

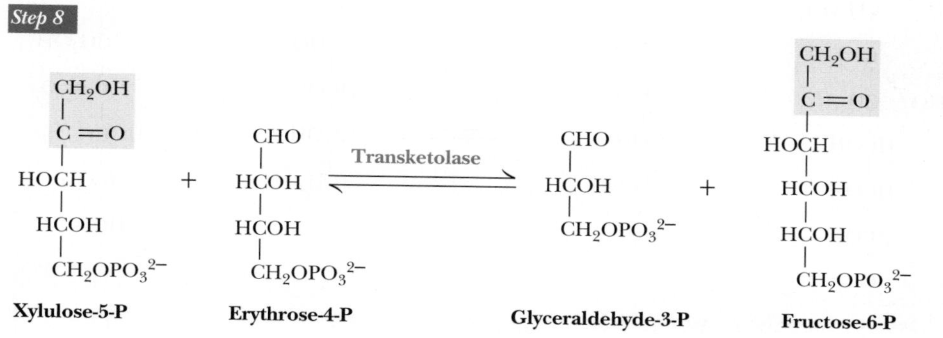

FIGURE 22.29 The transketolase reaction of step 8 in the pentose phosphate pathway.

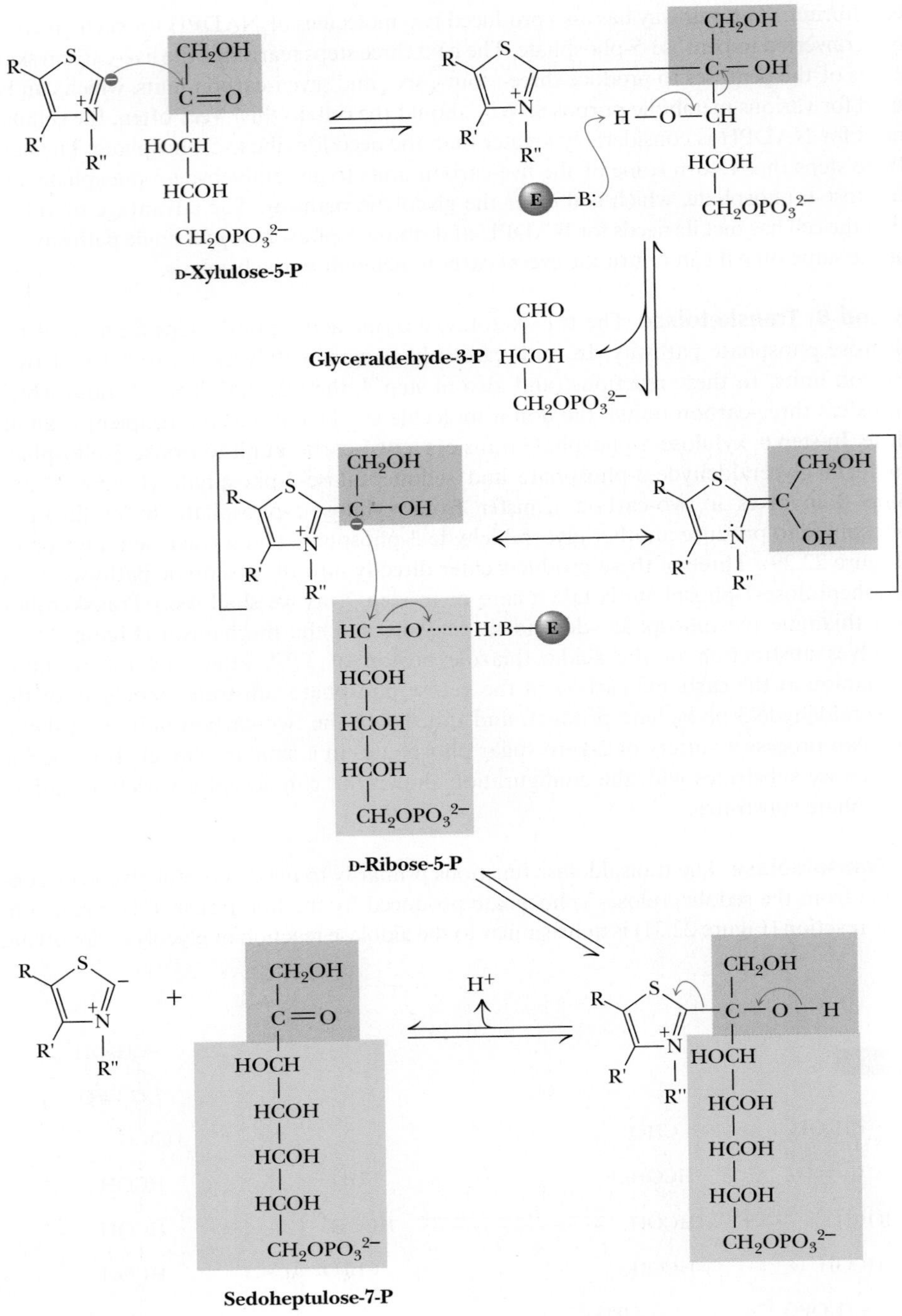

FIGURE 22.30 The mechanism of the TPP-dependent transketolase reaction. Ironically, the group transferred in the transketolase reaction might best be described as an aldol, whereas the transferred group in the transaldolase reaction is actually a ketol. Despite the irony, these names persist for historical reasons.

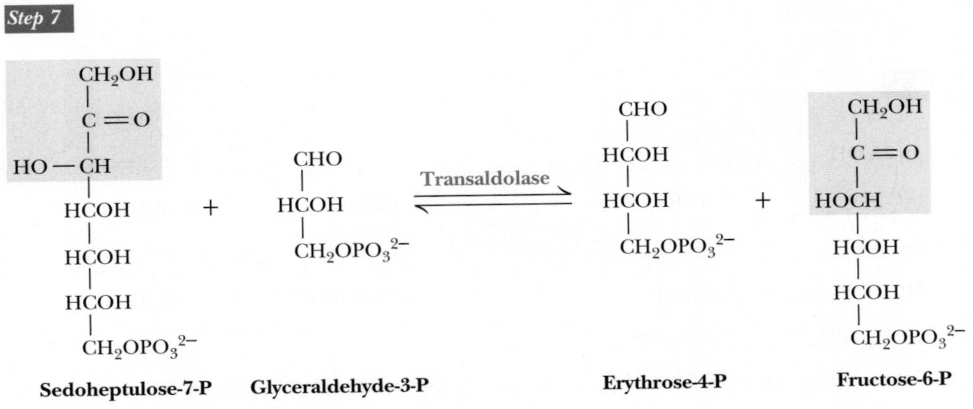

FIGURE 22.31 The transaldolase reaction.

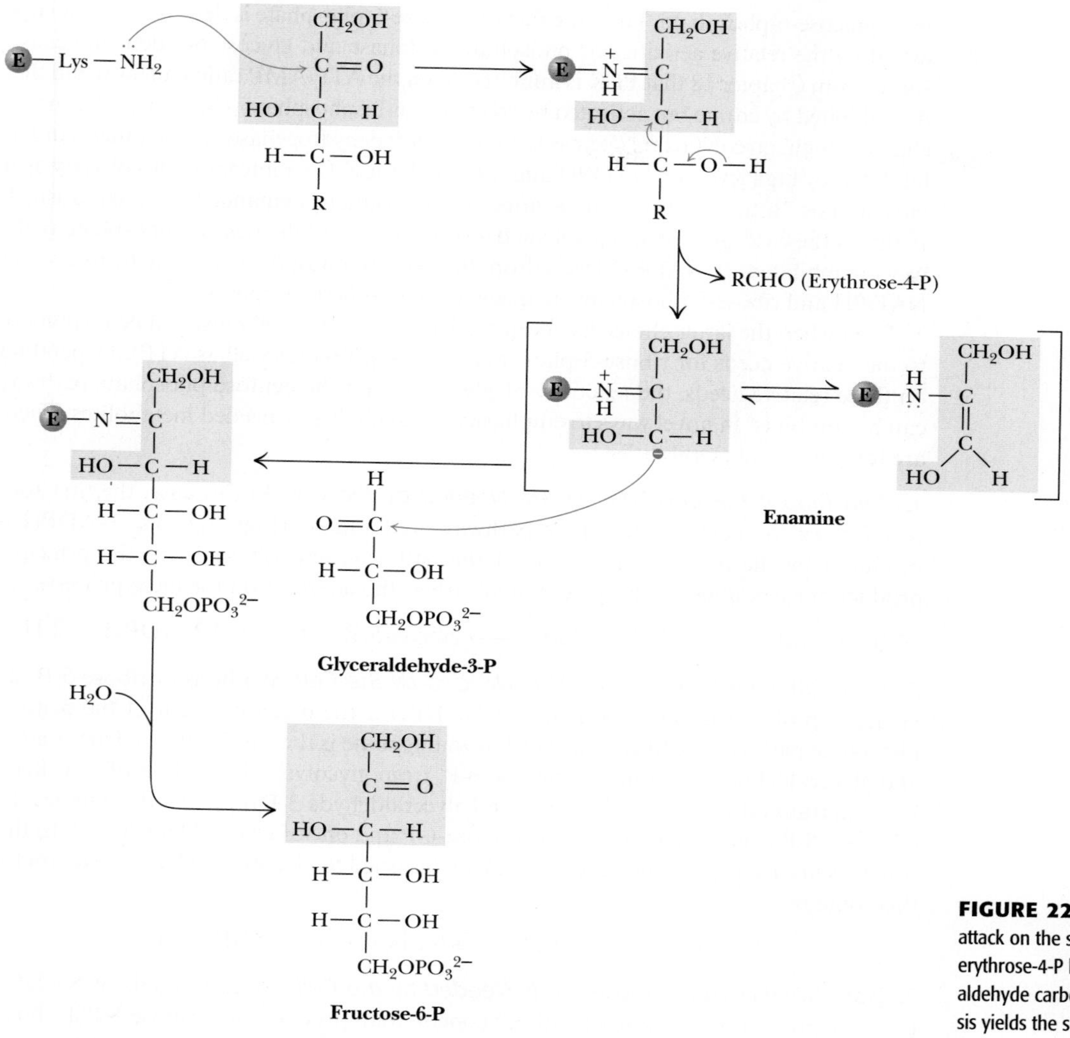

FIGURE 22.32 The transaldolase mechanism involves attack on the substrate by an active-site lysine. Departure of erythrose-4-P leaves the reactive enamine, which attacks the aldehyde carbon of glyceraldehyde-3-P. Schiff base hydrolysis yields the second product, fructose-6-P.

formation of a Schiff base intermediate between the sedoheptulose-7-phosphate and an active-site lysine residue (Figure 22.32). Elimination of the erythrose-4-phosphate product leaves an enamine of dihydroxyacetone, which remains stable at the active site (without imine hydrolysis) until the other substrate comes into position. Attack of the enamine carbanion at the carbonyl carbon of glyceraldehyde-3-phosphate is followed by hydrolysis of the Schiff base (imine) to yield the product fructose-6-phosphate.

Utilization of Glucose-6-P Depends on the Cell's Need for ATP, NADPH, and Ribose-5-P

It is clear that glucose-6-phosphate can be used as a substrate either for glycolysis or for the pentose phosphate pathway. The cell makes this choice on the basis of its relative needs for biosynthesis and for energy from metabolism. ATP can be produced in abundance if glucose-6-phosphate is channeled into glycolysis. On the other hand, if NADPH or ribose-5-phosphate is needed, glucose-6-phosphate can be directed to the pentose phosphate pathway. The molecular basis for this regulatory decision depends on the enzymes that metabolize glucose-6-phosphate in glycolysis and the pentose phosphate pathway. In glycolysis, phosphoglucoisomerase converts glucose-6-phosphate to fructose-6-phosphate, which is utilized by phosphofructokinase (a highly regulated enzyme) to produce fructose-1,6-bisphosphate. In the pentose phosphate pathway, glucose-6-phosphate dehydrogenase (also highly regulated) produces 6-phosphogluconolactone from glucose-6-phosphate. Thus, the fate of glucose-6-phosphate is determined to a large extent by the relative activities of phosphofructokinase and glucose-6-P dehydrogenase. Recall from Chapter 18 that PFK is inhibited when the ATP/AMP ratio increases and that it is inhibited by citrate but activated by fructose-2,6-bisphosphate. Thus, when the energy charge is high, glycolytic flux decreases. Glucose-6-P dehydrogenase, on the other hand, is inhibited by high levels of NADPH and also by the acyl-CoA intermediates of fatty acid biosynthesis. Both of these are indicators that biosynthetic demands have been satisfied. If that is the case, glucose-6-phosphate dehydrogenase and the pentose phosphate pathway are inhibited. If NADPH levels drop, the pentose phosphate pathway turns on and NADPH and ribose-5-phosphate are made for biosynthetic purposes.

Even when the latter choice has been made, however, the cell must still be responsive to the relative needs for ribose-5-phosphate and NADPH (as well as ATP). Depending on these relative needs, the reactions of glycolysis and the pentose phosphate pathway can be combined in novel ways to emphasize the synthesis of needed metabolites. There are four principal possibilities.

1. Both Ribose-5-P and NADPH Are Needed by the Cell In this case, the first four reactions of the pentose phosphate pathway predominate (Figure 22.33). NADPH is produced by the oxidative reactions of the pathway, and ribose-5-P is the principal product of carbon metabolism. As stated earlier, the net reaction for these processes is

$$\text{Glucose-6-P} + 2\ \text{NADP}^+ + \text{H}_2\text{O} \longrightarrow \text{ribose-5-P} + \text{CO}_2 + 2\ \text{NADPH} + 2\ \text{H}^+$$

2. More Ribose-5-P Than NADPH Is Needed by the Cell Synthesis of ribose-5-P can be accomplished without production of NADPH if the oxidative steps of the pentose phosphate pathway are bypassed. The key to this route is the withdrawal of fructose-6-P and glyceraldehyde-3-P, but not glucose-6-P, from glycolysis. The action of transketolase and transaldolase on fructose-6-P and glyceraldehyde-3-P produces three molecules of ribose-5-P from two molecules of fructose-6-P and one of glyceraldehyde-3-P. In this route, as in case 1, no carbon metabolites are returned to glycolysis. The net reaction for this route is

$$5\ \text{Glucose-6-P} + \text{ATP} \longrightarrow 6\ \text{ribose-5-P} + \text{ADP} + \text{H}^+$$

3. More NADPH Than Ribose-5-P Is Needed by the Cell Large amounts of NADPH can be supplied for biosynthesis without concomitant production of ribose-5-P if ribose-5-P produced in the pentose phosphate pathway is recycled to produce glycolytic intermediates. This alternative involves a complex interplay between the transketolase and transaldolase reactions to convert ribulose-5-P to fructose-6-P and glyceraldehyde-3-P, which can be recycled to glucose-6-P via gluconeogenesis. The net reaction for this process is

$$6\ \text{Glucose-6-P} + 12\ \text{NADP}^+ + 6\ \text{H}_2\text{O} \longrightarrow$$
$$6\ \text{ribulose-5-P} + 6\ \text{CO}_2 + 12\ \text{NADPH} + 12\ \text{H}^+$$
$$6\ \text{Ribulose-5-P} \longrightarrow 5\ \text{glucose-6-P} + \text{P}_i$$

Net: $\text{Glucose-6-P} + 12\ \text{NADP}^+ + 6\ \text{H}_2\text{O} \longrightarrow 6\ \text{CO}_2 + 12\ \text{NADPH} + 12\ \text{H}^+ + \text{P}_i$

Note that in this scheme, the six hexose sugars have been converted to six pentose sugars with release of six molecules of CO_2, and the six pentoses are reconverted to five glucose molecules.

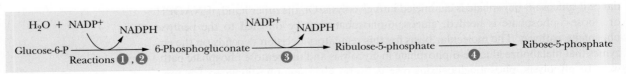

FIGURE 22.33 When biosynthetic demands dictate, the first four reactions of the pentose phosphate pathway predominate and the principal products are ribose-5-P and NADPH.

CRITICAL DEVELOPMENTS IN BIOCHEMISTRY

Integrating the Warburg Effect—ATP Consumption Promotes Cancer Metabolism

How and why do cells need to fine-tune the production of NADPH, ribose-5-phosphate, and ATP, as described on page 751–752? The Warburg effect, described in Chapter 18, is a good example of these choices. It is exhibited by cancer cells, which metabolize glucose differently than normal, differentiated cells. Normal cells generally have low rates of glycolysis, followed by oxidation of pyruvate in the mitochondria, leading to efficient generation of ATP. Rapidly proliferating cancer cells, on the other hand, use glycolytic intermediates for the synthesis of macromolecules and must balance their ATP requirements and biosynthetic needs. Moreover, cancer cells must fine-tune their metabolism in order to choose between synthesizing NADPH, ribose-5-phosphate, or both (page 751–752). Tumor cells also face another problem: If too much ATP is produced by aerobic glycolysis, feedback inhibition of phosphofructokinase may limit glucose utilization altogether.

How do tumor cells manage these needs? One piece of the puzzle has been provided by Xiaodong Wang and coworkers, who have shown that ENTPD5, a UDPase in the endoplasmic reticulum, is associated with ATP consumption in tumor cells. UMP made in this reaction is transported into the cytosol, where it is converted back to UDP, then UTP, and then to UDP-glucose. These reactions are coupled to adenylate kinase (see figure) with net production of AMP. Consumption of ATP and production of AMP stimulates phosphofructokinase, so that glycolysis proceeds to make glyceraldehyde-3-phosphate, which can, together with fructose-6-phosphate, yield ribose-5-phosphate in the pentose phosphate pathway (option 2 on page 752). Cancer cells use this strategy, together with other regulatory controls (green and red in the figure) to fine-tune the production of NADPH, ribose-5-phosphate, and ATP.

Can cancer cells afford to sacrifice so much ATP in this way? Compare the possible "yields" from glucose with the needs of a proliferating cell. One glucose can generate up to 34 ATPs, or 2 NADPH plus about 28 ATP (if shunted through the pentose phosphate pathway), or six carbons for biosynthesis. Compare these choices with the requirements for fatty acid synthesis (Chapter 24): One palmitate requires 16 carbons, 7 molecules of ATP, and 28 electrons from 14 NADPH. Thus one glucose provides nearly five times the amount of ATP required to make a palmitate, but only 1/7 of the NADPH required. Clearly, proliferating cells must devote much of their available glucose to NADPH production and biosynthesis, yet still provide sufficient ATP for cellular energy needs.

References

Fang, M., Shen, Z., Zhao, L., Chen, S., Mak, T. W., and Wang, X., 2010. The ER UDPase ENTPD5 promotes protein N-glycosylation, the Warburg effect, and proliferation in the PTEN pathway. *Cell* **143**:711–724.

Israelson, W. J., and Vander Heiden, M. G., 2010. ATP consumption promotes cancer metabolism. *Cell* **24**:669–671.

Vander Heiden, M. G., Cantley, L. C., and Thompson, C. B., 2009. Understanding the Warburg effect: The metabolic requirements of cell proliferation. *Science* **324**:1029–1033.

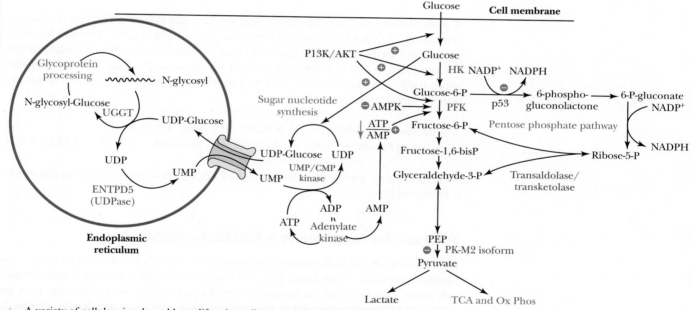

A variety of cellular signals enable proliferating cells to modulate the production of NADPH, ribose-5-phosphate, and ATP. ENTPD5 expression is enhanced in tumors and promotes ATP consumption, stimulating aerobic glycolysis. The protein kinases P13K and AKT stimulate glucose uptake, as well as hexokinase and phosphofructokinase. AMP kinase (AMPK) inhibits PFK, whereas p53 blocks synthesis of NADPH by inhibition of glucose-6-phosphate dehydrogenase.

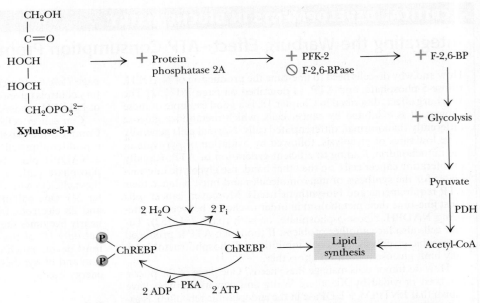

FIGURE 22.34 Xylulose-5-phosphate is a regulator of multiple metabolic pathways. Activation of PP2A triggers dephosphorylation of the bifunctional PFK-2/F2,6-BPase, which raises fructose-2,6-BP levels, in turn activating glycolysis and inhibiting gluconeogenesis. Simultaneously, ChREBP is dephosphorylated, leading to elevated expression of genes encoding enzymes for lipogenesis. These effects combine to trigger lipid biosynthesis in response to a carbohydrate-rich meal.

4. Both NADPH and ATP Are Needed by the Cell, but Ribose-5-P Is Not Under some conditions, both NADPH and ATP must be provided in the cell. This can be accomplished in a series of reactions similar to case 3 if the fructose-6-P and glyceraldehyde-3-P produced in this way proceed through glycolysis to produce ATP and pyruvate, which itself can yield even more ATP by continuing on to the TCA cycle. The net reaction for this alternative is

$$3 \text{ Glucose-6-P} + 5 \text{ NAD}^+ + 6 \text{ NADP}^+ + 8 \text{ ADP} + 5 \text{ P}_i \longrightarrow$$
$$5 \text{ pyruvate} + 3 \text{ CO}_2 + 5 \text{ NADH} + 6 \text{ NADPH} + 8 \text{ ATP} + 2 \text{ H}_2\text{O} + 8 \text{ H}^+$$

Note that, except for the three molecules of CO_2, all the other carbon from glucose-6-P is recovered in pyruvate.

Xylulose-5-Phosphate Is a Metabolic Regulator

In addition to its role in the pentose phosphate pathway, xylulose-5-phosphate serves as a signaling molecule. When blood [glucose] rises (for example, following a carbohydrate-rich meal), glycolysis and the pentose phosphate pathway are activated in the liver, and xylulose-5-phosphate produced in the latter pathway stimulates **protein phosphatase 2A (PP2A),** which dephosphorylates the bifunctional enzyme PFK-2/F-2,6-BPase (Figure 22.34, and see Figure 22.10). Increases in [fructose-2,6-bisphosphate] stimulate glycolysis and inhibit gluconeogenesis. At the same time, PP2A dephosphorylates **carbohydrate-responsive element-binding protein (ChREBP),** a transcription factor that activates expression of liver genes for lipid synthesis. These effects are a powerful combination. Increased glycolysis produces substantial amounts of acetyl-CoA, the principal substrate for lipid synthesis. The pentose phosphate pathway produces NADPH, the source of electrons for lipid biosynthesis. Elevated expression of the appropriate genes sets the stage for lipid biosynthesis in the liver, an important consequence of ingestion of carbohydrates.

SUMMARY

22.1 What Is Gluconeogenesis, and How Does It Operate? Gluconeogenesis is the generation *(genesis)* of new *(neo)* glucose. In addition to pyruvate and lactate, other noncarbohydrate precursors can be used as substrates for gluconeogenesis in animals, including most of the amino acids, as well as glycerol and all the TCA cycle intermediates. On the other hand, fatty acids are not substrates for gluconeogenesis in animals. Lysine and leucine are the only amino acids that are not substrates for gluconeogenesis. These amino acids produce only acetyl-CoA upon degradation. Acetyl-CoA can be a substrate for gluconeogenesis in plants when the glyoxylate cycle is operating. The major sites of gluconeogenesis are the liver and kidneys, which account for about 90% and 10% of the body's gluconeogenic activity, respectively.

22.2 How Is Gluconeogenesis Regulated? Glycolysis and gluconeogenesis are under reciprocal control, so glycolysis is inhibited when gluconeogenesis is active, and vice versa. When the energy status of the cell is low, glucose is rapidly degraded to produce needed energy. When the energy status is high, pyruvate and other metabolites are utilized for synthesis (and storage) of glucose. The three sites of regulation in the gluconeogenic pathway are glucose-6-phosphatase, fructose-1,6-bisphosphatase, and the pyruvate carboxylase–PEP carboxykinase pair, respectively. These are the three most appropriate sites of regulation in gluconeogenesis. Glucose-6-phosphatase is under substrate-level control by glucose-6-phosphate. Acetyl-CoA allosterically activates pyruvate carboxylase. Fructose-1,6-bisphosphatase is inhibited by AMP and activated by citrate. Fructose-2,6-bisphosphate is a powerful inhibitor of fructose-1,6-bisphosphatase.

22.3 How Are Glycogen and Starch Catabolized in Animals? Virtually 100% of digestible food is absorbed and metabolized. Digestive breakdown of starch is an unregulated process. On the other hand, tissue glycogen represents an important reservoir of potential energy, and the reactions involved in its degradation and synthesis are carefully controlled and regulated. Glycogen reserves in liver and muscle tissue are stored in the cytosol as granules exhibiting a molecular weight range from 6×10^6 to 1600×10^6. These granular aggregates contain the enzymes required to synthesize and catabolize the glycogen, as well as all the enzymes of glycolysis. The principal enzyme of glycogen catabolism is glycogen phosphorylase, a highly regulated enzyme. The glycogen phosphorylase reaction involves phosphorolysis at a nonreducing end of a glycogen polymer.

22.4 How Is Glycogen Synthesized? Luis Leloir, a biochemist in Argentina, showed in the 1950s that glycogen synthesis depended upon sugar nucleotides. The glycogen polymer is built around a tiny protein core. The first glucose residue is covalently joined to the protein glycogenin via an acetal linkage to a tyrosine–OH group on the protein. Sugar units are added to the glycogen polymer by the action of glycogen synthase. The reaction involves transfer of a glucosyl unit from UDP–glucose to the C-4 hydroxyl group at a nonreducing end of a glycogen strand. The mechanism proceeds by cleavage of the C—O bond between the glucose moiety and the β-phosphate of UDP–glucose, leaving an oxonium ion intermediate, which is rapidly attacked by the C-4 hydroxyl oxygen of a terminal glucose unit on glycogen.

22.5 How Is Glycogen Metabolism Controlled? Activation of glycogen phosphorylase is tightly linked to inhibition of glycogen synthase, and vice versa. Regulation involves both allosteric control and covalent modification, with the latter being under hormonal control. Glycogen synthase is also regulated by covalent modification. Storage and utilization of tissue glycogen are regulated by hormones, including insulin, glucagon, epinephrine, and the glucocorticoids. Insulin stimulates glycogen synthesis and inhibits glycogen breakdown in liver and muscle, whereas glucagon and epinephrine stimulate glycogen breakdown.

22.6 Can Glucose Provide Electrons for Biosynthesis? The pentose phosphate pathway is a collection of eight reactions that provide NADPH for biosynthetic processes and ribose-5-phosphate for nucleic acid synthesis. Several metabolites of the pentose phosphate pathway can also be shuttled into glycolysis. Utilization of glucose-6-P in the pentose phosphate pathway depends on the cell's need for ATP, NADPH, and ribose-5-P.

FOUNDATIONAL BIOCHEMISTRY Things You Should Know After Reading Chapter 22.

- The essential features of gluconeogenesis, including its substrates and its localization within tissues of the body.

- The reactions of gluconeogenesis that are different from glycolysis and how they drive the pathway.

- The function, mechanism, and compartmentation of pyruvate carboxylase.

- The essential features of PEP carboxykinase and fructose-1,6-bisphosphatase.

- The function, locus, and mechanism of action of glucose-6-phosphatase.

- The loci and interactions of glucose transporters in the plasma membrane and endoplasmic reticulum.

- The mechanism of action of gluconeogenesis inhibitors in diabetes therapy.

- The manner in which lactate is recycled from muscle to liver and the effects of this shuttle on the $[NAD^+]/[NADH]$ ratio in both tissues.

- The essential features of regulation of gluconeogenesis.

- The action of PFK-2 and the allosteric effector fructose-2,6-bisphosphate.

- The processes by which glycogen is synthesized and broken down.

- The essential features of regulation of glycogen metabolism.

- The systemic effects of glucagon and epinephrine.

- The essential features of the pentose phosphate pathway.

- The differences between the oxidative and nonoxidative reactions of the pentose phosphate pathway.

- The mechanisms of the transaldolase and transketolase reactions.

- The means by which the pentose phosphate pathway can be used to satisfy the cell's needs for ATP, NADPH, and ribose-5-phosphate.

- The function of xylulose-5-phosphate as a metabolic regulator.

- The manner in which ectonucleoside triphosphate diphosphohydrolase 5 (ENTPD5) modulates ATP consumption and favors aerobic glycolysis in cancer cells.

- The manner in which glucose flux through the hexosamine biosynthetic pathway regulates growth factor receptor glycosylation.

PROBLEMS

Answers to all problems are at the end of this book. Detailed solutions are available in the *Student Solutions Manual, Study Guide, and Problems Book.*

1. **Assessing the Overall Equation for Gluconeogenesis** Consider the balanced equation for gluconeogenesis in Section 22.1. Account for each of the components of this equation and the indicated stoichiometry.

2. **Assessing the Energetics of Gluconeogenesis** (Integrates with Chapters 3 and 18.) Calculate $\Delta G°'$ and ΔG for gluconeogenesis in the erythrocyte, using data on page 724 and in Tables 3.3, 18.1, and 18.2 (assume NAD⁺/NADH = 20, [GTP] = [ATP], and [GDP] = [ADP]). See how closely your values match those in Section 22.1.

3. **Understanding the Regulation of the Fructose-1,6-bisphosphatase Reaction** Use the data of Figure 22.9 to calculate the percent inhibition of fructose-1,6-bisphosphatase by 25 mM fructose-2,6-bisphosphate when fructose-1,6-bisphosphate is (a) 25 mM and (b) 100 mM.

4. **Understanding the Energetics of the Glycogen Synthase Reaction** (Integrates with Chapter 3.) Suggest an explanation for the exergonic nature of the glycogen synthase reaction ($\Delta G°'$ = -13.3 kJ/mol). Consult Chapter 3 to review the energetics of high-energy phosphate compounds if necessary.

5. **Determining the Rate of Energy Consumption by Muscles During Exercise** Using the values in Table 23.1 for body glycogen content and the data in part b of the illustration for A Deeper Look (page 740), calculate the rate of energy consumption by muscles in heavy exercise (in J/sec). Use the data for fast-twitch muscle.

6. **Assessing the Action of Sodium Borohydride on the Pentose Phosphate Pathway** Which reactions of the pentose phosphate pathway would be inhibited by NaBH₄? Why?

7. **Determining the Extent of Branching in Glycogen** (Integrates with Chapter 7.) Imagine a glycogen molecule with 8000 glucose residues. If branches occur every eight residues, how many reducing ends does the molecule have? If branches occur every 12 residues, how many reducing ends does it have? How many nonreducing ends does it have in each of these cases?

8. **Describing the Regulation of Gluconeogenesis and Glycogen Metabolism** Explain the effects of each of the following on the rates of gluconeogenesis and glycogen metabolism:
   a. Increasing the concentration of tissue fructose-1,6-bisphosphate
   b. Increasing the concentration of blood glucose
   c. Increasing the concentration of blood insulin
   d. Increasing the amount of blood glucagon
   e. Decreasing levels of tissue ATP
   f. Increasing the concentration of tissue AMP
   g. Decreasing the concentration of fructose-6-phosphate

9. **Assessing the Energetics of the Glycogen Phosphorylase Reaction** (Integrates with Chapters 3 and 15.) The free energy change of the glycogen phosphorylase reaction is $\Delta G°'$ = $+3.1$ kJ/mol. If $[P_i]$ = 1 mM, what is the concentration of glucose-1-P when this reaction is at equilibrium?

10. **Understanding Enzyme Mechanisms Related to Pyruvate Carboxylase** Based on the mechanism for pyruvate carboxylase (Figure 22.3), write reasonable mechanisms for the reactions that follow:

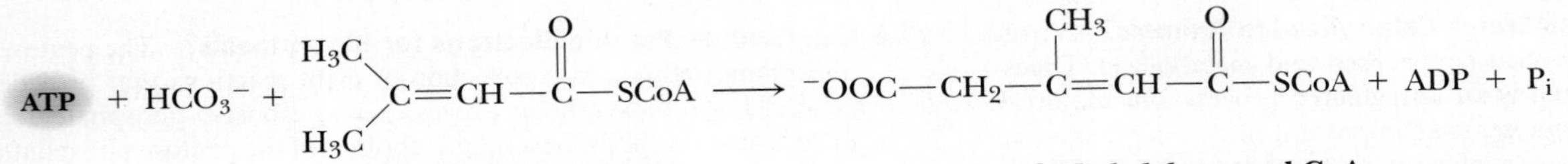

β-Methylcrotonyl-CoA   **β-Methylglutaconyl-CoA**

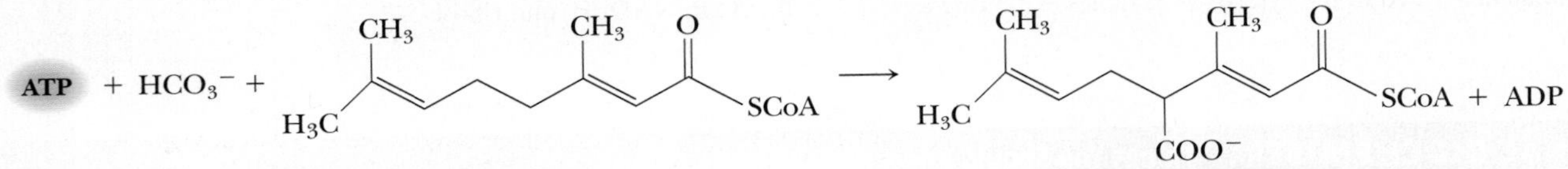

Geranyl-CoA   **γ-Carboxygeranyl-CoA**

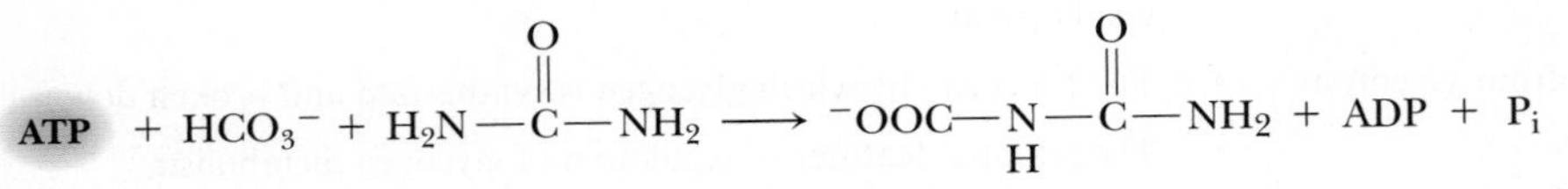

Urea   **N-Carboxyurea**

Transcarboxylase

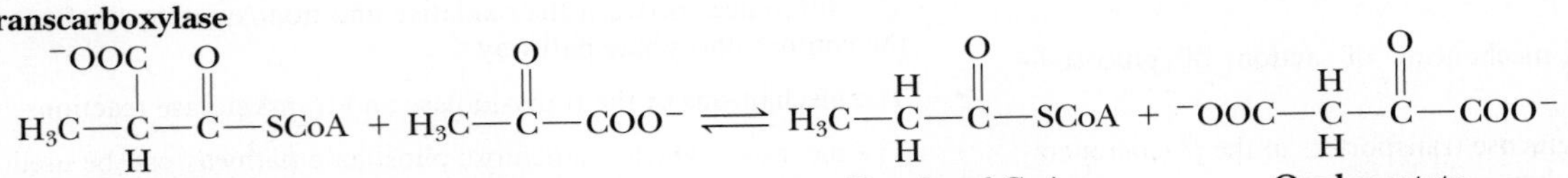

Methylmalonyl-CoA   **Pyruvate**   **Propionyl-CoA**   **Oxaloacetate**

11. Understanding the Mechanisms of Reactions Related to Transketolase The mechanistic chemistry of the acetolactate synthase and phosphoketolase reactions (shown here) is similar to that of the transketolase reaction (Figure 22.30). Write suitable mechanisms for these reactions.

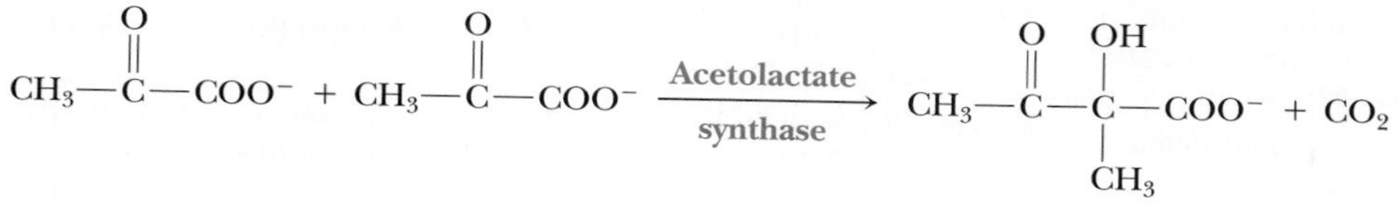

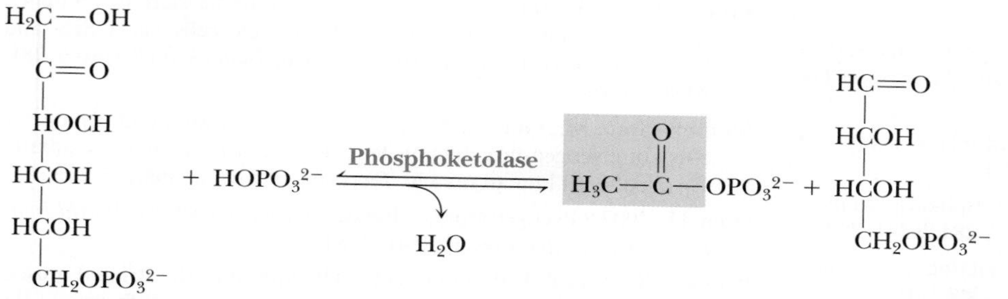

$$CH_3-\overset{\overset{\displaystyle O}{\|}}{C}-COO^- + CH_3-\overset{\overset{\displaystyle O}{\|}}{C}-COO^- \xrightarrow[\text{synthase}]{\text{Acetolactate}} CH_3-\overset{\overset{\displaystyle O}{\|}}{C}-\overset{\overset{\displaystyle OH}{|}}{\underset{\underset{\displaystyle CH_3}{|}}{C}}-COO^- + CO_2$$

Fructose-6-P   Acetyl-P   Erythrose-4-P

12. Understanding the Mechanisms of Action of a Popular Diabetes Medication Metaglip is a prescribed preparation (from Bristol-Myers Squibb) for treatment of type 2 diabetes. It consists of metformin (see Human Biochemistry, page 726) together with glipizide. The actions of metformin and glipizide are said to be complementary. Suggest a mechanism for the action of glipizide.

13. Assessing the Mechanism of Action of Aldose Reductase Inhibitors Study the structures of tolrestat and epalrestat in the Human Biochemistry box on page 748, and suggest a mechanism of action for these inhibitors of aldose reductase.

14. Understanding a Version of the Pentose Phosphate Pathway That Produces Mainly Ribose-5-P Based on the discussion on page 752, draw a diagram to show how several steps in the pentose phosphate pathway can be bypassed to produce large amounts of ribose-5-phosphate. Begin your diagram with fructose-6-phosphate.

15. Assessing the Labeling Patterns in Versions of the Pentose Phosphate Pathway Consider the diagram you constructed in problem 14. Which carbon atoms in ribose-5-phosphate are derived from carbon atoms in positions 1, 3, and 6 of fructose-6-phosphate?

16. Understanding a Version of the Pentose Phosphate Pathway That Produces Mainly NADPH As described on pages 752 to 754, the pentose phosphate pathway may be used to produce large amounts of NADPH without significant net production of ribose-5-phosphate. Draw a diagram, beginning with glucose-6-phosphate, to show how this may be accomplished.

17. Understanding a Version of the Pentose Phosphate Pathway That Produces Mainly NADPH and ATP The discussion on page 754 explains that the pentose phosphate pathway and the glycolytic pathway can be combined to provide both NADPH and ATP (as well as some NADH) without net ribose-5-phosphate synthesis. Draw a diagram to show how this may be accomplished.

18. Assessing the Labeling Patterns in Versions of the Pentose Phosphate Pathway Consider the pathway diagram you constructed in problem 17. What is the fate of carbon from positions 2 and 4 of glucose-6-phosphate after one pass through the pathway?

19. Understanding the Initial Reaction in Formation of Glycogen Glycogenin catalyzes the first reaction in the synthesis of a glycogen particle, with Tyr[194] of glycogenin (page 732) combining with a glucose unit (provided by UDP-glucose) to produce a tyrosyl glucose. Write a mechanism to show how this reaction could occur.

Preparing for the MCAT® Exam

20. Assessing the Biochemistry of Exercise Study the graphs in the Deeper Look box (page 740), and explain the timing of the provision of energy from different metabolic sources during periods of heavy exercise.

21. Understanding the Chemistry and Energetics of Creatine Phosphate (Integrates with Chapters 3 and 14.) What is the structure of creatine phosphate? Write reactions to indicate how it stores and provides energy for exercise.

ActiveModels Problems

22. Examine the ActiveModel for fructose-1,6-bisphosphatase, then describe the role of Arg-22 in the conformational change between the R and T states of this enzyme.

23. Using the ActiveModel for aldose reductase, describe the structure of the TIM barrel motif and the structure and location of the active site.

FURTHER READING

ENTPD5, ATP Consumption, and Cancer Metabolism

Fang, M., Shen, Z., Huang, S., Zhao, L., et al., 2010. The ER UDPase ENTPD5 promotes protein N-glycosylation, the Warburg effect, and proliferation in the PTEN pathway. *Cell* **143:**711–724.

Israelsen, W. J., and Vander Heiden, M. G., ATP consumption promotes cancer metabolism. *Cell* **143:**669–671.

Määttänen, P., Gehring, K., Bergeron, J. J. M., and Thomas, D. Y., 2010. Protein quality control in the ER: The recognition of misfolded proteins. *Seminars in Cell and Developmental Biology* **21:**500–511.

Schrag, J. D., Procopio, D. O., Cygler, M., Thomas, D. Y., and Bergeron, J. J. M., 2003. Lectin control of protein folding and sorting in the secretory pathway. *Trends in Biochemical Sciences* **28:**49–57.

Exercise Physiology

Gupta, A., and Gupta, V., 2010. Metabolic syndrome: What are the risks for humans? *BioScience Trends* **4**:204–212.

Horton, E. S., and Terjung, R. L., 1988. *Exercise, Nutrition and Energy Metabolism.* New York: Macmillan.

LeBrasseur, N. K., Walsh, K., and Arany, Z., 2011. Metabolic benefits of resistance training and fast glycolytic skeletal muscle. *American Journal of Physiology, Endocrinology and Metabolism* **300**:E3–E10.

Rhoades, R., and Pflanzer, R., 1992. *Human Physiology.* Philadelphia: Saunders College Publishing.

Gluconeogenesis

Dzugaj, A., 2006. Localization and regulation of muscle fructose-1,6-bisphosphatase, the key enzyme of glyconeogenesis. *Advances in Enzyme Regulation* **46**:51–71.

Hers, H.-G., and Hue, L., 1983. Gluconeogenesis and related aspects of glycolysis. *Annual Review of Biochemistry* **52**:617–653.

Hutton, J. C., and O'Brien, R. M., 2009. Glucose-6-phosphatase catalytic subunit gene family. *Journal of Biological Chemistry* **284**:29241–29245.

Jitrapakdee, S., and Wallace, J. C., 1999. Structure, function, and regulation of pyruvate carboxylase. *Biochemical Journal* **348**:1–16.

Jitrapakdee, S., Maurice, M. S., Rayment, I., Cleland, W.W., et al., 2008. Structure, mechanism and regulation of pyruvate carboxylase. *Biochemical Journal* **413**:369–387.

Kondo, S., Nakajima, Y., et al., 2007. Structure of the biotin carboxylase domain of pyruvate carboxylase from *Bacillus thermodenitrificans.* *Acta Crystallographica D* **63**:885–890.

Miller, R. A., and Birnbaum, M. J., 2010. An energetic tale of AMPK-independent effects of metformin. *The Journal of Clinical Investigation* **120**:2267–2270.

Nuttall, F. Q., Ngo, A., and Gannon, M. C., 2008. Regulation of hepatic glucose production and the rate of gluconeogenesis in humans: Is the rate of gluconeogenesis contant? *Diabetes/Metabolism Research and Reviews* **24**:438–458.

Previs, S. F., Brunengraber, D. Z., and Brunengraber, H., 2009. Is there glucose production outside of the liver and kidney? *Annual Review of Biochemistry* **29**:43–57.

Glucose Homeostasis

Dalsgaard, M. K., and Secher, N. H., 2007. The brain at work: A cerebral metabolic manifestation of central fatigue? *Journal of Neuroscience Research* **85**:3334–3339.

Feinman, R. D., and Fine, E. J., 2007. Nonequilibrium thermodynamics and energy efficiency in weight loss diets. *Theoretical Biology and Medical Modelling* **4**:1–13.

Heppner, K. M., Tong, J., Kirchner, H., Nass, R., and Tschop, M. H., 2011. The ghrelin O-acyltransferase-ghrelin system: A novel regulator of metabolism. *Current Opinions in Endocrinology, Diabetes, and Obesity* **18**:50–55.

Huang, S., and Czech, M. P., 2007. The GLUT4 glucose transporter. *Cell Metabolism* **5**:237–252.

Peters, A., 2011. The selfish brain: Competition for energy resources. *American Journal of Human Biology* **23**:29–34.

Glycogen Metabolism

Chou, J. Y., Jun, H. S., and Mansfield, B. C., 2011. Glycogen storage disease type I and G6Pase-β deficiency: Etiology and therapy. *Nature Reviews of Endocrinology* **6**:676–688.

Foster, J. D., and Nordlie, R. C., 2002. The biochemistry and molecular biology of the glucose-6-phosphatase system. *Experimental Biology and Medicine* **227**:601–608.

Horcajada, C., Guinovart, J. J., et al., 2006. Crystal structure of an archaeal glycogen synthase. *Journal of Biological Chemistry* **281**:2923–2931.

Hurley, T. D., Stout, S., et al., 2005. Requirements for catalysis in mammalian glycogenin. *Journal of Biological Chemistry* **280**:23892–23899.

Jope, R. S., Yuskaitis, C., et al., 2007. Glycogen synthase kinase-3 (GSK3): Inflammation, diseases, and therapeutics. *Neurochemical Research* **32**:577–595.

Kerkela, R., Woulfe, K., et al., 2007. Glycogen synthase kinase-3β: Actively inhibiting hypertrophy. *Trends in Cardiovascular Medicine* **17**: 91–96.

Kotova, O., Al-Khalili, L., et al., 2006. Cardiotonic steroids stimulate glycogen synthesis in human skeletal muscle cells via a Src- and ERK1/2-dependent mechanism. *Journal of Biological Chemistry* **281**: 20085–20094.

Montori-Grau, M., Guitart, M., et al., 2007. Expression and glycogenic effect of glycogen-targeting protein phosphatase 1 regulatory subunit G_L in cultured human muscle. *Biochemical Journal* **405**:107–113.

Ozen, H., 2007. Glycogen storage diseases: New perspectives. *World Journal of Gastroenterology* **13**:2541–2553.

Paterson, J., Kelsall, I. R., et al., 2008. Disruption of the striated muscle glycogen-targeting subunit of protein phosphatase 1: Influence of the genetic background. *Journal of Molecular Endocrinology* **40**: 47–59.

Saeed, Y. A., and Barger, S. W., 2007. Glycogen synthase kinase-3 in neurodegeneration and neuroprotection: Lessons from lithium. *Current Alzheimer Research* **4**:21–31.

O-GlcNAcylation

Hart, G. W., Slawson, C., Ramirez-Correa, G., and Lagerlof, O., 2011. Cross talk between O-GlcNAcylation and phosphorylation: Roles in signaling, transcription, and chronic disease. *Annual Review of Biochemistry* **80**:6.1–6.34.

Slawson, C., Copeland, R. J., and Hart, G. W., 2010. O-GlcNAc signaling: A metabolic link between diabetes and cancer? *Trends in Biochemical Sciences* **35**:547–555.

Stanley, P., and Okajima, T., 2010. Roles of glycosylation in Notch signaling. *Current Topics in Developmental Biology* **92**:131–164.

Regulation of Gluconeogenesis

Alves, G., and Sola-Penna, M., 2003. Epinephrine modulates cellular distribution of muscle phosphofructokinase. *Molecular Genetics and Metabolism* **78**:302–306.

Arden, C., Hampson, L., et al., 2008. A role for PFK-2/FBPase-2, as distinct from fructose-2,6-bisphosphate, in regulation of insulin secretion in pancreatic β-cells. *Biochemical Journal* **411**:41–51.

Moller, D. E., 2001. New drug targets for type 2 diabetes and the metabolic syndrome. *Nature* **414**:821–827.

Newsholme, E. A., and Leech, A. R., 1983. *Biochemistry for the Medical Sciences.* New York: John Wiley and Sons.

Newsholme, E. A., and Leech, A. R., 1983b. Substrate cycles: Their role in improving sensitivity in metabolic control. *Trends in Biochemical Sciences* **9**:277–280.

Rider, M., Bertrand, L., et al., 2004. 6-Phosphofructo-2-kinase/fructose-2,6-bisphosphatase: Head-to-head with a bifunctional enzyme that controls glycolysis. *Biochemical Journal* **381**:561–579.

Van Schaftingen, E., and Hers, H. G., 1981. Inhibition of fructose-1,6-bisphosphatase by fructose-2,6-bisphosphate. *Proceedings of the National Academy of Sciences U.S.A.* **78**:2861–2863.

Yabaluri, N., and Bashyam, M. D., 2010. Hormonal regulation of gluconeogenic gene transcription in the liver. *Journal of Bioscience* **35**:473–484.

Signaling and Carbohydrate Metabolism

Beckerman, R., and Prives, C., 2010. Transcriptional regulation by p53. *Cold Spring Harbor Perspectives in Biology* **2**:1–18.

Bensaad, K., and Vousden, K. H., 2007. p53: New roles in metabolism. *Trends in Cell Biology* **17**:286–291.

Gottlieb, E., 2010. p53 guards the metabolic pathway less travelled. *Nature Cell Biology* **13**:195–197.

Habegger, K. M., Heppner, K. M., Geary, N., Bartness, T. J., et al., 2010. The metabolic actions of glucagon revisited. *Nature Reviews of Endocrinology* **6**:689–697.

Heppner, K. M., Habegger, K. M., Day, J., Pfluger, P. T., et al., 2010. Glucagon regulation of energy metabolism. *Physiology and Behavior* **100**:545–548.

Lee, J. T., and Gu, W., 2010. The multiple levels of regulation by p53 ubiquitination. *Cell Death and Differentiation* **17**:86–92.

Li, H., and Jogl, G., 2009. Structural and biochemical studies of TIGAR (*TP53*-induced glycolysis and apoptosis regulator). *Journal of Biological Chemistry* **284**:1748–1754.

Metallo, C. M., and Vander Heiden, M. G., 2010. Metabolism strikes back: Metabolic flux regulates cell signaling. *Genes and Development* **24**:2717–2722.

Vander Heiden, M. G., Cantley, L. C., and Thompson, C. B., 2009. Understanding the Warburg effect: The metabolic requirements of cell proliferation. *Science* **324**:1029–1033.

Zhu, Y., and Prives, C., 2009. p53 and metabolism: The GAMT connection. *Molecular Cell* **36**:351–352.

Fatty Acid Catabolism

◄ The hummingbird's tremendous capacity to store and use fatty acids enables it to make migratory journeys of remarkable distances.

....

The fat is in the fire.

John Heywood
Proverbs (1497–1580)

KEY QUESTIONS

23.1 How Are Fats Mobilized from Dietary Intake and Adipose Tissue?

23.2 How Are Fatty Acids Broken Down?

23.3 How Are Odd-Carbon Fatty Acids Oxidized?

23.4 How Are Unsaturated Fatty Acids Oxidized?

23.5 Are There Other Ways to Oxidize Fatty Acids?

23.6 What Are Ketone Bodies, and What Role Do They Play in Metabolism?

ESSENTIAL QUESTIONS

Fatty acids represent the principal form of stored energy for many organisms. There are two important advantages to storing energy in the form of fatty acids. (1) The carbon in fatty acids (mostly —CH_2— groups) is almost completely reduced compared to the carbon in other simple biomolecules (sugars, amino acids). Therefore, oxidation of fatty acids will yield more energy (in the form of ATP) than any other form of carbon. (2) Fatty acids are not generally as hydrated as monosaccharides and polysaccharides are, and thus they can pack more closely in storage tissues.

How are fatty acids catabolized, and how is their inherent energy captured by organisms?

23.1 How Are Fats Mobilized from Dietary Intake and Adipose Tissue?

Modern Diets Are Often High in Fat

Fatty acids are acquired readily in the diet and can also be made from carbohydrates and the carbon skeletons of amino acids. Fatty acids provide 30% to 60% of the calories in the diets of most Americans. For our caveman and cavewoman ancestors, the figure was probably closer to 20%. Dairy products were apparently not part of their diet, and the meat they consumed (from fast-moving animals) was low in fat. In contrast, modern domesticated cows and pigs are actually bred for high fat content (and better taste). However, woolly mammoth burgers and saber-toothed tiger steaks are hard to find these days—even in the gourmet sections of grocery stores—and so, by default, we consume (and metabolize) large quantities of fatty acids.

Keren Su/Photodisc/Jupiter Images

ÖWL Online homework and a Student Self Assessment for this chapter may be assigned in OWL.

Triacylglycerols Are a Major Form of Stored Energy in Animals

Although some of the fat in our diets is in the form of phospholipids, triacylglycerols are a major source of fatty acids. Triacylglycerols are also our principal stored energy reserve. As shown in Table 23.1, the energy available in stores of fat in the average person far exceeds the energy available from protein, glycogen, and glucose. Overall, fat accounts for approximately 83% of available energy, partly because more fat is stored than protein and carbohydrate and partly because of the substantially higher energy yield per gram for fat compared with protein and carbohydrate. Complete combustion of fat yields about 37 kJ/g, compared with about 16 to 17 kJ/g for sugars, glycogen, and amino acids. In animals, fat is stored mainly as triacylglycerols in specialized cells called **adipocytes** or **adipose cells**. As shown in Figure 23.1, triacylglycerols, aggregated to form large globules, occupy most of the volume of adipose cells. Much smaller amounts of triacylglycerols are stored as small, aggregated globules in muscle tissue.

Hormones Trigger the Release of Fatty Acids from Adipose Tissue

The pathways for liberation of fatty acids from triacylglycerols, either from adipose cells or from the diet, are shown in Figures 23.2 and 23.3. Fatty acids are mobilized from adipocytes in response to hormone messengers such as adrenaline, glucagon, and adrenocorticotropic hormone (ACTH). These signal molecules bind to receptors on the plasma membrane of adipose cells and lead to the activation of adenylyl cyclase, which forms cyclic AMP from ATP. (Second messengers and hormonal signaling are discussed in Chapter 32.) In adipose cells, cAMP activates protein kinase A, which phosphorylates and activates a **triacylglycerol lipase** (also termed **hormone-sensitive lipase**) that hydrolyzes a fatty acid from C-1 or C-3 of triacylglycerols. Subsequent actions of **diacylglycerol lipase** and **monoacylglycerol lipase** yield fatty acids and glycerol. The cell then releases the fatty acids into the blood, where they are bound to **serum albumin** (the most abundant protein in blood serum). Serum albumin transports free fatty acids to sites of utilization.

Degradation of Dietary Triacylglycerols Occurs Primarily in the Duodenum

Dietary triacylglycerols are degraded to a small extent (via fatty acid release) by lipases in the low-pH environment of the stomach, but mostly they pass untouched into the duodenum. Alkaline pancreatic juice secreted into the duodenum (Figure 23.3a) raises the pH of the digestive mixture, allowing hydrolysis of the triacylglycerols by pancreatic lipase and by nonspecific esterases, which hydrolyze the fatty acid

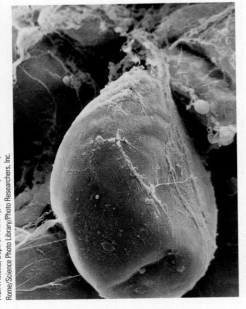

FIGURE 23.1 Scanning electron micrograph of an adipose cell (fat cell). Globules of triacylglycerols occupy most of the volume of such cells.

Prof. P. Motta, Dept. of Anatomy, University "La Sapienza," Rome/Science Photo Library/Photo Researchers, Inc.

TABLE 23.1	Stored Metabolic Fuel in a 70-kg Person		
Constituent	Energy (kJ/g dry weight)	Dry Weight (g)	Available Energy (kJ)
Fat (adipose tissue)	37	15,000	555,000
Protein (muscle)	17	6,000	102,000
Glycogen (muscle)	16	120	1,920
Glycogen (liver)	16	70	1,120
Glucose (extracellular fluid)	16	20	320
Total			660,360

Sources: Owen, O. E., and Reichard, G. A., Jr., 1971. Fuels consumed by man: The interplay between carbohydrates and fatty acids. *Progress in Biochemistry and Pharmacology* **6**:177; and Newsholme, E. A., and Leech, A. R., 1983. *Biochemistry for the Medical Sciences.* New York: Wiley.

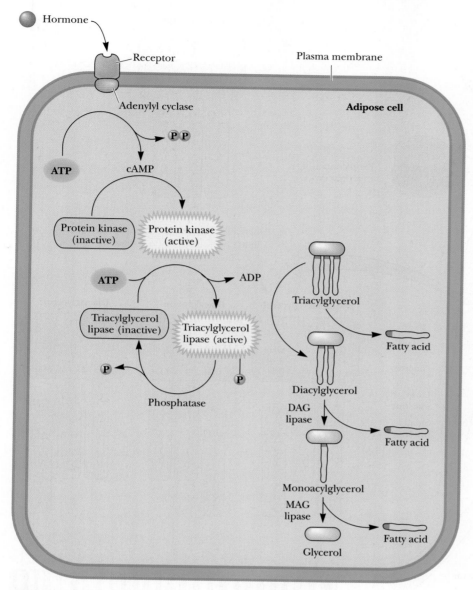

FIGURE 23.2 Liberation of fatty acids from triacylglycerols in adipose tissue is hormone-dependent.

ester linkages. Pancreatic lipase cleaves fatty acids from the C-1 and C-3 positions of triacylglycerols, and other lipases and esterases attack the C-2 position (Figure 23.3b). These processes depend upon the presence of **bile salts,** a family of carboxylic acid salts with steroid backbones (see also Chapter 24). These agents act as detergents to emulsify the triacylglycerols and facilitate the hydrolytic activity of the lipases and esterases. Short-chain fatty acids (ten or fewer carbons) released in this way are absorbed directly into the villi of the intestinal mucosa, whereas long-chain fatty acids, which are less soluble, form mixed micelles with bile salts and are carried in this fashion to the surfaces of the epithelial cells that cover the villi (Figure 23.4). The fatty acids pass into the epithelial cells, where they are condensed with glycerol to form new triacylglycerols. These triacylglycerols aggregate with lipoproteins to form particles called **chylomicrons,** which are then transported into the lymphatic system and on to the bloodstream, where they circulate to the liver, lungs, heart, muscles, and other organs (see Chapter 24). At these sites, the triacylglycerols are hydrolyzed to release fatty acids, which can then be oxidized in a highly exergonic metabolic pathway known as *β-oxidation.*

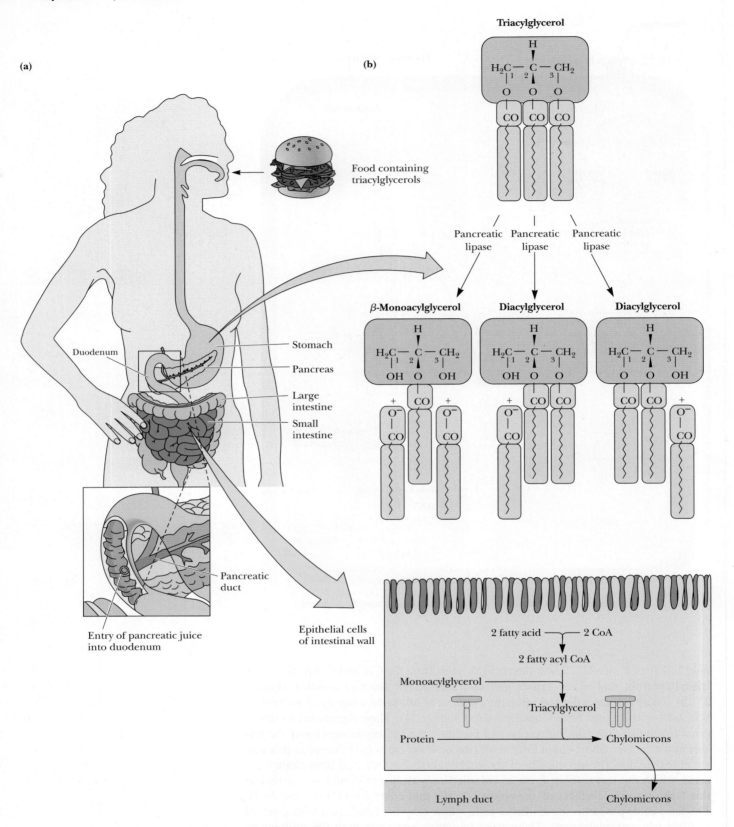

(a)

(b)

FIGURE 23.3 (a) The pancreatic duct secretes digestive fluids into the duodenum, the first portion of the small intestine. **(b)** Hydrolysis of triacylglycerols by pancreatic and intestinal lipases. Pancreatic lipases cleave fatty acids at the C-1 and C-3 positions. Resulting monoacylglycerols with fatty acids at C-2 are hydrolyzed by intestinal lipases. Fatty acids and monoacylglycerols are absorbed through the intestinal wall and assembled into lipoprotein aggregates termed chylomicrons (discussed in Chapter 24).

HUMAN BIOCHEMISTRY

Serum Albumin—Tramp Steamer of the Bloodstream

Human serum albumin, at a concentration of 40 mg/mL, is the most abundant protein in the blood serum. It is a single polypeptide of 585 amino acids (66.4 kD); the absence of any glycosylations makes it novel among extracellular proteins. Serum albumin has three principal functions: (1) It provides 80% of the osmotic pressure of blood plasma; (2) As the major macromolecular anion in the blood, it has important buffering capacity; (3) It is the major transport vehicle of hydrophobic substances. In this latter role, serum albumin merits the sobriquet "tramp steamer of the bloodstream" because of its ability to nonspecifically bind and transport numerous hydrophobic substances, including drugs taken for pharmacological benefit. ("Tramp steamers" are ships that operate on an as-needed basis, hauling any sort of cargo from place to place.) As much as 90% of some drugs are bound by serum albumin.

The solubility of free fatty acids in blood plasma is limited, amounting to only 0.1 n*M.* However, the concentration of fatty acids, as bound to serum albumin, approaches 1 m*M.* Serum albumin possesses 11 distinct fatty acid-binding sites.

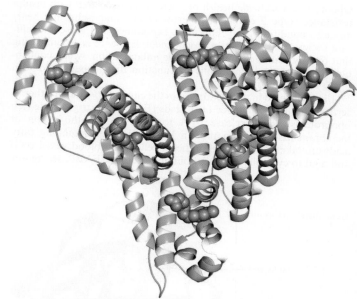

The heart-shaped tertiary structure of human serum albumin, with 11 molecules of decanoic acid bound (pdb id = 1E7F).

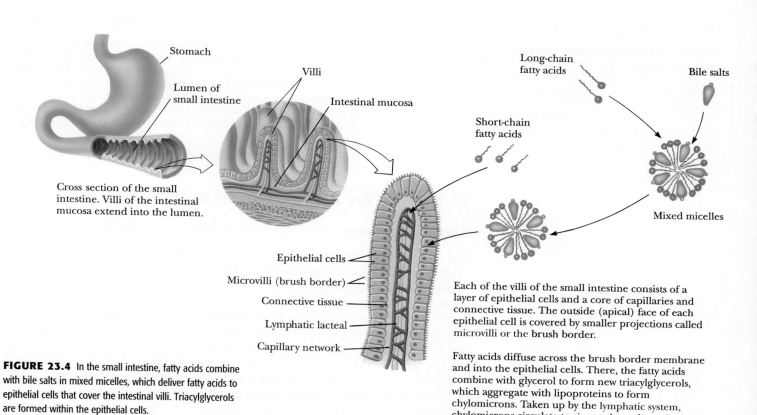

FIGURE 23.4 In the small intestine, fatty acids combine with bile salts in mixed micelles, which deliver fatty acids to epithelial cells that cover the intestinal villi. Triacylglycerols are formed within the epithelial cells.

Each of the villi of the small intestine consists of a layer of epithelial cells and a core of capillaries and connective tissue. The outside (apical) face of each epithelial cell is covered by smaller projections called microvilli or the brush border.

Fatty acids diffuse across the brush border membrane and into the epithelial cells. There, the fatty acids combine with glycerol to form new triacylglycerols, which aggregate with lipoproteins to form chylomicrons. Taken up by the lymphatic system, chylomicrons circulate to tissues throughout the body.

The Biochemistry of Obesity

Obesity is a worldwide health problem. It is associated with insulin resistance, type 2 diabetes, atherosclerosis, and ischaemic heart disease—conditions that reduce life expectancy and present significant societal and economic consequences. Evidence is accumulating that obesity causes chronic low-grade inflammation, which contributes to the systemic metabolic dysfunction that is associated with obesity-linked disorders.

Fat consumed in the diet traverses the circulatory system either as serum albumin-bound free fatty acids (FFA) or in triglyceride-rich lipoproteins, which release FFA to cells through the action of membrane-associated lipoprotein lipases (LPL). Circulating fatty acids are taken up by the endothelial cells lining blood vessel walls and bind to cytosolic fatty acid-binding proteins (FABP, see figure)

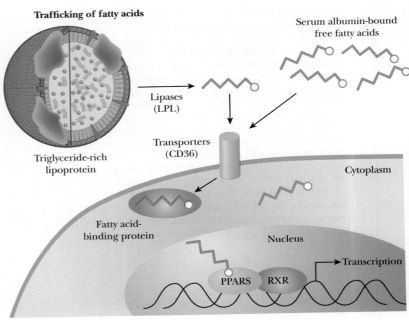

which determine their destination and fate, such as serving as ligands for transcription factors in the nucleus. For example, **peroxisome proliferator-activated receptors (PPARs)** regulate the expression of genes that control fatty acid oxidation and triglyceride metabolism (figure). Another transcription factor, **NHR-80,** regulates expression of desaturases, enzymes that convert saturated fatty acids into unsaturated fatty acids. Jerone Goudeau, Hugo Aguilaniu, and colleagues have shown that over-expression of NHR-80 in the roundworm *C. elegans* extends the organism's life span significantly.

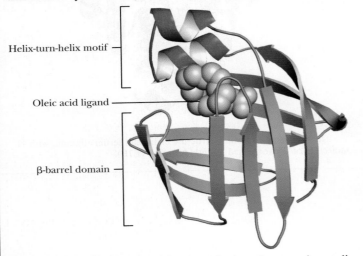

Accumulating evidence indicates that high levels of circulating FFA lead to altered deposition and accumulation of triglycerides in liver, muscle, and adipose tissue. In each of these tissues, altered fat accumulation provokes a cascade of signals that leads to stress and inflammatory responses and metabolic dysfunction. *In this context, adipose tissue is not only involved in energy storage but also functions as an endocrine organ that plays a key role in integration of system metabolism by secreting a variety of signaling proteins.*

Proteins secreted by adipose tissue are collectively called **adipokines.** These secreted proteins have either pro-inflammatory or anti-inflammatory activities. Altered production and/or secretion of these adipokines contributes to obesity-linked complications. The expression of adipokines varies depending on the site of the adipose tissue depot (see figure, facing page). The two most abundant depots are visceral and subcutaneous adipose tissues, but each of the adipose depots in the body produces a unique collection of adipokines. The various adipokines are produced either by adipose

23.2 How Are Fatty Acids Broken Down?

Knoop Elucidated the Essential Feature of β-Oxidation

The earliest clue to the secret of fatty acid oxidation and breakdown came in the early 1900s, when Franz Knoop carried out experiments in which he fed modified fatty acids to dogs. Knoop's experiments showed that fatty acids must be degraded by *oxidation at the β-carbon* (Figure 23.5), followed by cleavage of the C_α—C_β bond. Repetition of this process yielded two-carbon units, which Knoop assumed must be acetate. Much later, Albert Lehninger showed that this degradative process took place in the mitochondria, and F. Lynen and E. Reichart showed that the two-carbon unit released is *acetyl-CoA,* not free acetate. Because the entire process begins with oxidation of the carbon that is "β" to the carboxyl carbon, the process has come to be known as **β-oxidation.**

In mammalian cells, β-oxidation take place primarily in mitochondria, but a similar pathway occurs in peroxisomes. In yeast and other lower eukaryotes, β-oxidation is confined exclusively to peroxisomes. Mitochondrial β-oxidation provides energy to

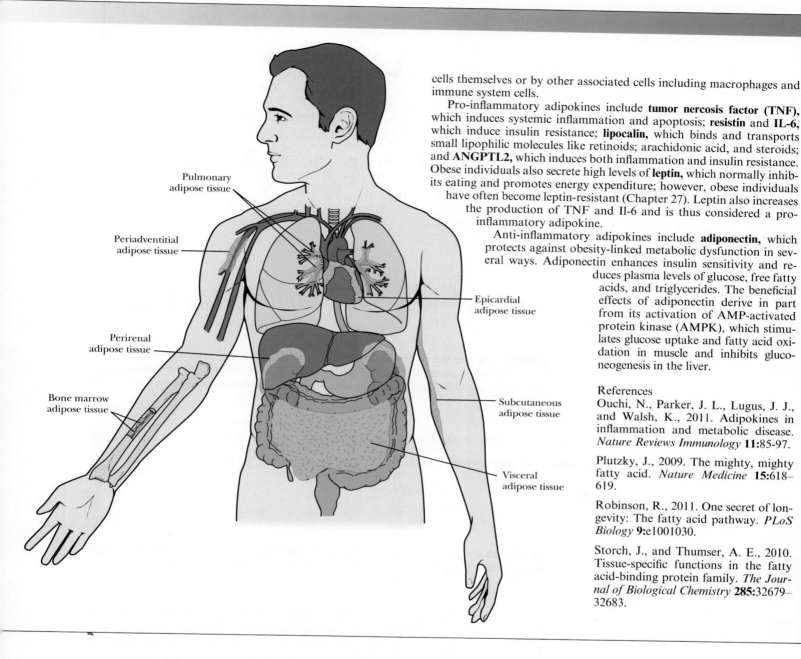

cells themselves or by other associated cells including macrophages and immune system cells.

Pro-inflammatory adipokines include **tumor nercosis factor (TNF),** which induces systemic inflammation and apoptosis; **resistin** and **IL-6,** which induce insulin resistance; **lipocalin,** which binds and transports small lipophilic molecules like retinoids; arachidonic acid, and steroids; and **ANGPTL2,** which induces both inflammation and insulin resistance. Obese individuals also secrete high levels of **leptin,** which normally inhibits eating and promotes energy expenditure; however, obese individuals have often become leptin-resistant (Chapter 27). Leptin also increases the production of TNF and Il-6 and is thus considered a pro-inflammatory adipokine.

Anti-inflammatory adipokines include **adiponectin,** which protects against obesity-linked metabolic dysfunction in several ways. Adiponectin enhances insulin sensitivity and reduces plasma levels of glucose, free fatty acids, and triglycerides. The beneficial effects of adiponectin derive in part from its activation of AMP-activated protein kinase (AMPK), which stimulates glucose uptake and fatty acid oxidation in muscle and inhibits gluconeogenesis in the liver.

References

Ouchi, N., Parker, J. L., Lugus, J. J., and Walsh, K., 2011. Adipokines in inflammation and metabolic disease. *Nature Reviews Immunology* **11:**85-97.

Plutzky, J., 2009. The mighty, mighty fatty acid. *Nature Medicine* **15:**618–619.

Robinson, R., 2011. One secret of longevity: The fatty acid pathway. *PLoS Biology* **9:**e1001030.

Storch, J., and Thumser, A. E., 2010. Tissue-specific functions in the fatty acid-binding protein family. *The Journal of Biological Chemistry* **285:**32679–32683.

the organism (Figure 23.6), whereas peroxisomal β-oxidation is responsible for shortening long-chain fatty acids that are poor substrates for mitochondrial β-oxidation. Such shortened fatty acids then become substrates for mitochondrial β-oxidation.

Coenzyme A Activates Fatty Acids for Degradation

The process of β-oxidation begins with the formation of a thiol ester bond between the fatty acid and the thiol group of coenzyme A. This reaction, shown in Figure 23.6, is catalyzed by **acyl-CoA synthetase,** which is also called **acyl-CoA ligase** or **fatty acid thiokinase.** This condensation with CoA activates the fatty acid for reaction in the β-oxidation pathway. For long-chain fatty acids, this reaction normally occurs at the outer mitochondrial membrane in higher eukaryotes before entry of the fatty acid into the mitochondrion, but it may also occur at the surface of the endoplasmic reticulum. Short- and medium-length fatty acids undergo this activating reaction in the mitochondria. In all cases, the reaction is

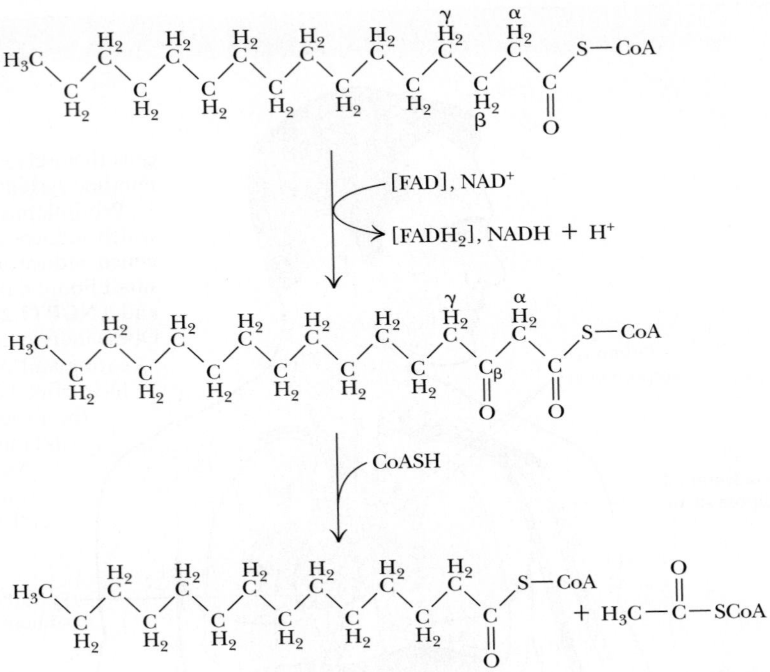

FIGURE 23.5 Fatty acids are degraded by repeated cycles of oxidation at the β-carbon and cleavage of the C_α—C_β bond to yield acetate units, in the form of acetyl-CoA. Each cycle of β-oxidation yields four electrons, captured as $FADH_2$ and NADH, which drive electron transport and oxidative phosphorylation pathways to produce ATP.

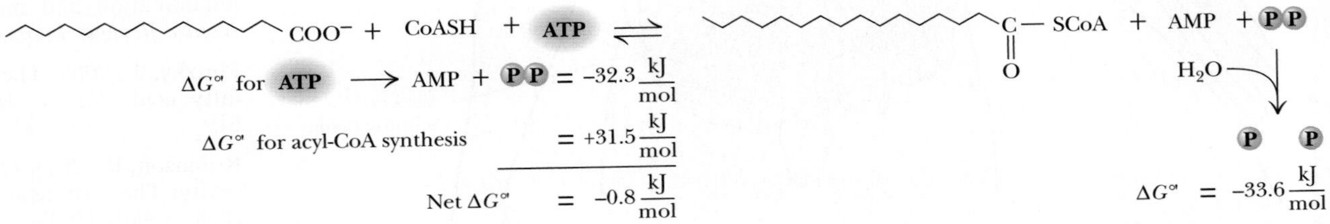

FIGURE 23.6 The acyl-CoA synthetase reaction activates fatty acids for β-oxidation. The reaction is driven by hydrolysis of ATP to AMP and pyrophosphate and by the subsequent hydrolysis of pyrophosphate.

accompanied by the hydrolysis of ATP to form AMP and pyrophosphate. As shown in Figure 23.6, the overall reaction has a net $\Delta G°'$ of about -0.8 kJ/mol, so the reaction is favorable but easily reversible. However, there is more to the story. As we have seen in several similar cases, the pyrophosphate produced in this reaction is rapidly hydrolyzed by inorganic pyrophosphatase to two molecules of phosphate, with a net $\Delta G°'$ of about -33.6 kJ/mol. Thus, pyrophosphate is maintained at a low concentration in the cell (usually less than 10 μM), and the synthetase reaction is strongly promoted. The mechanism of the acyl-CoA synthetase reaction is shown in Figure 23.7 and involves attack of the fatty acid carboxylate on ATP to form an *acyladenylate intermediate*, which is subsequently attacked by CoA, forming a fatty acyl-CoA thioester.

Carnitine Carries Fatty Acyl Groups Across the Inner Mitochondrial Membrane

All of the other enzymes of the β-oxidation pathway are located in the mitochondrial matrix. Short-chain fatty acids, as already mentioned, are transported into the matrix as free acids and form the acyl-CoA derivatives there. However, long-chain fatty acyl-CoA derivatives cannot be transported into the matrix directly. These long-chain derivatives

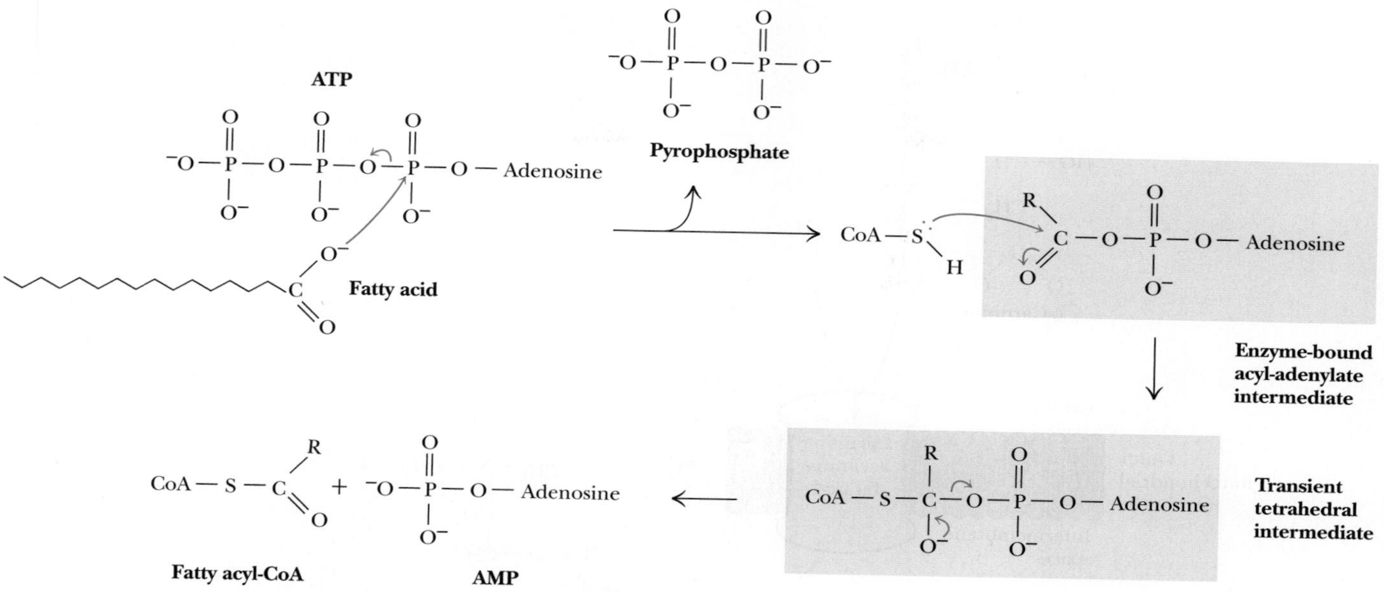

FIGURE 23.7 The mechanism of the acyl-CoA synthetase reaction involves fatty acid carboxylate attack on ATP to form an acyl-adenylate intermediate. The fatty acyl CoA thioester product is formed by CoA attack on this intermediate.

must first be converted to *acylcarnitine* derivatives, as shown in Figure 23.8. **Carnitine acyltransferase I,** associated with the outer mitochondrial membrane, catalyzes the formation of the *O*-acylcarnitine, which is then transported across the inner membrane by a **translocase.** At this point, the acylcarnitine is passed to **carnitine acyltransferase II** on the matrix side of the inner membrane, which transfers the fatty acyl group back to CoA to re-form the fatty acyl-CoA, leaving free carnitine, which can return across the membrane via the translocase.

Several additional points should be made. First, although oxygen esters usually have lower group-transfer potentials than thiol esters, the O—acyl bonds in acylcarnitines have high group-transfer potentials, and the transesterification reactions mediated by the acyltransferases have equilibrium constants close to 1. Second, note that eukaryotic cells maintain separate pools of CoA in the mitochondria and in the cytosol. The cytosolic pool is utilized principally in fatty acid biosynthesis (see Chapter 24), and the mitochondrial pool is important in the oxidation of fatty acids and pyruvate, as well as some amino acids.

β-Oxidation Involves a Repeated Sequence of Four Reactions

For saturated fatty acids, the process of β-oxidation involves a recurring cycle of four steps, as shown in Figure 23.9. The overall strategy in the first three steps is to create a carbonyl group on the β-carbon by oxidizing the C_α—C_β bond to form an olefin, with subsequent hydration and oxidation. In essence, this cycle is directly analogous to the sequence of reactions converting succinate to oxaloacetate in the TCA cycle. The fourth reaction of the cycle cleaves the β-keto ester in a reverse Claisen condensation, producing an acetate unit and leaving a fatty acid chain that is two carbons shorter than it began. (Recall from Chapter 19 that Claisen condensations involve attack by a nucleophilic agent on a carbonyl carbon to yield a β-keto acid.)

A Family of Acyl-CoA Dehydrogenases Carries Out the First Reaction of β-Oxidation The enzymes of mitochondrial β-oxidation are organized in two functional systems: a membrane-bound complex that is specific for long-chain fatty acids (14 carbons and longer) and a family of soluble enzymes in the matrix that is specific for short- and medium-chain fatty acids (Figure 23.10). As a fatty acyl chain is shortened in successive

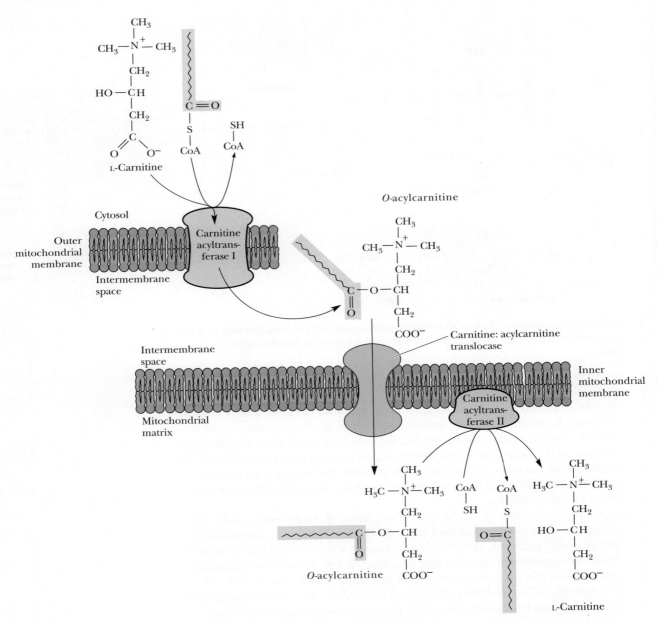

FIGURE 23.8 The formation of acylcarnitines and their transport across the inner mitochondrial membrane. The process involves the coordinated actions of carnitine acyltransferases on both the inner and outer mitochondrial membranes and of a translocase that shuttles O-acylcarnitines across the inner membrane.

cycles of β-oxidation, it moves from the membrane-bound complex to the family of soluble matrix enzymes. The first reaction of the β-oxidation cycle is catalyzed by one of four **acyl-CoA dehydrogenases.** These include the **very long-chain acyl-CoA dehydrogenase (VLCAD),** as well as acyl-CoA dehydrogenases specific for long-chain **(LCAD),** medium-chain **(MCAD),** and short-chain **(SCAD)** substrates. VLCAD is a membrane-bound homodimer of 67-kD subunits (Figure 23.11), whereas the soluble LCAD, MCAD, and SCAD are homotetramers of 40- to 45-kD subunits.

All acyl-CoA dehydrogenases carry noncovalently (but tightly) bound FAD, which is reduced during the oxidation of the fatty acid. As shown in Figure 23.12, FADH$_2$ transfers its electrons to an **electron transfer flavoprotein (ETF).** Reduced ETF is reoxidized by a specific oxidoreductase (an iron–sulfur protein), which in turn sends the electrons on to the electron-transport chain at the level of coenzyme Q. Recall from Chapter 20 that mitochondrial oxidation of FAD in this way eventually results in the net formation

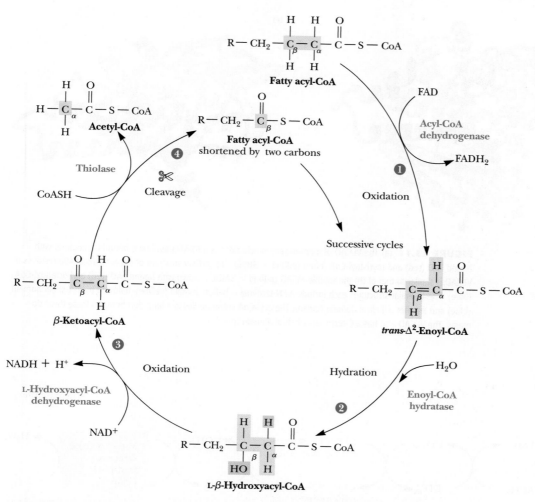

FIGURE 23.9 The β-oxidation of saturated fatty acids involves a cycle of four enzyme-catalyzed reactions. Each cycle produces single molecules of FADH₂, NADH, and acetyl-CoA, consumes a water, and yields a fatty acid shortened by two carbons. (The delta [Δ] symbol connotes a double bond, and its superscript indicates the lower-numbered carbon involved.)

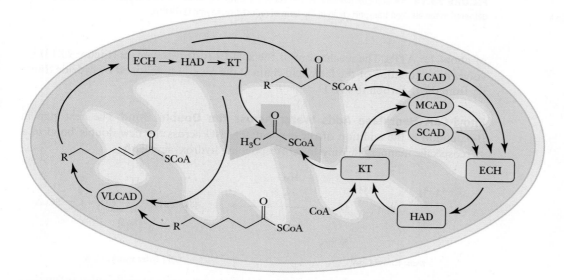

FIGURE 23.10 Very long-chain fatty acids proceed through several cycles of β-oxidation *(left)* via membrane-bound enzymes in mitochondria. A membrane-bound multifunctional complex includes the enoyl-CoA hydratase (ECH), hydroxyacyl-CoA dehydrogenase (HAD), and ketoacyl thiolase (KT) activities. As chains shorten progressively, they become substrates for the separate, soluble enzymes of β-oxidation *(right)*.

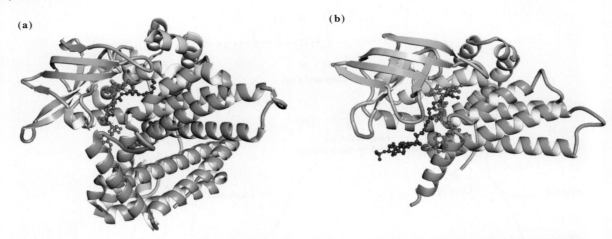

FIGURE 23.11 **(a)** The VLCAD of mammalian mitochondria is a 67-kD membrane-bound homodimer with bound FAD (red) and myristoyl-CoA (blue) (pdb id = 3B96). The tertiary structure of the N-terminal 400 residues of VLCAD is similar to that of **(b)** the soluble MCAD (pdb id = 3MDE), shown with bound FAD (red) and octanoyl-CoA (blue). These similar structures each include an N-terminal α-helical domain (yellow), followed by a β-sheet domain (blue) and another α-helical domain (green). The acyl-CoA substrate lies in a long cleft between these three domains. The VLCAD also has a C-terminal α-helical domain (purple).

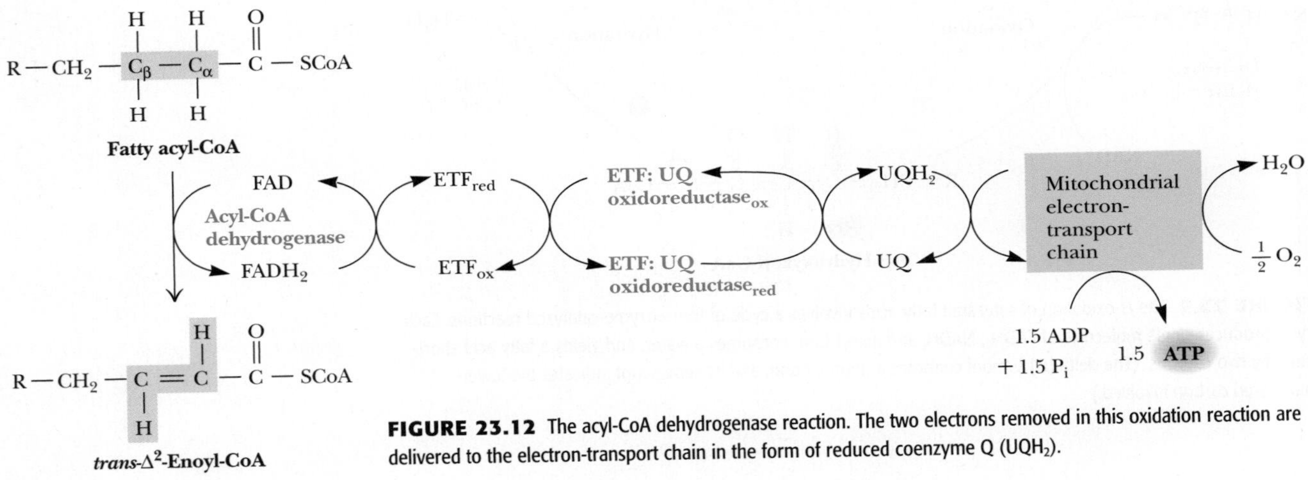

FIGURE 23.12 The acyl-CoA dehydrogenase reaction. The two electrons removed in this oxidation reaction are delivered to the electron-transport chain in the form of reduced coenzyme Q (UQH$_2$).

of about 1.5 ATPs. The mechanism of the acyl-CoA dehydrogenase (Figure 23.13) involves deprotonation of the fatty acid chain at the α-carbon, followed by hydride transfer from the β-carbon to FAD.

Enoyl-CoA Hydratase Adds Water Across the Double Bond The next step in β-oxidation is the addition of the elements of H$_2$O across the new double bond in a stereospecific manner, yielding the corresponding hydroxyacyl-CoA.

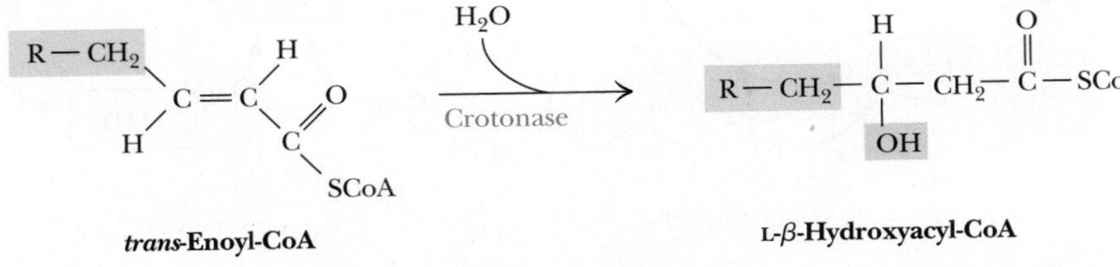

The reaction is catalyzed by **enoyl-CoA hydratase** (Figure 23.14). A number of different enoyl-CoA hydratase activities have been detected in various tissues. Also called **crotonases,** these enzymes specifically convert *trans*-enoyl-CoA derivatives to L-β-hydroxyacyl-CoA. Enoyl-CoA hydratases will also metabolize *cis*-enoyl-CoA (at slower rates) to give

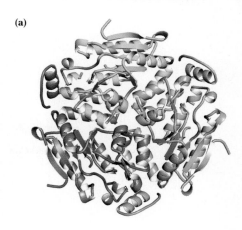

(a)

FIGURE 23.13 The mechanism of acyl-CoA dehydrogenase. Removal of a proton from the α-C is followed by hydride transfer from the β-carbon to FAD.

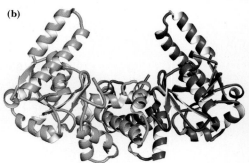

(b)

specifically D-β-hydroxyacyl-CoA. In addition, there is a novel enoyl-CoA hydratase that converts *trans*-enoyl-CoA to D-β-hydroxyacyl-CoA.

L-Hydroxyacyl-CoA Dehydrogenase Oxidizes the β-Hydroxyl Group The third reaction of this cycle is the oxidation of the hydroxyl group at the β-position to produce a β-ketoacyl-CoA derivative. This second oxidation reaction is catalyzed by L-**hydroxyacyl-CoA dehydrogenase,** an enzyme that requires NAD$^+$ as a coenzyme (see Figure 23.9).

L-β-**Hydroxyacyl-CoA** → β-**Ketoacyl-CoA**

(NAD$^+$ → NADH + H$^+$)

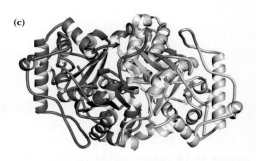

(c)

NADH produced in this reaction represents metabolic energy. Each NADH produced in mitochondria by this reaction drives the synthesis of 2.5 molecules of ATP in the electron-transport pathway. L-Hydroxyacyl-CoA dehydrogenase shows absolute specificity for the L-hydroxyacyl isomer of the substrate. (D-Hydroxyacyl isomers, which arise mainly from the metabolism of unsaturated fatty acids, are handled differently.)

β-Ketoacyl-CoA Intermediates Are Cleaved in the Thiolase Reaction The final step in the β-oxidation cycle is the cleavage of the β-ketoacyl-CoA. This reaction, catalyzed by ketoacyl **thiolase** (also known as β-**ketothiolase,** Figure 23.9), involves the attack of a cysteine thiolate from the enzyme on the β-carbonyl carbon, followed by cleavage to give the enolate of acetyl-CoA and an enzyme-thioester intermediate (Figure 23.15). Subsequent attack by the thiol group of a second CoA and departure of the cysteine thiolate yields a new (shorter) acyl-CoA. If the reaction in Figure 23.15 is read in reverse, it is easy to see that it is a Claisen condensation—an attack of the enolate anion of acetyl-CoA on a thioester. Despite the formation of a second thioester, this reaction has a very favorable K_{eq}, and it drives the three previous reactions of β-oxidation.

FIGURE 23.14 Structures of mitochondrial **(a)** enoyl-CoA hydratase trimer (pdb id = 2DUB), **(b)** hydroxyacyl-CoA dehydrogenase dimer (pdb id = 1F0Y), and **(c)** ketoacyl thiolase dimer (pdb id = 1AFW).

Repetition of the β-Oxidation Cycle Yields a Succession of Acetate Units

In essence, this series of four reactions has yielded a fatty acid (as a CoA ester) that has been shortened by two carbons and one molecule of acetyl-CoA. The shortened fatty acyl-CoA can now go through another β-oxidation cycle, as shown in Figure 23.9. Repetition of this cycle with a fatty acid with an even number of carbons eventually yields two molecules of acetyl-CoA in the final step. Complete β-oxidation of palmitic acid yields eight molecules of acetyl-CoA as well as seven molecules of [FADH$_2$] and seven molecules of NADH (Figure 23.16 and Table 23.2). The acetyl-CoA can be further metabolized in the

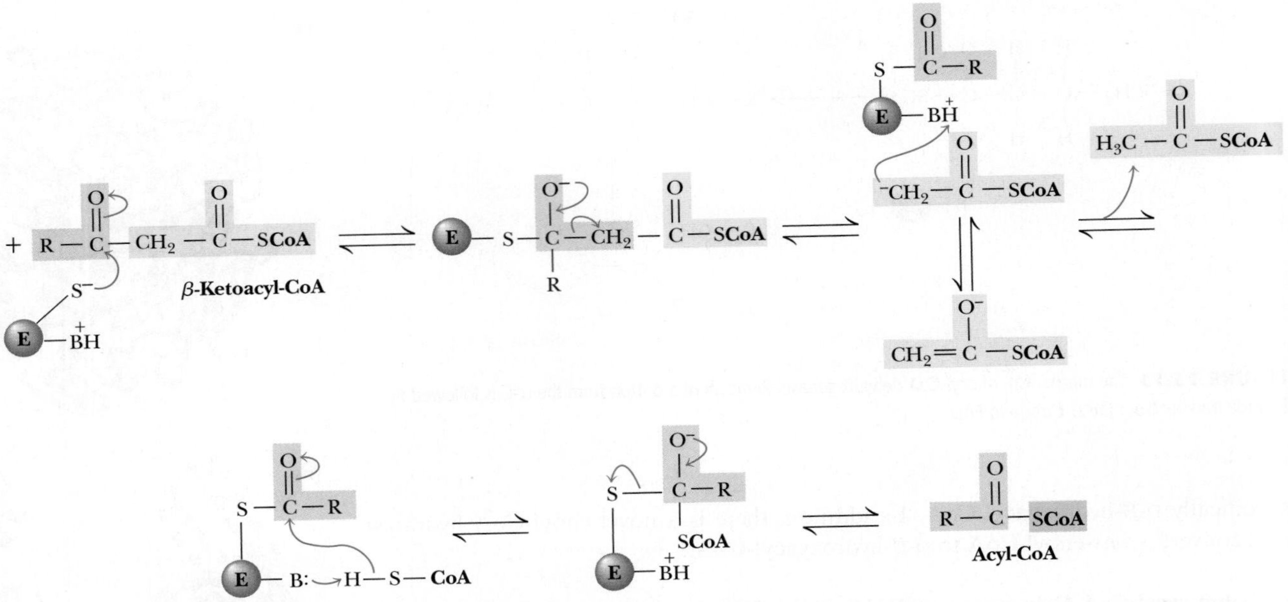

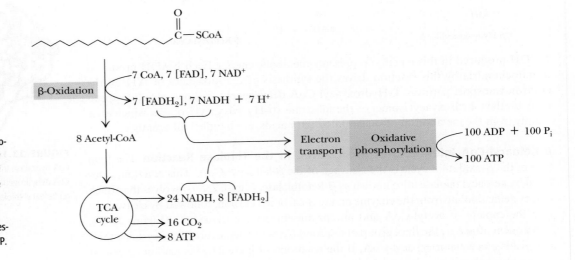

FIGURE 23.15 The mechanism of the thiolase reaction. Attack by an enzyme cysteine thiolate group at the β-carbonyl carbon produces a tetrahedral intermediate, which decomposes with departure of acetyl-CoA, leaving an enzyme thioester intermediate. Attack by the thiol group of a second CoA yields a new (shortened) acyl-CoA.

FIGURE 23.16 Reduced coenzymes produced by β-oxidation and TCA cycle activity provide electrons that drive the synthesis of ATP in oxidative phosphorylation. Complete oxidation of palmitoyl-CoA yields a total of 108 ATP. Subtracting the 2 ATP equivalents consumed in forming the original CoA thioester, oxidation of palmitate produces 106 ATP.

TABLE 23.2	Equations for the Complete Oxidation of Palmitoyl-CoA to CO_2 and H_2O		
Equation		**ATP Yield**	**Free Energy Yield (kJ/mol)**
$CH_3(CH_2)_{14}CO\text{-}CoA + 7\,[FAD] + 7\,H_2O + 7\,NAD^+ + 7\,CoA \longrightarrow 8\,CH_3CO\text{-}CoA + 7\,[FADH_2] + 7\,NADH + 7\,H^+$			
$7\,[FADH_2] + 10.5\,P_i + 10.5\,ADP + 3.5\,O_2 \longrightarrow 7\,[FAD] + 17.5\,H_2O + 10.5\,ATP$		10.5	320
$7\,NADH + 7\,H^+ + 17.5\,P_i + 17.5\,ADP + 3.5\,O_2 \longrightarrow 7\,NAD^+ + 24.5\,H_2O + 17.5\,ATP$		17.5	534
$8\,\text{Acetyl-CoA} + 16\,O_2 + 80\,ADP + 80\,P_i \longrightarrow 8\,CoA + 88\,H_2O + 16\,CO_2 + 80\,ATP$		80	2440
$CH_3(CH_2)_{14}CO\text{-}CoA + 108\,P_i + 108\,ADP + 23\,O_2 \longrightarrow 108\,ATP + 16\,CO_2 + 123\,H_2O + CoA$		108	3294
Energetic "cost" of forming palmitoyl-CoA from palmitate and CoA		−2	−61
	Total	106	3233

HUMAN BIOCHEMISTRY

Exercise Can Reverse the Consequences of Metabolic Syndrome

Metabolic syndrome is a combination of disorders that increase the risk of diabetes and cardiovascular disease. The hallmarks of metabolic syndrome include high blood pressure, elevated serum triglycerides, reduced serum high-density lipoprotein (HDL), insulin resistance, and obesity. The prevalence of these conditions is increasing in the United States. By most estimates, more than 30% of Americans are obese, and rising obesity has contributed to an epidemic of type 2 diabetes. Insights into how the body deals with high fat and high sugar diets are emerging from a variety of studies, and evidence points clearly to the benefits of exercise and dietary restriction. Endurance training (such as distance running) and resistance training (with weights) are both beneficial. Endurance training increases the mass of slow-twitch muscle fibers, resistance training builds fast-twitch muscle fibers, and both types of exercise reduce body fat, but in quite different ways.

Slow-twitch muscles depend on fatty acid oxidation and TCA cycle activity to support long periods of exercise and are termed *oxidative.* Effects of endurance training include increased expression of (1) **peroxisome proliferator-activated receptor δ,** a transcription factor that builds slow-twitch muscle fiber, and (2) insulin receptors and glucose transporters. Fast-twitch muscles are adapted for short bursts of energy, which can be supplied by glycolysis and thus are termed *glycolytic.* Effects of resistance training include activation of metabolic processes by the serine/threonine kinase **Akt1.** Induction of the Akt1 pathway results in growth of fast-twitch skeletal muscle fibers and subsequent effects on several other organs. These include increased fat uptake and oxidation by the liver and heart, reduction of adipose (fat cell) mass, and reduced blood glucose and insulin levels.

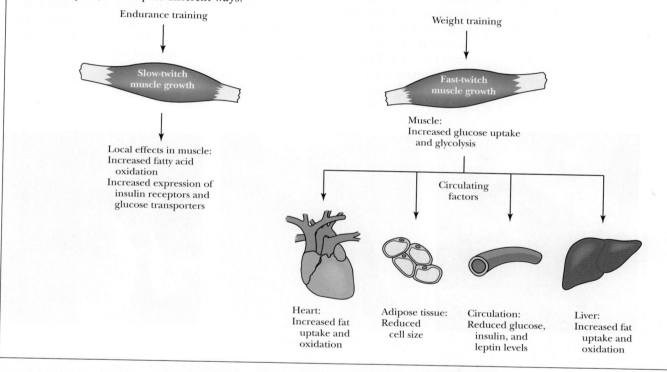

TCA cycle (as we have already seen). Alternatively, acetyl-CoA can also be used as a substrate in amino acid biosynthesis (see Chapter 25). As noted in Chapter 22, however, acetyl-CoA cannot be used as a substrate for gluconeogenesis.

Complete β-Oxidation of One Palmitic Acid Yields 106 Molecules of ATP

If the acetyl-CoA is directed entirely to the TCA cycle in mitochondria, it can eventually generate approximately ten high-energy phosphate bonds—that is, ten molecules of ATP synthesized from ADP (Table 23.2). Including the ATP formed from [FADH$_2$] and NADH, complete β-oxidation of a molecule of palmitoyl-CoA in mitochondria yields 108 molecules of ATP. Subtracting the two high-energy bonds needed to form palmitoyl-CoA, the substrate for β-oxidation, one concludes that β-oxidation of a molecule of palmitic acid yields 106 molecules of ATP. The $\Delta G°'$ for complete combustion of palmitate to CO_2 is -9790 kJ/mol. The hydrolytic energy embodied in 106 ATPs is 106×30.5 kJ/mol $= 3233$ kJ/mol, so the overall efficiency of β-oxidation under standard-state conditions is approximately 33%. The large energy yield from fatty acid oxidation is a reflection of the highly reduced state of the carbon in fatty acids. Sugars, in which the carbon is already partially oxidized, produce less energy, carbon for carbon, than do fatty acids. The breakdown of fatty acids is regulated by a variety of metabolites and hormones. Details of this regulation are described in Chapter 24, following a discussion of fatty acid synthesis.

Migratory Birds Travel Long Distances on Energy from Fatty Acid Oxidation

Because they represent the most highly concentrated form of stored biological energy, fatty acids are the metabolic fuel of choice for sustaining the incredibly long flights of many migratory birds. Although some birds migrate over landmasses and eat frequently, other species fly long distances without stopping to eat. The American golden plover (Figure 23.17) flies directly from Alaska to Hawaii, a 3300-km flight requiring 35 hours (at an average speed of nearly 60 miles/hr) and more than 250,000 wing beats! The ruby-throated hummingbird, which winters in Central America and nests in eastern North America, often flies nonstop across the Gulf of Mexico. These and similar birds accomplish these prodigious feats by storing large amounts of fatty acids (as triacylglycerols) in the days before their migratory flights. The percentage of dry-weight body fat in these birds may be as high as 70% when migration begins (compared with values of 30% and less for nonmigratory birds).

(a) Gerbil

(b) Ruby-throated hummingbird

(c) Golden plover

(d) Orca

(e) Camels

FIGURE 23.17 Animals whose existence is strongly dependent on fatty acid oxidation: **(a)** gerbil, **(b)** ruby-throated hummingbird, **(c)** golden plover, **(d)** orca (killer whale), and **(e)** camels.

Fatty Acid Oxidation Is an Important Source of Metabolic Water for Some Animals

Large amounts of metabolic water are generated by β-oxidation (123 H_2O per palmitoyl-CoA, see Table 23.2). For certain animals—including desert animals (such as gerbils) and killer whales (which do not drink seawater)—the oxidation of fatty acids can be a significant source of dietary water. A striking example is the camel (Figure 23.17), whose hump is essentially a large deposit of fat. Metabolism of fatty acids from this store provides needed water (as well as metabolic energy) during periods when drinking water is not available. It might well be said that "the ship of the desert" sails on its own metabolic water!

23.3 | How Are Odd-Carbon Fatty Acids Oxidized?

β-Oxidation of Odd-Carbon Fatty Acids Yields Propionyl-CoA

Fatty acids with odd numbers of carbon atoms are rare in mammals but fairly common in plants and marine organisms. Humans and animals whose diets include these food sources metabolize odd-carbon fatty acids via the β-oxidation pathway. The final product of β-oxidation in this case is the three-carbon propionyl-CoA instead of acetyl-CoA. Three specialized enzymes then carry out the reactions that convert propionyl-CoA to succinyl-CoA, a TCA cycle intermediate. (Because propionyl-CoA is a degradation product of methionine, valine, and isoleucine, this sequence of reactions is also important in amino acid catabolism, as we shall see in Chapter 25.) The pathway involves an initial carboxylation at the α-carbon of propionyl-CoA to produce D-methylmalonyl-CoA (Figure 23.18). The reaction is catalyzed by a biotin-dependent enzyme, **propionyl-CoA carboxylase.** The mechanism involves ATP-driven carboxylation of biotin at N_1, followed by nucleophilic attack by the α-carbanion of propionyl-CoA in a stereospecific manner.

D-Methylmalonyl-CoA, the product of this reaction, is converted to the L-isomer by **methylmalonyl-CoA epimerase** (Figure 23.18). (This enzyme has often and incorrectly been called "methylmalonyl-CoA racemase." It is not a racemase because the CoA moiety contains five other asymmetric centers.) The epimerase reaction involves a carbanion at the α-position formed via a reversible dissociation of the acidic α-proton (Figure 23.19). The L-isomer is the substrate for methylmalonyl-CoA mutase. Methylmalonyl-CoA epimerase is an impressive catalyst. The pK_a for the proton that must dissociate to initiate this reaction is approximately 21! If binding of a proton to the α-anion is diffusion

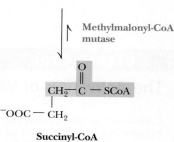

FIGURE 23.18 The conversion of propionyl-CoA (formed from β-oxidation of odd-carbon fatty acids) to succinyl-CoA is carried out by a trio of enzymes, as shown. Succinyl-CoA can enter the TCA cycle.

FIGURE 23.19 The methylmalonyl-CoA epimerase mechanism involves a resonance-stabilized carbanion at the α-position.

limited, with $k_{on} = 10^9 \ M^{-1} \ sec^{-1}$, then the initial proton dissociation must be rate limiting and the rate constant must be

$$k_{off} = K_a \cdot k_{on} = (10^{-21} \ M) \cdot (10^9 \ M^{-1} \ sec^{-1}) = 10^{-12} \ sec^{-1}$$

The turnover number of methylmalonyl-CoA epimerase is $100 \ sec^{-1}$, and thus the enzyme enhances the reaction rate by a factor of 10^{14}.

A B$_{12}$-Catalyzed Rearrangement Yields Succinyl-CoA from L-Methylmalonyl-CoA

The third reaction, catalyzed by **methylmalonyl-CoA mutase,** is quite unusual because it involves a migration of the carbonyl-CoA group from one carbon to its neighbor (Figure 23.18). The mutase reaction is vitamin B$_{12}$–dependent and begins with **homolytic cleavage** of the Co^{3+}—C bond in 5′-deoxyadenosylcobalamin, reducing the cobalt to Co^{2+} (see A Deeper Look, below). Transfer of a hydrogen atom from the substrate to the deoxyadenosyl group produces a methylmalonyl-CoA radical, which then can undergo a classic B$_{12}$-catalyzed rearrangement to yield a succinyl-CoA radical. Hydrogen transfer from the deoxyadenosyl group yields succinyl-CoA and regenerates the B$_{12}$ coenzyme (see problem 16 at the end of the chapter).

Net Oxidation of Succinyl-CoA Requires Conversion to Acetyl-CoA

Succinyl-CoA derived from propionyl-CoA can enter the TCA cycle. Oxidation of succinate to oxaloacetate provides a substrate for glucose synthesis. Thus, although the acetate units produced in β-oxidation cannot be utilized in gluconeogenesis by animals,

---

▶ A DEEPER LOOK

The Activation of Vitamin B$_{12}$

Conversion of inactive vitamin B$_{12}$ to active *5′-deoxyadenosylcobalamin* involves three steps (see accompanying figure). Two flavoprotein reductases sequentially convert Co^{3+} in cyanocobalamin to the Co^{2+} state and then to the Co$^+$ state. Co$^+$ is an extremely powerful nucleophile. It attacks the C-5′ carbon of ATP as shown, expelling the triphosphate anion to form 5′-deoxyadenosylcobalamin.

Because two electrons from Co$^+$ are donated to the Co—carbon bond, the oxidation state of cobalt reverts to Co^{3+} in the active coenzyme. This is one of only two known adenosyl transfers (that is, nucleophilic attack on the ribose 5′-carbon of ATP) in biological systems. (The other is the formation of *S*-adenosylmethionine; see Chapter 25.)

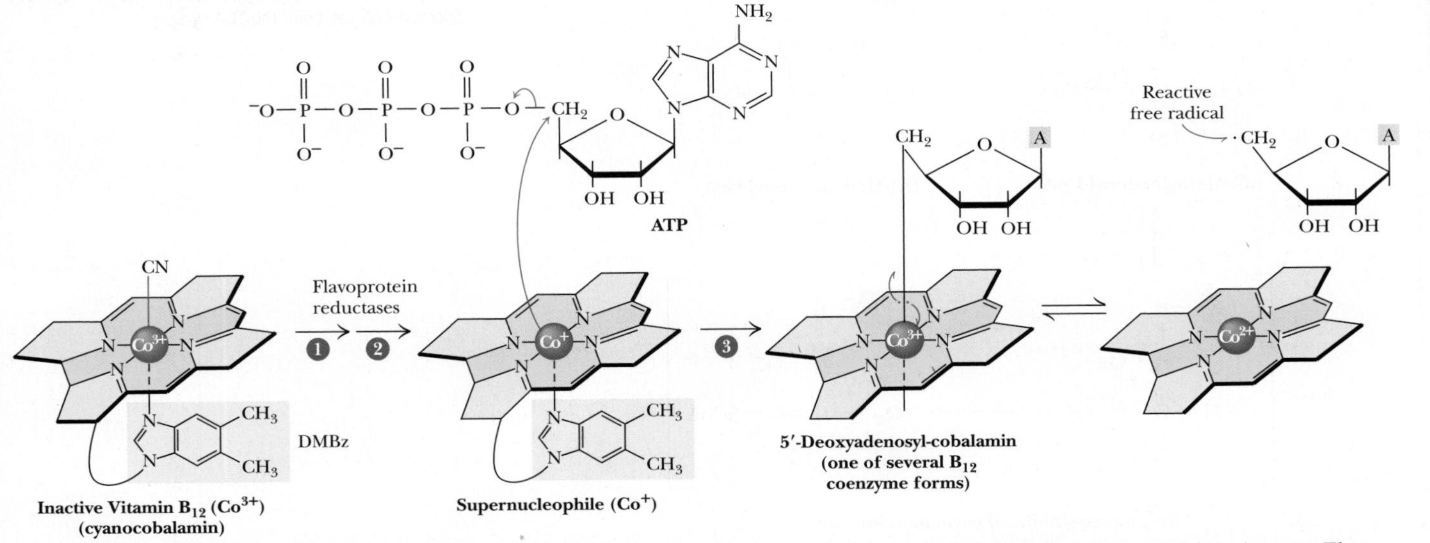

Formation of the active coenzyme 5′-deoxyadenosylcobalamin from inactive vitamin B$_{12}$ is initiated by the action of flavoprotein reductases. The resulting Co$^+$ species, dubbed a supernucleophile, attacks the 5′-carbon of ATP in an unusual adenosyl transfer. Homolytic cleavage of the Co^{3+}–C bond produces a reactive free radical that facilitates rearrangements such as that in the methylmalonyl-CoA mutase reaction.

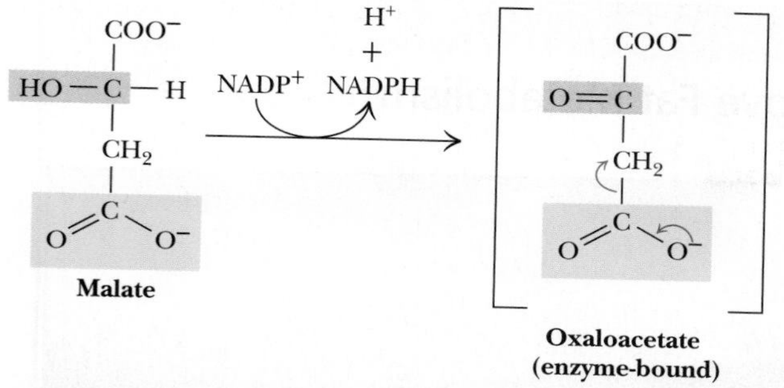

FIGURE 23.20 The malic enzyme reaction proceeds by oxidation of malate to oxaloacetate, followed by decarboxylation to yield pyruvate.

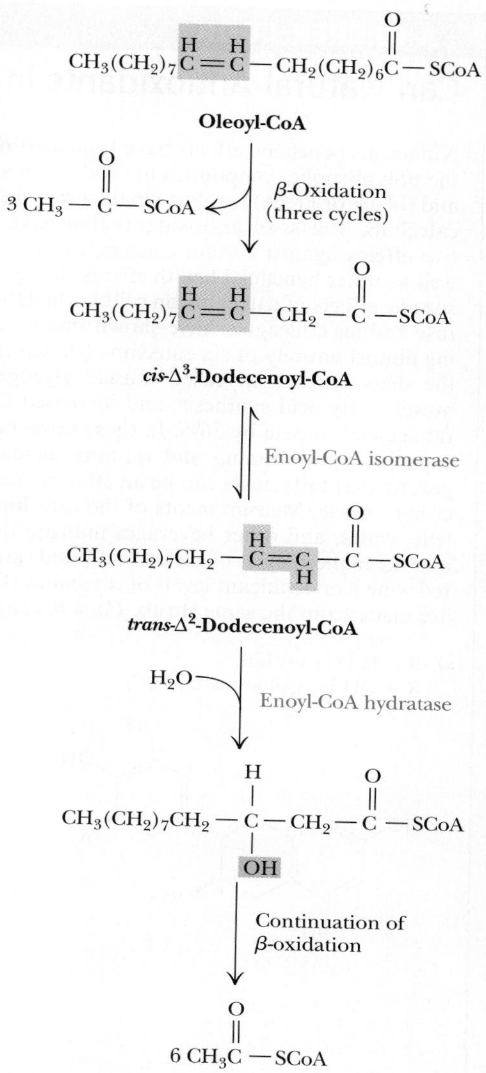

FIGURE 23.21 β-Oxidation of unsaturated fatty acids. In the case of oleoyl-CoA, three β-oxidation cycles produce three molecules of acetyl-CoA and leave *cis*-Δ^3-dodecenoyl-CoA. Rearrangement of enoyl-CoA isomerase gives the *trans*-Δ^2 species, which then proceeds normally through the β-oxidation pathway.

the occasional propionate produced from oxidation of odd-carbon fatty acids *can* be used for sugar synthesis. Alternatively, succinate introduced to the TCA cycle from odd-carbon fatty acid oxidation may be oxidized to CO_2. However, all of the four-carbon intermediates in the TCA cycle are regenerated in the cycle and thus should be viewed as catalytic species. Net consumption of succinyl-CoA thus does not occur directly in the TCA cycle. Rather, the succinyl-CoA generated from β-oxidation of odd-carbon fatty acids must be converted to pyruvate and then to acetyl-CoA (which is completely oxidized in the TCA cycle). To follow this latter route, succinyl-CoA entering the TCA cycle must be first converted to malate in the usual way and then transported from the mitochondrial matrix to the cytosol, where it is oxidatively decarboxylated to pyruvate and CO_2 by **malic enzyme,** as shown in Figure 23.20. Pyruvate can then be transported back to the mitochondrial matrix, where it enters the TCA cycle via pyruvate dehydrogenase. Note that malic enzyme plays a role in fatty acid synthesis (see Figure 24.1).

23.4 How Are Unsaturated Fatty Acids Oxidized?

An Isomerase and a Reductase Facilitate the β-Oxidation of Unsaturated Fatty Acids

Unsaturated fatty acids are also catabolized by β-oxidation, but two additional mitochondrial enzymes—an isomerase and a novel reductase—are required to handle the *cis* double bonds of naturally occurring fatty acids. As an example, consider the breakdown of oleic acid, an 18-carbon chain with a double bond at the 9,10-position. The reactions of β-oxidation proceed normally through three cycles, producing three molecules of acetyl-CoA and leaving the degradation product *cis*-Δ^3-dodecenoyl-CoA, shown in Figure 23.21. This intermediate is not a substrate for acyl-CoA dehydrogenase. With a double bond at the 3,4-position, it is not possible to form another double bond at the 2,3- (or β-) position. As shown in Figure 23.21, this problem is solved by **enoyl-CoA isomerase,** an enzyme that rearranges this *cis*-Δ^3 double bond to a *trans*-Δ^2 double bond. This latter species can proceed through the normal route of β-oxidation.

Degradation of Polyunsaturated Fatty Acids Requires 2,4-Dienoyl-CoA Reductase

Polyunsaturated fatty acids pose a slightly more complicated situation for the cell. Consider, for example, the case of linoleic acid shown in Figure 23.22. As with oleic acid, β-oxidation proceeds through three cycles, and enoyl-CoA isomerase converts the *cis*-Δ^3 double bond to a *trans*-Δ^2 double bond to permit one more round of β-oxidation. What results this time, however, is a *cis*-Δ^4 enoyl-CoA, which is converted normally by acyl-CoA dehydrogenase to a *trans*-Δ^2, *cis*-Δ^4 species. This, however, is a poor substrate for

> **A DEEPER LOOK**

Can Natural Antioxidants in Certain Foods Improve Fat Metabolism?

Numerous beneficial effects have been attributed in recent years to the polyphenolic compounds in foods such as chocolate, red wine, and (black or green) tea. Principal among these compounds are the catechins, a class of antioxidants that have demonstrated protective effects against certain cancers, obesity, and heart disease, as well as other beneficial health effects. Is it possible to demonstrate specific effects of catechins on cellular metabolism? Takatoshi Murase and his colleagues have shown that green tea extract, consisting almost entirely of six catechins (shown in the figure), added to the diets of rats increased muscle glycogen content, decreased muscle fatty acid synthesis, and increased fatty acid oxidation by rat skeletal muscle by 36%. In these same rats, exercise endurance capacity for swimming and running increased significantly, suggesting that fatty acids can be an effective source of energy in exercising muscle. Measurements of the catechin content of a range of teas, wines, and other beverages indicate that the broadest range and the highest levels of these compounds are found in tea, but that red wine has significant levels of catechins. (Green tea and black tea are made from the same shrub, *Camellia sinensis*.)

© iStockphoto.com/Michel de Nijs

▲ Tea plants

(a) R = H (+)-catechin,
R = OH (+)-gallocatechin (GC)

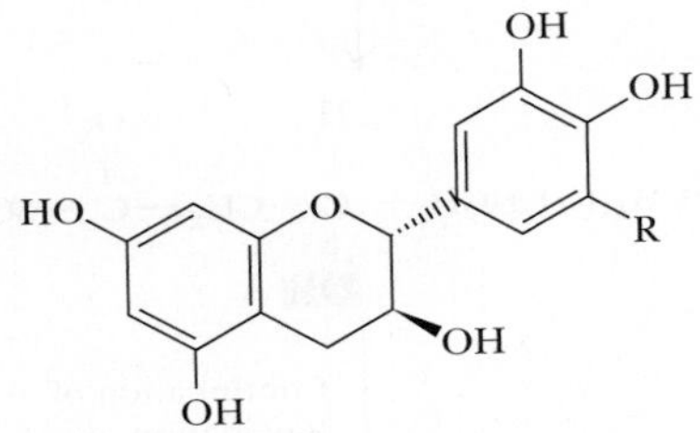

(b) R = H (−)-epicatechin,
R = OH (−)-epigallocatechin (EGC)

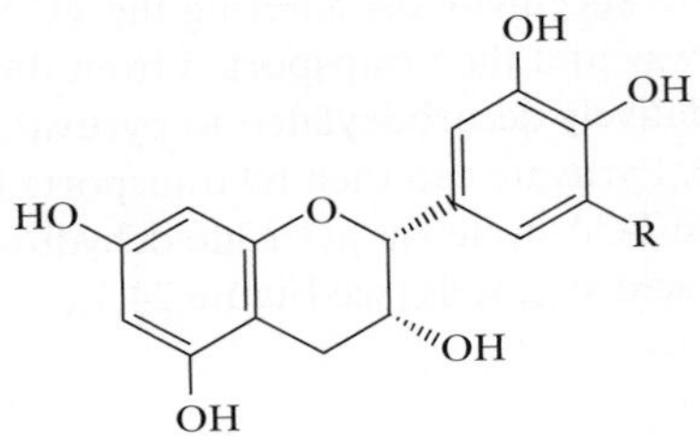

(c) R = H (−)-epicatechin gallate (ECg),
R = OH (−)-epigallocatechin gallate (EGCg)

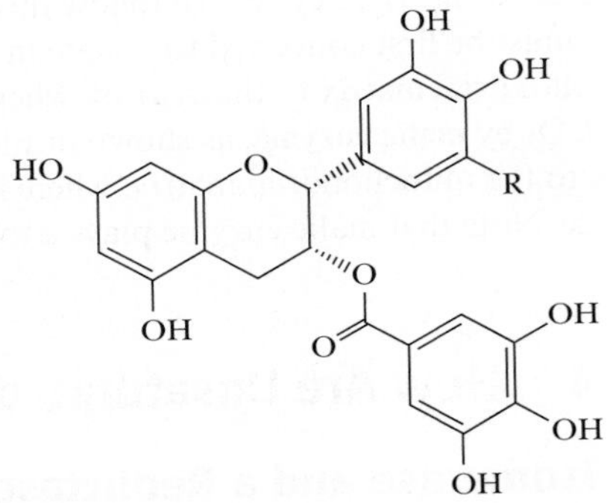

the enoyl-CoA hydratase. This problem is solved by **2,4-dienoyl-CoA reductase,** the product of which depends on the organism. The mammalian form of this enzyme produces a *trans*-Δ^3 enoyl product, as shown in Figure 23.22; this enoyl product can be converted by an enoyl-CoA isomerase to the *trans*-Δ^2 enoyl-CoA, which can then proceed normally through the β-oxidation pathway. *Escherichia coli* possesses a 2,4-dienoyl-CoA reductase that reduces the double bond at the 4,5-position to yield the *trans*-Δ^2 enoyl-CoA product in a single step.

23.5 : Are There Other Ways to Oxidize Fatty Acids?

Peroxisomal β-Oxidation Requires FAD-Dependent Acyl-CoA Oxidase

Although β-oxidation in mitochondria[1] is the principal pathway of fatty acid catabolism, several other minor pathways play important roles in fat catabolism. For example, organelles other than mitochondria, including *peroxisomes* and *glyoxysomes,*

[1]β-Oxidation does not occur significantly in plant mitochondria. Most β-oxidation in plants occurs in peroxisomes.

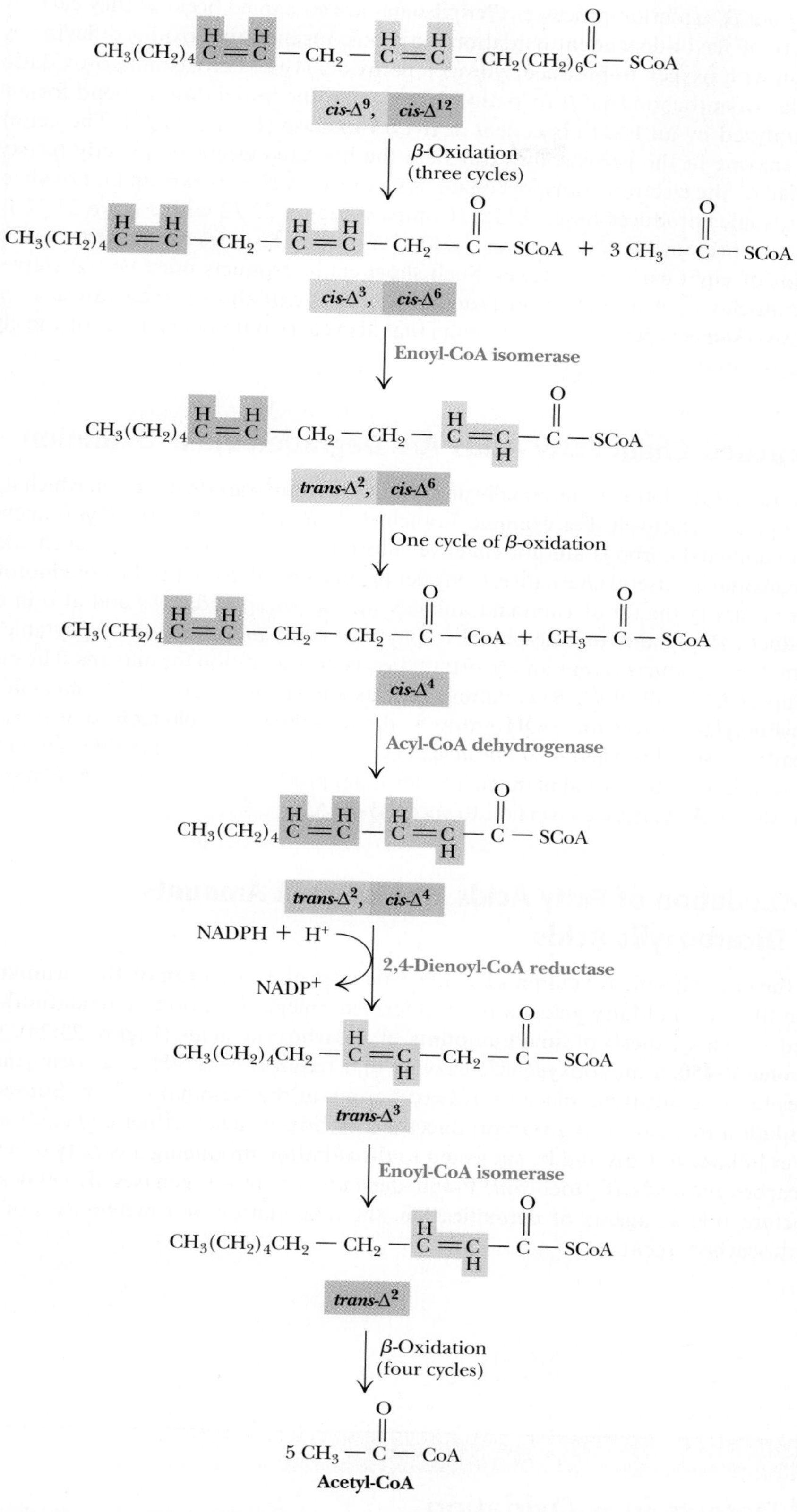

FIGURE 23.22 The oxidation pathway for polyunsaturated fatty acids in mammals, illustrated for linoleic acid. Three cycles of β-oxidation on linoleoyl-CoA yield the cis-Δ^3, cis-Δ^6 intermediate, which is converted to a $trans$-Δ^2, cis-Δ^6 intermediate. An additional round of β-oxidation gives cis-Δ^4 enoyl-CoA, which is oxidized to the $trans$-Δ^2, cis-Δ^4 species by acyl-CoA dehydrogenase. The subsequent action of 2,4-dienoyl-CoA reductase yields the $trans$-Δ^3 product, which is converted by enoyl-CoA isomerase to the $trans$-Δ^2 form. Normal β-oxidation then produces five molecules of acetyl-CoA.

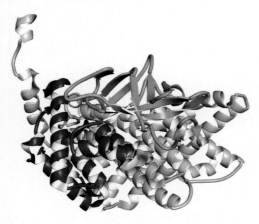

Structure of acyl-CoA oxidase from rat liver (pdb id = 1IS2). The enzyme is a homodimer, each monomer consisting of an N-terminal α-helical domain (blue), a β-sheet domain (orange), and a C-terminal α-helical domain (violet).

Fatty acyl-CoA

[FAD] → Acyl-CoA Oxidase → [FADH$_2$]

H$_2$O$_2$

O$_2$

$trans$-Δ^2-Enoyl-CoA

H$_2$O + $\frac{1}{2}$O$_2$

FIGURE 23.23 The acyl-CoA oxidase reaction in peroxisomes. Electrons captured as FADH$_2$ are used to produce the hydrogen peroxide required for degradative processes in peroxisomes and thus are not available for eventual generation of ATP. (Compare this reaction with the acyl-CoA dehydrogenase of mitochondria shown in Figure 23.12.)

carry out β-oxidation processes. **Peroxisomes** are so named because they carry out a variety of flavin-dependent oxidation reactions, regenerating oxidized flavins by reaction with oxygen to produce hydrogen peroxide, H$_2$O$_2$. Peroxisomal β-oxidation is similar to mitochondrial β-oxidation, except that the initial double-bond formation is catalyzed by an FAD-dependent **acyl-CoA oxidase** (Figure 23.23). The action of this enzyme in the peroxisomes transfers the liberated electrons directly to oxygen instead of the electron-transport chain. As a result, each two-carbon unit oxidized in peroxisomes produces fewer ATPs. (Compare Figure 23.23 with Figure 23.12.) The enzymes responsible for fatty acid oxidation in peroxisomes are inactive with acyl chains of eight carbons or fewer. Such short-chain products must be transferred to the mitochondria for further breakdown. Similar β-oxidation enzymes are also found in **glyoxysomes**—peroxisomes in plants that also carry out the reactions of the glyoxylate pathway.

Branched-Chain Fatty Acids Are Degraded Via α-Oxidation

Although β-oxidation is universally important, there are some instances in which it cannot operate effectively. For example, branched-chain fatty acids with alkyl branches at odd-numbered carbons are not effective substrates for β-oxidation. For such species, **α-oxidation** is a useful alternative. Consider **phytol,** a breakdown product of chlorophyll that occurs in the fat of ruminant animals such as sheep and cows and also in dairy products. Ruminants oxidize phytol to phytanic acid, and digestion of phytanic acid from dairy products is thus an important dietary consideration for humans. The methyl group at C-3 will block β-oxidation, but, as shown in Figure 23.24, **phytanic acid α-hydroxylase** places an —OH group at the α-carbon, and **phytanic acid α-oxidase** decarboxylates it to yield *pristanic acid.* The CoA ester of this metabolite can undergo β-oxidation in the normal manner. The terminal product, isobutyryl-CoA, can be sent into the TCA cycle by conversion to succinyl-CoA.

ω-Oxidation of Fatty Acids Yields Small Amounts of Dicarboxylic Acids

In the endoplasmic reticulum of eukaryotic cells, the oxidation of the terminal carbon of a normal fatty acid—a process termed omega oxidation (ω-oxidation)—can lead to the synthesis of small amounts of dicarboxylic acids (Figure 23.25). **Cytochrome P-450,** a monooxygenase enzyme that requires NADPH as a coenzyme and uses O$_2$ as a substrate, places a hydroxyl group at the terminal carbon. Subsequent oxidation to a carboxyl group produces a dicarboxylic acid. Either end can form an ester linkage to CoA and be subjected to β-oxidation, producing a variety of smaller dicarboxylic acids. (Cytochrome P-450–dependent monooxygenases also play an important role as agents of **detoxification,** the degradation and metabolism of toxic hydrocarbon agents.)

> **HUMAN BIOCHEMISTRY**

Refsum's Disease Is a Result of Defects in α-Oxidation

The α-oxidation pathway is defective in **Refsum's disease,** an inherited metabolic disorder that results in defective night vision, tremors, and other neurologic abnormalities. These symptoms are caused by accumulation of phytanic acid in the body. Treatment of Refsum's disease requires a diet free of chlorophyll, the precursor of phytanic acid. This regimen is difficult to implement because all green vegetables and even meat from plant-eating animals, such as cows, pigs, and poultry, must be excluded from the diet.

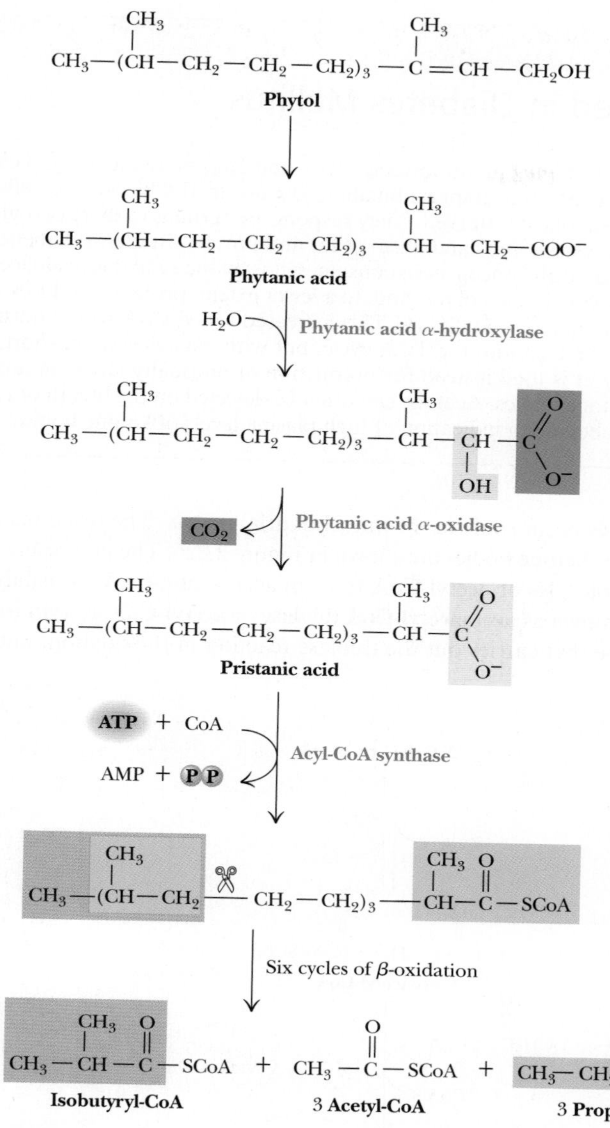

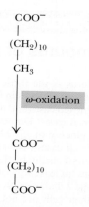

FIGURE 23.25 Dicarboxylic acids can be formed by oxidation of the omega carbon of fatty acids in a cytochrome P-450–dependent reaction.

FIGURE 23.24 Branched-chain fatty acids are oxidized by α-oxidation, as shown for phytanic acid. The product of the phytanic acid oxidase, pristanic acid, is a suitable substrate for normal β-oxidation. Isobutyryl-CoA and propionyl-CoA can both be converted to succinyl-CoA, which can enter the TCA cycle.

23.6 What Are Ketone Bodies, and What Role Do They Play in Metabolism?

Ketone Bodies Are a Significant Source of Fuel and Energy for Certain Tissues

Most of the acetyl-CoA produced by the oxidation of fatty acids in liver mitochondria undergoes further oxidation in the TCA cycle, as stated earlier. However, some of this acetyl-CoA is converted to three important metabolites: **acetone, acetoacetate,** and **β-hydroxybutyrate.** The process is known as **ketogenesis,** and these three metabolites are traditionally known as **ketone bodies,** despite the fact that β-hydroxybutyrate does not contain a ketone function. These three metabolites are synthesized primarily in the liver but are important sources of fuel and energy for many tissues, including brain, heart, and skeletal muscle. The brain, for example, normally uses glucose as its source of metabolic energy. However, during periods of starvation, ketone bodies may be the major energy source for the brain. Acetoacetate and 3-hydroxybutyrate are normal substrates for kidney cortex and for heart muscle.

HUMAN BIOCHEMISTRY

Large Amounts of Ketone Bodies Are Produced in Diabetes Mellitus

Diabetes mellitus is the most common endocrine disease and the third leading cause of death in the United States, with approximately 6 million diagnosed cases and an estimated 4 million more borderline but undiagnosed cases. Diabetes is characterized by an abnormally high level of glucose in the blood. In **type 1 diabetes** (representing 10% or less of all cases), elevated blood glucose results from inadequate secretion of insulin by the islets of Langerhans in the pancreas. **Type 2 diabetes** (at least 90% of all cases) results from an insensitivity to insulin. Type 2 diabetics produce normal or even elevated levels of insulin, but their cells are not responsive to insulin, often due to a shortage of insulin receptors (see Chapter 32). In both cases, trans-

port of glucose into muscle, liver, and adipose tissue is significantly reduced, and despite abundant glucose in the blood, the cells are metabolically starved. They respond by turning to increased gluconeogenesis and catabolism of fat and protein. In type 1 diabetes, increased gluconeogenesis consumes most of the available oxaloacetate, but breakdown of fat (and, to a lesser extent, protein) produces large amounts of acetyl-CoA. This increased acetyl-CoA would normally be directed into the TCA cycle, but with oxaloacetate in short supply, it is used instead for production of unusually large amounts of ketone bodies. Acetone can often be detected on the breath of type 1 diabetics, an indication of high plasma levels of ketone bodies.

Ketone body synthesis occurs only in the mitochondrial matrix. The reactions responsible for the formation of ketone bodies are shown in Figure 23.26. The first reaction—the condensation of two molecules of acetyl-CoA to form acetoacetyl-CoA—is catalyzed by thiolase, which is also known as **acetoacetyl-CoA thiolase** or **acetyl-CoA acetyltransferase.** This is the same enzyme that carries out the thiolase reaction in β-oxidation, but here it

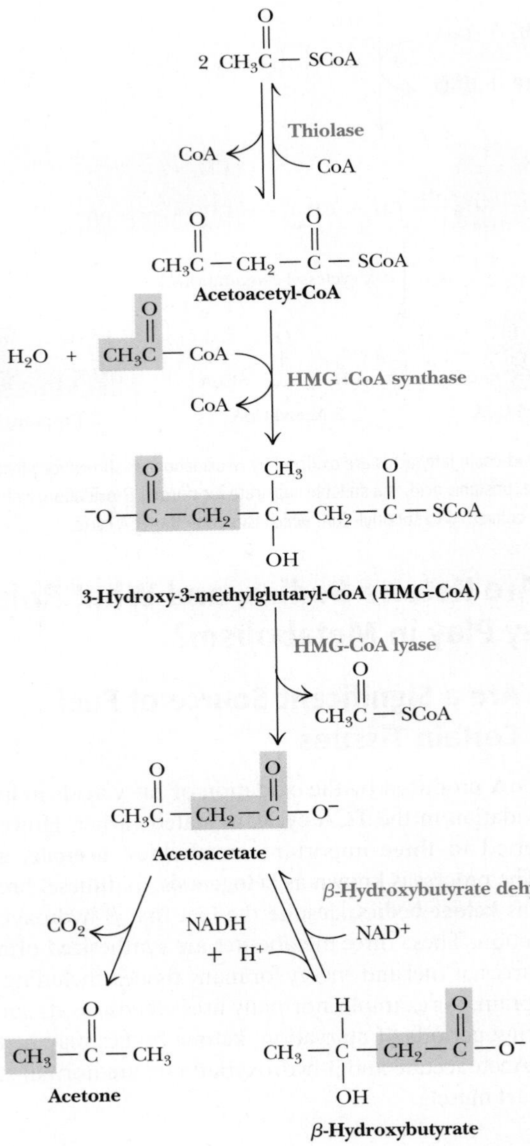

FIGURE 23.26 The formation of ketone bodies, synthesized primarily in liver mitochondria.

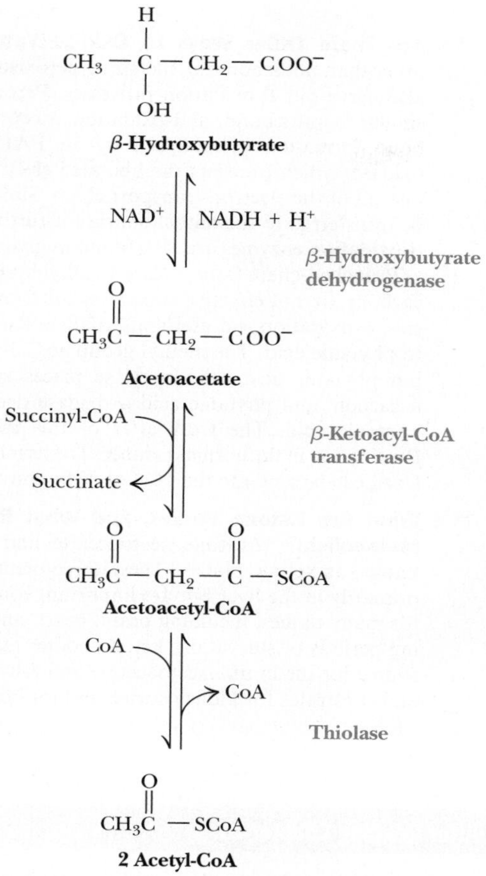

$$CH_3 - \overset{\overset{\displaystyle H}{|}}{\underset{\underset{\displaystyle OH}{|}}{C}} - CH_2 - COO^-$$

β-Hydroxybutyrate

NAD$^+$ ⇅ NADH + H$^+$

β-Hydroxybutyrate dehydrogenase

$$CH_3\overset{\overset{\displaystyle O}{\|}}{C} - CH_2 - COO^-$$

Acetoacetate

Succinyl-CoA ⤵

Succinate ⤸

β-Ketoacyl-CoA transferase

$$CH_3\overset{\overset{\displaystyle O}{\|}}{C} - CH_2 - \overset{\overset{\displaystyle O}{\|}}{C} - SCoA$$

Acetoacetyl-CoA

CoA ⤵

⤸ CoA

Thiolase

$$CH_3\overset{\overset{\displaystyle O}{\|}}{C} - SCoA$$

2 Acetyl-CoA

FIGURE 23.27 Reconversion of ketone bodies to acetyl-CoA in the mitochondria of many tissues (other than liver) provides significant metabolic energy.

runs in reverse. The second reaction adds another molecule of acetyl-CoA to give *3-hydroxy-3-methylglutaryl-CoA,* commonly abbreviated HMG-CoA. These two mitochondrial matrix reactions are analogous to the first two steps in cholesterol biosynthesis, a cytosolic process, as we shall see in Chapter 24. HMG-CoA is converted to acetoacetate and acetyl-CoA by the action of **HMG-CoA lyase** in a mixed aldol-Claisen ester cleavage reaction. This reaction is mechanistically similar to the reverse of the citrate synthase reaction in the TCA cycle. A membrane-bound enzyme, **β-hydroxybutyrate dehydrogenase,** then can reduce acetoacetate to β-hydroxybutyrate.

Acetoacetate and β-hydroxybutyrate are transported through the blood from liver to target organs and tissues, where they are converted to acetyl-CoA (Figure 23.27). *Ketone bodies are easily transportable forms of fatty acids that move through the circulatory system without the need for complexation with serum albumin and other fatty acid-binding proteins.*

SUMMARY

⚙️OWL Login to OWL to develop problem-solving skills and complete online homework assigned by your professor. •

23.1 How Are Fats Mobilized from Dietary Intake and Adipose Tissue? Triacylglycerols are a major source of fatty acids in the diet, and they are also our principal stored energy reserve in adipose tissue. Hormone messengers such as adrenaline, glucagon, and ACTH bind to receptors on the plasma membrane of adipose cells and lead to the activation of a triacylglycerol lipase that hydrolyzes a fatty acid from C-1 or C-3 of triacylglycerols. Subsequent actions of diacylglycerol lipase and monoacylglycerol lipase yield fatty acids and glycerol. The cell then releases the fatty acids into the blood, where they are carried to sites of utilization. Dietary triacylglycerols are degraded by lipases and esterases in the stomach and duodenum. Pancreatic lipase cleaves fatty acids from the C-1 and C-3 positions

of triacylglycerols, and other lipases and esterases attack the C-2 position. Bile salts act as detergents to emulsify the triacylglycerols and facilitate the hydrolytic activity of the lipases and esterases.

23.2 How Are Fatty Acids Broken Down? The process of β-oxidation begins with the formation of a thiol ester bond between the fatty acid and the thiol group of coenzyme A, catalyzed by acyl-CoA synthetase. The enzymes of the β-oxidation pathway are located in the mitochondrial matrix. Short-chain fatty acids are transported into the matrix as free acids and form the acyl-CoA derivatives there. However, long-chain fatty acyl-CoA derivatives must first be converted to acylcarnitine derivatives, which are transported across the inner membrane by a translocase. On the matrix side of the inner membrane, a second acyl carnitine transferase reforms the fatty acyl-CoA. The process of β-oxidation involves a recurring cycle of four steps. A double bond is formed, water is added across the double bond, and the resulting alcohol is oxidized to a carbonyl group. The fourth reaction of the cycle cleaves the resulting β-keto ester, producing an acetate unit and leaving a fatty acid chain that is two carbons shorter.

23.3 How Are Odd-Carbon Fatty Acids Oxidized?
Humans and animals metabolize odd-carbon fatty acids via the β-oxidation pathway, with the final product being propionyl-CoA. Three specialized enzymes then convert propionyl-CoA to succinyl-CoA, a TCA cycle intermediate. The pathway involves an initial ATP-dependent carboxylation (by propionyl-CoA carboxylase) at the α-carbon of propionyl-CoA to produce D-methylmalonyl-CoA, which is converted to the L-isomer by methylmalonyl-CoA epimerase. The L-isomer is the substrate for methylmalonyl-CoA mutase, which catalyzes a migration of a carbonyl-CoA group from one carbon to its neighbor, yielding succinyl-CoA.

23.4 How Are Unsaturated Fatty Acids Oxidized?
Two additional mitochondrial enzymes—an isomerase and a novel reductase—are required to handle the *cis*-double bonds of naturally occurring fatty acids. Consider the breakdown of oleic acid. The reactions of β-oxidation proceed normally through three cycles, producing three molecules of acetyl-CoA and leaving the product *cis*-Δ^3-dodecenoyl-CoA. This intermediate is not a substrate for acyl-CoA dehydrogenase. Instead, enoyl-CoA isomerase rearranges the *cis*-Δ^3 double bond to a *trans*-Δ^2 double bond, which can proceed through the normal route of β-oxidation.

23.5 Are There Other Ways to Oxidize Fatty Acids?
Organelles other than mitochondria, including peroxisomes and glyoxysomes, also carry out β-oxidation processes. Peroxisomal β-oxidation is similar to mitochondrial β-oxidation, except that the initial double-bond formation is catalyzed by an FAD-dependent acyl-CoA oxidase, which transfers the liberated electrons directly to oxygen instead of the electron-transport chain. Short-chain products must be transferred to the mitochondria for further breakdown. Similar β-oxidation enzymes are also found in glyoxysomes.

Branched-chain fatty acids with alkyl branches at odd-numbered carbons are not effective substrates for β-oxidation. For such species, α-oxidation is a useful alternative. Ruminants oxidize phytol to phytanic acid. The methyl group at C-3 will block β-oxidation, but phytanic acid α-hydroxylase places an —OH group at the α-carbon, and phytanic acid α-oxidase decarboxylates it to yield pristanic acid. The CoA ester of this metabolite can undergo β-oxidation in the normal manner. The terminal product, isobutyryl-CoA, can be sent into the TCA cycle by conversion to succinyl-CoA.

23.6 What Are Ketone Bodies, and What Role Do They Play in Metabolism?
Acetone, acetoacetate, and β-hydroxybutyrate are known as ketone bodies. These three metabolites are synthesized primarily in the liver but are important sources of fuel and energy for many tissues, including brain, heart, and skeletal muscle. During periods of starvation, ketone bodies may be the major energy source for the brain. Acetoacetate and 3-hydroxybutyrate are normal substrates for kidney cortex and for heart muscle.

FOUNDATIONAL BIOCHEMISTRY Things You Should Know After Reading Chapter 23.

- The essential features of fat mobilization from adipose tissue and dietary sources.

- How Franz Knoop elucidated the essence of β-oxidation.

- The mechanism by which acyl-CoA synthetases activate fatty acids for degradation.

- The essential features of formation and transport of acylcarnitines across the inner mitochondrial membrane.

- The mitochondrial organization of the enzymes involved in β-oxidation.

- The essential features of the chemistry of β-oxidation.

- The mechanism of the acyl-CoA dehydrogenase.

- The stereochemistry of formation and oxidation of hydroxyacyl-CoAs.

- The mechanism and energetics of the thiolase reaction.

- The stoichiometry of β-oxidation of palmitic acid.

- The reactions that produce and process propionyl-CoA from odd-carbon fatty acids.

- The mechanism of the methylmalonyl-CoA epimerase.

- The chemical and catalytic properties of vitamin B_{12}.

- The mechanism of oxidative decarboxylation by malic enzyme.

- The β-oxidation of unsaturated fatty acids.

- The unique reactions involved in peroxisomal β-oxidation of fatty acids.

- The manner in which branched-chain fatty acids are degraded in α-oxidation.

- The essential features of ω-oxidation.

- The essential features of ketogenesis and the identities of the ketone bodies.

- The reactions that form and interconvert the ketone bodies.

PROBLEMS

Answers to all problems are at the end of this book. Detailed solutions are available in the *Student Solutions Manual, Study Guide, and Problems Book*.

1. **Determining the Amount of Water Produced from Fatty Acid Oxidation** Calculate the volume of metabolic water available to a camel through fatty acid oxidation if it carries 30 pounds of triacylglycerol in its hump.

2. **Determining the Amount of ATP Produced from Fatty Acid Oxidation** Calculate the approximate number of ATP molecules that can be obtained from the oxidation of *cis*-11-heptadecenoic acid to CO_2 and water.

3. **Assessing the Mechanism of Oxidation of Phytanic Acid** Phytanic acid, the product of chlorophyll that causes problems for individuals with Refsum's disease, is 3,7,11,15-tetramethyl hexadecanoic acid. Suggest a route for its oxidation that is consistent with what you have learned in this chapter. (*Hint:* The methyl group at C-3 effectively blocks hydroxylation and normal β-oxidation. You may wish to initiate breakdown in some other way.)

4. **Examining the Labeling of Glucose from ^{14}C-labeled Acetate** Even though acetate units, such as those obtained from fatty acid oxidation, cannot be used for net synthesis of carbohydrate in animals, labeled carbon from ^{14}C-labeled acetate can be found in

newly synthesized glucose (for example, in liver glycogen) in animal tracer studies. Explain how this can be. Which carbons of glucose would you expect to be the first to be labeled by ^{14}C-labeled acetate?

5. **Assessing the Fatty Acid-Binding Capability of Serum Albumin** Human serum albumin (66.4 kD) is a soluble protein present in blood at 0.75 mM or so. Among other functions, albumin acts as the major transport vehicle for fatty acids in the circulation, carrying fatty acids from storage sites in adipose tissue to their sites of oxidation in liver and muscle. The albumin molecule has up to 11 distinct binding sites. Consult the biochemical literature to learn about the fatty acid–binding sites of albumin. Where are they located on the protein? What are their relative affinities? (Two suitable references with which to begin your study are Bhattacharya, A. A., Grüne, T., and Curry, S., 2000. Crystallographic analysis reveals common modes of binding of medium- and long-chain fatty acids in human serum albumin. *Journal of Molecular Biology* **303:**721–732; and Simard, J. R., Zunszain, P. A., Ha, C-E., et al., 2005. Locating high-affinity fatty acid-binding sites on albumin by X-ray crystallography and NMR spectroscopy. *Proceedings of the National Academy of Sciences U.S.A.* **102:**17958–17963.)

6. **Understanding the Implications of Storing Energy in the Form of Fat Versus Protein or Carbohydrate** Overweight individuals who diet to lose weight often view fat in negative ways because adipose tissue is the repository of excess caloric intake. However, the "weighty" consequences might be even worse if excess calories were stored in other forms. Consider a person who is 10 pounds "overweight," and estimate how much more he or she would weigh if excess energy were stored in the form of carbohydrate instead of fat.

7. **Understanding the Consequences of Vitamin B$_{12}$ Deficiency** What would be the consequences of a deficiency in vitamin B$_{12}$ for fatty acid oxidation? What metabolic intermediates might accumulate?

8. **Balancing the Equations for Fatty Acid Oxidation** Write properly balanced chemical equations for the oxidation to CO_2 and water of (a) myristic acid, (b) stearic acid, (c) α-linolenic acid, and (d) arachidonic acid.

9. **Tracing Tritium Incorporation into Acetate from Fatty Acid Oxidation** How many tritium atoms are incorporated into acetate if a molecule of palmitic acid is oxidized in 100% tritiated water?

10. **Understanding the Consequences of Carnitine Deficiency** What would be the consequences of a carnitine deficiency for fatty acid oxidation?

11. **Assessing Energy Consumption by a Migrating Bird** The ruby-throated hummingbird flies 500 miles nonstop across the Gulf of Mexico. The flight takes 10 hours at 50 mph. The hummingbird weighs about 4 grams at the start of the flight and about 2.7 grams at the end. Assuming that all the lost weight is fat burned for the flight, calculate the total energy required by the hummingbird in this prodigious flight. Does anything about the results of this calculation strike you as unusual?

12. **Understanding Human Energy Consumption During Exercise** Energy production in animals is related to oxygen consumption. The ruby-throated hummingbird consumes about 250 mL of oxygen per hour during its migration across the Gulf of Mexico. Use this number and the data in problem 11 to determine a conversion factor for energy expended per liter of oxygen consumed. If a human being consumes 12.7 kcal/min while running 8-minute miles, how long could a human run on the energy that the hummingbird consumes in its trans-Gulf flight? How many 8-minute miles would a person have to run to lose 1 pound of body fat?

13. **Understanding the Mechanism of the HMG-CoA Synthase Reaction** Write a reasonable mechanism for the HMG-CoA synthase reaction shown in Figure 23.26.

14. **Understanding the Mechanism of the Methylmalonyl-CoA Mutase Reaction** The methylmalonyl-CoA mutase reaction (see Figure 23.18) involves vitamin B$_{12}$ as a coenzyme. Write a reasonable mechanism for this reaction. (Hint: The reaction begins with

abstraction of a hydrogen atom—that is, a proton plus an electron—from the substrate by vitamin B$_{12}$. Consider the chemistry shown in A Deeper Look: The Activation of Vitamin B$_{12}$ as you write your mechanism.)

15. **Assessing the Oxidation States of Cobalt in the Methylmalonyl-CoA Mutase Reaction** Discuss the changes of the oxidation state of cobalt in the course of the methylmalonyl-CoA mutase reaction. Why do they occur as you suggested in your answer to problem 14?

16. **Extending the Mechanism of Methylmalonyl-CoA Mutase to Similar Reactions** Based on the mechanism for the methylmalonyl-CoA mutase (see problem 14), write reasonable mechanisms for the following reactions shown.

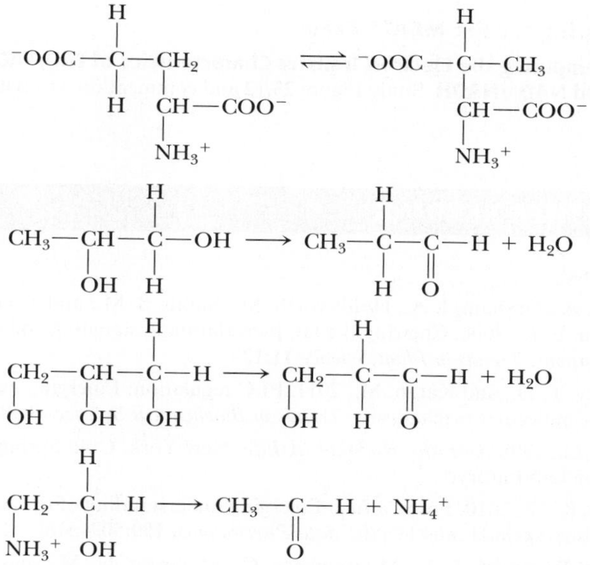

$$^-OOC-\overset{\overset{\displaystyle H}{|}}{\underset{\underset{\displaystyle H}{|}}{C}}-\overset{CH_2}{\underset{\underset{\displaystyle NH_3^+}{|}}{\underset{\displaystyle CH-COO^-}{|}}} \rightleftharpoons {}^-OOC-\overset{\overset{\displaystyle H}{|}}{\underset{\underset{\displaystyle NH_3^+}{|}}{C}}-\overset{CH_3}{\underset{\displaystyle CH-COO^-}{|}}$$

$$CH_3-\overset{\overset{\displaystyle H}{|}}{\underset{\underset{\displaystyle OH}{|}}{C}}-\overset{\overset{\displaystyle H}{|}}{\underset{\displaystyle H}{C}}-OH \longrightarrow CH_3-\overset{\overset{\displaystyle H}{|}}{\underset{\underset{\displaystyle H}{|}}{C}}-\overset{C}{\underset{\displaystyle O}{\|}}-H + H_2O$$

$$CH_2-\overset{\overset{\displaystyle H}{|}}{\underset{\underset{\displaystyle OH}{|}}{C}}-\overset{\overset{\displaystyle H}{|}}{\underset{\underset{\displaystyle OH}{|}}{C}}-H \longrightarrow CH_2-\overset{\overset{\displaystyle H}{|}}{\underset{\underset{\displaystyle OH}{|}}{C}}-\overset{C}{\underset{\displaystyle O}{\|}}-H + H_2O$$

$$CH_2-\overset{\overset{\displaystyle H}{|}}{\underset{\underset{\displaystyle OH}{|}}{\underset{\displaystyle NH_3^+}{}}}{C}-H \longrightarrow CH_3-\overset{C}{\underset{\displaystyle O}{\|}}-H + NH_4^+$$

17. **Understanding the Toxicity of Unripened Akee Fruit** A popular dish in the Caribbean islands consists of "akee and salt fish." Unripened akee fruit (from the akee tree, native to West Africa but brought to the Caribbean by African slaves) is quite poisonous. The unripened fruit contains hypoglycin (structure shown below), a metabolite that serves as a substrate for acyl-CoA dehydrogenase. However, the product of this reaction irreversibly inhibits the acyl-CoA dehydrogenase by reacting covalently with FAD on the enzyme. Consumption of unripened akee fruit can lead to vomiting and, in severe cases, convulsions, coma, and death. Write a reaction scheme to show the product of the acyl-CoA dehydrogenase reaction that reacts with FAD.

The akee tree.

$$H_2C=\overset{\overset{\displaystyle CH_2}{\diagup\diagdown}}{C}-CH-CH_2-\overset{\overset{\displaystyle H}{|}}{\underset{\underset{\displaystyle NH_3^+}{|}}{C}}-COO^-$$

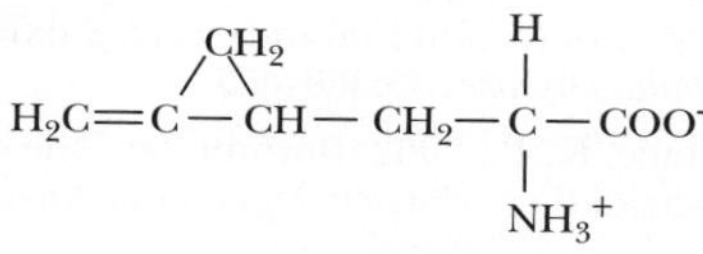

Hypoglycin A

18. **Exploring Substrate Channeling in Multi-Functional Enzyme Aggregates** In mammalian mitochondria, three of the enzymes in the membrane-bound β-oxidation system are components of a multifunctional enzyme (MFE). Similarly, glyoxysomes (in plants), peroxisomes (in most species), and Gram-negative bacteria carry out several reactions of β-oxidation via MFEs. In all these cases, the components of an MFE must cooperate to carry or "channel" the acyl-CoA substrate from one active site to the next in a cyclic fashion. The structure of the complete MFE from *Pseudomonas fragi* provides insights into how this channeling might occur. Consult the journal article that describes this structure (Ishikawa, M., Tsuchiya, D., et al., 2004. *EMBO Journal* **23**:2745–2754), and explain in your own words how this substrate channeling occurs.

Preparing for the MCAT® Exam

19. **Comparing the Electron Transfer Characteristics of FAD/FADH₂ and NAD⁺/NADH** Study Figure 23.12 and comment on why nature uses FAD/FADH₂ as a cofactor in the acyl-CoA dehydrogenase reaction rather than NAD⁺/NADH.

20. **Understanding a Ubiquitous Series of Metabolic Reactions** Study Figure 23.9. Where else in metabolism have you seen the chemical strategy and logic of the β-oxidation pathway? Why is it that these two pathways are carrying out the same chemistry?

ActiveModels Problem

21. Using the ActiveModel for enoyl-CoA dehydratase, give an example of a case in which conserved residues in slightly different positions can change the catalytic rate of reaction.

FURTHER READING

General

Baker, A., Graham, I. A., Holdsworth, M., Smith, S. M., and Theodoulou, F. L., 2006. Chewing the fat: β-oxidation in signaling and development. *Trends in Plant Science* **11**:124–132.

Bunney, T. D., and Katan, M., 2011. PLC regulation: Emerging pictures for molecular mechanisms. *Trends in Biochemical Sciences* **36**:88–96.

Ferry, G., 1998. *Dorothy Hodgkin: A Life.* New York: Cold Spring Harbor Laboratory.

Frayn, K. N., 2010. Fat as a fuel: Emerging understanding of the adipose tissue–skeletal muscle axis. *Acta Physiologica* **199**:509–518.

Fu, Z., Runquist, J. A., Montgomery, C., Miziorko, H. M., and Kim, J.-J. P., 2010. Functional insights into human HMG-CoA lyase from structures of acyl-CoA-containing ternary complexes. *Journal of Biological Chemistry* **285**:26341–26349.

Kleiner, S., Nguyen-Tran, V., Huang, X., Spiegelman, B., and We, Z., 2009. PPARδ agonism activates fatty acid oxidation via PGC-1α but does not increase mitochondrial gene expression and function. *The Journal of Biological Chemistry* **284**:18624–18633.

Lomb, D. J., Laurent, G., and Haigis, M. C., 2010. Sirtuins regulate key aspects of lipid metabolism. *Biochimica et Biophysica Acta* **1804**:1652–1657.

Scriver, C. R., Beaudet, A. L., et al., 1995. *The Metabolic and Molecular Bases of Inherited Disease,* 7th ed. New York: McGraw-Hill.

Steinberg, G. R., 2009. Role of the AMP-activated protein kinase in regulating fatty acid metabolism during exercise. *Applied Physiology, Nutrition and Metabolism* **34**:315–322.

Thomson, D. M., and Winder, W. W., 2009. AMP-activated protein kinase control of fat metabolism in skeletal muscle. *Acta Physiologica* **196**:147–154.

Vance, D. E., and Vance, J. E., 2008. *Biochemistry of Lipids, Lipoproteins and Membranes,* 5th ed. Amsterdam: Elsevier.

β-Oxidation

Henriques, B. J., Olsen, R. K., Bross, P., and Gomes, C. M., 2010. Emerging roles for riboflavin in functional rescue of mitochondrial oxidation flavoenzymes. *Current Medicinal Chemistry* **17**:3842–3854.

Houten, S. M., and Wanders, R. J. A., 2010. A general introduction to the biochemistry of mitochondrial fatty acid β-oxidation. *Journal of Inherited Metabolic Diseases* **33**:469–477.

Kim, J.-J., and Battaile, K. P., 2002. Burning fat: The structural basis of fatty acid beta-oxidation. *Current Opinion in Structural Biology* **12**:721–728.

Wanders, R. J. A., Ruiter, J. P. N., Ijist, L., Waterham, H. R., and Houten, S. M., 2010. The enzymology of mitochondrial fatty acid beta-oxidation and its application to follow-up analysis of positive neonatal screening results. *Journal of Inherited Metabolic Diseases* **33**:479–494.

Acyl-CoA Dehydrogenases

Ghisla, S., and Thorpe, C., 2004. Acyl-CoA dehydrogenases: A mechanistic overview. *European Journal of Biochemistry* **271**:494–508.

He, M., Pei, Z., Mohsen, A., Watkins, P., Murdoch, G., et al., 2011. Identification and characterization of new long chain acyl-CoA dehydrogenase. *Molecular Genetics and Metabolism* **102**:418–429.

Jethva, R., Bennett, M. J., and Vockley, J., 2008. Short-chain acyl-coenzyme A dehydrogenase deficiency. *Molecular Genetics and Metabolism* **95**:195–200.

Kim, J.-J., and Miura, R., 2004. Acyl-CoA dehydrogenases and acyl-CoA oxidases. *European Journal of Biochemistry* **271**:483–493.

McAndrew, R. P., Wang, Y., et al., 2008. Structural basis for substrate fatty-acyl chain specificity: Crystal structure of human very-long-chain acyl-CoA dehydrogenase. *Journal of Biological Chemistry* **283**:9435–9443.

Rao, K. S., Albro, M., et al., 2006. Kinetic mechanism of glutaryl-CoA dehydrogenase. *Biochemistry* **45**:15853–15861.

Regulation and Defects of Fatty Acid Oxidation

Gregersen, N., Bross, P., et al., 2004. Genetic defects in fatty acid beta-oxidation and acyl-CoA dehydrogenases. *European Journal of Biochemistry* **271**:470–482.

Oey, N. A., Den Boer, M. E. J., et al., 2005. Long-chain fatty acid oxidation during early human development. *Pediatric Research* **57**:755–759.

Spiekerkoetter, U., Bastin, J., Gillingham, M., Morris, A., et al., 2010. Current issues regarding treatment of mitochondrial fatty acid oxidation disorders. *Journal of Inherited Metabolic Diseases* **33**:555–561.

Spiekerkoetter, U., and Wood, P. A., 2010. Mitochondrial fatty acid oxidation disorders: Pathophysiological studies in mouse models. *Journal of Inherited Metabolic Diseases* **33**:539–546.

Branched-Chain and Unsaturated Fatty Acid Oxidation

Lloyd, M. D., Darley, D. J., Wierzbicki, A. S., and Threadgill, M. D., 2008. α-Methylacyl-CoA racemase—an 'obscure' metabolic enzyme takes centre stage. *The FEBS Journal* **275**:1089–1102.

Ran-Ressler, R. R., Sim, D., O'Donnell-Megaro, A. M., Bauman, D. E., et al., 2011. Branched chain fatty acid content of United States retail milk and implications for dietary intake. *Lipids* **46**:569–576.

Stepanov, I., Baeten, V., Abbas, O., Colman, E., et al., 2010. Analysis of milk odd- and branched-chain fatty acids using Fourier transform (FT)-Raman spectroscopy. *Journal of Agricultural and Food Chemistry* **58**:10804–10811.

Peroxisomes and Phytanic Acid Oxidation

Grimaldi, P. A., 2007. Peroxisome proliferator-activated receptors as sensors of fatty acids and derivatives. *Cell and Molecular Life Science* **64**: 2459–2464.

Hellgren, L. I., 2010. Phytanic acid—an overlooked bioactive fatty acid in dairy fat? *Annals of the New York Academy of Sciences* **1190**:42–49.

Poirer, Y., Antonenkov, V. D., Glumoff, T., and Hiltunen, J. K., 2006. Peroxisomal β-oxidation—a metabolic pathway with multiple functions. *Biochimica et Biophysica Acta* **1763**:1413–1426.

Schrader, M., and Fahimi, H. D., 2008. The peroxisome: Still a mysterious organelle. *Histochemistry and Cell Biology* **129**:421–440.

Van Veldhoven, P. P., 2010. Biochemistry and genetics of inherited disorders of peroxisomal fatty acid metabolism. *Journal of Lipid Research* **51**:2863–2895.

Wanders, R. J. A., and Waterham, H. R., 2006. Biochemistry of mammalian peroxisomes revisited. *Annual Review of Biochemistry* **75**:295–332.

Ylianttila, M. S., Pursiainen, N. V., et al., 2006. Crystal structure of yeast peroxisomal multifunctional enzyme: Structural basis for substrate specificity of (3R)-hydroxyacyl-CoA dehydrogenase units. *Journal of Molecular Biology* **358**:1286–1295.

Metabolic Syndrome

Deng, Y., and Scherer, P. E., 2010. Adipokines as novel biomarkers and regulators of the metabolic syndrome. *Annals of the New York Academy of Sciences* **1212**:E1–E19.

Gehart, H., Kumpf, S., Ittner, A., and Ricci, R., 2010. MAPK signaling in cellular metabolism: Stress or wellness. *EMBO Reports* **11**:834–840.

Rasmusson, A. M., Schnurr, P. P., Zukowska, Z., Scioli, E., and Forman, D. E., 2010. Adaptation to extreme stress: Post-traumatic stress disorder, neuropeptide Y and metabolic syndrome. *Experimental Biology and Medicine* **235**:1150–1162.

Multifunctional Enzyme Complexes

Gregersen, N., and Olsen, R. K. J., 2010. Disease mechanisms and protein structures in fatty acid oxidation defects. *Journal of Inherited Metabolic Diseases* **33**:547–553.

Ishikawa, M., Tsuchiya, D., et al., 2004. Structural basis for channeling mechanism of a fatty acid β-oxidation multienzyme complex. *EMBO Journal* **23**:2745–2754.

Carnitine Biosynthesis

Strijbis, K., van Roermund, C. W. T., Hardy, G. P., van den Burg, J., et al., 2009. Identification and characterization of a complete carnitine biosynthesis pathway in *Candida albicans*. *The FASEB Journal* **23**:2349–2359.

Strijbis, K., Vaz, F. M., and Distel, B., 2010. Enzymology of the carnitine biosynthesis pathway. *IUBMB Life* **62**:357–362.

Tang, L., Bai, L., Wang, W., and Jiang, T., 2010. Crystal structure of the carnitine transporter and insights into the antiport mechanism. *Nature Structural & Molecular Biology* **17**:492–497.

Fatty Acid-Binding Proteins

Atshaves, B. P., Martin, G. G., Hostetler, H. A., McIntosh, A. L., et al., 2010. Liver fatty acid-binding protein and obesity. *Journal of Nutritional Biochemistry* **21**:1015–1032.

Catalá, A., 2010. The function of very long chain polysaturated fatty acids in the pineal gland. *Biochimica et Biophysica Acta* **1801**:95–99.

Plutzky, J., 2009. The mighty, mighty fatty acid. *Nature Medicine* **15**:618–619.

Storch, J., and Thumser, A. E., 2010. Tissue-specific functions in the fatty acid-binding protein family. *Journal of Biological Chemistry* **285**:32679–32683.

Lipid Biosynthesis

© Alissa Crandall/CORBIS

◀ Walruses basking on the beach.

To everything there is a season, and a time for every purpose under heaven…A time to break down and a time to build up.

Ecclesiastes 3:1–3

KEY QUESTIONS

24.1 How Are Fatty Acids Synthesized?

24.2 How Are Complex Lipids Synthesized?

24.3 How Are Eicosanoids Synthesized, and What Are Their Functions?

24.4 How Is Cholesterol Synthesized?

24.5 How Are Lipids Transported Throughout the Body?

24.6 How Are Bile Acids Biosynthesized?

24.7 How Are Steroid Hormones Synthesized and Utilized?

ESSENTIAL QUESTION

We turn now to the biosynthesis of lipid structures. We begin with a discussion of the biosynthesis of fatty acids, stressing the basic pathways, additional means of elongation, mechanisms for the introduction of double bonds, and regulation of fatty acid synthesis. Sections then follow on the biosynthesis of glycerophospholipids, sphingolipids, eicosanoids, and cholesterol. The transport of lipids through the body in lipoprotein complexes is described, and the chapter closes with discussions of the biosynthesis of bile salts and steroid hormones.

What are the pathways of lipid synthesis in biological systems?

We have already seen several cases in which the synthesis of a class of biomolecules is conducted differently from degradation (glycolysis versus gluconeogenesis and glycogen or starch breakdown versus polysaccharide synthesis, for example). Likewise, the synthesis of fatty acids and other lipid components is different from their degradation. Fatty acid synthesis involves a set of reactions that follow a strategy different in several ways from the corresponding degradative process:

1. Intermediates in fatty acid synthesis are linked covalently to the sulfhydryl groups of special proteins, the **acyl carrier proteins** (ACPs). In contrast, fatty acid breakdown intermediates are bound to the —SH group of coenzyme A.
2. Fatty acid synthesis occurs in the cytosol, whereas fatty acid degradation takes place in mitochondria.
3. In animals, the enzymes of fatty acid synthesis are components of one long polypeptide chain, the **fatty acid synthase,** whereas no similar association exists for the degradative enzymes. (Plants and bacteria employ separate enzymes to carry out the biosynthetic reactions.)
4. The coenzyme for the oxidation–reduction reactions of fatty acid synthesis is $NADP^+/NADPH$, whereas degradation involves the $NAD^+/NADH$ couple.

ʘWL Online homework and a Student Self Assessment for this chapter may be assigned in OWL.

24.1 : How Are Fatty Acids Synthesized?

Formation of Malonyl-CoA Activates Acetate Units for Fatty Acid Synthesis

The design strategy for fatty acid synthesis is this:

1. Fatty acid chains are constructed by the addition of two-carbon units derived from *acetyl-CoA*.
2. The acetate units are activated by formation of *malonyl-CoA* (at the expense of ATP).
3. The addition of two-carbon units to the growing chain is driven by decarboxylation of malonyl-CoA.
4. The elongation reactions are repeated until the growing chain reaches 16 carbons in length (palmitic acid).
5. Other enzymes then add double bonds and additional carbon units to the chain.

Fatty Acid Biosynthesis Depends on the Reductive Power of NADPH

The net reaction for the formation of palmitate from acetyl-CoA is

$$\text{Acetyl-CoA} + 7 \text{ malonyl-CoA}^- + 14 \text{ NADPH} + 13 \text{ H}^+ + \text{H}_2\text{O} \longrightarrow$$
$$\text{palmitate}^- + 7 \text{ HCO}_3^- + 8 \text{ CoASH} + 14 \text{ NADP}^+$$

(Levels of free fatty acids are very low in the typical cell. The palmitate made in this process is rapidly converted to CoA esters in preparation for the formation of triacylglycerols and phospholipids.)

Cells Must Provide Cytosolic Acetyl-CoA and Reducing Power for Fatty Acid Synthesis

Eukaryotic cells face a dilemma in providing suitable amounts of substrate for fatty acid synthesis. Sufficient quantities of acetyl-CoA, malonyl-CoA, and NADPH must be generated *in the cytosol* for fatty acid synthesis. Malonyl-CoA is made by carboxylation of acetyl-CoA, so the problem reduces to generating sufficient acetyl-CoA and NADPH.

There are three principal sources of acetyl-CoA (Figure 24.1):

1. Amino acid degradation produces cytosolic acetyl-CoA.
2. Fatty acid oxidation produces mitochondrial acetyl-CoA.
3. Glycolysis yields cytosolic pyruvate, which (after transport into the mitochondria) is converted to acetyl-CoA by pyruvate dehydrogenase.

The acetyl-CoA derived from amino acid degradation is normally insufficient for fatty acid biosynthesis, and the acetyl-CoA produced by pyruvate dehydrogenase and by fatty acid oxidation cannot cross the mitochondrial membrane to participate directly in fatty acid synthesis. Instead, acetyl-CoA is linked with oxaloacetate to form citrate, which is transported from the mitochondrial matrix to the cytosol (Figure 24.1). Here it can be converted back into acetyl-CoA and oxaloacetate by **ATP–citrate lyase.** In this manner, mitochondrial acetyl-CoA becomes the substrate for cytosolic fatty acid synthesis. (Oxaloacetate returns to the mitochondria in the form of either pyruvate or malate, which is then reconverted to acetyl-CoA and oxaloacetate, respectively.)

NADPH can be produced in the pentose phosphate pathway as well as by malic enzyme (Figure 24.1). Reducing equivalents (electrons) derived from glycolysis in the form of NADH can be transformed into NADPH by the combined action of malate dehydrogenase and malic enzyme:

$$\text{Oxaloacetate} + \text{NADH} + \text{H}^+ \longrightarrow \text{malate} + \text{NAD}^+$$
$$\text{Malate} + \text{NADP}^+ \longrightarrow \text{pyruvate} + \text{CO}_2 + \text{NADPH} + \text{H}^+$$

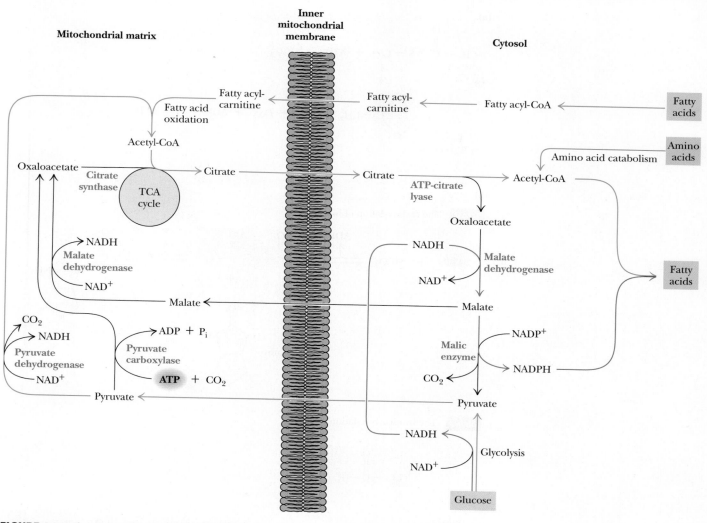

FIGURE 24.1 The citrate–malate–pyruvate shuttle provides cytosolic acetate units and some reducing equivalents (electrons) for fatty acid synthesis. The shuttle collects carbon substrates, primarily from glycolysis but also from fatty acid oxidation and amino acid catabolism. Pathways that provide carbon for fatty acid synthesis are shown in blue; pathways that supply electrons for fatty acid synthesis are shown in red.

How many of the 14 NADPH needed to form one palmitate (see equation on page 792) can be made in this way? The answer depends on the status of malate. Every citrate entering the cytosol produces one acetyl-CoA and one malate (Figure 24.1). Every malate oxidized by malic enzyme produces one NADPH, at the expense of a decarboxylation to pyruvate. Thus, when malate is oxidized, one NADPH is produced for every acetyl-CoA. Conversion of 8 acetyl-CoA units to one palmitate would then be accompanied by production of 8 NADPH. (The other 6 NADPH required, as shown in the equation on page 792, would be provided by the pentose phosphate pathway.) On the other hand, for every malate returned to the mitochondria, one NADPH fewer is produced.

Acetate Units Are Committed to Fatty Acid Synthesis by Formation of Malonyl-CoA

Rittenberg and Bloch showed in the late 1940s that acetate units are the building blocks of fatty acids. Their work, together with the discovery by Salih Wakil that bicarbonate is required for fatty acid biosynthesis, eventually made clear that this pathway involves synthesis of *malonyl-CoA*. The carboxylation of acetyl-CoA to form malonyl-CoA is essentially irreversible and is the **committed step** in the synthesis of fatty acids

(a)

$$CH_3 - \overset{\overset{\displaystyle O}{\|}}{C} - S - CoA \ + \ \textbf{ATP} \ + \ HCO_3^-$$

$$\overset{\overset{\displaystyle O}{\|}}{\underset{-O}{C}} - CH_2 - \overset{\overset{\displaystyle O}{\|}}{C} - S - CoA \ + \ ADP \ + \ P_i \ + \ H^+$$

(b)

Step 1 The carboxylation of biotin

ATP + HCO₃⁻ ⇌ (ADP) ⇌ ⁻O—C(=O)—O—P(=O)(O⁻)O⁻ ⇌ (Pᵢ) ⁻O—C(=O)—N-carboxybiotin-Lys

Biotin

Step 2 The transcarboxylation reaction

H₂C—C(=O)—SCoA

⇌ H₂C—C(=O)—SCoA with COO⁻ + HN NH (biotin)-Lys

FIGURE 24.2 **(a)** The acetyl-CoA carboxylase reaction produces malonyl-CoA for fatty acid synthesis. **(b)** A mechanism for the acetyl-CoA carboxylase reaction. Bicarbonate is activated for carboxylation reactions by formation of N-carboxybiotin. ATP drives the reaction forward, with transient formation of a carbonylphosphate intermediate **(Step 1)**. In a typical biotin-dependent reaction, nucleophilic attack by the acetyl-CoA carbanion on the carboxyl carbon of N-carboxybiotin—a transcarboxylation—yields the carboxylated product **(Step 2)**.

(Figure 24.2). The reaction is catalyzed by **acetyl-CoA carboxylase,** which contains a biotin prosthetic group. This carboxylase is the only enzyme of fatty acid synthesis in animals that is not part of the multienzyme complex called fatty acid synthase.

Acetyl-CoA Carboxylase Is Biotin-Dependent and Displays Ping-Pong Kinetics

The biotin prosthetic group of acetyl-CoA carboxylase is covalently linked to the ε-amino group of an active-site lysine in a manner similar to pyruvate carboxylase (see Figure 22.2). The reaction mechanism is also analogous to that of pyruvate carboxylase (see Figure 22.3): ATP-driven carboxylation of biotin is followed by transfer of the

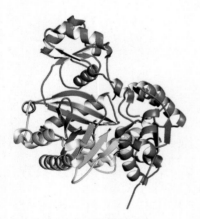

Biotin carboxylase
domain of human acetyl-CoA
carboxylase 2 (pdb id = 2HJW)

The biotin carboxylase domain from human acetyl-CoA carboxylase 2, with the A-subdomain in blue, the B-subdomain in red, the A–B linker in green, and the C-subdomain in yellow.

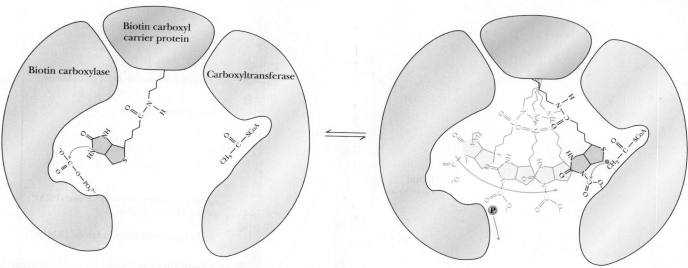

FIGURE 24.3 In the acetyl-CoA carboxylase reaction, the biotin ring, on its flexible tether, acquires carboxyl groups from carbonylphosphate on the biotin carboxylase subunit and transfers them to acyl-CoA molecules on the carboxyltransferase subunits. Colors of the domains correspond to those in Figure 24.4.

activated CO_2 to acetyl-CoA to form malonyl-CoA. The enzyme from *Escherichia coli* has three subunits: (1) a **biotin carboxyl carrier protein** (a dimer of 22.5-kD subunits); (2) **biotin carboxylase** (a dimer of 51-kD subunits), which adds CO_2 to the prosthetic group; and (3) **carboxyltransferase** (an $\alpha_2\beta_2$ tetramer with 30- and 35-kD subunits), which transfers the activated CO_2 unit to acetyl-CoA. The long, flexible biotin–lysine chain (biocytin) enables the activated carboxyl group to be carried between the biotin carboxylase and the carboxyltransferase (Figure 24.3).

Acetyl-CoA Carboxylase in Animals Is a Multifunctional Protein

In animals, acetyl-CoA carboxylase (ACC) is a filamentous polymer (4 to 8×10^6 D) composed of 265-kD protomers. Each of these subunits contains the biotin carboxyl carrier moiety, biotin carboxylase, and carboxyltransferase activities, as well as allosteric regulatory sites. Animal ACC is thus a multifunctional protein. The polymeric form is active, but the 265-kD protomers are inactive. The activity of ACC is thus dependent upon the position of the equilibrium between these two forms:

$$\text{Inactive protomers} \rightleftharpoons \text{active polymer}$$

Because this enzyme catalyzes the committed step in fatty acid biosynthesis, it is carefully regulated. *Palmitoyl-CoA,* the final product of fatty acid biosynthesis, shifts the equilibrium toward the inactive protomers, whereas *citrate,* an important allosteric activator of this enzyme, shifts the equilibrium toward the active polymeric form of the enzyme. Acetyl-CoA carboxylase shows the kinetic behavior of a Monod–Wyman–Changeux *V*-system[1] allosteric enzyme in which allosteric effectors shift the T/R equilibrium between active R conformers and inactive T conformers.

Phosphorylation of ACC Modulates Activation by Citrate and Inhibition by Palmitoyl-CoA

The regulatory effects of citrate and palmitoyl-CoA are dependent on the phosphorylation state of acetyl-CoA carboxylase. The animal enzyme is phosphorylated at eight to ten sites on each enzyme subunit (Figure 24.4). Some of these sites are regulatory, whereas

[1]The MWC *V*-system is an allosteric model characterized by (1) S binds equally well to R and T, but only R is active (T is inactive), and (2) effectors (A binds only to R; I binds only to T) regulate activity by shifting the relative R vs. T distribution.

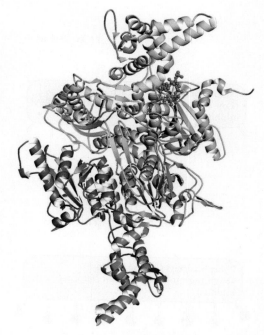

Carboxyltransferase domain of yeast acetyl-CoA carboxylase (pdb id = 1OD2)

The carboxyltransferase domain dimer of acetyl-CoA carboxylase-1 from *Saccharomyces cerevisiae.* The N- and C-subdomains of one monomer are cyan and yellow, whereas those of the other monomer are purple and green. CoA is shown as a ball-and-stick model in one subunit.

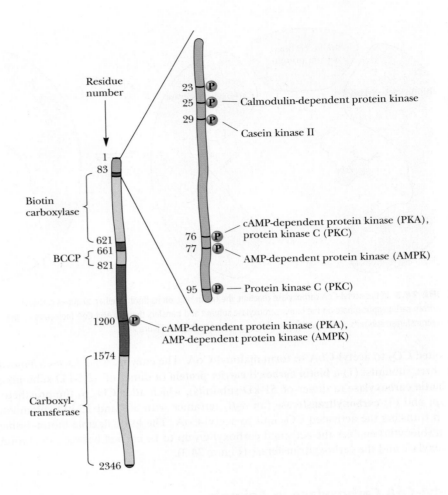

FIGURE 24.4 Schematic of the acetyl-CoA carboxylase polypeptide, with domains and phosphorylation sites indicated, along with the protein kinases responsible. Phosphorylation at Ser¹²⁰⁰ is primarily responsible for decreasing the affinity for citrate.

others are "silent" and have no effect on enzyme activity. Unphosphorylated acetyl-CoA carboxylase binds citrate with high affinity and thus is active at very low citrate concentrations (Figure 24.5). Phosphorylation of the regulatory sites decreases the affinity of the enzyme for citrate, and in this case, high levels of citrate are required to activate the carboxylase. The inhibition by fatty acyl-CoAs operates in a similar but opposite manner. Thus, low levels of fatty acyl-CoA inhibit the phosphorylated carboxylase, but the dephosphoenzyme is inhibited only by high levels of fatty acyl-CoA. Specific phosphatases act to dephosphorylate ACC, thereby increasing the sensitivity to citrate.

Acyl Carrier Proteins Carry the Intermediates in Fatty Acid Synthesis

The basic building blocks of fatty acid synthesis are acetyl and malonyl groups, but they are not transferred directly from CoA to the growing fatty acid chain. Rather, they are first passed to ACP. This protein consists (in *E. coli*) of a single polypeptide chain of 77 residues to which is attached (on a serine residue) a **phosphopantetheine group,** the same group that forms the "business end" of coenzyme A. Thus, ACP is a somewhat larger version of coenzyme A, specialized for use in fatty acid biosynthesis (Figure 24.6).

In Some Organisms, Fatty Acid Synthesis Takes Place in Multienzyme Complexes

The enzymes that catalyze formation of acetyl-ACP and malonyl-ACP and the subsequent reactions of fatty acid synthesis are organized quite differently in different organisms. Fatty acid synthesis in mammals occurs on homodimeric **fatty acyl synthase I**

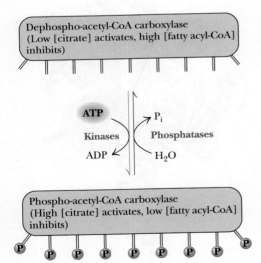

FIGURE 24.5 The activity of acetyl-CoA carboxylase is modulated by phosphorylation and dephosphorylation. The dephospho form of the enzyme is activated by low [citrate] and inhibited only by high levels of fatty acyl-CoA. In contrast, the phosphorylated form of the enzyme is activated only by high levels of citrate but is very sensitive to inhibition by fatty acyl-CoA.

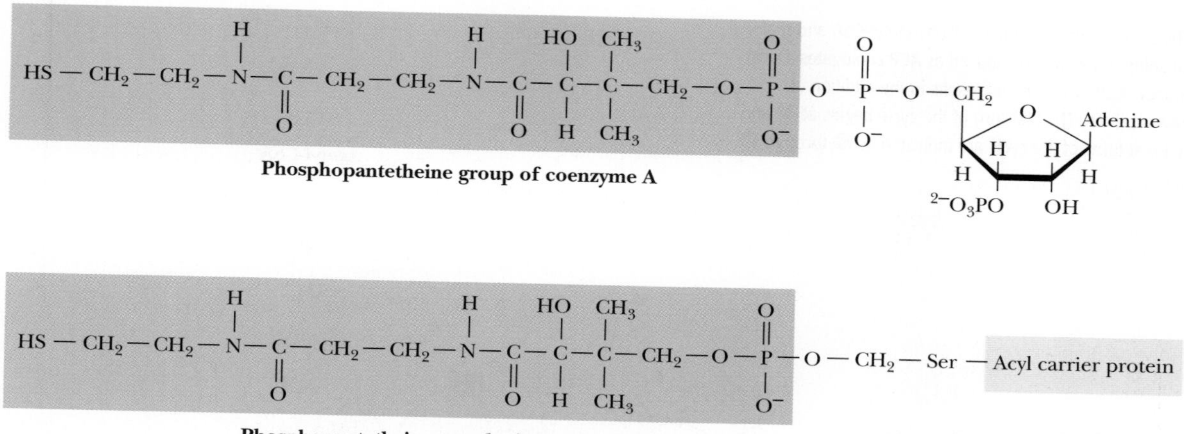

Phosphopantetheine group of coenzyme A

Phosphopantetheine prosthetic group of ACP

FIGURE 24.6 Fatty acids are conjugated both to coenzyme A and to acyl carrier protein through the sulfhydryl of phosphopantetheine prosthetic groups.

(FAS I), each 270-kD polypeptide of which contains all reaction centers required to produce a fatty acid. In lower eukaryotes, such as yeast and fungi, the enzymatic activities of FAS are distributed on two multifunctional peptide chains, which form 2.6-megadalton $\alpha_6\beta_6$ complexes. In plants, most bacteria, and parasites, the enzymes of fatty acid synthesis are separated and independent, and this collection of enzymes is referred to as **fatty acyl synthase II (FAS II)**.

The individual steps in the elongation of the fatty acid chain are quite similar across all organisms. The mammalian pathway (Figure 24.7) is a cycle of elongation that involves six enzyme activities. The elongation cycle is initiated by transfer of the acyl moiety of acetyl-CoA to the acyl carrier protein by the **malonyl-CoA–acetyl-CoA-ACP transacylase (MAT)**, which also transfers the malonyl group of malonyl-CoA to ACP.

Decarboxylation Drives the Condensation of Acetyl-CoA and Malonyl-CoA

The **β-ketoacyl-ACP synthase (KS)** catalyzes the decarboxylative condensation of the acyl group with malonyl-ACP to produce a β-ketoacyl-ACP intermediate (acetoacetyl-ACP in the first cycle). The mechanism (Figure 24.8) begins with acetyl group transfer to MAT, followed with attack by the ACP thiol sulfur to form an acetyl-ACP. The acetyl group is transferred to a cysteine sulfur on KS, freeing the ACP thiol to acquire the malonyl group. In the condensation reaction that follows, decarboxylation of the malonyl group creates a transient, highly nucleophilic carbanion that can attack the acetate group.

The net reaction for each turn of this cycle (see Figure 24.7) is addition of a two-carbon unit to the acyl group. *Why is the three-carbon malonyl group used here as a two-carbon donor?* The answer is that this is yet another example of a decarboxylation driving

> **A DEEPER LOOK**

Choosing the Best Organism for the Experiment

The selection of a suitable and relevant organism is an important part of any biochemical investigation. The studies that revealed the secrets of fatty acid synthesis are a good case in point.

The paradigm for fatty acid synthesis in plants has been the avocado, which has one of the highest fatty acid contents in the plant kingdom. Early animal studies centered primarily on pigeons, which are easily bred and handled and which possess high levels of fats in their tissues. Other animals, richer in fatty tissues, might be even more attractive but more challenging to maintain. Grizzly bears, for example, carry very large fat reserves but are difficult to work with in the lab!

FIGURE 24.7 The pathway of palmitate synthesis from acetyl-CoA and malonyl-CoA. Acetyl and malonyl building blocks are introduced as ACP conjugates. Decarboxylation drives the β-ketoacyl-ACP synthase and results in the addition of two-carbon units to the growing chain. The first turn of the cycle begins at ❶ and goes to butyryl-ACP; subsequent turns of the cycle are indicated as ❷ through ❻.

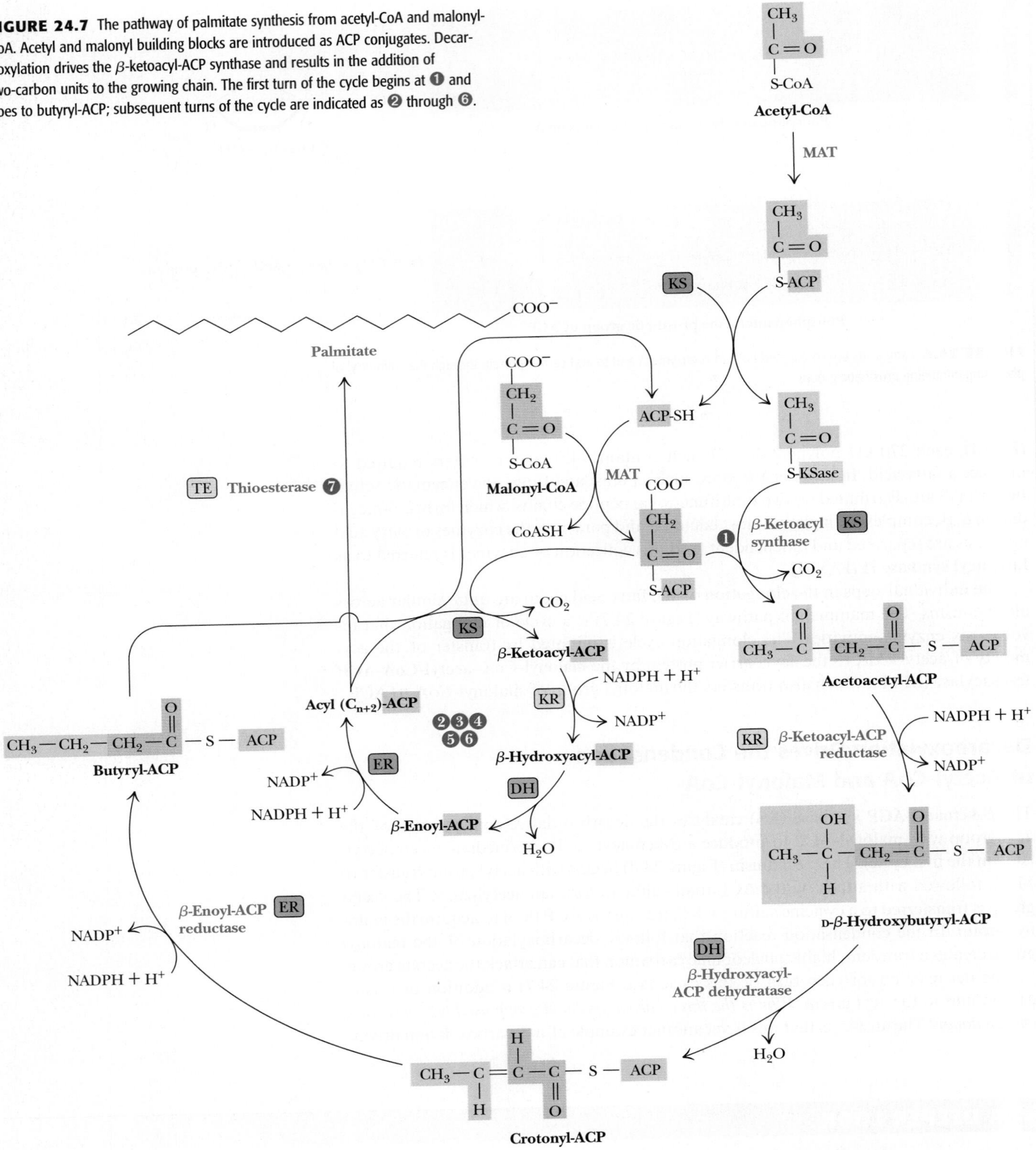

a desired but otherwise thermodynamically unfavorable reaction. The decarboxylation that accompanies the reaction with malonyl-ACP drives the synthesis of acetoacetyl-ACP. Note that hydrolysis of ATP drove the carboxylation of acetyl-CoA to form malonyl-ACP, so, indirectly, ATP is responsible for the condensation reaction to form acetoacetyl-ACP. Malonyl-CoA can be viewed as a form of stored energy for driving fatty acid synthesis.

It is also worth noting that the carbon of the carboxyl group that was added to drive this reaction is the one removed by the condensing enzyme. Thus, all the carbons of

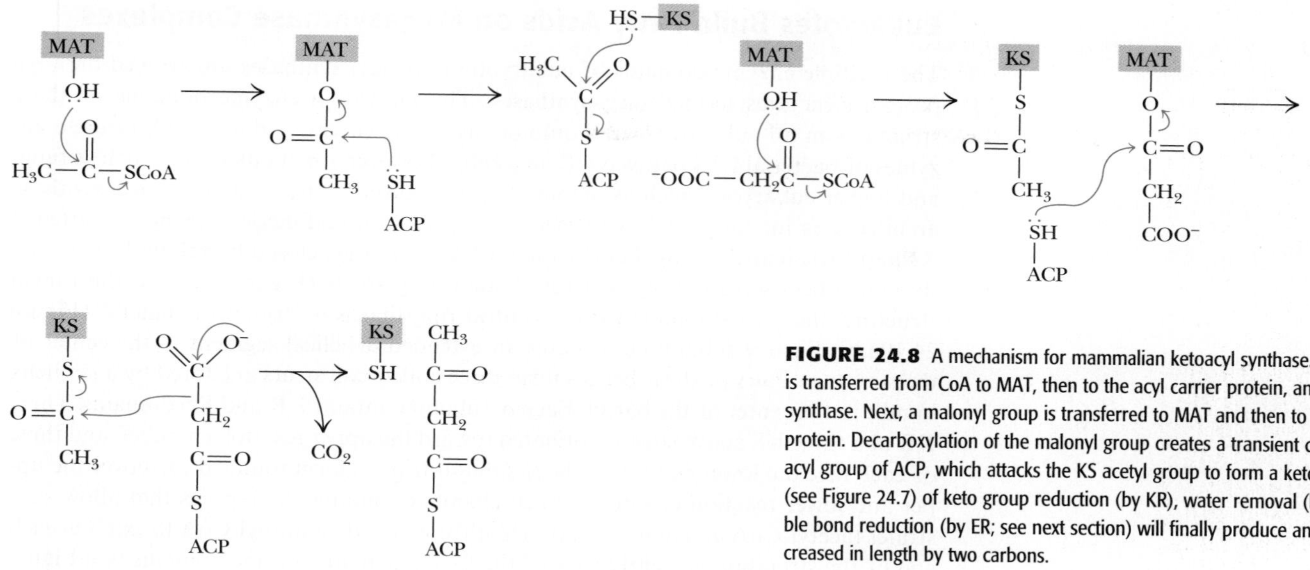

FIGURE 24.8 A mechanism for mammalian ketoacyl synthase. An acetyl group is transferred from CoA to MAT, then to the acyl carrier protein, and then to ketoacyl synthase. Next, a malonyl group is transferred to MAT and then to the acyl carrier protein. Decarboxylation of the malonyl group creates a transient carbanion on the acyl group of ACP, which attacks the KS acetyl group to form a ketoacyl-ACP. A cycle (see Figure 24.7) of keto group reduction (by KR), water removal (by DH), and double bond reduction (by ER; see next section) will finally produce an acyl group increased in length by two carbons.

acetoacetyl-ACP (*and* of the fatty acids to be made) are derived from acetate units of acetyl-CoA.

Reduction of the β-Carbonyl Group Follows a Now-Familiar Route

The next three steps—reduction of the β-carbonyl group by **β-ketoacyl-ACP reductase (KR)** to form a β-alcohol, then dehydration by **β-hydroxyacyl-ACP dehydratase (DH)** and reduction by **2,3-*trans*-enoyl-ACP reductase (ER)** to saturate the chain (see Figure 24.7)—look very similar to the fatty acid degradation pathway in reverse. However, there are two crucial differences between fatty acid biosynthesis and fatty acid oxidation (besides the fact that different enzymes are involved): First, the alcohol formed in biosynthesis has the D-configuration rather than the L-form seen in catabolism; second, the reducing coenzyme is NADPH, whereas NAD$^+$ and FAD are the oxidants in the catabolic pathway.

The net result of the first turn of the biosynthetic cycle is the synthesis of a four-carbon unit, a butyryl group, from two smaller building blocks. In the next cycle of the process, this butyryl-ACP condenses with another malonyl-ACP to make a six-carbon β-ketoacyl-ACP and CO$_2$. Subsequent reduction to a β-alcohol, dehydration, and another reduction yield a six-carbon saturated acyl-ACP. This cycle continues with the net addition of a two-carbon unit in each turn until the chain is 16 carbons long (see Figure 24.7). The KS cannot accommodate larger substrates, so the reaction cycle ends with a 16-carbon chain. Hydrolysis of the C$_{16}$-acyl-ACP yields a palmitic acid and the free ACP.

In the end, seven malonyl-CoA molecules and one acetyl-CoA yield a palmitate (shown here as palmitoyl-CoA):

$$\text{Acetyl-CoA} + 7 \text{ malonyl-CoA}^- + 14 \text{ NADPH} + 14 \text{ H}^+ \longrightarrow$$
$$\text{palmitoyl-CoA} + 7 \text{ HCO}_3^- + 14 \text{ NADP}^+ + 7 \text{ CoASH}$$

The formation of seven malonyl-CoA molecules requires

$$7 \text{ Acetyl-CoA} + 7 \text{ HCO}_3^- + 7 \text{ ATP}^{4-} \longrightarrow$$
$$7 \text{ malonyl-CoA}^- + 7 \text{ ADP}^{3-} + 7 \text{ P}_i^{2-} + 7 \text{ H}^+$$

Thus, the overall reaction of acetyl-CoA to yield palmitic acid is

$$8 \text{ Acetyl-CoA} + 7 \text{ ATP}^{4-} + 14 \text{ NADPH} + 7 \text{ H}^+ \longrightarrow$$
$$\text{palmitoyl-CoA} + 14 \text{ NADP}^+ + 7 \text{ CoASH} + 7 \text{ ADP}^{3-} + 7 \text{ P}_i^{2-}$$

Note: These equations are stoichiometric and are charge balanced. See problem 1 at the end of the chapter for practice in balancing these equations.

Eukaryotes Build Fatty Acids on Megasynthase Complexes

The multiple enzyme domains of eukaryotic fatty acyl synthases are arrayed on large protein structures termed **megasynthases.** The individual enzyme domains of these structures in all eukaryotes are homologous to the corresponding small, discrete enzymes of bacterial FAS pathways. Remarkably, however, lower eukaryotes such as fungi and higher eukaryotes such as mammals have evolved entirely different megasynthase architectures for fatty acid synthesis. Mammalian homodimeric FAS has a flattened X-shape, whereas the fungal dodecameric FAS is a large, closed barrel, with two reaction chambers separated by equatorial stabilizing struts (Figure 24.9). In the fungal structure, the six α-subunits form a central ring that is a "trimer of dimers" (Figure 24.10a,b). Each α-subunit contributes an extended α-helical segment to the center of the structure. Pairs of these helices form three coiled-coil struts anchored by a six-helix bundle in the center of the barrel. Each α-subunit contains KR and KS domains. Three KR and three KS active sites are oriented toward the upper reaction chamber, and three of each face the lower chamber. The β-subunit trimers form rounded caps over the upper and lower reaction chambers. Each chamber contains three pores that allow substrates (acetyl-CoA and malonyl-CoA) to diffuse in and palmitoyl-CoA to exit. On each end of the structure, the active sites of the four β-subunit enzyme domains (see Figure 24.9) are oriented toward the interior of the reaction chamber. Three ACP domains in each chamber shuttle growing acyl chains from site to site during the catalytic cycle. Each ACP is tethered by two flexible linker peptides, which facilitate its site-to-site movement (Figure 24.10c). The phosphopantetheine arm on each ACP can extend outward to reach into active sites or may retract to insert its acyl chain in a protective hydrophobic cavity during intersite transport.

The homodimeric mammalian FAS contains all six functional enzyme domains on each subunit (Figures 24.9 and 24.11). In the X-shaped dimer, three of the domains—including KS, ER, and DH—form dimeric structures, whereas the KR and MAT

AT: Acetyl transferase
MPT: Malonyl/palmitoyl transferase
MAT: Malonyl-CoA–acetyl-CoA-ACP
     transacylase
TE: Thioesterase
ACP: Acyl carrier protein
PPT: Phosphopantetheinyl
     transferase
KR: β-Ketoacyl reductase
KS: β-Ketoacyl synthase
ER: β-Enoyl reductase
DH: Dehydratase

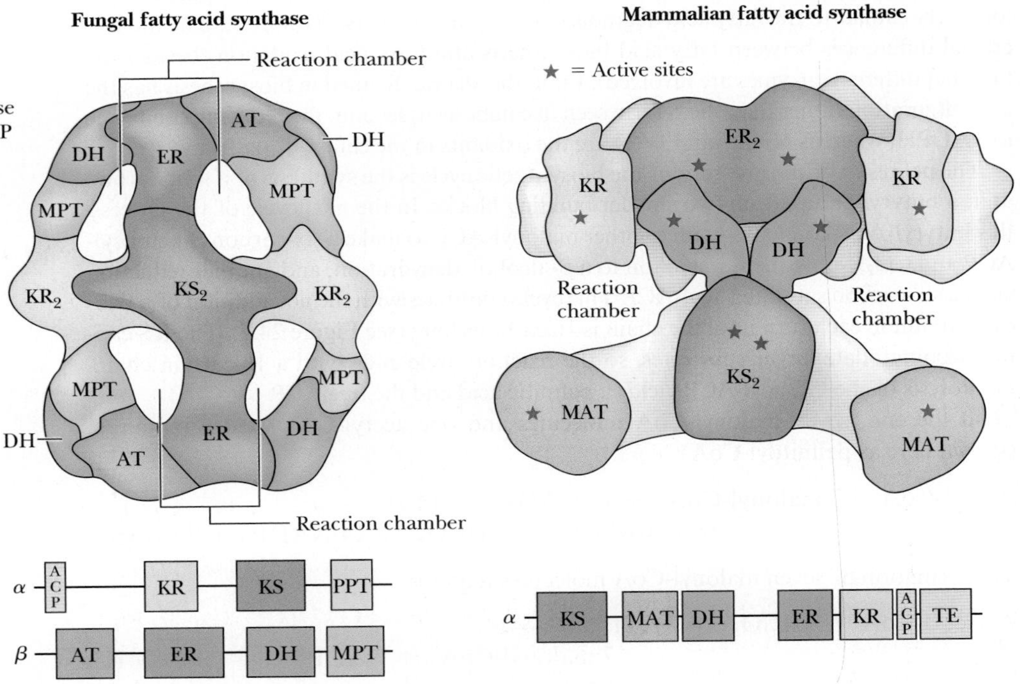

FIGURE 24.9 Organization of enzyme functions on two eukaryotic fatty acid synthases. (left) Fungal FAS is a closed barrel 260 Å high and 230 Å wide. (right) Mammalian (pig) FAS is an asymmetric X-shape 210 Å high, 180 Å wide, and 90 Å deep. The arrangement of functional domains along the FAS polypeptides is shown at the bottom of the figure. KS domains form dimers in both structures. KR domains form dimers in the fungal enzyme, whereas ER and DH domains form dimers in the mammalian complex.

(a)

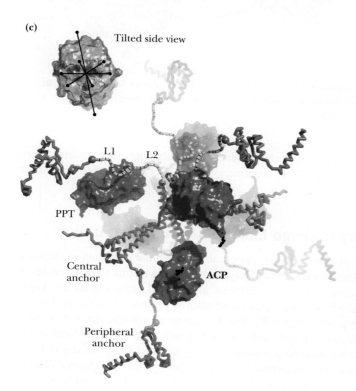

Upper reaction chamber

Lower reaction chamber

(b)

Central wheel

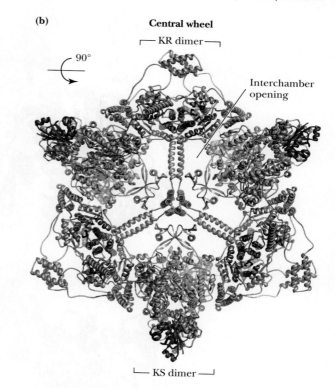

KR dimer

Interchamber opening

90°

KS dimer

(c)

Tilted side view

L1 L2

PPT

Central anchor

ACP

Peripheral anchor

FIGURE 24.10 **(a)** FAS from *S. cerevisiae* possesses two trimeric reaction chambers separated by equatorial stabilizing struts. ACP domains are located at the equatorial base of each reaction chamber, close to the catalytic site of KS. **(b)** The equatorial base of this structure is an α_6 trimer of dimers, with alternating KS and KR domains. The central stabilizing struts are α-helical extensions of the functional domains arranged around the outside of the ring. A central six-helix bundle stabilizes the structure. **(c)** The ACP domains (red) are tethered on flexible linkers (yellow) so that they can move from one active site to the next in the catalytic cycle. Comparison of the ACP domains of *S. cerevisiae* and *E. coli* reveals that the ACP domains probably extend the acyl-phosphopantetheine group to active sites but then retract the acyl group into a hydrophobic cleft while moving from one site to the next (pdb = 2UV8; images courtesy of Marc Leibundgut and Nenad Ban, ETH [Zurich]).

domains are separated and lie near the ends of the extended "arms." The arms form reaction chambers on either side of the structure. The flexible ACP domains do not appear in this structure (probably because they are not fixed in any one position in the crystals used for the structural studies). However, since it follows the KR domain in the polypeptide sequence, the ACP domain probably lies at the end of each KR arm, where it can rotate to interact with the adjacent active sites.

In both the fungal and the mammalian FAS structures, the close association of enzymic domains within one large complex permits efficient transfer of intermediates from

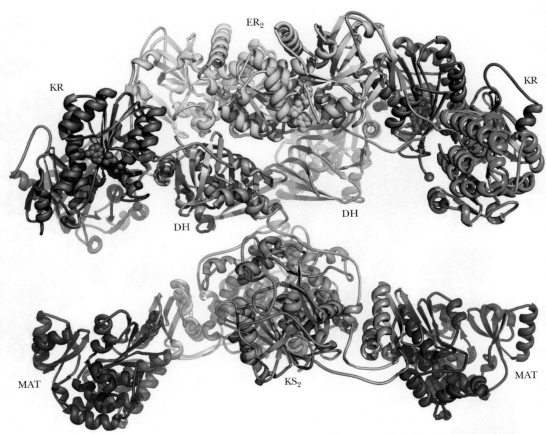

FIGURE 24.11 Structural studies reveal that mammalian FAS homodimer is X-shaped. The ACP domains are probably located adjacent to the KR domains at the ends of the arms (pdb id = 2VZ8; image courtesy of Marc Leibundgut and Nenad Ban, ETH [Zurich]).

one active site to the next. In addition, the presence of all these enzyme domains on one or two polypeptides allows the cell to coordinate synthesis of all enzymes needed for fatty acid synthesis.

C₁₆ Fatty Acids May Undergo Elongation and Unsaturation

Additional Elongation As seen already, palmitate is the primary product of the fatty acid synthase. Cells synthesize many other fatty acids. Shorter chains are easily made if the chain is released before reaching 16 carbons in length. Longer chains are made through special elongation reactions, which occur both in the mitochondria and at the surface of the endoplasmic reticulum (ER). The ER reactions are actually quite similar to those we have just discussed: addition of two-carbon units at the carboxyl end of the chain by means of oxidative decarboxylations involving malonyl-CoA. As was the case for the fatty acid synthase, this decarboxylation provides the thermodynamic driving force for the condensation reaction. The mitochondrial reactions involve addition (and subsequent reduction) of acetyl units. These reactions (Figure 24.12) are essentially a reversal of fatty acid oxidation, with the exception that NADPH is utilized in the saturation of the double bond, instead of FADH₂.

Introduction of a Single *cis* Double Bond Both prokaryotes and eukaryotes are capable of introducing a single *cis* double bond in a newly synthesized fatty acid. Bacteria such as *E. coli* carry out this process in an O₂-independent pathway, whereas eukaryotes have adopted an O₂-dependent pathway. There is a fundamental chemical difference between the two. The O₂-dependent reaction can occur anywhere in the fatty acid chain, with no (additional) need to activate the desired bond toward dehydrogenation. However, in the absence of O₂, some other means must be found to activate the

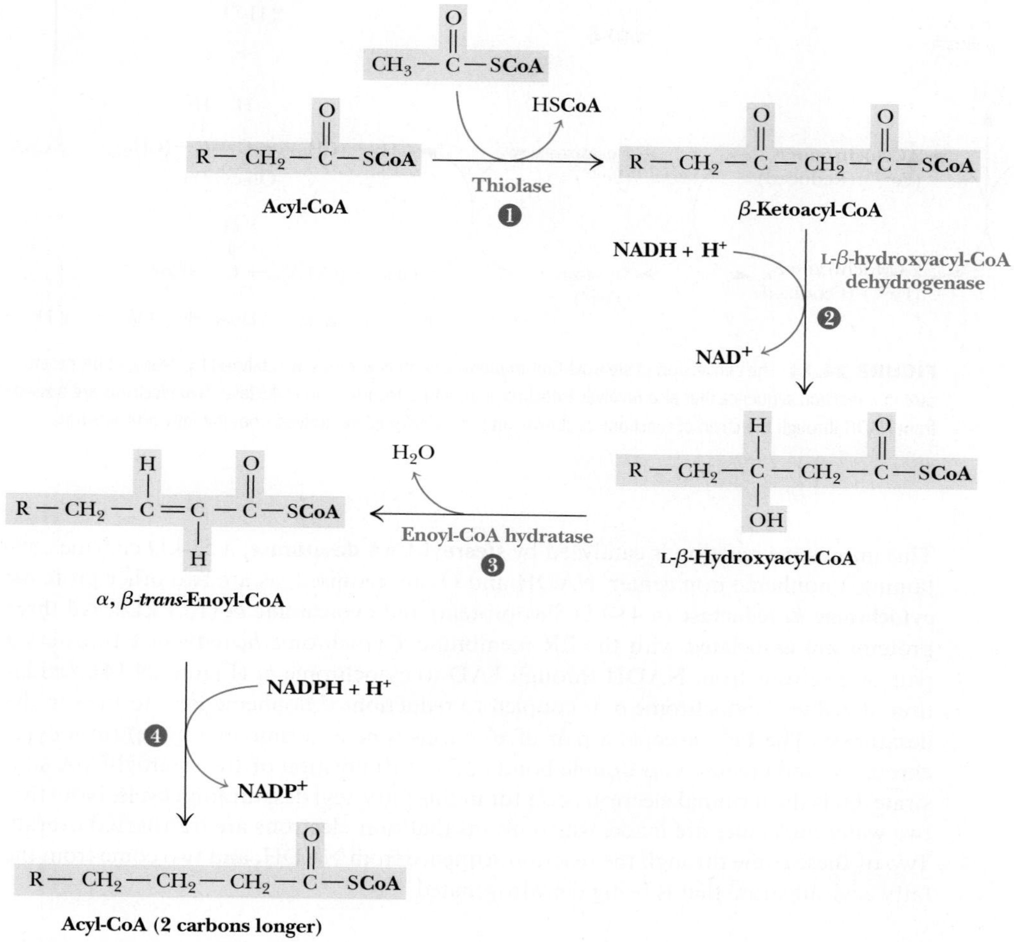

FIGURE 24.12 **(1)** Elongation of fatty acids in mitochondria is initiated by the thiolase reaction. The β-ketoacyl intermediate thus formed undergoes the same three reactions (in reverse order) that are the basis of β-oxidation of fatty acids. **(2)** Reduction of the β-keto group is followed by **(3)** dehydration to form a double bond. **(4)** Reduction of the double bond yields a fatty acyl-CoA that is elongated by two carbons. Note that the reducing coenzyme for the second step is NADH, whereas the reductant for the fourth step is NADPH.

bond in question. Thus, in the bacterial reaction, dehydrogenation occurs while the bond of interest is still near the β-carbonyl or β-hydroxy group and the thioester group at the end of the chain.

In *E. coli,* the biosynthesis of a monounsaturated fatty acid begins with four normal cycles of elongation to form a ten-carbon intermediate, β-hydroxydecanoyl-ACP (Figure 24.13). At this point, **β-hydroxydecanoyl thioester dehydrase** forms a double bond β,γ to the thioester and in the *cis* configuration. This is followed by three rounds of the normal elongation reactions to form *palmitoleoyl-ACP.* Elongation may terminate at this point or may be followed by additional biosynthetic events. The principal unsaturated fatty acid in *E. coli, cis-vaccenic acid,* is formed by an additional elongation step, using palmitoleoyl-ACP as a substrate.

Unsaturation Reactions Occur in Eukaryotes in the Middle of an Aliphatic Chain

The addition of double bonds to fatty acids in eukaryotes does not occur until the fatty acyl chain has reached its full length (usually 16 to 18 carbons). Dehydrogenation of stearoyl-CoA occurs in the middle of the chain, despite the absence of any useful functional group on the chain to facilitate activation:

$$CH_3—(CH_2)_{16}CO—SCoA \longrightarrow CH_3—(CH_2)_7CH=CH(CH_2)_7CO—SCoA$$

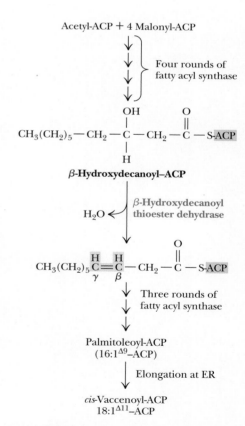

FIGURE 24.13 Double bonds are introduced into the growing fatty acid chain in *E. coli* by specific dehydrases. Palmitoleoyl-ACP is synthesized by a sequence of reactions involving four rounds of chain elongation, followed by double bond insertion by β-hydroxydecanoyl thioester dehydrase and three additional elongation steps. Another elongation cycle produces *cis*-vaccenic acid.

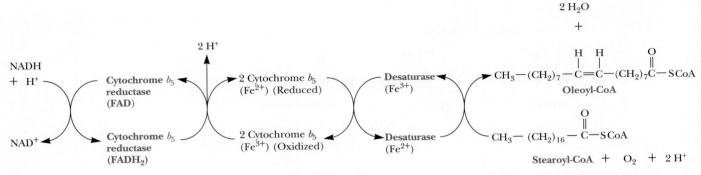

FIGURE 24.14 The conversion of stearoyl-CoA to oleoyl-CoA in eukaryotes is catalyzed by stearoyl-CoA desaturase in a reaction sequence that also involves cytochrome b_5 and cytochrome b_5 reductase. Two electrons are passed from NADH through the chain of reactions as shown, and two electrons are derived from the fatty acyl substrate.

This impressive reaction is catalyzed by **stearoyl-CoA desaturase,** a 53-kD enzyme containing a nonheme iron center. NADH and O_2 are required, as are two other proteins: **cytochrome b_5 reductase** (a 43-kD flavoprotein) and **cytochrome b_5** (16.7 kD). All three proteins are associated with the ER membrane. Cytochrome b_5 reductase transfers a pair of electrons from NADH through FAD to cytochrome b_5 (Figure 24.14). Oxidation of reduced cytochrome b_5 is coupled to reduction of nonheme Fe^{3+} to Fe^{2+} in the desaturase. The Fe^{3+} accepts a pair of electrons (one at a time in a cycle) from cytochrome b_5 and creates a *cis* double bond at the 9,10-position of the stearoyl-CoA substrate. O_2 is the terminal electron acceptor in this fatty acyl desaturation cycle. Note that two water molecules are made, which means that four electrons are transferred overall. Two of these come through the reaction sequence from NADH, and two come from the fatty acyl substrate that is being dehydrogenated.

The Unsaturation Reaction May Be Followed by Chain Elongation

Additional chain elongation can occur following this single desaturation reaction. The oleoyl-CoA produced can be elongated by two carbons to form a 20:1 *cis*-Δ^{11} fatty acyl-CoA. If the starting fatty acid is palmitate, reactions similar to the preceding scheme yield palmitoleoyl-CoA (16:1 *cis*-Δ^9), which subsequently can be elongated to yield *cis*-vaccenic acid (18:1 *cis*-Δ^{11}). Similarly, C_{16} and C_{18} fatty acids can be elongated to yield C_{22} and C_{24} unsaturated fatty acids, such as are often found in sphingolipids.

Mammals Cannot Synthesize Most Polyunsaturated Fatty Acids

Organisms differ with respect to formation, processing, and utilization of polyunsaturated fatty acids. *E. coli,* for example, does not have any polyunsaturated fatty acids. Eukaryotes *do* synthesize a variety of polyunsaturated fatty acids, certain organisms more than others. For example, plants manufacture double bonds between the Δ^9 and the methyl end of the chain, but mammals cannot. Plants readily desaturate oleic acid at the 12-position (to give linoleic acid) or at both the 12- and 15-positions (producing linolenic acid). Mammals require polyunsaturated fatty acids but must acquire them in their diet. As such, these fatty acids are referred to as **essential fatty acids.** On the other hand, mammals *can* introduce double bonds between the double bond at the 8- or 9-position and the carboxyl group. Enzyme complexes in the ER desaturate the 5-position, provided a double bond exists at the 8-position, and form a double bond at the 6-position if one already exists at the 9-position. Thus, oleate can be unsaturated at the 6,7-position to give an 18:2 *cis*-Δ^6,Δ^9 fatty acid.

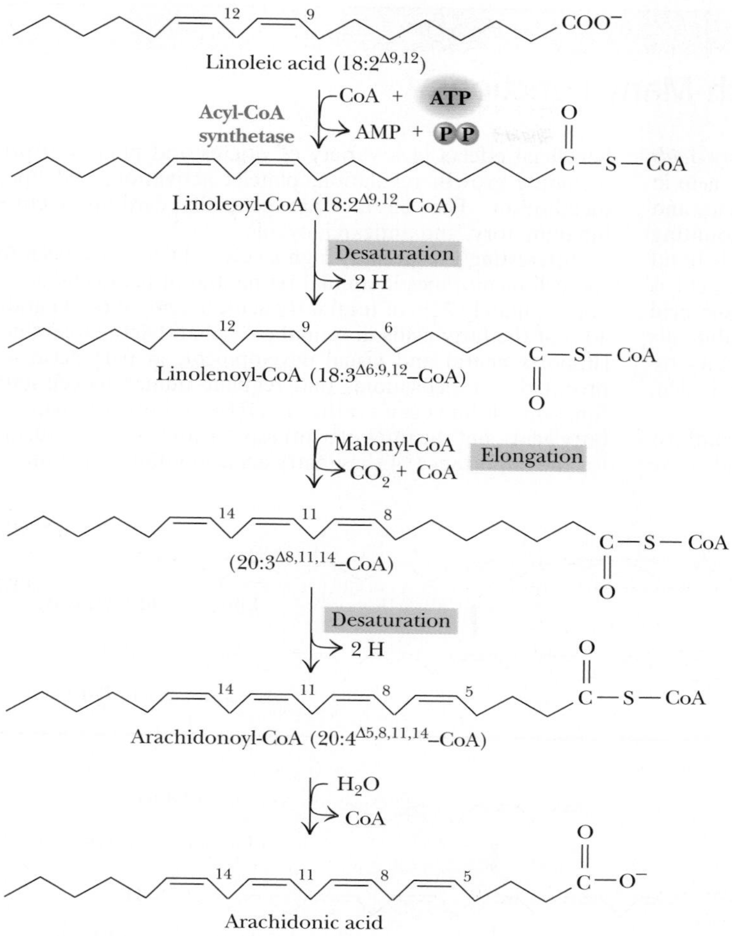

FIGURE 24.15 Arachidonic acid is synthesized from linoleic acid in eukaryotes. This is the means by which animals synthesize fatty acids with double bonds at positions other than C-9.

Arachidonic Acid Is Synthesized from Linoleic Acid by Mammals

Mammals can add additional double bonds to unsaturated fatty acids in their diets. Their ability to make arachidonic acid from linoleic acid is one example (Figure 24.15). This fatty acid is the precursor for prostaglandins and other biologically active derivatives such as leukotrienes. Synthesis involves formation of a linoleoyl ester of CoA from dietary linoleic acid, followed by introduction of a double bond at the 6-position. The triply unsaturated product is then elongated (by malonyl-CoA with a decarboxylation step) to yield a 20-carbon fatty acid with double bonds at the 8-, 11-, and 14-positions. A second desaturation reaction at the 5-position, followed by a reverse **acyl-CoA synthetase** reaction (see Chapter 23), liberates the product, a 20-carbon fatty acid with double bonds at the 5-, 8-, 11-, and 14-positions.

Regulatory Control of Fatty Acid Metabolism Is an Interplay of Allosteric Modifiers and Phosphorylation–Dephosphorylation Cycles

The control and regulation of fatty acid synthesis is intimately related to regulation of fatty acid breakdown, glycolysis, and the TCA cycle. Acetyl-CoA is an important intermediate metabolite in all these processes. In these terms, it is easy to appreciate the interlocking relationships in Figure 24.16. Malonyl-CoA can act to prevent fatty acyl-CoA derivatives from entering the mitochondria by inhibiting the carnitine

HUMAN BIOCHEMISTRY

ω3 and ω6—Essential Fatty Acids with Many Functions

Linoleic acid and α-linolenic acid are termed *essential fatty acids* because animals cannot synthesize them and must acquire them in their diet. Linoleic acid is the precursor of arachidonic acid, and both of these are referred to as **ω6 fatty acids** because, counting from the end (omega, ω) carbon of the chain, the first double bond is at the sixth position (see figure). Similarly, α-linolenic acid is the precursor of **eicosapentaenoic acid** and **docosahexaenoic acid (DHA),** and these three are termed **ω3 fatty acids.** Vegetable oils are rich in linoleic acid and thus satisfy the body's ω6 dietary requirements, whereas fish oils (for example, cod, herring, menhaden, and salmon) are rich in ω3 fatty acids.

The ω6 fatty acids are precursors of prostaglandins, thromboxanes, and leukotrienes (see Section 24.3). The ω3 fatty acids have

beneficial effects in a variety of organs and biological processes, including growth regulation, platelet activation, and lipoprotein metabolism. The ω3 fats are generally cardioprotective, anti-inflammatory, and anticarcinogenic.

Interestingly, especially high levels of DHA have been found in rod cell membranes in animal retina and in neural tissue. DHA is approximately 22% of total fatty acids in animal retina and 35% to 40% of the fatty acids in retinal phosphatidylethanolamine. DHA supports neural and visual development, in part because it is a precursor for eicosanoids that regulate numerous cell and organ functions. Infants can synthesize DHA and other polyunsaturated fatty acids, but the rates of synthesis are low. Strong evidence exists for the importance of these fatty acids in infant nutrition.

▶ The "ω" numbering system, where the position of the first double bond relative to the methyl (ω) group is indicated (red box), is useful because the ω double bond position is retained during chain elongation and desaturation.

ω6 Essential fatty acid

COOH

Linoleic acid (18:2 ω6)

COOH

Arachidonic acid (20:4 ω6)

ω3 Essential fatty acid

COOH

α-Linolenic acid (18:3 ω3)

COOH

Eicosapentaenoic acid (EPA; 20:5 ω3)

COOH

Docosahexaenoic acid (DHA; 22:6 ω3)

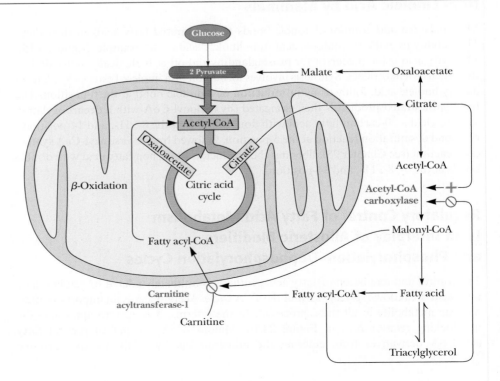

FIGURE 24.16 Regulation of fatty acid synthesis and fatty acid oxidation are coupled as shown. Malonyl-CoA, produced during fatty acid synthesis, inhibits the uptake of fatty acylcarnitine (and thus fatty acid oxidation) by mitochondria. When fatty acyl-CoA levels rise, fatty acid synthesis is inhibited and fatty acid oxidation activity increases. Rising citrate levels (which reflect an abundance of acetyl-CoA) similarly signal the initiation of fatty acid synthesis.

acyltransferase of the outer mitochondrial membrane that initiates this transport. In this way, when fatty acid synthesis is turned on (as signaled by higher levels of malonyl-CoA), β-oxidation is inhibited. As we pointed out earlier, citrate is an important allosteric activator of acetyl-CoA carboxylase, and fatty acyl-CoAs are inhibitors. The degree of inhibition is proportional to the chain length of the fatty acyl-CoA; longer chains show a higher affinity for the allosteric inhibition site on acetyl-CoA carboxylase. Palmitoyl-CoA, stearoyl-CoA, and arachidyl-CoA are the most potent inhibitors of the carboxylase.

Hormonal Signals Regulate ACC and Fatty Acid Biosynthesis

As described earlier, citrate activation and palmitoyl-CoA inhibition of acetyl-CoA carboxylase are strongly dependent on the phosphorylation state of the enzyme. This provides a crucial connection to hormonal regulation. Many of the enzymes that act to phosphorylate acetyl-CoA carboxylase (see Figure 24.4) are controlled by hormonal signals. Glucagon is a good example (Figure 24.17). As noted in Chapter 22, glucagon binding to membrane receptors activates an intracellular cascade involving activation of adenylyl cyclase. Cyclic AMP produced by the cyclase activates a protein kinase, which then phosphorylates acetyl-CoA carboxylase. Unless citrate levels are high, phosphorylation causes inhibition of fatty acid biosynthesis. The carboxylase (and fatty acid synthesis) can be reactivated by a specific phosphatase, which dephosphorylates the

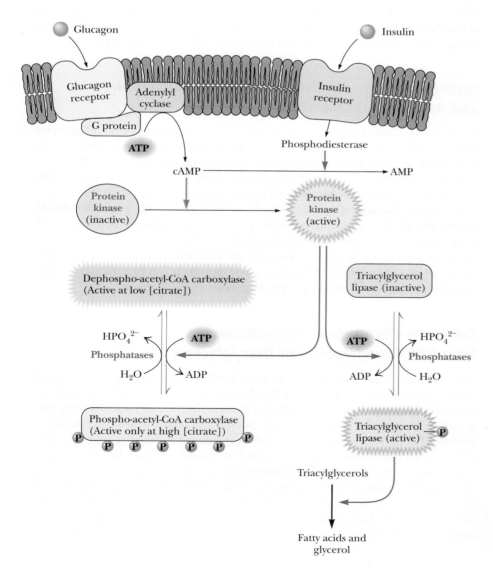

FIGURE 24.17 Hormonal signals regulate fatty acid synthesis, primarily through actions on acetyl-CoA carboxylase. Availability of fatty acids also depends upon hormonal activation of triacylglycerol lipase.

carboxylase. Also indicated in Figure 24.17 is the simultaneous activation by glucagon of triacylglycerol lipases, which hydrolyze triacylglycerols, releasing fatty acids for β-oxidation. Both the inactivation of acetyl-CoA carboxylase and the activation of triacylglycerol lipase are counteracted by insulin, whose receptor acts to stimulate a phosphodiesterase that converts cAMP to AMP.

24.2 How Are Complex Lipids Synthesized?

Complex lipids consist of backbone structures to which fatty acids are covalently bound. Principal classes include the **glycerolipids,** for which glycerol is the backbone, and **sphingolipids,** which are built on a sphingosine backbone. The two major classes of glycerolipids are **glycerophospholipids** and **triacylglycerols.** The **phospholipids,** which include both glycerophospholipids and sphingomyelins, are crucial components of membrane structure. They are also precursors of hormones such as the *eicosanoids* (for example, *prostaglandins*) and signal molecules (such as the breakdown products of *phosphatidylinositol*).

Different organisms possess greatly different complements of lipids and therefore invoke somewhat different lipid biosynthetic pathways. For example, sphingolipids and triacylglycerols are produced only in eukaryotes. In contrast, bacteria usually have rather simple lipid compositions. Phosphatidylethanolamine accounts for at least 75% of the phospholipids in *E. coli,* with phosphatidylglycerol and cardiolipin accounting for most of the rest. *E. coli* membranes possess no phosphatidylcholine, phosphatidylinositol, sphingolipids, or cholesterol. On the other hand, some bacteria (such as *Pseudomonas*) can synthesize phosphatidylcholine, for example. In this section and the one following, we consider some of the pathways for the synthesis of glycerolipids, sphingolipids, and the eicosanoids, which are derived from phospholipids.

Glycerolipids Are Synthesized by Phosphorylation and Acylation of Glycerol

A common pathway operates in nearly all organisms for the synthesis of **phosphatidic acid,** the precursor to other glycerolipids. **Glycerokinase** catalyzes the phosphorylation of glycerol to form glycerol-3-phosphate, which is then acylated at both the 1- and 2-positions to yield phosphatidic acid (Figure 24.18). The first acylation, at position 1, is catalyzed by **glycerol-3-phosphate acyltransferase,** an enzyme that in most organisms is specific for saturated fatty acyl groups. Eukaryotic systems can also utilize **dihydroxyacetone phosphate** as a starting point for synthesis of phosphatidic acid (Figure 24.18). Again a specific acyltransferase adds the first acyl chain, followed by reduction of the backbone keto group by **acyldihydroxyacetone phosphate reductase,** using NADPH as the reductant. Alternatively, dihydroxyacetone phosphate can be reduced to glycerol-3-phosphate by **glycerol-3-phosphate dehydrogenase.**

Eukaryotes Synthesize Glycerolipids from CDP-Diacylglycerol or Diacylglycerol

In eukaryotes, phosphatidic acid is converted directly either to diacylglycerol or to *cytidine diphosphodiacylglycerol* (or simply *CDP-diacylglycerol;* Figure 24.19). From these two precursors, all other glycerophospholipids in eukaryotes are derived. Diacylglycerol is a precursor for synthesis of triacylglycerol, phosphatidylethanolamine, and phosphatidylcholine. Triacylglycerol is synthesized mainly in adipose tissue, liver, and intestines and serves as the principal energy storage molecule in eukaryotes. Triacylglycerol biosynthesis in liver and adipose tissue occurs via **diacylglycerol acyltransferase,** an enzyme bound to the cytoplasmic face of the ER. A different route is used, however, in intestines. Recall (see Figure 23.3) that triacylglycerols from the diet are broken down to 2-monoacylglycerols by specific lipases. Acyltransferases then acylate 2-monoacylglycerol to produce new triacylglycerols (Figure 24.20).

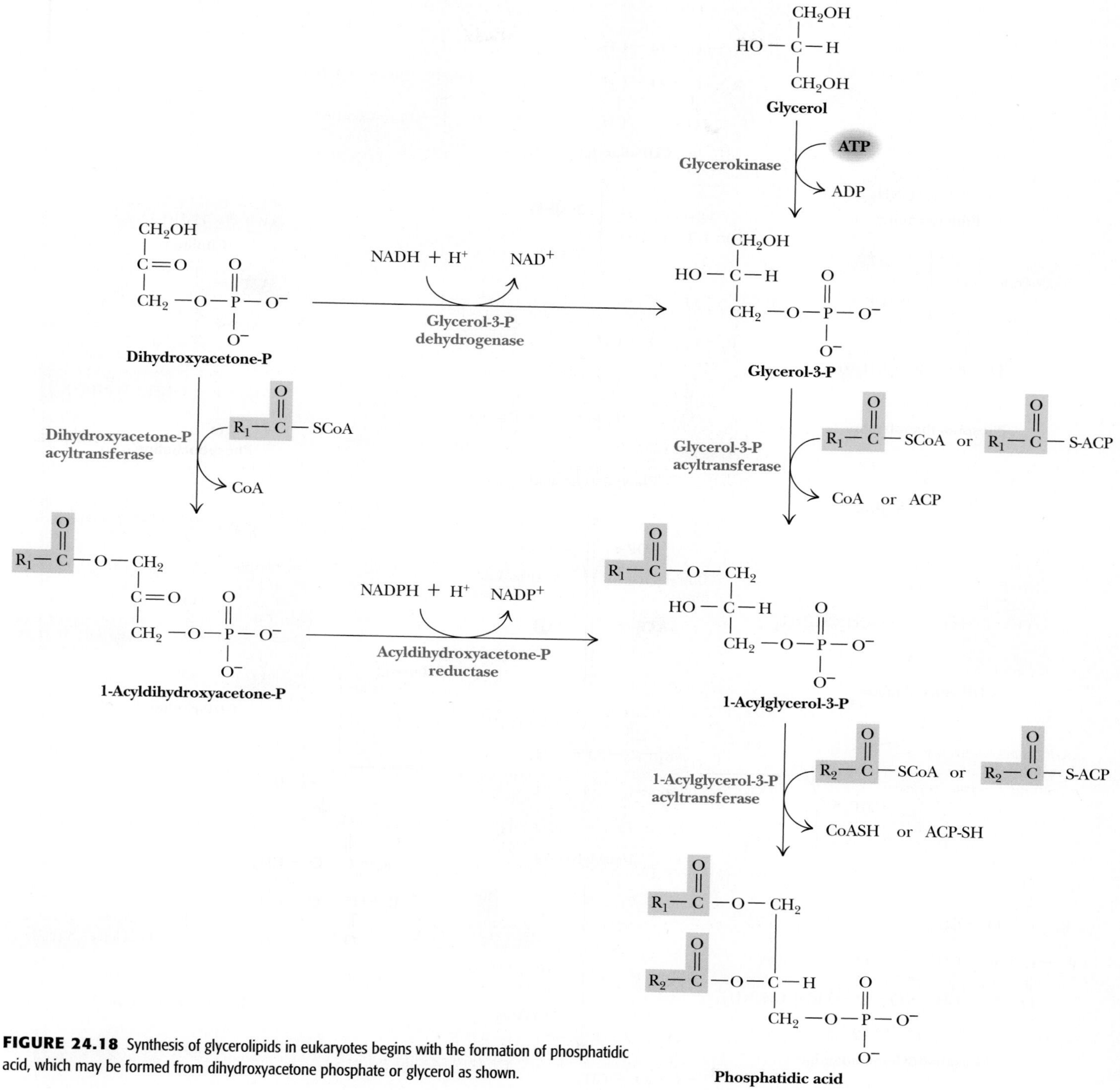

FIGURE 24.18 Synthesis of glycerolipids in eukaryotes begins with the formation of phosphatidic acid, which may be formed from dihydroxyacetone phosphate or glycerol as shown.

Phosphatidylethanolamine Is Synthesized from Diacylglycerol and CDP-Ethanolamine

Phosphatidylethanolamine synthesis begins with phosphorylation of ethanolamine to form phosphoethanolamine (see Figure 24.19). The next reaction involves transfer of a cytidylyl group from CTP to form CDP-ethanolamine and pyrophosphate. As always, PP$_i$ hydrolysis drives this reaction forward. A specific **phosphoethanolamine transferase** then links phosphoethanolamine to the diacylglycerol backbone. Biosynthesis of phosphatidylcholine is entirely analogous because animals can synthesize it directly. All of the choline utilized in this pathway must be acquired from the diet. On the other hand,

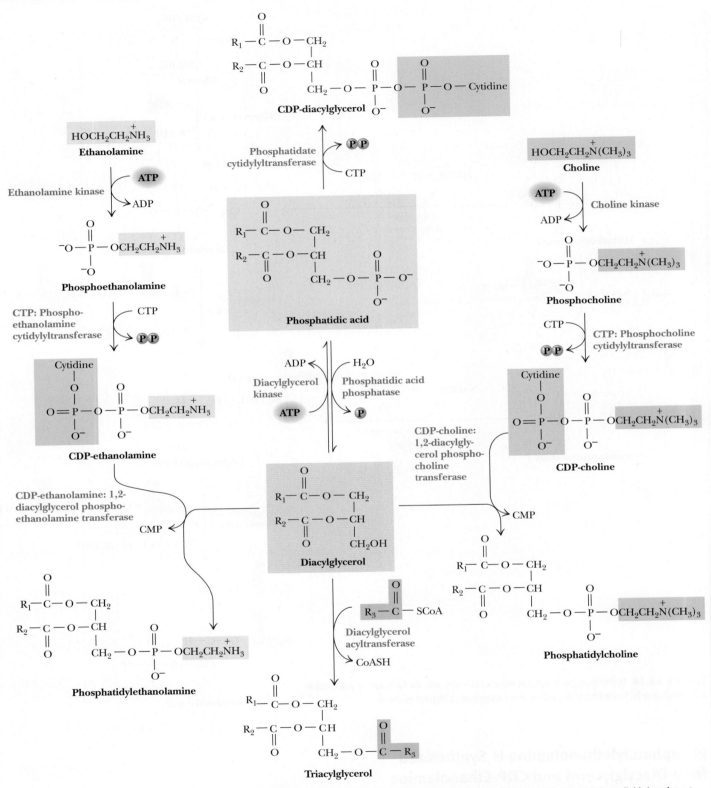

FIGURE 24.19 Diacylglycerol and CDP-diacylglycerol are the principal precursors of glycerolipids in eukaryotes. Phosphatidylethanolamine and phosphatidylcholine are formed by reaction of diacylglycerol with CDP-ethanolamine or CDP-choline, respectively.

yeast, certain bacteria, and animal livers can convert phosphatidylethanolamine to phosphatidylcholine by methylation reactions involving *S*-adenosylmethionine (see Chapter 25).

Exchange of Ethanolamine for Serine Converts Phosphatidylethanolamine to Phosphatidylserine

Mammals synthesize phosphatidylserine (PS) in a calcium ion–dependent reaction involving aminoalcohol exchange (Figure 24.21). The enzyme catalyzing this reaction is associated with the ER and will accept phosphatidylethanolamine (PE) and other phospholipid substrates. A mitochondrial **PS decarboxylase** can subsequently convert PS to PE. No other pathway converting serine to ethanolamine has been found.

Eukaryotes Synthesize Other Phospholipids Via CDP-Diacylglycerol

Eukaryotes also use CDP-diacylglycerol, derived from phosphatidic acid, as a precursor for several other important phospholipids, including phosphatidylinositol (PI), phosphatidylglycerol (PG), and cardiolipin (Figure 24.22). PI accounts for only about 2% to 8% of the lipids in most animal membranes, but breakdown products of PI, including inositol-1,4,5-trisphosphate and diacylglycerol, are second messengers in a vast array of cellular signaling processes.

Dihydroxyacetone Phosphate Is a Precursor to the Plasmalogens

Certain glycerophospholipids possess alkyl or alkenyl ether groups at the 1-position in place of an acyl ester group. These glyceroether phospholipids are synthesized from dihydroxyacetone phosphate (Figure 24.23). Acylation of dihydroxyacetone phosphate

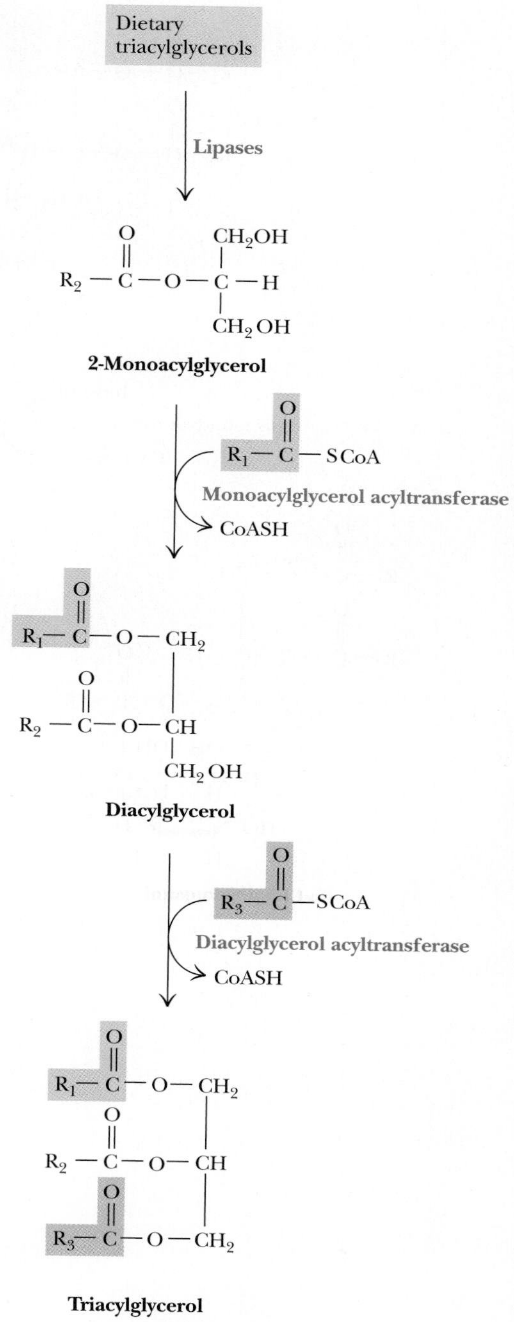

FIGURE 24.20 Triacylglycerols are formed primarily by the action of acyltransferases on monoacylglycerol and diacylglycerol.

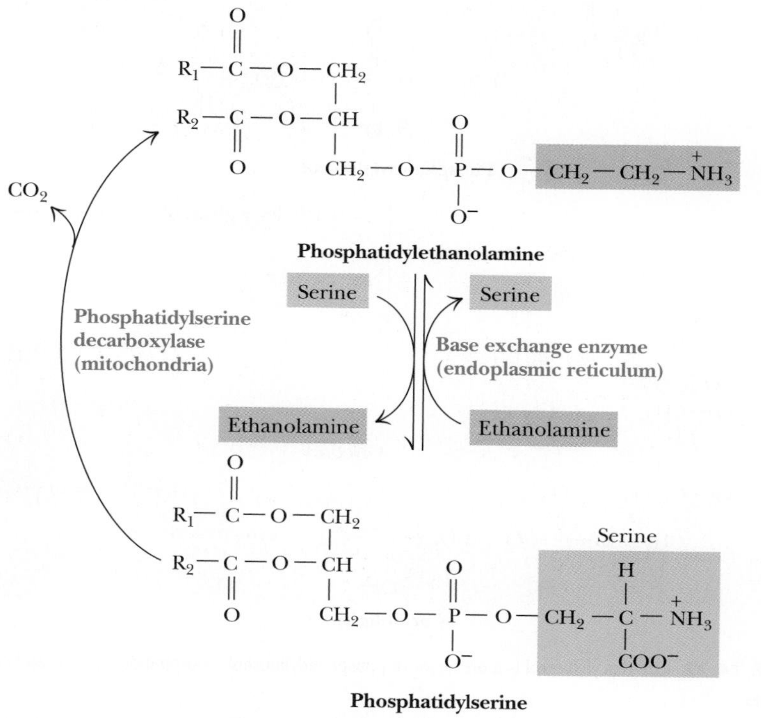

FIGURE 24.21 The interconversion of phosphatidylethanolamine and phosphatidylserine in mammals.

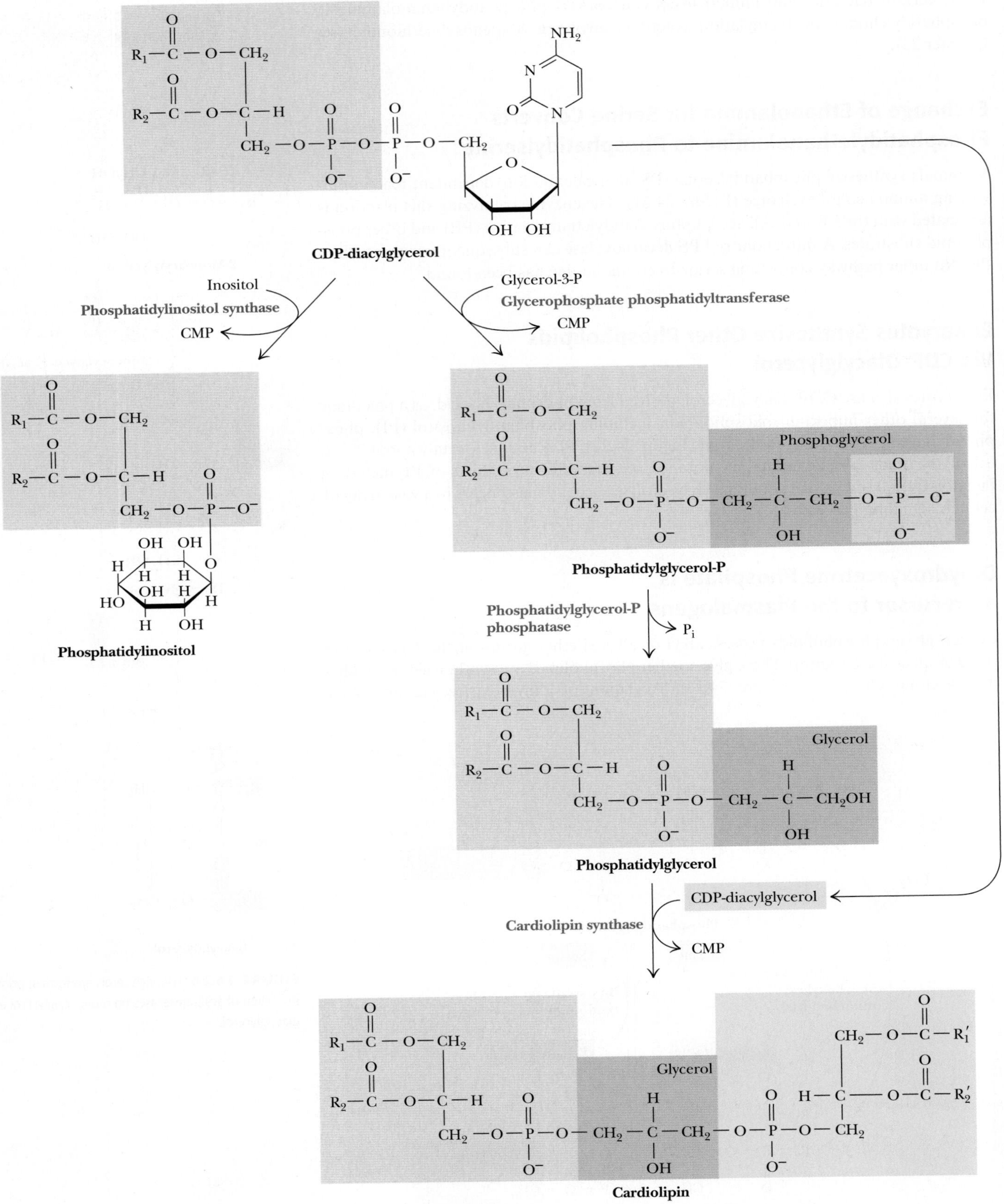

FIGURE 24.22 CDP-diacylglycerol is a precursor of phosphatidylinositol, phosphatidylglycerol, and cardiolipin in eukaryotes.

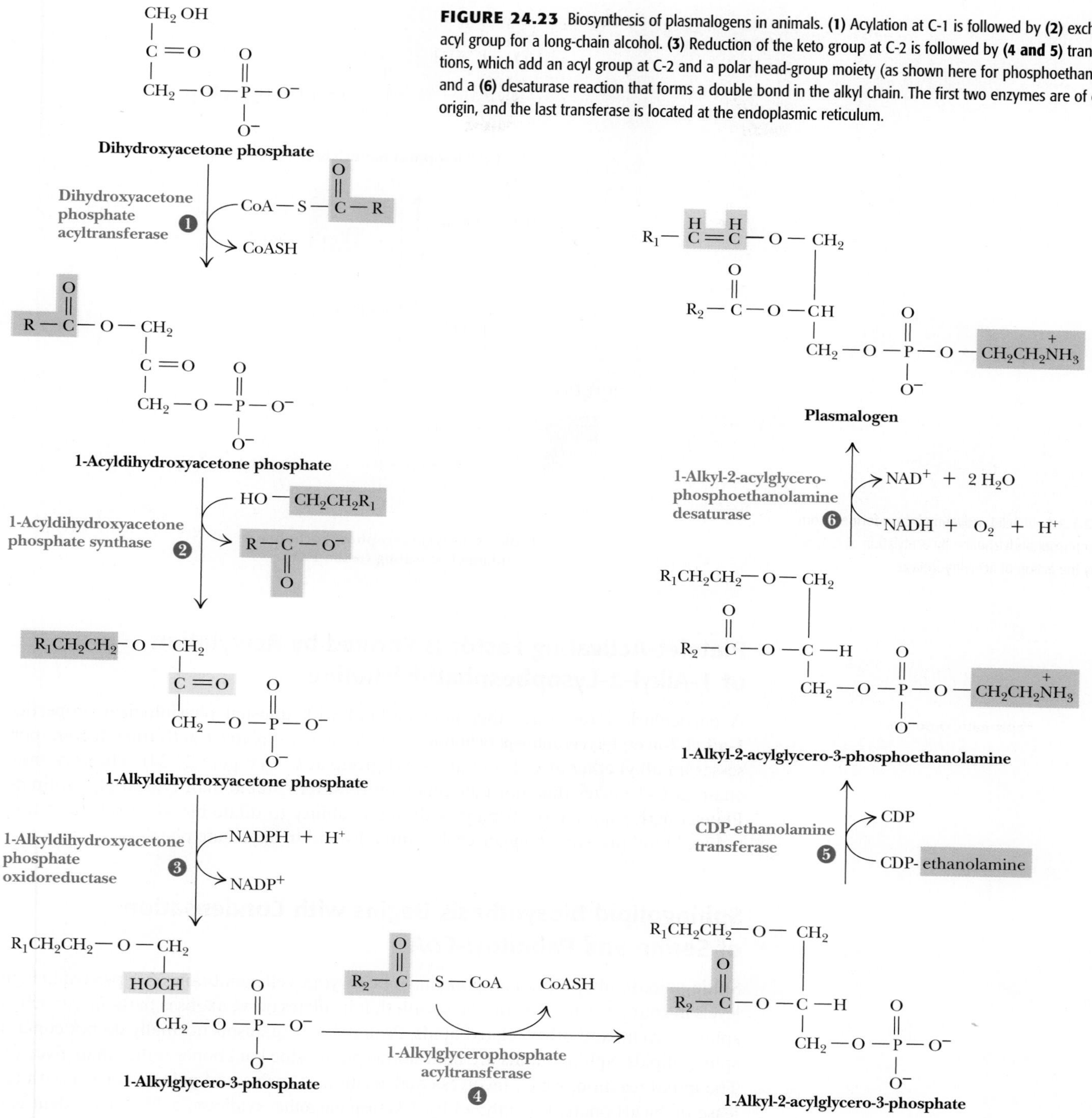

FIGURE 24.23 Biosynthesis of plasmalogens in animals. **(1)** Acylation at C-1 is followed by **(2)** exchange of the acyl group for a long-chain alcohol. **(3)** Reduction of the keto group at C-2 is followed by **(4 and 5)** transferase reactions, which add an acyl group at C-2 and a polar head-group moiety (as shown here for phosphoethanolamine), and a **(6)** desaturase reaction that forms a double bond in the alkyl chain. The first two enzymes are of cytoplasmic origin, and the last transferase is located at the endoplasmic reticulum.

(DHAP) is followed by an exchange reaction, in which the acyl group is removed as a carboxylic acid and a long-chain alcohol adds to the 1-position. This long-chain alcohol is derived from the corresponding acyl-CoA by means of an acyl-CoA reductase reaction involving oxidation of two molecules of NADH. The *2-keto* group of the DHAP backbone is then reduced to an alcohol, followed by acylation. The resulting 1-alkyl-2-acylglycero-3-phosphate can react in a manner similar to phosphatidic acid to produce ether analogs of phosphatidylcholine, phosphatidylethanolamine, and so forth (Figure 24.23). In addition, specific **desaturase** enzymes associated with the ER can desaturate the alkyl ether chains of these lipids as shown. The products, which contain α,β-unsaturated ether-linked chains at the C-1 position, are **plasmalogens;** they are abundant in cardiac tissue and in the central nervous system. The desaturases catalyzing these reactions are distinct from but similar to those that introduce unsaturations in fatty acyl-CoAs.

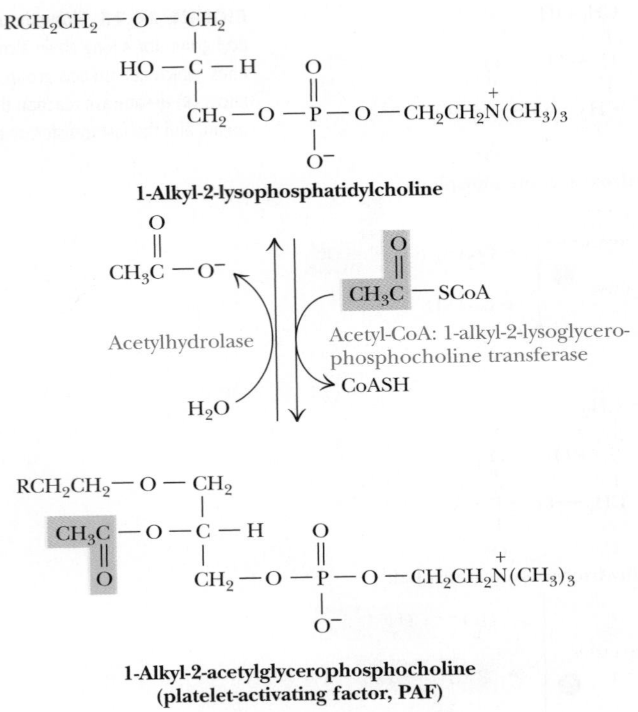

FIGURE 24.24 Platelet-activating factor, formed from 1-alkyl-2-lysophosphatidylcholine by acetylation at C-2, is degraded by the action of acetylhydrolase.

1-Alkyl-2-lysophosphatidylcholine

Acetylhydrolase

Acetyl-CoA: 1-alkyl-2-lysoglycero-phosphocholine transferase

1-Alkyl-2-acetylglycerophosphocholine
(platelet-activating factor, PAF)

Platelet-Activating Factor Is Formed by Acetylation of 1-Alkyl-2-Lysophosphatidylcholine

A particularly interesting ether phospholipid with unusual physiological properties, **1-alkyl-2-acetylglycerophosphocholine**, also known as **platelet-activating factor**, possesses an alkyl ether at C-1 and an acetyl group at C-2 (Figure 24.24). The very short chain at C-2 makes this molecule much more water soluble than typical glycerolipids. Platelet-activating factor displays a dramatic ability to dilate blood vessels (and thus reduce blood pressure in hypertensive animals) and to aggregate platelets.

Sphingolipid Biosynthesis Begins with Condensation of Serine and Palmitoyl-CoA

Sphingolipids, ubiquitous components of eukaryotic cell membranes, are present at high levels in neural tissues. The myelin sheath that insulates nerve axons is particularly rich in sphingomyelin and other related lipids. Prokaryotic organisms normally do not contain sphingolipids. Sphingolipids are built upon sphingosine backbones rather than glycerol. The initial reaction, which involves condensation of serine and palmitoyl-CoA with release of bicarbonate, is catalyzed by **3-ketosphinganine synthase,** a PLP-dependent enzyme (Figure 24.25). Reduction of the ketone product to form sphinganine is catalyzed by **3-ketosphinganine reductase,** with NADPH as a reactant. In the next step, sphinganine is acylated to form *N*-acyl sphinganine, which is then desaturated to form **ceramide.** Sphingosine itself does not appear to be an intermediate in this pathway in mammals.

Ceramide Is the Precursor for Other Sphingolipids and Cerebrosides

Ceramide is the building block for all other sphingolipids. **Sphingomyelin,** for example, is produced by transfer of phosphocholine from phosphatidylcholine (Figure 24.26). Glycosylation of ceramide by sugar nucleotides yields **cerebrosides,** such as galactosylceramide,

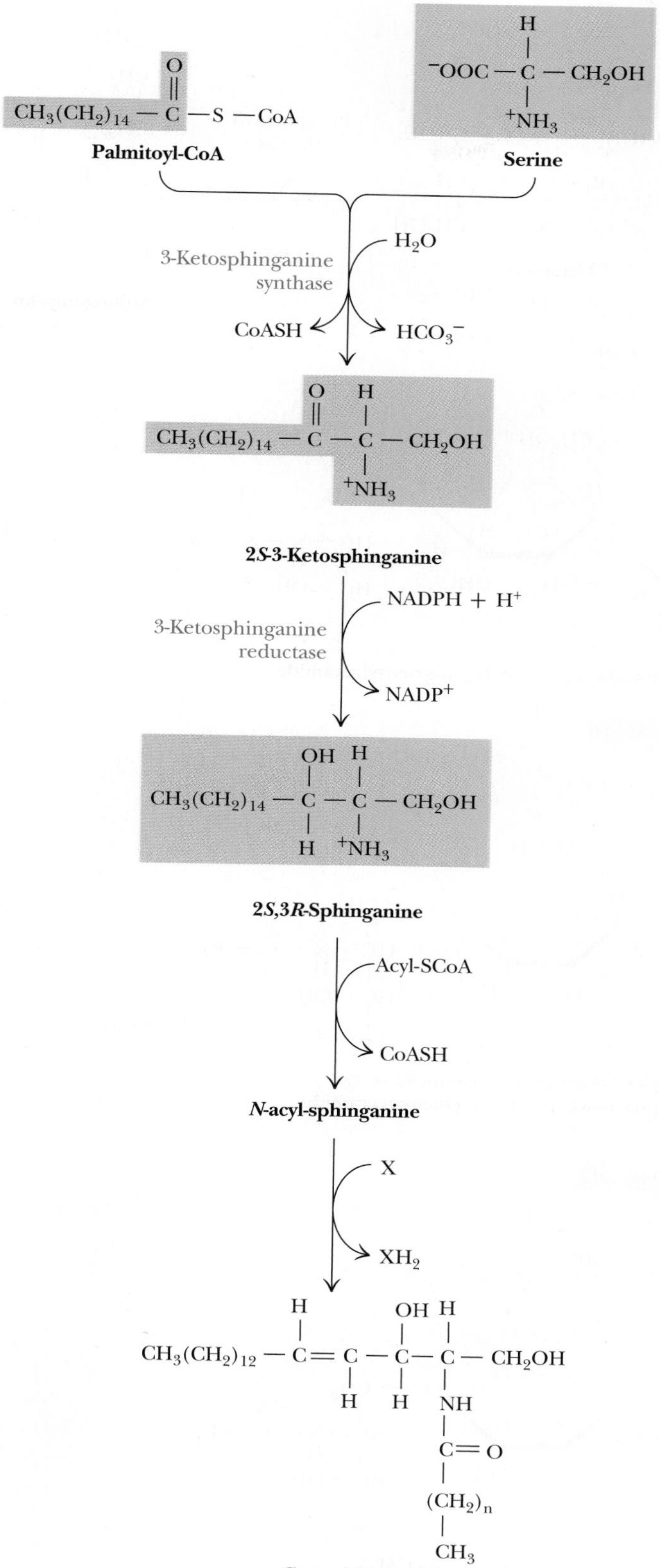

FIGURE 24.25 Biosynthesis of sphingolipids in animals begins with the 3-ketosphinganine synthase reaction, a PLP-dependent condensation of palmitoyl-CoA and serine. Subsequent reduction of the keto group, acylation, and desaturation (via reduction of an electron acceptor, X) form ceramide, the precursor of other sphingolipids.

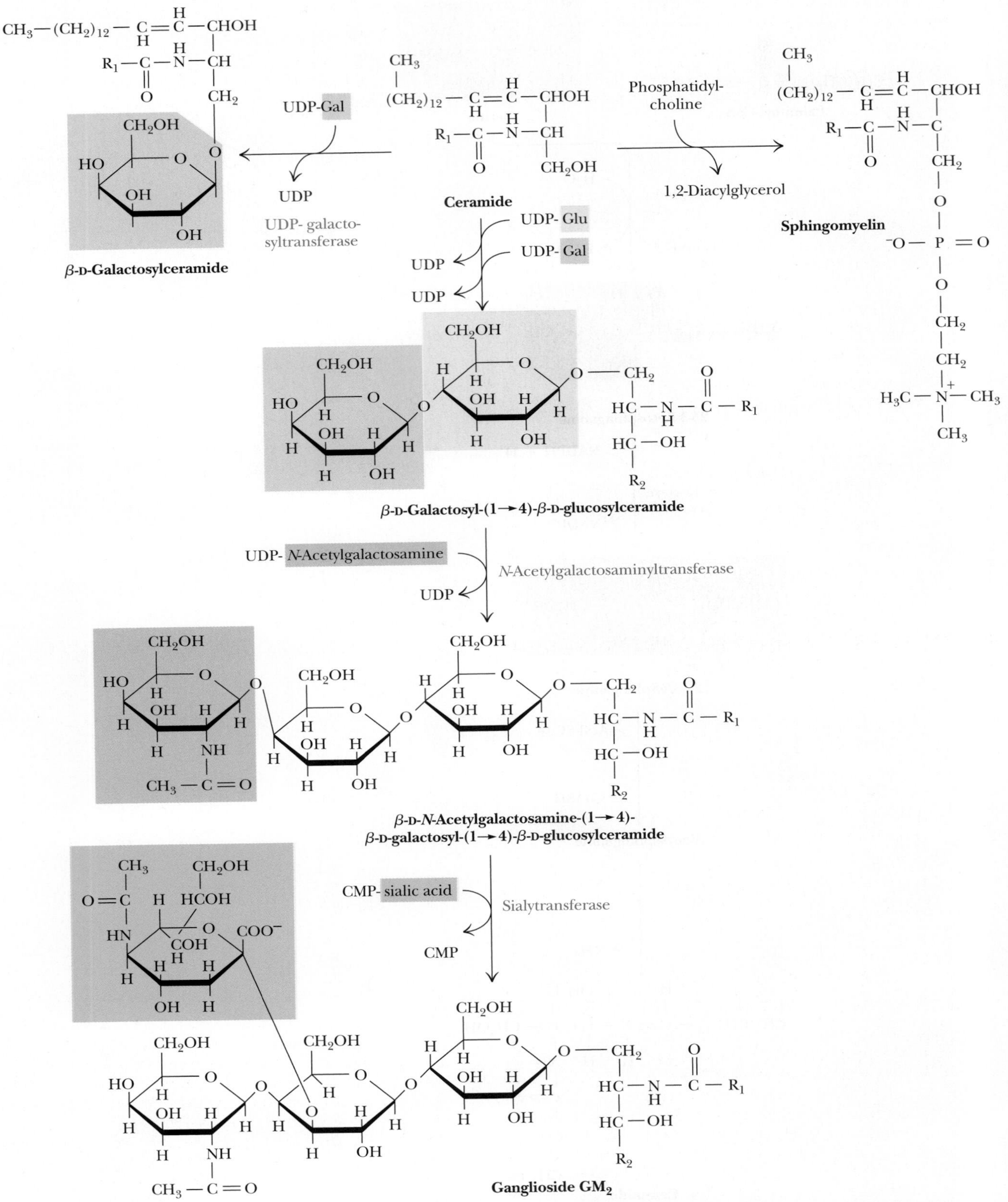

FIGURE 24.26 Glycosylceramides (such as galactosylceramide), gangliosides, and sphingomyelins are synthesized from ceramide in animals.

which makes up about 15% of the lipids of myelin sheath structures. Cerebrosides that contain one or more sialic acid (*N*-acetylneuraminic acid) moieties are called **gangliosides.** Several dozen gangliosides have been characterized, and the general form of the biosynthetic pathway is illustrated for the case of ganglioside GM_2 (Figure 24.26). Sugar units are added to the developing ganglioside from nucleotide derivatives, including UDP–*N*-acetylglucosamine, UDP–galactose, and UDP–glucose.

24.3 How Are Eicosanoids Synthesized, and What Are Their Functions?

Eicosanoids, so named because they are all derived from 20-carbon fatty acids, are ubiquitous breakdown products of phospholipids. In response to appropriate stimuli, cells activate the breakdown of selected phospholipids (Figure 24.27). Phospholipase A_2 (see Chapter 8) selectively cleaves fatty acids from the C-2 position of phospholipids. Often these are unsaturated fatty acids, among which is arachidonic acid. Arachidonic acid may also be released from phospholipids by the combined actions of phospholipase C (which yields diacylglycerols) and diacylglycerol lipase (which releases fatty acids).

Eicosanoids Are Local Hormones

Animal cells can modify arachidonic acid and other polyunsaturated fatty acids, in processes often involving cyclization and oxygenation, to produce so-called local hormones that (1) exert their effects at very low concentrations and (2) usually act near their sites of synthesis. These substances include the **prostaglandins (PG)** (Figure 24.27) as well as **thromboxanes (Tx), leukotrienes,** and other **hydroxyeicosanoic acids.** Thromboxanes, discovered in blood platelets *(thrombocytes),* are cyclic ethers (TxB_2 is actually a hemiacetal; see Figure 24.27) with a hydroxyl group at C-15.

Prostaglandins Are Formed from Arachidonate by Oxidation and Cyclization

All prostaglandins are cyclopentanoic acids derived from arachidonic acid. The biosynthesis of prostaglandins is initiated by an enzyme associated with the ER, called **prostaglandin endoperoxide H synthase (PGHS),** also known as **cyclooxygenase (COX).** The enzyme catalyzes simultaneous oxidation and cyclization of arachidonic acid. The enzyme is viewed as having two distinct activities, COX and peroxidase (POX), as shown in Figure 24.28.

> ## A DEEPER LOOK
>
> ## The Discovery of Prostaglandins
>
> The name *prostaglandin* was given to this class of compounds by Ulf von Euler, their discoverer, in Sweden in the 1930s. He extracted fluids containing these components from human semen. Because he thought they originated in the prostate gland, he named them prostaglandins. Actually, they were synthesized in the seminal vesicles, and it is now known that similar substances are synthesized in most animal tissues (both male and female). Von Euler observed that injection of these substances into animals caused smooth muscle contraction and dramatic lowering of blood pressure.
>
> Von Euler (and others) soon found that it is difficult to analyze and characterize these obviously interesting compounds because they are present at extremely low levels. Prostaglandin $E_2\alpha$, or $PGE_2\alpha$, is present in human serum at a level of less than $10^{-14} M$! In addition, they often have half-lives of only 30 seconds to a few
>
> minutes, not lasting long enough to be easily identified. Moreover, most animal tissues upon dissection and homogenization rapidly synthesize and degrade a variety of these substances, so the amounts obtained in isolation procedures are extremely sensitive to the methods used and highly variable even when procedures are carefully controlled. Sune Bergström, Bengt Samuelsson, and their colleagues described the first structural determinations of prostaglandins in the late 1950s. In the early 1960s, dramatic advances in laboratory techniques, such as NMR spectroscopy and mass spectrometry, made further characterization possible. Von Euler received the Nobel Prize for Physiology or Medicine in 1970, and Bergström, Samuelsson, and John Vane shared the Nobel for Physiology or Medicine in 1982.

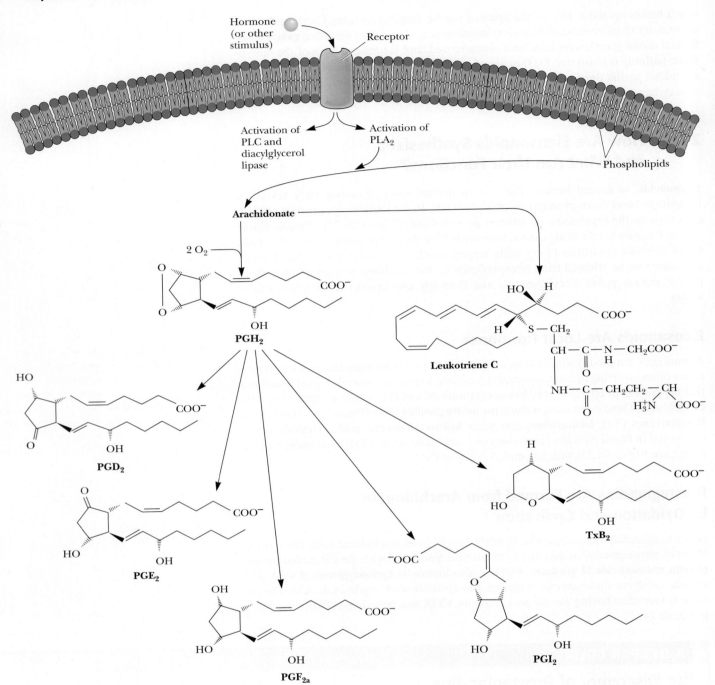

FIGURE 24.27 Arachidonic acid, derived from breakdown of phospholipids (PL), is the precursor of prostaglandins, thromboxanes, and leukotrienes. The letters used to name the prostaglandins are assigned on the basis of similarities in structure and physical properties. The class denoted PGE, for example, consists of β-hydroxyketones that are soluble in ether, whereas PGF denotes 1,3-diols that are soluble in phosphate buffer. PGA denotes prostaglandins possessing α,β-unsaturated ketones. The number following the letters refers to the number of carbon–carbon double bonds. Thus, PGE$_2$ contains two double bonds.

A Variety of Stimuli Trigger Arachidonate Release and Eicosanoid Synthesis

The release of arachidonate and the synthesis or interconversion of eicosanoids can be initiated by a variety of stimuli, including histamine, hormones such as epinephrine and bradykinin, proteases such as thrombin, and even serum albumin. An important mechanism of arachidonate release and eicosanoid synthesis involves tissue injury and inflammation. When tissue damage or injury occurs, special inflammatory cells, **monocytes** and **neutrophils,** invade the injured tissue and interact with the resident cells (such as smooth muscle cells and fibroblasts). *This interaction typically leads to arachidonate release and eicosanoid synthesis.* Examples of tissue injury in which eicosanoid synthesis has been characterized include heart attack (myocardial infarction), rheumatoid arthritis, and ulcerative colitis.

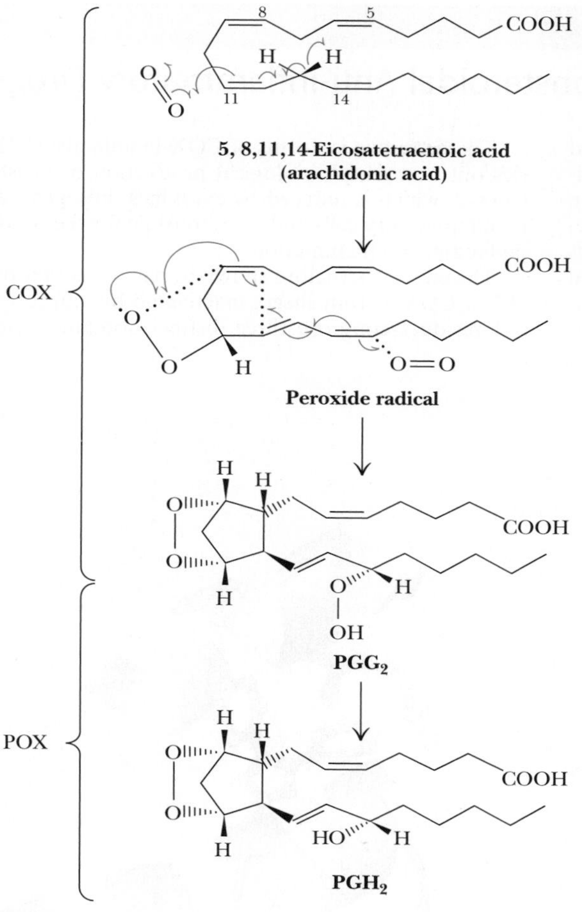

5, 8,11,14-Eicosatetraenoic acid
(arachidonic acid)

Peroxide radical

PGG₂

PGH₂

COX

POX

FIGURE 24.28 Prostaglandin endoperoxide H synthase (PGHS), the enzyme that converts arachidonic acid to prostaglandin PGH₂, possesses two distinct activities: cyclooxygenase (COX) and a glutathione-dependent hydroperoxidase (POX). The mechanism of the reaction begins with hydrogen atom abstraction by a tyrosine radical on the enzyme, followed by rearrangement to cyclize and incorporate two oxygen molecules. Reduction of the peroxide at C15 completes the reaction. COX is the site of action of aspirin and other analgesic agents.

"Take Two Aspirin and . . ." Inhibit Your Prostaglandin Synthesis

In 1971, biochemist John Vane was the first to show that **aspirin** (acetylsalicylate; Figure 24.29) exerts most of its effects by inhibiting the biosynthesis of prostaglandins. Its site of action is PGHS. COX activity is destroyed when aspirin *O*-acetylates Ser^{530} on the enzyme. From this you may begin to infer something about how prostaglandins (and aspirin) function. Prostaglandins are known to enhance inflammation in animal tissues. Aspirin exerts its powerful anti-inflammatory effect by inhibiting this first step in their synthesis. Aspirin does not have any measurable effect on the peroxidase activity of the

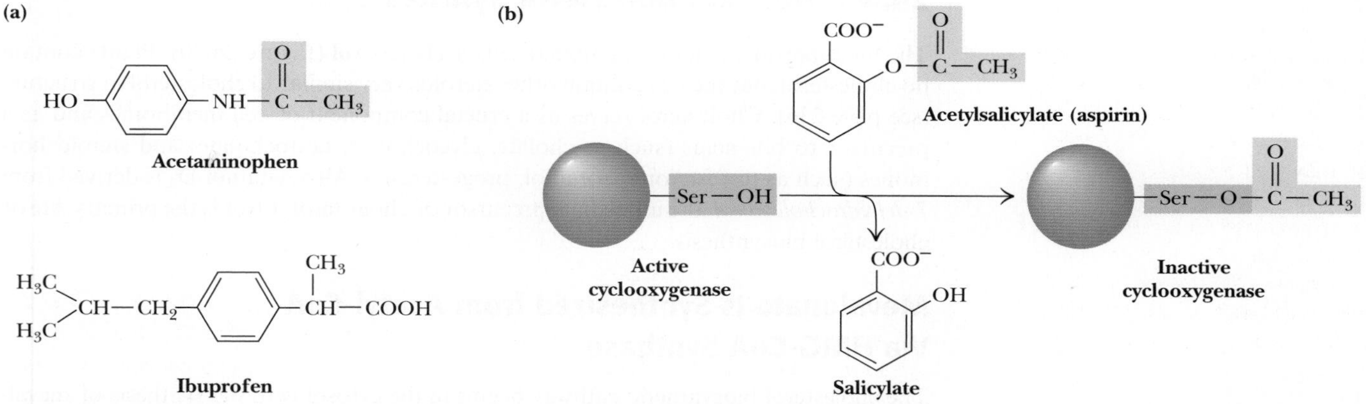

(a)

Acetaminophen

Ibuprofen

(b)

Acetylsalicylate (aspirin)

Active cyclooxygenase

Salicylate

Inactive cyclooxygenase

FIGURE 24.29 **(a)** The structures of several common analgesic agents. Acetaminophen is marketed under the trade name Tylenol. Ibuprofen is sold as Motrin, Nuprin, and Advil. **(b)** Acetylsalicylate (aspirin) inhibits the COX activity of endoperoxide synthase via acetylation (covalent modification) of Ser^{530}.

The Molecular Basis for the Action of Nonsteroidal Anti-inflammatory Drugs

Prostaglandins are potent mediators of inflammation. The first and committed step in the production of prostaglandins from arachidonic acid is the bis-oxygenation of arachidonate to prostaglandin PGG_2. This is followed by reduction to PGH_2 in a peroxidase reaction. Both these reactions are catalyzed by PGHS or COX. This enzyme is inhibited by the family of drugs known as nonsteroidal anti-inflammatory drugs, or NSAIDs. Aspirin, ibuprofen, flurbiprofen, and acetaminophen (trade name Tylenol) are all NSAIDs.

There are two isoforms of COX in animals: COX-1, which carries out normal, physiological production of prostaglandins, and COX-2, which is induced by cytokines, mitogens, and endotoxins in inflammatory cells and is responsible for the production of prostaglandins in inflammation.

The enzyme structure shown in panel a is that of residues 33 to 583 of COX-1 from sheep, inactivated by ibuprofen (cyan). These 551 residues comprise three distinct domains. The first of these,

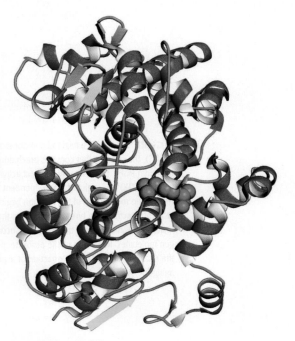

(a) pdb id = 1EQG

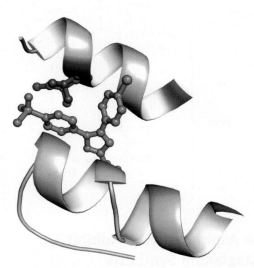

(b) Superposition of pdb id = 1EQG and 1CX2

synthase. Other nonsteroidal anti-inflammatory agents, such as **ibuprofen** (Figure 24.29) and phenylbutazone, inhibit COX by competing at the active site with arachidonate or with the peroxyacid intermediate (PGG_2, as in Figure 24.28). See A Deeper Look above.

24.4 | How Is Cholesterol Synthesized?

The most prevalent steroid in animal cells is **cholesterol** (Figure 24.30). Plants contain no cholesterol, but they *do* contain other steroids very similar to cholesterol in structure (see page 236). Cholesterol serves as a crucial component of cell membranes and as a precursor to bile acids (such as cholate, glycocholate, taurocholate) and steroid hormones (such as testosterone, estradiol, progesterone). Also, vitamin D_3 is derived from *7-dehydrocholesterol,* the immediate precursor of cholesterol. Liver is the primary site of cholesterol biosynthesis.

Mevalonate Is Synthesized from Acetyl-CoA Via HMG-CoA Synthase

The cholesterol biosynthetic pathway begins in the cytosol with the synthesis of mevalonate from acetyl-CoA (Figure 24.31). The first step is the **β-ketothiolase**–catalyzed Claisen condensation of two molecules of acetyl-CoA to form acetoacetyl-CoA. In the

residues 33 to 72, is a small, compact module that is similar to epidermal growth factor. The second domain, composed of residues 73 to 116, forms a right-handed spiral of four α-helical segments. These α-helical segments form a membrane-binding motif. The helical segments are amphipathic, with most of the hydrophobic residues facing away from the protein, where they can interact with a lipid bilayer. The third domain of the COX enzyme, the catalytic domain, is a globular structure that contains both the COX and the peroxidase active sites.

The COX active site lies at the end of a long, narrow, hydrophobic tunnel or channel. Three of the α-helices of the membrane-binding domain lie at the entrance to this tunnel. The walls of the tunnel are defined by four α-helices, formed by residues 106 to 123, 325 to 353, 379 to 384, and 520 to 535.

The COX-1 structure shown in panel a has a molecule of ibuprofen bound in the tunnel. Deep in the tunnel, at the far end, lies Tyr[385], a catalytically important residue. Heme-dependent peroxidase activity is implicated in the formation of a proposed Tyr[385] radical, which is required for COX activity. Aspirin and other NSAIDs block the synthesis of prostaglandins by filling and blocking the tunnel, preventing the migration of arachidonic acid to Tyr[385] in the active site at the back of the tunnel.

Why do the new "COX-2 inhibitors" bind to (and inhibit) COX-2 but not COX-1? A single amino acid substitution makes all the difference. Panel b shows an overlay of COX-1 (1EQG) and COX-2 (1CX2) structures. COX-2 has a valine (blue) at position 523, which leaves room for binding of a Celebrex-like inhibitor (orange). On the other hand, COX-1 has bulkier isoleucine (red) at position 523, which prevents binding of the inhibitor.

COX-2 inhibitors were introduced as pain medications in 1997, and by 2004 nearly half of the 100 million prescriptions written annually for NSAIDs in the United States were COX-2 inhibitors. However, several COX-2 inhibitors were taken off the U.S. market in late 2004 and early 2005, when their use was linked to heart attacks and strokes in a small percentage of users. Since that time, prescriptions for COX-2 inhibitors have dropped by 65%. Interestingly, although COX-2 inhibitors were originally intended to alleviate pain without the risk of adverse gastrointestinal effects, less than 5% of patients that used COX-2 prescriptions at the peak of their popularity were at high risk for these adverse effects.

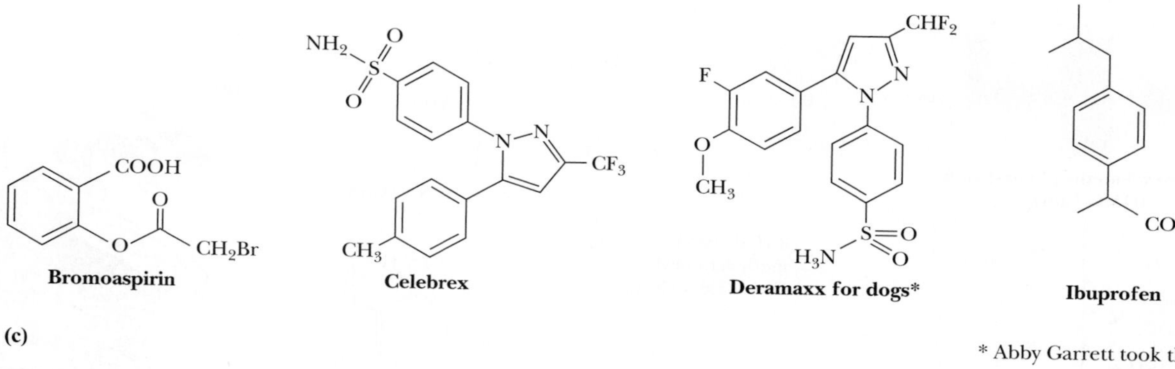

Bromoaspirin **Celebrex** **Deramaxx for dogs*** **Ibuprofen**

(c)

* Abby Garrett took this.

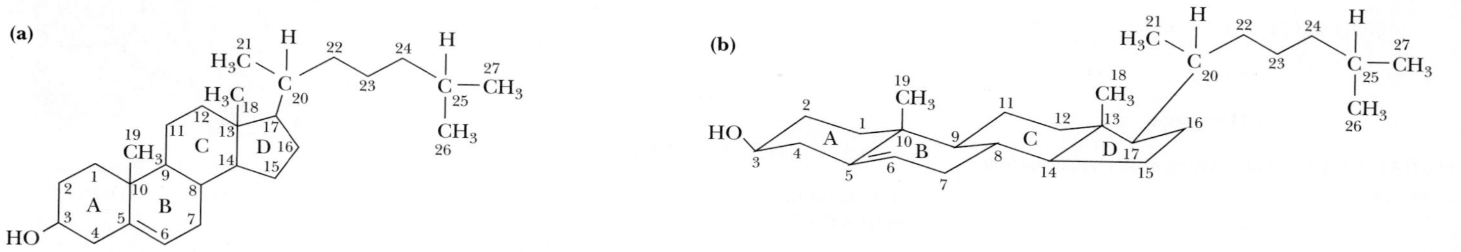

FIGURE 24.30 The structure of cholesterol, drawn (a) in the traditional planar motif and (b) in a form that more accurately describes the conformation of the ring system.

next reaction, acetyl-CoA and acetoacetyl-CoA join to form *3-hydroxy-3-methylglutaryl-CoA,* which is abbreviated *HMG-CoA.* The reaction, a second Claisen condensation, is catalyzed by **HMG-CoA synthase.** The third step in the pathway is the rate-limiting step in cholesterol biosynthesis. Here, HMG-CoA undergoes two NADPH-dependent reductions to produce *3R-mevalonate* (Figure 24.32). The reaction is catalyzed by **HMG-CoA reductase,** a 97-kD glycoprotein that spans the ER membrane with its active site facing the cytosol. As the rate-limiting step, HMG-CoA reductase is the principal site of regulation in cholesterol synthesis.

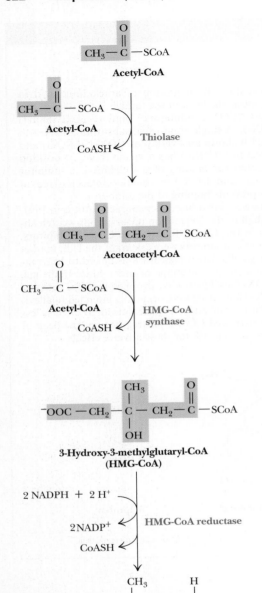

FIGURE 24.31 The biosynthesis of 3*R*-mevalonate from acetyl-CoA.

Three different regulatory mechanisms are involved:

1. Phosphorylation by cAMP-dependent protein kinases inactivates the reductase. This inactivation can be reversed by two specific phosphatases (Figure 24.33).
2. Degradation of HMG-CoA reductase. This enzyme has a half-life of only 3 hours, and the half-life itself depends on cholesterol levels: High [cholesterol] means a short half-life for HMG-CoA reductase.
3. Gene expression. Cholesterol levels control the amount of mRNA. If [cholesterol] is high, levels of mRNA coding for the reductase are reduced. If [cholesterol] is low, more mRNA is made. (Regulation of gene expression is discussed in Chapter 29.)

A Thiolase Brainteaser Asks Why Thiolase Can't Be Used in Fatty Acid Synthesis

If acetate units can be condensed by the thiolase reaction to yield acetoacetate in the first step of cholesterol synthesis, why couldn't this same reaction also be used in fatty acid synthesis, avoiding all the complexity of the fatty acyl synthase? The answer is that the thiolase reaction is more or less reversible but slightly favors the cleavage reaction. In the cholesterol synthesis pathway, subsequent reactions, including HMG-CoA reductase and the following kinase reactions, pull the thiolase-catalyzed condensation forward. However, in the case of fatty acid synthesis, a succession of eight thiolase condensations would be distinctly unfavorable from an energetic perspective. Given the

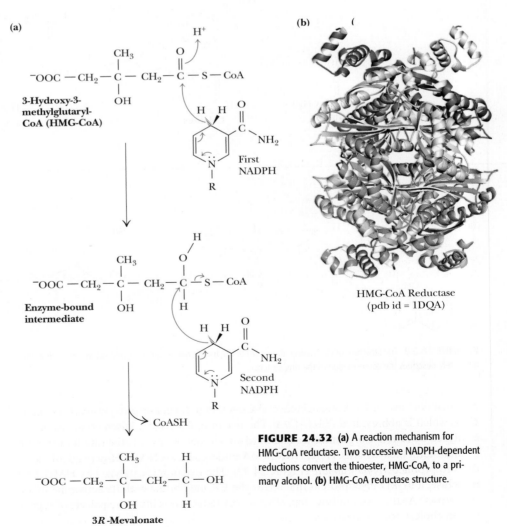

FIGURE 24.32 **(a)** A reaction mechanism for HMG-CoA reductase. Two successive NADPH-dependent reductions convert the thioester, HMG-CoA, to a primary alcohol. **(b)** HMG-CoA reductase structure.

HMG-CoA Reductase
(pdb id = 1DQA)

> ## CRITICAL DEVELOPMENTS IN BIOCHEMISTRY

The Long Search for the Route of Cholesterol Biosynthesis

Heilbron, Kamm, and Owens suggested as early as 1926 that squalene is a precursor of cholesterol. That same year, H. J. Channon demonstrated that animals fed squalene from shark oil produced more cholesterol in their tissues. Bloch and Rittenberg showed in the 1940s that a significant amount of the carbon in the tetracyclic moiety and in the aliphatic side chain of cholesterol was derived from acetate. In 1934, Sir Robert Robinson suggested a scheme for the cyclization of squalene to form cholesterol before the biosynthetic link between acetate and squalene was understood. Squalene is actually a polymer of isoprene units, and Bonner and Arreguin suggested in 1949 that three acetate units could join to form five-carbon *isoprene* units (see figure a).

In 1952, Konrad Bloch and Robert Langdon showed conclusively that labeled squalene is synthesized rapidly from labeled acetate and also that cholesterol is derived from squalene. Langdon, a graduate student of Bloch's, performed the critical experiments in Bloch's laboratory at the University of Chicago while Bloch spent the summer in Bermuda attempting to demonstrate that radioactively labeled squalene would be converted to cholesterol in shark livers. As Bloch himself admitted, "All I was able to learn was that sharks of manageable length are very difficult to catch and their oily livers impossible to slice" (Bloch, 1987).

In 1953, Bloch, together with the eminent organic chemist R. B. Woodward, proposed a new scheme (see figure b) for the cyclization of squalene. (Together with Fyodor Lynen, Bloch received the Nobel Prize in Medicine or Physiology in 1964 for his work.) The picture was nearly complete, but one crucial question remained: How could isoprene be the intermediate in the transformation of

acetate into squalene? In 1956, Karl Folkers and his colleagues at Merck Sharpe & Dohme isolated mevalonic acid and also showed that mevalonate was the precursor of isoprene units. The search for the remaining details (described in the text) made the biosynthesis of cholesterol one of the most enduring and challenging bioorganic problems of the 1940s, 1950s, and 1960s. Even today, several of the enzyme mechanisms remain poorly understood.

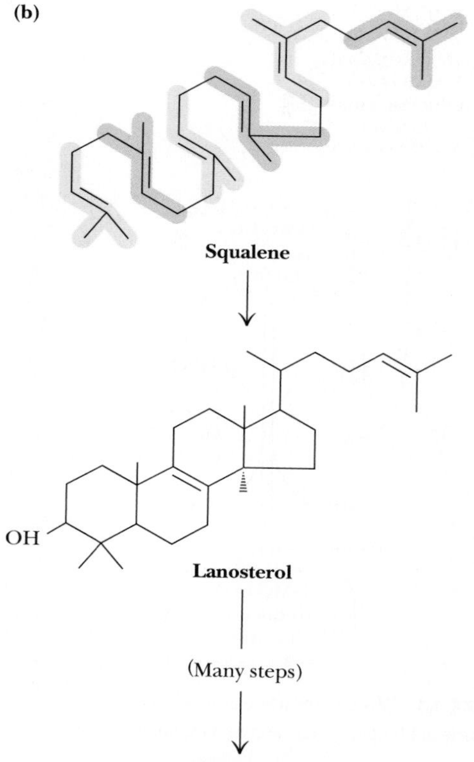

(a) An isoprene unit and a scheme for head-to-tail linking of isoprene units. (b) The cyclization of squalene to form lanosterol, as proposed by Bloch and Woodward.

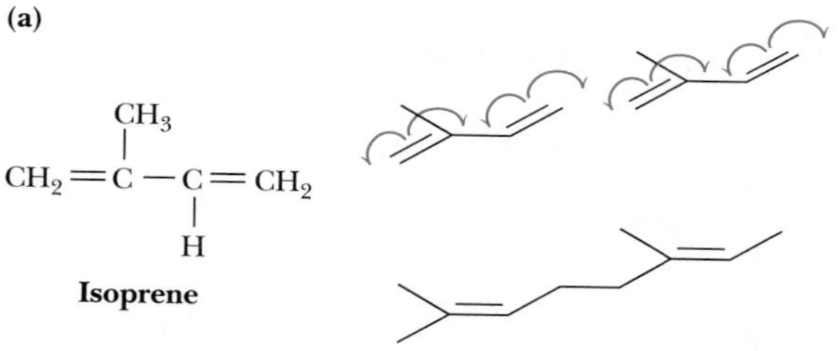

necessity of repeated reactions in fatty acid synthesis, it makes better energetic sense to use a reaction that is favorable in the desired direction.

Squalene Is Synthesized from Mevalonate

The biosynthesis of squalene involves conversion of mevalonate to two key 5-carbon intermediates, *isopentenyl pyrophosphate* and *dimethylallyl pyrophosphate,* which join to yield *farnesyl pyrophosphate* and then squalene. A series of four reactions converts mevalonate to isopentenyl pyrophosphate and then to dimethylallyl pyrophosphate (Figure 24.34). The first three steps each consume an ATP, two for the purpose of forming a pyrophosphate at the 5-position and the third to drive the decarboxylation and double bond formation in the third step. **Pyrophosphomevalonate decarboxylase** phosphorylates the 3-hydroxyl group, and this is followed by *trans* elimination of the phosphate and carboxyl groups to form the double bond in isopentenyl pyrophosphate. Isomerization of the double bond yields the dimethylallyl pyrophosphate. Condensation of these two 5-carbon intermediates produces *geranyl pyrophosphate;* addition of another 5-carbon

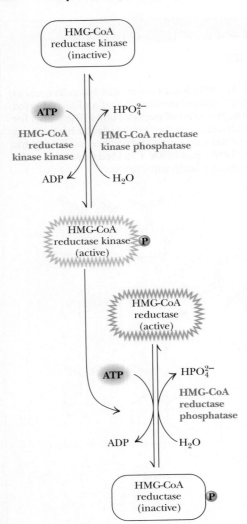

FIGURE 24.33 HMG-CoA reductase activity is modulated by a cycle of phosphorylation and dephosphorylation.

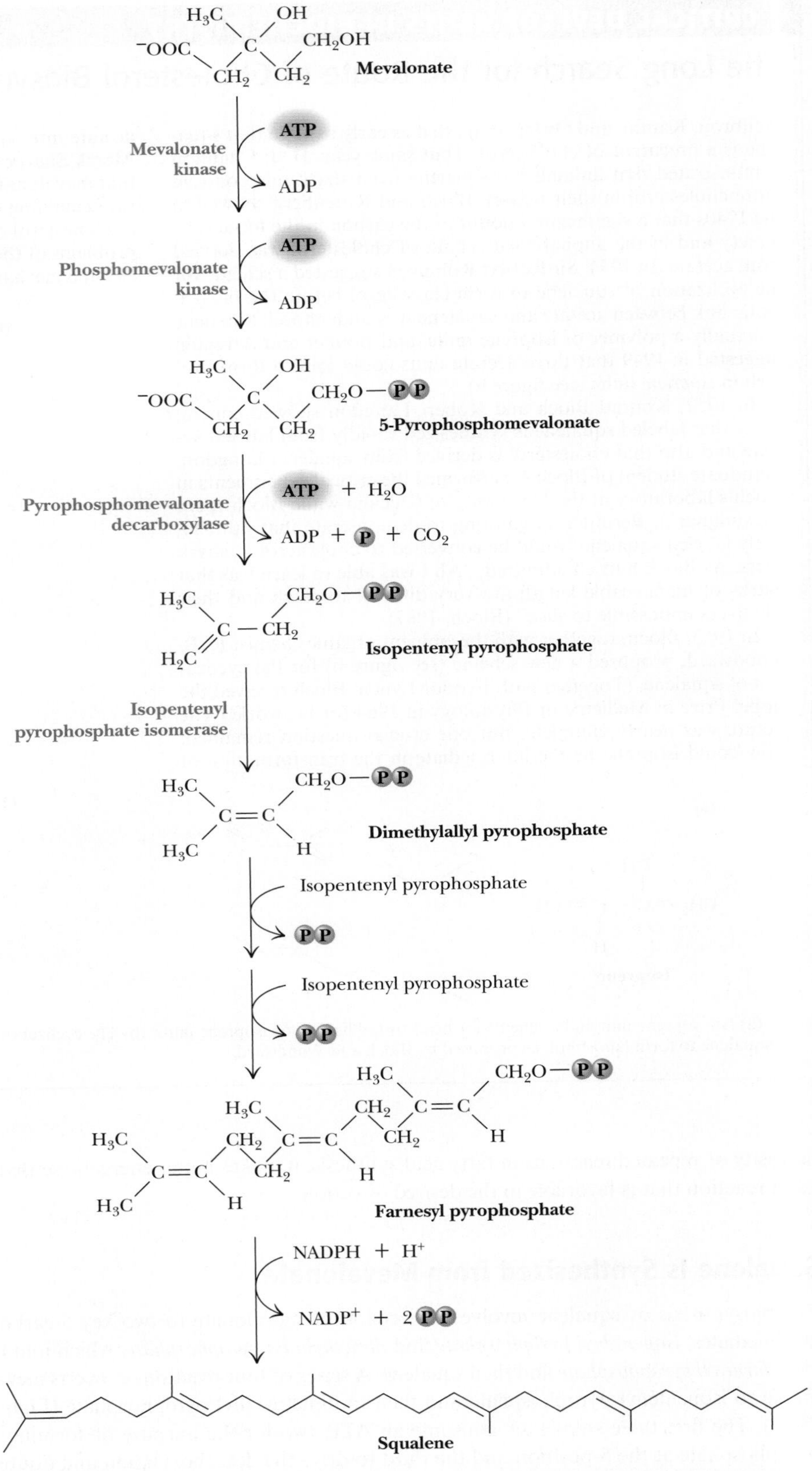

Mevalonate

Mevalonate kinase   ATP → ADP

Phosphomevalonate kinase   ATP → ADP

5-Pyrophosphomevalonate

Pyrophosphomevalonate decarboxylase   ATP + H_2O → ADP + ⓟ + CO_2

Isopentenyl pyrophosphate

Isopentenyl pyrophosphate isomerase

Dimethylallyl pyrophosphate

Isopentenyl pyrophosphate → ⓟⓟ

Isopentenyl pyrophosphate → ⓟⓟ

Farnesyl pyrophosphate

NADPH + H^+ → $NADP^+$ + 2 ⓟⓟ

Squalene

FIGURE 24.34 The conversion of mevalonate to squalene. In the last step, two farnesyl-PP condense to form squalene.

HUMAN BIOCHEMISTRY

Statins Lower Serum Cholesterol Levels

Chemists and biochemists have long sought a means of lowering serum cholesterol levels to reduce the risk of heart attack and cardiovascular disease. Because HMG-CoA reductase is the rate-limiting step in cholesterol biosynthesis, this enzyme is a likely drug target. **Mevinolin,** also known as **lovastatin** (see accompanying figure), was isolated from a strain of *Aspergillus terreus* and developed at Merck Sharpe & Dohme for this purpose. It is now a widely prescribed cholesterol-lowering drug. Dramatic reductions of serum cholesterol are observed at dosages of 20 to 80 mg per day.

Lovastatin is administered as an inactive lactone. After oral ingestion, it is hydrolyzed to the active **mevinolinic acid,** a competitive inhibitor of the reductase with a K_I of 0.6 nM. Mevinolinic acid is thought to behave as a transition-state analog (see Chapter 14) of the tetrahedral intermediate formed in the HMG-CoA reductase reaction (see figure).

Derivatives of lovastatin have been found to be even more potent in cholesterol-lowering trials. **Synvinolin** lowers serum cholesterol levels at much lower dosages than lovastatin. Lipitor, shown bound at the active site of HMG-CoA reductase, is the most-prescribed drug in the world, with annual sales of $10.7 billion.

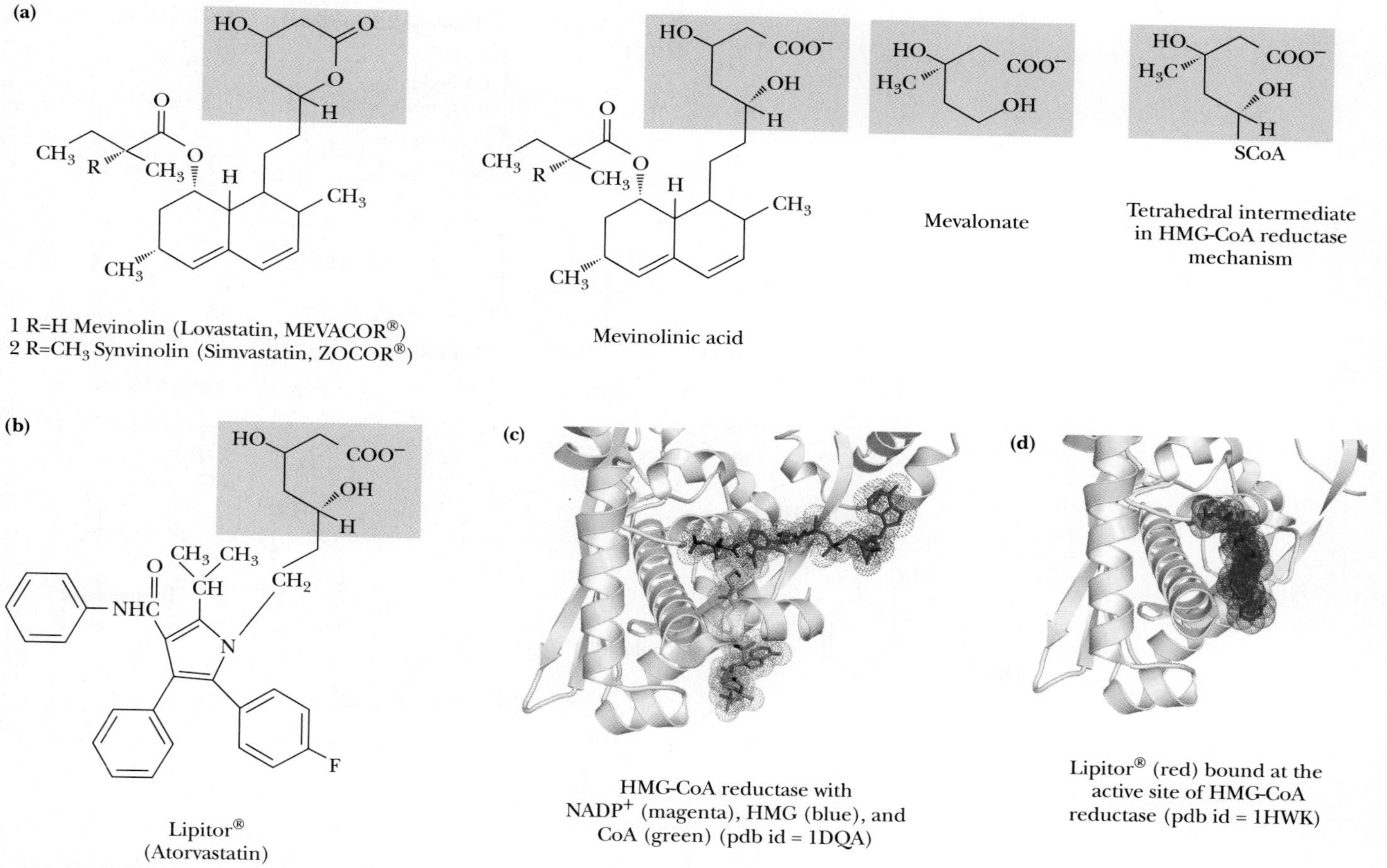

(a)

1 R=H Mevinolin (Lovastatin, MEVACOR®)
2 R=CH₃ Synvinolin (Simvastatin, ZOCOR®)

Mevinolinic acid

Mevalonate

Tetrahedral intermediate in HMG-CoA reductase mechanism

(b)

Lipitor®
(Atorvastatin)

(c)

HMG-CoA reductase with NADP⁺ (magenta), HMG (blue), and CoA (green) (pdb id = 1DQA)

(d)

Lipitor® (red) bound at the active site of HMG-CoA reductase (pdb id = 1HWK)

The structures of **(a)** (inactive) lovastatin, (active) mevinolinic acid, mevalonate, and **(b)** Lipitor (atorvastatin). **(c)** HMG-CoA reductase with NADP⁺, HMG, and CoA. **(d)** Lipitor bound at the HMG-CoA reductase active site.

isopentenyl group gives *farnesyl pyrophosphate*. Both steps in the production of farnesyl pyrophosphate occur with release of pyrophosphate, hydrolysis of which drives these reactions forward. Note too that the linkage of isoprene units to form farnesyl pyrophosphate occurs in a head-to-tail fashion. This is the general rule in biosynthesis of molecules involving isoprene linkages. The next step—the joining of two farnesyl pyrophosphates to produce squalene—is a "tail-to-tail" condensation and represents an important exception to the general rule.

Squalene monooxygenase, an enzyme bound to the ER, converts squalene to *squalene-2,3-epoxide* (Figure 24.35). This reaction employs FAD and NADPH as coenzymes and requires O_2 as well as a cytosolic protein called **soluble protein activator.** A

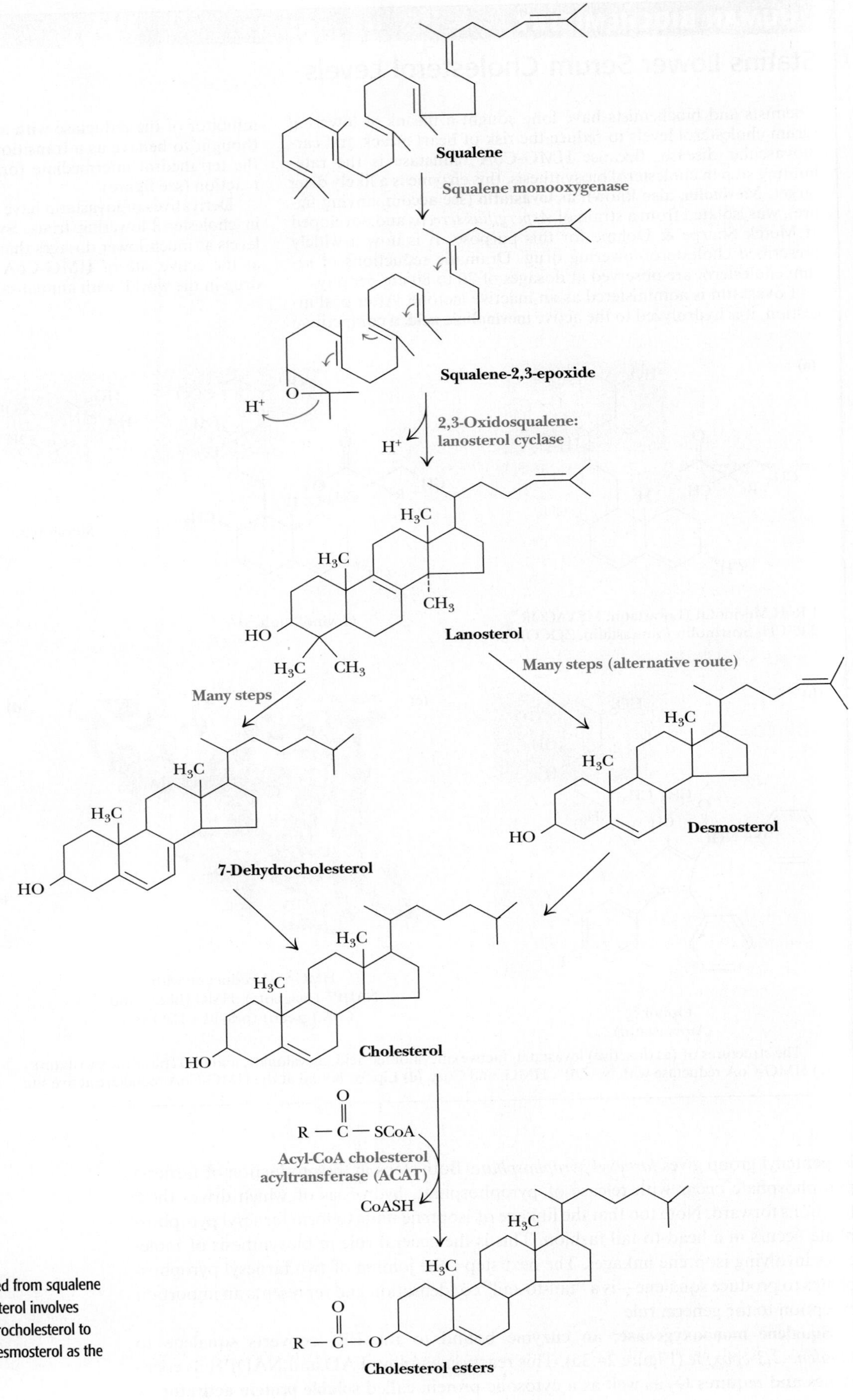

FIGURE 24.35 Cholesterol is synthesized from squalene via lanosterol. The primary route from lanosterol involves 20 steps, the last of which converts 7-dehydrocholesterol to cholesterol. An alternative route produces desmosterol as the penultimate intermediate.

second ER membrane enzyme, **2,3-oxidosqualene lanosterol cyclase,** catalyzes the second reaction, which involves a succession of 1,2 shifts of hydride ions and methyl groups.

Conversion of Lanosterol to Cholesterol Requires 20 Additional Steps

Although lanosterol may appear similar to cholesterol in structure, another 20 steps are required to convert lanosterol to cholesterol (Figure 24.35). The enzymes responsible for this are all associated with the ER. The primary pathway involves *7-dehydrocholesterol* as the penultimate intermediate. An alternative pathway, also composed of many steps, produces the intermediate *desmosterol.* Reduction of the double bond at C-24 yields cholesterol. Cholesterol esters—a principal form of circulating cholesterol—are synthesized by **acyl-CoA : cholesterol acyltransferases (ACAT)** on the cytoplasmic face of the ER.

24.5 How Are Lipids Transported Throughout the Body?

When most lipids circulate in the body, they do so in the form of **lipoprotein complexes.** Simple, unesterified fatty acids are merely bound to serum albumin and other proteins in blood plasma, but phospholipids, triacylglycerols, cholesterol, and cholesterol esters are all transported in the form of lipoproteins. At various sites in the body, lipoproteins interact with specific receptors and enzymes that transfer or modify their lipid cargoes. It is now customary to classify lipoproteins according to their densities (Table 24.1). The densities are related to the relative amounts of lipid and protein in the complexes. Because most proteins have densities of about 1.3 to 1.4 g/mL, and lipid aggregates usually possess densities of about 0.8 g/mL, the more protein and the less lipid in a complex, the denser the lipoprotein. Thus, there are **high-density lipoproteins (HDLs), low-density lipoproteins (LDLs), intermediate-density lipoproteins (IDLs), very-low-density lipoproteins (VLDLs),** and also **chylomicrons.** Chylomicrons have the lowest protein-to-lipid ratio and thus are the lowest-density lipoproteins. They are also the largest.

Lipoprotein Complexes Transport Triacylglycerols and Cholesterol Esters

HDL and VLDL are assembled primarily in the ER of the liver (with smaller amounts produced in the intestine), whereas chylomicrons form in the intestine. LDL is not synthesized directly but rather is made from VLDL. LDL appears to be the major circulatory complex for cholesterol and cholesterol esters. The primary task of chylomicrons is to transport triacylglycerols. Despite all this, it is extremely important to note that each of these lipoprotein classes contains some of each type of lipid. The relative amounts of HDL and LDL are important in the disposition of cholesterol in the body and in the

TABLE 24.1	Composition and Properties of Human Lipoproteins					
Lipoprotein Class	Density (g/mL)	Diameter (nm)	Composition (% dry weight)			
			Protein	Cholesterol	Phospholipid	Triacylglycerol
HDL	1.063–1.21	5–15	33	30	29	8
LDL	1.019–1.063	18–28	25	50	21	4
IDL	1.006–1.019	25–50	18	29	22	31
VLDL	0.95–1.006	30–80	10	22	18	50
Chylomicrons	<0.95	100–500	1–2	8	7	84

Adapted from Brown, M., and Goldstein, J., 1987. In Braunwald, E., et al., eds., *Harrison's Principles of Internal Medicine,* 11th ed. New York: McGraw-Hill; and Vance, D., and Vance, J., eds., 1985. *Biochemistry of Lipids and Membranes.* Menlo Park, CA: Benjamin/Cummings.

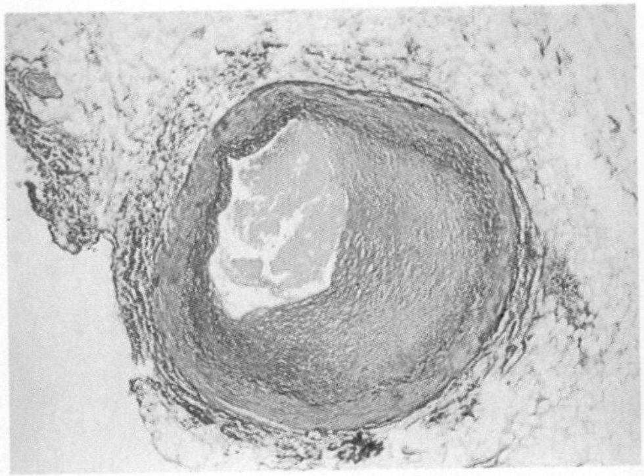

FIGURE 24.36 Photograph of an arterial plaque. The view is into the artery (orange), with the plaque shown in yellow at the back.

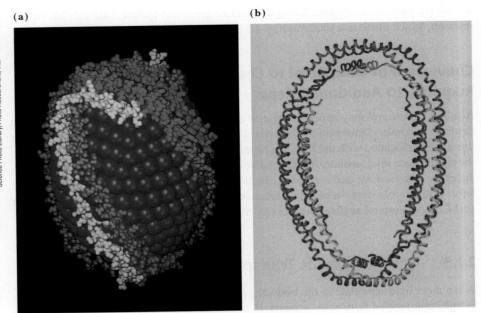

FIGURE 24.37 A model for the structure of a typical lipoprotein. **(a)** A core of cholesterol and cholesteryl esters is surrounded by a phospholipid (monolayer) membrane. Apolipoprotein A-I is modeled here as a long amphipathic α-helix, with the nonpolar face of the helix embedded in the hydrophobic core of the lipid particle and the polar face of the helix exposed to solvent. **(b)** A ribbon diagram of apolipoprotein A-I. (Adapted from Borhani, D. W., Rogers, D. P., Engler, J. A., and Brouillette, C. G., 1997. Crystal structure of truncated human apolipoprotein A-I suggests a lipid-bound conformation. *Proceedings of the National Academy of Sciences* **94:**12291–12296.)

development of arterial plaques (Figure 24.36). The structures of the various lipoproteins are approximately similar, and they consist of a core of mobile triacylglycerols or cholesterol esters surrounded by a single layer of phospholipid, into which is inserted a mixture of cholesterol and proteins (Figure 24.37). Note that the phospholipids are oriented with their polar head groups facing outward to interact with solvent water and that the phospholipids thus shield the hydrophobic lipids inside from the solvent water outside. The proteins also function as recognition sites for the various lipoprotein receptors throughout the body. A number of different apoproteins have been identified in lipoproteins (Table 24.2), and others may exist as well. The apoproteins have an

TABLE 24.2	Apoproteins of Human Lipoproteins		
Apoprotein	M_r	Concentration in Plasma (mg/100 mL)	Distribution
A-1	28,300	90–120	Principal protein in HDL
A-2	8,700	30–50	Occurs as dimer mainly in HDL
B-48	240,000	<5	Found only in chylomicrons
B100	500,000	80–100	Principal protein in LDL
C-1	7,000	4–7	Found in chylomicrons, VLDL, HDL
C-2	8,800	3–8	Found in chylomicrons, VLDL, HDL
C-3	8,800	8–15	Found in chylomicrons, VLDL, IDL, HDL
D	32,500	8–10	Found in HDL
E	34,100	3–6	Found in chylomicrons, VLDL, IDL, HDL

Adapted from Brown, M., and Goldstein, J., 1987. In Braunwald, E., et al., eds., *Harrison's Principles of Internal Medicine,* 11th ed. New York: McGraw-Hill; and Vance, D., and Vance, J., eds., 1985. *Biochemistry of Lipids and Membranes,* Menlo Park, CA: Benjamin/Cummings.

abundance of hydrophobic amino acid residues, as is appropriate for interactions with lipids. A **cholesterol ester transfer protein** also associates with lipoproteins.

Lipoproteins in Circulation Are Progressively Degraded by Lipoprotein Lipase

The livers and intestines of animals are the primary sources of circulating lipids. Chylomicrons carry triacylglycerol and cholesterol esters from the intestines to other tissues, and VLDLs carry lipid from liver, as shown in Figure 24.38. At various target sites, particularly in the capillaries of muscle and adipose cells, these particles are degraded by **lipoprotein lipase,** which hydrolyzes triacylglycerols. Lipase action causes progressive loss of triacylglycerol (and apoprotein) and makes the lipoproteins smaller. This process gradually converts VLDL particles to IDL and then LDL particles, which are either returned to the liver for reprocessing or redirected to adipose tissues and adrenal glands. Every 24 hours, nearly half of all serum LDL is removed from circulation in this way. The LDL binds to specific LDL receptors, which cluster in domains of the plasma membrane known as **coated pits** (discussed in subsequent paragraphs). These domains eventually invaginate to form **coated vesicles** (Figure 24.39), which pinch off from the plasma membrane and form **endosomes** (literally "bodies inside" the cell). In the low pH environment of the endosome, the LDL particles dissociate from their receptors. The endosomes then fuse with lysosomes, and the LDLs are degraded by **lysosomal acid lipases.**

HDLs have much longer life spans in the body (5 to 6 days) than other lipoproteins. Newly formed HDL contains virtually no cholesterol ester. However, over time, cholesterol esters are accumulated through the action of **lecithin:cholesterol acyltransferase (LCAT),** a 59-kD glycoprotein associated with HDLs. Another associated protein, **cholesterol ester transfer protein,** transfers some of these esters to VLDL and LDL. Alternatively, HDLs function to return cholesterol and cholesterol esters to the liver. This latter process apparently explains the correlation between high HDL levels and reduced

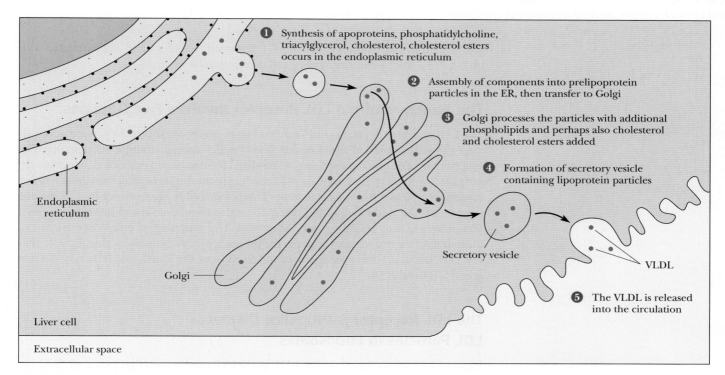

FIGURE 24.38 Lipoprotein components are synthesized predominantly in the ER of liver cells. Following assembly of lipoprotein particles (red dots) in the ER and processing in the Golgi, lipoproteins are packaged in secretory vesicles for export from the cell (via exocytosis) and released into the circulatory system.

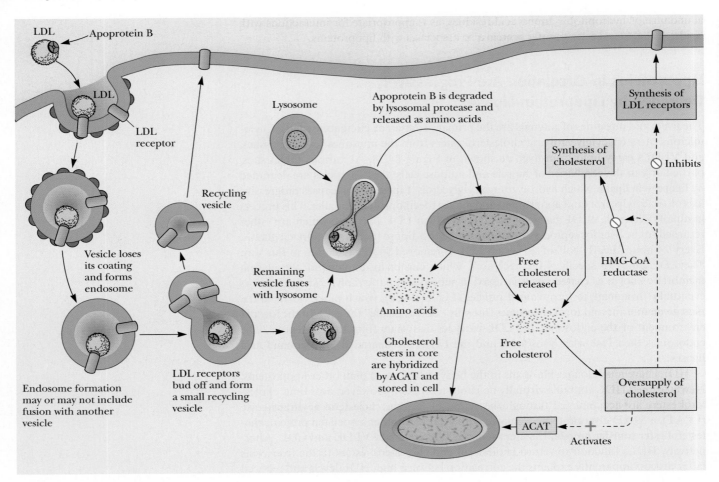

FIGURE 24.39 Endocytosis and degradation of lipoprotein particles. (ACAT is acyl-CoA cholesterol acyltransferase.)

risk of cardiovascular disease. (High LDL levels, on the other hand, are correlated with an *increased* risk of coronary artery and cardiovascular disease.)

The Structure of the LDL Receptor Involves Five Domains

The LDL receptor in plasma membranes (Figure 24.40) consists of 839 amino acid residues and is composed of five domains, two of which contain multiple subdomains. The N-terminal LDL-binding domain (292 residues) contains seven cysteine-rich repeats, denoted R1 to R7. The next segment (417 residues) contains three **epidermal growth factor repeats,** as well as a **β-propellor** module. This is followed in the sequence by a 58-residue segment of O-linked oligosaccharides, a 22-residue membrane-spanning segment, and a 50-residue segment extending into the cytosol. The clustering of receptors prior to the formation of coated vesicles requires the presence of this cytosolic segment. Note that the LDL particle binds specifically to the receptor at the fourth and fifth cysteine-rich repeats (R4 and R5).

The LDL Receptor β-Propellor Displaces LDL Particles in Endosomes

Figure 24.39 shows the release of LDL particles in endosomes that pinch off from the plasma membrane when cells take up LDLs. What molecular events trigger the release of LDL particles? A collaboration by three Nobel laureates has provided an answer. Johann Deisenhofer, Michael Brown, and Joseph Goldstein have determined the structure of the extracellular domain of the LDL receptor at pH 5.3, the typical pH inside endosomes. At this low pH, the receptor polypeptide is folded back on itself, with the β-propellor domain associated with R4 and R5, the two repeats that normally bind the LDL particle (Figure 24.41). The implication is that the β-propellor displaces the LDL particle in the lower pH environment of the endosome. What residues at the interface between the propellor and the R4 and R5 repeats act as the pH sensors? Three histidines at the propellor–R4/R5 interface—His[190], His[562], and His[586]—are the likely pH-sensing residues. His[190] lies at the tip of a loop on R5, whereas His[562] and His[586] are on the surface of the propellor domain (Figure 24.41). These three His residues form a cluster at the three-way junction between R4, R5, and the β-propellor.

Defects in Lipoprotein Metabolism Can Lead to Elevated Serum Cholesterol

The mechanism of LDL metabolism and the various defects that can occur therein have been studied extensively by Michael Brown and Joseph Goldstein, who received the Nobel Prize in Physiology or Medicine in 1985. **Familial hypercholesterolemia** is the term given to a variety of inherited metabolic defects that lead to greatly elevated levels of serum cholesterol, much of it in the form of LDL particles. The general genetic defect responsible for familial hypercholesterolemia is the absence or dysfunction of LDL receptors in the body. Only about half the normal level of LDL receptors is found in heterozygous individuals (persons carrying one normal gene and one defective gene). Homozygotes (with two copies of the defective gene) have few, if any, functional LDL receptors. In such cases, LDLs (and cholesterol) cannot be absorbed, and plasma levels of LDL (and cholesterol) are very high. Typical heterozygotes display serum cholesterol levels of 300 to 400 mg/dL, but homozygotes carry serum cholesterol levels of 600 to 800 mg/dL or even higher. There are two possible causes of an absence of LDL receptors—either receptor synthesis does not occur at all, or the newly synthesized protein does not successfully reach the plasma membrane due to faulty processing in the Golgi or faulty transport to the plasma membrane. Even when LDL receptors are made and reach the plasma membrane, they may fail to function for two reasons. They may be unable to form clusters competent in coated pit formation because of folding or sequence anomalies in the carboxy-terminal domain, or they may be unable to bind LDL because of sequence or folding anomalies in the LDL-binding domain.

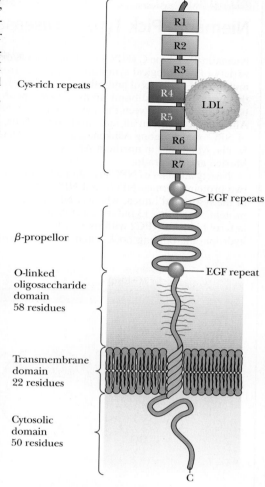

FIGURE 24.40 The structure of the LDL receptor. The amino-terminal binding domain is responsible for recognition and binding of LDL apoprotein. The B-100 apolipoprotein of the LDL particle is presumed to bind to the fourth and fifth cysteine-rich repeats (R4 and R5). The O-linked oligosaccharide-rich domain may act as a molecular spacer, raising the binding domain above the glycocalyx. The cytosolic domain is required for aggregation of LDL receptors during endocytosis.

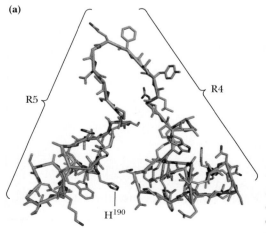

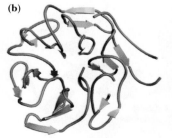

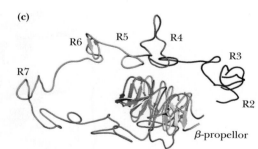

FIGURE 24.41 **(a)** The cysteine-rich repeats R4 and R5 are the site of LDL particle binding. Each repeat contains two loops connected by three disulfide bonds. The second loop in each repeat carries acidic residues (Asp and Glu) and forms a Ca²⁺-binding site. **(b)** The β-propellor domain of the LDL receptor. **(c)** The structure of the LDL receptor extracellular domain at pH 5.3 (similar to that found in endosomes). The β-propellor domain is associated with cysteine-rich domains R4 and R5. A cluster of histidines (His[190], His[562], and His[586]) and a variety of hydrophobic and charged interactions mediate the interaction (pdb id = 1N7D).

Niemann–Pick Type C Disease—A Hydrophobic Handoff Fumbled

Niemann-Pick type C (NPC) is a lipid storage disease that causes a variety of neurological symptoms and arises from abnormal accumulation of cholesterol and other lipids in the lysosomes of cells in the liver, spleen, and brain. In most cases, neurological symptoms begin appearing between the ages of 4 and 10. NPC is always fatal. Although it occurs in all races and populations, a higher incidence of NPC occurs among Ashkenazi Jews, French Canadians in Nova Scotia, Maghrebs in northern Africa, and Hispanic people in New Mexico and Colorado.

Nearly all cases of NPC are caused by mutations in the genes for two proteins, termed NPC1 and NPC2. NPC2 is a soluble protein in the lysosomal lumen, whereas NPC1 is a lysosomal membrane protein containing 13 putative membrane-spanning domains. Cholesterol binds to NPC2 with its 8-carbon side chain buried within a hydrophobic binding pocket and the 3β-hydroxyl group exposed

on the surface (see figure). In contrast, NPC1 binds cholesterol in the opposite orientation, with the 3β-hydroxyl deep in the binding site and the side chain partially exposed.

Michael Brown and Joseph Goldstein and co-workers have identified three residues adjacent to the cholesterol-binding site on NPC2 that are not required for cholesterol binding but are essential for the transfer of cholesterol from NPC2 to NPC1. In collaboration with Johann Deisenhofer, they have also identified 11 residues on NPC1 that surround (but are not part of) the cholesterol-binding site. These residues are essential for cholesterol transfer from NPC2 to NPC1 and also for restoring the export of cholesterol from lysosomes. Brown and Goldstein have proposed a "hydrophobic handoff" model for the interaction of NPC1 and NPC2 that enables cholesterol transfer between these proteins and that is essential for export of cholesterol from the lysosomes.

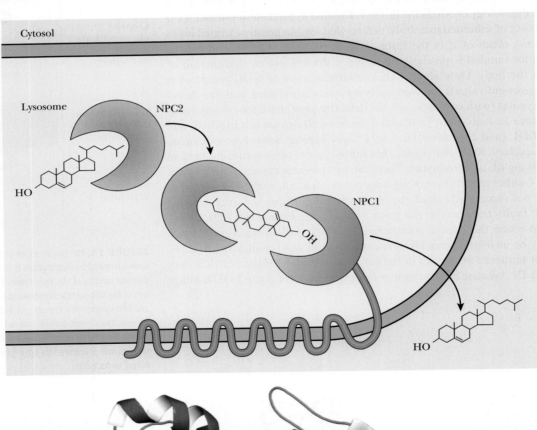

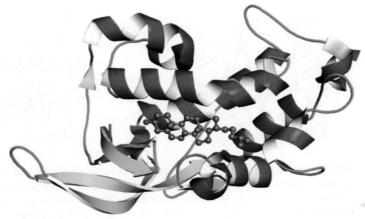

◀ NPC1 with cholesterol (blue) (pbd id = 3GKI)

24.6 | How Are Bile Acids Biosynthesized?

Bile acids, which exist mainly as **bile salts,** are polar carboxylic acid derivatives of cholesterol that are important in the digestion of food, especially the solubilization of ingested fats. The Na$^+$ and K$^+$ salts of *glycocholic acid* and *taurocholic acid* are the principal bile salts (Figure 24.42). Glycocholate and taurocholate are conjugates of *cholic acid* with glycine and taurine, respectively. Because they contain both nonpolar and polar domains, these bile salt conjugates are highly effective as detergents. These substances are made in the liver, stored in the gallbladder, and secreted as needed into the intestines.

The formation of bile salts represents the major pathway for cholesterol degradation. The first step involves hydroxylation at C-7 (Figure 24.42). **7α-Hydroxylase,** which catalyzes the reaction, is a mixed-function oxidase involving *cytochrome P-450.* **Mixed-function oxidases** use O$_2$ as substrate. One oxygen atom goes to hydroxylate the substrate while the other is reduced to water (Figure 24.43). The function of cytochrome P-450 is to activate O$_2$ for the hydroxylation reaction. Such hydroxylations are quite common in the synthetic routes for cholesterol, bile acids, and steroid hormones and also in detoxification pathways for aromatic compounds. Several of these are considered in the next section. 7α-Hydroxycholesterol is the precursor for cholic acid.

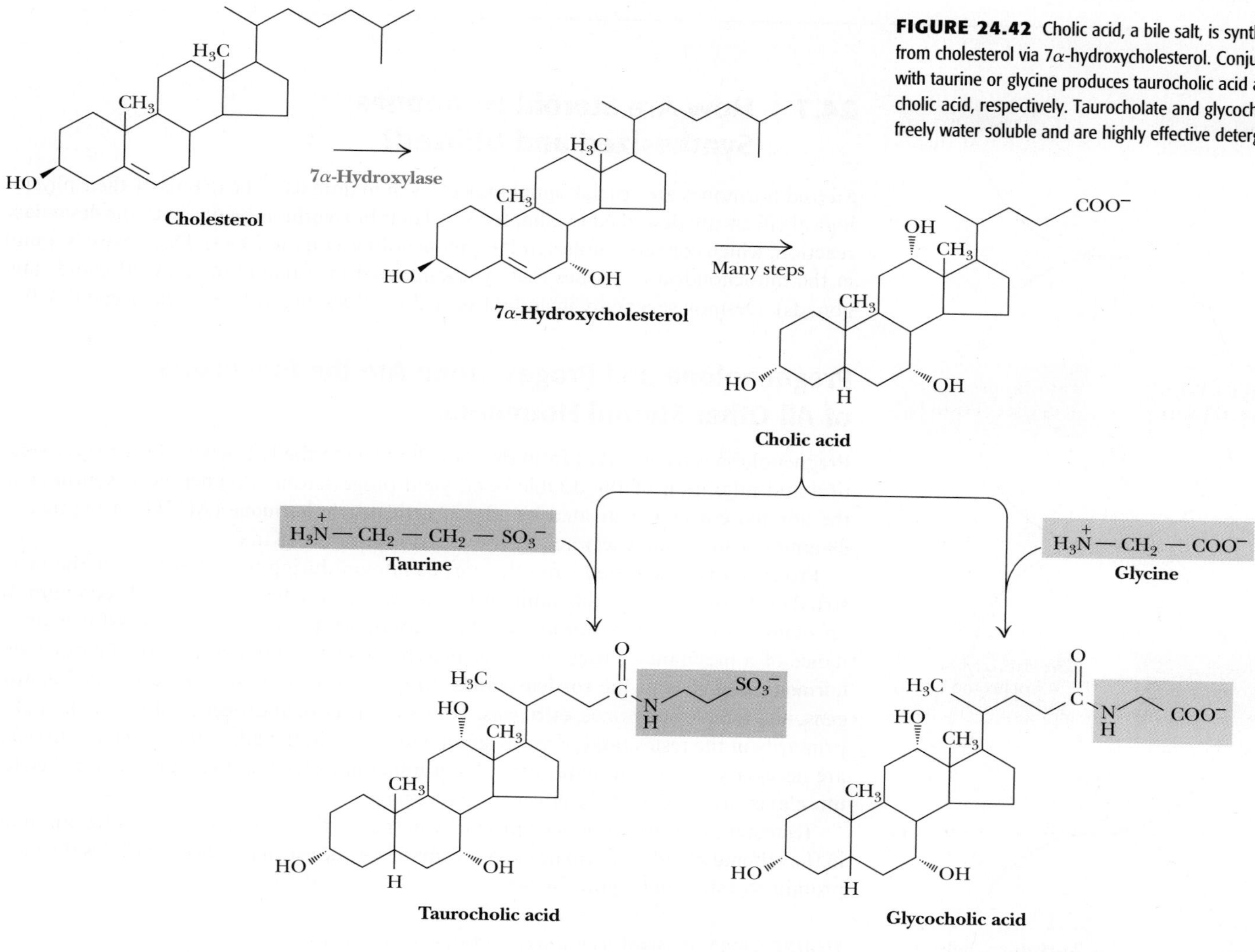

FIGURE 24.42 Cholic acid, a bile salt, is synthesized from cholesterol via 7α-hydroxycholesterol. Conjugation with taurine or glycine produces taurocholic acid and glycocholic acid, respectively. Taurocholate and glycocholate are freely water soluble and are highly effective detergents.

Steroid 5α-Reductase—A Factor in Male Baldness, Prostatic Hyperplasia, and Prostate Cancer

An enzyme that metabolizes testosterone may be involved in the benign conditions of male-pattern baldness (also known as *androgenic alopecia*) and benign prostatic hyperplasia (prostate gland enlargement), as well as potentially fatal prostate cancers. Steroid 5α-reductases are membrane-bound enzymes that catalyze the NADPH-dependent reduction of testosterone to dihydrotestosterone (DHT) (see accompanying figure). Two isoforms of 5α-reductase have been identified. In humans, the type I enzyme predominates in the sebaceous glands of skin and liver, whereas type II is most abundant in the prostate, seminal vesicles, liver, and epididymis. DHT is a contributory factor in male baldness and prostatic hyperplasia, and it has also been shown to act as a mitogen (a stimulator of cell division). For these reasons, 5α-reductase inhibitors are potential candidates for treatment of these human conditions.

Finasteride (see figure) is a specific inhibitor of type II 5α-reductase. It has been used clinically for treatment of benign prostatic hyperplasia, and it is also marketed under the trade name Propecia by Merck as a treatment for male baldness. Type II 5α-reductase inhibitors may also be potential therapeutic agents for treatment of prostate cancer. Somatic mutations occur in the gene for type II 5α-reductase during prostate cancer progression.

Because type I 5α-reductase is the predominant form of the enzyme in human scalp, the mechanism of finasteride's promotion of hair growth in men with androgenic alopecia has been uncertain. However, scientists at Merck have shown that whereas type I 5α-reductase predominates in sebaceous ducts of the skin, type II 5α-reductase is the only form of the enzyme present in hair follicles. Thus, finasteride's therapeutic effects may arise from inhibition of the type II enzyme in the hair follicle itself.

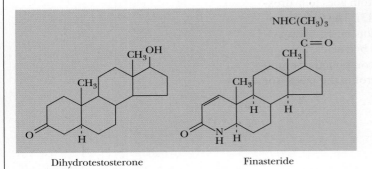

Dihydrotestosterone   Finasteride

24.7 How Are Steroid Hormones Synthesized and Utilized?

Steroid hormones are crucial signal molecules in mammals. (The details of their physiological effects are described in Chapter 32.) Their biosynthesis begins with the **desmolase** reaction, which converts cholesterol to pregnenolone (Figure 24.44). Desmolase is found in the mitochondria of tissues that synthesize steroids (mainly the adrenal glands and gonads). Desmolase activity includes two hydroxylases and utilizes cytochrome P-450.

Pregnenolone and Progesterone Are the Precursors of All Other Steroid Hormones

Pregnenolone is transported from the mitochondria to the ER, where a hydroxyl oxidation and migration of the double bond yield progesterone. Pregnenolone synthesis in the adrenal cortex is activated by **adrenocorticotropic hormone** (ACTH), a peptide of 39 amino acid residues secreted by the anterior pituitary gland.

Progesterone is secreted from the corpus luteum during the latter half of the menstrual cycle and prepares the lining of the uterus for attachment of a fertilized ovum. If an ovum attaches, progesterone secretion continues to ensure the successful maintenance of a pregnancy. Progesterone is also the precursor for synthesis of the other **sex hormone steroids** and the **corticosteroids.** Male sex hormone steroids are called **androgens,** and female hormones, **estrogens.** Testosterone is an androgen synthesized in males primarily in the testes (and in much smaller amounts in the adrenal cortex). Androgens are necessary for sperm maturation. Even nongonadal tissue (liver, brain, and skeletal muscle) is susceptible to the effects of androgens.

Testosterone is also produced primarily in the ovaries (and in much smaller amounts in the adrenal glands) of females as a precursor for the estrogens. β-Estradiol is the most prominent estrogen (Figure 24.44).

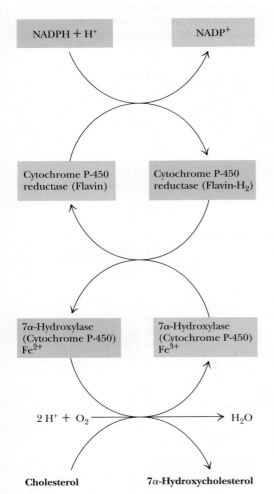

FIGURE 24.43 The mixed-function oxidase activity of 7α-hydroxylase.

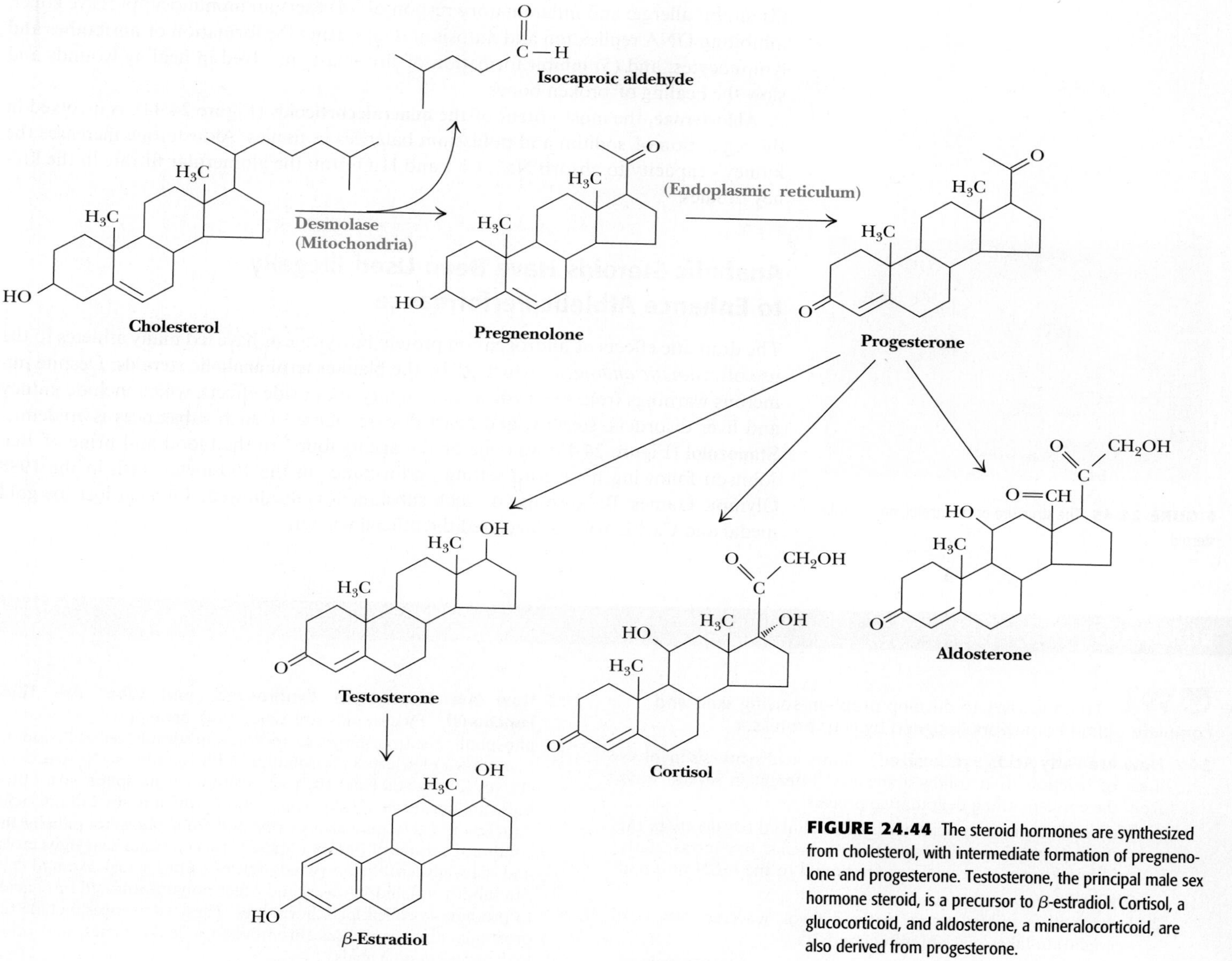

FIGURE 24.44 The steroid hormones are synthesized from cholesterol, with intermediate formation of pregnenolone and progesterone. Testosterone, the principal male sex hormone steroid, is a precursor to β-estradiol. Cortisol, a glucocorticoid, and aldosterone, a mineralocorticoid, are also derived from progesterone.

Steroid Hormones Modulate Transcription in the Nucleus

Steroid hormones act in a different manner from most hormones we have considered. In many cases, they do not bind to plasma membrane receptors but rather pass easily across the plasma membrane. Steroids may bind directly to receptors in the nucleus or may bind to cytosolic steroid hormone receptors, which then enter the nucleus. In the nucleus, the hormone-receptor complex binds directly to specific nucleotide sequences in DNA, increasing transcription of DNA to RNA (see Chapters 29 and 32).

Cortisol and Other Corticosteroids Regulate a Variety of Body Processes

Corticosteroids, including the *glucocorticoids* and *mineralocorticoids,* are made by the cortex of the adrenal glands on top of the kidneys. **Cortisol** (Figure 24.44) is representative of the **glucocorticoids,** a class of compounds that (1) stimulate gluconeogenesis and glycogen synthesis in liver (by promoting the synthesis of PEP carboxykinase, fructose-1,6-bisphosphatase, glucose-6-phosphatase, and glycogen synthase); (2) inhibit protein synthesis and stimulate protein degradation in peripheral tissues such as muscle;

(3) inhibit allergic and inflammatory responses; (4) exert an immunosuppressive effect, inhibiting DNA replication and mitosis and repressing the formation of antibodies and lymphocytes; and (5) inhibit formation of fibroblasts involved in healing wounds and slow the healing of broken bones.

Aldosterone, the most potent of the **mineralocorticoids** (Figure 24.44), is involved in the regulation of sodium and potassium balances in tissues. Aldosterone increases the kidney's capacity to absorb Na^+, Cl^-, and H_2O from the glomerular filtrate in the kidney tubules.

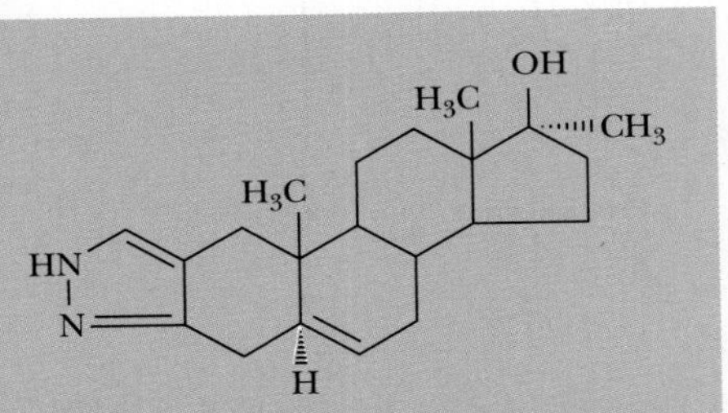

Stanozolol

FIGURE 24.45 The structure of stanozolol, an anabolic steroid.

Anabolic Steroids Have Been Used Illegally to Enhance Athletic Performance

The dramatic effects of androgens on protein biosynthesis have led many athletes to the use of *synthetic androgens,* which go by the blanket term **anabolic steroids.** Despite numerous warnings from the medical community about side effects, which include kidney and liver disorders, sterility, and heart disease, abuse of such substances is epidemic. Stanozolol (Figure 24.45) was one of the agents found in the blood and urine of Ben Johnson following his record-setting performance in the 100-meter dash in the 1988 Olympic Games. Because use of such substances is disallowed, Johnson lost his gold medal and Carl Lewis was declared the official winner.

SUMMARY

⊌WL Login to OWL to develop problem-solving skills and complete online homework assigned by your professor.

24.1 How Are Fatty Acids Synthesized? Fatty acid synthesis involves a set of reactions that follow a strategy different in several ways from the corresponding degradative process:

1. Intermediates in fatty acid synthesis are linked covalently to the sulfhydryl groups of the acyl carrier proteins. In contrast, fatty acid breakdown intermediates are bound to the —SH group of coenzyme A.

2. Fatty acid synthesis occurs in the cytosol, whereas fatty acid degradation takes place in mitochondria.

3. In animals, the enzymes of fatty acid synthesis are components of one long polypeptide chain, the fatty acid synthase, whereas no similar association exists for the degradative enzymes.

4. The coenzyme for the oxidation–reduction reactions of fatty acid synthesis is $NADP^+/NADPH$, whereas degradation involves the $NAD^+/NADH$ couple.

24.2 How Are Complex Lipids Synthesized? A common pathway operates in nearly all organisms for the synthesis of phosphatidic acid, the precursor to other glycerolipids. Glycerokinase catalyzes the phosphorylation of glycerol to form glycerol-3-phosphate, which is then acylated at both the 1- and 2-positions to yield phosphatidic acid. In eukaryotes, phosphatidic acid is converted directly either to diacylglycerol or to cytidine diphosphodiacylglycerol (or simply CDP-diacylglycerol). From these two precursors, all other glycerophospholipids in eukaryotes are derived. Phosphatidylethanolamine synthesis begins with phosphorylation of ethanolamine to form phosphoethanolamine. The next reaction involves transfer of a cytidylyl group from CTP to form CDP-ethanolamine and pyrophosphate. Diacylglycerol then displaces CMP to form phosphatidylethanolamine. Eukaryotes also use CDP-diacylglycerol, derived from phosphatidic acid, as a precursor for several other important phospholipids, including phosphatidylinositol (PI), phosphatidylglycerol (PG), and cardiolipin.

24.3 How Are Eicosanoids Synthesized, and What Are Their Functions? Eicosanoids are ubiquitous breakdown products of phospholipids. In response to appropriate stimuli, cells activate the breakdown of selected phospholipids. Phospholipase A_2 selectively cleaves fatty acids from the C-2 position of phospholipids. Often these are unsaturated fatty acids, among which is arachidonic acid. Arachidonic acid may also be released from phospholipids by the combined actions of phospholipase C (which yields diacylglycerols) and diacylglycerol lipase (which releases fatty acids). Animal cells can modify arachidonic acid and other polyunsaturated fatty acids to produce so-called local hormones. These substances include the prostaglandins, as well as thromboxanes, leukotrienes, and other hydroxyeicosanoic acids.

24.4 How Is Cholesterol Synthesized? The cholesterol biosynthetic pathway begins in the cytosol with the synthesis of mevalonate from acetyl-CoA. The first step is the condensation of two molecules of acetyl-CoA to form acetoacetyl-CoA. In the next reaction, acetyl-CoA and acetoacetyl-CoA join to form 3-hydroxy-3-methylglutaryl-CoA, which is abbreviated HMG-CoA, in a reaction catalyzed by HMG-CoA synthase. The third step in the pathway is the rate-limiting step in cholesterol biosynthesis; HMG-CoA undergoes two NADPH-dependent reductions to produce 3R-mevalonate. Biosynthesis of squalene involves conversion of mevalonate to isopentenyl pyrophosphate and dimethylallyl pyrophosphate. Condensation of these two 5-carbon intermediates produces geranyl pyrophosphate; addition of another 5-carbon isopentenyl group gives farnesyl pyrophosphate. Both steps in the production of farnesyl pyrophosphate occur with release of pyrophosphate, hydrolysis of which drives these reactions forward. Two farnesyl pyrophosphates join to produce squalene. Squalene monooxygenase converts squalene to squalene-2,3-epoxide. A second ER membrane enzyme produces lanosterol, and another 20 steps are required to convert lanosterol to cholesterol.

24.5 How Are Lipids Transported Throughout the Body? Most lipids circulate in the body in the form of lipoprotein complexes. Simple, unesterified fatty acids are merely bound to serum albumin and other proteins in blood plasma, but phospholipids, triacylglycerols, cholesterol, and cholesterol esters are all transported in the form of lipoproteins. At various sites in the body, lipoproteins interact with specific receptors and enzymes that transfer or modify their lipid cargoes.

24.6 How Are Bile Acids Biosynthesized? The formation of bile salts represents the major pathway for cholesterol degradation. The first step involves hydroxylation at C-7. 7α-Hydroxylase is a mixed-function oxidase involving cytochrome P-450. Mixed-function oxidases use O_2 as substrate. One oxygen atom goes to hydroxylate the substrate while the other is reduced to water. The function of cytochrome P-450 is to activate O_2 for the hydroxylation reaction. Such hydroxylations are quite common in the synthetic routes for cholesterol, bile acids, and steroid hormones and also in detoxification pathways for aromatic compounds.

24.7 How Are Steroid Hormones Synthesized and Utilized? Biosynthesis of steroid hormones begins with the desmolase reaction, which converts cholesterol to pregnenolone. Desmolase activity includes two hydroxylases and utilizes cytochrome P-450. Pregnenolone is transported from the mitochondria to the ER, where a hydroxyl oxidation and migration of the double bond yield progesterone. Progesterone is also the precursor for synthesis of the sex hormone steroids and the corticosteroids. Testosterone is an androgen synthesized in males primarily in the testes. β-Estradiol is the most prominent estrogen. Aldosterone, the most potent of the mineralocorticoids, is involved in the regulation of sodium and potassium balances in tissues.

FOUNDATIONAL BIOCHEMISTRY Things You Should Know After Reading Chapter 24.

- The role of acyl carrier proteins in fatty acid synthesis.
- How formation of malonyl-CoA activates acetate for fatty acid synthesis.
- The role of the reductive power of NADPH in fatty acid synthesis.
- How cells provide acetate units and NADPH for fatty acid synthesis.
- Conversion to malonyl-CoA commits acetate units to fatty acid synthesis.
- The mechanism of acetyl-CoA carboxylase and the role of biotin in the reaction.
- How the subunits of acetyl-CoA carboxylase work together to carry out malonyl-CoA synthesis.
- The roles of citrate, palmitoyl-CoA and phosphorylation in regulation of ACC.
- How acyl carrier proteins are similar to coenzyme A.
- The pathway of palmitate synthesis from acetyl-CoA and malonyl-CoA.
- The role of decarboxylation in driving condensation of acetyl-CoA and malonyl-CoA.
- The mechanism of the ketoacyl synthase reaction.
- The nature of the megasynthase complexes that orchestrate fatty acid synthesis.
- How C_{16} fatty acids are elongated and unsaturated.
- How polyunsaturated fatty acids are derived in mammals.
- The role of ω3 and ω6 fatty acids in human health.
- The roles of allosteric modifiers, hormones, and phosphorylation in regulation of fatty acid synthesis.
- The roles of phosphorylation and acylation in synthesis of glycerolipids.

- The role of CTP (and CDP-linked moieties) in synthesis of complex lipids.
- The pathways for synthesis of PC, PE, PS, PA, and triacylglycerols.
- The role of dihydroxyacetone in synthesis of plasmalogens.
- How condensation of serine and palmitoyl-CoA initiates synthesis of sphingolipids.
- The role of ceramide as a precursor for other sphingolipids and cerebrosides.
- The actions of prostaglandins as local hormones.
- How arachidonate oxidation and cyclization initiates synthesis of prostaglandins.
- How inhibition of cyclooxygenase by aspirin and other nonsteroidal anti-inflammatory drugs (NSAIDs) blocks inflammation.
- The route from acetate to mevalonate, squalene, and cholesterol.
- The role of lipoproteins in lipid transport in mammals.
- The pathways of lipoprotein formation and degradation in the human body.
- The essential features of the LDL receptor structure.
- The mechanism by which changes in pH regulate structure and function of the LDL receptor.
- How defects in lipoprotein metabolism lead to elevated serum cholesterol.
- How bile acids are synthesized from cholesterol.
- How steroid hormones are synthesized and utilized in the body.
- How anabolic steroids have been used to enhance athletic performance.

PROBLEMS

Answers to all problems are at the end of this book. Detailed solutions are available in the *Student Solutions Manual, Study Guide, and Problems Book.*

1. Explaining the Stoichiometry of Fatty Acid Synthesis Carefully count and account for each of the atoms and charges in the equations for the synthesis of palmitoyl-CoA, the synthesis of malonyl-CoA,

and the overall reaction for the synthesis of palmitoyl-CoA from acetyl-CoA.

2. Tracing Carbon Atom Incorporation in Fatty Acids (Integrates with Chapters 18 and 19.) Use the relationships shown in Figure 24.1 to determine which carbons of glucose will be incorporated into palmitic acid. Consider the cases of both citrate that is immediately exported to the cytosol following its synthesis and citrate that enters the TCA cycle.

3. **Modeling the Regulation of Acetyl-CoA Carboxylase** Based on the information presented in the text and in Figures 24.4 and 24.5, suggest a model for the regulation of acetyl-CoA carboxylase. Consider the possible roles of subunit interactions, phosphorylation, and conformation changes in your model.

4. **Estimating Active Site Separation in Animal Fatty Acid Synthase** Consider the role of the pantothenic acid groups in animal FAS and the size of the pantothenic acid group itself, and estimate a maximal separation between the malonyl transferase and the β-ketoacyl-ACP synthase active sites.

5. **Assessing the Stoichiometry of Stearoyl-CoA Desaturase** Carefully study the reaction mechanism for the stearoyl-CoA desaturase in Figure 24.14, and account for all of the electrons flowing through the reactions shown. Also account for all of the hydrogen and oxygen atoms involved in this reaction, and convince yourself that the stoichiometry is correct as shown.

6. **Understanding the Stoichiometry of Glycerophospholipid Synthesis** Write a balanced, stoichiometric reaction for the synthesis of phosphatidylethanolamine from glycerol, fatty acyl-CoA, and ethanolamine. Make an estimate of the $\Delta G^{\circ\prime}$ for the overall process.

7. **Understanding the Stoichiometry of Cholesterol Synthesis** Write a balanced, stoichiometric reaction for the synthesis of cholesterol from acetyl-CoA.

8. **Tracing Carbon Atom Incorporation in Cholesterol** Trace each of the carbon atoms of mevalonate through the synthesis of cholesterol, and determine the source (that is, the position in the mevalonate structure) of each carbon in the final structure.

9. **Understanding the Functional Role of O-linked Oligosaccharides in the LDL Receptor** Suggest a structural or functional role for the O-linked saccharide domain in the LDL receptor (Figure 24.40).

10. **Assessing the Consequences of Low Cellular Levels of CTP** Identify the lipid synthetic pathways that would be affected by abnormally low levels of CTP.

11. **Determining the ATP Requirements for Fatty Acid Synthesis** Determine the number of ATP equivalents needed to form palmitic acid from acetyl-CoA. (Assume for this calculation that each NADPH is worth 3.5 ATPs.)

12. **Understanding the Mechanism of the 3-ketosphinganine Synthase Reaction** Write a reasonable mechanism for the 3-ketosphinganine synthase reaction, remembering that it is a pyridoxal phosphate–dependent reaction.

13. **Understanding the Involvement of FAD in the Conversion of Stearic Acid to Oleic Acid** Why is the involvement of FAD important in the conversion of stearic acid to oleic acid?

14. **Understanding the Mechanism of the HMG-CoA Synthase Reaction** Write a suitable mechanism for the HMG-CoA synthase reaction. What is the chemistry that drives this condensation reaction?

15. **Understanding the Mechanism of the β-Ketoacyl-ACP Synthase Reaction** Write a suitable reaction mechanism for the β-ketoacyl-

ACP synthase, showing how the involvement of malonyl-CoA drives this reaction forward.

16. **Studying the Structure of the Fatty Acyl Synthase** In the FAS megasynthase structures, the multiple functional sites must lie within reach of the ACP and its bound acyl group substrates. Examine the mammalian FAS structure (see Figure 24.11), and determine the distances between the various functional sites. You could approach this problem either by using a molecular modeling program (such as PyMol at *www.pymol.org*) or by consulting appropriate references (the following end-of-chapter reference is a good place to start: Maier, T., Leibundgut, M., and Ban, N., 2008. *Science* **321**:1315–1322). You should convince yourself, with quantitative arguments, that the intersite distances can be traversed appropriately by the ACP group.

17. **Modeling the LDL Receptor** In the LDL receptor structure shown in Figure 24.41c, the β-propellor interaction with domains R4 and R5 is partly stabilized by salt bridges between acidic residues on R4 and R5 that also coordinate Ca^{2+} ions. Use a molecular modeling program or consult the literature to identify at least two such interactions. Two suitable references are Beglova, N., and Blacklow, S. C., 2005. *Trends in Biochemical Sciences* **30**:309–316; and Rudenko, G., Henry, L., et al., 2002. *Science* **298**:2353–2358.

18. **Understanding the Function of the LDL Receptor** Insights into the function of LDL receptors in displacing LDL particles in endosomes have come from an unlikely source: a study of LDL receptor binding by a human rhinovirus HRV2 (a common cold virus). Consult suitable references to learn how this study provided support for the model of LDL particle displacement presented in this chapter. Good references are Blacklow, S. C., 2004. *Nature Structural and Molecular Biology* **11**:388–390; Verdaguer, N., Fita, I., et al., 2004. *Nature Structural and Molecular Biology* **11**: 429–434; and Beglova, N., and Blacklow, S. C., 2005. *Trends in Biochemical Sciences* **30**:309–316.

Preparing for the MCAT® Exam

19. **Comparing the Reaction of Fatty Acid Synthesis and the Tricarboxylic Acid Cycle** Consider the synthesis of linoleic acid from palmitic acid, and identify a series of three consecutive reactions that embody chemistry similar to three reactions in the tricarboxylic acid cycle.

20. **Writing an Equation to Describe the Synthesis of Behenic Acid** Rewrite the equation in Section 24.1 to describe the synthesis of behenic acid (see Table 8.1).

ActiveModels Problems

21. Using the ActiveModel for Niemann-Pick C1, describe the structure of the sterol-binding pocket, and explain the interaction involved in cholesterol binding.

22. Using the ActiveModel for HMG-CoA reductase (HMGR), explain how HMGR is regulated. Why do researchers seek ways to inhibit HMGR?

23. Using the ActiveModel for cyclooxygenase-2, explain the mechanisms by which NSAIDs inhibit this enzyme. Discuss several examples.

FURTHER READING

General

Chun, J., 2007. The sources of a lipid conundrum. *Science* **316**:208–210.

Lusis, A., and Pajukanta, P., 2008. A treasure trove for lipoprotein biology. *Nature Genetics* **40**:129–130.

Ohlrogge, J., and Browse, J., 1995. Lipid biosynthesis. *Plant Cell* **7**:957–970.

Plutzky, J., 2011. The PPAR-RXR transcriptional complex in the vasculature: Energy in the balance. *Circulation Research* **108**:1002–1016.

Ruby, M. A., Goldenson, B., Orasanu, G., Johnston, T. P., et al., 2010. VLDL hydrolysis by LPL activates PPAR-α through generation of unbound fatty acids. *Journal of Lipid Research* **51**:2275–2281.

Vance, D. E., and Vance, J. E., 2008. *Biochemistry of Lipids, Lipoproteins and Membranes,* 5th ed. Amsterdam: Elsevier.

Vance, J. E., 2010. Transfer of cholesterol by the NPC team. *Cell Metabolism* **12**:105–106.

Wang. M. L., Motamed, M., Infante, R. E., Abi-Mosley, L., et al., 2010. *Cell Metabolism* **12**:168–173.

Acetyl-CoA Carboxylase

Bilder, P., Lightle, S., et al., 2006. The structure of the carboxyltransferase component of acetyl-CoA carboxylase reveals a zinc-binding motif unique to the bacterial enzyme. *Biochemistry* **45**:1712–1722.

Cho, Y. S., Lee, J. I., et al., 2007. Crystal structure of the biotin carboxylase domain of human acetyl-CoA carboxylase 2. *Proteins* **70**:268–272.

Cho, Y. S., Lee, J., Shin, D., Kim, H. T., et al., 2010. Molecular mechanism for the regulation of human ACC2 through phosphorylation by AMPK. *Biochemical and Biophysical Research Communications* **391**:187–192.

Jump, D. B., Torres-Gonzalez, M., and Olson, L. K., 2011. Soraphen A, an inhibitor of acetyl CoA carboxylase activity, interferes with fatty acid elongation. *Biochemical Pharmacology* **81**:649–660.

Munday, M. R., 2002. Regulation of mammalian acetyl-CoA carboxylase. *Biochemical Society Transactions* **30**:1059–1064.

Saggerson, D., 2008. Malonyl-CoA, a key signaling molecule in mammalian cells. *Annual Review of Nutrition* **28**:253–272.

Tong, L., and Harwood, H. J., Jr., 2006. Acetyl-coenzyme A carboxylases: Versatile targets for drug discovery. *Journal of Cellular Biochemistry* **99**:1476–1488.

Zhang, H., Tweel, B., et al., 2004a. Crystal structure of the carboxyltransferase domain of acetyl-coenzyme A carboxylase in complex with CP-640186. *Structure* **12**:1683–1691.

Zhang, H., Yang, Z., et al., 2003. Crystal structure of the carboxyltransferase domain of acetyl-coenzyme A carboxylase. *Science* **299**:2064–2067.

Fatty Acid Metabolism

Asturias, F. J., Chadick, J. Z., et al., 2005. Structure and molecular organization of mammalian fatty acid synthase. *Nature Structural and Molecular Biology* **12**:225–232.

Guillou, H., Zadravec, D., Martin, P. G. P., and Jacobsson, A., 2010. The key roles of elongases and desaturases in mammalian fatty acid metabolism: Insights from transgenic mice. *Progress in Lipid Research* **49**:186–199.

Green, C. D., Ozguden-Akkoc, C. G., Wang, Y., Jump, D. B., and Olson, L. K., 2010. Role of fatty acid elongases in determination of *de novo* synthesized monounsaturated fatty acid species. *Journal of Lipid Research* **51**:1871–1877.

Hiltunen, J. K., Chen, Z., Haapalainen, A. M., Wierenga, R. K., and Kastaniotis, A. J., 2010. Mitochondrial fatty acid synthesis—an adopted set of enzymes making a pathway of major importance for the cellular metabolism. *Progress in Lipid Research* **49**:27–45.

Jenni, S., Leibundgut, M., et al., 2007. Structure of fungal fatty acid synthase and implications for iterative substrate shuttling. *Science* **316**:254–261.

Jump, D. B., 2009. Mammalian fatty acid elongases. *Methods in Molecular Biology* **579**:375–389.

Kresge, N., Simoni, R. D., et al., 2006. Salih Wakil's elucidation of the animal fatty acid synthetase complex architecture. *Journal of Biological Chemistry* **281**:e5–e7.

Leibundgut, M., Jenni, S., et al., 2007. Structural basis for substrate delivery by acyl carrier protein in the yeast fatty acid synthase. *Science* **316**:288–290.

Leibundgut, M., Maier, T., Jenni, S., and Ban, N., 2008. The multienzyme architecture of eukaryotic fatty acid synthases. *Current Opinion in Structural Biology* **18**:714–725.

Maier, T., Jenni, S., and Ban, N., 2006. Architecture of mammalian fatty acid synthase at 4.5Å resolution. *Science* **311**:1258–1262.

Maier, T., Leibundgut, M., and Ban, N., 2008. The crystal structure of a mammalian fatty acid synthase. *Science* **321**:1315–1322.

Maier, T., Leibundgut, M., Boehringer, D., and Ban, N., 2010. Structure and function of eukaryotic fatty acid synthases. *Quarterly Reviews of Biophysics* **43**:373–422.

Meier, J. L., and Burkart, M. D., 2009. The chemical biology of modular biosynthetic enzymes. *Chemical Society Reviews* **38**:2012–2045.

Riezman, H., 2007. The long and short of fatty acid synthesis. *Cell* **130**: 587–588.

Tafesse, F. G., and Holthuis, J. C. M., 2010. A brake on lipid synthesis. *Nature* **463**:1028–1029.

White, S. W., Zheng, J., et al., 2005. The structural biology of type II fatty acid biosynthesis. *Annual Review of Biochemistry* **74**:791–831.

Function and Synthesis of Eicosanoids and Essential Fatty Acids

Grosser, T., Fries, S., et al., 2006. Biological basis for the cardiovascular consequences of COX-2 inhibition: Therapeutic challenges and opportunities. *Journal of Clinical Nutrition* **116**:4–15.

Hunter, W. N., 2007. The non-mevalonate pathway of isoprenoid precursor biosynthesis. *Journal of Biological Chemistry* **282**:21573–21577.

Kresge, N., Simoni, R. D., et al., 2006. The prostaglandins, Sune Bergström and Bengt Samuelsson. *Journal of Biological Chemistry* **281**:e9–e11.

Lands, W. E., 1991. Biosynthesis of prostaglandins. *Annual Review of Nutrition* **11**:41–60.

Newcomer, M. E., and Gilbert, N. C., 2010. Location, location, location: Compartmentalization of early events in leukotriene biosynthesis. *The Journal of Biological Chemistry* **285**:25109–25114.

Rouzer, C. A., and Marnett, L. J., 2008. Non-redundant functions of cyclooxygenases: Oxygenation of endocannabinoids. *The Journal of Biological Chemistry* **283**:8065–8069.

Smith, W. L., 2007. Nutritionally essential fatty acids and biologically indispensable cyclooxygenases. *Trends in Biochemical Sciences* **33**:27–37.

Sugimoto, Y., and Narumiya, S., 2007. Prostaglandin E receptors. *Journal of Biological Chemistry* **282**:11613–11617.

Wang, D., and DuBois, R. N., 2010. Eicosanoids and cancer. *Nature Reviews of Cancer* **10**:181–193.

Phospholipid and Triacylglycerol Synthesis

Cao, Y., 2010. Adipose tissue angiogenesis as a therapeutic target for obesity and metabolic diseases. *Nature Reviews Drug Discovery* **9**:107–115.

Carman, G. M., and Henry, S. A., 1989. Phospholipid biosynthesis in yeast. *Annual Review of Biochemistry* **58**:635–669.

Carman, G. M., and Henry, S. A., 2007. Phosphatidic acid plays a central role in the transcriptional regulation of glycerophospholipid synthesis in *Saccharomyces cerevisiae*. *Journal of Biological Chemistry* **282**: 37293–37297.

Hermansson, M., Hokynar, K., and Somerharju, P., 2011. Mechanisms of glycerophospholipid homeostasis in mammalian cells. *Progress in Lipid Research* **50**:240–257.

Kohlwein, S. D., 2010. Triacylglycerol homeostasis: Insights from yeast. *The Journal of Biological Chemistry* **285**:15663–15667.

Lumeng, C. N., Maillard, I., and Saltiel, A. R., 2009. T-ing up inflammation in fat. *Nature* **15**:846–847.

Sorger, D., and Daum, G., 2003. Triacylglycerol biosynthesis in yeast. *Applied Microbiology and Biotechnology* **61**:289–299.

Tafesse, F. G., Ternes, P., et al., 2006. The multigenic sphingomyelin synthase family. *Journal of Biological Chemistry* **281**:29421–29425.

Vance, D. E., Li, Z., et al., 2007. Hepatic phosphatidylethanolamine N-methyltransferase, unexpected roles in animal biochemistry and physiology. *Journal of Biological Chemistry* **282**:33237–33241.

Watkins, P. A., 2008. Very long-chain acyl-CoA synthetases. *Journal of Biological Chemistry* **283**:1773–1777.

Structure and Function of Lipoproteins and Their Receptors

Beglova, N., and Blacklow, S. C., 2005. The LDL receptor: How acid pulls the trigger. *Trends in Biochemical Sciences* **30**:309–316.

Davidson, W. S., and Thompson, T. B., 2007. The structure of apolipoprotein A-1 in high density lipoproteins. *Journal of Biological Chemistry* **282**:22249–22253.

Holla, Ø., Laerdahl, J. K., Strøm, T. B., Tveten, K., et al., 2011. Removal of acidic residues of the prodomain of PCSK9 increases its activity towards the LDL receptor. *Biochemical and Biophysical Research Communications* **406**:234–238.

Holla, Ø., Tveten, K., Cameron, J., Berge, K. E., and Leren, T. P., 2010. Recycling of the low density lipoprotein receptor by PCSK9 is not mediated by residues of the cytoplasmic domain. *Molecular Genetics and Metabolism* **101**:76–80.

Johs, A., Hammel, M., et al., 2006. Modular structure of solubilized human apolipoprotein B-100. *Journal of Biological Chemistry* **281**:19732–19739.

Rudenko, G., Henry, L., et al., 2002. Structure of the LDL receptor extracellular domain at endosomal pH. *Science* **298**:2353–2358.

Verdaguer, N., Fita, I., Reithmayer, M., Moser, R., and Blaas, D., 2004. X-ray structure of a minor group human rhinovirus bound to a fragment of its cellular receptor protein. *Nature Structural and Molecular Biology* **11**:429–434.

Von Eckardstein, A., and Sibler, R. A., 2011. Possible contributions of lipoproteins and cholesterol to the pathogenesis of diabetes mellitus type 2. *Current Opinion in Lipidology* **22**:26–32.

Yamamoto, T., Lu, C., and Ryan, R. O., 2011. A two-step binding model of PCSK9 interaction with the low density lipoprotein receptor. *The Journal of Biological Chemistry* **286**:5464–5470.

Cholesterol Metabolism

Allahverdian, S., and Grancis, G. A., 2010. Cholesterol homeostasis and high-density lipoprotein formation in arterial smooth muscle cells. *Trends in Cardiovascular Medicine* **20**:96–102.

Bloch, K., 1965. The biological synthesis of cholesterol. *Science* **150**:19–28.

Bloch, K., 1987. Summing up. *Annual Review of Biochemistry* **56**:1–19.

Brown, M. S., and Goldstein, J. L., 2006. Lowering LDL: Not only how low, but how long? *Science* **311**:1721–1723.

Bruckert, E., and Rosenbaum, D., 2011. Lowering LDL-cholesterol through diet: Potential role in the statin era. *Current Opinion in Lipidology* **22**:43–48.

Gimpl, G., Burger, K., et al., 2002. A closer look at the cholesterol sensor. *Trends in Biochemical Sciences* **27**:596–599.

Goldstein, J. L., and Brown, M. S., 2001. The cholesterol quartet. *Science* **292**:1310–1312.

Horvat, S., McWhir, J., and Rozman, D., 2011. Defects in cholesterol synthesis genes in mouse and in humans: Lessons for drug development and safer treatments. *Drug Metabolism Reviews* **43**:69–90.

Hsieh, F.-L., Chang, T.-H., Ko, T.-P., and Wang, A. H.-J., 2011. Structure and mechanism of an Arabidopsis medium/long-chain-length prenyl pyrophosphate synthase. *Plant Physiology* **155**:1079–1090.

Istvan, E. S., and Deisenhofer, J., 2001. Structural mechanism for statin inhibition of HMG-CoA reductase. *Science* **292**:1160–1164.

Kresge, N., Simoni, R. D., et al., 2006. 30 years of cholesterol metabolism: The work of Michael Brown and Joseph Goldstein. *Journal of Biological Chemistry* **281**:e25–e28.

Lou, K.-J., 2009. The secreted secret of PCSK9. *Science-Business Exchange* **2**:1–3.

Sato, R., 2010. Sterol metabolism and SREBP activation. *Archives of Biochemistry and Biophysics* **501**:177–181.

Nitrogen Acquisition and Amino Acid Metabolism

◀ Soybeans. Only plants and certain microorganisms are able to transform the oxidized, inorganic forms of nitrogen available in the inanimate environment into reduced, biologically useful forms. Soybean plants can meet their nitrogen requirements both by assimilating nitrate and, in symbiosis with bacteria, fixing N_2.

I was determined to know beans.

Henry David Thoreau (1817–1862)
The Writings of Henry David Thoreau, vol. 2,
p. 178, Houghton Mifflin (1906)

KEY QUESTIONS

25.1 Which Metabolic Pathways Allow Organisms to Live on Inorganic Forms of Nitrogen?

25.2 What Is the Metabolic Fate of Ammonium?

25.3 What Regulatory Mechanisms Act on *Escherichia coli* Glutamine Synthetase?

25.4 How Do Organisms Synthesize Amino Acids?

25.5 How Does Amino Acid Catabolism Lead into Pathways of Energy Production?

ESSENTIAL QUESTIONS

Nitrogen is an essential nutrient for all cells. Amino acids provide nitrogen for the synthesis of other nitrogen-containing biomolecules. Excess amino acids in the diet can be converted into α-keto acids and used for energy production.

What are the biochemical pathways that form ammonium from inorganic nitrogen compounds prevalent in the inanimate environment? How is ammonium incorporated into organic compounds? How are amino acids synthesized and degraded?

Amino acids and nucleotides, as well as their polymeric forms (proteins and nucleic acids), are nitrogen-containing molecules upon which cell structure and function rely. How do these various organic forms of nitrogen arise? As we look at these compounds, an obvious feature is that nitrogen atoms are typically bound to carbon and/or hydrogen atoms. That is, the nitrogen atom is in a reduced state. On the other hand, the prevalent forms of nitrogen in the environment are inorganic and oxidized; N_2 (dinitrogen gas) and NO_3^- (nitrate ions) being the most abundant species. The two principal routes for nitrogen acquisition from the inanimate environment, nitrate assimilation and nitrogen fixation, lead to formation of ammonium ions (NH_4^+). Reactions that incorporate NH_4^+ into organic linkage (the reactions of ammonium assimilation) follow. Among these, glutamine synthetase merits particular attention because it conveys several important lessons in metabolic regulation. This chapter presents the pathways of amino acid biosynthesis and degradation; those involving the sulfur-containing amino acids provide an opportunity to introduce aspects of sulfur metabolism.

25.1 | Which Metabolic Pathways Allow Organisms to Live on Inorganic Forms of Nitrogen?

Nitrogen Is Cycled Between Organisms and the Inanimate Environment

Nitrogen acquisition by biological systems is accompanied by its reduction to ammonium ion (NH_4^+) and the incorporation of NH_4^+ into organic linkage as amino or amido groups (Figure 25.1). The reduction of NO_3^- to NH_4^+ occurs in green plants,

OWL Online homework and a Student Self Assessment for this chapter may be assigned in OWL.

FIGURE 25.1 The nitrogen cycle. Organic nitrogenous compounds are formed by the incorporation of NH_4^+ into carbon skeletons. Note that denitrification and nitrogen fixation are anaerobic processes.

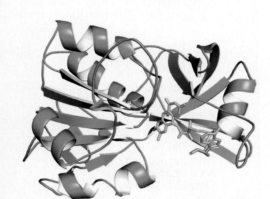

Corn nitrate reductase cytochrome domain (FAD shown in yellow) (pdb id = 1CNE)

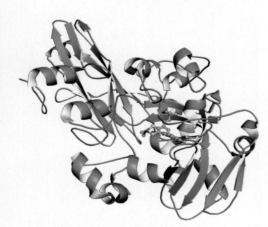

Fungal nitrate reductase molybdenum cofactor domain (Mo cofactor in gold) (pdb id = 2BIH)

Nitrate reductase has two structural domains, the cytochrome *b* domain (*top*) and the molybdenum cofactor domain (*bottom*).

various fungi, and certain bacteria in a two-step metabolic pathway known as **nitrate assimilation.** The formation of NH_4^+ from N_2 gas is termed **nitrogen fixation.** N_2 fixation is an exclusively prokaryotic process, although bacteria in symbiotic association with certain green plants also carry out nitrogen fixation. No animals are capable of either nitrogen fixation or nitrate assimilation, so they are totally dependent on plants and microorganisms for the synthesis of organic nitrogenous compounds, such as amino acids and proteins, to satisfy their requirements for this essential element.

Animals release excess nitrogen in a reduced form, either as NH_4^+ or as organic nitrogenous compounds such as urea. The release of N occurs both during life and as a consequence of microbial decomposition following death. Various bacteria return the reduced forms of nitrogen back to the environment by oxidizing them. The oxidation of NH_4^+ to NO_3^- by **nitrifying bacteria,** a group of chemoautotrophs, provides the sole source of chemical energy for the life of these microbes. Nitrate nitrogen also returns to the atmosphere as N_2 as a result of the metabolic activity of **denitrifying bacteria.** These bacteria are capable of using NO_3^- and similar oxidized, inorganic forms of nitrogen as electron acceptors in place of O_2 in energy-producing pathways. The NO_3^- is reduced ultimately to *dinitrogen* (N_2). These bacteria thus deplete the levels of *combined nitrogen,* that is, N joined with other elements in chemical compounds. Combined nitrogen is important as natural fertilizer. However, the denitrifying activity of bacteria is exploited in water treatment plants to reduce the load of combined nitrogen that might otherwise enter lakes, streams, and bays.

Nitrate Assimilation Is the Principal Pathway for Ammonium Biosynthesis

Nitrate assimilation occurs in two steps: the two-electron reduction of nitrate to nitrite, catalyzed by **nitrate reductase** (Equation 25.1), followed by the six-electron reduction of nitrite to ammonium, catalyzed by **nitrite reductase** (Equation 25.2).

$$(1) \; NO_3^- + 2\,H^+ + 2\,e^- \longrightarrow NO_2^- + H_2O \tag{25.1}$$

$$(2) \; NO_2^- + 8\,H^+ + 6\,e^- \longrightarrow NH_4^+ + 2\,H_2O \tag{25.2}$$

Nitrate assimilation is the predominant means by which green plants, algae, and many microorganisms acquire nitrogen. The pathway of nitrate assimilation accounts for more than 99% of the inorganic nitrogen (nitrate or N_2) assimilated into organisms.

Nitrate Reductase Contains Cytochrome b_{557} and Molybdenum Cofactor

$$\begin{array}{ccc} NADH & & NO_3^- \\ \Big\rangle & [-SH \to FAD \to \text{cytochrome } b_{557} \to MoCo] & \Big\langle \\ NADH^+ & & NO_2^- \end{array} \tag{25.3}$$

(a)

(b)

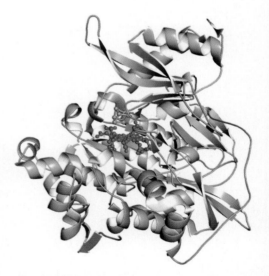

FIGURE 25.2 The novel prosthetic groups of nitrate reductase and nitrite reductase. **(a)** The molybdenum cofactor of nitrate reductase. The fifth ligand for the Mo atom is the sulfur atom of a conserved cysteine residue in the nitrate reductase. **(b)** Siroheme, an essential prosthetic group of nitrite reductase. Siroheme is novel among hemes in having eight carboxylate-containing side chains. These carboxylate groups may act as H^+ donors during the reduction of NO_2^- to NH_4^+.

A pair of electrons is transferred from NADH via enzyme-associated sulfhydryl groups, FAD, cytochrome b_{557}, and **MoCo** (an essential molybdenum-containing cofactor) to nitrate, reducing it to nitrite. The brackets [ ] denote the protein-bound prosthetic groups that constitute an e^- transport chain between NADH and nitrate. Nitrate reductases typically are cytosolic 220-kD dimeric proteins. The structure of the molybdenum cofactor (MoCo) is shown in Figure 25.2a. Molybdenum cofactor is necessary for both nitrate reductase activity and the assembly of nitrate reductase subunits into the active dimeric holoenzyme form. Molybdenum cofactor is also an essential cofactor for a variety of enzymes that catalyze hydroxylase-type reactions, including xanthine dehydrogenase, aldehyde oxidase, and sulfite oxidase.

Nitrite Reductase Contains Siroheme Six electrons are required to reduce NO_2^- to NH_4^+. Nitrite reductases in photosynthetic organisms obtain these electrons from six molecules of photosynthetically reduced ferredoxin (Fd_{red}).

$$\text{Light} \rightarrow 6\ Fd_{red} \quad\qquad NO_2^-$$
$$[(4Fe\text{-}4S) \rightarrow \text{siroheme}] \qquad\qquad (25.4)$$
$$6\ Fd_{ox} \qquad\qquad NH_4^+$$

Photosynthetic nitrite reductases are 63-kD monomeric proteins having a tetranuclear iron–sulfur cluster and a novel heme, termed **siroheme,** as prosthetic groups. The [4Fe–4S] cluster and the siroheme act as a coupled e^- transfer center. Nitrite binds directly to siroheme, providing the sixth ligand, much as O_2 binds to the heme of hemoglobin. Nitrite is reduced to ammonium while liganded to siroheme. The structure of siroheme is shown in Figure 25.2b.

In higher plants, nitrite reductase is found in chloroplasts, where it has ready access to its primary reductant, photosynthetically reduced ferredoxin. Microbial nitrite reductases closely resemble nitrate reductases in having essential —SH groups and FAD prosthetic groups to couple enzyme-mediated NADPH oxidation to nitrite reduction (Figure 25.3).

Spinach nitrate reductase (iron–sulfur cluster in gold, siroheme in red) (pdb id = 2AKJ)

Organisms Gain Access to Atmospheric N_2 Via the Pathway of Nitrogen Fixation

Nitrogen fixation involves the reduction of nitrogen gas (N_2) via an enzyme system found only in prokaryotic cells. The heart of the nitrogen fixation process is the enzyme known as **nitrogenase,** which catalyzes the reaction

$$N_2 + 10\ H^+ + 8\ e^- \longrightarrow 2\ NH_4^+ + H_2 \qquad (25.5)$$

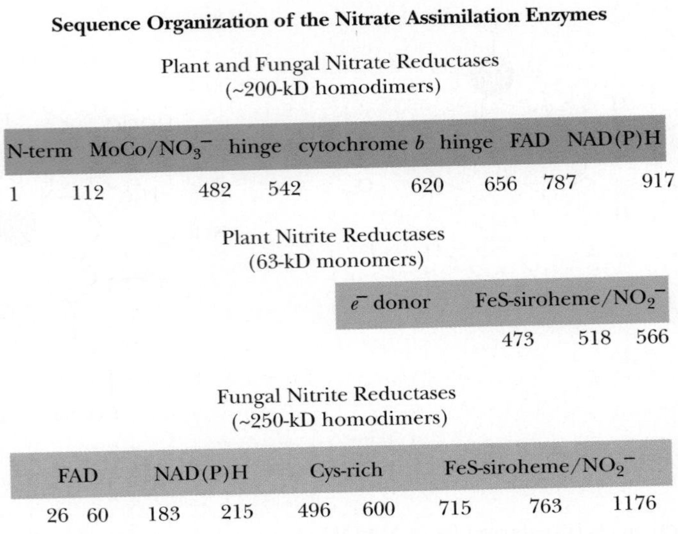

Sequence Organization of the Nitrate Assimilation Enzymes

Plant and Fungal Nitrate Reductases
(~200-kD homodimers)

N-term	MoCo/NO$_3^-$	hinge	cytochrome b	hinge	FAD	NAD(P)H
1	112	482	542	620	656 787	917

Plant Nitrite Reductases
(63-kD monomers)

e^- donor	FeS-siroheme/NO$_2^-$
	473 518 566

Fungal Nitrite Reductases
(~250-kD homodimers)

FAD	NAD(P)H	Cys-rich	FeS-siroheme/NO$_2^-$
26 60	183 215	496 600 715	763 1176

FIGURE 25.3 Domain organization within the enzymes of nitrate assimilation. The numbers denote residue number along the amino acid sequence of the proteins.

Note that an obligatory reduction of two protons to hydrogen gas accompanies the biological reduction of N_2 to ammonia. Less than 1% of the inorganic N incorporated into organic compounds by organisms can be attributed to nitrogen fixation; however, this process is the only way that organisms can tap into the enormous reservoir of N_2 in the atmosphere.

Although nitrogen fixation is exclusively prokaryotic, N_2-fixing bacteria may be either free-living or living as symbionts with higher plants. For example, *Rhizobia* are bacteria that fix nitrogen in symbiotic association with soybeans and other leguminous plants. Because nitrogen in a metabolically useful form is often the limiting nutrient for plant growth, such symbiotic associations can be an important factor in plant growth and agriculture.

Despite the wide diversity of bacteria in which nitrogen fixation takes place, all N_2-fixing systems are nearly identical and all have four fundamental requirements: (1) the enzyme *nitrogenase;* (2) a strong reductant, such as reduced ferredoxin; (3) ATP; and (4) O_2-free conditions. In addition, several modes of regulation act to control nitrogen fixation.

The Nitrogenase Complex Is Composed of Two Metalloproteins Two metalloproteins constitute the nitrogenase complex: the **Fe-protein** or nitrogenase reductase and the **MoFe-protein,** which is another name for nitrogenase. **Nitrogenase reductase** is a 60-kD homodimer possessing a single [4Fe–4S] cluster as a prosthetic group. Nitrogenase reductase is extremely O_2 sensitive. Nitrogenase reductase binds MgATP and hydrolyzes two ATPs per electron transferred during nitrogen fixation. Because reduction of N_2 to $2 NH_4^+ + H_2$ requires 8 electrons, 16 ATPs are consumed per N_2 reduced.

This ATP requirement seems paradoxical because the reaction is thermodynamically favorable: The $\mathscr{E}_o'$ for the reaction ($N_2 + 8 e^- + 10 H^+ \rightarrow 2 NH_4^+ + H_2$) is -0.314 V. Ferredoxin, the most common e^- donor for nitrogen fixation, has an $\mathscr{E}_o'$ that is more negative (see Table 20.1). The solution to the paradox is found in the very strong bonding between the two N atoms in N_2 (Figure 25.4). Substantial energy input is needed to overcome this large activation energy and break the N≡N triple bond. In this biological system, the energy is provided by ATP.

Nitrogenase, the MoFe-protein, is a 240-kD $\alpha_2\beta_2$-type heterotetramer. An $\alpha\beta$-dimer serves as the functional unit, and each $\alpha\beta$-dimer contains two types of metal centers: an unusual 8Fe-7S center known as the **P-cluster** (Figure 25.5a) and the novel 7Fe-1Mo-9S cluster known as the **FeMo-cofactor** (Figure 25.5b). Nitrogenase under unusual circumstances may contain an **iron:vanadium cofactor** instead of the molybdenum-containing one. Like nitrogenase reductase, nitrogenase is very oxygen labile.

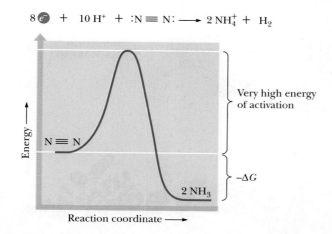

$$8\,\underline{e^-} + 10\,H^+ + :N\!\equiv\!N: \longrightarrow 2\,NH_4^+ + H_2$$

FIGURE 25.4 The triple bond in N_2 must be broken during nitrogen fixation.

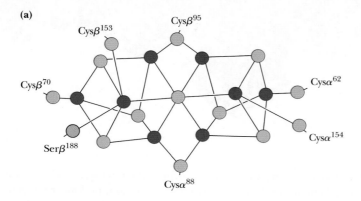

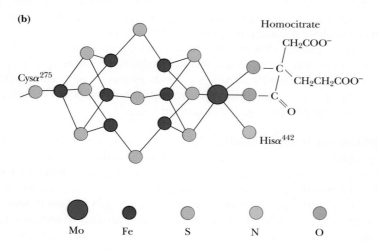

FIGURE 25.5 Structures of the two types of metal clusters found in nitrogenase. **(a)** The P-cluster consists of two Fe_4S_3 clusters that share an S atom. **(b)** The FeMo-cofactor contains 1 Mo, 7 Fe, and 9 S atoms. Homocitrate provides two oxo ligands to the Mo atom. (Adapted from Leigh, G. J., 1995. The mechanism of dinitrogen reduction by molybdenum nitrogenases. *European Journal of Biochemistry* **229**:14–20.)

The Nitrogenase Reaction In the nitrogenase reaction (Figure 25.6), electrons from reduced ferredoxin pass to nitrogenase reductase, which serves as electron donor to nitrogenase, the enzyme that actually catalyzes N_2 fixation. Electron transfer from nitrogenase reductase to nitrogenase takes place through docking of nitrogenase reductase with an $\alpha\beta$-subunit pair of nitrogenase (Figure 25.7). Nitrogenase reductase transfers e^- to nitrogenase one electron at a time. N_2 is bound within the FeMo-cofactor metal cluster until all electrons and protons are added; no free intermediates, such as $HN\!=\!NH$ or $H_2N\!-\!NH_2$ are released. Electron transfer takes place in the following sequence: Fe-protein $\rightarrow$ P-cluster $\rightarrow$ FeMo-cofactor $\rightarrow$ N_2. ATP hydrolysis is coupled

FIGURE 25.6 The nitrogenase reaction. Depending on the bacterium, electrons for N_2 reduction may come from light, NADH, hydrogen gas, or pyruvate. The primary e^- donor for the nitrogenase system is reduced ferredoxin.

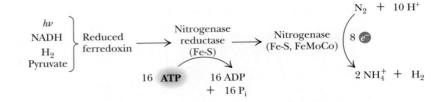

FIGURE 25.7 Ribbon diagram of nitrogenase reductase (the Fe-protein, blue) : nitrogenase (FeMo protein, green) complex. The Fe-protein iron-sulfur cluster is shown in yellow, bound ADP in orange. The nitrogenase FeMo cofactor is shown in cyan, the P-cluster in red (pdb id = 1N2C).

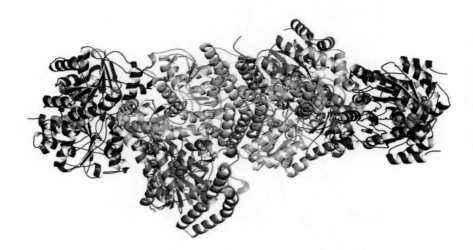

to the transfer of an electron from the Fe-protein to the P-cluster. ATP hydrolysis leads to conformational change in the nitrogenase reductase, so it no longer binds to nitrogenase. The ADP : oxidized nitrogenase reductase complex dissociates, making way for another ATP : reduced nitrogenase reductase complex to bind to nitrogenase. Interestingly, nitrogenase reductase is a member of the G-protein family; G proteins are molecular switches whose operation is driven by NTP hydrolysis.

Nitrogenase is a rather slow enzyme: Its optimal rate of e^- transfer is about 12 e^- pairs per second per enzyme molecule; that is, it reduces only three molecules of nitrogen gas per second. Because its activity is so weak, nitrogen-fixing cells maintain large amounts of nitrogenase so that their requirements for reduced N can be met. As much as 5% of the cellular protein may be nitrogenase.

The Regulation of Nitrogen Fixation To a first approximation, two regulatory controls are paramount (Figure 25.8): (1) ADP inhibits the activity of nitrogenase; thus, as the ATP/ADP ratio drops, nitrogen fixation is blocked. (2) NH_4^+ represses the expression of the *nif* genes, the genes that encode the proteins of the nitrogen-fixing system. To date, some 20 *nif* genes have been identified with the nitrogen fixation process. Repression of *nif* gene expression by ammonium, the primary product of nitrogen fixation, is an efficient and effective way of shutting down N_2 fixation when its end product is not needed. In addition, in some systems, covalent modification of nitrogenase reductase leads to its inactivation. Inactivation occurs when Arg[101] of nitrogenase reductase receives an ADP-ribosyl group donated by NAD^+.

25.2 What Is the Metabolic Fate of Ammonium?

Given the prevalence of N atoms in cellular components, it is surprising that only three enzymatic reactions introduce ammonium into organic molecules. Of these three, *glutamate dehydrogenase* and *glutamine synthetase* are responsible for most of the ammonium assimilated into carbon compounds. The third, *carbamoyl-phosphate synthetase I*, is a mitochondrial enzyme that participates in the urea cycle.

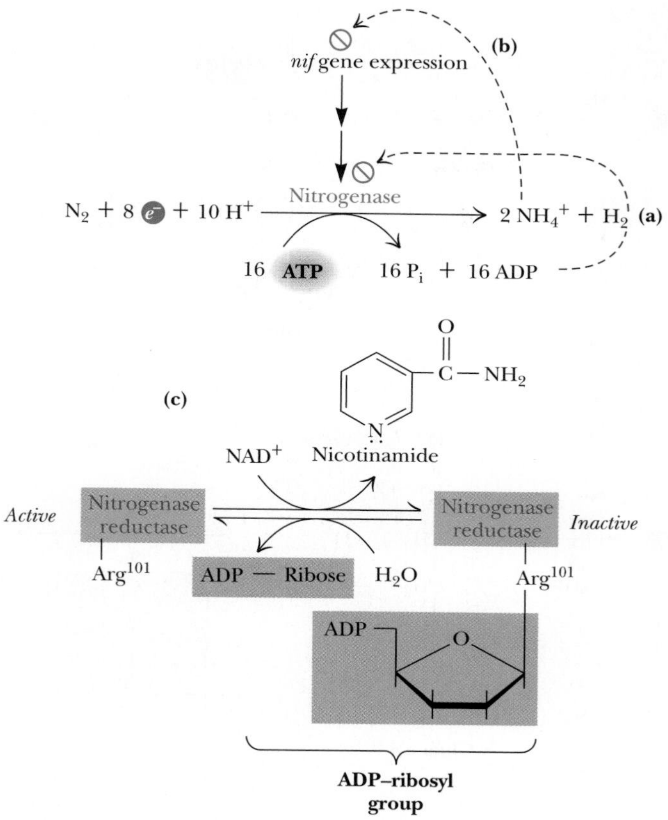

FIGURE 25.8 Regulation of nitrogen fixation. **(a)** ADP inhibits nitrogenase activity. **(b)** NH_4^+ represses *nif* gene expression. **(c)** In some organisms, the nitrogenase complex is regulated by covalent modification. ADP–ribosylation of nitrogenase reductase leads to its inactivation.

Glutamate dehydrogenase (GDH) catalyzes the reductive amination of α-ketoglutarate to yield glutamate. Reduced pyridine nucleotides (NADH or NADPH) provide the reducing power:

$$NH_4^+ + \alpha\text{-ketoglutarate} + NADPH + H^+ \longrightarrow$$
$$\text{glutamate} + NADP^+ + H_2O \qquad (25.6)$$

This reaction provides an important interface between nitrogen metabolism and cellular pathways of carbon and energy metabolism because α-ketoglutarate is a citric acid cycle intermediate. In vertebrates, GDH is an α_6-type multimeric enzyme localized in the mitochondrial matrix that uses NADPH as electron donor when operating in the biosynthetic direction (the direction of glutamate synthesis) (Figure 25.9). When GDH acts in the catabolic direction to generate α-ketoglutarate from glutamate, the binding of $NADP^+$, the electron acceptor for glutamate oxidation, shows negative cooperativity (see Chapter 15). The catabolic activity is allosterically activated by ADP and inhibited by GTP. Interestingly, mammalian GDH plays a prominent role in amino acid catabolism, because it is the only mammalian enzyme that can carry out the oxidative deamination of an amino acid.

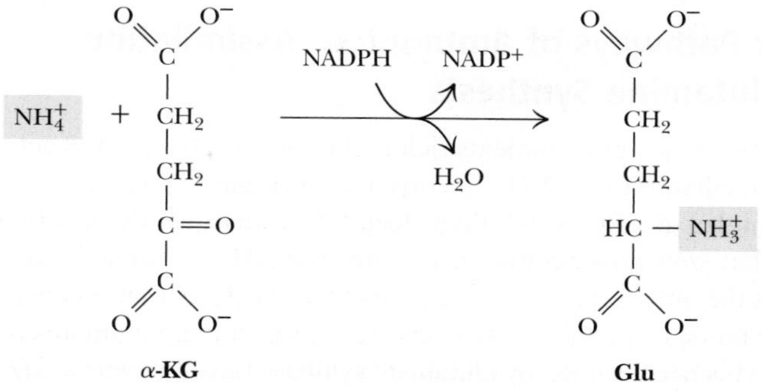

FIGURE 25.9 The glutamate dehydrogenase reaction.

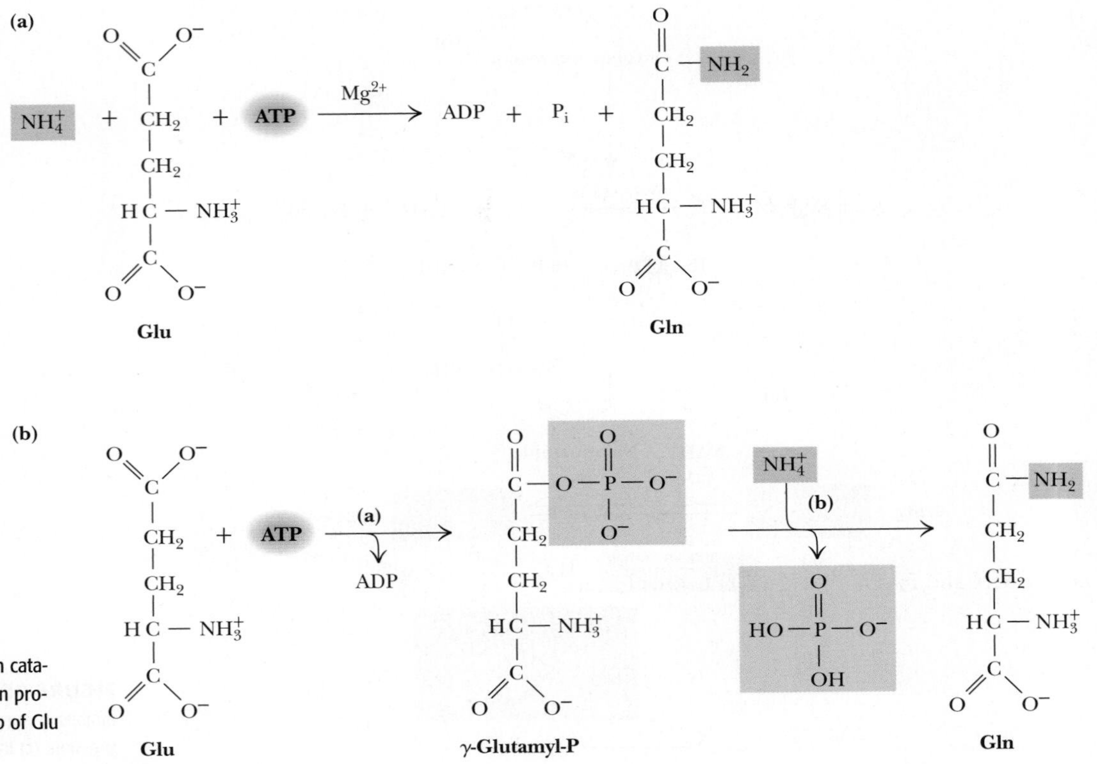

FIGURE 25.10 **(a)** The enzymatic reaction catalyzed by glutamine synthetase. **(b)** The reaction proceeds by (a) activation of the γ-carboxyl group of Glu by ATP, followed by (b) amidation by NH_4^+.

Glutamine synthetase (GS) catalyzes the ATP-dependent amidation of the γ-carboxyl group of glutamate to form glutamine (Figure 25.10). The reaction proceeds via a γ-glutamyl-phosphate intermediate, and GS activity depends on the presence of divalent cations such as Mg^{2+}. **Glutamine** is a major N donor in the biosynthesis of many organic N compounds such as purines, pyrimidines, and other amino acids, and GS activity is tightly regulated, as we shall soon see. The amide-N of glutamine provides the nitrogen atom in these biosyntheses. Glutamine is the most abundant amino acid in humans.

Carbamoyl-phosphate synthetase I, the third enzyme capable of using ammonium to form an N-containing organic compound, catalyzes an early step in the urea cycle. Two ATPs are consumed, one in the activation of HCO_3^- for reaction with ammonium and the other in the phosphorylation of the carbamate formed (see also Figure 25.22):

$$NH_4^+ + HCO_3^- + 2\ ATP \rightarrow H_2N\!-\!\overset{\overset{\displaystyle O}{\|}}{C}\!-\!O\!-\!PO_3^{2-} + 2\ ADP + P_i + 2\ H^+ \quad (25.7)$$

N-acetylglutamate is an essential allosteric activator for this enzyme. Both carbamoyl-P and *N*-acetylglutamate are precursors to arginine synthesis and intermediates in the urea cycle.

The Major Pathways of Ammonium Assimilation Lead to Glutamine Synthesis

In organisms that enjoy environments rich in nitrogen, GDH and GS acting in sequence furnish the principal route of NH_4^+ incorporation (Figure 25.11). However, GDH has a significantly higher K_m for NH_4^+ than does GS. Consequently, in organisms such as green plants that grow under conditions where little NH_4^+ is available, GDH is not effective and GS is the only NH_4^+-assimilative reaction. Such a situation creates the need for an alternative mode of glutamate synthesis to replenish the glutamate consumed by the GS reaction. This need is filled by **glutamate synthase** (also known as *GOGAT,* the acronym for the other name of this enzyme—**glutamate:oxo-glutarate amino-transferase**).

(a) NH$_4^+$ + α-ketoglutarate + NADPH $\xrightarrow{\text{GDH}}$ glutamate + NADP$^+$ + H$_2$O

(b) Glutamate + NH$_4^+$ + ATP $\xrightarrow{\text{GS}}$ glutamine + ADP + P$_i$

SUM: 2 NH$_4^+$ + α-ketoglutarate + NADPH + ATP $\longrightarrow$ glutamine + NADP$^+$ + ADP + P$_i$ + H$_2$O

FIGURE 25.11 The GDH/GS pathway of ammonium assimilation. The sum of these reactions is the conversion of 1 α-ketoglutarate to 1 glutamine at the expense of 2 NH$_4^+$, 1 ATP, and 1 NADPH.

Glutamate synthase catalyzes the reductive amination of α-ketoglutarate using the amide-N of glutamine as the N donor:

$$\text{Reductant} + \alpha\text{-KG} + \text{Gln} \longrightarrow 2\ \text{Glu} + \text{oxidized reductant} \qquad (25.8)$$

Two glutamates are formed—one from amination of α-ketoglutarate and the other from deamidation of Gln (Figure 25.12). These glutamates can now serve as ammonium acceptors for glutamine synthesis by GS. Organisms variously use NADH, NADPH, or reduced ferredoxin as reductant. Glutamate synthases are typically large, complex proteins; in *Escherichia coli*, GOGAT is an 800-kD flavoprotein containing both FMN and FAD, as well as [4Fe–4S] clusters.

Together, GS and GOGAT constitute a second pathway of ammonium assimilation, in which GS is the only NH$_4^+$-fixing step; the role of GOGAT is to regenerate glutamate (Figure 25.13). Note that this pathway consumes 2 equivalents of ATP and 1 NADPH (or similar reductant) per pair of N atoms introduced into Gln, in contrast to the

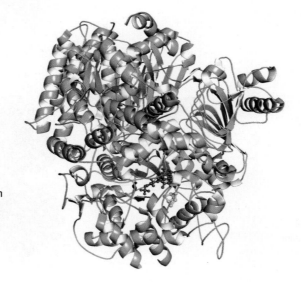

NADH (yeast, *N. crassa*) + H$^+$
NADPH (*E. coli*) + H$^+$ or } + α-KG + Gln $\longrightarrow$ 2 Glu +
2 H$^+$ + 2 reduced ferredoxin (plants)

NAD$^+$
NADP$^+$ or
2 oxidized ferredoxin

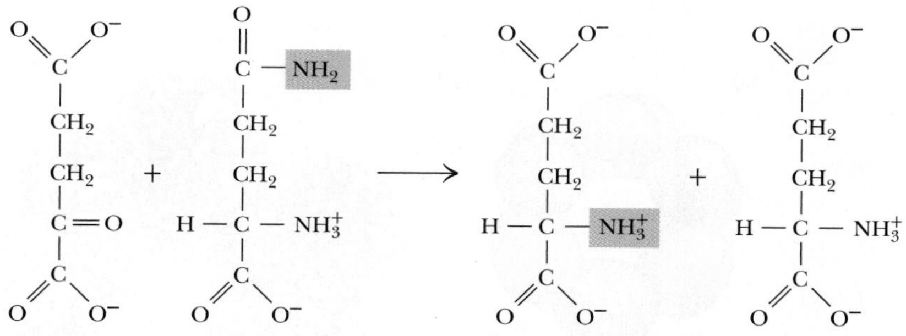

α-KG   Gln   Glu   Glu

FIGURE 25.12 The glutamate synthase reaction (*above*), showing the reductants exploited by different organisms in this reductive amination reaction. Structure of glutamate synthase (*right*) (pdb id = 1LM1). FAD is shown in blue, the Fe-S cluster in yellow.

FIGURE 25.13 The GS/GOGAT pathway of ammonium assimilation. The sum of these reactions results in the conversion of 1 α-ketoglutarate to 1 glutamine at the expense of 2 ATP and 1 NADPH.

(a) $2\ NH_4^+ + 2\ ATP + 2\ \text{glutamate} \xrightarrow{\ GS\ } 2\ \text{glutamine} + 2\ ADP + 2\ P_i$

(b) $NADPH + \alpha\text{-ketoglutarate} + \text{glutamine} \xrightarrow{\ GOGAT\ } 2\ \text{glutamate} + NADP^+$

SUM: $2\ NH_4^+ + \alpha\text{-ketoglutarate} + NADPH + 2\ ATP \longrightarrow \text{glutamine} + NADP^+ + 2\ ADP + 2\ P_i$

GDH/GS pathway, in which only 1 ATP and 1 NADPH are consumed per pair of NH_4^+ fixed. Clearly, coping with a nitrogen-limited environment has its cost.

25.3 What Regulatory Mechanisms Act on *Escherichia coli* Glutamine Synthetase?

As indicated earlier, glutamine plays a pivotal role in nitrogen metabolism by donating its amide nitrogen to the biosynthesis of many important organic N compounds. Consistent with its metabolic importance, in prokaryotic cells such as *E. coli,* GS is regulated at three different levels:

1. Its activity is regulated allosterically by *feedback inhibition.*
2. GS is interconverted between active and inactive forms by *covalent modification.*
3. Cellular amounts of GS are carefully controlled at the level of *gene expression* and *protein synthesis.*

Eukaryotic versions of glutamine synthetase show none of these regulatory features.

E. coli GS is a 600-kD dodecamer (α_{12}-type subunit organization) of identical 52-kD monomers (each monomer contains 468 amino acid residues). These monomers are arranged as a stack of two hexagons (Figure 25.14). The active sites are located at subunit

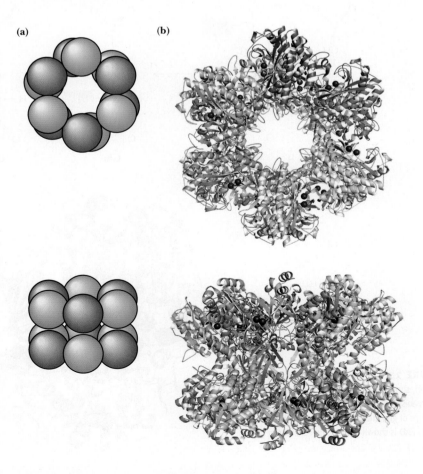

(a)   **(b)**

FIGURE 25.14 The subunit organization of bacterial glutamine synthetase. **(a)** Schematic; **(b)** molecular structure (note the pairs of metal ions [dark blue] that define the active sites) (pdb id = 1FPY).

interfaces within the hexagons; these active sites are recognizable in the X-ray crystal-lographic structure by the pair of divalent cations that occupy them. Adjacent subunits contribute to each active site, thus accounting for the fact that GS monomers are cata-lytically inactive.

Glutamine Synthetase Is Allosterically Regulated

Nine distinct feedback inhibitors (Gly, Ala, Ser, His, Trp, CTP, AMP, carbamoyl-P, and glucosamine-6-P) act on GS. Gly, Ala, and Ser are key indicators of amino acid metabolism in the cell; each of the other six compounds represents an end product of a biosynthetic pathway dependent on Gln (Figure 25.15). AMP competes with ATP for binding at the ATP substrate site. Gly, Ala, and Ser compete with Glu for binding at the active site. Carbamoyl-P binds at a site that overlaps both the Glu site and the site occupied by the γ-PO$_4$ of ATP.

Glutamine Synthetase Is Regulated by Covalent Modification

Each GS subunit can be adenylylated at a specific tyrosine residue (Tyr[397]) in an ATP-dependent reaction (Figure 25.16). Adenylylation inactivates GS. If we define n as the average number of adenylyl groups per GS molecule, GS activity is inversely proportional to n. The number n varies from 0 (no adenylyl groups) to 12 (every subunit in each GS molecule is adenylylated). Adenylylation of GS is catalyzed by the bifunctional

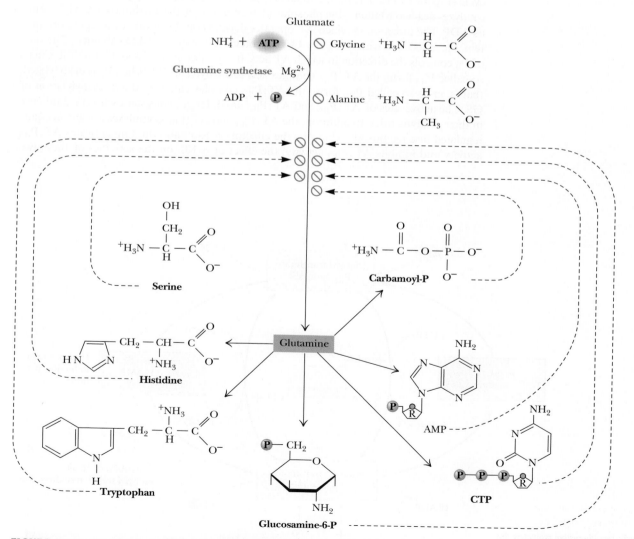

FIGURE 25.15 The allosteric regulation of glutamine synthetase activity by feedback inhibition.

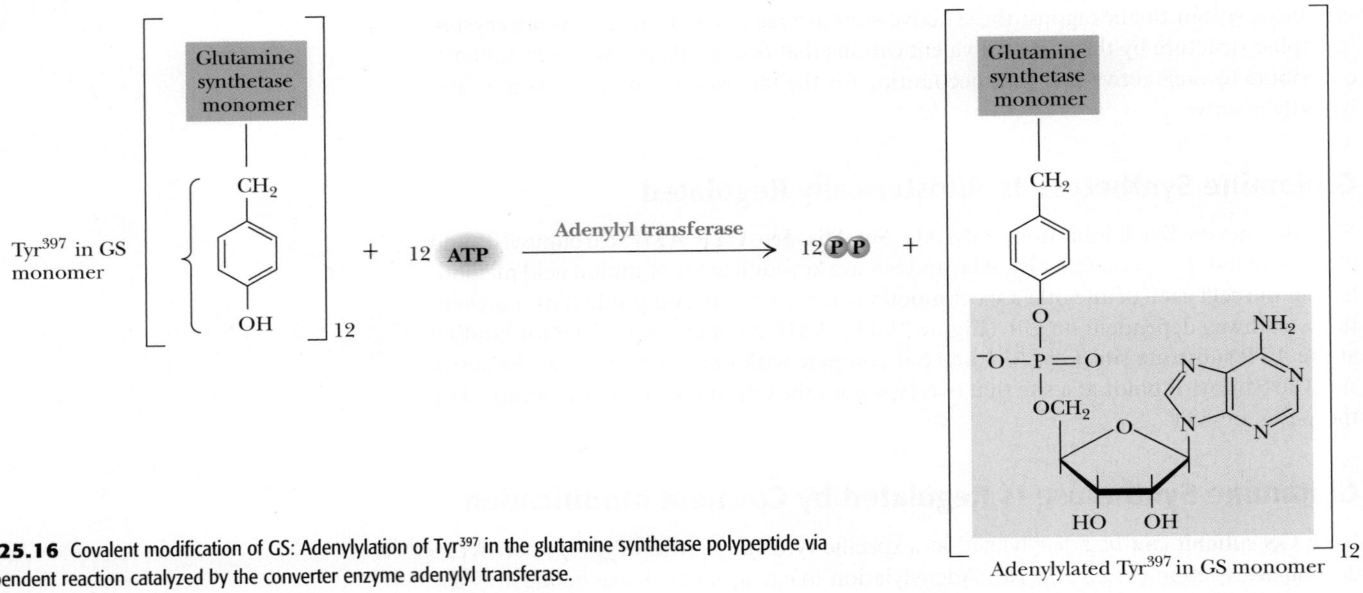

FIGURE 25.16 Covalent modification of GS: Adenylylation of Tyr^{397} in the glutamine synthetase polypeptide via an ATP-dependent reaction catalyzed by the converter enzyme adenylyl transferase.

converter enzyme **ATP:GS:adenylyl transferase,** or simply *adenylyl transferase* **(AT).** However, whether or not this covalent modification occurs is determined by a highly regulated cycle (Figure 26.17). The bifunctional AT not only catalyzes adenylylation of GS, it also catalyzes **deadenylylation**—the phosphorolytic removal of the Tyr-linked adenylyl groups as ADP. The direction in which AT operates depends on the nature of a regulatory protein, P_{II}, associated with it. P_{II} is a 44-kD protein (tetramer of 11-kD subunits): The state of P_{II} controls the direction in which AT acts. If P_{II} is not uridylylated, that is, if it is in its so-called P_{IIA} form, the AT:P_{IIA} complex acts to adenylylate GS. When P_{II} is uridylylated, that is, in its so-called P_{IID} form, the AT:P_{IID} complex catalyzes the deadenylylation of GS. The active sites of AT:P_{IIA} and AT:P_{IID} are different, consistent with the difference in their catalytic roles. In addition, the AT:P_{IIA} and AT:P_{IID} complexes are allosterically regulated in a reciprocal fashion by the effectors α-KG and Gln. Gln activates AT:P_{IIA} activity and inhibits AT:P_{IID} activity; the effect of α-KG on the activities of these two

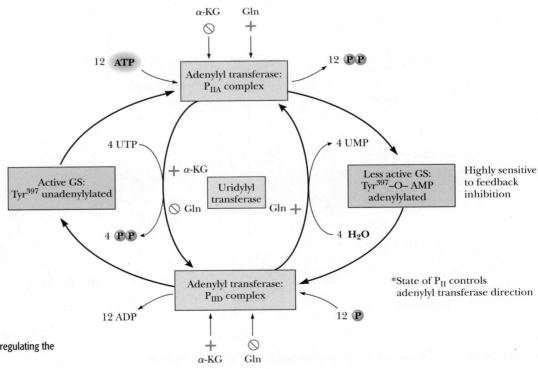

FIGURE 25.17 The cyclic cascade system regulating the covalent modification of GS.

complexes is diametrically opposite (Figure 25.17). Further, Gln favors conversion of P_{IID} to P_{IIA}, whereas α-ketoglutarate favors the P_{IID} over the P_{IIA} form.

Clearly, the determining factor regarding the degree of adenylylation, *n*, and hence the relative activity of GS, is the [Gln]/[α-KG] ratio. A high [Gln] level signals cellular nitrogen sufficiency, and GS becomes adenylylated and inactivated. In contrast, a high [α-KG] level is an indication of nitrogen limitation and a need for ammonium fixation by GS. Note that Gln affects all four reactions of the cyclic cascade system, whereas α-KG affects only three. Thus, Gln is ultimately the key effector in this system.

Glutamine Synthetase Is Regulated Through Gene Expression

The gene that encodes the GS subunit in *E. coli* is designated *GlnA*. The *GlnA* gene is actively transcribed to yield GS mRNA for translation and synthesis of GS protein only if a *specific transcriptional enhancer,* **NR$_I$,** is in its phosphorylated form, NR$_I$-P. In turn, NR$_I$ is phosphorylated in an ATP-dependent reaction catalyzed by **NR$_{II}$,** a protein kinase (Figure 25.18). However, if NR$_{II}$ is complexed with P_{IIA}, it acts not as a kinase but as a phosphatase, and the transcriptionally active form of NR$_I$, namely NR$_I$-P, is converted back to NR$_I$ with the result that *GlnA* transcription halts. Recall from the foregoing discussion that a high [Gln]/[α-KG] ratio favors P_{IIA} at the expense of P_{IID}. Under such conditions, GS gene expression is not necessary.

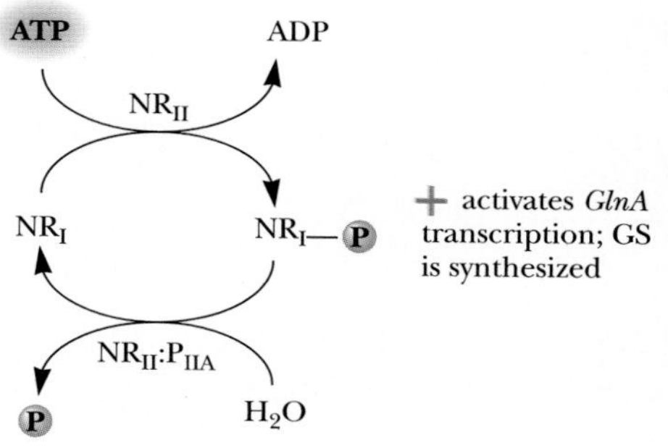

Glutamine in the Human Body

Glutamine is not among the ten most commonly occurring amino acids in proteins (see Figure 5.16), but it is the most abundant amino acid in human body fluids and tissues. The abundance of glutamine reflects its prominent anabolic role as the source of N atoms in the synthesis of purines and pyrimidines (see Chapter 26), amino sugars (Chapter 18), and other nitrogen-containing biological molecules.

Mammalian cells grown in culture rely on glucose and glutamine as their principal substrates for energy production, generation of reducing equivalents in the form of NADPH, and the carbon and nitrogen atoms for anabolic growth. Donation of the amide group of glutamine during nucleotide biosynthesis or the creation of other N-containing biomolecules creates glutamate. Oxidative deamination of Glu by GDH or transfer of its α-amino group through transamination (see Figure 25.19) leads to

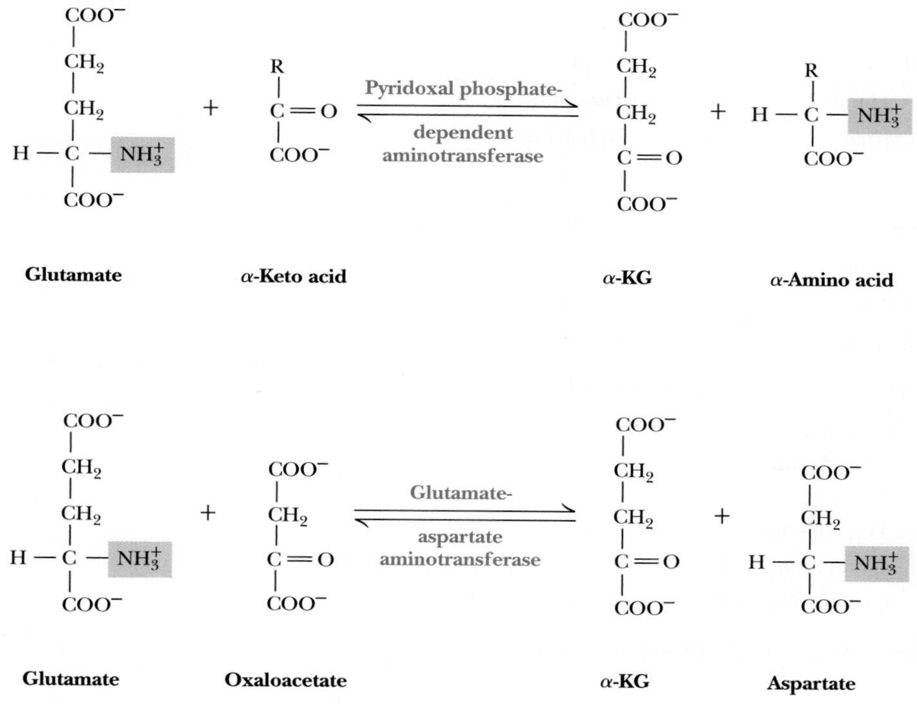

FIGURE 25.18 Transcriptional regulation of *GlnA* expression through the reversible phosphorylation of NR$_I$.

Glutamate α-Keto acid $\xrightarrow{\text{Pyridoxal phosphate-dependent aminotransferase}}$ α-KG α-Amino acid

Glutamate Oxaloacetate $\xrightarrow{\text{Glutamate-aspartate aminotransferase}}$ α-KG Aspartate

FIGURE 25.19 Glutamate-dependent transamination of α-keto acid carbon skeletons is a primary mechanism for amino acid synthesis. The transamination of oxaloacetate by glutamate to yield aspartate and α-ketoglutarate is a prime example.

the formation of α-ketoglutarate, thus replenishing a key citric acid cycle intermediate. Thus, carbon for biosynthesis is provided by glutamine through the amphibolic nature of the citric acid cycle.

Glutamine in the Brain Glutamine is by far the most abundant amino acid in the central nervous system. Glutamine synthetase in astrocytes, a star-shaped type of glial cell, aids in meeting the demand for Gln. (Glial cells are cells that support and nourish neurons.) Gln serves as the precursor for both glutamate and γ-aminobutyrate, two neurotransmitters that are synthesized in neurons. Excessive accumulation of Gln in the central nervous system can interfere with brain function.

25.4 How Do Organisms Synthesize Amino Acids?

Organisms show substantial differences in their capacity to synthesize the 20 amino acids common to proteins. Typically, plants and microorganisms can form all of their nitrogenous metabolites, including all of the amino acids, from inorganic forms of N such as NH_4^+ and NO_3^-. In these organisms, the α-amino group for all amino acids is derived from glutamate, usually via transamination of the corresponding α-keto acid analog of the amino acid (Figure 25.19). In many cases, amino acid biosynthesis is thus a matter of synthesizing the appropriate α-keto acid carbon skeleton, followed by transamination with Glu. The amino acids can be classified according to the source of intermediates for the α-keto acid biosynthesis (Table 25.1). For example, the amino acids Glu, Gln, Pro, and Arg (and, in some instances, Lys) are all members of the α-ketoglutarate family because they are all derived from the citric acid cycle intermediate α-ketoglutarate. We return to this classification scheme later when we discuss the individual biosynthetic pathways.

TABLE 25.1	The Grouping of Amino Acids into Families According to the Metabolic Intermediates That Serve as Their Progenitors	
α-Ketoglutarate Family	**Oxaloacetate Family**	
Glutamate	Aspartate	
Glutamine	Asparagine	
Proline	Methionine	
Arginine	Threonine	
Lysine*	Isoleucine	
	Lysine*	
Pyruvate Family	**3-Phosphoglycerate Family**	
Alanine	Serine	
Valine	Glycine	
Leucine	Cysteine	
Phosphoenolpyruvate and Erythrose-4-P Family		
The aromatic amino acids		
Phenylalanine		
Tyrosine		
Tryptophan		
The remaining amino acid, *histidine,* is derived from PRPP (5-phosphoribosyl-1-pyrophosphate) and ATP.		

*Different organisms use different precursors to synthesize lysine.

HUMAN BIOCHEMISTRY

Human Dietary Requirements for Amino Acids

Humans can synthesize only 10 of the 20 common amino acids (see table below); the others must be obtained in the diet. Those that can be synthesized are classified as **nonessential,** meaning it is not essential that these amino acids be part of the diet. In effect, humans can synthesize the α-keto acid analogs of nonessential amino acids and form the amino acids by transamination. In contrast, humans are incapable of constructing the carbon skeletons of **essential** amino acids, so they must rely on dietary sources for these essential metabolites. Excess dietary amino acids cannot be stored for future use, nor are they excreted unused. Instead, they are converted to common metabolic intermediates that can be either oxidized by the citric acid cycle to generate metabolic energy or used to form glucose (see Section 25.5).

Since autotrophic cells (and many prokaryotic cells) synthesize all 20 amino acids, several questions arise regarding human dietary requirements for amino acids. First, why is it that humans lack the ability to do what other organisms can do? The answer is that, over evolutionary time, human diets provided adequate amounts of those amino acids classified now as "essential." Thus, the loss of the metabolic pathways for synthesis of "essential" amino acids did not impair the fitness of humans. That is, synthesizing "essential" amino acids became superfluous, and no evolutionary pressure operated on humans to retain the genes for these pathways. A second question now emerges: Are there significant differences between synthesis of amino acids humans can make (the so-called nonessential amino acids) and those they can't? The table on the right summarizes amino acid biosynthesis from citric acid cycle

intermediates in terms of the number of reactions needed to make an amino acid from a TCA cycle intermediate.* Nonessential amino acids are shown in blue; essential amino acids in red. Two conclusions stand out: Nonessential amino acids require fewer reaction steps for synthesis than essential amino acids, and nonessential amino acids tend to be more abundantly represented in proteins than essential amino acids. Thus, evolutionary loss was not random: The biosynthetic pathways lost were those for amino acids requiring the most reaction steps.

*From Srinivasan, V., Morowitz, H., and Smith, E., 2007. Essential amino acids, from LUCA to LUCY. *Complexity* **13**:8–9.

Essential and Nonessential Amino Acids in Humans

Essential	Nonessential
Arginine*	Alanine
Histidine*	Asparagine
Isoleucine	Aspartate
Leucine	Cysteine
Lysine	Glutamate
Methionine	Glutamine
Phenylalanine	Glycine
Threonine	Proline
Tryptophan	Serine
Valine	Tyrosine†

*Arginine and histidine are essential in the diets of juveniles, not adults.
†Tyrosine is classified as nonessential only because it is readily formed from essential phenylalanine.

Nonessential Amino Acids Require Fewer Reactions for Synthesis

	Amino Acid	Reaction Steps	Mole % in Proteins†
1	Alanine	1	7.9
2	Aspartic acid	1	5.3
3	Glutamic acid	1	6.7
4	Asparagine	2	4.1
5	Glutamine	2	4.0
6	Serine	5	6.9
7	Glycine	6	6.9
8	Proline	6	4.8
9	Cysteine	7	1.5
10	Threonine	6	5.4
11	Valine	9	6.7
12	Isoleucine	13	5.9
13	Leucine	14	10.0
14	Lysine	14	5.9
15	Methionine	17	2.4
16	Arginine	24	5.4
17	Histidine	27	2.3
18	Phenylalanine	29	4.0
19	Tyrosine‡	30	3.0
20	Tryptophan	33	1.1

†Mole percentages are taken from amino acid representations among proteins in the Swiss-Prot protein knowledgebase: ca.expasy.org/sprot.
‡Note that "nonessential" tyrosine can be made only from "essential" phenylalanine.

Amino Acids Are Formed from α-Keto Acids by Transamination

Transamination involves transfer of an α-amino group from an amino acid to the α-keto position of an α-keto acid (Figure 25.19). In the process, the amino donor becomes an α-keto acid while the α-keto acid acceptor becomes an α-amino acid:

$$\text{Amino acid}_1 + \alpha\text{-keto acid}_2 \longrightarrow \alpha\text{-keto acid}_1 + \text{amino acid}_2 \qquad (25.9)$$

A DEEPER LOOK

The Mechanism of the Aminotransferase (Transamination) Reaction

The aminotransferase (transamination) reaction is a workhorse in biological systems. It provides a general means for exchange of nitrogen between amino acids and α-keto acids. This vital reaction is catalyzed by pyridoxal phosphate (PLP). The mechanism involves loss of the C_α proton, followed by an aldimine–ketimine tautomerization—literally a "flip-flop" of the Schiff base double bond from the pyridoxal aldehyde carbon to the α-carbon of the amino acid substrate.

This is followed by hydrolysis of the ketimine intermediate to yield the product α-keto acid. Left in the active site is a pyridoxamine phosphate intermediate, which combines with another (substrate) α-keto acid to form a second ketimine, which rearranges to form an aldimine, followed by release as an amino acid. Transaldiminization with a lysine at the active site completes the reaction.

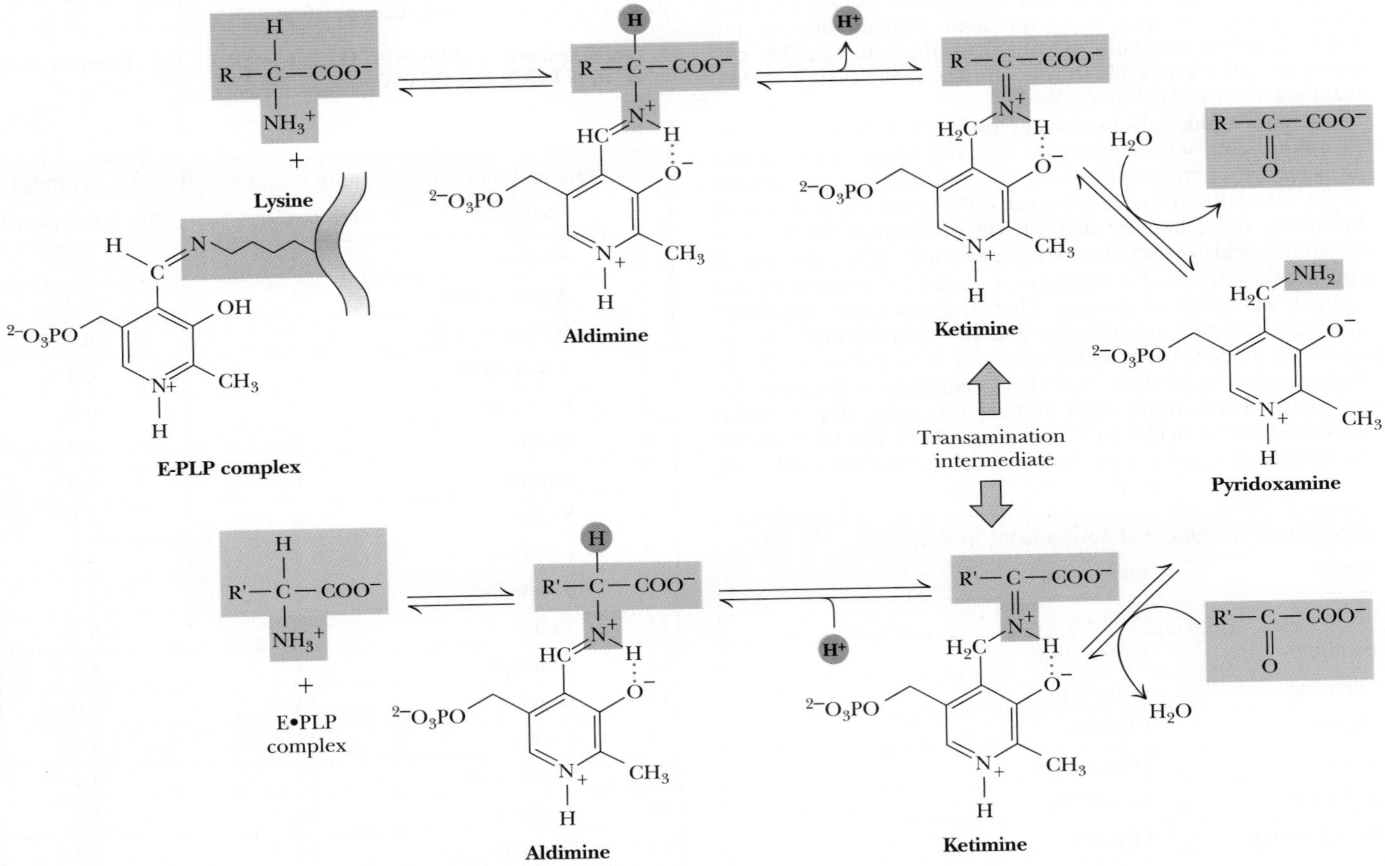

The mechanism of PLP-catalyzed transamination reactions.

The predominant amino acid/α-keto acid pair in these reactions is glutamate/α-ketoglutarate, with the net effect that glutamate is the primary amino donor for the synthesis of amino acids. Transamination reactions are catalyzed by **aminotransferases** (the preferred name for enzymes formerly termed *transaminases*). Aminotransferases are named according to their amino acid substrates, as in **glutamate–aspartate aminotransferase**. Aminotransferases are prime examples of enzymes that catalyze double displacement (ping-pong)–type bisubstrate reactions (see Figure 13.23).

The Pathways of Amino Acid Biosynthesis Can Be Organized into Families

As indicated in Table 25.1, the amino acids can be grouped into families on the basis of the metabolic intermediates that serve as their precursors.

The α-Ketoglutarate Family of Amino Acids Includes Glu, Gln, Pro, Arg, and Lys

Amino acids derived from α-ketoglutarate include glutamate (Glu), glutamine (Gln), proline (Pro), arginine (Arg), and in fungi and protista such as *Euglena,* lysine (Lys). The routes for Glu and Gln synthesis were described when we considered pathways of ammonium assimilation.

Proline is derived from glutamate via a series of four reactions involving activation, then reduction, of the γ-carboxyl group to an aldehyde *(glutamate-5-semialdehyde),* which spontaneously cyclizes to yield the internal Schiff base, *Δ^1-pyrroline-5-carboxylate* (Figure 25.20). NADPH-dependent reduction of the pyrroline double bond gives proline.

Arginine biosynthesis involves enzymatic steps that are also part of the **urea cycle,** a metabolic pathway that allows certain animals (including humans) to excrete any excess nitrogen arising from overconsumption of protein. Net synthesis of arginine depends on the formation of **ornithine.** Interestingly, ornithine is derived from glutamate via a reaction pathway reminiscent of the proline biosynthetic pathway (Figure 25.21). Glutamate is first *N*-acetylated in an acetyl-CoA–dependent reaction to yield *N-acetylglutamate* (Figure 25.21). An ATP-dependent phosphorylation of *N*-acetylglutamate to give *N-acetylglutamate-5-phosphate* primes this substrate for a reduced pyridine nucleotide-dependent reduction to the semialdehyde. *N*-acetylglutamate-5-semialdehyde then is aminated by a glutamate-dependent aminotransferase, giving *N-acetylornithine,* which is deacylated to ornithine. In mammals, ornithine is made from glutamate via a pathway that does not involve an *N*-acetyl block.

Ornithine has three metabolic roles: (1) to serve as a precursor to arginine, (2) to function as an intermediate in the urea cycle, and (3) to act as an intermediate in Arg degradation. In any case, the δ-NH_3^+ of ornithine is carbamoylated in a reaction catalyzed by **ornithine transcarbamoylase.** The carbamoyl group is derived from carbamoyl-P synthesized by **carbamoyl-phosphate synthetase I (CPS-I).** CPS-I is the mitochondrial CPS isozyme; it uses two ATPs in catalyzing the formation of carbamoyl-P from NH_3 and HCO_3^-, one to activate the bicarbonate ion and the other to phosphorylate the carbamate arising from carboxyphosphate through its reaction with the ammonium ion (Figure 25.22). CPS-I represents the committed step in the urea cycle, and CPS-I is

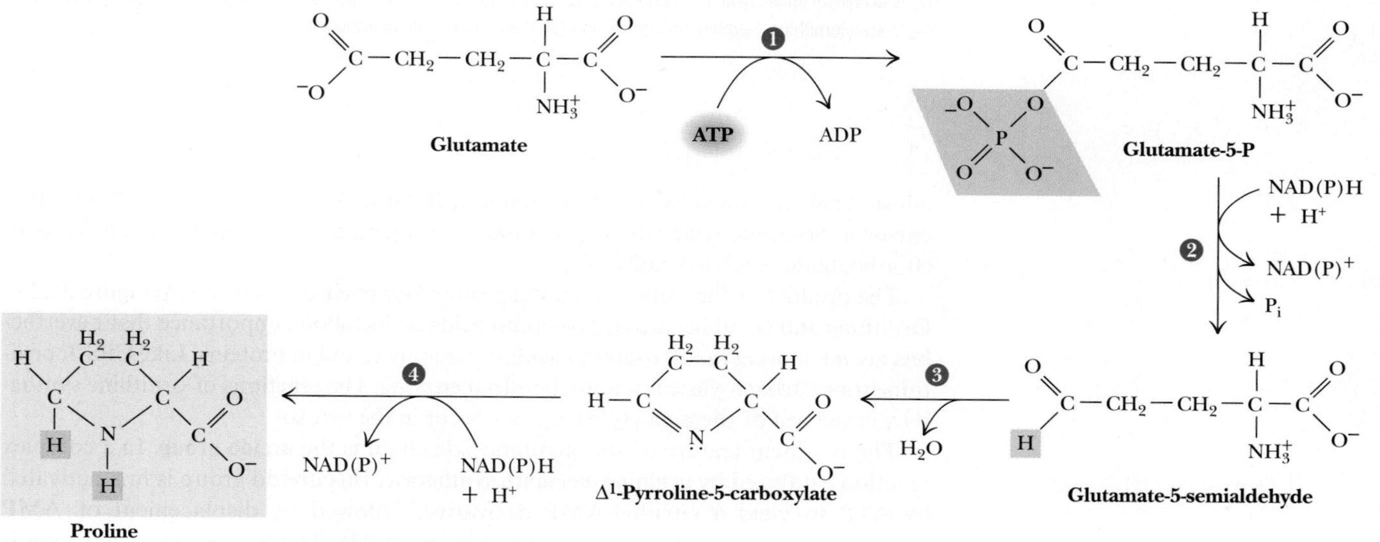

FIGURE 25.20 The pathway of proline biosynthesis from glutamate. The enzymes are **(1)** γ-glutamyl kinase, **(2)** glutamate-5-semialdehyde dehydrogenase, and **(4)** Δ^1-pyrroline-5-carboxylate reductase; reaction **(3)** occurs nonenzymatically.

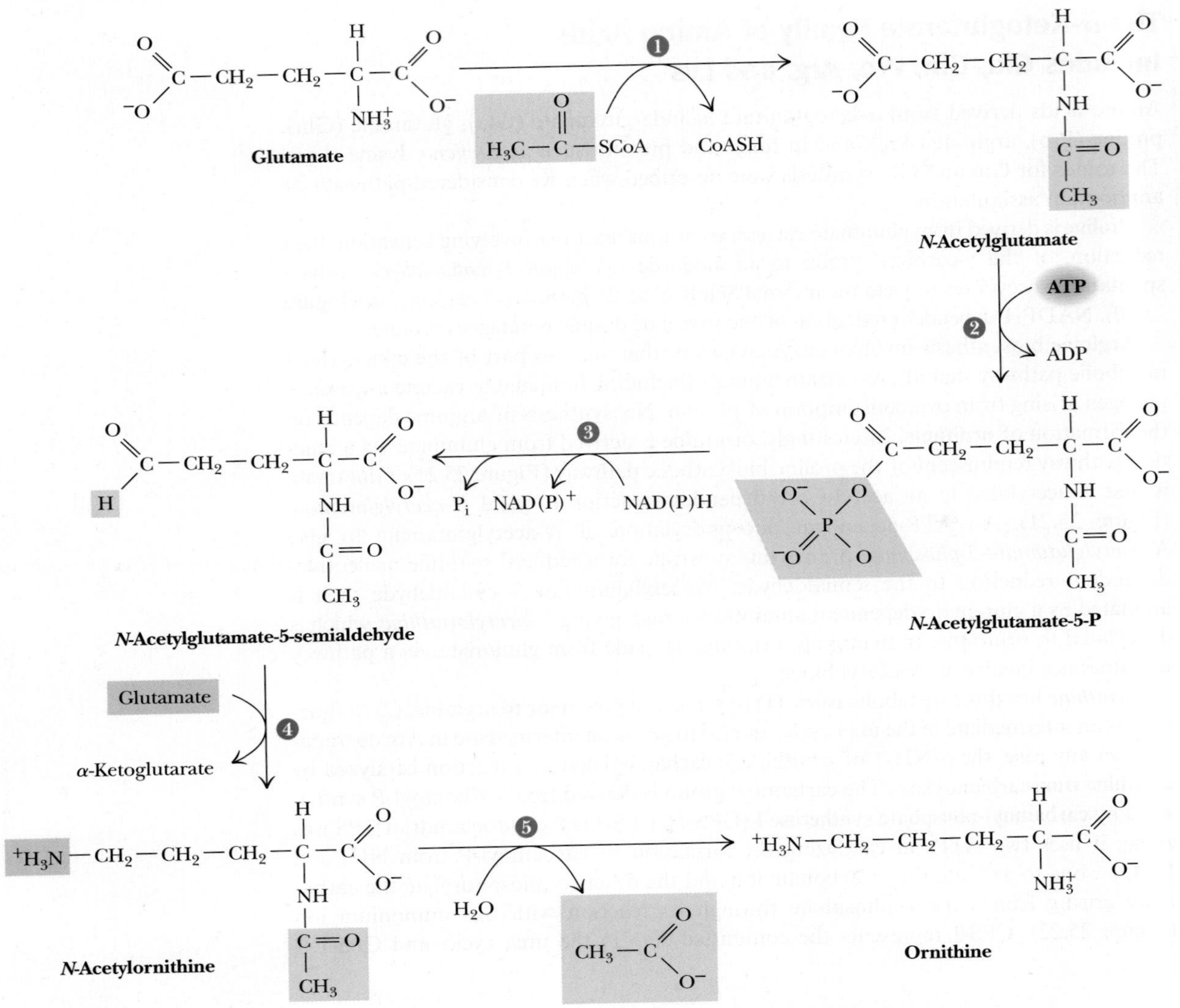

FIGURE 25.21 The bacterial pathway of ornithine biosynthesis from glutamate. The enzymes are
(1) N-acetylglutamate synthase, **(2)** N-acetylglutamate kinase, **(3)** N-acetylglutamate-5-semialdehyde dehydrogenase,
(4) N-acetylornithine δ-aminotransferase, and **(5)** N-acetylornithine deacetylase.

allosterically activated by *N*-acetylglutamate. Because *N*-acetylglutamate is both a precursor to ornithine synthesis and essential to the operation of the urea cycle, it serves to coordinate these related pathways.

The product of the ornithine transcarbamoylase reaction is *citrulline* (Figure 25.23). Ornithine and citrulline are two α-amino acids of metabolic importance that nevertheless are *not* among the 20 α-amino acids commonly found in proteins. Like CPS-I, ornithine transcarbamoylase is a mitochondrial enzyme. The reactions of ornithine synthesis and the rest of the urea cycle enzymes occur in the cytosol.

The pertinent feature of the citrulline side chain is the **ureido group.** In a complex reaction catalyzed by **argininosuccinate synthetase,** this ureido group is first activated by ATP to yield a citrullyl-AMP derivative, followed by displacement of AMP by aspartate to give *argininosuccinate* (Figure 25.23). The formation of arginine is then accomplished by **argininosuccinase,** which catalyzes the nonhydrolytic elimination of fumarate from argininosuccinate. This reaction completes the biosynthesis of Arg.

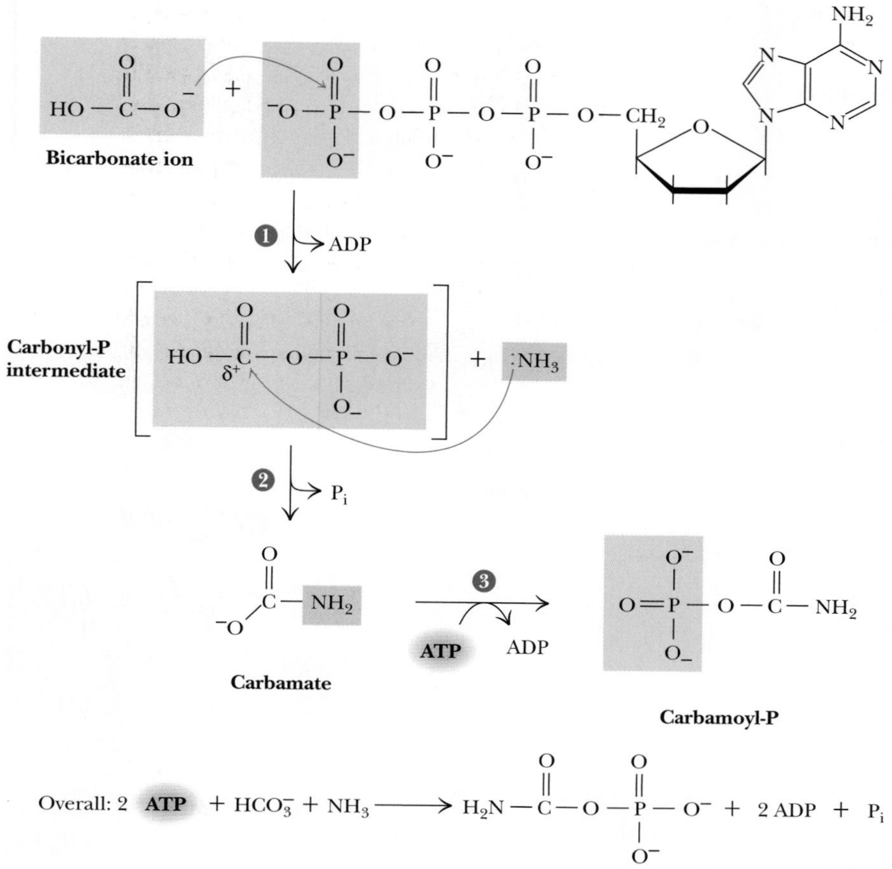

FIGURE 25.22 The mechanism of action of CPS-I.

The Urea Cycle Acts to Excrete Excess N Through Arg Breakdown

The carbon skeleton of arginine is derived principally from α-ketoglutarate, but the N and C atoms composing the **guanidino group** (Figure 25.23) of the Arg side chain come from NH_4^+, HCO_3^- (as carbamoyl-P), and the α-NH_2 groups of glutamate and aspartate. The circle of the urea cycle is closed when ornithine is regenerated from Arg by the **arginase**-catalyzed hydrolysis of arginine. Urea is the other product of this reaction and lends its name to the cycle. Note that the N atoms in urea arose from NH_4^+ and the Asp amino group. In terrestrial vertebrates, urea synthesis is required to excrete excess nitrogen generated by increased amino acid catabolism—for example, following dietary consumption of more than adequate amounts of protein. Urea formation is basically confined to the liver.

A healthy adult human male eating a typical American diet will consume about 100 g of protein per day. Because such an individual will remain in *nitrogen balance* (neither increasing or decreasing his net protein levels), his body must dispose of about 1 mole of excess N derived from the amino acids in this dietary protein. Glutamate is key to this process because of the position of glutamate dehydrogenase at the interface of amino acid and carbohydrate metabolism and the importance of glutamate to the urea cycle.

Increases in amino acid catabolism lead to elevated glutamate levels and a rise in *N*-acetylglutamate, the allosteric activator of CPS-I. Stimulation of CPS-I raises overall urea cycle activity because activities of the remaining enzymes of the cycle simply respond to increased substrate availability. Removal of potentially toxic NH_4^+ by CPS-I is another important aspect of this regulation. The urea cycle is linked to the citric acid

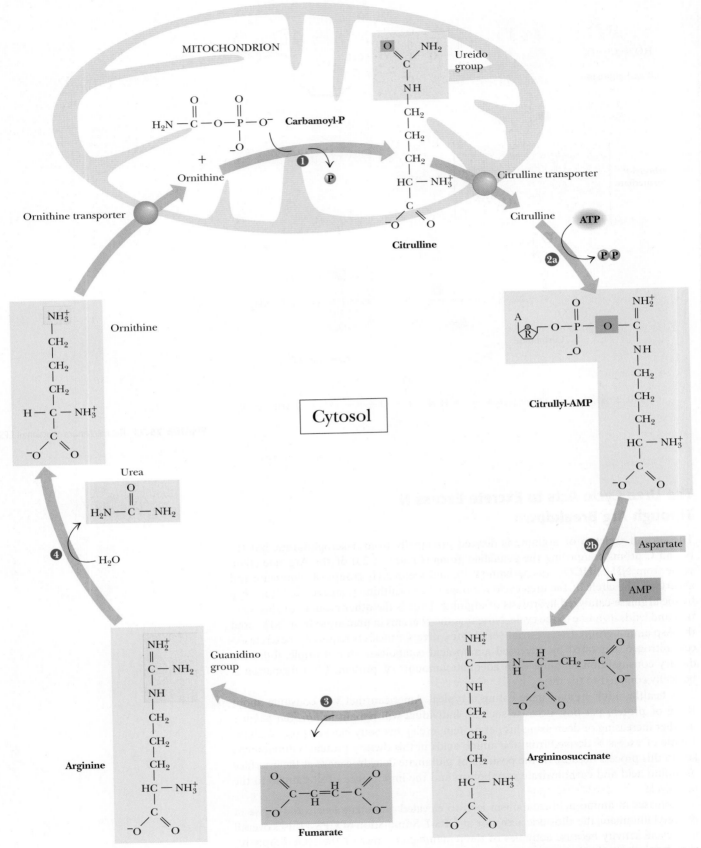

FIGURE 25.23 The urea cycle series of reactions: The enzymes are (1) ornithine transcarbamoylase (OTCase), (2a and 2b) argininosuccinate synthetase, (3) argininosuccinase, and (4) arginase.

> ▶ **A DEEPER LOOK**
>
> # The Urea Cycle as Both an Ammonium and a Bicarbonate Disposal Mechanism
>
> Excretion of excess NH_4^+ in the innocuous form of urea has traditionally been viewed as the physiological role of the urea cycle. However, the urea cycle also provides a mechanism for the excretion of excess HCO_3^- arising principally from α-carboxyl groups generated during the catabolism of α-amino acids. The following equations illustrate this property:
>
> (1) $HCO_3^- + 2\ NH_4^+ \longrightarrow H_2NCONH_2 + 2\ H_2O + H^+$
> (2) $HCO_3^- + H^+ \longrightarrow H_2O + CO_2$
> *Sum:*
> $2\ HCO_3^- + 2\ NH_4^+ \longrightarrow H_2NCONH_2 + CO_2 + 3\ H_2O$
>
> That is, *2* moles of HCO_3^- are eliminated in the synthesis of each mole of urea: One is incorporated into the product, urea (reaction
>
> 1), and the second is simply protonated and dehydrated to form CO_2 (reaction 2), which is easily excreted. *One interpretation of the preceding is that these coupled reactions allow a weak acid (NH_4^+) to protonate the conjugate base of a stronger acid (HCO_3^-).* At first glance, this protonation would appear thermodynamically unfavorable, but recall that in the urea cycle, 4 equivalents of ATP are consumed per equivalent of urea synthesized: 2 ATPs in the synthesis of carbamoyl-P, and 2 more as 1 ATP is converted to AMP + PP_i in the synthesis of argininosuccinate from citrulline (Figure 25.23). If this interpretation is correct, *the urea cycle may be considered an ATP-driven proton pump that transfers H^+ ions from NH_4^+ to HCO_3^- against a thermodynamic barrier. In the process, the potentially toxic waste products, ammonium and bicarbonate, are rendered innocuous and excreted.*

cycle through *fumarate,* a by-product of the action of *argininosuccinase* (Figure 25.23, reaction 3).

Lysine biosynthesis in some fungi and in the protist *Euglena* also stems from α-ketoglutarate, making lysine a member of the α-ketoglutarate family of amino acids in these organisms. (As we shall see, the other organisms capable of lysine synthesis—namely, bacteria, other fungi, algae, and green plants—rely on aspartate [derived from oxaloacetate] as a precursor.) To make lysine from α-ketoglutarate requires a lengthening of the carbon skeleton by one CH_2 unit to yield *α-ketoadipate* (Figure 25.24). This addition is accomplished by a series of reactions reminiscent of the initial stages of the citric acid cycle. First, a two-carbon acetyl-CoA unit is added to the α-carbon of α-ketoglutarate to form *homocitrate.* Then, in a reaction sequence like that catalyzed by aconitase, *homoisocitrate* is formed from homocitrate. Oxidative decarboxylation (as in isocitrate dehydrogenase) removes one carbon (the original α-carboxyl group of α-ketoglutarate), leaving *α-ketoadipate.* A glutamate-dependent aminotransferase enzyme then aminates α-ketoadipate to give *α-aminoadipate.* Next, the δ-COO^- group is activated in an ATP-dependent adenylylation reaction, priming this δ-COO^- group for reduction to an aldehyde by NADPH. *α-Aminoadipic-6-semialdehyde* is then reductively aminated by addition of glutamate to its aldehydic carbon in an NADPH-dependent reaction leading to the formation of *saccharopine.* Oxidative cleavage of saccharopine by way of an NAD^+-dependent dehydrogenase activity yields α-ketoglutarate and *lysine.* This pathway is known as the **α-aminoadipic acid pathway** of lysine biosynthesis. Interestingly, lysine degradation in animals leads to formation of α-aminoadipate by a reverse series of reactions identical to those occurring along the last steps of this biosynthetic pathway.

The Oxaloacetate Family of Amino Acids Includes Asp, Asn, Lys, Met, Thr, and Ile

The members of the oxaloacetate family of amino acids include aspartate (Asp), asparagine (Asn), lysine (via the diaminopimelic acid pathway), methionine (Met), threonine (Thr), and isoleucine (Ile).

Aspartate is formed from the citric acid cycle intermediate oxaloacetate by transfer of an amino group from glutamate via a PLP-dependent aminotransferase reaction (Figure 25.25). Like glutamate synthesis from α-ketoglutarate, aspartate synthesis is a drain on the citric acid cycle. As we already saw, the Asp amino group serves as the N donor in the conversion of citrulline to arginine. In Chapter 26, we shall see that this $-NH_2$ is also the source of one of the N atoms of the purine ring system during nucleotide biosynthesis,

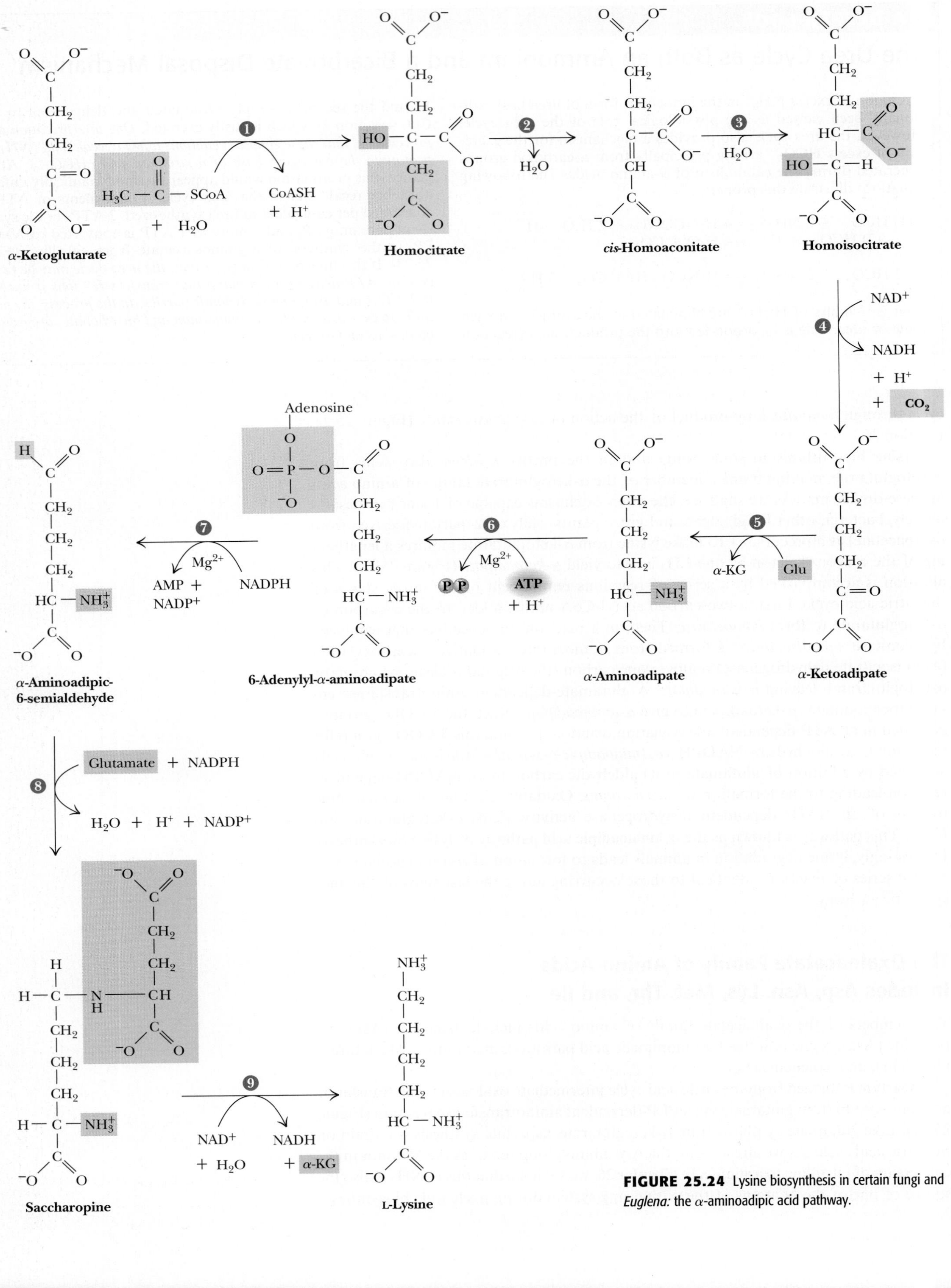

FIGURE 25.24 Lysine biosynthesis in certain fungi and *Euglena*: the α-aminoadipic acid pathway.

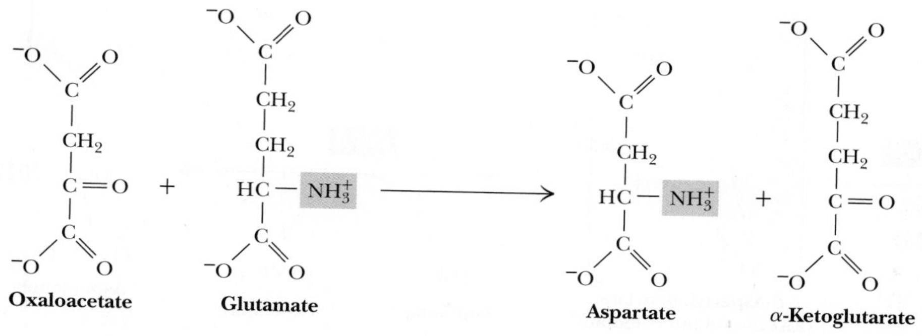

FIGURE 25.25 Aspartate biosynthesis via transamination of oxaloacetate by glutamate.

as well as the C-6-amino-group of the major purine adenine. In addition, the entire aspartate molecule is used in the biosynthesis of pyrimidine nucleotides.

Asparagine is formed by amidation of the β-carboxyl group of aspartate. In bacteria, in analogy with glutamine synthesis, the nitrogen added in this amidation comes directly from NH_4^+. In other organisms, **asparagine synthetase** catalyzes the ATP-dependent transfer of the amido-N of glutamine to aspartate to yield glutamate, AMP, PP_i, and asparagine (Figure 25.26).

Threonine, methionine, and **lysine** biosynthesis in bacteria proceeds from the common precursor aspartate, which is converted first to *aspartyl-β-phosphate* and then to *β-aspartyl-semialdehyde*. The first reaction is an ATP-dependent phosphorylation catalyzed by **aspartokinase** (Figure 25.27, reaction 1). In *E. coli,* there are three isozymes of aspartokinase, designated **aspartokinases I, II,** and **III.** Each of these isozymes is uniquely controlled by one of the three end-product amino acids. Form I is feedback-inhibited by threonine and form III, by lysine. Form II is not feedback-inhibited, but its synthesis is repressed by methionine.

β-Aspartyl-semialdehyde is formed via NADPH-dependent reduction of aspartyl-β-phosphate in a reaction catalyzed by **β-aspartyl-semialdehyde dehydrogenase** (Figure 25.27, reaction 2). From here, the pathway of lysine synthesis diverges. The methyl carbon of pyruvate is condensed with β-aspartyl-semialdehyde, and H_2O is eliminated to yield the cyclic compound *2,3-dihydropicolinate* (Figure 25.27, reaction 10). Thus, lysine synthesized by this pathway must be considered a member of both the oxaloacetate and

---

HUMAN BIOCHEMISTRY

Asparagine and Leukemia

Leukemia is a cancer of the bone marrow that affects the production of normal lymphocytes (white blood cells). **Acute lymphoblastic leukemia (ALL)** and **acute myeloblastic leukemia (AML)** are caused by overproduction of immature lymphocytes. Both normal and malignant lymphocytes are highly dependent on the uptake of asparagine from the blood for growth. Administration of *E. coli* asparaginase, an enzyme that converts Asn to Asp and NH_4^+ (page 880), is one chemotherapeutical approach to treat childhood ALL and some forms of AML, but patients often develop resistance to treatment with this "foreign" protein. Inhibition of asparagine synthetase presents an alternative way to deprive malignant lymphocytes of essential Asn, and asparagine synthetase inhibitors might offer a clinical strategy for treating asparaginase-resistant leukemias. The adenylated sulfoximine shown in the figure is an analog of the aspartyl-AMP intermediate formed in the asparagine synthetase reaction (Figure 25.26). In vitro, this compound inhibits asparagine synthetase at nanomolar concentrations. The polarity of

this substance limits its ability to cross cell membranes and thus its use in chemotherapy. Hopefully, "second-generation" compounds based on this structure's affinity for asparagine synthetase will lead to the development of useful drugs to treat these leukemias.

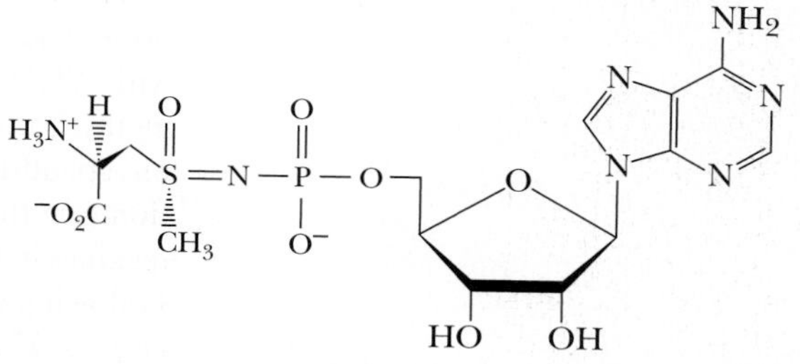

An adenylated sulfoximine

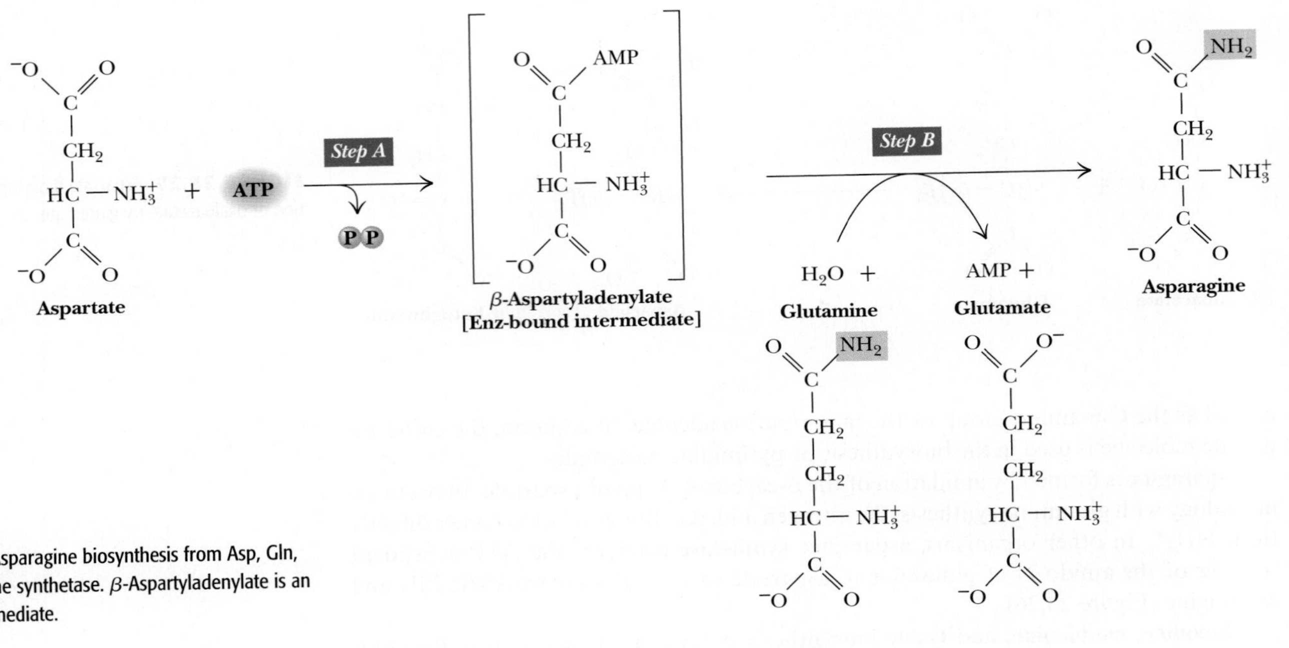

FIGURE 25.26 Asparagine biosynthesis from Asp, Gln, and ATP by asparagine synthetase. β-Aspartyladenylate is an enzyme-bound intermediate.

the pyruvate families of amino acids. Lysine is a feedback inhibitor of this branch-point enzyme. *Dihydropicolinate* is then reduced in an NADPH-dependent reaction to Δ¹-*piperidine*-2,6-*dicarboxylate* (Figure 25.27, reaction 11). A series of reactions, including a hydrolytic opening of the piperidine ring, a succinylation, a glutamate-dependent amination, and the hydrolytic removal of succinate, results in the formation of the symmetric L,L-α,ε-*diaminopimelate* (Figure 25.27, reactions 12 through 14). Epimerization of this intermediate to the *meso* form, followed by decarboxylation, yields the end product *lysine* (Figure 25.27, reactions 15 and 16). Because this pathway proceeds through the symmetric L,L-α,ε-*diaminopimelate,* one-half of the CO_2 evolved in the terminal decarboxylase step is derived from the carboxyl group of pyruvate and one-half from the α-carboxyl of Asp.

The other metabolic branch diverging from β-aspartyl-semialdehyde leads to *threonine and methionine* via *homoserine,* an analog of serine that is formed by the NADPH-dependent reduction of β-aspartyl-semialdehyde (Figure 25.27, reaction 3) catalyzed by **homoserine dehydrogenase.** From homoserine, the biosynthetic pathways leading to methionine and threonine separate. To form **methionine,** the —OH group of homoserine is first succinylated by **homoserine acyltransferase** (Figure 25.27, reaction 6). Methionine is a feedback inhibitor of this enzyme. The succinyl group of *O-succinylhomoserine* is then displaced by cysteine to yield *cystathionine* (Figure 25.27, reaction 7). The sulfur atom in methionine is contributed by a cysteine sulfhydryl. *Cystathionine* is then split to give pyruvate, NH_4^+, and *homocysteine,* a nonprotein amino acid whose side chain is one —CH_2— group longer than that of Cys (Figure 25.27, reaction 8). Methylation of the homocysteine —SH via methyl transfer from the methyl donor, N^5-methyl-THF (see Chapter 26) gives methionine (Figure 25.27, reaction 9).

In passing, it is important to note the role of methionine itself in methylation reactions. The enzyme **S-adenosylmethionine synthase** catalyzes the reaction of methionine with ATP to form *S-adenosylmethionine,* or SAM (Figure 25.28). SAM is a substrate of methyltransferases in a variety of methyl-donor reactions, such as the formation of phosphatidylcholine from phosphatidylethanolamine (see Figure 8.6), DNA methylation (see the Human Biochemistry box on page 349), and methylation of Arg and Lys residues in the regulation of DNA: histone interactions in chromatin (see Figure 29.30). Following decarboxylation, SAM is a source of propylamine for the synthesis of polyamines (Figure 25.28).

The remaining amino acids of the oxaloacetate family are threonine and isoleucine. **Threonine,** like methionine, is synthesized from homoserine. Indeed, homoserine is the

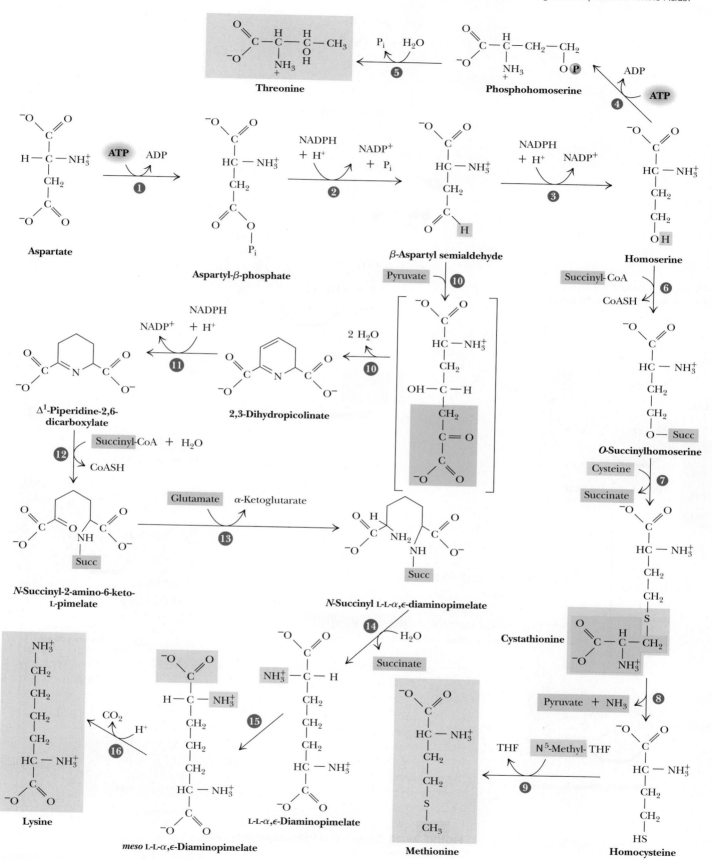

FIGURE 25.27 Biosynthesis of threonine, methionine, and lysine, members of the aspartate family of amino acids.

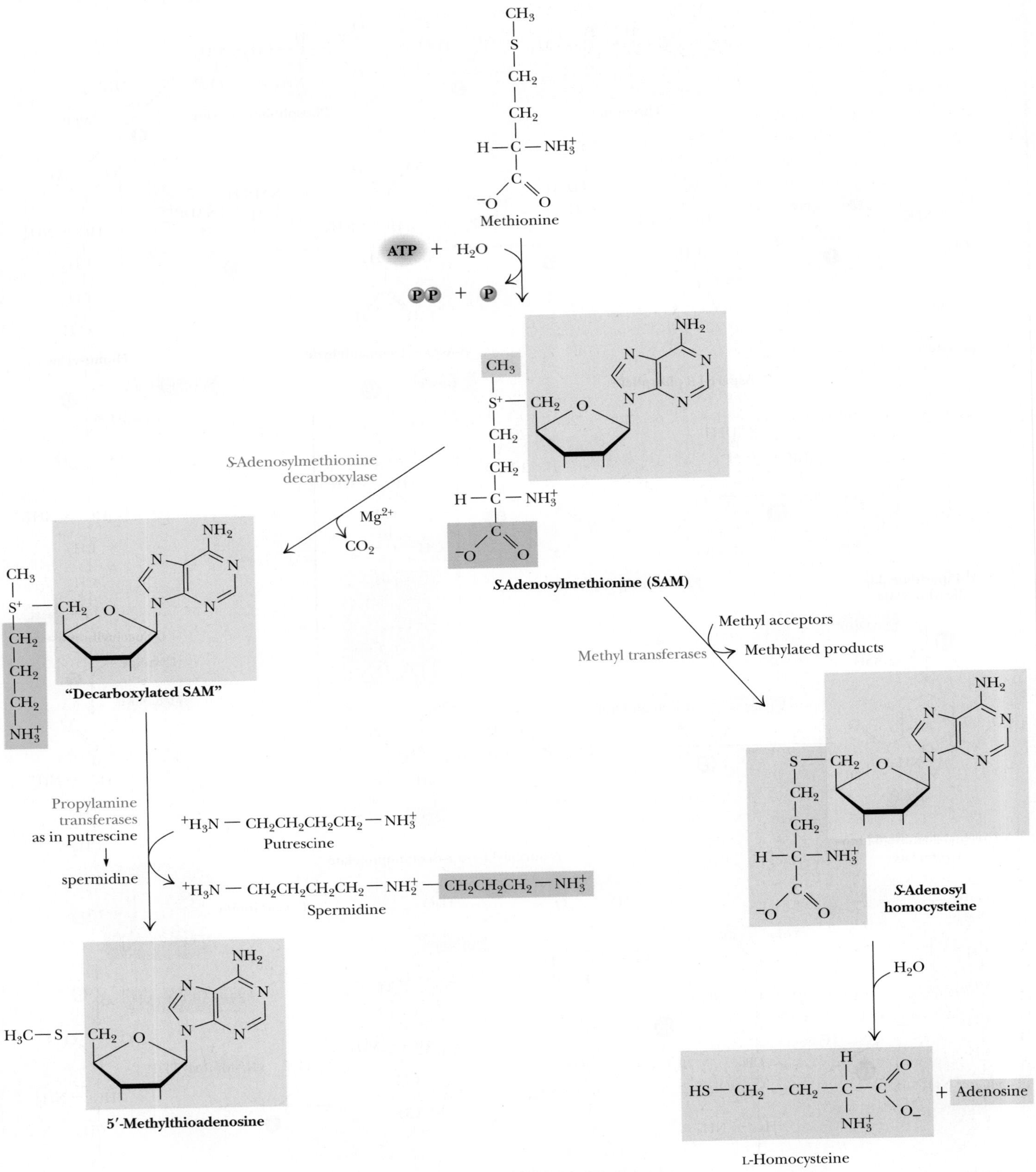

FIGURE 25.28 The synthesis of *S*-adenosylmethionine (SAM).

primary alcohol analog of the secondary alcohol Thr. To move this —OH from C-4 to C-3 requires activation of the hydroxyl through ATP-dependent phosphorylation by **homoserine kinase** (see Figure 25.27, reaction 4). As the first reaction unique to Thr biosynthesis, homoserine kinase is feedback inhibited by threonine. The last step is catalyzed by **threonine synthase,** a PLP-dependent enzyme (see Figure 25.27, reaction 5).

Isoleucine is included in the oxaloacetate family of amino acids because four of its six carbons derive from Asp (via threonine) and only two come from pyruvate. Nevertheless, four of the five enzymes necessary for isoleucine synthesis are common to the pathway for biosynthesis of valine, so discussion of isoleucine synthesis is presented under the biosynthesis of the pyruvate family of amino acids.

The Pyruvate Family of Amino Acids Includes Ala, Val, and Leu

The pyruvate family of amino acids includes alanine (Ala), valine (Val), and leucine (Leu). Transamination of pyruvate, with glutamate as amino donor, gives **alanine.** Because these transamination reactions are readily reversible, alanine degradation occurs via the reverse route, with α-ketoglutarate serving as amino acceptor.

Transamination of pyruvate to alanine is a reaction found in virtually all organisms, but valine, leucine, and isoleucine are essential amino acids, and as such, they are not synthesized in mammals. The pathways of **valine** and **isoleucine** synthesis can be considered together because a set of four enzymes is common to the last four steps of both pathways (Figure 25.29). Both pathways begin with an α-keto acid. Isoleucine can be considered a structural analog of valine that has one extra —CH_2— unit, and its α-keto acid precursor, namely, *α-ketobutyrate,* is one carbon longer than the valine precursor, pyruvate. Interestingly, α-ketobutyrate is formed from threonine by **threonine deaminase** (Figure 25.29, reaction 1). This PLP-dependent enzyme (also known as *threonine dehydratase* or *serine dehydratase*) is feedback-inhibited by isoleucine, the end product. Note that part of the carbon skeleton for Ile comes from Asp by way of Thr. From here on, the Val and Ile pathways employ the same set of enzymes. The first reaction involves the generation of hydroxyethyl-thiamine pyrophosphate from pyruvate in a reaction analogous to those catalyzed by transketolase and the pyruvate dehydrogenase complex. The two-carbon hydroxyethyl group is transferred from TPP to the respective keto acid acceptor by **acetohydroxy acid synthase** (acetolactate synthase) to give *α-acetolactate* or *α-aceto-α-hydroxybutyrate* (Figure 25.29, reaction 2). NAD(P)H-dependent reduction of these α-keto hydroxy acids yields the dihydroxy acids *α,β-dihydroxyisovalerate* and *α,β-dihydroxy-β-methylvalerate* (Figure 25.29, reaction 3). Dehydration of each of these dihydroxy acids by **dihydroxy acid dehydratase** gives the appropriate α-keto acid carbon skeletons *α-ketoisovalerate* and *α-keto-β-methylvalerate* (Figure 25.29, reaction 4). Transamination by the **branched-chain amino acid aminotransferase** yields Val or Ile, respectively (Figure 25.29, reaction 5).

Leucine synthesis depends on these reactions as well, because α-ketoisovalerate is a precursor common to both Val and Leu (Figure 25.30). Although Val and Leu differ by only a single —CH_2— in their respective side chains, the carboxyl group of α-ketoisovalerate first picks up *two* carbons from acetyl-CoA to give *α-isopropylmalate* in a reaction catalyzed by **isopropylmalate synthase;** the enzyme is sensitive to feedback inhibition by Leu (Figure 25.30, reaction 1). **Isopropylmalate dehydratase** (Figure 29.30, reaction 2) converts the α-isomer to the β-form, which undergoes an NAD$^+$-dependent oxidative decarboxylation by **isopropylmalate dehydrogenase** (Figure 29.30, reaction 3), so the carboxyl group of α-ketoisovalerate is lost as CO_2. Amination of α-ketoisocaproate by **leucine aminotransferase** (Figure 29.30, reaction 4) gives Leu.

The 3-Phosphoglycerate Family of Amino Acids Includes Ser, Gly, and Cys

Serine, glycine, and cysteine are derived from the glycolytic intermediate 3-phosphoglycerate. The diversion of 3-PG from glycolysis is achieved via **3-phosphoglycerate dehydrogenase** (Figure 25.31, reaction 1). This NAD$^+$-dependent oxidation of 3-PG

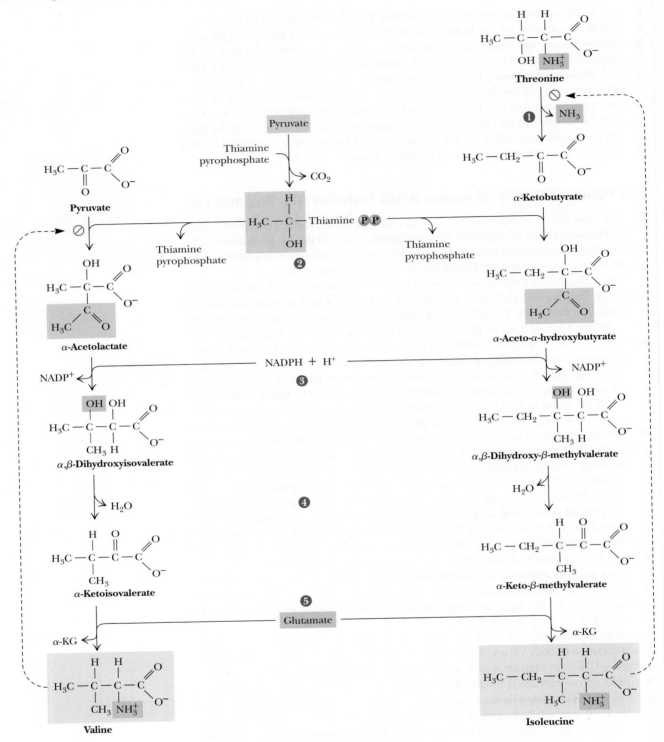

FIGURE 25.29 Biosynthesis of valine and isoleucine.

yields *3-phosphohydroxypyruvate*—which, as an α-keto acid, is a substrate for transamination by glutamate to give *3-phosphoserine* (Figure 25.31, reaction 2). **Serine phosphatase** then generates **serine** (Figure 25.31, reaction 3). Serine inhibits the first enzyme, 3-PG dehydrogenase, and thereby feedback-regulates its own synthesis.

Glycine is made from serine via two related enzymatic processes. In the first, **serine hydroxymethyltransferase,** a PLP-dependent enzyme, catalyzes the transfer of the serine β-carbon to tetrahydrofolate (THF), the principal agent of one-carbon metabolism

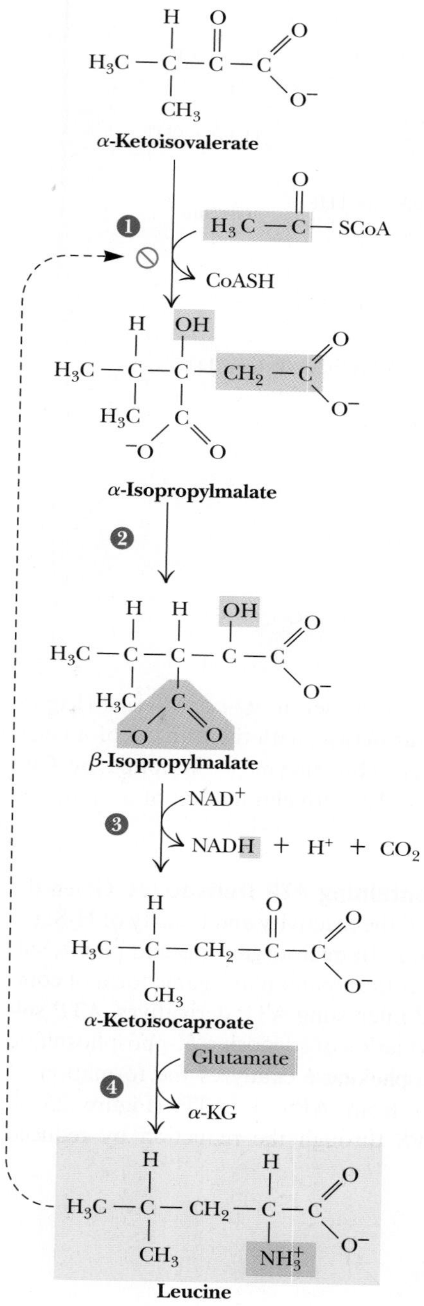

FIGURE 25.30 Biosynthesis of leucine.

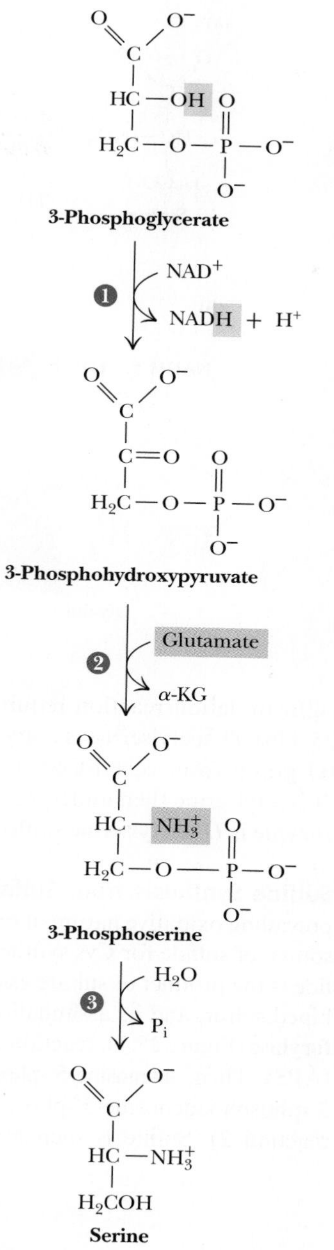

FIGURE 25.31 Biosynthesis of serine from 3-phosphoglycerate.

(Figure 25.32a). Glycine and N^5,N^{10}-methylene-THF are the products. In addition, glycine can be synthesized by a reversal of the **glycine oxidase** reaction (Figure 25.32b). Here, glycine is formed when N^5,N^{10}-methylene-THF condenses with NH_4^+ and CO_2. Via this route, the β-carbon of serine becomes part of glycine. The conversion of serine to glycine is a prominent means of generating one-carbon derivatives of THF, which are so important for the biosynthesis of purines and the C-5 methyl group of thymine (a pyrimidine, see Chapter 26), as well as the amino acid methionine. Glycine itself contributes to both purine and heme synthesis.

Cysteine synthesis is accomplished by sulfhydryl transfer to serine (Figure 25.33). In some bacteria, H_2S condenses directly with serine via a PLP-dependent enzyme-catalyzed reaction (Figure 25.33a), but in most microorganisms and green plants, the

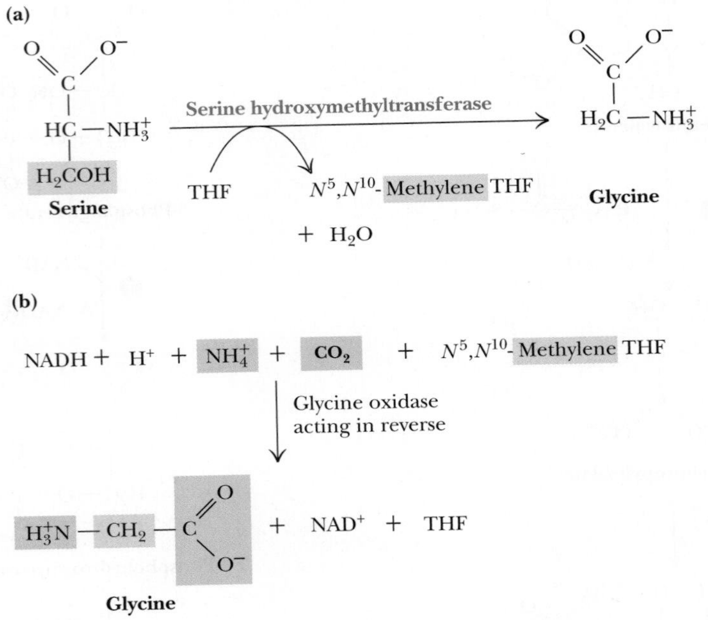

(a)

Serine hydroxymethyltransferase

Serine → Glycine

THF N^5,N^{10}-Methylene THF

$+$ H_2O

(b)

$NADH + H^+ + NH_4^+ + CO_2 + N^5,N^{10}$-Methylene THF

Glycine oxidase
acting in reverse

Glycine $+$ NAD^+ $+$ THF

FIGURE 25.32 Biosynthesis of glycine from serine
(a) via serine hydroxymethyltransferase and **(b)** via glycine
oxidase.

sulfhydrylation reaction requires an activated form of serine, *O-acetylserine* (Figure 25.33b). *O*-acetylserine is made by **serine acetyltransferase,** with the transfer of an acetyl group from acetyl-CoA to the —OH of Ser. This enzyme is inhibited by Cys. *O*-Acetylserine then undergoes sulfhydrylation by H_2S with elimination of acetate; the enzyme is ***O*-acetylserine sulfhydrylase.**

Sulfide Synthesis from Sulfate Involves S-Containing ATP Derivatives Given the prevailing oxidative nature of our environment and the reactivity and toxicity of H_2S, the source of sulfide for Cys synthesis merits discussion. In microorganisms and plants, sulfide is the product of sulfate assimilation. Sulfate is the common inorganic form of combined sulfur, and its assimilation involves several interesting ATP derivatives. **ATP sulfurylase** (Figure 25.34, reaction 1) catalyzes the formation of adenosine-5′-phosphosulfate (APS). Then, **adenosine-5′-phosphosulfate-3′-phosphokinase** catalyzes the formation of 3′-phosphoadenosine-5′-phosphosulfate (PAPS) from APS + ATP (Figure 25.34, reaction 2). Sulfite is then liberated from PAPS through the reduction by reduced

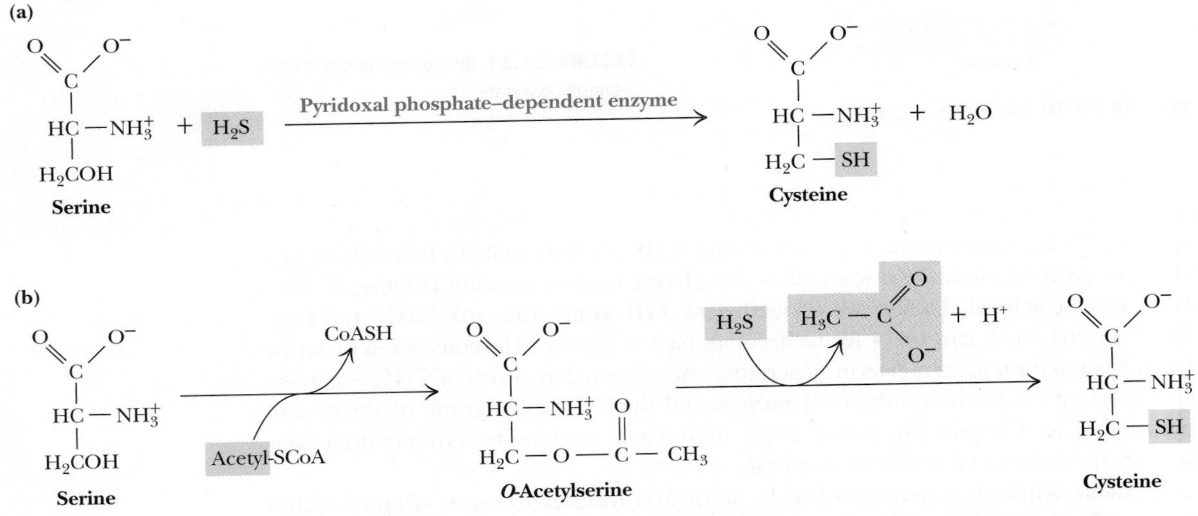

(a)

Serine $+$ H_2S → Cysteine $+$ H_2O

Pyridoxal phosphate–dependent enzyme

(b)

Serine → O-Acetylserine → Cysteine

CoASH

Acetyl-SCoA

H_2S H_3C—C $+$ H^+

FIGURE 25.33 Cysteine biosynthesis. **(a)** Direct sulfhydrylation of serine by H_2S. **(b)** H_2S-dependent sulfhydrylation of *O*-acetylserine.

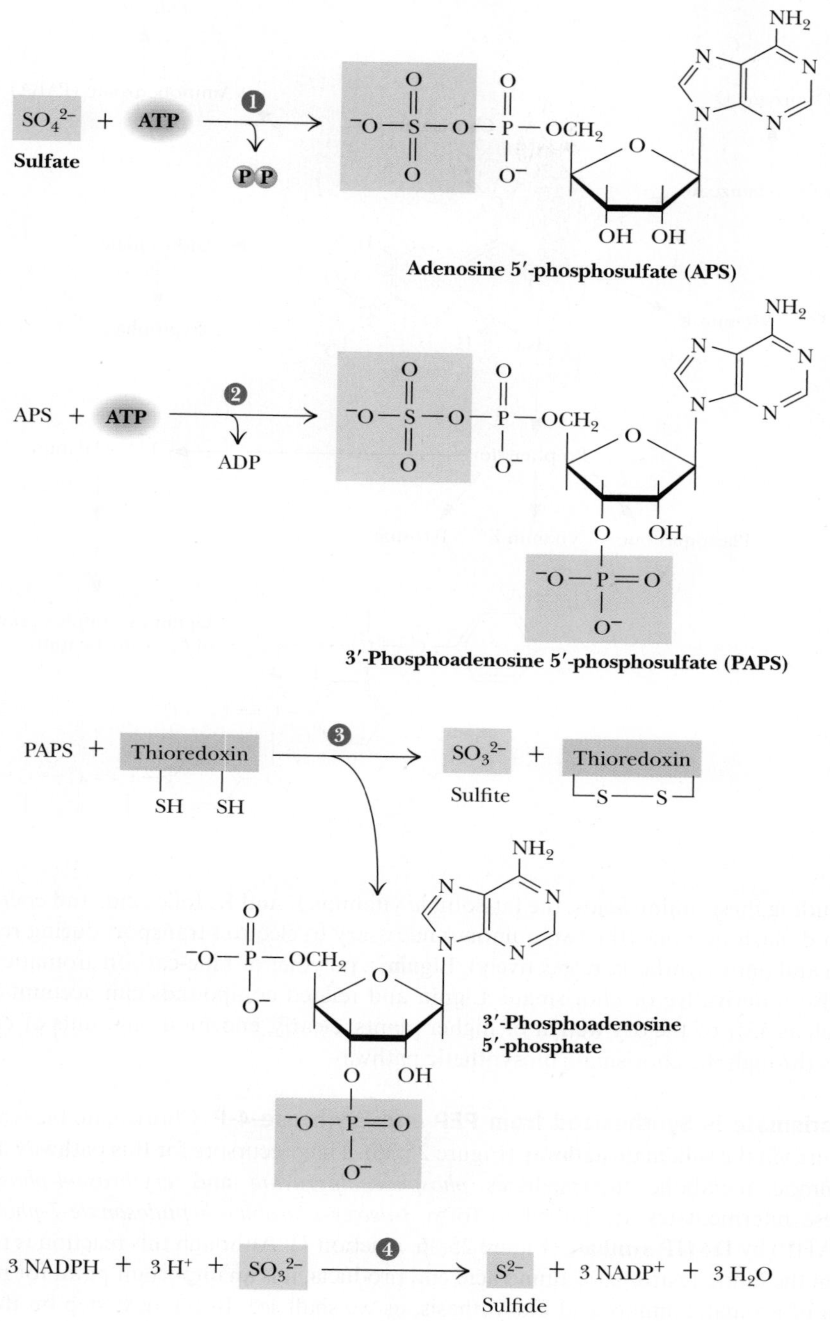

FIGURE 25.34 Sulfate assimilation and the generation of sulfide for synthesis of organic S compounds.

thioredoxin, leaving 3′-phosphoadenosine-5′-phosphate as a product (Figure 25.34, reaction 3). Sulfite (SO_3^-) is then reduced to sulfide (S_2^-) in a multielectron transfer reaction catalyzed by **sulfite oxidase** (Figure 25.34, reaction 4); NADPH is the electron donor. Sulfite reductase, like nitrite reductase, possesses siroheme as a prosthetic group (see Figure 25.2). 3′-Phosphoadenosine-5′-phosphosulfate is not only an intermediate in sulfate assimilation; it also serves as the substrate for synthesis of sulfate esters, such as the sulfated polysaccharides found in the glycocalyx of animal cells.

The Aromatic Amino Acids Are Synthesized from Chorismate

The aromatic amino acids, phenylalanine, tyrosine, and tryptophan, are derived from a shared pathway that has **chorismic acid** (Figure 25.35) as a key intermediate. Indeed, chorismate is common to the synthesis of cellular compounds having benzene rings,

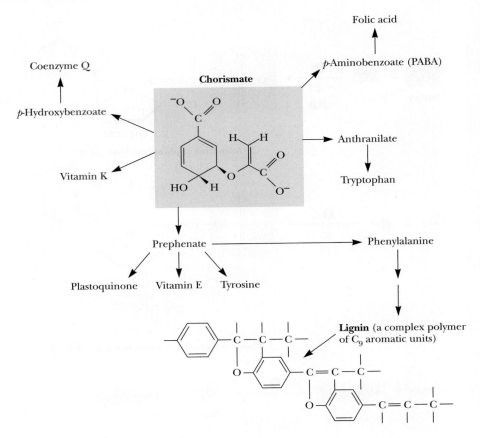

FIGURE 25.35 Some of the aromatic compounds derived from chorismate.

including these amino acids, the fat-soluble vitamins E and K, folic acid, and coenzyme Q and plastoquinone (the two quinones necessary to electron transport during respiration and photosynthesis, respectively). **Lignin,** a polymer of nine-carbon aromatic units, is also a derivative of chorismate. Lignin and related compounds can account for as much as 35% of the dry weight of higher plants; clearly, enormous amounts of carbon pass through the chorismate biosynthetic pathway.

Chorismate Is Synthesized from PEP and Erythrose-4-P Chorismate biosynthesis occurs via the **shikimate pathway** (Figure 25.36). The precursors for this pathway are the common metabolic intermediates *phosphoenolpyruvate* and *erythrose-4-phosphate*. These intermediates are linked to form *3-deoxy-D-arabino-heptulosonate-7-phosphate* (DAHP) by **DAHP synthase** (Figure 25.36, reaction 1). Although this reaction is remote from the ultimate aromatic amino acid end products, it is an important point for regulation of aromatic amino acid biosynthesis, as we shall see. In the next step on the way to chorismate, DAHP is cyclized to form a six-membered saturated ring compound, *5-dehydroquinate* (Figure 25.36, reaction 2), in a reaction catalyzed by **dehydroquinate synthase** (NAD$^+$ is a coenzyme in this reaction but is not modified by it). A sequence of reactions ensues that introduces unsaturations into the ring through dehydration (Figure 25.36, reaction 3, **5-dehydroquinate dehydratase**) and reduction (reaction 4, **shikimate dehydrogenase**), yielding *shikimate*. Phosphorylation of shikimate by **shikimate kinase** (reaction 5), then addition of PEP by **3-enolpyruvylshikimate-5-phosphate synthase** (reaction 6), followed by **chorismate synthase** (reaction 7), gives *chorismate*. Thus, two equivalents of PEP are needed to form chorismate from erythrose-4-P.

Phenylalanine and Tyrosine At chorismate, the pathway separates into three branches, each leading specifically to one of the aromatic amino acids. The branches leading to phenylalanine and tyrosine both pass through *prephenate* (Figure 25.37). In some organisms, such as *E. coli,* the branches are truly distinct because prephenate does not occur as a free intermediate but rather remains bound to the bifunctional enzyme

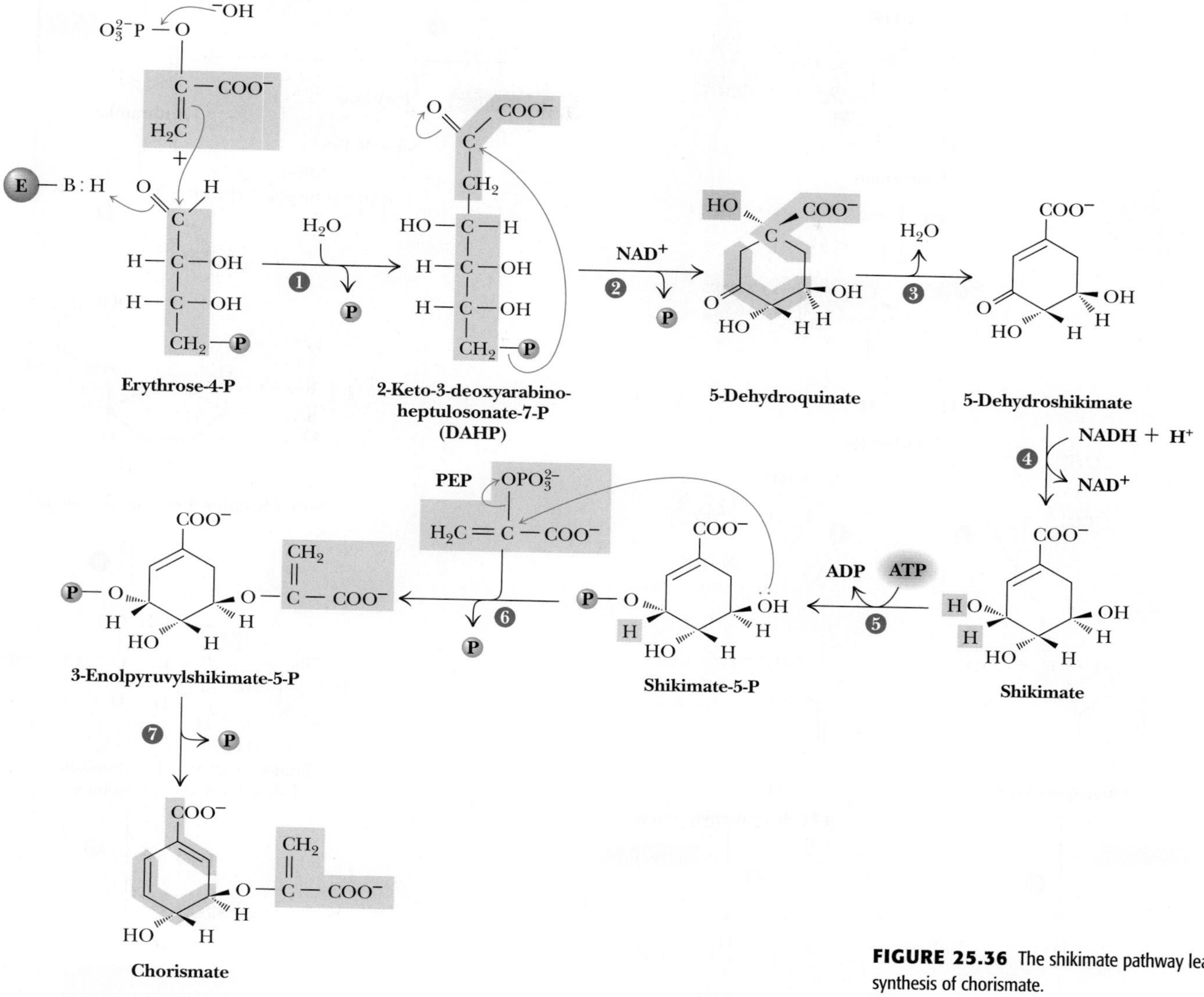

FIGURE 25.36 The shikimate pathway leading to the synthesis of chorismate.

that catalyzes the first two reactions after chorismate. In any case, **chorismate mutase** is the first reaction leading to Phe or Tyr (Figure 25.37, reaction 1). In the Phe branch, the —OH group *para* to the prephenate carboxyl is removed by **prephenate dehydratase** (Figure 25.37, reaction 2). In the Tyr branch, this —OH is retained; instead, an oxidative decarboxylation of prephenate catalyzed by **prephenate dehydrogenase** (Figure 25.37, reaction 4) yields 4-hydroxyphenylpyruvate. Glutamate-dependent aminotransferases (**phenylalanine aminotransferase** [Figure 25.27, reaction 3] and **tyrosine aminotransferase** [reaction 5]) introduce the amino groups into the two α-keto acids, *phenylpyruvate* and *4-hydroxyphenylpyruvate,* to give Phe and Tyr, respectively. Some mammals can synthesize Tyr from Phe obtained in the diet via **phenylalanine-4-monooxygenase** (also known as **phenylalanine hydroxylase**), using O_2 and *tetrahydrobiopterin,* an analog of tetrahydrofolic acid, as co-substrates (Figure 25.38).

Tryptophan The pathway of tryptophan synthesis is perhaps the most thoroughly studied of any biosynthetic sequence, particularly in terms of its genetic organization and expression. Synthesis of Trp from chorismate requires six steps (see Figure 25.37). In most microorganisms, the first enzyme, **anthranilate synthase** (see Figure 25.37, reaction 6), is an $\alpha_2\beta_2$-type protein, with the β-subunit acting in a glutamine–amidotransferase role to provide the —NH_2 group of anthranilate. Or, given high levels of NH_4^+, the

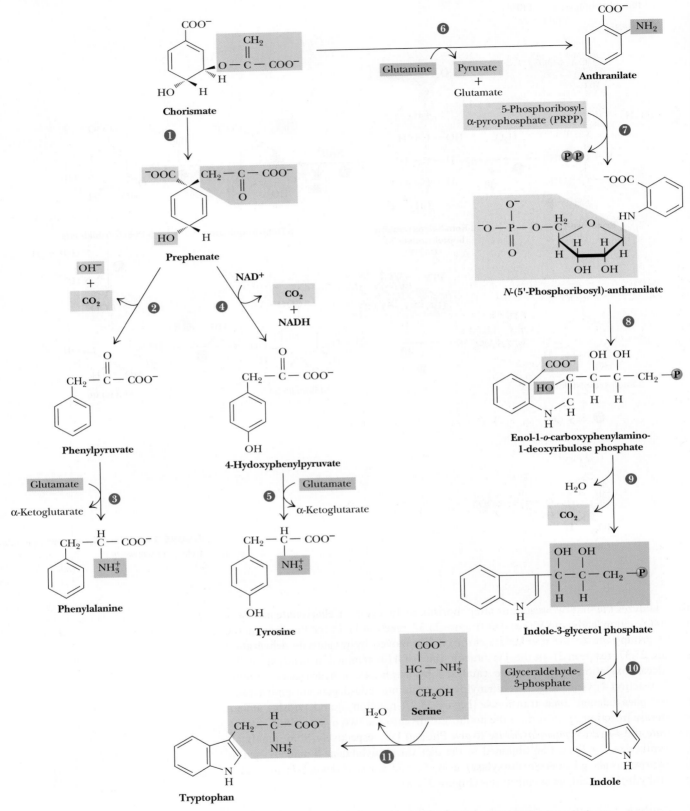

FIGURE 25.37 The biosynthesis of phenylalanine, tyrosine, and tryptophan from chorismate.

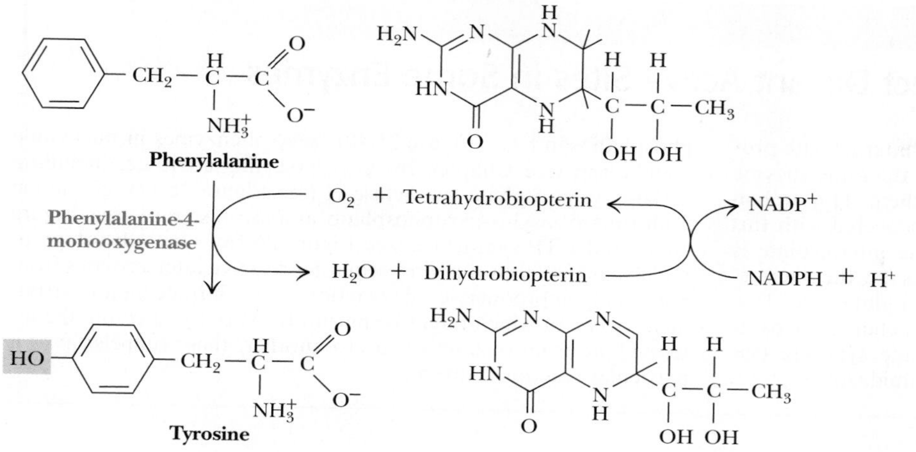

FIGURE 25.38 The formation of tyrosine from phenylalanine.

α-subunit can carry out the formation of anthranilate directly by a process in which the activity of the β-subunit is unnecessary. Furthermore, in certain enteric bacteria, such as *E. coli* and *Salmonella typhimurium,* the second reaction of the pathway, the **phosphoribosyl-anthranilate transferase** reaction (see Figure 25.37, reaction 7), is an activity catalyzed by the α-subunit of anthranilate synthase. *PRPP (5-phosphoribosyl-1-pyrophosphate),* the substrate of this reaction, is also a precursor for purine biosynthesis (see Chapter 26). *Phosphoribosyl-anthranilate* then undergoes a rearrangement wherein the ribose moiety is isomerized to the ribulosyl form in *enol-1-(o-carboxyphenylamino)-1-deoxyribulose-5-phosphate* by *N*-(5'-**phosphoribosyl)-anthranilate isomerase** (see Figure 25.37, reaction 8). Decarboxylation and ring closure ensue to yield the indole nucleus as *indole-3-glycerol phosphate* (**indole-3-glycerol phosphate synthase,** reaction 9). The final

> **A DEEPER LOOK**

Amino Acid Biosynthesis Inhibitors as Herbicides

Unlike animals, plants can synthesize all 20 of the common amino acids. Inhibitors acting specifically on the plant enzymes that are capable of carrying out the biosynthesis of the "essential" amino acids (that is, enzymes that animals lack) have been developed. These substances appear to be ideal for use as herbicides because they should show no effect on animals. **Glyphosate,** sold commercially as *RoundUp,* is a PEP analog that acts as an uncompetitive inhibitor of 3-enolpyruvylshikimate-5-P synthase (Figure 25.36). **Sulfmeturon methyl,** a sulfonylurea herbicide that inhibits *aceto-*

hydroxy acid synthase, an enzyme common to Val, Leu, and Ile biosynthesis (Figure 25.29), is the active ingredient in *Oust.* **Amino-triazole,** sold as *Amitrole,* blocks His biosynthesis by inhibiting *imidazole glycerol-P dehydratase* (Figure 25.40). **PPT (phosphi-nothricin)** is a potent inhibitor of *glutamine synthetase.* Although Gln is a nonessential amino acid and glutamine synthetase is a ubiquitous enzyme, PPT is relatively safe for animals because it does not cross the blood–brain barrier and is rapidly cleared by the kidneys.

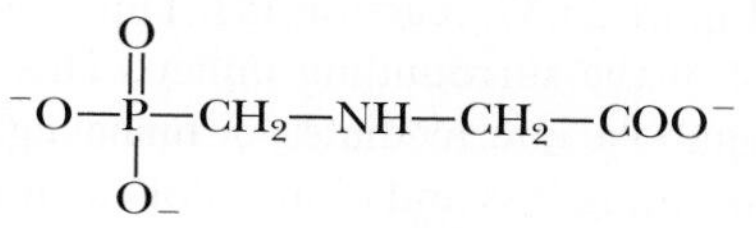

Glyphosate

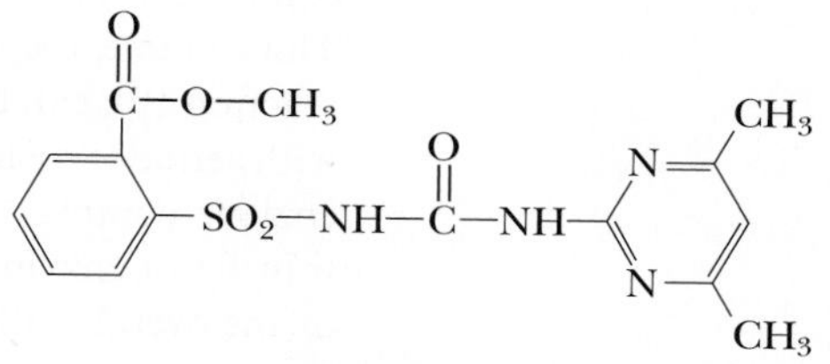

Sulfmeturon methyl

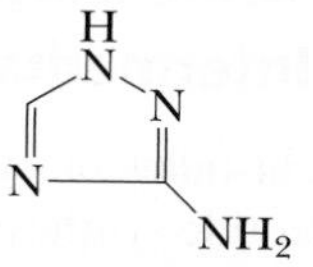

Aminotriazole

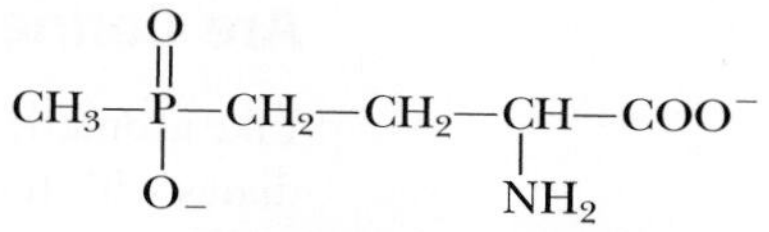

DL-Phosphinothricin (PPT)

> **A DEEPER LOOK**

Intramolecular Tunnels Connect Distant Active Sites in Some Enzymes

Molecular tunneling is the transfer of a reaction intermediate produced at one active site to another active site in the same enzyme through an intramolecular tunnel that connects them. Tryptophan synthase (Figure 25.39) was the first enzyme discovered with this structural feature. For tryptophan synthase, the intermediate is indole, but for most of these enzymes studied thus far, the intermediate transferred is ammonia (NH_3) derived from glutamine. Two such enzymes have been presented earlier in this chapter: asparagine synthetase (see Figure 25.26) and glutamate synthase (see Figure 25.12). Another will soon be considered: imidazole glycerol

phosphate synthase (Figure 25.40). Several enzymes in nucleotide metabolism (see Chapter 26) also have this attribute, including carbamoyl phosphate synthetase II (see Figure 26.14), glutamine 5-phosphoribosyl-α-pyrophosphate amidotransferase (see Figure 26.3), and CTP synthetase (see Figure 26.16). One advantage of molecular tunnels is that they sequester reactive intermediates from potentially unproductive side reactions in the intracellular environment (some of which might be harmful). Also, by directing the intermediate from one active site to another, these tunnels favor a particular reaction sequence.

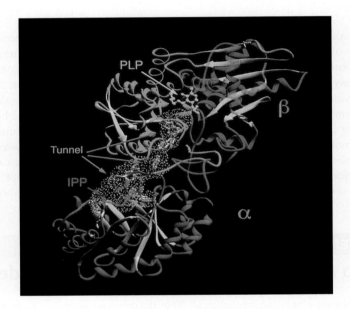

FIGURE 25.39 Tryptophan synthase ribbon diagram. The α-subunit is in blue, and the β-subunit is in orange (N-terminal domain) and red (C-terminal domain). The tunnel connecting them is outlined by the yellow dot surface and is shown with several indole molecules (green) packed in head-to-tail fashion. The labels "IPP" and "PLP" point to the active sites of the α- and β-subunits, respectively, in which a competitive inhibitor (indole propanol phosphate, IPP) and the coenzyme PLP are bound. (Adapted from Hyde, C. C., et al., 1988. Three-dimensional structure of the tryptophan synthase multienzyme complex from *Salmonella typhimurium. Journal of Biological Chemistry* **263**:17857–17871.)

two reactions (10 and 11 in Figure 25.37) are both catalyzed by **tryptophan synthase**, an $\alpha_2\beta_2$-type protein. The α-subunit cleaves indoleglycerol-3-phosphate to form indole and 3-glycerol phosphate. The indole is then passed to the β-subunit, which adds serine in a PLP-dependent reaction.

X-ray crystallographic analysis of *S. typhimurium* tryptophan synthase shows that the active sites of the α- and β-subunits are separated from each other by 2.5 nm but are connected by a hydrophobic tunnel wide enough to accommodate indole (Figure 25.39). Thus, indole, the product of the reaction catalyzed by the α-subunit (see Figure 25.37, reaction 10), can be transferred directly to the β-subunit, which catalyzes condensation with serine to yield Trp (see Figure 25.37, reaction 11). Thus, indole is not lost from the enzyme complex and diluted in the surrounding milieu. This phenomenon of direct transfer of enzyme-bound metabolic intermediates, or **tunneling,** increases the efficiency of the overall pathway by preventing loss and dilution of the intermediate.

Histidine Biosynthesis and Purine Biosynthesis Are Connected by Common Intermediates

Like aromatic amino acid biosynthesis, **histidine** biosynthesis shares metabolic intermediates with the pathway of purine nucleotide synthesis. The pathway involves ten separate steps, the first being an unusual reaction that links ATP and PRPP (Figure 25.40). Five carbon atoms from PRPP and one from ATP end up in histidine. Step 5 involves

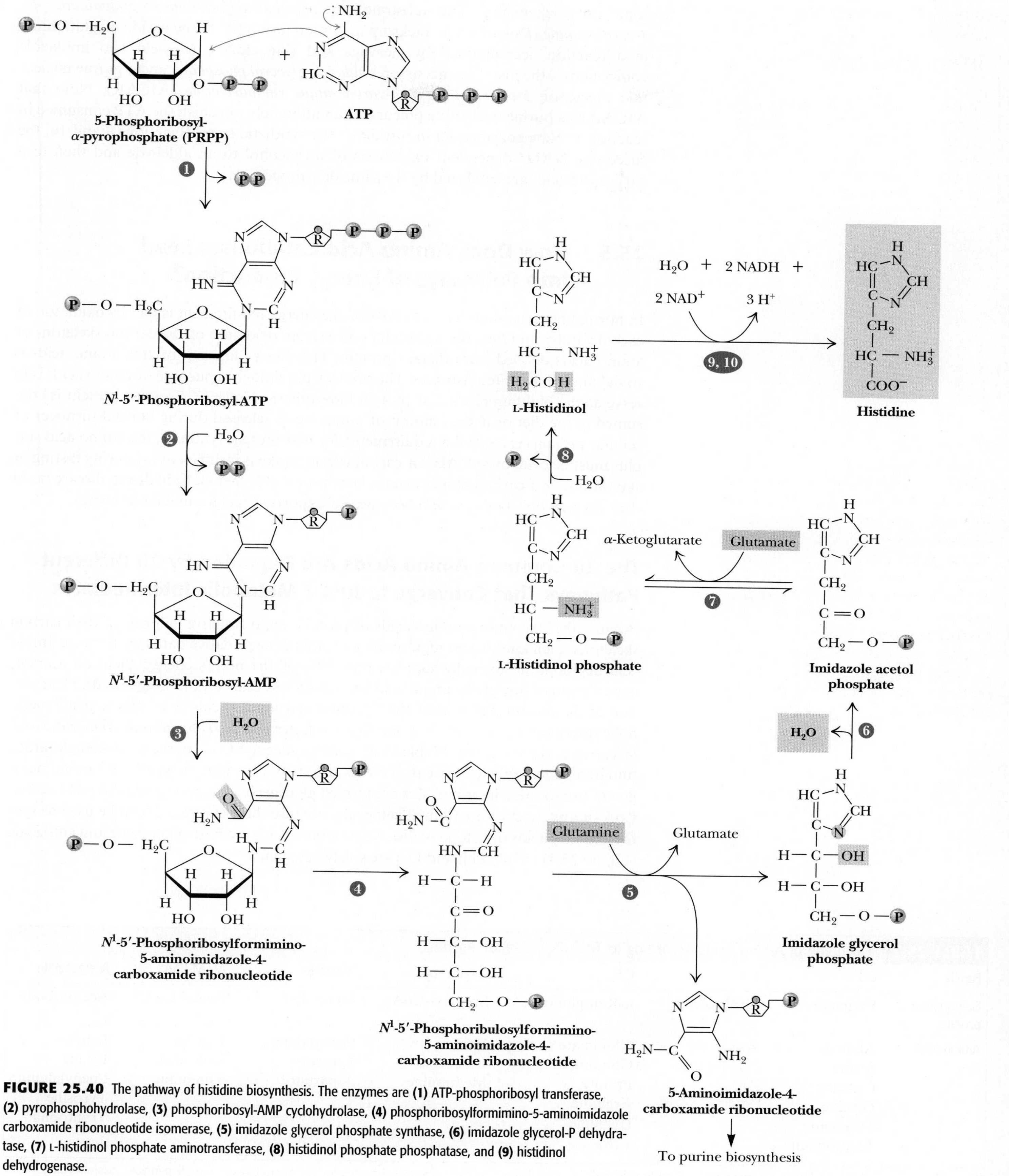

FIGURE 25.40 The pathway of histidine biosynthesis. The enzymes are **(1)** ATP-phosphoribosyl transferase, **(2)** pyrophosphohydrolase, **(3)** phosphoribosyl-AMP cyclohydrolase, **(4)** phosphoribosylformimino-5-aminoimidazole carboxamide ribonucleotide isomerase, **(5)** imidazole glycerol phosphate synthase, **(6)** imidazole glycerol-P dehydratase, **(7)** L-histidinol phosphate aminotransferase, **(8)** histidinol phosphate phosphatase, and **(9)** histidinol dehydrogenase.

some novel chemistry: The substrate, *phosphoribulosylformimino-5-aminoimidazole-4-carboxamide ribonucleotide,* picks up an amino group (from the amide of glutamine) in a reaction accompanied by cleavage and ring closure to yield two imidazole compounds—the histidine precursor, *imidazole glycerol phosphate,* and a purine nucleotide precursor, *5-aminoimidazole-4-carboxamide ribonucleotide* (AICAR). Note that AICAR as a purine nucleotide precursor can ultimately replenish the ATP consumed in reaction 1. Nine enzymes act in histidine's ten synthetic steps. Reactions 9 and 10, the successive NAD$^+$-dependent oxidations of an alcohol to an aldehyde and then to a carboxylic acid, are catalyzed by the same dehydrogenase.

25.5 How Does Amino Acid Catabolism Lead into Pathways of Energy Production?

In normal human adults, close to 90% of the energy requirement is met by oxidation of carbohydrates and fats; the remainder comes from oxidation of the carbon skeletons of amino acids derived from dietary protein. The dietary amount of free amino acids is trivial under most circumstances. The primary physiological purpose of amino acids is to serve as the building blocks for protein biosynthesis. However, if excess protein is consumed in the diet or if the amount of amino acids released during normal turnover of cellular proteins exceeds the requirements for new protein synthesis, the amino acid surplus must be catabolized. Also, if carbohydrate intake is insufficient (as during fasting or starvation) or if carbohydrates cannot be appropriately metabolized due to disease (as in *diabetes mellitus*), body protein becomes an important fuel for metabolic energy.

The 20 Common Amino Acids Are Degraded by 20 Different Pathways That Converge to Just 7 Metabolic Intermediates

Because the 20 common amino acids of proteins are distinctive in terms of their carbon skeletons, each amino acid requires its own unique degradative pathway. Because amino acid degradation normally supplies only 10% of the body's energy, then, on average, degradation of any given amino acid will satisfy less than 1% of energy needs. Degradation of the carbon skeletons of the 20 common α-amino acids converges to just 7 metabolic intermediates: *acetyl-CoA, succinyl-CoA, pyruvate, α-ketoglutarate, fumarate, oxaloacetate,* and *acetoacetate* (Table 25.2). Because succinyl-CoA, pyruvate, α-ketoglutarate, fumarate, and oxaloacetate can serve as precursors for glucose synthesis, amino acids giving rise to these intermediates are termed **glucogenic.** Those degraded to yield acetyl-CoA or acetoacetate are termed **ketogenic,** because these substances can be used to synthesize fatty acids or ketone bodies. Some amino acids are both glucogenic and ketogenic (Figure 25.41). Only Leu and Lys are solely ketogenic.

TABLE 25.2	Grouping of Amino Acids According to Their Degradation Product*						
Family	C-3	C-4	C-5	Succinyl-CoA	Fumarate	Acetyl-CoA	Acetoacetate
Degradation product	Pyruvate	Oxaloacetate	α-Ketoglutarate	Succinyl-CoA	Fumarate	Acetyl-CoA	Acetoacetate
Amino acids	Alanine Serine Cysteine Glycine Threonine Tryptophan	Aspartate Asparagine	Glutamate Glutamine Proline Arginine Histidine	Valine Isoleucine Methionine	Phenylalanine Tyrosine Aspartate‡	Leucine Isoleucine Threonine	Leucine Lysine Phenylalanine Tyrosine Tryptophan

*Five of the amino acid degradation product families lead to metabolic intermediates that can serve as substrates for gluconeogenesis, namely pyruvate, oxaloacetate, α-ketoglutarate, succinyl-CoA, and fumarate. The other two families lead to acetyl-CoA or acetoacetate, and these intermediates cannot be used as substrates for gluconeogenesis. Thus, amino acids in these degradation families contribute to ketone body formation.
‡Aspartate is readily deaminated to form oxaloacetate, but aspartate also leads to fumarate by virtue of the urea cycle (Figure 25.3, reaction 3) and reactions in the purine biosynthetic pathway (see Figure 26.3, reaction 9, and Figure 26.5).

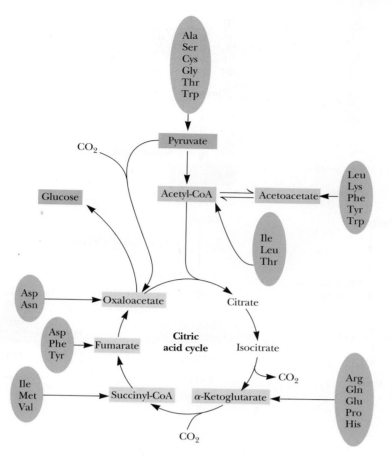

FIGURE 25.41 Metabolic degradation of the common amino acids. Glucogenic amino acids are shown in pink, ketogenic in blue.

The C-3 Family of Amino Acids: Alanine, Serine, and Cysteine The carbon skeletons of alanine, serine, and cysteine all converge to *pyruvate* (Figure 25.42). Transamination of alanine yields pyruvate:

$$\text{Alanine} + \alpha\text{-ketoglutarate} \rightleftharpoons \text{pyruvate} + \text{glutamate} \qquad (25.10)$$

Deamination of serine by **serine dehydratase** also yields pyruvate. Cysteine is converted to pyruvate via a number of paths.

The carbon skeletons of three other amino acids also become pyruvate. *Glycine* is convertible to serine and thus to pyruvate. The three carbon atoms of *tryptophan* that are not part of its indole ring appear as alanine (and, hence, pyruvate) upon Trp degradation. *Threonine* by one of its degradation routes is cleaved to glycine and acetaldehyde. The glycine is then converted to pyruvate via serine; the acetaldehyde is oxidized to acetyl-CoA (Figure 25.42).

The C-4 Family of Amino Acids: Aspartate and Asparagine Transamination of aspartate gives *oxaloacetate*:

$$\text{Aspartate} + \alpha\text{-ketoglutarate} \rightleftharpoons \text{oxaloacetate} + \text{glutamate} \qquad (25.11)$$

▶ **A DEEPER LOOK**

Histidine—A Clue to Understanding Early Evolution?

Histidine residues in the active sites of enzymes often act directly in the enzyme's catalytic mechanism. Catalytic participation by the imidazole group of His and the presence of imidazole as part of the purine ring system support a current speculation that life before the full evolution of protein molecules must have been RNA based. This notion correlates with the discovery that RNA molecules can have catalytic activity, an idea captured in the term ribozyme (see Chapter 13).

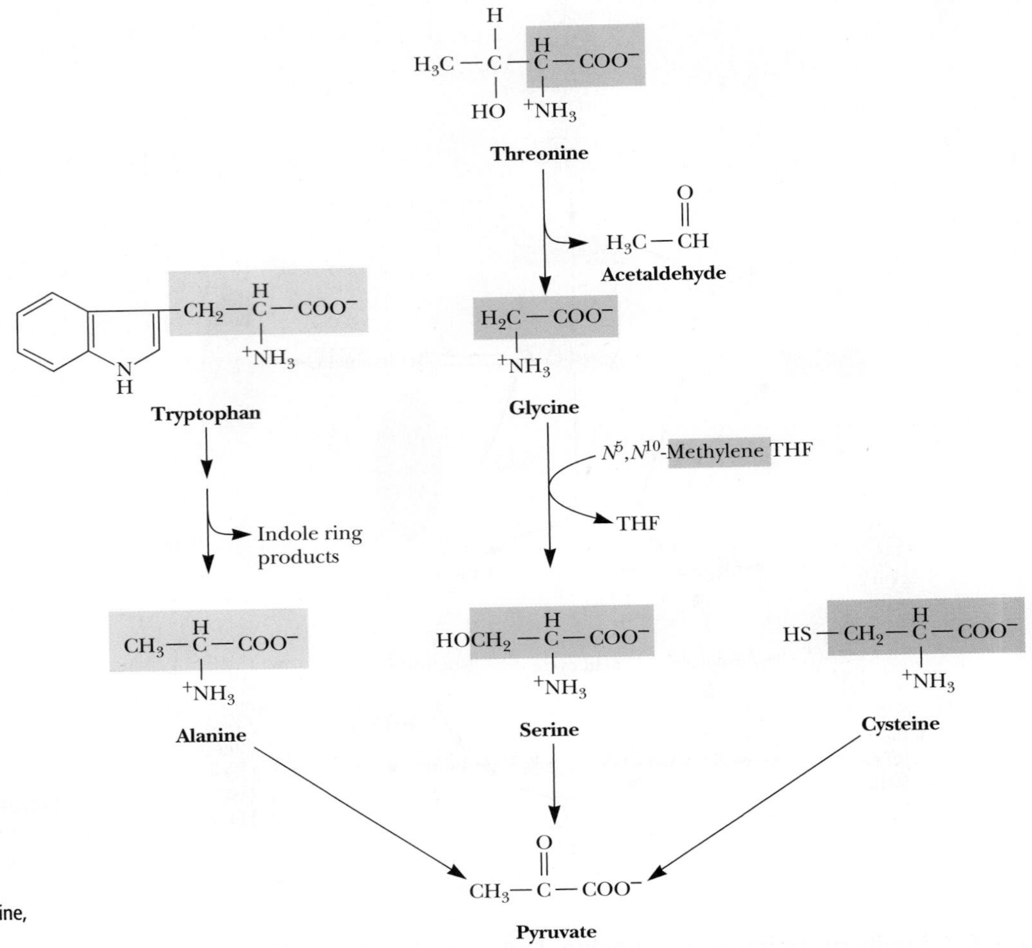

FIGURE 25.42 Formation of pyruvate from alanine, serine, cysteine, glycine, tryptophan, or threonine.

Hydrolysis of asparagine by **asparaginase** yields aspartate and NH_4^+. Alternatively, participation of aspartate in the urea cycle and purine biosynthesis converts its carbon skeleton to a different citric acid cycle intermediate, namely, *fumarate* (Figure 25.23).

The C-5 Family of Amino Acids Is Converted to α-Ketoglutarate Via Glutamate

The five-carbon citric acid cycle intermediate α-ketoglutarate is always a product of transamination reactions involving *glutamate*. Thus, glutamate *and* any amino acid convertible to glutamate are classified within the C-5 family. These amino acids include *glutamine, proline, arginine,* and *histidine* (Figure 25.43).

Degradation of Valine, Isoleucine, and Methionine Leads to Succinyl-CoA

The breakdown of valine, isoleucine, and methionine converges at *propionyl-CoA* (Figure 25.44). Methionine first becomes *S*-adenosylmethionine and then homocysteine (see Figure 25.28). The carboxyl groups from all three are lost as CO_2. The two distal carbon atoms of isoleucine become acetyl-CoA. Propionyl-CoA subsequently is converted to methylmalonyl-CoA and thence to *succinyl-CoA* via the same reactions mediating the oxidation of fatty acids that have odd numbers of carbon atoms (see Chapter 23).

Leucine Is Degraded to Acetyl-CoA and Acetoacetate

Leucine is one of only two purely ketogenic amino acids; the other is lysine. Deamination of leucine via a transamination reaction yields α-*ketoisocaproate,* which is oxidatively decarboxylated to *isovaleryl-CoA* (Figure 25.45). Subsequent reactions, one of which is a biotin-dependent carboxylation, give *β-hydroxy-β-methylglutaryl-CoA,* which is then cleaved to yield *acetyl-CoA* and *acetoacetate,* a ketone body (see Figure 23.26). Neither of these products is convertible to glucose.

> **A DEEPER LOOK**

The Serine Dehydratase Reaction—A β-Elimination

The degradation of serine to pyruvate (see Figure 25.42) is an example of a pyridoxal phosphate–catalyzed β-elimination reaction. β-Eliminations mediated by PLP yield products that have undergone a two-electron oxidation at C_α. Serine is thus oxidized to pyruvate, with release of ammonium ion (see accompanying figure). At first, this looks like a transaminase half-reaction, but there is an important difference. In each transaminase half-reaction, PLP undergoes a net two-electron reduction or oxidation (depending on the direction), whereas β-eliminations occur with no net oxidation or reduction of PLP. Note too that the aminoacrylate released from PLP is unstable in aqueous solution. It rapidly tautomerizes to the preferred imine form, which is spontaneously hydrolyzed to yield the α-keto acid product—pyruvate in this case.

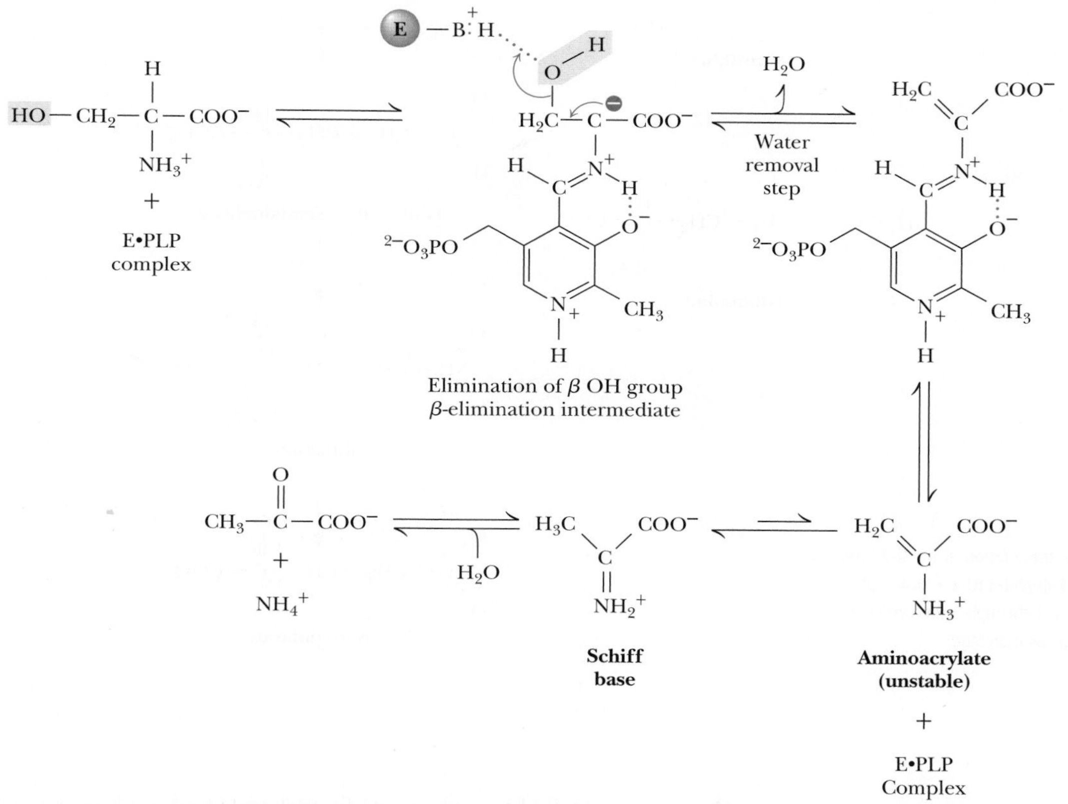

The serine dehydratase reaction mechanism—an example of a PLP-dependent β-elimination reaction.

Degradation of the Branched-Chain Amino Acids Follows a Common Pathway

Valine, leucine, and isoleucine are the three amino acids with branched aliphatic side chains, and all three are essential amino acids. Unlike the other 17 amino acids, which are broken down principally in the liver, Val, Ile, and Leu are also degraded in adipose tissue and skeletal muscle. The first three reactions in their catabolism are identical (Figure 25.46). These reactions, a transamination, followed by an oxidative decarboxylation to form a coenzyme A derivative, and then another dehydrogenation, are localized in the mitochondrial matrix. The liver has low levels of the first enzyme, branched-chain amino acid aminotransferase, thus explaining branched-chain amino acid degradation in other tissues. The oxidative decarboxylation is carried out by the branched-chain α-keto acid dehydrogenase (BCKDH), an enzyme complex with striking similarity to the mitochondrial pyruvate dehydrogenase and α-ketoglutarate dehydrogenase complexes discussed in Chapter 19 (see Figure 19.4 and Table 19.2). Hereditary defects in BCKDH leads to **maple sugar urine disease**. The α-keto acid derivatives of these amino acids accumulate in the blood and ultimately, in the urine, where their presence causes

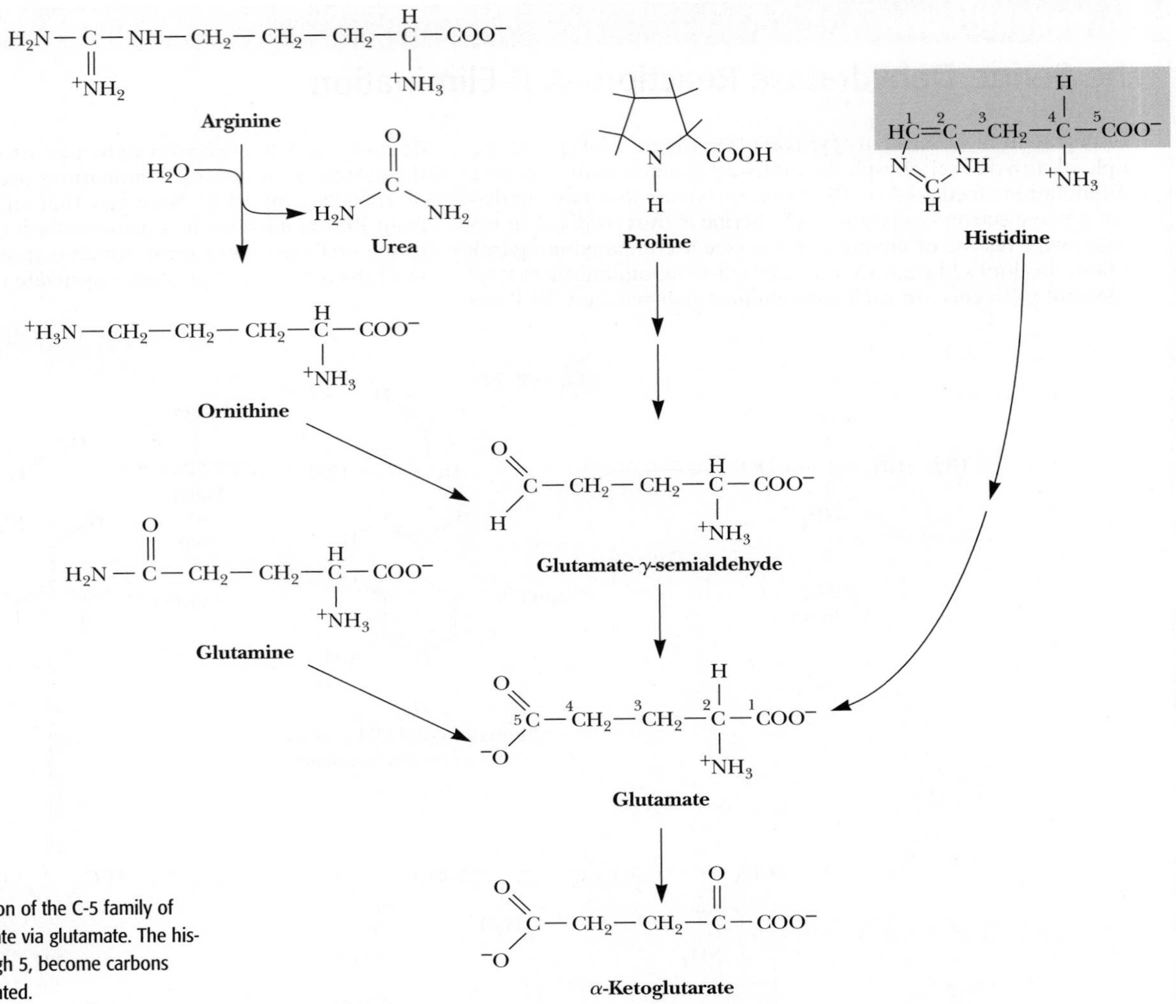

FIGURE 25.43 The degradation of the C-5 family of amino acids leads to α-ketoglutarate via glutamate. The histidine carbons, numbered 1 through 5, become carbons 1 through 5 of glutamate, as indicated.

the urine to smell like maple syrup. Such defects lead to abnormal development, mental retardation, and shortened lifespan, even if dietary intake of branched-chain amino acids is restricted.

Lysine Degradation Lysine degradation proceeds by several pathways, but the *saccharopine pathway* found in liver predominates (Figure 25.47). This degradative route proceeds backward along the lysine biosynthetic pathway through saccharopine and α-aminoadipate to α-ketoadipate (see Figure 25.24). Next, α-ketoadipate undergoes oxidative decarboxylation to *glutaryl-CoA,* which is then transformed into *acetoacetyl-CoA* and ultimately into the ketone body, *acetoacetate.*

Phenylalanine and Tyrosine Are Degraded to Acetoacetate and Fumarate The first reaction in phenylalanine degradation is the hydroxylation reaction of *tyrosine* biosynthesis (see Figure 25.38). Both these amino acids thus share a common degradative pathway. Transamination of Tyr yields the α-keto acid *p-hydroxyphenylpyruvate* (Figure 25.48, reaction 1). *p*-**Hydroxyphenylpyruvate dioxygenase,** a vitamin C–dependent enzyme, then carries out a ring hydroxylation–oxidative decarboxylation to yield homogentisate (Figure 25.48, reaction 2). Ring opening and isomerization (Figure 25.48, reactions 3 and 4) give *4-fumaryl-acetoacetate,* which is hydrolyzed to *acetoacetate* and *fumarate* (reaction 6).

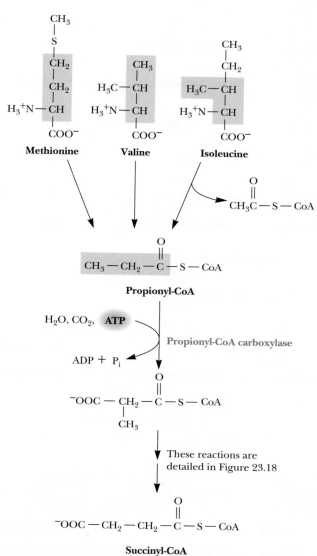

FIGURE 25.44 Valine, isoleucine, and methionine are converted via propionyl-CoA to succinyl-CoA for entry into the citric acid cycle. The shaded carbon atoms of the three amino acids give rise to propionyl-CoA.

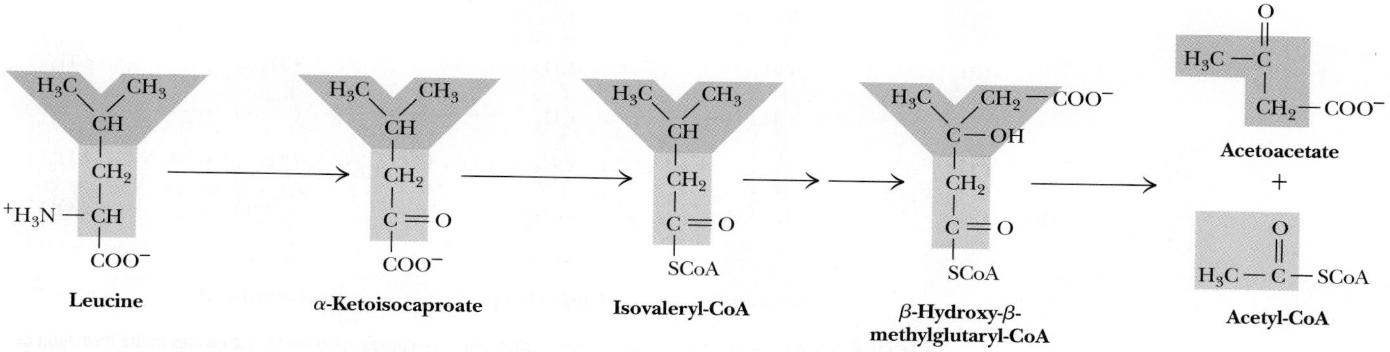

Leucine → **α-Ketoisocaproate** → **Isovaleryl-CoA** → **β-Hydroxy-β-methylglutaryl-CoA** → **Acetoacetate** + **Acetyl-CoA**

FIGURE 25.45 Leucine is degraded to acetyl-CoA and acetoacetate.

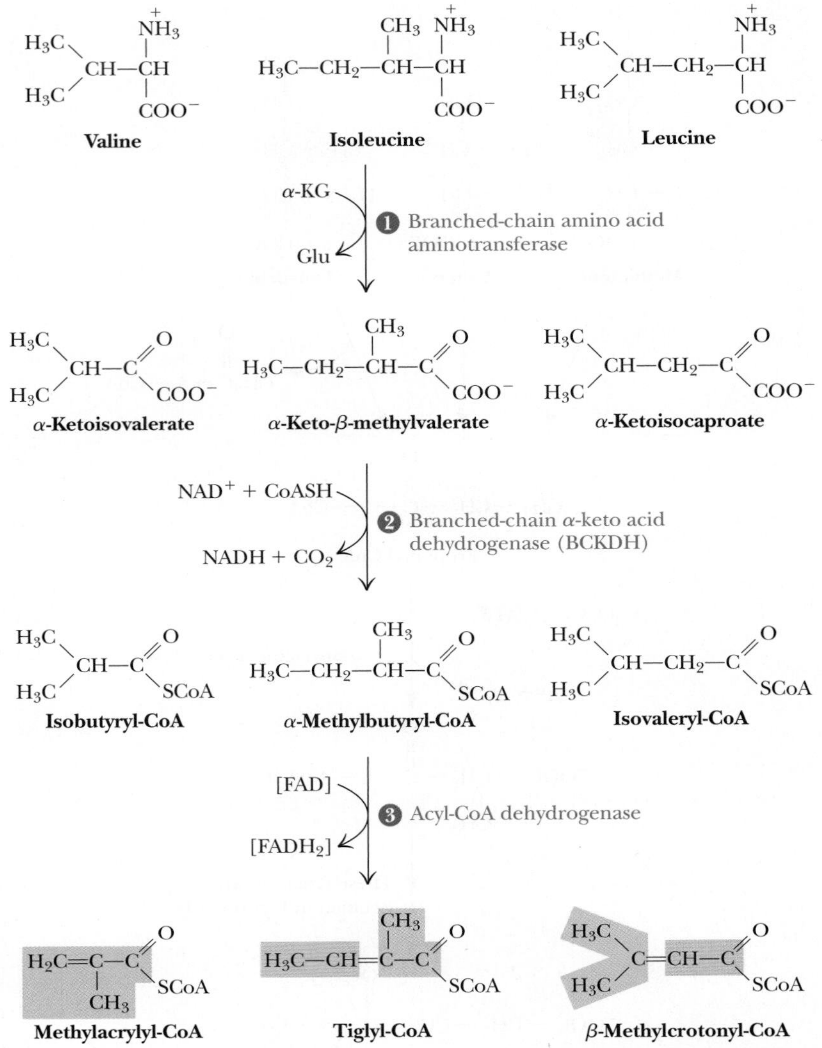

FIGURE 25.46 The first three steps for degradation of valine, isoleucine, and leucine follow a common pathway. Reaction 1 is catalyzed by the branched-chain amino acid aminotransferase; reaction 2 is catalyzed by the branched-chain α-keto acid dehydrogenase (BCKDH); and reaction 3 is catalyzed by the acyl-CoA dehydrogenase of the fatty acid β-oxidation pathway (see Figure 23.9, reaction 1). The carbons leading to succinyl-CoA via propionyl-CoA are shown within blue-shaded regions; those leading to acetyl-CoA in orange-shaded regions; and those to acetoacetate in green-shaded regions.

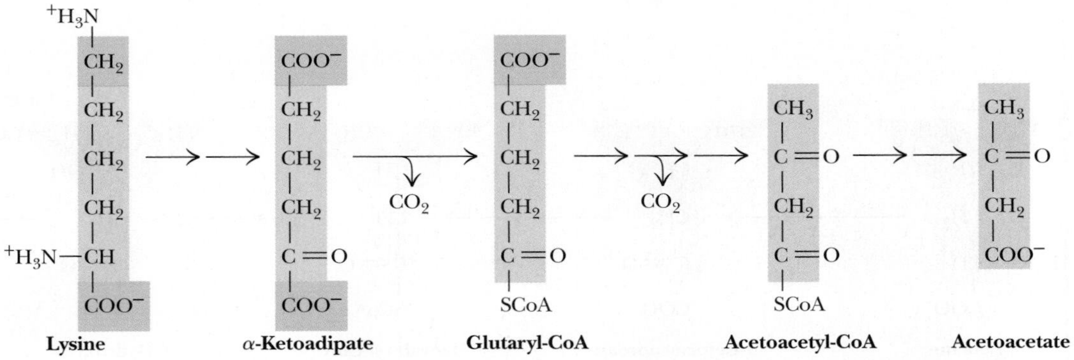

FIGURE 25.47 Lysine degradation via the saccharopine, α-ketoadipate pathway culminates in the formation of the ketone body, acetoacetate.

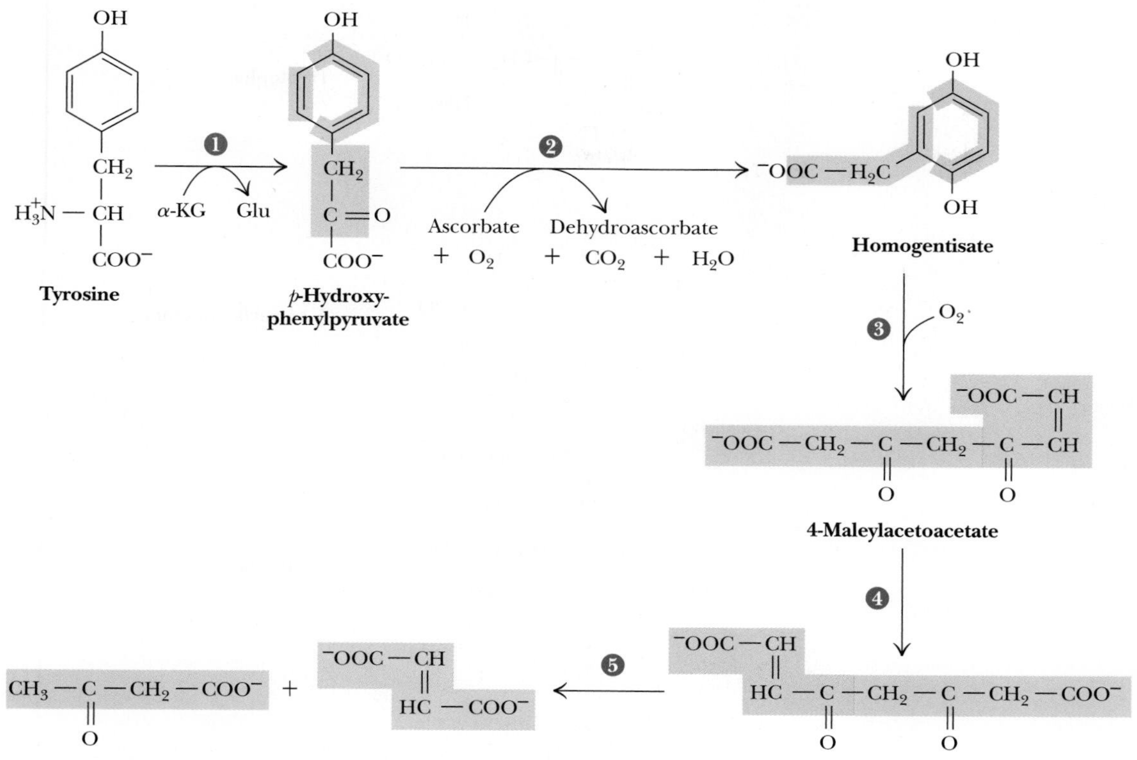

FIGURE 25.48 Phenylalanine and tyrosine degradation.

Tryptophan Degradation Tryptophan is the least abundant amino acid, so its catabolism to produce cellular energy is insignificant. However, tryptophan is a crucial precursor for synthesis of a variety of important substances, including the neurotransmitter **serotonin** (5-hydroxytryptamine) and the hormone **melatonin** (*N*-acetyl-5-methoxytryptamine). As noted earlier in the section on the degradation of the C-3 family of amino acids, degradation of the nonindole carbons of tryptophan yields pyruvate. The *indole ring* of Trp is converted by a series of reactions to α-ketoadipate and ultimately *acetoacetate* by the same reactions shown in Figure 25.47 for lysine. The first

---

Hereditary Defects in Phe Catabolism Underlie Alkaptonuria and Phenylketonuria

Alkaptonuria and phenylketonuria are two human genetic diseases arising from specific enzyme defects in phenylalanine degradation. **Alkaptonuria** is characterized by urinary excretion of large amounts of homogentisate and results from a deficiency in **homogentisate dioxygenase** (step 3, Figure 25.48). Air oxidation of homogentisate causes urine to turn dark on standing, but the only malady suffered by carriers of this disease is a tendency toward arthritis later in life.

In contrast, **phenylketonurics,** whose urine contains excessive *phenylpyruvate* (see accompanying figure), suffer severe mental retardation if the defect is not recognized immediately after birth and treated by putting the victim on a diet low in phenylalanine. These individuals are deficient in phenylalanine hydroxylase (Figure 25.38), and the excess Phe that accumulates is transaminated to phenylpyruvate and excreted.

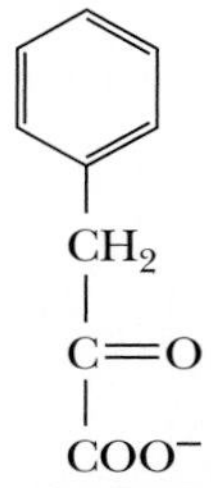

Phenylpyruvate

The structure of phenylpyruvate.

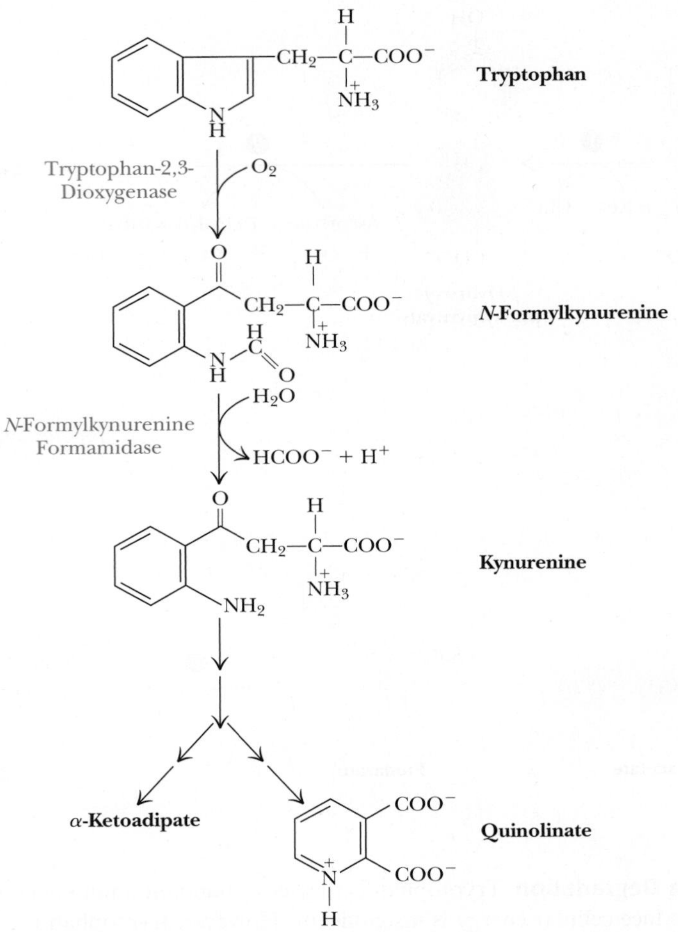

FIGURE 25.49 In the initial reaction of tryptophan degradation, tryptophan is oxidized by tryptophan-2,3-dioxygenase to yield *N*-formylkynurenine. Kynurenine is then formed by the action of *N*-formylkynurenine formamidase. The metabolite quinolinate can be produced from tryptophan if kynurenine is not metabolized to α-ketoadipate.

reaction in tryptophan degradation is catalyzed by **tryptophan-2,3-dioxygenase** and results in the formation of **N-formyl-kynurenine**; and then **kynurenine** (Figure 25.49). If not degraded to α-ketoadipate, kynurenine leads to the formation of a wide range of secondary metabolites, including **quinolinate** (2,3-pyridinedicarboxylate), which is a precursor to the B vitamin, **niacin**, and thus, to NAD^+ and $NADP^+$ synthesis. Interestingly, quinolinate is also a neurotoxic substance. Activation of the kynurenine pathway has been implicated as a potential cause of the neurodegeneration seen in Alzheimer's disease patients.

Animals Differ in the Form of Nitrogen That They Excrete

Animals often enjoy a dietary surplus of nitrogen. Excess nitrogen liberated upon metabolic degradation of amino acids is excreted by animals in three different ways, in accord with the availability of water. Aquatic animals simply release free ammonia to the surrounding water; such animals are termed **ammonotelic** (from the Greek *telos,* meaning "end"). On the other hand, terrestrial and aerial species employ mechanisms that convert ammonium to less toxic waste compounds that require little H_2O for excretion. Many terrestrial vertebrates, including humans, are **ureotelic,** meaning that they excrete excess N as **urea,** a highly water-soluble nonionic substance. Urea is formed by ureoteles via the urea cycle (see Figure 25.23). The **uricotelic** organisms are those animals using the third means of N excretion, conversion to **uric acid,** a rather insoluble purine analog. Birds and reptiles are uricoteles. Uric acid metabolism is discussed in the next chapter. Some animals can switch from ammonotelic to ureotelic to uricotelic metabolism, depending on water availability.

SUMMARY

25.1 Which Metabolic Pathways Allow Organisms to Live on Inorganic Forms of Nitrogen?

Nitrogen, an element essential to life, occurs in the environment principally as atmospheric N_2 and as NO_3^- ions in solution in soils and water. The metabolic pathways of nitrogen fixation and nitrate assimilation reduce these oxidized forms of nitrogen to the metabolically useful form, ammonium. Nitrate assimilation is a two-enzyme pathway: nitrate reductase and nitrite reductase. Nitrate reductase is a molybdenum cofactor-dependent flavohemoprotein. Nitrite reductase catalyzes the six-electron reduction of NO_2^- to NH_4^+ via a siroheme-dependent reaction. Nitrogen fixation is carried out by the nitrogenase system; biological reduction of N_2 to $2\ NH_4^+$ is an ATP-dependent eight-electron transfer reaction, with H_2 as an obligatory by-product. Nitrogenase is a metal-rich enzyme having an 8Fe-7S cluster as well as a 7Fe-1Mo-9S cluster known as the FeMo-cofactor. Nitrogenase is regulated in two ways: Its activity is inhibited by ADP, and its synthesis is repressed by NH_4^+.

25.2 What Is the Metabolic Fate of Ammonium?

Despite the great diversity of organic nitrogenous compounds found in cells, only a limited set of reactions incorporate ammonium ions into organic linkage: (1) glutamate dehydrogenase (GDH), (2) glutamine synthetase (GS), and (3) carbamoyl-P synthetase. Of these, the first two are quantitatively more important. Glutamate dehydrogenase, by adding NH_4^+ to the citric acid cycle intermediate α-ketoglutarate, sits at the interface between nitrogen metabolism and carbohydrate (and energy) metabolism. Glutamine synthetase catalyzes the ATP-dependent amidation of the γ-carboxyl group of Glu. Glutamine is the major donor of $-NH_2$ groups for the synthesis of many nitrogen-containing organic compounds, including purines, pyrimidines, and other amino acids. As such, its activity is tightly regulated. Most ammonium assimilation into organic linkage proceeds by one of two routes, depending on NH_4^+ availability: the GDH–GS route when ammonium is abundant and the glutamate synthase (GOGAT)–GS route when $[NH_4^+]$ is limiting.

25.3 What Regulatory Mechanisms Act on *Escherichia coli* Glutamine Synthetase?

Glutamine synthetase is a paradigm of enzyme regulation because its activity can be modulated at three different levels: (1) allosteric regulation by feedback inhibition, (2) covalent modification through adenylylation of Tyr^{397} in each of the 12 GS polypeptide chains, and (3) regulation of gene expression by the phosphorylated form of the transcriptional enhancer NR_I. Allosteric inhibitors of GS include five amino acids (Gly, Ser, Ala, His, and Trp), two nucleotides (one purine [AMP] and one pyrimidine [CTP]), one aminosugar (glucosamine-6-P), and carbamoyl-P. Adenylylation of GS converts it from a more active, allosterically unresponsive form to a less active, allosterically sensitive form. The ratio of adenylylated GS to deadenylylated GS is ultimately determined by the $[Gln]/[\alpha\text{-}KG]$ ratio, with a low ratio favoring the deadenylylated state and thus greater synthesis of glutamine.

25.4 How Do Organisms Synthesize Amino Acids?

In many cases, amino acid biosynthesis is a matter of synthesizing the appropriate α-keto acid carbon skeleton for the amino acid and then transaminating this α-keto acid using Glu as amino donor by action of an aminotransferase reaction. Mammals have retained the ability to synthesize the α-keto acid analog for 10 of the 20 common amino acids (the so-called nonessential amino acids), but the ability to make the other 10 α-keto acid analogs has been lost over evolutionary time, rendering these 10 amino acids as essential in the diet. The common amino acids can be grouped into families on the basis of the metabolic progenitor that serves as their precursor: The α-ketoglutarate family includes Glu, Gln, Pro, Arg, and (sometimes) Lys; the pyruvate family includes Ala, Val, and Leu; the oxaloacetate family includes Asp, Asn, Met, Thr, Ile, and (sometimes) Lys; the 3-phosphoglycerate family includes Ser, Gly, and Cys; and the PEP and erythrose-4-P family includes the aromatic amino acids Phe, Tyr, and Trp. Histidine is a special case—it is formed from PRPP and ATP. AICAR is a by-product.

25.5 How Does Amino Acid Catabolism Lead into Pathways of Energy Production?

The 20 common amino acids are degraded by 20 different pathways that converge to just 7 metabolic intermediates: pyruvate, acetyl-CoA, acetoacetate, oxaloacetate, α-ketoglutarate, succinyl-CoA, and fumarate. All seven of these compounds are intermediates in or readily feed into the pathways of energy production (citric acid cycle and oxidative phosphorylation). Five of them can serve as substrates for gluconeogenesis, namely pyruvate, oxaloacetate, α-ketoglutarate, succinyl-CoA, and fumarate. The other two, acetyl-CoA and acetoacetate, cannot. Amino acids leading to these products contribute to ketone body formation.

FOUNDATIONAL BIOCHEMISTRY Things You Should Know After Reading Chapter 25.

- Nitrogen is cycled between organisms and the inanimate environment.

- The two principal forms of inorganic nitrogen in the environment are N_2 and nitrate.

- Nitrate assimilation is the principal pathway for ammonium biosynthesis and nitrogen acquisition in the biosphere.

- Nitrate assimilation is a two-step pathway consisting of nitrate reductase and nitrite reductase.

- Organisms gain access to atmospheric N_2 via the pathway of nitrogen fixation.

- The nitrogenase enzyme complex that reduces N_2 to $2\ NH_4^+$ is composed of two metalloproteins.

- Ammonium is introduced into organic linkage principally through three enzymatic reactions: glutamate dehydrogenase, glutamine synthetase, and carbamoyl-P synthetase I.

- The major pathways of ammonium assimilation, the GDH/GS route and the GOGAT/GS route, both lead to glutamine synthesis.

- *E. coli* glutamine synthetase is regulated allosterically by feedback inhibition through the action of 9 metabolites that are key indicators of cellular nitrogen status.

- Glutamine synthetase is regulated by covalent modification through adenylylation of a Tyr side chain.

- Glutamine synthetase levels are regulated through controls on the expression of *GlnA*, the gene encoding the GS subunit.

- Glutamine is the most abundant amino acid in human body fluids and the brain.

- Amino acids are formed from α-keto acids by transamination.

- Aminotransferases (transaminases) are pyridoxal-P-dependent enzymes.

- Amino acids can be grouped into families according to the metabolic intermediate that serves as their progenitor.

- The amino acid progenitors are α-ketoglutarate, oxaloacetate, pyruvate, 3-phosphoglycerate, PEP, and erythrose-4-P.

- Humans can synthesize only 10 of the 20 common amino acids.

- Those amino acids humans cannot synthesize are called essential amino acids.

- With only one minor exception, the pathways for essential amino acid synthesis contain more steps than the corresponding pathways for nonessential amino acid synthesis.

- The α-ketoglutarate family of amino acids includes glutamate, glutamine, proline, arginine, and lysine.

- The urea cycle acts to synthesize Arg and to excrete excess N through synthesis of urea, which is disposed of in the urine.

- The urea cycle can be viewed as both an ammonium and a bicarbonate disposal mechanism.

- The oxaloacetate family of amino acids includes Asp, Asn, Lys, Met, Thr, and Ile.

- Methionine is the methyl donor in a variety of methylation reactions, including DNA methylation and histone modification.

- The pyruvate family of amino acids includes Ala, Val, and Leu.

- The 3-phosphoglycerate family of amino acids includes Ser, Gly, and Cys.

- The —SH group of cysteine comes from H_2S.

- H_2S synthesis from sulfate involves S-containing ATP derivatives.

- The aromatic amino acids are synthesized from chorismate.

- Chorismate is synthesized from PEP and erythrose-4-P.

- Chorismate is also a precursor for coenzyme Q, plastoquinone, vitamin E, and lignin.

- Anthranilate is an intermediate between chorismate and Trp in the tryptophan biosynthetic pathway.

- The pathways of essential amino acid biosynthesis are key targets of herbicides.

- Histidine biosynthesis and purine biosynthesis are connected by common intermediates.

- Dietary protein is an important source of fuel for energy production.

- The 20 common amino acids are degraded by 20 different pathways that converge to just 7 metabolic intermediates: α-ketoglutarate, oxaloacetate, pyruvate, succinyl-CoA, fumarate, acetyl-CoA, and acetoacetate.

- Those amino acids that lead to α-ketoglutarate, oxaloacetate, pyruvate, succinyl-CoA, or fumarate are considered glucogenic amino acids because these intermediates can serve as substrates for gluconeogenesis.

- Those amino acids that lead to acetyl-CoA or acetoacetate are ketogenic amino acids because they lead to the formation of ketone bodies.

- Only Leu and Lys are solely ketogenic.

- Ala, Ser, and Cys are degraded to pyruvate, as is Gly.

- Thr is degraded to pyruvate and acetyl-CoA.

- Asp and Asn are degraded to oxaloacetate.

- Glu, Gln, Pro, Arg, and His are degraded to α-ketoglutarate.

- Degradation of Val, Ile, and Met leads to succinyl-CoA.

- Leu is degraded to acetyl-CoA and acetoacetate.

- Degradation of the branched-chain amino acids (Val, Leu, Ile) follows a common pathway, a key reaction of which is the branched-chain α-keto acid dehydrogenase.

- Lys is degraded to acetoacetate.

- Phe and Tyr are degraded to fumarate and acetoacetate.

- Hereditary defects in Phe catabolism lead to alkaptonuria or phenylketonuria.

- Trp is degraded to pyruvate and acetoacetate.

- Trp is a precursor to the neurotransmitter serotonin, the hormone melatonin, and quinolinate, from which the B vitamin niacin can be formed.

PROBLEMS

Answers to all problems are at the end of this book. Detailed solutions are available in the *Student Solutions Manual, Study Guide, and Problems Book.*

1. **The Oxidation States of Nitrogen** What is the oxidation number of N in nitrate, nitrite, NO, N_2O, and N_2?

2. **The Energetic Cost of Nitrogen Acquisition from the Environment** How many ATP equivalents are consumed per N atom of ammonium formed by (a) the nitrate assimilation pathway and (b) the nitrogen fixation pathway? (Assume for this problem NADH, NADPH, and reduced ferredoxin are each worth 3 ATPs.)

3. **Regulation of Glutamine Synthetase by Covalent Modification** Suppose at certain specific metabolite concentrations in vivo the cyclic cascade regulating *E. coli* glutamine synthetase has reached a dynamic equilibrium where the average state of GS adenylylation is poised at $n = 6$. Predict what change in n will occur if:
   a. [ATP] increases.
   b. P_{IIA}/P_{IID} increases.
   c. $[\alpha\text{-KG}]/[\text{Gln}]$ increases.
   d. $[P_i]$ decreases.

4. **The Energetic Cost of Nitrogen Excretion via the Urea Cycle** How many ATP equivalents are consumed in the production of 1 equivalent of urea by the urea cycle?

5. **Why a High-Protein Diet Requires Greater Water Intake** Why are persons on a high-protein diet (such as the Atkins diet) advised to drink lots of water?

6. **The Energetic Cost of Lysine Synthesis** How many ATP equivalents are consumed in the biosynthesis of lysine from aspartate by the pathway shown in Figure 25.27?

7. **PEP as a Precursor for Aromatic Amino Acid Biosynthesis** If PEP labeled with ^{14}C in the 2-position serves as the precursor to chorismate synthesis, which C atom in chorismate is radioactive?

8. **Gluconeogenesis from Aspartate** (Integrates with Chapter 22.) Write a balanced equation for the synthesis of glucose (by gluconeogenesis) from aspartate.

9. **Synthesis of Amino Acids from α-Keto Acids** For each of the 20 common amino acids, give the name of the enzyme that catalyzes the reaction providing its α-amino group.

10. **The Essential Coenzyme in Amino Acid Metabolism** Which vitamin is central in amino acid metabolism? Why?

11. **Vitamins and Coenzymes in Homocysteine Metabolism** Vitamins B_6, B_{12}, and folate may be recommended for individuals with high blood serum levels of homocysteine (a condition called *hyperhomocysteinemia*). How might these vitamins ameliorate hyperhomocysteinemia?

12. **Vitamins and Coenzymes in Branched-Chain Amino Acid Degradation** (Integrates with Chapter 19.) On the basis of the following information, predict a reaction mechanism for the mammalian branched-chain α-keto acid dehydrogenase complex (the BCKAD complex). This complex carries out the oxidative decarboxylation of the α-keto acids derived from valine, leucine, and isoleucine.

   a. One form of maple syrup urine disease responds well to administration of thiamine.

   b. Lipoic acid is an essential coenzyme.

   c. The enzyme complex contains a flavoprotein.

13. **Dietary Concerns in Phenylketonuria** People with phenylketonuria must avoid foods containing the low-calorie sweetener *Aspartame*, also known as *NutraSweet*. Find the structure of *Aspartame* in the Merck Index (or other scientific source), and state why these people must avoid this substance.

14. **How the Herbicide *RoundUp* Affects Plant Metabolism** Glyphosate (otherwise known as *RoundUp*) is an analog of PEP. It acts as a noncompetitive inhibitor of 3-enolpyruvylshikimate-5-P synthase; it has the following structure in its fully protonated state:

$$HOOC—CH_2—NH—CH_2—PO_3H_2$$

Consult Figures 25.35 and 25.36, and construct a list of the diverse metabolic consequences that might be experienced by a plant cell exposed to glyphosate.

15. **Glycine Synthesis from Glucose** (Integrates with Chapter 18.) When cells convert glucose to glycine, which carbon atoms of glucose are represented in glycine?

16. **A Deficiency on 3-Phosphoglycerate Dehydrogenase Can Affect Amino Acid Metabolism** Although serine is a nonessential amino acid, serine deficiency syndrome has been observed in humans. One such form of the syndrome is traceable to a deficiency in 3-phosphoglycerate dehydrogenase (see Figure 25.31). Individuals with this syndrome not only are serine-deficient but also are impaired in their ability to synthesize another common amino acid, as well as a class of lipids. Describe why.

17. **Use the Protein Data Bank to Explore the Structure and Function of Glutamate Synthase** Go to www.pdb.org, and examine the pdb file 1LM1 for glutamate synthase. Find its iron–sulfur cluster and FMN prosthetic group. Discover how this enzyme is organized into an N-terminal domain that functions in ammonia removal from glutamine (the glutaminase domain) and the α-ketoglutarate–binding site near the Fe/S and flavin prosthetic groups. Consult van den Heuvel, R. H. H., et al., 2002. Structural studies on the synchronization of catalytic centers in glutamate synthase. *Journal of Biological Chemistry* **277:**24579–24583, to see how these two sites are connected by a tunnel for passage of ammonia from glutamine to α-ketoglutarate.

18. **The Energy Cost of Using Dietary Protein as Food** The thermic effect of food is a term used to describe the energy cost of processing the food we eat, digesting it, and either turning it into precursors for needed biosynthesis, usable energy in the form of ATP, or storing the excess intake as fat. The thermic effect is usually approximated at 10% of the total calories consumed, but the thermic effect of fat is only 2% to 3% of total fat calories and the thermic effect of protein is 30% or more of calories consumed as protein. Why do you suppose dietary protein has a much higher thermic effect than either dietary carbohydrate or fat?

Preparing for the MCAT® Exam

19. From the dodecameric (α_{12}) structure of glutamine synthetase shown in Figure 25.14, predict the relative enzymatic activity of GS monomers (isolated α-subunits).

20. Consider the synthesis and degradation of tyrosine as shown in Figures 25.37, 25.38, and 25.48 to determine where the carbon atoms in PEP and erythrose-4-P would end up in acetoacetate and fumarate.

ActiveModels Problems

21. Examine the ActiveModel for glutamine synthetase, and describe the formation of the GS dodecamer.

22. Using the ActiveModel for glutamate oxaloacetate transaminase 1, explain how this enzyme accomplishes a "ping-pong" mechanism for its catalyzed reaction.

FURTHER READING

Nitrate Assimilation and Nitrogen Fixation

Brewin, A. J., and Legocki, A. B., 1996. Biological nitrogen fixation for sustainable agriculture. *Trends in Microbiology* **4:**476–477.

Burgess, B. K., and Lowe, D. J., 1996. Mechanism of molybdenum nitrogenase. *Chemical Reviews* **96:**2983–3011.

Campbell, W. H., and Kinghorn, J. R., 1990. Functional domains of assimilatory nitrate reductases and nitrite reductases. *Trends in Biochemical Sciences* **15:**315–319.

Crawford, N. M., and Arst, H. N., Jr., 1993. The molecular genetics of nitrate assimilation in fungi and plants. *Annual Review of Genetics* **27:**115–146.

Lin, J. T., and Stewart, V., 1998. Nitrate assimilation in bacteria. *Advances in Microbial Physiology* **39:**1–30.

Mortenson, L. E., Seefeldt, L. C., Morgan, T. V., and Bolin, J. T., 1993. The role of metal clusters and MgATP in nitrogenase catalysis. *Advances in Enzymology* **67:**299–374.

Peters, J. W., and Szilagyi, R. K., 2006. Exploring new frontiers of nitrogenase structure and mechanism. *Current Opinion in Chemical Biology* **10:**101–108.

Rees, D. C., et al., 2005. Structural basis of nitrogen fixation. *Philosophical Transactions of the Royal Society A* **363:**971–984.

Schwarz G., Mendel, R. R., and Ribbe, M. U., 2009. Molybdenum cofactors, enzymes and pathways. *Nature* **460:**839–847.

Seefeldt, L. C., Hoffman, B. M., and Dean. D. R., 2009. Mechanism of Mo-dependent nitrogenase. *Annual Review of Biochemistry* **78:**701–722.

Wray, J. L., and Kinghorn, J. R., 1989. *Molecular and Genetic Aspects of Nitrate Assimilation.* New York: Oxford Science.

Glutamate Dehydrogenase and Glutamine Synthetase

Hart, Y., Madar, D., Yuan, J., Bren, A., et al., 2011. Robust control of nitrogen assimilation by a bifunctional enzyme in E. coli. *Molecular Cell* **41:**117–127.

Hudson, R. C., and Daniel, R. M., 1993. L-Glutamate dehydrogenases: Distribution, properties, and mechanism. *Comparative Biochemistry* **106B:**767–792.

Liaw, S-H., and Eisenberg, D. S., 1995. Discovery of the ammonium substrate site on glutamine synthetase, a third cation binding site. *Protein Science* **4:**2358–2365.

Liaw, S-H., Pan, C., and Eisenberg, D. S., 1993. Feedback inhibition of fully unadenylylated glutamine synthetase from *Salmonella typhimurium* by glycine, alanine, and serine. *Proceedings of the National Academy of Sciences U.S.A.* **90:**4996–5000.

Morris, S. M., Jr., 2002. Regulation of enzymes of the urea cycle and arginine metabolism. *Annual Review of Nutrition* **22**:87–105.

Mutalik, V. K., Shah, P., and Venkatesh, K.V., 2003. Allosteric interactions and bifunctionality make the response of glutamine synthetase cascade system of *Escherichia coli* robust and ultrasensitive. *Journal of Biological Chemistry* **278**:26327–26332.

Smith, T. J., and Stanley, C. A., 2008. Untangling the glutamate dehydrogenase allosteric nightmare. *Trends in Biochemical Sciences* **33**:555–564.

Stadtman, E. R., 2001. The story of glutamine synthetase regulation. *Journal of Biological Chemistry* **276**:44357–44364.

Vanoni, M. A., and Curti, B., 2008. Structure–function studies of glutamate synthases: A class of self-regulated iron-sulfur flavoenzymes essential for nitrogen assimilation. *IUBMB Life* **63**:287–300.

Wise, D. R., and Thompson, C. B., 2010. Glutamine addiction: A new therapeutic target in cancer. *Trends in Biochemical Sciences* **35**:427–433.

The Urea Cycle

Atkinson, D. E., and Camien, M. N., 1982. The role of urea synthesis in the removal of metabolic bicarbonate and the regulation of blood pH. *Current Topics in Cellular Regulation* **21**:261–302.

Amino Acid Metabolism

Bender, D. A., 1985. *Amino Acid Metabolism.* New York: Wiley.

Richards, N. G. J., and Kilberg, M. S., 2006. Asparagine synthetase chemotherapy. *Annual Review of Biochemistry* **75**:629–654.

Srere, P. A., 1987. Complexes of sequential metabolic enzymes. *Annual Review of Biochemistry* **56**:89–124.

Wagenmakers, A. J., 1998. Protein and amino acid metabolism in human muscle. *Advances in Experimental Medicine and Biology* **441**:307–319.

Clinical Disorders in Amino Acid Metabolism

Boushey, C. J., et al., 1995. A quantitative assessment of plasma homocysteine as a risk factor for vascular disease. *Journal of the American Medical Association* **274**:1049–1057.

Fernandez-Canon, J. M., et al., 1996. The molecular basis of alkaptonuria. *Nature Genetics* **14**:19–24.

De Koning, T. J., and Klomp, L., 2004. Serine-deficiency syndromes. *Current Opinion in Neurology* **17**:197–204.

Gong, C. Y., Li, Z., Wang, H. M., Chen, H. L., et al., 2011. Targeting the kynurenine pathway as potential to prevent and treat Alzheimer's disease. *Medical Hypotheses* **77**:383–385.

Seriver, C. R., et al., 1995. *The Metabolic and Molecular Bases of Inherited disease,* 7th ed. New York: McGraw-Hill.

Amino Acid Biosynthesis Inhibitors as Herbicides

Kishore, G. M., and Shah, D. M., 1988. Amino acid biosynthesis inhibitors as herbicides. *Annual Review of Biochemistry* **57**:627–663.

Tunneling in Enzyme-Catalyzed Reactions

Weeks, A., Lund, L., and Raushel, F. M., 2006. Tunneling intermediates in enzyme-catalyzed reactions. *Current Opinion in Chemical Biology* **10**:465–472.

Synthesis and Degradation of Nucleotides

© Adam Woolfitt/CORBIS

◄ Pigeon drinking at Gaia Fountain, Siena, Italy. The basic features of purine biosynthesis were elucidated initially from metabolic studies of nitrogen metabolism in pigeons. Pigeons excrete excess N as uric acid, a purine analog.

Guano, a substance found on some coasts frequented by sea birds, is composed chiefly of the birds' partially decomposed excrement. . . . The name for the purine guanine derives from the abundance of this base in guano.

J. C. Nesbit
On Agricultural Chemistry and the Nature
and Properties of Peruvian Guano (1850)

KEY QUESTIONS

26.1 Can Cells Synthesize Nucleotides?
26.2 How Do Cells Synthesize Purines?
26.3 Can Cells Salvage Purines?
26.4 How Are Purines Degraded?
26.5 How Do Cells Synthesize Pyrimidines?
26.6 How Are Pyrimidines Degraded?
26.7 How Do Cells Form the Deoxyribonucleotides That Are Necessary for DNA Synthesis?
26.8 How Are Thymine Nucleotides Synthesized?

ESSENTIAL QUESTION

Virtually all cells are capable of synthesizing purine and pyrimidine nucleotides. These compounds then serve as essential intermediates in metabolism and as the building blocks for DNA and RNA synthesis.

How do cells synthesize purines and pyrimidines?

Nucleotides are ubiquitous constituents of life, actively participating in the majority of biochemical reactions. Recall that ATP is the "energy currency" of the cell, that uracil nucleotide derivatives of carbohydrates are common intermediates in cellular transformations of carbohydrates (see Chapter 22), and that biosynthesis of phospholipids proceeds via cytosine nucleotide derivatives (see Chapter 24). In Chapter 30, we will see that GTP serves as the immediate energy source driving the endergonic reactions of protein synthesis. Many of the coenzymes (such as coenzyme A, NAD, NADP, and FAD) are derivatives of nucleotides. Nucleotides also act in metabolic regulation, as in the response of key enzymes of intermediary metabolism to the relative concentrations of AMP, ADP, and ATP (PFK is a prime example here; see also Chapter 18). Furthermore, cyclic derivatives of purine nucleotides such as cAMP and cGMP have no other role in metabolism than regulation. Last but not least, nucleotides are the monomeric units of nucleic acids. Deoxynucleoside triphosphates (dNTPs) and nucleoside triphosphates (NTPs) serve as the immediate substrates for the biosynthesis of DNA and RNA, respectively (see Part 4).

26.1 Can Cells Synthesize Nucleotides?

Nearly all organisms can make the purine and pyrimidine nucleotides via so-called de novo biosynthetic pathways. (*De novo* means "anew"; a less literal but more apt translation might be "from scratch" because de novo pathways are metabolic sequences that form complex end products from rather simple precursors.) Many organisms also have salvage pathways to recover purine and pyrimidine compounds obtained in the diet or

OWL Online homework and a Student Self Assessment for this chapter may be assigned in OWL.

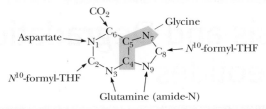

FIGURE 26.1 Nitrogen waste is excreted by birds principally as the purine analog, uric acid.

FIGURE 26.2 The metabolic origin of the nine atoms in the purine ring system.

released during nucleic acid turnover and degradation. Whereas the ribose of nucleotides can be catabolized to generate energy, the nitrogenous bases do *not* serve as energy sources; their catabolism does not lead to products used by pathways of energy conservation. Compared to slowly dividing cells, rapidly proliferating cells synthesize larger amounts of DNA and RNA per unit time. To meet the increased demand for nucleic acid synthesis, substantially greater quantities of nucleotides must be produced. The pathways of nucleotide biosynthesis thus become attractive targets for the clinical control of rapidly dividing cells such as cancers or infectious bacteria. Many antibiotics and anticancer drugs are inhibitors of purine or pyrimidine nucleotide biosynthesis.

26.2 How Do Cells Synthesize Purines?

Substantial insight into the de novo pathway for purine biosynthesis was provided in 1948 by John Buchanan, who cleverly exploited the fact that birds excrete excess nitrogen principally in the form of uric acid, a water-insoluble purine analog. Buchanan fed isotopically labeled compounds to pigeons and then examined the distribution of the labeled atoms in *uric acid* (Figure 26.1). By tracing the metabolic source of the various atoms in this end product, he showed that the nine atoms of the purine ring system (Figure 26.2) are contributed by aspartic acid (N-1), glutamine (N-3 and N-9), glycine (C-4, C-5, and N-7), CO_2 (C-6), and THF one-carbon derivatives (C-2 and C-8). THF is tetrahydrofolate, a coenzyme serving as a one-carbon transfer agent, not only in purine ring synthesis but also in amino acid metabolism (see Figures 25.27 and 25.32) and in synthesis of the pyrimidine thymine (see Figure 26.26). The formation and function of THF is summarized in A Deeper Look on pages 984–985.

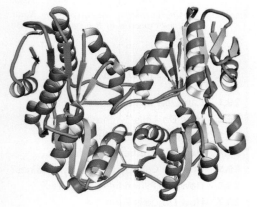

Human PRPP synthetase
(pdb id = 2H06).

IMP Is the Immediate Precursor to GMP and AMP

The de novo synthesis of purines occurs in an interesting manner: The atoms forming the purine ring are successively added to *ribose-5-phosphate;* thus, purines are directly synthesized as nucleotide derivatives by assembling the atoms that comprise the purine ring system directly on the ribose. Beginning with ribose-5-P and ending with IMP, the pathway consists of 11 enzymatic steps. In step 1, ribose-5-phosphate is activated via the direct transfer of a pyrophosphoryl group from ATP to C-1 of the ribose, yielding *5-phosphoribosyl-α-pyrophosphate (PRPP)* (Figure 26.3). The enzyme is **ribose-5-phosphate pyrophosphokinase** (also known as **PRPP synthetase**). PRPP is the limiting substance in purine biosynthesis. The two major purine nucleoside diphosphates, ADP and GDP, are negative effectors of ribose-5-phosphate pyrophosphokinase. However, because PRPP serves additional metabolic needs, the next reaction is actually the committed step in the pathway.

Step 2 (Figure 26.3) is catalyzed by **glutamine phosphoribosyl pyrophosphate amidotransferase.** The anomeric carbon atom of the substrate PRPP is in the α-configuration; the product is a β-glycoside (recall that all the biologically important nucleotides are β-glycosides). The N atom of this *N*-glycoside becomes N-9 of the nine-membered purine ring; it is the first atom added in the construction of this ring. Glutamine phosphoribosyl pyrophosphate amidotransferase is subject to feedback inhibition by GMP, GDP, and GTP as well as AMP, ADP, and ATP. The G series of nucleotides interacts at

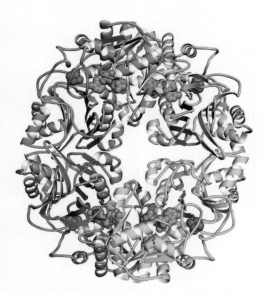

B. subtilis phosphoribosyl pyrophosphate amidotransferase (iron clusters in red, AMP in orange) (pdb id = 1GPH).

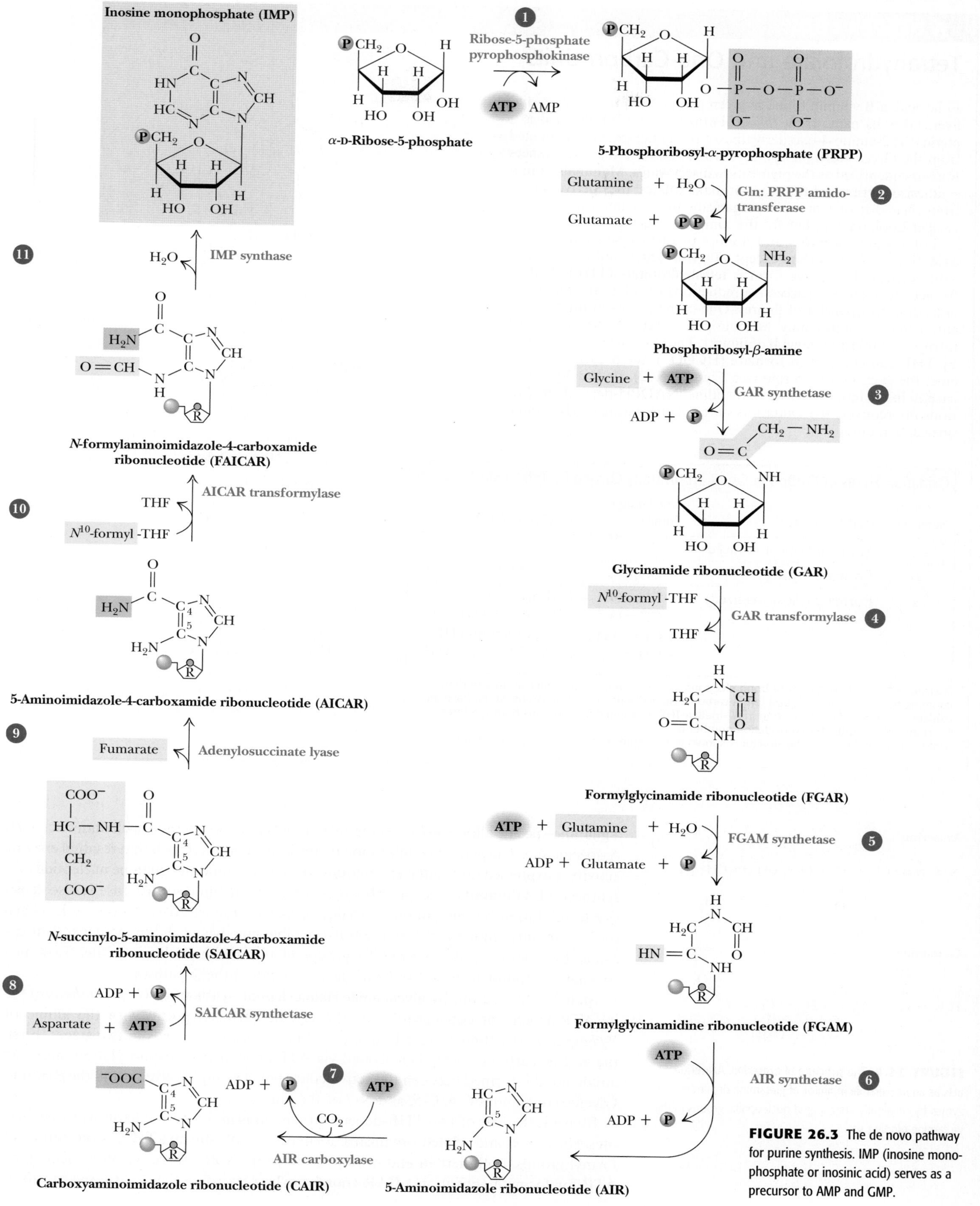

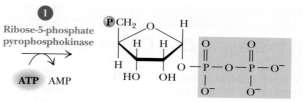

Inosine monophosphate (IMP)

α-D-Ribose-5-phosphate

Ribose-5-phosphate pyrophosphokinase ①

ATP AMP

5-Phosphoribosyl-α-pyrophosphate (PRPP)

Glutamine + H_2O

Gln: PRPP amido-transferase ②

Glutamate + PP

Phosphoribosyl-β-amine

Glycine + ATP

GAR synthetase ③

ADP + P

① IMP synthase
H_2O

N-formylaminoimidazole-4-carboxamide ribonucleotide (FAICAR)

Glycinamide ribonucleotide (GAR)

N^{10}-formyl -THF

GAR transformylase ④

THF

⑩ AICAR transformylase
THF
N^{10}-formyl -THF

5-Aminoimidazole-4-carboxamide ribonucleotide (AICAR)

Formylglycinamide ribonucleotide (FGAR)

ATP + Glutamine + H_2O

FGAM synthetase ⑤

ADP + Glutamate + P

⑨ Adenylosuccinate lyase
Fumarate

N-succinylo-5-aminoimidazole-4-carboxamide ribonucleotide (SAICAR)

Formylglycinamidine ribonucleotide (FGAM)

ATP

AIR synthetase ⑥

ADP + P

⑧ SAICAR synthetase
ADP + P
Aspartate + ATP

ADP + P ⑦ ATP

CO_2

AIR carboxylase

Carboxyaminoimidazole ribonucleotide (CAIR)

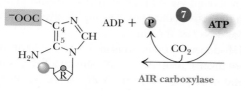

5-Aminoimidazole ribonucleotide (AIR)

FIGURE 26.3 The de novo pathway for purine synthesis. IMP (inosine monophosphate or inosinic acid) serves as a precursor to AMP and GMP.

> **A DEEPER LOOK**

Tetrahydrofolate and One-Carbon Units

Folic acid, a B vitamin found in green plants, fresh fruits, yeast, and liver, takes its name from *folium,* Latin for "leaf." Folic acid is a **pterin** (the 2-amino-4-oxo derivative of pteridine); pterins are named from the Greek word *pté ryj,* for "wing," because these substances were first identified as the pigments in insect wings. Mammals cannot synthesize pterins and thus cannot make folates; they derive folates from their diet or from microorganisms in their intestines. (See A Deeper Look on page 896 for the complete structure of folate.)

Folates are acceptors and donors of one-carbon units for all oxidation levels of carbon except CO_2 (for which biotin is the relevant carrier). The active form is **tetrahydrofolate (THF).** THF is formed through two successive reductions of folate by *dihydrofolate reductase* (panel a of figure). One-carbon units in three different oxidation states may be bound to THF at the N^5 or N^{10} nitrogens (table and panel b of figure). The one-carbon unit carried by THF can come from formate ($HCOO^-$), the α-carbon of glycine, the β-carbon of serine (see Figure 25.32), or the 3-position carbon in the imidazole ring of histidine. NADPH-dependent reactions interconvert the oxidation states of the various THF-bound one-carbon units.

(a)

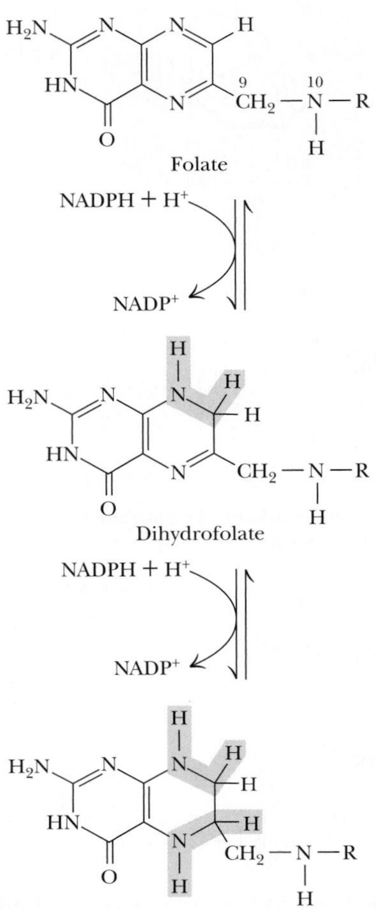

Folate

Dihydrofolate

Oxidation States of Carbon in One-Carbon Units Carried by Tetrahydrofolate

Oxidation Number*	Oxidation Level	One-Carbon Form[†]	Tetrahydrofolate Form
−2	Methanol (most reduced)	—CH_3	N^5-Methyl-THF
0	Formaldehyde	—CH_2—	N^5,N^{10}-Methylene-THF
2	Formate (most oxidized)	—CH=O	N^5-Formyl-THF
		—CH=O	N^{10}-Formyl-THF
		—CH=NH	N^5-Formimino-THF
		—CH=	N^5,N^{10}-Methenyl-THF

*Calculated by assigning valence bond electrons to the more electronegative atom and then counting the charge on the quasi ion. A carbon assigned four valence electrons would have an oxidation number of 0. The carbon in N^5-methyl-THF is assigned six electrons from the three C—H bonds and thus has an oxidation number of −2.
[†]Note: All vacant bonds in the structures shown are to atoms more electronegative than C.

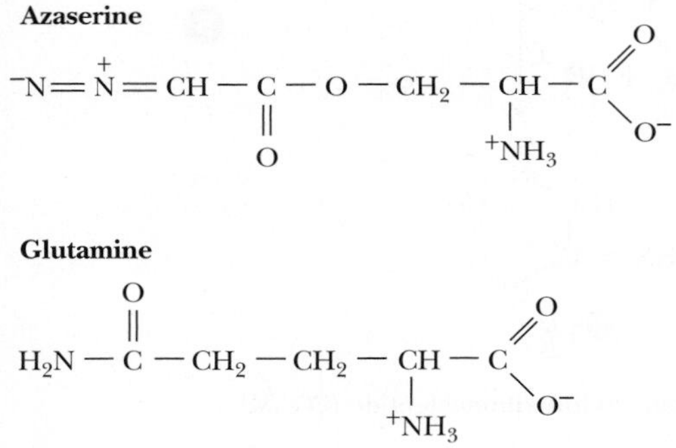

Azaserine

Glutamine

FIGURE 26.4 The structure of azaserine. Azaserine acts as an irreversible inhibitor of glutamine-dependent enzymes by covalently attaching to nucleophilic groups in the glutamine-binding site.

a guanine-specific allosteric site on the enzyme, whereas the adenine nucleotides act at an A-specific site. The pattern of inhibition by these nucleotides is such that residual enzyme activity is expressed until sufficient amounts of both adenine and guanine nucleotides are synthesized. Glutamine phosphoribosyl pyrophosphate amidotransferase is also sensitive to inhibition by the glutamine analog **azaserine** (Figure 26.4). Azaserine has been used as an antitumor agent because it irreversibly inactivates glutamine-dependent enzymes by reacting with nucleophilic groups at the glutamine-binding site. Two such enzymes are found at steps 2 and 5 of the purine biosynthetic pathway.

Step 3 is carried out by **glycinamide ribonucleotide synthetase** *(GAR synthetase)* via its ATP-dependent condensation of the glycine carboxyl group with the amine of *5-phosphoribosyl-β-amine* (see Figure 26.3). The reaction proceeds in two stages. First, the glycine carboxyl group is activated via ATP-dependent phosphorylation. Next, an amide bond is formed between the activated carboxyl group of glycine and the β-amine. Glycine contributes C-4, C-5, and N-7 of the purine.

Step 4 is the first of two THF-dependent reactions in the purine pathway of eukaryotes. (In *E. coli* and related organisms, formate, not N^{10}-formyl-THF, is the source of formyl groups, both here in and in step 10. In prokaryotes, these reactions depend on ATP for formate activation.) **GAR transformylase** transfers the N^{10}-formyl group of

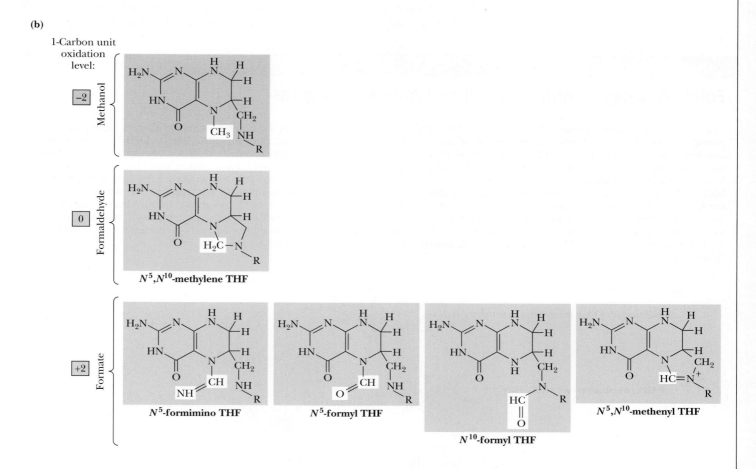

(b)

1-Carbon unit oxidation level:

N^5,N^{10}-methylene THF

N^5-formimino THF

N^5-formyl THF

N^{10}-formyl THF

N^5,N^{10}-methenyl THF

N^{10}-formyl-THF to the free amino group of GAR to yield *α-N-formylglycinamide ribonucleotide (FGAR)*. Thus, C-8 of the purine is "formyl-ly" introduced. Although all of the atoms of the imidazole portion of the purine ring are now present, the ring is not closed until Reaction 6.

Step 5 is catalyzed by **FGAR amidotransferase** (also known as *FGAM synthetase*). ATP-dependent transfer of the glutamine amido group to the C-4-carbonyl of FGAR yields *formylglycinamidine ribonucleotide (FGAM)*. The imino-N becomes N-3 of the purine.

Step 6 is an ATP-dependent dehydration that leads to formation of the imidazole ring. ATP is used to phosphorylate the oxygen atom of the formyl group, activating it for the ring closure step that follows. Because the product is *5-aminoimidazole ribonucleotide, or AIR,* this enzyme is called **AIR synthetase.** In higher organisms, the enzymatic activities for steps 3, 4, and 6 (GAR synthetase, GAR transformylase, and AIR synthetase) reside on a single multifunctional polypeptide.

In step 7, carbon dioxide is added at the C-4 position of the imidazole ring by **AIR carboxylase** in an ATP-dependent reaction; the carbon of CO_2 will become C-6 of the purine ring. The product is *carboxyaminoimidazole ribonucleotide (CAIR)*.

In step 8, the amino-N of aspartate provides N-1 through linkage to the C-6 carboxyl function of CAIR. ATP hydrolysis drives the condensation of Asp with CAIR.

The product is *N-succinylo-5-aminoimidazole-4-carboxamide ribonucleotide (SAICAR)*. **SAICAR synthetase** catalyzes the reaction. The enzymatic activities for steps 7 and 8 reside on a single, bifunctional polypeptide in eukaryotes.

Step 9 removes the four carbons of Asp as fumarate in a nonhydrolytic cleavage. The product is *5-aminoimidazole-4-carboxamide ribonucleotide (AICAR);* the enzyme is

---

HUMAN BIOCHEMISTRY

Folate Analogs as Antimicrobial and Anticancer Agents

The dependence of de novo purine biosynthesis on folic acid compounds at steps 4 and 10 means that antagonists of folic acid metabolism indirectly inhibit purine formation and, in turn, nucleic acid synthesis, cell growth, and cell division. Clearly, rapidly dividing cells such as malignancies or infective bacteria are more susceptible to these antagonists than slower-growing normal cells. Among the folic acid antagonists are *sulfonamides* (see accompanying figure). Folic acid is a vitamin for animals and is obtained in the diet. In contrast, bacteria synthesize folic acid from precursors, including *p-aminobenzoic acid (PABA),* and thus are more susceptible to sulfonamides than are animal cells.

Formation of THF, the functional folate form, depends on reduction of folate (and dihydrofolate or DHF) by dihydrofolate reductase, or DHFR (see A Deeper Look on page 894). Methotrexate (amethopterin), aminopterin, and trimethoprim are three analogs of folic acid. The first two have been used in cancer chemotherapy and the treatment of autoimmune disorders. Each binds to DHFR with about 1000-fold greater affinity than folate or DHF, thus acting as a virtually irreversible inhibitor of THF formation. Trimethoprim acts more effectively on bacterial DHFR and is prescribed for infections of the urinary tract.

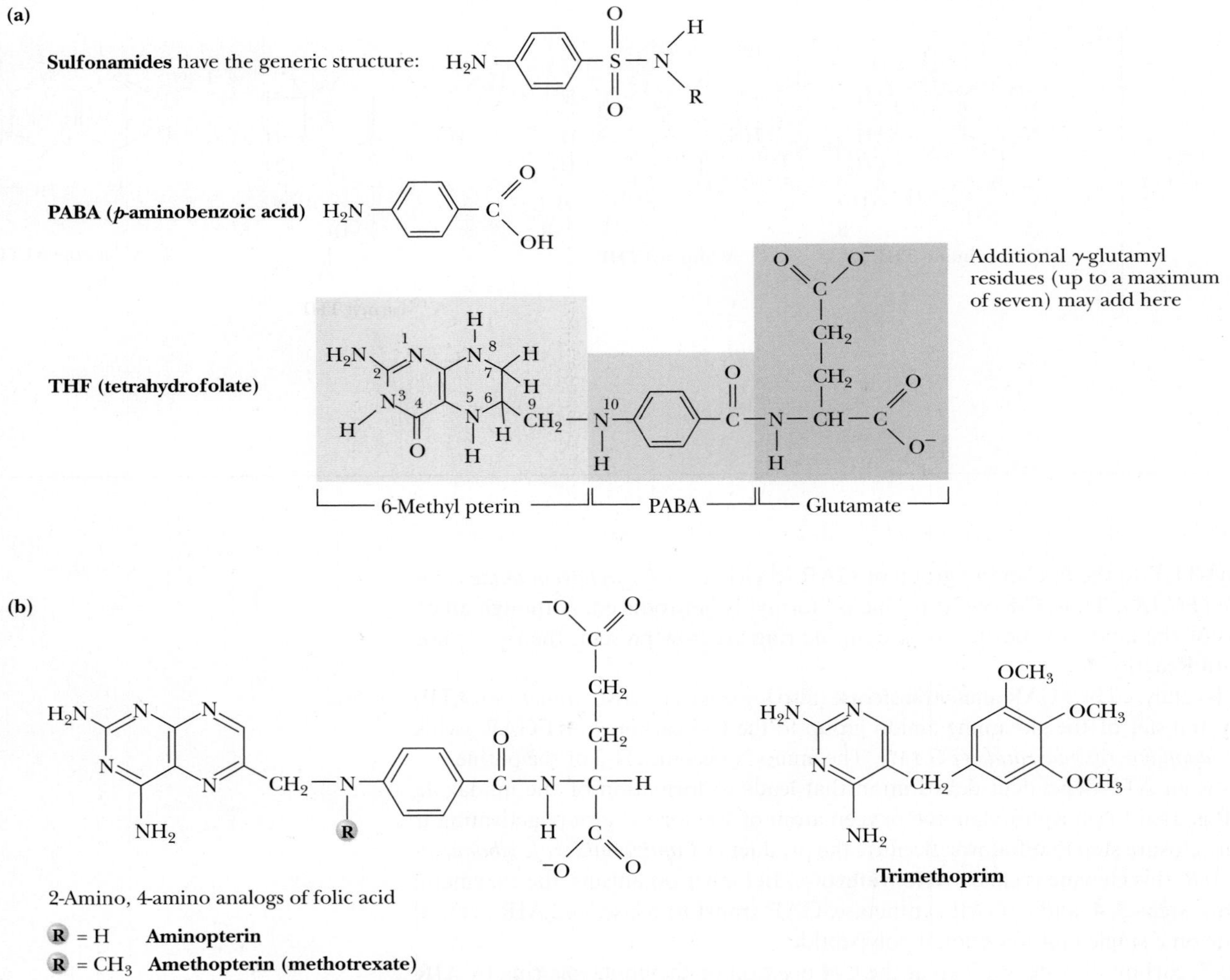

(a)

Sulfonamides have the generic structure:

PABA (*p*-aminobenzoic acid)

THF (tetrahydrofolate)

Additional γ-glutamyl residues (up to a maximum of seven) may add here

6-Methyl pterin — PABA — Glutamate

(b)

2-Amino, 4-amino analogs of folic acid

R = H   **Aminopterin**

R = CH₃   **Amethopterin (methotrexate)**

Trimethoprim

(a) Sulfa drugs, or sulfonamides, owe their antibiotic properties to their similarity to *p*-aminobenzoate (PABA), an important precursor in folic acid synthesis. Sulfonamides block folic acid formation by competing with PABA. **(b)** Precursors and analogs of folic acid employed as antimetabolites include methotrexate, aminopterin, and trimethoprim, as well as sulfonamides.

adenylosuccinase *(adenylosuccinate lyase)*. Adenylosuccinase acts again in that part of the purine pathway leading from IMP to AMP and takes its name from this latter reaction (see following). AICAR is also a by-product of the histidine biosynthetic pathway (see Chapter 25), but because ATP is the precursor to AICAR in that pathway, no net purine synthesis is achieved.

Step 10 adds the formyl carbon of N^{10}-formyl-THF as the ninth and last atom necessary for forming the purine nucleus. The enzyme is called **AICAR transformylase;** the products are THF and *N-formylaminoimidazole-4-carboxamide ribonucleotide (FAICAR)*.

Step 11 involves dehydration and ring closure to form the purine nucleotide **IMP (inosine-5′-monophosphate** or **inosinic acid);** this completes the initial phase of purine biosynthesis. The enzyme is **IMP cyclohydrolase** (also known as *IMP synthase* and *inosinicase*). Unlike step 6, this ring closure does not require ATP. In bacteria and eukaryotes, but not archaea, the enzymatic activities catalyzing steps 10 and 11 (AICAR transformylase and inosinicase) activities reside on a bifunctional polypeptide that dimerizes.

Thus, in higher organisms, 6 of the 11 enzymatic activities for purine synthesis are localized to 3 polypeptides: The enzymes for steps 3, 4, and 6 are found on a trifunctional polypeptide, and those for steps 7 and 8 and for 10 and 11 are found on bifunctional polypeptides. In total then, 7 proteins achieve the 11 steps necessary for IMP formation, starting from ribose 5-P. Benkovic and his colleagues have shown in a cultured human cell line that these 7 proteins cluster together in the cytosol to form multiprotein complexes when the cells are purine-depleted. This association is dynamic, because when the cell medium is supplemented with purines, the clustered proteins disperse. Benkovic and his colleagues have called these clusters *purinosomes*, since they represent the agents of de novo purine biosynthesis.

Note that 6 ATPs are required in the purine biosynthetic pathway from ribose-5-phosphate to IMP: one each at steps 1, 3, 5, 6, 7, and 8. However, 7 high-energy phosphate bonds (equal to 7 ATP equivalents) are consumed because α-PRPP formation in Reaction 1 followed by PP_i release in Reaction 2 represents the loss of 2 ATP equivalents.

AMP and GMP Are Synthesized from IMP

IMP is the precursor to both AMP and GMP. These major purine nucleotides are formed via distinct two-step metabolic pathways that diverge from IMP. The branch leading to AMP (adenosine 5′-monophosphate) involves the displacement of the 6-O group of inosine with aspartate (Figure 26.5) in a GTP-dependent reaction, followed by the nonhydrolytic removal of the four-carbon skeleton of Asp as fumarate; the Asp amino group remains as the 6-amino group of AMP. **Adenylosuccinate synthetase** and **adenylosuccinase** are the two enzymes. Recall that adenylosuccinase also acted at step 9 in the pathway from ribose-5-phosphate to IMP. Fumarate production provides a connection between purine synthesis and the citric acid cycle.

The formation of GMP from IMP requires oxidation at C-2 of the purine ring, followed by a glutamine-dependent amidotransferase reaction that replaces the oxygen on C-2 with an amino group to yield *2-amino,6-oxy purine nucleoside monophosphate,* or as this compound is commonly known, *guanosine monophosphate* (Figure 26.5). The enzymes in the GMP branch are **IMP dehydrogenase** and **GMP synthetase.** Note that, starting from ribose-5-phosphate, 8 ATP equivalents are consumed in the synthesis of AMP and 9 in the synthesis of GMP.

The Purine Biosynthetic Pathway Is Regulated at Several Steps

The regulatory network that controls purine synthesis is schematically represented in Figure 26.6. To recapitulate, the purine biosynthetic pathway from ribose-5-phosphate to IMP is allosterically regulated at the first two steps. Ribose-5-phosphate pyrophosphokinase, although not the committed step in purine synthesis, is subject to feedback inhibition by ADP and GDP. The enzyme catalyzing the next step, glutamine phosphoribosyl pyrophosphate amidotransferase, has two allosteric sites, one where the "A" series of nucleoside phosphates (AMP, ADP, and ATP) binds and feedback-inhibits, and another

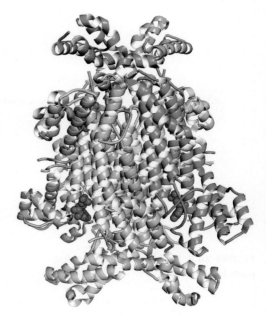

Human adenylosuccinate lyase (AMP in red) (pdb id =2J91)

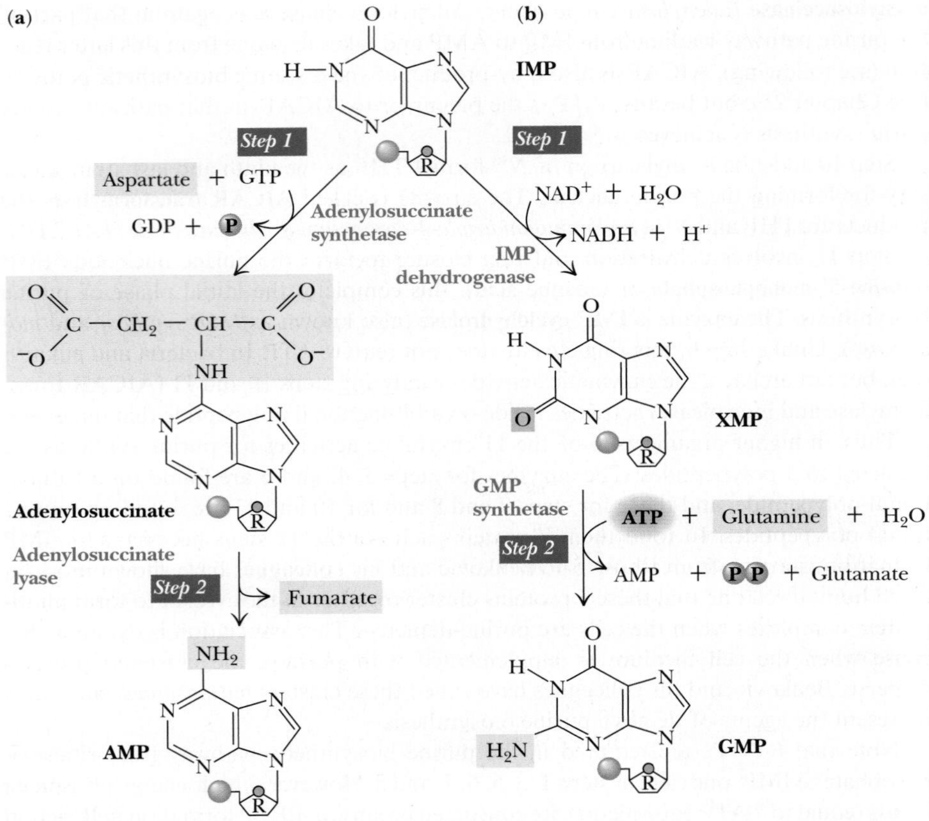

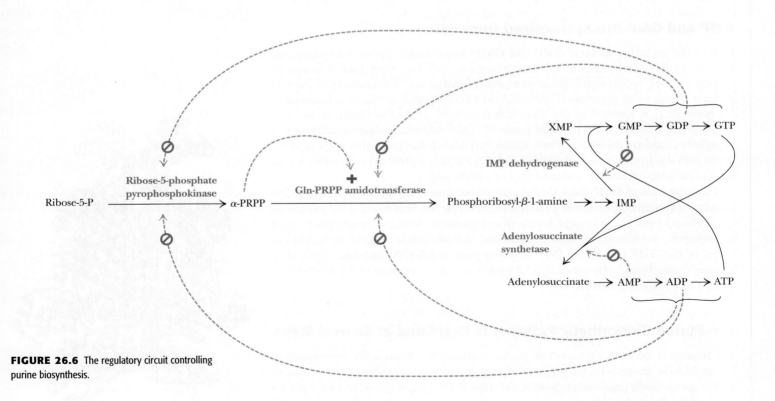

FIGURE 26.5 The synthesis of AMP and GMP from IMP.
(a) AMP synthesis: The two reactions of AMP synthesis
mimic steps 8 and 9 in the purine pathway leading to IMP.
(b) GMP synthesis.

FIGURE 26.6 The regulatory circuit controlling
purine biosynthesis.

where the corresponding "G" series binds and inhibits. Furthermore, PRPP is a "feed-forward" activator of this enzyme. Thus, the rate of IMP formation by this pathway is governed by the levels of the final end products, the adenine and guanine nucleotides.

The purine pathway splits at IMP. The first enzyme in the AMP branch, adenylosuc-cinate synthetase, is competitively inhibited by AMP. Its counterpart in the GMP branch, IMP dehydrogenase, is inhibited in a similar fashion by GMP. Thus, the fate of IMP is determined by the relative levels of AMP and GMP, so any deficiency in the amount of either of the principal purine nucleotides is self-correcting. This reciprocity of regulation is an effective mechanism for balancing the formation of AMP and GMP to satisfy cellular needs. Note also that reciprocity is even manifested at the level of energy input: GTP provides the energy to drive AMP synthesis, whereas ATP serves this role in GMP synthesis (Figure 26.6).

ATP-Dependent Kinases Form Nucleoside Diphosphates and Triphosphates from the Nucleoside Monophosphates

The products of de novo purine biosynthesis are the nucleoside monophosphates AMP and GMP. These nucleotides are converted by successive phosphorylation reactions into their metabolically prominent triphosphate forms, ATP and GTP. The first phosphory-lation, to give the nucleoside diphosphate forms, is carried out by two base-specific, ATP-dependent kinases, **adenylate kinase** and **guanylate kinase.**

$$\text{Adenylate kinase:} \quad \text{AMP} + \text{ATP} \longrightarrow 2 \text{ ADP}$$
$$\text{Guanylate kinase:} \quad \text{GMP} + \text{ATP} \longrightarrow \text{GDP} + \text{ADP}$$

These nucleoside monophosphate kinases also act on deoxynucleoside monophosphates to give dADP or dGDP.

Oxidative phosphorylation (see Chapter 20) is primarily responsible for the conversion of ADP into ATP. ATP then serves as the phosphoryl donor for synthesis of the other nucleoside triphosphates from their corresponding NDPs in a reaction catalyzed by **nucleoside diphosphate kinase,** a nonspecific enzyme. For example,

$$\text{GDP} + \text{ATP} \rightleftharpoons \text{GTP} + \text{ADP}$$

Because this enzymatic reaction is readily reversible and nonspecific with respect to both phosphoryl acceptor and donor, in effect any NDP can be phosphorylated by any NTP. The preponderance of ATP over all other nucleoside triphosphates means that, in quantitative terms, it is the principal nucleoside diphosphate kinase substrate. The enzyme does not discriminate between the ribose moieties of nucleotides and thus functions in phosphoryl transfers involving deoxy-NDPs and deoxy-NTPs as well.

26.3 | Can Cells Salvage Purines?

Nucleic acid turnover (synthesis and degradation) is an ongoing metabolic process in most cells. Messenger RNA in particular is actively synthesized and degraded. These degradative processes can lead to the release of free purines in the form of adenine, guanine, and hypoxanthine (the base in IMP). Purines represent a metabolic investment by cells. So-called salvage pathways exist to recover them in useful form. Also, purine salvage is the only way to form purine nucleotides in some parasitic organisms. Salvage reactions involve resynthesis of nucleotides from bases via **phosphoribosyltransferases.**

$$\text{Base} + \text{PRPP} \rightleftharpoons \text{nucleoside-5'-phosphate} + \text{PP}_i$$

The subsequent hydrolysis of PP_i to inorganic phosphate by pyrophosphatases renders the phosphoribosyltransferase reaction effectively irreversible.

The purine phosphoribosyltransferases are **adenine phosphoribosyltransferase (APRT),** which mediates AMP formation, and **hypoxanthine-guanine phosphoribosyltransferase (HGPRT),** which can act on either hypoxanthine to form IMP or guanine to form GMP (Figure 26.7).

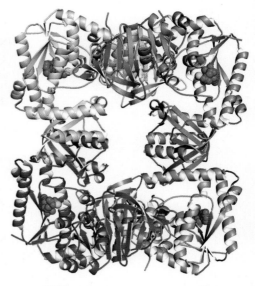

GMP synthetase tetramer with Mg^{2+} (yellow) adjacent to AMP (red) (pdb id = 1GPM).

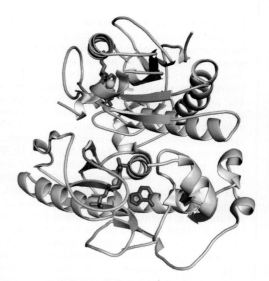

Human HGPRT (Mg^{2+} in yellow, α-PRPP in blue, purine analog in red) (pdb id = 1D6N).

FIGURE 26.7 Purine salvage by the HGPRT reaction.

26.4 How Are Purines Degraded?

Because nucleic acids are ubiquitous in cellular material, significant amounts are ingested in the diet. Nucleic acids are degraded in the digestive tract to nucleotides by various nucleases and phosphodiesterases. Nucleotides are then converted to nucleosides by base-specific nucleotidases and nonspecific phosphatases.

$$NMP + H_2O \longrightarrow nucleoside + P_i$$

HUMAN BIOCHEMISTRY

Lesch-Nyhan Syndrome—HGPRT Deficiency Leads to a Severe Clinical Disorder

The symptoms of **Lesch-Nyhan syndrome** are tragic: a crippling, gouty arthritis due to excessive uric acid accumulation (uric acid is a purine degradation product, discussed in the next section) and, worse, severe malfunctions in the nervous system that lead to mental retardation, spasticity, aggressive behavior, and self-mutilation. Lesch-Nyhan syndrome results from a complete deficiency in HGPRT activity. The structural gene for HGPRT is located on the X chromosome, and the disease is a congenital, recessive, sex-linked trait manifested only in males. The severe consequences of HGPRT deficiency argue that purine salvage has greater metabolic importance than simply the energy-saving recovery of bases. Although HGPRT might seem to play a minor role in purine metabolism, its absence has profound consequences: De novo purine biosynthesis is dramatically increased, and uric acid levels in the blood are elevated. Presumably, these changes ensue because lack of consumption of PRPP by HGPRT elevates its availability for glutamine-PRPP amidotransferase, enhancing overall de novo purine synthesis and, ultimately, uric acid production (see accompanying figure). The dramatically elevated uric acid levels lead to the particular neurological aberrations characteristic of the syndrome. Fortunately, deficiencies in HGPRT activity in fetal cells can be detected following amniocentesis. However, no medication ameliorates the neurological and behavioral consequences of this disease.

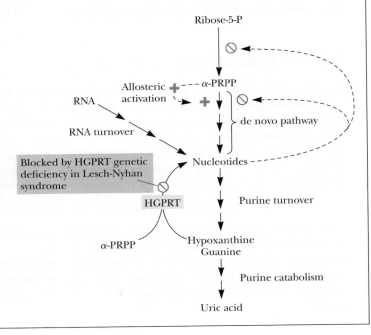

Nucleosides are hydrolyzed by nucleosidases or nucleoside phosphorylases to release the purine base:

$$\text{Nucleoside} + H_2O \xrightarrow{\textit{nucleosidase}} \text{base} + \text{ribose}$$

$$\text{Nucleoside} + P_i \xrightarrow{\textit{nucleoside phosphorylase}} \text{base} + \text{ribose-1-P}$$

The pentoses liberated in these reactions provide the only source of metabolic energy available from purine nucleotide degradation.

Feeding experiments using radioactively labeled nucleic acids as metabolic tracers have demonstrated that little of the nucleotide ingested in the diet is incorporated into cellular nucleic acids. Dietary purines are converted to uric acid (see following discussion) in the gut and excreted, and pyrimidine nucleosides are inefficiently absorbed into the bloodstream. These findings confirm the de novo pathways of nucleotide biosynthesis as the primary source of nucleic acid precursors. Ingested bases are, for the most part, excreted. Nevertheless, cellular nucleic acids do undergo degradation in the course of the continuous recycling of cellular constituents.

The Major Pathways of Purine Catabolism Lead to Uric Acid

The major pathways of purine catabolism in animals lead to uric acid formation (Figure 26.8). The various nucleotides are first converted to nucleosides by **intracellular nucleotidases.** These nucleotidases are under strict metabolic regulation to ensure that their substrates, which act as intermediates in many vital processes, are not depleted below critical levels. Nucleosides are then degraded by the enzyme **purine nucleoside phosphorylase (PNP)** to release the purine base and ribose-1-P. Note that neither adenosine nor deoxyadenosine is a substrate for PNP. Instead, these nucleosides are first converted to inosine by **adenosine deaminase.** The PNP products are merged into *xanthine* by **guanine**

HUMAN BIOCHEMISTRY

Severe Combined Immunodeficiency Syndrome—A Lack of Adenosine Deaminase Is One Cause of This Inherited Disease

Severe combined immunodeficiency syndrome, or **SCID,** is a group of related inherited disorders characterized by the lack of an immune response to infectious disease. This immunological insufficiency is attributable to the inability of B and T lymphocytes to proliferate and produce antibodies in response to an antigenic challenge. About 30% of SCID patients suffer from a deficiency in the enzyme *adenosine deaminase (ADA).* ADA deficiency is also implicated in a variety of other diseases, including AIDS, anemia, and various lymphomas and leukemias. **Gene therapy,** *the repair of a genetic deficiency through introduction of a functional recombinant version of the gene,* has been attempted on individuals with SCID due to a defective ADA gene (see Section 12.6). ADA is a

Zn^{2+}-dependent enzyme, and Zn^{2+} deficiency can also lead to reduced immune function.

In the absence of ADA, deoxyadenosine is not deaminated to deoxyinosine as normal (see Figure 26.8). Instead, it is salvaged by a nucleoside kinase, which converts it to dAMP. Phosphorylation of dAMP leads to dATP, a potent feedback inhibitor of deoxynucleotide biosynthesis (discussed later in this chapter). Without deoxyribonucleotides, DNA cannot be replicated and cells cannot divide (see accompanying figure). Rapidly proliferating cell types such as lymphocytes are particularly susceptible if DNA synthesis is impaired.

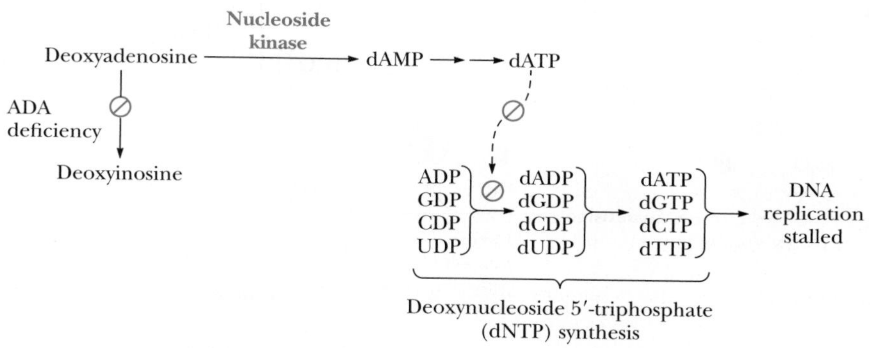

Deoxynucleoside 5′-triphosphate (dNTP) synthesis

deaminase and **xanthine oxidase,** and xanthine is then oxidized to uric acid by this latter enzyme.

The Purine Nucleoside Cycle in Skeletal Muscle Serves as an Anaplerotic Pathway

Deamination of AMP to IMP by **AMP deaminase** (Figure 26.8) followed by resynthesis of AMP from IMP by the de novo purine pathway enzymes, *adenylosuccinate synthetase* and *adenylosuccinate lyase,* constitutes a purine nucleoside cycle (Figure 26.9). This cycle has the net effect of converting aspartate to fumarate plus NH_4^+. Although this cycle might seem like senseless energy consumption, it plays an important role in energy metabolism in skeletal muscle: The fumarate that it generates replenishes the levels of

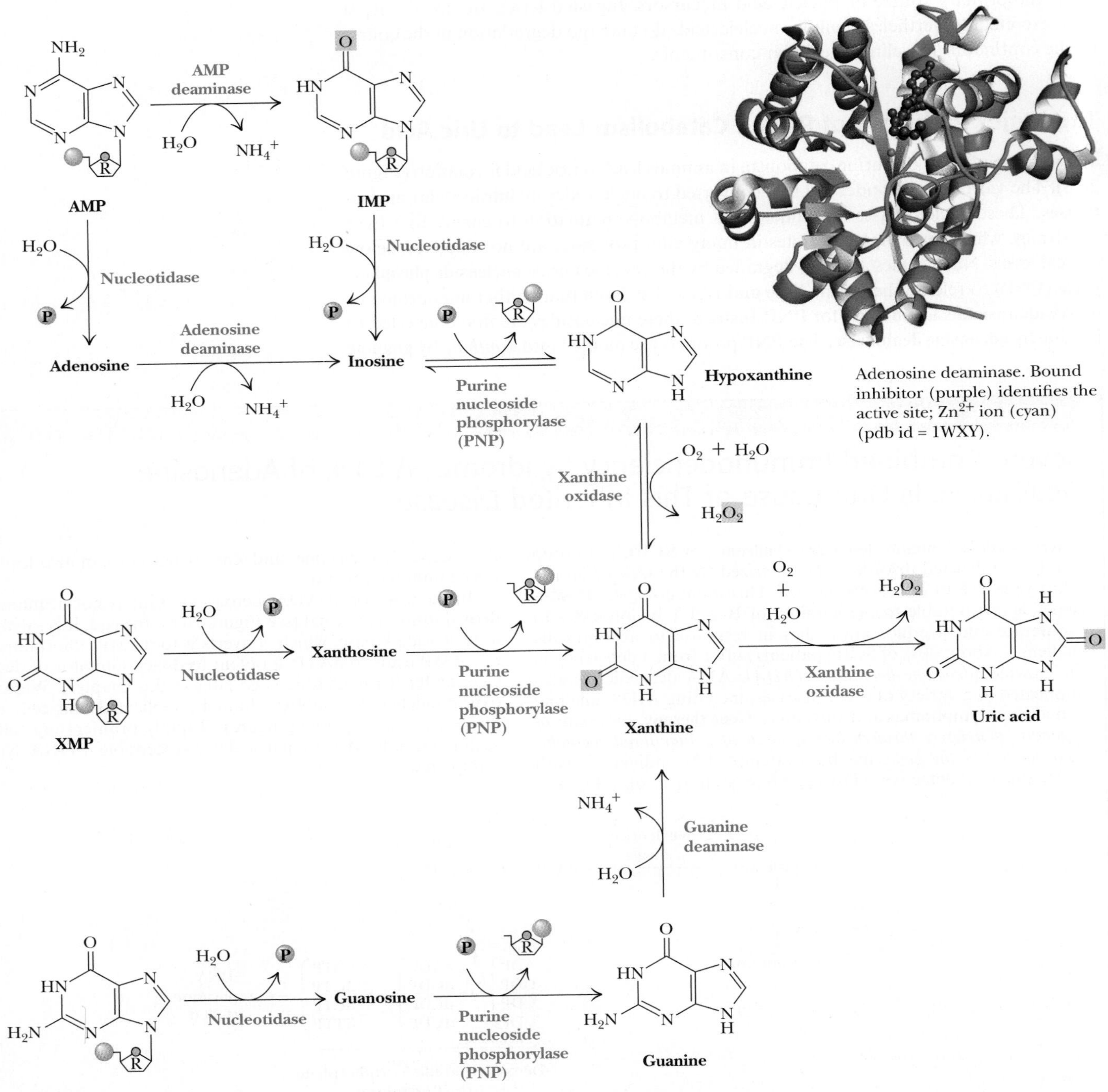

Adenosine deaminase. Bound inhibitor (purple) identifies the active site; Zn^{2+} ion (cyan) (pdb id = 1WXY).

FIGURE 26.8 The major pathways for purine catabolism.

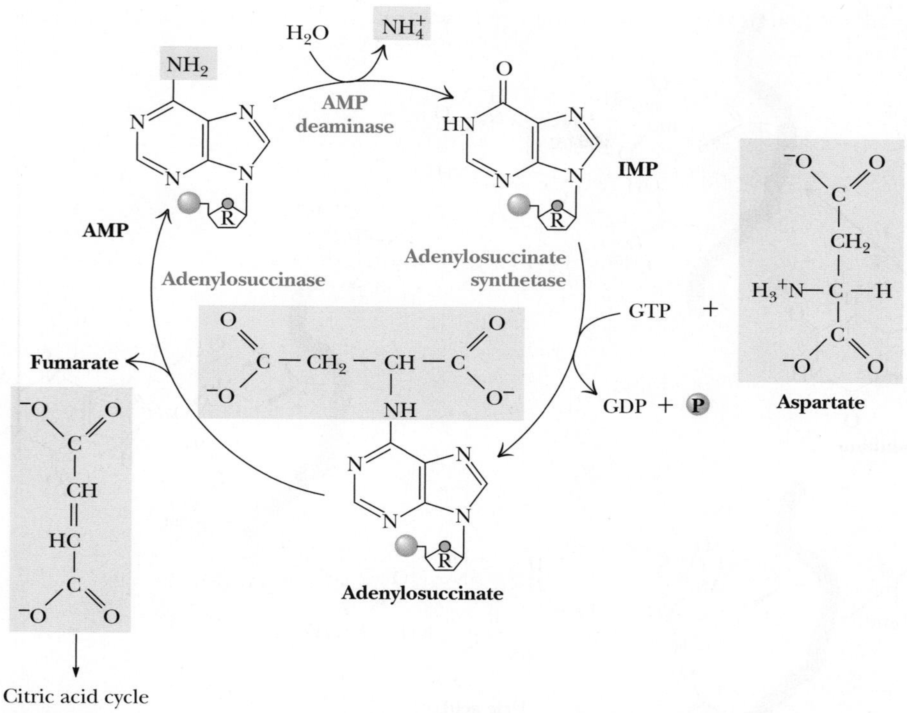

FIGURE 26.9 The purine nucleoside cycle for anaplerotic replenishment of citric acid cycle intermediates in skeletal muscle.

citric acid cycle intermediates lost in amphibolic side reactions (see Chapter 19). Skeletal muscle lacks the usual complement of anaplerotic enzymes and relies on enhanced levels of AMP deaminase, adenylosuccinate synthetase, and adenylosuccinate lyase to compensate.

Xanthine Oxidase

Xanthine oxidase (Figure 26.8) is present in large amounts in liver, intestinal mucosa, and milk. It oxidizes hypoxanthine to xanthine and xanthine to uric acid. Xanthine oxidase is a rather indiscriminate enzyme, using molecular oxygen to oxidize a wide variety of purines, pteridines, and aldehydes, producing H_2O_2 as a product. Xanthine oxidase possesses FAD, nonheme Fe-S centers, and *molybdenum cofactor* (a molybdenum-containing pterin complex very similar to the one shown in Figure 25.2) as electron-transferring prosthetic groups. Its mechanism of action is diagrammed in Figure 26.10. In humans and other primates, uric acid is the end product of purine catabolism and is excreted in the urine. Birds, terrestrial reptiles, and many insects also excrete uric acid, but in these organisms, uric acid represents the major nitrogen excretory compound because, unlike mammals, they do not also produce urea (see Chapter 25). Instead, the catabolism of all nitrogenous compounds, including amino acids, is channeled into uric acid. This route of nitrogen catabolism allows these animals to conserve water by excreting crystals of uric acid in pastelike solid form.

Gout Is a Disease Caused by an Excess of Uric Acid

Gout is the clinical term describing the physiological consequences accompanying excessive uric acid accumulation in body fluids. Uric acid and urate salts are rather insoluble in water and tend to precipitate from solution if produced in excess. The most common symptom of gout is arthritic pain in the joints as a result of urate deposition in cartilaginous tissue. The joint of the big toe is particularly susceptible. Urate crystals may also appear as kidney stones and lead to painful obstruction of the urinary tract.

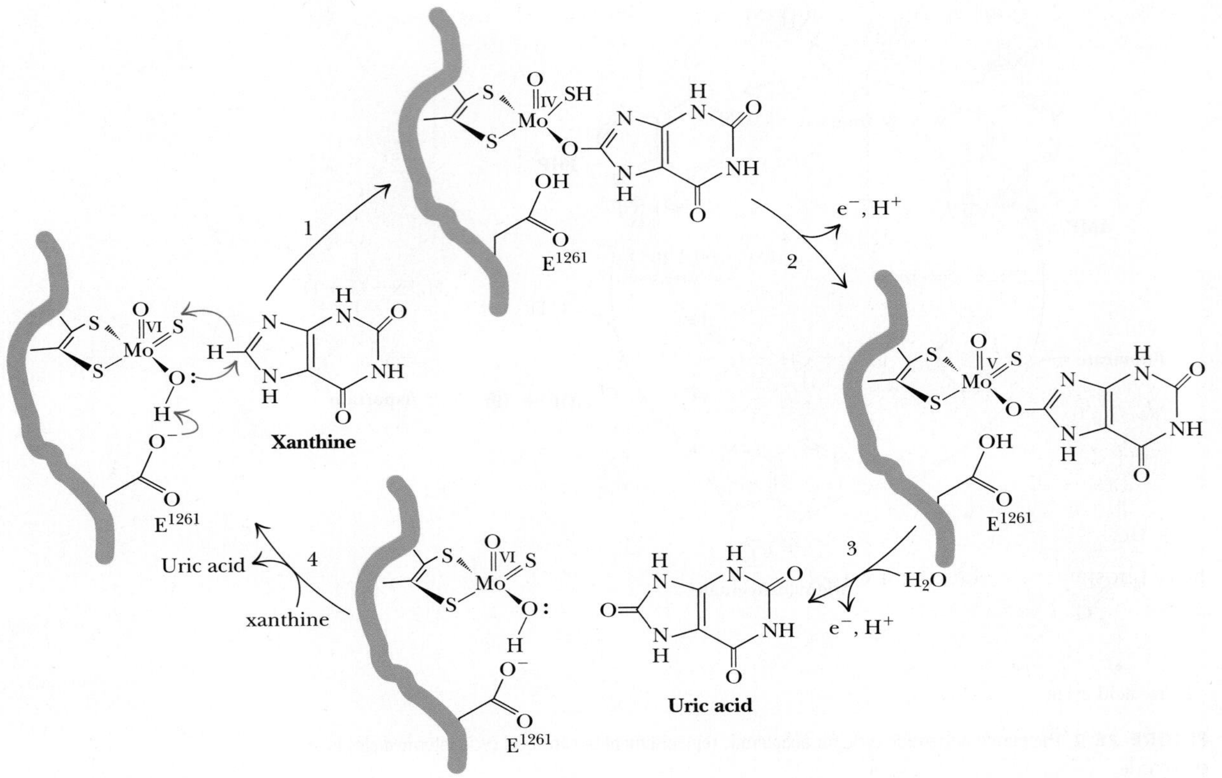

FIGURE 26.10 Xanthine oxidase catalyzes a hydroxylase-type mechanism. (1) Nucleophilic attack by the Mo^{IV}-OH on the xanthine C-8 carbon and concomitant hydride transfer from C-8 to Mo=S reduces Mo^{VI} to Mo^{IV}. (2, 3) Electron transfer from the molybdenum cofactor to the enzyme's Fe-S and FAD prosthetic groups and H$^+$ dissociations from the Mo-cofactor and Glu^{1261}, along with OH$^-$ displacement of oxidized substrate (uric acid) regenerates the Mo^{IV} form of the cofactor for another reaction cycle (4). Adapted from Figure 1 in Cao, H., Pauff, J. M., and Hille, R., 2010. Substrate orientation and catalytic specificity in the action of xanthine oxidase. The sequential hydroxylation of hypoxanthine to uric acid. *Journal of Biological Chemistry* **285:**28044–28053.

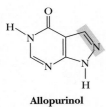

Allopurinol

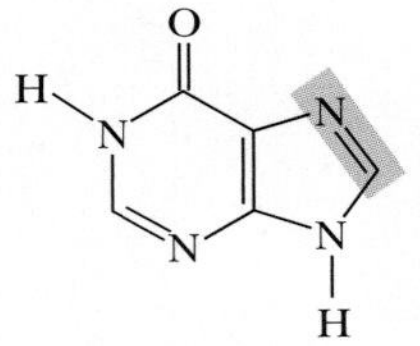

Hypoxanthine

FIGURE 26.11 Allopurinol, an analog of hypoxanthine, is a potent inhibitor of xanthine oxidase.

Hyperuricemia, chronic elevation of blood uric acid levels, occurs in about 3% of the population as a consequence of impaired excretion of uric acid or overproduction of purines. Purine-rich foods such as caviar (fish eggs rich in nucleic acids) may exacerbate the condition. The biochemical causes of gout are varied. However, a common treatment is *allopurinol* (Figure 26.11). This hypoxanthine analog binds tightly to xanthine oxidase, thereby inhibiting its activity and preventing uric acid formation. Hypoxanthine and xanthine do not accumulate to harmful concentrations because they are more soluble and thus more easily excreted.

Different Animals Oxidize Uric Acid to Form Various Excretory Products

The subsequent metabolism of uric acid in organisms that don't excrete it is shown in Figure 26.12. In molluscs and in mammals other than primates, uric acid is oxidized by **urate oxidase** to *allantoin* and excreted. In bony fishes (teleosts), uric acid degradation proceeds through yet another step wherein allantoin is hydrolyzed to *allantoic acid* by **allantoinase** before excretion. Cartilaginous fish (sharks and rays) and amphibians further degrade allantoic acid via the enzyme **allantoicase** to liberate glyoxylic acid and 2 equivalents of *urea*. Even simpler animals, such as most marine invertebrates (crustacea and so forth), use **urease** to hydrolyze urea to CO$_2$ and ammonia. In contrast to animals that must rid themselves of potentially harmful nitrogen waste products, microorganisms often are limited in growth by nitrogen availability. Many possess an identical pathway of uric acid degradation, using it instead to liberate NH_3 from uric acid so that it can be assimilated into organic-N compounds essential to their survival.

26.5 | How Do Cells Synthesize Pyrimidines?

In contrast to purines, pyrimidines are not synthesized as nucleotide derivatives. Instead, the pyrimidine ring system is constructed before a ribose-5-P moiety is attached. Also, only two precursors, carbamoyl-P and aspartate, contribute atoms to the six-membered pyrimidine ring (Figure 26.13), compared to seven precursors for the nine purine atoms.

Mammals have two enzymes for carbamoyl phosphate synthesis. Carbamoyl phosphate for pyrimidine biosynthesis is formed by **carbamoyl phosphate synthetase II (CPS-II),** a cytosolic enzyme. Recall that carbamoyl phosphate synthetase I is a mitochondrial enzyme dedicated to the urea cycle and arginine biosynthesis (see Figures 25.22 and 25.23). The substrates of carbamoyl phosphate synthetase II are HCO_3^-, H_2O, glutamine, and two ATPs (Figure 26.14). Thus, the N atom in carbamoyl phosphate made by CPS-II comes from the amide group of glutamine. The first ATP is used to form carboxy phosphate, an activated form of CO_2 (step 1, Figure 26.14). The glutamine amide displaces phosphate from carboxy phosphate to give carbamate (step 2). Phosphorylation of carbamate by the second ATP in the overall reaction yields carbamoyl phosphate (step 3). Because carbamoyl phosphate made by CPS-II in mammals has no fate other than incorporation into pyrimidines, mammalian CPS-II can be viewed as the committed step in the pyrimidine de novo pathway (Figure 26.15, step 1). Bacteria and plants have but one CPS, and its carbamoyl phosphate product is incorporated into arginine as well as pyrimidines. Thus, the committed step in bacterial pyrimidine synthesis is the next reaction, which is mediated by **aspartate transcarbamoylase (ATCase).**

ATCase catalyzes the condensation of carbamoyl phosphate with aspartate to form carbamoyl aspartate (Figure 26.15, step 2). No ATP input is required at this step because carbamoyl phosphate represents an "activated" carbamoyl group.

Step 3 of pyrimidine synthesis involves ring closure and dehydration via linkage of the $-NH_2$ group introduced by carbamoyl phosphate with the former β-COO^- of aspartate; this reaction is mediated by the enzyme **dihydroorotase.** The product of the reaction is *dihydroorotate (DHO),* a six-membered ring compound. Dihydroorotate is not a true pyrimidine, but its oxidation yields *orotate,* which is. This oxidation (step 4) is catalyzed by **dihydroorotate dehydrogenase.** Bacterial dihydroorotate dehydrogenases are NAD^+-linked flavoproteins which are somewhat unusual in possessing both FAD and FMN; these enzymes also have nonheme Fe-S centers as additional redox prosthetic groups. The eukaryotic version of dihydroorotate dehydrogenase is a protein component of the inner mitochondrial membrane; its immediate e^- acceptor is a quinone, and oxidation of the reduced quinone by the mitochondrial e^- transport chain can drive ATP synthesis via oxidative phosphorylation. At this stage, ribose-5-phosphate is joined to N-1 of orotate in appropriate *N-β-glycosidic* configuration, giving the pyrimidine nucleotide *orotidine-5'-monophosphate,* or *OMP* (step 5, Figure 26.15). The ribose phosphate donor is PRPP; the enzyme is **orotate phosphoribosyltransferase.** The next reaction is catalyzed by **OMP decarboxylase.** Decarboxylation of OMP gives *UMP (uridine-5'-monophosphate,* or *uridylic acid),* one of the two common pyrimidine ribonucleotides.

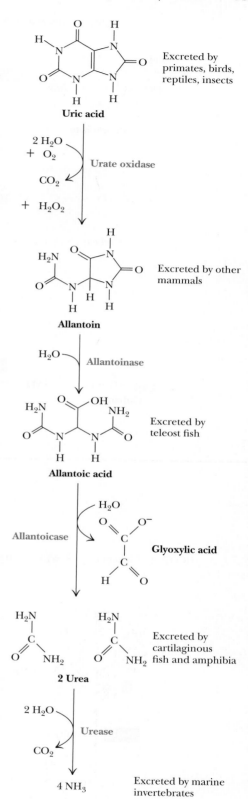

FIGURE 26.12 The catabolism of uric acid to allantoin, allantoic acid, urea, or ammonia in various animals.

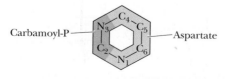

FIGURE 26.13 The metabolic origin of the six atoms of the pyrimidine ring.

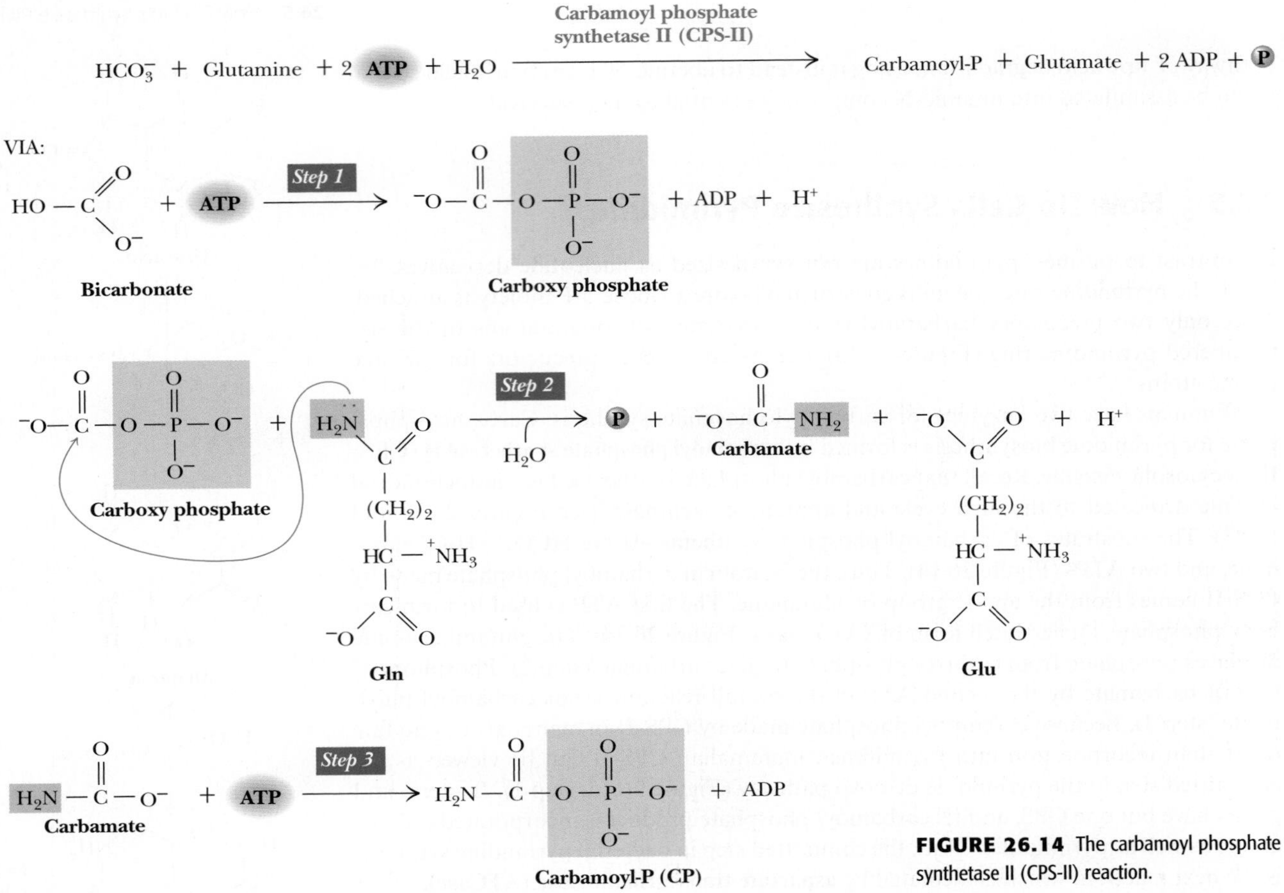

$$\text{HCO}_3^- + \text{Glutamine} + 2\ \textbf{ATP} + \text{H}_2\text{O} \xrightarrow{\text{Carbamoyl phosphate synthetase II (CPS-II)}} \text{Carbamoyl-P} + \text{Glutamate} + 2\ \text{ADP} + \textbf{P}$$

VIA:

FIGURE 26.14 The carbamoyl phosphate synthetase II (CPS-II) reaction.

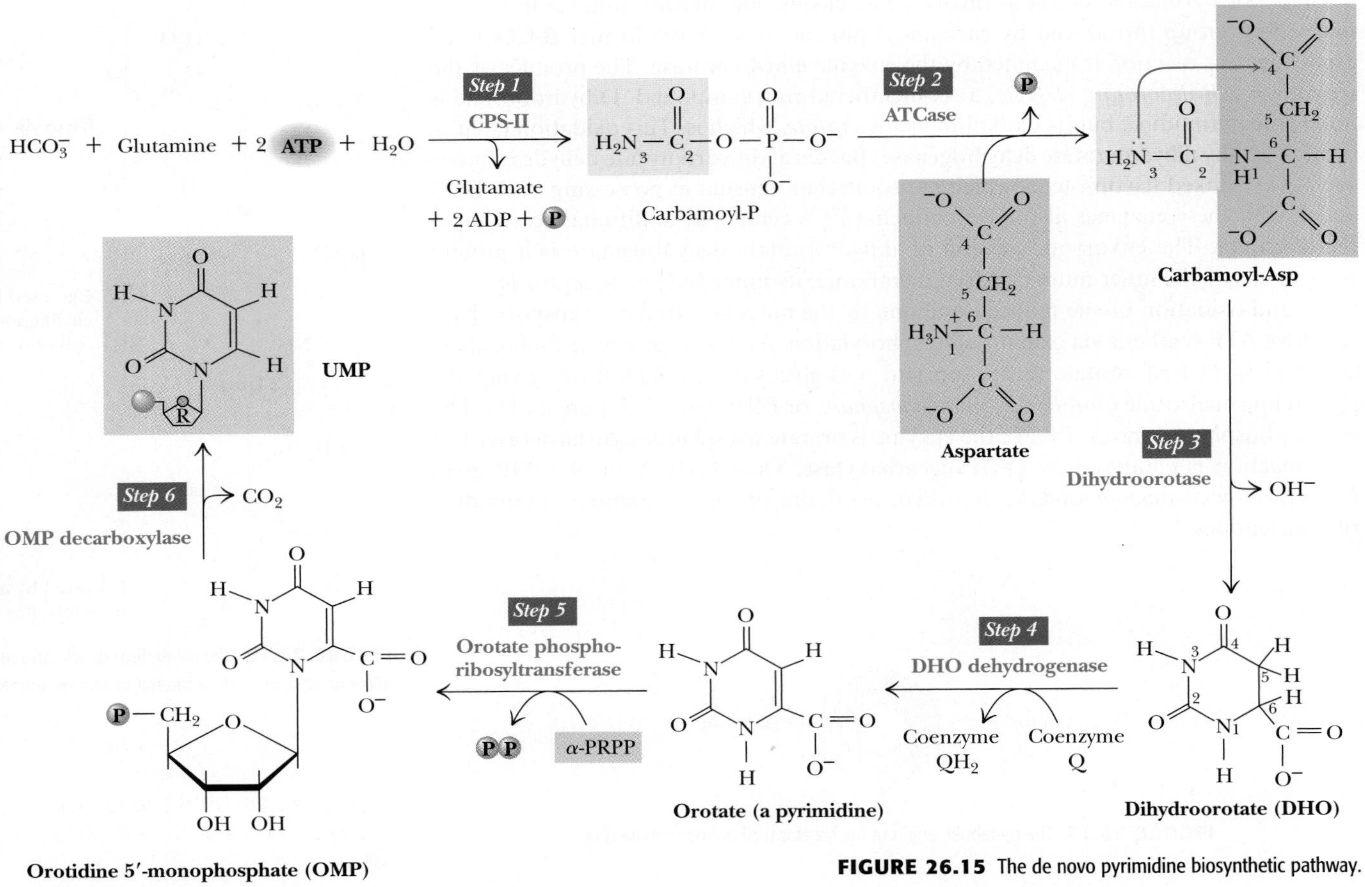

FIGURE 26.15 The de novo pyrimidine biosynthetic pathway.

"Metabolic Channeling" by Multifunctional Enzymes of Mammalian Pyrimidine Biosynthesis

CPS-II (Figure 26.15), glutamine 5-phosphoribosyl-α-pyrophosphate amidotransferase (see Figure 26.3), and CTP synthetase (Figure 26.16) are enzymes with metabolite tunnels (see A Deeper Look on page 876). In a subtle distinction, **metabolic channeling** refers to the transfer of metabolites between different enzymic sites on a **multifunctional polypeptide** chain.

In bacteria, the six enzymes of de novo pyrimidine biosynthesis exist as distinct proteins, each independently catalyzing its specific step in the overall pathway. In contrast, in mammals, the six enzymatic activities are distributed among only three proteins, two of which are multifunctional polypeptides. The first three steps of pyrimidine synthesis, CPS-II, aspartate transcarbamoylase, and dihydroorotase, are all localized on a single 210-kD cytosolic polypeptide. This multifunctional enzyme is the product of a solitary gene, yet it is equipped with the active sites for all three enzymatic activities. Step 4 in Figure 26.15 is catalyzed by DHO dehydrogenase, a separate enzyme associated with the outer surface of the inner mitochondrial membrane, but the enzymatic activities mediating steps 5 and 6, namely, orotate phosphoribosyltransferase and OMP decarboxylase in mammals, are also found on a single cytosolic polypeptide known as **UMP synthase.**

The purine biosynthetic pathway in higher organisms also provides examples of metabolic channeling. Recall that steps 3, 4, and 6 of de novo purine synthesis are catalyzed by three enzymatic activities localized on a single multifunctional polypeptide, and steps 7 and 8 and steps 10 and 11 by respective bifunctional polypeptides (see Figure 26.3). Fatty acid biosynthesis in animals is another instance of this mode of metabolic organization (see Figure 24.9).

Such multifunctional enzymes confer an advantage: The product of one reaction in a pathway is the substrate for the next. In multifunctional enzymes, such products remain bound and are channeled directly to the next active site, rather than dissociated into the surrounding medium for diffusion to the next enzyme. This metabolic

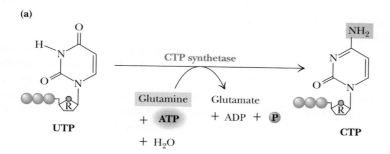

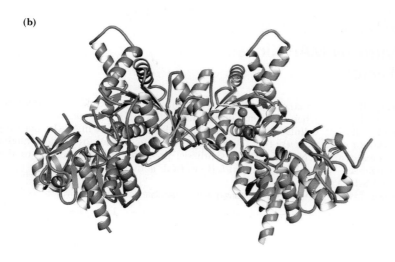

FIGURE 26.16 (a) CTP synthesis from UTP. (b) Structure of CTP synthetase dimer with bound Mg²⁺ ions (pdb id = 1S1M).

channeling is more efficient for a variety of reasons: Transit time for movement from one active site to the next is shortened, substrates are not diluted into the solvent phase, chemically reactive intermediates are protected from decomposition in the aqueous milieu, no pools of intermediates accumulate, and intermediates are shielded from interactions with other enzymes that might metabolize them. The end result is that metabolic channeling can increase the catalytic efficiency of a reaction sequence by as much as two orders of magnitude, while minimizing the overall concentration of the metabolic intermediates involved.

UMP Synthesis Leads to Formation of the Two Most Prominent Ribonucleotides—UTP and CTP

The two prominent pyrimidine ribonucleotide products are derived from UMP via the same unbranched pathway. First, UDP is formed from UMP via an ATP-dependent *nucleoside monophosphate kinase.*

$$UMP + ATP \rightleftharpoons UDP + ADP$$

Then, UTP is formed by *nucleoside diphosphate kinase.*

$$UDP + ATP \rightleftharpoons UTP + ADP$$

Amination of UTP at the 6-position gives CTP. The enzyme **CTP synthetase** is a glutamine amidotransferase (Figure 26.16). ATP hydrolysis provides the energy to drive the reaction.

Pyrimidine Biosynthesis Is Regulated at ATCase in Bacteria and at CPS-II in Animals

Pyrimidine biosynthesis in bacteria is allosterically regulated at aspartate transcarbamoylase (ATCase). *Escherichia coli* ATCase is feedback-inhibited by the end product, CTP. ATP, which can be viewed as a signal of both energy availability and purine sufficiency, is an allosteric activator of ATCase. CTP and ATP compete for a common allosteric site on the enzyme. In many bacteria, UTP, not CTP, acts as the ATCase feedback inhibitor.

In animals, CPS-II catalyzes the committed step in pyrimidine synthesis and serves as the focal point for allosteric regulation. UDP and UTP are feedback inhibitors of CPS-II, whereas PRPP and ATP are allosteric activators. With the exception of ATP, none of these compounds are substrates of CPS-II or of either of the two other enzymic activities residing with it on the trifunctional polypeptide. Figure 26.17 compares the regulatory circuits governing pyrimidine synthesis in bacteria and animals.

HUMAN BIOCHEMISTRY

Mammalian CPS-II Is Activated In Vitro by MAP Kinase and In Vivo by Epidermal Growth Factor

The rate-limiting step in mammalian de novo pyrimidine synthesis is catalyzed by CPS-II, and proliferating cells require lots of pyrimidine nucleotides for growth and cell division. Normally, CPS-II is feedback-inhibited by UTP, but in vitro phosphorylation of CPS-II by **MAP kinase** *(mitogen-activated protein kinase)* creates a covalently modified (phosphorylated) CPS-II that is no longer sensitive to UTP inhibition. Furthermore, phosphorylated CPS-II is more responsive to PRPP activation. Both of these responses favor enhanced pyrimidine biosynthesis. This regulation occurs in vivo when **epidermal growth factor (EGF)**, a **mitogen,** initiates an intracellular cascade of reactions that culminates in MAP kinase activation. By this action, pyrimidine nucleotides are made available for RNA and DNA synthesis, processes that are central to the cell proliferation that follows EGF activation.

A **mitogen** is a hormone that stimulates mitosis (cell division).

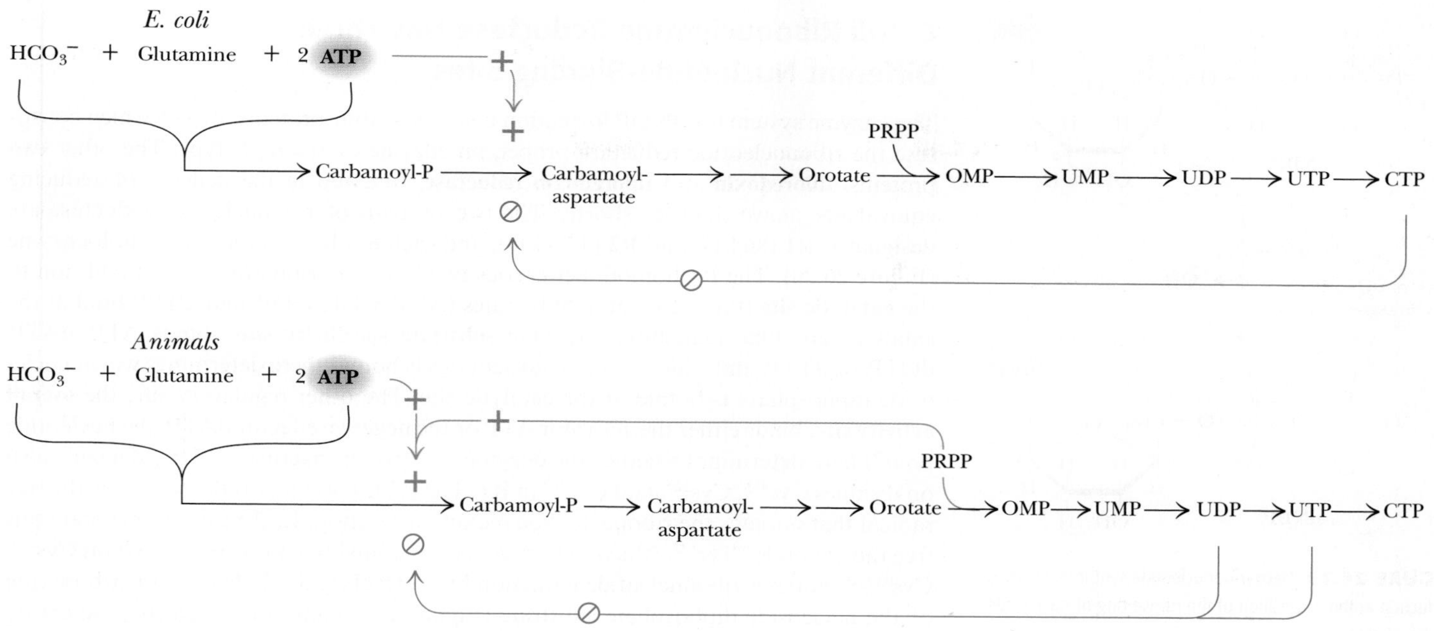

E. coli

HCO_3^- + Glutamine + 2 **ATP** ⟶

Carbamoyl-P ⟶ Carbamoyl-aspartate ⟶ ⟶ Orotate ⟶ OMP ⟶ UMP ⟶ UDP ⟶ UTP ⟶ CTP

Animals

HCO_3^- + Glutamine + 2 **ATP**

Carbamoyl-P ⟶ Carbamoyl-aspartate ⟶ ⟶ Orotate ⟶ OMP ⟶ UMP ⟶ UDP ⟶ UTP ⟶ CTP

FIGURE 26.17 A comparison of the regulatory circuits that control pyrimidine synthesis in *E. coli* and animals.

26.6 How Are Pyrimidines Degraded?

In some organisms, free pyrimidines, like purines, are salvaged and recycled to form nucleotides via phosphoribosyltransferase reactions similar to those discussed earlier. In humans, however, pyrimidines are recycled from nucleosides, but free pyrimidine bases are not salvaged. Pyrimidine catabolism results in degradation of the pyrimidine ring to products reminiscent of the original substrates, aspartate, CO_2, and ammonia (Figure 26.18). β-Alanine can be recycled into the synthesis of coenzyme A. Catabolism of the pyrimidine base, thymine (5-methyluracil), yields β-aminoisobutyric acid instead of β-alanine.

Pathways presented thus far in this chapter account for the synthesis of the four principal ribonucleotides: ATP, GTP, UTP, and CTP. These compounds serve important coenzymic functions in metabolism and are the immediate precursors for ribonucleic acid (RNA) synthesis. Roughly 90% of the total nucleic acid in cells is RNA, with the remainder being deoxyribonucleic acid (DNA). DNA differs from RNA in being a polymer of deoxyribonucleotides, one of which is deoxythymidylic acid. We now turn to the synthesis of these compounds.

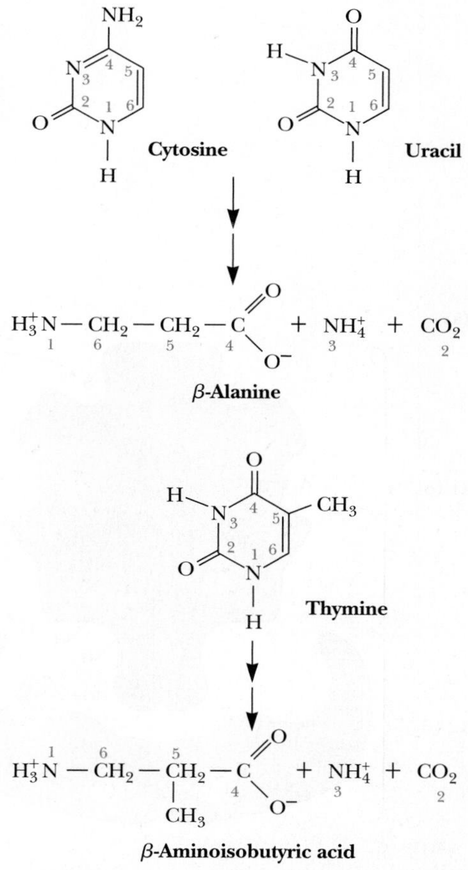

26.7 How Do Cells Form the Deoxyribonucleotides That Are Necessary for DNA Synthesis?

The deoxyribonucleotides have only one metabolic purpose: to serve as precursors for DNA synthesis. In most organisms, ribonucleoside diphosphates (NDPs) are the substrates for deoxyribonucleotide formation. Reduction at the 2′-position of the ribose ring in NDPs produces 2′-deoxy forms of these nucleotides (Figure 26.19). This reaction involves replacement of the 2′-OH by a hydride ion (H:⁻) and is catalyzed by an enzyme known as **ribonucleotide reductase.** Enzymatic ribonucleotide reduction involves a free radical mechanism, and three classes of ribonucleotide reductases are known, differing from each other in their mechanisms of free radical generation. Class I enzymes, found in *E. coli* and virtually all eukaryotes, are Fe-dependent and generate the required free radical on a specific tyrosyl side chain.

Cytosine

Uracil

$H_3^+N — CH_2 — CH_2 — C(=O)O^-$ + NH_4^+ + CO_2

β-Alanine

Thymine

$H_3^+N — CH_2 — CH_2 — C(=O)O^-$ with CH₃

β-Aminoisobutyric acid

FIGURE 26.18 Pyrimidine degradation.

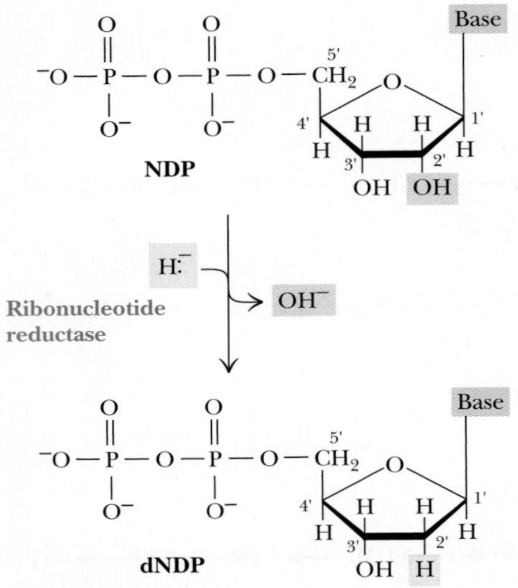

NDP

H:⁻ → OH⁻

Ribonucleotide
reductase

dNDP

FIGURE 26.19 Deoxyribonucleotide synthesis involves reduction at the 2'-position of the ribose ring of nucleoside diphosphates.

E. coli Ribonucleotide Reductase Has Three Different Nucleotide-Binding Sites

The enzyme system for dNDP formation consists of four proteins, two of which constitute the ribonucleotide reductase proper, an enzyme of the $\alpha_2\beta_2$ type. The other two proteins, **thioredoxin** and **thioredoxin reductase,** function in the delivery of reducing equivalents, as we shall see shortly. The two proteins of ribonucleotide reductase are designated **R1** (86 kD) and **R2** (43.5 kD), and each is a homodimer in the holoenzyme (Figure 26.20). The R1 homodimer carries two types of regulatory sites in addition to the **catalytic site** (the active site). Substrates (ADP, CDP, GDP, and UDP) bind at the catalytic site. One regulatory site—the **substrate specificity site**—binds ATP, dATP, dGTP, or dTTP, and which of these nucleotides is bound there determines which nucleoside diphosphate is bound at the catalytic site. The other regulatory site, the **overall activity site,** binds either the activator ATP or the negative effector dATP; the nucleotide bound here determines whether the enzyme is active or inactive. Activity depends also on residues Cys^{439}, Cys^{225}, and Cys^{462} in R1. The R2 homodimer is the source of the free radical that initiates the ribonucleotide reductase reaction. Its 2 Fe atoms generate this free radical on R2 Tyr^{122}, which in turn generates a thiyl free radical (Cys-S·) on Cys^{439}. Cys^{439}-S· initiates ribonucleotide reduction by abstracting the 3'-H from the ribose ring of the nucleoside diphosphate substrate (Figure 26.21) and forming a free radical on C-3'. Subsequent dehydration forms the deoxyribonucleotide product.

Thioredoxin Provides the Reducing Power for Ribonucleotide Reductase

NADPH is the ultimate source of reducing equivalents for ribonucleotide reduction, but the immediate source is reduced **thioredoxin,** a small (12-kD) protein with reactive Cys-sulfhydryl groups situated near one another in the sequence Cys-Gly-Pro-Cys. These Cys residues are able to undergo reversible oxidation–reduction between (—S—S—) and (—SH HS—) and, in their reduced form, serve as primary electron donors to regenerate

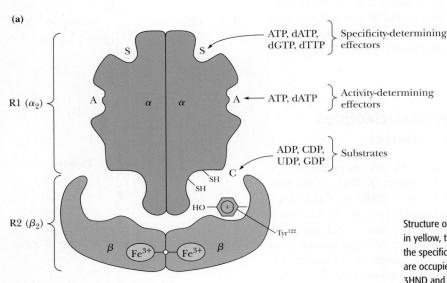

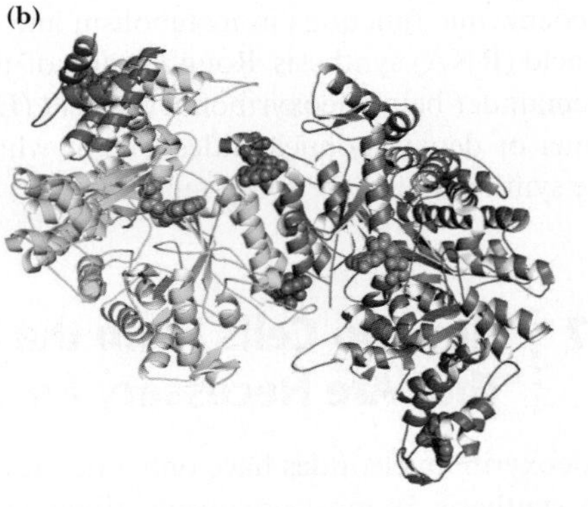

(a)

R1 (α_2)

R2 (β_2)

ATP, dATP, dGTP, dTTP } Specificity-determining effectors

ATP, dATP } Activity-determining effectors

ADP, CDP, UDP, GDP } Substrates

Tyr^{122}

(b)

Structure of the human ribonucleotide reductase R1 dimer. One monomer is shown in yellow, the other in cyan. The catalytic, or C, sites are occupied by GDP (orange); the specificity, or S, sites are occupied by TTP (green); and the activity, or A, sites are occupied by ATP (purple). This molecular graphic is a composite of pdb id = 3HND and pdb id = 3HNE. *Adapted from Figure 1 in Fairman, J. W., Wijerathna, S. R., Ahmad, M. F., Xu. H., et al., 2011. Structural basis for regulation of human ribonucleotide reductase by nucleotide-induced oligomerization.* Nature Structural and Molecular Biology **18:**316–323.

FIGURE 26.20 *E. coli* ribonucleotide reductase.

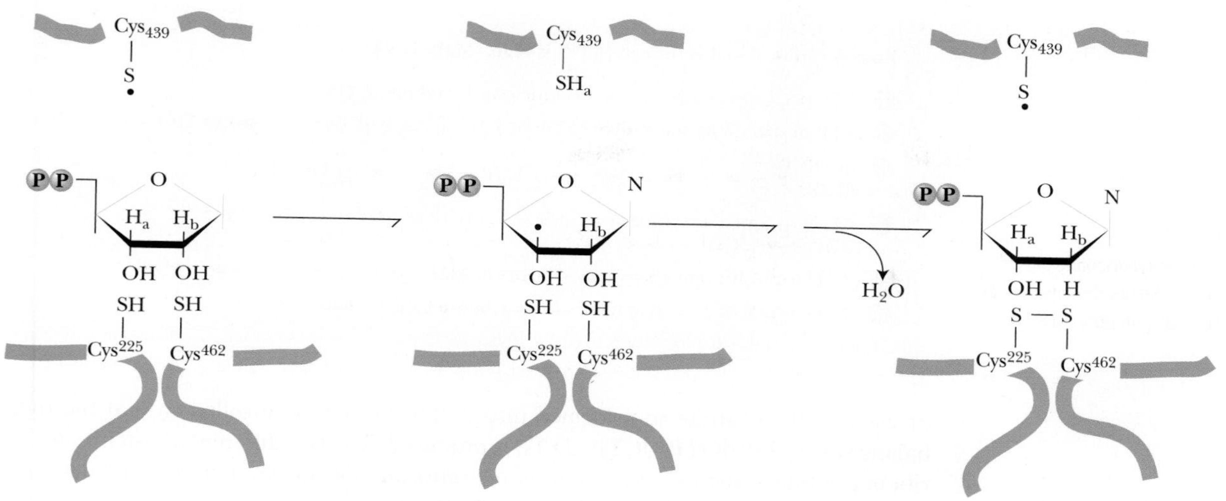

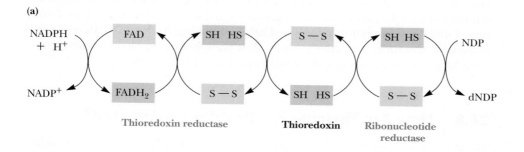

FIGURE 26.21 The free radical mechanism of ribonucleotide reduction. H_a designates the C-3' hydrogen, and H_b, the C-2' hydrogen atom.

the reactive —SH pair of the ribonucleotide reductase active site (Figure 26.21). In turn, the sulfhydryls of thioredoxin must be restored to the (—SH HS—) state for another catalytic cycle. **Thioredoxin reductase,** an α_2-type enzyme composed of 58-kD flavoprotein subunits, mediates the NADPH-dependent reduction of thioredoxin (Figure 26.22). Thioredoxin functions in a number of metabolic roles besides deoxyribonucleotide synthesis, the common denominator of which is reversible sulfide : sulfhydryl transitions.

The substrates for ribonucleotide reductase are CDP, UDP, GDP, and ADP, and the corresponding products are dCDP, dUDP, dGDP, and dADP. Because CDP is not an intermediate in pyrimidine nucleotide synthesis, it must arise by dephosphorylation of CTP, for instance, via nucleoside diphosphate kinase action. Although uridine nucleotides do not occur in DNA, UDP is a substrate. The formation of dUDP is justified because it is a precursor to dTTP, a necessary substrate for DNA synthesis (see following discussion).

Both the Specificity and the Catalytic Activity of Ribonucleotide Reductase Are Regulated by Nucleotide Binding

Ribonucleotide reductase activity must be modulated in two ways in order to maintain an appropriate balance of the four deoxynucleotides essential to DNA synthesis, namely, dATP, dGTP, dCTP, and dTTP. First, the overall activity of the enzyme must be turned on and off in response to the need for dNTPs. Second, the relative amounts

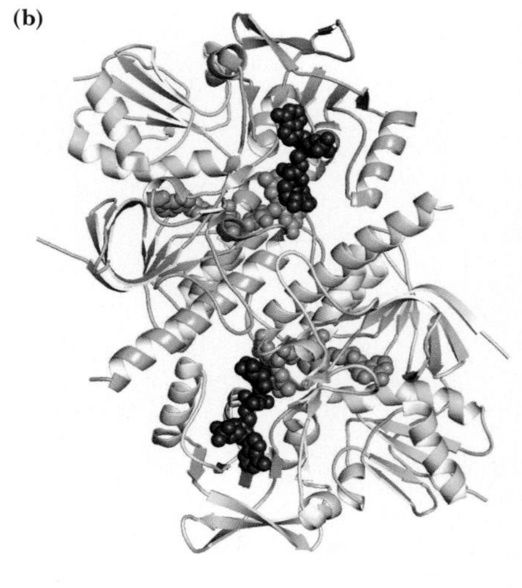

(a)

NADPH + H⁺ → FAD → SH HS → S—S → SH HS → NDP

NADP⁺ → FADH₂ → S—S → SH HS → S—S → dNDP

Thioredoxin reductase   Thioredoxin   Ribonucleotide reductase

(b)

FIGURE 26.22 **(a)** The (—S—S—)/(—SH HS—) oxidation–reduction cycle involving ribonucleotide reductase, thioredoxin, thioredoxin reductase, and NADPH. **(b)** Thioredoxin reductase (α_2 dimer) with bound ligands: FAD (gold) and NADP analog (blue). Each α-subunit is associated with a thioredoxin molecule (yellow) (pdb id = 1F6M).

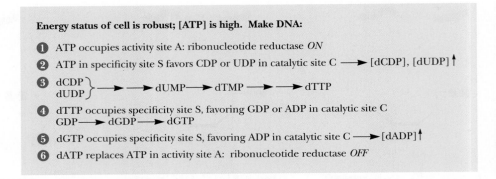

FIGURE 26.23 Regulation of deoxynucleotide biosynthesis: the rationale for the various affinities displayed by the two nucleotide-binding regulatory sites on ribonucleotide reductase.

of each NDP substrate transformed into dNDP must be controlled so that the right balance of dATP:dGTP:dCTP:dTTP is produced. The two different allosteric sites on ribonucleotide reductase, *discrete from the substrate-binding catalytic site,* are designed to serve these purposes. As noted previously, these two allosteric sites are designated the *overall activity site* and the *substrate specificity site.* Only ATP and dATP are able to bind at the *overall activity site.* ATP is an allosteric activator and dATP is an allosteric inhibitor, and they compete for the same site. If ATP is bound, the enzyme is active, whereas if its deoxy counterpart, dATP, occupies this site, the enzyme is inactive. Although ATP is more abundant than dATP, dATP binds to ribonucleotide reductase with 100-fold greater affinity. When dATP binds, R1 dimers aggregate to form inactive hexamers; ATP binding reverses this oligomerization.

The second allosteric site, the *substrate specificity site,* can bind either ATP, dTTP, dGTP, or dATP, and the substrate specificity of the enzyme is determined by which of these nucleotides occupies this site. If ATP is in the *substrate specificity site,* ribonucleotide reductase preferentially binds pyrimidine nucleotides (UDP or CDP) at its active site and reduces them to dUDP and dCDP. With dTTP in the specificity-determining site, GDP is the preferred substrate. When dGTP binds to the specificity site, ADP becomes the favored substrate for reduction. The rationale for these varying affinities is as follows (Figure 26.23): High [ATP] is consistent with cell growth and division and, consequently, the need for DNA synthesis. Thus, ATP binds in the *overall activity site* of ribonucleotide reductase, turning it on and promoting production of dNTPs for DNA synthesis. Under these conditions, ATP is also likely to occupy the *substrate specificity site,* so UDP and CDP are bound at the *catalytic site* and reduced to dUDP and dCDP. Both of these pyrimidine deoxynucleoside diphosphates are precursors to dTTP. Thus, elevation of dUDP and dCDP levels leads to an increase in [dTTP]. High dTTP levels increase the likelihood that it will occupy the *substrate specificity site,* in which case GDP becomes the preferred substrate and dGTP levels rise. Upon dGTP association with the *substrate specificity site,* ADP is the favored substrate, leading to ADP reduction and the eventual accumulation of dATP. Binding of dATP to the *overall activity site* then shuts the enzyme down. In summary, the relative affinities of the three classes of nucleotide-binding sites in ribonucleotide reductase for the various substrates, activators, and inhibitors are such that the formation of dNDPs proceeds in an orderly and balanced fashion. As these dNDPs are formed in amounts consistent with cellular needs, their phosphorylation by nucleoside diphosphate kinases produces dNTPs, the actual substrates of DNA synthesis.

26.8 How Are Thymine Nucleotides Synthesized?

The synthesis of thymine nucleotides proceeds from other pyrimidine deoxyribonucleotides. Cells have no requirement for free thymine ribonucleotides and do not synthesize them. Small amounts of thymine ribonucleotides do occur in tRNA (tRNA is notable for having unusual nucleotides), but these Ts arise via methylation of U residues already incorporated into the tRNA. Both dUDP and dCDP can lead to formation of dUMP, the immediate precursor for dTMP synthesis (Figure 26.24). Interestingly, formation of dUMP from dUDP passes through dUTP, which is then cleaved by **dUTPase,** a pyrophosphatase that removes PP$_i$ from dUTP. The action of dUTPase prevents dUTP from serving as a substrate in DNA synthesis. An alternative route to dUMP formation starts with dCDP, which is dephosphorylated to dCMP and then deaminated by **dCMP deaminase** (Figure 26.25), leaving dUMP. dCMP deaminase provides a second point for allosteric regulation of dNTP synthesis; it is allosterically activated by dCTP and feedback-inhibited by dTTP. Of the four dNTPs, only dCTP does not interact with either of the regulatory sites on ribonucleotide reductase (see Figure 26.20). Instead, it acts upon dCMP deaminase.

From dUDP: dUDP ⟶ dUTP ⟶ dUMP ⟶ dTMP
From dCDP: dCDP ⟶ dCMP ⟶ dUMP ⟶ dTMP

FIGURE 26.24 Pathways of dTMP synthesis. dTMP production is dependent on dUMP formation from dCDP and dUDP.

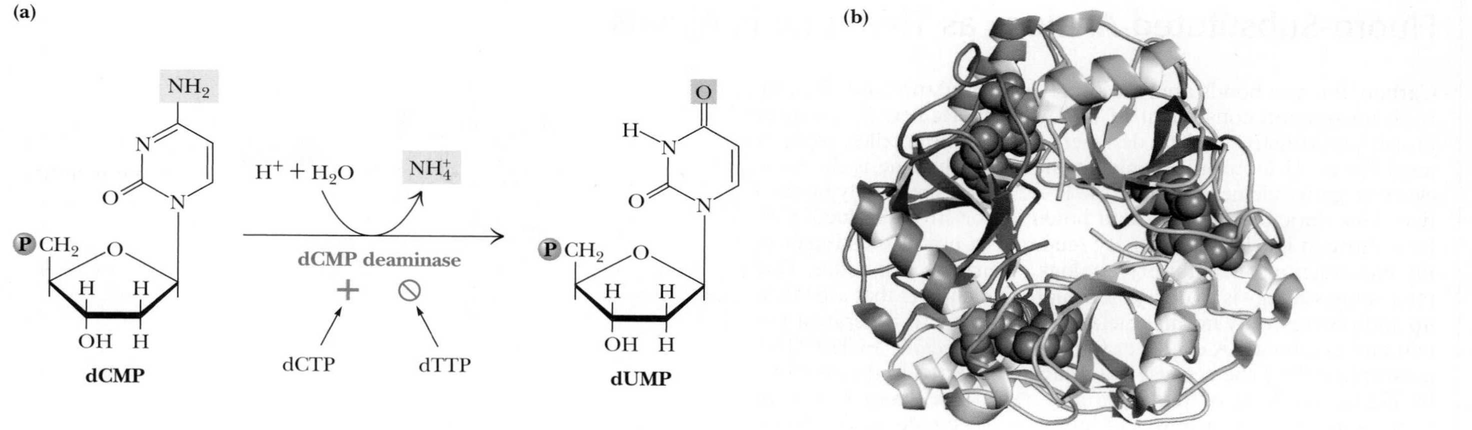

(a)

(b)

FIGURE 26.25 (a) The dCMP deaminase reaction. (b) Trimeric dCMP deaminase. Each chain has a bound dCTP molecule (purple) and a Mg^{2+} ion (orange) (pdb id = 1XS4).

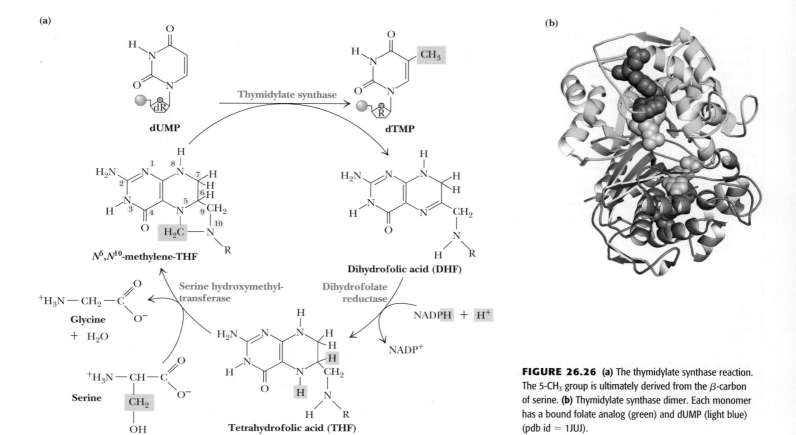

FIGURE 26.26 (a) The thymidylate synthase reaction. The 5-CH_3 group is ultimately derived from the β-carbon of serine. (b) Thymidylate synthase dimer. Each monomer has a bound folate analog (green) and dUMP (light blue) (pdb id = 1JUJ).

Synthesis of dTMP from dUMP is catalyzed by **thymidylate synthase** (Figure 26.26). This enzyme methylates dUMP at the 5-position to create dTMP; the methyl donor is the one-carbon folic acid derivative N^5,N^{10}-methylene-THF. The reaction is actually a reductive methylation in which the one-carbon unit is transferred at the methylene level of reduction and then reduced to the methyl level. The THF cofactor is oxidized at the expense of methylene reduction to yield DHF. DHFR then reduces DHF back to THF for service again as a one-carbon vehicle (see panel a of the figure in A Deeper Look on page 894). Thymidylate synthase sits at a junction connecting

A DEEPER LOOK

Fluoro-Substituted Analogs as Therapeutic Agents

Carbon–fluorine bonds are exceedingly rare in nature, and fluorine is an uncommon constituent of biological molecules. F has three properties attractive to drug designers: (1) It is the smallest replacement for an H atom in organic synthesis, (2) fluorine is the most electronegative element, and (3) the F—C bond is relatively unreactive. This steric compactness and potential for strong inductive effects through its electronegativity renders F a useful substituent in the construction of inhibitory analogs of enzyme substrates. One interesting strategy is to devise fluorinated precursors that are taken up and processed by normal metabolic pathways to generate a potent antimetabolite. A classic example is *fluoroacetate*. FCH_2COO^- is exceptionally toxic because it is readily converted to fluorocitrate by citrate synthase of the citric acid cycle (see Chapter 19). In turn, fluorocitrate is a powerful inhibitor of aconitase. The metabolic transformation of an otherwise innocuous compound into a poisonous derivative is termed **lethal synthesis.** *5-Fluorouracil* and *5-fluorocytosine* are also examples of this strategy (see Human Biochemistry on page 914).

Unlike hydrogen, which is often abstracted from substrates as H^+, electronegative fluorine cannot be readily eliminated as the corresponding F^+. Thus, enzyme inhibitors can be fashioned in which F replaces H at positions where catalysis involves H removal as H^+. Thymidylate synthase catalyzes removal of H from dUMP as H^+ through a covalent catalysis mechanism. A thiol group on this enzyme normally attacks the 6-position of the uracil moiety of 2'-deoxyuridylic acid so that C-5 can act as a carbanion in attack on the methylene carbon of N^5,N^{10}-methylene-THF (see accompanying figure). Regeneration of free enzyme then occurs through loss of the C-5 H atom as H^+ and dissociation of product dTMP. If F replaces H at C-5 as in 2'-deoxy-5-fluorouridylate (FdUMP), the enzyme is immobilized in a very stable ternary [enzyme : FdUMP : methylene-THF] complex and effectively inactivated. Enzyme inhibitors like FdUMP whose adverse properties are elicited only through direct participation in the catalytic cycle are variously called **mechanism-based inhibitors, suicide substrates,** or **Trojan horse substrates.**

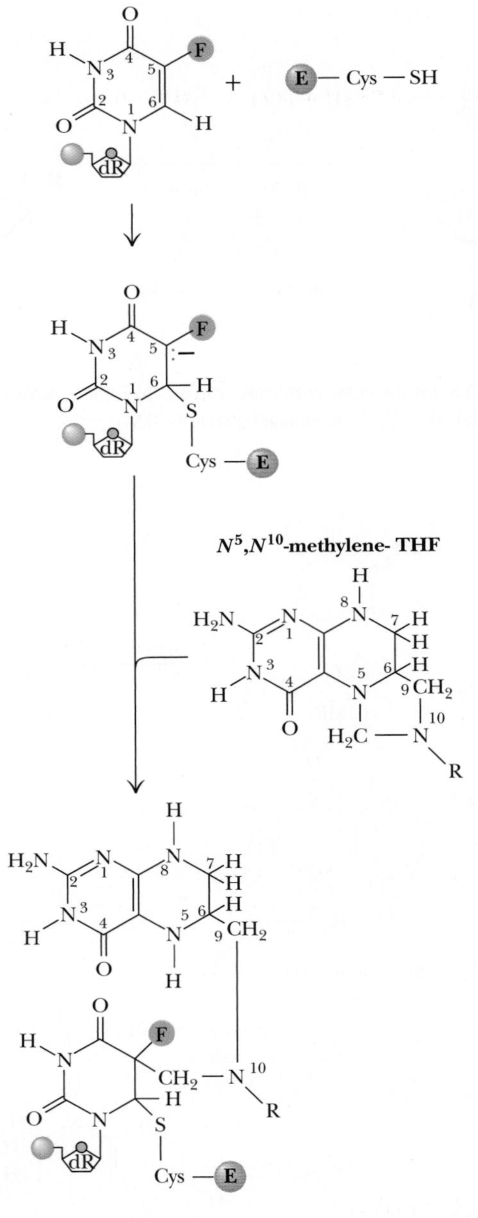

▶ The effect of the 5-fluoro substitution on the mechanism of action of thymidylate synthase. An enzyme thiol group (from a Cys side chain) ordinarily attacks the 6-position of dUMP so that C-5 can react as a carbanion with N^5,N^{10}-methylene-THF. Normally, free enzyme is regenerated following release of the hydrogen at C-5 as a proton. Because release of fluorine as F^+ cannot occur, the ternary (three-part) complex of [enzyme : fluorouridylate : methylene-THF] is stable and persists, preventing enzyme turnover. (The N^5,N^{10}-methylene-THF structure is given in abbreviated form.)

Ternary complex

dNTP synthesis with folate metabolism. It has become a preferred target for inhibitors designed to disrupt DNA synthesis. An indirect approach is to employ folic acid precursors or analogs as antimetabolites of dTMP synthesis (see panel b of the figure in A Deeper Look on page 896). Purine synthesis is affected as well because it is also dependent on THF (see Figure 26.3).

HUMAN BIOCHEMISTRY

Fluoro-Substituted Pyrimidines in Cancer Chemotherapy, Fungal Infections, and Malaria

5-Fluorouracil (5-FU; see part a of the figure) is a thymine analog. It is converted in vivo to *5'-fluorouridylate* by a PRPP-dependent phosphoribosyltransferase and passes through the reactions of dNTP synthesis, culminating ultimately as *2'-deoxy-5-fluorouridylic acid,* a potent inhibitor of dTMP synthase (see A Deeper Look on page 913). 5-FU is used as a chemotherapeutic agent in the treatment of human cancers. Similarly, *5-fluorocytosine* (see part b) is used as an antifungal drug because fungi, unlike mammals, can convert it to *2'-deoxy-5-fluorouridylate.* Furthermore, malarial parasites can use exogenous orotate to make pyrimidines for nucleic acid synthesis, whereas mammals cannot. Thus, *5-fluoroorotate* (see part c) is an effective antimalarial drug because it is selectively toxic to these parasites.

5-Fluorouracil   **5-Fluorocytosine**

5-Fluoroorotate

SUMMARY

OWL Login to OWL to develop problem-solving skills and complete online homework assigned by your professor.

26.1 Can Cells Synthesize Nucleotides? Nucleotides are ubiquitous constituents of life, and nearly all cells are capable of synthesizing them "from scratch" via de novo pathways. Rapidly proliferating cells must make lots of purine and pyrimidine nucleotides to satisfy demands for DNA and RNA synthesis. Nucleotide biosynthetic pathways are attractive targets for the clinical control of rapidly dividing cells such as cancers or infectious bacteria. Many antibiotics and anticancer drugs are inhibitors of purine or pyrimidine nucleotide biosynthesis.

26.2 How Do Cells Synthesize Purines? The nine atoms of the purine ring system are derived from aspartate (N-1), glutamine (N-3 and N-9), glycine (C-4, C-5, and N-7), CO_2 (C-6), and THF one-carbon derivatives (C-2 and C-8). The atoms of the purine ring are successively added to *ribose-5-phosphate,* so purines begin as nucleotide derivatives through assembly of the purine ring system directly on the ribose-P. Because purine biosynthesis depends on folic acid de-

rivatives, it is sensitive to inhibition by folate analogs. Distinct, two-step metabolic pathways diverge from IMP, one leading to AMP and the other to GMP. Purine biosynthesis is regulated at several stages: Reaction 1 (ribose-5-phosphate pyrophosphokinase) is feedback-inhibited by ADP and GDP; the enzyme catalyzing reaction 2 (glutamine phosphoribosyl pyrophosphate amidotransferase) has two inhibitory allosteric sites, one where adenine nucleotides bind and another where guanine nucleotides bind. PRPP is a "feed-forward" activator of this enzyme. In higher organisms, the 11 reactions leading to IMP synthesis from ribose-5-P are accomplished by seven proteins, one of which is trifunctional and two of which are bifunctional. The first reaction in the conversion of IMP to AMP involves adenylosuccinate synthetase, which is inhibited by AMP; the first step in the conversion of IMP to GMP is catalyzed by IMP dehydrogenase and is inhibited by GMP. ATP-dependent kinases form nucleoside diphosphates and triphosphates from AMP and GMP.

26.3 Can Cells Salvage Purines? Purine ring systems represent a metabolic investment by cells, and salvage pathways exist to recover them when degradation of nucleic acids releases free purines in the form of adenine, guanine, and hypoxanthine (the base in IMP). Hypoxanthine-guanine phosphoribosyltransferase (HGPRT) acts on either hypoxanthine to form IMP or guanine to form GMP; an absence of HGPRT is the basis of Lesch-Nyhan syndrome.

26.4 How Are Purines Degraded? Dietary nucleic acids are digested to nucleotides by various nucleases and phosphodiesterases, the nucleotides are converted to nucleosides by base-specific nucleotidases and nonspecific phosphatases, and then nucleosides are hydrolyzed to release the purine base. Only the pentoses of nucleotides serve as sources of metabolic energy. In humans, the purine ring is oxidized to uric acid by xanthine oxidase and excreted. Gout occurs when bodily fluids accumulate an excess of uric acid. Skeletal muscle operates a purine nucleoside cycle as an anaplerotic pathway.

26.5 How Do Cells Synthesize Pyrimidines? In contrast to formation of the purine ring system, the pyrimidine ring system is completed before a ribose-5-P moiety is attached. Only two precursors, carbamoyl-P and aspartate, contribute atoms to the six-membered pyrimidine ring. The first step in humans is catalyzed by CPS-II. ATCase then links carbamoyl-P with aspartate. Subsequent reactions close the ring and oxidize it before adding ribose-5-P, using α-PRPP as donor. Decarboxylation gives UMP. In mammals, the six enzymatic activities of pyrimidine biosynthesis are distributed among only three proteins, two of which are multifunctional polypeptides. Purine and pyrimidine synthesis in mammals are two prominent examples of metabolic channeling. UMP leads to UTP, the substrate for formation of CTP via CTP synthetase. Regulation of pyrimidine synthesis in animals occurs at CPS-II. UDP and UTP are feedback inhibitors, whereas PRPP and ATP are allosteric activators. In bacteria, regulation acts at ATCase through feedback inhibition by CTP (or UTP) and activation by ATP.

26.6 How Are Pyrimidines Degraded? Degradation of the pyrimidine ring generates β-alanine, CO_2, and ammonia. In humans, pyrimidines are recycled from nucleosides, but free pyrimidine bases are not salvaged.

26.7 How Do Cells Form the Deoxyribonucleotides That Are Necessary for DNA Synthesis? 2′-Deoxyribonucleotides are formed from ribonucleotides through reduction at the 2′-position of the ribose ring in NDPs. The reaction, catalyzed by ribonucleotide reductase, involves a free radical mechanism that replaces the 2′-OH by a hydride ion $(H:^-)$. Thioredoxin provides the reducing power for ribonucleotide reduction. Class Ia ribonucleotide reductases have three different nucleotide-binding sites: the catalytic site (or active site), which binds substrates (ADP, CDP, GDP, and UDP); the substrate specificity site, which can bind ATP, dATP, dGTP, or dTTP; and the overall activity site, which binds either the activator ATP or the negative effector dATP. The relative affinities of the three classes of nucleotide binding sites in ribonucleotide reductase for the various substrates, activators, and inhibitors are such that the various dNDPs are formed in amounts consistent with cellular needs.

26.8 How Are Thymine Nucleotides Synthesized? Both dUDP and dCDP can lead to formation of dUMP, the immediate precursor for dTMP synthesis. Formation of dTMP from dUMP is catalyzed by thymidylate synthase through reductive methylation of dUMP at the 5-position. The methyl donor is the one-carbon folic acid derivative N^5,N^{10}-methylene-THF. Fluoro-substituted pyrimidine analogs such as 5-fluorouracil (5-FU), 5-fluorocytosine, and 5-fluoroorotate can be converted to FdUMP, which inhibits thymidylate synthase. These fluoro compounds have found a range of therapeutic uses in treating diseases from cancer to malaria.

FOUNDATIONAL BIOCHEMISTRY Things You Should Know After Reading Chapter 26.

- Nucleotides are key participants in metabolism and essential building blocks for RNA and DNA synthesis.

- Nearly all organisms can carry out de novo synthesis of purine and pyrimidine nucleotides.

- The nine atoms that compose the purine ring come from five different precursors: glutamine, glycine, CO_2, aspartate, and N^{10}-formyl-THF.

- The atoms that form the purine ring are assembled on a ribose-5-P platform.

- THF, the active coenzyme form of the B vitamin folate, is a coenzyme in one-carbon transfer reactions.

- Folate analogs are used as drugs in cancer chemotherapy and as antimicrobial agents.

- IMP is the first purine nucleotide made by the de novo pathway of purine synthesis.

- Only seven proteins are needed to carry out the 11 reactions of purine biosynthesis, because 7 of these reactions are catalyzed by multifunctional polypeptides.

- AMP and GMP are synthesized from IMP by distinct two-step metabolic pathways.

- The purine biosynthetic pathway is regulated through feedback inhibition of ribose-5-P pyrophosphokinase (reaction 1) and glutamine-PPRP amidotransferase (reaction 2).

- ATP-dependent kinases form NDPs and NTPs from NMPs.

- Purine salvage pathways capture purines released in catabolism and relink them to ribose-5-P through the action of purine phosphoribosyltransferases.

- Some parasitic organisms rely on purine salvage to obtain purine nucleotides.

- The major pathways of purine catabolism lead to uric acid.

- The purine nucleoside cycle in skeletal muscle acts as an anaplerotic route to replenish citric acid cycle intermediates.

- Xanthine oxidase is a molybdo-enzyme that oxidizes hypoxanthine and xanthine to uric acid.

- Gout is a disease caused by excess uric acid.

- Different animals oxidize uric acid to form various excretory products.

- Pyrimidines are synthesized de novo from carbamoyl-P and aspartate.

- In contrast to purine biosynthesis, pyrimidines are synthesized as the free base, orotate, and then ribose-5-P is added, using PRPP as the substrate.

- The pyrimidine biosynthetic pathway in mammals is a prime example of metabolic channeling by multifunctional enzymes.

- UMP is the precursor to both UTP and CTP.

- Pyrimidine biosynthesis is regulated at carbamoyl-P synthetase II in mammals and at aspartate transcarbamoylase in bacteria.

- Pyrimidines are degraded to form β-alanine, NH_3, and CO_2. The carbons of β-alanine are the ones contributed by aspartate during pyrimidine synthesis.

- Deoxyribonucleotide synthesis involves reduction at the 2'-position of the ribose ring of NDPs; the reducing power comes from NADPH.

- Deoxyribonucleotides are synthesized through the action of ribonucleotide reductase.

- Eukaryotic ribonucleotide reductase has three nucleotide-binding sites: the catalytic (or C) site, the specificity (or S) site, and the activity (or A) site.

- The S site and the A site of ribonucleotide reductase are allosteric sites.

- Thioredoxin, reduced by NADPH through the action of thioredoxin reductase, is the direct reductant for ribonucleotide reductase.

- Reduction of NDPs to dNDPs by ribonucleotide reductase proceeds via a free radical mechanism.

- ATP is an allosteric activator of ribonucleotide reductase and dATP is an allosteric inhibitor.

- The substrate for dTMP synthesis is dUMP.

- Two dNDPs lead to formation of dUMP: dUDP and dCDP.

- Thymidylate synthase catalyzes the formation of dTMP from dUMP, using N^5, N^{10}-methylene-THF as methyl donor.

- 5-Fluoro-substituted pyrimidines are useful drugs in the treatment of cancer, fungal infections, and malaria.

PROBLEMS

Answers to all problems are at the end of this book. Detailed solutions are available in the *Student Solutions Manual, Study Guide, and Problems Book*.

1. **Metabolic Origin of the Atoms in Purine and Pyrimidine Rings** Draw the purine and pyrimidine ring structures, indicating the metabolic source of each atom in the rings.

2. **The Energy Cost of Nucleotide Biosynthesis** Starting from glutamine, aspartate, glycine, CO_2, and N^{10}-formyl-THF, how many ATP equivalents are expended in the synthesis of (a) ATP, (b) GTP, (c) UTP, and (d) CTP?

3. **Allosteric Regulation of Purine and Pyrimidine Biosynthesis** Illustrate the key points of regulation in (a) the biosynthesis of IMP, AMP, and GMP; (b) *E. coli* pyrimidine biosynthesis; and (c) mammalian pyrimidine biosynthesis.

4. **Inhibition of Purine and Pyrimidine Metabolism by Pharmacological Agents** Indicate which reactions of purine or pyrimidine metabolism are affected by the inhibitors (a) azaserine, (b) methotrexate, (c) sulfonamides, (d) allopurinol, and (e) 5-fluorouracil.

5. **The Metabolic Role of dUDP** Since dUTP is not a normal component of DNA, why do you suppose ribonucleotide reductase has the capacity to convert UDP to dUDP?

6. **Allosteric Regulation of Ribonucleotide Reductase by ATP and Deoxynucleotides** Describe the underlying rationale for the regulatory effects exerted on ribonucleotide reductase by ATP, dATP, dTTP, and dGTP.

7. **Nucleotide Catabolism as a Source of Metabolic Energy** (Integrates with Chapters 18–20 and 22.) By what pathway(s) does the ribose released upon nucleotide degradation enter intermediary metabolism and become converted to cellular energy? How many ATP equivalents can be recovered from one equivalent of ribose?

8. **Convergence of Purine and Histidine Biosynthesis** (Integrates with Chapter 25.) At which steps does the purine biosynthetic pathway resemble the pathway for biosynthesis of the amino acid histidine?

9. **The Mechanism of Action of Purine Biosynthetic Enzymes** Write reasonable chemical mechanisms for steps 6, 8, and 9 in purine biosynthesis (see Figure 26.3).

10. **The Anaplerotic Purine Nucleoside Cycle of Skeletal Muscle** Write a balanced equation for the conversion of aspartate to fumarate by the purine nucleoside cycle in skeletal muscle.

11. **The Catabolism of Uric Acid** Write a balanced equation for the oxidation of uric acid to glyoxylic acid, CO_2, and NH_3, showing each step in the process and naming all of the enzymes involved.

12. **The Allosteric Kinetics of Aspartate Transcarbamoylase** (Integrates with Chapter 15.) *E. coli* aspartate transcarbamoylase (ATCase) displays classic allosteric behavior. This $\alpha_6\beta_6$ enzyme is activated by ATP and feedback-inhibited by CTP. In analogy with the behavior of glycogen phosphorylase shown in Figure 15.15, illustrate the allosteric *v* versus [aspartate] curves for ATCase (a) in the absence of effectors, (b) in the presence of CTP, and (c) in the presence of ATP.

13. **The Functional Organization of a Heteromeric Allosteric Enzyme** (Integrates with Chapter 15.) Unlike its allosteric counterpart glycogen phosphorylase (an α_2 enzyme), *E. coli* ATCase has a heteromeric ($\alpha_6\beta_6$) organization. The α-subunits bind aspartate and are considered catalytic subunits, whereas the β-subunits bind CTP or ATP and are considered regulatory subunits. How would you describe the subunit organization of ATCase from a functional point of view?

14. **The Energy Cost of dTTP Synthesis** (Integrates with Chapter 20.) Starting from HCO_3^-, glutamine, aspartate, and ribose-5-P, how many ATP equivalents are consumed in the synthesis of dTTP in a eukaryotic cell, assuming dihydroorotate oxidation is coupled to oxidative phosphorylation? How does this result compare with the ATP costs of purine nucleotide biosynthesis calculated in problem 2?

15. **The Chemistry of dTMP Synthesis and the Substrate-Binding Site of Thymidylate Synthase** Write a *balanced* equation for the synthesis of dTMP from UMP and N^5,N^{10}-methylene-THF. Thymidylate synthase has four active-site arginine residues (Arg^{23}, Arg$^{178'}$, Arg$^{179'}$, and Arg^{218}) involved in substrate binding. Postulate a role for the side chains of these Arg residues.

16. **Representatives of the PRT Family of Enzymes in Purine Synthesis** Enzymes that bind phosphoribosyl-5-phosphate (PRPP) have a common structural fold, the PRT fold, which unites them as a structural family. PRT here refers to the phosphoribosyl transferase activity displayed by some family members. Typically, in such reactions, PP_i is displaced from PRPP by a nitrogen-containing nucleophile. Several such reactions occur in purine metabolism. Identify two such reactions where the enzyme involved is likely to be a PRT family member.

17. **Explore the Structure of DHFR Reductase** The crystal structure of *E. coli* dihydrofolate reductase (DFR) with NADP$^+$ and folate bound can be found in the Protein Data Bank (www.rcsb.org/pdb) as file 7DFR. Go to this website, enter "7DFR" in the search line, and click on one of the viewers under "Display options" when the 7DFR page comes up. Explore the graphic of the DFR structure to visualize how its substrates are bound. (If you hold down the left button on your mouse and move the cursor over the image, you can rotate the structure to view it from different perspectives.) Note in particular the spatial relationship between the nicotinamide ring of NADP$^+$ and the pterin ring of folate. Do you now have a better appreciation for how this enzyme works? Note also the location of polar groups on the two substrates in relation to the DFR structure.

18. **Explore the Structure of Aspartate Transcarbamoylase, an Allosteric Enzyme** *E. coli* aspartate transcarbamoylase is an allosteric enzyme (see problem 12) composed of six catalytic (C) subunits and six regulatory (R) subunits. Protein Data Bank file 1RAA shows one-third of the ATCase holoenzyme (two C subunits and two R subunits; CTP molecules are bound to the R subunits). Explore this structure using one of the display options. What secondary structural motif dominates the R-subunit structure? Protein Data Bank file 2IPO also shows one-third of the ATCase

holoenzyme (two C subunits and two R subunits), but in this structure molecules of the substrate analog *N*-(2-phosphonoacetyl)-L-asparagine are bound to the C subunits. Explore this structure using one of the display options. Note the distance separating the ATCase active site from its allosteric site. Interpret what you see in terms of the Monod–Wyman–Changeux model for allosteric regulation (see Chapter 15). Which of these structures corresponds to the MWC R-state, and which corresponds to the T-state?

Preparing for the MCAT® Exam

19. Examine Figure 26.6, and predict the relative rates of the regulated reactions in the purine biosynthetic pathway from ribose-5-P to GMP and AMP under conditions in which GMP levels are very high.

20. Decide from Figures 18.1, 25.31, 26.26, and the Deeper Look box on page 895 which carbon atom(s) in glucose would be most likely to end up as the 5-CH$_3$ carbon in dTMP.

ActiveModels Problems

21. Using the ActiveModel for CTP synthetase, describe how glutamine binding occurs, and explain how His[57] regulates ammonium diffusion.

22. Using the ActiveModel for HGPRT, discuss the putative function of the flexible loop. Describe the difference between the "loop open" and "loop closed" conformations.

FURTHER READING

Purine Metabolism

An, S., Kumar, R., Sheets, E. D., and Benkovic, S. J., 2008. Reversible compartmentalization of de novo purine biosynthetic complexes in living cells. *Science* **320:**103–106.

Cao, H., Pauff, J. M., and Hille, R., 2010. Substrate orientation and catalytic specificity in the action of xanthine oxidase: The sequential hydroxylation of hypoxanthine to uric acid. *Journal of Biological Chemistry* **285:**28044–28053.

Kisker, C., Schindelin, H., and Rees, D. C., 1997. Molybdenum-containing enzymes: Structure and mechanism. *Annual Review of Biochemistry* **66:**233–267.

Mueller, E. J., et al., 1994. N^5-carboxyaminoimidazole ribonucleotide: Evidence for a new intermediate and two new enzymatic activities in the de novo purine biosynthetic pathway of *Escherichia coli*. *Biochemistry* **33:**2269–2278.

Watts, R. W. E., 1983. Some regulatory and integrative aspects of purine nucleotide synthesis and its control: An overview. *Advances in Enzyme Regulation* **21:**33–51.

Zhang, Y., Morar, M., and Ealick, S. E., 2008. Structural biology of the purine biosynthetic pathway. *Cell and Molecular Life Sciences* **65:**3699–3724.

Pyrimidine Metabolism

Connolly, G. P., and Duley, J. A., 1999. Uridine and its nucleotides: Biological actions, therapeutical potentials. *Trends in Pharmacological Sciences* **20:**218–225.

Graves, L. M., et al., 2000. Regulation of carbamoyl phosphate synthetase by MAP kinase. *Nature* **403:**328–331.

Jones, M. E., 1980. Pyrimidine nucleotide biosynthesis in animals: Genes, enzymes and regulation of UMP biosynthesis. *Annual Review of Biochemistry* **49:**253–279.

Metabolic Disorders of Purine and Pyrimidine Metabolism

Löffler, M., et al., 2005. Pyrimidine pathways in health and disease. *Trends in Molecular Medicine* **11:**430–437.

Nyhan, W. L., 2005. Disorders of purine and pyrimidine metabolism. *Molecular Genetics and Metabolism* **86:**25–33.

Scriver, C. R., et al., 1995. *The Metabolic Bases of Inherited Disease*, 7th ed. New York: McGraw-Hill.

Metabolic Channeling

Benkovic, S. J., 1984. The transformylase enzymes in de novo purine biosynthesis. *Trends in Biochemical Sciences* **9:**320–322.

Conrado, R. J., Varner, J. D., and DeLisa, M. P., 2008. Engineering the spatial organization of metabolic enzymes: Mimicking nature's synergy. *Current Opinion in Biotechnology* **19:**492–499.

Henikoff, S., 1987. Multifunctional polypeptides for purine de novo synthesis. *BioEssays* **6:**8–13.

Huang, X., Holden, H. M., and Raushel, F. M., 2001. Channeling of substrates and intermediates in enzyme-catalyzed reactions. *Annual Review of Biochemistry* **70:**149–180.

Srere, P. A., 1987. Complexes of sequential metabolic enzymes. *Annual Review of Biochemistry* **56:**89–124.

Deoxyribonucleotide Biosynthesis

Carreras, C. W., and Santi, D. V., 1995. The catalytic mechanism and structure of thymidylate synthase. *Annual Review of Biochemistry* **64:**721–762.

Fairman, J. W., Wijerathna, S. R., Ahmad, M. F., Xu, H., et al., 2011. Structural basis for regulation of human ribonucleotide reductase by nucleotide-induced oligomerization. *Nature Structural and Molecular Biology* **18:**316–323.

Frey, P. A., 2001. Radical mechanisms of enzymatic catalysis. *Annual Review of Biochemistry* **70:**121–148.

Herrick, J., and Sciavi, B., 2007. Ribonucleotide reductase and the regulation of DNA replication: An old story and an ancient heritage. *Molecular Microbiology* **63:**22–34.

Jordan, A., and Reichard, P., 1998. Ribonucleotide reductases. *Annual Review of Biochemistry* **67:**71–98.

Licht, S., Gerfen, G. J., and Stubbe, J., 1996. Thiyl radicals in ribonucleotide reductases. *Science* **271:**477–481.

Marsh, E. N. G., 1995. A radical approach to enzyme catalysis. *BioEssays* **17:**431–441.

Stubbe, J., Ge, J., and Yee, C. S., 2001. The evolution of ribonucleotide reduction revisited. *Trends in Biochemical Sciences* **26:**93–99.

Inhibitors of Purine, Pyrimidine, and Deoxyribonucleotide Biosynthesis as Therapeutic Agents

Abeles, R. H., and Alston, T. A., 1990. Enzyme inhibition by fluoro compounds. *Journal of Biological Chemistry* **265:**16705–16708.

Galmarini, C. M., Mackey, J. R., and Dumontet, C., 2002. Nucleoside analogues and nucleobases in cancer treatment. *Lancet Oncology* **3:**415–424.

Hitchings, G. H., 1992. Antagonists of nucleic acid derivatives as medicinal agents. *Annual Review of Pharmacology and Toxicology* **32:**1–9.

Park, B. K., Kitteringham, N. R., and O'Neill, P. M., 2001. Metabolism of fluorine-containing drugs. *Annual Review of Pharmacology and Toxicology* **41:**443–470.

Zrenner, R., et al., 2006. Pyrimidine and purine biosynthesis and degradation in plants. *Annual Review of Plant Biology* **57:**805–836.

Metabolic Integration and Organ Specialization

PNC/Brand X Pictures/Jupiter Images

◄ The Washington, D.C., Metro map. The coordinated flow of passengers along different transit lines is an apt metaphor for metabolic regulation.

Study of an enzyme, a reaction, or a sequence can be biologically relevant only if its position in the hierarchy of function is kept in mind.

Daniel E. Atkinson
Cellular Energy Metabolism and Its Regulation (1977)

KEY QUESTIONS

27.1 Can Systems Analysis Simplify the Complexity of Metabolism?

27.2 What Underlying Principle Relates ATP Coupling to the Thermodynamics of Metabolism?

27.3 Is There a Good Index of Cellular Energy Status?

27.4 How Is Overall Energy Balance Regulated in Cells?

27.5 How Is Metabolism Integrated in a Multicellular Organism?

27.6 What Regulates Our Eating Behavior?

27.7 Can You Really Live Longer by Eating Less?

ESSENTIAL QUESTIONS

Cells are systems in a dynamic steady state, maintained by a constant flux of nutrients that serve as energy sources or as raw material for the maintenance of cellular structures. Catabolism and anabolism are ongoing, concomitant processes.

What principles underlie the integration of catabolism and energy production with anabolism and energy consumption? How is metabolism integrated in complex organisms with multiple organ systems?

In the preceding chapters, we have explored the major metabolic pathways—glycolysis, the citric acid cycle, electron transport and oxidative phosphorylation, photosynthesis, gluconeogenesis, fatty acid oxidation, lipid biosynthesis, amino acid metabolism, and nucleotide metabolism. Several of these pathways are catabolic and serve to generate chemical energy useful to the cell; others are anabolic and use this energy to drive the synthesis of essential biomolecules. Despite their opposing purposes, these reactions typically occur at the same time as nutrient molecules are broken down to provide the building blocks and energy for ongoing biosynthesis. Cells maintain a dynamic steady state through processes that involve considerable metabolic flux. The metabolism that takes place in just a single cell is so complex that it defies detailed quantitative description. However, an appreciation of overall relationships can be achieved by stepping back and considering intermediary metabolism at a systems level of organization.

27.1 | Can Systems Analysis Simplify the Complexity of Metabolism?

The metabolism of a typical heterotrophic cell can be portrayed by a schematic diagram consisting of just three interconnected functional blocks: (1) catabolism, (2) anabolism, and (3) macromolecular synthesis and growth (Figure 27.1).

OWL Online homework and a Student Self Assessment for this chapter may be assigned in OWL.

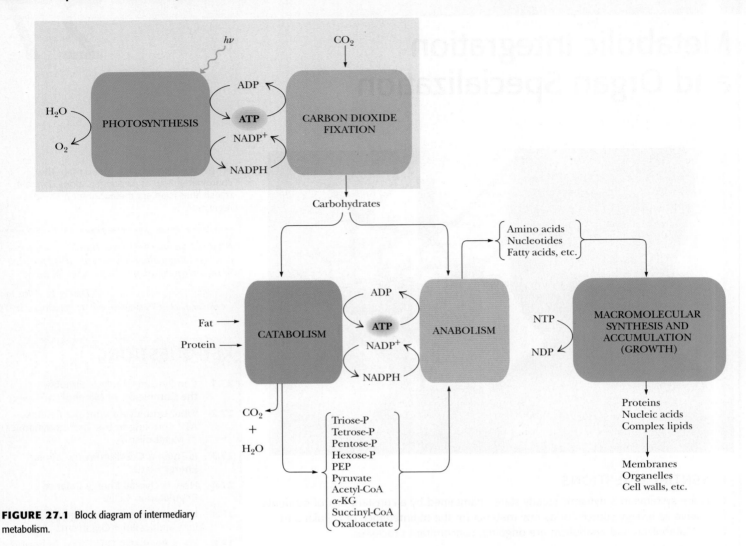

FIGURE 27.1 Block diagram of intermediary metabolism.

1. Catabolism Energy-yielding nutrients are oxidized to CO_2 and H_2O in catabolism, and most of the electrons liberated are passed to oxygen via an electron-transport pathway coupled to oxidative phosphorylation, resulting in the formation of ATP. Some electrons go to reduce $NADP^+$ to NADPH, the source of reducing power for anabolism. Glycolysis, the citric acid cycle, electron transport and oxidative phosphorylation, and the pentose phosphate pathway are the principal pathways within this block. The metabolic intermediates in these pathways also serve as substrates for processes within the anabolic block.

2. Anabolism The biosynthetic reactions that form the many cellular molecules collectively comprise anabolism. For thermodynamic reasons, the chemistry of anabolism is more complex than that of catabolism (that is, it takes more energy [and often more steps] to synthesize a molecule than can be produced from its degradation). Metabolic intermediates derived from glycolysis and the citric acid cycle are the precursors for this synthesis, with NADPH supplying the reducing power and ATP, the coupling energy.

3. Macromolecular Synthesis and Growth The organic molecules produced in anabolism are the building blocks for creation of macromolecules. Like anabolism, macromolecular synthesis is driven by energy from ATP, although indirectly in some cases: GTP is the principal energy source for protein synthesis, CTP for phospholipid synthesis, and UTP for polysaccharide synthesis. However, keep in mind that ATP is the principal phosphoryl donor for formation of GTP, CTP, and UTP from GDP, CDP, and UDP, respectively. Macromolecules are the agents of biological function and information—

proteins, nucleic acids, lipids that self-assemble into membranes, and so on. Growth can be represented as cellular accumulation of macromolecules and the partitioning of these materials of function and information into daughter cells in the process of cell division.

Only a Few Intermediates Interconnect the Major Metabolic Systems

Despite the complexity of processes taking place within each block, the connections between blocks involve only a limited number of substances. Just ten or so kinds of catabolic intermediates from glycolysis, the pentose phosphate pathway, and the citric acid cycle serve as the raw material for most of anabolism: four kinds of sugar phosphates (triose-P, tetrose-P, pentose-P, and hexose-P), three α-keto acids (pyruvate, oxaloacetate, and α-ketoglutarate), two coenzyme A derivatives (acetyl-CoA and succinyl-CoA), and PEP (phosphoenolpyruvate).

ATP and NADPH Couple Anabolism and Catabolism

Metabolic intermediates are consumed by anabolic reactions and must be continuously replaced by catabolic processes. In contrast, the energy-rich compounds ATP and NADPH are recycled rather than replaced. When these substances are used in biosynthesis, the products are ADP and $NADP^+$, and ATP and NADPH are regenerated during catabolism. ATP and NADPH are unique in that they are the only compounds whose purpose is to couple the energy-yielding processes of catabolism to the energy-consuming reactions of anabolism. Certainly, other coupling agents serve essential roles in metabolism. For example, NADH and [FADH$_2$] participate in the transfer of electrons from substrates to O_2 during oxidative phosphorylation. However, these reactions are solely catabolic, and the functions of NADH and [FADH$_2$] are fulfilled within the block called catabolism.

Phototrophs Have an Additional Metabolic System—The Photochemical Apparatus

The systems in Figure 27.1 reviewed thus far are representative only of metabolism as it exists in aerobic heterotrophs. The photosynthetic production of ATP and NADPH in photoautotrophic organisms entails a fourth block, the photochemical system (Figure 27.1). This block consumes H_2O and releases O_2. When this fourth block operates, energy production within the catabolic block can be largely eliminated. Yet another block, one to account for the fixation of carbon dioxide into carbohydrates, is also required for photoautotrophs. The inputs to this fifth block are the products of the photochemical system (ATP and NADPH) and CO_2 derived from the environment. The carbohydrate products of this block may enter catabolism, but not primarily for energy production. In photoautotrophs, carbohydrates are fed into catabolism to generate the metabolic intermediates needed to supply the block of anabolism. Although these diagrams are oversimplifications of the total metabolic processes in heterotrophic or phototrophic cells, they are useful illustrations of functional relationships between the major metabolic subdivisions. This general pattern provides an overall perspective on metabolism, making its purpose easier to understand.

27.2 What Underlying Principle Relates ATP Coupling to the Thermodynamics of Metabolism?

Virtually every metabolic pathway either consumes or produces ATP. The amount of ATP involved—that is, the stoichiometry of ATP synthesis or hydrolysis—lies at the heart of metabolic relationships. The overall thermodynamic efficiency of any metabolic sequence, be it catabolic or anabolic, is determined by **ATP coupling.** In every case, *the*

Stoichiometry is the measurement of the amounts of chemical elements and molecules involved in chemical reactions (from the Greek *stoicheion*, meaning "element," and *metria*, meaning "measure").

overall reaction mediated by any metabolic pathway is energetically favorable because of its particular ATP stoichiometry. In the highly exergonic reactions of catabolism, much of the energy released is captured in ATP synthesis. In turn, the thermodynamically unfavorable reactions of anabolism are driven by energy released upon ATP hydrolysis.

To illustrate this principle, we must first consider the three types of stoichiometries. The first two are fixed by the laws of chemistry, but the third is unique to living systems and reveals a fundamental difference between the inanimate world of chemistry and physics and the world of biological function, as shaped by evolution—that is, the world of living organisms. The fundamental difference is the stoichiometry of ATP coupling.

1. Reaction Stoichiometry This is simple chemical stoichiometry—the number of each kind of atom in any chemical reaction remains the same, and thus equal numbers must be present on both sides of the equation. This requirement holds even for a process as complex as cellular respiration:

$$C_6H_{12}O_6 + 6\ O_2 \longrightarrow 6\ CO_2 + 6\ H_2O$$

The six carbons in glucose appear as $6\ CO_2$, the 12 H of glucose appear as the 12 H in six molecules of water, and the 18 oxygens are distributed between CO_2 and H_2O.

2. Obligate Coupling Stoichiometry Cellular respiration is an oxidation–reduction process, and the oxidation of glucose is coupled to the reduction of NAD^+ and [FAD]. (Brackets here denote that the relevant FAD is covalently linked to succinate dehydrogenase; see Chapter 20). The NADH and [$FADH_2$] thus formed are oxidized in the electron-transport pathway:

(a) $C_6H_{12}O_6 + 10\ NAD^+ + 2\ [FAD] + 6\ H_2O \longrightarrow$
$$6\ CO_2 + 10\ NADH + 10\ H^+ + 2\ [FADH_2]$$

(b) $10\ NADH + 10\ H^+ + 2\ [FADH_2] + 6\ O_2 \longrightarrow 12\ H_2O + 10\ NAD^+ + 2\ [FAD]$

Sequence (a) accounts for the oxidation of glucose via glycolysis and the citric acid cycle. Sequence (b) is the overall equation for electron transport per glucose. The stoichiometry of coupling by the biological e^- carriers, NAD^+ and FAD, is fixed by the chemistry of electron transfer; each of the coenzymes serves as an e^- pair acceptor. Reduction of each O atom takes an e^- pair. Metabolism must obey these facts of chemistry: Biological oxidation of glucose releases 12 e^- pairs, creating a requirement for 12 equivalents of e^- pair acceptors, which transfer the electrons to 12 O atoms. By evolutionary chance, NAD^+/NADH and FAD/$FADH_2$ carry these electrons, but the stoichiometry is fixed by the chemistry.

3. Evolved Coupling Stoichiometries The participation of ATP is fundamentally different from the role played by pyridine nucleotides and flavins. The stoichiometry of adenine nucleotides in metabolic sequences is not fixed by chemical necessity. Instead, the "stoichiometries" we observe are the consequences of evolutionary design. The overall equation for cellular respiration,[1] including the coupled formation of ATP by oxidative phosphorylation, is

$$C_6H_{12}O_6 + 6\ O_2 + 38\ ADP + 38\ P_i \longrightarrow 6\ CO_2 + 38\ ATP + 44\ H_2O$$

The "stoichiometry" of ATP formation, $38\ ADP + 38\ P_i \longrightarrow 38\ ATP + 38\ H_2O$, cannot be predicted from any chemical considerations. The value of 38 ATP is an end result of biological adaptation. It is a trait that evolved through interactions between chemistry, heredity, and the environment over the course of evolution. Like any evolved character, ATP stoichiometry is the result of compromise. The final trait is one particularly suited to the fitness of the organism.

The number 38 is not magical. Recall that in eukaryotes, the consensus value for the net yield of ATP per glucose is 32 to 34, not 38 (see Table 20.4). Also, the value of 38 was

---

[1]This overall equation for cellular respiration is for the reaction within an uncompartmentalized (prokaryotic) cell. In eukaryotes, where much of the cellular respiration is compartmentalized within mitochondria, mitochondrial ADP/ATP exchange imposes a metabolic cost on the proton gradient of 1 H^+ per ATP, so the overall yield of ATP per glucose is 32, not 38.

established a long time ago in evolution, when the prevailing atmospheric conditions and the competitive situation were undoubtedly very different from those today. The significance of this number is that it provides a high yield of ATP for each glucose molecule, yet the yield is still low enough that essentially all of the glucose is metabolized.

ATP Coupling Stoichiometry Determines the K_{eq} for Metabolic Sequences

The fundamental biological purpose of ATP as an energy-coupling agent is to drive thermodynamically unfavorable reactions. In effect, the energy release accompanying ATP hydrolysis is transmitted to the unfavorable reaction so that the overall free energy change for the coupled process is negative (that is, favorable). The involvement of ATP serves to alter the free energy change for a reaction; or to put it another way, the role of ATP is to change the equilibrium ratio of [reactants] to [products] for a reaction. (See the A Deeper Look box on pages 69–70.)

Another way of viewing these relationships is to note that, at equilibrium, the concentrations of ADP and P_i will be vastly greater than that of ATP because $\Delta G^{\circ\prime}$ for ATP hydrolysis is a large negative number.[2] However, the cell where this reaction is at equilibrium is a dead cell. Living cells break down energy-yielding nutrient molecules to generate ATP. It is significant to note that the catabolism of glucose by glycolysis requires the investment of 2 ATP/glucose before any energy yields are realized. Further, fatty acid oxidation depends on fatty acid activation by acyl-CoA synthetase, a reaction that consumes two ATP equivalents. Thus, a cell deficient in ATP is doomed. Nevertheless, catabolism of energy-yielding nutrients proceeds with an overall decrease in free energy, and much of this energy is captured in ATP synthesis. Kinetic controls over the rates of the catabolic pathways are designed to ensure that the $[ATP]/([ADP][P_i])$ ratio is maintained very high. *The cell, by employing kinetic controls over the rates of metabolic pathways, maintains a very high $[ATP]/([ADP][P_i])$ ratio so that ATP hydrolysis can serve as the driving force for virtually all biochemical events.*

ATP Has Two Metabolic Roles

The role of ATP in metabolism is twofold:

1. It serves in a stoichiometric role to establish large equilibrium constants for metabolic conversions and to render metabolic sequences thermodynamically favorable. This is the role referred to when we call ATP the *energy currency* of the cell.
2. ATP also serves as an important allosteric effector in the kinetic regulation of metabolism. Its concentration (relative to those of ADP and AMP) is an index of the energy status of the cell and determines the rates of regulatory enzymes situated at key points in metabolism, such as PFK in glycolysis and FBPase in gluconeogenesis.

27.3 Is There a Good Index of Cellular Energy Status?

Energy transduction and energy storage in the *adenylate system*—ATP, ADP, and AMP—lie at the very heart of metabolism. The amount of ATP a cell uses per minute is roughly equivalent to the steady-state amount of ATP it contains. Thus, the metabolic lifetime of an ATP molecule is brief. ATP, ADP, and AMP are all important effectors in exerting kinetic control on regulatory enzymes situated at key points in metabolism, so uncontrolled changes in their concentrations could have drastic consequences. The regulation of metabolism by adenylates in turn requires close control of the relative concentrations of ATP, ADP, and AMP. Some ATP-consuming reactions produce ADP;

---

[2]Since $\Delta G^{\circ\prime} = -30.5$ kJ/mol, ln $K_{eq} = 12.3$. So $K_{eq} = 2.2 \times 10^5$. Choosing starting conditions of $[ATP] = 8$ mM, $[ADP] = 8$ mM, and $[P_i] = 1$ mM, we can assume that, at equilibrium, $[ATP]$ has fallen to some insignificant value x, $[ADP]$ = approximately 16 mM, and $[P_i]$ = approximately 9 mM. The concentration of ATP at equilibrium, x, then calculates to be about 1 nM.

PFK and hexokinase are examples. Others lead to the formation of AMP, as in fatty acid activation by acyl-CoA synthetases:

$$\text{Fatty acid} + \text{ATP} + \text{coenzyme A} \longrightarrow \text{AMP} + \text{PP}_i + \text{fatty acyl-CoA}$$

Adenylate Kinase Interconverts ATP, ADP, and AMP

Adenylate kinase (see Chapter 18), by catalyzing the reversible phosphorylation of AMP by ATP, provides a direct connection among all three members of the adenylate pool:

$$\text{ATP} + \text{AMP} \rightleftharpoons 2\,\text{ADP}$$

The free energy of hydrolysis of a phosphoanhydride bond is essentially the same in ADP and ATP (see Chapter 3), and the standard free energy change for this reaction is close to zero.

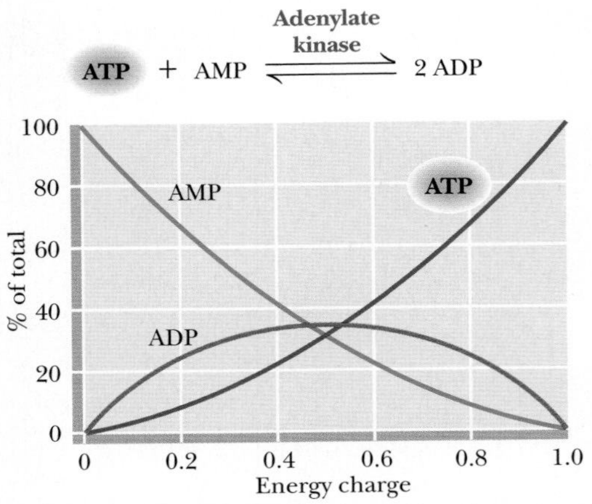

Energy Charge Relates the ATP Levels to the Total Adenine Nucleotide Pool

The role of the adenylate system is to provide phosphoryl groups at high group-transfer potential in order to drive thermodynamically unfavorable reactions. The capacity of the adenylate system to fulfill this role depends on how fully charged it is with phosphoric anhydrides. Energy charge is an index of this capacity:

$$\text{Energy charge} = \frac{1}{2}\left(\frac{2\,[\text{ATP}] + [\text{ADP}]}{[\text{ATP}] + [\text{ADP}] + [\text{AMP}]}\right)$$

The denominator represents the total adenylate pool ([ATP] + [ADP] + [AMP]); the numerator is the number of phosphoric anhydride bonds in the pool, two for each ATP and one for each ADP. The factor $\frac{1}{2}$ normalizes the equation so that energy charge, or E.C., has the range 0 to 1.0. If all the adenylate is in the form of ATP, E.C. = 1.0, and the potential for phosphoryl transfer is maximal. At the other extreme, if AMP is the only adenylate form present, E.C. = 0. It is reasonable to assume that the adenylate kinase reaction is never far from equilibrium in the cell. Then the relative amounts of the three adenine nucleotides are fixed by the energy charge. Figure 27.2 shows the relative changes in the concentrations of the adenylates as energy charge varies from 0 to 1.0.

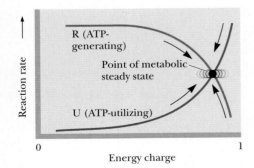

Adenylate kinase

$$\text{ATP} + \text{AMP} \rightleftharpoons 2\,\text{ADP}$$

FIGURE 27.2 Relative concentrations of AMP, ADP, and ATP as a function of energy charge. (This graph was constructed assuming that the adenylate kinase reaction is at equilibrium and that $\Delta G^{\circ\prime}$ for the reaction is -473 J/mol; $K_{eq} = 1.2$.)

Key Enzymes Are Regulated by Energy Charge

Regulatory enzymes typically respond in reciprocal fashion to adenine nucleotides. For example, PFK is stimulated by AMP and inhibited by ATP. If the activities of various regulatory enzymes are examined in vitro as a function of energy charge, an interesting relationship appears. Regulatory enzymes in energy-producing catabolic pathways show greater activity at low energy charge, but the activity falls off abruptly as E.C. approaches 1.0. In contrast, regulatory enzymes of anabolic sequences are not very active at low energy charge, but their activities increase exponentially as E.C. nears 1.0. These contrasting responses are termed **R,** for ATP-regenerating, and **U,** for ATP-utilizing (Figure 27.3). Regulatory enzymes such as PFK and pyruvate kinase in glycolysis follow the **R** response curve as E.C. is varied. Note that PFK itself is an ATP-utilizing enzyme, using ATP to phosphorylate fructose-6-phosphate to yield fructose-1,6-bisphosphate. Nevertheless, because PFK acts physiologically as the valve controlling the flux of carbohydrate down the catabolic pathways of cellular respiration that lead to ATP regeneration, it responds as an **"R"** enzyme to energy charge. Regulatory enzymes in anabolic pathways, such as acetyl-CoA carboxylase, which initiates fatty acid biosynthesis, respond as **"U"** enzymes.

The overall purposes of the **R** and **U** pathways are diametrically opposite in terms of ATP involvement. Note in Figure 27.3 that the **R** and **U** curves intersect at a rather high E.C. value. As E.C. increases past this point, **R** activities decline precipitously and

FIGURE 27.3 Responses of regulatory enzymes to variation in energy charge.

U activities rise. That is, when E.C. is very high, biosynthesis is accelerated while catabolism diminishes. The consequence of these effects is that ATP is used up faster than it is regenerated, and so E.C. begins to fall. As E.C. drops below the point of intersection, **R** processes are favored over **U**. Then, ATP is generated faster than it is consumed, and E.C. rises again. The net result is that the value of energy charge oscillates about a point of **steady state** (Figure 27.3). The experimental results obtained from careful measurement of the relative amounts of AMP, ADP, and ATP in living cells reveals that normal cells have an energy charge in the neighborhood of 0.85 to 0.88. Maintenance of this steady-state value is one criterion of cell health and normalcy.

Phosphorylation Potential Is a Measure of Relative ATP Levels

Because energy charge is maintained at a relatively constant value in normal cells, it is not an informative index of cellular capacity to carry out phosphorylation reactions. The relative concentrations of ATP, ADP, and P_i do provide such information, and a function called **phosphorylation potential** has been defined in terms of these concentrations:

$$ADP + P_i \rightleftharpoons ATP + H_2O$$

Phosphorylation potential, Γ, is equal to $[ATP]/([ADP][P_i])$.

Note that this expression includes a term for the concentration of inorganic phosphate. $[P_i]$ has substantial influence on the thermodynamics of ATP hydrolysis. In contrast with energy charge, phosphorylation potential varies over a significant range as the actual proportions of ATP, ADP, and P_i in cells vary in response to metabolic state. Γ ranges from 200 to 800 M^{-1}, or more, with higher levels signifying more ATP and correspondingly greater phosphorylation potential.

27.4 How Is Overall Energy Balance Regulated in Cells?

AMP-activated protein kinase (AMPK) is the cellular energy sensor. Metabolic inputs to this sensor determine whether its output, protein kinase activity, takes place. When cellular energy levels are high, as signaled by high ATP concentrations, AMPK is inactive. When cellular energy levels are depleted, as signaled by high [AMP], AMPK is allosterically activated and phosphorylates many targets controlling cellular energy production and consumption. Recall that, due to the nature of the adenylate kinase equilibrium (see page 585), AMP levels increase exponentially as ATP levels decrease. AMP is an allosteric activator of AMPK, whereas ATP at high levels acts as an allosteric inhibitor by displacing AMP from the allosteric site. Thus, competition between AMP and ATP for binding to the AMPK allosteric sites determines the activity of AMPK. Activation of AMPK (1) sets in motion catabolic pathways leading to ATP synthesis and (2) shuts down pathways that consume ATP energy, such as biosynthesis and cell growth.

AMPK is an $\alpha\beta\gamma$-heterotrimer (Figure 27.4). The α-subunit is the catalytic subunit; it has an N-terminal Ser/Thr protein kinase domain and a C-terminal $\beta\gamma$-binding domain. The β-subunit has at its C-terminus an $\alpha\gamma$-binding domain. The γ-subunit is the regulatory subunit; it has a pair of allosteric sites where either AMP or ATP binds. These sites are located toward its C-terminus in the form of four CBS domains (so named for their homology to *cystathionine-β-synthase*). These CBS domains act in pairs to form structures known as Bateman modules. The Bateman modules provide the binding sites for the allosteric ligands, AMP and ATP. AMP binding to these sites is highly cooperative, such that binding of AMP to one module markedly enhances AMP binding at the other. This cooperativity renders AMPK exquisitely sensitive to changes in AMP concentration.

AMP binding to AMPK increases its protein kinase activity by more than 1000-fold. The underlying mechanism involves a **pseudosubstrate sequence** (see the Protein Kinases: Target Recognition and Intrasteric Control section, pages 489–490) within CBS domain 2 that fits into the α-subunit catalytic site. When AMP binds to the Bateman modules, conformational changes in the γ-subunit displace the pseudosubstrate sequence

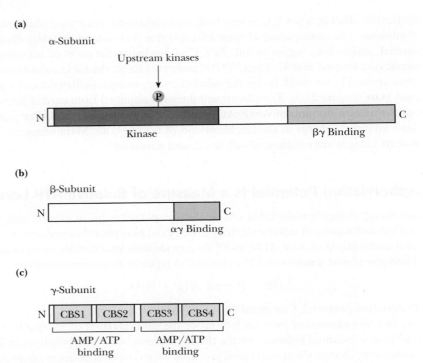

(a)

α-Subunit

Upstream kinases

N | Kinase | βγ Binding | C

(b)

β-Subunit

N | | αγ Binding | C

(c)

γ-Subunit

N | CBS1 | CBS2 | CBS3 | CBS4 | C

AMP/ATP binding   AMP/ATP binding

FIGURE 27.4 Domain structure of the AMP-activated protein kinase subunits. (Adapted from Figure 1 in Hardie, D. G., Hawley, S. A., and Scott, J. W., 2006. AMP-activated protein kinase: Development of the energy sensor concept. *Journal of Physiology* **574**:7–15.)

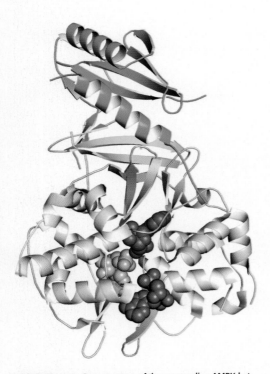

FIGURE 27.5 Core structure of the mammalian AMPK heterotrimer. The α-subunit is green, the β-subunit is yellow, and the γ-subunit is white. The γ-subunit has three AMP-binding sites; the three molecules of bound AMP are shown, one in red, one in pink. ATP and ADP can compete with AMP for binding at the AMP sites shown as occupied by red and pink AMP molecules. The molecule of AMP shown in blue cannot be exchanged for ATP or ADP (pdb id = 2V8Q).

from the kinase catalytic site, freeing it to act. The structural relationships between the AMPK subunits can be seen in the core structure of the mammalian αβγ complex (Figure 27.5).

Actually, AMP activates AMPK in two ways: First, it is an allosteric activator; second, AMP binding favors phosphorylation of Thr^{172} within the α-subunit kinase domain. Phosphorylation of Thr^{172} is necessary for α-subunit protein kinase activity and is achieved by several upstream protein kinases (such as LKB1 and CamKKβ), which are themselves constitutively active (that is, always active). Whether AMPK remains phosphorylated depends on the activity of phosphoprotein phosphatases, and AMP exerts its favorable effects on Thr^{172} phosphorylation by inhibiting its dephosphorylation by these phosphatases. Thr^{172} lies within the **activation loop** of the kinase; activation loops are common features of protein kinases whose activation requires phosphorylation by other protein kinases. Both of these favorable actions by AMP are reversed if ATP displaces AMP from the allosteric site.

In actuality, AMPK is acting as an energy charge sensor, because the two exchangeable adenine nucleotide-binding sites on the γ-subunit can bind either AMP, ADP, or ATP. AMP and ATP exert allosteric effects, but ADP does not. However, ADP occupation of these sites prevents dephosphorylation of Thr^{172}, whereas ATP favors Thr^{172} dephosphorylation. Thus, ADP binding to the γ-subunit protects AMPK from dephosphorylation, which would inactivate it. In its ADP-bound, Thr^{172}-phosphorylated state, AMPK is poised to undergo allosteric activation during energy stress, when AMP levels rise to a level where AMP displaces ADP from the allosteric sites. Since the intracellular concentrations of ADP normally exceed those of AMP, competition between ADP and ATP for binding to the γ-subunit is the likely determinant of the state of Thr^{172} phosporylation.

AMPK Targets Key Enzymes in Energy Production and Consumption

Activation of AMPK leads to phosphorylation of many key enzymes in energy metabolism. Those involved in energy production that are activated upon phosphorylation by AMPK include phosphofructokinase-2 (PFK-2; see Chapter 22). In contrast to protein kinase A phosphorylation of PFK-2, AMPK phosphorylation of liver PFK-2 enhances fructose-2,6-bisphosphate synthesis, which in turn stimulates glycolysis. Enzymes involved in energy consumption that are down-regulated upon phosphorylation by AMPK include glycogen synthase (see Chapter 22), acetyl-CoA carboxylase (which catalyzes the committed step in fatty acid biosynthesis; see Chapter 24), and 3-hydroxy-3-methylglutaryl-CoA reductase, which carries out the key regulatory reaction in cholesterol biosynthesis (see Chapter 24). Further, AMPK phosphorylation of various transcription factors leads to diminished expression of genes encoding biosynthetic enzymes and elevated expression of catabolic genes. Activation of AMPK also enhances energy production by promoting mitochondrial biogenesis.

AMPK Controls Whole-Body Energy Homeostasis

Beyond these cellular effects, AMPK plays a central role in energy balance in multicellular organisms (Figure 27.6). AMPK in skeletal muscle is activated by hormones such as adiponectin and leptin, adipocyte-derived hormones that govern eating behavior and energy homeostasis (see section 27.6). Physical activity (exercise) also activates muscle AMPK. In turn, skeletal muscle AMPK activates glucose uptake, fatty acid oxidation, and mitochondrial biogenesis through its phosphorylation of metabolic enzymes and transcription factors that control expression of genes involved in energy production and consumption. AMPK's actions in the liver lead to lowered ATP (energy) consumption through down-regulation of fatty acid synthesis, cholesterol synthesis, and gluconeogenesis. **Metformin,** a widely used drug for the treatment of type 2 diabetes (page 726), lowers blood glucose levels through inhibition of liver gluconeogenesis; metformin achieves this result through activation of AMPK. AMPK blocks insulin secretion by pancreatic β-cells; insulin is a hormone that favors energy storage (glycogen and fat synthesis). AMPK is also a master regulator of eating behavior through its activity in the hypothalamus, the key center for regulation of food intake. These effects of AMPK are described in section 27.6.

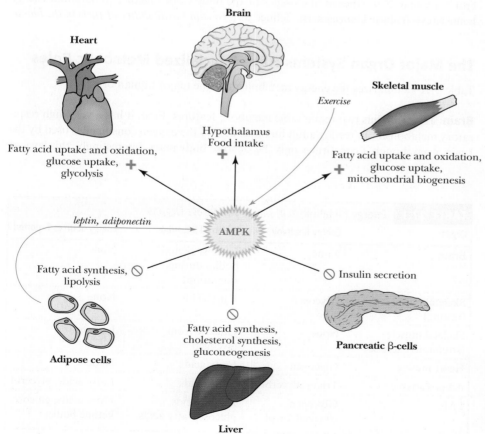

FIGURE 27.6 AMPK regulation of energy production and consumption in mammals. (Adapted from Figure 1 in Kahn, B. B., Alquier, T., Carling, D., and Hardie, D. G., 2005. AMP-activated protein kinase: Ancient energy gauge provides clues to modern understanding of metabolism. *Cell Metabolism* **1**:15–25.)

AMPK is also involved in regulating cellular growth through phosphorylation of a protein called **TSC2**. In turn, phosphorylation of TSC2 leads to inhibition of **mTORC1**, a complex containing **mTOR**. mTOR is a member of the phosphoinositide-3-kinase-related protein kinase family. When mTORC1 is active, it promotes protein synthesis, cell growth, and cell proliferation. Thus, inhibition of mTORC1 by AMPK conserves energy. (mTOR is an acronym for "mammalian Target of Rapamycin," so-named because it was first discovered through its ability to bind rapamycin. Rapamycin is an immunosuppressive drug.) We will return to a discussion of mTOR when we consider protein synthesis in eukaryotes in Chapter 30.

27.5 How Is Metabolism Integrated in a Multicellular Organism?

In complex multicellular organisms, organ systems have arisen to carry out specific physiological functions. Each organ expresses a repertoire of metabolic pathways that is consistent with its physiological purpose. Such specialization depends on coordination of metabolic responsibilities among organs so that the organism as a whole may thrive. Essentially all cells in animals have the set of enzymes common to the central pathways of intermediary metabolism, especially the enzymes involved in the formation of ATP and the synthesis of glycogen and lipid reserves. Nevertheless, organs differ in the metabolic fuels they prefer as substrates for energy production. Important differences also occur in the ways ATP is used to fulfill the organs' specialized metabolic functions. To illustrate these relationships, we will consider the metabolic interactions among the major organ systems found in humans: brain, skeletal muscle, heart, adipose tissue, and liver. In particular, the focus will be on energy metabolism in these organs (Figure 27.7). The major fuel depots in animals are *glycogen* in liver and muscle; *triacylglycerols* (fats) stored in adipose tissue; and *protein*, most of which is in skeletal muscle. In general, the order of preference for the use of these fuels is the order given: glycogen > triacylglycerol > protein. Nevertheless, the tissues of the body work together to maintain **energy homeostasis (caloric homeostasis),** defined as *a constant availability of fuels in the blood.*

The Major Organ Systems Have Specialized Metabolic Roles

Table 27.1 summarizes the energy metabolism of the major human organs.

Brain The brain has two remarkable metabolic features. First, it has a very high respiratory metabolism. In resting adult humans, 20% of the oxygen consumed is used by the brain, even though it constitutes only 2% or so of body mass. Interestingly, this level of

TABLE 27.1	Energy Metabolism in Major Vertebrate Organs		
Organ	Energy Reservoir	Preferred Substrate	Energy Sources Exported
Brain	None	Glucose (ketone bodies during starvation)	None
Skeletal muscle (resting)	Glycogen	Fatty acids	None
Skeletal muscle (strenuous exercise)	None	Glucose from glycogen	Lactate
Heart muscle	Glycogen	Fatty acids	None
Adipose tissue	Triacylglycerol	Fatty acids	Fatty acids, glycerol
Liver	Glycogen, triacylglycerol	Amino acids, glucose, fatty acids	Fatty acids, glucose, ketone bodies

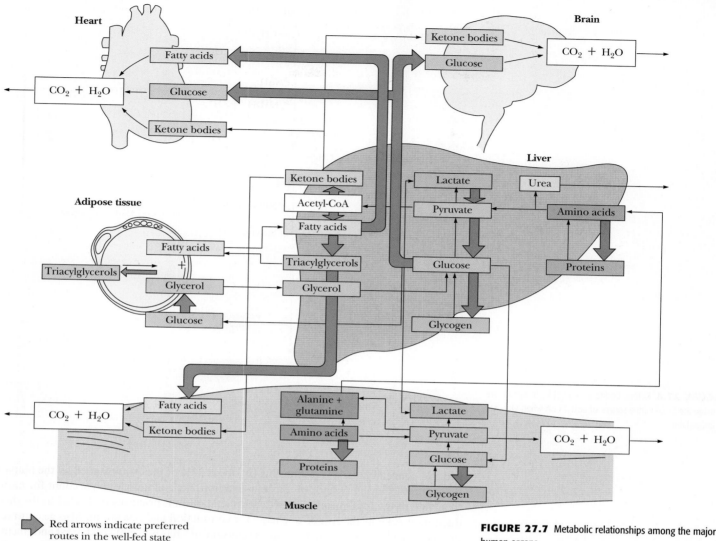

FIGURE 27.7 Metabolic relationships among the major human organs.

oxygen consumption is independent of mental activity, continuing even during sleep. Second, the brain is an organ with no significant fuel reserves—no glycogen, expendable protein, or fat (even in "fatheads"!). Normally, the brain uses only glucose as a fuel and is totally dependent on the blood for a continuous, incoming supply. Interruption of glucose supply for even brief periods of time (as in a stroke) can lead to irreversible losses in brain function. The brain uses glucose to carry out ATP synthesis via cellular respiration. High rates of ATP production are necessary to power the plasma membrane Na^+,K^+-ATPase so that the membrane potential essential for transmission of nerve impulses is maintained.

During prolonged fasting or starvation, the body's glycogen reserves are depleted. Under such conditions, the brain adapts to use β-hydroxybutyrate (Figure 27.8) as a source of fuel, converting it to acetyl-CoA for energy production via the citric acid cycle. β-Hydroxybutyrate (see Chapter 23) is formed from fatty acids in the liver. Although the brain cannot use free fatty acids or lipids directly from the blood as fuel, the conversion of these substances to β-hydroxybutyrate in the liver allows the brain to use body fat as a source of energy. The brain's other potential source of fuel during starvation is glucose obtained from gluconeogenesis in the liver (see Chapter 22), using the carbon skeletons of amino acids derived from muscle protein breakdown. The adaptation of the brain to use β-hydroxybutyrate from fat spares protein from degradation until lipid reserves are exhausted.

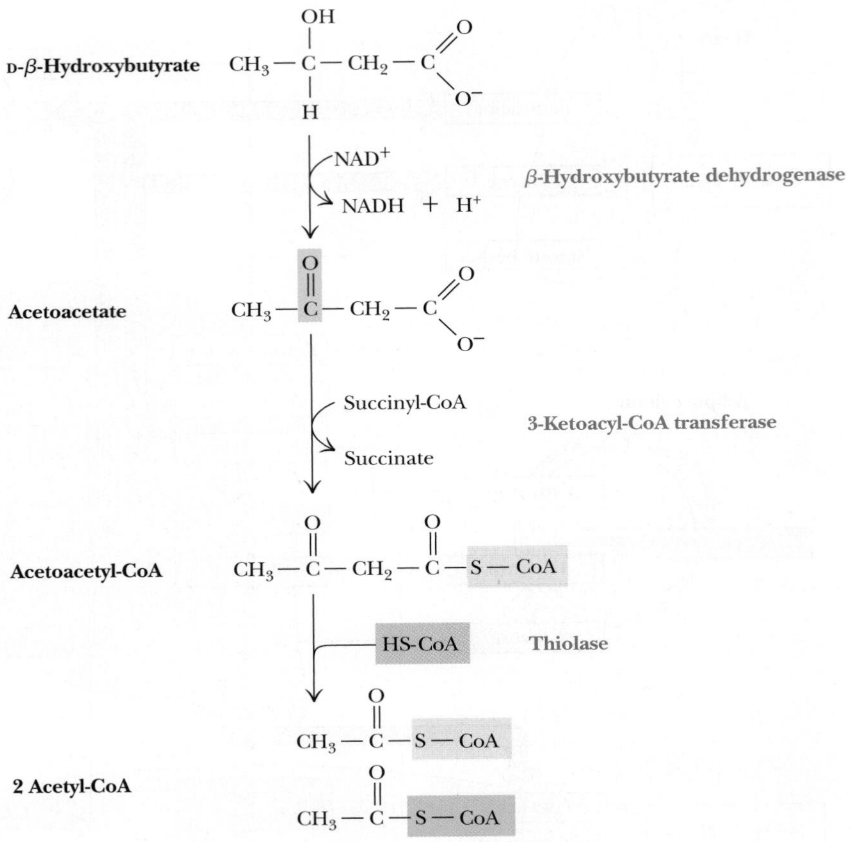

FIGURE 27.8 Ketone bodies such as β-hydroxybutyrate provide the brain with a source of acetyl-CoA when glucose is unavailable.

Muscle Skeletal muscle is responsible for about 30% of the O_2 consumed by the human body at rest. During periods of maximal exertion, skeletal muscle can account for more than 90% of the total metabolism. Muscle metabolism is primarily dedicated to the production of ATP as the source of energy for contraction and relaxation. Muscle contraction occurs when a motor nerve impulse causes Ca^{2+} release from specialized endomembrane compartments (the transverse tubules and sarcoplasmic reticulum). Ca^{2+} floods the *sarcoplasm* (the term denoting the cytosolic compartment of muscle cells), where it binds to **troponin C,** a regulatory protein, initiating a series of events that culminate in the sliding of myosin thick filaments along actin thin filaments. This mechanical movement is driven by energy released upon hydrolysis of ATP (see Chapter 16). The net result is that the muscle shortens. Relaxation occurs when the Ca^{2+} ions are pumped back into the sarcoplasmic reticulum by the action of a Ca^{2+}-transporting membrane ATPase. Two Ca^{2+} ions are translocated per ATP hydrolyzed. The amount of ATP used during relaxation is almost as much as that consumed during contraction.

Because muscle contraction is an intermittent process that occurs upon demand, muscle metabolism is designed for a demand response. Muscle at rest uses free fatty acids, glucose, or ketone bodies as fuel and produces ATP via oxidative phosphorylation. Resting muscle also contains about 2% glycogen and about 0.08% phosphocreatine by weight (Figure 27.9). When ATP is used to drive muscle contraction, the ADP formed can be reconverted to ATP by *creatine kinase* at the expense of phosphocreatine. Muscle phosphocreatine can generate enough ATP to power about 4 seconds of exertion. During strenuous exertion, such as a 100-meter sprint, once the phosphocreatine is depleted, muscle relies solely on its glycogen reserves, making the ATP for contraction via glycolysis. In contrast with the citric acid cycle and oxidative phosphorylation pathways, glycolysis is capable of explosive bursts of activity, and the flux of glucose-6-phosphate through this pathway can increase 2000-fold almost instantaneously. The triggers for this activation are Ca^{2+} and the "fight or flight" hormone *epinephrine* (see Chapters 22 and 32). Little interorgan cooperation occurs during strenuous (anaerobic) exercise.

FIGURE 27.9 Phosphocreatine serves as a reservoir of ATP-synthesizing potential.

$$\Delta G^{\circ\prime} = -13 \text{ kJ/mol}$$

Phosphocreatine + ADP → Creatine + ATP (Creatine kinase, Mg²⁺)

Muscle fatigue is the inability of a muscle to maintain power output. During maximum exertion, the onset of fatigue takes only 20 seconds or so. Fatigue is not the result of exhaustion of the glycogen reserves, nor is it a consequence of lactate accumulation in the muscle. Instead, it is caused by a decline in intramuscular pH as protons are generated during glycolysis. (The overall conversion of glucose to 2 lactate in glycolysis is accompanied by the release of 2 H⁺.) The pH may fall as low as 6.4. It is likely that the decline in PFK activity at low pH leads to a lowered flux of hexose through glycolysis and inadequate ATP levels, causing a feeling of fatigue. One benefit of PFK inhibition is that the ATP remaining is not consumed in the PFK reaction, thereby sparing the cell from the more serious consequences of losing all of its ATP.

During fasting or excessive activity, skeletal muscle protein is degraded to amino acids so that their carbon skeletons can be used as fuel. Many of the skeletons are converted to pyruvate, which can be transaminated back into alanine for export via the circulation (Figure 27.10). Alanine is carried to the liver, which in turn deaminates it back into pyruvate so that it can serve as a substrate for gluconeogenesis. Although muscle protein can be mobilized as an energy source, it is not efficient for an organism to consume its muscle and lower its overall fitness for survival. Muscle protein represents a fuel of last resort.

HUMAN BIOCHEMISTRY

Athletic Performance Enhancement with Creatine Supplements?

The creatine pool in a 70-kg (154-lb) human body is about 120 grams. This pool includes dietary creatine (from meat) and creatine synthesized by the human body from its precursors (arginine, glycine, and methionine). Of this creatine, 95% is stored in the skeletal and smooth muscles, about 70% of which is in the form of phosphocreatine. Supplementing the diet with 20 to 30 grams of creatine per day for 4 to 21 days can increase the muscle creatine pool by as much as 50% in someone with a previously low creatine level. Thereafter, supplements of 2 grams per day will maintain elevated creatine stores. Studies indicate that creatine supplementation gives some improvement in athletic performance during high-intensity, short-duration events (such as weight lifting), but no benefit in endurance events (such as distance running). The distinction makes sense in light of phosphocreatine's role as the substrate that creatine kinase uses to regenerate ATP from ADP. Intense muscular activity quickly (less than 2 seconds) exhausts ATP supplies; [phosphocreatine]ₘᵤₛcₗₑ is sufficient to restore ATP levels for a few extra seconds, but no more. The U.S. Food and Drug Administration advises consumers to consult with their doctors before using creatine as a dietary supplement.

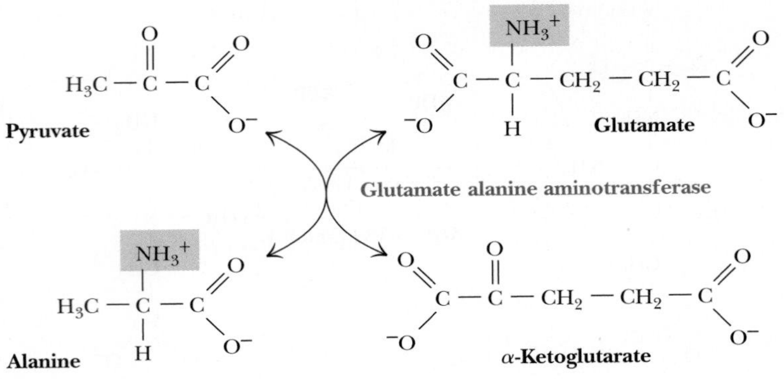

FIGURE 27.10 The transamination of pyruvate to alanine by glutamate : alanine aminotransferase.

Heart In contrast with the intermittent work of skeletal muscle, the activity of heart muscle is constant and rhythmic. The range of activity in heart is also much less than that in muscle. Consequently, the heart functions as a completely aerobic organ and, as such, is very rich in mitochondria. Roughly half the cytoplasmic volume of heart muscle cells is occupied by mitochondria. Under normal working conditions, the heart prefers fatty acids as fuel, oxidizing acetyl-CoA units via the citric acid cycle and producing ATP for contraction via oxidative phosphorylation. Heart tissue has minimal energy reserves: a small amount of phosphocreatine and limited quantities of glycogen. As a result, the heart must be continually nourished with oxygen and free fatty acids, glucose, or ketone bodies as fuel.

Adipose Tissue Adipose tissue is an amorphous tissue that is widely distributed about the body—around blood vessels, in the abdominal cavity and mammary glands, and most prevalently, as deposits under the skin. Long considered merely a storage depot for fat, adipose tissue is now appreciated as an endocrine organ responsible for secretion of a variety of hormones that govern eating behavior and caloric homeostasis. It consists principally of cells known as adipocytes that no longer replicate. However, adipocytes can increase in number as adipocyte precursor cells divide, and obese individuals tend to have more of them. As much as 65% of the weight of adipose tissue is triacylglycerol that is stored in adipocytes, essentially as oil droplets. The average 70-kg man has enough caloric reserve stored as fat to sustain a 6000 kJ/day rate of energy production for 3 months, which is adequate for survival, assuming no serious metabolic aberrations (such as nitrogen, mineral, or vitamin deficiencies). Despite their role as energy storage depots, adipocytes have a high rate of metabolic activity, synthesizing and breaking down triacylglycerol so that the average turnover time for a triacylglycerol molecule is just a few days. Adipocytes actively carry out cellular respiration,

HUMAN BIOCHEMISTRY

Fat-Free Mice—A Snack Food for Pampered Pets? No, A Model for One Form of Diabetes

Scientists at the National Institutes of Health have created transgenic mice that lack white adipose tissue throughout their lifetimes. These mice were created by blocking the normal differentiation of stem cells into adipocytes so that essentially no white adipose tissue can be formed in these animals. These "fat-free" mice have double the food intake and five times the water intake of normal mice. Fat-free mice also show decreased physical activity and must be kept warm on little heating pads to survive, because they lack insulating fat. They are also diabetic, with three times normal blood glucose and triacylglycerol levels and only 5% of normal leptin levels; they die prematurely. Like type 2 diabetic patients, fat-free mice have

markedly elevated insulin levels (50–400 times normal) but are unresponsive to insulin. These mice serve as an excellent model for the disease *lipoatrophic diabetes,* an inherited disease characterized by the absence of adipose tissue and severe diabetic symptoms. Indeed, transplantation of adipose tissue into these fat-free mice cured their diabetes. As the major organ for triacylglycerol storage, white adipose tissue helps control energy homeostasis (food intake and energy expenditure) via the release of leptin and other hormonelike substances (see the discussion on page 935). Clearly, absence of adipose tissue has widespread, harmful consequences for metabolism.

transforming glucose to energy via glycolysis, the citric acid cycle, and oxidative phosphorylation. If glucose levels in the diet are high, glucose is converted to acetyl-CoA for fatty acid synthesis. However, under most conditions, free fatty acids for triacylglycerol synthesis are obtained from the liver. Because adipocytes lack glycerol kinase, they cannot recycle the glycerol of triacylglycerol but rather depend on glycolytic conversion of glucose to dihydroxyacetone-3-phosphate (DHAP) and the reduction of DHAP to glycerol-3-phosphate for triacylglycerol biosynthesis. Adipocytes also require glucose to feed the pentose phosphate pathway for NADPH production.

Glucose plays a pivotal role for adipocytes. If glucose levels are adequate, glycerol-3-phosphate is formed in glycolysis and the free fatty acids liberated in triacylglycerol breakdown are re-esterified to glycerol to re-form triacylglycerols. However, if glucose levels are low, [glycerol-3-phosphate] falls and free fatty acids are released to the bloodstream (see Chapter 23).

"Brown Fat" A specialized type of adipose tissue, so-called **brown fat,** is found in newborns and hibernating animals. The abundance of mitochondria, which are rich in cytochromes, is responsible for the brown color of this fat. As usual, these mitochondria are very active in electron transport–driven proton translocation, but these particular mitochondria contain in their inner membranes a protein, **thermogenin,** also known as *uncoupling protein 1* (see Chapter 20), that creates a passive proton channel, permitting the H$^+$ ions to reenter the mitochondrial matrix without generating ATP. Instead, the energy of oxidation is dissipated as heat. Indeed, brown fat is specialized to oxidize fatty acids for heat production rather than ATP synthesis.

Liver The liver serves as the major metabolic processing center in vertebrates. Except for dietary triacylglycerols, which are metabolized principally by adipose tissue, most of the incoming nutrients that pass through the intestinal tract are routed via the portal vein to the liver for processing and distribution. Much of the liver's activity centers around conversions involving glucose-6-phosphate (Figure 27.11). Glucose-6-phosphate can be converted to glycogen, released as blood glucose, used to generate NADPH and pentoses via the pentose phosphate cycle, or catabolized to acetyl-CoA for fatty acid synthesis or for energy production via oxidative phosphorylation. Most of the

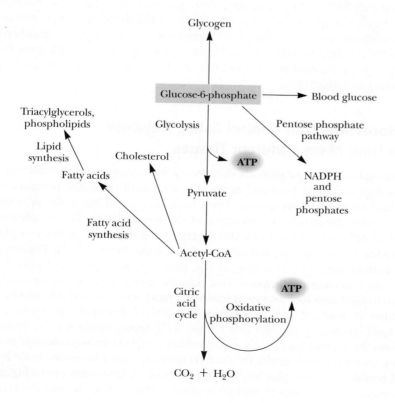

FIGURE 27.11 Metabolic conversions of glucose-6-phosphate in the liver.

liver glucose-6-phosphate arises from dietary carbohydrate, from degradation of glycogen reserves, or from muscle lactate that enters the gluconeogenic pathway.

The liver plays an important regulatory role in metabolism by buffering the level of blood glucose. Liver has two enzymes for glucose phosphorylation: hexokinase and glucokinase (type-IV hexokinase). Unlike hexokinase, glucokinase has a low affinity for glucose. Its K_m for glucose is high, on the order of 10 mM. When blood glucose levels are high, glucokinase activity augments hexokinase in phosphorylating glucose as an initial step leading to its storage in glycogen. The major metabolic hormones—epinephrine, glucagon, and insulin—all influence glucose metabolism in the liver to keep blood glucose levels relatively constant (see Chapters 22 and 32).

The liver is a major center for fatty acid turnover. When the demand for metabolic energy is high, triacylglycerols are broken down and fatty acids are degraded in the liver to acetyl-CoA to form ketone bodies, which are exported to the heart, brain, and other tissues. If energy demands are low, fatty acids are incorporated into triacylglycerols that are carried to adipose tissue for deposition as fat. Cholesterol is also synthesized in the liver from two-carbon units derived from acetyl-CoA.

In addition to these central functions in carbohydrate and fat-based energy metabolism, the liver serves other purposes. For example, the liver can use amino acids as metabolic fuels. Amino acids are first converted to their corresponding α-keto acids by aminotransferases. The amino group is excreted after incorporation into urea in the urea cycle. The carbon skeletons of glucogenic amino acids can be used for glucose synthesis, whereas those of ketogenic amino acids appear in ketone bodies (see Figure 25.41). The liver is also the principal detoxification organ in the body. The endoplasmic reticulum of liver cells is rich in enzymes that convert biologically active substances such as hormones, poisons, and drugs into less harmful byproducts.

Liver disease leads to serious metabolic derangements, particularly in amino acid metabolism. In cirrhosis, the liver becomes defective in converting NH_4^+ to urea for excretion, and blood levels of NH_4^+ rise. Ammonia is toxic to the central nervous system, and coma ensues.

27.6　What Regulates Our Eating Behavior?

Approximately 65% of Americans are overweight, and one in three Americans is clinically obese (overweight by 20% or more). Obesity is the single most important cause of type 2 (adult-onset insulin-independent) diabetes. Research into the regulatory controls that govern our feeding behavior has become a medical urgency with great financial incentives, given the epidemic proportions of obesity and widespread preoccupation with dieting and weight loss.

The Hormones That Control Eating Behavior Come from Many Different Tissues

Appetite and weight regulation are determined by a complex neuroendocrine system that involves hormones produced in the stomach, small intestines, pancreas, adipose tissue, and central nervous system. These hormones act in the brain, principally on neurons within the arcuate nucleus region of the hypothalamus. The arcuate nucleus is an anatomically distinct brain area that functions in homeostasis of body weight, body temperature, blood pressure, and other vital functions (Figure 27.12). The neurons respond to these signals by activating, or not, pathways involved in eating (food intake) and energy expenditure. Hormones that regulate eating behavior can be divided into short-term regulators that determine individual meals and long-term regulators that act to stabilize the levels of body fat deposits. Two subsets of neurons are involved: (1) the **NPY/AgRP**-producing neurons that release *NPY* **(neuropeptide Y),** the protein that stimulates the neurons that trigger eating behavior, and (2) the **melanocortin**-producing neurons, whose products inhibit the neurons initiating eating behavior. *AgRP* is **agouti-related peptide,** a protein that blocks the activity of melanocortin-producing neurons. *Melanocortins* are a group of peptide hormones that includes **α- and β-melanocyte–**

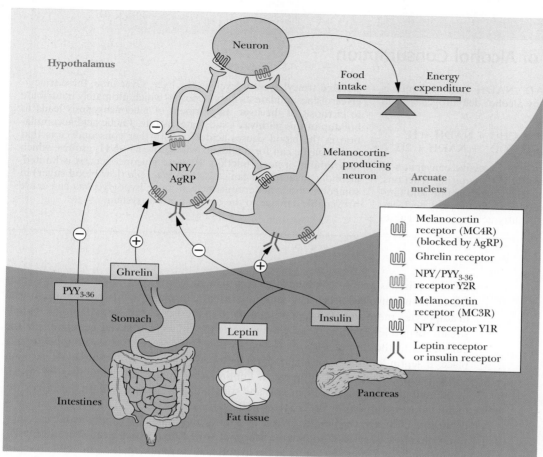

FIGURE 27.12 The regulatory pathways that control eating. (Adapted from Figure 1 in Schwartz, M. W., and Morton, G. J., 2002. Obesity: Keeping hunger at bay. *Nature* **418**:595–597.)

stimulating hormones (α-MSH and β-MSH). Melanocortins act on melanocortin receptors (MCRs), which are members of the 7-TMS G-protein–coupled receptor (GPCR) family of membrane receptors; MCRs trigger cellular responses through adenylyl cyclase activation (see Chapters 15 and 32).

Ghrelin and Cholecystokinin Are Short-Term Regulators of Eating Behavior

Short-term regulators of eating include **ghrelin** and **cholecystokinin.** Ghrelin is an appetite-stimulating peptide hormone produced in the stomach. Production of ghrelin is maximal when the stomach is empty, but ghrelin levels fall quickly once food is consumed. Cholecystokinin is a peptide hormone released from the gastrointestinal tract during eating. In contrast to ghrelin, cholecystokinin signals satiety (the sense of fullness) and tends to curtail further eating. Together, ghrelin and cholecystokinin constitute a meal-to-meal control system that regulates the onset and end of eating behavior. The activity of this control system is also modulated by the long-term regulators.

Insulin and Leptin Are Long-Term Regulators of Eating Behavior

Long-term regulators include **insulin** and **leptin,** both of which inhibit eating and promote energy expenditure. Insulin is produced in the β-cells of the pancreas when blood glucose levels rise. A major role of insulin is to stimulate glucose uptake from the blood into muscle, fat, and other tissues. Blood insulin levels correlate with body fat amounts. Insulin also stimulates fat cells to make leptin. Leptin (from the Greek word *lepto,* meaning "thin") is a 16-kD, 146–amino acid residue protein produced principally in adipocytes (fat

The Metabolic Effects of Alcohol Consumption

Ethanol metabolism alters the NAD$^+$/NADH ratio. Ethanol is metabolized to acetate in the liver by alcohol dehydrogenase and aldehyde dehydrogenase:

$$CH_3CH_2OH + NAD^+ \rightleftharpoons CH_3CHO + NADH + H^+$$
$$CH_3CHO + NAD^+ + H_2O \rightleftharpoons CH_3COO^- + NADH + 2H^+$$

Acetate is then converted to acetyl-CoA. Excessive conversion of available NAD$^+$ to NADH impairs NAD$^+$-requiring reactions, such as the citric acid cycle, gluconeogenesis from lactate, and fatty acid oxidation. Accumulation of acetyl-CoA favors fatty acid synthesis, which, along with blockage of fatty acid oxidation, causes elevated triacylglycerol levels in the liver. Over time, these triacylglycerols accumulate as fatty deposits, which ultimately contribute to cirrhosis of the liver. Impairment of gluconeogenesis leads to buildup of this pathway's substrate, lactate. Lactic acid accumulation in the blood causes acidosis. A further consequence is that acetaldehyde can form adducts with protein —NH$_2$ groups, which may inhibit protein function. Because gluconeogenesis is limited, alcohol consumption can cause *hypoglycemia* (low blood sugar) in someone who is undernourished. In turn, hypoglycemia can cause irreversible damage to the central nervous system.

cells). Leptin has a four-helix bundle tertiary structure similar to that of cytokines (protein hormones involved in cell–cell communication). Normally, as fat deposits accumulate in adipocytes, more and more leptin is produced in these cells and spewed into the bloodstream. Leptin levels in the blood communicate the status of triacylglycerol levels in the adipocytes to the central nervous system so that appropriate changes in appetite take place. If leptin levels are low ("starvation"), appetite increases; if leptin levels are high ("overfeeding"), appetite is suppressed. Leptin also regulates fat metabolism in adipocytes, inhibiting fatty acid biosynthesis and stimulating fat metabolism. In the latter case, leptin induces synthesis of the enzymes in the fatty acid oxidation pathway and increases expression of *uncoupling protein 2 (UCP2)*, a mitochondrial protein that uncouples oxidation from phosphorylation so that the energy of oxidation is lost as heat (thermogenesis). Leptin binding to leptin receptors in the hypothalamus inhibits release of NPY. Because NPY is a potent *orexic* (appetite-stimulating) peptide hormone, leptin is therefore an *anorexic* (appetite-suppressing) agent. Functional leptin receptors are also essential for pituitary function, growth hormone secretion, and normal puberty. When body fat stores decline, the circulating levels of leptin and insulin also decline. Hypothalamic neurons sense this decline and act to increase appetite to restore body fat levels.

Intermediate regulation of eating behavior is accomplished by the gut hormone **PYY$_{3-36}$**. PYY$_{3-36}$ is produced in endocrine cells found in distal regions of the small intestine, areas that receive ingested food some time after a meal is eaten. PYY$_{3-36}$ inhibits eating for many hours after a meal by acting on the NPY/AgRP-producing neurons in the arcuate nucleus. Clearly, the regulatory controls that govern eating are complex and layered. Some believe that defects in these controls are common and biased in favor of overeating, an advantageous evolutionary strategy that may have unforeseen consequences in these bountiful times.

To this point, our focus has been on hormonal controls influencing energy homeostasis and food intake. Superimposed on these regulatory mechanisms are neural controls originating in those cognitive centers of the brain involved in learning, memory, socialization, and decision-making processes. For example, hedonistic ("pleasure-seeking") mechanisms triggered by seeing, smelling, tasting, or even imagining food can override the normal hormonal regulation of eating behavior that is driven by nutrient levels in the blood. Clearly, eating or not is determined by a very complex interplay between energy status and the emotional and/or rational disposition of the brain. Our lack of understanding of these phenomena hinders our treatment of the obesity epidemic.

AMPK Mediates Many of the Hypothalamic Responses to These Hormones

The actions of leptin, ghrelin, and NPY converge at AMPK. Leptin inhibits AMPK activity in the arcuate nucleus, and this inhibition underlies the anorexic effects of leptin. Leptin action on AMPK depends on a particular melanocortin receptor type known as the **melanocortin-4 receptor (MC4R).** On the other hand, ghrelin and NPY activate hypothalamic AMPK, which stimulates food intake and leads, over time, to increased body weight. The effects of AMPK in the hypothalamus that lead to alterations in eating behavior may be mediated through changes in malonyl-CoA levels. Malonyl-CoA is believed to be a key indicator of energy levels in hypothalamic neurons. Low [malonyl-CoA] in hypothalamic neurons is associated with increased food intake, and elevated malonyl-CoA levels are associated with suppression of eating. The inhibition of acetyl-CoA carboxylase (and thus, malonyl-CoA synthesis) as a result of phosphorylation by AMPK plays an important part in the regulation of our eating behavior.

27.7 Can You Really Live Longer by Eating Less?

Caloric Restriction Leads to Longevity

Nutritional studies published in 1935 showed that rats fed a low-calorie, but balanced and nutritious diet lived nearly twice as long as rats with unlimited access to food (2.4 years versus 1.3 years). Subsequent research over the ensuing 75 years has shown that the relationship between diet and longevity is a general one for organisms from yeast to mammals. To achieve this effect of **caloric restriction (CR),** animals are given a level of food that amounts to 60% to 70% of what they would eat if they were allowed free access to food. In animals, CR results in lower blood glucose levels, declines in glycogen and fat stores, enhanced responsiveness to insulin, lower body temperature, and diminished reproductive capacity. The extended life span given by CR offers a definite evolutionary advantage: Any animal that could slow the aging process and postpone reproduction in times of food scarcity and then resume reproduction when food became available would out-compete animals without such ability.

Another remarkable feature of CR is that it diminishes the likelihood for development of many age-related diseases, such as cancer, diabetes, and atherosclerosis. Is this benefit simply the result of lowered caloric intake, or does CR lead to significant regulatory changes that affect many aspects of an organism's physiology? The answer to this and other questions is emerging from vigorous research efforts toward understanding CR.

Mutations in the *SIR2* Gene Decrease Life Span

Many important clues came from genetic investigations. Deletion of a gene termed *SIR2* (for silent information regulator 2) abolished the ability of CR to lengthen life span in yeast and roundworms, implicating the *SIR2* gene product in longevity. *SIR2* originally was discovered through its ability to silence the transcription of genes that encode rRNA. *SIR2*-related genes are found in some prokaryotes and virtually all eukaryotes, including yeast, nematodes, fruit flies, and humans. Humans have seven such genes, designated *SIRT1* through *SIRT7*, SIRT being an abbreviation of **sirtuin,** the generic name for proteins encoded by *SIR2* genes. SIRTs differ in their cellular localization, with SIRT1 cycling between the cytosol and nucleus, and SIRTs 3, 4, and 5 confined to the mitochondria. SIRT1 is the best studied of the human sirtuins. Sirtuins are NAD^+-dependent protein deacetylases (NAD^+-dependent KDACs) (Figure 27.13). Cleavage of acetyl groups from proteins is an exergonic reaction, as is cleavage of the N-glycosidic bond in NAD^+ ($\Delta G°' = -34$ kJ/mol). Thus, involvement of NAD^+ in the reaction is not a thermodynamic necessity. However, NAD^+ involvement does couple the reaction to an important signal of metabolic status, namely, the NAD^+/NADH ratio. Furthermore, both nicotinamide and NADH are potent inhibitors of the deacetylase reaction. Thus, the NAD^+/NADH ratio controls sirtuin protein deacetylase activity, so oxidative metabolism, which drives conversion of NADH to NAD^+, enhances sirtuin activity. One adaptive response found with CR is increased mitochondrial biogenesis in liver, fat, and muscle, which is a response that would raise the NAD^+/NADH ratio.

Sirtuin-catalyzed removal of acetyl groups from lysine residues of histones, the core proteins of nucleosomes, would allow the nucleosomes to interact more strongly with DNA, making transcription more difficult (see Chapter 29 for a discussion of the

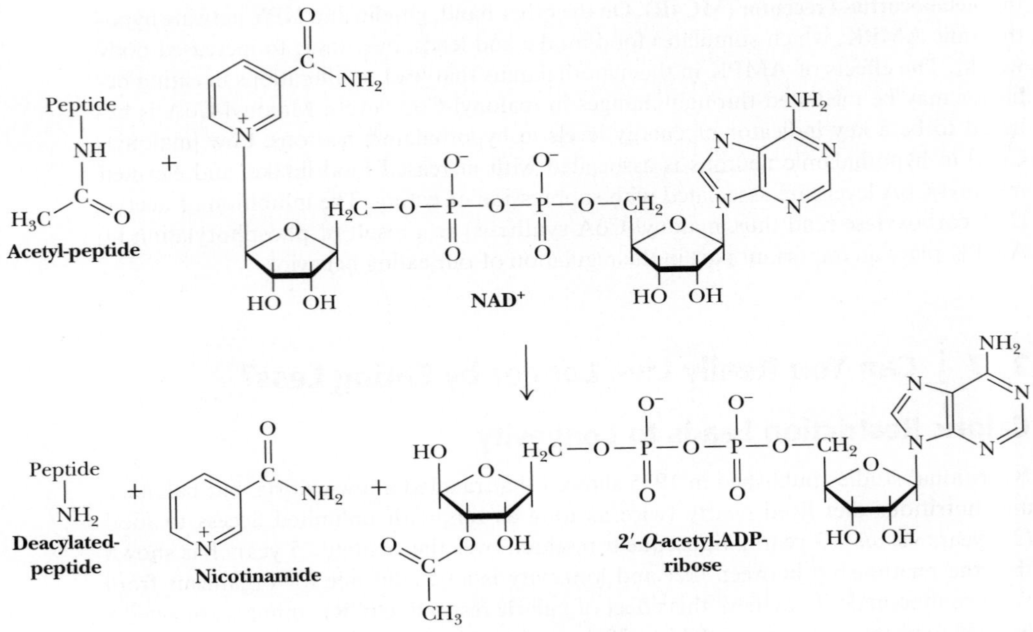

FIGURE 27.13 The NAD$^+$-dependent protein deacetylase reaction of sirtuins. Acetylated peptides include N-acetyl lysine side chains of histone H3 and H4 and acetylated p53 (p53 is the protein product of the *p53* tumor suppressor gene).

relationship between histone acetylation and transcriptional activity of genes). Perhaps more significantly, SIRT-catalyzed removal of acetyl groups from K residues in transcription factors and metabolic enzymes is an effective mechanism for transcriptional and metabolic regulation in response to changes in the NAD$^+$/NADH ratio. Almost all of the enzymes of energy metabolism are acetylated and thus potentially regulated by SIRT deacetylation.

SIRT1 Is a Key Regulator in Caloric Restriction

As a key regulator in CR, the human sirtuin protein SIRT1 connects nutrient availability to the expression of metabolic genes. A striking feature of CR is the loss of fat stores and reduction in **white adipose tissue (WAT)**. SIRT1 participates in the transcriptional regulation of adipogenesis through interaction with **PPARγ** (peroxisome proliferator-activator receptor-γ), a nuclear hormone receptor that activates transcription of genes involved in adipogenesis and fat storage. SIRT1 binding to two PPARγ corepressors, NCoR and SMRT, prevents transcription of genes controlled by PPARγ, leading to loss of fat stores. Because adipose tissue functions as an endocrine organ, this loss of fat has significant hormonal consequences for energy metabolism.

In liver, SIRT1 interacts with and deacetylates **PGC-1α** (peroxisome proliferator-activator receptor-γ coactivator-1), a transcriptional regulator of genes involved in energy metabolism. Thus, CR leads to increased transcription of the genes encoding the enzymes of gluconeogenesis and repression of genes encoding glycolytic enzymes. Acting in these roles, SIRT1 connects nutrient availability to the regulation of major pathways of energy storage (glycogen and fat) and fuel utilization.

Resveratrol, a Compound Found in Red Wine, Is a Potent Activator of Sirtuin Activity

Resveratrol (*trans*-3,4′,5-trihydroxystilbene, Figure 27.14) is a **phytoalexin**. Phytoalexins are compounds produced by plants in response to stress, injury, or fungal infection. Resveratrol is abundant in wine grape skins as a result of common environmental stresses, such as infection by *Botrytis cinerea*, a fungus important in making certain

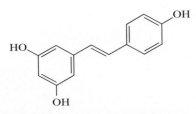

Resveratrol (*trans*-3,4′,5-trihydroxystilbene)

FIGURE 27.14 Resveratrol, a phytoalexin, is a member of the polyphenol class of natural products. As a polyphenol, resveratrol is a good free-radical scavenger, which may account for its cancer preventative properties.

| TABLE 27.2 | Caloric restriction and SIRT1 activators reverse many of the clinical signs of metabolic syndrome observed in mice that were maintained on a high-fat diet |

Metabolic Syndrome	High-Fat Diet	Caloric Restriction	High-Fat Diet + Resveratrol	High-Fat Diet + SIR1720
Body fat	↑	↓	↓	↓
LDL cholesterol	↑	↓	No change	↓
HDL cholesterol	↓	↑	No change	No change
Triglycerides	↑	↓	No change	↓
Glucose tolerance	↓	↑	↑	↑

Adapted from Table 2 in Lomb, D. J., Laurent, G., and Haigi, M. C., 2010. Sirtuins regulate key aspects of lipid metabolism. *Biochimica et Biophysica Acta* **1804**:1652–1657.

wines. Because the skins are retained when grapes are processed to make red wines, red wine is an excellent source of resveratrol. Resveratrol might be the basis of the French paradox—the fact that the French people enjoy longevity and relative freedom from heart disease despite a high-fat diet. When resveratrol is given to yeast cultures, round-worms *(Caenorhabditis elegans),* or fruit flies *(Drosophila melanogaster),* it has the same life-extending effects as CR. Resveratrol increased the replicative life span (the number of times a cell can divide before dying) of yeast by 70%. Resveratrol activates SIRT1 NAD^+-dependent deacetylase activity. Furthermore, resveratrol activates AMPK in the brain and in cultured neurons. Other activators of SIRT1 are known. SIR1720, a synthetic compound, is an even more potent activator of SIRT1 than resveratrol. Unlike resveratrol, SIR1720 does not directly activate AMPK. However, SIR1720 indirectly activates AMPK, because SIRT1 activation leads to increased energy expenditure, and thus a decrease in energy charge, which in turn activates AMPK.

Because the effects of CR and resveratrol on longevity are not additive, a reasonable conclusion is that they operate via a common mechanism. If this is so, then if you want to enjoy longevity, the appropriate advice would be "Eat less or drink red wine," not "Eat less and drink red wine"!

AMPK, SIRT1, Caloric Restriction, and Metabolic Syndrome While a detailed description of the relationships between AMPK, SIRT1, and CR awaits further research, certain generalizations are clear: AMPK is a key regulator of metabolism through its role as the sensor of energy status. The NAD^+-dependent deacetylase SIRT1 becomes activated when nutrients are limiting, as in CR. Thus, both AMPK and SIRT1 respond to nutrient limitation. AMPK responds by down-regulating anabolic processes and activating ATP-producing catabolic processes. AMPK accomplishes these ends through phosphorylation of metabolic enzymes and transcription factors. The effects of SIRT1 are achieved through its deacetylation of metabolic enzymes and transcription factors, and depending on the transcription factor, gene expression may be enhanced or diminished. AMPK and SIRT1 functions are coordinated to ensure that the cellular responses to changing conditions of nutrient availability are appropriate.

Metabolic syndrome is a suite of symptoms that precede the development of type 2 diabetes; it is a reasonable surrogate for the exact opposite of CR. The symptoms include high blood pressure, elevated blood triglyceride levels, low blood HDL-cholesterol levels, high blood glucose levels, and central obesity (the accumulation of body fat around the waist). The relationships between the altered metabolism seen in metabolic syndrome and the effects of CR, resveratrol, and SIR1720 are intriguing (Table 27.2). Caloric restriction reversed all of the symptoms of metabolic syndrome, and resveratrol and SIR1720 showed significant changes in several. A deeper understanding of the relationships between nutrient intake and changes in metabolic regulation seems promising.

SUMMARY

27.1 Can Systems Analysis Simplify the Complexity of Metabolism?
Cells are in a dynamic steady state maintained by considerable metabolic flux. The metabolism going on in even a single cell is so complex that it defies meaningful quantitative description. Nevertheless, overall relationships become more obvious by a systems analysis approach to intermediary metabolism. The metabolism of a typical heterotrophic cell can be summarized in three interconnected functional blocks: (1) catabolism, (2) anabolism, and (3) macromolecular synthesis and growth. Phototrophic cells require a fourth block: photosynthesis. Only a few metabolic intermediates connect these systems, and ATP and NADPH serve as the carriers of chemical energy and reducing power, respectively, between these various blocks.

27.2 What Underlying Principle Relates ATP Coupling to the Thermodynamics of Metabolism?
ATP coupling determines the thermodynamics of metabolic sequences. The ATP coupling stoichiometry cannot be predicted from chemical considerations; instead, it is a quantity selected by evolution and the fundamental need for metabolic sequences to be emphatically favorable from a thermodynamic perspective. Catabolic sequences generate ATP with an overall favorable K_{eq} (and hence, a negative ΔG) and anabolic sequences consume this energy with an overall favorable K_{eq}, even though such sequences may span the same starting and end points (as in fatty acid oxidation of palmitoyl-CoA to 8 acetyl-CoA versus synthesis of palmitoyl-CoA from 8 acetyl-CoA). ATP has two metabolic roles: a stoichiometric role in rendering metabolic sequences thermodynamically favorable and a regulatory role as an allosteric effector.

27.3 Is There a Good Index of Cellular Energy Status?
The level of phosphoric anhydride bonds in the adenylate system of ATP, ADP, and AMP can be expressed in terms of the energy charge equation:

$$\text{Energy charge} = \frac{1}{2}\left(\frac{2\,[\text{ATP}] + [\text{ADP}]}{[\text{ATP}] + [\text{ADP}] + [\text{AMP}]}\right)$$

More revealing of the potential for an ATP-dependent reaction to occur is Γ, the phosphorylation potential: $\Gamma = [\text{ATP}]/([\text{ADP}][\text{P}_i])$.

27.4 How Is Overall Energy Balance Regulated in Cells?
AMP-activated protein kinase (AMPK) is the cellular energy sensor. When cellular energy levels are high, as signaled by high ATP concentrations, AMPK is inactive. When cellular energy levels are depleted, as signaled by high [AMP], AMPK is allosterically activated and phosphorylates many enzymes involved in cellular energy production and consumption. Competition between AMP and ATP for binding to the AMPK allosteric sites determines AMPK activity. AMPK activation leads to elevated energy production and diminished energy consumption. AMPK is an $\alpha\beta\gamma$-heterotrimer. The α-subunit is the catalytic subunit. AMP or ATP binding to the γ-subunit determines AMPK activity, with AMP binding increasing activity by more than 1000-fold.

Activation of AMPK leads to phosphorylation of key enzymes in energy metabolism, activating those involved in ATP production and inactivating those in ATP consumption. In addition, AMPK phosphorylation of various transcription factors leads to elevated expression of catabolic genes and diminished expression of genes encoding biosynthetic enzymes. Beyond these cellular effects, AMPK plays a central role in energy balance in multicellular organisms because its activity is responsive to hormones that govern eating behavior and energy homeostasis.

27.5 How Is Metabolism Integrated in a Multicellular Organism?
Organ systems in complex multicellular organisms carry out specific physiological functions, with each expressing those metabolic pathways appropriate to its physiological purpose. Essentially all cells in animals carry out the central pathways of intermediary metabolism, especially the reactions of ATP synthesis. Nevertheless, organs differ in the metabolic fuels they prefer as substrates for energy production. The major fuel depots in animals are glycogen in liver and muscle; triacylglycerols (fats) in adipose tissue; and protein, most of which is in skeletal muscle. The order of preference for the use of these fuels is glycogen > triacylglycerol > protein. Nevertheless, the tissues of the body work together to maintain caloric homeostasis: Constant availability of fuels in the blood. The major organ systems have specialized metabolic roles within the organism.

The brain has a strong reliance on glucose as fuel. Muscle at rest primarily relies on fatty acids, but under conditions of strenuous contraction when O_2 is limiting, muscle shifts to glycogen as its primary fuel. The heart is a completely aerobic organ, rich in mitochondria, with a preference for fatty acids as fuel under normal operating conditions. The liver is the body's metabolic processing center, taking in nutrients and sending out products such as glucose, fatty acids, and ketone bodies. Adipose tissue takes up glucose and, to a lesser extent, fatty acids for the synthesis and storage of triacylglycerols.

27.6 What Regulates Our Eating Behavior?
Appetite and weight regulation are governed by hormones produced in the stomach, small intestines, pancreas, adipose tissue, and central nervous system. These hormones act on neurons within the arcuate nucleus region of the hypothalamus that control pathways involved in eating (food intake) and energy expenditure. Hormones that regulate eating behavior can be divided into short-term regulators that determine individual meals and long-term regulators that act to stabilize the levels of body fat deposits. Short-term regulators of eating include ghrelin and cholecystokinin. Ghrelin is produced when the stomach is empty, but ghrelin levels fall quickly once food is consumed. Cholecystokinin is released from the gastrointestinal tract during eating. In contrast to ghrelin, cholecystokinin signals satiety (the sense of fullness) and tends to curtail further eating. Together, ghrelin and cholecystokinin constitute a meal-to-meal control system that regulates the onset and end of eating behavior.

Long-term regulators include insulin and leptin, both of which inhibit eating and promote energy expenditure. Blood insulin levels correlate with body fat amounts. Insulin also stimulates fat cells to make leptin. Leptin is produced principally in adipocytes. As fat accumulates in adipocytes, more leptin is released into the bloodstream to communicate the status of adipocyte fat to the central nervous system. If leptin levels are low ("starvation"), appetite increases; if leptin levels are high ("overfeeding"), appetite is suppressed. Leptin binding to its receptors in the hypothalamus inhibits release of NPY. NPY is a potent orexic hormone; therefore, leptin is an anorexic agent. When body fat stores decline, the circulating levels of leptin and insulin also decline. Hypothalamic neurons sense this decline and act to increase appetite to restore body fat levels.

Intermediate regulation of eating behavior is accomplished by the gut hormone PYY_{3-36}. Produced in distal regions of the intestines, PYY_{3-36} delays eating for many hours after a meal by inhibiting the NPY/AgRP-producing neurons in the arcuate nucleus. The regulatory controls that govern eating are complex and layered. Superimposed on these biochemical mechanisms are neural controls arising from cognitive centers of the brain that can override metabolic regulation.

27.7 Can You Really Live Longer by Eating Less?
Caloric restriction (CR) prolongs the longevity of organisms from yeast to mammals. CR results in lower blood glucose levels, decline in glycogen and fat

stores, enhanced responsiveness to insulin, lower body temperature, and diminished reproductive capacity. CR also diminishes the likelihood for development of many age-related diseases, such as cancer, diabetes, and atherosclerosis. Genetic investigations revealed that mutations in the SIR2 gene abolish the extension of life span by CR. The human SIR2 gene equivalent is SIRT1. SIRT genes encode sirtuins, a family of NAD^+-dependent protein deacetylases. The NAD^+/NADH ratio controls sirtuin protein deacetylase activity, so oxidative metabolism, which drives conversion of NADH to NAD^+, enhances sirtuin action. CR increases mitochondrial biogenesis in liver, fat, and muscle, a response that would raise the NAD^+/NADH ratio. SIRT1 is a key regulator in CR. The physiological responses caused by CR are the result of a tightly regulated program that connects nutrient availability to energy metabolism. Almost all metabolic enzymes are acetylated and thus potentially regulated by SIRT1. A striking feature of CR is the loss of fat stores and reduction in white adipose tissue. SIRT1 can block the action of PPARγ, a nuclear hormone receptor that activates transcription of genes involved in adipogenesis and fat storage. The consequence is loss of fat stores. Because adipose tissue functions as an endocrine organ, this loss of fat has significant hormonal consequences for energy metabolism. In liver, SIRT1 interacts with and deacetylates PGC-1, a transcriptional regulator of glucose production. Transcription of genes encoding the enzymes of gluconeogenesis and repression of genes encoding glycolytic enzymes are increased upon PGC-1 deacetylation. Thus, SIRT1 connects nutrient availability to the regulation of major pathways of energy storage and fuel use.

Resveratrol, a phytoalexin, has the same life-extending effects as CR. Resveratrol activates both SIRT1 activity and brain AMPK, a key energy sensor. The influences of resveratrol on longevity may arise through its effects on caloric homeostasis.

Thus, both AMPK and SIRT1 respond to nutrient limitation. AMPK down-regulates anabolic processes and activates ATP-producing catabolic processes through protein phosphorylation. SIRT1 acts by deacetylating metabolic enzymes and transcription factors. AMPK and SIRT1 give cells the ability to respond to changing conditions of nutrient availability.

FOUNDATIONAL BIOCHEMISTRY   Things You Should Know After Reading Chapter 27.

- Systems analysis can simplify the complexity of metabolism.

- The metabolism of a heterotrophic cell can be portrayed as 3 interconnected functional blocks: catabolism, anabolism, and macromolecular synthesis/growth.

- Only a few intermediates connect the systems of catabolism and anabolism: 4 sugar-Ps, 3 α-keto acids, 2 coenzyme A derivatives, and PEP.

- ATP and NADPH couple catabolism and anabolism.

- Phototrophs have a fourth functional block consisting of photosynthesis and CO_2 fixation.

- In phototrophs, ATP and NAPH are produced by photosynthesis.

- Virtually every metabolic pathway either produces or consumes ATP.

- The overall thermodynamic efficiency of any metabolic sequence is determined by its ATP stoichiometry.

- Catabolic pathways release energy that is captured as ATP.

- The thermodynamically unfavorable reactions of anabolism are driven by ATP hydrolysis.

- Obligate coupling stoichiometry: Chemical considerations fix the stoichiometry of electron-pair carriers such as NADH and [FADH$_2$] in metabolic pathways.

- The stoichiometry of ATP coupling is not fixed by chemistry; instead, it is an evolved characteristic of living systems.

- ATP coupling stoichiometry determines the K_{eq} for a metabolic sequence.

- ATP has two metabolic roles: (1) ATP serves a stoichiometric role that renders metabolic pathways emphatically favorable, and (2) ATP has an allosteric role in the kinetic regulation of metabolism.

- Energy charge is a concept used to quantify cellular energy status.

- Energy charge is based on the ratio of the sum of the phosphoric anhydride bonds in ADP and ATP divided by the total adenine nucleotide pool.

- The adenylate kinase reaction connects the three principal adenine nucleotides: AMP, ADP, and ATP.

- Because of the reciprocal relationship between adenine nucleotide regulation of catabolic and anabolic pathways, the energy charge of cells is tightly regulated so that it remains fairly constant at high energy charge.

- Phosphorylation potential (the ratio of [ATP]/[ADP][P$_i$]) changes as ATP levels change.

- AMP-activated protein kinase (AMPK) is the cellular energy sensor.

- ATP is an allosteric inhibitor of AMPK; AMP is an allosteric activator of AMPK.

- When cellular energy status is high ([ATP] is high), AMPK is inactive.

- When cellular energy status is low ([AMP] is high), AMPK is activated.

- AMPK targets key enzymes in energy-producing and energy-consuming pathways, turning on those that lead to ATP synthesis and turning off those that consume ATP.

- AMPK also is involved in regulating cellular growth through phosphorylation of a protein called TSC2 that inhibits mTORC1, a complex that activates protein synthesis.

- Metabolism is integrated in multicellular organisms.

- The tissues of the body work together to maintain energy homeostasis.

- The organ systems of multicellular organisms have specialized metabolic roles.

- The brain has very high respiratory metabolism, accounting for 20% of the total oxygen consumption by the body.

- The brain has no stored fuel, so it depends on the bloodstream for a constant supply of glucose.

- Under conditions of starvation, the brain can use ketone bodies instead of glucose for fuel.

- Muscle is responsible for about 30% of O_2 consumption at rest, and more than 90% during strenuous exercise.

- Almost as much ATP is used to pump Ca^{2+} ions back into the sarcoplasmic reticulum as was used in muscle contraction.

- Creatine kinase allows muscle to use phosphocreatine to restore ATP levels that have been depleted.

- During fasting or starvation, muscle protein is degraded to form α-keto acids that can be used as energy sources or in gluconeogenesis.

- Heart muscle prefers fatty acids as fuel.

- Adipose tissue is not merely a repository for triglycerides; it is also an important endocrine organ, secreting hormones such as leptin that regulate eating behavior and caloric homeostasis.

- Brown fat is adipose tissue rich in mitochondria.

- Thermogenesis is the metabolic role of brown fat: The mitochondria in brown fat have uncoupling proteins that dissipate the proton gradient as heat.

- The liver is the major metabolic processing center in animals.

- The liver is responsible for glucose homeostasis through maintenance of constant blood [glucose].

- The liver is the principal site of fatty acid turnover.

- The liver can use amino acids as metabolic fuel, or for gluconeogenesis, or for ketone body synthesis.

- The hormones that control eating behavior come from many different organs—fat, stomach, pancreas, intestines, and the central nervous system.

- Two subsets of neurons located in the arcuate nucleus of the hypothalamus regulate eating behavior: the NPY/AgRP-producing neurons that release NPY, a hormone that triggers eating, and the melanocortin-producing neurons whose products inhibit eating behavior.

- Ghrelin and cholecystokinin are short-term regulators of eating behavior.

- Leptin and insulin are long-term regulators of eating behavior.

- The gut hormone PPY_{3-36} is an intermediate regulator of eating behavior.

- AMPK mediates many of the hypothalamic responses to these hormones.

- Superimposed on these biochemical mechanisms are neuronal controls arising from cognitive centers in the brain that can override metabolic regulation.

- Caloric restriction (CR) leads to longevity.

- Mutations in the yeast *SIR2* gene decrease life span.

- The human homologs of *SIR2* are the *SIRT* genes; humans have 7 *SIRT* genes.

- The *SIRT* gene products are sirtuins: NAD^+-dependent protein deacetylases (NAD^+-dependent KDACs).

- Almost all metabolic enzymes are acetylated and thus potentially regulated by the sirtuins.

- SIRT1 deacetylates metabolic enzymes, histones, and certain transcription factors.

- The metabolic role of SIRT1 is to integrate metabolism and gene expression with the availability of calories.

- SIRT1 is a key regulator in caloric restriction.

- Thus, both AMPK and SIRT1 respond to nutrient limitation.

- AMPK accomplishes these ends through phosphorylation of metabolic enzymes and transcription factors.

- SIRT1 activity is regulated by the NAD^+/NADH ratio and, thus, oxidative metabolism.

- The effects of SIRT1 are achieved through its deacetylation of metabolic enzymes and transcription factors.

- AMPK and SIRT1 functions are coordinated so that cellular responses to changing nutrient availability are appropriate.

- Resveratrol, a phytoalexin produced in grapes, is a potent activator of sirtuin activity and, indirectly, AMPK.

PROBLEMS

Answers to all problems are at the end of this book. Detailed solutions are available in the *Student Solutions Manual, Study Guide, and Problems Book.*

1. **The Energetics of Opposing Metabolic Sequences** (Integrates with Chapters 3, 18, and 22.) The conversion of PEP to pyruvate by pyruvate kinase (glycolysis) and the reverse reaction to form PEP from pyruvate by pyruvate carboxylase and PEP carboxykinase (gluconeogenesis) represent a so-called substrate cycle. The direction of net conversion is determined by the relative concentrations of allosteric regulators that exert kinetic control over pyruvate kinase, pyruvate carboxylase, and PEP carboxykinase. Recall that the last step in glycolysis is catalyzed by pyruvate kinase:

$$PEP + ADP \rightleftharpoons pyruvate + ATP$$

The standard free energy change is −31.7 kJ/mol.

a. Calculate the equilibrium constant for this reaction.

b. If [ATP] = [ADP], by what factor must [pyruvate] exceed [PEP] for this reaction to proceed in the reverse direction?

The reversal of this reaction in eukaryotic cells is essential to gluconeogenesis and proceeds in two steps, each requiring an equivalent of nucleoside triphosphate energy:

Pyruvate carboxylase
$$Pyruvate + CO_2 + ATP \longrightarrow oxaloacetate + ADP + P_i$$

PEP carboxykinase
$$Oxaloacetate + GTP \longrightarrow PEP + CO_2 + GDP$$

Net: $Pyruvate + ATP + GTP \longrightarrow PEP + ADP + GDP + P_i$

c. The $\Delta G°'$ for the overall reaction is +0.8 kJ/mol. What is the value of K_{eq}?

d. Assuming [ATP] = [ADP], [GTP] = [GDP], and $[P_i]$ = 1 mM when this reaction reaches equilibrium, what is the ratio of [PEP]/[pyruvate]?

e. Are both directions in the substrate cycle likely to be strongly favored under physiological conditions?

2. **Calculating Energy Charge and Phosphorylation Potential** (Integrates with Chapter 3.) Assume the following intracellular concentrations in muscle tissue: [ATP] = 8 mM, [ADP] = 0.9 mM, [AMP] = 0.04 mM, $[P_i]$ = 8 mM. What is the *energy charge* in muscle? What is the *phosphorylation potential?*

3. **How Long Do ATP and Phosphocreatine Supplies Last During Muscle Contraction?** Strenuous muscle exertion (as in the 100-meter dash) rapidly depletes ATP levels. How long will 8 mM ATP last if 1 gram of muscle consumes 300 μmol of ATP per minute? (Assume muscle is 70% water.) Muscle contains phosphocreatine as a reserve of phosphorylation potential. Assuming [phosphocreatine] = 40 mM, [creatine] = 4 mM, and $\Delta G°'$ (phosphocreatine + $H_2O \rightleftharpoons$ creatine + P_i) = −43.3 kJ/mol, how low must [ATP] become before it can be replenished by the reaction: phosphocreatine + ADP $\rightleftharpoons$ ATP + creatine? [Remember, $\Delta G°'$ *(ATP hydrolysis)* = −30.5 kJ/mol.]

4. **What is the Metabolic Value of NADPH in ATP Equivalents?** (Integrates with Chapter 20.) The standard reduction potentials for the (NAD^+/NADH) and ($NADP^+$/NADPH) couples are identical, namely, −320 mV. Assuming the in vivo concentration ratios

$NAD^+/NADH = 20$ and $NADP^+/NADPH = 0.1$, what is ΔG for the following reaction?

$$NADPH + NAD^+ \rightleftharpoons NADP^+ + NADH$$

Assuming standard state conditions for the reaction, $ADP + P_i \longrightarrow ATP + H_2O$, calculate how many ATP equivalents can be formed from $ADP + P_i$ by the energy released in this reaction.

5. **Adenylate Kinase Is an Amplifier** (Integrates with Chapter 3.) Assume the total intracellular pool of adenylates (ATP + ADP + AMP) = 8 mM, 90% of which is ATP. What are [ADP] and [AMP] if the adenylate kinase reaction is at equilibrium? Suppose [ATP] drops suddenly by 10%. What are the concentrations now for ADP and AMP, assuming that the adenylate kinase reaction is at equilibrium? By what factor has the AMP concentration changed?

6. **Regulating the Flux Through Metabolic Pathways** (Integrates with Chapters 18 and 22.) The reactions catalyzed by PFK and FBPase constitute another substrate cycle. PFK is AMP activated; FBPase is AMP inhibited. In muscle, the maximal activity of PFK (mmol of substrate transformed per minute) is ten times greater than FBPase activity. If the increase in [AMP] described in problem 5 raised PFK activity from 10% to 90% of its maximal value but lowered FBPase activity from 90% to 10% of its maximal value, by what factor is the flux of fructose-6-P through the glycolytic pathway changed? (*Hint:* Let PFK maximal activity = 10, FBPase maximal activity = 1; calculate the relative activities of the two enzymes at low [AMP] and at high [AMP]; let J, the flux of F-6-P through the substrate cycle under any condition, equal the velocity of the PFK reaction *minus* the velocity of the FBPase reaction.)

7. **The Effects of Leptin on Fat Metabolism** (Integrates with Chapters 23 and 24.) Leptin not only induces synthesis of fatty acid oxidation enzymes and uncoupling protein 2 in adipocytes, but it also causes inhibition of acetyl-CoA carboxylase, resulting in a decline in fatty acid biosynthesis. This effect on acetyl-CoA carboxylase, as an additional consequence, enhances fatty acid oxidation. Explain how leptin-induced inhibition of acetyl-CoA carboxylase might promote fatty acid oxidation.

8. **Ethanol as a Source of Metabolic Energy** (Integrates with Chapters 19 and 20.) Acetate produced in ethanol metabolism can be transformed into acetyl-CoA by the acetyl thiokinase reaction:

$$Acetate + ATP + CoASH \longrightarrow acetyl\text{-}CoA + AMP + PP_i$$

Acetyl-CoA then can enter the citric acid cycle and undergo oxidation to 2 CO_2. How many ATP equivalents can be generated in a liver cell from the oxidation of one molecule of ethanol to 2 CO_2 by this route, assuming oxidative phosphorylation is part of the process? (Assume all reactions prior to acetyl-CoA entering the citric acid cycle occur outside the mitochondrion.) Per carbon atom, which is a better metabolic fuel, ethanol or glucose? That is, how many ATP equivalents per carbon atom are generated by combustion of glucose versus ethanol to CO_2?

9. **ATP Coupling Coefficients and the Thermodynamics of Opposing Metabolic Sequences I** (Integrates with Chapter 23.) Assuming each NADH is worth 3 ATP, each $FADH_2$ is worth 2 ATP, and each NADPH is worth 4 ATP: How many ATP equivalents are produced when one molecule of palmitoyl-CoA is oxidized to 8 molecules of acetyl-CoA by the fatty acid β-oxidation pathway? How many ATP equivalents are consumed when 8 molecules of acetyl-CoA are transformed into one molecule of palmitoyl-CoA by the fatty acid biosynthetic pathway? Can both of these metabolic sequences be metabolically favorable at the same time if ΔG for ATP synthesis is +50 kJ/mol?

10. **ATP Coupling Coefficients and the Thermodynamics of Opposing Metabolic Sequences II** (Integrates with Chapters 18–21.) If each NADH is worth 3 ATP, each $FADH_2$ is worth 2 ATP, and each NADPH is worth 4 ATP, calculate the equilibrium constant for cellular respiration, assuming synthesis of each ATP costs 50 kJ/mol

of energy. Calculate the equilibrium constant for CO_2 fixation under the same conditions, except here ATP will be hydrolyzed to $ADP + P_i$ with the release of 50 kJ/mol. Comment on whether these reactions are thermodynamically favorable under such conditions.

11. **Metabolic Inhibitors as a Diabetes Treatment Strategy** (Integrates with Chapter 22.) In type 2 diabetics, glucose production in the liver is not appropriately regulated, so glucose is overproduced. One strategy to treat this disease focuses on the development of drugs targeted against regulated steps in glycogenolysis and gluconeogenesis, the pathways by which liver produces glucose for release into the blood. Which enzymes would you select for as potential targets for such drugs?

12. **Drug Targets to Regulate Eating Behavior** As chief scientist for drug development at PhatFarmaceuticals, Inc., you want to create a series of new diet drugs. You have a grand plan to design drugs that might limit production of some hormones or promote the production of others. Which hormones are on your "limit production" list and which are on your "raise levels" list?

13. **Leptin Injections as an Obesity Therapy** The existence of leptin was revealed when the *ob/ob* genetically obese strain of mice was discovered. These mice have a defective leptin gene. Predict the effects of daily leptin injections into *ob/ob* mice on food intake, fatty acid oxidation, and body weight. Similar clinical trials have been conducted on humans, with limited success. Suggest a reason why this therapy might not be a miracle cure for overweight individuals.

14. **Hormones That Regulate Eating Behavior** Would it be appropriate to call neuropeptide Y (NPY) the obesity-promoting hormone? What would be the phenotype of a mouse whose melanocortin-producing neurons failed to produce melanocortin? What would be the phenotype of a mouse lacking a functional MC3R gene? What would be the phenotype of a mouse lacking a functional leptin receptor gene?

15. **Consequences of Alcohol Consumption on the $NAD^+/NADH$ Ratio** The Human Biochemistry box, The Metabolic Effects of Alcohol Consumption, points out that ethanol is metabolized to acetate in the liver by alcohol dehydrogenase and aldehyde dehydrogenase:

$$CH_3CH_2OH + NAD^+ \rightleftharpoons CH_3CHO + NADH + H^+$$
$$CH_3CHO + NAD^+ + H_2O \rightleftharpoons CH_3COO^- + NADH + 2H^+$$

These reactions alter the $NAD^+/NADH$ ratio in liver cells. From your knowledge of glycolysis, gluconeogenesis, and fatty acid oxidation, what might be the effect of an altered $NAD^+/NADH$ ratio on these pathways? What is the basis of this effect?

16. **The Phenotype of a T172D Mutation in AMPK** A T172D mutant of the AMPK is locked in a permanently active state. Explain.

17. **Malonyl-CoA as a Key Indicator of Nutrient Availability**
   a. Some scientists support the "malonyl-CoA hypothesis," which suggests that malonyl-CoA is a key indicator of nutrient availability and the brain uses its abundance to assess whole-body energy homeostasis. Others have pointed out that malonyl-CoA is a significant inhibitor of carnitine acyltransferase-1 (see Figure 24.16). Thus, malonyl-CoA may be influencing the levels of another metabolite whose concentration is more important as a signal of energy status. What metabolite might that be?
   b. Another test of the malonyl-CoA hypothesis was conducted through the creation of a transgenic strain of mice that lacked functional hypothalamic fatty acid synthase (see Chapter 24). Predict the effect of this genetic modification on cellular malonyl-CoA levels in the hypothalamus, the eating behavior of these transgenic mice, their body fat content, and their physical activity levels. Defend your predictions.

18. **Exploring Leptin's Mechanism of Action**
   a. Leptin was discovered when a congenitally obese strain of mice (*ob/ob* mice) was found to lack both copies of a gene encoding a peptide hormone produced mainly by adipose tissue. The peptide

hormone was named leptin. Leptin is an anorexic (appetite-suppressing) agent; its absence leads to obesity. Propose an experiment to test these ideas.

b. A second strain of obese mice (*db/db* mice) produces leptin in abundance but fails to respond to it. Assuming the *db* mutation leads to loss of function in a protein, what protein is likely to be nonfunctional or absent? How might you test your idea?

Preparing for the MCAT® Exam

19. Consult Figure 27.7, and answer the following questions: Which organs use both fatty acids and glucose as a fuel in the well-fed state, which rely mostly on glucose, which rely mostly on fatty acids, which one never uses fatty acids, and which one produces lactate.

20. Figure 27.3 illustrates the response of R (ATP-regenerating) and U (ATP-utilizing) enzymes to energy charge.

a. Would hexokinase be an R enzyme or a U enzyme? Would glutamine:PRPP amidotransferase, the second enzyme in purine biosynthesis, be an R enzyme or a U enzyme?

b. If energy charge = 0.5: Is the activity of hexokinase high or low? Is ribose-5-P pyrophosphokinase activity high or low?

c. If energy charge = 0.95: Is the activity of hexokinase high or low? Is ribose-5-P pyrophosphokinase activity high or low?

ActiveModels Problem

21. Examine the ActiveModel for resistin, and describe the formation of the resistin hexamer.

FURTHER READING

Systems Analysis of Metabolism

Brand, M. D., and Curtis, R. K., 2002. Simplifying metabolic complexity. *Biochemical Society Transactions* **30**:25–30.

ATP Coupling and the Thermodynamics of Metabolism

Atkinson, D. F., 1977. *Cellular Energy Metabolism and Its Regulation.* New York: Academic Press.

Newsholme, E. A., Challiss, R. A. J., and Crabtree, B., 1984. Substrate cycles: Their role in improving sensitivity in metabolic control. *Trends in Biochemical Sciences* **9**:277–280.

Newsholme, E. A., and Leech, A. R., 1983. *Biochemistry for the Medical Sciences.* New York: John Wiley & Sons.

AMP-Activated Protein Kinase

Dunlop, E. A., and Tee, A. R., 2009. Mammalian target of rapamycin complex 1: Signalling inputs, substrates and feedback mechanisms. *Cellular Signalling* **21**:827–835.

Hardie, D. G., 2007. AMP-activated/SNF 1 protein kinases: Conserved guardians of cellular energy. *Nature Reviews Cell Molecular Biology* **8**:774–785.

Hardie, D. G., Hawley, S. A., and Scott, J. W., 2007. AMP-activated protein kinase: Development of the energy sensor concept. *Journal of Physiology* **574**:7–15.

McGee, S. L., and Hargreaves, M., 2008. AMPK and transcriptional regulation. *Frontiers in Bioscience* **13**:3022–3033.

Oakhill, J. S., Steel, R., Chen, Z. P., Scott, J. W., et al., 2011. AMPK is a direct adenylate charge-regulated protein kinase. *Science* **332**:1433–1435.

Steinberg, G. R., and Kemp, B. E., 2009. AMPK in health and disease. *Physiological Reviews* **89**:8105–1078.

Xiao, B., Heath, R., Salu, P., Leiper, F. C., et al., 2007. Structural basis for AMP binding to mammalian AMP-activated protein kinase. *Nature* **449**:496–500.

Xiao, B., Sanders, M. J., Underwood, E., Heath, R., et al., 2011. Structure of mammalian AMPK and its regulation by ADP. *Nature* **472**:230–233.

Metabolic Relationships Between Organ Systems

Harris, R., and Crabb, D. W., 1997. Metabolic interrelationships. In *Textbook of Biochemistry with Clinical Correlations,* 4th ed., Devlin, T. M., ed. New York: Wiley-Liss.

Sugden, M. C., Holness, M. J., and Palmer, T. N., 1989. Fuel selection and carbon flux during the starved-to-fed transition. *Biochemical Journal* **263**:313–323.

Creatine as a Nutritional Supplement

Ekblom, B., 1999. Effects of creatine supplementation on performance. *American Journal of Sports Medicine* **24**:S-38.

Kreider, R., 1998. Creatine supplementation: Analysis of ergogenic value, medical safety, and concerns. *Journal of Exercise Physiology* **1**, an international onnline journal available at http://www.css.edu/users/tboone2/asep/jan3.htm

Fat-Free Mice

Gavrilova, O., et al., 2000. Surgical implantation of adipose tissue reverses diabetes in lipoatrophic mice. *Journal of Clinical Investigation* **105**:271–278.

Moitra, J., et al., 1998. Life without white fat: A transgenic mouse. *Genes and Development* **12**:3168–3181.

Leptin and Hormonal Regulation of Eating Behavior

Barinaga, M., 1995. "Obese" protein slims mice. *Science* **269**:475–476, and references therein.

Buettner, C., 2007. Does FASing out new fat in the hypothalamus make you slim? *Cell Metabolism* **6**:249–251.

Clement, K., et al., 1998. A mutation in the human leptin receptor gene causes obesity and pituitary dysfunction. *Nature* **392**:398–401.

Coll, A. P., Farooqi, S., and O'Rahilly, S., 2007. The hormonal control of food intake. *Cell* **129**:251–262.

Saper, C. B., Chou, T. C., and Elmquist, J. K., 2002. The need to feed: Homeostatic and hedonic control of eating. *Neuron* **36**:199–211.

Schwartz, M. W., and Morton, G. J., 2002. Obesity: Keeping hunger at bay. *Nature* **418**:595–597.

Vaisse, C., et al., 2000. Melanocortin-4 receptor mutations are a frequent and heterogeneous cause of morbid obesity. *Journal of Clinical Investigation* **106**:253–262.

Zhou, Y-T., et al., 1997. Induction by leptin of uncoupling protein-2 and enzymes of fatty acid oxidation. *Proceedings of the National Academy of Sciences U.S.A.* **94**:6386–6390.

Caloric Restriction, Metabolic Syndrome, and Longevity

Baur, J. A., 2010. Biochemical effects of SIRT1 activators. *Biochimica et Biophysica Acta* **1804**:1626–1634.

Dasgupta, B., and Milbrandt, J., 2007. Resveratrol stimulates AMP kinase activity in neurons. *Proceedings of the National Academy of Science U.S.A.* **104**:7217–7222.

Denu, J. M., 2005. The Sir2 family of protein deacetylases. *Current Opinion in Chemical Biology* **9**:431–440.

Fulco, M., and Sartorelli, V., 2008. Comparing and contrasting the role of AMPK and SIRT1 in metabolic tissues. *Cell Cycle* **7**:3669–3679.

Guan, K.-L., and Xiong, Y., 2011. Regulation of intermediary metabolism by protein acetylation. *Trends in Biochemical Sciences* **36:**108–116.

Guarente, L., and Picard, F., 2005. Caloric restriction—The SIR2 connection. *Cell* **120:**473–482.

Hirschey, M. D., Shimazu, T., Jing, E., Greuter, C. A., et al., 2011. SIRT3 deficiency and mitochondrial hyperacetylation accelerate development of the metabolic syndrome. *Molecular Cell* **44:**1–14.

Hou, X., Xu, S., Maitland-Toolan, K. A., Sato, K., et al., 2008. SIRT1 regulates hepatocyte lipid metabolism through activating AMP-activated protein kinase. *Journal of Biological Chemistry* **283:**20015–20026.

Michan, S., and Sinclair, D., 2007. Sirtuins in mammals: Insights into their biological function. *Biochemical Journal* **404:**1–13.

Milne, J. C., Lambert, P. D., Schenk, S., Carney, D. P., et al., 2007. Small molecule activators of SIRT1 as therapeutics for the treatment of type 2 diabetes. *Nature* **450:**712–716.

Moynihan, K. A., and Imai, S-I., 2006. SIRT1 as a key regulator orchestrating the response to caloric restriction. *Drug Discovery Today: Disease Mechanisms* **3:**11–17.

Qui, X., Brown, K. V., Moran, Y., and Chen, D., 2010. Sirtuin regulation in caloric restriction. *Biochimica et Biophysica Acta* **1804:**1576–1583.

Ruderman, N. B., Xu, J., Nelson, L., Cacicedo, J. M., et al., 2010. AMPK and SIRT1: A long-standing partnership? *American Journal of Physiology. Endocrinology and Metabolism* **298:**E751–E760.

DNA Metabolism: Replication, Recombination, and Repair

28

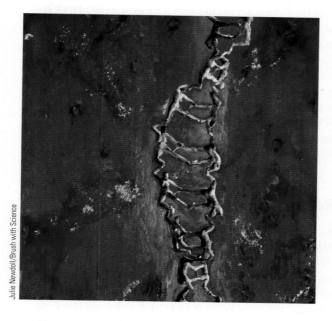

Julie Newdoll/Brush with Science

◀ Julie Newdoll's painting "Dawn of the Double Helix" composes the DNA duplex as human figures. Her theme in this painting is "Life Forms: The basic structures that make our existence possible."

Heredity

I am the family face;
Flesh perishes, I live on,
Projecting trait and trace
Through time to times anon,
And leaping from place to place
Over oblivion.

The years-heired feature that can
In curve and voice and eye
Despise the human span
Of durance—that is I;
The eternal thing in man,
That heeds no call to die.

Thomas Hardy
(in *Moments of Vision and*
Miscellaneous Verses, 1917)

ESSENTIAL QUESTIONS

DNA is the physical repository of genetic information in the cell and the material of heredity that is passed on to progeny.

How is this genetic information in the form of DNA replicated, how is the information rearranged, and how is its integrity maintained in the face of damage?

Heredity, which we can define generally as the tendency of an organism to possess the characteristics of its parent(s), is clearly evident throughout nature and since the dawn of history has served to justify the classification of organisms according to shared similarities. The basis of heredity, however, was a mystery. Early in the 20th century, geneticists demonstrated that **genes,** the elements or units carrying and transferring inherited characteristics from parent to offspring, are contained within the nuclei of cells in association with the chromosomes. Yet the chemical identity of genes remained unknown, and genetics was an abstract science. Even the realization that chromosomes are composed of proteins and nucleic acids did little to define the molecular nature of the gene because, at the time, no one understood either of these substances.

The material of heredity must have certain properties. It must be very stable so that genetic information can be stored in it and transmitted countless times to subsequent generations. It must be capable of precise copying or replication so that its information is not lost or altered. And, although stable, it must also be subject to change in order to account, in the short term, for the appearance of mutant forms and, in the long term, for evolution. DNA is the material of heredity.

KEY QUESTIONS

28.1 How Is DNA Replicated?

28.2 What Are the Functions of DNA Polymerases?

28.3 Why Are There So Many DNA Polymerases?

28.4 How Is DNA Replicated in Eukaryotic Cells?

28.5 How Are the Ends of Chromosomes Replicated?

28.6 How Are RNA Genomes Replicated?

28.7 How Is the Genetic Information Rearranged by Genetic Recombination?

28.8 Can DNA Be Repaired?

28.9 What Is the Molecular Basis of Mutation?

Special Focus: Gene Rearrangements and Immunology—Is It Possible to Generate Protein Diversity Using Genetic Recombination?

ƆWL Online homework and a Student Self Assessment for this chapter may be assigned in OWL.

DNA Metabolism

The chapters that precede this one have reviewed the major pathways of intermediary metabolism, pointing out the synthesis and breakdown of the major classes of cellular molecules. When we consider DNA from the perspective of metabolism, its role as the hereditary substance focuses our inquiry. Aside from its modest ribose content, DNA is not an important source of nutritional energy, and DNA catabolism contributes little to metabolism in most cells. In contrast, DNA synthesis is a central aspect of cellular existence. DNA is synthesized by copying existing DNA in a manner that reproduces its information content in the products of the reaction through a process called **DNA replication.** As the genetic material, DNA persists over the lifetime of the cell and appears essentially unaltered in subsequent generations. Occasionally, its information is shuffled, as regions of DNA are exchanged between chromosomes, a process known as **recombination.** Along with other kinds of mutation, recombination creates genetic variation upon which evolution can act. As a chemical, DNA is susceptible to reactions that alter its bases or disrupt its sugar-phosphate backbone. Because its effective lifetime exceeds the lifetime of cells, **DNA repair** is necessary to preserve the fidelity of its genetic information. This chapter addresses these dominant features of DNA metabolism: replication, recombination, and repair.

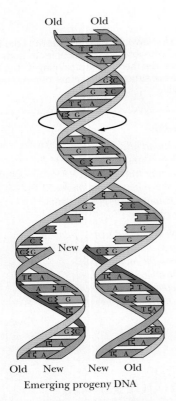

FIGURE 28.1 DNA replication: Strand separation followed by the copying of each strand.

A **template** is something whose edge is shaped in a particular way so that it can serve as a guide in making a similar object with a corresponding contour. In DNA replication, each of the original strands acts as a template by specifying the order of nucleotides in the newly synthesized strand according to the base-pairing rules of Watson and Crick.

28.1 How Is DNA Replicated?

Transfer of genetic information from generation to generation requires the faithful reproduction of the parental DNA. DNA reproduction produces two identical copies of the original DNA through DNA replication. The mechanism for DNA replication is *strand separation followed by the copying of each strand.* In the process, each separated strand acts as a **template** for the synthesis of a new complementary strand whose nucleotide sequence is fixed by the base-pairing rules Watson and Crick proposed (see Chapter 10). Strand separation is achieved by untwisting the double helix (Figure 28.1). Base pairing then dictates the proper sequence of nucleotide addition to achieve an accurate replication of each original strand. Thus, each original strand ends up paired with a new complementary partner, and two identical double-stranded DNA molecules are formed from one. This mode of DNA replication is referred to as **semiconservative** because one of the two original strands is conserved in each of the two progeny molecules.

DNA Replication Is Bidirectional

Replication of DNA molecules begins at one or more specific regions called the **origin(s) of replication** and, excepting certain bacteriophage chromosomes and plasmids, proceeds in both directions from this origin (Figure 28.2). For example, replication of *E. coli* DNA begins at *oriC,* a unique 245-bp chromosomal site that contains 11 GATC tetranucleotide sequences along its length. From *oriC,* replication advances in both directions around the circular chromosome. That is, bidirectional replication involves two **replication forks** that move in opposite directions. Bidirectional replication predicts that, if radioactively labeled nucleotides are provided as substrates for new DNA synthesis, both replication forks will become radioactively labeled. The experiment illustrated in Figure 28.2(b) confirms this prediction.

Replication Requires Unwinding of the DNA Helix

Semiconservative replication depends on unwinding the DNA double helix to expose single-stranded templates to polymerase action. For a double helix to unwind, it must either rotate about its axis (while the ends of its strands are held fixed), or positive supercoils must be introduced, one for each turn of the helix unwound (see Chapter 11). If the chromosome is circular, as in *E. coli,* only the latter alternative is possible. Because DNA replication in *E. coli* proceeds at a rate approaching 1000 nucleotides per second

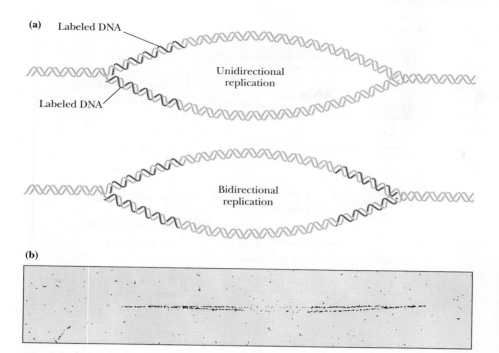

(a)

Labeled DNA

Unidirectional replication

Labeled DNA

Bidirectional replication

(b)

FIGURE 28.2 Bidirectional replication. **(a)** Comparison of labeling during unidirectional versus bidirectional replication. **(b)** An autoradiogram of *E. coli* chromosome replication in the presence of radioactive thymidine confirms bidirectional replication. (Photo courtesy of David M. Prescott, University of Colorado.)

and there are about 10 bp per helical turn, the chromosome would accumulate 100 positive supercoils per second! In effect, the DNA would become too tightly supercoiled to allow unwinding of the strands.

DNA gyrase, a Type II topoisomerase, acts to overcome the torsional stress imposed upon unwinding; DNA gyrase introduces negative supercoils at the expense of ATP hydrolysis. The unwinding reaction is driven by **helicases** (see also Chapter 16), a class of proteins that catalyze the ATP-dependent unwinding of DNA double helices. Unlike topoisomerases that alter the linking number of dsDNA through phosphodiester bond breakage and reunion (see Chapter 11), helicases simply disrupt the hydrogen bonds that hold the two strands of duplex DNA together. A helicase molecule requires a single-stranded region for binding. It then moves along the single strand, unwinding the double-stranded DNA in an ATP-dependent process. **SSB (single-stranded DNA-binding protein)** binds to the unwound strands, preventing their re-annealing. At least ten distinct DNA helicases involved in different aspects of DNA and RNA metabolism have been found in *E. coli* alone. **DnaB** is the DNA helicase acting in *E. coli* DNA replication. DnaB helicase assembles as a hexameric (α_6) "doughnut"-shaped protein ring, with DNA passing through its hole.

DNA Replication Is Semidiscontinuous

As shown in Figure 28.2, both parental DNA strands are replicated at each advancing replication fork. The enzyme that carries out DNA replication is **DNA polymerase.** This enzyme uses single-stranded DNA (ssDNA) as a template and makes a complementary strand by polymerizing deoxynucleotides in the order specified by their base pairing with bases in the template. DNA polymerases synthesize DNA only in a 5′→3′ direction, reading the antiparallel template strand in a 3′→5′ sense. A dilemma arises: How does DNA polymerase copy the parent strand that runs in the 5′→3′ direction at the replication fork? It turns out that *replication is semidiscontinuous* (Figure 28.3): As the DNA helix is unwound during its replication, the 3′→5′ strand (as defined by the direction that the replication fork is moving) can be copied continuously by DNA polymerase synthesizing in the 5′→3′ direction behind the replication fork. The other parental strand is copied only when a sufficient stretch of its sequence has been exposed for DNA polymerase to read it in the 3′→5′ sense. Thus, one parental strand is copied continuously to give a newly synthesized copy, called the **leading strand,** at each replication

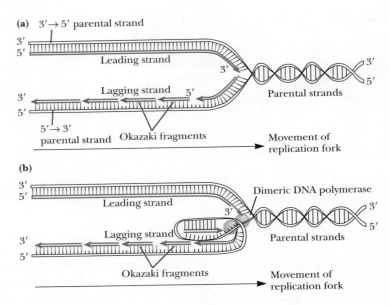

FIGURE 28.3 The semidiscontinuous model for DNA replication. Newly synthesized DNA is shown as red. **(a)** Leading- and lagging-strand synthesis. Note that the leading strand is synthesized continuously in the 5'→3' direction as the DNA polymerase reads along the 3'→5' parental template for the leading strand in the 3'→5' direction. Since DNA polymerases can only move along any template in a 3'→5' direction, the other parental strand, the one labeled "5'→3'" is copied, by necessity, in a discontinuous fashion to form short, 5'→3' fragments (so-called "lagging strand" synthesis). **(b)** Synthesis of both strands carried out by a dimeric DNA polymerase situated at the replication fork. Because DNA polymerase must read the template strand in the 3'→5' direction, the 5'→3' parental strand must wrap around in trombone fashion.

fork. The other parental strand is copied in an intermittent, or discontinuous, mode to yield a set of fragments 1000 to 2000 nucleotides in length, called the **Okazaki fragments** (Figure 28.3a). These fragments are then joined to form an intact **lagging strand.** Because both strands are synthesized in concert by a dimeric DNA polymerase situated at the replication fork, the 5'→3' parental strand must wrap around in *trombone fashion* so that the unit of dimeric DNA polymerase replicating it can move along it in the 3'→5' direction (Figure 28.3b). Overall, each of the two DNA duplexes produced in DNA replication contains one "old" and one "new" DNA strand, and half of the new strand was formed by leading strand synthesis and the other half by lagging strand synthesis.

The Biochemical Evidence for Semidiscontinuous DNA Replication

In 1968, Tsuneko and Reiji Okazaki provided biochemical verification of the semidiscontinuous pattern of DNA replication just described. The Okazakis exposed a rapidly dividing *E. coli* culture to ³H-labeled thymidine for 30 seconds, quickly collected the cells, and found that half of the label incorporated into nucleic acid appeared in short ssDNA chains just 1000 to 2000 nucleotides in length. (The other half of the radioactivity was recovered in very large DNA molecules.) Subsequent experiments demonstrated that with time, the newly synthesized short ssDNA **Okazaki fragments** became covalently joined to form longer polynucleotide chains, in accord with a semidiscontinuous mode of replication. The generality of this mode of replication has been corroborated with electron micrographs of DNA undergoing replication in eukaryotic cells.

28.2 What Are the Functions of DNA Polymerases?

The enzymes that replicate DNA are called DNA polymerases. All DNA polymerases, whether from prokaryotic or eukaryotic sources, share the same fundamental catalytic mechanism:

1. The incoming base is selected within the DNA polymerase active site, as determined by Watson–Crick geometric interactions with the corresponding base in the template strand.
2. Chain growth is in the 5'→3' direction and is antiparallel to the template strand.
3. DNA polymerases cannot initiate DNA synthesis de novo—all require a primer oligonucleotide with a free 3'-OH to build upon.

Biochemical Characterization of DNA Polymerases

DNA polymerases can catalyze the synthesis of DNA in vitro if provided with all four deoxynucleoside-5′-triphosphates (dATP, dTTP, dCTP, dGTP), a template DNA strand to copy, *and* a **primer**. A primer is essential because DNA polymerases can elongate only preexisting chains; they cannot join two deoxyribonucleoside-5′-phosphates together to make the initial phosphodiester bond. The primer base pairs with the template DNA, forming a short, double-stranded region. This primer must possess a free 3′-OH end to which an incoming deoxynucleoside monophosphate is added. In vivo, a short RNA strand is invariably the primer; it is synthesized by a DNA-dependent RNA polymerase activity known as **primase**. DNA replication begins when DNA polymerase binds at the 3′-end of the primer. The appropriate one of the four dNTPs is selected as substrate, pyrophosphate (PP_i) is released, and the dNMP is linked to the 3′-OH of the primer chain through formation of a phosphoester bond (Figure 28.4). The deoxynucleotide selected as substrate is chosen through its geometric fit with the template base to form a Watson–Crick base pair. As DNA polymerase catalyzes the successive addition of deoxynucleotide units to the 3′-end of the primer, the chain is elongated in the 5′→3′ direction, forming a polynucleotide sequence that is antiparallel and complementary to the template. Different DNA polymerases proceed along the template strand, synthesizing a complementary strand of varying length before they "fall off" (dissociate from) the template. The degree to which a particular DNA polymerase remains associated with the template through successive cycles of nucleotide addition is referred to as its **processivity**. Some DNA polymerases, such as those whose primary function is DNA repair rather than DNA replication, are only modestly processive, synthesizing a DNA strand only 3 to 200 bases long before falling off. On the other hand, replicative DNA polymerases are extraordinarily processive. The *E. coli* replicative DNA polymerase, known as DNA polymerase III holoenzyme, can replicate an entire strand of the *E. coli* chromosome (4.6 megabases) without falling off!

E. coli Cells Have Several Different DNA Polymerases

Table 28.1 compares the properties of the principal DNA polymerases in *E. coli*. These enzymes are nicknamed **pol** and numbered **I** through **V** in order of their discovery. DNA polymerases I, II, and V function principally in DNA repair; **DNA polymerase III** is the chief DNA-replicating enzyme of *E. coli*. Only 40 or so copies of this enzyme are present per cell.

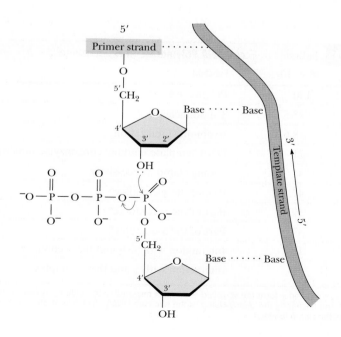

FIGURE 28.4 The chain elongation reaction catalyzed by DNA polymerase. The 3′-OH carries out a nucleophilic attack on the α-phosphoryl group of the incoming dNTP to form a phosphoester bond, and PP_i is released. The subsequent hydrolysis of PP_i by inorganic pyrophosphatase renders the reaction effectively irreversible.

TABLE 28.1	Properties of the DNA Polymerases of *E. coli*		
Property	Pol I	Pol II	Pol III (core)*
Mass (kD)	103	88	130(α), 27.5(ϵ), 8.6(θ)
Molecules/cell	400		40
Turnover number[†]	20	40	1000
Polymerization 5'$\longrightarrow$3'	Yes	Yes	Yes
Exonuclease 3'$\longrightarrow$5'	Yes	Yes	Yes
Exonuclease 5'$\longrightarrow$3'	Yes	No	No

*α-, ϵ-, and θ-subunits.
[†]Nucleotides polymerized at 37°C/second/molecule of enzyme.

E. coli DNA Polymerase III Holoenzyme Replicates the *E. coli* Chromosome

In its holoenzyme form, DNA polymerase III is the enzyme responsible for replication of the *E. coli* chromosome. The simplest form of DNA polymerase III showing any DNA-synthesizing activity in vitro, "core" DNA polymerase III, is 165 kD in size and consists of three polypeptides: α (130 kD), ϵ (27.5 kD), and θ (8.6 kD). In vivo, core DNA polymerase III functions as part of a multisubunit complex, the **DNA polymerase III holoenzyme,** which is composed of ten different kinds of subunits (Table 28.2). The various auxiliary subunits increase both the polymerase activity of the core enzyme and its processivity. DNA polymerase III holoenzyme synthesizes DNA strands at a speed of nearly 1 kb/sec. DNA polymerase III holoenzyme is organized in the following way: Two core ($\alpha\epsilon\theta$) DNA polymerase III units and one γ-**complex** are attached to DnaB helicase via two τ-subunits to form a structure known as **DNA polymerase III*.** (A γ-complex consists of one copy each of the γ-subunit, the δ-subunit, and the δ'-subunit.) In turn, each core polymerase within DNA polymerase III* binds to a β-subunit dimer to create **DNA polymerase III holoenzyme,** a 17-subunit (($\alpha\epsilon\theta$)$_2\beta_2\tau_2\gamma\delta\delta'\chi\psi$) complex (Figure 28.5). The γ-complex is responsible for assembly of the DNA polymerase III holoenzyme complex onto DNA. The γ-complex of the holoenzyme acts as a **clamp loader** by catalyzing the ATP-dependent transfer of a pair of β-subunits to each strand of the DNA template. Each β-subunit dimer forms a closed ring around a DNA strand and acts as a tight clamp that can slide along the DNA (Figure 28.6). Each β_2-**sliding clamp** tethers a core

TABLE 28.2	Subunits of *E. coli* DNA Polymerase III Holoenzyme	
Subunit	Mass (kD)	Function
α	130	Polymerase
ϵ	27.5	3'-Exonuclease
θ	8.6	ϵ-subunit stabilization
τ	71	DNA template binding; core enzyme dimerization
β	41	Sliding clamp, processivity
γ	47.5	Part of the γ-complex*
δ	39	Part of the γ-complex*
δ'	37	Part of the γ-complex*
χ	17	Interaction with SSB and the γ-complex
ψ	15	Interaction with χ and the γ-complex

*Subunits τ, γ, δ, δ', χ, and ψ form the so-called γ-complex responsible for adding β-subunits (the sliding clamp) to DNA and anchoring the sliding clamp to the two core DNA polymerase III structures. The γ-complex is referred to as the clamp loader.

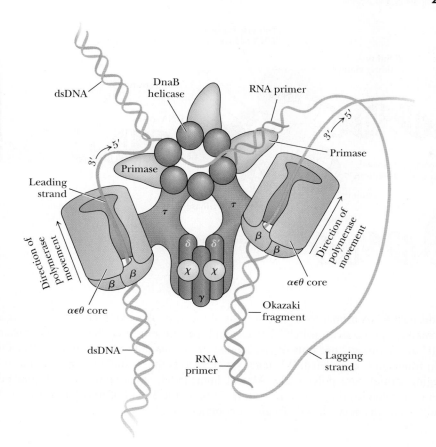

FIGURE 28.5 DNA polymerase III holoenzyme is a dimeric polymerase. One unit of polymerase synthesizes the leading strand, and the other synthesizes the lagging strand. Because DNA synthesis always proceeds in the 5′→3′ direction as the template strand is read in the 3′→5′ direction, lagging-strand synthesis must take place on a looped-out template. Lagging-strand synthesis requires repeated priming. Primase bound to the DnaB helicase carries out this function, periodically forming new RNA primers on the lagging strand. All single-stranded regions of DNA are coated with SSB (not shown).

polymerase to the template, accounting for the great processivity of the DNA polymerase holoenzyme. This complex can replicate an entire strand of the *E. coli* genome (more than 4.6 megabases) without dissociating. Compare this to the processivity of DNA polymerase I, which is only 20 bases!

The core polymerase synthesizing the lagging strand must release from the DNA template when synthesis of an Okazaki fragment is completed and rejoin the template

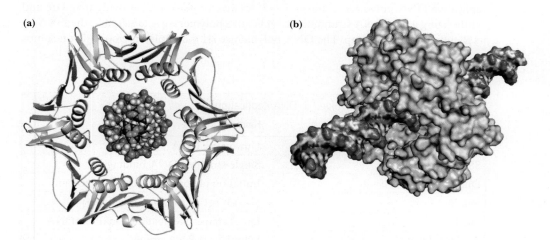

FIGURE 28.6 **(a)** Ribbon diagram of the β-subunit dimer of the DNA polymerase III holoenzyme on B-DNA, viewed down the axis of the DNA. One monomer of the β-subunit dimer is colored blue and the other yellow. The centrally located DNA is multicolored. **(b)** Space-filling model of the β-subunit dimer of the DNA polymerase III holoenzyme on B-DNA. The hole formed by the β-subunits (diameter ≈ 3.5 nm) is large enough to easily accommodate DNA (diameter ≈ 2.5 nm) with no steric repulsion (pdb id = 2POL). The rest of polymerase III holoenzyme ("core" polymerase + γ-complex) associates with this sliding clamp to form the replicative polymerase (not shown).

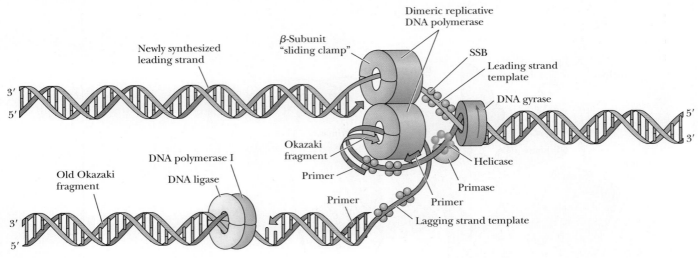

FIGURE 28.7 General features of a replication fork. The DNA duplex is unwound by the action of DNA gyrase and helicase, and the single strands are coated with SSB (ssDNA-binding protein). Primase periodically primes synthesis on the lagging strand. Each half of the dimeric replicative polymerase is a "core" polymerase bound to its template strand by a β-subunit sliding clamp. DNA polymerase I and DNA ligase act downstream on the lagging strand to remove RNA primers, replace them with DNA, and ligate the Okazaki fragments.

at the next RNA primer to begin synthesis of the next Okazaki fragment. The τ-subunit serves as a "processivity switch" that accomplishes this purpose. The τ-subunit is usually "off" and is turned "on" only on the lagging strand *and* only when synthesis of an Okazaki fragment is completed. When activated, τ ejects the β_2-sliding clamp bound to the lagging strand core polymerase. Almost immediately, the lagging strand core polymerase is reloaded onto a new β_2-sliding clamp at the 3'-end of next RNA primer, and synthesis of the next Okazaki fragment commences.

A DNA Polymerase III Holoenzyme Sits at Each Replication Fork

We now can present a snapshot of the enzymatic apparatus assembled at a replication fork (Figure 28.7 and Table 28.3). DNA gyrase (topoisomerase) and DnaB helicase unwind the DNA double helix, and the unwound, single-stranded regions of DNA are maintained through interaction with SSB. **Primase (DnaG)** synthesizes an RNA primer on the lagging strand; the leading strand, which needs priming only once, was primed when replication was initiated. The lagging strand template is looped around, and each replicative DNA polymerase moves 5'→3' relative to its strand, copying template and synthesizing a new DNA strand. Each replicative polymerase is tethered to the DNA by its β-subunit sliding clamp. The DNA polymerase III γ-complex periodically unclamps

TABLE 28.3	Proteins Involved in DNA Replication in *E. coli*
Protein	**Function**
DNA gyrase	Unwinding DNA
SSB	Single-stranded DNA binding
DnaA	Initiation factor; origin-binding protein
DnaB	5'→3' helicase (DNA unwinding)
DnaC	DnaB chaperone; loading DnaB on DNA
Primase (DnaG)	Synthesis of RNA primer
DNA polymerase III holoenzyme	Elongation (DNA synthesis)
DNA polymerase I	Excises RNA primer, fills in with DNA
DNA ligase	Covalently links Okazaki fragments
Tus	Termination

and then reclamps β-subunits on the lagging strand as the primer for each new Okazaki fragment is encountered. Downstream on the lagging strand, DNA polymerase I excises the RNA primer and replaces it with DNA, and DNA ligase seals the remaining nick.

DNA Polymerase I Removes the RNA Primers and Fills in the Gaps

E. coli DNA polymerase I, the first DNA polymerase discovered (in 1957 by Arthur Kornberg), acts to remove the RNA primers and replace them with DNA. DNA polymerase I also acts in DNA repair. It fulfills these roles because, in addition to its 5′→3′ polymerase activity, *E. coli* DNA polymerase I has two other catalytic functions: a *3′→5′ exonuclease (3′-exonuclease)* activity and a *5′→3′ exonuclease (5′-exonuclease)* activity. The three distinct catalytic activities of DNA polymerase I reside in separate active sites in the enzyme. The 5′-exonuclease activity is not a commonly found property of DNA polymerases. This 5′-exonuclease activity is the mechanism for removal of the RNA primers.

E. coli DNA Polymerase I Is Its Own Proofreader

The exonuclease activities of *E. coli* DNA polymerase I are functions that enhance the accuracy of DNA replication. The 3′-exonuclease activity removes nucleotides from the 3′-end of the growing chain (Figure 28.8), an action that negates the action of the polymerase activity. Its purpose, however, is to remove incorrect (mismatched) bases. Although the 3′-exonuclease works slowly compared to the polymerase, the polymerase cannot elongate an improperly base-paired primer terminus. Thus, the relatively slow 3′-exonuclease has time to act and remove the mispaired nucleotide. Therefore, the polymerase active site is a proofreader, and the 3′-exonuclease activity is an editor. This check on the accuracy of base pairing enhances the overall precision of the process.

DNA Ligase Seals the Nicks Between Okazaki Fragments

DNA ligase (see Chapter 12) seals nicks in double-stranded DNA where a 3′-OH and a 5′-phosphate are juxtaposed. This enzyme is responsible for joining Okazaki fragments together to make the lagging strand a covalently contiguous polynucleotide chain.

DNA Replication Terminates at the *Ter* Region

Located diametrically opposite from *oriC* on the *E. coli* circular map is a terminus region, the **Ter,** or *t,* locus. The oppositely moving replication forks meet here, and replication is terminated. The *Ter* region contains a number of short DNA sequences, with a consensus core element 5′-GTGTGTTGT. These *Ter* sequences act as terminators; clusters of three or four *Ter* sequences are organized into two sets inversely oriented with

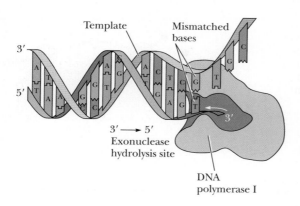

FIGURE 28.8 The 3′→5′ exonuclease activity of DNA polymerase I removes nucleotides from the 3′-end of the growing DNA chain. The newly synthesized strand is shown in purple.

respect to one another. One set blocks the clockwise-moving replication fork, and its inverted counterpart blocks the counterclockwise-moving replication fork. Termination requires binding of a specific replication termination protein, **Tus protein,** to *Ter.* Tus protein is a **contrahelicase.** That is, Tus protein prevents the DNA duplex from unwinding by blocking progression of the replication fork and inhibiting the ATP-dependent DnaB helicase activity. Final synthesis of both duplexes is completed.

DNA Polymerases Are Immobilized in Replication Factories

Most drawings of DNA replication (such as Figure 28.8) suggest that the DNA polymerases are tracking along the DNA, like locomotives along train tracks, synthesizing DNA as they go. Experimental evidence, however, favors the view that the DNA polymerases are immobilized, either via attachment to the cell membrane in prokaryotic cells or to the nuclear matrix in eukaryotic cells. All the associated proteins of DNA replication, as well as proteins necessary to hold DNA polymerase at its fixed location, constitute **replication factories.** The DNA is then fed through the DNA polymerases within the replication factory, much like tape is fed past the heads of a tape player, with all four strands of newly replicated DNA looping out from this fixed structure (Figure 28.9).

DNA Polymerase I Also Plays a Role in Double-Strand Break Repair DNA polymerase I can bind to double-stranded DNA where there is a break or gap in one of the strands. It then uses its 5′-exonuclease activity to remove mononucleotides and oligonucleotides, degrading the broken strand from its exposed 5′-end as it moves in a 5′→3 direction. One function of the 5′-exonuclease is to remove distorted (mispaired) segments lying in the path of an advancing polymerase. Indeed, the biological roles of DNA polymerase I depend on its ability to bind at nicks (single-stranded breaks) in dsDNA and move in the 5′→3′ direction, removing successive nucleotides with its 5-exonucleolytic activity. (This overall process is known as **nick translation** because the nick is translated [that is, moved] down the DNA.) This 5′-exonuclease activity plays an important role in primer removal during DNA replication, as we saw on page 955. The

---

> **A DEEPER LOOK**

A Mechanism for All Polymerases

Thomas A. Steitz of Yale University has suggested that biosynthesis of nucleic acids proceeds by an enzymatic mechanism that is universal among all DNA *and* RNA polymerases. His suggestion is based on structural studies indicating that DNA polymerases use a "two-metal-ion" mechanism to catalyze nucleotide addition during elongation of a growing polynucleotide chain (see accompanying figure). The incoming nucleotide has two Mg^{2+} ions (shown here as Me^{2+}, a generic representation for divalent cations) coordinated to its phosphate groups. These metal ions interact with two aspartate residues that are highly conserved in DNA (and RNA) polymerases. These residues in phage T7 DNA polymerase are D705 and D882. One metal ion, designated A, interacts with the O atom of the free 3′-OH group on the polynucleotide chain, lowering its affinity for its hydrogen. This interaction promotes nucleophilic attack of the 3′-O on the phosphorus atom in the α-phosphate of the incoming nucleotide. The second metal ion (B in the figure) assists departure of the product pyrophosphate group from the incoming nucleotide. Together, the two metal ions stabilize the pentacovalent transition state on the α-phosphorus atom.

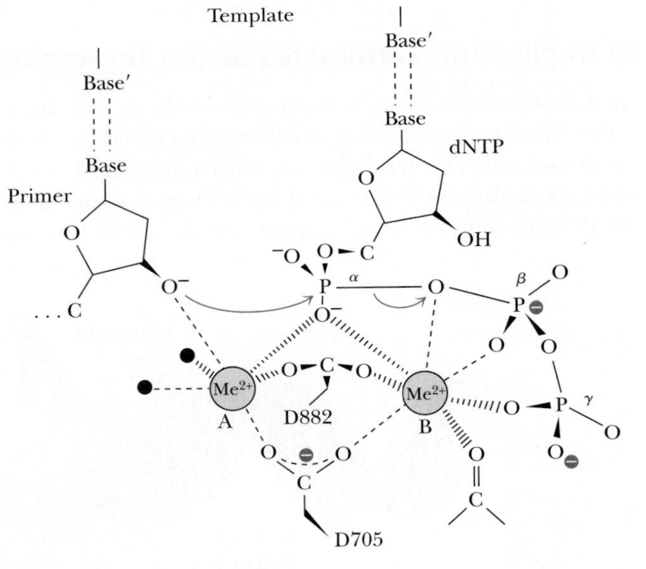

Adapted from Steitz, T., 1998. A mechanism for all polymerases. *Nature* **391:**231–232. (See also Doublié, S., et al., 1998. Crystal structure of bacteriophage T7 DNA replication complex at 2.2 Å resolution. *Nature* **391:**251–258; and Kiefer, J. R., et al., 1998. Visualizing DNA replication in a catalytically active *Bacillus* DNA polymerase crystal. *Nature* **391:**304–307.)

second *E. coli* DNA polymerase, **DNA polymerase II,** is a DNA repair enzyme expressed in response to DNA damage.

28.3 | Why Are There So Many DNA Polymerases?

Cells Have Different Versions of DNA Polymerase, Each for a Particular Purpose

A host of different DNA polymerases have been discovered, and even simple bacteria such as *Escherichia coli* have 7 distinct DNA polymerases. Yeast (*Saccharomyces cerevesiae*) has 9, and 16 have been identified in humans. Based on sequence homology, polymerases can be grouped into seven different families. The families differ in terms of the biological function served by family members. For example, family A includes DNA polymerases principally involved in DNA repair (*E. coli* DNA polymerase I, mammalian DNA-repair DNA polymerase θ, and mitochondrial DNA polymerase γ, which carries out mitochondrial DNA replication as well as repair). Family B polymerases include the eukaryotic DNA polymerases predominantly involved in replication of chromosomal DNA (*E. coli* DNA polymerase II and the mammalian replicative DNA polymerases α, δ, and ϵ, as well as the eukaryotic DNA-repair DNA polymerase ζ). Family C has the bacterial DNA-replicating enzymes (including *E. coli* DNA polymerase III). Members of families X and Y act in DNA repair pathways. Family X members include the eukaryotic DNA polymerases β, σ, λ, and μ, and others. Family Y comprises the translesion DNA polymerases, low-fidelity enzymes that allow the replication fork to bypass chemically damaged regions of DNA and continue on. Family Y includes eukaryotic DNA polymerases ι, κ, and η (η is also known as RevI). Family RT designates the DNA polymerases of retroviruses (such as HIV) and the telomerases that renew the ends of eukaryotic chromosomes. RT polymerases are novel in that they use RNA as the template.

The Common Architecture of DNA Polymerases

Despite sequence variation, the various DNA polymerase structures more or less follow a common architectural pattern that is reminiscent of a right hand, with distinct structural domains referred to as **fingers, palm,** and **thumb** (Figure 28.10). The active site of the polymerase, where deoxynucleotide addition to the growing chain is catalyzed, is located in the crevice within the palm domain that lies between the fingers and thumb domains. The fingers domain acts in deoxynucleotide recognition and binding, and the

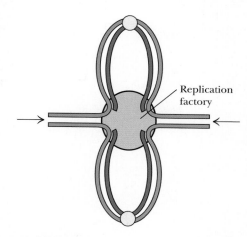

FIGURE 28.9 A replication factory "fixed" to a cellular substructure extrudes loops of newly synthesized DNA as parental DNA duplex is fed in from the sides. Parental DNA strands are green; newly synthesized strands are blue; small circles indicate origins of replication.

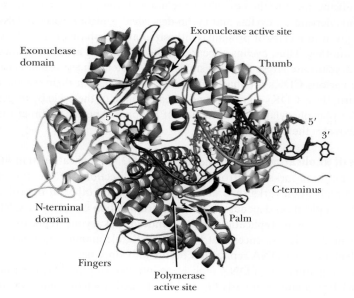

FIGURE 28.10 A structural paradigm for DNA polymerases, bacteriophage RB69 DNA polymerase. Ternary complex formed between the RB69 DNA polymerase, DNA, and dNTP. The N-terminal domain of the protein (residues 1–108 and 340–382) is in yellow, the exonuclease domain (residues 109–339) is in red, the palm (residues 383–468 and 573–729) is magenta, the fingers (residues 469–572) are blue, and the thumb (residues 730–903) is green. The DNA is given in stick representation, with the primer in gold and the template in blue-purple. A dNTP substrate (red) is shown at the active site, as are the two Ca^{2+} ions (light blue spheres that are necessary for the "two-metal-ion" DNA polymerase catalytic mechanism). Note also the calcium ion (blue sphere) at the exonuclease active site. (pdb id = 1IG9. Adapted from Figure 1 in Franklin, M. C., Wang, J., and Steitz, T. A., 2001. Structure of the replicating complex of a Pol α family DNA polymerase. *Cell* **105**:657–667.)

thumb is responsible for DNA binding, in the following manner: When the DNA polymerase binds to template-primer duplex DNA, its thumb domain closes around the DNA so that the DNA is bound in a groove formed by the thumb and palm. A dNTP substrate is then selected by the polymerase, and dNTP binding induces a conformational change in the fingers, which now rotate toward the polymerase active site in the palm. Catalysis ensues and a dNMP is added to the 3'-end of the growing primer strand; pyrophosphate is released, and the polymerase translocates one base farther along the template strand. In essence, all DNA polymerases are molecular motors that synthesize DNA, using dNTP substrates to add dNMP units to the primer strand, as they move along the template strand, reading its base sequence.

28.4　How Is DNA Replicated in Eukaryotic Cells?

DNA replication in eukaryotic cells shows strong parallels with prokaryotic DNA replication, but it is vastly more complex. First, eukaryotic DNA is organized into chromosomes that are compartmentalized within the nucleus. Furthermore, these chromosomes must be duplicated with high fidelity once (and only once!) each cell cycle. For example, in a dividing human cell, a carefully choreographed replication of 6 billion bp of DNA distributed among 46 chromosomes occurs. The events associated with cell growth and division in eukaryotic cells fall into a general sequence having four distinct phases: M, G_1, S, and G_2 (Figure 28.11). (Fully differentiated cells in multicellular organisms typically enter a quiescent, nondividing state after G_1 that is called G_0. Such cells may remain in G_0 indefinitely.) Eukaryotic cells have solved the problem of replicating their enormous genomes in the few hours allotted to the S phase by initiating DNA replication at multiple origins of replication distributed along each chromosome. Depending on the organism and cell type, replication origins are DNA regions 0.5 to 2 kbp in size that occur every 3 to 300 kbp (for example, an average human chromosome has several hundred replication origins). Since eukaryotic DNA replication proceeds concomitantly throughout the genome, each eukaryotic chromosome must contain many units of replication, called **replicons.**

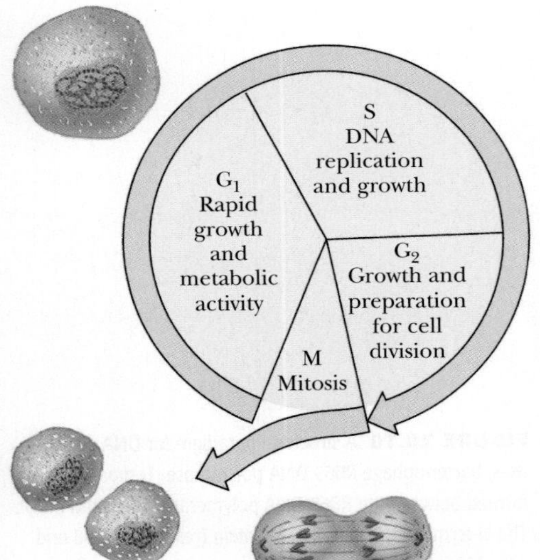

The Cell Cycle Controls the Timing of DNA Replication

Checkpoints, Cyclins, and CDKs Progression through the cell cycle is regulated through a series of **checkpoints** that control whether the cell continues into the next phase. These checkpoints are situated to ensure that *all* the necessary steps in each phase of the cycle have been satisfactorily completed before the next phase is entered. If conditions for advancement to the next phase are not met, the cycle is arrested until they are. Checkpoints depend on **cyclins** and **cyclin-dependent protein kinases (CDKs).** *Cyclin* is the name given to a class of proteins synthesized at one phase of the cell cycle and degraded at another. Thus, cyclins appear and then disappear at specific times during the cell cycle. Cyclins are larger than the small CDK protein kinase subunits to which they bind. The various CDKs are inactive unless complexed with their specific cyclin partners. In turn, these CDKs control events at each phase of the cycle by targeting specific proteins for phosphorylation. Destruction of the phase-specific cyclin at the end of the phase inactivates the CDK.

Initiation of Replication Eukaryotic cells initiate DNA replication at multiple origins, and two replication forks arise from each origin. Bidirectional DNA replication ensues, as the two replication forks then move away from each other in opposite directions. Initiation of replication depends on the **origin recognition complex,** or **ORC,** a protein complex that binds to replication origins. Indeed, eukaryotic replication origins are defined as nucleotide sequences that bind ORC. Stable maintenance of the eukaryotic genome demands that DNA replication occurs only once per cell cycle. This demand is met by dividing initiation of DNA replication into two steps: (1) the **licensing of replication origins** during late M or early G_1, and (2) the **activation of replication** at the origins

FIGURE 28.11 The eukaryotic cell cycle. The stages of mitosis and cell division define the M phase (*M* for *mitosis*). G_1 (*G* for *gap,* not growth) is typically the longest part of the cell cycle; G_1 is characterized by rapid growth and metabolic activity. Cells that are quiescent, that is, not growing and dividing (such as neurons), are said to be in G_0. The S phase is the time of DNA synthesis. S is followed by G_2, a relatively short period of growth when the cell prepares for cell division. Cell cycle times vary from less than 24 hours (rapidly dividing cells such as the epithelial cells lining the mouth and gut) to hundreds of days.

during S phase through the action of two protein kinases, **Cdc7-Dbf4** and **S-CDK** (the S-phase cyclin-dependent protein kinase).

Licensing involves the highly regulated assembly of **prereplication complexes (pre-RCs)** on origins of replication. Early in G_1 (just after M), the ORC (a heterohexameric complex of Orc1-6) serves as a "landing pad" for proteins essential to replication control. Binding of these proteins to ORC establishes a pre-RC, but only within this window of opportunity during G_1. Yeast, a simple eukaryote, provides an informative model: ORC binds to origins and recruits Cdc6 (in its phosphorylated form, Cdc6p), Cdt1, and the MCM proteins (Figure 28.12). The MCM proteins are a family of six related proteins. MCM proteins are also known as **replication licensing factors,** because they "license," or permit, DNA replication to occur. The MCM proteins assemble as hexameric helicases that render the chromosomes competent for replication. Two MCM complexes are active within each origin, one for each replication fork. The pre-RC therefore consists of ORC, Cdc6, Cdt1, the MCM complexes, and other proteins. The **Mcm2-7 helicase** is a double hexamer that uses ATP hydrolysis to move along duplex DNA and separate the strands. Each hexamer becomes an active **CMG helicase** upon binding Cdc45 and **GINS,** a protein that binds single-stranded DNA and recruits primase. CMG opens the DNA duplex and then the two helicase hexamers, each with a single strand at its core, move apart, leaving single-stranded DNA behind them, ready for replication. Cdc6 and Cdt1 are the **replicator activator proteins.** Cdc6 is degraded following replication initiation, thereby precluding the possibility for errant replication initiation events until after mitosis, when Cdc6 accumulates again.

DNA replication is the defining characteristic of the S phase of the cell cycle. The switch from G_1 to S is triggered by phosphorylation events carried out by S-CDK and Cdc7-Dbf4. Phosphorylation of the MCM proteins and binding of Cdc45 activates the helicase activity of MCM (Figure 28.12). Phosphorylation of **Sld2** and **Sld3,** a pair of proteins that interact with Dbp11, leads to the recruitment of DNA polymerase to the replication origins. (The protein acronyms refer to DNA polymerase-binding [Dpb] and proteins encoded by genes that give a lethal phenotype when mutated in Dpb11 mutant strains of yeast [Sld, or synthetic-lethal with Dpb11]. Collectively, these proteins are referred to as the "11-3-2 complex"). The actions of S-CDK and Cdc7-Dbf4 trigger bidirectional DNA replication from each origin, with the two diverging MCM complexes serving as helicases. Each helicase unwinds the duplex DNA to provide single-stranded templates for the DNA polymerases that follow.

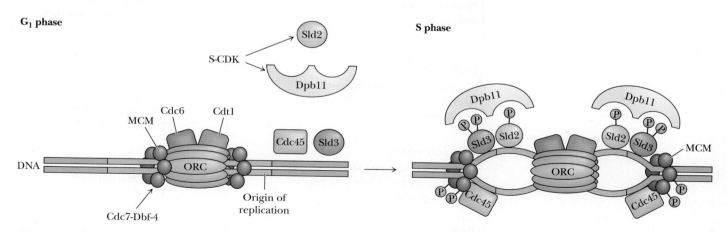

FIGURE 28.12 The initiation of DNA replication in eukaryotic cells. Binding of the pre-RC to origins of replication is followed by loading of MCM hexameric helicases, phosphorylation reactions mediated by S-CDK and Cdc7-Dbf4, and the binding of the 11-3-2 complex. The passage of the cell from G_1 to S is defined by these events. DNA polymerases are recruited to the origins of replication, where they can access single-stranded regions of DNA and initiate DNA synthesis.

Proteins of the Prereplication Complex Are AAA+ ATPase Family Members

Cdc6, the Orc proteins, and MCM proteins are members of the **AAA+ ATPase family,** a group of proteins characterized by sequence and structural homology, ATPase activity, and a general function as molecular chaperones. The binding of both ORC and Cdc6 to chromatin in the process of pre-RC assembly (Figure 28.12) is ATP-dependent. ORC with ATP bound can bind to origins of replication and recruit Cdc6 and Cdt1 to form a pre-RC in which both ORC and Cdc6 have bound ATP. Cdc6 is required for the recruitment of the MCM proteins, specifically, a complex of MCM2–7 with ATP-dependent helicase activity. To establish the pre-RC, MCM2–7 must be stably associated with the origin, and this stability is achieved following ATP hydrolysis, first by Cdc6 and then by ORC.

Geminin Provides Another Control Over Replication Initiation

Geminin is a protein that provides another level of regulatory control over DNA replication. Geminin inhibits DNA replication by preventing the incorporation of MCM complexes into the pre-RC. Geminin is active during S, G_2, and M phases, but its destruction during mitosis permits replication initiation in G_1. Geminin binds to Cdt1, preventing it from recruiting MCM proteins to the pre-RC. Geminin exists as a parallel coiled-coil homodimer. It interacts with Cdt1 in two ways: It has an array of glutamate residues on its surface that interact with positive charges on Cdt1, and an adjacent region on geminin interacts independently with the N-terminal region of Cdt1.

Eukaryotic Cells Also Contain a Number of Different DNA Polymerases

At least 19 different DNA polymerases have been described in eukaryotic cells thus far. These various polymerases have been assigned Greek letters in the order of their discovery (Table 28.4 lists the principal ones). Multiple DNA polymerases participate in

TABLE 28.4 Biochemical Properties of the Principal Human DNA Polymerases		
Polymerase Localization and Function	Subunits (mass in kD)	Subunit Function
DNA polymerase α		
Nuclear; initiation of nuclear DNA replication	180	Catalytic subunit
	68	Protein–protein interactions
	55	Primase
	48	Primase
DNA polymerase δ		
Nuclear; principal polymerase in leading- and lagging-strand synthesis; highly processive	125	Catalytic subunit
	66	Structural
	50	Interaction with PCNA
	12	Protein–protein interactions
DNA polymerase ε		
Nuclear; leading- and lagging-strand synthesis, sensor of DNA damage checkpoint control	261	Catalytic subunit
	59	Multimerization
	17	Protein–protein interactions
	12	Protein–protein interactions
DNA polymerase γ		
Mitochondria; mitochondrial DNA replication	140	Catalytic subunit
	55	Processivity
DNA polymerase β		
DNA repair	39	

leading- and lagging-strand synthesis, but three—α, δ, and ϵ—share the major burden. **DNA polymerase α** has an associated primase subunit and functions in the initiation of nuclear DNA replication. Given a template, it not only synthesizes an RNA primer of about 10 nucleotides, but it then adds 20 to 30 deoxynucleotides to extend the chain in the 5′→3′ direction. **DNA polymerase δ,** a heterotetrameric enzyme, is the principal DNA polymerase in eukaryotic DNA replication. It interacts with **PCNA** protein (PCNA stands for proliferating cell nuclear antigen). Through its association with PCNA, DNA polymerase δ carries out highly processive DNA synthesis. PCNA is the eukaryotic counterpart of the *E. coli* β_2-sliding clamp; it clamps DNA polymerase δ to the DNA. Like β_2, PCNA encircles the double helix, but in contrast to the prokaryotic β_2-sliding clamp, PCNA is a homotrimer, not a homodimer (Figure 28.13; see also Figure 28.7). **DNA polymerase ϵ** also plays a major role in DNA replication. It has an acidic C-terminal extension lacking in other DNA polymerases, and evidence suggests that this domain is a sensor for DNA damage checkpoint control, halting DNA replication until the damage is repaired. **DNA polymerase γ** is the DNA-replicating enzyme of mitochondria; **DNA polymerase β** functions in DNA repair. The more recently discovered eukaryotic DNA polymerases (including ζ, η, ι, κ, and Rev1) are novel in that they are more error-prone, resulting in lower fidelity of DNA replication. Nevertheless, they have the important ability to function in DNA replication and repair when damaged regions of DNA are encountered.

Other proteins involved in eukaryotic DNA replication include **replication protein A (RPA),** an ssDNA-binding protein that is the eukaryotic counterpart of SSB, and **replication factor C (RFC).** RFC loads the PCNA sliding clamp onto replicating DNA, thus acting as the eukaryotic counterpart of the prokaryotic γ-complex.

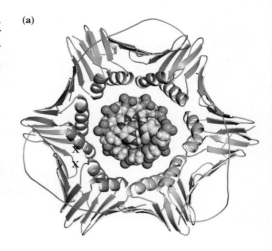

(a)

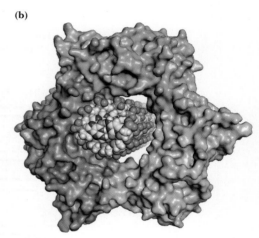

(b)

FIGURE 28.13 Structure of the human PCNA homotrimer. **(a)** Ribbon representation of the PCNA trimer (pdb id = 1AXC) with an axial view of a B-form DNA duplex in its center. The molecular mass of each PCNA monomer is 37 kD. **(b)** Molecular surface of the PCNA trimer-DNA complex.

28.5 | How Are the Ends of Chromosomes Replicated?

Eukaryotic chromosomes are linear. The ends of chromosomes have specialized structures known as **telomeres.** The telomeres of virtually all eukaryotic chromosomes consist of short (5–8 bp), tandemly repeated, GC-rich nucleotide sequences that form protective caps 1 to 12 kbp long at the ends of chromosomal DNA. For example, the telomeres of human germline cells (sperm and eggs) contain between 1000 and 1700 copies of the hexameric repeat TTAGGG (Figure 28.14a). Telomeres aid in chromosome maintenance and stability by protecting against DNA degradation or rearrangement. DNA polymerases cannot replicate the extreme 5′-ends of chromosomes because these enzymes require a template and a primer and replicate only in the 5′→3′ direction. Thus, lagging strand synthesis at the 3′-ends of chromosomes is primed by RNA primase to form Okazaki fragments, but these RNA primers are subsequently removed, resulting in gaps in the progeny 5′-terminal strands at each end of the chromosome after each round of replication ("primer gap"; see Figure 28.14).

Telomeres are added to the ends of chromosomal DNA by an RNA-dependent DNA polymerase known as **telomerase.** It maintains chromosome length by restoring telomeres at the 3′-ends of chromosomes. Telomerase is a specialized reverse transcriptase containing a catalytic subunit, referred to as **TERT,** for **TE**lomerase **R**everse **T**ranscriptase, and

Telomeres—A Timely End to Chromosomes?

Mammalian cells in culture undergo only 50 or so cell divisions before they die. Somatic cells are known to lack telomerase activity, and thus, they inevitably lose bits of their telomeres with each cell division. Telomerase activity is missing because the telomerase–reverse transcriptase gene (the *TRT* gene) is switched off. This fact has led to a telomere theory of cell aging, which suggests that cells senesce and die when their telomeres are gone. In support of this notion, a team of biologists headed by Calvin B. Harley at Geron Corporation used recombinant DNA techniques to express the catalytic subunit of human telomerase in skin cells in culture and observed that such cells divide 40 times more after cells lacking this treatment have become senescent. These results, although controversial, may have relevance to the aging process.

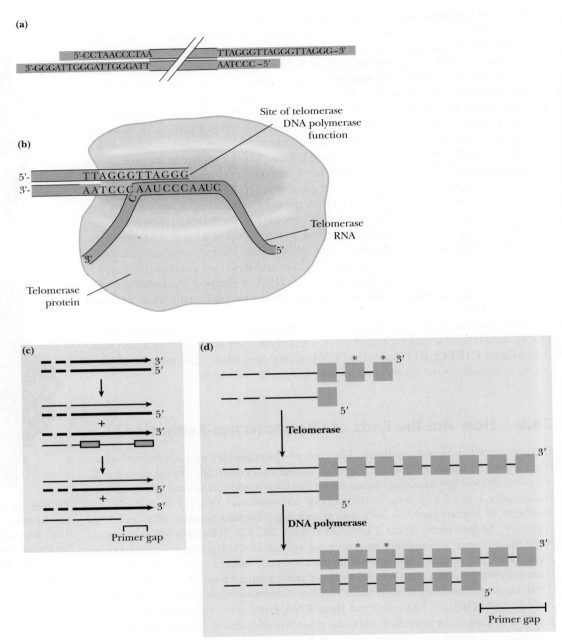

FIGURE 28.14 Telomere replication. **(a)** Telomeres on human chromosomes. TTAGGG tandem repeats are found at the 3′-ends of the DNA strands, paired with the complementary sequence 3′-AATCCC-5′ on the other DNA strand. Thus, a G-rich region is created at the 3′-end of each DNA strand, and a C-rich region is created at the 5′-end of each DNA strand. Typically, at each end of the chromosome, the G-rich strand protrudes 12 to 16 nucleotides beyond its complementary C-rich strand. **(b)** The ribonucleic acid of human telomerase serves as the template for the DNA polymerase activity of telomerase. Nucleotides 46 to 56 of this RNA are CUAA**CCCUAA**C and provide the template function for the telomerase-catalyzed addition of TTAGGG units to the 3′-end of a DNA strand. **(c)** In replication of the lagging strand, short RNA primers are added (pink) and extended by DNA polymerase. When the RNA primer at the 5′-end of each strand is removed, there is no nucleotide sequence to read in the next round of DNA replication. The result is a gap (primer gap) at the 5′-end of each strand (only one end of a chromosome is shown in this figure). **(d)** Asterisks indicate sequences at the 3′-end that cannot be copied by conventional DNA replication. Synthesis of telomeric DNA by telomerase extends the 5′-ends of DNA strands, allowing the strands to be copied by normal DNA replication.

an RNA component that serves as the template for telomere synthesis, referred to variously as **RT**, for RNA Template, or **TER**, for TEmplate-containing telomerase **RNA**. In human telomerase, the RNA part is 451 nucleotides long. Its template sequence is CUAACCCUAAC (base-pairs with TTAGGG). Telomerase uses the 3′-end of the DNA as a primer and adds successive TTAGGG repeats to it, employing its RNA as template over and over again (Figure 28.14). The G-rich ends of telomeres can arrange into four-stranded G-quadruplex structures (see Figure 11.19).

28.6 | How Are RNA Genomes Replicated?

Many viruses have genomes composed of RNA, not DNA. How then is the information in these RNA genomes replicated? In 1964, Howard Temin noted that inhibitors of DNA synthesis prevented infection of cells in culture by RNA tumor viruses such as avian sarcoma virus. On the basis of this observation, Temin proposed that DNA is an intermediate in the replication of such viruses; that is, *an RNA tumor virus can use viral RNA as the template for DNA synthesis.*

RNA viral chromosome ⟶ DNA intermediate ⟶ RNA viral chromosome

In 1970, Temin and David Baltimore independently discovered a viral enzyme capable of mediating such a process, namely, an **RNA-directed DNA polymerase** or, as it is usually called, **reverse transcriptase.** All RNA tumor viruses contain such an enzyme within their virions (viral particles). RNA viruses that replicate their RNA genomes via a DNA intermediate are classified as **retroviruses.**

Like other DNA and RNA polymerases, reverse transcriptase synthesizes polynucleotides in the 5′→3′ direction, and like all DNA polymerases, reverse transcriptase requires a primer. Interestingly, the primer is a specific tRNA molecule captured by the virion from the host cell in which it was produced. The 3′-end of the tRNA is base-paired with the viral RNA template at the site where DNA synthesis initiates, and its free 3′-OH accepts the initial deoxynucleotide once transcription commences. Reverse transcriptase then transcribes the RNA template into a complementary DNA (cDNA) strand to form a double-stranded DNA:RNA hybrid.

The Enzymatic Activities of Reverse Transcriptases

Reverse transcriptases possess three enzymatic activities, all of which are essential to viral replication:

1. *RNA-directed DNA polymerase activity,* for which the enzyme is named (see Figure 12.10).
2. *RNase H activity.* Recall that RNase H is a nuclease that specifically degrades RNA chains in DNA:RNA hybrids (see Figure 12.10). The RNase H function of reverse transcriptase is an exonuclease activity that degrades the template genomic RNA and also removes the priming tRNA after DNA synthesis is completed.

> **A DEEPER LOOK**
>
> ## RNA as Genetic Material
>
> Whereas the genetic material of cells is dsDNA, virtually all plant viruses, several bacteriophages, and many animal viruses have genomes consisting of RNA. In most cases, this RNA is single stranded. Viruses with single-stranded genomes use the single strand as a template for synthesis of a complementary strand, which can then serve as template in replicating the original strand. **Retroviruses** are an interesting group of eukaryotic viruses with single-stranded RNA genomes that replicate through a dsDNA intermediate. Furthermore, the life cycle of retroviruses includes an obligatory step in which the dsDNA is inserted into the host cell genome in a transposition event. Retroviruses are responsible for many diseases, including tumors and other disorders. **HIV-1,** the **human immunodeficiency virus** that causes **AIDS,** is a retrovirus.

(a)

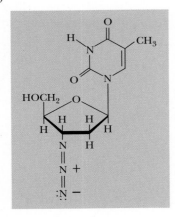

(b)

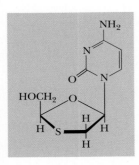

FIGURE 28.15 The structures of AZT (3′-azido-2′,3′-dideoxythymidine) and 3TC (2′,3′-dideoxy-3′-thiacytidine). These nucleosides are phosphorylated in vivo to form deoxynucleoside-5′-triphosphate substrate analogs for HIV reverse transcriptase.

3. *DNA-directed DNA polymerase activity.* This activity replicates the ssDNA remaining after RNase H degradation of the viral genome, yielding a DNA duplex. This DNA duplex directs the remainder of the viral infection process or becomes integrated into the host chromosome, where it can lie dormant for many years as a **provirus.** Activation of the provirus restores the infectious state.

HIV reverse transcriptase is of great clinical interest because it is the enzyme for replication of the AIDS virus. DNA synthesis by HIV reverse transcriptase is blocked by nucleotide analogs such as AZT and 3TC (Figure 28.15). HIV reverse transcriptase incorporates these analogs into growing DNA chains in place of dTMP (in the case of AZT) or dCMP (in the case of 3TC). Once incorporated, these analogs block further chain elongation because they lack a 3′-OH where the next incoming dNTP can be added. HIV reverse transcriptase is error-prone: It incorporates the wrong base at a frequency of 1 per 2000 to 4000 nucleotides polymerized. This high error rate during replication of the HIV genome means that the virus is ever changing, a feature that makes it difficult to devise an effective vaccine.

28.7 How Is the Genetic Information Rearranged by Genetic Recombination?

Genetic recombination is the natural process by which genetic information is rearranged to form new associations. For example, compared to their parents, progeny may have new combinations of traits because of genetic recombination. At the molecular level, genetic recombination is the exchange (or incorporation) of one DNA sequence with (or into) another. When recombination involves reaction between very similar sequences (homologous sequences) of DNA, the process is called **homologous recombination.** When very different nucleotide sequences recombine, it's **nonhomologous recombination. Transposition**—the enzymatic insertion of a *transposon* (a mobile segment of DNA, see pages 973–974) into a new location in the genome—and nonhomologous recombination (incorporation of a DNA segment whose sequence differs greatly from the DNA at the point of insertion) are two types of recombination that play a significant evolutionary role. Nonhomologous recombination occurs at a low frequency in all cells and serves as a powerful genetic force that reshapes the genomes of all organisms. Homologous recombination involves an exchange of DNA sequences between homologous chromosomes, resulting in the arrangement of genes into new combinations. Homologous recombination is generally used to fix the DNA so that information is not lost. For example, large lesions in DNA are repaired via recombination of the damaged chromosome with a homologous chromosome.

The process underlying homologous recombination is termed **general recombination** because the enzymatic machinery that mediates the exchange can use essentially any pair of homologous DNA sequences as substrates. Homologous recombination occurs in all organisms and is particularly prevalent during the production of gametes (meiosis) in diploid organisms. In higher animals—that is, those with immune systems—recombination also occurs in the DNA of somatic cells responsible for expressing proteins of the immune response, such as the immunoglobulins. This **somatic recombination** rearranges the immunoglobulin genes, dramatically increasing the potential diversity of immunoglobulins available from a fixed amount of genetic information. Homologous recombination can also occur in bacteria. Indeed, even viral chromosomes undergo recombination. For example, if two mutant viral particles simultaneously infect a host cell, a recombination event between the two viral genomes can lead to the formation of a virus chromosome that is wild-type.

General Recombination Requires Breakage and Reunion of DNA Strands

Recombination occurs by the breakage and reunion of DNA strands so that a physical exchange of parts takes place. Matthew Meselson and J. J. Weigle demonstrated this in 1961 by coinfecting *E. coli* with two genetically distinct bacteriophage λ strains, one of

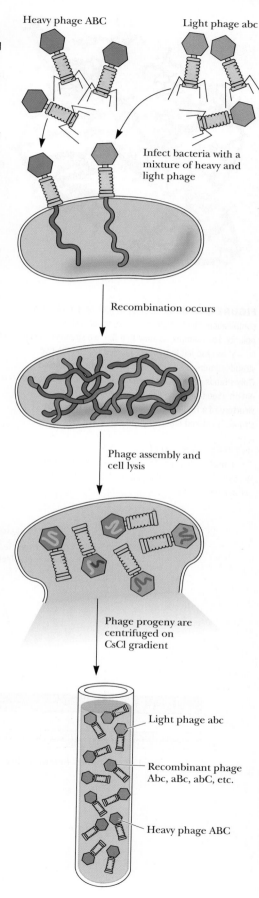

Infect bacteria with a mixture of heavy and light phage

Recombination occurs

Phage assembly and cell lysis

Phage progeny are centrifuged on CsCl gradient

Light phage abc

Recombinant phage Abc, aBc, abC, etc.

Heavy phage ABC

► **FIGURE 28.16** Meselson and Weigle's experiment. Density-labeled, "heavy" phage, symbolized as ABC phage, was used to coinfect bacteria along with "light" phage, the abc phage. The progeny from the infection were collected and subjected to CsCl density gradient centrifugation. Parental-type ABC and abc phage were well separated in the gradient, but recombinant phage (ABc, Abc, aBc, aBC, and so on) were distributed diffusely between the two parental bands because they contained chromosomes constituted from fragments of both "heavy" and "light" DNA.

which had been density-labeled by growth in ^{13}C- and ^{15}N-containing media (Figure 28.16). The phage progeny were recovered and separated by CsCl density gradient centrifugation. Phage particles that displayed recombinant genotypes were distributed throughout the gradient while parental (nonrecombinant) genotypes were found within discrete "heavy" and "light" bands in the density gradient. The results showed that recombinant phage contained DNA derived in varying proportions from both parents. The obvious explanation is that these recombinant DNAs arose via the breakage and rejoining of DNA molecules.

A second important observation made during this type of experiment was that some of the plaques formed by the phage progeny contained phage of two different genotypes, even though each plaque was caused by a single phage infecting one bacterium. Therefore, some infecting phage chromosomes must have contained a region of **heteroduplex DNA,** duplex DNA in which a part of each strand is contributed by a different parent (Figure 28.17).

Homologous Recombination Proceeds According to the Holliday Model

In 1964, Robin Holliday proposed a model for homologous recombination in fungi that has been proven to be a generally valid depiction of homologous recombination in bacteriophages, viruses, prokaryotes, and eukaryotes (Figure 28.18). The two homologous DNA duplexes are first juxtaposed so that their sequences are aligned. This process of **chromosome pairing** is called **synapsis** (Figure 28.18a). Holliday suggested that recombination begins with the introduction of single-stranded nicks in the DNA at homologous sites on the two paired chromosomes (Figure 28.18b). The two duplexes partially unwind, and the free, single-stranded end of one duplex begins to base-pair with its nearly complementary, single-stranded region along the intact strand in the other duplex, and vice versa (Figure 28.18c). This **strand invasion** is followed by ligation of the free ends from different duplexes to create a cross-stranded intermediate known as a **Holliday junction** (Figure 28.18d). The cross-stranded junction can now migrate in either

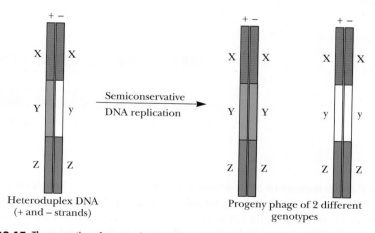

Heteroduplex DNA (+ and − strands)

Semiconservative DNA replication

Progeny phage of 2 different genotypes

FIGURE 28.17 The generation of progeny bacteriophage of two different genotypes from a single phage particle carrying a heteroduplex DNA region within its chromosome. The heteroduplex DNA is composed of one strand that is genotypically XYZ (the + strand), and the other strand that is genotypically XyZ (the − strand). That is, the genotype of the two parental strands for gene Y is different (one is Y, the other y).

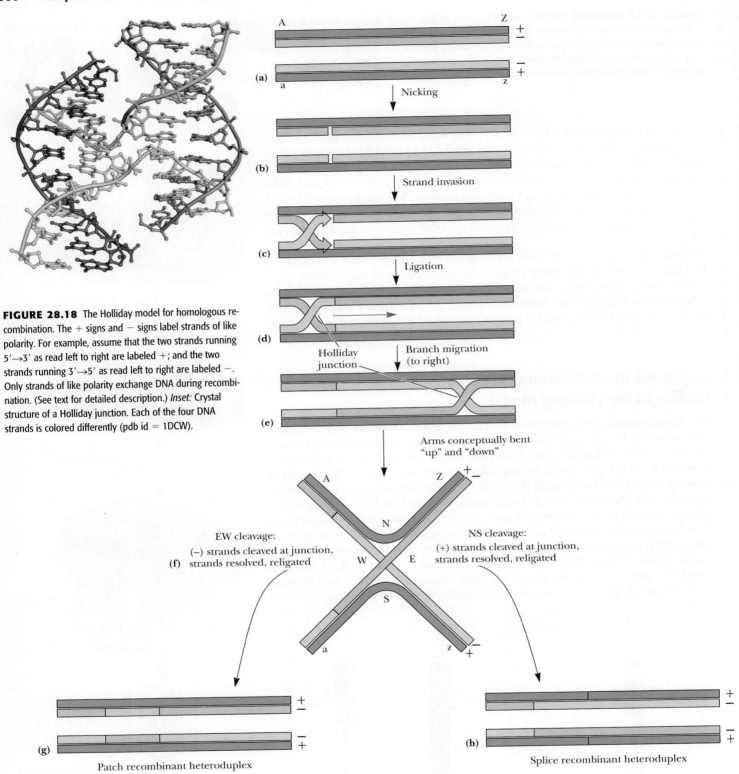

FIGURE 28.18 The Holliday model for homologous recombination. The + signs and − signs label strands of like polarity. For example, assume that the two strands running 5′→3′ as read left to right are labeled +; and the two strands running 3′→5′ as read left to right are labeled −. Only strands of like polarity exchange DNA during recombination. (See text for detailed description.) *Inset:* Crystal structure of a Holliday junction. Each of the four DNA strands is colored differently (pdb id = 1DCW).

direction **(branch migration)** by unwinding and rewinding of the two duplexes (Figure 28.18e). Branch migration results in **strand exchange;** heteroduplex regions of varying length are possible. In order for the joint molecule formed by strand exchange to be resolved into two DNA duplex molecules, another pair of nicks must be introduced. Resolution can be represented best if the duplexes are drawn with the chromosome arms bent "up" or "down" to give a planar representation (Figure 28.18f). Nicks then take place, either at E and W, that is, in the − strands that were originally nicked (see Figure 28.18b), or at N and S, that is, in the + strands (the strands not previously nicked). Duplex resolution is most easily kept straight by remembering that + strands are complementary to − strands and any resultant duplex must have one of each. Nicks made in the strands originally nicked lead to DNA duplexes in which one strand of each remains intact. Although these duplexes contain heteroduplex regions, they are not recombinant for the markers (AZ, az) that flank the heteroduplex region; such heteroduplexes are called **patch recombinants** (Figure 28.18g). Nicks introduced into the two strands not previously nicked yield DNA molecules that are both heteroduplex and recombinant for the markers A/a and Z/z; these heteroduplexes are termed **splice recombinants** (Figure 28.18h). Although this Holliday model explains the outcome of recombination, it provides no mechanistic explanation for the strand exchange reactions and other molecular details of the process.

The Enzymes of General Recombination Include RecA, RecBCD, RuvA, RuvB, and RuvC

To illustrate recombination mechanisms, we focus on general recombination as it occurs in *E. coli*. The principal players in the process are the **RecBCD** enzyme complex, which initiates recombination; the **RecA** protein, which binds single-stranded DNA, forming a nucleoprotein filament capable of strand invasion and homologous pairing; and the **RuvA, RuvB,** and **RuvC** proteins, which drive branch migration and process the Holliday junction into recombinant products. Eukaryotic homologs of these prokaryotic recombination proteins have been identified, indicating that the fundamental process of general recombination is conserved across all organisms.

The RecBCD Enzyme Complex Unwinds dsDNA and Cleaves Its Single Strands

The RecBCD complex is composed of the proteins **RecB** (140 kD; 1180 amino acids), **RecC** (130 kD; 1122 amino acids), and **RecD** (67 kD; 608 amino acids). This multifunctional enzyme complex has both helicase and nuclease activity and initiates recombination by attaching to the end of a DNA duplex (or at a double-stranded break in the DNA) and using its ATP-dependent helicase function to unwind the dsDNA (Figure 28.19a). RecB and RecD are helicases powered by ATP hydrolysis; each consumes an ATP per base pair of DNA traversed. The RecD motor is faster than the RecB motor and leads the way. The greater speed of RecD causes the DNA to loop out between RecD and RecB. The RecB subunit contains the nuclease domain. As RecBCD progresses along unwinding the duplex, the RecB nuclease activity cleaves both of the newly formed single strands (although the strand that provided the 3′-end at the RecBCD entry site is cut more frequently than the 5′-terminal strand [Figure 28.19b]). SSB (and some RecA protein) readily binds to the emerging single strands. Sooner or later, RecBCD encounters a particular nucleotide sequence, a so-called **Chi** (or χ) site, characterized by the sequence **5′-GCTGGTGG-3′**. These χ sites are recombinational "hot spots"; 1009 χ sites have been identified in the *E. coli* genome (on average, about one every 4.5 kb of DNA). When a χ sequence is encountered by a RecBCD complex approaching its 3′-side (the ..G-3′-side), RecBCD cleaves the χ-bearing DNA strand four to six bases to the 3′ side of χ (Figure 28.19c). RecBCD flips so that the RecB motor leads the way, and the RecBCD complex no longer expresses nuclease activity against the 3′-terminal strand, but nuclease activity against the 5′-terminal strand increases (Figure 28.19d).

Resuming its helicase function, RecBCD unwinds the dsDNA, and collectively these processes generate an ssDNA tail bearing a χ site at its 3′-terminal end. This ssDNA may reach several kilobases in length. RecA protein now binds to the 3′-terminal strand to form a **nucleoprotein filament** (Figure 28.19e). This filament is active in pairing and strand invasion with a homologous region in another dsDNA molecule.

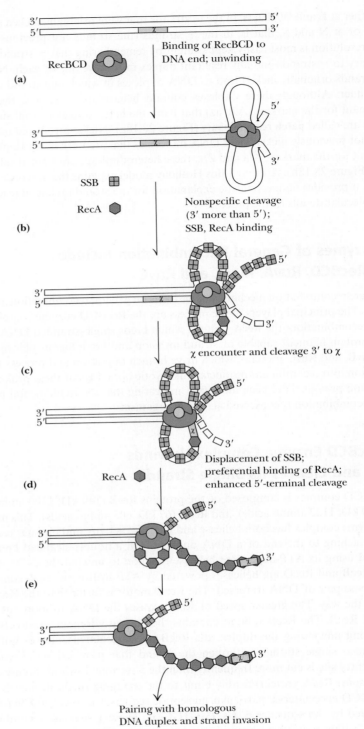

FIGURE 28.19 Model of RecBCD-dependent initiation of recombination. **(a)** RecBCD binds to a duplex DNA end, and its helicase activity begins to unwind the DNA double helix. "Rabbit ears" of ssDNA loop out from RecBCD because the rate of DNA unwinding exceeds the rate of ssDNA release by RecBCD. **(b)** As it unwinds the DNA, SSB (and some RecA) bind to the single-stranded regions; the RecBCD endonuclease activity randomly cleaves the ssDNA, showing a greater tendency to cut the 3'-terminal strand rather than the 5'-terminal strand. **(c)** When RecBCD encounters a properly oriented χ site, the 3'-terminal strand is cleaved just below the 3'-end of χ. **(d)** RecBCD now directs the binding of RecA to the 3'-terminal strand, as RecBCD endonuclease activity now acts more often on the 5'-terminal strand. **(e)** A nucleoprotein filament consisting of RecA-coated 3'-strand ssDNA is formed. This nucleoprotein filament is capable of homologous pairing with a dsDNA and strand invasion. (Adapted from Figure 2 in Eggleston, A. K., and West, S. C., 1996. Exchanging partners: Recombination in *E. coli. Trends in Genetics* **12**:20–25; and Figure 3 in Eggleston, A. K., and West, S. C., 1997. Recombination initiation: Easy as A, B, C, D ….χ? *Current Biology* **7**:R745–R749.)

The RecA Protein Can Bind ssDNA and Then Interact with Duplex DNA

The **RecA** protein, or **recombinase,** is a multifunctional protein that acts in general recombination (Figure 28.20a). RecA mediates the ATP-dependent **DNA strand exchange reaction** leading to formation of a Holliday junction (Figure 28.18b–f). In the presence of ATP and ssDNA, RecA forms a right-handed helical filament having six monomers per turn, with each monomer spanning about three nucleotides of DNA. The RecA nucleoprotein filament serves as a scaffold upon which the events of recombination take

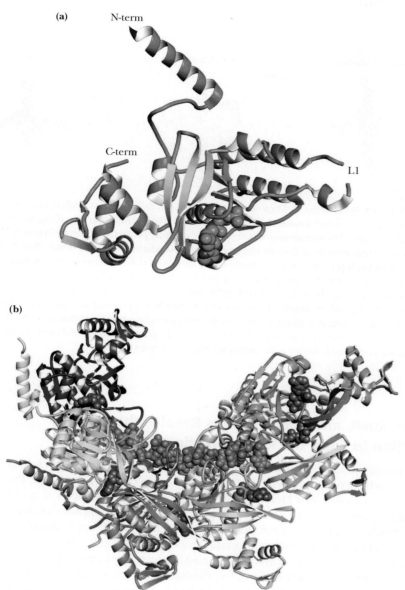

(a) N-term

C-term

L1

(b)

FIGURE 28.20 **(a)** RecA, a 352-residue, 38-kD protein, with dATP in the ATP-binding site (pdb id = 2ODN). **(b)** Structure of a fused RecA hexamer with bound ssDNA (as poly(dT)$_{18}$, magenta). Each of the six RecA units is a different color, and each has a bound ADP molecule (marine blue). The adenine nucleotide-binding sites lie at the interfaces between RecA units (pdb id = 3CMU).

place. This filament has a deep spiral groove large enough to encompass three strands of DNA. Although RecA-bound ssDNA is relatively stretched and underwound, with about 18.5 nucleotides per turn, its local conformation resembles B-DNA (Figure 28.20b). Thus, when dsDNA adds to the complex, the ssDNA is poised to search for the sequence homology that leads to strand exchange.

Procession of strand separation of dsDNA and the re-pairing into hybrid strands along the DNA duplex initiates **strand invasion** (Figure 28.21b). Strand invasion drives the displacement of the homologous DNA strand from the DNA duplex and its replacement with the ssDNA strand, a process known as **single-strand assimilation** (or **single-strand uptake**). Strand assimilation does not occur if there is no sequence homology between the ssDNA and the invaded DNA duplex. The DNA strand displaced by the invading 3′-terminal ssDNA is free to anneal with the 5′-terminal strand in the original DNA, a step that is also mediated by RecA protein and SSB (Figure 28.21c). The result is a Holliday junction, the classic intermediate in genetic recombination.

▷ **FIGURE 28.21** Model for homologous recombination as promoted by RecA enzyme. **(a)** RecA protein (and SSB) aid strand invasion of the 3′-ssDNA into a homologous DNA duplex, **(b)** forming a D-loop. **(c)** The D-loop strand that has been displaced by strand invasion pairs with its complementary strand in the original duplex to form a Holliday junction as strand invasion continues.

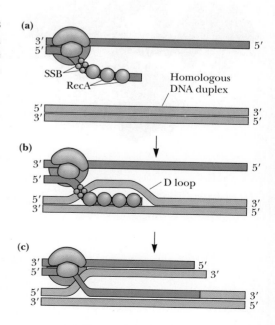

(a)
3′
5′ ⟶ 5′
SSB
RecA
Homologous DNA duplex
5′ ⟶ 3′
3′ ⟶ 5′

(b)
3′ ⟶ 5′
5′
D loop
5′ ⟶ 3′
3′ ⟶ 5′

(c)
3′ ⟶ 5′
5′ ⟶ 3′
5′ ⟶ 3′
3′ ⟶ 5′

(a)

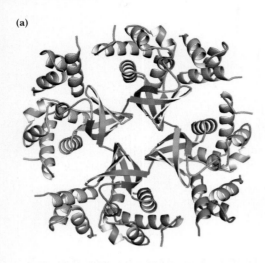

(b)

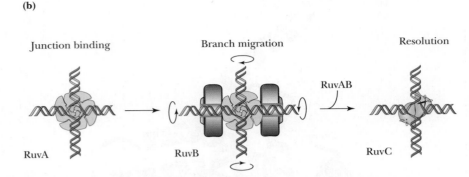

Junction binding

Branch migration

Resolution

RuvA

RuvB

RuvAB

RuvC

(c)

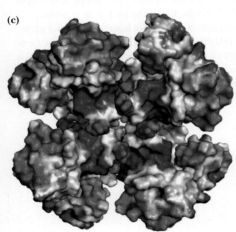

(d)

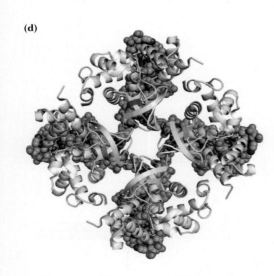

FIGURE 28.22 Model for the resolution of a Holliday junction in *E. coli* by the RuvA, RuvB, and RuvC proteins. **(a)** Ribbon diagram of the RuvA tetramer. RuvA monomers have an overall L shape (pdb id = 1CUK). **(b)** Model for RuvA/RuvB action. (left) The RuvA tetramer fits snugly within the Holliday junction point. (center) Oppositely facing RuvB hexameric rings assemble on the heteroduplexes, with the DNA passing through their centers. These RuvB hexamers act as motors to promote branch migration by driving the passage of the DNA duplexes through themselves. (right) Binding of RuvC at the Holliday junction and strand scission by its nuclease activity. The locations of the RuvC active sites are indicated by the scissors. **(c)** Charge distribution on the concave surface of an RuvA tetramer. Blue indicates positive charge and red, negative charge. Note the four red (negatively charged) pins at its center. **(d)** Structural model for the interaction of (RuvA)$_4$ with the hypothesized square-planar Holliday junction center (pdb id = 1C7Y). (Adapted from Figures 1, 2, and 3 in Rafferty, J. B., et al., 1996. Crystal structure of DNA recombination protein RuvA and a model for its binding to the Holliday junction. *Science* **274**:415–421.)

RuvA, RuvB, and RuvC Proteins Resolve the Holliday Junction to Form the Recombination Products

The Holliday junction is processed into recombination products by **RuvA** (203 amino acids), **RuvB** (336 amino acids), and **RuvC** (173 amino acids). Specifically, RuvA and RuvB work together as a Holliday junction–specific helicase complex that dissociates the RecA filament and catalyzes branch migration. An RuvA tetramer (Figure 28.22a) fits precisely within the junction point (Figure 28.22b), which has a square-planar geometry, and this RuvA tetramer mediates the assembly of RuvB around opposite arms of the DNA junction. The RuvB protein binds to form two oppositely oriented, hexameric [(RuvB)$_6$] ring structures encircling the dsDNAs, one on each side of the Holliday junction. Rotation of the dsDNAs by the RuvB hexameric rings pulls the dsDNAs through (RuvB)$_6$ and unwinds the DNA strands across the "spool" of RuvA, which threads the separated single strands into newly forming hybrid (recombinant) duplexes (Figure 28.22b). The RuvA tetramer is a disclike structure, one face of which (Figure 28.22c) has four negatively charged central pins, each contributed by a pair of residues (Glu^{55} and Asp^{56}) from each RuvA monomer. These four pins fit neatly into the hole at the center of the Holliday junction. The negatively charged sugar–phosphate backbones of the four DNA duplexes of the Holliday junction are threaded along grooves in the positively charged RuvA face, with the negatively charged central pins appropriately situated to transiently separate the dsDNA molecules into their component single strands through repulsive electrostatic interactions with the phosphate backbones of the DNA. The separated strands of each parental duplex are then channeled into grooves in the RuvA face, where they are led into hydrogen-bonding interactions with bases contributed by strands of the other parental DNA to form the two daughter hybrid duplexes flowing out from the RuvAB complex (Figure 28.22b). Figure 28.22d illustrates a model for the RuvA tetramer with the square-planar Holliday junction.

Depending on how the strands in the Holliday junction are cleaved and resolved, patch or splice recombinant duplexes result (Figure 28.18g and h). RuvC is an endonuclease that resolves Holliday junctions into heteroduplex recombinant products (**RuvC resolvase**). An RuvC dimer binds at the Holliday junction and cuts pairs of DNA

The Three R's of Genomic Manipulation: Replication, Recombination, and Repair

DNA replication, recombination, and repair have traditionally been treated as separate aspects of DNA metabolism. However, scientists have come to realize that DNA replication is an essential component of both DNA recombination and DNA repair processes. Furthermore, recombination mechanisms play an absolutely vital role in restarting replication forks that become halted at breaks or other lesions in the DNA strands. If a double-stranded break (DSB) or a nick in just one of the DNA strands (called a *single-stranded gap, or SSG*) is present in the DNA undergoing replication, the replication fork stalls and the replication complex dissociates *(replication fork "collapse")*. Significantly, the whole process of homologous recombination can initiate only at SSGs or DSBs, and establishment of homologous recombination at such sites can rescue DNA replication. This **recombination-dependent replication (RDR)** has the interesting property of initiating DNA replication at sites other than the *oriC* site, and thus RDR is an important mechanism for restarting DNA replication if the replication fork is disrupted for any reason. As might be expected from the close relationships between replication, recombination, and repair, many of the same proteins are involved in all three, and all three must be viewed as essential processes in the perpetuation of the genome.

strands of similar polarity (Figure 28.22b); whether a patch or a splice recombinant results depends on which DNA pair is cleaved.

RuvB hexamers are AAA$^+$-ATPase-type, DNA-driving molecular motors; similar motors act during DNA replication to propel strand separation and initiate DNA synthesis. Thus, the RuvABC system for processing Holliday junctions may represent a general paradigm for DNA manipulation in all cells.

"Knockout" Mice: A Method to Investigate the Essentiality of a Gene

Homologous recombination can be used to replace a gene with an inactivated equivalent of itself. Inactivation is accomplished by inserting a foreign gene, such as *neo,* a gene encoding resistance to the drug *G418,* within one of the exons of a copy of the gene of interest. Homologous recombination between the *neo*-bearing transgene and DNA in wild-type mouse embryonic stem (ES) cells replaces the target gene with the inactive transgene (see accompanying figure). ES cells in which homologous recombination has occurred will be resistant to G418, and such cells can be selected. These re-combinant ES cells can then be injected into early-stage mouse embryos, where they have a chance of becoming the germline cells of the newborn mouse. If they do, an inactivated target gene is then present in the gametes of this mouse. Mating between male and female mice with inactive target genes yields a generation of homozygous "knockout" mice—mice lacking a functional copy of the targeted gene. Characterization of these knockout mice reveals which physiological functions the gene directs.

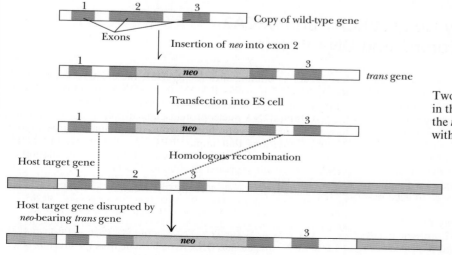

Two exchange events between homologous sequences in the trans *gene* and host target gene, one upstream of the *neo* cassette and one below, lead to insertion of *neo* within the host gene.

Recombination-Dependent Replication Restarts DNA Replication at Stalled Replication Forks

It is likely that most replication forks that begin at the *E. coli oriC* initiation sites (or analogous initiation sites in eukaryotes) are derailed by nicks or more extensive DNA damage lying downstream. However, DNA replication can be reinitiated (and genome replication can be completed) following **replication fork restart.** Recombinational repair of stalled replication forks requires the action of enzymes from every aspect of DNA metabolism: replication, recombination, and repair. The initial steps in restoration of a replication fork depend on the recombination proteins RecA and RecBCD and the formation of a D-loop (Figure 28.21). The *E. coli* protein **PriA** recognizes and binds with high affinity to D-loops. Once bound, PriA coordinates resumption of DNA replication by recruiting DnaB helicase to the D-loop and reestablishing a replication fork complete with two copies of the replicative DNA polymerase. The ability of RecA to mediate recombinational repair of stalled replication forks is undoubtedly the reason for the presence of RecA-type proteins in virtually all organisms. Indeed, the recombination system, a feature common to all cells, evolved to carry out this essential repair purpose. (A diagram illustrating recombination-dependent restart of a stalled replication fork is shown in the section on DNA repair [Section 28.8, Figure 28.26].)

Homologous Recombination in Eukaryotes Helps to Prevent Cancer

As in prokaryotes, eukaryotic homologous recombination eliminates double-stranded breaks and other deleterious lesions from chromosomes. Since the accumulation of lesions is a necessary step in the development of cancer, eukaryotic homologous recombination is crucial to the prevention of cancers in humans. As in prokaryotes, eukaryotic recombination is mediated by recombinases. The eukaryotic counterpart to RecA is **Rad51,** an ATP-dependent, single-stranded DNA-binding protein. Rad51 assembles on ssDNA in a manner similar to RecA, except Rad51 is present in solution as a heptamer, not a hexamer. The ssDNA:Rad51 nucleoprotein complex in eukaryotes is referred to as a **presynaptic filament.** Rad51 assembles on ssDNA, forming a right-handed helical nucleoprotein complex that can extend for thousands of bases. Each helical turn of the filament contains 18 or 19 bases of DNA and 6 Rad51 monomers. Thus, Rad51 holds the DNA in a highly extended conformation. Assembly of the Rad51:DNA filament is dependent on proteins that mediate the Rad51:DNA interaction, including BRCA2, the

---

HUMAN BIOCHEMISTRY

The Breast Cancer Susceptibility Genes BRCA1 and BRCA2 Are Involved in DNA Damage Control and DNA Repair

Mutations in the BRCA1 and BRCA2 genes cause increased likelihood of breast, ovarian, and other cancers. The BRCA1 protein functions in regulation of the cell cycle in response to DNA damage control. Phosphorylation of BRCA1 by DNA damage-response proteins controls the expression, phosphorylation, and cellular localization of specific cyclin-CDKs involved in the cell cycle G_2/M checkpoint and the onset of mitosis. Activation of these cyclin-CDKs leads to arrest of the cell cycle in G_2 until the damage to DNA is repaired. Mutations in BRCA1 that impair its function allow the cell cycle to enter mitosis and DNA damage to accumulate, raising the risk of cancer.

The BRCA2 protein participates in the pathway for DNA repair by homologous recombination. The BRCA2 protein (3418 amino acids) is a very large protein with 8 conserved sequence motifs of about 30 amino acids each, known as the *BRC repeats.* These

repeats act as binding sites for RAD51, the eukaryotic analog of RecA. BRCA2 transfers RAD51 to a ssDNA strand coated with RPA (the eukaryotic counterpart of SSB), allowing formation of the RAD51–ssDNA nucleoprotein filament that is an essential intermediate in eukaryotic homologous recombination (see Figure 28.20b). Mutations that impair BRCA2 function prevent DNA repair by homologous recombination, leading to accumulation of DNA damage and a greater likelihood of cancer.

Source: Yarden, R. I., et al., 2003. BRCA1 regulates the G_2/M checkpoint by activating Chk1 kinase upon DNA damage. *Nature Genetics* **30:**285–289; Simon, N., Powell, S. M., Willers, H., and Xia, F., 2003. BRCA2 keeps Rad51 in line: High-fidelity homologous recombination prevents breast and ovarian cancer? *Molecular Cell* **10:**1262–1263; and Pellegrini, L., et al., 2002. Insights into DNA recombination from the structure of a RAD51-BRCA2 complex. *Nature* **420:**287–293.

protein product of the BRCA2 gene whose mutation may lead to breast cancer. BRCA2 and other mediators, such as Rad52, are needed because **RPA,** the eukaryotic counterpart of the prokaryotic DNA replication protein **SSB,** competes with Rad51 for binding to ssDNA; the mediators displace RPA, allowing Rad51 to bind. Strand invasion and D-loop formation require other proteins, including Hop2-Mnd1. **Rad54** acts in the ATP hydrolysis-driven strand exchange reaction that leads to formation of a Holliday junction. Specific endonucleases then cleave the DNA strands to resolve the Holliday junction. BRCA2 can also mediate the binding of RAD51 to DNA regions that have become single-stranded as a result of DNA damage or DNA replication problems.

Transposons Are DNA Sequences That Can Move from Place to Place in the Genome

In 1950, Barbara McClintock reported the results of her studies on an **activator gene** in maize (*Zea mays,* or as it's usually called, corn) that was recognizable principally by its ability to cause mutations in a second gene. Activator genes were thus an internal source of mutation. A most puzzling property was their ability to move relatively freely about the genome. Scientists had labored to establish that chromosomes consisted of genes arrayed in a fixed order, so most geneticists viewed as incredible this idea of genes moving around. The recognition that McClintock so richly deserved for her explanation of this novel phenomenon had to await verification by molecular biologists. In 1983, Barbara McClintock was finally awarded the Nobel Prize in Physiology or Medicine. By this time, it was appreciated that many organisms, from bacteria to humans, possessed similar "jumping genes" able to move from one site to another in the genome. This mobility led to their designation as **mobile elements, transposable elements,** or, simply, **transposons.**

Transposons are segments of DNA that are moved enzymatically from place to place in the genome (Figure 28.23). That is, their location within the DNA is unstable. Transposons range in size from several hundred base pairs to more than 8 kbp. Transposons contain a gene encoding an enzyme necessary for insertion into a chromosome and for the remobilization of the transposon to different locations. These movements are termed **transposition events.** The smallest transposons are called **insertion sequences,** or **ISs,** signifying their ability to insert apparently at random in the genome. Insertion into a new site can cause a mutation if a gene or regulatory region at the site is disrupted. Because transposition events can move genes to new places or lead to the duplication of existing genes, transposition is a major force in evolution.

28.8 Can DNA Be Repaired?

Biological macromolecules are susceptible to chemical alterations that arise from environmental damage or errors during synthesis. For RNAs, proteins, or other cellular molecules, most consequences of such damage are avoided by replacement of these molecules through normal turnover (synthesis and degradation). However, the integrity of DNA is vital to cell survival and reproduction. Its information content must be protected over the life span of the cell and preserved from generation to generation. Safeguards include (1) high-fidelity replication systems and (2) repair systems that correct DNA damage that might alter its information content. DNA is the only molecule that, if damaged, is repaired by the cell. Usually, accurate repair is possible because the information content of duplex DNA is inherently redundant; the nucleotide sequence in one strand is directly related to the sequence in the other. DNA damage may arise from endogenous processes or from exogenous agents, such as UV light, ionizing radiation, or mutagenic chemicals. The most common forms of endogenous DNA damage arise from chemical reactions (oxidation, alkylation, or deamination of bases) or loss of bases due to cleavage of *N*-glycosidic bonds. Exogenous agents can damage DNA in a variety of ways, including UV-induced free-radical generation and crosslinking of adjacent pyrimidines, breakage of the polynucleotide backbone by ionizing radiation, and base modifications through chemical reactions. Cells have extraordinarily diverse and effective systems to repair these lesions in DNA so that the genetic information is not lost or

FIGURE 28.23 The typical transposon has inverted nucleotide-sequence repeats at its termini, represented here as the 12-bp sequence ACGTACGTACGT **(a)**. It acts at a target sequence (shown here as the sequence CATGC) within host DNA by creating a staggered cut **(b)** whose protruding single-stranded ends are then ligated to the transposon **(c)**. The gaps at the target site are then filled in, and the filled-in strands are ligated **(d)**. Transposon insertion thus generates direct repeats of the target site in the host DNA, and these direct repeats flank the inserted transposon.

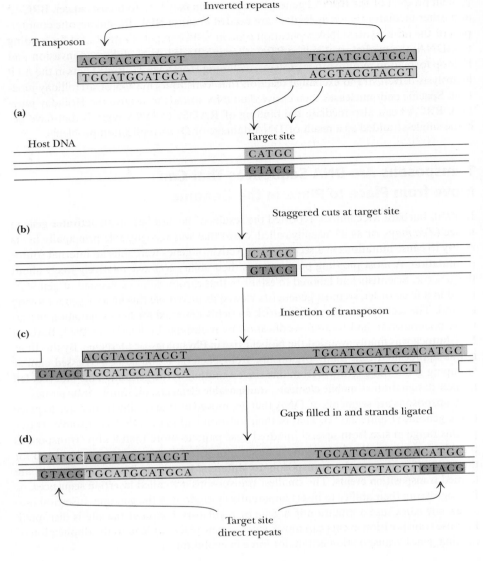

FIGURE 28.24 DSB repair through nonhomologous DNA end joining (NHEJ). Ku70/80 binds the ends and recruits a set of proteins that juxtaposes the broken ends. Processing of the ends to generate proper substrates for DNA ligase IV then occurs, followed by DNA-ligase-mediated end joining. (Adapted from Figure 4 in Helleday, T., Lo, J., van Gent, D. C., and Engelward, B. P., 2007. DNA double-strand break repair: From mechanistic understanding to cancer treatment. *DNA Repair* **6**:923–935.

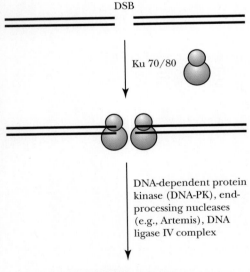

altered. The human genome has 150 or so genes associated with DNA repair. DNA repair systems include direct reversal damage repair, single-strand damage repair, double-stranded break (DSB) repair, and translesion DNA synthesis.

Chemical reactions that reverse the damage, returning the DNA to its proper state, are called **direct reversal** repair systems. Examples include methyltransferases to remove methyl groups from chemically modified bases and photolyase, which repairs thymine dimers, as discussed in a later section (see Figure 28.27). **Single-strand damage repair** relies on the intact complementary strand to guide repair. Systems repairing this sort of DNA damage include **mismatch repair (MMR), base excision repair (BER),** and **nucleotide excision repair (NER).** Such systems will be described further on.

DSBs are a particular threat to genome stability, because lost sequence information cannot be recovered from the same DNA double helix. Because the threat is so severe, cells have several systems to repair DSBs. The simplest way to repair a DSB is to rejoin the broken strands through **nonhomologous DNA end-joining,** or **NHEJ** (Figure 28.24). This repair can occur any time in the cell cycle. A key problem in NHEJ is to keep the ends near one another so that the fragments can be linked together again. A heterodimeric protein, Ku70/80, binds the DNA ends and recruits a set of proteins that juxtaposes the damaged ends, repairs them, and religates them without the use of a repair template. (Some proteins involved in repairing DSBs also function in immunoglobulin gene rearrangements; see Figure 28.38.) The lack of a proper template for NHEJ means that it is error-prone.

DSBs that arise as the cell progresses from S into G_2 during the cell cycle can be repaired through homologous recombination (Figure 28.25). The intact sister chromatid of the damaged DNA duplex guides the process. Processing of the DSB creates single-stranded tails that become substrates for RecA-mediated nucleoprotein filament formation and homology recognition within the sister chromatid. DNA synthesis propagated from the D-loop and migration of the Holliday junction, followed by strand cleavage within the Holliday junction by resolvase, either by cleaving the crossed strands (Figure 28.25(g), left) or the non-crossed strands of the Holliday junctions (Figure 28.25(g), right).

What is effectively a DSB can arise during DNA replication if DNA damage causes the replication fork to stall (Figure 28.26). Suppose a lesion in the leading strand (circle, Figure 28.26b) causes leading-strand synthesis to stall. Continued lagging-strand synthesis results in a single-stranded region on the leading-strand template. Leading-strand

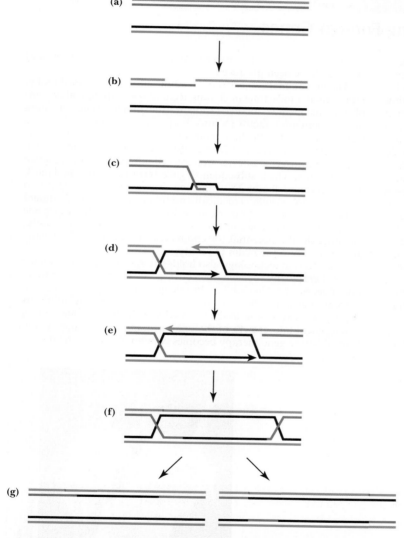

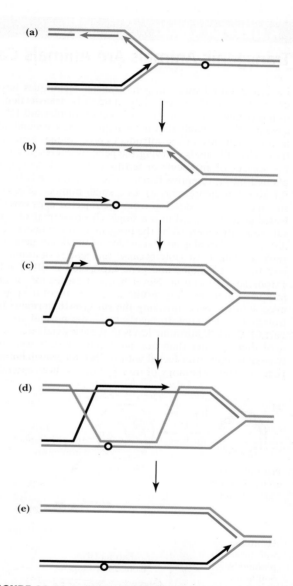

FIGURE 28.25 DSB repair through homologous DNA recombination. The orange–red pair of lines symbolizes the double-stranded DNA with a DSB; the black–blue pair represents the sister chromatid. Homologous recombination creates a D-loop **(c)**, and sister chromatid-directed DNA replication restores the information content of the damaged duplex **(d–f)**. Depending on how the Holliday junctions are resolved, the products **(g)** are either (left) noncrossover or (right) crossover recombinants. (Adapted from Figure 1 in Hiom, K., 2010. Coping with DNA double strand breaks. *DNA Repair* **9**:1256–1263.)

FIGURE 28.26 Restarting a stalled replication fork through homologous DNA recombination. A lesion in the DNA is symbolized by a circle; in this case, the lesion is in the leading-strand template **(a)**. Leading-strand synthesis halts because of the lesion **(b)**. Lagging-strand synthesis (red) continues, and the Okazaki fragments are ligated **(c)**. When the leading strand invades the new DNA duplex formed by lagging-strand synthesis, a D-loop is formed and strand exchange occurs. Using the lagging strand as a template, synthesis of the leading strand (black) resumes **(d)**, and the replication fork is re-established **(e)**. (Adapted from Figure 3 in Wyman, C., and Kanaar, R., 2006. DNA double-strand break repair: All's well that ends well. *Annual Review of Genetics* **40**:363–383.)

invasion of the new lagging-strand duplex (mediated by RecA) creates a D-loop (Figure 28.26c), and leading-strand synthesis directed by the lagging-strand template (Figure 29.26d) reestablishes a competent replication fork (Figure 28.26e). The site of the lesion is repaired later, usually by nucleotide excision repair (see following section). Homologous recombination-mediated restarting of replication forks is believed to be the evolutionary driving force for the emergence of homologous recombination. Even if the NHEJ or homologous recombination repair systems fail to act, the genome may be preserved if an "error-prone" mode of replication allows the lesion to be bypassed. Such **translesion DNA synthesis** is more of a tolerance mechanism than a repair mechanism because it allows replication without necessarily repairing the damage. Specialized

---

> **A DEEPER LOOK**

Transgenic Animals Are Animals Carrying Foreign Genes

Experimental advances in gene transfer techniques have made it possible to introduce genes into animals by **transfection.** Transfection is defined as the uptake or injection of plasmid DNA into recipient cells. Animals that have acquired new genetic information as a consequence of the introduction of foreign genes are termed **transgenic.** Plasmids carrying the gene of interest are injected into the nucleus of an oocyte or fertilized egg, and the egg is then implanted into a receptive female. The technique has been perfected for mice (see figure, part a). In a small number of cases—10% or so—the mice that develop from the injected eggs carry the transfected gene integrated into a single chromosomal site. The gene is subsequently inherited by the progeny of the transfected animal as if it were a normal gene. Expression of the donor gene in the transgenic animals is variable because the gene is randomly integrated into the host genome and gene expression is often influenced by chromosomal location. Nevertheless, transfection of animals has produced some startling results, as in the case of the transfection of mice with the gene encoding the **rat growth hormone (rGH).** The transgenic mice grew to nearly twice the normal size (see figure, part b). Growth hormone levels in these animals were several hundred times greater than normal. Similar results were obtained in transgenic mice transfected with the **human growth hormone (hGH)** gene. The biotechnology of transfection has been extended to farm animals, and transgenic chickens, cows, pigs, rabbits, sheep, and even fish have been produced.

The first animal cloned from an adult cell, a sheep named Dolly, represented a milestone in cloning technology. Subsequent accomplishments include incorporation of the human gene encoding blood coagulation factor IX into sheep. Fetal sheep fibroblast cells were transfected with the human factor IX gene, nuclei from the transfected cells were transferred into sheep oocytes lacking nuclei, and these transgenic oocytes were placed in the uterus of receptive female sheep, which subsequently gave birth to transgenic lambs. The introduced factor IX transgene was specifically designed so that factor IX protein, a medically useful product for the treatment of hemophiliacs, would be expressed in the milk of the transgenic sheep. Similar successes in cows, which produce much more milk, has brought the potential for commercial production of virtually any protein into the realm of reality.

Transfection technology also holds promise as a mechanism for "gene therapy" by replacing defective genes in animals with functional genes (see Chapter 12). Problems concerning delivery, integration and regulation of the transfected gene, including its appropriate expression in the right cells at the proper time during development and growth of the organism, must be brought under control before gene therapy becomes commonplace in humans.

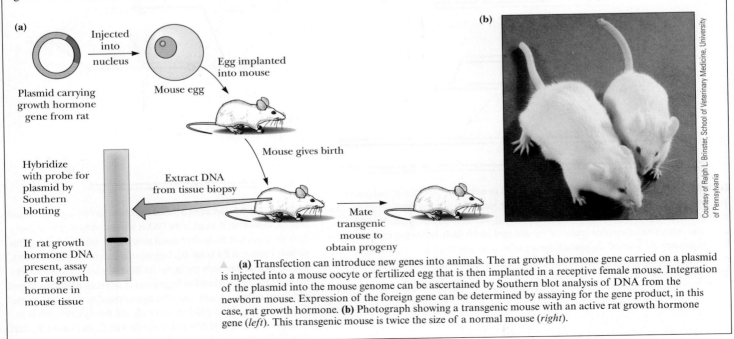

(a) Transfection can introduce new genes into animals. The rat growth hormone gene carried on a plasmid is injected into a mouse oocyte or fertilized egg that is then implanted in a receptive female mouse. Integration of the plasmid into the mouse genome can be ascertained by Southern blot analysis of DNA from the newborn mouse. Expression of the foreign gene can be determined by assaying for the gene product, in this case, rat growth hormone. **(b)** Photograph showing a transgenic mouse with an active rat growth hormone gene (*left*). This transgenic mouse is twice the size of a normal mouse (*right*).

translesion DNA polymerases, such as DNA polymerase IV in *E. coli* and DNA polymerase η (η = eta) in humans, substitute for the replicative DNA polymerase in this process. Although translesion DNA polymerases tend to be error-prone, they have the advantage of allowing DNA replication to continue.

Human DNA replication has an error rate of about three base-pair mistakes during copying of the 6 billion base pairs in the diploid human genome. The low error rate is due to those DNA repair systems that review and edit the newly replicated DNA. Furthermore, about 10^4 bases (mostly purines) are lost per cell per day from spontaneous breakdown in human DNA; repair systems must replace these bases to maintain the fidelity of the encoded information.

Mismatch Repair Corrects Errors Introduced During DNA Replication

The **mismatch repair system** corrects errors introduced when DNA is replicated. It scans newly synthesized DNA for mispaired bases, excises the mismatched region, and then replaces it by DNA polymerase–mediated local replication. The key to such replacement is to know which base of the mismatched pair is correct.

The *E. coli* **methyl-directed pathway** of mismatch repair relies on methylation patterns in the DNA to determine which strand is the newly synthesized one and which one was the parental (template) strand. DNA methylation, often an identifying and characteristic feature of a prokaryote's DNA, occurs just after DNA replication. During methylation, methyl groups are added to certain bases along the new DNA strand. However, a window of opportunity exists between the end of replication and the start of methylation, when only the parental strand of a dsDNA is methylated. This window in time provides an opportunity for the mismatch repair system to review the dsDNA for mismatched bases that arose as a consequence of replication errors. By definition, the newly synthesized strand is the one containing the error, and the methylated strand is the one having the correct nucleotide sequence. When the methyl-directed mismatch repair system encounters a mismatched base pair, it searches along the DNA—through thousands of base pairs if necessary—until it finds a methylated base.

The system identifies the strand bearing the methylated base as parental, assumes its sequence is the correct one, and replaces the entire stretch of nucleotides within the new strand from this recognition point to and including the mismatched base. Mismatch repair does this by using an endonuclease to cut the new, unmethylated strand and an exonuclease to remove the mismatched bases, creating a gap in the newly synthesized strand. DNA polymerase III holoenzyme then fills in the gap, using the methylated strand as template. Finally, DNA ligase reseals the strand.

Damage to DNA by UV Light or Chemical Modification Can Also Be Repaired

Repair of Pyrimidine Dimers Formed by UV Light UV irradiation promotes the formation of covalent bonds between adjacent thymine residues in a DNA strand, creating a cyclobutyl ring (Figure 28.27). Because the C—C bonds in this ring are shorter than the normal 0.34-nm base stacking in B-DNA, the DNA is distorted at this spot and is no longer a proper template for either replication or transcription. **Photolyase** (also called **photoreactivating enzyme**), a flavin- and pterin-dependent enzyme, binds at the dimer and uses the energy of visible light to break the cyclobutyl ring, restoring the pyrimidines to their original form.

Excision Repair Replacement of chemically damaged or modified bases occurs via two fundamental excision repair systems—**base excision** and **nucleotide excision**. *Base excision repair* acts on single bases that have been damaged through oxidation or other chemical modifications during normal cellular processes. The damaged base is removed by **DNA glycosylase,** which cleaves the glycosidic bond, creating an AP site where the

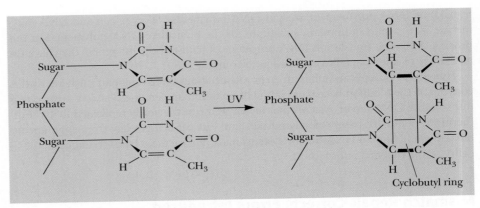

FIGURE 28.27 UV irradiation causes dimerization of adjacent thymine bases. A cyclobutyl ring is formed between carbons 5 and 6 of the pyrimidine rings. Normal base pairing is disrupted by the presence of such dimers.

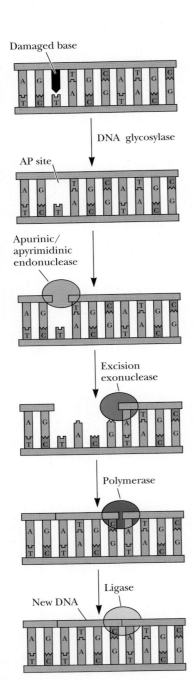

FIGURE 28.28 Base excision repair. A damaged base (■) is excised from the sugar-phosphate backbone by DNA glycosylase, creating an AP site. Then, an apurinic/apyrimidinic endonuclease severs the DNA strand, and an excision nuclease removes the AP site and several nucleotides. DNA polymerase I and DNA ligase then repair the gap.

sugar–phosphate backbone is intact but a purine *(apurinic site)* or a pyrimidine *(apyrimidinic site)* is missing. An **AP endonuclease** then cleaves the backbone, an exonuclease removes the deoxyribose-P and a number of additional residues, and the gap is repaired by DNA polymerase and DNA ligase (Figure 28.28). The information of the complementary strand is used to dictate which bases are added in refilling the gap. In *E. coli,* DNA polymerase I binds at the gap and moves in the 5′→3′ direction, removing nucleotides with its 5′-exonuclease activity. The 5′→3′ DNA polymerase activity of DNA polymerase I fills in the sequence behind the 5′-exonuclease action. No net synthesis of DNA results, but this action of DNA polymerase I "edits out" sections of damaged DNA. Excision is coordinated with 5′→3′ polymerase-catalyzed replacement of the damaged nucleotides so that DNA of the right sequence is restored.

Nucleotide excision repair recognizes and repairs larger regions of damaged DNA than base excision repair. The nucleotide excision repair system cuts the sugar–phosphate backbone of a DNA strand in two places, one on each side of the lesion, and removes the region. The region removed in prokaryotic nucleotide excision repair spans 12 or 13 nucleotides; in eukaryotic excision repair, an oligonucleotide stretch 27 to 29 units long is removed. The resultant gap is then filled in using DNA polymerase (DNA polymerase I in prokaryotes or DNA polymerase δ or ε and PCNA plus RFC in eukaryotes), and the sugar–phosphate backbone is covalently closed by DNA ligase.

In mammalian cells, nucleotide excision repair is the main pathway for removal of carcinogenic (cancer-causing) lesions caused by sunlight or other mutagenic agents. Such lesions are recognized by **XPA** protein, named for *xeroderma pigmentosum,* an inherited human syndrome whose victims suffer serious skin lesions if exposed to sunlight. At sites recognized by XPA, a multiprotein endonuclease is assembled and the damaged strand is cleaved and repaired.

28.9 What Is the Molecular Basis of Mutation?

Genes are normally transmitted unchanged from generation to generation, owing to the great precision and fidelity with which genes are copied during chromosome duplication. However, on rare occasions, genetically heritable changes **(mutations)** occur and result in altered forms. Most mutated genes function less effectively than the unaltered, wild-type allele, but occasionally, mutations arise that give the organism a selective advantage. When this occurs, they may be propagated to many offspring. Together with recombination, mutation provides for genetic variability within species and, ultimately, the evolution of new species.

Mutations change the sequence of bases in DNA, either by the substitution of one base pair for another (so-called **point mutations**) or by the insertion or deletion of one or more base pairs (**insertions** and **deletions**).

Point Mutations Arise by Inappropriate Base-Pairing

Point mutations arise when a base pairs with an inappropriate partner. The two possible kinds of point mutations are **transitions,** in which one purine (or pyrimidine) is replaced by another, as in A⟶G (or T⟶C), and **transversions,** in which a purine is substituted for a pyrimidine, or vice versa.

Point mutations arise by the pairing of bases with inappropriate partners during DNA replication, by the introduction of base analogs into DNA, or by chemical mutagens. Bases may rarely mispair (Figure 28.29), either because of their tautomeric properties (see Chapter 10) or because of other influences. Even in mispairing, the $C_1'–C_1'$ distances between bases must still be close to that of a Watson–Crick base pair (11 nm or so; see Figure 11.6) to maintain the mismatched base pair in the double helix. In tautomerization, for example, an amino group ($-NH_2$), usually an H-bond donor, can tautomerize to an imino form ($=NH$) and become an H-bond acceptor. Or a keto group (C=O), normally an H-bond acceptor, can tautomerize to an enol C—OH, an H-bond donor. Proofreading mechanisms operating during DNA replication catch most mispairings. The frequency of spontaneous mutation in prokaryotes and eukaryotes (including humans) is about 10^{-8} per base pair per generation.

Mutations Can Be Induced by Base Analogs

Base analogs that become incorporated into DNA can induce mutations through changes in base-pairing possibilities. Two examples are **5-bromouracil (5-BU)** and **2-aminopurine (2-AP).** 5-Bromouracil is a thymine analog and becomes inserted into DNA at sites normally occupied by T; its 5-Br group sterically resembles thymine's 5-methyl group. However, because 5-BU frequently assumes the enol tautomeric form

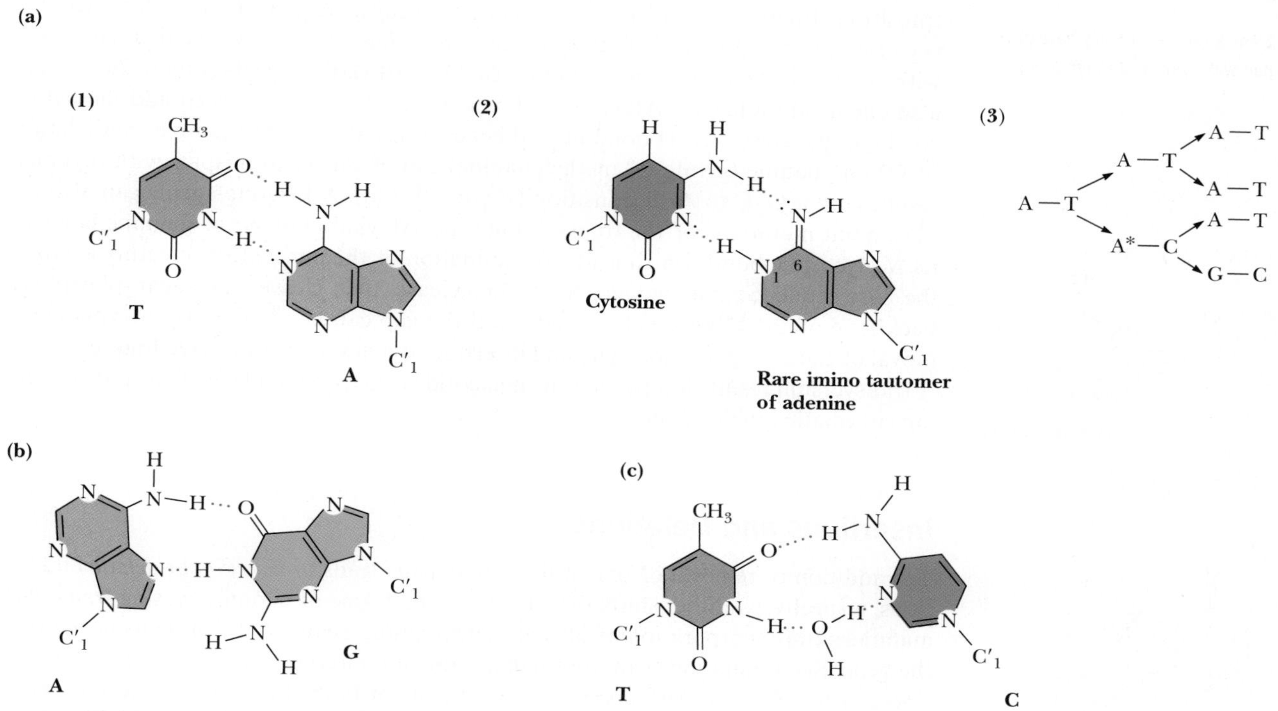

(a)

(b)

(c)

FIGURE 28.29 Point mutations due to base mispairings. **(a)** An example based on tautomeric properties. The rare imino tautomer of adenine base pairs with cytosine rather than thymine. **(1)** The normal A-T base pair. **(2)** The A*–C base pair is possible for the adenine tautomer in which a proton has been transferred from the 6-NH₂ of adenine to N-1. **(3)** Pairing of C with the imino tautomer of A (A*) leads to a transition mutation (A–T to G–C) appearing in the next generation. **(b)** A in the syn conformation pairing with G (G is in the usual anti conformation). **(c)** T and C form a base pair by H-bonding interactions mediated by a water molecule.

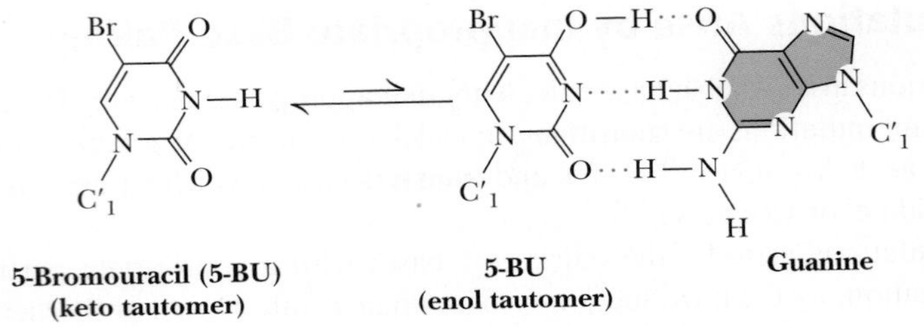

FIGURE 28.30 5-Bromouracil usually favors the keto tautomer that mimics the base-pairing properties of thymine, but it frequently shifts to the enol form, whereupon it can base-pair with guanine, causing a T–A to C–G transition.

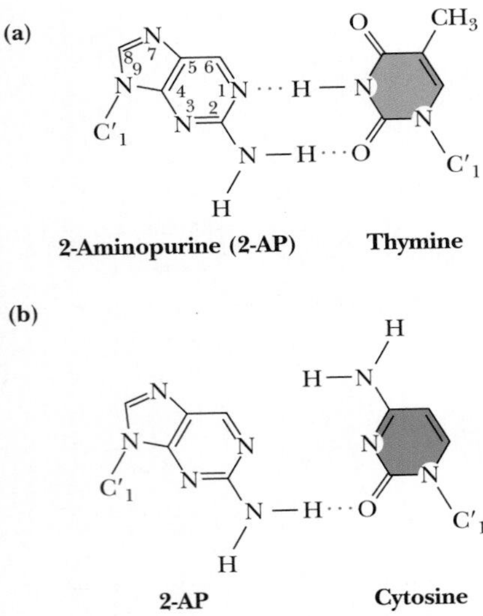

FIGURE 28.31 (a) 2-Aminopurine normally base-pairs with T but (b) may also pair with cytosine through a single hydrogen bond.

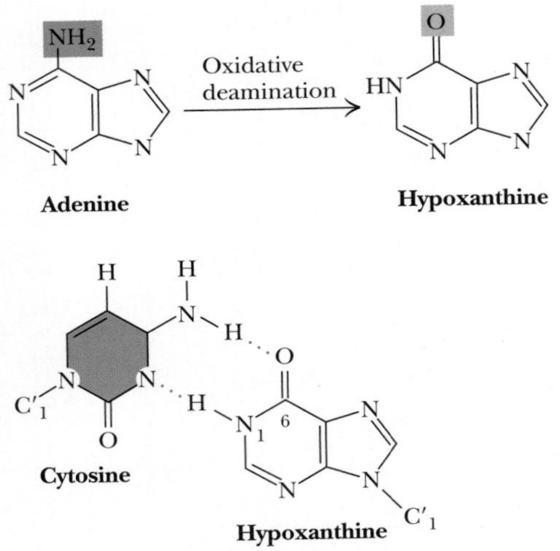

FIGURE 28.32 Oxidative deamination of adenine in DNA yields hypoxanthine, which base-pairs with cytosine, resulting in an A–T to G–C transition.

and pairs with G instead of A, a point mutation of the transition type may be induced (Figure 28.30). Less often, 5-BU is inserted into DNA at cytosine sites, not T sites. Then, if it base-pairs in its keto form, mimicking T, a C–G to T–A transition ensues. The adenine analog, 2-aminopurine (recall that adenine is 6-aminopurine) normally behaves like A and base-pairs with T. However, 2-AP can form a single H bond of sufficient stability with cytosine (Figure 28.31) that occasionally C replaces T in DNA replicating in the presence of 2-AP. Hypoxanthine (Figure 28.32) is an adenine analog that arises in situ in DNA through oxidative deamination of A. Hypoxanthine base-pairs with cytosine, creating an A–T to G–C transition.

Chemical Mutagens React with the Bases in DNA

Chemical mutagens are agents that chemically modify bases so that their base-pairing characteristics are altered. For instance, *nitrous acid* (HNO_2) causes the oxidative deamination of primary amine groups in adenine and cytosine. Oxidative deamination of cytosine yields uracil, which base-pairs the way T does and gives a C–G to T–A transition (Figure 28.33a). *Hydroxylamine* specifically causes C–G to T–A transitions because it reacts specifically with cytosine, converting it to a derivative that base-pairs with adenine instead of guanine (Figure 28.33c). **Alkylating agents** (Figure 28.33e) are also chemical mutagens. Alkylation of reactive sites on the bases to add methyl or ethyl groups alters their H bonding and hence base pairing. For example, methylation of O^6 on guanine (giving O^6-methylguanine) causes this G to mispair with thymine, resulting in a G–C to A–T transition (Figure 28.33d). Alkylating agents can also induce point mutations of the transversion type. Alkylation of N^7 of guanine labilizes its *N*-glycosidic bond, which leads to elimination of the purine ring, creating a gap in the base sequence. An enzyme, AP endonuclease, then cleaves the sugar–phosphate backbone of the DNA on the 5′-side, and the gap can be repaired by enzymatic removal of the 5′-sugar–phosphate and insertion of a new nucleotide (see Figure 28.28). A transversion results if a pyrimidine nucleotide is inserted in place of the purine during enzymatic repair of this gap.

Insertions and Deletions

The addition or removal of one or more base pairs leads to *insertion* or *deletion* mutations, respectively. Either shifts the triplet reading frame of codons, causing **frameshift mutations** (misincorporation of all subsequent amino acids) in the protein encoded by the gene. Such mutations can arise if flat, aromatic molecules such as *acridine orange* insert themselves between successive bases in one or both strands of the double helix. This insertion or, more aptly, **intercalation,** doubles the distance between the bases as measured along the helix axis (see Figure 11.12). This distortion of the DNA results in inappropriate insertion or deletion of bases when the DNA is replicated. Disruptions that arise from the insertion of a transposon within a gene also fall in this category of mutation (see Figure 28.23).

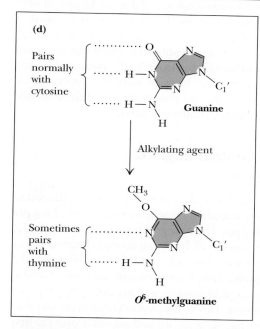

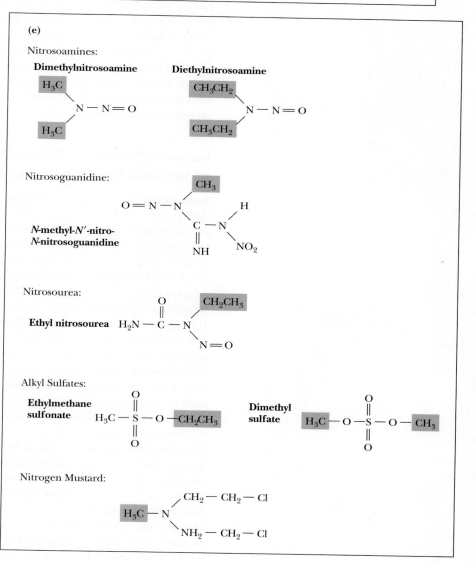

FIGURE 28.33 Chemical mutagens. **(a)** HNO₂ (nitrous acid) converts cytosine to uracil and adenine to hypoxanthine. **(b)** Nitrosoamines, organic compounds that react to form nitrous acid, also lead to the oxidative deamination of A and C. **(c)** Hydroxylamine (NH₂OH) reacts with cytosine, converting it to a derivative that base-pairs with adenine instead of guanine. The result is a C–G to T–A transition. **(d)** Alkylation of G residues to give O^6-methylguanine, which base-pairs with T. **(e)** Alkylating agents include nitrosoamines, nitrosoguanidines, nitrosoureas, alkyl sulfates, and nitrogen mustards. Note that nitrosoamines are mutagenic in two ways: They can react to yield HNO₂, or they can act as alkylating agents. The nitrosoguanidine, N-methyl-N′-nitro-N-nitrosoguanidine, is a very potent mutagen used in laboratories to induce mutations in experimental organisms such as *Drosophila melanogaster*. Ethylmethane sulfonate (EMS) and dimethyl sulfate are also favorite mutagens among geneticists.

SPECIAL FOCUS

Gene Rearrangements and Immunology—Is It Possible to Generate Protein Diversity Using Genetic Recombination?

Animals have evolved a way to exploit genetic recombination in order to generate protein diversity. This development was crucial to the evolution of the immune system. For example, the immunoglobulin genes are a highly evolved system for maximizing protein diversity from a finite amount of genetic information. This diversity is essential for gaining immunity to the great variety of infectious organisms and foreign substances that cause disease.

Cells Active in the Immune Response Are Capable of Gene Rearrangement

Only vertebrates show an immune response. If a foreign substance, called an **antigen,** gains entry to the bloodstream of a vertebrate, the animal responds via a protective system called the *immune response.* The immune response involves production of proteins capable of recognizing and destroying the antigen. This response is mounted by certain white blood cells—the **B-** and **T-cell lymphocytes** and the **macrophages.** B cells are so named because they mature in the bone marrow; T cells mature in the thymus gland. Each of these cell types is capable of gene rearrangement as a mechanism for producing proteins essential to the immune response. **Antibodies,** which can recognize and bind antigens, are immunoglobulin proteins secreted from B cells. Because antigens can be almost anything, the immune response must have an incredible repertoire of structural recognition. Thus, vertebrates must have the potential to produce immunoglobulins of great diversity in order to recognize virtually any antigen.

Immunoglobulin G Molecules Contain Regions of Variable Amino Acid Sequence

Immunoglobulin G (IgG or **γ-globulin)** is the major class of antibody molecules found circulating in the bloodstream. IgG is a very abundant protein, amounting to 12 mg per mL of serum. It is a 150-kD $\alpha_2\beta_2$-type tetramer. The α or H (for *heavy*) chain is 50 kD; the β or L (for *light*) chain is 25 kD. A preparation of IgG from serum is heterogeneous in terms of the amino acid sequences represented in its L and H chains. However, the IgG L and H chains produced from any given B lymphocyte are homogeneous in amino acid sequence. L chains consist of 214 amino acid residues and are organized into two roughly equal segments: the V_L and C_L regions. The V_L designation reflects the fact that L chains isolated from serum IgG show variations in amino acid sequence over the first 108 residues, V_L symbolizing this "variable" region of the L polypeptide. The amino acid sequence for residues 109 to 214 of the L polypeptide is constant, as represented by its designation as the "constant light," or C_L, region. The heavy, or H, chains consist of 446 amino acid residues. Like L chains, the amino acid sequence for the first 108 residues of H polypeptides is variable, ergo its designation as the V_H region, while residues 109 to 446 are constant in amino acid sequence. This "constant heavy" region consists of three quite equivalent domains of homology designated C_H1, C_H2, and C_H3. Each L chain has two intrachain disulfide bonds: one in the V_L region and the other in the C_L region. The C-terminal amino acid in L chains is cysteine, and it forms an interchain disulfide bond to a neighboring H chain. Each H chain has four intrachain disulfide bonds, one in each of the four regions. Figure 28.34 presents a diagram of IgG organization. Within the variable regions of the L and H chains, certain positions are **hypervariable** with regard to amino acid composition. These hypervariable residues occur at positions 24 to 34, 50 to 55, and 89 to 96 in the L chains and at positions 31 to 35, 50 to 65, 81 to 85, and 91 to 102 in the H chains. The hypervariable regions are also called **complementarity-determining regions,** or **CDRs,** because it is these regions that form the

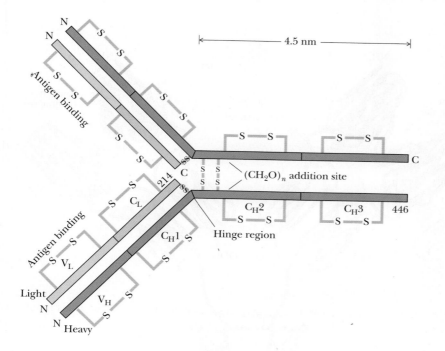

FIGURE 28.34 Diagram of the organization of the IgG molecule. Two identical L chains are joined with two identical H chains. Each L chain is held to an H chain via an interchain disulfide bond. The variable regions (purple) of the four polypeptides lie at the ends of the arms of the Y-shaped molecule. These regions are responsible for the antigen recognition function of the antibody molecules. The actual antigen-binding site is constituted from hypervariable residues within the V_L and V_H regions. For purposes of illustration, some features are shown on only one or the other L chain or H chain, but all features are common to both chains.

structural site that is complementary to some part of an antigen's structure, providing the basis for antibody:antigen recognition.

In the immunoglobulin genes, the arrangement of exons correlates with protein structure. In terms of its tertiary structure, the IgG molecule is composed of 12 discrete *collapsed β-barrel domains*. Within each domain, alternating β-strands are antiparallel to one another, a pattern known by the name *Greek key motif*. The characteristic structure of this domain is referred to as the **immunoglobulin fold** (Figure 28.35). Each of IgG's two heavy chains contributes four of these domains and each of its light chains contributes two. The four *variable-region* domains (one on each chain) are encoded by multiple exons, but the eight constant-region domains are each the product of a single exon. All of these *constant-region* exons are derived from a single ancestral exon encoding an immunoglobulin fold. The major variable-region exon probably derives from this ancestral exon also. Contemporary immunoglobulin genes are a consequence of multiple duplications of the ancestral exon.

The discovery of variability in amino acid sequence in otherwise identical polypeptide chains was surprising and almost heretical to protein chemists. For geneticists, it presented a genuine enigma. They noted that mammals, which can make millions of different antibodies, don't have millions of different antibody genes. How can the mammalian genome encode the diversity seen in L and H chains?

The Immunoglobulin Genes Undergo Gene Rearrangement

The answer to the enigma of immunoglobulin sequence diversity is found in the organization of the immunoglobulin genes. The genetic information for an immunoglobulin polypeptide chain is scattered among multiple gene segments along a chromosome in germline cells (sperm and eggs). During vertebrate development and the formation of B lymphocytes, these segments are brought together and assembled by **DNA rearrangement** (that is, genetic recombination) into complete genes. DNA rearrangement, or **gene reorganization**, provides a mechanism for generating a variety of protein isoforms from a limited number of genes. DNA rearrangement occurs in only a few genes, namely, those encoding the antigen-binding proteins of the immune response—the immunoglobulins and the T-cell receptors. The gene segments encoding the amino-terminal portion of the immunoglobulin polypeptides are also unusually susceptible to mutation events.

(a)

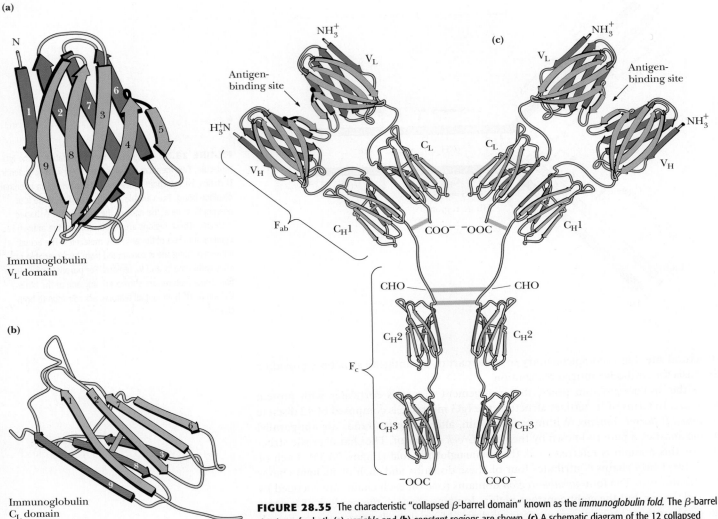

Immunoglobulin
V_L domain

(b)

Immunoglobulin
C_L domain
($\approx C_H$ domains)

FIGURE 28.35 The characteristic "collapsed β-barrel domain" known as the *immunoglobulin fold*. The β-barrel structures for both **(a)** *variable* and **(b)** *constant regions* are shown. **(c)** A schematic diagram of the 12 collapsed β-barrel domains that make up an IgG molecule. CHO indicates the carbohydrate addition site; F_{ab} denotes one of the two antigen-binding fragments of IgG, and F_c, the proteolytic fragment consisting of the pairs of C_H2 and C_H3 domains.

The result is a population of B cells whose antibody-encoding genes collectively show great sequence diversity even though a given cell can make only a limited set of immunoglobulin chains. Hence, at least one cell among the B-cell population likely will be capable of producing an antibody that will specifically recognize a particular antigen.

DNA Rearrangements Assemble an L-Chain Gene by Combining Three Separate Genes

The organization of various immunoglobulin gene segments in the human genome is shown in Figure 28.36. L-chain variable-region genes are assembled from two kinds of **germline genes: V_L** and **J_L** (*J* stands for *joining*). In mammals, there are two different families of **L-chain genes:** the **κ,** or **kappa, gene family** and the **λ,** or **lambda, gene family;** each family has V and J members. These families are on different chromosomes. Humans have 40 functional V genes and 5 functional J genes for the κ light chains and 31 V genes and 4 J genes for the λ light chain. The V and J genes lie upstream from the single **C_κ gene** that encodes the L-chain constant region. Each V_κ gene has its own L_κ segment for encoding

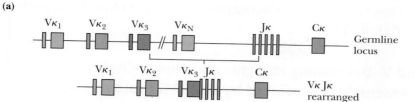

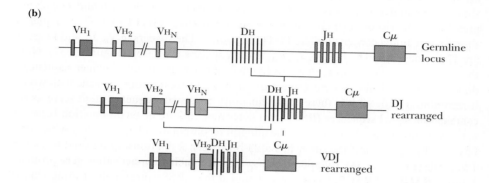

FIGURE 28.36 Organization of human immunoglobulin gene segments. Green, orange, blue, or purple colors indicate the exons of a particular V_L or V_H gene. **(a)** L-chain gene assembly: During B-lymphocyte maturation in the bone marrow, one of the 40 V genes combines with one of the 5 J genes and is joined with a C gene. During the recombination process, the intervening DNA between the gene segments is deleted (see Figure 28.38). These rearrangements occur by a mostly random process, giving rise to many possible light-chain sequences from each gene family. **(b)** H-chain gene assembly: H chains are encoded by V, D, J, and C genes. In H-chain gene rearrangements, a D gene joins with a J gene and then one of the V genes adds to the DJ assembly. (Adapted from Figure 2b and c in Nossal, G. J. V., 2003. The double helix and immunology. *Nature* **421**:440–444.)

the L-chain leader peptide that targets the L chain to the endoplasmic reticulum for IgG assembly and secretion. (This leader peptide is cleaved once the L chain reaches the ER lumen.) The λ family of L-chain genes is organized similarly (Figure 28.36). In different mature B-lymphocyte cells, V_κ and J_κ genes have joined in different combinations, and along with the C–V_κ gene, form complete L–V_κ chains with a variety of V_κ regions. However, any given B lymphocyte expresses only one V_κ–J_κ combination. Construction of the mature B-lymphocyte L-chain gene has occurred by DNA rearrangements that combine three genes (L–$V_{\kappa,\lambda}$, $J_{\kappa,\lambda}$, $C_{\kappa,\lambda}$) to make one polypeptide!

DNA Rearrangements Assemble an H-Chain Gene by Combining Four Separate Genes

The first 98 amino acids of the 108-residue, H-chain variable region are encoded by a **V_H gene.** Each V_H gene has an accompanying L_H gene that encodes its essential leader peptide. It is estimated that there are 200 to 1000 V_H genes and that they can be subdivided into eight distinct families based on nucleotide sequence homology. The members of a particular V_H family are grouped together on the chromosome, separated from one another by 10 to 20 bp. In assembling a mature H-chain gene, a V_H gene is joined to a **D gene** (*D* for *diversity*), which encodes amino acids 99 to 113 of the H chain. These amino acids comprise the core of the third CDR in the variable region of H chains. The V_H–D gene assemblage is linked in turn to a **J_H gene,** which encodes the remaining part of the variable region of the H chain. The V_H, D, and J_H genes are grouped in three separate clusters on the same chromosome. The four J_H genes lie 7 kb upstream of the eight C genes, the closest of which is C_μ. Any of four **C genes** may encode the constant region of IgG H chains: $C_{\gamma 1}$, $C_{\gamma 2a}$, $C_{\gamma 2b}$, and $C_{\gamma 3}$. Each C gene is composed of multiple exons (only C_μ is shown in Figure 28.36, none of the other C genes). Ten to 20 D genes are found 1 to 80 kb farther upstream. The V_H genes lie even farther upstream. In B lymphocytes, the variable region of an H-chain gene is composed of one each of the L_H–V_H genes, a D gene, and a J_H gene joined head to tail. Because the H-chain variable region is encoded in three genes and the joinings can occur in various combinations, the H chains have a greater potential for diversity than the L-chain variable regions that are assembled from just two genes (for example, L_κ–V_κ and J_κ). In making H-chain genes, four genes have been brought together and reorganized by DNA rearrangement to produce a single polypeptide!

V–J and V–D–J Joining in Light- and Heavy-Chain Gene Assembly Is Mediated by the RAG Proteins

Specific nucleotide sequences adjacent to the various variable-region genes suggest a mechanism in which these sequences act as joining signals. All germline *V* and *D* genes are followed by a trio of segments consisting of a consensus CACAGTG heptamer, a short spacer, and a consensus ACAAAAACA nonamer. Similarly, all *D* and *J* genes are preceded by a trio of segments, in this case, a consensus GGTTTTTGT nonamer, a short spacer, and a consensus CACTGTG heptamer. The spacers in V_κ, J_λ, and D genes are 12 bp long, whereas the V_λ, J_κ, V_H, and J_H spacers are 23 bp (Figure 28.37). Note that the conserved consensus elements (namely, the heptamer and nonamer sequences) downstream of a gene are complementary to those upstream from the gene with which it recombines. Indeed, it is these complementary consensus sequences that serve as **recombination signal sequences (RSSs)** and determine the site of recombination between variable-region genes. Functionally meaningful recombination happens only where one has a 12-bp spacer and the other has a 23-bp spacer, a relationship referred to as the 12/23 rule (Figure 28.38). Lymphoid cell-specific **recombination-activating gene** proteins 1 and 2 (**RAG1** and **RAG2**) recognize and bind at these RSSs, presumably through looping out of the 12- and 23-bp spacers and alignment of the homologous heptamer and nonamer regions (Figure 28.39). RAG1 and RAG2 together function as the **V(D)J recombinase**. RAG1/RAG2 action cleaves and processes the ends of the V and J DNA, producing what is effectively a double-stranded break (DSB). Proteins involved in NHEJ-type repair of DSBs (Ku70/80, DNA-PK, and DNA ligase) bind at the DSB and religate the DNA to create a recombinant immunoglobulin gene (Figure 28.38).

Imprecise Joining of Immunoglobulin Genes Creates New Coding Arrangements

Joining of the ends of the immunoglobulin-coding regions during gene reorganization is somewhat imprecise. This imprecision actually leads to even greater antibody diversity because new coding arrangements result. Position 96 in κ chains is typically encoded by the first triplet in the J_κ element. Most κ chains have one of four amino acids here, depending on which J_κ gene was recruited in gene assembly. However, occasionally only the second and third bases or just the third base of the codon for position 96 is contributed by the J_κ gene, with the other one or two nucleotides supplied by the V_κ

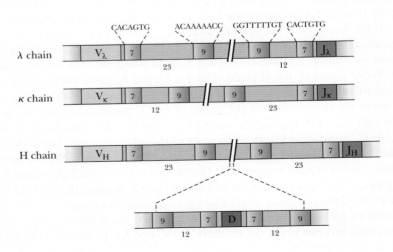

FIGURE 28.37 Consensus elements are located above and below germline variable-region genes that recombine to form genes encoding immunoglobulin chains. These consensus elements are complementary and are arranged in a heptamer-nonamer, 12- to 23-bp spacer pattern. (Adapted from Tonewaga, S., 1983. Somatic generation of antibody diversity. *Nature* **302**:575.)

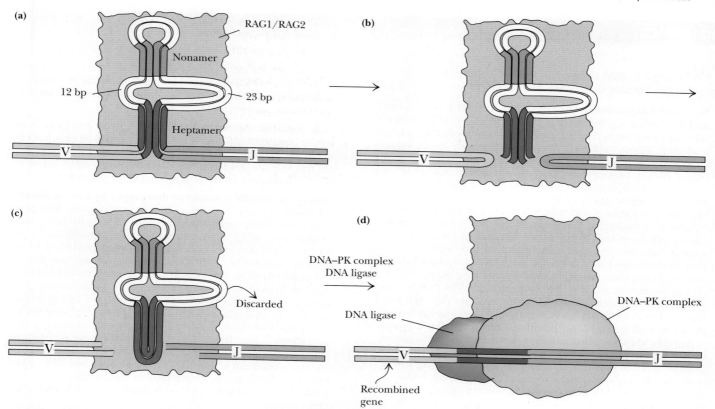

FIGURE 28.38 Model for V(D)J recombination. A RAG1 : RAG2 complex is assembled on DNA in the region of re-combination signal sequences **(a)**, and this complex introduces double-stranded breaks in the DNA at the borders of protein-coding sequences and the recombination signal sequences **(b)**. The products of RAG1 : RAG2 DNA cleavage are novel: The DNA bearing the recombination signal sequences has blunt ends, whereas the coding DNA has hairpin ends. That is, the two strands of the V and J coding DNA segments are covalently joined as a result of transesterifica-tion reactions catalyzed by RAG1 : RAG2. To complete the recombination process, the two RSS ends are precisely joined to make a covalently closed circular dsDNA, but the V and J coding ends undergo further processing **(c)**. Coding-end processing involves opening of the V and J hairpins and the addition or removal of nucleotides. This pro-cessing means that joining of the V and J coding ends is imprecise, providing an additional means for introducing antibody diversity. Finally, the V and J coding segments are then joined to create a recombinant immunoglobulin-encoding gene **(d)**. The processing and joining reactions require RAG1 : RAG2, DNA-dependent protein kinase (DNA-PK)–Ku70, Ku80, and DNA ligase. (Adapted from Figure 1 in Weaver, D. T., and Alt, F. W., 1997. From RAGs to stitches. *Nature* **388**:428–429.)

segment (Figure 28.39). So, the precise point where recombination occurs during gene reorganization can vary over several nucleotides, creating even more diversity.

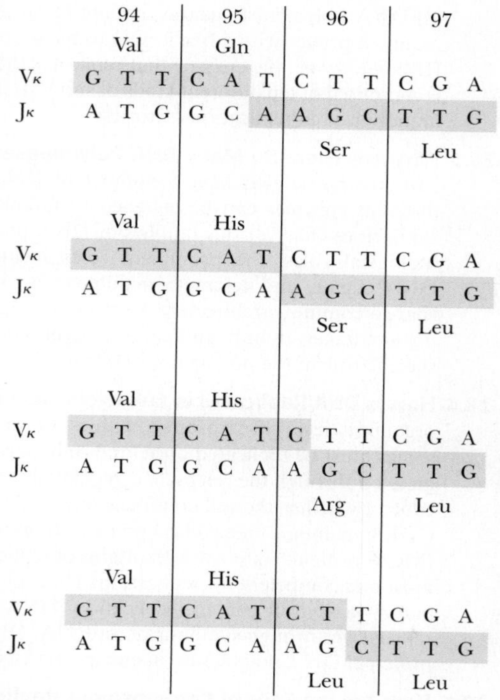

Antibody Diversity Is Due to Immunoglobulin Gene Rearrangements

Taking as an example the mouse with perhaps 300 V_κ genes, 4 J_κ genes, 200 V_H genes, 12 D genes, and 4 J_H genes, the number of possible combinations is given by $300 \times 4 \times 200 \times 12 \times 4$. Thus, more than 10^7 different antibody molecules can be cre-ated from roughly 500 or so different mouse variable-region genes. Including the possi-bility for V_κ–J_κ joinings occurring within codons adds to this diversity, as does the high rate of somatic mutation associated with the variable-region genes. (Somatic mutations are mutations that arise in diploid cells and are transmitted to the progeny of these cells within the organism, but not to the offspring of the organism.) Clearly, gene rearrange-ment is a powerful mechanism for dramatically enhancing the protein-coding potential of genetic information.

FIGURE 28.39 Recombination between the V_κ and J_κ genes can vary by several nucleotides, giving rise to varia-tions in amino acid sequence, and hence, diversity in immu-noglobulin L chains.

SUMMARY

⊘WL Login to OWL to develop problem-solving skills and complete online homework assigned by your professor.

28.1 How Is DNA Replicated? DNA replication is accomplished through strand separation and the copying of each strand. Strand separation is achieved by untwisting the double helix. Each separated strand acts as a template for the synthesis of a new complementary strand whose nucleotide sequence is fixed by Watson–Crick base-pairing rules. Base pairing then dictates an accurate replication of the original DNA double helix. DNA replication follows a semiconservative mechanism where each original strand is copied to yield a complete complementary strand and these paired strands, one old and one new, remain together as a duplex DNA molecule. Replication begins at specific regions called origins of replication and proceeds in both directions. Bidirectional replication involves two replication forks, which move in opposite directions. Helicases unwind the double helix, and DNA gyrases act to overcome torsional stress by introducing negative supercoils at the expense of ATP hydrolysis. Because DNA polymerases synthesize DNA only in a 5'→3' direction, replication is semidiscontinuous: The 3'→5' strand can be copied continuously by DNA polymerase proceeding in the 5'→3' direction. The other parental strand is copied only when a sufficient stretch of its sequence has been exposed for DNA polymerase to move along it in the 5'→3' mode. Thus, one parental strand is copied continuously to form the leading strand, while the other parental strand is copied in an intermittent, or discontinuous, mode to yield a set of Okazaki fragments that are joined later to give the lagging strand.

28.2 What Are the Functions of DNA Polymerases? All DNA polymerases share the following properties: (1) The incoming base is selected within the DNA polymerase active site through base-pairing with the corresponding base in the template strand, (2) chain growth is in the 5'→3' direction antiparallel to the template strand, and (3) DNA polymerases cannot initiate DNA synthesis de novo—all require a primer with a free 3'-OH to build upon. DNA polymerase III holoenzyme, the enzyme that replicates the E. coli chromosome, is composed of ten different kinds of subunits. DNA polymerases are immobilized in replication factories.

28.3 Why Are There So Many DNA Polymerases? Both prokaryotic and eukaryotic cells have a number of DNA polymerases. These different enzymes can be assigned to families based on sequence similarities. The various families of DNA polymerases fill different biological roles; the prominent roles include DNA replication, DNA repair, and telomere maintenance. All DNA polymerases share a common architecture resembling a right hand, composed of distinct finger, thumb, and palm structural domains, each serving a specific role in the polymerase reaction.

28.4 How Is DNA Replicated in Eukaryotic Cells? Eukaryotic DNA is organized into chromosomes within the nucleus. These chromosomes must be replicated once (and only once!) each cell cycle. Progression through the cell cycle is regulated through checkpoints that control whether the cell continues into the next phase. Cyclins and CDKs maintain these checkpoints. Replication licensing factors (MCM proteins) interact with origins of replication and render chromosomes competent for replication. Three DNA polymerases—α, δ, and ε—carry out genome replication. DNA polymerase α initiates replication through synthesis of an RNA. DNA polymerase δ is the principal DNA polymerase in eukaryotic DNA replication.

28.5 How Are the Ends of Chromosomes Replicated? Telomeres are short, tandemly repeated, G-rich nucleotide sequences that form protective caps on the chromosome ends. DNA polymerases cannot replicate the extreme 5'-ends of chromosomes, but a special polymerase called telomerase maintains telomere length. Telomerase is a ribonucleoprotein, and its RNA component serves as template for telomere synthesis.

28.6 How Are RNA Genomes Replicated? Many viruses have genomes composed of RNA, not DNA. DNA may be an intermediate in the replication of such viruses; that is, viral RNA serves as the template for DNA synthesis. This reaction is catalyzed by reverse transcriptase, an RNA-dependent DNA polymerase.

28.7 How Is the Genetic Information Rearranged by Genetic Recombination? Genetic recombination is the exchange (or incorporation) of one DNA sequence with (or into) another. Recombination between very similar DNA sequences is called homologous recombination. Homologous recombination proceeds according to the Holliday model. The RecBCD enzyme complex unwinds dsDNA and cleaves its single strands. RecA protein acts in recombination to catalyze the ATP-dependent DNA strand exchange reaction. Procession of strand separation and re-pairing into hybrid strands along the DNA duplex initiates strand invasion, displacing the homologous DNA strand from the DNA duplex and replacing it with the ssDNA strand. RuvA, RuvB, and RuvC resolve the Holliday junction to form the recombination products. DNA replication is an essential component of both DNA recombination and DNA repair processes. Furthermore, recombination mechanisms can restart replication forks that have halted at breaks or other lesions in the DNA strands.

Transposons are mobile DNA segments ranging in size from several hundred base pairs to more than 8 kbp that move enzymatically from place to place in the genome.

28.8 Can DNA Be Repaired? Repair systems correct damage to DNA in order to maintain its information content. The most common forms of damage are (1) replication errors, (2) deletions or insertions, (3) UV-induced alterations, (4) DNA strand breaks, and (5) covalent crosslinking of strands. Cells have extraordinarily diverse and effective DNA repair systems to deal with these problems, some of which are also involved in DNA replication and recombination. When repair systems fail, the genome may still be preserved if an "error-prone" mode of replication allows the lesion to be bypassed.

28.9 What Is the Molecular Basis of Mutation? Mutations change the sequence of bases in DNA, either by the substitution of one base pair for another (point mutations) or by the insertion or deletion of one or more base pairs. Point mutations arise by the pairing of bases with inappropriate partners during DNA replication, by the introduction of base analogs into DNA, or by chemical mutagens. Chemical mutagens are agents that chemically modify bases so that their base-pairing characteristics are altered.

Special Focus: Gene Rearrangements and Immunology—Is It Possible to Generate Protein Diversity Using Genetic Recombination? Animals have evolved a way to exploit genetic recombination in order to generate protein diversity. The immunoglobulin genes are a highly evolved system for maximizing protein diversity from a finite amount of genetic information. Cells active in the immune response are capable of gene rearrangements. The antibody diversity found in IgG molecules is a prime example of proteins produced via gene rearrangements. IgG L-chain genes are created by combining three separate genes, and H-chain genes by combining four. V–J and V–D–J joining in L- and H-chain gene assembly is mediated by RAG proteins.

- DNA is the material of heredity.

- The three Rs of DNA metabolism—replication, recombination, and repair—account for the transmission, rearrangement, and stability of genetic information.

- DNA replication is semiconservative: The DNA strands are separated and then the nucleotide sequence of each strand is copied by base-pairing.

- DNA replication is bidirectional, accomplished by two replication forks moving in opposite directions.

- DNA gyrases and helicases unwind the DNA double helix so the individual strands can be copied.

- Single-stranded DNA-binding protein (SSB) prevents DNA reannealing and protects the single strands from nuclease attack.

- DNA replication is semidiscontinuous: One parental strand, the $3' \rightarrow 5'$ strand, is copied in a continuous manner to form the leading strand. The other, the $5' \rightarrow 3'$ strand, is copied in an intermittent fashion to form the lagging strand.

- The products of lagging-strand synthesis are several kilobases in length and are called Okazaki fragments.

- DNA polymerases select the proper base for replication of a strand through Watson-Crick geometric interactions.

- DNA synthesis by DNA polymerase always proceeds in the $5' \rightarrow 3'$ direction, copying the template strand in the $3' \rightarrow 5'$ direction.

- DNA polymerases require a primer strand with a free 3'-OH to build upon.

- Prokaryotes and eukaryotes have multiple DNA polymerases, each serving specific functions in DNA replication and repair.

- *E. coli* DNA polymerase III holoenzyme replicates the *E. coli* chromosome.

- *E. coli* DNA polymerase III holoenzyme consists of 10 different kinds of subunits: 2 core ($\alpha\epsilon\theta$) DNA polymerase III units, one clamp loader (γ−complex), and a β_2-sliding clamp to tether the holoenzyme to DNA.

- Each *E. coli* DNA polymerase III holoenzyme is associated with a hexameric DnaB helicase that unwinds the DNA duplex and 3 primase subunits that synthesize RNA primers.

- *E. coli* DNA polymerase III holoenzyme is highly processive: It can replicate the entire *E. coli* chromosome without dissociating from the DNA.

- A DNA polymerase III holoenzyme sits at each replication fork.

- *E. coli* DNA polymerase I removes the RNA primers and fills in the gaps.

- *E. coli* DNA polymerase I is its own proofreader.

- *E. coli* DNA polymerase I activity is important in double-strand break repair (DSB repair).

- *E. coli* DNA ligase seals the links between Okazaki fragments.

- *E. coli* DNA replication terminates at the *Ter* region.

- DNA polymerases are localized in replication factories.

- DNA polymerases across all organisms have a common architecture and mechanism of action.

- DNA polymerases can be organized into families based on their sequence homology.

- In eukaryotic cells, the cell cycle determines the time of DNA replication through the action of checkpoints, CDKs and cyclins.

- Initiation of replication in eukaryotes depends on ORC (the origin of replication complex), the MCM2-7 double hexameric helicases (replication licensing factors), Cdc6, Cdt1, and other proteins.

- The MCM proteins, the Orc proteins, and Cdc6 are AAA+ ATP family members.

- Geminin inhibits DNA replication by preventing MCM recruitment to the prereplication complex.

- DNA polymerase δ is the principal replicative DNA polymerase of the eukaryotic nuclear genome.

- Telomerase, an RNA-dependent DNA polymerase, replicates telomeres, the structures at the ends of eukaryotic chromosomes.

- Telomerase consists of a DNA polymerase catalytic subunit (TERT) and an RNA subunit (RT or TER) that serves as the template for RNA-dependent synthesis of new telomeres.

- RNA genomes are replicated by reverse transcriptase, an RNA-dependent DNA polymerase.

- Reverse transcriptases possess RNA-directed DNA polymerase activity, RNase H activity, and DNA-directed DNA polymerase activity.

- Genetic information is rearranged by recombination.

- Recombination between very similar DNA sequences is called homologous recombination or general recombination.

- Recombination between dissimilar DNA sequences is called nonhomologous recombination.

- General recombination requires the breakage and reunion of DNA strands.

- General recombination can lead to formation of heteroduplex DNA, where part of each DNA strand is contributed by a different parental strand.

- Homologous recombination proceeds according to the Holliday model, with formation of a Holliday junction.

- Resolution of the Holliday junction leads to formation of patch recombinants or splice recombinants.

- The enzymes of general recombination include RecA, RecBCD, RuvA, RuvB, and RuvC.

- The RecBCD complex unwinds double-stranded DNA and cleaves its single strands.

- RecA can bind single-stranded DNA and then interact with double-stranded DNA.

- RecA mediates the ATP-dependent strand exchange reaction leading to formation of a Holliday junction.

- RecA binding to ssDNA leads to formation of a right-handed helical nucleoprotein filament in which the ssDNA is stretched and underwound.

- RuvA, RuvB, and RuvC resolve the Holliday junction to form recombination products.

- Recombination-dependent DNA replication restarts DNA replication at stalled replication forks.

- Transposons are DNA sequences that can move from place to place in the genome by recombination.

- DNA is susceptible to damage by chemical reactions, UV light, or ionizing radiation.

- DNA can be repaired by a number of different mechanisms, including mismatch repair (MMR), base excision repair (BER), and double-strand break repair (DSB repair).

- Mismatch repair corrects errors introduced during DNA replication.

- Pyrimidine dimers formed by UV light are repaired by photolyase.

- Base excision repair involves removal of the damaged base by DNA glycosylase, cleavage of the backbone by AP endonuclease, and gap repair by DNA polymerase.

- Nucleotide excision repair recognizes and removes larger regions of damaged DNA than BER.

- Point mutations in DNA arise by inappropriate base-pairing.

- Mutations can be induced by base analogs.

- Chemical mutagens react with bases in DNA.

- Insertions or deletions lead to frameshift mutations.

- Protein diversity can be generated through genetic recombination.

- Antibody molecules such as immunoglobulins G are an excellent example of protein diversity through genetic recombination.

- The variable regions of immunoglobulin G light and heavy chains are regions of variable amino acid sequence created by genetic recombination within the immunoglobulin genes.

- V-J and D-J joining in light- and heavy-chain gene assembly is mediated by the RAG proteins.

- Imprecise joining of IgG genes creates new coding arrangements.

PROBLEMS

Answers to all problems are at the end of this book. Detailed solutions are available in the *Student Solutions Manual, Study Guide, and Problems Book*.

1. **Semiconservative or Conservative DNA Replication** If ^{15}N-labeled *E. coli* DNA has a density of 1.724 g/mL, ^{14}N-labeled DNA has a density of 1.710 g/mL, and *E. coli* cells grown for many generations on ^{14}NH$_4^+$ as a nitrogen source are transferred to media containing ^{15}NH$_4^+$ as the sole N-source, (a) what will be the density of the DNA after one generation, assuming replication is semiconservative? (b) Suppose replication took place by a conservative mechanism in which the parental strands remained together and the two progeny strands were paired. Design an experiment that could distinguish between semiconservative and conservative modes of replication.

2. **The Enzymatic Activities of DNA Polymerase I** (a) What are the respective roles of the 5′-exonuclease and 3′-exonuclease activities of DNA polymerase I? (b) What might be a feature of an *E. coli* strain that lacked DNA polymerase I 3′-exonuclease activity?

3. **Multiple Replication Forks in E. coli I** Assuming DNA replication proceeds at a rate of 750 base pairs per second, calculate how long it will take to replicate the entire *E. coli* genome. Under optimal conditions, *E. coli* cells divide every 20 minutes. What is the minimal number of replication forks per *E. coli* chromosome in order to sustain such a rate of cell division?

4. **Multiple Replication Forks in E. coli II** On the basis of Figure 28.2, draw a simple diagram illustrating replication of the circular *E. coli* chromosome (a) at an early stage, (b) when one-third completed, (c) when two-thirds completed, and (d) when almost finished, assuming the initiation of replication at *oriC* has occurred only once. Then, draw a diagram showing the *E. coli* chromosome in problem 3 where the *E. coli* cell is dividing every 20 minutes.

5. **Molecules of DNA Polymerase III per Cell vs. Growth Rate** It is estimated that there are forty molecules of DNA polymerase III per *E. coli* cell. Is it likely that the growth rate of *E. coli* is limited by DNA polymerase III availability?

6. **Number of Okazaki Fragments in E. coli and Human DNA Replication** Approximately how many Okazaki fragments are synthesized in the course of replicating an *E. coli* chromosome? How many in replicating an "average" human chromosome?

7. **The Roles of Helicases and Gyrases** How do DNA gyrases and helicases differ in their respective functions and modes of action?

8. **Human Genome Replication Rate** Assume DNA replication proceeds at a rate of 100 base pairs per second in human cells and origins of replication occur every 300 kbp. Assume also that DNA polymerase III is highly processive and only 2 molecules of DNA polymerase III are needed per replication fork. How long would it take to replicate the entire diploid human genome? How many molecules of DNA polymerase does each cell need to carry out this task?

9. **Heteroduplex DNA Formation in Recombination** From the information in Figures 28.17 and 28.18, diagram the recombinational event leading to the formation of a heteroduplex DNA region within a bacteriophage chromosome.

10. **Homologous Recombination, Heteroduplex DNA, and Mismatch Repair** Homologous recombination in *E. coli* leads to the formation of regions of heteroduplex DNA. By definition, such regions contain mismatched bases. Why doesn't the mismatch repair system of *E. coli* eliminate these mismatches?

11. **RecA:DNA Nucleoprotein Helix/B-DNA Comparison** If RecA protein unwinds duplex DNA so that there are about 18.6 bp per turn, what is the change in $\Delta\phi$, the helical twist of DNA, compared to its value in B-DNA?

12. **Resolution of a Holliday Junction by Resolvase** Diagram a Holliday junction between two duplex DNA molecules, and show how the action of resolvase might give rise to either patch- or splice recombinant DNA molecules.

13. **Chemical Mutagenesis of DNA Bases** Show the nucleotide sequence changes that might arise in a dsDNA (coding strand segment GCTA) upon mutagenesis with (a) HNO$_2$, (b) bromouracil, and (c) 2-aminopurine.

14. **Transposons as Mutagens** Transposons are mutagenic agents. Why?

15. **Recombination in Immunoglobulin Genes** If recombination between a V_κ and J_κ gene formed a CCA codon at codon 95 (Figure 28.39), which amino acid would appear at this position?

16. **Helicase Unwinding of the E. coli Chromosome** Hexameric helicases, such as DnaB, the MCM proteins, and papilloma virus E1 helicase (illustrated in Figures 16.23–16.25), unwind DNA by passing one strand of the DNA duplex through the central pore, using a mechanism based on ATP-dependent binding interactions with the bases of that strand. The genome of *E. coli* K12 consists of 4,686,137 nucleotides. Assuming that DnaB functions like papilloma virus E1 helicase, from the information given in Chapter 16 on ATP-coupled DNA unwinding, calculate how many molecules of ATP would be needed to completely unwind the *E. coli* K12 chromosome.

17. **Translesion DNA Polymerase IV Experiment** Asako Furukohri, Myron F. Goodman, and Hisaji Maki wanted to discover how the translesion DNA polymerase IV takes over from DNA polymerase III at a stalled replication fork (see *Journal of Biological Chemistry* **283:**11260–11269, 2008). They showed that DNA polymerase IV could displace DNA polymerase III from a stalled replication fork formed in an in vitro system containing DNA, DNA polymerase III, the β-clamp, and SSB. Devise your own experiment to show how such displacement might be demonstrated. (Hint: Assume that you have protein identification tools that allow you to distinguish easily between DNA polymerase III and DNA polymerase IV.)

18. **Functional Consequences of Y-Family DNA Polymerase Structure** The eukaryotic translesion DNA polymerases fall into the Y family of DNA polymerases. Structural studies reveal that their fingers and thumb domains are small and stubby (see Figure 28.10). In addition, Y-family polymerase active sites are more open and less constrained where base pairing leads to selection of a dNTP

substrate for the polymerase reaction. Discuss the relevance of these structural differences. Would you expect Y-family polymerases to have 3′-exonuclease activity? Explain your answer.

Preparing for the MCAT® Exam

19. Figure 28.11 depicts the eukaryotic cell cycle. Many cell types "exit" the cell cycle and don't divide for prolonged periods, a state termed G_0; some, for example neurons, never divide again.

   a. In what stage of the cell cycle do you suppose a cell might be when it exits the cell cycle and enters G_0?

   b. The cell cycle is controlled by checkpoints, cyclins, and CDKs. Describe how biochemical events involving cyclins and CDKs might control passage of a dividing cell through the cell cycle.

20. Figure 28.39 gives some examples of recombination in IgG codons 95 and 96, as specified by the V_κ and J_κ genes. List the codon possibilities and the amino acids encoded if recombination occurred in codon 97. Which of these possibilities is less desirable?

ActiveModels Problems

21. Using the ActiveModel for reverse transcriptase, describe the function of the two magnesium ions in the active site.

22. Using the ActiveModel for human prion protein, give an example of how significant change in secondary structure affects protein folding. Can a change in protein folding lead to disease?

FURTHER READING

General

Holliday, R., 1964. A mechanism for gene conversion in fungi. *Genetic Research* **5**:282–304. The classic model for the mechanism of DNA strand exchange during homologous recombination.

Kornberg, A., 2005. *DNA Replication,* 2nd ed. New York: Macmillan. A comprehensive detailed account of the enzymology of DNA metabolism, including replication, recombination, repair, and more.

Krebs, J., Goldstein, E. S., and Kilpatrick, S. T., 2009. Lewin's Genes X. Sudbury, MA: Jones and Bartlett. A contemporary genetics text that seeks to explain heredity in terms of molecular structures.

Meselson, M., and Stahl, F. W., 1958. The replication of DNA in *Escherichia coli. Proceedings of the National Academy of Sciences U.S.A.* **44**:671–682. The classic paper showing that DNA replication is semiconservative.

Ogawa, T., and Okazaki, T., 1980. Discontinuous DNA replication. *Annual Review of Biochemistry* **49**:421–457. Okazaki fragments and their implications for the mechanism of DNA replication.

Palmiter, R. D., et al., 1982. Dramatic growth of mice that develop from eggs microinjected with metallothionein-growth hormone fusion genes. *Nature* **300**:611–615.

DNA Replication

Baker, T. A., and Bell, S. P., 1998. Polymerases and the replisome: Machines within machines. *Cell* **92**:295–305.

Bebenek, K., and Kunkel, T. A., 2004. Functions of DNA polymerases. *Advances in Protein Chemistry* **69**:137–165.

Bell, S. P., and Dutta, A., 2002. DNA replication in eukaryotic cells. *Annual Review of Biochemistry* **71**:333–374.

Berdis, A. J., 2009. Mechanisms of DNA polymerases. *Chemical Reviews* **109**:2862–2879.

Berdis, A. J., 2010. DNA polymerases: Perfect enzymes for an imperfect world. *Biochimica et Biophysica Acta* **1804**:1029–1031. Issue 5 of volume 1804 of *Biochimica et Biophysica Acta* contains 18 articles dedicated to the structure and function of DNA polymerases.

Blow, J. J., and Dutta, A., 2005. Preventing re-replication of chromosomal DNA. *Nature Reviews Molecular Cell Biology* **6**:476–486.

Botchan, M., 2007. A switch for S phase. *Nature* **445**:272–274.

Costa, A., Ilves, I., Tamberg, N., Petojevic, T., et al., 2011. The structural basis for MCM2-7 helicase activation by GINS and Cdc45. *Nature Structural and Molecular Biology* **18**:471–479.

Cvetic, C. A., and Walter, J. C., 2006. Getting a grip on licensing: Mechanism of stable MCM2-7 loading onto replication origins. *Cell* **21**: 143–148.

Franklin, M. C., Wang, J., and Steitz, T. A., 2001. Structure of the replicating complex of a pol α family DNA polymerase. *Cell* **105**:657–667.

Frick, D. N., and Richardson, C. C., 2001. DNA primases. *Annual Review of Biochemistry* **70**:39–80.

Goodman, M. F., 2002. Error-prone repair DNA polymerases in prokaryotes and eukaryotes. *Annual Review of Biochemistry* **71**:17–50.

Hamdan, S. M., and van Oijen, A. M., 2010. Timing, coordination, and rhythm: Acrobatics at the DNA replication fork. *Journal of Biological Chemistry* **285**:18979–18983.

Hübscher, U., Maga, G., and Spadari, S., 2002. Eukaryotic DNA polymerases. *Annual Review of Biochemistry* **71**:133–163.

Keck, J. L., 2000. Structure of the RNA polymerase domain of the *E. coli* primase. *Science* **287**:2482–2486.

Kool, E. T., 2002. Active site tightness and substrate fit in DNA replication. *Annual Review of Biochemistry* **71**:191–219.

Langston, L. D., Indiani, C., and O'Donnell, M., 2009. Whither the replisome: Emerging perspectives on the dynamic nature of the DNA replication machinery. *Cell Cycle* **8**:2686–2691.

Leu, F. P., Georgescu, R., and O'Donnell, M., 2003. Mechanism of the *E. coli* τ processivity switch during lagging-strand synthesis. *Molecular Cell* **11**:315–327.

Machida, Y. J., and Dutta, A., 2005. Cellular checkpoint mechanisms monitoring proper initiation of DNA replication. *Journal of Biological Chemistry* **280**:6253–6256.

Marians, K. J., 2008. Understanding how the replisome works. *Nature Structural and Molecular Biology* **15**:125–127.

McHenry, C., 2003. Chromosomal replicases as asymmetric dimers: Studies of subunit arrangement and functional consequences. *Molecular Microbiology* **49**:1157–1165.

Pomerantz, R. T., and O'Donnell, M., 2007. Replisome mechanics: Insights into a twin DNA polymerase machine. *Trends in Microbiology* **15**:156–164.

Randell, J. C. W., Bowers, J. L., Rodriguez, H. K., and Bell, S. P., 2006. Sequential ATP hydrolysis by Cdc6 and ORC directs loading of the Mcm2-7 helicase. *Molecular Cell* **21**:29–39.

Rothwell, P. J., and Waksman, G., 2005. Structure and mechanism of DNA polymerases. *Advances in Protein Chemistry* **71**:401–440.

Steitz, T. A., 1998. A mechanism for all polymerases. *Nature* **391**:231–232.

Tye, B. K., 1999. MCM proteins in DNA replication. *Annual Review of Biochemistry* **68**:649–686.

Protein Rings in DNA Metabolism

Hingorani, M. M., and O'Donnell, M., 2000. A tale of toroids in DNA metabolism. *Nature Reviews Molecular Cell Biology* **1**:22–30.

Wyman, C., and Botchan, M., 1995. DNA replication: A familiar ring to DNA polymerase processivity. *Current Biology* **5**:334–337.

Telomerase

Blackburn, E. H., 1992. Telomerases. *Annual Review of Biochemistry* **61:** 113–129.

Collins, K., 1999. Ciliate telomerase biochemistry. *Annual Review of Biochemistry* **68:**187–218.

Kim, N. W., 1994. Specific association of human telomerase activity with immortal cells and cancer. *Science* **266:**2011–2015.

Nakamura, T. M., et al., 1997. Telomerase catalytic subunit homologs from fission yeast and human. *Science* **277:**955–959.

Theimer, C. A., and Feigon, J., 2010. Structure and function of telomerase RNA. *Current Opinion in Structural Biology* **16:**307–318.

Wyatt, H. D. M., West, S. C., and Beattie, T. L., 2010. InTERTpreting telomerase structure and function. *Nucleic Acids Research* **38:**5609–5622.

Recombination

Alberts, B., 2003. DNA replication and recombination. *Nature* **421:**431–435.

Anderson, D. G., and Kowalczykowski, S. C., 1997. The translocating RecBCD enzyme stimulates recombination by directing RecA protein onto ssDNA in a χ-regulated manner. *Cell* **90:**77–86.

Baumann, P., and West, S. C., 1998. Role of the human RAD51 protein in homologous recombination and double-stranded-break repair. *Trends in Biochemical Sciences* **23:**247–252.

Beernink, H. T. H., and Morrical, S. W., 1999. RMPs: Recombination/replication proteins. *Trends in Biochemical Sciences* **24:**385–389.

Chen, Z., Yang, H., and Pavletich, N. P., 2008. Mechanism of homologous recombination from the RecA-ssDNA/dsDNA structures. *Nature* **453:**489–494.

Cox, M. M., 2007. Motoring along with the bacterial RecA protein. *Nature Reviews Molecular Cell Biology* **8:**127–138.

Haber, J. E., 1999. DNA recombination: The replication connection. *Trends in Biochemical Sciences* **24:**271–275.

Hays, F. A., Watson, J., and Ho, P. S., 2003. Caution! DNA crossing: Crystal structures of Holliday junctions. *Journal of Biological Chemistry* **278:**49663–49666.

Kowalczykowski, S. C., 2000. Initiation of genetic recombination and recombination-dependent replication. *Trends in Biochemical Sciences* **25:**156–165.

Krishna, R., Prabu, J. R., Manjunath, G. P., Datta, S., et al., 2007. Snapshots of RecA protein involving movement of the C-domain and different conformations of the DNA-binding loops: Crystallographic and comparative analysis of 11 structures of *Mycobacterium smegmatis* RecA. *Journal of Molecular Biology* **367:**1130–1144.

Lovett, S. T., 2003. Connecting replication and recombination. *Molecular Cell* **11:**554–556.

Lusetti, S. L., and Cox, M. M., 2002. The bacterial RecA protein and the recombinational DNA repair of replication forks. *Annual Review of Biochemistry* **71:**71–100.

Rafferty, J. B., et al., 1996. Crystal structure of DNA recombination protein RuvA and a model for its binding to the Holliday junction. *Science* **274:**415–421.

San Filippo, J., Sung, P., and Klein, H., 2008. Mechanisms of eukaryotic homologous recombination. *Annual Review of Biochemistry* **77:**229–257.

Taylor, A. F., and Smith, G. R., 2003. RecBCD enzyme is a DNA helicase with fast and slow motors of opposite polarity. *Nature* **423:**889–893. See also, Dillingham, M. S., Spies, M., and Kowalczykowski, S. C., 2003. RecBCD is a bipolar DNA helicase. *Nature* **423:**893–897.

Wigley, D. B., 2007. RecBCD: The supercar of DNA repair. *Cell* **131:**651–653.

Yamada, K., Ariyoshi, M., and Morikawa, K., 2004. Three-dimensional structural views of branch migration and resolution in DNA homologous recombination. *Current Opinion in Structural Biology* **14:**130–137.

Transposons

Lambowitz, A. M., and Belfort, M., 1993. Introns as mobile genetic elements. *Annual Review of Biochemistry* **62:**587–622.

Stellwagen, A. E., and Craig, N. L., 1998. Mobile DNA elements: Controlling transposition with ATP-dependent molecular switches. *Trends in Biochemical Sciences* **23:**486–490.

V(D)J Recombination and the Immunoglobulin Genes

Gellert, M., 2002. V(D)J recombination: RAG proteins, repair factors, and regulation. *Annual Review of Biochemistry* **71:**101–132.

Hiom, K., and Gellert, M., 1997. A stable RAG1-RAG2-DNA complex that is active in V(D)J cleavage. *Cell* **88:**65–72.

Lewis, S. M., and Wu, G. E., 1997. The origins of V(D)J recombination. *Cell* **88:**159–162.

Nossal, G. J. V., 2003. The double helix and immunology. *Nature* **421:**440–444.

Transgenic Animals

Morgan, R. A., and Anderson, W. F., 1993. Human gene therapy. *Annual Review of Biochemistry* **62:**192–217.

Schnieke, A. E., et al., 1997. Human factor IX transgenic sheep produced by transfer of nuclei from transfected fetal fibroblasts. *Science* **278:**2130–2133.

Wilmut, I., et al., 1997. Viable offspring derived from fetal and adult mammalian cells. *Nature* **385:**810–818. See also, Campbell, K. H. S., et al., 1996. Sheep cloned by nuclear transfer from a cultured cell line. *Nature* **380:**64–66.

Repair

Bartek, J., and Lukas, J., 2003. Damage alert. *Nature* **421:**486–488.

Friedberg, E. C., 2003. DNA damage and repair. *Nature* **421:**436–440.

Friedberg, E. C., Walker, G. C., and Siede, W., 1995. *DNA Repair and Mutagenesis.* Washington, DC: ASM Press.

Helleday, T., Lo, J., van Gent, D.C., and Engelward, B. P., 2007. DNA double-strand break repair: From mechanistic understanding to cancer treatment. *DNA Repair* **9:**1256–1263.

Hiom, K., 2010. Coping with DNA double strand breaks. *DNA Repair* **9:** 1256–1263.

Marians, K. J., 2000. PriA-directed replication fork restart in *Escherichia coli. Trends in Biochemical Sciences* **25:**185–189.

McCollough, A. K., et al., 1999. Initiation of base excision repair: Glycosylase mechanisms and structures. *Annual Review of Biochemistry* **68:**255–285.

Michel, B., 2000. Replication fork arrest and DNA recombination. *Trends in Biochemical Sciences* **25:**173–178.

Modrich, P., and Lahue, R., 1996. Mismatch repair in replication fidelity, genetic recombination, and cancer biology. *Annual Review of Biochemistry* **65:**101–133.

Mol, C. D., et al., 1999. DNA repair mechanisms for the recognition and removal of damaged DNA bases. *Annual Review of Biophysics and Biomolecular Structure* **28:**101–128.

Morgan, A. R., 1993. Base mismatches and mutagenesis: How important is tautomerism? *Trends in Biochemical Sciences* **18:**160–163.

Parikh, S. S., et al., 1999. Envisioning the molecular choreography of DNA base excision repair. *Current Opinion in Structural Biology* **9:**37–47.

Sancar, A., 1994. Mechanisms of DNA excision repair. *Science* **266:** 1954–1956.

Transcription and the Regulation of Gene Expression

British Museum, UK/Bridgeman Art Library

Now that we have all this useful information, it would be nice to do something with it.

From the *Unix Programmer's Manual*

KEY QUESTIONS

29.1 How Are Genes Transcribed in Prokaryotes?

29.2 How Is Transcription Regulated in Prokaryotes?

29.3 How Are Genes Transcribed in Eukaryotes?

29.4 How Do Gene Regulatory Proteins Recognize Specific DNA Sequences?

29.5 How Are Eukaryotic Transcripts Processed and Delivered to the Ribosomes for Translation?

29.6 Can Gene Expression Be Regulated Once the Transcript Has Been Synthesized?

29.7 Can We Propose a Unified Theory of Gene Expression?

ESSENTIAL QUESTION

Expression of the information encoded in DNA depends on transcription of that information into RNA.

How are the genes of prokaryotes and eukaryotes transcribed to form RNA products that can be translated into proteins?

All cells contain three major classes of RNA—messenger RNA (mRNA), ribosomal RNA (rRNA), and transfer RNA (tRNA)—and all participate in protein synthesis (see Chapters 10 and 30). Further, all RNAs are synthesized from DNA templates by **DNA-dependent RNA polymerases** in the process known as **transcription.** However, only mRNAs direct the synthesis of proteins. Protein synthesis occurs via the process of **translation,** wherein the instructions encoded in the sequence of bases in mRNA are translated into a specific amino acid sequence by ribosomes, the "workbenches" of polypeptide synthesis (see Chapter 30).

Transcription is tightly regulated in all cells. In prokaryotes, only 3% or so of the genes are undergoing transcription at any given time. The metabolic conditions and the growth status of the cell dictate which gene products are needed at any moment. Similarly, differentiated eukaryotic cells express only a small percentage of their genes in fulfilling their biological functions, not the full genetic potential encoded in their chromosomes.

29.1 | How Are Genes Transcribed in Prokaryotes?

In prokaryotes, virtually all RNA is synthesized by a single species of DNA-dependent RNA polymerase. The only exception is the short RNA primers formed by primase during DNA replication. Like DNA polymerases, RNA polymerase uses ribonucleoside

ⓄWL Online homework and a Student Self Assessment for this chapter may be assigned in OWL.

► A DEEPER LOOK

Conventions Used in Expressing the Sequences of Nucleic Acids and Proteins

Certain conventions are useful in tracing the course of information transfer from DNA to protein. The strand of duplex DNA that is read by RNA polymerase is termed the **template strand.** Thus, the strand that is not read is the **nontemplate strand.** Because the template strand is read by the RNA polymerase moving 3′→5′ along it, the RNA product, called the **transcript,** grows in the 5′→3′ direction (see accompanying figure). Note that the nontemplate strand has a nucleotide sequence and direction identical to those of the RNA transcript, except that the transcript has U residues in place of T. Portions of the RNA transcript will eventually be translated into the amino acid sequence of a protein (see Chapter 30) by a process in which successive triplets of bases (termed **codons**), read 5′→3′, specify a particular amino acid. Polypeptide chains are synthesized in the N—→C direction, and the 5′-end of mRNA encodes the N-terminus of the protein.

By convention, when the order of nucleotides in DNA is shown as a single strand, it is the 5′→3′ sequence of nucleotides in the nontemplate strand that is presented. Consequently, if convention is followed, DNA sequences are written in terms that correspond directly to mRNA sequences, which correspond in turn to the amino acid sequences of proteins as read beginning with the N-terminus.

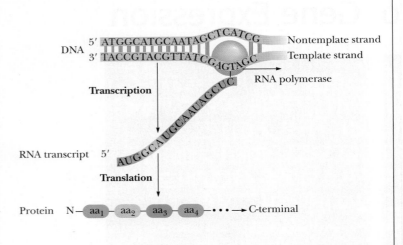

5′-triphosphates (ATP, GTP, CTP, and UTP, represented generically as NTPs) as substrates, linking them in an order specified by base pairing with a DNA template:

$$n \text{ NTP} \longrightarrow (\text{NMP})_n + n \text{ PP}_i$$

The enzyme moves along a DNA strand in the 3′→5′ direction, joining the 5′-phosphate of an incoming ribonucleotide to the 3′-OH of the previous residue. Thus, the RNA chain grows 5′→3′ during transcription, just as DNA chains do during replication. Subsequent hydrolysis of PP_i to inorganic phosphate by the pyrophosphatases present in all cells removes the product PP_i, thus making the polymerase reaction thermodynamically favorable.

Prokaryotic RNA Polymerases Use Their Sigma Subunits to Identify Sites Where Transcription Begins

Transcription is initiated in prokaryotes by **RNA polymerase holoenzyme,** a complex multimeric protein (about 400 kD) large enough to be visible in the electron microscope. Its subunit composition is $\alpha_2\beta\beta'\sigma$. After two α-subunits (35 kD each) dimerize, one recruits β and the other β' to form the clawlike core polymerase structure (Figure 29.1). The two largest subunits, β' (171 kD) and β (124 kD), perform most of the enzymatic functions. The β-subunit forms most of the upper jaw of the claw and contains the catalytic Mg^{2+}-binding site; β' forms the lower jaw. DNA passes through a 2.7-nm channel between the jaws of the claw. Nucleotide substrates reach the catalytic center through a secondary channel entering on the back side of the enzyme. Binding of the σ-subunit to β' allows the RNA polymerase to recognize different DNA sequences that act as **promoters.** A number of related proteins, **the sigma (σ) factors,** can serve as the σ-subunit. Promoters are nucleotide sequences that identify the location of *transcription start sites,* where transcription begins. Both β and β' contribute to formation of the catalytic site for RNA synthesis. Dissociation of the σ-subunit from the holoenzyme leaves the **core polymerase** ($\alpha_2\beta\beta'$), which can transcribe DNA into RNA but is unable to recognize promoters and initiate transcription. Bacteriophage T7 expresses a simpler (monomeric)

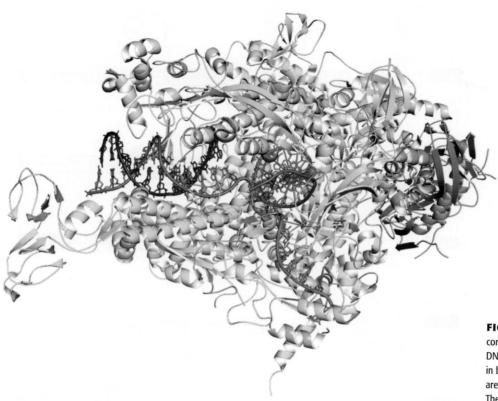

FIGURE 29.1 Structure of the *Thermus thermophilus* core RNA polymerase $\alpha_2\beta\beta'$ (pdb id = 2O5I). The template DNA strand is shown in green, the nontemplate DNA strand in blue, and the RNA transcript in hot pink. The two α chains are orange, the β chain is cyan, and the β' chain is yellow. The active-site Mg^{2+} is shown as a red sphere.

RNA polymerase (Figure 29.2) that shares the functional characteristics of prokaryotic RNA polymerases.

The Process of Transcription Has Four Stages

Transcription can be divided into four stages: (1) binding of RNA polymerase holoenzyme at promoter sites, (2) initiation of polymerization, (3) chain elongation, and (4) chain termination.

Binding of RNA Polymerase to Template DNA The process of transcription begins when the σ-subunit of RNA polymerase recognizes a promoter sequence (Figure 29.3), and RNA polymerase holoenzyme and the promoter form a **closed promoter complex** (Figure 29.3, Step 2). This stage in RNA polymerase:DNA interaction is referred to as the *closed* promoter complex because the dsDNA has not yet been "opened" (unwound) so that the RNA polymerase can read the base sequence of the DNA template strand and transcribe it into a complementary RNA sequence.

Once the closed promoter complex is established, the RNA polymerase holoenzyme unwinds about 14 base pairs of DNA (base pairs located at positions -12 to $+2$, relative to the transcription start site; see later discussion), forming the very stable **open promoter complex** (Figure 29.3, Step 3). Promoter sequences can be identified in vitro by **DNA footprinting**: RNA polymerase holoenzyme is bound to a putative promoter sequence in a DNA duplex, and the DNA:protein complex is treated with DNase I. DNase I cleaves the DNA at sites not protected by bound protein, and the set of DNA fragments left after DNase I digestion reveals the promoter (by definition, the promoter is the RNA polymerase holoenzyme binding site).

RNA polymerase binding typically protects a nucleotide sequence spanning the region from -70 to $+20$, where the $+1$ position is defined as the **transcription start site**: that base in the nontemplate DNA strand that is identical with the first base in the RNA transcript. The next base, $+2$, specifies the second base in the transcript. Nontemplate

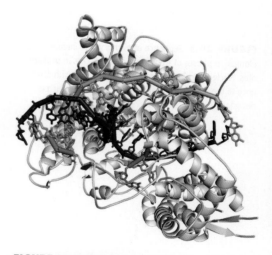

FIGURE 29.2 Bacteriophage T7 RNA polymerase (pdb id = 1MSW) in the act of transcription. T7 RNA polymerase is a 99-kD monomeric protein. The DNA is shown entering the enzyme from the upper right. The template strand is green, the nontemplate strand is blue, the RNA transcript is hot pink.

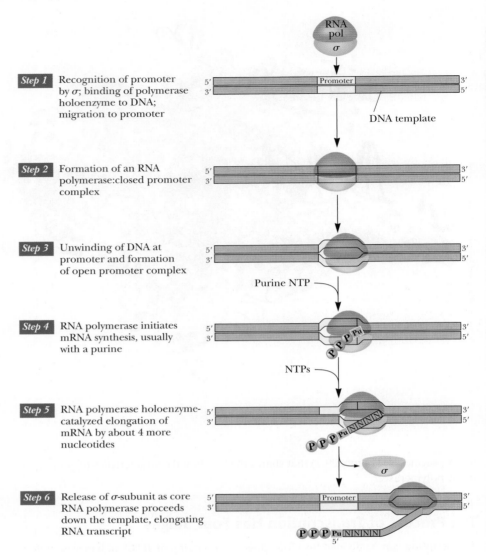

FIGURE 29.3 Sequence of events in the initiation and elongation phases of transcription as it occurs in prokaryotes. Nucleotides in this region are numbered with reference to the base at the transcription start site, which is designated +1.

Step 1 Recognition of promoter by σ; binding of polymerase holoenzyme to DNA; migration to promoter

DNA template

Step 2 Formation of an RNA polymerase:closed promoter complex

Step 3 Unwinding of DNA at promoter and formation of open promoter complex

Purine NTP

Step 4 RNA polymerase initiates mRNA synthesis, usually with a purine

NTPs

Step 5 RNA polymerase holoenzyme-catalyzed elongation of mRNA by about 4 more nucleotides

Step 6 Release of σ-subunit as core RNA polymerase proceeds down the template, elongating RNA transcript

A **consensus sequence** can be defined as the bases that appear with highest frequency at each position when a series of sequences believed to have common function is compared.

strand bases in the 5′, or "minus," direction from the **transcript start site** are numbered −1, −2, and so on. (Note that there is no zero.) Nontemplate nucleotides in the "minus" direction are said to lie **upstream** of the transcription start site, whereas nucleotides in the 3′, or "plus," direction are **downstream** of the transcription start site. The transcript start site on the template strand is usually a pyrimidine, so most transcripts begin with a purine. RNA polymerase binding protects 90 bp of DNA, equivalent to a distance of 30 nm along B-DNA. Because RNA polymerase is only 16 nm in its longest dimension, the DNA must be wrapped around the enzyme.

Properties of Prokaryotic Promoters Promoters recognized by the principal σ factor, σ^{70}, serve as the paradigm for prokaryotic promoters. These promoters vary in size from 20 to 200 bp but typically consist of a 40-bp region located on the 5′-side of the transcription start site. Within the promoter are two **consensus sequence elements.** These two elements are the **Pribnow box**[1] near −10, whose consensus sequence is the hexameric TATAAT, and a sequence in the **−35 region** containing the hexameric consensus TTGACA (Figure 29.4). The Pribnow box and the −35 region are separated by about 17 bp of nonconserved sequence. RNA polymerase holoenzyme uses its σ-subunit to bind to the conserved sequences, and the more closely the −35 region sequence corresponds to its consensus sequence, the greater is the efficiency of transcription of the

[1]Named for David Pribnow, who, along with David Hogness, first recognized the importance of this sequence element in transcription.

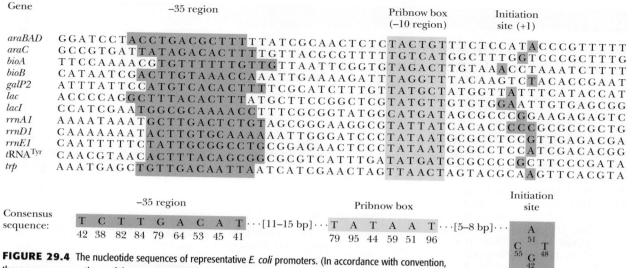

FIGURE 29.4 The nucleotide sequences of representative *E. coli* promoters. (In accordance with convention, these sequences are those of the nontemplate strand where RNA polymerase binds.) Consensus sequences for the −35 region, the Pribnow box, and the initiation site are shown at the bottom. The numbers represent the percent occurrence of the indicated base. (Note: The −35 region is only roughly 35 nucleotides from the transcription start site; the Pribnow box [the −10 region] likewise is located at approximately position −10.) In this figure, sequences are aligned relative to the Pribnow box.

gene. The highly expressed *rrn* genes in *E. coli* that encode ribosomal RNA (rRNA) have a third sequence element in their promoters, the **upstream element** (**UP** element), located about 20 bp immediately upstream of the −35 region. (Transcription from the *rrn* genes accounts for more than 60% of total RNA synthesis in rapidly growing *E. coli* cells.) Whereas the σ-subunit recognizes the −10 and −35 elements, the C-terminal domains (CTD) of the α-subunits of RNA polymerase recognize and bind the UP element.

In order for transcription to begin, the DNA duplex must be "opened" so that RNA polymerase has access to single-stranded template. The efficiency of initiation is inversely proportional to the melting temperature, T_m, in the Pribnow box, suggesting that the A : T-rich nature of this region is aptly suited for easy "melting" of the DNA duplex and creation of the open promoter complex (see Figure 29.3). Negative supercoiling facilitates transcription initiation by favoring DNA unwinding.

The RNA polymerase σ-subunit is directly involved in melting the dsDNA. Interaction of the σ-subunit with the nontemplate strand maintains the open complex formed between RNA polymerase and promoter DNA, with the σ-subunit acting as a sequence-specific single-stranded DNA-binding protein. Association of the σ-subunit with the nontemplate strand stabilizes the open promoter complex and leaves the bases along the template strand available to the catalytic site of the RNA polymerase.

Initiation of Polymerization Unlike DNA polymerase, RNA polymerase does not require a primer: It can catalyze the de novo synthesis of polynucleotides. RNA polymerase has two binding sites for NTPs: the initiation site and the elongation site. The first nucleotide binds at the initiation site, base-pairing with the +1 base exposed within the *open promoter complex* (see Figure 29.3, Step 4). The second incoming nucleotide binds at the elongation site, base-pairing with the +2 base. The ribonucleotides are then united when the 3′-O of the first nucleotide makes a phosphoester bond with the α-phosphorus atom of the second nucleotide, and PP$_i$ is eliminated. Note that the 5′-end of the transcript starts out with a triphosphate attached to it. Movement of RNA polymerase along the template strand *(translocation)* to the next base prepares the RNA polymerase to add the next nucleotide (see Figure 29.3, Step 5). Once an oligonucleotide 9 to 12 residues long has been formed, the σ-subunit dissociates from RNA polymerase, signaling the completion of initiation (see Figure 29.3, Step 6). The core RNA polymerase is highly processive and goes on to synthesize the remainder of the mRNA. As the core RNA polymerase progresses, advancing the 3′-end of the RNA chain, the

DNA Footprinting—Identifying the Nucleotide Sequence in DNA Where a Protein Binds

DNA footprinting is a widely used technique to identify the nucleotide sequence within DNA where a specific protein binds, such as the **promoter** sequence(s) bound by RNA polymerase holoenzyme. In this technique, the protein is incubated with a labeled (*) DNA fragment containing the nucleotide sequence where the protein is believed to bind. (The DNA fragment is labeled at only one end.) Then, a DNA cleaving agent, such as DNase I, is added to the solution containing the DNA:protein complex. DNase I cleaves the DNA backbone in exposed regions—that is, wherever the presence of the DNA-binding protein does not prevent DNase I from binding. A control solution containing naked DNA (a sample of the same labeled DNA fragment with no DNA-binding protein added) is also treated with DNase I. When these DNase I digests are analyzed by gel electrophoresis, a difference is found between the set of labeled fragments from the DNA:protein complex and the set from naked DNA. The absence of certain fragments in the digest of the DNA:protein complex reveals the location of the protein-binding site on the DNA (see accompanying figure).

Adapted from Rhodes, D., and Fairall, L., 1997. Analysis of sequence-specific DNA-binding proteins. In *Protein Function: A Practical Approach,* Creighton, T. E., ed., Oxford: IRL Press at Oxford University Press.

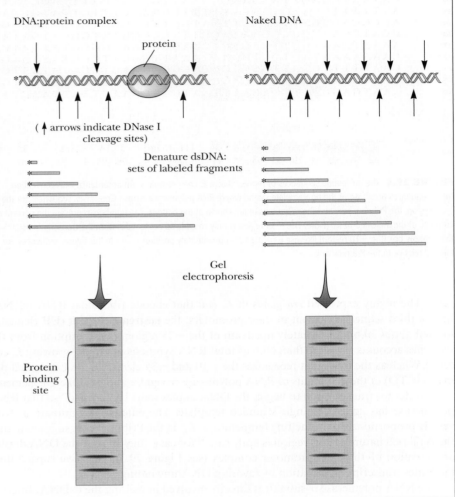

DNA duplex is unwound just ahead of it. About 12 base pairs of the growing RNA remain base-paired to the DNA template at any time, with the RNA strand becoming displaced as the DNA duplex rewinds behind the advancing RNA polymerase.

Chain Elongation   Elongation of the RNA transcript is catalyzed by the *core polymerase,* because once a short oligonucleotide chain has been synthesized, the σ-subunit dissociates. The accuracy of transcription is such that about once every 10^4 nucleotides, an error is made and the wrong base is inserted. Because many transcripts are made per gene and most transcripts are smaller than 10 kb, this error rate is acceptable.

Two possibilities can be envisioned for the course of the new RNA chain. In one, the RNA chain is wrapped around the DNA as the RNA polymerase follows the template strand around the axis of the DNA duplex, but this possibility seems unlikely due to its potential for tangling the nucleic acid strands (Figure 29.5a). In reality, transcription involves supercoiling of the DNA, so positive supercoils are created ahead of the transcription bubble and negative supercoils are created behind it (Figure 29.5b). To prevent torsional stress from inhibiting transcription, gyrases introduce negative supercoils (and thereby remove positive supercoils) ahead of RNA polymerase, and

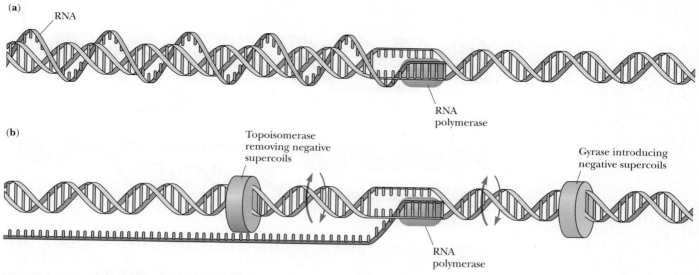

FIGURE 29.5 Supercoiling versus transcription. **(a)** If the RNA polymerase followed the template strand around the axis of the DNA duplex, no supercoiling of the DNA would occur but the RNA chain would be wrapped around the double helix once every 10 bp. This possibility seems unlikely because it would be difficult to untangle the transcript from the DNA duplex. **(b)** Alternatively, gyrases and topoisomerases could remove the torsional stresses induced by transcription.

topoisomerases remove negative supercoils behind the DNA segment undergoing transcription (Figure 29.5b).

Chain Termination Two types of transcription termination mechanisms operate in bacteria: one that is dependent on a specific protein called **rho termination factor** (for the Greek symbol, ρ) and another, **intrinsic termination**, that is not. Intrinsic termination is determined by specific sequences in the DNA called **termination sites.** These sites are not indicated by particular bases showing where transcription halts. Instead, these sites consist of three structural features whose base-pairing possibilities lead to termination:

1. Inverted repeats, which are typically G:C-rich, so a stable **stem-loop structure** can form in the transcript via intrachain base-pairing (Figure 29.6)
2. A nonrepeating segment that punctuates the inverted repeats
3. A run of 6 to 8 As in the DNA template, coding for Us in the transcript

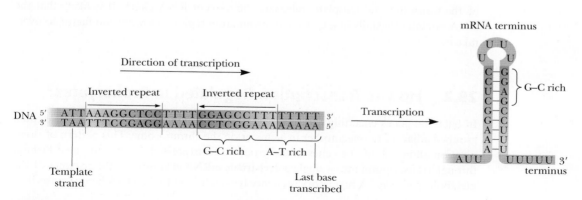

FIGURE 29.6 The termination site for the *E. coli trp* operon (the *trp* operon encodes the enzymes of tryptophan biosynthesis). The inverted repeats give rise to a stem-loop, or "hairpin," structure ending in a series of U residues.

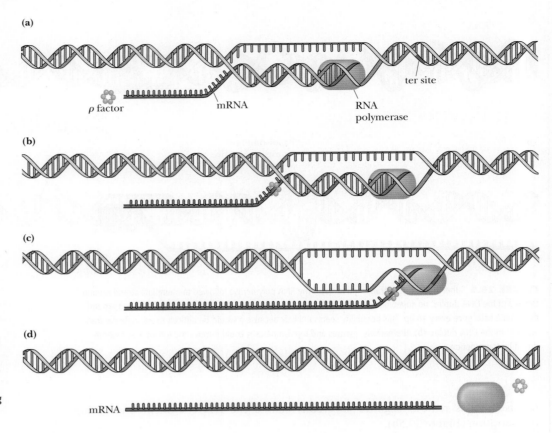

(a)

ρ factor mRNA RNA polymerase ter site

(b)

(c)

(d)

mRNA

FIGURE 29.7 The rho factor mechanism of transcription termination. Rho factor **(a)** attaches to a recognition site on mRNA and **(b)** moves along it behind RNA polymerase. **(c)** When RNA polymerase pauses at the termination site, rho factor unwinds the DNA : RNA hybrid in the transcription bubble, **(d)** releasing the nascent mRNA.

Termination then occurs as follows: A G : C-rich stem-loop structure, or "hairpin," forms in the transcript. The hairpin apparently causes the RNA polymerase to pause, whereupon the A : U base pairs between the transcript and the DNA template strand are displaced through formation of somewhat more stable A : T base pairs between the template and nontemplate strands of the DNA. The result is spontaneous dissociation of the nascent transcript from DNA.

The alternative mechanism of termination—factor-dependent termination—is less common and mechanistically more complex. Rho factor is an ATP-dependent hexameric helicase (50-kD subunits) that catalyzes the unwinding of RNA : DNA hybrid duplexes (or RNA : RNA duplexes). The rho factor recognizes and binds to C-rich regions in the RNA transcript. These regions must be unoccupied by translating ribosomes for rho factor to bind. Once bound, rho factor advances in the $5' \rightarrow 3'$ direction until it reaches the transcription bubble (Figure 29.7). There it catalyzes the unwinding of the transcript and template, releasing the nascent RNA chain. It is likely that the RNA polymerase stalls in a G : C-rich termination region, allowing rho factor to overtake it.

29.2 How Is Transcription Regulated in Prokaryotes?

In bacteria, genes encoding the enzymes of a particular metabolic pathway are often grouped adjacent to one another in a cluster on the chromosome. This pattern of organization allows all of the genes in the group to be expressed in a coordinated fashion through transcription into a **single polycistronic mRNA** encoding all the enzymes of the metabolic pathway.[2] A regulatory sequence lying adjacent to the DNA being transcribed

[2]A **polycistronic mRNA** is a single RNA transcript that encodes more than one polypeptide. *Cistron* is a genetic term for a DNA region representing a protein: *Cistron* and *gene* are essentially equivalent terms.

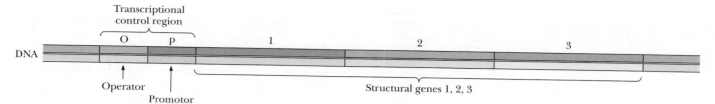

FIGURE 29.8 The general organization of operons. Operons consist of transcriptional control regions and a set of related structural genes, all organized in a contiguous linear array along the chromosome. The transcriptional control regions are the promoter and the operator, which lie next to, or overlap, each other, upstream from the structural genes they control. Operators may lie at various positions relative to the promoter, either upstream or downstream. Expression of the operon is determined by access of RNA polymerase to the promoter, and occupancy of the operator by regulatory proteins influences this access. Induction activates transcription from the promoter; repression prevents it.

determines whether transcription takes place. This sequence is termed the **operator** (Figure 29.8). The operator is located next to a promoter. Interaction of a **regulatory protein** with the operator controls transcription of the gene cluster by controlling access of RNA polymerase to the promoter.[3] Such co-expressed gene clusters, together with the operator and promoter sequences that control their transcription, are called **operons.**

Transcription of Operons Is Controlled by Induction and Repression

In prokaryotes, gene expression is often responsive to small molecules serving as signals of the nutritional or environmental conditions confronting the cell. Increased synthesis of enzymes in response to the presence of a particular substrate is termed **induction.** For example, lactose (Figure 29.9) can serve as both carbon and energy source for *E. coli.* Metabolism of lactose depends on hydrolysis into its component sugars, glucose and galactose, by the enzyme **β-galactosidase.** In the absence of lactose, *E. coli* cells contain very little β-galactosidase (less than 5 molecules per cell). However, lactose availability *induces* the synthesis of β-galactosidase by activating transcription of the *lac* **operon.** One of the genes in the *lac* operon, *lacZ,* is the structural gene for β-galactosidase. When its synthesis is fully induced, β-galactosidase can amount to almost 10% of the total soluble protein in *E. coli.* When lactose is removed from the culture, synthesis of β-galactosidase halts.

The alternative to induction—namely *decreased* synthesis of enzymes in response to a specific metabolite—is termed **repression.** For example, the enzymes of tryptophan biosynthesis in *E. coli* are encoded in the *trp* **operon.** If sufficient Trp is available to the growing bacterial culture, the *trp* operon is not transcribed, so the Trp biosynthetic enzymes are not made; that is, their synthesis is *repressed.* Repression of the *trp* operon in the presence of Trp is an eminently logical control mechanism: If the end product of the pathway is present, why waste cellular resources making unneeded enzymes?

Induction and repression are two faces of the same phenomenon. In induction, a substrate activates enzyme synthesis. Substrates capable of activating synthesis of the enzymes that metabolize them are called **co-inducers,** or, often, simply **inducers.** Some substrate analogs can induce enzyme synthesis even though the enzymes are incapable of metabolizing them. These analogs are called **gratuitous inducers.** A number of thiogalactosides, such as **IPTG** (isopropyl β-thiogalactoside, Figure 29.10), are excellent gratuitous inducers of β-galactosidase activity in *E. coli.* In repression, a metabolite, typically an end product, depresses synthesis of its own biosynthetic enzymes. Such metabolites are called **co-repressors.**

Lactose
(O-β-D-galactopyranosyl (1 → 4) β-D-glucopyranose)

FIGURE 29.9 The structure of lactose, a β-galactoside.

Isopropyl β-thiogalactoside (IPTG)

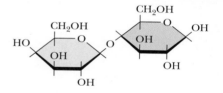

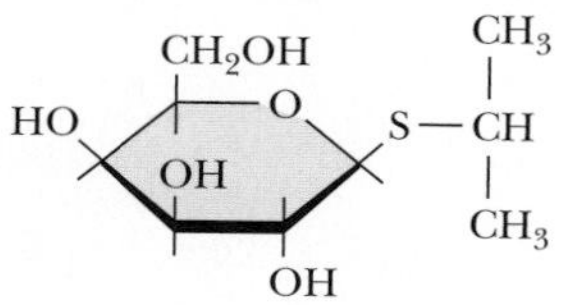

FIGURE 29.10 The structure of IPTG (isopropyl β-thiogalactoside).

[3]Although this is the paradigm for prokaryotic gene regulation, it must be emphasized that many prokaryotic genes do not contain operators and are regulated in ways that do not involve protein : operator interactions.

		p_{lacI}		lacI	p_{lac}, O		lacZ		lacY	lacA

DNA										
bp mRNA			1080		82		3069		1251	609
Polypeptide	Amino acids	360					1023		417	203
	kD	38.6					116.4		46.5	22.7
Protein	Structure	Tetramer					Tetramer		Membrane protein	Dimer
	kD	154.4					465		46.5	45.4
Function		Repressor					β-Galactosidase		Permease	Trans-acetylase

FIGURE 29.11 The *lac* operon. The operon consists of two transcription units. In one unit, there are three structural genes, *lacZ*, *lacY*, and *lacA*, under control of the promoter, p_{lac} and the operator *O*. In the other unit, there is a regulator gene, *lacI*, with its own promoter, p_{lacI}. *lacI* encodes a 360-residue, 38.6-kD polypeptide that forms a tetrameric *lac* repressor protein. *lacZ* encodes β-galactosidase, a tetrameric enzyme of 116-kD subunits. *lacY* is the β-galactoside permease structural gene, a 46.5-kD integral membrane protein active in β-galactoside transport into the cell. The remaining structural gene encodes a lactose transacetylase, whose function might allow excess lactose to be transported out of the cell by a multidrug resistance permease that exports acetylated substrates.

The *lac* Operon Serves as a Paradigm of Operons

In 1961, François Jacob and Jacques Monod proposed the **operon hypothesis** to account for the coordinate regulation of related metabolic enzymes. The operon was considered to be the unit of gene expression, consisting of two classes of genes: the **structural genes** for the enzymes and **regulatory genes** that control expression of the structural genes. The two kinds of genes could be distinguished by mutation. Mutations in a structural gene would abolish one particular enzymatic activity, but mutations in a regulatory gene would affect all of the different enzymes under its control. Mutations of both kinds were known in *E. coli* for lactose metabolism. Bacteria with mutations in either the *lacZ* gene or the *lacY* gene (Figure 29.11) could no longer metabolize lactose—the *lacZ* mutants (*lacZ*⁻ strains) because β-galactosidase activity was absent, the *lacY* mutants because lactose was no longer transported into the cell. Other mutations defined another gene, the *lacI* gene. *lacI* mutants were different because they both expressed β-galactosidase activity and immediately transported lactose, *without prior exposure to an inducer*. That is, a single mutation led to the expression of lactose metabolic functions independently of inducer. Expression of genes independently of regulation is termed **constitutive expression.** Thus, *lacI* had the properties of a regulatory gene. The *lac* operon includes the regulatory gene *lacI*; its promoter *p;* and three structural genes, *lacZ, lacY,* and *lacA,* with their own promoter p_{lac} and operator *O* (Figure 29.11).

lac Repressor Is a Negative Regulator of the *lac* Operon

The structural genes of the *lac* operon are controlled by **negative regulation.** That is, they are transcribed to give an mRNA unless turned off by the *lacI* gene product. This gene product is the ***lac* repressor,** a tetrameric protein (Figure 29.12). (Note that the language can be misleading: *Inducers* and *co-repressors* are small molecules/metabolites; *repressors* are proteins.) The *lac* repressor has an N-terminal DNA-binding domain; the rest of the protein functions in inducer binding and tetramer formation. In the absence of an inducer, *lac* repressor blocks *lac* gene expression by binding to the operator DNA site upstream from the *lac* structural genes. The *lac* operator is a palindromic DNA sequence (Figure 29.13). **Palindromes,** or "inverted repeats," provide a twofold, or dyad, symmetry, a structural feature common at sites in DNA where proteins specifically bind. Despite the presence of *lac* repressor, RNA polymerase can still initiate transcription at the promoter *(p_{lac}),* but *lac* repressor blocks elongation of transcription, so initiation is aborted. In *lacI* mutants, the *lac* repressor is absent or defective in binding to operator DNA, *lac* gene transcription is not blocked, and the *lac* operon is constitutively

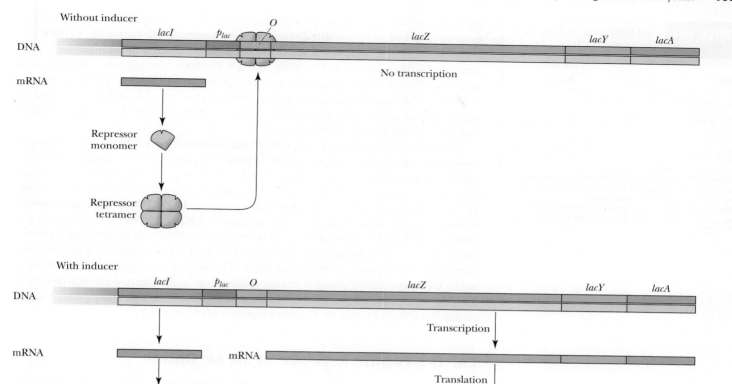

FIGURE 29.12 The mode of action of *lac* repressor. The structure of the *lac* repressor tetramer with bound IPTG (purple) is also shown (pdb id = 1LBH).

expressed. Note that *lacI* is normally expressed constitutively from its promoter, so *lac* repressor protein is always available to fill its regulatory role. About ten molecules of *lac* repressor are present in an *E. coli* cell.

Derepression of the *lac* operon occurs when appropriate β-galactosides occupy the inducer site on *lac* repressor, causing a conformational change in the protein that lowers

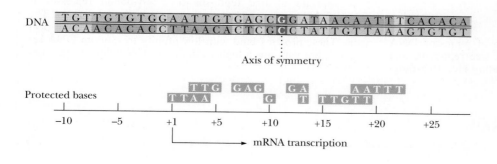

FIGURE 29.13 The nucleotide sequence of the *lac* operator. This sequence comprises 36 bp showing nearly palindromic symmetry. The inverted repeats that constitute this approximate twofold symmetry are shaded in rose. The bases are numbered relative to the +1 start site for transcription. The G:C base pair at position +11 represents the axis of symmetry. In vitro studies show that bound *lac* repressor protects a 26-bp region from −5 to +21 against nuclease digestion. Bases that interact with bound *lac* repressor are indicated below the operator. Note the symmetry of protection at +1 through +4 TTAA to +18 through +21 AATT.

TABLE 29.1	The Affinity of *lac* Repressor for DNA*	
DNA	Repressor	Repressor + Inducer
lac operator	$2 \times 10^{13} \ M^{-1}$	$2 \times 10^{10} \ M^{-1}$
All other DNA	$2 \times 10^6 \ M^{-1}$	$2 \times 10^6 \ M^{-1}$
Specificity†	10^7	10^4

*Values for repressor:DNA binding are given as association constants, K_A, for the formation of DNA:repressor complex from DNA and repressor.
†Specificity is defined as the ratio (K_A for repressor binding to operator DNA)/(K_A for repressor binding to random DNA).

the repressor's affinity for operator DNA. As a tetramer, *lac* repressor has four inducer binding sites, and its response to inducer shows cooperative allosteric effects. Thus, as a consequence of the "inducer"-induced conformational change, the inducer:*lac* repressor complex dissociates from the DNA, and RNA polymerase transcribes the structural genes (see Figure 29.12). Induction reverses rapidly: *lac* mRNA has a half-life of only 3 minutes, and once the inducer is used up through metabolism by the enzymes, free *lac* repressor reassociates with the operator DNA, transcription of the operon is halted, and any residual *lac* mRNA is degraded.

In the absence of inducer, *lac* repressor binds nonspecifically to duplex DNA with an association constant, K_A, of $2 \times 10^6 \ M^{-1}$ (Table 29.1) and to the *lac* operator DNA sequence with much higher affinity, $K_A = 2 \times 10^{13} \ M^{-1}$. Thus, *lac* repressor binds 10^7 times better to *lac* operator DNA than to any random DNA sequence. IPTG binds to *lac* repressor with an association constant of about $10^6 \ M^{-1}$. The IPTG:*lac* repressor complex binds to operator DNA with an association constant, $K_A = 2 \times 10^{10} \ M^{-1}$. Although this affinity is high, it is 3 orders of magnitude *less* than the affinity of inducer-free repressor for *lac* operator. There is no difference in the affinity of free *lac* repressor and *lac* repressor with IPTG bound for nonoperator DNA. The *lac* repressor apparently acts by binding to DNA and sliding along it, testing sequences in a one-dimensional search until it finds the *lac* operator. The *lac* repressor then binds there with high affinity until inducer causes this affinity to drop by 3 orders of magnitude (Table 29.1).

CAP Is a Positive Regulator of the *lac* Operon

Transcription by RNA polymerase from some promoters proceeds with low efficiency unless assisted by an accessory protein that acts as a *positive regulator*. One such protein is **CAP,** or **catabolite activator protein.** Its name derives from the phenomenon of catabolite repression in *E. coli*. Catabolite repression is a global control that coordinates gene expression with the total physiological state of the cell: As long as glucose is available, *E. coli* catabolizes it in preference to any other energy source, such as lactose or galactose. Catabolite repression ensures that the operons necessary for metabolism of these alternative energy sources, that is, the *lac* and *gal* operons, remain repressed until the supply of glucose is exhausted. Catabolite repression overrides the influence of any inducers that might be present.

> **A DEEPER LOOK**

Quantitative Evaluation of *lac* Repressor:DNA Interactions

The affinity of *lac* repressor for random DNA ensures that essentially all repressor is DNA bound. Assume that *E. coli* DNA has a single specific *lac* operator site for repressor binding and 4.64×10^6 base pairs. Assume also that any nucleotide sequence even one base out of phase with the operator constitutes a nonspecific binding site. Thus, there are 4.64×10^6 nonspecific sites for repressor binding.

The binding of repressor to DNA is given by the association constant, K_A:

$$K_A = \frac{[\text{repressor}:\text{DNA}]}{[\text{repressor}][\text{DNA}]}$$

where [repressor:DNA] is the concentration of repressor:DNA complex, [repressor] is the concentration of free repressor, and [DNA] is the concentration of nonspecific binding sites. Rearranging gives the following:

$$\frac{[\text{repressor}]}{[\text{repressor}:\text{DNA}]} = \frac{1}{K_A[\text{DNA}]}$$

If the number of nonspecific binding sites is 4.64×10^6, there are $(4.64 \times 10^6)/(6.022 \times 10^{23}) = 0.77 \times 10^{-17}$ moles of binding sites contained in the volume of a bacterial cell (roughly 10^{-15} liters). Therefore, [DNA] = $(0.77 \times 10^{-17})/(10^{-15}) = 0.77 \times 10^{-2} \ M$. Since $K_A = 2 \times 10^6 \ M^{-1}$ (Table 29.1),

$$\frac{[\text{repressor}]}{[\text{repressor}:\text{DNA}]} = \frac{1}{(2 \times 10^6)(0.77 \times 10^{-2})} = \frac{1}{(1.54 \times 10^4)}$$

So, the ratio of free repressor to DNA-bound repressor is 6.5×10^{-5}. *Less than 0.01% of repressor is not bound to DNA!* The behavior of *lac* repressor is characteristic of DNA-binding proteins. These proteins bind with low affinity to random DNA sequences, but with much higher affinity to their unique target sites (Table 29.1).

Catabolite repression is maintained until the *E. coli* cells become starved of glucose. Glucose starvation leads to activation of adenylyl cyclase, and the cells begin to make cAMP. (In contrast, glucose uptake is accompanied by deactivation of adenylyl cyclase.) The action of CAP as a positive regulator is cAMP-dependent. cAMP is a small-molecule inducer for CAP, and cAMP binding enhances CAP's affinity for DNA. CAP, also referred to as **CRP** (for **cAMP receptor protein**), is a dimer of identical 22.5-kD polypeptides. The N-terminal domains bind cAMP; the C-terminal domains constitute the DNA-binding site. Two molecules of cAMP are bound per dimer. The CAP–(cAMP)$_2$ complex binds to specific target sites near the promoters of operons (Figure 29.14). Binding of CAP–(cAMP)$_2$ to DNA causes the DNA to bend more than 80° (Figure 29.15). This CAP-induced DNA bending near the promoter assists RNA polymerase holoenzyme binding and closed promoter complex formation. Contacts made between the CAP–(cAMP)$_2$ complex and the α-subunit of RNA polymerase holoenzyme activate transcription.

Negative and Positive Control Systems Are Fundamentally Different

Negative and positive control systems operate in fundamentally different ways (although in some instances both govern the expression of the same gene). Genes under negative control are transcribed unless they are turned off by the presence of a repressor protein. Often, transcription activation is merely the release from negative control. In contrast, genes under positive control are expressed only if an active regulator protein is present. The *lac* operon illustrates these differences. The action of *lac* repressor is negative. It binds to operator DNA and blocks transcription; expression of the operon occurs only when this negative control is lifted through the release of the repressor. In contrast, regulation of the *lac* operon by CAP is positive: Transcription of the operon by RNA polymerase is stimulated by CAP's action as a positive regulator.

Operons can also be classified as **inducible, repressible,** or both, depending on how they respond to the small molecules that mediate their expression. Repressible operons are expressed only in the absence of their co-repressors. Inducible operons are transcribed only in the presence of small-molecule co-inducers (Figure 29.16).

The *araBAD* Operon Is Both Positively and Negatively Controlled by *AraC*

E. coli can use the plant pentose L-arabinose as sole source of carbon and energy. Arabinose is metabolized via conversion to D-xylulose-5-P (a pentose phosphate pathway intermediate and transketolase substrate [see Chapter 22]) by three enzymes encoded in

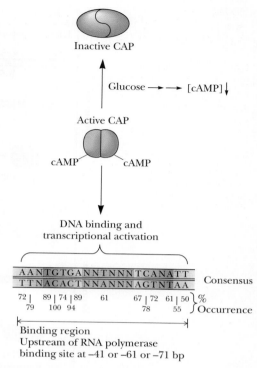

FIGURE 29.14 The mechanism of catabolite repression and CAP action. Glucose instigates catabolite repression by lowering cAMP levels. cAMP is necessary for CAP binding near promoters of operons whose gene products are involved in the metabolism of alternative energy sources such as lactose, galactose, and arabinose. The binding sites for the CAP–(cAMP)$_2$ complex are consensus DNA sequences containing the conserved pentamer TGTGA and a less well conserved inverted repeat, TCANA (where N is any nucleotide).

FIGURE 29.15 Binding of CAP–(cAMP)$_2$ induces a severe bend in DNA about the center of dyad symmetry at the CAP-binding site. The CAP dimer with two molecules of cAMP bound interacts with 27 to 30 base pairs of duplex DNA. Two α-helices of the CAP dimer insert into the major groove of the DNA at the dyad-symmetric CAP-binding site. The two cAMP molecules bound by the CAP dimer are indicated in yellow. Binding of CAP–(cAMP)$_2$ to its specific DNA site involves H bonding and ionic interactions between protein functional groups and DNA phosphates, as well as H-bonding interactions in the DNA major groove between amino acid side chains of CAP and DNA base pairs (pdb id = 1CGP).

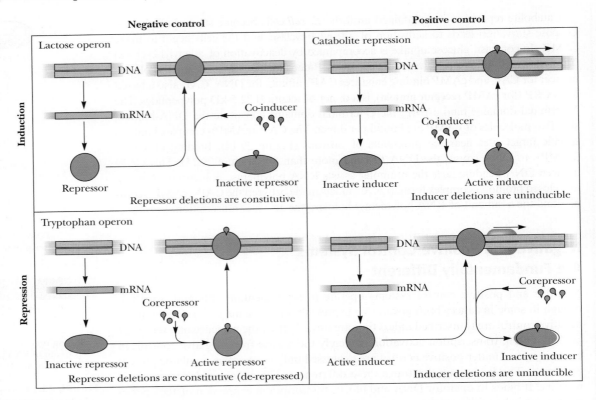

Negative control

Lactose operon

Positive control

Catabolite repression

Repressor deletions are constitutive

Inducer deletions are uninducible

Tryptophan operon

Repressor deletions are constitutive (de-repressed)

Inducer deletions are uninducible

FIGURE 29.16 Control circuits governing the expression of genes. These circuits can be either negative or positive, inducible or repressible.

the ***araBAD* operon.** Transcription of this operon is regulated by both catabolite repression and arabinose-mediated induction. CAP functions in catabolite repression; arabinose induction is achieved via the product of the *araC* gene, which lies next to the *araBAD* operon on the *E. coli* chromosome. The *araC* gene product, the protein **AraC**,[4] is a 292-residue protein consisting of an N-terminal domain (residues 1 to 170) that binds arabinose and acts as a dimerization motif and a C-terminal (residues 178 to 292) DNA-binding domain. Regulation of *araBAD* by AraC is novel in that it acts both negatively and positively. The *ara* operon has three binding sites for AraC: $araO_1$, located at nucleotides -106 to -144 relative to the *araBAD* transcription start site; $araO_2$ (spanning positions -265 to -294); and *araI*, the *araBAD* promoter. The *araI* site consists of two "half-sites"; $araI_1$ (nucleotides -56 to -78) and $araI_2$ (-35 to -51). (The $araO_1$ site contributes minimally to *ara* operon regulation.)

The details of *araBAD* regulation are as follows: When AraC protein levels are low, the *araC* gene is transcribed from its promoter p_c (adjacent to $araO_1$) by RNA polymerase (Figure 29.17). *araC* is transcribed in the direction away from *araBAD*. When cAMP levels are low and arabinose is absent, an AraC protein dimer binds to two sites, $araO_2$ and the $araI_1$ half-site, forming a DNA loop between them and restricting transcription of *araBAD* (Figure 29.17). In the presence of L-arabinose, the monomer of AraC bound to the $araO_2$ site is released from that site; it then associates with the unoccupied *araI* half-site, $araI_2$. L-Arabinose thus behaves as an allosteric effector that alters the conformation of AraC. In the arabinose-liganded conformation, the AraC dimer interacts with CAP–$(cAMP)_2$ to activate transcription by RNA polymerase. Thus, AraC protein is both a repressor and an activator.

Positive control of the *araBAD* operon occurs in the presence of L-arabinose *and* cAMP. Arabinose binding by AraC protein causes the release of $araO_2$, opening of the DNA loop, and association of AraC with $araI_2$. CAP–$(cAMP)_2$ binds at a site between $araO_1$ and *araI*, and together the AraC–(arabinose)$_2$ and CAP–$(cAMP)_2$ complexes

[4]Proteins are often named for the genes encoding them. By convention, the name of the protein is capitalized but not italicized.

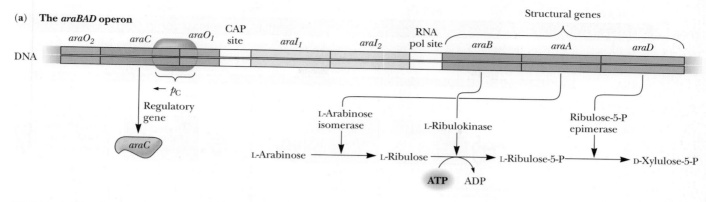

(a) **The *araBAD* operon**

(b) **Low [cAMP], no L-arabinose**

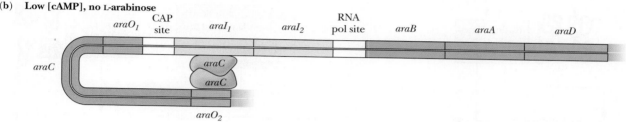

(c) **High [cAMP], L-arabinose present**

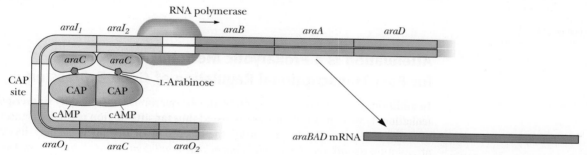

FIGURE 29.17 Regulation of the *araBAD* operon by the combined action of CAP and AraC protein.

influence RNA polymerase through protein:protein interactions to create an active transcription initiation complex. Supercoiling-induced DNA looping may promote protein:protein interactions between DNA-binding proteins by bringing them into juxtaposition.

The *trp* Operon Is Regulated Through a Co-Repressor–Mediated Negative Control Circuit

The *trp* operon of *E. coli* (and *S. typhimurium*) encodes the five polypeptides, *trpE* through *trpA* (Figure 29.18), that assemble into the three enzymes catalyzing tryptophan synthesis from chorismate (see Chapter 25). Expression of the *trp* operon is under the control of **Trp repressor,** a dimer of 108-residue polypeptide chains. When tryptophan is plentiful, Trp repressor binds two molecules of tryptophan and associates with the *trp* operator that is located within the *trp* promoter. Trp repressor binding excludes RNA polymerase from the promoter, preventing transcription of the *trp* operon. When Trp becomes limiting, repression is lifted because Trp repressor lacking bound Trp (Trp apo-repressor) has a lowered affinity for the *trp* promoter. Thus, the behavior of Trp repressor corresponds to a co-repressor–mediated, negative control circuit (see Figure 29.16). Trp repressor not only is encoded by the *trpR* operon but also regulates expression of the *trpR* operon. This is an example of **autogenous regulation (autoregulation):** regulation of gene expression by the product of the gene.

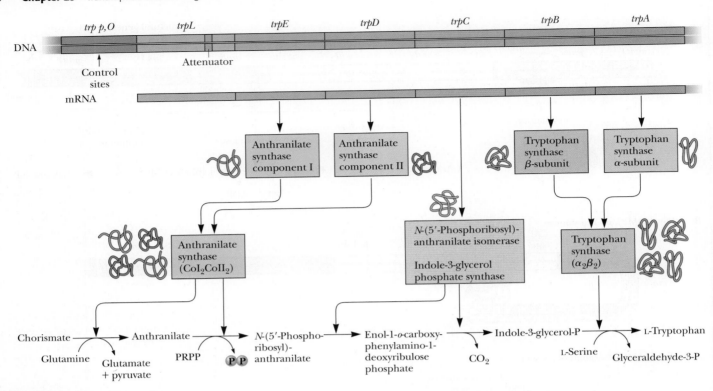

FIGURE 29.18 The *trp* operon of *E. coli.*

Attenuation Is a Prokaryotic Mechanism for Post-Transcriptional Regulation of Gene Expression

In addition to repression, expression of the *trp* operon is controlled by **transcription attenuation.** Unlike the mechanisms discussed thus far, attenuation regulates transcription after it has begun. Charles Yanofsky, the discoverer of this phenomenon, has defined attenuation as *any regulatory mechanism that manipulates transcription termination or transcription pausing to regulate gene transcription downstream.* In prokaryotes, transcription and translation (see Chapters 10 and 30) are coupled (see Figure 10.20), and the translating ribosome is affected by the formation and persistence of secondary structures in the mRNA. In many operons encoding enzymes of amino acid biosynthesis, a transcribed 150- to 300-bp leader region is positioned between the promoter and the first major structural gene. These regions encode a short leader peptide containing **multiple codons** for the pertinent amino acid. For example, the leader peptide of the *leu* operon has four leucine codons, the *trp* operon has two tandem tryptophan codons, and so forth (Figure 29.19). Translation of these codons depends on an adequate supply of the relevant aminoacyl-tRNA, which in turn rests on the availability of the amino acid.

Operon	Amino acid Sequence
his	Met — Thr — Arg — Val — Gln — Phe — Lys — His — His — His — His — His — His — His — Pro — Asp
ilv	Met — Thr — Ala — Leu — Leu — Arg — Val — Ile — Ser — Leu — Val — Val — Ile — Ser — Val — Val — Val — Ile — Ile — Ile — Pro — Pro — Cys — Gly — Ala — Ala — Leu — Gly — Arg — Gly — Lys — Ala
leu	Met — Ser — His — Ile — Val — Arg — Phe — Thr — Gly — Leu — Leu — Leu — Leu — Asn — Ala — Phe — Ile — Val — Arg — Gly — Arg — Pro — Val — Gly — Gly — Ile — Gln — His
pheA	Met — Lys — His — Ile — Pro — Phe — Phe — Phe — Ala — Phe — Phe — Phe — Thr — Phe — Pro
thr	Met — Lys — Arg — Ile — Ser — Thr — Thr — Ile — Thr — Thr — Thr — Ile — Thr — Ile — Thr — Thr — Gln — Asn — Gly — Ala — Gly
trp	Met — Lys — Ala — Ile — Phe — Val — Leu — Lys — Gly — Trp — Trp — Arg — Thr — Ser

FIGURE 29.19 Amino acid sequences of leader peptides in various amino acid biosynthetic operons regulated by attenuation. Color indicates amino acids synthesized in the pathway catalyzed by the operon's gene products. (The *ilv* operon encodes enzymes of isoleucine, leucine, and valine biosynthesis.)

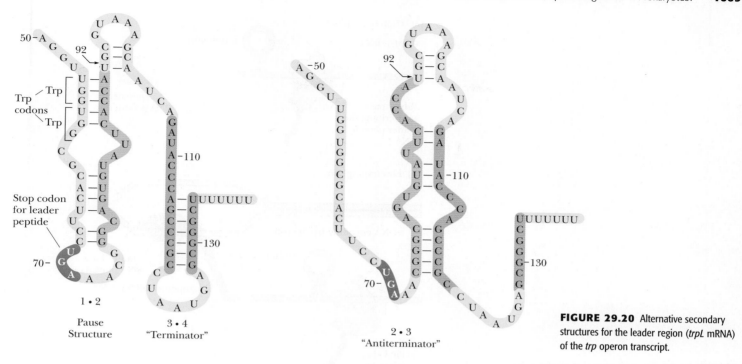

FIGURE 29.20 Alternative secondary structures for the leader region (*trpL* mRNA) of the *trp* operon transcript.

When tryptophan is scarce, the entire *trp* operon from *trpL* to *trpA* is transcribed to give a polycistronic mRNA. But, as [Trp] increases, more and more of the *trp* transcripts consist of only a 140-nucleotide fragment corresponding to the 5′-end of *trpL*. Tryptophan availability is causing premature termination of *trp* transcription, that is, transcription attenuation. Although attenuation occurs when tryptophan is abundant, attenuation is blocked when levels of tryptophan are low and little tryptophanyl-tRNA is available. The secondary structure of the 160-bp leader region transcript is the principal control element in transcription attenuation (Figure 29.20). This RNA segment includes the coding region for the 14-residue leader peptide. Three critical base-paired hairpins can form in this RNA: the **1:2 pause** structure, the **3:4 terminator,** and the **2:3 antiterminator.** Obviously, the 1:2 pause, 3:4 terminator, and the 2:3 antiterminator represent mutually exclusive alternatives. A significant feature of this coding region is the tandem UGG tryptophan codons.

Transcription of the *trp* operon by RNA polymerase begins and progresses until position 92 is reached, whereupon the 1:2 hairpin is formed, causing RNA polymerase to pause in its elongation cycle. While RNA polymerase is paused, a ribosome begins to translate the leader region of the transcript. Translation by the ribosome releases the paused RNA polymerase, and transcription continues, with RNA polymerase and the ribosome moving in unison. As long as tryptophan is plentiful enough that tryptophanyl-tRNA^Trp is not limiting, the ribosome is not delayed at the two tryptophan codons and follows closely behind RNA polymerase, translating the message soon after it is transcribed. The presence of the ribosome atop segment 2 blocks formation of the 2:3 antiterminator hairpin, allowing the alternative 3:4 terminator hairpin to form (Figure 29.21). Stable hairpin structures followed by a run of Us are features typical of *rho*-independent transcription termination signals, so the RNA polymerase perceives this hairpin as a transcription stop signal and transcription is terminated at this point. On the other hand, a paucity of tryptophan and, hence, low availability of tryptophanyl-tRNA^Trp causes the ribosome to stall on segment 1. This leaves segment 2 free to pair with segment 3 and to form the 2:3 antiterminator hairpin in the transcript. Because this hairpin precludes formation of the 3:4 terminator, termination is prevented and the entire operon is transcribed. Thus, transcription attenuation is determined by the availability of tryptophanyl-tRNA^Trp and its transitory influence over the formation of alternative secondary structures in the mRNA.

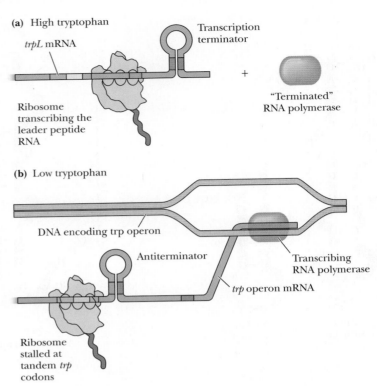

(a) High tryptophan

trpL mRNA

Transcription terminator

Ribosome transcribing the leader peptide RNA

"Terminated" RNA polymerase

(b) Low tryptophan

DNA encoding trp operon

Antiterminator

Transcribing RNA polymerase

trp operon mRNA

Ribosome stalled at tandem *trp* codons

FIGURE 29.21 The mechanism of attenuation in the *trp* operon.

DNA:Protein Interactions and Protein:Protein Interactions Are Essential to Transcription Regulation

Quite a variety of control mechanisms regulate transcription in prokaryotes. Several organizing principles materialize. First, **DNA:protein interactions** are a central feature in transcriptional control, and the DNA sites where regulatory proteins bind commonly display at least partial dyad symmetry or inverted repeats. Furthermore, DNA-binding proteins themselves are generally even-numbered oligomers (for example, dimers, tetramers) that have an innate, twofold rotational symmetry. Second, **protein:protein interactions** are an essential component of transcriptional activation. We see this latter feature in the activation of RNA polymerase by CAP–(cAMP)$_2$, for example. Third, the regulator proteins receive cues that signal the status of the environment (for example, Trp, lactose, cAMP) and act to communicate this information to the genome, typically via the medium of conformational changes and DNA:protein interactions.

Proteins That Activate Transcription Work Through Protein:Protein Contacts with RNA Polymerase

Although transcriptional control is governed by a variety of mechanisms, an underlying principle of transcriptional activation has emerged. Transcriptional activation can take place when a **transcriptional activator** protein [such as CAP–(cAMP)$_2$] bound to DNA makes protein:protein contacts with RNA polymerase, and the degree of transcriptional activation is proportional to the strength of the protein:protein interaction. Generally speaking, a nucleotide sequence that provides a binding site for a DNA-binding protein can serve as an **activator site** if the DNA-binding protein bound there can interact with promoter-bound RNA polymerase. These interactions can involve either the α-, β-, β'-, or σ-subunits of RNA polymerase. Moreover, if the DNA-bound transcriptional activator makes contacts with two different components of RNA polymerase, a synergistic effect takes place such that transcription is markedly elevated. Thus, transcriptional activation at specific genes relies on the presence of one or more activator sites where one or more transcriptional activator proteins can bind and make contacts

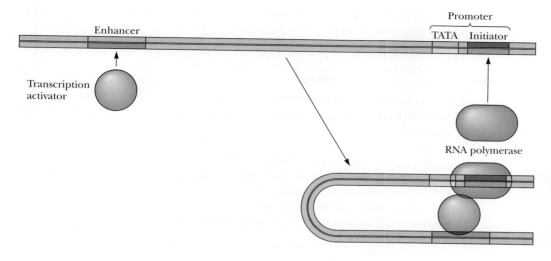

FIGURE 29.22 Formation of a DNA loop delivers DNA-bound transcriptional activator to RNA polymerase positioned at the promoter. Protein:protein interactions between the transcriptional activator and RNA polymerase activate transcription.

with RNA polymerase bound at the promoter of the gene. Indeed, transcriptional activators may facilitate the recruitment and binding of RNA polymerase to the promoter. This general principle applies to transcriptional activation in both prokaryotic and eukaryotic cells. In eukaryotes, transcriptional activators typically have discrete domains of protein structure dedicated to DNA binding (DB domains) and transcriptional activation (TA domains).

DNA Looping Allows Multiple DNA-Binding Proteins to Interact with One Another

Because transcription must respond to a variety of regulatory signals, multiple proteins are essential for appropriate regulation of gene expression. These regulatory proteins are the **sensors** of cellular circumstances, and they communicate this information to the genome by binding at specific nucleotide sequences. However, DNA is virtually a one-dimensional polymer, and there is little space for a lot of proteins to bind at (or even near) a transcription initiation site. DNA looping permits additional proteins to convene at the initiation site and to exert their influence on creating and activating an RNA polymerase initiation complex (Figure 29.22). The number of participants in transcriptional regulation is greatly expanded by DNA looping. It is important to note that, although illustrations on a printed page are two-dimensional, DNA looping occurs in the three dimensions of space. If you conceptualize DNA looping as three-dimensional, it becomes apparent that many different proteins, each bound to a DNA site remote from the others, could convene in three-dimensional space to recruit RNA polymerase to a promoter and execute exquisite control over its transcriptional activity.

29.3 How Are Genes Transcribed in Eukaryotes?

Although the mechanism of transcription in prokaryotes and eukaryotes is fundamentally similar, transcription is substantially more complicated in eukaryotes. The significant difference is that the DNA of eukaryotes is wrapped around histones to form nucleosomes, and the nucleosomes are further organized into chromatin (see Chapter 11). *Nucleosomes repress gene expression.* Nucleosomes control gene expression by controlling access of the transcriptional apparatus to genes. Two classes of transcriptional co-regulators are necessary to overcome nucleosome repression: (1) enzymes that covalently modify the nucleosome histone proteins and thereby loosen histone:DNA interactions and (2) ATP-dependent chromatin-remodeling complexes. However, gene activation depends not only on relief from nucleosome repression but also on interaction of RNA polymerase with the promoter. Only those genes activated by specific positive regulatory mechanisms are transcribed. A general understanding of transcription in eukaryotes rests on the following topics:

- The three classes of RNA polymerase in eukaryotes: RNA polymerases I, II, and III
- The structure and function of RNA polymerase II, the mRNA-synthesizing RNA polymerase
- Transcription regulation in eukaryotes, including:
  - General features of gene regulatory sequences: promoters, enhancers, and response elements
  - Transcription initiation by RNA polymerase II
    - The general transcription factors (GTFs)
    - Alleviating the repression due to nucleosomes
      - Histone acetyl transferases (HATs)
      - Chromatin-remodeling complexes
- A general model for eukaryotic gene activation, based on the preceding

We turn now to a review of these various features of eukaryotic transcription.

Eukaryotes Have Three Classes of RNA Polymerases

Eukaryotic cells have three classes of RNA polymerase, each of which synthesizes a different class of RNA. All three enzymes are found in the nucleus. **RNA polymerase I** is localized to the nucleolus and transcribes the major ribosomal RNA genes. **RNA polymerase II** transcribes protein-encoding genes, and thus it is responsible for the synthesis of mRNA. **RNA polymerase III** transcribes tRNA genes, the ribosomal RNA genes encoding 5S rRNA, and a variety of other small RNAs, including several involved in mRNA processing and protein transport.

All three RNA polymerase types are large, complex multimeric proteins (500 to 700 kD), consisting of ten or more types of subunits. Although the three differ in overall subunit composition, they have several smaller subunits in common. Furthermore, all possess two large subunits (each 140 kD or greater) having sequence similarity to the large β- and β'-subunits of *E. coli* RNA polymerase, indicating that the fundamental catalytic site of RNA polymerase is conserved among its various forms.

In addition to their different functions, the three classes of RNA polymerase can be distinguished by their sensitivity to **α-amanitin** (Figure 29.23), a bicyclic octapeptide produced by the poisonous mushroom *Amanita phalloides* (the "destroying angel" mushroom). α-Amanitin blocks RNA chain elongation. Although RNA polymerase I is resistant to this compound, RNA polymerase II is very sensitive and RNA polymerase III is less sensitive.

The existence of three classes of RNA polymerases acting on three distinct sets of genes implies that at least three categories of promoters exist to maintain this specificity. Eukaryotic promoters are very different from prokaryotic promoters. All three eukaryotic RNA polymerases interact with their promoters via so-called **transcription factors**—DNA-binding proteins that recognize and accurately initiate transcription at specific promoter sequences. For RNA polymerase I, its templates are the rRNA genes. Ribosomal RNA genes are present in multiple copies. Optimal expression of these genes requires the first 150 nucleotides in the immediate 5′-upstream region.

RNA polymerase III interacts with transcription factors **TFIIIA, TFIIIB,** and **TFIIIC.** Interestingly, TFIIIA and/or TFIIIC bind to specific recognition sequences that in some instances are located *within* the coding regions of the genes, not in the 5′-untranscribed region upstream from the transcription start site. TFIIIB associates with TFIIIA or TFIIIC already bound to the DNA. RNA polymerase III then binds to TFIIIB to establish an initiation complex.

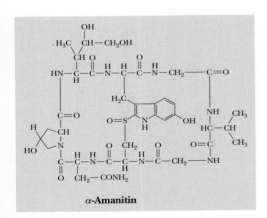

α-Amanitin

FIGURE 29.23 The structure of α-amanitin, one of a series of toxic compounds known as amatoxins that are found in the mushroom *Amanita phalloides.*

RNA Polymerase II Transcribes Protein-Coding Genes

As the enzyme responsible for the regulated synthesis of mRNA, RNA polymerase II has aroused greater interest than RNA polymerases I and III. RNA polymerase II must be capable of transcribing a great diversity of genes, yet it must carry out its function at

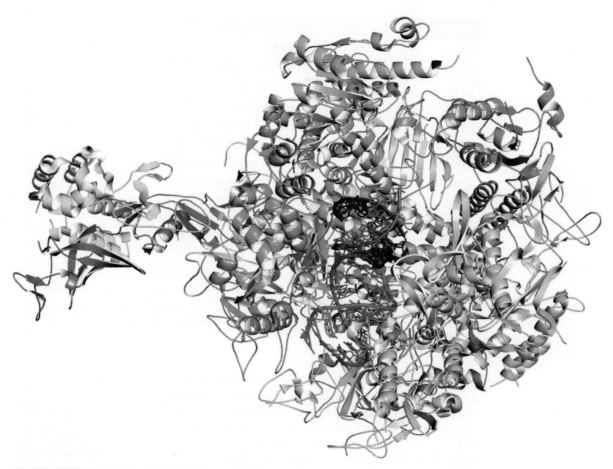

FIGURE 29.24 Structure of the yeast RNA polymerase (pdb id = 1Y77). The template DNA strand is shown in green, the nontemplate DNA strand in blue. The RNA transcript (hot pink) is emerging at the bottom of the structure. RPB1, the largest polypeptide chain, is shown in orange, its C-terminal domain (CTD) is to the upper right. The active-site Mg^{2+} is shown as a red sphere. The atoms of a ribonucleotide substrate analog (GMPCP) are shown as dark blue spheres.

any moment only on those genes whose products are appropriate to the needs of the cell in its ever-changing metabolism and growth. The RNA polymerase II enzymes from humans and yeast *(Saccharomyces cerevisiae)* have been extensively characterized, and the structure of the yeast enzyme has been solved by x-ray crystallography (Figure 29.24). Strong homology between yeast and human RNA polymerase II subunits suggests that the yeast RNA polymerase II is an excellent model for human RNA polymerase II. The yeast RNA polymerase II consists of 12 different polypeptides, designated RPB1 through RPB12 and ranging in size from 192 to 8 kD (Table 29.2). RPB3, RPB4, and RPB7 are unique to RNA polymerase II, whereas RPB5, RPB6, RPB8, and RPB10 are common to all three eukaryotic RNA polymerases.

RNA polymerases adopt a clawlike structure, grasping the DNA duplex as shown in Figure 29.24 for yeast RNA polymerase II. The DNA strands are unwound and separated, and the template strand enters the active site, where template-directed NTP substrate selection and NMP addition to the growing RNA transcript occur. NTP substrates access the active site through a channel below the active site. The DNA:RNA hybrid duplex exits at a right angle from the active site. As the hybrid duplex emerges from the protein structure, the RNA transcript is separated from the DNA template by RPB1 residue Phe[252], which acts as a wedge, splitting the position −10 RNA:DNA base pair (the base pair located 10 bases from the active site). The template strand is now free to re-anneal with the nontemplate strand to re-establish the dsDNA structure.

The RPB1 subunit has an unusual structural feature not found in prokaryotes: Its **C-terminal domain (CTD)** contains 27 repeats of the amino acid sequence YSPTSPS.

TABLE 29.2	Yeast* RNA Polymerase II Subunits		
Subunit	Size (kD)	Features	Prokaryotic Homolog
RPB1†	192	YSPTSPS CTD	β'
RPB2	139	NTP binding	β
RPB3	35	Core assembly	α
RPB4	25	Promoter recognition	σ
RPB5	25	In polymerases I, II, and III	
RPB6	18	In polymerases I, II, and III	
RPB7	19	Unique to polymerase II	
RPB8	17	In polymerases I, II, and III	
RPB9	14	Nonessential	
RPB10	8	In polymerases I, II, and III	
RPB11	14		
RPB12	8	In polymerases I, II, and III	

*A very similar RNA polymerase II can be isolated from human cells.
†RPB stands for RNA polymerase B; RNA polymerases I, II, and III are sometimes called RPA, RPB, and RPC.
Adapted from Myer, V. E., and Young, R. A., 1998. RNA polymerase II holoenzymes and subcomplexes. *The Journal of Biological Chemistry* **273**:27757–27760.

(The analogous subunit in RNA polymerase II enzymes of other eukaryotes has this heptapeptide tandemly repeated as many as 52 times.) Note that the side chains of 5 of the 7 residues in this repeat have —OH groups, endowing the CTD with considerable hydrophilicity *and* multiple sites for phosphorylation. A number of CTD kinases have been described, targeting different residues at different stages of the transcription process. The CTD domain may project more than 50 nm from the surface of RNA polymerase II. The phosphorylation possibilities represented by the YSPTSPS repeats within the CTD domain indicate the existence of a phosphorylation code that regulates mRNA synthesis and processing. In addition, an arginine residue (R^{810}) within the CTD can be methylated. Reversible methylation provides a selection mechanism for determining specific gene expression.

The CTD is essential to RNA polymerase II function. Only RNA polymerase II, whose CTD is not phosphorylated, can initiate transcription. However, transcription elongation proceeds only after protein phosphorylation within the CTD, suggesting that phosphorylation triggers the conversion of an initiation complex into an elongation complex. Such a mechanism would allow protein phosphorylation to regulate gene expression. Following termination of transcription, a phosphatase recycles RNA polymerase II to its unphosphorylated form. The CTD also plays a prominent role in orchestrating subsequent events in the transcription process. A multitude of additional proteins are essential to the formation of a translatable mRNA from the primary RNA polymerase II transcript; these proteins (described later in this chapter) include 5′-capping enzymes, splicing factors, and 3′-polyadenylylation complexes. Recruitment of these proteins is dependent upon phosphorylation of Ser residues at positions 2 and 5 in the CTD heptapeptide repeat. Phosphorylation of these Ser residues is also a prerequisite for interactions between the CTD and the histone methyltransferases capable of remodeling nucleosomes into a transcriptionally permissive state.

The Regulation of Gene Expression Is More Complex in Eukaryotes

Not only metabolic activity and cell division but also complex patterns of embryonic development and cell differentiation must be coordinated through the regulation of gene expression. All this coordinated regulation takes place in cells where the relative quantity (and diversity) of DNA is very great: A typical mammalian cell has 1500 times as much DNA as an *E. coli* cell. The structural genes of eukaryotes are rarely organized in clusters akin to operons. Each eukaryotic gene typically possesses a discrete set of regulatory sequences appropriate to the requirements for regulating its transcription. Certain of these sequences provide sites of interaction for general transcription factors, whereas others endow the gene with great specificity in expression by providing targets for specific transcription factors.

Gene Regulatory Sequences in Eukaryotes Include Promoters, Enhancers, and Response Elements

RNA polymerase II promoters commonly consist of two separate sequence features: the **core** element, near the transcription start site, where **general transcription factors (GTFs)** bind, and more distantly located **regulatory elements,** known variously as **enhancers** or **silencers.** These regulatory sequences are recognized by specific DNA-binding proteins that activate transcription above basal levels (*enhancers* bind *transcriptional activators*) or repress transcription (*silencers* bind *repressors*). The site of transcription initiation, called **initiator (Inr),** has the consensus sequence $(Py)_2CA(Py)_5$ (two pyrimidines, then CA, then five pyrimidines) located between positions -3 and $+5$, where $+1$ is the transcription start site. The core region often consists of a **TATA box** (a TATAAA consensus element) indicating the transcription start site; the TATA motif is usually located at position -25 (Figure 29.25). As in prokaryotic promoters, the A:T-rich nature of the TATA box facilitates promoter "melting" and access of the replication apparatus to the template strand. Genes that lack a TATA often have a **downstream promoter element (DPE)** centered on the -30 region. Other regulatory elements include short nucleotide sequences (sometimes called **response elements**) found near the promoter (the *promoter-proximal region*) that can bind certain specific transcription factors, such as proteins that trigger expression of a related set of genes in response to some physiological signal (hormone) or challenge (temperature shock).

Promoters The promoters of eukaryotic genes encoding proteins can be quite complex and variable, but they typically contain modules of short conserved sequences, such as the TATA box, the CAAT box, and the GC box. Sets of such modules embedded in the upstream region collectively define the promoter. The presence of a CAAT box, usually located around -80 relative to the transcription start site, signifies a strong promoter. One or more copies of the sequence GGGCGG or its complement (referred to as the GC box) have been found upstream from the transcription start sites of "housekeeping genes." Housekeeping genes encode proteins commonly present in all cells and essential to normal function; such genes are typically transcribed at more or less steady levels. Figure 29.26 depicts the promoter regions of several representative eukaryotic genes. Table 29.3 lists transcription factors that bind to respective modules. These transcription factors typically behave as positive regulatory proteins essential to transcriptional activation by RNA polymerase II at these promoters.

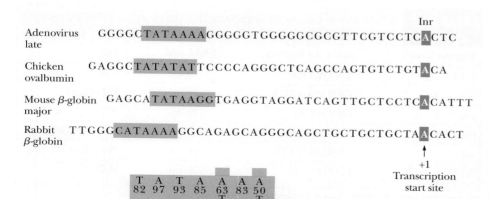

FIGURE 29.25 The Inr and TATA box in selected eukaryotic genes. The consensus sequence of a number of such promoters is presented in the lower part of the figure, the numbers giving the percent occurrence of various bases at the positions indicated.

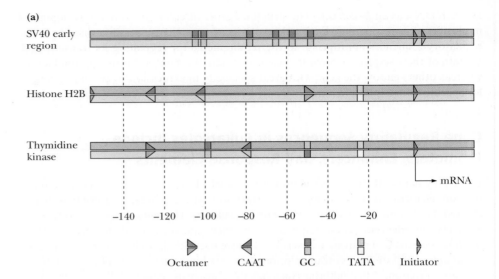

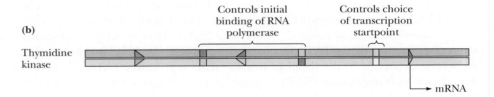

FIGURE 29.26 Promoter regions of several representative eukaryotic genes. **(a)** The SV40 early genes, the histone H2B gene, and the thymidine kinase gene. Note that these promoters contain different combinations of the various modules. In **(b),** the function of the modules within the thymidine kinase gene is shown.

Enhancers Eukaryotic genes have, in addition to promoters, regulatory sequences known as **enhancers.** Enhancers (also called **upstream activation sequences,** or **UASs**) assist initiation. Enhancers differ from promoters in two fundamental ways. First, the location of enhancers relative to the transcription start site is not fixed. Enhancers may be several thousand nucleotides away from the promoter, and they act to enhance transcription initiation even if positioned *downstream* from the gene. Second, enhancer sequences are *bidirectional* in that they function in either orientation. That is, enhancers can be removed and then reinserted in the reverse sequence orientation without impairing their function. Like promoters, enhancers represent modules of consensus sequence. Enhancers are "promiscuous," because they stimulate transcription from any promoter that happens to be in their vicinity. Nevertheless, *enhancer function is dependent on recognition by a specific transcription factor.* A specific transcription factor bound at an enhancer element stimulates transcription by interacting with an RNA polymerase II situated near a promoter.

TABLE 29.3	A Selection of Consensus Sequences That Define Various RNA Polymerase II Promoter Modules and the Transcription Factors That Bind to Them		
Sequence Module	Consensus Sequence	DNA Bound	Factor
TATA box	TATAAAA	~10 bp	TBP
CAAT box	GGCCAATCT	~22 bp	CTF/NF1
GC box	GGGCGG	~20 bp	SP1
Octamer	ATTTGCAT	~20 bp	Oct-1 or Oct-2
κB	GGGACTTTCC	~10 bp	NFκB or H2 TF1
ATF	GTGACGT	~20 bp	ATF

Adapted from Lewin, B., 1994. *Genes V.* Cambridge, MA: Cell Press.

TABLE 29.4	Response Elements That Identify Genes Coordinately Regulated in Response to Particular Physiological Challenges				
Physiological Challenge	Response Element	Consensus Sequence	DNA Bound	Factor	Size (kD)
Heat shock	HSE	CNNGAANNTCCNNG	27 bp	HSTF	93
Glucocorticoid	GRE	TGGTACAAATGTTCT	20 bp	Receptor	94
Cadmium	MRE	CGNCCCGGNCNC			
Phorbol ester	TRE	TGACTCA	22 bp	AP1	39
Serum	SRE	CCATATTAGG	20 bp	SRF	52

Adapted from Lewin, B., 1994. *Genes V.* Cambridge, MA: Cell Press.

Response Elements Promoter modules in genes responsive to common regulation are termed **response elements.** Examples include the **heat shock element (HSE),** the **glucocorticoid response element (GRE),** and the **metal response element (MRE).** These various elements are found in the promoter regions of genes whose transcription is activated in response to a sudden increase in temperature (heat shock), glucocorticoid hormones, or toxic heavy metals, respectively (Table 29.4). HSE sequences are recognized by a specific transcription factor, **HSTF** (for **heat shock transcription factor**). HSEs are located about 15 bp upstream from the transcription start site of a variety of genes whose expression is dramatically enhanced in response to elevated temperature. Similarly, the response to steroid hormones depends on the presence of a GRE positioned 250 bp upstream of the transcription start point. Activation of the **steroid receptor** (a specific transcription factor) at a GRE occurs when certain steroids bind to the steroid receptor.

Many genes are subject to multiple regulatory influences. Regulation of such genes is achieved through the presence of an array of different regulatory elements. The **metallothionein** gene is a good example (Figure 29.27). Metallothionein is a metal-binding protein that protects cells against metal toxicity by binding excess amounts of heavy metals and removing them from the cell. This protein is always present at low levels, but its concentration increases in response to heavy metal ions such as cadmium or in response to glucocorticoid hormones. The metallothionein gene promoter consists of two general promoter elements, namely, a TATA box and a GC box, two basal-level enhancers, four MREs, and one GRE. These elements function independently of one another; any one is able to activate transcription of the gene.

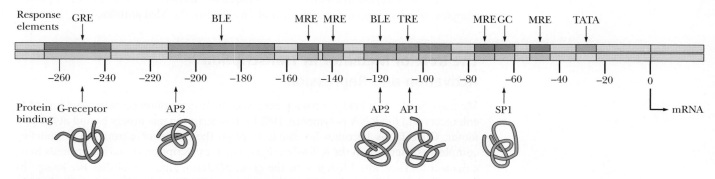

FIGURE 29.27 The metallothionein gene possesses several constitutive elements in its promoter (the TATA and GC boxes) as well as specific response elements such as MREs and a GRE. The BLEs are elements involved in basal level expression (constitutive expression). TRE is a tumor response element activated in the presence of tumor-promoting phorbol esters such as TPA (tetradecanoyl phorbol acetate).

TABLE 29.5		General Transcription Initiation Factors from Human Cells
Factor	Number of Subunits	Function
TFIID		
TBP	1	Core promoter recognition (TATA); platform for recruitment of TFIIA, TFIIB, and TAFs
TAFs	14	Core promoter recognition (non-TATA elements); positive and negative regulatory functions; HAT (histone acetyltransferase) activity
TFIIA	3	Stabilization of TBP binding; stabilization of TAF–DNA interactions
TFIIB	1	Binds TBP, RNA Polymerase II, and promoter DNA; determines transcription start site.
TFIIF	2	Promoter targeting of polymerase II; destabilization of nonspecific RNA polymerase II–DNA interactions
RNA pol II	12	Enzymatic synthesis of RNA; TFIIE recruitment
TFIIE	2	TFIIH recruitment; modulation of TFIIH helicase, ATPase, and kinase activities; promoter melting
TFIIH	9	Promoter melting using helicase activity; promoter clearance via CTD phosphorylation (2 subunits of TFHII are a cyclin:CDK pair)
TFIIS	34	Initiation complex stabilization, elongation

Adapted from Table 1 in Roeder, R. G., 1996. The role of general initiation factors in transcription by RNA polymerase II. *Trends in Biochemical Sciences* **21**:327–335; and Reese, J. C., 2003. Basal transcription factors. *Current Opinion in Genetics and Development* **13**:114–118; and Table 1 in Hahn, S., 2004. Structure and mechanism of the RNA polymerase II transcription machinery. *Nature Structural and Molecular Biology* **11**:394–403.

Transcription Initiation by RNA Polymerase II Requires TBP and the GTFs

Transcription initiation in eukaryotes proceeds through assembly of a **preinitiation complex (PIC)** at promoter DNA. The PIC consists of RNA polymerase II, and a set of general transcription factors (**TFIIA, TFIIB, TFIID, TFIIE, TFIIF, TFIIH, and TFIIS;** Table 29.5). **TFIID** consists of the **TBP** (the **T**ATA **b**inding **p**rotein), which directly recognizes the TATA box within the core promoter and a set of **T**BP-**a**ssociated **f**actors **(TAFs).** TBP is the TATA box-binding protein that recognizes the promoter core, making contacts with the DNA minor groove and distorting and bending the DNA so that sequences upstream and downstream of the TAT box come into closer proximity (Figure 29.28a). **TFIIA** stimulates transcription by stabilizing the interaction of TFIID and the TATA box. **TFIIB** recruits the PIC to the promoter, positioning the RNA polymerase active site over the promoter. **TFIIE** and **TFIIF** enhance various aspects of promoter:PIC interaction. The helicase activity of **TFIIH** aids in creating the transcription bubble. The transition from a state where the PIC is situated over the promoter (the "closed" initiation complex) occurs when TFIIB opens the promoter. The assemblage of RNA polymerase II and these other proteins at the promoter is now termed the **initiation complex.** However, transcription initiation in eukaryotes also requires **Mediator,** a 1.5 megadalton protein complex, which in humans is composed of 26 subunits, the **Med proteins.**

The Role of Mediator in Transcription Activation and Repression

Mediator serves as a bridge between gene-specific transcription co-activators bound to enhancers and the RNA polymerase II/GTF transcription machinery bound at the promoter. Once DNA becomes accessible through the action of chromatin remodeling complexes (discussed in the following sections), a transcription co-activator binds to an enhancer and recruits Mediator to the gene. Mediator then establishes the bridge by promoting assembly of the PIC at the promoter. Mediator has been described as the ultimate regulator of transcription since it integrates and communicates information from enhancers and transcription co-activators to RNA polymerase at the promoter.

Mediator is a somewhat crescent-shaped structure, with a head, middle, and tail. The tail section recognizes and binds the transcription co-activator. Both the head region

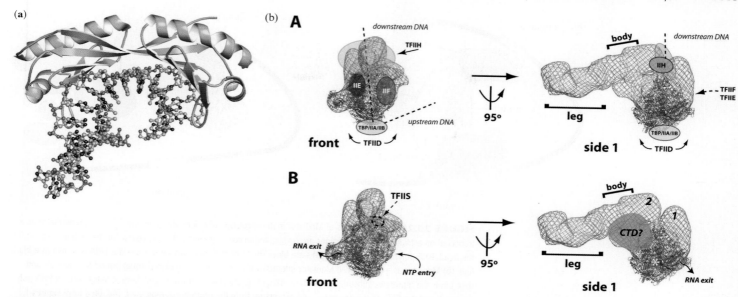

FIGURE 29.28 Transcription initiation. **(a)** Model of the TATA-binding protein (TBP) in complex with a DNA TATA sequence (pdb id = 1YTB). The saddle-shaped TBP (gold) is unusual in that it binds in the minor groove of DNA, sitting on the DNA like a saddle on a horse. TBP binding pries open the minor groove, creating a 100° bend in the DNA axis and unwinding the DNA within the TATA sequence. The other subunits of the TFIID complex (see Table 29.5) sit on TBP, like a "cowboy on a saddle." All known eukaryotic genes (those lacking a TATA box as well as those transcribed by RNA polymerase I or III) rely on TBP. **(b) A** Front and side views of the overall Mediator–RNA polymerase II–TFIIF complex, as obtained through cryo-electron microscopy, are shown as blue mesh. The location of RNA polymerase II is shown in red. The locations of TBP, TFIID, TFIIA, TFIIB, TFIIF, TFIIE, and TFIIH are indicated. The likely path of upstream and downstream promoter DNA is shown by the dashed line. **B** The binding site for TFIIS is shown (front view), along with the NTP substrate entry path and the RNA exit site. The probable location of the RNA polymerase II CTD is shown in green, and the asterisk indicates where the CTD extends from the enzyme. *Adapted from Figure 5 in Bernecky, C., Grob, P., Ebmeier, C. C., Nogales, E. and Taatjes, D. J., 2011. Molecular architecture of the human Mediator-RNA polymerase II-TFIIF assembly.* PLoS Biology **9***:e1000603. doi:10.1371/journal.pbio.1000603. Part (b) courtesy of Dylan J. Taatjes.*

and the middle region of Mediator bind RNA polymerase II through interactions with its CTD, with the middle region also associating with general transcription factor TFIIE (Figure 29.29a).

Two of the Mediator subunits are a cyclin-CDK pair **(CycC/CDK8),** and they act to phosphorylate S5 in the CTD YSPTSPS heptapeptide repeat. Mediator is a global regulator essential to transcription of virtually every RNA polymerase II-dependent gene. Mediator is even required for basal transcription of these genes. Mediator also displays HAT activity, which may aid in exposing promoters and subsequent binding of GTFs and RNA polymerase II. Taken together, the results indicate that Mediator provides a scaffold for assembly of the **PIC** at the promoter just prior to transcription initiation. Mediator apparently acts as the gateway by which RNA polymerase II gains access to the promoter. As a protein kinase, Mediator also acts in phosphorylation of the unphosphorylated RNA polymerase CTD, thereby prompting its transition into the elongation phase of transcription.

Mediator as a Repressor of Transcription

Mediator apparently acts in repression of transcription as well as activation. How can Mediator serve two opposing regulatory functions? Perhaps the explanation lies in the ability of a Mediator subcomplex (the MED12 and MED13 proteins, plus CycC/CDK8) to interact with co-repressor bound to a silencer (see Figure 29.29). This

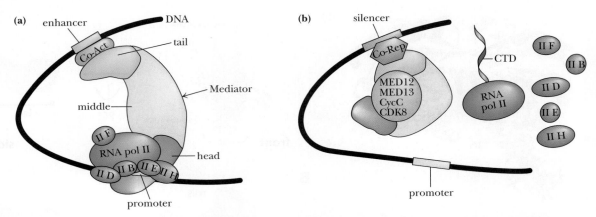

FIGURE 29.29 Simple models of Mediator in the regulation of eukaryotic gene transcription. **(a)** Mediator as a transcription activator. Mediator regions are highlighted in color: green for the tail, yellow for the middle, and red for the head. RNA polymerase II and the GTFs are blue. The transcription co-activator is orange. DNA is shown as a black line. **(b)** Mediator as a repressor. If Mediator interacts with a co-repressor (colored aqua) bound to a silencer and then binds the repressive subcomplex (MED12 : MED13 : CycC/CDK8), shown in pink here, it fails to recruit RNA polymerase II and the GTFs to the promoter, and expression from the promoter is repressed. (Adapted from Figures 1 and 2 in Björklund, S., and Gustafson, C. M., 2005. The yeast Mediator complex and its regulation. *Trends in Biochemical Sciences* **30**:240–244.)

Mediator subcomplex is a repressive module that stabilizes Mediator in a conformation that cannot recruit RNA polymerase II to the promoter (Figure 29.29b).

Chromatin-Remodeling Complexes and Histone-Modifying Enzymes Alleviate the Repression Due to Nucleosomes

The central structural unit of nucleosomes, the histone "core octamer" (see Figure 11.28), is constructed from the eight *histone-fold protein domains* of the eight various histone monomers comprising the octamer. Successive histone octamers are linked via histone H1, which is not part of the octamer. Each histone monomer in the core octamer has an unstructured N-terminal tail that extends outside the core octamer. Interactions between histone tails contributed by core histones in adjacent nucleosomes are an important influence in establishing the higher orders of chromatin organization. Activation of eukaryotic transcription is dependent on *two* sets of circumstances: (1) relief from the repression imposed by chromatin structure and (2) interaction of RNA polymerase II with the promoter and transcription regulatory proteins. Relief from repression requires factors that can reorganize the chromatin and then alter the nucleosomes so that promoters become accessible to the transcriptional machinery. Two sets of factors are important: **chromatin-remodeling complexes** that mediate ATP-dependent conformational (noncovalent) changes in nucleosome structure and **histone-modifying enzymes** that introduce covalent modifications into the N-terminal tails of the histone core octamer. Chromatin remodeling and histone modification are closely linked processes.

Chromatin-Remodeling Complexes Are Nucleic Acid–Stimulated Multisubunit ATPases

Chromatin-remodeling complexes are huge (1 megadalton) assemblies containing ATP-dependent enzymes that loosen the DNA : protein interactions in nucleosomes by sliding, ejecting, inserting, or otherwise restructuring core octamers. In the process, about 50 bp of DNA are "peeled" away from the edge of the nucleosome, creating a "bulge" that allows RNA polymerase II, GTFs, and other transcription factors to access the DNA. Chromatin-remodeling complexes contain proteins of the **SNF2** family of **DEAD/H-box-**containing, nucleic acid–stimulated ATPases. (DEAD signifies the Asp-

Glu-Ala-Asp tetrapeptide signature sequence of this protein family; in some of these proteins, a histidine residue (H) replaces the second D in the box.) SNF2 family members include SWI2 types with **bromodomains** that bind acetylated lysines, ISWI types that have separate domains for histone tail and linker DNA binding, CHD types with **chromodomains** that bind methylated lysines, and INO80 types with DNA-binding domains.

During the elongation phase of transcription, RNA polymerase II must pass by each nucleosome, and such passage is believed to result in the loss and regain of an H2A/H2B dimer from the core octamer. **FACT** (for **fac**ilitates **ch**romatin **t**ranscription) is a heterodimer that acts on some nucleosomes to catalyze H2A/H2B removal and readdition; its activity is markedly increased by **ubiquitination** (see Chapter 31) of H2B Lys^{120}.

Covalent Modification of Histones

Chromatin is also remodeled through the action of enzymes that covalently modify side chains on histones within the core octamer. These modifications either diminish DNA:histone associations through disruption of electrostatic interactions or introduce substitutions that can recruit binding of new protein participants through protein–protein interactions.

Initial events in transcriptional activation include acetyl-CoA–dependent acetylation of ε-amino groups on lysine residues in histone tails by **histone acetyltransferases (HATs)** (Figure 29.30). The histone transacetylases responsible are essential components of several megadalton-size complexes known to be required for transcription co-activation (*co-activation* in the sense that they are required along with RNA polymerase II and other components of the transcriptional apparatus). Examples of such complexes include the **TFIID** (some of whose TAF$_{II}$s have HAT activity), the **SAGA complex** (which also contains TAF$_{II}$s), and the **ADA complex**. *N*-Acetylation suppresses the positive charge in histone tails, diminishing their interaction with the negatively charged DNA.

Phosphorylation of Ser residues and methylation of Lys residues in histone tails also contribute to transcription regulation (Figure 29.30). Attachment of small proteins to histone C-terminal lysine residues through ubiquitination and **sumoylation** (see Chapter 31) are two additional forms of covalent modification found in nucleosomes. Collectively,

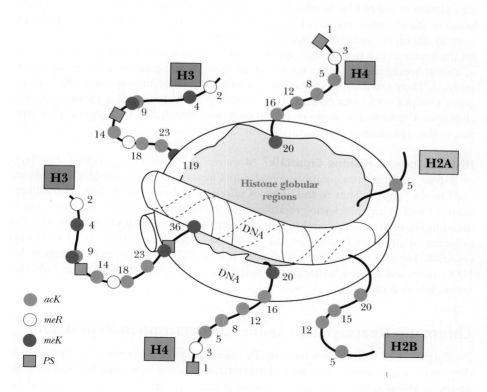

FIGURE 29.30 A schematic diagram of the nucleosome illustrating the various covalent modifications on the N-terminal tails of histones. *acK* = acetylated lysine residue; *meK* = methylated lysine residue; *meR* = methylated arginine residue; *PS* = phosphorylated serine residue. The numbers indicate the positions of the amino acids in the amino acid sequences. Note the prevalence of modifiable sites, particularly acetylatable lysines, on the N-terminal tails of histones H2B, H3, and H4.

these modifications create binding sites for proteins that modulate chromatin structure, such as the chromatin-remodeling complexes with bromodomains that interact specifically with acetylated lysine residues and chromodomains that bind to methylated lysine residues. A "histone code" has emerged.

Covalent Modification of Histones Forms the Basis of the Histone Code

A code based on histone-tail covalent modifications determines gene expression through selective recruitment of proteins. Proteins that cause chromatin compaction (heterochromatin formation) lead to repression; proteins giving easier access to DNA through relaxation of histone:DNA interactions favor the possibility of gene expression.

The prominent forms of histone covalent modification are lysine acetylation, lysine methylation, serine phosphorylation, lysine ubiquitination, and lysine sumoylation. The lysine residue at position 9 (K9) in the histone H3 amino acid sequence is methylated in heterochromatin, the compacted, repressed state of chromatin. In contrast, lysine 4 (K4) of histone H3 typically is methylated in chromatin where gene expression is active. Different proteins are recruited to these two methylated forms of histone H3. Methylated K9 recruits **heterochromatin protein 1 (HP1),** which binds via its chromodomain. On the other hand, methylated K4 binds **CHD1,** a chromatin remodeling protein with two chromodomains. (CHD1 is an acronym for **c**hromodomain, **h**elicase, **D**NA-binding.) Ubiquitination of Lys^{120} in the C-terminal tail of H2B favors methylation (and thus transcription activation), whereas ubiquitination of Lys^{119} in the C-terminal tail of H2A favors repression. Sumoylation of Lys residues tends to repress transcription; apparently, sumoylation antagonizes acetylation.

Methylation and Phosphorylation Act as a Binary Switch in the Histone Code

As cells enter mitosis, the chromatin becomes condensed and histone H3 is not only methylated at K9 but also phosphorylated at the adjacent serine residue, S10. S10 phosphorylation triggers the dissociation of HP1 from the heterochromatin. Thus, phosphorylation of the residue neighboring K9 trumps HP1 binding. Similarly, phosphorylation of the threonine residue (T3) neighboring K4 in the histone H3 tail evicts CHD1 from its site on the methylated K4. Apparently, lysine methylation is the "on" position for the binary switch that recruits specific proteins to histone tails, and phosphorylation at a neighboring residue turns the switch to the "off" position by ejecting the bound proteins. There are at least 16 instances of serines or threonines immediately flanking lysine residues in the four histones that constitute the histone core octamer of the nucleosome. The methylation-phosphorylation binary switch may be a general phenomenon in the regulation of chromatin dynamics.

Histone Code or Histone Crosstalk? More recent research suggests that the effects of histone modifications are more complex and nuanced than the concept of a histone code might suggest. That is, the order of addition and removal of histone modifications, as well as any preexisting modifications that might be present, are important in determining the transcriptional fate of the gene associated with the nucleosome. The influence of one modification on the likelihood of another is referred to as **histone crosstalk.** For example, phosphorylation of histone H3 S10 by different kinases at different times can affect whether or not acetylation of histone H3 occurs, and whether transcription ensues.

Chromatin Deacetylation Leads to Transcription Repression

Deacetylation of histones is a biologically relevant matter, and enzyme complexes that carry out such reactions have been characterized. Known as **histone deacetylase complexes,** or **HDACs,** they catalyze the removal of acetyl groups from lysine residues along the histone tails, restoring the chromatin to a repressed state. Beyond these effects on transcription, histone modifications determine whether significant cellular events involving DNA allocation through mitosis and meiosis may occur.

Nucleosome Alteration and Interaction of RNA Polymerase II with the Promoter Are Essential Features in Eukaryotic Gene Activation

Gene activation (the initiation of transcription) can thus be viewed as a process requiring two principal steps: (1) alterations in nucleosomes (and thus, chromatin) that relieve the general repressed state imposed by chromatin structure, followed by (2) the interaction of RNA polymerase II and the GTFs with the promoter. **Transcription activators** (proteins that bind to enhancers and response elements) initiate the process by recruiting chromatin-altering proteins (the *chromatin-remodeling complexes* and *histone-modifying enzymes* described previously). Once these alterations have occurred, promoter DNA is accessible to TBP:TFIID, the other GTFs, and RNA polymerase II. Transcription activation, however, requires communication between RNA polymerase II and the *transcription activator* for transcription to take place. Mediator fulfills this function. Mediator interacts with *both* the transcription activator *and* the CTD of RNA polymerase II. This Mediator bridge provides an essential interface for communication between enhancers and promoters, triggering RNA polymerase II to begin transcription. A general model for transcription initiation is shown in Figure 29.31. Once transcription begins, Mediator is replaced by another complex called **Elongator.** *Elongator* has HAT subunits whose activity remodels downstream nucleosomes as RNA polymerase II progresses along the chromatin-associated DNA.

The interactions described thus far emphasize regulation of gene expression at the level of RNA polymerase II recruitment to promoters. However, whole-genome analyses show that, for many genes, RNA polymerase II is already situated at promoters and appears to be paused there, awaiting signals that will activate the elongation phase of transcription. Thus, the expression of many genes may be regulated at the level of transcription elongation. What distinguishes between these possibilities? It seems that RNA polymerase pausing occurs more often on highly regulated genes because the promoters of these genes might otherwise be blocked by nucleosomes. Polymerase pausing is a mechanism to outcompete nucleosome occlusion of the promoter, so that the gene is poised for expression, as decided by regulatory signals. On the other hand, the promoters

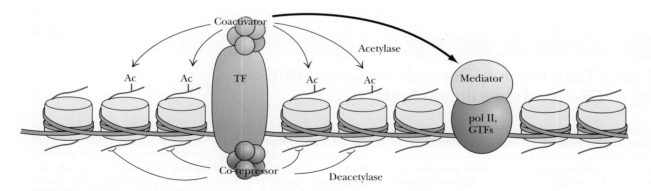

FIGURE 29.31 A model for the transcriptional regulation of eukaryotic genes. The DNA is a green ribbon wrapped around disclike nucleosomes. A specific transcription factor (TF, pink) is bound to a regulatory element (either an enhancer or silencer). RNA polymerase II and its associated GTFs (blue) are bound at the promoter. The N-terminal tails of histones are shown as wavy lines (blue) emanating from the nucleosome discs. A specific transcription factor that is a transcription activator stimulates transcription through interaction with a co-activator whose HAT activity renders the DNA more accessible *and* through interactions with the Mediator complex associated with RNA polymerase II. A specific transcription factor that is a repressor interacts with a co-repressor that has HDAC activity that deacetylates histones, restructuring the nucleosomes into a repressed state. (From Figure 1 in Kornberg, R. D., 1999. Eukaryotic transcription control. *Trends in Biochemical Sciences* **24:**M46–M49.)

of genes that are constitutively expressed, such as housekeeping genes, are relatively nucleosome-free.

Beyond these considerations, various mechanisms regulate gene expression through events that take place subsequent to transcription. **Post-transcriptional gene regulation** mediated by microRNAs (see Section 29.6), as well as **alternative RNA splicing** and nucleotide changes introduced through **RNA editing** (as described in this chapter), are mechanisms targeting transcripts. **Post-translational modifications** of proteins also play a major role in the regulation of gene expression, as assessed at the level of biological activity (see Chapter 30).

A SINE of the Times

An interesting twist on transcription regulation comes from the discovery that certain noncoding RNAs (ncRNAs) act as transcription factors through direct binding to RNA polymerase II. For example, ncRNA B2 in mice and Alu RNA in humans are RNAs encoded within **short interdispersed elements (SINEs).** SINEs are abundant within animal DNA and were once considered "junk" DNA because they lack protein-coding properties. Ala RNA or ncRNA B2 blocks promoter-bound RNA polymerase II by interfering with transcription initiation.

29.4 : How Do Gene Regulatory Proteins Recognize Specific DNA Sequences?

Proteins that recognize nucleic acids do so by the basic rule of macromolecular recognition. That is, such proteins present a three-dimensional shape or contour that is structurally and chemically complementary to the surface of a DNA sequence. When the two molecules come into contact, the numerous atomic interactions that underlie recognition and binding can take place. Nucleotide sequence–specific recognition by the protein involves a set of atomic contacts with the bases and the sugar–phosphate backbone. Hydrogen bonding is critical for recognition, with amino acid side chains providing most of the critical contacts with DNA. Protein contacts with the bases of DNA usually occur within the major groove (but not always). Protein contacts with the DNA backbone involve both H bonds and salt bridges with electronegative oxygen atoms of the phosphodiester linkages. Structural studies on regulatory proteins that bind to specific DNA sequences have revealed that roughly 80% of such proteins can be assigned to one of three principal classes based on their possession of one of three kinds of

---

HUMAN BIOCHEMISTRY

Storage of Long-Term Memory Depends on Gene Expression Activated by CREB-Type Transcription Factors

Learning can be defined as the process whereby new information is acquired and *memory* as the process by which this information is retained. Short-term memory (which lasts minutes or hours) requires only the covalent modification of preexisting proteins, but long-term memory (which lasts days, weeks, or a lifetime) depends on gene expression, protein synthesis, and the establishment of new neuronal connections.

The macromolecular synthesis underlying long-term memory storage requires **cAMP-r**esponse **e**lement-**b**inding **(CREB)** protein–related transcription factors and the activation of cAMP-dependent gene expression. Serotonin (5-hydroxytryptamine, or 5-HT, a hormone implicated in learning and memory) acting on neurons promotes cAMP synthesis, which in turn stimulates protein kinase A to phosphorylate CREB protein–related transcription factors that activate transcription of cAMP-inducible genes. These genes are characterized by the presence of **CRE** (**c**AMP **r**esponse **e**lement) consensus sequences containing the 8-bp TGACGTCA palindrome. CREB transcription factors are *bZIP*-type proteins (see later discussion). These exciting findings opened a new arena in molecular biology, the molecular biology of **cognition.** Eric Kandel was awarded the 2000 Nobel Prize in Physiology or Medicine for, among other things, his discovery of the role of CREB-type transcription factors in long-term memory storage.

Cognition is the act or process of knowing; the acquisition of knowledge.

small, distinctive structural motifs: the **helix-turn-helix** (or **HTH**), the **zinc finger** (or **Zn-finger**), and the **leucine zipper-basic region** (or **bZIP**). The latter two motifs are found only in DNA-binding proteins from eukaryotic organisms.

In addition to their DNA-binding domains, these proteins commonly possess other structural domains that function in protein:protein recognitions essential to oligomerization (for example, dimer formation), DNA looping, transcriptional activation, and signal reception (for example, effector binding).

α-Helices Fit Snugly into the Major Groove of B-DNA

A recurring structural feature in DNA-binding proteins is the presence of α-helical segments that fit directly into the major groove of B-form DNA. The diameter of an α-helix (including its side chains) is about 1.2 nm. The dimensions of the major groove in B-DNA are 1.2 nm wide by 0.6 to 0.8 nm deep. Thus, one side of an α-helix can fit snugly into the major groove. Although examples of β-sheet DNA recognition elements in proteins are known, the α-helix and B-form DNA are the predominant structures involved in protein:DNA interactions. Significantly, proteins can recognize specific sites in "normal" B-DNA; the DNA need not assume any unusual, alternative conformation (such as Z-DNA).

Proteins with the Helix-Turn-Helix Motif Use One Helix to Recognize DNA

The HTH motif is a protein structural domain consisting of two successive α-helices separated by a sharp β-turn (Figure 29.32). Within this domain, the α-helix situated more toward the C-terminal end of the protein, the so-called **helix 3,** is the DNA recognition helix; it fits nicely into the major groove, with several of its side chains touching DNA base pairs. **Helix 2,** the helix at the beginning of the HTH motif, creates a stable structural domain through hydrophobic interactions with helix 3 that locks helix 3 into its DNA interface. Proteins with HTH motifs bind to DNA as dimers. In the dimer, the two helix 3 cylinders are antiparallel to each other, such that their N⟶C orientations match the inverted relationship of nucleotide sequence in the dyad-symmetric DNA-binding site. An example is **Antp.** Antp is a member of a family of eukaryotic proteins involved in the regulation of early embryonic development that have in common an amino acid sequence element known as the **homeobox**[5] **domain.** The homeobox is a DNA motif that encodes a related 60–amino acid sequence (the homeobox domain) found among proteins of virtually every eukaryote, from yeast to man. Embedded within the homeobox domain is an HTH motif. Homeobox domain proteins act as **sequence-specific transcription factors.** Typically, the homeobox portion comprises only 10% or so of the protein's mass, with the remainder of the protein serving in protein:protein interactions essential to transcription regulation. Other DNA-binding proteins with HTH motifs are *lac* repressor, *trp* repressor, and the C-terminal domain of CAP.

How Does the Recognition Helix Recognize Its Specific DNA-Binding Site? The edges of base pairs in dsDNA present a pattern of hydrogen-bond donor and acceptor groups within the major and minor grooves, but only the pattern displayed on the major-groove side is distinctive for each of the four base pairs A:T, T:A, C:G, and G:C. (You can get an idea of this by inspecting the structures of the base pairs in Figure 11.6.) Thus, the base-pair edges in the major groove act as a **recognition matrix** identifiable through H bonding with a specific protein, so it is not necessary to melt the base pairs to read the base sequence. Although formation of such H bonds is very important in DNA:protein recognition, other interactions also play a significant role. For example, the C-5-methyl groups unique to thymine residues are nonpolar "knobs" projecting into the major groove.

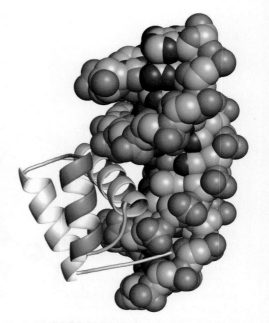

FIGURE 29.32 An HTH motif protein: *Antp* monomer bound to DNA. Helix 3 (yellow) is locked into the major groove of the DNA by helix 2 (magenta) (pdb id = 9ANT).

[5]*Homeo* derives from homeotic genes, a set of genes originally discovered in the fruit fly *Drosophila melanogaster* through their involvement in the specification of body parts during development.

Proteins Also Recognize DNA via "Shape Readout" **Shape readout,** also known as indirect readout, is the term for the ability of a protein to indirectly recognize a particular nucleotide sequence by recognizing local conformational variations resulting from the effects that base sequence has on DNA structure. Superficially, the B-form structure of DNA appears to be a uniform cylinder. Nevertheless, the conformation of DNA over a short distance along its circumference varies subtly according to local base sequence. That is, base sequences generate unique contours that proteins can recognize. Because these contours arise from the base sequence, the DNA-binding protein "indirectly reads out" the base sequence through interactions with the DNA backbone. In the *E. coli* Trp repressor : *trp* operator DNA complex, the Trp repressor engages in 30 specific hydrogen bonds to the DNA: 28 involve phosphate groups in the backbone; only 2 are to bases. Thus, some sequence-specific DNA-binding proteins are able to recognize an overall DNA conformation caused by the specific DNA sequence.

Some Proteins Bind to DNA via Zn-Finger Motifs

There are many classes of Zn-finger motifs. The prototype Zn-finger is a structural feature formed by a pair of Cys residues separated by 2 residues, then a run of 12 amino acids, and finally a pair of His residues separated by 3 residues ($Cys-x_2-Cys-x_{12}-His-x_3-His$). This motif may be repeated as many as 13 times over the primary structure of a Zn-finger protein. Each repeat coordinates a zinc ion via its 2 Cys and 2 His residues (Figure 29.33). The 12 or so residues separating the Cys and His coordination sites are looped out and form a distinct DNA interaction module, the so-called Zn-finger. When Zn-finger proteins associate with DNA, each Zn-finger binds in the major groove and interacts with about five nucleotides, adjacent fingers interacting with contiguous stretches of DNA. Many DNA-binding proteins with this motif have been identified. In all cases, the finger motif is repeated at least two times, with at least a 7– to 8–amino acid linker between Cys/Cys and His/His sites. Proteins with this general pattern are assigned to the C_2H_2 **class** of Zn-finger proteins to distinguish them from proteins bearing another kind of Zn-finger, the C_x **type,** which includes the C_4 and C_6 Zn-finger proteins. The C_x proteins have a variable number of Cys residues available for Zn **chelation.** For example, the vertebrate steroid receptors have two sets of Cys residues, one with four conserved cysteines (C_4) and the other with five (C_5).

Chelation is from the Greek word *chele,* meaning "claw"; it refers to the binding of a metal ion to two or more nonmetallic atoms in the same molecule.

Some DNA-Binding Proteins Use a Basic Region-Leucine Zipper *(bZIP)* Motif

bZIP is a structural motif characterizing the third major class of sequence-specific, DNA-binding proteins. This motif was first recognized by Steve McKnight in **C/EBP,** a heat-stable, DNA-binding protein isolated from rat liver nuclei that binds to both

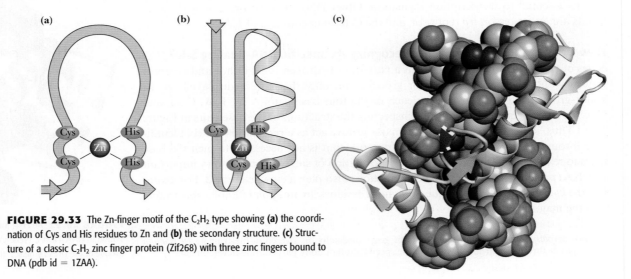

FIGURE 29.33 The Zn-finger motif of the C_2H_2 type showing **(a)** the coordination of Cys and His residues to Zn and **(b)** the secondary structure. **(c)** Structure of a classic C_2H_2 zinc finger protein (Zif268) with three zinc fingers bound to DNA (pdb id = 1ZAA).

CCAAT promoter elements and certain enhancer core elements.[6] The DNA-binding domain of C/EBP was localized to the C-terminal region of the protein. This region shows a notable absence of Pro residues, suggesting it might be arrayed in an α-helix. Within this region are two clusters of basic residues: A and B. Further along is a 28-residue sequence. When this latter region is displayed end-to-end down the axis of a hypothetical α-helix, beginning at Leu[315], an amphipathic cylinder is generated, similar to the one shown in Figure 6.22. One side of this amphipathic helix consists principally of hydrophobic residues (particularly leucines), whereas the other side has an array of negatively and positively charged side chains (Asp, Glu, Arg, and Lys), as well as many uncharged polar side chains (glutamines, threonines, and serines).

The Zipper Motif of *bZIP* Proteins Operates Through Intersubunit Interaction of Leucine Side Chains

The leucine zipper motif arises from the periodic repetition of leucine residues within this helical region. The periodicity causes the Leu side chains to protrude from the same side of the helical cylinder, where they can enter into hydrophobic interactions with a similar set of Leu side chains extending from a matching helix in a second polypeptide. These hydrophobic interactions establish a stable noncovalent linkage, fostering dimerization of the two polypeptides (as shown in Figure 29.34). The leucine zipper is not a DNA-binding domain. Instead, it functions in protein dimerization. Leucine zippers have been found in other mammalian transcriptional regulatory proteins, including *Myc, Fos,* and *Jun.*

The Basic Region of *bZIP* Proteins Provides the DNA-Binding Motif

The actual DNA contact surface of *bZIP* proteins is contributed by a 16-residue segment that ends exactly 7 residues before the first Leu residue of the Leu zipper. This DNA contact region is rich in basic residues and, hence, is referred to as the **basic region.** Two *bZIP* polypeptides join via a Leu zipper to form a Y-shaped molecule in which the stem of the Y corresponds to a coiled pair of α-helices held by the leucine zipper. The arms of the Y are the respective basic regions of each polypeptide; they act as a linked set of DNA contact surfaces (Figure 29.34). The dimer interacts with a DNA target site by situating the fork of the Y at the center of the dyad-symmetric DNA sequence. The two arms of the Y can then track along the major groove of the DNA in opposite directions, reading the specific recognition sequence (Figure 29.35). An interesting aspect of *bZIP* proteins is that the two polypeptides need not be identical (Figure 29.35). Heterodimers can form, provided both polypeptides possess a leucine zipper region. An important consequence of heterodimer formation is that the DNA target site need not be a palindromic sequence. The respective basic regions of the two different *bZIP* polypeptides (for example, *Fos* and *Jun*) can track along the major groove reading two different base sequences. Heterodimer formation expands enormously the DNA recognition and regulatory possibilities of this set of proteins.

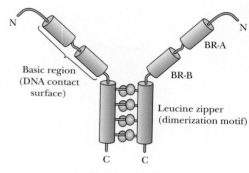

FIGURE 29.34 Model for a dimeric *bZIP* protein. Two *bZIP* polypeptides dimerize to form a Y-shaped molecule. The stem of the Y is the Leu zipper, and it holds the two polypeptides together. Each arm of the Y is the basic region from one polypeptide. Each arm is composed of two α-helical segments: BR-A and BR-B (basic regions A and B).

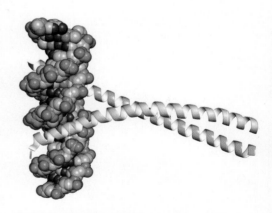

FIGURE 29.35 Model for the heterodimeric *bZIP* transcription factor *c-Fos:c-Jun* bound to a DNA oligomer containing the AP-1 consensus target sequence TGACTCA (pdb id = 1FOS).

29.5 | How Are Eukaryotic Transcripts Processed and Delivered to the Ribosomes for Translation?

Transcription and translation are concomitant processes in prokaryotes, but in eukaryotes, the two processes are spatially separated (see Chapter 10). *Transcription occurs on DNA in the nucleus, and translation occurs on ribosomes in the cytoplasm.* Consequently, transcripts must be transported from the nucleus to the cytosol to be translated. On the way, these transcripts undergo **processing:** alterations that convert the newly synthesized RNAs, or *primary transcripts,* into mature messenger RNAs. Also, unlike prokaryotes,

[6]The acronym *C/EBP* designates this protein as a "CCAAT and enhancer-binding protein."

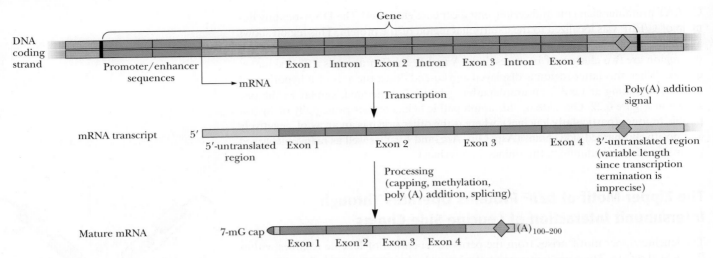

FIGURE 29.36 The organization of split eukaryotic genes.

in which many mRNAs encode more than one polypeptide (that is, they are polycistronic), eukaryotic mRNAs encode only one polypeptide (that is, they are exclusively monocistronic).

Eukaryotic Genes Are Split Genes

Most genes in higher eukaryotes are split into coding regions, called **exons,**[7] and noncoding regions, called **introns** (Figure 29.36; see also Figure 10.20). Introns are the intervening nucleotide sequences that are removed from the primary transcript when it is processed into a mature RNA. Gene expression in eukaryotes entails not only transcription but also the *processing of primary transcripts* to yield the mature RNA molecules we classify as mRNAs, tRNAs, rRNAs, and so forth.

The Organization of Exons and Introns in Split Genes Is Both Diverse and Conserved

Split genes occur in an incredible variety of interruptions and sizes. The yeast **actin gene** is a simple example, having only a single 309-bp intron that separates the nucleotides encoding the first 3 amino acids from those encoding the remaining 350 or so amino acids in the protein. The chicken **ovalbumin gene** is composed of 8 exons and 7 introns. The two **vitellogenin genes** of the African clawed toad *Xenopus laevis* are both spread over more than 21 kbp of DNA; their primary transcripts consist of just 6 kb of message that is punctuated by 33 introns. The chicken **pro α-2 collagen gene** has a length of about 40 kbp; the coding regions constitute only 5 kb distributed over 51 exons within the primary transcript. The exons are quite small, ranging from 45 to 249 bases in size.

Clearly, the mechanism by which introns are removed and multiple exons are spliced together to generate a continuous, translatable mRNA must be both precise and complex. If one base too many or too few is excised during splicing, the coding sequence in the mRNA will be disrupted. The mammalian **DHFR (dihydrofolate reductase) gene** is split into 6 exons spread over more than 31 kbp of DNA. The 6 exons are spliced together to give a 6-kb mRNA (Figure 29.37). Note that, in three different mammalian species, the size and position of the exons are essentially the same but that the lengths

---

[7]Although the term *exon* is commonly used to refer to the protein-coding regions of an interrupted or split gene, a more precise definition would specify exons as sequences that are represented in mature RNA molecules. This definition encompasses not only protein-coding genes but also the genes for various RNAs (such as tRNAs or rRNAs) from which intervening sequences must be excised in order to generate the mature gene product.

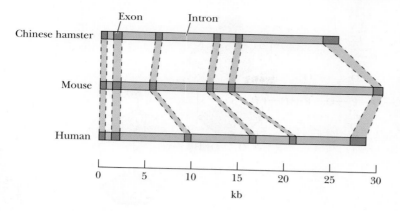

FIGURE 29.37 The organization of the mammalian DHFR gene in three representative species. Note that the exons are much shorter than the introns. Note also that the exon pattern is more highly conserved than the intron pattern.

of the corresponding introns vary considerably. Indeed, the lengths of introns in vertebrate genes range from a minimum of about 60 bases to more than 10,000 bp. Many introns have nonsense codons in all three reading frames and thus are untranslatable. Introns are found in the genes of mitochondria and chloroplasts as well as in nuclear genes. Although introns have been observed in archaea and even bacteriophage T4, none are known in the genomes of bacteria.

Post-Transcriptional Processing of Messenger RNA Precursors Involves Capping, Methylation, Polyadenylylation, and Splicing

Capping and Methylation of Eukaryotic mRNAs The protein-coding genes of eukaryotes are transcribed by RNA polymerase II to form primary transcripts or **pre-mRNAs** that serve as precursors to mRNA. As a population, these RNA molecules are very large and their nucleotide sequences are very heterogeneous because they represent the transcripts of many different genes, hence the designation **heterogeneous nuclear RNA**, or **hnRNA**. Shortly after transcription of hnRNA is initiated, the 5′-end of the growing transcript is capped by addition of a guanylyl residue. This reaction is catalyzed by the nuclear enzyme **guanylyl transferase** using GTP as substrate (Figure 29.38). The **cap structure** is methylated at the 7-position of the G residue. Additional methylations may occur at the 2′-O positions of the two nucleosides following the 7-methyl-G cap and at the 6-amino group of a first base adenine (Figure 29.39).

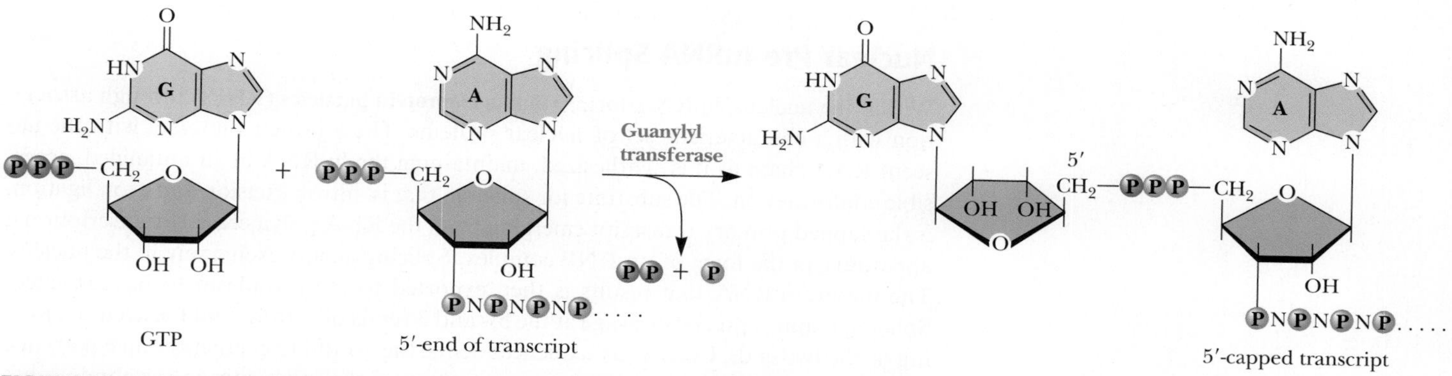

FIGURE 29.38 The capping of eukaryotic pre-mRNAs. Guanylyl transferase catalyzes the addition of a guanylyl residue (G_p) derived from GTP to the 5′-end of the growing transcript, which has a 5′-triphosphate group already there. In the process, pyrophosphate *(pp)* is liberated from GTP and the terminal phosphate (p) is removed from the transcript. Gppp + pppApNpNpNp... ⟶ GpppApNpNpNp... + *pp* + p (A is often the initial nucleotide in the primary transcript).

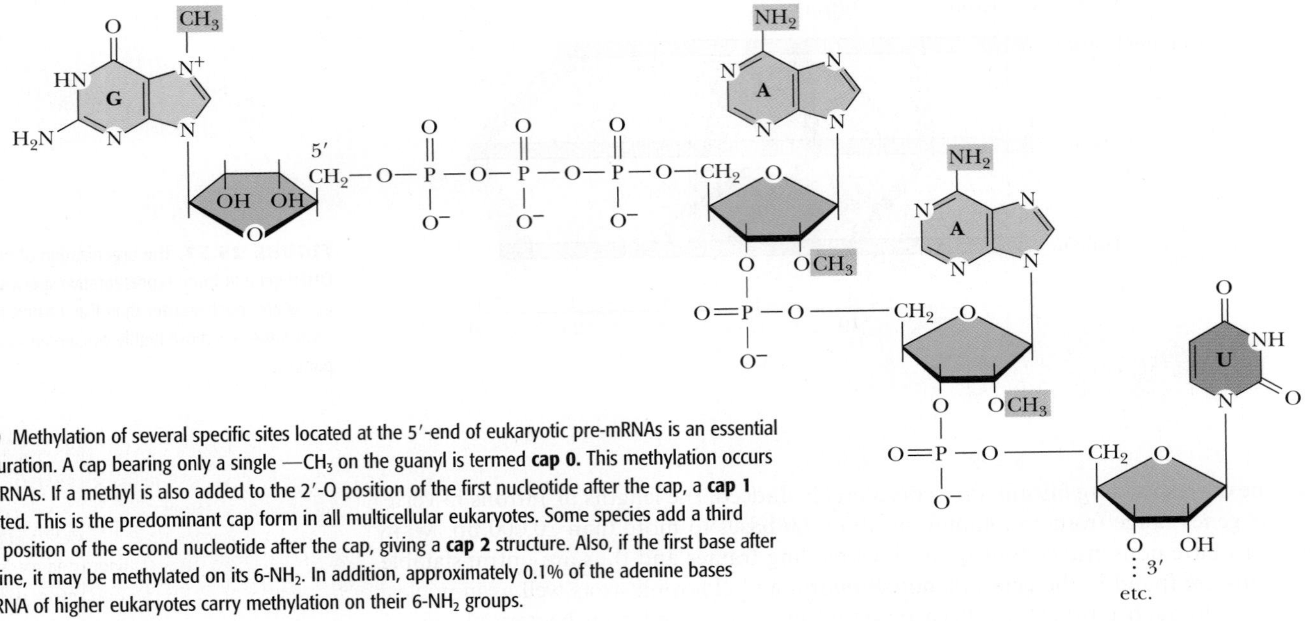

FIGURE 29.39 Methylation of several specific sites located at the 5'-end of eukaryotic pre-mRNAs is an essential step in mRNA maturation. A cap bearing only a single —CH_3 on the guanyl is termed **cap 0**. This methylation occurs in all eukaryotic mRNAs. If a methyl is also added to the 2'-O position of the first nucleotide after the cap, a **cap 1** structure is generated. This is the predominant cap form in all multicellular eukaryotes. Some species add a third —CH_3 to the 2'-O position of the second nucleotide after the cap, giving a **cap 2** structure. Also, if the first base after the cap is an adenine, it may be methylated on its 6-NH_2. In addition, approximately 0.1% of the adenine bases throughout the mRNA of higher eukaryotes carry methylation on their 6-NH_2 groups.

3'-Polyadenylylation of Eukaryotic mRNAs Transcription by RNA polymerase II typically continues past the 3'-end of the mature messenger RNA. Primary transcripts show heterogeneity in sequence at their 3'-ends, indicating that the precise point where termination occurs is nonspecific. However, termination does not normally occur until RNA polymerase II has transcribed past a consensus AAUAAA sequence known as the **polyadenylylation signal.**

Most eukaryotic mRNAs have 100 to 200 adenine residues attached at their 3'-end, the **poly(A) tail.** (Histone mRNAs are the only common mRNAs that lack poly(A) tails.) These A residues are not encoded in the DNA but are added post-transcriptionally by the enzyme **poly(A) polymerase,** using ATP as a substrate. The consensus AAUAAA is not itself the poly(A) addition site; instead, it defines the position where poly(A) addition occurs (Figure 29.40). The consensus AAUAAA is found 10 to 35 nucleotides upstream from where the nascent primary transcript is cleaved by an endonuclease to generate a new 3'-OH end. This end is where the poly(A) tail is added. The processing events of mRNA capping, poly(A) addition, and splicing of the primary transcript create the mature mRNA. Interestingly, both the guanylyl transferase that adds the 5'-cap structure and the enzymes that process the 3'-end of the transcript and add the poly (A) tail are anchored to RNA polymerase II via interactions with its RPB1 CTD.

Nuclear Pre-mRNA Splicing

Within the nucleus, hnRNA forms **ribonucleoprotein particles (RNPs)** through association with a characteristic set of nuclear proteins. These proteins interact with the nascent RNA chain as it is synthesized, maintaining the hnRNA in an untangled, accessible conformation. The substrate for splicing, that is, intron excision and exon ligation, is the capped primary transcript emerging from the RNA polymerase II transcriptional apparatus, in the form of an RNP complex. Splicing occurs exclusively in the nucleus. The mature mRNA that results is then exported to the cytoplasm to be translated. Splicing requires precise cleavage at the 5'- and 3'-ends of introns and the accurate joining of the two ends. Consensus sequences define the exon/intron junctions in eukaryotic mRNA precursors, as indicated from an analysis of the splice sites in vertebrate genes (Figure 29.41). Note that the sequences GU and AG are found at the 5'- and 3'-ends, respectively, of introns in pre-mRNAs from higher eukaryotes. In addition to the splice junctions, a conserved sequence within the intron, the **branch site,** is also essential to pre-mRNA splicing. The site lies 18 to 40 nucleotides upstream from the 3'-splice site

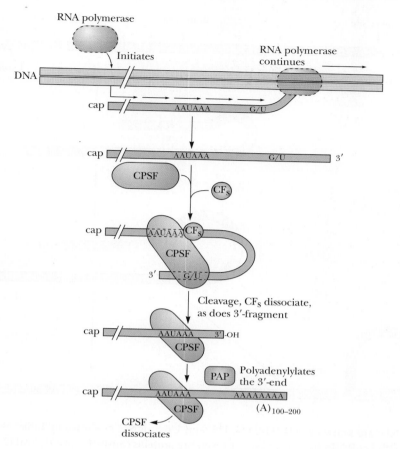

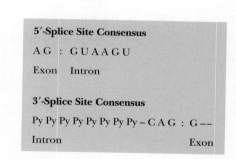

FIGURE 29.40 Poly(A) addition to the 3′-ends of transcripts occurs 10 to 35 nucleotides downstream from a consensus AAUAAA sequence, defined as the *polyadenylylation signal.* CPSF *(cleavage and polyadenylylation specificity factor)* binds to this signal sequence and mediates looping of the 3′-end of the transcript through interactions with a G/U-rich sequence even further downstream. Cleavage factors *(CFs)* then bind and bring about the endonucleolytic cleavage of the transcript to create a new 3′-end 10 to 35 nucleotides downstream from the polyadenylylation signal. Poly(A) polymerase (PAP) then successively adds 200 to 250 adenylyl residues to the new 3′-end. (RNA polymerase II is also a significant part of the polyadenylylation complex at the 3′-end of the transcript, but for simplicity in illustration, its presence is not shown in the lower part of the figure.)

and is represented in higher eukaryotes by the consensus sequence YNYRAY, where Y is any pyrimidine, R is any purine, and N is any nucleotide.

The Splicing Reaction Proceeds via Formation of a Lariat Intermediate

The mechanism for splicing nuclear mRNA precursors is shown in Figure 29.42. A covalently closed loop of RNA, the **lariat,** is formed by attachment of the 5′-phosphate group of the intron's invariant 5′-G to the 2′-OH at the invariant branch site A to form a 2′-5′ phosphodiester bond. Note that lariat formation creates an unusual branched nucleic acid. The lariat structure is excised when the 3′-OH of the consensus G at the 3′-end of the 5′ exon (Exon 1, Figure 29.42) covalently joins with the 5′-phosphate at the 5′-end of the 3′ exon (Exon 2). The reactions that occur are transesterification reactions where an OH group reacts with a phosphoester bond, displacing an —OH to form a new phosphoester link. Because the reactions lead to no net change in the number of phosphodiester linkages, no energy input (for example, as ATP) is needed. The lariat product is unstable; the 2′-5′ phosphodiester branch is quickly cleaved to give a linear excised intron that is rapidly degraded in the nucleus.

Splicing Depends on snRNPs

The hnRNA (pre-mRNA) substrate is not the only RNP complex involved in the splicing process. Splicing also depends on a unique set of small nuclear ribonucleoprotein particles called **snRNPs** (pronounced "snurps"). In higher eukaryotes, each snRNP consists of a small RNA molecule 100 to 200 nucleotides long and about 10 different proteins. The specific RNAs of SnRNPs are uridine-rich, and the five prominent

> **5′-Splice Site Consensus**
>
> A G : G U A A G U
>
> Exon   Intron
>
> **3′-Splice Site Consensus**
>
> Py Py Py Py Py Py Py Py – C A G : G – –
>
> Intron   Exon

FIGURE 29.41 Consensus sequences at the splice sites in vertebrate genes.

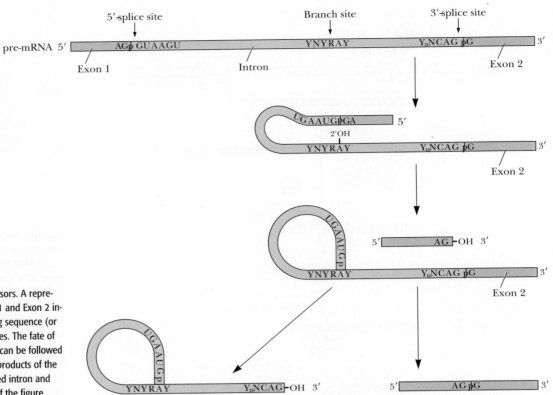

FIGURE 29.42 Splicing of mRNA precursors. A representative precursor mRNA is depicted. Exon 1 and Exon 2 indicate two exons separated by an intervening sequence (or intron) with consensus 5′, 3′, and branch sites. The fate of the phosphates at the 5′- and 3′-splice sites can be followed by tracing the fate of the respective *p*s. The products of the splicing reaction, the lariat form of the excised intron and the united exons, are shown at the bottom of the figure.

TABLE 29.6	The snRNPs Found in Spliceosomes	
snRNP	Length (nt)	Splicing Target
U1	165	5′ splice
U2	189	Branch
U4	145	5′ splice, recruitment of branch point to 5′-splice site
U5	115	
U6	106	

SnRNPs are known as U1, U2, U4, U5, and U6 SnRNP, as shown in Table 29.6. Four of the five snRNPs have a core set of seven **Sm proteins** (SmB/B′, SmD1, SmD2, SmD3, SmE, SmF, and SmG). The fifth, U1 SnRNP, has a related set of seven proteins that are designated LSm (for "like Sm"). Each snRNA has a single-stranded, U-rich RNA sequence, the **Sm site,** whose nucleotide sequence is AUUUUG. The Sm proteins assemble as a very stable heteroheptameric ring around the Sm site, in the order SmE-SmG-SmD3-SmB-SmD1-SmD2-SmF. Thus, the snRNA passes through the central hole in the Sm protein heteroheptamer (Figure 29.43). Each of the snRNPs also binds

FIGURE 29.43 Structure of the core domain of the U4 SnRNP. The U4 snRNA (orange) passes through the central hole in the heteroheptad Sm protein complex, SmG-SmD3-SmB-SmD1-SmD2-SmF. Each of the seven Sm proteins is a different color. *Adapted from Figure 1 in Leung, A. K. W., Nagai, K., and Li, J. 2011. Structure of the spliceosomal U4 snRNP core domain and its implication for snRNP biogenesis.* Nature **473:**536-539 (pdb id = 2Y9A).

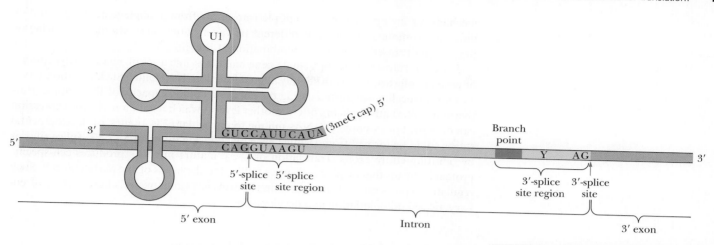

FIGURE 29.44 Mammalian U1 snRNA can be arranged in a secondary structure where its 5′-end is single-stranded and can base-pair with the consensus 5′-splice site of the intron. (Adapted from Figure 1 in Rosbash, M., and Seraphin, B., 1991. Who's on first? The U1 snRNP-5′ splice site interaction and splicing. *Trends in Biochemical Sciences* **16**:187.)

its own set of unique proteins to complete its ribonucleoprotein structure. These major snRNP species are very abundant, present at greater than 100,000 copies per nucleus. In the U1 SnRNP, the U1 snRNA folds into a secondary structure that leaves the 11 nucleotides at its 5′-end single-stranded. The 5′-end of U1 snRNA is complementary to the consensus sequence at the 5′-splice junction of the pre-mRNA (Figure 29.44), as is a region at the 5′-end of U6 snRNA. The U2 SnRNP snRNA is complementary to the consensus branch site sequence.

snRNPs Form the Spliceosome

Splicing occurs when the various snRNPs come together with the pre-mRNA to form a multicomponent complex called the **spliceosome.** The spliceosome is a large complex, roughly equivalent to a ribosome in size, and its assembly requires ATP. Assembly of the spliceosome begins with the binding of U1 snRNP at the 5′-splice site of the pre-mRNA (Figure 29.45). Each subsequent step in spliceosome assembly requires ATP-dependent RNA rearrangements catalyzed by spliceosomal **DEAD-box ATPases/helicases.** The branch-point sequence (UACUAAC in yeast) binds U2 snRNP, and then the triple snRNP complex of U4/U6·U5 replaces U1 at the 5′-splice site. The substitution of base-pairing interactions between U1 and the pre-mRNA 5′-splice site by base-pairing between U6 and the 5′-splice site is just one of the many RNA rearrangements that accompany the splicing reaction. Base-pairings between U6 and U2 RNA bring the 5′-splice site and the branch point RNA sequences into proximity. Interactions between U2 and U6 lead to release of U4 snRNP. The spliceosome is now activated for catalysis: A transesterification reaction involving the 2′-O of the invariant A residue in the branch-point sequence displaces the 5′-exon from the intron, creating the lariat intermediate. The free 3′-O of the 5′-exon now triggers a second transesterification reaction through attack on the P atom at the 3′-exon splice site. This second reaction joins the two exons and releases the intron as a lariat structure. In addition to the snRNPs, a number of proteins with RNA-annealing functions as well as proteins with ATP-dependent RNA-unwinding activity participate in spliceosome function. The spliceosome is thus a dynamic structure that uses the pre-mRNA as a template for assembly, carries out its transesterification reactions, and then disassembles when the splicing reaction is over.

Alternative RNA Splicing Creates Protein Isoforms

In one mode of splicing, every intron is removed and every exon is incorporated into the mature RNA without exception. This type of splicing, termed **constitutive splicing,** results in a single form of mature mRNA from the primary transcript. However, many eukaryotic genes can give rise to multiple forms of mature RNA transcripts. The

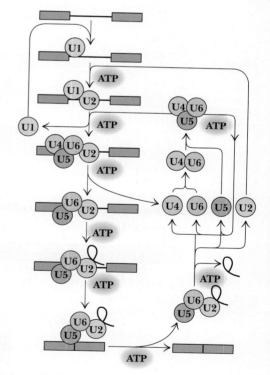

FIGURE 29.45 Events in spliceosome assembly. U1 snRNP binds at the 5′-splice site, followed by the association of U2 snRNP with the UACUAA*C branch-point sequence. The triple U4/U6-U5 snRNP complex replaces U1 at the 5′-splice site and directs the juxtaposition of the branch-point sequence with the 5′-splice site, whereupon U4 snRNP is released. Lariat formation occurs, freeing the 3′-end of the 5′-exon to join with the 5′-end of the 3′-exon, followed by exon ligation. U2, U5, and U6 snRNPs dissociate from the lariat following exon ligation.

mechanisms for production of multiple transcripts from a single gene include use of different promoters, selection of different polyadenylation sites, **alternative splicing** of the primary transcript, or even a combination of the three.

Different transcripts from a single gene make possible a set of related polypeptides, or **protein isoforms,** each with slightly altered functional capability. Such variation serves as a useful mechanism for increasing the apparent coding capacity of the genome. Furthermore, alternative splicing offers another level at which regulation of gene expression can operate. For example, mRNAs unique to particular cells, tissues, or developmental stages could be formed from a single gene by choosing different 5'- or 3'-splice sites or by omitting entire exons. Translation of these mature mRNAs produces cell-specific protein isoforms that display properties tailored to the needs of the particular cell. Such regulated expression of distinct protein isoforms is a fundamental characteristic of eukaryotic cell differentiation and development.

Fast Skeletal Muscle Troponin T Isoforms Are an Example of Alternative Splicing

In addition to many other instances, alternative splicing is a prevalent mechanism for generating protein isoforms from the genes encoding muscle proteins (see Chapter 16), allowing distinctive isoforms aptly suited to the function of each muscle. An impressive manifestation of alternative splicing is seen in the expression possibilities for the rat **fast skeletal muscle troponin T gene** (Figure 29.46). This gene consists of 18 exons, 11 of which are found in all mature mRNAs (exons 1 through 3, 9 through 15, and 18) and thus are **constitutive.** Five exons, those numbered 4 through 8, are **combinatorial** in that they may be individually included or excluded, in any combination, in the mature mRNA. Two exons, 16 and 17, are mutually exclusive: One or the other is always present, but never both. Sixty-four different mature mRNAs can be generated from the primary transcript of this gene by alternative splicing. Because each exon represents a cassette of genetic information encoding a segment of protein, alternative splicing is a versatile way to introduce functional variation within a common protein theme.

Fast skeletal troponin T gene and spliced mRNAs

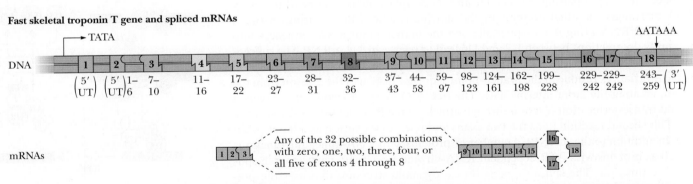

FIGURE 29.46 Organization of the fast skeletal muscle troponin T gene and the 64 possible mRNAs that can be generated from it. Exons are constitutive (yellow), combinatorial (green), or mutually exclusive (blue or orange). Exon 1 is composed of 5'-untranslated (UT) sequences, and Exon 18 includes the polyadenylation site (AATAAA) and 3'-UT sequences. The TATA box indicates the transcription start site. The amino acid residues encoded by each exon are indicated below. Many exon:intron junctions fall between codons. The "sawtooth" exon boundaries indicate that the splice site falls between the first and second nucleotides of a codon, the "concave/convex" exon boundaries indicate that the splice site falls between the second and third nucleotides of a codon, and flush boundaries between codons signify that the splice site falls between intact codons. Each mRNA includes all constitutive exons, 1 of the 32 possible combinations of Exons 4 to 8, and either Exon 16 or 17.

Inteins—Bizarre Parasitic Genetic Elements Encoding a Protein-Splicing Activity

Inteins are parasitic genetic elements found within protein-coding regions of genes. These selfish DNA elements are transcribed and translated along with the flanking host gene sequences. The typical intein protein consists of two domains: One domain is capable of self-catalyzed **protein splicing;** the other is an endonuclease that mediates the insertion of the intein nucleotide sequence into host genes. After the full protein is synthesized, the intein catalyzes excision of itself from the host protein *and* ligation of adjacent host polypeptide regions to form the functional protein that the host gene encodes. These adjacent polypeptides are termed **exteins ("external proteins")** to distinguish them from the intein ("internal protein"). Inteins have been found across all domains of life—archaea, bacteria, and eukaryotes—although thus far only in unicellular organisms. Inteins vary in size from about 130 to 600 amino acid residues. The protein splicing function of inteins is found in its N-terminal and its C-terminal regions; the endonuclease function that carries out parasitic insertion of the intein sequence into host genes is found in the central part of the intein. Splicing of the protein is an intramolecular process that liberates the intein sequence and ligates the host protein sequences (see accompanying figure).

Inteins are usually found as inserts in highly conserved host genes that have essential functions, such as genes encoding DNA or RNA polymerases, proton-translocating ATPases, or other vital metabolic enzymes. Their location in such genes means that removal of the intein via deletion or genetic rearrangement is more difficult. The endonuclease activity of the intein recognizes a 14- to 40–base-pair sequence in a potential host gene and cleaves the DNA there. During repair of the double-stranded DNA break, the intein gene is copied into the cleavage site, thereby establishing the parasitic genetic element in the host gene.

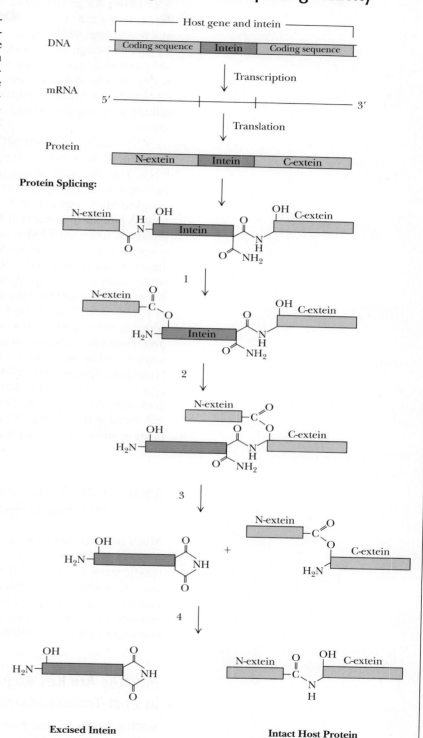

▶ Transcription and translation of the combined intein-host gene flanking sequences leads to synthesis of a fused intein–extein protein. The intein splices itself out when (1) the C-terminal residue of the N-extein is shifted to the O (or S) atom of a neighboring intein Ser (or Cys) residue, (2) the N-extein C-terminal carbonyl undergoes nucleophilic attack by the O (or S) atom of a Ser (or Cys) residue at the end of the C-extein in a transesterification reaction that creates a branched protein intermediate, (3) cyclization of the intein C-terminal asparagine residue excises the intein, and (4) the two exteins are properly united via a peptide bond when the N-extein C-terminus spontaneously shifts to the C-extein N-terminus to form an intact host protein.

Adapted from Paulus, H., 2000. Protein splicing and related forms of protein autoprocessing. *Annual Review of Biochemistry* **69:**447–496; and Gogarten, J. P., et al., 2002. Inteins: Structure, function, and evolution. *Annual Review of Microbiology* **56:**263–287.

RNA Editing: Another Mechanism That Increases the Diversity of Genomic Information

RNA editing is a process that changes one or more nucleotides in an RNA transcript by deaminating a base, either A→I (adenine to inosine, through deamination at the 6-position in a purine ring) or C→U (cytosine to uracil, through deamination at the 4-position in a pyrimidine ring). These changes alter the coding possibilities in a transcript, because I will pair with G (not U as A does) and U will pair with A (not G as C does). RNA editing has the potential to increase protein diversity by (1) altering amino acid coding possibilities, (2) introducing premature stop codons, or (3) changing splice sites in a transcript. If RNA splicing is cutting-and-pasting, then these single-base changes are aptly termed RNA editing.

A-to-I editing is carried out by **adenosine deaminases** that act on RNA (the **ADAR** family of RNA-editing enzymes). ADARs act only on double-stranded regions of RNA. Typically, such regions form when an exon region containing an A to be edited base-pairs with a complementary base sequence in an intron known as the **editing site complementary sequence,** or **ECS.** ADARs are abundant in the nervous system of animals. A prominent example of RNA editing occurs in transcripts encoding mammalian glutamate receptors (GluRs; see Chapter 32). Deamination of the GluR-B gene transcript changes a glutamine codon CAG to CIG, which is read by the translational machinery as an arginine codon (CGG), dramatically altering the conductance properties of the membrane receptor produced from the edited transcript, as compared to the receptor produced from the unedited transcript.

C-to-I editing is carried out on single-stranded regions of transcripts by an **editosome** core structure consisting of a cytosine deaminase and an adapter protein that brings the deaminase and the transcript together. A prominent example of C-to-I editing targets a single C residue in a 14-kb transcript encoding the 4536-residue apolipoprotein B100 protein (see Chapter 24). ApoB RNA editing changes codon 2153 (a CAA [glutamine] codon) to a UAA stop codon, which leads to a shortened protein product, ApoB48, consisting of the N-terminal 48% of apoB100. In humans, apoB100 is made in the liver and found in liver-derived VLDL serum lipoprotein complexes. In contrast, apoB48 is made in intestinal cells and found in intestinal-derived lipid complexes.

29.6 Can Gene Expression Be Regulated Once the Transcript Has Been Synthesized?

Much emphasis has been placed on the regulation of gene expression through mechanisms at the gene promoter that determine whether or not a gene will be transcribed. In reality, many protein-coding genes are transcribed into RNA but their transcripts are never translated to form proteins. Indeed, post-transcriptional regulation of gene expression is a pervasive form of regulation, affecting virtually every cellular process from metabolism to development and cancer. MicroRNAs (miRNAs; see Chapter 10) are important players in post-transcriptional gene regulation.

miRNAs Are Key Regulators in Post-Transcriptional Gene Regulation

miRNA Synthesis and Processing miRNAs are a large family of small (~21 nucleotides), noncoding RNAs found in animals, plants, and protists. At least 800 miRNAs are found in mammals and they are predicted to target the expression of about 60% of all protein-coding genes. miRNAs are synthesized mainly by RNA polymerase II transcription. Their coding sequences are found in intergenic regions of DNA, in the introns of protein-coding genes, and even in exons. miRNAs are synthesized as **pri-miRNAs,** primary transcripts hundreds or thousands of nucleotides long that are processed in the nucleus by **Drosha,** an RNase III family member, with the help of **Pasha** (a dsRNA-binding protein also called **DGCR8**), to form **pre-miRNAs,** which are 70-nucleotide stem-loop structures. Pre-miRNAs are exported to the cytosol, where **Dicer** (Chapter 12) processes it to form an approximately 20-nucleotide-long miRNA/miRNA* duplex (the * denotes the strand that is discarded). The miRNA strand represents the mature miRNA.

miRNA Mechanism of Action The mature miRNA is incorporated into a miRNA-induced silencing complex **(miRNA-RISC)** through interaction with **AGO2** (Figure 29.47). AGO2 contains three domains: PAZ, MID (which recognizes the 5′-end of the miRNA), and KIWI, the site of an RNase H-like enzyme activity. The 5′-terminal end of the miRNA binds in a deep pocket of AGO2 at the MID-KIWI junction. The KIWI domain also interacts with **GW182,** a 182-kD protein having a glycine- and tryptophan-rich (GW) N-terminal domain and an RNA-recognition motif (RRM) at its C-terminus. AGO2 and GW182 are the primary proteins in the miRNA-RISC, but they recruit additional proteins, such as **PABP,** poly(A)-binding protein (see Figure 30.29), a protein that binds to the poly(A) tail of mRNAs. miRNA-RISC mediates the interaction of the miRNA with its mRNA targets.

In most cases, miRNAs target the 3′-untranslated region (3′-UTR) of the mRNAs they regulate. As the key site for post-transcriptional gene regulation, the 3′-UTR can be viewed as the post-transcriptional equivalent of the gene promoter. It is here that the macromolecular interactions underlying post-transcriptional regulatory decisions take place. In almost every case, the miRNA base-pairs imperfectly with its target sequence in the 3′-UTR of the mRNAs it regulates. Effective mRNA targeting requires base-pairing of nucleotides 2-8 of the miRNA with the mRNA. Nucleotides 2-8 are termed the **seed region.** Because the RNase function of AGO2 cleaves only perfectly paired RNA duplexes, miRNA-RISC interaction with the mRNA does not result directly in RNA degradation. Further, because the miRNA-mRNA base-pairing is not perfect, miRNAs are promiscuous. They can interact with the 3′-UTRs of multiple mRNA targets; keep in mind that these 3′-UTRs are very extensive, 700 nucleotides or more in human mRNAs, affording many opportunities for miRNA interactions.

miRNA-RISC Blocks Gene Expression in Two Ways A key aspect of eukaryotic mRNA translation is 'closing the loop': the formation of an mRNA circle maintained through RNA:protein interactions (see Figure 30.29). Closing the loop is achieved through translation factors binding at the capped 5′-end of the mRNA and then

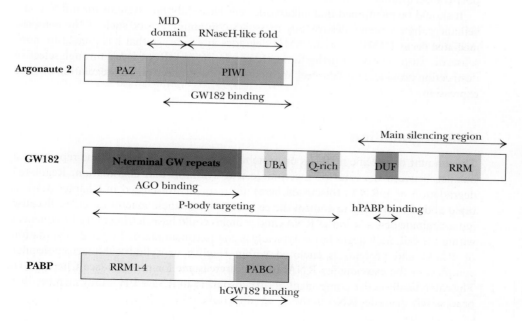

FIGURE 29.47 Domain organization of human Argonaute 2 (AGO2), GW182, and poly(A)-binding protein (PABP). The AGO2 PAZ (Piwi-Argonaute-Zwilli) domain is found in other proteins involved in post-transcriptional gene silencing. The GW-repeat domain of GW182 interacts with AGO2. The C-terminal GW182 domain has an RNA recognition motif (RRM). PABP has four RRM domains and a conserved C-terminal domain, PABC. Mammalian PABP binds directly to the silencing region of human GW182 proteins via PABC. (Adapted from Figure 1 in Fabian, M. R., Sonenberg, N., and Filipowicz, W., 2010. Regulation of mRNA stability and translation by miRNAs. *Annual Review of Biochemistry* **79**:351–379.)

interacting with PABP bound to the poly(A) tail of the mRNA. Because of the proximity of the 5'- and 3'-ends of the mRNA when the loop is closed, miRNA-RISC binding at the 3'-UTR can interfere with recruitment of ribosomes to translate the mRNA, a mode of miRNA action referred to as **translation repression.** In addition, miRNA-RISC complexes destabilize mRNAs through deadenylylation at their 3'-ends, decapping at their 5'-ends, and subsequent 5'→3' exonucleolytic degradation. The miRNA-RISC complex disrupts the closed loop through competing interactions with PABP, exposing the mRNA to decapping enzymes and exonucleases. Thus, the second mode of miRNA action is promoting **mRNA decay.**

29.7 Can We Propose a Unified Theory of Gene Expression?

The stages of eukaryotic gene expression—from transcriptional activation, transcription, transcript processing, nuclear export of mRNA, to translation—have traditionally been presented as a linear series of events, that is, as a pathway of discrete, independent steps. However, it now is clear that each stage is part of a continuous process, with physical and functional connections between the various transcriptional and processing machineries. This realization led George Orphanides and Danny Reinberg to propose a "unified theory of gene expression." The principal tenet of this theory is that eukaryotic gene expression is a continuous process, from transcription through processing and protein synthesis: DNA→RNA→protein (Figure 29.48). Furthermore, regulation occurs at multiple levels in this continuous process, in a coordinated fashion. Tom Maniatis and Robin Reed provide additional support for this theory by pointing out that eukaryotic gene expression depends on an interacting network of multicomponent protein machines—nucleosomes, HATs, and the remodeling apparatus; RNA polymerase II and its associated factors, which include capping, splicing, and polyadenylylation enzymes; and the proteins involved in mRNA export to the cytoplasm for translation on ribosomes, the topic of the next chapter. Translation is not inevitable. If a noncoding RNA base-pairs with the mRNA, translation will be thwarted. For example, base pairing with a microRNA will result in gene silencing (see Chapter 10). Base pairing with a small interfering RNA (siRNA, Chapter 10) leads to gene knockdown by RNAi through mRNA destruction by the RISC protein complex (see Figure 12.20). Recall that gene silencing is a post-transcriptional regulatory mechanism that prevents translation of an mRNA, whereas RNAi carries out post-transcriptional destruction of mRNAs.

It should be mentioned that eukaryotic cells have elaborate systems for mRNA surveillance; these systems destroy any messages containing errors, such as the **nonsense mediated decay (NMD)** system. NMD degrades any message that has premature nonsense, or "stop," codons. Furthermore, the regulation of mRNA levels through selective destruction provides another mechanism for the post-transcriptional regulation of gene expression.

RNA Degradation

The amount of specific mRNAs or proteins present in a cell at any time represents a balance between the rates of macromolecular synthesis and degradation. Regulated degradation of mRNAs (discussed here) and proteins (discussed in Chapter 31) is a rapid and effective way to control the cellular levels of these macromolecules. Because indiscriminate degradation of RNAs and proteins could have detrimental consequences within the cell, such degradation typically is compartmentalized. Targeted degradation of RNAs and proteins is enclosed within ringlike or cylindrical macromolecular complexes—the **exosome** for RNA and the **proteasome** for proteins (see Chapter 31). The catalytically active component of an exosome is an **RNase PH** family member that processively degrades RNA in the 3'→5' direction.

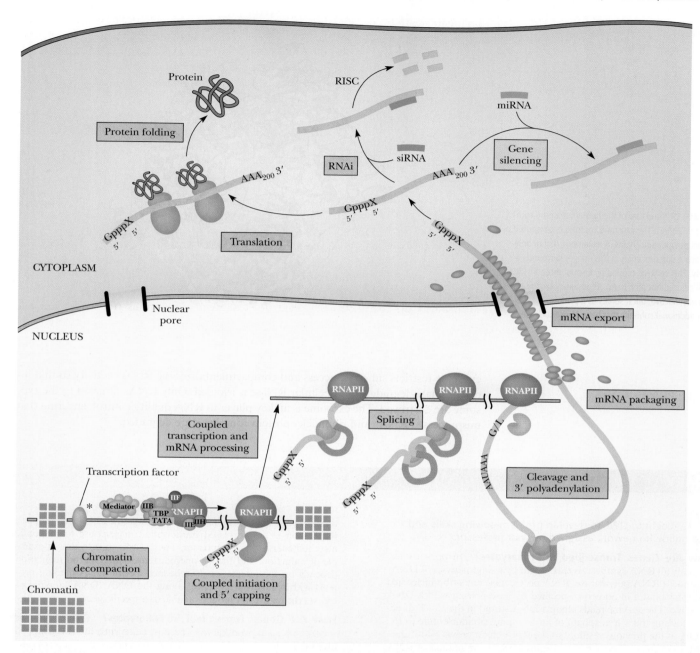

FIGURE 29.48 A unified theory of gene expression. Each step in gene expression, from transcription to translation, is but a stage within a continuous process. Each stage is physically and functionally connected to the next, ensuring that all steps proceed in an appropriate fashion and overall regulation of gene expression is tightly integrated. (Adapted from Figure 2 in Orphanides, G., and Reinberg, D., 2002. A unified theory of gene expression. *Cell* **108**:439–451.)

Exosomes have a fundamental structural pattern: a ring of six subunits surrounding a central cavity, with one or more of the subunits having RNase PH activity. Superimposed on top of the hexameric RNase PH ring are three different proteins with **S1 RNA-binding domains** (RBD proteins), as shown in Figure 29.49. Collectively, these nine proteins constitute the exosome complex. Depending on the cellular compartment—nucleus, nucleolus, or cytosol—other proteins associate with the exosome complex. These proteins determine the substrates degraded—rRNA, mRNA, tRNA, or any of the variety of small RNAs. RNAs to be degraded are threaded into the central cavity. The architecture of the

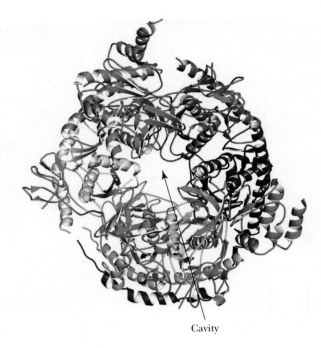

FIGURE 29.49 Structure of the human exosome complex (pdb id = 2NN6). The exosome complex is composed of nine different polypeptide chains. A hexameric ring of polypeptides (each a different color in this image) surrounds a central cavity. This cavity is capped by a set of three S1 RBD proteins (all colored hot pink here). The human exosome core is catalytic inactive, but it serves as a platform for the association of additional subunits that have 3'-exonuclease activity.

Cavity

exosome restricts substrate access and compartmentalizes the RNase activity so that indiscriminate degradation of cellular RNase is avoided. Only RNAs targeted to the exosome are destroyed. The exosome is an key player in RNA quality control, ensuring that misprocessed RNAs and ribonucleoprotein complexes are degraded.

SUMMARY

OWL Login to OWL to develop problem-solving skills and complete online homework assigned by your professor.

29.1 How Are Genes Transcribed in Prokaryotes? In prokaryotes, virtually all RNA synthesis is carried out by a single species of DNA-dependent RNA polymerase. RNA polymerase links ribonucleoside 5'-triphosphates in an order specified by base pairing with a DNA template. The enzyme reads along a DNA strand in the 3'→5' direction, joining the 5'-phosphate of an incoming ribonucleotide to the 3'-OH of the previous residue, so the RNA chain grows 5'→3' during transcription. Transcription begins when the σ-subunit of RNA polymerase recognizes a promoter and forms a complex with it. Next, the RNA polymerase holoenzyme unwinds about 14 base pairs of DNA, and transcription commences. Once an oligonucleotide 9 to 12 residues long has been formed, the σ-subunit dissociates. The core RNA polymerase is highly processive and goes on to synthesize the remainder of the mRNA. Prokaryotes have two types of transcription termination mechanisms: one that is dependent on ρ termination factor protein and another that depends on dissociation of the mRNA through reestablishment of DNA base pairs.

29.2 How Is Transcription Regulated in Prokaryotes? Bacterial genes encoding a common metabolic pathway are often grouped adjacent to one another in an operon, allowing all of the genes to be expressed in a coordinated fashion. The operator, a regulatory sequence adjacent to the structural genes, determines whether transcription takes place. The operator is located next to a promoter. Interaction of a regulatory protein with the operator controls transcription. Small molecules act as signals of the nutritional or environmental conditions. These small molecules interact with operator-binding regulatory proteins and determine whether transcription occurs. Induction is the increased synthesis of enzymes in response to a small molecule called a co-inducer. Repression is the decreased transcription in response to a specific metabolite termed a co-repressor. The *lac* operon provides an example of induction, and the *trp* operon, repression. Operon regulation depends on the interaction of sequence-specific DNA-binding proteins with regulatory sequences along the DNA. DNA looping increases the regulatory input available to a specific gene.

29.3 How Are Genes Transcribed in Eukaryotes? Transcription is more complicated in eukaryotes because eukaryotic DNA is wrapped around histones to form nucleosomes, and nucleosomes repress gene expression by limiting access of the transcriptional apparatus to genes. Two classes of transcriptional co-regulators are necessary to overcome nucleosome repression: (1) histone-modifying enzymes, such as HATs, and (2) ATP-dependent chromatin-remodeling complexes. Gene activation also requires interaction of RNA polymerase with the promoter. RNA polymerase II consists of 12 different polypeptides. The largest, RPB1, has a C-terminal domain (CTD) containing multiple repeats of the heptapeptide sequence YSPTSPS; 5 of these 7 residues can be phosphorylated by protein kinases. The CTD orchestrates events in the transcription process. RNA polymerase II promoters commonly consist of the core element, near the transcription start site, where general transcription factors bind, and more distantly located regulatory elements, known as enhancers or silencers. These regulatory sequences are recognized by specific DNA-binding proteins that activate transcription above basal levels. A eukaryotic transcription initiation complex consists of RNA polymerase II, five general transcription factors, and Mediator. The CTD of RNA polymerase II anchors the polymerase to Mediator, allowing RNA polymerase II to communicate with transcriptional activators bound at sites distal from the promoter.

29.4 How Do Gene Regulatory Proteins Recognize Specific DNA Sequences? Proteins that recognize nucleic acids present a three-dimensional shape or contour that is structurally and chemically complementary to the surface of a DNA sequence. Nucleotide sequence–specific recognition by the protein involves a set of atomic contacts with the bases and the sugar–phosphate backbone. Protein contacts with the bases of DNA usually occur within the major groove (but not always). Roughly 80% of DNA-binding regulatory proteins fall into one of three principal classes based on distinctive structural motifs: the helix-turn-helix (or HTH), the zinc finger (or Zn-finger), and the leucine zipper-basic region (or *bZIP*). In addition to their DNA-binding domains, these proteins also have protein:protein recognition domains essential to oligomerization, DNA looping, transcriptional activation, and signal reception (effector binding).

29.5 How Are Eukaryotic Transcripts Processed and Delivered to the Ribosomes for Translation? In eukaryotes, primary transcripts must be processed to form mature messenger RNAs and exported from the nucleus to the cytosol for translation. Shortly after transcription initiation, the 5′-end of the growing transcript is capped with a guanylyl residue that is then methylated at the 7-position. Additional methylations may occur at the 2′-O positions of the next two nucleotides and at the 6-amino group of a first adenine. Transcription termination does not normally occur until RNA polymerase II has transcribed past the polyadenylation signal. Most eukaryotic mRNAs have a poly(A) tails consisting of 100 to 200 adenine residues at their 3′-end, added post-transcriptionally by poly(A) polymerase. Most eukaryotic genes are split genes, subdivided into coding regions, called exons, and noncoding regions, called introns. Intron excision and exon ligation, a process called splicing, also occurs in the nucleus. Splicing is mediated by the spliceosome, which is assembled from a set of small nuclear ribonucleoprotein particles called snRNPs. Splicing requires precise cleavage at the 5′- and 3′-ends of introns and the accurate joining of the two exons. Exon/intron junctions are defined by consensus sequences recognized by the spliceosome. In addition, a conserved sequence within the intron, the branch site, is also essential to splicing. The splicing reaction involves formation of a lariat intermediate through attachment of the 5′-phosphate group of the intron's invariant 5′-G to the 2′-OH at the invariant branch site A to form a 2′-5′ phosphodiester bond. The lariat structure is excised when the exons are ligated. In constitutive splicing, every intron is removed and every exon is incorporated into the mature RNA without exception. However, alternative splicing can give rise to different transcripts from a single gene, making possible a set of protein isoforms, each with slightly altered functional capability. Fast skeletal muscle troponin T isoforms are an example of alternative splicing.

29.6 Can Gene Expression Be Regulated Once a Transcript Has Been Synthesized? Post-transcriptional gene regulation is pervasive, affecting virtually every cellular process. miRNAs are key regulators in post-transcriptional gene regulation. miRNAs are transcribed from intergenic regions of DNA, introns, and even exons as pri-miRNAs that are processed to form mature miRNAs. miRNAs act as part of miRNA-RISC complexes that include AGO2 and GW182 proteins. The miRNA in an miRNA-RISC imperfectly base-pairs with sequences in the 3′-UTR region of mRNAs. miRNA binding to an mRNA leads to translation repression or mRNA decay.

29.7 Can We Propose a Unified Theory of Gene Expression? Each stage in eukaryotic transcription is part of a continuous process, with physical and functional connections between the various transcriptional and processing machineries. These multicomponent protein machines are organized into an interacting network, and regulation occurs in a coordinated fashion at multiple levels in the continuous process. Eukaryotic cells also have elaborate systems for mRNA surveillance. Not all protein-coding transcripts are translated. Gene silencing or RNAi may intervene to prevent translation of mature RNAs. Degradation of misprocessed or otherwise aberrant RNA molecules by the exosome complex prevents these defective RNAs from interfering with cellular processes.

FOUNDATIONAL BIOCHEMISTRY **Things You Should Know After Reading Chapter 29.**

- All RNAs are synthesized from DNA templates by DNA-directed RNA polymerases.

- The synthesis of RNA from DNA templates is called transcription.

- Prokaryotes have a single version of RNA polymerase.

- RNA is synthesized in the 5′→3′ direction, as RNA polymerase moves along the DNA in a 3′→5′ direction.

- The strand of DNA that is transcribed by RNA polymerase is called the template strand.

- The nucleotide sequence of the RNA transcript is identical to the nucleotide sequence of the nontemplate DNA strand (except with U instead of T).

- Prokaryotic RNA polymerase has an $\alpha_2\beta\beta'\sigma$ subunit composition.

- The σ-subunit recognizes gene promoters, which identify the transcription start site for prokaryotic genes.

- Transcription has four stages:
  RNA polymerase holoenzyme binding to DNA at promoter sites
  Initiation of polymerization
  Elongation by the $\alpha_2\beta\beta'$ RNA polymerase core
  Termination of polymerization

- Prokaryotic promoters contain two consensus sequence elements: the Pribnow box and the -35 region.

- Initiation of polymerization begins formation of an open promoter complex.

- Two transcription termination mechanisms operate in prokaryotes: rho (ρ)-mediated termination and intrinsic termination at specific termination sites.

- Prokaryotic genes are often grouped together in operons.

- Transcription of operons is controlled through induction and repression.

- Operons are units of gene expression that consist of structural genes and regulatory genes.

- *lac* repressor is a negative regulator of the *lac* operon.

- CAP is a positive regulator of the *lac* operon.

- Negative and positive control systems are fundamentally different.

- The *araBAD* operon is both positively and negatively controlled by AraC.

- The *trp* operon is repressed if the Trp repressor protein binds tryptophan.

- Attenuation, a prokaryotic mechanism for post-transcriptional gene regulation, is achieved through formation of alternative secondary structures in the *trp* operon transcript.

- DNA–protein interactions and protein–protein interactions are essential in transcription regulation.

- Proteins that activate transcription exert their effects through protein–protein contacts with RNA polymerase.

- DNA looping allows multiple DNA-binding proteins to convene at the promoter.

- Eukaryotes have three RNA polymerases:
    RNA polymerase I transcribes rRNA genes.
    RNA polymerase II transcribes protein-coding genes.
    RNA polymerase III transcribes tRNA genes and other small RNA genes, including 5S rRNA.

- RNA polymerase II is composed of 12 different subunits.

- The C-terminal domain (CTD) of the RPB1 subunit of RNA polymerase II contains multiple repeats of the amino acid sequence YSPTSPS.

- The CTD multiple YSPTSPS repeats provide multiple phosphorylation sites.

- A "phosphorylation" code controls RNA polymerase II, regulating mRNA synthesis and processing.

- The regulation of gene expression is more complex in eukaryotes compared to prokaryotes.

- RNA polymerase II promoters consist of a core element near the transcription start site where GTFs bind and more distantly located enhancers or silencers where transcriptional activators or repressors bind.

- Inr is the site of transcription initiation.

- Typical eukaryotic promoters contain modules of short, conserved sequences, such as the TATA box, CAAT box, and GC box.

- The location of enhancers (also called upstream activating sequences) is not fixed relative to the promoter.

- Specific transcription factors bind at enhancers and regulate transcription.

- Transcription initiation by RNA polymerase II requires TBP and the GTFs.

- Formation of the pre-initiation complex (PIC) and promoter opening establishes an initiation complex.

- Transcription initiation also depends on interactions between RNA polymerase II and Mediator.

- Mediator also functions in repression of transcription.

- Chromatin-remodeling complexes and histone-modifying enzymes alleviate repression due to nucleosomes.

- Chromatin-remodeling complexes are huge, ATP-dependent enzymes that loosen DNA–protein interactions in nucleosomes.

- The various covalent modifications of histones can modulate histone–DNA interactions and create binding sites for recruitment of additional proteins.

- A histone code based on the different possible covalent modifications can determine gene regulation.

- Methylation and phosphorylation act as a binary switch in the histone code.

- Chromatin deacetylation leads to gene repression.

- Gene regulatory proteins recognize specific DNA sequences through DNA–protein interactions that may or may not involve direct readout of base sequence.

- The three prominent motifs in DNA-binding proteins include the HTH (helix-turn-helix), the Zn finger, and the *bZIP*.

- Eukaryotic genes are split into coding regions (exons) and noncoding regions (introns).

- Post-transcriptional processing of eukaryotic genes includes capping, methylation, 3′-polyadenylylation, and splicing.

- Intron excision by the splicing reaction proceeds through formation of a lariat intermediate.

- Splicing depends on a set of snRNPs, each of which contains in its core a specific 100-200-nucleotide U-rich RNA chain and a heteroheptamer of Sm proteins.

- Spliceosomes also contain DEAD-box ATPases/helicases that rearrange transcript secondary structure so that the splicing reaction can occur.

- Alternative splicing dramatically increases the number of protein isoforms possible from a single transcript.

- RNA editing through A→I or C→U deamination reactions also increases the diversity of proteins encoded by a transcript.

- Post-transcriptional regulation of gene expression affects virtually all cellular processes.

- miRNAs are key regulators in post-transcriptional gene expression.

- miRNAs are transcribed from intergenic regions of DNA, introns, and even exons.

- miRNAs act as part of miRNA-RISC complexes.

- The miRNAs in miRNA-RISC complexes base-pair imperfectly with seqences in the 3′-UTR of mRNA.

- miRNAs lead to either transcription repression or mRNA decay.

- A "unified theory of gene expression" describes the events through the sequential participation of multicomponent molecular machines: chromatin decompaction, regulated transcription through RNA polymerase II-Mediator interactions, capping, methylation, polyadenylylation, splicing, mRNA packaging and export from the nucleus, and either translation by ribosomes, destruction by RNAi, or silencing by miRNAs in the cytosol.

- The RNA exosome degrades RNAs.

PROBLEMS

Answers to all problems are at the end of this book. Detailed solutions are available in the *Student Solutions Manual, Study Guide, and Problems Book*.

1. **Template–Transcript Relationships** The 5′-end of an mRNA has the sequence

    ...AGAUCCGUAUGGCGAUCUCGACGAAGACUC-
         CUAGGGAAUCC...

What is the nucleotide sequence of the DNA template strand from which it was transcribed? If this mRNA is translated beginning with the first AUG codon in its sequence, what is the N-terminal amino acid sequence of the protein it encodes? (See Table 30.1 for the genetic code.)

2. **The Events in Transcription Initiation** Describe the sequence of events involved in the initiation of transcription by *E. coli* RNA

polymerase. Include in your description those features a gene must have for proper recognition and transcription by RNA polymerase.

3. Substrate Binding by RNA Polymerase RNA polymerase has two binding sites for ribonucleoside triphosphates: the initiation site and the elongation site. The initiation site has a greater K_m for NTPs than the elongation site. Suggest what possible significance this fact might have for the control of transcription in cells.

4. Comparison of Prokaryotic and Eukaryotic Transcription Make a list of the ways that transcription in eukaryotes differs from transcription in prokaryotes.

5. DNA Recognition by DNA-Binding Proteins DNA-binding proteins may recognize specific DNA regions either by reading the base sequence or by "shape readout." How do these two modes of protein:DNA recognition differ?

6. The Size of Gene Promoters (Integrates with Chapter 11.) The metallothionein promoter is illustrated in Figure 29.27. How long is this promoter, in nm? How many turns of B-DNA are found in this length of DNA? How many nucleosomes (approximately) would be bound to this much DNA? (Consult Chapter 11 to review the properties of nucleosomes.)

7. Heterodimer Formation by *bZIP* Proteins Describe why the ability of *bZIP* proteins to form heterodimers increases the repertoire of genes whose transcription might be responsive to regulation by these proteins.

8. Alternative Splicing Possibilities Suppose exon 17 were deleted from the fast skeletal muscle troponin T gene (Figure 29.46). How many different mRNAs could now be generated by alternative splicing? Suppose that exon 7 in a wild-type troponin T gene were duplicated. How many different mRNAs might be generated from a transcript of this new gene by alternative splicing?

9. Histone Modifications and Histone–DNA Interactions Figure 29.30 illustrates some of the various covalent modifications that occur on histone tails. How might each of these modifications influence DNA:histone interactions?

10. Inducer Binding to *lac* Repressor (Integrates with Chapter 15.) Predict from Figure 29.12 whether the interaction of *lac* repressor with inducer might be cooperative. Would it be advantageous for inducer to show cooperative binding to *lac* repressor? Why?

11. Post-transcriptional Modification of Eukaryotic mRNAs What might be the advantages of capping, methylation, and polyadenylylation of eukaryotic mRNAs?

12. The RNA Polymerase Mechanism of Action (Integrates with Chapter 28.) Figure 29.24 shows only one Mg^{2+} ion in the RNA polymerase II active site; more recent studies reveal the presence of two. Why is the presence of two Mg^{2+} ions significant?

13. Chromatin Remodeling Exposes DNA (Integrates with Chapter 11.) The SWI/SNF chromatin-remodeling complex peels about 50 bp from the nucleosome. Assuming B-form DNA, how long is this DNA segment? In forming nucleosomes, DNA is wrapped in turns about the histone core octamer. What fraction of a DNA turn around the core octamer does 50 bp of DNA comprise? How does 50 bp of DNA compare to the typical size of eukaryotic promoter modules and response elements?

14. The Lariat Intermediate in RNA Splicing Draw the structures that comprise the lariat branch point formed during mRNA splicing: the invariant A, its 5′-R neighbor, its 3′-Y neighbor, and its 2′-G neighbor.

15. HTH DNA-Binding Proteins and B-DNA Dimensions (Integrates with Chapters 6 and 11.) The α-helices in HTH (helix-turn-helix motif) DNA-binding proteins are formed from 7– or 8–amino acid residues. What is the overall length of these α-helices? How does their length compare with the diameter of B-form DNA?

16. Exploring the Structure of RNA Polymerase Bacteriophage T7 RNA polymerase bound to two DNA strands and an RNA strand, as shown in pdb file 1MSW, provides a glimpse of transcription. View this structure at *www.pdb.org* to visualize how the template DNA strand is separated from the nontemplate strand and transcribed into an RNA strand. Which Phe residue of the enzyme plays a significant role in DNA strand separation? In which domain of the polymerase is this Phe located? (You might wish to consult Yin, Y. W., and Steitz, T. A., 2002. Structural basis for the transition from initiation to elongation transcription in T7 RNA polymerase. *Science* **298**:1387–1395 to confirm your answer.)

17. α-Amanitin Inhibition of RNA Polymerase II RNA polymerase II is inhibited by α-amanitin. This mushroom-derived toxin has no effect on the enzyme's affinity for NTP substrates, but it dramatically slows polymerase translocation along the DNA. Go to *www.pdb.org* to view pdb file 1K83, which is the structure of RNA polymerase II with bound α-amanitin. Locate α-amanitin within this structure, and discuss why its position is consistent with its mode of inhibition.

18. Exploring the Structure of a DNA-Bound *bZIP* Transcription Factor C/EBPβ is a *bZIP* transcription factor in neuronal differentiation, learning and memory process, and other neuronal and glial functions. The structure of the *bZIP* domain of C/EBPβ bound to DNA is shown in pdb file 1GU4. Explore this structure to discover the leucine zipper dimerization domain and the DNA-binding basic regions. On the left side of the *www.pdb.org* 1GU4 page under "Display Files," click "pdb file" to see the atom-by-atom coordinates in the three-dimensional structure (scroll down past "Remarks" to find this information). Toward the end of this series, find the amino acid sequence of the C/EBPβ domain used in this study. Within this amino acid sequence, find the leucine residues of the leucine zipper and the basic residues in the DNA-binding basic region.

Preparing for the MCAT® Exam

19. Figure 29.15 highlights in red the DNA phosphate oxygen atoms. Some of them interact with catabolite activator protein (CAP). What kind of interactions do you suppose predominate, and what kinds of CAP amino acid side chains might be involved in these interactions?

20. Chromatin decompaction is a preliminary step in gene expression (Figure 29.48). How is chromatin decompacted?

FURTHER READING

Transcription in Prokaryotes

Busby, S., and Ebright, R. H., 1994. Promoter structure, promoter recognition, and transcription activation in prokaryotes. *Cell* **79**:743–746.

Campbell, E. A., Pavlova, O., Zenkin, N., Leon, F., et al., 2005. Structural, functional, and genetic analysis of sorangicin inhibition of bacterial RNA polymerase. *EMBO Journal* **24**:674–682.

Nudler, E., 2009. RNA polymerase active center. *Annual Review of Biochemistry* **78**:335–361.

Yin, Y. W., and Steitz, T. A., 2002. Structural basis for the transition from initiation to elongation transcription in T7 RNA polymerase. *Science* **298**:1387–1395.

Regulation of Transcription in Prokaryotes

Berg, O. G., and von Hippel, P. H., 1988. Selection of DNA binding sites by regulatory proteins. *Trends in Biochemical Sciences* **13**:207–211.

Dover, S. L., et al., 1997. Activation of prokaryotic transcription through arbitrary protein–protein contacts. *Nature* **386**:627–630.

Jacob, F., and Monod, J., 1961. Genetic regulatory mechanisms in the synthesis of proteins. *Journal of Molecular Biology* **3**:318–356.

Matthews, K. S., 1992. DNA looping. *Microbiological Reviews* **56**:123–136.

Platt, T., 1998. RNA structure in transcription elongation, termination, and antitermination. In *RNA Structure and Function,* Simons, R. W., and Grunberg-Monago, M., eds., pp. 541–574. Cold Spring Harbor, NY: Cold Spring Harbor Press.

Schleif, R., 1992. DNA looping. *Annual Review of Biochemistry* **61**:199–223.

Transcription in Eukaryotes

Burley, S., 1998. X-ray crystallographic studies of eukaryotic transcription factors. *Cold Spring Harbor Symposium on Quantitative Biology* **LXIII**:33–40.

Burley, S. K., and Roeder, R. G., 1996. Biochemistry and structural biology of transcription factor IID (TFIID). *Annual Review of Biochemistry* **65**:769–799.

Conaway, R. C., and Conaway, J. W., 1999. Transcription elongation and human disease. *Annual Review of Biochemistry* **68**:301–319.

Cramer, D., 2006. Recent structural studies of RNA polymerases II and III. *Biochemical Society Transactions* **34**:1058–1061.

Eichner, J., Chen, H.T., Warfield, L., and Hahn, S., 2010. Position of the general transcription factor TFIIF within the RNA polymerase II transcription preinitiation complex. *The EMBO Journal* **29**:706–716.

Haag, J. R., Pikaard, C. S., 2007. RNA polymerase I: A multifunctional molecular machine. *Cell* **131**:1224–1225.

Hager, G. L., McNally, J. G., and Misteli, T., 2009. Transcription dynamics. *Molecular Cell* **35**:741–753.

Kettenberger, H., Armache, K. J., and Cramer, P., 2004. Complete RNA polymerase II elongation complex structure and its interactions with NTP and TFIIS. *Molecular Cell* **16**:955–965.

Kim, H., Erickson. B., Luo, W., Seward, D., et al., 2010. Gene-specific RNA polymerase II phosphorylation and the CTD code. *Nature Structural & Molecular Biology* **17**:1279–1286.

Kostrewa, D., Zeller, M. E., Armache, K.-J., Seizl, M., et al., 2009. RNA polymerase II-TFIIB structure and mechanism of transcription initiation. *Nature* **462**:323–330.

Kornberg, R. D., 1998. Mechanism and regulation of yeast RNA polymerase II transcription. *Cold Spring Harbor Symposium on Quantitative Biology* **LXIII**:229–232.

Liu, X., Bushnell, D. A., Wang, D., Calero, G., and Kornberg, R. D., 2010. Structure of an RNA polymerase II-TFIIB complex and the transcription initiation mechanism. *Science* **327**:206–209.

Saunders, A., Core, L. J., and Lis, J. T., 2006. Breaking barriers to transcription elongation. *Nature Reviews Molecular Cell Biology* **7**:557–567.

Selth, L. A., Sigurdsson, S., and Svejstrup, J. Q., 2010. Transcription elongation by RNA polymerase II. *Annual Review of Biochemistry* **79**:271–293.

Wang, D., Bushnell, D. A., Westover, K. D., Kaplan, C. D., and Kornberg, R. D., 2006. Structural basis of transcription: Role of the trigger loop in substrate specificity and catalysis. *Cell* **127**:941–954.

Westover, K. D., Bushnell, D. A., and Kornberg, R. D., 2004. Structural basis of transcription: Nucleotide selection by rotation in the RNA polymerase II active center. *Cell* **119**:481–489.

Westover, K. D., Bushnell, D. A., and Kornberg, R. D., 2004. Structural basis of transcription: Separation of RNA from DNA by RNA polymerase II. *Science* **303**:1014–1016.

Regulation of Transcription in Eukaryotes

Amaral, P. P., Dinger, M. E., Mercer, T. R., and Mattick, J. S., 2008. The eukaryotic genome as an RNA machine. *Science* **319**:1787–1789.

Amoutzias, G. D., Robertson, D. L., Van de Peer, Y., and Oliver, S. G., 2008. Choose your partners: Dimerization in eukaryotic transcription factors. *Trends in Biochemical Sciences* **33**:220–229.

Bailey, C. H., Bartsch, D., and Kandel, E. R., 1996. Toward a molecular definition of long-term memory storage. *Proceedings of the National Academy of Sciences U.S.A.* **93**:13445–13452.

Björklund, S., et al., 1999. Global transcription regulators of eukaryotes. *Cell* **96**:759–767.

Carey, M., and Smale, S. T., 2000. *Transcriptional Regulation in Eukaryotes: Concepts, Strategies, and Techniques.* New York: Cold Spring Harbor Laboratory Press.

Core, L. J., and Lis, J. T., 2008. Transcriptional regulation through promoter-proximal pausing of RNA polymerase II. *Science* **319**:1791–1792.

Hobart, O., 2004. Common logic of transcription factor and microRNA action. *Trends in Biological Sciences* **29**:462–468.

Makeyev, E. V., and Maniatis, T., 2008. Multilevel regulation of gene expression by microRNAs. *Science* **319**:1789–1790.

Margaritis, T., and Holstege, F. C. P., 2008. Poised RNA polymerase II gives pause for thought. *Cell* **133**:581–584.

Maston, G. A., Evans, S. K., and Green, M. R., 2006. Transcriptional regulatory elements in the human genome. *Annual Review of Genomics and Human Genetics* **7**:29–59.

Moore, M. J., 2002. Nuclear RNA turnover. *Cell* **108**:431–434.

Severinov, K., 2000. RNA polymerase structure-function: Insights into points of transcriptional regulation. *Current Opinion in Microbiology* **3**:118–125.

Shamovsky, I., and Nudler, E., 2008. Modular RNA heats up. *Molecular Cell* **29**:415–417.

Struhl, K., 1999. Fundamentally different logic of gene regulation in prokaryotes and eukaryotes. *Cell* **98**:1–4.

Tuch, B. B., Ki, H., and Johnson, A. D., 2008. Evolution of eukaryotic transcription circuits. *Science* **319**:1797–1799.

Tully, T., 1997. Regulation of gene expression and its role in long-term memory and synaptic plasticity. *Proceedings of the National Academy of Sciences U.S.A.* **94**:4239–4241.

Utley, R. T., et al., 1998. Transcriptional activators direct histone acetyltransferase complexes to promoters. *Nature* **394**:498–502.

Mediator

Bernecky, C., Grob, P., Ebmeier, C. C., Nogales, E., and Taatjes, D. J., 2011. Molecular architecture of the human Mediator-RNA polymerase II-TFIIF assembly. *PLoS Biology* **9**:e1000603. doi:10.1371/journal.pbio.1000603.

Biddick, R., and Young, E. T., 2005. Yeast Mediator and its role in transcription regulation. *Comptes Rendus Biologies* **328**:773–782.

Björklund, S., and Gustafson, C. M., 2005. The yeast Mediator complex and its regulation. *Trends in Biochemical Sciences* **30**:240–244.

Imasaki, T., Calero, G., Cai, G., Tsai, K-L., et al., 2011. Architecture of the Mediator head module. *Nature* **475**:240–243.

Kornberg, R. D., 2005. Mediator and the mechanism of transcription activation. *Trends in Biochemical Sciences* **30**:235–239. This article introduces a series of reviews on Mediators that highlights the May 2005 issue of this journal (*Trends in Biochemical Sciences* **30**:235–271).

Sims, R. J., III, Rojas, L. A., Beck, D., Bonasio, R., et al., 2011. The C-terminal domain of RNA polymerase II is modified by site-specific methylation. *Science* **332**:99–103.

Soutourina, J., Wydau, S., Ambroise, Y., Boschiero, C., and Werner, M., 2011. Direct interaction between RNA polymerase II and Mediator required for transcription in vivo. *Science* **331**:1451–1454.

The Histone Code

Allis, D. C., Jenuwin, T., Reinberg, D., and Caparros, M-L., 2006. *Epigenetics.* New York: Cold Spring Harbor Laboratory Press.

Eisenberg, J. C., and Elgin, C. R., 2005. Antagonizing the neighbors. *Nature* **438:**1090–1091.

Goldberg, A. D., Allis, C. D., and Bernstein, E., 2007. Epigenetics: A landscape takes shape. *Cell* **128:**635–638.

Lee, J.-S., Smith, E., and Shilatifard, A., 2010. The language of histone crosstalk. *Cell* **142:**682–685.

Shahbazian, M. D., and Grunstein, M., 2007. Functions of site-specific histone acetylation and deacetylation. *Annual Review of Biochemistry* **76:**75–100.

Nucleosome Structure and Gene Expression

Armstrong, J. A., 2007. Negotiating the nucleosome: Factors that allow RNA polymerase II to elongate through chromatin. *Biochemistry and Cell Biology* **85:**426–434.

Boeger, H., et al., 2003. Nucleosomes unfold completely at a transcriptionally active promoter. *Molecular Cell* **11:**1587–1598.

Brown, C. E., et al., 2000. The many HATs of transcriptional coactivators. *Trends in Biochemical Sciences* **25:**15–19.

Clapier, C. R., and Cairns, B. R., 2009. The biology of chromatin remodeling complexes. *Annual Review of Biochemistry* **78:**273–304.

Fan, H. Y., et al., 2003. Distinct strategies to make nucleosomal DNA accessible. *Molecular Cell* **11:**1311–1322.

Felsenfeld, G., and Groudine, M., 2003. Controlling the double helix. *Nature* **421:**448–453.

Gilchrist, D., Dos Santos, G., Fargo, D. C., Xie, B., et al., 2010. Pausing of RNA Polymerase II disrupts DNA-specified nucleosome organization. *Cell* **143:**540–551.

Hampsey, M., and Reinberg, D., 2003. Tails of intrigue: Phosphorylation of RNA polymerase II mediates histone methylation. *Cell* **113:**429–432.

Iñigues-Llhî, J. A., 2006. For a healthy histone code, a little SUMO in the tail keeps the acetyl away. *ACS Chemical Biology* **1:**204–206.

Kassabov, S. R., et al., 2003. SWI/SNF unwraps, slides, and rewraps the nucleosome. *Molecular Cell* **11:**391–403.

Kornberg, R. D., and Lorch, Y., 1999. Twenty-five years of the nucleosome, fundamental particle of the eukaryotic chromosome. *Cell* **98:**285–294.

Ng, H. H., and Bird, A., 2000. Histone deacylases: Silencers for hire. *Trends in Biochemical Sciences* **25:**121–126.

Osley, M. A., 2006. Regulation of histone H2A and H2B ubiquitylation. *Briefings in Functional Genomics and Proteomics* **5:**179–189.

Reinberg, D., and Sims, R. J. III, 2006. deFACTo nucleosome dynamics. *Journal of Biological Chemistry* **281:**23297–23301.

Van Vugt, J. J. F. A., Ranes, M., Campsteijn, C., and Logie, C., 2007. The ins and outs of ATP-dependent chromatin remodeling in budding yeast: Biophysical and proteomic perspectives. *Biochimica et Biophysica Acta* **1769:**153–171.

Workman, J. L., ed., 2003. *Protein Complexes That Modify Chromatin.* New York: Springer.

Wu, C., et al., 1998. ATP-dependent remodeling of chromatin. *Cold Spring Harbor Symposium on Quantitative Biology* **LXIII:**525–534.

Yamada, K., Frouws, T. D., Angst, B., Fitzgerald, D. J., et al., 2011. Structure and mechanism of the chromatin remodelling factor ISW1a. *Nature* **472:**448–455.

Zaman, Z., et al., 1998. Gene transcription by recruitment. *Cold Spring Harbor Symposium on Quantitative Biology* **LXIII:**167–171.

Zlatanova, J., et al., 2000. Linker histone binding and displacement: Versatile mechanism for transcriptional regulation. *FASEB Journal* **14:**1697–1704.

DNA-Binding Gene Regulatory Proteins

Berg, J. M., and Shi, Y., 1996. The galvanization of biology: A growing appreciation for the roles of zinc. *Science* **271:**1081–1085.

Edmondson, D. G., and Olson, E. N., 1993. Helix-loop-helix proteins as regulators of muscle-specific transcription. *Journal of Biological Chemistry* **268:**755–758.

Glover, J. N. M., and Harrison, S. C., 1995. Crystal structure of the heterodimeric *bZIP* transcription factor *c-Fos-c-Jun* bound to DNA. *Nature* **373:**257–261.

Landschulz, W. H., Johnson, P. F., and McKnight, S. L., 1988. The leucine zipper: A hypothetical structure common to a new class of DNA-binding proteins. *Science* **240:**1759–1764.

Pabo, C. O., and Sauer, R. T., 1992. Transcription factors: Structural families and principles of DNA recognition. *Annual Review of Biochemistry* **61:**1053–1095.

Patikoglou, G., and Burley, S. K., 1997. Eukaryotic transcription factor–DNA complexes. *Annual Review of Biophysics and Biomolecular Structure* **26:**289–325.

Rohs, R., Jin, X., West, S. M., Joshi, R., et al., 2010. Origins of specificity in protein-DNA recognition. *Annual Review of Biochemistry* **79:**233–269.

Vinson, C. R., Sigler, P. B., and McKnight, S. L., 1989. Scissors-grip model for DNA recognition by a family of leucine zipper proteins. *Science* **246:**911–916.

von Hippel, P. H., 2007. From "simple" DNA–protein interactions to the macromolecular machines of gene expression. *Annual Review of Biophysics and Structural Biology* **36:**79–105.

Processing of Eukaryotic Transcripts

Breitbart, R. E., Andreadis, A., and Nadal-Ginard, B., 1987. Alternative splicing: A ubiquitous mechanism for the generation of multiple protein isoforms from single genes. *Annual Review of Biochemistry* **56:**467–495.

Egecioglu, D. E., and Chanfreau, G., 2011. Proofreading and spellchecking: A two-tier strategy for pre-mRNA splicing quality control. *RNA* **17:**383–389.

Hocine, S., Singer, R. H., and Grünwald, D., 2011. RNA processing and export. *Cold Spring Harbor Perspectives in Biology* **2:**2a000752.

Kramer, A., 1996. The structure and function of proteins involved in mammalian pre-mRNA splicing. *Annual Review of Biochemistry* **65:**367–409.

Leff, S. E., Rosenfeld, M. G., and Evans, R. M., 1986. Complex transcriptional units: Diversity in gene expression by alternative RNA processing. *Annual Review of Biochemistry* **55:**1091–1117.

Leung, A. K. W., Nagai, K., and Li, J., 2011. Structure of the spliceosomal U4 snRNP core domain and its implication for snRNP biogenesis. *Nature* **473:**536–539.

Maeder, C., and Guthrei, C., 2008. Modifications target spliceosome dynamics. *Nature Structural Biology* **15:**426–428.

Sachs, A., and Wahle, E., 1993. Poly (A) tail metabolism and function in eukaryotes. *Journal of Biological Chemistry* **268:**22955–22958.

Sharp, P. A., 1987. Splicing of messenger RNA precursors. *Science* **235:**766–771.

Sims, R. J. III, Milhouse, S., Chen, C.-F., Lewis, B. A., et al., 2007. Recognition of trimethylated histone H3 lysine 4 facilitates the recruitment of transcription post-initiation factors and pre-mRNA splicing. *Molecular Cell* **28:**665–676.

Staley, J. P., and Guthrie, C., 1998. Mechanical devices of the spliceosome: Motors, clocks, springs, and things. *Cell* **92:**315–326.

RNA Editing

Blanc, V., and Davidson, N. O., 2003. C-to-U RNA editing: Mechanisms leading to genetic diversity. *Journal of Biological Chemistry* **278:**1395–1398.

Hoopengardner, B., 2006. Adenosine-to-inosine RNA editing: Perspectives and predictions. *Mini-Reviews in Medicinal Chemistry* **6**:1213–1216.

Maas, S., Rich, A., and Nishikura, K., 2003. A-to-I RNA editing: Recent news and residual mysteries. *Journal of Biological Chemistry* **278**:1391–1394.

Nishikura, K., 2010. Functions and regulation of RNA editing by ADAR deaminases. *Annual Review of Biochemistry* **79**:321–349.

Samuel, C. E., 2003. RNA editing minireview series. *Journal of Biological Chemistry* **278**:1389–1390.

Post-Transcriptional Gene Regulation

Fabian, M. R., Sonenberg, N., and Filipowicz, W., 2010. Regulation of mRNA stability and translation by miRNAs. *Annual Review of Biochemistry* **79**:351–379.

Krol, J., Loedige, I., and Filipowicz, W., 2010. The widespread regulation of microRNA biogenesis, function and decay. *Nature Reviews Genetics* **11**:597–610.

Salmena, L., Poliseno, L., Tay, Y., Kats, L., and Pandolfi, P. P., 2011. A ceRNA hypothesis: The Rosetta Stone of a hidden RNA language? *Cell* **146**:353–358.

A Unified Theory of Gene Expression

Maniatis, T., and Reed, R., 2002. An extensive network of coupling among gene expression machines. *Nature* **416**:499–506.

Narliker, G. J., Fan, H.-Y., and Kingston, R. E., 2002. Cooperation between complexes that regulate chromatin structure and transcription. *Cell* **108**:475–487.

Orphanides, G., and Reinberg, D., 2002. A unified theory of gene expression. *Cell* **108**:439–451.

Schreiber, S. L., and Bernstein, B. E., 2002. Signaling network model of chromatin. *Cell* **111**:771–778.

Woychik, N. A., and Hampsey, M., 2002. The RNA polymerase II machinery: Structure illuminates function. *Cell* **108**:439–451.

RNA Degradation

Liu, Q., Greimann, J. C., and Lima, C. D., 2006. Reconstitution, activities, and structure of the eukaryotic RNA exosome. *Cell* **127**:1223–1237.

Lorentzen, E., and Conti, E., 2006. The exosome and proteasome: Nanocompartments for degradation. *Cell* **125**:651–654.

Pruijn, G., 2005. Doughnuts dealing with RNA. *Nature Structural and Molecular Biology* **12**:562–564.

Schmid, M., and Jensen, T. H., 2008. The exosome: A multipurpose RNA-decay machine. *Trends in Biochemical Sciences* **33**:501–510.

Protein Synthesis

◀ The Maya encoded their history in hieroglyphs carved on stelae and temples like these ruins in Yucatan Peninsula, Mexico.

....

We are a spectacular, splendid manifestation of life. We have language and can build metaphors as skillfully and precisely as ribosomes make proteins. We have affection. We have genes for usefulness, and usefulness is about as close to a "common goal" of nature as I can guess at. And finally, and perhaps best of all, we have music.

Lewis Thomas (1913–1994)
"The Youngest and Brightest Thing Around"
in *The Medusa and the Snail* (1979)

KEY QUESTIONS

30.1 What Is the Genetic Code?

30.2 How Is an Amino Acid Matched with Its Proper tRNA?

30.3 What Are the Rules in Codon–Anticodon Pairing?

30.4 What Is the Structure of Ribosomes, and How Are They Assembled?

30.5 What Are the Mechanics of mRNA Translation?

30.6 How Are Proteins Synthesized in Eukaryotic Cells?

ESSENTIAL QUESTION

Ribosomes synthesize proteins by reading the nucleotide sequence of mRNAs and polymerizing amino acids in an N→C direction.

How is the nucleotide sequence of an mRNA molecule translated into the amino acid sequence of a protein molecule?

We turn now to the problem of how the sequence of nucleotides in an mRNA molecule is translated into the specific amino acid sequence of a protein. The problem raises both informational and mechanical questions. First, what is the **genetic code** that allows the information specified in a sequence of bases to be translated into the amino acid sequence of a polypeptide? That is, how is the 4-letter language of nucleic acids translated into the 20-letter language of proteins? Implicit in this question is a mechanistic problem: It is easy to see how base pairing establishes a one-to-one correspondence that allows the template-directed synthesis of polynucleotide chains in the processes of replication and transcription. However, there is no obvious chemical affinity between the purine and pyrimidine bases and the 20 different amino acids. Nor is there any obvious structural or stereochemical connection between polynucleotides and amino acids that might guide the translation of information.

Francis Crick reasoned that **adapter molecules** must bridge this information gap. These adapter molecules must interact specifically with both nucleic acids (mRNAs) and amino acids. At least 20 different adapter molecules would be needed, at least one for each amino acid. The various adapter molecules would be able to read the genetic code in an mRNA template and align the amino acids according to the template's directions so that they could be polymerized into a unique polypeptide. Transfer RNAs (tRNAs; Figure 30.1) are the adapter molecules (see Chapter 10). Amino acids are attached to the 3'-OH at the 3'-CCA end of tRNAs as aminoacyl esters. The formation

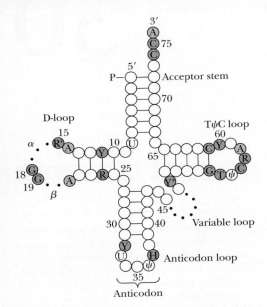

FIGURE 30.1 Generalized secondary structure of tRNA molecules. Circles represent nucleotides in the tRNA sequence. The numbers given indicate the standardized numbering system for tRNAs (which differ in total number of nucleotides). Dots indicate places where the number of nucleotides may vary in different tRNA species. All tRNAs have the invariant 3-base sequence CCA at their 3'-ends. Recall from Chapter 10 that tRNA molecules often have modified or unusual bases.

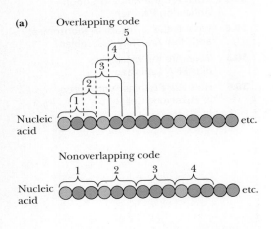

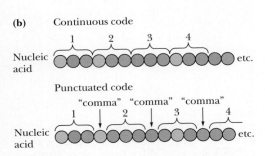

FIGURE 30.2 (a) An overlapping versus a nonoverlapping code. (b) A continuous versus a punctuated code.

of these aminoacyl-tRNAs, so-called **charged tRNAs,** is catalyzed by specific **aminoacyl-tRNA synthetases.** There is one of these enzymes for each of the 20 amino acids and each aminoacyl-tRNA synthetase loads its amino acid only onto tRNAs designed to carry it. In turn, these tRNAs specifically recognize unique sequences of bases in the mRNA through complementary base pairing.

30.1 | What Is the Genetic Code?

Once it was realized that the sequence of bases in a gene specified the sequence of amino acids in a protein, various possibilities for such a genetic code were considered. How many bases are necessary to specify each amino acid? Is the code overlapping or nonoverlapping (Figure 30.2)? Is the code punctuated or continuous? Mathematical considerations favored a triplet of bases as the minimal code word, or **codon,** for each amino acid: A doublet code based on pairs of the four possible bases, A, C, G, and U, has $4^2 = 16$ unique arrangements, an insufficient number to encode the 20 amino acids. A triplet code of four bases has $4^3 = 64$ possible code words, more than enough for the task.

The Genetic Code Is a Triplet Code

The genetic code is a triplet code read continuously from a fixed starting point in each mRNA. Specifically, it is defined by the following:

1. A group of three bases codes for one amino acid.
2. The code is not overlapping.
3. The base sequence is read from a fixed starting point without punctuation. That is, the mRNA sequences contain no "commas" signifying appropriate groupings of triplets. If the reading frame is displaced by one base, it remains shifted throughout the subsequent message; no "commas" are present to restore the "correct" frame.
4. The code is **degenerate,** meaning that, in most cases, each amino acid can be coded by any of several triplets. Recall that a triplet code yields 64 codons for 20 amino acids. Most codons (61 of 64) code for some amino acid.

Codons Specify Amino Acids

The complete translation of the genetic code is presented in Table 30.1. Codons, like other nucleotide sequences, are read $5' \rightarrow 3'$. Codons represent triplets of bases in mRNA or, replacing U with T, triplets along the nontranscribed (nontemplate) strand of DNA.

Several noteworthy features characterize the genetic code:

1. *All the codons have meaning.* Of the 64 codons, 61 specify particular amino acids. The remaining 3—UAA, UAG, and UGA—specify no amino acid and, thus, they are **nonsense codons.** Nonsense codons serve as **termination codons;** they are "stop" signals indicating that the end of the protein has been reached.
2. *The genetic code is unambiguous.* Each of the 61 "sense" codons encodes only one amino acid.
3. *The genetic code is degenerate.* With the exception of Met and Trp, every amino acid is coded by more than one codon. Several—Arg, Leu, and Ser—are represented by six different codons. Codons coding for the same amino acid are called **synonymous codons.**
4. *Codons representing the same amino acid or chemically similar amino acids tend to be similar in sequence.* Often the third base in a codon is irrelevant, so, for example, all four codons in the GGX family specify Gly, and the UCX family specifies Ser (Table 30.1). This feature is known as **third-base degeneracy.** Note also that codons with a pyrimidine as second base likely encode amino acids with hydrophobic side chains, and codons with a purine in the second-base position typically specify polar or charged amino acids. The two negatively charged amino acids, Asp and Glu, are

TABLE 30.1 The Genetic Code

First Position (5'-end)	Second Position				Third Position (3'-end)
	U	C	A	G	
U	UUU Phe	UCU Ser	UAU Tyr	UGU Cys	U
	UUC Phe	UCC Ser	UAC Tyr	UGC Cys	C
	UUA Leu	UCA Ser	UAA Stop	UGA Stop	A
	UUG Leu	UCG Ser	UAG Stop	UGG Trp	G
C	CUU Leu	CCU Pro	CAU His	CGU Arg	U
	CUC Leu	CCC Pro	CAC His	CGC Arg	C
	CUA Leu	CCA Pro	CAA Gln	CGA Arg	A
	CUG Leu	CCG Pro	CAG Gln	CGG Arg	G
A	AUU Ile	ACU Thr	AAU Asn	AGU Ser	U
	AUC Ile	ACC Thr	AAC Asn	AGC Ser	C
	AUA Ile	ACA Thr	AAA Lys	AGA Arg	A
	AUG Met*	ACG Thr	AAG Lys	AGG Arg	G
G	GUU Val	GCU Ala	GAU Asp	GGU Gly	U
	GUC Val	GCC Ala	GAC Asp	GGC Gly	C
	GUA Val	GCA Ala	GAA Glu	GGA Gly	A
	GUG Val	GCG Ala	GAG Glu	GGG Gly	G

Third-Base Degeneracy Is Color-Coded

	Third-Base Relationship	Third Bases with Same Meaning	Number of Codons
	Third base irrelevant	U, C, A, G	32 (8 families)
	Purines	A or G	12 (6 pairs)
	Pyrimidines	U or C	14 (7 pairs)
	Three out of four	U, C, A	3 (AUX = Ile)
	Unique definitions	G only	2 (AUG = Met) (UGG = Trp)
	Unique definition	A only	1 (UGA = Stop)

*AUG signals translation initiation as well as coding for Met residues.

> **A DEEPER LOOK**

Natural and Unnatural Variations in the Standard Genetic Code

The genomes of some lower eukaryotes, prokaryotes, and mitochondria show some exceptions to the standard genetic code (Table 30.1) in codon assignments. The phenomenon is more common in mitochondria. For example, the termination codon UGA codes for tryptophan in mitochondria from various animals, protozoans, and fungi. AUA, normally an Ile codon, codes for methionine in some animal and fungal mitochondrial genomes, and AGA (an Arg codon) is a termination codon in vertebrate mitochondria but is a Ser codon in fruit fly mitochondria. Mitochondria in several species of yeast use the CUX codons to specify Thr instead of Leu. Some yeast and algal mitochondria use CGG, normally an Arg codon, as a stop codon.

Less common are genomic codon variations within the genomes of prokaryotic and eukaryotic cells. Among the lower eukaryotes, certain ciliated protozoans (*Tetrahymena* and *Paramecium*) use UAA and UGA as glutamine codons rather than stop codons. Instances in prokaryotes include use of the stop codon UGA to specify Trp by *Mycoplasma*. Perhaps most interesting is the use of some UGA codons by both prokaryotes and eukaryotes (including humans) to specify **selenocysteine (Sec),** an analog of cysteine in which the sulfur atom is replaced by a selenium atom. Indeed, the identification of Sec residues in proteins from bacteria, archaea, and eukaryotes has led some people to nominate Sec as the 21st amino acid! Sec formation requires a novel Sec-specific tRNA known as tRNA[Sec]. This tRNA[Sec] is loaded with a Ser residue by seryl-tRNA synthetase, the aminoacyl-tRNA synthetase for serine. Then, in an ATP-dependent process, the Ser-O is replaced by Se. Translation of UGA codons by selenocysteinyl-tRNA[Sec] depends on the presence of specific stem-loop secondary structures in the mRNA called **SECIS elements** that recode the UGA codon from "stop" to "Sec." SECIS elements are recognized by specific proteins that recruit selenocysteinyl-tRNA[Sec] to the UGA codon during protein synthesis. Most selenoproteins (proteins containing selenocysteine) are involved in oxidation–reduction reactions, and Sec participates directly in the catalytic mechanism. Sec provides a more reactive functionality than Cys.

$$H-Se-CH_2-\underset{\underset{NH_3^+}{|}}{\overset{\overset{H}{|}}{C}}-COO^-$$

Selenocysteine (Sec)

It is also possible now to manipulate the genetic code, tRNAs, and aminoacyl-tRNA synthetases to introduce amino acids never before imagined into proteins of interest. By this methodology, bacterial, yeast, and mammalian cells have been engineered so that genetically encoded unnatural amino acids with desired physical, chemical, or biological properties can be placed at any position in a protein. More than 70 such unnatural amino acids, each with a chemically unique side chain, can be incorporated into proteins using a unique codon and corresponding tRNA–aminoacyl-tRNA synthetase pair. (*See Wang L., Xie J., and Schultz, P. G., 2006. Expanding the genetic code.* Annual Review of Biophysics and Biomolecular Structure **35:**225–249 *and Liu, C. C., and Schultz, P. G., 2010. Adding new chemistry to the genetic code.* Annual Review of Biochemistry **79:**413–444.)

encoded by GAX codons; GA–pyrimidine gives Asp and GA–purine specifies Glu. The consequence of these similarities is that mutations are less likely to be harmful because single base changes in a codon will result either in no change or in a substitution with an amino acid similar to the original amino acid. The degeneracy of the code is evolution's buffer against mutational disruption.

5. *The genetic code is "universal."* Although certain minor exceptions in codon usage occur (see A Deeper Look box on page 1049), the more striking feature of the code is its universality: Codon assignments are virtually the same throughout all organisms—archaea, bacteria, and eukaryotes. This conformity means that all extant organisms use the same genetic code, providing strong evidence that they all evolved from a common primordial ancestor.

30.2 How Is an Amino Acid Matched with Its Proper tRNA?

Codon recognition is achieved by aminoacyl-tRNAs. In order for accurate translation to occur, the appropriate aminoacyl-tRNA must "read" the codon through base pairing via its **anticodon loop** (see Chapter 11). Once an aminoacyl-tRNA has been synthesized, the amino acid part makes no contribution to accurate translation of the mRNA. That is, the amino acid is passively chauffeured by its tRNA and becomes inserted into a growing peptide chain following codon–anticodon recognition between the mRNA and tRNA.

Aminoacyl-tRNA Synthetases Interpret the Second Genetic Code

A **second genetic code** must exist, the code by which each aminoacyl-tRNA synthetase matches up its amino acid with tRNAs that can interact with codons specifying its amino acid. To interpret this second genetic code, an aminoacyl-tRNA synthetase discriminates between the 20 amino acids and the many tRNAs and uniquely picks out its proper substrates—one specific amino acid and the tRNA(s) appropriate to it—from among the more than 400 possible combinations. The appropriate tRNAs are those having anticodons that can base-pair with the codons specifying the particular amino acid. It is imperative that the proper amino acids be loaded onto the various tRNAs so that the mRNA can be translated with fidelity. Although the primary genetic code is key to understanding the central dogma of molecular biology on how DNA encodes proteins, the second genetic code is just as crucial to the fidelity of information transfer.

Eukaryotic cells and some bacteria have 20 different aminoacyl-tRNA synthetases, one for each amino acid. Each of these enzymes catalyzes ATP-dependent attachment of its specific amino acid to the 3'-end of its **cognate tRNA molecules** (Figure 30.3). The products of the reaction are an aminoacyl-tRNA, AMP, and PP$_i$. Ever-present pyrophosphatases quickly hydrolyze the pyrophosphate product to give 2 P$_i$. This highly exergonic reaction provides the overall thermodynamic driving force for aminoacyl-tRNA synthesis. The aminoacyl-tRNA synthetase reaction serves two purposes:

1. It activates the amino acid so that it will readily react to form a peptide bond.
2. It bridges the information gap between amino acids and codons.

The underlying mechanisms of molecular recognition used by each aminoacyl-tRNA synthetase to bring the proper amino acid to its cognate tRNA are the embodiment of the second genetic code. An exception to these rules is the fact that most bacteria and archaea do not have specific aminoacyl-tRNA synthetases for Asn and/or Gln. Instead, the Asp- and Glu-tRNA synthetases are used to load Asp and Glu onto tRNA^{Asn} and tRNA^{Gln}, respectively. Then these aminoacyl groups are amidated by tRNA-dependent amidotransferases to yield Asn-tRNA^{Asn} and Gln-tRNA^{Gln} for protein synthesis.

Most aminoacylated-tRNAs are destined to deliver their amino acids to the translational apparatus for use in protein synthesis. However, aminoacylated-tRNAs also serve as

Cognate kindred; in this sense, cognate refers to those tRNAs having anticodons that can read one or more of the codons that specify one particular amino acid.

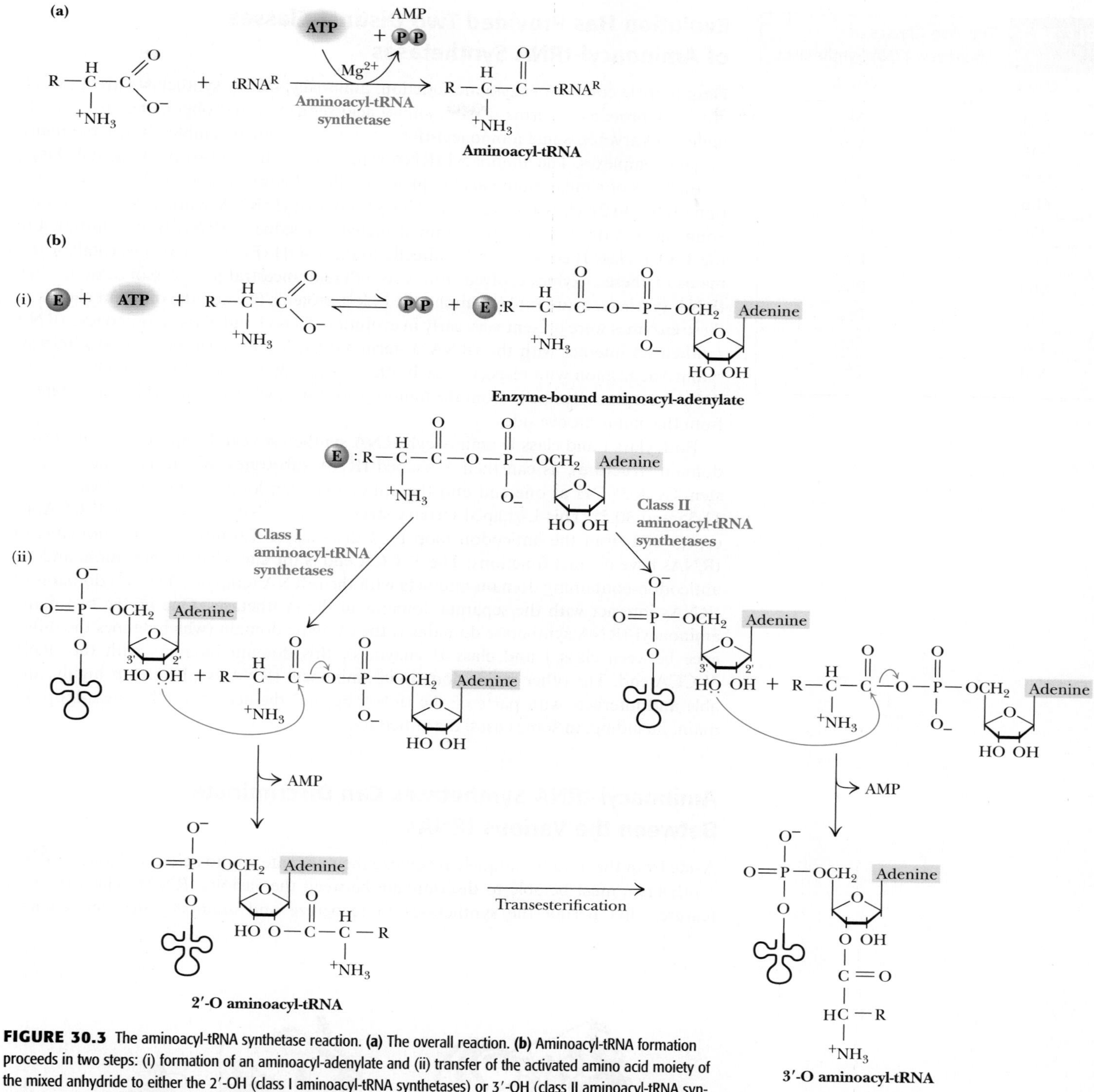

FIGURE 30.3 The aminoacyl-tRNA synthetase reaction. **(a)** The overall reaction. **(b)** Aminoacyl-tRNA formation proceeds in two steps: (i) formation of an aminoacyl-adenylate and (ii) transfer of the activated amino acid moiety of the mixed anhydride to either the 2′-OH (class I aminoacyl-tRNA synthetases) or 3′-OH (class II aminoacyl-tRNA synthetases) of the ribose on the terminal adenylic acid at the 3′-CCA terminus common to all tRNAs. Those aminoacyl-tRNAs formed as 2′-aminoacyl esters undergo a transesterification that moves the aminoacyl function to the 3′-O of tRNA. Only the 3′-esters are substrates for protein synthesis.

aminoacyl donors for a number of other purposes. For example, aminoacyl-tRNAs are involved in tagging the N-termini of proteins for degradation (see Figure 31.9), in the delivery of amino acids for peptidoglycan synthesis (see Figure 7.29), as substrates for amino acid addition to membrane lipids (as in the formation of lysyl-phosphatidylglycerol, thereby dramatically altering the charge of this membrane lipid), and as substrates in antibiotic biosynthesis (e.g., the synthesis of valanimycin).

TABLE 30.2	The Two Classes of Aminoacyl-tRNA Synthetases	
Class I		**Class II**
Arg		Ala
Cys		Asn
Gln		Asp
Glu		Gly
Ile		His
Leu		Lys
Met		Phe
Trp		Pro
Tyr		Ser
Val		Thr

Evolution Has Provided Two Distinct Classes of Aminoacyl-tRNA Synthetases

Despite their common enzymatic function, aminoacyl-tRNA synthetases are a diverse group of proteins in terms of size, amino acid sequence, and oligomeric structure. In higher eukaryotes, some aminoacyl-tRNA synthetases are assembled into large multi-protein complexes. The aminoacyl-tRNA synthetases fall into two fundamental classes on the basis of similar amino acid sequence motifs, oligomeric state, and acylation function (Table 30.2): class I and class II. Class I aminoacyl-tRNA synthetases first add the amino acid to the 2'-OH of the terminal adenylate residue of tRNA before shifting it to the 3'-OH; class II enzymes add it directly to the 3'-OH (Figure 30.3). The catalytic domains of these enzymes evolved from two different ancestral predecessors. Aminoacyl-tRNA synthetases are ranked among the oldest proteins because the different classes of these enzymes were present very early in evolution. Class I and class II aminoacyl-tRNA synthetases interact with the tRNA 3'-terminal CCA and acceptor stem in a mirror-symmetric fashion with respect to each other (Figure 30.4). Class I enzymes bind to the tRNA acceptor stem helix from the minor-groove side, whereas class II enzymes bind it from the major-groove side.

Both class I and class II aminoacyl-tRNA synthetases can be approximated as two-domain structures, as can their L-shaped tRNA substrates, which have the acceptor stem/CCA-3'-OH at one end and the anticodon stem-loop at the other (see Figures 11.35 and 30.5). This L-shaped tertiary structure of tRNAs separates the 3'-CCA acceptor end from the anticodon loop by a distance of 7.6 nm. The two domains of tRNAs have distinct functions: The 3'-CCA end is the site of aminoacylation, and the anticodon-containing domain interacts with the mRNA template. The two domains of tRNAs interact with the separate domains in the synthetases. One of the two major aminoacyl-tRNA synthetase domains is the catalytic domain (which defines the difference between class I and class II enzymes); this domain interacts with the tRNA 3'-CCA end. The other major domain in aminoacyl-tRNA synthetases is highly variable and interacts with parts of the tRNA beyond the acceptor-TΨC stem-loop domain, including, in some cases, the anticodon.

Aminoacyl-tRNA Synthetases Can Discriminate Between the Various tRNAs

Aside from the need to uniquely recognize their cognate amino acids, aminoacyl-tRNA synthetases must be able to discriminate between the various tRNAs. The structural features that permit the synthetases to recognize and aminoacylate their cognate

FIGURE 30.4 Mirror-symmetric interactions of class I versus class II aminoacyl-tRNA synthetases with their tRNA substrates. The two different classes of aminoacyl-tRNA synthetases bind to opposite faces of tRNA molecules. On the left is the structure of the class I glutaminyl-tRNA^Gln synthetase : tRNA^Gln complex with a bound active-site inhibitor (orange) (pdb id = 1EUQ). On the right is the structure of the class II threonyl-tRNA^Thr synthetase : tRNA^Thr complex with AMP (red) at the active site (pdb id = 1QF6). The relative orientation of the tRNA is the same in both structures, with the 3'-CCA end of the tRNA pointed away from the viewer.

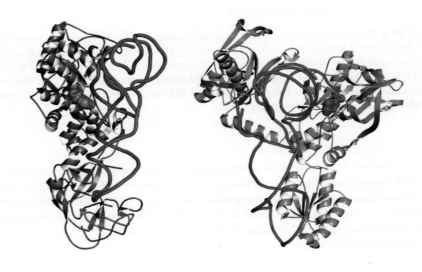

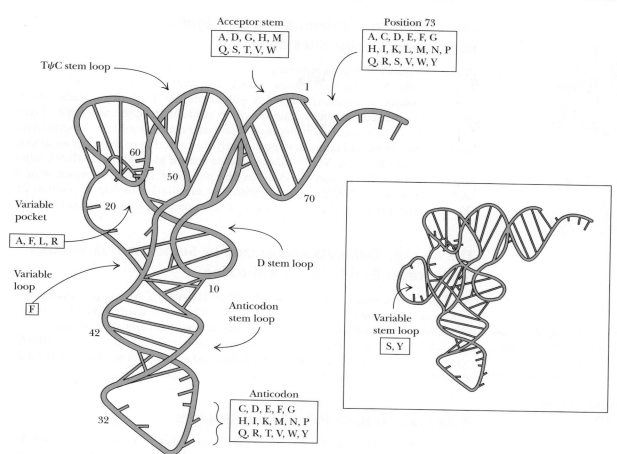

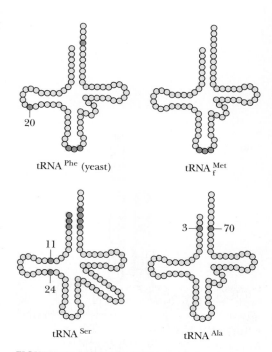

FIGURE 30.5 Ribbon diagram of tRNA tertiary structure. Numbers represent the consensus nucleotide sequence (see Figure 30.1). The locations of nucleotides recognized by the various aminoacyl-tRNA synthetases are indicated; shown within the boxes are one-letter designations of the amino acids whose respective aminoacyl-tRNA synthetases interact at the discriminator base (position 73), acceptor stem, variable pocket and/or loop, or anticodon. The inset shows additional recognition sites in those tRNAs having a variable loop that forms a stem-loop structure. (Adapted from Figure 2 in Saks, M. E., Sampson, J. R., and Abelson, J. N., 1994. The transfer RNA problem: A search for rules. *Science* **263**:191–197.)

tRNA(s) are *not* universal. That is, a common set of rules does not govern tRNA recognition by these enzymes. Most surprising is the fact that the recognition features are not limited to the anticodon and, in some instances, do not even include the anticodon. For most tRNAs, a set of sequence elements is recognized by its specific aminoacyl-tRNA synthetase, rather than a single distinctive nucleotide or base pair. These elements include one or more of the following: (1) at least one base in the anticodon; (2) one or more of the three base pairs in the acceptor stem; and (3) the base at canonical position 73 (the unpaired base preceding the CCA end), referred to as the **discriminator base** because this base is invariant in the tRNAs for a particular amino acid. Figure 30.5 presents a ribbon diagram of a tRNA molecule showing the location of nucleotides that contribute to specific recognition by the respective aminoacyl-tRNA synthetases for each of the 20 amino acids. Interestingly, the same set of tRNA features that serves as positive determinants for binding and aminoacylation of the tRNA by its cognate aminoacyl-tRNA synthetase may act as negative determinants that prohibit binding and aminoacylation by other (noncognate) aminoacyl-tRNA synthetases. Because no common set of rules exists, the second genetic code is an **operational code** based on aminoacyl-tRNA synthetase recognition of varying sequence and structural features in the different tRNA molecules during the operation of aminoacyl-tRNA synthesis. Some examples of this code are given in Figure 30.6.

FIGURE 30.6 Major identity elements in four tRNA species. Each base in the tRNA is represented by a circle. Numbered filled circles indicate positions of identity elements within the tRNA that are recognized by its specific aminoacyl-tRNA synthetase. (Adapted from Schulman, L. H., and Abelson, J., 1988. Recent excitement in understanding transfer RNA identity. *Science* **240**:1591–1592.)

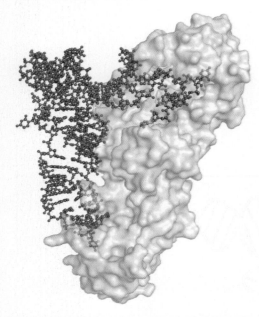

FIGURE 30.7 Structure of *E. coli* glutaminyl-tRNA^Gln synthetase complexed with tRNA^Gln and ATP (pdb id = 1GSG). The protein:tRNA contact region extends along one side of the entire length of this extended protein. The acceptor stem of the tRNA and the ATP *(green)* fit into a cleft at the top of the protein in this view. The enzyme also interacts extensively with the anticodon (lower tip of tRNA^Gln).

Escherichia coli Glutaminyl-tRNA^Gln Synthetase Recognizes Specific Sites on tRNA^Gln

E. coli glutaminyl-tRNA^Gln synthetase, a class I enzyme, provides a good illustration of aminoacyl-tRNA synthetase:cognate tRNA interactions. This glutaminyl-tRNA^Gln synthetase shares a continuous interaction with its cognate tRNA that extends from the anticodon to the acceptor stem along the entire inside of the L-shaped tRNA (Figure 30.7). Specific recognition elements include enzyme contacts with the discriminator base, acceptor stem, and anticodon, particularly the central U in the CUG anticodon. The carboxylate group of Asp[235] makes sequence-specific H bonds in the tRNA minor groove with the 2-NH_2 group of G3 in the base pair G3:C70 of the acceptor stem. A mutant glutaminyl-tRNA^Gln synthetase with Asn substituted for Asp at position 235 shows relaxed specificity; that is, it now will acylate noncognate tRNAs with Gln.

The Identity Elements Recognized by Some Aminoacyl-tRNA Synthetases Reside in the Anticodon

Alteration of the anticodons of either tRNA^Trp or tRNA^Val to CAU, the anticodon for the methionine codon AUG, transforms each of the tRNAs into a substrate for methionyl-tRNA synthetase, and they are loaded with methionine. Similarly, reversing the methionine CAU anticodon of tRNA^Met to UAC transforms it into a substrate for valyl-tRNA^Val synthetase. Clearly, methionyl-tRNA synthetase and valyl-tRNA synthetase rely on the anticodon in selecting tRNAs for loading.

A Single G:U Base Pair Defines tRNA^Ala s

The noncanonical base pair, G3:U70, is the singular feature by which alanyl-tRNA^Ala synthetase recognizes tRNAs as its substrates. All tRNA^Ala representatives, from archaea to eukaryotes, possess this G3:U70 acceptor stem base pair. Altering this unusual G3:U70 base pair of tRNA^Ala to G:C, A:U, or even U:G abolishes its ability to be aminoacylated with alanine. On the other hand, provided the G3:U70 base pair is present, alanyl-tRNA^Ala synthetase aminoacylates a 24-nucleotide stem-loop analog of tRNA^Ala (Figure 30.8). The key feature of the G3:U70 base pair is the 2-NH_2 group of G3. In the RNA A-form double-helical structure adopted by the tRNA acceptor stem,

(a)

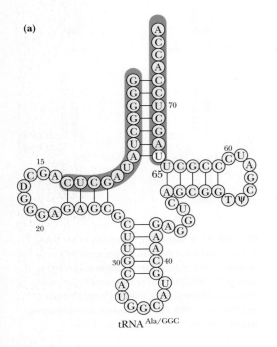

tRNA ^Ala/GGC

(b)

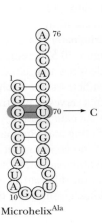

Microhelix^Ala

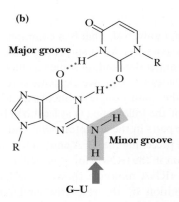

FIGURE 30.8 **(a)** A microhelix analog of tRNA^Ala is aminoacylated by alanyl-tRNA^Ala synthetase, provided it has the characteristic tRNA^Ala G3:U70 acceptor stem base pair. The microhelix^Ala consists of nucleotides 1 through 13 of tRNA^Ala/GGC connected directly to 66 through 76 to re-create the tRNA^Ala 7-bp acceptor stem. **(b)** The unbonded 2-NH_2 group of the G3 in the G3:U70 base pair would lie within the minor groove of an A-form RNA double helix. [(a) Adapted from Schimmel, P., 1989. Parameters for molecular recognition of transfer RNAs. *Biochemistry* **28**:2747–2759.]

the G3 2-NH$_2$ group is exposed in the minor groove of the helix, and if the G3 pairing partner is a U, this 2-NH$_2$ group lacks an H-bonding partner (Figure 30.8). Thus, an unpaired G 2-amino group at the right place in a tRNA acceptor stem marks a tRNA for aminoacylation by alanyl-tRNA^{Ala} synthetase.

30.3 : What Are the Rules in Codon–Anticodon Pairing?

Protein synthesis depends on the codon-directed binding of the proper aminoacyl-tRNAs so that the right amino acids are sequentially aligned according to the specifications of the mRNA undergoing translation. This alignment is achieved via codon–anticodon pairing in antiparallel orientation (Figure 30.9). However, considerable degeneracy exists in the genetic code at the third position. Conceivably, this degeneracy could be handled in either of two ways: (1) Codon–anticodon recognition could be highly specific so that a complementary anticodon is required for each codon, or (2) fewer than 61 anticodons could be used for the "sense" codons if certain allowances were made in the base-pairing rules. Then, some anticodons could recognize more than one codon. As early as 1965, it was known that poly(U) bound *all* Phe-tRNA^{Phe} molecules even though UUC is also a Phe codon. The phenylalanine-specific tRNAs could recognize either UUU or UUC. Also, one particular yeast tRNA^{Ala} was able to bind to three codons: GCU, GCC, and GCA.

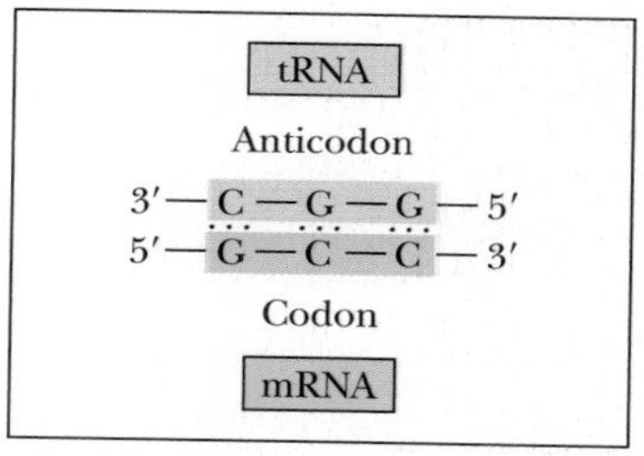

Francis Crick Proposed the "Wobble" Hypothesis for Codon–Anticodon Pairing

Francis Crick considered these results and tested alternative base-pairing possibilities by model building. He hypothesized that the first two bases of the codon and the last two bases of the anticodon form canonical Watson–Crick A:U or G:C base pairs, but pairing between the third base of the codon and the first base of the anticodon follows less stringent rules. That is, a certain amount of play, or **wobble**, might be allowed in base pairing at this position. The third base of the codon is sometimes referred to as the **wobble position.**

Crick's investigations suggested a set of rules for pairing between the third base of the codon and the first base of the anticodon (Table 30.3). The wobble rules indicate that a first-base anticodon U could recognize either an A or G in the codon third-base position; first-base anticodon G might recognize either U or C in the third-base position of the codon; and first-base anticodon I^1 might interact with U, C, or A in the codon third position (Figure 30.10).[2]

The wobble rules also predict that four-codon families (like Pro or Thr), where any of the four bases may be in the third position, require at least two different tRNAs. However, all members of the set of tRNAs specific for a particular amino acid—termed **isoacceptor tRNAs**—are served by one aminoacyl-tRNA synthetase.

Some Codons Are Used More Than Others

Because more than one codon exists for most amino acids, the possibility for variation in codon usage arises. Indeed, variation in codon usage accommodates the fact that the DNA of different organisms varies in relative A:T/G:C content. Nevertheless, even in organisms of average base composition, codon usage may be biased. Table 30.4 gives some examples from *E. coli* and humans reflecting the nonrandom usage of codons. Of more than 109,000 Leu codons tabulated in a set of human genes, CUG was used in excess of 48,000 times, CUC more than 23,000 times, but UUA just 6000 times.

FIGURE 30.9 Codon–anticodon pairing. Complementary trinucleotide sequence elements align in antiparallel fashion.

TABLE 30.3	Base-Pairing Possibilities at the Third Position of the Codon
Base on the Anticodon	Bases Recognized on the Codon
U	A, G
C	G
A	U
G	U, C
I	U, C, A

Adapted from Crick, F. H. C., 1966. Codon–anticodon pairing: The wobble hypothesis. *Journal of Molecular Biology* **19:**548–555.

---

^{1}I is inosine (6-OH purine).
[2]Thus, the first base of the anticodon indicates whether the tRNA can read one, two, or three different codons: Anticodons beginning with A or C read only one codon, those beginning with G or U read two, and anticodons beginning with I can read three codons.

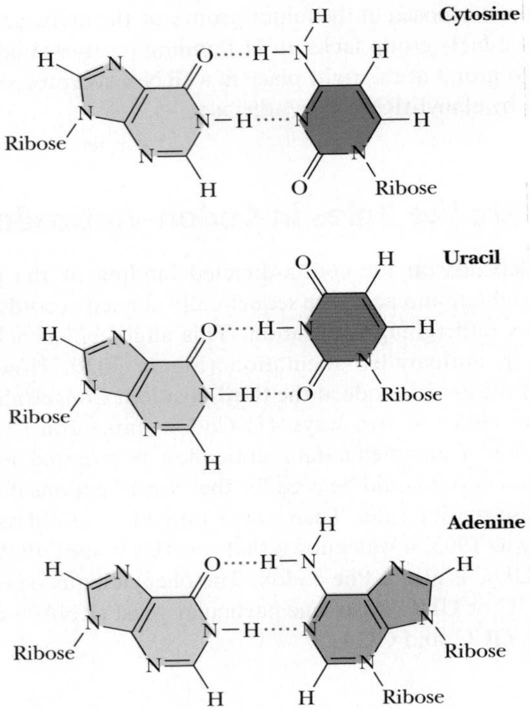

Cytosine

Uracil

Adenine

FIGURE 30.10 Pairing of anticodon inosine (I, *left*) with C, U, or A as the codon third base. Note that I is in the keto tautomeric form.

TABLE 30.4	Representative Examples of Codon Usage in *E. coli* and Human Genes

The results are expressed as frequency of occurrence of a codon per 1000 codons tabulated in 1562 *E. coli* genes and 2681 human genes, respectively. (Because *E. coli* and human proteins differ somewhat in amino acid composition, the frequencies for a particular amino acid do not correspond exactly between the two species.)

Amino Acid	Codon	*E. coli* Gene Frequency/1000	Human Gene Frequency/1000
Leu	CUA	3.2	6.1
	CUC	9.9	20.1
	CUG	54.6	42.1
	CUU	10.2	10.8
	UUA	10.9	5.4
	UUG	11.5	11.1
Pro	CCA	8.2	15.4
	CCC	4.3	20.6
	CCG	23.8	6.8
	CCU	6.6	16.1
Ala	GCA	15.6	14.4
	GCC	34.4	29.7
	GCG	32.9	7.2
	GCU	13.4	18.9
Lys	AAA	36.5	21.9
	AAG	12.0	35.2
Glu	GAA	43.5	26.4
	GAG	19.2	41.6

Adapted from Wada, K., et al., 1992. Codon usage tabulated from Genbank genetic sequence data. *Nucleic Acids Research* **20**:2111–2118.

The occurrence of codons in *E. coli* mRNAs correlates well with the relative abundance of the tRNAs that read them. Preferred codons are represented by the most abundant isoacceptor tRNAs. Furthermore, mRNAs for proteins that are synthesized in abundance tend to employ preferred codons. Rare tRNAs correspond to rarely used codons, and messages containing such codons might experience delays in translation.

Nonsense Suppression Occurs When Suppressor tRNAs Read Nonsense Codons

Mutations that alter a sense codon to one of the three nonsense codons—UAA, UAG, or UGA—result in premature termination of protein synthesis and the release of truncated (incomplete) polypeptides. Geneticists found that second mutations elsewhere in the genome were able to *suppress* the effects of nonsense mutations so that the organism survived, a phenomenon termed **nonsense suppression.** The molecular basis for nonsense suppression was a mystery until it was realized that **suppressors** were mutations in tRNA genes that altered the anticodon so that the mutant tRNA could now read a particular "stop" codon and insert an amino acid. For example, alteration of the anticodon of a tRNA^{Tyr} from GUA to CUA allows this tRNA to read the *amber* stop codon, UAG, and insert Tyr. (The nonsense codons are named *amber* [UAG], *ochre* [UAA], and *opal* [UGA]). **Suppressor tRNAs** are typically generated from minor tRNA species within a set of isoacceptor tRNAs, so their recruitment to a new role via mutation does not involve loss of an essential tRNA; that is, the mutation is not particularly deleterious to the organism. A suppressor tRNA, as a mutant tRNA, may even carry and introduce an amino acid different from the one borne by the wild-type tRNA.

30.4 | What Is the Structure of Ribosomes, and How Are They Assembled?

Protein biosynthesis is achieved by the process of **translation.** Translation converts the language of genetic information embodied in the base sequence of a messenger RNA molecule into the amino acid sequence of a polypeptide chain. During translation, proteins are synthesized on ribosomes by linking amino acids together in the specific linear order stipulated by the sequence of codons in an mRNA. Ribosomes are the agents of protein synthesis.

Ribosomes are compact ribonucleoprotein particles found in the cytosol of all cells, as well as in the matrix of mitochondria and the stroma of chloroplasts. The general structure of ribosomes is described in Chapter 10; here we consider their structure in light of their function in synthesizing proteins. Ribosomes are mechanochemical systems that move along mRNA templates, orchestrating the interactions between successive codons and the corresponding anticodons presented by aminoacyl-tRNAs. As they align successive amino acids via codon–anticodon recognition, ribosomes also catalyze the formation of peptide bonds between the growing peptide chain and incoming amino acids.

Prokaryotic Ribosomes Are Composed of 30S and 50S Subunits

E. coli ribosomes are representative of the structural organization of the prokaryotic versions of these supramolecular protein-synthesizing machines (Table 30.5, see also Figure 10.22). The *E. coli* ribosome is a roughly globular particle with a diameter of 25 nm, a sedimentation coefficient of 70S, and a mass of about 2520 kD. It consists of two unequal subunits that dissociate from each other at Mg^{2+} concentrations below 1 m*M*. The smaller, or **30S,** subunit is composed of 21 different proteins and a single rRNA, **16S ribosomal RNA (rRNA).** The larger **50S** subunit consists of 31 different

TABLE 30.5	Structural Organization of *E. coli* Ribosomes		
	Ribosome	Small Subunit	Large Subunit
Sedimentation coefficient	70S	30S	50S
Mass (kD)	2520	930	1590
Major RNAs		16S = 1542 bases	23S = 2904 bases
Minor RNAs			5S = 120 bases
RNA mass (kD)	1664	560	1104
RNA proportion	66%	60%	70%
Protein number		21 polypeptides*	31 polypeptides†
Protein mass (kD)	857	370	487
Protein proportion	34%	40%	30%

*The S proteins
†The L proteins

proteins and two rRNAs: **23S rRNA** and **5S rRNA.** Ribosomes are roughly two-thirds RNA and one-third protein by mass. An *E. coli* cell contains around 20,000 ribosomes, constituting about 20% of the dry cell mass.

Prokaryotic Ribosomes Are Made from 50 Different Proteins and Three Different RNAs

Ribosomal Proteins There is one copy of each ribosomal protein per 70S ribosome, excepting protein L7/L12 (L7 and L12 have identical amino acid sequences and differ only in the degree of N-terminal acetylation). Only one protein is common to both the small and large subunit: S20 = L26. The largest ribosomal protein is S1 (557 residues, 61.2 kD); the smallest is L34 (46 residues, 5.4 kD). The sequences of ribosomal proteins share little similarity. These proteins are typically rich in the cationic amino acids Lys and Arg and have few aromatic amino acid residues, properties appropriate to proteins intended to interact strongly with polyanionic RNAs.

FIGURE 30.11 The seven ribosomal RNA operons in *E. coli.* Numerals to the right of the brackets indicate the number of species of tRNA encoded by each transcript.

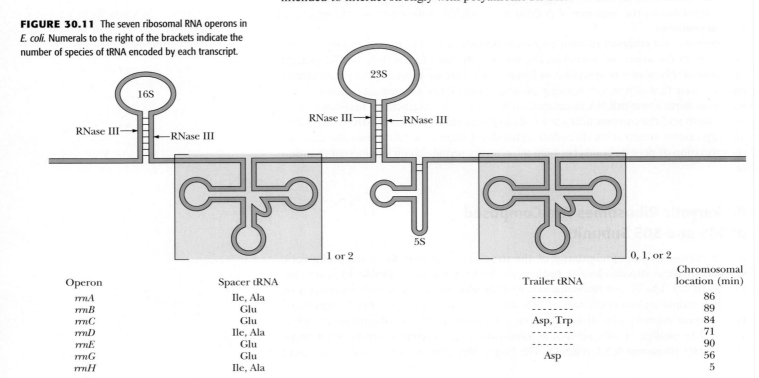

Operon	Spacer tRNA	Trailer tRNA	Chromosomal location (min)
rrnA	Ile, Ala	--------	86
rrnB	Glu	--------	89
rrnC	Glu	Asp, Trp	84
rrnD	Ile, Ala	--------	71
rrnE	Glu	--------	90
rrnG	Glu	Asp	56
rrnH	Ile, Ala		5

rRNAs The rRNAs of *E. coli* are encoded by a set of seven operons (Figure 30.11). Each of these operons is transcribed into a 30S rRNA precursor that includes several tRNAs. RNase III and other nucleases cleave these precursors to generate 23S, 16S, and 5S rRNA, as well as several tRNAs that are unique to each operon. Transcription of rRNA genes accounts for 80% to 90% of total cellular RNA synthesis. Ribosomal RNAs show extensive potential for intrachain hydrogen bonding and assume secondary structures reminiscent of tRNAs, although substantially more complex (see Figures 11.39 and 11.40). About two-thirds of rRNA is double-helical. Double-helical regions are punctuated by short, single-stranded stretches, generating hairpin conformations that dominate the molecule; four distinct domains can be discerned in the secondary structure of 16S rRNA and six in 23S rRNA. The three-dimensional structures of both the 30S and 50S ribosomal subunits show that the general shapes of the ribosomal subunits are determined by the conformation of the rRNA molecules within them. Figure 30.12 illustrates the three-dimensional structure of 16S rRNA within the 30S subunit. The overall form of the 30S structure is essentially that of the rRNA (compare Figures 30.12 and 30.13a). The same relationship is true for the 23S plus 5S rRNA and the large ribosomal subunit (compare Figures 11.40 and 30.13b). Ribosomal proteins serve a largely structural role in ribosomes; their primary function is to brace and stabilize the rRNA conformations within the ribosomal subunits.

Ribosomes Spontaneously Self-Assemble In Vitro

Ribosomal subunit self-assembly is one of the paradigms for the spontaneous formation of supramolecular complexes from their macromolecular components. If the individual proteins and rRNAs composing ribosomal subunits are mixed together in vitro under appropriate conditions of pH and ionic strength, spontaneous self-assembly into functionally competent subunits takes place without the intervention of any additional

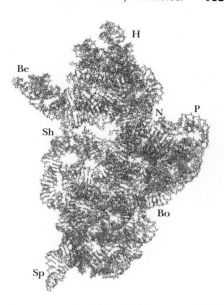

FIGURE 30.12 Tertiary structure of the 16S rRNA within the *Thermus thermophilus* 30S ribosomal subunit (pdb id = 2J02). This view is of the face that interacts with the 50S subunit (see Figures 30.13 and 30.15). H, head; Be, beak; N, neck; P, platform; Sh, shoulder; Sp, spur; Bo, body.

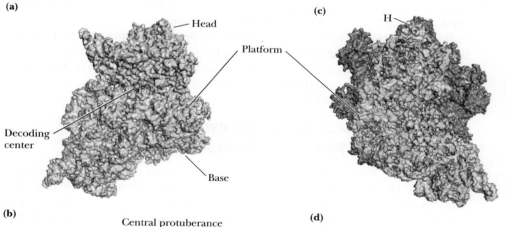

(a) Head, Platform, Decoding center, Base

(b) Central protuberance, L1, L7/L12, Peptidyl transferase center

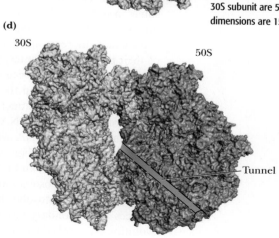

(c) H

(d) 30S, 50S, Tunnel

FIGURE 30.13 Structure of the *T. themophilus* ribosomal subunits and 70S ribosome, as deduced by X-ray crystallography. Prominent structural features are labeled. **(a)** 30S (pdb id = 2J02) and **(b)** 50S (pdb id = 2J03) subunits. These views show the sides of these two that form the interface between them when they come together to form a 70S subunit **(c)**. **(d)** is a side view of the 70S ribosome; the white area represents the region where mRNA and tRNAs are bound and peptide bond formation occurs. The tunnel through the 50S subunit that the growing peptide chain transits is shown as a dashed line. The approximate dimensions of the 30S subunit are 5.5 × 22 × 22 nm; the 50S subunit dimensions are 15 × 20 × 20 nm.

factors or chaperones. The rRNA acts as a scaffold upon which the various ribosomal proteins convene. Ribosomal proteins bind in a specified order.

Ribosomes Have a Characteristic Anatomy

Ribosomal subunits have a characteristic three-dimensional architecture that has been revealed by image reconstructions from cryoelectron microscopy, X-ray crystallography, and X-ray and neutron solution scattering. Such analyses provide images as depicted in Figure 30.13. The 30S, or small, subunit features a "head" and a "base," or "body," from which a "platform" projects. A cleft is defined by the spatial relationship between the head, base, and platform (Figure 30.13a). The mRNA passes across this cleft. The platform represents the central domain of the 30S subunit; it contains one-third of the 16S rRNA. This central domain binds mRNA and the anticodon stem-loop end of aminoacyl-tRNAs, providing the framework for decoding the genetic information in mRNA by mediating codon–anticodon recognition. As such, this central domain of the 30S subunit serves as the **decoding center.** This center is composed only of 16S rRNA; no ribosomal proteins are involved in decoding the message.

The 50S, or large, subunit is a mitt-like globular structure with three distinctive projections: a "central protuberance," the "stalk" containing protein L1, and a winglike ridge known as the "L7/L12 region" (Figure 30.13b). The large subunit binds the aminoacyl-acceptor ends of the tRNAs and is responsible for catalyzing formation of the peptide bond formed between successive amino acids in the polypeptide chain. This catalytic center, the **peptidyl transferase,** is located at the bottom of a deep cleft. From it, a 10-nm-long tunnel passes outward through the back of the large subunit.

The small and large subunits associate with each other in the manner shown in Figure 30.13c and d. The contacts between the 30S and 50S subunits are rather limited, and the subunit interface contains mostly rRNA, with relatively little contribution from ribosomal proteins. The decoding center in the 30S subunit is aligned somewhat with the peptidyl transferase and the tunnel in the large subunit, and the growing peptidyl chain is threaded through this tunnel as protein synthesis proceeds. Even though the ribosomal proteins are arranged peripherally around the rRNAs in ribosomes, rRNA occupies 30% to 40% of the ribosomal subunit surface areas.

Both subunits are involved in **translocation,** the process by which the mRNA moves through the ribosome, one codon at a time. (Although it is physically more likely for the mRNA to move through the ribosome, the descriptions of the events in protein synthesis that follow imply that the ribosome moves along the mRNA.)

The Cytosolic Ribosomes of Eukaryotes Are Larger Than Prokaryotic Ribosomes

Eukaryotic cells have ribosomes in their mitochondria (and chloroplasts) as well as in the cytosol. The mitochondrial and chloroplastic ribosomes resemble prokaryotic ribosomes in size, overall organization, structure, and function, a fact reflecting the prokaryotic origins of these organelles. Although eukaryotic cytosolic ribosomes are larger and considerably more complex, they retain the "core" structural and functional properties of their prokaryotic counterparts, confirming that the fundamental ribosome organization and operation has been conserved across evolutionary time. The rRNA genes of eukaryotes are present in the form of several hundred tandem clusters; these clusters define the **nucleolus,** a distinct region of the nucleus where these clusters are located and where transcription of rRNA genes occurs. As in prokaryotes, 80% to 90% of eukaryotic transcription is rRNA synthesis. Higher eukaryotes have more complex ribosomes than lower eukaryotes. For example, the yeast cytosolic ribosomes have major rRNAs of 3392 (large subunit) and 1799 nucleotides (small subunit); the major rRNAs of mammalian cytosolic ribosomes are 4718 and 1874 nucleotides, respectively. Table 30.6 lists the properties of cytosolic ribosomes in a mammal, the rat. Comparison of base sequences and secondary structures of rRNAs from different organisms suggests that evolution has worked to conserve the secondary structure of these molecules, although

TABLE 30.6	Structural Organization of Mammalian (Rat Liver) Cytosolic Ribosomes		
	Ribosome	Small Subunit	Large Subunit
Sedimentation coefficient	80S	40S	60S
Mass (kD)	4220	1400	2820
Major RNAs		18S = 1874 bases	28S = 4718 bases
Minor RNAs			5.8S = 160 bases 5S = 120 bases
RNA mass (kD)	2520	700	1820
RNA proportion	60%	50%	65%
Protein number		33 polypeptides	49 polypeptides
Protein mass (kD)	1700	700	1000
Protein proportion	40%	50%	35%

not necessarily the nucleotide sequences creating such structure. That is, the retention of a base pair at a particular location seems more important than whether the base pair is G : C or A : U.

30.5 : What Are the Mechanics of mRNA Translation?

In translating an mRNA, a ribosome must move along it in the 5′→3′ direction, recruiting aminoacyl-tRNAs whose anticodons match up with successive codons and joining amino acids in peptide bonds in a polymerization process that forms a particular protein. Like chemical polymerization processes, protein biosynthesis in all cells is characterized by three distinct phases: initiation, elongation, and termination. At each stage, the energy driving the assembly process is provided by GTP hydrolysis, and specific soluble protein factors participate in the events. These soluble proteins are often **G-protein family** members that use the energy released upon hydrolysis of bound GTP to fuel switchlike conformational changes. Such conformational changes are at the heart of the mechanical steps necessary to move a ribosome along an mRNA and to deliver an aminoacyl-tRNA into appropriate register with a codon.

Initiation involves binding of mRNA by the small ribosomal subunit, followed by association of a particular **initiator aminoacyl-tRNA** that recognizes the first codon. This codon often lies within the first 30 nucleotides or so of mRNA spanned by the small subunit. The large ribosomal subunit then joins the initiation complex, preparing it for the elongation stage.

Elongation includes the synthesis of all peptide bonds from the first to the last. The ribosome remains associated with the mRNA throughout elongation, moving along it and translating its message into an amino acid sequence. This is accomplished via a repetitive cycle of events in which successive aminoacyl-tRNAs are added to the ribosome : mRNA complex as directed by codon binding, the 50S subunit catalyzes peptide bond formation, and the polypeptide chain grows by one amino acid at a time.

Three tRNA molecules may be associated with the ribosome : mRNA complex at any moment. Each lies in a distinct site (Figure 30.14). The **A**, or **acceptor, site** is the attachment site for an incoming aminoacyl-tRNA. The **P**, or **peptidyl, site** is occupied by peptidyl-tRNA, the tRNA carrying the growing polypeptide chain. The elongation reaction transfers the peptide chain from the peptidyl-tRNA in the P site to the aminoacyl-tRNA in the A site. This transfer occurs through covalent attachment of the α-amino group of the aminoacyl-tRNA to the α-carboxyl group of the peptidyl-tRNA, forming a new peptide bond. The ribosome now moves (translocates) precisely one codon further along the mRNA. Accompanying this translocation is the movement of the new, longer peptidyl-tRNA from the A site into the P site. The A site, left vacant by this translocation, can accept the next incoming aminoacyl-tRNA. The **E**, or **exit, site,** is transiently occupied by the

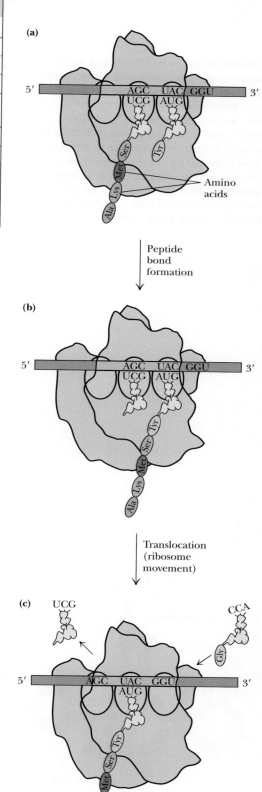

FIGURE 30.14 The basic steps in protein synthesis. The ribosome has three distinct binding sites for tRNA: the A, or acceptor, site; the P, or peptidyl, site; and the E, or exit, site.

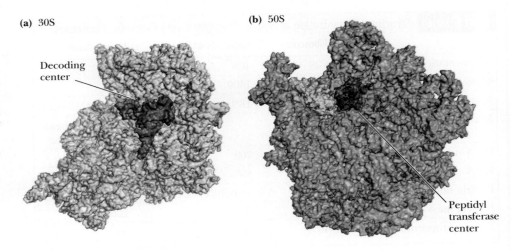

(a) 30S **(b)** 50S

Decoding center

Peptidyl transferase center

FIGURE 30.15 The three tRNA-binding sites on ribosomes. The view shows the ribosomal surfaces that form the interface between the 30S **(a)** and 50S **(b)** subunits in a 70S ribosome (as if a 70S ribosome has been "opened" like a book to expose facing "pages"). The A *(green)*, P *(blue)*, and E *(yellow)* sites are occupied by tRNAs. The decoding center on the 30S subunit (a) lies behind the top of the tRNAs in the A and P sites (which is where the anticodon ends of the tRNAs are located). The peptidyl transferase center on the 50S subunit (b) lies at the lower tips (acceptor ends) of the A- and P-site tRNAs. (a: pdb id = 2J02; b: pdb id = 2J03.)

"unloaded," or deacylated, tRNA, which has lost its peptidyl chain through the peptidyl transferase reaction. These events are summarized in Figure 30.14. The contributions made to each of the three tRNA-binding sites by each ribosomal subunit are shown in Figure 30.15.

Termination is triggered when the ribosome reaches a "stop" codon on the mRNA. At this point, the polypeptide chain is released and the ribosomal subunits dissociate from the mRNA.

Protein synthesis proceeds rapidly. In vigorously growing bacteria, about 20 amino acid residues are added to a growing polypeptide chain each second. So an average protein molecule of about 300 amino acid residues is synthesized in only 15 seconds. Eukaryotic protein synthesis is only about 10% as fast. Protein synthesis is also highly accurate: An inappropriate amino acid is incorporated only once in every 10^4 codons. We focus first on protein synthesis in *E. coli,* the system for which we know the most.

Peptide Chain Initiation in Prokaryotes Requires a G-Protein Family Member

The components required for peptide chain initiation include (1) mRNA; (2) 30S and 50S ribosomal subunits; (3) a set of proteins known as **initiation factors;** (4) GTP; and (5) a specific charged tRNA, **f-Met-tRNA$_i$^{fMet}.** A discussion of the properties of these components and their interaction follows.

Initiator tRNA tRNA$_i$^{fMet} is a particular tRNA for reading an AUG (or GUG, or even UUG) codon that signals the start site, or N-terminus, of a polypeptide chain; the subscript $_i$ signifies "initiation." This tRNA$_i$^{fMet} does not read internal AUG codons, so it does not participate in chain elongation. Instead, that role is filled by another methionine-specific tRNA, referred to as tRNA^{Met}, which cannot replace tRNA$_i$^{fMet} in peptide chain initiation. (However, both of these tRNAs are loaded with Met by the same methionyl-tRNA synthetase.) The structure of *E. coli* tRNA$_i$^{fMet} has several distinguishing features (Figure 30.16). Collectively, these features identify this tRNA as essential to initiation and inappropriate for chain elongation.

The synthesis of all *E. coli* polypeptides begins with the incorporation of a modified methionine residue, *N*-formyl-Met, as N-terminal amino acid. However, in about half of the *E. coli* proteins, this Met residue is removed once the growing polypeptide is ten or so residues long; as a consequence, many mature proteins in *E. coli* lack N-terminal Met.

The methionine contributed in peptide chain initiation by tRNA$_i$^{fMet} is unique in that its amino group has been formylated. This reaction is catalyzed by a specific enzyme, **methionyl-tRNA$_i$^{fMet} formyl transferase** (Figure 30.17). Note that the addition of the formyl group to the α-amino group of Met creates an N-terminal block resembling a peptidyl grouping. That is, the initiating Met is transformed into a minimal analog of a peptidyl chain.

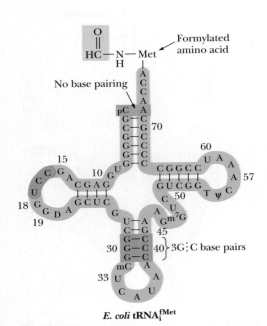

Formylated amino acid

No base pairing

E. coli tRNA$_i$^{fMet}

FIGURE 30.16 The secondary structure of *E. coli N*-formyl-methionyl-tRNA$_i$^{fMet}. The features distinguishing it from noninitiator tRNAs are highlighted.

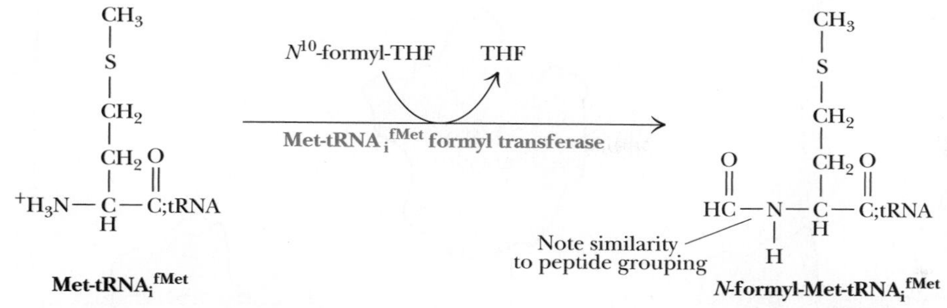

FIGURE 30.17 Methionyl-tRNA$_i^{fMet}$ formyl transferase catalyzes the transformylation of methionyl-tRNA$_i^{fMet}$ using N^{10}-formyl-THF as formyl donor. The tRNA for reading Met codons within a protein (tRNA^{Met}) is not a substrate for this transformylase.

mRNA Recognition and Alignment In order for the mRNA to be translated accurately, its sequence of codons must be brought into proper register with the translational apparatus. Recognition of translation initiation sequences on mRNAs involves the 16S rRNA component of the 30S ribosomal subunit. Base pairing between a pyrimidine-rich sequence at the 3′-end of 16S rRNA and complementary purine-rich tracts at the 5′-end of prokaryotic mRNAs positions the 30S ribosomal subunit in proper alignment with an initiation codon on the mRNA. The purine-rich mRNA sequence, the **ribosome-binding site,** is often called the **Shine–Dalgarno sequence** in honor of its discoverers. Figure 30.18 shows various Shine–Dalgarno sequences found in prokaryotic mRNAs, along with the complementary 3′-tract on *E. coli* 16S rRNA. The 3′-end of 16S rRNA resides in the "head" region of the 30S small subunit, and the double helical duplex formed by the Shine-Dalgarno sequence and the 16S rRNA 3′-end fits snugly in a chamber between the 30S subunit "head" and "platform" domains.

Initiation Factors Initiation involves interaction of the **initiation factors (IFs)** with GTP, *N*-formyl-Met-tRNA$_i^{fMet}$, mRNA, and the 30S subunit to give a **30S initiation complex** to which the 50S subunit then adds to form a **70S initiation complex.** The initiation factors are soluble proteins required for assembly of proper initiation complexes. Their properties are summarized in Table 30.7.

Events in Initiation Initiation begins when a 30S subunit : (IF-3 : IF-1) complex binds mRNA and a complex of IF-2, GTP, and f-Met-tRNA$_i^{fMet}$. The sequence of events is summarized in Figure 30.19. Although IF-3 is absolutely essential for mRNA binding

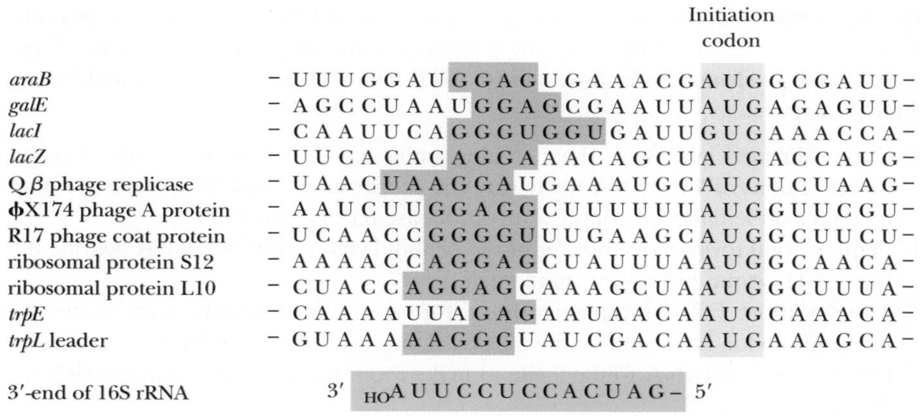

FIGURE 30.18 Various Shine–Dalgarno sequences recognized by *E. coli* ribosomes. These sequences lie about ten nucleotides upstream from their respective AUG initiation codon and are complementary to the UCCU core sequence element of *E. coli* 16S rRNA. G:U as well as canonical G:C and A:U base pairs are involved here.

TABLE 30.7			Properties of *E. coli* Initiation Factors
Factor	Mass (kD)	Molecules/ Ribosome	Function
IF-1	9	0.15	Binds to 30S A site and prevents tRNA binding
IF-2	97		G-protein that binds fMet-tRNA$_i^{fMet}$; interacts with IF-1
IF-3	23	0.25	Binds to 30S E site; prevents 50S binding

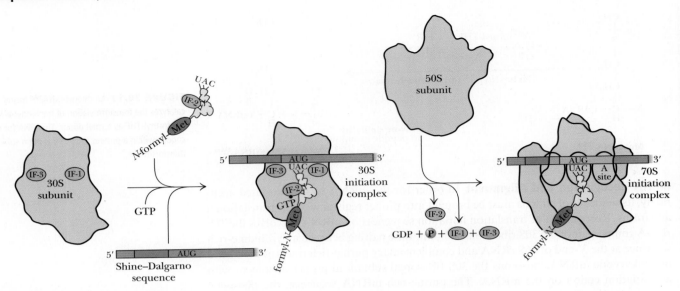

FIGURE 30.19 The sequence of events in peptide chain initiation.

by the 30S subunit, it is not involved in locating the proper translation initiation site on the message. The presence of IF-3 on 30S subunits also prevents them from reassociating with 50S subunits. IF-3 must dissociate before the 50S subunit will associate with the mRNA:30S subunit complex.

IF-2 delivers the initiator f-Met-tRNA$_i^{fMet}$ in a GTP-dependent process. Apparently, the 30S subunit is aligned with the mRNA such that the initiation codon is situated within the "30S part" of the P site. Upon binding, f-Met-tRNA$_i^{fMet}$ enters this 30S portion of the P site. GTP hydrolysis is necessary to form an active 70S ribosome. GTP hydrolysis is triggered when the 50S subunit joins and is accompanied by IF-1 and IF-2 release. The A site of the *70S initiation complex* is ready to accept an incoming aminoacyl-tRNA; the 70S ribosome is poised to begin chain elongation.

Peptide Chain Elongation Requires Two G-Protein Family Members

The requirements for peptide chain elongation are (1) an mRNA:70S ribosome:peptidyl-tRNA complex (peptidyl-tRNA in the P site), (2) aminoacyl-tRNAs, (3) a set of proteins known as **elongation factors,** and (4) GTP. Chain elongation can be divided into three principal steps:

1. Codon-directed binding of the incoming aminoacyl-tRNA at the A site. Decoding center regions of 16S rRNA make sure the proper aminoacyl-tRNA is in the A site by direct surveillance of codon–anticodon base pairing geometry.
2. Peptide bond formation: transfer of the peptidyl chain from the tRNA bearing it to the —NH$_2$ group of the new amino acid.
3. Translocation: Following peptide bond formation, the ribosome translocates precisely one codon further along the mRNA. Translocation also shifts the "one-residue-longer" peptidyl-tRNA to the P site to make room for the next aminoacyl-tRNA at the A site.

The Elongation Cycle

The properties of the soluble proteins essential to peptide chain elongation are summarized in Table 30.8. These proteins are present in large quantities, reflecting the great importance of protein synthesis to cell vitality. For example, **elongation factor Tu (EF-Tu)** is the most abundant protein in *E. coli,* accounting for 5% of total cellular protein. EF-Tu and EF-G, along with IF-2, are all members of the G-protein family of GTPases (see Figures 15.20 and 32.24). As such, the conformation of the GTP-bound and GDP-bound

TABLE 30.8	Properties of *E. coli* Elongation Factors		
Factor	Mass (kD)	Molecules/ Cell	Function
EF-Tu	43	70,000	G protein that binds aminoacyl-tRNA and delivers it to the A site
EF-Ts	74	10,000	Guanine-nucleotide exchange factor (GEF) that replaces GDP on EF-Tu with GTP
EF-G	77	20,000	G protein that promotes translocation of mRNA

forms of these proteins are significantly different, allowing them to act as conformational switches that, in turn, can trigger dramatic conformational changes in the ribosome to which they are bound. These ribosomal conformational changes lie at the heart of the mechanical processes of protein synthesis by a ribosome translocating along an mRNA.

Aminoacyl-tRNA Binding

EF-Tu binds aminoacyl-tRNA and GTP. There is only one EF-Tu species serving all the different aminoacyl-tRNAs, and aminoacyl-tRNAs are accessible to the A site of active 70S ribosomes only in the form of aminoacyl-tRNA : EF-Tu : GTP complexes. The ternary aa-tRNA : EF-Tu : GTP complex binds to the L7/L12 stalk region of the 50S ribosomal subunit. Once correct base pairing between codon and anticodon has been established within the A site, the GTP is hydrolyzed to GDP and P_i by EF-Tu. Because the sites of codon–anticodon recognition and GTP hydrolysis are 7.5 nm apart, conformational changes in the ribosome must convey the information of cognate tRNA recognition to the EF-Tu GTPase site. Indeed, conformational adjustments in the 50S ribosome driven by codon–anticodon recognition allow the phosphate group of residue A2662 in 23S rRNA to activate the intrinsic GTPase activity of EF-Tu through interactions with His[84] in EF-Tu. The region of the 50S ribosomal subunit where 23S rRNA A2662 is located is near where EF-Tu binds and is referred to as the **GTPase-activating center,** or **GAC.** As a result of these events, the aminoacyl end of the tRNA is properly oriented in the peptidyl transferase site of the 50S subunit, and the EF-Tu molecule is released as a EF-Tu : GDP complex (Figure 30.20).

Elongation factor Ts (EF-Ts) is a **guanine-nucleotide exchange factor (GEF)** that catalyzes the recycling of EF-Tu by mediating the displacement of GDP and its replacement by GTP. EF-Ts forms a transient complex with EF-Tu by displacing GDP, whereupon GTP displaces EF-Ts from EF-Tu (Figure 30.20).

The Decoding Center: A 16S rRNA Function Analysis of the structures of the 70S ribosome : tRNA complexes and isolated 30S subunits has revealed the decoding center in the 30S subunit. This decoding center, where anticodon loops of the A- and P-site tRNAs and the codons of the mRNA are matched up, is primarily a property of 16S rRNA. Figure 30.21 reveals the location of the decoding center and highlights the interface between an mRNA UUU codon, the GAA anticodon of a cognate tRNA[Phe], and the elements of the ribosome. 16S rRNA nucleotides A1493, A1492, G530, C518, and C1054 interact extensively with the minor groove of the UUU : AAG duplex; residues Ser[50] and Pro[48] of ribosomal protein S12 also participate (Figure 30.21b). A1492 and A1493 are part of 16S rRNA helix 44; G530 and C518 are from 16S helix 18. Triggered conformational changes within these 16S rRNA regions are key to codon–anticodon recognition.

Peptidyl Transfer Peptidyl transfer, or **transpeptidation,** is the central reaction of protein synthesis, the actual peptide bond–forming step. No energy input (for example, in the form of ATP) is needed; the ester bond linking the peptidyl moiety to tRNA is intrinsically reactive. As noted earlier, *peptidyl transferase,* the activity catalyzing peptide

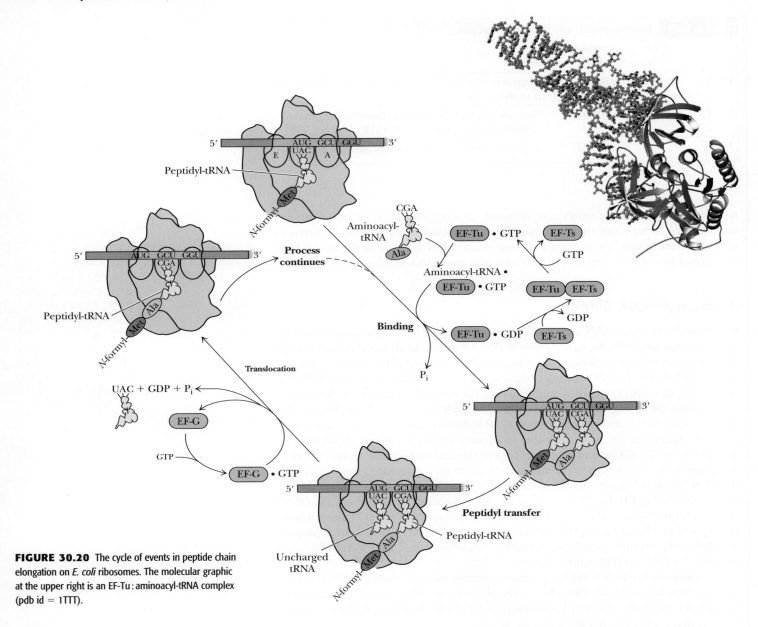

FIGURE 30.20 The cycle of events in peptide chain elongation on *E. coli* ribosomes. The molecular graphic at the upper right is an EF-Tu : aminoacyl-tRNA complex (pdb id = 1TTT).

bond formation, is associated with the 50S ribosomal subunit. Indeed, this reaction is a property of the 23S rRNA in the 50S subunit.

23S rRNA Is the Peptidyl Transferase Enzyme Peptide bond formation is catalyzed by the large rRNA in the large ribosomal subunit (e.g., the 23S rRNA in prokaryotic 50S ribosomal subunits) through its **peptidyl transferase center (PTC).** No ribosomal proteins lie within 1.5 nm of the PTC; thus, none can participate in the reaction mechanism. In kinetic terms, the catalytic power of ribosomes is modest: A ribosome, through the rRNA comprising its PTC, accelerates the peptide-bond forming reaction rate 4×10^6-fold over the uncatalyzed reaction.

The Catalytic Power of the Ribosome Comes from Tight Binding of Its Peptidyl-tRNA and Aminoacyl-tRNA Substrates Structurally, the PTC can be described as a funnel-shaped active-site crater where the 3′-acceptor ends of the peptidyl-tRNA and aminoacyl-tRNA meet. This crater lies directly above the entrance to the peptide exit tunnel. Bases at the 3′-ends of the tRNAs base-pair with bases in the PTC. G2251 and

(a)

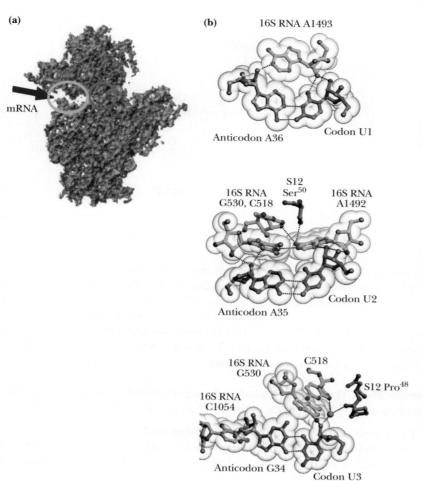

(b)

16S RNA A1493

Anticodon A36   Codon U1

16S RNA   S12   16S RNA
G530, C518   Ser[50]   A1492

Anticodon A35   Codon U2

16S RNA   C518
G530

16S RNA   S12 Pro[48]
C1054

Anticodon G34   Codon U3

FIGURE 30.21 The decoding center of the 30S ribosomal subunit is composed only of 16S rRNA. **(a)** The 30S subunit, as viewed from the 50S subunit. The circle *(cyan)* shows the latch structure of the 30S subunit that encircles and encloses the mRNA. The mRNA enters along the path indicated by the arrow and follows a groove along this face of the subunit. The location of the decoding center is indicated by the *red* circle. (Adapted from Figure 3 in Schluenzen, F., et al., 2000. Structure of the functionally activated small ribosomal subunit at 3.3 Å resolution. *Cell* **102:**615–623.) **(b)** 30S ribosomal subunit interactions with the codon-anticodon duplex during cognate tRNA recognition. (Adapted from Figure 2 in Ogle, J. M., and Ramakrishnan, V., 2005. Structural insights into translational fidelity. *Annual Review of Biochemistry* **74:**129–177.)

G2252 of the PTC (Figure 30.22) base-pair with C75 and C74, respectively, of the P-site tRNA. PTC G2553 forms a base pair with C75 of the A-site tRNA. Significantly, PTC C2501 interacts with A76 of the P-site tRNA, and nearby U2506 interacts with A76 of the A-site tRNA. These interactions and others lead to tight binding of the substrates by the ribosome in a manner such that the reactive groups (aminoacyl and peptidyl) are juxtaposed and properly oriented for reaction to occur. Collectively, functional groups

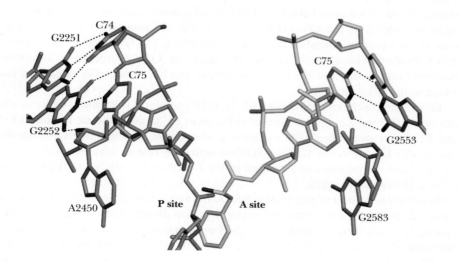

C74
G2251
C75
C75
G2252
G2553
A2450   **P site**   **A site**
G2583

FIGURE 30.22 The peptidyl transferase active site: Base-pairing between cytosine residues of A-site tRNA *(yellow)* and P-site tRNA *(orange)* with 23S rRNA bases *(pale green)* (pdb id = 1VQN). The α-amino group of the aminoacyl group on the aminoacyl-tRNA *(blue)* is positioned for the attack on the ester carbonyl carbon of the peptidyl group of the P-site tRNA *(green)*. (Adapted from Figure 2 in Beringer, M., and Rodnina, M. V., 2007. The ribosomal peptidyl transferase. *Molecular Cell* **26:**311–321.)

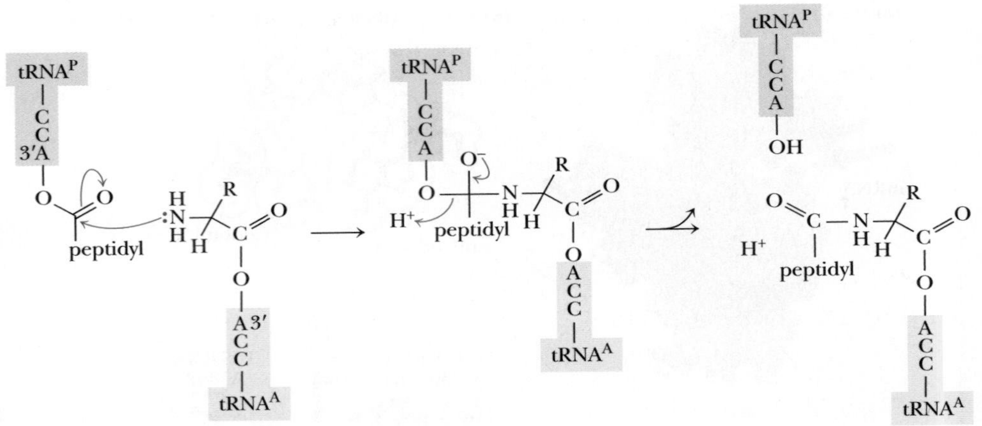

FIGURE 30.23 The protein synthesis reaction proceeds via deprotonation of the α-amino group of the aminoacyl-tRNA, followed by nucleophilic attack of the α-NH$_2$ on the peptidyl-tRNA carbonyl carbon to form a highly polar tetrahedral intermediate. Proton transfer to the 3'-O of the tRNA that contributed the peptidyl chain leads to release of this tRNA in the deacylated form and formation of a "one-residue-longer"-peptidyl-tRNA.

provided by the PTC, the peptidyl-tRNA, and the aminoacyl-tRNA establish a network of H bonds and electrostatic interactions that stabilize the highly polar transition state (Figure 30.23). In effect, interactions of the tRNAs with the rRNA of the PTC create a highly organized environment that facilitates peptide bond synthesis. Ribosome-catalyzed protein synthesis relies more on proximity and orientation of substrates than on chemical catalysis. The role of the PTC in protein synthesis is to align the tRNA-linked substrates so that the reaction is facilitated.

Translocation   Three things remain to be accomplished in order to return the active 70S ribosome:mRNA complex to the starting point in the elongation cycle:

1. The deacylated tRNA must be removed from the P site.
2. The peptidyl-tRNA must be moved (translocated) from the A site to the P site.
3. The ribosome must move one codon down the mRNA so that the next codon is positioned in the A site.

The precise events in translocation are still being resolved, but several distinct steps are clear. The acceptor ends (the aminoacylated ends) of both A- and P-site tRNAs interact with the PTC of the 50S subunit. Because the growing peptidyl chain doesn't move during peptidyl transfer, the acceptor end of the A-site aminoacyl-tRNA must move into the P site as its aminoacyl function picks up the peptidyl chain. At the same time, the acceptor end of the deacylated P-site tRNA is shunted into the E site (see Figure 30.15). Then, the mRNA and the anticodon ends of tRNAs move together with respect to the 30S subunit so that the mRNA is passively dragged one codon further through the ribosome. With this movement, the anticodon end of the now one-residue-longer peptidyl-tRNA goes from the A site of the 30S subunit to the P site. Concomitantly, the anticodon end of the deacylated tRNA is moved into the E site. These ratchetlike movements of the 30S subunit relative to the 50S subunit are catalyzed by the translocation protein **elongation factor G (EF-G),** which, as a G-protein family GTPase-driven conformational switch, couples the energy of GTP hydrolysis to movement. Note that translocation of the mRNA relative to the 30S subunit will deliver the next codon to the 30S A site.

EF-G binds to the ribosome as an EF-G:GTP complex. GTP hydrolysis is essential not only for translocation but also for subsequent EF-G dissociation. Because EF-G and EF-Tu compete for a common binding site on the ribosome (the **factor-binding center,** also known as the **GAC** or **GTPase-activating center**) adjacent to the A site, EF-G release is a prerequisite for return of the 70S ribosome:mRNA to the beginning point in the elongation cycle.

In this simple model of peptidyl transfer and translocation, the ends of both tRNAs move relative to the two ribosomal subunits in two discrete steps, the acceptor ends moving first and then the anticodon ends. Furthermore, the readjustments needed to reposition the ribosomal subunits relative to the mRNA and to one another imply that the 30S and 50S subunits must move relative to one another in ratchetlike fashion. *This model provides a convincing explanation for why ribosomes are universally organized into a two-subunit structure: The small and large subunits must move relative to each other in a rachetlike fashion, as opposed to moving as a unit, in order to carry out the process of translation.*

GTP Hydrolysis Fuels the Conformational Changes That Drive Ribosomal Functions

Two GTPs are hydrolyzed for each amino acid residue incorporated into peptide during chain elongation: one upon EF-Tu–mediated binding of aa-tRNA and one more in translocation. The role of GTP (with EF-Tu as well as EF-G) is mechanical, in analogy with the role of ATP in driving muscle contraction (see Chapter 16). GTP binding induces conformational changes in ribosomal components that actively engage these components in the mechanics of protein synthesis; subsequent GTP hydrolysis followed by GDP and Pi release relax the system back to the initial conformational state so that another turn in

> ## A DEEPER LOOK

Molecular Mimicry—The Structures of EF-Tu : Aminoacyl-tRNA, EF-G, and RF-3

EF-Tu and EF-G compete for binding to ribosomes. EF-Tu has the unique capacity to recognize and bind any aminoacyl-tRNA and deliver it to the ribosome in a GTP-dependent reaction. EF-G catalyzes GTP-dependent translocation. The structure of the EF-Tu:tRNA complex is remarkably similar to the structure of EF-G (EF-Tu: tRNA is shown on the left in the figure; EF-G is in the center). The EF-Tu:tRNA structure is *Thermus aquaticus* EF-Tu:Phe-tRNA[Phe] complexed with GMPCP (purple); GMPCP is a nonhydrolyzable analog of GTP (pdb id = 1TTT). The EF-G structure has GDP bound (purple) (pdb id = 1DAR). Note that parts of the EF-G structure mimic the structure of the tRNA molecule. Both EF-Tu and EF-G bind to the factor-binding center, which is located on the L7/L12 side of 50S ribosomal subunit. Part of this center has an associated GTPase function that plays an integral role in GTP hydrolysis and release of these factors.

Nature has extended this mimicry with RF-3, a ribosome-binding protein essential to the termination phase of protein synthesis. RF-3 is a GTP-binding protein that looks like a tRNA (the structure of RF-3 with bound GDP (purple) is shown on the right in the figure) (pdb id = 2H5E).

One view of early evolution suggests that RNA was the primordial macromolecule, fulfilling all biological functions, including those of catalysis and information storage that are now assumed for the most part by proteins and DNA. The mimicry of EF-Tu: tRNA by EF-G may represent a fossil of early macromolecular evolution when the proteins first began to take over some functions of RNA by mimicking shapes known to work as RNAs.

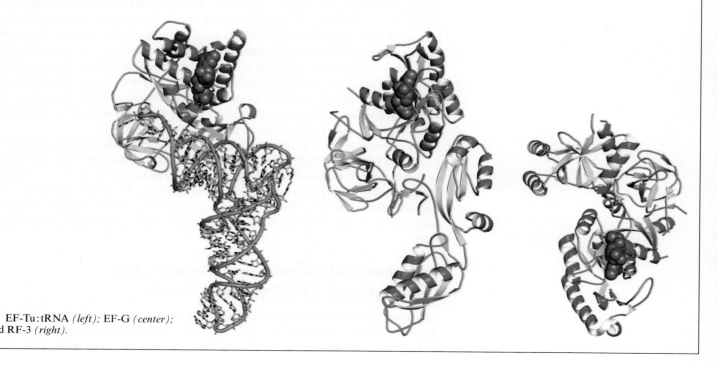

> EF-Tu:tRNA *(left)*; EF-G *(center)*; and RF-3 *(right)*.

the cycle can take place. The energy expenditure for protein synthesis is at least four high-energy phosphoric anhydride bonds per amino acid. In addition to the two provided by GTP, two from ATP are expended in amino acid activation via aminoacyl-tRNA synthesis (see Figure 30.3).

Peptide Chain Termination Requires Yet Another G-Protein Family Member

The elongation cycle of polypeptide synthesis continues until the 70S ribosome encounters a "stop" codon. At this point, polypeptidyl-tRNA occupies the P site and the arrival of a "stop" or nonsense codon in the A site signals that the end of the polypeptide chain has been reached (Figure 30.24) These nonsense codons are not "read" by any "terminator tRNAs" but instead are recognized by specific proteins known as **release factors,** so named because they promote polypeptide release from the ribosome. The release factors bind at the A site. **RF-1** recognizes UAA and UAG, whereas **RF-2** recognizes UAA and UGA. RF-1 and RF-2 are members of the guanine nucleotide exchange factor (GEF) family of proteins, which includes EF-Ts. Like EF-G, RF-1 and RF-2 interact well with the ribosomal A-site structure. Furthermore, these release factors "read" the nonsense codons through specific tripeptide sequences that serve as the RF protein equivalent of the tRNA anticodon loop. There is about one molecule each of RF-1 and RF-2 per 50 ribosomes. Ribosomal binding of RF-1 or RF-2 is competitive with EF-G. RF-1 or RF-2 recruit a third release factor, **RF-3,** complexed with GTP; this protein is a structural mimic of tRNA. RF-3 is the fourth G-protein family member (the other three are IF-2, EF-Tu, and EF-G) involved in protein synthesis. All share the same ribosomal binding site. The state of the peptidyl-tRNA in the P site determines which is bound and, importantly, the progression of protein synthesis through initiation, elongation, and termination.

The presence of release factors with a nonsense codon in the A site creates a **70S ribosome : RF-1** (or **RF-2) : RF-3-GTP : termination signal** complex that transforms the ribosomal peptidyl transferase into a hydrolase. That is, instead of catalyzing the transfer of the polypeptidyl chain from a polypeptidyl-tRNA to an acceptor aminoacyl-tRNA, the peptidyl transferase hydrolyzes the ester bond linking the polypeptidyl chain to its tRNA carrier. Both RF-1 and RF-2 have a conserved GGQ sequence located in a loop that enters the PTC. This GGQ sequence faces A76 of the P-site tRNA; the closest 23S rRNA residues are A2451 and A2602. RF-1 and RF-2 promote conformational adjustments that expose the peptidyl-tRNA ester to attack by a water molecule. The conserved Gln in the release factor GGQ motif positions this hydrolytic water molecule through hydrogen-bonding interactions. The peptidyl transferase transfers the polypeptidyl chain to this water molecule instead of an aminoacyl-tRNA. Peptide release is followed by expulsion of RF-1 (or RF-2) from the ribosome (Figure 30.24). This leaves a ribosome : mRNA : P-site tRNA complex that must be disassembled by a protein **ribosome recycling factor (RRF)** with the help of EF-G. The structure of RRF resembles the letter "L"; thus, this protein is also structurally akin to a tRNA.

We can now recount the central role played by GTP in protein synthesis. IF-2, EF-Tu, EF-G, and RF-3 are all GTP-binding proteins, and all are part of the G-protein superfamily (whose name is derived from the heterotrimeric G proteins that function in transmembrane signaling pathways, as in Figure 15.20). IF-2, EF-Tu, EF-G, and RF-3 interact with the same site on the 50S subunit, the *factor-binding center,* in the 50S cleft. This factor-binding center activates the GTPase activity of these factors, once they become bound.

The Ribosomal Subunits Cycle Between 70S Complexes and a Pool of Free Subunits

Ribosomal subunits cycle rapidly through protein synthesis. In actively growing bacteria, 80% of the ribosomes are engaged in protein synthesis at any instant. Once a polypeptide chain is synthesized and the nascent polypeptide chain is released, the 70S

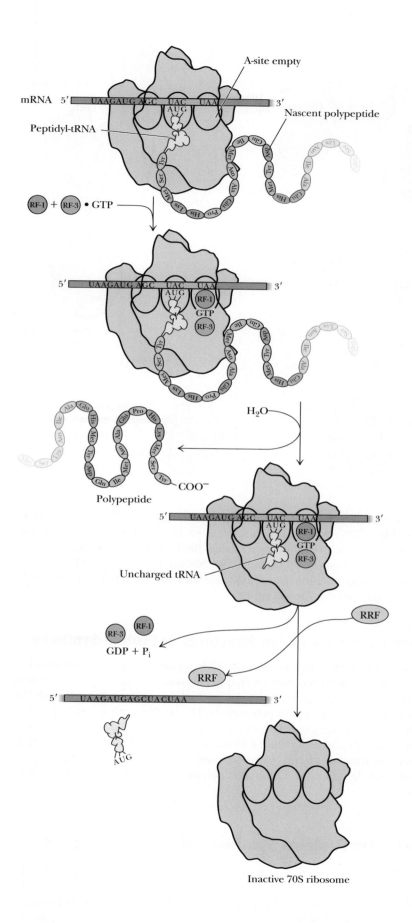

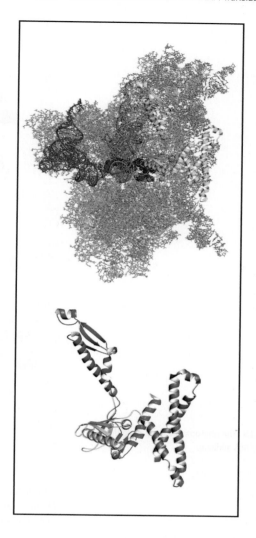

FIGURE 30.24 The events in peptide chain termination.
Inset: Upper: Ribosome:RF-2 termination complex. The view here shows only the 30S subunit:tRNA complex, as if it were folded down and away from the 50S subunit, opening the ribosome as if you were opening your car's glove compartment door. RF-2 is shown in yellow, the P-site tRNA in orange, the E-site tRNA in red, mRNA in dark blue, the 30S proteins in cyan, and the 30S ribosomal subunit 16S rRNA in light purple. Lower: The ribosome-bound conformation of RF-2, with its GGQ motif, is shown alone as the lower image in the inset. These structures are derived from pdb id = 3F1E. *Adapted from Korostelev, A., Asahara, H., Lancaster, L., Laurberg, M., et al., 2008. Crystal structure of a translation termination complex formed with release factor RF2. Proceedings of the National Academy of Sciences, USA* **105**:*19684-19689.*

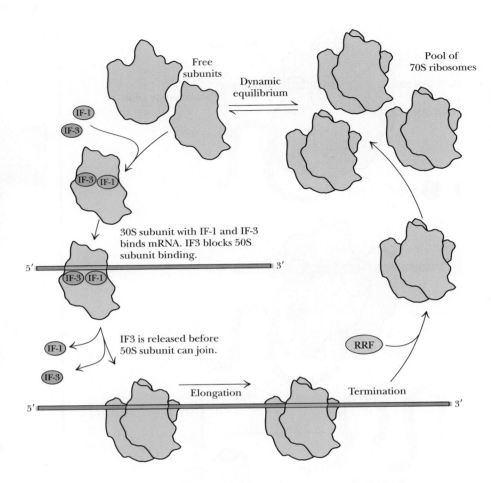

FIGURE 30.25 The ribosome life cycle. Note that IF-3 is released prior to 50S addition.

ribosome dissociates from the mRNA and separates into free 30S and 50S subunits (Figure 30.25). Intact 70S ribosomes are inactive in protein synthesis because only free 30S subunits can interact with the initiation factors. Binding of initiation factor IF-3 by 30S subunits and interaction of 30S subunits with 50S subunits are mutually exclusive. 30S subunits with bound initiation factors associate with mRNA, but 50S subunit addition requires IF-3 release from the 30S subunit.

Polyribosomes Are the Active Structures of Protein Synthesis

Active protein-synthesizing units consist of an mRNA with several ribosomes attached to it. Such structures are **polyribosomes,** or, simply, **polysomes** (Figure 30.26). All protein synthesis occurs on polysomes. In the polysome, each ribosome is traversing the mRNA and independently translating it into polypeptide. The further a ribosome has moved along the mRNA, the greater the length of its associated polypeptide product. In prokaryotes, as many as ten ribosomes may be found in a polysome. Ultimately, as many as 300 ribosomes may translate an mRNA, so as many as 300 enzyme molecules may be produced from a single transcript. Eukaryotic polysomes typically contain fewer than 10 ribosomes.

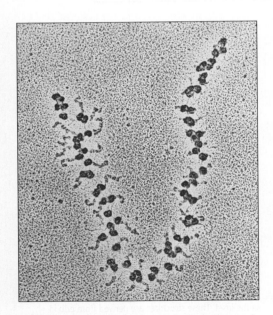

FIGURE 30.26 Electron micrograph of polysomes: multiple ribosomes translating the same mRNA. (From Francke, C., et al., 1982. Electron microscopic visualization of a discrete class of giant translation units in salivary gland cells of *Chironomus tentans. The EMBO Journal* **1**:59–62. Photo courtesy of Oscar L. Miller, Jr., University of Virginia.)

30.6 How Are Proteins Synthesized in Eukaryotic Cells?

Eukaryotic mRNAs are characterized by two post-transcriptional modifications: the 5′-terminal ^{7}**methyl-GTP cap** and the 3′-terminal **poly(A) tail** (Figure 30.27). The ^{7}methyl-GTP cap is essential for mRNA binding by eukaryotic ribosomes and also enhances the stability of these mRNAs by preventing their degradation by 5′-exonucleases. The poly(A) tail enhances both the stability and translational efficiency of eukaryotic

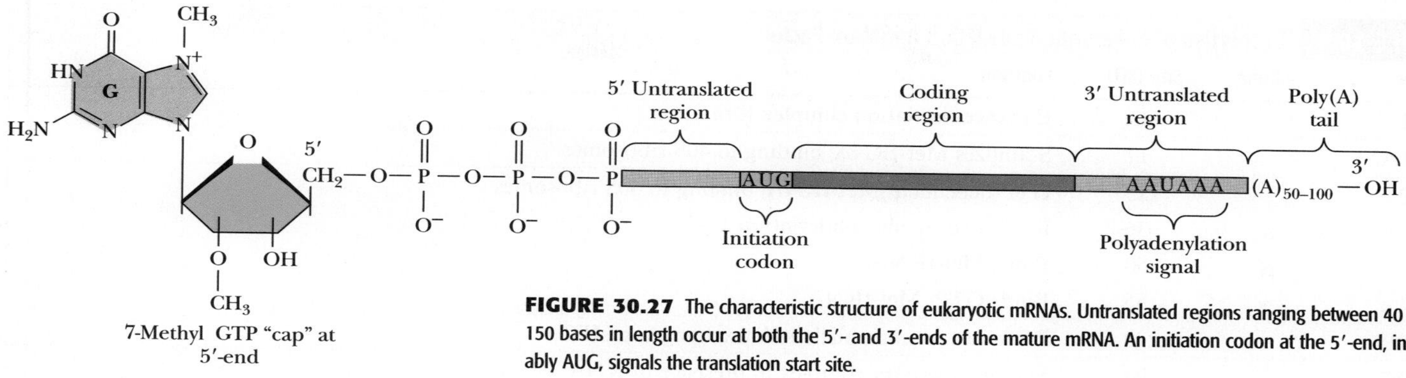

FIGURE 30.27 The characteristic structure of eukaryotic mRNAs. Untranslated regions ranging between 40 and 150 bases in length occur at both the 5'- and 3'-ends of the mature mRNA. An initiation codon at the 5'-end, invariably AUG, signals the translation start site.

mRNAs. The Shine–Dalgarno sequences found at the 5'-end of prokaryotic mRNAs are absent in eukaryotic mRNAs.

Peptide Chain Initiation in Eukaryotes

The events in eukaryotic peptide chain initiation are summarized in Figure 30.28, and the properties of **eukaryotic initiation factors,** symbolized **eIFs,** are presented in Table 30.9. As might be expected, eukaryotic protein synthesis is considerably more complex

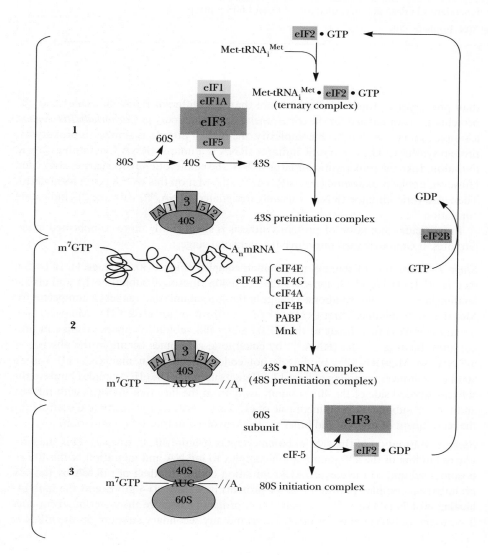

FIGURE 30.28 The three stages in the initiation of translation in eukaryotic cells. See Table 30.9 for a description of the functions of the eukaryotic initiation factors (eIFs).

TABLE 30.9		Properties of Eukaryotic Translation Initiation Factors	
Factor	Subunit	Size (kD)	Function
eIF1		15	Enhances initiation complex formation
eIF1A		17	Stabilizes Met-tRNA$_i$ binding to 40S ribosomes
eIF2		125	GTP-dependent Met-tRNA$_i$ binding to 40S ribosomes
	α	36	Regulated by phosphorylation
	β	50	Binds Met-tRNA$_i$
	γ	55	Binds GTP, Met-tRNA$_i$
eIF2B		270	Promotes guanine nucleotide exchange on eIF2
eIF2C		94	Stabilizes ternary complex in presence of RNA
eIF3		800	Promotes Met-tRNA$_i$ and mRNA binding
eIF4F		243	Binds to mRNA caps and poly(A) tails; consists of eIF4A, eIF4E, and eIF4G; RNA helicase activity unwinds mRNA 2° structure
eIF4A		46	Binds RNA; ATP-dependent RNA helicase; promotes mRNA binding to 40S ribosomes
eIF4E		24	Binds to 5′-terminal ^{7}methyl-GTP cap on mRNA
eIF4G		173	A scaffolding protein that associates the mRNA with ribosome-bound eIF3, binds to PABP
eIF4B		80	Binds mRNA; promotes RNA helicase activity and mRNA binding to 40S ribosomes
eIF4H		25	Acts with eIF4B to stimulate eIF4A helicase activity
eIF5		49	Promotes GTPase of eIF2, ejection of eIF2 and eIF3
eIF5B		175	Ribosome-dependent GTPase activity; mediates 40S and 60S joining
eIF6			Dissociates 80S; binds to 60S

than prokaryotic protein synthesis. More than 100 different RNA molecules and 200 proteins are required just for the core translational machinery in *Caenorhabditis elegans,* a simple animal. Despite such complexity, the overall process is similar to prokaryotic protein synthesis. The eukaryotic initiator tRNA is a unique tRNA functioning only in initiation. Like the prokaryotic initiator tRNA, the eukaryotic version carries only Met. However, unlike prokaryotic f-Met-tRNA$_i^{fMet}$, the Met on this tRNA is not formylated. The eukaryotic initiator tRNA is usually designated **tRNA$_i^{Met}$,** with the "i" indicating "initiation."

In particular, initiation of protein synthesis is significantly more complicated in eukaryotes. It can be divided into three fundamental stages:

Stage 1: Formation of the **43S preinitiation complex** (Figure 30.28, stage 1). Initiation factors eIF1, eIF1A, eIF3, and eIF5 bind to a 40S ribosomal subunit. eIF1 and eIF1A binding induces a conformational change in the 40S subunit that renders it competent for Met-tRNA$_i^{Met}$ binding. Met-tRNA$_i^{Met}$ (in the form of an eIF2:GTP:Met-tRNA$_i^{Met}$ ternary complex) then binds to the eIF1/1A/3/5:40S subunit complex. (Unlike in prokaryotes, binding of Met-tRNA$_i^{Met}$ by eukaryotic ribosomes occurs in the absence of mRNA, so Met-tRNA$_i^{Met}$ binding is not codon-directed.) In mammals, eIF3 is an 800-kD multimer containing at least 13 different subunits. The eIF3 complex binds to the solvent-exposed side of the 40S subunit, away from the face that interacts with mRNA and the 60S subunit. A prominent role of eIF3 is to serve as a platform and scaffold for the recruitment of mRNA and other proteins involved in translation initiation.

Stage 2: Formation of the **48S initiation complex** (Figure 30.28, stage 2). This stage involves binding of the 43S preinitiation complex to mRNA and migration of the 40S ribosomal subunit to the correct AUG initiation codon. Binding of mRNA by the 43S preinitiation complex requires a set of proteins including **eIF4 group** and the **(poly)A-binding protein (PABP).** Collectively, these proteins recognize the 5′-terminal cap and 3′-terminal poly(A) tail of an mRNA, unwind any secondary structure in the mRNA,

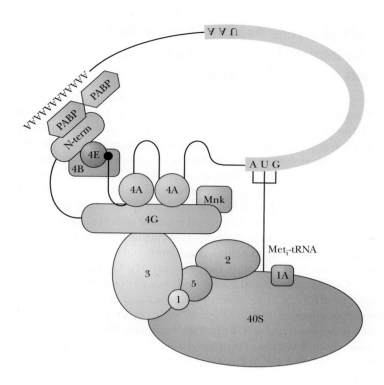

FIGURE 30.29 The 48S initiation complex. mRNA is shown as a black line, its 5′-cap as a black circle, and its coding region as a yellow bar beginning with AUG and ending with UAA; downstream lies the poly(A) tail. The initiator Met-tRNA$_i$ is shown as a pitchfork bound to the AUG initiation codon. The various initiation factors are shown by their numerical and letter codes. Mnk is a stress- and mitogen-activated protein kinase that phosphorylates eIF4E, increasing its affinity for the 5′-cap. (Adapted from Figure 2 in Rhoads, R., Dinkova, T. D., and Korneeva, N. L., 2006. Mechanism and regulation of translation in *C. elegans. Worm-Book* **28**:1–18.)

and transfer the mRNA to the 43S preinitiation complex. The eIF4 group includes eIF4B and eIF4F. eIF4F is a trimeric complex consisting of eIF4A (an ATP-dependent RNA helicase), eIF4E (which binds the 5′-terminal ^{7}methyl-GTP of mRNAs), and eIF4G. Because eIF4G interacts with PABP (Figure 30.29), eIF4G serves as the bridge between the cap-binding eIF4E, the poly(A) tail of the mRNA, and the 40S subunit (through interaction with eIF3). These interactions between the 5′-terminal ^{7}methyl-GTP cap and the 3′-poly(A) tail initiate scanning of the 40S subunit in search of an AUG codon. Both eIF1 and eIF1A are necessary for opening the mRNA-binding channel of the 40S subunit so that scanning in search of an initiation codon can take place.

eIF4E, the mRNA cap-binding protein of eIF4F, represents a key regulatory element in eukaryotic translation. eIF4F binding to the cap structure is necessary for association of eIF4B and formation of the 48S preinitiation complex. Translation is inhibited when the eIF4E subunit of eIF4F binds with **4EBP** (the eIF4E binding protein). Growth factors stimulate protein synthesis by causing the phosphorylation of 4EBP. For example, mammalian target of rapamycin complex 1 (mTORC-1)-mediated phosphorylation of 4EBP causes it to release bound eIF4E, which then can bind to the 5′-cap of an mRNA, triggering translation initiation. Amino acid levels, glucocorticosteroid hormones, and resistance exercise all affect protein synthesis through effects on mTORC-1.

Note that recruitment of both the 5′-cap and 3′-poly(A) tail of the mRNA to eIF4G (Figure 30.29) brings the two ends of the mRNA together, "closing the circle." These interactions ensure that only appropriately processed transcripts are translated. Further, "closing the circle" helps to protect the mRNA from exonucleolytic degradation.

Stage 3: Formation of the **80S initiation complex.** When the 48S preinitiation complex stops at an AUG codon, GTP hydrolysis in the eIF2:Met-tRNA$_i^{Met}$ ternary complex causes ejection of the initiation factors bound to the 40S ribosomal subunit. EIF5, in conjunction with eIF5B, acts here by stimulating the GTPase activity of eIF2. Ejection of the eIFs is followed by 60S subunit association to form the 80S initiation complex, whereupon translation begins (Figure 30.28, stage 3). eIF2:GDP is recycled to eIF2:GTP by eIF2B (eIF2B is a *guanine nucleotide exchange factor*).

Control of Eukaryotic Peptide Chain Initiation Is One Mechanism for Post-Transcriptional Regulation of Gene Expression

Regulation of gene expression can be exerted post-transcriptionally through control of mRNA translation, as indicated by the effects of mTORC-1 mentioned earlier. Phosphorylation/dephosphorylation of translational components is a dominant mechanism for control of protein synthesis. Several initiation factors—eIF2α, eIF2B, eIF4E, eIF4G, 40S ribosomal protein S6, and two eukaryotic elongation factors, eEF1 and eEF2 (see following)—have been identified as targets of regulatory controls. Modification of some factors affects the rate of mRNA translation; modification of others affects which mRNAs are selected for translation. Peptide chain initiation, the initial phase of the synthetic process, is the optimal place for such control. Phosphorylation of S6 facilitates initiation of protein synthesis, resulting in a shift of the ribosomal population from inactive ribosomes to actively translating polysomes. S6 phosphorylation is stimulated by serum growth factors (see Chapter 32). On the other hand, the phosphorylation of some translational components inhibits protein synthesis. For example, the α-subunit of eIF2 can be reversibly phosphorylated at a specific Ser residue by an eIF2α kinase/phosphatase system (Figure 30.30). Four different eIF2α kinases are known; each is responsive to specific metabolic signals. Phosphorylation of eIF2α inhibits peptide chain initiation. Phosphorylation of eIF2 by **heme-regulated inhibitor (HRI,** the heme-inhibited eIF2α kinase) is an important control governing globin synthesis in reticulocytes. If heme for hemoglobin synthesis becomes limiting in these cells, eIF2α is phosphorylated, so globin mRNA is not translated and chains are not synthesized. Availability of heme inhibits HRI, leading to resumption of protein synthesis upon phosphatase-mediated removal of the phosphate group from the Ser residue.

Peptide Chain Elongation in Eukaryotes Resembles the Prokaryotic Process

Once translation initiation is completed, elongation factors take over the role of translation. Eukaryotic peptide elongation occurs in very similar fashion to the process in prokaryotes. An incoming aminoacyl-tRNA enters the ribosomal A site while peptidyl-tRNA occupies the P site. Peptidyl transfer then occurs, followed by translocation of the ribosome one codon further along the mRNA. Two elongation factors, eEF1 and eEF2, mediate the elongation steps. eEF1 consists of two components: eEF1A and eEF1B. eEF1A is the eukaryotic counterpart of EF-Tu; it serves as the aminoacyl-tRNA binding factor and requires GTP. eEF1B is the eukaryotic equivalent of prokaryotic EF-Ts; it catalyzes the exchange of bound GDP on eEF1:GDP for GTP so that active eEF1:GTP can be regenerated. EF2 is the eukaryotic translocation factor. eEF2 (like its prokaryotic kin, EF-G) binds GTP, and GTP hydrolysis accompanies translocation. Elongation continues until a stop codon is reached, signaling termination.

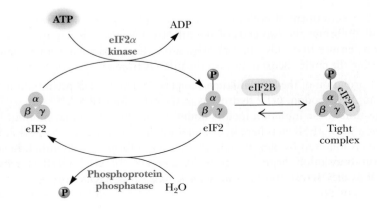

FIGURE 30.30 Control of eIF2 functions through reversible phosphorylation of a Ser residue on its α-subunit. The phosphorylated form of eIF2 (eIF2–P) enters a tight complex with eIF2B and is unavailable for initiation.

Diphtheria Toxin ADP-Ribosylates eEF2

Diphtheria arises from infection by *Corynebacterium diphtheriae* bacteria carrying bacteriophage *corynephage β*. Diphtheria toxin is a phage-encoded enzyme secreted by these bacteria that is capable of inactivating a number of GTP-dependent enzymes through covalent attachment of an ADP-ribosyl moiety derived from NAD^+. That is, diphtheria toxin is an NAD^+-dependent ADP-ribosylase. One target of diphtheria toxin is the eukaryotic translocation factor, eEF2. This protein has a modified His residue known as diphthamide. Diphthamide is generated post-translationally on eEF2;

its biological function is unknown. (EF-G of prokaryotes lacks this unusual modification and is not susceptible to diphtheria toxin.) Diphtheria toxin specifically ADP-ribosylates an imidazole-N within the diphthamide moiety of eEF2 (see accompanying figure). ADP-ribosylated eEF2 retains the ability to bind GTP but is unable to function in protein synthesis. Because diphtheria toxin is an enzyme and can act catalytically to modify many molecules of its target protein, just a few micrograms suffice to cause death.

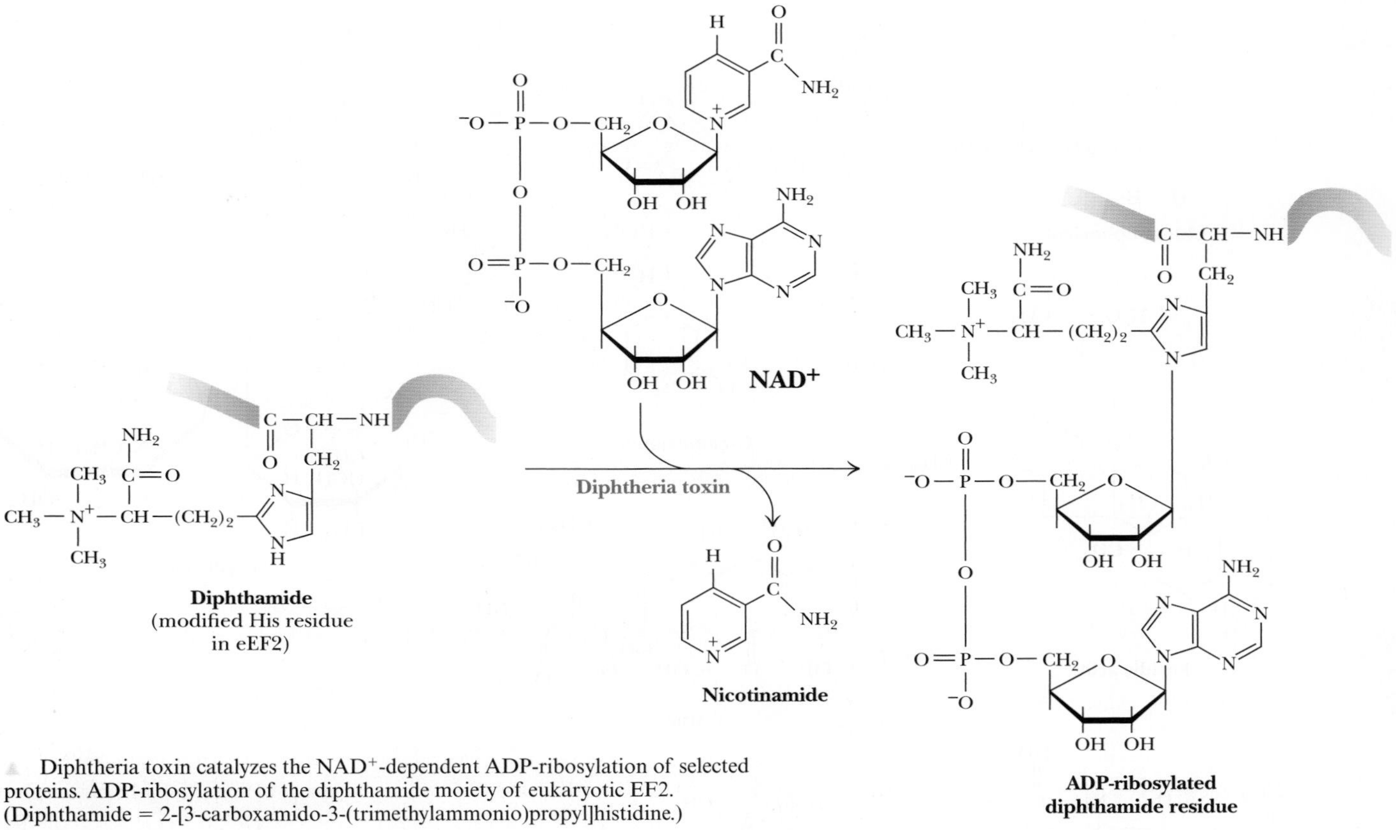

Diphtheria toxin catalyzes the NAD^+-dependent ADP-ribosylation of selected proteins. ADP-ribosylation of the diphthamide moiety of eukaryotic EF2. (Diphthamide = 2-[3-carboxamido-3-(trimethylammonio)propyl]histidine.)

Eukaryotic Peptide Chain Termination Requires Just One Release Factor

Whereas prokaryotic termination involves three different release factors (RFs), just one RF is sufficient for eukaryotic termination. Eukaryotic RF binding to the ribosomal A site is GTP-dependent, and RF:GTP binds at this site when it is occupied by a termination codon. Then, hydrolysis of the peptidyl-tRNA ester bond, hydrolysis of GTP, release of nascent polypeptide and deacylated tRNA, and ribosome dissociation from mRNA ensue.

Inhibitors of Protein Synthesis

Protein synthesis inhibitors have served two major, and perhaps complementary, purposes. First, they have been very useful scientifically in elucidating the biochemical mechanisms of protein synthesis. Second, some of these inhibitors affect prokaryotic but not eukaryotic protein synthesis, and thus, are medically important antibiotics. Table 30.10 is a partial list of these inhibitors and their mode of action. The structures of some of these compounds are given in Figure 30.31.

As indicated in Table 30.10, many antibiotics inhibit protein synthesis by binding to key rRNA components. The two principal sites for antibiotic binding are the 16S rRNA of the decoding center within the small ribosomal subunit and the PTC of the 23S rRNA in the large subunit. More antibiotics are targeted to the PTC than to the decoding center.

Why is rRNA such an optimal target for antibiotics? Generally speaking, it is difficult for organisms to defend themselves against agents that bind rRNA rather than protein. First of all, there are many rRNA genes, so mutation of one gene to antibiotic resistance (loss in antibiotic binding, for instance) would affect few rRNA molecules among the many. Further, four bases comprise RNA (as opposed to 20 amino acids in

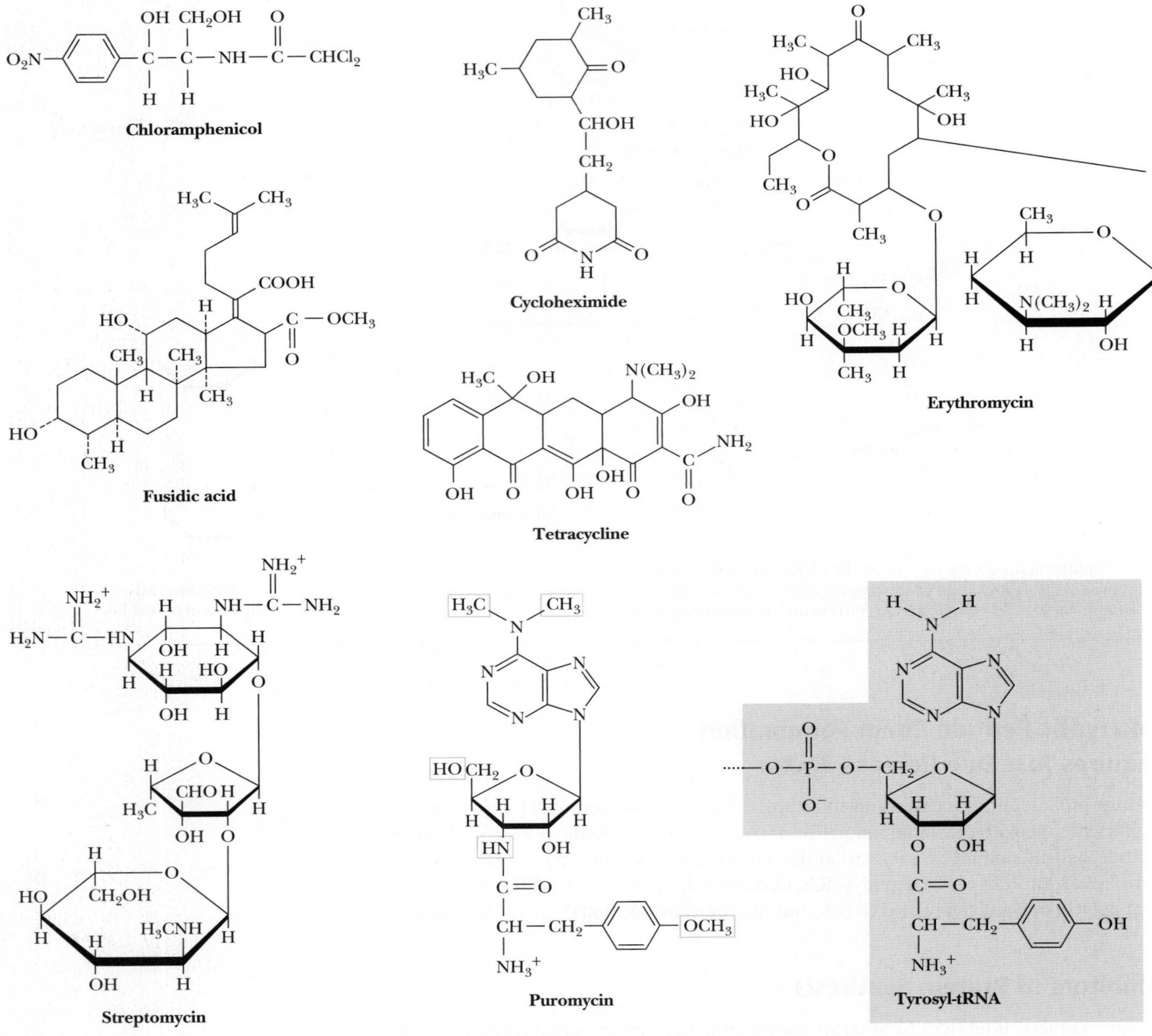

FIGURE 30.31 The structures of various antibiotics that act as protein synthesis inhibitors. Puromycin mimics the structure of aminoacyl-tRNA in that it resembles the 3′-terminus of a Tyr-tRNA (shaded box).

TABLE 30.10	Some Protein Synthesis Inhibitors	
Inhibitor	Cells Inhibited	Mode of Action
Initiation		
Aurintricarboxylic acid	Prokaryotic	Prevents IF binding to 30S subunit
Kasugamycin	Prokaryotic	Inhibits fMet-tRNA$_i$^{fMet} binding
Streptomycin	Prokaryotic	Prevents formation of initiation complexes
Elongation: Aminoacyl-tRNA Binding		
Tetracycline	Prokaryotic	Inhibits aminoacyl-tRNA binding at A site
Streptomycin	Prokaryotic	Codon misreading, insertion of improper amino acid
Kirromycin	Prokaryotic	Binds to EF-Tu, preventing conformational switch from EF-Tu:GTP to EF-Tu:GDP
Elongation: Peptide Bond Formation		
Sparsomycin	Prokaryotic	Peptidyl transferase inhibitor
Chloramphenicol	Prokaryotic	Binds to 50S subunit, blocks the A site and inhibits peptidyl transferase activity
Clindamycin	Prokaryotic	Binds to 50S subunit, overlapping the A and P sites and blocking peptidyl transferase activity
Erythromycin	Prokaryotic	Blocks the 50S subunit tunnel, causing premature peptidyl-tRNA dissociation
Elongation: Translocation		
Fusidic acid	Both	Inhibits EF-G:GDP dissociation from ribosome
Thiostrepton	Prokaryotic	Inhibits ribosome-dependent EF-Tu and EF-G GTPase activity
Diphtheria toxin	Eukaryotic	Inactivates eEF-2 through ADP-ribosylation
Cycloheximide	Eukaryotic	Inhibits translocation of peptidyl-tRNA
Premature Termination		
Puromycin	Both	Aminoacyl-tRNA analog, binds at A site and acts as peptidyl acceptor, aborting peptide elongation
Ribosome Inactivation		
Ricin	Eukaryotic	Catalytic inactivation of 28S rRNA via *N*-glycosidase action on A^{4256}

proteins), so there are fewer possibilities for modification of the rRNA so that the RNA retains its function but loses its ability to bind the antibiotic.

The Decoding Site Is a Target of Aminoglycoside Antibiotics Aminoglycoside antibiotics are a series of compounds with a **2-deoxy-streptamine (2-DOS)** core structure (Figure 30.32a) that is central to their action. These antibiotics bind to an unpaired adenine residue (A^{1408}) within the 16S rRNA decoding site (that part of the 16S rRNA that decodes the mRNA through interactions with the anticodon of the aminoacyl-

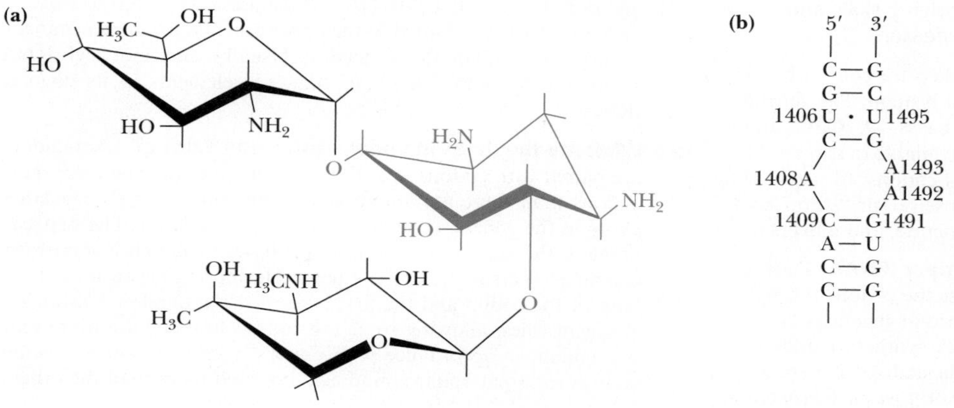

FIGURE 30.32 (a) Structure of geneticin, a representative aminoglycoside antibiotic. Note the characteristic 2-deoxystreptamine (2-DOS) core structure, in red. (b) The base sequence of the small RNA loop within the 16S rRNA decoding center. Note that unpaired adenine residues 1408, 1492, and 1493 constitute the internal loop structure. (Adapted from Figure 1 in Hermann, T., 2005. Drugs targeting the ribosome. *Current Opinion in Structural Biology* **15**:355–366.)

tRNA in the A site). A^{1408} lies within a small internal rRNA loop (Figure 30.32b) that is part of the decoding center. The sugar moiety attached to the 4-position of the 2-DOS core forms H bonds with A^{1408} that limit the flexibility of A^{1492} and A^{1493} on the other side of the small RNA loop; A^{1492} and A^{1493} interact with codon nucleotides (see Figure 30.21b). This flexibility loss results in errors in translation fidelity, where the code is misread, the wrong amino acid is inserted in the growing polypeptide chain, and the protein being made is nonfunctional. Aminoglycoside antibiotics have been useful probes for decoding center properties.

Many Antibiotics Target the PTC and the Peptide Exit Tunnel A chemically diverse series of antibiotics targets the PTC and/or the adjacent peptide exit tunnel. Among these, the **macrolide antibiotics** are one of the clinically most important classes; **erythromycin** (see Figure 30.31), a broadly prescribed macrolide, is representative of this class. (The newer, semisynthetic ketolide drugs are analogs of the macrolide class and act in a similar manner.) X-ray crystallographic analyses of ribosome–macrolide complexes show the macrolide bound within the peptide exit tunnel, 1 to 1.5 nm below the PTC. Macrolide binding plugs the exit tunnel, preventing the movement of the growing peptide chain through the tunnel. Protein synthesis is aborted and the peptidyl-tRNA eventually dissociates from the ribosome. Substituents bound at the 5-position of the macrolide extend outward from the peptide exit tunnel toward the PTC. (The 5-position in macrolides is the position where the N(CH$_3$)$_2$-containing saccharide ring of erythromycin is attached; see Figure 30.31). All direct interactions of macrolide antibiotics with the ribosome involve 23S rRNA, not ribosomal proteins. Both macrolide and ketolide antibiotics adopt a conformation within the ribosome that projects a polar face into the peptide exit tunnel and a nonpolar face toward the tunnel wall. Hydrophobic interactions between the lactone-ring methyl groups and the tunnel wall contribute significantly to the binding energy of these drugs. Residue A^{2058} of the prokaryotic 23S rRNA lies at the entrance to the peptide exit tunnel and favors macrolide antibiotic binding. The major large-subunit rRNA of ribosomes found in the cytosol and mitochondria of eukaryotic cells have G at the equivalent position and are unaffected by macrolides.

The most common effect of antibiotics that interact with the PTC is to occupy space within this center, such that the amino acid or peptidyl chain linked at the 3′-end of a tRNA cannot be positioned properly for the peptide-bond forming reaction. This mode of inhibition is consistent with the catalytic role of PTC in precisely orienting the substrates so that the peptide bond-forming reaction can occur. This effect is more common for aminoacyl-tRNAs in the A site, although some drugs can bridge the A and P sites and affect both aminoacyl-tRNA and peptidyl-tRNA orientation. Ribosomes with long peptidyl chains attached to the tRNA in the P site are less susceptible to macrolide antibiotics.

SUMMARY

⊙WL Login to OWL to develop problem-solving skills and complete online homework assigned by your professor.

30.1 What Is the Genetic Code? The genetic code is the code of bases that specifies the sequence of amino acids in a protein. The genetic code is a triplet code. Given the four RNA bases—A, C, G, and U—a total of $4^3 = 64$ three-letter codons are available to specify the 20 amino acids found in proteins. Of these 64 codons, 61 are used for amino acids, and the remaining 3 are nonsense, or "stop," codons. The genetic code is unambiguous, degenerate, and universal.

30.2 How Is an Amino Acid Matched with Its Proper tRNA? During protein synthesis, aminoacyl-tRNAs recognize the codons through base pairing using their anticodon loops. A second genetic code exists, the code by which each aminoacyl-tRNA synthetase adds its amino acid to tRNAs that can interact with the codons that specify its amino acid. A common set of rules does not govern tRNA rec-

ognition by aminoacyl-tRNA synthetases. The tRNA features recognized are not limited to the anticodon and, in some instances, do not even include the anticodon. Usually, an aminoacyl-tRNA synthetase recognizes a set of sequence elements in its cognate tRNAs.

30.3 What Are the Rules In Codon–Anticodon Pairing? Anticodons are paired with codons in antiparallel orientation. There are more codons than there are amino acids, and considerable degeneracy exists in the genetic code at the third base position. The first two bases of the codon and the last two bases of the anticodon form canonical Watson–Crick base pairs, but pairing between the third base of the codon and the first base of the anticodon follows less stringent rules, allowing some anticodons to recognize more than one codon, in accordance with Crick's wobble hypothesis. Some codons for a particular amino acid are used more than the others.

Nonsense suppression occurs when suppressor tRNAs read nonsense codons.

30.4 What Is the Structure of Ribosomes, and How Are They Assembled?

Ribosomes are ribonucleoprotein particles that act as mechanochemical systems in protein synthesis. They move along mRNA templates, orchestrating the interactions between successive codons and the corresponding anticodons presented by aminoacyl-tRNAs. Ribosomes catalyze the formation of peptide bonds. Prokaryotic ribosomes consist of two subunits, 30S and 50S, which are composed of 50 different proteins and 3 rRNAs—16S, 23S, and 5S. The general shapes of the ribosomal subunits are determined by their rRNA molecules; ribosomal proteins serve a largely structural role in ribosomes. Ribosomes spontaneously self-assemble in vitro. The 30S subunit provides the decoding center that matches up the tRNA anticodons with the mRNA codons. The 50S subunit has the peptidyl transferase center that catalyzes peptide bond formation. This center consists solely of 23S rRNA; the ribosome is a ribozyme. Eukaryotic cytosolic ribosomes are larger than prokaryotic ribosomes.

30.5 What Are the Mechanics of mRNA Translation?

Ribosomes move along the mRNA in the $5' \rightarrow 3'$ direction, recruiting aminoacyl-tRNAs whose anticodons match up with successive codons and joining amino acids in peptide bonds in a polymerization process that forms a particular protein. Protein synthesis proceeds in three distinct phases: initiation, elongation, and termination. Elongation involves two steps: peptide bond formation and translocation of the ribosome one codon further along the mRNA. At each stage, energy is provided by GTP hydrolysis, and specific soluble protein factors participate. Many of these soluble proteins are G-protein family members. Initiation involves binding of mRNA by the small ribosomal subunit, followed by binding of fMet-tRNA$_i^{fMet}$ that recognizes the first codon. Elongation is accomplished via a repetitive cycle in which successive aminoacyl-tRNAs add to the ribosome: mRNA complex as directed by codon binding, the 50S subunit catalyzes peptide bond formation, and the polypeptide chain grows by one amino acid at a time. Ribosomes have three tRNA-binding sites: the A site, where incoming aminoacyl-tRNAs bind; the P site, where the growing peptidyl-tRNA chain is bound; and the E site, where deacylated tRNAs exit the ribosome. Termination occurs when the ribosome encounters a stop codon in the mRNA. Polysomes are the active structures in protein synthesis.

30.6 How Are Proteins Synthesized in Eukaryotic Cells?

The process of protein synthesis in eukaryotes strongly resembles that in prokaryotes, but the events are more complicated. Eukaryotic mRNAs have 5'-terminal ^{7}methyl G caps and 3'-polyadenylylated tails. Initiation of eukaryotic protein synthesis involves three stages and multiple proteins. This complexity offers many opportunities for regulation, and eukaryotic cells employ a variety of mechanisms for post-transcriptional regulation of gene expression. Many antibiotics are specific inhibitors of prokaryotic protein synthesis, making them particularly useful for the treatment of bacterial infections and diseases.

FOUNDATIONAL BIOCHEMISTRY Things You Should Know After Reading Chapter 30.

- The genetic code allows the information expressed in a base sequence to be translated into the amino acid sequence of a protein.

- tRNAs are adapter molecules that bridge the gap between nucleotide sequence and amino acid sequence.

- The genetic code is a triplet code.

- Codons specify amino acids.

- Of the 64 codons possible from the four bases (A, C, G, U), 61 specify an amino acid (the sense codons) and three are termination signals (stop codons or nonsense codons).

- The genetic code is unambiguous, degenerate, and universal.

- An amino acid is matched with its proper tRNA by its specific aminoacyl-tRNA synthetase.

- Cells have 20 different aminoacyl-tRNA synthetases.

- The code by which each aminoacyl-tRNA matches up its amino acid with appropriate tRNAs constitutes a "second" genetic code.

- Aminoacyl-tRNA synthetases interpret the second genetic code.

- Evolution has provided two distinct classes of aminoacyl-tRNA synthetases.

- Recognition of mRNA codons by tRNA anticodons is through antiparallel base-pairing.

- The wobble hypothesis points out that the third base of the codon and the first base of the anticodon follow less stringent base-pairing rules.

- Some codons are used more than others.

- Nonsense suppression occurs when suppressor tRNAs read nonsense codons.

- Ribosomes are the agents by which mRNAs are translated into proteins.

- Prokaryotic ribosomes are composed of 30S and 50S subunits.

- Prokaryotic ribosomes contain 50 different proteins and 3 different rRNAs.

- Ribosomes can spontaneously assemble in vitro from a mixture of their component proteins and rRNAs.

- Ribosomes have a characteristic anatomy.

- The central domain of the 30S ribosomal subunit is the site of the decoding center.

- The defining reaction of protein synthesis is performed by peptidyl transferase.

- Peptidyl transferase is a catalytic activity of the large rRNA of the large ribosomal subunit.

- Cytosolic ribosomes of eukaryotes are considerably larger and more complex than prokaryotic ribosomes.

- In translation, a ribosome moves along an mRNA in the $5' \rightarrow 3'$ direction, synthesizing a protein in the N→C direction.

- Initiation of translation involves mRNA binding by the small ribosomal subunit, following by association of the initiator Met-tRNA$_i^{Met}$ that recognizes the first codon, then recruitment of a large ribosomal subunit to form the 70S initiation complex, which is composed of 30 + 50S ribosomal subunits: mRNA: f-Met-tRNA$_i^{Met}$.

- Elongation is the phase of translation that forms every peptide bond, from the first to the last.

- Elongation is a repetitive series of: codon-directed binding of an aminoacyl-tRNA, peptidyl transfer from the peptidyl-tRNA to the amino group of the aminoacyl-tRNA, and translocation of the ribosome one codon further along the mRNA.

- Three tRNAs may be associated with the ribosome:mRNA complex at any moment: one in the A (or acceptor) site, one in the P (or peptidyl) site, and one in the E (or exit) site.

- Termination is triggered when the ribosome reaches a termination codon on the mRNA.

- Peptide chain initiation requires: mRNA, 30S and 50S ribosomal subunits, initiation factors, GTP, and f-Met-tRNA$_i^{Met}$.

- In prokaryotes, recognition of the mRNA and proper alignment for translation depends on interaction between the Shine-Dalgarno sequence at the 5′-end of the mRNA and a complementary sequence near the 3′-end of the small ribosomal subunit 16S rRNA.

- Three initiation factors (IF-1, IF-2, and IF-3) participate in initiation of translation.

- IF-2, a G-protein family member, delivers the initiator f-Met-tRNA$_i^{Met}$ to the 30S ribosomal subunit:mRNA complex.

- GTP hydrolysis by IF-2 is triggered when a 50S ribosomal subunit joins the 30S:mRNA:f-Met-tRNA$_i^{Met}$ complex.

- Three elongation factors are required: EF-Tu, EF-Ts, and EF-G.

- The peptide chain elongation cycle has three principal steps:
  1. Codon-directed binding of an incoming aminoacyl-tRNA at the A site.
  2. Peptide bond formation by the 50S subunit peptidyl transferase activity through transfer of the peptidyl chain from the P-site tRNA to the amino group of the aminoacyl-tRNA in the A site.
  3. Translocation of the ribosome one codon further along the mRNA, accompanied by movement of the now longer peptidyl-tRNA from the A site to the P site.

- The decoding center that ensures proper codon:anticodon recognition is a 16S rRNA function.

- EF-Tu is a G-protein family member that mediates delivery of the aminoacyl-tRNA to the A site.

- EF-Ts is a guanine nucleotide exchange factor (GEF) that exchanges GDP for GTP on Ef-Tu.

- 23S rRNA is the peptidyl transferase enzyme.

- Catalysis of peptide bond formation relies on tight binding of the aminoacyl-tRNA and the peptidyl-tRNA by the ribosome.

- The translocation phase of protein synthesis has three principal steps:
  1. The deacylated tRNA must be moved from the P site to the E site.
  2. The newly extended peptidyl-tRNA must be moved from the A site to the P site.
  3. Concomitant with these events, the ribosome must translocate one codon further along the mRNA.

- EF-G acts as a GTP-driven molecular switch that mediates ribosome movement along the mRNA.

- EF-Tu and EF-G compete for binding at a common site on the ribosome, called the factor-binding center or GTPase-activating center (GAC).

- Peptide chain termination requires yet another G-protein family member, RF-3.

- Molecular mimicry is seen in the structures of EF-TU:tRNA complex, EF-G, and RF-3.

- Termination occurs when release factors recognize a termination codon.

- The ribosomal subunits cycle between 70S complexes and free ribosomal subunits.

- Polysomes are the active structures of protein synthesis.

- Protein synthesis in eukaryotes is more complex than in prokaryotes.

- More than a dozen eukaryotic initiation factors, symbolized as eIFs, are required.

- Eukaryotic initiation proceeds through three stages:
  1. Formation of the 43S preinitiation complex, composed of a set of the eIFs, the 40S ribosomal subunit, and eIF2:GTP:Met-tRNA$_i^{Met}$.
  2. Formation of the 48S initiation complex, which includes the 43S preinitiation complex, mRNA, and additional eIFs.
  3. Formation of the 80S initiation complex, with the initiator Met-tRNA$_i^{Met}$ atop the initiation codon through addition of the 60S large ribosomal subunit.

- Control of eukaryotic peptide chain initiation is an important mechanism for post-transcriptional regulation of gene expression.

- Eukaryotic peptide chain elongation resembles the prokaryotic process.

- Eukaryotic peptide chain termination requires just one release factor.

- Because eukaryotic and prokaryotic protein synthesis is different, some inhibitors of protein synthesis serve as medically important antibiotics.

- The decoding center is a target of aminoglycoside antibiotics.

- Many antibiotics target the peptidyl transferase center and the peptide exit tunnel.

PROBLEMS

Answers to all problems are at the end of this book. Detailed solutions are available in the *Student Solutions Manual, Study Guide, and Problems Book.*

1. **Translating Nucleotide Sequences into Amino Acid Sequences** (Integrates with Chapter 12.) The following sequence represents part of the nucleotide sequence of a cloned cDNA:

   ...CAATACGAAGCAATCCCGCGACTAGACCTTAAC...

   Can you reach an unambiguous conclusion from these data about the partial amino acid sequence of the protein encoded by this cDNA?

2. **Nucleotide Sequences, Possible Codons, and Amino Acid Specification** A random (AG) copolymer was synthesized using a mixture of 5 parts adenine nucleotide to 1 part guanine nucleotide as substrate. If this random copolymer is used as an mRNA in a cell-free protein synthesis system, which amino acids will be incorporated into the polypeptide product? What will be the relative abundances of these amino acids in the product?

3. **The Second Genetic Code** Review the evidence establishing that aminoacyl-tRNA synthetases bridge the information gap between amino acids and codons. Indicate the various levels of specificity possessed by aminoacyl-tRNA synthetases that are essential for high-fidelity translation of messenger RNA molecules.

4. **Codon–Anticodon Recognition: Base-Pairing Possibilities** (Integrates with Chapter 11.) Draw base-pair structures for (a) a G:C base pair, (b) a C:G base pair, (c) a G:U base pair, and (d) a U:G base pair. Note how these various base pairs differ in the potential hydrogen-bonding patterns they present within the major groove and minor groove of a double-helical nucleic acid.

5. **Consequences of the Wobble Hypothesis** Point out why Crick's wobble hypothesis would allow fewer than 61 anticodons to be used to translate the 61 sense codons. How might "wobble" tend to accelerate the rate of translation?

6. **Sense-to-Nonsense Codon Mutations** How many codons can mutate to become nonsense codons through a single base change? Which amino acids do they encode?

7. **Amino Acid Possibilities in Nonsense Suppression** Nonsense suppression occurs when a suppressor mutant arises that reads a nonsense codon and inserts an amino acid, as if the nonsense codon were actually a sense codon. Which amino acids do you think are most likely to be incorporated by nonsense suppressor mutants?

8. **Prokaryotic vs. Eukaryotic Protein Synthesis** Why do you suppose eukaryotic protein synthesis is only 10% as fast as prokaryotic protein synthesis?

9. **Capacity of the Ribosomal Peptide Exit Tunnel** If the tunnel through the large ribosomal subunit is 10 nm long, how many amino acid residues might be contained within it? (Assume that the growing polypeptide chain is in an extended β-sheet–like conformation.)

10. **The Consequences of Ribosome Complexity** Eukaryotic ribosomes are larger and more complex than prokaryotic ribosomes. What advantages and disadvantages might this greater ribosomal complexity bring to a eukaryotic cell?

11. **Ribosomes as Two-Subunit Structures** What ideas can you suggest to explain why ribosomes invariably exist as two-subunit structures, instead of a larger, single-subunit entity?

12. **tRNA$_i^{Met}$ Recognition and Translation Initiation** How do prokaryotic cells determine whether a particular methionyl-tRNA^{Met} is intended to initiate protein synthesis or to deliver a Met residue for internal incorporation into a polypeptide chain? How do the Met codons for these two different purposes differ? How do eukaryotic cells handle these problems?

13. **The Function of the Shine-Dalgarno Sequence** What is the Shine–Dalgarno sequence? What does it do? The efficiency of protein synthesis initiation may vary by as much as 100-fold for different mRNAs. How might the Shine–Dalgarno sequence be responsible for this difference?

14. **Reconciling Peptide Bond Formation and tRNA Movements** In the protein synthesis elongation events described under the section on translocation, which of the following seems the most apt account of the peptidyl transfer reaction: (a) The peptidyl-tRNA delivers its peptide chain to the newly arrived aminoacyl-tRNA situated in the A site, or (b) the aminoacyl end of the aminoacyl-tRNA moves toward the P site to accept the peptidyl chain? Which of these two scenarios makes more sense to you? Why?

15. **Why Translation Factors are G Proteins** (Integrates with Chapter 15.) Why might you suspect that the elongation factors EF-Tu and EF-Ts are evolutionarily related to the G proteins of membrane signal transduction pathways described in Chapter 15?

16. **The Energetic Cost of Peptide Elongation** How many ATP equivalents are consumed for each amino acid added to an elongating polypeptide chain during the process of protein synthesis?

17. **Exploring the Structure of the 30S Ribosomal Subunit** Go to *www.pdb.org* and bring up PDB file 1GIX, which shows the 30S ribosomal subunit, the three tRNAs, and mRNA. In the box on the right titled "Biological Assembly," click "More Images," and then scroll down to look at the Interactive View. By moving your cursor over the image, you can rotate it to view it from any perspective.

    a. How are the ribosomal proteins represented in the image?

    b. How is the 16S rRNA portrayed?

    c. Rotate the image to see how the tRNAs stick out from the structure. Which end of the tRNA is sticking out?

    d. Where will these ends of the tRNAs lie when the 50S subunit binds to this complex?

18. **Exploring the Structure of the 50S Ribosomal Subunit** Go back to *www.pdb.org* and bring up PDB file 1FFK, which shows the 50S ribosomal subunit. In the box titled "Biological Assembly," click "More Images," and scroll down to look at the Interactive View. Right-click the image to discover more information and tools.

    a. How many atoms are represented in this structure?

    b. Are the bases of the nucleotides visible anywhere in the structure?

    c. Can you find double helical regions of RNA?

    d. Right-click and, under "Select," select all proteins. Right-click again and select "Render," then "Scheme," and then "CPK Space-fill" to highlight the ribosomal proteins. Go back and cancel the protein selection. Then select "Nucleic," and render nucleic acid in "CPK Spacefill." Which macromolecular species seems to predominate the structure?

19. **Exploring the Structure of the Peptide Chain Termination Complex** Go again to *www.pdb.org* and bring up the PDB file 3F1E. Scroll down and view the designations for the various components in the structure, as given in the **Molecular Description** box (note, for example, that the mRNA is designated "chain V"). In the box on the right, click on "view in Jmol". View the overall structure, then scroll down below the Jmol box and under "PDP," click on the various 3F1E selections to highlight RF-2 in the Jmol image. Within the Jmol image, right click to explore the structure further. Clicking on "model 1/2" allows you to select the 30S or 50S subunits. Other possibilities allow you to delve further.

Preparing for the MCAT® Exam

20. Review the list of Shine–Dalgarno sequences in Figure 30.18 and select the one that will interact best with the 3′-end of *E. coli* 16S rRNA.

21. Chloramphenicol (Figure 30.31) inhibits the peptidyl transferase activity of the 50S ribosomal subunit. The 50S peptidyl transferase active site consists solely of functionalities provided by the 23S rRNA. What sorts of interactions do you think take place when chloramphenicol binds to the peptidyl transferase center? Which groups on chloramphenicol might be involved in these interactions?

FURTHER READING

General

Cech, T. R., Atkins, J. F., and Gesteland, R. F., 2005. *The RNA World*, 3rd ed. Cold Spring Harbor, NY: Cold Spring Harbor Laboratory Press.

Krebs, J., Goldstein, E. S., and Kilpatrick, S. T., 2009. *Lewin's Genes X*. Sudbury, MA: Jones and Bartlett.

The Genetic Code

Cedergren, R., and Miramontes, P., 1996. The puzzling origin of the genetic code. *Trends in Biochemical Sciences* **21**:199–200.

Huttenhofer, A., and Bock, A., 1998. RNA structures involved in seleno-protein synthesis. In *RNA Structure and Function*, Simons, R. W.,

and Grunberg-Monago, M., eds., pp. 603–639. Cold Spring Harbor, NY: Cold Spring Harbor Laboratory Press.

Khorana, H. G., et al., 1966. Polynucleotide synthesis and the genetic code. *Cold Spring Harbor Symposium on Quantitative Biology* **31**:39–49. The use of synthetic polyribonucleotides in elucidating the genetic code.

Knight, R. D., et al., 1999. Selection, history, and chemistry: Three faces of the genetic code. *Trends in Biochemical Sciences* **24**:241–247.

Liu, C. C., and Schultz, P. G., 2010. Adding new chemistries to the genetic code. *Annual Review of Biochemistry* **79**:413–444.

Nirenberg, M. W., and Leder, P., 1964. RNA codewords and protein synthesis. *Science* **145**:1399–1407.

Nirenberg, M. W., and Matthaei, J. H., 1961. The dependence of cell-free protein synthesis in *E. coli* upon naturally occurring or synthetic polyribonucleotides. *Proceedings of the National Academy of Sciences U.S.A.* **47:**1588–1602.

Wang, L., Xie, J., and Schultz, P. G., 2006. Expanding the genetic code. *Annual Review of Biophysics and Biomolecular Structure* **35:**225–249.

Aminoacylation of tRNAs and the Second Genetic Code

Arnez, J. G., and Moras, D., 1997. Structural and functional considerations of the aminoacylation reaction. *Trends in Biochemical Sciences* **22:**211–216.

Banarjee, R., Chen, S., Dare, K., Gilreath, M., et al., 2009. tRNAs: Cellular barcodes for amino acids. *FEBS Letters* **584:**387–395.

Carter, C. W., Jr., 1993. Cognition, mechanism, and evolutionary relationships in aminoacyl-tRNA synthetases. *Annual Review of Biochemistry* **62:**715–748.

Hale, S. P., et al., 1997. Discrete determinants in transfer RNA for editing and aminoacylation. *Science* **276:**1250–1252.

Ibba, M., Curnow, A. W., and Söll, D., 1997. Aminoacyl-tRNA synthesis: Divergent routes to a common goal. *Trends in Biochemical Sciences* **22:**39–42.

Normanly, J., and Abelson, J., 1989. tRNA identity. *Annual Review of Biochemistry* **58:**1029–1049. Review of the structural features of tRNA that are recognized by aminoacyl-tRNA synthetases.

Park, S. G., Ewalt, K. L., and Kim, S., 2005. Function expansion of aminoacyl-tRNA synthetases and their interacting factors: New perspectives on housekeepers. *Trends in Biochemical Sciences* **30:**569–574.

Perona, J. J., and Hou, Y. M., 2007. Indirect readout of tRNA for aminoacylation. *Biochemistry* **46:**10419–10432.

Schimmel, P., and Schmidt, E., 1995. Making connections: RNA-dependent amino acid recognition. *Trends in Biochemical Sciences* **20:**1–2.

Sheppard, K., Yuan, J., Hohn, M. J., Jester, B., Devine, K. M., and Söll, D., 2008. From one amino acid to another: tRNA-dependent amino acid biosynthesis. *Nucleic Acids Research* **36:**1813–1825.

Codon–Anticodon Recognition

Crick, F. H. C., 1966. Codon–anticodon pairing: The wobble hypothesis. *Journal of Molecular Biology* **19:**548–555. Crick's original paper on wobble interactions between tRNAs and mRNA.

Crick, F. H. C., et al., 1961. General nature of the genetic code for proteins. *Nature* **192:**1227–1232. An insightful paper on insertion/deletion mutants providing convincing genetic arguments that the genetic code was a triplet code, read continuously from a fixed starting point. This genetic study foresaw the nature of the genetic code, as later substantiated by biochemical results.

Ribosome Structure and Function

Ban, N., et al., 2000. The complete atomic structure of the large ribosomal subunit at 2.4 Å resolution. *Science* **289:**905–920.

Ben-Shem, A., Jenner, L., Yusupova, G., and Yusupova, M., 2010. Crystal structure of the eukaryotic ribosome. *Science* **330:**1203–1209.

Carter, A. P., et al., 2000. Functional insights from the structure of the 30S ribosomal subunit and its interactions with antibiotics. *Nature* **407:**340–348.

Cate, J. H., et al., 1999. X-ray crystal structure of 70S functional ribosomal complexes. *Science* **285:**2095–2104.

Frank, J., and Gonzalez, R. L., Jr., 2010. Structure and dynamics of a processive Brownian motion. *Annual Review of Biochemistry* **79:**381–412.

Kaminishi, T., Wilson, D. N., Takemoto, C., Harms, J. M., et al., 2007. A snapshot of the 30S ribosomal subunit capturing mRNA via the Shine-Dalgarno interaction. *Structure* **15:**289–297.

Korostelev, A., and Noller, H. F., 2007. The ribosome in focus: New structures bring new insights. *Trends in Biochemical Sciences* **32:**434–441.

Moore, P. B., and Steitz, T. A., 2002. The involvement of RNA in ribosome function. *Nature* **418:**229–235.

Ogle, J. M., Carter, A. P., and Ramakrishnan, V., 2003. Insights into the decoding mechanism from recent ribosome structures. *Trends in Biochemical Sciences* **28:**259–266.

Rabl, J., Leibundgut, M., Ataide, S. F., Haag, A., and Ban, N., 2011. Crystal structure of the eukaryotic 40S ribosomal subunit in complex with initiation factor 1. *Science* **331:**730–736.

Ramakrishnan, V., 2010. The eukaryotic ribosome. *Science* **331:**681–682.

Spahn, C. M. T., et al., 2001. Structure of the 80S ribosome from *Saccharomyces cerevisiae*–tRNA-ribosome and subunit–subunit interactions. *Cell* **107:**373–386.

Stark, H., et al., 2002. Ribosome interactions of aminoacyl-tRNA and elongation factor Tu in the codon-recognition complex. *Nature Structural Biology* **9:**849–854.

Tenson, T., and Ehrenberg, M., 2002. Regulatory nascent peptides in the ribosomal tunnel. *Cell* **108:**591–594.

Valle, M., et al., 2002. Locking and unlocking of ribosomal motions. *Cell* **114:**123–134.

Voorhees, R. M., Schmeing, T. M., Kelley, A. C., and Ramakrishnan, V., 2010. The mechanism for activation of GTP hydrolysis on the ribosome. *Science* **330:**835–838.

The Ribosome Is a Ribozyme

Cech, T. R., 2000. The ribosome is a ribozyme. *Science* **289:**878–879.

Green, R., Samaha, R. R., and Noller, H. F., 1997. Mutations at nucleotides G2251 and U2585 of 23 S rRNA perturb the peptidyl transferase center of the ribosome. *Journal of Molecular Biology* **266:**40–50.

Green, R., Switzer, C., and Noller, H. F., 1998. Ribosome-catalyzed peptide-bond formation with an A-site substrate covalently linked to 23S ribosomal RNA. *Science* **280:**286–289.

Protein Synthesis: Initiation, Elongation, and Termination Factors

Allen, G. S., Zavialov, A., Gursky, R., Ehrenberg, M., and Frank, J., 2005. The cryo-EM structure of a translation initiation complex from *Escherichia coli. Cell* **121:**703–712.

Beringer, M., 2008. Modulating the activity of the peptidyl transferase center of the ribosome. *RNA* **14:**795–801.

Beringer, M., and Rodnina, M. V., 2007. The ribosomal peptidyl transferase. *Molecular Cell* **26:**311–321.

Bieling, P., Beringer, M., Adio, S., and Rodnina, M. V., 2006. Peptide bond formation does not involve acid–base catalysis by ribosomal residues. *Nature Structural and Molecular Biology* **13:**423–428.

Clark, B. F. C., and Nyborg, J., 1997. The ternary complex of EF-Tu and its role in protein synthesis. *Current Opinion in Structural Biology* **7:**110–116.

Clark, B. F. C., et al., eds., 1996. Prokaryotic and eukaryotic translation factors. *Biochimie* **78:**1119–1122.

Dever, T. E., 1999. Translation initiation: Adept at adapting. *Trends in Biochemical Sciences* **24:**398–403.

Ehrenberg, M., and Tenson, T., 2002. A new beginning to the end of translation. *Nature Structural Biology* **9:**85–87.

Nissen, P., et al., 1995. Crystal structure of the ternary complex of Phe-tRNA^Phe, Ef-Tu, and a GTP analog. *Science* **270:**1464–1472.

Ogle, J. M., and Ramakrishnan, R., 2005. Structural insights into translational fidelity. *Annual Review of Biochemistry* **74:**129–177.

Poole, E. S., Askarian-Amiri, M. E., Major, L. L., McCaughan, K. K., et al., 2003. Molecular mimicry in the decoding of translational stop signals. *Progress in Nucleic Acids Research and Molecular Biology* **74:**83–121.

Voss, N. R., Gerstein, M., Steitz, T. A., and Moore, P. B., 2006. The geometry of the ribosomal polypeptide exit tunnel. *Journal of Molecular Biology* **360:**893–906.

Zavialov, A. V., and Ehrenberg, M., 2003. Peptidyl-tRNA regulates the GTPase activity of translation factors. *Cell* **114:**113–122.

Eukaryotic Protein Synthesis

Dinman, J. D., 2009. The eukaryotic ribosome: Current status and challenges. *Journal of Biological Chemistry* **284:**11761–11765.

Gingras, A-C., et al., 1999. eIF-4 initiation factors: Effectors of mRNA recruitment to ribosomes and regulators of translation. *Annual Review of Biochemistry* **68:**913–963.

Hinnebush, A. G., 2006. eIF3: A versatile scaffold for translation initiation complexes. *Trends in Biochemical Sciences* **31:**553–562.

Kimball, S. R., and Jefferson, L. S., 2010. Control of translation initiation through integration of signals generated by hormones, nutrients, and exercise. *Journal of Biological Chemistry* **285:**29027–29032.

Kolitz, S. E., and Lorsch, J. R., 2010. Eukaryotic initiator tRNA: Finely tuned and ready for action. *FEBS Letters* **584:**396–404.

Lorsch, J. R., and Dever, T. E., 2010. Molecular view of 43S complex formation and start site selection in eukaryotic translation initiation. *Journal of Biological Chemistry* **285:**21203–21207.

Matsuo, H., et al. 1997. Structure of translation factor eIF4E bound to 7mGDP and interaction with 4E-binding protein. *Nature Structural Biology* **4:**717–724.

Merrick, W. C., 2010. Eukaryotic protein synthesis: Still a mystery. *Journal of Biological Chemistry* **285:**21197–21201.

Pain, V. M., 1996. Initiation of protein synthesis in eukaryotic cells. *European Journal of Biochemistry* **236:**747–771.

Rhoads, R. E., 1999. Signal transduction pathways that regulate eukaryotic protein synthesis. *Journal of Biological Chemistry* **274:**30337–30340.

Rhoads, R. E., 2009. eIF4E: New family members, new binding partners, new roles. *Journal of Biological Chemistry* **284:**16711–16715.

Rhoads, R., Dinkova, T. D., and Komeeva, N. L., 2006. Mechanism and regulation of translation in *C. elegans*. *WormBook* **28:**1–18.

Sachs, A. B., and Varani, G., 2000. Eukaryotic translation initiation: There are two sides (at least) to every story. *Nature Structural Biology* **7:**356–361.

Samuel, C. E., 1993. The eIF-2a protein kinases, regulators of translation in eukaryotes from yeast to humans. *Journal of Biological Chemistry* **268:**7603–7606.

Tarun, S. Z., Jr., et al., 1997. Translation factor eIF4G mediates in vitro poly(A) tail dependent translation. *Proceedings of the National Academy of Sciences U.S.A.* **94:**9046–9051.

Protein Synthesis Inhibitors

Endo, Y., et al., 1987. The mechanism of action of ricin and related toxic lectins on eukaryotic ribosomes. The site and the characteristics of the modification in 28S ribosomal RNA caused by the toxins. *Journal of Biological Chemistry* **262:**5908–5912.

Hermann, T., 2005. Drugs targeting the ribosome. *Current Opinion in Structural Biology* **15:**355–366.

Polacek, N., and Mankin, A. S., 2005. The ribosomal peptidyl transferase center: Structure, function, evolution, inhibition. *Critical Reviews in Biochemistry and Molecular Biology* **40:**285–311.

Schlünzen, F., et al., 2000. Structural basis for the interaction of antibiotics with the peptidyl transferase center in eubacteria. *Nature* **413:**814–821.

Yonath, A., 2005. Antibiotics targeting ribosomes: Resistance, selectivity, synergism, and cellular regulation. *Annual Review of Biochemistry* **74:**649–679.

Completing the Protein Life Cycle: Folding, Processing, and Degradation

Gabriel Vong

◀ Vong's Flying Crane. *Origami*—the Asian art of paper folding—arose in China almost 2000 years ago when paper was rare and expensive and the folded shape added special meaning. Protein folding, like origami, takes a functionless form and creates a structure with unique identity and purpose.

* *

Life is a process of becoming, a combination of states we have to go through.

Anais Nin (1903–1977)

KEY QUESTIONS

31.1 How Do Newly Synthesized Proteins Fold?

31.2 How Are Proteins Processed Following Translation?

31.3 How Do Proteins Find Their Proper Place in the Cell?

31.4 How Does Protein Degradation Regulate Cellular Levels of Specific Proteins?

ESSENTIAL QUESTION

Proteins are the agents of biological function. Protein turnover (synthesis and decay) is a fundamental aspect of each protein's natural history.

How are newly synthesized polypeptide chains transformed into mature, active proteins, and how are undesired proteins removed from cells?

The human genome apparently contains about 20,500 genes, but some estimates suggest that the total number of proteins in the human proteome may approach 1 million. What processes introduce such dramatically increased variation into the products of protein-encoding genes? We've reviewed (or will soon cover) many of these processes; a partial list (with examples) includes:

1. Gene rearrangements (immunoglobulin G)
2. Alternative splicing (fast skeletal muscle troponin T)
3. RNA editing (apolipoprotein B)
4. Proteolytic processing (chymotrypsinogen or prepro-opiomelanocortin; see Chapter 32)
5. Isozymes (lactate dehydrogenase)
6. Protein sharing (the glycolytic enzyme enolase is identical to τ-crystallin in the eye)
7. Protein–protein interactions at many levels (oligomerization, supramolecular complexes, assembly of signaling pathway protein complexes upon scaffold proteins)
8. Covalent modifications of many kinds (phosphorylation or glycosylation, with multisite phosphorylation or variable degrees of glycosylation, to name just two of the dozens of possibilities)

Thus, the nascent polypeptide emerging from a ribosome is not yet the agent of biological function that is its destiny. First, the polypeptide must fold into its native tertiary structure. This complicated process is made even more so by the crowding of macromolecules

OWL Online homework and a Student Self Assessment for this chapter may be assigned in OWL.

Nascent means "undergoing the process of being born" or, in the molecular sense, "newly synthesized."

inside a typical cell (see Figure 31.3d). Even then, seldom is the nascent, folded protein in its final functional state. Proteins often undergo various proteolytic processing reactions and covalent modifications as steps in their maturation to functional molecules. Finally, at the end of their usefulness, damaged by chemical reactions or denatured due to partial unfolding, they are degraded. In addition, some proteins are targeted for early destruction as part of regulatory programs that carefully control available amounts of particular proteins. Damaged or misfolded proteins are a serious hazard; accumulation of protein aggregates can be a cause of human disease, including the prion diseases and diseases of amyloid accumulation, such as Alzheimer's, Parkinson's, or Huntington's disease.

31.1 | How Do Newly Synthesized Proteins Fold?

As Christian Anfinsen pointed out 40 years ago, the information for folding each protein into its unique three-dimensional architecture resides within its amino acid sequence or primary structure (see Chapter 6). Proteins begin to fold even before their synthesis by ribosomes is completed (Figure 31.1a). However, the cytosolic environment is a very crowded place, with effective protein concentrations as high as 0.3 grams/mL. Macromolecular crowding enhances the likelihood of nonspecific protein association and aggregation. The primary driving force for protein folding is the burial of hydrophobic side chains away from the aqueous solvent and reduction in solvent-accessible surface area

---

Alzheimer's, Parkinson's, and Huntington's Disease Are Late-Onset Neurodegenerative Disorders Caused by the Accumulation of Protein Deposits

As noted in Chapter 6, protein misfolding problems can cause disease by a variety of mechanisms. For example, protein aggregates can impair cell function. **Amyloid plaques** (so named because they resemble the intracellular starch, or amyloid, deposits found in plant cells) and **neurofibrillary tangles (NFTs)** are proteinaceous deposits found in the brains of individuals suffering from any of several neurogenerative diseases. In each case, the protein is different. In Alzheimer's, disease is caused both by extracellular amyloid deposits composed of proteolytic fragments of the amyloid precursor protein (APP) termed **amyloid-β (Aβ)** and intracellular NFTs composed of the **microtubule-binding protein tau (τ).** Aβ is a peptide 39 to 43 amino acids long that polymerizes to form long, highly ordered, insoluble fibrils consisting of a hydrogen-bonded parallel β-sheet structure in which identical residues on adjacent chains are aligned directly, in register (see accompanying figure). Why Aβ aggregates in some people but not others is not clear. In Parkinson's, the culprit is NFTs composed of polymeric τ; no amyloid plaques are evident. In Huntington's disease, the protein deposits occur as nuclear inclusions composed of polyglutamine (polyQ) aggregates. PolyQ aggregates arise from mutant forms of **huntingtin,** a protein that characteristically has a stretch of glutamine residues close to its N-terminus. Huntingtin is a 3144-residue protein encoded by the *ITI5* gene, which has 67 exons. Exon 1 encodes the polyglutamine region. Individuals whose huntingtin gene has fewer that 35 CAG (glutamine codon) repeats never develop the disease; those with 40 or more always develop the disease within a normal lifetime. The nuclear inclusions in Huntington's disease are huntingtin-derived polyglutamine fragments that have aggregated to form β-sheet–containing amyloid fibrils.

Impairment of cellular function by proteinaceous deposits may be a general phenomenon. In vitro experiments have demonstrated

that aggregates of proteins not associated with disease can be cytotoxic, and the ability to form amyloid deposits is a general property of proteins. The evolution of chaperones to assist protein folding and proteasomes to destroy improperly folded proteins may have been driven by the necessity to prevent protein aggregation.

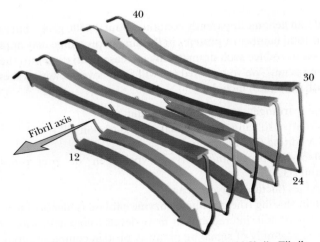

A model for the Aβ$_{1-40}$ structural unit in β-amyloid fibrils. Fibrils contain β-strands perpendicular to the fibril axis, with interstrand hydrogen bonding parallel to the fiber axis. The top face of the β-sheet is hydrophobic and presumably interacts with neighboring Aβ molecules in fibril formation. (Figure adapted from Figure 1 in Thompson, L. K., 2003. Unraveling the secrets of Alzheimer's β-amyloid fibrils. *Proceedings of the National Academy of Sciences, U.S.A.* **100:**383–385.)

(see Chapter 6). The folded protein typically has a buried hydrophobic core and a hydrophilic surface. Protein aggregation is typically driven by hydrophobic interactions, so burial of hydrophobic regions through folding is a crucial factor in preventing aggregation. To evade such problems, nascent proteins are often assisted in folding by a family of helper proteins known as **molecular chaperones** (see Chapter 6) because, like the chaperones at a prom, their purpose is to prevent inappropriate liaisons. Chaperones also serve to shepherd proteins to their ultimate cellular destinations. Also, mature proteins that have become partially unfolded may be rescued by chaperone-assisted refolding.

Chaperones Help Some Proteins Fold

A number of chaperone systems are found in all cells. Many of the proteins in these systems are designated by the acronym **Hsp** (for **heat shock protein**) and a number indicating their relative mass in kilodaltons (as in Hsp60). Hsps were originally observed as abundant proteins in cells given brief exposure to high temperature (42°C or so). The principal Hsp chaperones are **Hsp70, Hsp60** (the **chaperonins**), and **Hsp90.** In general, proteins whose folding is chaperone-dependent pass down a pathway in which Hsp70 acts first on the newly synthesized protein and then passes the partially folded intermediate to a chaperonin for completion of folding.

Nascent polypeptide chains exiting the large ribosomal subunits are met by ribosome-associated chaperones (**TF,** or **Trigger Factor,** in *Escherichia coli;* **NAC [nascent chain-associated complex]** in eukaryotes). In *E. coli,* the 50S ribosomal protein L23, which is situated at the peptide exit tunnel, serves as the docking site for TF, directly linking protein synthesis with chaperone-assisted protein folding. TF and NAC mediate transfer of the emerging nascent polypeptide chain to the Hsp70 class of chaperones, although many proteins do not require this step for proper folding.

Hsp70 Chaperones Bind to Hydrophobic Regions of Extended Polypeptides

In Hsp70-assisted folding, proteins of the Hsp70 class bind to nascent polypeptide chains while they are still on ribosomes (Figure 31.1b). Hsp70 (known as **DnaK** in *E. coli*) recognizes exposed, extended regions of polypeptides that are rich in hydrophobic residues. By interacting with these regions, Hsp70 prevents nonproductive associations and keeps the polypeptide in an unfolded (or partially folded) state until productive folding interactions can occur. Completion of folding requires release of the protein from Hsp70; release is energy-dependent and is driven by ATP hydrolysis.

Hsp70 proteins such as DnaK consist of two domains: a 44-kD N-terminal ATP-binding domain and an 18-kD central domain that binds polypeptides with exposed hydrophobic regions (Figure 31.2a). The DnaK : ATP complex receives an unfolded (or partially folded) polypeptide chain from **DnaJ** (Figure 31.2b). DnaJ is an **Hsp40** family member. Interaction of DnaK with DnaJ triggers the ATPase activity of DnaK; the DnaK : ADP complex forms a stable complex with the unfolded polypeptide, preventing its aggregation with other proteins. A third protein, **GrpE,** catalyzes nucleotide exchange on DnaK, replacing ADP with ATP, which converts DnaK back to a conformational form having low affinity for its polypeptide substrate. Release of the polypeptide gives it the opportunity to fold. Multiple cycles of interaction with DnaK (or Hsp70) give rise to partially folded intermediates or, in some cases, completely folded proteins. The partially folded intermediates may be passed along to the Hsp60/chaperonin system for completion of folding (Figure 31.1c).

The GroES–GroEL Complex of *E. coli* Is an Hsp60 Chaperonin

The Hsp60 class of chaperones, also known as **chaperonins,** assists some partially folded proteins to complete folding after their release from ribosomes. Chaperonins sequester partially folded molecules from one another (and from extraneous interactions), allowing

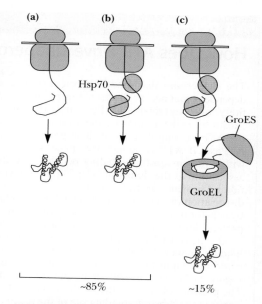

FIGURE 31.1 Protein folding pathways. **(a)** Chaperone-independent folding. The protein folds as it is synthesized on the ribosome (*green*) (or shortly thereafter). **(b)** Hsp70-assisted protein folding. Hsp70 (*gray*) binds to nascent polypeptide chains as they are synthesized and assists their folding. **(c)** Folding assisted by Hsp70 and chaperonin complexes. The chaperonin complex in *E. coli* is GroES–GroEL. The majority of proteins fold by pathways (a) or (b). (Adapted from Figure 2 in Netzer, W. J., and Hartl, F. U., 1998. Protein folding in the cytosol: Chaperonin-dependent and -independent mechanisms. *Trends in Biochemical Sciences* **23:**68–73; and Figure 2 in Hartl, F. U., and Hayer-Hartl, M., 2002. Molecular chaperones in the cytosol: From nascent chain to folded protein. *Science* **295:**1852–1858.)

A DEEPER LOOK

How Does ATP Drive Chaperone-Mediated Protein Folding?

The chaperones that mediate protein folding do so in an ATP-dependent manner, as illustrated in Figures 31.2 and 31.3. The affinity of chaperones for their unfolded or misfolded protein substrates is determined by the nature of the nucleotide bound by the ATP-binding domain of these proteins, which functions as an ATPase. If ATP is bound, the chaperone adopts a conformation with an open substrate-binding pocket. ATP increases the rate of association of the chaperone Hsp70 (DnaK) with an unfolded peptide or protein substrate by 100-fold, but it increases the rate of dissociation of the unfolded protein from the chaperone even more, by a factor of 1000. Overall, the chaperone's affinity for an unfolded protein substrate decreases 10-fold (or more) when it binds ATP.

On the other hand, if the substrate-binding site on the peptide-binding domain of DnaK is occupied by an unfolded protein substrate in conjunction with binding of the co-chaperone (DnaJ in Figure 31.2), ATP hydrolysis by the ATPase domain is triggered. The presence of ADP in the ATP-binding (ATPase) domain causes a shift in the substrate-binding site of the peptide-binding domain to a closed conformation, high-affinity state. Thus, ATP-dependent chaperones cycle between two stable conformational states, just like allosteric proteins. Bound ATP favors the open conformation for the protein substrate-binding site, and ADP favors the closed conformation. (When ADP is released and no nucleotide occupies the ATP-binding site of the ATPase domain, the peptide-binding site remains in the closed, high-affinity conformation; see Figure 31.2). What is the underlying mechanism that controls these ATP-regulated conformational changes?

The ATP-Dependent Allosteric Regulation of Hsp70 Chaperones Is Controlled by a Proline Switch

Clearly, the two domains of Hsp70 (DnaK)—the peptide-binding domain and the ATP-binding (or ATPase) domain—communicate with each other, because the nature of the nucleotide bound to the ATP domain determines the affinity of the peptide-binding domain for unfolded substrates. Markus Vogel, Bernd Bukau, and Matthais Mayer of the Center for Molecular Biology at the University of Heidelberg (Germany) argue that four distinct elements are needed for communication between these separate domains: an **ATP sensor** (which must include residues within the ATP-binding site), a **transducer** (to communicate the presence of ATP to the distant peptide-binding site), a **lever** (operated by the ATP-binding domain to exert its effect on the distant peptide-binding domain), and a **switch** that controls the lever (the switch is needed to lock the protein in either the open conformation or the closed conformation so that either alternative conformation is stable).

The switch that controls the conformational transitions of Hsp70 involves two universally conserved residues in the ATPase domain of Hsp70 family members, a proline (Pro^{143}) and a surface-exposed arginine (Arg^{151}; panel a of the figure). Pro^{143} is the switch, and Arg^{151} is a relay for the lever. Replacement of either of these residues by amino acid substitutions disrupts and/or destabilizes the switch. Other nearby residues, Glu^{171} and Lys^{70}, function as ATP sensors. It is believed that Lys^{70} also serves as the nucleophile that initiates ATP hydrolysis through attack on the γ-phosphate of ATP. Arg^{151} acts as a relay between Pro^{143}, events occurring during ATP hydrolysis, and the peptide-binding domain (panels b–d of the figure). When ATP binding is sensed by Lys^{70} and Glu^{171} (panel b), Pro^{143} is shifted, which causes Arg^{151} to move in the direction of the peptide-binding domain of DnaK (panel c). In turn, this protein-binding domain assumes the open, low-affinity conformation. The interaction of DnaK with an unfolded protein substrate and co-chaperone DnaJ moves Arg^{151} back toward Pro^{143}, which causes Lys^{70} and Glu^{171} to initiate ATP hydrolysis (panel d). ADP now occupies the nucleotide-binding site of the ATPase domain, and the protein-binding domain of DnaK is locked in the closed, high-affinity conformation.

The consequence of these events is that DnaK cycles between binding and releasing unfolded (or partially folded) proteins, fueled by ATP hydrolysis within the ATPase domain. In effect, ATP binding and hydrolysis drive DnaK from an open, low-affinity conformational state to a closed, high-affinity conformational state. When unfolded (or partially folded) proteins are not held by DnaK, they have the opportunity to fold so that any solvent-accessible hydrophobic surfaces they might still retain are buried. Once a protein has adopted a stable folded state, it lacks exposed hydrophobic surfaces and thus escapes the cycle of binding and release by DnaK. More generally, Hsp70 provides an elegant example of protein conformational transitions based on the binding of ATP versus ADP at an effector site, with the added dimension that the effector site in this case is also an ATPase.

folding to proceed in a protected environment. This protected environment is sometimes referred to as an **"Anfinsen cage"** because it provides an enclosed space where proteins fold spontaneously, free from the possibility of aggregation with other proteins. Chaperonins are large, cylindrical protein complexes formed from two stacked rings of subunits. The chaperonins have been organized into two groups, I and II, on the basis of their source and structure. Group I chaperonins are found in bacteria, group II in archaea and eukaryotes. The group I chaperonin in *E. coli* is the **GroES–GroEL complex** (Figure 31.1c). GroEL is made of two stacked seven-membered rings of 60-kD subunits that form a cylindrical α_{14} oligomer 15 nm high and 14 nm wide (Figure 31.3). Each GroEL ring has a 5-nm central cavity where folding can take place. This cavity can accommodate proteins up to 60 kD in size. GroES, sometimes referred to as a **co-chaperonin,** consists of a single seven-membered ring of 10-kD subunits that sits like a dome on one end of GroEL (Figure 31.3). The end of GroEL where GroES is sitting is referred to as the *apical end.* Each GroEL subunit has two structural domains: an equatorial domain that binds ATP and interacts with neighbors in the other α_7 ring and an apical domain with hydrophobic residues that can interact with hydrophobic regions on partially folded proteins. The apical

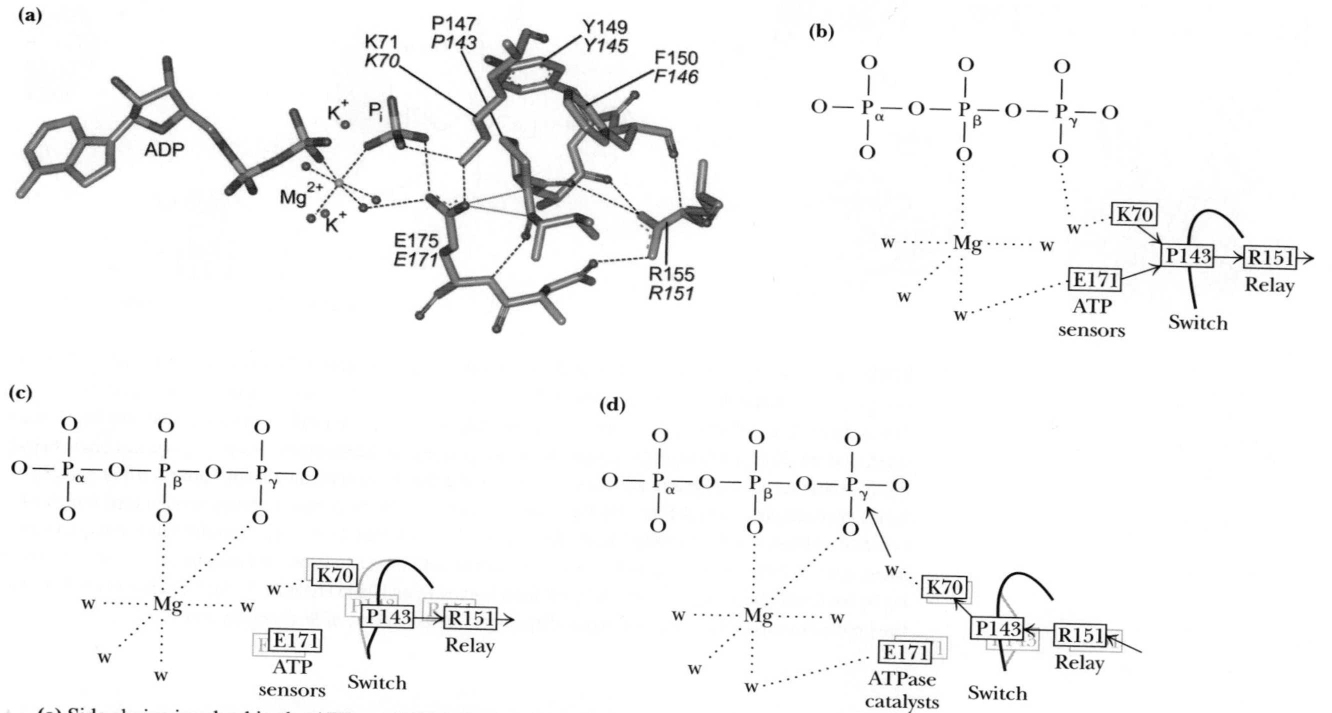

(a) Side chains involved in the ATP- and ADP-dependent allosteric regulation of the Hsp70 family member DnaK. Lower numbers indicate DnaK amino acid residues; upper numbers indicate the corresponding residues in Hsc70 (the human counterpart to DnaK). This illustration highlights residues in or near the ATP-binding site of the ATPase domain. The site is occupied by ADP, P_i, one Mg^{2+}, and two K^+ ions. See Figure 31.2a for the location of this site within the ATPase domain. Carbon atoms are gray; oxygen, red; nitrogen, blue; and phosphorus, yellow. P143 sits at the center of an H-bonded network of residues, which includes K70, Y145, F146, R151, and E171. **(b–d)** The mechanism of ATP- and ADP-dependent allosteric transitions in Hsp70. Events in the ATP-binding site of the ATPase domain are communicated to the protein-binding site of the peptide-binding domain through K70 and E171, which act as ATP sensors; P143, which acts as the switch; and R151, which is part of the lever that relays the events from the ATPase domain to the other domain. (a: Adapted from Figure 1 in Vogel, M., Bukau, B., and Mayer, M. P., 2006. Allosteric regulation of Hsp70 chaperones by a proline switch. *Molecular Cell* **21:**359–367. Courtesy of Bernd Bukau and Matthias Mayer. b–d: Adapted from Figure 6 in Vogel, M., Bukau, B., and Mayer, M. P., 2006. Allosteric regulation of Hsp70 chaperones by a proline switch. *Molecular Cell* **21:**359–367.)

domain hydrophobic patches face the interior of the central cavity. An unfolded (or partially folded) protein binds to the apical patches and is delivered to the central cavity of the upper α_7 ring (Figure 31.3c). ATP binding to the subunits of the upper α_7 ring causes rapid (<100 msec), forced unfolding of the substrate protein, followed by two events that occur on a slower time scale (~1 sec): (1) GroES is recruited to GroEL, and (2) the α-subunits undergo a conformational change that buries their hydrophobic patches. The α-subunits now present a hydrophilic surface to the central cavity. This change displaces the bound partially folded polypeptide into the sheltered hydrophilic environment of the central cavity, where it can fold, free from danger of aggregation with other proteins. GroES also promotes ATP hydrolysis (Figure 31.3c). The GroEL:ADP:GroES complex dissociates when ATP binds to the subunits of the other (lower) α_7 ring. Dissociation of GroES allows the partially folded (or folded) protein to escape from GroEL. If the protein has achieved its native conformation, its hydrophobic residues will be buried in its core and the hydrophobic patches on the α_7 rings will have no affinity for it. On the other hand, if the protein is only partially folded, it may be bound again, gaining access to the Anfinsen cage of the α_7 ring and another cycle of folding. The folding of rhodanese, a 33-kD

(a) Domain organization and structure of the Hsp70 family member, DnaK

(b) DnaK mechanism of action

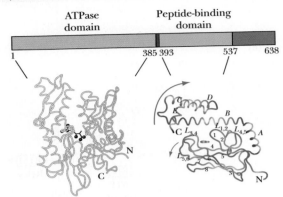

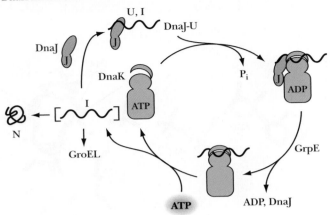

FIGURE 31.2 Structure and function of DnaK: **(a)** Domain organization and structure of the Hsp70 family member, DnaK. The ribbon diagram on the lower left is the ATP-binding domain of the DnaK analog, bovine Hsc70; bound ADP is shown as a stick diagram (purple). The ribbon diagram on the lower right is the polypeptide-binding domain of DnaK. The small blue ovals highlight the position of the polypeptide substrate; the protein regions that bind the polypeptide substrate are blue-green. **(b)** DnaK mechanism of action: DnaJ binds an unfolded protein (U) or partially folded intermediate (I) and delivers it to the DnaK:ATP complex. The nucleotide exchange protein GrpE replaces ADP with ATP on DnaK and the partially folded intermediate ([I]) is released. I has several possible fates: It may fold into the native state, N; it may undergo another cycle of interaction with DnaJ and DnaK; or it may become a substrate for folding by the GroEL chaperonin system. (Adapted from Figures 1a and 2a in Frydman, J., 2001. Folding of newly translated proteins in vivo: The role of molecular chaperones. *Annual Review of Biochemistry* **70**:603–647.)

protein, requires the hydrolysis of about 130 equivalents of ATP. The extreme crowding of macromolecules in the typical cell, which could interfere with the folding of nascent proteins, justifies the need for elaborate structures like GroES-GroEL, and also the substantial investment of ATP energy in the folding process (Figure 31.3d).

The group II chaperonin and eukaryotic analog of GroEL, **CCT** (also called **TriC**) is also a double-ring structure, but each ring consists of eight different subunits that vary in size from 50 to 60 kD. Furthermore, group II chaperonins lack a GroES counterpart. **Prefoldin** (also known as **GimC**), a hexameric protein composed of six subunits from two related classes (two α and four β), can serve as a co-chaperone for CCT, much as GroES does for GroEL. However, prefoldin also acts like an Hsp70 protein because it binds unfolded polypeptide chains emerging from ribosomes and delivers them to CCT. Prefoldin resembles a jellyfish, with six tentacle-like coiled coils extending from a barrel-shaped body. The ends of the tentacles have hydrophobic patches for binding unfolded proteins.

Prior to substrate protein binding, CCT exists in a partly open state. ATP binding opens the ring even more, a state in which prefoldin delivers the substrate protein. ATP hydrolysis closes the chamber and drives the folding process. ATP-induced conformation changes that promote protein folding propagate from one subunit to the next around the ring structure.

The Eukaryotic Hsp90 Chaperone System Acts on Proteins of Signal Transduction Pathways

Hsp90 constitutes 1% to 2% of the total cytosolic proteins of eukaryotes, its abundance reflecting its importance. Like other Hsp chaperones, its action depends on cyclic binding and hydrolysis of ATP. Conformational regulation of signal transduction molecules seems to be a major purpose of Hsp90. Receptor tyrosine kinases, soluble tyrosine kinases, and steroid hormone receptors are some of the signal transduction molecules (see Chapter 32) that must associate with Hsp90 in order to become fully competent; proteins fitting this description are called Hsp90 "client proteins." The maturation of Hsp70 client proteins requires other proteins as well, and together with Hsp90, these proteins come together to

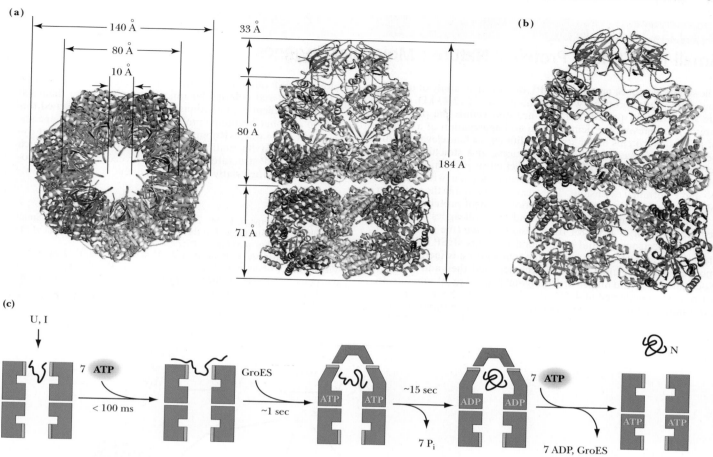

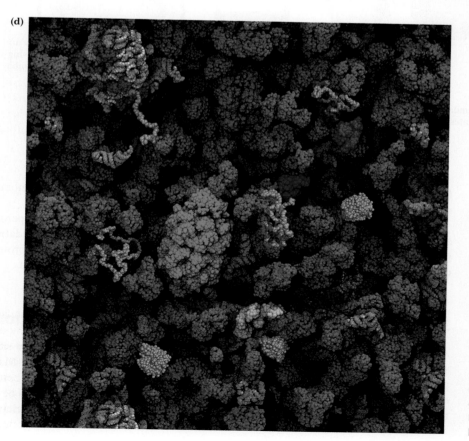

FIGURE 31.3 Structure and function of the GroES–GroEL complex. **(a)** Structure and overall dimensions of GroES–GroEL (top view, left; side view, right) (pdb id = 1AON). **(b)** Section through the center of the complex to reveal the central cavity. **(c)** Model of the GroEL cylinder (blue) in action. An unfolded (U) or partially folded (I) polypeptide binds to hydrophobic patches on the apical ring of α_7-subunits, followed by ATP binding, forced protein unfolding, and GroES (red) association. **(d)** Depiction of a protein folding reaction in the cytoplasm of an *E. coli* cell, emphasizing macromolecular crowding. The cytoplasmic components are present at their known concentrations. Proteins shown include a new protein undergoing folding (orange), chaperone structures (GroEL in green, DnaK in red, and Trigger Factor in yellow), ribosomes (purple), and RNA (salmon). Image courtesy of A. Elcock and L. Gierasch.

A DEEPER LOOK

Small Heat Shock Proteins: Nature's Molecular Sponges

The **small heat-shock proteins (sHSP)** are a diverse family of ubiquitous intracellular proteins, with subunit masses of 12–43 kD. sHSPs are important in protein maintenance and cellular responses to stress. They prevent the denaturation and aggregation of a variety of unrelated proteins under stress conditions, such as elevated temperature or reduced or oxidized conditions, at a stoichiometry as low as one subunit of sHSP to one target client protein. The small HSPs have a 100-residue core α-crystallin domain with variable N- and C-terminal extensions. These extensions enable the sHSPs to oligomerize and also mediate recognition of client proteins.

Carol Robinson, Justin Benesch, and their colleagues have studied the behavior of HSP18.1 from *Pisum sativum* (the garden pea) by mass spectrometry and have found that this sHSP exists as a dodecamer at 22°C. However, as the temperature is increased, dimers escape from some of the dodecamers and then add to other dodecamers to form higher-order oligomers of 13 to 20 subunits. In vitro assays conducted at 42°C (the classic "heat shock" temperature) show that monomers of a model client protein (firefly luciferase) add to sHSP oligomers to form heteromultimeric assemblages that reorganize and enlarge dynamically to yield more than 300 different sHSP:client combinations! Lila Gierasch has likened this process to a sponge becoming moist and swelling so that it accommodates yet more water. Robinson and Benesch conclude that dynamic equilibrium fluctuations of quaternary structure in such ensembles of sHSP structures underlie the protection of client proteins as they seek their native folded structures.

References

Eyles, S. J., and Gierasch, L. M., 2010. Nature's molecular sponges: Small heat shock proteins grow into their chaperone roles. *Proceedings of the National Academy of Sciences* **107**:2727–2728.

Stengel, F., Baldwin, A. J., Painter, A. J., Jaya, N., Bashu, E., Kay, L. E., Vierling, E., Robinson, C. V., and Benesch, J. L. P., 2010. Quarternary dynamics and plasticity underlie small heat shock protein chaperone function. *Proceedings of the National Academy of Sciences U.S.A.* **107**:2007–2012.

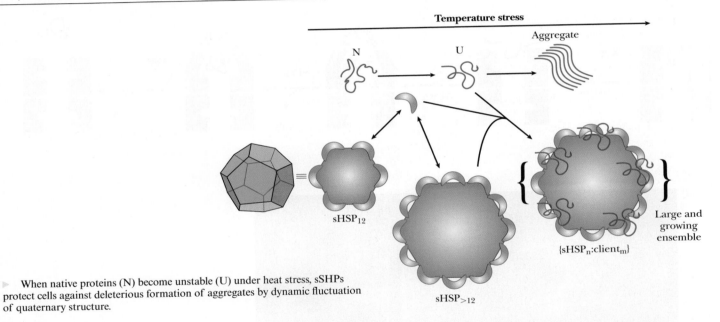

When native proteins (N) become unstable (U) under heat stress, sSHPs protect cells against deleterious formation of aggregates by dynamic fluctuation of quaternary structure.

form an assembly that has been called a **foldosome.** CFTR (cystic fibrosis transmembrane regulator), telomerase, and nitric oxide synthase are also Hsp90-dependent.

Association of nascent polypeptide chains with proteins of the various chaperone systems commits them to a folding pathway, redirecting them away from degradation pathways that would otherwise eliminate them from the cell. However, if these protein chains fail to fold, they are recognized as non-native and targeted for destruction.

31.2 | How Are Proteins Processed Following Translation?

Aside from these folding events, release of the completed polypeptide from the ribosome is not necessarily the final step in the covalent construction of a protein. Many proteins must undergo covalent alterations before they become functional. In the course of these **post-translational modifications,** the primary structure of a protein may be altered, and/or novel derivations may be introduced into its amino acid side chains.

Hundreds of different amino acid variations have been described in proteins, virtually all arising post-translationally. The list of such modifications is very large; some are rather commonplace, whereas others are peculiar to a single protein. The diphthamide moiety in elongation factor eEF-2 is one example of an amino acid modification (see the Human Biochemistry box on page 1077 in Chapter 30); the fluorescent group of green fluorescent protein (GFP; see Chapter 4) is another. In addition, common chemical groups such as carbohydrates and lipids may be covalently attached to a protein during its maturation. Phosphorylation, acetylation, and methylation of proteins are common mechanisms for regulating protein function. Interestingly, many proteins are modified in multiple ways, and many post-translational modifications act in combinations—a phenomenon termed **cross-talk.** (The majority of proteins in cells can be phosphorylated on one or more residues. A survey of some of the more prominent chemical groups conjugated to proteins is given in Chapter 5.) To put a number on the significance of post-translational modifications, we have seen that the number of human proteins is estimated to exceed the number of human genes (20,000 or so) by more than an order of magnitude.

Proteolytic Cleavage Is the Most Common Form of Post-Translational Processing

Proteolytic cleavage, as the most prevalent form of protein post-translational modification, merits special attention. The very occurrence of proteolysis as a processing mechanism seems strange: Why join a number of amino acids in sequence and then eliminate some of them? Three reasons can be cited. First, diversity can be introduced where none exists. For example, a simple form of proteolysis, enzymatic removal of N-terminal Met residues, occurs in many proteins. **Met-aminopeptidase,** by removing the invariant Met initiating all polypeptide chains, introduces diversity at N-termini. Second, proteolysis serves as an activation mechanism so that expression of the biological activity of a protein can be delayed until appropriate. A number of metabolically active proteins, including digestive enzymes and hormones, are synthesized as larger inactive precursors termed **pro-proteins** that are activated through proteolysis (see **zymogens,** Chapter 15). The N-terminal pro-sequence on such proteins may act as an intramolecular chaperone to ensure correct folding of the active site. Third, proteolysis is involved in the targeting of proteins to their proper destinations in the cell, a process known as **protein translocation.**

31.3 : How Do Proteins Find Their Proper Place in the Cell?

Proteins are targeted to their proper cellular locations by **signal sequences:** Proteins destined for service in membranous organelles or for export from the cell are synthesized in precursor form carrying an N-terminal stretch of amino acid residues, or **leader peptide,** that serves as a *signal sequence.* In effect, signal sequences serve as "zip codes" for sorting and dispatching proteins to their proper compartments. Thus, the information specifying the correct cellular localization of a protein is found within its structural gene. Once the protein is routed to its destination, the signal sequence is often, but not always, proteolytically clipped from the protein by a signal sequence-specific endopeptidase called a **signal peptidase.**

Proteins Are Delivered to the Proper Cellular Compartment by Translocation

Protein translocation is the name given to the process whereby proteins are inserted into membranes or delivered across membranes. Protein translocation occurs in all cells. Newly synthesized chains of membrane proteins or secretory proteins are targeted to the plasma membrane (in prokaryotes) or the endoplasmic reticulum (in eukaryotes) by their signal sequences. In addition to the ER, a number of eukaryotic membrane systems are competent in protein translocation, including the membranes of the nucleus, mitochondria, chloroplasts, and peroxisomes. Several common features characterize protein translocation systems:

1. Proteins to be translocated are made as preproteins containing contiguous blocks of amino acid sequence that act as organelle-specific sorting signals.

2. **Signal recognition particles (SRPs)** recognize the presence of a nascent protein chain in the ribosomal exit tunnel and, together with **signal receptors (SRs),** deliver the nascent chain to the membrane. If the nascent sequence emerging from the ribosome is a signal sequence, it is delivered to a specific membrane protein complex, the *translocon,* that mediates protein integration into the membrane or protein translocation across the membrane.

3. **Translocons** are selectively permeable protein-conducting channels that catalyze movement of the proteins across the membrane, and metabolic energy in the form of ATP, GTP, or a membrane potential is essential. In eukaryotes, ATP-dependent chaperone proteins within the membrane compartment usually associate with the entering polypeptide and provide the energy for translocation. Proteins destined for membrane integration contain amino acid sequences that act as **stop-transfer signals,** allowing diffusion of transmembrane segments into the bilayer.

4. Preproteins are maintained in a loosely folded, translocation-competent conformation through interaction with molecular chaperones.

Prokaryotic Proteins Destined for Translocation Are Synthesized as Preproteins

Gram-negative bacteria typically have four compartments: cytoplasm, plasma (or inner) membrane, periplasmic space (or periplasm), and outer membrane. Most proteins destined for any location other than the cytoplasm are synthesized with amino-terminal leader sequences 16 to 26 amino acid residues long. These leader sequences, or *signal sequences,* consist of a basic N-terminal region, a central domain of 7 to 13 hydrophobic residues, and a nonhelical C-terminal region (Figure 31.4). The conserved features of the last part of the leader, the C-terminal region, include a helix-breaking Gly or Pro residue and amino acids with small side chains located one and three residues before the proteolytic cleavage site. Unlike the basic N-terminal and nonpolar central regions, the C-terminal features are not essential for translocation but instead serve as recognition signals for the **leader peptidase,** which removes the leader sequence. The exact amino acid sequence of the leader peptide is unimportant. Nonpolar residues in the center and a few Lys residues at the amino terminus are sufficient for successful translocation. The functions of leader peptides are to retard the folding of the preprotein so that molecular chaperones have a chance to interact with it and to provide recognition signals for the translocation machinery and leader peptidase.

Eukaryotic Proteins Are Routed to Their Proper Destinations by Protein Sorting and Translocation

Eukaryotic cells are characterized by many membrane-bounded compartments. In general, signal sequences targeting proteins to their appropriate compartments are located at the N-terminus as *cleavable presequences,* although many proteins have N-terminal localization signals that are not cleaved and others have internal targeting sequences

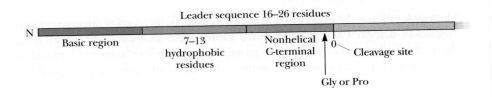

FIGURE 31.4 General features of the N-terminal signal sequences on *E. coli* proteins destined for translocation: a basic N-terminal region, a central apolar domain, and a nonhelical C-terminal region.

that may or may not be cleaved. Proteolytic removal of the leader sequences is also catalyzed by specialized proteases, but removal is not essential to translocation. No sequence similarity is found among the targeting signals for each compartment. Thus, the targeting information resides in more generalized features of the leader sequences such as charge distribution, relative polarity, and secondary structure. For example, proteins destined for secretion enter the lumen of the endoplasmic reticulum (ER) and reach the plasma membrane via a series of vesicles that traverse the endomembrane system. Recognition by the ER depends on an N-terminal amino acid sequence that contains one or more basic amino acids followed by a run of 6 to 12 hydrophobic amino acids. An example is serum albumin, which is synthesized in precursor form (**preproalbumin**) having a M*K*WVTF**LLLLLFISGSAFSR** N-terminal signal sequence. The italicized K highlights the basic residue in the sequence, and the bold residues denote a continuous stretch of (mostly) hydrophobic residues. A signal peptidase in the ER removes the signal sequence by cleaving the preproprotein between the S and R.

The Synthesis of Secretory Proteins and Many Membrane Proteins Is Coupled to Translocation Across the ER Membrane

The signals recognized by the ER translocation system are virtually indistinguishable from bacterial signal sequences; indeed, the two are interchangeable in vitro. In addition, the translocon systems in prokaryotes and eukaryotes are highly analogous. In higher eukaryotes, translation and translocation of many proteins destined for processing via the ER are tightly coupled. Translocation across the ER occurs co-translationally (that is, as the protein is being translated on the ribosome). As the N-terminal sequence of a protein undergoing synthesis enters the exit tunnel of the ribosome, it is detected by a **signal recognition particle (SRP;** Figure 31.5). SRP is a 325-kD nucleoprotein assembly that contains six polypeptides and a 300-nucleotide **7S RNA. SRP54,** a 54-kD subunit of SRP and a G-protein family member, recognizes the nascent protein's signal sequence, and SRP binding of the signal sequence causes the ribosome to cease translation. This arrest prevents release of the growing protein into the cytosol before it reaches the ER and its intended translocation. The SRP–ribosome complex is referred to as the **RNC–SRP (ribosome nascent chain∶SRP complex).**

Interaction Between the RNC–SRP and the SR Delivers the RNC to the Membrane

The RNC–SRP is then directed to the cytosolic face of the ER, where it binds to the signal receptor (SR), an $\alpha\beta$ heterodimeric protein. The 70-kD α-subunit is anchored to the membrane by the transmembrane β-subunit; both subunits are G-protein family members, and both have bound GTP. When SRP54 docks with SRα, the RNC–SRP becomes membrane associated (Figure 31.5). If the nascent chain emerging from the

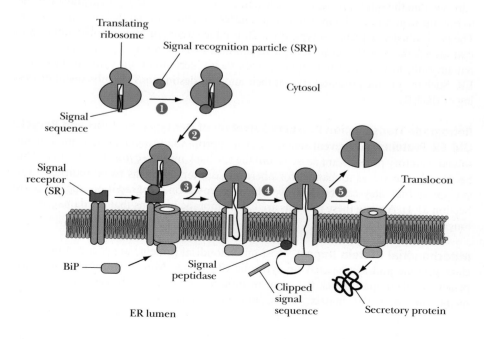

FIGURE 31.5 Synthesis of a eukaryotic secretory protein and its translocation into the endoplasmic reticulum. **(1)** The signal recognition particle (SRP, *red*) recognizes the signal sequence emerging from a translating ribosome (ribosome nascent complex [RNC], *gray*). **(2)** The RNC-SRP interacts with the signal receptor (SR, *purple*) and is transferred to the translocon *(pink)*. **(3)** Release of the SRP and alignment of the peptide exit tunnel of the RNC with the protein-conducting channel of the translocon stimulates the ribosome to resume translation. **(4)** The membrane-associated signal peptidase *(purple circle)* clips off the N-terminal signal sequence, and BiP (the ER lumen Hsp70 chaperone, *blue*) binds the nascent chain mediating its folding into its native conformation. **(5)** Following dissociation of the ribosome, BiP plugs the translocon channel. Not shown are subsequent secretory protein maturation events, such as glycosylation. (Adapted from Figures 1a and 2a in Frydman, J., 2001. Folding of newly translated proteins in vivo: The role of molecular chaperones. *Annual Review of Biochemistry* **70**:603–647.)

ribosome is not a signal sequence, the RNC is released from the SRP and the membrane. If the nascent chain emerging from the ribosome is a signal sequence, the complex remains intact, and SRP54 and SRα function together as reciprocal GTPase-activating proteins. GTP hydrolysis causes the dissociation of SRP from SR and transfer of the RNC to the translocon.

The Ribosome and the Translocon Form a Common Conduit for Transfer of the Nascent Protein Through the ER Membrane and into the Lumen Through interactions with the translocon, the ribosome resumes protein synthesis, delivering its growing polypeptide through the ER membrane. The peptide exit tunnel of the large ribosomal subunit and the protein-conducting channel of the translocon are aligned with one another, forming a continuous conduit from the peptidyl transferase center of the ribosome to the ER lumen.

The mammalian translocon is a complex, multifunctional entity that has as its core the **Sec61 complex,** a heterotrimeric complex of membrane proteins, and a unique fourth subunit, **TRAM,** that is required for insertion of nascent integral membrane proteins into the membrane. The 53-kD α-subunit of Sec61p has ten membrane-spanning segments, whereas the β- and γ-subunits are single TMS proteins. Sec61α forms the transmembrane protein-conducting channel through which the nascent polypeptide is transported into the ER lumen (Figure 31.5). The pore size of Sec61p is very dynamic, ranging from about 0.6 to 6 nm in diameter. Thus, a great variety of protein structures could be accommodated easily within the translocon. This flexibility allows the Sec61p translocon complex to function in post-translational translocation (translocation of completely formed proteins) as well as co-translational translocation.

As the protein is threaded through the Sec61p channel into the lumen, an Hsp70 chaperone family member called **BiP** binds to it and mediates proper folding. BiP function, like that of other Hsp70 proteins, is ATP-dependent, and ATP-dependent protein folding provides the driving force for translocation of the polypeptide into the lumen. When the ribosome dissociates from the translocon, BiP serves as a plug to block the protein-conducting channel, preventing ions and other substances from moving between the ER lumen and the cytosol.

A Signal Peptidase Within the ER Lumen Clips Off the Signal Peptide Soon after it enters the ER lumen, the signal peptide is clipped off by membrane-bound **signal peptidase** (also called *leader peptidase*), which is a complex of five proteins. Other modifying enzymes within the lumen introduce additional post-translational alterations into the polypeptide, such as glycosylation with specific carbohydrate residues. ER-processed proteins destined for secretion from the cell or inclusion in vesicles such as lysosomes end up contained within the soluble phase of the ER lumen. On the other hand, polypeptides destined to become membrane proteins carry **stop-transfer** sequences within their mature domains. The stop-transfer sequence is typically a 20-residue stretch of hydrophobic amino acids that arrests the passage across the ER membrane. Proteins with stop-transfer sequences remain embedded in the ER membrane with their C-termini on the cytosolic face of the ER. Such membrane proteins arrive at their intended destinations via subsequent processing of the ER.

Retrograde Translocation Prevents Secretion of Damaged Proteins and Recycles Old ER Proteins To prevent secretion of inappropriate proteins, fragmented or misfolded secretory proteins are passed from the ER back into the cytosol via Sec61p. Thus, Sec61p also serves as a channel for aberrant secretory proteins to be returned to the cytosol so that they can be destroyed by the proteasome degradation apparatus (see Section 31.4). Among these proteins are ER membrane proteins that are damaged or no longer needed.

Mitochondrial Protein Import Most mitochondrial proteins are encoded by the nuclear genome and synthesized on cytosolic ribosomes. Mitochondria consist of four principal subcompartments: the outer membrane, the intermembrane space, the inner membrane, and the matrix. Thus, not only must mitochondrial proteins find mitochondria, they must gain access to the proper subcompartment; and once there, they must attain a functionally active conformation. As a consequence, mitochondria possess multiple preprotein translocons and chaperones. Similar considerations apply to protein import to chloroplasts, organelles with five principal subcompartments (outer membrane, intermembrane space, inner/thylakoid membrane, stroma, and thylakoid lumen; see Chapter 21).

Signal sequences on nuclear-encoded proteins destined for the mitochondria are N-terminal cleavable presequences 10 to 70 residues long. These mitochondrial presequences lack contiguous hydrophobic regions. Instead, they have positively charged and hydroxy amino acid residues spread along their entire length. These sequences form **amphipathic α-helices** (Figure 31.6) with basic residues on one side of the helix and uncharged and hydrophobic residues on the other; that is, mitochondrial presequences are positively charged amphipathic sequences. In general, mitochondrial targeting sequences share no sequence homology. Once synthesized, mitochondrial preproteins are retained in an unfolded state with their target sequences exposed, through association with Hsp70 molecular chaperones. Import involves binding of a preprotein to the **mitochondrial outer membrane translocon (TOM)** (Figure 31.7). If the protein is destined to be an outer mitochondrial membrane protein, it is transferred from the TOM to the **sorting and assembly complex (SAM)** and inserted in the outer membrane. If it is an integral protein of the inner mitochondrial membrane, it traverses the TOM complex, enters the intermembrane space, and is taken up by the **inner mitochondrial membrane translocon (TIM22)** and inserted into the inner membrane. On the other hand, if it is destined to be a mitochondrial matrix protein, a different TIM complex, **TIM23,** binds the preprotein and threads it across the inner mitochondrial membrane into the matrix. Chloroplasts have **TOCs** (translocon outer chloroplast membrane) and **TICs** (translocon inner chloroplast membrane) for these purposes.

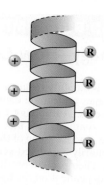

FIGURE 31.6 Structure of an amphipathic α-helix having basic (+) residues on one side and uncharged and hydrophobic (R) residues on the other.

31.4 How Does Protein Degradation Regulate Cellular Levels of Specific Proteins?

Cellular proteins are in a dynamic state of turnover, with the relative rates of protein synthesis and protein degradation ultimately determining the amount of protein present at any point in time. In many instances, transcriptional regulation determines the concentrations of specific proteins expressed within cells, with protein degradation playing a minor role. In other instances, the amounts of key enzymes and regulatory proteins, such as cyclins and transcription factors, are controlled via selective protein degradation. In addition, abnormal proteins arising from biosynthetic errors or post-

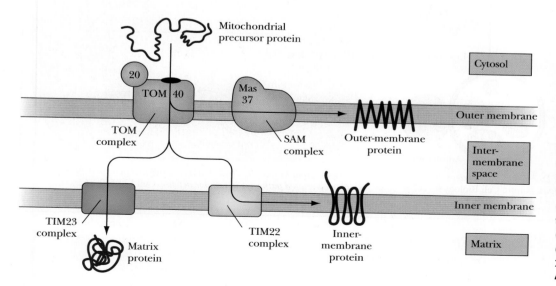

FIGURE 31.7 Translocation of mitochondrial preproteins involves distinct translocons. All mitochondrial proteins must interact with the outer mitochondrial membrane (TOM). From there, depending on their destiny, they are (1) passed to the SAM complex if they are integral proteins of the outer mitochondrial membrane or (2) traverse the TOM and enter the intermembrane space, where they are taken up by either TIM22 or TIM23, depending on whether they are integral membrane proteins of the inner mitochondrial membrane (TIM22) or mitochondrial matrix proteins (TIM23). (Adapted from Figure 1 in Mihara, K., 2003. Moving inside membranes. *Nature* **424:**505–506.)

HUMAN BIOCHEMISTRY

Autophagy—How Cells Recycle Their Materials

Autophagy is a tightly regulated catabolic process whereby cells use their lysosomes to degrade their own macromolecular and organellar constituents. Autophagy functions to maintain a balance between the synthesis of macromolecules and organelles, their degradation, and recycling of their component materials. More than 30 evolutionarily conserved genes, the **ATG genes**, are involved in autophagy. Their protein products are designated **Atg proteins.** Autophagy is dependent on the hydrolytic capacities of acidic hydrolyases in lysosomes to achieve its ends. Autophagy begins when a volume of cytoplasm or an organelle (such as a mitochondrion) becomes separated from the rest of the cell by a cup-shaped membranous structure called the **phagophore** (Figure a). The phagophore derives from a **pre-autophagosomal structure (PAS),** a dynamic assembly of Atg proteins with associated lipids that functions as a membrane-forming apparatus. Closure of the phagophore around this sequestered material gives rise to the double-membrane-bounded **autophagosome.** Fusion of the autophagosome membrane with the membrane of a lysosome creates a hybrid vesicle, the **autolysosome** (sometimes called an **autophagolysosome**). Within the autolysosome, the contents delivered by the autophagosome are broken down so that their constituent amino acids, nucleotides, and lipids can be recycled to meet the cell's need for building blocks to create new macromolecules and organelles.

Three forms of autophagy are recognized: **macroautophagy** (just described), **microautophagy** (wherein small blebs of cytoplasm are engulfed directly by a lysosome), and **chaperone-mediated autoph-agy (CMA),** in which proteins targeted by Hsc (heat shock cognate)-containing chaperone/cochaperone complexes are taken up directly by lysosomes. The focus here is on **macroautophagy,** the principal form.

Although autophagy is an ongoing process of homeostasis in normal cells, it becomes crucial when extracellular nutrients become limiting (Figure b). Recycling cellular components through autophagy allows the cell to maintain energy production and macromolecular synthesis despite a paucity of nutrients. As noted in Chapter 27, when mTORC1 (a protein complex that serves as a nutrient and energy sensor) is active, it promotes protein synthesis, cell growth, and cell proliferation. mTORC1 is a negative regulator of autophagy, acting to limit its expression when nutrients are in abundance. When mTORC1 is inactivated upon nutrient depletion (starvation), a particular Atg protein, Atg13, is dephosphorylated, a triggering event that culminates in activation of autophagosome formation.

Autophagy as a source of cellular nutrients can be only a short-term solution. Interestingly, starvation-induced autophagy in cultured mammalian AgRP neurons derived from hypothalamus causes an intracellular increase in free fatty acids (FFA). In turn, these FFA promote a rise in AgRP levels. Expression of AgRP along with NPY is known to promote eating behavior (see Figure 27.12). Thus, upon starvation, autophagy provides a short-term fix and a long-term solution to nutrient limitation. Autophagy is also implicated in immunity, inflammation, and cellular defenses against microbial infection.

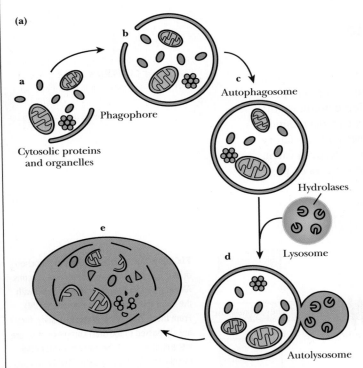

(a)

b

a

Cytosolic proteins and organelles

Phagophore

c

Autophagosome

Hydrolases

Lysosome

e

d

Autolysosome

▲ *Adapted from Xie, Z., and Klionsky, D. J., 2007. Autophagosome formation: Core machinery and adaptations.* Nature Cell Biology **9:**1102–1109.

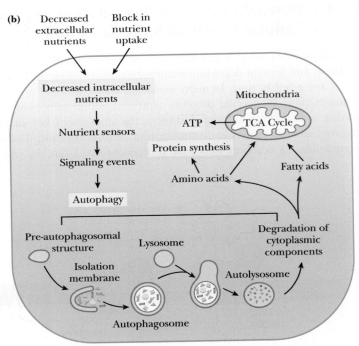

(b)

Decreased extracellular nutrients

Block in nutrient uptake

Decreased intracellular nutrients

Mitochondria

Nutrient sensors

ATP

TCA Cycle

Protein synthesis

Signaling events

Fatty acids

Amino acids

Autophagy

Pre-autophagosomal structure

Lysosome

Degradation of cytoplasmic components

Isolation membrane

Autolysosome

Autophagosome

▲ *Adapted from Levine, B., 2005. Eating oneself and uninvited guests: Autophagey-related pathways in cellular defense.* Cell **120:**159–162.

synthetic damage must be destroyed to prevent the deleterious consequences of their buildup. The elimination of proteins typically follows first-order kinetics, with half-lives ($t_{1/2}$) of different proteins ranging from several minutes to many days. A single, random proteolytic break introduced into the polypeptide backbone of a protein is believed sufficient to trigger its rapid disappearance because no partially degraded proteins are normally observed in cells.

Protein degradation poses a real hazard to cellular processes. To control this hazard, protein degradation is compartmentalized, either in macromolecular structures known as **proteasomes** or in degradative organelles such as lysosomes. Protein degradation within lysosomes is largely nonselective; selection occurs during lysosomal uptake. Proteasomes are found in eukaryotic as well as prokaryotic cells. The proteasome is a functionally and structurally sophisticated counterpart to the ribosome. Regulation of protein levels via degradation is an essential cellular mechanism. Regulation by degradation is both rapid and irreversible.

Eukaryotic Proteins Are Targeted for Proteasome Destruction by the Ubiquitin Pathway

Ubiquitination is the most common mechanism to label a protein for proteasome degradation in eukaryotes. **Ubiquitin** is a highly conserved, 76-residue (8.5-kD) polypeptide widespread in eukaryotes. Proteins are condemned to degradation through ligation to ubiquitin. Three proteins in addition to ubiquitin are involved in the ligation process: **E1, E2,** and **E3** (Figure 31.8). E1 is the **ubiquitin-activating enzyme** (105-kD dimer). It becomes attached via a thioester bond to the C-terminal Gly residue of ubiquitin through ATP-driven formation of an activated ubiquitin-adenylate intermediate. Ubiquitin is then transferred from E1 to an SH group on E2, the **ubiquitin-carrier protein.** (E2 is actually a family of at least seven different small proteins, several of which are heat shock proteins; there is also a variety of E3 proteins.) In protein degradation, E2-S ~ ubiquitin transfers ubiquitin to free amino groups on proteins selected by E3 (180 kD), the **ubiquitin-protein ligase.** Upon binding a protein substrate, E3 catalyzes the transfer of ubiquitin from E2-S ~ ubiquitin to free amino groups (usually Lys ϵ-NH$_2$)

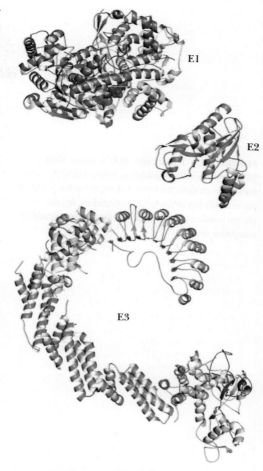

(Top) A ubiquitin : E1 heterodimer complex (pdb id = 1R4N; ubiquitin is shown in blue). **(Middle)** A ubiquitin : E2 complex (pdb id = 1FXT; E2 is shown in orange, ubiquitin in blue). **(Bottom)** The clamp-shaped E3 heteromultimer (pdb id = 1LDK and 1FQV). The target protein is bound between the jaws of the clamp.

E1 : Ubiquitin-activating enzyme

① Ubiquitin—C(=O)—O⁻ (C-term. Gly) + **ATP** → **E1** → **P P** + Ubiquitin—C(=O)—AMP Ubiquitinyl-acyladenylate

② Ubiquitin—C(=O)—AMP + **E1**—SH → AMP + **E1**—S—C(=O)—Ubiquitin (Thioester)

E2 : Ubiquitin-carrier protein

E1—S—C(=O)—Ubiquitin + **E2**—SH → **E1**—SH + **E2**—S—C(=O)—Ubiquitin

E3 : Ligase

E3 + Protein (substrate) → **E3** : Protein

E3 : Protein + **E2**—S—C(=O)—Ubiquitin → **E2**—SH + **E3** + Protein—Ubiquitin

FIGURE 31.8 Enzymatic reactions in the ligation of ubiquitin to proteins. Ubiquitin is attached to selected proteins via isopeptide bonds formed between the ubiquitin carboxy-terminus and free amino groups (α-NH$_2$ terminus, Lys ϵ-NH$_2$ side chains) on the protein.

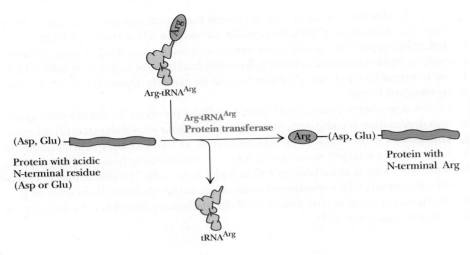

FIGURE 31.9 Proteins with acidic N-termini show a tRNA requirement for degradation. Arginyl-tRNA^{Arg}: protein transferase catalyzes the transfer of Arg to the free α-NH$_2$ of proteins with Asp or Glu N-terminal residues. Arg-tRNA^{Arg}: protein transferase serves as part of the protein degradation recognition system.

on the protein. More than one ubiquitin may be attached to a protein substrate, and tandemly linked chains of ubiquitin also occur via *isopeptide bonds* between the C-terminal glycine residue of one ubiquitin and the ϵ-amino of Lys residues in another. Ubiquitin has seven lysine residues, at positions 6, 11, 27, 29, 33, 48, and 63. Only isopeptide linkages to K^{11}, K^{29}, K^{48}, and K^{63} have been found, with the K^{48}-type being most common as a degradation signal.

E3 plays a central role in recognizing and selecting proteins for degradation. E3 selects proteins by the nature of the N-terminal amino acid. Proteins must have a free α-amino terminus to be susceptible. Proteins having either Met, Ser, Ala, Thr, Val, Gly, or Cys at the amino terminus are resistant to the ubiquitin-mediated degradation pathway. However, proteins having Arg, Lys, His, Phe, Tyr, Trp, Leu, Asn, Gln, Asp, or Glu N-termini have half-lives of only 2 to 30 minutes.

Interestingly, proteins with acidic N-termini (Asp or Glu) show a tRNA requirement for degradation (Figure 31.9). Transfer of Arg from Arg-tRNA to the N-terminus of these proteins alters their N-terminus from acidic to basic, rendering the protein susceptible to E3. It is also interesting that Met is less likely to be cleaved from the N-terminus if the next amino acid in the chain is one particularly susceptible to ubiquitin-mediated degradation.

Most proteins with susceptible N-terminal residues are *not* normal intracellular proteins but tend to be secreted proteins in which the susceptible residue has been exposed by action of a signal peptidase. Perhaps part of the function of the N-terminal recognition system is to recognize and remove from the cytosol any invading "foreign" or secreted proteins.

Other proteins targeted for ubiquitin ligation and proteasome degradation contain **PEST sequences**—short, highly conserved sequence elements rich in proline (P), glutamate (E), serine (S), and threonine (T) residues.

Proteins Targeted for Destruction Are Degraded by Proteasomes

Proteasomes are large oligomeric structures enclosing a central cavity where proteolysis takes place. The 20S proteasome from the archaeon *Thermoplasma acidophilum* is a 700-kD barrel-shaped structure composed of two different kinds of polypeptide chains, α and β, arranged to form four stacked rings of $\alpha_7\beta_7\beta_7\alpha_7$-subunit organization. The barrel is about 15 nm in height and 11 nm in diameter, and it contains a three-part central cavity (Figure 31.10a). The proteolytic sites of the 20S proteasome are found within this cavity. Access to the cavity is controlled through a 1.3-nm opening formed by the outer α_7 rings. These rings are believed to unfold proteins destined for degradation and transport them into the central cavity. The β-subunits possess the proteolytic activity. Proteolysis occurs when the β-subunit N-terminal threonine side-chain O atom makes

(a)

(b)

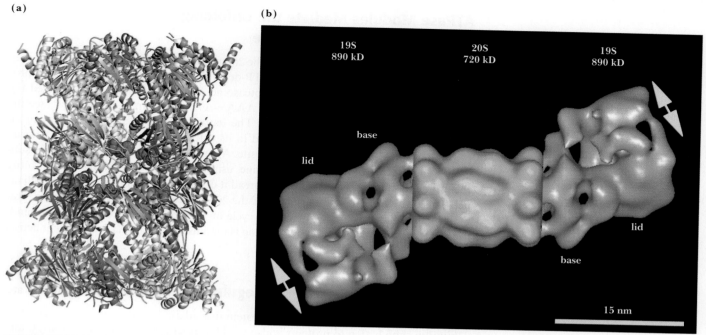

FIGURE 31.10 The structure of the 26S proteasome. **(a)** The yeast *(Saccharomyces cerevisiae)* 20S proteasome core with bortezomib bound (red) (pdb id = 2F16). Bortezomib is the first therapeutic proteasome inhibitor used in humans. It is approved in the United States for treatment of relapsed multiple myeloma and mantle cell lymphoma. **(b)** Composite model of the 26S proteasome. The 20S proteasome core is shown in yellow; the 19S regulator (19S cap) structures are in blue. (Adapted from Figure 5 in Voges, D., Zwickl, P., and Baumeister, W., 1999. The 26S proteasome: A molecular machine designed for controlled proteolysis. *Annual Review of Biochemistry* **68:**1015–1068.)

nucleophilic attack on the carbonyl-C of a peptide bond in the target protein. The products of proteasome degradation are oligopeptides seven to nine residues long.

Eukaryotic cells contain two forms of proteasomes: the **20S proteasome,** and its larger counterpart, the **26S proteasome.** The eukaryotic 26S proteasome is a 45-nm-long structure composed of a 20S proteasome plus two additional substructures known as **19S regulators** (also called **19S caps** or **PA700** [for **p**roteasome **a**ctivator-**700** kD]) (Figure 31.10b). Overall, the 26S proteasome (approximately 2.5 megadaltons) has 2 copies each of 32 to 34 distinct subunits, 14 in the 20S core and 18 to 20 in the cap structures. Unlike the archaeal 20S proteasome, the eukaryotic 20S core structure contains seven different kinds of α-subunits and seven different kinds of β-subunits. Interestingly, only three of the seven different β-subunits have protease active sites. The 26S proteasome forms when the 19S regulators dock to the two outer α_7 rings of the 20S proteasome cylinder. Many of the 19S regulator subunits have ATPase activity. Replacement of certain 19S regulator subunits with others changes the specificity of the proteasome. The 19S regulators cause the proteolytic function of the 20S proteasome to become ATP-dependent and specific for ubiquitinylated proteins as substrates. That is, these 19S caps act as regulatory complexes for the recognition and selection of ubiquitinylated proteins for degradation by the 20S proteasome core (Figure 31.11). The 26S proteasome shows a preference for proteins having four or more ubiquitin molecules attached to them. The 19S regulators also carry out the unfolding and transport of ubiquitinylated protein substrates into the proteolytic central cavity.

The 19S regulators consist of two parts: the base and the lid. The base subcomplex connects to the 20S proteasome and contains the six ATPase subunits that unfold proteasome substrates. These subunits are members of the **AAA+ family of ATPases** (**A**TPases **a**ssociated with various cellular **a**ctivities); AAA+-ATPases are an evolutionarily ancient family of proteins involved in a variety of cellular functions requiring energy-dependent unfolding, disassembly, and remodeling of proteins. The lid subcomplex acts as a cap on the base subcomplex and one of its subunits functions in recognition and ubiquitin-chain processing of proteasome protein substrates.

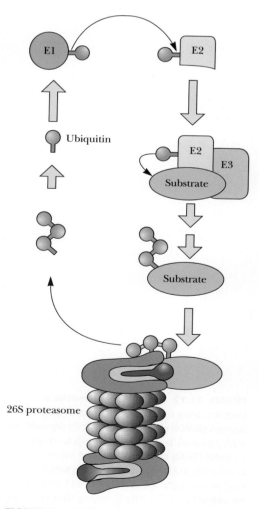

FIGURE 31.11 Diagram of the ubiquitin-proteasome degradation pathway. Pink "lollipop" structures symbolize ubiquitin molecules. (Adapted from Figure 1 in Hilt, W., and Wolf, D. H., 1996. Proteasomes: Destruction as a program. *Trends in Biochemical Sciences* **21:**96–102.)

ATPase Modules Mediate the Unfolding of Proteins in the Proteasome

The base of the 19S regulators that cap the 26S proteasome consists of a hexameric ring of AAA+-ATPases that mediate the ATP-dependent unfolding of ubiquitinated proteins targeted for destruction in the proteasome. Structural studies have revealed the presence of loops extending from these AAA+-ATPase subunits; these loops face the central channel of the hexameric ring. The loops are in an "up" position when an AAA+-ATPase subunit has ATP bound in its active site, but they shift to a "down" position when ADP occupies the active site. Apparently, these loops bind protein substrates and act like the levers of a machine, using the energy of ATP hydrolysis to tug the protein into an unfolded state and thread it down through the narrow central channel. This channel leads into the cavity of the 20S proteasome, where proteolytic degradation takes place. The AAA+-ATPase cycle of ATP binding, protein substrate binding, ATP hydrolysis, and protein unfolding is reminiscent of the ATP-dependent action of Hsp70 chaperones.

Ubiquitination Is a General Regulatory Protein Modification

Protein ubiquitination is a signal for protein degradation, as described on page 1101. Ubiquitin conjugation to proteins is also used for other purposes in cells. Nondegradative functions for ubiquitination include roles in chromatin remodeling, DNA repair, transcription, signal transduction, endocytosis, spliceosome assembly, and sorting of proteins to specific organelles and cell structures. In addition, cells possess a variety of protein modifiers attached to target proteins by processes similar to the ubiquitin pathway, as described in the following section.

Small Ubiquitin-Like Protein Modifiers Are Post-transcriptional Regulators

Small ubiquitin-like protein modifiers (SUMOs) are a highly conserved family of proteins found in all eukaryotic cells. Like ubiquitin, SUMO family members are covalently ligated to lysine residues in target proteins by a three-enzyme conjugating system (Figure 31.12). SUMO proteins share only limited homology to ubiquitin, and sumoylated proteins are not targeted for destruction. Instead, sumoylation alters the ability of the modified protein to interact with other proteins. This ability to change protein–protein interactions is believed to be the biological purpose of SUMO proteins.

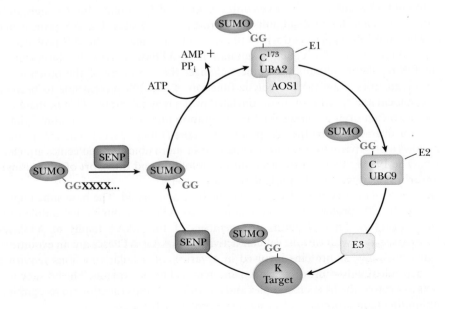

FIGURE 31.12 The mechanism of reversible sumoylation. Before conjugation, small ubiquitin-like protein modifiers (SUMOs) need to be proteolytically processed, removing anywhere from 2 to 11 amino acids to reveal the C-terminal Gly–Gly motif. SUMOs are then activated by the E1 enzyme in an ATP-dependent reaction to form a thioester bond between SUMO and E1. SUMO is then transferred to the catalytic Cys residue of the E2 enzyme, Ubc9. Finally, an "isopeptide bond" is formed, between the C-terminal Gly of SUMO and a Lys residue on the substrate protein, through the action of an E3 enzyme. The SUMO-specific protease SENP can deconjugate SUMO from target proteins.

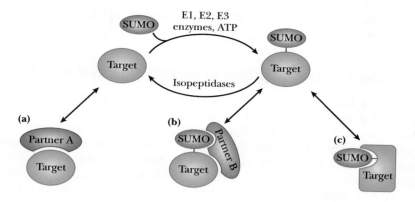

FIGURE 31.13 The molecular consequences of sumoylation. The process can affect a modified protein in three ways: **(a)** Sumoylation can interfere with the interaction between a target protein and its binding partner. **(b)** Sumoylation can provide a new binding site for an interacting partner. **(c)** Sumoylation can induce a conformational change in the modified target protein.

Sumoylation can have three general consequences for modified proteins (Figure 31.13): (1) sumoylation can interfere with the interactions between the target and its partner so that the interaction can occur only in the absence of sumoylation; (2) sumoylation can create a binding site for an interacting partner protein; and (3) sumoylation can induce a conformational change in the modified target protein, altering its interactions with partner proteins. The regulatory opportunities associated with sumoylation are significant for many cellular functions, including transcriptional regulation, chromosome organization, nuclear transport, and signal transduction (see Chapter 32).

Many E2 enzymes participate in ubiquitination processes, but the only known SUMO E2 enzyme is Ubc9 (Figure 31.14a). Ubc9 recognizes a ψKXD/E consensus sequence in proteins destined for sumoylation. In this sequence, ψ is an aliphatic branched amino acid (such as Leu), K is the lysine to which SUMO is conjugated, and X is any amino acid, with an acidic D or E completing the sequence. Recognition by Ubc9 is possible only if the consensus sequence is in a relatively unstructured part of a target protein or if it is part of an extended loop, as in RanGAP1 (Figure 31.14a). In the Ubc9–RanGAP1 complex, Leu^{525} of RanGAP1 is in van der Waals contact with several non-polar Ubc9 residues, whereas RanGAP1 Lys^{526} lies in a hydrophobic groove of Ubc9, juxtaposed with the catalytic Cys^{93} (Figure 31.14b). The pK_a of Lys^{526} is lowered in this

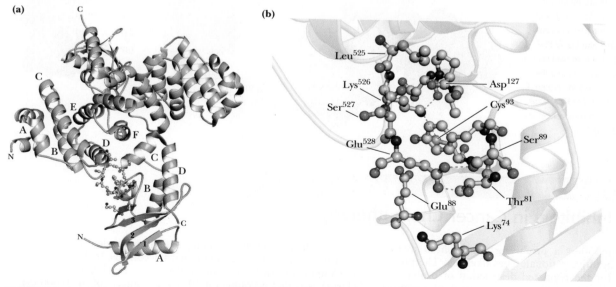

FIGURE 31.14 **(a)** The complex formed by the E2 enzyme, Ubc9 *(yellow)*, and a target protein, RanGAP1 *(blue)*. **(b)** In the Ubc9–RanGAP1 complex, the exposed loop of RanGAP1 lies in the binding pocket of Ubc9. The exposed loop contains the consensus sequence for sumoylation (ψKXD/E), including Leu^{525}, which is surrounded by hydrophobic residues from Ubc9; Lys^{526}, which is coordinated by Asp^{127} and Cys^{93} of Ubc9; and Glu^{528}, which is coordinated by Ubc9 Thr^{81}.

complex, activating the lysine amino group for nucleophilic attack to form the SUMO–target protein conjugate.

HtrA Proteases Also Function in Protein Quality Control

The discussion thus far has stressed the importance of protein quality control to cellular health. **HtrA proteases** are a class of proteins involved in quality control that combine the dual functions of chaperones and proteasomes. (The acronym **Htr** comes from "**h**igh **t**emperature **r**equirement" because *E. coli* strains bearing mutations in HtrA genes do not grow at elevated temperatures.) In addition to their novel ability to be either chaperones or proteases, HtrA proteases are the only known protein quality control factor that is not ATP-dependent. Prokaryotic HtrA proteases act as chaperones at low temperatures (20°C) where they have negligible protease activity. However, as the temperature increases, these proteins switch from a chaperone function to a protease function to remove misfolded or unfolded proteins from the cell. With this functional duality, HtrA proteases have the potential to mediate quality control through protein triage (see A Deeper Look box on page 1107).

The *E. coli* HtrA protein **DegP** is the best characterized HtrA protease. DegP is localized in the *E. coli* periplasmic space, where it oversees quality control of proteins involved in the cell envelope. It is a 448-residue protein containing a central protease domain with a classic Ser protease Asp-His-Ser catalytic triad (see Chapter 14) and two C-terminal **PDZ domains.** These domains are structural modules involved in protein–protein interactions, and they recognize and bind selectively to the C-terminal three or four residues of target proteins. Like other quality control systems, HtrA proteases have a central cavity where proteolysis occurs (Figure 31.15); the height of this cavity is 1.5 nm, which excludes folded proteins from access to the proteolytic sites. Thus, HtrA proteases can act only on misfolded proteins. As we have seen, limited access to

FIGURE 31.15 The HtrA protease structure. **(a)** A trimer of DegP subunits represents the HtrA functional unit. The different domains are color-coded: The protease domain is green, PDZ domain 1 (PDZ1) is yellow, and PDZ domain 2 (PDZ2) is orange. Protease active sites are highlighted in blue. The trimer has somewhat of a funnel shape, with the protease in the center and the PDZ domains on the rim. **(b)** Two HtrA trimers come together to form a hexameric structure in which the two protease domains form a rigid molecular cage *(blue)* and the six PDZ domains are like tentacles *(red)* that both bind protein substrate targets and control lateral access into the protease cavity. (Adapted from Figure 3 in Clausen, T., Southan, C., and Ehrmann, M., 2002. The HtrA family of proteases: Implications for protein composition and cell fate. *Molecular Cell* **10**:443–455.)

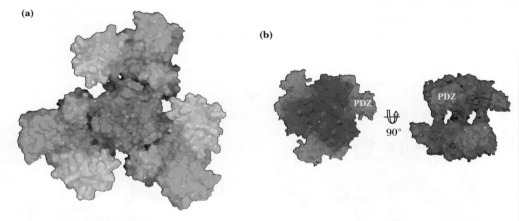

(a)

(b)

PDZ

PDZ

90°

HUMAN BIOCHEMISTRY

Proteasome Inhibitors in Cancer Chemotherapy

Proteasome inhibition offers a promising approach to treating cancer. The counterintuitive rationale goes like this: The proteasome is responsible for the regulated destruction of proteins involved in cell cycle progression and the control of apoptosis (programmed cell death). Inhibition of proteasome function leads to cell cycle arrest and apoptosis. In clinical trials, proteasome inhibitors have retarded cancer progression by interfering with the programmed degradation of regulatory proteins, causing cancer

cells to self-destruct. Bortezomib, a small-molecule proteasome inhibitor developed by Millenium Pharmaceuticals, Inc., has received FDA approval for the treatment of multiple myeloma, a cancer of plasma cells that accounts for 10% of all cancers of the blood (see Figure 31.10a).

Source: Adams, J., 2003. The proteasome: Structure, function, and role in the cell. *Cancer Treatment Reviews Supplement* **1**:3–9.

▶ **A DEEPER LOOK**

Protein Triage—A Model for Quality Control

Triage is a medical term for the sorting of patients according to their need for (and their likelihood to benefit from) medical treatment. Sue Wickner, Michael Maurizi, and Susan Gottesman have pointed out that cells control the quality of their proteins through a system of triage based on chaperones and the ubiquitination-proteasome degradation pathway. These systems recognize non-native proteins (proteins that are only partially folded, misfolded, incorrectly modified, damaged, or in an inappropriate compartment). Depending on the severity of its damage, a non-native protein is directed to chaperones for refolding or targeted for destruction by a proteasome (see accompanying figure).

Adapted from Figures 2 and 3 in Wickner, S., Maurizi, M., and Gottesman, S., 1999. Posttranslational quality control: Folding, refolding, and degrading proteins. *Science* **286:**1888–1893.

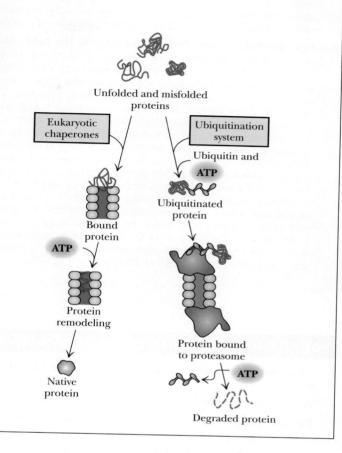

proteolytic sites is an important regulatory feature of quality control proteases; only proteins targeted for destruction have access to such sites. The PDZ domains of the HtrA proteases act as gatekeepers, determining access of protein substrates to the proteolytic centers. Human HtrA proteases are implicated in stress response pathways. Human HtrA1 is expressed at higher levels in osteoarthritis and aging and lower levels in ovarian cancer and melanoma. Secreted human Htr1 may be involved in degradation of extracellular matrix proteins involved in arthritis as well as tumor progression and invasion.

SUMMARY

⬤**WL** Login to OWL to develop problem-solving skills and complete online homework assigned by your professor.

31.1 How Do Newly Synthesized Proteins Fold? Most proteins fold spontaneously, as anticipated by Anfinsen, whose experiments suggested that all of the information necessary for a polypeptide chain to assume its active, folded conformation resides in its amino acid sequence. However, some proteins depend on molecular chaperones to achieve their folded conformation within the crowded intracellular environment. Hsp70 chaperones are ATP-dependent proteins that bind to exposed hydrophobic regions of polypeptides, preventing nonproductive associations with other proteins and keeping the protein in an unfolded state until productive folding steps can take place. Hsp60 chaperones such as GroEL–GroES are ATP-dependent cylin-

drical chaperonins that provide a protected central cavity or "Anfinsen cage," where partially folded proteins can fold spontaneously, free from the danger of nonspecific hydrophobic interactions with other unfolded protein chains. Hsp90 chaperones assist a subset of "client proteins" involved in signal transduction pathways in assuming their active conformations.

31.2 How Are Proteins Processed Following Translation? Nascent proteins are seldom produced in their final, functional form. Maturation typically involves proteolytic cleavage and may require posttranslational modification, such as phosphorylation, glycosylation, or other covalent substitutions. Removal of nascent N-terminal methionine residues is a common form of protein processing by proteolysis. The number of human proteins is believed to exceed the number of human genes (20,000 or so) by an order of magnitude or more. The great number of proteins available from a fixed set of genes is attributed to a variety of processes, including alternative splicing of mRNAs and post-translational modification of proteins.

31.3 How Do Proteins Find Their Proper Place in the Cell?

Most proteins destined for compartments other than the cytosol are synthesized with N-terminal signal sequences that target them to their proper destinations. These signal sequences are recognized by signal recognition particles as they emerge from translating ribosomes. The signal recognition particle interacts with a membrane-bound signal receptor, delivering the translating ribosome to a translocon, a multimeric integral membrane protein structure having at its core the Sec61p complex. The translocon transfers the growing protein chain across the membrane (or in the case of nascent integral membrane proteins, inserts the protein into the membrane). Signal peptidases within the membrane compartment clip off the signal sequence. Other post-translational modifications may follow. Mitochondrial protein import and membrane insertion are mediated by specific translocon complexes in the outer mitochondrial membrane called TOM and SAM. Proteins destined for the inner mitochondrial membrane or mitochondrial matrix must interact with inner mitochondrial translocons (either TIM23 or TIM22) as well as TOM.

31.4 How Does Protein Degradation Regulate Cellular Levels of Specific Proteins?

Protein degradation is potentially hazardous to cells because cell function depends on active proteins. Therefore, protein degradation is compartmentalized in lysosomes or in proteasomes. Proteins targeted for destruction are selected by ubiquitination. A set of three enzymes (E1, E2, and E3) mediate transfer of ubiquitin to free —NH_2 groups in targeted proteins. Proteins with charged or hydrophobic residues at their N-termini are particularly susceptible to ubiquitination and destruction. The ubiquitin moieties are recognized by 19S cap structures found at either end of 26S proteasomes. Protein degradation occurs when, in an ATP-dependent process, the ubiquitinated protein is unfolded and threaded into the central cavity of the cylindrical $\alpha_7\beta_7\beta_7\alpha_7$ 20S part of the 26S proteasome. The β-subunits possess protease active sites that chop the protein substrate into short (seven- to nine-residue) fragments; the ubiquitin moieties are recycled. Linkage of SUMO (small ubiquitin-like protein modifiers) to target proteins has the ability to alter their protein-protein interactions. HtrA proteases also function in protein quality control. HtrA proteins are novel in two aspects: Unlike other chaperones and proteasomes, they are ATP-independent; and unlike the others, they have dual chaperone and protease activities and switch between these two functions in response to stress conditions.

FOUNDATIONAL BIOCHEMISTRY   Things You Should Know After Reading Chapter 31.

- The variety of processes that modify the structure and function of proteins.
- The role that chaperones play in protein folding.
- The involvement of protein aggregates in Alzheimer's, Parkinson's, and Huntington's diseases.
- Why chaperones were first called heat shock proteins.
- The structure and function of the Hsp70 chaperone and the manner in which ATP drives protein folding.
- The structure and mechanism of action of the GroES-GroEL chaperone.
- The function of the eukaryotic Hsp90 chaperone.
- The functional significance of post-translational modifications of proteins.
- The involvement of signal sequences and leader peptides in targeting proteins to specific locations in the cell.
- The essential features of protein translocation, including the roles of signal recognition particles, signal receptors, and translocons, and the role of proteolytic processing.

- The differences between translocation and routing for prokaryotic versus eukaryotic proteins.
- The manner in which secretory protein synthesis is coupled to ER membrane translocation.
- The role of RNC-SRP complexes in eukaryotic protein translocation.
- The structural and functional characteristics of the eukaryotic translocon.
- The essential features of mitochondrial protein import.
- The essential features of the ubiquitin pathway for protein degradation.
- The structural and functional characteristics of the proteasome.
- The manner in which AAA+-ATPases mediate unfolding of proteins at the proteasome.
- The essential features of the SUMO pathway for regulation of protein–protein interactions.
- How HtrA proteases function in protein quality control.

PROBLEMS

Answers to all problems are at the end of this book. Detailed solutions are available in the *Student Solutions Manual, Study Guide, and Problems Book.*

1. **Assessing the ATP Cost of Protein Synthesis** (Integrates with Chapter 30.) Human rhodanese (33 kD) consists of 296 amino acid residues. Approximately how many ATP equivalents are consumed in the synthesis of the rhodanese polypeptide chain from its constituent amino acids *and* the folding of this chain into an active tertiary structure?

2. **Understanding the Consequences of Proteolysis** A single proteolytic break in a polypeptide chain of a native protein is often sufficient to initiate its total degradation. What does this fact suggest to you regarding the structural consequences of proteolytic nicks in proteins?

3. **Understanding the Relevance of Chaperones in Protein Folding** Protein molecules, like all molecules, can be characterized in terms of general properties such as size, shape, charge, solubility/hydrophobicity. Consider the influence of each of these general features on the likelihood of whether folding of a particular protein

will require chaperone assistance or not. Be specific regarding just Hsp70 chaperones or Hsp70 chaperones *and* Hsp60 chaperonins.

4. **Assessing the Chaperone-Assisted Folding of Multidomain Proteins** Many multidomain proteins apparently do not require chaperones to attain the fully folded conformations. Suggest a rational scenario for chaperone-independent folding of such proteins.

5. **Assessing the Dimensions and Capacity of GroEL** The GroEL ring has a 5-nm central cavity. Calculate the maximum molecular weight for a spherical protein that can just fit in this cavity, assuming the density of the protein is 1.25 g/mL.

6. **Understanding the Possible Phosphorylated States of Acetyl-CoA Carboxylase** (Integrates with Chapter 24.) Acetyl-CoA carboxylase has at least seven possible phosphorylation sites (residues 23, 25, 29, 76, 77, 95, and 1200) in its 2345-residue polypeptide (see Figure 24.4). How many different covalently modified forms of acetyl-CoA carboxylase protein are possible if there are seven phosphorylation sites?

7. **Comparing the Mechanisms of Action of EF-Tu/EF-Ts and DnaK/GrpE** (Integrates with Chapter 30.) In what ways are the mechanisms of action of EF-Tu/EF-Ts and DnaK/GrpE similar? What mechanistic functions do the ribosome A-site and DnaJ have in common?

8. **Understanding the Structure and Function of Signal Sequences and Signal Peptidases** The amino acid sequence deduced from the nucleotide sequence of a newly discovered human gene begins: MRSLLILVLCFLPLAALGK…. Is this a signal sequence? If so, where does the signal peptidase act on it? What can you surmise about the intended destination of this protein?

9. **Understanding the Mechanism of Transport of Integral Membrane Proteins** Not only is the Sec61p translocon complex essential for translocation of proteins into the ER lumen, it also mediates the incorporation of integral membrane proteins into the ER membrane. The mechanism for integration is triggered by stop-transfer signals that cause a pause in translocation. Figure 31.5 shows the translocon as a closed cylinder spanning the membrane. Suggest a mechanism for lateral transfer of an integral membrane protein from the protein-conducting channel of the translocon into the hydrophobic phase of the ER membrane.

10. **Assessing the Dimensions and Capacity of Sec61p** The Sec61p core complex of the translocon has a highly dynamic pore whose internal diameter varies from 0.6 to 6 nm. In post-translational translocation, folded proteins can move across the ER membrane through this pore. What is the molecular weight of a spherical protein that would just fit through a 6-nm pore? (Adopt the same assumptions used in problem 5.)

11. **Assessing the Dimensions and Capacity of the Protein-Conducting Translocon** (Integrates with Chapters 6, 9, and 30.) During co-translational translocation, the peptide tunnel running from the peptidyl transferase center of the large ribosomal subunit and the protein-conducting channel are aligned. If the tunnel through the ribosomal subunit is 10 nm and the translocon channel has the same length as the thickness of a phospholipid bilayer, what is the minimum number of amino acid residues sequestered in this common conduit?

12. **Understanding the Nature of the Ubiquitin-Ubiquitin Linkage** Draw the structure of the isopeptide bond formed between Gly^{76} of one ubiquitin molecule and Lys^{48} of another ubiquitin molecule.

13. **Understanding the Nature of E3 Ubiquitin Protein Ligation** Assign the 20 amino acids to either of two groups based on their

susceptibility to ubiquitin ligation by E3 ubiquitin protein ligase. Can you discern any common attributes among the amino acids in the less susceptible versus the more susceptible group?

14. **Assessing the Consequences of Proteasome Inhibition** Lactacystin is a *Streptomyces* natural product that acts as an irreversible inhibitor of 26S proteasome β-subunit catalytic activity by covalent attachment to N-terminal threonine —OH groups. Predict the effects of lactacystin on cell cycle progression.

15. **Understanding the Dual Functionality of the HtrA Protease** HtrA proteases are dual-function chaperone-protease protein quality control systems. The protease activity of HtrA proteases depends on a proper spatial relationship between the Asp-His-Ser catalytic triad. Propose a mechanism for the temperature-induced switch of HtrA proteases from chaperone function to protease function.

16. **Understanding the Scope and Diversity of Post-Translational Protein Modification** As described in this chapter, the most common post-translational modifications of proteins are proteolysis, phosphorylation, methylation, acetylation, and linkage with ubiquitin and SUMO proteins. Carry out a Web search to identify at least eight other post-translational modifications and the amino acid residues involved in these modifications.

17. **Understanding the Nature and Functional Utility of Fluorescence Resonance Energy Transfer (FRET)** Fluorescence resonance energy transfer (FRET) is a spectroscopic technique that can be used to provide certain details of the conformation of biomolecules. Look up FRET on the Web or in an introductory text on FRET uses in biochemistry, and explain how FRET could be used to observe conformational changes in proteins bound to chaperonins such as GroEL. A good article on FRET in protein folding and dynamics can be found here: Haas, E., 2005. The study of protein folding and dynamics by determination of intramolecular distance distributions and their fluctuations using ensemble and single-molecule FRET measurements. *ChemPhysChem* **6**:858–870. Studies of GroEL using FRET analysis include the following: Sharma, S., et al., 2008. Monitoring protein conformation along the pathway of chaperonin-assisted folding. *Cell* **133**:142–153; and Lin, Z., et al., 2008. GroEL stimulates protein folding through forced unfolding. *Nature Structural and Molecular Biology* **15**:303–311.

18. **Understanding the Nature and Functions of the Phosphodegron** The crosstalk between phosphorylation and ubiquitination in protein degradation processes is encapsulated in the concept of the "phosphodegron." What is a phosphodegron, and how does phosphorylation serve as a recognition signal for protein degradation? (A good reference on the phosphodegron and crosstalk between phosphorylation and ubiquitination is Hunter, T., 2007. The age of crosstalk: Phosphorylation, ubiquitination, and beyond. *Molecular Cell* **28**:730–738.)

Preparing for the MCAT® Exam

19. **Assessing the Consequences of a Common Post-Translational Modification** A common post-translational modification is removal of the universal N-terminal methionine in many proteins by Met-aminopeptidase. How might Met removal affect the half-life of the protein?

20. **Designing a Sequence for a Peptide with Amphipathic α-Helical Secondary Structure** Figure 31.6 shows the generalized amphipathic α-helix structure found as an N-terminal presequence on a nuclear-encoded mitochondrial protein. Write out a 20-residue-long amino acid sequence that would give rise to such an amphipathic α-helical secondary structure.

FURTHER READING

Protein-Folding Diseases

Aguzzi, A., and O'Connor, T., 2010. Protein aggregation diseases: Pathogenicity and therapeutic perspectives. *Nature Review Drug Discovery* **9:**237–248.

Basaiawmoit, R. V., and Rattan, S. I., 2010. Cellular stress and protein misfolding during aging. *Methods in Molecular Biology* **648:**107–117.

Bowman, G. R., Voelz, V. A., and Pande, V. S., 2011. Taming the complexity of protein folding. *Current Opinion in Structural Biology* **21:**4–11.

Herczenik, E., and Gebbink, M. F., 2008. Molecular and cellular aspects of protein misfolding and disease. *FASEB Journal* **22:**2115–2133.

Lyubchenko, Y. L., Kim. B.-H., Krasnoslobodtsev, A. V., and Yu, J., 2010. Nanoimaging for protein misfolding diseases. *Wiley Interdisciplinary Reviews: Nanomedicine and Nanobiotechnology* **2:**526–543.

Naiki, H., and Nagai, Y., 2009. Molecular pathogenesis of protein misfolding diseases: Pathological molecular environments verses quality control systems against misfolded proteins. *Journal of Biochemistry (Tokyo)* **146:**751–756.

Silva, J. L., Vieira, T. C., Gomes, M. P., Bom, A. P., et al., 2010. Ligand binding and hydration in protein misfolding: Insights from studies of prion and p53 tumor suppressor proteins. *Accounts of Chemical Research* **43:**271–279.

Soto, C., Estrada, L., et al., 2006. Amyloids, prions, and the inherent infectious nature of misfolded protein aggregates. *Trends in Biochemical Sciences* **31:**150–156.

Straub, J. E., and Thirumalai, D., 2010. Principles governing oligomer formation in amyloidogenic peptides. *Current Opinion in Structural Biology* **20:**187–195.

Welberg, L., 2010. A protective role for prions. *Nature Reviews Neuroscience* **11:**151.

Chaperone-Assisted Protein Folding

Clare, D. K., Bakkes, P. J., van Heerikhuizen, H., van der Vies, S. M., and Saibil, H. R., 2009. Chaperonin complex with a newly folded protein encapsulated in the folding chamber. *Nature* **457:**107–111.

Eyles, S. J., and Gierasch, L. M., 2010. Nature's molecular sponges: Small heat shock proteins grow into their chaperone roles. *Proceedings of the National Academy of Sciences U.S.A.* **107:**2727–2728.

Fedyukina, D. V., and Cavagnero, S., 2011. Protein folding at the exit tunnel. *Annual Review of Biophysics* **40:**337–359.

Gershenson, A., and Gierasch, L. M., Protein folding in the cell: Challenges and progress. *Current Opinion in Structural Biology* **21:**32–41.

Hartl, F. U., and Hartl, M., 2009. Converging concepts of protein folding *in vitro* and *in vivo*. *Nature Structural and Molecular Biology* **16:**574–581.

Hebert, D. N., and Gierash, L. M., 2009. The molecular dating game: An antibody heavy chain hangs loose with a chaperone while waiting for its life partner. *Molecular Cell* **34:**635–636.

Hoffman, A., Bukau, B., and Kramer, G., 2010. Structure and function of the molecular chaperone Trigger Factor. *Biochimica et Biophysica Acta* **1803:**650–661.

Hong, J., and Gierasch, L. M., 2010. Macromolecular crowding remodels the energy landscape of a protein by favoring a more compact unfolded state. *Journal of the American Chemical Society* **132:**10445–10452.

Liu, Y., Hu, Y., Li, X., Niu, L., and Teng, M., 2010. The crystal structure of the human nascent polypeptide-associated complex domain reveals a nucleic acid-binding region on the NACA subunit. *Biochemistry* **49:**2890–2896.

Makareeva, E., Aviles, N. A., and Leiken, S., 2011. Chaperoning osteogenesis: New protein-folding disease paradigms. *Trends in Cellular Biology* **21:**168–176.

Mayer, M. P., 2010. Gymnastics of molecular chaperones. *Molecular Cell* **39:**321–331.

Neckers, L., Tsutsumi, S., and Mollapour, M., 2009. Visualizing the twists and turns of a molecular chaperone. *Nature Structural and Molecular Biology* **16:**235–236.

O'Brien, E. P., Christodoulou, J., Vendruscolo, M., and Dobson, C. M., 2011. New scenarios of protein folding can occur on the ribosome. *Journal of the American Chemical Society* **133:**513–526.

Pech, M., Spreter, T., Beckmann, R., and Beatrix, B., 2010. Dual binding mode of the nascent polypeptide-associated complex reveals a novel universal adapter site on the ribosome. *The Journal of Biological Chemistry* **285:**19679–19687.

Richter, K., Haslbeck, M., and Buchner, J., 2010. The heat shock response: Life on the verge of death. *Molecular Cell* **40:**253–266.

Schlect, R., Erbse, A. H., Bukau, B., and Mayer, M. P., 2011. Mechanics of Hsp70 chaperones enables differential interaction with client proteins. *Nature Structural and Molecular Biology* **18:**345–352.

Schulz, E. C., Dickmanns, A., Urlaub, H., Schmitt, A., et al., 2010. Crystal structure of an intramolecular chaperone mediating triple-β-helix folding. *Nature Structural and Molecular Biology* **17:**210–216.

Stengel, F., Baldwin, A. J., Painter, A. J., Jaya, N., et al., 2010. *Proceedings of the National Academy of Sciences U.S.A* **107:**2027–2012.

Vabulas, R. M., Raychaudhuri, S., Hayer-Hartl, M., and Hartl, F. U., 2010. Protein folding in the cytoplasm and the heat shock response. *Cold Spring Harbor Perspectives on Biology* **2:**1–18.

Wang, L., Zhang, W., Wang, L., Zhang, X. C., et al., 2010. Crystal structures of NAC domains of human nascent polypeptide-associated complex (NAC) and its αNAC subunit. *Protein and Cell* **1:**406–416.

Zhang, G., and Ignatova, Z., 2011. Folding at the birth of the nascent chain: Coordinating translation with co-translational folding. *Current Opinion in Structral Biology* **21:**25–31.

Protein Translocation

Ataide, S. F., Schmitz, N., Shen, K., Ke, A., et al., 2011. The crystal structure of the signal recognition particle in complex with its receptor. *Science* **331:**881–886.

Bolender, N., Sickmann, A., et al., 2008. Multiple pathways for sorting mitochondrial precursor proteins. *EMBO Reports* **9:**42–49.

Bornemann, T., Jockel, J., et al., 2008. Signal sequence-independent membrane targeting of ribosomes containing short nascent peptides within the exit tunnel. *Nature Structural and Molecular Biology* **15:**494–499.

Cheng, Z., 2010. Protein translocation through the Sec61/SecY channel. *Bioscience Reports* **30:**201–207.

Cross, B. C. S., Sinning, I., Luirink, J., and High, S., 2009. Delivering proteins for export from the cytosol. *Nature Reviews Molecular Cell Biology* **10:**255–264.

Mandon, E. C., Trueman, S. F., and Gilmore, R., 2010. Translocation of proteins through the Sec61 and SecYEG channels. *Current Opinion in Cell Biology* **21:**501–507.

Mast, F. D., Fagarasanu, A., and Rachubinski, R., 2010. The peroxisomal protein importomer: A bunch of transients with expanding waistlines. *Nature Cell Biology* **12:**203–205

Meinecke, M., Cizmowski, C., Schliebs, W., Krüger, V., et al., 2010. The peroxisomal importomer constitutes a large and highly dynamic pore. *Nature Cell Biology* **12:**273–278.

Schleiff, E. and Becker, T., 2011. Common ground for protein translocation: Access control for mitochondria and chloroplasts. *Nature Reviews Molecular Cell Biology* **12:**48–59.

Tsukazaki, T., Mori, H., Fukai, S., Ishitani, R., et al., 2008. Conformational transition of Sec machinery inferred from bacterial SecYE structures. *Nature* **455**:988–992.

Zimmerman, R., Eyrisch, S., Ahmad, M., and Helms, V., 2011. Protein translocation across the ER membrane. *Biochimica et Biophysica Acta* **1808**:912–924.

Autophagy

Cuervo, A. M., 2011. Autophagy's top chef. *Science* **332**:1392–1393.

Kaushik, S., Rodriguez-Navarro, J. A., Arias, E., Kiffin, R., et al., 2011. Autophagy in hypothalamic AgRP neurons regulates food intake and energy balance. *Cell Metabolism* **14**:173–183.

Kettern, N., Dreiseidler, M., Tawo, R., and Höhfeld, J., 2010. Chaperone-assisted degradation: Multiple paths to destruction. *Biological Chemistry* **391**:481–489.

Kraft, C., Peter, M., and Hofmann, K., 2010. Autophagy: Ubiquitin-mediated recognition. *Nature Cell Biology* **12**:836–841.

Kroemer, G., Marino, G., and Levine, B., 2010. Autophagy and the integrated stress response. *Molecular Cell* **40**:280–293.

Lamark, T., and Johansen, T., 2010. Autophagy: Links with the proteasome. *Current Opinion in Cell Biology* **22**:192–198.

Levine, B., 2005. Eating oneself and uninvited guests: Autophagy-related pathways in cellular defense. *Cell* **120**:159–162.

Levine, B., Mizushima, N., and Virgin, H. W., 2011. Autophagy in immunity and inflammation. *Nature* **469**:323–335.

Li, W., Yang, Q., and Mao, Z., 2011. Chaperone-mediated autophagy: Machinery, regulation and biological consequences. *Cellular and Molecular Life Sciences* **68**:749–763.

Mizushima, N., and Komatsu, M., 2011. Autophagy: Renovation of cells and tissues. *Cell* **147**:728–741.

Nakatogawa, H., Suzuki, K., Kamada, Y., and Ohsumi, Y., 2009. Dynamics and diversity in autophagy mechanisms: Lessons from yeast. *Nature Reviews Molecular Cell Biology* **10**:458–467.

Orenstein, S. J., and Cuervo, A. M., 2010. Chaperone-mediated autophagy: Molecular mechanisms and physiological relevance. *Seminars in Cell & Development Biology* **21**:719–726.

Rabinowitz, J. D., and White, E., 2010. Autophagy and metabolism. *Science* **330**:1344–1348.

Wong, E., and Cuervo, A. M., 2011. Integration of clearance mechanisms: The proteasome and autophagy. *Cold Spring Harbor Perspectives in Biology* **2**:1–19.

Xie, Z., and Klionsky, D. J., 2007. Autophagosome formation: Core machinery and adaptations. *Nature Cell Biology* **9**:1102–1109.

Yang, Z. and Klionsky, D. J., 2010. Mammalian autophagy: Core molecular machinery and signaling regulation. *Current Opinion in Cell Biology* **22**:124–131.

Ubiquitin Selection of Proteins for Degradation

Bellare, P., Small, E. C., et al., 2008. A role for ubiquitin in the spliceosome assembly pathway. *Nature Structural and Molecular Biology* **15**:444–451.

Dikic, I., Wakatsuki, S., and Walters, K. J., 2009. Ubiquitin-binding domains—from structures to functions. *Nature Reviews Molecular Cell Biology* **10**:659–671.

Geoffroy, M.-C., and Hay, R. T., 2009. An additional role for SUMO in ubiquitin-mediated proteolysis. *Nature Reviews Molecular Cell Biology* **10**:564–568.

Hurley, J. H., and Stenmark, H., 2011. Molecular mechanics of ubiquitin-dependent membrane traffic. *Annual Review of Biophysics* **40**:119–142.

Komander, D., Clague, M. J., and Urbe, S., 2009. Breaking the chains: Structure and function of the deubiquitinases. *Nature Reviews Molecular Cell Biology* **10**:550–563.

Matta-Camacho, E., Kozlov, G., Li, F. F., and Gehring, K., 2010. Structural basis of substrate recognition and specificity in the N-end rule pathway. *Nature Structural & Molecular Biology* **17**:1182–1187.

Schulman, B. A., and Harper, J. W., 2009. Ubiquitin-like protein activation by E2 enzymes: The apex for downstream signaling pathways. *Nature Reviews Molecular Cell Biology* **10**:31–331.

Ye, Y., and Rape, M., 2009. Building ubiquitin chains: E2 enzymes at work. *Nature Reviews Molecular Cell Biology* **10**:755–764.

Proteasome-Mediated Protein Degradation

Bedford, L., Lowe, J., Dick, L. R., Mayer, R. J., and Brownell, J. E., 2011. Ubiquitin-like protein conjugation and the ubiquitin-proteasome system as drug targets. *Nature Reviews Drug Discovery* **10**:29–46.

Bedford, L., Paine, S., Sheppard, P. W., Mayer, R. J., and Roelofs, J., 2010. Assembly, structure, and function of the 26S proteasome. *Trends in Cell Biology* **20**:391-401.

Breusing, N., and Grune, T., 2008. Regulation of proteasome-mediated protein degradation during oxidative stress and aging. *Biological Chemistry* **389**:203–209.

Clague, M. J., and Urbe, S., 2011. Ubiquitin: Same molecule, different degradation pathways. *Cell* **143**:682–685.

Ruschak, A. M., Religa, T. L., Breuer, S., Witt, S., and Kay, L. E., 2010. The proteasome antechamber maintains substrates in an unfolded state. *Nature* **467**:868–871.

Stadtmueller, B. M., and Hill, C. P., 2011. Proteasome activators. *Molecular Cell* **41**:8–19.

Xie, Y., 2010. Structure, assembly and homeostatic regulation of the 26S proteasome. *Journal of Molecular Cell Biology* **2**:308–317.

HtrA Proteases

Clausen, T., Kaiser, M., Huber, R., and Ehrmann, M., 2011. HTRA proteases: Regulated proteolysis in protein quality control. *Nature Reviews Molecular Cell Biology* **12**:152–162.

Kim, D. Y., and Kim, K. K., 2005. Structure and function of HtrA family proteins, the key players in protein quality control. *Journal of Biochemistry and Molecular Biology* **38**:266–274.

Sohn, J., Grant, R. A., et al., 2007. Allosteric activation of DegS, a stress sensor PDZ protease. *Cell* **131**:572–583.

Walle, L. V., Lamkanfi, M., et al., 2008. The mitochondrial serine protease HtrA2/Omi: An overview. *Cell Death and Differentiation* **15**:453–460.

Post-translational Modification by Sumoylation

Anckar, J., and Sistonen, L., 2007. SUMO: Getting it on. *Biochemical Society Transactions* **35**:1409–1413.

Gareau, J. R., and Lima, C. D., 2010. The SUMO pathway: Emerging mechanisms that shape specificity, conjugation and recognition. *Nature Reviews Molecular Cell Biology* **11**:861–871.

Geiss-Friedlander, R., and Melchior, F., 2007. Concepts in sumoylation: A decade on. *Nature Reviews Molecular Cell Biology* **8**:947–956.

Karaca, E., Tozluoglu, M., Nussinov, R., and Halilioglu, T., 2011. Alternative allosteric mechanisms can regulate the substrate and E2 in SUMO conjugation. *Journal of Molecular Biology* **406**:620–630.

Sarge, K. D., and Park-Sarge, O. K., 2009. Sumoylation and human disease pathogenesis. *Trends in Biochemical Sciences* **34**:200–205.

Sarge, K. D., and Parke-Sarge, O. K., 2011. SUMO and its role in human disease. *International Review of Cell and Molecular Biology* **288**:167–183.

Wilkinson, K. A., and Henley, J. M., 2010. Mechanisms, regulation and consequences of protein SUMOylation. *Biochemical Journal* **438**:133–145.

The Reception and Transmission of Extracellular Information

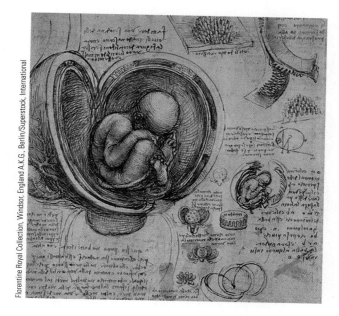

Florentine Royal Collection, Windsor, England A.K.G., Berlin/Superstock, International

"How little we know, how much to discover,
What chemical forces flow from lover to lover.
How little we understand what touches off that
*   tingle,*
That sudden explosion when two tingles
*   intermingle.*
Who cares to define what chemistry this is?
Who cares with your lips on mine
How ignorant bliss is,
So long as you kiss me and the world around us
*   shatters?*
How little it matters how little we know"

"How Little We Know"

by P. Springer and C. Leigh
(Excerpt from "How Little We Know" by P. Springer and C. Leigh as recorded by Frank Sinatra, April 5, 1956. Capitol Records, Inc.)

KEY QUESTIONS

32.1 What Are Hormones?

32.2 What Is Signal Transduction?

32.3 How Do Signal-Transducing Receptors Respond to the Hormonal Message?

32.4 How Are Receptor Signals Transduced?

32.5 How Do Effectors Convert the Signals to Actions in the Cell?

32.6 How Are Signaling Pathways Organized and Integrated?

32.7 How Do Neurotransmission Pathways Control the Function of Sensory Systems?

ESSENTIAL QUESTIONS

Higher life forms must have molecular mechanisms for detecting environmental information as well as mechanisms that allow for communication at the cell and tissue levels. Sensory systems detect and integrate physical and chemical information from the environment and pass this information along by the process of **neurotransmission.** Control and coordination of processes at the cell and tissue levels are achieved not only by neurotransmission but also by chemical signals in the form of **hormones** that are secreted by one set of cells to direct the activity of other cells.

What are the mechanisms of information transfer that comprise the molecular basis of hormone action? How do excitable membranes transduce the signals of neurotransmission and sensory systems?

Hormones are secreted by certain cells, usually located in glands, and travel, either by simple diffusion or circulation in the bloodstream, to specific target cells (Figure 32.1). As we shall see, some hormones bind to specialized receptors on the plasma membrane and induce responses within the cell without themselves entering the target cell (Figure 32.2). Other hormones actually enter the target cell and interact with specific receptors there. By these mechanisms, hormones:

• Regulate the metabolic processes of various organs and tissues
• Facilitate and control growth, differentiation, reproductive activities, learning, and memory
• Help the organism cope with changing conditions and stresses in its environment

All these effects of hormonal signals are either **metabolic responses** or **gene expression responses** (Figure 32.2).

OWL Online homework and a Student Self Assessment for this chapter may be assigned in OWL.

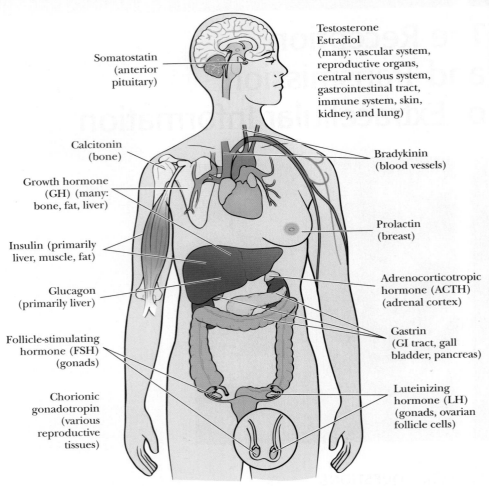

FIGURE 32.1 The sites of synthesis and action of a few of the polypeptide and steroid hormones. Hormones typically circulate at low concentrations (1 nM or less) and bind with high affinity to their receptor proteins.

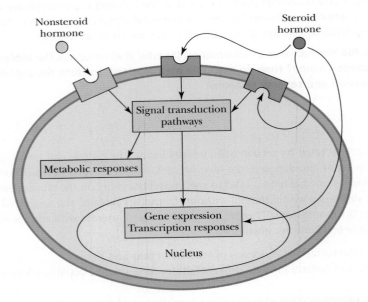

FIGURE 32.2 Nonsteroid hormones bind exclusively to plasma membrane receptors, which mediate the cellular responses to the hormone. Steroid hormones exert their effects either by binding to plasma membrane receptors or by diffusing to the nucleus, where they modulate transcriptional events. Intracellular responses to hormone binding include metabolic changes and alterations of gene expression.

32.1 : What Are Hormones?

Many different chemical species act as hormones. **Steroid hormones,** all derived from cholesterol, regulate metabolism, salt and water balances, inflammatory processes, and sexual function. Several hormones are **amino acid derivatives.** Among these are *epinephrine* and *norepinephrine* (which regulate smooth muscle contraction and relaxation, blood pressure, cardiac rate, and the processes of lipolysis and glycogenolysis) and the *thyroid hormones* (which stimulate metabolism). **Peptide hormones** are a large group of hormones that regulate processes in all body tissues, including the release of yet other hormones.

Hormones and other signal molecules in biological systems bind with very high affinities to their receptors, displaying K_D values in the range of 10^{-12} to 10^{-6} M. The hormones are produced at concentrations equivalent to or slightly above these K_D values. Once hormonal effects have been induced, the hormone is usually rapidly metabolized.

Steroid Hormones Act in Two Ways

The steroid hormones include the glucocorticoids (cortisol and corticosterone), the mineralocorticoids (aldosterone), and the sex hormones (progesterone and testosterone, for example) (Figure 32.1; see Chapter 24 for the details of their synthesis). The steroid hormones exert their effects in two ways: First, by entering cells and migrating to the nucleus, steroid hormones act as transcription regulators, modulating gene expression. These effects of the steroid hormones occur on time scales of hours and involve synthesis of new proteins. Steroids can also act at the cell membrane, directly regulating ligand-gated ion channels and perhaps other processes. These latter processes take place very rapidly, on time scales of seconds and minutes.

Polypeptide Hormones Share Similarities of Synthesis and Processing

The largest class of hormones in vertebrate organisms is that of the **polypeptide hormones** (Figure 32.1). One of the first polypeptide hormones to be discovered, **insulin,** was described by Banting and Best in 1921. Insulin, a secretion of the pancreas, controls glucose utilization and promotes the synthesis of proteins, fatty acids, and glycogen. Insulin, which is typical of the **secreted polypeptide hormones,** is discussed in detail in Chapters 5, 15, and 22.

Many other polypeptide hormones are produced and processed in a manner similar to that of insulin. Three unifying features of their synthesis and cellular processing should be noted. First, all secreted polypeptide hormones are originally synthesized with a signal sequence, which facilitates their eventual routing to secretory granules, and thence to the extracellular milieu. Second, peptide hormones are usually synthesized from mRNA as inactive precursors, termed **preprohormones,** which become activated by proteolysis. Third, a single polypeptide precursor or preprohormone may produce several different peptide hormones by suitable proteolytic processing.

An impressive example of the production of many hormone products from a single precursor is the case of **prepro-opiomelanocortin,** a 250-residue precursor peptide synthesized in the pituitary gland. A cascade of proteolytic steps produces, as the name implies, a natural *opi*ate substance **(endorphin)** and several other hormones (Figure 32.3). Endorphins and other opiatelike hormones are produced by the body in response to systemic stress. These substances probably contribute to the "runner's high" that marathon runners describe.

32.2 : What Is Signal Transduction?

Hormonal regulation depends on the transduction of the hormonal signal across the plasma membrane to specific intracellular sites, particularly the nucleus. **Signal transduction** consists of a stepwise progression of signaling stages: receptor⟶transducers⟶effectors. The receptor perceives the signal, transducers relay the signal, and the effectors convert the signal into an intracellular response.

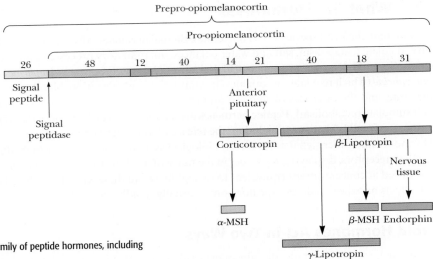

FIGURE 32.3 The conversion of prepro-opiomelanocortin to a family of peptide hormones, including corticotropin, β- and γ-lipotropin, α- and β-MSH, and endorphin.

What are the characteristics of signal transduction systems? Regardless of the organism or the tissue, certain features appear to be universal, or nearly so:

- Transmembrane communication of hormonal signals by **receptor proteins**
- Clustering of membrane receptors and their ligands in large aggregates called **signalsomes**
- Reversible **covalent modifications** that change the function of certain proteins and lipids (including phosphorylation, methylation, acetylation, ubiquitylation, hydroxylation, and cleavage)
- **Protein interaction domains** that selectively recognize specific structural motifs and bind them with high affinity and specificity
- **Second messengers** that bind to specific targets, changing their activity and behavior
- Intracellular **signaling pathways,** often involving a series of enzymes (such as protein kinases), that link receptors to their downstream functional targets
- Cooperativity
- Spatial and temporal control of signals and messengers
- Integration of signals
- Converging and diverging networks
- Signal amplification
- Desensitization and adaptation

Many Signaling Pathways Involve Enzyme Cascades

Signaling pathways must operate with speed and precision, facilitating the accurate relay of intracellular signals to specific targets. But how does this happen? Enzyme cascades are one answer to this question. Enzymes can produce (or modify) a large number of molecules rapidly and specifically. Many enzymes of signaling cascades are protein kinases and protein phosphatases, and many steps in signaling pathways involve phosphorylation of serine, threonine, and tyrosine residues on target proteins. The complexity of signal transduction is thus manifested in the estimates that the human genome contains about a thousand protein kinase and protein phosphatase genes. Enzyme cascades act like a series of amplifiers, dramatically increasing the magnitude of the intracellular response available from a very small amount of hormone (for an example, see the discussion of the phosphorylase cascade on page 741).

Signaling Pathways Connect Membrane Interactions with Events in the Nucleus

The complete pathway from hormone binding at the plasma membrane to modulation of transcription in the nucleus is understood for many signaling pathways. Figure 32.4 shows a complete signal transduction pathway that connects receptor tyrosine kinases, the Ras

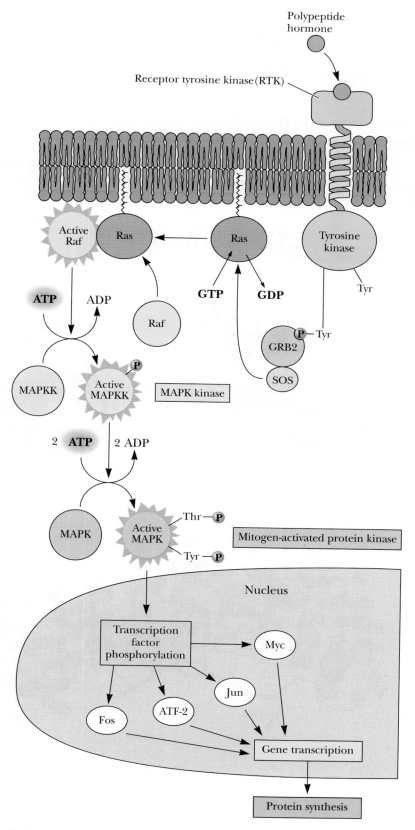

FIGURE 32.4 A complete signal transduction pathway that connects a hormone receptor with transcription events in the nucleus. Many similar pathways have been characterized.

Shawn Li and colleagues have proposed that the diverse specificities of PIDs are achieved either by variability in binding-site residues or in the surface loops adjacent to the binding site itself.
Kaneko, T., Sidhu, S. S., and Li, S. C., 2010. Evolving specificity from variability for protein interaction domains. Trends in Biochemical Sciences **36**:183–190.

GTPase, cytoplasmic Raf, and two other protein kinases with transcription factors that alter gene expression in the nucleus. This pathway represents just one component of a complex signaling network that involves many other proteins and signaling factors. The existence of nearly 4000 human genes devoted to signal transduction portends a complex and interwoven network of signaling interactions in nearly all human cells.[1]

Signaling Pathways Depend on Multiple Molecular Interactions

Each step in a signaling pathway depends on one or more molecular interactions. For example, a protein may bind to a small molecule, a peptide, or even another protein (Figure 32.5). In fact, many signaling proteins are constructed in a cassettelike fashion

[1]The American Association for the Advancement of Science oversees a consortium of researchers who have established a Web site—the Signal Transduction Knowledge Environment (STKE). This site is an up-to-date and ongoing compilation of information about cell signaling and signal transduction. The URL is *http://www.stke.org.*

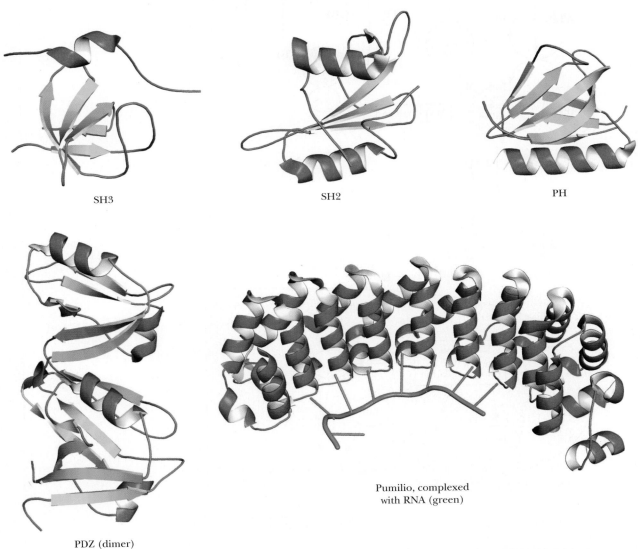

SH3

SH2

PH

PDZ (dimer)

Pumilio, complexed with RNA (green)

FIGURE 32.5 Five of the protein modules found in cell signaling proteins. The binding specificity of these modules is shown in Figure 32.6a. SH3 domains bind to proline-rich peptides; SH2 domains bind to phosphorylated tyrosine residues; PH domains bind to phosphoinositides (such as IP₃); PDZ domains bind to the terminal four or five residues of a target protein; pumilio domains bind to segments of RNA. A given protein may have a number of these protein modules, giving it the ability to interact with multiple partners.

(a)

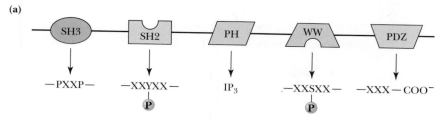

(b)

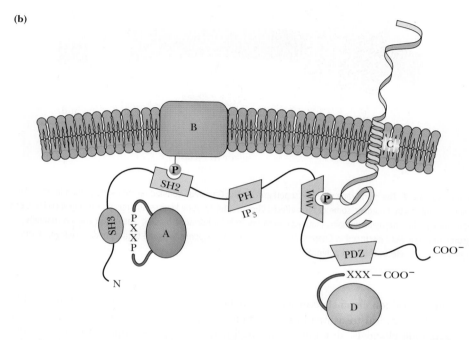

FIGURE 32.6 **(a)** Many signaling proteins consist of combinations of protein modules, each with a specific binding or enzymatic function. **(b)** Such multiply-interacting proteins can act as scaffolds that direct the assembly of large signaling complexes, termed signalsomes.

with several distinct modules, termed **protein interaction domains (PIDs),** each of which mediates a particular interaction (Figure 32.6a). A protein with multiple modules can interact with several binding targets at once to create a complex, called a **signalsome,** with multiple signaling capabilities (Figure 32.6b). Several signaling systems described in this chapter involve events at a signalsome.

32.3 How Do Signal-Transducing Receptors Respond to the Hormonal Message?

All receptors that mediate transmembrane signaling processes fit into one of three **receptor superfamilies** (Figure 32.7):

1. The **G-protein–coupled receptors** (see Section 32.4) are integral membrane proteins with an extracellular recognition site for ligands and an intracellular recognition site for a **GTP-binding protein** (see following discussion).

2. The **single-transmembrane segment (1-TMS) catalytic receptors** are proteins with only a single transmembrane segment and substantial globular domains on both the extracellular and the intracellular faces of the membrane. The extracellular domain is the ligand recognition site, and the intracellular catalytic domain is either a **tyrosine kinase** or a **guanylyl cyclase.**

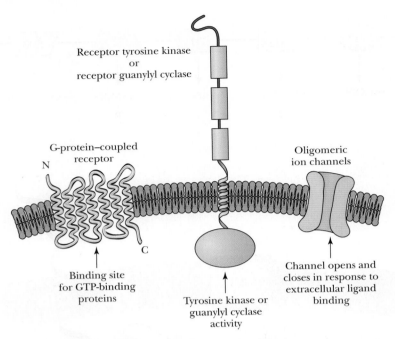

FIGURE 32.7 The membrane receptor superfamilies. The G-protein–coupled receptors are named for the GTP-binding proteins that mediate some of their effects. The receptor tyrosine kinases and receptor guanylyl cyclases contain intracellular enzymatic domains that respond to extracellular hormone binding. The oligomeric ion channels (some of which were discussed in Chapter 9) open and close in response to ligand binding and/or changes of the transmembrane electrochemical potential.

3. **Oligomeric ion channels** consist of associations of protein subunits, each of which contains several transmembrane segments. These oligomeric structures are **ligand-gated ion channels.** Binding of the specific ligand typically opens the ion channel. The ligands for these ion channels are neurotransmitters.

The G-Protein–Coupled Receptors Are 7-TMS Integral Membrane Proteins

There are 600 genes for non-olfactory GPCRs in the human genome, but alternative splicing produces between 1000 and 2000 active GPCRs in human cells. Louis Luttrell and colleagues have shown that synthesis of GPCRs and trafficking to the plasma membrane are regulated by RNA signals. *Tholanikunnel, B. G., Joseph, K., Kandasamy, K., Baldys, A., et al., 2010. Novel mechanisms in the regulation of G protein-coupled receptor trafficking to the plasma membrane. The Journal of Biological Chemistry 285:33816–33825.*

The G-protein–coupled receptors (GPCRs) have primary and secondary structure similar to that of bacteriorhodopsin (see Chapter 9), with seven transmembrane α-helical segments; they are thus known as *7-transmembrane segment* (7-TMS) proteins. Rhodopsin and the β-adrenergic receptors, for which epinephrine is a ligand, are good examples (Figure 32.8). The site for binding of cationic catecholamines to the adrenergic receptors is located within the hydrophobic core of the receptor.

Binding of hormone to a GPCR induces a conformation change that activates a GTP-binding protein, also known as a G protein (discussed in Section 32.4). Activated G proteins trigger a variety of cellular effects, including activation of adenylyl and guanylyl cyclases (which produce cAMP and cGMP from ATP and GTP), activation of phospholipases (which produce second messengers from phospholipids) and activation of Ca^{2+} and K^+ channels (which leads to elevation of intracellular $[Ca^{2+}]$ and $[K^+]$). (All of these effects are described in Section 32.4.)

The Single TMS Receptors Are Guanylyl Cyclases or Tyrosine Kinases

Receptor proteins that span the plasma membrane with a single helical transmembrane segment—either **receptor tyrosine kinase (RTK)** or **receptor guanylyl cyclase (RGC)**—possess an external ligand recognition site and an internal domain with enzyme activity.

(a) β_2-Adrenergic receptor

(b)

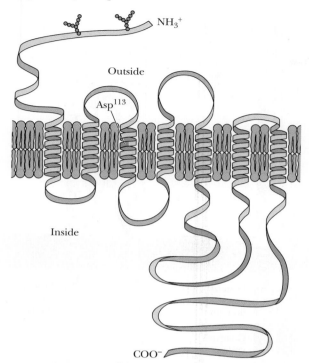

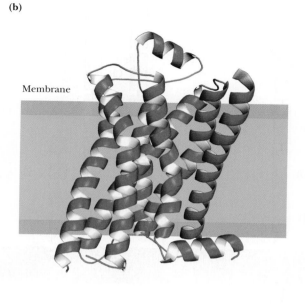

FIGURE 32.8 (a) The arrangement of the β_2-adrenergic receptor in the membrane. Substitution of Asp[113] in the third hydrophobic domain of the β-adrenergic receptor with an Asn or Gln by site-directed mutagenesis results in a dramatic decrease in affinity of the receptor for both agonists and antagonists. Significantly, this Asp residue is conserved in all GPCRs that bind biogenic amines but is absent in receptors whose ligands are not amines. Asp[113] appears to be the counterion of the amine moiety of adrenergic ligands. (b) The structure of a β_2-adrenergic receptor (pdb id = 2RH1). The flexible third intracellular loop and C-terminal segment are not shown. (From Figure 2 from Palczewski, K., et al., 2000. Crystal structure of rhodopsin: A G-protein-coupled receptor. *Science* **289**:739–745.)

Each of these enzyme activities is manifested in two different cellular forms. Thus, guanylyl cyclase activity is found in both membrane-bound receptors and in soluble, cytoplasmic proteins. Tyrosine kinase activity, on the other hand, is exhibited by two different types of membrane proteins: The RTKs are integral transmembrane proteins, whereas the non-RTKs are peripheral, lipid-anchored proteins.

RTKs and RGCs Are Membrane-Associated Allosteric Enzymes

The binding of polypeptide hormones and growth factors to RTKs and RGCs activates the intracellular enzyme activity of these proteins. These catalytic receptors are composed of three domains (Figure 32.9): an extracellular receptor-binding domain (which may itself include several subdomains), a transmembrane domain consisting of a single transmembrane α-helix, and an intracellular domain. This intracellular portion includes a tyrosine kinase or guanylyl cyclase domain that mediates the biological response to the hormone or growth factor via its catalytic activity and a regulatory domain that contains multiple phosphorylation sites. The human genome contains at least 58 different RTKs, which can be grouped into about 20 families on the basis of their kinase domain sequences and the various extracellular subdomains. The **epidermal growth factor (EGF) receptor** and the **insulin receptor** are representative of this class of receptor proteins.

Given that the extracellular and intracellular domains of RTKs and RGCs are joined by only a single transmembrane helical segment, how does extracellular hormone

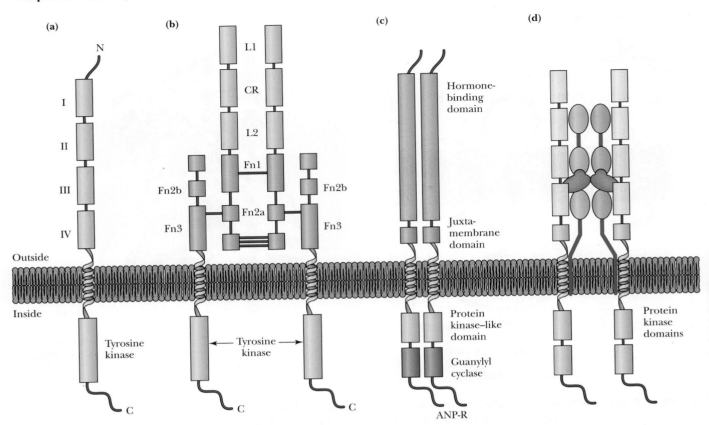

FIGURE 32.9 (a) The EGF receptor and (b) the insulin receptor are receptor tyrosine kinases. EGF receptors are activated by ligand-induced dimerization, whereas the insulin receptor is a glycoprotein composed of two kinds of subunits in an $\alpha_2\beta_2$ tetramer stabilized by disulfide bonds. The extracellular portions of the EGF and insulin receptors consist of multiple modules or domains. Fn refers to a series of fibronectin domains numbered 1, 2a, 2b, and 3. (c) The atrial natriuretic peptide receptor is a receptor guanylyl cyclase with a large extracellular hormone-binding domain and two intracellular domains, and activation typically involves ligand-induced dimerization. (d) The growth factor receptor Ret is a receptor tyrosine kinase that mediates the effects of neuronal growth factors known as neurotrophins. The Ret receptor domains (yellow) requires a GPI-anchored coreceptor (blue) for binding of ligands (pink).

binding activate intracellular enzyme activity? How is the signal transduced? As shown in Figure 32.10, signal transduction occurs by hormone-induced oligomeric association of receptors. Hormone binding triggers a conformational change in the *extracellular* domain, which induces oligomeric association. Oligomeric association allows adjacent *cytoplasmic* domains to interact, leading to phosphorylation of the cytoplasmic domains and stimulation of cytoplasmic enzyme activity. By virtue of these ligand-induced conformation changes and oligomeric interactions, RTKs and RGCs are **membrane-associated allosteric enzymes.**

The EGF Receptor Is Activated by Ligand-Induced Dimerization

Human EGF is a 53-amino acid peptide that stimulates proliferation of epithelial cells (cells that cover a surface or line a cavity in biological tissues). The human EGF receptor is a 1186–amino acid RTK. The extracellular domain, where EGF binds, contains 622 amino acids and is divided into four domains (Figure 32.9). Domains I and III are similar β-helical barrels, whereas domains II and IV are long, slender, cysteine-rich domains stabilized by disulfide bonds. Domain II is characterized by a β-hairpin structure that protrudes from the middle of the domain (Figure 32.11a).

Prior to EGF binding, EGF receptors exist as inactive monomers in the plasma membrane. In this state, the four domains are folded so that domains II and IV lie parallel, with the β-hairpin of domain II in contact with domain IV (Figure 32.11a). Binding

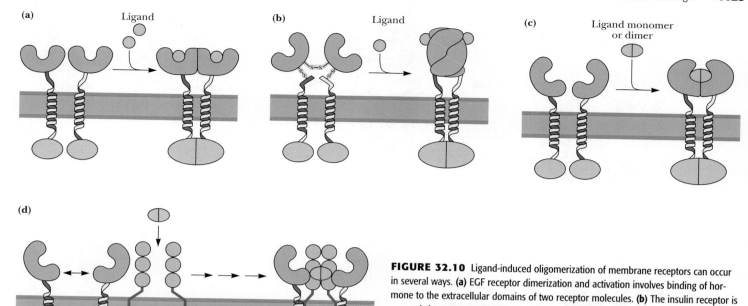

FIGURE 32.10 Ligand-induced oligomerization of membrane receptors can occur in several ways. **(a)** EGF receptor dimerization and activation involves binding of hormone to the extracellular domains of two receptor molecules. **(b)** The insulin receptor is a preexisting tetramer. Two insulin-binding sites are created by association of the extracellular domains. **(c)** RGCs bind a single hormone ligand at the dimer interface. (Ligand dimers could presumably act in a similar manner). **(d)** Neurotrophin activation of Ret involves association of two molecules of the Ret kinase and two molecules of a GPI-anchored coreceptor protein.

of EGF to domain I induces a conformation change that rotates domains I and II so that the bound EGF is brought into contact with domain III and the β-hairpin is extended away from the rest of the structure. Pairs of such receptor monomers then dimerize by mutual association of the β-hairpin structures (Figure 32.11b). *The next event is the critical step in transmembrane signal transduction.* Dimerization of the extracellular domains brings the intracellular domains together, activating the tyrosine kinase activity. Thus, EGF-induced dimerization allows an extracellular signal (EGF) to exert an intracellular response.

EGF Receptor Activation Forms an Asymmetric Tyrosine Kinase Dimer

The tyrosine kinase domain of the EGF receptor consists of an N-terminal domain built around a twisted β-sheet and a C-terminal domain that is primarily α-helical (Figure 32.12a). In the inactive, monomeric EGF receptor, the active site of the tyrosine kinase domain is blocked by a 30-residue loop of the protein (residues 831–860) that is termed the **activation loop.** Extracellular dimer formation by the EGF receptor appears to promote formation of an asymmetric dimer of the intracellular tyrosine kinase domains (Figure 32.12b), with the C-terminal lobe of one kinase domain juxtaposed with the N-terminal lobe of the other kinase domain. In this asymmetric dimer, one monomer is inactive but acts as an activator of kinase activity in the other monomer. Conversion of the kinase domain from its inactive state to the active conformation involves rotation of the activation loop out of the active site, making room for a peptide substrate to enter the site (Figure 32.12c).

The activated tyrosine kinase domain of the EGF receptor can phosphorylate several tyrosine residues at its own C-terminus (Figure 32.13), a process termed **autophosphorylation.** These phosphorylated tyrosines are binding sites for a variety of other signaling proteins that contain phosphotyrosine-binding SH2 domains (see Figures 32.5 and 32.6). Each of these SH2-domain–containing proteins can initiate several signal transduction cascades.

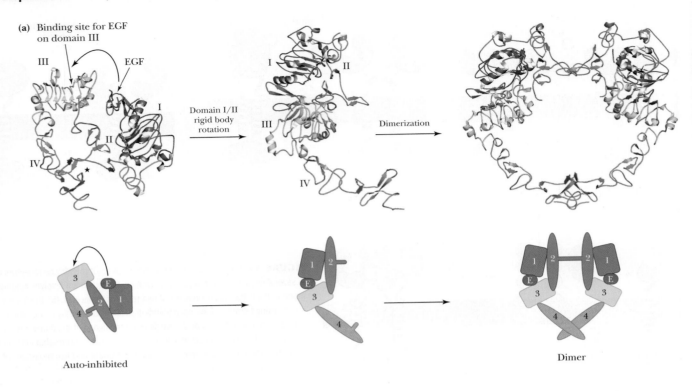

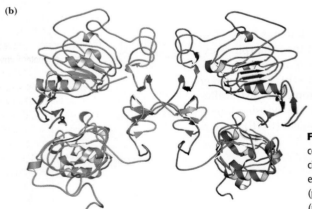

FIGURE 32.11 (a) In the absence of hormone, the extracellular domain of the EGF receptor is folded with the hairpin loop and dimerization interface on domain II (green) associated with domain IV (orange). Binding of EGF (red) induces a conformation change that exposes the dimerization interface and promotes formation of the activated dimer complex (pdb id = 1NQL). (b) Dimerization involves dove-tailing of the domain II hairpin loops (pdb id = 1IVO).

The Insulin Receptor Mediates Several Signaling Pathways

Insulin, a small heterodimeric peptide (see Figure 5.8), is the most potent anabolic hormone known. It regulates blood glucose levels, and it promotes the synthesis and storage of carbohydrates, proteins, and lipids. Abnormalities of insulin production, action, or both lead to diabetes, as well as other serious health issues. Insulin binding to receptors in liver, muscle, and other tissues triggers multiple signaling pathways, and insulin action is responsible for a variety of cellular effects.

The insulin receptor is an RTK that catalyzes the phosphorylation of tyrosine residues of several intracellular substrates, including the **insulin receptor substrate (IRS)** proteins, and several other proteins known as **Gab-1, Shc,** and **APS** (Figure 32.14). Each of these phosphorylated substrates binds a particular family of proteins containing SH2 domains (see Figures 32.5 and 32.6). These SH2 proteins interact specifically with phosphotyrosine-containing IRS sequences. Each of these signaling pathways can be confined to distinct cellular locations and can proceed with a different time course, thus providing spatial and temporal dimensions to insulin action in cells.

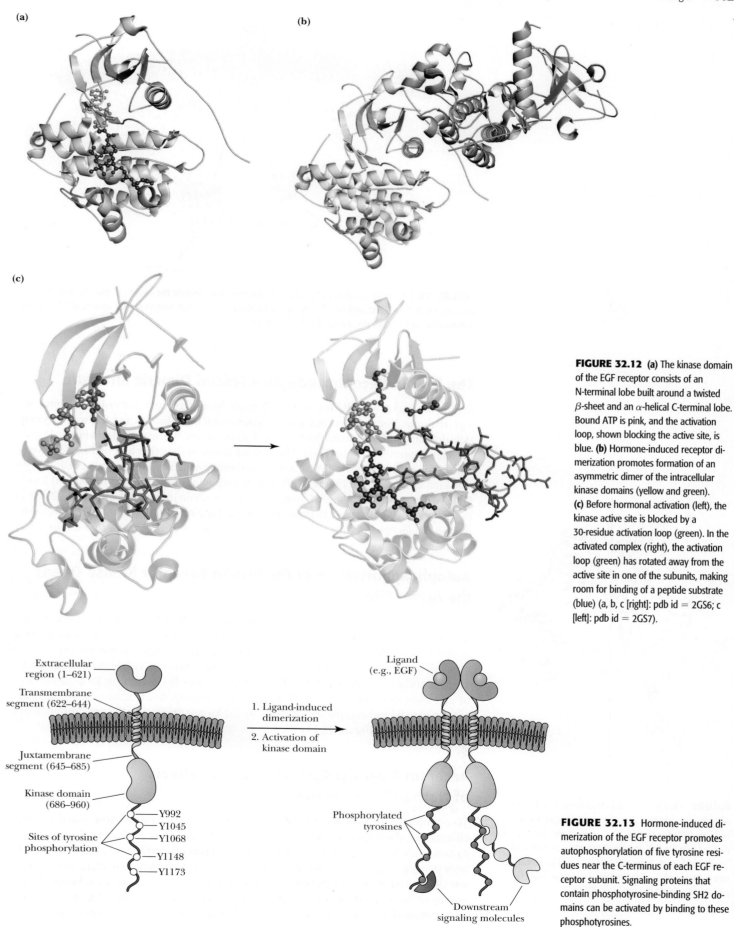

(a)

(b)

(c)

FIGURE 32.12 **(a)** The kinase domain of the EGF receptor consists of an N-terminal lobe built around a twisted β-sheet and an α-helical C-terminal lobe. Bound ATP is pink, and the activation loop, shown blocking the active site, is blue. **(b)** Hormone-induced receptor dimerization promotes formation of an asymmetric dimer of the intracellular kinase domains (yellow and green). **(c)** Before hormonal activation (left), the kinase active site is blocked by a 30-residue activation loop (green). In the activated complex (right), the activation loop (green) has rotated away from the active site in one of the subunits, making room for binding of a peptide substrate (blue) (a, b, c [right]: pdb id = 2GS6; c [left]: pdb id = 2GS7).

Extracellular region (1–621)

Transmembrane segment (622–644)

Juxtamembrane segment (645–685)

Kinase domain (686–960)

Sites of tyrosine phosphorylation

Y992
Y1045
Y1068
Y1148
Y1173

1. Ligand-induced dimerization

2. Activation of kinase domain

Ligand (e.g., EGF)

Phosphorylated tyrosines

Downstream signaling molecules

FIGURE 32.13 Hormone-induced dimerization of the EGF receptor promotes autophosphorylation of five tyrosine residues near the C-terminus of each EGF receptor subunit. Signaling proteins that contain phosphotyrosine-binding SH2 domains can be activated by binding to these phosphotyrosines.

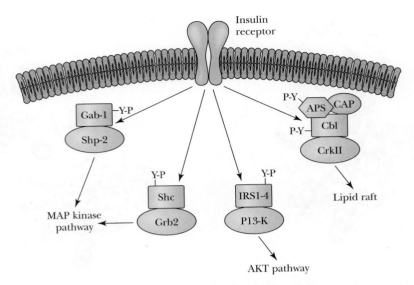

FIGURE 32.14 Substrates of the insulin receptor tyrosine kinase include the insulin receptor substrate (IRS) proteins, as well as Gab-1, Shc, and APS. The phosphorylated substrates in turn bind to several families of SH2 domain–containing proteins, activating several signaling pathways.

The Insulin Receptor Adopts a Folded Dimeric Structure

Unlike the majority of RTKs, which are single-chain receptors, the insulin receptor is an $\alpha_2\beta_2$ tetramer. The α-chain contains two leucine-rich domains (L1 and L2) with a cysteine-rich domain (CR) between them, as well as an intact fibronectin domain (Fn1) and a partial fibronectin domain (Fn2). The β-chain contains the other half of Fn2, followed by a third fibronectin domain (Fn3), a transmembrane α-helix, and (inside the cell) the tyrosine kinase domain (see Figure 32.9). The extracellular domain (the ectodomain) forms a **folded dimer,** with the L1 and L2 domains of one α-chain juxtaposed with the Fn2 domain of the other α/β chain pair, to create the insulin-binding site (Figure 32.15). Thus, *each of the two insulin-binding sites of the ectodomain consists of portions of both α-chains.*

Autophosphorylation of the Insulin Receptor Kinase Opens the Active Site

Binding of insulin to its receptor activates the tyrosine kinase activity of the intracellular domains. Like the EGF receptor kinase, the insulin receptor kinase contains an activation loop that lies across the kinase active site, excluding substrate peptides and thus inhibiting the kinase. Phosphorylation of three tyrosine residues on the activation loop—another case of autophosphorylation—causes the activation loop to move out of and away from the active site (Figure 32.16). This opens the active site so that target proteins are bound and phosphorylated by the kinase, triggering the appropriate signaling pathways (see Figure 32.14).

Receptor Guanylyl Cyclases Mediate Effects of Natriuretic Hormones

When you eat a salty meal, your body secretes hormones that protect you from the harmful effects of excess salt intake. When your heart senses that blood volume is too great, it sends signals to the kidneys to excrete NaCl and water (processes termed natriuresis and diuresis, respectively). These are just two examples of the action of **natriuretic hormones,** which allow tissues and organs to communicate with one another to regulate the volumes of blood and other body fluids and the osmotic effects of Na^+, K^+, Cl^-, and other ions. **Guanylin** and **uroguanylin** are produced in the intestines after ingestion of a salty meal and are

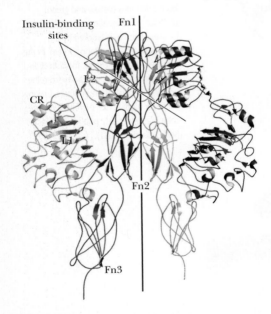

FIGURE 32.15 The ectodomain formed by the insulin receptor is constructed from the two α/β units. These two units are folded so that the L1 domain of one α-chain is juxtaposed with the Fn1 domain of the other α-chain. Each insulin-binding site is created by the L1 and L2 domains of one α-chain and the Fn2 domain of the other α/β chain pair (see Figure 32.9) (pdb id = 2DTG).

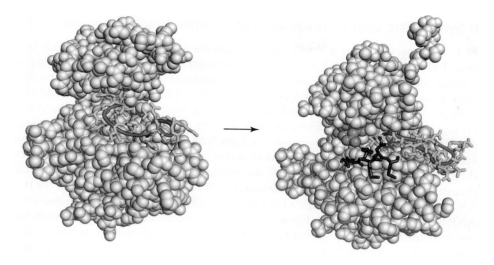

FIGURE 32.16 In its inactive state, the insulin receptor tyrosine kinase is inhibited by an activation loop (yellow and red), which prevents substrate access to the active site (left—pdb id = 1IRK). Autophosphorylation of three tyrosine residues on the activation loop induces a conformation change that rotates the loop out of the active site (right—pdb id = 1IR3), permitting access by insulin receptor substrates (blue) and ATP (ATP analog in orange).

secreted into the intestinal lumen. Binding of guanylin and uroguanylin to receptor guanylyl cyclases on cell membranes lining the lumen activates the intracellular guanylyl cyclase (Figure 32.17a). cGMP (see Figure 10.12) produced in this reaction is a second messenger that inhibits Na^+ uptake from the lumen and activates Cl^- export into the lumen. The result is a beneficial enhanced excretion of NaCl and water. (This effect can be carried too far, however. **Heat-stable enterotoxin (ST)** produced by *E. coli*—with a sequence similar to guanylin and uroguanylin—causes violent diarrhea.) **Atrial natriuretic peptide (ANP)** and **brain natriuretic peptide (BNP,** so named because it was discovered first in the brain) are both produced primarily in the heart. When the heart muscle is stressed and stretched by increased blood volume, the heart secretes ANP and BNP into the blood. In the kidneys, ANP and BNP bind to RGCs in kidney tubules, activating intracellular guanylyl cyclase and producing cGMP (Figure 32.17b). As in the intestines, cGMP inhibits Na^+ uptake, once again stimulating excretion of salt and water and reducing blood volume.

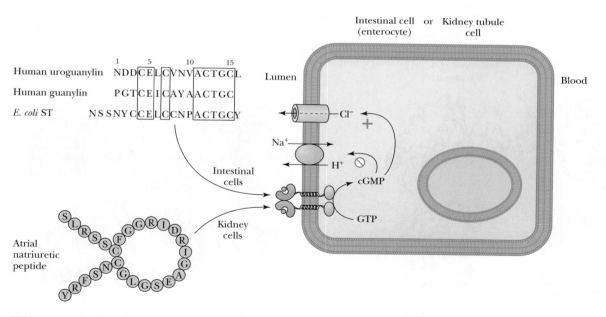

FIGURE 32.17 **(a)** Ingestion of a salty meal triggers excretion of NaCl and water in the intestines. Binding of guanylin and uroguanylin to RGC on cell membranes lining the lumen activates the intracellular guanylyl cyclase. cGMP produced by the cyclase inhibits Na^+ uptake from the lumen and activates Cl^- export into the lumen. **(b)** Atrial natriuretic peptide (ANP) protects the heart and circulatory system from the deleterious effects of increased blood volume. When the heart muscle is stressed and stretched, the heart secretes ANP. ANP binding to RGCs in kidney tubules activates intracellular guanylyl cyclase, producing cGMP. Inhibition of Na^+ uptake in the kidneys stimulates excretion of salt and water, reducing blood volume.

A Symmetric Dimer Binds an Asymmetric Peptide Ligand

RGC monomers associate as dimers in the membrane, even in the absence of their hormone ligands (Figure 32.18a). A dimeric receptor complex is activated by the binding of a single polypeptide hormone. This raises two important questions about the mechanism of action of these receptors: (1) How does an asymmetric hormone ligand (for example, ANP) bind to a symmetric homodimeric receptor? And (2) how does hormone binding to its extracellular dimeric receptor activate the intracellular guanylyl cyclase domain? Answers to these provocative questions have been provided by Kunio Misono and his colleagues, who determined the structure of the ANP receptor–ANP complex (Figure 32.18b). Remarkably, there is no significant intramolecular conformational change in either of the ANP receptor monomers. However, upon ANP binding, the two ANP receptor molecules undergo a twist motion in the membrane in order to insert the ANP hormone between them (Figure 32.19). The hormone lies like a disc between the two receptor subunits. The hormone-induced twist of the extracellular domains suggests a mechanism for activation of guanylyl cyclase activity across the membrane. Rotation of the transmembrane helices upon hormone binding could reorient the two intracellular domains, giving rise to guanylyl cyclase activity.

Nonreceptor Tyrosine Kinases Are Typified by pp60^{src}

The first tyrosine kinases to be discovered were associated with **viral transforming proteins.** These proteins, produced by **oncogenic viruses,** enable the virus to *transform* animal cells, that is, to convert them to the cancerous state. A prime example is the tyrosine kinase expressed by the **src gene** of **Rous** or **avian sarcoma virus.** The protein product of this gene is **pp60^{v-src}** (the abbreviation refers to phosphoprotein, 60 kD, viral origin, sarcoma-causing). The v-src gene was derived from the avian proto-oncogenic gene c-src during the original formation of the virus. The cellular proto-oncogene homolog of pp60^{v-src} is referred to as pp60^{c-src}. pp60^{v-src} is a 526-residue peripheral membrane protein. It undergoes two post-translational modifications: First, the amino group of the NH$_2$-terminal glycine is modified by the covalent attachment of a **myristoyl** group (this modification is required for membrane association of the kinase; see Figure 32.20). Then Ser^{17} and Tyr^{416} are phosphorylated. The phosphorylation at Tyr^{416}, which increases kinase activity twofold to threefold, appears to be an autophosphorylation. On the other hand, phosphorylation at Tyr^{527} is inhibitory and is catalyzed by another

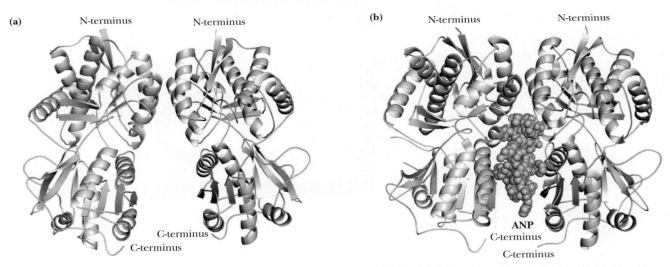

(a) N-terminus   N-terminus   **(b)** N-terminus   N-terminus

C-terminus
C-terminus

ANP
C-terminus

C-terminus

FIGURE 32.18 Activation of the ANP receptor involves binding of an asymmetric ligand at the interface of two identical receptor domains. Comparison of the structures of the receptor domain in the absence **(a)** and presence **(b)** of ANP reveals no significant intramolecular conformational change in either of the receptor monomers. ANP binding induces a twist of the two ANP receptor molecules, allowing the ANP hormone to insert between them. (a) PDB file provided by Kunio Misono, University of Nevada, Reno; (b) (pdb id = 1T34).

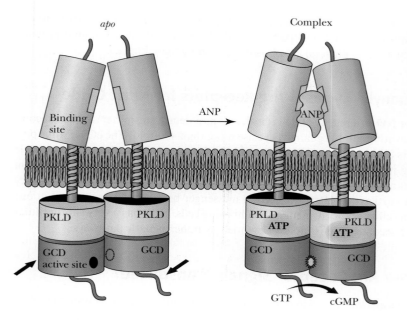

(b)

FIGURE 32.19 The rotation mechanism proposed by Misono and colleagues for transmembrane signaling by the ANP receptor. ANP binding causes a twist of the two extracellular domains, leading to rotation of the two intracellular domains and activation of guanylyl cyclase activity. PKLD is a protein-kinase-like domain, and GCD is a guanylyl cyclase domain. (Adapted from Misono, K., Ogawa, H., Qiu, Y., and Ogata, C., 2005. Structural studies of the natriuretic peptide receptor: A novel hormone-induced rotation mechanism for transmembrane signal transduction. *Peptides* **26**:957–968.)

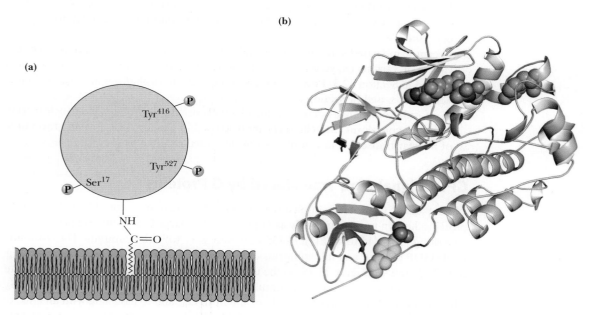

FIGURE 32.20 **(a)** The soluble tyrosine kinase pp60$^{v\text{-}src}$ is anchored to the plasma membrane via an N-terminal myristoyl group. **(b)** The structure of protein tyrosine kinase pp60$^{c\text{-}src}$, showing AMP–PNP in the active site (blue, green, red), Tyr^{416} (orange), and Tyr^{527} (yellow). Tyr^{527} is phosphorylated (purple).

▶ A DEEPER LOOK

Nitric Oxide, Nitroglycerin, and Alfred Nobel

NO · is the active agent released by **nitroglycerin** (see accompanying figure), a powerful drug that ameliorates the symptoms of heart attacks and **angina pectoris** (chest pain due to coronary artery disease) by causing the dilation of coronary arteries. Nitroglycerin is also the active agent in dynamite. Ironically, Alfred Nobel, the inventor of dynamite who also endowed the Nobel prizes, himself suffered from angina pectoris. In a letter to a friend in 1885, Nobel wrote, "It sounds like the irony of fate that I should be ordered by my doctor to take nitroglycerin internally."

$$CH_2 — CH — CH_2$$

The structure of nitroglycerin, a potent vasodilator.

kinase known as CSK. The significance of nonreceptor tyrosine kinase activity to cell growth and transformation is only partially understood, but 1% of all cellular proteins (many of which are also kinases) are phosphorylated by these kinases.

Soluble Guanylyl Cyclases Are Receptors for Nitric Oxide

Nitric oxide, or **NO ·**, a reactive free radical, acts as a neurotransmitter and as a second messenger, activating soluble guanylyl cyclase more than 400-fold. The cGMP thus produced also acts as a second messenger, inducing relaxation of vascular smooth muscle and mediating penile erection. As a dissolved gas, NO · is capable of rapid diffusion across membranes in the absence of any apparent carrier mechanism. This property makes NO · a particularly attractive second messenger because NO · generated in one cell can exert its effects quickly in many neighboring cells. NO · has a very short cellular half-life (1 to 5 seconds) and is rapidly degraded by nonenzymatic pathways.

32.4   How Are Receptor Signals Transduced?

Receptor signals are *transduced* in one of three ways to initiate actions inside the cell:

1. Exchange of GDP for GTP by GTP-binding proteins (G proteins), which leads to generation of *second messengers,* including cAMP, phospholipid breakdown products, and Ca^{2+}.
2. Receptor-mediated activation of phosphorylation cascades that in turn trigger activation of various enzymes. This is the action of the receptor tyrosine kinases described in Section 32.3. Protein kinases and protein phosphatases acting as effectors will be discussed in Section 32.5.
3. Conformation changes that open ion channels or recruit proteins into nuclear transcription complexes. Ion channels are discussed in Section 32.7, and the formation of nuclear transcription complexes was described in Chapter 29.

GPCR Signals Are Transduced by G Proteins

The signals of G-protein–coupled receptors (GPCRs) are transduced by GTP-binding proteins, known more commonly as G proteins. The large G proteins are heterotrimers consisting of α- (45 to 47 kD), β- (35 kD), and γ- (7 to 9 kD) subunits. The α-subunit binds GDP or GTP and has an intrinsic, slow GTPase activity. The $G_{\alpha\beta\gamma}$ complex (Figure 32.21, and see Figure 15.19) in the unactivated state has GDP at the nucleotide site. Binding of hormone to receptor stimulates a rapid exchange of GTP for GDP on G_α. The binding of GTP causes G_α to dissociate from $G_{\beta\gamma}$ and to associate with an effector protein such as adenylyl cyclase (Figure 32.22). *Binding of $G_\alpha(GTP)$ activates adenylyl cyclase.* The adenylyl cyclase actively synthesizes cAMP as long as G_α(GTP) remains bound to it. However, the intrinsic GTPase activity of G_α eventually hydrolyzes GTP to GDP, leading to dissociation of G_α(GDP) from adenylyl cyclase and reassociation with the $G_{\beta\gamma}$ dimer, regenerating the inactive heterotrimeric $G_{\alpha\beta\gamma}$ complex.

The hormone-activated GPCR is a **guanine–nucleotide exchange factor (GEF)**—promoting the exchange of GDP with GTP on the G protein—in a manner entirely similar to the interaction of EF-Ts with EF-Tu(GDP) (pages 1065–1066). By contrast, the $G_{\beta\gamma}$ complex, which normally acts to inhibit the spontaneous release of GDP from G_α (in the inactivated state of the GPCR), is termed a **guanine–nucleotide dissociation inhibitor (GDI).** Other proteins may also behave as GEFs and GDIs; their actions are discussed in Section 32.5.

Two stages of amplification occur in the G-protein–mediated hormone response. First, a single hormone-receptor complex can activate many G proteins before the hormone dissociates from the receptor. Second, and more obvious, the G_α-activated adenylyl cyclase synthesizes many cAMP molecules. Thus, the binding of hormone to a very small number of membrane receptors stimulates a large increase in concentration of cAMP within the cell. The hormone receptor, G protein, and cyclase constitute a complete hormone **signal transduction unit.**

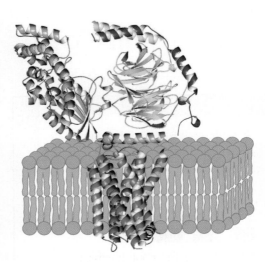

FIGURE 32.21 A heterotrimeric G protein (α–pink, β–yellow, γ–blue) docked with a β_2-adrenergic receptor (green) (pdb id = 2RH1 and 1GOT).

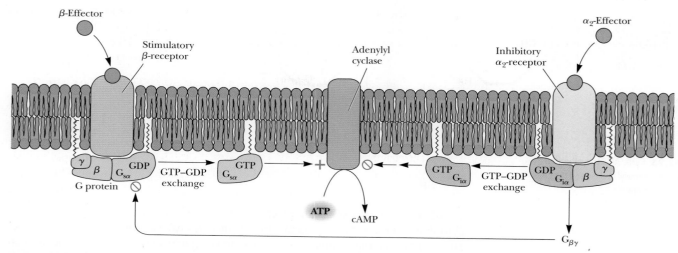

FIGURE 32.22 Adenylyl cyclase activity is modulated by the interplay of stimulatory (G_s) and inhibitory (G_i) G proteins. Binding of hormones to β_1- and β_2-adrenergic receptors activates adenylyl cyclase via G_s, whereas hormone binding to α_2-adrenergic receptors leads to the inhibition of adenylyl cyclase. Inhibition may occur by direct inhibition of cyclase activity by $G_{i\alpha}$ or by binding of $G_{i\beta\gamma}$ to $G_{s\alpha}$.

Hormone-receptor–mediated processes regulated by G proteins may be stimulatory or inhibitory. Each hormone receptor interacts specifically with either a stimulatory G protein, denoted G_s, or an inhibitory G protein, denoted G_i (Figure 32.22).

Cyclic AMP Is a Second Messenger

Cyclic AMP (denoted *cAMP*) was identified in 1956 by Earl Sutherland, who termed cAMP a **second messenger** because it is the intracellular response provoked by binding of hormone (the first messenger) to its receptor. Since Sutherland's discovery of cAMP, many other second messengers have been identified (Table 32.1). The concentrations of second messengers in cells are carefully regulated. Synthesis or release of a second messenger is followed quickly by degradation or removal from the cytosol. Following its synthesis by adenylyl cyclase, cAMP is broken down to 5′-AMP by phosphodiesterase (Figure 32.23).

TABLE 32.1	Intracellular Second Messengers*	
Messenger	Source	Effect
cAMP	Adenylyl cyclase	Activates protein kinases
cGMP	Guanylyl cyclase	Activates protein kinases, regulates ion channels, regulates phosphodiesterases
Ca^{2+}	Ion channels in ER and plasma membrane	Activates protein kinases, activates Ca^{2+}-modulated proteins
IP_3	PLC action on PI	Activates Ca^{2+} channels
DAG	PLC action on PI	Activates protein kinase C
Phosphatidic acid	Membrane component and product of PLD	Activates Ca^{2+} channels, inhibits adenylyl cyclase
Ceramide	PLC action on sphingomyelin	Activates protein kinases
Anandamide	Action of phospholipases and other enzymes on glycerophospholipids	Activation or inhibition of neuronal circuits
Hydrogen sulfide (H_2S)	Action of cystathionine γ lyase on cysteine	Regulation of blood vessels and heart and inflammatory responses
Carbon monoxide (CO)	Action of heme oxygenase in heme catabolism	Cryoprotective and homeostatic functions
Nitric oxide (NO·)	NO synthase	Activates guanylyl cyclase, relaxes smooth muscle
Cyclic ADP-ribose	cADP-ribose synthase	Activates Ca^{2+} channels

*IP_3 is inositol-1,4,5-trisphosphate; PI is phosphatidylinositol; DAG is diacylglycerol; PLC is phospholipase C; PLD is phospholipase D (see Figure 32.26).

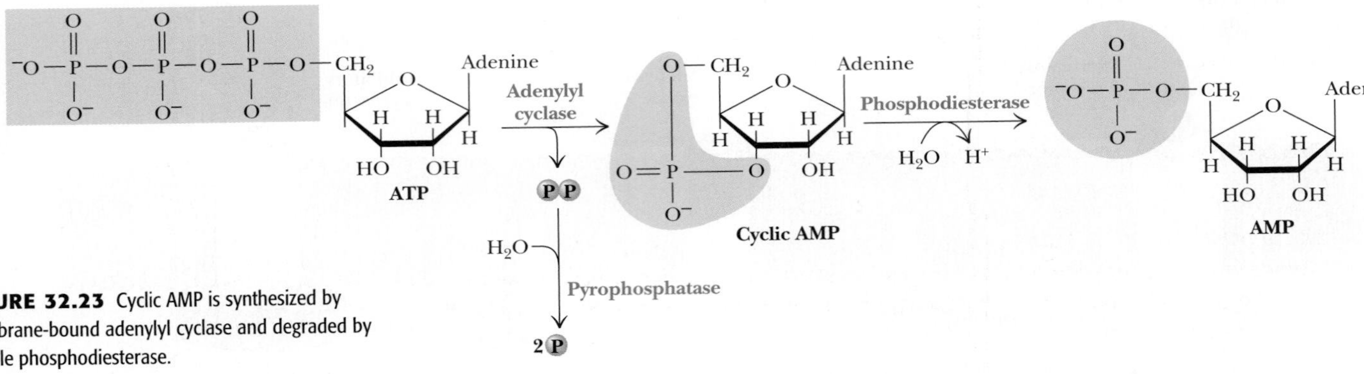

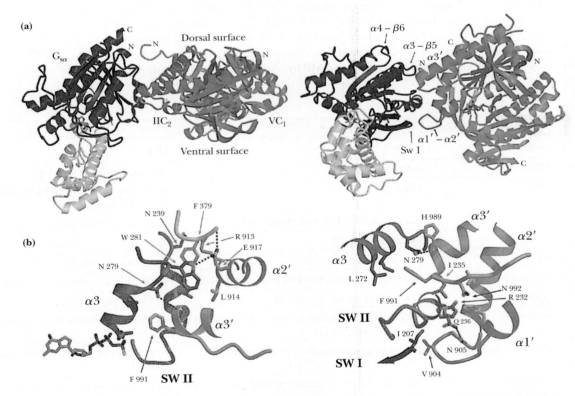

FIGURE 32.23 Cyclic AMP is synthesized by membrane-bound adenylyl cyclase and degraded by soluble phosphodiesterase.

Miles Houslay and others have shown that spatial and temporal cAMP concentration gradients modulate the actions of this second messenger. These gradients lie between the sites of cAMP synthesis (by adenylyl cyclase) and the sites of degradation (by specific phosphodiesterases).
Houslay, M., 2010. Underpinning compartmentalized cAMP signaling through targeted cAMP breakdown. Trends in Biochemical Sciences **36**:91–100.

Adenylyl cyclase (AC) is an integral membrane enzyme. Its catalytic domain, on the cytoplasmic face of the plasma membrane, includes two subdomains, denoted VC_1 and IIC_2. Binding of the α-subunit of G_s (denoted $G_{s\alpha}$) activates the AC catalytic domain.

Alfred Gilman, Stephen Sprang, and co-workers have determined the structure of a complex of $G_{s\alpha}$ (with bound GTP) with the cytoplasmic domains (VC_1 and IIC_2) of adenylyl cyclase (Figure 32.24). The $G_{s\alpha}$ complex binds to a cleft at one corner of the C_2 domain, and the surface of $G_{s\alpha}$-GTP that contacts adenylyl cyclase is the same surface that binds the $G_{\beta\gamma}$ dimer. The catalytic site, where ATP is converted to cyclic AMP, is far removed from the bound G protein.

cAMP Activates Protein Kinase A

All second messengers exert their cellular effects by binding to one or more target molecules. cAMP produced by adenylyl cyclase activates a protein kinase, which is thus known as *cAMP-dependent protein kinase*. Protein kinase A, as this enzyme is also known, activates many other cellular proteins by phosphorylation. The activation of

FIGURE 32.24 (a) Two views of the complex of $G_{s\alpha}$ with the VC_1–IIC_2 catalytic domain of adenylyl cyclase and $G_{s\alpha}$. **(b)** Details of the $G_{s\alpha}$ complex in the same orientation as the structures in (a). SW-I and SW-II are "switch regions," whose conformations differ greatly depending on whether GTP or GDP is bound. (Courtesy of Alfred Gilman, University of Texas Southwestern Medical Center.)

protein kinase A by cAMP and regulation of the enzyme by intrasteric control was described in detail in Chapter 15. The structure of protein kinase A has served as a paradigm for understanding many related protein kinases (see Figure 15.9).

Ras and Other Small GTP-Binding Proteins Are Proto-Oncogene Products

GTP-binding proteins are implicated in growth control mechanisms in higher organisms. Certain tumor virus genomes contain genes encoding 21-kD proteins that bind GTP and show regions of homology with other G proteins. The first of these genes to be identified was found in *ra*t *s*arcoma virus and was dubbed the ***ras* gene.** Genes implicated in tumor formation are known as **oncogenes;** they are often mutated versions of normal, noncancerous genes involved in growth regulation, so-called **proto-oncogenes.** The normal, cellular Ras protein is a GTP-binding protein that functions in a manner similar to that of other G proteins described previously, activating metabolic processes when GTP is bound and becoming inactive when GTP is hydrolyzed to GDP. The GTPase activity of the normal Ras p21 is very low, as is appropriate for a G protein that regulates long-term effects like growth and differentiation. A specific **GTPase-activating protein (GAP)** increases the GTPase activity of the Ras protein. Mutant (oncogenic) Ras proteins have severely impaired GTPase activity, which apparently causes serious alterations of cellular growth and metabolism in tumor cells. The conformations of Ras proteins (Figure 32.25) in complexes with GDP are different from the corresponding complexes with GTP analogs such as GMP–PNP (a nonhydrolyzable analog of GTP in which the β-P and γ-P are linked by N rather than by O). Two regions of the Ras structure change conformation upon GTP hydrolysis. These conformation changes mediate the interactions of Ras with other proteins, termed **effectors.**

G Proteins Are Universal Signal Transducers

A given G protein can be activated by several different hormone-receptor complexes. For example, either glucagon or epinephrine, binding to their distinctive receptor proteins, can activate the same species of G protein in liver cells. The effects are additive, and combined stimulation by glucagon and epinephrine leads to higher cytoplasmic concentrations of cAMP than activation by either hormone alone.

G proteins are a universal means of signal transduction in higher organisms, activating many hormone-receptor–initiated cellular processes in addition to adenylyl cyclase. Such processes include, but are not limited to, activation of phospholipases C and A_2 and the opening or closing of transmembrane channels for K^+, Na^+, and Ca^{2+} in brain, muscle, heart, and other organs (Table 32.2). G proteins are integral components of

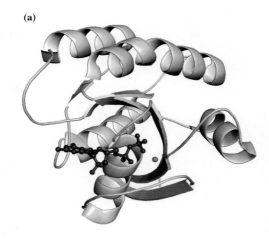

(a)

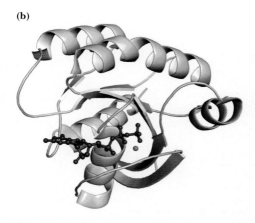

(b)

FIGURE 32.25 The structure of Ras complexed with **(a)** GDP (pdb id = 1LF5) and **(b)** GMP–PNP (pdb id = 1LF0). The Ras p21–GMP–PNP complex is the active conformation of this protein. A Mg^{2+} ion (red) is shown in both structures.

TABLE 32.2	G Proteins and Their Physiological Effects			
G Protein	Location	Stimulus	Effector	Effect
G_s	Liver	Epinephrine, glucagon	Adenylyl cyclase	Glycogen breakdown
G_s	Adipose tissue	Epinephrine, glucagon	Adenylyl cyclase	Fat breakdown
G_s	Kidney	Antidiuretic hormone	Adenylyl cyclase	Conservation of water
G_i	Heart muscle	Acetylcholine	Potassium channel	Decreased heart rate and pumping force
G_i/G_o	Brain neurons	Enkephalins, endorphins, opioids	Adenylyl cyclase, potassium channels, calcium channels	Changes in neuron electrical activity
G_q	Smooth muscle cells in blood vessels	Angiotensin	Phospholipase C	Muscle contraction, blood pressure elevation
G_{olf}	Neuroepithelial cells in the nose	Odorant molecules	Adenylyl cyclase	Odorant detection
Transducin (G_t)	Retinal rod and cone cells of the eye	Light detection	cGMP phosphodiesterase	Light detection (vision)

sensory pathways such as vision and olfaction. More than 100 different GPCRs and at least 21 distinct G proteins are known. At least a dozen different G-protein effectors have been identified, including a variety of enzymes and ion channels.

Specific Phospholipases Release Second Messengers

A diverse array of second messengers are generated by breakdown of membrane phospholipids. Binding of certain hormones and growth factors to their respective receptors triggers a sequence of events that can lead to the activation of **specific phospholipases.** The action of these phospholipases on membrane lipids produces the second messengers shown in Figure 32.26.

Inositol Phospholipid Breakdown Yields Inositol-1,4,5-Trisphosphate and Diacylglycerol

Breakdown of **phosphatidylinositol (PI)** and its derivatives by **phospholipase C** produces a family of second messengers. In the best-understood pathway, successive phosphorylations of PI produce **phosphatidylinositol-4-P (PIP)** and **phosphatidylinositol-4,5-bisphosphate (PIP_2)**. Four isozymes of phospholipase C (denoted α, β, γ, and δ) hydrolyze PI, PIP, and PIP_2. Hydrolysis of PIP_2 by phospholipase C yields the second messenger **inositol-1,4,5-trisphosphate (IP_3)**, as well as another second messenger, **diacylglycerol (DAG)** (Figure 32.27). IP_3 is water soluble and diffuses to intracellular organelles where release of Ca^{2+} is activated. DAG, on the other hand, is lipophilic and remains in the plasma membrane, where it activates a Ca^{2+}-dependent protein kinase known as **protein kinase C** (see following discussion).

Activation of Phospholipase C Is Mediated by G Proteins or by Tyrosine Kinases

Phospholipase C-β, C-γ, and C-δ are all Ca^{2+}-dependent, but the different phospholipase C isozymes are activated by different intracellular events. Phospholipase C-β is stimulated by G proteins (Figure 32.28). On the other hand, phospholipase C-γ is

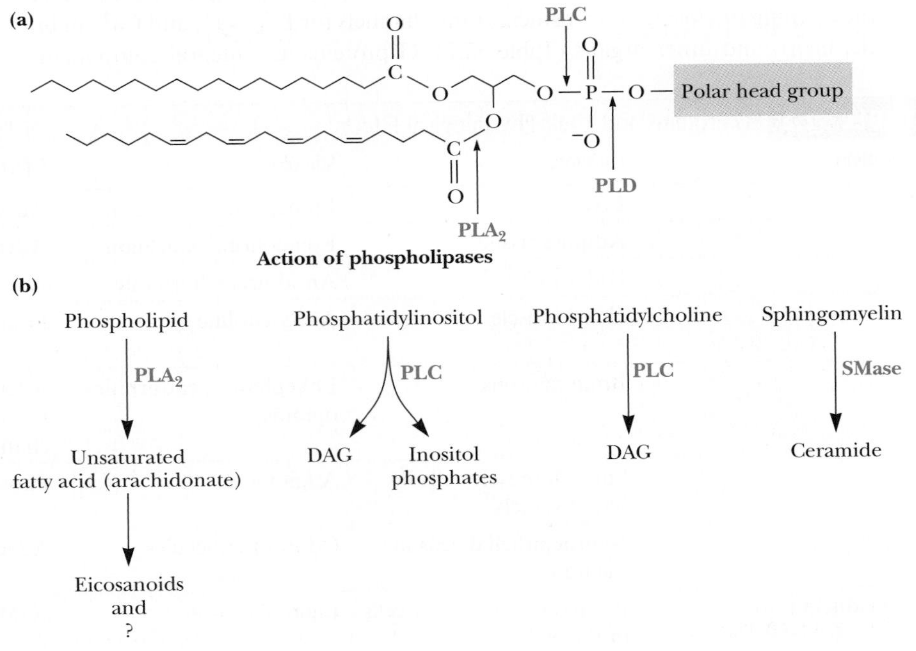

FIGURE 32.26 (a) The general action of phospholipase A_2 (PLA₂), phospholipase C (PLC), and phospholipase D (PLD). (b) The synthesis of second messengers from phospholipids by the action of phospholipases and sphingomyelinase (SMase), which for example cleaves a choline sphingomyelin to yield phosphocholine and ceramide.

activated by **receptor tyrosine kinases** (Figure 32.29). The domain organization of phospholipase C-β and C-γ is shown in Figure 32.30. The X and Y domains of phospholipase C-β and C-γ are highly homologous, and both of these domains are required for phospholipase C activity. The other domains of these isozymes confer specificity for G-protein activation or tyrosine kinase activation.

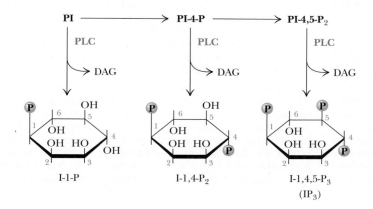

FIGURE 32.27 The family of second messengers produced by phosphorylation and breakdown of phosphatidylinositol. PLC action instigates a bifurcating pathway culminating in two distinct and independent second messengers: DAG and IP$_3$.

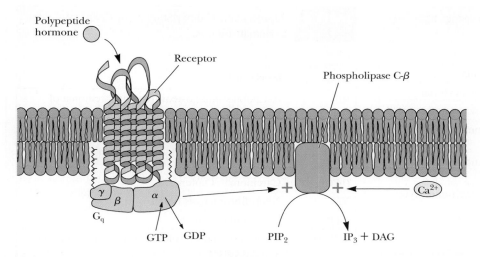

FIGURE 32.28 Phospholipase C-β is activated specifically by G$_q$, a GTP-binding protein, and also by Ca^{2+}.

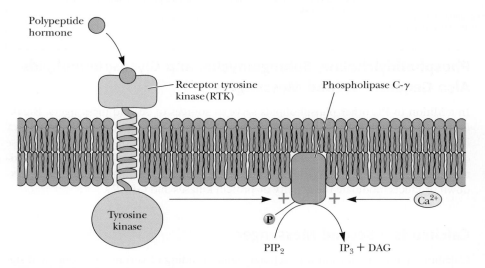

FIGURE 32.29 Phospholipase C-γ is activated upon phosphorylation by receptor tyrosine kinases and by Ca^{2+}.

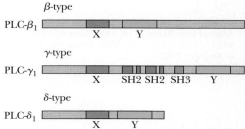

FIGURE 32.30 The amino acid sequences of phospholipase C isozymes β, γ, and δ share two homologous domains, denoted X and Y. The sequence of γ-isozyme contains src homology domains, denoted SH2 and SH3. SH2 domains (approximately 100 residues in length) interact with phosphotyrosine-containing proteins (such as RTKs), whereas SH3 domains mediate interactions with Pro-rich sequences. (Adapted from Dennis, E., Rhee, S., Gillah, M., and Hannun, E., 1991. Role of phospholipases in generating lipid second messengers in signal transduction. *The FASEB Journal* 5:2068–2077.)

HUMAN BIOCHEMISTRY

Cancer, Oncogenes, and Tumor Suppressor Genes

The disease state known as **cancer** is the uncontrolled growth and proliferation of one or more cell types in the body. Control of cell growth and division is an incredibly complex process, involving the signal-transducing proteins (and small molecules) described in this chapter and many others like them. The genes that give rise to these growth-controlling proteins are of two distinct types:

1. **Oncogenes:** These genes code for proteins that are capable of stimulating cell growth and division. In normal tissues and organisms, such growth-stimulating proteins are regulated so that growth is appropriately limited. However, mutations in these genes may result in loss of growth regulation, leading to uncontrolled cell proliferation and tumor development. These mutant genes are known as *oncogenes* because they induce the oncogenic state—cancer. The normal versions of these genes are termed **proto-oncogenes;** proto-oncogenes are essential for normal cell growth and differentiation. Oncogenes are *dominant* because mutation of only one of the cell's two copies of the gene can lead to tumor formation. Table A lists a few of the known oncogenes (more than 60 are now known).

2. **Tumor suppressor genes:** These genes code for proteins whose normal function is to *turn off* cell growth. A mutation in one of these growth-limiting genes may result in a protein product that has lost its growth-limiting ability. Since the normal products suppress tumor growth, the genes are known as *tumor suppressor genes*. Because both cellular copies of a tumor suppressor gene must be mutated to foil its growth-limiting action, these genes are *recessive* in nature. Table B presents several recognized tumor suppressor genes.

Careful molecular analysis of cancerous tissue has shown that tumor development may result from mutations in several proto-oncogenes or tumor suppressor genes. The implication is that *there is redundancy in cellular growth regulation.* Many (if not all) tumors are either the result of interactions of two or more oncogene products or arise from simultaneous mutations in a proto-oncogene and both copies of a tumor suppressor gene. Cells have thus evolved with overlapping growth-control mechanisms. When one is compromised by mutation, others take over.

TABLE A	A Representative List of Proto-Oncogenes Implicated in Human Tumors
Proto-Oncogene	Neoplasm(s)
Abl	Chronic myelogenous leukemia
ErbB-1	Squamous cell carcinoma; astrocytoma
ErbB-2 (Neu)	Adenocarcinoma of breast, ovary, and stomach
Myc	Burkitt's lymphoma; carcinoma of lung, breast, and cervix
H-*Ras*	Carcinoma of colon, lung, and pancreas; melanoma
N-*Ras*	Carcinoma of genitourinary tract and thyroid; melanoma
Ros	Astrocytoma
Src	Carcinoma of colon
Jun *Fos*	Several

Adapted from Bishop, J. M., 1991. Molecular themes in oncogenesis. *Cell* **64:**235–248; Croce, C. M., 2008. Oncogenes and cancer. *New England Journal of Medicine* **358:**502–511.

TABLE B	Representative Tumor Suppressor Genes Implicated in Human Tumors
Tumor Suppressor Gene	Neoplasm(s)
RB1	Retinoblastoma; osteosarcoma; carcinoma of breast, bladder, and lung
p53	Astrocytoma; carcinoma of breast, colon, and lung; osteosarcoma
WT1	Wilms' tumor
DCC	Carcinoma of colon
NF1	Neurofibromatosis type 1
FAP	Carcinoma of colon
MEN-1	Tumors of parathyroid, pancreas, pituitary, and adrenal cortex

Adapted from Bishop, J. M., 1991. Molecular themes in oncogenesis. *Cell* **64:**235–248, and Sherr, C. J., 2004. Principles of tumor suppression. *Cell* **116:**235–246.

Phosphatidylcholine, Sphingomyelin, and Glycosphingolipids Also Generate Second Messengers

Anandamide, formed by the action of several phospholipases from PC and PE (Chapter 8 and Table 32.1), is an intracellular messenger in many cells. Mauro Maccarrone and colleagues have proposed that anandamide for such purposes is stored in lipid droplets (adiposomes) and other intracellular reservoirs. *Maccarrone, M., Dainese, E., and Oddi, S., 2010. Intracellular trafficking of anandamide: New concepts for signaling.* Trends in Biochemical Sciences **36:**601–608.

In addition to PI, other phospholipids serve as sources of second messengers. Breakdown of phosphatidylcholine by phospholipases yields a variety of second messengers, including DAG, phosphatidic acid, and prostaglandins. The action of **sphingomyelinase** on sphingomyelin produces **ceramide,** which stimulates **ceramide-activated protein kinase.** Similarly, gangliosides (such as ganglioside G_{M3}; see Chapter 8) and their breakdown products modulate the activity of protein kinases and GPCRs.

Calcium Is a Second Messenger

Calcium ion is an important intracellular signal. Binding of certain hormones and signal molecules to plasma membrane receptors can cause transient increases in cytoplasmic Ca^{2+} levels, which in turn can activate a wide variety of enzymatic processes, including smooth muscle contraction, exocytosis, and glycogen metabolism. (Most of these activation processes depend on special Ca^{2+}-binding proteins discussed in the following section.) Cytoplasmic $[Ca^{2+}]$ can be increased in two ways (Figure 32.31). As mentioned briefly earlier, cAMP can activate the opening of plasma membrane Ca^{2+} channels, allowing extracellular Ca^{2+} to stream in. On the other hand, cells also contain intracellular reservoirs of Ca^{2+}, within the endoplasmic reticulum and **calciosomes,** small membrane vesicles that are similar in some ways to muscle sarcoplasmic reticulum. These special intracellular Ca^{2+} stores are *not* released by cAMP. They respond to IP_3, a second messenger derived from PI.

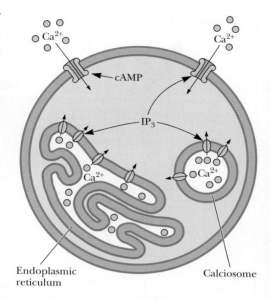

FIGURE 32.31 Cytosolic $[Ca^{2+}]$ increases occur via the opening of Ca^{2+} channels in the membranes of calciosomes, the endoplasmic reticulum, and the plasma membrane.

Intracellular Calcium-Binding Proteins Mediate the Calcium Signal

Given the central importance of Ca^{2+} as an intracellular messenger, it should not be surprising that complex mechanisms exist in cells to manage and control Ca^{2+}. When Ca^{2+} signals are generated by cAMP, IP_3, and other agents, these signals are translated into the desired intracellular responses by **calcium-binding proteins,** which in turn regulate many cellular processes (Figure 32.32). One of these, protein kinase C, is described in Section 32.5. The other important Ca^{2+}-binding proteins can, for the most part, be divided into two groups on the basis of structure and function: (1) the **calcium-modulated proteins,** including **calmodulin, parvalbumin, troponin C,** and many others, all of which have in common a structural feature called the **EF hand** (Figure 32.33), and (2) the **annexin proteins,** a family of homologous proteins that interact with membranes and phospholipids in a Ca^{2+}-dependent manner.

More than 170 calcium-modulated proteins are known. All possess a characteristic peptide domain consisting of a short α-helix, a loop of 12 amino acids, and a second α-helix (Figure 32.33). Robert Kretsinger at the University of Virginia initially discovered this pattern in parvalbumin, a protein first identified in the carp fish and later in neurons possessing a high firing rate and a high oxidative metabolism. Kretsinger lettered the six helices of parvalbumin A through F. He noticed that the E and F helices, joined by a loop, resembled the thumb and forefinger of a right hand and named this structure the *EF hand,* a name in common use today to identify the helix-loop-helix motif in calcium-binding proteins. In the EF hand, Ca^{2+} is coordinated by six carboxyl oxygens contributed by a glutamate and three aspartates, by a carbonyl oxygen from a peptide bond, and by the oxygen of a coordinated water molecule. The EF hand was subsequently identified in calmodulin, troponin C, and **calbindin-9K.** Most of the known EF-hand proteins possess two or more (as many as eight) EF-hand domains, usually arranged so that two EF-hand domains may directly contact each other.

Calmodulin Target Proteins Possess a Basic Amphiphilic Helix

The conformations of EF-hand proteins change dramatically upon binding of Ca^{2+} ions. This change promotes binding of the EF-hand protein with its target protein(s). For example, calmodulin (CaM), a 148-residue protein found in many cell types, modulates the

HUMAN BIOCHEMISTRY

PI Metabolism and the Pharmacology of Li^+

An intriguing aspect of the phosphoinositide story is the specific action of lithium ion, Li^+, on several steps of PI metabolism. Lithium salts have been used in the treatment of *manic-depressive illnesses* for more than 30 years, but the mechanism of lithium's therapeutic effects had been unclear. Recently, however, several reactions in the phosphatidylinositol degradation pathway have been shown to be sensitive to Li^+ ion. For example, Li^+ is an uncompetitive inhibitor of *myo*-inositol monophosphatase (see Chapter 13). Li^+ levels similar to those used in treatment of manic illness thus lead to the accumulation of several key intermediates. This story is far from complete, and many new insights into phosphoinositide metabolism and the effects of Li^+ can be anticipated.

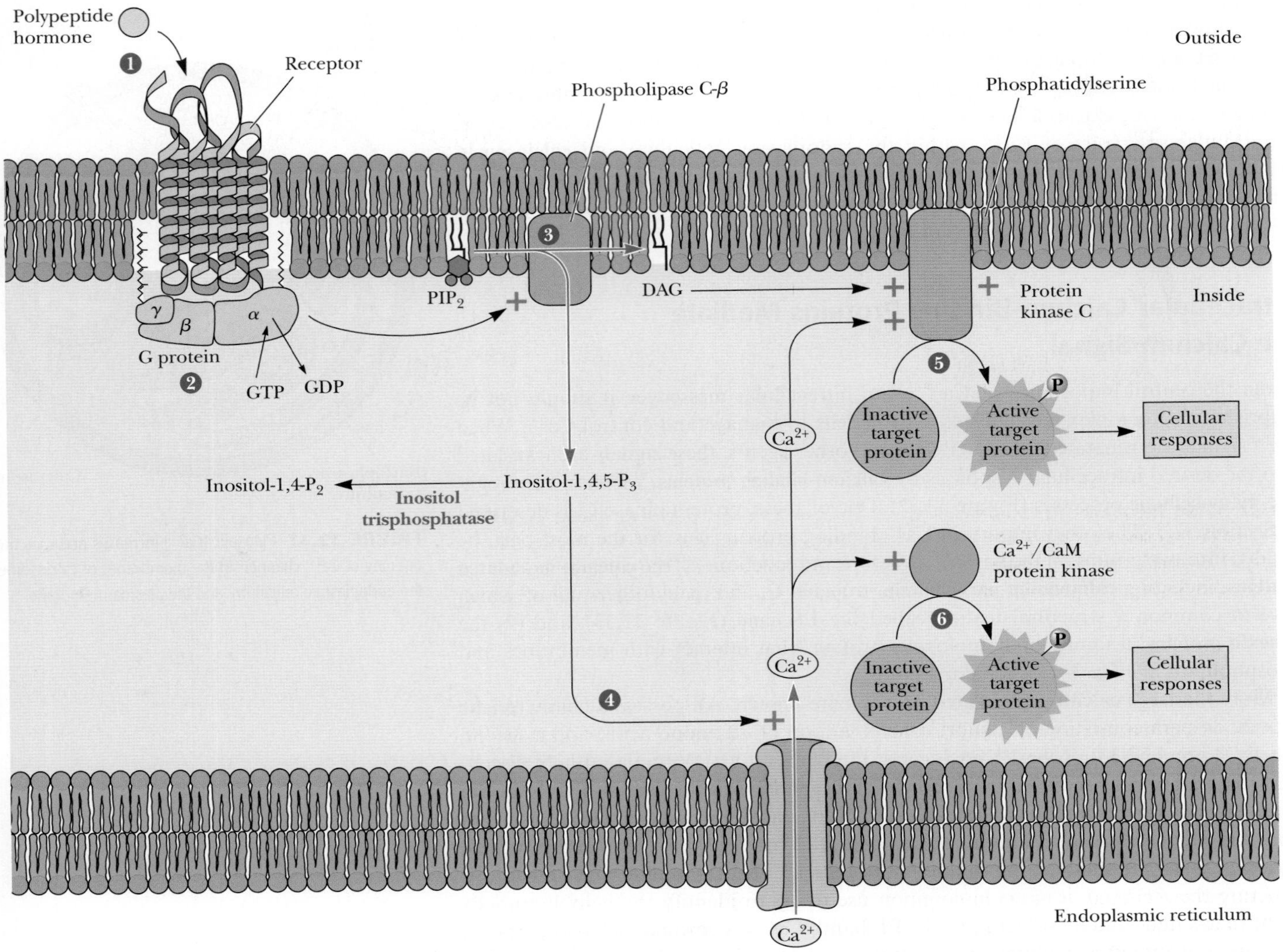

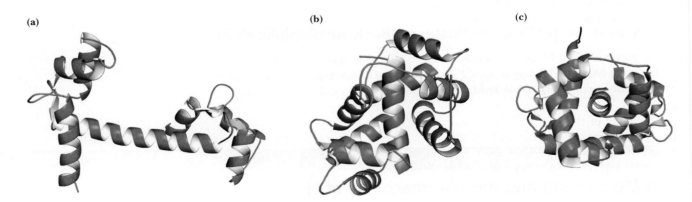

FIGURE 32.32 IP$_3$-mediated signal transduction pathways. Increased [Ca^{2+}] activates protein kinases, which phosphorylate target proteins. Ca^{2+}/CaM represents calci-calmodulin (Ca^{2+} complexed with the regulatory protein calmodulin).

(a)   **(b)**   **(c)**

FIGURE 32.33 (a) Structure of uncomplexed calmodulin (pdb id = 1LKJ). Calmodulin, with four Ca^{2+}-binding domains, forms a dumbbell-shaped structure with two globular domains joined by an extended, central helix. Each globular domain juxtaposes two Ca^{2+}-binding EF-hand domains. An intriguing feature of these EF-hand domains is their nearly identical three-dimensional structure despite a relatively low degree of sequence homology (only 25% in some cases). **(b, c)** Complex of calmodulin (red) with a peptide from myosin light chain kinase (blue); (b) side view; (c) top view (pdb id = 1QTX).

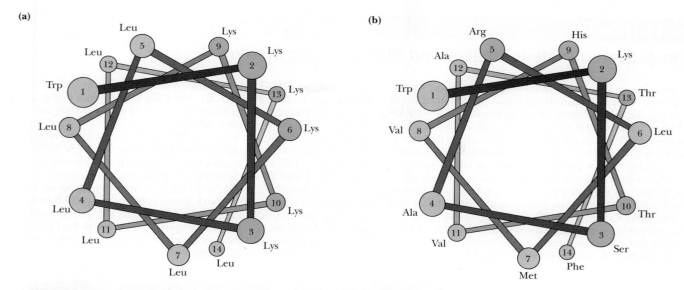

FIGURE 32.34 Helical wheel representations of **(a)** a model calmodulin-binding peptide, Ac-WKKLLKLLKKLLKL-CONH$_2$, and **(b)** the calmodulin-binding domain of spectrin. Positively charged and polar residues are indicated in green, and hydrophobic residues are orange. (Adapted from O'Neil, K., and DeGrado, W., 1990. How calmodulin binds its targets: Sequence independent recognition of amphiphilic α-helices. *Trends in Biochemical Sciences* **15**:59–64.)

activities of a large number of target proteins, including Ca^{2+}-ATPases, protein kinases, phosphodiesterases, and NAD$^+$ kinase. CaM binds to these and to many other proteins with extremely high affinities (K_D values typically in the high picomolar to low nanomolar range). All CaM target proteins possess a **b**asic **a**mphiphilic **a**lpha helix (a **Baa helix**), to which CaM binds specifically and with high affinity. Viewed end-on, in the so-called **helical wheel** representation (Figure 32.34), a Baa helix has mostly hydrophobic residues on one face; basic residues are collected on the opposite face. However, the Baa helices of CaM target proteins, although conforming to the model, show extreme variability in sequence. How does CaM, itself a highly conserved protein, accommodate such variety of sequence and structure? Each globular domain consists of a large hydrophobic surface flanked by regions of highly negative electrostatic potential—a surface suitable for interacting with a Baa helix. The long central helix joining the two globular regions behaves as a long, flexible tether. When the target protein is bound, the two globular domains fold together (Figure 32.33b). The flexible nature of the tethering helix allows the two globular domains to adjust their orientation synergistically for maximal binding of the target protein or peptide.

32.5 How Do Effectors Convert the Signals to Actions in the Cell?

Transduction of the hormonal signal leads to activation of **effectors**—usually protein kinases and protein phosphatases—that elicit a variety of actions that regulate discrete cellular functions. Of the thousands of mammalian kinases and phosphatases, the structures and functions of a few are representative.

Protein Kinase A Is a Paradigm of Kinases

Most protein kinases share a common catalytic core first characterized in protein kinase A (PKA), the enzyme that phosphorylates phosphorylase kinase (see Figure 15.9 and Figure 15.18). The active site of the catalytic subunit of PKA in a ternary complex with MnAMP–PNP and a pseudosubstrate inhibitor peptide, as shown in Figure 32.35,

> ## A DEEPER LOOK

Mitogen-Activated Protein Kinases and Phosphorelay Systems

In multicellular organisms, many physiological processes, including mitosis, gene expression, metabolism, and programmed death of cells, are regulated by a family of **mitogen-activated protein kinases (MAPKs).** (A mitogen is any agent that induces cell division, that is, mitosis.) MAPKs phosphorylate specific serines and threonines of target protein substrates, and these phosphorylation events function as switches to turn on or off the activity of the substrate proteins. These "substrates" may be other protein kinases, phospholipases, transcription factors, and cytoskeletal proteins. Protein phosphatases reverse the process, removing the phosphates that were added by MAPKs.

MAPKs are part of a **phosphorelay system** composed of three kinases that are activated in sequence (see accompanying figure). In such systems, the MAPK itself is phosphorylated by a MAPK kinase (denoted MKK), which is itself phosphorylated by a MAPK kinase kinase (denoted MKKK). MKKKs have distinct domains and motifs that respond to different cellular stimuli, and they have other domains that recognize specific MKKs. The same kinds of specificities regulate the action of MKKs. These specificities are accounted for in the classification of four subfamilies of MAPKs: One group is that of the extracellular signal-regulated kinases, notably ERK1 and ERK2; another is the c-Jun-amino-terminal kinases, including JNK, JNK1, and JNK2; a third group depends on the ERK5 kinase; and the fourth group involves the p38 kinases, including p38α, p38β, p38γ, and p38δ. There are undoubtedly other MAPK families yet to be discovered. As shown in the figure, these phosphorelay systems link a variety of stimuli to substrates that affect many cellular functions.

▶ MAPK phosphorelay systems. The left column is a general model. The four columns to the right show the four known MAPK phosphorelay families.

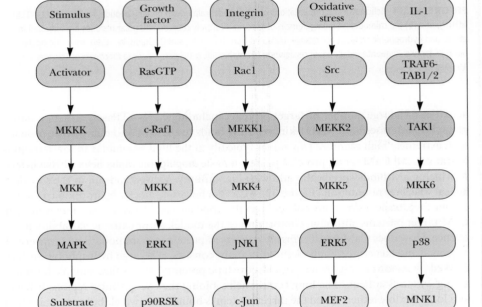

From Johnson, G. L., and Lapadat, R., 2002. Mitogen-activated protein kinase pathways mediated by ERK, JNK, and p38 protein kinases. *Science* **298:** 1911–1912.

includes a glycine-rich β-strand that acts as a flap over the triphosphate moiety of the bound nucleotide. A conserved residue, Asp[166], is the catalytic base that deprotonates the Ser/Thr-OH during phosphorylation, and Lys[168] stabilizes the transition state of the reaction. Three Glu residues on the enzyme are involved in recognition of the pseudosubstrate inhibitor peptide.

Protein Kinase C Is a Family of Isozymes

The enzymes called *protein kinase C* are actually a family of similar enzymes—isozymes—that encompass three subclasses. The "conventional PKCs," α, βI, βII, and γ, are regulated by Ca^{2+}, diacylglycerol (DAG), and phosphatidylserine (PS). Because Ca^{2+} levels increase in the cell in response to IP_3, the activation of conventional PKCs depends on both of the second messengers released by the hydrolysis of PIP_2. The "novel PKCs," δ, ϵ, θ, and η, are Ca^{2+}-independent but are regulated by DAG and PS. PKCs ζ, ι, and λ are

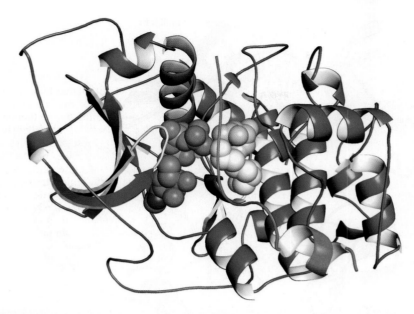

FIGURE 32.35 The structure of the catalytic subunit of PKA in a ternary complex with MnAMP–PNP and a pseudosubstrate inhibitor peptide (pink). A glycine-rich β-strand acts as a flap over the triphosphate moiety of the bound nucleotide. The glycine-rich flap that covers the ATP-binding site is shown in white, AMP–PNP is bound in the ATP site, Asp[166] is shown in beige, and Lys[168] is shown below in white (pdb id = 1ATP).

termed "atypical" and are activated by PS alone. These various cofactor requirements are imparted by subdomains represented in the conventional PKC polypeptide sequence. Conventional PKCs are comprised (Figure 32.36) of four conserved domains (C1–C4) and five variable regions (V1–V5). Domain C1 is a *pseudosubstrate sequence* that regulates the kinase by *intrasteric control* (see pages 489-490), C2 is a Ca^{2+}-binding domain, C3 is the ATP-binding domain, and C4 binds peptide substrates.

PKC phosphorylates serine and threonine residues on a wide range of protein substrates. A role for protein kinase C in cellular growth and division is demonstrated through its strong activation by **phorbol esters** (Figure 32.37). These compounds, from the seeds of *Croton tiglium,* are **tumor promoters**—agents that do not themselves cause tumorigenesis but that potentiate the effects of carcinogens. The phorbol esters mimic DAG, bind to the regulatory pseudosubstrate domain of the enzyme, and activate protein kinase C.

Protein Tyrosine Kinase pp60[c-src] Is Regulated by Phosphorylation/Dephosphorylation

The structure of protein tyrosine kinase pp60[c-src] (see Figure 32.20) consists of an N-terminal "unique domain," an SH2 domain, an SH3 domain, and a kinase domain that includes a small lobe comprised mainly of a twisted β-sheet and a large lobe that is predominantly α-helical (Figure 32.38; see also Section 32.2). Phosphorylation of Tyr[527]

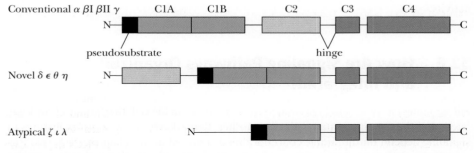

FIGURE 32.36 The primary structures of the PKC isozymes. Conserved domains C1–C4 are indicated. Variable regions are shown as simple lines.

12-*O*-Tetradecanoylphorbol-13-acetate

FIGURE 32.37 The structure of a phorbol ester. Long-chain fatty acids predominate at the 12-position, whereas acetate is usually found at the 13-position.

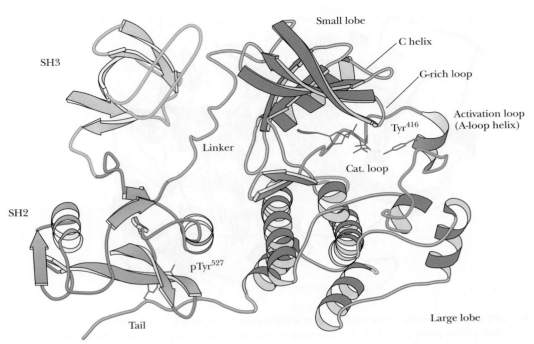

FIGURE 32.38 A ribbon diagram showing the structure of protein tyrosine kinase pp60[c-src] with bound AMP–PNP. (Image kindly provided by Stephen C. Harrison.)

in the SH2 domain inhibits tyrosine kinase activity by drawing an "activation loop" into the active site, blocking ATP and/or substrate binding. Dephosphorylation of Tyr^{527} induces a conformation change that removes the activation loop from the active site, permitting autophosphorylation of Tyr^{416}, which stimulates tyrosine kinase activity.

Protein Tyrosine Phosphatase SHP-2 Is a Nonreceptor Tyrosine Phosphatase

The human phosphatase SHP-2 is a cytosolic nonreceptor tyrosine phosphatase. It comprises two SH2 domains, a catalytic phosphatase domain, and a C-terminal domain. The SH2 domains enable the enzyme to bind to its target substrates, and they also regulate the phosphatase activity. The catalytic domain of SHP-2 consists of nine α-helices and a ten-stranded mixed β-sheet that wraps around one of the helices (Figure 32.39). The other eight helices pack together on the opposite side of the β-sheet.

The N-terminal SH2 domain regulates phosphatase activity by binding to the phosphatase domain and directly blocking the active site. When a target peptide containing a phosphotyrosine group binds to the SH2 domain, a conformation change causes this domain to dissociate from the catalytic domain, exposing the active site and allowing peptide substrate to bind. Binding of phosphotyrosine-containing peptide to the second SH2 domain provides additional activation. Target peptides with two phosphotyrosines (one to bind to each SH2 domain) provide maximal activation of the phosphatase activity.

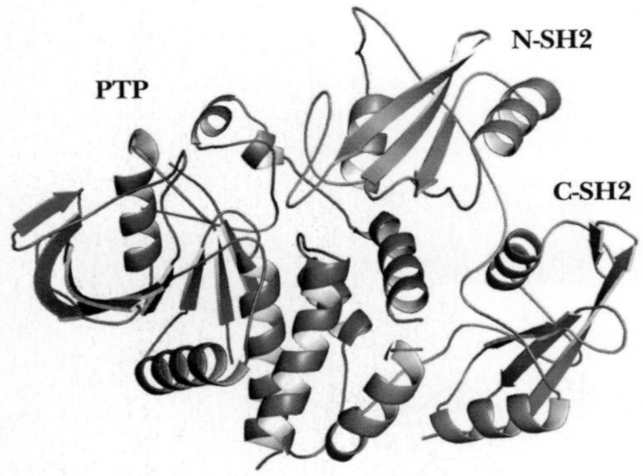

FIGURE 32.39 A ribbon diagram showing the structure of protein tyrosine phosphatase SHP-2 in its autoinhibited, closed conformation. The N- and C-terminal SH2 domains are orange and purple, respectively. The catalytic domain is blue (pdb id = 2SHP).

32.6 How Are Signaling Pathways Organized and Integrated?

All signaling pathways are organized in time and space in the cell, they are carefully regulated, and they are integrated with one another. Remarkably, these complex features of signaling depend on the simple concepts already covered in this chapter: PIDs (see Figures 32.5 and 32.6) modulate and control the association of signaling molecules with one another, often in large signalsomes; signaling molecules are switched on and off by covalent modifications such as phosphorylations; and signaling often involves amplification and cooperative effects. The multifaceted behavior of GPCRs serves as a paradigm for the organization and integration of signaling pathways and is the focus of this section.

GPCRs Can Signal Through G-Protein–Independent Pathways

The classic GPCR signaling pathway involves coupling to heterotrimeric G proteins, but GPCRs can interact with other effector molecules as well. The cellular src kinase (page 1128) can be activated directly by the β_2-adrenergic receptor (Figure 32.40), leading to activation of a MAPK pathway (see A Deeper Look box, page 1140). Activated src phosphorylates Tyr[317] on the Shc adaptor protein, promoting binding by an SH2 domain of Grb2. Grb2 in turn binds to and activates Sos1, which activates Ras. Ras then activates a kinase cascade (Figure 32.40).

The **Janus protein kinase (JAK)** and its **associated transducers and activators of transcription (STAT)** constitute the **JAK/STAT** signaling pathway (Figure 32.41). For example, Jana Stankova and her colleagues have shown that binding of the platelet-activating factor (see Figure 8.8b) to the platelet-activating factor receptor (a GPCR) directly stimulates phosphorylation of **tyk2,** a Janus kinase. The activations of src (described earlier) and tyk2 raise the intriguing possibility that GPCR action may represent an alternate means of activating these signaling pathways (Figures 32.40 and 32.41).

G-Protein Signaling Is Modulated by RGS/GAPs

The signal-transducing effects of G proteins persist as long as the bound GTP is not hydrolyzed. However, G proteins, such as Ras p21 and $G_{s\alpha}$, are themselves weak GTPases. For example, Ras p21 hydrolyzes GTP with a rate constant of only 0.02 min^{-1}. If Ras p21 and $G_{s\alpha}$ were efficient enzymes, the GTP-bound state would be short-lived and G-protein–mediated signaling would be ineffective.

How can G proteins be switched off if they are inherently poor GTPases? The answer is provided by **regulators of G-protein signaling (RGS),** which act as **GTPase-activating proteins (GAPs).** RGS/GAPs elicit dramatic increases in GTPase activity when bound to G proteins. For example, RGS/GAPs increase the GTPase activity of Ras p21 by a factor of 100,000 and accelerate $G_{s\alpha}$-catalyzed GTP hydrolysis nearly 100-fold. RGS/GAPs induce conformation changes in the switch domains of their G protein targets and stabilize the transition state of the GTPase reaction (Figure 32.42).

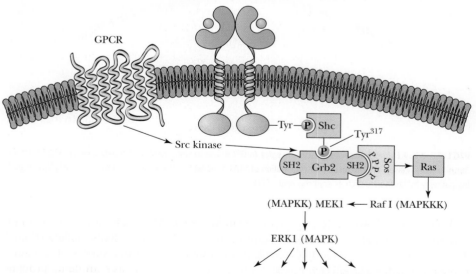

FIGURE 32.40 GPCRs can initiate cellular signaling pathways without involvement of G proteins. Under certain conditions, binding to GPCRs can induce autophosphorylation of the src kinase. Activated src can then bind Shc and phosphorylate Tyr[317], which promotes binding of Grb2 and initiation of a Ras-dependent kinase cascade.

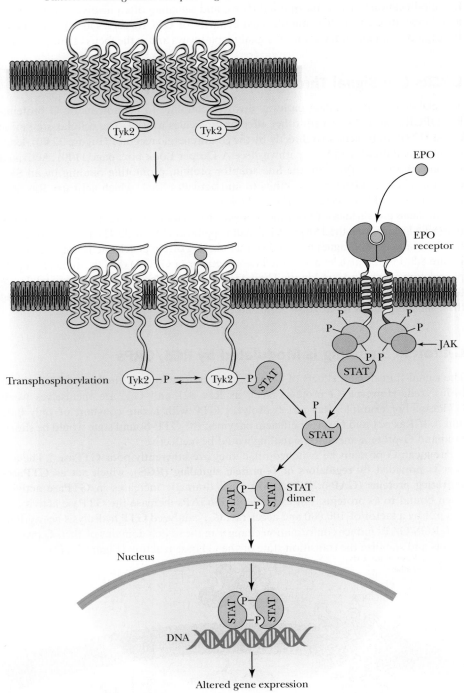

Platelet-activating factor receptor oligomer

FIGURE 32.41 Platelet-activating factor receptor (a GPCR) can trigger autophosphorylation of the tyk2 kinase (a Janus protein kinase). Tyk2-induced phosphorylation of STAT essentially mimics the activation of the JAK/STAT signaling pathway by hormones such as erythropoietin (EPO).

Interestingly, however, RGS proteins are more than GAPs. All RGS proteins contain several other signaling modules in addition to the 120-residue **RGS module** (Figure 32.43). These additional modules enable RGS proteins to bind to a variety of signaling proteins, to behave as effector molecules themselves, and to act as scaffolding proteins in the formation of signalsome complexes. RGS proteins interact directly, for example, with adenylyl cyclase, phospholipase C-β, cGMP phosphodiesterase, guanylyl cyclase, Ca^{2+} channels, and potassium channels (Figure 32.44). These RGS-mediated interactions enable "cross-talk" between many signaling pathways.

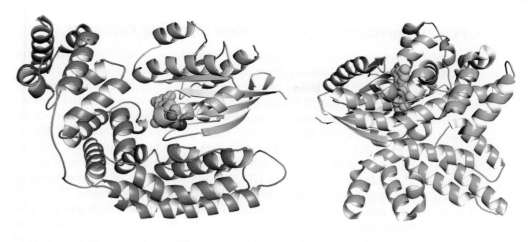

FIGURE 32.42 **(a)** A fragment of an RGS/GAP (brown) bound to Ras p21 (blue). GAPs increase the GTPase activity of Ras p21 by a factor of 100,000 (pdb id = 1WQ1). **(b)** An RGS/GAP (pink) bound to $G_{i\alpha}$ (blue). GDP is shown in yellow (pdb id = 2IK8).

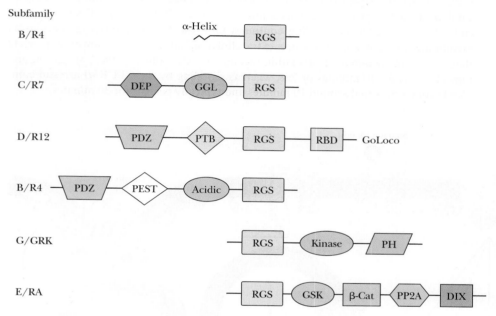

FIGURE 32.43 The RGS proteins (mammalian cells contain more than 30) are classified in subfamilies. Representative members of six of those subfamilies are shown. In addition to the RGS module of 121 residues, each RGS protein contains other motifs and modules that define its functionality. Many of those shown are discussed elsewhere in Chapter 32. (Adapted from Bansal, G., Druey, K. M., and Xie, Z., 2007. R4 RGS proteins: Regulation of G-protein signaling and beyond. *Pharmacology and Therapeutics* **116:**473–495.)

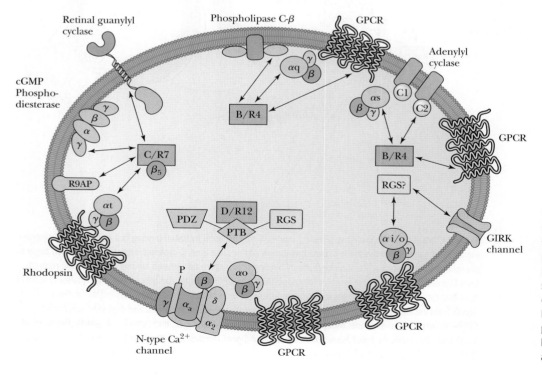

FIGURE 32.44 Several interactions between RGS proteins and G-protein effectors. RGS proteins interact directly with adenylyl cyclase, phospholipase C-β, cGMP phosphodiesterase, and retinyl guanylyl cyclase, as well as with the potassium channel known as GIRK and certain calcium channels. (Adapted from Abramow-Newerly, M., Roy, A. A., Nunn, C., and Chidiac, P., 2006. RGS proteins have a signaling complex: Interactions between RGS proteins and GPCRs, effectors, and auxiliary proteins. *Cellular Signaling* **18:**579–591.)

GPCR Desensitization Leads to New Signaling Pathways

Activation of GPCRs by hormones and other extracellular signals leads to heterotrimeric G protein binding, which triggers a variety of intracellular signals as shown. Importantly, the activated GPCR also binds two other classes of molecules: a family of protein kinases known as **GPCR kinases (GRKs)** and a family of adaptor and scaffolding proteins known as **β-arrestins.** On one level, these two protein families work together to desensitize the GPCRs, "arresting" the G-protein activation by GPCRs. On another level, the GRKs and β-arrestins act together as a molecular switch, directing GPCRs to a distinctly different role in cell signaling. GRK phosphorylation of several sites on the C-terminal sequence of the GPCR promotes binding of β-arrestin (Figure 32.45) to form a signalsome assembly. Binding of β-arrestin has two effects: The GPCR is no longer able to interact with and activate G proteins, and the GPCR is targeted for **clathrin-mediated endocytosis** (Figure 32.45). Following formation of the **endosome,** the GPCR–β-arrestin signalsome enters an **arrestin signaling** mode, recruiting a variety of catalytically active proteins, such as components of the Raf-MEK-ERK signaling cascade. Arrestin-mediated signaling continues until the GPCR returns to the plasma membrane, where it waits for a new extracellular signal. It is important to appreciate that G-protein signaling occurs within seconds of extracellular GPCR activation and typically lasts for 10 minutes or less, whereas signaling by the GPCR–β-arrestin complex becomes maximal around 10 minutes and typically persists for 30 minutes.

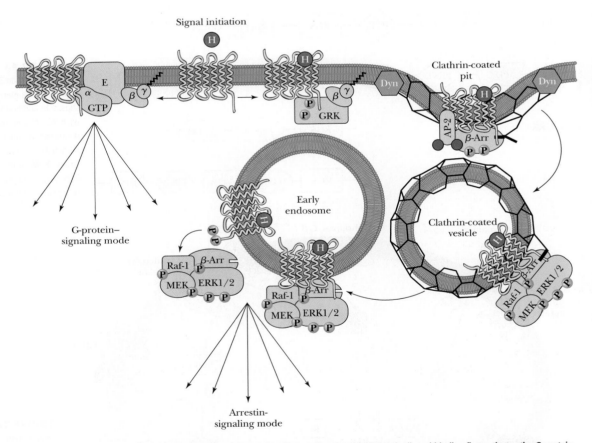

FIGURE 32.45 GPCRs have two signaling modes. Signal initiation by ligand binding first activates the G-protein–mediated signaling pathways at the plasma membrane. At the same time, G-protein receptor kinases (GRKs) begin to phosphorylate the receptor, creating high-affinity binding sites for arrestin. Arrestin binding uncouples the receptor from G proteins and targets the receptor for endocytosis. The GPCR–arrestin complex serves as a signalsome scaffold, recruiting catalytically active proteins such as components of the Raf-MEK-ERK cascade. This initiates a distinct set of signals from the endosome-bound signalsome. Endosomes are eventually processed and degraded, returning the GPCRs to the plasma membrane. (Adapted from Figure 1 of Gesty-Palmer, D., and Luttrell, L. M., 2008. Heptahelical terpsichory. Who calls the tune? *Journal of Receptors and Signal Transduction* **28**:39–58.)

> **A DEEPER LOOK**

Whimsical Names for Proteins and Genes

The study of cell signaling and the identification of hundreds of new signaling proteins provided an unprecedented creative opportunity for cell biologists and geneticists in the naming of these proteins. In the early days of molecular biology, such names were typically arcane abbreviations and acronyms. One such case is the family of **14-3-3 proteins,** named for the migration patterns of these proteins on DEAE-cellulose chromatography and starch-gel electrophoresis. In the 1970s, a few creative scientists suggested whimsical names for newly discovered genes, such as *seuenless,* named in reference to R7, one of the eight photoreceptor cells in the compound eye of *Drosophila,* the common fruit fly. What began as a trickle became a torrent of whimsical names for proteins and genes. *Seuenless* was followed by *bride of seuenless* (*boss,* a ligand of seuenless), and *son of seuenless* (*sos,* first isolated in a genetic screen of the seuenless receptor tyrosine kinase pathway in *Drosophila*). The *hedgehog (hh)* genes, including *sonic hedgehog (Shh),* play critical roles in the development and patterning of animal embryonic tissues but were named for a popular video game.

The accompanying table lists a few notable examples of whimsically named genes and gene products, many of which were first identified in *Drosophila.*

Name	Role or Function
Armadillo	Plakoglobin = β-catenin
Bag of marbles	Novel protein involved in oogenesis and spermatogenesis
Bullwinkle	Oocyte protein
Cactus	Signaling protein—IkB homolog
Cheap date	Alcohol sensitivity
Chickadee	Profilin homolog—regulation of actin cytoskeleton
Corkscrew	A protein tyrosine phosphatase
Dachshund	Nuclear protein required for cell fate determination
Dishevelled	Novel cytoplasmic protein in the wingless pathway
Dunce	A cAMP phosphodiesterase
Hopscotch	A Janus family tyrosine kinase
Naked	A segment polarity gene
Reaper	A death-domain protein functioning in apoptosis
Rutabaga	A Ca^{2+}/calmodulin-dependent protein kinase
Shark	A tyrosine kinase SH2-nonreceptor
Yak	Literally "yet another kinase"

Receptor Responses Can Be Coordinated by Transactivation

Hormone and signaling receptors and their pathways do not operate in isolation. Rather, one or more of the second messengers and effectors of a given pathway can often activate (or inhibit) another pathway. For example, binding of **insulin-like growth factor 1 (IGF-1)** to its own RTK signals enhanced synthesis of a peptide known as **RANTES.** RANTES is secreted into the extracellular medium, where it binds to a GPCR known as the **CCR5 receptor** (Figure 32.46a). As an extracellular signal, RANTES initiates a signaling pathway that induces cell migration, a part of the cell's response to inflammation caused by infection. This is an example of transactivation of a GPCR by a RTK.

The reverse can happen as well. GPCRs can transactivate RTKs by a variety of mechanisms. For example, in certain neurons in the hypothalamus, stimulation of **α_1-adrenergic receptors** triggers a G-protein–mediated pathway that activates a matrix metalloproteinase. Metalloproteinase action releases heparin-binding EGF-like growth factor (HB-EGF) in the extracellular matrix. Binding of HB-EGF to the EGF receptor initiates a classic RTK-activated signaling pathway (Figure 32.46b).

Signals from Multiple Pathways Can Be Integrated

A cell can be exposed simultaneously to multiple, potentially contradictory signals in the form of soluble hormones and ligands anchored to adjacent cells or the extracellular matrix. Cells must have mechanisms for sorting these various signals into a defined response. The **Rsk1** protein serine/threonine kinases exhibit such behavior, integrating several signals to achieve full activation. Rsk1 has two protein kinase domains (Figure 32.47), of which the N-terminal domain phosphorylates downstream targets. This N-terminal

(a) Transcriptional regulation of GPCR ligand synthesis

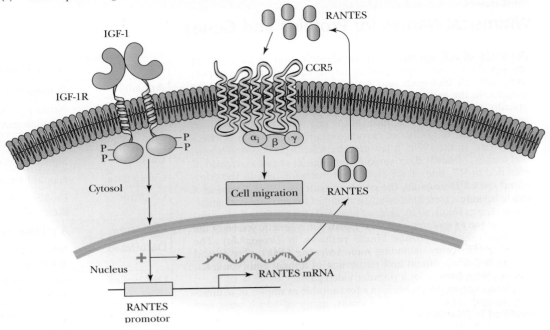

(b)

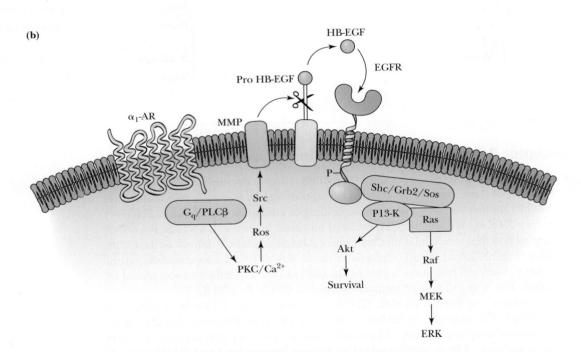

FIGURE 32.46 **(a)** An example of transactivation of cell signaling pathways. Binding of IGF-1 to RTK (in human breast cancer cells) initiates a signaling pathway that leads to an enhanced synthesis of RANTES (a peptide). RANTES is secreted into the extracellular medium, where it can bind a GPCR known as CCR5. CCR5 activation triggers G-protein–mediated signaling pathways. **(b)** Transactivation that is the reverse of that in panel a. Activation of the α_1-adrenergic receptor activates phospholipase C-β, which triggers metalloproteinase (MMP) activation. MPP cleaves a precursor of heparin-binding EGF-like growth factor (HB-EGF). Binding of HB-EGF to the EGF receptor initiates an RTK signaling pathway.

kinase domain is controlled by multiple inputs, including from the C-terminal domain. The Erk MAPK binds to a docking site at the C-terminus of Rsk1, phosphorylating sites in the linker region between the two kinase domains and in the C-terminal domain, all of which are essential for activation. Full activation, however, also requires phosphorylation of the N-terminal kinase domain by the PIP_3-stimulated PDK1 protein kinase. Rsk1 activation thus requires inputs from both the Erk MAPK pathway and the PIP$_3$ pathway (Figure 32.47).

32.7 How Do Neurotransmission Pathways Control the Function of Sensory Systems?

The survival of higher organisms is predicated on the ability to respond rapidly to sensory input from physical signals (sights, sounds) and chemical cues (smells). The responses to such stimuli may include muscle movements and many forms of intercellular communication. Hormones (as described earlier in this chapter) can move through an organism only at speeds determined by the circulatory system. In most higher organisms, a faster means of communication is crucial. Nerve impulses, which can be propagated at speeds up to 100 m/sec, provide a means of intercellular signaling that is fast enough to encompass sensory recognition, movement, and other physiological functions and behaviors in higher animals. The generation and transmission of nerve impulses in vertebrates is mediated by an extremely complicated neural network that connects every part of the organism with the brain—itself an interconnected array of as many as 10^{12} cells.

Despite their complexity and diversity, the nervous systems of animals all possess common features and common mechanisms. Physical or chemical stimuli are recognized by specialized **receptor proteins** in the membranes of **excitable cells.** Conformational changes in the receptor protein result in a change in enzyme activity or a change in the permeability of the membrane. These changes are then propagated throughout the cell or from cell to cell in specific and reversible ways to carry information through the organism. This section describes the characteristics of excitable cells and the mechanisms by which these cells carry information at high speeds through an organism.

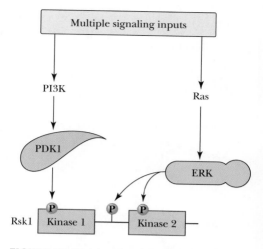

FIGURE 32.47 Integration of signaling pathways. Activation of a ribosomal serine/threonine kinase known as Rsk1 requires phosphorylation by two protein kinases: a phosphoinositide-dependent kinase (PDK1) and a mitogen-activated protein kinase (ERK; see the box on page 1140). Thus, both phosphoinositide-mediated and Ras-mediated pathways must be active to activate Rsk1.

Nerve Impulses Are Carried by Neurons

Neurons and **neuroglia** (or **glial cells**) are cell types unique to nervous systems. The reception and transmission of nerve impulses are carried out by neurons (Figure 32.48), whereas glial cells serve protective and supportive functions. (*Neuroglia* could be translated as "nerve glue.") Glial cells differ from neurons in several ways. Glial cells do not possess axons or synapses, and they retain the ability to divide throughout their life spans. Glial cells outnumber neurons by at least 10 to 1 in most animals.

Neurons are distinguished from other cell types by their long cytoplasmic extensions or projections, called **processes.** Most neurons consist of three distinct regions (see Figure 32.48): the **cell body** (called the **soma**), the **axon,** and the **dendrites.** The axon ends in small structures called **synaptic terminals, synaptic knobs,** or **synaptic bulbs.** Dendrites are short, highly branched structures emanating from the cell body that receive neural impulses and transmit them to the cell body. The space between a synaptic knob on one neuron and a dendrite ending of an adjacent neuron is the **synapse** or **synaptic cleft.**

Three kinds of neurons are found in higher organisms: sensory neurons, interneurons, and motor neurons. **Sensory neurons** acquire sensory signals, either directly or from specific receptor cells, and pass this information along to either **interneurons** or **motor neurons.** Interneurons simply pass signals from one neuron to another, whereas motor neurons pass signals from other neurons to muscle cells, thereby inducing muscle movement (motor activity).

Ion Gradients Are the Source of Electrical Potentials in Neurons

The impulses that are carried along axons, as signals pass from neuron to neuron, are electrical in nature. *These electrical signals occur as transient changes in the electrical potential differences (voltages) across the membranes of neurons (and other cells). Such potentials are generated by ion gradients.* The cytoplasm of a neuron at rest is low in Na$^+$

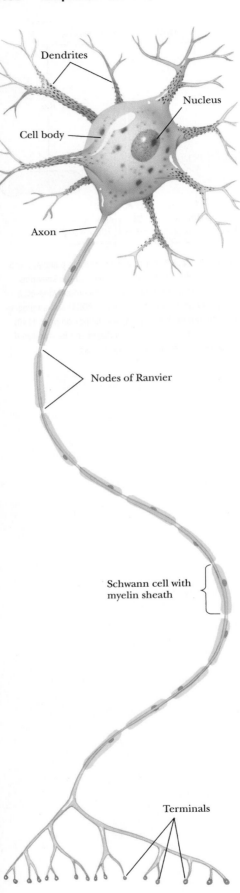

Dendrites

Nucleus

Cell body

Axon

Nodes of Ranvier

Schwann cell with myelin sheath

Terminals

◀ **FIGURE 32.48** The structure of a mammalian motor neuron. The nucleus and most other organelles are contained in the cell body. One long axon and many shorter dendrites project from the body. The dendrites receive signals from other neurons and conduct them to the cell body. The axon transmits signals from this cell to other cells via the synaptic knobs. Glial cells called Schwann cells envelop the axon in layers of an insulating myelin membrane. Although glial cells lie in proximity to neurons in most cases, no specific connections (such as gap junctions, for example) connect glial cells and neurons. However, gap junctions can exist between adjacent glial cells.

and Cl^- and high in K^+, relative to the extracellular fluid (Figure 32.49). These gradients are generated by the Na^+,K^+-ATPase (see Chapter 9). A resting neuron exhibits a potential difference of approximately -60 mV (that is, negative inside).

Action Potentials Carry the Neural Message

Nerve impulses, also called **action potentials,** are transient changes in the membrane potential that move rapidly along nerve cells. Action potentials are created when the membrane is locally **depolarized** by approximately 20 mV from the resting value of about -60 mV to a new value of approximately -40 mV. This small change is enough

(a)

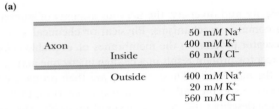

Axon		50 mM Na⁺
	Inside	400 mM K⁺
		60 mM Cl⁻
	Outside	400 mM Na⁺
		20 mM K⁺
		560 mM Cl⁻

(b) (c)

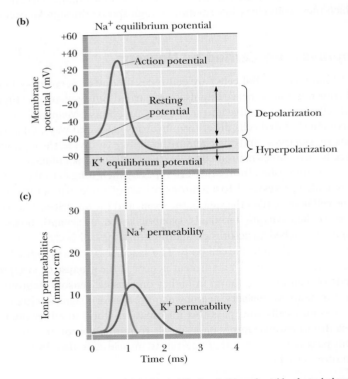

FIGURE 32.49 **(a)** The concentrations of Na^+, K^+, and Cl^- ions inside and outside of a typical resting mammalian axon are shown. Assuming relative permeabilities for K^+, Na^+, and Cl^- are 1, 0.04, and 0.45, respectively, the Goldman equation yields a membrane potential of -60 mV. (See problem 14, page 1167.) **(b** and **c)** The time dependence of an action potential, compared with the ionic permeabilities of Na^+ and K^+. **(b)** The rapid rise in membrane potential from -60 mV to slightly more than $+30$ mV is referred to as a "depolarization." This depolarization is caused **(c)** by a sudden increase in the permeability of Na^+. As the Na^+ permeability decreases, K^+ permeability is increased and the membrane potential drops, eventually falling below the resting potential—a state of "hyperpolarization"—followed by a slow return to the resting potential. (Adapted from Hodgkin, A., and Huxley, A., 1952. A quantitative description of membrane current and its application to conduction and excitation in nerve. *Journal of Physiology* **117**:500–544.)

to have a dramatic effect on specific proteins in the axon membrane called **voltage-gated ion channels.** These proteins are ion channels that are specific either for Na^+ or K^+. These ion channels are normally closed at the resting potential of -60 mV. When the potential difference rises to -40 mV, the "gates" of the Na^+ channels are opened and Na^+ ions begin to flow into the cell. As Na^+ enters the cell, the membrane potential continues to increase and additional Na^+ channels are opened (Figure 32.49). The potential rises to more than $+30$ mV. At this point, Na^+ influx slows and stops. As the Na^+ channels close, K^+ channels begin to open and K^+ ions stream out of the cell, returning the membrane potential to negative values. The potential eventually overshoots its resting value a bit. At this point, K^+ channels close, and the resting potential is eventually restored by action of the Na^+,K^+-ATPase and the other channels. Alan Hodgkin and Andrew Huxley originally observed these transient increases and decreases, first in Na^+ permeability and then in K^+ permeability. For this and related work, Hodgkin and Huxley, along with J. C. Eccles, won the Nobel Prize in Physiology or Medicine in 1963.

The Action Potential Is Mediated by the Flow of Na^+ and K^+ Ions

These changes in potential in one part of the axon are rapidly passed along the axonal membrane (Figure 32.50). The sodium ions that rush into the cell in one localized region actually diffuse farther along the axon, raising the Na^+ concentration *and* depolarizing the membrane, causing Na^+ gates to open in that adjacent region of the axon. In this way, the action potential moves down the axon in wavelike fashion. This simple process has several very dramatic properties:

1. Action potentials propagate very rapidly—up to and sometimes exceeding 100 m/sec.
2. The action potential is not attenuated (diminished in intensity) as a function of distance transmitted.

The input of energy all the way along an axon—in the form of ion gradients maintained by Na^+,K^+-ATPase—ensures that the shape and intensity of the action potential are maintained over long distances. The action potential has an all-or-none character. There are no gradations of amplitude; a given neuron is either at rest (with a polarized

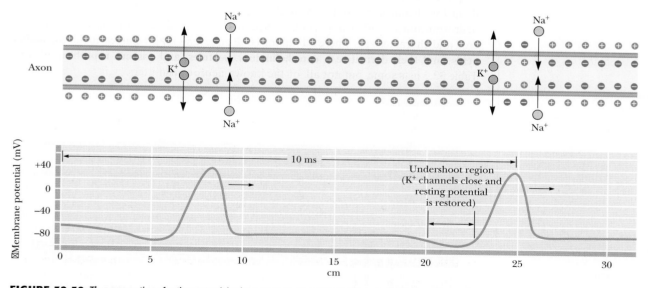

FIGURE 32.50 The propagation of action potentials along an axon. Figure 32.49 shows the time dependence of an action potential at a discrete point on the axon. This figure shows how the membrane potential varies along the axon as an action potential is propagated. (For this reason, the shape of the action potential is the apparent reverse of that shown in Figure 32.49.) At the leading edge of the action potential, membrane depolarization causes Na^+ channels to open briefly. As the potential moves along the axon, the Na^+ channels close and K^+ channels open, leading to a drop in potential and the onset of hyperpolarization. When the resting potential is restored, another action potential can be initiated.

membrane) or is conducting a nerve impulse (with a reversed polarization). Because nerve impulses display no variation in amplitude, the size of the action potential is not important in processing signals in the nervous system. *Instead, it is the number of action potential firings and the frequency of firing that carry specific information.*

The action potential is a delicately orchestrated interplay between the Na^+,K^+-ATPase and the voltage-gated Na^+ and K^+ channels that is initiated by a stimulus at the postsynaptic membrane. The density and distribution of Na^+ channels along the axon are different for myelinated and unmyelinated axons (Figure 32.51). In unmyelinated axons, Na^+ channels are uniformly distributed, although they are few in number—approximately 20 channels per μm^2. On the other hand, in myelinated axons, Na^+ channels are **clustered** at the nodes of Ranvier. In these latter regions, they occur with a density of approximately 10,000 per μm^2. (Ion channel structure and function were discussed in Chapter 9.)

Neurons Communicate at the Synapse

How are neuronal signals passed from one neuron to the next? Neurons are juxtaposed at the synapse. The space between the two neurons is called the **synaptic cleft.** The number of synapses in which any given neuron is involved varies greatly. There may be as few as one synapse per postsynaptic cell (in the midbrain) to many thousands per cell. Typically, 10,000 synaptic knobs may impinge on a single spinal motor neuron, with 8000 on the dendrites and 2000 on the soma or cell body. The ratio of synapses to neurons in the human forebrain is approximately 40,000 to 1!

Synapses are actually quite specialized structures, and several different types exist. *A minority of synapses in mammals, termed **electrical synapses,** are characterized by a very small gap—approximately 2 nm—between the **presynaptic cell** (which delivers the signal) and the **postsynaptic cell** (which receives the signal).* At electrical synapses, the arrival of an action potential on the presynaptic membrane leads directly to depolarization of the postsynaptic membrane, initiating a new action potential in the postsynaptic cell. However, most synaptic clefts are much wider—on the order of 20 to 50 nm. In these, an action potential in the presynaptic membrane causes secretion of a chemical substance—called a **neurotransmitter**—by the presynaptic cell. This substance binds to receptors on the postsynaptic cell, initiating a new action potential. Synapses of this type are thus **chemical synapses.**

Different synapses utilize specific neurotransmitters. The **cholinergic synapse,** a paradigm for chemical transmission mechanisms at synapses, employs acetylcholine as a neurotransmitter. Other important neurotransmitters and receptors have been discovered and characterized. These all fall into one of several major classes, including **amino acids** (and their derivatives), **catecholamines, peptides,** and **gaseous neurotransmitters.** Table 32.3 lists some, but not all, of the known neurotransmitters.

TABLE 32.3	Families of Neurotransmitters
Cholinergic Agents	
Acetylcholine	
Catecholamines	
Norepinephrine (noradrenaline)	
Epinephrine (adrenaline)	
L-Dopa	
Dopamine	
Octopamine	
Amino Acids (and Derivatives)	
γ-Aminobutyric acid (GABA)	
Alanine	
Aspartate	
Cystathione	
Glycine	
Glutamate	
Histamine	
Proline	
Serotonin	
Taurine	
Tyrosine	
Peptide Neurotransmitters	
Cholecystokinin	
Enkephalins and endorphins	
Gastrin	
Gonadotropin	
Neurotensin	
Oxytocin	
Secretin	
Somatostatin	
Substance P	
Thyrotropin releasing factor	
Vasopressin	
Vasoactive intestinal peptide (VIP)	
Gaseous Neurotransmitters	
Carbon monoxide (CO)	
Nitric oxide (NO)	

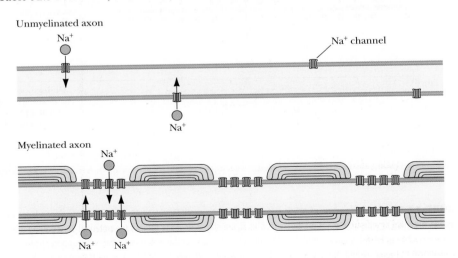

FIGURE 32.51 Na^+ channels are infrequently and randomly distributed in unmyelinated nerve. In myelinated axons, Na^+ channels are clustered in large numbers in the nodes of Ranvier, between the regions surrounded by myelin sheath structures.

Communication at Cholinergic Synapses Depends upon Acetylcholine

In **cholinergic synapses,** small **synaptic vesicles** inside the synaptic knobs contain large amounts of acetylcholine (approximately 10,000 molecules per vesicle; Figure 32.52). When the membrane of the synaptic knob is stimulated by an arriving action potential, special **voltage-gated Ca²⁺ channels** open and Ca²⁺ ions stream into the synaptic knob, causing the acetylcholine-containing vesicles to attach to and fuse with the knob membrane. The vesicles open, spilling acetylcholine into the synaptic cleft. Binding of acetylcholine to specific **acetylcholine receptors** in the postsynaptic membrane causes opening of ion channels and the creation of a new action potential in the postsynaptic neuron.

A variety of toxins can alter or affect this process. The anaerobic bacterium *Clostridium botulinum,* which causes botulism poisoning, produces several toxic proteins that strongly inhibit acetylcholine release. The black widow spider, *Lactrodectus mactans,* produces a venom protein, **α-latrotoxin,** that stimulates abnormal release of acetylcholine at the neuromuscular junction. The bite of the black widow causes pain, nausea, and mild paralysis of the diaphragm but is rarely fatal.

There Are Two Classes of Acetylcholine Receptors

Two different acetylcholine receptors are found in postsynaptic membranes. They were originally distinguished by their responses to **muscarine,** a toxic alkaloid in toadstools, and **nicotine** (Figure 32.53). The **nicotinic receptors** are cation channels in postsynaptic membranes, and the **muscarinic receptors** are transmembrane proteins that interact with G proteins. The receptors in sympathetic ganglia and those in motor endplates of skeletal muscle are nicotinic receptors. Nicotine locks the ion channels of these receptors in their open conformation. Muscarinic receptors are found in smooth muscle and in glands. Muscarine mimics the effect of acetylcholine on these latter receptors.

The nicotinic acetylcholine receptor is a 290-kD transmembrane glycoprotein consisting of a ring of four homologous subunits (α, β, γ, and δ) in the order αγαβδ

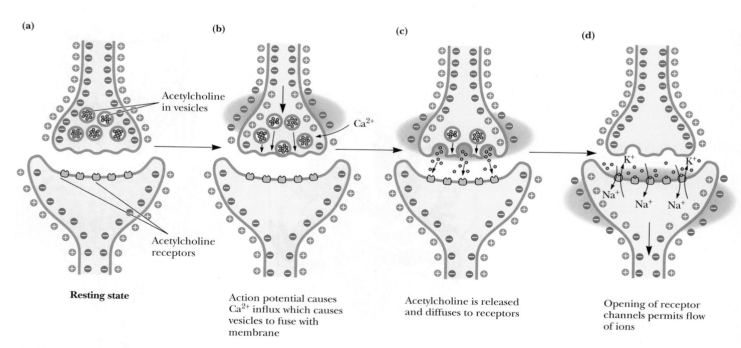

| (a) | (b) | (c) | (d) |

Acetylcholine in vesicles

Acetylcholine receptors

Ca²⁺

K⁺   K⁺

Na⁺   Na⁺   Na⁺

Resting state

Action potential causes Ca²⁺ influx which causes vesicles to fuse with membrane

Acetylcholine is released and diffuses to receptors

Opening of receptor channels permits flow of ions

FIGURE 32.52 Cell–cell communication at the synapse **(a)** is mediated by neurotransmitters such as acetylcholine, produced from choline by choline acetyltransferase. The arrival of an action potential at the synaptic knob **(b)** opens Ca²⁺ channels in the presynaptic membrane. Influx of Ca²⁺ induces the fusion of acetylcholine-containing vesicles with the plasma membrane and release of acetylcholine into the synaptic cleft **(c)**. Binding of acetylcholine to receptors in the postsynaptic membrane opens Na⁺ channels **(d)**. The influx of Na⁺ depolarizes the postsynaptic membrane, generating a new action potential.

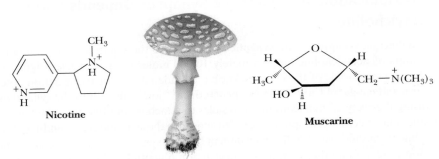

Nicotiana tabacum

Amanita muscaria

FIGURE 32.53 Two types of acetylcholine receptors are known. Nicotinic acetylcholine receptors are locked in their open conformation by nicotine. Obtained from tobacco plants, nicotine is named for Jean Nicot, French ambassador to Portugal, who sent tobacco seeds to France in 1550 for cultivation. Muscarinic acetylcholine receptors are stimulated by muscarine, obtained from the intensely poisonous mushroom, *Amanita muscaria.*

(Figure 32.54a). The receptor is shaped like an elongated (160 Å) funnel, with a large extracellular ligand-binding domain, a membrane-spanning pore, and a smaller intracellular domain. Acetylcholine binds to the two α-subunits at sites that lie 40 Å from the membrane surface.

The Nicotinic Acetylcholine Receptor Is a Ligand-Gated Ion Channel

The nicotinic acetylcholine receptor functions as a **ligand-gated ion channel,** and on the basis of its structure, it is also an **oligomeric ion channel.** When acetylcholine (the ligand) binds to this receptor, a conformational change opens the channel, which is equally permeable to Na^+ and K^+. Na^+ rushes in while K^+ streams out, but because the Na^+ gradient

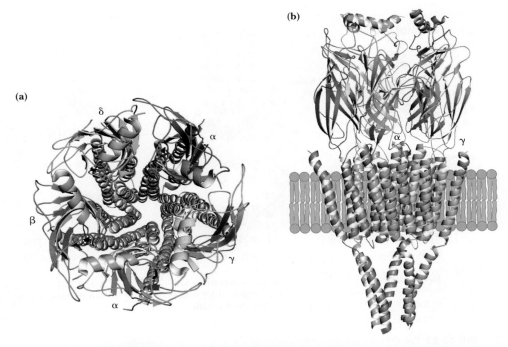

FIGURE 32.54 The nicotinic acetylcholine receptor is an elongated funnel constructed from homologous subunits named α, β, γ, and δ. The pentameric channel includes two copies of the α-subunit. The extracellular domain of each subunit is a β-barrel, whereas the transmembrane and intracellular domains are α-helical. **(a)** Top view; **(b)** side view (pdb id = 2BG9).

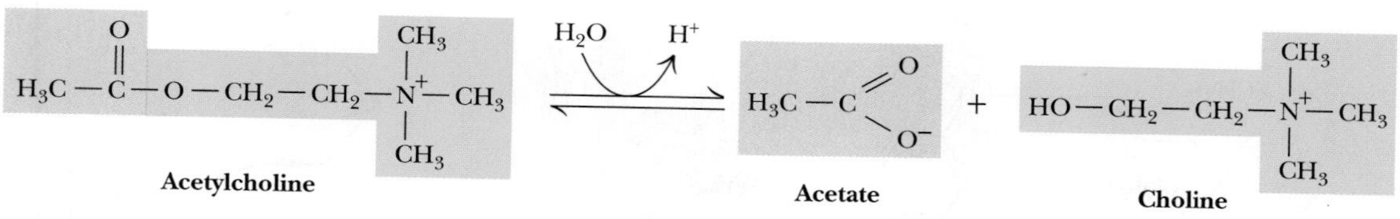

FIGURE 32.55 Acetylcholine is degraded to acetate and choline by acetylcholinesterase, a serine protease.

across this membrane is steeper than that of K^+, the Na^+ influx greatly exceeds the K^+ efflux. The influx of Na^+ depolarizes the postsynaptic membrane, initiating an action potential in the adjacent membrane. After a few milliseconds, the channel closes, even though acetylcholine remains bound to the receptor. At this point, the channel will remain closed until the concentration of acetylcholine in the synaptic cleft drops to about 10 nM.

Acetylcholinesterase Degrades Acetylcholine in the Synaptic Cleft

Following every synaptic signal transmission, the synapse must be readied for the arrival of another action potential. Several things must happen very quickly. First, the acetylcholine left in the synaptic cleft must be rapidly degraded to resensitize the acetylcholine receptor and to restore the excitability of the postsynaptic membrane. This reaction is catalyzed by **acetylcholinesterase** (Figure 32.55).

When [acetylcholine] has decreased to low levels, acetylcholine dissociates from the receptor, which thereby regains its ability to open in a ligand-dependent manner. Second, the synaptic vesicles must be reformed from the presynaptic membrane by endocytosis (Figure 32.56) and then must be restocked with acetylcholine.

This occurs through the action of an ATP-driven H^+ pump and an **acetylcholine transport protein.** The H^+ pump in this case is a member of the family of **V-type ATPases.** It uses the free energy of ATP hydrolysis to create an H^+ gradient across the vesicle membrane. This gradient is used by the acetylcholine transport protein to drive acetylcholine into the vesicle, as shown in Figure 32.56.

Antagonists of the nicotinic acetylcholine receptor are particularly potent neurotoxins. These agents, which bind to the receptor and prevent opening of the ion channel, include **d-tubocurarine**, the active agent in the South American arrow poison **curare,** and several small proteins from poisonous snakes. These latter agents include **cobratoxin** from cobra venom, and **α-bungarotoxin,** from *Bungarus multicinctus,* a snake common in Taiwan (Figure 32.57).

Muscarinic Receptor Function Is Mediated by G Proteins

There are several different types of muscarinic acetylcholine receptors, with different structures and different apparent functions in synaptic transmission. However, certain structural and functional features are shared by this class of receptors. Muscarinic receptors are 70-kD glycoproteins and are members of the GPCR family.

Activation of muscarinic receptors (by binding of acetylcholine) results in several G-protein–mediated effects, including the inhibition of **adenylyl cyclase,** the stimulation of **phospholipase C,** and the opening of K^+ channels. Many antagonists for muscarinic acetylcholine receptors are known, including **atropine** from *Atropa belladonna,* the deadly nightshade plant whose berries are sweet and tasty but highly poisonous (Figure 32.57).

Both the nicotinic and muscarinic acetylcholine receptors are sensitive to certain agents that inactivate acetylcholinesterase itself. Acetylcholinesterase is a serine esterase similar to trypsin and chymotrypsin (see Chapter 14). The reactive serine at the active site of such enzymes is a vulnerable target for organophosphorus inhibitors (Figure 32.58). **DIPF** and related agents form stable covalent complexes with the active-site serine, irreversibly blocking the enzyme. **Malathion** is a commonly used insecticide, and **sarin** and

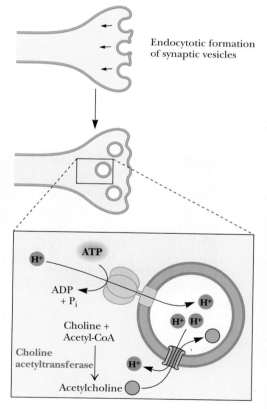

FIGURE 32.56 Following a synaptic transmission event, acetylcholine is repackaged in vesicles in a multistep process. Synaptic vesicles are formed by endocytosis, and acetylcholine is synthesized by choline acetyltransferase. A proton gradient is established across the vesicle membrane by an H^+-transport ATPase, and a proton–acetylcholine transport protein transports acetylcholine into the synaptic vesicles, exchanging acetylcholine for protons in an electrically neutral antiport process.

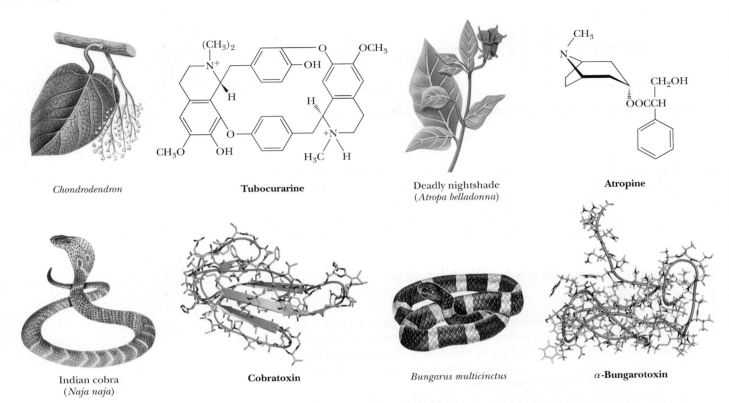

Chondrodendron **Tubocurarine** Deadly nightshade (*Atropa belladonna*) **Atropine**

Indian cobra (*Naja naja*) **Cobratoxin** *Bungarus multicinctus* **α-Bungarotoxin**

FIGURE 32.57 Tubocurarine, obtained from the plant *Chondrodendron tomentosum,* is the active agent in "tube curare," named for the bamboo tubes in which it is kept by South American tribal hunters. Atropine is produced by *Atropa belladonna,* the poisonous deadly nightshade. The species name, which means "beautiful woman," derives from the use of atropine in years past by Italian women to dilate their pupils. Atropine is still used for pupil dilation in eye exams by ophthalmologists. Cobratoxin and α-bungarotoxin are produced by the cobra *(Naja naja)* and the banded krait snake *(Bungarus multicinctus),* respectively.

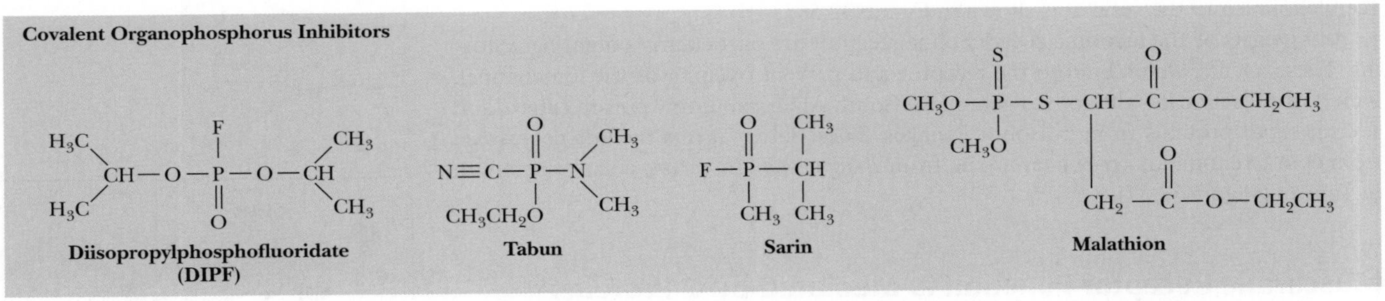

Covalent Organophosphorus Inhibitors

Diisopropylphosphofluoridate (DIPF) **Tabun** **Sarin** **Malathion**

FIGURE 32.58 Covalent inhibitors of acetylcholinesterase include DIFP, the nerve gases tabun and sarin, and the insecticide malathion.

tabun are nerve gases used in chemical warfare. All these agents effectively block nerve impulses, stop breathing, and cause death by suffocation.

Other Neurotransmitters Can Act Within Synaptic Junctions

Synaptic junctions that use amino acids, catecholamines, and peptides (see Table 32.3) appear to operate much the way the cholinergic synapses do. Presynaptic vesicles release their contents into the synaptic cleft, where the neurotransmitter substance can bind to specific receptors on the postsynaptic membrane to induce a conformational change and elicit a particular response. Some of these neurotransmitters are **excitatory** in nature and

> **A DEEPER LOOK**

Tetrodotoxin and Saxitoxin Are Na⁺ Channel Toxins

Tetrodotoxin and **saxitoxin** are highly specific blockers of Na⁺ channels and bind with very high affinity ($K_D < 1$ nM). This unique specificity and affinity have made it possible to use radioactive forms of these toxins to purify Na⁺ channels and map their distribution on axons. Tetrodotoxin is found in the skin and several internal organs of puffer fish, also known as blowfish or swellfish, members of the family *Tetraodontidae,* which react to danger by inflating themselves with water or air to nearly spherical (and often comical) shapes (see accompanying figure). Although tetrodotoxin poisoning can easily be fatal, puffer fish are delicacies in Japan, where they are served in a dish called fugu. For this purpose, the puffer fish must be cleaned and prepared by specially trained chefs. Saxitoxin is made by *Gonyaulax catenella* and *G. tamarensis,* two species of marine dinoflagellates or plankton that are responsible for "red tides" that cause massive fish kills. Saxitoxin is concentrated by certain species of mussels, scallops, and other shellfish that are exposed to red tides. Consumption of these shellfish by animals, including humans, can be fatal.

(a)

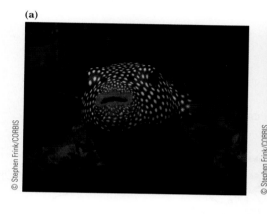

(b)

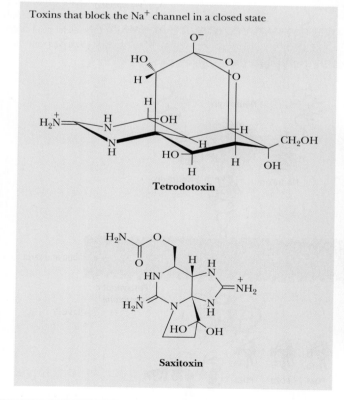

▲ Tetrodotoxin is found in puffer fish, which are prepared and served in Japan as fugu. **(a)** The puffer fish on the left is unexpanded; the one on the right is inflated. **(b)** Structures of tetrodotoxin and saxitoxin.

HUMAN BIOCHEMISTRY

Neurexins and Neuroligins—the Scaffolding of Learning and Memory

How do we learn, remember, and forget? And how do our brains change in response to new experiences? Answers to these questions are being revealed by new research at the crossroads of cell biology and biochemistry. The human brain contains about 100 billion neurons that are interconnected by more than 100 trillion synapses—a total of approximately 100,000 miles of biological wiring. Synapses are the fundamental unit of neural circuits. Each of the 20 billion neurons of the neocortex have, on average, 7,000 synaptic connections to other neurons, with some neurons making up to 100,000 synaptic connections to other neurons. Remarkably, it is estimated that each neuron is able to contact any other neuron with no more than six interneuronal connections. Thus, any two neurons are no more than six connections from one another.

Learning and the acquisition of long-term memory involve **plasticity**—i.e., the ability to change the structures and functions of synaptic connections. Plasticity may include the remodeling and strengthening of existing synapses as well as the formation of new synapses. In all cases, synaptic function requires that postsynaptic neurotransmitter receptors on the dendrite be precisely aligned opposite chemically matched presynaptic vesicles in the axon. This

References:

Choi, Y.-B., Li, H.-L., Kassabov, S. R., Jin, I., Puthanveettil, S. V., Karl, K. A., Lu, Y., Kim, J.-H., Bailey, C. H., and Kandel, E. R., 2011. Neurexin-neuroligin transsynaptic interaction mediates learning-related synaptic remodeling and long-term facilitation in *Aplysia. Neuron* **70:**468–481.

Siddiqui, T. J., and Craig, A. M., 2011. Synaptic organizing complexes. *Current Opinion in Neurobiology* **21:**132–143.

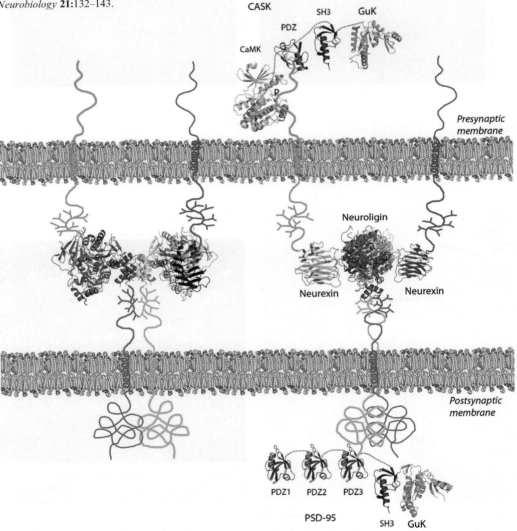

Trans-synaptic complexes of scaffolding proteins anchor and stabilize the synapse and, by virtue of their associated intracellular functional domains, also activate a variety of signaling processes that regulate synaptic function in both neurons. *Adapted with permission from Südhof, T., 2008. Neuroligins and neurexins link synaptic function to cognitive function.* Nature **455:***903–911.*

alignment is achieved with a complex scaffolding network composed of presynaptic and postsynaptic partner proteins that associate across the cleft. The best understood of such partners are presynaptic **neurexins** and postsynaptic **neuroligins** (figure). Neurexins were first identified as the receptors for α-latrotoxin, the venom protein of the black widow spider. The mammalian genome contains only three neurexin genes, but extensive alternative splicing at five positions generates thousands of neurexin isoforms. Neurexins partner with a variety of postsynaptic scaffolding proteins besides the neuroligins, and some neurexin-partner associations involve soluble secreted "bridge" proteins such as the 35 kD **cerebellin precursor protein,**

Cbln1. Emerging research also is identifying other presynapse-postsynapse scaffolding protein pairs, including the presynaptic **LAR protein tyrosine phosphatase receptor** and its partner the postsynaptic **netrin G ligand-3**.

The involvement of trans-synaptic scaffold protein interactions such as these in learning and memory is becoming increasingly apparent. For example, Nobelist Eric Kandel and his colleagues have shown recently that expression of an autism-linked variant of neuroligin in *Aplysia* (large, shell-less gastropod molluscs commonly known as sea slugs) blocks both intermediate and long-term memory formation while having no effect on basal synaptic transmission. In other words, neuroligin defects can lead to "sluggish" learning.

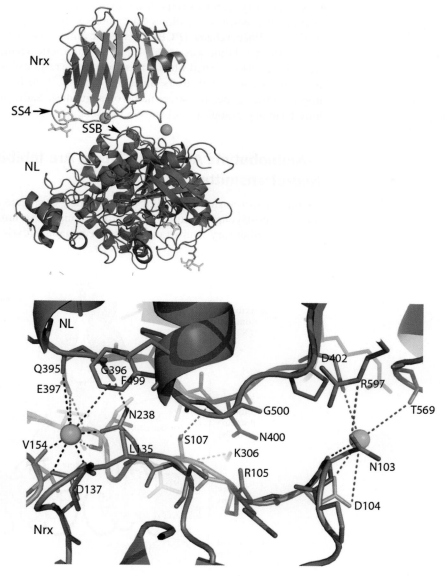

The complex of neurexin-1β with neuroligin-1. The top image shows the complete complex; the bottom image is a close-up of the neurexin-1β:neuroligin-1 interface. (pdb id=3BIW) *From Arac, D., Boucard, A. A., Ozkan, E., Strop, P., Newell, E., Südhof, T., and Brunger, A. T., 2007. Structures of neuroligin-1 and the neuroligin-1/neurexin-1β complex reveal specific protein-protein and protein-Ca²⁺ interactions.* Neuron **56:**992–1003.

stimulate postsynaptic neurons to transmit impulses, whereas others are **inhibitory** and prevent the postsynaptic neuron from carrying other signals. Just as acetylcholine acts on both nicotinic and muscarinic receptors, so most of the known neurotransmitters act on several (and in some cases, many) different kinds of receptors. Biochemists are just beginning to understand the sophistication and complexity of neuronal signal transmission.

Glutamate and Aspartate Are Excitatory Amino Acid Neurotransmitters

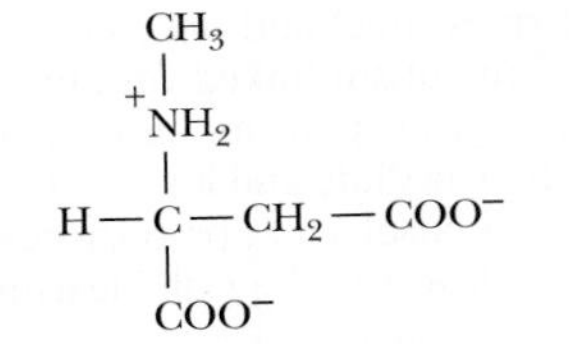

The common amino acids glutamate and aspartate act as neurotransmitters. Like acetylcholine, glutamate and aspartate are excitatory and stimulate receptors on the postsynaptic membrane to transmit a nerve impulse. No enzymes that degrade glutamate exist in the extracellular space, so glutamate must be cleared by the high-affinity presynaptic and glial transporters—a process called **reuptake.**

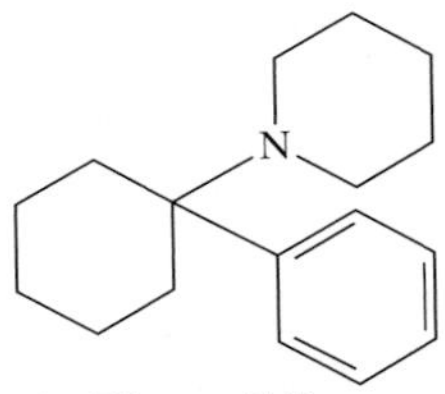

At least five subclasses of glutamate receptors are known. The best understood of these excitatory receptors is the N-methyl-D-aspartate (NMDA) receptor, a ligand-gated channel that, when open, allows Ca^{2+} and Na^+ to flow into the cell and K^+ to flow out of the cell. **Phencyclidine (PCP)** is a specific antagonist of the NMDA receptor (Figure 32.59). Phencyclidine was once used as an anesthetic agent, but legitimate human use was quickly discontinued when it was found to be responsible for bizarre psychotic reactions and behavior in its users. Since this time, PCP has been used illegally as a hallucinogenic drug, under the street name of **angel dust.** Sadly, it has caused many serious, long-term psychological problems in its users.

γ-Aminobutyric Acid and Glycine Are Inhibitory Neurotransmitters

Certain neurotransmitters, acting through their conjugate postsynaptic receptors, inhibit the postsynaptic neuron from propagating nerve impulses from other neurons. Two such inhibitory neurotransmitters are **γ-aminobutyric acid (GABA)** and **glycine.**

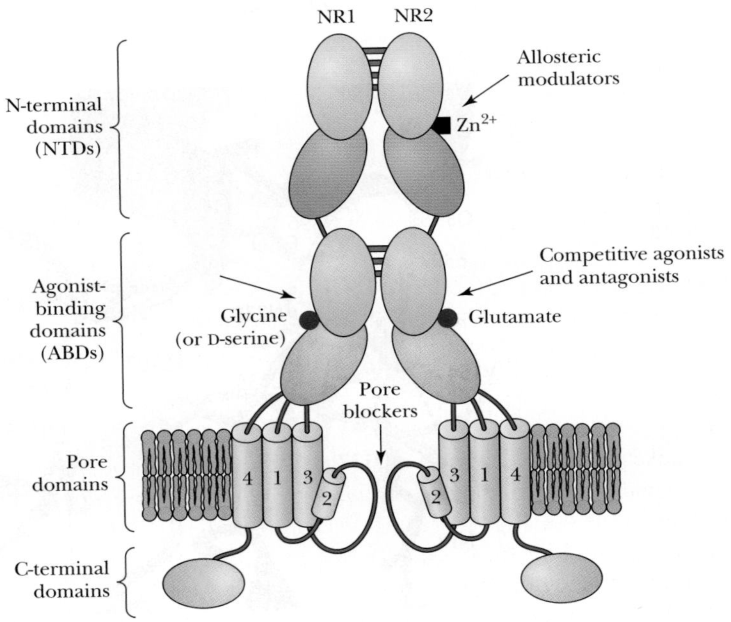

N-Methyl-D-aspartate (NMDA)

Phencyclidine

FIGURE 32.59 NMDA receptors assemble as tetramers, with two NR1 subunits and two NR2 subunits. (For clarity, only one of the NR1–NR2 pairs is shown.) The extracellular portion of each subunit consists of an N-terminal domain (NTD) and an agonist-binding domain (ABD). Red lines indicate that stabilizing interactions occur between these domains. NMDA receptors are Na^+ and Ca^{2+} channels. They are stimulated by NMDA, inhibited by phencyclidine, and regulated by Zn^{2+} and glycine.

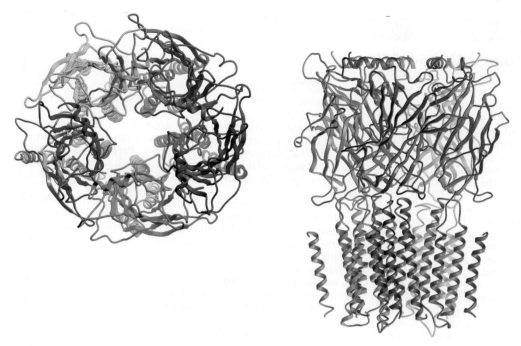

FIGURE 32.60 GABA (γ-aminobutyric acid) and glycine are inhibitory neurotransmitters that activate chloride channels. Influx of Cl^- causes a hyperpolarization of the postsynaptic membrane. GABA receptors are similar in many respects to nicotinic acetylcholine receptors, with an $\alpha_2\beta\gamma\delta$ stoichiometry. **(a)** Top view; **(b)** side view. (Images courtesy of Donald Weaver, University of Nova Scotia, and Valerie Campagna-Slater, University of Toronto.)

FIGURE 32.61 Glutamate is converted to GABA by glutamate decarboxylase. GABA is degraded by the action of GABA–glutamate transaminase and succinate semialdehyde dehydrogenase to produce succinate.

These agents make postsynaptic membranes permeable to chloride ions and cause a net influx of Cl^-, which in turn causes **hyperpolarization** of the postsynaptic membrane (making the membrane potential more negative). Hyperpolarization of a neuron effectively raises the threshold for the onset of action potentials in that neuron, making the neuron resistant to stimulation by excitatory neurotransmitters. These effects are mediated by the GABA and glycine receptors, which are ligand-gated chloride channels (Figure 32.60). GABA is derived by a decarboxylation of glutamate (Figure 32.61) and appears to operate mainly in the brain, whereas glycine acts primarily in the spinal cord. The glycine receptor has a specific affinity for the convulsive alkaloid **strychnine** (Figure 32.62). The effects of ethanol on the brain arise in part from the opening of GABA receptor Cl^- channels.

Strychnine

FIGURE 32.62 Glycine receptors are distinguished by their unique affinity for strychnine.

The Biochemistry of Neurological Disorders

Defects in catecholamine processing are responsible for the symptoms of many neurological disorders, including clinical depression (which involves norepinephrine [NE]) and parkinsonism (involving dopamine [DA]). Once these neurotransmitters have bound to and elicited responses from postsynaptic membranes, they must be efficiently cleared from the synaptic cleft (see accompanying figure, part a). Clearing can occur by several mechanisms. NE and DA transport or reuptake proteins exist both in the presynaptic membrane and in nearby glial cell membranes. On the other hand, catecholamine neurotransmitters can be metabolized and inactivated by two enzymes: **catechol-*O*-methyl-transferase** in the synaptic cleft and **monoamine oxidase** in the mitochondria (see figure, part b). Catecholamines transported back into the presynaptic neuron are accumulated in synaptic vesicles by the same H^+-ATPase/H^+-ligand exchange mechanism described for glutamate. Clinical depression has been treated by two different strategies. **Monoamine oxidase inhibitors** act as antidepressants by increasing levels of catecholamines in the brain. Another class of antidepressants, the **tricyclics,** such as desipramine (see figure, part c), act on several classes of neurotransmitter reuptake transporters and facilitate

more prolonged stimulation of postsynaptic receptors. **Prozac** is a more specific reuptake inhibitor and acts only on serotonin reuptake transporters.

Parkinsonism is characterized by degeneration of dopaminergic neurons, as well as consequent overproduction of postsynaptic dopamine receptors. In recent years, Parkinson's patients have been treated with dopamine agonists such as **bromocriptine** (see figure, part d) to counter the degeneration of dopamine neurons.

Catecholaminergic neurons are involved in many other interesting pharmacological phenomena. For example, **reserpine** (see figure, part e), an alkaloid from a climbing shrub of India, is a powerful sedative that depletes the level of brain monoamines by inhibiting the H^+–monoamine exchange protein in the membranes of synaptic vesicles.

Cocaine (see figure, part f), a highly addictive drug, binds with high affinity and specificity to reuptake transporters for the monoamine neurotransmitters in presynaptic membranes. Thus, at least one of the pharmacological effects of cocaine is to prolong the synaptic effects of these neurotransmitters.

(a)

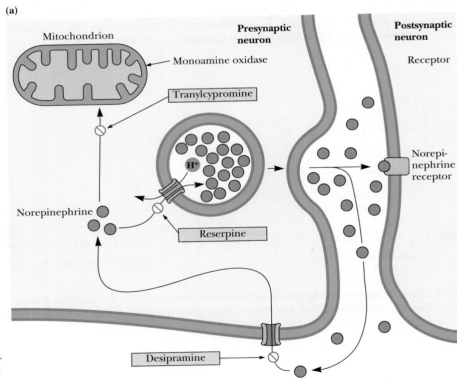

▶ (a) The pathway for reuptake and vesicular repackaging of the catecholamine neurotransmitters. The sites of action of desipramine, tranylcypromine, and reserpine are indicated.

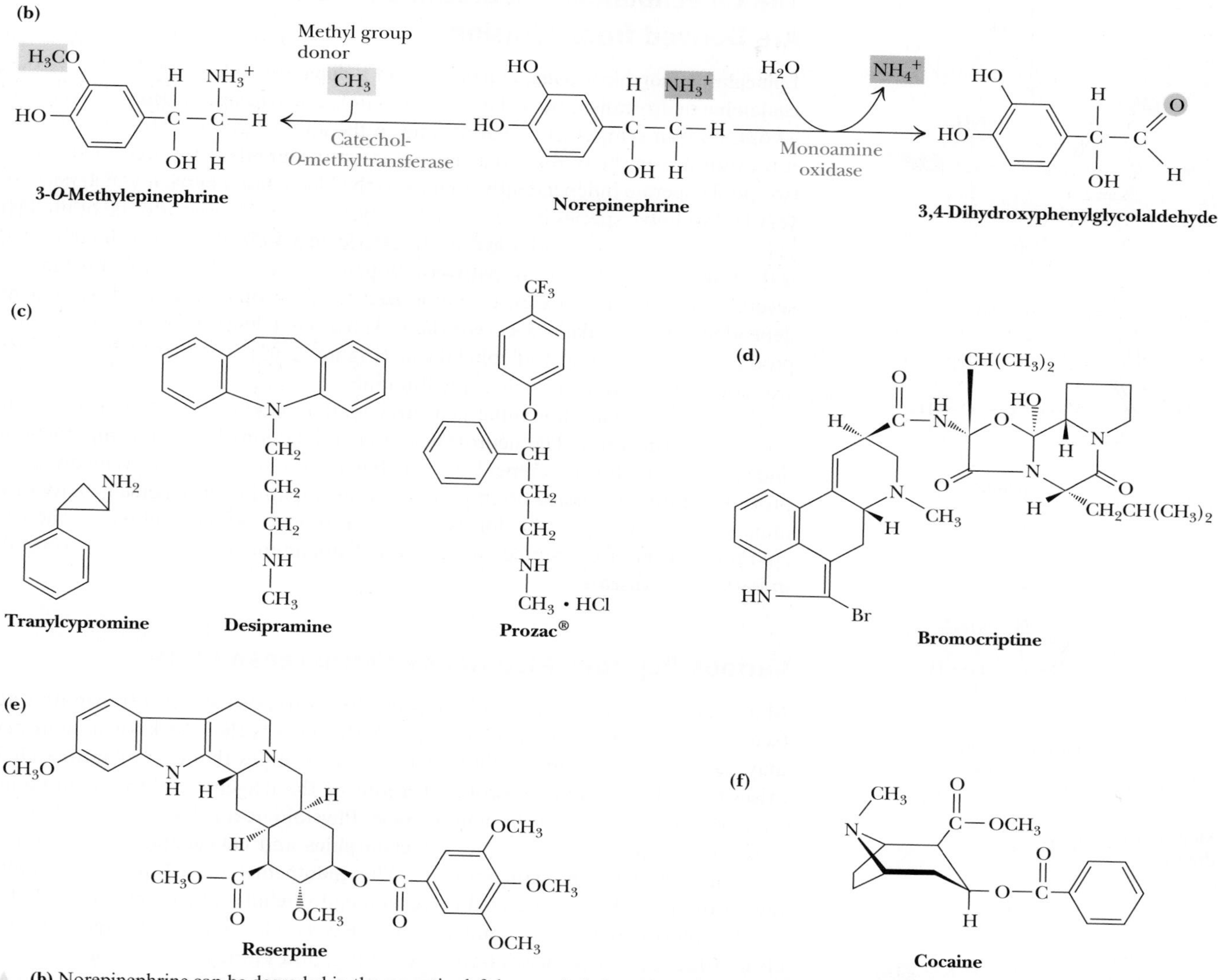

(b)

3-O-Methylepinephrine ← Catechol-O-methyltransferase ← **Norepinephrine** → Monoamine oxidase → **3,4-Dihydroxyphenylglycolaldehyde**

Methyl group donor

(c)

Tranylcypromine   **Desipramine**   **Prozac®**

(d)

Bromocriptine

(e)

Reserpine

(f)

Cocaine

(b) Norepinephrine can be degraded in the synaptic cleft by catechol-O-methyltransferase or in the mitochondria of presynaptic neurons by monoamine oxidase. **(c)** The structures of tranylcypromine, desipramine, and Prozac. **(d)** The structure of bromocriptine. **(e)** The structure of reserpine. **(f)** The structure of cocaine.

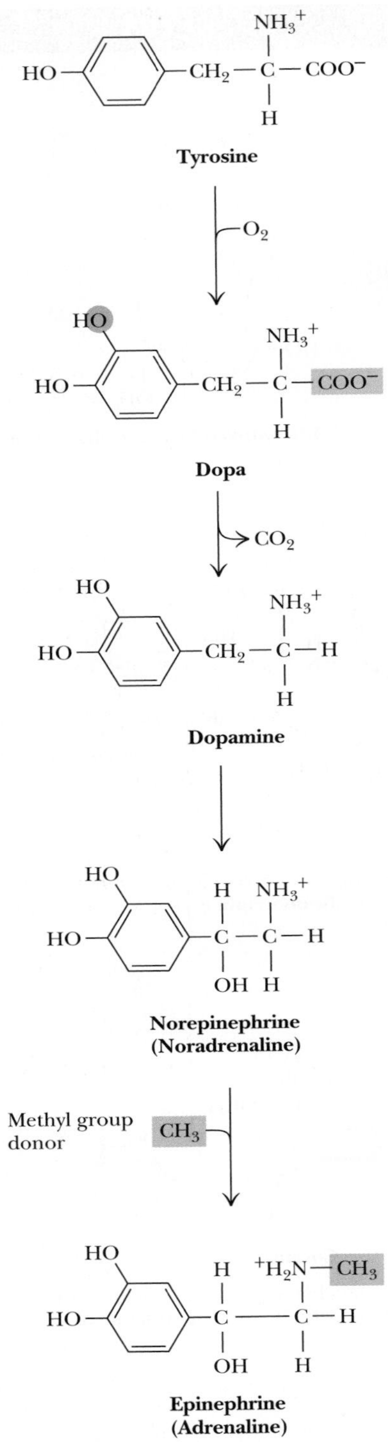

FIGURE 32.63 The pathway for the synthesis of catecholamine neurotransmitters. Dopa, dopamine, noradrenaline, and adrenaline are synthesized sequentially from tyrosine.

The Catecholamine Neurotransmitters Are Derived from Tyrosine

Epinephrine, norepinephrine, dopamine, and L-**dopa** are collectively known as the **catecholamine** neurotransmitters. These compounds are synthesized from tyrosine (Figure 32.63), both in sympathetic neurons and in the adrenal glands. They function as neurotransmitters in the brain and as hormones in the circulatory system. However, these two pools operate independently, thanks to the **blood–brain barrier,** which permits only very hydrophobic species in the circulatory system to cross over into the brain. Hydroxylation of tyrosine (by **tyrosine hydroxylase**) to form **3,4-*d*ihydr*oxy*phenyl*a*lanine** (L-dopa) is the rate-limiting step in this pathway. Dopamine, a crucial catecholamine involved in several neurological diseases, is synthesized from L-dopa by a pyridoxal phosphate-dependent enzyme, **dopa decarboxylase.** Subsequent hydroxylation and methylation produce norepinephrine and epinephrine (Figure 32.63). The methyl group in the final reaction is supplied by *S*-adenosylmethionine.

Each of these catecholamine neurotransmitters is known to play a unique role in synaptic transmission. The neurotransmitter in junctions between sympathetic nerves and smooth muscle is norepinephrine. On the other hand, dopamine is involved in other processes. Either excessive brain production of dopamine or hypersensitivity of dopamine receptors is responsible for psychotic symptoms and schizophrenia, whereas lowered production of dopamine and the loss of dopamine neurons are important factors in Parkinson's disease.

Various Peptides Also Act as Neurotransmitters

Many relatively small peptides have been shown to possess neurotransmitter activity (see Table 32.3). One of the challenges of this field is that the known neuropeptides may represent a very small subset of the neuropeptides that exist. Another challenge arises from the small in vivo concentrations of these agents and the small number of receptors that are present in neural tissue. Physiological roles for most of these peptides are complex. For example, the **endorphins** and **enkephalins** are natural opioid substances and potent pain relievers. The **endothelins** are a family of homologous regulatory peptides, synthesized by certain endothelial and epithelial cells that act on nearby smooth muscle and connective tissue cells. They induce or affect smooth muscle contraction; vasoconstriction; heart, lung, and kidney function; and mitogenesis and tissue remodeling. **Vasoactive intestinal peptide (VIP)** produces a G protein–adenylyl cyclase–mediated increase in cAMP, which in turn triggers a variety of protein phosphorylation cascades, one of which leads to conversion of phosphorylase *b* to phosphorylase *a,* stimulating glycogenolysis. Moreover, VIP has synergistic effects with other neurotransmitters, such as norepinephrine. In addition to increasing cAMP levels through β-adrenergic receptors, norepinephrine acting at α_1-adrenergic receptors markedly stimulates the increases in cAMP elicited by VIP. Many other effects have also been observed. For example, injection of VIP increases rapid eye movement (REM) sleep and decreases waking time in rats. VIP receptors exist in regions of the central nervous system involved in sleep modulation.

SUMMARY

OWL Login to OWL to develop problem-solving skills and complete online homework assigned by your professor.

32.1 What Are Hormones? Many different chemical species act as hormones. Steroid hormones, all derived from cholesterol, regulate metabolism, salt and water balances, inflammatory processes, and sexual function. Several hormones are amino acid derivatives. Among these are epinephrine and norepinephrine (which regulate smooth muscle contraction and relaxation, blood pressure, cardiac rate, and the processes of lipolysis and glycogenolysis) and the thyroid hormones (which stimulate metabolism). Peptide hormones are a large group of hormones that appear to regulate processes in all body tissues, including the release of yet other hormones. Hormones and other signal molecules in biological systems bind with very high affinities to their receptors, displaying K_D values in the range of 10^{-12} to 10^{-6} M. Hormones are produced at concentrations equivalent to or slightly above these K_D values. Once hormonal effects have been induced, the hormone is usually rapidly metabolized.

32.2 What Is Signal Transduction? Hormonal regulation depends upon the transduction of the hormonal signal across the plasma membrane to specific intracellular sites, particularly the nucleus. Signal transduction pathways consist of a stepwise progression of signaling stages: receptor→transducer→effector. The receptor perceives the signal, transducers relay the signal, and the effectors convert the signal into an intracellular response. Often, effector action involves a series of steps, each of which is mediated by an enzyme, and each of these enzymes can be considered as an amplifier in a pathway connecting the hormonal signal to its intracellular targets.

32.3 How Do Signal-Transducing Receptors Respond to the Hormonal Message? Steroid hormones may either bind to plasma membrane receptors or exert their effects within target cells, entering the cell and migrating to their sites of action via specific cytoplasmic receptor proteins. The nonsteroid hormones, which act by binding to outward-facing plasma membrane receptors, initiate signal transduction pathways that mobilize various second messengers—cyclic nucleotides, Ca^{2+} ions, and other substances—that activate or inhibit enzymes or cascades of enzymes in very specific ways.

32.4 How Are Receptor Signals Transduced? Receptor signals are transduced in one of three ways, to initiate actions inside the cell:

1. Exchange of GDP for GTP on GTP-binding proteins (G proteins), which in turn leads to generation of second messengers, including cAMP, phospholipid breakdown products, and Ca^{2+}.

2. Receptor-mediated activation of phosphorylation cascades that, in turn, trigger activation of various enzymes.

3. Conformation changes that open ion channels or recruit proteins into nuclear transcription complexes.

32.5 How Do Effectors Convert the Signals to Actions in the Cell? Transduction of the hormonal signal leads to activation of effectors—usually protein kinases and protein phosphatases—that elicit a variety of actions that regulate discrete cellular functions. Of the thousands of mammalian kinases and phosphatases, the structures and functions of a few are representative, including protein kinase A (PKA), protein kinase C (PKC), and protein tyrosine phosphatase SHP-2.

32.6 How Are Signaling Pathways Organized and Integrated? All signaling pathways are organized in time and space in the cell, they are carefully regulated, and they are integrated with one another. PIDs modulate and control the association of signaling molecules with one another, often in large signalsomes; signaling molecules are switched on and off by covalent modifications such as phosphorylations; and signaling often involves amplification and cooperative effects. GPCR signaling can occur through G-protein-independent pathways and is sometimes modulated by RGS/GAPs. Responses of signaling receptors can be coordinated by transactivation, and signals from multiple pathways can be integrated.

32.7 How Do Neurotransmission Pathways Control the Function of Sensory Systems? Nerve impulses, which can be propagated at speeds up to 100 m/sec, provide a means of intercellular signaling that is fast enough to encompass sensory recognition, movement, and other physiological functions and behaviors in higher animals. The generation and transmission of nerve impulses in vertebrates is mediated by an incredibly complicated neural network that connects every part of the organism with the brain—itself an interconnected array of as many as 10^{12} cells. Despite their complexity and diversity, the nervous systems of animals all possess common features and common mechanisms. Physical or chemical stimuli are recognized by specialized receptor proteins in the membranes of excitable cells. Conformational changes in the receptor protein result in a change in enzyme activity or a change in the permeability of the membrane. These changes are then propagated throughout the cell or from cell to cell in specific and reversible ways to carry information through the organism.

FOUNDATIONAL BIOCHEMISTRY Things You Should Know After Reading Chapter 32.

- The functional differences between steroid hormones, amino acid derivatives, and peptide hormones.

- The two different functions of steroid hormones.

- The essential features of polypeptide hormone synthesis, processing, and action.

- The twelve essential features of signal transduction.

- The role of enzymatic cascades in signaling pathways.

- The connection between membrane interactions and nuclear effects in signal transduction.

- The critical role of protein interaction domains and the signalsome in signaling pathways.

- The manner in which signal-transducing receptors respond to hormonal messages.

- The structural differences between GPCRs, guanylyl cyclases, and tyrosine kinases.

- The allosteric behaviors of RTKs and RGCs.

- The role of ligand-induced dimerization in activation of epidermal growth factor-like receptors.

- The asymmetric nature of the tyrosine kinase dimer in EGF.

- The folded dimeric nature of the insulin receptor.

- The manner in which autophosphorylation opens the active site of the insulin receptor.

- The mechanism of guanylyl cyclases in mediating the effects of natriuretic hormones.

- How a symmetric dimer of the atrial natriuretic receptor binds an asymmetric peptide ligand.

- The essential structural and functional features of pp60^{src}.

- The manner in which nitric oxide activates guanylyl cyclase.

- The three ways in which receptor signals initiate actions inside the cell.

- How G proteins (large and small) transduce the signals of GPCRs.

- The manner in which cAMP is synthesized and degraded, acting as a second messenger.

- The manner in which phospholipases produce second messengers.

- The role of calcium (and calcium oscillations) as a second messenger.

- The manner in which protein kinases and phosphatases and other effectors convert cell signals into actions in the cell.

- How signaling pathways are organized and integrated.

- How GPCRs signal in G-protein-independent ways.

- The manner in which nerve impulses are carried by neurons.

- How ion gradients produce electrical potentials that carry the neural message.

- The mechanism of synaptic communication.

- How acetylcholine is synthesized and degraded, and how it acts as a neurotransmitter.

- How muscarinic receptor function is mediated by G proteins.

- The action of glutamate and aspartate as excitatory neurotransmitters.

- The action of γ-aminobutyric acid and glycine as inhibitory neurotransmitters.

- How dopamine, norepinephrine, and epinephrine are derived from tyrosine.

- The action of peptides such as endorphins and enkephalins as neurotransmitters.

PROBLEMS

Answers to all problems are at the end of this book. Detailed solutions are available in the Student Solutions Manual, Study Guide, and Problems Book.

1. **Assessing the Features and Physiology of Hormones** Compare and contrast the features and physiological advantages of each of the major classes of hormones, including the steroid hormones, polypeptide hormones, and the amino acid–derived hormones.

2. **Understanding the Features and Physiology of the Second Messengers** Compare and contrast the features and physiological advantages of each of the known classes of second messengers.

3. **Are there Gaseous Second Messengers Besides NO?** Nitric oxide may be merely the first of a new class of gaseous second messenger/neurotransmitter molecules. Based on your knowledge of the molecular action of nitric oxide, suggest another gaseous molecule that might act as a second messenger, and propose a molecular function for it.

4. **Assessing the Action of an Antibiotic** Herbimycin A is an antibiotic that inhibits tyrosine kinase activity by binding to SH groups of cysteine in the *src* gene tyrosine kinase and other similar tyrosine kinases. What effect might it have on normal rat kidney cells that have been transformed by Rous sarcoma virus? Can you think of other effects you might expect for this interesting antibiotic?

5. **Proposing Uses for Monoclonal Antibodies Against Phosphotyrosine** Monoclonal antibodies that recognize phosphotyrosine are commercially available. How could such an antibody be used in studies of cell signaling pathways and mechanisms?

6. **Are there Functions of Hormone Receptors Other Than Signal Amplification?** Explain and comment on this statement: The main function of hormone receptors is that of signal amplification.

7. **Determining the Concentration of a Neurotransmitter in a Synaptic Vesicle** Synaptic vesicles are approximately 40 nm in outside diameter, and each vesicle contains about 10,000 acetylcholine molecules. Calculate the concentration of acetylcholine in a synaptic vesicle.

8. **Imagining a Model for Neurotransmitter Release** GTPγS is a nonhydrolyzable analog of GTP. Experiments with squid giant axon synapses reveal that injection of GTγS into the presynaptic end (terminal) of the neuron inhibits neurotransmitter release (slowly and irreversibly). The calcium signals produced by presynaptic action potentials and the number of synaptic vesicles docking on the presynaptic membrane are unchanged by GTPγS. Propose a model for neurotransmitter release that accounts for all of these observations.

9. **Assessing the Binding of a Hormone to Its Receptor** A typical hormone binds to its receptor with an affinity (K_D) of approximately 1×10^{-9} M. Consider an in vitro (test-tube) system in which the total hormone concentration is approximately 1 nM and the total concentration of receptor sites is 0.1 nM. What fraction of the receptor sites is bound with hormone? If the concentration of receptors is decreased to 0.033 nM, what fraction of receptor is bound with the hormone?

10. **Assessing the Effects of Statins on Steroid Hormones** (Integrates with Chapter 24.) All steroid hormones are synthesized in the human body from cholesterol. What is the consequence for steroid hormones and their action from taking a "statin" drug, such as Zocor, which blocks the synthesis of cholesterol in the body? Are steroid hormone functions compromised by statin action?

11. **Reconciling the Apparent Preference for Helices in Transmembrane Proteins** Given that β-strands provide a more genetically economical way for a polypeptide to cross a membrane, why has nature chosen α-helices as the membrane-spanning motif for G-protein–coupled receptors? That is, what other reasons can you imagine to justify the use of α-helices?

12. **Understanding Reaction Mechanisms for Adenylyl Cyclase and Phosphodiesterases** Write simple reaction mechanisms for the formation of cAMP from ATP by adenylyl cyclase and for the breakdown of cAMP to 5'-AMP by phosphodiesterases.

13. **Determining the Equilibrium Transmembrane Potential for K$^+$ and Na$^+$** (Integrates with Chapter 9.) Consider the data in Figure 32.49a. Recast Equation 9.2 to derive a form from which you could calculate the equilibrium electrochemical potential at which no net flow of potassium would occur. This is the Nernst equation. Calculate the equilibrium potential for K$^+$ and also for Na$^+$, assuming $T = 37°C$.

14. Calculating Transmembrane Potential Using the Goldman Equation
The calculation of the actual transmembrane potential difference for a neuron is accomplished with the Goldman equation:

$$\Delta\psi = \frac{RT}{\mathcal{F}} \ln\left(\frac{\Sigma P_C[C]_{outside} + \Sigma P_A[A]_{inside}}{\Sigma P_C[C]_{inside} + \Sigma P_A[A]_{outside}}\right)$$

where [C] and [A] are the cation and anion concentrations, respectively, and P_C and P_A are the respective permeability coefficients of cations and anions.

Assume relative permeabilities for K^+, Na^+, and Cl^- of 1, 0.04, and 0.45, respectively, and use this equation to calculate the actual transmembrane potential difference for the neuron whose ionic concentrations are those given in Figure 32.49a.

15. Understanding the Movement of K⁺, Na⁺, and Cl⁻ During an Action Potential
Use the information in problems 13 and 14, together with Figure 32.50, to discuss the behavior of potassium, sodium, and chloride ions as an action potential propagates along an axon.

16. Assessing Signal Amplification in a Signaling Pathway
Review the cell signaling pathway shown in Figure 32.4. With the rest of the chapter as context, discuss all the steps of this pathway that involve signal amplification.

17. Understanding the Consequences of a Ras Mutation on a Signaling Pathway
In the cell signaling pathway shown in Figure 32.4, what would be the effect if Ras were mutated so that it had no GTPase activity?

18. Understanding the Action of GPCRs in G-Protein-Independent Pathways
One of the topics discussed in this chapter is the ability of GPCRs to exert signaling effects without the involvement of G proteins. Using the pathway shown in Figure 32.40, and considering everything you have learned in this chapter, suggest some reasons that would explain why this G-protein–independent signaling was difficult to verify experimentally.

Preparing for the MCAT® Exam

19. Assessing the Action of a Popular Pesticide
Malathion (Figure 32.58) is one of the secrets behind the near-complete eradication of the boll weevil from cotton fields in the United States. For most of the 20th century, boll weevils wreaked havoc on the economy of states from Texas to the Carolinas. When boll weevils attacked cotton fields in a farming community, the destruction of cotton plants meant loss of jobs for farm workers, bankruptcies for farm owners, and resulting hardship for the entire community. Relentless application of malathion to cotton crops and fields has turned the tide, however, and agriculture experts expect that boll weevils will be completely gone from cotton fields within a few years. Remarkably, malathion-resistant boll weevils have not emerged despite years of this pesticide's use. Consider the structure and chemistry of malathion, and suggest what you would expect to be the ecological consequences of chronic malathion application to cotton fields.

20. Assessing the Structural Requirements for a Regulatory Protein in a Signaling Network
Consult the excellent review article "Assembly of Cell Regulatory Systems Through Protein Interaction Domains" (*Science* **300**:445–452, 2003, by Pawson and Nash), and discuss the structural requirements for a regulatory protein operating in a signaling network.

ActiveModels Problems

21. Using the ActiveModel for Brca2 (the protein encoded by the BRCA2 gene associated with breast cancer susceptibility), discuss how a mutation in a BRC motif could cause breast cancer.

22. Using the ActiveModel for c-Abl kinase, explain how Gleevec functions as a therapy for chronic myelogenous patients.

FURTHER READING

Signal Transduction and Signaling Pathways

Gesty-Palmer, D., and Luttrell, L. M., 2008. Heptahelical terpsichory. Who calls the tune? *Journal of Receptors and Signal Transduction* **28**:39–58.

Housley, M. D., 2010. Underpinning compartmentalized cAMP signaling through targeted cAMP breakdown. *Trends in Biochemical Sciences* **35**:91–100.

Kearns, J. D., and Hoffman, A., 2009. Integrating computational and biochemical studies to explore mechanisms in NF-κB signaling. *The Journal of Biological Chemistry* **284**:5439–5443.

Lee, J. H., Budanov, A. V., Park, E. J., Birse, R., et al., 2010. Sestrin as a feedback inhibitor of TOR that prevents age-related pathologies. *Science* **327**:1223–1228.

Lefkowitz, R. J., and Shenoy, S. K., 2005. Transduction of receptor signals by β-arrestins. *Science* **308**:512–517.

Lim, W., 2010. Designing customized cell signaling circuits. *Nature Reviews Molecular and Cell Biology* **11**:393–403.

Luttrell, L. M., 2005. Composition and function of G protein–coupled receptor signalsomes controlling mitogen-activated protein kinase activity. *Journal of Molecular Neuroscience* **26**:253–264.

Maccarrone, M., Dainese, E., and Oddi, S., 2010. Intracellular trafficking of anandamide: New concepts for signaling. *Trends in Biochemical Sciences* **35**:601–608.

Maizels, R. M., and Allen, J. E., 2011. Eosinophils forestall obesity. *Science* **332**:186–187.

Raaijmakers, J. H., and Bos, J. L., 2009. Specificity in Ras and Rap signaling. *The Journal of Biological Chemistry* **284**:10995–10999.

Sudol, M., and Harvey, K. F., 2010. Modularity in the Hippo signaling pathway. *Trends in Biochemical Sciences* **35**:627–633.

Topisirovic, I., and Sonenberg, N., 2010. Burn out or fade away? *Science* **327**:1210–1211.

Wagner, E. F., and Nebreda, A. R., 2009. Signal integration by JNK and p38 MAPK pathways in cancer development. *Nature Reviews Cancer* **9**:537–549.

Wimmer, R. and Baccarini, M., 2010. Partner exchange: Protein–protein interactions in the Raf pathway. **35**:660–668.

Worthington, J. J., Klementowicza, J. E., and Travis, M. A., 2011. TGFβ: A sleeping giant awoken by integrins. *Trends in Biochemical Sciences* **36**:47–54.

Zdanov, A., and Wlodawer, A., 2008. A new look at cytokine signaling. *Cell* **132**:179–181.

Scaffolding, Adaptor Proteins, and Signal Integration

Deribe, Y. L., Pawson, T., and Dikic, I., 2011. Post-translational modifications in signal integration. *Nature Structural and Molecular Biology* **17**:666–672.

Good, M. C., Zalatan, J. G., and Lim, W. A., 2011. Scaffold proteins: Hubs for controlling the flow of cellular information. *Science* **332**:680–686.

Groves, J. T., and Kuriyan, J., 2010. Molecular mechanisms in signal transduction at the membrane. *Nature Structural and Molecular Biology* **17**:659–665.

Haucke, V., Neher, E., and Sigrist, S. J., 2011. Protein scaffolds in the coupling of synaptic exocytosis and endocytosis. *Nature Reviews Neuroscience* **12**:127–138.

Huang, R., Martinez-Ferrando, I., and Cole, P. A., 2010. Enhanced interrogation: Emerging strategies for cell signaling inhibition. *Nature Structural and Molecular Biology* **17**:646–649.

Jaillais, Y., and Chory, J., 2010. Unraveling the paradoxes of plant hormone signaling integration. *Nature Structural and Molecular Biology* **17:**642–645.

Pawson, C. T., and Scott, J. D., 2010. Signal intergration through blending, bolstering, and bifurcating of intracellular information. *Nature Structural and Molecular Biology* **17:**653–658.

Scott, J. D., and Pawson, T., 2009. Cell signaling in space and time: Where proteins come together and when they're apart. *Science* **326:** 1220–1224.

Membrane Receptors

Cherezov, V., Rosenbaum, D. M., et al., 2007. High-resolution crystal structure of an engineered human β_2-adrenergic G protein–coupled receptor. *Science* **318:**1258–1265.

Huang, C., and Tesmer, J. J. G., 2011. Recognition in the face of diversity: Interactions of hetertrimeric G proteins and G protein-coupled receptor (GPCR) kinases with activated GPCRs. *The Journal of Biological Chemistry* **286:**7715–7721.

Lawrence, M. C., McKern, N. M., and Ward, C. W., 2007. Insulin receptor structure and its implications for the IGF-1 receptor. *Current Opinion in Structural Biology* **17:**699–705.

Linderman, J. J., 2009. Modeling of G-protein-coupled receptor signaling pathways. *The Journal of Biological Chemistry* **284:**5427–5431.

Luttrell, L. M., and Gesty-Palmer, D., 2010. Beyond desensitization: Physiological relevance of arrestin-dependent signaling. *Pharmacological Reviews* **62:**305–330.

McKern, N. M., Lawrence, M. C., et al., 2006. Structure of the insulin receptor ectodomain reveals a folded-over conformation. *Nature* **443:** 218–221.

Michino, M., Abola, E., GPCR Dock 2008 participants, Brooks, C. L., Dixon, J. S., Moult, J., and Stevens, R. C., 2009. Community-wide assessment of GPCR structure modeling and ligand docking: GPCR Dock 2008. *Nature Reviews Drug Discovery* **8:**455–463.

Misono, K., Ogawa, H., et al., 2005. Structural studies of the natriuretic peptide receptor: A novel hormone-induced rotation mechanism for transmembrane signal transduction. *Peptides* **26:**957–968.

Nakagawa, T., 2010. The biochemistry, ultrastructure, and subunit assembly mechanism of AMPA receptors. *Molecular Neurobiology* **42:**161–184.

Palczewski, K., 2010. Oligomeric forms of G protein-coupled receptors (GPCRs). *Trends in Biochemical Sciences* **35:**595–600.

Peeters, M. C., van Westen, G. J. P., and Ijzerman, A. P., 2011. Importance of the extracellular loops in G-protein-coupled receptors for ligand recognition and receptor activation. *Trends in Pharmacological Sciences* **32:**35–42.

Rajagopol, S., Kim, J., Ahn, S., Craig, S., et al., 2010. β-arrestin- but not G protein-mediated signaling by the "decoy" receptor CXCR7. *Proceedings of the National Academy of Sciences, U.S.A.* **107:**628–632.

Rajagopal, S., Rajagopal, K., and Lefkowitz, R. J., 2010. Teaching old receptors new tricks: Biasing seven-transmembrane receptors. *Nature Reviews Drug Discovery* **9:**373–386.

Rasmussen, S. G. F., Choi, H.-J., et al., 2007. Crystal structure of the human β_2-adrenergic G-protein–coupled receptor. *Nature* **450:**383–388.

Rasmussen, S. G. F., Choi, H.-J., Fung, J. J., Pardon, E., et al., 2011. Structure of a nanobody-stabilized active state of the β_2 adrenoceptor. *Nature* **469:**175–181.

Ritter, S. L., and Hall, R. A., 2010. Fine-tuning of GPCR activity by receptor-interacting proteins. *Nature Reviews Molecular Cell Biology* **10:**819–830.

Rosenbaum, D. M., Zhang, C., Lyons, J. A., Holl, R., et al., 2011. Structure and function of an irreversible agonist-β_2 adrenoceptor complex. *Nature* **469:**236–240.

Rozenfeld, R., and Devi, L. A., 2011. Exploring a role for heteromerization in GPCR signaling specificity. *Biochemical Journal* **433:**11–18.

Sprang, S. R., 2007. A receptor unlocked. *Nature* **450:**355–356.

Tholanikunnel, B. G., Joseph, K., Kandasamy, K., Baldys, A., et al., 2010. Novel mechanisms in the regulation of G protein-coupled receptor trafficking to the plasma membrane. *The Journal of Biological Chemistry* **285:**33816–33825.

Vilardaga, J.-P., Agnati, L. F., Fuxe, K., and Ciruela, F., 2010. G-protein-coupled receptor heteromer dynamics. *Journal of Cell Science* **123:**4215-4220.

Warne, T., Moukhametzianov, R., Baker, J. G., Nehme, R., et al., 2011. The structural basis for agonist and partial agonist action on a β_1-adrenergic receptor. *Nature* **469:**241-245.

Xu, F., Wu, H., Katritch, V., Han, H. W., et al., 2011. Structure of an agonist-bound human A_{2A} adenosine receptor. *Science* **332:**322–327.

Young, M. T., 2010. P2X receptors: Dawn of the post-structure era. *Trends in Biochemical Sciences* **35:**83–90.

G Proteins

Abramow-Newerly, M., Roy, A. A., et al., 2006. RGS proteins have a signalling complex: Interactions between RGS proteins and GPCRs, effectors, and auxiliary proteins. *Cellular Signalling* **18:**579–591.

Bansal, G., Druey, K. M., et al., 2007. R4 RGS proteins: Regulation of G-protein signaling and beyond. *Pharmacology and Therapeutics* **116:** 473–495.

Cabrera-Vera, T. M., Vanhauwe, J., et al., 2003. Insights into G protein structure, function, and regulation. *Endocrine Reviews* **24:**765–781.

Kimple, R. J., Willard, F. S., et al., 2002. The GoLoCo motif: Heralding a new tango between G protein signaling and cell division. *Molecular Interventions* **2:**88–100.

Mirshahi, T., Mittal, V., et al., 2002. Distinct sites on G protein β–γ subunits regulate different effector functions. *Journal of Biological Chemistry* **277:**36345–36350.

Oldham, W. M., and Hamm, H. E., 2008. Heterotrimeric G protein activation by G-protein–coupled receptors. *Nature* **9:**60–71.

Siderovski, D. P., and Willard, F. S., 2005. The GAPs, GEFs, and GDIs of heterotrimeric G-protein α-subunits. *International Journal of Biological Sciences* **1:**51–66.

Willard, F. S., Kimple, R. J., et al., 2004. Return of the GDI: The GoLoCo motif in cell division. *Annual Review of Biochemistry* **73:**925–951.

Willars, G. B., 2006. Mammalian RGS proteins: Multifunctional regulators of cellular signalling. *Seminars in Cell and Developmental Biology* **17:**363–376.

Second Messengers

Bunney, T. D., and Katan, M., 2010. PLC regulation: Emerging pictures for molecular mechanisms. *Trends in Biochemical Sciences* **35:**88–96.

Demuro, A., Parker, I., and Stutzmann, G. E., 2010. Calcium signaling and amyloid toxicity in Alzheimer disease. *The Journal of Biological Chemistry* **285:**12463–12468.

Hill, B. G., Dranka, B. P., Bailey, S. M., Lancaster, J. R., Jr., Darley-Usmar, V. M., 2010. What part of NO don't you understand? Some answers to the cardinal questions in nitric oxide biology. *The Journal of Biological Chemistry* **285:**19699–19704.

Li, L., Rose, P., and Moore, P. K., 2011. Hydrogen sulfide and cell signaling. *Annual Review of Pharmacology and Toxicology* **51:**169–187.

Motterlini, R., and Otterbein, L. E., 2010. The therapeutic potential of carbon monoxide. *Nature Reviews Drug Discovery* **9:**728–743.

Parekh, A. B., 2011. Decoding cytosolic Ca^{2+} oscillations. *Trends in Biochemical Sciences* **36:**78–87.

Protein Kinases and Protein Phosphatases

Bollen, M., Peti, W., Ragusa, M. J., and Beullens, M., 2010. The extended PP1 toolkit: Designed to create specificity. *Trends in Biochemical Sciences* **35**:450-458.

Julien, S. G., Dube, N., Hardy, S., and Tremblay, M. L., 2011. Inside the human cancer tyrosine phosphatome. *Nature Reviews Cancer* **11**:35–49.

Park, S. W., Zhou, Y., Lee, J., Lu, A., et al., 2010. The regulatory subunits of PI3K, p85α and p85β, interact with XBP-1 and increase its nuclear translocation. *Nature Medicine* **16**:429-437.

Pawson, T., and Kofler, M., 2010. Kinome signaling through regulated protein-protein interaction in normal and cancer cells. *Current Opinions in Cell Biology* **21**:147–153.

Rosse, C., Linch, M., Kermorgant, S., Cameron, A. J. M., et al., 2010. PKC and the control of localized signal dynamic. *Nature Reviews Molecular Cell Biology* **11**:103–112.

Sabio, G., and Davis, R. J., 2010. cJun NH$_2$-terminal kinase 1 (JNK1): Roles in metabolic regulation of resistance. *Trends in Biochemical Sciences* **45**:490–496.

Taylor, S. S., and Kornev, A. P., 2011. Protein kinases: Evolution of dynamic regulatory proteins. *Trends in Biochemical Sciences* **35**:65–77.

Winnay, J. N., Boucher, J., Mori, M. A., Ueki, K., and Kahn, C. R., 2010. A regulatory subunit of phosphoinositide 3-kinase increases the nuclear accumulation of X-box-binding protein-1 to modulate the unfolded protein response. *Nature Medicine* **16**:438-446.

Wong, K.-K., Engelman, J. A., and Cantley, L. C., 2010. Targeting the PI3K signaling pathway in cancer. *Current Opinion in Genetics and Development* **20**:87–90.

Wu, D., and Pan, W., 2010. GSK3: A multifaceted kinase in Wnt signaling. *Trends in Biochemical Sciences* **35**:161–168.

Yip, S.-C., Saha, S., and Chernoff, J., 2010. PTP1B: A double agent in metabolism and oncogenesis. *Trends in Biochemical Sciences* **35**:442–449.

Steroid Hormones

Filardo, E., Quinn, J., et al., 2007. Activation of the novel estrogen receptor G protein–coupled receptor 30 (GPR30) at the plasma membrane. *Endocrinology* **148**:3236–3245.

Kampa, M., and Castanas, E., 2006. Membrane steroid receptor signaling in normal and neoplastic cells. *Molecular and Cellular Endocrinology* **246**:76–82.

Prossnitz, E. R., Arterburn, J. B., et al., 2008. Estrogen signaling through the transmembrane G protein–coupled receptor GPR30. *Annual Review of Physiology* **70**:165–190.

Prossnitz, E. R., Sklar, L. A., et al., 2008. GPR30: A novel therapeutic target in estrogen-related disease. *Trends in Pharmacological Sciences* **29**: 116–123.

Raz, L., Khan, M. M., et al., 2008. Rapid estrogen signaling in the brain. *Neurosignals* **16**:140–153.

Zhao, C., Dahlman-Wright, K., and Gustafsson, J.-A., 2010. Estrogen signaling via estrogen receptor β. *The Journal of Biological Chemistry* **285**:39575–39579.

Neurotransmission

Arnadottir, J., and Chalfie, M., 2010. Eukaryotic mechanosensitive channels. *Annual Review of Biophysics* **39**:111–137.

Baenziger, J. E., and Corringer, P.-J., 2011. 3D structure and allosteric modulation of the transmembrane domain of pentameric ligand-gated ion challels. *Neuropharmacology* **60**:116–135.

Biel, M., 2009. Cyclic nucleotide-regulated cation channels. *The Journal of Biological Chemistry* **284**:9017–9021.

Chalfie, M., 2009. Neurosensory mechanotransduction. *Nature Reviews Molecular Cell Biology* **10**:44–52.

Corringer, P.-J., Baaden M., Bocquet, N., Delarue, M., et al., 2010. Atomic structure and dynamics of pentameric ligand-gated ion channels: New insight from bacterial homologues. *The Journal of Physiology* **588**:565–572.

Cukkemane, A., Seifert, R., Kaupp, U. B., 2011. Cooperative and uncooperative cyclic-nucleotide-gated ion channels. *Trends in Biochemical Sciences* **36**:55–64.

Fritsch, S., Ivanov, I., Wang, H., and Cheng, X., 2011. Ion selectivity mechanism in a bacterial pentameric ligand-gated ion channel. *Biophysical Journal* **100**:390–398.

Shipston, M. J., 2011. Ion channel regulation by protein palmitoylation. *The Journal of Biological Chemistry* **286**:8709–8719.

Unwin, N., 2005. Refined structure of the nicotinic acetylcholine receptor at 4 Å resolution. *Journal of Molecular Biology* **346**:967–989.

Zhou, H.-X., and McCammon, A., 2010. The gates of ion channels and enzymes. *Trends in Biochemical Sciences* **35**:179–185.

Detailed answers to the end-of-chapter problems and additional problems to solve are available in the *Student Solutions Manual, Study Guide and Problems Book* by David Jemiolo and Steven Theg. For additional information see the listing for this book in the Preface.

CHAPTER 1

1. Because bacteria (compared with humans) have simple nutritional requirements, their cells obviously contain enzyme systems that allow them to convert rudimentary precursors (even inorganic substances such as NH_4^+, NO_3^-, N_2, and CO_2) into complex biomolecules—proteins, nucleic acids, polysaccharides, and complex lipids. On the other hand, animals have an assortment of different cell types designed for specific physiological functions; these cells possess a correspondingly greater repertoire of complex biomolecules to accomplish their intricate physiology.

2. Consult Figures 1.20 to 1.22 to confirm your answer.

3. a. Laid end to end, 250 *E. coli* cells would span the head of a pin.
   b. The volume of an *E. coli* cell is about 10^{-15} L.
   c. The surface area of an *E. coli* cell is about 6.3×10^{-12} m². Its surface-to-volume ratio is 6.3×10^6 m⁻¹.
   d. 600,000 molecules.
   e. 1.7 nM.
   f. Because we can calculate the volume of one ribosome to be 4.2×10^{-24} m³ (or 4.2×10^{-21} L), 15,000 ribosomes would occupy 6.3×10^{-17} L, or 6.3% of the total cell volume.
   g. Because the *E. coli* chromosome contains 4600 kilobase pairs (4.6×10^6 bp) of DNA, its total length would be 1.6 mm—approximately 800 times the length of an *E. coli* cell. This DNA would encode 4300 different proteins, each 360 amino acids long.

4. a. The volume of a single mitochondrion is about 4.2×10^{-16} L (about 40% the volume calculated for an *E. coli* cell in problem 3).
   b. A mitochondrion would contain on average fewer than eight molecules of oxaloacetate.

5. a. Laid end to end, 25 liver cells would span the head of a pin.
   b. The volume of a liver cell is about 8×10^{-12} L (8000 times the volume of an *E. coli* cell).
   c. The surface area of a liver cell is 2.4×10^{-9} m²; its surface-to-volume ratio is 3×10^5 m⁻¹, or about 0.05 (1/20) that of an *E. coli* cell. Cells with lower surface-to-volume ratios are limited in their exchange of materials with the environment.
   d. The number of base pairs in the DNA of a liver cell is 6×10^9 bp, which would amount to a total DNA length of 2 m (or 6 feet of DNA!) contained within a cell that is only 20 μm on a side. Maximal information content of liver-cell DNA =

3×10^9 bp, which, expressed in proteins 400 amino acids in length, could encode 2.5×10^6 proteins.

6. The amino acid side chains of proteins provide a range of shapes, polarity, and chemical features that allow a protein to be tailored to fit almost any possible molecular surface in a complementary way.

7. Biopolymers may be informational molecules because they are constructed of different monomeric units ("letters") joined head to tail in a particular order ("words, sentences"). Polysaccharides are often linear polymers composed of only one (or two repeating) monosaccharide unit(s) and thus display little information content. Polysaccharides with a variety of monosaccharide units may convey information through specific recognition by other biomolecules. Also, most monosaccharide units are typically capable of forming branched polysaccharide structures that are potentially very rich in information content (as in cell surface molecules that act as the unique labels displayed by different cell types in multicellular organisms).

8. Molecular recognition is based on structural complementarity. If complementary interactions involved covalent bonds (strong forces), stable structures would be formed that would be less responsive to the continually changing dynamic interactions that characterize living processes.

9. Two carbon atoms interacting through van der Waals forces are 0.34 nm apart; two carbon atoms joined in a covalent bond are 0.154 nm apart.

10. Slight changes in temperature, pH, ionic concentrations, and so forth may be sufficient to disrupt weak forces (H bonds, ionic bonds, van der Waals interactions, hydrophobic interactions).

11. Living systems are maintained by a continuous flow of matter and energy through them. Despite the ongoing transformations of matter and energy by these highly organized, dynamic systems, no overt changes seem to occur in them: They are in a *steady* state.

12. The fraction of the *M. genitalium* genes encoding proteins = 0.925. Genes not encoding proteins encode RNA molecules. (0.925)(580,074 base pairs) = 536,820 base pairs devoted to protein-coding genes. Since 3 base pairs specify an amino acid in a protein, 369 amino acids are found in the average *M. genitalium* protein. If each amino acid contributes on average 120 D to the mass of a protein, the mass of an average *M. genitalium* protein is 44,280 D.

13. (0.925)(206) = 191 proteins. Assuming its genes are the same size as *M. genitalium*, the minimal genome would be 228,480 base pairs.

14. Given 1109 nucleotides (or base pairs) per gene, the minimal virus, with a 3500-nucleotide genome, would have only 3 genes; the maximal virus, with a 280,000-bp genome, would have 252 genes.

15. Fate of proteins synthesized by the rough ER:
    a. Membrane proteins would enter the ER membrane, and, as part of the membrane, be passed to the Golgi, from which vesicles depart and fuse with the plasma membrane.
    b. A secreted protein would enter the ER lumen and be transferred as a luminal protein to the Golgi, from which vesicles depart. When the vesicle fuses with the plasma membrane, the protein would be deposited outside the cell.

16. Increasing kinetic energy increases the motions of molecules and raises their average energy, which means that the difference between the energy to disrupt a weak force between two molecules and the energy of the weak force is smaller. Thus, increases in kinetic energy may break the weak forces between molecules.

17. Informational polymers must have "sense" or direction, and they must be composed of more than one kind of monomer unit.

CHAPTER 2

1. a. 3.3; b. 9.85; c. 5.7; d. 12.5; e. 4.4; f. 6.97.

2. a. 1.26 mM; b. 0.25 mM; c. 4×10^{-12} M; d. 2×10^{-4} M; e. 3.16×10^{-10} M; f. 1.26×10^{-7} M (0.126 μM).

3. a. $[H^+] = 2.51 \times 10^{-5}$ M; b. $K_a = 3.13 \times 10^{-8}$ M; pK_a = 7.5.

4. a. pH = 2.38; b. pH = 4.23.

5. Combine 187 mL of 0.1 M acetic acid with 813 mL of 0.1 M sodium acetate.

6. $[HPO_4^{-2}]/[H_2PO_4^-]$ = 0.398.

7. Combine 555.7 mL of 0.1 M Na$_3$PO$_4$ with 444.3 mL of 0.1 M H$_3$PO$_4$. Final concentrations of ions will be $[H_2PO_4^-]$ = 0.0333 M; $[HPO_4^{2-}]$ = 0.0667 M; $[Na^+]$ = 0.1667 M; $[H^+]$ = 3.16×10^{-8} M.

8. a. Fraction of H$_3$PO$_4$: @pH 0 = 0.993; @pH 2 = 0.58; @pH 4 = 0.01; negligible @pH 6.
   b. Fraction of H$_2$PO$_4^-$: @pH 0 = 0.007; @pH 2 = 0.41; @pH 4 = 0.986; @pH 6 = 0.94; @pH 8 = 0.14; negligible @pH 10.
   c. Fraction of HPO$_4^{2-}$: negligible @pH 0, 2, and 4; @pH 6 = 0.06; @pH 8 = 0.86; @pH 10 ≈ 1.0; @pH 12 = 0.72.
   d. Fraction of PO$_4^{3-}$: negligible at any pH <10; @pH 12 = 0.28.

9. At pH 5.2, $[H_3A]$ = 4.33×10^{-5} M; $[H_2A^-]$ = 0.0051 M; $[HA^{2-}]$ = 0.014 M; $[A^{3-}]$ = 0.0009 M.

10. a. pH = 7.02; $[H_2PO_4^-]$ = 0.0200 M; $[HPO_4^{2-}]$ = 0.0133 M.
    b. pH = 7.38; $[H_2PO_4^-]$ = 0.0133 M; $[HPO_4^{2-}]$ = 0.0200 M.

11. $[H_2CO_3]$ = 2.2 μM; $[CO_{2(d)}]$ = 0.75 mM. When $[HCO_3^-]$ = 15 mM and $[CO_{2(d)}]$ = 3 mM, pH = 6.8.

12. Titration of the fully protonated form of anserine will require the addition of three equivalents of OH$^-$. The pK_a values lie at 2.64 (COOH); 7.04 (imidazole-N$^+$H); and 9.49 (NH$_3^+$). Its isoelectric point lies midway between pK_2 and pK_3, so pH$_I$ = 8.265. To prepare 1 L of 0.04 M anserine buffer, add 164 mL of 0.1 M HCl to 400 mL of 0.1 M anserine at its isoelectric point, and make up to 1 L final volume.

13. Add 410 mL of 0.1 M NaOH to 250 mL of 0.1 M HEPES in its fully protonated form and make up to 1 L final volume.

14. 166.7 g/mole.

15. Add 193 mL of water and 307 mL of 0.1 M HCl to 500 mL of 0.1 M triethanolamine.

16. Combine 200 mL of 0.1 M Tris-H$^+$ with 732 mL water and 68 mL of 0.1 M NaOH.

17. a.

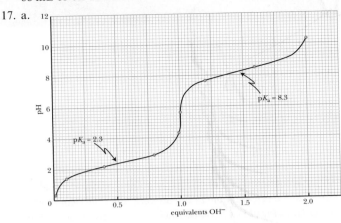

b.

CH$_2$CH$_2$OH
|
:NCH$_2$COO$^-$
|
CH$_2$CH$_2$OH

c. The relevant pK_a for this calculation is 8.3. Combine 400 mL of 0.1 M bicine at its pH$_I$ (pH 5.3) with 155 mL of 0.1 M NaOH and 345 mL of water.

d. The relevant pK_a for this calculation is 2.3. The concentration of fully protonated form of bicine at pH 7.5 is 2.18×10^{-7} M.

18. 0.063 M

19. 5.2 mM

20.

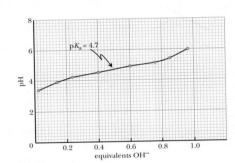

21. 1.53 nmol/mL · sec

22. A drop in blood pH would occur.

23. c.

CHAPTER 3

1. K_{eq} = 613 M; $\Delta G°$ = -15.9 kJ/mol.

2. $\Delta G°$ = 1.69 kJ/mol at 20°C; $\Delta G°$ = -5.80 kJ/mol at 30°C. $\Delta S°$ = 0.75 kJ/mol · K.

3. ΔG = -24.8 kJ/mol.

4. State functions are quantities that depend on the state of the system and not on the path or process taken to reach that state. Volume, pressure, and temperature are state functions. Heat and all forms of work, such as mechanical work and electrical work, are not state functions.

5. $\Delta G°'$ = $\Delta G°$ $-$ 39.5 n (in kJ/mol), where n is the number of H$^+$ produced in any process. So $\Delta G°$ = $\Delta G°'$ + 39.5 n = -30.5 kJ/mol + 39.5(1) kJ/mol.

$\Delta G°$ = 9.0 kJ/mol at 1 M [H$^+$].

6. a. $K_{eq}(AC) = (0.02 \times 1000) = 20$.

 b. $\Delta G°(AB) = 10.1$ kJ/mol.

 $\Delta G°(BC) = -17.8$ kJ/mol.

 $\Delta G°(AC) = -7.7$ kJ/mol.

 $K_{eq} = 20$.

7. See *The Student Solutions Manual, Study Guide and Problems Book* for resonance structures.

8. $K_{eq} = [Cr][P_i]/[CrP][H_2O]$.

 $K_{eq} = 3.89 \times 10^7$.

9. CrP in the amount of 135.3 moles would be required per day to provide 5860 kJ energy. This corresponds to 17,730 g of CrP per day. With a body content of 20 g CrP, each molecule would recycle 886 times per day.

 Similarly, 637 moles of glycerol-3-P, or 108,300 g of glycerol-3-P, would be required. Each molecule would recycle 5410 times/day.

10. $\Delta G = -46.1$ kJ/mol.

11. The hexokinase reaction is a sum of the reactions for hydrolysis of ATP and phosphorylation of glucose:

$$ATP + H_2O \rightleftharpoons ADP + P_i$$
$$\underline{Glucose + P_i \rightleftharpoons G\text{-}6\text{-}P + H_2O}$$
$$Glucose + ATP \rightleftharpoons G\text{-}6\text{-}P + ADP$$

 The free energy change for the hexokinase reaction thus can be obtained by summing the free energy changes for the first two reactions listed here.

 $\Delta G°'$ for hexokinase $= -30.5$ kJ/mol $+ 13.9$ kJ/mol $=$

 -16.6 kJ/mol

12. Comparing the acetyl group of acetoacetyl-CoA and the methyl group of acetyl-CoA, it is reasonable to suggest that the acetyl group is more electron-withdrawing in nature. For this reason, it tends to destabilize the thiol ester of acetoacetyl-CoA, and the free energy of hydrolysis of acetoacetyl-CoA should be somewhat larger than that of acetyl-CoA. In fact, $\Delta G°' = -43.9$ kJ/mol, compared with -31.5 kJ/mol for acetyl-CoA.

13. Carbamoyl phosphate should have a somewhat larger free energy of hydrolysis than acetyl phosphate, at least in part because of greater opportunities for resonance stabilization in the products. In fact, the free energy of hydrolysis of carbamoyl phosphate is -51.5 kJ/mol, compared with -43.3 kJ/mol for acetyl phosphate.

14. The denaturation of chymotrypsinogen is spontaneous at 58°C because the $\Delta G°$ at this temperature is negative (at approximately -2.8 kJ/mol). The native and denatured forms are in equilibrium at approximately 56.6°C.

15. The positive values for ΔC_p for the protein denaturations described in Table 3.1 reflect an increase in motion of the peptide chain in the denatured state. This increased motion provides new ways to store heat energy.

16. $K_{eq} = 2.04 \times 10^{-2}$ at 298K; $K_{eq} = 9$ at 320K.

 $\Delta H°' = 219$ kJ/mol

 $\Delta G°' = 9.64$ kJ/mol

 $\Delta S°' = 702$ J/mol $\cdot$ K

17. The value of $\Delta G°$ is determined at standard state, which includes 1M [H$^+$]. Using Equation 3.23, a value for $\Delta G°$ of -3.36 kJ/mol can be calculated. This value applies at pH 2, 7, and 12 because, of course, $\Delta G°$ is pH-independent. However, Equation 3.23 can also be used to determine that $\Delta G°'$ is -14.77 kJ/mol at pH 2

and -71.82 kJ/mol at pH 12. Finally, $\Delta G°'$ for enolase is not pH-dependent because the enolase reaction (see Figure 3.11) neither consumes nor produces protons. $\Delta G° = \Delta G°' = 1.8$ kJ/mol.

18. The magnitude of $\Delta G°'$ for ATP hydrolysis is sufficiently great that it provides ample energy to drive the conversion of A to B, even though the reaction is unfavorable. The result is that the equilibrium ratio of B to A is more than 10^8 greater when the reaction is coupled to ATP hydrolysis. If A→B were coupled instead to 1,3-bisphosphoglycerate hydrolysis, the ratio of B to A would be even greater, since the $\Delta G°'$ for hydrolysis of 1,3-bisphosphoglycerate is substantially greater than that of ATP. Using the concentrations for 1,3-BPG and 3-PG in Table 18.2, and repeating the calculations on pages 69–70 yields a ratio of $[B_{eq}]/[A_{eq}] = 4.14 \times 10^9$, an even greater ratio than that calculated on pages 69–70.

19. This exercise is left to the student. Use Figure 3.8 as a guide.

20. Without pyrophosphate cleavage, the acyl-CoA synthetase reaction is only slightly favorable, with a $\Delta G°'$ of 0.8 kJ/mol. With pyrophosphate cleavage included, the net $\Delta G°'$ for the reaction is -33.6 kJ/mol, a far more favorable value.

CHAPTER 4

1. Structures for glycine, aspartate, leucine, isoleucine, methionine, and threonine are presented in Figure 4.3.

2. Asparagine = Asn = N.
 Arginine = Arg = R.
 Cysteine = Cys = C.
 Lysine = Lys = K.
 Proline = Pro = P.
 Tyrosine = Tyr = Y.
 Tryptophan = Trp = W.

3. **Alanine dissociation:**

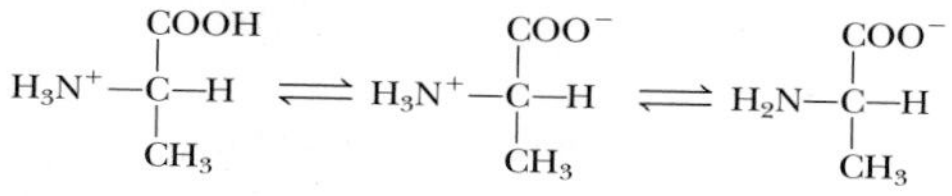

 Glutamate dissociation:

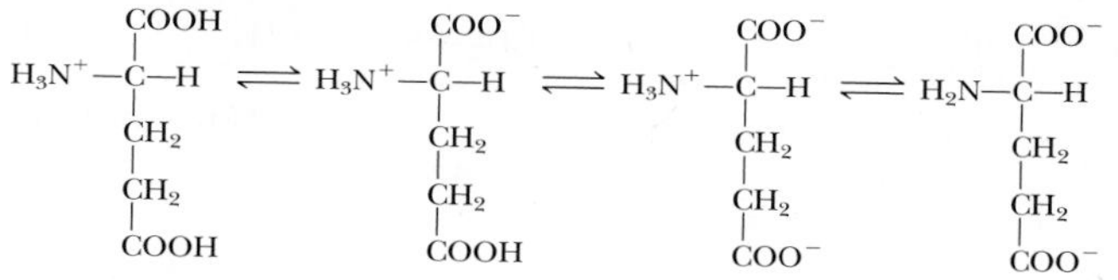

 Histidine dissociation:

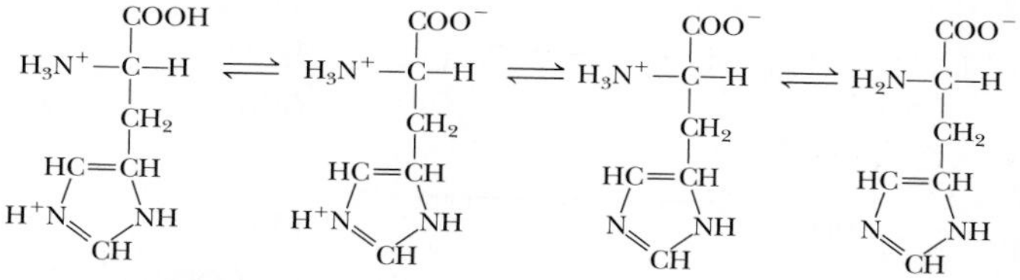

 Lysine dissociation:

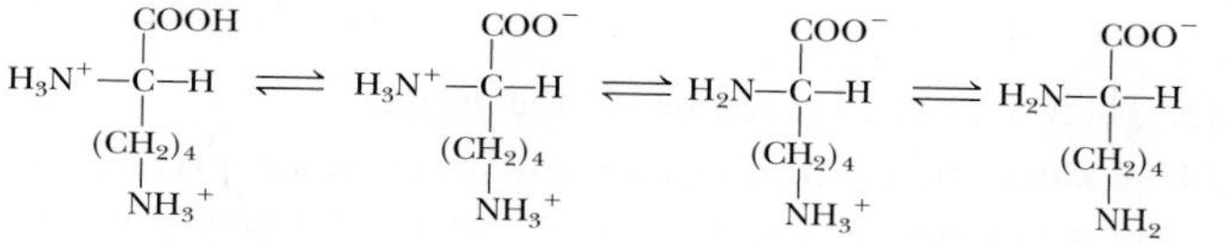

 Phenylalanine dissociation:

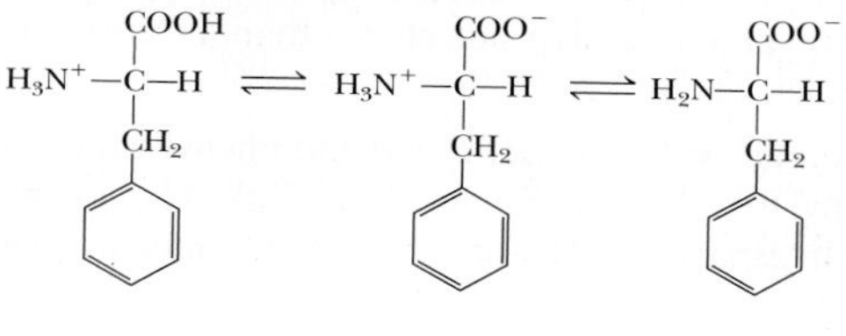

4. The proximity of the α-carboxyl group lowers the pK_a of the α-amino group.

5.

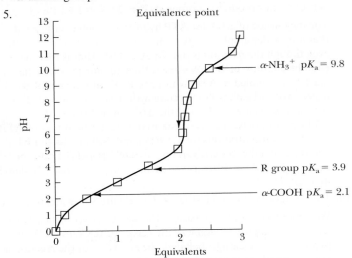

6. Denoting the four histidine species as His^{2+}, His^+, His^0, and His^-, the concentrations are:

pH 2: $[His^{2+}] = 0.097\ M$, $[His^+] = 0.153\ M$, $[His^0] = 1.53 \times 10^{-5}\ M$, $[His^-] = 9.6 \times 10^{-13}\ M$.

pH 6.4: $[His^{2+}] = 1.78 \times 10^{-4}\ M$, $[His^+] = 0.071\ M$, $[His^0] = 0.179\ M$, $[His^-] = 2.8 \times 10^{-4}\ M$.

pH 9.3: $[His^{2+}] = 1.75 \times 10^{-12}\ M$, $[His^+] = 5.5 \times 10^{-5}\ M$, $[His^0] = 0.111\ M$, $[His^-] = 0.139\ M$.

7. $pH = pK_a + \log(2/1) = 4.3 + 0.3 = 4.6$.

The γ-carboxyl group of glutamic acid is 2/3 dissociated at $pH = 4.6$.

8. $pH = pK_a + \log(1/4) = 10.5 + (-0.6) = 9.9$.

9. a. The pH of a 0.3 M leucine hydrochloride solution is approximately 1.46.
   b. The pH of a 0.3 M sodium leucinate solution is approximately 11.5.
   c. The pH of a 0.3 M solution of isoelectric leucine is approximately 6.05.

10. The sequence of reactions shown would demonstrate that L(−)-serine is related stereochemically to L(−)-glyceraldehyde:

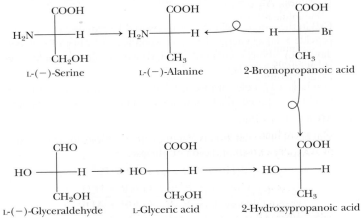

Straight arrows indicate reactions that occur with retention of configuration. Looped arrows indicate inversion of configuration. (*From Kopple, K. D., 1966. Peptides and Amino Acids. New York: Benjamin Co.*)

11. Cystine (disulfide-linked cysteine) has two chiral carbons, the two α-carbons of the cysteine moieties. Each chiral center can exist in two forms, so there are four stereoisomers of cystine. However, it is impossible to distinguish the difference between L-cysteine/D-cysteine and D-cysteine/L-cysteine conjugates. So three distinct isomers are formed:

12.

13. There are eight Tyr residues in the protein.

14. A water is removed from each amino acid when it is incorporated into a protein, so the "molecular weight" of a residue is lower by 18 units. Also, most proteins have relatively more small side chains (Gly, Ala) and fewer Trp side chains than a statistical average would predict.

15. Aspartame is composed of aspartic acid and phenylalanine, with a carboxymethyl cap. These amino acids are linked by a peptide (amide) bond in aspartame. Heating can cleave amide linkages.

For this reason, drinks such as coffee and hot chocolate must be consumed relatively quickly after preparation. Aspartame kept hot for several hours is quite bitter tasting (based on the experience of one of the authors).

16. Phenylketonuria is an autosomal recessive genetic disease caused by a deficiency or absence of the enzyme phenylalanine hydroxylase (PAH), an essential enzyme that converts phenylalanine to tyrosine. Without sufficient PAH activity, phenylalanine accumulates and is converted to phenylpyruvate (which can be detected in the urine). Without treatment, phenylketonurics eventually experience progressive mental retardation and seizures. Phenylketonuria can be controlled by eliminating phenylalanine from the diet, and phenylketonurics should be advised not to use aspartame.

17. The process for distinguishing R- and S-configurations of chiral molecules is described in the text on page 91. Enzymes discriminate between isomers of chiral molecules thanks to the asymmetric arrangement of amino acid residues in the enzyme active site.

18. Appropriate ranges for buffering:

Alanine—1.4–3.4, 8.7–10.7
Histidine—0.8–2.8, 5.0–7.0, 8.2–10.2
Aspartic acid—1.1–4.9, 8.8–10.8
Lysine—1.2–3.2, 8–11.5

19. With pK_a of 8.3, cysteine would make a useful buffer except that cystine, the disulfide of cysteine, can form readily in this pH range. For every cystine formed, two cysteine-SH groups are eliminated, making the buffering capacity of cysteine of limited usefulness. Also, the cysteine sulfhydryl group is the most potent nucleophile among the side chains of the 20 common amino acids.

20. L-threonine is (2S, 3R)-threonine.
D-threonine is (2R, 3S)-threonine.
L-allothreonine is (2S, 3S)-threonine.
D-allothreonine is (2R, 3R)-threonine.

CHAPTER 5

1. Nitrate reductase is a dimer (2 Mo/240,000 M_r).

2. Phe-Asp-Tyr-Met-Leu-Met-Lys.

3. Ser-Glu-Tyr-Arg-Lys-Lys-Phe-Met-Asn-Pro.

4. Ala-Arg-Met-Tyr-Asn-Ala-Val-Tyr or Asn-Ala-Val-Tyr-Ala-Arg-Met-Tyr sequences both fit the results. (That is, in one-letter code, either *ARMYNAVY* or *NAVYARMY*.)

5. Gly-Arg-Lys-Trp-Met-Tyr-Arg-Phe.

6. Actually, there are four possible sequences: NIGIRVIA, GINIRVIA, VIRNIGIA, and, of course, VIRGINIA.

7. Gly-Trp-Arg-Met-Tyr-Lys-Gly-Pro.

8.
Leu-Met-Cys-Val-Tyr-Arg-Cys-Gly-Pro.

9. Alanine, attached to a solid-phase matrix via its α-carboxyl group, is reacted with diisopropylcarbodiimide-activated lysine. Both the α-amino and ϵ-amino groups of the lysine must be blocked with 9-fluorenyl-methoxycarbonyl (Fmoc) groups. To add leucine to Lys-Ala to form a linear tripeptide, precautions must be taken to prevent the incoming Leu α-carboxyl group from reacting inappropriately with the Lys ϵ-amino group instead of the Lys α-amino group.

10. Bovine pancreatic RNase A; *Neurospora crassa* NADPH-nitrite reductase

11. The mass of the myoglobin chain is calculated to be 16,947 ± 1 daltons.

12. Unlike any amino acid side chain, the phosphate group (or more appropriately, the phosphoryl group) bears two equivalents of negative charge at physiological pH. Furthermore, replacing an H atom on an S, T, or Y side chain with a phosphoryl group introduces a very bulky substituent into the protein structure where none existed before.

13. A graph of ν versus $[L]$ reveals that at $\nu = 0.5$, $[L] = K_D = 2.4$ mM.

14. IRS-1 has 1242 amino acids. Its average molecular mass is 131,590.97. The amino acid sequence of the tryptic peptide of IRS-1 of mass 1741.9629 is LNSEAAAVVLQLMNIR. The sequence of the tryptic fragment containing the SHPTP-2 site is LCGAAGGLENGLNYIDLDLVK.

15. Nucleophilic attack by the hydroxyl O of the active-site serine on the carbonyl carbon of a peptide bond.

16. a. Amino acid changes in mutant hemoglobins that appear on the surface of the folded globin chains may affect quaternary structure.

b. Amino acid substitutions on the surface on the quaternary hemoglobin structure that create hydrophobic patches might lead to polymerization. Such amino acids would include all of the hydrophobic amino acids.

CHAPTER 6

1. The central rod domain of keratin is composed of distorted α-helices, with 3.6 residues per turn, but a pitch of 0.51 nm, compared with 0.54 nm for a true α-helix.

(0.51 nm/turn)(312 residues)/(3.6 residues/turn) =
$$44.2 \text{ nm} = 442 \text{ Å}.$$

For an α-helix, the length would be:

(0.54 nm/turn)(312 residues)/(3.6 residues/turn) =
$$46.8 \text{ nm} = 468 \text{ Å}.$$

The distance between residues is 0.347 nm for antiparallel β-sheets and 0.325 nm for parallel β-sheets. So 312 residues of antiparallel β-sheet amount to 1083 Å and 312 residues of parallel β-sheet amount to 1014 Å.

2. The collagen helix has 3.3 residues per turn and 0.29 nm per residue, or 0.96 nm/turn. Then:

(4 in/year)(2.54 cm/in)(10^7 nm/cm)/(0.96 nm/turn) =
$$1.06 \times 10^8 \text{ turns/year.}$$

$(1.06 \times 10^8$ turns/year)(1 year/365 days)(1 day/24 hours)
$$(1 \text{ hour/60 minutes}) = 201 \text{ turns/minute.}$$

3. **Asp:** The ionizable carboxyl can participate in ionic and hydrogen bonds. Hydrophobic and van der Waals interactions are negligible.

Leu: The leucine side chain does not participate in hydrogen bonds or ionic bonds, but it will participate in hydrophobic and van der Waals interactions.

Tyr: The phenolic hydroxyl of tyrosine, with a relatively high pK_a, will participate in ionic bonds only at high pH but can both donate and accept hydrogen bonds. Uncharged tyrosine is capable of hydrophobic interactions. The relatively large size of the tyrosine side chain will permit substantial van der Waals interactions.

His: The imidazole side chain of histidine can act as both an acceptor and donor of hydrogen bonds and, when protonated, can participate in ionic bonds. Van der Waals interactions are expected, but hydrophobic interactions are less likely in most cases.

4. As an imino acid, proline has a secondary nitrogen with only one hydrogen. In a peptide bond, this nitrogen possesses no hydrogens and, thus, cannot function as a hydrogen-bond donor in α-helices. More importantly, ϕ, the angle of rotation around the C_α-N bond, is fixed in Pro. On the other hand, proline stabilizes the *cis*-configuration of a peptide bond and thus, is well suited to β-turns, which require the *cis*-configuration.

5. For a right-handed crossover, moving in the N-terminal to C-terminal direction, the crossover moves in a clockwise direction when viewed from the C-terminal side toward the N-terminal side. The reverse is true for a left-handed crossover; that is, movement from N-terminus to C-terminus is accompanied by counterclockwise rotation.

6. The Ramachandran plot reveals allowable values of ϕ and ψ for α-helix and β-sheet formation. The plots consider steric hindrance and will be somewhat specific for individual amino acids. For example, peptide bonds containing glycine can adopt a much wider range of ϕ and ψ angles than can peptide bonds containing tryptophan.

7. The protein appears to be a tetramer of four 60-kD subunits. Each of the 60-kD subunits in turn is a heterodimer of two peptides, one of 34 kD and one of 26 kD, joined by at least one disulfide bond.

8. Hydrophobic interactions frequently play a major role in subunit–subunit interactions. The surfaces that participate in subunit–subunit interactions in the B_4 tetramer are likely to possess larger numbers of hydrophobic residues than the corresponding surfaces of protein A.

9. The length is given by (53 residues)$\times$(0.15 nm run/residue) = 7.95 nm. The number of turns in the helix is given by (53 residues)/(3.6 residues/turn) = 14.7 turns. There are 49 hydrogen bonds in this helix.

10. Glycines are essential components of tight turns (β-turns) and, thus, are often essential for maintenance of protein structure.

11. Asp, Glu, Ser, Thr, His, and perhaps also Asn, Gln, Cys, Arg, Lys.

12. The ability of poly-Glu to form α-helices requires that the glutamate carboxyls be protonated. Deprotonation produces a polyanionic peptide that is not amenable to helix formation.

13. A coiled-coil formed from α-helices with 3.5 residues per turn would form a symmetrical seven-residue-repeating structure that would place the first and fourth residues of the seven-residue repeat at the same positions about the helix axis in every seven-residue repeat. This would allow the two helices of a coiled-coil structure to lie side by side with no twist about the coiled-coil axis. Such a structure would probably not be as stable as the twisted structure of coiled coils formed from α-helices with 3.6 residues per turn.

14. Linear motifs are peptide sequences that are short and conserved. These blocks occur within a larger protein and act as recognition modules. A sequence containing a phosphorylatable tyrosine and a proline-rich sequence are two examples. Linear motifs are different from domains in the sense that the linear sequence rather than a folded domain is the recognizable unit. Consult the references provided in this question for additional information.

15. The student should consult the references provided to answer this question.

16. The work of Dorothee Kern has provided evidence that conformational transitions in proteins may be assisted by transient interactions resulting from internal protein motions. Consult the reference provided for more information on this interesting phenomenon.

17. Born in Pocahontas, Virginia, Herman Branson received a B.S. from Virginia State University in 1936 and a Ph.D. in Physics from the University of Cincinnati in 1939. He became a professor of physics and chemistry at Howard University in 1941 and spent 27 years on that faculty. In 1948, he took a sabbatical leave in the laboratory of Linus Pauling at the California Institute of Technology. Pauling asked him to do a mathematical analysis of all helical structures for peptide chains that would fit the structural constraints determined previously by Pauling and Robert Corey. The first (and most revolutionary) paper on the alpha helix was submitted to Proceedings of the National Academy of Sciences on Pauling's 50th birthday (2/28/1951) by Pauling,

Corey, and Branson. An excellent account of this story is provided by David Eisenberg (*Proceedings of the National Academy of Sciences* **100**:11207–11210 [2003]).

18. a. The third sequence would place hydrophobic residues on both sides of a β-strand and thus could be found in a parallel β-sheet.
    b. The second sequence would place hydrophobic residues on just one side of a β-strand and thus could be found in an antiparallel β-sheet.
    c. The sixth sequence consists of GPX repeats (where X is any amino acid) and thus could be part of a tropocollagen molecule.
    d. The first sequence consists of seven-residue repeats, with first and fourth residues hydrophobic, and thus could be part of a coiled coil structure.

19. The solution to this exercise is to be completed by the student.

20. $\Delta G°' = -34.23$ kJ/mol, a number that corresponds to one to two H bonds.

CHAPTER 7

1. See structures in *The Student Solutions Manual, Study Guide and Problems Book*.

2. See structures and titration curve in *The Student Solutions Manual, Study Guide and Problems Book*.

3. Glycated hemoglobin can be separated from ordinary hemoglobin on the basis of charge difference (by ion-exchange chromatography, high-performance liquid chromatography [HPLC] electrophoresis, and isoelectric focusing) or on the basis of structural difference (by affinity chromatography).

4. The systematic name for trehalose is α-D-glucopyranosyl-$(1 \longrightarrow 1)$-α-D-glucopyranoside.

   Trehalose is not a reducing sugar. Both anomeric carbons are occupied in the disaccharide linkage.

5. See structures in *The Student Solutions Manual, Study Guide and Problems Book*.

6. A sample that is 0.69 g α-D-glucose/mL and 0.31 g β-D-glucose/mL will produce a specific rotation of 83°.

7. See structures in *The Student Solutions Manual, Study Guide and Problems Book*.

8. A 0.2 g sample of amylopectin corresponds to 0.2 g/162 g/mole or 1.23×10^{-3} mole glucose residues; 50 μmole is 0.04 of the total sample or 4% of the residues. Methylation of such a sample should yield 1,2,3,6-tetramethylglucose for the glucose residues on the reducing ends of the sample. The amylopectin sample contains 1.2×10^{18} reducing ends.

9. There are several target sites for trypsin and chymotrypsin in the extracellular sequence of glycophorin, and it would be reasonable to expect that access to these sites by trypsin and chymotrypsin would be restricted by the presence of oligosaccharides in the extracellular domain of glycophorin.

10. Energy yield upon combustion (whether by metabolic pathways or other reactions) depends on the oxidation level. Carbohydrate and protein are at approximately the same oxidation level, and both of these are significantly less than that of fat.

11. This mechanism could involve either an S_N1 or S_N2 mechanism. In the former, protonation of the bridging oxygen would result in dissociation to produce a carbo-cation intermediate, which could be attacked by the phosphate nucleophile. The observation of retention of configuration at the anomeric carbon favors this

mechanism. An S_N2 mechanism would presumably involve water attack at the anomeric carbon, with dissociation of the oxygen of the carbohydrate chain. This would be followed by S_N2 attack by phosphate. Two S_N2 attacks would result in retention of configuration at the anomeric carbon atom.

12. Laetrile contains a cyanide group. Breakdown of laetrile and release of this cyanide function in the body would be highly toxic.

13. Chondroitin and glucosamine are amino sugar components of cartilage and connective tissue. Dietary supplement with these substances could help replenish the cartilage matrix proteoglycan, relieving pain and restoring the proper function of connective tissue structures.

14. Two of the sugar units in stachyose are glucose and fructose, and the bond joining them is easily cleaved by stomach enzymes. However, the other two sugar units in stachyose are galactoses in $\beta(1 \longrightarrow 6)$-linkages, which are not broken down by human enzymes. The result is that stachyose loses a fructose in the stomach but the resulting trisaccharide passes into the intestines, where bacterial enzymes degrade it, producing intestinal gas in the process. Beano contains an enzyme that hydrolyzes $\beta(1 \longrightarrow 6)$-galactose linkages. Taking several Beano tablets before a meal of beans or legumes facilitates complete breakdown of stachyose and other related oligosaccharides in the stomach, avoiding the production of intestinal gas. The active enzyme in Beano is referred to as a $\beta(1 \longrightarrow 6)$-galactosidase.

15. β-D-Fructofuranosyl-O-α-D-galactopyranosyl-$(1 \longrightarrow 6)$-O-α-D-galactopyranosyl-$(1 \longrightarrow 6)$-α-D-glucopyranoside

16. Starch phosphorylase cleaves glucose units, one at a time, from starch chains until a $(1 \longrightarrow 6)$-branch is encountered. Thus, limit dextrins are amylopectin fragments with a glucose in $(1 \longrightarrow 6)$-linkage at each nonreducing end. The mechanism of the reaction that cleaves these glucose units is essentially the same as that for starch phosphorylase, except that it occurs at a glucose unit that is $(1 \longrightarrow 6)$-linked.

17. In the beer-making process, the mash fermented by yeast contains starch, which is partially broken down by the amylase from the malt to produce limit dextrins (see problem 16). These limit dextrins add significant caloric content to regular beers. Joseph Owades used the enzyme amyloglucosidase, which hydrolyzes both $(1 \longrightarrow 4)$ and $(1 \longrightarrow 6)$ linkages in starch, thus breaking down the limit dextrins to simple glucose, which is fermented normally. (This raises the alcohol content of the beer above what normally would be produced, and water typically is added to adjust the alcohol content.)

18. As described in problem 14, Beano is a $\beta(1 \longrightarrow 6)$-galactosidase. This enzyme can also hydrolyze $\alpha(1 \longrightarrow 6)$-glucose linkages, although somewhat more slowly than for its natural substrate. Thus, Beano can slowly convert the limit dextrins produced in the beer-brewing process into simple glucose units, much like the enzyme used by Joseph Owades (see problem 17). Amateur brewers have learned to use Beano to make their own light beer, often referred to as Beano beer.

19. Assuming each glucose unit contributes 0.55 nm to the growing cellulose polymer (and the plant length), a growth rate of 1 foot per day corresponds to a rate for cellulose synthase of 6,400 glucose units per second.

20. Basic amino acid side chains (Arg, His, Lys) in antithrombin III present positively charged side chains for ionic interactions with sulfate functions on heparin; H-bond donating amino acid side chains could form H bonds with O atoms in —OH groups on the heparin carbohydrate residues; H-bond accepting amino acid side chains could form H bonds with H atoms in —OH groups of heparin.

21. Because these glycosaminoglycans are rich in hydroxyl groups, amine groups, and anionic functions (carboxylates, sulfates), they interact strongly with water. The heavily hydrated proteoglycans formed from these glycosaminoglycans are reversibly dehydrated in response to the pressure imposed on a joint during normal body movements. This dehydration has a cushioning effect on the joint; when the pressure is relieved, the glycosaminoglycans are spontaneously rehydrated due to their affinity for water.

CHAPTER 8

1. Because the question specifically asks for triacylglycerols that contain stearic acid *and* arachidonic acid, we can discount the triacylglycerols that contain only stearic or only arachidonic acid. In this case, there are six possibilities:

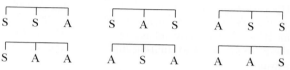

2. See *The Student Solutions Manual, Study Guide and Problems Book* for a discussion.

3. a. Phosphatidylethanolamine and phosphatidylserine have a net positive charge at low pH.
   b. Phosphatidic acid, phosphatidylglycerol, phosphatidylinositol, phosphatidylserine, and diphosphatidylglycerol normally carry a net negative charge.
   c. Phosphatidylethanolamine and phosphatidylcholine carry a net zero charge at neutral pH.

4. Diets high in cholesterol contribute to heart disease and stroke. On the other hand, plant sterols bind to cholesterol receptors in the intestines but are not taken up by the cells containing these receptors. There is substantial evidence that a diet that includes plant sterols can reduce serum cholesterol levels significantly.

5. Former Interior Secretary James Watt was well known during the Reagan administration for his comments on several occasions that trees caused and produced air pollution. As noted in this chapter, it is, of course, true that trees emit isoprenes that are the cause of the blue–gray haze that is common in still air in mid- to late-summer in the eastern United States. However, these isoprene compounds are not significantly toxic to living things, and it would be misleading to call them pollutants.

6. Louis L'Amour clearly knew his biochemistry. His protagonist knew that fat carries a higher energy content than protein or carbohydrate. If meals are going to be scarce (as for a person living in the wild and on the run), it is wise to consume fat rather than protein or carbohydrate. The same reasoning applies for migratory birds in the weeks preceding their long flights.

7. Phospholipase A_2 from snake bites operates without regulation or control, progressively breaking down cell membranes. Phospholipase A_2 action is under precise control and is carefully regulated to produce just the right amounts of required cell signals.

8. The lethal dose for 50% of animals is referred to as the LD_{50}. The LD_{50} for dogs is approximately 3 mg/kg of body weight. Thus, for a 40-lb dog (18.2 kg), consumption of approximately 55 mg of warfarin would be lethal for 50% of animals.

9. See the *Student Solution Guide* and *www.cengage.com/login* for the solutions to this problem.

10. Humans require approximately 2000 kcal per day. Seal blubber is predominantly triglycerides, which yield approximately 9 kcal per gram. This means that a typical human would need to consume 222 grams of seal blubber per day (about half a pound) in order to obtain all of his or her calories from this energy source.

11. Results for this problem will depend on the particular cookies chosen by the student.

12. The only structural differences between cholesterol and stigmasterol are the double bond of stigmasterol at C_{22}–C_{23}, and the ethyl group at C_{24}.

13. Androgens mediate the development of sexual characteristics and sexual function in animals. Glucocorticoids participate in the control of carbohydrate, protein, and lipid metabolism. Mineralocorticoids regulate salt (Na^+, K^+, and Cl^-) balances in tissues.

14. The answers to this question will depend on the household products chosen and the isoprene substances identified.

15. Hydroxide ions (in lye) catalyze the breakdown of triglycerides to produce fatty acid salts such as sodium stearate and sodium palmitate and leaving glycerol as a by-product. Micelle formation by these "soaps" leads to emulsification of fats and other nonpolar substances.

16. Amphipathic phosphatidylcholine from egg yolk exerts a detergentlike action on mixtures of vegetable oil and water. Micelles made primarily of egg phosphatidylcholine emulsify the vegetable oil, forming a stable suspension that persists indefinitely. (Thus, mayonnaise does not need to be mixed or shaken before use, whereas oil and vinegar mixtures do.) The micellar particles of mayonnaise are large enough to scatter visible light, so mayonnaise appears milky white, whereas the separated layers of oil and water in containers of oil and vinegar appear clear.

17. Stanol esters function in cholesterol reduction by binding to cholesterol receptors in the intestines. However, each serving of stanol esters consumed only blocks a fraction of all intestinal receptors. Regular consumption of stanol esters eventually blocks all or most available receptors. Binding of stanol esters is tight and long lasting, but not indefinite, so stanol ester consumption must be continued to maintain the beneficial results. The graph on page 250 shows that reductions of serum cholesterol levels can approach 15%; given that the typical diet accounts for about 15% of total serum cholesterol, it may be concluded that stanol esters are highly effective in preventing uptake of dietary cholesterol.

18. Serum cholesterol is partly derived from diet and partly from synthesis in the liver. Stanol esters prevent uptake of dietary cholesterol but do not affect synthesis in the liver. Cholesterol-lowering drugs, on the other hand, block cholesterol synthesis but have no effect on uptake from diet. Thus, the effects of these two classes of cholesterol-lowering agents are additive.

19. Research has shown that consumption of substantial quantities of plant fats can lead to significant reduction of serum cholesterol. The content of so-called phytosterols varies depending on the source. However, a sterol- or stanol-fortified spread like Benecol probably provides the highest concentration of dietary agents for cholesterol lowering.

20. Tetrahydrogestrinone (THG) is synthesized by the catalytic hydrogenation of gestrinone (which has an acetylene group in place of the ethyl group on the D-ring of tetrahydrogestrinone). Patrick Arnold, known as the "father of prohormones," is credited with the synthesis of THG from gestrinone and the promotion of THG as an anabolic steroid in preparations such as The Clear and The Cream. These substances were marketed aggressively to top athletes in several sports by Arnold and Victor Conte of the Bay Area Laboratory Co-Operative. THG (at that time) was undetectable by existing laboratory analyses, and its use was not specifically prohibited in athletic competitions by the World Anti-Doping Code. Use of THG has since been prohibited by nearly all sports regulatory agencies, and effective methods for its detection have been developed. Numerous athletes have by now admitted to use of THG or have been convicted for perjury in denial of its use, and many have been stripped of medals and awards for athletic accomplishments aided by THG.

21. Most obviously, a diet of triglycerides (from the blubber of seals, the polar bears' favorite food) provides high energy and material that can be reprocessed into other triglycerides and used as insulation under the skin. Less obvious is the need of the polar bear to stay warm and conserve water. (Polar bears cannot afford to eat snow or ice [they are too cold], and they cannot drink seawater [it is too salty].) The polar bear is adapted to conserve body water and stay warm, and thus it does not urinate for months at a time. (Urination would give up both water and heat.) To achieve this, it must consume little or no nitrogen because a diet rich in nitrogen would require urination and defecation. Triglycerides contain no nitrogen and are thus ideal food for the adult polar bear. Juvenile polar bears, which have not yet reached their full adult body size, must consume protein in order to make their own proteins. Once the bear reaches its full size, it changes its diet and no longer consumes protein!

22. Snake venom phospholipase A_2 cleaves fatty acids from phospholipids at the C-2 position. The fatty acids behave as detergents and form micelles (see Chapter 9) that can remove lipids and proteins from the membrane and disrupt membrane structure, causing pores and eventually rupturing the cell itself.

CHAPTER 9

1. Glycerophospholipids with an unsaturated chain at the C-1 position and a saturated chain at the C-2 position are rare to nonexistent. Glycerophospholipids with two unsaturated chains, or with a saturated chain at C-1 and an unsaturated chain at C-2, are commonly found in biomembranes.

2. The phospholipid/protein molar ratio in purple patches of *H. halobium* is 10.8.

3. See *The Student Solutions Manual, Study Guide and Problems Book* for plots of sucrose solution density versus percent by weight and by volume. The plot in terms of percent by weight exhibits a greater curve because less water is required to form a solution that is, for example, 10% by weight (10 g sucrose/100 g total) than to form a solution that is 10% by volume (10 g sucrose/100 mL solution).

4. $r = (4Dt)^{1/2}$.

   According to this equation, a phospholipid with $D = 1 \times 10^{-8}$ cm^2 will move approximately 200 nm in 10 milliseconds.

5. Fibronectin: For $t = 10$ msec, $r = 1.67 \times 10^{-7}$ cm $= 1.67$ nm. Rhodopsin: $r = 110$ nm.

   All else being equal, the value of D is roughly proportional to $(M_r)^{-1/3}$. Molecular weights of rhodopsin and fibronectin are 40,000 and 460,000, respectively. The ratio of diffusion coefficients is thus expected to be $(40,000)^{-1/3}/(460,000)^{-1/3} = 2.3$.

   On the other hand, the values given for rhodopsin and fibronectin give an actual ratio of 4286. The explanation is that fibronectin is anchored in the membrane via interactions with cytoskeletal proteins, and its diffusion is severely restricted compared with that of rhodopsin.

6. a. Divalent cations increase T_m.
   b. Cholesterol broadens the phase transition without significantly changing T_m.
   c. Distearoylphosphatidylserine should increase T_m, due to increased chain length and also to the favorable interactions between the more negative PS head group and the more positive PC head groups.
   d. Dioleoylphosphatidylcholine, with unsaturated fatty acid chains, will decrease T_m.
   e. Integral proteins will broaden the transition and could raise or lower T_m, depending on the nature of the protein.

7. $\Delta G = RT \ln([C_2]/[C_1])$.

   $\Delta G = +4.0$ kJ/mol.

8. $\Delta G = RT \ln([C_{out}]/[C_{in}]) + Z\mathscr{F}\Delta\psi$.

   $\Delta G = +4.08$ kJ/mol -2.89 kJ/mol.

   $\Delta G = +1.19$ kJ/mol.

   The unfavorable concentration gradient thus overcomes the favorable electrical potential, and the outward movement of Na^+ is not thermodynamically favored.

9. One could solve this problem by going to the trouble of plotting the data in v vs. [S], $1/v$ vs. $1/$[S], or [S]/v vs. [S] plots, but it is simpler to examine the value of [S]/v at each value of [S]. The Hanes–Woolf plot makes clear that [S]/v should be constant for all [S] for the case of passive diffusion. In the present case, [S]/v is a constant value of 0.0588 (L/min)$^{-1}$. It is thus easy to recognize that this problem describes a system that permits passive diffusion of histidine.

10. This is a two-part problem. First calculate the energy available from ATP hydrolysis under the stated conditions; then use the answer to calculate the maximal internal fructose concentration against which fructose can be transported by coupling to ATP hydrolysis. Using a value of -30.5 kJ/mol for the $\Delta G°'$ of ATP and the indicated concentrations of ATP, ADP, and P_i, one finds that the ΔG for ATP hydrolysis under these conditions (and at 298 K) is -52.0 kJ/mol. Putting the value of $+52.0$ kJ/mol into Equation 9.1 and solving for C_2 yields a value for the maximum possible internal fructose concentration of 1300 M! Thus, ATP hydrolysis could (theoretically) drive fructose transport against internal fructose concentrations up to this value. (In fact, this value is vastly in excess of the limit of fructose solubility.)

11. Each of the transport systems described can be inhibited (with varying degrees of specificity). Inhibition of the rhamnose transport system by one or more of these agents would be consistent with involvement of one of these transport systems with rhamnose transport. Thus, nonhydrolyzable ATP analogs should inhibit ATP-dependent transport systems, ouabain should specifically block Na^+ (and K^+) transport, uncouplers should inhibit proton gradient–dependent systems, and fluoride should inhibit the PTS system (via inhibition of enolase).

12. N-myristoyl lipid anchors are found linked only to N-terminal Gly residues. Only the peptide in (e) of this problem contains an N-terminal Gly residue.

13. Only the peptide in (a) possesses a CAAX sequence (where C = Cys, A = aliphatic (Ala, Val, Leu, Ile), and X = any amino acid).

14. The hydropathy plot of a soluble protein should show no substantial stretches of hydrophobic residues, except for the signal sequence.

15. Prolines are destabilizing to α-helices, and a short helix with a proline would not be likely to be stable. Where a proline occurs in an α-helix, the helix is bent or kinked. Helices with a proline kink tend to be longer than average, presumably because longer helices are more stable and more able to tolerate the loss of H bonds in the kink region.

16. The structural consequences of a proline-induced kink in an α-helix are described in detail in the references provided in this problem.

17. Porin proteins typically consist of 18-stranded β-barrels. With 9 to 11 residues per strand, about 180–200 residues would be required to form the barrel, with a roughly equivalent number required to form the loops between strands. (The maltoporin chains from E. coli consists of about 420 residues.) The transmembrane domain of Wza consists of eight α-helical segments with about 25 residues per helix, about 200 residues in all. Thus, the number of residues needed to create the transmembrane pore in these two proteins is about the same, even though the number of residues per membrane spanning segment is less for a β-strand (9–11) than for an α-helix (21–25).

18. From Figure 9.29, the area within a typical "fenced" area is about 0.3 μm $\times$ 0.3 μm, or 9×10^6 Å^2. Dividing by 60 Å^2, it appears that there are about 150,000 phospholipids in a monolayer-fenced area in the membrane (assuming a membrane of pure phospholipid, with no cholesterol, etc.).

19. The lysine side chain N is about 5.5 Å from the α-carbon (about 1.1 Å per bond). If a Lys side chain was reoriented toward the membrane center, the maximum change in position relative to the membrane center would be 10–11 Å (assuming the position of the α-carbon did not change). Figure 9.15 indicates that the energy of a Lys side chain would change by 4kT over the 15-Å distance from the membrane surface to the membrane center. A movement of 10 Å thus would correspond to 2/3 of 4kT, or 8kT/3—almost twice the average translation kinetic energy of a molecule in the gas phase (3kT/2).

20. In a hydrophobic environment, the aspartate carboxyl group will be more stable in the protonated (uncharged) form than in the deprotonated (charged) form.

21. In a hydrophobic environment, the side chains of lysine and arginine would be more stable in their deprotonated (uncharged) forms than in the protonated (charged) forms. As a result, the pK_a values of these residues would be lowered significantly in a hydrophobic environment.

22. Based on the discussion in problems 20 and 21 in this chapter, it would be reasonable to imagine that light-induced conformation changes could alter the pK_a values of the proton-transferring moieties in the protein. For example, a conformation change that made the environment around an Asp carboxyl more polar would reduce the pK_a of that group, promoting dissociation (i.e., proton release). An appropriate sequence of such changes could accomplish transmembrane proton transport.

23. Point (b)—that proteins can be anchored to the membrane via covalent links to lipid molecules—was not part of the Singer–Nicolson fluid mosaic model.

CHAPTER 10

1.

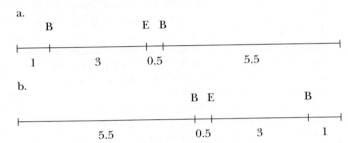

2. See Figure 10.15.

3. $f_A = 0.29$; $f_G = 0.22$; $f_C = 0.25$; $f_T = 0.28$.

4. The number of A and T residues $= 3.14 \times 10^9$ each; the number of G and C residues $= 2.68 \times 10^9$ each.

5. 5'-TAGTGACAGTTGCGAT-3'.

6. 5'-ATCGCAACTGTCACTA-3'.

7. 5'-TACGGTCTAAGCTGA-3'.

8. There are two possibilities, a and b. (E = *Eco*RI site; B = *Bam*HI site.) Note that b is the reverse of a.

a.

```
      B              E  B
|-----+---------------+--+----------------|
  1      3        0.5         5.5
```

b.

```
                   B  E          B
|------------------+--+----------+--------|
       5.5          0.5      3       1
```

9. a. GGATCCCGGGTCGACTGCAG;
   b. GTCGACCCGGGATCCTGCAG.

   *Sma*I products: a. GGATCCC and GGGTCGACTGCAG.
   b. GTCGACCC and GGGATCCTGCAG.

10. Synthesis of a polynucleotide 100 residues long requires formation of 99 phosphodiester bonds. $\Delta G^{\circ\prime}$ for phosphodiester synthesis (assuming it is the same magnitude but opposite sign as that for phosphoric anhydride cleavage of an NTP to give NMP + PP$_i$) is +32.3 kJ/mol. $\Delta G^{\circ\prime}_{overall} = (99)(+32.3) = 3198$ kJ/mol.

11. a. Hydrogen bonding and van der Waals interactions between amino acid side chains and DNA, and ionic interactions of amino acid side chains with the nucleic acid backbone phosphate groups. Double-helical DNA does not present hydrophobic regions for interaction with proteins because of base-pair stacking.
    b. Proteins can recognize specific base sequences if they can fit within the major or minor groove of DNA and "read" the H-bonding pattern presented by the edges of the bases in the groove. The dimensions of an α-helix are such that it fits snugly within the major groove of B-DNA; then, depending on the amino acid sequence of the protein, the side chains displayed on the circumference of the α-helix have the potential to form H bonds with H-bonding functions provided by the bases.

12. a. The restriction endonuclease must be able to recognize a specific nucleotide sequence in the DNA that has twofold rotational symmetry.
    b. In order to read a base sequence within DNA, either the restriction endonuclease must interact with the bases by direct access, for example, by binding in the major groove, or the restriction endonuclease must be able to "read" the base sequence by indirect means, for example, if the base sequence imparts some local variation in the cylindrical surface of the DNA that the enzyme might recognize.
    c. The restriction endonuclease must be able to cleave both DNA strands, often in a staggered fashion.
    d. An obvious solution to the requirements listed in (a) and (c) would be a homodimeric subunit organization for restriction endonucleases.

13. a. The ribose group of nucleosides greatly increases the water solubility of the base.
    b. Ribose has fewer hydroxyl groups and thus less likelihood to undergo unwanted side reactions.
    c. The absence of the 2-OH in 2-deoxyribose leads to a polynucleotide sugar–phosphate backbone that is more stable because it is not susceptible to alkaline hydrolysis.

14. a. Phosphate groups bear a negative electrical charge at neutral pH.
    b. Cleavage of phosphoric anhydride bonds is strongly exergonic and can provide the thermodynamic driving force for diverse metabolic reactions.
    c. Cleavage of phosphoric anhydride bonds is strongly exergonic and can provide the thermodynamic driving force for phosphodiester bond formation in polynucleotide synthesis.

15. Once in every 4.4×10^{12} nucleotides.

16. A DNA sequence 16 nucleotides long.

17. The strategy for protein sequencing by Edman degradation, as described on pages 87 and 114, could be adapted, replacing Edman's reagent with snake venom phosphodiesterase acting on an immobilized nucleotide sequence.

18. 3.59×10^{12} D.

19. The use of bases as "information symbols" in metabolism allows the cell to allocate portions of its phosphorylation potential (as total NTP) to dedicated tasks, as in GTP for protein synthesis, CTP for phospholipid synthesis, UTP for carbohydrate synthesis, with ATP serving the central role. Enzymes in these pathways selectively bind the proper nucleotide through recognition of the particular base.

20. The most prominent structural feature of the DNA double helix in terms of structural complementarity is found in the canonical base-pairing of A with T and G with C. Base pairing is essential to DNA replication (preservation of the genetic information) and transcription (expression of genetic information).

CHAPTER 11

1.

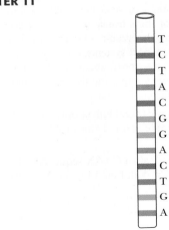

2. Original nucleotide: 5'-GATAGCGCAAAGATCAACCTT.

3. a. 10.5 base pairs per turn; b. $\Delta\phi = 34.3°$; c (true repeat) = 6.72 nm.

4. 27.3 nm; 122 base pairs.

5. 4.35 μm.

6. $L_0 = 160$. If $W = -12$, $L = T + W = 160 + (-12) = 148$. $\sigma = \Delta L/L_0 = -12/160 = -0.075$.

7. For 1 turn of B-DNA (10 base pairs): $L_B = 1.0 + W_B$.

For Z-DNA, 10 base pairs can form only 10/12 turn (0.833 turn), and $L_Z = 0.833 + W_Z$.

For the transition B-DNA to Z-DNA, strands are not broken, so $L_B = L_Z$; that is, $1.0 + W_B = 0.833 = W_Z$, or $W_Z - W_B = +0.167$.

(In going from B-DNA to Z-DNA, the change in W, the number of supercoils, is positive. This result means that, if B-DNA contains negative supercoils, their number will be reduced in Z-DNA. Thus, all else being equal, negative supercoils favor the B $\longrightarrow$ Z transition.)

8. 6×10^9 bp/200 bp = 3×10^7 nucleosomes.

The length of B-DNA 6×10^9 bp long = $(0.34 \text{ nm})(6 \times 10^9)$ = 2.04×10^9 nm (more than 2 meters!). The height of 3×10^7 nucleosomes = $(6 \text{ nm})(3 \times 10^7)$ = 18×10^7 nm (0.18 meter).

9. From Figure 11.33: Similarly shaded regions indicate complementary sequences joined via intrastrand hydrogen bonds:

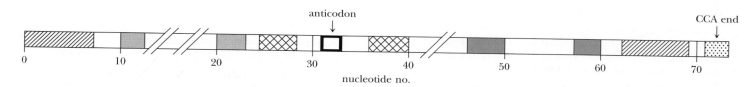

10. Increasing order of T_m: yeast < human < salmon < wheat < E. coli.

11. In 0.2 M Na$^+$, $T_m(°C) = 69.3 + 0.41(\%G + C)$:

Rats (%G + C) = 40%, $T_m = 69.3 + 0.41(40) = 85.7°C$

Mice (%G + C) = 44%, $T_m = 69.3 + 0.41(44) = 87.3°C$

Because mouse DNA differs in GC content from rat DNA, they could be separated by isopycnic centrifugation in a CsCl gradient.

12. GC content = 0.714 (from Table 10.1 and equations used in problem 3, Chapter 10). $\rho = 1.660 + 0.098(GC) = 1.730$ g/mL.

13. See Figure 11.35 and compare your structure with the base pair formed between G18 and ψ55 in yeast phenylalanine tRNA, which has a single H bond between the 2-amino group of G (2-NH$_2$ $\cdots$ O$=$) and the O atom at position 4 in ψ.

14. Assuming the plasmid is in the B-DNA conformation, where each pair contributes 0.34 nm to the length of the molecule, the circumference of a perfect circle formed from pBR322 would be (0.34 nm/bp)(4363 bp) = 1483 nm = 1.48 μm.

The E. coli K12 chromosome laid out as a perfect circle would have a circumference of (0.34 nm/bp)(4,639,000 bp) = 1,577,260 nm = 1.58 mm.

The diameter of a B-DNA molecule is about 2.4 nm (see Table 11.1). For pBR322, the length/diameter ratio would be 1480/2.4 = 616.7. For the E. coli K12 chromosome, the length/diameter ratio would be 1,577,260/2.4 = 657,192.

15. (a) Z-DNA; (b) cruciform; (c) triplex DNA; (d) tRNA; (e) type-II restriction endonuclease site.

16. Rely on the features in Figure 11.33 to draw a tRNA cloverleaf with the anticodon positioned at nucleotides 34–36: CUG.

17. Erythromycin is within the palm of the mitten.

18. (a) β-globin; (b) hexosaminidase A (α-polypeptide); (c) insulin receptor; (d) neutral amino acid transporter.

19. The DNA in such a thermophilic organism would be subjected to very high temperatures that might denature the DNA. Because G:C base pairs are more heat stable than A:T base pairs, one might expect DNA from a thermophilic organism to have a high G + C content.

20. DNA is the material of heredity; that is, genetic information. This information is encoded in the sequence of bases in DNA, so this is its most important structural feature. The double-stranded nature of DNA and the complementary base sequence of the two strands are also crucial structural features in the transmission of genetic information through DNA replication. Beyond these points, one might cite the structural features of DNA that impart stability to the double helix and structural aspects that render DNA less susceptible to degradation.

CHAPTER 12

1. Linear and circular DNA molecules consisting of one or more copies of just the genomic DNA fragment; linear and circular DNA molecules consisting of one or more copies of just the vector DNA; linear and circular DNA molecules containing one or more copies of both the genomic DNA fragment and plasmid DNA.

2.

-GAATTCCCGGGGATCCTCTAGAGTCGACCTGCAGGCATGC-			
GAATTC	GGATCC	GTCGAC	GCATGC
EcoRI	*BamHI*	*SalI*	*SphI*
	CCCGGG	TCTAGA	CTGCAG
	SmaI	*XbaI*	*PstI*

3. a. AAGCTTGAGCTCGAGATCTAGATCGAT

*Hind*III   *Xho*I   *Xba*I

    *Sac*I   *Bg*III   *Cla*I

b.

Vector: *Hind*III: 5'-A.......-gap-.........CGAT-3': *Cla*I
                 3'-TTCGA....-gap-..........TA-5'

Fragment: *Hind*III: 5'-AGCTT(NNNN-etc-NNNN)AT-3'
                      3'-A(NNNN-etc-NNNN)TAGC-5'

4. N = 3480.

5. N = 10.4 million.

6. 5'-ATGCCGTAGTCGATCAT and
5'-ATGCTATCTGTCCTATG.

7. -Thr-Met-Ile-Thr-Asn-Ser-Pro-*Asp-Pro-Phe-Ile-His-Arg-Arg-Ala-Gly-Ile-Pro-Lys-Arg-Arg-Pro*...

   The junction between β-galactosidase and the insert amino acid sequence is between Pro and Asp, so the first amino acid encoded by the insert is Asp. (The polylinker itself codes for Asp just at the *Bam*HI site, but in constructing the fusion, this Asp and all of this downstream section of polylinker DNA is displaced to a position after the end of the insert.)

8. 5′(G)AATTCNGGNATGCAYCCNGGNAAR$_C$TT$_N$^{Y}GCN<u>AGY</u>TGG-TTYGTNGGGAATTCN-

   (*Note:* The underlined triplet AGY represents the middle Ser residue. Ser codons are either AGY or TCN [where Y = pyrimidine and N = any base]; AGY was selected here so that the mutagenesis of this codon to a Cys codon [TGY] would involve only an A ⟶ T change in the nucleotide sequence.)

   Because the middle Ser residue lies nearer to the 3′-end of this *Eco*RI fragment, the mutant primer for PCR amplification should encompass this end. That is, it should be the primer for the 3 ⟶ 5′ strand of the *Eco*RI fragment:

   5′-NNNGAATTCCCN$_c$ACR$_c$AACCAR$_c$CAN$_c$GC-3′

   where the mutated Ser ⟶ Cys triplet is underlined, NNN = several extra bases at the 5′-end of the primer to place the *Eco*RI site internal, N$_c$ = the nucleotide complementary to the nucleotide at this position in the 5′→3′ strand, and R$_c$ = the pyrimidine complementary to the purine at this position in the 5′→3′ strand.

9. The number of sequence possibilities for a polymer is given by x^y, where x is the number of different monomer types and y is the number of monomers in the oligomers. Thus, for RNA oligomers 15 nucleotides long: $x^y = 4^{15} = 1,073,741,824$. For pentapeptides, $x^y = 20^5 = 3,200,000$.

10. See Figure 12.17. You would need a *GAL4*-deficient yeast strain expressing a fusion protein composed of the *GAL4* DB domain fused with a cytoskeletal protein (such as actin or tubulin) to serve as the bait. These *GAL4*⁻ cells would be transformed with a cDNA library of human epithelial proteins constructed so that the human proteins were expressed as *GAL4* TA fusion proteins; these fused proteins are the target. Interaction of a target protein with the cytoskeletal protein fused to the *GAL4* DB domain will lead to *GAL4*-driven expression of the *lacZ* gene, whose presence can be revealed by testing for β-galactosidase activity.

11. See Figure 12.9 for isolation of mRNA and Figure 12.10 for preparation of a cDNA library from mRNA. The mRNA would be isolated, and the cDNA libraries would be prepared from two batches of yeast cells, one grown aerobically and one grown anaerobically.

12. See Figure 12.11. Differently labeled single-stranded cDNA from separate libraries prepared from aerobically versus anaerobically grown yeast cells would be hybridized with a DNA microarray (gene chip) of yeast genes.

13. An antibody against hexokinase A could be used to screen the yeast cDNA library to identify a yeast colony expressing this protein (see discussion on page 394). Once a sample of the putative yeast hexokinase A protein was isolated, its identity could be confirmed by peptide mass fingerprinting using mass spectrometry (see page 121).

14. The experiment in problem 12 identified the cDNA clone for the protein, but this clone will not have the regulatory elements (promoter) necessary for the experiment at hand. Using the cDNA clone as a probe, the genomic clone for this *nox* gene could be identified in a yeast genomic library. Cloning and sequencing a genomic clone would identify putative promoter regions that could be fused to the coding region of green fluorescent protein (GFP) to create a reporter gene construct (see page 396) that would "light up" more and more as oxygen levels declined.

15. Go to the NCBI site map at *http://www.ncbi.nlm.nih.gov/Sitemap/index.html*. In the alphabetical index, click on Genomes and Maps to go to the Entrez Genome site at *http://www.ncbi.nlm.nih.gov/Sitemap/index.html#Genomes*.

    Under Organism Collections, Entrez Genomes section, click on the links for bacteria, archea, and eukaryotes to see:

    Number of bacterial genomes completed at *http://www.ncbi.nlm.nih.gov/PMGifs/Genomes/eub_g.html*

    Number of archeal genomes completed at *http://www.ncbi.nlm.nih.gov/PMGifs/Genomes/a_g.html*

    Completed eukaryotic genomes accessible at *http://www.ncbi.nlm.nih.gov/PMGifs/Genomes/euk_g.html*

16. Insertion of DNA at any of the following restriction sites would render cells harboring the recombinant plasmid ampicillin-sensitive: *Ssp*I, *Sca*I, *Pvu*I, *Pst*I, and *Ppa*I.

17. The respective codons are Trp = UGG; Cys = UGC and UGU. Thus, changing only the third base G in the Trp codon to a pyrimidine (either C or U) would yield a Cys codon. In this case, there are now 6 differences in codon possibilities, so $2^6 = 64$ different oligonucleotides must be synthesized.

CHAPTER 13

1. $v/V_{max} = 0.8$.

2. $v = 91\ \mu\text{mol/mL} \cdot \text{sec}$.

3. $K_s = 1.43 \times 10^{-5}\ M$; $K_m = 3 \times 10^{-4}\ M$. Because k_2 is 20 times greater than k_{-1}, the system behaves like a steady-state system.

4. If the data are graphed as double-reciprocal Lineweaver–Burk plots:

   a. $V_{max} = 51\ \mu\text{mol/mL} \cdot \text{sec}$ and $K_m = 3.2\ \text{m}M$.

   b. Inhibitor (2) shows competitive inhibition with a $K_I = 2.13\ \text{m}M$. Inhibitor (3) shows noncompetitive inhibition with a $K_I = 4\ \text{m}M$.

5.

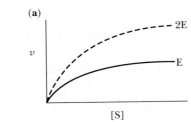

(a)

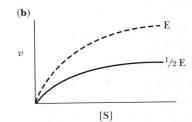

(b)

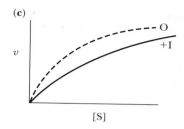

(c)

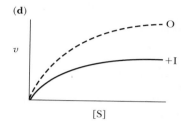

(d)

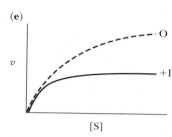

(e)

In (a), V_{max} doubles, but K_m is constant.

In (b), V_{max} halves, but K_m is constant.

In (c), V_{max} is constant, but the apparent K_m increases.

In (d), V_{max} decreases, but K_m is constant.

In (e), V_{max} decreases, K_m decreases, but the ratio K_m/V_{max} is constant.

6. a. The slope is given by $K_m{}^B/V_{max}(K_S{}^A/[A] + 1)$.
   b. y-intercept = $((K_m{}^A/[A]) + 1)1/V_{max}$.
   c. The horizontal and vertical coordinates of the point of intersection are $1/[B] = -K_m{}^A/K_S{}^A K_m{}^B$ and $1/v = 1/V_{max}(1 - (K_m{}^A/K_S{}^A))$.

7. Top left: (1) Competitive inhibition (I competes with S for binding to E). (2) I binds to and forms a complex with S.

   Top right: (1) Pure noncompetitive inhibition. (2) Random, single-displacement bisubstrate reaction, where A doesn't affect B binding, and vice versa. (Other possibilities include [3] Irreversible inhibition of E by I; [4] $1/v$ vs. $1/[S]$ plot at two different concentrations of enzyme, E.)

   Bottom left: (1) Mixed noncompetitive inhibition. (2) Ordered single-displacement bisubstrate mechanism.

   Bottom right: (1) Uncompetitive inhibition. (2) Double-displacement (ping-pong) bisubstrate mechanism.

8. Clancy must drink 694 mL of wine, or about one 750-mL bottle.

9. a. $K_S = 5\ \mu M$; b. $K_m = 30\ \mu M$; c. $k_{cat} = 5 \times 10^3\ sec^{-1}$; d. $k_{cat}/K_m = 1.67 \times 10^8\ M^{-1}sec^{-1}$; e. Yes, because k_{cat}/K_m approaches the limiting value of $10^9\ M^{-1}sec^{-1}$; f. $V_{max} = 10^{-5}\ mol/mL \cdot sec$; g. $[S] = 90\ \mu M$; h. V_{max} would equal $2 \times 10^{-5}\ mol/mL \cdot sec$, but $K_m = 30\ \mu M$, as before.

10. a. $V_{max} = 132\ \mu mol\ mL^{-1}\ sec^{-1}$; b. $k_{cat} = 44,000\ sec^{-1}$; c. $k_{cat}/K_m = 24.4 \times 10^8\ M^{-1}\ sec^{-1}$; d. Yes! k_{cat}/K_m actually exceeds the theoretical limit of $10^9\ M^{-1}\ sec^{-1}$ in this problem; e. The rate at which E encounters S; the ultimate limit is the rate of diffusion of S.

11. a. $V_{max} = 1.6\ \mu mol\ mL^{-1}\ sec^{-1}$; b. $v = 1.45\ \mu mol\ mL^{-1}\ sec^{-1}$; c. $k_{cat}/K_m = 1.6 \times 10^8\ M^{-1}\ sec^{-1}$; d. Yes.

12. a. $V_{max} = 6\ \mu mol\ mL^{-1}\ sec^{-1}$; b. $k_{cat} = 1.2 \times 10^6\ sec^{-1}$; c. $k_{cat}/K_m = 1 \times 10^8\ M^{-1}\ sec^{-1}$; d. Yes! k_{cat}/K_m approaches the theoretical limit of $10^9\ M^{-1}\ sec^{-1}$; e. The rate at which E encounters S; the ultimate limit is the rate of diffusion of S.

13. a. $V_{max} = 64\ \mu mol\ mL^{-1}\ sec^{-1}$; b. $k_{cat} = 1.28 \times 10^4\ sec^{-1}$; c. $k_{cat}/K_m = 1.4 \times 10^8\ M^{-1}\ sec^{-1}$; d. Yes! k_{cat}/K_m approaches the theoretical limit of $10^9\ M^{-1}\ sec^{-1}$.

14. a. $V_{max} = 120\ mmol\ mL^{-1}\ sec^{-1}$; b. $v = 104.7\ mmol\ mL^{-1}\ sec^{-1}$; c. $k_{cat}/K_m = 3.64 \times 10^8\ M^{-1}\ sec^{-1}$; d. Yes! k_{cat}/K_m approaches the theoretical limit of $10^9\ M^{-1}\ sec^{-1}$.

15. a. Starting from $V_{max}{}^f = k_2[E_T]$ and $V_{max}{}^r = k_{-1}[E_T]$ for the maximal rates of the forward and reverse reactions respectively, and the Michaelis constants $K_m{}^S = (k_{-1} + k_2)/k_1$ and $K_m{}^P = (k_{-1} + k_2)/k_{-2}$ for S and P, respectively,

    $$v = (V_{max}{}^f\ [S]/K_m{}^S) - V_{max}{}^r[P]/K_m{}^P)/(1 + [S]/K_m{}^S + [P]/K_m{}^P)$$

    b. At equilibrium, $v = 0$ and $K_{eq} = [P]/[S] = V_{max}{}^f\ K_m{}^P/V_{max}{}^r K_m{}^S$.

16. a. S is the preferred substrate; its K_m is smaller than the K_m for T, so a lower [S] will give $v = V_{max}/2$, compared with [T].
    b. k_{cat}/K_m defines catalytic efficiency. k_{cat}/K_m for S = $2 \times 10^7\ M^{-1}sec^{-1}$; k_{cat}/K_m for T = $4 \times 10^7\ M^{-1}sec^{-1}$, so the enzyme is a more efficient catalysis with T as substrate.

17. a. Because the enzyme shows maximal activity at or below 40°C, it seems more like a mammalian enzyme than a plant enzyme, which would be expected to have a broader temperature optimum because plants experience a broader range of temperatures.

    b. An enzyme from a thermophilic bacterium growing at 80°C would show an activity versus temperature profile similar to this one but shifted much farther to the right.

CHAPTER 14

1. a. Nucleophilic attack by an imidazole nitrogen of His[57] on the —CH_2— carbon of the chloromethyl group of TPCK covalently inactivates chymotrypsin. (See *The Student Solutions Manual, Study Guide and Problems Book* for structures.)
   b. TPCK is specific for chymotrypsin because the phenyl ring of the phenylalanine residue interacts effectively with the binding pocket of the chymotrypsin active site. This positions the chloromethyl group to react with His[57].
   c. Replacement of the phenylalanine residue of TPCK with arginine or lysine produces reagents that are specific for trypsin.

2. a. The structures proposed by Craik et al., 1987 (*Science* **237**:905–907) are shown here. (If you look up this reference, note that the letters A and B of the figure legend for Figure 3 of this article actually refer to parts B and A, respectively. Reverse either the letters in the figure or the letters in the figure legend and it will make sense.)

B 57 D102N mutant

b. Asn^{102} of the mutant enzyme can serve only as a hydrogen-bond donor to His^{57}. It is unable to act as a hydrogen-bond acceptor, as aspartate does in native trypsin. As a result, His^{57} is unable to act as a general base in transferring a proton from Ser^{195}. This presumably accounts for the diminished activity of the mutant trypsin.

3. a. The usual explanation for the inhibitory properties of pepstatin is that the central amino acid, statine, mimics the tetrahedral amide hydrate transition state of a good pepsin substrate with its unique hydroxyl group.

   b. Pepsin and other aspartic proteases prefer to cleave peptide chains between a pair of hydrophobic residues, whereas HIV-1 protease preferentially cleaves a Tyr-Pro amide bond. Because pepstatin more closely fits the profile of a pepsin substrate, we would surmise that it is a better inhibitor of pepsin than of HIV-1 protease. In fact, pepstatin is a potent inhibitor of pepsin ($K_I < 1$ nm) but only a moderately good inhibitor of HIV-1 protease, with a K_I of about 1 μM.

4. The enzyme-catalyzed rate is given by:

$$v = k_e[ES] = k_e{}'[EX^\ddagger]$$

$$K_e{}^\ddagger = \frac{[EX^\ddagger]}{[ES]}$$

$$\Delta G_e{}^\ddagger = -RT \ln K_e{}^\ddagger$$

$$K_e{}^\ddagger = e^{-\Delta G_e{}^\ddagger/RT}$$

$$[EX^\ddagger] = K_e{}^\ddagger[ES] = e^{-\Delta G_e{}^\ddagger/RT}[ES]$$

So

$$k_e[ES] = k_e{}'e^{-\Delta G_e{}^\ddagger/RT}[ES]$$

or

$$k_e = k_e{}'e^{-\Delta G_e{}^\ddagger/RT}$$

Similarly, for the uncatalyzed reaction:

$$k_u = k_u{}'e^{-\Delta G_u{}^\ddagger/RT}$$

Assuming that $k_u{}' \cong k_e{}'$

Then

$$\frac{k_e}{k_u} = \frac{e^{-\Delta G_e{}^\ddagger/RT}}{e^{-\Delta G_u{}^\ddagger/RT}}$$

$$\frac{k_e}{k_u} = e^{(\Delta G_u{}^\ddagger - \Delta G_e{}^\ddagger)/RT}$$

5. This problem is solved best by using the equation derived in problem 4.

$$\frac{k_e}{k_u} = e^{(\Delta G_u{}^\ddagger - \Delta G_e{}^\ddagger)/RT}$$

Using this equation, we can show that the difference in activation energies for the uncatalyzed and catalyzed hydrolysis reactions ($\Delta G_u - \Delta G_c$) is 92 kJ/mol.

6. Trypsin catalyzes the conversion of chymotrypsinogen to π-chymotrypsin, and chymotrypsin itself catalyzes the conversion of π-chymotrypsin to α-chymotrypsin.

7. The mechanism suggested by Lipscomb is a general base pathway in which Glu^{270} promotes the attack of water on the carbonyl carbon of the substrate:

Carboxypeptidase

8. Using the equation derived in problem 4, it is possible to calculate the ratio k_e/k_u as 1.86×10^{12}.

9. If the concentration of free enzyme is equal to the concentration of enzyme–ligand complex, the concentration of ligand would be 1×10^{-27} M. This corresponds to 1.67×10^3 liters per molecule.

10. $\Delta G = -154$ kJ/mol. This value is intermediate between noncovalent forces (H bonds are typically 10–30 kJ/mol) and covalent bonds (300–400 kJ/mol).

11. Assuming that the rate of gluconeogenesis would be equal to the slowest step in the pathway, with $k = 2 \times 10^{-20}$/sec, and assuming a cellular concentration of fructose-1,6-bisphosphatase of 0.031 (see Table 18.2), the rate of sugar synthesis would be 2×10^{-20}/sec $\times$ 0.031 mM, or 6.2×10^{-22} mM/sec. Assuming a total human cell volume of 40 L, this corresponds to 2.48×10^{-23} moles glucose/sec. Assuming 30 ATP per glucose (under cellular conditions) and 50 kJ/mole of ATP hydrolyzed, we find that the rate of energy production is $(2.48 \times 10^{-23}) \times$ (30 ATP/glucose) $\times$ (30.5 kJ/mole), or 2.269×10^{-20} kJ/sec. Converting to kilocalories and years, we find that the time to synthesize the needed 480 kilocalories with an uncatalyzed reaction would be 2.8×10^{15} years, or roughly 200,000 times the lifetime of the universe so far.

12. The correct answer is c. The stomach is a very acidic environment, whereas the small intestine is slightly alkaline. The lower part of the figure shows that enzyme X has optimal activity near pH 2, whereas enzyme Y works best at a pH near 8.

13. The correct answer is a. The two enzymes have nonoverlapping pH ranges, so it is highly unlikely that they could operate in the same place at the same time.

14. The correct answer is b. Only enzyme A has a temperature range that encompasses human body temperature (37°C).

15. The correct answer is d. The activities of the two enzymes overlap between 40° and 50°C.

16. The correct answer is c. We have no information on the pH behavior of enzymes A and B, nor on the behavior of X and Y as a function of temperature. The only answer that is appropriate to the data shown is c.

17. The only possible answer is b, because a "leveling off" implies that all the enzyme is saturated with S.

18. The correct answer is c. In order to bring the substrate into the transition state, an enzyme must enjoy environmental conditions that favor catalysis. There is no activity apparent for enzyme Y below pH 5.5.

19. Serine and aspartic proteases both exhibit general acid–base catalysis, but only the serine proteases employ covalent catalysis. The steps that lead to peptide bond cleavage are similar in the two mechanisms: Proton abstraction forms a nucleophilic species that attacks the carbonyl carbon of the bond to be cleaved, and then protonation of the amide nitrogen facilitates bond cleavage.

Observation of burst kinetics with model substrates (esters) could provide evidence for a covalent mechanism. X-ray crystallography and NMR could provide evidence for involvement of low-barrier hydrogen bonds. (See Kidd, R. E., 1999. *Protein Science* **8:**410–417.)

CHAPTER 15

1. a. As [P] rises, the rate of P formation shows an apparent decline, as enzyme-catalyzed conversion of P $\longrightarrow$ S becomes more likely.
   b. Availability of substrates and cofactors.
   c. Changes in [enzyme] due to enzyme synthesis and degradation.
   d. Covalent modification.
   e. Allosteric regulation.
   f. Specialized controls, such as zymogen activation, isozyme variability, and modulator protein influences.

2. Proteolytic enzymes have the potential to degrade the proteins of the cell in which they are synthesized. Synthesis of these enzymes as zymogens is a way of delaying expression of their activity to the appropriate time and place.

3. Monod, Wyman, Changeux allosteric system:

Lineweaver–Burk plot

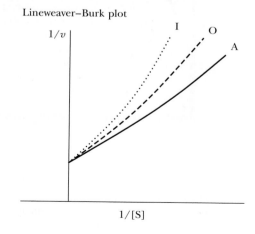

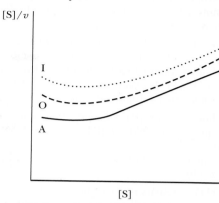

Hanes–Woolf plot

4. Using the curves for negative cooperativity as shown in Figure 15.8 as a guide:

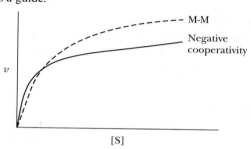

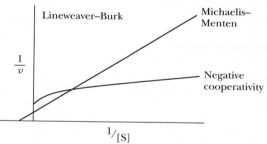

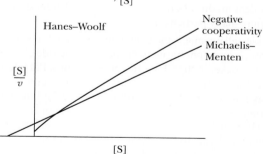

5. For $n = 2.8$, $Y_{lungs} = 0.98$ and $Y_{capillaries} = 0.77$.
   For $n = 1.0$, $Y_{lungs} = 0.79$ and $Y_{capillaries} = 0.61$.

Thus, with an n of 2.8 and a P_{50} of 26 torr, hemoglobin becomes almost fully saturated with O_2 in the lungs and drops to 77% saturation in resting tissue, a change of 21%. If n of 1.0 and a P_{50} of 26 torr, hemoglobin would become only 79% saturated with O_2 in the lungs and would drop to 61% in resting tissue, a change of 18%. The difference in hemoglobin O_2 saturation conditions between the values for $n = 2.8$ and 1.0 (21% − 18%, or 3% saturation) seems small, but note that the potential for O_2 delivery (98% saturation versus 79% saturation) is large and becomes crucial when pO_2 in actively metabolizing tissue falls below 40 torr.

6. More glycogen phosphorylase will be in the glycogen phosphorylase *a* (more active) form, but caffeine promotes the less active T conformation of glycogen phosphorylase.

7. Over time, stored erythrocytes will metabolize 2,3-BPG via the pathway of glycolysis. If [BPG] drops, hemoglobin may bind O_2 with such great affinity that it will not be released to the tissues (see Figure 15.29). The patient receiving a transfusion of [BPG]-depleted blood may actually suffocate.

8. a. By definition, when $[P_i] = K_{0.5}$, $v = 0.5\ V_{max}$.
   b. In the presence of AMP, at $[P_i] = K_{0.5}$, $v = 0.85\ V_{max}$ (from Figure 15.15c).
   c. In the presence of ATP, at $[P_i] = K_{0.5}$, $v = 0.12\ V_{max}$ (from Figure 15.15b).

9. If G_α–GTPase activity is inactivated, the interaction between G_α and adenylyl cyclase will be persistent, adenylyl cyclase will be active, [cAMP] will rise, and glycogen levels will fall because glycogen phosphorylase will be predominantly in the active, phosphorylated *a* form.

10. An excess of a negatively cooperative allosteric inhibitor could never completely shut down the enzyme. Because the enzyme leads to several essential products, inhibition by one product might starve the cell for the others.

11. a. -R(R/K)X(S*/T*)-
    b. -KRKQIAVRGL-

12. Ligand binding is the basis of allosteric regulation, and allosteric effectors are common metabolites whose concentrations reflect prevailing cellular conditions. Through reversible binding of such ligands, enzymatic activity can be adjusted to the momentary needs of the cell. On the other hand, allosteric regulation is inevitably determined by the amounts of allosteric effectors at any moment, which can be disadvantageous. Covalent modification, like allosteric regulation, is also rapid and reversible because the converter enzymes act catalytically. Furthermore, covalent modification allows cells to escape allosteric regulation by covalently locking the modified enzyme in an active (or inactive) state, regardless of effector concentrations. One disadvantage is that covalent modification systems are often elaborate cascades that require many participants.

13. Sickle-cell anemia is the consequence of Hb S polymerization through hydrophobic contacts between the side chain of Valβ6 and a pocket in the EF corner of β-subunits. Potential drugs might target this interaction directly or indirectly. For example, a drug might compete with Valβ6 side chain for binding in the EF corner. Alternatively, a useful drug might alter the conformation of Hb S such that the EF corner was no longer accessible or accommodating to the Valβ6 side chain. Another possibility might be to create drugs that deter the polymerization process in other ways through alterations in the surface properties between Hb S molecules.

14. Nitric oxide is covalently attached to Cys93β. Thus, the interaction is not reversible binding, as in allosteric regulation. On the other hand, the reaction of NO · with this cysteine residue is apparently spontaneous, and no converter enzyme is needed to add or remove it. Thus, the regulation of covalent modification

that is afforded by converter enzyme involvement is obviated. The Hb:NO · interaction illustrates that nature does not always neatly fit the definitions that we create.

15. Pro: Lactate is a metabolic indicator of the need for oxygen; it binds to Mb at a distinct site (the allosteric site?), and it lowers Mb's affinity for O_2.

    Con: Mb is a monomeric protein. Traditionally, allosteric phenomena have been considered the realm of oligomeric proteins. How then is Mb "allosteric"? Since a ligand-induced conformational change in a monomeric protein can affect binding of "substrate" (i.e., O_2), the definition of allostery may need to be broadened.

16.

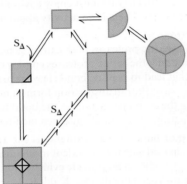

The wedge-shaped protein monomers (red) assemble into trimers, but the alternative conformation for the monomer (square, green) forms tetramers. The substrate or allosteric regulator (yellow) binds only to the square conformation, and its binding prevents the square from adopting the wedge conformation. Thus, if S or the allosteric regulator is present, equilibrium favors a greater population of square tetramers among the morpheein ensemble at the expense of round trimers.

17.

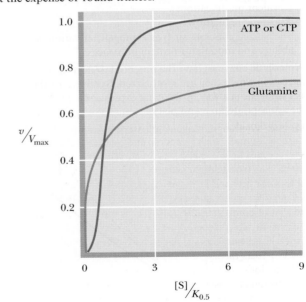

18. Negative cooperativity in NAD^+ binding to glyceraldehyde-3-P dehydrogenase:

Where affinity for NADH is

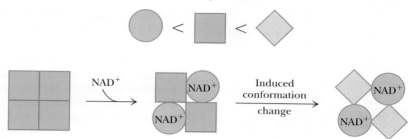

19. Mass spectrometry is a prominent tool in proteomics. Proteomics would reveal the presence of acetylated forms of metabolic enzymes.

20. Hyperventilation results in decreased blood pCO_2, which in turn leads to an increase in pH. Thus, the affinity of Hb for O_2 increases (less O_2 will be released to tissues). Hypoventilation has exactly the opposite effects on pCO_2 and pH; thus, hypoventilation leads to a diminished affinity of Hb for O_2 (more O_2 will be released to the tissues).

21. When hormone disappears, the hormone:receptor complex dissociates, the GTPase activity of the G_α subunit cleaves bound GTP to GDP + P_i, and the affinity of G_α–GDP for adenylyl cyclase is low, so it dissociates and adenylyl cyclase is no longer activated. Residual cAMP will be converted to 5'-AMP by phosphodiesterase, and the catalytic subunits of protein kinase A will be bound again by the regulatory subunits, whose cAMP ligands have dissociated. When protein kinase A becomes inactive, phosphorylase kinase will revert to the unphosphorylated, inactive form through loss of phosphoryl groups. Phosphoprotein phosphatase 1 will act on glycogen phosphorylase *a*, removing the phosphoryl group from Ser^{14} and thereby converting glycogen phosphorylase to the less active, allosterically regulated *b* form.

CHAPTER 16

1. The pronghorn antelope is truly a remarkable animal, with numerous specially evolved anatomical and molecular features. These include a large windpipe (to draw in more oxygen and exhale more carbon dioxide), lungs that are three times the size of those of comparable animals (such as goats), and lung alveoli with five times the surface area so that oxygen can diffuse more rapidly into the capillaries. The blood contains larger numbers of red blood cells and thus more hemoglobin. The skeletal and heart muscles are likewise adapted for speed and endurance. The heart is three times the size of that of comparable animals and pumps a proportionally larger volume of blood per contraction. Significantly, the muscles contain much larger numbers of energy-producing mitochondria, and the muscle fibers themselves are shorter and thus designed for faster contractions. All these characteristics enable the pronghorn antelope to run at a speed nearly twice the top speed of a thoroughbred racehorse and to sustain such speed for up to 1 hour.

2. Refer to Figure 16.9. The step in which the myosin head conformation change occurs, is the step that should be blocked by β,γ-methylene-ATP, because hydrolysis of ATP should occur in this step and β,γ-methylene-ATP is nonhydrolyzable.

3. Phosphocreatine is synthesized from creatine (via creatine kinase) primarily in muscle mitochondria (where ATP is readily generated) and then transported to the sarcoplasm, where it can act as an ATP buffer. The creatine kinase reaction in the sarcoplasm yields the product creatine, which is transported back into the mitochondria to complete the cycle. Like a number of other mitochondrial proteins, the expression of mitochondrial creatine kinase is directed by mitochondrial DNA, whereas sarcoplasmic creatine kinase is derived from information encoded in nuclear DNA.

4. Note in step 3 of Figure 16.9 that it is ATP that stimulates dissociation of myosin heads from the actin filaments—the dissociation of the cross-bridge complex. When ATP levels decline (as happens rapidly after death), large numbers of myosin heads are unable to dissociate from actin and the muscle becomes stiff and unable to relax.

5. The skeletal muscles of the average adult male have a cross-sectional area of approximately 35,000 cm^2. The gluteus maximus muscles represent approximately 300 cm^2 of this total. Assuming 4 kg of maximal tension per square centimeter of cross-sectional area, one calculates a total tension for the gluteus maximus of 1200 kg (as stated in the problem). The same calculation shows that the total tension that could be developed by all the muscles in the body is 140,000 kg (or 154 tons)!

6. Taking 55,000 g/mol divided by 6.02×10^{23}/mol and by 1.3 g/mL, one obtains a volume of 7.03×10^{-20} mL. Assume a sphere and use the volume of a sphere ($V = (4/3)\pi r^3$) to obtain a radius of 25.6 Å. The diameter of the tubulin dimer (see Figure 16.12) is 8 nm, or 80 Å, and two times the radius we calculated here is approximately 51 Å, a reasonable value by comparison.

7. A liver cell is 20,000 nm long, which would correspond to 5000 tubulin monomers if using the value of Figure 16.12, and about 7800 tubulin monomers using the value of 25.6 Å calculated in problem 6.

8. 4 inches (length of the giant axon) = 10.16 cm. Movement at 2 to 5 μ/sec would correspond to a time of 5.6 to 14 hours to traverse this distance.

9. 14 nm is 140 Å, which would be approximately 93 residues of a coiled coil.

10. Using Equation 9.1, one can calculate a ΔG of 17,100 J/mol for this calcium gradient.

11. Using Equation 3.13, one can calculate a cellular ΔG of −48,600 J/mol for ATP hydrolysis.

12. 17,100 J are required to transport 1 mole of calcium ions. Two moles of calcium would require 34,200 J, and three would cost 51,300 J. Thus, the gradient of problem 10 would provide enough energy to drive the transport of two calcium ions per ATP hydrolyzed.

13. Energy (or work) = force × distance. If 1 ATP is hydrolyzed per step and the cellular value of $\Delta G^{\circ\prime}$ for ATP hydrolysis is

−50 kJ/mol, then the calculation of force exerted by a motor is force = 50 kJ/mol/(step size). The SI unit of force is the newton, and 1 newton · meter = 1 J.

For the kinesin-1 motor, 50 kJ/mol/(8×10^{-9} m) = 6.25×10^{12} newtons.

For the myosin-V motor, 50 kJ/mol/(36×10^{-9} m) = 1.39×10^{12} newtons.

For the dynein motor, 50 kJ/mol/(28×10^{-9} m) = 1.78×10^{12} newtons.

14. Perhaps the simplest way to view this problem is to consider the potential energy change for lifting a 10-kg weight 0.4 m. E = mgh = (10 kg) · (9.8 m/sec^2) · (0.4 m) = 39.2 J. Now, if 1 ATP is expended per myosin step along an actin filament, 39.2 J/(50 kJ/mol) = 7.8×10^{-4} mol.

Then, (7.8×10^{-4} mol) · (6.02×10^{23}) = 4.72×10^{20} molecules of ATP expended, one per step. So there must be 4.72×10^{20} myosin steps along actin. However, the stepping process is almost certainly not 100% efficient. As shown in the box on page 526, a step size of 11 nm against a force of 4 pN would correspond to an energy expended per myosin per step of 4.4×10^{-20} J. With this assumed energy expended per step, we can calculate 39 J/(4.4×10^{-20} J) = 8.86×10^{20} steps total.

If we take the step size value for skeletal myosin from the box on page 526 as 11 nm, one myosin head would have to take 0.4 m/11×10^{-9} m or 3.64×10^7 steps to raise the weight a distance of 0.4 m. Taking (8.86×10^{20} steps total)/(3.64×10^7 steps per myosin) = 2.43×10^{13} myosin heads involved per 11-nm step.

15. All these are smooth muscle except for the diaphragm. (See *http://chanteur.net/contrib/index.htm#http://chanteur.net/contrib/cJMdiaph.htm* for an explanation of the common misconception that the diaphragm is smooth muscle.)

16. 1, a; 2, e; 3, d; 4, c; 5, b.

17. This exercise is left to the student and will presumably be different for every student.

18. The 80 Å step of the kinesin motor is a 0.008 μm step. To cover 10.16 cm would require 12.7 million steps.

19. The correct answer is d, although each answer is reasonable. ATP is needed for several processes involved in muscle relaxation. Salt imbalances can also prevent normal muscle function, and interrupted blood flow could prevent efficient delivery of oxygen needed for ATP production during cellular respiration.

20. The correct answer is a. Understanding the inheritance pattern of sex-linked traits is essential. Males always inherit sex-linked traits (on the X chromosome) from their mothers. In addition, sex-linked traits are much more commonly expressed in males because they have only one X chromosome.

CHAPTER 17

1. 6.5×10^{12} (6.5 trillion) people.

2. Consult Table 17.2.

3. O_2, H_2O, and CO_2.

4. See Section 17.2.

5. Consult Figure 17.5.

6. Consult *Corresponding Pathways of Catabolism and Anabolism Differ in Important Ways*, p. 561. See also Figure 17.8.

7. See *The ATP Cycle*, p. 562; *NAD$^+$ Collects Electrons Released in Catabolism*, p. 562; and *NADPH Provides the Reducing Power for Anabolic Processes*, p. 563.

8. In terms of quickness of response, the order is allosteric regulation > covalent modification > enzyme synthesis and degradation. See *The Student Solutions Manual, Study Guide and Problems Book* for further discussions.

9. See *Metabolic Pathways Are Compartmentalized within Cells*, p. 569, and discussions in *The Student Solutions Manual, Study Guide and Problems Book*.

10. Many answers are possible here. Some examples: Large numbers of metabolites in tissues and fluids; great diversity of structure of biomolecules; need for analytical methods that can detect and distinguish many different metabolites, often at very low concentrations; need to understand relationships between certain metabolites; time and cost required to analyze and quantitate many metabolites in a tissue or fluid.

11. Mass spectrometry provides great sensitivity for detection of metabolites. NMR offers a variety of methods for resolving and discriminating metabolites in complex mixtures. Other comparisons and contrasts are discussed in the references at the end of the chapter.

12. See Figure 13.23; The Coenzymes of the Pyruvate Dehydrogenase Complex, pages 614–616, and the Activation of Vitamin B$_{12}$, page 778.

13. The genome is the entire hereditary information in an organism, as encoded in its DNA or (for some viruses) RNA. The transcriptome is the set of all messenger RNA molecules (transcripts) produced in one cell or a population of cells under a defined set of conditions. The proteome is the entire complement of proteins produced by a genome, cell, tissue, or organism under a defined set of conditions. The metabolome is the complete set of low molecular weight metabolites present in an organism or excreted by it under a given set of physiological conditions.

14. A mechanism for liver alcohol dehydrogenase:

15. TCA cycle: in mitochondria; converts acetate units to CO_2 plus NADH and FADH$_2$.

Glycolysis: in cytosol; converts glucose to pyruvate.

Oxidative phosphorylation: in mitochondria; uses electrons (from NADH and FADH$_2$) to produce ATP.

Fatty acid synthesis: in cytosol; uses acetate units to synthesize fatty acids.

16. Most discussions of ocean sequestration of CO_2 are devoted to capture of CO_2 by phytoplankton or injection of CO_2 deep in the ocean. However, the box on page 553 offers another possibility—carbon sequestration in the shells of corals, mollusks, and crustaceans. This approach may be feasible. However, there are indications that these classes of sea organisms are already suffering from global warming and from pollution of ocean water.

17. ^{32}P and ^{35}S both decay via beta particle emission. A beta particle is merely an electron emitted by a neutron in the nucleus. Thus, beta decay does not affect the atomic mass, but it does convert a

neutron to a proton. Thus, beta emission changes ^{32}P to ^{32}S and converts ^{35}S to ^{35}Cl:

$$^{32}P \longrightarrow \beta^- + 32S, \ t_{1/2} = 14.3 \text{ days.}$$

$$^{35}S \longrightarrow \beta^- + 35Cl, \ t_{1/2} = 87.1 \text{ days.}$$

The decay equation is a first-order decay equation:

$$A/A_0 = e^{-0.693/t_{1/2}}$$

Thus, after 100 days of decay, the fraction of ^{32}P remaining would be 0.786% and the fraction of ^{35}S remaining would be 45%.

18. The correct answer is b. Obligate anaerobes can survive only in the absence of oxygen.

19. The correct answer is c, mass spectrometry.

CHAPTER 18

1. a. Phosphoglucoisomerase, fructose bisphosphate aldolase, triose phosphate isomerase, glyceraldehyde-3-P dehydrogenase, phosphoglycerate mutase, and enolase.
   b. Hexokinase/glucokinase, phosphofructokinase, phosphoglycerate kinase, pyruvate kinase, and lactate dehydrogenase.
   c. Hexokinase and phosphofructokinase.
   d. Phosphoglycerate kinase and pyruvate kinase.
   e. According to Equation 3.13, reactions in which the number of reactant molecules differs from the number of product molecules exhibit a strong dependence on concentration. That is, such reactions are extremely sensitive to changes in concentration. Using this criterion, we can predict from Table 19.1 that the free energy changes of the fructose bisphosphate aldolase and glyceraldehyde-3-P dehydrogenase reactions will be strongly influenced by changes in concentration.
   f. Reactions that occur with ΔG near zero operate at or near equilibrium. See Table 18.1 and Figure 18.22.

2. The carboxyl carbon of pyruvate derives from carbons 3 and 4 of glucose. The keto carbon of pyruvate derives from carbons 2 and 5 of glucose. The methyl carbon of pyruvate is obtained from carbons 1 and 6 of glucose.

3. Increased [ATP], [citrate], or [glucose-6-phosphate] inhibits glycolysis. Increased [AMP], [fructose-1,6-bisphosphate], or [fructose-2,6-bisphosphate] stimulates glycolysis.

4. See *The Student Solutions Manual, Study Guide and Problems Book* for discussion.

5. The mechanisms for fructose bisphosphate aldolase and glyceraldehyde-3-P dehydrogenase are shown in Figures 18.12 and 18.13, respectively.

6. The relevant reactions of galactose metabolism are shown in Figure 18.24. See *The Student Solutions Manual, Study Guide and Problems Book* for mechanisms.

7. Iodoacetic acid would be expected to alkylate the reactive active-site cysteine that is vital to the glyceraldehyde-3-P dehydrogenase reaction. This alkylation would irreversibly inactivate the enzyme.

8. Ignoring the possibility that ^{32}P might be incorporated into ATP, the only reaction of glycolysis that utilizes P_i is glyceraldehyde-3-P dehydrogenase, which converts glyceraldehyde-3-P to 1,3-bisphosphoglycerate. $^{32}P_i$ would label the phosphate at carbon 1 of 1,3-bisphosphoglycerate. The label will be lost in the next reaction, and no other glycolytic intermediates will be directly labeled. (Once the label is incorporated into ATP, it will also show up in glucose-6-P, fructose-6-P, and fructose-1, 6-bisphosphate.)

9. The sucrose phosphorylase reaction leads to glucose-6-P without the need for the hexokinase reaction. Direct production offers the obvious advantage of saving a molecule of ATP.

10. All of the kinases involved in glycolysis, as well as enolase, are activated by Mg^{2+} ion. A Mg^{2+} deficiency could lead to reduced activity for some or all of these enzymes. However, other systemic effects of a Mg^{2+} deficiency might cause even more serious problems.

11. a. 7.5.
    b. 0.0266 mM.

12. +1.08 kJ/mol.

13. a. −13.8 kJ/mol.
    b. $K_{eq} = 194,850$.
    c. [Pyr]/[PEP] = 24,356.

14. a. −13.8 kJ/mol.
    b. $K_{eq} = 211$.
    c. [FBP]/[F-6-P] = 528.

15. a. −19.1 kJ/mol.
    b. $K_{eq} = 1651$.
    c. [ATP]/[ADP] = 13.8.

16. $\Delta G°' = -33.9$ kJ/mol.

17. An 8% increase in [ATP] changes the [AMP] concentration to 20 μM (from 5 μM).

18.

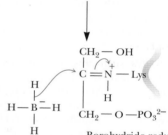

Formation of Schiff base intermediate

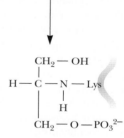

Borohydride reduction of Schiff base intermediate

Stable (trapped) E–S derivative

Degradation of enzyme (acid hydrolysis)

N^6-dihydroxypropyl-L-lysine

19.

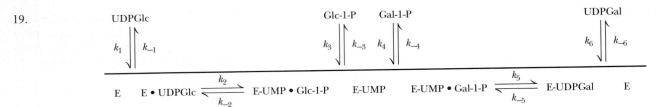

20. Pyruvate kinase deficiency primarily affects red blood cells, which lack mitochondria and can produce ATP only via glycolysis. Absence of pyruvate kinase reduces the production of ATP by glycolysis in red cells, which in turn reduces the activity of the Na,K-ATPase (see Chapter 9). This results in alterations of electrolyte (Na^+ and K^+) concentrations, which leads to cellular distortion, rigidity, and dehydration. Compromised red blood cells are destroyed by the spleen and liver.

21. The best answer is d. PFK is more active at low ATP than at high ATP, and F-2,6-bisP activates the enzyme.

22. Hexokinase is inhibited by high concentrations of glucose-6-P, so glycolysis would probably stop at such a high level of glucose-6-P. Also, the increased concentration of glucose-6-P would change the cellular ΔG of the hexokinase reaction to approximately $-22,000$ J/mol.

CHAPTER 19

1. Glutamate enters the TCA cycle via a transamination to form α-ketoglutarate. The γ-carbon of glutamate entering the cycle is equivalent to the methyl carbon of an entering acetate. Thus, no radioactivity from such a label would be lost in the first or second cycle, but in each subsequent cycle, 50% of the total label would be lost (see Figure 19.15).

2. NAD^+ activates pyruvate dehydrogenase and isocitrate dehydrogenase and thus would increase TCA cycle activity. If NAD^+ increases at the expense of NADH, the resulting decrease in NADH likewise would activate the cycle by stimulating citrate synthase and α-ketoglutarate dehydrogenase. ATP inhibits pyruvate dehydrogenase, citrate synthase, and isocitrate dehydrogenase; reducing the ATP concentration thus would activate the cycle. Isocitrate is not a regulator of the cycle, but increasing its concentration would mimic an increase in acetate flux through the cycle and increase overall cycle activity.

3. For most enzymes that are regulated by phosphorylation, the covalent binding of a phosphate group at a distant site induces a conformation change at the active site that either activates or inhibits the enzyme activity. On the other hand, X-ray crystallographic studies reveal that the phosphorylated and unphosphorylated forms of isocitrate dehydrogenase share identical structures with only small (and probably insignificant) conformation changes at Ser^{113}, the locus of phosphorylation. What phosphorylation *does* do is block isocitrate binding (with no effect on the binding affinity of $NADP^+$). As shown in Figure 2 of the paper cited in the problem (Barford, D., 1991. Molecular mechanisms for the control of enzymic activity by protein phosphorylation. *Biochimica et Biophysica Acta* **1133**:55–62), the γ-carboxyl group of bound isocitrate forms a hydrogen bond with the hydroxyl group of Ser^{113}. Phosphorylation apparently prevents isocitrate binding by a combination of a loss of the crucial H bond

between substrate and enzyme and by repulsive electrostatic and steric effects.

4. A mechanism for the first step of the α-ketoglutarate dehydrogenase reaction:

TPP α-KG

Covalent TPP intermediate

5. Aconitase is inhibited by fluorocitrate, the product of citrate synthase action on fluoroacetate. In a tissue where inhibition has occurred, all TCA cycle metabolites should be reduced in concentration. Fluorocitrate would replace citrate, and the concentrations of isocitrate and all subsequent metabolites would be reduced because of aconitase inhibition.

6. $FADH_2$ is colorless, but FAD is yellow, with a maximal absorbance at 450 nm. Succinate dehydrogenase could be conveniently assayed by measuring the decrease in absorbance in a solution of the flavoenzyme and succinate.

7. The central (C-3) carbon of citrate is reduced, and an adjacent carbon is oxidized in the aconitase reaction. The carbon bearing the hydroxyl group is obviously oxidized by isocitrate dehydrogenase. In the α-ketoglutarate dehydrogenase reaction, the departing carbon atom and the carbon adjacent to it are both oxidized. Both of the —CH_2— carbons of succinate are oxidized in the succinate dehydrogenase reaction. Of these four molecules, all but citrate undergo a net oxidation in the TCA cycle.

8. Several TCA metabolite analogs are known, including malonate, an analog of succinate, and 3-nitro-2-S-hydroxypropionate, the anion of which is a transition-state analog for the fumarase reaction.

Malonate (succinate analog)   3-Nitro-2-S-hydroxypropionate   Transition-state analog for fumarase

9. A mechanism for pyruvate decarboxylase:

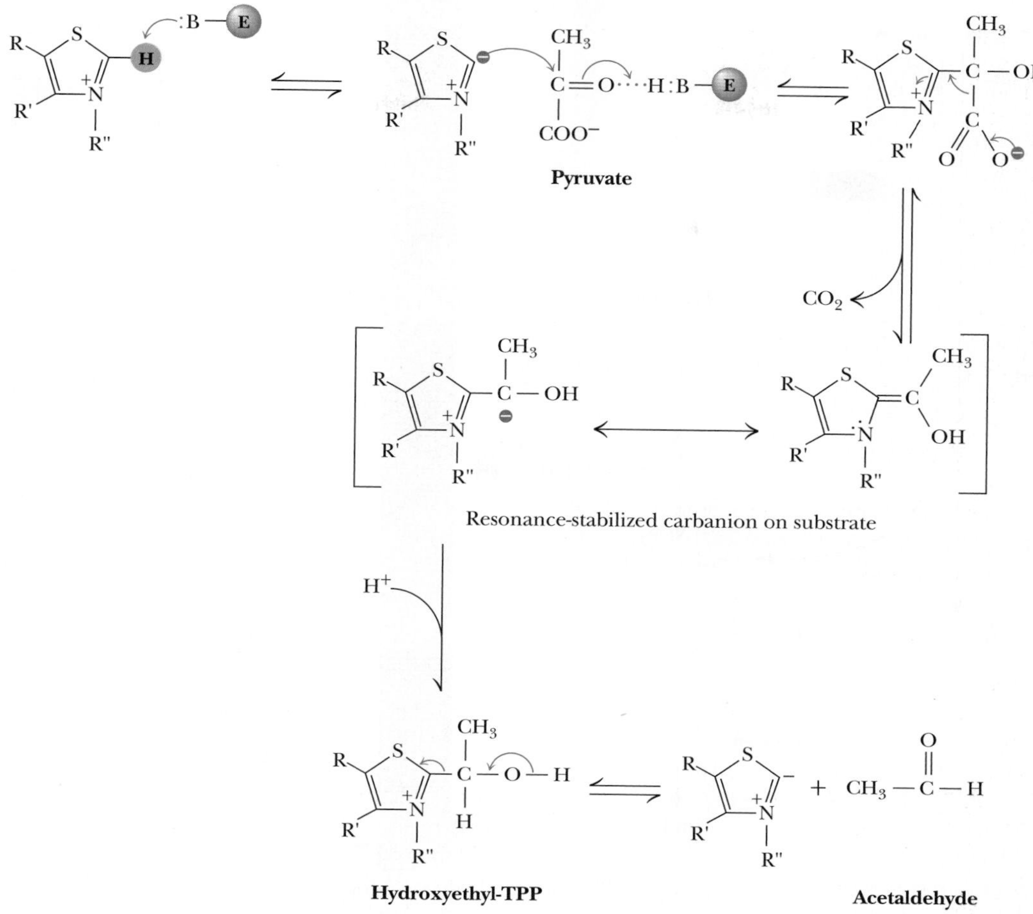

Pyruvate

CO_2

Resonance-stabilized carbanion on substrate

H^+

Hydroxyethyl-TPP

Acetaldehyde

$CH_3 - C - H$ (with O double bond)

10. [isocitrate]/[citrate] = 0.1. When [isocitrate] = 0.03 mM, [citrate] = 0.3 mM.

11. $^{14}CO_2$ incorporation into TCA via the pyruvate carboxylase reaction would label the —CH_2—COOH carboxyl carbon in oxaloacetate. When this entered the TCA cycle, the labeled carbon would survive only to the α-ketoglutarate dehydrogenase reaction, where it would be eliminated as $^{14}CO_2$.

12. ^{14}C incorporated in a reversed TCA cycle would label the two carboxyl carbons of oxaloacetate in the first pass through the cycle. One of these would be eliminated in its second pass through the cycle. The other would persist for more than two cycles and would be eliminated slowly as methyl carbons in acetyl-CoA in the reversed citrate synthase reaction.

13. The labeling pattern would be the same as if methyl-labeled acetyl-CoA was fed to the conventional TCA cycle (see Figure 19.15).

14.

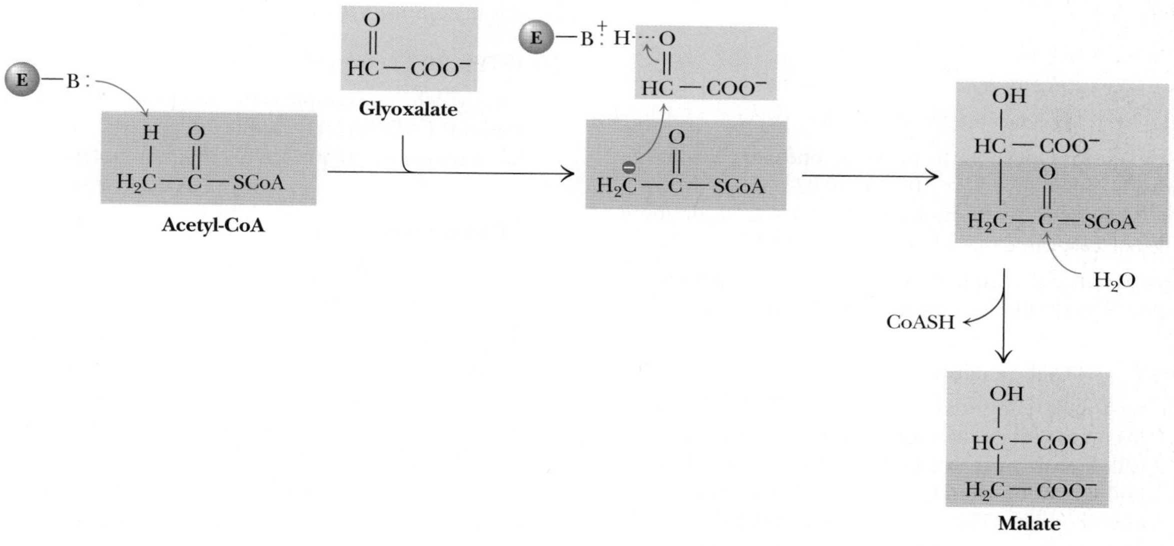

Glyoxalate

Acetyl-CoA

Malate

CoASH

H_2O

15.

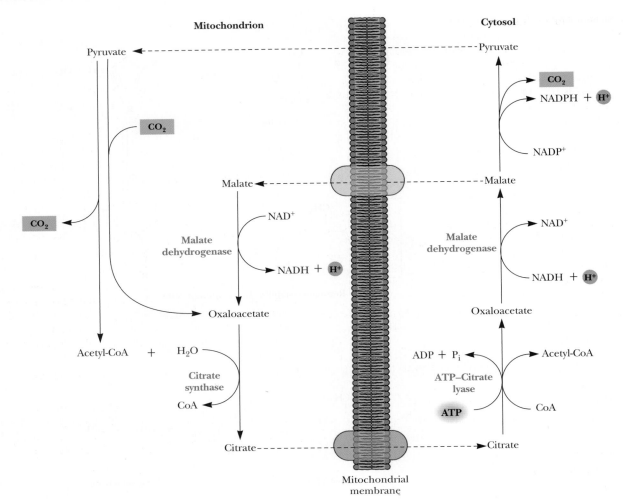

16. For the malate dehydrogenase reaction,

$$\text{Malate} + \text{NAD}^+ \rightleftharpoons \text{oxaloacetate} + \text{NADH} + \text{H}^+$$

with the value of $\Delta G^{\circ\prime}$ being $+30$ kJ/mol. Then

$$\Delta G^{\circ\prime} = -RT \ln K_{eq}$$

$$= -(8.314 \text{ J/mol} \cdot \text{K})(298)\ln\left(\frac{[1]x}{[20][2.2 \times 10^{-4}]}\right)$$

$$\frac{-30{,}000 \text{ J/mol}}{2478 \text{ J/mol}} = \ln(x/4.4 \times 10^{-3})$$

$$-12.1 = \ln(x/4.4 \times 10^{-3})$$

$$x = -RT \ln K_{eq}$$

$$x = [\text{oxaloacetate}] = 0.024 \ \mu\text{M}$$

Using the dimensions given in the problem, one can calculate a mitochondrial volume of 1.57×10^{-15} L. $(0.024 \times 10^{-6} \text{ M}) \cdot (1.57 \times 10^{-15} \text{ L}) \cdot (6.02 \times 10^{23} \text{ molecules/mole}) = 22.7$, or about 23 molecules of OAA in a mitochondrion ("pOAA" = 7.62).

17. This exercise is left to the student. Review of the calculation of oxidation numbers should be a prerequisite for answering this problem.

18. This exercise is left to the student.

19. 2R, 3R-fluorocitrate is converted by aconitase to 2-fluoro-*cis*-aconitate. This intermediate then rotates 180 degrees in the active site. Addition of hydroxide at the C4 position is followed by double bond migration from C3-C4 to C2-C3, to form 4-hydroxy-*trans*-aconitate. This product remains tightly bound at the aconitase active site, inactivating the enzyme. Studies by

Lauble, et al. (The reaction of fluorocitrate with aconitase and the crystal structure of the enzyme-inhibitor complex. *Proceedings of the National Academy of Sciences U.S.A.* **93**:13699–13703 (1996)) have shown that the inhibitory product can be displaced by a 10^6-fold excess of isocitrate.

20. Only eight ATPs would be generated in the succinyl-CoA synthetase (and nucleoside diphosphate kinase) reaction of the TCA cycle itself.

21. d is false. Succinyl-CoA is an inhibitor of citrate synthase.

CHAPTER 20

1. The cytochrome couple is the acceptor, and the donor is the (bound) FAD/FADH$_2$ couple, because the cytochrome couple has a higher (more positive) reduction potential.

$$\Delta \mathscr{E}_o{}' = 0.254 \text{ V} - 0.02 \text{ V*}$$

*This is a typical value for enzyme-bound [FAD].

$$\Delta G^{\circ\prime} = -n\mathscr{F}\Delta\mathscr{E}_o{}' = -43.4 \text{ kJ/mol}.$$

2. $\Delta \mathscr{E}_o{}' = -0.03 \text{ V}$; $\Delta G^{\circ\prime} = 5790 \text{ J/mol}$.

3. This situation is analogous to that described in Section 3.3. The net result of the reduction of NAD$^+$ is that a proton is consumed. The effect on the calculation of free energy change is similar to that described in Equation 3.26. Adding an appropriate term to Equation 20.12 yields:

$$\mathscr{E} = \mathscr{E}_o{}' - RT \ln[\text{H}^+] + (RT/n\mathscr{F} \ln([\text{ox}]/[\text{red}])$$

4. Cyanide acts primarily via binding to cytochrome a_3, and the amount of cytochrome a_3 in the body is much lower than the

amount of hemoglobin. Nitrite anion is an effective antidote for cyanide poisoning because of its unique ability to oxidize ferrohemoglobin to ferrihemoglobin, a form of hemoglobin that competes very effectively with cytochrome a_3 for cyanide. The amount of ferrohemoglobin needed to neutralize an otherwise lethal dose of cyanide is small compared with the total amount of hemoglobin in the body. Even though a small amount of hemoglobin is sacrificed by sequestering the cyanide in this manner, a "lethal dose" of cyanide can be neutralized in this manner without adversely affecting oxygen transport.

5. You should advise the wealthy investor that she should decline this request for financial backing. Uncouplers can indeed produce dramatic weight loss, but they can also cause death. Dinitrophenol was actually marketed for weight loss at one time, but sales were discontinued when the potentially fatal nature of such "therapy" was fully appreciated. Weight loss is best achieved by simply making sure that the number of calories of food eaten is less than the number of calories metabolized. A person who metabolizes about 2000 kJ (about 500 kcal) more than he or she consumes every day will lose about a pound in about 8 days.)

6. The calculation in Section 20.4 assumes the same conditions as in this problem (1 pH unit gradient across the membrane and an electrical potential difference of 0.18 V). The transport of 1 mole of H^+ across such a membrane generates 23.3 kJ of energy. For three moles of H^+, the energy yield is 69.9 kJ. Then, from Equation 3.13, we have:

$$69{,}900 \text{ J/mol} = 30{,}500 \text{ J/mol} + RT \ln([ATP]/[ADP][P_i])$$
$$39{,}400 \text{ J/mol} = RT \ln([ATP]/[ADP][P_i])$$
$$[ATP]/[ADP][P_i] = 4.36 \times 10^6 \ M^{-1} \text{ at } 37°C.$$

In the absence of the proton gradient, this same ratio is $7.25 \times 10^{-6} \ M^{-1}$ at equilibrium!

7. The succinate/fumarate redox couple has the highest (that is, most positive) reduction potential of any of the couples in glycolysis and the TCA cycle. Thus, oxidation of succinate by NAD^+ would be very unfavorable in the thermodynamic sense.

Oxidation of succinate by NAD^+:

$$\Delta \mathcal{E}_o' \approx -0.35 \text{ V}; \Delta G°' \approx 67{,}500 \text{ J/mol}.$$

On the other hand, oxidation of succinate is quite feasible using enzyme-bound FAD.

Oxidation of succinate by [FAD]:

$\Delta \mathcal{E}_o' \approx 0$, $\Delta G°' \approx 0$, depending on the exact reduction potential for bound FAD.

8. a. -73.3 kJ/mol.
   b. $K_{eq} = 7.1 \times 10^{12}$.
   c. [ATP]/[ADP] = 19.7.

9. a. -151.5 kJ/mol.
   b. $K_{eq} = 3.54 \times 10^{26}$.
   c. [ATP]/[ADP] = 8.8.

10. a. -219 kJ/mol.
    b. $K_{eq} = 2.63 \times 10^{38}$.
    c. [ATP]/[ADP] = 3340

11. a. -217 kJ/mol.
    b. $K_{eq} = 1.08 \times 10^{38}$.
    c. [ATP]/[ADP] = 3340.

12. a. -26 kJ/mol.
    b. $K_{eq} = 3.68 \times 10^4$.
    c. $[NAD^+]/[NADH] = 1.48$.

13. a. $K_{eq} = 4.8 \times 10^3$.
    b. $+48$ kJ/mol.
    c. [ATP]/[ADP] = 2.3.

14. This answer should be based on the information on pages 652–655 and should be in the student's own words.

15. This answer should be based on the information on pages 655–658 and should be in the student's own words.

16. Several mechanisms are possible. See, for example, the mechanisms in the following references: Pryor, W. A., and Porter, N. A., 1990. Suggested mechanisms for the production of 4-hydroxy-2-nonenal from the autoxidation of polyunsaturated fatty acids. *Free Radical Biology and Medicine* **8**:541–543; and Niki, E., Yoshida, Y., Saito, Y., and Noguchi, N., 2005. Lipid peroxidation: Mechanisms, inhibition, and biological effects. *Biochemical and Biophysical Research Communications* **338**:668–676.

17. A calculation similar to that in problem 16 in Chapter 19 yields about 15 H^+ in a typical mitochondrion.

18. In the succinate dehydrogenase reaction, a two-electron transfer reaction (the conversion of succinate to fumarate) must be coupled to iron–sulfur centers, which can transfer only one electron at a time. FAD can carry on both one-electron and two-electron transfers and thus is ideally suited for this reaction.

19. The answer is a. The mitochondrial ATPase reaction proceeds via nucleophilic S_N2-type substitution.

20. At 298 K, doubling the proton gradient would change ΔG by 1717 J/mole. The mitochondrial membrane potential would have to change by approximately 18 mV. The student should decide whether this change should be positive or negative.

CHAPTER 21

1. Efficiency = $\Delta G°'/(\text{light energy input}) = n\mathcal{F}\Delta\mathcal{E}_o'/(Nhc/\lambda)$ = 0.564.

2. $\Delta G°' = -n\mathcal{F}\Delta\mathcal{E}_o'$, so $\Delta G°'/-n\mathcal{F} = \Delta\mathcal{E}_o'$. $n = 4$ for 2 $H_2O \longrightarrow 4 e^- + 4 H^+ + O_2$. $\Delta\mathcal{E}_o' = (\mathcal{E}_o'(\text{primary oxidant}) - \mathcal{E}_o'(\frac{1}{2} O_2/H_2O)) = +0.0648$ V. Thus, $\mathcal{E}_o'(\text{primary oxidant}) = 0.881$ V.

3. ΔG for ATP synthesis = $+50{,}336$ J/mol. $\Delta\mathcal{E}_o' = -0.34$ V.

4. Reduced plastoquinone = PQH_2; oxidized plastoquinone = PQ; Reduced plastocyanin = $PC(Cu^+)$; plastocyanin = $PC(Cu^{2+})$:

$$PQH_2 + 2 H^+_{stroma} + 2 PC(Cu^{2+}) \longrightarrow$$
$$PQ + 2 PC(Cu^+) + 4 H^+_{thylakoid\ lumen}$$

5. Noncyclic photophosphorylation: 3 ATP would be synthesized from 8 quanta, which equates to 2.67 $h\nu$/ATP. Cyclic photophosphorylation theoretically yields 2 $H^+/h\nu$, so if 14 H^+ yield 3 ATP, 7 $h\nu$ yield 3 ATP, which equates to 2.33 $h\nu$/ATP.

6. In mitochondria, H^+ translocation leads to a decline in intermembrane space pH and hence cytosolic pH, because the outer mitochondrial membrane is permeable to protons. Thus, the eukaryotic cytosol pH is at risk from mitochondrial proton translocation. So, mitochondria rely on a greater membrane potential ($\Delta\psi$) and a smaller ΔpH to achieve the same proton-motive force. Because proton translocation in eukaryotic photosynthesis deposits H^+ into the thylakoid lumen, the cytosol does not experience any pH change and ΔpH is not a problem. Moreover, the light-induced efflux of Mg^{2+} from the lumen diminishes $\Delta\psi$ across the thylakoid membrane, so a greater contribution of ΔpH to the proton-motive force is warranted.

7. Replacement of Tyr by Phe would greatly diminish the possibility of e^- transfer between water and P680$^+$, limiting the ability of P680$^+$ to regain an e^- and return to the P680 ground state.

8. Assuming 12 c-subunits means 12 H$^+$ are needed to drive one turn of the c-subunit rotor and the synthesis of 3 ATP by the CF$_1$ part of the ATP synthase. If the *R. viridis* cytochrome bc_1 complex drives the translocation of 2 H$^+$/e^-, then 2 $h\upsilon$ gives 4 H$^+$, and 6 $h\upsilon$ gives 12 H$^+$ and thus 3 ATP (thus, 2 $h\upsilon$ yield 1 ATP).

9. $\Delta G° = -n\mathcal{F}\Delta\mathcal{E}_o' = -2(96,485 \text{ J/volt} \cdot \text{mol})$
($\mathcal{E}_o'(1,3\text{-BPG/Gal3P}) - \mathcal{E}_o'(\text{NADP}^+/\text{NADPH})) =$
$-(192,970 \text{ J/volt} \cdot \text{mol})(-0.29 \text{ V} - (-0.32 \text{ V})) = -192,970$
$(0.03) \text{ J/mol} = -5,789 \text{ J/mol}.$

10. Use the first eight reactions in Table 21.1 to show that:
$$\text{CO}_2 + \text{Ru-1,5-BP} + 3 \text{ H}_2\text{O} + 2 \text{ ATP} + 2 \text{ NADPH} + 2 \text{ H}^+ \longrightarrow$$
$$\text{glucose} + 2 \text{ ADP} + 4 \text{ P}_i + 2 \text{ NADP}^+$$

11. Radioactivity will be found in C-1 of 3-phosphoglycerate; C-3 and C-4 of glucose; C-1 and C-2 of erythrose-4-P; C-3, C-4, and C-5 of sedoheptulose-1,7-bisP; and C-1, C-2, and C-3 of ribose-5-P.

12. Light induces three effects in chloroplasts: (1) pH increase in the stroma, (2) generation of reducing power (as ferredoxin), and (3) Mg^{2+} efflux from the thylakoid lumen. Key enzymes in the Calvin–Benson CO$_2$ fixation pathway are activated by one or more of these effects. In addition, rubisco activase is activated indirectly by light, and, in turn, activates rubisco.

13. The following series of reactions accomplishes the conversion of 2-phosphoglycolate to 3-phosphoglycerate:

1. 2 phosphoglycolate + 2 H$_2$O $\longrightarrow$ 2 glycolate + 2 P$_i$

2. 2 glycolate + 2 O$_2$ $\longrightarrow$ 2 glyoxylate + 2 H$_2$O$_2$

3. 2 glyoxylate + 2 serine $\longrightarrow$ 2 hydroxypyruvate + 2 glycine

4. 2 glycine $\longrightarrow$ serine + CO$_2$ + NH$_3$

5. hydroxypyruvate + glutamate $\longrightarrow$ serine + α-ketoglutarate (an aminotransferase reaction)

6. α-ketoglutarate + NH$_3$ + NADH$^+$ + H$^+$ $\longrightarrow$ glutamate + H$_2$O + NAD$^+$ (the glutamate dehydrogenase reaction)

7. hydroxypyruvate + NADH$^+$ + H$^+$ $\longrightarrow$ glycerate + NAD$^+$

8. glycerate + ATP $\longrightarrow$ 3-phosphoglycerate + ADP

Net: 2 phosphoglycolate + 2 NADH$^+$ + 2 H$^+$ + ATP + 2 O$_2$ + H$_2$O $\longrightarrow$ 3-phosphoglycerate + CO$_2$ + 2 P$_i$ + ADP + 2 H$_2$O$_2$ + 2 NAD$^+$

14. Considering the reactions involving water separately:

1. ATP synthesis: 18 ADP + 18 P$_i$ $\longrightarrow$ 18 ATP + 18 H$_2$O

2. NADP$^+$ reduction and the photolysis of water:
12 H$_2$O + 12 NADP$^+$ $\longrightarrow$ 12 NADPH + 12 H$^+$ + 6 O$_2$

3. Overall reaction for hexose synthesis:
6 CO$_2$ + 12 NADPH + 12 H$^+$ + 18 ATP + 12 H$_2$O $\longrightarrow$ glucose + 12 NADP$^+$ + 6 O$_2$ + 18 ADP + 18 P$_i$

Net: 6 CO$_2$ + 6 H$_2$O $\longrightarrow$ glucose + 6 O$_2$

Of the 12 waters consumed in O$_2$ production in reaction 2 and the 12 waters consumed in the reactions of the Calvin–Benson cycle (reaction 3), 18 are restored by H$_2$O release in phosphoric anhydride bond formation in reaction 1.

15. a. If the number of c-subunits increases, the H$^+$/ATP ratio will increase.

b. If the number of c-subunits increases, Δp can be smaller. (More protons moving down a shallower Δp will yield the same overall ΔG for ATP synthesis.)

16. a. 870-nm light: $\Delta G = -192$ kJ/mol
b. 700-nm light: $\Delta G = -259$ kJ/mol
c. 680-nm light: $\Delta G = -269$ kJ/mol

17. $\Delta G = +35.1$ kJ

18. $\Delta G = -950$ kJ

19. Ideally, accessory light-harvesting pigments would absorb visible light of wavelengths that the chlorophylls do not absorb, that is, light in the 470–620 nm wavelength range.

20.

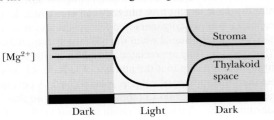

$[\text{Mg}^{2+}]$ — Stroma — Thylakoid space — Dark — Light — Dark

CHAPTER 22

1. The reactions that contribute to the equation on page 725 are:
$$2 \text{ Pyruvate} + 2 \text{ H}^+ + 2 \text{ H}_2\text{O} \longrightarrow \text{glucose} + \text{O}_2$$
$$2 \text{ NADH} + 2 \text{ H}^+ + \text{O}_2 \longrightarrow 2 \text{ NAD}^+ + 2 \text{ H}_2\text{O}$$
$$4 \text{ ATP} + 4 \text{ H}_2\text{O} \longrightarrow 4 \text{ ADP} + 4 \text{ P}_i$$
$$2 \text{ GTP} + 2 \text{ H}_2\text{O} \longrightarrow 2 \text{ GDP} + 2 \text{ P}_i$$

Summing these four reactions produces the equation on page 725.

2. This problem essentially involves consideration of the three unique steps of gluconeogenesis. The conversion of PEP to pyruvate was shown in Chapter 18 to have a $\Delta G°'$ of -31.7 kJ/mol. For the conversion of pyruvate to PEP, we need only add the conversion of a GTP to GDP + P$_i$ (equivalent to ATP to ADP + P$_i$) to the reverse reaction. Thus:

Pyruvate $\longrightarrow$ PEP: $\Delta G°' = +31.7$ kJ/mol $- 30.5$ kJ/mol $= 1.2$ kJ/mol

Then, using Equation 3.13:
$\Delta G = 1.2$ kJ/mol $+ RT \ln ([\text{PEP}][\text{ADP}]^2[\text{Pi}]/[\text{Py}][\text{ATP}]^2)$
ΔG (in erythrocytes) $= -31.96$ kJ/mol

In the case of the fructose-1,6-bisphosphatase reaction, $\Delta G°' = -16.3$ kJ/mol (see Equation 18.6).

$\Delta G = -16.7$ kJ/mol $+ RT \ln ([\text{F-6-P}][\text{P}_i]/[\text{F-1,6-BP}])$
ΔG (in erythrocytes) $= -36.5$ kJ/mol

For the glucose-6-phosphatase reaction, $\Delta G°' = -13.9$ kJ/mol (see Table 3.3). $\Delta G = -13.9$ kJ/mol $+ RT \ln ([\text{Glu}][\text{P}_i]/[\text{G-6-P}]) = -21.1$ kJ/mol. From these ΔG values and those in Table 18.1, ΔG for gluconeogenesis $= -87.7$ kJ/mol.

3. Inhibition by 25 μM fructose-2,6-bisphosphate is approximately 94% at 25 μM fructose-1,6-bisphosphate and approximately 44% at 100 μM fructose-1,6-bisphosphate.

4. The hydrolysis of UDP–glucose to UDP and glucose is characterized by a $\Delta G°'$ of -31.9 kJ/mol. This is more than sufficient to overcome the energetic cost of synthesizing a new glycosidic bond in a glycogen molecule. The net $\Delta G°'$ for the glycogen synthase reaction is -13.3 kJ/mol.

5. According to Table 24.1, a 70-kg person possesses 1920 kJ of muscle glycogen. Without knowing how much of this is in fast-

twitch muscle, we can simply use the fast-twitch data from A Deeper Look (page 740) to calculate a rate of energy consumption. The plot on page 740 shows that glycogen supplies are exhausted after 60 minutes of heavy exercise. Ignoring the curvature of the plot, 1920 kJ of energy consumed in 60 minutes corresponds to an energy consumption rate of 533 J/sec.

6. Although other inhibitory processes might also occur, enzymes with mechanisms involving formation of Schiff base intermediates with active-site lysine residues are likely to be inhibited by sodium borohydride (see Chapter 18, problem 18). The transaldolase reaction of the pentose phosphate pathway involves this type of active-site intermediate and would be expected to be inhibited by sodium borohydride.

7. Glycogen molecules do not have any free reducing ends, regardless of the size of the molecule. If branching occurs every 8 residues and each arm of the branch has 8 residues (or 16 per branch point), a glycogen molecule with 8000 residues would have about 500 ends. If branching occurs every 12 residues, a glycogen molecule with 8000 residues would have about 334 ends.

8. a. Increased fructose-1,6-bisphosphate would activate pyruvate kinase, stimulating glycolysis.
   b. Increased blood glucose would decrease gluconeogenesis and increase glycogen synthesis.
   c. Increased blood insulin inhibits gluconeogenesis and stimulates glycogen synthesis.
   d. Increased blood glucagon inhibits glycogen synthesis and stimulates glycogen breakdown.
   e. Because ATP inhibits both phosphofructokinase and pyruvate kinase, and because its level reflects the energy status of the cell, a decrease in tissue ATP would have the effect of stimulating glycolysis.
   f. Increasing AMP would have the same effect as decreasing ATP—stimulation of glycolysis and inhibition of gluconeogenesis.
   g. Fructose-6-phosphate is not a regulatory molecule and decreases in its concentration would not markedly affect either glycolysis or gluconeogenesis (ignoring any effects due to decreased [G-6-P] as a consequence of decreased [F-6-P]).

9. At 298 K, assuming roughly equal concentrations of glycogen molecules of different lengths, the glucose-1-P concentration would be about 0.286 mM.

10. All four of these reactions begin with the formation of N-carboxybiotin:

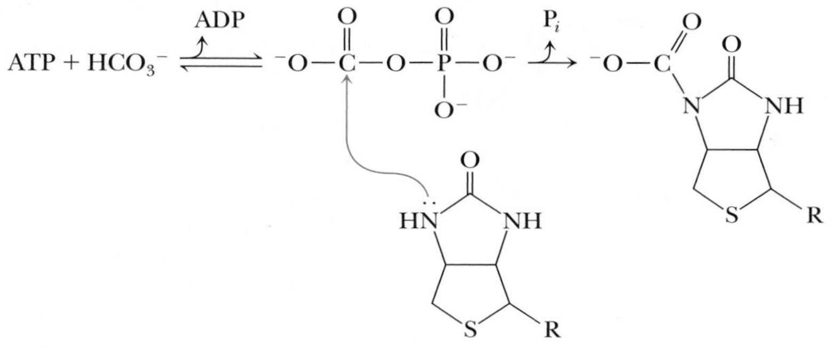

The carboxylation of β-methylcrotonyl-CoA occurs as follows:

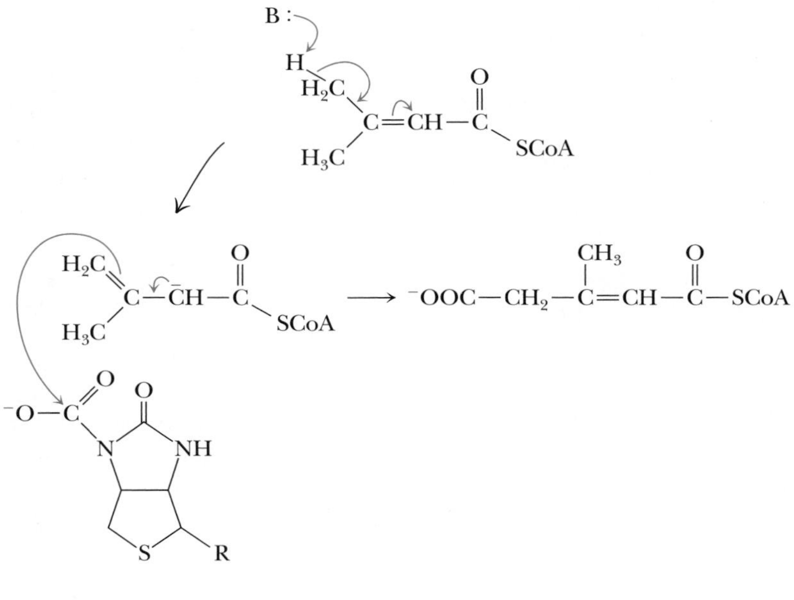

The carboxylations of geranyl-CoA and urea are shown below:

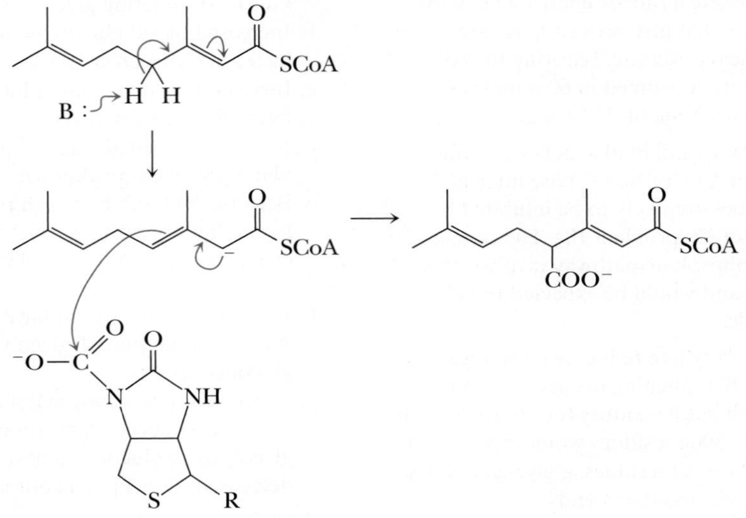

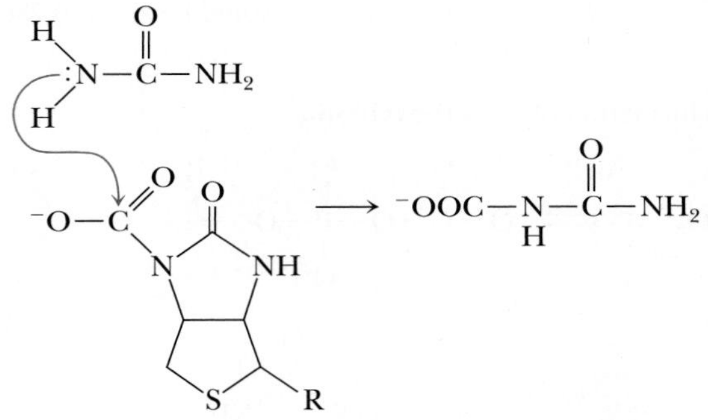

The mechanism of the carboxyltransferase reaction is a combination of the pyruvate carboxylase mechanism (shown in Figure 22.3) and a reverse propionyl-CoA carboxylase reaction.

11. a. The mechanism of the acetolactate synthase reaction.

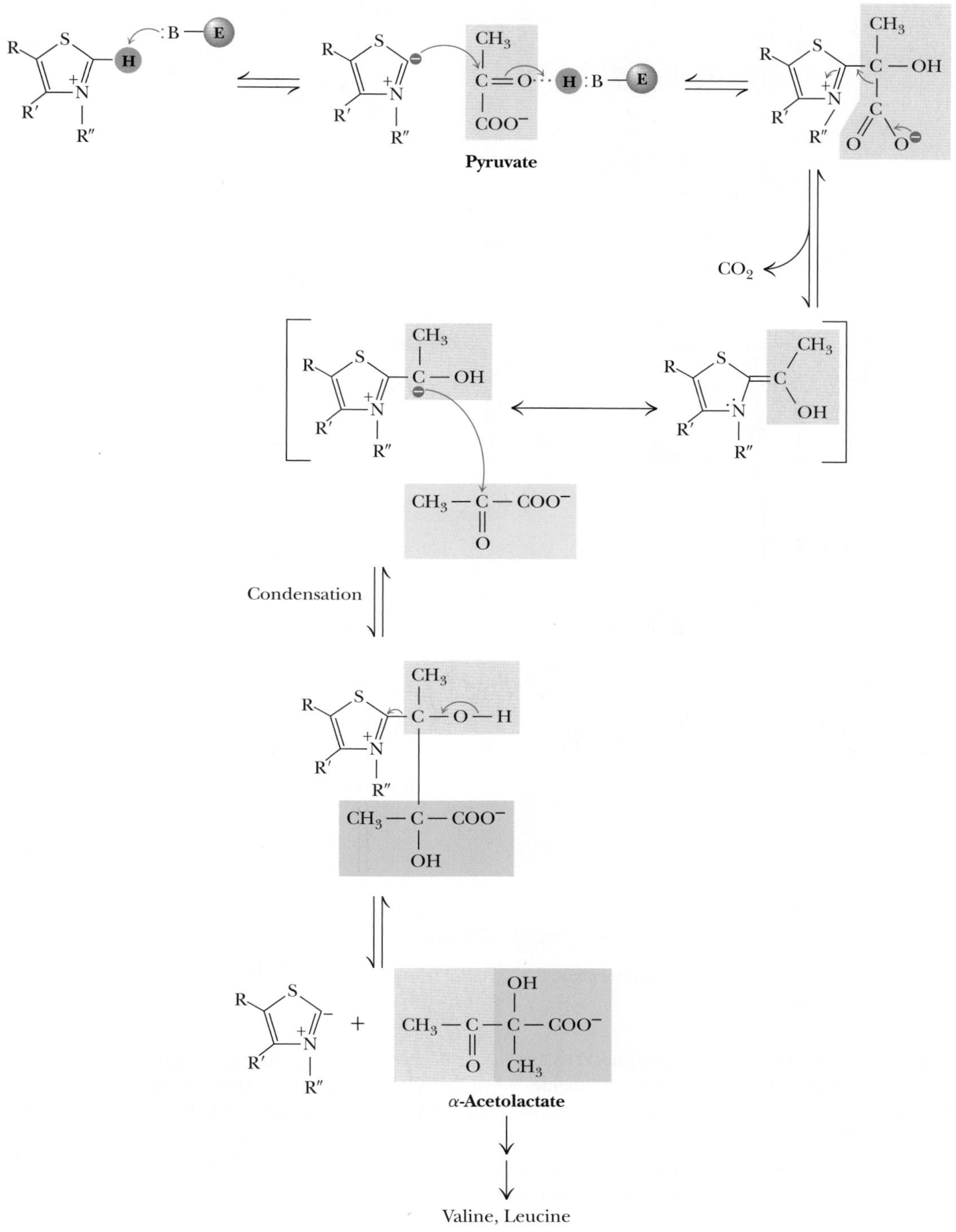

α-**Acetolactate**

Valine, Leucine

b. The phosphoketolase reaction.

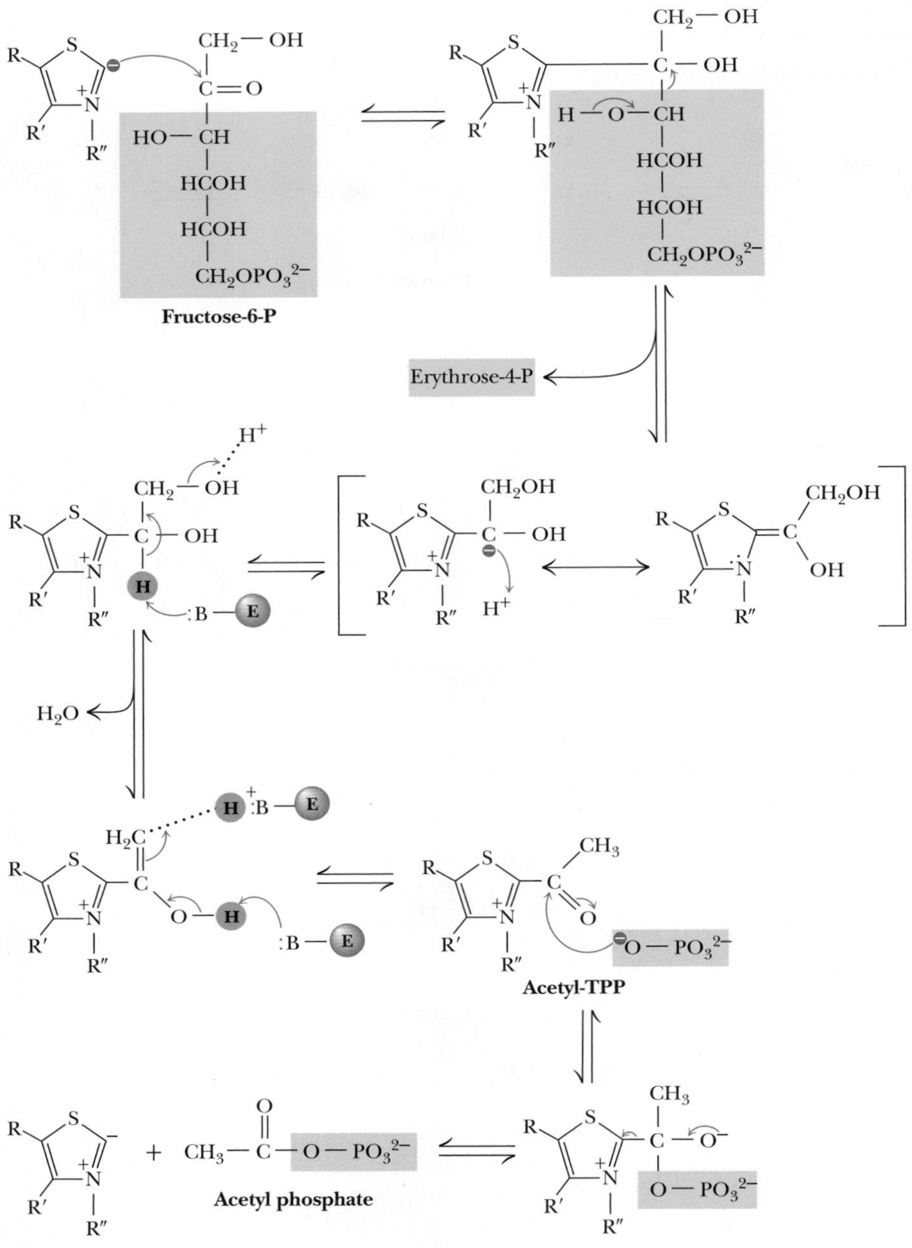

Fructose-6-P

Erythrose-4-P

Acetyl-TPP

Acetyl phosphate

12. Metformin both stimulates glucose uptake by peripheral tissues and enhances the binding of insulin to its receptors. Glipizide complements the actions of metformin by stimulating increased insulin secretion by the pancreas.

13. Epalrestat and tolrestat do not resemble the transition state for aldose reductase, and evidence from studies by Franklin Prender-gast and others (Ehrig, T., Bohren, K., Prendergast, F., and Gabbay, K., 1994. Mechanism of aldose reductase inhibition: Binding of NADP$^+$/NADPH and alrestatin-like inhibitors. *Biochemistry* **33:**7157–7165) show that these inhibitors probably bind to the enzyme:NADP$^+$ complex at a site other than the active site.

14.

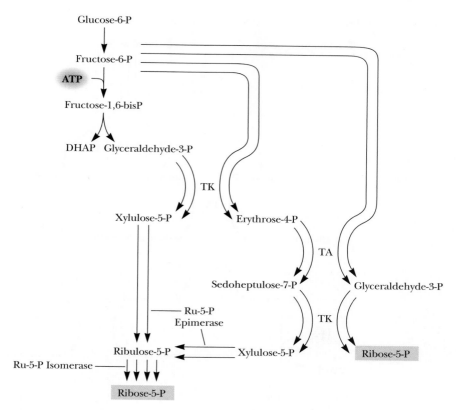

15. Carbons 1, 2, and 5 of ribose-5-P will be labeled.

16.

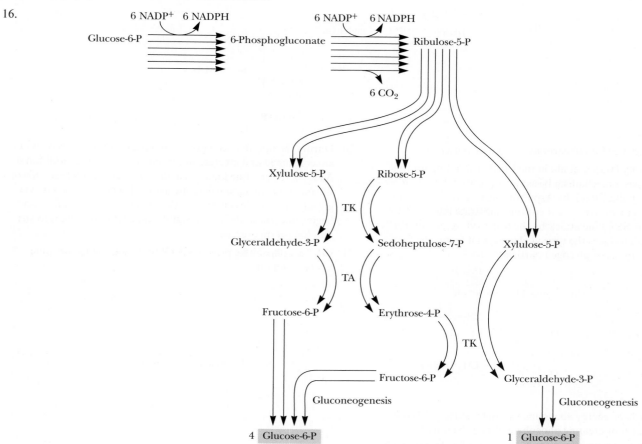

17.

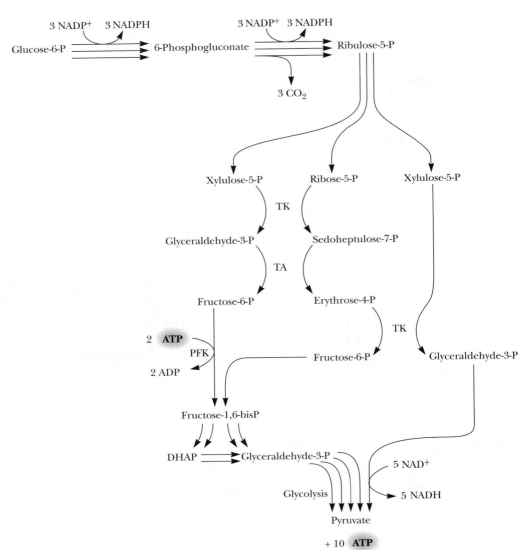

18. Carbons 2 and 4 end up in carbons 1 and 3 of pyruvate.

19. The reference by Hurley, et al., in the Further Reading for this chapter suggests a mechanism based on structural data on glycogenin. Nucleophilic attack by Asp[162] on the C-1 of glucose (in UDP-glucose) forms a covalent enzyme-substrate intermediate. Subsequent nucleophilic attack by the hydroxyl oxygen of Tyr[194] of glycogenin produces the tyrosyl glucose that forms the foundation for synthesis of glycogen particles.

20. During the first few seconds of exercise, existing stores of ATP are consumed and creatine phosphate then provides additional ATP via the creatine kinase reaction. For the next 90 seconds or so, anaerobic metabolism (conversion of glucose to lactate via glycolysis) provides energy. At this point, aerobic metabolism begins in earnest, delivering significant energy resources to sustain long-term exercise.

21. Creatine phosphate provides ATP to muscle via the creatine kinase reaction:

The creatine kinase reaction

$$^-O-\overset{\overset{O}{\|}}{\underset{\underset{-O}{}}{P}}-\overset{H}{\underset{}{N}}-\overset{\overset{CH_3}{|}}{\underset{\underset{H_2N^+}{}}{C}}-N-CH_2-COO^- + ATP \longrightarrow \overset{H_2N}{\underset{\overset{+}{H_2N}}{}}C-\overset{\overset{CH_3}{|}}{N}-CH_2-COO^- + ADP$$

During periods of energy abundance, muscles store ATP equivalents in the form of creatine phosphate. When the muscles demand energy, creatine kinase runs in the reverse direction, converting creatine phosphate to ATP.

CHAPTER 23

1. Assuming that all fatty acid chains in the triacylglycerol are palmitic acid, the fatty acid content of the triacylglycerol is 95% of the total weight. On the basis of this assumption, one can calculate that 30 lb of triacylglycerol will yield 118.7 L of water.

2. 11-*cis*-Heptadecenoic acid is metabolized by means of seven cycles of β-oxidation, leaving a propionyl-CoA as the final product. However, the fifth cycle bypasses the acyl-CoA dehydrogenase reaction because a *cis*-double bond is already present at the proper position. Thus, β-oxidation produces 7 NADH (= 17.5 ATP), 6 FADH$_2$ (= 9 ATP), and 7 acetyl-CoA (= 70 ATP), for a total of 96.5 ATP. Propionyl-CoA is converted to succinyl-CoA (with expenditure of 1 ATP), which can be converted to oxaloacetate in the TCA cycle (with production of 1 GTP, 1 FADH$_2$, and 1 NADH). Oxaloacetate can be converted to pyruvate (with no net ATP formed or consumed), and pyruvate can be metabolized in the TCA cycle (producing 1 GTP, 1 FADH$_2$, and 4 NADH). The net for these conversions of propionate is 16.5 ATP. Together with the results of β-oxidation, the total ATP yield for the oxidation of one molecule of 11-*cis*-heptadecenoic acid is 113 ATP.

3. Instead of invoking hydroxylation and β-oxidation, the best strategy for oxidation of phytanic acid is α-hydroxylation, which places a hydroxyl group at C-2. This facilitates oxidative α-decarboxylation, and the resulting acid can react with CoA to form a CoA ester. This product then undergoes six cycles of β-oxidation. In addition to CO_2, the products of this pathway are three molecules of acetyl-CoA, three molecules of propionyl-CoA, and one molecule of 2-methylpropionyl-CoA.

4. Although acetate units cannot be used for net carbohydrate synthesis, oxaloacetate can enter the gluconeogenesis pathway in the PEP carboxykinase reaction. (For this purpose, it must be converted to malate for transport to the cytosol.) Acetate labeled at the carboxyl carbon will first label (equally) the C-3 and C-4 positions of newly formed glucose. Acetate labeled at the methyl carbon will label (equally) the C-1, C-2, C-5, and C-6 positions of newly formed glucose.

5. This exercise is left to the student, in consultation with the references suggested in the problem.

6. Fat is capable of storing more energy (37 kJ/g) than carbohydrate (16 kJ/g). Ten pounds of fat contains $10 \times 454 \times 37 = 167,980$ kJ of energy. This same amount of energy would require $167,980/16 = 10,499$ g, or 23 lb, of stored carbohydrate.

7. The enzyme methylmalonyl-CoA mutase, which catalyzes the third step in the conversion of propionyl-CoA to succinyl-CoA, is B$_{12}$-dependent. If a deficiency in this vitamin occurs, and if large amounts of odd-carbon fatty acids were ingested in the diet, L-methymalonyl-CoA could accumulate.

8. a. Myristic acid:
$$CH_3(CH_2)_{12}COOH + 92\ P_i + 92\ ADP + 20\ O_2 \longrightarrow$$
$$92\ ATP + 14\ CO_2 + 106\ H_2O$$

b. Stearic acid:
$$CH_3(CH_2)_{16}COOH + 120\ P_i + 120\ ADP + 26\ O_2 \longrightarrow$$
$$120\ ATP + 18\ CO_2 + 138\ H_2O$$

c. α-Linolenic acid:
$$C_{17}H_{29}COOH + 113.5\ P_i + 113.5\ ADP + 24.5\ O_2 \longrightarrow$$
$$113.5\ ATP + 18\ CO_2 + 128.5\ H_2O$$

d. Arachidonic acid:
$$C_{19}H_{31}COOH + 125\ P_i + 125\ ADP + 27\ O_2 \longrightarrow$$
$$125\ ATP + 20\ CO_2 + 141\ H_2O$$

9. During the hydration step, the elements of water are added across the double bond. Also, the proton transferred to the acetyl-CoA carbanion in the thiolase reaction is derived from the solvent, so each acetyl-CoA released by the enzyme would probably contain two tritiums. Seven tritiated acetyl-CoAs thus would derive from each molecule of palmitoyl-CoA metabolized, each with two tritiums at C-2.

10. A carnitine deficiency would presumably result in defective or limited transport of fatty acids into the mitochondrial matrix and reduced rates of fatty acid oxidation.

11. 1.3 grams $\times$ 37 kJ/gram = 48,000 J. This at first seems like a remarkably small amount of energy to sustain the hummingbird during a 500-mile flight at 50 mph. However, keep in mind that the hummingbird weighs only 3 to 4 grams, and see problem 12 following.

12. 48,000 J in 10 hours is 4800 J/hr. If the hummingbird consumes 250 mL per hour during migration, this means that it is consuming 4800/250 or 19.2 J/mL of oxygen consumed. If a human consumes 12.7 kcal/min while running, this is equivalent to $12,700 \times 4.184$ J/cal = 53.1 kJ/min; 48,000 J/53,100 J/min = 0.9 minute. So a human could run for only less than a minute on the energy consumed by the hummingbird (only 3 to 4 g) on a 500-mile flight. A typical person would have to run about 40 miles to lose 1 lb of fat.

13. A mechanism for the HMG-CoA synthase reaction:

14.

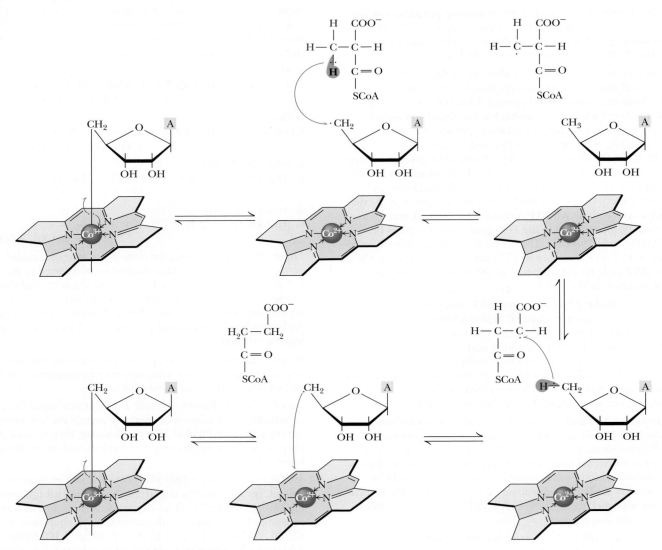

15. The changes in oxidation state of cobalt in the course of any
B$_{12}$-dependent reaction arise from homolytic cleavages of the
Co^{3+}—C bonds involved. Moreover, in the Co^{3+}—C bonds
shown in a typical B$_{12}$-dependent reaction, the two electrons of
the bond are not ascribed to the Co^{3+} atom in the calculation of
the Co oxidation state. Thus, when the Co—C bond is cleaved
homolytically, one electron reverts to the Co, changing its
oxidation state from 3+ to 2+.

16. See the following mechanisms.

α-Methyleneglutarate mutase

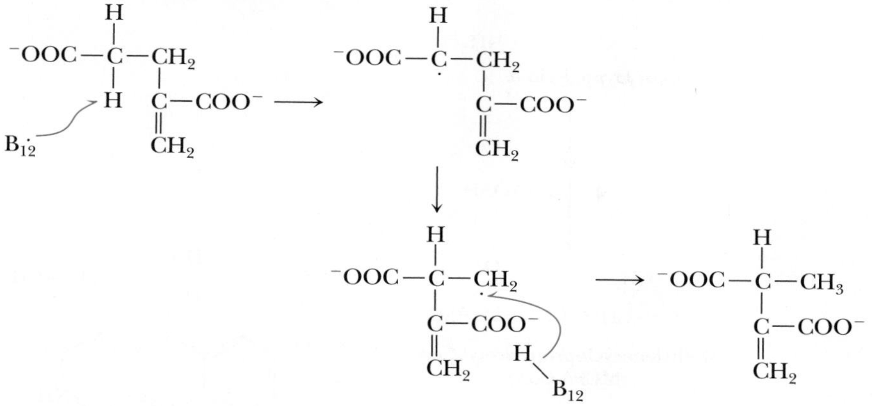

$$^-OOC-\underset{\underset{H}{|}}{\overset{\overset{H}{|}}{C}}-CH_2 \quad \underset{\overset{|}{C}-COO^-}{} \longrightarrow \quad ^-OOC-\underset{\overset{|}{C}-COO^-}{\overset{\overset{H}{|}}{C}}-CH_2$$

$$B_{12} \qquad CH_2 \qquad\qquad CH_2$$

$$^-OOC-\underset{\underset{C-COO^-}{|}}{\overset{\overset{H}{|}}{C}}-CH_2 \longrightarrow {}^-OOC-\underset{\underset{C-COO^-}{|}}{\overset{\overset{H}{|}}{C}}-CH_3$$

$$CH_2 \quad H \quad B_{12} \qquad CH_2$$

Diol dehydrase

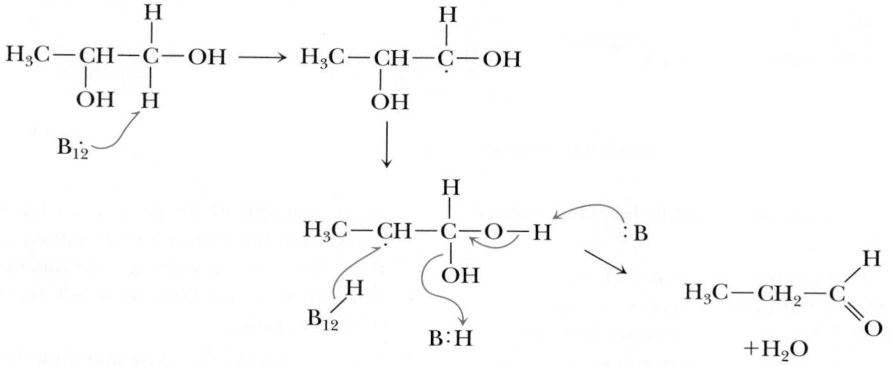

$$H_3C-\underset{\underset{OH}{|}}{CH}-\underset{\underset{H}{|}}{\overset{\overset{H}{|}}{C}}-OH \longrightarrow H_3C-\underset{\underset{OH}{|}}{CH}-\overset{\overset{H}{|}}{\underset{\cdot}{C}}-OH$$

$$B_{12}$$

$$H_3C-CH-\underset{\underset{OH}{|}}{\overset{\overset{H}{|}}{C}}-O-H \qquad :B$$

$$H \qquad\qquad B:H \qquad H_3C-CH_2-\overset{\overset{H}{|}}{\underset{\parallel}{C}}$$
$$B_{12} \qquad\qquad\qquad\qquad O$$
$$+H_2O$$

Glycerol dehydrase

$$\underset{OH}{CH_2}-\underset{OH}{CH}-\underset{OH}{\overset{\overset{H}{|}}{C}}-H \longrightarrow \underset{OH}{CH_2}-\underset{OH}{CH}-\underset{OH}{\overset{\overset{H}{|}}{C}}\cdot$$

$$\cdot B_{12}$$

$$\underset{OH}{CH_2}-\underset{\cdot}{CH}-\overset{\overset{H}{|}}{\underset{O-H}{C}}-OH \quad \overset{H:B}{\longleftarrow} \longrightarrow \underset{OH}{CH_2}-CH_2-\overset{\overset{H}{|}}{\underset{\parallel}{C}}$$

$$B_{12}-H \quad :B \qquad\qquad O$$
$$+H_2O$$

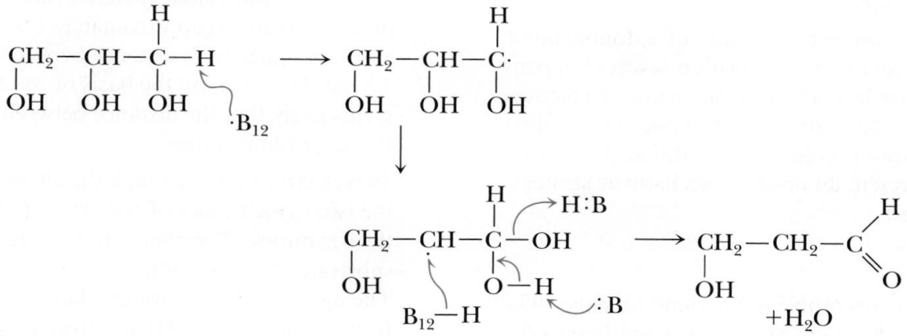

Ethanolamine ammonia-lyase

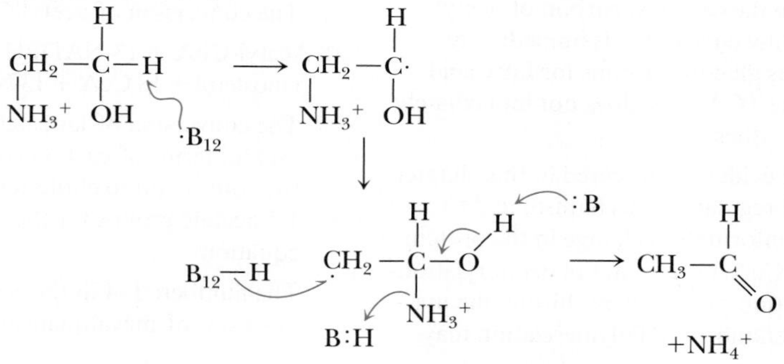

$$\underset{NH_3^+}{CH_2}-\underset{OH}{\overset{\overset{H}{|}}{C}}-H \longrightarrow \underset{NH_3^+}{CH_2}-\underset{OH}{\overset{\overset{H}{|}}{C}}\cdot$$

$$\cdot B_{12}$$

$$B_{12}-H \quad \underset{\cdot}{CH_2}-\overset{\overset{H}{|}}{\underset{NH_3^+}{C}}-O \quad \overset{H}{\underset{}{}} \quad :B$$

$$B:H \qquad\qquad\qquad CH_3-\overset{\overset{H}{|}}{\underset{\parallel}{C}}$$
$$O$$
$$+NH_4^+$$

17.

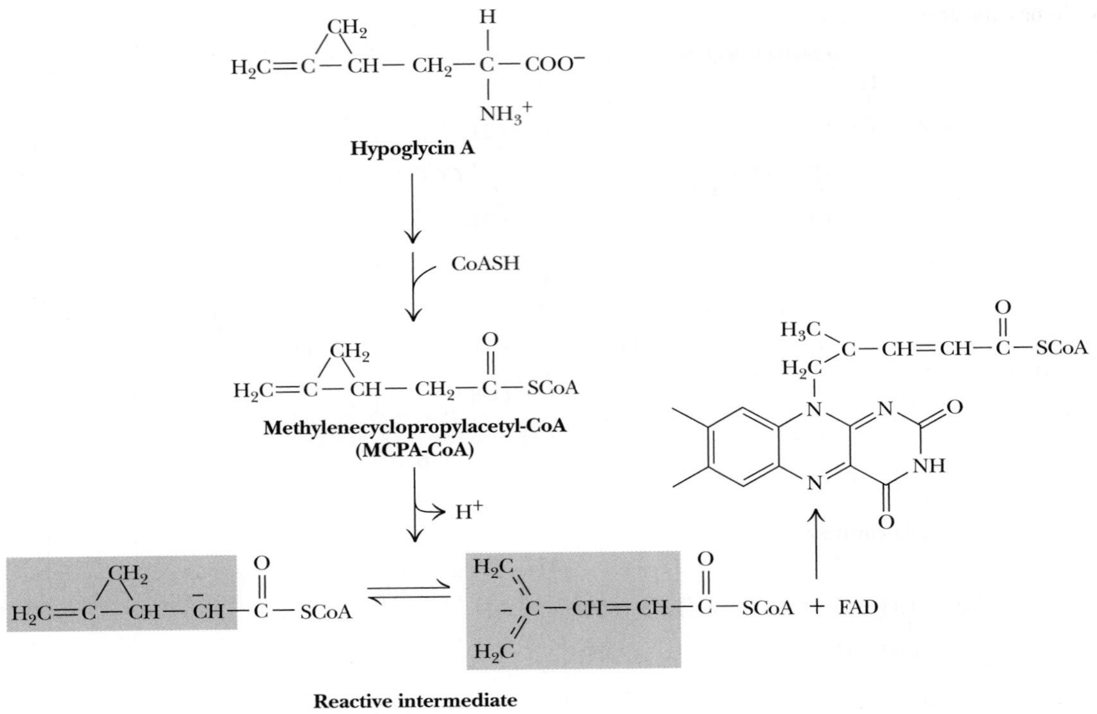

$$H_2C=\overset{\overset{\displaystyle CH_2}{\diagup\!\!\!\diagdown}}{C}-CH-CH_2-\overset{\overset{\displaystyle H}{|}}{\underset{\underset{\displaystyle NH_3^+}{|}}{C}}-COO^-$$

Hypoglycin A

↓ CoASH

$$H_2C=\overset{\overset{\displaystyle CH_2}{\diagup\!\!\!\diagdown}}{C}-CH-CH_2-\overset{\overset{\displaystyle O}{||}}{C}-SCoA$$

**Methylenecyclopropylacetyl-CoA
(MCPA-CoA)**

↓ H⁺

$$H_2C=\overset{\overset{\displaystyle CH_2}{\diagup\!\!\!\diagdown}}{C}-CH-\overset{-}{C}H-\overset{\overset{\displaystyle O}{||}}{C}-SCoA \rightleftharpoons \overset{H_2C}{\underset{H_2C}{>}}C-CH=CH-\overset{\overset{\displaystyle O}{||}}{C}-SCoA + FAD$$

Reactive intermediate

18. This exercise is left to the student and should be based on the reference provided in the problem.

19. It may be presumed that the oxidation of the acyl chain is accomplished via a two-electron transfer, whereas the steps involved in reoxidation of FADH$_2$ by ETF are one-electron transfers. FAD/FADH$_2$ can participate both in one-electron and two-electron transfers, whereas NAD$^+$/NADH can participate only in two-electron transfers.

20. The sequence of reactions involving creation of a double bond, then hydration across it, followed by oxidation is what happens to succinate in the TCA cycle. (This same sequence of reactions is also employed in fatty acid synthesis and in both the catabolism and anabolism of amino acids.) Clearly, this sequence of three reactions must represent an optimal mechanistic strategy for the chemistry achieved.

CHAPTER 24

1. The equations needed for this problem are found on page 799. See *The Student Solutions Manual, Study Guide and Problems Book* for details.

2. Carbons C-1 and C-6 of glucose become the methyl carbons of acetyl-CoA that is the substrate for fatty acid synthesis. Carbons C-2 and C-5 of glucose become the carboxyl carbon of acetyl-CoA for fatty acid synthesis. Only citrate that is immediately exported to the cytosol provides glucose carbons for fatty acid synthesis. Citrate that enters the TCA cycle does not immediately provide carbon for fatty acid synthesis.

3. A suitable model, based on the evidence presented in this chapter, would be that the fundamental regulatory mechanism in ACC is a polymerization-dependent conformation change in the protein. All other effectors—palmitoyl-CoA, citrate, and phosphorylation-dephosphorylation—may function primarily by shifting the inactive protomer-active polymer equilibrium. Polymerization may bring domains of the protomer (that is, bicarbonate-, acetyl-CoA-, and biotin-binding domains) closer together or may bring these domains on separate protomers close to each other. See *The Student Solutions Manual, Study Guide and Problems Book* for further details.

4. The phosphopantetheine may function, at least to some extent, as a flexible "arm" to carry acyl groups between the malonyl transferase and ketoacyl-ACP synthase active sites. The phosphopantetheine is approximately 1.9 nm in length, setting an absolute upper-limit distance between these active sites of 3.8 nm. However, on the basis of modeling considerations, it seems likely that the distance between these sites is smaller than this upper-limit value.

5. Two electrons pass through the chain from NADH to FAD to the two cytochromes of the cytochrome b_5 reductase and then to the desaturase. Together with two electrons from the fatty acyl substrate, these electrons reduce an O_2 to two molecules of water. The hydrogen for the waters that are formed in this way comes from the substrate (2H) and from two protons from solution.

6. Ethanolamine + glycerol + 2 fatty acyl-CoA + 2 ATP + CTP + $H_2O \longrightarrow$ phosphatidylethanolamine + 2 ADP + 2 CoA + CMP + PP$_i$ + P$_i$

7. The conversion of acetyl-CoA to lanosterol can be written as:

18 Acetyl-CoA + 13 NADPH + 13 H$^+$ + 18 ATP + 0.5 O$_2 \longrightarrow$ lanosterol + 18 CoA + 13 NADP$^+$ + 18 ADP + 6 P$_i$ + 6 PP$_i$ + CO$_2$

The conversion of lanosterol to cholesterol is complicated; however, in terms of carbon counting, three carbons are lost in the conversion to cholesterol. This process might be viewed as 1.5 acetate groups for the purpose of completing the balanced equation.

8. The numbers 1–4 in the cholesterol structure indicate the carbon positions of mevalonate as shown (note that the numbering

shown here is not based on the systematic numbering of mevalonate):

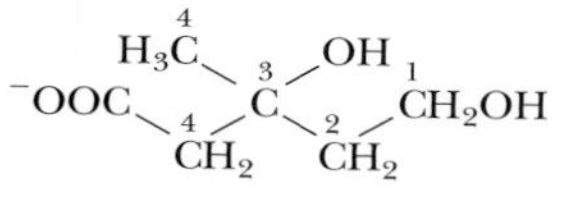

Mevalonate 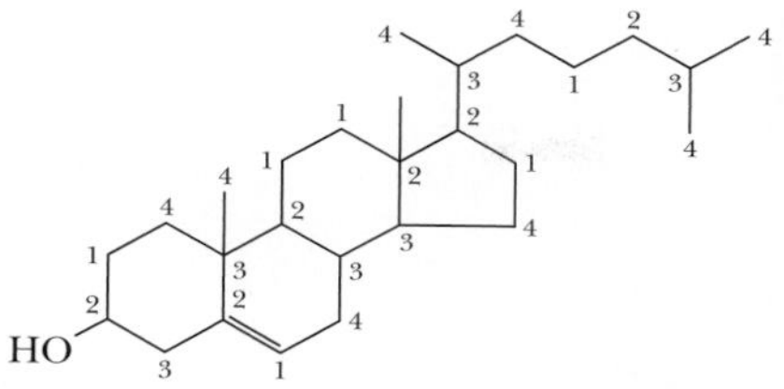   Cholesterol

9. The O-linked saccharide domain of the LDL receptor probably functions to extend the receptor domain away from the cell sur-face and above the glycocalyx coat so that the receptor can recognize circulating lipoproteins.

10. As shown in Figures 24.19 and 24.22, the syntheses (in eukaryotes) of phosphatidylcholine, phosphatidylethanolamine, phosphatidylinositol, and phosphatidylglycerol are dependent upon CTP. A CTP deficiency would be likely to affect all these synthetic pathways.

11. It "costs" 1 ATP to form a malonyl-CoA. Each cycle of the fatty acyl synthase consumes 2 NADPH molecules, each worth 3.5 ATP. Thus, each of the seven cycles required to form a palmitic acid consumes 8 ATPs. A total of 56 ATPs are consumed to synthesize one molecule of palmitic acid.

12. The mechanism of the 3-ketosphinganine synthase reaction is shown in the following figure.

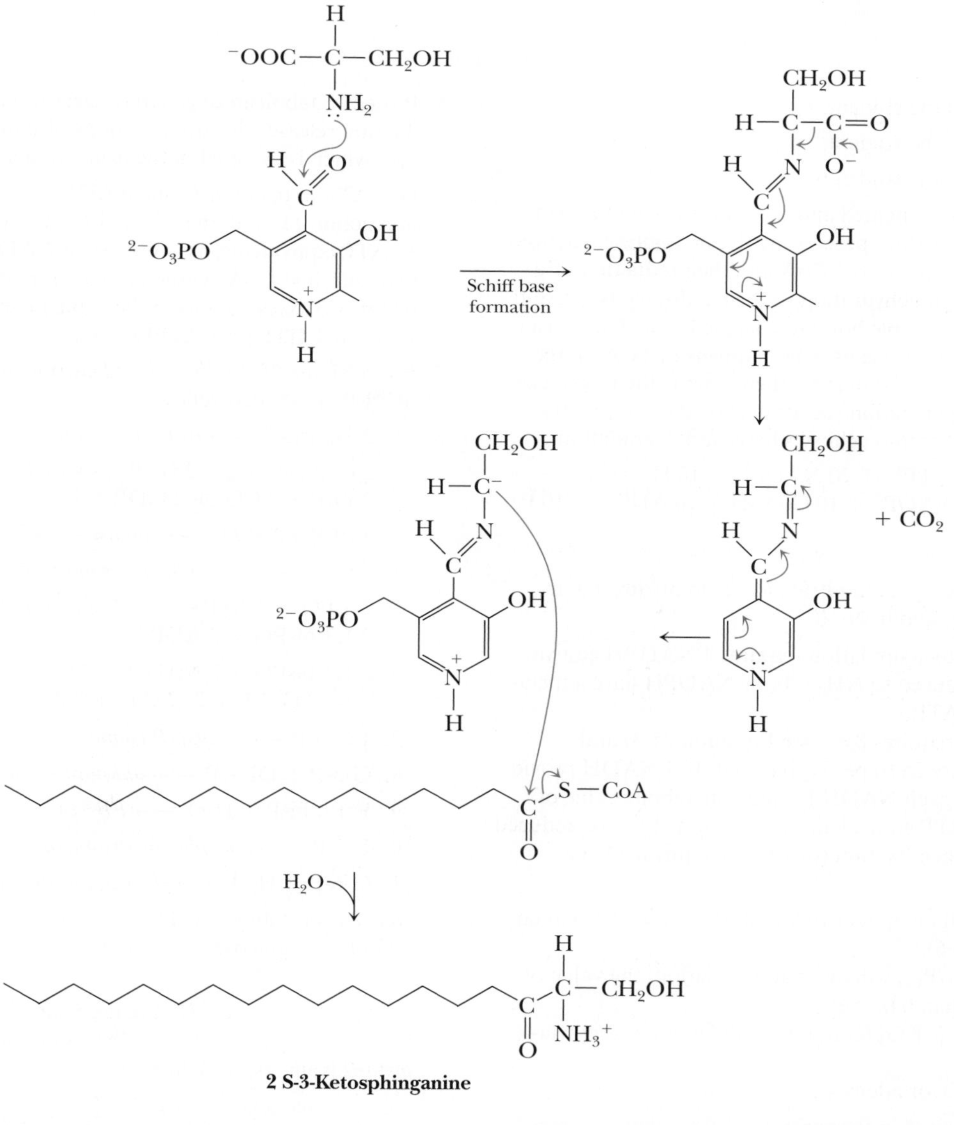

2 S-3-Ketosphinganine

13. FAD is required for the eukaryotic reaction that converts stearic acid to oleic acid because the oxidation of a single bond to a double bond in stearic acid involves NADH and thus requires a two-electron transfer, whereas the reoxidation of $FADH_2$ to FAD is accomplished by electron transfer to cytochrome b_5. Cyto-chromes are capable of one-electron transfers only, so $FADH_2$/FAD is required because it can participate both in one-electron and two-electron transfers.

14. See Chapter 23, problem 13, for a mechanism for the HMG-CoA synthase reaction.

15. A mechanism for the β-ketoacyl ACP synthase reaction:

β-Ketoacyl ACP synthase

Acetoacetyl – ACP

16. This exercise is left to the student.

17. This exercise is left to the student.

18. This exercise is left to the student.

19. Palmitic acid must be elongated and then unsaturated to form linoleic acid. The elongation process involves a thiolase reaction to add two carbons to palmitoyl-CoA and then reduction of a carbonyl to a hydroxyl, dehydration to form a double bond, and then reduction of the double bond to a single bond. These last three reactions are the reverse of what happens in TCA, in the conversion of succinate to fumarate, then malate, then oxaloacetate. These same three reactions occur in β-oxidation, in fatty acid synthesis, and in amino acid synthesis and degradation.

20. 11 Acetyl-CoA + 10 ATP^{4-} + 20 NADPH + 10 H$^+$ ⟶ behenoyl-CoA + 20 NADP$^+$ + 10 CoASH + 10 ADP^{3-} + 10 P$_i^{2-}$

CHAPTER 25

1. The oxidation number of N in nitrate is +5; in nitrite, +3; in NO, +2; in N$_2$O, +1; and in N$_2$, 0.

2. a. Assume that nitrate assimilation requires 4 NADPH equivalents per NO$_3^-$ reduced to NH$_4^+$. Four NADPH have a metabolic value of 14 ATP.
   b. Nitrogen fixation requires 8 e^- (see Equation 25.3) and 16 ATPs (see Figure 25.6) per N$_2$ reduced. If 4 NADH provide the requisite 8 e^-, each NADH having a metabolic value of 3 ATPs, then 28 ATP equivalents are consumed per N$_2$ reduced in biological nitrogen fixation (or 14 ATP equivalents per NH$_4^+$ formed).

3. a. [ATP] increase will favor adenylylation; the value of n will be greater than 6 (n>6).
   b. An increase in P$_{IIA}$/P$_{IID}$ will favor adenylylation; the value of n will be greater than 6 (n >6).
   c. An increase in the [αKG]/[Gln] ratio will favor deadenylylation; n <6.
   d. [P$_i$] decrease will favor adenylylation; n >6.

4. Two ATPs are consumed in the carbamoyl-P synthetase-I reaction, and 2 phosphoric anhydride bonds (equal to 2 ATP equivalents) are expended in the argininosuccinate synthetase reaction. Thus, 4 ATP equivalents are consumed in the urea cycle, as 1 urea and 1 fumarate are formed from 1 CO$_2$, 1 NH$_3$, and 1 aspartate.

5. Protein catabolism to generate carbon skeletons for energy production releases the amino groups of amino acids as excess nitrogen, which is excreted in the urine, principally as urea.

6. One ATP in reaction 1, one NADPH in reaction 2, one NADPH in reaction 11, and one succinyl-CoA in reaction 12 add up to 10 ATP equivalents, assuming each NADPH is worth 4 ATPs. (The succinyl-CoA synthetase reaction of the citric acid cycle [see p. 623] fixes the metabolic value of succinyl-CoA versus succinate at 1 GTP [= 1 ATP].)

7. From Figure 25.36: ^{14}C-labeled carbon atoms derived from ^{14}C-2 of PEP are shaded yellow.

8. 1. 2 aspartate ⟶ *transamination* ⟶ 2 oxaloacetate.
   2. 2 oxaloacetate + 2 GTP ⟶ *PEP carboxykinase* ⟶ 2 PEP + 2 CO$_2$ + 2 GDP.
   3. 2 PEP + 2 H$_2$O ⟶ *enolase* ⟶ 2 2-PG.
   4. 2 2-PG ⟶ *phosphoglyceromutase* ⟶ 2 3-PG.
   5. 2 3-PG + 2 ATP ⟶ *3-P glycerate kinase* ⟶ 2 1,3-bisPG + 2 ADP.
   6. 2 1,3-bisPG + 2 NADH + 2 H$^+$ ⟶ *G-3-P dehydrogenase* ⟶ 2 G-3-P + 2 NAD$^+$ + 2 P$_i$.
   7. 1 G-3-P ⟶ *triose-P isomerase* ⟶ 1 DHAP.
   8. G-3-P + DHAP ⟶ *aldolase* ⟶ fructose-1,6-bisP.
   9. F-1,6-bisP + H$_2$O ⟶ *FBPase* ⟶ F-6-P + P$_i$.
   10. F-6-P ⟶ *phosphoglucoisomerase* ⟶ G-6-P.
   11. G-6-P + H$_2$O ⟶ *glucose phosphatase* ⟶ glucose + P$_i$.

   Net: 2 aspartate + 2 GTP + 2 ATP + 2 NADH + 6 H$^+$ + 4 H$_2$O ⟶ glucose + 2 CO$_2$ + 2 GDP + 2 ADP + 4 P$_i$ + 2 NAD$^+$

   (Note that 4 of the 6 H$^+$ are necessary to balance the charge on the 4 carboxylate groups of the 2 OAA and the loss of amino groups from Asp is ignored.)

   (As a consequence of reaction [1], 2 α-keto acids [for example, α-ketoglutarate] will receive amino groups to become 2 α-amino acids [for example, glutamate].)

9. Alanine: glutamate:pyruvate aminotransferase.

   Arginine: from glutamate via ornithine, so it's glutamate dehydrogenase.

   Aspartate: glutamate:oxaloacetate aminotransferase.

Asparagine: from aspartate, so it's the glutamate:oxaloacetate aminotransferase.

Cysteine: cysteine is formed from serine, so it's glutamate via 3-phosphoserine aminotransferase.

Glutamate: glutamate dehydrogenase.

Glutamine: glutamine synthetase.

Glycine: glycine is formed from serine, so it's glutamate via 3-phosphoserine aminotransferase.

Histidine: from glutamate via L-histidinol phosphate aminotransferase.

Isoleucine: glutamate:α-keto-β-methylvalerate aminotransferase.

Leucine: glutamate:α-ketoisocaproate aminotransferase.

Lysine: from glutamate via saccharopine formation by a glutamate-dependent NADPH dehydrogenase; in bacteria, lysine is synthesized from aspartate, so it's glutamate:oxaloacetate aminotransferase.

Methionine: from aspartate, so it's the glutamate:oxaloacetate aminotransferase.

Phenylalanine: glutamate:phenylpyruvate aminotransferase (= phenylalanine aminotransferase).

Proline: from glutamine, so it's glutamate dehydrogenase.

Serine: glutamate via 3-phosphoserine aminotransferase.

Threonine: from aspartate, so it's glutamate:oxaloacetate aminotransferase.

Tryptophan: from serine via tryptophan synthase, so its α-amino group comes from serine, which gets its amino group from glutamate via 3-phosphoserine aminotransferase.

Tyrosine: glutamate:4-hydroxyphenylpyruvate aminotransferase (= tyrosine aminotransferase).

Valine: glutamate:α-ketoisovalerate aminotransferase.

10. Pyridoxal (vitamin B_6), because it is the precursor to pyridoxal-P, the key coenzyme in aminotransferase reactions, as well as other aspects of amino acid metabolism.

11. The conversion of homocysteine to methionine is folate-dependent; dietary folate absorption is dependent on vitamin B_{12} for removal of methyl groups added to folate during digestion; finally, the α-amino group of homocysteine formed in the methione biosynthetic pathway comes from aspartate via the pyridoxal-P–dependent glutamate:oxaloacetate aminotransferase.

12. The BCKAD complex is structurally and functionally analogous to the pyruvate dehydrogenase complex and the α-ketoglutarate dehydrogenase complexes. See Figure 19.4; the reaction mechanism for the pyruvate dehydrogenase complex is essentially the same as that of the BCKAD complex.

13. Aspartame is a N-α-L-aspartyl-L-phenylalanine-1-methyl ester that is broken down in the digestive tract and phenylalanine is released. Phenylalanine is the substance that phenylketonurics must avoid.

14. Glyphosate inhibits 3-enolpyruvylshikimate-5-P synthase, an essential enzyme in the biosynthesis of chorismate. Not only is chorismate the precursor for synthesis of the aromatic amino acids Phe, Tyr, and Trp, it is also the precursor for formation of lignin, a major structural component in plant cell walls, as well as other essential substances such as folate, coenzyme Q, plastoquinone, and vitamins E and K.

15. Glycine is formed from serine, which is formed from 3-phosphoglycerate. Carbons 3 and 4, 2 and 5, and 1 and 6 of glucose contribute carbons 1, 2, and 3 of 3-phosphoglycerate, respectively. The β-carbon of serine derives from the 3-C in 3-phosphoglycerate, the C_α-carbon atom in serine comes from 3-phosphoglycerate C-2, and the carboxyl-C in serine comes from C-1 of 3-PG. It is these latter two carbons of 3-PG that are found in glycine; that is, C-1 and C-2 of 3-PG; C-1 of 3-PG came from C-3 and C-4 in glucose, and C-2 of 3-PG came from C-2 and C-5 in glucose.

16. Serine is an important precursor for glycine synthesis.

17. Consult *www.pdb.org* to view pdb id = 1LM1.

18. The liver converts amino acids to glucose to provide a source of energy for other cells, such as nerve and red blood cells. Conversion of amino acids to glucose by gluconeogenesis requires energy; for example, six ATP equivalents are needed to form glucose from aspartic acid.

19. GS monomers are inactive because GS active sites require elements of protein structure contributed by adjacent subunits in the GS_{12} dodecamer.

20. Four of the six carbons in the ring of Tyr come from erythrose-4-P and the remaining two from PEP; the carboxyl-C, C_α, and C_β of Tyr come from PEP. Degradation of Tyr to acetoacetate + fumarate yields acetoacetate composed from the C_α and C_β plus 2 C atoms from the ring, which are either both from E-4-P or one each from E-4-P and PEP, depending on the orientation of the ring (rotation about the C_β-ring C bond). Fumarate will be composed from 2 C atoms from PEP and 2 from E-4-P *or* 3 C atoms from E-4-P and 1 from PEP.

CHAPTER 26

1. See Figure 26.2 (purines) and Figure 26.13 (pyrimidines).

2. Assume ribose-5-P is available.

   Purine synthesis: 2 ATP equivalents in the ribose-5-P pyrophosphokinase reaction, 1 in the GAR synthetase reaction, 1 in the FGAM synthetase reaction, 1 in the AIR carboxylase reaction, 1 in the CAIR synthetase reaction, and 1 in the SACAIR synthetase reaction yields IMP, the precursor common to ATP and GTP. Net: 7 ATP equivalents.
   a. *ATP:* 1 GTP (an ATP equivalent) is consumed in converting IMP to AMP; 2 more ATP equivalents are needed to convert AMP to ATP. Overall, ATP synthesis from ribose-5-P onward requires 10 ATP equivalents.
   b. *GTP:* 2 high-energy phosphoric anhydride bonds from ATP, but 1 NADH is produced in converting IMP to GMP; 2 more ATP equivalents are needed to convert GMP to GTP. Overall, GTP synthesis from ribose-5-P onward requires 8 ATP equivalents.
   Pyrimidine synthesis: Starting from HCO_3^- and Gln, 2 ATP equivalents are consumed by CPS-II, and an NADH equivalent is produced in forming orotate. OMP synthesis from ribose-5-P plus orotate requires conversion of ribose-5-P to PRPP at a cost of 2 ATP equivalents. Thus, the net ATP investment in UMP synthesis is just 1 ATP.
   c. *UTP:* Formation of UTP from UMP requires 2 ATP equivalents. Net ATP equivalents in UTP biosynthesis = 3.
   d. *CTP:* CTP biosynthesis from UTP by CTP synthetase consumes 1 ATP equivalent. Overall ATP investment in CTP synthesis = 4 ATP equivalents.

3. a. See Figure 26.6.
   b. See Figure 26.17.
   c. See Figure 26.17.

4. a. Azaserine inhibits glutamine-dependent enzymes, as in steps 2 and 5 of IMP synthesis (glutamine:PRPP amidotransferase and FGAM synthetase), as well as GMP synthetase (step 2, Figure 26.5), and CTP synthetase (Figure 26.16).

   b. Methotrexate, an analog of folic acid, antagonizes THF-dependent processes, such as steps 4 and 10 (GAR transformylase and AICAR transformylase) in purine biosynthesis (Figure 26.3), and the thymidylate synthase reaction (Figure 26.26) of pyrimidine metabolism.

   c. Sulfonamides are analogs of *p*-aminobenzoic acid (PABA). Like methotrexate, sulfonamides antagonize THF formation. Thus, sulfonamides affect nucleotide biosynthesis at the same sites as methotrexate, but only in organisms such as prokaryotes that synthesize their THF from simple precursors such as PABA.

   d. Allopurinol is an inhibitor of xanthine oxidase (Figure 26.10).

   e. 5-Fluorouracil inhibits the thymidylate synthase reaction (Figure 26.26).

5. UDP, via conversion to dUDP (Figure 26.24), is ultimately a precursor to dTTP, which is essential to DNA synthesis.

6. See Figure 26.23.

7. Ribose, as ribose-5-P, is released during nucleotide catabolism (as in Figure 26.8). Ribose-5-P is catabolized via the pentose phosphate pathway and glycolysis to form pyruvate, which enters the citric acid cycle. From Chapter 22, problem 17, note that 3 ribose-5-P (in the form of 1 ribose-5-P and 2 xylulose-5-P) would give a net consumption of 2 ATPs and a net production of 8 ATPs and 5 NADH (= 15 ATPs), when converted to 5 pyruvate. If each pyruvate is worth 15 ATP equivalents (as in a prokaryotic cell), the overall yield of ATP from 3 ribose-5-P is 75 ATP + 23 ATP − 2 ATP = 96 ATP. Net yield per ribose-5-P is thus 32 ATP equivalents.

8. Comparing Figures 26.3 and 25.40, note that AICAR (5-aminoimidazole-4-carboxamide ribonucleotide) is a common intermediate in both pathways. It is a product of step 5 of histidine biosynthesis (Figure 25.40) and step 9 of purine biosynthesis (Figure 26.3). Thus, formation of AICAR as a by-product of histidine biosynthesis from PRPP and ATP by-passes the first nine steps in purine synthesis. However, cells require greater quantities of purine than of histidine, and these nine reactions of purine synthesis are essential in satisfying cellular needs for purines.

9. See the following figure.

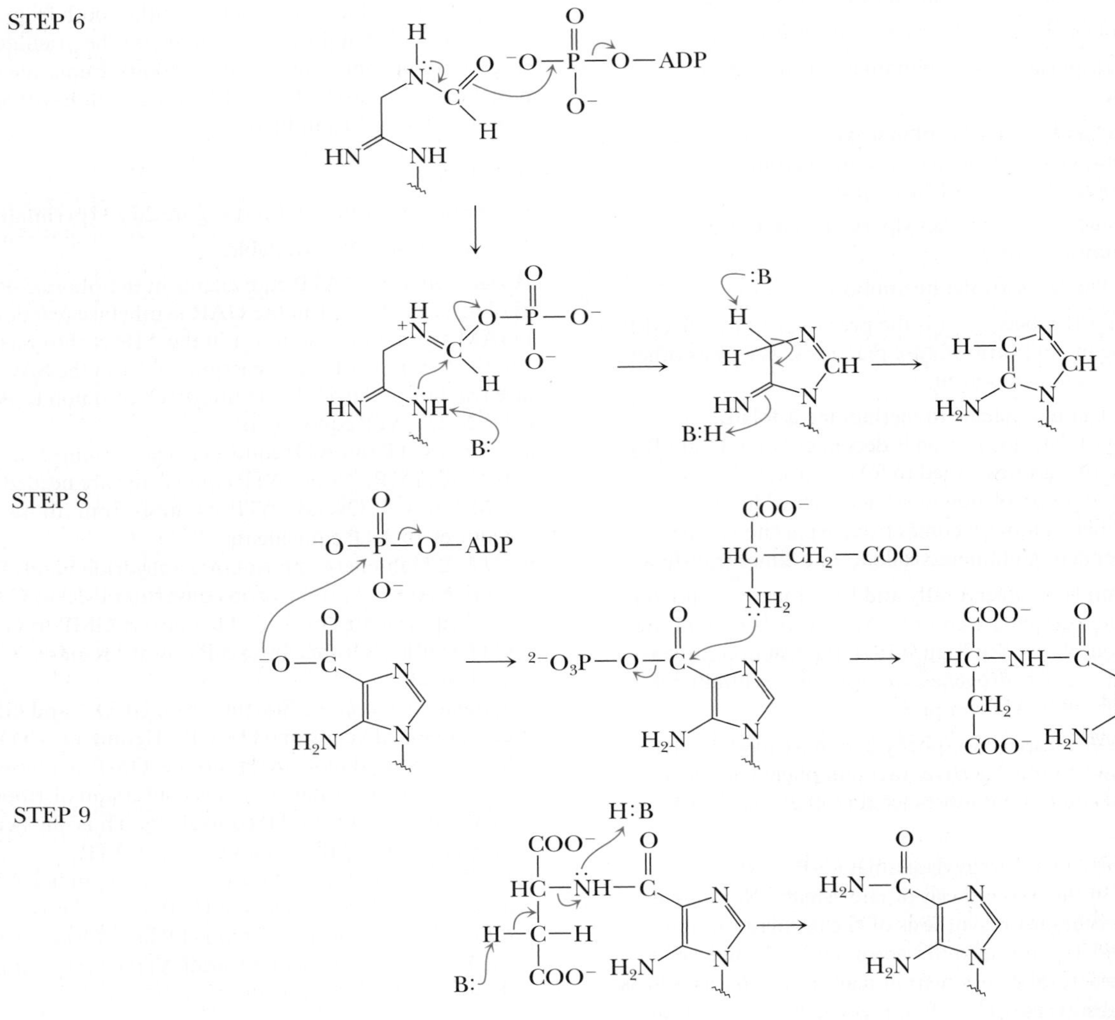

10. Aspartate + GTP^{4-} + H_2O $\longrightarrow$
$$\text{fumarate} + NH_4^+ + GDP^{3-} + P_i^{2-}$$

11. 1. Uric acid + 2 H_2O + O_2 $\longrightarrow$ *urate oxidase* $\longrightarrow$ allantoin + H_2O_2 + CO_2

2. Allantoin + H_2O $\longrightarrow$ *allantoinase* $\longrightarrow$ allantoic acid + glyoxylic acid

3. Allantoic acid + H_2O $\longrightarrow$ *allantoicase* $\longrightarrow$ 2 urea

4. 2 Urea + 2 H_2O $\longrightarrow$ *urease* $\longrightarrow$ 2 CO_2 + 2 HN_3

12.

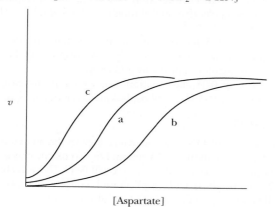

[Aspartate]

13. ATCase (subunit organization $\alpha_6\beta_6$) consists of 6 functional units, each composed of 1 α-subunit and 1 β-subunit, best described as $(\alpha\beta)_6$.

14. From these starting materials to UTP in a eukaryotic cell, where mitochondrial oxidation of dihydroorotate via a coenzyme Q–linked flavoprotein would yield 1.5 ATPs, has a net cost of 2.5 ATPs. From UTP to dTTP would require an ATP at the CTP synthetase step; recovery of an ATP in the nucleoside diphosphate kinase reaction ADP + CTP → CDP + ATP; an NADPH (= 4 ATP equivalents) in converting CDP → dCDP, recovery of an ATP equivalent in the deoxycytidylate kinase reaction dCDP + ADP → dCMP + ATP, deamination of dCMP to dUMP by dCMP deaminase; conversion of dUMP to dTMP by thymidylate synthase and then 2 ATP equivalents to form dTTP. Net: 7.5 ATP equivalents versus 10 ATP for ATP synthesis via the purine biosynthetic pathway.

15. 1. UMP + ATP $\longrightarrow$ UDP + ADP

2. UDP + NADPH $\longrightarrow$ dUDP + $NADP^+$ + OH^-

3. dUDP + ATP $\longrightarrow$ dUTP + ADP

4. dUTP + H_2O $\longrightarrow$ dUMP + PP_i

5. PP_i + H_2O $\longrightarrow$ 2 P_i

6. dUMP + N^5,N^{10}-methylene-THF $\longrightarrow$ dTMP + DHF

Net: UMP + N^5,N^{10}-methylene-THF + 2 ATP + NADPH + 2 H_2O $\longrightarrow$ dTMP + DHF + 2 ADP + 2 P_i + OH^- + $NADP^+$

R^{23}, $R^{178'}$, $R^{179'}$ and R^{218} in thymidylate synthase are hydrogen-bond donors to the phosphate group of the substrate, dUMP. See Kawase, S., Cho, S. W., Rozell, J., and Stroud, R. M., 2000. Replacement set mutagenesis of the four phosphate-binding arginine residues of thymidylate synthase. *Protein Engineering* **13**:557–563. (The other substrate, N^5,N^{10}-methylene-THF bears negative charge on its glutamate units, and without knowledge of thymidylate synthase structure and function, would also be a possibility.)

16. Possibilities include glutamine phosphoribosyl pyrophosphate amidotransferase (see Figure 26.3), adenine phosphoribosyl transferase and hypoxanthine–guanine phosphoribosyl transferase (see Figure 26.7), ribose-5-phosphate pyrophosphokinase (see Figure 26.3), and enzymes of the pyrimidine salvage pathway.

17. Explore pdb id = 7DFR. Note the proximity of the 4-position of the nicotinamide ring (the hydride donor) to the 7-position of folate (the hydride acceptor). Note that polar groups on the substrates are solvent-exposed.

18. View pdb id = 1RAA to see that the CTP-binding sites on the R-subunits are dominated by β-strands. View pdb id = 1RAA to see the ATCase substrate-binding sites (evident from the location of the substrate analog *N*-2-(phosphonoacetyl)-L-asparagine). The allosteric (CTP-binding) sites and the active sites are at opposite ends of the protein. Thus, CTP binding is communicated to the active site via ligand-induced conformational changes extending across the entire protein. R-state ATCase is represented by pdb id = 2IPO; T-state by pdb id = 1RAA.

19. If GMP levels are high (but GDP or AMP is not), ribose-5-P pyrophosphokinase will not be inhibited, glutamine-PRPP amidotransferase will be roughly 50% inhibited due to GMP binding at the G nucleotide allosteric site, and IMP dehydrogenase will be blocked, so IMP is directed toward AMP synthesis to give a balanced amount of A versus G nucleotides.

20. Carbons 1 and 6.

CHAPTER 27

1. a. K_{eq} = 360,333.

b. Assuming [ATP] = [ADP], [pyruvate] must be greater than 360,333 [PEP] for the reaction to proceed in reverse.

c. K_{eq} = 0.724.

d. [PEP]/[pyruvate] = 724.

e. Yes. Both reactions will be favorable as long as [PEP]/[pyruvate] falls between 0.0000028 and 724.

2. Energy charge = 0.945; phosphorylation potential = 1111 M^{-1}.

3. 8 mM ATP will last 1.12 sec. Because the equilibrium constant for creatine-P + ADP → Cr + ATP is 175, [ATP] must be less than 1750 [ADP] for the reaction creatine-P + ADP → Cr + ATP to proceed to the right when [Cr-P] = 40 mM and [Cr] = 4 mM.

4. From Equation 3.48, $\mathscr{E}$($NADP^+$/NADPH) = −0.350 V; $\mathscr{E}$(NAD^+/NADH) = −0.281V; thus, $\Delta\mathscr{E}$ = 0.069 V, and ΔG = −13,316 J/mol. If an ATP "costs" 50 kJ/mol, this reaction can produce about 0.27 ATP equivalent at these concentrations of NAD^+, NADH, $NADP^+$, and NADPH.

5. Assume that the K_{eq} for ATP + AMP → 2 ADP = 1.2 (see legend to Figure 27.2). When [ATP] = 7.2 mM, [ADP] = 0.737 mM and [AMP] = 0.063 mM. When [ATP] decreases by 10% to 6.48 mM, [ADP] + [AMP] = 1.52 mM and thus [ADP] = 1.30 mM and [AMP] = 0.22 mM. A 10% decrease in [ATP] has resulted in a 0.22/0.063 = 3.5-fold increase in [AMP].

6. J (the flux of F-6-P through the substrate cycle) at low [AMP] = 0.1; J at high [AMP] = 8.9. Therefore, the flux of F-6-P through the cycle has increased 89-fold.

7. Inhibition of acetyl-CoA carboxylase will lower the concentration of malonyl-CoA. Because malonyl-CoA inhibits uptake of fatty acids (see Figure 24.16), decreases in [malonyl-CoA] favor fatty acid synthesis.

8. Ethanol oxidation to acetate yields 2 NADH in the cytosol, each worth 1.5 ATPs. The acetyl thiokinase reaction consumes 2 ATP equivalents in converting acetate to acetyl-CoA (due to ATP → AMP + PP_i). Combustion of acetyl-CoA → 2 CO_2 in a liver cell yields a net of 10 ATPs. Therefore, the net yield of ATP from ethanol in a liver cell is 11 ATPs. Glucose → 6 CO_2 in a liver cell yields 30 ATPs (see Table 20.4) or 5 ATP/C atom. Ethanol → 2 CO_2 gives 11 ATPs or 5.5 ATP/C atom.

9. Palmitoyl-CoA → 8 acetyl-CoA yields 7 NADH and 7 [FADH₂] = 21 + 14 = 35. Eight acetyl-CoA → palmitoyl-CoA requires 14 NADPH + 7 ATP = −63 ATP (negative sign denotes ATP consumed). The palmitoyl-CoA ⇌ 8 acetyl-CoA conversion is favorable in both directions provided the free energy release is more than 1750 kJ/mol in the catabolic (acetyl-CoA forming) direction (the energy necessary to produce 35 ATP equivalents at a cost of 50 kJ/mol each) and less than 3150 kJ/mol in the anabolic (palmitoyl-CoA forming) direction (the energy released from 63 ATP equivalents).

10. Cellular respiration releases 2870 kJ/mol of glucose under standard-state conditions (and about the same amount under cellular conditions). Assuming the cell is a bacterial cell where cellular respiration produces 38 ATPs, the total cost of ATP synthesis is 38(50) = 1900 kJ/mol. Thus, the overall free energy change = −2870 + 1900 = −970 kJ/mol. Because $\Delta G°' = -RT \ln K_{eq}$, $K_{eq} = e^{391.4447} = 10^{168}$. This is a very large number! It will be even larger for a cell where ATP yields per glucose are less.

Carbon dioxide fixation leads to glucose synthesis at the cost of 12 NADPH and 18 ATP = 66 ATP equivalents. At 50 kJ/mol, the energy investment from ATP is −3300 kJ/mol and the value of a glucose is +2870 kJ/mol. Thus, $\Delta G°' = -430$ kJ/mol = $-RT \ln K_{eq}$. $K_{eq} = e^{173.52} = 10^{75}$, which is also a very large number!

11. Glycogen phosphorylase, phosphorylase kinase, glycogen synthase, PFK-1, PFK-2, FBPase, and glucose-6-P phosphatase are liver enzymes that act specifically in glycogenolysis or gluconeogenesis; these enzymes are potential targets to control blood glucose levels.

12. From Figure 27.12: The "limit production" list would include NPY, AgRP, and ghrelin; the "raise levels" list would include cholecystokinin, leptin, insulin, melanocortins, and PYY₃₋₃₆.

13. Leptin injection in *ob/ob* mice decreases food intake, raises fatty acid oxidation levels, and lowers body weight. Obese humans are not usually defective in leptin production. Indeed, because leptin is produced in adipocytes, obese individuals may already have high levels of leptin, so leptin injection has limited success.

14. No, although NPY is an orexic agent, it is not necessarily an obesity-promoting hormone. Mice deficient in melanocortin production, melanocortin receptors, or leptin receptors would have an obese phenotype because all these agents act in appetite-suppressing pathways.

15. Alcohol consumption lowers the NAD⁺/NADH ratio, which limits glycolysis because glyceraldehyde-3-P dehydrogenase requires NAD⁺, stimulates gluconeogenesis because NADH favors glyceraldehyde-3-P dehydrogenase working in the glucose synthesis direction, and limits fatty acid oxidation because β-oxidation requires NAD⁺. NADH is also a negative regulator of several citric acid cycle enzymes.

16. The T172D mutant of AMPK mimics the phosphorylated form in having negative charge on the side chain at position 172.

17. a. Fatty acyl-CoA.
   b. Transgenic mice lacking functional hypothalamic fatty acid synthase should have elevated [malonyl-CoA]. If hypothalamic levels of malonyl-CoA are high, eating should diminish, as should body fat content. Assuming physical activity is proportional to signals of nutrient availability, physical activity should increase.

18. a. Leptin-deficient mice should respond to regular leptin injections by losing weight.
   b. The leptin receptor. Assay wild-type and *db/db* mutant cells for leptin binding ability.

19. The heart is the principal organ using both glucose and fatty acids in the well-fed state; the brain relies mostly on glucose (although adipose tissue also relies on glucose for glycerol production); muscle relies mostly on fatty acids; the brain never uses fatty acids; and muscle produces lactate, which is converted in the liver into glucose.

20. a. Hexokinase is an R enzyme because it is a glycolytic enzyme. Glutamine:PRPP amidotransferase is not an ATP-dependent enzyme, so it is neither an R or a U enzyme. It responds to the adenine nucleotide pool through feedback inhibition by AMP, ADP, and ATP.
   b. If E.C. = 0.5, hexokinase activity is high and ribose-5-P pyrophosphokinase activity is likely to be low because R-5-P pyrophosphokinase is probably a U enzyme.
   c. At E.C. = 0.95, the situation reverses and hexokinase activity is low and R-5-P pyrophosphokinase activity is high.

CHAPTER 28

1. The density of DNA can be determined by density gradient ultracentrifugation. (a) Assuming DNA replication is semiconservative, the density of the DNA will be (1.724 + 1.710)/2 = 1.717 g/mL. (b) If DNA replication is conservative, two bands will be seen upon density gradient ultracentrifugation of the DNA after one generation of growth on ¹⁵N-containing media: a band at density = 1.724 g/mL representing the progeny dsDNA and a parental band at 1.710 g/mL.

2. The 5′-exonuclease activity of DNA polymerase I removes mispaired segments of DNA sequence that lie in the path of the advancing polymerase. Its biological role is to remove mispaired bases during DNA repair. The 3′-exonuclease activity acts as a proofreader to see whether the base just added by the polymerase activity is properly base-paired with the template. If not (that is, if it is an improper base with respect to the template), the 3′-exonuclease removes it and the polymerase activity can try once more to insert the proper base. An *E. coli* strain lacking DNA polymerase I 3′-exonuclease activity would show a high rate of spontaneous mutation.

3. Assume that the polymerization rate achieved by each half of the DNA polymerase III homodimer is 750 nucleotides per sec. The entire *E. coli* genome consists of 4.64 × 10⁶ bp. At a rate of 750 bp/sec per DNA polymerase III homodimer (one at each replication fork), DNA replication would take almost 3100 sec (51.7 min; 0.86 hr). When *E. coli* is dividing at a rate of once every 20 min, *E. coli* cells must be replicating DNA at the rate of 4.64 × 10⁶ bp per 20 min (2.32 × 10⁵ bp/min or 3867 bp/sec). To achieve this rate of replication would require initiation of DNA replication once every 20 min at *ori*, and a minimum of 3933/1500 = 2.57 replication bubbles per *E. coli* chromosome, or 5.14 replication forks.

4.

The *E. coli* chromosome replicating once every 20 min (3 replication bubbles, 6 replication forks)

5. If there are 40 molecules of DNA polymerase III (as DNA polymerase III homodimers) per *E. coli* cell and *E. coli* growing at its maximum rate has about 5 replication forks per chromosome, DNA polymerase III availability is sufficient to sustain growth at this rate.

6. Okazaki fragments are 1000 to 2000 nucleotides in length. Because DNA replication in *E. coli* generates Okazaki fragments whose total length must be 4.64×10^6 nucleotides, a total of 2300 to 4700 Okazaki fragments must be synthesized. Consider in comparison a human cell carrying out DNA replication. The haploid human genome is 3×10^9 bp, but most cells are diploid (6×10^9 bp). Collectively, the Okazaki fragments must total 6×10^9 nucleotides in length, distributed over 3 to 6 million separate fragments (assuming they are 1–2 kb in length).

7. DNA gyrases are ATP-dependent topoisomerases that introduce negative supercoils into DNA. DNA gyrases change the linking number, L, of double-helical DNA by breaking the sugar–phosphate backbone of its DNA strands and then religating them (see Figure 11.25). In contrast, helicases are ATP-dependent enzymes that disrupt the hydrogen bonds between base pairs that hold double-helical DNA together. Helicases move along dsDNA, leaving ssDNA in their wake.

8. A diploid human cell contains 6×10^9 bp of DNA. One origin of replication every 300 kbp gives 20,000 origins of replication in 6×10^9 bp. If DNA replication proceeds at a rate of 100 bp/sec at each of 40,000 replication forks (2 per origin), 6×10^9 bp could be replicated in 1500 sec (25 min). To provide 2 molecules of DNA polymerase per replication fork, a cell would require 80,000 molecules of this enzyme.

9. For purposes of illustration, consider how the heteroduplex bacteriophage chromosome in Figure 28.17 might have arisen:

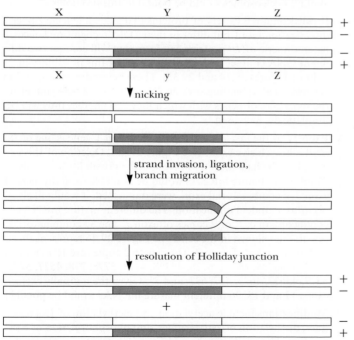

10. The mismatch repair system of *E. coli* relies on DNA methylation to distinguish which DNA strand is "correct" and which is "mismatched"; the unmethylated strand (the newly synthesized one that has not had sufficient time to become methylated is, by definition, the mismatched one). Homologous recombination involves DNA duplexes at similar stages of methylation, neither of which would be interpreted as "mismatched," and thus the mismatch repair system does not act.

11. B-DNA normally has a helical twist of about 10 bp/turn, so the rotation per residue (base pair), $\Delta\phi$, is 36°. If the DNA is unwound to 18.6 bp/turn, $\Delta\phi$ becomes 19.4°. Thus, the change in $\Delta\phi$ is 16.6°.

12. Consider the two DNA duplexes, *WC* and *wc*, respectively, displayed as "ladders" in the following figure. The gap at the right denotes the initial cleavage, in this case of the *W* and *w* strands. Cleavage by resolvase at the arrows labeled 1 and ligation of like strands (*W* with *w* and *C* with *c*) will yield patch recombinants; cleavage at the arrows labeled 2 and ligation of like strands gives splice recombinants.

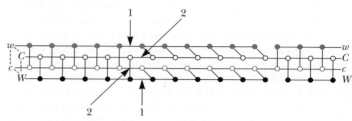

13. The DNA is: GCTA
                    CGAT

    a. HNO$_2$ causes deamination of C and A. Deamination of C yields U, which pairs the way T does, giving:

         GTTA and ACTA and (rarely) ATTA

         CAAT   TGAT   TAAT

    Deamination of A yields I, which pairs as G does, giving:

         GCTG and GCCA and (rarely) GCCG

         CGAC   CGGT   CGGC

    b. Bromouracil usually replaces T, and pairs the way C does:

         GCCA and GCTG and (rarely) GCCG

         CGGT   CGAC   CGGC

    Less often, bromouracil replaces C, and mimics T in its base-pairing:

         GTTA and ACAT and (rarely) ATTA

         CAAT   TGTA   TAAT

    c. 2-Aminopurine replaces A and normally base-pairs with T but may also pair with C:

         GCTG and GCCA and (rarely) GCCG

         CGAC   CGGT   CGGC

14. Transposons are mobile genetic elements that can move from place to place in the genome. Insertion of a transposon within or near a gene may disrupt the gene or inactivate its expression.

15. Proline

16. Hexameric helicases unwind DNA at a cost of 1 ATP/nucleotide. The genome of *E. coli* K12 consists of 4,686,137 nucleotides, so 4,686,137 ATP equivalents would be needed to completely unwind the *E. coli* K12 chromosome.

17. Antibodies against specific proteins are commonly used protein identification tools. Armed with an antibody specific for DNA polymerase III, another antibody specific for DNA polymerase IV, and a procedure to separate replication forks (large macromolecular complexes) from soluble proteins (gel filtration, centrifugation?), addition of DNA polymerase IV should lead to association of IV with the replication fork and release of III.

IV should now be found in the large macromolecular complex and III among the soluble proteins.

18. Eukaryotic translesion DNA polymerases with small and stubby thumb and finger domains are error-prone because they lack the structural elements ensuring high selectivity for the proper dNTP substrate. Since 3′-exonuclease activity is a proofreading activity, translesion DNA polymerases would not need it.

19. a. Cells exit the cell at G_1 and enter G_0 because this stage is prior to DNA replication. Once DNA replication takes place, cells move on to mitosis.
    b. Progression through the cell cycle is regulated at checkpoints. Checkpoints, which control whether the cell continues into the next phase, ensure that steps necessary for successful completion of each phase of the cycle are satisfactorily accomplished. If conditions for advancement to the next phase are not met, the cycle is arrested until they are. Checkpoints depend on cyclins and cyclin-dependent protein kinases (CDKs). Cyclins are synthesized at one phase of the cell cycle and degraded at another. CDK interaction with a specific cyclin is essential for the protein kinase activity of the CDK. In turn, these CDKs control events at each phase of the cycle by targeting specific proteins for activation (or inactivation) through phosphorylation. Destruction of the phase-specific cyclin at the end of the phase inactivates the CDK.

20. Recombination in codon 97 suggests that the V_κ gene provides codons 94–96, which are GTT CAT CTT and specify Val-His-Leu. The fourth codon is either TTG (Leu), CTG (Leu), or CGG (Arg). CGG (Arg) would be the change. However, the desirability of a change cannot be anticipated.

CHAPTER 29

1. The first AUG codon is underlined:

   5′-AGAUCCGU<u>AUG</u>GCGAUCUCGACGAAGACUCCUA
   GGGAAUCC...

   Reading from the AUG codon, the amino acid sequence of the protein is:

   (N-term)Met-Ala-Ile-Ser-Thr-Lys-Thr-Pro-Arg-Glu-Ser-...

2. See Figure 29.3 for a summary of the events in transcription initiation by *E. coli* RNA polymerase. A gene must have a promoter region for proper recognition and transcription by RNA polymerase. The promoter region (where RNA polymerase binds) typically consists of 40 bp to the 5′-side of the transcription start site. The promoter is characterized by two consensus sequence elements: a hexameric TTGACA consensus element in the −35 region and a Pribnow box (consensus sequence TATAAT) in the −10 region (see Figure 29.4).

3. The fact that the initiator site on RNA polymerase for nucleotide binding has a higher K_m for NTP than the elongation site ensures that initiation of mRNA synthesis will not begin unless the available concentration of NTP is sufficient for complete synthesis of an mRNA.

4. First, transcription is compartmentalized within the nucleus of eukaryotes. Furthermore, in contrast to prokaryotes (which have only a single kind of RNA polymerase), eukaryotes possess three distinct RNA polymerases—I, II, and III—acting on three distinct sets of genes (see Section 29.3). The subunit organization of these eukaryotic RNA polymerases is more complex than that of their prokaryotic counterparts. The three sets of genes are recognized by their respective RNA polymerases because they possess distinctive categories of promoters. In turn, all three polymerases interact with their promoters via particular transcription factors that recognize promoter elements within their respective genes. Protein-coding eukaryotic genes, in analogy with prokaryotic genes, have two consensus sequence elements within their promoters, a TATA box (consensus sequence TATAAA) located in the −25 region, and *Inr*, an initiator element that encompasses the transcription start site. In addition, eukaryotic promoters often contain additional short, conserved sequence modules such as enhancers and response elements for appropriate regulation of transcription.

   Transcription termination differs as well as prokaryotes and eukaryotes. In prokaryotes, two transcription basic termination mechanisms occur: *rho* protein-dependent termination and DNA-encoded termination sites. Transcription in eukaryotes tends to be imprecise, occurring downstream from consensus AAUAAA sequences known as poly(A) addition sites. An important aspect of eukaryotic transcription is post-transcriptional processing of mRNA (see Section 29.5).

5. When DNA-binding proteins "read" specific base sequences in DNA, they do so by recognizing a specific matrix of H-bond donors and acceptors displayed within the major groove of B-DNA by the edges of the bases. When DNA-binding proteins recognize a specific DNA region by "shape readout," the protein is discerning local conformational variations in the cylindrical surface of the DNA double helix. These conformational variations are a consequence of unique sequence information contributed by the base pairs that make up this region of the DNA.

6. The metallothionine promoter is about 265 bp long. Because each bp of B-DNA contributes 0.34 nm to the length of DNA, this promoter is about 90 nm long, covering 26.5 turns of B-DNA (10 bp/turn). Each nucleosome spans about 146 bp, so about 2 nucleosomes would be bound to this promoter.

7. If cells contain a variety of proteins that share a leucine zipper dimerization motif but differ in their DNA recognition helices, heterodimeric *bZIP* proteins can form that will contain two different DNA contact basic regions (consider this possibility in light of Figures 29.34 and 29.35). These heterodimers are no longer restricted to binding sites on DNA that are dyad-symmetric; the repertoire of genes with which they can interact thus will be dramatically increased.

8. Exon 17 of the fast skeletal muscle troponin T gene is one of the mutually exclusive exons (see Figure 29.46). Deletion of this exon would cut by half the alternative splicing possibilities, so only 32 different mature mRNAs would be available from this gene. If exon 7, a combinatorial exon, were duplicated, it would likely occur as a tandem duplication. The different mature mRNAs could contain 0, 1, or 2 copies of exon 7. The 32 combinatorial possibilities shown in the center of Figure 29.46 represent those with 0 or 1 copy of exon 7. Those with 2 copies are 16 in number: 456778, 56778, 46778, 6778, 45778, 4778, 5778, 778, 4577, 577, 5677, 45677, 4677, 477, 677, and 77. Given the mutually exclusive exons 17 and 18, 96 different mature mRNAs would be possible.

9. Modifications include acetylation and methylation of Lys residues, methylation of Arg residues, and phosphorylation of Ser residues. The positively charged Lys residues interact electrostatically with phosphate groups on the DNA backbone, so these interactions would be diminished through acetylation. Methylation of Lys and Arg residues introduces bulky aliphatic groups that have the potential to cause steric hindrances between the DNA backbone and the histone tails. Such methylations (as well as the other modifications discussed here) may also establish

recognition sites for chromatin-binding proteins that could modulate DNA:histone interactions. Phosphorylation of Ser residues would lead to electrostatic repulsion between the DNA backbone and the histone tails. Collectively, these modifications weaken histone:DNA interactions.

10. Because the *lac* repressor is tetrameric and each subunit has an inducer-binding site, it seems likely that inducer binding is cooperative. Cooperative binding of inducer would be advantageous because the slope of the *lac* operon expression versus [inducer] would be steeper over a narrower [inducer] range.

11. Capping of the 5′-end and polyadenylation of the 3′-end of mRNA protect the RNA from both 5′- and 3′-exonucleolytic degradation. Methylations at the 5′-end enhance the interaction between cap-binding protein eIF4E (see Figure 30.29), thus increasing the likelihood that the mRNA will be translated. Note also in Figure 30.29 that polyadenylylation provides a protein-binding site for proper assembly of the 40S initiation complex essential to translation.

12. In the A Deeper Look box on page 956, the argument is made that polymerases have a common enzymatic mechanism for nucleotide addition to a growing polynucleotide chain based on two metal ions coordinated to the incoming nucleotide. These metal ions interact with two aspartate residues that are highly conserved in both DNA and RNA polymerases. The discovery of a second Mg^{2+} ion in the RNA polymerase II active site confirms to the universality of this model.

13. 50 bp of DNA is (50)(0.34 nm/bp) = 17 nm. Because 147 bp of DNA make 1.7 turns around the nucleosome (see Chapter 11) = 86.4 bp/turn of DNA, 50 bp = 0.58 turns, or slightly more than $\frac{1}{2}$ turn of DNA around the histone core octamer. Promoter and response modules are about 20 to 25 bp.

14.

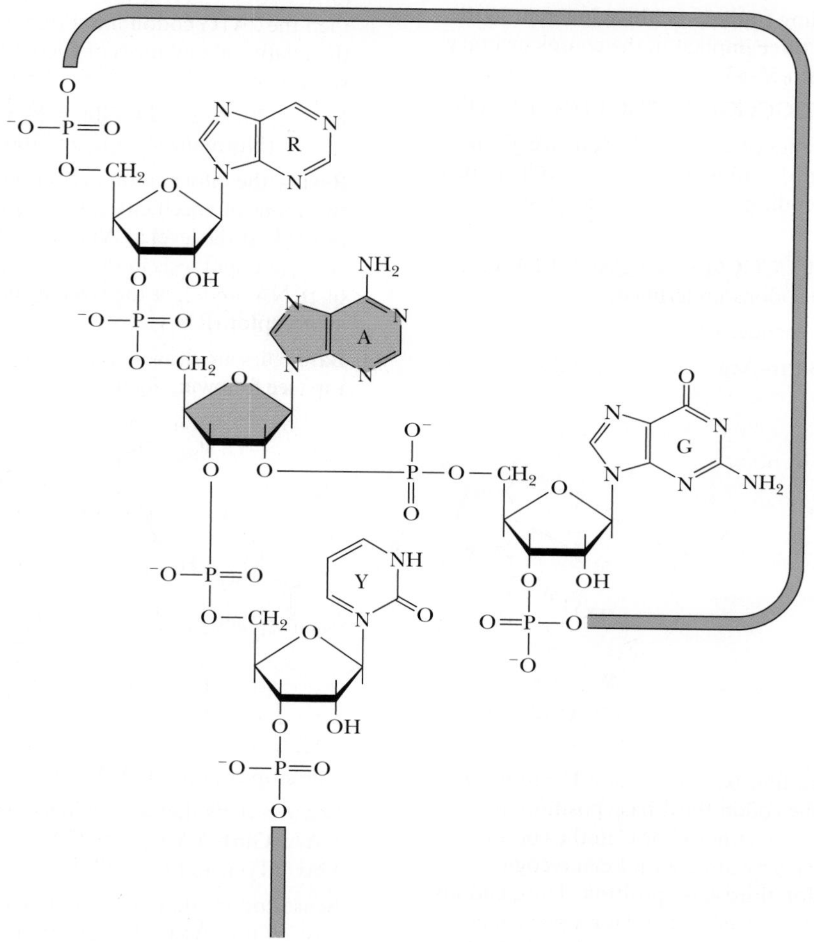

15. α-Helices composed of 7 to 8 residues would have a length of 1.05 to 1.2 nm. The overall diameter of B-DNA is about 2.4 nm, but the diameter at the bottom of major groove is significantly less (as a reference, the C_1'–C_1' distance between base-paired nucleotides in separate chains is 1.1 nm). A distance along the B-DNA helix axis of 1.1 nm would correspond to about 3 base pairs, so if such an α-helix were laid along the major groove, it could not make contacts with more than 3 base pairs (see also Figure 29.32).

16. Phe^{644} acts as a wedge to separate the template and nontemplate strands from one another; Phe^{644} is located at the end of the Y-helix.

17. α-Amanitin binds in the large cleft between the two largest RNA polymerase II subunits, in particular to an α-helix, the "bridge" helix, running across this cleft. Its position in the cleft slows inhibits translocation of DNA and RNA and the entry of the next template nucleotide into the active site.

18. The two protein chains in pdb id = 1GU4 are identical through positions 268 (D) to 333 (Q). Leucines are found at positions 297,

304, 306, 313, 320, 324, 327, and 330. Those at positions 306, 313, 320, and 327 are seven residues apart, as in a leucine zipper. Basic residues (K and R) precede these leucines, occurring at positions 268, 275, 277, 280, 286, 287, 289, 291, 293, 295, and 302.

19. A protein such as CAP that interacts with phosphate groups in the DNA backbone does so via electrostatic interactions based on positively charged Arg and Lys side chains.

20. Chromatin is decompacted by HATs and chromatin remodeling complexes. Histone acetylation by HATs disrupts chromatin structure (see problem 9) and chromatin remodeling complexes then mediate ATP-dependent conformational changes in the chromatin that peels about 50 bp of DNA from the core octamer, exposing the DNA for access by the transcriptional machinery.

CHAPTER 30

1. The cDNA sequence as presented:

    CAATACGAAGCAATCCCGCGACTAGACCTTAAC...

    represents six potential reading frames, the three inherent in the sequence as written and the three implicit in the complementary DNA strand, which as written $5' \rightarrow 3'$, is:

    ...GTTAAGGTCTAGTCGCGGGATTGCTTCGTATTG

    Two of the three reading frames of the cDNA sequence given contain stop codons. The third reading frame is a so-called open reading frame (a stretch of coding sequence devoid of stop codons):

    <u>CAA</u>TAC<u>GAAG</u>CA<u>ATCCCG</u>C<u>GA</u>CTA<u>GAC</u>CTT<u>AAC</u>...
    (alternate codons underlined)

    The amino acid sequence it encodes is:

    Gln-Tyr-Glu-Ala-Ile-Pro-Arg-Leu-Asp-Leu-Asn....

Two of the three reading frames of the complementary DNA sequence also contain stop codons. The third may be an open reading frame:

...G<u>TTA</u>AGG<u>TCT</u>AGT<u>CGC</u>GGG<u>ATT</u>GCT<u>TCG</u>TAT<u>TG</u>
(alternate codons underlined)

An unambiguous conclusion about the partial amino acid sequence of this cDNA cannot be reached.

2. A random (AG) copolymer would contain varying amounts of the following codons:

    AAA AAG AGA GAA AGG GAG GGA GGG

    codons for Lys Lys Arg Glu Arg Glu Gly Gly, respectively.

    Therefore, the random (AG) copolymer would direct the synthesis of a polypeptide consisting of Lys, Arg, Glu, and Gly. The relative frequencies of the various codons are a function of the probability that a base will occur in a codon. For example, if the A/G ratio is 5/1, the ratio of AAA/AAG is $(5 \times 5 \times 5)/(5 \times 5 \times 1) = 5/1$. If the 3A codon is assigned a value of 100, then the 2A1G codon has a frequency of 20. From this analysis, the relative abundances of these amino acids in the polypeptide should be:

    Lys = 120; Arg = 24; Glu = 24; Gly = 4.8
    (Normalized to Lys = 100: Arg = 20; Glu = 20; Gly = 4)

3. Review the information in Section 30.2, noting in particular the two levels of specificity exhibited by aminoacyl-tRNA synthetases (1: at the level at ATP $\rightleftharpoons$ PP$_i$ exchange and aminoacyl adenylate synthesis in the presence of amino acid and the absence of tRNA; and 2: at the level of loading the aminoacyl group on an acceptor tRNA).

4. Base pairs are drawn such that the B-DNA major groove is at the top (see following figure).

G:C   C:G   G:U   U:G

5. The wobble rules state that a first-base anticodon U can recognize either an A or a G in the codon third-base position; first-base anticodon G can recognize either U or C in the codon third-base position; and first-base anticodon I can recognize either U, C, or A in the codon third-base position. Thus, codons with third-base A or G, which are degenerate for a particular amino acid (Table 30.3 reveals that all codons with third-base purines are degenerate, except those for Met and Trp), could be served by single tRNA species with first-base anticodon U. More emphatically, codons with third-base pyrimidines (C or U) are always degenerate and could be served by single tRNA species with first-base anticodon G. Wobble involving first-base anticodon I further minimizes the number of tRNAs needed to translate the 61 sense codons. Wobble tends to accelerate the rate of translation because the noncanonical base pairs formed between bases in the third position of codons and bases occupying the first-base wobble position of anticodons are less stable. As a consequence, the codon:anticodon interaction is more transient.

6. The stop codons are UAA, UAG, and UGA.

    Sense codons that are a single-base change from UAA include CAA (Gln), AAA (Lys), GAA (Glu), UUA (Leu), UCA (Ser), UAU (Tyr), and UAC (Tyr).

    Sense codons that are a single-base change from UAG include CAG (Gln), AAG (Lys), GAG (Glu), UCG (Ser), UUG (Leu), UGG (Trp), UAU (Tyr), and UAC (Tyr).

    Sense codons that are a single-base change from UGA include CGA (Arg), AGA (Arg), GGA (Gly), UUA (Leu), UCA (Ser), UGU (Cys), UGC (Cys), and UGG (Trp).

    That is, 19 of the 61 sense codons are just a single base change from a nonsense codon.

7. The list of amino acids in problem 6 is a good place to start in considering the answer to this question. Amino acid codons in which the codon base (the wobble position) is but a single base change from a nonsense codon are the more likely among this list

because pairing is less stringent at this position. These include UAU (Tyr) and UAC (Tyr) for nonsense codons UAA and UAG, and UGU (Cys), UGC (Cys), and UGG (Trp) for nonsense codon UGA.

8. The more obvious answer to this question is that eukaryotic ribosomes are larger, more complex, and hence slower than prokaryotic ribosomes. In addition, initiation of translation requires a greater number of initiation factors in eukaryotes than in prokaryotes. It is also worth noting that eukaryotic cells, in contrast to prokaryotic cells, are typically under less selective pressure to multiply rapidly.

9. Each amino acid of a protein in an extended β-sheet–like conformation contributes about 0.35 nm to its length. 10 nm ÷ 0.35 nm per residue = 28.6 amino acids.

10. Larger, more complex ribosomes offer greater advantages in terms of their potential to respond to the input of regulatory influences, to have greater accuracy in translation, and to enter into interactions with subcellular structures. Their larger size may slow the rate of translation, which in some instances may be a disadvantage.

11. The universal organization of ribosomes as two-subunit structures in all cells—archaea, eubacteria, and eukaryotes—suggests that such an organization is fundamental to ribosome function. Translocation along mRNA, aminoacyl-tRNA binding, peptidyl transfer, and deacylated-tRNA release are processes that require repetitious uncoupling of physical interactions between the large and small ribosomal subunits.

12. Prokaryotic cells rely on N-formyl-Met-tRNA$_i^{fMet}$ to initiate protein synthesis. The tRNA$_i^{fMet}$ molecule has a number of distinctive features, not found in noninitiator tRNAs, that earmark it for its role in translation initiation (see Figure 30.16). Furthermore, N-formyl-Met-tRNA$_i^{fMet}$ interacts only with initiator codons (AUG or, less commonly, GUG). A second tRNA^{Met}, designated tRNA$_m^{Met}$, serves to deliver methionyl residues as directed by internal AUG codons. Both tRNA$_i^{fMet}$ and tRNA$_m^{Met}$ are loaded with methionine by the same methionyl-tRNA synthetase, and AUG is the Met codon, both in initiation and elongation. AUG initiation codons are distinctive in that they are situated about 10 nucleotides downstream from the Shine–Dalgarno sequence at the 5'-end of mRNAs; this sequence determines the translation start site (see Figure 30.18). Eukaryotic cells also have two tRNA^{Met} species, one of which is a unique tRNA$_i^{Met}$ that functions only in translation initiation. Eukaryotic mRNAs lack a counterpart to the prokaryotic Shine–Dalgarno sequence; apparently the eukaryotic small ribosomal subunit binds to the 5'-end of a eukaryotic mRNA and scans along it until it encounters an AUG codon. This first AUG codon defines the eukaryotic translation start site.

13. The Shine–Dalgarno sequence is a purine-rich sequence element near the 5'-end of prokaryotic mRNAs (see Figure 30.18). It base-pairs with a complementary pyrimidine-rich region near the 3'-end of 16S rRNA, the rRNA component of the prokaryotic 30S ribosomal subunit. Base pairing between the Shine–Dalgarno sequence and 16S rRNA brings the translation start site of the mRNA into the P site on the prokaryotic ribosome. Because the nucleotide sequence of the Shine–Dalgarno element varies somewhat from mRNA to mRNA, whereas the pyrimidine-rich Shine–Dalgarno-binding sequence of 16S rRNA is invariant, different mRNAs vary in their affinity for binding to 30S ribosomal subunits. Those that bind with highest affinity are more likely to be translated.

14. The most apt account is b. Polypeptide chains typically contain hundreds of amino acid residues. Such chains, attached as a peptidyl group to a tRNA in the P site, would show significant inertia to movement, compared with an aminoacyl-tRNA in the A site.

15. Elongation factors EF-Tu and EF-Ts interact in a manner analogous to the GTP-binding G proteins of signal transduction pathways (see pages 496–498, as well as pages 1130–1133 and Figure 15.19). EF-Tu binds GTP and in the GTP-bound form delivers an aminoacyl-tRNA to the ribosome, whereupon the GTP is hydrolyzed to yield EF-Tu:GDP and P$_i$. EF-Ts mediates an exchange of the bound GDP on EF-Tu:GDP with free GTP, regenerating EF-Tu:GTP for another cycle of aminoacyl-tRNA delivery. The α-subunit of the heterotrimeric GTP-binding G proteins also binds GTP. G$_\alpha$ has an intrinsic GTPase activity and the GTP is eventually hydrolyzed to form G$_\alpha$:GDP and P$_i$. Guanine nucleotide exchange factors facilitate the exchange of bound GDP on G$_\alpha$ for GTP, in analogy with EF-Ts for EF-Tu. The amino acid sequences of EF-Tu and G proteins reveal that they share a common ancestry.

16. Four: two in the aminoacyl-tRNA synthetase reaction; one by EF-Tu; one by EF-G.

17. a. The proteins are shown as ribbons.
    b. The rRNA is shown as a strand (line).
    c. The tRNAs (shown with the sugar–phosphate backbone as a strand and the bases as ring structures) are oriented such that their acceptor stems extend away from the 30S subunit, in the direction from which a 50S subunit will add.
    d. The tRNA acceptor stems will lie within the peptidyl transferase center of the 50S subunit.

18. a. 64,281.
    b. If you use the cursor to rotate the structure, many bases around the periphery are easy to see (otherwise, the complexity of the structure makes it hard to see them).
    c. Many such regions are easy to find along the structure's periphery.
    d. RNA.

19. The problem is an exploration, with no discrete answer.

20. The Qβ phage replicase mRNA shows the most consecutive perfect matches (6) complementary to the Shine-Dalgarno sequence.

21. Chloramphenicol consists of an aromatic ring and several polar functions (a nitro group, two —OH groups, and an amine-N). The aromatic ring might intercalate between bases in the 23S rRNA peptidyl transferase site, the polar functions might form H bonds with nitrogenous bases or sugar —OH groups in the rRNA. The —NO$_2$ group is also rather bulky in addition to being polar, so it could disrupt peptidyl transferase in a number of ways.

CHAPTER 31

1. Human rhodanese has 296 amino acid residues. Its synthesis would require the involvement of 4 ATP equivalents per residue or 1184 ATP equivalents. Folding of rhodanese by the Hsp60 α_{14} complex consumes another 130 ATP equivalents. The total number of ATP equivalents expended in the synthesis and folding of rhodanese is approximately 1314.

2. Because a single proteolytic nick in a protein can doom it to total degradation, nicked proteins clearly are not tolerated by cells and are degraded quickly. Cells are virtually devoid of partially degraded protein fragments, which would be the obvious intermediates in protein degradation. The absence of such

intermediates and the rapid disappearance of nicked proteins from cells indicate that protein degradation is a rigorously selective, efficient cellular process.

3. *Hsp70:* Hsp70 binds to exposed hydrophobic regions of unfolded proteins. The Hsp70 domain involved in this binding is 18 kD in size and therefore would have a diameter of roughly 3 nm (see problem 5 for the math). Assuming the hydrophobic binding site of Hsp70 stretches across its diameter, it would interact with about 3 nm/0.35 nm = 8 or 9 amino acid residues. Multiple Hsp70 monomers could interact with longer hydrophobic stretches. So, for Hsp70 to interact with a polypeptide chain does not necessarily depend on the protein's size, but it does depend on its hydrophobicity (and absence of charged groups).

Hsp70 and Hsp60 chaperonins: Proteins that interact with both of these classes of chaperones not only must fit the description for Hsp70 targets but must also be small enough to access the Hsp60 chaperonin chamber, which has a diameter of about 5 nm. This restriction means that Hsp60 cannot interact with proteins more than roughly 50 kD in mass (see problem 5).

4. Many eukaryotic proteins have a multidomain or modular organization, where each module is composed of a contiguous sequence of amino acid residues that folds independently into a discrete domain of structure (see Figure 6.26 for examples of this type of sequence and structure organization). Such proteins would be ideal for co-translational folding: As each newly synthesized contiguous sequence emerges from the ribosome tunnel, it begins folding into its characteristic domain structure. The final, fully folded state of the complete protein would be achieved when the various domains assumed their proper spatial relationships to one another through hinge motions occurring at intradomain regions of the protein.

5. The maximal diameter for a spherical protein would be 5 nm (radius = 2.5 nm). The volume of this protein is given by $V = \frac{4}{3}\pi r^3 = 4/3(3.14)(2.5 \times 10^{-9}\,m)^3 = 65.4 \times 10^{-21}\,mL$. If its density is 1.25 g/mL, its mass would be $(1.25\,g)(65.4 \times 10^{-21}\,mL) = 81.8 \times 10^{-21}\,g/molecule$, so its molecular weight = $(6.023 \times 10^{23})(81.8 \times 10^{-21}\,g) = 492.5 \times 10^2\,g/mol = 49{,}250\,g/mol$, or about 50 kD.

6. There are $2^7 - 1$, or 127 different phosphorylated forms for a protein having 7 separate phosphorylation sites.

7. GrpE catalyzes nucleotide exchange on DnaK, replacing ADP with ATP, which converts DnaK back to a conformational form having low affinity for its polypeptide substrate. This change

leads to release of bound polypeptide, giving it the opportunity to fold, which is an important step in DnaK function. EF-Ts catalyzes nucleotide exchange on EF-Tu, converting it to the conformational form competent in aminoacyl-tRNA binding. Binding of the aminoacyl-tRNA:EF-Tu complex to the ribosome A site triggers the GTPase activity of EF-Tu as codon:anticodon recognition takes place and the aminoacyl-tRNA:EF-Tu complex conformationally adjusts to the A site. DnaJ delivers an unfolded polypeptide chain to DnaK, and its interaction with DnaK also triggers the ATPase activity of DnaK, whereupon the DnaK:ADP complex forms a stable complex with the unfolded polypeptide. In both instances, hydrolysis of bound nucleoside triphosphate is triggered and leads to conformational changes that stabilize the respective complexes.

8. Protein targeting information resides in more generalized features of the leader sequences such as charge distribution, relative polarity, and secondary structure, rather than amino acid sequence per se. Proteins destined for secretion have N-terminal amino acid sequence with one or more basic amino acids followed by a run of 6 to 12 hydrophobic amino acids. The sequence MRSLLILVLCFLPAALGK... has a basic residue (Arg) at position 2 and a run of 12 nonpolar residues (...LLILVLPLAALG...); thus, it appears to be a signal sequence for a secretory protein. Cleavage by the signal peptidase would occur between the G and K.

9. Translocation proceeds until a stop transfer signal associated with a hydrophobic transmembrane protein segment is recognized. The stop-transfer signal induces a pause in translocation, and the translocon changes its conformation such that the wall of this closed cylindrical structure either opens or exposes a hydrophobic path, allowing the transmembrane segment to diffuse laterally into the hydrophobic phase of the membrane.

10. The maximal diameter for a spherical protein would be 6 nm (radius = 3 nm). The volume of this protein is given by $V = \frac{4}{3}\pi r^3 = 4/3(3.14)(3 \times 10^{-9}\,m)^3 = 113 \times 10^{-21}\,mL$. If its density is 1.25 g/mL, its mass would be $(1.25\,g)(113 \times 10^{-21}\,mL) = 141 \times 10^{-21}\,g/molecule$, so its molecular weight = $(6.023 \times 10^{23})(141 \times 10^{-21}\,g) = 851 \times 10^2\,g/mol = 85{,}100\,g/mol$, or about 85 kD.

11. The thickness of a phospholipid bilayer is 5 nm, so the overall channel formed by a 50S ribosome and the translocon channel is 15 nm; 15 nm ÷ 0.35 nm per amino acid = about 43 amino acid residues.

12.

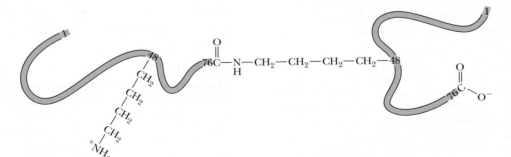

13. E3 ubiquitin protein ligase selects proteins by the nature of the N-terminal amino acid. Proteins with Met, Ser, Ala, Thr, Val, Gly, or Cys at the amino terminus are resistant to its action. Proteins having Arg, Lys, His, Phe, Tyr, Trp, Leu, Ile, Asn, Gln, Asp, or Glu at their N-terminus are susceptible. N-terminal Pro residues lack a free α-amino group, so such proteins are not susceptible.

14. The cell cycle relies on the cyclic synthesis and destruction of cell cycle regulatory proteins. Lactacystin inhibits cell cycle

progression by interfering with programmed destruction of proteins by proteasomes.

15. The temperature-induced switch could be based on a temperature-dependent conformational change in the HtrA protein that brings the His-Ser-Asp catalytic triad into the proper spatial relationship for protease function.

16. There are many other post-translational modifications, including acylation, alkylation, amidation (at a protein's C-terminus), biotinylation, formylation, glycosylation, isoprenylation, and oxidation. See also Table 5.5.

17. This exercise is left to the student.

18. A phosphodegron is defined as one or a series of phosphorylated residues on a substrate protein that interact directly with a protein–protein interaction domain in an E3 ubiquitin ligase, thereby linking the substrate to the ubiquitin conjugation machinery. Phosphorylation can affect the affinity of target protein binding; alternatively, phosphorylation can stimulate E2 activity.

19. It depends on which amino acid is penultimate. If Arg, Lys, His, Phe, Tyr, Trp, Leu, Ile, Asn, Gln, Asp, or Glu follow immediately after Met, the protein will have a short half-life after Met removal. If Ser, Ala, Thr, Val, Gly, Cys, Pro, or another Met follow, the protein should have a long half-life after N-terminal Met removal.

20. To array positively charged residues on one side of an α-helix and hydrophobic residues on the other requires a pattern with positively charged residues positioned every 3.6 residues and similarly for hydrophobic residues. A sequence with Arg or Lys at positions 1, 4, 7, 11, 14, and 18 and an array of hydrophobic residues at 2, 3, 5, 6, 8, 9, 10, 12, 13, 15, 16, 17, 19, and 20 would fit this pattern.

CHAPTER 32

1. Polypeptide hormones constitute a larger and structurally more diverse group of hormones than either the steroid or amino acid–derived hormones, and it thus might be concluded that the specificity of polypeptide hormone-receptor interactions, at least in certain cases, should be extremely high. The steroid hormones may act either by binding to receptors in the plasma membrane or by entering the cell and acting directly with proteins controlling gene expression, whereas polypeptides and amino acid–derived hormones act exclusively at the membrane surface. Amino acid–derived hormones can be rapidly interconverted in enzyme-catalyzed reactions that provide rapid responses to changing environmental stresses and conditions. See *The Student Solutions Manual, Study Guide and Problems Book* for additional information.

2. The cyclic nucleotides are highly specific in their action, because cyclic nucleotides play no metabolic roles in animals. Ca^{2+} ion has an advantage over many second messengers because it can be very rapidly "produced" by simple diffusion processes, with no enzymatic activity required. IP_3 and DAG, both released by the metabolism of phosphatidylinositol, form a novel pair of effectors that can act either separately or synergistically to produce a variety of physiological effects. DAG and phosphatidic acid share the unique property that they can be prepared from several different lipid precursors. Nitric oxide, a gaseous second messenger, requires no transport or translocation mechanisms and can diffuse rapidly to its target sites.

3. Nitric oxide functions primarily by binding to the heme prosthetic group of soluble guanylyl cyclase, activating the enzyme.

An agent that could bind in place of NO—but that does not activate guanylyl cyclase—could reverse the physiological effects of nitric oxide. Interestingly, carbon monoxide, which has long been known to bind effectively to heme groups, appears to function in this way. Solomon Snyder and his colleagues at Johns Hopkins University have shown that administration of CO to cells that have been stimulated with nitric oxide causes attenuation of the NO-induced effects.

4. Herbimycin, whose structure is shown in the figure below, reverses the transformation of cells by Rous sarcoma virus, presumably as a direct result of its inactivation of the viral tyrosine kinase. The manifestations of transformation (on rat kidney cells, for example) include rounded cell morphology, increased glucose uptake and glycolytic activity, and the ability to grow without being anchored to a physical support (termed *anchorage-independent growth*). Herbimycin reverses all these phenotypic changes. On the basis of these observations, one might predict that herbimycin might also inactivate tyrosine kinases that bear homology to the viral $pp60^{v-src}$ tyrosine kinase. This inactivation has in fact been observed, and herbimycin is used as a diagnostic tool for implicating tyrosine kinases in cell-signaling pathways.

5. The identification of phosphorylated tyrosine residues on cellular proteins is difficult. Quantities of phosphorylated proteins are generally extremely small, and to distinguish tyrosine phosphorylation from serine/threonine phosphorylation is tedious and laborious. On the other hand, monoclonal antibodies that recognize phosphotyrosine groups on protein provide a sensitive means of detecting and characterizing proteins with phosphorylated tyrosines, using, for example, Western blot methodology.

6. Hormones act at extremely low concentrations, but many of the metabolic consequences of hormonal activation (release of cyclic nucleotides, Ca^{2+} ions, DAG, etc., and the subsequent alterations of metabolic pathways) occur at and involve higher concentrations of the affected molecular species. As we have seen in this chapter, most of the known hormone receptors mediate hormonal signals by activating enzymes (adenylyl cyclase, phospholipases, protein kinases, and phosphatases). One activated enzyme can produce many thousands of product molecules before it is inactivated by cellular regulation pathways.

7. Vesicles with an outside diameter of 40 nm have an inside diameter of approximately 36 nm and an inside radius of 18 nm. The data correspond to a volume of 2.44×10^{-20} L. Then 10,000 molecules/6.02×10^{23} molecules/mole = 1.66×10^{-20} mole. The concentration of acetylcholine in the vesicle is thus 1.66/2.44 M or 0.68 M.

8. The evidence outlined in this problem points to a role for cAMP in fusion of synaptic vesicles with the presynaptic membrane and the release of neurotransmitters. GTPγS may activate an inhibitory G protein, releasing $G_{\alpha i}$(GTPγS), which inhibits adenylyl cyclase and prevents the formation of requisite cAMP.

9. In both cases, the fraction of receptor that is bound with hormone is approximately 50%.

10. Statin drugs inhibit the synthesis of new cholesterol in human subjects, but they have no effect on dietary intake of cholesterol. Dietary cholesterol accounts for approximately 2/3 of all body cholesterol. Moreover, synthesis of steroid hormones is not generally dependent on ambient levels of cholesterol in the body. Thus the taking of statin drugs would not be expected to influence the levels and functions of steroid hormones in the body.

11. β-Strands indeed provide more genetically economical means of traversing a membrane. However, the only significant classes of proteins that employ β-strands for membrane traversal are the porins and toxins such as hemolysin. These β-strand structures provide no easy way to communicate conformational changes across the membrane. On the other hand, bundles of α-helices are more able to convey and transmit conformational changes across a bilayer membrane.

12. The formation and breakdown of cAMP:

13. The point at which no ion flow occurs is the point of equilibrium that balances the chemical and electrical forces across a membrane. The Nernst equation is obtained by setting ΔG to zero in Equation 9.2. Rearrangement yields

$$RT \ln (C_2/C_1) = -Z\mathscr{F}\Delta\psi$$

Using the data in Figure 32.49, one can use this equation to yield an equilibrium potential for K^+ of -80 mV and an equilibrium potential for Na^+ of $+55.5$ mV.

14. Using the Goldman equation, one can calculate $\Delta\psi = -60$ mV, in agreement with values measured experimentally in neurons.

15. The resting potential in neurons is -60 mV. Local depolarization of the membrane causes the potential to rise approximately 20 mV to about -40 mV. This change causes the Na^+ channels to open and Na^+ ions begin to flow into the cell. This causes the potential to continue increasing to a value of about $+30$ mV. Because this value is close to the equilibrium potential for Na^+, the Na^+ channels begin to close and K^+ channels open. K^+ rushes out of the neuron, returning the membrane potential to a very negative value. There is a slight overshoot of the potential, and the K^+ channels close and the resting membrane potential is restored.

16. Pathway steps (in Figure 32.4) that involve amplification include the tyrosine kinase of the receptor protein, Raf, MAPKK, and MAPK.

17. If Ras were mutated so as to have no GTPase activity, it would activate Raf indefinitely, and the pathway would run as if hormones were continually activating the receptor tyrosine kinase.

18. There are several possible answers to this problem. One is that the original assays of kinase activities that reflected signaling effects were carried out on whole-cell preparations, in which the G-protein-dependent and G-protein-independent activities would appear as components of the overall activation. The availability of G-protein-uncoupled receptor mutants and ligands that could stimulate arrestin recruitment without G-protein activation made it possible to distinguish the G-protein-dependent and G-protein-independent effects. (See the Further Reading references in this chapter by Luttrell for additional discussion of these issues.)

19. Malathion has been used for years in the eradication of boll weevils, but it has not presented a serious toxicity problem in this program. Apparently, malathion is very toxic to insects, and relatively less toxic to humans, particularly in the low volume applications used for this purpose.

20. This exercise is left to the student, for obvious reasons.

Common Abbreviations Used by Biochemists

A	adenine or the amino acid alanine
Ab	antibody
Ag	antigen
Ac-CoA	acetyl-coenzyme A
ACh	acetylcholine
ACP	acyl carrier protein
ADH	alcohol dehydrogenase
ADP	adenosine diphosphate
AIDS	acquired immunodeficiency syndrome
AMP	adenosine monophosphate
ALA	d-aminolevulinic acid
ATCase	aspartate transcarbamoylase
atm	atmosphere
ATP	adenosine triphosphate
BChl	bacteriochlorophyll
bp	base pair
BPG	bisphosphoglycerate
BPheo	bacteriopheophytin
C	cytosine or the amino acid cysteine
cal	calorie
CaM	calmodulin
cAMP	cyclic 39,59-adenosine monophosphate
CAP	catabolite activator protein
cDNA	complementary DNA
CDP	cytidine diphosphate
CDR	complementarity-determining region
Chl	chlorophyll
CM	carboxymethyl
CMP	cytidine monophosphate
CoA or CoASH	coenzyme A
CoQ	coenzyme Q
cpm	counts per minute
CTP	cytidine triphosphate
D	dalton or the amino acid aspartate
d	deoxy
dd	dideoxy
DAG	diacylglycerol
DEAE	diethylaminoethyl
DHAP	dihydroxyacetone phosphate
DHF	dihydrofolate
DHFR	dihydrofolate reductase
DNP	dinitrophenol
Dopa	dihydroxyphenylalanine
DNA	deoxyribonucleic acid
ds	double-stranded
%	reduction potential
E4P	erythrose-4-phosphate
EF	elongation factor
EGF	epidermal growth factor
EPR	electron paramagnetic resonance
ER	endoplasmic reticulum
ESI-MS	electrospray ionization mass spectrometry
^	Faraday's constant
FAB	antibody molecule fragment that binds antigen
FAD	flavin adenine dinucleotide
FADH2	reduced flavin adenine dinucleotide
FBP	fructose-1,6-bisphosphate
FBPase	fructose bisphosphatase
Fd	ferredoxin
fMet	N-formyl-methionine
FMN	flavin mononucleotide
F-1-P	fructose-1-phosphate
F-6-P	fructose-6-phosphate
G	guanine or the amino acid glycine
G	Gibbs free energy
GABA	g-aminobutyric acid
Gal	galactose
GDP	guanosine diphosphate
GFP	green fluorescent protein
GLC	gas-liquid chromatography
Glc	glucose
GMP	guanosine monophosphate
G-1-P	glucose-1-phosphate
G-3-P	glyceraldehyde-3-phosphate
G-6-P	glucose-6-phosphate
G6PD	glucose-6-phosphate dehydrogenase
GS	glutamine synthetase
GSH	glutathione (reduced glutathione)
GSSG	glutathione disulfide (oxidized glutathione)
GTP	guanosine triphosphate
h	hour
h	Planck's constant
Hb	hemoglobin
HDL	high-density lipoprotein
HGPRT	hypoxanthine-guanine phosphoribosyltransferase
HIV	human immunodeficiency virus
HMG-CoA	hydroxymethylglutaryl-coenzyme A
hnRNA	heterogeneous nuclear RNA
HPLC	high-pressure (or high-performance) liquid chromatography
HX	hypoxanthine
Hyl	hydroxylysine
Hyp	hydroxyproline
I	inosine or the amino acid isoleucine
IDL	intermediate density lipoprotein
IF	initiation factor
IgG	immunoglobulin G
IMP	inosine monophosphate
IP3	inositol-1, 4, 5-trisphosphate
IPTG	isopropylthiogalactoside
IR	infrared
ITP	inosine triphosphate
J	joule
k	kilo (103)
k	rate constant
Km	Michaelis constant
kb	kilobases
kD	kilodaltons
L	liter
LDH	lactate dehydrogenase
LDL	low density lipoprotein
Lys	lysine
M	molar
m	milli (1023)
mL	milliliter
mm	millimeter
Man	mannose
Mb	myoglobin
miRNA	micro RNA
mol	mole
mRNA	messenger RNA
MS	mass spectrometry
mV	millivolt
N	Avogadro's number